# OLIN'S CONSTRUCTION

# OLIN'S CONSTRUCTION
## Principles, Materials, and Methods

**EIGHTH EDITION**

H. Leslie Simmons, RA, CSI

John Wiley & Sons, Inc.

Published by John Wiley & Sons, Inc., Hoboken, New Jersey
Published simultaneously in Canada

Limit of Liability/Disclaimer of Warranty: While the publisher and the author have used their best
efforts in preparing this book, they make no representations or warranties with respect to the
accuracy or completeness of the contents of this book and specifically disclaim any implied
warranties of merchantability or fitness for a particular purpose. No warranty may be created or
extended by sales representatives or written sales materials. The advice and strategies contained
herein may not be suitable for your situation. You should consult with a professional where
appropriate. Neither the publisher nor the author shall be liable for any loss of profit or any other
commercial damages, including but not limited to special, incidental, consequential, or other
damages.

For general information on our other products and services or for technical support, please contact
our Customer Care Department within the United States at (800) 762-2974, outside the United States
at (317) 572-3993 or fax (317) 572-4002.

Wiley also publishes its books in a variety of electronic formats. Some content that appears in print
may not be available in electronic books. For more information about Wiley products, visit our web
site at www.wiley.com.

*Library of Congress Cataloging-in-Publication Data:*

Simmons, H. Leslie
Olin's construction : principles, materials, and methods / H. Leslie Simmons. -- 8th ed.
    p.  cm.
Rev. ed. of: Construction : principles, materials, and methods. 7th ed. 2001.
ISBN-13: 978-0-471-71405-7
ISBN-10: 0-471-71405-4
1. Building.   I. Simmons, H. Leslie. Construction, principles, materials, and methods.  II. Title.
TH145.S513 2006
690--dc22

                                        2006028489

Printed in the United States of America

10  9  8  7  6  5  4  3  2  1

# Contents

# Preface

*Olin's Construction: Principles, Materials, and Methods* is a widely adopted course textbook in nearly 200 colleges and universities offering architectural and building technology curricula. It is also becoming a standard reference source in professional offices. The seventh edition has been translated into mainland Chinese for distribution in China.

This eighth edition is the result of almost 50 years of research and editorial effort costing several million dollars. During the book's first 26 years, much of this work was carried out in an ongoing membership education program initiated in 1964 by the United States League of Savings Institutions. The sixth edition was the result of that research and a major program carried out during the 3 years prior to its publication by its then publisher, Van Nostrand Reinhold. This program consisted of completely revising the data in the previous edition to reflect the many changes that had occurred in recent years and to bring those data up-to-date with latest industry standards. Furthermore, new data were added to introduce construction materials and methods not in general use when the previous edition was prepared. Materials and construction methods related to commercial construction, including high-rise buildings, were also added. Previous editions primarily covered residential construction. The sixth edition addressed materials and methods used in both.

The seventh edition was written to further develop the book as a modern teaching and research tool for residential, commercial, and institutional building construction. To this end, the seventh edition contained almost 1200 pages divided into 16 chapters. The organization and names of these chapters were similar to those used in the 1995 edition of *MasterFormat*™, which made the seventh edition easier to relate to the specifications and data filing formats then being used in the construction industry.

The eighth edition updates the data contained in the seventh edition and continues the book's development as a modern teaching and research tool for residential, commercial, and institutional building construction. To this end, the eighth edition again contains about 1200 pages. It is accompanied by a Web site that enhances the data contained in the print version. To make the book more useful to the teaching profession, an *Instructor's Manual* is available for use with the eighth edition.

In addition, the eighth edition has been extensively revised to bring it essentially into agreement with *MasterFormat* 2004. To that end, the book now contains 25 chapters, with chapter numbers and titles revised where necessary to bring them into conformance with *MasterFormat* 2004, with two exceptions. First, *MasterFormat* 2004 is divided into two groups, the "Procurement and Contracting Requirements Group" and the "Specifications Group." The eighth edition of *Olin's Construction: Principles, Materials, and Methods* does not divide data into these two groups. Rather, Chapter 1, "Design and Contracting Requirements," addresses some requirements from the "Procurement and Contracting Requirements Group" and many requirements not specifically covered by *MasterFormat* 2004, including building design, industry standards, codes, barrier-free design, sustainable building design, construction documents, bidding and negotiation, construction contract administration, and construction management.

The second exception relates to the numbering and naming of chapters within the "Specifications Group." The names and numbers of Chapters 2 through 14 exactly mirror those used in *MasterFormat* 2004. With a few exceptions, the subdivision of data in Chapters 2 through 14, the names associated with those data, and the order of presentation of those data are essentially the same as the names and order used in *MasterFormat* 2004. In the interest of continuity, the numbers associated with individual data may vary somewhat from those used *MasterFormat* 2004, and some headings that are separate from each other in *MasterFormat* 2004 have been combined in *Olin's Construction: Principles, Materials, and Methods*. In addition, some names have been altered to more correctly reflect the data contained in a particular section.

Many *MasterFormat* 2004 divisions 15 through 49 are reserved for future expansion of the system and therefore contain no data. It makes little sense to have book chapters that contain no data. Therefore, Chapters 15 through 25 of *Olin's Construction: Principles, Materials, and Methods* follow the order used in *MasterFormat* 2004, but the names and numbers used are different. For example: "Fire Suppression," which is Division 21 in *MasterFormat* 2004, is Chapter 15 in *Olin's Construction: Principles, Materials, and Methods;* Chapter 16, "Plumbing," is Division 22 in *MasterFormat* 2004; and so forth. A notation in the "Applicable *MasterFormat* 2004 sections of each chapter indicates the *MasterFormat* 2004 division that is applicable to that chapter.

*MasterFormat* 2004 moves subject matter from its location in the 1995 edition of *MasterFormat* to new locations. The eighth edition of *Olin's Construction: Principles, Materials, and Methods* also relocates these data to agree with *MasterFormat* 2004. For example, Earthwork was in Division 2 of the 1995 edition of *MasterFormat* and, accordingly, was in Chapter 2 of the seventh edition. In *MasterFormat* 2004, earthwork is in Division 31. Because there are no unused chapter numbers in the eighth edition for reasons discussed earlier, there is no Chapter 31. Earthwork is, therefore, moved to the corresponding consecutive number, which is Chapter 21.

In addition to the subjects mentioned earlier that are covered in Chapter 1 but are not listed in *MasterFormat* 2004, other subjects not included in *MasterFormat* 2004 are also included in the eighth edition. For example, Chapter 2 addresses land survey and description,

but *MasterFormat* 2004 does not include them. Several chapters also include a discussion of the manufacture of products and systems related to those chapters, and several chapters have a discussion of the background and history associated with the covered materials or systems. Chapter 24, "Properties of Materials," and Chapter 25, "The Metric System of Measurement," are not listed in *MasterFormat* 2004.

Construction managers have been associated with large construction projects for many years, but their concerns have been inadequately addressed in most university-level textbooks. Earlier editions of *Olin's Construction: Principles, Materials, and Methods* included coverage of most of the technical construction–related subjects that need to be addressed by construction managers. The eighth edition supplements those data with new Section 1.9, "Construction Management," to address construction management in far more depth than existed in earlier editions.

Many chapters contain a new section entitled "Sustainability Considerations." These sections supplement the data presented in Section 1.5, "Sustainable Building Design." The purpose of these sections is to alert the reader to the implications of the covered subjects for the sustainability of the structure and their effects on the natural environment.

Extensive rewriting has been done in the eighth edition in recognition of the advent of the *International Codes* and the demise of the previous three major model code groups. References to the earlier codes have been largely removed, and references to the *International Code®* have been substituted.

Conversely, many subjects addressed in *MasterFormat* 2004 are beyond the scope of this book and are not addressed here—for example, none of the subjects addressed in the *MasterFormat* 2004 "Process Equipment Subgroup" (material processing and handling equipment, pollution control equipment, electrical power generation, and so forth).

A list of applicable *MasterFormat* 2004 sections has been included as a subsection in each chapter. The numbers and titles listed there are from *MasterFormat* 2004, published by the Construction Specifications Institute (CSI) and Construction Specifications Canada (CSC), and are used with permission from CSI.

For those interested in a more in-depth explanation of *MasterFormat* and its use in the construction industry, contact:

The Construction Specifications Institute (CSI)
99 Canal Center Plaza, Suite 300
Alexandria, VA 22314
(800) 689-2900; (703) 684-0300
http://www.csinet.org

An attempt has been made in the eighth edition to cover every principle, material, and method used to design and construct both large and small buildings of most types. The information presented includes the background and history of the materials and systems described. In each case, materials and their manufacture are discussed first, followed by the methods of construction used to erect these materials.

The first five editions of this book were directed primarily the home construction portion of the construction industry. They included coverage, much of it extensive, of wood construction, masonry, interior and exterior finishes, heating, ventilating, and air conditioning, plumbing, electrical, and many other subjects as they related to small residential construction. The sixth edition included most of these subjects as well, but supplemented them with information about materials and methods used in the construction of other building types. Construction systems covered included those of both precast and cast-in-place concrete, steel, wood, and masonry. The discussions about mechanical and electrical systems were still limited to requirements for residential construction. Other subjects covered in earlier editions were expanded to address most kinds of commercial, larger residential, and institutional buildings. As a result, the sixth edition introduced materials and systems used in most types of building construction.

The sixth edition also covered doors and windows; glazed curtain walls; glazing; and finishes, including plaster, gypsum board, tile, terrazzo, acoustical ceilings and other acoustical treatment, wood flooring, resilient flooring, carpet, and painting and finishing.

Coverage was given to industry standards, codes, land survey and descriptions, properties of materials, barrier-free design, metrication, siteworks (including excavation, grading, shoring and sheeting, and other earthwork), and sound control in buildings.

The seventh edition covered most of the subjects addressed in the sixth edition but supplemented them with much new data. New subjects in the seventh edition included a discussions of the architect's role in building design and construction, including the development of construction documents, and the architect's responsibilities during the bidding and negotiation and construction phases of a building construction project. The sections on codes and standards have been expanded to include a discussion of their effect on building design. Additional subjects include glass-fiber reinforced concrete, concrete toppings, cement-based underlayment, glass unit masonry, miscellaneous metal fabrications, heavy timber framing, finish carpentry, wood siding, metal roofing, door hardware, stone flooring, resinous flooring, wall coverings, fire protection specialties, residential appliances, unit kitchens, elevators, and fire sprinkler systems. The discussions of mechanical systems have been greatly expanded to include information related to buildings other than residences. For example, there is coverage of the different types of HVAC systems, including heating systems (forced air, air-water, steam, water, electrical), components (boiler types, furnaces, controls, finned tube radiators, radiant heaters and panels, heat recovery systems, ducts, pipes, diffusers, grills, and registers), and fuels (coal, gas, electricity). There is also coverage of cooling systems including discussions of the refrigeration cycle, cooling components, heat pumps, direct refrigeration, and delivery systems (fan coil units, unit ventilators). The electrical discussion has been expanded to include commercial lighting and cable distribution systems.

The eighth edition includes the same subjects covered in the seventh edition, updating them when necessary. It also includes the new subjects discussed earlier, and adds materials and systems that did not exist when earlier editions were published.

Dual measurements in the metric system have been included throughout the book. To avoid changing copyrighted figures, however, metric measure has not been added to them. Instead, Figure 25.7-1 has been added to list all metric conversions that were not made in these figures.

Each chapter has a section called, "Additional Reading," that lists sources of additional information about the subjects discussed in that chapter.

Each chapter also contains a section entitled, "Acknowledgments and References" that lists the name and Web site address of the sources of data in that chapter.

The eighth edition also includes an extensive glossary of terms used in the book.

# A Tribute

For greater market recognition, this edition's title has been changed, giving greater prominence to my name. I am therefore prompted to give much deserved credit to my several collaborators, but especially the two original co-authors, John Schmidt and Walter Lewis, without whose vision, perseverance and expertise this book would not exist. In 1960 John and Walter sold the idea of the book to Norman Strunk, EVP, of the U.S. League of Savings Institutions, worked out the book's format and drafted the early sections of the work. The intent was to educate lending officers in saving and loan associations to encourage suitable construction quality based on knowledge and understanding, not blind reliance on the Federal Housing Administration Minimum Property Standards. I joined the team in 1965 and was privileged to have the mentorship and guidance of my colleagues as I gradually took over the tasks of writing and editing. Van Nostrand Reinhold acquired the book in 1994, and was fortunate to attract Leslie Simmons to author the 6th Edition. Now as the sole author, Mr. Simmons' careful research and extensive experience have widened the book's scope and greatly increased its appeal.

Harold B. Olin, FAIA

# Acknowledgments

It takes the help of many people and organizations to produce a book such as this. These include those who prepared previous editions. First and foremost of these are the original authors, Harold B. Olin, John L. Schmidt, and Walter H. Lewis, without whom this book would not exist. The preface for the fifth edition, written by the three authors, acknowledges the contributions of several persons: Research Associate Christina Farnsworth; Architectural Writer Richard Laya; and Production Editor Betsy Pavichevich.

Even though the original authors' material was, in most cases, extensively revised in the sixth edition and has been further revised in the seventh and eighth editions to bring it up-to-date with current materials, methods, and standards, the contribution of these original authors is to be commended.

I am indebted to the professionals, both practicing architects and teachers, who lent their time, energy, and expertise to review the sixth edition manuscript so that it would be more accurate and more useful as a text for students of architecture and allied fields. These include Harold B. Olin, AIA, the original primary editor of previous editions; Daniel W. Halpin, Ph.D., Professor and Head, Division of Construction Engineering and Management, Purdue University; Larry Grosse, Ph.D., Associate Professor, College of Architecture, Texas A&M University; and Terry L. Patterson, AIA, Professor, College of Architecture, University of Oklahoma, whose extensive, well-informed, and detailed comments contributed greatly to making this book as good as it could be. Professor Patterson was also the author of the sixth edi-

tion's *Student Workbook* and *Instructor's Manual.*

To determine the needs of the readers of *Construction: Principles, Materials, and Methods* for development of the seventh edition, the sixth edition was reviewed by then current users, including teaching professionals and architects in practice. Their comments were of immeasurable help in developing the seventh edition. Among them were the aforementioned Harold B. Olin and Terry Patterson; and Atilla Lawrence, University of Nevada, Department of Architecture; Dana S. Mosher, Professor of Building Construction Technology, NHCTC/ Manchester; Robert O. Segner, Jr., Professor of Construction Science, Department of Construction Science, College of Architecture, Texas A&M University; Khalid T. Al-Hamdouni, Rogers University School of Engineering; Maryrose McGowan, AIA, IIDA; and Tim Kirby, consultant to the construction industry.

I am also indebted to the professionals, both practicing architects and teachers, who lent their time, energy, and expertise to review the manuscript for the seventh edition so that it would be more accurate and more useful as a text for students of architecture and allied fields, and as a reference for professionals in practice. Their extensive, well-informed, and detailed comments contributed greatly to making this book a success. They include Harold B. Olin, William C. Mason, Owens Community College, Toledo, OH; Linda Swoboda, University of Nebraska; John Lebduska, Hope, NJ; Harold Tepper, New Jersey Institute of Technology; Felix Rospond, West Orange, NJ; and Anthony Abbate, Florida Atlantic University.

# Acknowledgments for the Eighth Edition

To determine the needs of the readers of *Olin's Construction: Principles, Materials, and Methods* for development of the eighth edition, the seventh edition was reviewed by current users, including teaching professionals and architects in practice. Their comments were of immeasurable help in developing the seventh edition. Among them were Thomas Mills, Virginia Technological Institute and State University; Anthony Ray, R.A. CSI, CDT, Valencia Community College West Campus, Orlando, FL; Terry L. Patterson of University of Oklahoma; Wen Andrews of J. Sargent Reynolds Community College; Richard Clarke of Georgia Southern University; Rives Taylor of the University of Houston; David Gunderson of Colorado State University; T. Michael Toole of Bucknell University; Tim Brockman of Bemidji State University; and James Dahlin of the University of Nebraska.

I gratefully acknowledge the contributions of the many manufacturers, trade and professional associations, standards-setting bodies, government agencies, periodicals, book publishers, and individuals who provided valuable research material and illustrations, as well as text reviews and comments. Without their help, this book would not have been possible. The names of most of them are listed in the "Acknowledgments and References" sections at the end of the chapters.

I would also like to thank my wife, Nancy, whose help and support have been invaluable.

Finally, I would like to express my thanks to the professionals at John Wiley & Sons, Inc., who lent their talents to this work. They include Vice President and Publisher Amanda Miller; Acquisitions Editor Paul Drougas; Assistant Development Editor Lauren LaFrance; Editorial Assistant Raheli Millman; Senior Production Editor Donna Conte, who ably guided the book through production; and Matrix Publishing Services, which typeset the book.

I would also like to thank Judith Joseph, who initially contacted me about revising the book. My sincere thanks to all those, both named and unnamed here, who participated in making this work better than it might otherwise have been.

H. Leslie Simmons, RA, CSI

# Disclaimer

The information in this book was derived from data published by trade associations, standards-setting organizations, manufacturers, and government organizations, and from statements made to the author by their representatives; model codes; and related books and periodicals. The author and publisher have exercised their best judgment in selecting data to be presented, have reported the recommendations of the sources consulted in good faith, and have made every reasonable effort to make the data presented accurate and authoritative. However, neither the author nor the publisher warrants the accuracy or completeness of the data nor assumes liability for their fitness for a particular purpose. Users are expected to apply their own professional knowledge and experience or consult with someone who has such knowledge and experience when using the data contained in this book. Users are also expected to consult the original sources of the data, obtain additional and updated information as needed, and seek expert advice when appropriate.

A few manufacturers and products are mentioned in this book. Such mention is intended merely to indicate the existence of such products and manufacturers, not to imply endorsement by the author or the publisher of the mentioned manufacturer or product. Nor is there any implied endorsement of other products made by a mentioned manufacturer, or any statement made by a mentioned manufacturer or associated in any way with any product, its accompanying literature, or advertising copy.

Literature and standards produced by various manufacturers, associations, and other organizations are mentioned as well. This does not imply that the mentioned item is the only available data or standard, or the best of its kind, or is the legal requirement or accepted standard in a particular locality. The author and publisher expect the reader to seek out other manufacturers, appropriate associations, and other organizations to ascertain whether they have similar literature, or will make similar data available, and to determine whether the data presented conform with code and other legal requirements in a particular locale.

# CHAPTER 1

# Design and Contracting Requirements

Introduction

Applicable *MasterFormat*™ Sections

Building Design

Industry Standards

Codes

Barrier-Free Design

Sustainable Building Design

Construction Documents

Bidding and Negotiation

Construction Contract Administration

Construction Management

Additional Reading

Acknowledgments and References

# Introduction

Many factors influence an architect's work related to building design and construction contract administration. In addition to architectural design, an architect must be aware of and conversant in site, structural, mechanical, and electrical design. He or she must also be aware of the legal constraints, such as codes, laws, and regulations, and of the many industry standards that influence design and construction. An architect must also be knowledgeable and conversant in the production of construction documents and must understand the means and methods used in constructing buildings. He or she must understand the construction process and be able to render an architect's services during the construction phase of a building project regardless of the construction contract type or employment by the owner of a construction manager. He or she must understand the financial constraints on building construction and be able to design within those constraints. And in all of these, an architect must not be just a jack-of-all-trades; he or she must be a master of them all.

This chapter covers facets of the building design and construction process that a professional must understand to be able to carry out an architect's responsibilities in the design and construction of buildings. The chapter also addresses the function of a construction manager in the construction

process and the architect's relationship to a construction manager. Chapters 2 through 23 address construction materials and methods of design and construction of which an architect must be knowledgeable. Chapter 24 addresses the fundamental properties of materials. Chapter 25 describes the metric system of measurement.

The first five parts of this chapter discuss some of the many factors affecting building design.

Sections 1.6, 1.7, and 1.8 discuss the services architects provide related to a building construction project. The American Institute of Architects (AIA) has divided an architect's services into the categories *basic* and *additional*.

*Basic* services are those included in standard services contracts developed by AIA and included in the architect's basic fee for services.

*Additional* services are optional and are performed only when agreed to by the architect and the owner, with additional compensation to the architect.

Following the flow of a project from conception to the completion of the warranty period (one year after construction completion), an architect's services can be broken down into *predesign* services, *design* services, *construction* services, *postconstruction* services, and *supplemental* services.

Predesign services are *additional* services. They include such acts as programming, existing facilities studies, project budgeting, and site analysis.

*Basic* services include *design* and *construction* services. Design services are further broken down into *schematic* design, *design development* (a further refinement of schematic design documents), and *construction documents* services.

*Construction* services include services performed during the *bidding and negotiation* phase and those performed during the *construction contract administration* phase.

*Postconstruction* services are additional services performed after substantial completion of the building. They include such acts as maintenance and operational programming, record drawings, start-up assistance, and warranty review.

*Supplemental* services are additional services. They include such items as renderings, models, life cycle cost analysis, quantity surveys, graphic design, and many others.

Section 1.9 addresses the function of a construction manager related to a construction project and a construction manager's relationships to the owner, the architect, and the contractor.

# Applicable *MasterFormat* Sections

The following *MasterFormat* 2004 Level 2 sections are applicable to this chapter.

# 1.1 Building Design

An architect's first and primary contractual responsibility related to building construction is design. Building design requires training, experience, an aesthetic sense, and an understanding of certain basic principles. Among these principles are (1) the objectives good design should strive for, (2) an architect's responsibilities related to design, (3) basic building use and shape types, and (4) available construction systems and methods.

## 1.1.1 DESIGN OBJECTIVES

An architect's primary design objective should be to produce buildings that serve their intended purpose and that permit the activities that take place in them to proceed with appropriate dispatch and ease. They should be efficient in their use and operation. In addition, commercial buildings should be capable of producing a profit.

An architect's buildings should be of good-quality construction, and should be able to be built at as low a cost as is practicable. An architect's designs should produce individual buildings that are aesthetically pleasing and that do not diminish the beauty of or reduce the quality of the natural environment around them. They should also produce the most practicable conservation of energy and the least practicable degradation of the environment.

### 1.1.1.1 Environmental Considerations

In addition to his or her responsibility to the public as defined by law and ethical considerations, an architect bears a responsibility to protect and maintain the environment. One factor in fulfilling this responsibility is to design buildings for sustainability, as discussed in Section 1.5. But protecting the environment goes far beyond designing green buildings. It also entails consideration of how a building works and how it fits into its environment. This concern must be considered not just for the present, but also throughout the life of the building.

A building should be designed so that it fits within its site and does not overpower the environment. Fitting is accomplished by placing and orienting building elements to take the best advantage of sun angles, site features, and prevailing weather patterns such as wind. Where practicable, earth-sheltered design and passive solar design can be used to reduce heating and cooling loads on a building. Refer to Section 17.6 for a discussion of solar heating and cooling.

Whenever possible, buildings should be sited so as to preserve as much of the existing vegetation and land features as is practicable. Means should also be provided to assure the protection of existing preservable vegetation and land features, such as wetlands and waterways, from damage during construction.

Where possible, construction waste should be reduced to the smallest amount possible, which can be aided by selecting materials that have little waste and by employing off-site prefabrication of building elements. Debris and waste should be recycled where possible, preferably on the construction site.

Most jurisdictions require the prevention of storm sewer water and groundwater pollution by prescribed methods of control. This should be accomplished whether or not it is required by law or code.

Designing buildings for energy conservation is discussed in Chapters 17 and 18; for sound reduction, refer to Chapter 13.

### 1.1.1.2 Occupant Health Considerations

It is important to consider the health of the occupants when designing a building. It is necessary, for instance, to design heating, cooling, and ventilating systems that will not introduce toxic gases into the building. Ductwork should be easy to clean internally and maintain. Design of ventilating systems is discussed in Chapter 17.

In creating a healthy environment in a building, it is important to select non-toxic building materials. Emanation of toxic compounds and gases from building materials can create a condition known as "sick building syndrome" and can cause serious conditions such as allergies and even cancer. In addition, products that are known to have high rates of outgassing of volatile organic compounds (VOCs) should be avoided. Even the materials necessary to clean some building products can lead to indoor air pollution.

Another factor in maintaining healthy building is the control of moisture intrusion and condensation, which can lead to the growth of mold and mildew. Conditions that create condensation problems are discussed in Chapter 17. Control of moisture and free water penetration are discussed in Chapter 7.

## 1.1.2 THE BUILDING CONSTRUCTION TEAM

Many organizations and individuals must work together to produce buildings. These include owners, design professionals, constructors, members of supporting professions and industries, and sometimes construction managers.

*Owners* are the architect's clients. They are not necessarily the users of a building, but they conceive, finance, and usually own the project.

A *design professional* is a person or organization that designs a construction project. The *prime design professional* is the one who is hired by the owner to lead the design team. In building construction projects, the prime professional is usually an architect. An engineer may be the prime design professional on some primarily engineering projects—for example, in the construction of bridges or in the major renovation of an existing heat-

ing, ventilating, and air conditioning system. In this section, it is assumed that the project being considered is a building and that the prime design professional is an architect. In other project types, the chores here delineated as the responsibility of the architect may fall to an engineer as the prime design professional for a particular project.

It is ordinarily the architect's responsibility to (1) determine the legal, financial, and other constraints on project design, (2) program and design the project, (3) produce contract documents, (4) provide professional services during the bidding or negotiation phase, and (5) provide construction contract administration services. For a residence or other small building, an architect may carry out these functions alone. Larger and more complicated buildings often present design problems that are beyond the expertise of most architects. For these more complicated building construction projects, an architect functions as a member, usually as the leader, of a team of design professionals that includes structural, mechanical, civil, and electrical engineers, and interior designers, who function as consultants to the architect.

The architect and each of the architect's consultants will design, produce construction documents, and provide construction contract administration for the one portion of a building's components that falls within his or her field of expertise. The architect coordi-

nates the activities of all design team members.

The construction process often also requires input from a second group of design professionals working as consultants, either to the owner directly or to one of the team members. These other professionals include, but are not limited to, those with special knowledge about schools, hospitals, food service facilities, laboratories, industrial complexes, computer systems, communication systems, furniture, specialized equipment, and many other components. The architect usually coordinates the activities of these other professionals.

*Constructors*, also called *builders*, are usually a group of organizations that together erect construction projects. They also provide most of the training of construction workers (Fig. 1.1-1). They consist of many types of *contractors*, including, but not limited to, *general contractors*, who oversee the work of, and usually hire, the others; and specialty, or *trade*, contractors, such as those who provide sitework, excavation, concrete, masonry, steel, carpentry, casework, moisture protection, doors, windows, finishes, specialties, equipment, and conveying, plumbing, electrical, and mechanical systems. Supporting these contractors are suppliers, who provide construction equipment, such as cranes,

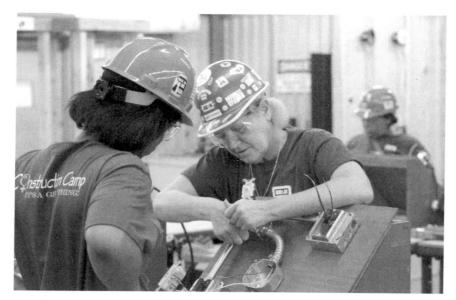

FIGURE 1.1-1 An instructor in a construction firm works with an electrical technician student. (Courtesy BE&K.)

and product suppliers who furnish the materials, products, systems, and equipment that go into a building.

Supporting professions and industries include, but are not limited to, construction managers (see Section 1.9); legal professionals; accountants; lenders and investors, who provide construction money and long-term loans that permit construction projects to be erected; insurance providers; testing and research agencies, which develop new products and test existing ones; and regulators, including code and law writers and enforcers, who control health and safety issues, aesthetics, environmental issues, zoning, utilities, financial institutions, and design professionals' licensing and practices.

### 1.1.3 BUILDING USE TYPES

Construction projects can be identified by their use: residential, commercial (stores, office buildings, etc.), institutional (hospitals, schools, jails, etc.), industrial (manufacturing, laboratories, etc.), and nonbuilding types (bridges, towers, etc.). From this point forward, this chapter addresses only those building construction projects for which an architect is the primary professional. In such projects, it is the architect's job to determine the design requirements specific to each use. For example, there is little resemblance between the requirements for a single-family residence and those of a hospital. There may be major differences even within a group. There are great differences, for example, between the requirements for a single-family residence and a high-rise apartment building. A local jail will probably bear little resemblance to a federal prison.

Some buildings are designed for *common use*, meaning that they have more than one use type in the same structure.

Street-front stores may have residential or office spaces above them. High-rise buildings may house commercial uses on the lower floors, office uses on intermediate floors, and apartments on the upper floors.

### 1.1.4 BUILDING SHAPE TYPES

Buildings take many forms and shapes, depending on their use, the materials used to build them, the needs and desires of the owner, the construction budget, the building's potential operating costs, and the designer's preferences. Buildings other than single-family residences and townhouses are so varied in size and shape as to make simplification of their types difficult (Fig. 1.1-2). However, some basic types and construction methods can be identified (Fig. 1.1-3).

The simplest building is a one-story, single-span, slab-on-grade structure with

**FIGURE 1.1-3** Basic building types for other than single-family residential buildings: (a) one-story; (b) one-story with basement; (c) one-story with multiple framing bays; (d–f) typical roof shapes; and (g) multistory. (Drawings by HLS.)

a flat roof (Fig. 1.1-3a). Similar buildings with basements are also commonly built (Fig. 1.1-3b). Single-story structures with more than one structural span (Fig. 1.1-3c), in which one or more intermediate rows of walls or columns supports the roof structure, enclose more space per unit of exterior wall cladding than do smaller buildings.

Single-story buildings may also have full or partial basements. The structural systems in buildings of this type may be concrete, masonry, steel-framed, or woodframed bearing walls with steel, concrete, or wood roof framing systems; steel, concrete, or wood interior and exterior columns with steel, concrete, or wood roof framing; or a combination of these systems. Foundations are usually poured concrete, but treated wood foundations are sometimes used (see Section 6.5). The roof of a single-story building may be either flat or any of a wide variety of shapes (Figs. 1.1-3d–f). Roof decks may be of wood, steel, or concrete. Basements may have either poured concrete or reinforced masonry walls, depending on the level of the earth against the wall and the height and hydrostatic head of adjacent underground water. Floors above basements may be steel-framed with a concrete or wood floor, concrete-framed with a concrete floor, steel-framed with a concrete floor, wood-framed with a wood floor, or a combination of these systems.

The same principles apply to multistory structures (Fig. 1.1-3g). The construction materials and structural systems in multistory structures and the height of such buildings are usually dictated by economic factors, such as land cost, but may be affected by codes and laws that restrict building height, land area coverage, or the materials that may be used. Fire codes, for example, may restrict the types of construction systems and the materials that may be left exposed. Many fire codes do not permit wood construction or the exposure of wood finishes on the exterior of buildings in certain locations.

Multistory buildings require less roof surface than single-story buildings with the same floor area. This results in a savings in the cost of roofing materials. In addition, multistory buildings require less land per unit of usable space. Because of their higher ratio of interior space to building shell area, they are also generally more energy efficient than single-story buildings. Except in rare instances, these advantages increase with the number of stories. The lower costs are somewhat offset by the increased costs for maintenance of the exterior surfaces of multistory buildings, the relatively high costs of materials that can be used there, and the increased cost of construction associated with moving materials to high levels and working with them far above ground level.

Low-rise multistory buildings may be of steel or concrete construction or a combination of these. Some even have masonry bearing walls. Steel columns and concrete floors are common. Foundations are usually poured concrete spread footings, although poor soil conditions sometimes dictate the use of piles or caissons.

High-rise buildings are usually framed in steel, with thin concrete floor slabs, because concrete structures of great height have heavier and larger framing members than steel structures, which reduces the amount of usable space and increases the cost of construction. Some recent very high buildings have been designed as a series of steel shells or tubes that extend for the entire height of the building; others have been designed using the same principles as tall radio and television towers. Foundations are either poured concrete footings or pads, piles, or caissons, depending on the soil conditions and the size and load imposed on the soil by the building.

Sometimes the desire to create a statement for ego-enhancing or advertising purposes affects the size, height, and appearance of a building. For example, a corporation may wish to use its headquarters building as a symbol or may just want to own the tallest, largest, or most spectacular building in town.

Multistory buildings need elevators or escalators to make their use practicable. In addition, in most types of uses, federal accessibility laws and rules make elevators or wheelchair lifts a legal requirement in every building that is not inherently accessible to the handicapped (see Section 1.4), which, of course, includes every multistory building. The additional cost of this vertical transportation must be considered in deciding whether to construct a multistory building.

The basic building types used in single-family and townhouse construction are easier to define. Figure 1.1-4 shows some common types. Most of these types are also used for buildings other than single-family homes or townhouses, however, so they should not be thought of for only these restricted applications. The most prevalent of these is a one-story building (Fig. 1.1-4a), because this type provides the most size and shape variations. These buildings may or may not contain a basement. Their roofs may be sloped, as shown, or flat.

One-and-one-half-story buildings, with or without basements (Fig. 1.1-44b), are sometimes used for housing. They offer more living space than single-story buildings with a minimum of additional cost. The second-floor space varies with the building size and roof slope. Light, ventilation, and a view can be provided by dormers.

One-and-one-half-story buildings are seldom built for other types of uses because their inherently small second-floor rooms, with their sloped ceilings, while adequate for sleeping rooms, often do not make satisfactory work spaces.

Two-story (Fig. 1.1-4c) or taller buildings, with or without basements, provide the maximum usable area at relatively low cost. Two- and three-story single-family houses and townhouses are common. These types of buildings can reduce construction costs, depending on the value of the land. When they must be accessible to the handicapped (see Section 1.4), buildings of more than one story require elevators, as described earlier for multistory buildings. Bilevel buildings (Fig. 1.1-4d) are well suited for single-family houses, townhouses, or small commercial buildings in hillside locations. They provide habitable space at both grade levels when connected with full flights of stairs. In certain types of uses, accessibility restrictions may require that elevators be included. This configuration can also be used for two different occupancies, such as an apartment on one level and a small store on the other. In this case, both levels can be easily made independently accessible to the handicapped. Roofs may be either sloped, as shown, or flat.

Split-level (Fig. 1.1-4e) and bilevel/split-entry (Fig. 1.1-4f) buildings are used mostly for single-family houses and townhouses. They are infrequently used for other purposes because of the difficulty of making them accessible to the handicapped. Split-level houses offer distinct separation of functions, either on three levels or four, including a base-

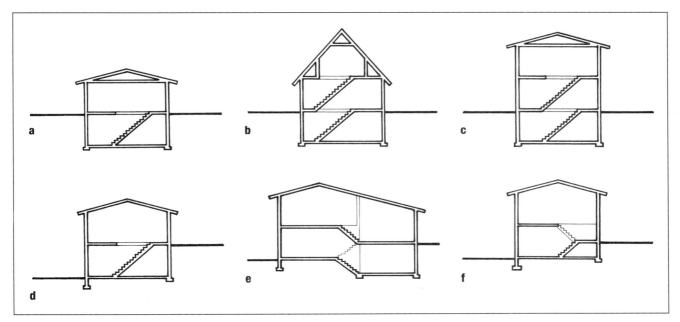

**FIGURE 1.1-4** Basic single-family residential building types: (a) single-story; (b) one-and-one-half-story; (c) two-story or higher; (d) bilevel; (e) split-level; and (f) bilevel/split-entry. (Drawings by HLS.)

ment. These are best suited for sloping lots. They offer numerous design possibilities but can have awkward proportions if not carefully designed.

Bilevel/split-entry buildings are also best suited to sloping lots. They are characterized by a split foyer between two full living levels. This configuration can provide either a sunken two-story (or more) house without a basement or a raised one-story (or more) house with a finished basement.

### 1.1.5 CONSTRUCTION SYSTEMS AND METHODS

The selection of construction methods and systems that produce the basic building types shown in Figures 1.1-3 and 1.1-4 is usually governed by three criteria: functional requirements, cost, and the desired appearance.

These basic criteria may require a consideration of climate, site topography, initial costs, maintenance costs, building codes, zoning ordinances or other laws, availability of materials and labor, builder resourcefulness and size, owner taste, local custom, and other factors.

The selection of methods and systems is further complicated by the thousands of materials, products, and construction system choices available, many of which are interdependent. Sometimes the rela-

tionship of these building elements to each other will create situations in which the construction method or system is the major influence on a building's design. For example, the selection of a dome as the means to roof a coliseum may dictate that the shape of the building be circular or near-circular. Conversely, design requirements may dictate the framing system. A dome, for example, may be a poor choice for roofing a theater because of the inherent acoustical difficulties of a dome and because a circular building may not be preferable for the kind of theater that is desired. The cost and availability of very large laminated wood (gluelam) structural elements, as compared with smaller gluelam units, may influence the width or even the shape of a church, or the way in which the gluelam units are fitted together to make a roof structure. The permissible span of the available wood decking may further influence the spacing of these gluelam units or the design of the roof structure and the placement of purlins.

Although there have been experiments with a few revolutionary construction systems since World War II, most new homes and many small commercial and institutional buildings in the United States are still built using conventional light-wood-platform framing (see Chapter 6), often with wood-truss-framed roofs. In many areas, the use of preassembled compo-

nents, such as those discussed in Chapter 6, is common. In the future, advanced industrialization techniques using new materials and methods may offer new construction forms far different from those typical today.

Other basic systems in use today include wood-post-and-beam framing and wood-pole construction (see Chapter 6); masonry bearing-wall construction (see Chapter 4), sometimes with concrete floors (see Chapter 3), often supported on metal framing or bar joists; concrete-framed construction; and structural steel-framed construction and light-gauge metal framing of walls and roofs (see Chapter 5).

In small construction, conventional wood framing still offers many advantages. As a complete construction system, it still is one of the most economical ways to build. The ease of working and fastening wood together with simple tools provides flexibility, which permits job changes without extensive re-engineering. Wood framing is still the basis for most building codes and labor practices and will probably remain so for some time to come. Conventional framing is adaptable to site fabrication by the smallest builder handling each member piece by piece, as well as to off-site fabrication of individual pieces into larger preassembled components that require additional manpower or machinery for erection.

### 1.1.6 THE FUTURE

Further industrialization, using more and larger prefinished and prefabricated components, appears essential to help offset the rising costs of land, labor, and materials. Off-site fabrication permits maximum utilization of labor and materials under factory-controlled conditions with little loss in on-site time owing to bad weather. Efficiency may be increased with the use of power tools and machinery; volume purchasing of materials and stockpiling of finished parts is possible; greater convenience for workers and better protection for finished materials is provided; and site erection of components can usually be accomplished more economically and in less time by semiskilled or even unskilled labor.

To save costs, mechanical components for small buildings have been developed that combine a furnace, air conditioner, water heater, and electric power panel in one package. Larger mechanical components include completely furnished kitchens and bathrooms. The concept of prefinishing complete rooms has been extended to prefabricating as much as half of a small building, such as a house, so that upon setting and joining two halves, an entire building is completed. Future developments may include assembling an entire building and completely finishing it prior to site placement.

Some future building construction methods will be highly sophisticated and closely integrated systems. For instance, integrated floor and ceiling systems available for use in commercial construction include structure, lighting, acoustical control, heating, cooling, and air distribution in a single system.

Components should be capable of satisfying varying design requirements; should permit simple modifications in the field in case of errors; and should be sized for ease of shipping, storage, and assembly. As component size increases, design and construction problems increase and design flexibility is lessened. The dimensions of large units are restricted to what can be transported physically and legally over the highways, and larger components usually require more manpower and larger erection equipment at the site.

Accordingly, the design, engineering, or selection of preassembled components requires judgments between size and flexibility. The most useful systems will combine the advantages of fully standardized factory-built modular units, which capitalize on the inherent savings resulting from repetitive production, and those that offer the design advantages of custom fabrication in the field.

Unfortunately, there are also certain disadvantages associated with prefabrication that have so far limited its use. For example, to be profitable, large components require a large market willing to accept a standardized design, which has not been forthcoming. In comparison, because they can be adapted to many building sizes, shapes, and designs, there is a huge market for prefabricated roof trusses, making them relatively inexpensive and readily available.

There are also potential disputes among construction trade unions and between trade unions and manufacturers about the right to do certain work. Union-member plumbers, for example, are not likely to be pleased when the plumbing piping and fixtures in a prefabricated building are installed by nonunion factory workers.

Other problems with prefabrication include consumer and builder resistance to prefabricated structures associated with the preconceived notion that prefabricated buildings will be shabbily constructed and look like house trailers. In fact, while the construction may actually be superior to that of stick-built units, the appearance possibilities are somewhat limited, and design variety is difficult to achieve.

A final deterrent to prefabrication is the lack of consistency among building codes. These differences can require slight, but costly, modifications in prefabricated units to comply with the codes in different jurisdictions.

Construction practices vary widely across the country. Conventional methods and systems are not usually engineered but are based on a combination of long-established custom, rules of thumb, and arbitrary building code requirements. These practices have resulted in buildings that have usually provided satisfactory performance over the years. However, when these practices are unduly conservative, as is often the case, they foster excessive waste and therefore higher cost. New performance standards for methods and systems are constantly being established. These design criteria, based on laboratory and field testing, can materially reduce overdesign, waste, and cost. The members of the building construction industry should continue to encourage the development of criteria based on performance rather than the requirement that a specific product or system be used. They should also take steps to further the design of methods and systems based on these performance criteria. In addition, the industry needs to find the means for quicker acceptance of innovations in the marketplace as they occur.

## 1.2 Industry Standards

The building construction industry is made up of so many diverse interest groups that it has not been possible to develop a single comprehensive set of criteria or standards acceptable to all concerned. In addition, groups that are more intimately involved in a product or construction system are generally best qualified to establish standards for them. The result is myriad organizations that establish standards.

Before beginning a discussion of construction industry standards, it is necessary to clarify some terms. There is much confusion in the industry about the use of the terms *specifications*, *standards*, and *codes*. Unfortunately, the three are often erroneously used interchangeably, which can lead to some confusion. Specifications are discussed in Section 1.6, codes in Section 1.3. In the sense that they are a detailed, precise description of a product or practice, some construction industry *standards* could be called *specifications*, and some are so called by their producers. Some of the manufacturers' data that designers, builders, and owners rely on are also specifications in the dictionary sense, and those data are sometimes called *specifications* by the manufacturers. But calling either of them *specifications* sometimes leads to confusion. Therefore, this book refers to construction industry standards as *standards* and manufacturers' data as *product descriptions* or *product literature*. Unless specifically modified in the text, the term *specification* is used in its narrow construction industry sense, as defined in Section 1.6.

## 1.2.1 TYPES, OBJECTIVES, AND USES OF STANDARDS

When selecting materials or determining the suitability of materials and methods, specifiers and builders refer to a variety of *industry standards*. These are also sometimes called *reference standards*.

### 1.2.1.1 Types of Standards

Some standards result from the efforts of manufacturers, professionals, and tradespeople to simplify and increase the efficiency of their work or to ensure a minimum level of quality. Other standards are the work of governmental agencies and other groups interested in establishing minimum levels of safety and performance. Therefore, standards take a variety of forms, depending on their source and purpose.

#### MATERIAL STANDARDS

Standards that define the properties of a material are called *material standards.* For example, these include standards for extruded aluminum bars, rods, shapes, and tubes or for a particular type of steel item. They usually specify the constituents of a material, its physical properties, and its performance under stress and varying climatic conditions.

#### PRODUCT STANDARDS

The requirements of a specific product, such as aluminum windows, are defined by *product standards*. These often define terms, classify constituent materials, and state acceptable thicknesses, lengths, and widths. They may also spell out the acceptable methods of joining separate materials, of fabricating various parts of a product, or of assembling systems.

#### DESIGN STANDARDS

Design standards define the requirements for sound design using a particular material, product, or system. They are published by such organizations as the American Concrete Institute (ACI), the Architectural Woodwork Institute (AWI), and the U.S. Department of Housing and Urban Development (HUD).

#### WORKMANSHIP STANDARDS

Workmanship standards are standards for installing materials, products, and systems. ASTM International (formerly the American Society for Testing and Materials) (ASTM) produces many of these.

#### TEST METHOD STANDARDS

Test method standards spell out acceptable criteria for testing materials and systems. Again, ASTM is a prime producer of standards of this type.

### 1.2.1.2 Objectives of Standards

Construction standards have two basic objectives: (1) to establish levels of quality that may be recognized by a user, specifier, approver, or buyer of a material, product, or system and (2) to standardize or simplify such variables as dimensions, varieties, and other characteristics of specific products so as to minimize variations in manufacture and use.

### 1.2.1.3 Uses of Standards

Construction standards are used by manufacturers, specifiers, consumers, communities, and others. Standards may be used and referred to either separately or within collections, such as in the HUD *Minimum Property Standards for Housing*. Standards may also be incorporated into municipal or state building codes by inclusion or reference. Such larger works may have broader objectives than to act only as construction standards. Codes, for example, are concerned basically with minimum acceptable standards of public health, safety, and welfare; the HUD *Minimum Property Standards for Housing* establishes minimum requirements of design and construction for the insurance of mortgage loans. Standards are often incorporated by reference into construction document specifications to help establish requirements for materials, equipment, finishes, and workmanship for a particular construction project.

## 1.2.2 PRIVATE INDUSTRY STANDARDS

Standards are established by two kinds of private industry organizations: trade associations and standards-setting and testing agencies.

### 1.2.2.1 Trade Associations

Manufacturers, tradespeople, and suppliers, working through *trade associations*, prepare most private industry standards. Some trade associations are called *societies* or *institutes*. A *trade association* is an organization of individual manufacturers or businesses engaged in the production, supply, or installation of materials or services of a similar nature. A basic function of a trade association is to promote the interests of its membership. Those interests generally are best served by the proper use of the groups' materials, products, and services. The proper use of building materials and methods is guided largely by the development of suitable levels of quality for their manufacture, use, and installation. Some of the most important activities of trade associations are directed toward research into the use and improvement of materials and methods, and toward formulating performance standards. In many instances, trade associations also sponsor programs of certification in which labels, seals, or other identifiable marks are placed on materials or products manufactured to particular standards.

#### STANDARDS-SETTING TRADE ASSOCIATIONS

Many trade associations engage in developing standards for materials and products their members manufacture and services they perform. APA–The Engineered Wood Association, is an example of a trade association active in developing standards. This association is supported by a large membership of manufacturers producing softwood plywood, oriented strand board, gluelam, I-joists, siding, laminated veneer lumber, and other engineered wood products. The production of various types and grades of these products requires the selection and classification of wood veneers and other components according to strength and appearance. For example, producing plywood from various veneers demands careful manufacturing control of such factors as moisture content and adhesive type. The completed product, properly assembled, bonded, and finished, must conform to a body of material standards, manufacturing procedures, and performance testing standards.

Engineered wood products manufacturers, through their own research and combined efforts within the APA, financially and technically support the research and development of criteria that the association formulates into industry standards. The members of the association then agree to produce products that conform to these standards.

## TRADE ASSOCIATIONS STANDARDS AS A BASE FOR OTHER STANDARDS

The standards developed by trade associations often are adopted by or used as a base from which other groups, such as the American National Standards Institute (ANSI), may develop standards. For example, industry standards may be promulgated as *ANSI standards*, which are often incorporated into the HUD *Minimum Property Standards for Housing* or building codes.

## CERTIFICATION

Many trade associations and other industry groups provide assurance that established standards have been met by materials and manufactured products. Certification of quality may take the form of *grademarks*, labels, or *seals*. For example, the APA maintains a continuing program of product testing during and after manufacture, with grade markings applied directly to plywood. A grademark is a visible statement that the appropriate APA product standard has been met.

The reliability of certifications issued by manufacturers or their associations varies. Some are excellent. Others are worthless. The most reliable certification is one issued by an *independent testing agency*.

## 1.2.2.2 Standards-Setting and Testing Agencies

In addition to trade associations, organizations have been established whose primary purpose is the setting of standards or the testing of materials and products to ensure that they comply with established standards. When a material or product is compliance tested by its manufacturer, its trade association, or an independent testing agency, a *testing standard* produced by one of the agencies discussed in this subsection is often used as the standard for the testing procedure.

## ASTM INTERNATIONAL

ASTM International (ASTM) is an international, privately financed, nonprofit, technical, scientific, and educational society. The objectives of the society are "the promotion of knowledge of the materials of engineering, and the standardization of specifications and methods of testing." ASTM membership consists of individual engineers, scientists, and educators, as well as organizational members, including companies, government

agencies, and universities. Technical committees formulate and recommend ASTM standards covering many types of materials; a number of administrative committees deal with publications, research, testing, consumer standards, and other activities. The society is supported mainly by membership dues, with some income from the sale of publications.

ASTM started in 1898 as the American Section of the International Society for Testing Materials. It was incorporated in 1902 and became the American Society for Testing Materials. In 1961, the name was changed to the American Society for Testing and Materials to emphasize its interest in basic information about materials. The name was recently again changed and is now ASTM International.

Two general categories of information and publications are available from ASTM: (1) *ASTM standards*, which include definitions of terms, materials standards, workmanship standards, and methods of test standards used throughout the industry, and (2) data dealing with research and testing of materials, including monthly and quarterly publications and technical publications that cover symposia and collections of data.

ASTM standards are designated by the initials ASTM, followed by a code number and the year of last revision. For example, ASTM C55-03 refers to ASTM Document C55, "Standard Specification for Concrete Brick," as last revised in 2003.

## AMERICAN NATIONAL STANDARDS INSTITUTE

American National Standards Institute (ANSI) is the name adopted by the United States of America Standards Institute (USASI) in October 1969. USASI was created in 1966 by the complete reorganization of the earlier standards organization, the American Standards Association (ASA). ASA was founded during World War I to prevent duplication and waste in war production. In 1918 five leading American engineering societies, including the American Society of Mechanical Engineers (ASME), the American Society of Civil Engineers (ASCE), and ASTM, and three departments of the federal government, Commerce, War, and Navy, formed the American Engineering Standards Committee to coordinate the development of national

standards. This committee was reorganized in 1928 into the American Standards Association, which was later incorporated into USASI and subsequently renamed ANSI.

More than 3000 American National Standards have been developed and approved under ANSI procedures. These standards apply in the fields of engineering, industry, safety, and consumer goods.

An American National Standard is designated by the code number and date of the last revision; for example, ICC/ANSI A117.1-2003, "Accessible and Usable Buildings and Facilities." Unlike ASTM, ANSI does not formulate its own standards or provide testing services. Instead, one part of ANSI, composed of national trade, professional, and scientific associations, establishes and maintains procedures for the approval of standards developed by other associations, agencies, or groups as American National Standards. In this way, a standard developed by a trade association, such as the American Architectural Manufacturers Association (AAMA), can become an American National Standard.

A second part of ANSI consists of representatives of industrial firms, labor, and government. The two parts work closely together to recommend areas of standardization deemed essential and to review standards.

ANSI is privately financed by voluntary membership dues and from the sale of the published American National Standards. These national standards are available for voluntary use and often are incorporated in regulations and codes.

## UNDERWRITERS LABORATORIES, INC.

Underwriters Laboratories, Inc. (UL) is chartered as a nonprofit organization to establish, maintain, and operate laboratories for the examination and testing of devices, systems, and materials. The stated objectives of UL are (1) to determine the relation of various devices, systems, and materials to life and property and (2) to ascertain, define, and publish standards, classifications, and specifications for materials, devices, products, equipment, constructions, methods, and systems affecting hazards to life and property.

UL, formed in 1894, was originally subsidized by stock insurance companies. Before the turn of the twentieth century, as new electrical devices and products came rapidly into the market, it became necessary to test and inspect them to ensure public safety. The National Board of Fire Underwriters (now the American Insurance Association) organized and sponsored UL to meet this demand.

UL became self-supporting in about 1916. To sustain its testing program, UL contracts with a product submitter for testing, reporting, and listing of devices, systems, or materials on a time and material basis. The cost of the inspection service is provided for either by an annual fee or by service charges for labels, depending on the type of service. Materials and products carrying UL labels and certificates must meet published standards of performance and manufacture and are subjected to UL inspection during manufacture.

Although primarily interested in public safety, UL's policy is to list and label only products that perform their intended function. If a product does not perform with reasonable efficiency, even though it may be perfectly safe, it does not qualify for a UL label. UL standards are designated by the initials UL, followed by a code number. No date of last revision is indicated by the number. For example, UL 70, "Septic Tanks, Bituminous Coated Metal," was issued in 2001.

## NAHB RESEARCH CENTER

The NAHB Research Center is a wholly owned subsidiary of the National Association of Home Builders. Its objectives are as follows:

1. To conduct and disseminate the results of research and development with respect to homes, apartments, and light commercial structures for the purpose of lowering the cost and improving the quality of buildings constructed by the U.S. homebuilding industry
2. To conduct, for itself or by contract for others, tests and investigations into and on materials, products, systems, and other matters related to the design, construction, or occupancy of homes and other buildings
3. To encourage lower construction costs and improved quality in the design and construction of residential and related structures
4. To provide a system for labeling materials and products, and to grant a seal or certificate of quality or similar device

The NAHB Research Center was founded in 1964 as an expansion of the NAHB Research Institute, which was founded in 1952. The Research Center's activities have included the design and construction of a number of "research houses" that incorporated new methods and materials. The Research Center also developed the successful TAMAP system, which marked the first time that industrial engineering techniques were used to improve productivity in the design and construction of new homes.

The Center has also carried out product, standards, and systems research, development, and evaluation studies for many homebuilding industry manufacturers and associations. The Center is thereby supported by its clients, including the National Association of Home Builders, a number of building industry manufacturers, trade associations, and other organizations. The Center's laboratories include a broad range of facilities that can conduct ASTM standard tests, vibration studies, acoustical measuring, and temperature-humidity control testing.

## NATIONAL FIRE PROTECTION ASSOCIATION

The National Fire Protection Association (NFPA) was organized in 1896 to promote the science and advance the methods of fire protection. NFPA is a nonprofit educational organization that publishes and distributes various publications on fire safety, including model codes, materials standards, and recommended practices. These technical materials, aimed at minimizing losses of life and property by fire, are prepared by NFPA Technical Committees and are adopted at the annual meeting of the Association. All are published as National Fire Codes, a 12-volume compilation of NFPA's official technical material. *The National Electrical Code* is Volume 3.

## THE AMERICAN SOCIETY OF HEATING, REFRIGERATING AND AIR CONDITIONING ENGINEERS

Since 1894, the American Society of Heating, Refrigerating and Air Conditioning Engineers (ASHRAE) and its predecessor societies have pursued their objective of advancing the arts and sciences of heating, ventilating, and air conditioning buildings. ASHRAE conducts an extensive research program, publishes meeting transactions, and establishes standards.

ASHRAE standards are established to assist industry and the public by offering a uniform method of testing for rating purposes, by suggesting safe practices in designing and installing heating, ventilating, and air conditioning equipment and systems, and by providing other information that may serve to guide the industry. The creation of ASHRAE standards is determined by need. Conformance is voluntary.

ASHRAE standards are updated on a 5-year cycle; each title is preceded by a hyphenated number. The digits before the hyphen are the standard's numerical designation; the digits after the hyphen are the year of approval, revision, or update. For example, ASHRAE Standard 90.1-2001 (SI edition)—"Energy Standard for Buildings Except Low-Rise Residential Buildings (IESNA cosponsored; ANSI approved; Continuous Maintenance Standard), SI Edition"—describes a standard of designation 90.1, approved in 2001.

ASHRAE has developed standards not only in the traditional areas of heating, ventilating, and air conditioning equipment, but also on such diverse subjects as fire safety in buildings, energy conservation, solar energy, pollution control, and ozone depletion.

### 1.2.3 FEDERAL GOVERNMENT STANDARDS

Many federal agencies either develop standards themselves or commission their development by other federal agencies or private-sector organizations.

#### 1.2.3.1 Department of Commerce

Manufacturers seek to encourage product acceptance and improve their own efficiency by establishing basic levels of quality for materials and products and by coordinating dimensions, terminology, and other variables such as type and style.

Manufacturers cannot legally agree to establish unreasonable standards that might rule out the success of individual competitors, nor can they engage in price-fixing agreements. Therefore, in recognition of the desirability of certain types of industry-supported standardization, the U.S. Department of Commerce provides for the development of Product Standards.

#### VOLUNTARY PRODUCT STANDARDS

Voluntary Product Standards (PS) are the modern replacement for the older Commercial Standards (CS) and Simplified Practice Recommendations (SPR), previously published by the Department of Commerce.

Voluntary PS are developed by manufacturers, distributors, and users in cooperation with the Standards Services Division of the National Institute of Standards and Technology (NIST). The purpose of a PS may be either (1) to establish standards of practice for sizes, dimensions, varieties, or other characteristics of a specific product or (2) to establish quality criteria, including standard methods of testing, rating, certifying, and labeling of the manufactured products.

The adoption and use of a PS are voluntary. However, when reference to a PS is made in contracts, labels, invoices, or advertising literature, the provisions of the standard are enforceable through usual legal channels as a part of the sales contract.

A PS usually originates with the manufacturing segment of the industry. The sponsors may be manufacturers, distributors, or users of the specific product. One of these three elements of industry (the *proponent*) submits to the Standards Services Division the necessary data to be used as the basis for developing a PS. The Division, by means of assembled conferences, letter referenda, or both, assists the sponsor group in arriving at a tentative standard of practice and thereafter refers it to the other elements of the same industry for approval or for constructive criticism that will be helpful in making necessary adjustments. The regular procedure of the Division ensures continuous servicing of each PS through review and revision whenever, in the opinion of the industry, changing conditions warrant such action.

A Voluntary PS is designated by the letters PS, followed by an identification number and the last two digits of the year of issuance or last revision. For example, PS 1-95 is the PS for "Construction and Industrial Plywood." It was originally issued in 1966 as the first PS. A list of PSs, currently available, a price list, and ordering instructions may be obtained from the Standards Services Division, NIST. Procedures for the development of Voluntary PS are available from the Standards Coordination and Conformity Group of NIST.

### 1.2.3.2 General Services Administration

The General Services Administration (GSA) of the U.S. government develops *Federal Standardization Documents*, including *Federal Specifications*, *Interim Federal Specifications*, and *Federal Standards*, through the cooperation of federal agencies and industry groups. The purpose of these documents is to standardize the variations and quality of materials and products being purchased by governmental agencies. Approximately 5600 Federal and Interim Federal Specifications have been developed by the GSA. The *Index of Federal Specifications, Standards and Commercial Item Descriptions*, which lists those documents alphabetically by title and numerically, may be purchased from the Superintendent of Documents, U.S. Government Printing Office, and is also available at the GSA Web site.

#### FEDERAL SPECIFICATIONS

In the pure sense of the dictionary definition, Federal Specifications can be called specifications. According to the definitions we are using in this book, however, and according to the standard practices of the construction industry, even on projects for the government, they are actually used as standards. They are not permitted, for example, as a part of a *project manual* for a building construction project, except by reference. It is not possible to enter Federal Specifications intact into a project manual, and editing them for this purpose is neither permissible nor desirable. However, since the GSA calls them specifications and this terminology is generally accepted in the building industry, we will accede to this convention.

A new Federal Specification is developed when a government procurement need arises, when a present specification becomes obsolete, or when revision is required for other reasons. Although Federal Specifications are gradually being replaced by industry standards, such as those promulgated by ASTM, Federal Specifications are still referenced in government guide specifications and are the only standards available for some products. In addition, Federal Specifications are still referenced by some product manufacturers even when industry standards are available. There may come a day when Federal Specifications are no longer used in the construction industry, but that day has not yet arrived.

The GSA may assign the development of a particular Federal Specification to a federal agency that has specialized technical competence and the necessary facilities. However, nationally recognized industry, technical society, and trade association standards, such as those by ASTM, are used and adopted in Federal Specifications to the maximum extent practicable.

Federal Specifications are designated by a letter and number code. For example, A-A-3130A Paint (For Application to Wet Surfaces) 13-Jun-2003 is the Federal Specification for a particular type of paint. The term Federal Specification is often abbreviated as Fed. Spec. or simply FS.

### 1.2.3.3 Military Agencies

The federal government is one of the world's largest buyers of equipment, materials, and supplies, with annual purchases in the billions of dollars. Various departments within the Department of Defense have developed specifications covering materials, products, and services used predominantly by military activities.

Military Specifications may be used by any interested civilian organization or specifier. Military Specifications are indexed by the title and code letter prefix MIL.

### 1.2.3.4 Department of Housing and Urban Development

The National Housing Act, enacted by Congress in 1934 and amended from time to time, created the Federal Housing Administration (FHA) to stimulate home construction by insuring mortgage loans. The functions of this agency were transferred by Congress in 1965 to the newly created Department of Housing and Urban Development (HUD), and FHA became part of this larger cabinet-level department.

The overall purpose of HUD is to assist in the sound development of the nation's communities and metropolitan areas. Encouragement of housing production through mortgage insurance and various subsidies has been one of HUD's chief objectives. Improvement in housing quality and in land planning standards has been another HUD objective mandated by Congress.

## FHA/HUD HOUSING PROGRAMS

FHA/HUD makes no loans, nor does it plan or build housing. It functions mainly as an insuring agency for mortgage loans made by private lenders, such as savings associations and commercial banks. For instance, through the Section 203(b) program, FHA/HUD encourages lenders to make loans with low down payments and long maturities on one- to four-family dwellings. The borrower pays an annual insurance premium of a small percentage of the average principal outstanding over the premium year. The Secretary of HUD sets the interest rate ceiling on FHA/HUD loans at a level required to meet market conditions. Another frequently used section, 221(d)(4), provides for mortgage insurance of new or rehabilitated low- or middle-income rental housing.

The traditional role of the FHA was transformed when the agency became the administrator of interest-rate subsidy and rent-supplement programs authorized by Congress since 1965. The Section 235 program combines insurance with interest assistance payments for owner-occupied homes. In addition to insuring the loan, FHA/HUD pays part of the interest the borrower owes the mortgage lender. Section 236 offers insurance and interest assistance for rental projects. Section 237 provides insurance on loans to borrowers with poor credit histories. Section 238 authorizes insurance for mortgage loans in high-risk situations, such as transitional urban areas, not covered by other programs.

## HUD MINIMUM PROPERTY STANDARDS

Because not all housing programs authorized by Congress involve mortgage insurance, not all of them are administered by FHA/HUD. For instance, Section 8 of the Housing Act of 1974 authorizes rental subsidies for leased low-income housing. The housing may be existing or new and may be financed either conventionally or with FHA/HUD mortgage insurance. Before 1973, FHA-insured private housing had to conform to the FHA *Minimum Property Standards*, and subsidized public housing was regulated by a different set of standards. With the adoption in that year of the HUD *Minimum Property Standards* (MPS), uniform standards became applicable to all HUD housing programs.

The MPS were intended to provide a sound technical basis for the planning and design of housing under the numerous programs of HUD. The standards described those characteristics in a property that would provide initial and continuing utility, durability, desirability, economy of maintenance, and a safe and healthful environment.

Environmental quality was considered throughout the MPS. As a general policy, property development was required to be consistent with the national program for conservation of energy and other natural resources. Care had to be exercised to avoid air, water, land, and noise pollution and other environmental hazards.

The MPS consisted of three volumes of mandatory standards: (1) *MPS for One- and Two-Family Dwellings*, HUD 4900.1; (2) *MPS for Multifamily Housing*, HUD 4910.1; and (3) *MPS for Care-Type Housing*, HUD 4920.1. Variations and exceptions for seasonal homes intended for other than year-round occupancy were listed in HUD 4900.1. Exceptions for elderly housing were listed in HUD 4900.1 and 4910.1. A fourth volume, the MPS *Manual of Acceptable Practices*, HUD 4930.1, contained advisory and illustrative material for the three volumes of mandatory standards.

Today these documents have been withdrawn and replaced by a single document, HUD 4910.1, *Minimum Property Standards for Housing*. This document is intended to supplement the requirements of the applicable local and state building codes and the *International Residential Code* (see Section 1.3.3.3).

## MATERIALS BULLETINS

The Architectural Standards Division of HUD issues *Use of Materials* bulletins for specific proprietary products or products that HUD engineers have investigated and found their performance to be acceptable. Each bulletin describes a product and its use and is issued to HUD field offices for guidance in determining the acceptance of the product.

The absence of a bulletin for a particular product does not preclude its use. Use of Materials bulletins are not intended to indicate approval, endorsement, or acceptance by HUD. Manufacturers of materials and products for which Use of Materials bulletins have been issued are not authorized to use them in any manner for sales promotion. Copies of Use of Materials bulletins are on file in HUD field offices but are not available for general distribution.

HUD has additional provisions for the review of special materials, products, and construction methods that it may be asked to insure. Design, materials, equipment, and construction methods other than those described in *Minimum Property Standards for Housing* are considered for use, provided that complete substantiating data satisfactory to HUD are submitted. Local HUD field offices are authorized to accept variations from the standards for specific cases, subject to conditions outlined in the standards.

Variations on an area or regional basis, or variations involving a substantial number of properties on a repetitive basis, are authorized only after consideration of recommendations by the HUD field office and approval by the Architectural Standards Division. Under certain conditions, some variations are established and published as Local Acceptable Standards (LAS) for a specific area.

### 1.2.4 EUROPEAN STANDARDS

In January 1993, 12 European countries joined economically into a *Single European Market* (SEM). In addition, the *European Community* (EC) is moving steadily toward the opening of borders and establishment of the free trade of ideas and goods between the member nations. These major changes in Europe will greatly affect trade and other relationships between the United States and the EC. They will also affect the U.S. building industry in many ways. One effect that will be greatly felt is the change in European standards. As a part of the establishment of SEM, the EC member states have deemed it necessary to unify each of their existing 12 separate groups of national standards into common European standards. The European Committee for Standarization (CEN) and the Committee for European Electrotechnical Standarization (CENELEC) have been charged with publishing standards for the EC member states.

The *International Organization for Standardization* (IOS) is a nongovernmental organization made up of representatives from the standards institutions of 91 countries. The United States is a member, represented by ANSI, but is not very influential there. IOS approves and publishes standards produced by its members. Most of those published by IOS are European standards.

The new EC standards will be based either on reconciling the differences between existing standards in the EC states or simply adoption of the IOS standards. The United States could exert greater influence in this area by increasing its involvement in IOS, but there seems to be little movement in that direction. Although development is in progress, single European standards for construction products do not yet exist. Until they do, EC member states will continue to enforce either their own standards or interim standards.

American firms doing work in Europe and U.S. products sold in Europe today have to comply with the standards of the country in which the work is being done. Eventually they will all have to comply with the unified EC standards. In the interim, there will be a morass of conflicting and possibly overlapping standards in the various countries. Some standards will be the old national ones, some will be the new EC standards, and others will be IOS standards. Probably the best first step of anyone contemplating working in the EC would be to contact the U.S. Department of Commerce Office of European Community Affairs for advice.

What eventual effect the new EC standards will have on U.S. standards is unknown at this point. There will probably be some changes in our standards relatively soon. Down the road, there may be extensive changes as the United States tries to compete economically with a unified Europe.

# 1.3  Codes

The planning, construction, location, and use of buildings are regulated by a variety of laws enacted by local, state, and federal governments. These statutes and ordinances include zoning, building, plumbing, electrical, and mechanical codes that are intended to protect the health, safety, and general welfare of the public. These codes incorporate many recognized construction industry standards (see Section 1.2), but they do not

necessarily contain criteria that ensure efficient, convenient, or adequately equipped buildings.

A *zoning code* (see Section 1.3.2) establishes requirements for land use.

A *building code* (see Section 1.3.3) establishes requirements for the construction and occupancy of buildings. It contains standards of performance and requirements for materials, methods, and systems. It also covers structural strength, fire resistance, adequate light and ventilation, egress, occupant safety, and other considerations determined by the design, construction, alteration, and demolition of buildings.

A collection of building requirements becomes a code when it is adopted by a municipality as a public ordinance or law. Local communities may write their own codes or may legally adopt other codes, such as state building codes or one of the model codes (see Section 1.3.3.3).

## 1.3.1  HISTORY

Laws controlling building construction are not new. The ancient code of the Babylonian emperor Hammurabi, dating back to about 1800 B.C., is often cited as the first recorded building code. It provided severe penalties for construction practices that violated the health or safety of citizens. For example, if a building collapsed, killing the occupants, its architects and builders were put to death. This ancient code, based on the idea that the strong should not injure the weak, sets the principle for today's construction regulations: The public has a right to be protected from the harmful acts of others.

Modern building regulation evolved over time, starting in the early nineteenth century with the adoption of fire laws in some large cities. These laws prohibited the construction of wooden buildings in congested areas. At about the same time, cities also began to adopt health regulations to improve the living conditions of the poor, which were the forerunners of today's minimum housing codes. Some courts, however, resisted the enactment of such laws as infringements on personal property rights.

As the validity of fire and health laws was slowly established, courts began to accept governmental control of all aspects of building construction involving the health and safety of the public. Today most states and cities, and a great many counties and towns, regulate the

planning, construction, and installation of building systems through a variety of laws and ordinances.

The right to regulate building construction constitutionally rests in the states, but before 1960 few state governments exercised that right. The states usually delegated to local governments the power to regulate buildings to protect public health, safety, and welfare. Today many states are taking an active role in building regulation. More than half of the states have now adopted some form of statewide building regulation concerned with the construction of industrialized buildings, mobile homes, and conventional construction.

Federal legislation in the areas of occupational health and safety, environmental protection, pollution controls, and consumer protection, coupled with increased state legislative activity, offer evidence that the building regulatory process will become ever more complex.

## 1.3.2  ZONING CODES

Zoning codes are developed, interpreted, enacted, and enforced by local jurisdictions (cities, counties, townships, etc.) to achieve the following:

- Promote the general welfare by ensuring adequate light, air, and convenience of access to buildings
- Provide for safety from fire, flood, and other dangers
- Reduce congestion in the public streets
- Create a convenient, attractive, and harmonious community
- Expedite adequate police, fire, and rescue protection
- Expedite emergency evacuation
- Provide adequate public facilities (schools, parks, playgrounds, etc.)
- Reduce encroachment on historic areas
- Protect against overcrowding of land
- Prevent undue density of population
- Encourage economic development

Zoning codes affect building design by controlling where building types may be located, both within their communities and on their particular sites. Zoning codes also restrict the size of buildings that can be built on a specific site and affect the shapes of buildings. Many highrise buildings owe their shapes to zoning setback rules that require higher floors to be farther from the property line.

A typical zoning code contains specific requirements for the type of use that is permitted in individual zones within the community. The naming of zones differs from community to community. The following are zones included in the *International Zoning Code:*

### Agricultural:

A-1, any so designated open space
A-2, any agricultural use
A-3, any public park land or other similar recreational use

### Residential:

R-1, single-family
R-2, single-family and two-family
R-3, apartment and multiple-family dwelling

### Commercial and Commercial/Residential:

C-1, general business
C-2, light commercial
C-3, amusement uses, wholesale and retail sales facilities
C-4, major commercial and manufacturing facilities
CR-1, uses permitted in C-1, plus some residential
CR-2, uses permitted in C-2, plus some residential

### Factory/Industrial:

FI-1, light manufacturing or industrial
FI-2, stadiums, arenas, and light industrial
FI-3, heavy industrial

Special districts may also be established to control requirements for special-use zones, such as historic zones; wetlands; areas where flood damage is likely; architectural districts, where design appearance is controlled; planned development districts, where special rules apply, such as allowance of increased occupant density in exchange for leaving green areas; and water runoff restricted districts established to protect lakes, rivers, and bays. Typical restrictions in a zone include requirements such as those for:

■ Permitted uses
■ Building height
■ Minimum lot size and street frontage length
■ Minimum floor area
■ Required yards (distance from a building to a property line in front, rear, and sides)
■ Building setback (distance from the building to lot lines at each height above the ground, which often increases in cities as the building grows taller)
■ Permitted types of accessory buildings and their permitted size and location
■ Off-street parking and loading requirements
■ Lighting of parking and other site areas
■ Visual clearance on corner lots
■ Swimming pools
■ Site fences and walls

## 1.3.3 BUILDING CODES

The construction industry is enmeshed in an extraordinary network of building codes that attempt to ensure that buildings and their environs will be safe. They generally accomplish that purpose; however, codes and code administration are criticized widely as restricting building progress by retarding the acceptance of new and improved uses of materials and methods and thereby unnecessarily increasing construction costs. In some instances, the adoption of improved codes has stimulated better building practices. But the existing complex and chaotic building code situation is recognized as one of the problems facing the building industry today because of the use of descriptive-type codes (see Section 1.3.3.1), the lack of code uniformity, the multiplicity of codes, the slow response of codes to change, and inadequate performance standards in the codes (see Section 1.3.3.3).

The major building officials organizations and numerous federal, state, and local groups are constantly working to improve codes and the code regulatory system. The latest efforts center on standardizing the major model codes (see Section 1.3.3.3) into a single code for each discipline. This has currently been accomplished by the introduction in 2003 of the International Building Codes (see Section 1.3.3.3).

Some so-called codes are actually standards because they are voluntary and have no status in law. Only when standards are adopted by legislative bodies and incorporated into law do they become codes. In one sense, the model codes discussed in Section 1.3.3.3 are actually standards that have been adopted in many, but not all, jurisdictions as codes. Other examples of documents called codes that actually are not codes include various ANSI and NFPA codes.

Standards that are called codes by their producers often fall in the design standards category (see Section 1.2.1) but may encompass one or more of the other types.

Because some documents called codes are universally recognized as such, we will accede to that convention in this book. Therefore, publications such as ANSI A17.1, "Safety Code for Elevators and Escalators," NFPA 101, "Life Safety Code," and "National Electrical Code," are here called codes.

### 1.3.3.1 The Effect of Building Codes on Design

Building codes affect building design in a number of ways. One way is by limiting the construction type and size for each building use. Figure 1.3-1 is a table from the 2006 edition of the International Building Code. Similar tables are included in all the model codes (see Section 1.3.3.3). The table shown lists 26 *groups*. These groups are defined and described in the body of the code and are partially shown in the following list. The list is not all-inclusive.

Groups A-1 through A-5—assembly occupancies, such as theaters, art galleries, restaurants, assembly halls, and so forth
Group B—business occupancies, including buildings for office, professional, or service-type uses
Group E—educational facilities
Groups F-1 and F-2—factories
Groups H-1 through H-5—high-hazard facilities, such as those where the manufacture or storage of materials that constitute a physical or health hazard takes place
Groups I-1 through I-4—institutional facilities, including those for supervised residential environments, hospitals and other medical facilities, and prisons, jails, and other correctional and detention facilities
Group M—mercantile facilities including buildings used for the display or sale of merchandise
Groups R-1 through R-4—residential occupancies, including single-family and multifamily dwellings, apartments, hotels, and assisted living facilities
Group S-1 and S-2—storage facilities
Group U—utility and miscellaneous facilities, such as barns, carports, aircraft hangers, buildings that are accessories to dwellings, greenhouses, and private garages

# TABLE 503
## ALLOWABLE HEIGHT AND BUILDING AREAS[a]
### Height limitations shown as stories and feet above grade plane.
### Area limitations as determined by the definition of "Area, building," per story

| GROUP | HGT(S) | TYPE I A | TYPE I B | TYPE II A | TYPE II B | TYPE III A | TYPE III B | TYPE IV HT | TYPE V A | TYPE V B |
|---|---|---|---|---|---|---|---|---|---|---|
| HGT(feet) | | UL | 160 | 65 | 55 | 65 | 55 | 65 | 50 | 40 |
| A-1 | S | UL | 5 | 3 | 2 | 3 | 2 | 3 | 2 | 1 |
| | A | UL | UL | 15,500 | 8,500 | 14,000 | 8,500 | 15,000 | 11,500 | 5,500 |
| A-2 | S | UL | 11 | 3 | 2 | 3 | 2 | 3 | 2 | 1 |
| | A | UL | UL | 15,500 | 9,500 | 14,000 | 9,500 | 15,000 | 11,500 | 6,000 |
| A-3 | S | UL | 11 | 3 | 2 | 3 | 2 | 3 | 2 | 1 |
| | A | UL | UL | 15,500 | 9,500 | 14,000 | 9,500 | 15,000 | 11,500 | 6,000 |
| A-4 | S | UL | 11 | 3 | 2 | 3 | 2 | 3 | 2 | 1 |
| | A | UL | UL | 15,500 | 9,500 | 14,000 | 9,500 | 15,000 | 11,500 | 6,000 |
| A-5 | S | UL | UL | UL | UL | UL | UL | UL | UL | UL |
| | A | UL | UL | UL | UL | UL | UL | UL | UL | UL |
| B | S | UL | 11 | 5 | 4 | 5 | 4 | 5 | 3 | 2 |
| | A | UL | UL | 37,500 | 23,000 | 28,500 | 19,000 | 36,000 | 18,000 | 9,000 |
| E | S | UL | 5 | 3 | 2 | 3 | 2 | 3 | 1 | 1 |
| | A | UL | UL | 26,500 | 14,500 | 23,500 | 14,500 | 25,500 | 18,500 | 9,500 |
| F-1 | S | UL | 11 | 4 | 2 | 3 | 2 | 4 | 2 | 1 |
| | A | UL | UL | 25,000 | 15,500 | 19,000 | 12,000 | 33,500 | 14,000 | 8,500 |
| F-2 | S | UL | 11 | 5 | 3 | 4 | 3 | 5 | 3 | 2 |
| | A | UL | UL | 37,500 | 23,000 | 28,500 | 18,000 | 50,500 | 21,000 | 13,000 |
| H-1 | S | 1 | 1 | 1 | 1 | 1 | 1 | 1 | 1 | NP |
| | A | 21,000 | 16,500 | 11,000 | 7,000 | 9,500 | 7,000 | 10,500 | 7,500 | NP |
| H-2[d] | S | UL | 3 | 2 | 1 | 2 | 1 | 2 | 1 | 1 |
| | A | 21,000 | 16,500 | 11,000 | 7,000 | 9,500 | 7,000 | 10,500 | 7,500 | 3,000 |
| H-3[d] | S | UL | 6 | 4 | 2 | 4 | 2 | 4 | 2 | 1 |
| | A | UL | 60,000 | 26,500 | 14,000 | 17,500 | 13,000 | 25,500 | 10,000 | 5,000 |
| H-4 | S | UL | 7 | 5 | 3 | 5 | 3 | 5 | 3 | 2 |
| | A | UL | UL | 37,500 | 17,500 | 28,500 | 17,500 | 36,000 | 18,000 | 6,500 |
| H-5 | S | 4 | 4 | 3 | 3 | 3 | 3 | 3 | 3 | 2 |
| | A | UL | UL | 37,500 | 23,000 | 28,500 | 19,000 | 36,000 | 18,000 | 9,000 |
| I-1 | S | UL | 9 | 4 | 3 | 4 | 3 | 4 | 3 | 2 |
| | A | UL | 55,000 | 19,000 | 10,000 | 16,500 | 10,000 | 18,000 | 10,500 | 4,500 |
| I-2 | S | UL | 4 | 2 | 1 | 1 | NP | 1 | 1 | NP |
| | A | UL | UL | 15,000 | 11,000 | 12,000 | NP | 12,000 | 9,500 | NP |
| I-3 | S | UL | 4 | 2 | 1 | 2 | 1 | 2 | 2 | 1 |
| | A | UL | UL | 15,000 | 10,000 | 10,500 | 7,500 | 12,000 | 7,500 | 5,000 |
| I-4 | S | UL | 5 | 3 | 2 | 3 | 2 | 3 | 1 | 1 |
| | A | UL | 60,500 | 26,500 | 13,000 | 23,500 | 13,000 | 25,500 | 18,500 | 9,000 |
| M | S | UL | 11 | 4 | 4 | 4 | 4 | 4 | 3 | 1 |
| | A | UL | UL | 21,500 | 12,500 | 18,500 | 12,500 | 20,500 | 14,000 | 9,000 |
| R-1 | S | UL | 11 | 4 | 4 | 4 | 4 | 4 | 3 | 2 |
| | A | UL | UL | 24,000 | 16,000 | 24,000 | 16,000 | 20,500 | 12,000 | 7,000 |
| R-2 | S | UL | 11 | 4 | 4 | 4 | 4 | 4 | 3 | 2 |
| | A | UL | UL | 24,000 | 16,000 | 24,000 | 16,000 | 20,500 | 12,000 | 7,000 |
| R-3 | S | UL | 11 | 4 | 4 | 4 | 4 | 4 | 3 | 3 |
| | A | UL | UL | UL | UL | UL | UL | UL | UL | UL |
| R-4 | S | UL | 11 | 4 | 4 | 4 | 4 | 4 | 3 | 2 |
| | A | UL | UL | 24,000 | 16,000 | 24,000 | 16,000 | 20,500 | 12,000 | 7,000 |
| S-1 | S | UL | 11 | 4 | 3 | 3 | 3 | 4 | 3 | 1 |
| | A | UL | 48,000 | 26,000 | 17,500 | 26,000 | 17,500 | 25,500 | 14,000 | 9,000 |
| S-2[b, c] | S | UL | 11 | 5 | 4 | 4 | 4 | 5 | 4 | 2 |
| | A | UL | 79,000 | 39,000 | 26,000 | 39,000 | 26,000 | 38,500 | 21,000 | 13,500 |
| U[c] | S | UL | 5 | 4 | 2 | 3 | 2 | 4 | 2 | 1 |
| | A | UL | 35,500 | 19,000 | 8,500 | 14,000 | 8,500 | 18,000 | 9,000 | 5,500 |

For SI: 1 foot = 304.8 mm, 1 square foot = 0.0929 m².

UL = Unlimited, NP = Not permitted.

a. See the following sections for general exceptions to Table 503:
   1. Section 504.2, Allowable height increase due to automatic sprinkler system installation.
   2. Section 506.2, Allowable area increase due to street frontage.
   3. Section 506.3, Allowable area increase due to automatic sprinkler system installation.
   4. Section 507, Unlimited area buildings.
b. For open parking structures, see Section 406.3.
c. For private garages, see Section 406.1.
d. See Section 415.5 for limitations.

**FIGURE 1.3-1** From the 2006 *International Building Code.* Copyright ICC; reprinted with permission.

Figure 1.3-1 shows height and area restrictions for each of these groups, based on each of nine types of construction. The types of construction are also defined in the text of the code and are summarized in the following list.

Types I and II, A and B—construction in which the building groups listed in Figure 1.3-1 are of noncombustible construction

Types III, A and B—construction in which the exterior walls of buildings in the groups listed in the table in Figure 1.3-1 are of noncombustible construction, but interior elements may not be necessarily of noncombustible construction

Type IV—construction in which the exterior walls of buildings in the groups listed in Figure 1.3-1 are of noncombustible construction, and interior elements are of solid wood (heavy timber) construction

Type V, A and B—construction in which the building groups listed in Figure 1.3-1 are of any construction type permitted by the code, including that in which the building is constructed of wood and other combustible materials

As an example of the use of Figure 1.3-1, suppose that an architect has been commissioned to design a building that falls in group E, educational facilities, that is required to have 40,000 sq ft (3716 m$^2$) of area. The table shows that if the building is Noncombustible Type lA construction, the area and height are not limited by code (UL). Therefore, the building could be entirely on a single floor or built on five floors of 8000 sq ft (743 m$^2$) each.

Our example building can be also constructed of Type IB or IIA, but not of Type IIB, because we need 40,000 sq ft (3716 m$^2$) and IIB permits only two floors of a maximum 14,500 sq ft (1347 m$^2$) each, for a total of only 29,000 sq ft (2694 m$^2$). Our building could also be built of Type IIIA construction but not IIIB, or of Type lV but not Type V.

Codes further affect building design by controlling materials and methods of construction. This effect is illustrated in Figure 1.3-2, another table from the 2006 International Building Code. In this table, structural elements are listed down the left side and types of construction across the top. These types of construction are the same as those used in Figure 1.3-1. The figures at the intersection of the lines

and columns are the hourly fire ratings of the assemblies in the left column for each construction type listed across the top of the table. For example, the floor construction in a Type lA building must have a 2-hour fire rating.

Building codes also restrict fire resistance ratings for exterior walls. Figure 1.3-3 is another table from the 2006 International Building Code. In this table, fire separation distances are listed in the far left column. Types of construction are listed in the second column. The rest of the columns list groups. The types of construction and groups are the same as those in Figure 3.3-1. To determine the required horizontal separation, enter the table knowing the building's type of construction and group and extend the combined line across the table to read the separation distance. Conversely, if the separation distance is known, the table will yield the required type of construction for the group into which the building falls.

Building codes also affect building design in other ways. Building codes are either descriptive or performance oriented. A *descriptive* (sometimes called *specification* or *prescription*) building code establishes construction requirements by reference to a particular material or method. For example, a code may require that exterior walls be built of 2 × 4 (50 × 100) wall studs spaced 16 in. (406.4 mm) on center and covered by 1-in. (25.4 mm)-thick board sheathing applied diagonally. A builder seeking to space studs at 24 in. (610 mm) on center, which may be structurally sound and safe, would be in violation of the specifically stated code requirements.

A *performance* code does not limit the selection of methods and systems to a single type, but establishes requirements for the performance of building elements. Such a code states design and engineering criteria without reference to specific materials or methods of construction. For example, an outside wall may be required to support loads (wind, dead, live, etc.) with specific defined values and to meet or exceed stated insulating and permeability requirements. Any system performing as required by the code would be acceptable, regardless of the materials or methods used.

True performance codes are idealistically excellent but impractical in use. To enforce them, local code administrators would have to be extraordinarily compe-

tent and equipped to interpret performance criteria and evaluate proposed methods, uses, and systems. Such people and equipment are rare. What's more, these decisions can be more readily challanged. A workable solution lies somewhere between descriptive and performance codes. Both types of codes should adequately provide for the acceptance of alternate methods and systems.

### 1.3.3.2 Building Code Enforcement

Local administration and enforcement of codes are usually done by a building inspector or engineer who has the authority to approve materials and methods that may not be directly referenced in the code. Qualified people are necessary to administer a building code properly. No matter how good a code may be, it must be enforced by someone who is experienced, informed, and objective. Builders complain about inspectors who do not understand construction and may thus be arbitrary and inconsistent. They are seldom upset by a careful and competent inspector who is consistent even though tough. A competent code enforcer knows when the letter of the code should prevail and when subjective interpretation should be made.

However, local code administrators, faced continually with difficult decisions, may well argue that their job is solely to check compliance. An often-drawn analogy is that a local policeman does not make judgments about whether a law is right or just, but is charged solely with determining compliance. Perhaps it is best that the determination of performance criteria and judgment as to whether methods and systems perform suitably must be made by technically qualified people and not by local code administrators. Members of the model code groups (see Section 1.3.3.3) are qualified to offer this type of service.

Model code groups, which are supported by building officials themselves, have performed a great service to the building industry. But model codes have not fully solved the problems of code uniformity. The reason is, partly, that even though most communities have adopted a model code, some have adopted them with modifications. The National Association of Home Builders (NAHB) has said that many of the changes related to housing made to model codes to adapt them to local conditions come from local

TABLE 601
## FIRE-RESISTANCE RATING REQUIREMENTS FOR BUILDING ELEMENTS (hours)

| BUILDING ELEMENT | TYPE I A | TYPE I B | TYPE II A^e | TYPE II B | TYPE III A^e | TYPE III B | TYPE IV HT | TYPE V A^e | TYPE V B |
|---|---|---|---|---|---|---|---|---|---|
| Structural frame[a] | $3^b$ | $2^b$ | 1 | 0 | 1 | 0 | HT | 1 | 0 |
| Bearing walls<br>Exterior[g]<br>Interior | 3<br>$3^b$ | 2<br>$2^b$ | 1<br>1 | 0<br>0 | 2<br>1 | 2<br>0 | 2<br>1/HT | 1<br>1 | 0<br>0 |
| Nonbearing walls and partitions<br>Exterior | See Table 602 | | | | | | | | |
| Nonbearing walls and partitions<br>Interior[f] | 0 | 0 | 0 | 0 | 0 | 0 | See Section 602.4.6 | 0 | 0 |
| Floor construction<br>Including supporting beams and joists | 2 | 2 | 1 | 0 | 1 | 0 | HT | 1 | 0 |
| Roof construction<br>Including supporting beams and joists | $1^{1}/_{2}{}^{c}$ | $1^{c,\,d}$ | $1^{c,\,d}$ | $0^{c,\,d}$ | $1^{c,\,d}$ | $0^{c,\,d}$ | HT | $1^{c,\,d}$ | 0 |

For SI: 1 foot = 304.8 mm.

a. The structural frame shall be considered to be the columns and the girders, beams, trusses and spandrels having direct connections to the columns and bracing members designed to carry gravity loads. The members of floor or roof panels which have no connection to the columns shall be considered secondary members and not a part of the structural frame.

b. Roof supports: Fire-resistance ratings of structural frame and bearing walls are permitted to be reduced by 1 hour where supporting a roof only.

c. Except in Group F-1, H, M and S-1 occupancies, fire protection of structural members shall not be required, including protection of roof framing and decking where every part of the roof construction is 20 feet or more above any floor immediately below. Fire-retardant-treated wood members shall be allowed to be used for such unprotected members.

d. In all occupancies, heavy timber shall be allowed where a 1-hour or less fire-resistance rating is required.

e. An approved automatic sprinkler system in accordance with Section 903.3.1.1 shall be allowed to be substituted for 1-hour fire-resistance-rated construction, provided such system is not otherwise required by other provisions of the code or used for an allowable area increase in accordance with Section 506.3 or an allowable height increase in accordance with Section 504.2. The 1-hour substitution for the fire resistance of exterior walls shall not be permitted.

f. Not less than the fire-resistance rating required by other sections of this code.

g. Not less than the fire-resistance rating based on fire separation distance (see Table 602).

**FIGURE 1.3-2**   From the 2006 *International Building Code.* Copyright ICC; reprinted with permission.


TABLE 602
## FIRE-RESISTANCE RATING REQUIREMENTS FOR EXTERIOR WALLS BASED ON FIRE SEPARATION DISTANCE[a, e]

| FIRE SEPARATION DISTANCE = X (feet) | TYPE OF CONSTRUCTION | OCCUPANCY GROUP H | OCCUPANCY GROUP F-1, M, S-1 | OCCUPANCY GROUP A, B, E, F-2, I, R, S-2, U[b] |
|---|---|---|---|---|
| X < 5[c] | All | 3 | 2 | 1 |
| 5 ≤ X < 10 | IA<br>Others | 3<br>2 | 2<br>1 | 1<br>1 |
| 10 ≤ X < 30 | IA, IB<br>IIB, VB<br>Others | 2<br>1<br>1 | 1<br>0<br>1 | $1^d$<br>0<br>$1^d$ |
| X ≥ 30 | All | 0 | 0 | 0 |

For SI: 1 foot = 304.8 mm.

a. Load-bearing exterior walls shall also comply with the fire-resistance rating requirements of Table 601.

b. For special requirements for Group U occupancies see Section 406.1.2.

c. See Section 705.1.1 for party walls.

d. Open parking garages complying with Section 406 shall not be required to have a fire-resistance rating.

e. The fire-resistance rating of an exterior wall is determined based upon the fire separation distance of the exterior wall and the story in which the wall is located.

**FIGURE 1.3-3**   From the 2006 *International Building Code.* Copyright ICC; reprinted with permission.

codes prepared 20 or more years earlier. Sometimes, progressive time- and money-saving requirements of a model code are revised so that antiquated provisions apply instead, simply to agree with local custom.

### 1.3.3.3 Model Codes

Problems with code development, use, and enforcement occur for several reasons. Because establishing and enforcing building codes are local functions, a designer or builder who works in more than one community is often faced with a frustrating variety of requirements. A product manufacturer must win acceptance by thousands of local building code administrators, instead of concerning itself only with performance and public acceptance. Much effort has been made to unify the codes used by communities, and considerable improvement has been gained through the local adoption of model codes.

If designers and builders were dealing with a single code, their problems would be greatly simplified. However, in addition to adhering to local building codes that tell them how a structure must be assembled, they have to satisfy a number of other codes covering a variety of subjects such as plumbing, electrical wiring, traffic, utilities, health and sanitation, land planning, building occupancy, access by the handicapped, and zoning.

The lack of code uniformity and the multiplicity of additional regulations often so complicate matters that buildings are designed and built with the use of very conservative and often expensive criteria and methods. Designers and builders alike are often unable to improve their techniques because of the restrictive nature of one or more of the applicable codes.

Codes are often criticized for failing to recognize new materials and methods. Judging new products on the basis of performance criteria must be performed by technically qualified and equipped organizations.

This function is one in which model code groups can exercise great leadership and provide a major stimulus to progress. These groups have responded reasonably to progressive change. However, they are subject to many pressures that deter progress, and they have sometimes been slow in acting on proposed changes. For example, it took almost 4 years for all of the major code groups to accept the NAHB's proposal to eliminate floor bridging in most home building situations, a proposition well documented through extensive research and testing.

As discussed in Section 1.2, no single group of building industry standards exists. In selecting materials or determining the suitability of materials and methods, a specifier and a builder make reference to a variety of trade association certification programs, grademarks and trademarks, UL labels, and many other construction standards. There are independent testing agencies, such as UL, and standards-setting bodies, such as ASTM, and some manufacturers have substantial commitments in research and testing facilities, but no single agency is presently recognized by all groups interested in building construction.

To help fill the need for workable codes, three major organizations of industry and professional groups and states have developed building codes that may be adopted into law for use in local communities. These codes are commonly referred to as *model codes*. The major model code groups are the Building Officials and Code Administrators International (BOCAI), the International Conference of Building Officials (ICBO), and the Southern Building Code Congress International (SBCCI). The major national model codes and their sponsors are listed in Figure 1.3-4.

Model codes have been widely accepted by local communities and now are used by more than 80% of those communities with a population of more than 10,000. The organizations that prepare model codes have also provided for the continuous updating necessary to include recommendations based on industry research and the development of new materials and methods. Other model codes that deal with specialized subjects not fully covered in the major model codes are also available. Most of these codes and their sponsors are listed in Figure 1.3-4. Many local communities have also adopted these codes.

### INTERNATIONAL CODE COUNCIL

For years, code developers and building industry practitioners have been talking about developing a single national building code, and these efforts may have actually reached a satisfactory conclusion. BOCA, SBCCI, and ICBO have joined together to create the International Code Council (ICC). ICC produces a series of international codes, which replace those codes produced by these three major model code groups. They also issue codes that replace other model codes, such as the One- and Two-Family Dwelling Code produced by the Council of American Building Officials (CABO). Other codes still are being produced (see Figure 1.3-4). Codes currently produced by ICC include the following *International Codes*™:

*International Building Code®*
*International Residential Code®*
*International Existing Building Code®*
*International Fire Code®*
*International Mechanical Code®*
*International Plumbing Code®*
*International Fuel Gas Code®*
*International Property Maintenance Code®*
*International Energy Conservation Code®*
*International Zoning Code®*
*International Private Sewage Disposal Code®*
*International Urban-Wildland Interface Code*™
*ICC Electrical Code®*
*ICC Performance Based Code*™

The first ICC codes appeared in 1995. The first edition of the *International Building Code* was published in 2000. These codes, like all others, must be adopted by local and state jurisdictions before they can have an effect. If history can be relied on, this will take some time. To date, the International Codes have been adopted in more than 75 jurisdictions: the *International Building Code* in 44 states and by the U.S. Department of Defense; the *International Residential Code* in 43 states plus Washington, D.C.; and the *International Fire Code* in 32 states. Although model codes are no longer being updated, they are discussed here because these older codes are still in effect in many jurisdictions. Sooner or later, all jurisdictions will need to adopt the ICC codes as the other model codes become obsolete. When this book refers to code-related data, the reference is to the ICC codes

### NATIONAL BUILDING CODE

The first model building code was introduced in 1905 by the National Board of Fire Underwriters, later called the American Insurance Association (AIA). This organization was concerned primarily with public protection against fire hazards, and the original code was designed to guide communities in setting up fire

**FIGURE 1.3-4** Model Building Codes and Their Sponsoring Organizations

| Building Codes | Sponsoring Organization |
|---|---|
| International Building Code | International Code Council |
| International Residential Code | International Code Council |
| International Existing Building Code | International Code Council |
| National Building Code | Building Officials and Code Administrators International, Inc. |
| Standard Building Code | Southern Building Code Congress International |
| One- and Two-Family Dwelling Code | Council of American Building Officials |

| Electrical Codes | |
|---|---|
| National Electrical Code | National Fire Protection Association |
| Electrical Code for One- and Two-Family Dwellings | National Fire Protection Association |

| Elevator Codes | |
|---|---|
| Safety Code for Elevators and Escalators | American Society of Mechanical Engineers |

| Fire Prevention Codes | |
|---|---|
| International Fire Code | International Code Council |
| National Fire Prevention Code | Building Officials and Code Administrators International, Inc. |
| Uniform Fire Code | International Conference of Building Officials |

| Housing Codes | |
|---|---|
| International Property Maintenance Code | International Code Council |
| Property Maintenance Code | Building Officials and Code Administrators International |
| Standard Housing Code | Southern Building Code Congress International |
| Uniform Housing Code | International Conference of Building Officials |

| Plumbing Codes | |
|---|---|
| International Plumbing Code | International Code Council |
| International Private Sewage Disposal Code | International Code Council |
| National Plumbing Code | Building Officials and Code Administrators International |
| National Standard Plumbing Code | National Association of Plumbing-Heating-Cooling Contractors |
| Standard Plumbing Code | Southern Building Code Congress International |
| Uniform Plumbing Code | International Association of Plumbing and Mechanical Officials |

| Mechanical Codes | |
|---|---|
| International Mechanical Code | International Code Council |
| International Fuel Gas Code | International Code Council |
| National Mechanical Code | Building Officials and Code Administrators International |
| Standard Gas Code | Southern Building Code Congress International |
| Standard Mechanical Code | Southern Building Code Congress International |

| Miscellaneous Codes | |
|---|---|
| International Energy Conservation Code | International Code Council |
| International Zoning Code | International Code Council |
| International Urban-Wildlife Interface Code | International Code Council |
| Boiler and Pressure Vessel Code | American Society of Mechanical Engineers |
| Flammable and Combustible Liquids Code | National Fire Protection Association |
| Safety Code for Mechanical Refrigeration | American Society of Heating, Refrigerating, and Air Conditioning Engineers, Inc. |
| Life Safety Code | National Fire Protection Association |

safety standards. With subsequent additions, however, its National Building Code laid the groundwork for the development of building codes throughout the country. AIA produced its last edition of the *National Building Code* in 1976. In 1950, BOCAI published the *Basic Building Code*. When AIA ceased publishing the *National Building Code*, BOCAI picked it up and, until recently, has published it since. BOCAI also published the other codes listed for it in Figure 1.3-4.

BOCAI is no longer updating these codes.

## STANDARD BUILDING CODE

SBCCI drafted the *Southern Standard Building Code* in 1945. Renamed the

*Standard Building Code*, this model code is designed to recognize the special problems, such as high winds, in the southern region of the country and is used extensively in this area. SBCCI also published the other codes listed for it in Figure 1.3-4. BOCAI is no longer updating these codes.

### UNIFORM BUILDING CODE

In 1927, ICBO (initially the Pacific Coast Building Officials Conference) published its *Uniform Building Code*. This code gained wide acceptance, particularly in the West. ICBO also published the other codes listed for it in Figure 1.3-4. It copublished the *Uniform Fire Code* and the *Uniform Fire Code Standards* with the Western Fire Chiefs Association. ICBO also jointly published the *Uniform Mechanical Code* with the International Association of Plumbing and Mechanical Officials. The most recent edition of these codes was published in 1997.

### NATIONAL STANDARD PLUMBING CODE

The *National Standard Plumbing Code* is sponsored by the Plumbing-Heating-Cooling Contractors Association (PHCC). The *National Standard Plumbing Code* covers the proper design, installation, and maintenance of plumbing systems according to principles of sanitation and safety, but not necessarily for efficiency, convenience, or adequacy for good service or future expansion of system use. Standards for materials and fixtures are based largely on industry standards such as Commercial Standards, American Standards, and ASTM standards.

### UNIFORM PLUMBING AND UNIFORM MECHANICAL CODES

The Western Plumbing Officials Association (now the International Association of Plumbing and Mechanical Officials [IAPMO]) established the first *Uniform Plumbing Code* in 1929 and continues to publish it today. IAPMO also publishes the *Uniform Mechanical Code*. These codes cover the design, installation, and maintenance of plumbing, heating, and air conditioning systems. Materials and equipment standards are based largely on other industry standards. These codes have gained widespread use in the western states, as well as in other communities across the country.

### NATIONAL ELECTRICAL CODE

The *National Electrical Code* is produced by the NFPA. Its purpose is to ensure the safeguarding of persons and of buildings and their contents from hazards arising from the use of electricity for light, heat, power, and other purposes. The code contains basic minimum provisions for safety which, with proper maintenance, will result in installations free from hazard, but not necessarily efficient, convenient, or adequate for good service or future expansion of electrical use. The *National Electrical Code* makes reference to many industry standards, such as ANSI/ASME A17.1, and to UL labels. It is revised periodically and is adopted by reference into many building codes and federal and state laws and regulations.

### LIFE SAFETY CODE

The *Life Safety Code* began in 1912 with a pamphlet entitled *Exit Drills in Factories, Schools, Department Stores and Theaters* and has evolved over the years to cover most building types. It is published by the NFPA to establish minimum requirements for a reasonable degree of safety from fire in structures. It addresses requirements for egress; design and construction that will prevent undue danger from fire, smoke, and gases; fire detection and alarm systems; and automatic sprinklers and other fire extinguishing systems. It includes requirements for building design and construction and for heating, ventilating, air conditioning, and electrical systems

### ONE- AND TWO-FAMILY DWELLING CODE

In 1971, the American Insurance Association, BOCA, ICBO, and SBCCI adopted a consensus code entitled the *One- and Two-Family Dwelling Code*. This model code is now published by the ICC as the 2006 *International Residential Code*.

The *International Residential Code* contains requirements for building planning and construction, including requirements for heating, cooling, and plumbing. HUD requires compliance with the *International Residential Code* in addition to its own *Minimum Property Standards for One- and Two-Family Dwellings* and local and state building codes for consideration for HUD/FHA loan guarantees.

Publication of this single national model code for one- and two-family dwellings constituted a significant step toward uniform minimum regulations in

the housing industry, but it still has not been recognized and adopted by every state and local government.

### 1.3.4 NATIONAL ENERGY CODES

In the United States, energy consumed in buildings amounts to about one-third of all energy consumption. Building energy consumption has been increasing rapidly for many years. The U.S. Department of Commerce's earlier estimate, that energy consumption would increase by 25% between 1984 and the year 2000, appears to have been fairly accurate. That increase represents an annual growth rate of 1.14% for housing and 1.66% for commercial use. Slowing the rate of increase through increased efficiency in building systems design can significantly reduce overall energy demand.

The *National Energy Plan* calls for substantial decreases in energy consumption through conservation. A major element of an energy conservation program is enforcement of thermal efficiency standards and insulation requirements in new and renovated buildings by state and local governments.

The National Conference of States on Building Codes and Standards (NCSBCS) in 1973 requested the National Bureau of Standards (NBS) (now the National Institute of Standards and Technology [NIST]) to develop standards that could be used by all states for energy-efficient design in new buildings. NBS completed these standards in 1974, but they were considered too complex for most state and local code enforcement officials to administer. As a result, NCSBCS requested ASHRAE to translate the NBS standards into enforceable language.

In 1975, ASHRAE published its Standard 90-75, "Energy Conservation in New Building Design." This standard was the first nationally recognized standard for energy-efficient design applicable to all building types. It was, however, written for use by design engineers. State and local practitioners still needed a document that could be administered through the traditional code enforcement system. In 1976, NCSBCS translated ASHRAE Standard 90-75 into the Model Code for Energy Conservation in New Building Construction. Under a grant from the Energy Research and Development Administration (now the Department of Energy), NCSBCS worked with building officials from all parts of the country who administer model codes so that the Model Code for Energy Conservation would be applicable to all geographic areas in the country and could be incorporated easily into existing codes. The NCSBCS Model Code for Energy Conservation in New Building Construction was published in January 1977.

The *Energy Policy and Conservation Act* (EPCA) became law in 1975. The law provided for substantial grants to those states that developed and implemented statewide energy conservation plans aimed at reducing statewide energy consumption 5% by 1980. State energy conservation plans had to include at least five elements: (1) mandatory lighting efficiency standards for public buildings, (2) programs to promote the availability and use of car pools and van pools, (3) mandatory standards and policies related to procurement practices of state and local governments, (4) a traffic law that permits a right turn on red after stopping, and (5) mandatory thermal energy standards and insulation requirements for new and renovated buildings.

In 1976, Congress passed the *Energy Conservation and Production Act* (ECPA). Title III of ECPA contained the following provisions: (1) HUD is directed to develop performance-oriented thermal energy efficiency standards, (2) HUD is to monitor the progress of state and local governments in implementing the standards, and (3) if a substantial amount of new housing is not constructed in conformance with the standards, HUD can recommend to the Congress that the standards be made mandatory. If Congress concurs:

1. All construction assisted by federal money will have to be consistent with the standards. This will include federal grants, loans, and loan guarantees and will apply to all savings and loan associations that are federally insured.
2. Local governments will not be required to incorporate the standards into their building codes. HUD and the states will be responsible for ensuring that all proposed construction is in conformance with the standards. However, local governments that elect to adopt the standards will be reimbursed for the cost of the certification.
3. If a local government incorporates the federal standards into its codes, then no further approvals will be required.

Money will be available to communities to assist them in incorporating and enforcing the federal standards.

Current requirements for insulation are addressed in Section 7.3. The currently accepted standards for energy conservation and thermal design criteria are discussed in Chapter 17.

### 1.3.5 CODE ADVANCEMENT

The growing complexity of building design and construction and the subsequent increase of regulatory controls have induced several organizations to improve the effectiveness of the regulatory process with better code documents, administration, and enforcement.

#### 1.3.5.1 National Institute of Building Sciences

The Housing and Community Development Act of 1974 authorized the creation of the National Institute of Building Sciences (NIBS). NIBS was initiated by the federal government with the assistance of the National Academy of Science and the National Academy of Engineering Research Council to help improve the way building construction is regulated. NIBS gives the United States a national center for (1) assembly, storage, and dissemination of technical data and related information on construction, (2) development and promulgation of nationally recognized performance criteria, standards, test methods, and other evaluating techniques, and (3) evaluation and prequalification of existing and new building techniques.

#### 1.3.5.2 Council of American Building Officials

In 1972 the governing bodies of BOCAI, ICBO, and SBCCI established the Council of American Building Officials (CABO). CABO's policies are determined by a board of trustees composed of representatives of the governing bodies of the member organizations.

Examples of CABO's work include (1) advancing the *One- and Two-Family Dwelling Code* as a recognized national standard through the procedures of ANSI, (2) sponsoring the development of model performance standards that will complement the requirements of model codes, (3) sponsoring a national research activity that provides a single approval

agency for manufacturers of building components, systems, and materials, (4) developing and adopting a model mechanical, plumbing, and fire prevention code as the standard for all three organizations, and (5) establishing ICC to implement a single group of international codes to replace all the model codes in current use.

#### 1.3.5.3 Associated Major City Building Officials

In 1972 building code officials from the nation's 20 largest cities created an organization called Associated Major City/County Building Officials (AMCBO). Its purpose is to exchange ideas, experience, and information on the growing complexities of code administration and enforcement in densely populated urban areas.

#### 1.3.5.4 National Conference of States on Building Codes and Standards

NCSBCS was formed in 1967. The organization is composed of state delegates designated by their respective governors to represent the 50 states in discussions and programs pertaining to building regulation; state and local building officials; various representatives of the building industry's design, manufacturing, and construction sectors; federal government officials; and consumers. It has affiliations with governmental agencies, such as NIST and the Consumer Product Safety Commission. Its members participate in the activities of many organizations, including CABO, BOCAI, ICBO, SBCCI, NFPA, ASHRAE, HUD, NIBS, the American Institute of Architects (AIA), ANSI, the Council of State Community Affairs Agencies (COSCAA), and ASTM.

NCSBCS was established to (1) provide a forum for states to discuss and develop solutions to code regulatory problems, (2) promote adoption and administration of uniform building codes and standards that regulate construction in and between states, (3) create an effective voice for state input on the committees of nationally recognized standards-writing organizations, (4) support comprehensive training and educational programs for code enforcement personnel, and (5) foster cooperation and encourage innovation between code administration officials and the design, manufacturing, and consumer interests affected by the code regulatory system.

# 1.4 Barrier-Free Design

As related to building design, a *physically handicapped* person is an individual who has a physical impairment, including impaired sensory, manual, or speaking abilities, that results in a functional limitation in gaining access to and using a building or facility.

Until recently, most of the buildings, facilities, transportation systems, and other structures and spaces in which we work, play, and live (the built environment) were designed for use by average, able-bodied, young adults who walk without aid or assistance (ambulatory). People who were older, younger, smaller, taller, or who had physical handicaps were at a disadvantage in most places. Many buildings were inaccessible to them, and some were even dangerous. Fortunately, this picture is rapidly changing.

Physical conditions that make a building or facility unsafe or confusing or that prevent physically handicapped people from using them are called *architectural barriers*. *Barrier-free*, or *accessible*, design eliminates or avoids creating such barriers. Barrier-free design is frequently thought of as a way to accommodate a few special people called the *handicapped*. The wheelchair symbol, used to designate public parking spaces, toilets, telephones, and water fountains, tends to foster the belief that barrier-free design offers accessibility mainly to those in wheelchairs. Actually, barrier-free design benefits everyone, because it makes facilities safer and more convenient to use.

Barrier-free design benefits not only the physically handicapped, but also (1) children who are physically and mentally immature, (2) pregnant women who have reduced agility, stamina, or balance, (3) people who care for children and must carry them, hold their hands, or maneuver their baby carriages, (4) older people who may suffer progressive degeneration of physical, perceptual, and mental faculties, (5) those disabled by size-related disorders such as giantism, dwarfism, or obesity, (6) able-bodied people who suffer temporary illness or injury, and (7) even able-bodied people who are carrying large packages.

Some people are born with a limited or unusable physical, mental, or sensory function (*disability*), but even a normally able-bodied person can become either temporarily or permanently disabled at any moment as the result of an illness or injury. *Disabilities* include visual impairment, hearing impairment, mental or perceptual impairment, confinement to a wheelchair, and coordination disability.

Nationwide, about 22% of those between the ages of 15 and 64 and about 59% of those 65 and older are disabled to some degree. That adds up to about 20% of the total population over 15 years old, or more than 37 million disabled people. Some sources put the number of disabled people at more than 43 million. It is most likely that the percentage of disabled people will rise in the future, because many people today survive illnesses and injuries that were once fatal. In fact, the average life span has increased dramatically during the twentieth century. In addition, the U.S. population as a whole is getting older. The number of people more than 65 years old is projected to grow from a little more than 10% of the total population in 1986 to more than 20% by 2040.

This section divides barrier-free design into four broad categories: (1) safety, (2) general accessibility criteria applicable to all buildings, (3) recommendations applicable to buildings for people with specific disabilities, and (4) requirements for making buildings adaptable for future use by the handicapped.

## 1.4.1 THE LAW AND APPLICABLE STANDARDS

In the late 1950s, the *President's Committee on Employment of the Handicapped* joined with consumer groups, such as the *National Easter Seal Society*, and standards makers to develop standards for making buildings accessible to handicapped persons. The result was ANSI A117.1, which was first published in 1961 by the American Standards Association (now ANSI). ANSI A117.1 quickly became, and remains today, a major standard for design for the handicapped. It is a voluntary standard, of course, having force only when adopted by a governing body.

In 1968, Congress passed the *Architectural Barriers Act*, which required that access for the handicapped be achieved in accordance with standards to be established by the GSA, HUD, and the Department of Defense (DOD). The rules they developed established ANSI A117.1

as the generic standard of access for buildings owned or leased by the federal government. Enforcement, however, was spotty at best. As a result, Congress passed the Rehabilitation Act of 1973, which established the *Architectural and Transportation Barriers Compliance Board* (ATBCB) to enforce the Architectural Barriers Act. The ATBCB issued a document called *ATBCB Minimum Guidelines and Requirements*.

The various states followed the federal government in passing laws requiring that buildings be made accessible to the handicapped. Unfortunately, many of them did not require compliance with ANSI A117.1 or with the *ATBCB Minimum Guidelines*, but rather developed their own standards. In an attempt to gain wider acceptance, ANSI expanded and reissued A117.1, but it still failed to gain universal acceptance. To make matters even more complicated, different federal agencies required compliance with different editions of ANSI A117.1.

In 1984, in the midst of all this confusion, GSA, HUD, DOD, and the U.S. Postal Service jointly issued the *Uniform Federal Accessibility Standard* (UFAS). Its purpose was to cause all federal agencies to follow the same technical requirements for complying with the Architectural Barriers Act. It was based on the *ATBCB Minimum Guidelines* but incorporated the basic requirements of ANSI A117.1 with some changes and additions.

As a result, various agencies of the federal government were requiring compliance with four different standards, two versions of ANSI A117.1, UFAS, and the *ATBCB Minimum Guidelines*. In some states, those four were being used in addition to the state's own standards. Some states actually require compliance with more than one standard. Even at the federal level, compliance with more than one was required by some agencies.

Gradually the two major standards have become the latest edition of ANSI A117.1, which at the time of this writing was ICC/ANSI A117.1-2003, "Standard on Accessible and Usable Buildings and Facilities," for private sector work and the UFAS for federal government work, although the others mentioned are still required by some state and local authorities. More than half of the states have adopted some version of ANSI A117.1 or incorporated it into their own standards. The International Codes, all model codes, and the NFPA 101 *Life Safety Code* incorporate or reference some edition of ANSI A117.1.

The *Federal Fair Housing Act of 1988, Title VIII of the 1968 Civil Rights Act*, prohibits housing discrimination based on a physical handicap in housing projects with four or more dwelling units. Basically, it says that everything in such dwelling units must be adaptable for accessibility by the handicapped, and all public areas serving such dwelling units must be accessible. Compliance must be in accord with the *Final Fair Housing Accessibility Guidelines* published in the Federal Register on March 6, 1991.

The *Americans with Disabilities Act* of 1990 prohibits discrimination on the basis of disability in places of employment, public service (transportation facilities), or in public accommodations (restaurants, hotels, theaters, doctors' offices, retail stores, museums, libraries, parks, private schools, day care centers, and others), and in telecommunications services. The standards developed by the ATBCB to implement this law differ slightly from and expand on ICC/ANSI A117.1 and the UFAS but may ultimately incorporate one or the other as the basis of the new standard.

The recommendations in this section come from many sources, but most of them comply with the requirements of ICC/ANSI A117.1, the *Uniform Federal Accessibility Standards*, the HUD *Minimum Property Standards for Housing*, and the HUD document *Adaptable Housing*. Keep in mind that there are some differences between these standards, and that state and local requirements may contain some other differences. In addition, these standards and the *Fair Housing Accessibility Guidelines* issued in conjunction with the *Federal Fair Housing Act of 1988* are subject to modification at any time. Therefore, when dealing with an actual project, whether for a federal or state agency or for a private sector owner, it is necessary to determine the specific requirements of the project and local, state, and federal laws and ordinances that dictate the requirements for accessibility.

Rules for providing accessibility have developed slowly and will no doubt continue to expand and change as more is understood about the needs of handicapped people. Until a single standard emerges, it is a mistake to follow any one standard alone, or any edition of a standard, before verifying the actual requirements. It is necessary also to remember that both the Federal Fair Housing Act of 1988 and the Americans with Disabilities Act of 1990 are civil rights laws rather than code compliance laws. Accordingly, it is best to comply with their spirit and not try to circumvent their intent. In the event of a compliance dispute, a court is likely to favor the intent of the law rather than the wording of the standards with which it requires compliance.

## 1.4.2 SAFETY

Many accidents in buildings can be traced to obvious causes, such as slippery floors, the lack of grab bars, or inappropriate or faulty stair railings. Accident prevention requires elimination of these obvious causes as well as other measures. Floors that are likely to get wet, such as those of approach walkways, stoops, entryways, corridors, toilet rooms, bathrooms, shower rooms, locker rooms, and kitchens, should have slip-resistant surfaces. Throw rugs and area rugs should not be used in public spaces; where they are appropriate, they should have a non-slip backing.

### 1.4.2.1 Doors and Sidelights

*Safety glazing* consisting of tempered glass, laminated glass, or plastic should be used in all glazed or glazing-insert doors and sidelights. Most building codes and regulations require such glazing.

Rounded door and jamb edges or resilient door edges minimize injury to fingers.

### 1.4.2.2 Stairs

To help reduce the many accidents that occur on stairs, (1) the number of risers in a series should be at least three, because people are careless on fewer, (2) treads and risers should be uniform, (3) treads should be no less than 11 in. (280 mm) wide, measured from riser to riser, and should have a slip-resistant surface, and (4) risers should be not more than 7 in. (180 mm) high and should be closed.

### NOSINGS

Nosings should project not more than $1\frac{1}{2}$ in. (38 mm), and their undersides should be rounded. The radius at the leading edge of the tread should not be greater than $1\frac{1}{2}$ in. (13 mm).

### HANDRAILS

Handrails should run continuously along both sides of a stairway, extend parallel to the floor at least 12 in. (305 mm) beyond the top and bottom of the staircase, and be free of protrusions that might snag clothing. Rails should be securely mounted at a height of 30 to 34 in. (815 to 865 mm) from the floor. If children regularly use the stairs, an extra railing should be mounted at a height of 24 in. (610 mm) from the floor. The space between balusters, if present, should be no more than 5 in. (127 mm) since wider spacings may allow a child's head to become trapped.

### LANDINGS

Because long flights of stairs can be tiring, a landing should be provided for flights with more than 16 risers.

### 1.4.2.3 Toilet Rooms, Bathrooms, Shower Rooms, Locker Rooms

Most accidents in toilet rooms, bathrooms, shower rooms, and locker rooms involve falls caused by slipping or burns caused by scalding.

### WATER

A person can receive a third-degree burn in just 2 seconds in 150°F (6°C) water, 6 seconds in 140°F (65.6°C) water, and 30 seconds in 130°F (54.4°C) water. Many people have limited sensitivity to heat and may not be able to adequately gauge the temperature of water in a lavatory, sink, tub, or running shower. Others may not be able to react quickly enough to a sudden surge of hot water.

To prevent scalding, the temperature of hot water should be regulated by (1) setting the temperature of the heated water to below 115°F (40.6°C), (2) providing a temperature-regulating device on tub fillers and shower heads, or (3) providing a temperature-regulating device on the hot water supply lines. The use of either individual fixture-mounted temperature controls or a temperature-regulating device on supply lines is preferred because other uses, such as dishwashing and clothes washing, require temperatures from 120°F (48.9°C) to 140°F (60°C). Fittings are available that can be set to maintain a specific water temperature either at the fixture or at the supply pipe.

In addition, exposed hot water supply and drain pipes should be wrapped with insulation to prevent contact burns.

## SHOWERS AND TUBS

Nonslip strips should be provided on the bottom surfaces of tubs and showers. Soap dishes, tub fillers, and controls should be recessed so that users cannot fall on them.

## GRAB BARS

Firmly mounted grab bars should be placed in and near tubs and showers and adjacent to water closets to help wheelchair users transfer from the chair to the fixture and to help those with other disabilities and the elderly get up and down and maintain their balance.The width of a grab bar gripping surface should be $1^1/_4$ to $1^1/_2$ in. (32 to 38 mm). The shape of the bar should allow a natural grip. There should be no protrusions or rough surfaces to catch clothing or cause injury.

Bars should be spaced with $1^1/_2$ in. (38 mm) clearance from the wall to allow proper grasping but prevent the arm from slipping behind the bar.

### 1.4.2.4 Kitchens

Kitchen accidents, such as burns and falls, can be reduced by proper planning.

## APPLIANCES

More accidents are related to cooktops and ranges than to any other kitchen appliance. To prevent such injury, cooktop and range *controls* should be front- or side-mounted so that a cook does not have to reach over hot burners to adjust them. When dwelling residents include small children, controls should be out of sight on a horizontal surface and should either have a protective lid or require an extra movement, such as pushing, before a control can be turned to activate a burner. Electric cooktops should be equipped with a light to indicate that burners are on.

## CABINETS

Reaching for cabinets above refrigerators requires standing on a stool, which presents a falling hazard. Reaching for cabinets over cooktops or stoves may cause burns. Therefore, placing cabinets over refrigerators and stoves is not desirable.

### 1.4.2.5 Protruding Objects

Objects, such as telephones, located between 27 and 80 in. (685 and 2030 mm) above a finished floor should not project more than 4 in. (100 mm) into a room or circulation path (see Section 1.4.3.1). Objects mounted at or below 27 in. (685 mm) may project any amount because they are detectable by a moving cane. Free-standing objects mounted between 27 and 80 in. (685 mm and 2030 mm) on posts or pylons may overhang a maximum of 12 in. (305 mm). No projection should reduce the clear width of an accessible route or maneuvering space.

When objects project into circulation spaces, a change in texture or a contrast in color of the floor surface can serve as a warning. Such warning devices should, however, be used with discretion, because overuse will diminish their effectiveness.

### 1.4.2.6 Lighting

Sufficient glare-free lighting should be available everywhere to meet the varying intensity needs of different tasks and to eliminate dark shadows that may conceal hazards. Changes in light intensity should be gradual so as to give the eyes time to adjust.

The paths people follow to get from one place to another in the built environment are called *circulation paths* (see Section 1.4.3.1). Good, glare-free lighting in circulation paths, especially those in potentially hazardous areas such as stairways, is particularly important in the creation of a safe environment.

### 1.4.3 GENERAL ACCESSIBILITY

An *accessible* site, building, facility, or portion thereof is one that complies with current standards and can be approached, entered, and used by physically handicapped people.

### 1.4.3.1 Accessible Route

A *circulation path* is an exterior or interior way of passage from one place to another for pedestrians, including, but not limited to, walks, hallways, courtyards, stairways, and stair landings.

An *accessible route* is a continuous, unobstructed circulation path connecting every accessible element and space in a facility that can be negotiated by a person with a severe disability using a wheelchair and that is also safe for and usable by people with other disabilities.

### 1.4.3.2 Site Planning

Barrier-free design should include the planning of accessible site facilities. Individuals with disabilities should have as much freedom of movement outside as inside.

Barrier-free buildings demand barrier-free routes of access, including (1) convenient parking, (2) accessible walks and curb ramps, (3) suitable paving, and (4) logical organization.

Direct routes should be provided into a building from public transportation, parking lots, and passenger loading zones.

## PARKING

Parking spaces and passenger loading zones for people with disabilities should be located as close as possible to an accessible entrance and should be identified by the *international symbol of accessibility* (Fig. 1.4-1). If an accessible entrance cannot be seen from the parking area, signs should direct users to it.

Accessible parking spaces should be at least 96 in. (2440 mm) wide and have an adjacent access aisle at least 60 in. (1525 mm) wide. Two adjacent spaces may share one access aisle. Passenger loading zones should have an access aisle at least 48 in. (1220 mm) wide and 20 ft (6100 mm) long, adjacent and parallel to the vehicle pull-up space.

Curbs between access aisles and walks should have *accessible curb cuts* and *curb ramps*, as described in a following section, "Curb Ramps."

**FIGURE 1.4-1** International symbol of accessibility.

## WALKS

Walks and other access paths should be at least 36 in. (915 mm) wide for single passage and 60 in. (1525 mm) wide for passing. Paths that are less than 60 in. (1525 mm) wide should have periodically spaced areas where a pedestrian or wheelchair user can move over to allow passing. The clear *passage space* should not be reduced by protruding objects.

Walks should be smooth and nonslip, with gaps of not more than $1/4$ in. (6 mm) and level changes of not more than $1/2$ in. (13 mm) (see Section 1.4.3.4). Textural changes should warn users of level changes, projecting objects, and other hazards.

## CURB RAMPS

Curbs define and separate walks from vehicular traffic areas. A *curb ramp* is a short ramp cutting through a curb or built into it to remove barriers for wheelchair travel. There should be a curb ramp wher-

ever an accessible route crosses a curb and the path changes level more than $1/2$ in. (13 mm).

Curb ramps should be at least 36 in. (915 mm) wide and should not project into vehicular traffic (Fig. 1.4-2c), unless there is a *safety zone* on the roadway. Their slope should be 1:12 or less.

If anyone must cross perpendicular to a curb ramp, the side should be flared at a maximum slope of 1:10 (see Fig. 1.4-2a) or protected by planting strips or some other nonwalking surface (see Fig. 1.4-2b).

Curb ramps should be of a different texture than the surrounding passage to warn visually handicapped users of hazards. This *textural warning* area should extend for the entire width and depth of the curb ramp

## PLANTING STRIPS

Figure 1.4-3 shows typical *planting strips*. These strips are a good way to keep protruding objects and street furniture from intruding on a clear accessible path.

Planting strips can be combined with benches to provide convenient resting places for the elderly or for those with limited stamina (see Fig. 1.4-3b). Hazards should be marked with *detectable warnings*, as recommended in Section 1.4.4.1.

## LIGHTING

Proper illumination is just as important outdoors as it is indoors. Weatherproof fixtures should provide a minimum of 5 footcandles (53.82 lx) of light on the travel surfaces at entrances, ramps, and steps. Weatherproof surface or post lights or floodlights should provide 1 footcandle (10.76 lx) and at least $1/2$ footcandle (5.38 lx) of light along paths of travel and in parking areas.

## ORGANIZATION AND ARCHITECTURAL HARMONY

Frequently used public spaces inside and outside a building should be easy to find and use. Elements that make buildings

a

b

**FIGURE 1.4-2** Curb ramps may either (a) start at the curb or walkway and project into the roadway or (b and c) be curb cuts.

**FIGURE 1.4-3** (a) Planting strips help organize and separate street furniture from the pedestrian path. (b) Rest areas may be integrated with planting strips to provide convenient stopping places outside the circulation path.

**FIGURE 1.4-4** An attractively landscaped, fully accessible space provides a pleasant meeting area for everyone.

accessible, such as ramps, curb ramps, and signs, should blend harmoniously with other building elements. Careful design and site development in new construction can often reduce the need to add special elements that single out the handicapped. For example, a single-level building with appropriately sized doors and grade-level entrances would require no ramps, elevators, platform lifts, or special entrances.

Landscaping also contributes to the site (Fig. 1.4-4). Care should be taken to choose plants that do not obstruct the view or leave excessive debris in walkways. Use of poisonous and thorny plants should be avoided.

### 1.4.3.3 Entries and Passageways

Entries should be at grade level or made accessible by means of elevators or lifts (see Section 14.1). Doorways should have a minimum clear opening of 32 in. (815 mm) with the door open 90 degrees, measured from the face of the door to the opposite stop.

Hallways and corridors should be at least 36 in. (915 mm) wide. Where a door opens into a corridor, the corridor should be 54 in. (1370 mm) wide. Corridors where wheelchair users must pass each other or where wheelchairs must turn around should be 60 in. (1525 mm) wide.

### 1.4.3.4 Ramps

Passageways and walks are said to *blend to a common level* when their various portions meet in such a way that there is no abrupt rise or drop in the surface. Level changes up to $^1/_2$ in. (6 mm) high need no edge treatment; those between $^1/_4$ in. (6 mm) and $^1/_2$ in. (13 mm) should be beveled with a slope no greater than 1:2. Level changes greater than $^1/_2$ in. (13 mm) require a ramp in order to be accessible by wheelchair users (Fig. 1.4-5) and to provide easier passage for those who must use crutches, canes, or walkers. Ramps are the primary means by which wheelchair users enter buildings that do not have level entryways.

A *ramp* is a walking surface in an accessible space that has a running slope greater than 1:20. A *running slope* is the slope of a pedestrian way that is parallel to the direction of travel. Conversely, a *cross slope* is the slope of a pedestrian way that is perpendicular to the direction of travel.

Ramp slopes should be as gradual as possible. The maximum slope should be 1:12, but more permissive ramp dimensions are sometimes allowed for existing construction where space limitations prevent the use of this preferred slope. The maximum rise for a ramp is 30 in. (760 mm); minimum clear width is 36 in. (915 mm).

Level *landings* at least as wide as the ramp itself and at least 60 in. (1525 mm) long should be provided at the top and bottom of each ramp. Landings are also necessary at intermediate levels where the rise of a ramp exceeds 30 in. (760 mm). If a ramp changes direction, the landing should be a minimum of 60 by 60 in. (1525 by 1525 mm). The width of the platform at a *switch-back ramp* should be at least as wide as the width of the two ramp portions plus the width of the space between them (Fig. 1.4-6).

The surfaces of ramps and landings should be *slip resistant* and should not collect water. Their edges should have

**FIGURE 1.4-5** Ramps can be made to blend with other building elements.

**FIGURE 1.4-6** Layouts and sizes of flat platforms at ramps. For clarity, railings are not shown.

---

curbs, walls, railings, or other protection to prevent people from falling.

*Handrails* should be provided on both sides of ramps that rise more than 6 in. (150 mm) or are longer than 72 in. (1830 mm). Handrails should (1) extend at least 12 in. (305 mm) beyond the top and bottom of a ramp, (2) be parallel with the ramp, (3) have a clear space of $1\frac{1}{2}$ in. (38 mm) between the handrail and any adjacent wall, and (4) be 30 to 34 in. (760 to 865 mm) above the ramp surfaces. Inside handrails on switch-back or *dogleg ramps* should be continuous.

Because many handicapped people find it easier to negotiate stairs than ramps, both should be provided in buildings with an above-ground-level entry.

### 1.4.3.5 Stairs

Stairs should comply with the requirements stated in Section 1.4.2. In addition, there should be *detectable warnings* (see "Texture" in Section 1.4.4.1) at least 36 in. (915 mm) wide at the top of every stair run. Stairs should have closed risers of uniform height, treads should be slip resistant, and nosings should not be square or abrupt.

Steps and landings should have sharp color contrasts to help people with limited sensitivity to light and dark and limited depth perception. It is also helpful if the edge of a stair contrasts with the rest of the tread.

### 1.4.3.6 Signage

*Signage* is defined as verbal, symbolic, and pictorial information presented in a graphic, two-dimensional format.

Informational and warning signs should be easy to find and to read; that is, (1) they should be positioned as close as possible and perpendicular to the path of travel, (2) the information on them should contrast with the background, and (3) they should be made of glare-free materials.

Signage should also have a consistent format or use *international symbols*.

### LOCATION AND VIEWING DISTANCE

Most people can see within an angle of 30 degrees to either side of the centerline of their faces without moving their heads. People with disabilities often have limited head movement or reduced peripheral vision. Therefore, signs should be positioned as close as possible and perpendicular to the path of travel. Signs close enough to touch should be positioned for easy hand detection by a standing person, 54 to 60 in. (1372 to 1524 mm) from the floor.

The distance from which a sign will be read determines the size of the letters, numbers, and symbols used on it.

### DESIGN AND PROPORTION OF LETTERS, NUMBERS, AND SYMBOLS

Letters and numbers should be of sans serif (block) type and should be sized and proportioned as recommended in Figure 1.4-7.

Characters on signs that must be read from a distance should be large enough to be read easily from the maximum viewing distance expected. Characters on signs close enough to touch should be raised at least $\frac{1}{32}$ in. (0.8 mm) from their background. They should also be at least $\frac{5}{8}$ in. (16 mm) tall, but no taller than 2 in. (50 mm), and have a stroke width of at least $\frac{1}{4}$ in. (6 mm) in order to be easily detectable by one finger (Fig. 1.4-8).

The raised characters on each sign close enough to touch must be accompanied by grade 2 Braille. *Braille* is a special raised touch alphabet for the blind that uses a cell of six dots. Unfortunately, only about 10% of those with visual impairments can read braille, but raised Arabic symbols can be read by anyone, regardless of visual handicap. To aid those who can read it, braille characters in the standard dot size and spacing may be added to the left of standard characters.

Borders, which may confuse a reader with a visual impairment, are not recommended on signs.

**FIGURE 1.4-7** (a) Proportions of letters expressed as a ratio; (b) dimensions of letters for signs close enough to be touched.

**FIGURE 1.4-8** A sign with raised lettering can be read by both those with normal sight and those with visual impairments.

### COLOR CONTRAST

The background and characters on a sign should contrast. Dark characters may be used on a light background, but light characters on a dark background are more easily readable. Surfaces should be glare free.

### 1.4.3.7 Surfaces

Floor surfaces should be nonslip. When practicable, they should be carpeted to add the advantage of noise reduction, which helps those who must use hearing aids.

Carpet should (1) be securely attached, (2) have either a firm backing or no backing, and (3) have a pile height not in excess of $1/2$ in. (13 mm) so that it remains usable by those in wheelchairs. Shag carpets should be avoided. Carpets should be of *antistatic* materials or treated to be static free, because static electricity interferes with hearing aid operation. Humidifiers may be used to help control static electricity. Exposed carpet edges should have carpet trim along the entire carpet length.

### 1.4.3.8 Electrical

Electrical outlets should be centered at least 15 in. (380 mm), and preferably 24 in. (610 mm), above the floor.

The best height for wall switches is centered 48 in. (1220 mm) above the floor, but they should never be placed higher than 54 in. (1370 mm) above the floor. Where persons in wheelchairs will approach from a forward position, as is the case when switches are located over counters with knee spaces, switches should be centered between 36 and 40 in.

(915 and 1016 mm) above the floor. In no case should switches and outlets over counters be placed more than 48 in. (1220 mm) above the floor.

Lighting should be as described in Sections 1.4.2 and 1.4.3.2.

### 1.4.4 SPECIFIC HANDICAPS

The kinds of specific handicaps discussed in this section include (1) *visual impairments*, (2) *hearing impairments*, (3) *mental and perceptual impairments*, (4) *disabilities* that require the use of a wheelchair, and (5) *coordination disabilities*.

Designs that *remove barriers* for people with one kind of disability sometimes *create barriers* for those with a different type of disability. For example, curb ramps (see Fig. 1.4-2) installed to permit easy street crossings for wheelchair users can cause a person with a visual impairment to wander unknowingly into automobile traffic. Both environmental needs can be met by providing a textured pavement that will act as a warning to those with visual impairments but will not impede wheelchair travel.

The recommendations in Sections 1.4.2 and 1.4.3 apply to spaces for those with specific handicaps, but the more demanding provisions discussed in this section are also necessary to accommodate specific disabilities.

### 1.4.4.1 Visual Impairment

Only about 10% of those with visual disabilities are totally blind. Legal blindness is usually defined as vision at or below 20/200. Some of those with visual disabilities have impaired peripheral vision; others cannot distinguish light and dark; some cannot perceive the full spectrum of colors.

Many people who have a visual impairment rely on memory instead of their eyes. They use a long cane, a guide dog, or other devices such as an *infrared detector* as aids for traveling to specific destinations. These destinations have often been explored with the help of someone with sight. Dwelling units may need little adaptation, because those with visual impairments become familiar with their surroundings.

Persons with visual impairments learn to rely to a certain extent on *tactile response* (touch). A *tactile object* is one that can be perceived using the sense of touch.

### SIGNAGE

The proper type of signage is a very important means of communication for visually handicapped people. This signage should follow the rules set down in Section 1.4.3.6.

### TEXTURE

A *cue* is a device that alerts a handicapped person to an upcoming condition. Cues may be *audible*, *visual*, or *textural*. A cue that consists of a standardized surface texture applied to or built into a walking surface or other element, to warn visually impaired people of a hazard in the path of travel, is called a *detectable warning*.

Detectable warnings help those with visual impairments to avoid hazards such as automobile traffic, level changes, and physical obstacles. Detectable warnings on walking surfaces include (1) exposed aggregate concrete, (2) rubber or plastic cushioned surfaces, (3) raised strips, and (4) grooves. Textures used for warning should contrast with surrounding surfaces but should be uniform in design within any building or site. Grooves should be used indoors only. Handles on doors leading to hazardous areas (such as spaces housing mechanical equipment) should be knurled or roughened as a warning.

### OBSTACLES

Benches and other obstacles in the direct path of travel should be surrounded by detectable warnings to aid those with visual impairments, unless the obstacles are encircled by other indicators such as curbs (Fig. 1.4-9).

**FIGURE 1.4-9** The textured surface around the planter and the bench serves as a warning to those with visual impairments.

## PROJECTIONS

Many people with severe visual impairment use white canes to aid their mobility. However, the cane technique is useful only if hazards are detectable. Refer to Section 1.4.2 for requirements related to projecting objects.

## SOUND

An *audible cue* is a sound or verbal alert. These are other means of helping those with visual impairments. Audible alarms should be incorporated into emergency warning systems. Alarm signals should be at least 15 decibels (dB) louder than the room's normal sound level but should not exceed 120 dB.

### 1.4.4.2 Hearing Impairment

There are more than 13,000,000 people in the United States with partial or total hearing loss. Just as sound compensates those with visual impairments for the loss of sight, visual cues help those with hearing impairments to make up for their loss.

### WARNING AND INDICATION LIGHTS

*Emergency signals*, such as fire and burglar alarms, should have *visual* as well as *audible warning systems*. These should be located so that the signal or its reflection is clearly visible and should flash with a frequency of not more than 5 cycles per second (Hz).

Elevator *directional lights* should be at least $2^1/_2$ in. (63 mm) in the smallest dimension and should be mounted at least 72 in. (1830 mm) above the floor. Refer to Section 14.1 for additional requirements.

### SOUND

Control of background noises and amplification of sound can help those with partial hearing loss to communicate.

For hearing aids to function properly, reduction of background noise, including reverberation, and control of sound frequencies are important. Use of ultrahigh-frequency sound security systems and low-cycle electric transformers should be avoided. Further information may be found in Section 13.1 and in Chapter 18.

Wheelchair-accessible telephones and 25% of all other telephones must be fitted with adjustable volume controls and other devices. Such telephones in public areas must be identified as such (Fig. 1.4-10a).

Text telephones (TTYs) are required in some locations. Where TTYs are re-

**FIGURE 1.4-10** (a) Sign identifying a volume-adjustable telephone for those with hearing impairments; (b) international TTY symbol. (From CD-ROM version 3.0 of Architectural Graphic Standards. Courtesy John Wiley & Sons, Inc.)

quired by the applicable standard (ICC/ANSI 117.1 or USAF) in public areas, they must be identified by a sign bearing the international TTY symbol (Fig. 1.4-10b).

### 1.4.4.3 Mental and Perceptual Impairment

Clear signage and good organization are especially important for those who are *mentally immature* or who have *perceptual* or *cognitive disabilities*.

If people must frequently ask for the location of a commonly used facility such as a washroom, that environment may have a barrier. Many individuals with disabilities are self-conscious and will not ask for help; if they feel uncomfortable in a facility, they often will not use it. Public facilities should be easy to locate and to utilize.

### SIGNAGE

It is especially important to those with mental impairments that signs be easy to locate and to read. The basic requirements for signs are discussed in Section 1.4.3.6. Many individuals with mental impairments who are unable to read learn to recognize information by word length. For example, *women* is a longer word than *men*. Use of the words *ladies* and *gentlemen* may confuse those familiar with the three- and five-letter words.

### ORGANIZATION

The location of specific rooms within an area should be clearly identified. If corridors in different areas are similar in appearance, an effort should be made to create *landmarks* using color or other decoration. Stairways and other circulation areas should be clearly identified or sited in highly conspicuous locations.

### 1.4.4.4 Wheelchair Users

The abilities of wheelchair users differ widely. Disability may affect only the legs or may involve the entire body. Pressure sores, muscle atrophy, and blood pooling are frequent problems. Paralysis reduces the muscle tone of the affected area. Barrier-free environments encourage mobility, and the exercise gained helps prevent additional medical problems.

### WHEELCHAIR ACCESS

Most adult-sized wheelchairs are of a standard size; some special wheelchairs may be larger. Wheelchairs need an accessible route from parking spaces or bus stops to the interior of a building and within the building to the final destination of the wheelchair user.

***Site Access*** Refer to Sections 1.4.3.2 and 1.4.3.4 for requirements.

***Elevators and Lifts*** Unless every space is accessible by level movement, or ramps or other vertical means of access are provided, elevators or platform lifts should be provided for wheelchair users. In some applications, *chair-type stair lifts* can be provided to assist persons who cannot readily climb stairs, when the stairs cannot be eliminated and elevators or platform lifts cannot be provided.

***Entries and Passageways*** Refer to Sections 1.4.2 and 1.4.3.3 for requirements for doors, hardware, and thresholds. Figure 1.4-11 shows the recommended minimum approach spaces necessary for convenient maneuvering of wheelchairs at doorways.

Wheelchair users frequently use the wheelchair itself to push a door open, scuffing or scratching the door. Therefore, the bottom 10 in. (255 mm) of hinged accessible doors should have a smooth, hard-surface *kickplate* (Fig. 1.4-12).

### ROOM SIZE AND ARRANGEMENT

*Accessible* spaces for wheelchair users should be on one level and have adequate circulation and maneuvering space. Windows and doors should be located so that furniture arrangements do not obstruct wheelchair circulation.

The minimum clear floor area that will accommodate a wheelchair is 30 by 48 in. (760 by 1220 mm). The minimum diame-

FIGURE 1.4-11 Minimum clear floor space at doors: (a) front approach, (b) hinge side approach, and (c) latch side approach.

FIGURE 1.4-12 An abrasion-resistant kickplate of stainless steel or hard plastic should protect the lower 10 in. (254 mm) of doors intended for wheelchair users.

ter needed for a wheelchair user to execute a 180 degree turn is 60 in. (1525 mm).

Figure 1.4-13 shows examples of accessible bedroom plans. In an accessible dwelling, at least two bedrooms should be on the main level.

## TOILET ROOMS, SHOWER ROOMS, AND BATHROOMS

Wheelchair users should be able to enter and use toilet, shower, and bathroom facilities without squeezing past obstructing fixtures or doorways.

*Clear Floor Space* Accessible toilet rooms, shower rooms, and bathrooms should have the recommended clearances around plumbing fixtures and controls. Ideally, an accessible facility should have a clear floor area 60 in. (1525 mm) in diameter so that a wheelchair user can make a 180 degree turn. The clear floor spaces of the fixtures, controls, and turning space may overlap.

*Doors* Doors should have a minimum clear opening of 32 in. (815 mm). Doors should not swing into the clear floor area of a fixture or obstruct entrance to the room. If a door must swing out, it should not obstruct passage through a corridor or any other circulation space.

*Water Closets* Water closets in an accessible public toilet room should be 17 to 19 in. (430 to 484 mm) high measured to the top of the toilet seat. In a dwelling, the seat height should be at least 15 in. (380 mm). The bowl should be centered 16 to 18 in. (405 to 455 mm) from each side wall.

*Lavatories* Lavatories should be mounted so as to provide at least 29 in. (735 mm) of knee space to the underside of the apron. Exposed water pipes that in-

trude into the knee space should be padded and insulated.

*Mirrors* Mirrors should be mounted with the bottom edge not higher than 40 in. (1015 mm) from the floor. Medicine cabinets should be mounted with a usable shelf not higher than 44 in. (1120 mm) above the floor.

*Bathtubs and Showers* Depending on the location of the bathroom door, a bathtub should have a clear floor space 30 to 48 in. (760 to 1220 mm) wide and 60 to 72 in. (1525 to 1905 mm) long. Showers should have a clear floor space at least 36 in. (915 mm) wide and 48 in. (1220 mm) long for *transfer stalls* and 60 in. (1525 mm) long for *roll-in stalls*.

Seats should be provided in both transfer shower stalls and tubs. In shower stalls the seat should extend the full depth of the stall on the wall opposite the controls.

A shower spray unit that can be used either as a fixed shower head or a handheld shower should be provided. The hose should be at least 60 in. (1525 mm) long.

Curbs in shower stalls 36 in. (915 mm) square should be no higher than 4 in. (100 mm). Stalls, such as those 30 by 60 in. (760 by 1525 mm), that allow a wheelchair to be rolled in should not have curbs, but may have a maximum $1/2$ in. (13 mm) threshold.

Enclosures on bathtubs and showers should not obstruct the use of controls or the transfer from wheelchairs to seats.

*Faucets* The controls on bathtubs, showers, and lavatories should be within comfortable reach of a wheelchair user. Controls such as handle and drain mechanisms should be operable with one hand and not require tight grasping or twisting.

Single bed

Double bed

FIGURE 1.4-13 Minimum dimensions for accessible bedrooms.

## STORAGE

A clear floor space at least 30 by 48 in. (760 by 1220 mm) should be provided near storage areas. Closet rods should be mounted not higher than 54 in. (1370 mm) above the floor.

## KITCHENS

Suitable kitchen appliances, proper countertop heights, and adequate circulation space are the primary requisites of a wheelchair user's kitchen.

*Clear Floor Space*  The minimum clear floor space between opposing base cabinets, countertops, appliances, and walls should be 40 in. (1015 mm), unless the kitchen is U-shaped. U-shaped kitchens should have 60 in. (1525 mm) of clear floor space between opposing cabinets (Fig. 1.4-14). Clear floor space of 30 by 48 in. (760 by 1220 mm) should be provided at appliances such as a cooktop, oven, refrigerator, and dishwasher.

*Controls*  Knobs, faucets, and appliance controls must be within comfortable reach and operable with one hand.

*Counters*  One 30 in. (760 mm) section of counter, intended for food preparation, should be adjustable to a range of heights, including at least three positions: 28, 32, and 36 in. (710, 815, and 915 mm).

There should be no cabinets under the adjustable counter, and it should have a clear knee space 30 in. (760 mm) wide by 19 in. (485 mm) deep over a finished

**FIGURE 1.4-14** Minimum clearances between opposing cabinets in (a) parallel and (b) U-shaped kitchens.

floor space. Counter thickness and supports should be 2 in. (50 mm) maximum over the necessary clear space. The underside of the counter should not have any sharp protrusions or rough surfaces.

*Appliances*  Refer to Section 11.1.2 for accessibility requirements for kitchen appliances.

*Cabinets*  Both base and wall cabinets should be reachable from a wheelchair. At least one continuous shelf in the cabinets above a work counter should not be higher than 48 in. (1220 mm). Upper cabinet doors should have pulls located as close to the bottom as possible; base cabinets should have pulls located as close to the top as possible. Because of the necessity for knee space, cabinets should not be built under counters. Instead, pantries should be used for storage. They should have storage below the 54 in. (1370 mm) height and a clear space of 30 by 48 in. (760 by 1220 mm).

Pull-out shelves in base cabinets, and lazy Susan rotating corner units are highly recommended.

*Sinks*  Sink heights should be adjustable to 28, 32, and 36 in. (710, 815, and 915 mm). The minimum width of a counter surrounding a sink should be 30 in. (760 mm). Plumbing should be roughed in for sinks mounted at the 28 in. (710 mm) height. Counter thickness and supports should be 2 in. (50 mm) maximum over the necessary clear space. A sink bowl should be not more than $6^1/_2$ in. (165 mm) deep; in a double-bowl sink only one of the bowls needs to be so shallow.

Faucets should be at the side of the sink and operated by a single lever control.

## ELECTRICAL

The location of electrical outlets and switches should be as described in Section 1.4.3.8. Lighting requirements are discussed in Section 1.4.2.

### 1.4.4.5  Coordination Disabilities

*Coordination disabilities* include balance, leg, upper limb, and flexibility impairments. Many people have problems with balance, stamina, and coordination, which do not confine them to a wheelchair but do limit their use of an environment. They may have difficulty walking and may need aids such as walkers

or crutches, which require approximately the same amount of space as wheelchairs. Others may have difficulty bending, sitting, kneeling, or rising; still others may find it difficult to push open doors or turn knobs and faucets.

The needs of those with coordination disabilities and the needs of the elderly frequently overlap. However, younger people with disabilities do not want to be associated or housed solely with the elderly.

## ENTRY AND CIRCULATION

People who use walking aids need the same amount of *circulation space* as those who use wheelchairs. Many people with coordination disabilities prefer stairs to ramps because ramps are more tiring to use. However, ramps may be beneficial to crutch and walker users who cannot negotiate stairs as easily. Ramps should be constructed according to the criteria established in Section 1.4.3.4. Wherever possible, both stairs and ramps should be provided.

## SURFACES

Unevenness, raised joints, and debris can be hazardous to those with walking disabilities. Surfaces should be kept free of debris; joints and level changes should be less than $1/_4$ in. (6 mm).

### 1.4.4.6  The Elderly

Accessibility recommendations for the elderly overlap those for wheelchair users and persons with coordination disabilities. Sensory losses accelerate at various ages. For example, hearing starts to diminish at 40; vision, at 50; taste, between 55 and 60; smell, after 70. The actual ages and degree of loss vary considerably.

The elderly have a higher rate of visual impairment than those in other age groups. Age impedes the eyes' ability to change focus, adjust to sudden changes in light intensity, and function at low levels of light intensity. For example, an 80-year-old may need three times as much light for comfortable reading as a 16-year-old. As it ages, the lens of the eye becomes increasingly rigid, yellow, and opaque. This process affects the ability to discern color, especially blues and greens, and to distinguish between colors such as pink and lavender. With aging, depth perception decreases and glare becomes a greater problem.

Aging also increases hearing loss, reducing the ability to hear sounds in general and high-frequency sounds in particular.

The cushioning of the foot degenerates with advancing age. This loss can cause pain and contributes to the shuffling gait characteristic of many elderly people (Fig. 1.4-15).

The mobility of the elderly varies from complete ambulatory freedom to wheelchair confinement. Entries and other access facilities should be barrier-free for elderly persons, including wheelchair users. Only the criteria that add to or differ from those established for wheelchair users will be mentioned in this section.

### SIGNAGE

Many elderly people suffer visual impairment, and others may be easily confused; signage and lighting should conform to the recommendations outlined in Sections 1.4.3.6 and 1.4.4.1.

### ROOM SIZE AND ARRANGEMENT

Recommended room sizes for the elderly are greater than the minimums specified for wheelchair users. This is partially because the elderly sometimes have coordination problems that make it more difficult for them to maneuver a wheelchair

Pressure point          Pressure point

**a**

Pressure point          Pressure point

**b**

**FIGURE 1.4-15** The foot of a younger person (a) contains more cushioning material at pressure points than that of an older person (b).

within the minimum spaces recommended for other wheelchair users.

The criteria for corridor widths and floor surfaces should follow those indicated in Sections 1.4.3.3 and 1.4.3.7, respectively.

### COLOR AND TONE

As mentioned earlier, changes in the eye lens caused by aging often affect color perception. It may be difficult for some elderly people to distinguish between the light tints of pastel colors such as blue, lavender, and pink, and sometimes even between dark colors such as navy, brown, and charcoal. Therefore, the selection of colors for elderly persons should emphasize objects against their background. Examples include light entry–dark door jamb and light floor–dark furniture.

### ELECTRICAL

Electrical outlet and switch locations and lighting requirements should be as described in Section 1.4.3.8.

### BATHROOMS

Because the elderly may have to use a wheelchair at times, bathroom requirements should be the same as for wheelchair users.

### KITCHENS

The requirements for kitchens used by the elderly vary. Some of the elderly are reasonably able-bodied; others use wheelchairs. Kitchens for the elderly generally should follow the recommendations for adaptable kitchens (see Section 1.4.5), thus permitting ready alterations to suit the specific needs of the occupants.

Kitchen storage should be accessible without endangering the safety of the elderly. Because not all of the elderly are wheelchair users, the use of wall cabinets is acceptable. However, these are not recommended above a refrigerator or cooktop. Pantries are an excellent means of providing safe and accessible storage. Base cabinets should have *pull-out* shelves and corner lazy Susans to make items in them more accessible.

### 1.4.5 ADAPTABILITY

Most people with disabilities resent being segregated into housing built specifically for the handicapped. They prefer *mainstreaming*, a process which (1) equips

people who have disabilities with the personal devices and adaptive skills needed to function effectively in the built environment and (2) removes physical barriers that prevent handicapped people from functioning like able-bodied individuals.

Current laws and regulations require that most types of buildings be made fully accessible to handicapped people. Other laws and regulations have been proposed that would mandate making new housing also fully accessible. When they take effect, the following discussion may become moot. Until then, however, one way to prevent the segregation of people who become disabled into housing prepared specifically for handicapped people is to build adaptable buildings. An *adaptable* building is one that can be modified to satisfy the needs of an occupant's physical limitations with ease and without excessive costs. The kinds of elements that may be modified or added in an adaptable house, for example, include, but are not limited, to kitchen counters, sinks, and grab bars.

Adaptability for wheelchair users requires the most modification and should therefore receive the greatest attention in the design process.

The recommendations in Sections 1.4.2 and 1.4.3 apply fully to adaptable buildings. In general, adaptability requires that the following design requirements be followed.

### 1.4.5.1 Entries

When practicable, entrances should be at grade level so that ramps or stairs will not be needed. When this is not practicable, an adaptable dwelling can be on a level accessible by a ramp or an elevator or platform lift.

### DOORS

Passage doors should have a minimum clear opening of 32 in. (815 mm) with the door open 90 degrees, measured from the face of the door to the opposite stop.

*Hardware* Door knobs, handles, pulls, latches, and locks should have a shape that is easy to hold with one hand and does not require tight grasping, pinching, or twisting. Extensions are available that adapt a regular door knob for people with manual disabilities such as those caused by arthritis.

*Thresholds* Thresholds on exterior sliding glass doors should not be higher than

$3/4$ in. (18 mm). A recessed track is preferred where suitable, such as for interior sliding doors. Thresholds on hinged passage doors should not exceed $1/2$ in. (13 mm) in height and should be beveled. Exterior doors should have weather stripping applied to the bottom edge of the door, because weather stripping applied to thresholds can present a tripping hazard (Fig. 1.4-16).

### 1.4.5.2 Room Size and Arrangement

A single-level dwelling is the preferred type for adaptability. Rooms should be large enough to permit convenient furniture arrangement and wheelchair maneuvering. Most living rooms, dining rooms, and bedrooms already meet the requirements for adaptability. Bathrooms and kitchens require the most modification.

### CORRIDORS

Hallways and corridors should be a minimum of 36 in. (915 mm) wide. If a door opens into a corridor, the passage width should be 54 in. (1370 mm); corridors where people must pass each other should be 60 in. (1525 mm) wide.

### BATHROOMS

An adaptable bathroom has the appearance of a conventional bathroom but with the layout and clearances suitable for later adaptation to the needs of those with disabilities.

***Clear Floor Space*** A typical bathroom should have a minimum width of 60 in. (1525 mm). Clear floor spaces around fixtures should be as shown in Figure 1.4-17.

***Doors*** Bathroom doors should have a minimum clear opening of 32 in. (815 mm); doors should not swing into the

**FIGURE 1.4-16** Thresholds.

**FIGURE 1.4-17** Recommended minimum dimensions for accessible bathrooms.

clear floor space around fixtures or obstruct entrance into the bathroom.

Bathroom doors can be rehung to open out in order to gain the necessary clear door space, but should not obstruct corridors or other circulation space outside the bathroom.

***Lavatories*** Lavatories should be mounted so as to provide at least 29 in. (735 mm) of clearance from the floor to the underside of the apron. The clear floor space should be at least 30 in. (760 mm) wide by 48 in. (1220 mm) deep; the lavatory basin should take no more than 19 in. (485 mm) of that depth, and pipes should not protrude into the knee space.

***Water Closets*** Water closet seats should be at least 15 in. (380 mm) above the floor. Clear floor space and structural wall reinforcement for grab bars should be provided.

***Mirrors*** The bottom edge of mirrors should be mounted not higher than 40 in. (1015 mm) above the floor. Medicine cabinets should be mounted with at least one usable shelf not higher than 44 in. (1120 mm) above the floor.

***Bathtubs and Showers*** Depending on the location of the bathroom door, bathtubs and showers should have adequate clear floor space to accommodate wheelchairs.

Structural wall support and floor space should be provided so that grab bars and a seat can be readily installed for a resident with a handicap. A flexible-hose shower unit that can be used as either a fixed or hand-held head should be provided. Shower stalls 36 by 36 in.

(915 by 915 mm) should have curbs not higher than 4 in. (100 mm). Stalls 30 by 60 in. (760 by 1525 mm) should not have curbs so that a wheelchair may be rolled in; a $1/2$ in. (13 mm) threshold is permissible, and the stall floor should be pitched slightly toward the drain to help avoid drainage problems.

***Grab Bars*** Structural reinforcement or other provisions should be provided near bathroom fixtures in the locations where grab bars would be installed.

### KITCHENS

Adjustable-height counters, sinks, and cooktops, and adequate circulation space are the key elements of an adaptable kitchen that allow a wheelchair user to maneuver freely within the space.

***Clear Floor Space*** Clear floor space requirements are the same as those for wheelchair users (see Section 1.4.4.4).

***Counters*** One section of counter 30 in. (760 mm) wide should be capable of being repositioned to heights of 28, 32, and 36 in. (710, 815, and 915 mm) to provide a comfortable area for food preparation. Cabinets may be mounted under the adjustable counter, but should be readily removable in order to provide the necessary knee space for a wheelchair user.

***Appliances*** Appliances in adaptable kitchens should fulfill the accessibility requirements indicated in Section 11.1.

***Cabinets*** Wall cabinets should have pulls located as close to the bottom as possible; base cabinets should have pulls located as close to the top as possible.

Cabinet pulls should be smooth and rounded, without any sharp edges or pointed projections. Because storage under adjustable spaces may have to be removed, pantry storage is very desirable. Other helpful devices are lazy Susans for corner cabinets, pull-out shelves in base cabinets, and cutting boards with mixing bowl cutouts.

*Sinks* The requirements for sinks are the same as those outlined in Section 1.4.4.4, except that cabinets may be mounted under the sink; however, they should be removable in order to provide the necessary knee space for a wheelchair user.

## FLOORS

Flooring should be *nonslip* even when wet.

## ELECTRICAL

Lighting should be as described in Section 1.4.2. Electrical outlet and switch locations should be as indicated in Section 1.4.3.8.

## COST

When practicable, new dwellings should be made adaptable, because the cost of making an adaptable dwelling suit the needs of a wheelchair user is small in comparison with the cost of retrofitting an existing conventional house. In addition, making a house adaptable does not in any way harm its use by the able-bodied.

# 1.5 Sustainable Building Design

Until relatively recently, architects and constructors have had a tendency to take little notice of the extensive use of materials, some of them nonrenewable, in their building projects. It seemed that the supply of wood, for example was inexhaustible. Electricity and fuel oil were relatively inexpensive. Architects tried to preserve the natural environment adjacent to their buildings as much as was practicable, but often took little notice of the environmental impact of the building materials they were specifying. The exhausting of irreplaceable materials in the making of building products was more or

less ignored, as were the production of waste as a result of the manufacturing of these products and the introduction of toxic by-products to the air, water, soils, and people. Only recently have we become more aware of the need to conserve both materials and energy in constructing our environment.

Recent studies have shown that for every 10.76 sq ft (1 m$^2$) of building 131,000 Btu (138.2 million J) of energy are expended, 815 lb (369.7 kg) of green house gasses are produced, and 132 lb (59.9 kg) of solid waste go into a landfill. These amounts do not include the effects of building operations after construction.

As our awareness has grown, we have struggled with the means to effect the kinds of changes necessary to preserve the environment without making buildings too expensive to build. Slowly, the terms *green buildings* and *sustainability* have come into vogue, but there have been no consensus definitions of them. Before architects could effectively design green buildings, there had to be a common standard of measurement for them. It was first necessary to define the problem in a way that was understandable to building professionals and the public alike.

The purpose of producing green or sustainable buildings is to meet the needs of the present society while preserving the ability of future generations to meet their own needs. To design sustainable buildings, it is necessary to think in terms of integrated whole-building design. Other than small residential and commercial projects, most buildings today are designed in several separate segments. Architects are the designers of the overall building; structural engineers make buildings stand up; and the mechanical, plumbing, and electrical services necessary are the responsibility of the architect's mechanical, plumbing, and electrical engineering consultants. Other consultants design other portions of a building. The architect coordinates all of their work. It is difficult to produce sustainable buildings using these methods. Buildings use too much power, their useful life span is reduced, and they are not easily brought up to date when new methods and systems arrive. As depletable materials become less plentiful, not only will the cost of producing building materials increase, the environment will become more toxic and less livable.

To remedy this situation, architects must incorporate resource conservation into their design philosophies, and architects, engineers, and constructors must provide environmental leadership in the building industry. They and governmental agencies together must raise consumer awareness of the benefits achieved by the production of green buildings, and they must stimulate competition among building product producers to provide products that are less stressful of the environment. Doing so will increase the availability of relatively inexpensive green building materials.

## 1.5.1 LEED®

The best current attempt to alleviate the lack of standards for green buildings is the LEED program developed by the U.S. Green Building Council (USGBC). With contributions from representatives of most segments of the building industry, the USGBC developed the LEED program, and continues to contribute to its evolution. *LEED* is an acronym for "Leadership in Energy and Environmental Design."

The LEED Green Building Rating System® is a voluntary, consensus-based national standard for developing high-performance, sustainable buildings. The LEED Project Checklist shown in Figure 1.5-1 demonstrates the extent of the measures necessary to provide sustainable design and the rating necessary to become a LEED-certified building.

The USGBC also certifies buildings for compliance with LEED guidelines. More information about the USGBC and LEED, and a copy of the LEED certification program, including the checklist reprinted in this section, are available at the USGBC Web site (http://www.usgbc.org).

## 1.5.2 LIFE CYCLE

To determine the sustainability or lack of sustainability of a building, it is necessary to assess the effect of the building on the environment during its entire life cycle. Factors that must be addressed include the following:

- The life cycle of each of the materials proposed for use in the building
- The most significant environmental risks associated with the building
- The opportunities available to provide for sustainability in the building and

# Sustainable Sites

14 Possible Points

| | | |
|---|---|---|
| **Y** Prereq 1 | **Erosion & Sedimentation Control** | Required |
| Y ? N Credit 1 | **Site Selection** | 1 |
| Y ? N Credit 2 | **Urban Redevelopment** | 1 |
| Y ? N Credit 3 | **Brownfield Redevelopment** | 1 |
| Y ? N Credit 4.1 | **Alternative Transportation**, Public Transportation Access | 1 |
| Y ? N Credit 4.2 | **Alternative Transportation**, Bicycle Storage & Changing Rooms | 1 |
| Y ? N Credit 4.3 | **Alternative Transportation**, Alternative Fuel Vehicles | 1 |
| Y ? N Credit 4.4 | **Alternative Transportation**, Parking Capacity | 1 |
| Y ? N Credit 5.1 | **Reduced Site Disturbance**, Protect or Restore Open Space | 1 |
| Y ? N Credit 5.2 | **Reduced Site Disturbance**, Development Footprint | 1 |
| Y ? N Credit 6.1 | **Stormwater Management**, Rate and Quantity | 1 |
| Y ? N Credit 6.2 | **Stormwater Management**, Treatment | 1 |
| Y ? N Credit 7.1 | **Heat Island Effect**, Non-Roof | 1 |
| Y ? N Credit 7.2 | **Heat Island Effect**, Roof | 1 |
| Y ? N Credit 8 | **Light Pollution Reduction** | 1 |

# Water Efficiency

5 Possible Points

| | | |
|---|---|---|
| Y ? N Credit 1.1 | **Water Efficient Landscaping**, Reduce by 50% | 1 |
| Y ? N Credit 1.2 | **Water Efficient Landscaping**, No Potable Use or No Irrigation | 1 |
| Y ? N Credit 2 | **Innovative Wastewater Technologies** | 1 |
| Y ? N Credit 3.1 | **Water Use Reduction**, 20% Reduction | 1 |
| Y ? N Credit 3.2 | **Water Use Reduction**, 30% Reduction | 1 |

# Energy & Atmosphere

17 Possible Points

| | | |
|---|---|---|
| **Y** Prereq 1 | **Fundamental Building Systems Commissioning** | Required |
| **Y** Prereq 2 | **Minimum Energy Performance** | Required |
| **Y** Prereq 3 | **CFC Reduction in HVAC&R Equipment** | Required |
| Y ? N Credit 1 | **Optimize Energy Performance** | 1–10 |
| Y ? N Credit 2.1 | **Renewable Energy**, 5% | 1 |
| Y ? N Credit 2.2 | **Renewable Energy**, 10% | 1 |
| Y ? N Credit 2.3 | **Renewable Energy**, 20% | 1 |
| Y ? N Credit 3 | **Additional Commissioning** | 1 |
| Y ? N Credit 4 | **Ozone Depletion** | 1 |
| Y ? N Credit 5 | **Measurement & Verification** | 1 |
| Y ? N Credit 6 | **Green Power** | 1 |

**FIGURE 1.5-1** LEED Project Checklist, Version 2.1. (Courtesy U.S. Green Building Council.)

# Materials & Resources

13 Possible Points

| | | | | |
|---|---|---|---|---|
| **Y** | | Prereq 1 | **Storage & Collection of Recyclables** | Required |
| Y ? N | | Credit 1.1 | **Building Reuse**, Maintain 75% of Existing Shell | 1 |
| Y ? N | | Credit 1.2 | **Building Reuse**, Maintain 100% of Shell | 1 |
| Y ? N | | Credit 1.3 | **Building Reuse**, Maintain 100% Shell & 50% Non-Shell | 1 |
| Y ? N | | Credit 2.1 | **Construction Waste Management**, Divert 50% | 1 |
| Y ? N | | Credit 2.2 | **Construction Waste Management**, Divert 75% | 1 |
| Y ? N | | Credit 3.1 | **Resource Reuse**, Specify 5% | 1 |
| Y ? N | | Credit 3.2 | **Resource Reuse**, Specify 10% | 1 |
| Y ? N | | Credit 4.1 | **Recycled Content**, Specify 5% (p.c. + $1/2$ p.i.) | 1 |
| Y ? N | | Credit 4.2 | **Recycled Content**, Specify 10% (p.c. + $1/2$ p.i.) | 1 |
| Y ? N | | Credit 5.1 | **Local/Regional Materials**, 20% Manufactured Locally | 1 |
| Y ? N | | Credit 5.2 | **Local/Regional Materials**, of 20% in MRc5.1, 50% Harvested Locally | 1 |
| Y ? N | | Credit 6 | **Rapidly Renewable Materials** | 1 |
| Y ? N | | Credit 7 | **Certified Wood** | 1 |

# Indoor Environmental Quality

15 Possible Points

| | | | |
|---|---|---|---|
| **Y** | Prereq 1 | **Minimum IAQ Performance** | Required |
| **Y** | Prereq 2 | **Environmental Tobacco Smoke** (ETS) **Control** | Required |
| Y ? N | Credit 1 | **Carbon Dioxide** ($CO_2$) **Monitoring** | 1 |
| Y ? N | Credit 2 | **Ventilation Effectiveness** | 1 |
| Y ? N | Credit 3.1 | **Construction IAQ Management Plan**, During Construction | 1 |
| Y ? N | Credit 3.2 | **Construction IAQ Management Plan**, Before Occupancy | 1 |
| Y ? N | Credit 4.1 | **Low-Emitting Materials**, Adhesives & Sealants | 1 |
| Y ? N | Credit 4.2 | **Low-Emitting Materials**, Paints | 1 |
| Y ? N | Credit 4.3 | **Low-Emitting Materials**, Carpet | 1 |
| Y ? N | Credit 4.4 | **Low-Emitting Materials**, Composite Wood | 1 |
| Y ? N | Credit 5 | **Indoor Chemical & Pollutant Source Control** | 1 |
| Y ? N | Credit 6.1 | **Controllability of Systems**, Perimeter | 1 |
| Y ? N | Credit 6.2 | **Controllability of Systems**, Non-Perimeter | 1 |
| Y ? N | Credit 7.1 | **Thermal Comfort**, Comply with ASHRAE 55-1992 | 1 |
| Y ? N | Credit 7.2 | **Thermal Comfort**, Permanent Monitoring System | 1 |
| Y ? N | Credit 8.1 | **Daylight & Views**, Daylight 75% of Spaces | 1 |
| Y ? N | Credit 8.2 | **Daylight & Views**, Views for 90% of Spaces | 1 |

# Innovation & Design Process

5 Possible Points

| | | | |
|---|---|---|---|
| Y ? N | Credit 1.1 | **Innovation in Design** | 1 |
| Y ? N | Credit 1.2 | **Innovation in Design** | 1 |
| Y ? N | Credit 1.3 | **Innovation in Design** | 1 |
| Y ? N | Credit 1.4 | **Innovation in Design** | 1 |
| Y ? N | Credit 2 | **LEED™ Accredited Professional** | 1 |

# Project Totals

69 Possible Points

**Certified** 26-32 points  **Silver** 33-38 points  **Gold** 39-51 points  **Platinum** 52-69 points

**FIGURE 1.5-1** *(Continued)*

how those are affected by construction cost and the operating cost of the completed building

- The client's attitude and resources

### 1.5.2.1 Building Materials Life Cycles

A material's life cycle begins with extraction of the raw materials used to manufacture it and ends when it is removed from the building and disposed of. This life cycle includes obtaining the raw materials, manufacturing the product, and its packaging, distribution, use, and ultimate reuse, recycling, or disposal. The environmental impact of a building material is affected by whether or not the material is renewable, the amount of toxic and waste materials that occur as a result of the conversion of the raw material into a building product, and the amount of energy the product consumes during its entire life cycle.

### RAW MATERIALS

To assess the life cycle of a building product, we must take into account the energy cost of obtaining the raw materials that make up that product and whether the raw materials are available locally, where the available manufacturing facilities are to convert the raw materials into a finished building product, and the toxicity of the raw materials. We must also take into account the energy that must be used to extract and refine the raw materials and the pollutants and other environmentally harmful by-products of that extraction.

### MANUFACTURE, PACKAGING, AND DISTRIBUTION

Another analysis that must be made when considering the sustainability of a building product is the expenditure of energy and the amount and type of harmful by-products associated with the manufacture, packaging, and distribution of the product. It is also necessary to ascertain whether the product is reusable, and if not, whether it or its raw material is extractable and recyclable and, if so, how many time it is recyclable before it becomes useless. Other factors to consider include whether the packaging is recyclable and whether it is manufactured with nonrenewable materials.

### THE CONSTRUCTION PROCESS

Sustainable building products must be examined to determine the difficulties associated with installing them in a building, including the energy used to install them and the waste and harmful by-products that may be produced during that installation. It is also necessary to take into account whether the waste produced by this installation is recyclable.

### BUILDING USE

The energy expended in using and maintaining a building product is another factor affecting its life cycle. Other considerations include the toxicity of materials required to clean the product, the amount of air pollution the product generates within the building, and how much water and other consumables are necessary for the use of the product.

### REUSE, RECYCLING, OR DISPOSAL

The final aspect of a building product's life cycle has to do with what happens to the product when it is removed from the building or the building is demolished. Considerations include whether the product can be reused, or its raw materials or the product itself can be recycled, or whether the product is not usable. If the product is recyclable, what is the energy expenditure to recycle it? If the product is not recyclable, how much damage will discarding it do to the environment?

### 1.5.3 DESIGNING FOR SUSTAINABILITY

Factors necessary to consider when designing sustainable building include, but are not limited to, the following:

- Select building materials and products with the most favorable life cycles.
- Take measures to reduce the impact on the environment of the materials and products selected.
- Design to reduce materials use by such measures as using the smallest building footprint that is practicable, configuring the building to reduce waste, and simplifying the building's geometry.
- Design for durability by using materials that have long life spans and configuring the design to prevent deterioration of materials and ensure easy replacement of components.
- Design for adaptability so that the building can be modified to take into account new systems and products or changing uses.

### 1.5.4 ADDITIONAL INFORMATION

Additional information related to sustainable building design occurs in most chapters of this book where relevant. In addition, environmental concerns and design consideration for building occupant safety are addressed in various chapters. Energy conservation is also addressed in several chapters, including, but not limited to, Chapters 11, 17, and 18.

# 1.6 Construction Documents

Architects and their consultants provide predesign and design services. In the predesign phase of their work, they may be engaged by the owner to provide such services as programming, existing facilities surveys, feasibility studies, and site analysis. In the design phase of their services, they provide schematic design, design development, and construction document services. Building predesign and design methods and the presentation of designs are beyond the scope of this book.

Construction documents include both drawn and written documents. Their presentations vary somewhat with different project delivery systems.

### 1.6.1 PROJECT DELIVERY SYSTEMS

There are three basic project delivery systems in general use today. The team responsible for project delivery remains the same, regardless of the delivery system used. They are the owner, the architect, and the constructor. The three delivery systems are *design-bid-build*, *construction management*, and *design-build*. There are several subcategories in the three basic systems. Documents will be organized differently for each of the three delivery systems and may also vary somewhat to accommodate variations within the systems.

### 1.6.1.1 Design-Bid-Build Delivery Systems

In the past, most construction projects were built using the design-bid-build system. In spite of the infusion of other delivery systems, the design-bid-build system is still used extensively today. There are several variations within the design-bid-build delivery system. In the first and most common of these, a project is built under a single prime contractor agreement. In such agreements, a single general contractor is the contractor of record who signs a construction agreement with the owner. There is only a single construction agreement. In this system, the architect prepares the construction documents and the constructor offers a bid to build the project for a fixed bid price.

In a variation of this system, the constructor agrees to build the project for a negotiated sum. In this system, the contractor may be asked to present a bid in competition with other constructors or may be preselected to enter into a negotiated agreement. In this system, the constructor is often selected during the design process and is involved in the building design to the extent of helping to control the building's cost.

In another variation, the constructor agrees to build the project under a fixed price plus a fee agreement. The price and fee may be bid for or negotiated. This method is usually used on projects where the scope of the work involved cannot be determined in advance.

Design-bid-build projects may also be built using more than one prime contractor. In this system, separate constructors are involved for the different disciplines involved, and there is a separate owner/contractror agreement with each prime contractor. That is, there is one agreement for a general works contractor, a separate agreement for a site works contractor, another for a mechanical contractor, still another for an electrical contractor, and so on. Such a project will require as many sets of construction documents as there are prime contractors, and each set may be presented in a slightly different order.

### 1.6.1.2 Construction Management Delivery Systems

In a *construction management* (CM) contract, the owner employs a firm, or an individual, as a construction manager to oversee the project. Many CM projects are also *fast-track* projects, in which the general contractor is brought on board during the design phase and construction documents are issued as needed for the contractor to price and bid subcontract work. The front-end documents (see Section 1.6.2.3), such as bidding requirements, contract forms, and contract condi-

tions, will be agreed to by the owner and the contractor before the other contract documents have been completely developed. Building costs are usually established as a guaranteed maximum price (GMP) with one or more contingency amounts to take care of unforseen conditions.

This method of delivery is discussed in some detail in Section 1.9.

### 1.6.1.3 Design-Build Delivery Systems

In a *design-build* system, a team of design professionals (architects and engineers) and one or more constructors are employed by an owner to deliver a complete functioning building ready for occupancy and use. In such projects, drawings are usually similar to those for a single-contract design-bid-build project, though they may be somewhat less detailed. The written construction documents will contain some different provisions, especially in the front-end portions where legal and contractual obligations are delineated

### 1.6.1.4 Fast-Track Delivery

Any of the three basic delivery systems may be used in fast-tract projects. In a *fast-tract* delivery system, the architect completes early portions of the construction documents, and those portions are then let to a constructor. The constructor builds those portions of the project while the architect continues to design the rest of the project. This system might to used to prepare the site for construction before the building design has been completed.

Foundation and structural framing drawings and specifications may be issued as a package before the rest of the drawings and specifications have been completed. This fast tracking often permits an accelerated construction schedule, lowering the cost of construction and financing. In some cases, fast tracking may even permit occupancy of portions of a project before the rest of the project has been completed. Unfortunately, these cost reductions are somewhat negated by errors that are inevitable in the fast-track method.

Ultimately, when all the drawings have been completed, they will be organized in the same format as they would be for a single prime contractor project. Specifications sections addressing procuring and contract provisions and general project requirements will differ from those of other

projects, but the rest of the specifications will be the same as for other project delivery systems.

### 1.6.2 DRAWINGS

AIA has defined *drawings* as construction documents that "show in graphic and quantitative form the extent, design location, relationships, and dimensions of the work to be done. They generally contain site and building plans, elevations, sections, details, schedules, and diagrams."

### 1.6.2.1 Organization

The organization of drawings differs from one architectural firm to another, but there are some generalities. Sitework drawings are sometimes issued as a separate set of drawings but often are placed at the beginning of the architectural drawings. The architectural drawings are usually placed first in a set, followed by structural, mechanical, plumbing, and electrical drawings, in that order. Drawings for other elements may be placed in various locations within a set of drawings. For example, furniture and equipment drawings may be placed immediately after the architectural drawings or may be issued as separate drawings attached at the end of the full set.

There is no universally accepted order within the disciplines. There have been numerous attempts to standardize the organization, but none has yet achieved a consensus. One possible order for architectural drawings places general notes first, followed by overall building plans, building cross sections and exterior elevations, floor plan blowups, interior elevations, reflected ceiling plans, vertical circulation (elevators, stairs, escalators, etc.), exterior wall sections and details, and, finally, interior sections and details. Drawings for the other disciplines follow similar arrangements, with notes first, plans next, then elevations, details, schedules, and diagrams.

### 1.6.2.2 Drawing Production Methods

Until relatively recently, drawings were produced manually on a drawing board using a T-square, parallel bar, or drafting instrument. Standard details and notes were sometimes repeated from project to project, and some firms issued them in a separate bound volume. Drawing sheets had standardized boarders and title blocks.

## OVERLAY DRAFTING

An advancement to the manual drafting system included the use of overlay drafting techniques, in which a base drawing was made manually and reproduced on Mylar™ or other plastic film. The film was drilled to fit over a pinned registration bar. Additional sheets were pinned over the first to produce equipment, furniture, mechanical, and electrical drawings. The floor plan was drawn only once. This method effected savings in time and, subsequently, in document production cost and tended to produce more accurate drawings.

## CAD

Overlay drafting techniques eventually led to a similar system of production using computers. Today, *computer-aided design and drafting* (CADD or simply CAD) has all but supplanted the earlier methods of drawing production. CAD systems are more expensive to implement than the earlier systems. They require the purchase of expensive hardware and software and extensive employee training. Even so, they are rapidly replacing earlier methods. They are increasingly required by owners, notably federal and state government agencies. Many architectural firms are now requiring new employees to be able to use CAD programs.

All CAD systems are based to some extent on the use of layers. A single CAD file can be used to show a floor plan, a reflected ceiling plan, an electrical power plan, a lighting plan, and more.

In the beginning, there were no recognized standards for organizing or naming CAD drawing files. Every architectural firm had to develop its own standards, which made it difficult for owners, constructors, and building officials to locate requirements in the CAD files, especially when they were dealing with more than one architect's documents. The first industry standard, *CAD Layer Guidelines*, was published in 1990 and updated in 1997 by the AIA. This document provides a standard means for coordinating the work of architects and their engineering and other consultants. It provides a standard for naming and organizing CAD layers and discipline code designations that allocate layers to specific disciplines. For example, architectural drawings are designated with an A, civil drawings with a C, electrical drawings with an E, equipment drawings with a Q, and so forth.

Combining the discipline code with the standard for naming layers gives a method for locating data within a set of CAD files for a building project that is the same from project to project and from design professional to design professional. For example, a CAD layer showing toilet partitions is named A-FLOR-TPTN. A layer showing exit lighting is called E-LITE-EXIT.

The current standard for CAD is the U.S. National CAD Standard (NCS). NCS coordinates the CAD publications of the AIA, the Construction Specifications Institute (CSI), and other CAD-related documents. It was developed using the expertise and input of many construction industry organizations. NCS has been adopted for use by several government agencies, including the GSA and the departments of the Navy and Air Force. It is also used by several large organizations, several states, and many architectural firms.

The aforementioned *CAD Layer Guidelines* are a major component of the NCS, as is the CSI's *Uniform Drawing System*. The *Uniform Drawing System* is intended to format and unify the organization of construction drawings, whether produced in the conventional drafting method or by CAD.

Early CAD packages were used primarily to develop construction contract documents in much the same way that was used to produce manually generated drawings. Many are still used in that manner today, but newer programs are capable of interactive functions that permit a change in one view to automatically change all other views. Change a plan, for example, and sections and elevations change automatically to reflect the plan change. Add a door in a perspective drawing, and the door is automatically added in plans and elevations. Change a wall section, and the elevations change automatically. Even more advanced programs allow a viewer to change the viewpoint of a computer-generated perspective drawing at will, allowing a user to, in effect, walk through a design, looking in every possible direction and from every possible altitude during the walk. Virtual reality systems have been demonstrated that permit a user to see a design in three dimensions and to move or resize elements by merely raising a hand. A window, for example, can be moved in a wall by a wave of the operator's hand. At some time in the future, such systems will revolutionize the design of buildings and render all present methods obsolete.

### 1.6.2.3 Drawing Reproduction

Traditionally drawings were drawn large and reproduced for use by designers and constructors in the full size in which they were drawn. Half-size prints were occasionally used, but they had to be made during the reproduction process, and maintaining a usable scale was difficult, although not impossible. Drawings are still reproduced in large size for use in architects' and constructors' offices and for bidding purposes. In recent years, however, half-size drawings have been used more frequently, for all purposes, but especially on the job. Half-size drawings are less unwieldy to use, especially in the field (Figure 1.6-1) and less costly to reproduce.

Today, most drawings for buildings are produced on a computer screen and stored in electronic form. They can easily be reproduced in any size desired. Even though drawings should not be scaled for construction purposes, it is still convenient to have drawings in a measurable scale. Careful selection of the scale to use for individual portions of the reproduction can make smaller drawings easier to read. In the past, scale notations on drawings were given at the full size scale. Now, scale notations on half-size drawings are usually given in the scale in which they will exist in the reproduction.

## 1.6.3 THE PROJECT MANUAL

A *project manual* is a single volume that contains all written requirements for a building construction project.

**FIGURE 1.6-1** Using a half-size print on the job. (Courtesy BE&K, Inc.)

### 1.6.3.1 Concept and Contents

A project manual provides a standardized format and location for all written project requirements. It includes bidding requirements, contract forms, conditions of the contract, information about alternates and allowances, and specifications. The number of possible requirements in project manuals is large, but not all are needed for every project. For example, one project may be of concrete construction and have no requirement for wood trusses. Another may be framed with wood and contain no structural steel.

### 1.6.3.2 The CSI

In 1948 a group of construction specifiers, frustrated with the chaotic specifications practices of that time, organized the CSI with the broad goal of improving construction specifications practices. CSI's current membership includes primarily specifiers, product manufacturers' representatives, and government agency representatives interested in specifications. Construction Specifications Canada (CSC) is a similar organization operating in Canada.

### MASTERFORMAT

Project manuals are complex and need a numbering system to make it possible to find individual requirements. Without a standard to follow, design firms tend to invent their own numbering systems and methods of organizing these materials. This makes it difficult for a contractor to find a particular requirement, thus increasing the possibility of errors in estimating that affect project cost, and errors in the field that affect the completed building. For instance, in a certain project, the requirements for structural steel may be in the third section of the manual, and in another project manual similar requirements may be in the twelfth section. To unify the location of the parts of a project manual, in 1963 CSI and CSC jointly produced a publication entitled *MasterFormat*.

*MasterFormat* has become the industry standard for naming and numbering the data in project manuals. It has been adopted by AIA in its publications and by most federal government agencies for their construction projects. It is also the format generally used in manufacturers' product literature. It is the basis of organization for Sweet's Catalog Files, McGraw-Hill's yearly collection of manufacturers' catalogs, which are a major source of product literature in most architects' offices, and in all CSI-sponsored publications, including another major source of manufacturers' data, *The Architect's First Source for Products*. It is also the basis for data filing in many design offices. More important to our purposes here, *MasterFormat* is used on the vast majority of private-sector and government building construction projects in the United States.

Because of its almost universal use in the U.S. construction industry, students and professionals need to become familiar with *MasterFormat*. The 1995 edition of *MasterFormat* divided the work associated with construction projects into 16 *divisions*. The chapters in the seventh edition of this book were named and numbered in the same way as *MasterFormat* divisions. In addition, the material there was roughly equivalent to the recommended material in the corresponding divisions of *MasterFormat*. *MasterFormat* 2004 divides its content into two *groups*: the "Procurement and Contracting Requirements Group," which contains the 00 documents in the 1995 edition, and the "Specifications Group," which contains the requirements in Divisions 1 through 16 of the 1995 edition and some added requirements. The "Specifications Group" is further divided in five subgroups: the "General Requirements Subgroup," which contains the requirements in Division 1 of the 1995 edition; the "Facility Construction Subgroup," which contains the requirements in Divisions 2 through 14 of the 1995 edition; the "Facility Services Subgroup," which contains the requirements in Divisions 15 and 16 of the 1995 edition; the "Site and Infrastructure Subgroup," which contains most of the requirements in Division 2 of the 1995 edition and some requirements not formerly included in *MasterFormat*; and the "Process Equipment Subgroup," which contains some requirements formerly in Divisions 15 and 16, but mostly new requirements that were not included in the 1995 edition.

In addition, the 2004 edition has drastically changed the naming and numbering and the subjects included in some divisions. Instead of the 16 divisions in the 1995 edition, the 2004 version contains 49 divisions. Some of the 49 divisions contain no requirements, but are instead reserved for additional requirements in the future. This edition of this book follows the intent of the 2004 edition of MasterFormat, with some necessary modifications. Except for Chapter 1, the chapters in the eighth edition of this book have the same names as the divisions, and the data they contain are roughly those included in the corresponding division of *MasterFormat* 2004. The numbers of Chapters 2 through 14 are the same as the corresponding division in *MasterFormat* 2004. The order and the names of the rest of the chapters are the same as those in MasterFormat 2004's divisions, but the chapter numbers differ from those of *MasterFormat* 2004's divisions because it makes no sense to have chapters in a textbook reserved for future use.

In addition, some divisions of *MasterFormat* 2004 are beyond the scope of this book and are thus not included here. These include divisions covering integrated automation, transportation, waterway and marine construction, and the entire "Process Equipment Group" of the edition of *MasterFormat* 2004.

Furthermore, because the order of sections in *MasterFormat* tends to change slightly from edition to edition, some of the subjects in the eighth edition of this book, as was true in the seventh edition, differ in location from the order used in *MasterFormat* 2004. Moreover, the limited size of the book precludes discussing every section listed in *MasterFormat* 2004.

It is reasonable to assume that over time the construction industry will adopt the 2004 edition of *MasterFormat* to the same extent that it has adopted earlier versions.

### UNIFORMAT™

*MasterFormat* is an organization system based on materials and their uses. CSI's Uniformat is an organization system based on systems and assemblies. *Uniformat* is intended not to replace *MasterFormat* but rather to supplement it.

*Uniformat* is intended for use in arranging project descriptions to show how a design meets the owner's requirements; developing preliminary cost estimates and construction cost models; filing information and drawing details; and even arranging project manuals for certain design-build delivery projects in lieu of using *MasterFormat*, especially when performance specifications are used.

*Uniformat* contains multiple levels of involvement. Level 1 divides a building into nine categories. The first, Project Description, is self-explanatory. The succeeding eight categories cover construction systems and assemblies. The first seven are given alphabetic designations from A to G. The designations are substructure (A), Shell (B), Interiors (C), Services (D), Equipment and Finishings (E), Special Construction and Demolition (F), and Building Sitework (G). The eighth category, Z, is called General. It includes contract requirements, general requirements, and cost estimates.

There are also levels 2, 3, 4, 5, and 6. Levels 2 and 3 add numeric designations to subcategories of the level 1 categories. An example would be A10, where A is level 1 and 10 is the level 2 designation, "Foundations." In Levels 4, 5, and 6, numbers designate subdivisions of the subdivisions in a manner similar to those in *MasterFormat*. In fact, sometimes level 4 numbers correspond exactly with *MasterFormat* numbers.

*Uniformat* and *MasterFormat* are most often used independently for different purposes, but sometimes it is advantageous to use them together. An example is in the production of cost data. They may also be both used in certain design-build projects. When using them together, it is best to use *MasterFormat* numbers for *Uniformat* levels 4, 5, and 6 designations.

### 1.6.3.3 Bidding and Contracting Requirements

*Bidding requirements* include invitations to bid, instructions to bidders, a listing of information available to bidders, and bid and bid bond forms. They are issued before bids on a project are offered and are not considered part of the contract for construction.

*Contracting requirements* documents are also issued to potential bidders before they offer bids, but completed forms and other contracting requirements documents are considered part of the construction contract. Contracting requirements documents include the owner/contractor agreement, bonds (performance and payment), certificates (insurance, conformance, etc.), the conditions of the contract (general and supplementary), and addenda and modifications issued prior to acceptance of bids.

The general conditions of the contract and standard contract documents have evolved over the years from the early be-

ginnings of AIA documents to today's complex and extensive interrelated documents. For example, a current standard form of the owner/contractor agreement produced by AIA is fully compatible with AIA A201, "General Conditions of the Contract for Construction."

Engineering organizations have also formed the Engineers Joint Contract Documents Committee (EJCDC), which has developed standard contract documents for use on engineering projects.

### 1.6.3.4 Specifications

*Specifications* constitute that portion of the written requirements for a building construction project that are contained in the divisions of a project manual. AIA has defined specifications as "written requirements for materials, equipment, construction systems, standards, and workmanship for the work as well as standards for the construction services required to produce the work."

Originally, specifications were just notes on the drawings. As the construction world became more complex, lawsuits proliferated, and many new materials and systems were introduced, the notes became too voluminous to put on the drawings. Architects and engineers began to type them and issue them as separate documents. Legal and procedural requirements now occupy as many pages as an entire specification set did in the 1950s.

For most construction projects today, specifications are included in a project manual and are organized according to *MasterFormat* 2004. Within each of the *MasterFormat* divisions, specifications are organized into level 2 sections, level 3 sections, and sometimes level 4 sections. Each level represents a more detailed subdivision of the work. In lieu of the five-digit numbering system used in the 1995 *MasterFormat* edition, sections in the 2004 edition are numbered with a six-digit system. Therefore, each level is represented by two digits instead of the older system's one digit. This increases the possible number of entries at each level 10-fold. As was true in previous editions, the first two digits are the division number and the remaining digits are the section number. For example, in Division 10, "Specialties," and a level 2 section 10 20 00, "Interior Specialties," and a level 3 section 10 21 00, "Compartments and Cubicles." Below that is a level 3

section 10 21 13, "Toilet Compartments," and a level 4 section 10.21.13.13, "Metal Toilet Compartments." Note that the level 4 number is the level 3 designation with a dot preceding the level 4 designation. A user may add level 4 designations as necessary.

Specifications should include clear and accurate descriptions of technical requirements related to materials, products, and services. They should also state a project's requirements for quality and the prescribed use of materials and methods to produce a desired product, system, application, or finish. And they should do all this in a form that is understandable and in a format in which the requirements can be easily found and clearly understood.

Recognizing that a clear and concise format, that was the same from job to job, was essential to producing clear specifications, shortly after the introduction of *MasterFormat*, CSI introduced a three-part section format and soon after that a recommended page format. In the section format, the data in each specifications section are divided into three *parts* as follows:

Part 1: General
Part 2: Products
Part 3: Execution

The content of each part is divided into articles, and the articles are divided into paragraphs.

The content and order of the data in each part have changed somewhat over the years, but the general intent is the same. Part 1, "General," includes requirements that affect the entire section and that are general in nature, such as requirements for allowances; unit prices; alternates and alternatives; submittals; quality assurance; delivery, storage, and handling; project and site conditions related to that section; sequencing; scheduling; warranties; and maintenance. Part 1 also includes definitions related to that section and cross references to other sections.

Part 2, "Products," includes the names of acceptable manufacturers; descriptions of required materials, manufactured units, equipment, and accessories; and requirements for mixes, shop fabrication and assembly, tolerances, and source quality control, including tests and performance verifications.

Part 3, "Execution," includes requirements for examination of conditions prior to installation; preparation for installation; and erection, installation, or application, as applicable. Part 3 also includes requirements for field quality control and testing; adjustments, cleaning, and protection of the installed work; and requirements for demonstration of proper operation of items installed under the section and requirements for scheduling and coordination of the work of this section with other work.

Each part is organized into articles and paragraphs using the following format:

Part 1 General
1.01 Article
    a. Paragraph
    b. Paragraph
      1. Subparagraph
1.02 Article
    a. Paragraph
    b. Paragraph
      1. Subparagraph

The CSI page format addresses the margins and page arrangement, including recommended indents for each level of articles and paragraphs. Most guide specifications (see Section 1.6.2.5) are already set up in the CSI page format or in an organization resembling it.

### 1.6.3.5 Guide Specifications

Nonproprietary guide specifications sets are commercially available today from organizations representing national associations. For example, Arcom Master Systems, sponsored by AIA, produces *MASTERSPEC®*; Construction Sciences Research Corporation, sponsored by CSI, produces *SPECTEXT®*. Commercial guides are also available from several private organizations, such as Kalin Associates and Building Systems Design, Inc. (BSD), which produces an automated guide set named *SpecLink®*. Most of these sets are available in either paper or electronic form, using various methods of delivery. An advantage of using such sets is that they are updated periodically and cover a wide range of materials and systems. In addition, they are produced by organizations with no vested interest in a particular product, which may make them more reliable in some ways. Some of these sets come with detailed instructions and recommendations for selecting one product type over another. Most of them are thoroughly researched and reasonably up-to-date.

Proprietary guide specifications are available from building product manufacturers by mail and at their Web sites. Some organizations offer direct links to an extensive list of manufacturers' Web sites where guide specifications can be accessed. In addition, guide specifications sections produced by manufacturers are available from a variety of organizations in sets. These sets are often less complete than those produced by the associations. Examples of organizations offering access to manufacturers' sites are ARCAT, Architects' First Source, and Sweet's Group.

Guide specifications are also produced by federal agencies, such as the GSA and the various military services. These are intended for use only on government projects but contain data that can be valuable to a specifier even on private-sector projects.

## 1.7 Bidding and Negotiation

Construction contracts are let based on the delivery system (see Section 1.6.1) to be used and the cost of construction.

### 1.7.1 BASIS OF CONSTRUCTION COST

The cost of construction may be determined on the basis of a fixed-price stipulated sum, a cost-plus-fee arrangement, unit prices, a guaranteed maximum price (GMP), or a combination of these.

In a fixed-price stipulated sum agreement, a constructor agrees to provide a building project for a fixed amount of money based on contract documents furnished by the architect. In a cost-plus-fee arrangement, the constructor agrees to construct the project and the owner agrees to pay the constructor's costs plus a fixed fee.

In a unit-price agreement, the cost is based on a price per unit of a material, such as cubic yards of topsoil or rock excavation. The unit costs include the actual cost of the material and an amount to cover the constructor's overhead and profit. Unit-price arrangements are frequently used with other types of pricing, such as fixed-price stipulated sum agreements. They are less frequently used as the basic agreement for an entire project.

A GMP agreement is exactly what it sounds like. Such agreements are usually used with design-build or CM projects where the design document are not complete when the project starts. GMPs are usually established at the time of completion of the design development phase of the architect's services. Thereafter, all concerned parties must ensure that the cost of the project does not exceed the GMP.

### 1.7.2 METHOD OF SELECTING A CONTRACTOR

Construction contracts are let based on dollar amounts and contract terms determined by either bidding or negotiation. *Bidding* is a process whereby a prime design professional engaged by an owner, and the design professional's consultants, prepare bidding documents and issue them to a group of constructors, who then submit the dollar amounts they will charge to build the building under the terms of the bidding documents. The prime design consultant may be an architect or engineer, depending on the project type. For example, for a bridge, a structural engineer may be the prime design professional. This section assumes that the project under consideration is a building and that the prime design professional is an architect.

*Negotiation* is a process whereby an architect engaged by an owner, and the architect's consultants, prepare negotiation documents; the owner, often with the advice of the architect, selects a potential constructor; and the architect issues the documents to the constructor, who then submits an offer to construct the project for a set number of dollars and within defined contract conditions. Then, either this offer is accepted or the offer and the terms are negotiated between the owner, with advice from the architect, and the constructor until an amicable arrangement is agreed upon.

#### 1.7.2.1 Bidding

SELECTING BIDDERS

There are several methods for selecting bidders for a building construction project. In most government projects, all licensed contractors must, by law, be permitted to offer a bid. The only way to limit bidders is by imposing requirements for performance and payment bonds. Contractors who are unable to perform the work because of financial or other difficulties will not be able to obtain the required bonds and can be disqualified on this basis.

Some private work contracts are handled in the same way as government projects, whereby bidding is open to all legitimate constructors, but in most cases constructors requested to bid are restricted in some way. In some instances, the owner will simply develop a list of constructors it wants to bid and exclude all others. Sometimes this list is a result of consultation between the owner and the architect, but not always.

For some projects, potential bidders are examined and *prequalified* to ascertain whether they are legitimate, reputable, licensed contractors who are large enough and financially stable enough to perform the work and are not disqualified by legal restrictions. This investigation is usually conducted by the architect, who forwards the results to the owner. The final decision on which constructors are permitted to bid lies solely with the owner. Prequalification is usually performed with the use of standard forms such as AIA's Document A305, "Contractor's Qualification Statement."

Some sophisticated owners develop and maintain lists of prequalified bidders and select from that list for each project. Government agencies are also permitted to prequalify contractors and develop a list of qualified bidders, but the failure to qualify must be based on defined restrictions, such as insufficient financial assets or company size.

### BIDDING PROCESS

For government projects and for private-sector projects in which the owner has elected to use completely *open* bidding, the owner or the architect places an advertisement for bids in one or more newspapers and in other publications deemed appropriate. This advertisement lists the pertinent project information, such as its name and address, a general description of the project, the type of contract, the date, time, and place of receiving bids, bid security requirements, and other relevant information.

When the bidders have been preselected, either the owner or the architect issues an invitation to bid to all bidders on the prequalified list. The invitation to

bid contains the same information as in an advertisement for bids.

### BIDDING DOCUMENTS

When a legitimate bidder responds to an advertisement for bids or an invitation to bid and requests bidding documents, the architect issues one or more sets of bidding documents to each requester. In addition to the advertisement or invitation to bid, the bidding documents issued initially include the following:

- Instructions to bidders, which explain bidding procedures and list the requirements of the bid, such as the requirements for bonds
- Bid forms, which are developed by the owner or the architect to keep the content of all bids in the same format, so as to make interpretation easier and to ensure that bids for required alternates, allowances, and unit prices are included
- Information related to bonds, including bid bonds required to ensure that a bidder will sign an agreement based on his or her bid or forfeit the bond amount, and performance and payment bonds that ensure completion of the signed contract and payment of the constructor's related debts
- Form of the owner/contractor agreement
- The conditions of the contract (general and supplementary)
- Drawings
- The project manual

Bidding documents also include addenda and other modifications to the previously issued documents that are issued prior to contract signing.

### 1.7.2.2 Negotiation

Bid projects are often negotiated with the lowest bidder, or several of the lowest bidders, to arrive at a final cost and determine changes to the original proposal. Conversely, in a *negotiated* project, a single constructor is selected and the contract sums and terms are negotiated between the constructor and the owner, usually with advice from the architect.

### SELECTING POTENTIAL CONSTRUCTORS

An owner may select a potential constructor by firsthand knowledge of the firm or by reputation, or may select several potential constructors and reduce the list to one by a prequalification process similar to that described for bid projects.

A negotiated contract may be a straight single-prime-contractor contract or may have multiple prime contractors. It may result in a fixed-price stipulated sum, cost-plus-fee, or unit price agreement. It may result in a construction management system with a GMP. It may also be used for design-build agreements.

### NEGOTIATION PROCESS

The constructor in a negotiated contract is sometimes engaged before the architect's and associated consulting engineer's design work has been completed. The advantage of this arrangement is that the constructor can be called on to issue advice related to systems and products as the design work progresses. This advice can be related to costs, availability, and interface problems that may be encountered in the field. The disadvantage is that materials and systems selections and design judgments are sometimes based on factors other than costs and ease of construction. Care must be taken to ensure that the architect maintains control of the use, safety, and aesthetic aspects of the project.

### NEGOTIATION DOCUMENTS

Negotiation documents are mostly the same as bidding documents. However, there is no advertisement, invitation to bid, or instructions to bidders. Additional requirements must sometimes be established, however, for subcontract bids. The conditions of the contract will also vary somewhat based on the delivery system to be used.

### 1.7.3 ARCHITECT'S RESPONSIBILITIES

In addition to the responsibilities already delineated, the architect should conduct public bid openings, unless the owner elects to do so. The architect should assist the owner in evaluating bids and negotiated contract terms. The architect should also advise the owner in regard to acceptance of bids and selection of a contractor, but the responsibility for these duties and for executing the owner/contractor agreement rests solely on the owner.

# 1.8 Construction Contract Administration

Some owners undervalue the potential contribution of their architects and consulting engineers during building construction and do not continue their services past the bidding and negotiation phase. Fortunately, most owners understand the advantages of these continued services and employ their design professionals until the construction has been completed. In some types of delivery systems, the design professionals' involvement during construction is essential. Obviously, in a design-build project, the design professionals are part of the design-build team and must continue with the project until completion.

Fast-track delivery systems also require the services of design professionals through the construction process. For the most part, in such projects the design and construction documents are not finished until a short time before the construction has been completed. Construction management projects are also frequently fast-track and need the services of design professionals throughout.

When engaged by owners to provide services during the construction phase of a building project, architects and engineers can provide the following services:

1. Hold preconstruction conferences to clarify procedures and establish the responsibilities of the various team members. Who the team members are will vary with the delivery system. At the least, there will be an owner, an architect, consulting engineers, and the general contractor.
2. Review and process the contractor submittals, such as shop drawings, samples, product data, mockups, test results, bonds, and record drawings.
3. Conduct periodic field observations to ensure that the work is proceeding in accordance with the contract documents, to issue field reports detailing these observations and noting the progress of the work, and to reject work found to be not in compliance with the contract documents.
4. Respond in writing to the contractor's questions concerning the intent of the contract documents.

5. Issue documents to make changes in the work. Minor changes can be made by a simple written order by the architect. Changes affecting project delivery time or costs are made by *change orders*. Change orders are sometimes initiated by the architect, who asks the contractor in writing to submit a *proposal request* for the desired changes. The contractor then responds with a proposal, indicating changes in contract time or cost associated with the requested change. In other cases, proposal requests are initiated by the contractor. In either case, the architect reviews the contractor's proposal and advises the owner whether to accept or reject it. Once the terms of the proposal request are acceptable to both the owner and the contractor, the architect issues a change order, which must then be signed by both the owner and the contractor. Signed change orders become a part of the contract for construction.
6. Attend regularly scheduled project progress meetings with representatives of the owner and the contractor.
7. Review and process the contractor's applications for payment. At the beginning of a project, the contractor submits to the architect a *schedule of values*, which is broken down into the various portions of the work. Periodically, usually once each month, the contractor submits to the architect an *application for payment*, indicating the progress of the work to date and the amount of the contract sum due that month. The application is broken down in exactly the same way as the schedule of values so that the two can be compared. The architect and his or her consultants visit the project site and compare the contractor's application with the work accomplished. The consultants report their findings to the architect, who then certifies payment by filling in, on the contractor's application for payment, the dollar amount due that month based on the architect's and consultants' observations. The amount certified may be the amount shown in the application or may be less, depending on the architect's judgment of the work actually completed. When a discrepancy is found, there is usually an attempt to have the contractor revise the proposal, but the certified amount is based ultimately on the architect's judgment, taking into account the consultants' opinions. The architect must submit written justification for payments certified in amounts less than those requested. The owner bears the final responsibility for payment and can pay more or less than is certified. Underpayment, however, can trigger a legal challenge by the contractor and should not be done without documented justification and consultation with an attorney.
8. Handle closeout procedures. When the contractor considers that the work is complete, it initiates *contract closeout*. To do this, the contractor submits a request for a review by the design professional pursuant to *substantial completion*. Substantial completion means that the owner can occupy the premises and use it for its intended purpose. If minor items remain to be completed, the contractor's request must include a list of these items, along with reasons for the incompleteness and the dollar value of each incomplete item. The contractor must also submit all closeout documents required by the contract documents, including, but not limited to, warranties, maintenance agreements, operating instructions, certificates of inspection and occupancy issued by the local jurisdiction, bonds, insurance certificates, and record documents, including drawings and a project manual marked to show all changes made during construction. The contractor must also submit maintenance and replacement stock and keys, and must instruct the owner's personnel in the operation of equipment as required by the contract documents.
9. Upon receipt of a request for a substantial completion inspection, the architect and the architect's consultants will inspect the project. If they consider the work not substantially complete, the architect will so notify the contractor with a list of observed incomplete work. The contractor must then correct the deficiencies and submit a new request for inspection. The architect and the architect's consultants will again visit the site and conduct a second inspection. When the

architect and the architect's consultants deem the work to be substantially complete, and after all required closeout materials have been received, the architect will issue a certificate of substantial completion. If items still remain incomplete at that time, the certificate will be accompanied by a list (punch list) of these remaining items and a date, usually 30 days, within which the remaining work must be completed.

10. Issue a *certificate of final completion.* When the items on the punch list have been completed and the architect is convinced that all contract work and the contractor's other obligations under the contract have been completed, the architect will issue a certificate of final completion. The contractor may then submit a request for final payment.

11. Review and certify the contractor's final request for payment. This will complete the architect's basic services under most construction contracts.

# 1.9 Construction Management

A construction manager usually acts as an adviser to the owner to determine probable construction costs during the design phase of a building construction project. The construction manager may or may not assist the owner during the bidding or negotiation phase (see Section 1.7). A construction manager may be involved in a project during the construction phase. In this case, the construction manager may be a separate entity or the prime contractor. In either case, the construction manager acts as an agent of the owner, coordinates the work of the prime contractor or multiple prime contractors, and oversees quality control for the project. Whether the construction manager is a separate entity or the prime general contractor, when representing the owner during construction, the construction manager usually bears the responsibility of ensuring that the construction project is completed on time and within its budget.

The architect sometimes functions as a construction manager. An AIA survey done in 2000 showed that 36% of the responding architectural firms offered CM services. More often, however, construction management services are performed by a professional other than the designing architect or by the prime contractor.

CM is a relatively recent innovation in the building construction industry that has gained a strong following. The owners of many major construction projects employ the services of a construction manager in some form. Colleges and universities are now offering degrees in CM. Some architects have expanded their services to include CM.

CM services may be performed by the designing architect, by the prime building contractor, or by a third party representing neither. A construction manager may perform services in one of three ways:

1. As an adviser. In this capacity, a construction manager will act as a consultant on cost management while the design is proceeding, but will have no involvement in the construction process.
2. As an agent of the owner. In this capacity, the construction manager will consult with the owner about cost matters during the design phase, and will be involved in selecting contractors and subcontractors and coordinating the work of the constructors during construction of the building. In this capacity, the construction manager is a consultant only and bears no financial responsibility for the project.
3. As a constructor. In this capacity, the construction manager may be involved in the design phase and also act as the prime contractor during the construction phase. This arrangement is usually done under a GMP agreement (see Section 1.7). A major problem with this arrangement is that the cost estimator (the construction manager), the quality control overseer (also the construction manager), and the constructor (prime building contractor) are one and the same. This can lead to inherent conflicts of interest. It does not, however, mean that such an arrangement should not be considered, so long as the owner is aware of the potential conflict.

Other conditions under which CM services may be provided are also possible.

Many CM projects are also done using fast-tract procedures (see Section 1.6). When this is done, a GMP agreement is usually employed. This combination allows the owner to establish a maximum cost while taking advantage of speedy delivery of the building and quality control during construction. Often the agreement between the construction manager/constructor and the owner stipulates a guaranteed completion date.

## 1.9.1 ADVANTAGES AND DISADVANTAGES OF THE CM APPROACH

### 1.9.1.1 Advantages

A CM delivery system offers an owner a single point responsibility and thus accountability for a construction project. It also can lead to increased construction and design quality and a significant cost savings to the owner. Properly done, CM should also assure a better chance of having the building completed within is construction schedule.

### 1.9.1.2 Disadvantages

There are inherent disadvantages of using a CM delivery system for building construction projects. These are related to the type of CM system used and the relationship of the construction manager to the owner and the project. When the CM services are performed by the constructor or a third party, there is the possibility that providing this additional layer of services will increase the cost to the owner, but there is no guarantee that this additional cost will result in a better project. As stated earlier, there is an inherent conflict of interest that may be difficult to overcome when the same firm estimates the cost of the project, influences which products will be used in the project, and then has to purchase those products. Sometimes the relationships between construction managers and the subcontractors they employ can also result in a conflict of interest.

The owner is somewhat better protected when the construction manager is also the designing architect. Unfortunately, there are drawbacks to this system as well. The owner may be concerned that the architect acting as the construction manager will protect the architect in his capacity as designer when a conflict arises during construction and an interpretation must be made. To alleviate this and other potential problems, an owner may feel the need to engage a full-time observer to oversee the construction manager. This is an increased cost to the owner.

## 1.9.2 CONSTRUCTION MANAGER QUALIFICATIONS

A construction manager must have the ability to perform as an administrator, must have an extensive background in the construction industry, and must have accumulated the knowledge of both design and construction necessary to act in this capacity. In addition, a construction manager must have the skills necessary to deal with people both on the professional level and in the building trades. The industry skills necessary are an understanding and ability in specifications writing, estimating, and record keeping. A construction manager will encounter many crises during the performance of the work and must be able to handle them without floundering. Competent administrative support is essential to anyone engaging in construction management.

A construction manager must have an extensive knowledge of the relationships between prime contractors, subcontractors, and product suppliers. An understanding of contracts and contractual arrangements is essential.

## 1.9.3 SCHEDULING

Construction managers are responsible for scheduling and controlling many construction-related activities. Several types of methods are used for scheduling these activities. The bar chart method and the critical path method are the ones most widely used.

### 1.9.3.1 Bar Chart Method

The *bar chart method* of scheduling is the oldest scheduling control device. In spite of the advent of computerized planning methods, bar charts are still the most widely used scheduling control method today. On projects with no construction manager, the prime contractor is usually required to submit a schedule of activities and to update this schedule periodically, usually monthly, to show the progress of the work. This submittal is required as a supplement to the contractor's application for payment. When there is a construction manager, he or she is often responsible for similar schedules.

Bar charts can take several forms, but the most common is a chart similar to that shown in Figure 1.9-1. The table is basically self-explanatory. The lighter gray bar represents the scheduled (SCH) beginning and end of the task. The darker gray bar represents the actual (ACT) beginning and end of the task. Sometimes the day of the month is indicated above each box. The number in the Weight column is the percentage that task represents of the entire project.

### 1.9.3.2 Critical Path Method

The *critical path method* (CPM) of scheduling is a computer-based, highly mathematical method that is frequently used on large or complex projects and is often required as a project's scheduling method, especially for government projects. A detailed discussion of CPM programing is beyond the scope of this book, but some basics are included here.

**FIGURE 1.9-1** A Construction Schedule Bar Chart*

| Description of Task | Weight % | | 2006 | | | | | | | | | | | | 2007 | | | | | | | |
|---|---|---|---|---|---|---|---|---|---|---|---|---|---|---|---|---|---|---|---|---|---|---|
| | | | J | F | M | A | M | J | J | A | S | O | N | D | J | B | M | A | M | J | J | A |
| 1. Strip site | 8 | SCH | | | | | | | | | | | | | | | | | | | | |
| | | ACT | | | | | | | | | | | | | | | | | | | | |
| 2. Footing excavation | 6 | SCH | | | | | | | | | | | | | | | | | | | | |
| | | ACT | | | | | | | | | | | | | | | | | | | | |
| 3. Place footings | 7 | SCH | | | | | | | | | | | | | | | | | | | | |
| | | ACT | | | | | | | | | | | | | | | | | | | | |
| 4. Foundation walls | 12 | SCH | | | | | | | | | | | | | | | | | | | | |
| | | ACT | | | | | | | | | | | | | | | | | | | | |
| 5. Column bases | 2 | SCH | | | | | | | | | | | | | | | | | | | | |
| | | ACT | | | | | | | | | | | | | | | | | | | | |
| 6. Steel columns on site | 1 | SCH | | | | | | | | | | | | | | | | | | | | |
| | | ACT | | | | | | | | | | | | | | | | | | | | |
| | | SCH | | | | | | | | | | | | | | | | | | | | |
| | | ACT | | | | | | | | | | | | | | | | | | | | |
| | | SCH | | | | | | | | | | | | | | | | | | | | |
| | | ACT | | | | | | | | | | | | | | | | | | | | |
| | | SCH | | | | | | | | | | | | | | | | | | | | |
| | | ACT | | | | | | | | | | | | | | | | | | | | |
| | | SCH | | | | | | | | | | | | | | | | | | | | |
| | | ACT | | | | | | | | | | | | | | | | | | | | |
| | | SCH | | | | | | | | | | | | | | | | | | | | |
| | | ACT | | | | | | | | | | | | | | | | | | | | |

*This type of chart is also known as a *Gantt chart,* because it is based on a chart originated by Henry L. Gantt.

CPM begins with the production of a detailed list showing each task that must be performed during a particular construction project. An identifying code, often an alphanumeric one, is assigned to each task. For example, a task might be C1, "Clear North Site."

Next, the relationships between the various tasks are identified. There are four possible relationships: (1) Task 2 bears no relationship to Task 1; (2) Task 2 cannot be started until Task 1 is partially completed; (3) Task 2 cannot be started until Task 1 has been completed; (4) Task 2 cannot be completed until Task 1 has been completed. Network diagrams and precedence diagrams are developed that show these relationships in graphic form. Network diagrams are spider-web-like diagrams. Precedence diagrams are charts. These diagrams show which task must be done before the next task can commence. For example, it is not possible to begin rough grading before clearing has been at least partially done. Obviously, some tasks will not have to wait on other tasks to proceed. For example, Task O1, "Order Topsoil," will not have to wait for Task C1, "Clear North Site," to be completed, or even started for that matter.

Once the relationship between tasks has been identified and charted, it is then necessary to determine a duration time (how long it will take) for each task. This duration time is added to the network diagram.

Once the duration time has been established, a schedule is developed to show the first date that a task can be started (*early start date*) and the first date that it can be completed (*early completion date*). The early completion date occurs on the date revealed by adding the task duration to the early start date. When a task cannot be started before another task has been completed, the early start date for the second task is the same as the early completion date for the first task.

When the schedule showing early start and early finish dates has been completed, a completion date will emerge. Once the completion date has been established, it is possible to work backward through the chart and determine the latest date that each task can be completed, as well as the latest date the each task can be completed without affecting the completion date of the project. For each task, the difference in days between the early

start date and the late start date or between the late finish date and the early finish date is called *float time*. This float time is the maximum number of days a task can be late without affecting the project completion date.

From the completed schedule, it is possible to determine the *critical path* for the project. A task that has the same late start and early start date or the same early finish and late finish date is on the critical path. It is essential that tasks that fall on the critical path be completed by their late finish date to avoid jeopardizing the project completion date.

CPM scheduling is time-consuming and costly. It will probably not be justified for small or simple projects unless the owner demands its use. CPM schedules are difficult, almost impossible, to prepare without the use of a computer. Fortunately, several organizations offer software for CPM schedules. These include, but are not limited to, AEC Software in Sterling, Virginia; Primavera Systems, Inc., in Bala Cynwyd, Pennsylvania, and Microsoft Corporation in Redwood, Washington.

### 1.9.4 PROJECT CASH FLOW

A construction manager involved in the construction phase of a project is usually responsible for keeping track of the progress of the project, and for advising the owner concerning periodic payments to the prime contractor and subcontractors. The prime contractor's monthly progress reports and applications for payment are reviewed by the construction manager before they are forwarded to the architect for review and comment. The construction manager is responsible for determining whether the completion percentages shown by the contract are accurate, and for returning the application for payment to the contractor for revision if such is necessary. An architect involved in the construction phase of a project is also responsible for reviewing these figures, but when a construction manager is involved, the architect reviews the figures after the construction manager has reviewed them and they are accepted or returned to the contractor for revision.

When a construction manager is involved in a construction project, the payment procedure can take more time than for a project where no construction manager is involved and may delay payments to the contractor. Usually the time period

from the contractor's submittal of a payment application to the required payment is increased in the construction agreement to a longer time than would otherwise be required to accommodate anticipated delays.

The general contractor's costs of operations are usually not the responsibility of the construction manager unless the construction manager is also the general contractor. In any event, control of construction cost is beyond the scope of this book.

### 1.9.5 OTHER CONSIDERATIONS

A construction manager is responsible for numerous other activities. When the construction manager is also the constructor, many of the construction manager's responsibilities are the same as those of a general contractor and are beyond the scope of this book.

In addition to the functions already mentioned, a construction manager is responsible for the same activities and to the same extent that the architect is responsible in a project where there is no construction manager. These responsibilities include observation and reporting of safety issues, inferior construction practices, and inferior completed work; violations of the construction agreement; failure to comply with the requirements of the contract document; review of contractor submittals; and most of the other architect's responsibilities delineated in Section 1.8, with the exception of duties that are clearly within the purview of the architect, such as issuing clarifications of the construction documents.

### 1.9.6 APPLICABLE AIA DOCUMENTS

AIA provides a number of documents to aid architects and owners in obtaining and developing agreements for construction management services. These are listed in Section 1.11.

# 1.10 Additional Reading

More information about the subjects discussed in this chapter can be found in the references listed in Section 1.11, and in the following publications.

Allen, Edward. 2004. *Fundamentals of Building Construction: Materials and Methods*, 4th ed. Hoboken, NJ: John Wiley & Sons, Inc. (www.wiley.com).

Allen, Edward and Joseph Iano. 2001. *The Architect's Studio Companion: Rules of Thumb for Preliminary Design*, 3rd ed. Hoboken, NJ: John Wiley & Sons, Inc. (www.wiley.com).

Antill, James M. and Ronald W. Woodhead. 1990. *Critical Path Methods in Construction Practice*, 4th ed. Hoboken, NJ: John Wiley & Sons, Inc. (www.wiley.com).

Bannister, Jay M. 1991. *Building Construction Inspection: A Guide for Architects*. Hoboken, NJ: John Wiley & Sons, Inc. (www.wiley.com).

Belcher, Osmund F. 2001. "Cost Control and Schedule Control Management." *The Construction Specifier* (June): 47 (www.csinet.org).

Construction Specifications Institute 2004. *The Project Resource Manual—CSI Manual of Practice*. Alexandria, VA: CSI (www.csinet.org).

————. *Masterformat 2004*. Alexandria, VA: CSI (www.csinet.org).

Construction Specifications Institute (CSI), American Institute of Architects (AIA), and National Institute of Building Sciences (NIBS). 2004. *National CAD Standard Version 3.0 Update from Version 2.0*. CSI (www.csinet.org), AIA (www.aia.org), and NIBS (www.nibs.org).

Dell'Isola, Michael D. 2001. "Contract Management: Know Your Options." *The Construction Specifier* (September): 38 (www.csinet.org).

Foss, Rebecca. 2001. "Sustainable Building." *The Construction Specifier* (July): 32 (www.csinet.org).

Haney, Craig K. 2002. "Construction Documents for Historic Projects." *The Construction Specifier* (October): 52 (www.csinet.org).

Hartmen, Douglas C. 2001. "Construction Administration—New Tools and Methods for Efficiency." *The Construction Specifier* (November): 56 (www.csinet.org).

Hentschel, Rick. 2001. "Closing the Deal—The Role of the Project Manager in the Justification Process." *The Construction Specifier* (November): 46 (www.csinet.org).

Halpin, Daniel W. and Ronald W. Woodhead. 1998. *Construction Management*. Hoboken, NJ: John Wiley & Sons, Inc. (www.wiley.com).

Ingardia, Michael P. and G. William Quatman. 2001. "Computer-Aided Design and Electronic Documents." *The Construction Specifier* (March): 45 (www.csinet.org).

International Code Council. 2003. *International Codes*. Falls Church, VA: ICC (www.iccsafe.org).

Kepler, Stephen and Judith Pinkelny Tibbs. 2000. "LEED: Good for Business, Good for the Environment." *The Construction Specifier* (July): 27 (www.csinet.org).

Langdon, Geoffrey Moore. 2001. "Architectural CADD." *The Construction Specifier* (May): 31 (www.csinet.org).

Liebing, Ralph W. 1999. *Architectural Working Drawings*, 4th ed. Hoboken, NJ: John Wiley & Sons, Inc. (www.wiley.com).

Mays, Patrick C. and B. J. Novitski. 1997. *Construction Administration: An Architect's Guide to Surviving Information Overload*. Hoboken, NJ: John Wiley & Sons, Inc. (www.wiley.com).

Nunnally, S. W. 2001. *Construction Methods and Management*. Upper Saddle River, NJ: Prentice Hall (www.vig.prenhall.com).

Ramsey, Charles George, Harold Reeve Sleeper, and John Ray Hoke, Jr. 2000. *Architectural Graphic Standards*, 10th ed. Hoboken, NJ: John Wiley & Sons, Inc. (www.wiley.com).

Reifsnider, Randal J. 2002. "National CAD Standard." *The Construction Specifier* (April): 39 (www.csinet.org).

Roberts, Greg. 2001. "The Green Team." *The Construction Specifier* (July): 47 (www.csinet.org).

Snyder, James. 1998. *Architectural Construction Drawings with AutoCAD® R14*. Hoboken, NJ: John Wiley & Sons, Inc. (www.wiley.com).

Stovall, Gerald L. 1995. "Project Contract Administration." *The Construction Specifier* (May): 149 (www.csinet.org).

Stratton, Peter A. 2001. "Universal Design." *The Construction Specifier* (February): 27 (www.csinet.org).

Templeton, Peter, Tom Dietsche, and Kris Price. 2001. "Green Buildings in Black and White." *The Construction Specifier* (July): 29 (www.csinet.org).

Wakita, Osamu and Richard M. Linde. 2002. *The Professional Practice of Architectural Working Drawings*, 3rd ed. Hoboken, NJ: John Wiley & Sons, Inc. (www.wiley.com).

Watson, Wayne. 2001. "Understanding and Classifying Performance Specifications." *The Construction Specifier* (August): 38 (www.csinet.org).

Wilson, Alex. 2001. "Identifying Green Building Products." *The Construction Specifier* (July): 24 (www.csinet.org).

Yee, Rendow. 2002. *Architectural Drawing: A Visual Compendium of Types and Methods*, 2nd ed. Hoboken, NJ: John Wiley & Sons, Inc. (www.wiley.com).

## 1.11 Acknowledgments and References

### ACKNOWLEDGMENTS

We gratefully acknowledge the assistance of the following organizations and individuals in preparing this chapter. We are also indebted to them for permission to use their illustrations when requested and for the use of their publications as references.

American Institute of Architects (AIA) (www.aia.org)

American National Standards Institute (ANSI) (www.ansi.org)

American Society of Heating, Refrigerating and Air Conditioning Engineers (ASHRAE) (www.ashrae.org)

ASTM International (ASTM) (www.astm.org)

BE&K, Inc. (www.bek.com)

Bethlehem Steel Corporation (www.bethsteel.com)

Capital Development Board of the State of Illinois (www.cdb.state.il.us)

Construction Specifications Institute (CSI) (www.csinet.org)

Geoffrey Lee Farnsworth, AIA

International Code Council (ICC) (www.iccsafe.org)

National Association of Home Builders (NAHB) Research Foundation (www.nahbrc.org)

National Association of Women in Construction (NAWCI) (www.nawic.org)

National Easter Seal Society (www.cdc.gov/nasd/orgs)

National Institute of Building Sciences (NIBS) (www.nibs.org)

National Institute of Standards and Technology (NIST) of the U.S. Department of Commerce (www.nist.gov)

Betsy A. Pavichevich

Plumbing-Heating-Cooling Contractors Association (PHCCA) (www.phccweb.org)

Associate Professor Edward Steinfeld

T & S Brass and Bronze Works, Inc. (www.tsbrass.com)

Underwriters Laboratories, Inc. (UL) (www.ul.com)

U.S. Architectural and Transportation Barriers Compliance Board (ATBCB) (www.access-board.gov)

U.S. Department of Housing and Urban Development (HUD) (www.hud.gov)

U.S. General Services Administration (GSA), Specifications Unit (www.usgs.gov)

U.S. Green Building Council (www.usgbc.org)

## REFERENCES

We would also like to thank the authors and publishers of the publications in the following list for their contribution to our research for this chapter.

American Institute of Architects. 2001. *The Architect's Handbook of Professional Practice*, 13th ed. and supplements. Hoboken, NJ. John Wiley & Sons, Inc. (www.wiley.com).

———. "Construction Management Checklist." This document is included in Chapter 18 of *The Architect's Handbook of Professional Practice*. Hoboken, NJ: John Wiley & Sons, Inc. (www.wiley.com).

American Institute of Architects (AIA). Washington, DC (www.aia.org). Documents:

A101, "Owner-Contractor Agreement Stipulated."

A101/CMa, "Owner-Contractor Agreement Stipulated Sum—Construction Manager-Adviser Edition."

A111, "Owner-Contractor Agreement Form—Cost of the Work Plus a Fee with or without a Guaranteed Maximum Price."

A121/CMa (AGC 565), "Standard Form of Agreement Between Owner and Construction Manager Where the Construction Manager Is Also the Constructor."

A131/CMa (AGC 566), "Standard Form of Agreement Between Owner and Construction Manager Where the Construction Manager Is Also the Constructor; and Where the Basis of Payment Is the Cost of the Work Plus a Fee and There Is No Guarantee of Cost."

A191, "Standard Form of Agreement Between Owner and Design/Builder."

A201 "General Conditions of the Contract for Construction."

A201/CMa, "General Conditions of the Contract for Construction—Construction Manager—Adviser Edition."

A305, "Contractor's Qualification Statement."

A310, "Bid Bond."

A312, "Performance and Payment Bond."

A491, "Standard Form of Agreement Between Design/Builder and Contractor.

A501, "Recommended Guide for Competitive Bidding Procedures and Contract Awards for Building Construction."

A511/CMa, "Guide for Supplementary Conditions—Construction Manager-Adviser Edition."

A701, "Instructions to Bidders."

B141/CMa, "Standard Form of Agreement Between Owner and Architect, Construction Manager—Adviser Edition."

B144/ARCG-CM, "Standard Form of Amendment for the Agreement Between the Owner and Architect—Construction Management Services."

B352, "Duties, Responsibilities, and Limitations of Authority of the Architect's Project Representative."

B801/CMa, "Standard Form of Agreement Between Owner and Construction Manager-Adviser."

B901, "Standard Form of Agreement Between Design/Builder and Architect."

G701, "Change Order."

G701/CMa, "Change Order—Construction Manager—Adviser Edition."

G702/G703, "Application and Certificate for Payment and Continuation Sheet."

G702/CMa, "Application and Certificate for Payment—Construction Manager-Adviser Edition."

G703/CMa, "Continuation Sheet."

G704, "Certificate of Substantial Completion."

G704/CMa, "Certificate of Substantial Completion—Construction Manager-Adviser Edition."

G706, "Contractor's Affidavit of Payment of Debts and Claims."

G707, "Consent of Surety to Final Payment."

G707A, "Consent of Surety to Reduction in or Partial Release of Retainage."

G709, "Proposal Request."

G714, "Construction Change Directive."

G714/CMa, "Construction Change Directive—Construction Manager—Adviser Edition."

G722/CMa, G723/CMa, "Project Application and Project Certificate for Payment and Project Application Summary."

American National Standards Institute. A117.1. 2003. *Accessible and Usable Buildings And Facilities*. New York: ANSI (www.ansi.org).

A156.10. 2003. *Power Operated Pedestrian Doors*.

A156.19. 2002. *Power Assist and Low Energy Power Operated Doors*.

American Society of Mechanical Engineers (ASME). 2004. *Safety Code for Elevators and Escalators*. Fairfield, NJ: ASME (www.asme.org).

*Architectural Graphic Standards*. See Ramsey et al. (2000).

Arcom Master Systems. *Masterspec*. Salt Lake City, UT (www.arcomnet.com). Basic Sections:

01210, "Allowances," 11/02 ed.

01290, "Payment Procedures," 11/02 ed.

01310, "Project Management and Coordination," 8/04 ed.

01330, "Submittal Procedures," 11/02 ed.

01352, "LEED Requirements," 2/03 ed.

01600, "Product Requirements," 11/02 ed.

01770 "Closeout Procedures," 11/02 ed.

01781. "Project Record Documents," 11/02 ed.

01782. "Operation and Maintenance Data," 11/02 ed.

01810. "General Commissioning Requirements," 3/03 ed.

"HVAC Commissioning Requirements," 3/03 ed.

Construction Specifications Institute. 2004. *The Project Resource Manual—CSI Manual of Practice*. Alexandria, VA: CSI (www.csinet.org).

———. *Masterformat 2004*. Alexandria, VA: CSI (www.csinet.org).

Construction Specifications Institute (CSI), American Institute of Architects AIA), and National Institute of Building Sciences (NIBS). 2004. *National CAD Standard Version 3.0 Update from Version 2.0*. CSI (www.csinet.org), AIA (www.aia.org), and NIBS (www.nibs.org).

Halpin, Daniel W. and Ronald W. Woodhead. 1998. *Construction Management*. Hoboken, NJ: John Wiley & Sons, Inc. (www.wiley.com).

International Code Council. (ICC) 2006. International Codes (see section 1.3.3.3 for a list of ICC codes).

Liebing, Ralph W. 1999. *Architectural Working Drawings*, 4th ed. Hoboken, NJ: John Wiley & Sons, Inc. (www.wiley.com).

Mays, Patrick C. and B. J. Novitski. 1997. *Construction Administration: An Architect's Guide to Surviving Information Overload*. Hoboken, NJ: John Wiley & Sons, Inc. (www.wiley.com).

National Fire Protection Association (NFPA). 2002. NFPA 70A, "Electrical Code for One- and Two-family Dwellings." Quincy. MA: NFPA (www.nfpa.org).

———. 2003. *NFPA 101®. Life Safety® Code*. Quincy, MA: NFPA (www.nfpa.org).

———. 2005. NFPA 70. "National Electrical Code." Quincy, MA: NFPA (www.nfpa.org).

Nunnally, S. W. 2004. *Construction Methods and Management*, 6th ed. Upper Saddle River, NJ: Pearson Education, Prentice Hall (www.vig.prenhall.com).

Ramsey, Charles George, Harold Reeve Sleeper, and John Ray Hoke, Jr. 2000. *Architectural Graphic Standards*, 10th ed. Hoboken, NJ: John Wiley & Sons, Inc. (www.wiley.com).

Rosen, Harold J. 1999. *Construction Specifications Writing: Principles and Procedures*, 4th ed. New York: John Wiley & Sons, Inc. (www.wiley.com).

Snyder, James. 1998. *Architectural Construction Drawings with AutoCAD®* R14. Hoboken, NJ: John Wiley & Sons, Inc. (www.wiley.com).

U.S. Department of Housing and Urban Development (HUD). 1988. *Uniform Federal Accessibility Standards (UFAS)*. Washington, DC: HUD (www.hud.gov).

———. 1994. *Minimum Property Standards for One- and Two-Family Dwellings*, HUD-4910.1. Washington, DC: HUD (www.hud.gov).

U.S. General Services Administration (GSA). *Index of Federal Specifications, Standards, and Commercial Item Descriptions*. Washington, DC: GSA (www.apps.fss.gsa.gov).

U.S. Green Building Council. 2003. "LEED NC Version 2.1 Rating System." Washington, DC: Green Building Council (www.usgbc.org.).

Wakita, Osamu and Richard M. Linde. 2002. *The Professional Practice of Architectural Working Drawings*, 3rd ed. Hoboken, NJ: John Wiley & Sons, Inc. (www.wiley.com).

Yee, Rendow. 2002. *Architectural Drawing: A Visual Compendium of Types and Methods*, 2nd ed. Hoboken, NJ: John Wiley & Sons, Inc. (www.wiley.com).

# 2

# Existing Conditions

# Introduction

This chapter discusses the handling of conditions that exist at a construction site before the work under a construction contract begins. Some of these elements are addressed in this chapter, including land survey and descriptions, demolition, and subsurface investigation. Some aspects of existing conditions at the site are touched on in Chapter 1 as part of the design parameters for a project. Other work involving conditions at a building site that exist before construction begins are beyond the scope of this book. These include assessment of existing conditions including the environment (air, water, and land), traffic, accessibility, and existing materials on the site; and, where an existing building is involved, investigation of the buildings, including, but not limited to, investigations related to waterproofing, water intrusion, roofing, existing building materials, necessary decontamination or removal of hazardous materials, and underground storage tank removal.

Below-grade site construction activities that take place after construction begins are addressed in Chapter 21. Above-grade site construction is addressed in Chapter 22. Water supply sources, sanitary drainage utilities, storm drainage utilities, and electrical distribution to the building are covered in Chapter 23.

# Sustainability Considerations

Demolition is by nature destructive of the environment. The process produces dust and waste products that often cannot be reused or recycled. The dust can be controlled by the judicious use of protective enclosures and water spraying or misting.

Some waste products are very harmful to people and the environment in general. Asbestos, for example, must be handled with great care, employing protective gear for workers and enclosures to prevent the escape of asbestos-laden dust. Lead, mercury chemical, and carcinogenic materials, such as polychlorinate biphenyl (PCB), must be similarly handled with great care to prevent harmful effects.

Materials and products that can be reused or recycled should be so handled.

As much care as is practicable should be employed to prevent damaging beyond use materials and products that would otherwise be reusable.

Materials that can be recycled, such as broken glass, concrete, stone, brick, and ceramic products, should be salvaged for eventual reuse. Refer to the applicable chapters of this book for additional comments concerning specific materials.

# Applicable *MasterFormat* Sections

The following *MasterFormat* 2004 Level 2 sections are applicable to this chapter.

02 21 00  Surveys
02 26 00  Hazardous Material
          Assessment
02 30 00  Subsurface Investigation
02 32 00  Geotechnical Investigations
02 41 00  Demolition
02 42 00  Removal and Salvage of
          Construction Materials
02 81 00  Transportation and Disposal of
          Hazardous Materials
02 82 00  Asbestos Remediation
02 83 00  Lead Remediation
02 84 00  Polychlorinate Biphenal
          Remediation

# 2.1 Land Surveys and Descriptions

The transfer of land ownership and the mortgaging of property require a legal description to locate and establish the boundaries of the affected land parcel. A legal description is prepared based on a *survey* of the land. A survey is the measure and marking of land, accompanied by maps and field notes to describe the measures and marks made in the field. The lengths and directions of boundary lines are established between reference points on the ground. The reference points may be natural marks, such as rivers, or they may be stone, iron, concrete, or other artificial markers set in the

Portions of the text in versions of this section published before the seventh edition were taken from publications by, and were used with permission of, The Chicago Title and Trust Company. The seventh and eighth editions have been extensively revised from earlier versions, and do not reflect the opinions of, and are not endorsed by The Chicago Title and Trust Company.

field by a surveyor. Natural points, noted as permanent, and permanent markers set in the field by surveyors are called *monuments*.

*Land descriptions* and *surveys* have been used for centuries. It is not possible to assign the birth of the science of *surveying* to a particular year or even country, but the earliest known records refer to the making of skillful measurements and calculations with respect to land. The Chinese, at an early date, and the ancient Egyptians practiced the art of surveying. In Egypt, it appears that it was necessary every spring to reestablish corners and boundary lines obliterated by floods of the Nile River.

Two general systems of survey are used in the United States: the *metes and bounds* system and the *rectangular system*.

### 2.1.1 METES AND BOUNDS SYSTEM

Survey by *metes and bounds* (Fig. 2.1-1) consists of beginning at a known point, *running out* the boundaries of the area by *courses* (directions) and *distances* (lengths), and fixing natural or artificial

*monuments* at the corners. The *place of beginning* (P.O.B.) must be a known point that can be readily identified. The point must be established and witnessed so that it can be located again with certainty if the monument which identifies it is destroyed or removed.

The metes and bounds system, one of the oldest known methods of surveying and describing land, was used in colonial days. Land being bought, claimed, or inherited was identified by known natural points such as trees and rivers, and by artificial points such as roads, structures, or other manmade marks. For example, a survey might start at the middle of a white pine stump, move north to a small white oak tree, and continue with a series of measures by chains and links to trees, stakes, and piles of stones, until it returned to the stump where it began. Early metes and bounds surveys and descriptions were often based on monuments that lacked permanence. The surveying of large and irregular tracts of land without regard to uniformity and the failure of surveyors to make their survey notes a matter of public record gave rise to frequent boundary line disputes and litigation.

Present-day surveys and descriptions may refer to government survey section lines and corners (see Section 2.1.2), in addition to permanent monuments placed by surveyors.

### 2.1.2 RECTANGULAR SYSTEM

After the Revolutionary War, the federal government found itself with vast tracts of undeveloped and uninhabited land, with few natural characteristics suitable for use as monuments in metes and bounds descriptions. A committee headed by Thomas Jefferson evolved a new standard system of describing land so that parcels could be located readily and permanently. This system, designated the *rectangular system* (also called the *government system*) of survey, divided land into a series of rectangles. The Continental Congress adopted this system 1785, and it is in use today in a majority of the 50 states.

#### 2.1.2.1 Meridians and Base Lines

The world is divided into a grid of imaginary north-south lines called *meridians of longitude* that extend from the North Pole to the South Pole (Fig. 2.1-2). The

**FIGURE 2.1-1** A typical metes and bounds survey plat. (Drawing by HLS.)

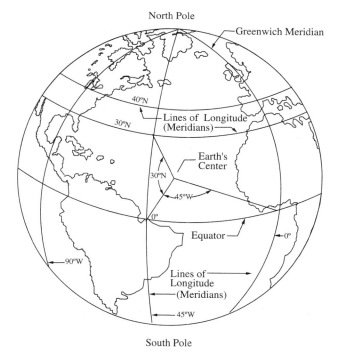

**FIGURE 2.1-2** The rectangular system of surveying is based on a grid of meridians and parallels measured from the Greenwich prime meridian and the equator, respectively. (Drawing by HLS.)

meridian that passes through the Royal Observatory at Greenwich, England, is designated as 0 (zero) degrees longitude and is called the *Greenwich*, or *prime*, *meridian*. Locations east or west of Greenwich are measured in degrees from the prime meridian.

The reference lines running around the globe from east to west are called *parallels of latitude*. The *equator* is designated as 0 (zero) degrees latitude. Locations that fall north or south of the equator are measured in degrees from the equator.

The rectangular system of survey is based on a series of 35 *principal meridians* and 32 *base lines*, as shown in Figures 2.1-3a and 2.1-3b.

### PRINCIPAL MERIDIANS

In establishing the rectangular system of survey, surveyors first selected a substantial landmark from which a start could be made in surveying an area of land. Usually a place was selected that could be identified readily, such as the mouth of a river, and whenever possible a monument was placed. From this point, the surveyors ran a line due north and south and designated it as the principal meridian for that particular state or area. The location of the principal meridian (its longitude) was fixed by a reading measured in degrees, minutes, and seconds (see the Glossary) west of the prime meridian.

As territories were opened and surveyed by the government, principal meridians were established for each area opened. Some of the principal meridians were referred to by number, such as the First, Second, and Third principal meridian. The Third principal meridian, for example, is the line located 89 degrees, 10 minutes, and 15 seconds west of Greenwich. It extends from the mouth of the Ohio River to the northern boundary of Illinois. Other principal meridians were named for the states, such as the Michigan Meridian, which covers the survey of that state.

### BASE LINES

After the principal meridian for a particular area was fixed, a point on the meridian was selected from which a line was run at right angles, due east and west. This line was designated as the base line for the area, and its location (its latitude) was established in degrees, minutes, and seconds north of the equator.

Sometimes the base line for a new territory was established by extending the base line of an adjoining territory. For example, the base line for both the Second and Third principal meridians are at 38 degrees, 28 minutes, and 20 seconds north latitude.

### CORRECTION LINES

Meridians meet at the North and South Poles. They are not parallel lines. An accurate measure of a midwestern area 6 miles square would show its north line to be about 50 ft shorter than its south line. To compensate for this convergence, additional east and west lines (*correction lines*), parallel to the base line, were located at intervals 24 miles north and south of the base line.

### GUIDE MERIDIANS

After the east-west correction lines were set, *guide meridians* running due north and south at 24-mile intervals on each side of the principal meridians were established. Guide meridians extend from the base line to the first correction line, and then from correction line to correction line. The guide meridians and the correction lines form squares approximately 24 miles on each side, as shown in Figure 2.1-4.

### TOWNSHIP AND RANGE LINES

The 24-mile squares were then divided into 16 smaller tracts by *township lines* and *range lines*. Township lines run east and west at 6-mile intervals. Range lines run north and south at 6-mile intervals. This crosshatching resulted in a grid of squares, called *townships*.

Two *reference numbers* were assigned to each square: a *township number* and a *range number*. Rows of squares (*tiers*) were numbered consecutively to the north or south of the base line. Thus, each square in the first row north of the base line is called Township 1 (one)

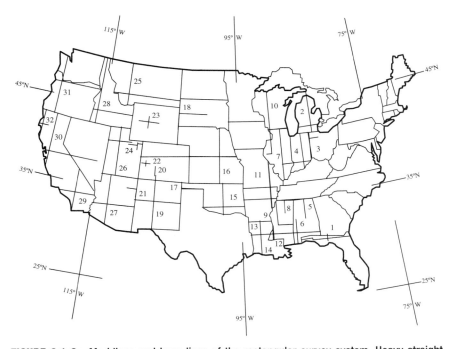

**FIGURE 2.1-3a** Meridians and base lines of the rectangular survey system. Heavy straight lines running from east to west on the map are base lines. Heavy straight lines running from north to south are principal meridians. The principal meridians are numbered. These numbers have been used only to identify the principal meridians on this map and to tie them to the accompanying table. These numbers are in no way related to the rectangular survey system. The names of the principal meridians identified by the numbers on the map and their longitudes are shown in Figure 2.1-3b. The latitude of the base line associated with each principal meridian is also shown there. The principal meridians governing Alaska are included in the table, but a map of Alaska has not been included. (Drawing by HLS.)

FIGURE 2.1-3b Meridians and Base Lines

| Map Number | Meridian Name | Governed States | Longitude of Meridian °  '  " | Latitude of Associated Base Line °  '  " |
|---|---|---|---|---|
| 1 | Tallahassee | FL | 84  16  42 | 30  28  00 |
| 2 | Michigan | MI, OH | 84  22  24 | 42  26  30 |
| 3 | First Principal | OH, IN | 84  48  50 | 41  00  00 |
| 4 | Second Principal | IL, IN | 86  28  00 | 38  28  20 |
| 5 | Huntsville | AL, MS | 86  34  45 | 35  00  00 |
| 6 | St. Stephens | AL, MS | 88  02  00 | 31  00  00 |
| 7 | Third Principal | IL | 89  10  15 | 38  28  20 |
| 8 | Chickasaw | MS | 89  15  00 | 34  59  00 |
| 9 | Choctaw | MS | 90  14  45 | 31  54  40 |
| 10 | Fourth Principal | IL | 90  28  45 | 40  00  30<br>42  30  00 |
| 11 | Fifth Principal | AR, IA, MN, MO, ND, SD | 91  03  42 | 34  44  00 |
| 12 | St. Helena | LA | 91  09  15 | 31  00  00 |
| 13 | Washington | WA | 91  09  15 | 31  00  00 |
| 14 | Louisiana | LA | 92  24  15 | 31  00  00 |
| 15 | Indian | OH | 97  14  30 | 34  30  00 |
| 16 | Sixth Principal | CO, KS, NE, SD, WY | 97  23  00 | 40  00  00 |
| 17 | Cimarron | OK | 103  00  00 | 36  30  00 |
| 18 | Black Hills | SD | 104  03  00 | 44  00  00 |
| 19 | New Mexico | NM | 106  52  41 | 34  15  25 |
| 20 | New Mexico | CO | 106  53  36 | |
| 21 | Navajo | AZ, NM | 108  32  45 | 35  45  00 |
| 22 | Ute | CO | 108  33  20 | 29  06  20 |
| 23 | Wind River | WY | 108  48  40 | 43  01  20 |
| 24 | Uinta | UT | 109  57  30 | 40  26  20 |
| 25 | Principal | MO | 111  38  50 | 45  46  48 |
| 26 | Salt Lake | UT | 111  54  00 | 40  46  04 |
| 27 | Gila and Salt River | AZ | 112  18  24 | 33  22  33 |
| 28 | Boise | ID | 116  24  15 | 43  22  31 |
| 29 | San Bernardino | CA | 116  56  15 | 34  07  10 |
| 30 | Mount Diablo | CA, NV | 121  54  48 | 37  51  30 |
| 31 | Willamette | OR, WA | 122  44  20 | 45  31  00 |
| 32 | Humbolt | CA | 124  07  11 | 40  25  04 |
| 33 | Copper River | AK | 145  18  42 | 61  49  11 |
| 34 | Fairbanks | AK | 147  38  33 | 64  51  49 |
| 35 | Seward | AK | 149  21  53 | 60  07  26 |

North (T.1N.); each in the second row, Township 2 (two) North (T.2N.), and so forth. Similarly, each square in the first row south of the base line is called Township 1 South; each in the second row, Township 2 South, and so on. In the same manner, rows of squares (*ranges*) were numbered consecutively to the east and west of the principal meridian. Thus, each square in the first row east of the principal meridian is called Range 1 East (R.1E.), those in the second row Range 2 East (R.2E.), and so forth, and on the west of the principal meridian, those in the first row were numbered Range 1 West, those in the second row, Range 2 West, and so on.

## SECTIONS

*Townships* were the smallest divisions of land provided for in the 1785 act that created the rectangular system of survey. In making early surveys, the outside boundaries of the townships were surveyed and monuments were placed at every mile on the township lines. It soon became apparent that a township 6 miles square was too large an area in which to describe and locate a given tract of land. In 1796, Congress passed an act directing that townships already surveyed be subdivided into 36 *sections*, each to be approximately 1 mile square containing about 640 acres. The sections were numbered consecutively from 1 to 36, beginning with number 1 in the northeast corner of the township and ending in the southeast corner with number 36, as shown in Figure 2.1-4b. A tract of land can be located within a square mile by giving the section number, the township number, and the range number.

## FRACTIONAL SECTIONS

A *fractional section* is one that is larger or smaller than others in a township. Fractional sections may result from geographical factors, such as land area bordering on lakes. They may also occur where the surveys of separately surveyed areas do not tie together perfectly, and sections result which are less than 1 mile square.

In addition, because of the convergence of the meridians, each township does not form a perfect square, and cannot be surveyed to form 36 identical sections in shape and area (Fig. 2.1-4c). Therefore, the excess or deficiency in

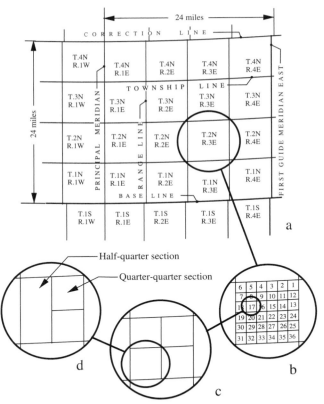

**FIGURE 2.1-4** Division of land in the rectangular system: (a) A square (24 mi on a side), shown with townships delineated. (b) A township (6 mi on a side); this is Township 2 North (T.2N.), Range 3 East (R.3E.) from the square shown in (a), with 36 sections delineated. (c) A section (640 acres) shown with quarter section divisions; this is Section 17 from the township shown in (b). (d) A quarter section (160 acres) shown with half-quarter (80 acres) and quarter-quarter sections (40 acres) delineated. In a legal description, the quarter-quarter section in (d) would be described as "the northeast quarter of the southwest quarter of Section 17, Township 2 North, Range 3 East of the First (Second, Third, New Mexico, etc., as applicable) Principal Meridian." (Drawing by HLS.)

south-to-north measurements is added to or deducted from the north portion of the sections in the north row in the township (Sections 1 through 6). The excess or shortage in the east-to-west measurements is added to or deducted from the west portion of the west row in the township (Sections 6, 7, 18, 19, 30, and 31). These adjusted sections are classified as *fractional sections*.

After fractional sections are divided into equal fractional parts so far as possible, remaining portions are divided into lots of approximately the same size. These *government lots* are numbered consecutively.

### DIVISION OF SECTIONS

Sections are also divided into still smaller units (Fig. 2.1-4d). In 1800, Congress directed the subdivision of sections into east and west halves of approximately

320 acres each by running lines north and south through the centers of sections. In 1805 regulations provided for *quarter sections* by running east and west lines through the centers of sections (see Fig. 2.1-4d). Congress later provided for the further division of quarter sections into *half-quarters* and finally *quarter-quarters*. These subdivisions were made by running east-west lines and north-south lines through the quarter sections. A quarter-quarter section of 40 acres is the smallest statutory division of regular sections in the rectangular survey system (Fig. 2.1-4d).

### 2.1.3 LEGAL DESCRIPTIONS

A *legal description* identifies, locates, and defines a specific piece of land to distinguish it from all other tracts of land. Legal descriptions in the United States may be

based on (1) the metes and bounds survey system, (2) the rectangular survey system, and (3) reference to recorded maps and plats. It is not uncommon to find a combination of these methods in the same description. For example, a lot in a subdivision may be described by reference to a recorded plat of a land, which in turn is described by a metes and bounds survey within the framework of the rectangular system (Fig. 2.1-5).

### 2.1.3.1 Metes and Bounds Description

A metes and bounds description consists of a series of statements describing each portion of the boundary around the parcel of land being described. Starting from a P.O.B., the direction and distances of the property lines are *called* (stated in the description), until the perimeter has been traced around the entire property, returning to the starting point. Each leg of the perimeter is described first by direction, then by length (see Fig. 2.1-5). The survey description must "close"; that is, if the courses and distances of the description are traced in order, one must return to the starting point.

All that certain lot or parcel of land with the improvements thereon and appurtenances thereto, belonging, lying, being, and situated in the City of Rightown, Virginia, being known, numbered, and designated as Lot No. six (6), in Block C, Section three (3), on a certain plat of Hunt Club View Hills, Section three (3) dated March 28, 1989, made by Simpton, Inc. Surveyors, and recorded in the Clerk's Office of the Circuit Court of the City of Rightown, Virginia, in Plat Book 2, at page 1.

Said property commencing at the North West corner of Section 8, thence South along the section line 12 feet; thence East 10 feet to the point of beginning; thence South 61 degrees, 5 minutes, 20 seconds East for 119.34 feet; thence South 31 degrees, 43 minutes, 40 seconds West for 100.72 feet; thence South 80 degrees, 47 minutes West for 140 feet to a point of tangency; thence along a curve convex Northwesterly and having a radius of 20 feet; thence North 28 degrees, 54 minutes, 40 seconds East, for a distance of 174.68 feet to the point of beginning.

**FIGURE 2.1-5** Legal description for the plat shown in Figure 2.1-1.

## DIRECTION

Directions (*courses*) that fall along major compass directions are so named. Courses other than due north, south, east, or west (cardinal points) are called *angular courses*. The direction of an angular course is stated in degrees, minutes, and seconds as an angular deviation east or west of due north or south (see Fig. 2.1-5). (There is one exception to this convention explained in the next section, "Prolongation.")

The direction is always written with the statement *north* or *south*, followed by the angle of eastward or westward deviation from north or south. For example, "north 30 degrees east" defines an angular course direction 30 degrees eastward of due north. A call "south 42 degrees 15 minutes west" defines a direction 42 degrees and 15 minutes westward of south.

## PROLONGATION

Sometimes course direction is described by stating the amount of deviation from the preceding course line. An angle is measured from an extension, a *prolongation*, of the preceding line, and might be stated as "thence northerly 30 degrees to the left of an extension of the last described line."

## CURVED LINES

A curved line is sometimes referred to in a description as "thence south 80 degrees east 100 feet to a point of tangency; thence along a curve convex northeasterly and having a radius of 30 feet; a distance of 40 feet." The point of tangency (the point of contact between the last described straight line and the curve) is the starting place of the curve. *Convex* refers to the outside of a curve; *concave*, if used, refers to the inside of a curve. "Convex northeasterly" means that the outside of the curve lies to the northeast. Figure 2.1-1 shows a curved property line with typical notation at the southwest corner of the property.

## DISTANCE

Distance is usually stated in feet and decimal parts of feet, and sometimes in feet and inches or fractions of feet. Very old surveys may state distances in chains, rods, and links. The measure is made from point to point along straight lines and along the curve of a curved line.

### 2.1.3.2 Rectangular System Descriptions

A tract of land in the rectangular survey system can be described by reference to its principal meridian, base line, township and range numbers, section number, and, if necessary, its location within the section (see Fig. 2.1-4).

Government lots are described by identifying the lot number and the quarter section in which the lot is located.

### 2.1.3.3 Reference to Recorded Plat

Curved street patterns often produce irregularly shaped blocks and lots. An entire subdivision may be described within the rectangular system or by a metes and bounds description, but it may be impractical to use these systems to describe individual blocks or lots. Therefore, many states provide for the legal description of property by the reference to plats properly recorded, generally in the office of the appropriate county.

A surveyor surveys the tract and prepares a *plat*. The tract or subdivision may be divided into blocks and then into lots. Lots and blocks may be given consecutive numbers or letters enabling identification of a particular area of land within the subdivision by a lot and block number. For further identification it is common practice to give the subdivision a name (see Fig. 2.1-5).

The surveyor's plat is captioned with the legal description of the land. It shows all boundary lines, all necessary monuments and dividing lines for blocks, lots, and streets, and the numbering and dimensions of each lot.

*Easements* and *restrictions* may also be indicated on a plat. Upon approval by the proper authorities (such as municipal or county zoning boards, planning or building commissions), a plat is recorded generally in the appropriate government recorder's office. When recorded, the lots in a subdivision may be easily identified by reference to lot and block numbers, the subdivision name, and the section, township, and range in which the entire subdivision boundaries are described within a rectangular survey or metes and bounds description.

### 2.1.3.4 Uncertainties

Mistakes, errors, and judgments in a survey may result in conflicts and uncertainties in the legal description of land. There may be differing opinions as to the precise location of the boundary lines of a tract of land, and the exact location of a point on the land can be subject to interpretation.

## LOCATION OF MARKERS

Through the years, monuments may have been moved or uprooted during earth-moving projects, such as road construction, or in farming. Relocating a missing corner may require measuring from monuments in existing surveys (which may also contain uncertainties) or from other references.

The disappearance of monuments constantly causes uncertainties in surveying. Metes and bounds surveys in which a monument was a tree or a post that no longer exists may require the surveyor to dig carefully in the probable vicinity of the lost monument in an attempt to find the decayed remains of the tree or post.

Roads described in an old survey may have been abandoned and the pavement removed. Even though title companies may have unrecorded evidence of such roads, surveying judgment is required to fix a line to substitute for the missing road.

Many other uncertainties are caused by conjecture as to the location of lakes and rivers, which have receded, dried up, or changed course and location over the years.

## CONVERGENCE

The original government surveys that divided land areas into 24-mile squares, each containing 16 townships, and subsequent divisions into sections and fractional sections made a practical but not perfect effort to adjust for the convergence of meridians of longitude as they travel north. Sections were marked by lines spaced at 1-mile intervals, running parallel to the range line closest to the principal meridian. This causes a divergence in direction which becomes progressively more pronounced in each successive tier of sections. Therefore, an ambiguity may result when a cardinal point is used as a course in a legal description. For instance, the call "thence north 150 feet to a point . . . ," unless marked by a monument, usually is interpreted as indicating a course which runs parallel with the east line of the section. If the section line deviates from true north, an uncertainty exists.

The measurement of distances called for in some legal descriptions is virtually impossible. For example, the distance 126.01 feet seems precise, but it is unlikely that two surveying crews measuring this distance would observe the same measurement. A distance of .01 ft is a little less than $1/_8$ in. Differences in the tautness of a measuring tape and changes in the length of the tape due to the effect of temperature fluctuations can easily negate the accuracy of such a measurement.

In surveying a tract, a surveyor must locate the boundaries of abutting tracts as shown in earlier surveys and reconcile the perimeter lines of the tract being surveyed with them. The surveyor must also obtain a legal description of the abutting tracts.

Section corners and other marks in the rectangular system sometimes were placed incorrectly in early surveys, which were based on lengths such as *chains* and *links*. Surveyors made field measurements with a metal chain made of 100 links. A chain was 66 ft long. The combined length of 80 chains was equal to 1 statute mile, or 5280 ft.

Weather changes expanded and contracted these chains, and wear and use caused them to stretch. Sometimes surveyors counted the revolutions of the wheel of the vehicle in which they rode to measure long, straight stretches. Distances also were "stepped off" by the measured stride. Even though these measurements were far from accurate, markers placed according to them are considered legal points of reference.

In addition to errors in distance, there are also deviations in the perimeter lines of a section from true east-west or north-south courses. In locating the south boundary of a tract described as the "north 10 acres of the north west quarter" of a section in which the north line deviates from a true east and west course, some judgment is required. The south line might be considered to be parallel with the north line of the section, or it might run due east and west. Again, the surveyor must search for physical evidence, such as a stake set by an earlier surveyor or a fence indicating a line of occupation, in order to decide which line was intended.

Courses may show an appreciable error in direction. For example, a metes and bounds course shown as "south 7 degrees west" in an earlier survey might be observed to be actually "south 6 degrees west," indicating that an error could have been made by an earlier surveyor or that faulty equipment had been used.

Mistakes, such as careless observations and inaccurate measurements, are random and unpredictable. Even the best surveying instruments are subject to a certain degree of error, but these errors are predictable and can be allowed for in observing measurements. Again, however, this calls for some human judgment.

### 2.1.3.5 Recorded Plats

When an individual lot is described by reference to a recorded subdivision plat, the plat together with all information and data shown thereon is part of the lot description (see Fig. 2.1.5). It is possible that the lot size may not be exactly as shown on the plat. The incorporation by reference is of the entire plat and all data appearing thereon. This includes not only dimensions and expressions of quantity and area, but also the notations of the courses and distances of the lines of the subdivision and of the identity and locations of the monuments marking the boundaries.

When this information is incorporated into a legal description, it is, in effect, a metes and bounds description and is subject to rules of interpretation, should any conflict or discrepancy be contained in the data of the plat. For example, monuments prevail over courses and distances, courses prevail over distances, and expressions of quantity or area are generally the least reliable.

It is possible that the overall recorded measurements of a block may not equal the recorded widths of the lots in the block. If such an error is found, the widths of some or all of the lots must be adjusted accordingly.

The systems of survey upon which legal descriptions are based are as precise as possible. There is always the possibility of uncertainties in separate surveys, and mistakes may occur in the writing and rewriting of legal descriptions. Descriptions may identify land parcels adequately, but surveys are necessary to actually locate boundary lines.

The complexities of resolving uncertainties in either land survey or legal land description are a matter of real estate law and interpretation and are beyond the scope of this book.

## 2.2 DEMOLITION

Demolition includes removal from the site of existing features such as buildings that would interfere with new construction. Movable structures, such as house trailers, may be simply towed away. Houses and other light buildings are sometimes removed from the site intact.

A major aspect of some projects is the demolition and removal of existing buildings or sections of existing buildings that are not to be a part of the new work. When structures are to be removed as a part of the general construction contract, the term *demolition* is often used to refer to that part of such structures that is above ground. Removing below-grade portions of structures is usually thought of as excavation work (see Section 21.3.2). However, when demolition is done under a separate contract, removing basement walls and slabs and building foundations is usually a part of the separate contract.

Demolition may be either *complete* or *selective*. In complete demolition, an existing structure is demolished completely, and the debris is removed from the site. Selective demolition consists of removing a portion of an existing structure, usually to make way for remodeling or an addition.

Demolition is potentially hazardous to the workers involved, to the public, to adjacent property, and to the environment. Therefore, extraordinary measures should be taken to protect all of them from such dangers. Most heavily populated jurisdictions have code and other legal requirements for public and property safety during demolition. There are usually restrictions on practices that affect the health and welfare of the workers and the public or cause damage to the environment. For example, there are regulations for the handling of hazardous materials such as asbestos, toxic chemicals, explosive materials, flammable materials, and radioactive waste.

Moreover, there are also often extensive requirements relative to the bracing and shoring of buildings during demolition. These restrictions are usually more stringent when a building has been damaged in a fire, a storm, or an earthquake. Restrictions are likely to be extensive regarding the use of explosives.

Even when safety and environmental restrictions are not mandated by code or laws, precautions should be observed. Sprinkling with water, temporary enclosures, and other appropriate means should be used to limit the spread of dust and dirt in the air. Pedestrian and vehicular traffic should be routed around the area of danger. Rodents and insects should be controlled before and during demolition using legal and environmentally sound means so that they are not driven into adjacent structures. These means should not damage the environment.

In selective demolition, there is usually a need to protect building portions that will not be demolished. Cutting and patching of existing materials is often required. Protective enclosures should be erected between areas to be demolished and areas to be let alone. People often occupy spaces adjacent to the area in which selective demolition will take place. They should be protected from danger and from exposure to fumes, dust, dirt, loud noises, and toxic materials by whatever means are appropriate: temporary dustproof walls, sound barriers, barricades, and covered passageways.

Portions of a building to remain should similarly be protected, and in addition should be protected from the weather by temporary walls and roofs, or by weatherproof tarpaulins or plastic sheeting.

When portions of a building undergoing selective demolition remain occupied, utility service interruptions should be kept to a minimum and should be done so that the occupants are not harmed, placed in danger, or caused significant discomfort. For example, fire sprinklers should not be interrupted during times when a building is occupied. Exits should be maintained at the number and size required by code. Electric, water, sewer, and other services should not be interrupted for extended periods while the building is occupied.

## 2.3 Subsurface Investigation

The accuracy of soil classification, and descriptions defining density, consistency, plasticity, bearing capacity, and other soil properties, depend on the skill of the individual making the observation and the extent of the controls under which the observations are made. As a result,

**FIGURE 2.3-1** Manual and power-driven hand augers for preliminary field investigation of soils. (Courtesy ELE International, Inc., Soiltest Products Division.)

a

b

**FIGURE 2.3-2** Test borings to determine soil properties are made with power-driven equipment and (a) a split-barrel sampler for coarse-grained soil or (b) a thin-walled sampler for fine-grained soils. (Courtesy ELE International, Inc., Soiltest Products Division.)

classification by field observation is subject to uncertainties and inaccuracies. The extensiveness of soil investigation will vary with building type, with local knowledge of subsurface conditions, and with conclusions drawn from preliminary investigations.

### 2.3.1 PRELIMINARY INVESTIGATION FOR SMALL BUILDINGS

For single-family homes and other small buildings, a preliminary investigation should be made to obtain general information about the nature and origin of existing soils. This investigation should determine the need for and the necessary extent of further investigation. In some cases, it can eliminate the expense of extensive soil boring and testing. A preliminary examination usually consists of at least one test boring at the site to determine the acceptability of the soil for ground-supported slabs and foundations. Such test borings can be made with simple tools (Fig. 2.3-1) to determine soil types and the extent of each to the depth usually necessary to predict the performance of small building foundations.

Obtaining the professional advice of a soils engineer experienced in a given locality may preclude the need for even a single test boring at a single-family home site. This is especially true if the engineer's advice is coupled with an examination of other structures in the immediate area for evidence of apparent settlement or expansion of the soil, provided that similar conditions of soil, topography, and proposed construction prevail.

No additional soil investigation is required for single-family home sites if the preliminary site investigation indicates that the site has nonproblem soils consisting of firm, nonexpansive material, that no surface water or groundwater problems exist, and that fill thickness greater than 3 ft (914 mm) will not be required.

When a preliminary investigation reveals problem soils at a single-family dwelling site, the procedures, test borings, and soil investigation procedures indicated in the following paragraphs for commercial, institutional, and other building types should be followed.

### 2.3.2 SUBSURFACE INVESTIGATION FOR MEDIUM-SIZED AND LARGE BUILDINGS

An engineering study should be conducted by a qualified soils engineer for every building other than a single-family dwelling. This study should be used to determine necessary design and construction procedures.

Depending on the extent of the project, this subsurface investigation may be made by digging test pits, driving sounding rods, or making test borings. Though limited in depth, test pits are effective investigative means when the footings will not go deeper than about 8 ft (2.44 m).

Load tests can be made at their bottoms, soil samples can be taken in them, and the water table can be measured easily after water fills the pit to the maximum level of the table.

Bearing capacities can be measured using standard penetration tests by driving rods using a standard hammer and counting the number of blows needed to drive the rod a given distance.

However, the most versatile method of obtaining soil test samples and determining bearing capacities involves using soil borings. Borings can be made much deeper into the soil than can either test pits or sounding rods.

**FIGURE 2.3-3** Relative Density of Coarse-Grained Soils

| Soil Description | Rule-of-Thumb Field Guide[a] | ASTM Method No. of Blows[b] | Relative Density (%) |
|---|---|---|---|
| Very loose | | 0–4 | 0–20 |
| Loose | Easily penetrated with $1/2$ in. (12.7 mm) reinforcing rod pushed by hand | 4–10 | 20–40 |
| Medium | Easily penetrated with $1/2$ in. (12.7 mm) reinforcing rod driven with 5 lb (2.27 kg) hammer | 10–30 | 40–60 |
| Dense | Penetrated 1 in. with $1/2$ in. (12.7 mm) reinforcing rod driven with 5 lb (2.27 kg) hammer | 30–50 | 60–80 |
| Very dense | Penetrated only a few inches with $1/2$ in. (12.7 mm) reinforcing rod driven with 5 lb (2.27 kg) hammer | Over 50 | 80–100 |

[a]Field guide intended as an example of one of many field procedures currently in use for indicating density and is not necessarily a preferred method.

[b]Number of blows required to penetrate to a depth of 1 ft (304.8 mm) as measured in accordance with ASTM D1586.

**FIGURE 2.3-4** Consistency of Undisturbed Cohesive Soils

| Soil Description | Rule-of-Thumb Field Guide[a] | Unconfined Comprehensive Strength, $q_u$[b] (MPa) |
|---|---|---|
| Very soft | Core (height = 2 × diameter) sags under own weight | 0.25 (23.94) |
| Soft | Can be pinched in two between thumb and forefinger | 0.25–0.50 (23.94–47.88) |
| Medium | Can be imprinted easily with fingers | 0.50–1.00 (47.88–95.76) |
| Stiff | Can be imprinted with considerable pressure from fingers | 1.00–2.00 (95.76–191.52) |
| Very stiff | Can be barely imprinted by pressure from fingers | 2.00–4.00 (191.52–383.04) |
| Hard | Cannot be imprinted by fingers | 4.00 (383.04) |

[a]Field guide is only an indication of soil consistency. Values of $q_u$ are given as basic values of consistency by which field classification can be verified and should not be used for design purposes without laboratory verification.

[b]$q_u$ is unconfined compressive strength in tons/sq ft (not the ultimate bearing capacity of the soil for design).

### 2.3.2.1 Density and Consistency Tests

The relative density of coarse-grained soils can be estimated using a standard penetration test, with a split-barrel sampler and power-driven equipment (Fig. 2.3-2). This test is not always necessary for single sites for one- or two-family residences, but it should be required for other types of buildings. The relative density may be estimated from the number of blows causing the sampler to penetrate 1 ft (304.8 mm) and is classed from very loose to very dense (Fig. 2.3-3). The consistency of clay (cohesive) soils should be determined by test from undisturbed samples using a thin-walled sampler (Fig. 2.3-4), or the assumed characteristics may not be valid. A direct numerical value of consistency determined by laboratory test is the load per unit of area that causes failure of an unconfined soil sample in a simple compression test. The value obtained is called the *unconfined compressive strength* ($q_u$) of the soil. Where CL, CH, OL, or OH soils (see Fig. 21.1-3) are encountered, this laboratory test should be performed to determine the unconfined compressive strength in order to select the type of slab used. The consistency of cohesive soils is given in Figure 2.3-4.

### 2.3.2.2 Tests for Other Properties

Soil samples should be tested in a soils laboratory to establish their other properties. These include the soil's particle sizes, plasticity, cohesiveness, shear strength, water content, shrinkage coefficient, compressive strength, and the amount that the soil can be expected to consolidate (compress) under load.

## 2.4 Additional Reading

More information about the subjects discussed in this chapter can be found in the references listed in Section 2.5 and in the following publications.

Bowles, Joseph E. 1992. *Engineering Properties of Soils and Their Measurement*, 4th ed. New York: McGraw-Hill (www.mcgraw-hill.com).

Day, Robert W. 2001. *Soil Testing Manual Procedures, Classification Data, and Sampling Practices*. New York: McGraw-Hill (www.mcgraw-hill.com).

## 2.5 Acknowledgments and References

### ACKNOWLEDGMENTS

We gratefully acknowledge the assistance of the following organizations and individuals in preparing this chapter. We are also indebted to them for permission to use their illustrations when requested and for the use of their publications as references.

ELE International, Soiltest Products Division (www.soiltest.com)

### REFERENCES

We would also like to thank the authors and publishers of the publications in the following list for their contribution to our research for this chapter.

*Architectural Graphic Standards. See* Ramsey et al. (2000).

ASTM International (ASTM). West Conshohoken, PA. (www.astm.org). Standards.

D427, "Test Method for Shrinkage Factors of Soils by the Mercury Method."

D698, "Standard Test Method for Laboratory Compaction Characteristics of Soil Using Standard Effort (12,400 ft-lbf/ft$^3$) (600 kN-m/m$^3$)."

D1557, "Test Method for Laboratory Compaction Characteristics of Soil Using Modified Effort (56,000 ft-lbf/ft$^3$ (2,700 kN-m/m$^3$))."

D1586, "Standard Test Method for Penetration Test and Split-Barrel Sampling of Soils."

D1883, "Standard Test Method for CBR (California Bearing Ratio) of Laboratory Compacted Soils."

D2487, "Standard Classification of Soils for Engineering Purposes (Unified Soil Classification System)."

D2488, "Standard Practice for Description and Identification of Soils (Visual-Manual Procedure)."

D4253, "Standard Test Methods for Maximum Index Density and Unit Weight of Soils Using a Vibratory Table."

D4254, "Standard Test Method for Minimum Index Density and Unit Weight of Soils and Calculation of Relative Density."

D4318, "Standard Test Method for Liquid Limit, Plastic Limit, and Plasticity Index of Soils."

Ramsey, Charles George, Harold Reeve Sleeper, and John Ray Hoke, Jr. 2000. *Architectural Graphic Standards*, 10th ed. Hoboken, NJ: John Wiley & Sons, Inc. (www.wiley.com).

# Concrete

# Introduction

Concrete is used in almost every type and size of architectural and engineering structure. Its major uses are for foundation walls and footings, building super-structures, floor slabs, stairs, sidewalks, paving, curbs and gutters, streets, high-ways, bridges, and for site structures such as storm water inlets and piping, septic tanks, and drainage swales. The evidence is that the use of concrete is increasing. This is partly due to the inherent fire re-sistance of concrete, which eliminates the expensive fireproofing required on steel structures. Ease of erection has increased the use of precast concrete units in all types of buildings. Concrete is an old ma-terial, but the technology associated with it continues to evolve. Fiber reinforce-ment is being specified increasingly to toughen concrete slabs and help control cracking, thus reducing the need for control joints. The use of glass-fiber-reinforced concrete (GFRC) for exterior cladding, column covers, and decorative panels is also increasing. Super plasticiz-ers are being used more and more to in-crease the fluidity of concrete, which makes pumping it to its final location more practicable.

The term *concrete of a suitable qual-ity* and similar phrases used in this book refer to concrete that will perform satis-factorily in its intended use. To do so, it must possess:

- Strength to carry superimposed loads
- Sufficient watertightness to prevent water penetration
- Durability to resist wear and weather
- Workability to ensure proper han-dling, placing, finishing, and curing

These properties can only be achieved with good materials and careful work-manship. Required surface treatments may range from smooth-troweled sur-faces for finished floors to highly deco-rative geometric patterns for use on ter-races, walls, and other surfaces. The plastic quality of concrete offers almost unlimited design possibilities of form, pattern, and texture. The in-place cost of concrete is three to six times the cost of its ingredients. The in-place cost and the long-term cost of maintenance are de-pendent largely on proper care in han-dling, placing, finishing, and curing. Skillful selection and combining of in-gredients in the mix are necessary to achieve the most economical concrete. The same factors also influence concrete quality. No concrete mix is ever better than the workmanship it receives in the field.

To produce concrete of suitable qual-ity, sound materials must be carefully se-lected, proportioned, and combined into the correct mixture with a consistency that can be worked into place to provide the required surface finish. The placed concrete must then be properly cured and protected.

Unless quality control measures are observed, undesirable properties may oc-cur in the hardened concrete, especially on the wearing surface. These undesir-able properties may lead to a soft or *dust-ing* surface, permeable concrete, crack-ing, or poor durability.

# Sustainability Considerations

All construction materials cause some damage to the environment, either inher-ently or during the obtaining of their raw materials, during conversion of their raw materials into a finished product, during their installation, over the useful life of the building, and when they are disposed of after the building in which they have been used is demolished. This section discusses some of the potential damage that may be caused by the use of rein-forced concrete and some ways to allevi-ate some of this potential damage. Most of the comments here apply as well to cast-in-place and precast concrete.

### Raw Materials

Concrete consumes the largest quantity of nonrenewable materials, such as port-land cement, aggregates, and water, of all the industries in the world. Obtaining these raw materials can damage a land-scape, leaving deep pits or strip-mined hillsides. It can also produce massive ero-sion, harmful runoff, and habitat loss and release dangerous pollutants into the environment. Concrete consumes large quantities of many materials other than its own raw materials. These include wood and wood panel products; plastics; and steel, aluminum, and other metals. Each of these has its own impacts on the environment. For discussions of each of their impacts on the environment, refer to the applicable chapters of this book.

Concrete consumes large quantities of water. The amount of water used can be reduced by increasing the amount of fly ash in the concrete, as discussed under "Manufacture of Products Used in Con-crete." Another way to reduce water use is by the employment of super plasticiz-ers and air entrainment, both of which produce concrete that uses less water.

Natural aggregates make up nearly three-quarters of the volume of concrete. The amounts of these can be reduced by substituting such waste materials as crushed concrete and recycled glass as a portion of the aggregate.

### Manufacture of Products Used in Concrete

The major cementing agent in concrete is portland cement. The manufacture of portland cement consumes more energy than the manufacture of any other con-struction material except for steel and aluminum. Portland cement production produces about 7% of all the carbon diox-ide, a greenhouse gas, released yearly into the atmosphere from all sources. The process also emits massive amounts of other gases and dust. The obvious solu-tion to this excessive use is to reduce the amount of portland cement used. This can be done by increasing the amount of other cementing agents, such as fly ash, wood ash, blast furnace slag, and rice husks, in concrete as a cementing agent. This not only allows the reduction of portland ce-ment in concrete mixes, it substitutes for the portland cement a waste by-product that might otherwise be discarded. Fly ash, for example, is a waste by-product always produced during the operation of coal-fired electricity generators. Studies have indicated that up to 50% of the portland cement used today in concrete can be replaced with fly ash.

### Production and Installation of Concrete

In addition to the water used in the concrete itself, potable water is also used to clean concrete transit mixing trucks, chutes, and other equipment. This water is almost always wasted. Reclaiming this water for other purposes would reduce the consumption necessary for a construction project.

Concrete transit mixing trucks often arrive at a project site carrying more concrete than is needed for a particular pour. Usually, this leftover concrete is dumped at the site and is then either carried away to a landfill or simply buried at the project site. Reclaiming this concrete for use as crushed aggregate or other purposes would produce a source of material for these uses and save space in the landfill.

Formwork consumes large quantities of wood, wood panel products, and metals. These can be reduced by the use of reusable forms.

Some form-release agents contain substantial amounts of volatile organic compounds (VOCs), which are harmful to the atmosphere. The obvious solution is the use of low-VOC form-release agents.

The use of insulated forms, although they are not reusable, saves time and produces walls with higher insulating values, which saves energy and reduces the size, initial cost, and operating cost of a heating and air conditioning system.

### During Useful Life

Properly designed and constructed concrete structures can have long useful lives, thus resulting in lower life cycle costs. This natural long life can be further enhanced by increasing the quality of the concrete used.

The design of a concrete building can be used to generate energy savings in it-self. For example, mass concrete can be used as a heat sink in passive solar heating design (see Section 17.10).

### Disposal

Concrete binds its raw materials together, which renders most of them impossible to recycle or reuse. For example, concrete provides a much more economical use of steel than does structural steel framing, but many of the elements in a steel-framed building can be taken apart and either reused or melted down and re-formed, while steel reinforcement in concrete is at best difficult and costly to extract. Nevertheless, reinforcing steel can and should be salvaged when possible.

Most concrete is disposed of by simply burying it at the construction site or by hauling it to a landfill. A better solution would be to salvage as much of the concrete waste as possible and have it crushed for use as aggregate or roadway sub-grades, or broken into larger fragments for use as riprap for the lining of waterways and to protect roadway embankments.

## Applicable *MasterFormat* Sections

The following *MasterFormat* 2004 Level 2 sections are applicable to this chapter.

03 11 00  Concrete Forming
03 15 00  Concrete Accessories
03 21 00  Reinforcing Steel
03 22 00  Welded Wire Fabric
            Reinforcing
03 23 00  Stressing Tendons
03 24 00  Fibrous Reinforcing
03 31 00  Structural Concrete
03 33 00  Architectural Concrete

03 34 00  Low Density Concrete
03 35 00  Concrete Finishing
03 37 00  Specially Placed Concrete
03 38 00  Post-Tensioned Concrete
03 39 00  Concrete Curing
03 41 00  Precast Structural Concrete
03 45 00  Precast Architectural Concrete
03 47 00  Site-Cast Concrete
03 48 00  Precast Concrete Specialties
03 49 00  Glass-Fiber-Reinforced
            Concrete

03 51 00  Cast Roof Decks
03 52 00  Lightweight Concrete Rood
            Insulation
03 53 00  Concrete Topping
03 54 00  Cast Underlayment
03 61 00  Cementitious Grouting
03 62 00  Non-Shrink Grouting
03 63 00  Epoxy Grouting
03 64 00  Injection Grouting
03 71 00  Mass Concrete for Raft
            Foundations

## 3.1 History

The Romans were using a form of concrete made with lime, broken stones, and sand to construct temples and other buildings as early as the third century B.C. At first, they left the surface of concrete rough and finished it with a form of stucco. Soon, they began to embed small stones in the surface to produce a decorative finish. Eventually, they incorpo-rated broken terra cotta roof tiles, embedding these at the surface with their smooth sides outward. This innovation eventually led to the manufacture of clay bricks.

Roman architecture was altered forever by the accidental discovery near Mt. Vesuvius of a native natural cement called *puzzolana*. The Romans combined this silica- and alumina-bearing material with crushed limestone to produce a slow-drying cement that was much stronger, harder, and more adhesive than the cements they had been making using lime. In addition, it was *hydraulic*, meaning that it would set both in air and under water. The name, *puzzolana*, now spelled *pozzolana*, is used today to describe a variety of siliceous or siliceous and aluminous materials used as a substitute for part of the portland cement in some modern portland cement concrete.

Better concrete soon led the Romans to use more of this material and to build larger structures, especially arches, vaults, and domes. Concrete then, as now, had no permanent form of its own before it set and needed some sort of form or mold to hold it in the desired shape. The Romans used wooden forms often supported by temporary support (*centering*). To conserve materials and labor, they invented reusable wood forms and centering that could be used repeatedly. They also used brick or stone to form the outer layers of a structure and filled the voids between the outer layers with concrete. They built impressive domed and vaulted structures using unreinforced concrete to house temples, baths, and other large spaces, some with spans as large as 150 ft (46 m).

With the collapse of the Roman Empire, concrete technology largely fell into disuse, and remained virtually unknown until the rebirth of learning during the Renaissance. In the fifteenth century, the 10-book series *De architectura*, written by the Roman architect and engineer Marcus Vitruvius Pollio in the first century B.C., became a popular object of study.

It was not until near the end of the eighteenth century, however, that research related to concrete technology resumed, and not until 1824 was the essential ingredient in modern concrete discovered. In that year, a brick mason named Joseph Aspdin, living in Leeds, England, patented a new type of cement. He made it by treating, heating (*calcining*), and cooling raw materials into small rock-like shapes called *clinker*, which consisted essentially of calcium silicates, and crushing them into a fine powder. Aspdin called his discovery *portland cement* because of its resemblance to a gray rock found on the Isle of Portland in the English Channel. Aspdin's cement was an overnight success. However, even though ample raw materials were present in the United States, the manufacture of portland cement did not begin in this country until 1872.

Concrete has many uses on its own, but modern concrete buildings are not possible using concrete alone. Concrete has high strength in compression but little tensile strength. Even though other forces are also involved, the primary forces acting on a beam or unsupported slab are those that tend to push it together (compression) at the top of the beam or slab and those that pull it apart

**a**

**b**

**FIGURE 3.1-1** Dashed lines show the beam shape (exaggerated) after a load is applied. (a) The unreinforced beam bottom will separate, leading to collapse. (b) The reinforced beam also deflects, but the magnitude is less. Concrete cracks at the bottom but does not collapse. (Drawing by HLS.)

(tension) at the bottom. Therefore, a concrete slab or beam of any appreciable length will pull apart at the bottom and collapse, especially if a weight (load) is applied to it (Fig. 3.1-1a). This problem can be solved by placing in the concrete unit a material, such as iron or steel, that is strong in tension. Such a *reinforced concrete* unit is capable of safely spanning large distances (Fig. 3.1-1b).

In the 1850s, the race was on to develop a practical method of embedding iron and steel in concrete shapes. Early experimentation involved not only beams, but also boats and other devices. In 1857, a French gardener named F. Joseph Monier patented a system for producing reinforced concrete. He used it to produce flower pots, water tanks, and garden bridges. His techniques were quickly adapted to building construction.

**a** The **concrete arch** is curved in one direction only and is simple and pleasing in appearance. The arch thickness is variable, being thinner at the center and gradually increasing in thickness toward the supports.

**b** The **barrel-shell** roof is also curved in one direction but is more complicated in construction and spans between rigid frame supports.

**c** The **dome roof** is a slab of double curvature.

**d** **Groined vaults** are curved cross vaults often used in large structures such as convention halls or airport buildings.

**FIGURE 3.1-2** (a) Concrete arch; (b) barrel shell; (c) dome; (d) groined vaults. (Courtesy Concrete Reinforcing Steel Institute.)

In 1857, the first application of reinforced concrete in the United States occurred in Port Chester, New York. By 1900, the use of reinforced concrete in buildings was widespread.

A reinforced concrete unit's strength can be increased by reducing tensile stresses in the concrete. This is done by stretching the reinforcing steel, which in turn places the concrete in compression before it is loaded. The scientific basis for modern *prestressing* of con-

FIGURE 3.1-6 Art and Architecture Building, Yale University, New Haven, Connecticut. (Architect: Paul Rudolph; drawing by HLS.)

crete was developed in the late 1920s by a French engineer named Eugene Freyssinet for use in long-span bridge construction. Prestressing is used extensively in precast concrete units. Prestressed, precast concrete requires the use of a high-strength concrete and special curing methods developed in the twentieth century.

In addition to more conventional structures built with columns, beams, and slabs, concrete is effective in erecting thin shell structures, some with very large unsupported spans. Because concrete is placed as a fluid, many configurations of such structures are possible. Figure 3.1-2 depicts some of the more common shapes.

Concrete has been used by such renowned architects as Frank Lloyd Wright, LeCorbusier, Pier Luigi Nervi, Felix Candela, Ero Saarinen, Paul Rudolph, and I.M. Pei. The sketches in Figures 3.1-3 through 3.1-6 depict some of their concrete masterpieces. Contemporary architects continue to use concrete today in both high-rise and smaller buildings (Figs. 3.1-7 to 3.1-9).

Advances in concrete technology continue at a rapid pace, with the relatively recent development of:

- Plastic and metal reusable forms, some that impart textured finishes to concrete
- *Flying forms* that permit rapid form erection and removal
- Advanced chemical formulas for portland cement that produce different cement formulations (high-early strength,

FIGURE 3.1-3 The Guggenheim Museum, New York. (Architect: Frank Lloyd Wright; drawing by HLS.)

FIGURE 3.1-4 The chapel of Notre Dame du Haut at Ronchamp, France. (Architect: LeCorbusier; drawing by HLS.)

FIGURE 3.1-5 TWA Terminal, John F. Kennedy Airport, New York. (Architect: Eero Saarinen; drawing by HLS.)

FIGURE 3.1-7 One Magnificent Mile, Chicago, Illinois. (Architect: Skidmore Owings and Merrill; photo courtesy Portland Cement Association.)

**FIGURE 3.1-8** Commissioners of Public Works Administrative Offices, Charleston, South Carolina. (Architect: Lucus Stubbs Pascullis Powell and Penny, Ltd.; photo courtesy Portland Cement Association.)

**FIGURE 3.1-9** Associates Center, Chicago, Illinois. (Architects: A. Epstein and Sons; photo courtesy Portland Cement Association.)

air-entrained, etc.) for different uses

- Admixtures to control the setting time, ease of delivery, and other characteristics of concrete mixes
- High-strength steel strands that make prestressing possible
- Steel and plastic fiber reinforcement that make slab surfaces stronger and reduce the need for joints
- Strong and light glass-fiber-reinforced concrete building cladding
- Machine delivery devices such as transit mixing trucks, batch mixing plants, and concrete pumps
- Machine placing and finishing devices, such as internal and form vibrators, motor-driven screeds, and mechanical riding trowels

Currently developing technology will in the future produce concrete that will:

- Be corrosion-resistant for use in bridges, parking garages, roads, and radiant floor heating systems.
- Conduct electricity. Currently, concrete acts as an insulator. Future concrete that conducts electricity may have far-reaching uses, including electrical transmission tower support structures and bridge decks that will dissipate lightning strikes.

## 3.2  Concrete Materials

Concrete is made by combining four materials into a mixture of basically two parts: aggregate and portland cement paste. The paste, composed of portland cement and water, binds the fine aggregate (sand) and coarse aggregate (gravel or crushed stone) into a rock-like mass when the paste hardens. Concrete hardening (*setting*) is not simply caused by evaporation or drying but, rather, results from a chemical reaction (*hydration*) between portland cement and water. The portland cement and fine and coarse aggregates used should conform to the requirements of the standards of ASTM International (ASTM) mentioned in this section and summarized in Figure 3.2-1.

Concrete is often reinforced with steel mesh or rods to increase its tensile strength (see Section 3.4)

### 3.2.1  PORTLAND CEMENT

Portland cement is a finely pulverized material consisting principally of compounds of lime, silica, alumina, and iron. It is manufactured from selected materials in closely controlled processes (Fig. 3.2-2). The process starts by mixing limestone or marl with other ingredients such

as clay, shale, or blast furnace slag in the proper proportions and then burning this mixture in a rotary kiln at a temperature of approximately 2700°F (1482°C) to fuse it into *clinker* (rock-like shapes of varying size averaging $^3/_4$ in. [19 mm] in diameter). The clinker is cooled and then pulverized together with a small amount of gypsum, which slows the setting time. The finished product, portland cement, is so fine that nearly all of it will pass through a sieve with 40,000 openings

**FIGURE 3.2-1** Concrete Materials

| Material | Standard |
|---|---|
| **Portland cement** | |
| Types I, II, III, IV, V | ASTM C150 |
| Types IA, IIA, IIIA | |
| (air entraining) | ASTM C150 |
| Pozzolan and blast | |
| furnace | ASTM C595 |
| **Ready-mixed concrete** | ASTM C94 |
| **Aggregates** | |
| Normal weight | ASTM C33 |
| Lightweight | ASTM C330 |
| **Admixtures** | |
| Air-entraining agents | ASTM C260 |
| Accelerators: | |
| calcium chloride | ASTM D98 |

(Courtesy Portland Cement Association.)

**FIGURE 3.2-2** Steps in the manufacture of portland cement showing both dry and wet processes. (Courtesy Portland Cement Association.)

per square inch (62 openings per square millimeter).

### 3.2.1.1 Types of Portland Cement

Portland cements are made to meet different physical and chemical requirements for specific purposes. ASTM C150 describes eight portland cement types: Type I, Normal; Type IA, Normal, Air-entraining; Type II, Moderate; Type IIA, Moderate, Air-entraining; Type III, High Early Strength; Type IIIA, High Early Strength, Air-entrain-

ing; Type IV, Low Heat of Hydration; Type V, Sulfate Resisting. These types are summarized in Figure 3.2-3.

#### TYPE I, NORMAL

Type I is a general-purpose portland cement suitable for most uses when the special properties of other types are not required. Type I portland cement should not be used where concrete will be in contact with high-sulfate soils or will be subject to high temperatures during hydration.

#### TYPE II, MODERATE

Type II portland cement is used where precaution against moderate sulfate attack is important, as in drainage structures where sulfate concentrations in the groundwater are higher than normal. Type II portland cement generates less heat of hydration and cures at a slower rate than Type I. This moderate heat of hydration reduces the temperature rise which is especially important when concrete is placed in warm weather in structures of considerable mass, such as in large piers or heavy retaining walls.

#### TYPE III, HIGH EARLY STRENGTH

Type III portland cement is used when high strengths are desired at a very early time, usually a week or less. It is used:

- When early form removal is desirable
- When concrete must be put into service quickly
- In cold weather to reduce the period required for protection against low temperatures
- When its use will secure high early strengths more satisfactorily or more

**FIGURE 3.2-3** Portland Cement Types and Uses

| ASTM C150 | Use | ASTM C175 |
|-----------|-----|-----------|
| Type I | In general construction when special properties of other types are not required | Type IA* |
| Type II | In general construction where moderate heat of hydration is required | Type IIA* |
| Type III | When high early strength is required | Type IIIA* |
| Type IV | When low heat of hydration is required | |
| Type V | When high sulfate resistance is required | |

*Air-entraining portland cement.
(Courtesy Portland Cement Association.)

economically than by using richer mixes of Type I portland cement

### TYPE IV, LOW HEAT

Type IV is a special portland cement for use where the amount and rate of heat generated during hydration must be kept to a minimum. The development of strength also occurs at a slower rate. Type IV is intended for use in large masses of concrete, such as in large gravity dams, where the temperature rise resulting from heat generated during hardening is a critical factor.

### TYPE V, SULFATE RESISTING

Type V is a special portland cement intended for use only in construction exposed to severe sulfate action, such as in some western states having soils and waters of high alkali content. It has a slower rate of strength gain than normal portland cement.

### AIR-ENTRAINING PORTLAND CEMENTS

Types IA, IIA, and IIIA correspond in composition to Types I, II and III, respectively. They produce concrete that is resistant to severe frost action and to scaling caused by chemicals used for snow and ice removal. In these portland cements, small quantities of air-entraining additions have been incorporated by grinding them in along with clinker during manufacture. These additions should conform to the requirements of ASTM C226. Concrete made with these portland cements contains tiny, well-distributed, and completely separated air bubbles. Air-entraining portland cements cost no more than the corresponding non-air-entraining types in ASTM C150.

### 3.2.1.2 White Portland Cement

White portland cement is manufactured to conform to ASTM C150, although it is not specifically mentioned there. It is used primarily in precast wall and facing panels, terrazzo, stucco, cement paint, tile grout, and decorative concrete. It is recommended wherever white or colored concrete or mortar is desired. It is made of selected raw materials containing negligible amounts of iron and manganese oxide, the substances that give concrete its gray color.

### 3.2.1.3 Portland Blast-Furnace Slag Cement

In portland blast-furnace slag cement, granulated and selected blast-furnace slag, which is obtained by rapidly chilling or quenching molten slag in water, steam, or air, is added to and ground along with portland cement. This type of portland cement should meet the requirements of ASTM C595, Type IS and Type IS-A, Air-entraining, which requires that the slag constituent be between 25% and 65% of the total weight of the portland cement.

Portland blast-furnace slag cement is usually available only in areas where it can be obtained economically, such as near blast-furnace sites.

### 3.2.1.4 Portland-Pozzolan Cement

In portland-pozzolan cement, pozzolan, which consists of siliceous (or siliceous and aluminous) material, is blended with ground portland cement. Typical pozzolan materials are fly ash, volcanic ash, and calcined shale and clay. This type of portland cement should meet the requirements of ASTM C595, Type IP and Type IP-A, Air-entraining, which requires that the pozzolan constituent be between 15% and 40% of the total weight of the portland cement. Concrete with pozzolan has a higher cured strength than does concrete containing only portland cement, aggregate, and water. Pozzolan also decreases the heat of hydration, which is important in mass concrete structures. Pozzolans should not be used, however, where high early strength is a requirement, because they tend to slow the curing process.

### 3.2.2 WATER

Mixing water should be clean and free from oil, alkali, and acid. In general, water that is fit to drink is suitable for use in concrete. However, water with excessive quantities of sulfates should be avoided even when it is potable.

### 3.2.3 AGGREGATES

Aggregates constitute 60% to 75% of the volume of concrete (Fig. 3.2-4). Therefore, both the cost and quality of a concrete are affected by the kinds of aggregates used in it. Aggregates should be obtained from reliable dealers whose materials are of a quality known to make suitable concrete. They should comply with the requirements of ASTM C33, which places limits on the allowable amounts of damaging substances and specifies requirements for the grading, strength, and soundness of normal-weight aggregate.

The most commonly used aggregates—sand, crushed stone, and air-cooled blast-furnace slag—produce normal-weight concrete weighing between 135 and 160 lb/ft³ (2163 and 2563 kg/m³). Aggregates of expanded shale, clay, slate, and slag are used to produce structural lightweight concrete weighing 85 to 115 lb/ft³ (1362 to 1842 kg/m³). Other lightweight materials, such as pumice, scoria, perlite, vermiculite, and diatomite, are used to produce insulating concretes ranging in weight from 15 to 90 lb/ft³ (240 to 1442 kg/m³) (Fig. 3.2-5). Both fine and coarse aggregates should be uniformly graded from their finest particles up to their largest. They should also be clean and free of loam, clay, and vegetable matter, because these foreign particles prevent portland cement paste from properly binding aggregate particles together. Con-

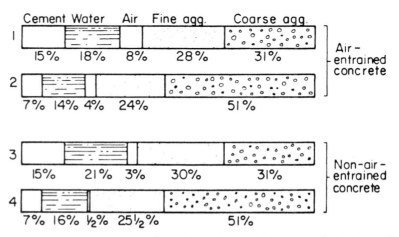

**FIGURE 3.2-4** Range in proportions of materials used in concrete, by absolute volume. Bars 1 and 3 represent rich mixes with small aggregates. Bars 2 and 4 represent lean mixes with large aggregates. (Courtesy Portland Cement Association.)

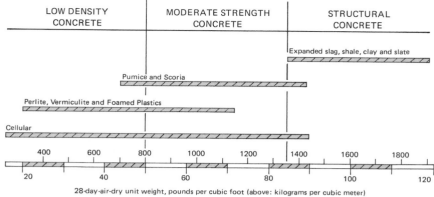

FIGURE 3.2-5 Concrete made with lightweight aggregate. (Courtesy Portland Cement Association.)

FIGURE 3.2-6 Cross section of concrete showing cement paste completely surrounding each aggregate particle and filling the spaces between particles. The quality of its paste largely controls the quality of a concrete. (Courtesy Portland Cement Association.)

crete containing these objectionable materials will be porous, have low resistance to weathering, and have low strength. It can also develop surface defects such as popouts.

In addition, the most economical concrete results when both the fine and coarse aggregates are uniformly graded, because this requires the least amount of portland cement paste to surround the aggregate and fill all the spaces between the particles (Fig. 3.2-6). Less paste means a reduction in shrinkage because most shrinkage occurs in the paste.

### 3.2.3.1 Fine Aggregate

Fine aggregate consists of sand or another suitable fine material. A good concrete sand will contain particles varying uniformly in size from very fine up to $1/4$ in. (6.4 mm) in diameter (Fig. 3.2-7). In well-graded sand, the finer particles help fill the spaces between the larger particles. Sufficient fine particles are necessary for good workability and smooth surfaces, but an excess of fines requires an increase in the amount of portland cement-water paste to achieve a desired strength, thus increasing the cost. Failure to increase the portland cement content when there is an excess of fines in the aggregate will lower the strength of a concrete. A poor concrete sand is shown in Figure 3.2-8.

FIGURE 3.2-7 Well-graded sand before and after separation into its various sizes. The particles vary from dust to about $1/4$ in. (6.4 mm) in diameter. (Courtesy Portland Cement Association.)

FIGURE 3.2-8 A poor concrete sand, before and after being separated into four sizes. This sand lacks particles larger than $1/16$ in. (1.6 mm). More cement is needed when fine sand is used. (Courtesy Portland Cement Association.)

**FIGURE 3.2-9** Gradings of Fine Aggregate for Concrete Slabs

| Sieve Size | Percent of Aggregate Passing Sieve | |
| --- | --- | --- |
| | Normal Weight | Lightweight |
| 3/8 in. (10 mm) | 100 | 100 |
| No. 4 | 95–100 | 85–100 |
| No. 8 | 80–90 | — |
| No. 16 | 50–75 | 40–80 |
| No. 30 | 30–50 | 30–65 |
| No. 50 | 10–20 | 10–35 |
| No. 100 | 2–5 | 0–5 |

(Courtesy Portland Cement Association.)

**FIGURE 3.2-11** The effect of aggregate size on required water content. (Courtesy Portland Cement Association.)

a

b

**FIGURE 3.2-10** (a) A well-graded coarse aggregate for concrete. Notice how the small pieces fit between the larger ones. (b) The same material separated, from left to right, into 1/4- to 3/8 in. (6.4 to 9.5 mm), 3/4 in. (19.1 mm), and 1 1/2 in. (38.1 mm) sizes. (Courtesy Portland Cement Association.)

Figure 3.2-9 shows a typical grading of fine aggregates for light concrete slabs on grade.

### 3.2.3.2 Coarse Aggregate

Coarse aggregate consists of gravel, crushed stone, or another suitable materials larger than 1/4 in. (6.4 mm) in diameter (Fig. 3.2-10). Coarse aggregate that is sound, hard, and durable is best suited for making concrete. Material that is soft or flaky or wears away rapidly is unsatisfactory.

In most concrete, normal-weight coarse aggregate sizes are limited to those between 3/8 in. (9.5 mm) and 1 1/2 in. (38.1 mm). Thin toppings and slabs may be limited to 3/8 in. (9.5 mm) aggregate. Coarse aggregate as large as 6 in. (152.4 mm) in diameter is often found in massive concrete structures, such as dams. Lightweight aggregate is usually limited to a maximum of 1 in. (25.4 mm). In

walls, the largest piece of any type of coarse aggregate should never be more than one-fifth of the overall wall thickness. The maximum size in slabs should be about one-third of the slab thickness. The largest piece of aggregate should never be larger than three-quarters of the width of the narrowest space through which the concrete will be required to pass during placing. This is usually the space between the reinforcing bars or between the bars and the forms.

Coarse aggregate is well graded when particles range uniformly from 3/8 in. (9.5 mm) up to the largest size that may be used on the kind of work to be done. Using the maximum allowable particle size usually results in less drying shrinkage and more economical concrete.

The size of the coarse aggregate used in concrete has a bearing on economy. More water is usually required for small aggregates than for large ones. The water required for a slump of 3 to 4 in. (76 to

100 mm) is shown in Figure 3.2-11 for a wide range of coarse aggregate sizes. For a given water-cement ratio (see Section 3.7.1), the amount of portland cement required decreases as the maximum size of coarse aggregate increases. However, the increased cost of obtaining and handling aggregates larger than 2 in. (50.8 mm) may offset the savings in portland cement.

### 3.2.4 ADMIXTURES

Admixtures are materials other than portland cement, aggregates, and water that are added to concrete either immediately before or during its mixing to alter the properties of the concrete in a variety of ways. For example, they can be used to:

■ Improve workability
■ Reduce separation of coarse and fine aggregates due to settling out of the heavier coarse aggregate
■ Entrain air

- Accelerate or retard setting and hardening

Concrete should be workable, finishable, strong, durable, watertight, and wear resistant. The following factors should be considered when deciding whether to use admixtures in a concrete mix:

- Changing aggregate gradation, portland cement type or amount, or mix proportions may be a better, surer, and more economical way to achieve the desired objectives than will incorporating an admixture.
- Many admixtures affect more than one property of concrete, sometimes hurting desirable properties.
- The effect of some admixtures is changed by mix wetness and richness, aggregate gradation, and the character and length of mixing.
- Some admixtures will not react the same way with all portland cements, even of the same type.

Accordingly, admixtures should be used with caution since specific effects that will result from their use can seldom be predicted accurately. In addition, two or more admixtures should not be used in the same concrete mix unless their combined effect is well known and will not produce an undesirable condition.

ASTM C494 classifies some chemical admixtures by function as follows:

Type A: Water-Reducing
Type B: Retarding
Type C: Accelerating
Type D: Water-Reducing and Retarding
Type E: Water-Reducing and Accelerating
Type F: High-Range Water-Reducing
Type G: High-Range Water-Reducing and Retarding

Of these, the most frequently used are water-reducing agents, retarders, accelerators, and air-entraining agents.

### 3.2.4.1 Water-Reducing Admixtures

Water-reducing admixtures, such as hydroxylated carboxylic acid, should be used in almost all concrete. They permit a lower water content, improve workability, and increase the efficiency of the portland cement in a mix, which lowers a concrete's cost relative to its performance.

High-range water-reducing admixtures (*super plasticizers*) are mostly used in concrete that is to be pumped. They produce a mix that flows easily, with no increase in its water content.

### 3.2.4.2 Retarders

Admixtures that have a retarding effect on the set of portland cement overcome the accelerating effect that temperature has on setting during hot weather and in large masses of concrete, and delay the early stiffening of concrete placed under difficult conditions. Retarder solutions are also sometimes applied directly to the surface of concrete to retard the set of a surface layer of mortar so that it can be readily removed by brushing, thus exposing the aggregate and producing textured surface effects.

Because most retarders also act as water reducers, they are frequently called *water-reducing retarders*. Retarders may also entrain some air in concrete. Many chemicals have a retarding influence on the normal setting time of portland cement. Some of these are variable in action, retarding the set of some portland cements and accelerating the set of others. Unless experience has been gained with a retarder to determine the extent of its effect on the setting time and other properties of a concrete, its use as an admixture should not be attempted without technical advice or advance experimentation with the portland cement and other concreting materials involved to determine its effect on the properties of the concrete.

### 3.2.4.3 Accelerators

Accelerators increase the rate of early strength development in concrete to:

- Reduce the time before finishing operations can be started
- Permit earlier removal of forms and screeds
- Reduce the period required for curing in certain types of work
- Advance the time when a structure can be placed in service
- Partially compensate for the slow gain in strength of a concrete even with proper protection during cold weather
- Reduce the period of protection required for initial and final set in making emergency repairs and other work

Under most conditions, commonly used accelerators cause an increase in the drying shrinkage of concrete. In many cases, it must be decided whether to use an admixture, increase the portland cement content, use high-early-strength portland cement, provide greater protection or a longer curing period, or use a combination of these.

Calcium chloride is used to accelerate the time of set and to increase the rate of strength gain. It should meet the requirements of ASTM D98. The amount used should not exceed 2% by weight of the portland cement used. Greater amounts may cause rapid stiffening, increase drying shrinkage, and corrode reinforcement. Calcium chloride should always be added in solution as part of the mixing water to ensure uniform distribution. If it is added in dry form, not all of the dry particles may completely dissolve during mixing. The undissolved lumps can cause popouts or dark spots in hardened concrete. Calcium chloride should never be used as an antifreeze. To lower the freezing point of concrete appreciably would require using so much calcium chloride that the concrete would be ruined. Instead, protective measures should be taken to prevent the concrete from freezing (see Section 3.10.3.1).

Calcium chloride or admixtures containing soluble chlorides should not be used in:

- Prestressed concrete because of possible corrosion hazards
- Concrete containing embedded aluminum, such as conduit, because serious corrosion can result, especially if the aluminum is in contact with embedded steel and the concrete is in a humid environment
- Concrete exposed to soils or water containing sulfates
- Floor slabs intended to receive dry-shake metallic finishes
- Hot weather generally

### 3.2.4.4 Air-Entraining Agents

Air-entrained concrete can be made either with admixtures added to the mix or with air-entraining portland cement, as listed in ASTM C150, which already contains the admixtures. Air-entraining admixtures produce many microscopic stable air bubbles in concrete. These materials should conform with the requirements of ASTM C260, which includes specifications and methods of testing. Such admixtures will, at little or no additional cost, improve a concrete's workability and durability, and produce a hardened concrete that is resistant to severe frost action and the effects of salt applied for snow and ice removal. Common air-entrainment admixtures include wood resin, sulfonated hydrocarbons, and fatty resinous acids. Because careful control is essential in producing air-entrained concrete, its use should be restricted to plant or transit-mixed batches.

Properly proportioned air-entrained concrete has better workability and contains less water per cubic yard (cubic meter) than non-air-entrained concrete of the same slump (see Section 3.7.3.1 for a discussion of slump tests). This results in a more solid, weather resistant, blemish-free surface. Air-entrained concrete can be handled and placed with less segregation of materials and less tendency to bleed (see Section 3.9.1.1). These properties indirectly aid in promoting durability by increasing the uniformity of the concrete.

In concrete made with air-entraining portland cement or admixtures containing air-entraining agents, the quantity of water in the mix must be adjusted to maintain the desired slump. Such mix designs and adjustments should be made only by a qualified technician.

Because of its proven record, both in the laboratory and in the field, and because it increases workability and durability, all concrete would benefit from entrained air regardless of exposure conditions. Generally, however, air entrainment is specified only for concrete that is exposed to cycles of freezing and thawing because, except in very lean and harsh mixtures, air entrainment reduces both compressive and flexural strengths by 2% to 6% for each 1% of air added. However, when freezing and thawing will be encountered, and when water tightness is necessary, air entrainment is essential. Whether by the use of air-entraining portland cement or an admixture, a proper percentage of entrained air in a concrete mix should be obtained, depending on the maximum size of aggregate, as shown in Figure 3.2-12.

### 3.2.4.5 Synthetic Fiber Reinforcement

Synthetic fiber reinforcement consists of glass, steel, or polypropylene fibers with a nominal fiber length of 2 in. (50 mm). It is sometimes added to concrete slabs to help prevent plastic and shrinkage cracking, and can be quite effective when properly used. Fiber reinforcement is not a substitute for wire mesh, how-

**FIGURE 3.2-12** Recommended Air-Entrainment Amounts for Various Maximum Coarse Aggregate Sizes

| Nominal Maximum Size Coarse Aggregate | Total Air Content Percentage by Volume of Concrete |
|---|---|
| ³⁄₈ in. | 6 to 10 |
| ¹⁄₂ in. | 5 to 9 |
| ³⁄₄ in. | 4 to 8 |
| 1 in. | 3½ to 6½ |
| 1½ in. | 3 to 6 |

(Courtesy Portland Cement Association.)

ever, and should not be used at all where service temperatures will exceed 300°F (150°C). Concentrations above 0.1% by volume are not cost-effective. The effective standard for synthetic fiber reinforcement is ASTM C1116.

## 3.3 Formwork

Because concrete is unable to maintain a particular shape before it sets, it must be placed in a form or mold. In building construction, this form or mold is called *formwork*. Architects usually select the form liners and inserts often used to impart special finishes or patterns to concrete. Architects and structural engineers, working together, also select special forms for ribbed slabs. But the design of the structural portions of formwork and the necessary external supports should be, and usually is, the responsibility of the building contractor. Codes require certain minimum provisions for formwork, and a particular circumstance may cause a structural engineer to place minimum limits on form design. The minimum location and number of slab form supports and shoring may be dictated, for example. Unfortunately, in some jurisdictions the architect is legally prevented from disclaiming responsibility for form design.

The discussion here is general in nature. Additional requirements for formwork are mentioned in the sections of this chapter where a particular type of concrete unit is discussed. For example, there are discussions about forming slabs, including lift-slab forming, in Section 3.13.4 and some additional requirements for precast concrete formwork in Section 3.14.

### 3.3.1 MATERIALS

Forms are made from wood, metal, and plastic. They can be divided into two classes: temporary forms and permanent forms.

#### 3.3.1.1 Temporary Forms

Most formwork is temporary for two reasons: (1) permanent forms would detract from the desired finish, and (2) it is more economical to reuse formwork.

Most temporary concrete forms are made from wood. The parts of the form in contact with the concrete are often high-grade lumber or plastic-overlaid plywood. Wood structural forms are also used to support metal and plastic pans and domes and metal- and plastic-form textured liners. Temporary forms for pan slabs and ribbed slabs (Section 3.13.4.1) are either steel or plastic.

Round column forms made of metal, molded glass-fiber-reinforced plastic, and multilayered paper or fiber are available. These are often plastic coated on the interior to improve the smoothness of the cast column and to eliminate the lines imparted by the spiral winding of the form.

An almost unlimited number of textures and patterns can be imparted to concrete using metal or plastic *form liners*. Most of these are proprietary items, but custom designs are also possible.

#### 3.3.1.2 Permanent Forms

Permanent corrugated steel forms are used to make supported concrete slabs. A supported slab rests on beams or columns rather than on earth (slab on grade). Permanent steel forms are made from corrugated sheet steel (see Section 3.13.4.5).

### 3.3.2 DESIGN AND CONSTRUCTION

The potential failure of formwork represents a great hazard to people working around and on it and to the property adjacent to or supported by it. Therefore, formwork must be designed and built using the same skills and practices that would be applied to a permanent structure. It must be strong enough to withstand the great pressures imposed by heavy, wet concrete and the additional stresses that result from placing and consolidating the concrete. In addition, formwork must be rigid enough to maintain the desired shape under these stresses. It must also remain tight so that cement paste and water are not lost from the con-

**FIGURE 3.3-1** Prefabricated formwork. (Courtesy the Burke Company.)

**FIGURE 3.3-2** Prefabricated formwork, some with bracing, some without. (Courtesy the Burke Company.)

crete during hydration. Such loss will affect the quality of the cured concrete.

Forms must be accessible so that concrete can be easily placed, moved in the form, and consolidated. They must also be designed so that they can be easily removed without damaging the concrete or the form. For example, there must be no undercuts, intrusions, or sharp edges left in the concrete, because these will either be broken off or will tear the forms apart as the forms are removed. Corners should be rounded off or beveled by inserting wood, rubber, or plastic chamfering strips in the forms.

Form surfaces must impart the desired degree of smoothness and texture to the concrete. Form surfaces that will be in contact with concrete should be coated with a wax, oil, or plastic compound form release (parting) agent to prevent the concrete from sticking to them. Where the concrete will be exposed to view, the parting agent should be nonstaining.

The following discussions assume that forms will be custom-built on the site. However, today prefabricated proprietary wood, steel, and aluminum systems are available for forming footings, columns, and walls (Fig. 3.3-1). These systems are similar in concept to the custom-built systems described here, although they may use different materi-

als for the various components. For example, a column form may use angle-shaped steel clamps for yokes instead of wood members or steel bands to hold the form together; footing forms may use corrugated steel sides rather than boards or plywood; wall forms may be framed with aluminum or steel studs instead of wood studs; and whales and braces may be metal as well, or may not be necessary at all (Fig. 3.3-2). When these types of systems are used, it is necessary to

follow the manufacturer's instructions with regard to erection, safety, and removal.

### 3.3.2.1 Forming Footings

Footings should be formed, unless they are in very stable soil and have been cut cleanly and accurately to the shape of the footing, which is a rare condition. Continuous footing may be formed using plywood, as shown in Figure 3.3-3, or using

**FIGURE 3.3-3** A typical footing form. (Drawing by HLS.)

wood boards. Small footing forms can be braced using stakes and spreaders as shown. On small footings, steel rods are sometimes used instead of stakes.

Large footings are framed in a similar manner, except that plywood or multiple boards are used for the side forms and these are braced by vertical boards (*cleats*) and continuous horizontal framing (*wales*) (Fig. 3.3-4). Tie wires are passed through the form from side to side and fastened to the cleats. Stakes may be used to prevent the formwork from spreading. Steel straps or lumber spreaders are sometimes necessary across the top of larger forms to prevent the side walls from spreading apart when the concrete is placed.

### 3.3.2.2 Forming Walls

Custom forms for walls are built at the site using plywood supported by studs, tied together by continuous wales, and braced diagonally to the ground (Fig. 3.3-5).

To erect a wall form, the inner liner and its framing are installed first and braced back to the ground. The reinforcing steel is placed next. Then the other liner and its framing are placed and braced (Fig. 3.3-6). As the formwork is erected, the wales are tied together using form ties to keep the form from spreading. There are two types of ties: snap ties and screw ties (Fig. 3.3-7). *Snap ties* are inserted through holes in the forms as the forms are erected. A washer on the tie rod at the interior side of each form surface maintains the width of the wall and prevents concrete from leaking out through the hole. The ends are then clamped at the wales with slotted tapered wedges. The rod is deformed just inside each washer so that when the forms are removed, the rod may be twisted until it breaks at the weakened spot. This process leaves a ragged hole in the concrete which can be filled with mortar.

The cones on *screw ties* have interior threads. These ties are inserted into the forms with the cones against the inner lining of the form. Threaded rods are then passed through holes in the form and screwed into the cones at each wall face. The cones prevent concrete from leaking out through the holes. The threaded rods are then locked into place at the wales by slotted tapered wedges or plates. When the forms are removed, the threaded rods are unscrewed from the cones and the cones are unscrewed from the tie, leaving a clean cone-shaped hole in the concrete which can be left open or filled with mortar or a sealant.

### 3.3.2.3 Forming Columns

Columns require a liner, bracing, and yokes, straps, or clamps to hold the form together (Fig. 3.3-8). The straps shown may be replaced by wood yokes held together with threaded rods and bolts or by a proprietary system using two overlapping angle brackets (Fig. 3.3-9). Larger columns may require additional rows of bracing inside the strapping or brackets.

### 3.3.2.4 Forming Beams and Girders

Beams and girders are formed with wood or plywood. One way to form light beams is with braces, kickers, and spreaders, as shown in Figure 3.3-10. The bottoms of these forms are supported on the shoring system. A deep beam or girder form is built with ties and diagonal braces much like a wall form. The bottom is supported on continuous wood cleats turned with the long edges horizontal (as shown) or vertical, depending on the size of the beam or girder. Forms for large beams and girders may also require additional bracing.

Plywood form walls; lumber may also be used

Steel strap or lumber spreaders sometimes necessary to prevent form from spreading when concrete is placed

Cleat

Wale

**FIGURE 3.3-4** One of several methods commonly used to form large column footings. (Drawing by HLS.)

**FIGURE 3.3-5** Typical wall forms. (Courtesy Portland Cement Association.)

**FIGURE 3.3-6** Detail of one type of wall form.
(Drawing by HLS.)

**FIGURE 3.3-7** Wall form ties: (a) a snap tie and (b) a screw tie. (Drawing by HLS.)

**FIGURE 3.3-8** A column form with steel straps. (Drawing by HLS.)

### 3.3.2.5 Forming Floors

Both slabs on grade and supported floors require forming.

#### SLABS ON GRADE

The edges of and openings through slabs on grade require forming. This is usually done using wood or metal forms staked to the ground, similar to the wall footing form shown in Figure 3.3-3.

#### SUPPORTED FLOORS

In general, beams and girders are formed first; then the slab is formed. Flat slabs are formed with plywood supported on joists. Pans and domes are supported on a wood frame of joists and plywood called *centering* (Fig. 3.3-11). Alternatively, the plywood beneath pans or domes is omitted and wide boards are placed across the joists at the location of each rib. The forms for beams and floors are supported by a system of temporary wood or metal beams or joists. These in turn are supported by adjustable-length steel columns called *shores*, or wood posts, or a series of adjustable ladder-like steel assemblages called *staging*. The carrier shown in Figure 3.3-11 is a part of this form support system. The entire support system for forms is called *shoring*.

### 3.3.2.6 Flying Forms

Flying forms are made by building a large section of form, as described earlier for centering, and supporting the entire sec-

**FIGURE 3.3-9** A column form with overlapping angle brackets. (Drawing by HLS.)

**FIGURE 3.3-10** A typical beam form. Heavier beams may need wall ties or diagonal braces back to the header. Additional framing members will also be necessary to support floor slab forms. Floor slab form framing will usually eliminate the need for spreaders. (Drawing by HLS.)

tion on deep steel trusses. This assembly can then be lifted by crane and positioned in the next location without dismantling it, thus reducing the cost and speeding construction.

### 3.3.2.7 Slip Forming

Slip forming is a method of continuously moving a form for vertical structures, such as elevator or stair shafts, upward

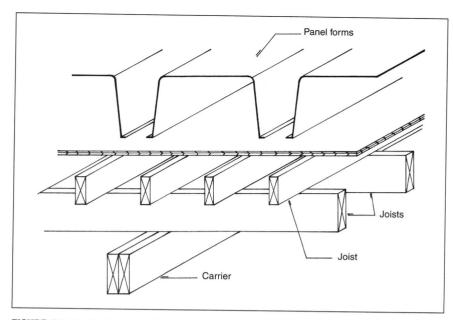

**FIGURE 3.3-11** Typical centering for a supported floor system. (Drawing by HLS.)

on jacks as new concrete is placed on top of the old.

### 3.3.3 FORM REMOVAL

Leaving forms in place as long as possible helps the curing process and protects freshly placed concrete from damage (see Section 3.10.2.1). However, it is sometimes desirable to remove forms from beams and slabs as soon as this can be done safely. For example, patching and repairing of formed surfaces should be done as early as possible, which requires removal of the forms. It is also often nec-

essary to remove forms quickly to permit their reuse.

In any case, forms should not be removed until the concrete has attained sufficient strength to ensure structural stability and to safely carry both the dead load and construction loads that may be imposed on it. Concrete should be hard enough that its surfaces will not be injured in any way when reasonable care is used in removing forms.

When their forms are removed, slabs, beams, and girders, although strong enough to support themselves, will often not be sufficiently strong for several more weeks to support the construction loads that will be imposed on them. This is especially true when additional concrete will be placed on higher floors,

whose forms will be at least partially supported by the concrete from which the forms have just been removed. So that work may continue uninterrupted, wood posts or adjustable metal columns called *reshores* are placed beneath slabs and beams from which the forms have been recently removed (see Fig. 3.13-12). These reshores are then left in place until the beams and slabs are able to support the imposed loads.

## 3.4 Reinforcement

Until *reinforced concrete* was developed, concrete could only be used to resist forces that tended to compress it (compressive loads) (Fig. 3.4-1). Concrete is strong in resisting *compressive* (pushing) loads, but its ability to resist *tensile* (pulling) forces is practically nonexistent. All slabs, beams, and girders, and even some columns and walls, are subject to both compressive loads (*compression*) and tensile loads (*tension*) caused by bending (see Fig. 3.1-1), drying shrinkage, temperature changes, and vertical and horizontal shear. The net result of vertical and horizontal shear forces is called *diagonal tension*. Tensile forces that result from bending will cause an unreinforced concrete beam or slab to crack along the bottom and collapse (see Fig. 3.1-1a). Diagonal tension will cause a crack at the support, which will act at an approximate 45 degree angle (Fig. 3.4-2). Diagonal tension at a column or other support is called *punching shear*.

The requirements in this section apply in principle to all reinforced concrete.

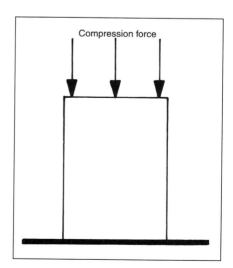

**FIGURE 3.4-1** Concrete is best at supporting compression loads. (Drawing by HLS.)

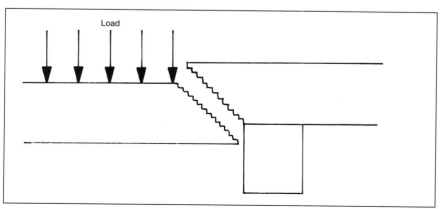

**FIGURE 3.4-2** Diagonal tension (shear) failure. (Drawing by HLS.)

Specific requirements for reinforcing steel in different types of slabs, beams, girders, walls, and columns are addressed in other sections of this chapter.

### 3.4.1 PURPOSE OF CONCRETE REINFORCEMENT

The concept of reinforced concrete is simple. Concrete resists compression forces; steel bars resist tensile forces and shear. When both are placed in the same structural member (slab, wall, beam, girder, or column), the steel will resist the tensile, bending, and shear forces that would otherwise tear the concrete apart, and the concrete will resist applied compression forces that would bend the steel.

The theory is simple; the practice is not. The location and amount of steel in a concrete structural component are critical and vary with the type of component, the loads that will be applied, whether the unit is prestressed or not, and whether the unit is cast in place or precast. For example, the reinforcement required in a simple (single-span) beam varies considerably from the reinforcement in a beam that is continuous across several spans.

Other complications also occur. For instance, because steel is also strong in compression, it is sometimes used to resist a portion of the applied compression loads. This often occurs in columns, and in beams and girders when limiting their depth is important.

In addition, steel is used in concrete even when it is not required to resist tensile and shear forces. For example, slabs on grade simply pass loads through them to the underlying base. Structural cracks in slabs on grade are due to a failure in the base and not in the concrete or its lack of reinforcement. However, slabs on grade do tend to crack as a result of drying shrinkage and when they are subjected to widely varying temperatures. Therefore, a layer of light welded wire fabric is often added to them. This is not structural reinforcement and should not be counted on to help a slab span depressions in the slab base. While it will not prevent cracking, it will hold cracks together so that they are less unsightly. Some practitioners believe that welded wire fabric is not necessary in slabs on grade and routinely omit it. The Portland Cement Association's *Concrete Floors on Ground* suggests that welded wire fabric may not be necessary when the slab is uniformly supported and joints are located close together. The spacing of joints that would permit deletion of fabric is not specified.

Reinforcing bars are also needed in other types of slabs to resist cracks caused by drying shrinkage and temperature changes. This steel is called *temperature steel*. When structural reinforcement occurs, it acts to resist these stresses, and temperature steel is not required. When structural reinforcement is placed in one direction only, temperature steel is required in the perpendicular direction (see Section 3.13.4).

The use of concrete and steel together is made possible by the chemical compatibility of the two materials. Concrete bonds very strongly to steel, and this bond is not broken by thermal movement, because the coefficients of expansion of steel and concrete are very similar. The alkalis in concrete protect steel from corroding.

### 3.4.2 MATERIALS

Steel reinforcement is either *structural reinforcement* or *temperature reinforcement*. Structural reinforcement resists primary tensile, shear, and compression forces. It consists mainly of steel bars, but welded wire fabric is also used in structural slabs and walls and in some precast concrete elements.

Temperature reinforcement holds together cracks that occur from drying shrinkage and from expansion and contraction due to temperature variations. It consists of light welded wire fabric in slabs on grade and of small bars in supported slabs. Synthetic fiber reinforcement is also sometimes added to concrete slabs to help prevent plastic and shrinkage cracking (see Section 3.2.4.5).

#### 3.4.2.1 Reinforcing Bars

Reinforcing bars are either plain or *deformed* (Fig. 3.4-3). Deformations improve the bond between steel and concrete. They are made in 11 standard sizes

(Fig. 3.4-4) and in several grades (Fig. 3.4-5). They should comply with ASTM A615 or A706. Most concrete reinforcement for building construction is made from Grade 40 or Grade 60, deformed new billet or ingot steel bars, conforming with ASTM A615.

Reinforcing bars typically have numbers and letters stamped on them to identify the producing mill, bar size, types of steel, and grade (Fig. 3.4-6).

#### 3.4.2.2 Welded Wire Fabric

Welded wire fabric is made by welding either smooth or deformed reinforcing bars into a square- or rectangular-shaped mesh. Welded wire fabric is designated by the size and spacing of the reinforcing bars (*wires*) used to make it. An example of the designation system is WWF 6 × 12-W16 × W8, which denotes welded wire fabric (WWF) with a 6 in. (152 mm) spacing between longitudinal wires and a 12 in. (305 mm) spacing between transverse wires, with smooth W16 longitudinal wires and smooth W8 transverse wires. The W represents plain (smooth) wire. The letter D substituted for the W would represent deformed wires. The number following the W (or D) is the cross-sectional area of the wire in hundredths of an inch. Figure 3.4-7 shows WWF styles designated by the Wire Reinforcement Institute as common stock styles. Figure 3.4-8 shows the nominal wire diameters and weights of steel associated with size numbers. The normal welded wire fabric used in light slabs on grade is WWF 6 × 6-W1.4 × W1.4.

Lighter gauges of welded wire fabric are available in rolls; heavier gauges are available in sheets (see Fig. 3.4-7).

#### 3.4.2.3 Accessories

Accessories are used to support structural steel and to make splices in it. The devices shown in Figure 3.4-9 are used to support reinforcing steel in forms.

**FIGURE 3.4-3** Typical deformed steel bars. (Drawing by HLS.)

**FIGURE 3.4-4** Deformed Bar Designation Numbers, Nominal Weights, Nominal Dimensions, and Deformation Requirements

| Bar Designation No.[b] | Nominal Weight, lb/ft | Nominal Dimensions[a] | | | Deformation Requirements, in. | | |
|---|---|---|---|---|---|---|---|
| | | Diameter, in. | Cross-Sectional Area, sq. in. | Perimeter, in. | Maximum Average Spacing | Minimum Average Height | Maximum Gap (Chord of 12½% of Nominal Perimeter) |
| 3 | 0.376 | 0.375 | 0.11 | 1.178 | 0.262 | 0.015 | 0.143 |
| 4 | 0.668 | 0.500 | 0.20 | 1.571 | 0.350 | 0.020 | 0.191 |
| 5 | 1.043 | 0.625 | 0.31 | 1.963 | 0.437 | 0.028 | 0.239 |
| 6 | 1.502 | 0.750 | 0.44 | 2.356 | 0.525 | 0.038 | 0.286 |
| 7 | 2.044 | 0.875 | 0.60 | 2.749 | 0.612 | 0.044 | 0.334 |
| 8 | 2.670 | 1.000 | 0.79 | 3.142 | 0.700 | 0.050 | 0.383 |
| 9 | 3.400 | 1.128 | 1.00 | 3.544 | 0.790 | 0.056 | 0.431 |
| 10 | 4.303 | 1.270 | 1.27 | 3.990 | 0.889 | 0.064 | 0.487 |
| 11 | 5.313 | 1.410 | 1.56 | 4.430 | 0.987 | 0.071 | 0.540 |
| 14 | 7.65 | 1.693 | 2.25 | 5.32 | 1.185 | 0.085 | 0.648 |
| 18 | 13.60 | 2.257 | 4.00 | 7.09 | 1.58 | 0.102 | 0.864 |

[a]The nominal dimensions of a deformed bar are equivalent to those of a plain round bar having the same weight per foot as the deformed bar.
[b]Bar numbers are based on the number of eighths of an inch included in the nominal diameter of the bars.
(Copyright ASTM; reprinted with permission.)

**FIGURE 3.4-5** ASTM Standards Covering Various Bars by Grade

| Material | ASTM Standard | Grade | Yield Strength, psi (MPa) | Bar Sizes Available |
|---|---|---|---|---|
| Billet or ingot steel | A615 | 40 | 40,000 (275.6) | 3–6 |
| | A615 | 60 | 60,000 (413.4) | 3–11, 14, 18 |
| | A615 | 75 | 75,000 (516.75) | 11, 14, 18 |
| Rail steel | A616 | 50 | 50,000 (344.5) | 3–11 |
| | A616 | 60 | 60,000 (413.4) | 3–11 |
| Axle steel | A617 | 40 | 40,000 (275.6) | 3–11 |
| | A617 | 60 | 60,000 (413.4) | 3–11 |

**FIGURE 3.4-6** Standard markings on steel bars. (Courtesy Concrete Reinforcing Steel Institute.)

Welded wire fabric in slabs on grade should be supported on similar devices having flat plates (*sand plates*) at their legs to prevent the legs from sinking into the slab base. The common practice of supporting reinforcement in slabs on grade using pieces of concrete or concrete bricks is discouraged because concrete often does not adhere to such materials. This produces cracks, which permits water to reach and rust the rein-

forcement, destroying its effectiveness. The plastic caps shown on the feet of the support devices in Figure 3.4-9 are used when the underside of the slab will be exposed and in other locations to prevent the supports from rusting and transferring this rust to the reinforcement.

In beams, girders, walls, and slabs, when bar space is not at a premium, splices are usually made by lapping the bars a prescribed number of bar diame-

ters and tying them together with wire to keep them in place until the concrete has set. Sometimes, in columns where space is limited, bars are butted end to end and connected using special devices made for the purpose or are welded together.

### 3.4.3 DESIGN AND PLACEMENT

The primary standards for reinforcing steel design, fabrication, and installation

**FIGURE 3.4-7** Common Stock Styles of Welded Wire Fabric

| Style Designation | | Steel Area, sq in./ft (mm²/m²) | | Approximate Weight, lb/100 sq ft (kg/m²) |
|---|---|---|---|---|
| New Designation (by W-number) | Old Designation (by Steel Wire Gauge) | Longitudinal | Transverse | |
| **Rolls** | | | | |
| 6 × 6–W1.4 × W1.4 | 6 × 6–10 × 10 | .03 (208.33) | .03 (208.33) | 21 (1.03) |
| 6 × 6–W2 × W2 | 6 × 6–8 × 8[a] | .04 (277.78) | .04 (277.78) | 29 (1.42) |
| 6 × 6–W2.9 × W2.9 | 6 × 6–6 × 6 | .06 (416.67) | .06 (416.67) | 42 (2.05) |
| 6 × 6–W4 × W4 | 6 × 6–4 × 4 | .08 (555.56) | .08 (555.56) | 58 (2.83) |
| 4 × 4–W1.4 × W1.4 | 4 × 4–10 × 10 | .04 (277.78) | .04 (277.78) | 31 (1.51) |
| 4 × 4–W2 × W2 | 4 × 4–8 × 8[a] | .06 (416.67) | .06 (416.67) | 43 (2.10) |
| 4 × 4–W2.9 × W2.9 | 4 × 4–6 × 6 | .09 (625) | .09 (625) | 62 (3.03) |
| 4 × 4–W4 × W4 | 4 × 4–4 × 4 | .12 (833.33) | .12 (833.33) | 86 (4.20) |
| **Sheets** | | | | |
| 6 × 6–W2.9 × W2.9 | 6 × 6–6 × 6 | .06 (416.67) | .06 (416.67) | 42 (2.05) |
| 6 × 6–W4 × W4 | 6 × 6–4 × 4 | .08 (555.56) | .08 (555.56) | 58 (2.83) |
| 6 × 6–W5.5 × W5.5 | 6 × 6–2 × 2[b] | .11 (763.89) | .11 (763.89) | 80 (3.11) |
| 4 × 4–W4 × W4 | 4 × 4–4 × 4 | .12 (833.33) | .12 (833.33) | 86 (4.20) |

[a]Exact W-number size for 8-gauge is W2.1.
[b]Exact W-number size for 2-gauge is W5.4.
(Courtesy Concrete Reinforcing Steel Institute.)

are the Concrete Reinforcing Steel Institute publication *Manual of Standard Practice* and the American Welding Society publications D1.4, "Structural Welding Code-Reinforcing Steel."

Reinforcing steel is usually fabricated in a mill or shop following shop drawings prepared by the fabricator from the structural engineer's drawings and approved by the structural engineer and the architect. The fabricated steel is bundled, tagged to indicate its location in the building, and delivered to the construction site. Details of *bend*s and *hooks* are standard (Figs. 3.4-10 and 3.4-11). Bundles are delivered to their place in the building by hand or crane. The reinforcement is then wired in place to maintain its location and alignment until the concrete has set. In some cases, sections of reinforcement, such as those for a column, are preassembled and delivered to the site in one piece. Such assemblies

**FIGURE 3.4-8** Sectional Area and Weight of Welded Wire Fabric

| Wire Size Number[a] | | Nominal Diameter,[b] in. | Nominal Weight, lb/lin ft | Area of Width for Various Spacings, sq in./ft | | | | | | |
|---|---|---|---|---|---|---|---|---|---|---|
| | | | | Center-to-Center Spacing | | | | | | |
| Plain | Deformed | | | 2 in. | 3 in. | 4 in. | 6 in. | 8 in. | 10 in. | 12 in. |
| W20 | D20 | 0.505 | .680 | 1.20 | .80 | .60 | .40 | .30 | .24 | .20 |
| W18 | D18 | 0.479 | .612 | 1.08 | .72 | .54 | .36 | .27 | .216 | .18 |
| W16 | D16 | 0.451 | .544 | .96 | .64 | .48 | .32 | .24 | .192 | .16 |
| W14 | D14 | 0.422 | .476 | .84 | .56 | .42 | .28 | .21 | .168 | .14 |
| W12 | D12 | 0.391 | .408 | .72 | .48 | .36 | .24 | .18 | .144 | .12 |
| W11 | D11 | 0.374 | .374 | .66 | .44 | .33 | .22 | .165 | .132 | .11 |
| W10.5 | | 0.366 | .357 | .63 | .42 | .315 | .21 | .157 | .126 | .105 |
| W10 | D10 | 0.357 | .340 | .60 | .40 | .30 | .20 | .15 | .12 | .10 |
| W9.5 | | 0.348 | .323 | .57 | .38 | .285 | .19 | .142 | .114 | .095 |
| W9 | D9 | 0.338 | .306 | .54 | .36 | .27 | .18 | .135 | .108 | .09 |
| W8.5 | | 0.329 | .289 | .51 | .34 | .255 | .17 | .127 | .102 | .085 |
| W8 | D8 | 0.319 | .272 | .48 | .32 | .24 | .16 | .12 | .096 | .08 |
| W7.5 | | 0.309 | .255 | .45 | .30 | .225 | .15 | .112 | .09 | .075 |
| W7 | D7 | 0.299 | .238 | .42 | .28 | .21 | .14 | .105 | .084 | .07 |
| W6.5 | | 0.288 | .221 | .39 | .26 | .195 | .13 | .097 | .078 | .065 |
| W6 | D6 | 0.276 | .204 | .36 | .24 | .18 | .12 | .09 | .072 | .06 |
| W5.5 | | 0.265 | .187 | .33 | .22 | .165 | .11 | .082 | .066 | .055 |
| W5 | D5 | 0.252 | .170 | .30 | .20 | .15 | .10 | .075 | .06 | .05 |
| W4.5 | | 0.239 | .153 | .27 | .18 | .135 | .09 | .067 | .054 | .045 |
| W4 | D4 | 0.226 | .136 | .24 | .16 | .12 | .08 | .06 | .048 | .04 |
| W3.5 | | 0.211 | .119 | .21 | .14 | .105 | .07 | .052 | .042 | .035 |
| W3 | | 0.195 | .102 | .18 | .12 | .09 | .06 | .045 | .036 | .03 |
| W2.9 | | 0.192 | .099 | .174 | .116 | .087 | .058 | .043 | .035 | .029 |
| W2.5 | | 0.178 | .085 | .15 | .10 | .075 | .05 | .037 | .03 | .025 |
| W2.1 | | 0.162 | .070 | .126 | .084 | .063 | .042 | .031 | .025 | .021 |
| W2 | | 0.160 | .068 | .12 | .08 | .06 | .04 | .03 | .024 | .02 |
| W1.5 | | 0.138 | .051 | .09 | .06 | .045 | .03 | .022 | .018 | .015 |
| W1.4 | | 0.134 | .048 | .084 | .056 | .042 | .028 | .021 | .017 | .014 |

*Note:* The above listing of smooth and deformed wire sizes represents wires normally selected to manufacture welded wire fabric styles to specific areas of reinforcement. Wire sizes other than those listed here, including larger sizes, may be available if the quantity required is sufficient to justify manufacture.
[a]The number following the prefix W or the prefix D identifies the cross-sectional area of the wire in hundredths of a square inch.
[b]The nominal diameter of a deformed wire is equivalent to the diameter of a smooth wire having the same weight per foot as the deformed wire.
(Courtesy Concrete Reinforcing Steel Institute.)

| SYMBOL | BAR SUPPORT ILLUSTRATION | BAR SUPPORT ILLUSTRATION PLASTIC CAPPED OR DIPPED | TYPE OF SUPPORT | TYPICAL SIZES |
|---|---|---|---|---|
| SB | | CAPPED | Slab Bolster | ¾, 1, 1½, and 2 inch heights in 5 ft. and 10 ft. lengths |
| SBU* | | | Slab Bolster Upper | Same as SB |
| BB | | CAPPED | Beam Bolster | 1, 1½, 2, over 2″ to 5″ heights in increments of ¼″ in lengths of 5 ft. |
| BBU* | | | Beam Bolster Upper | Same as BB |
| BC | | DIPPED | Individual Bar Chair | ¾, 1, 1½, and 1¾″ heights |
| JC | | DIPPED  DIPPED | Joist Chair | 4, 5, and 6 inch widths and ¾, 1 and 1½ inch heights |
| HC | | CAPPED | Individual High Chair | 2 to 15 inch heights in increments of ¼ inch |
| HCM* | | | High Chair for Metal Deck | 2 to 15 inch heights in increments of ¼ in. |
| CHC | | CAPPED | Continuous High Chair | Same as HC in 5 foot and 10 foot lengths |
| CHCU* | | | Continuous High Chair Upper | Same as CHC |
| CHCM* | | | Continuous High Chair for Metal Deck | Up to 5 inch heights in increments of ¼ in. |
| JCU** | | DIPPED | Joist Chair Upper | 14″ Span Heights −1″ thru +3½″ vary in ¼″ increments |
| CS | | | Continuous Support | 1½″ to 12″ in increments of ¼″ in lengths of 6′-8″ |

*Usually available in Class 3 only, except on special order.
**Usually available in Class 3 only, with upturned or end bearing legs.

**FIGURE 3.4-9** Typical types and sizes of wire bar supports. (Courtesy Concrete Reinforcing Steel Institute.)

**NOTES:**

1. All dimensions are out-to-out of bar except "A" and "G" on standard 180° and 135° hooks.
2. "J" dimension on 180° hooks to be shown only where necessary to restrict hook size, otherwise ACI standard hooks are to be used.
3. Where "J" is not shown, "J" will be kept equal to or less than "H" on Types 3, 5, and 22. Where "J" can exceed "H", it should be shown.
4. "H" dimension stirrups to be shown where necessary to fit within concrete.
5. Where bars are to be bent more accurately than standard fabricating tolerances, bending dimensions which require closer fabrication should have limits indicated.
6. Figures in circles show types.
7. For recommended diameter "D", of bends, hooks, etc., see tables, page 6-4.
8. Type S1-S6, S11, T1-T3, T5-T9 apply to bar sizes #3 through #8.
9. Unless otherwise noted, diameter D is the same for all bends and hooks on a bar (except for bend types 11 and 13).

Where slope differs from 45° dimensions "H" and "K" must be shown.

**ENLARGED VIEW SHOWING BAR BENDING DETAILS**

**FIGURE 3.4-10 Typical bar bends.** (Courtesy Concrete Reinforcing Steel Institute.)

All specific sizes recommended by CRSI below meet minimum requirements of ACI 318

RECOMMENDED END HOOKS
All Grades

D=Finished bend diameter

| Bar Size | D | 180° HOOKS | | 90° HOOKS |
|----------|-----|------|-------|-----------|
| | | A or G | J | A or G |
| # 3 | 2¼ | 5 | 3 | 6 |
| # 4 | 3 | 6 | 4 | 8 |
| # 5 | 3¾ | 7 | 5 | 10 |
| # 6 | 4½ | 8 | 6 | 1-0 |
| # 7 | 5¼ | 10 | 7 | 1-2 |
| # 8 | 6 | 11 | 8 | 1-4 |
| # 9 | 9½ | 1-3 | 11¾ | 1-7 |
| #10 | 10¾ | 1-5 | 1-1¼ | 1-10 |
| #11 | 12 | 1-7 | 1-2¾ | 2-0 |
| #14 | 18¼ | 2-3 | 1-9¾ | 2-7 |
| #18 | 24 | 3-0 | 2-4½ | 3-5 |

STIRRUP AND TIE HOOKS

135° SEISMIC STIRRUP/TIE HOOKS

STIRRUPS
(TIES SIMILAR)

STIRRUP AND TIE HOOK DIMENSIONS
Grades 40-50-60

| Bar Size | D (in.) | 90° Hook | 135° Hook | |
|----------|---------|----------|-----------|------|
| | | Hook A or G | Hook A or G | H Approx. |
| #3 | 1½ | 4 | 4 | 2½ |
| #4 | 2 | 4½ | 4½ | 3 |
| #5 | 2½ | 6 | 5½ | 3¾ |
| #6 | 4½ | 1-0 | 8 | 4½ |
| #7 | 5¼ | 1-2 | 9 | 5¼ |
| #8 | 6 | 1-4 | 10½ | 6 |

135° SEISMIC STIRRUP/TIE
HOOK DIMENSIONS
Grades 40-50-60

| Bar Size | D (in.) | 135° Hook | |
|----------|---------|-----------|------|
| | | Hook A or G | H Approx. |
| #3 | 1½ | 4¼ | 3 |
| #4 | 2 | 4½ | 3 |
| #5 | 2½ | 5½ | 3¾ |
| #6 | 4½ | 8 | 4½ |
| #7 | 5¼ | 9 | 5¼ |
| #8 | 6 | 10½ | 6 |

NOTES:
1. 180° hook J dimension (sizes #10, #11, #14 and #18), and A or G dimension (#14 and #18) have been revised to reflect test research using ASTM/ACI bend test criteria as a minimum.
2. Tables for Stirrup and Tie Hook dimensions have been expanded to include sizes #6, #7, and #8 to reflect current design practices.

**FIGURE 3.4-11 Standard hooks.** (Courtesy Concrete Reinforcing Steel Institute.)

may also be made up at the site on the ground and lifted into place by crane.

Steel should be accurately positioned, secured to prevent its displacement during the placing of concrete, and cleaned of rust and foreign matter before concrete is placed against it. Steel should be protected from possible harm by fire and corrosion by covering it with at least the thickness of concrete shown in Figure 3.4-12.

**FIGURE 3.4-12 Recommended Minimum Concrete Cover for Reinforcing Steel**

| Location and Use | Minimum Cover |
|------------------|---------------|
| **Against ground without forms** | 3 in. |
| **Exposed to weather or ground but placed in forms** | |
| Greater than ⅝-in.-diameter bars | 2 in. |
| Less than ⅝-in.-diameter bars | 1½ in. |
| **Slabs and walls (no exposure)** | ¾ in. |
| **Beams, girders, columns (no exposure)** | 1½ in. |

(Courtesy Portland Cement Association.)

Welded wire fabric should be installed in the largest practicable length. Adjoining pieces should be lapped at least one full mesh and tied together with wire to prevent movement during concrete placement.

# 3.5 Accessories

Many types of accessories are used in concrete construction. These include forming accessories (see Section 3.3), reinforcement accessories (see Section 3.4), anchors and fasteners, sleeves, joint inserts, reglets, and waterstops.

### 3.5.1 ANCHORS AND FASTENERS

Anchors and fasteners are used to anchor other materials, such as wood, stone, and masonry, to concrete. They range from simple threaded anchor bolts to elaborate adjustable inserts. Figures 3.5-1 through 3.5-6 show a few of the many types of anchors and fasteners available for use in concrete.

Figure 3.5-1 shows a simple anchor bolt. The bent portion turns out into the concrete to hold the bolt in place. Anchor bolts are used for fastening precast concrete, metal, and wood construction to concrete. Examples are precast concrete columns, steel column plates, and wood plates. The diameter and length of such

**a**

**b**

**FIGURE 3.5-2** Concrete inserts are installed in formwork so that they finish flush with the concrete: (a) shelf-angle insert and anchor; (b) two types of light-duty internally threaded inserts. (Drawing by HLS.)

bolts vary, depending on the loads that will be applied and the construction materials they will anchor.

Figure 3.5-2 shows two of many available concrete insert configurations. Figure 3.5-2a is a *shelf-angle insert* designed to receive a special bolt. Figure 3.5-2b shows two types of internally *threaded inserts* which receive a standard bolt. In each case, the insert is fastened to the inside face of the forms before concrete is placed. The shelf-angle insert shown is one of many available configurations, all of which work in essentially the same way. An anchor bolt with a special head designed to fit the insert is placed into the cross slot on the insert and dropped into the vertical slot. The device to be supported is slipped over the bolt and the washer and nut are installed and tightened, pulling the anchor head against the inside of the insert. With a threaded insert, a bolt is passed through the item to be supported into the threaded insert. Then a washer and nut are installed and the nut is tightened.

Figure 3.5-3 shows a dovetail slot and two of the many types of anchors that can be used in such a slot. These are used to anchor masonry and stone to concrete. The slot is formed from galvanized steel, copper, or stainless steel. The anchors may be galvanized steel or stainless steel. Dovetail slots are available with a felt or plastic foam filler to prevent intrusion of concrete and other materials, or with no filler. Dovetail slots are nailed to the inside of the form before the concrete is

**FIGURE 3.5-1** Threaded anchor bolt with nut and washer. (Drawing by HLS.)

Corrugated masonry anchor

Dovetail anchor slot

Adjustable masonry anchor

**FIGURE 3.5-3** Dovetail anchor slots are designed to receive anchors for stone and masonry. (Drawing by HLS.)

placed. The wedge-shaped ends of the anchors are slipped into the slot vertically and twisted to lock them in place.

Figure 3.5-4 shows two types of expansion anchors. Figure 3.5-4a is representative of the many types of heavy-duty expansion anchors available. They all work in essentially the same way. A hole is drilled in the concrete after it has set. The anchor is placed in the hole. The bolt furnished with the anchor is passed through the item to be supported and into the anchor shell. As the bolt is tightened,

FIGURE 3.5-6 Various types of anchor plates and angles. (Drawing by HLS.)

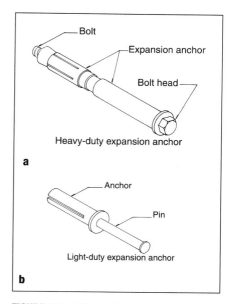

FIGURE 3.5-4 Expansion anchors. (Drawing by HLS.)

the expanding portion of the anchor is forced apart, wedging the anchor securely into the drilled hole.

Figure 3.5-4b shows one of many types of light-duty expansion anchors. The one shown has a *drive pin*; others have bolts or screws. Light-duty expansion anchors are placed in drilled holes in set concrete. A pin, screw, or bolt is passed through the item to be supported into the anchor. A drive pin is driven with a hammer. Screws are tightened with a screwdriver, bolts with a wrench. As the pin, screw, or bolt is forced into the anchor, the anchor expands, locking itself into the drilled hole.

Figure 3.5-5 shows two types of powder-actuated fasteners. These are driven into concrete by an exploding charge of gunpowder in a gun specially designed for the purpose. Figure 3.5-5a is a drive pin that is driven through the item to be supported into the concrete. Figure 3.5-5b is a stud, which has threads so that a supported item can be held in place with washers and nuts.

Figure 3.5-6 shows several configurations of anchor plates and angle clips used in concrete construction. These devices are fastened to the forms before concrete is placed with the anchors shown toward the concrete. Bolts may be welded to plate and angle clips to support or anchor other construction, or other steel shapes may be welded to them.

Other anchor types not shown include hammer- and powder-driven nails and screws.

### 3.5.2 SLEEVES

Metal, cardboard, and plastic sleeves are used to produce holes in concrete for the insertion of pipes, railings, and other items that pass through or are set into concrete. Openings for banks of several pipes, conduits, and the like are often formed with a wood box set into the form. Openings for single pipes are usually formed with galvanized steel pipes or cardboard or plastic tubes. When railing standards or other items that must be wedged in tightly for support are placed in sleeves, paper and concrete materials are removed; metal sleeves are left in place (Fig. 3.5-7).

### 3.5.3 JOINT INSERTS

Joint inserts are used to form control joints in slabs. Figure 3.5-8 shows a typical tongue-and-groove control joint insert for use in slabs on grade. These inserts are made of either plastic or galvanized steel. Similar inserts are used in supported slabs.

Similar inserts are also used to form keyways in concrete to permit bonding

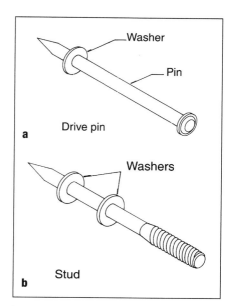

FIGURE 3.5-5 Powder-actuated fasteners. (Drawing by HLS.)

FIGURE 3.5-7 A sleeve for use in concrete. (Drawing by HLS.)

FIGURE 3.5-8 Control joint insert. (Drawing by HLS.)

of adjacent concrete or masonry (Fig. 3.5-9). Keyways are often placed in the tops of footings to key in the foundation walls, whether the wall is concrete or masonry (Fig. 3.5-10).

Plastic strips are used as inserts to create control joints. These may be just flat strips or flanged units that remain in place. Some of these have removable tops to permit installing sealant on top of the insert (Fig. 3.5-11).

Plastic caps are sometimes used over fibrous joint fillers to permit easier application of sealant in the joint (Fig. 3.5-12).

**FIGURE 3.5-9 Keyway insert.** (Drawing by HLS.)

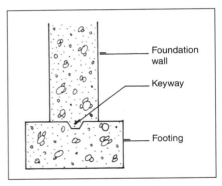

**FIGURE 3.5-10 Keyway used with a concrete wall.** (Drawing by HLS.)

**FIGURE 3.5-11 Control joint insert with removable plastic cap.** (Drawing by HLS.)

**FIGURE 3.5-12 Removable cap for joint filler.** (Drawing by HLS.)

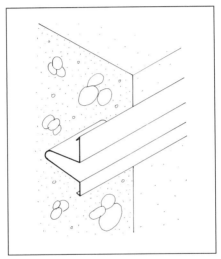

**FIGURE 3.5-13 A reglet.** (Drawing by HLS.)

### 3.5.4 REGLETS

Reglets are devices that form slots in concrete for the insertion of flashings. Figure 3.5-13 shows one reglet configuration. Others are available, but the general characteristics of all reglets are similar. They consist essentially of a slot into which flashing is inserted. They are available in galvanized steel, copper, stainless steel, and plastic.

### 3.5.5 WATERSTOPS

Waterstops are rubber or vinyl inserts designed to be placed in concrete joints to prevent water from penetrating the joint. There are many available configurations. A few of the more commonly used ones are shown in Figure 3.5-14.

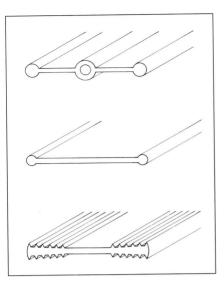

**FIGURE 3.5-14 Waterstops.** (Drawing by HLS.)

## 3.6 Joints

It is not possible to entirely prevent cracks in concrete, but good jointing practice will help to reduce their number and encourage them to occur in more acceptable locations. Joints in concrete work permit movement and volume changes; the handling, placing, and finishing of conveniently sized areas; and the separation of independent elements. The types of joints required in concrete include control joints, construction joints, isolation and separation joints, and building expansion joints.

The amount of thermal movement in concrete can be calculated using its approximate coefficient of expansion (and contraction) of 0.0000055 in./in./°F (0.00014 mm/mm/17.77°C). For example, a 100 ft (30.5 m) length of unrestrained concrete would expand 0.66 in. (16.76 mm) in length for each 100°F (37.8°C) rise in temperature. If the temperature were to drop 100°F (37.8°C), the same amount of contraction would occur. Concrete would be exposed to a 100°F (37.8°C) rise in temperature if it were placed in the spring when the temperature was 50°F (10°C) and then, in the summer, the temperature rose to 150°F (65.6°C).

However, concrete shortens about 0.72 in. (18.29 mm) per 100 ft (30.5 m) while drying from its saturated condition at placing to an average hardened state, with a moisture content equilibrium with air at 50% relative humidity. This shrinkage slightly exceeds the expansion caused by such an extreme increase in temperature as 100°F (37.8°C). Furthermore, in most outdoor applications, concrete reaches its maximum moisture content (and moisture expansion) during the season of low temperature (and thermal contraction). Thus, the volume changes due to moisture and temperature variations frequently offset each other. There is, therefore, no need for expansion joints in most concrete associated with buildings. The common reference to *expansion* is misleading because it implies that an increase in size after placement must be allowed for. There may be rare instances in very large structures and in highways or other large paved surfaces when extensive surfaces of concrete are subject to large temperature variations where sufficient expansion may occur to justify expansion joints. In most instances, however, concrete is at its greatest mass when it is placed.

Nevertheless, some practitioners in warm regions insist that expansion joints are needed and include them as a safety factor, especially in concrete paving. Where they are used, expansion joints are similar to the isolation joints discussed in Section 3.6.3.

### 3.6.1 CONTROL JOINTS

As concrete sets or hardens, its excess mixing water is lost through evaporation and hydration. This initial water loss produces shrinkage that is larger than subsequent increases in the size of the hardened mass resulting from an increase in temperature or moisture content.

Shrinkage as water dries from large areas of freshly placed concrete results in jagged, irregularly spaced cracks. By anticipating shrinkage and installing *control joints* to limit areas and control where cracking occurs, concrete work can be made more attractive, serviceable, and relatively free from unsightly random cracking. These joints provide a break or a reduction in slab thickness and thus create a weakened section that encourages cracking to occur at that location. When concrete shrinks, the cracks in these joints will open slightly, reducing the number of irregular and unsightly random cracks.

The maximum spacing between control joints depends on the concrete's thickness and shrinkage potential, the curing environment, and the absence or presence of reinforcement.

Control joints in sidewalks should be spaced at intervals generally equal to the width of the slab (Fig. 3.6-1), but should not be more than 6 ft (1800 mm) apart. Joints in driveways should be spaced generally equal to the width of the slab, but never more than 20 ft (6000 mm) apart. In large slab areas that do not contain structural reinforcement (Types I and II, as described in Section 3.12.2.1), control joints should divide the slab into approximately square panels, with a maximum spacings of 20 ft (6000 mm). Actual spacings depend on slab thickness (Fig. 3.6-2c). A rule of thumb for plain concrete slabs is that joint spacing in feet should not exceed two slab thicknesses in inches for unreinforced concrete made with coarse $^3/_4$ in. (19 mm) maximum aggregate. Thus, a 4-in. (100-mm)-thick plain slab would require control joints at intervals not to exceed 8 ft (2400 mm). Additional control joints should be provided at slab intersections (see Fig. 3.12-4).

Control joints are unnecessary in a slab that is to receive a floor finish, such as resilient flooring or carpeting, because the flooring will cover unsightly cracks that result from shrinkage. Control joints are usually unnecessary in structurally reinforced slabs (Types III and IV, as described in Section 3.12.2).

Control joints can be made by sawing a groove in hardened but not yet fully cured concrete, using a power saw, or by installing a keyed joint (Figure 3.6-2a,b). Sawed joints are approximately $^1/_8$-in. (3.2-mm)-wide grooves cut in the concrete to a depth equal to one-fourth of the total slab thickness, but not less than $^3/_4$ in. (19 mm), and at least equal to the maximum size of the aggregate. Tooled control joints should be of a similar minimum depth. However, grooves cut only for decorative purposes may be shallower.

It is good practice to use a straight board as a guide when making a groove in a concrete slab (see Fig. 3.6-1), because tooled joints should be straight and perpendicular to the edge of the slab. The same care must be taken in running joints as in edging, for a tooled joint can add to or detract from the appearance of the finished slab.

On large concrete flat surfaces, it may be more convenient to cut joints with an electric or gasoline-driven power saw fitted with an abrasive or diamond blade. Joints can be cut as soon as the concrete surface is firm enough not to be torn or damaged by the blade (within 4 to 12 hours) and before random shrinkage cracks form in the slab.

### 3.6.2 CONSTRUCTION JOINTS

Construction joints are necessary in large construction work because it is not practicable to handle, place, and finish a large area in one operation. When construction joints are installed, concrete can be conveniently and practically placed in several operations with no loss in the appearance or performance of the complete job. Although they may double as control or isolation joints, construction joints must allow no vertical movement in the completed floor and are thus often keyed joints, as shown in Figure 3.6-2b.

Plastic or rubber waterstops are inserted in construction joints below grade and in other locations to prevent water penetration through a concrete structural element (Fig. 3.6-3).

### 3.6.3 ISOLATION AND SEPARATION JOINTS

Isolation and separation joints are often necessary to separate concrete sections and prevent the bonding of one concrete section with another, or to separate a concrete section from another material or structural part so that one can move independently of the other. *Isolation joints* are sometimes called *expansion joints*.

Isolation and separation joints are usually formed by installing $^1/_8$-in. (3.2-mm)-thick, or slightly thicker, asphalt-

**FIGURE 3.6-1** Jointing using a jointing tool. (Courtesy Portland Cement Association.)

**FIGURE 3.6-2** (a) Several types of control joints for slabs on grade. (b) Keyed joint details for slabs on grade. Using too large a key may cause a crack parallel to the joint; if the key is too small, it may shear off, resulting in loss of transfer capacity. A horizontal saw cut permits the wood key strip to swell slightly without damaging the concrete and allows for easy removal of the strip. The joint should be filled with joint filler. (c) Isolation joint for an interior column. The column is boxed out so that the corners meet the control joints. This eliminates cracks radiating from the column corners. Circular fiberboard forms can be used instead of square wood forms. (d) Components of a slab on grade.

(All courtesy Portland Cement Association.)

impregnated fiber sheets in floors at columns, footings, the junctures between floors and walls, and anywhere else that adjacent surfaces are required to move independently. Figure 3.6-2c shows how isolation joints are formed around a column. Similar consideration should be given around the perimeter of a slab, as shown in Figure 3.6-2d. Figure 3.6-4 shows an isolation joint at a column.

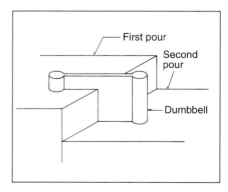

**FIGURE 3.6-3** Waterstop in a control joint.
(Drawing by HLS.)

**FIGURE 3.6-4** A construction joint used as an isolation joint. (Drawing by HLS.)

**FIGURE 3.6-5** Waterstop in a building expansion joint. (Drawing by HLS.)

### 3.6.4 BUILDING EXPANSION JOINTS

While expansion of the concrete itself is seldom a problem, expansion of an entire structure often requires the insertion of *building expansion joints* that pass through both structural elements and finishes. Concrete that crosses a building expansion joint must be interrupted. Various types of joint covers have been developed to cover such joints. In slabs and in walls, where water is likely to pass through a building expansion joint, waterstops are placed in the concrete (Fig. 3.6-5).

## 3.7 Mixtures and Mixing

The mixture of its ingredients will determine whether the consistency, workability, and finishability of plastic concrete will be in a desirable range. It will also determine whether the essential properties of hardened concrete (strength, durability, watertightness, and wear resistance) are obtained and whether the concrete will be economical.

In properly made concrete, every aggregate particle, no matter how large or small, is surrounded completely by portland cement and water paste, and the spaces between aggregate particles are all filled completely with paste. This is illustrated in Figure 3.2-6, which shows a section cut through hardened concrete.

Aggregates are considered inert materials, and the paste is the cementing medium that binds the aggregate particles into a solid mass. Therefore, the quality of a concrete is greatly dependent on the quality of its paste, and it is important that the paste itself have the required strength, durability, and watertightness.

### 3.7.1 WATER-CEMENT RATIO

A concrete mixture's water-cement ratio is the ratio by weight of water to portland cement in the mixture. This means pound (kg) of water per pound (kg) of cement (or ounce [g] per ounce [g]). A water-cement ratio of 0.31 could be thought of as 0.31 lb (kg) of water for each pound (kg) of cement. The units (pounds, ounces, kg) are not used, because they cancel out mathematically (pound/pound = 1). The cementing or binding properties of a cement paste are due to a chemical reaction between the cement and water (hydration) that requires time and favorable temperature and moisture conditions. The reaction takes place very rapidly at first and then more slowly for a long time under favorable conditions. Only a relatively small amount of water, about $3^1/_2$ gal (13.25 L) per 94 lb (42.64 kg) of portland cement, is actually required to complete the chemical reaction. This amount is equivalent to a water-cement ratio of 0.31. But a water-cement ratio ranging from 0.35 to 0.80 is usually maintained for the sake of workability and placeability. Another advantage of using more water is that more aggregate

can be used, with resulting economy. All the water that is not required for hydration eventually evaporates, leaving the small holes that account for the porosity of concrete.

For successful results, a proper proportion of water to portland cement is essential. If there is a high water-cement ratio, meaning that too much water has been added, a portland cement paste becomes thin or diluted and will be weak and porous when it hardens. Using too little water may result in a mix that cannot be properly placed and finished. Portland cement paste made with the correct amount of water has strong binding qualities and is watertight and durable when cured. If its portland cement paste and aggregates are strong and durable, a concrete will be strong and durable. The strength, durability, freeze-thaw resistance, watertightness, and wear resistance of a concrete are largely controlled by using a sufficiently low ratio of water to portland cement. Additional control can be gained by adding an adequate amount of entrained air.

The water-cement ratio selected should be the lowest value required to meet design requirements such as durability, strength, and impermeability. For instance, concrete that will be exposed to a combination of wet-dry and freeze-thaw cycling and de-icing chemicals requires the following for durability: (1) a low water-cement ratio, (2) air-entrainment, (3) suitable materials, (4) adequate curing, and (5) good construction practices.

Figure 3.7-1 is a guide for selecting the water-cement ratios for various exposure conditions.

#### 3.7.1.1 Strength

The compressive strength required of a particular concrete is dependent on several factors. One is the exposure condition (see Fig. 3.7-1); another is the applied load. Strengths as low as 2500 psi (17 MPa) may be sufficient in some small building foundations; columns in very tall buildings may employ concrete with a strength of 19,000 psi (1312 MPa). Even higher-strength concretes may be used in the future when technology permits their development. There is a direct link between the strength of concrete and its water-cement ratio. In fact, for given materials and handling conditions, strength is determined primarily by the water-cement ratio, as long as the mix-

**FIGURE 3.7-1** Recommended Concrete Mixtures for Various Exposure Conditions

| Concrete Construction Element | Exposure Condition[a] | Minimum Compressive Strength at 28 days,[b] Plant or Transit, psi | Practical Water-Cement Ratio by Weight | | Normal Maximum Coarse Aggregate Size,[e] in. | Minimum Cement[f] Content, lb/cu yd | Air-Entrainment by Volume, % | Slump, in. |
|---|---|---|---|---|---|---|---|---|
| | | | Non-Air-Entrained Concrete | Air-Entrained Concrete | | | | |
| Foundations, basement walls, and slabs not exposed to weather | Severe | 3000[c] | See footnote c | 0.55 | 1 | 564 | See footnote f | 5-in. maximum for hand methods of strike-off and consolidation; 3-in. maximum for mechanical strike-off and consolidation |
| | Moderate | | | | | | | |
| | Negligible | | | | | 520 | | |
| Foundations, basement walls, exterior walls, and other concrete work exposed to weather | Severe | 3500[c] | See footnote c | 0.45 | 1 | 564 | 6–8 | |
| | Moderate | 3000 | 0.58 | 0.50 | 1 | 520 | 5–7 | |
| | Negligible | 2500 | 0.67 | 0.55 | | | See footnote f | |
| Concrete exposed to sulfate attack | Severe | 5000[c] | See footnote d | See footnote d | 1 | 564 | 6–8 | |
| | Moderate | | | | | 520 | 5–7 | |
| | Negligible | 5000 | | | | | See footnote f | |
| Driveways, garage floors, walks, porches, patios, and stairs exposed to weather | Severe | 4000[c] | See footnote c | 0.45 | 1 | 564 | 6–8 | |
| | Moderate | 3000[c] | | 0.50 | 1 | 520 | 5–7 | |
| | Negligible | 2500 | 0.67 | 0.55 | | | See footnote f | |

[a]See Figure 3.7-8 for severity of exposure, regional weathering areas.

[b]With most materials, the water-cement ratios shown will provide average strengths greater than required.

[c]Use air-entrained concrete only.

[d]Proportions should be established by the trial batch method.

[e]Maximum size of coarse aggregate should not be larger than one-fifth the narrowest dimension between forms, nor larger than three-quarters the minimum clear spacing between reinforcing bars.

[f]2% to 3% air entrainment, to improve cohesiveness and reduce bleeding, is recommended but not essential.

(Courtesy Portland Cement Association.)

ture is plastic and workable. The age-strength relationships for Types I and III portland cements are shown in Figure 3.7-2.

For a given water-cement ratio, the strength at a certain age is practically fixed if the mixture is plastic and workable, the aggregates are strong, clean, and sound, and the concrete is properly cured. More water will result in less strength, and less water will result in greater strength.

After concrete has been placed, its strength continues to increase while moisture and optimum temperatures are present to promote hydration and as long as the evaporation of its water (*drying*) is prevented. When concrete is permitted to dry, the water that would cause hydration is gone, and the chemical reactions between the portland cement and water cease. This will result in lower strengths than would occur if the concrete were kept moist. Therefore, it is desirable to keep concrete constantly moist as long as possible after placing (Fig. 3.7-3). The temperature at which concrete is mixed and cured affects the rate of the chemical reactions between the portland ce-

ment and water. Seventy-five °F (23.9°C) is considered the normal curing temperature. Most specifiers require that the range be between 50° and 85°F (10° and 29.4°C). Some will permit temperatures as high as 90°F (32°C). At higher temperatures, concrete increases in strength faster than normal during the first few days of curing but much slower than normal during later periods (Fig. 3.7-4). The final strength of concrete cured at temperatures above 90°F (32°C) is lower than that of concrete cured at between 50° and 90°F (10° and 32°C). Concrete made and cured at 45°F (7.2°C) will gain strength slowly during curing and perhaps never reach the strength of concrete cured at between 50° and 90°F (10° and 32°C). As the temperature in concrete rises, the slump decreases. Adding water to maintain a desirable slump decreases the concrete's strength and adversely affects its other properties.

### 3.7.1.2 Durability

There is a direct link between a concrete's durability and its water-cement ra-

tio. The most destructive natural weathering force is freezing and thawing while concrete is wet or moist because of the expansion of water as it changes to ice. A portland cement paste made with a low water-cement ratio will be much more resistant to damage by freezing and thawing than one with a higher water-cement ratio. This is demonstrated by the concrete cubes in Figure 3.7-5, which were subjected to 70 cycles of freezing and thawing while saturated with water. Those in the upper portion, which were made with a water-cement ratio of 0.80, show much more disintegration than those in the lower portion, which were made with a water-cement ratio of 0.64. Furthermore, introducing entrained air into concrete will significantly increase its resistance to freezing and thawing damage and the application of de-icer salts.

### 3.7.1.3 Watertightness

For a concrete to be watertight and impervious, its portland cement paste must be formulated to produce this result. As

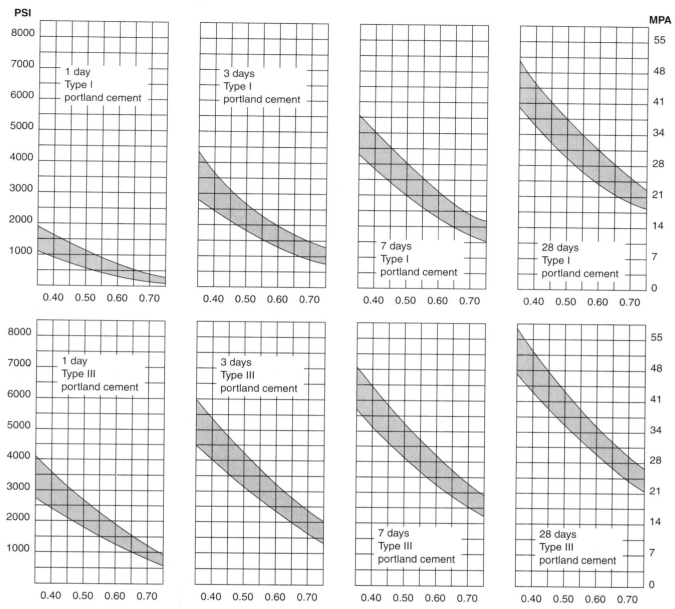

**FIGURE 3.7-2** The relationship between age and compressive strength for Types I and III portland cement concretes. A majority of the results of tests for compressive strength made by several laboratories, using a variety of materials complying with ASTM standards, fall within the shaded areas. (Drawing by HLS.)

**FIGURE 3.7-3** The strength of a concrete continues to increase as long as moisture is present to promote hydration of the cement. Notice that a resumption of moist curing after a period of drying also increases concrete strength. However, the test specimens used in this study were small compared with those of most concrete members. Because resaturation is difficult in the field, moist curing should be continuous. (Courtesy Portland Cement Association.)

FIGURE 3.7-4 The effect of temperature on the strength of a concrete. Other characteristics are also affected. (Courtesy Portland Cement Association.)

FIGURE 3.7-5 The durability of a concrete exposed to freezing and thawing is affected by the quality of its paste. The specimens at the bottom of this photograph were made with a lower water-cement ratio (0.64) than those in the upper part (0.80) and were, therefore, more resistant to damage. The same aggregate was used in all the specimens. (Courtesy Portland Cement Association.)

with strength and durability, there is a direct link between a concrete's watertightness and its water-cement ratio.

Tests show that a portland cement concrete's permeability depends on the amount of mixing water in its water paste and the extent to which the chemical reactions between the portland cement and water have progressed. The results of subjecting discs made with portland cement mortar, fine aggregate, and water to 20 psi (0.14 MPa) water pressure are shown in Figure 3.7-6. In these tests, mortar cured moist for 7 days had no leakage when made with a water-cement ratio of 0.50, but there was considerable leakage with mortars made with the higher water-cement ratios. Moreover, in

FIGURE 3.7-6 Effect of the water-cement ratio on watertightness. Leakage is reduced as the water-cement ratio is decreased and the curing period is increased. Specimens were 1 × 6 in. (25 × 150 mm) mortar discs. Pressure was 20 psi (0.14 MPa). Values are 48-hour averages. (Courtesy Portland Cement Association.)

each case, leakage lessened as the curing period was increased. Mortar discs with a water-cement ratio of 0.80 leaked, even when moist-cured for a month.

Air entrainment improves watertightness by allowing reduction of the water-cement ratio. To be watertight, concrete must also be free from cracks and honeycombing.

As a result of testing and experience in the field, definite recommendations can be made regarding the maximum amount of mixing water that should be used for various construction applications (see Fig. 3.7-1).

## 3.7.2 SELECTING A CONCRETE MIX

The best proportions for a concrete mix are the most economical combination of portland cement, aggregate, and other materials that will produce concrete of the required durability, abrasion resistance, and strength, and that can be placed and finished properly. For slab construction, the properties of the plastic concrete that affect placing and finishing are as important as the properties of the hardened concrete (abrasion resistance, durability, and strength). The plastic properties will greatly affect the quality of the top $1/16$ to $1/8$ in. (1.6 to 3.2 mm) of the slab surface. However, because these properties are not easily measured, there

is a tendency to emphasize the importance of more readily determined properties such as compressive strength.

Recommendations for slump, strength, maximum coarse aggregate size, and minimum portland cement content required for each size aggregate are given in Figure 3.7-7. The portland cement content of a mix should be properly proportioned to produce concrete that can be easily placed and finished and that will, in addition, possess adequate strength and durability. Mixtures with low portland cement content produce harsh concretes and slabs that bleed excess water to the surface, are difficult to finish, and have poor surface characteristics. Mixtures with excessive portland cement content can result in increased drying shrinkage.

When concrete mixtures are not established by tests before construction work begins, the concrete mixtures shown in Figure 3.7-1 may be used. It is first necessary to determine the exposure condition of the concrete from the weathering areas shown in Figure 3.7-8. Areas shown in the severe exposure range are subject to many freeze-thaw cycles per year and to de-icer chemicals; moderate-exposure areas are subject to few freeze-thaw cycles and no de-icer chemicals; negligible-exposure areas have no freeze-thaw cycles, and de-icer chemicals are not used.

When the exposure condition of a concrete construction element is known, it is possible to use Figure 3.7-1 to determine the recommended minimum compressive strength. It is also possible to determine the maximum permissible water-cement ratio necessary to obtain the specified minimum compressive strength. The minimum portland cement content is included in addition to the lowest maximum water-cement ratio. Minimum portland cement requirements ensure satisfactory finishability, improved water resistance, and suitable appearance of vertical surfaces. The percentage of air entrainment for various maximum aggregate sizes is also given; slumps should be in the ranges shown.

An example showing how to use Figure 3.7-1 follows. The proposed project is in the Chicago area, which is in a severe weathering region (see Fig. 3.7-8). According to Figure 3.7-1, the minimum compressive strength required for basement walls that will be exposed to weather is 3500 psi (24 MPa). Reading

**FIGURE 3.7-7** Recommended Mixtures for Light Slabs on Grade[a]

| Maximum Size of Coarse Aggregate,[b] in. (mm) | Minimum Cement Content, lb/cu yd (kg/m³) | Maximum Water/Cement by Weight | | Slump, in. (mm) | Probable Compressive Strength,[c] psi (MPa) |
| | | Non-Air-Entrained Concrete | Air-Entrained Concrete | | |
|---|---|---|---|---|---|
| 1½ (38.1) | 470 (278.84) | | | | |
| 1 (25.4) | 520 (308.10) | | | | |
| ¾ (19.1) | 540 (320.37) | 0.51 | 0.40 | 2–4 (50.8–101.6) | 3500 (24) |
| ½ (12.7) | 590 (350.03) | | | | |
| ⅜ (9.5) | 610 (361.90) | | | | |

[a]For both air-entrained and non-air-entrained concretes.

[b]Normal-weight aggregate; different mixes may be required for lightweight aggregate concretes.

[c]Strength refers to approximate compressive strength of cylinders made and tested according to applicable ASTM standards at 28 days that have been continuously moist cured.

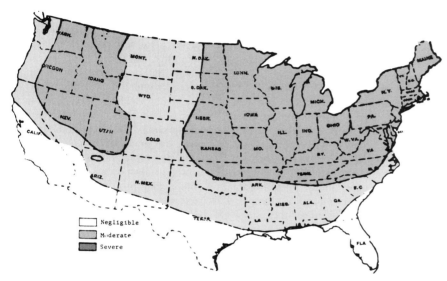

**FIGURE 3.7-8** Regional weathering areas. Areas are based upon the ASTM C62 weathering index map adjusted to conform with an experience survey by HUD and the Portland Cement Association. Local conditions may be more or less severe than those indicated by the regional classification.

across the table shows a practical water-cement ratio of 0.45, and 6% to 8% air entrained concrete. The maximum portland cement content likely to be required to achieve that strength at the recommended air-entrainment level is 564 lb/yd³ (334.6 kg/m³). Even though 2500 psi (17 MPa) concrete might satisfactorily support typical vertical and lateral loading in a residential basement wall, this lower-strength concrete would be less watertight than the concrete recommended here. A higher-strength concrete provides a margin of safety against the possibility of foundation cracking and leaking.

High-strength concrete, as might occur in tall building columns and in mass concrete for gravity dam structures, is produced by including additives such as silica fume, fly ash, and superplasticizers in the mix.

### 3.7.2.1 Allowance for Water in the Aggregate

When selecting the proper amount of water to add to a concrete mix to produce the desired water-cement ratio, it is necessary to allow for the water contained in most fine aggregates A simple test for determining whether sand is *damp*, *wet*, or *very wet* is to press some together in your hand. If the sand falls apart after your hand is opened, it is damp (Fig. 3.7-9a); if it forms a ball that holds its shape, it is wet (Fig. 3.7-9b); if the sand sparkles and leaves water on your hand, it is very wet (Fig. 3.7-9c).

The most practical procedure is to make trial mixes. On large jobs where many materials or unusual materials may be under consideration, or where strength is particularly important, the trials may involve full-size batches checked in the field. The first trial mix may be selected on the basis of experience or established relationships such as those given in Figure 3.7-1. This table has been developed from experience and data from several sources and indicates the suggested proportions of all the various ingredients which should result in good workability.

### 3.7.3 CONSISTENCY OF THE MIX

With a given amount of portland cement paste, more aggregate is used in stiff mixes than in more fluid mixes; consequently, stiff mixes are more economical in terms of cost of materials. Stiff mixes require more labor in placing, however, and when a mixture is too stiff, the additional cost of placing may offset savings in materials. Mixes should always be of a consistency and workability that will permit the concrete to be worked into the angles and corners of forms and around reinforcement without permitting the materials to segregate or excess free water to collect on the surface. Thin members and heavily reinforced members require more fluid mixes than large members or flat surfaces containing little reinforcing.

a

b

c

**FIGURE 3.7-9** (a) Damp sand; (b) wet sand; (c) very wet sand. (Courtesy Portland Cement Association.)

A *plastic* concrete is one that is readily molded and yet will change its form only slowly if the mold is removed. Mixes of plastic consistency are required for most concrete work. Concrete of such consistency does not crumble but flows sluggishly without segregation. Thus, neither very stiff, crumbly mixes nor very fluid, watery mixes are of plastic consistency (Fig. 3.7-10).

The ease or difficulty of placing concrete in a particular location is referred to as *workability*. A stiff but plastic mix with large aggregate would be workable in a large open form but not in a thin wall with closely spaced reinforcement.

Excessive slump and consequent bleeding and segregation are primary causes of poor concrete performance. For slabs, the slump may vary from 2 to 4 in. (50 to 100 mm) (see Fig. 3.7-7), which will produce slabs that can be easily placed and properly consolidated, yet will have a consistency not conducive to bleeding or segregation during placing and finishing. Concrete for slabs where the appearance is important, and concrete for ramps and sloping surfaces, should have a slump that does not exceed 3 in. (76 mm). Different batches of concrete placed in the same slab, ramp, or sloping surface should have approximately the same slump. When lightweight concrete is placed at slumps in excess of 3 in. (76 mm), the coarser lightweight aggregate may rise to the surface and excessive bleeding may also occur.

Reinforced foundation systems may have slumps as low as 1 in. (25.4 mm) but should not exceed 3 in. (76 mm). Concrete that will be struck off and consolidated by hand may have slumps of up to 5 in. (127 mm). The slump of concrete that will be struck off and consolidated using mechanical devices should be limited to 3 in. (76 mm). Concrete that contains high-range water-reducing admixtures may have slumps of up to 8 in. (203 mm) after the admixture has been added. But the slump before the admixture is added must be between 2 and 3 in. (50 and 76 mm). Slumps for other concrete should be not less than 1 in. (25.4 mm) or more than 4 in. (100 mm).

### 3.7.3.1 Slump Test

A slump test conforming with the requirements of ASTM C143 may be used as a rough measure of the consistency of concrete. This test is not a measure of workability, and it should not be used to compare mixes of entirely different proportions or containing different kinds of aggregate. Any change in slump on the job indicates that changes have been made in grading or proportioning the aggregate or in water content. The mix should be corrected immediately to get the proper consistency by adjusting the amounts and proportions of sand and coarse aggregate, with care taken not to change the water-cement ratio.

In a slump test, a test specimen is made in a mold called a *slump cone*, which is made of 16-gauge (1.6 mm) gal-

a

b

c

**FIGURE 3.7-10** (a) A concrete mixture with insufficient cement-sand mortar to fill the spaces between the coarse aggregate. This mixture will be difficult to handle and place, and will result in honeycombed surfaces and porous concrete. (b) A concrete mixture with the correct amount of cement-sand mortar. With light troweling, the spaces between aggregate are filled with mortar. Note the appearance on the edge of the pile. This is a good, workable mixture that will give the maximum yield of concrete. (c) A concrete mixture with too much cement-sand mortar. This mixture is plastic and workable and will produce smooth surfaces, but the yield of concrete will be low and uneconomical, and the concrete will probably be porous. (Courtesy Portland Cement Association.)

vanized metal to the shape shown in Figure 3.7-11a. The base and top are open. The mold is provided with foot pieces and handles, as shown.

To make a slump test, a concrete sample is taken just before its batch is placed in the forms. A slump cone is placed on a flat surface, such as a smooth plank or concrete slab, is held firmly in place by the tester standing on the foot pieces, and is filled with concrete to about one-third of its volume. Then the concrete is puddled with 25 strokes of a $^5/_8$ in. (15.9 mm) diameter rod about 24 in. (610 mm) long and bullet pointed at the lower end (Fig. 3.7-11b). The filling is completed in two more layers, each about one-third of the volume of the cone. Each new layer is rodded 25 times, with the rod penetrating into the underlying layer on each stroke. After the top layer has been rodded, it is struck off with a trowel so that the mold is filled completely. The mold is removed immediately after it has been struck off by gently raising it vertically.

The slump of the concrete is measured, as shown in Figure 3.7-11c, immediately after the cone is removed. If the top of the slump pile is 5 in. (127 mm) below the top of the cone, the slump for this concrete is 5 in. (127 mm).

### 3.7.3.2 Compression Test

Compression tests are made in accordance with ASTM C31 to determine whether a concrete has the specified compressive strength. In these tests, concrete specimens are tested at 7 days and at 28 days after they are taken to determine the concrete's rate of strength gain. Standard procedures provide for curing the specimens either in the laboratory or in the field. Laboratory curing gives an indication of the potential quality of the concrete when properly cured under ideal conditions. Field-cured specimens may give a more accurate interpretation of actual strength in the structure or slab, under the less than ideal curing conditions at a job site, but they offer no explanation as to whether lack of strength is due to error in proportioning, poor materials, or unfavorable curing conditions. Sometimes both methods are used, especially when the weather is unfavorable, in order to interpret the tests properly. When differences occur, the laboratory test results prevail.

Compressive strength test samples are taken at three or more regular intervals during discharge of a concrete batch. Samples should not be taken at the beginning or end of discharge. The batch of concrete thus sampled should be noted as to its location in the work, the air temperature, and unusual conditions at the time. The usual number of test specimens in a sample is four; one is tested at 7 days, two at 28 days, and the fourth is held in reserve in case additional testing is necessary. If both laboratory and field tests are required, duplicate samples are taken. A compressive test specimen is placed in a watertight cylindrical mold to prevent water loss. Standard cylindrical molds, like those shown in Figure 3.7-12, are 3 in. (76 mm) in diameter by 6 in. (152 mm) long, 4 in. (101.6 mm) in diameter by 8 in. (203 mm) long, and 6 in. (152.4 mm) in diameter by 12 in. (304.8 mm) long. The 6 in. by 12 in. (152.4 mm by 304.8-mm) cylinders are generally used when the coarse aggregate does not ex-

**FIGURE 3.7-12** Standard plastic test cylinders. Waxed cardboard cylinders with metal bottoms are also available. (Courtesy ELE International, Inc., Soil Test Products Division.)

a          b          c

**FIGURE 3.7-11** (a) Slump cone (mold) for the slump test. (Courtesy ELE International, Inc., Soil Test Products Division.) (b) Rodding the concrete in a slump cone to ensure complete filling. (Copyright ASTM, reprinted with permission.) (c) Measuring slump by the depression below a rod laid across the cone. This amount of slump indicates a medium-wet mixture. (Copyright ASTM, reprinted with permission.)

ceed 2 in. (50.8 mm) in nominal size. A mold is filled in three equal layers. Each layer is puddled with 25 strokes of a $^5/_8$-in. (15.9-mm)-diameter steel rod about 24 in. (610 mm) long with a bullet-pointed tip. Reinforcing bars and other tools should not be used as puddle rods. After the top layer has been rodded, the surface of the concrete is struck off with a trowel and covered to prevent evaporation.

The cured specimens are crushed in a laboratory testing machine to determine their strength (Fig. 3.7-13). If the strength of the specimens is too low, test cores should be cut from the concrete structural unit containing the suspect concrete. These should then be tested for compression. If this test indicates unsatisfactory strength, the concrete should be removed and new concrete of adequate strength should be placed.

### 3.7.4 MATERIALS MEASUREMENT

Producing uniform batches of concrete of proper proportions and consistency requires that the ingredients be measured accurately. The effects of the water-cement ratio on concrete quality make it necessary to measure all the ingredients,

including the water. A troublesome factor is the effect of the varying amounts of moisture nearly always present in aggregate, particularly in natural sand. The amount of free moisture introduced into the mixture with the aggregates must be determined and allowance made for it if accurate control is to be obtained.

#### 3.7.4.1 Measuring Portland Cement

On small projects where sacked portland cement is used, the batches of concrete should be of such size that only full sacks are used. If partial sacks are used, they should be weighed for each batch. It is not satisfactory to divide sacks of portland cement on the basis of volume. Bulk (unsacked) portland cement, which is used today for most concrete, should be measured by weight for each batch.

#### 3.7.4.2 Measuring Water

Dependable and accurate means for measuring mixing water are essential. Portable mixers should be equipped with water tanks and measuring devices that are accurate when properly operated. The measuring device most generally used operates on the principle of the siphon.

The tank is filled, and the desired amount of water is siphoned off. When the water level reaches the point at which the bottom of the siphon is set, the water is shut off automatically.

#### 3.7.4.3 Measuring Aggregates

Measuring aggregates by volume cannot be depended on except under the most careful supervision. A small amount of moisture in fine aggregate, which is nearly always present, causes the aggregate to bulk up (increase in volume), as indicated in Figure 3.7-14. The amount of bulking varies with the amount of moisture present and the grading; fine sands bulk more than coarse sands for a given amount of moisture, because a given quantity of smaller particles presents more surface to which water can adhere than does the same amount of aggregate with larger particles. Note in Figure 3.7-14 that a fine sand with a 5% moisture content increases in volume almost 40% over its dry volume. Sand often contains almost enough moisture to produce the maximum bulking. Small changes in moisture can cause large changes in the amount of bulking until all particles have been surrounded with water. After that, the addition of water causes a lesser degree of increased bulking. Therefore, the relationship between moisture and bulk is not linear. For these reasons, aggregates should always be measured by weight.

To reduce segregation of aggregates and to produce uniformity from batch to batch, it is desirable to weigh the coarse aggregate in two or more sizes, especially if the maximum size exceeds 1 in. (25.4 mm). Generally, the ratio of the maxi-

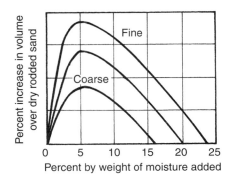

**FIGURE 3.7-14 Bulking in fine aggregate caused by surface moisture.** (Courtesy Portland Cement Association.)

**FIGURE 3.7-13 A concrete compression tester.** (Courtesy ELE International, Inc., Soil Test Products Division.)

mum size particle to the minimum size for coarse aggregate separations should not exceed 2:1 for materials larger than 1 in. (25.4 mm) and should not exceed 3:1 for finer materials. Therefore, $1^1/_2$-in. (38 mm) aggregate would be separated into two batches, one with $^1/_4$- to $^3/_4$-in. (6.4- to 19.1-mm)-sized aggregate and the other with $^3/_4$- to $1^1/_2$-in. (19.1 to 38 mm) sizes.

### 3.7.5 MIXING

The ingredients should be in the proper condition when they are charged into a mixer. Aggregates should have the correct moisture content, and portland cement should be free of lumps. If lumps are present which cannot be pulverized between the thumb and finger, the portland cement should not be used.

Concrete should be mixed thoroughly until its appearance, including its color, is uniform and its ingredients are uniformly distributed. The time required for thorough mixing depends on several factors. Specifications usually require a minimum of 1 minute mixing for mixtures of up to 1 cu yd (.765 m$^3$) capacity, with an increase of 15 seconds mixing for each $^1/_2$ cu yd (.38 m$^3$), or fraction thereof of additional capacity. In general, longer mixing gives more uniform results. The mixing period should be measured from the time all solid materials are in the mixer drum, provided that all of the water is added before a quarter of the mixing time has elapsed. Many mixers are equipped with timing devices, some of which can be set for a given mixing time and locked so that the batch cannot be discharged until the designated time has elapsed.

Mixers should not be loaded above their rated capacity and should be operated at approximately the speeds for which they are designed. If the blades of the mixer become worn or coated with hardened concrete, the mixing action will be less efficient. Badly worn blades should be replaced, and hardened concrete should be removed before each run of concrete. Under usual operating conditions, up to about 10% of the mixing water should be placed in the drum before the aggregates and portland cement are added. Water should then be added uniformly with the dry materials, leaving about 10% to be added after all other materials are in the drum. When heated water is used during cold weather, this or-

der of charging may require some modification to prevent flash setting of the portland cement. In this case, addition of the portland cement should be delayed until most of the aggregate and water have intermixed. The concrete should be discharged completely from a mixer before the mixer is recharged.

#### 3.7.5.1 Ready-Mixed Concrete

Most concrete used in building construction originates in a central ready-mix plant (Fig. 3.7-15). Ready-mixed concrete should comply with ASTM C94, which requires, among other things, that the hauling be done in agitator trucks or transit mixer trucks operated at agitator speed.

In some ready-mixed operations, the materials are dry-batched at the central plant and then mixed en route to the job in truck mixers. In others, the concrete is mixed in a stationary mixer at the central plant only sufficiently to intermingle the ingredients, generally about 30 seconds. The mixing is completed in a truck mixer en route to the job.

Truck mixers consist essentially of a mixer with a separate water tank and a water measuring device mounted on a truck chassis. They usually are made with capacities of 1 to 5 cu yd (.76 to 3.8 m$^3$). Agitator trucks are similar but do not have separate water tanks.

ASTM C94 requires that concrete must be delivered to and discharged from a truck mixer or agitator truck within $1^1/_2$ hours after water was added to the portland cement and aggregate.

#### 3.7.5.2 Remixing Concrete

The initial set of concrete does not usually take place until 2 or 3 hours after the portland cement is mixed with water, but fresh concrete that is left standing tends to dry out and stiffen somewhat before the portland cement sets. Such concrete may be remixed and used within $1^1/_2$ hours of the time water was added to the mix if, when remixed, it becomes sufficiently plastic to be compacted in the forms. However, adding water to make the mixture more workable (retempering) should never be permitted because it lowers the quality, just as would adding a larger amount of water in the original mixing.

## 3.8 Handling, Transporting, Placing, and Consolidating

Each step in handling, transporting, placing, and consolidating concrete should be

**FIGURE 3.7-15** Loading trucks at a ready-mix plant. (Courtesy Portland Cement Association.)

controlled carefully to maintain uniformity within each batch and from batch to batch so that the complete structure has uniform quality throughout. It is essential to avoid separation of coarse aggregate from mortar and of water from the other ingredients. Equipment for conveying concrete from mixer to place of final deposit should do so without material loss or separation so as to provide a continuous flow of concrete at the point of delivery. Segregation at the point of discharge from the mixer can be avoided by providing a down pipe at the end of the chute so that the concrete will drop vertically into the center of the receiving bucket, hopper, or form. Concrete is handled and transported by:

- Chutes
- Push buggies
- Buckets handled by cranes
- Pumping through a pipeline
- Pneumatically forcing through a hose (shotcrete)

The methods of handling and transporting, and the equipment used, should not place restrictions on the consistency of the concrete, which should be governed instead by the placing conditions. If placing conditions permit a stiff mix, the delivery equipment used should be designed and arranged to facilitate handling and transporting such a mix.

### 3.8.1 PREPARATION

Before concrete is placed, the surfaces to receive the concrete should be properly prepared, forms should be erected, and reinforcement should be installed.

#### 3.8.1.1 Requirements Common to All Concrete

Concrete should not be placed on a frozen subgrade. Snow, ice, and debris must be removed from within forms and excavations and from slab beds before concrete is placed. When concrete is to be placed on rock, loose material should be removed and cut faces should be nearly vertical or horizontal rather than sloping.

Chlorides should not be used in concrete in which dissimilar metals are embedded, such as reinforcing steel and heating ducts.

Reinforcing steel should be clean and free of loose rust and mill scale at the time concrete is placed. Hardened mortar should be removed from the steel.

#### 3.8.1.2 Slabs on Grade

Subgrades should be trimmed to specified elevations and uniformly compacted before concrete is placed. Except when a vapor retarder, waterproof membrane, or separator has been installed immediately below a slab, the subgrade should be moistened (Fig. 3.8-1) to prevent too rapid extraction of water from the concrete. However, at the time of placing, there should be no mud, soft spots, or free water standing where concrete will be placed.

The Portland Cement Association (PCA) recommends that a vapor retarder be placed under every concrete slab on grade where an impervious floor finish will be installed and in every other location where the passage of water vapor through a slab on grade is undesirable. However, a vapor retarder placed in contact with the concrete will cause excess moisture in the concrete to bleed to the surface rather than downward, increasing the number of capillaries in the concrete and contributing to weakening of the surface, more cracking, and higher permeability of the slab. Therefore, PCA recommends that a vapor retarder not be placed directly beneath a slab, but rather be covered by a 3-in. (76-mm)-thick layer of granular, self-draining material, such as sand. Refer to Section 21.3 for additional information.

When a vapor retarder is not used beneath a slab on grade, a separator of building paper or other material that will withstand handling and construction traffic is sometimes installed. Such separators are usually installed between the base course or fill and the slab to prevent the fines or paste of plastic concrete from seeping down into the base course or fill. However, if such a separator is complete

**FIGURE 3.8-1 Wetting the subgrade before concrete is placed.** (Courtesy Portland Cement Association.)

and impervious, it should not be placed directly in contact with the slab. If it is, it may act in the same way that a vapor retarder often acts, and contribute to warping (curling) of the slab and exacerbate drying shrinkage cracking.

Steel reinforcing, ductwork, heating pipes, and other items embedded in a slab should be set and supported at the proper elevation before concrete is placed.

#### 3.8.1.3 Formed Concrete

Forms should be tight, adequately braced, and constructed of material that will impart the desired texture to the finished concrete. Before concrete is placed in them, forms should be cleaned to remove dirt, nails, sawdust, wood chips, and other foreign materials. To facilitate form removal, forms should be treated with a release agent such as oil or lacquer. For architectural concrete, the agent should be a nonstaining lacquer or emulsified stearate. Wood forms should be moistened before concrete is placed, or they will absorb water from the concrete and swell.

### 3.8.2 PLACING

Concrete should be placed as nearly as practicable in its final position to avoid segregation resulting from rehandling or flowing. It should not be placed in large quantities at a given point and allowed to run or be worked over a long distance in the form. This practice results in segregation because the mortar tends to flow out ahead of the coarser material. It also results in sloping work planes and potentially weak bonding between successive concrete layers. In general, concrete should be placed in horizontal layers of uniform thickness, with each layer thoroughly compacted before the next is placed.

Layers should be 6 to 24 in. (150 to 610 mm) thick; thinner layers may be used in narrow and heavily reinforced forms. Each layer must be placed before the previous one stiffens.

#### 3.8.2.1 Placing Systems

Concrete is either delivered to the site by truck or, in small projects, mixed at the site and moved to its placement location by:

- Dumping directly from a truck's chute.
- Buggy.

along form faces, the order of placing should be as follows: on walls, start at ends and work toward the center; on slabs, start around the perimeter.

On flat surfaces, such as slabs, the placing of concrete should be started at the far end of the work so that each batch will be dumped against previously placed concrete, not away from it (Fig. 3.8-5). Concrete should not be dumped in separate piles and then leveled and worked together. If stone pockets occur, some of the excess large particles should be removed and distributed to areas where there is more mortar present to surround them.

In walls, the first batches should be placed at either end of the section; the placing should then progress toward the

**FIGURE 3.8-2** Placing concrete by a crane-delivered bucket. (Courtesy Portland Cement Association.)

■ Dumping from a truck's chute into a large bucket, which is then lifted by crane and deposited in its proper location (Fig. 3.8-2).
■ Dumping from a truck into a concrete pump which delivers the mix to its final location through a hose. Pumps are capable of delivering concrete over long distances and up many floors (Fig. 3.8-3).

Regardless of the method used to get concrete to its final location, it should not be dropped freely more than 4 ft (1200 mm) to prevent segregation. Nor should concrete be permitted to strike against the forms as it is being placed. Instead, drop chutes should be used to prevent spattering of mortar on reinforcement and forms (Fig. 3.8-4). If placement can be completed before the mortar dries, drop chutes may not be needed.

### 3.8.2.2 Order of Placing

To prevent water from collecting at the ends of forms, in corners of forms, or

CORRECT

*Separation is avoided by discharging concrete into hopper feeding into drop chute. This arrangement also keeps forms and steel clean until concrete covers them.*

INCORRECT

*Permitting concrete from chute or buggy to strike against form and ricochet on bars and form faces causes separation and honeycomb at the bottom.*

**FIGURE 3.8-4** Placing concrete in the top of a narrow form. (Courtesy Portland Cement Association.)

**FIGURE 3.8-3** Placing concrete by a pump hose. (Courtesy Portland Cement Association.)

**a**

CORRECT

*Concrete should be dumped into the face of previously placed concrete.*

INCORRECT

*Dumping concrete away from previously placed concrete causes separation.*

**b**

Chute

Baffle

CORRECT

*A baffle and drop at end of chute will avoid separation and concrete remains on slope.*

Chute

INCORRECT

*Discharging concrete from free end of chute onto a slope causes separation of rock which goes to bottom of slope. Velocity tends to carry concrete down the slope.*

**FIGURE 3.8-5** (a) Placing slab concrete; (b) placing concrete on a sloping surface. (Courtesy Portland Cement Association.)

center. This same procedure should be used for each layer.

### 3.8.2.3 Placing on Hardened Concrete

In order to secure a good bond and a watertight joint, when fresh concrete is placed on hardened concrete, the hardened concrete should be level, rough, clean, and moist, and some aggregate particles should be exposed. *Laitance* (soft layers of mortar) should be removed from the top surface of hardened concrete. An appreciable thickness of laitance indicates that segregation and bleeding occurred or that too much water was used in the concrete. Laitance can be prevented by using a stiffer mix or finer material, particularly in the upper part of the lift.

On floor slabs to be built in two courses, the top of the lower course may be broomed just before it sets with a stiff fiber or steel broom. The surface should be level, scored, and free of laitance. It must then be protected and thoroughly cleaned just before the second course is placed.

When new concrete is to be placed on old concrete, the old concrete must be roughened and cleaned thoroughly. In most cases, it is necessary to remove the entire surface to expose a new surface satisfactory for bonding.

Hardened concrete should be moistened thoroughly before new concrete is placed on it. Where concrete has dried out, it is necessary to saturate it for several days. There should be no pools of water, however, when the new concrete is placed.

Where concrete is to be placed on hardened concrete or brick, a layer of mortar should be placed on the substrate to provide a cushion against which the new concrete can be placed; this prevents stone pockets and ensures a tight joint. This mortar should be made with the same content as the concrete and should have a slump of about 6 in. (150 mm). It should be placed to a thickness of $1/2$ to 1 in. (13 to 25 mm) and should be well worked into the irregularities of the hard surface. In two-course floor construction, a coat of portland cement and water paste of the consistency of thick paint should be brushed into the hard surface just before the second course is placed.

### 3.8.2.4 Shotcrete

*Shotcrete* is pneumatically placed concrete, used primarily for swimming pools and other in-ground and aboveground free-form structures and for repairing damaged concrete. Its major advantage is that it does not require formwork even when placed on vertical surfaces. Shotcrete concrete is forced through a hose by compressed air onto earth, rock, existing concrete, or another layer of shotcrete.

Shotcrete is reinforced with bars and mesh, and sometimes with steel fiber reinforcement to minimize cracking. Finishes include the natural finish left by the placement gun; a screeded finish left by a sharp-edged cutting screed; a flash-coat finish produced by spraying on a final coat of shotcrete made with fine aggregate; and a float and trowel finish, which cannot be used on thin slabs.

### 3.8.3 CONSOLIDATING

Concrete should be compacted by a method appropriate to the material and its location to:

- Eliminate stone pockets and large air bubbles
- Consolidate each layer with that previously placed
- Completely embed reinforcing and fixtures
- Bring just enough fine material to the faces and top surfaces to produce the desired finish

### 3.8.3.1 Spading and Puddling

Medium- to high-slump (3 to 8 in. [76 to 200 mm]) concrete should be compacted and worked into place by spading or puddling. Spades or sticks, long enough to reach the bottom of the forms and thin enough to pass between the reinforcing steel and forms, should be used.

### 3.8.3.2 Vibrating

Low- to medium-slump (1 to 6 in. [25 to 150 mm]) concrete should be compacted using mechanical vibrators either within the concrete (internal vibrators) or on the top of the concrete or on the forms (external vibrators). Relatively thin slabs that are not heavily reinforced may be vibrated by mechanical beam or truss screeds. Thick slabs, slabs with structural reinforcement, and thicker concrete members, such as footings, walls, columns, beams, and girders, should be vibrated with internal vibrators (Fig. 3.8-6).

When concrete is vibrated, the internal friction between the coarse aggregate particles is temporarily eliminated and the concrete behaves as a liquid; it settles in the forms under the action of gravity, and the trapped air bubbles rise more easily to the surface. Friction is reestablished as soon as vibration stops. Therefore, mixes of lower water content or leaner mixes for a given water content can be used. If less water is used per unit vol-

**FIGURE 3.8-6** Using an internal vibrator. (Courtesy Portland Cement Association.)

ume of portland cement, the concrete will be of better quality; if less portland cement per unit volume of water is used, the concrete will be more economical. Vibrators should not be used to push or move concrete laterally over long distances within a form. The concrete should be deposited as near its final position as possible. It should be distributed in layers and then vibrated (Fig. 3.8-7). Some hand spading or puddling may be necessary along with the vibration to secure smooth surfaces and reduce pitting on formed surfaces.

Precautions should be taken not to overvibrate to the point that segregation

results, especially when the concrete has a higher slump than is necessary for vibration. Precaution and judgment must be used to be sure that complete consolidation is secured without segregation and that no areas are missed. Sufficient vibration usually is indicated by a line of mortar along the forms and by the submerging of the coarse aggregate particles in the mortar.

## 3.9 Finishing

Concrete may be finished in several ways, depending on its location, the concrete material used, and the esthetic effect desired. In this section, general finishing procedures are summarized first, followed by techniques used to produce special surface finishes.

### 3.9.1 FINISHING STANDARD-WEIGHT CONCRETE SLABS

The finishing of standard-weight concrete slabs proceeds through several steps in a defined order. These steps are screeding, leveling, edging, jointing, floating, troweling, and broom finishing or other surface texturing.

#### 3.9.1.1 Bleeding

Generally, the dry materials used in making quality concrete are heavier than water. Thus, shortly after placement, they have a tendency to settle to the bottom and displace the mixing water to the surface, which is called *bleeding*. This oc-

curs more readily when the concrete is not air-entrained.

If any operation is performed on the surface while bleed water is present, serious scaling, dusting, or crazing of the hardened concrete can result. This point cannot be overemphasized and is the basic rule for successful finishing of concrete surfaces. Section 3.9.6 contains a more detailed discussion of scaling, crazing, and dusting.

Placing, screeding, consolidating, and leveling of placed concrete must be performed before bleeding takes place. Therefore, concrete should not be placed faster than it can be spread, struck off, and leveled, and should not be allowed to remain in wheelbarrows, buggies, or buckets any longer than is absolutely necessary. It should be dumped and spread as soon as possible, and immediately struck off to the proper grade and leveled.

#### 3.9.1.2 Screeding

The surface of newly placed concrete is struck off (*screeded*) by moving a straightedge back and forth with a saw-like motion across the top of the forms and screeds. If mechanical vibrating equipment is used in the striking-off process, the need for leveling may be eliminated. Whether boards, vibratory screeds, or roller screeds are used, a small amount of concrete always should be kept ahead of the straightedge to fill in low spots and maintain a plane surface (Fig. 3.9-1).

Of all the placing and finishing operations, screeding the surface to a predetermined grade has the greatest effect on surface tolerances. Concrete should be struck off immediately after it is placed, and screed stakes used to establish surface elevation should be removed as the work progresses so that workers do not

**a**

CORRECT     INCORRECT

*Correct: Start placing at bottom of slope so that compaction is increased by weight of newly added concrete. Vibration consolidates the concrete.*

*Incorrect: When placing is begun at top of slope, the upper concrete tends to pull apart, especially when vibrated below because this starts flow and removes support from the concrete above.*

**b**

CORRECT     INCORRECT

*Correct: Start placing at bottom of slope so that compaction is increased by weight of newly added concrete. Vibration consolidates the concrete.*

*Incorrect: Haphazard random penetration of the vibrator at all angles and spacings without sufficient depth will not assure intimate combination of the two layers.*

**FIGURE 3.8-7** (a) Consolidation of concrete on slopes; (b) consolidation of concrete in multiple lifts. (Courtesy Portland Cement Association.)

**FIGURE 3.9-1** Striking off (screeding). (Courtesy Portland Cement Association.)

have to walk back into areas that have already been struck off.

### 3.9.1.3 Leveling

Leveling is the bringing of a concrete surface to true grade with enough mortar to produce the desired finish. After a concrete slab has been screeded, it should be immediately smoothed with a *darby* to level raised spots and fill depressions (Fig. 3.9-2) left after screeding. Long-handled floats, called *bull floats*, of either wood or metal, are sometimes used instead of darbies to smooth and level concrete surfaces (Fig. 3.9-3). Because it is hard to produce surfaces in plane near the edges of a slab using bull floats, darbies are sometimes needed after the bull floating has been done.

Leveling is sometimes called "darbying or bull floating," but this terminology requires that both terms be used or that the architect dictate which is to be used and which is inappropriate. The PCA in its literature uses the broader term *leveling*.

The purpose of leveling is to eliminate the ridges and voids left by screeding. In addition, it should slightly embed the coarse aggregate, thus preparing the surface for subsequent finishing op-

**FIGURE 3.9-2 Leveling using a darby.** (Courtesy Portland Cement Association.)

erations of edging, jointing, floating, and troweling.

A slight stiffening of the concrete is necessary after leveling before further finishing operations are started. No subsequent operations should be performed until the concrete will sustain foot pressure with only about $^1/_4$ in. (6 mm) indentation.

### 3.9.1.4 Edging

When all bleed water and water sheen have left the surface and the concrete has started to stiffen, other finishing operations such as *edging* may be started (Fig. 3.9-4). Edging rounds off the formed edge of a slab to prevent chipping or damage. An *edger* should be run back and forth until a finished edge is produced. All coarse aggregate particles should be covered, and the edger should not leave too deep an impression in the slab. If it does, the indentation may be difficult to remove with a subsequent finishing operation.

Usually, edging is not required for most interior slabs on grade, and is more commonly performed on sidewalks, driveways, and steps. An edger should not be used when a slab is to be finished with resilient flooring requiring a smooth, level subfloor. Edges at construction joints in the slab may be ground lightly with a silicon carbide stone to remove irregularities after forms are stripped and before the adjacent slab is placed.

### 3.9.1.5 Jointing

Except when joints will be later sawed, immediately following or during edging, premolded inserts are placed in concrete slabs to control cracking in the concrete as a result of shrinkage. Refer to Section 3.6 for a detailed discussion about joints.

### 3.9.1.6 Floating

After edging and hand jointing operations, a slab should be floated. Many variables, such as concrete temperature, air temperature, relative humidity, wind, and other factors, affect the process, making it difficult to set a definite time to begin floating. This knowledge comes only through job experience. In general, floating may be done when the water sheen has disappeared and the concrete will support the weight of the finisher.

The purpose of floating is to:

- Embed large aggregate just beneath the surface
- Remove slight imperfections, humps, and voids to produce a level or plane surface
- Consolidate mortar at the surface in preparation for other finishing operations
- Open the surface to permit excess moisture to escape

Aluminum or magnesium floats should be used, especially on air-entrained concrete. This type of metal float greatly reduces the amount of work to be done by the finisher because the float slides more readily over the concrete surface, has a good floating action, and forms a smoother surface texture than a wood float. A wood float tends to stick to and "tear" the concrete surface.

The marks left by edgers and jointers should be removed by floating (Figs. 3.9-5 and 3.9-6) unless such marks are desired for decorative purposes, in which case the edger or jointer should be rerun after the floating operation.

### 3.9.1.7 Troweling

Troweling is done on slabs that are to be left exposed or to receive thin finishes, such as resilient flooring, carpet, tile, or

**FIGURE 3.9-3 Leveling using a bull float.** (Courtesy Portland Cement Association.)

**FIGURE 3.9-4 Edging.** (Courtesy Portland Cement Association.)

**FIGURE 3.9-5 Hand floating.** (Courtesy Portland Cement Association.)

**FIGURE 3.9-6** Power floating. (Courtesy Portland Cement Association.)

paint. When troweling is required, the surface should be steel-troweled immediately after floating (Figs. 3.9-7 and 3.9-8). It is customary for a cement mason using hand tools to float and then steel-trowel an area before moving the knee boards. If necessary, tooled joints and edges should be rerun before and after troweling to maintain true lines, proper depths, and uniformity, and to remove kinks.

The purpose of troweling is to produce a smooth, hard surface. For the first troweling, whether by power or by hand, the trowel blade must be kept as flat against the surface as possible. If it is tilted or pitched at too great an angle, an objectionable *washboard* or *chatter* surface will result. For the first troweling, a new trowel is not recommended. An older trowel which has been broken in can be worked quite flat without the

**FIGURE 3.9-7** Hand troweling. (Courtesy Portland Cement Association.)

**FIGURE 3.9-8** Power troweling. Also available are self-propelled power trowels on which the operator rides in a seat. (Courtesy Portland Cement Association.)

edge digging into the concrete. The smoothness of a surface can be improved by timely additional trowelings. There should be a lapse of time between successive trowelings to permit the concrete to increase its set. As the surface stiffens, each successive troweling should be made by a smaller trowel tipped at a progressively higher angle so that sufficient pressure can be applied for proper finishing.

For exposed slabs, additional troweling increases the compaction of fines at the surface, giving greater density and better wear resistance. A second troweling is recommended even if the slab is to be finished with resilient flooring because it results in closer surface tolerances and a better surface to receive the flooring.

### 3.9.1.8 Broom Finishng

Steel-troweled concrete surfaces are very smooth and become slippery when wet. They can be slightly roughened to produce a nonslip surface by *brushing* or *brooming* them. A brushed surface is made by drawing a broom over the surface after steel troweling. Notice in Figure 3.9-9 that the worker is drawing the broom right over the edge joint, but the joint is not being marred. This indicates that the concrete surface has been steel-troweled properly and is hard enough. Other types of surface finishes are discussed in Section 3.9.4.

**FIGURE 3.9-9** Broom finishing. (Courtesy Portland Cement Association.)

### 3.9.2 FINISHING AIR-ENTRAINED CONCRETE SLABS

The microscopic air bubbles in air-entrained concrete tend to hold the ingredients, including water, in suspension. This type of concrete requires less mixing water and still has good workability with the same slump. Since there is less water and it is held in suspension, little or no bleeding occurs. There is, therefore, no need to wait for the evaporation of free water from the surface, and floating and troweling can and should be started as soon as the slab can support the finisher and equipment. Many horizontal surface defects and failures are caused by performing finishing operations while bleed water or excess surface moisture is present. Therefore, better results are generally obtained with air-entrained concrete.

As with other concrete, if floating is done by hand, an aluminum or magnesium float should be used. A wood float drags and greatly increases the amount of work necessary to achieve the same result. If floating is done with power equipment, the only major difference between the finishing procedures for air-entrained concrete and those for other concrete is that floating may be started sooner on air-entrained concrete.

### 3.9.3 FINISHING LIGHTWEIGHT STRUCTURAL CONCRETE SLABS

Finishing operations for slabs made with lightweight structural concrete containing coarse aggregates of expanded clay, shale, or slag vary somewhat from those used on slabs made with normal-weight concrete. When the surface of the concrete is worked, there is a tendency for the coarse lightweight aggregate rather than the mortar to rise to the surface. Lightweight concrete can be easily finished if the following precautions are observed: (1) the mix should be properly proportioned and should not be over- or undersanded in an attempt to meet unit weight requirements; (2) finishing should not be started too early, and the concrete should not be overworked or overvibrated. A well-proportioned mix can generally be placed, struck off, leveled, and floated with less effort than is necessary for normal-weight concrete. Excessive leveling and floating are principal causes of finishing problems because the heavier mortar is driven down and the coarse aggregate is brought to the surface.

### 3.9.4 SPECIAL SLAB FINISHES

Due to the plastic quality of concrete, many surface finishes can be applied. The surface may be scored or tooled with a jointer in decorative and geometric patterns. Some of the more common special finishes are discussed here.

#### 3.9.4.1 Exposed Aggregate

An exposed aggregate surface is often chosen for an area where a special textural effect is desired (Fig. 3.9-10). If the surface is ground and polished, it is suitable especially for such places as entrances, interior terraces, and courtyards.

Selection of aggregates is so important that test panels should be made before the job is started. Colorful gravel aggregate that is quite uniform in gradation and in sizes ranging from $1/2$ to $3/4$ in. (13 to 19 mm) is recommended. Flat, sliver-shaped particles or aggregate less than $1/2$ in. (13 mm) in diameter should be avoided, because they become dislodged during exposing operations. Exposing the aggregate used in ordinary concrete generally is unsatisfactory since this will not necessarily reveal a high percentage of coarse aggregates.

A five and one-half to six-sack concrete with a maximum 3 in. (76 mm) slump should be used. Immediately after the slab has been screeded and leveled, the selected aggregate should be scattered by hand and evenly distributed so that the entire surface is completely covered (Fig. 3.9-11a). The initial embedding of the aggregate is usually done by patting with a darby or the flat side of a 2 × 4 (50 × 100) (Fig. 3.9-11b). After the aggregate is thoroughly embedded and as soon as the concrete will support the weight of a mason on kneeboards, the surface should be hand floated with a float or darby (Fig. 3.9-11c). This operation should be performed so thoroughly that all aggregate is embedded entirely just beneath the surface. The grout should completely surround and slightly cover all aggregate, leaving no holes or openings in the surface.

Shortly after this floating, a reliable retarder may be sprayed or brushed over the surface following the manufacturer's recommendations. On small jobs, a retarder may not be necessary. Retarders generally are used on large jobs for better control of exposing operations. Where a retarder has been used, exposing the aggregate is usually done some hours later by brushing and hosing with water. However, the manufacturer's recommendations should be followed closely. Whether or not a retarder has been used, the proper time for exposing aggregate is quite critical. It should be done as soon as the grout covering the aggregate can be removed by simultaneously brushing and hosing (Fig. 3.9-11d), without overexposing or dislodging the aggregate. If during exposing it is necessary for masons to move about on the surface, kneeboards should be used gently. If possible, this should be avoided because of the risk of breaking the aggregate bond.

For interior areas and where a smooth surface is desired, no retarder is used and exposure of the aggregate is accomplished entirely by grinding. This may be followed by polishing, which will produce a surface similar to that of terrazzo.

a                                              b

c                                              d

**FIGURE 3.9-11** Producing an exposed aggregate surface: (a) decorative aggregate is distributed to cover the entire surface; (b) initial embedding of aggregate; (c) floating the surface to completely cover aggregate with grout; (d) exposing aggregate is done by simultaneously brushing and hosing with water. (Photos courtesy Portland Cement Association.)

**FIGURE 3.9-10** Exposed aggregate surface.
(Courtesy Portland Cement Association.)

Because the aggregate completely covers the surface, tooled joints in this type of work are impractical. Decorative or control joints are accomplished best by sawing. Control joints should be cut 4 to 12 hours after the slab is placed and should be at least one-fifth the depth of the slab. A small-radius edger should be used before and immediately after the aggregate has been embedded to provide a more decorative edge to the slab. Another method of providing control joints is to install permanent strips of redwood before placing concrete.

Another method of placement requires spreading a 1-in. (25.4-mm)-thick or thicker top course containing the special aggregate. As in all concrete work, exposed aggregate slabs should be cured thoroughly.

### 3.9.4.2 Dry-Shake Colored Surface

Dry-shake colored-concrete surfaces are used for interior areas, terraces, decorative walks, driveways, and other areas where a colored surface is desired. They are made by applying one of a number of proprietary dry-shake materials. Their basic ingredients are mineral oxide pigment, white portland cement, and specially graded silica sand or fine aggregate. Other pigments are inappropriate and should not be used. Selecting the ingredients, proportioning, and mixing of a dry-shake material on the job are not recommended, because it is difficult to control the proportions and the results cannot be predicted with accuracy. Field-mixed colors are likely to produce a mottled, uneven coloring.

After concrete has been struck off and leveled and free water and excess moisture have evaporated from its surface, the surface should be floated either by power or by hand. If it is done by hand, a magnesium or aluminum float should be used. Preliminary floating should be done before the dry-shake material is applied to bring up enough moisture for combining with the dry-shake material. Floating also removes ridges and depressions that may cause variations in color intensity. Immediately following floating, the dry-shake material is shaken evenly by hand over the surface, following the dry-shake material manufacturer's instructions. If the wrong amount of dry-shake material is applied in one spot, nonuniformity in color and possibly surface peeling will result.

Unless the manufacturer recommends otherwise, the first application of a colored dry-shake should use about two-thirds of the total amount needed. Amounts are usually specified in pounds per square foot (kg/m$^2$) of surface. In a few minutes, this dry material will absorb some moisture from the plastic concrete and should then be thoroughly floated into the surface, preferably with a power float. Immediately afterward, the balance of the dry-shake material should be distributed evenly over the surface, thoroughly floated again, and made a part of the surface, taking care that a uniform color is obtained.

Tooled edges and joints should be run both before and after the dry-shake application.

Shortly after final floating, the surface should be troweled. If done by hand, troweling should immediately follow final floating. After the first troweling, whether by power or by hand, there should be a lapse of time to allow the concrete to increase its set. The length of elapsed time should depend on the temperature and humidity present. Concrete may be troweled a second time to improve the texture and to produce a denser, harder surface.

For exterior surfaces, two trowelings are usually enough. Then a fine, soft-bristled pushbroom may be drawn over the surface to produce a roughened texture for better traction underfoot. For interior surfaces, a third hard troweling may be appropriate. This final troweling should be done by hand to eliminate washboard or trowel marks and to produce a smooth, hard, dense, wearing surface.

Dry-shake colored concrete, like other types of freshly placed concrete, must be cured thoroughly. After thorough curing and surface drying, interior surfaces may be given at least two coats of special concrete floor wax, which is available from various reliable manufacturers. It contains the same mineral oxide pigment used in the dry-shake. Care should be taken to avoid staining, such as by dirt or foot traffic, during the curing or drying period and before waxing.

### 3.9.4.3 Swirl Design

A swirl design, nonskid surface texture can be produced on a slab using a magnesium or aluminum float or a steel finishing trowel. When a float is used, the finish is called a *swirl float* finish; when

a trowel is used, it is called a *swirl trowel* finish.

After a concrete surface has been struck off, leveled, floated, and steel-troweled, it is ready to be given either a swirl float or a swirl trowel finish. A float should be worked flat on the surface in a semicircular or fan-like motion. Pressure applied on the float with this motion will give a rough-textured swirl design, as shown in Figure 3.9-12a.

With the same motion, using a steel trowel held flat, a cement mason can obtain a finer-textured swirl design on a concrete surface (Fig. 3.9-12b). Moist curing of the slab is the final operation.

### 3.9.4.4 Keystone Finish

A keystone finish produces a travertine-like texture. It can be used on a terrace, garden walk, driveway, deck around a swimming pool, and in any other location where an unusually decorative flat concrete surface is desired. It should not be used where it will be subjected to freezing weather.

a

b

FIGURE 3.9-12 Producing a swirl design on a concrete surface: (a) using an aluminum float; (b) using a steel finishing trowel. (Courtesy Portland Cement Association.)

After a concrete slab has been screeded, leveled, and edged in the usual manner, it is broomed with a stiff-bristled broom to ensure that the finish (mortar) coat will bond. A mortar coat is made by mixing one sack of white portland cement and 2 cu ft (0.57 m³) of sand with about $1/2$ lb (0.23 kg) of a color pigment. Yellow is usually selected to tint the mortar coat, but any mineral oxide color may be used. Care must be taken to keep the proportions exactly the same for all batches. Enough water is added to make a soupy mixture with the consistency of thick paint.

Mortar coat material is placed in pails and thrown vigorously on the slab with a stiff bristle brush called a *dash brush* to produce an uneven surface with ridges and depressions. The ridges should be about $1/4$ to $1/2$ in. (6.5 to 13 mm) high.

When the surface is hard enough to permit the use of kneeboards, it is troweled with a steel trowel to flatten the ridges, leaving it smooth in the high spots and rough or coarse-grained in the low ones. Many interesting textures can be produced, depending on the amount of troweling (Figs. 3.9-13a,b). A textured slab can then be scored into random geometrical designs before it cures.

### 3.9.5 FINISHING FORMED SURFACES

The way in which formed surfaces are finished depends on their exposure and the esthetic effect the designer is trying

a

b

**FIGURE 3.9-13** (a) Keystone or travertine finish; (b) random-scored keystone finish. (Courtesy Portland Cement Association.)

to achieve. Several classifications are used in the industry to describe the various finishes.

#### 3.9.5.1 Rough Form Finish

A *rough form finish* is left by the formwork in locations that will be concealed from view in the completed building. The surface is improved only by the repair of tie holes and defective areas; removal of fins left by concrete leaking out through joints in the forms; and chipping off or grinding down of projections larger than $1/4$ in. (6.4 mm) high.

#### 3.9.5.2 Smooth Form Finish

A *smooth form finish* is used in two instances. The first is when form liners are used to impart a pattern or texture to the concrete. The second is for concrete that will receive paint, waterproofing, damp-proofing, veneer plaster, or another thin coating or covering material. After the forms are removed, fins and projections are removed completely, and imperfections are repaired. The surface is made smooth, or of the texture imparted by the forms to adjacent areas, as applicable. Tie holes may be patched or left alone, depending on the designer's intent. Sometimes tie holes are used as a decorative element. *Screw ties* with cones are usually used when tie holes will not be filled (see Section 3.3.2.2).

#### 3.9.5.3 Smooth-Rubbed Finish

A *smooth-rubbed finish* is applied when a smooth form or form liner is used. This finish is applied by moistening the new concrete within 1 day after the forms have been removed and rubbing it with an abrasive, such as a carborundum brick. Cement grout is not used during this finishing process.

#### 3.9.5.4 Grout-Cleaned Finish

A *grout-cleaned finish* is also used on concrete that has been placed against a smooth form or form liner. It consists of applying to the surface a thick paint-consistency grout of 1 part portland cement and $1^1/2$ parts fine sand, mixed with equal parts of an acrylic or styrene butadiene bonding material and water. A blend of standard (gray) and white portland cement is used to achieve the color and tone desired by the designer. After the grout has been applied, the finish is completed by rubbing it with burlap.

### 3.9.6 SURFACE DEFECTS

There are many causes for concrete surface defects. Some of the major defects, their causes, and the construction techniques that should be used to prevent them are described here. Many defects result from improper curing or errors in mixing or finishing.

#### 3.9.6.1 Scaling

*Scaling* is the breaking away of the hardened concrete surface of a slab to a depth of about $1/6$ to $3/16$ in. (1.6 to 4.8 mm). It usually occurs at an early age of the slab.

Scaling of a slab may occur if it is subjected to cycles of freezing and thawing soon after the slab has been placed. A favorable temperature must be maintained long enough to prevent injury (see Section 3.10.3.1). Cycles of freezing and thawing, and applications of de-icing salts on non-air-entrained concrete, can also cause scaling. This is the reason that air-entrained concrete is recommended for all severe exposure conditions.

Performing a finishing operation while free excess water or bleed water is on the surface causes segregation of the surface fines (sand and portland cement) and brings a thin layer of neat portland cement, clay, and silt to the surface, leaving another layer of nearly clean, washed sand that is not bonded to the concrete under it. To prevent scaling from this cause, water should be allowed to evaporate from the surface or be forced to evaporate by fans or blower-type heaters. Alternatively, it should be removed by dragging a rubber garden hose over the surface before finishing operations begin.

#### 3.9.6.2 Crazing

*Crazing* is the occurrence of numerous fine hair cracks in the surface of a newly hardened slab due to surface shrinkage. These cracks form an overall pattern similar in appearance to that of a crushed eggshell.

Crazing can be the result of rapid surface drying usually caused by either high air temperatures, hot sun, or drying winds or a combination of these. It can be lessened by curing with water, because water will maintain or lower the concrete surface temperature (see Section 3.10.3.3).

Premature floating and troweling when there is an excess amount of moisture at the surface, or while the concrete is still too plastic, brings an excess amount of fines and moisture to the surface. Rapid loss of this moisture will cause shrinkage at the surface, which may result in crazing. Floating and troweling should be delayed until excess moisture has evaporated from the surface and the concrete has started its initial set. Excess moisture can be avoided by reducing the slump and using air-entrained concrete.

Overuse of a vibrating screed, darby, or bull float may also contribute to crazing by working an excess of mortar to the surface, which tends to cause additional surface shrinkage.

### 3.9.6.3 Dusting

*Dusting* is the appearance of a powdery material on the surface of a newly hardened concrete slab.

An excess of harmful fines (clay or silt) in a concrete mix with the sand and portland cement at the surface can result in dusting. This condition emphasizes the need to use clean, well-graded coarse and fine aggregates.

Premature troweling and floating mix excess surface water with surface fines, weakening the portland cement paste. Troweling should be delayed until all free water or excess moisture has disappeared and the concrete has started its initial set.

When carbon dioxide, as might be emitted from open portable fossil-fuel heaters and gasoline engines, comes in contact with the surface of plastic concrete, a reaction takes place that impairs proper hydration. Such fumes should be vented to the outside and sufficient fresh air ventilation provided.

Condensation sometimes occurs on a concrete surface before floating and troweling have been completed, usually in the spring and fall, when materials become cold due to low night temperatures. If possible, this condition should be anticipated, and the concrete should be heated or at least hot water should be used for mixing. If this is impossible, blower-type heaters should be used to lower the humidity directly over the slab, and fans should be used to increase air circulation. If heaters and fans are not available, windows and doors should be opened. When condensation is present, floating and initial troweling should be held to a minimum, and the concrete surface should not be given a second troweling.

Neat portland cement and mixtures of portland cement and fine sand should never be used as a dry-shake. Because condensation may occur for several hours while concrete is beginning to harden, emergency measures may have to be taken in order to finish a slab. A well-mixed dry mixture of 1 part portland cement and 1 part well-graded concrete sand may be evenly and lightly distributed over a concrete surface if it is followed at once by troweling. There should be no second troweling, because additional condensation may take place after the first troweling.

Winter-protection heaters may lower the relative humidity around concrete excessively and inhibit proper hydration of the portland cement. Water jackets should be placed on heaters to increase the relative humidity by evaporation, and moist curing methods should be employed. Heaters should be moved periodically so that no area will be subjected to an extreme or harmful amount of heat.

Proper curing for the correct length of time is essential. Concrete that is not cured properly will often be weak, and its surface will be easily worn by foot traffic.

# 3.10 Curing and Protection

Concrete curing is an important construction operation that is often neglected. Even when concrete has been properly mixed, carefully placed, and correctly finished, a poor job will result if proper curing techniques are not followed.

### 3.10.1 PURPOSE OF CURING

The strength and watertightness of concrete improve with age as long as conditions are favorable for continued *hydration* of its portland cement. Other qualities, such as weathering and resistance to freezing and thawing, are affected similarly. Improvement is rapid early and continues more slowly for an indefinite period as long as moisture and favorable temperatures are present.

Fresh concrete contains more than enough water for hydration, but under most job conditions much of this water is lost by evaporation unless certain precautions are taken. Hydration proceeds at a much slower rate as the temperature drops until it stops altogether at about

14°F ($-10°C$). However, at about 40°F (4.4°C), hydration becomes too slow for concrete work to proceed without special provisions to heat the ingredients so that the concrete will be at least 50°F (10°C) when it is placed. Therefore, concrete should be protected so that moisture is not lost during the early stages of hardening, and it should be kept at a temperature that will promote hydration and protect against injury from subsequent construction activities.

### 3.10.2 CURING METHODS

Concrete can be kept moist in a number of ways, including leaving its forms in place, sprinkling, ponding, and using moisture-retention covers or a seal coat that is applied as liquid and then hardens to form a thin membrane.

Curing methods that use water, such as sprinkling, ponding, and using wet coverings, are the most effective means and should be used whenever practicable. Although not usually as effective as curing with wet coverings, moisture-retention membranes are widely used because of their convenience. In any case, the most important consideration is that the surface of the concrete be kept uniformly wet or moist. Partial drying of any part of the surface of a slab can result in crazing and cracking. Curing compounds will affect the bonding properties of adhesives and should not be used on slabs that are to receive resilient flooring or other adhesive-applied finishes unless they are known to be compatible with the adhesive. Curing compounds are also not recommended for slabs that will be given a further surface treatment.

### 3.10.2.1 Leaving Forms in Place

Leaving forms in place is of great assistance in retaining moisture. In hot, dry weather, wood forms will dry out and should be kept moist by sprinkling. In all cases, exposed surfaces must be protected from moisture loss.

### 3.10.2.2 Sprinkling

When concrete is kept moist by sprinkling, the drying of surfaces between water applications must be prevented. Alternate cycles of wetting and drying are conducive to surface crazing or cracking. A fine spray of water applied continuously provides a more constant supply of moisture and is better than copious applications of water with periods of drying between theme.

### 3.10.2.3 Ponding

Ponding is sometimes used on flat surfaces, such as pavements, sidewalks, and floors. A small dam of earth or other water-retaining material is placed around the perimeter of a surface, and the enclosed area is kept flooded with water. Ponding provides a more constant condition than does sprinkling.

### 3.10.2.4 Curing Covers

Moisture-retaining covers such as wet sand, burlap, polyethylene film, or other materials are sometimes used. Care should be taken to cover the entire concrete surface, including exposed sides of members and the sides of pavements and sidewalks where the forms have been removed. Covers should be kept constantly moist enough to provide a film of moisture on the concrete surface.

#### WATERTIGHT COVERS

Watertight paper is used on floors and other horizontal areas. It should be non-staining and strong enough to withstand the wind and abrasive action of workers walking over it. Seams should be overlapped several inches and covered with glued tape. Polyethylene films often are used as watertight curing covers.

#### SEALING COMPOUNDS

Sealing curing compounds are available in black, red, colorless, and white pigmented coatings. Some of them can be applied in one coat, but two coats will give better results. The application should be made immediately after concrete has been finished. If there is a delay, the concrete should be kept moist until the application is made. In extremely hot weather, it is advisable to cover a slab with water for 12 hours before using curing compounds. On formed surfaces, such as beams and columns, the forms should be removed, the concrete sprayed lightly with water, and then a sealing compound applied. Where the membrane must be protected against traffic, it should be covered with at least 1 in. (25 mm) of sand or earth or by some other means. Such a protective cover should be placed not sooner than 24 hours after the sealing compound.

### 3.10.3 CURING TEMPERATURES

Temperature affects the rate of chemical reactions between portland cement and water. Consequently, temperature affects the rate at which concrete hardens, as well as its strength and other properties.

The temperatures and time periods in the following discussions are approximations to introduce the subject. The actual required temperatures and times vary with concrete thickness and air temperature. Refer to American Concrete Institute (ACI) publications ACI 305R and ACI 306 for more specific requirements.

### 3.10.3.1 Cold Weather Construction

In cold weather, it is often necessary to heat a concrete's ingredients, and either cover fresh concrete or provide a heated enclosure, or both. Hydration of portland cement causes some heat to be generated, and if this heat is retained, it raises the temperature of the concrete. In cold weather, concrete should have a temperature at the time of placing of between 50° and 70°F (10° and 21°C). In no case should its ingredients be heated to the point where the temperature of fresh concrete is above 70°F (21°C). If it is, built-in thermal stresses will result.

Therefore, when the temperature of the surrounding air is below 40°F, the water and aggregate used in a concrete should be heated so that the temperature of the mixed concrete at the time of placing is between 50° and 70°F (10° and 21°C). The temperature of the air adjacent to the placed concrete should then be kept at between 50° and 70°F (10° and 21°C) during the curing period. This may require covering the concrete with an insulating material, such as straw, or surrounding the concrete with a heated enclosure, or both, until sufficient time has passed for the concrete to remain undamaged by the prevailing conditions. This time varies with the thickness of the concrete, the air temperature, and other conditions, such as high winds. The normal range is above 70°F (21°C) for 3 days or above 50°F (10°C) for 5 days when normal portland cement (Type 1) is used; or above 70°F (21°C) for 2 days or 50°F (10°C) for 3 days when high-early-strength portland cement (Type III) is used. This range may vary, however, if high winds occur, and will have to be increased if the concrete has not properly cured within the specified times.

Fresh concrete should not be placed on a frozen subgrade. When such a subgrade

thaws, there is likely to be uneven settlement and cracking of the concrete member it supports. The insides of forms, and reinforcing steel and embedded fixtures, should be free of ice at the time concrete is placed. A thin layer of warm concrete should not be placed on cold, hardened concrete, for the thick upper layer will shrink as it cools and the lower layer will expand as it warms. Bond failure will result.

No dependence should be placed on salt or other chemicals to prevent freezing.

### 3.10.3.2 Mild Weather Construction

In relatively mild weather, that is, when the temperature is generally above 40° to 45°F (4.4° to 7.2°C) with only short periods below this range, heating only the mixing water will usually provide the desired temperatures in the concrete. The mixing water should be heated or other steps taken to raise the temperature of the concrete to between 50° and 70°F (10° and 21°C) at the time of placing.

In-place concrete should be kept at a favorable temperature long enough to avoid injury by exposure to the atmospheric temperature. In general, the air surrounding placed concrete made with normal portland cement (Type I) should be kept either at 70°F (21°C) or above for the first 3 days, or above 50°F (10°C) for the first 5 days. When high-early-strength concrete (Type III) is used, air temperatures should be kept at 70°F (21°C) for 2 days or 50°F (10°C) for 3 days. In mild weather, a covering of tarpaulins may be enough protection. A layer of straw covered with tarpaulins will protect against more severe conditions, but an enclosure of tarpaulins or other watertight material with artificial heat may be necessary to provide proper protection in many cases. To prevent rapid drying of concrete, especially near heating elements, it may be necessary to elevate these elements and protect concrete near them with sand kept wet constantly.

### 3.10.3.3 Hot Weather Construction

In extremely hot weather, extra care is required to avoid high temperatures in newly placed concrete and to prevent its rapid drying. Keeping aggregate stockpiles moistened with cool water will help to lower the temperature. In very hot weather, mixing water should be chilled by refrigeration or by using ice as part of the mixing water. The ice should be melted by the time the concrete leaves the mixer.

**FIGURE 3.10-1** Length of Curing

| Type of Portland Cement | Curing Time, days | Concrete Surface Temperature, °F (°C) |
|---|---|---|
| Type I | 5 | 70 (21) or higher |
|  | 7 | 50–70 (10–21) |
| Type II | 3 | 50 (10) |
| Other than Type I or Type II | 7 | 70 (21) |
|  | 14 | 50–70 (10–21) |

A subgrade on which concrete will be placed should be saturated some time in advance, and then sprinkled just before the placing operation, so that it will not absorb water from the mix. Wood forms should be wetted thoroughly if they have not been treated otherwise. There should be no delay in placing concrete, and it should be struck off and leveled at once. Temperature covers, such as burlap kept constantly wet, should be placed over fresh concrete immediately after it has been leveled. During final finishing, only a small section immediately ahead of the finishers should be uncovered. The finished surface should then be re-covered, and the cover should be kept wet. A delay in finishing air-entrained concrete in hot weather usually leads to the formation of a surface that is difficult to finish without leaving ridges.

Fresh concrete should be shaded as soon as possible after it has been finished, and moist curing should be started as quickly as it can be without marring the surface. Concrete surfaces should be kept constantly wet during the curing period.

### 3.10.4 LENGTH OF CURING

Sufficient curing time should always be allowed for concrete to develop adequate strength before it is loaded. Curing should be started as soon as it is possible to apply a curing medium without damaging the surface and should continue as long as is practicable. Some curing times and surface temperatures for normal mild weather conditions are summarized in Figure 3.10-1. However, during hot weather, curing should continue for at least 7 days, regardless of the type of concrete or the temperature. In mild weather, the curing time can sometimes be reduced to 2 days when high-early-strength portland cement (Type III) is used. Even this time may have to be increased, though, in very hot weather.

In any case, concrete temperature should not be allowed to fall below 50°F (10°C) at any time during curing. Rapid cooling may induce thermal cracking, especially during the first 12 to 24 hours.

# 3.11 Concrete Foundation Systems

A building's foundations form its *substructure* and support its *superstructure*, transmitting loads to the earth. They also resist *lateral* (horizontal) loads from the ground and the superstructure and provide anchorage for the superstructure against uplift (Fig. 3.11-1). A foundation system is composed of load-carrying elements which may include footings, caissons, piles, grade beams, walls, piers, columns, pilasters (Fig. 3.11-2), and the turned-down portion of a combination footing and floor system. Foundation systems are constructed of concrete, masonry, steel, wood, and combinations of these. Some common variations are shown in Figure 3.11-3.

Foundations take three basic forms: basements, crawl spaces, and slabs on grade. Figures 3.11-3a and 3.11-3b show a basement and a crawl space with spread footings, concrete or masonry walls, and steel joist floor framing. The floor may also be framed in wood, cast-in-place concrete, or precast concrete. Figures 3.11-3c and 3.11-3d show a similar basement or crawl space supported on piles and caissons (the floor framing has been omitted for clarity). Basements and crawl spaces provide underfloor space for easier installation of mechanical, electrical, and plumbing systems.

Slabs on grade may turn down at the edges to form foundations (Fig. 3.11-3e) or rest on foundation walls and footings (Fig. 3.11-3f). Slabs may also be supported on footings or grade beams resting on piles (Fig. 3.11-3g). Floating foundations (see Section 3.11.2.1) are also combination slab and foundation systems.

This section discusses concrete foundation systems. Other relevant chapters are:

- Chapter 4—masonry foundation walls, columns, piers, and pilasters

**FIGURE 3.11-1** Foundation walls must (1) resist vertical loads from the weight of the building and its occupants; (2) resist lateral loads from the soil, water, and other superimposed loads such as earth-moving equipment and earthquakes; and (3) anchor the building to its foundation.

**FIGURE 3.11-2** The substructure (foundations) of a building.

### 3.11.1.4 Other Loads

Some loads that affect foundations are difficult to categorize as purely live, dead, or wind loads. Sometimes they incorporate all three. They are usually calculated and accounted for in the design separately. Among these are lateral pressures exerted on foundation walls by the adjacent soil and water in that soil; multidirectional loads imparted by seismic disturbances (earthquakes); thrust loads transmitted into the foundations from domes, rigid frames, vaults, and arches; uplift caused by expansion due to freezing and subsequent *frost heave*; and uplift resulting from the tendency of a building with its foundations below the water table to float (*buoyancy*).

### 3.11.1.5 Settlement

Refer to Section 21.1.4 for a more detailed discussion of the causes and effects of settlement in different types of soil.

A building's substructure must distribute building loads so that settlement will be either negligible or uniform under all parts of the building. *Differential* (uneven) settlement under different parts of a foundation may cause substantial problems. Uneven settlement can occur when a foundation is out of plumb or not level, or when loads are not balanced on the substructure, and can cause such damage as cracks in foundation walls and in finished walls and ceilings, sloping floors, open joints in woodwork, and functional annoyances, such as binding doors and windows. When differential settlement is extreme, partial failure of the structural integrity of the building can result.

In time, all foundations built on soil will settle to some extent. The extent of settling depends on the compressibility of the underlying soil and the loads placed on it. Differential settlement can be controlled by designing footings so that they will transmit an equal load per unit of area to the soil.

The bearing capacity and compressibility of soils can be calculated by site and laboratory tests on actual soil samples (see Section 2.3). With a knowledge of bearing capacity and the magnitude of superimposed loads, designers can use engineering analysis to size foundations and predict settlement. Although engineering analysis is used for most commercial, institutional, industrial, and multifamily housing structures, this

- Chapter 6—wood foundations, additional precautions necessary to prevent termite damage, and foundation anchorage
- Chapter 7—waterproofing, dampproofing, and foundation wall thermal insulation
- Chapter 21—soil materials and earthwork needed to support foundations, special foundation and load-bearing elements (piles and cassions), surface water and groundwater control, thermal control, and termite protection

Spread foundations, such as footings and mat and raft foundations, resting on the soil receive loads from foundation walls, piers, pilasters, and columns and must be able to distribute them directly to the soil without exceeding its allowable bearing capacity. Piles and caissons receive loads from foundation walls, isolated columns, or grade beams and transfer them to the soil or into bedrock.

Basements are habitable spaces that lie mostly below the adjacent finished grade. Foundations form the substructure for buildings whether or not there is a basement. Buildings without basements include those with a crawl space or a slab on grade. Some such buildings have limited basement space, but many do not.

Many of the recommendations in this section are based on ANSI's A56.1, "Building Code Requirements for Excavation and Foundations," and the Underground Space Center's *Building Foundation Design Handbook*.

### 3.11.1  DESIGN

A foundation system should be economical and should carry all expected loads with a reasonable factor of safety. The design of foundation elements is based on assumptions about the kinds of loads the foundation must carry.

Loads associated with foundation design are either *vertical* or *lateral* (horizontal); *static* (nonmoving) or *dynamic* (moving); and *concentrated* at a point or *uniform* along a surface. These loads may be further divided into dead loads, live loads, wind loads, and other loads.

### 3.11.1.1  Dead Loads

Dead loads act constantly. They include the weight of the superstructure and of the foundation itself. Dead loads are calculated by summing the total weight of the materials used in the construction. Dead loads are usually assumed to be uniformly distributed over the length and thickness of a foundation wall.

### 3.11.1.2  Live Loads

Live loads are variable, periodic, and sometimes moving. They include, but are not limited to, the weight of snow or water on the roof, furniture, equipment, and people. The type and magnitude of live loads acting on a structure vary according to locality, occupancy, and other factors. Most building codes specify minimum live loads that may be used in sizing foundation walls, footings, and other elements of the substructure.

### 3.11.1.3  Wind Loads

Wind loads result from the intermittent effect of wind on the superstructure. They may act downward, laterally at any angle, or upward.

procedure is seldom followed in designing foundations for single-family dwellings or small buildings of other types. Instead, their footings are often sized using rules of thumb based on established practices and on the performance of other foundations in their locality. Even in these types of buildings, however, it is necessary to proportion footing sizes for equal settlements and to balance applied concentrated loads, such as those that will occur at a chimney or column.

### 3.11.2 FOUNDATION TYPES

Foundations are of two general types: (1) *spread foundations*, which distribute a building's loads directly to a sufficient area of soil to obtain adequate bearing capacity, and (2) *pile* and *caisson* foundations, which transmit a building's loads through soils that have inadequate bearing capacity for spread footings to deeper layers of soil or rock that have adequate bearing value.

Most foundations in single-family dwellings and other one- and two-story buildings are of the spread type. Many buildings of three or more floors are also supported on spread foundations. Piles and caissons are used when the soil-bearing capacity is insufficient to support the loads imposed on the soil by a building.

#### 3.11.2.1 Spread Foundations

Spread foundations support transmitting elements, such as walls, pilasters, columns, piers, or grade beams, on an abruptly enlarged base called a *footing*, *mat foundation*, or *raft foundation*. These, in turn, spread the load directly to the supporting soil.

#### FOOTINGS

Except on bedrock, or in the extremely rare instances when the bearing capacity of the soil is inadequate to support a foundation wall or pier alone, a footing is required to spread the loads on the supporting soil. Foundation walls built of masonry units bonded with mortar should have footings, regardless of the soil's bearing capacity. Refer to Section 21.1.3 for a discussion of the bearing capacities of various soil types.

Footings are classified according to the way in which they are loaded, shaped, and constructed. Several types are shown

**FIGURE 3.11-3** Basic foundation forms: (a) basement and (b) crawl space, both with spread footings, concrete or masonry walls, and steel-joist floor framing; (c) basement and (d) crawl space, both supported on piles and caissons; (e) slab on grade that turns down at the edges to form a combination slab-and-foundation system; (f) slab on grade that rests on foundation walls and footings; and (g) slab supported on foundation walls and footings that are in turn supported on piles. (Drawing by HLS.)

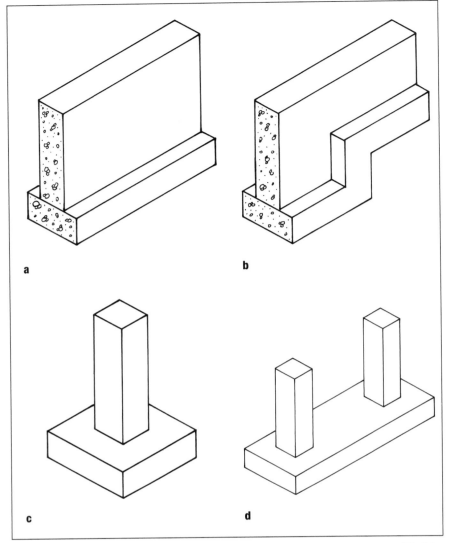

**FIGURE 3.11-4** Various types of footings spread building loads over the soil area. All of the types shown can be reinforced. (Drawing by HLS.)

in Figure 3.11-4. Unreinforced, or plain, footings carry light loads. Reinforced footings contain embedded steel to strengthen them in tension and in shear. Any footing type can be either plain or reinforced.

***Footing Material*** Footings have been made from many materials, including wood, stone, brick, and concrete masonry units. Most footings today, however, are monolithic cast-in-place concrete. Concrete for unreinforced footings in foundations should have a minimum compressive strength of 2000 psi (14 MPa) at 28 days. Higher strengths may be required when determined by engineering analysis. Sufficient time should be allowed for the strength of the concrete to develop before subjecting a footing to loads. In cold weather, concrete should be placed as soon after excavation as possible and must be protected from freezing until it cures (see Section 3.10.3).

When a large excavation has been carried deeper than desired, the depth should be refilled with concrete rather than with earth fill material. If it is necessary to excavate beneath a footing for a sewer, water line, or other small opening, the excavation may be backfilled with compacted stone or gravel or with concrete.

***Continuous Footings*** Continuous footings support foundation walls of either concrete or masonry (Fig. 3.11-4a). They also may be used to support a row of several metal or wood columns or masonry piers to minimize differential settlement. When a single footing supports more than one column or masonry pier, it is called a *combined footing* (Fig. 3.11-4d). Combined footings can also support combined piers for multiple columns (Figure 3.11-5).

***Stepped Footings*** Stepped footings change levels in stages to accommodate a sloping grade (Fig. 3.11-4b). Typically, the number of steps necessary to traverse a slope is minimized by making them large both vertically and horizontally. A few large steps are generally less expensive to form than are many small ones.

The vertical part of the step in a stepped footing should be of the same thickness and width as the horizontal footing and should be placed at the same time. If more than one step is necessary, the horizontal distance between steps should not be less than 2 ft (610 mm), and the vertical step should not exceed three-quarters of the horizontal distance between steps (see Fig. 3.11-4b).

***Isolated Footings*** Isolated footings are independent footings that receive the loads of free-standing columns or piers (Fig. 3.11-4c). Masonry piers are bonded to a footing with mortar. Steel and precast concrete columns are usually anchored to a footing with embedded bolts. Wood columns are frequently held in position on a footing by protruding steel pins embedded in the concrete.

***Footing Size*** Footings should be sized for existing soil conditions and the magnitude of dead and live loads. The usual rule-of-thumb size for normal residential footings resting on soil of average bearing capacity (2000 psf [9 765 kg/m²]) is that the width of the footing should equal twice the thickness of the wall, and the

**FIGURE 3.11-5** A combined pier for support of more than one column. (Courtesy BE&K, Inc.)

**FIGURE 3.11-6** Dimensions for Continuous Wall Footings

| Number of Floors | | Frame Construction | | Masonry or Veneer Construction | |
|---|---|---|---|---|---|
| | | $t''$ | $p''$ | $t''$ | $p''$ |
| **1** | Basement | 6" (150 mm) | 2" (50 mm) | 6" (150 mm) | 3" (75 mm) |
| | No Basement | 6" (150 mm) | 3" (75 mm) | 6" (150 mm) | 4" (100 mm) |
| **2** | Basement | 6" (150 mm) | 4" (100 mm) | 8" (200 mm) | 5" (125 mm) |
| | No Basement | 6" (150 mm) | 3" (75 mm) | 6" (150 mm) | 4" (100 mm) |

Based on average soil bearing value = 2000 P.S.F. (9.76 kg/m²).

**p** = Footing projection
**t** = Footing thickness

depth of the footing should equal the wall thickness. In other words, given a wall thickness *t*, the footing width will be 2*t* and its depth will be *t*. In the absence of specific building code requirements or engineering analysis, the footing sizes given in Figure 3.11-6 have been used successfully. However, their applicability to a particular project should be verified in every instance. In no case should the sizes be smaller than those shown in Figure 3.11-6. These sizes are based on a conventionally loaded wall footing and on soil of average bearing value (approximately 2000 psf [9 765 kg/m²] or greater). Footings should not be less than 6 in. (150 mm) thick and, unless transversely reinforced, their thickness should not be less than one and a half times the footing projection.

Footings for piers, columns, and masonry chimneys should be at least 8 in. (200 mm) thick and, unless reinforced, not less than one and a half times the footing projection. In addition, the minimum size for interior pier or column footings should be 24 × 24 in. (610 × 610 mm).

***Footing Location*** In most cases, footings should rest on undisturbed earth. Footings for small buildings are occasionally supported on *engineered fill* (see Section 21.3.4) when other types of foundations are inappropriate or would be more expensive than placing spread footings on engineered fill. In these cases, the footings must be designed by engineering analysis for these conditions. In addition, more than usual care must be taken in compacting and testing the en-

gineered fill to ensure that the necessary compaction has been achieved.

Footings should not be placed on frozen ground. As frozen soil thaws, excess water can transform the soil into liquid mud that is unusable to support loads. Footings should be located far enough below ground level to protect them from possible frost heaving caused by the freezing of soil moisture (see Section 21.1.9).

In areas where soils expand due to moisture changes, footings should be deep enough to be unaffected by seasonal moisture changes or should be specifically engineered for these conditions.

***Reinforced Footings*** In reinforced footings, steel bars are placed transversely (perpendicular to the wall) when loads must be spread over wider areas. When its projection exceeds two-thirds of its thickness, a footing should be reinforced transversely with steel as determined by engineering analysis. Footings are reinforced longitudinally when they must bridge weak spots such as excavations for service or sewer connections, and when the soil has low bearing capacity or varies in compressibility.

## MAT AND RAFT FOUNDATIONS

Mat and raft foundations (Fig. 3.11-7) are used over soils of low bearing capacity when other foundations would be inadequate. They are made of concrete and heavily reinforced with steel so that the entire foundation will act as a unit.

A *mat* foundation is a thickened slab that supports loads and transmits them as a single structural unit over the entire slab

and soil surface area. Mat foundations are used when the foundations for a building become so large, usually due to poor soil, that it is more economical to join them into a single unit than to cast them each separately.

A mat foundation becomes a *raft* or *floating* foundation when it is combined with basement walls and placed beneath a building in such a way that the weight of the soil removed to make the basement equals the weight of the building's superstructure and substructure combined. In such a structure, settlement is minimum even on low-strength soils because the building load on the soil is no more than the weight of the excavated soil.

### 3.11.2.2 Piles and Caissons

Piles and caissons (Section 21.5) are used as foundation support for buildings when the soil is not capable of supporting the

**FIGURE 3.11-7** Mat and raft foundations. (Drawing by HLS.)

loads that will be imposed by a building using spread foundations.

# 3.12 Concrete Slabs on Grade

The suitability of a particular slab-on-grade construction is generally influenced by the geographic location of the building, the building site, and the construction of the superstructure. Specific influences on the type and design of a slab include site preparation, topography, soil properties, climate, the type and support of the superstructure, and comfort factors such as thermal and moisture control. Producing satisfactory slabs requires properly designing the slabs, correctly selecting the concrete materials and mixtures, and carefully placing, finishing, and curing the concrete.

This section includes recommendations for selecting and constructing concrete slabs on grade. If care is exercised to control concrete quality and if the necessary precautions dictated by the characteristics of the site, superstructure, and geographic location are taken, slabs on grade are suitable for floors and combination foundation and floor systems in all parts of the United States. Unless specifically modified in this section, the provisions of Section 3.2 are applicable to concrete slabs on grade.

These recommendations are based on:

- Portland Cement Association (PCA): "Concrete Floors on Ground"
- American Concrete Institute International (ACI): ACI 302.1R, *Guide for Concrete Floor and Slab Construction*
- National Academy of Science: *Criteria for Selection and Design for Residential Slabs-on-Ground*
- Underground Space Center: *Building Foundation Design Handbook*

## 3.12.1 DESIGN

Slabs on grade may be either independent floors or combination foundation and floor systems. In independent floors, the slab is separate from the building's load-carrying elements, such as walls, columns, and footings, which transmit their loads directly to the supporting soil (Fig. 3.12-1a). In combination founda-

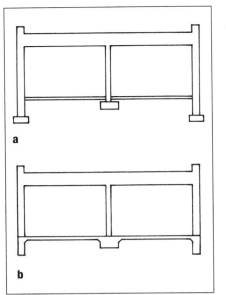

**FIGURE 3.12-1** (a) Ground-supported floor slab built independently of foundation walls; (b) ground-supported slab built integrally with the foundation grade beam. (Drawing by HLS.)

tion and floor systems, the slab is an integral part of the foundation, and carries superstructure loads and transmits them to the soil (Fig. 3.12-1b). Concrete foundation elements such as walls and footings which are common to many types of foundation systems are discussed in Section 3.11. Piles and caissons are addressed in Section 21.5. Masonry walls, columns, piers, and pilasters are covered in Chapter 4.

Slab-on-grade construction requirements are determined largely by site characteristics, such as drainage and soil type, and by the methods used to build the slab. Slabs may be ground-supported or supported structurally. They may be unreinforced (*plain*), contain welded wire fabric reinforcement, or be structurally reinforced using welded wire fabric and steel reinforcing bars. Four types of slabs are described in the following section.

Slab construction depends on the following factors:

- Slab type
- Site preparation, including preparation of the slab bed, grading, and backfilling
- Precautions to control ground and surface moisture
- Thermal control when required

The performance of slabs is dependent on the quality of concrete materials and workmanship.

## 3.12.2 SLAB TYPES

Concrete slabs for floors and combination foundation-and-floor systems can be classed according to construction as either:

- Ground-supported (slabs that rest directly on a slab bed consisting of undisturbed soil, compacted fill, or a base course) or
- Structurally supported (slabs supported independently of the ground by walls, piers, grade beams, or piles).

The appropriateness of a concrete slab floor or combination foundation-and-floor system should be based on soil properties, climate, functional and esthetic requirements, and economy.

### 3.12.2.1 Ground-Supported Slabs

Ground-supported slabs can be designated as Types I, II, or III.

#### TYPE I SLABS

A Type I slab is mostly unreinforced and is separated from the elements that support the superstructure (Fig. 3.12-2) so that it carries no superstructure loads. Superstructure loads are supported on foundations independently of the slab. A Type I slab has a nominal thickness of at least 4 in. (100 mm) and is cast directly on a dense or compacted slab bed. It is unreinforced but may contain welded wire fabric or bars in localized areas. Type I slabs should not be placed on a slab bed that can develop significant changes in volume over time. Differential movement in the slab bed may produce enough tensile stress in the concrete to crack the slab. Random cracking due to drying shrinkage is controlled by:

- Isolation joints
- Construction joints
- Limiting the area between control joints placed to induce cracking at preselected locations
- Fiber reinforcing

The controlling factors in the suitable performance of a Type I slab are:

- Selection of the correct slab bed
- Concrete quality
- Slab location
- Control joint spacing

**FIGURE 3.12-2 Type I slab.** (Drawing by HLS.)

- Grade
- Foundation wall
- 4-in. (100 mm) unreinforced cast-in-place concrete slab
- Porous fill
- Perimeter Insulation
- Footing

must be capable of supporting a Type I slab and be uniformly and adequately compacted to ensure that warping and tensile stresses that could contribute to cracking are not induced in the slab.

Type I slabs can be applied successfully to:

- Gravely soils of any density (Types GW, GP)
- Sandy soils (with or without silts) and clays (Types GM, GC, SW, SP, SM, SC)
- Silts (Types ML, MH), provided that they are dense or medium dense
- Loose sandy and silty soils if the soil can be compacted to a dense or medium-dense state to its entire depth before slab construction (see Fig. 3.12-3)

Refer to Section 21.1.1 for a discussion of soil types.

Type I slabs should be rectangular or square in shape (Fig. 3.12-4a). To control shrinkage cracking during drying and subsequent thermal volume changes, no dimension of a slab perimeter should exceed 32 ft (9.75 m). However, if a slab is to be left exposed, control joints spaced 15 to 20 ft (4.57 to 6.1 m) apart are recommended (Fig. 3.12-4b).

Irregularities in shape, such as when a nonrectangular T or L shape is required, need control joints to divide the slab into squares or rectangles (see Fig. 3.12-4b). If control joints are not provided, diagonal cracks are likely to occur where the slab shape changes.

Irregularities in thickness that reduce a slab to less than 4 in. (100 mm) in nom-

Because a Type I slab is unreinforced, except for specific localized areas, it possesses only compressive strength and cannot tolerate appreciable amounts of tension or warping. Such a slab may crack from shrinkage during drying, but when it is used under the conditions recommended in the following paragraphs, cracks that do occur should not open excessively or prove detrimental to the serviceability of the slab.

Type I slabs should be placed only on well-drained and properly graded coarse-grained soils (Fig. 3.12-3). Such soils are not affected significantly by climate or moisture content changes and should develop no appreciable volume change that would be detrimental to the slab. The soil

**FIGURE 3.12-3 Recommended Slab Bed for Slabs on Grade**

| Slab Type | Soil Type | Density | Plasticity[a] | Consistency, $q_u/w$[b] | Climatic Rating |
|---|---|---|---|---|---|
| TYPE I | GW, GP | All | — | — | All |
| | GM, GC, SW, SP, SM, SC, ML, MH | Dense or medium dense[c] | | | |
| TYPE II | GW, GP, GM, GC, SW, SP, SM, SC, ML, MH | All | — | — | All |
| | CL, CH, | — | PI < 15 | $q_u/w \geq 7.5$ | All |
| | OL, OH | — | PI > 15 | $q_u/w \geq 7.5$ | $C_w \geq 45$ |
| TYPE III | CL, CH, | — | PI > 15 | $q_u/w \geq 7.5$ | $C_w < 45$ |
| | OL, OH | — | — | $7.5 > q_u/w > 1.5$ | All |
| TYPE IV | CL, CH, OL, OH | — | — | $q_u/w < 1.5$ | All |
| | PT | — | — | All | |

[a]PI = Plasticity Index, determined in accordance with ASTM D4318.

[b]$q_u/w$ = unconfined compressive strength/ton/sq ft of undisturbed sample determined in accordance with ASTM D2166.

[c]May be loose if properly compacted to its entire depth.

32' = 9.75 m
75' = 22.86 m

**FIGURE 3.12-4** Maximum dimensions for Types I and II slabs.

inal thickness should not be permitted. Stresses caused by drying shrinkage or thermal change tend to cause cracking at these locations. When the upper surface of a slab must be lowered, such as to permit installation of ceramic tile or another floor finish, the uniform slab thickness should be maintained by lowering the underside of the slab beginning at least 24 in. (610 mm) away from the slab depression (Fig. 3.12-5a). In addition, when the vertical displacement is greater than $1^1/_2$ in. (38 mm), WWF 6 × 6-W2.9 × W2.9 welded wire fabric should be placed in the slab extending 25 in. (635 mm) on each side of the point of displacement (Fig. 3.12-5b). Heating coils or pipes should not be embedded in Type I slabs, because thermal stresses will be induced that the slab is unable to resist without cracking. Heating ducts may be embedded in Type I slabs if WWF 6 × 6-W2.9 × W2.9 welded wire fabric is placed to extend 19 in. (482 mm) on each side of the centerline of the duct or the slab edge, whichever is closer. Refer to Chapter 17 for additional discussion about embedding heating elements in Type I slabs, including additional reinforcement needed.

Superstructure loads should be supported directly on foundations independently of the slab, but Type I slabs can accommodate partition loads of up to 500 lb/lin ft (744 kg/m). Because a Type I slab is unreinforced, it cannot carry greater loads without the possibility of cracking. Where this loading must be exceeded, reinforcement or special construction should be provided to minimize the cracking which would result. Methods of supporting partitions and concentrated loads are illustrated in Figure 3.12-6.

Openings in Type I slabs should be kept to a minimum, because they can introduce stresses that are not uniform. When an opening 12 in. (305 mm) or wider must be provided for a column or other penetration, a slab should be reinforced with 25-in. (635-mm)-wide WWF 6 × 6-W1.4 × W1.4 welded wire fabric around the opening (Fig. 3.12-7).

## TYPE II SLABS

A Type II slab is also separated from the elements that support the superstructure (Fig. 3.12-8) but contains welded wire fabric or reinforcing bars in localized areas. The controlling factors in the suitable performance of a Type II slab include:

- Correct slab bed preparation
- Concrete quality
- Correct size and placement of the welded wire fabric reinforcement

Type II slabs are cast on a dense or compacted bed and contain welded wire fabric to hold drying shrinkage and thermal movement cracks together once they form. This reinforcement does not prevent cracking or add to the load-carrying capacity of a slab. Grading and loading are similar to those for Type I slabs. As with Type I slabs, the design of the slab is not influenced by the type of superstructure that is supported directly on independent foundations.

Because they contain welded wire fabric, Type II slabs can be cast onto some soils that are unsuitable for Type I slabs. However, Type II slabs cannot be expected to resist differential movement of the soil, because the amount of reinforcement provided is based on the premise that the entire slab will remain uniformly supported by the underlying soil.

Type II slabs can be constructed on all soils suitable for Type I slabs under all climatic conditions, and these soils can be of any density, including loose (see Fig. 3.12-3). In addition, Type II slabs can be applied to certain fine-grained soils (CL, OL, CH, OH) (see Section 21.1.1) under certain climatic conditions if the unconfined compressive strength

**FIGURE 3.12-5** (a) Method of maintaining minimum slab thickness; (b) location of welded wire fabric reinforcement. (Drawing by HLS.)

**FIGURE 3.12-6** (a) In Types I and II slabs, a heavy partition or concentrated load should be carried on a separate footing. (b,c) Alternative methods that may be used when the slab is designed using engineering methods and structural reinforcement. (Drawing by HLS.)

*In addition to normal slab reinforcement.

12″ = 305 mm
25″ = 635 mm

**FIGURE 3.12-7** Location of reinforcing around a hole in a Type I or II slab.

and plasticity index fall within the limits shown in Figure 3.12-3.

Type II slabs must be at least 4 in. (100 mm) in uniform thickness, and no dimension of the slab perimeter should exceed 75 ft (22.86 m) (see Fig. 3.12-4).

The same recommendations apply as for Type I slabs with respect to irregularities in thickness (see Fig. 3.12-5) and shape (see Fig. 3.12-4), with the added requirement that reinforcement be continuous across control joints or that keyed joints be provided to eliminate the possibility of vertical displacement between adjacent sections.

**FIGURE 3.12-8** Type II slab. (Drawing by HLS.)

Welded wire fabric reinforcement in Type II slabs provides control of crack size only. Cracks that are expected to occur should be held tightly closed by the tensile reinforcement and should not be objectionable. To control crack size, the slab should be provided with the minimum welded wire fabric reinforcement recommended in Figure 3.12-9. Reinforcement should be placed in the center of the slab, midway between top and bottom, and should be supported while the

**FIGURE 3.12-9** Minimum Fabric Reinforcement for Type II Slabs

| Maximum Slab Dimension, ft (m) | Wire Spacing, in. (mm) | Wire Gauge, number |
|---|---|---|
| Up to 45 (13.72) | 6 × 6 (150 × 150) | W10 × W10 |
| 45–60 (13.72–18.29) | 6 × 6 (150 × 150) | W8 × W8 |
| 60–75 (18.29–22.86) | 6 × 6 (150 × 150) | W6 × W6 |

**FIGURE 3.12-10 Type III slab.** (Drawing by HLS.)

concrete is being placed, preferably on metal chairs designed for the purpose. Edges should be lapped at least one mesh space. The common practice of laying the mesh on the bed and pulling it up using hooks (*hooking*) after the concrete has been placed is not recommended, because accurate, level placement of the mesh within the slab using this method is almost impossible. The practice of laying the mesh on top of a poured slab and embedding it by having workers walk on it is also discouraged for the same reasons.

Because the reinforcement in Type II slabs will hold together cracks that form due to temperature changes, heating pipes or coils and heating ducts can be embedded in them. However, the construction recommendations given in Chapter 17 should be followed.

Type II slabs are expected to deflect with slight soil movement; therefore, superstructure loads should be supported independently on foundations. However, like Type I slabs, Type II slabs can accommodate partition loads of up to 500 lb/lin ft (744 kg/m) and equivalent concentrated loads. An additional layer of reinforcement can be provided to extend at least 25 in. (635 mm) on each side of the partition or concentrated load to distribute added stresses (see Fig. 3.12-6). Heavy concentrated loads, such as masonry walls, should be supported on independent or structurally reinforced footings (see Figs. 3.12-6a and 3.12-6c).

The recommendations for Type I slabs relative to openings apply also to Type II slabs, except that the additional layer of reinforcing placed around the opening should be 6 × 6-W2.9 × W2.9 welded wire fabric. The additional reinforcing around the opening will prevent the concentration of stresses that are likely to crack the slab at these points (see Fig. 3.12-7).

## TYPE III SLABS

Type III slabs are combination foundation-and-floor systems (Fig. 3.12-10) that are structurally reinforced to make them usable over problem soils that undergo substantial volume changes with time and climate. The use of foundations with spread footings is not advisable on such soils. Type III slabs receive superstructure loads and transmit them to the foundation soil under the entire slab area. Figure 3.12-11 shows an example of a Type III slab.

Depending on its sensitivity to differential settlement, the superstructure imposes limits on the maximum tolerable deflection in a Type III slab. Because the superstructure is supported on the slab itself, walls and other load-bearing superstructure elements will deform when the slab deforms. Different materials tolerate deformations of differing intensities before developing undesirable defects such as cracking and distortions. For example, wood frame construction generally can sustain more deformation than masonry construction before warping and cracking create mechanical and esthetic problems. Comparatively brittle cementitious materials such as concrete, stucco, plaster, and mortar can tolerate only small deformations before cracking develops.

The objective of a Type III slab is to cause the foundation-and-floor system and the superstructure to act as one monolithic unit that restricts differential movements in both slab and superstructure. To ensure that the slab and the superstructure actually act in this manner, the slab must be designed with the necessary rigidity and strength. Therefore, in addition to concrete quality control, Type III slabs should be designed by a professional engineer in accordance with good engineering practice to provide the proper dimensions for stiffness and the correct amount and size of steel reinforcement for strength. The National Academy of Sciences's *Criteria for Selection and Design for Residential Slabs-on-Ground* is applicable.

Because a Type III slab is structurally reinforced, it can be used over certain problem soils such as highly compressible,

5'-0" = 1.5 m, 12'-0" = 3.6 m, 15'-0" = 4.5 m, 30'-0" = 9.1 m, 53'-0" = 16.2 m
1¼" = 29 mm, 3" = 76 mm, 4" = 100 mm, 5" = 125 mm, 6" = 150 mm, 10" = 250 mm, 12" = 300 mm, 18" = 450 mm

**FIGURE 3.12-11** An example of Type III slab construction. This design should not be used until it has been reviewed by a professional engineer.

highly expansive, and highly plastic soils. The unconfined compressive strength, plasticity index, and climatic ratings for soils on which Type III slabs should be used are given in Figure 3.12-3.

Highly compressible soils include those that are high in organic content (PT), such as peat, which are so compressible that they will not support ground-supported slabs.

Highly compressible clays are soils with sufficient moisture content to have a soft or very soft consistency or high organic content. Excessive settlement may result even under relatively light residential loading or from the weight of fill material.

Loose fine sands and silts can cause serious problems beneath a slab. Vibration of loose sands and silts may cause a decrease in the volume of soil voids with resultant settlement, depending on the amplitude and frequency of the vibration. When it is not practical to compact very loose sand and silt to a suitably dense

state for ground-supported slabs, a Type IV slab should be used.

Highly expansive and plastic soils are problem soils because they are greatly affected by climatic conditions. Any soil with a plasticity index greater than 15 may cause problems by expanding under exposure to severe climatic conditions. However, under favorable or intermediate conditions this material may be considered for Type II slabs.

A soil with a plasticity index of 30 or higher will react vigorously to changes in moisture content and will present the most difficult design problems because of the great volume change that can be expected. Loads associated with small buildings are not usually sufficient to control such expansion by loading. In any case, when such soils are encountered, professional advice should be sought.

Loose soils can become problem soils when they are heavily loaded and become saturated. When slab construction is desirable over these soils and the slab

must be heavily loaded but problems of groundwater or surface water exist, a Type III slab should be selected.

### 3.12.2.2 Structurally Supported Slabs

Type IV slabs are structurally reinforced and can be used over very poor soils such as highly expansive clay, which are extremely sensitive to moisture; over soils of negligible bearing capacity or that are high in organic material content; and where Type I, II, and III slabs would not be suitable. Type IV slabs require professional design based on engineering analysis.

Like Type III slabs, Type IV slabs receive and transmit superstructure loads to the soil, but unlike Type III slabs, they do not depend on soil for support. The slab bed functions as a form and is used only for temporary support of the slab until it develops sufficient strength to be supported on beams, piers, or intermedi-

**FIGURE 3.12-12 Type IV slab.** (Drawing by HLS.)

ate foundation walls. These foundation elements, in turn, are carried by piles or footings that distribute the loads to stable ground below the level of the slab (Fig. 3.12-12).

Type IV slabs can be used over very poor soils (see Fig. 3.12-3) which cannot be treated economically or practically to receive ground-supported slabs. These soils might include expansive clays,

**FIGURE 3.12-13 Construction details for a variation of a Type IV slab.**

highly organic soils, and very loose sands and silts. Foundation elements such as grade beams should be designed so that swelling of the soils beneath the slab will not lift them off the foundations or pull the foundations out of the ground.

A method used in some areas to prevent soil swelling from lifting grade beams is to cast the beam between its supports (piles, caissons, grade beams, etc.) on rectangular cardboard tube forms. These forms are folded from a flat sheet with a diagonal across the center. They are about 6 in. (150 mm) thick. When the earth swells, it crushes the cardboard instead of lifting the beam. Variations of Type IV slabs are sometimes used over unstable soils which may not necessarily be considered extreme problem soils (Fig. 3.12-13). Foundation piles, piers, or footings should be designed to penetrate to a depth providing adequate support below unstable soil.

### 3.12.3 MOISTURE CONTROL

Refer to Section 21.4 for a discussion of the control of groundwater and surface water, including provisions for slab beds and vapor retarders directly under slabs on grade. Subdrainage and foundation drainage systems are discussed in Chapter 23.

### 3.12.4 THERMAL CONTROL

For the most comfortable conditions, and to conserve energy, slab perimeters should be insulated in all except the warmest climatic regions. Whenever walls are insulated, slabs should be insulated. Slab insulation is discussed in Section 7.3.

Slabs are sometimes used to deliver heat throughout buildings. Slabs heated by embedded pipes or containing warm air ducts, and slabs in buildings for which summer cooling is to be provided, are discussed in Chapter 17.

### 3.12.5 SLAB FLATNESS AND LEVELNESS

*Flatness* is a measure of the degree to which the surface of a slab deviates from a plane. *Levelness* is a measure of the degree to which a slab deviates from horizontal. A slab can be flat (having no depressions or rises in its surface) and still not be level. For example, a slab that is flat within the allowable tolerances may have one corner that is higher than the others or one side that is lower than the opposite side. Conversely, a level slab may have depressions, rises, or pockets in its surface.

Until recently, slab surface tolerances were specified as a dimension between $^1/_8$ in. (3.2 mm) and $^1/_2$ in. (305 mm) in every 10 ft (3050 mm), depending on the activities that would take place where the slab was located and the finishes that would be applied. Historically, these deviations were measured using a 10-ft (305-mm)-long straight board, with its long edge vertical, placed across the slab in each primary direction and diagonally. The slab was to be within the specified tolerance dimension of the board along its entire length. This method was inaccurate, especially in measuring levelness.

A more accurate method is called the *F-curve* or *F-number* system. F-numbers are designated $F_F$ for flatness and $F_L$ for levelness. The numbers have no units. The method for determining a slab's F-numbers is defined in ASTM E1155. This system depends on marking multiple lines on a slab and measuring the deviation at 12 in. (305 mm) intervals along these lines. F-numbers are obtained mathematically from these measurements using formulas contained in ASTM E1155.

Several devices are used to determine the deviations. In one, a 10- or 12-ft (3 050- or 3 658-mm)-long highway straightedge is laid along the lines. The deviations are then measured down from the straightedge. This method produces an accurate reading when done properly but is time-consuming. Laser levels, optical levels, floor profilometers, and inclinometers give quick and accurate readings.

The degree of flatness and levelness increases as the F-numbers increase. Slabs requiring a closer tolerance must have higher F-numbers. The required F-numbers for a particular slab depend on the activities that will take place on it and the finish that will be applied. In general, the required $F_F$ is 15 and the required $F_L$ is 13 for a slab with a scratch finish, as is permitted on slabs that are to receive concrete floor toppings, mortar setting beds for tile or terrazzo, and other bonded cementitious finishes. Slabs with a float finish that are to receive membrane waterproofing, elastic roofing, or sand-bed terrazzo should have an $F_F$ of 18 and a $F_L$ of 15. Trowel finish slabs that are to receive resilient flooring, carpet, paint,

and other thin-film finishes should have an $F_F$ of 20 and an $F_L$ of 17.

$F_F$-numbers for slabs in buildings range from 15 to 100, $F_L$-numbers from 13 to 50. The higher numbers are required when extraordinary degrees of flatness and levelness are needed, such as in skating rinks.

# 3.13 Cast-in-Place Structural Concrete

The cast-in-place structural concrete discussed in this section includes walls, columns, piers, pilasters, beams, girders, and slabs. The discussion here related to the use and placement of reinforcing steel should be considered introductory only. The actual size and placement of bars, ties, stirrups, and accessories should be determined from structural analysis by a structural engineer. In addition, forming, installing reinforcement, and placing concrete are the responsibility of the contractor. The methods described here may not be followed on a particular project and may, in fact, be inappropriate under certain circumstances. For example, the procedures described here do not apply to mass concrete structures.

## 3.13.1 WALLS

Foundation walls, which rest on footings, must be able to support vertical loads from the superstructure; resist lateral (horizontal) loads from the ground and the superstructure; and provide anchorage for the superstructure against uplift. Wall construction should be delayed until the footing is sufficiently cured to support the loads. This time will vary, depending on the air temperature and the magnitude of the loads to be supported.

In addition to their structural functions, foundation walls must be durable, resist free water and water vapor penetration, provide a barrier against fire, and control air and heat flow. A foundation wall that encloses a habitable space must present a suitable appearance or be able to accept finishes, and must accommodate windows, doors, and other openings.

Reinforced concrete and concrete masonry are used extensively for foundation walls. Foundation walls for most commercial, institutional, and industrial build-

ings are designed using engineering principles based on theoretical analysis of the forces and stresses involved. Foundations for single- and two-family homes and other small buildings are more commonly designed using minimum requirements based on past experience and observation of foundations for such buildings or on similar requirements listed in the applicable building code. Such requirements are generally empirical, particularly in designing basement walls to resist lateral earth pressures. The usual minimum thickness for concrete foundation walls for small buildings is 10 in. (254 mm) for concrete walls and 12 in. (305 mm) for concrete masonry walls. However, these thicknesses should not be used until they have been verified as acceptable under local conditions. The type of soil and the height of the water table affect the lateral loads acting on foundation walls and must be taken into account.

Figure 3.13-1 shows a typical concrete foundation wall. The key shown on the top of the footing is sometimes omitted. The procedure for building such a wall is as follows: The footing is formed. Reinforcement, including dowels, is installed, and the concrete is placed and screeded. When the concrete in the footing has cured sufficiently (about 1 day), one side of the wall forms is erected and coated with a form-release agent. Vertical and horizontal reinforcing steel bars are placed in the wall and wired to the dowels and to each other. Figure 3.13-1 shows two layers of steel, but walls carrying lighter loads often require only one layer. The reinforcing ties shown are necessary only for walls that bear heavy vertical loads.

After the steel has been placed and inspected, wall ties are inserted through holes in the side of the form that is in place. The remaining side of the formwork is coated with form-release agent and slipped into place over the wall ties. When screw-type wall ties are used, the threaded sections are screwed into the cones. The wales and braces are then installed, and the wall ties are anchored to the wales. The formwork is inspected to ensure that it has been installed properly and that it is in the proper position, tight, and securely braced.

Foreign material is removed from the inside of the forms, the level to which the concrete will be placed is marked on the forms using surveying instruments, and concrete of the proper consistency is

Projecting rods to tie to
slab or wall above

Ties used in
walls carrying
heavy loads

Reinforcing rods

Dowels

Keyway

Footing reinforcing rods

**FIGURE 3.13-1** Typical concrete foundation wall. (Drawing by HLS.)

placed. Placing of the concrete is done by dumping it from a bucket or by hose from a concrete pump. In a low wall, the concrete may be placed directly from a transit-mix truck's chute. As the concrete is placed, it is consolidated using internal vibrators. When it reaches the line on the forms, the concrete is smoothed using hand floats. The top is then covered with canvas or sheet plastic to prevent drying, and the concrete is permitted to cure.

Several days later, the forms are removed, the exposed wall ties are twisted off (unscrewed in the case of screw ties), and finishing is begun.

Most cast-in-place concrete walls higher in a building are built in essentially the same way as foundation walls. Their vertical steel is tied to the steel projecting from below rather than to dowels. Exposed-to-view joints between levels of concrete in walls should be treated in some manner, such as with a grooved joint or other positive articulation. This is needed to prevent an unsightly condition that might occur because of slight

differences in the concrete or due to the difficulty of obtaining clean horizontal joints at such locations.

## 3.13.2 COLUMNS, PIERS, AND PILASTERS

Columns, piers, and pilasters act in compression to transmit loads to footings. Columns and piers are freestanding; pilasters are built integrally with the wall and also may function as stiffeners to provide necessary lateral support to a wall. Masonry piers and pilasters are discussed in Chapter 4.

Column, pier, and pilaster construction should be delayed until the footing is sufficiently cured to support the loads. This time will vary, depending on the air temperature and the magnitude of the loads to be supported.

Figure 3.13-2 shows the basic types of reinforcement in a concrete column. Piers and pilasters are similar. Several parts of this drawing have been stylized for reasons of clarity. For example, single ver-

tical bars are shown. When more steel is required, bundles containing several individual bars are often used in each location. Only four vertical bar locations are shown, but additional bar locations are often necessary. Eight bars are common, with one falling halfway between each of the corner bars. Sometimes many more locations are required. In the figure, vertical steel is shown in a rectangular pattern, but round patterns are also used, and either may be used in square or round columns. The reinforcement shown is that for a tied column. When the steel is in a circular pattern, circular ties are often used, but more expensive spiral ties are also sometimes used (Fig. 3.13-3) when the structural engineer deems their use justified. Spiral ties are shipped to the job site coiled like a spring. They are stretched and wired in place at the site.

The process for building a concrete column, pier, or pilaster is similar to that described for a wall. There are, however, some exceptions:

- Dowels are set to line up with the vertical steel.
- Reinforcement ties serve as horizontal reinforcement. The spacing of reinforcement ties is not necessarily the same throughout the height of a column, being doubled where splices occur and increased in number where loads so dictate. Their spacing and size are selected by the structural engineer.
- The reinforcement is often wired together in the horizontal position into a *column cage*, which is then hoisted into place by a crane.
- The reinforcement is placed and wired to the dowels before any formwork is installed. When space is limited within a column, bars may be spliced in line rather than being lapped.
- Form ties are not usually needed.

### 3.13.3 BEAMS AND GIRDERS

There are several types of beams and girders. *Grade beams* are located approximately at grade level. In contrast to a continuous footing, which is supported uniformly along its length, a grade beam receives building loads and transfers them to either spread, pile, or caisson foundations. Grade beams may take many forms. The examples in Figure 3.13-4 include (a) a reinforced concrete

**FIGURE 3.13-3** Typical spiral column reinforcement. (Drawing by HLS.)

**FIGURE 3.13-2** Typical tied column reinforcement. (Drawing by HLS.)

beam placed integrally with the concrete slab and supported on piles and (b) a pre-cast reinforced wall carried on footings and supported laterally by concrete or masonry piers.

*Superstructure beams* are located above grade level, where they serve to support floors and roofs. Beams are of three general types: simple beams, continuous beams, and beams that are part of a slab. Beams usually carry only uniform loads that act over the entire length of the beam. *Girders* too may be simple or continuous, and they also often carry a uniform load, but in addition they carry beams. Large transfer girders sometimes also carry columns.

#### 3.13.3.1 Simple Beams

*Simple beams* (Fig. 3.13-5) span only one bay from support to support. They are not often found in large concrete-framed buildings, but an understanding of them will be helpful in understanding the more complex beams and girders usually found in actual construction. As was discussed earlier when a uniform load, including its own weight, is applied to a simple beam, compression stresses are imposed in the top portion of the beam and tensile stresses occur in the lower portion (see Fig. 3.1-1a). Reinforcing steel is added near the bottom of the beam to absorb the tensile stresses (see Fig 3.1-1b). This re-inforcement is supported at the proper

height in the form by chairs or bolsters (see Fig. 3.4-10).

Tensile stresses in a simple beam are concentrated at the bottom of the beam in the center of its span and are strongest at the center. Near the ends of the beam, the stresses move diagonally upward, where they tend to cause the beam to crack diagonally. Although weaker than the stresses at the center of the beam, these diagonal stresses are still sufficient to cause damage. *Stirrups* are added and wire is tied to the bottom reinforcing bars near the supports to absorb this diagonal stress (Fig. 3.13-6). No stirrups are needed near the center of a simple beam. Stirrups may be U-shaped (*U-stirrups*), as shown in the main figure, or closed (*closed stirrup ties*, as shown in Fig. 2.13-6a), or capped (two-piece) ties, as shown in Figure 3.13-6b. Light reinforcing rods are placed near the top of a beam that contains stirrups to hold the stirrups in place until the concrete has set. When heavy loads are present and the depth of a beam is limited by factors other than structural considerations, compression steel may be added to the top of a beam. Then this compression steel is used to hold the stirrups in place.

Reinforcing steel dissipates the forces it absorbs into the concrete by means of

**a** GRADE BEAM SUPPORTED ON PILES      **b** GRADE BEAM SUPPORTED ON FOOTINGS

**FIGURE 3.13-4** Grade beams: (a) reinforced concrete beam and (b) precast reinforced wall.

**FIGURE 3.13-5 Simple beam.** (Drawing by HLS.)

its bond with the concrete until the stresses have been neutralized. However, when the reinforcement reaches the end of a beam, it typically has some remaining stress, but there is no more concrete. This remaining stress is dissipated by bending each bar into a U shape called a *hook*. Beams are formed, and reinforcement is installed and concrete is placed in them in a sequence similar to that used in walls. The major differences are that the steel reinforcement in beams is supported on chairs or bolsters resting on the bottom of the formwork, and wall ties are usually not required.

### 3.13.3.2 Continuous Beams

A continuous beam crosses more then one span (Fig. 3.13-7). The dashed lines in the figure show a greatly exaggerated view of the bending that tends to take place in such a beam. Reinforcing steel is added to resist this bending. These beams are formed, and steel and concrete are placed in them as described for simple beams. Some ways in which continuous beams are different from simple beams are as follows:

■ The reinforcement in a simple beam tends to be the same throughout. In a continuous beam, more cross-sectional area of reinforcement is required at the bottom of the beam near the center of each bay than is required to pass through the columns. This usually means fewer bottom bars at the columns.

■ Top steel is required in a continuous beam at every column to resist the tendency of the beam to bend upward at those locations.

■ Stirrups are required at each end of a continuous beam and, in addition, in the beam on both sides of each intermediate column.

■ Right-angle bends with the steel turning downward into the columns at the ends of a continuous beam may be used in lieu of hooks when there is sufficient room for them.

### 3.13.3.3 Beams That Are Part of Floor Slabs

Beams are often part of a concrete slab (Fig. 3.13-8). They do not act like simple beams or even continuous beams, but rather have some unique characteristics. For example, they act as part of the slab, rather than as separate elements, and they are affected by forces acting in the slab. Therefore, they are discussed in Section 3.13.4.

### 3.13.3.4 Girders

Girders are essentially large beams, and the principles discussed in Sections 3.13.3.1 and 3.13.3.2 apply to them as well. Their steel reinforcement must be specifically designed by a structural engineer, but they usually have both top and bottom steel reinforcement and heavier bars than do beams. Additional steel is required where concentrated loads are carried.

**a** Closed stirrup tie

**b** Capped (two-piece) stirrup tie

Hooks

Reinforcing bar
Stirrup
Chair or bolster
Form button

**FIGURE 3.13-6 Reinforcement and forming for a typical beam.** (Drawing by HLS.)

Column

Beam

**FIGURE 3.13-7 A continuous beam.** (Drawing by HLS.)

**FIGURE 3.13-8** Beam and floor slab working together. (Drawing by HLS.)

### 3.13.3.5 Post-Tensioning

Tensile forces near the bottoms of loaded beams tend to pull the concrete apart, making it crack (see Fig. 3.1-1). Even when reinforcing steel is added, these cracks are often large enough to be seen. Such cracks can be eliminated altogether by placing the concrete in compression before an external load is applied. If enough compression is applied and held there, when a load is applied the amount of compression will be lessened but will never disappear completely. Compressing the concrete in a beam will also permit it to carry the same loads with less concrete, making it lighter and less expensive.

Compression loads can be added to a beam by stretching the steel and locking it to the concrete at the ends. As the steel tries to return to its normal length, the resulting stress is transferred to the concrete. Two procedures are used to produce this result. *Prestressing* is done before concrete is placed around the steel. This process is best done under controlled conditions in a shop or factory and is very expensive to do anywhere else. Therefore, it is used almost exclusively in precast concrete work and is described in Section 3.14.

A process called *post-tensioning* is used to add permanent tension to the reinforcing steel in cast-in-place concrete and subsequently to place the concrete under compression. Essentially it is done using *tendons*, which are bundles of high-strength, cold-drawn steel wire strands or steel bars. The tendons are coated with oil or placed in a steel tube to prevent bonding between the concrete and the tendons. The concrete is placed and permitted to cure. Generally, the tendons are fastened to one end of the beam and stretched from the other end using hydraulic jacks. When the desired amount of tension has been achieved, the jack end of the tendon is secured to the concrete, usually with a bearing plate of some sort. When the strands are very long, they must be stretched from both ends simultaneously to ensure uniform tensioning.

Tendons can be placed level, as conventional reinforcing bars are placed. Post-tensioned beams are more efficient, however, if the tendons are placed in a shape that approximates the lines of tensile force in the beam. This is near the bottom of the beam in the center and sloping upward toward the top of the beam at the ends in a V shape.

Post-tensioned beams tend to shorten due to elastic compression, shrinkage, and creep. These movements must be accommodated in the building design and construction. Adjacent elements that would be affected by these changes should be built after the post-tensioning has been completed or they should be isolated from the post-tensioned unit. In addition, the amount of tension to be applied can sometimes be adjusted slightly upward to compensate at least partially for these movements.

### 3.13.4 STRUCTURAL SLABS

The term *structural slabs* in this book refers to slabs that are not cast against the earth (slabs on grade). Structural slabs fall into one of two categories: one-way slabs and two-way slabs.

#### 3.13.4.1 One-Way Slabs

*One-way slabs* span in one direction from support to support. There are two types of one-way slabs: *solid* and *ribbed* or *joist* slabs (Figs. 3.13-9 and 3.13-10).

**SOLID SLAB**

The solid slab shown in Figure 3.13-9 is cut to show its shape. Such slabs usually have beams, girders, or bearing walls on each of their sides (Fig 3.13-11).

The first step in constructing a one-way solid slab is to form the supporting walls and columns, place the concrete in them, and permit it to cure sufficiently to support the loads that will be imposed. The second step is to build the forms for the beams and girders and then the forms for the slabs. The forms should be supported on temporary supports and shores of sufficient strength and stability to resist the large loads that will be imposed by the forms and the concrete. When all the forms have been built, the side of the forms that will be in contact with concrete should be coated with form release agent and the steel should be placed in them, following the shop drawings.

Structural reinforcement in a one-way solid slab is similar to that described in Section 3.13.3.2 for continuous beams, with bottom bars generally and top bars over the beams and girders. The structural reinforcement in one-way slabs is placed parallel to the span (see Fig. 3.13-9). Temperature steel, which usually consists of number 3 bars, is placed perpendicular to the structural bars. Tempera-

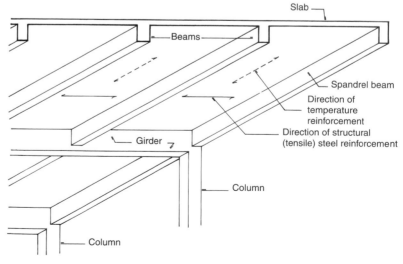

**FIGURE 3.13-9** A one-way solid slab. (Drawing by HLS.)

**FIGURE 3.13-10** A one-way ribbed slab. (Drawing by HLS.)

ture steel is not needed parallel to the span, because the structural bars there will serve to resist shrinkage and temperature cracks in that direction.

When the steel is in place and has been inspected, concrete may be placed into the forms and consolidated. The slab should then be cured and finished as described in Section 3.9. When the slab, beams, and girders are sufficiently strong to support themselves, the forms should be removed and reshores placed (Fig. 3.13-12). *Reshoring* consists of vertical metal or wood props designed to help the new concrete support construction loads for several additional weeks until it has cured sufficiently to stand alone. After

reshoring has been done, formed surfaces should be finished. If there is another floor above the one just completed, formwork should be moved up to it and the process repeated.

## RIBBED SLAB

One-way solid slabs become uneconomical when slab spans are long, because a slab must become thick and heavy to support the loads imposed and the steel must be placed lower in the slab. Eventually, the weight of the slab itself becomes a problem. The solution then is to eliminate part of the concrete. This is possible, because much of the concrete near

the bottom of a deep slab is not normally needed to resist loads. Most of the stress there is tensile, which is resisted by the steel. The solution then is to remove part of this nonworking concrete; one way is by forming the slab using metal or plastic *pans* (Fig. 3.13-13), which produces the effect of a thick slab using less concrete.

Standard pans come in two widths, 20 in. (508 mm) and 30 in. (762 mm). They range in depth from 8 in. (203 mm) to 20 in. (508 mm). Pans are tapered to permit their easy removal after the concrete has set. Tapered ends are used to increase the amount of concrete at the beams. The ribs formed between pans are called *joists*. Joist width can be varied by widening the spacing of the pans. Ribs are supported on beams. When these beams are kept at the same depth as the pans, they are called *joist bands*. Joists are often interrupted at midspan by a beam known as a *distribution rib*, which helps to distribute concentrated loads to several joists.

Ribbed one-way slabs are constructed by first building the supporting elements such as columns and walls and permitting them to cure sufficiently to support the loads that will be imposed. Then the supporting structure (centering) is built for the pans as described in Section 3.3.2.5. The centering decking forms the bottom of the joists. The pans should be placed on this centering and coated with a form release agent. The steel should then be placed.

After the steel has been installed and inspected, the concrete should be placed and the slab surface cured and finished as described in Section 3.9. When the ribbed one-way slab has set sufficiently to support its weight, the centering should be removed and the pans dropped from the concrete and moved to the next forming area. The formed surfaces should then be finished.

### 3.13.4.2 Two-Way Slabs

*Two-way slabs* span in two perpendicular directions at the same time. They are generally selected when the columns in a building can be located to create square or almost square bays. In this case, they are usually less expensive than comparable one-way slabs. It is possible to construct a two-way slab that is supported on all four sides by beams, similar to the one-way solid slabs described in Section 3.13.4.1, but they are seldom used. Two-

**FIGURE 3.13-11** Solid one-way slab framing. (Drawing by HLS.)

**FIGURE 3.13-12** Formwork is under construction on the top floor. On the floor directly below, the formwork is still in place. The formwork has been removed from the next lower floor, and reshoring has been placed. (Courtesy Portland Cement Association.)

Intermediate section          Tapered end form

**FIGURE 3.13-13** Typical one-way ribbed-slab pan. (Drawing by HLS.)

way slabs without beams are easier to form and thinner when the beam depth is taken into account.

There are essentially two types of two-way slabs in general use: *flat plate* and *waffle* slabs (Figs. 3.13-14 and 3.13-15). Two-way flat plate slabs are used on buildings with relatively short spans. When spans become longer, the same thing happens with two-way slabs that happens with one-way slabs: the slabs get thicker until they become uneconomical. The solution is similar. The two-way

equivalent of a pan slab is a *waffle slab*, which has ribs in both directions (Fig. 3.13-15).

### 3.13.4.3 Flat Plate Slabs

Flat plate slabs that carry light loads, such as those in office buildings, hotels, apartment buildings, and dormitories, usually have no thickened areas (see Fig. 3.13-14). Flat plate slabs in more heavily loaded storage and industrial buildings often have dropped panels at the columns

to resist the heavy shear loads associated with heavier loading (Fig. 3.13-16).

The first step in constructing a two-way flat plate slab is to form the supporting walls and columns, place the concrete in them, and permit it to cure sufficiently to support the loads that will be imposed. The second step is to build the forms for the dropped panels, when they occur, and then the forms for the slabs. The forms should be supported on temporary supports and shores of suffi-

**FIGURE 3.13-14** Two-way flat plate slab. (Drawing by HLS.)

**FIGURE 3.13-15 Waffle slab.** (Courtesy of Portland Cement Association.)

cient strength and stability to resist the large loads that will be imposed by the forms and the concrete. When all the forms have been built, the side of the forms that will be. in contact with concrete should be coated with form release agent, and the steel should be placed in them, following the shop drawings.

Structural reinforcement in each of the two perpendicular directions in a two-way flat slab varies, depending on its location relative to the columns. The slab is divided into zones, one between the columns called the *middle strips* and another called the *column strips*, which

lie along the column rows in both directions (Fig. 3.13-17). The least amount of reinforcement is placed in the middle strips where the loads are lightest. Additional steel and closer spacing of bars occur in the column strips to accommodate the higher bending stresses there. Steel is located near the top of the beam in the column strips where upward-bending stresses are present. Temperature steel is not needed in two-way slabs, because there is already steel in both directions. There is usually no additional steel in column caps when they are used. Rather, the additional con-

crete is relied on to resist the additional shear stress.

When the steel is in place and has been inspected, concrete may be placed into the forms and consolidated. The slab should then be cured and finished as described in Section 3.9. When the slab is sufficiently strong to support itself, the forms should be removed and reshores placed. The formed surfaces should then be finished. If there is another floor above the one just completed, formwork should be moved up and the process repeated.

A variation in two-way flat slab construction that eliminates most of the formwork is *lift-slab* construction. This is a process in which the slabs for the floors of a building occurring above the ground floor are cast on the ground floor in a stack and are then lifted into place using a system of hydraulic jacks. The slabs have cast-in sleeves that ride up the columns. When the lowest slab of a stack has reached its proper level, the sleeves are welded to the columns and the jacks are used to pull the remaining slabs up the column to the next level. When the last slab has been welded into place, the jacks are removed.

### 3.13.4.4 Waffle Slabs

Two-way waffle slabs are formed using metal or plastic pans called *domes* (Fig. 3.13-18). The ribs formed between domes are called *joists*. There are two basic sizes of standard domes. The smaller of the two forms joists is 5 in. (127 mm) wide at the bottom on 24 in. (610 mm) centers and 8 to 14 in. (203 to 356 mm) deep. The second standard dome size produces joists 6 in. (152 mm) wide at the bottom on 36 in. (914 mm) centers and 8 to 20 in. (203 to 508 mm) deep. Larger sizes are also available. Domes are tapered to permit their easy removal after the concrete has set. Some domes have compressed-air fittings to permit even easier removal.

Waffle slabs are most efficient when allowed to cantilever at the perimeter. When a cantilever is not desirable, a perimeter beam (*slab band*) must be placed there (Fig. 3.13-19). In addition, column *heads* are created around the columns to accommodate the shear stresses there (Fig. 3.13-19). Slab bands and heads are created by omitting domes.

Waffle slabs are constructed by first building the supporting elements, such as columns and walls, and permitting them to cure sufficiently to support the loads that will be imposed. Then the support-

**FIGURE 3.13-16 Two-way flat plate slab with dropped panels.** (Drawing by HLS.)

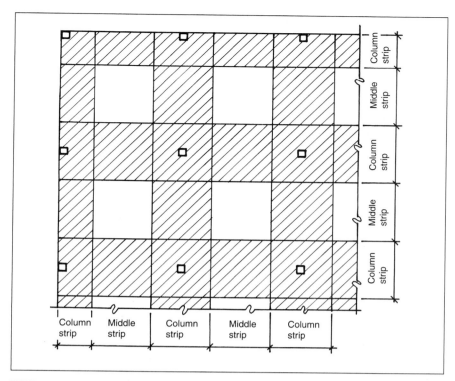

**FIGURE 3.13-17** Plan showing locations of middle strips and column strips in a two-way flat slab. (Drawing by HLS.)

**FIGURE 3.13-20** A concrete slab on a permanent steel form. (Drawing by HLS.)

steel decking acts as a form until the concrete sets. Such slabs are usually reinforced with welded wire fabric. This produces a finished lightweight deck. Such slabs are finished and cured as described in Section 3.9.

Heavier-gauge steel decking is also used with a concrete fill to produce *composite decking* (see Chapter 5).

### 3.13.4.6 Slab Flatness

The principles governing flatness requirements for slabs on grade discussed in Section 3.12.5 also apply to structural slabs.

### 3.13.4.7 Post-Tensioning

A process similar to that described in Section 3.13.3.5 for post-tensioning beams is sometimes used to post-tension flat-plate

ing structure (*centering*) is built for the domes as described in Section 3.3.2.5. The domes should be placed on this centering and coated with a form release agent. The steel should then be placed.

After the steel has been installed and inspected, the concrete should be placed and the slab surface cured and finished as described in Section 3.9. When the waffle slab has set sufficiently to support its weight, the centering should be removed and the domes dropped from the concrete and moved to the next forming area. The formed surfaces should then be finished.

### 3.13.4.5 Slabs on Permanent Forms

In steel structures, concrete slabs are often placed on a permanent corrugated steel sheet decking (Fig. 3.13-20). The

**FIGURE 3.13-18** A waffle slab dome. (Drawing by HLS.)

**FIGURE 3.13-19** Pan slab in plan showing slab bands and column heads. (Drawing by HLS.)

---

**CONCRETE** ■ Cast-in-Place Structural Concrete **133**

Main Bars One Direction                    Main Bars Two Directions

**FIGURE 3.13-21** Typical stair reinforcement. (Courtesy Concrete Reinforcing Steel Institute.)

concrete slabs. Post-tensioning concrete slabs allows their thickness to be reduced, which results in smaller floor-to-floor heights and less cladding of the building. The amount of steel required can be as little as one-third of that needed for a conventional concrete slab.

### 3.13.5 CONCRETE STAIRS

Concrete stairs are of two types. Each run of the first type consists of an inclined reinforced slab with integral risers and horizontal landings at the top and bottom (Fig. 3.13-21). The bottoms, sides, and risers are formed. The risers usually slope outward at the top. All concrete in such a stair is placed simultaneously. This type of stair usually occurs in concrete-framed buildings. Similar stairs are also used at entrances and site locations with the underneath supported by earth rather than by formwork.

The second type of concrete stair consists of steel pans supported on steel stringers. These pans are filled with concrete. This type of stair is most often used in steel-framed buildings.

## 3.14 Precast Concrete

Precast concrete can be either site cast or plant cast.

### 3.14.1 SITE-CAST UNITS

Concrete structural elements are sometimes precast at a job site and lifted or tilted into place by construction cranes. Any precast element can be cast in the field on a horizontal surface such as the ground or a concrete slab and lifted into place in the building. Job-cast units are not subject to the size limitations imposed on plant-cast units. Therefore, it is sometimes advantageous to precast large elements, such as transfer beams, on the ground. Site-cast precast units are subject to the same disadvantages as cast-in-place concrete relative to curing time and adverse weather conditions.

A common element precast in the field is the *tilt-up* slab. To save time and money by eliminating most of the usual formwork, concrete wall sections are cast on a poured-in-place concrete slab and tilted or lifted into place (Fig. 3.14-1).

The greatest use of tilt-up construction has been on single-story industrial buildings, but the method has also been used in other types of buildings.

Some additional reinforcement is necessary in field-precast elements to permit attachment of lines from a crane and to resist the stresses generated by lifting the units into place.

Joining of sections of tilt-up panels is accomplished in several ways. The space between panels may be filled with a poured-in-place concrete pilaster or column. Panels may be joined using welded-together steel angles cast into the tilt-up panels. Tilt-up panels may be supported by

**FIGURE 3.14-1** A tilt-up panel being lifted into place. (Courtesy the Burke Company.)

separate concrete columns, by concrete columns cast as a part of the tilt-up panel, or by structural steel columns. Panels may themselves be bearing walls, supporting a roof structure without columns.

## 3.14.2 PLANT-CAST UNITS

The cost of forming and placing concrete in the field, and the need to cope with unfavorable weather conditions while doing so, have led to *plant-cast* precast concrete elements that are produced in a factory environment. Units are cast at ground level in the plant, shipped to the construction site, and set in place using cranes in much the same way as structural steel is erected (Fig. 3.14-2).

Precasting concrete elements in a plant reduces their cost and makes it easier to control their quality. For example, concrete in beds in a plant can be easily consolidated by mechanical vibration, which results in fewer voids and better surfaces than can be achieved in concrete vibrated in the field. The ease of *prestressing* structural

units in an industrial plant allows the units to be made smaller. It is also possible in a plant to cure units using steam. Steam-cured precast concrete units made using high-early-strength portland cement can be removed from the casting bed within 24 hours after the concrete has been placed.

Plant-cast precast concrete units, which are heavy and bulky, must be shipped from plant to job site by truck. Legal limitations on the size of trucks and the use of roads limit the size of precast units that can be shipped. This influences the types of units that can be effectively precast in a plant. For these and other reasons, plant-cast precast concrete units are usually modular in nature and relatively small. For example, it is impractical to precast several bays of a two-way structural slab in a plant. Entire walls and even rooms have been produced, but most plant-cast precast concrete is made in relatively small modular units in a few standard shapes.

Once plant-cast precast concrete units arrive at a job site, they can be erected quickly in most kinds of weather (Fig.

3.14-3). There is no need to wait for them to cure or to provide the moisture conditions required to cure cast-in-place concrete.

Precast concrete units can be divided into two basic categories: *structural* precast concrete units and *architectural* precast concrete units.

## 3.14.3 STRUCTURAL PRECAST CONCRETE

Structural precast concrete units are so called because they form a part of the structural system of the building and carry loads other than their own weight.

Precast/Prestressed Concrete Institute (PCI) publication MNL-116 establishes standard practices for the manufacturer of structural precast concrete elements. PCI also administers a plant certification program.

### 3.14.3.1 Types of Units

Structural precast concrete elements include walls, beams and girders, columns, and floor and roof systems.

**FIGURE 3.14-2** A precast concrete beam being erected by crane. (Photo courtesy Precast/Prestressed Concrete Institute.)

**FIGURE 3.14-3** A precast concrete structure being erected in the snow. (Photo courtesy Precast/Prestressed Concrete Institute.)

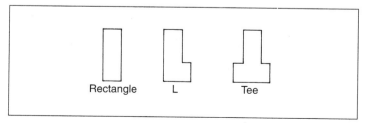

**FIGURE 3.14-4** Typical precast beam shapes. (Drawing by HLS.)

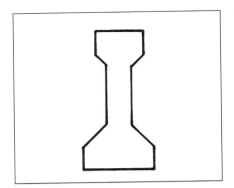

**FIGURE 3.14-5** AASHTO bridge girder shape. (Drawing by HLS.)

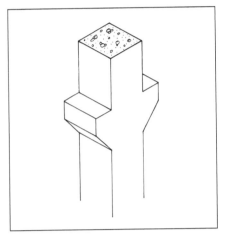

**FIGURE 3.14-6** Typical precast concrete haunched column. (Drawing by HLS.)

## WALLS

Structural precast walls are occasionally used, including when they are sandwich panels, but most precast wall panels are architectural concrete. Therefore, this category is discussed in Section 3.14.4.

## BEAMS, GIRDERS, AND COLUMNS

The usual shapes for precast beams are rectangular, L-shaped, and inverted-T-shaped (Fig. 3.14-4). The projections serve as support ledges for floor and roof systems.

Girders are usually L-shaped or inverted T-shaped, but heavier girders may be the American Association of State Highway and Transportation Officials (AASHTO) shape shown in Figure 3.14-5. These girders were originally used on bridges, but they work as well in buildings.

Precast columns are usually either square or rectangular. While most structural precast concrete is post-tensioned, columns often have conventional reinforcement. Columns may be either single-story units or cast in multistory sections with corbels to support beams and girders (Figs. 3.14-6 and 3.14-7).

## FLOOR AND ROOF ELEMENTS

Floor and roof elements rest on bearing walls, beams, or girders and are either left exposed to become the floor or roof slab or covered with a concrete topping (see Section 3.16).

Floor and roof elements may be either flat slab or T-shaped. Both types may be made from either normal-weight or lightweight concrete. A balance must be drawn in this selection between the decreased weight of lightweight concrete and its greater cost.

*Flat Slab Precast Elements*   Flat slab precast elements may be either solid or hollow-core. *Solid flat slab units* are simple flat concrete slabs (Fig. 3.14-8). They are used where spans are about 20 ft (6.1 m) maximum and maximum headroom is needed.

*Hollow-core flat slab units* are cast with round voids running their entire length (Fig. 3.14-9). They are appropriate for intermediate spans (about 50 ft [15.2 m] maximum) where solid flat units would be too thick and heavy. The principle is the same as that discussed in Section 3.13.4 as justification for selecting pan slabs instead of flat plate slabs. The cores eliminate the nonworking concrete, lightening the slab section. The steel is contained in the top and bottom of the cored slab.

Both solid and hollow-core slabs are relatively thin, which saves on floor-to-floor height. They can also be painted or given a thin textured coating and left exposed as a finished ceiling.

**FIGURE 3.14-8** Solid flat precast concrete slab. (Drawing by HLS.)

**FIGURE 3.14-9** Cored flat precast concrete slab. (Drawing by HLS.)

**FIGURE 3.14-7** Concrete-framed building under construction showing two-story columns and several types of beam haunches. (Photo courtesy Precast/Prestressed Concrete Institute.)

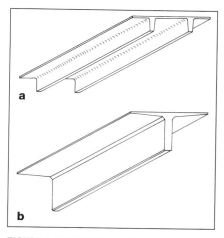

**FIGURE 3.14-10** Stemmed precast concrete elements: (a) double tee; (b) single tee. (Drawing by HLS.)

**FIGURE 3.14-11** Installing a precast hollow core flat slab section. (Photo courtesy Precast/Prestressed Concrete Institute.)

*T-Shaped Precast Elements* When spans are longer than can be accommodated by cored slabs, the cores are eliminated, the slab thinned, and stems added to give stiffness to the system. The result is T-shaped floor and roof units. When the units have a single stem, they are called *single tees*; two-stemmed elements are called *double tees* (Fig. 3.14-10).

Double tees can span as much as 100 ft (30.5 m). They can also be used on shorter spans where solid slabs and cored slabs will also work. Where headroom is not a problem and their appearance is desirable, double tees may be appropriate on short spans.

Single tees have the disadvantage of having to be braced during construction because they are subject to tipping until they are finally fixed in place. They are, therefore, seldom used except on long spans. Their normal spans of between 60 and 150 ft (18.3 and 45.7 m) require depths ranging from 36 to 60 in. (914 to 1500 mm), which is appropriate in few buildings.

Tees usually have steel weld plates cast into their edges so that adjacent units can be welded together to develop a diaphragm action in the entire system. This process helps to reduce differential deflection in adjacent units.

*Topping* Precast concrete floor and roof elements are usually finished rough and covered with a 2- to 4-in. (50.8- to 102-mm)-thick topping as described in Section 3.16.

#### 3.14.3.2 Manufacturer

Most structural precast concrete elements are manufactured in a plant on casting beds that range from 200 to 400 ft (61 to 122 m) in length. However, some are cast into molds. The process is essentially as follows.

Prestressing steel is placed in a casting bed and tensioned using hydraulic jacks. Bulkheads are placed in the bed to separate individual units. Bulkheads are sometimes omitted for solid and cored slabs, which are then sawed apart later. Core forms and opening forms are placed in the beds. Concrete is placed in the beds, consolidated by vibration, and leveled. The top surface is then finished as appropriate with floats or trowels. Steam or radiant heat is then used to cure the unit. In some plants, some units, especially flat slab units, are wet-cured for 7 days in lieu of steam curing.

Test cylinders are taken at the same time that the concrete is placed and cured. When the units are steam-cured, about 24 hours after the concrete is placed, the cylinders are tested to verify the strength of the concrete. When it is found to be satisfactory, the prestressing strands are cut, and the units are separated and removed from the casting bed.

#### 3.14.3.3 Erection

Precast structural concrete elements are handled and erected in much the same way as is structural steel. Units are hoisted into place by crane (Fig. 3.14-11). Some units are held in place by gravity, others by the welding together of inserts in the elements.

### CONNECTIONS

Precast concrete columns usually have a cast-in steel plate at the bottom. This plate is fastened to anchor bolts set into a concrete footing (Fig. 3.14-12). Column sections are joined using similar plates and bolts (Fig. 3.14-13).

Beams and girders are anchored to columns in a variety of ways. A typical system is the use of corbels to support beams and girders (see Figure 3.14-6).

Tees usually rest on L- or T-shaped beams or girders (Fig. 3.14-14). Tees are often, but not always, given a concrete topping (see Section 3.16).

Flat and cored slabs usually rest on beams or girders (Fig. 3.14-15). Such slabs are usually given a concrete topping (see Section 3.16).

**FIGURE 3.14-12** A typical precast concrete column base at the footing. Pockets are filled with nonshrink grout after anchor nuts are set. (Drawing by HLS.)

Upper column

Pocket

Base plate

Top plate

Shims

Anchor bolts with nuts in each corner

Lower column

**FIGURE 3.14-13** A typical bolted column-to-column connection. Pockets and spaces around shims are packed solid with non-shrink grout after anchor nuts are set. (Drawing by HLS.)

Bearing pads are usually inserted between concrete elements and other concrete elements supported by them so that the elements may move to accommodate thermal changes and other structural movement. Bearing pads are usually synthetic rubber for beams, girders, and tees, and plastic for slabs.

Precast structural elements must be held together to resist seismic and wind loads. This is sometimes accomplished by installing reinforcing rods in a topping or by welding together plates cast into the elements. However, the stems of tees are never fastened solidly to beams or columns because doing so would prevent them from expanding and contracting freely.

### 3.14.4 ARCHITECTURAL PRECAST CONCRETE

Architectural precast concrete units include panels and other members used as part of a building's exterior cladding. Most are not load-bearing, but some sandwich panels are load-bearing.

The PCI publication MNL-117 establishes standard practices for the manufacture of architectural precast concrete elements. PCI also administers a plant certification program. Most architectural precast concrete panels are made from standard-weight concrete and reinforced conventionally. However, glass-fiber reinforced concrete panels, which are enjoying increased use today, require no conventional steel reinforcement (see

Section 3.15). These panels are lighter than conventional panels and can have steel studs cast into them, thus providing a cavity for insulation. PCI MNL 128 establishes recommendations for glass-fiber-reinforced panels.

Architectural precast concrete panels are available in many finishes, ranging from flat, as-cast finishes to ornate articulation made using form liners, and many colors. PCI publication CTG lists more than 500 available colors and textures. Exposed aggregate, bush-hammered, and

abrasive blasted finishes are possible, as are panels faced with granite, marble, brick, and other materials.

Many panel configurations are possible, from flat panels to those with deep shadow lines and recesses. Several configurations of nonflat panels are available, including ribbed panels; panels with a configuration similar to that of double T structural panels; sculptured panels with cast-in moldings to simulate wood panels; and panels that have openings to frame windows, doors, and other openings. Ar-

**FIGURE 3.14-14** Tees resting on a beam and on masonry. (Photo courtesy Precast/Prestressed Concrete Institute.)

**FIGURE 3.14-15** Hollow-core flat slab resting on a beam. (Photo courtesy Precast/Prestressed Concrete Institute.)

chitectural precast concrete column and spandrel covers are also available.

Architectural precast sandwich panels are enjoying current popularity. *Sandwich panels* consist of one or more layers of precast concrete and a layer of foam insulation. They may be formed in any of the configurations used for other architectural precast panels, but most have either cored, flat, or T-shaped concrete portions (Fig. 3.14-16). They are available in the same finishes as other precast wall panels and are installed in essentially the same manner.

Architectural precast panels are usually installed by crane and fastened in place using cast-in angles and plates

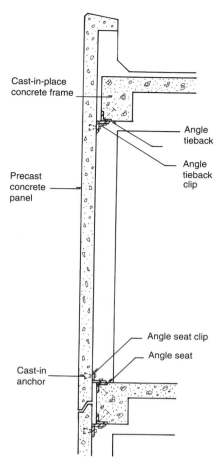

**FIGURE 3.14-17** Architectural precast cladding panel, supported by a cast-in-place concrete structural frame showing a typical method of supporting and anchoring panels. Finishes, flashing, caulking, roofing, and thermal and safing insulation have been omitted for clarity. Seat and tieback clips usually have slotted bolt holes to permit adjustment, and are shimmed as necessary and bolted to panel and to seat and tieback angles. Plates are sometimes used in place of seat and tieback angles. (Drawing by HLS.)

(Figs. 3.14-17 to 3.14-19). Joints are usually treated with sealants.

### 3.14.5 PRESTRESSING AND POST-TENSIONING

Most structural precast concrete units and some architectural precast concrete units are prestressed. *Prestressing* is done in precast concrete for the same reasons post-tensioning is done in cast-in-place beams and slabs (see Section 3.11.3.5). The process is also similar to the post-tensioning described in Section 3.11.3.5, except that the tendons are stretched be-

tween *abutments* before the concrete is placed around them. When the concrete has cured, the tension is released from the steel, which recoils, transferring the stresses in it to the concrete in the form of compression. This load transfer is done through the bond between the steel and the concrete. Bearing plates are not required.

Usually, precast units are cast end to end over continuous tendons. When the concrete has cured, the tendons are cut at both ends of each concrete element. This reduces the need for a separate set of costly abutments for each precast element.

## 3.15 Glass-Fiber-Reinforced Concrete

Glass-fiber-reinforced concrete (GFRC) is a plant-cast composite product based on portland cement reinforced with randomly dispersed glass fibers in lieu of reinforcing steel. It is used to manufacture components of exterior wall-cladding systems, such as panels, column covers, window wall units, and ancillary components of curtain wall systems, such as spandrels, mullions, fascia panels, and soffits. It is also used for mansards, sunscreens, and interior panels and column covers. GFRC units are not load-bearing building components.

GFRC weighs about 80% less than conventional precast concrete. This light weight makes it ideal for rehabilitation projects where increased weight is a problem. In new buildings, this light weight can result in significant cost savings in the structure compared to other forms of cladding, such as conventional precast concrete.

GFRC elements should comply with the requirements of PCI standards MNL 117 and MNL 128. Manufacturers should be members of PCI or participate in its plant certification program.

### 3.15.1 DESIGN FLEXIBILITY

GFRC is capable of almost unlimited design flexibility in terms of possible shapes. Deep reveals, complex curves, and even 90-degree corners are possible. Panels may be flat, flutted, sculptured, or grained to look like wood. GFRC is ideal for producing vertical and horizontal sun-

**a**

**b**

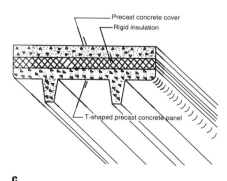

**c**

**FIGURE 3.14-16** Precast concrete insulated sandwich panels: (a) cored panel; (b) flat panel; (c) T-shaped ribbed panel. (Drawing by HLS.)

**FIGURE 3.14-18** A light precast panel bring installed. (Photo courtesy Precast/Prestressed Concrete Institute.)

**FIGURE 3.14-19** A heavyweight cored precast wall panel being installed. These panels are designed to carry the weight of the panels above. (Photo courtesy Precast/Prestressed Concrete Institute.)

shades and deeply recessed windows, which aids in reducing solar heat gain on the glass and thereby reduces the cost of mechanical cooling systems and their operation. GFRC can be placed in rubber molds made from existing building elements to produce new elements that exactly match an existing configuration.

A wide range of colors is possible, produced by colored cement in the matrix and colored aggregates. Exposed aggregate faces are also possible. Face mixes can be added that make the GFRC panel's appearance indistinguishable from conventional precast concrete, granite, or limestone, and it can be made to resemble metal panels.

GFRC can be faced with natural stone, veneer-thickness brick, tile, and architectural terra cotta. These materials are separated from the GFRC with a bond breaker to ensure that differential thermal movements in them does not damage the GFRC. However, some sources advise against using GFRC as a backup for clay masonry, due to past failures possibly attributable to such use.

### 3.15.2 MANUFACTURER

The matrix for GFRC panels is a mixture of portland cement, aggregates (mostly sand), glass fibers, water, and a water reducer or superplasticizer. Coloring agents, air-entraining agents, fly ash, and natural pozzolan are also used by some fabricators. The glass fibers constitute 4% to 5% of the matrix.

Most panels are manufactured by spraying the matrix on a mold. In small ornamental elements, the matrix is sometimes packed into a mold by hand.

Except on small panels, thickened sections or stiffeners are usually added to the back of a GFRC panel. Stiffeners may be steel studs or structural steel tubes or may be formed from the GFRC concrete material by spraying it over rib forms, such as polystyrene foam strips. The stiffeners, especially studs, provide a space for insulation, making high U-values possible and helping to decrease heating and cooling loads. They also provide space for electrical, mechanical, and telephone conduits.

GFRC will expand and contract up to $^1/_8$ in. (3.2 mm) in 10 ft (3000 mm) due to moisture and thermal effects. This movement must be accounted for in joint widths. Joints in GFRC panels are finished with polysulfide, polyurethane, or silicone sealants.

### 3.15.3 INSTALLATION

Quality control during installation is important if GFRC panel systems are to resist water penetration. Joints should provide for a double layer of sealant.

Anchors should be flexible to prevent them from applying stress to the GFRC. Fastening is done with bolts or welding.

Windows and doors should be attached to panel frames, not to the GFRC itself. If windows or doors would prohibit the GFRC from moving freely, a separate framing system should be used for them. Windows should not be permitted to bear on the GFRC material or be supported by it.

Alignment of panel faces may result in the backing studs being slightly out of plane or not plumb. It is therefore necessary to provide horizontal furring members attached to the embedded stiffener studs for attachment of interior finishes. However, shimming and installing such furring may be more expensive than installing a secondary row of studs to receive finishes.

## 3.16 Concrete Toppings

Concrete toppings are thin slabs of concrete placed on top of other concrete, waterproofing, or insulation. They are of two basic types: *bonded* and *unbonded.*

Most toppings are composed of portland cement, coarse aggregate, and fine aggregate mixed with water. Sometimes a plasticizing agent is used instead of the water or in addition to it. The aggregates may be either normal-weight or heavy-duty types. The latter are especially hard aggregates, such as granite, quartz, or corundum, which is used when abrasive resistance is needed. Aggregates are limited to a $1/2$ in. (13 mm) diameter maximum size, which is generally smaller than those used in normal concrete. However, aggregates in 3 in. (76 mm) thick toppings may be somewhat larger. Aggregates may be of selected colors if desired. Sometimes iron particles are added to the mix to improve surface wear and increase the compressive strength of the concrete.

Toppings may be machine- or ready-mixed. They are placed, cured, and finished as described for slabs in Sections 3.8 and 3.9. Toppings that contain specially selected aggregate are sometimes given a ground surface finish using grinding machines to expose the aggregate.

### 3.16.1 BONDED TOPPINGS

A *bonded topping* is a layer of concrete placed over an underlying concrete slab or precast concrete structural element (see Section 3.13). Bonded toppings may be placed at the same time as the structural slab or later.

Bonded toppings are used for several purposes. Over monolithic concrete slabs, they can create a special finish, such as an abrasive finish, where it would be impracticable, too expensive, or otherwise undesirable to include the aggregate necessary to produce this finish in the structural concrete. Unreinforced bonded toppings are also used over monolithic concrete in heavy industrial buildings where the movement of equipment may damage the finished floor. The toppings are placed after the equipment has been installed. Bonded toppings over monolithic concrete slabs are usually 1 to $1^{1}/_{2}$ in. (25 to 38 mm) thick and contain neither reinforcing steel nor welded wire fabric.

Bonded toppings are also used to make a level, smooth surface over structural precast concrete elements and to make a composite structural unit with them (see Section 3.14). Such toppings often contain a layer of welded wire fabric or steel reinforcing bars to help tie together a series of precast elements. Toppings over precast concrete elements are usually 2 to 4 in. (50 to 100 mm) thick.

### 3.16.2 UNBONDED TOPPINGS

*Unbonded* toppings are thin concrete slabs used to produce wearing surfaces over sheet membrane waterproofing or insulation (see Chapter 7). They are usually 3 in. (76 mm) thick but are sometimes thicker. They always contain a layer of welded wire fabric.

### 3.16.3 SLAB FLATNESS

The principles concerning flatness requirements for slabs on grade discussed in Section 3.12.5 also apply to toppings.

## 3.17 Structural Insulating Roof Decks

Two kinds of structural insulating roof decks are addressed here: poured-in-place gypsum concrete decks and prefabricated board decks. The material, gypsum, is described in Section 9.1.

### 3.17.1 POURED-IN-PLACE GYPSUM CONCRETE ROOF DECKS

A poured-in-place gypsum concrete roof deck is a reinforced gypsum concrete slab placed on permanent formboards that may or may not be supported by subpurlins.

#### 3.17.1.1 Materials

GYPSUM CONCRETE

Gypsum concrete mix is a factory-controlled mixture of gypsum and wood chips or mineral aggregate. It requires only the addition of water at the job site. A gypsum concrete slab, which weighs 35 to 55 lb/cu ft (561 to 881 kg/m$^3$), depending on the type and amount of aggregate, is poured to an average thickness of 2 to $3^{1}/_{2}$ in. (51 to 89 mm). Gypsum concrete slabs are reinforced with galvanized steel, welded wire mesh, or woven wire mesh to provide at least 0.26 sq in. (168 mm$^2$) of cross-sectional area of steel in each foot (305 mm) of slab width (Fig. 3.17-1).

a

b

**FIGURE 3.17-1** Gypsum concrete roof construction: (a) with subpurlins; (b) without subpurlins. (Courtesy Gypsum Association.)

## FORMBOARDS

Formboards are nonstructural components of a gypsum concrete roof deck and are usually not considered in the load-bearing capacity of the slab. They serve as a form and remain in place to provide a functional underside of the deck. Various formboards are available to provide fire resistance, light reflection, insulation, sound absorption, and economy.

Gypsum formboards, which are also called *roof boards*, are surfaced on the exposed face and longitudinal edges with glass fiber sheets or with calendered manila paper specially treated to resist fungus growth. The back face is surfaced with gray liner paper that is used on regular gypsum wallboard (see Section 9.5) or with glass fiber sheets. Gypsum formboards are usually $1/2$ or $5/8$ in. (12.7 or 15.9 mm) thick, 32 in. (812.8 mm) wide, and either 8 or 10 ft (2438 or 3048 mm) long, although other sizes may be available.

Wood-fiber, glass-fiber, and mineral-fiber boards, and various types of precast planks, are also used as forms for gypsum concrete. Their widths generally coincide with the standard 24, 32, and 48 in. (610, 313, and 1219 mm) nominal subpurlin spacing.

## SUBPURLINS

Steel subpurlins (rails, bulb tees, or other tee sections) are welded transversely to the primary framing to support formboards and roof slab. They span between main purlins, beams, or joists. Subpurlins also anchor the deck against uplift forces, restrict deck movement caused by temperature changes, and provide lateral bracing for the main roof purlins. Subpurlins, which vary in size, weight, and shape, are selected on the basis of the span between the main purlins or joists supporting them and the required safe load-bearing capacity of the deck.

When subpurlins are not used, the formboards rest directly on purlins, beams, or joists (Fig. 3.17-2).

### 3.17.1.2 Use and Construction

Gypsum concrete decks are used when noncombustible roof construction is required and other types are less practicable or more costly. They can be installed on virtually any roof shape, size, or configuration, even on those that are warped, sawtoothed, curved, or sloped. The primary framing can be wood, steel, or concrete. The light weight of gypsum concrete, compared with that of portland cement concrete, often allows the use of a lighter structural frame. Gypsum concrete sets rapidly, and the roof slab may be used as a working surface within 30 minutes after pouring.

Decks of gypsum concrete poured over gypsum- or mineral-fiber formboards are classified as noncombustible. When combustible wood-fiber formboards are used, there is generally an increase in insurance premiums. Gypsum concrete decks, 2 and $2^1/2$ in. (50.8 and 63.5 mm) thick on gypsum formboards without a suspended ceiling below, qualify for 1- and 2-hour fire resistance ratings, respectively. The materials used in gypsum concrete should conform with the requirements of ASTM C317; installation should conform with those of ASTM C956.

a

b

**FIGURE 3.17-3** Insulating deck: (a) lightweight perlite aggregate, (b) wood fiber.

### 3.17.2 PREFABRICATED STRUCTURAL INSULATING ROOF DECKS

Prefabricated structural insulating roof deck boards are designed to provide a structural deck as well as thermal insula-

**FIGURE 3.17-2** Subpurlins and reinforcing: (a) bulb tee and welded wire; (b) trussed tee and woven wire. (Courtesy Gypsum Association.)

tion and interior ceiling finish. They are used primarily on flat or sloping roofs with exposed beams. The underside of the decking is factory-finished in various designs or textures, some with acoustical value.

Two types of structural insulating roof decks are discussed here: lightweight aggregate boards and wood-fiber boards. These boards are available in a variety of thicknesses and sizes in either butt-joint or tongue-and-groove edge profiles (Fig. 3.17-3). Because the underside of the deck is also the interior finish, care must be taken not to damage the surface during erection. In areas with high interior humidities, such as bathrooms or laundries, adequate roof insulation and ventilation should be provided to prevent condensation.

# 3.18 Cement-Based Underlayment

An underlayment is a material placed beneath floor finishing materials, such as tile, resilient flooring, carpets, and liquid-applied flooring, to level and stabilize the substrate to receive floor finishes. Underlayments may be cement-based or board types. This section addresses two cement-based types: gypsum cement underlayment and cement-based, polymer modified, self-leveling underlayment. Other underlayment types are addressed in the sections covering the finish flooring materials to be applied over them.

## 3.18.1 GYPSUM CEMENTITIOUS UNDERLAYMENT

Gypsum-based floor underlayments have been available for more than 50 years. They were originally developed to replace conventional underlayments. They are used mostly in multifamily residences and commercial buildings. They have two major advantages over cement-based underlayments: they cost less, and they are lighter in weight.

Gypsum-based underlayments have two major disadvantages compared with cement-based underlayments. They are not as strong, and they are more susceptible to moisture damage. Early gypsum-based underlayments were limited in their use by a very low compressive strength

(100 psi [6.9 MPa]). Today's formulations have strengths of up to 2500 psi (17.2 MPa), which makes them useful in more applications, but they are still low in strength compared with cement-based underlayments, which have a strength of between 4100 and 5500 psi (28 and 38 MPa). The lower compressive strengths make the use of gypsum-based underlayments questionable where rolling loads or heavy concentrated loads will be present. They are also more susceptible to damage when the bond between applied floor coverings is greater that the internal strength of the underlayment.

Gypsum-based underlayment manufacturers do not recommend its use in locations where it will be subjected to moisture damage. Therefore, its use is questionable over cast-in-place concrete, which will continue to emit moisture vapor for a long time after it is placed. It should not be used where the adhesives used to apply a finish over it are water-based. Water from floor cleanings can also deteriorate gypsum-based underlayment. It should not be used as an exposed floor or on exterior locations, such as balconies. Its use below ground is limited to those situations where water intrusion is impossible. Where it will be exposed to water, such as in kitchens and baths, it must be covered by an impervious material, such as vinyl sheet flooring.

The advice of the flooring products manufacturer should be sought, however, regarding whether gypsum-based underlayment is acceptable beneath its product. Installation adhesives must be compatible with the underlayment, and the bond strength of the adhesive must be low. Resilient flooring manufacturers generally do not recommend the use of gypsum-based underlayments beneath their products.

### 3.18.1.1 Products

Gypsum-based underlayments are propriety products made with gypsum cement, sand aggregate, and water. They are self-leveling. Several types are available, and selection is based on project requirements and cost. Some gypsum-based underlayments are relatively low strength 1200 to 1500 psi (8.3 to 10.3 MPa). Others range from 1600 to 2500 psi (11 to 17.2 MPa). Their point loading resistance is similarly different, ranging from 550 to 2000 lb (250 to 910 kg) on a 1 in. (25.4 mm) disc.

### 3.18.1.2 Installation

Gypsum-based underlayment is often installed over wood-based substrates, such as plywood, oriented strand board, or wafer board, but may be installed over concrete that has been allowed to cure for the time recommended by the underlayment manufacturer and has been sealed with a primer recommended by the underlayment manufacturer. It can also be installed over extruded or expanded polystyrene foam and over sound-deadening pads.

The deflection in the substrates beneath gypsum-based underlayment must be limited to L/360. Greater deflection will cause the underlayment to crack.

Installation of a gypsum-based underlayment must be done in accordance with its manufacturer's recommendations. These will vary from manufacturer to manufacturer and from product to product. Probably the worst and most frequent problem is overwatering the mix, which lowers the underlayment's compressive strength and ability to resist point loads.

Another major problem occurs when finish flooring is installed before the underlayment has had adequate time to cure. Normally, a $3/4$-in. (19-mm)-thick underlayment will be ready to receive finish flooring in 5 to 7 days, but the actual time required will vary, depending on the ambient temperature and humidity. The underlayment should be tested as recommended by its manufacturer for its moisture content before flooring application.

## 3.18.2 CEMENT-BASED UNDERLAYMENTS

Cement-based underlayments are self-leveling, polymer-modified hydraulic cement products. They are primarily used to level and resurface existing cast-in-place and precast concrete, terrazzo, and tile substrates. They can also be used over vinyl asbestos tile, extruded or expanded polystyrene foam, and wood. They are not a finish flooring, nor are they structural in nature. They can serve as a substrate for wood, tile, carpet, and resilient finish flooring.

### 3.18.2.1 Products

Cement-based underlayments are proprietary products composed of a binder of portland cement or hydraulic or blended cement, as defined in ASTM C219, washed-gravel or sand aggregate, and water. Some manufacturers also recommend that some chemical additives be included in cement-based underlayment that will be used over certain substrates, such as cut-back adhesives, wood, or metal.

### 3.18.2.2 Installation

Cement-based underlayments are generally installed in thicknesses varying from $1/8$ to 6 in. (3 to 150 mm). Different manufacturers have different maximum thickness and filler requirements. Without a filler (sand) added, the limits vary from $1/2$ in. (12 mm) to $1^1/2$ in. (38 mm). Most manufacturers recommend that a filler be used at all thicknesses and require it when the thickness increases beyond the $1/2$ in. (12 mm) to $1^1/2$ in. (38 mm) maximums. Some manufacturers recommend the use of several layers of unfilled material to provide increased thicknesses.

The requirements for placing and curing concrete discussed elsewhere in this chapter apply to cement-based underlayments.

# 3.19 Additional Reading

More information about the subjects discussed in this chapter can be found in the references listed in Section 3.20 and in the following publications.

Ambrose, James. 1996. *Simplified Design of Concrete Structures*, 7th ed. Hoboken, NJ: John Wiley & Sons, Inc. (www.wiley.com).

Butt, Thomas K. 1994. "Concrete Slab Floor Flatness and Levelness Tolerances." *The Construction Specifier* (November): 30 (www.csinet.org).

DiPasquale, Raymond A. 2000. "Tilt-Up Construction." *The Construction Specifier* (December): 29 (www.csinet.org).

Dobrowolski, Joseph A. 1998. *Concrete Construction Handbook*. New York: McGraw-Hill (www.mcgraw-hill.com).

Ferguson, Phil M., John E Breen, and James O Jirsa. 1988. *Reinforced Concrete Fundamentals*, 5th ed. Hoboken, NJ: John Wiley & Sons, Inc. (www.wiley.com).

Gencarelli, Frank. 1992. "Fiber for Secondary Reinforcement." *The Construction Specifier* (December): 123 (www.csinet.org).

Gerwick, Ben C. 1997. *Construction of Prestressed Concrete Structures*, 2nd
ed. Hoboken, NJ, John Wiley & Sons, Inc. (www.wiley.com).

Hime, William G. 1995. "Avoiding Gypsum Concrete Failures." *The Construction Specifier* (July): 38 (www.csinet. org).

King, Stephen M. 1995. "Shrinkage-Compensating Concrete Floors." *The Construction Specifier* (July): 30 (www.csinet.org).

Lin, Tung Yen and Ned H Burns. 1981. *Design of Prestressed Concrete Structures*, 3rd ed. Hoboken, NJ: John Wiley & Sons, Inc. (www.wiley.com).

Mailvaganam, Noel P. 1993. "Concreting with Chemical Admixtures." The Construction Specifier (July): 106 (www.csinet.org).

McCormac, Jack C. 2000. *Design of Reinforced Concrete*, 5th ed. Hoboken, NJ: John Wiley & Sons, Inc. (www.wiley.com).

McDougle, Edwin A. 1995. "GFRC Comes of Age." *The Construction Specifier* (December): 46 (www.csinet.org).

Michaud, Pete 1996. "Design Guidelines for Concrete Anchoring." *The Construction Specifier* (December): 38 (www.csinet.org).

Nawy, Edward G. 2003. *Reinforced Concrete: A Fundamental Approach*, 5th ed. Upper Saddle River, NJ: Prentice Hall (www.vig.prenhall.com).

Nicastro, David H. 1994. "Cracking in GFRC." *The Construction Specifier* (September): 160 (www.csinet.org).

Nilson Arthur H. 1987. *Design of Prestressed Concrete*, 2nd ed. Hoboken, NJ: John Wiley & Sons, Inc. (www.wiley.com).

Nilson, Arthur H., David Darwin, and Charles W. Dolan. 2004. *Design of Concrete Structures*. New York: McGraw-Hill (www.mcgraw-hill.com).

Peurifoy, R. L. 1996. *Formwork for Concrete Structures*. New York: McGraw-Hill (www.mcgraw-hill.com).

Raeber, John A. 1991. "Architectural Precast Concrete Finishes." *The Construction Specifier* 12 (December): 29 (www.csinet.org).

———. 1994. "Glass-Fiber Reinforced Concrete." *The Construction Specifier* (October): 21 (www.csinet.org).

Robinson, Ralph C. 1997. "Architectural Precast Concrete Connections." *The Construction Specifier* (December): 32 (www.csinet.org).

Socha, Dennis, Edwin Jakacki, and William Rector. 1997. "Installation
Guidelines for Full-Thickness Gypsum Underlayments." *The Construction Specifier* (September): 73 (www.csinet.org).

Wang, Chu-Kia and Charles G. Salmon. 1998. *Reinforced Concrete Design*, 6th ed. Hoboken, NJ: John Wiley & Sons, Inc. (www.wiley.com).

# 3.20 Acknowledgments and References

## ACKNOWLEDGMENTS

We gratefully acknowledge the assistance of the following organizations and individuals in preparing this chapter. We are also indebted to them for permission to use their illustrations when requested and for the use of their publications as references.

American Concrete Institute (www.concrete.org)
ASTM International (www.astm.org)
Burke Company
Concrete Reinforcing Steel Institute (www.crsi.org)
ELE International, Inc. (www.soiltest.com)
Gypsum Association (www.gypsum.org)
Martin Fireproofing Corporation (www.martinfireproofing.com)
Martin Fireproofing Georgia, Inc.
Maxxon Corporation (www.maxxon.com)
National Academy of Sciences (NAS), Building Research Board (www.nas.edu)
National Association of Home Builders (NAHB) (www.nahbrc.com)
Portland Cement Association (PCA) (www.cement.org)
Precast/Prestressed Concrete Institute (PCI) (www.pci.org)
U.S. Department of Housing and Urban Development (HUD)(www.hud.gov)
Wire Reinforcement Institute (www.wirereinforcementinstitute.org)

## REFERENCES

We would also like to thank the authors and publishers of the publications in the following list for their contribution to our research for this chapter.

American Concrete Institute (ACI). Farmington Hills, MI (www.concrete.org).

———. 2002. ACI 211.1: *Standard Practice for Selecting Proportions for Normal, Heavy-weight, and Mass Concrete.*

———. 1999. ACI 212.3R: *Chemical Admixtures for Concrete.*

———. 1999. ACI 301: *Specifications for Structural Concrete for Buildings.*

———. 2004. ACI 302.1R: *Guide for Concrete Floor and Slab Construction.*

———. 1999. ACI 305R: *Hot Weather Concreting.*

———. 2002. ACI 306R, *Cold Weather Concreting.*

———. 2001. ACI 308R, *Guide to Curing Concrete.*

———. 1996. ACI 309R: *Guide for Consolidation of Concrete.*

———. 2002. ACI 318: *Building Code Requirements for Structural Concrete, and Commentary.*

———. 1999. ACI 332R: *Guide to Residential Cast-in-Place Concrete Construction.*

———. 2003. ACI 347R: *Guide to Formwork for Concrete.*

———. 2004. *Manual of Concrete Practice.*

American National Standards Institute (ANSI). ANSI A56.1, *Building Code Requirements for Excavations and Foundations.* New York: ANSI (www.ansi.org).

American Welding Society. 1998. D1.4, "Structural Welding Code-Reinforcing Steel." Miami, FL: AWS (www.aws.org).

Arcom Master Systems. *Masterspec®.* Salt Lake City, UT (www.arcomnet.com). Basic Sections:

03120, "Architectural Cast-In-Place Concrete Formwork," 8/91 ed.

03300, "Cast-in-place Concrete," 8/92 ed.

03350, "Concrete Toppings," 5/91 ed.

03355, "Special Concrete Finishes," 8/94 ed.

03361, "Shotcrete," 5/95 ed.

03410, "Structural Precast Concrete—Plant Cast," 2/94 ed.

03450, "Architectural Precast Concrete—Plant Cast," 8/94 ed.

03455, "Glass-Fiber Reinforced Concrete—Plant Cast, 5/92 ed.

03470, "Tilt-Up Precast Concrete," 5/91 ed.

03520, "Insulating Concrete Decks," 2/95 ed.

03531, "Cementitious Wood-Fiber Planks," 5/93 ed.

03542, "Cement-Based Underlayment," 8/97 ed.

ASTM International (ASTM). West Conshohocken, PA. (www.astm.org). Standards:

A82, "Standard Specification for Steel Wire, Plain, for Concrete Reinforcement."

A185, "Standard Specification for Steel Welded Wire, Fabric, Plain, for Concrete Reinforcement."

A416, "Standard Specification for Steel Strand, Uncoated Seven-Wire for Prestressed Concrete."

A497, "Standard Specification for Steel Welded Wire Fabric, Deformed, for Concrete Reinforcement."

A615, "Standard Specification for Deformed and Plain Billet-Steel Bars for Concrete Reinforcement."

A706/A706M, "Standard Specification for Low-Alloy Steel Deformed and Plain Bars for Concrete Reinforcement."

A767, "Standard Specification for Zinc-Coated (Galvanized) Steel Bars for Concrete Reinforcement."

A775, "Standard Specification for Epoxy-Coated Steel Reinforcing Bars."

C31, "Standard Practice for Making and Curing Concrete Test Specimens in the Field."

C33, "Standard Specification for Concrete Aggregates."

C39, "Standard Test Method for Compressive Strength of Cylindrical Concrete Specimens."

C94, "Standard Specification for Ready-Mixed Concrete."

C143, "Standard Test Method for Slump of Hydraulic Cement Concrete."

C150, "Standard Specification for Portland Cement."

C219, "Standard Terminology Relating to Hydraulic Cement."

C226, "Standard Specification for Air-Entraining Additions for Use in the Manufacture of Air-Entraining Hydraulic Cement."

C260, "Standard Specification for Air-Entraining Admixtures for Concrete."

C309, "Standard Specification for Liquid Membrane-Forming Compounds for Curing Concrete."

C317, "Standard Specification for Gypsum Concrete."

C330, "Standard Specification for Lightweight Aggregates for Structural Concrete."

C494, "Standard Specification for Chemical Admixtures for Concrete."

C595, "Standard Specification for Blended Hydraulic Cements."

C618, "Standard Specification for Coal Fly Ash and Raw or Calcined Natural Pozzolan for Use as a Mineral Admixture in Concrete."

C956, "Standard Specification for Installation of Cast-In-Place Re-inforced Gypsum Concrete."

C1116, "Standard Specification for Fiber-Reinforced Concrete and Shotcrete."

D98, "Standard Specification for Calcium Chloride."

D2166, "Standard Test Method for Unconfined Compressive Strength of Cohesive Soil."

D4318, " Standard Test Method for Liquid Limit, Plastic Limit, and Plasticity Index of Soils."

E1155, "Standard Test Method for Determining $F_F$." and $F_L$ Floor Levelness Numbers."

Concrete Reinforcing Steel Institute (CRSI) 1997. *Placing Reinforcing Bars.* Schaumburg, IL: CRSI (www.crsi.org).

———. 2001. *Manual of Standard Practice.* Schaumburg, IL: CRSI.

National Academy of Sciences. 1968. *Criteria for Selection and Design for Residential Slabs-on-Ground.* Washington, DC: NAS (www.nas.edu).

Portland Cement Association (PCA). Skokie, IL (www.cement.org).

———. 1965. "Moisture Migration—Concrete Slab-on-Ground Construction."

———. 1996. *Resurfacing Concrete Floors.*

———. 2001. *Concrete Floors on Ground.*

———. 2002. *Batching, Mixing, Transporting and Handling Concrete.*

———. 2003. *Concrete Slab Surface Defects: Causes, Prevention, Repair.*

———. 2002. *Design and Control of Concrete Mixtures,* 14th ed.

Precast/Prestressed Concrete Institute (PCI). Chicago (www.pci.org). 1999. MNL 116, *Manual for Quality Control.*

———. 1996. MNL 117, *Manual for Quality Control for Plants and*

*Production of Architectural Precast Products*, 3rd ed.

———. 1999. MNL 120, *PCI Design Handbook—Precast and Prestressed Concrete*, 5th ed.

———. 1989. MNL 122, *Architectural Precast Concrete* 2nd ed.

———. 1999. MNL 127, *Erectors' Manual: Standards and Guidelines for the Erection of Precast Concrete Products*.

———. 2001. MNL 128, *Recommended Practice for Glass Fiber Reinforced Concrete Panels*, 4th ed.

———. 1991. MNL 130, *Manual for Quality Control for Plants and Production of Glass Fiber Reinforced Concrete Panels*.

———. 2003. CTG, *Architectural Precast Concrete—Color and Texture Selection Guide*, 2nd ed.

Pruter, Walter F. 1987. "The Advantages of Perlite Concrete Roof Deck Systems." *The Construction Specifier* (November): 52–55 (www.csinet.org).

Ramsey, Charles George, Harold Reeve Sleeper, and John Ray Hoke, Jr. 2000. *Architectural Graphic Standards*, 10th ed. Hoboken, NJ: John Wiley & Sons, Inc. (www.wiley.com).

Underground Space Center. 1988. *Building Foundation Design Handbook*. Minneapolis: Underground Space Center (Now the Center for Sustainability Research) (www.cala.umn.edu).

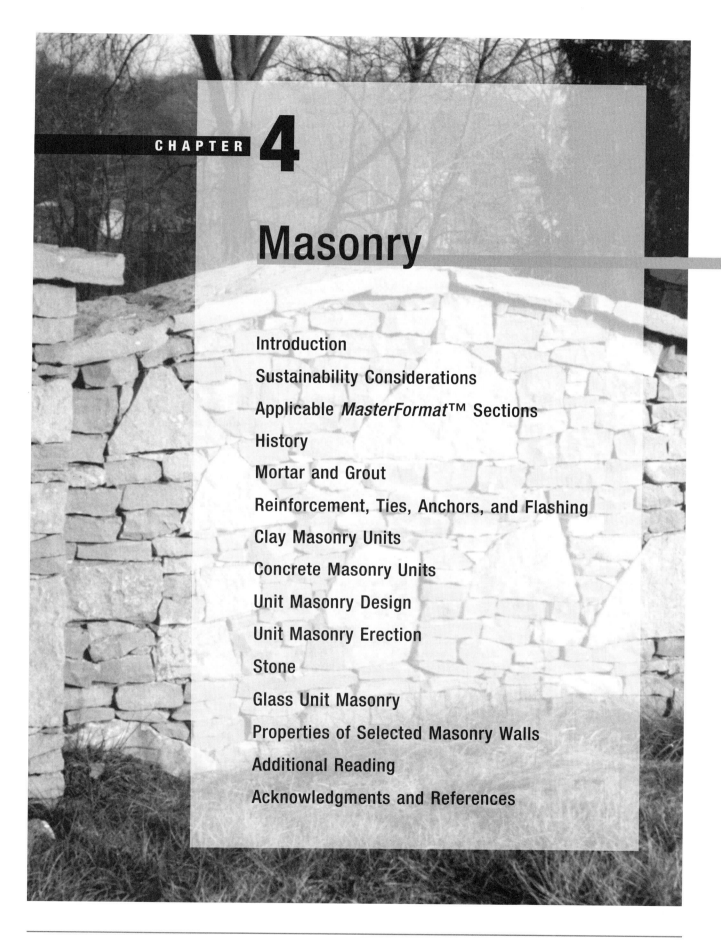

**CHAPTER 4**

# Masonry

# Introduction

Stone, clay masonry units, and concrete masonry units are all used extensively in construction today. Brick, which is one of the oldest manufactured building materials, is an example of clay masonry. Concrete masonry units include concrete block, a relatively new material. To take advantage of their compressive strength, durability, and fireproof character, these products are used primarily in wall assemblies, often appearing together. In addition, all three are also used as decorative screening. Stone and clay masonry units are also suitable for floor surfacing and exterior paving. Lightweight aggregates have made lighter concrete masonry units possible and have permitted the fabrication of larger masonry panels than earlier technology permitted. The development of high-bond-strength mortars has made prefabricated brick masonry panels practicable. Continuing improvement of sealants, adhesives, and masonry units will assure continued use of these time-proven masonry materials.

To serve their intended function, masonry walls must perform as they were designed to do. Exterior walls, for example, must possess at least four essential functional properties. They must be:

- Strong enough to carry applied loads
- Watertight enough to prevent groundwater and storm water penetration
- Durable so that they will not quickly erode due to weather or other atmospheric conditions
- Resistant to heat, sound, and fire transmission when these properties are required

Esthetic properties often influence the selection of masonry units that will be exposed to view. Appearance is affected by the following:

- Color of both the units and the mortar
- Texture of the units, glazes applied to the units, and texture created by the method of laying the units
- Pattern produced by the type of bond and the unit

Assuming that the design had been done correctly, the procedures for producing satisfactory masonry assemblies can be summarized as follows.

- Materials must be sound.
- Materials must be properly handled, stored, protected, and prepared.
- Mortar type must be the one best suited for assembly use and exposure.
- Mortar constituents must be proportioned and mixed correctly.
- Units must be laid with full head and bed joints.
- Joints must be tooled properly.
- Masonry must be protected from damage by the weather during construction.
- Water must be prevented from penetrating to the inside of the assembly after it has been built by properly located and installed flashing and weep holes.

This chapter describes materials and procedures that will help to produce masonry of suitable quality.

# Sustainability Considerations

Brick is one of the more sustainable building materials in general use today. The raw materials used, clay and shale, are among the earth's most abundant materials. The energy used to make brick, although more than 14,000 Btu per brick. is less than that used to produce concrete, steel, and many other materials. Since brick is inert, it does not pose any long-term environmental threats. And bricks can usually be recycled when the building they are used in is demolished.

Some of the same considerations apply to concrete masonry units that were discussed in Chapter 3 for reinforced concrete. The same materials are used in both reinforced concrete and concrete masonry units, and the same environmental effects occur from obtaining, refining, and using these materials. Both uses consume large quantities of water, but concrete masonry units are produced in a controlled environment that can result in less wasted water, such as the water used to clean mixing trucks, which is usually wasted.

Concrete masonry units are produced in reusable molds with far less wastage than usually occurs with the use of conventional forms in the field for homogeneous concrete. A major advantage of using concrete masonry units instead of homogeneous concrete occurs when the building in which they are used reaches the end of its useful life span. Unlike homogeneous concrete, concrete masonry units are readily salvageable.

# Applicable *MasterFormat* Sections

The following *Masterformat* 2004 Level 2 sections are applicable to this chapter.

| | | | |
|---|---|---|---|
| 04 21 00 | Clay Unit Masonry | 04 25 00 | Unit Masonry Panels |
| 04 22 00 | Concrete Unit Masonry | 04 27 00 | Multiple-Wythe Unit Masonry |
| 04 23 00 | Glass Unit Masonry | 04 40 00 | Stone Assemblies |
| 04 24 00 | Adobe Unit Masonry | 04 50 00 | Refractory Masonry |

# 4.1 History

Masonry structures today are made from brick and block, which together are called masonry units, or from stone. Masonry structures predate written history. The earliest structures were huts made from unshaped native field stones piled upon one another without mortar or other materials in the joints. Where stone was not available, sod or dried mud served the same purpose. Later, clay and silt were mixed with water and formed by hand into bricks. The spaces between these bricks were sometimes packed with mud to keep out the wind and rain and to make it easier to build level walls with irregular bricks. Later still, it was discovered that clay bricks placed in or adjacent to a fire became harder and more weather resistant. The Romans used this knowledge to build kilns to produce burned-clay roofing tiles and eventually burned-clay bricks.

The Mesopotamians were constructing stone and sun-dried brick buildings 4000 years before the birth of Christ. A thousand years later, the Egyptians began building temples and pyramids of cut stone by laboriously cutting the stone using bronze tools so that the pieces fit closely together.

Early stone buildings were limited in that openings and column spacings could be only as large as could be spanned with a single stone. Roofs were made of wood. The Babylonians built small arches over windows and other small openings, but the Romans were the first to build arches large enough to produce bridges and large buildings.

When it became possible to make stoneworking tools from iron, the art of stone building developed to a high order. The Greeks refined the process to produce fine details in stone. The Romans built, for the first time, buildings with large open spaces. They used arches and vaults to build temples, forum buildings, theaters, amphitheaters, baths, aqueducts, and homes.

The building of brick and stone structures without joint materials changed when the Etruscans developed a lime mortar that could be used to fill gaps between masonry units. Later, the Romans discovered how to make a hydraulic (that which will set underwater) cement by burning and grinding a type of volcanic rock. This led to stronger and more watertight stone and brick structures. This same discovery led to the expanded use of concrete.

The sun-baked bricks used extensively by ancient peoples began to disappear in sophisticated cultures after the Romans invented kilns. Clay could then be baked hard enough that its resistance to the elements increased dramatically.

The fall of the Roman Empire was followed by the emergence of the semioriental Byzantine Empire with its capital in Constantinople (Istanbul). The Byzantine Empire flourished for a 1000 years until it was overrun in 1453 by the Ottoman Turks. Byzantine architecture combined Roman arch forms with other shapes and added detail and color. The development of the *pendative* permitted the construction of a stone dome over a square space. The Romans had been limited to circular supporting shapes, such as in the Pantheon. A pendative is a spherical triangle whose lower point rests on a pier and whose curved upper edge, in combination with three other pendatives, forms a circle that can support a dome. Using this device, the Romans built large stone domes over square buildings and were able to construct buildings with several domes. The Byzantines also used broad arches and massive buttresses to add aisles and galleries to their buildings.

Beginning in the eleventh century, a 200-year-long revival, called the *Romanesque period*, saw the beginning of types of stone structures that would be refined in later periods. These included ribbed vaults, membered piers, and pier buttresses.

The *Gothic period,* which followed, saw the development of flying buttresses that could withstand the thrust from large vaulted and domed structures. This allowed construction of even larger and more complex structures and, when combined with vaults and piers, produced what has been called a "cage of stone." Walls could be replaced with windows, leading to the development of ornate stained glass windows, especially in the cathedrals of western Europe.

Beginning in the fourteenth century, the *Renaissance (Age of Discovery)* stirred an interest in the neoclassical. This was a period of great strides in the arts, humanities, and science.

In the late eighteenth and early nineteenth centuries (the "age of revivals"), architects actually measured Roman and Greek ruins, Gothic buildings, and other former styles and reproduced them as new buildings. However, little was done to further the science of building. In the late eighteenth century, the Industrial Revolution ushered in the modern era. Machines began to replace much of the handwork necessary to quarry and cut stone and to mold and fire bricks. Stone units became more uniform in shape and size. Bricks became more consistent in color, strength, and size.

Blocks made from cement and various aggregates appeared after the development of portland cement and portland cement concrete in the late eighteenth and early nineteenth centuries. They were cheaper and lighter than cut stone and larger than bricks, which reduced the time needed to lay them. They were more versatile than concrete in that they could be easily placed in small quantities and did not require formwork.

Until the development of the theory of elasticity in the nineteenth century, unit masonry and stone construction were based solely on experience. After that, masonry structures could be built using rational design based on calculated stresses.

Until the late nineteenth century, masonry and stone were the materials of choice throughout the world for constructing buildings, bridges, and viaducts. The advent of cast iron and steel structural elements quickly replaced masonry and stone as the major structural support elements in large buildings. Reinforced concrete had the same effect in smaller buildings and in bridges and other arched structures. The building of large unit masonry and stone arches reached its climax at about the turn of the twentieth century, when they were replaced by reinforced concrete. Concrete was less labor intensive, easier to form and place, and it could be made to span distances three to four times that possible in stone.

Although they have been largely replaced by steel and concrete as primary load-carrying elements in larger buildings, unit masonry and stone remain in great use today for cladding, partitions, and flooring. They are especially valuable where fire and weather resistance are required. Their varied appearance is used to create a variety of esthetic effects. They are adaptable to current standards and methods for insulating buildings to make them more comfortable and to increase their ability to conserve energy. Modern masonry cavity walls resist water penetration and provide space to house insulation. Masonry partitions are fire resistant and help to restrict sound penetration. New mortars and construction adhesives reduce the labor needed to place masonry walls, and some walls are built from prefabricated masonry panels and held together with adhesives.

Further advances are certain. In spite of innovations in metal and concrete construction techniques and materials, masonry and stone construction are likely to be with us for a very long time.

# 4.2 Mortar and Grout

Mortar is a combination of one or more cementitious materials (portland cement, lime, or masonry cement), a clean, well-graded aggregate, such as sand, and enough water to give the mixture a plastic, workable quality. The term *grout* refers to several different types of construction products, depending on the material and its use. Here we are concerned with only two types of grout. One type is used to fill voids in both reinforced and unreinforced masonry. This type is made from the same ingredients as is mortar, but is produced in a pourable consistency. The various ingredients must be proportioned properly to give the mortar or grout a good balance of essential properties. The second type of grout is a tile-setting grout used in filling the joints in unit masonry or stone paving and flooring. In this section, the term *grout* refers to the material used to fill voids unless specifically identified as the type used in paving and flooring joints.

Mortar is placed in the joints between individual masonry or stone units in a wall or other building element to cushion the units and to provide a level setting bed for them. It also seals the spaces between the units, compensates for size variations in the units, and provides esthetic quality by creating shadow lines and color effects. Grout is poured into the voids between masonry unit wythes and into the cores in hollow masonry units to produce solid filled and reinforced masonry walls. Mortar and grout also grip metal ties and reinforcement, helping them to act integrally with the other parts

of a wall. Tile-setting-type grout is used to fill the joints between brick pavers.

## 4.2.1 MORTAR AND GROUT PROPERTIES

Because mortar and grout are placed while plastic and then harden, they must have two sets of properties: (1) those present when they are in their plastic state and (2) those that result after they have hardened. Proper plastic properties and hardened properties are both necessary for a mortar or grout to be suitable for use in building construction. Both sets of properties affect a finished wall's strength, durability, and watertightness.

The basic constituents of most modern masonry mortars and grouts are portland cement, lime, sand, and water. Masonry cement also contains other ingredients (see Section 4.2.2.4). Sometimes additives are used in portland cement-lime mortars and grouts, even though some additives are harmful to the hardened mortar or grout. Since both the plastic and hardened properties of a mortar or grout depend largely on its ingredients, these should conform to the appropriate standards of ASTM International (ASTM) and must be properly proportioned. Because a mortar's or grout's properties also depend on the proportions of its ingredients, different mortar types (see Section 4.2.4.1) and grout mixes have different properties.

### 4.2.1.1 Plastic Properties

The properties of mortars and grouts while they are still plastic include *workability* and *water retention*, including *initial flow* and *flow after suction*.

#### WORKABILITY

A *workable* mortar is uniform, cohesive, and of a consistency that makes it usable by a mason. A mortar is workable when particles in the mix do not segregate and when it spreads easily, holds the weight of the units, makes alignment easy, clings to the vertical surfaces of masonry units, and easily extrudes from the mortar joints without dropping or smearing.

Water retention, flow, resistance to segregation, and other factors affect a mortar's workability. These, in turn, are affected by the properties of the mortar ingredients. This complex relationship makes quantitative estimates of worka-

bility difficult. There is, in fact, no standard laboratory test for measuring it.

Grout is designed to pour and therefore must flow freely without being too soupy. It also must be uniform.

#### WATER RETENTION

Water retention in a mortar or grout prevents rapid loss of water and a resultant loss of plasticity when the mortar or grout contacts a masonry unit with a high absorption rate. A high degree of water retention also prevents a mortar from bleeding when it comes into contact with a masonry unit that has a low absorption rate. *Bleeding* is a process in which water leaves the mortar and is deposited in a thin layer between the masonry unit and the mortar. When this happens, the unit is said to *float*. Such floating materially reduces bond.

The water retention of a mortar is the ratio of a plastic characteristic called *flow* immediately after *mixing* to the *flow* of the same mortar after *suction*. The flow of a mortar immediately after mixing (initial flow) is measured by a simple laboratory test (Fig. 4.2-1). A truncated cone of mortar 4 in. (100 mm) in diameter is placed on a testing apparatus called a *flow table*. The flow table is then mechanically raised and dropped 25 times in 15 seconds through a height of $1/2$ in. (12.7 mm).

**FIGURE 4.2-1** The laboratory test for measuring flow. (Courtesy Portland Cement Association.)

This movement causes the mortar to flow outward, increasing the diameter of the cone. The flow is the ratio of the final diameter to the original diameter of the cone, expressed as a percentage. For example, if the diameter of the mortar mass is 8 in. (200 mm) after the test, the flow is 100%. ASTM C270 requires an initial flow of 110%, plus or minus 5% for specimens tested for water retention or compressive strength.

A *flow-after-suction* test is then used to predict the theoretical flow of a mortar after it has lost water to an absorbent masonry unit. In this test, a sample of the same mortar previously tested for initial flow is placed over an absorbent filter in a perforated metal dish mounted over a vacuum device. Operation of the vacuum device removes some water from the mortar. This mortar is then removed from the suction device and tested for flow in the same way as the first test sample. The laboratory value of water retention is the ratio of flow after suction to initial flow, expressed as a percentage. The formula is $(A/B) \times 100$, where $A$ is the flow after suction and $B$ is the flow immediately after mixing. ASTM C270 requires a water retention of not less than 75%.

### 4.2.1.2 Hardened Properties

Among the properties of hardened mortar are *bond strength*, *durability*, *compressive strength*, *volume change*, *water tightness*, *rate of hardening*, and *mortar color*. The most important of these are bond strength and compressive strength.

#### BOND STRENGTH

Bond strength is perhaps the most important property of hardened mortar and an important characteristic of hardened grout, especially of grout used in conjunction with the type of terra cotta called *ceramic veneer* (see Section 4.7.7). Some variables that can affect bond strength are addressed in this section.

A mortar's or grout's other properties often determine its hardened bond strength. Properties that may affect bond strength include the type and amount of cementitious materials and the amount of water in the mix (see Sections 4.2.4.3 and 4.2.5), the water retention characteristics of the mortar or grout (see Section 4.2.1.1), the entrained-air content of the mortar or grout, and the mortar's or grout's compressive strength.

The amount of water in a mortar mix and its water retention affect its flow (see Section 4.2.1.1). The more water, the greater the flow. A mortar's tensile bond strength is a mechanical function rather than a chemical one. The mortar flows into pores in the masonry unit and interlocks with them. Therefore, bond strength always increases as the mortar's flow increases, because wetter mortar flows more readily into these pores. Adding lime to portland cement mortar increases its workability and therefore facilitates the interlocking (bonding) process. Too little water can also affect bonding (see Section 4.2.6).

When the air content of a mortar or grout exceeds 12%, there is an accompanying decrease in bonding.

The type of masonry unit in contact with a mortar or grout can affect its bond strength. Masonry unit characteristics that can affect the bond include surface texture, the suction of clay masonry units (see Section 4.7.1.2), and the moisture content of concrete masonry units (see Section 4.4.2.3).

Workmanship can also affect the bond strength of a mortar. For example, movement of a masonry unit after it has been laid can cause loss of bond, as can water on the surface of a clay masonry unit or a delay in laying masonry after the mortar has been placed. The air temperature and relative humidity during a mortar's curing period can also affect its bond strength.

#### DURABILITY

The durability of mortar is measured principally by its ability to resist water penetration. Water entering the mortar in a masonry joint, and thus the adjacent masonry, can damage the mortar and the adjacent masonry in several ways. Repeated cycles of freezing and thawing of such water under natural weather conditions will cause the surface of the mortar or masonry unit to pop off (*spall*). Water entering a mortar will dissolve salts in the mortar. These salts will then leach out into or onto the surface of adjacent masonry. Upon drying, these salts will crystallize. Salts that crystallize on the exterior face of a wall (*efflorescence*) are unsightly but do little harm. However, when the crystallization occurs within a wall (*subflorescence* or *cryptoflorescence*), considerable damage can result. Subsequent cycles of wetting and drying

cause larger and larger crystalline deposits to form until the deposits grow large enough to cause the surface of the mortar or masonry units to spall.

An increase in entrained-air content may increase the durability of mortar, but will substantially decrease its bond strength and adversely affect other desirable properties. Because portland cement-lime mortar is sufficiently durable without air entrainment, admixtures should not be used to increase its entrained-air content. Masonry cement mortars tend to be less durable, so air entrainment admixtures are used in them.

#### COMPRESSIVE STRENGTH

The compressive strength of a mortar or grout depends largely on the amount of portland cement in its mix, rising when the cement content is increased. Conversely, a larger flow (see Section 4.2.1.1) brought about by a rise in the water content of a mortar will decrease its compressive strength.

Since there are few reports of structural distress or failure due to low compressive strength problems, it is not necessary to use greater than moderate strength mortars in normal construction. Bond strength, workability, and water retention are more important and are generally given a higher priority when mortar is being selected. Conversely, using mortar with too high a strength can cause significant problems, especially in repointing older masonry. A high-strength mortar used in an existing joint may be stronger and less flexible than the existing mortar or even the masonry units. Normal thermal and other movements can cause the stronger mortar to tear away from the weaker mortar, causing cracks that may leak, and may even cause the masonry units to chip or crack. Therefore, mortar strength should never be higher than is necessary for the prevailing conditions, and in existing walls new mortar should match the existing mortar in composition and materials.

#### VOLUME CHANGE

Some people believe that mortar shrinkage can be extensive and can cause leaky walls. Actually, the maximum possible shrinkage in a properly tooled mortar joint containing sound mortar that has a good balance of all desirable properties is so small that any resultant crack cannot be seen with the naked eye.

## WATERTIGHTNESS

When a wall leaks significantly and there are no major holes in it, the leaks are usually through fine cracks between the mortar and the masonry or stone units, especially in the vertical joints. The watertightness of the masonry units themselves or of mortars commonly used today is seldom a factor in wall leaks.

## RATE OF HARDENING

The rate of hardening of mortar or grout is the speed at which it develops a resistance to indentation and crushing. Too rapid hardening may interfere with the use of a mortar by a mason or the pouring of a grout. Too slow hardening may impede the progress of the work or may subject a mortar or grout to early damage from frost action during winter. A well-defined, consistent rate of hardening allows a mason to tool joints at the same degree of hardness and thus obtain uniform joint color.

## MORTAR COLOR

Color is an inherent characteristic of hardened mortar. The materials needed to produce colored mortar are discussed in Section 4.2.3.4. Regardless of the color, color uniformity affects the overall appearance of masonry. Several factors other than the materials used can affect the uniformity of color and shade of a hardened mortar. They include atmospheric conditions, the moisture content of the masonry units, the admixtures used, the control exercised over the mixture, and the timing of the mason's tooling of the joints.

Controlling the mix from batch to batch and day to day is necessary if a mortar is to have the same color throughout. Careful measurement of materials and thorough mixing are important in providing uniformity.

The timing of joint tooling influences the shade of a hardened mortar. If the mason tools the joint when the mortar is relatively hard, a darker color results than if he tools the joint when the mortar is relatively soft. To ensure a uniform mortar color in the finished wall, the mason should tool all the joints at like degrees of mortar hardness (Fig. 4.2-2).

## 4.2.2 KINDS OF MORTAR AND GROUT

Four kinds of mortars and two kinds of grout are described in this section. The

**FIGURE 4.2-2** Joints should be tooled when the mortar is at a like degree of hardness to prevent variations in color. (Courtesy Portland Cement Association.)

mortars are sand-lime, portland cement, portland cement-lime, and masonry cement. All of the discussions about mortar elsewhere in this chapter refer to either portland cement-lime or masonry cement mortars, and they are the only ones covered in ASTM C270. Lime mortars and portland cement mortars are seldom used today except in small special cases, such as in patching a like mortar.

The two kinds of grout discussed here are portland cement grout and portland cement-lime grout. Lime mortars are not suitable for grout. Masonry cement is usually not permitted in grout.

### 4.2.2.1 Sand-Lime Mortars

Until relatively recently, all mortar was a mixture of lime, sand, and water. The lime was the cementing material. Sand-lime mortars harden at a slow, variable rate, develop low compressive strength, and have poor durability when subjected to freeze-thaw cycles. Therefore, the joints in many old masonry installations that were built using sand-lime mortar have deteriorated until there is almost no binder left. The sand can be raked from the joint with a finger.

However, sand-lime mortars have high workability and high water retention, and they are more flexible than portland cement mortars. Because lime hardens only when in contact with air, complete hardening of a lime mortar occurs very slowly over a long period of time. This is a disadvantage in terms of project progress. The advantage of this slow hardening is that even when cracks occur, they often heal spontaneously during the hardening process, which tends to help keep water penetration to a minimum.

Sand-lime mortar is seldom used today except in renovation work where an existing masonry was laid using a similar material.

### 4.2.2.2 Portland Cement Mortars and Grouts

Portland cement mortars and grouts (portland cement, sand, and water) harden quickly at a consistent rate and develop high compressive strengths. They also have good durability when subjected to freeze-thaw cycles, but their workability and water retention are low. Portland cement grouts are used in the installation of the kind of architectural terra cotta called *ceramic veneer* (see Section 4.7.7). They may also be used in other types of masonry but are not permitted in engineered unit masonry walls.

### 4.2.2.3 Portland Cement-Lime Mortars and Grouts

To combine the advantages of portland cement and lime and to compensate for their disadvantages, and because their mixture produces the most predictable mortars and grouts, portland cement-lime mortars and grouts (portland cement, lime, sand, and water) are widely used. In portland cement-lime mortar and grout, the portland cement contributes durability, high early strength, a consistent rate of hardening, and high compressive strength. Lime adds workability, water retention, and plasticity. Both contribute bond strength. Sand acts as a filler, making the mix economical, and adds strength. Water is the mixing vehicle. It creates plastic workability and initiates the cementing action.

### 4.2.2.4 Masonry Cement Mortars

Masonry cement mortars (masonry cement, sand, and water) are made using proprietary mortar mixes called *masonry cement*. Masonry cement manufacturers combine in one bag such ingredients as portland cement, portland blast-furnace slag cement, portland-pozzolan cement, natural cement, and slag cement. Masonry cement may also contain hydrated lime, limestone, chalk, calcareous shell, talc, or slag, and an air-entraining agent or other admixtures.

The principal advantages of using masonry cement mortar are that it is readily available, convenient to mix, and produces a mortar with generally good workability. The first of those advantages is somewhat reduced, however, by the availability of ready-mixed portland cement-lime mortars, which resemble ready-mixed concrete in that the ingredients are batched and mixed at a central location, and delivered to a project site ready to use.

The principal disadvantage of using masonry cement is that the standard for it, ASTM C91, does not limit the ingredients or their proportions, or control the types or quantities of admixtures a producer can include in it. Therefore, the performance of masonry cement is relatively unpredictable when compared with portland cement-lime mortars made in compliance with ASTM C270, which does limit the ingredients and admixtures in mortar. While it is true that laboratory test results furnished by individual manufacturers for specific masonry cements can serve as a guide to performance, masonry cements are not recommended for use in reinforced masonry or in masonry that may have to withstand lateral forces or heavy loads.

## 4.2.3 MORTAR AND GROUT MATERIALS

ASTM C270, which is the currently recognized standard for mortar for unit masonry work and stone, lists the required standards controlling the ingredients for mortar. ASTM C476 similarly lists the controlling standards for grout ingredients for use in engineered masonry. The listed standards are based on laboratory tests, field use, and many years of experience.

To comply with ASTM C270 and C476, mortar and grout materials must conform to the individual ASTM standards given there. Many of the numbers of those standards are repeated in this section. Some of them are:

- C5—Quicklime.
- C91—Masonry cement.
- C144—Aggregates (sand) for use in mortar.
- C150—Portland cement, Types I or II. Type III may also be used for cold-weather construction. Air-entraining portland cement is also covered, but should not be used in mortars.
- C207—Hydrated lime (Type S).
- C404—Aggregates for use in grout.

Some mortars and grouts, such as those used in terra cotta, for example, may not be required by a code or other governmental regulation, or recommended by an industry standard, to comply with ASTM standards. Nevertheless, it is prudent to require that their ingredients comply with the individual standards listed in ASTM C270 and C476.

### 4.2.3.1 Cementitious Materials

Cementitious materials used in mortar and grout include portland cement, masonry cement, quicklime, and hydrated lime.

#### PORTLAND CEMENT

ASTM C150 covers three portland cement types recommended for use in mortar and grout:

- Type I: Used when no special properties are required
- Type II: Used when moderate sulfate resistance or moderate heat of hydration is desired
- Type III: Used when high early strength is need, such as in cold weather

ASTM C150 also covers several air-entraining types of cement, but these should not be used in masonry mortar or grout because the amount of air entrained is often difficult to predict. Excess entrained air in a mortar reduces the bond between the mortar and masonry units laid with it.

#### MASONRY CEMENT

Masonry cements should conform to the requirements of ASTM C91. They should not be used in grout. Where their use is permitted in mortar, more satisfactory results may be obtained by requiring the use of a specific manufacturer's product which has proven to be acceptable for the intended use. As an alternative, masonry cements may be specified by listing the acceptable properties and limiting the ingredients. Admixtures should also be limited to those that are permitted in mortars by ASTM C270.

Masonry cements are available with color pigments added at the factory. The advantage of using them when colored mortar is desired is that it is much easier to obtain uniformity of color when the pigments are added to the mix in the plant rather than at the job site. Masonry cement is often selected for architectural projects for this very reason. Care must be taken, however, to ensure that the colored masonry cement selected will produce a mortar with the other properties required.

The air content of masonry cement is permitted by ASTM C91 to go as high as 22%. When masonry cement is permitted, however, its air content should be limited to 12%, since higher amounts often result in loss of bond strength.

#### QUICKLIME

Quicklime, which is essentially calcium oxide, CaO, should conform to ASTM C5. It must always be mixed with water (slaked) to the consistency of putty and stored for as long as 2 weeks before use. Though used extensively in the past, quicklime is used in that form today only in preservation work to match an existing mortar.

#### HYDRATED LIME

Hydrated lime must conform with ASTM C207. Hydrated lime is quicklime that has been slaked before packaging, converting the calcium oxide into calcium hydroxide, $Ca(OH)_2$. Hydrated lime, which can be mixed and used without delay, is much more convenient than quicklime. Hydrated lime is available in both S and N types, but only Type S is recommended for masonry mortar and grout, because ASTM C207 does not control the amount of unhydrated oxides in Type N or its plasticity. Air-entraining hydrated lime is also available, but is not suitable for masonry mortars or grout.

### 4.2.3.2 Aggregates

The only aggregate suitable for use in masonry mortars is material that complies with ASTM C144. The standard for aggregates for use in grout is ASTM C404. Either natural or manufactured aggregates may be used, but both must be clean, sound, and well graded. Manufactured aggregate may be ground stone such as marble, granite, or other sound stone of the desired color. The aggregate used has an important effect on the workability and durability of a mortar or grout.

#### AGGREGATE GRADATION FOR MASONRY- AND STONE-SETTING MORTAR

For the best workability and most economy, aggregate for general use should be graded uniformly to include all particle sizes from very fine up to $1/4$ in. (6.4 mm). Aggregates with less than 5% very fine particles produce harsh, hard-to-work mortars that need additional cement or lime to become workable. Aggregates deficient in large particles generally result in weaker mortars. Nevertheless, aggregates for joints narrower than $1/4$ in. (6.4 mm) wide should have 95% passing through a No. 16 sieve and 100% passing through a No. 8 sieve.

Many commercially available sands do not comply with the gradation limits set in ASTM C144 and should not be used in masonry mortar. Fortunately, the gradation of many commercial sands can be altered easily and inexpensively by adding fine or coarse sands as appropriate.

### AGGREGATE GRADATION FOR STONE JOINT-POINTING MORTAR

Aggregates for stone joint-pointing mortar should be very fine, with no granules retained on a No. 16 sieve.

### AGGREGATE GRADATION FOR GROUT

Grout for use in engineered unit masonry is divided in ASTM C476 into two categories, fine and coarse. The aggregate controlled by ASTM C404 is similarly graded, and there are several subgrades within each category. The gradation of the aggregate for a particular grout is determined by the size of the grouted space and the height of the pour.

The aggregate recommended for use in grout associated with ceramic veneer (see Section 4.4.4.3) is pea gravel. This pea gravel is usually defined as aggregate that is uniformly graded from $1/4$ in. (6.4 mm) in size down to that which can pass through a No. 8 sieve. It should comply with the other requirements of ASTM C404.

#### 4.2.3.3 Water

The water used in mortar and grout should be clean and free from deleterious amounts of acids, alkalis, and organic materials. Water that is fit to drink (potable) is generally suitable for use in mortar or grout unless it contains minerals that would be harmful to the mortar.

#### 4.2.3.4 Materials for Colored Mortar

Mortar may be colored either by controlling the color of the ingredients or by adding pigments to the mix.

### COLORED AGGREGATE MORTAR

The preferred method of obtaining colored mortar is by using colored aggregate and cement. This method will usually give permanent color and will not weaken the mortar. Either white or gray

portland cement may be used. The aggregates may be white or brown sand or ground granite, marble, limestone, or other stone. For example, white sand, ground limestone, or ground marble may be used with white portland cement and lime to produce white joints. The disadvantage of this method, of course, is the limit on the number of possible mortar colors.

### COLORED PIGMENTED MORTAR

The second method of producing colored mortar is to add pigments to the mix. Masonry cement is available with the pigments already added. Pigments must be added in the field for colored portland cement-lime mortars.

Mortar pigments must be fine enough to disperse throughout the mix, must be capable of imparting the desired color when used in permissible quantities, and must not react with other ingredients to the detriment of the mortar. It is usually possible to meet these requirements using metallic oxides, such as iron, manganese, and chromium oxides. Carbon black and ultramarine blue have also been used successfully. Organic pigments should be avoided, as should colors containing Prussian blue, cadmium lithopone, and zinc and lead chromates.

The minimum quantity of pigment that will produce the desired results should be used, because an excess may seriously impair strength and durability. The American Concrete Institute's (ACI) ACI 530.1/ASCE 6/TMS 602 limits the amount of mineral oxide pigments to 10% by weight of the cement in portland cement-lime mortar and 5% of the weight of masonry cement in masonry cement mortar. It limits carbon black to 2% of the weight of the cement in portland cement-mortar and 1% of the weight of the masonry cement in masonry cement mortar. It permits no other pigments. Therefore, if black mortar is desired, the best pigment to use is black iron oxide.

For best results, pigments should be premixed with portland cement in large, controlled quantities. Premixing large quantities will assure a more uniform color than can be obtained by mixing smaller batches at the project site.

#### 4.2.3.5 Admixtures

Some admixtures are permissible in mortar and grout. For example, color pigments are an admixture. Other admix-

tures, some of which are permitted in some mortars and grouts, include air-entraining agents, water-repellent agents, accelerators, retarders, and antifreeze compounds. ASTM C270 mentions the use of calcium chloride as an accelerator, but ACI 530.1/ASCE 6/TMS 602 specifically prohibits the use of admixtures containing chlorides in mortar and grout that are used in engineered masonry. Nonchloride accelerating admixtures are available, but if these are necessary, specific products should be selected that are known to give satisfactory results without damage to the properties of the mortar or grout.

ASTM C476 prohibits any kind of antifreeze compound in grout used in engineered masonry and permits other admixtures only with specific approval from the purchaser of the grout. All admixtures should be laboratory tested to ensure that they will not be harmful to a mortar or grout.

Except for latex additives and colored pigments, additives should not be used in mortar that will come into contact with stone. Some additives will discolor or otherwise harm stone.

### 4.2.4 MORTAR TYPES, USES, AND PROPORTIONS

Mortar types are defined in ASTM C270.

#### 4.2.4.1 Mortar Types

The four mortar types described in ASTM C270 (M, S, N, and O) are those recommended for use with unit masonry and stone. Of those, only Types M, S, and N are permitted in engineered unit masonry. The four types are as follows.

### TYPE M

Type M mortar is a high-strength mortar that has somewhat greater durability than other mortar types. It is specifically recommended for use in reinforced masonry and where high compressive strength is necessary. A typical use is in a reinforced masonry wall that is below grade, such as a foundation or retaining wall. It is also used extensively for walks, sewers, and manholes and to set stone in paving installations.

### TYPE S

Type S mortar is a medium-high-strength mortar recommended for use where bond and lateral strengths are more important than high compressive strength. Tensile bond strength between brick and Type S mortar approaches the maximum obtainable with cement-lime mortars. Type S mortar is recommended for use in reinforced masonry and in unreinforced masonry where maximum flexural strength is required. It is also used in setting beds for granite, marble, and slate in other than paving installations.

## TYPE N

Type N mortar is a medium-high-strength mortar recommended for general use in exposed masonry above grade where high compressive and lateral strengths are not required. It is specifically recommended for chimneys, parapet walls, and exterior walls subject to severe exposure. It is also used to set limestone other than in paving installations.

## TYPE O

Type O mortar is a low-strength mortar suitable only for general interior use in non-load-bearing masonry and for tuck pointing. It should never be used where it will be subjected to freezing conditions.

### 4.2.4.2 Mortar Uses

No single mortar type is best for all purposes. The basic rule to follow in selecting a mortar type is to never select one

that is stronger in compression than it needs to be to satisfy the structural requirements of the wall.

Selection of the proper mortar type for each unit masonry use should be based on the masonry type and its location in the building (Fig. 4.2-3). No mortar type rates highest in all desirable properties. Adjustments in the mix to improve one property are often made at the expense of others. For this reason, the properties of each mortar type should be evaluated and the type chosen that will best satisfy the end-use requirements. Figure 4.2-3 shows the recommended mortar types for six common construction applications. Additional information about selection of mortar types can be found in the Brick Industry Association's (BIA) *Technical Notes on Brick Construction*, specifically Technical Notes No. 8, "Mortars for Brick Masonry," 8A, "Standards Specifications for Portland Cement-lime Mortar for Brick Masonry, BIA Designation M1-88," and 8B, "Mortar for Brick Masonry—Selection and Controls."

In addition to the general uses suggested in Section 4.2.4.1, some specific cases should be mentioned. The following sections suggest the mortar types usually selected for the specific conditions indicated. Mortar types for brick and stone flooring and paving are addressed in Sections 9.7 and 9.8, respectively.

## CAVITY WALLS

In cavity walls where wind velocities will exceed 80 miles per hour, Type S mortars should be used. In locations where winds of lower velocity are expected, either Type S or Type N may be used.

## FACING TILE

For facing tile, either M, S, N, or O types are usually acceptable. ASTM C144 allows the use of sand particle sizes as large as $1/4$ in. (6.4 mm). However, facing tile is often laid with $1/4$-in. (6.4-mm)-wide joints. In that case, the aggregate used must pass through a No. 16 sieve, which allows aggregates to pass that are slightly less than $1/16$ in. (1.6 mm) in size.

## DIRT RESISTING MORTAR

For dirt resistance, and when medium resistance to staining is needed, aluminum tristerate, calcium stearate, or ammonium stearate is sometimes added to a masonry mortar or the mortar in ceramic veneer (see Section 4.7.7). When this is done, the amount should be limited to that dictated by code, if applicable, or to that recommended by the masonry unit manufacturer. In the absence of a code restriction or manufacturer's recommendation, the amount should be limited to 3% of the weight of the portland cement in the mortar.

Where maximum dirt resistance is desired, as for facing tile, for example, the mortar should be raked out to a depth of about $3/8$ in. (9.5 mm) and the joints pointed with a mortar consisting of 1 part portland cement, $1/8$ part lime, and 2 parts graded fine (80 mesh) sand, by volume, to which has been added aluminum tristerate, calcium stearate, or ammonium stearate in amounts equal to about 2% of the weight of the portland cement in the mortar.

Admixtures should not be used in mortar for engineered masonry unless specifically allowed by ACI 530/530.1, ASCE 6, TMS 602, and the applicable building code.

## TUCK-POINTING MORTAR

*Tuck pointing* is the refilling of an existing joint from which some of the mortar has been removed. Tuck-pointing mortar must be *prehydrated*. To prehydrate a mortar, the ingredients are thoroughly mixed without water and then mixed again, adding only enough water to produce a damp, unworkable compound that

**FIGURE 4.2-3** Recommended Mortar Types for Various Construction Applications

| Construction Application | Recommended Minimum ASTM Mortar Types | Order of Relative Importance of Principal Properties | | |
| --- | --- | --- | --- | --- |
| | | Plasticity[a] | Compressive Strength | Weather Resistance |
| Foundations, basements, walls, isolated piers[b] | M, S[c] | 3 | 2 | 1 |
| Exterior walls | S, N | 2 | 3 | 1 |
| Solid masonry unit veneer over wood frame | N | 2 | 3 | 1 |
| Interior walls, load bearing | S, N | 1 | 2 | 3 |
| Interior partitions, non-load-bearing | N, O | 1 | — | — |
| Reinforced masonry (columns, pilasters, walls, beams) | M, S[c] | 3 | 1 | 2 |

[a]Adequate workability and a minimum water retention (flow after suction of 70%) assumed for all mortars.
[b]Also any masonry wall subject to unusual lateral loads for earthquakes, hurricanes, etc.
[c]Only portland cement-lime Type S and M mortars.
(Courtesy Portland Cement Association.)

will retain its form when pressed into a ball. After 1 to 2 hours, enough water is added to make the mortar workable; the resulting mortar should be used immediately.

For best results, tuck-pointing mortar should have the same ingredients and mixture as the original mortar. When the original mortar type is unknown, prehydrated Type N mortar is often used.

### TWO DIFFERENT MORTARS IN THE SAME JOINT

Sometimes it is helpful to place two different mortars in the same joint. These mortars should be the same type, however, and masonry cement and portland cement-lime mortars should not be used together. Two different mortars might be installed, for example, where white, black, or colored mortar is desired. One option, of course, is to use the same mortar throughout. However, white, black, and colored mortars are usually more expensive than natural colored mortars, and their properties are not as consistent or predictable as those of mortars that do not contain pigments. An alternative is to lay the unit masonry in a normal, natural-colored mortar. After the units have been laid, rake the joints back to a depth of about $1/2$ in. (12.7 mm), and then point them with the desired white, black, or colored mortar.

### 4.2.4.3 Mortar Proportions

The ingredient proportions required by ASTM C270 for the mortar types listed in Section 4.2.4.1 are shown in Figure 4.2-4.

The mortar used in adhesion-type ceramic veneer is usually required to have a mixture consisting of 1 part portland cement, $1/2$ part lime putty or hydrated lime, and 4 parts sand, by volume.

### 4.2.5 GROUT TYPES, USES, AND PROPORTIONS

In ASTM C476 grout is classified as either fine or coarse. The only difference between the two is the gradation of the aggregate. They both contain 1 part portland cement, 0 to $1/10$ part hydrated lime or lime putty, and damp, loose, fine aggregate in an amount between two and one-quarter and three times the sum of the volumes of the cementitious materials. In addition, coarse grout contains coarse aggregate in an amount equal to between one and two times the sum of the volumes of the cementitious materials. Both grout types are used to fill reinforced masonry walls, tie beams, lintels, and other reinforced masonry items.

Grout for use in anchored-type ceramic veneer is usually required to have 1 part portland cement, 1 part sand, and 5 parts pea gravel.

### 4.2.6 MORTAR AND GROUT MIXING

Thorough mixing is important to the development of desirable properties in a mortar or grout. Because some properties begin to deteriorate soon after a mortar or grout is mixed, no more should be mixed than can be placed before hydration (curing) begins.

Water is essential to the development of bond strength in mortar (see Section 4.2.1.2). If a mortar contains too little water, the bond will be weak and spotty. Therefore, a mortar should be mixed with the maximum amount of water possible, so long as it does not become so soupy that it will not stay where placed, extrude from the joint under the weight of the masonry units, or otherwise become unusable by a mason.

Grout should be mixed with sufficient water to produce a slump of between 8 and 11 in. (200 and 280 mm). Retempering should not be necessary. Grout should be used within $1 1/2$ hours after water was first added to the mix. In hot climates, as are found in some parts of the southern United States, it may be necessary to drastically shorten this maximum time limit. ASTM C270 and C476 give the proportions of the ingredients in mortars and grouts in terms of parts by volume. Ingredients may be measured in the field either by volume or by weight. In either case, the proportions of the ingredients must be accurately controlled if a predictable mortar or grout is to be produced. Both ASTM C270 and C476 contain estimated weights per cubic foot of mortar ingredients.

Whether volume or weight measures are employed, containers of known measure must be used. Cement is usually added directly from the bag, for example, either in full or half bags, depending on the amount of mortar or grout needed. The materials must also be in the condition required by ASTM C270 and C476. For example, moisture in sand causes it to bulk or fluff up, which can make a cubic foot of wet sand contain a different amount of sand than is assumed by the design tables in ASTM C270 and C476. ASTM C270 and C476, therefore, both require that sand be in a damp, loose condition.

Oversanding of mortar happens often because the sand is measured by shovel rather than in a container having a known capacity. Oversanded mortar is harsh and unworkable, provides a weak bond with

**FIGURE 4.2-4** Mortar Proportions by Volume[a]

| Type[b] | Portland Cement | Hydrated Lime or Lime Putty | Masonry Cement | Maximum Damp Loose Aggregate[c] | Minimum Compressive Strength, 2-in. Cubes at 28 Days, psi (MPa) |
|---|---|---|---|---|---|
| M | 1 | $1/4$ | — | 3 | 2500 (17) |
| | or | | | | |
| | 1 | — | 1 | 6 | 2500 (17) |
| S | 1 | $1/2$ | — | $4 1/2$ | 1800 (12) |
| | or | | | | |
| | $1/2$ | — | 1 | $4 1/2$ | 1800 (12) |
| N | 1 | 1 | — | 6 | 750 (5) |
| | or | | | | |
| | — | — | 1 | 3 | 750 (5) |
| O | 1 | 2 | — | 9 | 350 (2.4) |

[a]The weight of 1 cu ft (0.028 m³) of materials used is considered to be: portland cement, 94 lb (42.64 kg), 1 bag; hydrated lime, 40 lb (18.14 kg); lime putty, 80 lb (36.27 kg); dry sand, 80 lb (36.27 kg); masonry cement, weight printed on bag.

[b]Shaded figures show proportions for portland cement-lime mortar. Mortars made with masonry cement are unshaded.

[c]The damp loose aggregate should be not less than $2 1/4$ times, or more than 3 times, the sum of the cementitious material used.

masonry units, and tends to erode. The problem is that the amount of sand a shovel will hold depends on the size of the shovel and the fatigue of the person holding it. Therefore, the measurement of sand solely by shovel should be prohibited. BIA suggests several accurate methods for measuring sand. Refer to BIA Technical Note 8B, "Mortar for Brick Masonry—Selection and Controls," for specifics.

### 4.2.6.1 Machine Mixing

Except on very small projects, mortar should always be mixed in a mechanical batch mixer designed for the purpose. Grout should always be mixed in a mechanical batch mixer, regardless of the size of the project. The size of the mixer needed will depend on the amount of mortar to be mixed.

When machine mixing is used, mortar and grout ingredients should be batched in small enough quantities to mix properly. Water should be added slowly until the proper consistency has been reached. After all the ingredients have been added, the mortar or grout should be mixed a minimum of 3 to a maximum of 5 minutes. The mixer drum should be completely empty before it is recharged with the next batch.

### 4.2.6.2 Hand Mixing

When mortar must be mixed by hand, the sand should first be spread in the mortar box, the cement and other cementitious materials should then be spread on top of the sand, and the ingredients should be mixed well with a hoe from both ends of the box. Next, about three-quarters of the required water should be added and mixed until the cementitious ingredients and aggregates are uniformly damp. Water should then be added in small amounts, and mixing should be continued until the mortar reaches a suitable workability. The batch should be allowed to stand for about 5 minutes and remixed thoroughly with the hoe without additional water.

### 4.2.6.3 Cold Weather Mixing

The temperature of a mortar or grout when placed should be between 70° and 120°F (21.1° and 48.9°C) when the outside temperature is below 40°F (4.4°C). Higher mortar temperatures may result in fast hardening, which can make it im-

possible to lay masonry units with that mortar. Even when the units can be laid, the mortar may have reduced compressive and bond strengths.

Heating the mixing water is one of the easiest methods of raising the temperature of a mortar or grout. Mixing water should not be heated above 160°F (71.1°C), however, because a flash set can occur when very hot water comes in contact with cement.

In freezing weather, the moisture that is almost always present in masonry sand will freeze. Therefore, sand must be heated slowly and carefully to prevent scorching in order to remove the ice.

The use of calcium chloride and other antifreeze admixtures to lower the freezing point of mortar or grout is not permitted by most modern codes and standards and many building codes. BIA also recommends against their use. The quantity of such materials necessary to lower the freezing point of mortar appreciably is so large that mortar strength and other desirable properties would be seriously impaired. Such materials not only hasten the corrosion of metal anchors and reinforcement, they weaken mortar, contribute to efflorescence, and may even cause the masonry to spall when the chemicals dissolve in water and then recrystallize.

For additional information concerning cold weather masonry construction, refer to BIA Technical Note No. 1, "All-Weather Construction."

### 4.2.7 RETEMPERING

When mortar is not used immediately after it is mixed, as is often the case, some of its water may evaporate. Adding water to such mortar, called *retempering* (Fig. 4.2-5), is sometimes prohibited or limited in extent by code, because it is thought to have detrimental effects. In some jurisdictions, particularly in the Deep South, retempering is not allowed by inspectors even when it is permitted by code. While it is true that retempering slightly reduces a mortar's compressive strength, failing to retemper mortar that needs it may result in a materially lowered bond strength. Therefore, retempering should be permitted to replace water lost by evaporation, provided that the mortar has not begun to set. Regardless of whether it has been retempered or not, no mortar should be placed more than

**FIGURE 4.2-5** Retempering of mortar. (Courtesy Portland Cement Association.)

$2\frac{1}{2}$ hours after its initial mixing. This time limit may need to be greatly reduced in hot weather due to the accelerated drying time caused by high temperatures. In no case should mortar be used after it has begun to set.

A change in the time between the spreading of a mortar and the placing of masonry units in it will affect the flow of the mortar, especially if the masonry units have high suction. The highest bond strength is achieved when this time interval is a minimum. Varying this interval to a large extent will result in a wall with varying tensile bond strengths.

## 4.3 Reinforcement, Ties, Anchors, and Flashing

Masonry and stone walls require various types of reinforcement, ties, anchors, and flashing, depending on the masonry or stone materials in the wall and its configuration and location.

### 4.3.1 JOINT REINFORCEMENT, TIES, AND ANCHORS FOR UNIT MASONRY

Many types of joint reinforcement, ties, and anchors are used in unit masonry construction. Some of them are shown in Figure 4.3-1. The use of these items is discussed and illustrated in Sections 4.7.2.2 and 4.7.2.3.

Metal joint reinforcement, ties, and anchoring devices in masonry or stone walls that are exposed to the weather tend to corrode, especially if they are exposed to salt-laden water. Four inches or more

**FIGURE 4.3-1** Joint reinforcement, ties, and anchoring devices.

of masonry cover may offer some protection, but all such devices should be of copper-coated steel, hot dip galvanized (zinc-coated) steel, or stainless steel. Even though stainless steel is more expensive than the other two, it offers the highest degree of safety and is essential in stone construction and preferable in unit masonry construction as well. The uses, types, and relative ASTM standards for masonry joint reinforcement, anchors, and ties are summarized in Figure 4.3-2.

The requirements for some types of joint reinforcement, ties, and anchors in unit masonry walls are as follows.

### 4.3.1.1 Joint Reinforcement

Joint reinforcement for use in horizontal courses of interior masonry and other masonry protected from the weather should be fabricated from zinc-coated, cold-drawn steel wire (ASTM A82) or Type 304 stainless steel wire (ASTM A580). Joint reinforcement for exterior walls should be fabricated from Type 304 stainless steel wire (ASTM A580). Joint reinforcement should consist of two smooth or deformed longitudinal wires, 9 gauge or larger, that are tied together with welded-on 12-gauge wires. The dis-

tance between weld contacts of the cross-wires should not exceed 6 in. (150 mm) for smooth longitudinal wire or 16 in. (406 mm) for deformed longitudinal

wire. The width of the joint reinforcement should be such that it can be embedded in a mortar joint with its wires not closer than $\frac{5}{8}$ in. (16 mm) to the ma-

**FIGURE 4.3-2** Joint Reinforcement, Ties, and Anchoring Device Materials and Uses

| Use | Material | ASTM Standard |
|---|---|---|
| Interior wall joint reinforcement | Copper-coated wire<br>Zinc-coated wire | B227<br>A116 |
| Exterior wall joint reinforcement | Stainless steel wire | A580, Type 304 |
| Reinforced masonry reinforcement | Billet steel bars<br>Epoxy-coated billet steel bars<br>Deformed wire<br>Welded wire fabric<br>    Plain<br>    Deformed | A615<br>A615 and A775<br>A496<br><br>A185<br>A497 |
| Interior wall anchors and ties | Galvanized carbon steel wire<br>Galvanized steel sheet | A82<br>A366 and A153 |
| Exterior wall anchors and ties | Stainless steel wire<br>Galvanized steel sheet<br>Stainless steel sheet | A580, Type 304<br>A366 and A153<br>A167, Type 304 |
| Rigid anchors | Heavy-thickness galvanized steel sheet<br>Plates and bars<br>    Steel<br>        Painted<br>        Galvanized<br>    Stainless steel | A635 and A653<br><br>A36<br>2 coats epoxy paint<br>A153<br>A666, Type 304 |

sonry face surface. The configuration of joint reinforcement may be either ladder design, truss design, or tab design (see Fig. 4.3-1).

### 4.3.1.2 Wire Mesh Ties

Wire mesh ties or hardware cloth are used to anchor non-load-bearing partitions to intersecting masonry. They should be made from at least 15-gauge zinc- or copper-coated steel wire, have a $1/2$ in. (12.7 mm) mesh, and be at least 12 in. (305 mm) long. Their width should be 1 in. (25.4 mm) less than the wall thickness.

### 4.3.1.3 Cavity Wall Ties

Ties for use in exterior cavity walls should be $3/16$-in. (4.8-mm)-diameter Type 304 stainless steel wire (ASTM A580) formed either in a rectangular shape not less than 4 in. (100 mm) wide (for use with hollow units having vertical cells) or in a Z shape with 2 in. (50.8 mm) legs (for use with solid units or hollow units with horizontal cells).

Ties should be long enough to permit the ends of the tie to be embedded in the outer face-shell mortar beds of hollow units or the center of mortar beds of solid units.

Joint reinforcement comprising 9-gauge or larger cross-wires may be used instead of metal ties, provided that the cross-wires are spaced to render strength and stiffness per square foot of wall area equivalent to that of $3/16$-in. (4.8-mm)-diameter Z ties used at the maximum spacing.

### 4.3.1.4 Faced Wall Ties

Ties used in faced walls should meet the same requirements given above for cavity wall ties.

### 4.3.1.5 Veneered Wall Ties, Wood Studs

The ties usually used for brick veneer over wood studs are 22-gauge, 6-in. (152.4-mm)-long, $7/8$-in. (22.2-mm) wide corrugated galvanized steel. These are

usually bent up and nailed through the sheathing into the underlying studs.

### 4.3.1.6 Veneered Wall Ties, Metal Studs

Ties for masonry veneer over metal studs should be triangular, flexible, stainless steel wire ties similar to those used to anchor masonry to steel structural members. The fixed portion of the ties should be screwed through the sheathing to the studs. The flexible triangular portion of the tie should then be threaded through the support and built into the masonry as it is erected.

### 4.3.1.7 Dovetail Anchors

Dovetail anchors for use with embedded slots or inserts in concrete should be 16-gauge, zinc-coated sheet steel, $7/8$ in. (22.2 mm) wide and long enough to permit the ends of the tie to be embedded in the outer face-shell mortar beds of hollow units or the center of mortar beds in solid units.

Anchors should be Stainless Steel

**strap anchors**
(Recommended $1/8''$ & $3/16'' \times 1''$ or $1 1/4''$ wide)

**rod anchors**
(Recommended $3/8'' \phi$ min.)

**dovetail anchors**
(Recommended $1/8''$ & $3/16'' \times 1''$ or $1 1/4''$ wide)

**miscellaneous anchors**

Epoxy Anchors

15°   15°

**expansion anchors**
follow manufacturer's recommendations on use.

**special anchors**
Bracket mounted to masonry or steel

**FIGURE 4.3-3** Stone anchors. (Courtesy Indiana Limestone Institute.)

#### 4.3.1.8 Rigid Steel Anchors

Rigid steel anchors for anchorage of load-bearing walls and fire walls should be $1^{1}/_{4}$ in. (6.4 mm) by $^{1}/_{4}$ in. (6.4 mm) steel with the ends turned up not less than 2 in. (50.8 mm).They should be not less than 16 in. (406.4 mm) long for 8-in. (200-mm)-thick walls and not less than 24 in. (610 mm) long for 12-in. (305-mm)-thick walls.

### 4.3.2 STRUCTURAL REINFORCEMENT FOR UNIT MASONRY WALLS

Structural reinforcement for unit masonry walls consists mainly of steel bars and welded wire mesh.

### 4.3.3 REINFORCEMENT, TIES, ANCHORS, AND SUPPORT SYSTEMS FOR STONE

Various types of reinforcement, ties, anchors, and grid support systems are used in stone construction, depending on the type of stone and its use. All such devices that come in contact with stone should be fabricated from AISI Type 304 stainless steel (ASTM A666). Devices and support components that are not in direct contact with stone may be fabricated from hot-dip galvanized steel (ASTM A36 for materials, ASTM A123 for galvanizing). Some of the various types of devices used are shown in Figure 4.3-3.

### 4.3.4 FLASHING

Flashing is installed at vulnerable locations in masonry to help prevent water from entering and to collect water that has entered the masonry and divert it through weep holes to the exterior (Fig. 4.3-4).

**FIGURE 4.3-4** Through-wall flashing in a concrete masonry wall and a rope-wick weep hole.

Flashings are formed from sheet metal, bituminous fabric membranes, plastic sheet materials, or a combination of these materials. Flashing materials are discussed in Section 7.9. Because replacement costs can exceed the initial installed costs, it is prudent to select a durable flashing material for the original installation. Recommendations for locating and installing flashing are presented in Section 4.7.2.8.

# 4.4 Clay Masonry Units

Clay masonry units are burned clay units used in buildings. They include brick, hollow clay tile, and architectural terra cotta. They do not include ceramic tile, flue linings, drain tile, or other ceramic products. Ceramic tile and related materials are discussed in Section 9.4. Solid clay masonry units, such as brick, and hollow clay masonry units, such as tile, are composed of burned clay, shale, fire clay, or a mixture of these, formed into the desired shape while plastic and then burned in a kiln.

*Terra cotta* is Italian for "cooked earth." It is composed of an aged clay to which admixtures and sand or pulverized, previously fired clay, called *grog*, are added. It is extruded or formed into the desired shape by hand packing and then fired at a high temperature to produce a material that is harder and more compact than most brick.

Both the raw materials used in making clay masonry and the way in which it is manufactured have a marked influence on the properties of the units in a finished wall. Therefore, a general understanding of both materials and methods is essential to the selection and use of brick and tile.

### 4.4.1 RAW MATERIALS

Clays used in making clay masonry units must have plasticity so that they can be shaped or molded when mixed with water. They must also have sufficient tensile strength to maintain their shape after they have been formed. In addition, their particles must fuse when they are subjected to high temperatures in a kiln. The three principal forms of clays used to

manufacture clay masonry units have similar chemical compositions but different physical characteristics. They are:

- *Surface clays*: sedimentary clays, found near the surface of the earth.
- *Shales*: clays that have been subjected to high pressure until they have hardened almost to the form of slate.
- *Fire clays*: clays found deeper in the ground than the other types and having refractory qualities, making them resistant to high temperatures and useful in such applications as fire brick. They usually have fewer impurities and more uniform chemical and physical properties than surface clays or shales.

All three are compounds of silica and alumina with varying amounts of metallic oxides and other impurities. The metallic oxides act as fluxes, which promote fusion at lower temperatures than would be possible without them. They also influence the color of finished units.

There are slight differences in the properties of any two finished clay masonry units, regardless of how hard their manufacturer tried to minimize variations. An attempt is made to standardize the properties of units by mixing clays from different locations in a pit, mixing clays from different pits, and making compensations during the manufacturing processes, but none of these is completely successful.

### 4.4.2 MANUFACTURE OF CLAY MASONRY UNITS

In general, the manufacturing process (Fig. 4.4-1) consists of mining (also called *winning*), preparing, and storing the raw materials; and forming, drying, glazing, burning, cooling, drawing, and storing the finished units.

#### 4.4.2.1 Mining (Winning), Preparation, and Storage

In the process of winning, raw clay is taken from either open pits or underground mines. Raw clay that does not contain large lumps or stones is immediately passed through an inclined vibrating screen to control its particle size and then stored.

The clay is crushed to break up lumps, and stones are removed. Then the clay is pulverized and mixed by large grinding wheels that revolve in a circular pan.

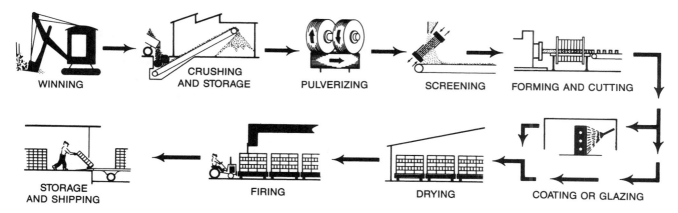

**FIGURE 4.4-1** Manufacturing flowchart. (Courtesy Brick Institute of America [renamed Brick Industry Association].)

WINNING → CRUSHING AND STORAGE → PULVERIZING → SCREENING → FORMING AND CUTTING

STORAGE AND SHIPPING ← FIRING ← DRYING ← COATING OR GLAZING

a

b

c

**FIGURE 4.4-2** (a) Mixing powdered clay with water. (b) Clay emerging from a de-airing machine. (c) Mixture being extruded through a die that molds it into a long, flat column. (Courtesy Brick Institute of America [renamed Brick Industry Association].)

Each of these wheels weighs between 4 and 8 tons (3629 to 7257 kg). This pulverized clay is then screened and stored for later use.

The clay from a single pit is often placed in several storage areas to permit later blending. Blending produces more uniformity in raw clay, helps to control the color of finished units, and permits some control over the suitability of the raw material for manufacturing a given type of unit.

### 4.4.2.2 Forming

The first step in the forming process, *tempering*, produces a homogeneous, plastic mass ready for molding. Tempering is usually achieved by adding water to the clay in a mixing chamber called a *pug mill*, which contains one or two revolving shafts with blades (Fig. 4.4-2a). Three methods are used to form clay masonry units: the stiff-mud, the soft-mud, and the dry-press processes.

### STIFF-MUD PROCESS

The stiff-mud process is the most commonly used method for forming brick and clay tile. All structural clay tiles and many bricks are made this way. A similar method is used for extruded terra cotta. In the stiff-mud process, clay is mixed with sufficient water (12% to 15% by weight) to produce the desired degree of plasticity. After thorough mixing (*pugging*), the tempered clay goes through a de-airing machine in which a partial vacuum is maintained. De-airing removes air, thus eliminating holes and bubbles, which improves the clay's plasticity and workability and results in greater strength in finished units (Fig. 4.4-2b).

After de-airing, the clay is forced (extruded) through a die, which produces a clay column of the desired shape (Fig. 4.4.2c). Here, various textures may be applied. The column then passes through an automatic cutter (Fig. 4.4-3). Die sizes and the spacing of cutter wires must be calculated to compensate for normal shrinkage during drying and burning.

After they are cut from the column, individual units proceed from the cutting table onto a continuous moving belt, where they are inspected. Acceptable units are removed from the belt as it passes and placed on dryer cars (Fig. 4.4-4). Imperfect units are returned to the pug mill.

**FIGURE 4.4-3** Forming and cutting a clay column in the stiff-mud process. (Courtesy Brick Institute of America [renamed Brick Industry Association].)

**FIGURE 4.4-4** Bricks on a belt. Core holes of various sizes are cut through the body of bricks as they are extruded from the die to make them lighter and enable them to burn more uniformly. (Courtesy Brick Institute of America [renamed Brick Industry Association].)

## SOFT-MUD PROCESS

The soft-mud process is used only for producing brick. It is particularly suitable for clays that contain too much natural water (20% to 30% by weight) to be sat-isfactorily extruded in the stiff-mud process. In the soft-mud process, raw clay is mixed with water, and the mixture is then formed into units in molds. To prevent this clay from sticking, the molds are lubricated with sand and water or just water. When sand is used, the bricks are called *sandstruck* (Fig. 4.4-5h); if water is used, they are called *waterstruck* (Fig. 4.4-5g).

a. Smooth

b. Matte-vertical markings

c. Matte-horizontal markings

d. Rugs

e. Barks

f. Sandmold

g. Waterstruck

h. Sandstruck

**FIGURE 4.4-5** Typical brick textures. Some of these may not be available in a some locations. (Courtesy Brick Institute of America [renamed Brick Industry Association].)

A method similar to the soft-mud process for forming brick has been used to make molded terra cotta shapes, except that a high water content is not a prerequisite.

### DRY-PRESS PROCESS

The dry-press process is used with clay that has very low plasticity. This clay is mixed with a minimum of water (up to 10% by weight) and then formed in steel molds under pressure of between 500 and 1500 psi (211 and 105.5 Mg/m$^2$).

### 4.4.2.3 Drying

When wet clay masonry units come from molding or cutting machines, they contain 7% to 30% moisture, depending on the forming method. Before brick or clay tile units are burned, most of this water is evaporated in dryer kilns at temperatures ranging from 100°F (37.8°C) to 300°F (148.9°C). Drying times vary from 24 to 48 hours, depending on the type of clay used in the units. During this time, heat and humidity must be regulated to avoid excessive cracking of the units.

Terra cotta units must be thoroughly dried before they are fired. Unless the units are bone dry at the time of firing, they will crack or explode.

### 4.4.2.4 Ceramic Glazing

Ceramic glazing is the process used to produce the finish on glazed facing tile, glazed facing brick, and glazed terra cotta. It is a specialized, controlled procedure in which glazes are sprayed onto masonry units before the units are fired. Spray glazes are composed of mineral ingredients that fuse into a glass-like coating at a given temperature. They are available in almost any color. There are two basic variations of the ceramic glazing process: *high-fired glazing* and *low-fired glazing*.

### HIGH-FIRED GLAZING

In the high-fired process, the glazes are sprayed on the masonry units sometimes before and sometimes after drying. The units are then kiln-fired at normal firing temperatures.

### LOW-FIRED GLAZING

Some glazed colors cannot be produced at high temperatures. When these are desired, low-fired glazing is used. In this process, the glazes are applied after the units have been burned to maturity and cooled. Then the units with the glazes applied are fired again at a lower temperature.

### 4.4.2.5 Burning and Cooling

*Burning* is the process of baking or firing dried clay masonry units in ovens called *kilns* to produce finished units. *Cooling* is the process of returning the masonry units to atmospheric temperature. The two chief types of kilns used are *periodic kilns* and *tunnel kilns* (Figs. 4.4-6a and 4.4-6b). Both types are fueled by either natural gas, oil, or coal.

Burning of brick and clay tile units requires 2 to 5 days, depending on the kiln used and other variables. The burning time for terra cotta may be a little shorter.

During burning and cooling, the temperatures in a kiln must be raised and lowered at controlled rates. To permit careful control of temperatures and their rates of change, kilns are equipped with devices to permit constant monitoring and regulating of their interior temperatures.

The burning process begins when dried masonry units are loaded into kilns. In periodic kilns (see Fig. 4.4-6a), the

a

b

**FIGURE 4.4-6** Until recently, bricks were fired in (a) periodic (beehive) kilns. Today, most bricks are fired in (b) tunnel kilns. (Courtesy Brick Institute of America [renamed Brick Industry Association].)

units are placed in the kiln in a pattern that permits free circulation of hot gases. In tunnel kilns (see Fig. 4.4-6b), the units are placed on special cars which pass through various temperature zones as they traverse the tunnel.

Burning may be divided into six stages: the evaporation of free water (water-smoking), dehydration, oxidation, vitrification, flashing, and cooling. The first four are associated with rising temperatures in the kiln. The actual temperatures used in the burning process will vary somewhat, depending on the type of clay. In general, for brick and clay tile units, water-smoking takes place at about 400°F (204.4°C), dehydration at between 300°F (148.9°C) and 1800°F (982°C), oxidation at between 1000°F (537.8°C) and 1800°F (982°C), and vitrification at between 1600°F (871°C) and 2400°F (1315.6°C).

To produce different colors and shades of colors, masonry units are sometimes subjected to a reducing atmosphere (one that contains insufficient oxygen for complete combustion) near the end of the burning process. The colors and shades produced by this so-called *flashing* will vary with the type of clay used in the units.

Cooling begins after the temperature in a kiln has reached the maximum needed to properly burn the units in it. The rate of cooling must be controlled, because it has a direct effect on the color of the units and because excessively rapid cooling will cause the masonry units to crack or check. In tunnel kilns, cooling seldom requires more than 48 hours, but in periodic kilns, proper cooling usually requires at least 48 hours and sometimes as much as 72 hours.

### 4.4.2.6 Drawing

*Drawing* is the process of unloading a kiln after cooling (Fig. 4.4-7). During drawing, units are sorted and graded. They are then taken to a storage yard or loaded directly onto rail cars or trucks for delivery.

### 4.4.3 PROPERTIES OF CLAY MASONRY UNITS

The properties of finished units depend on the characteristics of the raw materials and the effects of the manufacturing process.

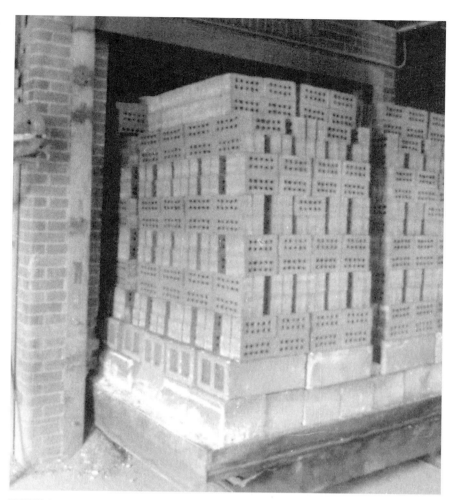

**FIGURE 4.4-7** Unloading a kiln. (Courtesy Brick Institute of America [renamed Brick Industry Association].)

#### 4.4.3.1 Compressive Strength

The clay, the method of manufacturing, and the degree of burning affect compressive strength. With some exceptions, the plastic clays used in the stiff-mud process have higher compressive strengths when burned than do the clays used in the soft-mud or dry-press processes. For a given clay and method of manufacture, higher compressive strengths are associated with higher burning temperatures. The compressive strength of brick varies from 1500 psi (1054.6 Mg/m$^2$) to more than 20,000 psi (14 061 Mg/m$^2$), due mainly to the wide variation in the properties of the clays used.

#### 4.4.3.2 Absorption

The degree of water absorption of a clay masonry unit also depends on the clay, manufacturing method, and temperature of burning. Plastic clays and higher burn-ing temperatures generally produce units having lower absorption. Generally, the stiff-mud process produces units with lower absorption than either the soft-mud or dry-press processes. A brick's initial rate of absorption, which is called *suction*, results from pores or small openings in the burned clay which act as capillaries to draw or suck water into the unit. Suction has little bearing on the transmission of free water through a brick, but it has an important effect on the bond between brick and mortar. Maximum bond strength and minimum water penetration are obtained when the suction of the brick at the time of laying does not exceed 20 g of water (0.7 oz) per minute. Bricks having suction of more than 20 g per minute should be wetted before they are laid to reduce the suction. Section 4.7.1.2 contains more details about acceptable degrees of absorption and describes a field test that will help determine suction.

### 4.4.3.3 Durability

The durability of clay masonry units is affected by pore size within the units and by the amount and completeness of fusion during burning. In general, higher burning temperatures produce harder clay masonry units.

A major weathering action that affects burned clay masonry units is alternate freezing and thawing in the presence of moisture. The ability of clay masonry units to withstand freeze-thaw cycles without disintegration is important in every cold climate, but is especially so where the annual average precipitation exceeds 20 in. (508 mm) of water and alternate freezing and thawing is common. Units produced from the same raw materials and by the same method of manufacture may have different resistances to damage by freeze-thaw cycles, depending on their compressive strengths and absorption. Units with high compressive strength and units with low absorption are usually the more resistant to damage during freeze-thaw cycles.

A second factor that affects the durability of clay masonry units is their soluble salts content. Such salts will dissolve when water is present and crystallize either on the surface of the masonry units or within them. Surface deposits (efflorescence), while unsightly, are not damaging. Internal crystallization (cryptoflorescence), however, will build up until they eventually cause the surface of the masonry units to spall. For this reason, brick that will be exposed to water in any form should be tested and rated *not effloresced* in accordance with ASTM C67.

### 4.4.3.4 Natural Color

Burned clays produce a wide range of colors, from pearl grays and creams through buff, and in a descending scale of reds to purple, maroon, and gun metal black. The clay's chemical composition, the burning temperature, and the method of burning control color. Terra cotta clays produce units that vary widely in color from red to brown to white.

Clays used in masonry units contain varying degrees of many metallic oxides. Of these, iron oxides probably have the greatest effect on color. Clay containing iron in almost any form will burn red when exposed to an oxidizing fire, because the iron combines with the oxygen to form ferrous oxide, which is the same as red rust. When clay containing iron

is burned in a reducing atmosphere (*flashed*), it will take on a purple cast.

Lighter colors are produced when a clay is underburned. This is the method used to produce salmon-colored brick from a clay that would ordinarily burn red. Underburned clay masonry is softer, more absorptive, and has a lower compressive strength than units that are burned completely.

Overburning of clay produces *clinker brick*, which is dark red to black when the clays used are those that produce red units when burned normally and dark speckled brown when the clay used produce buff units when burned normally. Clinker brick is ordinarily used only as an esthetic oddity.

With few exceptions, the natural colors of clay masonry units are mixtures of shades rather than pure colors. Therefore, the accepted practice in specifying color is to require that the color of the delivered units match an approved sample panel made up of several units representative of the color range expected.

By using applied coatings, such as glazes or nonlustrous finishes, and by introducing chemicals into the kiln that vaporize and combine with the clay, color effects can be produced in almost the complete range of the spectrum.

### 4.4.3.5 Texture

Textures result from the dies or molds used in forming masonry units. The pressure exerted by a steel die normally produces smooth units. Many textures may be applied, however, by mounting at the outlet end of a die an attachment that cuts, scratches, rolls, brushes, or otherwise roughens the surface of the uncut clay column as it leaves the die. The possible textures range from fine through medium to coarse. Some typical textures are smooth, matt with vertical or horizontal markings, rugs, barks, sandmold, waterstruck, and sandstruck (see Fig. 4.4-5).

### 4.4.3.6 Size Variation

The clays used to manufacture brick and clay tile shrink 4.5% to 15% during drying and burning. Terra cotta clay shrinks 6% to 7%. In general, this shrinkage can be compensated for by increasing the die size and adjusting the spacing of cutter wires so that the units approximate the desired final size after they are burned. Molds are simply made larger to compensate for shrinkage.

A more serious problem exists when units to be used together do not shrink the same amount. Units shrink at differing rates for several reasons. For example, units made with clays having different chemical composition may not be the same size. In spite of the manufacturer's attempts to limit it, this kind of variation will always occur to at least some extent. It is not possible to have every unit made from exactly the same clay. Large differences in size can be prevented, however, by careful monitoring to ensure that all units to be used together are made from similar clays.

The amount of shrinkage during burning increases with higher temperatures, which are necessary to produce darker shades of color. As a result, when a wide range of color tone is desired, some variation in size between the dark and light units is inevitable. Size differences can be reduced by selecting masonry units of consistent color throughout.

Manufacturers try to control shrinkage to obtain units of uniform size, but variations in raw materials and kiln temperatures make absolute uniformity impossible. Therefore, specifications should spell out permissible variations in size.

### 4.4.4 CLASSIFICATIONS OF CLAY MASONRY UNITS

Clay masonry units are classified (Fig. 4.4-8) as solid masonry units, hollow masonry units, and architectural terra cotta. The many individual products are qualified further as to size, grade, type, color, and texture.

Clay masonry units should conform to the current ASTM standards as follows:

Building (common) brick: ASTM C62
Facing brick, unglazed: ASTM C216
Facing brick, ceramic glazed: ASTM C126
Hollow brick: ASTM C652
Structural clay tile, load-bearing: ASTM C34
Structural clay tile, non-load-bearing: ASTM C56
Structural clay facing tile, unglazed facing tile: ASTM C212
Structural clay facing tile, ceramic glazed: ASTM C126
Architectural terra cotta: None

ASTM standards do not fix the required size, color, or texture of masonry units for a specific use, but rather cover a range of sizes and a wide variety of col-

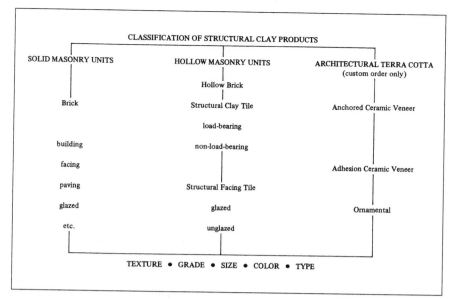

FIGURE 4.4-8 Classification of clay masonry units.

ors and textures. When construction documents require that masonry units be of a certain type and grade as defined in an ASTM standard, it is usually necessary to add specific requirements related to the units' size, color, and texture, and to the quality of workmanship required and the specific finished appearance desired. This can be done by requesting that these variables match approved samples.

### 4.4.4.1 Solid Clay Masonry Units

The term *brick* usually refers to a solid clay masonry unit. Most solid clay masonry units are either frogged, cored, or celled (Figure 4.4-9). Some special-use solid clay masonry units, such as pavers, are produced without frogs, cores, or cells, but such units are rarely used in walls. Frogs, cores, and cells are introduced to promote even drying during burning and to reduce the weight of the units. The number of cores may be relatively large and few in number, as shown in Figure 4.4-9, or as many as 21 small "pencil" cores may be used. Cells can also be smaller in size and larger in number than those shown, may vary in size and shape within a unit, and may be placed either vertically or horizontally within a unit. Cells are usually larger than cores, but in no case may the collective area of the cores or cells exceed 25% of the unit's total cross-sectional area. The applicable ASTM standards (see Section 4.4.4) dictate the size of frogs, the distance from cores and cells to the faces of

a

b

c

FIGURE 4.4-9 Solid clay masonry units (brick) having (a) frogs, (b) cores, and (c) cells. (Drawing by HLS.)

a unit, and the thickness of the webs between cores or cells. Brick pavers, which are used in both brick flooring and paving, are also solid clay masonry, but they are discussed in Section 9.7.

Brick has been produced in many shapes and sizes, but most brick comes in modular sizes having nominal dimensions based on a 4 in. (100 mm) module. Figures 4.4-10 and 4.4-11 list some of the many rectangular shapes available. Common designations and sizes of brick are shown in Figure 4.4-12. Not many manufactures produce all the sizes shown, so it is important to determine during the design phase which sizes are available in the locality of the project.

Special brick shapes are also available. The units used in some of the arches in Figure 4.6-34 are typical special shapes. Other special shapes include those for use as water tables, sills, stair treads, copings, jambs, corners, and angles, (Fig. 4.4-13). Many variations of the shapes shown are available. Most brick manufacturers will make special shapes if sufficient lead time is allowed. Some manufacturers will produce custom units in shapes designed by a building's design professionals in addition to the standard shapes usually available. Some manufacturers will produce brick sculpture.

The term *modular coordination* refers to a system of components sized to fit together easily with a minimum of cutting and fitting in the field. Coordinated assemblies often finish with an overall dimension in even units without fractions. Designing masonry units with modular coordination in mind tends to increase productivity during construction and can result in significant cost savings. Masonry units so designed are referred to as *modular units*. The nominal dimensions of a modular brick are equal to the manufactured (actual) dimensions plus the thickness of the mortar joint for which the unit is designed. In general, joint thicknesses in brick are either $3/8$ or $1/2$ in. (9.5 or 12.7 mm).

The manufactured height for modular brick that is designed to be laid with three courses in every 8 in. (203 mm) of height is about $2^1/4$ in. (57.2 mm). This is to ensure that the vertical coursing in an addition built with modular brick can be made to match the coursing in an existing building constructed with either modular or nonmodular brick. Those manufacturers who produce modular brick have agreed to keep the manufactured

FIGURE 4.4-10 Standard Nomenclature for Brick Sizes[a]

**Modular Brick Sizes**

| Unit Designation | Nominal Dimensions, in. | | | Joint Thickness,[b] in. | Specified Dimensions,[c] in. | | | Vertical Coursing |
|---|---|---|---|---|---|---|---|---|
| | w | h | l | | w | h | l | |
| Modular | 4 | 2⅔ | 8 | 3/8 | 3⅝ | 2¼ | 7⅝ | 3C = 8 in. |
| | | | | 1/2 | 3½ | 2¼ | 7½ | |
| Engineer modular | 4 | 3⅕ | 8 | 3/8 | 3⅝ | 2¾ | 7⅝ | 5C = 16 in. |
| | | | | 1/2 | 3½ | 2³/₁₆ | 7½ | |
| Closure modular | 4 | 4 | 8 | 3/8 | 3⅝ | 3⅝ | 7⅝ | 1C = 4 in. |
| | | | | 1/2 | 3½ | 3½ | 7½ | |
| Roman | 4 | 2 | 12 | 3/8 | 3⅝ | 1⅝ | 11⅝ | 2C = 4 in. |
| | | | | 1/2 | 3½ | 2¼ | 7½ | |
| Norman | 4 | 2⅔ | 12 | 3/8 | 3⅝ | 2¼ | 11⅝ | 3C = 8 in. |
| | | | | 1/2 | 3½ | 2¼ | 11½ | |
| Engineer Norman | 4 | 3⅕ | 12 | 3/8 | 3⅝ | 2¾ | 11⅝ | 5C = 16 in. |
| | | | | 1/2 | 3½ | 2¹³/₁₆ | 11½ | |
| Utility | 4 | 4 | 12 | 3/8 | 3⅝ | 3⅝ | 11⅝ | 1C = 4 in. |
| | | | | 1/2 | 3½ | 3½ | 11½ | |

**Nonmodular Brick Sizes**

| Unit Designation | Nominal Dimensions, in. | | | Joint Thickness,[b] in. | Specified Dimensions,[c] in. | | | Vertical Coursing |
|---|---|---|---|---|---|---|---|---|
| Standard | | | | 3/8 | 3⅝ | 2¼ | 8 | 3C = 8 in. |
| | | | | 1/2 | 3½ | 2¼ | 8 | |
| Engineer standard | | | | 3/8 | 3⅝ | 2¾ | 8 | 5C = 16 in. |
| | | | | 1/2 | 3½ | 2¹³/₁₆ | 8 | |
| Closure standard | | | | 3/8 | 3⅝ | 3⅝ | 8 | 1C = 4 in. |
| | | | | 1/2 | 3½ | 3½ | 8 | |
| King | | | | 3/8 | 3 | 2¾ | 9⅝ | 5C = 16 in. |
| | | | | 3 | 3 | 2⅝ | 9⅝ | |
| Queen | | | | 3/8 | 3 | 2¾ | 8 | 5C = 16 in. |

[a]1 in. = 25.4 mm; 1 ft = 0.3 m.
[b]Common joint sizes used with length and width dimensions. Joint thicknesses of bed joints vary based on vertical coursing and specified unit height.
[c]Specified dimensions may vary within this range from manufacturer to manufacturer.
(Courtesy Brick Institute of America [renamed Brick Industry Association].)

FIGURE 4.4-11 Other Brick Sizes[a]

**Modular Brick Sizes**

| Nominal Dimensions, in. | | | Joint Thickness,[b] in. | Specified Dimensions,[c] in. | | | Vertical Coursing |
|---|---|---|---|---|---|---|---|
| w | h | l | | w | h | l | |
| 4 | 6 | 8 | 3/8 | 3⅝ | 5⅝ | 7⅝ | 2C = 12 in. |
| | | | 1/2 | 3½ | 5½ | 7½ | |
| 4 | 8 | 8 | 3/8 | 3⅝ | 7⅝ | 7⅝ | 1C = 8 in. |
| | | | 1/2 | 3½ | 7½ | 7½ | |
| 6 | 3⅕ | 12 | 3/8 | 5⅝ | 2¾ | 11⅝ | 5C = 16 in. |
| | | | 1/2 | 5½ | 2¹³/₁₆ | 11½ | |
| 6 | 4 | 12 | 3/8 | 5⅝ | 3⅝ | 11⅝ | 1C = 4 in. |
| | | | 1/2 | 5½ | 3½ | 11½ | |
| 8 | 4 | 12 | 3/8 | 7⅝ | 3⅝ | 11⅝ | 1C = 4 in. |
| | | | 1/2 | 7½ | 3½ | 11½ | |
| 8 | 4 | 16 | 3/8 | 7⅝ | 3⅝ | 15⅝ | 1C = 4 in. |
| | | | 1/2 | 7½ | 3½ | 15½ | |

**Nonmodular Brick Sizes**

| Nominal Dimensions, in. | | | Joint Thickness,[b] in. | Specified Dimensions,[c] in. | | | Vertical Coursing |
|---|---|---|---|---|---|---|---|
| | | | 3/8 | 3 | 2¾ | 8⅝ | 5C = 16 in. |
| | | | 3 | 3 | 2⅝ | 8⅝ | |

[a]1 in. = 25.4 mm; 1 ft = 0.3 m.
[b]Common joint sizes used with length and width dimensions. Joint thicknesses of bed joints vary based on vertical coursing and specified unit height.
[c]Specified dimensions may vary within this range from manufacturer to manufacturer.
(Courtesy Brick Institute of America [renamed Brick Industry Association].)

face height 2¼ in. (57.2 mm), even when the length and thickness of their units are modular. They consider as minimal the differences in mortar bed thickness necessary to maintain the modular coursing of three courses in 8 in. (203 mm).

The brick unit dimensions in all the figures in this chapter are shown with the thickness first, followed by the face dimensions (height and length), which has become the standard in the industry. The unit dimensions of a brick are listed in this order, regardless of the positions in which they are laid (see Fig. 4.7-2). Specifications and drawings related to brick should follow this standard order of listing brick dimensions. For example, if the nominal dimensions of a Roman brick are $4 \times 2 \times 12$ in. $(100 \times 50 \times 300$ mm) and the units are designed to be laid with $3/8$ in. (9.5 mm) mortar joints, their actual size will be $3⅝ \times 1⅝ \times 11⅝$ in. $(92.1 \times 41.3 \times 295.3$ mm), with allowance, of course, for permissible size variations in manufacture.

#### 4.4.4.2 Hollow Clay Masonry Units

Hollow clay masonry units include hollow brick and hollow clay tile. Hollow clay masonry units contain either cells or cores (see Fig. 4.4-9). In hollow clay masonry units, the collective area of all cores and cells must exceed 25% of the gross cross-sectional area of the unit. The applicable ASTM standards (see Section 4.4.4) dictate the distance from cores and cells to the faces of a unit and the thickness of the webs between cores or cells. The size, number, and shape of cells and cores will vary greatly, depending on the size, shape, and use of the unit and the unit manufacturer's preferences.

The many types of hollow clay tile are categorized as either structural clay tile or structural facing tile. Tile designed to be laid with its cells in a horizontal plane is called *side construction* or *horizontal cell tile*; that designed to be laid with its cells placed vertically is called *end construction* or *vertical cell tile*.

Hollow brick is defined by ASTM C652 as a hollow clay masonry unit whose net cross-sectional area in every plane parallel to the bearing surface is not less than 60% of the gross cross-sectional area measured in the same plane. Hollow brick units are cored between 25% (the upper limit for solid brick units) and 40%. At the higher limit, they are only 60% solid.

**FIGURE 4.4-12** Designations for brick: (a) and (b) modular brick sizes (dimensions are nominal); (c) nonmodular brick sizes (dimensions shown are actual size). (Courtesy Brick Institute of America [renamed Brick Industry Association].)

Generally, hollow brick can be used where solid brick is used; it is the same size and looks the same on the outside as solid brick. However, solid brick design standards are not applicable to hollow brick construction. Instead, hollow brick structures should be individually designed according to accepted engineering practice.

Technological improvements in masonry construction and lower production, transportation, and installation costs have given hollow brick advantages over solid brick in both speed of erection and economy of construction. Unfortunately, hollow bricks are not readily available in some parts of the country.

Engineered brick masonry design and construction have created a trend toward larger brick units. Typical hollow brick units are shown in Figure 4.4-14. Their greater core area makes hollow bricks lighter and permits easier handling. Larger sizes permit faster, more economical construction. For example, three units, each 8 in. (200 mm) × 4 in. (100 mm) × 12 in. (300 mm), produce 1 sq ft (0.093 m²) of 8-in. (200-mm)-thick wall, while 13 standard-size brick units would be required.

Structural clay tile is available in both load-bearing and non-load-bearing units. A few of the available shapes and sizes are shown in Figure 4.4-15.

Structural facing tile is available either glazed or unglazed. Glazed tile is made from a high-grade, light-burning fire clay to which a ceramic glaze has been applied. During burning, the glaze fuses into a glass-like coating on the unit. Unglazed tile is produced from light- or dark-burning clays and shales and may have either a smooth or rough textural finish. The nominal modular sizes of structural facing tile are shown in Figure 4.4-16. Figure 4.4-15 shows a few common and special shapes currently produced.

### 4.4.4.3 Architectural Terra Cotta

Terra cotta has been used to make sewer pipes, fireproofing for metal structural elements, floor vaulting, and flue liners, but it is chiefly used today for copings and decorative cladding. It is made in practically every color possible and in a variety of shapes and forms. The surface may be flat, grooved, or raised and either smooth or textured. It may be glazed or natural.

One form of terra cotta used for exterior decoration is called *ceramic veneer*. Made in a process similar to that for extruded brick, it is limited to items and shapes that can be extruded. There are two types. The first, called *adhesion type*, is designed to be installed using mortar as one would install ceramic tile. The maximum thickness of this type is $1^1/_8$ in. (28.6 mm). The maximum face size is 540 sq in (0.348 m²), which usually results in pieces that are about 18 by 30 in.

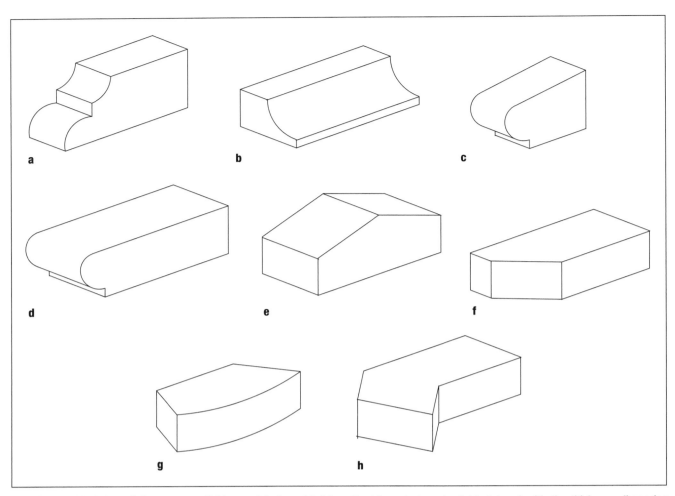

**FIGURE 4.4-13** A few of the many available special shaped brick units: (a) rowlock water table (placed with the thickness dimension vertical), (b) flat water table (placed in the direction of a stretcher), (c) sill (door, window, and other opening sills), (d) stair tread, (e) coping, (f) jamb (door, window, and other opening jambs), (g) convex raidal (concave raidals are also available) (used to build curved walls without field cutting of brick), and (h) angle (used where walls intersect at other than 90 degrees. (Drawing by HLS.)

4 × 2⅔ × 8   4 × 4 × 12   6 × 4 × 12   8 × 4 × 12   10 × 4 × 12
(1) 8 × 2⅔ × 12

**FIGURE 4.4-14** Typical hollow brick units. (Courtesy Brick Institute of America [renamed Brick Industry Association].)

(457.2 by 762 mm) or 20 by 27 in. (508 by 686 mm).

The second type of ceramic veneer is called *anchored type*. It is designed to be held in place by anchors tied to reinforcing rods installed in a thick bed of grout. This type of ceramic veneer can be made in larger panels than can the adhesion type. The panels have ribs or scoring on their backs, making their total thickness 2 to $2^{1}/_{2}$ in. (50.8 to 63.5 mm).

The second form of terra cotta in general use today is tile. The term *tile* is used here to designate those terra cotta shapes that are produced by casting. Earlier terra cotta tile often had an egg-crate structure on the back side to lighten the units. Some modern tiles have vertical voids similar to those in concrete block, while others are flat and ribbed on the back, similar to ceramic tile. Tiles may be surfacing units with repetitive patterns, or

**4" (100 mm) WALL THICKNESS**

12 X 12    8 X 8 or 12    10 2/3 X 12    5 1/3 X 12    5 1/3 X 12    5 1/3 X 12

**6" (150 mm) WALL THICKNESS**

12 X 12    12 X 12    8 X 12    8 X 12

**8" (200 mm) WALL THICKNESS**

12 X 12    12 X 12    8 X 12    5 1/3 X 12    8 X 8    5 1/3 X 12

8 X 12    5 1/3 X 12    10 2/3 X 12    6 2/3 X 12    8 X 12 or 16    8 X 12

**10" (250 mm) WALL THICKNESS**

5 1/3 or 8 X 12 or 16    12 X 12    12 X 12

**12" (305 mm) WALL THICKNESS**

12 X 12    12 X 12    8 X 12    8 X 12    8 X 12

**a**

5⅓″ = 135.46 mm; 6⅔″ = 161.32 mm; 8″ = 203 mm; 10⅔″ = 270.92 mm;
12″ = 305 mm; 16″ = 406 mm; 1¾″ = 44.5 mm; 2⅜″ = 60.3 mm;

Stretcher Scored or unscored    Soap

Soap    6" Stretcher Scored or unscored    8" Stretcher

Soap

Stretcher    Soap

Soap    Stretcher    Soap

**b**

3¾″ = 95.3 mm; 5¹⁄₁₆″ = 128.6 mm; 5¾″ = 146.1 mm;
7¾″ = 196.9 mm; 11¾″ = 298.5 mm; 15¾″ = 400.1 mm

**FIGURE 4.4-15** (a) Structural clay tile. (b) Structural facing tile.

they may be decorative elements with letters or numbers engraved in them. They may also be sculpted.

### 4.4.5 STANDARDS CONTROLLING CLAY MASONRY UNITS

Standards developed by ASTM for the many types and grades of brick and clay tile are widely used to control the quality of clay masonry units. Every architect's specification and every masonry contractor's order for brick, hollow brick, structural clay tile, and structural facing tile should include a requirement that the units delivered comply with the applicable ASTM standard. The standard should be specifically named.

ASTM standards classify clay masonry units in types, grades, and classes based on the weathering index (Fig. 4.4-

17), the appearance of the completed wall, and other factors. When ASTM standards are used to control masonry quality, it is essential that the grade, type, and class of the units be spelled out.

#### 4.4.5.1 Weathering Index

The weathering index is a measure of the potential effect of weathering on clay masonry units. The various regions shown in Figure 4.4-17 were established by multiplying the average annual number of freezing cycle days by the annual winter rainfall.

A freezing cycle day is a day when the air temperature moves above or below 32°F (0°C). The average annual number of freezing cycle days equals the difference between the mean numbers of days when the minimum temperature is 32°F (0°C) or below and the mean number of days when

the maximum temperature is 32°F (0°C) or below. Winter rainfall is the sum, in inches, of the mean monthly precipitation occurring between the first and last killing frosts in the fall and spring.

A severe weathering region has an index greater than 500; a moderate region, 50 to 500; and a negligible region, less than 50. Figure 4.4-17 shows regions where clay masonry units will be subjected to severe, moderate, and negligible weathering.

Masonry in contact with the earth is considered to be in a severe weathering condition, regardless of the climate.

#### 4.4.5.2 Summary of Physical Characteristics and Uses

A summary of the physical requirements and normal uses for various types of clay masonry units follows. The accompany-

## FIGURE 4.4-16 Nominal Modular Sizes of Structural Facing Tile[a]

| Thickness | Face Dimension in Wall | |
|---|---|---|
| | Height | Length |
| 2, 4, 6, and 8 in.<br>(50, 100, 150, and 200 mm) | 4 in.<br>(100 mm) | 8 and 12 in.<br>(200 and 300 mm) |
| 2, 4, 6, and 8 in.<br>(50, 100, 150, and 200 mm) | 5⅓ in.<br>(135 mm) | 8 and 12 in.<br>(200 and 300 mm) |
| 2, 4, 6, and 8 in.<br>(50, 100, 150, and 200 mm) | 6 in.<br>(150 mm) | 12 in.<br>(300 mm) |
| 2, 4, 6, and 8 in.<br>(50, 100, 150, and 200 mm) | 8 in.<br>(200 mm) | 12 and 16 in.<br>(300 and 400 mm) |

[a]Nominal sizes include the thickness of the mortar joint for all dimensions.

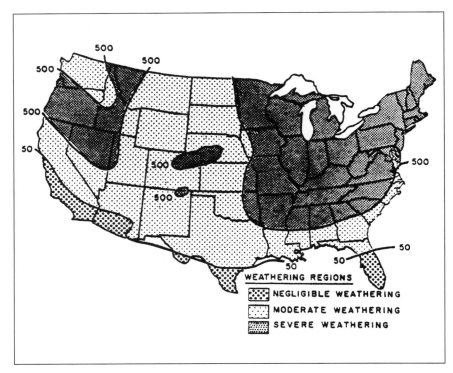

**FIGURE 4.4-17** U.S. weathering indexes. (Copyright ASTM; reprinted with permission.)

ing figures show more detail and guides to selecting the proper types, classes, and grades of units based on their use.

### BUILDING (COMMON) BRICK

ASTM C62 covers three grades, SW (severe weathering), MW (moderate weathering), and NW (negligible weathering), based on resistance to weathering (see Fig. 4.4-17). The physical requirements for these grades are given in Figure 4.4-18; the uses, in Figure 4.4-19.

### FACING BRICK

ASTM C216 covers two grades, SW and MW, based on resistance to weathering, and three types, FBX, FBS, and FBA, based on the appearance of the finished wall. FBX units are allowed small dimension tolerances, small distortion tolerances, little chippage, and small color range differences from unit to unit. In fact, they require such high precision and uniformity that they are seldom specified. FBS units require less precision and uniformity, and FBA units require even less than FBS units. The most common are FBS units. The physical requirements for facing brick are the same as for those building brick grades SW and MW (see Fig. 4.4-18). The uses of facing brick based both on weathering and appearance are shown in Figure 4.4-20.

### HOLLOW BRICK

ASTM C652 covers two grades, SW and MW, based on resistance to weathering, and four types, HBS, HBX, HBA, and HBB, based on the appearance of the finished wall. The physical requirements for hollow brick are the same as those for building brick grades SW and MW (see Fig. 4.4-18). The uses of hollow brick are shown in Figure 4.4-21.

## FIGURE 4.4-18 Physical Requirements for Solid Clay Masonry Units by Grade

| Designation | Minimum Compressive Strength (brick flatwise), Gross Area, psi | | Maximum Water Absorption by 5-Hour Boiling | | Maximum Saturation Coefficient[a] | |
|---|---|---|---|---|---|---|
| | Average of 5 Bricks | Individual | Average of 5 Bricks | Individual | Average of 5 Bricks | Individual |
| Grade SW | 3000 | 2500 | 17.0% | 20.0% | 0.78 | 0.80 |
| Grade MW | 2500 | 2200 | 22.0% | 25.0% | 0.88 | 0.90 |
| Grade NW | 1500 | 1250 | No limit | No limit | No limit | No limit |

[a]The saturation coefficient is the ratio of absorption by 24-hour submersion in cold water to that after 5-hour submersion in boiling water.

(Copyright ASTM; reprinted with permission.)

**FIGURE 4.4-19** Uses of Building (Common) Brick[a] (ASTM C62) by Grade

| Grade[b] (based on weathering index) | Uses |
|---|---|
| SW (severe weathering) | Foundations or other structures below grade or in contact with earth, especially when subjected to freezing temperatures |
| MW (moderate weathering) | Exterior walls and other exposed vertical masonry surfaces above grade |
| NW (negligible weathering) | All interior masonry or as backup |

[a]Units with core areas less than 25% of unit's gross cross-sectional area.
[b]See Figure 4.4-17.

**FIGURE 4.4-20** Uses of Facing Brick[a] (ASTM C216) by Grade and Type

| Grade[b] (based on weathering index) | Uses |
|---|---|
| SW (severe weathering) | Masonry in contact with earth, especially when subject to freezing temperatures |
| MW (moderate weathering) | Exterior walls and other exposed masonry above grade |

| Type (based on appearance of finished wall) | Uses |
|---|---|
| FBX | Masonry where high degree of mechanical perfection, minimum variation in color range, and minimum variation in size are required |
| FBS | Where wider color ranges and greater size variations are permissible than specified for FBX |
| FBA | Where architectural effects are desired resulting from nonuniformity in size, color, and texture of units |

[a]Units with core areas less than 25% of unit's total cross-sectional area.
[b]See Figure 4.4-17.

**FIGURE 4.4-21** Uses of Hollow Brick[a] (ASTM C652) by Grade and Type

| Grade[b] (based on weathering index) | Uses |
|---|---|
| SW (severe weathering) | Where a high and uniform degree of resistance to frost action and disintegration by weathering is desired and where brick may be frozen when permeated with water |
| MW (moderate weathering) | Where a moderate and somewhat nonuniform degree of resistance to frost action is permissible and where brick is unlikely to be permeated with water when exposed to below-freezing temperatures |

| Type (based on appearance of finished wall) | Uses |
|---|---|
| HBS | Exposed exterior and interior walls where wider color ranges and greater variations in size are permitted than specified for HBX |
| HBX | Exposed exterior and interior walls where narrow color range and minimum permissible variation in size are required |
| HBA | Where architectural effects are desired resulting from nonuniformity in size, color, and texture of units |
| HBB | Where color and texture are not a consideration and greater size variation is permitted than specified for HBX |

[a]Units with core area greater than 25% of unit's total cross-sectional area.
[b]See Figure 4.4-17.

## STRUCTURAL CLAY TILE, LOAD-BEARING

ASTM C34 covers two grades, LBX and LB, based on compressive strength and resistance to weathering. The physical requirements for load-bearing structural clay tile are given in Figure 4.4-22; the uses, in Figure 4.4-23.

## STRUCTURAL CLAY TILE, UNGLAZED FACING TILE

ASTM C212 covers two classes, standard and special duty, based on the thickness of the face shells, and two types, FTX and FTS, based on the appearance of the finished wall. The physical requirements for unglazed structural clay facing tile are given in Figure 4.4-24; the uses, in Figure 4.4-25.

## FACING BRICK AND STRUCTURAL CLAY FACING TILE, CERAMIC GLAZED

ASTM C126 covers ceramic glazed facing brick and other solid units, as well as facing tile having a finish consisting of a ceramic color glaze or a clear glaze. It includes two grades, S (select) for use with comparatively narrow mortar joints and SS (select sized or ground edge) for use where variation of the face dimension must be very small, and two types, I (single-faced units) for use where only one finished face will be exposed and II (two-faced units) for use where two opposite finished faces will be exposed

**FIGURE 4.4-22** Physical Requirements for Load-Bearing Structural Clay Tile

| Grade | Maximum Water Absorption[a] by 1-Hour Boiling, % | | Minimum Compressive Strength (based on gross area),[b] psi | | | |
| | Average of 5 Bricks | Individual | End Construction Tile | | Side Construction Tile | |
| | | | Average of 5 Bricks | Individual | Average of 5 Bricks | Individual |
|---|---|---|---|---|---|---|
| LBX | 16 | 19 | 1400 | 1000 | 700 | 500 |
| LB | 25 | 28 | 1000 | 700 | 700 | 500 |

[a]The range in percentage absorption for tile delivered to any one job should be not more than 12.

[b]Gross area of a unit should be determined by multiplying the horizontal face dimension of the unit, as placed in the wall, by its thickness.

(Copyright ASTM; reprinted with permission.)

**FIGURE 4.4-23** Uses of Load-Bearing Structural Clay Tile[a] (ASTM C34) by Grade

| Grade[b] (based on weathering index) | Uses |
|---|---|
| LBX | Masonry exposed to weathering and direct application of stucco |
| LB | Masonry not exposed to frost action or earth; may be used in exposed masonry if protected with a facing of 3 in. (76 mm) or more of other masonry |

[a]Units with core area greater than 25% of unit's total cross-sectional area.

[b]See Figure 4.4-19.

(Fig. 4.4-26). The physical requirements include minimum compressive strength, tests of finish (imperviousness, chemical resistance, and crazing), limitations on distortion, and limitations on variation in dimensions.

## ARCHITECTURAL TERRA COTTA

There is no ASTM standard for terra cotta. Terra cotta units of every type are custom products. Many stock dies are available, but in general, terra cotta sizes

**FIGURE 4.4-24** Physical Requirements for Unglazed Structural Clay Facing Tile

| Type | Maximum Water Absorption | | | |
| | By 24-Hour Submersion in Cold Water, % | | By 1-Hour Boiling, % | |
| | Average | Individual | Average | Individual |
|---|---|---|---|---|
| FTX | 7 | 9 | 9 | 11 |
| FTS | 13 | 16 | 16 | 19 |

| Class | Compressive Strength Based on Gross Area | | | |
| | End Construction Tile | | Side Construction Tile | |
| | Minimum Average of 5 Tests, psi | Individual Minimum psi | Minimum Average of 5 Tests, psi | Individual Minimum psi |
|---|---|---|---|---|
| Standard | 1400 | 1000 | 700 | 500 |
| Special duty | 2500 | 2000 | 1200 | 1000 |

(Copyright ASTM; reprinted with permission.)

**FIGURE 4.4-25** Uses of Unglazed Structural Clay Facing Tile[a] (ASTM C212) Based on Type

| Type | Uses |
|---|---|
| FTX | Exposed masonry: smooth face, low absorption, and resistance to staining. Has high degree of mechanical perfection, minimal variation in color and face dimensions |
| FTS | May be smooth or rough-textured, with moderate absorption and variation in dimensions, medium color range, and minor surface finish defects |

[a]Units with core area less than 25% of unit's total cross-sectional area.

**FIGURE 4.4-26** Uses of Ceramic Glazed Facing Tile and Brick[a] (ASTM C126) Based on Grade and Type

| Grade | Uses |
|---|---|
| S (select) | Masonry with narrow mortar joints ($\frac{1}{4}$ in./6.4 mm) |
| SS (select sized or ground edge) | Masonry where face dimension variation must be very small |

| Type | Uses |
|---|---|
| I (single-face units) | Masonry where only one finished face will be exposed |
| II (two-faced units) | Masonry where two opposite finished faces will be exposed |

[a]Units with core area greater than 25% of unit's total cross-sectional area.

and shapes are produced to meet specific project requirements.

# 4.5 Concrete Masonry Units

Concrete masonry units are molded concrete units used in building construction as an integral part of the structure, as facing for or filler panels between structural elements, and to construct partitions. Concrete masonry units include concrete brick, hollow concrete block, slump block, split-face block, and other special units. They do not include concrete splash blocks, manhole blocks, paving units, sills, or similar concrete products or cast stone.

Solid and hollow masonry units are made from a relatively dry mixture of portland cement (and sometimes other cementitious materials), aggregates, and water. Admixtures are also often used. The ingredients are thoroughly mixed, then molded into the desired shapes by compaction and vibration, both by machine. They are then cured under controlled conditions of moisture and temperature. After a period of aging during which the required strength, moisture content, and other desired properties are attained, the units are ready for use. Because the raw materials and manufacturing methods used influence both the strength and other properties of concrete masonry units in a finished wall, a general understanding of both will help in the selection and use of such units.

## 4.5.1 RAW MATERIALS

The raw materials used to manufacture concrete masonry units include cementitious materials, water, aggregates, and various admixtures.

### 4.5.1.1 Cementitious Materials

The primary cementitious material used in concrete masonry units is portland cement. Most of the portland cement used is normal portland cement without an air-entraining admixture. To a lesser extent, high-early-strength portland cement, and both normal and high-early-strength portland cements containing an air-entraining admixture, are also used. Units formed using high-early-strength cement have greater strength during the early stages of setting and curing than do their counterparts made with normal portland cement. They are also less likely to break during handling and delivery.

Portland blast-furnace slag cements, fly ash, silica flour, and other pozzolanic or hydraulic materials may also be used as part of the cementing medium. Pozzolanic materials, such as fly ash, have little or no cementitious value in themselves, but combine at ordinary temperatures with lime liberated from portland cement during hydration to form cementing compounds. This reaction is slow, which contributes to strength development after several months. Materials such as silica flour react similarly at the elevated temperatures used in high-pressure (autoclave) curing. Both fly ash and silica flour are used as a portion of the cementing material with portland cement. The correct proportioning of these materials depends on the curing temperatures and type of aggregate used.

Although its use is not common, natural cement can be used satisfactorily to replace up to 20% of the portland cement in concrete masonry units.

The types of cementitious materials used in concrete masonry units and the applicable ASTM standards covering them are shown in Figure 4.5-1.

**FIGURE 4.5-1** Standards Covering the Raw Materials Used in Concrete Masonry Units

| Cementitious Materials | |
|---|---|
| Portland cement | ASTM C150 |
| Portland blast-furnace cement | ASTM C595 |
| Fly ash | ASTM C618 |
| Hydrated lime | ASTM C207 –– Type S |
| Siliceous and pozzolanic | No ASTM standards. Siliceous and pozzolanic materials should be shown by test or experience not to be detrimental to the durability of the concrete. |
| **Aggregates** | |
| Normal weight | ASTM C33 |
| Lightweight | ASTM C331 |
| **Admixtures** | |
| Air-entraining agents | ASTM C260 |
| Accelerators and calcium chloride | ASTM D98 |
| Coloring pigments, integral water repellents, etc. | No ASTM standards. These admixtures should be previously established as suitable for use in concrete or be shown by test or experience not to be detrimental to the durability. |

### 4.5.1.2 Aggregates

Aggregates constitute about 90% of the weight of a concrete masonry unit, and therefore have an important effect on the desirable properties as well as the cost of the finished units. In general, local availability is the primary determinant in the use of any one type of aggregate. The applicable ASTM standards for aggregates are shown in Figure 4.5-1.

Aggregates are classified according to their weight as either dense or lightweight. Dense aggregates, which are also called *normal-weight* aggregates, include sand, gravel, crushed limestone, and air-cooled slag. Lightweight aggregates include expanded shale, clay, slag, coal cinders, pumice, and scoria. The effect of aggregate type on the weight, strength, and other characteristics of concrete block is shown in Figure 4.5-2. Desirable aggregate properties include:

- Toughness, hardness, and strength to resist impact, abrasion, and loading
- Durability to resist freezing and thawing, and expansion and contraction, resulting from moisture and temperature changes
- Uniform gradation of fine and coarse aggregate sizes to produce an eco-nomical, workable mixture and a uniform appearance—the maximum size of aggregate should not exceed one-third of the smallest section of a block's shell
- Cleanliness and lack of foreign particles, which would reduce strength or cause surface imperfections

### 4.5.1.3 Admixtures

Only a few of the numerous admixtures marketed for use in concrete mixes have been found beneficial or desirable in concrete masonry units. The admixtures used include air-entraining agents, accelerators, coloring pigments, integral water repellents, and a few others. The applicable ASTM standards are shown in Figure 4.5-1. Admixtures should be investigated individually to determine whether they are appropriate and to ensure that they will perform satisfactorily under the individual manufacturing plant conditions.

Air-entraining agents increase the plasticity and workability of a concrete mixture and distribute minute air bubbles uniformly throughout the hardened concrete, which increases its ability to withstand frost action. The addition of an air-entraining admixture to the mix for concrete masonry units usually permits greater compaction of the mix and produces a denser unit. In addition, the appearance of the units is improved, closer reproduction of the contours of the mold is obtained, and breakage of freshly molded units is reduced.

Water-repelling agents, such as metallic stearates, are used to a limited extent. They are somewhat effective in reducing the rate of absorption and capillarity of the units.

Accelerators, the most common type being calcium chloride, are used by some producers to speed up the hardening of units during cold weather operations.

Workability agents, many of which are of the air-entraining type, are sometimes used to correct the harshness of concrete mixes made with poorly graded aggregates.

### 4.5.2 MANUFACTURE

Block plants vary in size, layout, equipment, type of automation, and manufacturing details, but their basic operations are essentially alike. They include the steps shown in Figure 4.5-3 and discussed in the following paragraphs.

**FIGURE 4.5-2** Properties of Concrete Masonry Units Based on the Aggregates Used

| Aggregate (graded; ⅜ in. (9.5 mm) to 0) | | Weight, lb/cu ft (kg/m³) of Concrete | Weight, 8 in. × 8 in. × 16 in. (200 mm × 200 mm × 400 mm) Unit | Compressive Strength (gross area), psi[a] (MPa) | Water Absorption, lb/cu ft (kg/m³) of Concrete | Thermal Expansion Coefficient (per °F) × 10⁻⁶ |
|---|---|---|---|---|---|---|
| Type | Density (air-dry), lb/cu ft (kg/m³) | | | | | |
| **Normal weight** | | | | | | |
| Sand and gravel | 130–145 (2082.41–2242.59) | 135 (2162.5) | 40 (18.14) | 1200–1800 (8.3–12.4) | 7–10 (112.13–160.19) | 5.0 |
| Limestone | 120–140 (1922.22–2242.59) | 135 (2162.5) | 40 (18.14) | 1100–1800 (7.6–12.4) | 8–12 (128.15–192.22) | 5.0 |
| Air-cooled slag | 100–125 (1601.85–2002.31) | 120 (1922.22) | 35 (15.88) | 1100–1500 (7.6–10.3) | 9–13 (144.17–208.24) | 4.6 |
| **Lightweight** | | | | | | |
| Expanded shale | 75–90 (1201.39–1441.67) | 85 (1361.57) | 25 (11.34) | 1000–1500 (7–10.3) | 12–15 (192.22–240.29) | 4.5 |
| Expanded slag | 80–105 (1281.48–1681.94) | 95 (1521.76) | 28 (12.70) | 700–1200 (4.8–8.3) | 12–16 (192.22–256.3) | 4.0 |
| Cinders | 80–105 (1281.48–1681.94) | 95 (1521.76) | 28 (12.70) | 700–1000 (4.8–7) | 12–18 (192.22–288.33) | 4.5 |
| Pumice | 60–85 (961.11–1361.57) | 75 (1201.39) | 22 (9.98) | 700–900 (4.8–6.2) | 13–19 (208.24–304.35) | 4.0 |
| Scoria | 75–100 (1201.39–1601.85) | 95 (1521.76) | 28 (12.70) | 700–1200 (4.8–8.3) | 12–16 (192.22–256.3) | 4.0 |

[a]Multiply these values by 1.80 to obtain approximate corresponding values of strength of the concrete (strength of unit on net area).

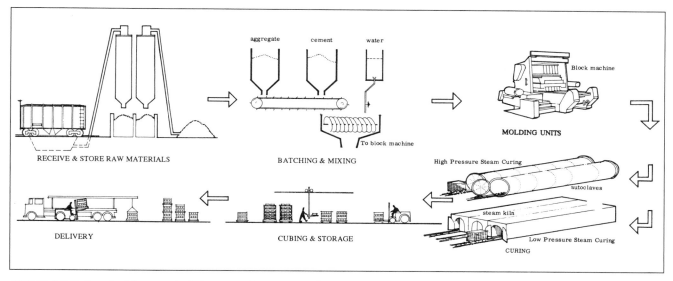

**FIGURE 4.5-3** Flowchart for manufacturing concrete masonry units.

### 4.5.2.1 Receiving Raw Materials

Materials are delivered in bulk to a plant by truck or rail and elevated into overhead steel storage bins above the batching and mixing area (Fig. 4.5-4). At some plants, cement is delivered in special tank trucks or railroad cars and forced by compressed air through pipes directly into the overhead cement storage bin. Different aggregates are handled and stored separately to permit proper blending and proportioning.

### 4.5.2.2 Batching and Mixing

Materials in the concrete mix are proportioned by weight with a *weigh batcher*. The weigh batcher is placed under the storage bins, and the correct amount of each material for a batch is discharged into the batcher. The weigh batcher is then positioned over the mixer and the materials are discharged into the mixer, where they are mixed with a controlled quantity of water for 6 to 8 minutes (Fig. 4.5-5). An electronic water meter senses the water contained in the aggregate and adjusts the amount of mixing water. Admixtures in liquid form may be added as part of the mixing water.

**FIGURE 4.5-4** Placing materials into an overhead storage bin above a batching and mixing machine.

**FIGURE 4.5-5** A weigh batcher positioned over a mixer.

**FIGURE 4.5-6** A molding machine.

### 4.5.2.3 Molding Units

The mixed batch is then discharged into a hopper above the molding machine. Controlled quantities of the mix are fed into the machine for molding into desired shapes and sizes. Concrete is consolidated in the mold by a combination of vibration and pressure (Fig. 4.5-6). Typically, three 8 × 8 × 16 in. (200 × 200 × 400 mm) concrete blocks or their equivalent in volume of concrete are molded per machine cycle at a speed of about 5 cycles per minute.

The freshly molded (*green*) concrete units emerge from the machines on steel pallets and are placed on a steel curing rack. This operation, called *off bearing*, may be done manually or with an automated rack loader. At this stage, the units are sufficiently dry to hold their shape outside the mold, but are fragile and can be broken by hand pressure.

### 4.5.2.4 Curing

Occasionally, where the climate is favorable, units are moist cured at normal temperatures of 70° to 100°F (21.1° to 37.8°C). Most units, however, are cured using either low-pressure or high-pressure steam.

#### LOW-PRESSURE STEAM CURING

Steam curing at atmospheric pressure accelerates the hardening process. This low-pressure steam curing, which is done in a kiln, accounts for most concrete masonry units produced in the United States. The main benefit of steam curing is rapid strength gain, which permits placing units in inventory within hours after they are molded. Two to 4 days after molding, the compressive strength of low-pressure steam-cured blocks will be 90% or more of the final ultimate strength. Moist-cured blocks cured at normal temperatures, on the other hand, develop strength slowly and have less than 40% of their ultimate strength 4 days after molding. Low-pressure steam curing produces units of lighter color than is usually obtained with moist curing.

### HIGH-PRESSURE STEAM CURING

Units cured by application of saturated steam in an autoclave (Fig. 4.5-7) under a pressure between 125 and 150 psi (460 and 553 kPa/mm$^2$) are called *autoclaved* units. Autoclaved units, which were widely available in the past, are not made in many parts of the country today, and are available there only at a high price due to high shipping costs.

The principal benefits of high-pressure steam curing are high early strength and greater stabilization against volume changes caused by varying moisture conditions. Within 24 hours after molding, usable units can be produced that have a permanent compressive strength equal to that of moist-cured units only after they have been cured for a full 28 days.

### 4.5.2.5 Cubing and Storage

When curing has been completed, the curing racks are conveyed to a *cubing station*, where the units are assembled into cubes, usually consisting of six layers of 15 to 18 blocks, 8 × 8 × 16 in. (200 × 200 × 400 mm) or an equivalent number of other sized units. Cubes are

**FIGURE 4.5-7** An autoclave.

stacked so that two low openings are left to accommodate the forks of a forklift. They are then secured together using steel bands and transported to the storage area, where they are placed in inventory in stacks that are three or four cubes high. Depending on the curing methods used, the units remain in inventory anywhere from a few days to several weeks before they are released for use.

### 4.5.2.6 Delivery

When units are ready to be released, they are loaded onto flatbed trucks by the cube using a forklift or other loading device. At a construction project's site, cubes of units are picked up one or two at a time by a mechanical unloader and set at convenient locations for use.

### 4.5.3 PHYSICAL PROPERTIES

The physical properties of concrete masonry units are determined largely by the properties of the hardened cement paste and the aggregate. In general, much of the technical knowledge relating to concrete is applicable to concrete masonry units. However, differences do occur as a result of variations in mix composition and consistency, the consolidation method used, textural requirements, method of curing, and other factors.

Concrete masonry units usually have a substantially lower cement factor than structural concrete. Between 2.75 and 4 sacks of cement are used per cubic yard in masonry units compared with between 4 and 7 bags in concrete. The water-cement ratio is much lower too, between 2 and 4 gal (7.57 and 15.14 L) per sack of cement in concrete masonry and between 5 and 7 gal (18.93 and 26.5 L) per sack in concrete. The aggregate in concrete masonry is graded finer than that used in concrete, with the largest size seldom exceeding $^3/_8$ in. (9.5 mm). Concrete masonry units are more often made with porous or lightweight materials than is concrete. Concrete masonry units contain a relatively large volume of interparticle void spaces (space between aggregate particles that are not filled with cement paste). They are also cured at higher temperatures than is cast-in-place structural concrete.

Typical ranges in compressive strength, water absorption, density, and other physical properties of commercial-grade hollow concrete masonry units are shown in Figure 4.5-2.

### 4.5.3.1 Compressive Strength

The strength of concrete masonry units is difficult to predict from mix data, because a simple water-cement ratio is not valid for harsh, dry mixes and because each type of aggregate exhibits different characteristics during mixing. Unlike cast-in-place concrete, in concrete masonry units the highest strengths are obtained from the wettest mixes. However, the maximum amount of mixing water is not usually added in order to cut down on breakage of freshly molded units during handling before curing. Dry mixes do not compact as well as wet ones and therefore have lower strengths. Through experimentation, each manufacturing plant develops mixes and techniques to produce units of optimum strength compatible with other desirable properties.

The compressive strengths shown in Figure 4.5-2 are based on the gross-bearing area of the concrete masonry unit, including core spaces. The compressive strength of the unit based on net area, excluding core space, is about 1.8 times the values shown. Solid face and veneer units, such as concrete brick and split block, when made with sand and gravel or crushed stone aggregate, will generally exceed 3000 psi (21 MPa) in compression.

As shown in Figure 4.5-2, units made with lightweight aggregates have lower compressive strengths than units made with sand and gravel, limestone, or air-cooled slag. Major influences on compressive strength are:

- Type and gradation of aggregate
- Type and amount of cementitious material
- Degree of compaction attained in molding the units
- Moisture content and temperature of the units at the time of testing

### 4.5.3.2 Tensile Strength, Flexural Strength, and Modulus of Elasticity

The tensile strength, flexural strength, and modulus of elasticity of concrete masonry units vary in proportion to their compressive strength. Tensile strength will range from 7% to 10% of the compressive strength, flexural strength from 15% to 20% of the compressive strength, and modulus of elasticity from 300 to 1200 times the compressive strength.

#### 4.5.3.3 Water Absorption

Absorption tests are used to determine the density of a concrete. Absorption is a measure of the number of pounds (kilograms) of water in each cubic foot (cubic meter) of concrete. In hollow concrete masonry units this measure varies over a wide range, from as little as 4 or 5 lb/cu ft (64.1 to 80.1 kg/m³) for the heaviest sand and gravel units to as much as 20 lb/cu ft (320.4 kg/m³) for the very porous, lightweight aggregate types. Solid units made with sand and gravel will normally have less than 7 lb/cu ft (112 kg/m³) absorption. The absorption of the aggregate, which may vary from 1% to 5% of the dry weight of aggregate for dense aggregates to 30% or more for the lightest aggregate, has a strong influence on the absorption of the units.

The porosity and pore structure of a concrete influence other properties, such as permeability, thermal conductivity, and sound absorption, but the influence is not always predictable, especially when comparing units made with different types of aggregates. High water absorption is not a desirable or purposely incorporated property, but it is accepted as a natural consequence when properties such as lightness in weight, higher sound absorption, or thermal insulation are desired.

A high initial rate of absorption (*suction rate*) results when concrete contains a large portion of relatively large, interconnected pores and voids. Units with a high suction rate combined with high absorption have high permeability to water, air, and sound and may also have less resistance to frost action. On the other hand, the unconnected, air-filled pores present in lightweight aggregate and in air-entrained cement paste impart the advantages due to porosity yet minimize permeability to water, air, and sound.

An abnormally high suction rate adversely affects the structural bond of mortar to units, but this problem can be minimized by using mortars with high water retention properties.

Concrete masonry intended for use in exterior walls and which will not be painted should have low absorption, and mortar joints should be tooled for weathertightness.

#### 4.5.3.4 Volume Changes

Concrete masonry units undergo small dimensional changes due to changes in temperature, changes in moisture content, and a chemical reaction called *carbonation*. Of these factors, only moisture changes can be conveniently controlled appreciably. This can be done by pre-shrinking the units and by adequately drying them before they are used.

Carbonation causes irreversible shrinkage in masonry units as a result of a chemical reaction within the concrete brought about when it absorbs carbon dioxide from the air. Although few test data exist, it appears that the magnitude of this change under certain conditions and over extended periods may approximate that due to changes in moisture content.

Concrete masonry units expand when heated and contract upon cooling. These volume changes are fully reversible. Units return to their original shape when the temperature returns to its original level. The coefficient of thermal expansion for concrete masonry units depends largely on the coefficient of the aggregate, since it comprises about 90% of the concrete volume. Coefficients of thermal expansion for concrete masonry units are shown in Figure 4.5-2. Moisture changes cause concrete masonry units to expand when wetted and contract when dried. Moisture volume changes that occur when units are new may not be fully reversible because of concrete's tendency to assume a permanent contraction during the first few cycles of wetting and drying. Later moisture volume changes are usually fully reversible, however.

An important contributor to the development of cracks in concrete masonry walls is the volume change in the units due to original shrinkage during drying. If the units are laid up in a wall before they have become dimensionally stable and have been allowed to shrink, tensile stresses will develop where the wall is restrained from shrinking, and cracking can be expected. Drying shrinkage can be greatly reduced with proper curing and drying so that when units are placed, their moisture content is in equilibrium with the surrounding air—that is, dried down to the average air-dry condition to which the finished wall will be exposed. For a given aggregate type, autoclave curing will sometimes reduce drying by as much as 50% compared with that of units cured by low-pressure steam or cured at normal temperatures.

The amount of residual shrinkage, or movement, a concrete unit will undergo due to moisture change after placement is the product of its shrinkage potential and its moisture loss after placement. In other words, residual shrinkage = shrinkage potential × moisture loss. Shrinkage potential depends on the manufacturing technique, raw materials used, age of the unit, and other factors. Moisture loss depends on the moisture content of the units when laid and the relative humidity and temperature of the drying environment. Although measurable in the laboratory, the small magnitudes make it impractical to measure shrinkage on a routine basis as a means of demonstrating compliance with specifications. Instead, shrinkage is controlled by limiting the moisture content of the units at the time of delivery.

Figure 4.5-8 shows the correlation between moisture content and drying environment, expressed in terms of relative humidity. Masonry units complying with the limits for moisture content shown will have a tolerable residual shrinkage, and cracking of the wall will be minimized. The linear shrinkage classifications shown depend on aggregate type and curing method. Figure 4.5-9 shows the average values of linear shrinkage. The term *normal weight* refers to units made using sand, gravel, limestone, and other dense aggregates. *Lightweight* refers to all com-

**FIGURE 4.5-8** Maximum Moisture Content of Type I Units

| Linear Shrinkage Classification,[a] % | Maximum Moisture Content, % of Total Absorption (average 5 units) | | |
|---|---|---|---|
| | Average Annual Relative Humidity,[b] % | | |
| | Over 75 | 75–50 | Under 50 |
| Up to 0.03 | 45 | 40 | 35 |
| 0.03 to 0.045 | 40 | 35 | 30 |
| 0.045 to 0.065 | 35 | 30 | 25 |

[a]As determined by ASTM C426 test for drying shrinkage of concrete within previous 12 months.

[b]As reported by the U.S. Weather Service nearest source of manufacture.

FIGURE 4.5-9 Linear Shrinkage of Concrete Masonry Units

| Aggregate Type | Steam-Curing Method | Average Linear Shrinkage,[a] % |
|---|---|---|
| Normal weight | High pressure | 0.019 |
| | Low pressure | 0.027 |
| Lightweight | High pressure | 0.023 |
| | Low pressure | 0.042 |
| Pumice | High pressure | 0.039 |
| | Low pressure | 0.063 |

[a]As determined by test, ASTM C426.

mon lightweight aggregate types except pumice. Aggregate classifications are not necessarily exclusive and may overlap since many lightweight blocks contain minor amounts of normal-weight aggregate.

### 4.5.3.5 Surface Texture

Texture may be greatly varied (Fig. 4.5-10) to satisfy esthetic requirements or to suit a desired physical requirement, such as a coarse texture to provide a mechanical bond for direct application of plaster. Either normal-weight or lightweight aggregate units can be given many different surface textures by controlling (1) the gradation of the aggregates, (2) the amount of water used, and (3) the degree of compaction at the time of molding. In addition, the surface of a unit may be ground to produce the exact texture desired (Fig. 4.5-11), and the face mold can also be changed to produce numerous contours in the face shell of the unit (Fig. 4.5-12).

Units with open surface textures offer some value in absorbing sound. Exposed concrete masonry walls built with ordinary commercial units will absorb between 18% and 68% of the incident sound. The textures shown in Figure 4.5-10 illustrate this range. Surface finishes, such as paint, tend to close the surface pores and thereby reduce the absorption significantly.

### 4.5.3.6 Color

The usual aggregates and portland cements used in commercial production produce units with a range of color from off-white to gunmetal gray and from light tan to brownish shades. A wider color range, including blues, browns, buffs,

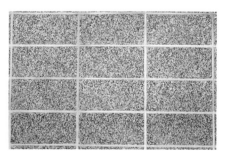

FIGURE 4.5-11 Ground-face units laid in stack bond.

FIGURE 4.5-12 The decorative concrete blocks shown are called *Shadowal*. They may be laid in other patterns to produce a variety of effects.

FIGURE 4.5-10 The surface texture of a concrete masonry unit can affect its insulating value and sound absorption characteristics. As shown here, surface textures may range from fine to coarse.

greens, reds, and slate grays, can be obtained using pure mineral oxide pigments singly or in combination, mixed integrally with the concrete before it is molded.

### 4.5.4 CLASSIFICATION OF CONCRETE MASONRY UNITS

Concrete masonry units can be classified as concrete brick, consisting of solid units (building brick and slump brick); concrete block, consisting of both solid units (load-bearing and non-load-bearing) and hollow units (load-bearing and non-load-bearing); and special units (split-faced block, faced block, and decorative block).

#### 4.5.4.1 Concrete Brick

This category includes concrete building brick and slump brick.

##### BUILDING BRICK

Building bricks are either completely solid or have a shallow depression called a *frog* (Fig. 4.5-13). The frog's purpose is to reduce weight and provide for a better mechanical bond when these units are laid in mortar. Concrete building bricks are modular units sized to finish 4 in. (101.6 mm) in width and 8 in. (203.2 mm) in length when laid with $^3/_8$-in. (9.5-mm)-thick mortar joints. The thickness varies, but in most units it is such that by slightly varying the bed joints, a mason can lay the units with three brick and three mortar joints in every 8 in. (203.2 mm) of wall height. Some manufacturers produce oversized jumbo (see Fig. 4.5-13) and double brick units.

##### SLUMP BRICK

The concrete mixture used in making slump brick has a consistency that causes the units to sag or *slump* when they are removed from the molds, resulting in irregularly faced units that vary in height, surface texture, and appearance (Fig. 4.5-14). The slump may occur from the effects of gravity alone or may be enhanced by a mechanical device that presses down on the units after they have been removed from the molds. Slump brick units are used when their special and unusual appearance is desired.

#### 4.5.4.2 Concrete Block

Concrete blocks are either solid load-bearing units, hollow load-bearing units, or hollow non-load-bearing units. Re-

**FIGURE 4.5-13** Typical shapes of concrete masonry units.

gardless of the grade or type of the unit (see Section 4.5.5) or whether it is load-bearing or non-load-bearing, a solid unit has a core area less than 25% of the total cross-sectional area of the unit, while a hollow unit has a core area greater than 25% of the total cross-sectional area of the unit.

Some typical shapes of concrete block are shown in Figure 4.5-13, but it is seldom possible to obtain all of them in a given locality. Most producers also make half-length and half-height units for use with the full-sized units they manufacture. Concrete blocks are made in modular sizes for use with $3/8$-in. (9.5-mm)-wide head and bed joints. Therefore an $8 \times 8 \times 16$ in. (200 $\times$ 200 $\times$ 400 mm)

block is actually only $7^5/8 \times 7^5/8 \times 15^5/8$ in. (193.7 $\times$ 193.7 $\times$ 396.9 mm). The typical dimensions and component parts of a three-core corner unit are shown in Figure 4.5-15. Face shells and webs increase in thickness from the bottom of the unit to the top to permit easy removal of the core form during manufacture and to provide greater bedding area for mortar in the bed joints. The end flanges of most units have $3/4$-in. (19.1-mm)-wide by $3/16$-in. (4.8-mm)-deep mortar grooves to increase the transverse strength in a wall laid up with them and to provide a vertical mortar key which helps create a weathertight joint.

Figure 4.5-15 shows a three-core unit, but units are also made with two cores. These offer some advantages over the three-core design. The principal advantage is the resulting graduated increase of the face shell thickness at the center web, which increases the strength of the unit in tension and reduces the tendency of units to crack due to drying shrinkage and temperature changes. This is particularly true when units are laid in running bond, since the head joints of the course above and below a block occur over the thickened center web rather than over an open core, as is the case with three-core blocks. Two-core blocks have three webs instead of four, which reduces heat conduction from 3% to 4%. They are also about 4 lb (1.8 kg) lighter in weight than the same-sized three-core block. And finally, they have more vertical space to accommodate reinforcement and conduit or other utilities.

The uses of some of the available types of concrete block follow.

**FIGURE 4.5-14** Slump blocks and slump bricks can produce bold, irregular wall textures.

$3/16'' = 4.76$ mm
$3/4'' = 19.1$ mm
$1^1/4'' = 31.8$ mm
$1^3/4'' = 44.45$ mm
$1'' = 25.4$ mm

**FIGURE 4.5-15** A typical $8 \times 8 \times 16$ in. (200 $\times$ 200 $\times$ 400 mm) concrete corner block.

**FIGURE 4.5-16** Split-face blocks have a rough, natural-stone-like texture.

- *Corner blocks* have one flush end for use in pilasters and piers and on exposed corners.
- *Bullnose blocks* have one or more small radius-rounded corners. They are used instead of square-edged corner units where a rounded corner is desirable.
- *Jamb or sash blocks* are used to facilitate the installation of windows, louvers, and other opening fillers.
- *Solid top blocks* have solid tops for use as a bearing surface in the finishing course of a foundation wall.
- *Header blocks* have a recess to receive the header unit in a masonry bonded wall.
- *Lintel blocks* are U- or W-shaped. They are used to construct horizontal lintels or beams.

- *Partition blocks* are usually 4 or 6 in. thick. They are used to build non-load-bearing partitions and walls.
- *Return (L) corner blocks* are used in 6-in. (150-mm), 10-in. (250-mm), and 12-in. (300-mm)-thick walls at corners to maintain a consistent horizontal coursing with the appearance of full-length units and half-length units without cutting of the corner units.
- *Control joint blocks* of various types are used to construct vertical shear-type control joints.

### 4.5.4.3 Special Units

Some special units (see Fig. 4.5-13) are stock units, while others are custom-made. Some of the types available follow.

- *Split-face blocks* are solid or hollow units that are produced in parallel pairs and fractured (split) lengthwise by machine after they have hardened to produce two units, each with a rough surface texture on one face (Fig. 4.5-16). Color variations are obtained by introducing mineral colors and by using aggregates of different gradation and color in the concrete mixture. The units are laid in a wall with the fractured faces exposed. Units are available in random sizes, which are often larger than brick sizes, so that broken joint patterns can be used.
- *Slump blocks* are similar to slump brick but larger (see Fig. 4.5-14).
- *Faced blocks* are units with ceramic glazed, plastic, polished, or ground faces. These facings are applied in separate operations on cured concrete masonry units. One type of glazed unit uses a glazing material which is a thermosetting, resinous binder combined with glass silica and color pigments or colored granules. This mixture, in a semiliquid state, is cast onto each block in individual molds. The units are then heat-treated to set the glaze and bind it integrally into the pores of the concrete unit.
- *Decorative blocks* are manufactured in many different types and configurations. Some have beveled face shell recesses. Others have surfaces with more than one plane (see Fig. 4.5-12). Still others have clear rectangular or curved holes through them. Different effects can be created by using the same block design in different pattern bonds. Standard units can also be used to produce decorative effects.
- *Chimney blocks* are designed for use with a flue lining in chimneys.
- *Pilaster blocks* are used in plain or reinforced pilasters and columns.

### 4.5.5 STANDARDS CONTROLLING CONCRETE MASONRY UNITS

The four types of concrete masonry units that are controlled by ASTM standards are listed with the applicable standard in Figure 4.5-17.

**FIGURE 4.5-17** Strength and Water Absorption Requirements for Concrete Masonry Units

| Unit Type and Standard | Grade[a] | Minimum Compressive Strength on Average Gross Area, psi | | Maximum Water Absorption for Units of Different Densities, lb/cu ft | | | |
|---|---|---|---|---|---|---|---|
| | | | | Lightweight Concrete | | Medium-Weight Concrete | Normal-Weight Concrete |
| | | Average of 3 Units | Individual Unit | Less than 85 lb/cu ft | Less than 105 lb/cu ft | 105 to 125 lb/cu ft | More than 125 lb/cu ft |
| Hollow load-bearing units, ASTM C90 | N-I, N-II | 1000 | 800 | — | 18 | 15 | 13 |
| | S-I, S-II | 700 | 600 | 20 | — | — | — |
| Solid load-bearing units, ASTM C90 | N-I, N-II | 1800 | 1500 | — | 18 | 15 | 13 |
| | S-I, S-II | 1200 | 1000 | 20 | — | — | — |
| Concrete building brick, ASTM C55 | N-I, N-II | 3500 | 3000 | 15 | 15 | 13 | 10 |
| | S-I, S-II | 2500 | 2000 | 18 | 18 | 15 | 13 |
| Hollow non-load-bearing block, ASTM C129 | — | 350 | 300 | — | — | — | — |

[a]Grades S-I and S-II are limited to walls not exposed to the weather and to exterior above-grade walls protected with weather-resistant coatings. Grades N-I and N-II may be used below grade as well as above; protective coatings are recommended below grade and where required above grade.

(Copyright ASTM; reprinted with permission.)

### 4.5.5.1 Grades

Concrete masonry units intended for load-bearing construction are classified in ASTM standards as Grade N and Grade S. Grade N units are suitable for general use. They may be used in exterior walls, whether below or above grade, that may be exposed to moisture or weather, and for interior walls and backup.

Grade S units are suitable for use only where they will not be exposed to the weather or other moisture. They may not be used below grade. They may be used in exterior walls only when a weather-protective coating is applied over them. They may be unprotected in walls that are not exposed to the weather or other moisture.

### 4.5.5.2 Types

Concrete building brick and both load-bearing and non-load-bearing concrete block can be manufactured with a controlled moisture content. *Moisture-controlled* units are designated by ASTM as Type I units. They are available in both Grades N and S. The two designations are written together (N-I or S-I). The moisture content of Type I units is regulated during manufacture to conform with the limits established by the applicable ASTM standard. Type II units, as designated in the ASTM standards, are not restricted to a definite moisture content.

The maximum moisture contents for Type I units based on various average annual relative humidities and the linear shrinkage classification of the masonry units are shown in Figure 4.5-8. The selection should be based on the local humidity conditions as reported by the U.S. Weather Service nearest the place where the masonry units are to be manufactured and the shrinkage classification of the particular masonry units used.

Units with limited moisture content should be used in most cases to minimize in-place shrinkage of masonry units and the wall cracks that result from it. As shown in Figure 4.5-8, when the construction is located in an area of low relative humidity and a low equilibrium moisture condition is expected, the maximum moisture content at the time of use should be lower than when the construction is located in a less arid region. A lower maximum moisture content is also required when the shrinkage potential of the masonry unit is greater. The shrinkage classification for different units can be determined from Figure 4.5-9.

In addition to regulating moisture content, ASTM standards establish requirements for the strength and water absorption of masonry units (see Fig. 4.5-17). The effect of water absorption and of other properties on masonry construction is described in Section 4.5.3.

## 4.6 Unit Masonry Design

Masonry walls provide excellent structural performance and offer long-term, low-maintenance finishes. They are a basic element of buildings of all kinds, from enormous skyscrapers to single-family dwellings. They are used as part of the structural bearing system in small and large structures and as part of the enclosing and finishing systems of many large and small steel-, concrete-, and wood-framed buildings. To be successful, masonry wall design, selection, and construction must take into consideration the required functional performance level, loading conditions, allowable costs, exposure, type of occupancy, and desired finished appearance. These factors are often influenced by the systems and finishes used in other construction elements.

To function adequately, a masonry wall must:

- Support the applied *dead load* (the weight of the wall and of the other building elements it carries, such as floors and roofs)
- Resist forces applied directly to the wall, such as those that result from wind and earthquakes
- Support *live loads* transmitted into the wall by other building elements that are supported by the wall, such as floors and roofs
- Be weathertight to adequately control the flow of heat, moisture, air, and water vapor
- Produce satisfactory levels of sound and visual privacy
- Have appropriate fire resistance
- Accommodate heating, air conditioning, electrical, and plumbing equipment
- Be suitable for the application of various finish materials
- Be economical
- Permit installation of doors, windows, louvers, and other opening closers

When requirements other than the above exist, the materials, design, and construction methods described in this chapter may be inadequate.

Wall systems must safely support superimposed vertical loads and resist horizontal and racking loads caused by lateral forces, such as wind and earthquake, and, where they are load-bearing, transmit all loads to the foundation. Masonry walls on small structures are rarely specifically engineered. Ordinarily, building codes and rules of thumb based on established practices determine the materials and methods used in their construction.

The costs of a masonry wall include in-place costs, maintenance costs, and building operating costs associated with the masonry. *In-place costs* are the sum of the costs of materials, labor, and overhead required to erect a wall. The selection of a particular wall type will affect other costs, such as that for installing insulation, vapor retarders, siding, and finishes, and for providing openings for windows and doors. The time and costs required to enclose a building and provide its heating, air conditioning, electrical, and plumbing systems may also be influenced by the wall system selected. The costs of maintaining the masonry and the costs of maintaining the heating, ventilating, and air conditioning systems that are attributable to a wall system, including heat loss or gain through the construction, have an important influence on the ultimate cost of a wall and should not be overlooked if long-term economy is to be achieved.

The esthetics of a building are influenced by the location, size, and type of doors, windows, and other openings that form part of the building's walls. The wall construction should also be compatible with the roof and floor systems used in the construction. Masonry walls are usually left exposed on their exterior side, but they may be finished with stucco or other coatings. Interior surfaces can be finished with applications such as plaster or gypsum board or can be left unfinished.

Many of the industry's currently recognized minimum requirements for masonry wall design are based on experience with and observation of the performance of masonry assemblies over many years. It naturally follows that building codes and published standards often require minimum wall thicknesses and maximum lateral support spacings, instead of demanding that walls be

**FIGURE 4.6-1** Masonry wall types.

designed using theoretical analysis of the forces and stresses involved.

Most of the recommendations for the design of masonry walls in this section are based on the provisions of ACI 530, ASCE 5, TMS 402, and ACI 530.1, ASCE 6, and TMS 602, which have been developed jointly by the American Concrete Institute (ACI), the American Society of Civil Engineers (ASCE), and The Masonry Society (TMS). These documents have been adopted by reference in the *International Building Code* published by the International Code Council. These standards consolidate and supersede several masonry codes which were the standards used for masonry for many years but which are no longer valid.

Section 4.2 of this book explains how to choose the correct mortar type for the use and exposure of a wall. Sections 4.4 and 4.5 explain how to select the right masonry units. Section 4.7 illustrates the techniques of proper masonry installation. This section (4.6) describes the various types of masonry wall assemblies

and their properties when built and relates many details essential to satisfactory design. Taken together, the sections of this chapter explain how to produce masonry construction of suitable quality.

The need for insulating masonry walls, and the materials and application methods generally used for that insulation, are discussed in Section 7.3.

## 4.6.1 WALL TYPES

Masonry walls may be classified as solid walls, cavity walls, veneered walls, and reinforced walls (Fig. 4.6-1). Each type may be constructed of either solid or hollow masonry units. Solid and cavity walls are built to be either load-bearing or non-load-bearing; veneered walls are non-load-bearing; reinforced walls are load-bearing. Figure 4.6-2 shows one load-bearing condition; Figure 4.6-3 shows another. No wall type is suitable for all exposures or loading conditions, nor is any one type appropriate for all localities or types of occupancy. The final selection of a wall

type should be made only after weighing the various requirements of function, economy, and appearance.

### 4.6.1.1 Solid Walls

A solid wall is built up of units laid close together, with all joints between units filled solidly with mortar. Solid walls may be built of either solid or hollow masonry units (or both in combination) in any required thickness and are used for either load-bearing or non-load-bearing construction. The structural bond of solid walls is provided by metal ties, masonry headers, or joint reinforcement, as shown in Figure 4.6-1. Walls that utilize masonry headers or have units overlapping in alternate courses (see Fig. 4.6-1G) are called *masonry bonded*. These are used in areas of slight weather exposure, for economical foundation walls, and for exterior load-bearing walls. There is a possibility that masonry headers may break due to wall movement and may not, therefore, provide as good a bond as was previously supposed.

**FIGURE 4.6-2** Many types of clay and concrete unit masonry load-bearing walls are possible. This drawing shows concrete masonry walls that have been reinforced to resist lateral forces.

FIGURE 4.6-3 This building has load-bearing masonry walls. The floor framing is supported directly on the masonry.

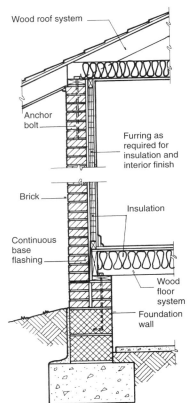

FIGURE 4.6-5 Section through a nominal 6-in. (150 mm)-thick load-bearing wall. (Drawing by HLS.)

Solid walls may consist of solid units of brick or concrete brick and block; hollow units of concrete block, hollow brick, or structural clay tile; or may be a composite (faced) construction consisting of facing and backup units of different materials bonded so that both facing and backup units are load-bearing.

Typical details of 8 in. (200 mm) solid brick walls and composite walls are shown in Figures 4.6-1H, 4.6-1I, and 4.6-1J. Solid and composite walls bonded with masonry headers are less resistant to rain penetration than metal-tied walls, and should be used only in areas where they will not be exposed to wind and rain (Fig. 4.6-4) (see also Section 4.6.3). The addition of furring to the inner wall reduces through-the-wall moisture penetration and permits use in areas of severe exposure. In areas subjected to hurricanes or wind-driven rains, the exterior surface of concrete block walls is usually stuccoed to resist rain penetration.

The *International Building Code* permits the use of exterior load-bearing 6-in. (150-mm)-thick masonry walls for one-story, single-family dwellings where the wall height does not exceed 9 ft (2743 mm) to the eaves or 15 ft (4572 mm) to the peak of the gable. Figure 4.6-5 is a typical wall section using SCR Brick. SCR Brick is a solid unit designed to form a nominal 6-in. (150-mm)-thick wall laid up in running bond with full $1/2$-in. (12.7-mm)-thick bed and head joints. The length and width of the SCR Brick make it possible to turn corners and maintain the half-bond. An air space, formed by 2 × 2 (50 × 50) wood furring strips, is flashed at the base of the wall to provide a barrier to moisture penetration. The furred space also permits installation of electrical wiring and outlets and provides room for wall insulation if it is required. Similar construction using solid or hollow concrete masonry units may also be used.

Where solid masonry walls are used, insulation and mechanical equipment are

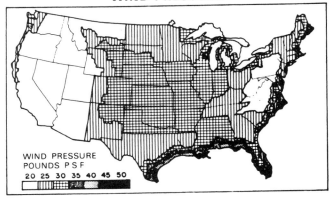

FIGURE 4.6-4 The weather exposure of masonry units and assemblies is based on precipitation and wind pressure. These maps show the annual precipitation rates and wind pressures that may be expected in the United States. (Courtesy Brick Institute of America [renamed Brick Industry Association].)

often installed within furred spaces on the interior side of the wall. On below-grade basement and foundation walls, insulation is often placed on the exterior side of the wall.

### 4.6.1.2 Cavity Walls

A cavity wall consists of two wythes of either solid or hollow masonry, each of which is at least 4 in. (100 mm) thick, separated by a continuous air space not less than 2 in. (50.8 mm) or more than $4^1/_2$ in. (114 mm) wide, and bonded together with metal ties or joint reinforcement (see Figs. 4.6-1D–F and 4.6-6). The exterior wythe may be brick, hollow brick, or concrete masonry units. The interior wythe may be brick, hollow brick, structural clay tile, or solid or hollow concrete masonry units (Fig 4.6-7). The exterior wythe is usually a nominal 4 in. (100 mm) thickness. The interior wythe may be 4, 6, or 8 in. (100, 150, or 200 mm), depending on the height and length of the wall and the loads to be carried. The nominal overall thickness is 10, 12, or 14 in. (250, 305, or 356 mm) when the

**FIGURE 4.6-7** Section through a non-load-bearing cavity wall on a concrete-framed building. (Drawing by HLS.)

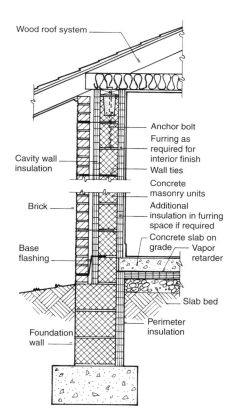

**FIGURE 4.6-6** Section through a nominal 12-in. (300 mm)-thick load-bearing cavity wall. (Drawing by HLS.)

air space is maintained at a nominal 2 in. (50.8 mm). The cavity may be left as all air space or insulated in a variety of ways.

A cavity offers two advantages in areas of severe exposure. The continuous air space provides insulation value to the wall and permits insulation to be installed within the wall to further reduce heat transfer (Fig. 4.6-8) (see also Section 7.3). It also acts as a barrier to moisture, eliminating the need for furring, since significant rain penetration to the interior is practically impossible if proper flashing and weep holes are installed. A cavity also may be used to conceal mechanical and electrical utilities (Fig. 4.6-9).

For a cavity wall to be effective, certain precautions must be observed during its construction. The cavity must be kept free of mortar droppings that could form

a bridge allowing moisture to penetrate to the interior face. Two methods for doing this are shown in Figure 4.6-10. Flashings and weep holes must be provided to drain the cavity of water that passes through the exterior wythe. Metal

**FIGURE 4.6-8** Cavity wall with rigid insulation in the cavity. (Courtesy DOW USA.)

**FIGURE 4.6-9** The cavity in this wall is being used to conceal plumbing pipes and electric wiring runs.

**FIGURE 4.6-10** This figure demonstrates two methods of keeping mortar out of a cavity. The two drawings on the left show the beveling of a mortar bed to prevent mortar from being squeezed into the cavity when the next unit is placed on it. The drawing on the right shows how a wood strip can be placed on each tier of metal ties or continuous joint reinforcement to catch mortar droppings. The wood strip is raised as the work progresses. (Courtesy Brick Institute of America [renamed Brick Industry Association].)

cavity wall ties, designed to resist both tension and compression, should be used to bond the exterior and interior wythes and should be solidly embedded in mortar so that the two wythes will act together to resist wind and other lateral forces. These ties may be separate items or may be part of continuous joint reinforcement.

A vapor retarder is unnecessary in a cavity wall when the cavity is insulated with fill materials that will not retain excessive moisture, such as water-repellent vermiculite or silicon-treated perlite, or with rigid board materials, such as foam glass and foamed plastics, that are at least 1 in. (25.4 mm) less thick than the cavity and are installed next to the inner wythe. In cavity walls, insulation and mechanical equipment are often installed between thicknesses of masonry or within furred spaces on the interior side of the wall.

### 4.6.1.3 Veneered Walls

Installing masonry units as a facing material (veneer), without using their load-bearing properties, is common in all building types. Figure 4.6-11 shows typical brick veneer over wood studs and metal studs in a single-story wood-framed structure. Figure 4.6-12 shows a two-story steel-framed building with a cavity wall consisting of brick veneer over insulated metal studs. Similar construction is also used in higher buildings and those framed with a concrete structural system. The facing of a veneered wall is attached but not bonded to the studs and does not act structurally with the load-bearing portion of the wall.

In masonry veneer construction, wood or metal framing carries the vertical loads, while the masonry performs

**FIGURE 4.6-11** Sections through brick veneer walls. Section (a) is through a wood stud wall. Section (b) is through a metal stud wall. Insulation has been omitted from the drawings for clarity.

**Anchor**

**Adjustable anchor**

**Anchor**

**2 in. (50 mm) min. air space**

**Building paper over sheathing**

**Brick**

**Flashing**

**Adjustable anchor**

**Building paper over sheathing**

**Anchor**

**Flashing**

**Mortar**

**Foundation wall**

**Roof decking**

**Roof framing**

**Interior finishes**

**Metal studs with insulation**

**Concrete slab on permanent form**

**Floor framing**

**Interior finish**

**Metal studs with insulation**

**Concrete slab on grade**

**Slab bed**

**Perimeter insulation**

**FIGURE 4.6-12** Section through a brick veneer wall on a steel-framed building. (Drawing by HLS.)

mostly as an exterior finish surface and provides a weather barrier. The space between the studs typically is used to contain insulation and to conceal heating, air conditioning, electrical, and plumbing lines.

Key factors in constructing brick veneer walls include correct selection of masonry materials, adequate anchorage to the backup wall, and proper installation of flashing and weep holes.

Since the veneer is isolated from the backup wall by an air space and exposed to the exterior, Severe Weathering (SW) Grade facing brick should be used. Since a masonry veneer carries only its own weight, Type N mortar should be used in it, consistent with the principle that mortar must be of the lowest strength that is compatible with structural requirements.

Refer to Section 4.7.2.2 for a description of the bonding together of brick veneer walls and a short discussion about the controversy surrounding the use of brick veneer with metal studs.

Flashing and weep holes to provide resistance to rain penetration should be located at the bottom of the wall and at all openings.

#### 4.6.1.4 Reinforced Walls

Following the principles used in reinforced concrete, steel reinforcement is sometimes placed in a masonry wall and embedded in grout to give the wall greatly increased resistance to forces that produce tensile, shear, and compressive stresses. The chief uses of reinforced masonry are as walls, columns, lintels, and bond beams (Fig. 4.6-13).

There are two ways to reinforce masonry walls. In the first method, which is sometimes called *reinforced hollow masonry*, the reinforcing and grout are placed in the cores of concrete masonry units (see Fig 4.6-1B) or hollow brick. In this method, concrete masonry units are laid with full face shell bed joints. The head joints are filled with mortar at least to a distance in from the face equal to the face shell thickness. Cross webs that are adjacent to cores that will be filled with grout should be filled solidly with mortar. Hollow brick should be laid with full head joints, but the cores must be left open in the bed joints. Both concrete masonry units and hollow brick must be laid and lapped with the cores lined up vertically to permit installation of reinforcing steel and proper filling with grout.

In the second method, which is called *reinforced solid masonry* or *reinforced grouted masonry*, the reinforcement and grout are placed in the space between two wythes of masonry (see Fig. 4.6-1C). In this method, the two masonry wythes are laid with completely filled bed and head joints. Cores and joints adjacent to spaces to be filled with grout should also be filled with mortar.

In both methods, care must be taken to prevent mortar droppings from entering the spaces to be filled with grout and to remove mortar fins from those spaces. Sometimes, cleanouts are placed in walls to be grouted so that foreign materials can be removed.

Reinforcing bars are placed vertically in walls and tied to horizontal bars. Continuous joint reinforcement is usually also placed in reinforced masonry walls.

The kind of grout used varies with the size of the spaces to be filled. When the least clear dimension of the space is $2\frac{1}{2}$ in. (63.5 mm) or less, fine grout (see Section 4.2.5) should be used. Coarse grout should be used when the spaces to be filled are larger. The height of pour of fine grout should be limited to 12 in. (305 mm), while coarse grout can be placed in layers as much as 4 ft (1219 mm) thick. Fine grout must be agitated or puddled to ensure that it fills the grout space completely. Coarse grout should be vibrated.

Refer to BIA's *Technical Notes on Brick Construction* No. 17, "Reinforced Brick Masonry: Part I," No. 17A, "Reinforced Brick Masonry—Materials and Construction," No. 17B, "Reinforced Brick Masonry—Beams," No. 17L, "Four-Inch RBM Curtain and Panel Walls," No. 17M, "Reinforced Brick Masonry Girders—Examples," and No. 41, "Hollow Brick Masonry" for additional information about reinforced masonry. Reinforced masonry design should be done in accordance with the requirements of ACI 530/531, ASCE 6, and TMS 602.

Insulation and mechanical equipment are usually installed in furred spaces on the interior side of reinforced masonry walls. Insulation is often placed on the exterior side of below-grade basement and foundation walls.

### 4.6.2 PATTERN BONDS

The pattern formed by the masonry units and mortar joints is called *pattern bond*. Pattern bond may result from the type of structural bond used or may be purely decorative. Structural bond is the method by which individual masonry units are interlocked together, using masonry headers or metal ties, to cause the entire wall assembly to act as a single structural unit.

#### 4.6.2.1 Brick Pattern Bonds

Traditional brick bond patterns still commonly used today are described briefly below and are illustrated in Figure 4.6-14.

##### RUNNING BOND

Running bond is the simplest of the basic pattern bonds in that there are no headers. Metal ties are used for structural bond. Running bond is used largely in cavity and veneer walls of solid units and often in concrete masonry unit and clay tile walls where masonry bonding may be accomplished by wider stretcher units (see Fig. 4.6-1G).

##### COMMON (AMERICAN) BOND

Common bond is similar to running bond except that there is a course of full-length

A-Continuous reinforced concrete bond beam. Lap bars at corners
B-Reinforced concrete studs tied to footing
C-Reinforcement in horizontal mortar joints
D-Reinforced concrete footing

**FIGURE 4.6-13** Typical methods of reinforcing concrete unit masonry walls.

RUNNING BOND     1/3 RUNNING BOND

6th Course Headers
COMMON BOND

6th Course Flemish Headers
COMMON BOND

Dutch Corner   English Corner
FLEMISH BOND

English Corner   Dutch Corner
ENGLISH BOND

STACK BOND

English Corner   Dutch Corner
ENGLISH CROSS OR DUTCH BOND

**FIGURE 4.6-14** Traditional brick pattern bonds. (Courtesy Brick Institute of America [renamed Brick Industry Association].)

headers at every fifth, sixth, or seventh course, depending on the structural bonding requirements. In some historic buildings, header courses were placed at even wider spacings. It is important in laying out every bond pattern that the corners be started correctly. For common bond, a three-quarter brick starts each header course at the corner.

### FLEMISH BOND

Flemish bond has alternate stretchers and headers in each course. Where the headers are not used for structural bonding, half-brick, called *blind headers*, may be used. There are two methods of starting the corners in a Flemish bond pattern. In the *Dutch* corner, a three-quarter brick starts each course. In the *English* corner, 2 in. (50.8 mm) or quarter-brick closures are used.

### ENGLISH BOND

In the English bond pattern, alternate courses are composed of headers and stretchers. The joints between stretchers in all courses line up vertically. Blind headers are used except in structural bonding courses.

### ENGLISH CROSS OR DUTCH BOND

The English cross or Dutch bond pattern is a variation of English bond. Vertical joints between the stretchers in alternate courses do not line up vertically, but center on the stretchers in the courses above and below.

### STACK BOND

In a stack bond pattern, all the vertical joints are aligned. Since units do not overlap, this pattern is usually bonded to

the backing with rigid steel ties. In large wall areas and in load-bearing construction, such walls should be reinforced using continuous joint reinforcement or steel rods placed in the horizontal mortar joints. If the vertical mortar joints are to align, the masonry units must be dimensionally accurate or prematched units must be carefully selected for size.

### CONTEMPORARY BONDS

Many traditional pattern bonds have been modified by projecting and recessing units or by leaving out units to form perforated walls and screens. A few contemporary brick bonds are shown in Figure 4.6-15. The effect of a particular pattern bond can be altered by varying the pattern formed by variations in the color and texture of individual units, by changing the joint types used, and by re-

cessing or projecting individual units, courses, or wythes.

The basket weave and diagonal bond patterns shown for concrete masonry units in Figure 4.6-16 are also used in brick paving.

### 4.6.2.2 Concrete Masonry Unit Pattern Bonds

Many bond patterns currently used with concrete masonry units are shown in Figure 4.6-16. The more commonly used ones are the running bond patterns in which the units overlap those in the next course by one-half unit and the vertical joints in every second row are in alignment. Although some of the patterns shown are decorative, they all have been shown by test to be suitable for load-bearing construction.

**FIGURE 4.6-15** Some other brick pattern bonds. (Courtesy Brick Institute of America [renamed Brick Industry Association].)

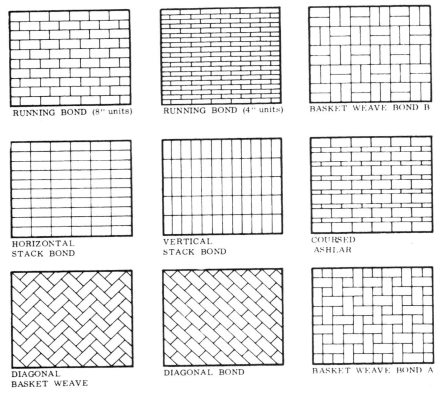

FIGURE 4.6-16 Concrete masonry unit pattern bonds.

*(Pattern labels, reading top-left to bottom-right:)*
RUNNING BOND (8" units) — RUNNING BOND (4" units) — BASKET WEAVE BOND B — HORIZONTAL STACK BOND — VERTICAL STACK BOND — COURSED ASHLAR — DIAGONAL BASKET WEAVE — DIAGONAL BOND — BASKET WEAVE BOND A

### 4.6.3 MOISTURE CONTROL

Refer to Section 7.1 for a general discussion about moisture control in buildings and some specific requirements for moisture control in foundations and basements with masonry walls. The principal sources of moisture in masonry are rain or melting snow penetration, capillary action from contact with free water or moisture in the ground, and condensation of vapor within the masonry. Other possible sources are faulty drainage and leaking plumbing systems.

Preventing moisture from entering a masonry wall is important to prevent the formation of efflorescence and cryptoflorescence and to prevent freeze-thaw cycle damage, both as discussed earlier (Sections 4.2.1.2 and 4.4.5.1). Excluding moisture from walls is also necessary to prevent corrosion of reinforcement and ties, and resultant damage to the wall and other building components.

Methods used to control rain and melting snow penetration into masonry walls include providing:

- Adequate flashing
- Proper tooling of mortar joints
- Parging and dampproofing
- Painting
- Filling joints between masonry and doors, windows, and other openings and penetrations with sealants
- A sufficient slope or wash to readily drain horizontal surfaces, such as those at sills and copings
- An overhang on sills and copings and the like, including drips to keep rain and melted snow away from the walls
- Adequate gutters and downspouts

Methods used to control groundwater penetration into masonry foundations, basements, and retaining walls include parging and dampproofing (see Sections 4.7.3 and 7.2), waterproofing (see Section 7.2), and sloping the exterior grade to drain water away from the walls.

Water vapor that may condense within uninsulated masonry can often be controlled by installing a vapor retarder on the warm side of the wall (see Section 7.1). A water-emulsion asphalt paint applied to the wall's surface is one method used to provide such a retarder. Masonry insulated with proper types of insulation may not require an additional vapor retarder.

### 4.6.4 CRACK CONTROL

Differential movements within masonry walls and conditions that prohibit necessary movement can produce stresses that may cause serious cracking unless precautions are taken to minimize them. Conditions that can generate stress concentrations large enough to cause cracking include:

- Expansion or contraction due to temperature changes
- Expansion or contraction due to changes in the moisture content of the units, especially of concrete masonry units
- Contraction in concrete masonry units due to carbonation (see Section 4.5.3.4)
- Structural movements caused by unequal foundation settlement or by the concentration of applied loads
- Concentration of stress that develops where doors, windows, and other openings occur
- Members built into the wall that locally restrain or prevent the wall from moving, as at the junction of floors, roofs, columns, and intersecting walls

The potential for cracking in masonry walls is exacerbated by the following:

- Dissimilar materials, as often occur in composite walls, expand and contract at different rates (metal lintels, windows, and door frames have substantially greater coefficients of expansion than does masonry, which causes relatively large differential movements).
- Exterior walls having a cold exterior surface and a warm interior surface tend to warp.
- The degree of restraint to movement imposed by roofs, floors, and intersecting walls and the effect of openings are difficult to estimate.

Expansion and control joints are built into masonry walls to reduce cracking caused by movement. The location of expansion and control joints cannot be established with mathematical accuracy because the many variables involved may act singly or in combination in a given case. Therefore, these joints are usually located using experience with and observation of existing buildings. The recommendations that follow have evolved in just that way.

#### 4.6.4.1 Expansion Joints for Clay Masonry

An unrestrained clay masonry wall 100 ft (30.48 m) long will tend to expand

$^3/_8$ in. (9.5 mm) in response to a temperature increase of 100°F (37.8°C). If this expansion is cumulative, it could cause the wall to extend beyond its foundation (Fig. 4.6-17a). When the wall contracts, its tensile strength is not sufficient to overcome frictional resistance and cracks will appear. Cumulative expansion causes failure in walls at offsets where the masonry in the offset at right angles to a long wall is placed in bending and shear by the expansion of the long wall (Fig. 4.6-17b).

Expansion joints are installed to provide a complete separation through the structure. Figure 4.6-18 shows several of the many methods used to build these joints. Expansion joints are expensive, may be difficult to maintain, and should be avoided when it is possible to do so without having the walls crack.

In general, expansion joints should be located at offsets when the wall expanding into the offset is 50 ft (15.24 m) or more in length and at junctions of walls in L-, T-, and U-shaped buildings. Recommended locations are shown in Figure 4.6-18. Figure 4.6-19 gives recommended spacings.

### 4.6.4.2 Control Joints for Concrete Masonry

Most initial movement in a concrete masonry wall is due to contraction from drying shrinkage. As a new masonry wall tries to contract, tensile forces occur that must be resisted by the tensile strength of the wall. The problem is made worse when a foundation, roof, or other restraining element offers resistance to the normal shrinkage of the masonry. A crack will occur when the forces that occur exceed the tensile strength of either the masonry units or the mortar.

Cracking due to shrinkage can be controlled by one or more of the following methods:

- Using moisture-controlled (Type I) concrete masonry units manufactured according to ASTM standards
- Installing horizontal steel reinforcement to increase the wall's tensile resistance to cracking and to minimize the width of cracks
- Building in control joints located to accommodate wall movements

Sometimes, more than one method must be used in the same wall.

**FIGURE 4.6-17** Cracks in masonry caused by movement. The crack in (a) is in a foundation wall. The one in (b) is in a building wall. (Courtesy Brick Institute of America [renamed Brick Industry Association].)

Control joints are continuous vertical joints built into concrete masonry walls as they are laid up, to minimize shrinkage cracks and to control cracks at locations where points of weakness or concentration of stresses is expected. Properly built control joints should prevent slight wall movements from cracking the wall, be themselves sealed against the weather, and stabilize the

**FIGURE 4.6-18** Expansion joint design and location in masonry walls. (Courtesy Brick Institute of America [renamed Brick Industry Association].)

**FIGURE 4.6-19** Recommended Spacing of Expansion Joints

| Outside Temperature Ranges[a] | Maximum Length of Wall[b] | | | |
| --- | --- | --- | --- | --- |
| | Unheated or Insulated | | Heated, Not Insulated | |
| | Solid, ft | Openings, ft[c] | Solid, ft | Openings, ft[c] |
| 100 and over | 200 | 100 | 250 | 125 |
| Less than 100 | 250 | 125 | 300 | 150 |

[a]The range from the lowest average temperature to the highest in degrees F.

[b]For insulated cavity or veneer walls, distances are approximately one-half the lengths shown.

[c]Opening: 20% or more of wall area.

(Courtesy Brick Institute of America [renamed Brick Industry Association].)

wall laterally across the joint by means of a shear key.

## LOCATION AND SPACING

The layout of control joints will depend on the following variables:

- Wall length and height.
- Wall type.
- Moisture content of units at the time of laying.
- Joint reinforcement spacing.
- Use of bond beams.
- Architectural details.
- Local experience concerning the need for control joints in the locality where the building is being built. In the absence of specific local data or practice experience, the locations and spacings of control joints shown in Figures 4.6-20 and 4.6-21 have been used successfully. Other spacings and locations may be necessary, however, in a particular project, especially if there are unusual circumstances.

Control joints should be placed in both load-bearing and non-load-bearing walls at changes in wall height or thickness, at the junction of walls with columns and pilasters, where openings occur, at wall intersections, and at return angles in L-, T-, and U-shaped buildings. In composite walls, where concrete masonry is used as a backup for other materials, control joints should be extended through the facing when the wall is laid with masonry bonding. Control joints do not need to extend through the facing material when it is bonded to the backup with metal ties.

When stucco or plaster is applied directly to masonry units, control joints should extend through the stucco or plaster.

Because the temperature- and moisture-volume change of concrete masonry below grade in foundation walls is usually quite low, and because it is difficult to construct a watertight joint in masonry below grade, control joints are not recommended for basement or foundation walls. The movements at a control joint in above-grade masonry may cause cracks to develop in the foundation wall below. To minimize this possibility, a continuous bond beam can be provided at the top of the foundation walls. Then the control joints in the above-grade wall can be stopped at the bond beam with no damage to the wall below it.

## CONSTRUCTING A CONTROL JOINT

Control joints should be laid up in mortar just like any other vertical joint. If the control joint is to be exposed to the weather or to view, a recess should be provided to receive a joint sealant by raking the mortar out to a depth of about $3/4$ in. (19 mm) after it has become quite stiff (Fig. 4.6-22). Sometimes a pressure gun and a thin, flat caulking trowel are used to force a knife-grade caulking compound into the joint (Fig. 4.6-23). In other cases, the methods discussed in Section 7.13 are used to apply an elastomeric sealant. To keep them as unnoticeable as possible, control joints should be built plumb and be of the same thickness as the other mortar joints.

One type of control joint can be built with regular stretcher block. Building paper or roofing felt is inserted in the end

**FIGURE 4.6-20** Typical control joint locations in concrete masonry unit walls and partitions.

**FIGURE 4.6-21** Maximum Spacings of Control Joints in Unreinforced Masonry[a]

| Panel Dimensions[b] | Vertical Spacing of Joint Reinforcements[c] | | | |
| --- | --- | --- | --- | --- |
| | None | 24 in. (600 mm) | 16 in. (400 mm) | 8 in. (200 mm) |
| Ratio of panel, length to height (L/H)[d] | 2 | 2½ | 3 | 4 |
| Panel length (L) not to exceed | 40 ft (12.9 m) | 45 ft (13.72 m) | 50 ft (15.24 m) | 60 ft (18.29 m) |

[a]Moisture controlled, Type I concrete masonry units assumed. If nonmoisture control units, reduce joint spacing by 50%; if wall is solid grouted, spacing should be 33% less.

[b]A panel is a wall element in one plane lying between (a) corners or wall ends, (b) control joints, or (c) a control joint and end wall.

[c]Continuous metal ties with a minimum of two No. 9 gauge longitudinal wires.

[d]Maximum panel length (L) is applicable regardless of height (H).

**FIGURE 4.6-22** The mortar in this control joint has been raked out to a depth of about $^3/_4$ in. preparatory to applying a sealant. (Courtesy Portland Cement Association.)

**FIGURE 4.6-23** This control joint is being sealed with a knife-grade caulking compound. (Courtesy Portland Cement Association.)

**FIGURE 4.6-24** This control joint is being built by inserting building paper in the end core of the block. Roofing felt may also have been used. (Courtesy Portland Cement Association.)

**FIGURE 4.6-25** A control joint with building paper or roofing felt and a mortar-filled core. (Courtesy Portland Cement Association.)

**FIGURE 4.6-26** Special control joint units provide lateral stability by means of tongue-and-grove shaped ends. (Courtesy Portland Cement Association.)

**FIGURE 4.6-27** Laying a half-length control joint unit. (Courtesy Portland Cement Association.)

core of full- and half-length block and extended the full height of the control joint. The core is then filled with mortar for lateral support (Fig. 4.6-24). The paper or felt, cut to convenient lengths and wide enough to extend across the joint, prevents the mortar from bonding on one side of the joint, thus permitting the control joint to function (Fig. 4.6-25). Control joint blocks with tongue-and-groove shaped ends, which are available in many areas, provide lateral support to the wall sections on each side of the control joint (Fig. 4.6-26). These control joint blocks are made in both full- and half-length units (Fig. 4.6-27).

Because control joints can be unsightly and require maintenance, they should be avoided when possible. Methods used to increase the tensile strength of a wall and to improve the dimensional stability of the masonry units used in it will reduce the number of control joints required, as will designing to eliminate the conditions that dictate their use.

Other methods are also available to increase the crack resistance of walls. For example, two-core blocks with thickened fillets and three-core blocks with thickened center cores have higher tensile strengths than standard units. Their use can increase the crack resistance of a wall about 50% and 25%, respectively, over standard three-core blocks. Another example is the inclusion of continuous joint reinforcement in a wall. As Figure 4.6-21 shows, when joint reinforcement is introduced, the spacing of control joints may be increased.

### 4.6.5 STRUCTURAL DESIGN OF MASONRY WALLS

Masonry walls are used both as bearing walls to support floor or roof structure and as non-load-bearing walls and parti-

tions. The same basic principles apply to the design of masonry walls, regardless of their use. Masonry should always be supported either on other masonry, on concrete, or on steel. Wood girders and other types of wood framing should never be used to support masonry.

#### 4.6.5.1 Compressive Strength

The strength at the point of failure, which is called the *ultimate compressive strength*, of masonry walls is closely related to the compressive strength of the masonry units themselves and of the mortar. The quality of workmanship, thickness of mortar joints, regularity of the units' bearing surfaces, and workability of the mortar are also important.

High compressive strength is developed by installing high-strength units using proper workmanship so that every joint is completely filled with mortar. The strengths used for designing purposes, which are called *allowable compressive stresses*, recommended for various wall types, unit strengths, and mortar types are given in Figure 4.6-28.

The principal loads producing compressive stresses in a wall are the weight of the wall itself and the weights of associated floors and roofs that are transferred into the wall, which are collec-

**FIGURE 4.6-28** Allowable Compressive Stresses in Unit Masonry

| Construction; Compressive Strength of Unit, Gross Area, psi | Allowable Compressive Stresses,[a] Gross Cross-Sectional Area, psi | |
|---|---|---|
| | Type M or S Mortar | Type N Mortar |
| Solid masonry of brick and other solid units of clay or shale; sand-lime or concrete bricks | | |
| 8000 or greater | 350 | 300 |
| 4500 | 225 | 200 |
| 2500 | 160 | 140 |
| 1500 | 115 | 100 |
| Grouted masonry of clay or shale; sand-lime or concrete | | |
| 4500 or greater | 225 | 200 |
| 2500 | 160 | 140 |
| 1500 | 115 | 100 |
| Solid masonry of solid concrete masonry units | | |
| 3000 or greater | 225 | 200 |
| 2000 | 160 | 140 |
| 1200 | 115 | 100 |
| Masonry of hollow load-bearing units | | |
| 2000 or greater | 140 | 120 |
| 1500 | 115 | 100 |
| 1000 | 75 | 70 |
| 700 | 60 | 55 |
| Hollow walls (noncomposite masonry bonded[b]) | | |
| Solid units | | |
| 2500 or greater | 160 | 140 |
| 1500 | 115 | 100 |
| Hollow units | 75 | 70 |
| Stone ashlar masonry | | |
| Granite | 720 | 640 |
| Limestone or marble | 450 | 400 |
| Sandstone or cast stone | 360 | 320 |
| Rubble stone masonry | | |
| Coarse, rough, or random | 120 | 100 |

[a]Linear interpolation for determining allowable stresses for masonry units having compressive strengths that are intermediate between those given in the table is permitted.

[b]Where floor and roof loads are carried upon one wythe, the gross cross-sectional area is that of the wythe under load; if both wythes are loaded, the gross cross-sectional area is that of the wall minus the area of the cavity between the wythes. Walls bonded with metal ties shall be considered as noncomposite walls unless collar joints are filled with mortar or grout.

(Courtesy American Concrete Institute.)

tively called the *dead load*; and the weights of occupants, furniture, and equipment, which are collectively called the *live load*. Both dead and live loads can be determined with reasonable accuracy, and once they have been determined, the resultant compressive stresses can be computed. Such stresses should not exceed the allowable stresses given in Figure 4.6-28 for the type of wall, kinds of units, and type of mortar being used. When composite walls or other structural masonry elements are composed of different kinds or types of units or mortars, the maximum compressive stress should not exceed the allowable stress for the weakest of the combinations of either the masonry unit(s) or the mortar type(s) of which the wall is composed. Fortunately, typical construction today does not require most masonry walls to carry large vertical loads. For example, the superimposed load on the foundation wall of a typical one-story residence rarely exceeds 15 psi (0.1 MPa). About 75% of the bricks produced today have compressive strengths of more than 4500 psi (31 MPa); concrete block, more than 1000 psi (7 MPa). Typical mortars used today have compressive strengths in the range of 750 psi (5 MPa) to 2500 psi (17 MPa). Therefore, it is evident that most masonry walls for one-story buildings have far more compressive strength than they need.

Tests have established the relationship between the compressive strength of walls and the masonry units in them. The compressive strength of walls constructed of hollow concrete masonry units is about 42% of the compressive strength of the units when they are laid with face-shell mortar bedding and 53% when they are laid with full mortar bedding (Fig. 4.6-29). So a wall built with full mortar bedding and concrete masonry units with compressive strengths of 1000 psi (7 MPa) would have a compressive wall strength of 0.53 × 1000, or 530 psi (3.7 MPa). Using an allowable compressive stress of 70 psi (0.48 MPa) (from Fig. 4.6-28), a wall having a strength of 530 psi (3.7 MPa) will have a safety factor of 530/70 (3.7/0.48), or about $7^1/_2$. A safety factor of 4 is ample for masonry walls.

### 4.6.5.2 Transverse Strength

The transverse strength of a wall is a measure of its *lateral stability*, which is its resistance to such lateral forces as earth pressures, as occur in foundation,

**FIGURE 4.6-29** The top drawing shows face-shell mortar bedding; the bottom, full mortar bedding. (Courtesy American Concrete Institute.)

basement, and retaining walls; wind pressures; and earthquake forces. Transverse loads induce tensile stresses in masonry that must be resisted. The lateral stability of masonry walls depends on the tensile strength of the bond between mortar and masonry units. Actions that increase tensile bond strength also increase transverse strength. These include using Type S mortar mixed with the maximum amount of water consistent with workability, maintaining the suction of clay masonry units below 20 g per minute, per 30 sq in (0.02 m²) (a simple field test to determine the suction of brick is described in Section 4.7.1.2), and making sure that both head and bed joints are completely full of mortar.

The lateral stability of a masonry wall is also increased by vertical loads which produce compressive stresses that offset tensile stresses caused by transverse loads. The addition of reinforcing steel also greatly increases the lateral strength of a masonry wall.

**FIGURE 4.6-30** One method of bracing masonry walls during construction. (Courtesy Portland Cement Association.)

### 4.6.5.3 Effect of Flashing on Wall Strength

The effect of flashing on wall strength depends on where mortar is placed and the bond of the mortar to both the flashing and the masonry. Through-wall flashing, such as base and head flashing, does not affect a wall's compressive strength but will reduce bending and shearing strengths. If flashing is placed directly on masonry with no mortar beneath it, the flexural strength at that point is zero. Fortunately, through-wall flashing usually occurs where bending resistance is least significant.

Test data suggest that when mortar is placed immediately above and below copper flashing, flexural strength is 30% to 70% of that of a comparable unflashed wall. There are no sufficient data that permit a similar generalization for shearing strength under similar conditions. Assuming that there is no mortar under a base flashing, a wall's shear resistance, its ability to withstand lateral forces, depends on friction. The wall still retains some shearing resistance, since the coefficient of friction between the flashing and the masonry is on the order of 0.25 to 0.50.

### 4.6.5.4 Lateral Support of Walls

Masonry walls require lateral support both while they are under construction and after they have been completed.

## LATERAL SUPPORT DURING CONSTRUCTION

Masonry walls are seldom designed to be freestanding, without support from columns, piers, or cross walls. Wind pressures can create four times as much bending stress in a freestanding wall as they do in a wall that is supported by other structural parts of the building. This stress often occurs at the bottom of the wall, where flashing and lack of bond in fresh mortar decrease the wall's strength

to resist tensile wind forces (see Section 4.6.5.3).

Bracing (Fig. 4.6-30) should be installed to resist wind pressure. Figure 4.6-31 shows commonly accepted limits on wall heights based on various peak wind velocities. Where bracing is used, the heights shown are the safe heights above the bracing. For example, to withstand 50 mph (80.47 km/hr) wind gusts, the freestanding, unbraced height of a 10-in. (250-mm)-thick wall weighing 67 psi

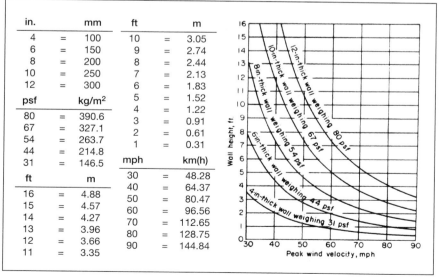

**FIGURE 4.6-31** Maximum unsupported height of nonreinforced, ungrouted concrete masonry walls during construction.

(0.46 MPa) or less should not be more than $7^1/_2$ ft (2286 mm) tall. The curves in the figure assume that the mortar has no tensile strength and the walls are free-standing, unreinforced, ungrouted concrete masonry. In cavity walls, the thickness is assumed to be two-thirds of the sum of the thickness of the two wythes. Table values may not be adequate under some circumstances.

Required bracing should be left in place until permanent bracing, such as that afforded by piers, cross walls, floors, and roof structures, has been placed and permanently braces the wall.

### LATERAL SUPPORT OF COMPLETED WALLS

Masonry walls must be permanently laterally supported or braced at certain intervals either by vertical or horizontal supports. When the limiting distance is length, vertical supports may be by columns, piers, pilasters, buttresses, or cross walls. When the limiting distance is height, horizontal supports may be floors, beams, or roofs. The spacing of these supports is primarily dependent on wall thickness. Possible spacing of supports and other lateral support requirements are summarized in Figure 4.6-32. Codes often require different spacings than those shown and should be consulted before limits are imposed. In walls of different classes of units or mortars, the ratio of wall height or length to wall thickness should not exceed that allowed for the weakest of the combinations of units and mortars in the member. Sufficient bonding and anchorage must be provided to transfer the loads from the wall to its supports.

If vertical supports are provided, there is no limit on the height of walls between floors or between floor and roof. When horizontal supports are provided, there is no limit on the length of a wall, and lateral supports in the form of columns, pilasters, or cross walls are not required. Neither the ratio of the unsupported wall height to wall thickness nor the ratio of unsupported wall length to wall thickness should be permitted to exceed the values given in Figure 4.6-32 for various wall types.

#### 4.6.5.5 Wall Thicknesses, Heights, and Lengths

Of course, it is possible to design every wall by theoretical analysis of the forces involved, and such an analysis is imperative when unusual conditions occur or when a code requires that it be done. However, theoretical analysis is seldom needed or required by code for houses or for small buildings of any type. Therefore, the design of masonry walls in most small projects is based on empirical requirements, as are the suggestions in this section.

In general, walls should not vary in thickness between their lateral supports. An exception is made, however, in residential buildings one or two stories in height, where vertical recesses may be built into 8-in. (200-mm)-thick walls if such a recess is not more than 4 in. (100 mm) deep and does not occupy more than 4 sq ft (2581 mm$^2$) of the wall's area.

**FIGURE 4.6-32** Lateral support requirements for completed masonry walls.

**FIGURE 4.6-33** Foundation Wall Thicknesses Based on Wall Types and Height of Unbalanced Fill

| Foundation Wall Construction | Maximum Height of Unbalanced Fill H (feet) | Minimum Wall Thickness T (inches) | |
|---|---|---|---|
| | | Frame Construction | Masonry or Veneer |
| Hollow Masonry | 3' (900 mm) | 8" (200 mm) | 8" (200 mm) |
| | 5' (1500 mm) | 8" (200 mm) | 8" (200 mm) |
| | 7' (2000 mm) | 12" (300 mm) | 10" (250 mm) |
| Solid Masonry | 3' (900 mm) | 6" (150 mm) | 8" (200 mm) |
| | 5' (1500 mm) | 8" (200 mm) | 8" (200 mm) |
| | 7' (2000 mm) | 10" (250 mm) | 8" (200 mm) |

## FOUNDATION WALLS

Unreinforced concrete masonry is used extensively for foundation walls, especially in single-family dwellings and other single-story buildings. Experience and observation have resulted in the development of minimum requirements governing their design. The requirements are empirical, particularly in designing basement walls to resist lateral earth pressures. Therefore, foundation design for small buildings is rarely based on theoretical analysis of the forces and stresses involved. In the absence of specific building code requirements or individual engineering analysis, the wall thicknesses shown in Figure 4.6-33 have been used when the height of unbalanced fill against the wall did not exceed the limits given in the figure and no other unusual circumstances existed. However, the figure does not take into account the presence of hydrostatic pressure against the wall due to groundwater. In addition, some practitioners believe that masonry foundation walls should not be less than 12 in. (305 mm) thick, regardless of the height of earth against them.

## OTHER WALLS

This section includes empirical design data for walls *other* than single-family residential foundations walls. These data should be verified with applicable code requirements and local experience in every case.

*Load-Bearing Walls* Load-bearing walls in residential buildings three stories or less in height may be 8 in. (200 mm) thick when not more than 35 ft (10.67 m) high if the roof is designed to impart no horizontal thrust.

Except where earthquake design is required, exterior walls in one-story residential buildings may be 6 in. (150 mm) thick when they are not more than 9 ft (2743 mm) high, the height to the peak of the gable does not exceed 15 ft (4572 mm), and girders and concentrated loads are supported on not less than 8 × 12 in. (200 × 305 mm) integral piers or pilasters.

Interior bearing walls in dwellings not more than one and a half stories or 20 ft (6.1 m) high may be 6 in. (150 mm) thick.

Load-bearing walls more than 35 ft (10.67 m) high should be at least 12 in. (305 mm) thick for the top 35 ft (10.67 m) of their height, and should increase 4 in. (100 mm) in thickness for each successive 35 ft (10.67 m).

Cavity walls built of solid masonry units and more than 25 ft (7.6 m) high above the support of the wall may be 10 in. (250 mm) thick. The facing and backing of cavity walls should each be at least 4 in. (100 mm) thick, and the cavity should not be less than 2 in. (50.8 mm) or more than $4^{1}/_{2}$ in. (114 mm) wide.

*Non-Load-Bearing Walls* Non-load-bearing walls on the exterior may be 4 in. (100 mm) less in thickness than is required for bearing walls but not less than 8 in. (200 mm) in thickness except where 6-in. (150-mm)-thick walls are permitted in one-story residences.

## 4.6.6 LINTELS AND ARCHES

The design of a structural member to support masonry over an opening should not be taken lightly. It is true that rule-of-thumb methods often result in overdesign. It is also true that arbitrary selection of steel shapes or reinforcing steel can result in inadequate sizing, causing structural cracks that lead to leaks and an unsightly appearance. Underdesigned supports may also fail completely, with disastrous results. Therefore, both masonry lintels and arches should be designed in accordance with engineering practice.

The dead weight of a masonry wall which must be supported over an opening can safely be assumed as the weight of a triangular section whose height is one-half the clear span of the opening (Fig. 4.6-34). Arching action of the masonry above the top of the opening may be counted on to support the additional weight. A concentrated load over an opening may be distributed along a wall length equal to the base of a triangle whose sides make an angle of 30 degrees with the vertical from the point of application of the load. The horizontal thrust resulting from an arching action must be resisted by sufficient mass in the adjoining wall.

Historically, various forms of arches have been used to span openings. Figure 4.6-35 shows some typical brick arch types. Arches built in contact with adjacent arches form a barrel vault. Domes may be thought of as arches rotated about their vertical centerlines.

**FIGURE 4.6-34** Assumptions for determining the loads imposed on lintels in masonry walls.

**FIGURE 4.6-35 Typical brick arches.** (Courtesy Brick Institute of America. [renamed Brick Industry Association].)

**FIGURE 4.6-36** Typical lintels for clay and concrete masonry.

Brick arch construction requires extensive labor, especially when the bricks are ground into wedge shapes with an abrasive stone to produce the curve. Most arches today are made using tapered joints between rectangular bricks to save the labor cost of producing wedge-shaped bricks. These are called *rough arches*. Most arches are built in solid walls because they are difficult to construct and even more difficult to flash properly in cavity walls and veneer walls. The arch must extend across the cavity to present a uniform appearance across the bottom of the arch. This not only blocks the cavity, it also ties the inner and outer wythes together, producing difficulties due to differential movement between the wythes.

In North America, where labor costs are high, arches have for the most part been replaced by lintels. The most common lintel for supporting masonry over openings in concrete unit masonry is a reinforced concrete masonry lintel. Such lintels may be rectangular concrete units made either in a shop or in the field from U-shaped lintel block, reinforcing steel, and grout (Fig. 4.6-36).

The most common method of supporting masonry over exterior openings is by steel angles or structural shapes (see Fig. 4.6-36). Using galvanized steel lintels will reduce the possibility of rusting but will slightly increase the initial cost of the lintels. Stainless steel lintels would eliminate maintenance and the possibility of failure due to corrosion altogether, and even though the initial, costs are higher, they may well be worth the difference.

When the steel reinforcement can be completely protected from the elements, reinforced masonry lintels have certain advantages over steel ones. Their maintenance costs are lower, because the steel does not require periodic painting. Even the initial costs are lower because less steel is required. However, if the wall around the lintel permits water to penetrate to the reinforcement, it will rust.

Lintels should bear on the wall not less than 4 in. (100 mm) at each end and should be of sufficient stiffness to carry the superimposed load without deflection of more than $1/360$ of the clear span.

### 4.6.7 PIERS AND PILASTERS

Piers are freestanding columns; pilasters are columns that are bonded to and built as integral parts of a wall. Both carry concentrated vertical loads. In addition, pi-

**FIGURE 4.6-37** Standard-time temperature curve (ASTM E119 Fire Test). (Copyright ASTM; reprinted with permission.)

lasters often provide necessary lateral support to walls.

Pier size is usually governed by code and is often a ratio of the height of the pier to its smallest cross-sectional dimension ($h/t$). The minimum pier section for solid units is usually 8 × 12 in. (200 × 305 mm), and for hollow units it is usually 8 × 16 in. (200 × 400 mm).

The unsupported height of a solid masonry pier should not exceed 10 times its smallest cross-sectional dimension. The same is true of piers constructed with hollow masonry units, provided that the cells are filled with Type M or Type S mortar (see Section 4.2.4.1). Piers constructed with unfilled hollow masonry units may have unsupported heights not exceeding four times their smallest dimension.

### 4.6.8 FIRE RESISTANCE

The fire resistance of walls, floors, and partitions is usually based on ASTM E119. The standard fire test does not measure the fire hazard in terms of actual performance in a real fire. Rather, it gives a comparison of the measure of performance between assemblies tested under similar conditions.

The test consists of exposing one side of a wall to a fire of controlled intensity measured by the standard time-temperature curve (Fig. 4.6-37). Immediately after firing, the hot face of the wall is subjected to a blast of water from a fire hose. The wall must withstand the fire and the hose stream without permitting flame or gases to pass through it, and the heat transmitted through the wall must be limited to the extent that there is less than a 250°F (121°C) rise in temperature. A load-bearing wall must also be able to resist its design stresses without failing. Within 72 hours after the tests have been completed, bearing walls must safely sustain loads twice those applied during the test. Such tests provide the basis for estimating the performance of selected construction types.

Although some walls have ultimate fire-resistance test periods of more than 10 hours, they are often rated lower than their ultimate fire-resistance test periods, because most building codes require ratings in multiples of 1 hour with a 4-hour maximum. Figure 4.6-38 gives ultimate fire resistance periods for solid brick load-bearing walls. Figure 4.6-39 lists the equivalent thickness of concrete masonry units required for different aggregate types to give the fire ratings found in model building codes (see Section 1.3).

### 4.6.9 FIREPLACES AND CHIMNEYS

The design and construction of an efficient, functional fireplace and chimney

**FIGURE 4.6-38** Ultimate Fire-Resistance Periods for Load-Bearing Clay and Shale Brick Walls

| Nominal Wall Thickness, in. | Wall Type | Ultimate Fire Resistance Period, hours | | | | |
|---|---|---|---|---|---|---|
| | | Incombustible Members Framed into Wall or No Framed-in Members | | | Combustible Members Framed into Wall | |
| | | No Plaster | Plaster[a] on One Side | Plaster[a] on Two Sides | No Plaster | Plaster on Exposed Side[a] |
| 4 | Solid | 1¼ | 1¾ | 2½ | — | — |
| 8 | Solid | 5 | 6 | 7 | 2 | 2½ |
| 12 | Solid[b] | 10 | 10 | 12 | 8 | 9 |
| 12 | Solid[c] | 12 | 13 | 15 | — | — |
| 10 | Cavity | 5 | 6 | 7 | 2 | 2½ |

[a]To achieve these ratings, each plastered wall face must have at least ½ in., 1:3 gypsum-sand plaster.
[b]Based on load failure.
[c]Based on temperature rise (for non-load-bearing walls).
(Courtesy Brick Institute of America [renamed Brick Industry Association].)

**FIGURE 4.6-39** Estimated Fire-Resistance Ratings for Concrete Masonry Unit Walls of Equivalent Thickness[a]

| Concrete Masonry Units | Members Framed into Wall or Partition | | | | | | |
|---|---|---|---|---|---|---|---|
| | None or Noncombustible, in. (mm)[b] | | | | | | |
| | 4 hours | 3 hours | 2 hours | 1½ hours | 1 hour | ¾ hour | ½ hour |
| Expanded slag or pumice aggregates | 4.7 (119.38) | 4.0 (101.6) | 3.2 (81.28) | 2.7 (68.58) | 2.1 (53.34) | 1.9 (48.26) | 1.5 (38.1) |
| Expanded clay or shale aggregates | 5.7 (144.78) | 4.8 (121.92) | 3.8 (96.52) | 3.3 (83.82) | 2.6 (66.04) | 2.2 (55.88) | 1.8 (45.72) |
| Limestone, cinders, or unexpanded slag aggregates | 5.9 (149.86) | 5.0 (127) | 4.0 (101.6) | 3.4 (86.36) | 2.7 (68.58) | 2.3 (58.42) | 1.9 (48.26) |
| Calcareous gravel aggregates | 6.2 (157.48) | 5.3 (134.62) | 4.2 (106.68) | 3.6 (91.44) | 2.8 (71.12) | 2.4 (60.96) | 2.0 (50.8) |
| Siliceous gravel aggregates | 6.7 (170.18) | 5.7 (144.78) | 4.5 (114.3) | 3.8 (96.52) | 3.0 (76.2) | 2.6 (66.04) | 2.1 (53.34) |

[a]Equivalent thickness is the solid thickness that would be obtained if the same amount of concrete contained in a hollow unit were recast without core holes.
[b]Where combustible members are framed into the wall, the wall must be of such thickness or be so constructed that the thickness of solid material between the end of each member and the opposite face of the wall, or between members set in from opposite sides, will not be less than 93% of the thickness shown in this table.

**FIGURE 4.6-40** Typical fireplace and chimney construction details. Hearths are often cantilevered concrete slabs rather than the arch-supported or floor-framing-supported structures shown. Modern codes and regulations often require fresh air intakes near a fireplace to provide combustion air.

requires adherence to some basic rules concerning dimensions and placement of the various component parts. A correctly designed fireplace will:

- Assure proper fuel combustion
- Deliver smoke and other combustion products up the chimney
- Radiate the maximum amount of heat into the room
- Afford simplicity and fire safety in construction

The first two of those objectives are closely related and depend mainly on the shape and dimensions of the combustion chamber, the proper location of the fireplace throat and the smoke shelf, and the ratio of the flue area to the area of the fireplace opening. The third objective depends on the dimensions of the combustion chamber, and the fourth depends on the size and shape of the masonry units and their ability to withstand high temperatures without warping, cracking, or deterioration. Typical details of fireplace construction are shown in Figure 4.6-40.

### 4.6.9.1 Separation from Other Construction

Fireplace walls should be separated from combustible construction as follows:

1. *Framing members:* 2 in. (50.8 mm) air space. This air space should be firestopped at floor level with extension of ceiling finish, strips of fire-resistant board, or other noncombustible material.
2. *Subfloor and flooring:* $^3/_4$ in. (19 mm) air space. Trim in contact with the back of the fireplace should be of non-combustible material or should be fire-retardant wood.
3. *Wall sheathing:* 2 in. (50.8 mm) air space.
4. *Furring strips:* 1 in. (25.4 mm) air space, except that furring may be applied directly to masonry at the sides or corners of the fireplace.

Combustible materials should not be placed within 8 in. (200 mm) of the top or side edges of a fireplace opening. When the floor construction or finish is combustible, an incombustible hearth should be installed. This hearth should extend at least 16 in. (400 mm) in front and 8 in. (200 mm) beyond each side of the fireplace opening when the opening is less than 6 sq ft (0.56 m$^2$) in area and 20 in. (508 mm) in front and 12 in. (305 mm) to the sides when the opening is larger.

Wood framing and furring members should not be placed closer than 2 in. (50.8 mm) from the walls of a chimney. The space between the chimney and floor members should be filled with an incombustible material providing a firestop to prevent the transfer of fire from below through the space.

### 4.6.9.2 Size and Shape

The size of a fireplace depends not only on esthetic considerations, but also on the size of the room to be heated. For a room with 300 sq ft (27.87 m$^2$) of floor area, a fireplace with an opening 30 to 36 in. (762 to 914 mm) wide is sufficient. Approximate dimensions for sizing a fireplace are given in Figure 4.6-41.

#### COMBUSTION CHAMBER

The shape of the combustion chamber influences both the draft and the amount of heat that will be radiated into the room. A slope at the back throws the flame forward and leads the gases with increasing velocity through the throat. Sloping the back and sides also helps radiate the maximum amount of heat into the room. The dimensions, wall angles, and slopes shown in Figures 4.6-40 and 4.6-41 may be varied slightly to correspond with brick coursing, but no major changes should be made in the dimensions shown.

The combustion chamber, unless it is a preformed metal one, should be lined with firebrick at least 2 in. (50.8 mm) thick laid in fireclay mortar. The back and end walls, including the thickness of the firebrick, should be at least 8 in. (200 mm) thick to support the weight of the chimney above.

#### THROAT

Because of its effect on draft, a fireplace's throat must be carefully designed.

**FIGURE 4.6-41** Fireplace Types and Dimensions

| Type | Opening Height h″ (mm) | Hearth Size w″ × d″ (mm) | Flue Size (standard modular) |
|---|---|---|---|
| Single Face | 29 (736.6) | 30 × 16 (762 × 406.4) | 12 × 12 (304.8 × 304.8) |
| | 29 (736.6) | 36 × 16 (914.4 × 406.4) | 12 × 12 (304.8 × 304.8) |
| | 29 (736.6) | 40 × 16 (1016 × 406.4) | 12 × 16 (304.8 × 406.4) |
| | 32 (812.8) | 48 × 18 (1219.2 × 461.5) | 16 × 16 (406.4 × 406.4) |
| Two Face (adjacent) | 26 (666.7) | 32 × 16 (812.8 × 406.4) | 12 × 16 (304.8 × 406.4) |
| | 29 (736.6) | 40 × 16 (1016 × 406.4) | 16 × 16 (406.4 × 406.4) |
| | 29 (736.6) | 48 × 20 (1219.2 × 508) | 16 × 16 (406.4 × 406.4) |
| Two Face (opposite) | 29 (736.6) | 32 × 28 (812.8 × 711.2) | 16 × 16 (406.4 × 406.4) |
| | 29 (736.6) | 36 × 28 (914.4 × 711.2) | 16 × 20 (406.4 × 508) |
| | 29 (736.6) | 40 × 28 (1016 × 711.2) | 16 × 20 (406.4 × 508) |
| Three Face (2 long, 1 short) | 27 (685.8) | 36 × 32 (914.4 × 812.8) | 20 × 20 (508 × 508) |
| | 27 (685.8) | 36 × 36 (914.4 × 914.4) | 20 × 20 (508 × 508) |
| | 27 (685.8) | 44 × 40 (1117.6 × 1016) | 20 × 20 (508 × 508) |

It should be not less than 8 in. (200 mm) above the highest point of the fireplace opening. The sloping back should extend to the same height and support the back of the damper.

A metal damper, preferably of a design in which the valve plate opens upward and toward the back, should be placed in the throat. It should extend the full width of the fireplace opening. When open, the damper's valve plate must form a barrier against downdrafts, and deflect them upward with the ascending column of smoke. When the fireplace is not in use, the damper should be kept closed to prevent heat loss from the room and to keep dirt out of the flue.

### SMOKE SHELF AND CHAMBER

The position of the smoke (downdraft) shelf is established by the location of the throat. The smoke shelf should be directly under the bottom of the flue and extend horizontally the full width of the throat. The space above the shelf is called the *smoke chamber*. The back wall of the chimney should be built straight, and the other three sides should be sloped uniformly toward the center to meet the bottom of the flue lining.

Metal linings are available for smoke chambers to give the chamber its proper form. They provide smooth surfaces and simplify masonry installation.

### FLUE

It is desirable to obtain relatively high velocities of flue gases and smoke through the throat and flue. The velocity is affected by both the area of the flue and the height of the chimney. The proper sizes of fireplace flues are given in Figure 4.6-41.

Each fireplace should have a separate flue with no other openings or connections. A flue lining should be supported on at least three sides by a ledge of protecting brick that finishes flush with the inside of the lining.

### HEARTH

A widely used practice in hearth construction is to form a cantilevered reinforced concrete underhearth, as shown in Figure 4.6-40. An alternative more often used in the past than today is to construct an arch of brick from the chimney base to the floor construction. The arch is then filled to a level surface and receives the finished hearth (see Fig. 4.6-40).

### EXTERIOR AIR SUPPLY

Some codes now require that each fireplace be provided with an exterior air intake to ensure proper combustion. The *International Residential Code* requires an air intake unless the room in which the fireplace is located is mechanically ventilated and the air pressure is neutral or positive. An exterior air intake should permit air to reach the fireplace either directly from the exterior or from an enclosed space open to the exterior, such as a crawl space or attic. The air intake should open into the side wall of the fireplace chamber or within 24 in. (406 mm) of the fireplace opening at or near the floor.

### 4.6.9.3 Fireplace Types

Several types of masonry fireplaces are used today, but the basic principles involved in their design and construction are the same. The single-faced fireplace is the most common and the oldest variety, and therefore most of the standard design information is based on this type.

Multifaced fireplaces, if used properly, may be effective and attractive. They present some problems concerning draft and opening size which usually must be solved on an individual basis. Certain standard sizes that have been found to work satisfactorily under most conditions are shown in Figure 4.6-41.

Most fireplaces are not energy efficient but can be made more so by using the following practices:

- Provide a source of outside air for fireplace combustion so that heated air is not drawn out of the house.
- Adjust the damper to suit the type of fire. For example, a slow-burning log may only need a 1 in. (25.4 mm) or 2 in. (50.8 mm) damper opening to provide adequate combustion without smoking.
- Provide a glass screen with an adjustable damper so that the damper in the throat of the fireplace can be left open while the entrance of combustion air is controlled by the damper in the glass screen. When the fireplace is not in use, the glass screen can be closed, preventing heated inside air from being wasted.

### 4.6.9.4 Chimneys

Because of the large mass and weight of a chimney, it is quite possible that its foundation will have to support a greater load per square foot than the surrounding foundations. Therefore, special attention must be given to a chimney's footing and the rest of the foundation's design so that the unit load (psi) on the soil under all portions of the building's foundation will be nearly the same. If not, there may be uneven settling of the foundation, which could result in cracking of masonry or other damage to the building.

The thickness of the masonry in the base of a chimney that contains only flues may be 4 in. (100 mm) thick, but if a fireplace is included, it must be at least 8 in. (200 mm) thick. The ash pit in the base should be fitted with a tight cast-iron door and a frame of the same material anchored securely approximately 16 in. (406 mm) above the floor. If there is no basement, the door should be located one or two courses above the grade line. A flue lining is not required 8 in. (200 mm) or more below the point where the lowest smoke pipe thimble or flue ring enters the flue.

Each flue must be built as a separate unit entirely free of other flues or openings. Joints in flue walls should be completely filled with mortar. In masonry chimneys with walls less than 8 in. (200 mm) thick, liners should be separated from the chimney wall and the space between the liner and masonry should be left unfilled; only enough mortar should be used to make a good joint and hold the liners in position. Exposed joints inside a flue should be struck smooth. The interior surface should not be plastered.

Although unlined fireplace chimneys with masonry walls at least 8 in. (200 mm) thick may perform adequately, the use of a fireclay flue liner is recommended. A flue lining provides a smooth interior surface which permits more efficient operation of the flue.

The tops of chimneys without flue linings may have to be rebuilt every few years because of the disintegrating effect of smoke and gases on the mortar. When rebuilding is necessary, a chimney must be taken down to a point where the mortar joints are solid and a new top built.

Each flue lining section should be set in place before the brick has been laid to the top of the lining section below, and then the brick should be built up around the just-placed lining to a level below the next section. Offsets and bends should be made by equally mitering both ends of abutting sections of lining. This prevents

**FIGURE 4.6-42** Smoke pipe connections to the chimney.

reduction of the flue area so that the same effective area is maintained for the entire height of a flue.

Smoke pipe connections should enter the side of a flue at a thimble or flue ring, which should be built as the work progresses. A metal smoke pipe should not extend beyond the inside face of the flue (Fig. 4.6-42). The top of a smoke pipe should not be less than 18 in. (457 mm) below the ceiling, and no wood or combustible materials should be placed within 6 in. (150 mm) of a thimble.

A flue lining should project at least 4 in. (100 mm) above the top course or capping of the chimney and should be surrounded with portland cement mortar at least 2 in. (50.8 mm) thick, finished with a straight or concave slope to direct the air currents upward at the top of the flue. This wash also serves to drain water from the top of the chimney and should be constructed with a drip to keep the walls dry and clean.

Unless the applicable code has more stringent requirements, chimneys should extend at least 2 ft (609 mm) above every part of the roof that lies within 10 ft (3048 mm) of the chimney and at least 3 ft (914 mm) above the roof at the point of penetration.

After a chimney has been completed and the masonry cured, a test of each flue should be conducted. The test is made by building a smudge fire at the top of the flue and, when smoke is pouring freely from the top, covering the top tightly. Any escaping smoke indicates openings that must be repaired. Since such repair is usually difficult and expensive, it is far more satisfactory and economical to ensure that construction is properly executed as the work progresses.

Smoke passages and chimney flues should be kept clean. The most efficient method of cleaning a chimney flue is with a weighted brush or bundle of rags lowered and raised from the top. An accumulation of soot and resins may cause a chimney fire whose sparks may ignite the roof. A chimney fire may also damage the chimney, which may in turn permit the passage of more fire and cause significant damage to the building.

# 4.7 Unit Masonry Erection

Selection of masonry materials is covered in Sections 4.2, 4.3, 4.4, and 4.5; design of masonry assemblies, in Section 4.6. This section addresses masonry unit and mortar preparation, unit laying, mortar bedding and jointing, joint finishing, and flashing installation. All these must all be done correctly if a completed masonry assembly is to be watertight, strong, and durable and present a satisfactory appearance.

## 4.7.1 PREPARATION OF MATERIALS

Protection during shipping, handling, and storage of masonry and associated materials and accessories is essential to prevent damage to them. Proper preparation is necessary before masonry and related materials and accessories are placed in a masonry assembly.

### 4.7.1.1 Storage and Protection of Materials

All materials that will be used in masonry assemblies should be stored under cover, off the ground, in a dry place, and protected from water and freezing. They should also be kept free of soil, ice, frost, and other contaminants. If they become contaminated during storage or handling, the contaminants must be removed before the materials are used.

It is common to place clay and concrete masonry units and cementitious mortar materials outside on skids or platforms and to cover them during inclement weather with tarpaulins, clear plastic film, or another suitable material. It is better to store these materials indoors when possible.

Structural clay tile units, either glazed or unglazed, and terra cotta units should be covered at all times. It is best to store these units indoors.

Sand and other aggregates are often not protected as carefully as they should be, which results in their becoming soaked with water or contaminated with soil or other foreign materials. Foreign materials, excess water, and ice must be removed before aggregates are used. If different aggregates become mixed together during storage, it is necessary to grade them again, which is difficult, or discard them, which is expensive.

Metal ties, anchors, and reinforcement must be kept dry and protected from contact with soil and away from conditions that would harm them. Before they are placed, these items must be free of loose rust and other coatings that would destroy or reduce the bond.

### 4.7.1.2 Clay Masonry

The initial rate of absorption (suction) of a brick is the amount of water in grams absorbed by 30 sq in. (0.02 m²) of surface when a brick is placed in water to a depth of $1/8$ in. (3.2 mm) for 1 minute, as tested in accordance with ASTM C67 (Fig. 4.7-1). In almost every case, mortar bonds best to brick whose suction is 5 to 20 g of water.

Bricks having a suction in excess of 20 g of water per minute should be wetted sufficiently so that the suction when the brick is laid does not exceed that amount. If high-suction bricks are not wetted, they will absorb water from the mortar before a bond has developed. Sprinkling brick to wet it will not produce the desired results. Rather, a hose stream should be played on each brick pile until water runs from all sides. Wetting should be done several hours before

**FIGURE 4.7-1** Apparatus for testing the suction of a brick. (Courtesy Brick Institute of America [renamed Brick Industry Association].)

### 4.7.1.4 Mortar

The strength and resistance of masonry walls to water penetration are dependent on the completeness and strength of the bond between the mortar and the masonry units. Mortar characteristics that affect bond strength are discussed in Section 4.2.1. Mortar preparation includes mixing the mortar and maintaining the proper water ratio until the mortar is used. Mixing is discussed in Section 4.2.6; retempering to control water content, in Section 4.2.7. Mortar application techniques that affect bond strength and water penetration resistance are discussed in Section 4.7.2.5.

### 4.7.2 INSTALLATION OF SOLID AND HOLLOW MASONRY UNITS

The installation of masonry units requires laying the units, bonding them together, and anchoring them to the structure. It also includes installing continuous joint reinforcement, building in bearings for supported structural members, mortar bedding and jointing, backparging, joint finishing, and installing flashing and weep holes.

### 4.7.2.1 Clay Masonry Positioning

Figure 4.7-2 shows the six positions in which solid and hollow clay masonry units are most often placed in a wall, but other positions are also used. For example, such units may be tilted (placed at an angle other than vertical) in any of the six indicated positions to create special effects and in arches.

Figure 4.7-3 shows brick units in their most frequently used positions in a wall and indicates the names of the various joints between units. Some other positions are also frequently used. Rowlock coursing, for example, is used in copings on retaining walls and other freestanding walls; as sills at window, door, and other openings (see Fig. 4.6-11); and as water tables. Slopped headers are also sometimes used as sills. Individual brick units may be flush with adjacent units or may project beyond the surrounding surface. Special shapes (see Section 4.4.1) are also used as copings, sills, water tables, stair treads, jambs, radials, angles, and corners.

The terms shown in Figures 4.7-2 and 4.7-3 are further defined in the Glossary.

Structural clay tile and structural facing tile are usually placed in the stretcher

the units are laid, and preferably the previous day. The surfaces of wetted brick must be allowed to dry before the brick is laid. Water on the surface of a brick will cause it to *float* on the mortar bed and prevent it from bonding with the mortar.

Conversely, if bricks with a low suction are wetted or if other bricks are wetted too much, they may become saturated. Saturated brick will not absorb enough moisture from the mortar and will float, which will usually not permit the formation of sufficient bond between the brick and the mortar. When brick suction exceeds 60 g of water at the time of laying, the bond may be extremely poor, regardless of the quality of workmanship or mortar type used.

An effective on-the-job method of determining whether a brick should be wetted before it is laid is to sprinkle drops of water on its flat side. If these drops are absorbed completely in less than 1 minute, the brick should be wetted. A more accurate method is to draw a circle on the flat side of a brick using a wax pencil and a quarter as a guide. With a medicine dropper, apply 20 drops of water to the surface of the brick inside the circle. If the water is completely absorbed in less than $1\frac{1}{2}$ minutes, the brick should be wetted before it is laid.

### 4.7.1.3 Concrete Unit Masonry

ACI 530/530.1, ASCE 6, and TMS 402 specifically prohibit the wetting of concrete masonry units. Therefore, care must be taken to keep these units dry on the job. They should be stockpiled on planks or other supports, free from contact with the ground, and covered for protection against wetting. Concrete units should never be wetted before or during laying in the wall.

The suction test described in Section 4.7.1.2 is not applicable to concrete masonry units. A concrete masonry unit with a moisture content less than 20% of the total amount it can absorb can draw water out of mortar on hot, dry days, but concrete masonry unit suction is usually insignificant because the units generally contain sufficient moisture to retard suction effects.

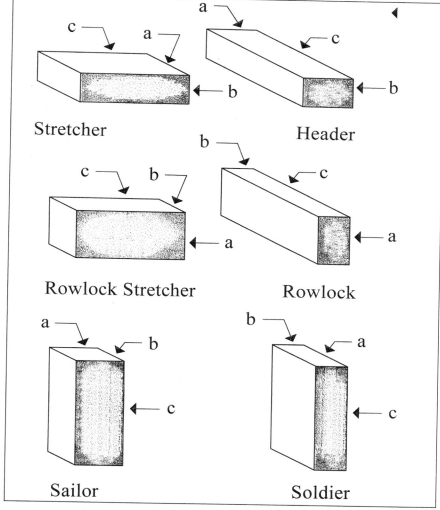

**FIGURE 4.7-2** The various positions in which bricks are most often placed in a wall. The shaded portion is the surface that is exposed. The unit dimensions (see Section 4.4.4.1) are shown by (a) thickness, (b) height, and (c) length. (Drawing by HLS.)

**FIGURE 4.7-3** The names of the three most frequently used positions in which masonry units are placed in walls and the names given to various joint locations. (Drawing by HLS.)

position. Copings and sills are usually special units made for the purpose.

Architectural terra cotta units are placed in the positions for which they are designed.

### 4.7.2.2 Concrete Masonry Positioning

Other positioning is occasionally used to create special effects, but most concrete masonry units are placed in either the stretcher or header position (the same as in Fig. 4.7-2), depending on the design of the wall. Copings are usually special units created for the purpose. Solid brick-sized units are often used as sills in concrete masonry walls.

### 4.7.2.3 Laying Masonry Units

Many general techniques are applicable to the laying of both solid and hollow masonry units, including clay brick, hollow clay brick, structural clay tile, concrete brick, and concrete block.

Masonry should be laid *plumb* (exactly vertical) and *true* (accurately placed) to lines. Each unit should be adjusted to its final position while the mortar is still soft and plastic. A mason should avoid overplumbing, tapping, pounding on, or otherwise attempting to realign units after they have been set into position (Figs 4.7-4a and 4.7-4b). Such movements will often break the bond between the unit and the mortar. When the bond has been broken in this manner it usually remains broken, regardless of a mason's attempts to reestablish it. When a unit must be adjusted after the mortar has started to harden, the units and the mortar should be removed and the units should be laid again using new mortar.

The highest bond strength is achieved when the time interval between the placing of a mortar bed and the laying of a masonry unit into it is kept to a minimum. In addition, a change in the time interval between the spreading of mortar and the placing of masonry units into that mortar will affect the flow of the mortar (see Section 4.2.1.1), especially if the masonry units have a high suction (see Section 4.7.1.2). Varying this interval to any large extent will result in a wall with varying tensile bond strengths.

Concrete masonry units and some other hollow masonry units have thicker face shells on the top than on the bottom, and should be laid with the thicker face

**a**

**b**

**FIGURE 4.7-4** Moving masonry units after they have been laid can result in cracks. Even hammering a brick back in line, as shown in (a), can break the bond. Cracks caused by re-aligning units often occur at corners, as shown in (b). (Courtesy Brick Institute of America [renamed Brick Industry Association].)

shell up to provide a larger mortar bedding area.

Anchors, accessories, flashings, and other items required to be built into masonry should be built in as the work progresses.

When fresh masonry joins masonry that is partially or totally set, the exposed surface of the set masonry should be cleaned, and in the case of clay masonry, wetted, to obtain the best possible bond with the new work.

A board marked with the courses (*story pole* or *course pole*) provides an accurate method of finding the top of the masonry for each course (Fig. 4.7-5). The same story pole should be used throughout the work.

### STARTING THE WORK

After locating the corners, a mason often spaces out the units dry (without mortar) to determine the extent to which they must be cut or have their joint sizes varied to accomplish accurate horizontal coursing (Fig. 4.7-6).

The mason then spreads a mortar bed and beds the units in it (see Section 4.7.2.5). The corner unit is laid first and carefully placed in its correct position.

After several units have been laid, a mason's level is used as a straight edge to verify that they are in correct alignment at the proper grade and are plumb. Corrections are made by tapping the units with a trowel handle (Fig. 4.7-7). The first course of masonry should be laid carefully to ensure that it is properly aligned, leveled, and plumbed. This will assist the mason in laying succeeding courses and in building a straight, true wall.

### LAYING UP THE CORNERS

After the first course has been laid, the corners of the wall are built before the rest of the wall is laid. The corners are started by laying up several courses higher than the center of the wall. Each

**FIGURE 4.7-5** Using a story pole. (Courtesy Portland Cement Association.)

**FIGURE 4.7-6** Spacing units at a corner without mortar before laying them. (Courtesy Portland Cement Association.)

course is stepped back by one-half unit. As each course is laid, it is checked to make sure that it is in alignment and plumb (Fig. 4.7-8) and level (Fig. 4.7-9). At the same time, each course is checked

**FIGURE 4.7-7** A mason verifying that the first course of blocks is in correct alignment at the proper grade, and plumb. (Courtesy Portland Cement Association.)

**FIGURE 4.7-8** Aligning units at a corner and checking for plumb. (Courtesy Portland Cement Association.)

**FIGURE 4.7-9** Checking to see whether units are level. (Courtesy Portland Cement Association.)

carefully with a level or straightedge to make certain that the faces of the units are all in the same plane so that the wall will be true and straight (Fig. 4.7-10).

## LAYING BETWEEN THE CORNERS

Before filling in the wall between the corners, a mason stretches a line from corner to corner for each course, then lays the top outside edge of the units to this line. In laying concrete masonry units to a mason's line, the way a mason handles and grips the units is important. Tipping the units slightly toward the mason will permit seeing the upper edge of the course below, thus enabling the placement of the block's lower edge directly over the course below (Fig. 4.7-11). By rolling the block slightly to a vertical position and shoving it against the adjacent block, the mason can lay it to the line with minimum adjustment.

As stated earlier, adjustments to the final position must be made while the mortar is soft and plastic. By tapping lightly with a trowel handle, a mason can level and adjust each unit to the mason's line. The use of a level between corners is necessary only to check the alignment of each unit with the face of the wall.

**FIGURE 4.7-10** Checking to see whether units are in line. (Courtesy Portland Cement Association.)

**FIGURE 4.7-11** A mason positions units between the corners by laying them to a line. (Courtesy Portland Cement Association.)

### 4.7.2.4 Bonding and Anchoring

The units in a single-wythe masonry wall are bonded together by the wall's pattern bond. Two methods are generally used to structurally bond together the several wythes of wall assemblies that have more than one wythe of masonry. One method (*masonry bonded*) consists of overlapping and interlocking the masonry units. The second (*metal tied*), which is the preferred method in most cases, bonds the masonry together using metal ties embedded in the mortar joints.

Masonry must also be anchored to other structural elements when it relies on them for support. Most such anchoring is done using metal anchors.

### MASONRY-BONDED SINGLE-WYTHE WALLS

The units in a single-wythe brick wall are bonded together by building the wall in the running bond pattern shown in Figure 4.6-14. Other bond patterns can be used in single-wythe walls, of course, but the headers must be cut to a length that matches the runner's thickness. This method is unsatisfactory when both sides of the wall are exposed because the finish of the cut units cannot be made to match the face texture of the brick. When a stack bond pattern is used, it is usually necessary to provide continuous joint reinforcement (see Section 4.7.2.3).

The units in a single-wythe concrete unit masonry wall are bonded together by building the walls in one of the patterns shown in Figure 4.6-16. Stack bond patterns require the use of continuous joint reinforcement (see Section 4.7.2.3).

Corners and intersections in a single-wythe brick or concrete unit masonry non-load-bearing wall or partition can be bonded by laying at least 50% of the units in an overlapping pattern. Each overlapping unit should bear at least 3 in. (76.2 mm) on the unit below.

Masonry-bonded, load-bearing, single-wythe brick and concrete unit masonry walls may be bonded at intersections by overlapping 50% of the units, with alternate units bearing at least 3 in. (76.2 mm) on the unit below. This method is not recommended by some experts, however, except in earthquake zones. In other areas, they prefer that such walls be masonry bonded only at corners, and be built up separately at other intersections and tied together there with metal connectors, as discussed later in this section.

### MASONRY-BONDED MULTIPLE-WYTHE WALLS

The bonding of the units in a masonry-bonded multiple-wythe wall is based on variations of the two traditional methods of bonding: (1) English bond, consisting of alternating courses of headers and stretchers, and (2) Flemish bond, consisting of alternating headers and stretchers in every course (see Fig. 4.6-14). In this regard, common bond is but a variation of English bond. English and common bonds are seldom used in this country in contemporary construction. Flemish bond sometimes appears in contemporary churches. Of course, any of the three may be needed to match existing construction.

Overlapping stretchers, laid parallel with the length of the wall, develop longitudinal bonding strength; headers laid across the width of the wall bond the wall transversely. Some types of masonry-bonded walls are shown in Figure 4.6-1, using brick (Fig. 4.6-1H), overlapping stretchers (Fig. 4.6-1G), brick and block or tile backup (Fig. 4.6-1J).

Masonry-bonded brick walls should be built so that not less than 4% of the wall's surface is composed of headers, the headers should extend not less than 4 in. (100 mm) into the backing, and they should be not more than 24 in. (610 mm) apart either vertically or horizontally. This means that there must be at least one header course in every seven brick courses. In multiple-wythe tile or block walls, structural bonding is achieved by overlapping stretcher units rather than by using header units (see Fig. 4.6-1G). The stretcher units may be bonded at vertical intervals not exceeding 34 in. (864 mm) when the bonding course can overlap one-half of the unit below and the lap is at least 4 in. (100 mm) (Fig. 4.7-12a). When these conditions cannot be met, the bond may be achieved by overlapping headers that are at least 50% greater in thickness than the units below, with the bond courses occurring at vertical intervals of 17 in. (432 mm) (Fig. 4.7-12b).

Unfortunately, water can penetrate a multiple-wythe masonry-bonded wall fairly easily for several reasons:

1. It is sometimes difficult to obtain full mortar joints between brick headers. This permits water to find its way to the inside face of the wall along the headers, which are continuous through the full wall thickness.

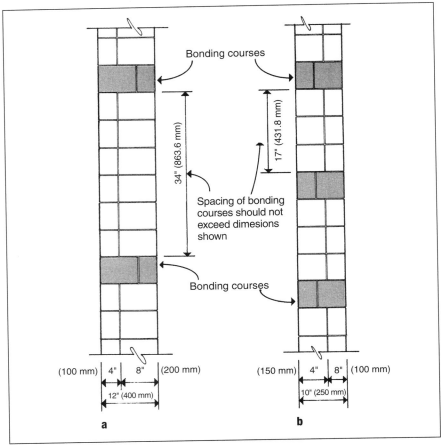

FIGURE 4.7-12 The overlapping stretcher method of bonding.

2. It is difficult to bond brick and larger, heavier backup units without impairing the bond between the headers and the mortar. If the backup unit is set on the header before the mortar has had time to stiffen, the header will settle unevenly and a crack may develop (Fig. 4.7-13). In a thicker wall, this problem can be solved by the use of header blocks (Fig. 4.7-14).

3. Perhaps the most important reason that water may find its way through a masonry-bonded wall is that there are differences in thermal, moisture, and other movements between different facing and backup materials. The differences are greatest between brick and concrete masonry units where movement of the concrete masonry unit backup, which shrinks as it dries, and the brick, which progressively expands over time, create an eccentric load on the brick headers. This tends to rupture the bond between the headers and the mortar at the external face. The flexibility inherent in metal ties negates this disadvantage.

## METAL-TIED SINGLE-WYTHE WALLS

The units in the field of metal-tied single-wythe walls are not bonded using metal ties. Metal ties are used, however, where such walls abut other walls.

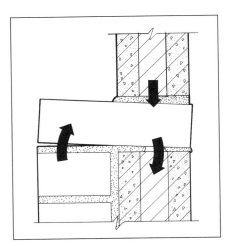

FIGURE 4.7-13 The effect of a heavy backup unit on a masonry header. (Courtesy Portland Cement Association.)

When a single-wythe, load-bearing brick or concrete masonry unit wall is not masonry bonded to walls it abuts, except at the corners, it should terminate at the faces of these other walls in control joints. For lateral support, bearing walls should be tied to walls they intersect with rigid steel tie bars 1/4 in. (6.4 mm) thick, 1 1/4 in. (31.8 mm) wide, and 28 in. (711 mm) long, with a 2 in. (50.8 mm) right angle bend on each end (Fig. 4.7-15). Tie bars should be placed not more than 4 ft (1219 mm) apart vertically. The bends at the ends of the tie bars should be embedded in cores filled with mortar (Fig. 4.7-16).

Single-wythe, non-load-bearing masonry unit walls or partitions that are 6 in. (152 mm) or less in thickness may be tied to walls they intersect using strips of metal lath or 1/4 in. (6.4 mm) mesh galvanized hardware cloth placed across the

FIGURE 4.7-14 A mason laying a header course in a 12 in. (300 mm) solid brick and block wall using header block. (Courtesy Portland Cement Association.)

FIGURE 4.7-15 A metal bar anchor being installed to tie a bearing wall to an intersecting wall. Metal lath was used to support the mortar filling the core where the turned-down end of the bar is being inserted. (Courtesy Portland Cement Association.)

**FIGURE 4.7-16** The core into which the upturned end of the metal tie bar shown in Figure 4.7-13 projects is here being filled with mortar. (Courtesy Portland Cement Association.)

**FIGURE 4.7-17** A strip of galvanized hardware cloth being used to tie a non-load-bearing wall to an intersecting wall. (Courtesy Portland Cement Association.)

**FIGURE 4.7-18** This hardware cloth strip was left protruding from the completed wall at the right and is now being bedded into a joint in an intersecting wall. (Courtesy Portland Cement Association.)

joint between the two walls (Fig. 4.7-17). These metal strips should be placed in alternate courses (not more than 16 in. [406 mm] apart vertically) in the wall. When one wall is constructed first, the metal strips should be built into this wall and later embedded into the mortar joint in the second wall as it is built (Fig. 4.7-18). Corrugated metal tie anchors may be used in the same way. Where the two walls meet, a control joint should be provided if the joint is exposed to view.

## METAL-TIED MULTIPLE-WYTHE WALLS

Both multiple-wythe solid and cavity walls can be structurally bonded using metal ties (see Fig. 4.6-1). For exterior walls, especially where weather resistance is important or the physical characteristics of the facing and backing materials are different, bonding with metal ties is better than the masonry-bonded method. This is because metal-tied walls have greater resistance to moisture penetration and the ties allow slight differential movement between the facing and backing units, which may help to relieve stress and reduce cracking.

The usual requirements for metal-tie bonding are for $3/16$-in. (4.8-mm)-diameter steel ties, or metal ties of equivalent stiffness and strength, to be spaced so that at least one tie occurs in every $4\frac{1}{2}$ sq ft (0.42 m²) of wall area. Except where different spacings are required by code or local practice, and unless another spacing is recommended later in this section for a specific wall type or condition, in solid, composite, and faced walls the maximum vertical distance between ties should not exceed 18 in. (457 mm) and the maximum horizontal distance should not exceed 36 in. (914 mm) (Fig. 4.7-19). Ties in alternate courses should be staggered (see Fig. 4.7-19). Ties should be embedded in horizontal joints of the facing and backing. Additional bonding ties should be provided at openings. These should be spaced not more than 3 ft (914 mm) apart around the perimeter and within 12 in. (305 mm) of the openings and, in addition, on both sides of expansion and control joints.

General and specific recommendations for some types of walls and conditions are summarized below. Notice that some conditions require masonry bonding or a combination of masonry bonding and metal ties.

*Reinforced Masonry Walls* The units in a single-wythe hollow reinforced masonry wall are bonded together by their pattern bond and by the reinforcing steel and the grout in their cores (see Section 4.6.1.4).

The units in the separate wythes of reinforced solid masonry walls are bonded together by their pattern bond (see Figs. 4.6-14 and 4.6-16). The wythes are structurally bonded together by the reinforcing steel and grout in the cavity (*collar joint*) between the wythes of the wall (see Section 4.6.1.4).

*Faced Walls* The facing should be bonded to the backing with metal ties.

*Cavity Walls* Cavity walls should be bonded with metal ties. Their spacings should not exceed 24 in. (610 mm) vertically or 36 in. (914 mm) horizontally.

*Brick-Veneered Walls* In wood frame construction (see Fig. 4.6-11a), brick veneer is often anchored to the backup wall by nailing corrosion-resistant corrugated metal ties through the sheathing to the

**FIGURE 4.7-19** Wall tie locations in multiple-wythe walls.

studs. This type of construction has been used for many years, but recently some experts have begun to question its longevity. Failure may result from rusting of the ties due to water reaching them from leaks or from condensation due to high humidity. Professional guidance should be sought before electing to use this method of construction, and local practice should be followed.

There is some controversy about whether brick veneer should be used over metal studs (see Fig. 4.6-11b). There are more questions about using this combination on tall structures than about its use on one- and two-story residences and other small buildings. Its supporters claim that brick veneer over metal studs will give satisfactory long-term results if the proper materials are selected and are applied correctly. Detractors claim that brick veneer applied over metal studs will eventually crack, regardless of the materials used or the configuration of the wall. This will allow water to enter the wall and corrode the metal supports, resulting in veneer failure.

The key elements in the controversy are the required stiffness of the studs, the type of anchors to use, and whether any available anchoring system will perform satisfactorily in the long run. The most authoritative current advice about the methods needed to construct such a brick-veneered wall are offered by the BIA. However, before deciding to use brick veneer over metal studs, it would be wise to contact industry sources, such as BIA, the Portland Cement Association (PCA), the Metal Lath/Steel Framing Association, and the manufacturers of the products that will be used, and obtain their latest recommendations concerning this controversial use of brick. It is also necessary to determine if this type of construction is permitted by the applicable codes and the owner. There has been some movement by some federal agencies and some states to ban its use altogether.

### ANCHORAGE

Masonry assemblies must be securely anchored to elements that are used for their vertical and horizontal support. These include steel and concrete columns and beams, concrete walls, concrete and masonry bond beams, and floors, other walls, and roofs. This anchoring is usually done with flexible metal anchors that permit at least some differential movement.

Exterior walls facing or abutting concrete members should be anchored by dovetailed anchors inserted in slots set into the concrete (see Fig. 4.3-1). These anchors should be spaced not more than 18 in. (457 mm) vertically and 24 in. (610 mm) horizontally.

### 4.7.2.5 Continuous Joint Reinforcement

Continuous joint reinforcement is often used today to control cracks in masonry walls (see Figs. 4.3-1, 4.6-1, and 4.6-9). When joint reinforcement is used, it often replaces and does the same job usually done by metal wall ties. In wythes laid in a stacked bond pattern, continuous joint reinforcement is necessary to bond the units together. Continuous joint reinforcement is also sometimes used in lieu of horizontal reinforcement in reinforced masonry walls.

Continuous joint reinforcement should be installed in every masonry wall and partition that is more than 20 ft (6.1 m) long, and is usually installed in every wall regardless of length.

Using continuous joint reinforcement not only reduces the amount of cracking associated with thermal movements, it also permits greater spacing of control joints (see Fig. 4.6-22). Joint reinforcement should be embedded in horizontal mortar joints so that it is at least $5/8$ in. (15.9 mm) from the masonry surface, and should be continuous except through control and expansion joints. Joint reinforcement sections should be lapped at least 6 in. (152 mm), and the lap should contain at least one cross wire of each section.

Continuous joint reinforcement should be installed in most cases at not more than 16 in. (406 mm) on center vertically. In reinforced walls, continuous reinforcement is sometimes placed at closer spacings. In parapet walls and foundation walls, reinforcement is often placed as close together as every 8 in. (203 mm). Local codes sometimes govern this placement. To compensate for loss of reinforcement at wall openings, joint reinforcement should be installed in the first and second bed joints immediately above and below openings and should extend at least 24 in. (610 mm) beyond the ends of sills and lintels.

### 4.7.2.6 Masonry Support for Structural Members

The bearing course of masonry under structural members such as floor joists,

concrete slabs, beams, and lintels should be at least 8-in. (203-mm)-thick solid masonry or hollow masonry with the cells filled solidly with mortar or fine grout.

### 4.7.2.7 Mortar Bedding and Jointing

Laboratory tests and field observations of masonry indicate that good masonry construction demands that the joints be completely filled with mortar as the units are laid.

Partially filling joints results in leaky walls, reduces the strength of masonry to between 50% and 60%, and contributes to disintegration and cracking when water penetrates the wall and expands as it freezes.

### BRICK

Clay and concrete brick and other solid masonry units should generally be bedded and jointed in similar ways. Every head, bed, and collar joint in them should be filled solidly with mortar.

Mortar joints in contemporary brick walls usually vary from $1/4$ in. (6.4 mm) to $1/2$ in. (12.7 mm) in thickness, depending on the size of the brick and the bonding. Thicker joints are sometimes used, however.

*Bed Joints*  The mortar in bed joints in brick should be spread to a uniform thickness and furrowed only slightly (Fig. 4.7-20). As the brick is laid, excess mortar will fill the furrow and ensure a full bed joint. If the bed is too thin or the furrow is too deep, there will not be enough mortar to fill the furrow completely, and the resulting void will enable water entering an opening in the mortar or between the mortar and the brick to penetrate the wall (Fig. 4.7-21).

To ensure a good bond, mortar should not be spread too far ahead of laying block or brick, or it will stiffen and lose its plas-

**FIGURE 4.7-20** In brick, the mortar furrow in the bed joint should be shallow. (Courtesy Brick Institute of America [renamed Brick Industry Association].)

ticity (Fig. 4.7-22). This is especially true in hot weather when absorbent bricks are being laid. One way to determine if a bond is developing is to remove a freshly laid brick from the wall (Fig. 4.7-23). If the mortar is soft and plastic, it will stick to the brick placed on top of the bed as well as to the brick on which the bed is spread, and an adequate bond will be attained.

***Head Joints*** Head joints are more vulnerable to water penetration than bed joints, because the weight of the wall above tends to close cracks in bed joints. Therefore, every head joint in both the facing brick and backup units should be completely filled with mortar. One method is to spread the mortar on the end of a brick before it is laid (Fig. 4.7-24). When the brick is shoved into place, mortar should squeeze out at the sides and top, indicating that the head joint is completely filled (Fig. 4.7-25). An alternate method is to throw plenty of mortar on the end of the brick already in place (Fig. 4.7-26), giving the desired result (see Fig. 4.7-25) when the next brick is shoved into place. A third method is to place a full trowel of mortar on the wall (Fig. 4.7-27) and shove the next brick into this deep mortar bed, which will squeeze mortar out at the sides and top of the joint (Fig. 4.7-28).

Dabs of mortar spotted on both corners of the brick (Fig. 4.7-29) will not completely fill the head joints (Fig. 4.7-30). Attempting to fill the joints from above after the brick is placed (Fig. 4.7-31), which is called *slushing*, cannot be relied upon to fill voids left in head joints (Fig. 4.7-32).

**FIGURE 4.7-21** The furrow in the bed joint shown here is too deep. (Courtesy Brick Institute of America [renamed Brick Industry Association].)

**FIGURE 4.7-22** Mortar should be spread only a few units ahead of the masonry. (Courtesy Brick Institute of America [renamed Brick Industry Association].)

**FIGURE 4.7-23** This mason has removed a newly laid brick to determine if an adequate bond is being achieved. (Courtesy Brick Institute of America [renamed Brick Industry Association].)

**FIGURE 4.7-24** Loading a brick before laying it. (Courtesy Brick Institute of America [renamed Brick Industry Association].)

**FIGURE 4.7-25** Mortar oozing out indicates that the joint has been filled completely. (Courtesy Brick Institute of America [renamed Brick Industry Association].)

**FIGURE 4.7-26** Buttering the end of a brick that has already been laid. (Courtesy Brick Institute of America [renamed Brick Industry Association].)

**FIGURE 4.7-27** Throwing a full trowel of mortar on an already installed brick to receive the next unit. (Courtesy Brick Institute of America [renamed Brick Industry Association].)

**FIGURE 4.7-28** A brick being shoved into the mortar shown being thrown on in Figure 4.7-27. (Courtesy Brick Institute of America [renamed Brick Industry Association].)

**FIGURE 4.7-29** Spotting the corners of a brick. This will produce unsatisfactory head joints and must not be permitted. (Courtesy Brick Institute of America [renamed Brick Industry Association].)

**FIGURE 4.7-30** The result of spotting the corners, as shown in Figure 4.7-29. Note that the head joint is not filled completely. (Courtesy Brick Institute of America [renamed Brick Industry Association].)

**FIGURE 4.7-31** Attempting to fill a head joint by slushing. (Courtesy Brick Institute of America [renamed Brick Industry Association].)

**FIGURE 4.7-32** A freshly laid brick has been removed after slushing (see Figure 4.7-31). Note that the head joint was not adequately filled. (Courtesy Brick Institute of America [renamed Brick Industry Association].)

Cross joints in header courses must also be completely filled with mortar. This is accomplished by spreading mortar over the entire side of the header brick before it is placed and then shoving it into place to fill all voids (Fig. 4.7-33).

*Closers* The last brick in a course is called a *closer*. Before laying a closer, plenty of mortar should be placed both on the brick already in place (Fig. 4.7-34) and

**FIGURE 4.7-33** Filling a cross joint in a header course with mortar. The side of a header brick is buttered (top) and shoved into place (bottom). (Courtesy Brick Institute of America [renamed Brick Industry Association].)

**FIGURE 4.7-34** Placing mortar on a brick that is already in place. (Courtesy Brick Institute of America [renamed Brick Industry Association].)

**FIGURE 4.7-35** A closer unit being buttered. (Courtesy Brick Institute of America [renamed Brick Industry Association].)

on both sides or ends, as applicable, of the closer (Fig. 4.7-35). The closer should then be laid without disturbing the brick already in place (Fig. 4.7-36).

### CONCRETE MASONRY UNITS

Mortar joints for concrete masonry units should be $3/_8$ in. (9.5 mm) thick.

*Bed Joints* For the first course of both solid and hollow units, a full, thick mortar bed should be spread on the foundation or other support under the entire unit, including the cells, and furrowed with a trowel to ensure that there will be enough mortar along the bottom edge of the face shells and webs of the block to completely fill the joint (Fig. 4.7-37).

In columns, piers, pilasters, and where adjacent cells will be filled with grout, the face shells and webs should be set in a full and complete bed of mortar. This is called *full mortar bedding* (Fig. 4.7-38).

**FIGURE 4.7-36** The closer unit shown being buttered in Figure 4.7-35 is here being placed in the wall. (Courtesy Brick Institute of America [renamed Brick Industry Association].)

**FIGURE 4.7-37** A furrowed bed joint and three blocks with their face shells buttered prior to laying. Note the full mortar bedding. (Courtesy Portland Cement Association.)

**FIGURE 4.7-38** Full mortar bedding. (Courtesy Portland Cement Association.)

Solid units should be laid in full mortar beds, regardless of their use or location. In all other concrete unit masonry work, mortar is usually applied only to the horizontal face shells of the units. This is called *face-shell mortar bedding* (Fig. 4.7-39).

***Head Joints*** In vertical joints in concrete block, only the ends of the face shells are bedded in mortar. This bedding should, however, be complete and solid for the entire depth of the face shell.

Mortar for vertical joints in block can be applied to the vertical face shells of either the block to be placed or the block previously laid. Some masons butter both edges to ensure filled joints. By placing three or four blocks on end, a mason can butter the vertical face shells of them in one operation (see Fig. 4.7-37). Each block is then brought to its final position and pushed downward into the mortar bed and against the previously laid blocks, thus producing full vertical mortar joints (Fig. 4.7-40).

***Closers*** The last unit in a course is called a *closer*. Before laying closers, plenty of mortar should be placed both on the unit already in place and on both ends of the closer. The closer should then be laid without disturbing the unit already in place (Fig. 4.7-41).

**STRUCTURAL CLAY TILE AND OTHER HOLLOW MASONRY UNITS**

In general, the requirements for hollow concrete masonry units apply equally well to hollow brick, structural clay tile, and other hollow masonry units. There are several exceptions. Vertical-cell structural clay tile units should be laid with divided (not full) head joints, and when they are exposed to the weather, care should be taken to prevent continuous mortar joints through the wall.

**FIGURE 4.7-39** Face-shell mortar bedding. (Courtesy Portland Cement Association.)

**FIGURE 4.7-40** Positioning a block before setting it in place. Note the full vertical joints between the blocks that are already in place. (Courtesy Portland Cement Association.)

Horizontal-cell structural clay tile units that are more than 4 in. (100 mm) thick should be laid with divided bed joints. Head joints of horizontal-cell units should be placed on both sides of the tile and sufficient mortar used so that excess mortar will be squeezed out of the joints as the units are placed in position.

**4.7.2.8 Backparging**

The vertical, longitudinal joints (*collar joints*) between the wythes of multiwythe masonry walls should be completely filled by parging either the face of the backing or the back of the facing to help prevent water leakage through the wall (Fig. 4.7-42). Mortar extruding into a collar joint should be cut flush before it hardens. Parging placed over hardened mortar will not bond to it and may result in a leaky wall (Fig. 4.7-43). Filling collar joints from above (slushing) will not fill the joint completely and should not be permitted (Fig. 4.7-44). For best results, parging should be done as the walls are laid up. Careful workmanship is necessary because the pressure required to

**FIGURE 4.7-41** Setting a closer block. (Courtesy Portland Cement Association.)

**FIGURE 4.7-42** Backparging a collar joint. (Courtesy Portland Cement Association.)

**FIGURE 4.7-43** Mortar joints that were cut flush before they hardened. (Courtesy Portland Cement Association.)

**FIGURE 4.7-44** A freshly laid wythe of masonry has been removed to show the spotty and incomplete backparging that resulted from slushing a collar joint. (Courtesy Portland Cement Association.)

spread a parge coat of mortar on newly laid units can break the bond between the units and their mortar joints, producing cracks that may contribute to future leakage.

### 4.7.2.9 Joint Finishing

When masonry walls leak, the water often enters through improperly sealed cracks and voids around openings, through flashings between the wall and the roof, or through cracks where the wall intersects with other materials or construction. Even when water actually penetrates a wall, the leak is almost never through the masonry units themselves. Most of the time, water enters a masonry wall through cracks and voids between the masonry units and the mortar in the joints. Many such leaks can be prevented by properly finishing the joints. Mortar joint finishes fall loosely into three types: troweled joints, tooled joints, and sealant-filled joints.

Regardless of the joint finishing type used, care should be taken to prevent smearing mortar or sealants onto the finished surface of masonry units that will be exposed to view. Hardened, embedded mortar often cannot be removed. Mortar droppings that adhere to a wall should be allowed to dry before their removal is attempted. Removal should be done using a trowel so that the mortar will not become embedded in the face of the unit.

#### TROWELED JOINTS

In a troweled joint, excess mortar is simply cut off (*struck*) with a trowel, and no further finishing is used. Joints should be struck flush on surfaces that will be parged, plastered, stuccoed, or covered with other masonry.

A *weathered* joint (Fig. 4.7-45, Col. 3) will shed water, but producing this type of joint does not force enough mortar against the brick above to form a watertight joint. Therefore, it should not be used where it will be exposed to the weather.

A flush joint (see Fig. 4.7-45, Col. 4) produces an uncompacted joint with a hairline crack where the mortar is pulled away from the masonry by the cutting action of a trowel (Fig. 4.7-46).

A struck joint (see Fig. 4.7-45, Col. 5) and its kin, a raked joint (Col. 6), which is made by removing the surface of the mortar while it is still soft, both leave a shelf on which water can collect. Raked joints are popular with some architects in spite of their drawbacks, however, because they produce crisp, dark joint lines.

#### TOOLED JOINTS

Because tooled joints provide excellent protection against water penetration, they should be used in every wall that is exposed to the weather, except where sealant-filled joints are required. Joints in below-grade walls that will not be parged should also be tooled. For example, slab on grade and crawl space foundations need not be parged when moisture penetration is not a problem, and if they are not, then those joints should be tooled. Most exposed interior joints are also tooled to give them a finished appearance, even though water penetration is not usually a problem on the interior.

The most common tooled joints used today are either concave (see Fig. 4.7-45, Col. 1) or V-shaped (Col. 2). Many other tooled joint shapes have been used historically, however, and may be encountered in renovation work. To make a tooled joint, a special tool is used to compress and shape the mortar. The tooling operation compacts the mortar and forces it tightly against the masonry on each side of the joint (Fig. 4.7-47). Tooling should be done as soon as a section of the wall has been laid and the mortar has become "thumbprint" hard. Every joint should be tooled at the same degree of mortar hardness so that the joints will all be the same color and shade, and will be uniform in appearance with clean, sharp lines. The jointer for tooling joints should be slightly larger than the joint so that complete contact is made along the edges of the masonry unit, thus compressing and sealing the surface of the mortar, and upturned on the ends to prevent gouging the mortar (Figs. 4.7-48a and 4.7-48b).

#### SEALANT-FILLED JOINTS

Joints between masonry and the perimeter of door and window frames, louvers,

**FIGURE 4.7-46** A mortar joint being cut flush. (Courtesy Brick Institute of America [renamed Brick Industry Association].)

**FIGURE 4.7-47** A mortar joint both before and after tooling. (Courtesy Brick Institute of America [renamed Brick Industry Association].)

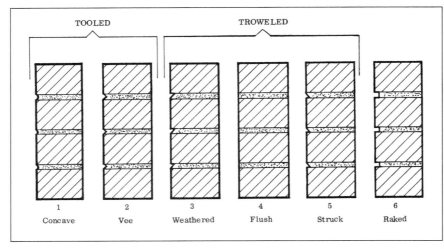

**FIGURE 4.7-45** Types of mortar joints.

a

b

**FIGURE 4.7-48** Here the joints are being tooled (a) in a concrete masonry unit wall with a concave jointer (Courtesy Portland Cement Association) and (b) in a brick wall with a V-jointer. (Courtesy Brick Institute of America [renamed Brick Industry Association].)

and other penetrations, and joints where masonry abuts other materials, should be sealant-filled joints. Expansion and control joints are also usually sealant filled. Where joints to be sealant filled contain mortar or grout, they should be cleaned out to a uniform depth of at least $^3/_4$ in. (19.1 mm) and then filled solidly with an elastic sealant compound forced into place with a pressure gun. Sealant-filled joints should not be less than $^1/_4$ in. (6.4 mm) wide. Sealant materials and methods of installation are discussed and illustrated in Section 7.13.

### 4.7.2.10 Flashing

The general principles, materials, and methods used to install flashings are discussed in detail in Section 7.9. Wall flashing in masonry is included here, however, because, unlike many other flashing installations, it is usually installed by the mason as the masonry is erected, and its installation is generally thought of in the industry as part of the masonry work. Because most of the recommendations about flashing in Section 7.9 apply equally well to the flashing discussed here, those requirements have not been repeated here.

Except in areas of very slight exposure to precipitation (see Fig. 4.6-4) where local experience supports omitting it, flashing should be installed at the following vulnerable locations (Figs. 4.7-49 and 4.7-50):

- At the base of a wall
- Under the sills of wall openings
- Over the heads of wall openings
- At floor lines
- Beneath copings
- Where masonry penetrates a roof surface

#### PURPOSE OF FLASHING

Masonry wall flashing (Fig. 4.7-51) is installed to collect moisture that may penetrate the wall and divert it through weep holes to the exterior. The amount of exposure of masonry to rain, which varies greatly in the United States, is a basic consideration when designing masonry walls to resist moisture penetration. Exposure may be defined roughly in terms of the resultant wind pressures and precipitation maps shown in Figure 4.6-4 as either severe, moderate, or slight. *Severe* exposure indicates annual precipitation of 30 in. (762 mm) or more, with a wind pressure of 30 psf (146.5 kg/m$^2$) or higher. *Moderate* exposure means annual precipitation of 30 in. (762 mm) or more and a wind pressure of 20 to 25 psf (97.65 to 122.1 kg/m$^2$). *Slight* exposure means annual precipitation of less than 30 in. (762 mm) and a wind pressure of 20 to 25 psf (97.65 to 122.1 kg/m$^2$) or annual precipitation of less than 20 in. (483 mm).

The following recommendations apply to areas of severe exposure such as the Gulf Coast, Atlantic Seaboard, and Great Lakes regions, and to areas of moderate exposures which are typical in the Midwest. Where exposure is slight, completely concealed flashing such as lintel or sill flashing is often eliminated, and in some areas even exposed flashing is reduced to a minimum.

#### FLASHING LOCATIONS

The vulnerable locations in masonry where wall flashing is necessary are described below (see Figs. 4.7-49 and 4.7-50). Flashing should be provided under horizontal masonry surfaces such as sills and copings, at intersections of masonry walls with horizontal surfaces such as the roof and a chimney, over the heads of openings such as doors and windows,

and frequently at floor lines, depending on the type of construction.

Base flashing and damp checks are used at the base of a wall. Base flashing diverts to the exterior moisture that enters the wall above the location of the flashing. A damp check stops the upward capillary travel of ground moisture. If installed properly, metal shields used for termite protection of wood joists (see Sections 21.3.9 and 6.4.2) may double for these purposes.

Sill flashing should be placed under and behind every sill. The ends of sill flashing should extend beyond the sides of an opening and turn up at least 1 in. (25.4 mm) into the wall. Sills should slope and project from the wall to drain water away from the building and to prevent staining. When the undersides of sills do not slope, a drip notch should be provided or the flashing may be extended and bent down to form a drip.

Head flashing should be placed over all openings except those completely protected by overhanging projections. At steel lintels, the flashing should be placed under and behind the facing material with its outer edge bent down over the lintel to form a drip.

#### FLASHING INSTALLATION

When flashing is to be laid on or against masonry, the surface of the masonry must be smooth and free from projections that would puncture the flashing material or otherwise destroy its effectiveness. Through-wall flashing should be placed on a thin mortar bed, and another thin mortar bed should be placed on top of the flashing to receive and bond the next masonry course. Metal flashing should be deformed to help lock the mortar in place.

Flashing seams must be overlapped and thoroughly bonded or sealed to prevent water penetration. Although most sheet metal flashing materials can be soldered, lockslip joints are required at intervals to permit thermal expansion and contraction. Many plastics can be permanently and effectively joined by heat or an appropriate adhesive. The elastic pliability of plastic flashings eliminates the need for expansion and contraction seams.

### 4.7.2.11 Weep Holes

The water collected by flashing must be drained to the outside. Concealed flashing in tooled mortar joints is not

**FIGURE 4.7-49** Details showing flashing installations in masonry assemblies.

**FIGURE 4.7-50** Wall flashing locations in masonry construction. (Drawing by HLS.)

Labels in figure (top to bottom): COPING — Stone coping, Flashing, Cap flashing, Roofing and base flashing; ROOF — Flashing; Flashing; FLOOR — Flashing; Flashing — Ceiling; OPENING HEAD — Flashing; OPENING SILL — Flashing; Brick veneer; BOTTOM OF WALL — Flashing — Concrete slab; Foundation wall — Insulation

**FIGURE 4.7-51** Through-wall flashing. (Courtesy Portland Cement Association.)

self-draining without weep holes. Actually, such flashing acts as a trap to collect water, and by concentrating it in one place can do more harm than good. Therefore, except at window sill flashing where little water collects, weep holes should be provided in the head joint immediately above all flashing.

Weep holes may be formed in several ways, and while no one method provides better weep holes than the others, some have a few disadvantages. Weep holes may be formed by simply omitting the mortar in the head joints at the desired intervals, or by inserting oiled rods or pins and then removing them when the mortar is ready for tooling. Weep holes may also be made by inserting plastic or metal tubes, or metal or plastic vents in the head joints and leaving them in place. Another way is to insert short lengths of sash cord and leave them in place to disintegrate later. Using these methods, weep holes should be placed at 24 in. (610 mm) intervals horizontally. Unfortunately, these methods all produce open weep holes that provide access for insects and, especially when placed over lintel flashing, may contribute to wall staining. Care must be taken to prevent open weep holes from being clogged with mortar droppings, dirt, or other foreign materials.

Closed weep holes can be made by inserting wicks in the head joints (see Fig. 4.3-4). These wicks are made of $1/4$ in. (6.4 mm) glass fiber rope or a similar inorganic material.

Another way is to insert $1/2$ in. (12.7 mm) glass fiber insulation in open head joints. Since they do not drain as well as open joints, closed weep holes should be spaced every 16 in. (406 mm). Care must be taken to ensure that mortar droppings do not block the holes behind the wick or insulation and that the materials in the holes do not themselves become clogged with mortar or dirt.

### 4.7.3 PARGING AND DAMPPROOFING

When the groundwater table and subsoil conditions do not cause water to build up hydrostatic pressure against masonry foundation and basement walls below grade, but moisture is present in the soil, penetration of that moisture can usually be controlled by parging and dampproofing the wall. Dampproofing is a coating applied to resist moisture penetration due to capillary action. It is discussed in Section 7.2. When hydrostatic pressures may develop, concrete foundation walls are normally used rather than masonry and water penetration is controlled by waterproofing. Waterproofing is a continuous watertight membrane capable of resisting water under pressure (see Section 7.2).

Parging consists of Type M mortar or portland cement plaster applied to the exterior face of a below-grade wall and troweled to a smooth, dense surface (Fig. 4.7-52). Portland cement mortar applied above grade is called *stucco* (see Section 9.2). Parging should be thickened and turned out onto the footing to form a cove (see Fig. 4.7-52b) and should extend upward from the footing to a line about 6 in. (152 mm) above finished grade. Its top should be beveled to form a wash.

Parging should be applied in two coats. The first coat should be roughened when partially set, hardened for 24 hours, then moistened before the second coat is applied. Parging should be moist cured for at least 48 hours and dry before dampproofing is applied over it. In single-family residential buildings, parging is sometimes applied in only one $^1/_4$-in. (6.4-mm)-thick coat and is often applied in two $^1/_4$-in. (6.4-mm)-thick coats to a completed thickness of $^1/_2$ in. (12.7 mm). In larger projects, two $^3/_8$-in. (9.5-mm)-thick coats are usually applied for a total thickness of $^3/_4$ in. (19.1 mm). The latter method produces a better job and should be used even on small projects. A single $^1/_4$-in. (6.4-mm)-thick coat is much too thin to form a credible parging. Even $^1/_2$ in. (12.7 mm) is too thin; the additional cost of the extra $^1/_4$ in. (6.4 mm) of parging is small when compared with the benefits. One problem with thin coats of parging is that walls are seldom straight and never smooth. The parging tends to average the projections and depressions. Over projections in a wall, a thin coat of parging tends to get even thinner. Very thin coats may flake off the wall when they are subjected to wall movement. When pieces of parging fall off, the covering dampproofing goes with them.

Bituminous dampproofing products (see Section 7.2) may be applied either over parging or directly on the masonry to increase resistance to moisture penetration. Parging is sometimes omitted when the wall to be dampproofed is smooth and the soil is well drained and dry. Dampproofing of interior masonry walls is also discussed in Section 7.2.

### 4.7.4 PROTECTION DURING CONSTRUCTION

Masonry under construction must be protected from rain and snow and, in cold weather, from freezing. Methods of providing this protection are discussed in this section. Lateral support of walls during construction is covered in Section 4.6.5.4.

#### 4.7.4.1 Protection from Rain and Snow

Unprotected, partially finished masonry that becomes saturated with water during a rain- or snowstorm may take months to dry out, and during this drying period efflorescence may appear on the surface (see Section 4.7.5.2).

At the end of each day or shutdown period, when there is a threat of precipitation, partially completed masonry should be protected by covering it with securely weighted-down canvas or another strong waterproof membrane. This covering should overhang at least 2 ft (610 mm) on each side of a wall and be securely anchored (Fig. 4.7-53).

#### 4.7.4.2 Cold Weather Construction

As the air temperature falls, mason productivity and masonry strength are affected. The chemical reaction between portland cement and water (hydration), slows significantly at temperatures below 40°F (4.4°C). Below 32°F (0°C), hydra-

a

b

**FIGURE 4.7-52** (a) Parging on a basement wall; (b) a wall section showing parging and dampproofing. Note that the parging turns out onto the footing to form a cove. (Courtesy Portland Cement Association.)

**FIGURE 4.7-53** The top of a masonry wall covered with a tarp to protect it from the elements and to aid in curing. (Courtesy Portland Cement Association.)

tion proceeds only when sufficient liquid water is available. Hydration ceases altogether at about 14°F ($-10$°C). Therefore, masonry should not be laid when the air temperature is 40°F (4.4°C) and falling or lower, unless the proper precautions are taken.

Procedures for mortar mixing in cold weather are addressed in Section 4.2.6.3. Figure 4.7-54 summarizes the other precautions required in cold weather. The basic principle is that at air temperatures below 40°F (4.4°C), masonry materials must be heated to achieve satisfactory masonry assemblies. Masonry construction may continue even when temperatures drop below freezing, but only when the proper measures are taken to make the materials act as they would at temperatures above 40°F (4.4°C). For example, heated mortar materials have performance characteristics identical to those they have in the normal temperature range.

It is also very important that mortar and masonry units not be laid on frozen materials, and that the masonry units be at a temperature of at least 20°F ($-6.7$°C)

and contain no visible frost or ice at the time of laying. Stored masonry units must be protected and warmed if necessary. When wetting of brick is necessary (see Section 4.7.1.2) in cold weather, the water used must be warmed.

Masonry must be protected from precipitation and winds in excess of 15 mph (24 km/hr) and maintained at above-freezing temperatures during construction and for at least 24 hours after construction. Even when temperatures are above freezing, if they are below 40°F (4.4°C) completed masonry must be kept covered and protected from rain and snow for at least 24 hours after construction.

## 4.7.5 POINTING AND CLEANING MASONRY

Exposed masonry should be protected against staining by covering it, and excess mortar should be wiped off the surface as the work progresses. When the masonry work has been otherwise completed, the joints should be pointed and the entire masonry should be carefully

cleaned. Large particles of mortar should be removed using a putty knife or chisel, and the masonry should be cleaned of dirt, dust, efflorescence, stains, and other foreign materials.

### 4.7.5.1 Pointing

Except for weep holes, holes in exposed masonry should be *pointed* (filled), and defective joints should be cut out and re-pointed with fresh mortar.

Holes left by nails or line pins should be pointed immediately while the adjacent mortar is green enough to allow adequate bond with the patch. Otherwise, water entering through nail holes may penetrate the wall (Fig. 4.7-55).

### 4.7.5.2 Cleaning Clay Masonry

Unglazed brick and tile should be cleaned to remove general dirt and grime and stains. Cleaning should not be attempted until at least 48 hours after completion of construction.

### GENERAL CLEANING

Unglazed brick and tile on which no efflorescence or other stain appears should be first cleaned with stiff brushes and clean water. Soil not removed by that cleaning should be scrubbed using stiff brushes, clean water, and soap powder or another mild solution recommended by the manufacturer of the masonry unit.

Only after such mild cleaning has been unsuccessful should stronger measures be used. With the concurrence of the masonry unit manufacturer, stubborn dirt may be removed with a mild solution of acid similar to that suggested in the later subsection entitled "Mortar Stains." The surface should be thoroughly wetted with clear water and then scrubbed with the

**FIGURE 4.7-54** Recommended Procedures for Masonry Construction in Cold Weather[a]

| Work Day Temperature | Construction Requirement[b] | Recommended Protection[b] |
| --- | --- | --- |
| All temperatures | Normal masonry procedures. | Masonry materials should be received, stored, and protected in ways that prevent water damage. Cover tops of newly laid masonry with plastic or canvas to prevent water damage. |
| 40° to 32°F (4.4 to 0°C) | Heat mixing water to produce mortar temperatures between 40° and 120°F (4.4 and 48.9°C). Never place masonry on frozen, snowy, or ice-covered surfaces. | Cover walls and materials with plastic or canvas to prevent wetting and freezing. |
| 32° to 25°F (0 to $-3.9$°C) | Heat mixing water and sand to produce mortar temperatures between 40°F and 120°F (4.4 and 48.9°C). | Provide windbreaks for wind velocities over 15 mph (24.14 km/h). Cover walls to prevent freezing. Maintain masonry above 32°F (0°C) for 24 hours using auxiliary heat for insulating blankets. |
| 25° to 20°F ($-3.9$ to $-6.7$°C) | Mortar on boards should be maintained above 40°F (4.4°C). Masonry units below 20°F ($-6.7$°C) must be heated above 20°F ($-6.7$°C). | |
| 20°F and below ($-6.7$°C) | Heat mixing water and sand to produce mortar temperatures between 40° and 120°F (4.4 and 48.9°C). | Provide enclosures and supply heat to maintain masonry enclosure above 32°F (0°C) for 24 hours. |

[a]For unenclosed construction sites. These recommendations do not apply where the complete work area, including materials storage space, is enclosed and heated.

[b]Each requirement is in addition to the one above it.

**FIGURE 4.7-55** Nail and line-pin holes should be filled as soon as possible. (Courtesy Brick Institute of America [renamed Brick Industry Association].)

acid, followed immediately by a thorough rinsing with clear water.

The exposed surfaces of glazed brick, glazed structural clay tile, and terra cotta should be washed with detergent powder and warm water, applied with a scrubbing brush, and then rinsed thoroughly with clear water. Metal cleaning tools and brushes, acid solutions, and abrasive materials should not be used.

Regardless of the cleaning solution used, cleaning should begin at the top and proceed downward. The wall should be wetted with water before a cleaning solution is applied, and adjacent construction should be protected. Stiff fiber brushes should be used. Only a small area should be cleaned at one time. Cleaning should be followed immediately by thorough rinsing with clean water.

### EFFLORESCENCE AND OTHER STAINS

Large amounts of soluble salts in mortar and masonry units can cause efflorescence or lime run to appear on the face of the masonry if excessive water penetrates the masonry. The most common form of efflorescence is a white powdery deposit of alkali salts and carbonates (Fig. 4.7-56), but not all white deposits on masonry are efflorescence and salts of other colors sometimes appear as well. For example, a white stain may be *lime run*, as described below, or a silicate deposit called *scumming* or *white scum*. Other stains on brick masonry may be so-called *green stain* caused by vanadium salts or by brown, tan, or gray manganese dioxide salts. These water-soluble salts are originally present in either the mortar or the masonry units. They are carried to the surface by movement of water that has entered the wall. The dissolved salts are left on the face of the masonry when the water evaporates.

White efflorescence will often be washed away through normal weathering, especially if it has resulted from initial water in a wall that is dry after it has been completed. When necessary, it can usually be removed by dry brushing alone. If that measure fails, a cleaning with water and stiff brushes will usually do the job. In severe cases, a solution of 1 part muriatic acid to 12 parts water may be used.

White scum is very difficult and sometimes impossible to remove. Green stain is also difficult to remove. Magnesium stains, while not difficult to remove, do require special techniques. All such stains should be removed in the manner and using the materials recommended by the manufacturer of the masonry units that are affected. Different methods may be needed for light-colored masonry than those used for darker masonry units. In the absence of specific recommendations regarding removal of stains from brick masonry, the BIA's Technical Notes 20, "Cleaning Brick Masonry," 23, "Efflorescence—Causes and Mechanisms, Part I of II," and 23A, "Efflorescence—Prevention and Control, Part II of II" offer excellent guidance.

The condition called *lime run*, which is actually a deposit of carbonate, presents as a vertical white, crusty area that appears to have flowed from a hole in a mortar joint. It is often mistaken for efflorescence. Despite its name, lime run is not associated with the lime in a mortar, but is thought to be a deposit of one or more calcium compounds that have been dissolved in a water stream that came through the joint and then reacted with carbon dioxide in the air. Lime run can be removed using a weak solution of hydrochloric acid. The condition will return, however, unless the water source is removed.

### MORTAR STAINS

When possible, mortar stains should be removed from brick by scrubbing it with detergent and warm water. Harsher cleaning methods should not be used on buff, gray, or other light-colored bricks without the concurrence of their manufacturer. Unless specifically not recommended by the manufacturer, when milder methods have failed, mortar may be removed from darker-colored bricks using a mild acid solution.

When use of acid is unavoidable, a solution of hydrochloric acid and water (*muriatic acid*) is often used. Muriatic acid should be further diluted until it is no stronger than 1 part muriatic acid to 9 parts water. Even a solution this dilute should not be used until the mortar in the wall to be cleaned is well hardened and at least 7 days old. Acid solutions should always be tested in an inconspicuous location on a wall to ensure against adverse effects.

Lumps of mortar should be removed from terra cotta using sharpened wood paddles. Acid should never be used on terra cotta.

### 4.7.5.3 Cleaning Concrete Unit Masonry

Because hardened, embedded mortar smears can never be removed from it and because concrete masonry should not be cleaned with an acid wash, mortar should never be smeared on the surface of an exposed unit. Mortar droppings should be

**FIGURE 4.7-56 A bad case of efflorescence.** (Courtesy Brick Institute of America [renamed Brick Industry Association].)

**FIGURE 4.7-57** Final brushing. (Courtesy Portland Cement Association.)

allowed to dry before removal is attempted. Trying to remove wet mortar will probably result in smearing of the mortar into the surface of the unit, from where it cannot be removed. However, most dry, hard mortar can be removed with a trowel. A final brushing will then remove practically all the remaining mortar (Fig. 4.7-57)

### 4.7.6 INSTALLATION OF ARCHITECTURAL TERRA COTTA

Historically, architectural terra cotta was tied to a building's structure with metal anchors. Then it was backfilled to make it more fireproof and to protect the concealed anchors. Modern-day terra cotta is applied in two ways. The first method is used to install adhesion-type ceramic veneer and terra cotta tile (see Section 4.4.4.3). It is similar to the method used to install ceramic tile. In this method, the wall is dampened and the terra cotta units are soaked in water for 1 hour. Then a brush coat of portland cement and water is applied to the back of the units. Mortar (see Section 4.2) is then applied to both the units and the wall, and the units are pressed into the mortar. The units are set plumb and level with even $1/4$-in. (6.4-mm)-wide mortar joints, and then supported by wood wedges placed in the joints until the mortar hardens. The joints are then tooled or raked back $1/2$ in. (12.7 mm) deep, pointed, and tooled.

The second method of installing terra cotta is used to install anchored-type ceramic veneer. In this method, the ceramic veneer may be mounted with or without vertical pencil rods, depending on the design and size of the units and the design of the supporting elements. Pencil rods are $1/4$-in. (6.4-mm)-diameter steel reinforcing rods. When they are used, they are mounted vertically and held in place by passing them through the loops of dowel anchors attached to the supporting construction. The ceramic veneer units are set in place with wood wedges in the joints to support their weight. Wire anchors are installed in slots in the units and tied to the pencil rods. When rods are not used, the anchors are tied to anchors set into or attached to the supporting construction. After one horizontal course has been laid, the space between the units and the supporting construction is filled with the grout described in Section 4.2. The installation is completed by removing the wood wedges, pointing the joints, and cleaning the units.

## 4.8 Stone

Stone played an important part in the lives of people long before recorded history. Undoubtedly, early humans used naturally occurring stones as both tools and weapons. About half a million years ago, our predecessors began to flake chips off flint and other stone to make more useful implements and weapons. Early dwellings were in natural stone caves. Where these were not available, shelter was made from wood and animal skins, which were relatively portable, a necessity for nomadic peoples. With the development of farming came more permanent settlements and the associated need for permanent dwellings. Where loose stones were available, which was often the case as land was cleared for farming, these were piled up to form crude stone huts.

Shaped stone was not used extensively until iron stoneworking tools were developed. Before the Etruscans introduced lime mortar, cut stone buildings were made by accurately cutting and fitting stones together with no joint filler, which was time-consuming, tedious work. The use of mortar joints reduced the need for extreme accuracy in stone cutting and speeded the construction process. When the Romans developed hydraulic mortar, much stronger stone structures could be made and much longer spans built.

Stone was gradually replaced to a large extent in smaller and residential buildings by lighter, easier-to-handle brick and, later, concrete masonry units. However, stone remained in extensive use in decorative and ceremonial buildings, such as cathedrals and large public buildings (Fig. 4.8-1), and as cladding on large and small commercial buildings.

**FIGURE 4.8-1** National Archives Building, Washington, DC. (Architect: John Russell Pope. Photo courtesy Indiana Limestone Company, Inc.)

**FIGURE 4.8-2 U.S. Courthouse, Hammond, Indiana.** (Architect: Pei, Cobb, Freed, and Partners. Photographer George James. Photo courtesy Indiana Limestone Company, Inc.)

In the eighteenth and nineteenth centuries, lighter steel and concrete frameworks virtually replaced mass stone as the structural support elements in buildings. Stone structures cannot compete with these newer materials either in terms of economy or speed of construction. The large structures possible with these other materials simply cannot be built in stone.

Mass stone is still sometimes used, but most stone today is hung on concrete or steel structural framing. Massive stone structures, and stone bridges and aqueducts, are a thing of the past. However, stone still remains in great use as a modern building material (Fig. 4.8-2). It is used in many forms, including the following:

*Fieldstone* is regular or irregular stone units that are gathered from stream beds and fields. It is most often used for fences and site walls, for landscaping, and sometimes as riprap.

*Rubble* is rough-cut or blasted stone used in walls (Fig. 4.8-3).

*Flagstone* is thin slabs of stone that is used primarily as flooring or paving.

*Dimension stone* is stone that is cut with square corners. It may be thin slabs, large or small blocks, or sculptured shapes.

*Cut stone* consists of large pieces of dimension stone.

*Ashlar* is small rectangular blocks of dimension stone (Fig. 4.8-4).

*Crushed stone* is made by crushing stone from the quarry. It is used as fill beneath building slabs and paving; drainage fill; riprap; as aggregate in concrete, terrazzo, and asphalt; and in landscaping.

*River gravel* is relatively small, smooth stone gathered from stream beds. It is

**FIGURE 4.8-3 A limestone rubble wall.** (Photo Courtesy Indiana Limestone Institute.)

used in locations similar to those for crushed stone.

*Stone dust* results from quarrying, milling, and finishing of stone. It is used as mulch and for other landscaping purposes.

### 4.8.1 MATERIALS

The types of natural stone used most often today—limestone, marble, granite, and slate—are addressed here. Other types of stones are also used, however. These include sandstone, bluestone, soapstone, and many locally available stones.

#### 4.8.1.1 Limestone

Limestone is sedimentary rock composed of either calcium carbonate (calcite) or dolomite, which is a mixture of calcium and magnesium or a mixture of calcite and dolomite. It is classified as either *oolitic* or *dolomitic*. Oolitic limestone is

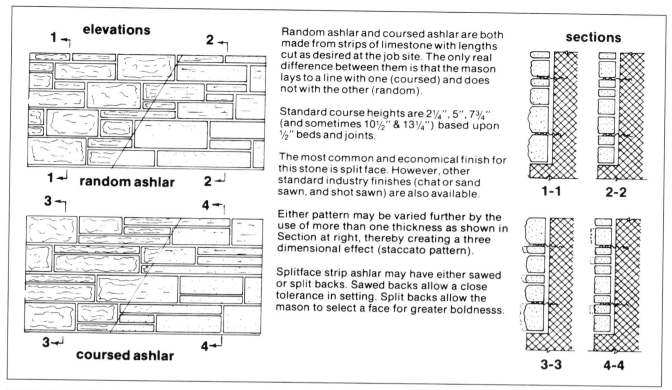

**FIGURE 4.8-4 Ashlar stone patterns.** (Courtesy Indiana Limestone Institute.)

quarried mostly in Indiana and dolomitic limestone mostly in Minnesota, but both are also found in other locations. Dolomitic limestone is more crystalline and stronger than oolitic limestone, and some of it will take a polish, whereas oolitic limestone will not. Finishes for oolitic limestone include smooth finishes and textured and coarse finishes such as plucked, split-faced, tooled, chat-sawed, and rock-faced. Finishes for dolomitic limestone include honed, sandblasted, and polished smooth. Sometimes different finishes are used together.

Most limestone ranges in color from a light creamy buff to a brownish buff or from silver gray to bluish gray, but some is green. Stone primarily of one of these colors may contain various spots and streaks of the others. Full color blend limestone contains both buff and gray colors either in the same stone or in adjacent stones. Limestone ranges in texture from very fine-grained to coarse-grained with pit holes, reedy formations, honeycomb, iron spots, and travertine-like formations.

Limestone is not suitable for use in acidic environments or where chemicals will be used to melt ice and snow, nor should it be cleaned with harsh chemical cleaners. Acids and de-icing and cleaning chemicals will damage the stone.

The industry standard for limestone dimension stone is ASTM C568.

### 4.8.1.2 Marble

Marble is metamorphic stone composed principally of calcite and dolomite (limestone) that has been recrystallized for a long time by heat and pressure. Marble is classified into four categories: I Calcite, II Dolomite, III Serpentine, and IV Travertine. It comes in so many colors and with such a variety of veining that it is necessary to specify it by its source or by requiring that the material supplied match a sample furnished by the architect.

Many types of stone are classified as marbles. Among them are onyx, alabaster, some gypsums, and some marble and granite mixtures. The marble category known as *travertine* is a favorite of many architects due to its distinctive appearance distinguished by its color and its irregularly shaped surface pores. Like other marbles, travertine began as limestone, which through geological shifting descended in the earth. Heated by the earth's inner core, water in underground *aquifers* rises as steam and hot pressurized water to form hot mud baths, such as "Old Faithful." This rising hot water dissolves the limestone and brings with it granules from below, forming mud beds on the surface. In time, the mud beds cool and crystallize into solid travertine.

A requirement of marble is that it be capable of taking a polish, and interior marble is often so finished. However, highly polished finishes are inappropriate on traffic surfaces and on the exterior, where weathering will dull them. Typical finishes used on exterior surfaces are honed, sandblasted, abrasive, axed, hammered, rock-faced, rough-sawn, and tooled. The industry standard for marble dimension stone is ASTM C503.

### 4.8.1.3 Granite

Granite is granular igneous rock consisting mostly of quartz or feldspar. It is hard, strong, durable, and almost impervious to water. True granite ranges in color from pink to light or dark gray. Other igneous stones that do not strictly meet the chemical and mineralogical requirements for granite are nevertheless sold as granite. These come in black, brown, green, and buff. Because there are so many possible colors and grain char-

**FIGURE 4.8-5** A diamond belt saw being drawn through limestone bedrock. The parallel lines are cuts already made by the saw. (Courtesy Indiana Limestone Company, Inc.)

acteristics in granite, the best way to specify it is either by its specific source or as a match for samples furnished by the architect.

**FIGURE 4.8-6** Turning a cut in the quarry. (Courtesy Indiana Limestone Company, Inc.)

**FIGURE 4.8-7** A quarry crew drilling holes in a turned-down cut using pneumatic drills preparatory to splitting it into blocks. (Courtesy Indiana Limestone Company, Inc.)

Common granite finishes include mirror-like polished, honed, thermal, split-faced, and jet-honed. It can be used under severe weathering conditions and in contact with the ground.

The industry standard for granite dimension stone is ASTM C615.

**FIGURE 4.8-8** In the foreground are blocks tagged and stored for shipment to the mill. In the background, blocks are being loaded on a flatbed truck for shipment to the mill. In the far background, the cut steps in the bedrock are clearly shown where stone has been removed. (Courtesy Indiana Limestone Company, Inc.)

### 4.8.1.4 Slate

Slate is a microcrystalline metamorphic rock formed originally from clay. It consists of thin plates that can be easily split into sheets. It is used primarily for paving and roofing. Slate comes in red, purple, green, blue, and black. Normal finishes are *natural cleft*, which is the face left by splitting the slate, honed, and *sand-rubbed*, in which the texture left by splitting is completely removed to provide a smooth surface.

The industry standard for slate dimension stone is ASTM C629.

### 4.8.2 MANUFACTURE

Stone for building is obtained by *quarrying* and prepared for use by *milling* and finishing.

### 4.8.2.1 Quarrying and Milling

*Quarrying* is the removal of the stone from the ground. The methods of quarrying differ, depending on the type of stone and the use to which it will be put. Stone that will be used for rubble may be blasted out of the ground or broken off using picks and crowbars.

The methods of quarrying dimension stone vary with the type of stone and may vary with the finished stone required. The methods described here are the normal ones in use today.

A stone supplier may vary these methods when necessary to provide a finished stone that differs from the normal. In addition, stone quarrying technology has been changing rapidly in recent years, and the methods in use when you read this text may no longer be the same as those described here.

*Milling* is the process of cutting quarried stone splits into blocks and slabs and further refining them into the final shapes that will be used in a building.

## LIMESTONE QUARRYING

Generally, a diamond belt saw is drawn through the stone at intervals of about 4 ft 6 in. (1372 mm), cutting the bedrock into sections about 15 ft (4500 mm) deep and 50 ft (15.24 m) long (Fig. 4.8-5). Pneumatic air bags are inserted into the sawed joint and inflated, tilting a section away from the bedrock (Fig. 4.8-6). Holes are then drilled in the turned section parallel and perpendicular to the grain (Fig. 4.8-7). The cut sections are then broken into 8-ft (2400-mm)-long strips called *blocks*. Traditionally, workers inserted wedges into these holes and tapped them intermittently with hammers

until the stone split. Today, the holes are drilled deeper and devices called *rock splitters* are inserted into the holes and activated, splitting the rock. The rock splitter method is quicker and less labor intensive than the older method. Using this method, the stone is separated into blocks about $4 \times 8 \times 3^1/_2$ ft ($1200 \times 2400 \times 1050$ mm) or smaller. These blocks are tagged to signify their final use and stored until transported to the mill (Fig. 4.8-8).

In the milling process, blocks are cut into smaller blocks or slabs in a shop using several types of saws and splitters (Fig. 4.8-9). Limestone is not usually cut

a

b

c

**FIGURE 4.8-9** (a,b) Two of the various types of saws used in milling limestone blocks into smaller blocks, slabs, and other shapes. (c) A splitter used to create split-face pieces. (Courtesy Indiana Limestone Company, Inc.)

into blocks or slabs that are less than 2 in. (50 mm) thick. The maximum slab size is dictated by the thickness. Thicker limestone can be handled in large sheets.

Most stone blocks and slabs are too heavy to be moved and placed by hand. Cranes and other mechanical lifting devices are used extensively in the quarry (see Fig. 4.8-8), at the mill (Fig. 4.8-10), and at the building site. Individual finished stones must have the proper holes and sinkages to permit them to be lifted without damage (Fig. 4.8-11) and to be placed without interference or damage to mortar or other bed joints.

Natural stone contains moisture (quarry sap) when quarried. When the moisture dries and the stone stabilizes, it is said to be *seasoned*. Stone is often cut and used before seasoning is complete, but until it is seasoned, which can take as long as a year, stone will not reach its final color and should not be coated with water repellants of any kind.

## MARBLE QUARRYING

Marble quarrying is similar to the methods described for quarrying limestone. Marble can be used in thicknesses as small as $3/4$ in. (19 mm).

## GRANITE QUARRYING

Because granite is much harder than limestone and marble, the methods used to quarry it differ substantially from those used to quarry limestone and marble. Some types of granite can be successfully quarried using a wire cutter, in which a large diamond-studded wire is drawn through the stone at a high speed, cutting the stone from the bedrock. Most granite, however, is quarried by drilling, blasting, or the application of extreme heat using a device called a *jet burner*, which produces a flame by forcing compressed air through a stream of fuel oil. Thermal stresses induced by the jet burner cause the granite to split away (spall) from the bedrock.

Granite can be cut and installed in sheets as thin as $3/8$ in. (9.5 mm), but thicker stones are advisable.

a

b

c

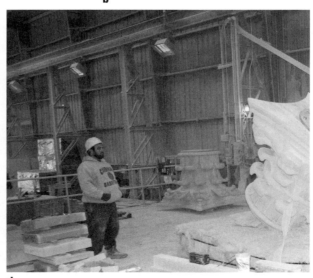

d

**FIGURE 4.8-10** Loading and handling of stone at the mill; (a) overhead crane at a mill, (b) forklift being used to lift a stack of limestone slabs, (c) lifting a slab section, (d) lifting a carved column cap using an overhead crane. (Courtesy Indiana Limestone Company, Inc.)

The supplier of Cut Indiana Limestone will provide holes and sinkages for both anchoring and lifting. However, it is the responsibility of the general contractor or erector to correlate the types of equipment he plans to use with the matching types of lifting holes to be provided by the supplier. Where no arrangements are made, the supplier will provide those sinkages, if any, which suit his own handling requirements. Do not use lewis devices in stones under 3½″ thick.

NOTE: ILI has included this page for illustrative purposes only. This information is not intended as a guide to lifting procedures. Commercial chain, clamps, pins and other devices are available. Their manufacturers will provide instructions for their safe and proper use.

**Slings**

Damage
Incorrect

Too-short slings cause damage at stone edges. Loads lifted by such slings tend to be unstable.

Correct

**Lewis Pins**

**Box Lewis**

**Stone Clamp**

On textured finishes, a ribbed pressure pad evenly distributes the clamp pressure to the smooth surface.

Lifting Pulley allows self-leveling of stone.

Sometimes two clamps can handle heavy stones better.

Pressure Pad

Tightening Screw

Button into Stone Back

Chain retains ribbed pressure pad from falling when loosened.

**Note:** Pins, box lewises and other devices attached to or near the tops of stones should not be used to raise stone panels to vertical position from the horizontal. They may, however, be used in conjunction with slings or other devices which support the main weight of panels while turning them to the position for installation. **In contemporary stone design and usage, lewis devices are unreliable and dangerous.** ILI recommends against their use by persons unfamiliar with safe practices.

**FIGURE 4.8-11** Some methods of lifting large cut limestone units. Other devices are also used. (*Indiana Limestone Handbook*, 19th ed. p. 37, reprinted with permission of Indiana Limestone Institute, Inc.)

### 4.8.3 FINISHES

Cut stones are finished by grinding, polishing, or other treatment to produce the desired finish (Fig. 4.8-12). Final cutting is often done using a diamond saw. Cylindrical shapes are turned on a lath. Articulation is sculpted by machine or by hand, depending on the final configuration (Fig. 4.8-13). Intricate detailing is possible in stone (Fig. 4.8-14). Each stone is marked with a unique code so that its correct location in the building can be ascertained when the stone is delivered to the construction site.

Finishes applied on stone include the following:

*Chat-sawed*: Chat-sawed surfaces are coarse pebbled surfaces made by using a coarse abrasive during sawing.

*Flamed*: A flamed (thermal) finish results from the application of a direct flame at high temperatures (with a blow torch). This usually is seen in granites and some limestones. Most stones cannot withstand this treatment.

*Hand-tooled*: A hand-tooled surface is any surface made using hand tools.

*Honed*. Stones having a honed finish are smooth, like polished ones, but are nonreflective. Honed finishes are achieved by rubbing by hand or machine using abrasives.

*Plucked*: A plucked surface has a rough finish made by rough-planing the surface and breaking out (plucking) small particles.

*Polished*: Polished stone has a mirror-like surface. Only crystallized stone can take a polish. Essentially, at a microscopic level, polishing is putting a facet on each crystal, much as a jeweler puts a facet on a diamond. The result is a surface that allows light to

**FIGURE 4.8-13** An artisan sculpting limestone using a hand-held power tool. (Courtesy Indiana Limestone Company, Inc.)

a

c

b

**FIGURE 4.8-12** Some of the many machines used to provide finishes on stone: (a) a computer-driven device that can do sawing, routing, and carving to preset patterns; (b) this machine houses a diamond-surfaced drum used to create a smooth surface on slabs; (c) profiling a shape using a planer. (Courtesy Indiana Limestone Company, Inc.)

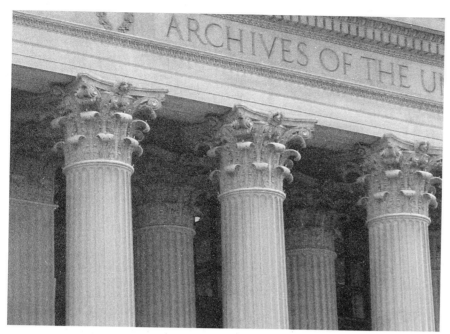

**FIGURE 4.8-14** Carved capitals on columns at the National Archives Building, Washington, DC. (Architect: John Russell Pope. Photo courtesy Indiana Limestone Company, Inc.)

reflect in and out of the stone in a parallel way, enhancing the visible light and color and giving the appearance of depth. A polished finish does not affect the porosity of the stone.

*Rock-face*: A rock-face surface is coarse and irregular. It is sawed on the top and bottom, and the face is exposed by splitting and is then dressed by machine.

*Rubbed or fine-rubbed*: Rubbed finishes are achieved by rubbing with abrasives, such as sand or grit. A rubbed finish has occasional light trails or scratches. A fine rubbed finish is smooth but has no sheen.

*Rusticated*: A rusticated stone face has a deep joint effect cut by machine and sometimes finished by hand.

*Sandblasted:* Sandblasted stone has a smooth mat surface.

*Sawcut*: A sawcut finish is simply the rough sawcut surface of the stone with no other treatment applied. Sawcut material is seldom used where it will be visible in a completed building, but tumbled marble is often made from sawcut material; if not tumbled enough, it will show circular saw marks.

*Sculptured*: A sculptured finish is a reproduction of a sculptor's model carved in the stone by hand.

*Split-face*: A split-face finish is a rough, irregular surface made by sawing the stone at the top and bottom and splitting it.

### 4.8.4 DESIGN AND CONSTRUCTION

Stone may be laid in mortar beds and supported by the stone below or supported from the structure by metal. It may also be adhered to backing panels of concrete or metal. It may be used to build solid walls, arches, and vaults; as a facing over masonry; or as cladding. As cladding, it may be supported by the structure or mounted on a metal framework.

#### 4.8.4.1 Traditional Stone Setting

Stone structures may be built in the traditional method using mortar beds either as a complete wall or as a facing for masonry. Finished joints may be either mortar or sealant filled. Traditional mortar bed setting methods are used today primarily to build stone retaining walls, planters, and the like; for installing stone copings and trim in masonry walls; and in restoration work on existing stone structures.

Stones laid using traditional mortar bed methods are placed either as rubble or as ashlar. Rubble consists of natural stone shapes that may be laid in either random or coursed patterns (see Fig. 4.8-3).

Ashlar is cut stone with squared corners. It may be laid in either random or coursed patterns (see Fig. 4.8-4). Large blocks with square corners are called *dimension stone*.

The methods used are first to place a setting bed, then set the stone in this bed. The joints are raked back. When the bed has set, the joints are either filled with mortar and tooled or filled with a sealant.

#### 4.8.4.2 Mechanical Fastening

Stone cladding may be set on a field-erected grid system of stone anchors, fasteners, and metal struts and braces (Figs. 4.8-15, 4.8-16, and 4.8-17). Very thin stone cladding may also be installed as panels in aluminum frames. Thin stone cladding panels may be applied as a facing on concrete panels. They may also be preassembled into panels on a prefabricated metal support system that is then lifted into place and mechanically attached to metal frames or to the structure. Panels may have insulation sprayed on their interior surfaces. Joints in mechanically fastened stone cladding are usually sealant filled.

Stone cladding may be of almost any thickness, but the trend today is toward using stone that is 2 in. (50.8 mm) thick or less. The performance requirements for stone cladding are determined in much the same way as those for aluminum curtain walls or concrete cladding. Thin stone is not likely to perform as well as thicker stone. It is more subject to water penetration; to damage due to chemical weathering because of its thinness; to damage due to relatively light impact forces, such as those from wind-driven rain or sleet; and to bowing from internal stresses that are less likely to cause damage in thicker stone. Therefore, greater safety factors must be used when selecting and installing thin stone veneer. However, the savings resulting from the reduction in the cost of the stone material, the ease of erecting lighter stone, and the reduction of the load on the structure often make the use of thin stone cladding worthwhile.

## 4.9 Glass Unit Masonry

Glass block, which enjoyed great popularity in the mid-1900s and then almost disappeared from the market, has returned to favor. Glass blocks are available in many sizes and shapes. They are

**FIGURE 4.8-15** Multistory stone cladding anchored to a steel frame without masonry backup. Three different connection systems are shown.
(Drawings courtesy Indiana Limestone Institute.)

Adapter Channel Welded to Steel

Built into Masonry

1″ Minimum
2″ Preferred

Grout Anchor in C.M.U.

S.S. Split Anchor

Shim as Req'd.

Plate with Bar

Expansion Anchor

S.S. Split Anchor

Bond Beam

Shims

Adjustable

Adjustable

Gusseted Plate with Dowels

alt. "c"

alt. "c"

Anchors shown are at Horiz. Joints only.

**Note:** It is permissible to pierce blocks to insert anchors. Block must be dammed and well-filled with grout.

**FIGURE 4.8-16** Some of the many possible ways to anchor multistory stone cladding with masonry backup to a steel structural system. The number of types of anchors in a single building should be the fewest practicable. Panels should be supported at least at each floor level by a positive structural element (angle, plate). (Drawings courtesy Indiana Limestone Institute.)

**FIGURE 4.8-17** Anchoring multistory stone cladding with masonry backup to a concrete frame is similar to anchoring it to a steel frame (see Figure 4.8-16), and the same principles apply. (Drawings courtesy Indiana Limestone Institute.)

**FIGURE 4.9-1** A typical glass block unit.
(Drawing by HLS.)

made with clear, textured, reflective, and heat-absorbing glass and may be either solid or hollow. They can be installed in the traditional way using mortar in the joints, in systems using aluminum frames, or in frameless systems held together with sealant.

Glass blocks are used in both interior and exterior locations. They may be placed vertically in the conventional manner or horizontally in luminous floors and skylights. They provide a partially transparent or translucent surface, depending on the block selected.

Glass block units are not themselves fire rated, but some building codes permit their use in fire-rated assemblies in certain limited cases. For example, they can be used in 45-minute assemblies in some locations when they do not exceed 15% of the surface area. They can be used in 90-minute rated masonry walls, with certain exceptions and restrictions. It is necessary to verify their use in fire-rated assemblies with code requirements, but they should not be summarily excluded from such use without checking.

Glass block units can be used to build both straight and curved walls.

### 4.9.1 MATERIALS

#### 4.9.1.1 Glass Block

Glass blocks (Fig. 4.9-1) are available as hollow units and as solid blocks. Hollow units are made by joining two molded units together, with a partial vacuum between them. The face thickness of blocks varies. A wide variety of patterns are available. Patterns are usually, but not always, formed on the inner surfaces of hollow units, with the outer surfaces kept flat and smooth. Decorative units may have ceramic faces fused onto the glass block.

Solar reflective units are coated on one or both outer faces with a reflective oxide.

Both single-cavity and double-cavity blocks are available. Double cavities are formed by inserting a glass fiber insert between the two halves of the block, forming two parallel air spaces. Ambient heat transmission control is good in single-cavity hollow blocks and is further enhanced in double-cavity blocks.

**FIGURE 4.9-2** Glass block accessories: (a) joint reinforcement; (b) panel anchor; (c) metal channels. (Drawing by HLS.)

**FIGURE 4.9-3** Typical exterior wall glass block installation with mortar joints and metal channels. (Drawing by HLS.)

**FACE BRICK** (2¼ in. × 8 in.)

| | |
|---|---|
| Fire Rating (hr.): | 1 |
| Sound Transmission Class (dB): | 45 |
| U-value (Btu/sq. ft. hr., °F): | .76 |
| Wall Weight (lb./sq. ft.): | 38 |

**COMMON BRICK** (2⅛ in. × 8 in.)

| | |
|---|---|
| Fire Rating (hr.): | 1 |
| Sound Transmission Class (dB): | 34 |
| U-value (Btu/sq. ft. hr., °F): | .76 |
| Wall Weight (lb./sq. ft.): | 38 |

**CONCRETE BRICK** (2¼ in. × 7⅝ in.)

| | |
|---|---|
| Fire Rating (hr.): | 1 |
| Sound Transmission Class (dB): | 37 |
| U-value (Btu/sq. ft. hr., °F): | .67 |
| Wall Weight (lb./sq. ft.): | 35 |

**HEAVYWEIGHT CONCRETE "SLUMP" BLOCK** (4 in. × 16 in. Face Size)

| | |
|---|---|
| Fire Rating (hr.): | 1 |
| Sound Transmission Class (dB): | 36 |
| U-value (Btu/sq. ft. hr., °F): | .46 |
| Wall Weight (lb./sq. ft.): | 33 |

**GLAZED TILE** (6 TC Series and 6 TCD Series—Tooled 2 Sides)

| | |
|---|---|
| Fire Rating (hr.): | −1 |
| Sound Transmission Class (dB): | 32 |
| U-value (Btu/sq. ft. hr., °F): | .40 |
| Wall Weight (lb./sq. ft.): | 27 |

**LANNON STONE**

| | |
|---|---|
| Fire Rating (hr.): | 1 |
| Sound Transmission Class (dB): | 46 |
| U-value (Btu/sq. ft. hr., °F): | .68 |
| Wall Weight (lb./sq. ft.): | 50 |

**LIGHTWEIGHT CONCRETE BLOCK**

| | |
|---|---|
| Fire Rating (hr.): | 1 |
| Sound Transmission Class (dB): | 40 |
| U-value (Btu/sq. ft. hr., °F): | 43 |
| Wall Weight (lb./sq. ft.): | 24 |

**CLAY PARTITION TILE** (12 in. × 12 in. Face Size) (Trowel Cut Joint)

| | |
|---|---|
| Fire Rating (hr.): | 1 |
| Sound Transmission Class (dB): | 28 |
| U-value (Btu/sq. ft. hr., °F): | .55 |
| Wall Weight (lb./sq. ft.): | 21 |

**GYPSUM PARTITION TILE** (12 in. × 30 in. Face Size) (Trowel Cut Joint)

| | |
|---|---|
| Fire Rating (hr.): | 4 |
| Sound Transmission Class (dB): | 42 |
| U-value (Btu/sq. ft. hr., °F): | N.A. |
| Wall Weight (lb./sq. ft.): | 26 |

**GLASS BLOCK*** (8 in. × 8 in.) (Tooled 2 Sides)

| | |
|---|---|
| Fire Rating (hr.): | – |
| Sound Transmission Class (dB): | 27 |
| U-value (Btu/sq. ft. hr., °F): | .56 |
| Wall Weight (lb./sq. ft.): | 20 |

*6 in. × 6 in. and 12 in. × 12 in. Glass Block are also available.

| in. | | mm |
|---|---|---|
| 2⅛ | = | 53.98 |
| 2¼ | = | 57.15 |
| 4 | = | 100 |
| 6 | = | 150 |
| 8 | = | 200 |
| 12 | = | 300 |
| 16 | = | 400 |

| U-value<br>Btu/ft.² × h. | | k-value<br>W/m² × k |
|---|---|---|
| 0.40 | = | 2.27 |
| 0.43 | = | 2.44 |
| 0.46 | = | 2.61 |
| 0.55 | = | 3.12 |
| 0.56 | = | 3.18 |
| 0.67 | = | 3.80 |

| U-value<br>Btu/ft.² × h. | | k-value<br>W/m² × k |
|---|---|---|
| 0.68 | = | 3.86 |
| 0.76 | = | 4.32 |
| **lb./sq. ft.** | | **kg/m²** |
| 20 | = | 97.65 |
| 21 | = | 102.53 |
| 24 | = | 117.17 |

| lb./sq. ft. | | kg/m² |
|---|---|---|
| 26 | = | 126.94 |
| 27 | = | 131.82 |
| 33 | = | 161.12 |
| 35 | = | 170.88 |
| 38 | = | 185.53 |
| 50 | = | 244.12 |

**FIGURE 4.10-1** Properties of 4 in. (100 mm) walls.

#### 4.9.1.2 Mortar

Mortar for glass block installation is Type S portland cement-lime mortar in accordance with ASTM C270. For use in exterior panels, a waterproofing admixture is included. Colored pigment mortars or colored aggregates are sometimes used.

#### 4.9.1.3 Accessories

Accessories for mortar installations include panel reinforcement (Fig. 4.9-2a), panel anchors (Fig. 4.9-2b), expansion strips, spacers, and aluminum- or steel-channel perimeter chases (Fig. 4.9-2c). These are usually proprietary products of the glass block manufacturers or are recommended by the glass block manufacturers for use with their products. Sealants and backers for such installations are those recommended by the glass block manufacturers.

Accessories for framed systems are provided by or recommended by the system's manufacturer. They include separators, aluminum framing systems, and plastic foam tape.

### 4.9.2 INSTALLATION

The following are general recommendations. The various manufacturers' instructions vary, and they must be followed when installing glass block.

#### 4.9.2.1 Installation with Mortar Joints

The traditional method for installing glass block assemblies is to lay the blocks with both horizontal and vertical mortar joints between the blocks (Fig. 4.9-3). Units may be anchored to other con-

struction at the head and jambs using sheet metal channels, as shown in Figure 4.9-3, but blocks may also be anchored at jambs using either:

- Panel anchors, consisting of perforated steel strips (see Fig. 4.9-2b)
- Dovetail wire ties into dovetail slots in concrete
- Trapezoidal ties welded to steel columns

Blocks may also be anchored at the head using bolts extending down into the joints from above. This method is most often used in interior applications.

Horizontal panel reinforcement, consisting of parallel wires joined by cross wires in a ladder fashion, is installed in the joints at about 16 in. (406 mm) intervals, depending on unit size and thickness. A compressible glass fiber or plastic foam expansion strip is installed between the glass blocks and adjacent construction to permit expansion in the blocks.

General recommendations for mortar joints are similar to those for other types of masonry. Some, but not all, glass block manufacturers recommend that joints be raked out and pointed with mortar. All manufacturers recommend that joints at adjacent materials be finished with a sealant. Some manufacturers recommend that all joints be finished with a sealant.

Joint width is usually $1/4$ in. (6.4 mm), but $3/8$ in. (9.5 mm) joints are sometimes used.

#### 4.9.2.2 Other Installation Methods

*Framed* and *frameless* glass block systems should be installed in accordance with their manufacturer's recommendations. In framed (grid) systems, the glass

block is set into an aluminum grid designed and sized to accept the block. The grid encases each block. A plastic foam filler is placed between the block and the aluminum grid members to permit expansion and contraction. Joints are then sealed with a silicone sealant.

A frameless system has an overall frame like a window frame. There is no frame between the individual glass blocks, as there is in a framed system. Blocks are set into the overall frame. A glass block sealant is applied to the vertical edges of each block, and a vertical separator is installed. When a course of blocks has been installed, a horizontal separator is placed across it to act as a separator and as joint reinforcement. This procedure is repeated until the entire panel is in place. Sealant is applied as the components are installed, leaving a watertight structure.

## 4.10 Properties of Selected Masonry Walls

Figures 4.10-1 through 4.10-6 contain data that permit comparison of several masonry wall properties. The figures show various wall assemblies and thicknesses and, for each, give the wall's fire rating, measured in hours; sound transmission class (STC), measured in decibels; heat transmission, measured by the wall's U-value in Btu/sq ft hr°F (K-value in W/m²hr°C); and weight, measured in psf (kg/m²).

**BRICK, THRU WALL UNITS**
($3\frac{1}{2}$ in. × $11\frac{1}{2}$ in. Face Size)

| | |
|---|---|
| Fire Rating (hr.): | 2+ |
| Sound Transmission Class (dB): | 41 |
| U-value (Btu/sq. ft. hr., °F): | .66 |
| Wall Weight (lb./sq. ft.): | 42 |

| in. | | mm |
|---|---|---|
| $3\frac{1}{2}$ | = | 88.9 |
| 6 | = | 150 |
| $11\frac{1}{2}$ | = | 292.1 |

| U-value Btu/ft.$^2$ × h. | | k-value W/m$^2$ × k |
|---|---|---|
| 0.38 | = | 2.16 |
| 0.66 | = | 3.75 |

**LIGHTWEIGHT CONCRETE BLOCK**

| | |
|---|---|
| Fire Rating (hr.): | 1 |
| Sound Transmission Class (dB): | 45 |
| U-value (Btu/sq. ft. hr., °F): | .38 |
| Wall Weight (lb./sq. ft.): | 26 |

| lb./sq. ft. | | kg/m$^2$ |
|---|---|---|
| 26 | = | 126.94 |
| 42 | = | 205.06 |

**FIGURE 4.10-2** Properties of 6 in. (150 mm) walls.

**4 IN. FACE BRICK AND 4 IN. LIGHTWEIGHT CONCRETE BLOCK**

| | |
|---|---|
| Fire Rating (hr.): | 3+ |
| Sound Transmission Class (dB): | 48 |
| U-value (Btu/sq. ft. hr., °F): | .33 |
| Wall Weight (lb./sq. ft.): | 53 |

**4 IN. COMMON BRICK AND 4 IN. LIGHTWEIGHT CONCRETE BLOCK**

| | |
|---|---|
| Fire Rating (hr.): | 3 |
| Sound Transmission Class (dB): | 48 |
| U-value (Btu/sq. ft. hr., °F): | .33 |
| Wall Weight (lb./sq. ft.): | 53 |

**4 IN. FACE BRICK AND 4 IN. FACE BRICK** (Tooled 2 Sides)

| | |
|---|---|
| Fire Rating (hr.): | 4 |
| Sound Transmission Class (dB): | 49 |
| U-value (Btu/sq. ft. hr., °F): | .54 |
| Wall Weight (lb./sq. ft.): | 62 |

**LIGHTWEIGHT CONCRETE BLOCK** (Tooled 2 Sides)

| | |
|---|---|
| Fire Rating (hr.): | 3 |
| Sound Transmission Class (dB): | 46 |
| U-value (Btu/sq. ft. hr., °F): | .35 |
| Wall Weight (lb./sq. ft.): | 32 |

**BRICK, THRU WALL UNITS**
(3½ in. × 11½ in. Face Size)

| | |
|---|---|
| Fire Rating (hr.): | 3 |
| Sound Transmission Class (dB): | 50 |
| U-value (Btu/sq. ft. hr., °F): | .47 |
| Wall Weight (lb./sq. ft.): | 61 |

**HEAVYWEIGHT CONCRETE "SLUMP" BLOCK** (4 in. × 16 in. Face Size)

| | |
|---|---|
| Fire Rating (hr.): | 3 |
| Sound Transmission Class (dB): | 44 |
| U-value (Btu/sq. ft. hr., °F): | .38 |
| Wall Weight (lb./sq. ft.): | 45 |

| in. | | mm |
|---|---|---|
| 3½ | = | 88.9 |
| 4 | = | 100 |
| 16 | = | 400 |

| U-value Btu/ft.² × h. | | k-value W/m² × k |
|---|---|---|
| 0.33 | = | 1.87 |
| 0.35 | = | 1.99 |
| 0.38 | = | 2.16 |
| 0.47 | = | 2.67 |
| 0.54 | = | 3.07 |

| lb./sq. ft. | | kg/m² |
|---|---|---|
| 32 | = | 156.24 |
| 45 | = | 219.70 |
| 53 | = | 258.77 |
| 61 | = | 279.83 |
| 62 | = | 302.71 |

**FIGURE 4.10-3** Properties of 8 in. (200 mm) walls.

**4 IN. FACE BRICK AND 6 IN. LIGHTWEIGHT CONCRETE BLOCK**

| | |
|---|---|
| Fire Rating (hr.): | 4 |
| Sound Transmission Class (dB): | 50 |
| U-value (Btu/sq. ft. hr., °F): | .32 |
| Wall Weight (lb./sq. ft.): | 62 |

**LIGHTWEIGHT CONCRETE BLOCK** (Tooled 2 Sides)

| | |
|---|---|
| Fire Rating (hr.): | 3 |
| Sound Transmission Class (dB): | 48 |
| U-value (Btu/sq. ft. hr., °F): | .33 |
| Wall Weight (lb./sq. ft.): | 39 |

**4 IN. COMMON BRICK AND 6 IN. LIGHTWEIGHT CONCRETE BLOCK**

| | |
|---|---|
| Fire Rating (hr.): | 4 |
| Sound Transmission Class (dB): | 50 |
| U-value (Btu/sq. ft. hr., °F): | .32 |
| Wall Weight (lb./sq. ft.): | 62 |

**FLUTED (4 FLUTES) HEAVYWEIGHT CONCRETE BLOCK** (Tooled 2 Sides)

| | |
|---|---|
| Fire Rating (hr.): | 3 |
| Sound Transmission Class (dB): | 48 |
| U-value (Btu/sq. ft. hr., °F): | .33 |
| Wall Weight (lb./sq. ft.): | 39 |

| in. | | mm |
|---|---|---|
| 4 | = | 100 |
| 6 | = | 150 |
| 10 | = | 250 |

| U-value Btu/ft.² × h. | | k-value W/m² × k |
|---|---|---|
| 0.32 | = | 1.82 |
| 0.33 | = | 1.87 |
| 0.49 | = | 2.78 |

| lb./sq. ft. | | kg/m² |
|---|---|---|
| 39 | = | 190.41 |
| 61 | = | 297.83 |
| 62 | = | 302.71 |

**FIGURE 4.10-4** Properties of 10 in. (250 mm) walls.

**4 IN. FACE BRICK, 2 IN. AIR SPACE, AND 4 IN. LIGHTWEIGHT CONCRETE BLOCK**

| | |
|---|---|
| Fire Rating (hr.): | 3 |
| Sound Transmission Class (dB): | 48 |
| U-value (Btu/sq. ft. hr., °F): | .27 |
| Wall Weight (lb./sq. ft.): | 53 |

**4 IN. FACE BRICK, 2 IN. URETHANE INSULATION, AND 4 IN. LIGHTWEIGHT CONCRETE BLOCK**

| | |
|---|---|
| Fire Rating (hr.): | 3 |
| Sound Transmission Class (dB): | 49 |
| U-value (Btu/sq. ft. hr., °F): | .05 |
| Wall Weight (lb./sq. ft.): | 54 |

**4 IN. FACE BRICK, 2 IN. AIR SPACE, AND 4 IN. FACE BRICK**

| | |
|---|---|
| Fire Rating (hr.): | 4 |
| Sound Transmission Class (dB): | 51 |
| U-value (Btu/sq. ft. hr., °F): | .37 |
| Wall Weight (lb./sq. ft.): | 64 |

**4 IN. COMMON BRICK, 2 IN. AIR SPACE, AND 4 IN. LIGHTWEIGHT CONCRETE BLOCK**

| | |
|---|---|
| Fire Rating (hr.): | 3 |
| Sound Transmission Class (dB): | 48 |
| U-value (Btu/sq. ft. hr., °F): | .27 |
| Wall Weight (lb./sq. ft.): | 53 |

| in. | | mm |
|---|---|---|
| 2 | = | 50.8 |
| 4 | = | 100 |
| 10 | = | 250 |

| U-value Btu/ft.$^2 \times$ h. | | k-value W/m$^2 \times$ k |
|---|---|---|
| 0.05 | = | 0.28 |
| 0.27 | = | 1.53 |
| 0.37 | = | 2.10 |

| lb./sq. ft. | | kg/m$^2$ |
|---|---|---|
| 53 | = | 258.77 |
| 54 | = | 263.65 |
| 64 | = | 312.48 |

**FIGURE 4.10-5** Properties of 10 in. (250 mm) cavity walls.

**4 IN. FACE BRICK AND 8 IN. LIGHTWEIGHT CONCRETE BLOCK**

| | |
|---|---|
| Fire Rating (hr.): | 4 |
| Sound Transmission Class (dB): | 50+ |
| U-value (Btu/sq. ft. hr., °F): | .29 |
| Wall Weight (lb./sq. ft.): | 72 |

**LIGHTWEIGHT CONCRETE BLOCK**

| | |
|---|---|
| Fire Rating (hr.): | 3–4 |
| Sound Transmission Class (dB): | 49 |
| U-value (Btu/sq. ft. hr., °F): | .31 |
| Wall Weight (lb./sq. ft.): | 47 |

**4 IN. COMMON BRICK AND 8 IN. LIGHTWEIGHT CONCRETE BLOCK**

| | |
|---|---|
| Fire Rating (hr.): | 4 |
| Sound Transmission Class (dB): | 50+ |
| U-value (Btu/sq. ft. hr., °F): | .29 |
| Wall Weight (lb./sq. ft.): | 72 |

**FLUTED (4 FLUTES) HEAVYWEIGHT CONCRETE BLOCK**

| | |
|---|---|
| Fire Rating (hr.): | 3–4 |
| Sound Transmission Class (dB): | 54 |
| U-value (Btu/sq. ft. hr., °F): | .47 |
| Wall Weight (lb./sq. ft.): | 73 |

| in. | | mm |
|---|---|---|
| 4 | = | 100 |
| 8 | = | 200 |
| 12 | = | 300 |

| U-value Btu/ft.$^2 \times$ h. | | k-value W/m$^2 \times$ k |
|---|---|---|
| 0.29 | = | 1.65 |
| 0.31 | = | 1.76 |
| 0.47 | = | 2.67 |

| lb./sq. ft. | | kg/m$^2$ |
|---|---|---|
| 47 | = | 229.47 |
| 72 | = | 351.53 |
| 73 | = | 356.42 |

**FIGURE 4.10-6** Properties of 12 in. (300 mm) walls.

## 4.11 Additional Reading

More information about the subjects discussed in this chapter can be found in the references listed in Section 4.12 and in the following publications.

Ambrose, James. 1997 *Simplified Design of Masonry Structures*. Hoboken NJ: John Wiley & Sons, Inc. (www.wiley.com).

Beall, Christine. 2004. "Proper Masonry Details Can Beat the Mold Menace" *The Construction Specifier* (April): 94 (www.csinet.org).

Borchelt, J. Gregg 2001. "Brick-Related Standards." *The Construction Specifier* (August): 20 (www.csinet.org).

Boyer, David W. 2002. "Chemical or Abrasive Masonry Cleaning." *The Construction Specifier* (September): 44 (www.csinet.org).

Brick Industry Association (BIA). *Technical Notes on Brick Construction*. Reston, VA: BIA (www.bia.org).

Chacon, Mark A. 1999. *Architectural Stone: Fabrication, Installation, and Selection*. Hoboken NJ: John Wiley & Sons, Inc. (www.wiley.com).

Clift, Charles D. 2001. "Limestone Cladding." *The Construction Specifier* (May): 27 (www.csinet.org).

Dairymple, Gerald A. 2005. "Designing for Differential Movement–Clay Brick Veneer Over Wood Frame Construction." *The Construction Specifier* (April): 44 (www.csinet.org).

Filloramo, Richard. 2005. "Buildable Masonry Details." *The Construction Specifier* (April): 26 (www.csinet.org).

Haney, Craig K. 2001. "Restoration of Historic Brick Masonry." *The Construction Specifier* (March): 37 (www.csinet.org).

Heyman, Jacques. 1997. *The Stone Skeleton: Structural Engineering of Masonry Architecture*. New York: Cambridge University Press (www.uk.cambridge.org).

Nadel, Barbara A. 2000. "Designing Successful Marble Projects." *The Construction Specifier* (March): 23 (www.csinet.org).

Solinski, Joseph P. 2002. "Thin Stone Veneer." *The Construction Specifier* (May): 100 (www.csinet.org).

## 4.12 Acknowledgments and References

### ACKNOWLEDGMENTS

We gratefully acknowledge the assistance in preparing this chapter of the following organizations and individuals. We are also indebted to them for permission to use their illustrations when requested and for the use of their publications as references.

A.A. Wire Products Company

American Concrete Institute (ACI) (www.aci-int.org)

American National Standards Institute (ANSI) (www.ansi.org)

American Slate Company (www.americanslate.com)

Ancor Granite Tile Inc. (www.ancor.ca)

ASTM International (ASTM) (www.astm.org)

Athenian Marble (www.athenianmarble.net)

Brick Industry Association (BIA) (www.bia.org)

Buckingham-Virginia Slate Corporation (bvslate.com)

Burlington Natstone Inc. (www.burlingtonstone.co.uk)

Building Research Board of the National Academy of Sciences (NAS) (www4.nationalacademies.org)

Camara Slate, Inc. (www.cqmaraslate.com)

Cold Spring Granite Co. (www.coldspringgranit.com)

Dow Corning Corporation (www.dowcorning.com)

Facing Tile Institute (FTI)

Georgia Marble Company (www.georgiamarble.com)

Greek Trade Commission (www.hepogreektrade.com)

Indiana Limestone Company, Inc. (ILC) (www.indianalimestonecompany.com)

Indiana Limestone Institute of America (ILI) (www.iliai.com)

Marble Institute of America (MIA) (www.marble-institute.com)

National Building Granite Quarries Association, Inc. (NBGQA) (www.nbgqa.com)

National Concrete Masonry Association (NCMA) (www.ncma.org)

Portland Cement Association (PCA) (www.cement.org)

Rock of Ages Corporation (www.rockofages.com)

Solnhofen Natural Stone, Inc. (www.solnhofen.com)

Structural Slate Company (www.structuralslate.com)

U.S. Department of Housing and Urban Development (HUD) (www.hud.gov)

### REFERENCES

We would also like to thank the authors and publishers of the publications in the following list for their contribution to our research for this chapter.

American Concrete Institute (ACI) (www.concrete.org). 2002. ACI 530/530.1, *Building Code Requirements for Masonry Structures and Specs for Masonry Structures and Commentaries*. Farmington Hills, MI: ACI.

American Concrete Institute (ACI) (www.concrete.org), American Society of Civil Engineers (AISC) (www.asce.org), and The Masonry Society (TMS) (www.masonrysociety.org). 1992. ACI 530.1-92/ASCE 6-92/TMS 602.92, with selected ASTM references. *Reference Manual: Specifications for Masonry Structures*. Farmington Hills, MI: ACI, ASCE, TMS.

American Iron and Steel Institute (AISI). Washington, DC (www.steel.org).

———. 2002. CF02-1, "Cold-Formed Steel Framing Design Guide."

———. 2003. CF03-1, "Steel Stud Brick Veneer Design Guide."

———. 2001. SG02-1 and SG02-2, "North American Specification for the Design of Cold-Formed Steel Structural Members and the Commentary."

———. 2004. SG02-3, "Standard for Cold-Formed Steel Framing—General Provisions."

———. 2004. SG02-4, "Standard for Cold-Formed Steel Framing—Prescriptive Method for one and Two Family Dwellings."

———. 2002. SG03, *Cold-Formed Steel Design Manual*.

American National Standards Institute (ANSI). New York (www.ansi.org). Standards:

————. 1988. A10.20, "Safety Requirements, Ceramic Tile, Terrazzo, Marble."

————. 1999. A108.1, "Installation of Ceramic Tile."

*Architectural Graphic Standards. See* Ramsey et al. (2000).

*Architectural Technology.* 1984. "Rediscovering Terra Cotta: A Glowing Recommendation." *Architectural Technology*, Spring: 12–13.

————. 1986. "Technical Tips: Slope the Sills." *Architectural Technology*, May/June: 66.

————. 1986. "Technical Tips: Tension Bars Save Masonry." *Architectural Technology*, May/June: 66–67.

Arcom Master Systems. Salt Lake City, UT (www.arcomnet.com). *Masterspec®.* Basic Sections:

04720, "Cast Stone," 11/02 ed.

04810, "Unit Masonry Assemblies," 8/02 ed.

04815, "Glass Unit Masonry Assemblies," 8/03 ed.

04851, "Dimension Stone Cladding," 2/04 ed.

04860, "Stone Veneer Assemblies," 11/99 ed.

04900, "Masonry Restoration and Cleaning," 8/00 ed.

04901, " Clay Masonry Restoration and Cleaning," 8/00 ed.

04902, "Stone Restoration and Cleaning," 8/00 ed.

07620, "Sheet Metal Flashing and Trim," 11/00 ed.

Beall, Christine. 1988. "Controlling Moisture Movement in Masonry Walls." *The Construction Specifier* (June): 36–47 (www.csinet.org).

————. 1988. "The Types of Masonry Mortar." *The Magazine of Masonry Construction*, 1(4): 164–166.

————. 1989. "Weatherproof Masonry Walls." *Architecture* (March): 133–136 (www.architecturemag.com).

————. 1989. "The Many Differences Between Concrete and Masonry Technology." *Architecture* (April): 123–124 (www.architecturemag.com).

————. 1989. "Preventing Problems with Thin Brick Veneers." *The Construction Specifier* (September): 68–78 (www.csinet.org).

————. 2004. *Masonry Design and Detailing.* New York: McGraw-Hill (www.mcgraw-hill.com).

ASTM International (ASTM) (www.astm.org). 1998. *Annual Book of ASTM Standards, Vol. 4.01, Construction, (Cement, Lime, Gypsum).* West Conshohocken, Pa: ASTM.

ASTM International (ASTM). West Conshohocken, PA (www.astm.org). Standards:

A36/A 36M, "Standard Specification for Carbon Structural Steel."

A82, "Standard Specification for Steel Wire, Plain, for Concrete Reinforcement."

A116, "Standard Specification for Metallic-Coated Steel Woven Wire Fence Fabric."

A123/A 123M, "Standard Specification for Zinc (Hot-Dip Galvanized) Coatings on Iron and Steel Products."

A153/A 153M, "Standard Specification for Zinc Coating (Hot-Dip) on Iron and Steel Hardware."

A167, "Standard Specification for Stainless and Heat-Resisting Chromium-Nickel Steel Plate, Sheet, and Strip."

A185, "Standard Specification for Steel Welded Wire Reinforcement, Plain, for Concrete."

A496, "Standard Specification for Steel Wire, Deformed, for Concrete Reinforcement."

A497, "Standard Specification for Welded Wire, Reinforcement, Deformed, for Concrete."

A510, "Standard Specification for General Requirements for Wire Rods and Coarse Round Wire, Carbon Steel."

A580/A 580M, "Standard Specification for Stainless Steel Wire."

A615/A 615M, "Standard Specification for Deformed and Plain Billet-Steel Bars for Concrete Reinforcement."

A635/A 635M, "Standard Specification for Steel, Sheet and Strip, Heavy-Thickness Coils, Hot-Rolled, Carbon, Structural, High-Strength Low-Alloy, and High-Strength Low-Alloy with Improved Formability, General Requirements for."

A653/A653M, "Standard Specification for Steel Sheet, Zinc-Coated (Galvanized) or Zinc-Iron Alloy-Coated (Galvannealed) by the Hot-Dip Process."

A666, "Standard Specification for Annealed or Cold-Rolled Austenitic Stainless Steel Sheet, Strip, Plate, and Flat Bar."

A775/A775M, "Standard Specification for Epoxy-Coated Steel Reinforcing Bars."

B101, "Standard Specification for Lead-Coated Copper Sheet and Strip for Building Construction."

B152, "Standard Specification for Copper Sheet, Strip, Plate, and Rolled Bar."

B227, "Standard Specification for Hard-Drawn Copper-Clad Steel Wire."

C5, "Standard Specification for Quicklime for Structural Purposes."

C33, "Standard Specifications for Concrete Aggregates."

C34, "Standard Specification for Structural Clay Load-Bearing Wall Tile."

C55, "Standard Specification for Concrete Brick."

C56, "Standard Specification for Structural Clay Non-Load-Bearing Tile."

C62, "Standard Specification for Building Brick (Solid Masonry Units Made from Clay or Shale)."

C67, "Standard Test Method for Sampling and Testing Brick and Structural Clay Tile."

C90, "Standard Specification for Load-Bearing Concrete Masonry Units."

C91, "Standard Specification for Masonry Cement."

C97, "Standard Test Method for Absorption and Bulk Specific Gravity of Dimension Stone."

C99, "Standard Test Method for Modulus of Rupture of Dimension Stone."

C119, "Standard Terminology Relating to Dimension Stone."

C126, "Standard Specification for Ceramic Glazed Structural Clay Facing Tile, Facing Brick, and Solid Masonry Units."

C129, "Standard Specification for Nonloadbearing Concrete Masonry Units."

C144, "Standard Specification for Aggregate for Masonry Mortar."

C150, "Standard Specification for Portland Cement."

C170, Test Method for Compressive Strength of Dimension Stone."

C206, "Standard Specification for Finishing Hydrated Lime."

C207, "Standard Specification for Hydrated Lime for Masonry Purposes."

C212, "Standard Specification for Structural Clay Facing Tile."

C216, "Standard Specification for Facing Brick (Solid Masonry Units Made from Clay or Shale)."

C260, "Standard Specification for Air-Entraining Admixtures for Concrete."

C270, "Standard Specification for Mortar for Unit Masonry."

C331, "Standard Specification for Lightweight Aggregates for Concrete Masonry Units."

C404, "Standard Specification for Aggregate for Masonry Grout."

C426, "Standard Test Method for Linear Drying Shrinkage of Concrete Masonry Units."

C476, "Standard Specification for Grout for Masonry."

C503, "Standard Specification for Marble Dimension Stone (Exterior)."

C568, "Standard Specification for Limestone Dimension Stone."

C595, "Standard Specification for Blended Hydraulic Cements."

C615, "Standard Specification for Granite Dimension Stone."

C618, "Standard Specification for Coal Fly Ash and Raw or Calcined Natural Pozzolan for Use in Concrete."

C629, "Standard Specification for Slate Dimension Stone."

C645, "Standard Specifications for Nonstructural Steel Framing Members."

C652, "Standard Specification for Hollow Brick (Hollow Masonry Units Made from Clay or Shale)."

C880, "Standard Test Method for Flexural Strength of Dimension Stone."

C902, "Standard Specification for Pedestrian and Light Traffic Paving Brick."

C955, "Standard Specifications for Load-Bearing (Transverse and Axial) Steel Studs, Runners (Track), and Bracing or Bridging, for Screw Application of Gypsum Panel Products and Metal Plaster Bases."

C1007, "Standard Specification for Installation of Load-Bearing (Transverse and Axial) Steel Studs and Related Accessories."

C1242, "Standard Guide for Design, Selection, and Installation of Dimension Stone Anchoring Systems."

D98, "Standard Specification for Calcium Chloride."

E90, "Standard Test Method for Laboratory Measurement of Airborne-Sound Transmission Loss of Building Partitions and Elements."

E96, "Standard Test Methods for Water Vapor Transmission of Materials."

E119, "Standard Methods for Fire Tests of Building Construction and Materials."

E413, "Classification for Rating Sound Insulation."

E514, "Standard Test Method for Water Penetration and Leakage Through Masonry."

Borchelt, J. ed. 1982. *STP 778—Masonry: Materials, Properties, and Performance.* West Conshohocken, PA: ASTM (www.astm.org).

Brick Industry Association (BIA). *Technical Notes on Brick Construction.* Reston, VA: BIA (www.bia.org).

Catani, Mario J. 1985. "Protection of Embedded Steel in Masonry." *The Construction Specifier* (January): 62–68 (www.csinet.org).

———. 1987. "Connecting Masonry with Ties and Anchors." *The Construction Specifier* (August): 108–113 (www.csinet.org).

———. 1989. "Setting New Standards for Masonry." *The Construction Specifier* (September): 108–114 (www.csinet.org).

Clifton, J. R., ed. 1986. *STP 935—Cleaning Stone and Masonry.* West Conshohocken, PA: ASTM (www.astm.org).

Grimm, Clayford T. 1985. "Clay Brick/Concrete Masonry Wall Design Checklist." *The Construction Specifier* (August): 67–69 (www.csinet.org).

———. 1989. "What They Say About Brick Veneer Over Steel Studs." *The Construction Specifier* (September): 76–77 (www.csinet.org).

Grogan, John C. 1983. "Water Permeance in Brick Masonry." *The Construction Specifier* (December): 36–38 (www.csinet.org).

Grogan, John C. and John T. Conway, eds. 1985. *STP 871—Masonry Research Application and Problems.* West Conshohocken, PA: ASTM (www.astm.org).

Harris, Harry A., ed. 1988. *STP 992—Masonry: Materials, Design, Construction, and Maintenance."* West

Conshohocken, PA: ASTM (www.astm.org).

Indiana Limestone Institute of America, Inc. (ILI). Bedford, IN (www.iliai.com).

———. *The Contractors Handbook on Indiana Limestone*, 7th ed.

———. "The Finishing Touch."

———. *Indiana Limestone Handbook*, 21st ed.

———. "Specifications for Indiana Limestone."

———. "ILI Technote on Safety Factors."

Marble Institute of America, Inc. 2006. *Dimension Stone Design Manual VI.* Cleveland: Marble Institute of America (www.marble-institute.com).

Masonry Society, The (TMS). 2003. *Masonry Designers' Guide.* Boulder, CO: TMS (www.masonrysociety.org).

National Building Granite Quarries Association. *Specifications for Architectural Granite.* Barre, VT: National Building Granite Quarries Association (www.nbgqa.com).

National Concrete Masonry Association (NCMA). 2005. (wwwncma.org). Herndon, VA.

———. "The Structural Role of Joint Reinforcement in Concrete Masonry."

———. 2002. "Building Code Requirements Masonry Structures."

———. 2002. *Annotated & Construction Details for CM Design.*

Panarese, W., S. Kosmatka, and F. Randall. 1991. *Concrete Masonry Handbook for Architects, Engineers, and Builders*, 5th ed. Skokie, IL: Portland Cement Association (www.cement.org).

Portland Cement Association (PCA). 1991. *Water Penetration Tests of Masonry Walls.* Skokie, IL: PCA. (www.cement.org)

Raeber, John A. 1991. "Brick Veneer in Earthquake Country." *The Construction Specifier* (January): 23–24 (www.csinet.org)

Ramsey, Charles George, Harold Reeve Sleeper, and John Ray Hoke, Jr. 2000. *Architectural Graphic Standards*, 10th ed. Hoboken, NJ: John Wiley & Sons, Inc. (www.wiley.com).

Rhault, Robert. 1991. "Masonry Cleaning." *The Construction Specifier* (January): 53–61 (www.csinet.org).

Smith, Baird M. 1984. *Moisture Problems in Historic Masonry Walls: Di-*

agnosis and Treatment. Technical Report. Washington, DC: Technical Preservation Services, U.S. Department of the Interior (www.doi.gov).

Specialty Steel Industry of North America (SSINA). Stainless Steel for: Wall Ties, Stone Anchors, Masonry Fastening Systems. Washington, DC: SSINA (www.ssina.com).

Steel Stud Manufacturers Association (SSMA). Chicago (www.ssma.com). 2001.

———. Product Technical Information.

———. Cold Formed Steel Details.

Tiller, deTeel Patterson. 1979. The Preservation of Historic Glazed Architectural Terra Cotta. Preservation Brief #7: Washington, DC: U.S. Department of the Interior, National Park Service, Preservation Assistance Division, Technical Preservation Services (www.nps.gov).

Underground Space Center. 1988. Building Foundation Design Handbook. Minneapolis: Underground Space Center.

CHAPTER **5**

# Metals

# Introduction

Metals have been an essential part of human development since our beginnings. The materials for very early stone structures were cut and shaped laboriously using tools made of flint or other hard stone. The first known metals were those found in nature, such as gold and silver. Later, other metals, probably initially discovered accidentally, were obtained by melting ores. The first relatively hard metals produced were mostly copper alloys, such as brass and bronze. These metals were vastly superior to flint and other stone for some purposes. They are still used today in building construction. Copper is used in roofing, flashing, piping, wiring, and in the manufacture of equipment. Copper alloys are used to manufacture doors and their frames, windows, door and window hardware, cladding panels, plumbing fixtures, and for many decorative purposes.

The knowledge that metals could be obtained by heating ores naturally led to a search for other metals. The discovery of iron came fairly early because the process of making it was relatively easy. Iron and its modern derivative, steel, are used extensively in buildings today.

It is impossible to construct a modern building containing no metal products. Products made from iron, steel, aluminum, copper, bronze, lead, alloys of these metals, and other metals are contained in a great many building components, ranging from those forming the underlying structure to those in the shell. Masonry structures contain metal reinforcements, anchors, and fasteners. Concrete structures are reinforced with steel. Wood framing is held together with metal nails, lag bolts, and screws. Finishes are hung on metal furring and framing.

In addition to the use of steel and iron in structural systems, ferrous metals are used in many other building products. For example, ferrous metals used in sitework applications include fencing components, barricades, drains, manhole covers, guard rails, corrugated pipe for drainage, and many others.

Ferrous metal components are used in roofing, siding, cladding, and flashing. Both structural and nonstructural ferrous standing seam metal roofing (see Section 7.10) have a long history of use in several styles. These products offer a broad choice of finishes and coatings for durability, texture, and color.

Aluminum, which was not produced until the nineteenth century, is another common metal in most of today's buildings. It is used to make cladding panels, storefronts, curtain walls, doors and their frames, windows, siding, roofing, flashing, wiring, piping, railings, hardware, and many other products.

Both steel and aluminum siding, in traditional clapboard designs as well as vertical and other patterns, are increasingly popular products (see Section 7.7).

Wall panel systems (cladding), often in combination with lightweight steel framing, are used in many types of buildings today.

Stainless steel is used for flashing, and carbon steel sheets are used in composite flashing (see Section 7.9).

Many types and styles of metal exterior and interior doors are available. Steel doors (see Section 8.1) provide fire safety, security, durability, and dimensional stability. Modern exterior swinging steel doors contain an insulating core surrounded by steel skins and sides. Designs range from smooth flush faces to variations with glazing, moldings, textured surfaces, and vinyl laminations.

Metal windows (see Section 8.6) offer a complete range of designs for residential, commercial, and institutional uses.

Steel products used in interior finishes include non-load-bearing studs and furring components (see Section 9.1), as well as acoustical ceilings and wall panels (see Sections 9.5 and 9.6). Non-load-bearing studs and furring members provide noncombustible, dimensionally stable framing and furring for gypsum wallboard, plaster, and other interior surfaces.

Steel acoustical ceilings are found in many types of buildings, sometimes as an integral part of a steel deck floor and ceiling system.

Metals are used to make ornamental building components, such as railings and handrails, panels, and ornamentation.

The following are a few of the many other steel products used in buildings:

- Steel pan (concrete filled) stairs
- Spiral stairs
- Fixed ladders
- Prefabricated fireplaces and flues
- Laboratory, kitchen, and other cabinets
- Unit kitchens
- Roof accessories (skylights, stacks, ventilators, roof hatches, etc.)
- Loose lintels

Aluminum and steel are also used in the following products:

- Gratings and stair treads
- Rainware (flashing, gutters, and downspouts)
- Roof accessories (skylights, stacks, ventilators, roof hatches, etc.)
- Gratings
- Floor plates
- Sidewalk hatches
- Access doors
- Loose lintels
- Various castings

Iron and steel are used extensively in plumbing fixtures and piping; hot water and steam piping; radiators, valves, and accessories; and in heating and cooling ductwork.

Plumbing ware includes both cast iron and sheet steel products, such as tubs, shower and tub enclosures, toilets, bidets, and sinks. These are often finished in porcelain enamel. Stainless steel is widely used for kitchen, laboratory, and bar sinks.

Modern mechanical and electrical systems cannot function without metals. Sheet steel ductwork, galvanized for corrosion resistance, is part of most heating and cooling systems that employ moving air. Such systems generally have fans, vents, louvers, and other sheet steel accessories. (See Chapter 17 for additional discussion.)

Electrical systems (see Chapter 18) have many steel and iron elements, including panel and junction boxes, conduit, outlet boxes and outlets, and many others.

# Sustainability Considerations

The mining of raw materials and the manufacture of metal products consume large amounts of energy and contribute to disruption to and damage of the environment. Therefore, sustainability is a major consideration when deciding whether to use steel, aluminum, and other metals in any particular circumstance.

## Steel

It takes more than 19,000 Btus to produce 1 lb (0.45 kg) of steel from ore. The process consumes about 3200 lb (1451.5 kg) of ore, 300 lb (136 kg) of limestone, and 900 lb (408.2 kg) of coke for each ton of steel produced. The coke, in turn, is made from coal, with its own cost in energy of production. Some of the alloying metals used to produce the various types of steel are running low. High-grade iron ore is also being rapidly depleted. This depletion increases the cost of obtaining high-quality raw materials.

The manufacture of steel products produces large quantities of pollutants that must be captured to prevent their release into the environment. The production of a ton of steel releases more than 4500 lb (2041.2 kg) of gaseous pollutants, and leaves behind 50 lb (22.67 kg) of dust and 600 lb (272.16 kg) of slag that must be disposed of.

On the plus side, steel is one of the world's most recycled materials. According to the Steel Recycling Institute (SRI), in 2003 almost 71% of all steel was recycled. In construction-related steel, 96% of beams and plates and 60% of rebars were recycled. This amounts to more than 60 million tons (66.14 metric tons) of scrap a year. This recycling reduces the amount of energy used to produce a ton of steel by 75% when compared to what would be required if no scrap were used. In fact, more steel is recycled than paper, aluminum, glass, and plastic combined. The recycling of steel scrap also saves landfill space and natural resources.

Unfortunately, however, even with the current extensive reuse of steel to make new steel products, the depletion of quality raw materials will require even more recycling and development of different production methods for steel products if the current rate of use of steel is to continue.

Steel is strong and durable and permits the development of relatively lightweight structures. These qualities have long led to the use of steel to construct tall buildings. In fact, without steel, there would be no really tall buildings. There is a current trend toward using lightweight steel structures in residences and other small buildings. In determining the advantages or disadvantages of using light-gauge steel framing instead of the traditional wood framing, it is necessary to study the relative environmental cost of wood verses steel framing. The same applies to producing steel for lightweight structures that were discussed earlier in this section for steel in general. There are two major disadvantages of using lightweight steel framing instead of wood framing. The first is that wood is a renewable resource; steel is not. The second is the high thermal conductivity of the metal framing when compared with that of wood. This deficiency can be partly corrected in the design of the structure and the judicious use of foam insulation. Unfortunately, the fasteners used are metal as well and conduct heat. Refer to Chapter 6 for a discussion of sustainability considerations relative to wood and wood framing.

Steel framing enjoys some advantages over wood framing, but they are mostly related to cost and ease of construction and less to environmental concerns. Steel framing is easier to handle on the job and easier to erect. Using steel framing results in less scrap to dispose of on a job site. It does present some hazards to personnel, such as sharp edges and more difficult and less safe cutting procedures.

## Aluminum

Most aluminum is used in the United States in three major industries: 31% in the transportation industry, 23% for packaging, and 14% in construction.

To produce primary aluminum, bauxite ore is refined in an electrical process. The cost of the electricity used is more than $2 billion year, which is about one-third of the total production cost of primary aluminum. The industry is also a large consumer of natural gas. The production of aluminum products consumes 548 trillion Btus (548.17 trillion kJ) annually. Of this, 507.16 trillion Btus (535.08 trillion kJ) are used to produce 8 billion lb (3.63 billion kg) of aluminum made from ore and 41.3 trillion Btus for 8.2 billion lb (3.72 billion kg) for recycled secondary uses. The energy used to produce a pound of aluminum is about six time that needed to produce a pound of steel. Therefore, energy efficiency and energy management must be prime objectives for the aluminum industry.

The aluminum industry has been a leader in conservation of natural resources. Of the 10.69 million metric tons of aluminum produced in 2001, 33% was recycled aluminum. Recycling of aluminum saves almost 95% of the energy needed to produce aluminum from bauxite ore and removes about 95% of the emissions associated with making new aluminum from ore. Recycling also saves. More than two-thirds of all of the aluminum ever produced is still in use through continuous aluminum recycling.

Aluminum is a high conductor of heat. Therefore, aluminum components that are exposed to exterior conditions must be separated from interior components by thermal-break materials. Refer to Chapters 7 and 8 for additional discussion of this requirement.

## Metal Finishes

Factory-applied finishing materials that produce airborne pollutants offer the same hazards as those discussed in Chapter 9 for similar field-applied finishes. However, pollutants released during application of liquid finishes in the shop or factory are easier to contain and eliminate than those from similar materials applied in the field. Such application in the shop or factory may more easily be done inside an enclosure constructed for the purpose. Powder coatings, which produce no airborne pollutants, are generally better environmentally than brush- or spray-applied materials.

# Applicable *MasterFormat* Sections

The following *MasterFormat* 2004 Level 2 sections are applicable to this chapter.

05 12 00  Structural Steel Framing
05 14 00  Structural Aluminum Framing
05 21 00  Steel Joist Framing
05 31 00  Steel Decking
05 33 00  Aluminum Decking
05 34 00  Acoustical Metal Decking

05 35 00  Raceway Decking Assemblies
05 36 00  Composite Metal Decking
05 41 00  Structural Metal Stud Framing
05 42 00  Cold-Formed Metal Joist Framing
05 44 00  Cold-Formed Metal Trusses
05 45 00  Metal Support Assemblies
05 51 00  Metal Stairs

05 52 00  Metal Railings
05 53 00  Metal Gratings
05 54 00  Metal Floor Plates
05 55 00  Metal Stair Treads and Nosings
05 56 00  Metal Castings
05 58 00  Formed Metal Fabrications
05 59 00  Metal Specialties

## 5.1 Iron and Steel Materials and Products

The chief ingredient of the metals known as iron and steel is the chemical element iron (*ferrum*), from which iron-based alloys get the generic name *ferrous metals*. Although about 5% of the earth's crust is composed of iron, it is rarely found in its metallic form in nature. It is almost always combined with oxygen (iron oxide) and sulfur silicon and is mixed with other minerals. It is found in rock, gravel, sand, clay, and mud. In this form it is mined as iron ore, which is the chief source of iron. Iron- and steelmaking involve extracting iron from ore, combining it with carbon and other elements, and then forming it into shapes.

Pure iron is soft, ductile, easily shaped, and relatively weak. However, *wrought iron*, which is nearly pure, is strong enough for many construction purposes. In fact, it has been used for many bridges and buildings and as the structural frame for the Eiffel Tower. The many irons and steels used in modern construction and manufacture get their strength and hardness mainly from the element carbon. Most steels have the best combination of properties when carbon is present in amounts less than 1.2%. Ferrous metals containing substantially larger amounts of carbon become so hard and brittle that they cannot be readily shaped by either hot or cold working methods. These metals are formed by being cast into molds and are referred to as *cast iron*.

### 5.1.1 HISTORY

Iron was discovered at different times in different parts of the world. Iron implements that are more than 4000 years old have been found in Egypt; others in China are more than 2700 years old. By the time of Rameses II, who reigned from 1292 to 1225 B.C., iron was being widely used in Egypt. In the following 1000 years, the use of iron spread throughout most of the known world. Iron was not used extensively in western Europe, however, until around 100 B.C.

Fairly high-quality steel was made in ancient Syria and in Toledo, in what is now Spain, but the methods used were lost, and the steelmaking processes used between the beginning of the Dark Ages and the 1730s could not control the impurities and carbon content sufficiently to produce metal with the properties of modern steel. The so-called blister steel they produced was nothing more than carburized wrought iron. Benjamin Huntsman changed that situation in the 1730s with his invention of the crucible process, which permits the melting of blister steel into a homogeneous mass, thus producing steel in the modern sense.

Huntsman's crucible process, which produced steel of very high quality, was quite labor-intensive and yielded only small quantities of steel with each charging. In addition, modern rolling, drawing, and heat-treating processes that improve the mechanical properties of steel were unknown. Therefore, the earliest ferrous metals used in construction were essentially wrought iron and cast iron, shaped by hand working or by casting of hot metal into molds.

The first successful North American ironworks was built in Saugus, Massachusetts, in the late 1640s. Its earliest products were cast iron cooking pots and wrought iron nails. Nails were so expensive that old buildings were often burned so that the nails could be recovered.

The first high-tonnage steelmaking process was discovered almost simultaneously by an American, William Kelly, and an Englishman, Sir Henry Bessemer. The Civil War interrupted Kelly's work and slowed the development of commercial steelmaking here until after 1864. This gave Bessemer time to develop the process and make it practical. Therefore, the process and the furnace used in it were given his name. In the Bessemer (pneumatic) process, air was forced through molten iron contained in a pear-shaped vessel. As the air rose through the iron, it burned out impurities. The addition of manganese caused chemical changes which converted the iron to steel.

After 1864, the Bessemer process and the discovery of abundant iron ore deposits in the Lake Superior region made the United States a world leader in steel production. The modern revolution in construction, industry, commerce, transportation, and communications is due in large part to the economical production of iron and steel.

The name *steel* covers a variety of ferrous metals of different chemical compositions and properties. These properties are affected by the alloying elements included, the production processes used, and the methods of manufacture.

### 5.1.2 RAW MATERIALS

In this section, the term *raw material* is used to designate all materials used to make iron and steel up to the point of producing metal ingots used for product manufacture.

The general steps in the mining and processing of raw materials, smelting them into pig iron, and refining that into steel ingots are illustrated in Figure 5.1-1. Iron- and steelmaking starts with the sep-

**FIGURE 5.1-1** Flowchart of operations in the production of steel ingot. (Courtesy American Iron and Steel Institute.)

aration of the metal iron from its oxide, which is the way iron appears in iron ores. This is done by heating iron ore while it is in contact with coke, a form of carbon. The carbon then combines with both the oxygen present and the iron in the ore, forming a volatile gas and a carbon-rich crude iron called *pig iron.* The term *pig iron* comes from the shape of the trenches into which the molten iron from early furnaces ran. The iron was poured into a main trench and ran along it into smaller connected trenches where it hardened into ingots. These smaller trenches, lined up against the main trench, looked to some like baby pigs lined up against their mother for suckling.

The primary raw materials used to make pig iron are iron ore, fuel (coke), air, flux, and refractories. In steelmaking, pig iron and metal scrap replace iron ore as the primary source of iron, and electricity may replace fossil fuel as a source of heat. In some processes, gaseous oxygen either supplements or completely replaces ordinary combustion air. The emphasis in steelmaking is to remove some or most of the carbon and other impurities from pig iron.

### 5.1.2.1 Mining and Processing

The primary raw materials—iron ores, fuels, fluxes, and refractories—are often mined in areas that are far from iron- and steelmaking plants. To reduce shipping costs, raw materials are sometimes processed near the mines to increase their yield and reduce their bulk. They are then shipped by rail and water (Fig. 5.1-2) to steel mills or plants for direct use or further processing.

Mills and plants are typically located along navigable water routes, which permits low-cost transportation of bulk materials and provides an abundant supply of the fresh water that is essential in steelmaking.

### IRON ORES

Iron ore is mined either by underground or open pit methods (Fig. 5.1-3). The iron ore of the Lake Superior region dominates steel production; other deposits are mined in Alabama, New Jersey, New York, California, and a few other states. Iron ore is also imported, with Canada providing more than half of all imports. The most common iron minerals are

**FIGURE 5.1-2** Ore and limestone are often delivered to a steel mill (background) by boat. (Courtesy Republic Steel Corporation.)

**FIGURE 5.1-3** Typical terraced effect of open pit ore mining. (Courtesy American Iron and Steel Institute.)

hematite ($FE_2O_3$), siderite ($FeCo_3$), and pyrite ($FeS$), which have iron contents ranging as high as 70%. Limonite may be considered as magnetite plus water; *taconite* and *jasper* are names referring to a wide variety of iron-bearing rocks with iron contents ranging between 22% and 30%.

Most ore mined in the Lake Superior region is taconite, which lies close to the surface but is so hard that superheated flames must be used to drill holes for explosive charges. After it has been blasted out of the ground, taconite is processed by *beneficiation*, a treatment that increases its iron content per ton of raw material. Ores with iron contents near 50% are usually shipped as is. Ores like taconite, with a 23% to 25% iron content, can be improved to nearly 66% iron content by beneficiation. Richer ores and beneficiated taconite can be further improved by direct reduction to an iron content of more than 90%.

***Beneficiation*** The beneficiation of ores generally consists of (1) grinding and concentrating them to increase the iron content and (2) *agglomeration*, which increases the ore's particles to a size suitable for blast furnace use. Two forms of agglomeration are pelletizing, used for jasper and taconite, and sintering, used to recycle iron ore dust.

*Pelletizing* by both magnetic separation and concentration is usually done near mine sites to avoid transporting large tonnages of non-iron-bearing minerals. In one pelletizing process, magnetic taconite is crushed into a fine powder with the consistency of flour. The powder is passed through a magnetic separator which removes the iron particles. The iron powder then goes to a balling drum, where it is mixed with coal dust and a binder material to form pellets with 60% iron content. The pellets, each about 1 in. (25.4 mm) in diameter, are baked into a hard finish in a kiln or furnace.

In another pelletizing process, high-grade nonmagnetic ores are crushed and then concentrated in flotation cells. Flotation cells contain liquids in which the heavier iron-bearing particles sink to the bottom of the cell while the lighter particles remain suspended and are skimmed away. After filtering, the iron-bearing particles are formed into pellets and baked hard.

*Sintering* is a process by which iron ore dust, reclaimed from blast furnace gases and other manufacturing processes, is fused with coke and fluxes into a clinker-like product containing 52% to 65% iron.

***Direct Reduction*** Direct reduction includes various processes in which ore is concentrated to a high iron content comparable to that of molten iron from a blast furnace. Some ores are made into pellets, others into briquettes. Direct reduced ores with over 90% iron contents can bypass traditional blast furnace processing and be charged directly into electric steelmaking and basic oxygen furnaces.

## FUELS

Of the three main fuels used in the iron and steel industry—coal, oil, and natural gas—coal supplies more than 65% of the industry's heat and energy requirements.

The chief fuel used in a blast furnace is coke, which is produced in coking ovens from selected types of bituminous coal (Fig. 5.1-4). Most coal of coking quality is mined in West Virginia, Pennsylvania, Kentucky, and Alabama. In addition to furnishing heat and carbon, coke acts as a reducing element to separate iron from its oxide in the ores to make pig iron.

## FLUXES

*Fluxes* are minerals that have an affinity for the impurities in iron ore or pig iron. They combine with these impurities and separate them from the molten metal in a furnace by forming a liquid slag. Some fluxes are used mainly to separate the impurities, others to make the slag more fluid, thus floating it to the surface of the metal sooner. Depending on the affinity of a flux to combine chemically with certain impurities, it is classed as either *basic* or *acid*. Limestone and dolomite are examples of basic fluxes; sand, gravel, and quartz rock are acid fluxes. Fluorspar is a neutral flux used mainly to make the slag more fluid.

## REFRACTORIES

*Refractories* are nonmetallic materials with superior heat and abrasion resistance used as linings of steelmaking furnaces, flues, and vessels. Like fluxes, refractories may be *acid* or *basic*, depending on their chemical reaction with impurities. Theoretically, acid and basic refractories must be used with like fluxes to minimize the chemical reaction between them, but in practice this rule is not always followed.

A common acid refractory material is *ganister*, a form of quartzite rock obtained from quarries and mines in Pennsylvania, Wisconsin, Alabama, Utah, and California. The most common basic refractory material is *magnesia*, which is obtained from the mineral magnesite and from sea water (Fig. 5.1-5).

## ALLOY ORES

The chemistry of steel requires many elements for alloying purposes, such as manganese, silicon, nickel, chromium,

**FIGURE 5.1-4** Hot coke is forced from one of a battery of slot-shaped ovens into a quenching car for cooling with water. (Courtesy Bethlehem Steel Corporation.)

**FIGURE 5.1-5** Magnesia is extracted from seawater or brines for use as a basic refractory in steel furnaces. (Courtesy American Iron and Steel Institute.)

and molybdenum. Although the United States is a major steel producer, it depends on imports to furnish many of these elements. Most alloys are imported as ores and processed into useful forms called *ferroalloys*.

*Manganese*  All steel contains manganese, both as a scavenger to remove oxygen from molten metal and as an alloying element to improve hardness and wear resistance. An average of 12 lb (5.44 kg) of manganese goes into each ton (907 kg) of steel.

*Chromium*  Added in amounts of 1% to 5%, chromium provides stainless steels with resistance to heat, rust, and corrosion. To other alloy steels and cast iron, it provides hardness and wear resistance. Chromium is also important as a coating metal.

*Nickel*  Cryogenic steels (those produced at low temperatures) use nickel to increase their hardness. Steels containing nickel are especially suitable for case hardening, the process of surface hardening by absorption of carbon at high temperatures.

*Silicon*  In steelmaking, silicon serves as a deoxidizer and as an alloying element. When used in large percentages, silicon enables cast irons to withstand highly corrosive acids.

*Tungsten*  Tungsten provides heat resistance and is used in cutting tools of high-speed steel and to make tungsten carbide, an abrasion-resistant material almost as hard as diamond. It is also used as a welded-on deposit on parts exposed to extreme wear.

*Molybdenum*  One of the more versatile alloying elements, molybdenum increases the hardness and corrosion resistance of steels.

*Vanadium*  Vanadium is often used in conjunction with other alloying elements: with chromium in steel springs, with manganese in special plates and other structural forms, and with molybdenum in certain high-temperature steels. Small amounts (often measured in hundreds of a percent) provide the desired effect in construction steels.

*Boron*  Known widely as a cleansing agent, boron is used as a hardening agent in steel. Minute quantities are added to obtain the desired effect, but these must be protected by other deoxidizers or must be added to a completely deoxidized molten metal bath.

### AIR AND OXYGEN

Air and oxygen can be considered as much raw materials as the specially prepared iron ore, coke, and lime consumed in steel production. Air must be forced into furnaces, heated or cooled, and cleaned to control air pollution. Oxygen is a product of special plants that also produce nitrogen and argon from ordinary air. It takes about 4 tons (3630 kg) of air to make 1 ton (970 kg) of iron.

### WATER

Each ton (907 kg) of raw steel produced requires 35,000 gal (132 489 L) of water, mostly to cool closed systems. Often, this water must be cleaned both before and after use.

### 5.1.2.2 Iron Making

*Smelting* is a metallurgical operation by which metal ore is heated to separate the metal in it from impurities with which the metal may be chemically combined or physically mixed. The separation of chemically combined iron and oxygen is called *reduction*. Pig iron is made in a *blast furnace* by smelting, which extracts the iron by reduction of its oxide and by physically separating it from other impurities.

### BLAST FURNACE

A *blast furnace* is a cylindrical structure, approximately 120 ft (36.58 m) tall, lined with refractory brick and encased in a

steel shell (Fig. 5.1-6). A *charge* consisting of ore, coke, and limestone is loaded at the top, and a blast of hot air is introduced near the bottom. A charge consists of about $1\frac{1}{2}$ tons (1361 kg) of ore, $\frac{3}{4}$ ton (680 kg) of coke, and $\frac{1}{4}$ ton (227 kg) of limestone for each ton (907 kg) of iron to be produced. Oxygen in the air combines with carbon in the hot coke, causing combustion and producing heat. The charge is smelted by the heat and, as the gases of combustion pass through the hot ore, another chemical reaction takes place which frees the iron from its oxide. The hot combustion gases are cleaned in dust catchers and recycled through brick checkerworks (stoves) to heat the incoming air blast (Fig. 5.1-7). A blast furnace is operated 24 hours a day.

The impurities in the ore and the flux combine with oxygen to form a floating slag 4 to 5 ft (1220 to 1500 mm) deep which is tapped from the furnace at about 2-hour intervals. The molten crude iron, which contains about 93% iron and 7% extraneous elements, is tapped at a lower level at about 3-hour intervals. This is then conveyed in ladle cars to *mixers* (holding furnaces), where the liquid iron is mixed to equalize its chemical composition and kept hot pending further refinement.

Most of this molten iron is sent to open hearth or basic oxygen steelmaking facilities. Some goes to a casting machine, where it solidifies into molds, known as *pigs* (ingots). Most iron from pig casting (*pig iron*) is made to rigid specifications which vary widely according to its end use. Pig iron that is intended to be sold as an end product is called *merchant* pig iron. End products made from pig iron vary from soil pipe to fancy dutch ovens. A merchant pig iron producer may have several hundred piles of different grades of iron and may blend a dozen or more grades of iron ore to arrive at a desired chemical composition. While 30,000 cu ft (850 m³) of solids weighing just under 1000 tons (907.184 Mg) are being processed in a furnace, the chemical composition of the mix is controlled so that the contents of carbon, silicon, sulfur, phosphorus, and manganese are maintained to within fractions of 1% of the desired amounts.

### FOUNDRY

A small percentage of pig iron is remelted in *foundries* and formed into special

Iron ore, coke and limestone are the principal materials used in making iron.

IRON ORE STOCKPILE

STOCK HOUSE receives raw materials which are carried up the inclined ramp (skip hoist) and charged into the blast furnace.

STOVES pre-heat air for the blast furnace.

BLAST FURNACE

DUST CATCHER

MOLTEN IRON

LADLE CAR

MOLTEN SLAG for disposal.

**FIGURE 5.1-6** Diagrammatic illustration of a blast furnace and its auxiliaries. (Courtesy American Iron and Steel Institute.)

shapes by casting in sand or loam molds. This process is called *iron founding*. Since no major change in chemical composition takes place when pig iron is cast, it is most often melted together with scrap metal and alloying elements before it is cast in order to impart desirable properties. The carbon content of cast iron

**FIGURE 5.1-7** Overall view of a blast furnace and its auxiliaries. A battery of cylindrical checkerworks appears at the center left. (Courtesy American Iron and Steel Institute.)

therefore is somewhat lower than that of pig iron, ranging between 1.5% and 4%.

The types of cast iron produced are *gray, white, malleable, chilled, alloyed,* and *nodular*. Only the first three are important in construction, and only these will be discussed here.

***Gray and White Cast Iron*** When molten cast iron solidifies, carbon in the iron may remain chemically combined in the form of iron carbide or may separate out as graphite. *White cast iron* contains carbon mainly in the chemically combined form (iron carbide), which makes this metal hard and nonductile. In *gray cast iron*, the carbon is chiefly in the form of graphite flakes, which impart the characteristic gray fracture and promote machinability and resistance to wear.

***Malleable Cast Iron*** White cast iron may be made somewhat softer and more ductile, though not strictly malleable, by *annealing* (controlled heating and cool-

ing). The carbon in the iron carbide is converted to a more dispersed graphite form, which provides greater ductility than does the flaky graphite of gray iron or the carbide of white iron.

**DIRECT REDUCTION**

There are direct reduction processes reported to be capable of bypassing the blast furnace. These include processes using solid reductants, processes using gaseous reductants, and direct steel processes. However, none of these processes is ready to replace efficient blast furnaces.

**5.1.2.3 Steelmaking**

Steelmaking requires lowering the carbon content of pig iron, controlling impurities in the pig iron and in other raw materials, and adding alloying elements to obtain desired properties.

Carbon is beneficial in steel when it is contained within certain ranges that vary with the type of steel. Carbon contents up to about 1.2% of the total material increase hardness and strength, but this increase is less when the carbon content is between 0.5% and 1.2% of the total. Because brittleness increases and the degree of ductility decreases as the carbon content rises, lower contents are used in most steels.

Uncontrolled or excessive amounts of carbon, phosphorus, sulfur, manganese, and silicon are classed as impurities. Pig iron usually contains these elements as impurities originating from the raw materials of the blast furnace. Up to certain limits or within specified ranges, the effect of these elements on steel may be beneficial, and steelmaking processes attempt to regulate their contents within these limits or ranges. Phosphorus and sulfur impart some desirable properties, but must be limited to less than 0.05% in most steels due to their adverse effects on *malleability* (ability to be shaped). Silicon and manganese are strengthening elements desired in ordinary steels in the range of 0.01% to 0.35% and 0.20% to 2.00%, respectively, and may be added when raw materials contain insufficient quantities. All of these elements are present in controlled amounts in most steels. These and other elements that occur naturally are called *residual* elements; when deliberately added, they are called *alloying* elements.

Steelmaking processes generally start with hot pig iron delivered in ladles by overhead cranes to the steel furnace, where scrap metal and fluxes are added. After the melting operation, the hot steel is tapped from the furnace into other ladles and moved to awaiting molds for casting into ingots or to a strand casting machine. The chief steelmaking processes are *open hearth*, *electric*, and *pneumatic* (Fig. 5.1-8).

## OPEN HEARTH PROCESS

The majority of open hearth furnaces are of the basic type, which are capable of controlling the phosphorus present in pig iron. Both ordinary carbon grades and high-grade alloy steels are usually produced by this process.

The name *open hearth* is applied because the saucer-shaped hearth, or floor of the furnace, is exposed to the sweep of the flames. The furnace itself is a masonry structure lined with refractory materials arranged on two levels: the upper

**FIGURE 5.1-8** Hot pig iron may be converted to steel in one of the steelmaking furnaces shown or in induction furnaces or basic oxygen furnaces. (Courtesy American Iron and Steel Institute.)

level for charging the raw materials, the lower for tapping the molten steel and slag by gravity. Furnaces may hold 50 to 600 tons (45.36 to 544.3 Mg) and produce 25 to 50 tons (22.68 to 45.36 Mg) of steel per hour. The charge generally consists of molten pig iron, scrap metal, and fluxes (Fig. 5.1-9).

The fuel (gas, oil, tar, or a combination of these mixed with hot air) is blown through ports at the ends of the furnace to produce combustion and heat over the hearth. As in a blast furnace, the hot gases of combustion are recycled through checkerworks, which preheat the incoming air. In addition, gaseous oxygen is usually introduced to speed the melting operation. Refinement of the metal results when the impurities and carbon in the pig iron and the oxygen and fluxes in the mix react to each other chemically to

**FIGURE 5.1-9** An open hearth furnace. Hot pig iron is being poured from a ladle. A charging machine (foreground) adds scrap metal and limestone flux. Open hearth furnaces are seldom used today. (Courtesy American Iron and Steel Institute.)

form a slag that floats to the surface. The molten steel flows through a hole in the bottom of the hearth into ladles. When desired, the carbon content may be increased by adding pig iron in the furnace or coke or coal in the ladle. Oxidizable *ferroalloys* (iron-based compounds rich in desired alloying elements) may also be added in the ladle.

## ELECTRIC PROCESSES

Most high-grade steels, such as alloy, stainless, and heat-resisting steel, are produced in electric furnaces capable of developing the high temperatures and reducing conditions necessary for melting these metals. Here electrical resistance and arc radiation rather than combustion are used to produce heat, and air or oxygen is not necessary for combustion. High-purity oxygen is introduced at appropriate times to oxidize impurities, which results in better control of the oxidation process and a smaller loss of expensive alloying elements.

The charge usually consists of scrap metal, mill scale, fluxes, and often alloying elements. Electric furnaces are mostly of the basic type and produce low-phosphorus steel. Many grades of basic steel can be produced at a rate of 30 to 40 tons (27.2 to 36.3 Mg) per hour in an electric arc furnace, which may hold up to 300 tons (272.2 Mg). Special grades involving valuable alloying elements are generally produced in smaller quantities, often in an electric induction furnace.

***Electric Arc Furnace*** The body of an electric arc furnace is a circular steel shell resembling a huge kettle, lined with refractory materials and mounted so that it may be tilted to pour off the molten steel. Three or more cylindrical electrodes project into the furnace from above. When high-voltage current is passed through the electrodes, an electric arc is produced between them and the charge, generating heat. Many of the impurities are oxidized during the melting operation and float to the surface as slag, allowing the steel to be poured off from under it (Fig. 5.1-10).

***Induction Furnace*** An induction furnace consists of a magnesia pot, insulated by other refractory materials and surrounded by windings of copper tubing (Fig. 5.1-11). When high-voltage alternating electric current is passed through the copper tubing around the pot, an in-duction current is generated in the metal charge of the furnace. The resistance of the metal to the induced current creates the heat for the melting operation.

## PNEUMATIC PROCESS

Steelmaking operations that use high-speed oxygen or air to oxidize impurities in the charge without fuel are called *pneumatic processes*. Their chief advantages are shortened heat time and high output. With furnace capacities of 300 tons (272.2 Mg) and tap-to-tap heat cycles of 45 minutes, it is possible to produce steel at a rate 6 to 10 times that of open hearth and electric arc furnaces. The pneumatic process, which takes place in a *basic oxygen furnace*, is the predominant method of producing steel in the United States.

In a basic oxygen furnace, which is a pear-shaped steel vessel lined with refractory materials (Fig. 5.1-12), a jet of high-purity oxygen is directed onto molten metal. The charge, consisting of molten pig iron, some scrap, and fluxes, is added from the top. Larger amounts of hot pig iron and smaller amounts of scrap metal are used in the oxygen furnace than in other steel furnaces.

**FIGURE 5.1-10** An electric arc furnace. (Courtesy Lukens Inc.)

**FIGURE 5.1-11** A 1500 ton (1360.78 Mg) channel induction furnace. (Courtesy Ajax Magnethermic Corporation.)

**FIGURE 5.1-12** A basic oxygen furnace. A ladle is pouring hot pig iron into the mouth of a tilted furnace. (Courtesy American Iron and Steel Institute.)

No fuel is required in this process, because the heat of the large mass of molten pig iron is sufficient to start a chemical reaction between the oxygen, the carbon, and the other impurities. Oxidation, in turn, produces sufficient heat to continue the process without the introduction of fuel.

If the as-tapped metal has a lower carbon or manganese content than is required, *ferromanganese* (a manganese-rich ferroalloy containing carbon) may be added in the ladle. The carbon gives added strength and hardness, while the manganese, in addition to imparting strength, assures proper ductility at elevated temperatures by combining with excess sulfur.

***Steel Casting*** No matter which steel-making process is used, steel must be formed into a specific shape before further processing. To do this, it is either poured into *ingot molds* or put through a *strand casting machine.*

***Ingots*** Raw steel from a furnace is typically *teemed* (poured) from a ladle into case iron ingot molds, which may be more than 8 ft (2400 mm) high and 3 ft (915 mm) in diameter. These molds are coated on the inside with a releasing agent that prevents splashed metal from marring the ingot's surface quality. Often, stripped ingots are moved to furnaces called *soaking pits* while the outside is solid but the interior is still molten.

***Strand Casting*** Strand casting produces a continuous ribbon of steel rather than separate molds. It can result in semifinished solid products, thus bypassing many of the steps necessary to produce ingots. At some point, the continuous ribbon of steel is cut into desired lengths (Fig. 5.1-13).

### 5.1.3 PROPERTIES

The properties of various irons and steels are closely related to their chemical compositions and grain structure, that is, the size and arrangement of the crystalline microscopic particles making up the metal. In the initial crude refining processes, the most important factors influencing grain structure are the carbon and its chemical reaction with iron as it cools. Further refining processes, alloying elements, heat treatments, and hot and cold working operations also affect grain structure. In ordinary steels, carbon and the other residual elements are important. In higher grades of steel, alloying elements contribute special properties to the metal.

The essential physical property that distinguishes steel from iron is its malleability as it comes from the furnace and is initially cast. Other unique properties of steel are its ability to resist high stresses by deforming without breaking (toughness) and its ability to be hardened by hot and cold working. The effects of the several residual elements originating in the raw materials and of the added alloying elements are outlined in Figure 5.1-14.

#### 5.1.3.1 Cast Iron

Only slight variations in properties or in the grain structure of the metal occur in cast iron as a result of the reaction between its carbon and its iron. Most ordinary white and gray cast irons, therefore, like pig iron, are hard and brittle, and have relatively high compressive strength and low tensile strength. However, with the proper combination of selected pig irons, scrap metals, and alloying elements, cast iron can be made with increased hardness, toughness, and corrosion and wear resistance. Casting also

1. Molten steel pours from a ladle into a reservoir called a tundish.

2. The metal flows out the bottom of the tundish at a carefully regulated rate into the mold, which is moving up and down to prevent the hot metal from sticking. The interior of the mold is hollow—just the size, in width and thickness, of the slab to be formed. Lining the walls are pipes through which water flows, chilling the metal. A thin shell of steel begins to solidify around the molten metal.

3. The gradually solidifying slab moves down through the secondary cooling zone. A series of rollers support the slab and gradually turn it into a horizontal position. Sprays of water under high pressure cool and harden the metal still further.

4. The ribbon of steel moves on to a level table.

5. A flame-cutting torch slices down through the metal. When the slab is cut off, it is carried on rollers to a cooling bed. The entire trip from the ladle has taken less than one-half hour.

Hot steel is transported rapidly by ladle from electric and open hearth furnaces to the casting unit and is fed into the tundish.

Hot Metal Ladle

The refractory-lined tundish controls the flow and distribution of metal into the molds.

Operator's console

In the water-cooled mold the steel begins to solidify. A solid shell is formed.

Gantry service crane

Roller aprons and secondary cooling

Solidifying steel enters the secondary cooling zone. Cooling is accomplished by direct water spray. Roller aprons are arranged to guide and support the strands and simultaneously take up the ferrostatic pressure exerted by the liquid metal core upon the strand shell.

Traveling Slab Cut-off Torch

Slabs are cut into predetermined lengths and removed by roller tables.

Roller Leveler

Here rolls withdraw and level the strands.

Slab Run Out Table

**FIGURE 5.1-13** Strand, or continuous, casting. (Courtesy American Iron and Steel Institute.)

permits the formation of complex and massive shapes not readily produced by machining or rolling.

Drainage, vent, and waste pipe, ornamental railings, lampposts, handrail brackets, stair treads, manhole covers, and concrete inserts for anchors are familiar cast iron products. In sanitary ware involving enameling and baking at elevated temperatures, cast iron's resistance to warping and cracking at high heat is an important advantage.

*Malleable cast iron* has improved breakage resistance as a result of its increased toughness and ductility. It is also easier to weld than gray or white cast iron. It is used extensively for small articles of door hardware, concrete inserts, brackets, flanges, and anchors. Special *alloy cast irons*, containing between 1% and 5% alloying elements, and *high-strength cast irons*, which contain molybdenum, chromium, and nickel, are also produced for special uses. *Nodular cast iron*, which is produced in this country by adding magnesium, has greater strength, ductility, toughness, machin-

ability, and corrosion resistance than other cast irons.

### 5.1.3.2 Carbon Steels

*Carbon steels* are those in which the residual elements, such as carbon, manganese, phosphorus, sulfur, and silicon, are controlled, but in which no alloying elements are added to achieve special properties. They contain up to 1.2% carbon, and other elements are controlled within specified limits or ranges. More than 90% of the steel manufactured into finished products is carbon steel.

### CLASSIFICATION AND USES

In the past, steels were classified according to their carbon content as follows:

Soft steel: 0.20% maximum
Mild steel: 0.15% to 0.25%
Medium steel: 0.25% to 0.45%
Hard steel: 0.45% to 0.85%
Spring steel: 0.85% to 1.15%

These broad classifications became less useful as the need for precisely com-

pounded steel types increased. Today, the American Iron and Steel Institute (AISI) and ASTM International(ASTM) designations are more commonly used in manufacturing and construction.

*AISI Designations* The AISI designation system assigns code numbers that indicate the chemical composition and the steelmaking process of various types of steel. For carbon steels, each four-digit code number specifies the permissible maximums or ranges for residual elements, including carbon, manganese, phosphorus, sulfur, and sometimes silicon. Some types of alloy steels are designated by five numerals.

The first two digits in the AISI system indicate the type of steel (carbon or alloy) (Fig. 5.1-15). The last two digits of the four-numeral series and the last three digits of the five-numeral series indicate the approximate mean of the carbon range. For example, in Series Designation 1035, 35 represents a carbon range of 0.32% to 0.38%. It is necessary, however, to deviate from this system and to

**FIGURE 5.1-14** Effects of Alloying Elements

| Element | Common Content | Effects |
|---|---|---|
| Carbon | Up to 0.90% | Increases hardness, tensile strength, and responsiveness to heat treatment with corresponding increases in strength and hardness |
| | Over 0.90% | Increases hardness and brittleness; over 1.2% causes loss of malleability |
| Manganese | 0.50% to 2.0% | Imparts strength and responsiveness to heat treatment; promotes hardness, uniformity of internal grain structure |
| Silicon | Up to 2.50% | Same general effects as manganese |
| Sulfur | Up to 0.050% | Maintained below this content to retain malleability at high temperatures, which is reduced with increased content |
| | 0.05% to 3.0% | Improves machinability |
| Phosphorus | Up to 0.05% | Increases strength and corrosion resistance, but is maintained below this content to retain malleability and weldability at room temperature |
| Aluminum | Variable | Promotes small grain size and uniformity of internal grain structure in the as-cast metal or during heat treatment |
| Copper | Up to 0.25% | Increases strength and corrosion resistance |
| Lead | 0.15% to 0.35% | Improves machinability without detrimental effect on mechanical properties |
| Chromium | 0.50% to 1.50% | In alloy steels, increases responsiveness to heat treatment and hardenability |
| | 4.0% to 12% | In heat-resisting steels, causes retention of mechanical properties at high temperatures |
| | Over 12% | Increases corrosion resistance and hardness |
| Nickel | 1.0% to 4.0% | In alloy steels, increases strength, toughness, and impact resistance |
| | Up to 27.0% | In stainless steels, improves performance at elevated temperatures and prevents work hardening |
| Molybdenum | 0.10% to 0.40% | In alloy steels, increases toughness and hardenability |
| | Up to 4.0% | In stainless steels, increases corrosion resistance and strength retention at high temperatures |
| Tungsten | 17% to 20% | In tool steels, promotes hardness at high cutting temperatures; in stainless steels, smaller amounts ensure strength retention at high temperatures |
| Vanadium | 0.15% to 0.20% | Promotes small grain size and uniformity of internal grain structure in the as-cast metal or during heat treatment; improves resistance to thermal fatigue shock |
| Tellurium | Up to .05% | Improves machinability when added to leaded steels |
| Titanium | Variable | Prevents loss of effective chromium through carbide precipitation in 18/8 stainless steels |
| Cobalt | 17.0% to 36.0% | Increases magnetic properties of alloy steels; in smaller amounts, promotes strength at high temperatures in heat-resisting steels |

(Courtesy American Iron and Steel Institute.)

interpolate numbers in the case of some carbon ranges, and for variations in manganese, sulfur, or other elements with the same carbon range.

***ASTM Designations*** Structural steels are commonly specified according to ASTM standards. The carbon steels typically used in building construction conform to ASTM standards A6, A36, A53, A82, A185, A283, A497, A500, A501, A615, A663, A675, A767, and A996. The steel most often used for construction purposes is that complying with ASTM A36.

ASTM standards indicate certain partial chemical compositions and specified minimums for strength and ductility. However, they regulate only those chemical elements which directly affect the fabrication and erection of steel, making it possible to produce proprietary steels of different chemical compositions conforming with the required ASTM performance standards.

***Unified Numbering System*** Alloy and carbon steels have been assigned a designation in the Unified Numbering System (UNS) for Metals and Alloys. The UNS was established in 1975 by ASTM and the Society of Automotive Engineers (SAE). These designations appear in ASTM E527 and SAE Standard J1086.

A UNS number consists of a single letter prefix followed by five digits. The prefix *E* designates steels made by the basic electric furnace process with special practices; the absence of a prefix indicates open hearth steels. The prefix *M* designates a series of merchant-quality steels; *G* indicates standard alloy or carbon steels; and *H* indicates standard hardenability steels. The first four digits correspond to AISI, ASTM, or SAE steel designations, and the last digit usually relates to an additional chemical requirement such as lead or boron. The use of 6 as the last digit designates steels which are made by the basic electric furnace process with special practices. An example of the UNS is "Series Designation E52100," which is the same steel as the 1.45% chromium steel (52) shown in Figure 5.1-15, made by the basic electric furnace process (E), which has a carbon content between 0.98% and 1.10% (100).

**FIGURE 5.1-15** AISI Designations

| Series Designation[a] | Type and Approximate Percentages of Identifying Elements |
|---|---|
| **Alloy Steels** | |
| 13xx | Manganese 1.75 |
| 40xx | Molybdenum 0.20 or 0.25; or molybdenum 0.25 and sulfur 0.042 |
| 41xx | Chromium 0.50, 0.80, or 0.95; molybdenum 0.12, 0.20, or 0.30 |
| 43xx | Nickel 1.83, chromium 0.50 or 0.80, molybdenum 0.05 |
| 46xx | Nickel 0.85 or 1.83, molybdenum 0.20 or 0.25 |
| 47xx | Nickel 1.05, chromium 0.45, molybdenum 0.20 or 0.35 |
| 48xx | Nickel 3.50, molybdenum 0.25 |
| 51xx | Chromium 0.80, 0.88, 0.93, 0.95, or 1.00 |
| 51xxx | Chromium 1.03 |
| 52xxx | Chromium 1.45 |
| 61xx | Chromium 0.60 or 0.95, vanadium 0.13 or min. 0.15 |
| 86xx | Nickel 0.55, chromium 0.50, molybdenum 0.20 |
| 87xx | Nickel 0.55, chromium 0.50, molybdenum 0.25 |
| 88xx | Nickel 0.55, chromium 0.50, molybdenum 0.35 |
| 92xx | Silicon 2.00; or silicon 1.40 and chromium 0.70 |
| 50Bxx | Chromium 0.28 or 0.50 |
| 51Bxx | Chromium 0.80 |
| 81Bxx | Nickel 0.30, chromium 0.45, molybdenum 0.12 |
| 94Bxx | Nickel 0.45, chromium 0.40, molybdenum 0.12 |
| **Carbon Steels** | |
| 10xx | Nonresulfurized, manganese 1.00% maximum |
| 11xx | Resulfurized |
| 12xx | Rephosphorized and resulfurized |
| 15xx | Nonresulfurized, manganese maximum over 1.00% |

[a]B denotes boron steel.

(Courtesy American Iron and Steel Institute.)

***Common Uses*** Carbon steels are used for most products in the building industry, from structural shapes, concrete reinforcing bars, sheets, plates, and pipes to the smallest items of builders' hardware. Carbon steels generally possess adequate strength, hardness, stiffness, malleability, and weldability but poor resistance to corrosion.

Occasionally, copper and other elements may be added to improve atmospheric corrosion resistance, and by varying the chemical compositions, heat treatments, and subsequent hot and cold working operations, other desirable properties can be enhanced.

### 5.1.3.3 Alloy Steels

*Alloy steel* is so classified when the content of alloying elements exceeds certain limits or when a definite range of alloying elements is specified. However, the term *alloy steel* does not cover stainless steel, tool steel, specialty steel, or some other classifications, even though those steels contain specified amounts of alloying elements.

### CLASSIFICATION AND USES

Alloy steels are generally grouped by the element or combination of elements used to obtain desired properties. Manganese-molybdenum steels, for example, are known for hardenability and resistance to fatigue; silicon-manganese steels for high relative strength and shock resistance; nickel steels for toughness and impact and corrosion resistance.

***AISI Designations*** The AISI has developed a numerical designation system for identifying alloy steels. These steels are designated by a four-digit code, from 1200 to 9800, according to chemical composition (see Fig. 5.1-15). There are specific allowable ranges for carbon, manganese, silicon, nickel, chromium, and molybdenum; limits for sulfur and phosphorus are also usually specified. The prefix *E* is used to designate electric furnace steel; the absence of a prefix indicates open hearth steels. The suffix *H* is sometimes used to identify steels of controlled hardenability; the letters *B* or *BV* between the second and third digits indicate boron or boron-vanadium alloys, respectively.

These steels are grouped in series according to the most important alloying element(s), identified by the first two digits; the last two digits indicate the approximate carbon content. For example, AISI Type 4023 steel indicates a molybdenum alloy steel (40) with an average carbon content of 0.23% (23) made in an open hearth furnace (no *E* prefix).

***Common Uses*** Alloy steels have properties essentially similar to or exceeding those of carbon steels. In general, particular properties are improved by certain elements. For example, a high silicon content gives steel excellent magnetic permeability and low core loss, making it useful in motors, generators, and transformers; cobalt improves magnetic properties for such uses as permanent magnets in electrical apparatus and sealing gaskets in modern refrigerators; nickel imparts great toughness, which is useful in rock drilling and air hammer equipment.

Because of their higher cost, alloy steels are not often used in construction materials. However, a particular group known as *heat-treated constructional alloy steels* have been rolled in a limited number of structural shapes and used successfully in large structures where unusually large loads or temperatures are encountered. These steels are capable of developing very high strengths (ultimate strength of up to 135,000 psi [930 MPa], yield point of up to 100,000 psi [689 MPa]) and show promise of increased structural use.

### 5.1.3.4 High-Strength, Low-Alloy Steels

*High-strength, low-alloy steels* comprise a group with chemical compositions specially developed to impart better mechanical properties and greater resistance to atmospheric corrosion than are obtainable from conventional carbon structural steels.

## CLASSIFICATION AND USES

AISI classifies high-strength, low-alloy steels according to chemical composition. ASTM publishes recommended practices and standards for them.

***Structural Steels*** Among the ASTM standards for high-strength, low-alloy structural steels are A242, A572, and A588.

***Sheet Steel*** The most widely produced and ordered high-strength, low-alloy sheet steel products are:

- *Conventional high-strength, low-alloy steel* (ASTM A1011), which is produced to various strength levels where atmospheric corrosion resistance and maximum formability are not a requirement.
- *Improved atmospheric corrosion resistance high-strength, low-alloy sheet steel* (ASTM A606), which is produced to 45 or 50 kips per square inch (ksi) (310.3 or 344.7 MPa) minimum yield points (1 kip equals 1000 lb [453.6 kg]). The types of steel with enhanced corrosion resistance include type 2 with a corrosion resistance twice, and Type 4 with a corrosion resistance at least four times, that of plain carbon steel.
- *Improved formability high-strength, low-alloy steel* (ASTM A1011), which is produced with special forming properties to various strength levels where maximum formability is required. This type is generally furnished for making parts too difficult for the fabricating properties of the other two types.

In general, the minimum yield points for these steels vary within a range of 45 to 80 ksi (310.3 to 551.6 MPa), and their minimum tensile strength varies from 55 to 90 ksi (379 to 620.5 MPa).

***Steel Bars*** The most frequently specified high-strength, low-alloy steel bars have a minimum yield point ranging from 42 to 80 ksi (289.4 to 551.5 MPa) and a minimum tensile strength that varies between 60 and 95 ksi (413.7 and 655 MPa).

## COMMON USES

High-strength, low-alloy steels are commonly used where high strength in relation to weight is important. The steels listed in the previous subsections are often used in bridge and high-rise building construction. These and other high-strength steels also are used extensively in various transportation vehicles, heavy construction equipment, industrial containers, and wherever resistance to corrosion and impact damage is important.

Some proprietary low-alloy steels can be made with increased corrosion resistance when intended for exposed architectural use. These steels, sometimes called *weathering steels*, form natural rust-colored, self-healing oxide coatings that inhibit further corrosion when left unpainted. This rusting may not stop, however, if certain pollutants are present in the atmosphere. They are used increasingly in commercial and industrial construction but are not used extensively in residential work.

Unfortunately, the rust-colored oxide coating that forms to protect weathering steels will wash off if water runs across it. Special detailing is necessary to prevent this runoff from staining adjacent materials.

## 5.1.3.5 Stainless and Heat-Resisting Steel

The outstanding characteristics of the two groups of steels referred to as *stainless* and *heat-resisting* are implied in their names. Stainless steels possess excellent corrosion resistance at varying temperature ranges. Heat-resisting steels retain their essential physical and mechanical properties at elevated temperatures. Stainless and heat-resisting steels account for about 1% of total steel production.

Chromium is the alloying element mainly responsible for corrosion-resisting and heat-resisting properties, although other elements, such as nickel, manganese, and molybdenum, also contribute to these properties. The stainless qualities of steel are derived from a self-healing chromium oxide that forms a transparent skin and prevents further oxidation. Steels with less than 4% chromium are regarded as alloy steels. Heat-resisting steels normally contain 4% to 12% chromium and stainless steels more than 12%.

## CLASSIFICATION AND USES

Stainless and heat-resisting steels are classified according to chromium content and internal grain structure into three major groups—*martensitic, ferritic,* and *austenitic*—and a minor group under the martensitic group, *precipitation hardening.*

***Martensitic Steels*** Heat-resisting and stainless steels with a chromium content between 4% and 18% are known as *martensitic*. These may be hardened by heat treatment and are therefore sometimes referred to as *hardenable chromium alloys*. They also have adequate cold-forming characteristics and are satisfactory for hot working or forging, but must be slowly cooled or annealed after forging to prevent cracking. They are easily welded but require subsequent annealing or tempering to restore ductility.

***Ferritic Steel*** Stainless steels that contain between 12% and 27% chromium are called *ferritic*. These steels do not respond readily to heat treatment and are sometimes referred to as *nonhardenable chromium alloys*, even though their hardness may be increased somewhat by cold working. Ductility may be improved by hot or cold working followed by annealing. These steels are not welded as easily as are the martensitic steels, but they have greater corrosion resistance. Fair ductility after annealing allows product fabrication by a variety of processes such as cold forming, spinning, and light drawing. Ferritic steels possess low strength at elevated temperatures and a low coefficient of thermal expansion.

***Austenitic Steel*** The chromium content of austenitic steels ranges between 16% and 26%. They also contain between 3.5% and 22% nickel. These steels are characterized by a greater ability to harden as a result of cold working than the ferritic types. Austenitic steels include both nickel-chromium and nickel-chromium-manganese groups of alloys. Both groups provide superior resistance to corrosion and are suited to many fabrication techniques. They have high ductility, which is important for deep drawing and forging operations, are easily welded, and have good oxidation resistance at elevated temperatures. A relatively high coefficient of thermal expansion is characteristic of these steels and must be considered in the design of high-temperature equipment. They are also more susceptible to corrosive attack by deoxidizing sulfur gases than other stainless steels.

**Precipitation Hardening Steels** Precipitation hardening steels are iron-chromium-nickel alloys with additional elements, which are hardenable by solution treating and aging. These steels combine high strength, hardness, and corrosion resistance.

**Industry Designations** Heat-resisting and stainless steels are identified by Specialty Steel Industry of North America (SSINA) and ASTM and SAE designations. SSINA designations follow the old AISI system in which heat-resisting and stainless steels are grouped into five series bearing number designations from 200 to 500.

The six-digit UNS system mentioned earlier applies as well to stainless steel. Stainless steels are identified in the UNS with an *S* followed by five digits.

**Common Uses** There are many applications for stainless steels and a wide variety of alloys from which to choose.

*Martensitic chromium steels* are suitable for applications requiring high strength, hardness, and resistance to abrasion, such as steam and gas turbine parts, bearings, and cutlery. SSINA Type 410 (basic type) is one of the most widely used alloys in this category. *Ferritic chromium steels* are used mostly for automotive trim, applications involving nitric acid, high-temperature service requiring resistance to scaling, and uses which call for low thermal expansion. SSINA Types 405 and 430 steels are the most economical and widely used alloys in this category. Type 430 is sometimes used in building construction for gutters and downspouts, architectural trim, sills, and column covers. It is also used for kitchen applications such as range hoods, range tops, and appliance fronts.

*Austenitic nickel-chromium steels* are the most numerous and popular category of stainless steels. They are sometimes referred to as *18-8 alloys* because several of the most common types (302, 303, and 304) contain approximately 18% chromium and 8% nickel. Among these, Types 302 (the basic 18-8 grade) and 304 are used most often for such building applications as fascias, curtain walls, storefronts, doors and windows, column covers, and railings. Their relative chemical inertness and finish-retention qualities also make them useful in food preparation and surgical equipment, hospital clean rooms, laboratories,

and for such familiar kitchen applications as sinks, countertops, and appliances.

In addition to a variety of automotive and transportation equipment uses, Type 301 steel is sometimes used for structural members because it work-hardens rapidly to high strength and toughness. Because it has good cold forming properties, it is used for making roof drainage products.

Type 316 is the most resistant to attack by salt spray and corrosive industrial fumes, and is therefore used in seacoast and industrial area where protection from such exposures is essential.

*Austenitic nickel-chromium manganese steels* Types 201 and 202 are similar to their nickel-chromium counter parts, Types 301 and 302, and may be used interchangeably with them. In addition to automotive and food storage equipment, they are used for flatware, cooking utensils, kitchen sinks, and countertops.

## 5.1.4 MANUFACTURE

Steel products are manufactured by casting in foundries and by forging, extruding, rolling, and cold drawing in mills. The products of the foundry are referred to as *cast products*; all others are *wrought products*. Wrought products account for more than 90% of the total steel production, and by far the largest part of that is produced by hot rolling and cold finishing. Steel mill products are regarded as semifinished when in the form of blooms,

billets, and slabs, and as finished when in a form suitable for further fabrication into components or for use directly in construction. Figure 5.1-16 shows the variety of finished mill products produced at steel mills.

### 5.1.4.1 Cast Steel Products

Castings are generally made by pouring molten steel into sand molds. Before it is cast, pig iron may be further refined and melted with scrap metal and ferroalloys at the foundry to obtain desired casting and mechanical properties.

Casting is used in preference to other forming operations when desired shapes are of a size or complexity not readily obtained by rolling or machining. Complex and massive shapes possessing great strength and impact resistance at high and low temperatures usually are cast (Fig. 5.1-17). Other cast steel products used in construction are concrete inserts, anchors, and items of hardware. Iron and steel castings contribute many million of tons (kg) to a broad range of uses in a typical year. Of this, most are in gray and malleable cast iron, while some is in steel. Of the cast steel products, approximately two-thirds are carbon steels; the rest are alloy and special steels.

The transportation industry relies on castings for many engine and car body parts, as does the excavating and mining equipment industry. Castings are also used extensively in many parts of steel-making furnaces and rolling mills.

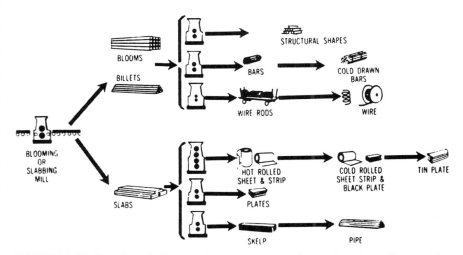

**FIGURE 5.1-16** Operations in the manufacturing sequence for steel products. Blooms, billets, slabs, bars, plates, and pipes are shown as stacks of individual units. (Courtesy American Iron and Steel Institute.)

**FIGURE 5.1-17** Steel castings may range in weight from a few ounces to several hundred tons, as in this bridge saddle. (Courtesy Bethlehem Steel Corporation.)

## EXTRUDING

Semifinished shapes may be converted into lengths of uniform cross section by *extruding*. In this process, an advancing ram (Fig. 5.1-20) forces preheated plastic metal through a tough, heat-resistant die of the desired profile (Fig. 5.1-21). Presses vary in size from a small 700 ton (635 Mg), $2^1/_2$ in. (63.5 mm) maximum circle die press designed for fast extrusion speeds of up to 25 ft/sec (7.62 m/s) to 12,000 ton (1088.6 Mg) installations capable of extruding sections of up to 21 in. (533.4 mm) in diameter at slower speeds. Stainless steel is generally extruded in shapes of up to $6^1/_2$ in. (165.1 mm) in diameter.

Extruding can produce more complex sections with better surface characteristics than can be produced by rolling. It can be used to shape certain alloys without undesirable residual effects (such as excessive hardness) and is more economical than other forming methods for small quantities. However, most steel extrusions are limited to 35 ft (10.67 m) in length and 21 in. (533.4 mm) in diameter.

## HOT ROLLING AND COLD FINISHING

By far the largest proportion of wrought steel products are manufactured by hot rolling and cold finishing.

Hot rolling is used to make semifinished shapes as well as some finished products. Hot rolling may be done as part of the continuous casting process on a strand caster or by using hot ingots made by traditional steelmaking processes.

In a hot-rolling process, hot steel passes through a system of rolls which gradually imparts rectangular bloom, billet, or slab shapes. These are cooled, cleaned of surface irregularities, inspected, and reheated for further rolling. Semifinished shapes may be hot rolled

### 5.1.4.2 Wrought Steel Products

A few steel products that are complex and have an irregular cross section are made by *forging*. Linear products of constant cross section, such as bars, tubes, and rods, and complex shapes may be made by *extruding*. Less complex and larger linear forms, and sheets, strips, rods, bars, and most structural shapes used in construction are produced by *hot rolling* and *cold finishing* or by hot rolling without cold finishing.

### FORGING

*Forging* is a method of forming hot metal into desired shapes by pressing it between heat-resistant dies. When metal is forged, its toughness, strength, and ductility increase significantly along the lines of flow (Fig. 5.1-18), which is an important advantage of this forming technique. The metal to be forged is preheated to between 2200°F (1204°C) and 2400°F (1315.6°C) in special furnaces, and in some operations the dies are heated as well. The two major types of forging are *open-die* and *closed-die*.

***Open-Die Forging*** In open-die forging, a large press squeezes (rather than strikes) steel between two flat surfaces.

Temperature must be carefully controlled, which requires frequent reheating of the steel between shaping actions. Some hydraulic open-die forging presses may accept ingots weighing several hundred tons (Fig. 5.1-19).

***Closed-Die Forging*** In closed-die forging, a hammer pounds a section of steel between carved dies until it reaches the desired shape. Closed-die forging most often uses steam hammers. Drop hammers develop forging force by the fall of a heavy weight or ram. In double-action hammers, the speed of the falling weight is usually given additional impetus.

**FIGURE 5.1-18** Continuity of flow lines in forged metal structure results in greater strength and toughness. (Courtesy American Iron and Steel Institute.)

FIGURE 5.1-19 A hydraulic press, which may exert a force of up to 18,000 tons (16 329.3 Mg), is particularly suitable for forging high-strength steels. (Courtesy Kaiser Aluminum and Chemical Corporation.)

FIGURE 5.1-20 Hot steel is forced through a lubricated flow extrusion press. (Courtesy American Iron and Steel Institute.)

FIGURE 5.1-21 Bars and tubing can be extruded into a variety of open and hollow shapes. (Courtesy American Iron and Steel Institute.)

directly into finished hot-rolled products such as structural shapes, plates, sheets, strip, and bars; or hot rolling may be used as an intermediate step before cold finishing, as in the manufacture of bar, sheet, strip, and wire. The hot rolling of finished products is sometimes referred to as *hot finishing*.

Hot-rolled products intended for cold finishing are *descaled* (cleaned of surface oxide scale) usually by *pickling*, which involves passing the steel through sulfuric or hydrochloric acid, followed by rinsing with hot and cold water, steam drying, and, usually, oiling.

Cold finishing consists of cold rolling, cold reduction, and cold drawing of previously hot-rolled, descaled shapes. *Cold rolling* involves passing metal at room temperature through sets of rolls to impart the desired shape, finish, and mechanical properties. *Cold reduction* is a form of cold rolling that drastically reduces the thickness of a flat product (sheet, strip, blackplate, etc.) and improves its strength, surface finish, and flatness. Cold reduction often hardens the metal excessively, so that it must be annealed to soften it and further *temper* (skin) rolled to obtain the proper strength and stiffness and the desired surface texture.

*Cold drawing* is used to make smaller or more complex bar shapes from hot-rolled annealed bars by pulling the bars through a hard, abrasion-resistant die. Wire is similarly cold drawn from wire rods. This operation results in improved machinability and strength, smoother finish, and greater dimensional accuracy.

***Slabs, Blooms, and Billets*** Slabs, blooms, and billets are semifinished shapes from which other products are made. *Slabs* are flat rectangular shapes with a width more than twice the thickness, which is usually less than 10 in. (254 mm). *Blooms* are rectangular shapes generally larger than 36 sq in. (23 226 mm²). *Billets* are less than 36 sq in. (23 226 mm²), more nearly square in cross section, and longer.

The type of semifinished shape rolled depends on the intended final use. Slabs are commonly used in the rolling of flat products such as plates, sheets, strip, and skelp; blooms in the rolling of rails, tube rounds, and structural shapes; and billets for making bars, tube rounds, and wire rods (Fig. 5.1-22). Skelp, tube rounds, and wire rods are also semifinished shapes. Skelp is used to make welded pipe, tube rounds for piercing into seamless pipe, and wire rods for drawing into wire.

**FIGURE 5.1-22** Intermediate mill processes used to convert semifinished steel shapes into finished mill products. (Courtesy American Iron and Steel Institute.)

*Sheet and Strip* Much of the steel shipped annually is the product of continuous sheet mills. The definition of these and other flat-rolled products is complex and varies, depending on whether they are stainless or carbon steel. Generally, the dimensions of sheet and strip lie between those of bars and plates

(Fig. 5.1-23). For carbon steels, the method of rolling (whether hot or cold) is also a factor.

*Hot-rolled sheet and strip* are manufactured from semifinished slabs in sheet and strip mills. Preheated slabs are squeezed and shaped by progressive hot rolling until the final desired thickness and width are obtained. Sheets and strip emerge from the roller tables at speeds of up to 3500 ft/min (1066.8 m/min) and are wound into coils for shipment or for transfer to other departments for further working.

Modern cold reducing mills may receive hot-rolled steel about as thick as a half dollar and three-quarters of a mile (1207 m) long; 2 minutes later, it will be the thickness of two playing cards and more than 2 miles (3220 m) long.

*Cold-reduced sheets and strip* are made by cold reducing previously hot-rolled and pickled flat products. Much sheet and strip are cold-rolled to meet consumer requirements for thickness, surface finish, or mechanical properties. Cold-reduced products may also be annealed and temper-rolled or zinc or tin coated.

*Blackplate* is the classification for cold-reduced products thinner than 0.0141 in. (0.3556 mm) and wider than 12 in. (305 mm) (see Fig. 5.1-23).

*Plates* Flat-rolled finished products thicker than sheet and strip, produced

**FIGURE 5.1-23** General Classification of Flat-Rolled Steel Products by Size[a]

| Carbon Steel, Hot Rolled | | | | | |
|---|---|---|---|---|---|
| | Width | | | | |
| Thickness | To 3$\frac{1}{2}$ in. incl. | Over 3$\frac{1}{2}$ in. to 6 in. incl. | Over 6 in. to 8 in. incl. | Over 8 in. to 12 in. incl. | Over 12 in. to 48 in. incl. | Over 48 in. |
| 0.2300 in. & thicker | Bar | Bar | Bar | Plate | Plate | Plate |
| 0.2299 in. to 0.2031 in. | Bar | Bar | Strip | Strip | Sheet | Plate |
| 0.2030 in. to 0.1800 in. | Strip | Strip | Strip | Strip | Sheet | Plate |
| 0.1799 in. to 0.0449 in. | Strip | Strip | Strip | Strip | Sheet | Sheet |
| 0.0448 in. to 0.0344 in. | Strip | Strip | | | | |
| 0.0343 in. to 0.0255 in. | Strip | | | | | |

| Carbon Steel, Cold Rolled | | | | | |
|---|---|---|---|---|---|
| | Width | | | | |
| Thickness | To $\frac{15}{32}$ in. incl. | | Over $\frac{15}{32}$ in. to 12 in. incl. | | Over 12 in. to 23$\frac{15}{16}$ in. incl. | | Over 23$\frac{15}{16}$ in. |
| 0.2500 in. & thicker | Bar | | Bar | | Strip[c] Sheet[e] | Sheet |
| 0.2499 in. to 0.0142 in. | Flat Wire[b] | Strip | Strip | Sheet[d] | Strip[c] Sheet[e] | Sheet |
| 0.0141 in. & thinner | Flat Wire[b] | Strip | Strip | | Strip Blackplate[e] | Blackplate |

*(continues on p. 264)*

FIGURE 5.1-23 *(Continued)*

**Stainless Steel, Hot and Cold Rolled**

| Thickness | Width | | | |
|---|---|---|---|---|
| | ¼ in. to ⅜ in. | ⅜ in. to 10 in. | 10 in. to 24 in. | 24 in. & Over |
| ³⁄₁₆ in. & over | Bar[f] | Bar | Plate | Plate |
| ⅛ in. to ³⁄₁₆ in. | Flat Wire[g] | Bar[g] | Strip | Sheet[h] |
| Up to ⅛ in. | Flat Wire | Strip | Strip | Sheet[h] |

[a]Dimensional ranges for flat-rolled alloy steels are customarily slightly different from those for carbon or stainless steel.

[b]When the material has rolled or prepared edges.

[c]When special temper, edge, finish, or single-strand rolling is specified.

[d]Cut lengths or sheet coils slit from wider coils (with resulting No. 3 edges), in thicknesses 0.0142 in. to 0.0821 in. inclusive, and widths 2 in. to 12 in. carbon 0.20% maximum.

[e]When no special temper, edge, or finish (other than Dull or Luster) is specified.

[f]Only hot rolled.

[g]May also be rolled as strip.

[h]In polished finishes Nos. 3, 4, 6, 7 and 8.

(Courtesy American Iron and Steel Institute.) (Refer to Figure 25.7-1 for metric conversions.)

**FIGURE 5.1-24** A hot plate is rolled back and forth in a reversing mill until the desired thickness is achieved. (Courtesy American Iron and Steel Institute.)

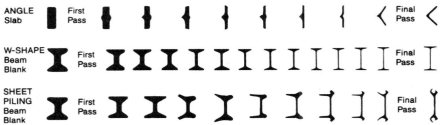

**FIGURE 5.1-25** Successive steps in rolling structural shapes. Hot blanks are roughly shaped by one set of rolls, then move onto other rolls which impart the final shape. (Courtesy American Iron and Steel Institute.)

**FIGURE 5.1-26** Some Common Structural Steel Designations

| Shape | Designation | Example (as used on drawings)[a] | |
|---|---|---|---|
| Wide flange | W | W24 × 62 | 24 is depth, 62 is weight |
| Angle | L | L5 × 3 × ¼ | 5 and 3 are leg lengths ¼ is leg thickness |
| Channel | C | C10 × 15.3 | 10 is depth, 15.3 is weight |
| Structural tee | WT | WT12 × 52 | 12 is depth, 52 is weight |

[a]Dimensions are in inches; weights are in pounds per linear foot.

from slabs or slab ingots by hot rolling, are called *plates* (Fig. 5.1-24 [see also Fig. 5.1-23]). The two principal classes of plates are *sheared* and *universal*, depending on the type of mill on which they are rolled. Sheared plate mills have only horizontal rolls and produce plates with uneven edges and ends, which must later be flame cut or sheared straight to the desired size. Universal plate mills have vertical rolls which are capable of producing smooth, straight edges so that only the ends must be trimmed. Lighter plates may also be rolled on some sheet and strip mills from slabs produced in blooming or slabbing mills.

***Structural Shapes*** Structural members, such as W-shaped (wide flange) sections, angles, channels, tees, zees, and piling, are hot rolled from blooms or billets. As many as 40 passes through grooved rolls may be required to produce a desired shape (Fig. 5.1-25). The first sets of rolls drastically reduce the cross section and impart the approximate shape; the finishing rolls gradually form the final shape and, for deep sections such as channels and angles, bend the legs into position. Wide flange shapes are rolled on universal mills which have vertical rolls capa-

ble of shaping the vertical surfaces. Different weights of a shape are produced by varying the spacing of the rollers, wider spacings producing heavier sections.

Structural shapes are identified by standard designations (Fig. 5.1-26). Wide-flange shapes are available in a great many sizes, a few of which are shown in Figure 5.1-27. Those used for beams and girders tend to be relatively tall with narrow flanges; those for columns and piles tend to be more square in shape.

Angles are available with legs of equal or unequal length (Fig. 5.1-28).

**FIGURE 5.1-27** A Few of the Many Available Wide-Flange (W) Shapes[a]

| Designation | $A$ (in.²) | $d$ (in.) | $t_w$ (in.) | $t_w/2$ (in.) | $b_f$ (in.) | $t_f$ (in.) | $T$ (in.) | $k$ (in.) | $k_1$ (in.) | Wt per Ft (lb) | $b_f/2t_f$ | $F'_y$ (ksi) | $d/t_w$ | $F''_y$ (ksi) | $r_T$ (in.) | $d/A_f$ | $I_x$ (in.⁴) | $S_x$ (in.³) | $r_x$ (in.) | $I_y$ (in.⁴) | $S_y$ (in.³) | $r_y$ (in.) | $J$ (in.⁴) | $Z_x$ (in.³) | $Z_y$ (in.³) |
|---|---|---|---|---|---|---|---|---|---|---|---|---|---|---|---|---|---|---|---|---|---|---|---|---|---|
| W18×119 | 35.1 | 18.97 (19) | 0.655 (⅝) | ⁵/₁₆ | 11.265 (11¼) | 1.060 (1¹/₁₆) | 15½ | 1¾ | ¹⁵/₁₆ | 119 | 5.3 | — | 29.0 | — | 3.02 | 1.59 | 2190 | 231 | 7.90 | 253 | 44.9 | 2.69 | 10.6 | 261 | 69.1 |
| ×106 | 31.1 | 18.73 (18¾) | 0.590 (⁹/₁₆) | ⁵/₁₆ | 11.200 (11¼) | 0.940 (¹⁵/₁₆) | 15½ | 1⅝ | ¹⁵/₁₆ | 106 | 6.0 | — | 31.7 | — | 3.00 | 1.78 | 1910 | 204 | 7.84 | 220 | 39.4 | 2.66 | 7.48 | 230 | 60.5 |
| ×97 | 28.5 | 18.59 (18⅝) | 0.535 (⁹/₁₆) | ⁵/₁₆ | 11.145 (11⅛) | 0.870 (⅞) | 15½ | 1⁹/₁₆ | ⅞ | 97 | 6.4 | — | 34.7 | 54.7 | 2.99 | 1.92 | 1750 | 188 | 7.82 | 201 | 36.1 | 2.65 | 5.86 | 211 | 55.3 |
| ×86 | 25.3 | 18.39 (18⅜) | 0.480 (½) | ¼ | 11.090 (11⅛) | 0.770 (¾) | 15½ | 1⁷/₁₆ | ⅞ | 86 | 7.2 | — | 38.3 | 45.0 | 2.97 | 2.15 | 1530 | 166 | 7.77 | 175 | 31.6 | 2.63 | 4.10 | 186 | 48.4 |
| ×76 | 22.3 | 18.21 (18¼) | 0.425 (⁷/₁₆) | ¼ | 11.035 (11) | 0.680 (¹¹/₁₆) | 15½ | 1⅜ | ¹³/₁₆ | 76 | 8.1 | 64.2 | 42.8 | 36.0 | 2.95 | 2.43 | 1330 | 146 | 7.73 | 152 | 27.6 | 2.61 | 2.83 | 163 | 42.2 |
| W18×71 | 20.8 | 18.47 (18½) | 0.495 (½) | ¼ | 7.635 (7⅝) | 0.810 (¹³/₁₆) | 15½ | 1½ | ⅞ | 71 | 4.7 | — | 37.3 | 47.4 | 1.98 | 2.99 | 1170 | 127 | 7.50 | 60.3 | 15.8 | 1.70 | 3.48 | 145 | 24.7 |
| ×65 | 19.1 | 18.35 (18⅜) | 0.450 (⁷/₁₆) | ¼ | 7.590 (7⅝) | 0.750 (¾) | 15½ | 1⁷/₁₆ | ⅞ | 65 | 5.1 | — | 40.8 | 39.7 | 1.97 | 3.22 | 1070 | 117 | 7.49 | 54.8 | 14.4 | 1.69 | 2.73 | 133 | 22.5 |
| ×60 | 17.6 | 18.24 (18¼) | 0.415 (⁷/₁₆) | ¼ | 7.555 (7½) | 0.695 (¹¹/₁₆) | 15½ | 1⅜ | ¹³/₁₆ | 60 | 5.4 | — | 44.0 | 34.2 | 1.96 | 3.47 | 984 | 108 | 7.47 | 50.1 | 13.3 | 1.69 | 2.17 | 123 | 20.6 |
| ×55 | 16.2 | 18.11 (18⅛) | 0.390 (⅜) | ³/₁₆ | 7.530 (7½) | 0.630 (⅝) | 15½ | 1⁵/₁₆ | ¹³/₁₆ | 55 | 6.0 | — | 46.4 | 30.6 | 1.95 | 3.82 | 890 | 98.3 | 7.41 | 44.9 | 11.9 | 1.67 | 1.66 | 112 | 18.5 |
| ×50 | 14.7 | 17.99 (18) | 0.355 (⅜) | ³/₁₆ | 7.495 (7½) | 0.570 (⁹/₁₆) | 15½ | 1¼ | ¹³/₁₆ | 50 | 6.6 | — | 50.7 | 25.7 | 1.94 | 4.21 | 800 | 88.9 | 7.38 | 40.1 | 10.7 | 1.65 | 1.24 | 101 | 16.6 |
| W18×46 | 13.5 | 18.06 (18) | 0.360 (⅜) | ³/₁₆ | 6.060 (6) | 0.605 (⅝) | 15½ | 1¼ | ¹³/₁₆ | 46 | 5.0 | — | 50.2 | 26.2 | 1.54 | 4.93 | 712 | 78.8 | 7.25 | 22.5 | 7.43 | 1.92 | 1.22 | 90.7 | 11.7 |
| ×40 | 11.8 | 17.90 (17⅞) | 0.315 (⁵/₁₆) | ³/₁₆ | 6.015 (6) | 0.525 (½) | 15½ | 1³/₁₆ | ¹³/₁₆ | 40 | 5.7 | — | 56.8 | 20.5 | 1.52 | 5.67 | 612 | 68.4 | 7.21 | 19.1 | 6.35 | 1.27 | 0.81 | 78.4 | 9.95 |
| ×35 | 10.3 | 17.70 (17¾) | 0.300 (⁵/₁₆) | ³/₁₆ | 6.000 (6) | 0.425 (⁷/₁₆) | 15½ | 1⅛ | ¾ | 35 | 7.1 | — | 59.0 | 19.0 | 1.49 | 6.94 | 510 | 57.6 | 7.04 | 15.3 | 5.12 | 1.22 | 0.51 | 66.5 | 8.06 |
| W16×100 | 29.4 | 16.97 (17) | 0.585 (⁹/₁₆) | ⁵/₁₆ | 10.425 (10⅜) | 0.985 (1) | 13⅝ | 1¹¹/₁₆ | ¹⁵/₁₆ | 100 | 5.3 | — | 29.0 | — | 2.81 | 1.65 | 1490 | 175 | 7.10 | 186 | 35.7 | 2.51 | 7.73 | 198 | 54.9 |
| ×89 | 26.2 | 16.75 (16¾) | 0.525 (½) | ¼ | 10.365 (10⅜) | 0.875 (⅞) | 13⅝ | 1⁹/₁₆ | ⅞ | 89 | 5.0 | — | 31.9 | 64.9 | 2.79 | 1.85 | 1300 | 155 | 7.05 | 163 | 31.4 | 2.49 | 5.45 | 175 | 48.1 |
| ×77 | 22.6 | 16.52 (16½) | 0.455 (⁷/₁₆) | ¼ | 10.295 (10¼) | 0.760 (¾) | 13⅝ | 1⁷/₁₆ | ⅞ | 77 | 6.8 | — | 36.3 | 50.1 | 2.77 | 2.11 | 1110 | 134 | 7.00 | 138 | 26.9 | 2.47 | 3.57 | 150 | 41.1 |
| ×67 | 19.7 | 16.33 (16⅜) | 0.395 (⅜) | ³/₁₆ | 10.235 (10¼) | 0.665 (¹¹/₁₆) | 13⅝ | 1⅜ | ¹³/₁₆ | 67 | 7.7 | — | 41.3 | 38.6 | 2.75 | 2.40 | 954 | 117 | 6.96 | 119 | 23.2 | 2.46 | 2.39 | 130 | 35.5 |
| W16×57 | 16.8 | 16.43 (16⅜) | 0.430 (⁷/₁₆) | ¼ | 7.120 (7⅛) | 0.715 (¹¹/₁₆) | 13⅝ | 1⅜ | ⅞ | 57 | 5.0 | — | 38.2 | 45.2 | 1.86 | 3.23 | 758 | 92.2 | 6.72 | 43.1 | 12.1 | 1.60 | 2.22 | 105 | 18.9 |
| ×50 | 14.7 | 16.26 (16¼) | 0.380 (⅜) | ³/₁₆ | 7.070 (7⅛) | 0.630 (⅝) | 13⅝ | 1⁵/₁₆ | ¹³/₁₆ | 50 | 5.6 | — | 42.8 | 36.1 | 1.84 | 3.65 | 659 | 81.0 | 6.68 | 37.2 | 10.5 | 1.59 | 1.52 | 92.0 | 16.3 |
| ×45 | 13.3 | 16.13 (16⅛) | 0.345 (⅜) | ³/₁₆ | 7.035 (7) | 0.565 (⁹/₁₆) | 13⅝ | 1¼ | ¹³/₁₆ | 45 | 6.2 | — | 46.8 | 30.2 | 1.83 | 4.06 | 586 | 72.7 | 6.65 | 32.8 | 9.34 | 1.57 | 1.11 | 82.3 | 14.5 |
| ×40 | 11.8 | 16.01 (16) | 0.305 (⁵/₁₆) | ³/₁₆ | 6.995 (7) | 0.505 (½) | 13⅝ | 1³/₁₆ | ¹³/₁₆ | 40 | 6.9 | — | 52.5 | 24.0 | 1.82 | 4.53 | 518 | 64.7 | 6.63 | 28.9 | 8.25 | 1.57 | 0.79 | 72.9 | 12.7 |
| ×36 | 10.6 | 15.86 (15⅞) | 0.295 (⁵/₁₆) | ³/₁₆ | 6.985 (7) | 0.430 (⁷/₁₆) | 13⅝ | 1⅛ | ¾ | 36 | 8.1 | 64.0 | 53.8 | 22.9 | 1.79 | 5.28 | 448 | 56.5 | 6.51 | 24.5 | 7.00 | 1.52 | 0.54 | 64.0 | 10.8 |
| W16×31 | 9.12 | 15.88 (15⅞) | 0.275 (¼) | ⅛ | 5.525 (5½) | 0.440 (⁷/₁₆) | 13⅝ | 1⅛ | ¾ | 31 | 6.3 | — | 57.7 | 19.8 | 1.39 | 6.53 | 375 | 47.2 | 6.41 | 12.4 | 4.49 | 1.17 | 0.46 | 54.0 | 7.03 |
| ×26 | 7.68 | 15.69 (15¾) | 0.250 (¼) | ⅛ | 5.500 (5½) | 0.345 (⅜) | 13⅝ | 1¹/₁₆ | ¾ | 26 | 8.0 | — | 62.8 | 16.8 | 1.36 | 8.27 | 301 | 38.4 | 6.26 | 9.59 | 3.49 | 1.12 | 0.26 | 44.2 | 5.48 |

[a]The shapes listed here are only a few of the many available. This table shows every detail needed by a designer.

(Courtesy American Institute of Steel Construction, Inc.)

## FIGURE 5.1-28  Some of the Available Structural Steel Angles[a]

| Size and Thickness | k | Weight per Foot | Area | Axis X-X | | | | Axis Y-Y | | | | Axis Z-Z | |
|---|---|---|---|---|---|---|---|---|---|---|---|---|---|
| | | | | $I$ | $S$ | $r$ | $y$ | $I$ | $S$ | $r$ | $x$ | $r$ | |
| in. | in. | lb | in.² | in.⁴ | in.³ | in. | in. | in.⁴ | in.³ | in. | in. | in. | Tan $\alpha$ |
| L9 × 4 × 5/8 | 1⅛ | 26.3 | 7.73 | 64.9 | 11.5 | 2.90 | 3.36 | 8.32 | 2.65 | 1.04 | 0.858 | .847 | 0.216 |
| 9/16 | 1 1/16 | 23.8 | 7.00 | 59.1 | 10.4 | 2.91 | 3.33 | 7.63 | 2.41 | 1.04 | 0.834 | .850 | 0.218 |
| 1/2 | 1 | 21.3 | 6.25 | 53.2 | 9.34 | 2.92 | 3.31 | 6.92 | 2.17 | 1.05 | 0.810 | .854 | 0.220 |
| L8 × 8 × 1⅛ | 1¾ | 56.9 | 16.7 | 98.0 | 17.5 | 2.42 | 2.41 | 98.0 | 17.5 | 2.42 | 2.41 | 1.56 | 1.000 |
| 1 | 1⅝ | 51.0 | 15.0 | 89.0 | 15.8 | 2.44 | 2.37 | 89.0 | 15.8 | 2.44 | 2.37 | 1.56 | 1.000 |
| 7/8 | 1½ | 45.0 | 13.2 | 79.6 | 14.0 | 2.45 | 2.32 | 79.6 | 14.0 | 2.45 | 2.32 | 1.57 | 1.000 |
| 3/4 | 1⅜ | 38.9 | 11.4 | 69.7 | 12.2 | 2.47 | 2.28 | 69.7 | 12.2 | 2.47 | 2.28 | 1.58 | 1.000 |
| 5/8 | 1¼ | 32.7 | 9.61 | 59.4 | 10.3 | 2.49 | 2.23 | 59.4 | 10.3 | 2.49 | 2.23 | 1.58 | 1.000 |
| 9/16 | 1 3/16 | 29.6 | 8.68 | 54.1 | 9.34 | 2.50 | 2.21 | 54.1 | 9.34 | 2.50 | 2.21 | 1.59 | 1.000 |
| 1/2 | 1⅛ | 26.4 | 7.75 | 48.6 | 8.36 | 2.50 | 2.19 | 48.6 | 8.36 | 2.50 | 2.19 | 1.59 | 1.000 |
| L8 × 6 × 1 | 1½ | 44.2 | 13.0 | 80.8 | 15.1 | 2.49 | 2.65 | 38.8 | 8.92 | 1.73 | 1.65 | 1.28 | 0.543 |
| 7/8 | 1⅜ | 39.1 | 11.5 | 72.3 | 13.4 | 2.51 | 2.61 | 34.9 | 7.94 | 1.74 | 1.61 | 1.28 | 0.547 |
| 3/4 | 1¼ | 33.8 | 9.94 | 63.4 | 11.7 | 2.53 | 2.56 | 30.7 | 6.92 | 1.76 | 1.56 | 1.29 | 0.551 |
| 5/8 | 1⅛ | 28.5 | 8.36 | 54.1 | 9.87 | 2.54 | 2.52 | 26.3 | 5.88 | 1.77 | 1.52 | 1.29 | 0.554 |
| 9/16 | 1 1/16 | 25.7 | 7.56 | 49.3 | 8.95 | 2.55 | 2.50 | 24.0 | 5.34 | 1.78 | 1.50 | 1.30 | 0.556 |
| 1/2 | 1 | 23.0 | 6.75 | 44.3 | 8.02 | 2.56 | 2.47 | 21.7 | 4.79 | 1.79 | 1.47 | 1.30 | 0.558 |
| 7/16 | 15/16 | 20.2 | 5.93 | 39.2 | 7.07 | 2.57 | 2.45 | 19.3 | 4.23 | 1.80 | 1.45 | 1.31 | 0.560 |
| L8 × 4 × 1 | 1½ | 37.4 | 11.0 | 69.6 | 14.1 | 2.52 | 3.05 | 11.6 | 3.94 | 1.03 | 1.05 | 0.846 | 0.247 |
| 3/4 | 1¼ | 28.7 | 8.44 | 54.9 | 10.9 | 2.55 | 2.95 | 9.36 | 3.07 | 1.05 | 0.953 | 0.852 | 0.258 |
| 9/16 | 1 1/16 | 21.9 | 6.43 | 42.8 | 8.35 | 2.58 | 2.88 | 7.43 | 2.38 | 1.07 | 0.882 | 0.861 | 0.265 |
| 1/2 | 1 | 19.6 | 5.75 | 38.5 | 7.49 | 2.59 | 2.86 | 6.74 | 2.15 | 1.08 | 0.859 | 0.865 | 0.267 |
| L7 × 4 × 3/4 | 1¼ | 26.2 | 7.69 | 37.8 | 8.42 | 2.22 | 2.51 | 9.05 | 3.03 | 1.09 | 1.01 | 0.860 | 0.324 |
| 5/8 | 1⅛ | 22.1 | 6.48 | 32.4 | 7.14 | 2.24 | 2.46 | 7.84 | 2.58 | 1.10 | 0.963 | 0.865 | 0.329 |
| 1/2 | 1 | 17.9 | 5.25 | 26.7 | 5.81 | 2.25 | 2.42 | 6.53 | 2.12 | 1.11 | 0.917 | 0.872 | 0.335 |
| 3/8 | 7/8 | 13.6 | 3.98 | 20.6 | 4.44 | 2.27 | 2.37 | 5.10 | 1.63 | 1.13 | 0.870 | 0.880 | 0.340 |
| L6 × 6 × 1 | 1½ | 37.4 | 11.0 | 35.5 | 8.57 | 1.80 | 1.86 | 35.5 | 8.57 | 1.80 | 1.86 | 1.17 | 1.000 |
| 7/8 | 1⅜ | 33.1 | 9.73 | 31.9 | 7.63 | 1.81 | 1.82 | 31.9 | 7.63 | 1.81 | 1.82 | 1.17 | 1.000 |
| 3/4 | 1¼ | 28.7 | 8.44 | 28.2 | 6.66 | 1.83 | 1.78 | 28.2 | 6.66 | 1.83 | 1.78 | 1.17 | 1.000 |
| 5/8 | 1⅛ | 24.2 | 7.11 | 24.2 | 5.66 | 1.84 | 1.73 | 24.2 | 5.66 | 1.84 | 1.73 | 1.18 | 1.000 |
| 9/16 | 1 1/16 | 21.9 | 6.43 | 22.1 | 5.14 | 1.85 | 1.71 | 22.1 | 5.14 | 1.85 | 1.71 | 1.18 | 1.000 |
| 1/2 | 1 | 19.6 | 5.75 | 19.9 | 4.61 | 1.86 | 1.68 | 19.9 | 4.61 | 1.86 | 1.68 | 1.18 | 1.000 |
| 7/16 | 15/16 | 17.2 | 5.06 | 17.7 | 4.08 | 1.87 | 1.66 | 17.7 | 4.08 | 1.87 | 1.66 | 1.19 | 1.000 |
| 3/8 | 7/8 | 14.9 | 4.36 | 15.4 | 3.53 | 1.88 | 1.64 | 15.4 | 3.53 | 1.88 | 1.64 | 1.19 | 1.000 |
| 5/16 | 13/16 | 12.4 | 3.65 | 13.0 | 2.97 | 1.89 | 1.62 | 13.0 | 2.97 | 1.89 | 1.62 | 1.20 | 1.000 |
| L6 × 4 × 7/8 | 1⅜ | 27.2 | 7.98 | 27.7 | 7.15 | 1.86 | 2.12 | 9.75 | 3.39 | 1.11 | 1.12 | 0.875 | 0.421 |
| 3/4 | 1¼ | 23.6 | 6.94 | 24.5 | 6.25 | 1.88 | 2.08 | 8.68 | 2.97 | 1.12 | 1.08 | 0.860 | 0.428 |
| 5/8 | 1⅛ | 20.0 | 5.86 | 21.1 | 5.31 | 1.90 | 2.03 | 7.52 | 2.54 | 1.13 | 1.03 | 0.864 | 0.435 |
| 9/16 | 1 1/16 | 18.1 | 5.31 | 19.3 | 4.83 | 1.90 | 2.01 | 6.91 | 2.31 | 1.14 | 1.01 | 0.866 | 0.438 |
| 1/2 | 1 | 16.2 | 4.75 | 17.4 | 4.33 | 1.91 | 1.99 | 6.27 | 2.08 | 1.15 | 0.987 | 0.870 | 0.440 |
| 7/16 | 15/16 | 14.3 | 4.18 | 15.5 | 3.83 | 1.92 | 1.96 | 5.60 | 1.85 | 1.16 | 0.964 | 0.873 | 0.443 |
| 3/8 | 7/8 | 12.3 | 3.61 | 13.5 | 3.32 | 1.93 | 1.94 | 4.90 | 1.60 | 1.17 | 0.941 | 0.877 | 0.446 |
| 5/16 | 13/16 | 10.3 | 3.03 | 11.4 | 2.79 | 1.94 | 1.92 | 4.18 | 1.35 | 1.17 | 0.918 | 0.882 | 0.448 |
| L6 × 3½ × 1/2 | 1 | 15.3 | 4.50 | 16.6 | 4.24 | 1.92 | 2.08 | 4.25 | 1.59 | 0.972 | 0.833 | 0.759 | 0.344 |
| 3/8 | 7/8 | 11.7 | 3.42 | 12.9 | 3.24 | 1.94 | 2.04 | 3.34 | 1.23 | 0.988 | 0.787 | 0.767 | 0.350 |
| 5/16 | 13/16 | 9.8 | 2.87 | 10.9 | 2.73 | 1.95 | 2.01 | 2.85 | 1.04 | 0.996 | 0.763 | 0.772 | 0.352 |
| L5 × 5 × 7/8 | 1⅜ | 27.2 | 7.98 | 17.8 | 5.17 | 1.49 | 1.57 | 17.8 | 5.17 | 1.49 | 1.57 | 0.973 | 1.000 |
| 3/4 | 1¼ | 23.6 | 6.94 | 15.7 | 4.53 | 1.51 | 1.52 | 15.7 | 4.53 | 1.51 | 1.52 | 0.975 | 1.000 |
| 5/8 | 1⅛ | 20.0 | 5.86 | 13.6 | 3.86 | 1.52 | 1.48 | 13.6 | 3.86 | 1.52 | 1.48 | 0.978 | 1.000 |
| 1/2 | 1 | 16.2 | 4.75 | 11.3 | 3.16 | 1.54 | 1.43 | 11.3 | 3.16 | 1.54 | 1.43 | 0.983 | 1.000 |
| 7/16 | 15/16 | 14.3 | 4.18 | 10.0 | 2.79 | 1.55 | 1.41 | 10.0 | 2.79 | 1.55 | 1.41 | 0.986 | 1.000 |
| 3/8 | 7/8 | 12.3 | 3.61 | 8.74 | 2.42 | 1.56 | 1.39 | 8.74 | 2.42 | 1.56 | 1.39 | 0.990 | 1.000 |
| 5/16 | 13/16 | 10.3 | 3.03 | 7.42 | 2.04 | 1.57 | 1.37 | 7.42 | 2.04 | 1.57 | 1.37 | 0.994 | 1.000 |

[a]Angles in shaded rows may not be readily available. Availability is subject to rolling accumulation and geographical location, and should be checked with material suppliers.

(Courtesy American Institute of Steel Construction, Inc.)

Among their many uses are as structural lintels over masonry openings, supports for precast concrete and stone cladding elements, floor opening framing, concrete slab opening edging, structure cross bracing, brackets used to connect structural beams to girders and columns, components of trusses, and supports for finishes furring.

Channels have many of the same applications as angles. They are also used as light beams, wall framing, and purlins. Tees are also used as light beams and in such applications as purlins in roof framing and as support for cementitious and plank decking.

Each structural shape is marked at the mill with its shape designation (*W, S, L, C, WT*), mill identification, and a code that identifies the batch of steel from which it was produced. This marking provides a convenient means of tracing a shape back to its source in the event of its failure. A certificate showing the chemical analysis of the batch of steel accompanies steel shapes from the mill to the fabricator to display compliance with the required specifications.

*Bars*  Bars may be round, square, hexagonal, or multifaceted long shapes, generally thicker than wire. They are produced by hot rolling and cold drawing. In hot rolling, between 10 and 15 roll passes, taking less than 2 minutes, are required to change a square hot billet to a bar. Bars more than $3/4$ in. (19.1 mm) in diameter are usually sheared into lengths 16 to 20 ft (4.88 to 6.1 m) long. Smaller bars are typically wound in coils for ease of handling and shipping. Carbon steel bars may be galvanized or treated with other corrosion-resistant coatings. Flat rectangular shapes (flats), which are narrower than sheet and strip, are also known as *bars* (see Fig. 5.1-23) and may be either hot or cold rolled.

Hot-rolled bars may be cold drawn to further reduce their cross section and produce complex shapes (Fig. 5.1-29) of closer dimensional tolerances and smoother finishes. Improved mechanical properties, such as greater tensile strength and hardness, also result from cold drawing. The degree of reduction depends on the type of metal used and the desired mechanical properties of the end product.

*Wire*  Cold-drawn products of round, square, or multifaceted cross section and smaller than bars are called *wire*. Round

**FIGURE 5.1-29** Irregular as well as round and rectangular bars can be easily cold drawn. (Courtesy American Iron and Steel Institute.)

wire is drawn from 0.005 in. (0.127 mm) up to 1 in. (25.4 mm) in diameter. Flat wire is a cold-rolled product, rectangular in shape and generally narrower than bars (see Fig. 5.1-23).

Semifinished wire rods of low-carbon steel may be drawn directly into wire by *angle draft drawing*. For finer wire sizes or harder steels, the rods must be softened by annealing before the continuous *multiple draft drawing* (Fig. 5.1-30) through a series of dies. Wire may also be annealed after drawing to make extremely soft *annealed wire*.

Cold-drawn wire of carbon steel can be produced with very high tensile strengths ranging up to more than 500,000 psi (351 535 $Mg/m^2$). Corrosion resistance

**FIGURE 5.1-30** Fine wire can be made by drawing rods through successively smaller die openings. (Courtesy American Iron and Steel Institute.)

can be achieved by galvanizing or coating with other corrosion-resistant metals.

*Tubular Products*  Tubular products are long, hollow metal products of round, oval, square, rectangular, or multifaceted cross section. Tubular products can be either *pipe* or *tubing*, the terms being applied somewhat interchangeably. Round pipe in a wide range of sizes and square or rectangular structural tubing are widely employed in building construction. Tubular products which are mechanically worked and produced on *welded* and *seamless* mills are referred to as *wrought pipe* to distinguish them from cast pipe. Wrought tubular products may be marketed in the hot- or cold-finished condition resulting from a mill operation as *black pipe*, or as *galvanized pipe* when hot dipped in zinc.

*Welded pipe* is made chiefly by the butt weld and electric weld processes. In the *butt weld* process, a strip of preheated, semifinished skelp is continuously formed into a tubular shape between concave rolls, and the edges are joined by mechanical pressure during rolling (Fig. 5.1-31a). Pipe sizes between $1/8$ in. (3.2 mm) and 4 in. (101.6 mm) are commonly produced by this process.

The majority of steel pipe in this country is produced by *electric weld* processes. Preheating of the metal is not required, and localized heat necessary for joining the edges is generated during the welding process itself. Resistance welding is used to make *electric-resistance welded* (ERW) tubular products of up to 20 in. (508 mm) in diameter. In this process, a strip of cold metal is formed into a tubular shape by rolling, and the edges are joined by heat and pressure (Fig. 5.1-31b). Welding heat is generated in the seam by the resistance of the metal to an electric current introduced by a wheel-shaped roller acting as the electrode. For larger-diameter and thick-walled pipe, one of several electric welding processes may be used. In most, steel strip or plates are die formed cold by several successive forming operations consisting of crimping (bending the edges), U forming, and O forming. The last forming operations result in a nearly round shape with edges touching and ready for welding. Submerged arc welding and metal inert-gas-shielded arc (MIG) welding are often used for this purpose. The final shape and diameter are obtained by expanding

## FROM SKELP TO BUTT WELD PIPE

(continuous rolled)

TRAVELING SAW
CUTS PIPE TO
LENGTH

a

## FROM SKELP TO ELECTRIC WELD PIPE

DC

WATER

CUTTING
TOOLS

b

**FIGURE 5.1-31** (a) The butt weld process can produce pipes up to 4 in. (101.6 mm) in diameter. (b) The electric weld process can produce pipes up to 8 in. (203.2 mm) in diameter. (Courtesy American Iron and Steel Institute.)

## FROM BILLET TO SEAMLESS PIPE

SIZING
MANDREL

END VIEW

a   CENTER PUNCHED

## FROM PIPE TO COLD DRAWN TUBING

MANDREL

DIE

LUBRICANT

MANDREL

CHAIN PULLS GRIP DRAWING
PIPE THROUGH DIE

DIE

b

**FIGURE 5.1-32** (a) Most seamless carbon steel pipe is made by hot piercing and extruding. (b) Stainless and alloy steel tubing is cold drawn. (Courtesy American Iron and Steel Institute.)

the welded shell with hydraulic power against a retaining jacket.

*Seamless pipe and tubing* are made by hot piercing and hot extrusion. Piercing is accomplished by continuously feeding a hot billet between pairs of tapered rolls that are skewed to apply uneven pressure, which causes a pilot hole to form in the center of the billet as it rolls between them. A piercing point enlarges the pilot hole, and a sizing mandrel shapes the final opening size and wall thickness (Fig. 5.1-32a). Most seamless products are made by hot working, although some seamless tubing is made by cold drawing a prepierced billet through a die over an internal mandrel (Fig. 5.1-32b). Stainless and alloy steel tubing up to 11 in. (279.4 mm) in diameter are often made in this manner. Hot piercing methods can produce seamless pipe up to about 26 in. (660.4 mm) in diameter.

Hot extrusion produces a tubular shape by forcing hot predrilled billets through a die over an internal mandrel. This method is used primarily to make small sizes of stainless or high-alloy pipe for cold reducing. It is also the most economical method of producing complex tubular shapes in small quantities.

### 5.1.4.3 Fabrication

Most steels can be fabricated by hot forming, cold forming, and machining. Some forming methods may be the same as those previously described; however, *fabrication* as used here implies reworking the products of the mills and foundries in a fabricating plant. Cold-forming techniques are by far the more common ones used in the manufacture of products for building construction.

### HOT FORMING

Hot-forming operations are used on metals too thick or too hard to be formed cold. These operations include (1) forging (described earlier); (2) high-temperature forming (electroforming), which uses heat generated by the resistance of the workpiece to an electric current passed through it by the forming roll; and (3) high-energy rate forming, a method of hot (sometimes cold) shaping of materials by the sudden impact of a small explosive charge, a shock wave generated by an electric spark in a fluid, or sudden release of a compressed gas through a system of valves.

### COLD FORMING

Cold-forming operations are performed on most flat-rolled and tubular products by shaping, bending, or drawing the metal beyond its yield point so that a permanent set is achieved. The most common cold-forming methods are *roller*, *stretch*, *shear*, *brake forming*, and *deep drawing*.

**Roller Forming** A variety of shapes (Fig. 5.1-33) are formed by passing sheet and strip through a series of roller dies which gradually impart the desired shape (Fig. 5.1-34). Many types of corrugated and ribbed decking and wall panels are made in this manner.

**Stretch Forming** In stretch forming, light-gauge flat products are shaped over form blocks into shallow curved shapes by pressure. The steps in stretch forming a particular shape are illustrated in Figure 5.1-35.

**FIGURE 5.1-33** Sheet and strip may be roller formed into a variety of shapes. (Courtesy American Iron and Steel Institute.)

**FIGURE 5.1-34** Wall and roofing panels typically are roller formed from galvanized or stainless steel sheets. (Courtesy American Iron and Steel Institute.)

**FIGURE 5.1-35** In stretch forming, an oversized piece of sheet metal is gripped (a) and bent over a form block (b) to produce the desired contour (c) and then is trimmed to the final shape (d). (Courtesy American Iron and Steel Institute.)

*Shear Forming* In shear forming, flat round blanks are formed into curved surface such as cylinders, domes, and cones by turning the workpiece on a mandrel and shaping it over a block with external pressure. *Spinning* is a type of shear forming in which the deformation and reduction in original blank thickness are not quite as severe (Fig. 5.1-36). Tank heads, television tube cones, and stainless steel railing base flanges and caps are examples.

*Brake Forming* Brake forming is a common method of fabricating sheet metal components for construction. In this method, sheet steel is shaped on a press brake by squeezing it between a punch and a die block in one or several consecutive operations (Fig. 5.1-37). Heating and air conditioning ducts, steel roofing products, and many of the sections used in architectural stainless steel storefront construction are fabricated by this method.

*Deep Drawing* Flat, thin-gauge blanks are often formed in a draw press between an operating punch and a stationary die. Many household utensils, such as pots and pans, and commercial vessels and

**FIGURE 5.1-36** In spinning, a round blank is clamped on a form block (a) and spun while pressure is applied (b), gradually imparting the shape of the form block (c). Then it is trimmed to the desired profile (d). (Courtesy American Iron and Steel Institute.)

**FIGURE 5.1-37** Both carbon steel and stainless steel can be bent sharply in a press brake to form various products. (Courtesy American Iron and Steel Institute.)

containers are made this way. Stainless steel can be drawn to various degrees, but the austenitic group lends itself most readily to deep-drawing operations.

## MACHINING

Most steels can be milled, sawed, drilled, punched, sheared, tapped, and reamed with appropriate equipment. Tool steels containing tungsten, molybdenum, and chromium that have been subjected to tempering and other heat treatments are used for this purpose. Sulfur and lead improve the machining properties of carbon steels and are added to steels requiring extensive machining. Among the stainless steels, the ferritic alloys are the easiest to machine.

# 5.2 Aluminum Materials and Products

Aluminum was first detected in 1807 as an ingredient of common clay, but it was not until 1825 that the first metallic aluminum was produced, and not until 1845 was aluminum powder successfully transformed into solid particles. In 1852, due to costly production methods, pure aluminum sold for $545 per pound and was used only for the finest jewelry and tableware.

The 1886 discovery that metallic aluminum could be produced by dissolving alumina (aluminum oxide) in molten cryolite and then passing an electric current through the solution gave birth to the modern aluminum industry.

Today, buildings and other construction comprise a large market for aluminum in the United States. Products that use large amounts include windows, doors, storefronts, building cladding systems, screens, siding, and mobile homes.

A unique combination of properties makes aluminum one of the most versatile engineering and construction materials. It is light in weight, yet some of its alloys have strengths greater than that of structural steel. It has high resistance to corrosion and will not rust red to stain adjacent surfaces or discolor products.

It has no toxic action, and possesses high electrical and thermal conductivities and high reflectivity to both heat and light. Aluminum can be cast or worked into a large variety of shapes and accepts many surface finishes. These characteristics give aluminum its versatility, and usually two or more of them come into play.

One of aluminum's most important attributes is that it functions as designed and maintains its finish over extended periods. For example, an aluminum cap installed on the top of the Washington Monument in 1884 remains in good condition. Many large and small buildings have been clad with aluminum components, and millions of homes and other small buildings have been sided with aluminum during the past 70 years.

As with other materials, however, aluminum's properties must be understood to ensure its proper use. This section discusses the production of aluminum and aluminum alloys and their various properties, as well as methods by which aluminum products are manufactured. Individual aluminum products are discussed in Chapters 7, 8, and 9.

## 5.2.1 RAW MATERIAL

Aluminum forms 8% of the earth's crust, making it the most plentiful metal, but it is never found free in nature. Unlike gold and silver, aluminum is always combined chemically with other elements. Though nearly all common rocks and clays contain some aluminum, it has not been historically practical to extract aluminum unless the raw ore contained at least 45% aluminum oxide. This percentage may change from time to time, however, as economic conditions vary.

Such ores are called *bauxites*, after the town of Les Baux in France where one of the first bauxite ore deposits was found. Most of the world's bauxite deposits are outside the United States, making it necessary for the United States to import most of its supply. About 95% of the bauxite mined in the United States comes from Arkansas, and the remainder from small, lower-grade deposits in Georgia and Alabama.

Work is now underway in the United States to develop methods of using domestic ores, such as kaolin and alunite, as replacements for imported bauxite. The U.S. Bureau of Mines, which is coordinating the project, believes that the technology exists to extract aluminum commercially from these domestic ores.

### 5.2.1.1 Mining

Strip or open pit mining methods are generally used to obtain aluminum ore, which is crushed, washed, screened, ground, and then dried in kilns (large revolving steel drums) at temperatures of up to 250°F (121.11°C). The general steps in the mining, refining, and reduction of aluminum are shown in Figure 5.2-1.

### 5.2.1.2 Refining

Commercial bauxite refining is done by the Bayer process, which separates aluminum oxide (alumina) from unwanted minerals.

Dried, ground bauxite is mixed with a solution of caustic soda (sodium hydroxide) which dissolves the alumina to form sodium aluminate. The silica in the bauxite reacts and precipitates out of solution. Iron oxide and other impurities are not affected chemically and, being solids, settle out.

This *green liquor* is now highly supersaturated sodium aluminate. Previously prepared hydrated alumina crystals are added to the solution. These form larger crystals, which gradually settle out. After being washed to remove remaining traces of impurities, the concentrated aluminum hydrate crystals are roasted at temperatures of more than 2000°F (1093.3°C), which drives off the water. The resulting alumina is a fine white powder similar to sugar in appearance and consistency.

### 5.2.1.3 Reduction

*Reduction* is the electrolytic separation of aluminum from its oxide, alumina. The general reduction procedure is described in Figure 5.2-1.

This process employs a carbon-lined vessel (pot) containing molten cryolite (sodium aluminum fluoride) in which alumina is dissolved. Metallic aluminum is separated from the alumina by passing an electric current through this solution at a temperature of about 1775°F (963.33°C).

Molten aluminum collects at the bottom of the pot and is tapped or siphoned off into large ladles. It is then poured into molds to form ingots or transferred to holding furnaces for alloying. Aluminum, as it comes from the pots, is about 99.5% pure. By additional refining, superpurity aluminum, which is about 99.99% pure, can be produced.

### 5.2.1.4 Recycling

Recycled aluminum, which has already had a useful life in a product and is used again, is playing an increasingly important role (see Fig. 5.2-1). Recycling is important because it saves 95% of the energy required to make aluminum from ore. Ingots that come from recycling are called *secondary ingots*.

Old aluminum windows, doors, and siding are among the many products that are being brought to scrap dealers for recycling.

### 5.2.1.5 Aluminum Alloys

High-purity aluminum has a tensile strength of about 7000 psi (4921.5 Mg/m²) and is relatively soft and ductile. Most applications, however, require greater strength and other properties achieved by the addition of other elements to produce various aluminum *alloys*. Aluminum alloys can be divided into two general categories: those whose strength characteristics may be improved by heat treatment (*heat-treatable* alloys) and those whose strength may be improved by mechanical methods of cold working (*non-heat-treatable* alloys).

#### HEAT-TREATABLE ALLOYS

The initial strength of heat-treatable alloys is improved by the addition of alloying elements such as copper, magnesium, zinc, and silicon. Since these elements (singly or in various combinations) show increasingly solid solubility in aluminum with increasing temperature, it is possible to subject them to thermal treatments which will impart further strengthening.

The first step, called *solution heat treatment*, is an elevated-temperature process. This is followed by rapid quenching, usually in water, which momentarily "freezes" the structure and for a short time renders the alloy very workable. It is at this stage that some fabricators retain this more workable structure by storing the alloys at below-freezing temperatures until they are ready to be formed, At room or elevated temperatures the alloys are not stable after quenching, and after a period of several days at room temperature, the alloys are considerably stronger. This is called *natural aging* or *room temperature precipitation*. Some alloys, particularly those containing magnesium and silicon or magnesium and zinc, continue to

**ALLOYING & CASTING**

EXTRUSION BILLETS

ALLOY INGOTS

ROLLING INGOTS

CASTING MOLDS

HOLDING AND ALLOYING FURNACE

**ALUMINUM PRODUCTS**

SKYSCRAPERS

WHEELS

CANS

FOOD TRAYS

CANOES

AIRPLANES

SPACE SHUTTLE

TENNIS RACKETS

WINDOWS/SIDING

**RECYCLING**

ALLOYING FURNACE

USED ALUMINUM

RECYCLED ALUMINUM PRODUCTS

© Copyright 1982 Aluminum Association/Jerline Learning Systems, Inc.

**SMELTING**

ALUMINA

TO REDUCTION

BUS BAR

CARBON ANODE

CRYOLITE BATH

MOLTEN ALUMINUM

CARBON LINING

STEEL SHELL

SIPHON

CRUCIBLE

**HALL-HEROULT REDUCTION PROCESS**

$$2Al_2O_3 + 3C \xrightarrow[Na_3AlF_6]{E} 4Al + 3CO_2$$

**MINING & REFINING**

STORAGE

SPRAY WATER

COVERED HOPPER CARS

CRUSHED BAUXITE

MINING BAUXITE

CRUSHER

CLAY, SILICA WASTE

WASTE TREATMENT

DRYING KILN

SODA ASH

CRUSHED LIME

CRUSHED BAUXITE

MIXER

SETTLING TANK

DIGESTER

PRESSURE REDUCER

FILTER

COOLING TOWER

PRECIPITATOR

WASTE TREATMENT

THICKENER

ROTARY CALCINATING KILN

COVERED HOPPER CARS

FILTER

the Aluminum Association ®
818 Connecticut Avenue, N.W. Washington, D.C. 20006

**FIGURE 5.2-1** The mining, refining, and reduction of aluminum. (Courtesy Aluminum Association.)

age-harden for long periods of time at room temperature.

By heating the alloy at a controlled temperature, even further strengthening is possible and properties are stabilized. This process is called *artificial aging* or *precipitation hardening*. The highest strengths are obtained by the proper combination of solution heat treatment, quenching, cold working, and artificial aging.

## NON-HEAT-TREATABLE ALLOYS

Non-heat-treatable (*common*) aluminum alloys have alloying elements that do not show an increase in strength with heat treatment. The initial strength of non-heat-treatable alloys depends on the hardening effects of elements such as manganese, silicon, iron, and magnesium, singly or in various combinations. These alloys can be strengthened by *strain-hardening*, *cold rolling*, and other mechanical working.

## ANNEALING

*Annealing* is a thermal treatment that is applicable to both heat-treatable and non-heat-treatable alloys and to cast and wrought products. It is done by elevating the temperature of the metal to between 600°F (315.56°C) and 800°F (426.7°C) and then cooling it slowly. The purpose of this treatment is to relieve internal stresses and to return the alloy to its softest and most ductile condition. Annealing is usually performed to condition the metal for severe forming operations. Subsequent working processes increase the mechanical properties of non-heat-treatable alloys. The mechanical properties of heat-treatable alloys may be increased by appropriate thermal treatments after forming operations have been completed.

## ALLOY DESIGNATION SYSTEMS

Aluminum products fall into two general groups: *wrought* products and *cast* products. Alloys and temper are designated by numerical systems for both cast and wrought products. Figure 5.2-2 shows some typical uses for various alloys.

***Temper Designations*** A notation or temper designation is used following the alloy (Fig 5.2-3). The letter *H* is used for non-heat-treatable and the letter *T* for heat-treatable alloys. Some temper designations apply only to wrought products and others only to cast products, but most apply to both. Additional digits after the

**FIGURE 5.2-2** Examples of Aluminum Alloys Used for Some Building Applications

| Application | Alloy Number | Product |
|---|---|---|
| Acoustic ceiling | 3003-H14 | Sheet |
| Air duct | 3003-H14, 1100-H16, Duct stock | Sheet |
| Awning | 3003-H14 | Sheet |
| Builder's hardware | 5050-0 | Sheet |
| | B443.0, 356.0 | Casting |
| | 6063-T5 | Extrusion |
| Curtain wall | 3003-H14, Alclad 4043, Alclad 1235 | Sheet |
| | B443.0 | Casting |
| | 6063-T5, 4043-F | Extrusion |
| Deck (corrugated) | 3003-H14 | Sheet |
| Door frame | 6063-T5, 6063-T6 | Extrusion |
| Fascia plate | 6063-T42 | Extrusion |
| | 3003-H14 | Sheet |
| Flashing | 3003-0 | Sheet |
| Flue lining | 1100-H16 | Sheet |
| Garage door | 3003-H14 | Sheet |
| Grating | 6061-T6 | Bar |
| Gravel stop | 6063-T5, 6063-T42 | Extrusion |
| Grille | B443.0 | Sand casting |
| | 3003-H14 | Sheet |
| | 6063-T5 | Extrusion |
| Gutter | 3003-H14 | Sheet |
| Hood | 3003-H14 | Sheet |
| Insulation | 1235 | Foil |
| Kick plate | 3003-H18 | Sheet |
| Letters | B443.0, 514.0 | Sand casting |
| | 3003-H14 | Sheet |
| Louver | 3003-H14 | Sheet |
| | 6063-T5 | Extrusion |
| Mullion | 3003-H14 | Sheet |
| | 6063-T5, 6063-T6 | Extrusion |
| Nails | 6061-T913 | Wire |
| Railing | 6063-T5, 6061-T83 | Extrusion |
| Relief panel (sculptural) | B443.0 | Casting |
| Roofing | Special roofing alloys | Sheet |
| Screen | Alclad 5056-H392 | Wire |
| Screws | 2024-T4, 2017-T4 | Screw machine stock |
| Shade screening | 5052-H38 | Sheet |
| Shingle | 3003-H14 | Sheet |
| Siding | 3003-H14 | Sheet |
| Structural shape | 2014-T6, 6061-T6 | Extrusion |
| | 6061-T6, 2014-T6 | Rolled shape |
| Termite shield | 3003-H14 | Sheet |
| Terrazzo strip | 3003-H14 | Sheet |
| Threshold | 6063-T5 | Extrusion |
| Vapor barrier | 1235-0 | Foil |
| Venetian blind | 5052-H18 | Sheet |
| Weatherstrip | 5052-H38 | Sheet |
| Window | 6063-T5, 6063-T6, 6061-T6 | Extrusion |

**FIGURE 5.2-3** Aluminum Association's Temper Designations

### Temper Designations

| | | |
|---|---|---|
| -F | As fabricated | |
| -O | Annealed | |
| -H | Strain hardened (wrought only) | |
| | -H1 | Plus one or more digits[a]: strain hardened only |
| | -H2 | Plus one or more digits[a]: strain hardened and partially annealed |
| | -H3 | Plus one or more digits[a]: strain hardened, then stabilized |
| | -H4 | Plus one or more digits[a]: strain hardened and lacquered or painted |
| -W | Solution heat treated—unstable temper | |
| -T | Thermally treated to produce stable tempers other than -F, -O, or -H | |
| | -T1 | Cooled from an elevated temperature shaping process and naturally aged to a substantially stable condition |
| | -T2 | Cooled from an elevated temperature shaping process, cold worked, and naturally aged to a substantially stable condition |
| | -T3 | Solution heat treated, cold worked, and naturally aged to a substantially stable condition |
| | -T4 | Solution heat treated and naturally aged to a substantially stable condition |
| | -T5 | Cooled from an elevated temperature shaping process, then artificially aged |
| | -T6 | Solution heat treated, then artificially aged |
| | -T7 | Solution heat treated and stabilized |
| | -T8 | Solution heat treated, cold worked, then artificially aged |
| | -T9 | Solution heat treated, artificially aged, then cold worked |
| | -T10 | Cooled from an elevated temperature shaping process, cold worked, then artificially aged |

[a]Second digit indicates final degree of strain hardening, i.e., 2 is ¼ hard, 6 is ¾ hard, 8 is full hard.

The temper letter *H* is used for non-heat-treatable alloys, *T* for heat-treatable alloys.

*T* or *H* (H1, H2, T1, T2, etc.) indicate variations in the basic treatment.

***Wrought Alloys*** Almost all aluminum products used in construction are made from wrought alloys, which are designated by a four-digit index system (Fig. 5.2-4). The first digit of the number indicates an alloy group or series.

The 1000 Series designates aluminum of 99% or higher purity. These alloys have excellent corrosion resistance, high thermal and electrical conductivity, low mechanical properties, and excellent workability. Moderate increase in strength may be obtained by strain hardening. This group includes alloys (1235) used in electrical and chemical fields and alu-

minum foils for insulation and vapor retarders; alloy 1100 is used for sheet metal work.

In the 2000 Series, copper is the principal alloying element. These alloys require heat treatment to obtain optimum properties, and in the heat-treated condition their mechanical properties are similar to and may exceed those of mild steel.

Sometimes artificial aging is employed to increase strength properties. The alloys in the 2000 Series do not have as good corrosion resistance as most other aluminum alloys, and in sheet form are usually clad (see Section 5.9.8.1) with high-purity alloy or a magnesium-silicon alloy of the 6000 Series. Alloy 2024 is used for fasteners; alloy 2014 for structural shapes.

The 3000 Series alloys have manganese as the major alloying element. They are usually non-heat-treatable. Because only a limited percentage of manganese (up to about 1.5%) can be effectively added to aluminum, it is used as a major element in only a few alloys.

Alloy 3003 is a general-purpose alloy used for moderate-strength applications requiring good workability. In sheet form, it is used for air ducts, acoustical ceiling pans, awnings, corrugated decks, garage doors, flashings, gutters, siding, shingles, and other products. The major alloying element in the 4000 Series is silicon, which substantially lowers the melting point of the alloy without producing brittleness. Most alloys in this series are non-heat-treatable. Those containing appreciable amounts of silicon become dark gray when anodic oxide finishes are applied and hence are used for architectural applications.

The 5000 Series alloys use magnesium as the major alloying element. Magnesium is one of the most effective and widely used alloying elements, and when it is used as the major alloying element (or with manganese), the result is a moderate-to high-strength, non-heat-treatable alloy. Magnesium is more effective than manganese as a hardener and can be added in considerably greater quantities. These alloys have especially good resistance to corrosion in marine atmospheres and sea water. Shade screening, venetian blinds, and weatherstripping are often made from alloy 5052.

The 6000 Series alloys are heat-treatable alloys containing silicon and magnesium. They are not as strong as most alloys of the 2000 or 7000 Series, but they

**FIGURE 5.2-4** The Aluminum Association's Four-Digit Number System for Designating Wrought Aluminum Alloys

X   X   X   X

**First Digit**

**Major Alloying Elements**

2 xxx — Copper
3 xxx — Manganese
4 xxx — Silicon
5 xxx — Magnesium
6 xxx — Magnesium and silicon
7 xxx — Zinc
8 xxx — Other elements
9 xxx — Unused series

**Second Digit**

**Modification of Alloy**

x0xx — Original alloy
x1xx
x2xx
x3xx
x4xx
x5xx  } Alloy modifications
x6xx
x7xx
x8xx
x9xx

**Third and Fourth Digits**

**Arbitrary Numbers**

Used to identify alloys within the series

possess good formability and corrosion resistance with medium strength. They may be formed in the solution heat-treated temper (T4 temper) and then reach higher strengths by artificial aging (T6 temper). This series is the one most commonly used in products for residential construction. For example, 6061 is used in structural shapes, nails, and bars; 6063 in window and door frames, hardware, louvers, tubing, and curtain walls.

The 7000 Series alloys have zinc as the major alloying element, which, when combined with a smaller percentage of magnesium, results in heat-treatable alloys of very high strength. Usually, other elements, such as copper and chromium, are also added in small quantities. Among the highest-strength alloys available is 7075. It is often used in aircraft structures and other high-stressed parts.

Experimental alloys are also designated with the groups described above but carry the prefix *X*, which is dropped when the alloy becomes standard.

*Cast alloy* designations use a four-digit system similar to that for wrought alloys, shown in Figure 5.2-4.

## 5.2.2 PROPERTIES

Aluminum has a specific gravity of about 2.70, a density of about 170 lb/cu ft (1.71 kg/m$^3$), and weighs about 0.1 lb/cu in. (276.8 mg/m$^3$), compared with 0.28 16/cu in. (775.04 mg/m$^3$) for iron and 0.32 lb/cu in. (885.76 mg/m$^3$) for copper.

Pure aluminum melts at about 1200°F (648.89°C). Melting points for various alloys range from 900°F (482.2°C) to 1250°F (676.67°C).

Aluminum and aluminum alloys have relatively high coefficients of thermal expansion. For structural calculation, the coefficient of expansion is assumed to be $13 \times 10^{-6}$. This means, for example, that an unrestrained piece of aluminum 18 ft (5.29 m) long will expand or contract 0.259 in. (6.5786 mm) (about $^1/_4$ in. [6.4 mm]) in a 100°F (37.8°C) temperature change. The two common metals used as electric conductors are copper and aluminum. The electrical conductivity of EC grade (electric-conductor) aluminum is about 62% that of the International Annealed Copper Standard. However, with a specific gravity of less than one-third that of copper, 1 lb (0.45 kg) of aluminum will conduct twice as much electricity as 1 lb (0.45 kg) of copper. Unfortunately, aluminum is stiffer than copper, so connections are more difficult to make. This stiffness can result in inadequate and therefore unsafe connections.

The high thermal conductivity of aluminum makes it well fitted for use wherever a transfer of thermal energy from one medium to another is desired. However, high thermal conductivity is generally not required in most building material applications and, in fact, is often a problem in such components as window frames and in storefronts and curtain wall sections, where *thermal breaks* (other nonconducting materials) are often introduced to reduce the flow of heat through the aluminum.

Aluminum is an excellent reflector of radiant energy throughout the entire range of wavelengths, from ultraviolet through the visible spectrum to infrared (heat) waves. Aluminum's light reflectivity of more than 80% accounts for its wide use in lighting fixtures, and its high heat reflectivity and low emissivity have led to its use as reflective insulation and improved performance in such applications as paint, roofing, and ductwork.

Aluminum's nonmagnetic properties make it useful for electrical shielding purposes in electrical equipment, and its nonsparking characteristics make it an excellent material for use around inflammable or explosive substances.

The ease with which aluminum may be fabricated is one of its most important properties. The metal can be cast, rolled to many thicknesses (down to foil thinner than paper), extruded, stamped, drawn, spun, roll formed, hammered, or forged; it can be turned, milled, bored, or machined at maximum speeds.

Almost any method of joining is applicable to aluminum, and a wide variety of mechanical aluminum fasteners simplifies the assembly of products (see Section 5.3).

### 5.2.2.1 Strength

Although pure aluminum has a tensile strength of only about 7000 psi (4921.5 Mg/m$^2$), by working the metal, as by cold rolling, its strength can be about doubled. Much larger increases in strength can be obtained by *alloying* aluminum with small percentages of one or more other metals such as manganese, silicon, copper, magnesium, or zinc. Like pure aluminum, the alloys are also made stronger by cold working, and heat-treatable alloys are further strengthened and hardened by heat treatments so that tensile strengths approaching 100,000 psi (703 070 Mg/m$^2$) are possible (Fig. 5.2-5). However, such high-strength aluminums are seldom used in building construction products.

Strength decreases at elevated temperatures, although some alloys retain good strength at temperatures from 400°F (204.4°C) to 500°F (260°C). At subzero temperatures, strength increases without loss of ductility so that aluminum is a particularly useful metal for low-temperature applications.

#### 5.2.2.2 Corrosion Resistance

Aluminum's high resistance to corrosion is due to the very thin, inert surface film of aluminum oxide which forms rapidly and naturally in air. After it reaches a thickness of about one ten-millionth of an inch, this film effectively halts further atmospheric oxidation of the metal and thus protects the surface. If this natural oxide film is broken, as by a scratch, a new protective film forms immediately. The oxide film increases in thickness with temperature and remains protective of the underlying aluminum even at the melting point.

It may be necessary to protect aluminum in certain severely corrosive environments by protective coatings such as organic paints, by cladding, or by increasing the thickness and effectiveness of the oxide film by anodizing (see Section 5.9).

#### WEATHERING

The extent of corrosion due to weathering depends on the type and extent of contamination of the surrounding atmosphere. More corrosive elements are present in industrial and marine areas than in rural areas.

In tests and observations (in part under the direction of the National Institute of Standards and Technology [NIST]), made over a 10-year period in different atmospheric conditions on alloys 1100, 3003, and 3004, the depth of penetration of atmospheric corrosion was less than one mil (1/1000 in. [0.0254 mm]) in rural areas, about 4 mils (0.1016 mm) in industrial areas, and up to 6 mils (0.1524 mm) in seacoast locations.

#### GALVANIC CORROSION

When two dissimilar metals are connected by a solution which conducts elec-

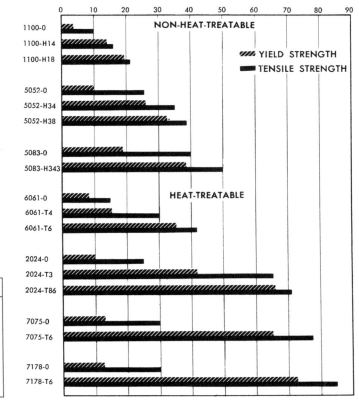

**FIGURE 5.2-5** Mechanical properties of representative aluminum alloys.

similar metal joint may be sealed with compatible building mastic or caulking.

A dangerous source of electrolytic attack may be water drainage from metals such as copper. Therefore, aluminum gutters on buildings with copper flashings must be protected by painting either the copper flashing or both the copper and aluminum.

## DIRECT CHEMICAL ATTACK

Some chemicals will dissolve metals on contact. For example, solutions of strong alkalis, sulfuric acid, hydrochloric acid, carbonates, and fluorides will attack aluminum, because they can dissolve or penetrate its protective oxide coating. Some chemical attack may also occur when aluminum is placed in contact with wet alkaline materials such as mortar, concrete, and plaster. Therefore, surface protection of aluminum products is often necessary during construction. Where aluminum is buried in concrete, it is usually not necessary to paint the parts. However, if chlorides are present from additives or other sources in steel-reinforced concrete, the aluminum parts should be

tricity (an electrolyte), *galvanic corrosion* may occur. As in a battery, an electric current is created, and one metal corrodes away while the other is being plated. Galvanic attack can occur when moisture condenses from the air and contaminating elements act as an electrolyte. The threat of such corrosion is greatest in industrial areas, because the atmospheric moisture is more contaminated, and in coastal areas because of the salt and moisture in the air.

The severity of galvanic attack varies, depending on how far apart the two metals are in the galvanic table (Fig. 5.2-6). For example, coupling mild steel to aluminum produces less corrosion than does coupling copper to aluminum under identical exposure conditions. The metal that is higher in the potential table is sacrificial and corrodes. Thus, zinc provides cathodic protection to aluminum and aluminum alloys. That is, when zinc and aluminum are in the presence of an electrolyte, the zinc usually corrodes. Thus, zinc-plated steel parts are commonly

used with aluminum to prevent the harmful effect of galvanic action on the aluminum.

Cadmium-plated steel in contact with aluminum alloys is also satisfactory. Since there is no appreciable difference in the electrolytic potential of these two metals, practically no galvanic action occurs.

Aluminum is compatible with some stainless steels, chromium, zinc, and small areas of white bronze. Where permanent contact with other metals cannot be avoided, the risk of galvanic corrosion can be greatly reduced by painting the other metal and the aluminum at the contact area with zinc chromate, followed by one coat of a nonleaded paint such as aluminum paint or of a heavy-bodied bituminous paint. When the dissimilar metal cannot be painted, the aluminum may be given the insulating treatment. As an alternative, a strip of plastic or a similar insulator may be placed between the two metals.

In severely corrosive atmospheres and in high-moisture areas, the edges of a dis-

**FIGURE 5.2-6** Electrolytic Solution Potentials[a]

| Metal or Alloy | Potential in Volts |
|---|---|
| Magnesium | −1.73 |
| Zinc | −1.00 |
| 520.0-T4 alloy (cast) | −0.92 |
| 7072 (wrought) | −0.96 |
| 514.0 alloy (cast) | −0.87 |
| Pure aluminum and 5052-0, 1100-0, 1100-H18, 3003-H18, 6061-T6 alloys (wrought) and 43 alloy (cast) | −0.83 |
| Cadmium | −0.82 |
| 356.0-T4 alloy (cast) | −0.81 |
| 2017-T4, 2024-T4 alloys (wrought) | −0.68 |
| Mild steel | −0.58 |
| Lead | −0.55 |
| Tin | −0.49 |
| Brass (60–40) | −0.28 |
| Copper | −0.20 |
| Stainless steel (18–8) | −0.15 |

[a]Measured in a 5.85% sodium chloride solution containing 0.3% hydrogen peroxide. The values vary somewhat depending on the particular lot of material and on the surface preparation employed. Coupling metals with dissimilar electrolytic potentials in an electrolyte results in galvanic corrosion of the metal with the highest negative potential. Degree of attack generally increases as potential difference increases.

coated with bituminous paint or another suitable coating.

Aluminum in contact with concrete, masonry, or other absorbent materials under even intermittently wet conditions should be protected with a coating of bituminous paint, zinc chromate primer, or a separating layer of plastic or other gasketing material. Creosote and tar coatings should not be used, however, because of their acid contents. When wet concrete or another alkaline material may splash against aluminum during construction, the aluminum surfaces should be protected. A coating of clear acrylic-type lacquer or a suitable strippable coating may be used when maintenance of appearance is important.

Although aluminum is inert when buried in many types of soil, some soils and the presence of stray electrical currents in the ground may cause corrosion. Therefore, buried aluminum pipe may have to be coated or otherwise protected.

### 5.2.3 MANUFACTURE

*Wrought* mill products make up the majority of aluminum products used in the building industry. They include plate, sheet, foil, bar, rod wire, extruded sections, structural shapes, tube, pipe, and forging stock. The mechanical properties of such materials are developed by alloying, hot working, cold working, or heat treatment, or by combinations of these processes.

*Cast* products include sand castings, permanent mold castings, and die castings. These are produced in a foundry by pouring molten metal into molds of a desired shape. The mechanical properties of some castings may be improved by heat treatment.

The general categories of wrought and cast products are outlined in Figure 5.2-7.

#### 5.2.3.1 Wrought Products

Wrought mill products are those shaped by hot and cold rolling, cold drawing, extruding, and forging.

##### HOT AND COLD ROLLING

Hot rolling is performed in mills, in rolling machines through which an ingot, heated to soften it, is passed until it is reduced to the desired shape. Hot rolling can be continued until a thickness of about 0.125 in. (3.175 mm) is obtained. Then, further thinning is usually accomplished by cold rolling. Cold-rolled material has a better finish and, in the case of non-heat-treatable alloys, increased strength and hardness.

*Plate* Plate has a minimum thickness of 0.250 in. (6.40 mm). To make it, a slab of aluminum is hot rolled in a reversing mill (acting like a rolling pin) until its thickness has been reduced to 1 to 3 in. (25.2 to 76.2 mm). Further rolling operations are then performed in a hot mill.

*Sheet* Sheet is material with a thickness between 0.006 and 0.249 in. (0.1524 and 6.3246 mm). It is rolled in a hot mill to about 0.125 in. (3.2 mm) thick, then cold rolled for additional thinning and to give it more strength and a better finish.

*Coiled sheet* is rolled to its finish gauge and width in a continuous strip up to several hundred feet in length and wound into coils. *Flat sheet* is sheared to proper lengths from coiled sheet. It may be stretcher leveled for improved flatness.

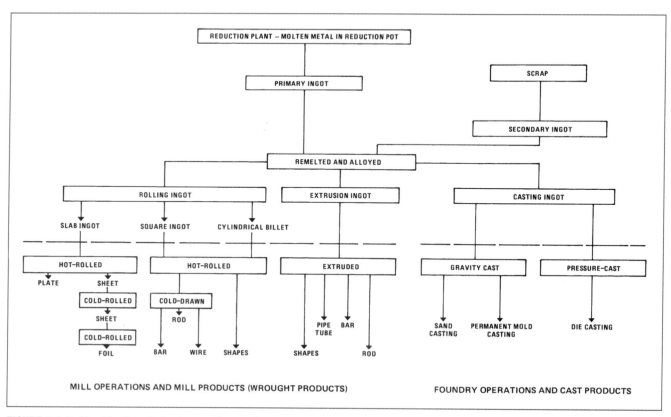

**FIGURE 5.2-7** Classification of aluminum mill operations used to manufacture wrought products and foundry operations used to make cast products.

*Brazing sheet* is a special type of sheet which has a cladding that melts at a lower temperature than the core. Assemblies of parts made from this sheet can be brazed together by subjecting them to a temperature of about 1000°F (537.78°C), which does not melt the core alloy but only the cladding, joining the surfaces.

*Aluminum foil* has a thickness of less than 0.006 in. (0.1524 mm). It is made by rolling sheet in cold-rolling mills and is usually supplied in coil form. Its most common applications in the building industry are for thermal insulation and vapor retarders.

**Structural Shapes**  Most structural aluminum shapes are extruded, but some are hot rolled by a series of specially shaped rolls. Complicated structural shapes require many roller passes to prevent the undue straining that is possible when too much force is applied in a single pass. Because they require considerable metal and a high degree of machining and finishing to get the desired contour, rolls are expensive. Therefore, lower tooling costs often make extruded structural aluminum shapes less expensive than rolled ones. Finishing operations include roller or stretch straightening, and when required, the rolled shapes may be drawn through dies to improve the surface finish and to obtain closer dimensional tolerances.

Standard rolled structural shapes, usually supplied in 2014 and 6061 alloys, include angles, channels, and beams.

Hot rolling is used to produce rod and bar shapes. Cold-finished rod and bar are produced by hot rolling to a size larger than specified and then cold drawing through a die for a better finish and closer dimensional tolerances. Bar is commercially available as square (round or square-edged), square-edge rectangular, and hexagonal.

Rod, bar, and other sections suitable for further shaping by impact or pressure in a die are called *forging stock*. Forging rod is nearly always rolled, but forging bar may be rolled or extruded, and forging shapes are usually extruded.

Redraw rod is typically ³/₈ in. (9.5 mm) in diameter. This rod is suitable for further drawing to produce wire.

## COLD DRAWING

Drawn products are formed by pulling (drawing) metal through a die. The purpose of cold drawing is to obtain close dimensional tolerances and a better surface finish.

Wire, which must be kept within close tolerances, is cold drawn by pulling redraw rod (³/₈ in. [9.5 mm] diameter) through a series of progressively smaller dies.

Hollow tube sections are drawn in a process similar to drawing bar and rod, except that the tube is threaded over a mandrel (shaper) positioned in the center of the die. When the tube is drawn, close tolerances both inside and out are maintained.

## EXTRUDING

Extruding is a process in which a heated aluminum ingot is pushed by hydraulic pressure through a die opening. The extruded aluminum assumes the same size and cross section as the die opening. Extruded products include shapes of varying cross section, rod, bar, and seamless tube. Aluminum window sashes, storefront sections, and expansion joint covers are examples of extruded shapes.

Shapes that cannot be extruded as a single section because of dimensional limitations or complexity of cross section can be obtained by fitting (mating) more than one section together. The extruding process provides great flexibility of design shape. The economical manufacture of practically any shape is possible due to the relatively low cost of dies. Close tolerances in manufacture can be obtained, and small parts of desired shapes can be sliced from a longer extrusion, often minimizing or eliminating machining. Extrusions are straighter, dimensionally more accurate, and have a better surface finish than rolled shapes.

## 5.2.3.2 Cast Products

Cast products are manufactured in a foundry. They include sand castings, permanent mold castings, and die castings.

## SAND CASTINGS

Molds made of sand are lowest in cost and permit maximum flexibility in design changes, although labor costs are higher than for other castings. Sand castings require heavier section thicknesses than other methods, and tolerances are not as close. They are used for spandrel panels, decorative panels, lighting standard bases, and general fittings.

## PERMANENT-MOLD CASTINGS

Permanent-mold casting employs metal molds with a metal core. It is used where a large number of castings from the same mold are required. If a sand core is used, the casting is called a *semipermanent* mold casting. Molten metal is poured into the mold by gravity.

## DIE CASTINGS

Die castings are permanent mold castings in which metal is forced into the mold under pressure. Die castings have smoother surfaces than permanent mold castings and permit greater repetitive use of the molds. Die casting is generally used for the mass production of small items such as handles, latches, and other door and window fittings.

The main advantages of permanent mold and die castings over sand castings are higher strength, better surface appearance, and closer dimensional tolerances.

### 5.2.3.3 Fabrication

Fabrication includes forming and machining operations performed on aluminum products and castings after they leave the mill or foundry.

## FORMING

Aluminum, one of the most workable common metals, can be fabricated into a variety of shapes by bending, stretching, rolling, drawing, and forging. Formability varies greatly, depending on the aluminum alloy and temper.

**Bending**  Temper and thickness greatly affect the minimum bend radius that can be used in forming. Certain aluminum alloys can be cold formed to a sharp 90-degree bend; others must be formed on stipulated radii.

**Stretch Forming**  Sheets are stretched to flatten them or to form them into special shapes. The most suitable alloys for stretch forming are those with high elongation (ability to stretch without fracturing). The advantage of stretch forming is that there is hardly any tendency of the material to assume its original shape after forming stresses are removed.

**Roll Forming**  Products are roll formed by passing strips through a series of roller dies (Fig. 5.2-8). Each successive pair of rolls shapes the metal slightly—only a small deformation is made by any single

**FIGURE 5.2-8** A series of cold rolls form corrugated decking from embossed sheet. (Courtesy Kaiser Aluminum and Chemical Corporation.)

**FIGURE 5.2-9** An 8000 ton (7257.47 Mg) hydraulic press produces forgings for use as architectural panels. (Courtesy Kaiser Aluminum and Chemical Corporation.)

pair of rolls. No straightening is required since roll-formed products are ordinarily straight and free from distortion.

*Forging* High-strength aluminum parts are often produced by forging (Fig. 5.2-9). Most forgings are made by hammering or pressing heated aluminum between closed steel dies. Press forging forces the metal into a die cavity of the desired shape under mechanical or hydraulic pressure. Hammer forging produces the same results by repeated blows from a drop hammer. Aluminum forgings may be heat treated after fabrication.

Since forging refines and provides desirable directional characteristics to the grain structure of the metal, improving strength and fatigue resistance, it is used to make various types of door hardware where strength is important.

*Drawing* Tubes and wire can be made by drawing aluminum alloy through a die in lieu of pushing it, as in the extrusion process.

## MACHINING

The machining characteristics of aluminum and its alloys, especially of wrought products, are excellent. Sawing, drilling, tapping, and threading methods vary little from those employed for other metals.

# 5.3 Joints and Connections in Metal Assemblies

Joints and connections in metal assemblies can be made with *mechanical fastenings*, *mechanically formed joints*, *welding*, *brazing*, *soldering*, and *adhesive bonding*.

## 5.3.1 MECHANICAL FASTENINGS

Mechanical fastenings include a variety of permanent and semipermanent types. They are used to connect metal to metal and metal to other building materials. Permanent mechanical fastening types (welding, rivets, etc.) involve permanent deformation, so the components may be disassembled only by destruction or cutting of the fastener or the component. Semipermanent fastenings (screws, bolts, and nails) allow disassembly without permanent damage to either the component or the fastener.

Bolting and welding are the most significant joining methods for steel structural components. Riveting is almost never employed for that purpose today and is used only occasionally in shop-fabricated steel structural components. One of the disadvantages of riveting is the high noise level of riveting machines.

Mechanical fastening of aluminum and most other nonferrous metals is generally done with bolts, screws, pop rivets, or nails or by stitching.

### 5.3.1.1 Riveting

#### STRUCTURAL STEEL

Riveting in structural steel and other thick metals involves the joining of components by means of a rivet with a prefabricated head on one end and deformation of the shank to form a head on the other end. Structural rivets are generally deformed by a pneumatic or hydraulic riveter while red hot. As the rivet cools it shortens in length, drawing the components together. The resulting compressive and frictional forces between the contact surfaces are important in resisting tensile and shear stresses developed in structural joints.

#### ALUMINUM

Riveting in aluminum assemblies is a method of permanent mechanical fastening in which a headed pin (rivet) inserted through the pieces being joined is mechanically deformed so that a head is formed on both sides of the work, tightly gripping the pieces together. Riveting is probably the most widely used aluminum fastening method (particularly in the aircraft industry) and makes use of a great variety of rivet shapes, sizes, and assembling techniques.

Aluminum rivets are typically used in joining aluminum for a neater appearance and a uniform finish, avoidance of galvanic corrosion, low cost, and light weight.

Aluminum rivets are available in standard designs and sizes with round, brazier, flat, and countersunk heads.

Rivets can be made in practically any alloy. The following alloys meet most needs: 1100, 2024, 2117, 5056, and 6053. Alloys 1100 (commercially pure aluminum) and 5056 are non-heat-treatable. The others are heat-treatable. In general, rivets to be used with parts of a given alloy should match that alloy in mechanical and physical properties.

Although standard rivets meet most requirements, many special types are available, including tubular and other designs for blind riveting (where the work is accessible from only one side).

### 5.3.1.2 Bolting

*Bolting*, a semipermanent mechanical method of fastening with nuts, bolts, and washers, is seldom used within aluminum and other nonferrous metal components, but is sometimes used to fasten such assemblies to supporting construction.

Bolting and welding (see Section 5.3.3) are the primary methods of fastening used in structural steel assemblies. Bolting is easier to accomplish in the field than welding, especially in locations where access is difficult and when adverse weather conditions exist. Bolted connections work best when it is necessary for them to resist only shearing forces. When moment forces also occur, bolted joints often become complex and bulky compared with welded joints. Bolts are sometimes used to hold joints together for welding.

Specified tensile values must be developed in bolts by tightening them with calibrated torque wrenches, by turning the nuts a specified number of turns from a snug-tight position, or using a special power tool that locks onto the tip of the bolt and breaks it off when the nut is sufficiently tight, thus preventing further tightening. Proper tightening is important to develop frictional resistance in a joint to counteract shearing forces. Keeping the threaded part of a bolt out of the structural parts being joined is also essential in developing the full strength of the connection.

Nuts and bolts are made in sizes ranging up to 4 in. (102 mm) in diameter and with several types of heads (square, round, hexagonal, countersunk, etc.). Most bolts have square or hexagonal heads, as do most nuts, which allows the use of wrenches. Although bolts are available in both carbon and alloy steels, using high-strength bolts for most major structural applications is becoming more prevalent (Fig. 5.3-1). Ordinary carbon steel bolts are used in securing many steel components in small building construction, such as sill plates to foundations and beams to columns and to each other.

Bolting is also used to fasten components together and to supports in steel fabrications, such as stairs, gratings, railings, ladders, and fire escapes, and in decorative metal assemblies.

### 5.3.1.3 Welded Studs

*Studs* are steel rods. They may be straight or have one end bent at a 90-degree or other angle to the straight portion. Special studs are also available that have one end configured as a round or rectangular eye-bolt or as a j-bolt. Straight studs may be unthreaded or have one end externally or internally threaded.

The use of studs constitutes a semipermanent method of fastening most often used when nonstructural materials must be attached to steel, as in the application of architectural finishes to a structural framework. Welded studs are also used in composite structural design as a means of transferring stresses between structural steel and reinforced concrete to produce combined resistance to stresses. In each of these cases, the plain end of a stud is arc welded by a gun-shaped unit to a steel component. When a threaded stud is employed, the supported item is installed over the stud and a nut is applied to the stud's threaded end (Fig. 5.3-2). Refer to Section 5.6 for a discussion of studs used in composite concrete and steel floor decking.

### 5.3.1.4 Powder-Actuated Fastening

Powder-actuated fastenings are permanent or semipermanent fastenings that employ a gun powered by a small explosive charge to drive a pin into hard and tough materials. In permanent fastenings, a specially hardened drive pin is used to connect components in a fashion similar to nailing. Often, however, a pointed stud, threaded on one end, is driven into

**FIGURE 5.3-3** Powder-actuated fasteners often are used to secure steel members to concrete and masonry surfaces.

one of the components, and assembly is completed with a nut. The availability of inexpensive guns and a variety of stud sizes and shapes make this a useful fastening procedure in all types of construction (Fig. 5.3-3).

### 5.3.1.5 Stitching

Joining of various metals to other metals and other materials by wire sewing is called *stitching*. Aluminum's relatively low density and softness make it very suitable for stitching. Stitching requires no backing material, prepunching, or drilling. The work need not be clamped or the metal cleaned. Inspection is simple since it can be determined visually whether a stitch has penetrated the work and is properly clinched. Stitching is used in assembling aluminum garage doors, on sheet metal such as ductwork and blower installations, and in joining aluminum to rubber, fabrics, and other materials.

### 5.3.1.6 Threaded Fasteners and Nails

The most familiar semipermanent fastening devices are screws, bolts, and nails. Threaded fasteners include a multitude of screw and bolt types. Nails are plain (smooth), threaded, or grooved.

Threaded fasteners and nails for use in steel products are manufactured from carbon, alloy, and stainless steels in many shapes and sizes. Their uses include ordinary household applications, assemblies for product fabrication, and for the joining of trim components in construction. Fasteners used in steel should be the same metal as the material being fastened or a material that is compatible with the metal being fastened. This compatibility

**FIGURE 5.3-1** Air-impact wrenches speed high-strength bolting of steel connections in modern buildings. (Courtesy American Iron and Steel Institute.)

**FIGURE 5.3-2** Special stud welding equipment permits fast and efficient joining of many materials to steel. (Courtesy TRW Nelson Co.)

is sometimes achieved by using coated fasteners, such as those coated with zinc (see Section 5.9). Fasteners used in steel siding and roofing are discussed in Chapter 7.

Threaded fasteners are widely used in assembling aluminum products. Such fasteners should be aluminum or another metal compatible with aluminum, such as stainless steel.

Screws for use with aluminum are usually made from alloy 2024-T4 and are manufactured in practically any head type, size, and standard thread (Fig. 5.3-4). Bolts, nuts, and washers are usually made from alloy 2024-T4, 6061-T6, or 7075-T73. They are manufactured in practically any head style (typically hexagonal), size, and standard machine thread.

Because of their freedom from corrosion and staining, aluminum nails are popular fasteners for roofing and siding made of aluminum or of nonmetallic materials. These nails are manufactured from alloys 6061 and 5056, which, in this form, have tensile strengths in excess of 60,000 psi (42 184.2 Mg/m$^2$).

Threaded fasteners and nails used in other metals should be of a metal that is compatible with the material being fastened.

### 5.3.1.7 Expansion Bolts

Expansion bolts are semipermanent fasteners generally used to join materials to concrete and masonry surfaces. They are also referred to as *cinch anchors*, or *expansion anchors*, and by many proprietary names. These devices consist of two or more units that develop holding power in concrete or masonry through wedge action and friction with the walls of a pilot hole by expansion. This action is produced by striking or turning the threaded machine bolt insert. Expansion bolts act in a manner similar to that of screws used with lead, fiber, or plastic shields but usually develop much greater withdrawal strengths (Fig. 5.3-5).

### 5.3.1.8 Adhesive-Held Fasteners

Adhesive-held fasteners are designed to be held in place by the action of an adhesive. A device consisting of two glass containers, each holding one component of a two-part adhesive, is inserted into a hole drilled in concrete. The desired fastener is driven into the hole, breaking both parts of the glass container and permitting the components of the adhesive to mix and begin curing. When the adhesive has fully cured, the fastener is held securely in place.

### 5.3.1.9 Special Fasteners

Many other special fasteners and rivets are available for blind fastening and other connections. Blind fasteners include pop rivets, explosive rivets, and others where only one side of the work is accessible. Other special fasteners, such as speed nuts and quick-release fasteners, are also commonly used.

### 5.3.2 MECHANICALLY FORMED JOINTS

Joints in thin sheets are often made without the use of fasteners. Instead, a part of the material itself is used to make the connection. Connections obtained by forming or cutting the parts to be joined in such manner that they interlock are called *mechanically formed joints*. Where high strength is of primary importance, other joining methods will usually be preferred. But the building industry offers many applications, such as roof work, ducts, gutters, downspouts, and railings, where mechanically formed joints are quite satisfactory (Fig. 5.3-6). Two common methods for making mechanically formed joints are staking and lock-seam joining.

*Staking* is a method of joining flat materials by foldover tabs inserted through matching slots in the elements to be joined. It is often used in product fabrication. For example, a staked joint is commonly used in assembling extruded aluminum members. Tabs or prongs formed at the end of one member are in-

**FIGURE 5.3-4** Extruded aluminum window and door frame sections are often connected using screw fasteners. A common method is to thread the screws into continuous slots.

**FIGURE 5.3-5** Self-drilling expansion anchors speed joining of steel to concrete. (Courtesy ITW Ramset/Red Head.)

**FIGURE 5.3-6** Mechanically formed slip joints are used to assemble aluminum gutters and downspouts.

serted through mating holes in the member to be joined and are twisted, bent over, or spun down (Fig. 5.3-7).

*Lock-seam joining* is a mechanical method of joining sheet metals in both shop and field by bending and interlocking the elements (Fig. 5.3-8). This is a common method of fabricating air ducts for heating and air conditioning, as well as roofing and roof drainage components.

### 5.3.3 WELDING, BRAZING, AND SOLDERING

Welding, brazing, and soldering are methods of forming a metallurgical bond

**FIGURE 5.3-7** Staked joints are often used in assembling aluminum frame members of windows and doors.

**FIGURE 5.3-8** Typical methods of lock-seam joining of sheet metal components in product manufacture. (Courtesy American Iron and Steel Institute.)

between two or more components of a metal assembly. The selection of a specific process for a particular job depends upon many variables, including conditions encountered in service, the shape and thickness of parts, volume of production, and the alloys employed.

Welded joints are made by melting the edges of the components to be joined and adding filler metal if required. Brazing and soldering generally employ nonferrous filler metals which have melting points below those of the base metals and do not require melting of the base metal to effect a bond. When the filler metal has a melting point below 800°F (426.7°C), the joining method is considered soldering; if above, it is brazing. Brazed connections can be as strong as welded joints, depending on service requirements. The brazing filler metal flows into a joint of any configuration by capillary attraction. Soldering and brazing are most often used in product manufacture; welding is used both in manufacturing and in structural applications.

Welding, brazing, and soldering are dangerous to both people and property. The most obvious hazard is fire. While these processes are being performed, it is necessary to protect adjacent combustible materials from the heat generated. A number of dangers to people are also created by these processes. Welding arcs are bright enough to produce permanent eye damage to both welders and onlookers. Those not directly involved in the process should be prohibited from entering the area where welding is being done. Steps should be taken to protect the eyes of those who must be in the vicinity. These processes, particularly welding, also produce sparks of hot materials that can burn through clothing and skin. Welders must wear specialized protective clothing (Fig. 5.3-9) and head covering. Vision panels in their headgear must be glazed with materials specifically designed to block out damaging rays. Their clothing must be flame resistant. Their hands must be protected with heavy gloves. Bystanders who must be in the immediate vicinity of the welding process must be similarly protected.

#### 5.3.3.1 Ferrous Metals

Steel product manufacturers employ many different welding, brazing, and soldering processes. The following discussion, however, is limited to common

**FIGURE 5.3-9** Typical protective gear worn by welders. (Courtesy National Association of Women in Construction.)

techniques used in building construction.

Structural joints in building frames may be either welded or bolted (see Section 5.3.1.2). The decision about whether to use welding or bolting to form a particular joint rests with the structural designer. Welding is probably best for most shop joints because control of the surrounding environment is easier than it is in the field, and welding is usually more economical in the shop than bolting. For example, computer-driven robots can sometimes be used for shop welding, reducing labor costs. When joints must resist both shear and moment forces, welding usually produces cleaner and less complicated joints.

The thickness and number of welds in a particular joint are decided by the structural designer based on the forces at the joint.

#### WELDING TECHNIQUES

The majority of welding of steel components in building construction can be classified into two main processes, *arc* and *resistance*, both of which are *electric* welding processes. *Arc welding* depends on the heat of an electric arc established between an electrode and the components to be welded. It is most commonly used in the field for connecting steel structural components. Steel decking is also welded in place using arc welding.

Welding is an exacting process requiring care and skill. Welders should therefore be well trained and periodically tested to ensure their competence. In addition, each weld should be inspected to ensure that it was properly made. These inspections may be done visually for minor conditions, but for structural joints more sophisticated inspection methods are usually employed. These include radiographic, ultrasonic, magnetic particle, and dye-penetrant testing, each of which will detect hidden flaws and voids within welds. The decision as to which type of testing is appropriate rests with the structural designer.

*Resistance welding* is generally a shop process in which the components are butted together and heat is generated in the seam by the resistance of the metal to the passage of an electric current. An electrode in contact with the component introduces the current. Welded wire fabric is usually produced using resistance welding.

Oxygen and nitrogen in the air have a detrimental effect on hot metal in a weld area. Most arc processes *shield* the weld area from atmospheric contamination with inert gases or granular fusible materials. *Shielded metal arc welding* (also known as *manual metal arc* or *coated stick electrode welding*) is the most common arc process (Fig. 5.3-10a). This process uses a consumable metal stick electrode that is gradually melted by the arc and deposits filler metal in the weld. A chemical coating on the electrode releases an inert gas which forms a shielding envelope around the weld area. This process depends on manual replacement of the stick electrode as it is consumed.

In *semiautomatic* processes using self-advancing wire electrodes, the wire comes off a spool at a predetermined rate and goes through a hand-held welding gun. The two chief processes of this type are *metal inert-gas shielded arc* (MIG), or *gas-shielded metal arc* welding (Fig. 5.3-10b), and *submerged arc* welding (Fig. 5.3-10c). In MIG, an inert gas or mixture of gases (argon, helium, a combination of both, or carbon dioxide) from an external supply is directed through a gun on the weld area to shield it from atmospheric contaminants. In submerged arc welding, protection is achieved by burying the arc and the immediate weld

a

b

c

**FIGURE 5.3-10** (a) Shielded metal arc welding. (b) MIG welding. (c) Submerged arc welding. (Courtesy American Iron and Steel Institute.)

area under a blanket of fusible granular material. Arc heat melts the material to form a layer of protective slag over the molten metal. The slag and loose material are removed after welding has been completed. In MIG, filler metal is provided by a consumable electrode; in submerged arc welding, either a consumable electrode or a separate filler rod may be used.

The most common welded joint types are shown in Figure 5.3-11. Exposed welds

are often ground smooth if the welds will permit it. The welding processes with various types of steel are outlined in Figure 5.3-12. Symbols are used on drawings to denote the types of welds required (Fig. 5.3-13).

*Spot welding* is a type of resistance welding in which the components are joined by a series of spot welds, rather than a continuous line of weldment as in other techniques. It is produced by the resistance heating of a small area between two cylindrical electrodes under pressure.

*Seam welding* is actually a series of overlapping spot welds produced under pressure by a pair of disc electrodes, as in welded pipe manufacture.

*Projection welding* is also similar to spot welding but makes use of preapplied projective embossments on one of the components to localize the heat and effect of a weld in the desired spot.

### EFFECTS OF WELDING ON STEEL

Because heat is an essential element of all welding techniques and because metals respond to heat exposure to varying degrees, some steels are more weldable than others. The degree of weldability is related to the sensitivity of the metal to heat, as reflected in the soundness of the weld and the degree of undesirable side effects produced by the operation, including:

- Cracks in the base and weld metals
- Decreases in the base metal's physical properties (strength, ductility, corrosion resistance, etc.)
- Finished surface oxidation

The most important element affecting the weldability of steel is its carbon content. Low-carbon steels (less than 0.15%) are the most weldable; mild steels (0.15% to 0.30% carbon) are next in weldability and can be welded by most processes up to a section thickness of $1/2$ in. (12.7 mm). Thicker sections require greater input of heat to effect a weld, and this may have detrimental results unless proper procedures and care are used. All ASTM structural steels have a carbon content of less than 0.35% and, therefore, can be welded.

Although all stainless steels can be welded by some method, austenitic steels are the most weldable. Almost any commercial welding method can be used to join these steels without problems of

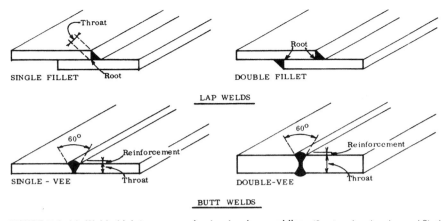

**FIGURE 5.3-11** Welded joints common in structural assemblies. (Courtesy American Iron and Steel Institute.)

**FIGURE 5.3-12** Welding Processes Used to Join Various Types of Steel

| Type of Steel | Metal-Shielded Arc | Submerged Arc | Metal-Inert Gas-Shielded (MIG) Arc | Spot, Seam, and Projection |
|---|---|---|---|---|
| Low Carbon | | | | |
| Sheets | Common | Common | Common | Common |
| Plates and bars | Common | Common | Common | Occasional |
| Medium Carbon[a] | | | | |
| Sheets | Common | Common | Common | Common |
| Plates and bars | Common | Common | Common | Rare |
| High Carbon[a] | | | | |
| Sheets | Common | Occasional | Occasional | Common |
| Plates and bars | Common | Occasional | Occasional | Not used |
| Low Alloy[a] | | | | |
| Sheets | Common | Common | Common | Occasional |
| Plates and bars | Common | Common | Common | Rare |
| Stainless[a] | | | | |
| Sheets | Common | Occasional | Common | Common |
| Plates and bars | Common | Common | Common | Rare |

[a]Preheating and/or post-heating (stress relieving) structures may be required in the weld zones, to prevent embrittlement and excessive residual stresses on certain carbon and low alloy steels, and to stabilize in order to prevent corrosion on stainless steels.

(Courtesy American Iron and Steel Institute.)

soundness and porosity, or loss of toughness or ductility. However, the high rate of thermal expansion in austenitic steels must be considered in order to control distortion. Another hazard is the reduction of corrosion resistance in the weld area, because some of the chromium becomes ineffective through reaction with carbon. This often occurs when stainless steels are exposed to high welding temperatures over extended periods. This effect can be minimized by limiting heat input, promoting heat dissipation, and selecting low-heat resistance welding methods. Annealing of the finished assembly followed by fast cooling also restores corrosion resistance.

### 5.3.3.2 Nonferrous Metals

The choice of filler metal alloy for welding nonferrous metals depends on the metal being joined, joint design, service conditions, surface finish, welding process, the alloy or alloy combinations to be joined, and the desired appearance. Therefore, filler alloys vary considerably and must be selected for each particular circumstance.

Many welding processes are suitable for use with aluminum alloys. Some of the more important processes are discussed here.

### GAS WELDING

The earliest welding process used on aluminum was gas welding. In gas welding, oxygen, acetylene, oxyhydrogen, or another gas flame is the heat source. The gas flame melts the parent metal, and a flux-coated filler rod is used to complete the weld. The disadvantages of gas welding include:

- Necessity to remove the flux residue, a potential source of corrosion or weld defects in multipass welds
- Distortion, which is greater than in arc welding
- A wide heat-affected zone
- Slow welding speeds

These conditions generally result in higher costs compared with gas-shielded arc methods.

Aluminum about $1/32$ in. (0.794 mm) to 1 in. (25.4 mm) thick may be gas welded. Heavier material is seldom gas welded, because heat dissipation is so rapid that it is difficult to apply sufficient heat to melt the metal with a torch.

Gas welding finds its greatest use today where few or occasional welds are required. It is also the obvious choice in the field when electric power lines or portable generators for arc welding are not accessible.

### ARC WELDING

Most forms of arc welding depend on the production of heat from high-voltage electric flow (arc) from an electrode to the parent metal.

### INERT-GAS-SHIELDED ARC WELDING

Of all the fusion welding processes, the inert-gas-shielded arc processes are by far the most important and most widely used. In the tungsten inert gas (TIG) arc process, an arc is drawn between the work and a nonconsumable tungsten electrode; the arc is shielded by a flow of inert argon or helium. Filler metal is added manually.

In MIG arc welding, the arc in inert gas is drawn between an automatically fed filler wire and the work. Fluxes, used to dissolve the aluminum oxide surface film on the metal in other welding processes, are not required because of the action of the arc and the inert gas shield.

**BASIC WELD SYMBOLS**

| BACK | FILLET | PLUG OR SLOT | GROOVE OR BUTT ||||||||
|------|--------|--------------|-----------------|---|------|---|---|--------|-------------|
|      |        |              | SQUARE | V | BEVEL | U | J | FLARE V | FLARE BEVEL |

**SUPPLEMENTARY WELD SYMBOLS**

| BACKING | SPACER | WELD ALL AROUND | FIELD WELD | CONTOUR ||
|---------|--------|-----------------|------------|---------|--------|
|         |        |                 |            | FLUSH | CONVEX |

**STANDARD LOCATION OF ELEMENTS OF A WELDING SYMBOL**

Finish symbol

Contour symbol

Root opening, depth of filling for plug and slot welds

Effective throat

Depth of preparation or size in inches

Reference line

Specification, process or other reference

Tail (omitted when reference is not used)

Basic weld symbol or detail reference

Groove angle or included angle of countersink for plug welds

Length of weld in inches

Pitch (c. to c. spacing) of welds in inches

Field weld symbol

Weld-all-around symbol

F
A
R
S(E)
(Both sides) (Arrow side) (Other side)
L @ P
T

Arrow connects reference line to arrow side of joint. Use break as A or B to signify that arrow is pointing to the grooved member in bevel or J-grooved joints.

A       B

Note:

Size, weld symbol, length of weld and spacing must read in that order from left to right along the reference line. Neither orientation of reference line or location of the arrow alter this rule.

The perpendicular leg of weld symbols must be at left.

Arrow and Other Side welds are of all the same size unless otherwise shown. Dimensions of fillet welds must be shown on both the Arrow Side and the Other Side Symbol.

The point of the field weld symbol must point toward the tail.

Symbols apply between abrupt changes in direction of welding unless governed by the "all around" symbol or otherwise dimensioned.

These symbols do not explicitly provide for the case that frequently occurs in structural work, where duplicate material (such as stiffeners) occurs on the far side of a web or gusset plate. The fabricating industry has adopted the convention that when the billing of the detail material discloses the existence of a member on the far side as well as on the near side, the welding shown for the near side shall be duplicated on the far side.

**FIGURE 5.3-13 Standard welding symbols.** (Courtesy American Institute of Steel Construction, Inc.)

These processes are applicable to all welding positions. Thinner gauges are commonly welded by TIG. MIG is particularly appropriate for automatic welding; TIG is also appropriate there when filler metal is not required. Since welding speeds are high, the effect of heat input on strength is minimized.

## RESISTANCE WELDING

In resistance welding, an electric current is conducted through the two parts to be joined. Heat is generated at the juncture of the parts as a result of resistance to the passage of heavy current. When pressure is applied to the heated metal, fusion occurs. Resistance welding has practically no effect on the alloy temper since the weld is made so quickly that the adjoining area does not heat appreciably.

*Flash welding* is a type of resistance welding used to butt weld shapes, such as tubes, extruded shapes, rods, bars, and wire. Typical is the miter joining of aluminum window sections. In general, flash welding is used where technically practical and where mass production is involved.

*Spot and seam welding* are resistance welding methods especially important in fabricating high-strength heat-treated sheet alloys which can be so joined with practically no loss in strength. Spot welding is widely used to replace riveting, joining sheets at intervals as required. *Seam welding* is merely spot welding with the spots spaced so closely that they overlap, producing an airtight joint where desired.

## BRAZING

Brazing is a welding process using a filler metal that is nonferrous, or an alloy with a melting point higher than 800°F (426.7°C) but lower than that of the materials being joined.

In brazing, the molten metal is called *brazing filler metal*; in soldering, it is called *solder*.

Brazing is employed to join millions of aluminum parts each year, which range in thickness from foil to heavy plate and castings. It is used for joints that, because of their complexity of contour, limited access, or high ratio of length to part size, are unsuitable for welding.

Filler alloys for brazing aluminum include such aluminum-silicon alloys as 4047 and 4145. They have liquidus temperatures much closer to the solidus temperature of the base metal than filler alloys for brazing most other metals. Consequently, temperature control is critical, and only those aluminum alloys with relatively high melting points are brazeable. These include alloys 1050, 1100, 3003, 6063, and the self-aging foundry alloy Alcan 432.

Although alloy 6061 can be brazed, its melting range is very close to that of alloy 4047, the commonest brazing filler alloy. Alloys with high magnesium content are not considered brazeable because of poor wetting by the filler alloys. Flux is essential to remove the oxide film.

Aluminum may be brazed by several methods, differing mainly in the way that heat is applied to the work. These include torch brazing, furnace brazing, and salt bath brazing.

*Torch brazing* is employed where the joints are accessible both for heating with the torch flame and for feeding filler metal into the joint.

In *furnace brazing*, the entire work assembly is held in a furnace for 1 to 5 minutes. Due to the capillary action of a brazing alloy, the furnace brazing process is particularly useful for inaccessible joints.

In the *salt bath process* (also known as *dip brazing*), the work is submerged in molten salt which performs both as the heating medium and as the flux. Time in the salt bath varies from 30 to 180 seconds.

Whatever the process, it is essential that surfaces be thoroughly cleaned before they are brazed. For alloys 1100 and 3003, degreasing is usually sufficient; for other alloys, an etching treatment is recommended.

Under some conditions, brazed joints develop strengths comparable to those of welded joints, make more economical use of filler metal, can be made faster, and have a neater appearance (requiring little finishing work, if any). In addition, brazing makes possible the joining of extremely thin sections. In general, the corrosion resistance of brazed joints is comparable to that of welded joints.

## SOLDERING

Filler metals (*solders*) with melting points under 800°F (426.7°C) are used in soldering. Difficulties in soldering aluminum are caused by the great affinity of aluminum for oxygen and the lack of bond between solder and the natural oxide film present on aluminum. Ordinarily, a suitable flux is applied to dissolve the oxide and prevent it from reforming, or otherwise reacting, so that the solder may bond directly to the underlying aluminum. Another way to get below the oxide film is to mechanically abrade it down through an overlying layer of molten solder, using a soldering iron, a wire brush, or another tool.

When soldering aluminum to other metals, the aluminum surface should be *tinned* (precoated with solder) and then joined to the other metal with solder and flux. If any difficulty is encountered, the other metal surface should also be tinned first.

The essentials of soldering aluminum are to remove the surface oxide and to use a filler metal which will not encourage galvanic corrosion. Therefore, appropriate fluxes, solders with higher melting points, pure zinc, and zinc-aluminum (95-5) are used where corrosion is a factor. Where corrosion resistance is less important, lower-melting zinc-tin and lead-tin alloy solders are used.

In corrosive atmospheres, lacquer coatings may be used to improve the corrosion resistance of soldered aluminum.

### 5.3.4 ADHESIVE BONDING
#### 5.3.4.1 Ferrous Metals

Many decorative and protective laminated coatings are applied to ferrous metal base materials using adhesives. Adhesive bonding is also used to secure thin carbon, galvanized, and stainless steel sheets and foils to nonferrous materials that provide the required rigidity, flatness, strength, or insulating qualities.

Carbon steel surfacing adds abrasion and indentation resistance to softer core materials. Steel-clad honeycomb doors and steel bonded to plywood panels for column and radiator covers are common examples; galvanized or vitreous enamel-coated steel sheets, bonded with rigid urethane foam to form sandwich panels, are often used in building construction.

Stainless steel sheets also are bonded to rigid or insulating materials in the manufacture of many products with corrosion-resistant finishes. Very thin stainless steel foil is laminated to kraft paper for use in packaging where additional strength and impermeability to vapor are important considerations.

The chief advantages of adhesive bonding are uniform distribution of stresses, absence of surface fasteners, and economy of mass production. Ease of joining small parts not accessible for welding, and low weight in relation to strength, make this a valuable structural joining method in the aircraft industry. However, the high initial cost of specialized equipment, low peel and creep (flow under stress) strengths, and sensitivity to certain temperatures (characteristic of low-cost adhesives) at present limit their use in building construction mainly to the manufacture of panelized products.

The most common adhesives used in the lamination of steel to other materials are epoxies, phenolic resins, rubber polymers modified by resins, polyvinyl acetates, and chemically cured neoprenes. The most promising adhesives for laminating steel are the polyurethanes which, with epoxies and phenolics, withstand considerable loads at a greater temperature range and come closest to the concept of *structural* adhesives. These adhesives often possess high peel strength, chemical inertness, resistance to wetting

action, and shear and tensile strengths greater than those of the material bonded. Under certain conditions, they may be used for joining structural members in building construction.

Adhesives are available in a variety of forms, such as pressure-sensitive tapes, films, liquids, pastes, and solids. They can be applied manually with ordinary building tools or with specialized mass production equipment that sprays, rolls, or flows the adhesives onto component surfaces. Some adhesives acquire permanent mechanical properties when heated (thermosetting); others set at moderate temperatures but can be softened repeatedly at high temperatures (thermoplastics); still others depend on the chemical reaction between two agents to acquire their permanent characteristics.

Although many adhesives can perform well under a given set of service conditions, there is no universal adhesive and each must be selected to meet the anticipated operating requirements.

#### 5.3.4.2 Nonferrous Metals

Adhesive bonding utilizes adhesives for bonding aluminum to aluminum or other materials. Design possibilities are limited only by extremes of stress, temperature, corrosive environment, and cost. A special consideration in designing joints for the use of adhesives is that peeling stresses should be kept as low as possible. The pretreatment of aluminum used in adhesive bonding is concerned mainly with cleaning the surface. Where the adhesive bond must develop the greatest strengths, a chromic-sulfuric acid dip is often used as a chemical cleaning method.

Adhesives may be applied by any of several techniques. Roller coating is particularly effective for high-speed application of a thin coat of adhesive, as in the production of laminates. Various building materials are adhesively laminated with aluminum foil to increase their thermal insulation properties.

For more limited areas, adhesives may be applied by brush, blade, or spray. Certain adhesives are adapted to mass production practice by manufacturing them in film form. The film is cut into sheets, ribbons, or special shapes which follow the contours of specific joint areas and are assembled with the metal parts before curing.

Curing processes for adhesives vary considerably in type and form. The cur-

**FIGURE 5.4-1** Large-span atrium spaceframe supporting tension fabric cladding, Hyatt Regency Hotel, Burlingame California. (Architect: Hornberger Worstell and Associates; photograph courtesy Mero Structures, Inc.)

ing time ranges from 3 seconds to over 76 hours, depending on the type of adhesives and whether or not heat and pressure are applied to the assembled joint. Either heat, pressure, or both are used to obtain the highest bond strengths.

As is true of adhesives used with ferrous metals, there is no universal adhesive for use with nonferrous metals. The nature of the material joined and the ultimate conditions of service determine the type of adhesive to be used. Therefore, many proprietary adhesives are developed for varying bonding processes. Natural and synthetic rubbers, polyvinyl acetates, phenolic blends, and epoxies are among the commonly used adhesive bases.

## 5.4 Structural Steel Framing

Many low-rise and medium-rise buildings and all very tall buildings are built with structural steel frames. In such construction, structural steel shapes are erected and fastened together using various methods to form a rigid framework from which are hung floors and exterior walls. Floors are supported either on steel structural members or on steel joists supported by structural framing.

Conventional steel-framed structures are built from hot-rolled beams, columns, and girders, combined with bar joists (see Section 5.5) and steel decking (see Section 5.6). Sometimes these components are combined into complete preengineered buildings (see Section 5.4.1.5). Small structures and parts of larger structures carrying light loads are sometimes erected using lightweight cold-formed framing (see Section 5.7).

**FIGURE 5.4-2** Part of a designer's steel framing plan. Squares with X's are elevator shafts. Column sizes and types are usually shown in tables. (Drawing by HLS.)

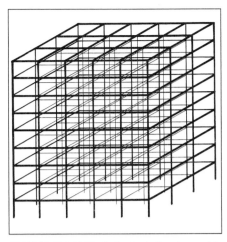

**FIGURE 5.4-3** Typical steel framing grid. (Drawing by HLS.)

Long spans are achieved in steel using many different types of systems. Conventional framing can be stretched using plate girders and castellated beams. Roofs may be framed with long-span trusses. Special areas, such as arenas, stadium roofs, and airline and train terminals, may be spanned with domes, space frames (Fig. 5.4-1), or steel cable systems.

## 5.4.1 CONVENTIONAL STEEL FRAMING SYSTEMS

Steel building frames may be designed by either architects or structural engineers. However, architects seldom design frames for large buildings or space frames, steel cable roof systems, or trusses.

### 5.4.1.1 The Design Process

The designer produces a set of working drawings (Fig. 5.4-2) and specifications detailing the type of steel, basic requirements for fasteners, the location and sizes of the basic elements of the frame (columns, girders, beams, etc.), the basic dimensions of the frame (column centers, floor heights, etc.), and unusual erection details.

Using the designer's drawings and specifications as a guide, a fabricator, employed by a building contractor or a structural steel subcontractor, produces a set of shop drawings showing the same basic information as on the designer's drawings, but adding information about joints, member lengths, and all other de-

**FIGURE 5.4-4** Details of shear connections. (Courtesy American Institute of Steel Construction, Inc.)

tails necessary to fabricate and erect the frame. The shop drawings also show details of the methods and bolts necessary to anchor the frame to the foundations.

These shop drawings are then sent by the contractor to the designer for review. If revisions are necessary, the designer marks the drawings and returns them to the contractor for revision and resubmittal. This process continues until the designer finds the shop drawings acceptable. The designer then places a stamp on the drawings stating that they are acceptable and returns them to the contractor for return to the fabricator.

### 5.4.1.2 Conventional Steel Framing

The structural frame for a conventional steel-framed building is composed of columns, girders, and beams fastened to-

**FIGURE 5.4-5** Details of moment connections. (Courtesy American Institute of Steel Construction, Inc.)

gether to form a three-dimensional grid (Fig. 5.4-3).

Under normal conditions, the joints in conventional structural steel building grids are subjected to both shear and moment forces. Unless some means is employed to resist these forces, the frame will be inherently unstable and subject to wracking and ultimate collapse. The American Institute of Steel Construction (AISC) lists three methods for solving these inherent difficulties as types of steel frame construction. In Type 1, connections must be made sufficiently rigid to resist all forces that will occur without significant change in the angle between framing members (beams, girders, columns). In Type 2, diagonal bracing or shear panels are introduced to resist moment forces. Thus, most joints need resist only shear forces. Type 3 is a semirigid system in which the joints are sufficiently rigid to resist the applicable moment forces, but not as rigid as those required in Type 1.

Joints that are required to resist only shear are usually just bolted (Fig. 5.4-4). Joints required to resist both moment and shear forces can be fastened using bolts alone, but such joints are both complicated and expensive. Therefore, joints resisting both shear and moment forces are either welded or, more frequently, both bolted and welded (Fig. 5.4-5). Such joints are more expensive to produce than those designed to resist only shear forces. For this reason, an attempt is usually made, especially on large buildings, to reduce the number of moment joints in a steel frame.

Such a reduction in the number of moment joints in a building is often accomplished by making part of the frame rigid and tying the remainder of the frame to the rigid portions by means of rigid floor systems, such as steel decking with a concrete topping. Such a floor system must be designed structurally as diaphragms, meaning that it acts as a rigid unit to distribute horizontal forces acting on the building to the rigid portion of the frame, which is, in turn, capable of resisting those forces and transmitting them harmlessly to the ground.

The rigid portion of a building frame may be either in the perimeter or in the core. The core is that portion of a building that houses elevator and stair towers and service facilities such as toilets. The amount of the frame that must be made rigid and its location are functions of the

design and are the responsibility of the structural designer.

The three basic ways of creating a rigid core or perimeter of a steel building frame are diagonal bracing, shear panels, and moment connections. Most of the rest of the frame can usually be fastened with shear-only connections, although there are some exceptions. For example, cantilever beams usually require moment connections.

Diagonal bracing consists simply of steel members jointing opposite corners of a structural bay (Fig. 5.4-6) and thus forming rigid triangles within the bay. Because the triangular structure itself cannot distort, moment joints are not necessary to attach the bracing to the frame. Shear panels often consist of rigid concrete walls but are sometimes constructed of braced steel framing. In either case, they must resist the forces imposed on them without distorting.

Moment connections can be used to form a rigid portion of the frame, with the same result as that obtained with diagonal bracing or shear panels. While structural design and safety must be considered, the final decision about the method used to stabilize a steel-framed building is often based on which method is least expensive.

### 5.4.1.3 Fabrication

After receipt of shop drawings bearing the designer's stamp of acceptance, a fabricator produces full-sized templets of wood or cardboard to assist in the positioning of bolt holes and cutouts and begins the fabrication process.

Fabrication consists of measuring and cutting beams, columns, girders, base plates, and other components to the proper length and configuration; punching or boring holes for bolts; welding on stiffeners, connection devices such as angles, and smaller base plates; beveling of joints for welding; flattening and squaring of bearing surfaces for a tight fit; cambering (curving slightly upward) of beams and girders where the design requires it; and assembly of large components, such as plate girders, trusses, and built-up columns.

### 5.4.1.4 Erection

The various components of most steelframed structure are fabricated in the shop, delivered to the construction site as individual components (beams, girders,

joists, decking, etc.) and fastened together in the field (Figs. 5.4-7 and 5.4-8).

The procedures and equipment used in the erection of steel-framed buildings are the responsibility of the *erector*. The actual work is done by *ironworkers* (sometimes called *steelworkers*) (Fig. 5.4-9). The designer should require that layout

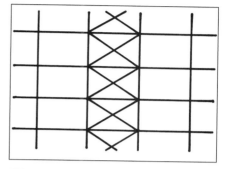

**FIGURE 5.4-6** Diagonal bracing. (Drawing by HLS.)

**FIGURE 5.4-7** An ironworker high on a steel frame building. (Courtesy Bethlehem Steel Corporation.)

**FIGURE 5.4-8** A complicated steel frame under construction. (Courtesy Bethlehem Steel Corporation.)

**a**          **b**

**FIGURE 5.4-9** (a) Ironworkers wrestling a steel column into place. (b) Ironworkers welding steel members together. (Courtesy Bethlehem Steel Corporation.)

and positioning of the steel, positioning of anchor bolts, and establishing and checking of bearing levels and other elevations be done by a licensed surveyor. The erection should be accomplished in accordance with the requirements and within the tolerances established by AISC, unless the structural designer specifically specifies or approves deviations from AISC requirements.

Structural steel is placed in position in the field using cranes or other lifting devices. On fairly low buildings, ground-mounted cranes are used to erect the entire structure (Fig. 5.4-10). On taller structures, either guy derricks (Fig. 5.4-11) or climbing cranes (Fig. 5.4-12), both of which are mounted on the building, are employed.

**FIGURE 5.4-10** Steel being erected using a ground-mounted crane, First Indiana Plaza, Indianapolis, Indiana. (Architect: CSO Architects, Inc.; design architect, 3DI International; photograph courtesy of the Vulcraft Division of Nucor Corporation.)

**FIGURE 5.4-11** Guy derricks are lifted onto the structure by a ground-mounted crane when the building has reached the practical limit of the crane. A guy derrick rests on the structure and is stabilized by guys fastened to the structure. Components of guy derricks are moved upward to a new level as the frame progresses. To move a guy derrick upward, the boom is disconnected from the mast, temporarily supported on the structure, and guyed to stabilize it. The mast is then lifted by the boom to a higher level, where it is supported on the structure and stabilized. The crane is then reassembled. (Courtesy Bethlehem Steel Corporation.)

The general process usually followed in erecting a steel structure begins with the placement of anchor bolts and leveling plates and the placement of grout beneath the leveling plates by the general contractor. Leveling plates are often omitted, however, especially for larger columns.

Each of the lowest-level columns is then transported to its proper position and lowered by crane so that the anchor bolts pass through the corresponding holes in the base plate. On multistory buildings, the lowest (first-*tier*) columns are usually two stories high.

After the first-tier columns have been bolted in place and are plumb, the beams and girders for the first two floors are placed and fastened in place (Fig. 5.4-13).

The next-tier columns, which are also usually two stories in height, are erected and the beams and girders for the next two floors are installed (Fig. 5.4-14). This process continues until the entire structure has been completed (Fig. 5.4-15).

### 5.4.1.5 Preengineered Buildings

Preengineered buildings are primarily used as industrial facilities but may also be used as commercial buildings. They are built in a shop or factory from conventional rolled steel framing elements (beams, angles, tees, plates, etc.), open-web steel joists or joist-girders, steel decking, and sheet steel roofing and siding.

Such preengineered buildings are then shipped to the site broken down

FIGURE 5.4-12 A climbing crane rises through an opening in the structure such as an elevator shaft. Additional sections are added at the top of the crane as the structure grows. (Courtesy Bethlehem Steel Corporation.)

FIGURE 5.4-14 An ironworker guiding a beam into place. (Courtesy Bethlehem Steel Corporation.)

FIGURE 5.4-13 Installing beams. (Courtesy Bethlehem Steel Corporation.)

FIGURE 5.4-15 Erecting a portion of a sloping steel structure. (Courtesy Bethlehem Steel Corporation.

into units and fastened together in the field to form a complete building frame and shell.

# 5.5 Metal Joists, Girders, and Trusses

Whether the building is framed with concrete or steel, or built with supporting walls of masonry, the floors and roofs in many low-rise and medium-rise buildings, and in some high-rise buildings, are supported with steel joists and girders. Where spans are large or the roof is sloped, the roof of such a building is often supported by trusses. When the building is framed with structural steel, the steel is erected and braced; then the joists, girders, or trusses are set onto the framing members and welded or bolted in place. When the building is framed with concrete, the concrete supporting beams are erected to the proper elevation braced, steel bearing plates or angles are installed on the beams, and the joists, girders, or trusses are set on them. When the building is of masonry construction, the masonry supporting walls are erected to the proper elevation and topped with a bond beam or solid masonry, steel bearing plates or angles are installed on the masonry, and the joists, girders, or trusses are set on them.

## 5.5.1 STEEL JOISTS AND JOIST GIRDERS

Standard open web steel joists (Fig. 5.5-1) are used for floor and roof framing mostly in masonry or steel-framed buildings (Fig. 5.5-2). Joist girders are open web trusses designed for longer spans and as primary framing members in lieu of beams. Both are usually built from high-strength steel to standard sizes. Selection is made by a structural designer from tables produced by the Steel Joists Institute (SJI).

Joists are classified as K-Series, LH-Series, and DLH-Series. K-Series joists range in depth from 8 to 30 in. (203.2 to 762 mm) in 2 in. (50.8 mm) increments. They are designed to span up to 60 ft (18.29 m). Except where their ends abut over steel beams where bolts and welding can be used to provide additional support, K-Series joists must be supported at least $2^1/_2$ in. (63.5 mm) on steel bearing plates.

LH-Series (longspan) joists are available in depths from 18 to 48 in. (457.2 to 1219.2 mm). The deeper ones can span up to 96 ft (29.26 m). LH-Series joists must be supported by at least 4 in. (101.6 mm) of steel beam or 6 in. (152.4 mm) of steel bearing plate when resting on concrete or masonry.

DLH-Series (deep longspan) joists range in depth from 56 to 72 in. (1422.4 to 1828.8 mm). Some can span as much as 144 ft (43.89 m). Bearing requirements

**FIGURE 5.5-2** Bar joist being placed on joists girders. (Photograph courtesy of the Vulcraft Division of Nucor Corporation.)

for DLH-Series joists are the same as those for LH-Series joists.

Joists are usually spaced about 2 ft (609.6 mm) apart, but spacings as wide as 10 ft (3048 mm) sometimes occur when the floor or roof decking can accommodate them.

Joists usually require bridging to prevent them from twisting or overturning. This can be accomplished by horizontal angles or rods running across the bottom chords or by diagonal steel bracing from joist to joist. Bridging is usually welded in place.

## 5.5.2 STEEL TRUSSES

Steel trusses are manufactured from steel structural shapes to produce units that are lighter in weight than comparable solid steel construction, such as welded steel plate girders. Most trusses form roof structures rather than floors.

The structures known as *space frames* (see Fig. 5.4-1) are actually three-dimensional trusses designed to span in two directions simultaneously.

# 5.6 Steel Decking

Many types of products are used as roof decks on steel-framed buildings. Examples include the poured-in-place gypsum concrete decks and the lightweight aggregate board and wood fiber board decks

**FIGURE 5.5-1** Bar joist. (Drawing by HLS.)

described in Chapter 3. A major type of roof decking used on steel-framed buildings is corrugated steel decking, either with or without a concrete topping (Fig. 5.6-1).

Concrete slabs are often used as floor decking in steel-framed buildings. Sometimes, such slabs are formed using temporary formwork, as described in Chapter 3. More frequently, floor slabs are placed over corrugated steel decking.

Many different configurations of corrugated steel decking are available, depending on their use and span. The *Design Manual for Composite Decks, Form Decks, and Roof Decks* of the Steel Deck Institute (SDI) spells out the minimum requirements for each type.

Steel decking is usually fastened in place by welding through it to the supporting steel. In thin metal decking, and where greater strength is required, welding washers are used. Occasionally, decking is fastened in place with mechanical fasteners. When a deck must act as a diaphragm, its panels are sometimes fastened together at the laps between panels with screws or welds. The method of attachment to be used and whether deck laps need fastening are the responsibility of the structural designer. The important consideration is to ensure that the deck and its fastenings will adequately resist the forces that may act on them.

### 5.6.1 STEEL ROOF DECKING

Steel roof decking is made from 22-, 20-, 18-, and 16-gauge (0.85-, 1.0-, 1.3-, and 1.6-mm) steel in depths from $1\frac{1}{2}$ in. (38.1 mm) to $7\frac{1}{2}$ in. (190.5 mm) (see Fig. 5.3-19). It can span from 4 ft 6 in. (1371.6 mm) for the lightest decking to as much as 30 ft (9.14 m) for heavy, deep cellular decking. Rib openings in $1\frac{1}{2}$ in. (38.1 mm) roof deck are either 1 in. (25.45 mm) wide (*narrow rib*), $1\frac{3}{4}$ in. (44.45 mm) wide (*intermediate rib*), or $2\frac{1}{2}$ in. (63.5 mm) wide (*wide rib*). Corrugated sheet steel roof decks are either *open flute* (Fig. 5.6-2a) or cellular (Fig. 5.6-2b), meaning that they have a sheet steel plate affixed to the bottom of the decking. *Acoustical roof decking* is also available. Some of it is of the open-flute type, and some is of the cellular type. The open-flute type has openings in the vertical webs. Openings in the cellular type are in the bottom plate. The concealed spaces in acoustical roof decking are filled with sound-absorptive matter, usually glass fiber insulation.

### 5.6.2 STEEL FLOOR DECKING

Corrugated steel floor decking is available in three different types: *form decking*, *composite decking*, and *cellular composite decking*.

*Form decking* is used as a permanent form for a concrete slab (Fig. 5.6-3a). It varies in thickness from $\frac{1}{2}$ in. (12.7 mm) to 2 in. (50.8 mm). Spans range from 1 to 2 ft (304.8 to 610 mm) for light $\frac{1}{2}$ in. (12.7 mm) form decking to as much as 12 ft (3.66 m) for heavy 2-in. (50.8-mm)-thick decking with 6 in. (152.4 mm) rib spacings. Form decking adds no strength to the slab. It simply reduces the amount of work involved in forming and eliminates the need to strip the forms from a slab.

*Composite decking* is designed not only to support a concrete slab while it is being placed, but also to bond with it, providing tensile reinforcement for the com-

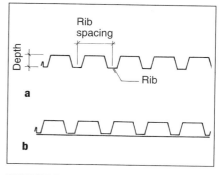

**FIGURE 5.6-2** Roof deck: (a) open flute type; (b) cellular type. (Drawing by HLS.)

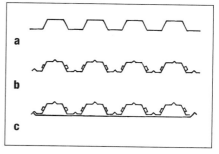

**FIGURE 5.6-3** Floor decks: (a) noncomposite form deck; (b) composite deck. The indentations and projections act to produce a bond with the concrete slab as it is placed. The configuration of these indentations and projections varies from manufacturer to manufacturer. Some composite decking has steel rods welded across the top of the flutes. (c) Composite cellular deck. Many configurations are available. (Drawing by HLS.)

**FIGURE 5.6-1** Steel deck being placed on steel joists and joists girders. (Courtesy of the Vulcraft Division of Nucor Corporation.)

posite structure (Fig. 5.6-3b). The resulting longer slab spans and greater strength produce a more economical structure than does noncomposite construction. Sometimes the composite nature of the system is further enhanced by welding *shear connectors* to the tops of supporting beams, thus bonding the concrete slab to the steel structural members as well as to the decking (Figs. 5.6-4 and 5.6-5).

*Cellular composite decking* works in essentially the same way as does noncellular composite decking (Fig. 5.6-3c). In addition, cellular decking can be designed to house electrical wiring or to provide acoustical treatment in the same way that cellular roof decking does.

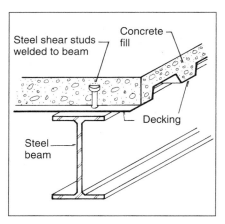

**FIGURE 5.6-4** Composite construction shear studs. (Drawing by HLS.)

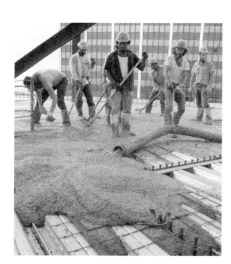

**FIGURE 5.6-5** Concrete being placed on a composite steel deck. (Courtesy of the Vulcraft Division of Nucor Corporation.)

# 5.7 Cold-Formed Metal Framing

Cold-formed steel framing, which is also called *light-gauge metal framing* and *lightweight steel framing*, is equivalent to wood structural framing for partitions, walls, floors, and roofs. Cold-formed steel framing is so called because the components employed are cold formed from galvanized coil and sheet steel stock by brake forming (see Section 5.1.4.3) and punching. This section covers cold-formed steel studs, joists, rafters, trusses, and accessories that are capable of carrying loads other than themselves and their finishes. These provide a nonflammable, dimensionally stable structural system for light loads. Although compatible with conventional stud or bearing wall construction, lightweight steel framing also introduces the possibility of longer spans and large-scale prefabrication (Fig. 5.7-1). Cold-formed steel studs used primarily for non-load-bearing partitions and furring runners, hat channels, and z-shaped furring are addressed in Section 9.1.

## 5.7.1 DESIGN

Unlike structural steel framing, cold-formed steel framing is sometimes not designed by a project's structural engineer. In such a case, carrying loads, per-

**FIGURE 5.7-1** A prefabricated steel stud wall assembly can often be lifted into place by one or two workers. (Courtesy American Iron and Steel Institute.)

missible deflections, element spacings, and other characteristics are specified in the construction documents, and the cold-formed steel manufacturer has the system designed by a structural engineer. The manufacturer is required to submit design criteria, an engineering analysis, construction details, and shop drawings showing the system in detail for approval by the architect and the project's structural engineer. Cold-formed steel framing systems and their components are generally required to be in compliance with ASTM standards A1003, C955, and C1007. System design and installation should be in accordance with AISI standards CF02, CF03, the SG02 series, and SG03.

Required spacings and depths for studs, joists, and rafters are often specified in the construction documents as part of the design requirements. A change in these criteria by the manufacturer may necessitate redesign of items being supported by the cold-formed framing. Fortunately, the necessary strength (load-carrying potential) of members can usually be obtained by increasing the thickness (gauge) of the metal without changing the depth or spacing of the members.

Allowable span tables for cold-formed steel floor joists, roof joists, and rafters are available and can be used in a building's design. Stud height tables are also available and can be used in establishing stud depths for design purposes. The International Code Council's *International Residential Code* includes tables for studs, rafters, and roof framing members that can be used for residential construction without resort to design by a structural engineer where this code has been adopted by a particular jurisdiction.

## 5.7.2 COMPONENTS

Cold-formed structural systems are made up of framing components and auxiliary components (accessories).

### 5.7.2.1 Framing Components

The major framing components in a cold-formed steel structural system are called *C-members* (also known as *C-studs*) and *channels* (also known as *channel studs*). C-members have $1/2$ in. (12.77 mm) lips on the flanges (Fig. 5.7-2a), and channel members do not have lips (Fig. 5.7-2b). Both shapes are used as component parts of rafters, joists, headers, and trusses.

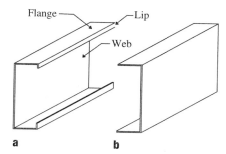

**FIGURE 5.7-2** (a) C-member with lips; (b) channel. (Drawing by HLS.)

Studs are usually C-members. Members used for studs, rafters, and joists are punched to permit the passage of pipes and wiring. Members used for tracks and rim joists are usually not punched.

Track components, which are used at the tops and bottoms of studs, as part of built-up joists and built-up rim members, and in other locations where they are appropriate, are slightly larger than elements with lips to permit studs and other framing members to fit easily, though snugly, within them.

Standard stud lengths are 10 ft (3.048 m), but they and other cold-formed steel framing components are available in lengths of up to 40 ft (12.19 m).

Cold-formed C-members are available in stock sizes with depths of $2^1/_2$ in. (63.5 mm), $3^1/_2$ (88.8 mm), $3^5/_8$ in. (92.08 mm), 4 in. (101.6 mm), $5^1/_2$ in. (139.7 mm), 6 in. (152.4 mm), 8 in. (203.32 mm), 10 in. (254 mm), 12 in. (304.8 mm), and 14 in. (355.6 mm).

C-members flange sizes are $1^1/_4$ in. (31.8 mm), $1^3/_8$ in. (34.93 mm), $1^5/_8$ in. (41.28 mm), 2 in. (50.8 mm), $2^1/_2$ in. (63.5 mm), and 3 in. (76.2 mm).

Channel member leg heights are: 1 in. (25.4 mm), $1^1/_4$ in. (31.8 mm), $1^1/_2$ in. (38.1 mm), 2 in. (50.8 mm), and 3 in. (76.2 mm).

Standard C-member and channel members gauges are 20 (33 mils), 18 (43 mils), 16 (54 mils), 14 (68 mils), and 12 (97 mils).

The standard yield strength for all components is 33 ksi (227 MPa), but 50 ksi (345 MPa) is available on request.

The standard galvanized coating is G-60 in accordance with ASTM A653 or the equivalent.

The most common sizes used for studs are those that match wood framing members used for the same purpose.

Custom sizes, lengths, and coatings are available.

### 5.7.2.2 Accessories

Many types of accessory components are needed to complete a cold-formed framing system. These include furring channels and runner channels similar to those described in Section 9.1. Needed accessories also include various sheet metal plates, angles, web stiffeners, clips, hangers, and fasteners.

### 5.7.2.3 Precut Joists and Rafters

Cold-formed steel joists and rafters are often cut from stock members in the field, but C-shaped joists that are precut to exact sizes are also readily available. Joists may be made up from stock C-members or channel members or a combination of the two used together (Fig. 5.7-3).

Available precut joist sizes include 8 in. (203.32 mm), 10 in. (254 mm), 12 in. (304.8 mm), and 14 in. (355.6 mm).

Gauges of joist members are 18 (43 mils), 16 (54 mils), 14 (68 mils), and 12 (97 mils). Flange widths include $1^5/_8$ in. (41.28 mm), 2 in. (50.8 mm), $2^1/_2$ in. (63.5 mm), and 3 in. (76.2 mm).

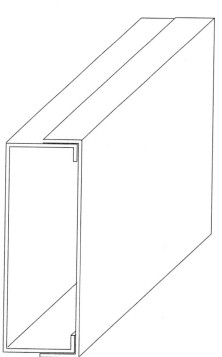

**FIGURE 5.7-3** Nested C-members and channel members are often used to provide the necessary strength for rafters, ceiling joists, headers, rim joists, and other structural elements. (Drawing by HLS.)

### 5.7.2.4 Trusses and Girders

Cold-formed steel trusses and girders may be assembled in the field from cut sections of unpunched stud and floor framing components or they may be factory fabricated and delivered to the site complete with all components necessary for installation, such as valley components, truss-to-truss and truss-to-bearing connection devices, special connections, truss bracing and installation bracing products, permanent bracing components, and required hip, ridge, and valley plates. Such trusses can be designed to span as much as 100 ft (30.48 m). Factory-prefabricated, preengineered, cold-formed steel trusses and girders can be UL rated. In any case, trusses and girders should be designed in accordance with AISI SG02-5, "Standard for Cold-Formed Steel Framing-Truss Design." The *International Residential Code* requires that ANSI SG02-5 be followed.

Trusses and girders are assembled using mechanical fasteners. Specially designed systemic and hurricane ties are used to attach trusses and girders to the building frame and to anchor them against forces generated by these phenomena.

### 5.7.3 INSTALLATION

Cold-formed steel framing components are assembled by welding or with screws. However, welding is limited to components of 18 gauge (0.0428 in. [1.22 mm]) or thicker. Studs fit within runner tracks and are screwed to the tracks (Fig. 5.7-4).

Joists are either seated on a runner supported by studs (Fig. 5.7-5) or on beams or walls, or are hung from headers or girders (Fig. 5.7-6). Conditions at

**FIGURE 5.7-4** Runner tracks are slightly larger than studs to permit the studs to fit within the flanges of the runners. This characteristic also permits the nesting shown in Figure 5.7-3. (Drawing by HLS.)

**FIGURE 5.7-5** Joists bearing on stud framing are placed directly over load-bearing studs located at the same spacing as the joists. Such bearing locations usually require insertion of web stiffeners into the flanges of the joist members. Such stiffeners are generally fastened with screws through both of the overlapping joists with four screws. Stiffeners are also fastened to the runner with two screws in each stiffener. Studs are screwed to the runner channels with a screw through each flange. Stiffeners are also usually required where a single joist is continuous over more than one span. Sometimes clip angles or bent stiffeners are used to accomplish fastening to the runner channel. The condition at supporting concrete or steel beams is similar. (Drawing by HLS.)

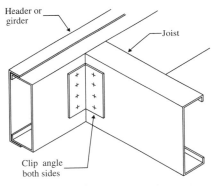

**FIGURE 5.7-6** One method of hanging joists from a header or girder. The clip angles are screwed to the joist and to the header or girder. Joist are also sometimes hung using hangers similar to those used in wood framing. (Drawing by HLS.)

the building's perimeter are similar (Fig. 5.7-7).

When possible, floor joists, ceiling joists, and rafters are placed directly above and in alignment with the studs supporting them (see Fig. 5.7-7). When such an arrangement is not possible, distribution members are installed to trans-

**FIGURE 5.7-7** Framing at the building perimeter is similar to that at an interior support with the addition of rim joists. Depending on load requirements, the rim joist may be a single member, as shown here, or a built-up (nested) member made up of a C-member and a channel member. Sometimes two C-members or two or more nested members may be required, depending on the loads involved. (Drawing by HLS.)

fer the loads to the supporting studs. Studs are also aligned with the floor framing members below them. In multi-floor framing, studs are aligned vertically from foundation to roof.

Roof framing members and trusses are generally supported on tracks supported by studs (Fig. 5.7-8).

Framing around floor openings in cold-formed steel framing is similar to that in wood framing, except that framing members are single or nested cold-formed members. Openings in walls in cold-formed steel-framed buildings are framed in a manner that is also similar to that used in wood framing. C-member jack studs and king studs are used at the jambs. Headers are built up from C-members and runners. The configuration, size, and gauge of members, and the number of components in opening headers, depend on the span of the opening and the loads to be carried by the header.

Joists, rafters, and trusses in cold-formed steel structures must be stabilized with cross bracing or continuous lateral bracing in a manner similar to that used in wood-framed structures. In addition, cold-formed steel-framed buildings must be properly braced to resist racking under wind and seismic loads. This can be accomplished with diagonal strap bracing fastened by screws or welds to studs and plates. Lateral steel bracing can be used, to resist stud rotation and minor axis bending under wind, seismic, and axial loads, especially during construction.

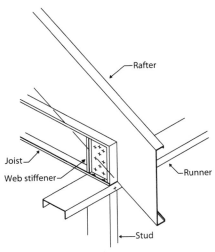

**FIGURE 5.7-8** Typical roof framing support. The rafter is supported by screws through the web stiffener and the ceiling framing joists. Loads will sometimes require additional supports and anchorage. Trusses are supported in a similar fashion, with the addition of clip angles fastened to the truss and the runner to resist uplift from wind-related forces. Where it is not possible to place roof framing members directly over a load-bearing stud, it is often necessary to install distribution members within the runner track to distribute the roof load to the load-bearing studs. (Drawing by HLS.)

## 5.8 Other Metals

Other metals used in buildings include copper and its alloys brass and bronze, zinc, lead, chromium, nickel, tin, and gold.

### 5.8.1 COPPER AND ITS ALLOYS

Copper is probably the world's second most used metal, following the various alloys of iron, including steel. Copper occurs in nature in its metallic state, which probably explains its use as early as 5000 B.C. Today, it is mostly obtained by refining the ores cuprite, which is copper oxide; malachite, which is basic copper carbonate; or chalcocite or chalopyrite, both of which are copper sulfide. The process is a complicated one, requiring roasting, smelting, and converting the ore, and then refining the resulting metal by electrolysis.

Newly refined copper is about 99.95% pure. It has a reddish yellow color and is soft, malleable, ductile, relatively strong, and a very good conductor of electricity. Copper does not corrode in dry air but will oxidize when heated. In normal atmosphere it oxidizes slowly, forming a thin surface coating of verdigris, which is similar to the ore malachite.

Copper has many uses, including making copper sulfate for printing, electroplating, and to purify public water supplies. It is also used in tanning leather, in manufacturing pigments, insecticides, and dyes for cotton, and even as an astringent. Copper is also used extensively in making coins, cooking utensils, art objects, wire, and various sheet metal items, such as gutters, downspouts, and roofing. One of the primary uses of copper is in the making of various alloys, including the two most interesting to those in the building industry, brass and bronze.

### 5.8.1.1 The Copper Alloys

There are more than 1100 recognized alloys of copper. Of these, more than half are alloys of copper and zinc or copper and tin. The former has traditionally been known as *brass*, the latter as *bronze*. Today, most copper alloys used for other than utilitarian purposes in building and related construction are called *bronze*, even though the many that contain zinc are really brass. Bronze was almost certainly the first industrial alloy. In some places in the world, copper and tin ore are found naturally intermixed, which probably explains the appearance of bronze as early as 3500 B.C. Brass followed much later, but even it has been in use since about 50 B.C.

### ALLOY DESIGNATIONS

The Copper Development Association (CDA) has a numbering system for copper alloys. CDA is also the administrator of the UNS, which was developed jointly by ASTM and SAE. The CDA system consists of three digits. The UNS adds the prefix *C* to the three digits and a 00 suffix. The alloy 280 (*Muntz metal*) in the CDA system is C28000 in the UNS.

### COMMON ALLOYS

The CDA classifies the most common copper alloys used in the construction industry into three groups. The A group contains copper, including Alloy 110, which is 99.9% copper, and Alloy 122, which is 0.02% phosphorus.

The B group contains the common brasses and extruded *architectural bronze*. It includes Alloy 220 (*commercial brass*), which is 10% zinc; Alloy 230 (*red brass*), which is 15% zinc; Alloy 260 (*cartridge brass*), which is 30% zinc; Alloy 280 (*Muntz metal*), which is 40% zinc; and Alloy 385 (*architectural bronze*), which is 40% zinc and 3% lead.

The C group contains the so-called *white bronzes*. It includes Alloy 651 (*low-silicon bronze*), which has between 1.5% and 3% silicon; Alloy 655 (*high-silicon bronze*), which has 3% silicon; Alloy 745 (*nickel silver*), which has 25% zinc and 10% nickel; and Alloy 796 (*leaded nickel silver*), which has 42% zinc, 10% nickel, 2% manganese, and 1% lead.

*Casting alloys* are special alloys made specifically for casting. They fall in the 800 and 900 series of CDA alloy designations.

Other alloys are used also, of course, such as when color matching is required. The CDA's *Copper in Architecture Design Handbook* includes a chart showing which alloys match others in color. When a nonstructural metal item contains several different shapes or types of components, such as sheets, pipes, extrusions, fasteners, castings, and so forth, the different parts are most likely made from different alloys, even though the colors match.

### 5.8.1.2 Copper Alloy Products Production

Many copper alloy products are used in producing nonstructural metal items for use in buildings and related construction. Figure 5.8-1 is a partial list of such products showing one possible series of color-matched alloys that might be used.

### FORMING METHODS

The production of final copper alloy items for use in buildings starts with the forming or fabricating of copper alloy materials into useful products, such as those listed in Figure 5.8-1. Four basic methods are generally used to form these products: casting, rolling, extruding, and drawing.

*Casting*   Intricate shapes are often made using the casting method. Castings may later be drilled, milled, or otherwise machined to produce necessary openings and close fits with adjacent materials.

*Rolling*   Many copper alloy items used in buildings and related construction are at some point in their development run through a hot rolling mill. Many of them are then passed through a cold-rolling process to make their surfaces brighter and smoother. Items produced by the rolling process include plates, sheets, shapes, rods, bars, and forging stock.

*Extruding*   Many copper alloy products used in buildings and related construction are produced by extrusion. Extruded items include structural shapes, rods, bars, seamless tubes and pipes, and the kinds of special shapes often used in nonstructural metal items. Extrusions are superior in virtually every way to rolled products. Among other characteristics, they are straighter, more accurate in dimension, and smoother. They are, however, limited to a diagonal cross section of 6 in. (152.4 mm) and a thickness of about $1/8$ in. (3.2 mm).

*Drawing*   Tubes and wire can be made by drawing copper alloy through a die in lieu of pushing it, as in the extrusion process.

**FIGURE 5.8-1** Copper Alloy Products That Are Compatible in Color[a]

| Product | ASTM Standard | CDA Alloy Number and Name |
|---|---|---|
| Extrusions | B455 | 385, architectural bronze |
| Sheets, plates, and bars | B370 | 280, Muntz metal |
| Round tubing | B135 | 230, red brass |
| Castings | B271 | 857 |
| Sand castings | B584 | 857 |
| Wire and rods | —[b] | 280, Muntz metal |

[a]The alloys listed represent only a single group of materials that generally match in color and are by no means the only alloys used for the products listed. Refer to the CDA publication *Copper Brass Bronze Design Handbook, Architectural Applications* for an indication of other alloys used for the purposes listed.

[b]No current universally recognized standard for alloy 280 wire rods.

## FABRICATION

Some rolled, extruded, and drawn products are used exactly as they come from the mill to fabricate items used in buildings. Others require only additional finishing. Many other components, however, are made by fabricating these products into usable shapes. The methods employed include additional roll forming, forging, bending, break forming, explosive forming, and spinning. Not all methods are applicable to all alloys. Refer to the CDA's *Copper in Architecture Design Handbook* for an indication of the alloys that can be formed by each method.

*Roll Forming* The rolled products mentioned earlier are passed through an additional roll-forming process to make the final product. Roll forming at this stage consists of passing sheets through a series of shaped rollers to produce the desired shapes.

*Forging* Forging is done using one of four methods. The first two methods require closed steel dies.

1. In *hammering*, the copper alloy is hammered into the die cavity by repeated blows of a *drop hammer*.
2. In *pressing*, the metal is forced into the die cavity under hydraulic pressure.
3. In *hydroforming*, a punch or formed steel die is forced under hydraulic pressure against the metal, which is backed up by rubber.
4. In *stamping*, the metal is bent, shaped, cut, indented, embossed, coined, or formed by means of a press or hammer that is faced with a shaped die.

*Bending* Many copper alloy sheet metal and plate components used in buildings, such as door frames and the like, and many roofing-related copper items are shaped by bending.

*Brake Forming* Brake forming is probably the most often used method for shaping copper alloy sheets used in buildings. Many copper alloy components used in nonstructural metal items, shapes such as door frames and the like, and many roofing-related copper alloy items are shaped by brake forming. In essence, brake forming consists of squeezing sheet metal between a punch and a die block. Several consecutive op-erations are usually required to make a final shape.

*Explosive Forming* High-energy rate forming is done by forcing metal against a single die using a shock wave generated by detonation of small explosive charges.

*Spinning* A type of shear forming, spinning, is done by clamping the sheet metal to a form block and spinning the block and the metal against a roller, which gradually forces the metal to fit the shape of the form block. Spin forming is done using either power or hand tools.

### 5.8.2 ZINC

Zinc is a bluish-white metal that is plastic and highly corrosion resistant. It is not decorative, however, and is usually given an applied finish, mostly paint and other organic coatings.

Nonstructural metal items whose strength is not paramount are sometimes made from zinc. Most zinc is used, however, as an alloying metal or as a coating for other metals. Refer to Section 5.9.4.1 for a discussion about zinc coatings, such as galvanizing on iron and steel. Another major use of zinc is as an alloying metal. It is alloyed with copper to make brass and with aluminum to make high-strength, heat-treatable alloys. It is also added to magnesium-aluminum alloys to increase their ductility.

Zinc is also a pigment in some paints, particularly in primer paints for galvanized surfaces and in paints for rusted steel surfaces.

### 5.8.3 LEAD

The heavy metal, lead, is soft and weak due to its ductility. Its properties, however, especially its ductility, its resistance to corrosion, and its density, which make it able to stop x-ray penetration, cause it to be used extensively in the construction industry.

Lead can be easily rolled into sheets and extruded to make pipes. It is used extensively in the chemical industry because of its resistance to corrosion by many chemicals, including sulfuric acid. Its chief use in buildings is in sheet form as radiation protection, sound barriers, and baffles, and as a coating for other metals to provide corrosion protection. Sheet lead is also used in roofing and as-sociated rainwear, but lead-coated copper is a more common roofing material.

Some paints contain lead, especially primer coatings on steel. This use may diminish in the future because of the fear of lead poisoning, especially of children, but lead paints will be found on existing metals for many years, even if their use is eventually banned altogether.

### 5.8.4 CHROMIUM

Chromium, which is very hard and highly corrosion resistant, is mostly used as a coating on steel or other metals and as an alloying metal for steel. All stainless steel, for example, contains chromium.

Chromium is so hard that it resists polishing. Therefore, to produce the common highly polished chrome-plated surfaces, such as those on plumbing fixtures, it must be deposited over a highly polished steel surface, usually with an intermediate layer of nickel.

### 5.8.5 NICKEL

The primary use of nickel in buildings is as an alloying metal. *Monel metal*, for example, is an alloy of nickel with 31% copper. *Nickel silver* is an alloy of nickel and brass. Perhaps half of the nickel produced in this country ends up in steel alloys.

At one time, nickel was a popular finish coating for steel. Today, it is more likely to be used as an undercoating over steel for the more popular chromium.

### 5.8.6 TIN

Probably the best-known use of tin in the building industry today is as an alloying metal with copper to make bronze. Tin is also used as a coating for steel and iron. In earlier days, much roofing was made from tin plate. Alloys of tin and other metals, such as zinc, are also used for coatings on steel. Solder contains tin.

Tin is used less than zinc as a protective coating for steel, because tin is much rarer than zinc, especially in North America, and because tin does not protect steel from corrosion as effectively as zinc does.

### 5.8.7 GOLD

Gold is seldom used in the building industry except to gild decorative items. It is worth mentioning here, however, because there are many, especially older buildings on which gold leaf was used extensively. Gilding with gold is not as expensive as it might appear because of the metal's great malleability. Gold leaf is only $\frac{1}{250,000}$ in. (0.000 101 6 mm) thick. At that thickness, 1 oz (28.3495 g) of gold will cover 120 sq ft (11.15 m²) of surface.

## 5.9 Metal Finishes

Some metals are used in construction applications with their natural finish, but most metals used in construction have an applied finish of some sort. This section discusses the various natural and applied finishes generally used on metal construction products.

### 5.9.1 FINISH CLASSIFICATIONS

Finishes for ferrous metals can be generally classified as *mechanical finishes*, *chemical finishes*, and *coatings*, and these are the designations used here.

In its *Metal Finishes Manual*, the National Association of Architectural Metal Manufacturers (NAAMM) specifies the requirements for three types of basic finishes for aluminum and copper. The Aluminum Association (AA) has adopted the NAAMM metal finish types, terminology, and descriptions and has added a few of its own. AA's terminology is applicable only to aluminum surfaces. The NAAMM types are *mechanical finishes*, *chemical finishes*, and *coatings*. Within these types there are many finishes.

The discussion of each nonferrous metal finish type in this section includes the NAAMM designation of the finishes in that type. In the complete system, all the appropriate designations are used together when describing a particular finish. For example, the finish M31-M34-07× has a fine satin (M31), hand-rubbed (M34) directional textured mechanical finish and a clear thermoset organic coating (07×). To make the designation complete, the organic coating, represented by the x, must be completely described, including its characteristics and manufacturer.

### 5.9.2 MECHANICAL FINISHES ON FERROUS METALS

*Mechanical* finishes on ferrous metals include those left by the manufacturing process and those created by grinding, polishing, sandblasting, and rolling. They include the as-rolled and as-drawn natural mill finishes, and those imparted by further grinding, polishing, or patterning.

#### 5.9.2.1 Corrosive Ferrous Metals

Ordinary cast iron, most carbon steels, and some alloy steels oxidize under ordinary atmospheric conditions. Elevated temperatures and moisture accelerate this oxidation, which is known as *atmospheric corrosion*. The resulting surface scale (rust) deteriorates the metal surface and, if not prevented, progresses until the metal reverts to the oxide state found in nature. The chief function of protective coatings used on iron and carbon steel is to inhibit corrosion. Not all ferrous metals require equal protection against progressive corrosion. Copper-bearing steels, some cast irons, and some high-strength, low-alloy steels acquire excellent corrosion resistance from a tight surface oxide that inhibits further corrosion. Sometimes the appearance of the natural patina is objectionable, and finish coatings are applied on these steels for decorative purposes. Decorative and protective coatings are used less often on stainless steels because of their superior corrosion resistance and attractive natural finishes.

The *hot-rolled mill finish* (black, as-rolled finish) on carbon steel is characterized by a mill scale and rust powder which is sometimes acceptable as a base for organic coatings applied by brush. For spray painting, this rust powder must be removed by sandblasting or wire brushing.

*Cold-finished* surfaces are usually smoother and contain little or no surface corrosion, but must be degreased of oil coatings and sometimes roughened to provide a good base for organic coatings. Steel products that will not be coated may be sent to a *temper mill,* which rolls the steel to the desired flatness and surface quality. Cold-reduced sheet and strip have a natural mill finish acceptable for many uses, and as a base for subsequent galvanizing, tinning, or terne coating. *Black pipe* has a natural dark finish often left uncoated. Stainless steel sheet is produced in a variety of patterned finishes (Fig. 5.9-1) imparted by roll forming for greater stiffness and decorative effect, and to reduce differential light reflection that emphasizes the waviness of sheet surfaces in thin, flat products.

#### 5.9.2.2 Stainless Steel Sheet and Strip

AISI has developed standard designations for mechanical finishes on stainless steel (Fig. 5.9-2). Because stainless steels are inherently corrosion resistant and can be made strong and rigid in thin sheets, these products are the ones most often not protected by coatings. They are most often either left with their natural cold-rolled finish, such as Nos. 1, 2D, and 2B sheet finishes, or given a polished finish, such as Nos. 3, 4, 6, 7, or 8 (see Fig. 5.9-2). An exception is terne-coated stainless steel, which is used in roofing.

#### 5.9.2.3 Stainless Bars, Pipes, and Tubing

Stainless bars, pipes, and tubing are manufactured in several neutral and polished finishes similar to those of sheets and strip described in the preceding subsection. In industrial applications, they may be used in one of the natural finishes; for architectural work where uniformity of

**FIGURE 5.9-1** Steel sheets can be stiffened and decorated with patterned finishes. (Photographs furnished by Rigidized Metals Corp.)

**FIGURE 5.9-2** Common Stainless Sheet Steel Mechanical Finishes

| | Designation | Description | Uses |
|---|---|---|---|
| **Natural Finishes** (unpolished) | No. 1 Finish | Rough dull surface resulting from hot rolling, annealing, and descaling | Generally available on heavy gauge sheets; not used in architectural applications |
| | No. 2D Finish (No. 1 Strip Finish) | Smooth dull surface resulting from cold rolling, annealing, and descaling | Some architectural applications requiring low luster, as in roofing and drainage products; suitable for cold forming and further polishing |
| | No. 2B Finish (No. 2 Strip Finish) | Bright, smooth cold-rolled surface produced either by highly polished rolls or bright annealing | General-purpose natural finish for transportation equipment; limited use in curtain walls and storefronts; used for further polishing |
| **Polished Finishes** | No. 3 Finish | A surface produced by grinding and polishing with slightly finer abrasives than for finish No. 4 | General-purpose natural finish for transportation equipment; limited use in curtain walls and storefronts; used for further polishing |
| | No. 4 Finish | Bright, polished surface produced by grinding and polishing either at the mill or fabricator plant | Most common for architectural uses; directional grit line permits blending of fabricated or field joints with mill-polished surface |
| | No. 6 Finish | Dull matte finish produced by brushing No. 4 finish | Architectural applications where softer, less reflective finish is desired |
| | No. 7 Finish | A highly reflective finish produced by further polishing and buffing | Chiefly in architectural applications |
| | No. 8 Finish | Mirrorlike finish produced by polishing with buffing rouges until surface is free of grit lines | Infrequent in architectural applications |

(Courtesy American Iron and Steel Institute.)

appearance is important, polished finishes are generally preferred.

### 5.9.3 MECHANICAL FINISHES ON NONFERROUS METALS

NAAMM divides the mechanical finishes on nonferrous metals into five categories: *as fabricated, buffed, directional textured, nondirectional textured,* and *patterned.* Within these types are many finishes. Each of the first four categories is given a designation that consists of the letter *M* (mechanical) followed by a two-digit number. The first digit corresponds to the four types of finish. Thus, as fabricated finishes are designated M1; buffed, M2; and directional textured, M3. The second digit designates the specific finish within that type.

The patterned type is designated M4x. When an x appears in a designation, it is necessary to follow that designation with a description, usually of a proprietary finish.

#### 5.9.3.1 As Fabricated Mechanical Finishes

An *as fabricated finish* is simply the natural finish imparted by the casting, rolling, or extruding process. It is the finish a copper alloy has when it leaves the initial manufacturing stage. It is not the final finish after the material has been fabricated into a metal item. The term *as fabricated finish* corresponds to the term *mill finish* used for iron and steel.

#### ALUMINUM

NAAMM divides as fabricated mechanical finishes on aluminum into four classes:

1. Unspecified (M10)
2. Specular as fabricated (M11)
3. Nonspecular as fabricated (M12)
4. Other (M1x)

An *unspecified* as fabricated finish is the natural finish imparted by casting, hot rolling, cold rolling with unpolished rollers, or extruding. Grinding with a dry abrasive wheel may be used to remove surface roughness from castings.

A *specular as fabricated* finish is a mirror-like finish produced by cold rolling with polished rollers. A specular finish may be applied to one or both sides, but it cannot be applied to castings, extrusions, or forgings.

A *nonspecular as fabricated* finish is more uniform than an unspecified finish but lacks the mirror-like quality of a specular as fabricated finish.

The *other* category applies to finishes that do not fit into one of the previous categories.

The surfaces of many aluminum products require no further finishing. The natural oxide film makes protective coatings unnecessary under ordinary conditions. However, finishes are frequently applied in order to (1) change or improve the appearance for decorative purposes; (2) improve its abrasion or corrosion resistance; or (3) provide a base for subsequent surface treatment such as painting.

Natural finishes and other common surface treatments are briefly described in this section and summarized in Figure 5.9-3.

Natural finishes on aluminum can be classified as uncontrolled or controlled.

***Uncontrolled Natural Finishes*** Hot-rolled products show a certain amount of discoloration or darkening, while cold-finished products have a whiter, brighter surface. Extruded products have an intermediate appearance closer to that of cold-finished items and show traces of longitudinal striations (*die lines*) caused by the extrusion process. Castings vary little in their surface appearance, except that die castings and permanent mold castings are smoother than sand castings.

***Controlled Natural Finishes*** Natural finishes of cold-rolled sheet may be controlled by the degree of polishing given to the rollers of the rolling mill as follows:

**FIGURE 5.9-3** Principal aluminum finishing processes.

■ *Mill finish*—uncontrolled, slightly variable, not entirely free from stains or oil

■ *One-side bright mill finish*—moderate degree of brightness on one side; uncontrolled on the reverse side, possibly with a dull, nonuniform appearance

■ *Standard bright finish*—relatively bright, uniform appearance on both sides; somewhat less lustrous than the standard one-side bright sheet

■ *Standard one-side bright sheet*—uniformly bright, lustrous surface on one side; uncontrolled on the reverse side, possibly with a dull, nonuniform appearance

*Embossed sheet* is obtained by passing a mill finished sheet through a pair of rolls with a matched design or between a design roll and a plain roll. Embossed surfaces increase the stiffness and tend to hide wear and scratches. Patterns include wood grain, leather grain, fluted, stucco, diamond, notched, ribbed, bark, and pebble designs (Fig. 5.9-4).

### COPPER

NAAMM divides mechanical finishes on copper also into four classes:

1. Unspecified (M10)
2. Specular as fabricated (M11)
3. Matte finish as fabricated (M12)
4. Other (M1x)

An *unspecified* as fabricated finish is the natural finish imparted by casting, hot rolling, cold rolling with unpolished rollers, or extruding. Grinding with a dry abrasive wheel may be used to remove surface roughness from castings.

A *specular as fabricated* finish is a mirror-like finish produced by cold rolling with polished rollers. A specular finish may be applied to one or both sides but cannot be applied to castings, extrusions, or forgings.

A *matte as fabricated* finish is a dull finish produced by annealing copper alloy items that have been produced by hot

**FIGURE 5.9-4** Aluminum sheet and plate can be embossed to produce a variety of surface textures.

rolling, cold rolling, cold rolling with unpolished rollers, casting, or extruding.

The *other* category applies to finishes that do not fit into one of the previous categories. Most copper alloys with an *as fabricated* finish will display some imperfections, which must be taken into account when deciding whether to use them without further finishing. Castings, for example, have a rough, matte finish. Sand castings are rougher than die castings. Hot-rolled copper alloy items have a dull surface, may be darker than cast, cold-rolled, or extruded copper alloy items, and usually have some random discolorations as well. Extrusions are usually the same color as cold-rolled material but are often left with parallel striations (*die lines*). Most imperfections tend to become worse in appearance when the metal is further formed and assembled into metal items.

Cold-rolled copper alloy items usually come closer to their final desired appearance than those produced by the other methods. The process permits some control by varying the amount of polishing used on the rollers. The smoother the rollers are, the closer the cold-rolled surface will be to the desirable finish. Even cold-rolled copper alloy items, however, may have stains, a coat of oil, or both.

### 5.9.3.2 Buffed Mechanical Finishes

Nonferrous metal finishes created by a process of grinding, polishing, and buffing are called *buffed finishes*. NAAMM lists buffed finishes on both aluminum and copper in four categories:

1. Unspecified (M20)
2. Smooth specular (M21)
3. Specular (M22)
4. Other (M2x)

An *unspecified* buffed finish is optional with the finisher.

A *smooth specular* buffed finish, which is the brightest and most lustrous mechanical finish that can be developed, is produced by successively grinding, polishing, and buffing. Buffing with a fine abrasive, results in a highly lustrous surface. Polishing with even finer abrasives and very light pressure can result in a mirror-like surface.

A *specular* buffed finish is produced by the same methods as a smooth specular buffed finish, but is not as smooth.

The *other* category is reserved for special finishes that do not fit into one of the previous categories.

The M21 category is the one most used on aluminum.

### 5.9.3.3 Directional Textured Mechanical Finishes

NAAMM's standard designation system recognizes eight finish types in the directional textured finish category for aluminum and copper:

1. Unspecified (M30)
2. Fine satin (M31)
3. Medium satin (M32)
4. Coarse satin (M33)
5. Hand rubbed (M34)
6. Brushed (M35)
7. Uniform (M36) (This category does not apply to aluminum surfaces.)
8. Other (M3x)

An *unspecified* directional textured finish is optional with the metal's finisher.

The three *satin* finishes represent different degrees of fineness. They are all produced by wheel or belt polishing with fine wire bristles, very fine abrasives, or a muslin or felt wheel to produce a soft, smooth sheet. *Spin finishing*, in which an abrasive cloth is held against the metal and rotated, followed by stainless steel wool or emery polishing, produces a bright finish with a fine pattern of concentric circles.

*Hand rubbed* is an expensive finish type that is produced by rubbing the metal with a fine brass brush or a non-woven abrasive mesh pad, abrasive cloths, or stainless steel wool. It is seldom used for general finishing but is often used for final touch-up on the other satin finishes.

*Brushed* is a directional finish characterized by fine or coarse parallel scratches that are produced by power-driven rotating stainless steel, nickel steel, or brass wire brushes, sander heads, impregnated plastic discs, and other means. Whether a brushed finish is fine or coarse depends on the size of the wire bristles used.

A *uniform* finish is produced by a single pass of a number 80 grit belt.

The *other* category is reserved for special finishes that do not fit into one of the previous seven categories.

*Highlighting* provides a two-tone effect for relief surfaces. Elevated areas are masked and recessed areas are sandblasted. Then the masking is reversed and the elevated areas are given a highlight effect by painting or polishing.

The finish categories most often used on aluminum are M32 and M34. Those most often used on copper are M32, M33, M34, and M36.

### 5.9.3.4 Nondirectional Textured Mechanical Finishes

Finishes that are designated as *nondirectional textured* finishes are produced by abrasive blasting with various combinations of pressure, steel shot, and sand to produce matte and coarse surfaces. Most of these finishes are matte in appearance with various degrees of roughness. They are not widely used on metal items for building construction. Their most extensive use is on castings. They should not be used on thin metal ($^1/_4$ in. [6.4 mm] or less) because they tend to distort the metal. Rough textures produced by blasting may be given a lacquer coat or may be anodized to minimize the collection of dust.

NAAMM's standard designation system recognizes nine finish types in the nondirectional textured finish category for nonferrous metals:

1. Unspecified (M40)
2. Extra fine matte (M41) (This category does not apply to copper.)
3. Fine matte (M42)
4. Medium matte (M43)
5. Coarse matte (M44)
6. Fine shot blast (M45)
7. Medium shot blast (M46)
8. Coarse shot blast (M47)
9. Other (M4x)

An *unspecified* nondirectional textured finish is optional with the finisher.

The categories with *matte* in their names are produced by blasting with silica sand or aluminum oxide. Those with *shot blast* in their names are produced by steel shot blasting.

The *other* category is reserved for special finishes that do not fit into one of the previous eight categories.

Combinations are sometimes used on surfaces that have relief in which the incised areas are abrasive blasted while the high areas are masked. Then the high areas are polished.

The finish categories most often used on aluminum are M42, M43, M44, and M46. The one most often used on copper is M42.

### 5.9.3.5 Patterned Mechanical Finishes

Thin aluminum and copper alloy sheets can be given a patterned finish by rolling as fabricated sheet between rollers that have been shaped to the desired design. There is no specific standard category for patterned finishes.

### 5.9.3.6 Electropolishing of Aluminum

High-purity aluminum may be treated by electropolishing to obtain a light reflectivity of about 85%. The metal surface is anodically smoothed to a highly polished surface suitable for use in aluminum light fixture reflectors.

Electropolished reflecting surfaces are usually given a final anodizing treatment for surface protection. Final reflectivity depends on the purity of the aluminum, the thickness of the oxide film, and the particular procedure employed.

### 5.9.3.7 Mechanical Cleaning of Nonferrous Metal

Bare new aluminum that is to receive an organic coating should be washed with mineral spirits or terpentine, then allowed to weather for 1 month or roughened with stainless steel wool.

Bare copper and copper alloy material that is to receive an organic coating should be washed with mineral spirits. Then stains, mill scale, and other foreign materials should be removed by sanding.

## 5.9.4 CHEMICAL FINISHES FOR AND CLEANING OF FERROUS METALS

*Chemical* finishes are used on ferrous metals chiefly as cleaning and conditioning steps for other coating applications. *Pickling*, which removes oxide scale, and other operations for removing lubricants in product manufacture are examples. *Conversion coatings* are sometimes used as decorative finishes on carbon and stainless steels.

### 5.9.4.1 Carbon Steel Conversion Coatings

Conversion coatings are mostly chemical treatments (as with an acid phosphate solution) which *convert* the chemical nature of the surface film to improve its bonding properties with paint and other applied coatings. In one decorative finish, textured steel is zinc plated and dyed using a chromate conversion coating, followed by buffing. Buffing causes the bright zinc to stand out in relief, highlighted by the color coating in the recesses.

### 5.9.4.2 Stainless Conversion Coatings

Surface blackening, a common conversion coating, on stainless steel, produces a decorative finish ranging from a bluish hue to dark brown or dead black. It is produced by encouraging the growth of a thin oxide, either by controlled heat treatment or by dipping in a hot chemical oxidizing bath.

## 5.9.5 CHEMICAL FINISHES FOR AND CLEANING OF ALUMINUM

Many methods of treating aluminum surfaces employ the reaction of various chemicals on the metal. Some of the most common ones are discussed here. NAAMM designates four types of chemical finishes for use on aluminum: *nonetched cleaned*, *etched*, *brightened*, and *conversion coatings*.

### 5.9.5.1 Nonetched Cleaned Aluminum

NAAMM divides nonetched cleaned finishes on aluminum into four categories:

1. Unspecified C10
2. Degreased (C11)
3. Chemically cleaned (C12)
4. Other (C1x)

An *unspecified* nonetched cleaned finish is optional with the finisher.

*Degreasing*, which is also called *vapor degreasing*, consists of exposing the aluminum to the vapors of a chlorinated solvent in a machine designed for the purpose.

*Chemical cleaning* is sometimes called *inhibited* because the chemicals used are inhibited to prevent them from etching the aluminum. Chemical cleaning may be used as the sole treatment for aluminum, but it is more often a second step following degreasing.

The *other* category is reserved for special finishes that do not fit into one of the previous three categories.

### 5.9.5.2 Etched Aluminum

Etched surfaces are common in aluminum production. Etching is sometimes given a lacquer finish with no further pretreatment and is often used as a pretreatment for anodizing. An excellent bond for paint can be achieved by etching the surface with a phosphoric acid solution. An etched finish is produced by dipping, washing, or otherwise exposing aluminum to an alkili or acid solution. Different chemicals produce different degrees of finish. NAAMM divides etched finishes on aluminum into five categories:

1. Unspecified (C20)
2. Fine matte (C21)
3. Medium matte (C22)
4. Coarse matte (C23)
5. Other (C2x)

An *unspecified* etched finish is optional with the finisher.

The three *matte* finishes represent different degrees of fineness.

The *other* category is reserved for special finishes that do not fit into one of the previous four categories.

A decorative, silvery white finish, also known as *frosted* finish, may be produced by etching the metal with caustic soda (sodium hydroxide) or another chemical. Several methods permit partial or *design* etching to produce patterns of varying finish and colored background.

### 5.9.5.3 Brightened Aluminum

A mirror-bright surface finish can be provided on aluminum by immersing the parts in a hot chemical solution, followed by subsequent rinsing. Depending on the alloy and its mechanical preparation, reflectivity values of up to 85% may be obtained. NAAMM divides brightened finishes on aluminum into four categories:

1. Unspecified (C30)
2. Highly specular (C31)
3. Diffuse bright (C32)
4. Other (C3x)

An *unspecified* brightened finish is optional with the finisher.

The *highly specular* and *diffuse* bright categories refer to different levels of brightening.

The *other* category is reserved for special finishes that do not fit into one of the previous three categories.

### 5.9.5.4 Conversion Coatings on Aluminum

Although chemical conversion coatings are generally used on aluminum to prepare the surface for painting, some types may also be used as a final finish.

Since the natural oxide film on aluminum surfaces may not provide a suitable bond for paints and other organic coatings and laminates, its chemical nature is often converted (altered) to improve adhesion. *Conversion films*, which are also called *conversion coatings*, are generally applied by the use of phosphate or chromate solutions, many of which are proprietary. One of the simplest methods for paint surface preparation is to etch the surface with a phosphoric acid solution, thus providing a good mechanical bond. NAAMM divides conversions coating used on aluminum into seven categories:

1. Unspecified (C40)
2. Acid chromate-fluoride (C41)
3. Acid chromate-fluoride-phosphate (C42)
4. Alkaline chromate (C43)
5. Nonchromate (C44)
6. Nonrinsed chromate (C45)
7. Other (C4x)

An *unspecified* conversion coating is optional with the finisher.

The conversion coatings in the categories from C41 to C45 are produced by proprietary chemicals, each of which produces a different color. Some are clear, some green, and some gray. Sometimes these colors are permitted to stay as the aluminum's final color.

The *other* category is reserved for special finishes that do not fit into one of the previous six categories.

### 5.9.5.5 Chemical Oxidizing of Aluminum

Chemical oxide films have improved resistance to corrosion. As a base for painting, they provide additional protection in the event that the finish is scratched or broken. They are more economical and easier to apply than anodic finishes (see Section 5.9.8.4), which also produce oxide films, but are thinner and softer than anodic finishes. Chemical oxide coatings may be dyed, but anodic films are preferred if a superior color quality is desired.

### 5.9.6 CHEMICAL FINISHES FOR AND CLEANING OF COPPER AND COPPER ALLOYS

Chemicals are used on copper alloys for two primary purposes: to clean the copper alloys of foreign matter and to change the metal's color and provide a final finish. The two chemical finishes we are interested in are identified by NAAMM as *nonetched cleaned* and *conversion coatings*. Some other chemical treatments are used on copper alloys, but they are not used extensively on building construction items.

#### 5.9.6.1 Nonetched Cleaned Copper and Copper Alloys

NAAMM divides nonetched cleaned copper into four categories:

1. Unspecified (C10)
2. Degreased (C11)
3. Cleaned (C12)
4. Other (C1x)

An *unspecified* nonetched cleaned finish is optional with the finisher.

*Degreasing*, which is also called *vapor degreasing*, consists of exposing the metal to the vapors of a chlorinated solvent in a machine designed for the purpose.

*Chemical cleaning* is sometimes called *inhibited* because the chemicals used are inhibited to prevent them from etching the aluminum. Chemical cleaning may be used as the sole treatment for aluminum, but it is more often a second step following degreasing.

The *other* category is reserved for special finishes that do not fit into one of the previous three categories.

#### 5.9.6.2 Conversion Coatings on Copper and Copper Alloys

Conversion coatings are used to change the color of a copper alloy and provide a final finish for the metal. As a copper alloy ages it oxidizes, changing its color and producing a coating that not only changes the appearance of the material, but also protects it from further oxidization. Conversion coatings are not always completely successful in duplicating by accelerated chemical means the natural appearance and protective coating that result from aging. The coatings produced are oxides or sulfides of the metal.

Two basic types of conversion coatings are commonly used: those that produce a *patina* (verde antique) finish and those that produce an oxidized finish known as *statuary bronze*.

NAAMM divides the more commonly used conversion coatings for copper into eight categories:

1. Ammonium chloride (C50)
2. Cuprous chloride-hydrochloric acid (C51)
3. Ammonium sulfate (C52)
4. Carbonate (C53)
5. Oxide (C54)
6. Sulfide (C55)
7. Selenide (C56)
8. Other (C1x)

C50, C51, C52, and C53 produce a *patina*. C54, C55, and C56 produce a *statuary* finish. The chemicals that produce a patina are chlorides, which are more difficult to control than *statuary bronze* finishes. Patinas often have variations in color, especially over large surface areas, and sometimes fail to adhere to the metal. They are also likely to stain adjacent materials. Oxidized (*statuary*) finishes are somewhat more stable than patinas, and the color is easier to control. They come in three tones: *light*, *medium*, and *dark*. A range of color should be expected, however, even within a single tone group.

The *other* category is reserved for conversion coatings that do not fit into one of the previous seven categories.

### 5.9.7 ORGANIC COATINGS ON METALS

*Organic coatings* are a group of coatings that have been classified by NAAMM as such. They can be classified in two groups: *paint* and *other organic coatings*. Air-drying, opaque, organic coatings, including air-drying enamels, are classified as paint. Paint can be applied in the shop or in the field. Finish coats are usually applied in the field, but prime coats may also be applied in the shop or factory. The *other organic coatings* classification includes baking finishes, high-performance coatings, and the types of clear coatings used on aluminum and other nonferrous metals. Field-applied paint coatings are addressed in Chapter 9. The discussion here relates to organic coatings applied in the shop or factory.

#### 5.9.7.1 Application Methods

Paints and other organic coatings are applied in the shop or factory by hand or by machine. Methods include brushing, rolling, spraying, dipping, flow coating, or roller coating. Baked enamel coatings are achieved by coating the metal with an organic coating and then causing the coating to cure by application of elevated temperatures. Baked finish organic coatings dry either by evaporation of their solvent or by a chemical reaction. In either case, increased temperatures are applied. When the applied temperature is below 200°F (93.3°C), the process is called *forced drying*. *Baking* occurs at temperatures between 250°F (121.1°C) and 600°F (315.6°C), depending on the coating material.

Powder coatings are applied by covering the surface with a fine powder of epoxy, polyester, acrylic, or another resin and fusing the powder to the metal by heating them to temperatures between 300° and 400°F (148.9° and 204.4°C).

*Roller coating* is a coil-coating process that is frequently used to coat sheet metals that will be assembled into panel products. Roller coating enables the painting of sheet at line speeds of up to 600 ft/min (182.88 m/min).

First, the bare coiled sheet is passed through a series of cleaning and surface preparation baths to provide a surface to which paint will strongly adhere; then it is roller-coated with an organic paint; finally, it is passed through an oven where the paint is baked hard. The resulting finish has an excellent combination of hardness, ductility, adhesion, and gloss and can undergo further fabrication without harm.

Castings and extrusions are commonly sprayed. Electrostatic spraying, in which electrically charged coating particles are attracted to the surface by maintaining it at the opposite charge, provides a more uniform finish, with less waste, than conventional spraying.

#### 5.9.7.2 Paint and Other Organic Coatings on Ferrous Metal

Many steel products are factory finished with baked enamel or powder coatings.

Ferrous metal products that are not to be factory finished are usually coated in the shop or factory with a *primer* to form a base coat for the finish coat(s) that will be applied in the field and to protect the metal until the finish coat has been applied. Several materials have achieved success in this capacity. They include red-lead mixed pigment paints, red-lead iron oxide, basic lead silico-chromate-base iron oxide, alkyd varnish, alkyd paint, linseed oil, and raw linseed oil paints. These materials are usually applied at a rate that will yield a final thickness of about 2 mils (0.051 mm). Where steel will be in contact with masonry or concrete, and on surfaces that will be inaccessible after the steel is in place, it is usual to apply two coats. The primer must be compatible with the metal being coated and with the finish coat, and, of course, must comply with legal and code requirements. This provision limits some applications. For example, in most jurisdictions, paint containing lead cannot be used where it will be accessible to children.

### 5.9.7.3 Paint and Other Organic Coatings on Aluminum

Organic coatings on aluminum include opaque coatings and clear coatings.

#### OPAQUE ORGANIC COATINGS

A substantial volume of aluminum is painted for both decorative and protective purposes. Several important applications are siding, cladding panels, doors, door and window frames, curtain wall components, and storefront components.

Although other factory-applied organic finishes are also used, most factory-applied permanent organic finishes on aluminum are either baked enamel or fluorocarbon polymers.

The trend in painting aluminum is toward continuous processing, with pretreatment, coating, and curing integrated into a single operation.

#### CLEAR ORGANIC COATINGS

Waxes and lacquers applied to aluminum surfaces provide some weather protection and make cleaning easier. Aluminum so treated can usually be kept clean by wiping with a damp cloth or rinsing with clear water.

Clear lacquer, applied by a fabricator or contractor, provides the most common type of clear coating for aluminum. Methacrylate lacquers, which are com-monly used, are resistant to alkalis such as lime mortar, plaster, and concrete and provide some protection against abrasion. Under normal conditions, a methacrylate coating lasts for several years and weathers by chalking rather than by unsightly peeling. Methacrylate lacquer should not be used on color-anodized aluminum, however, because lacquer never fully adheres to an anodized coating. Where adhesion is incomplete, the lacquer will appear white. On clear anodizing, this effect is largely unnoticeable. On a color-anodized surface, however, the white spots produce a blotchy, unappealing effect.

Liquid waxes or wax base cleaners offer some protection against atmospheric weathering and retard grime accumulation on aluminum surfaces. However, after repeated waxings, surfaces may tend to yellow. If necessary, the wax may be removed with a solvent or mild abrasive cleaner before rewaxing.

*Strippable coatings* are rather expensive coatings that are usually applied by the fabricator to protect surfaces from the normal hazards encountered during shipping and construction. They are ideal for finer finishes which may be scratched or damaged more easily. Strippable coatings should be left in place until the final cleanup operation. Aluminum with such coatings should not be stored in places exposed to sunlight, since the coatings may get "tacky" and difficult to remove. If they cannot be easily removed, a solvent should be used.

Various types of tape may be used for protection against abrasion or staining on smaller areas. They may be applied either by the fabricator or by the contractor.

### 5.9.8 INORGANIC FINISHES ON METAL

Inorganic finishes include all finishes on metals that are not classified by NAAMM as organic coatings. They include metallic, vitreous, laminated, and anodic finishes.

### 5.9.8.1 Metallic Finishes

Some metals are plated or coated with another metal or another alloy of the same metal. These metallic (metal-on-metal) finishes are used on a variety of metals. The base metal may be either ferrous or nonferrous, but the coating or plating metal is usually nonferrous. In most cases, the two metals are bonded together, but remain separate metals, and do not form an alloy. In an *alloy*, the constituent materials are melted together to form a single material. For example, the various types of steel are all alloys of iron and other materials. Bronze is an alloy of copper.

### PURPOSE

Metallic finishes are applied to metals for two purposes: to protect the base metal and to decorate it. Some metallic finishes, such as galvanizing and aluminum plating on steel, are primarily intended only to protect the metal and do not serve to decorate it, although they may be left as the final finish in inconspicuous locations. When decoration is desired, those kinds of metallic finishes are usually given a finish coat of another material. Other metallic finishes, such as chromium plating, both decorate and protect the underlying metal.

Metal coatings on nonferrous metals are used mostly to give the underlying metal a desired finish. Some, however, also protect the underlying metal. Chromium is much harder and more scratch resistant, for example, than either aluminum or copper, which it is often used to coat. Some metallic coatings protect other materials from the coated metals. Lead-coated copper, for example, prevents water runoff from the copper from staining, corroding, or otherwise damaging adjacent materials.

### METHODS

Metallic coatings provide protection against corrosive elements, either by (1) acting as a *sacrificial* metal (one that is purposely permitted to corrode), as in the case of zinc coatings, or by (2) being relatively inert and corrosion resistant (thus protecting the underlying base metal), as in a nickel or chromium coating.

There are six basic methods of coating metal with other metals: hot dipping, electroplating, spraying (metallizing), cladding, alloying (cementation), and fusion welding.

The most common methods used in the building industry to coat steel with other metals are the *hot-dip* and *electroplating* processes. *Galvanizing* (zinc coating) and *aluminizing* (aluminum coating) are the most common coatings used on steel.

The most common method used to coat nonferrous metals with metal for use in the building industry is electroplating, although some dipping is also used.

***Hot Dipping*** In the *hot-dipping* process the underlying metal is immersed in a molten bath of the coating metal. Zinc, aluminum, tin, and terne are applied to steel, for example, by the hot-dipping process. The most prevalent method is continuous hot-dip galvanizing (Fig. 5.9-5), which accounts for more than 10 times as much galvanized steel as does the electrolytic process. The hot-dipping process, which results in heavier coatings than the electrolytic process, is the chief method of galvanizing tubular and flat-rolled products. Hot-dipped galvanized sheet and strip can be made with zinc coatings of designation G90 (0.90 oz/sq ft [0.275 kg/m$^2$]) which are used frequently in architectural applications. Heavier coatings of approximately 2 oz/sq ft (0.61 kg/m$^2$) are made for severely corrosive applications. These coating weights are the combined weights on both surfaces.

Hot-dipped *aluminized* sheets are usually coated with an aluminum-silicon (5% to 11%) coating known as *type 1*. Aluminum-silicon coatings provide excellent resistance to atmospheric and high-temperature (up to 1250°F [676.67°C]) corrosion and are extensively used in the automotive industry.

In a process called *zincating*, aluminum plate is immersed in a zincate bath to coat it with a thin film of zinc. Zincating is generally used to prepare aluminum for electroplating. The standard for zincating on aluminum is ASTM B253.

***Electroplating*** *Electroplating* employs an electric current and an electrolytic solution to deposit metallic coatings on steel or iron (Fig. 5.9-6). Tin, zinc, cadmium, aluminum, and nickel are often deposited on steel, aluminum, and copper alloy by the electroplating process. Chromium and copper are similarly deposited on nickel which has been deposited on steel or aluminum. Chromium is also deposited over nickel that has been deposited over copper over steel.

Electroplating is the chief method of making modern tin plate and the second most important process for making galvanized products. Lighter coatings of 0.10 oz/sq ft (0.031 kg/m$^2$) (for minimum corrosion resistance) and greater variations in coating thickness are possible by electroplating. The lighter coatings are

**FIGURE 5.9-5** A continuous hot-dip galvanizing unit. (Courtesy American Iron and Steel Institute.)

**FIGURE 5.9-6** Steps in the production of tinplate and galvanized products by the electrolytic process. The looping pits shown are necessary to ensure that the proper tension is applied to the metal throughout the process. (Courtesy American Iron and Steel Institute.)

normally recommended on surfaces to be painted or to be used in mildly corrosive exposures.

***Metallizing*** *Metallizing* consists of spraying molten metal against the surface to be coated. Most metal coatings may be applied to steel and nonferrous metals by spraying, but probably the most frequent use of that method is to apply zinc and aluminum to steel. This is the only method, other than touchup galvanizing repair by brush, that may be effectively used in field applications. The coating adheres to the base metal by a combination of mechanical interlocking and metallurgical bonding.

***Cladding*** *Cladding* produces bimetallic "sandwich" products consisting usually of a carbon steel, low-alloy steel, or aluminum core covered with a thin sheet layer of coating metal. The cladding process may be accomplished by dipping, rolling, electrolytic deposit, or electrowelding, depending on the metal that forms the cladding, but in all cases the cladding metal and the steel remain separate materials.

Steel is covered with copper, stainless steel, and aluminum by the cladding process. Stainless steel and aluminum coatings are applied by hot rolling under controlled, nonoxidizing conditions. Copper may also be used for cladding steel plates by hot dipping or electroplating. An aluminum alloy or other metal coating may be bonded to an aluminum product to increase its corrosion-resistant properties, to provide a special surface appearance, or in preparation for additional surface treatment. Most of this cladding is done on sheet and plate products, but some tube, rod, and wire products are also clad. Aluminum clad with high-purity aluminum or a non-heat-treatable alloy to improve its surface corrosion resistance is designated by its four-digit alloy number (see Section 5.2) preceded by the word *Alclad*. For example, Alclad 3004 is a common sheet and plate material used for building applications.

*Alloying*   *Cementation* is the forming of an alloy between a coating metal and the coated metal.

*Fusion Welding*   *Fusion welding* is, as it sounds, the fusing of the cladding metal to the base metal by welding.

## TYPES

There are many types of metal-coated metal used in building construction, including those discussed in the following subsections.

*Metallic Finishes on Ferrous Metals* The metals most often used for protecting carbon steels are zinc, tin, terne metal, aluminum, cadmium, chromium, and nickel. These coating metals are most commonly applied by hot dipping, electroplating, metallizing, or cladding. Galvanizing (coating with zinc) is usually done by hot dipping or electroplating. Aluminum coatings are generally applied by hot dipping. Tin plating is coating with tin, either by hot dipping or electroplating. Bonderizing on galvanized sheets and galvannealing are light metallic coatings with a suitable surface for paint coatings.

*Zinc-based paint pretreatments* are common on steel to improve the bond of subsequent paint finishes.

*Bonderizing* is the most common process for producing a suitable paint base on fabricated products. Zinc-coated products are dipped in a hot phosphate solution, resulting in a thin (0.02 to 0.03 oz/sq ft [0.006 to 0.009 kg/m$^2$]) crystalline surface film of zinc phosphate. Steel fasteners and hardware items are sometimes electroplated with cadmium to provide an electrolytic separation between the steel and other materials to prevent galvanic corrosion. Since cadmium is toxic, cadmium-plated steel items should not be used where they will come in contact with people, especially children, or food. *Cadmium-plated steel* should comply with the requirements of ASTM B766.

Brass-plated steel is sometimes used as door hardware and similar items. Terne is an alloy of lead and tin, usually in a ratio of 4 parts lead to 1 part tin. Terneplate, which is the "terne" used in older buildings, is sheet iron or steel coated with a layer of terne. Today's terne-coated steel (TCS) is copper-bearing steel or stainless steel coated with terne. Terneplate and TCS are used mostly in roofing but may occasionally be found in other nonstructural metal items.

Copper-coated steel sheets are sometimes used in flashing and other sheet metal applications.

*Metallic Finishes on Aluminum*   Chromium and copper plating are the most common and easiest-to-apply *electroplated* finishes for aluminum. Chromium is readily plated on aluminum either after zincating or after copper, brass, or nickel plating. Lustrous chrome plate is easy to obtain by buffing. Chromium plates are generally applied for decorative effect or resistance to abrasion.

Aluminum may be copper plated after a zincate treatment or by other methods, including anodizing in phosphoric acid. Copper may also be electroplated after first precleaning and zincating the aluminum. Copper plates on aluminum are used where soldering operations are required, for certain electrical requirements, for a base for chromium or silver plating, and for decorative items.

*Silver plating* decreases electrical resistance and is sometimes used for electrical bus-bar connections. Tin and brass plates are especially good as a base for soldering. Zinc and cadmium plates provide corrosion protection. Chemically treated nickel plates yield a black decorative finish.

*Metallic Finishes Used on More Than One Type of Metal*   Some metallic coatings are applied over more than one type of base metal. For example, chromium is used to plate steel, aluminum, and copper. Iron, carbon steel, and stainless steel are all chromium plated to produce products used in buildings. Applicable standards include ASTM B177, B254, B320, and B650.

Chromium is plated onto aluminum after zincating or after plating the aluminum with copper, brass, or nickel.

Copper can be best plated with chromium after a coating of nickel has been applied. Chromium-plated copper has many uses in the construction industry, especially on hardware, toilet accessories, and fasteners.

Nickel plating is seldom used alone in the building industry. It is sometimes used on steel as an intermediate coat beneath chromium or copper plating. Nickel plating on copper is used primarily as a base for chromium plating.

*Lead-Coated Copper*   Sheets of lead-coated copper are used extensively in roofing and flashing applications and in other nonstructural metal items, especially when water running across bare copper would be harmful to adjacent materials. Lead-coated copper is produced by electroplating or hot dipping. It may be on both sides of the sheet or on one side only. The standard for lead-coated copper is ASTM B101, which lists two types and two classes. Type I denotes application by dipping. A Type II coating is electrodeposited. Class A Standard requires a lead coating weight of 6 to 7$\frac{1}{2}$ lb/100 sq ft (2.722 to 3.402 kg/9.29 m$^2$) of surface. Class B Heavy requires 10 to 15 lb/100 sq ft (4.536 to 6.804 kg/9.29 m$^2$) The weights given above are the weights of lead on each side of the sheet. Thus, the total weight of lead on a Class A Standard sheet with lead on both sides, which is the way B101 lists it, is 12 to 15 lb/100 sq ft (5.443 to 6.804 kg/9.29 m$^2$) of sheet.

*Gold Plating (Gilding)*   *Gilding* with gold is an ancient art, at least 4000 years old, that is still practiced today. *Gold plating* is the application of gold leaf to wood, plaster, metal, glass, or another material to simulate the look of solid gold. It does not include the application of gold paint. There are some imitation golds (called *Dutch leaf*) that are used to simulate real gold leaf, but that material tarnishes (oxidizes) even when protected by varnish. Pure gold needs no protection, since gold, for all practical purposes, does not oxidize. Even extremely thin high-caret gold leaf, when properly maintained and protected, will withstand most climates and pollution levels for between 30 and 90 years.

The purity of gold is measured by its weight in carats. Pure gold weighs 24 carets. Golds of lesser caret weight are alloys that include silver, copper, or both. Gold leaf can weigh as much as $23^1/_2$ carets, but most is lighter, because pure gold does not wear as well as its lower-caret alloys.

Gold leaf is available today in four forms. *Loose leaf* comes in loose sheets for use on interior surfaces only. *Patent leaf* comes in books of sheets adhered to backing paper. It is the form used for most architectural gilding, especially on exteriors. *Surface leaf* is used where close inspection is not critical, since it is made from scrap gold and has flaws and discolorations. Finally, *glass leaf* is used on signs and glass. Among these forms there are many qualities and grades.

There are essentially three methods of gilding in general use today, although there are others. The three are *oil* (mordant) *gilding*, *water gilding*, and *glass gilding*.

Different gilders have their own ways of using each method. In essence, however, both the *oil* and *water* methods are started by applying several coats of a material known as *gesso*, which is a mixture of calcium carbonate (*gilders' whiting*) and glue. In oil gilding, the gesso is then coated with shellac or marine varnish. The gold leaf is next laid into a coat of oil sizing, which acts as an adhesive. In water gilding, the gold leaf is laid over the gesso in a wet mixture of alcohol and water called *gilders' liquor*. If the gold leaf is to be burnished, a layer of a special clay, called *burnishing clay*, is laid just ahead of the gold. In glass gilding, an *oil size* is sometimes used, but usually the gold is held in place by a *size* that is a mixture of gelatin and water.

### 5.9.8.2 Vitreous Finishes on Metals

Vitreous finishes are fused-on glassy inorganic coatings that are bonded to metal by fusion at high temperatures. Although several types of vitreous coatings are used on ferrous metal, including glass-on-steel linings applied to hot water tanks and process piping, *porcelain enamel* is the type most common in architectural applications. Common uses are porcelain enamel on cast iron or steel in bath tubs, sinks, and other sanitary ware. Porcelain-enameled building panels are used in commercial and institutional buildings.

Porcelain enamels have inert abrasion- and corrosion-resistant properties similar to those of glass. They can be produced in a great variety of lightfast colors and textures. Porcelain enamel is applied in one or two coats to sheets that have been prepared by etching. In the two-coat system, the first enamel coat contains adherence promoters, such as cobalt or nickel oxides, and is applied by spraying, dipping, or flow coating followed by firing at high temperatures. The second (and sometimes third) coat contains the coloring elements, and is applied and fired in a similar manner. Coating thickness ranges between 4 and 20 mils (0.1 and 0.51 mm), though durability is not necessarily a function of thickness; specifically, where flexure is a possibility, thinner or fewer coatings may be preferred.

Porcelain enamel is also used extensively as a finish on aluminum. A porcelain enamel coating significantly improves the existing properties of aluminum, increasing durability and corrosion resistance. Practically any texture and lightfast color is available.

### 5.9.8.3 Laminated Finishes on Metals

Laminated finishes involve the adhesive application of inert plastic films to metal.

#### LAMINATED FINISHES ON FERROUS METALS

Laminated coatings are thin (2 to 12 mils [0.05 to 0.3 mm] thick), tough plastic films applied with thermosetting (heat-hardening) adhesives to steel products. The most common films used are polyvinyl chloride (PVC) and polyvinyl fluoride (PVF), both of which possess good abrasion resistance and sufficient tenacity, toughness, and flexibility to withstand severe forming operations. They have excellent resistance to weathering, good color retention, and resistance to attack by a variety of chemical agents. Although color selections are more limited than they are in porcelain enamels, the potential color variations are substantial.

#### LAMINATED FINISHES ON ALUMINUM

For special use and design applications, many other materials may be laminated to aluminum surfaces. For example, PVC and vinyl films are used to provide a non-metallic color and a long-lasting finish

for exterior applications such as aluminum siding. The film is adhesive bonded to the aluminum surfaces which have been suitably prepared, usually by application of a chemical conversion coating.

### 5.9.8.4 Anodic Coatings on Aluminum

Anodized electrolytic oxide coatings are readily applied to aluminum alloys and are widely used. Thinner anodic "flash" films may be used as a prepaint treatment.

The distinction between a coating and a film is thickness. A film is less than 0.1 mil (0.0254 mm); any thicker covering is a coating.

Ordinarily, clear anodic coatings on architectural products are preceded by either a chemical etch or some form of mechanical finish, or both. Etching is the most economical and common prefinish, imparting a frosted appearance which helps to conceal variations in the mill finish and die lines on extrusions. Mechanical finishes also leave a pleasing texture which is retained after anodizing.

If neither etching nor mechanical finishing is specified, the aluminum still must be chemically cleaned before it is anodized to ensure satisfactory results. Contaminants deposited on the metal surfaces during fabrication processes may affect the anodic coating and should be completely removed before anodizing.

Anodizing consists of immersing a product in an acid solution (*electrolyte*) and passing an electric current, usually of the direct type, between the metal and the solution, with the aluminum serving as the anode. This results in the controlled formation of a relatively thin but durable oxide or coating on the surface of the metal.

In general, such coatings are many times thicker than the natural oxide film and may be transparent, translucent, or opaque, depending on the alloys and the electrolyte used. Anodic coatings do not change the surface texture of the metal, but they greatly increase resistance to corrosion and provide increased resistance to abrasion.

Several anodizing processes are commonly used. The principal differences between these processes are the type of solution used for the electrolyte, the voltage and current required, and the temperature of the bath.

The *sulfuric acid process*, because of its relatively low cost, is most widely used for architectural anodizing. It produces comparatively thick, transparent, and absorptive coatings which are suitable for receiving coloring dyes and pigments (see the following subsection). Thin films may be produced by this process, which are excellent pretreatments for paint, enamel, and lacquer coatings.

## COLOR ANODIZING

There are three methods of color anodizing in common use: sulfuric acid anodizing and dyeing, integral-color anodizing, and sulfuric acid anodizing and electrolytically deposited color (Fig. 5.9-7).

**FIGURE 5.9-7** Distribution of coloring elements in the anodic film of the three main color anodizing processes.

*Sulfuric Acid Anodizing and Conventional Dyeing* In this method, the work is rinsed after it leaves the anodizing bath, and is dipped in a bath containing either an organic dye or an inorganic pigment. To set the dye, the work is dipped in a nickel acetate solution and is sealed (usually in boiling deionized water) to produce a clear film that provides minimum interference with, and optimum lightfastness of, the color,

A wide range of colors is possible with this process, but not all of them are satisfactory for outdoor service. As indicated in Figure 5.9-7, the color is concentrated in the top third of the anodic film; therefore, it is necessary to adjust dyeing conditions (time, temperature, dye concentration, and pH) to get similar colors with different film thicknesses.

*Integral-Color Anodizing* In this method, the coloring material is spread uniformly throughout the anodic film. Integral-color anodic films may be produced by one of two methods or by a combination of both. The first method makes use of an alloy with constituents that color the anodic film as it grows. The second method makes use of an electrolyte that yields a colored precipitate during anodizing, again coloring the anodic film as it grows.

A combination of these two methods increases the range of possible colors. The colors produced by integral-color anodizing are very lightfast but, as with sulfuric acid anodizing and electrolytically deposited color, the depth of the color varies with the anodic film thickness. In addition, alloy is a major variable; some colors are restricted to specific alloys.

*Sulfuric Acid Anodizing and Electrolytically Deposited Color* This method concentrates the color at the base of the anodic film (see Fig. 5.9-7). Consequently, the color receives maximum protection from sunlight and weathering.

In most applications, the pores in the anodic film are sealed off to increase resistance to staining and corrosion and to improve color durability. Because most applications require a clean film, immersion of the work in boiling deionized water with a pH of 5.5 to 6.5 is the usual method of sealing. When maximum corrosion resistance is required, potassium dichromate is added to the water.

The thickness of anodic films and coatings ranges from less than 0.1 mil (0.0254 mm) in primer films for receiving organic finishes, to 5.0 mils (0.127 mm) for some of the engineering hardcoats.

For architectural applications, the thickness and weight of coatings are of prime importance; no coating thinner than 0.4 mil (0.1016 mm) should be used even for interior purposes. Coatings for exterior parts which will be subject to contaminated atmospheres, or will not receive regular maintenance care, should have at least a 0.7 mil (0.01778 mm) thickness and a minimum coating weight of 27 mg/sq in. (27 mg/645.16 mm$^2$).

## CLASSIFICATION OF ANODIC COATINGS

The Aluminum Association classifies the various anodic finishes into four general types: General Anodic Coatings, Protective and Decorative Coatings, Architectural Class I Coatings, and Architectural Class II Coatings. The first two classifications apply to general industrial work. Therefore, only the last two are discussed here.

*Architectural Class II Coatings* These coatings range in thickness from 0.4 to 0.7 mil (0.1016 to 0.01778 mm), with corresponding weights of 17 mg/sq in. (17 mg/645.16 mm$^2$) to 27 mg/sq in. (27 mg/645.16 mm$^2$). These are appropriate for interior items not subject to excessive wear or abrasion, and for exterior items such as storefronts and entrances which are regularly cleaned and maintained.

*Architectural Class I Anodic Coatings* These coatings are more than 0.7 mil (0.01778 mm) thick, weigh more than 27 mg/sq in. (27 mg/645.16 mm$^2$), and include the hardest and most durable anodic coatings available. They are suitable for exterior items which receive no regular maintenance care and for interior architectural items subject to normal wear.

The Aluminum Association divides anodic coatings on aluminum into five groups:

1. Clear (natural) (A31)
2. Integral color (A32)
3. Impregnated color (A33
4. Electrolytically deposited color (A34)
5. Other (A3x)

*Clear* anodic coatings are usually produced by the sulfuric acid process.

*Integral color* coatings are mostly proprietary products that are manufactured to suit the user's requirements. *Hardcoat* coatings fit into this category.

*Impregnated color* coatings are produced by the sulfuric acid process, and are then dyed or colored using mineral spirits.

*Electrolytically deposited* color coatings are produced by the sulfuric acid process combined with electrolytic depositing of pigment.

The *other* category is reserved for anodic coatings that do not fit into one of the previous categories. Specifying this category requires a complete description of the finish that is required.

# 5.10 Additional Reading

More information about the subjects discussed in this chapter can be found in the references listed in Section 5.11 and in the following publications.

Allen, Edward. 2004. *Fundamentals of Building Construction: Materials and Methods*, 4th ed. Hoboken NJ: John Wiley & Sons, Inc. (www.wiley.com).

Ambrose, James. 1994. *Design of Building Trusses*. Hoboken NJ: John Wiley & Sons, Inc. (www.wiley.com).

————. 1995. *Simplified Design of Building Structures*, Hoboken NJ: John Wiley & Sons, Inc. (www.wiley.com).

————. 1997. *Simplified Design of Steel Structures*, Hoboken NJ: John Wiley & Sons, Inc. (www.wiley.com).

Amon, Rene, David Fanella, Bruce Knobloch, and Atanu Mazumder. 1994. "Making Connections." *The Construction Specifier* (April): 120 (www.csinet.org).

Crall, Chris P. 2001. "Metal Buildings." *The Construction Specifier* (July): 55 (www.csinet.org).

Di Girolamo, Edward. 2001. "Specifications for Light Steel Framing—Only as Strong as Its Weakest Link." *The Construction Specifier* (October): 39 (www.csinet.org).

Drisko, Richard W. 2001 "Specifying for Corrosion Control Using Coatings." *The Construction Specifier* (October): 45 (www.csinet.org).

Englekirk, Robert. 1994. *Steel Structures: Controlling Behavior Through*

*Design*. Hoboken NJ: John Wiley & Sons, Inc. (www.wiley.com).

Houska, Catherine. 2003. "Stainless Steel Selection." *The Construction Specifier* (January): 32 (www.csinet. org).

International Code Council. 2003. *International Residential Code®*. Falls Church, VA: ICC (www.iccsafe.org).

Nemchock, Gary. 2002. "Metal for a New Era–Expanding Possibilities with Titanium." *The Construction Specifier* (June): 70 (www.csinet.org).

Newman, Alexander. 2004. *Metal Building Systems; Design and Specifications*. New York: McGraw-Hill (www.mcgraw-hill.com).

Ramsey, Charles George, Harold Reeve Sleeper, and John Ray Hoke, Jr. 2000. *Architectural Graphic Standards*, 10th ed. Hoboken, NJ: John Wiley & Sons, Inc. (www.wiley.com).

Scharff, Robert. 1996. *Residential Steel Framing Handbook*. New York: McGraw-Hill (www.mcgraw-hill.com).

Smith, J.C. 1996. *Structural Steel Design: LRFD Approach*, 2nd ed. Hoboken NJ: John Wiley & Sons, Inc. (www.wiley.com).

Underwood, James R. and Michele Chiuini. 1998. *Structural Design: A Practical Guide for Architects*. Hoboken NJ: John Wiley & Sons, Inc. (www.wiley.com).

Williamson, Tom, and Marilyn H. LeMoine. 2005 "Building a Case for Wood and Sustainability Design." *The Construction Specifier* (March): 64 (www.csi.net.org).

Yu, Wei-Wen. 2000. *Cold-Formed Steel Design*, 3rd ed. Hoboken NJ: John Wiley & Sons, Inc. (www.wiley.com).

Zahner, L. William, 1995. *Architectural Metals: A Guide to Selection, Specification, and Performance*. Hoboken NJ: John Wiley & Sons, Inc. (www.wiley.com).

# 5.11 Acknowledgments and References

## ACKNOWLEDGMENTS

We gratefully acknowledge the assistance of the following organizations and individuals in preparing this chapter. We are also indebted to them for permission to use their illustrations when requested and for the use of their publications as references.

ALCOA Building Products (www.alcoa.com)

Allegheny Ludlum Steel Corporation (www.alleghenyludlum.com)

Aluminum Association (AA) (www.aluminum.org)

American Architectural Manufacturers Association (AAMA) (www.aamanet.org)

American Institute of Steel Construction (AISC) (www.aisc.org)

American Iron and Steel Institute (AISI) (www.steel.org)

American Welding Society (AWS) (www.aws.org)

ASTM International (ASTM) (www.astm.org)

Bethlehem Steel Corporation (www.bethsteel.com)

Copper Development Association (CDA) (www.copper.org)

International Code Council (ICC) (www.iccsafe.org)

ITW Ramset/Redhead (www.ramset-redhead.com)

Kaiser Aluminum and Chemical Corporation (www.kaiseral.com)

Lukens Inc.

*Modern Metals Magazine* (www.ffjournal.net)

National Association of Architectural Metal Manufacturers (NAAMM) (www.naamm.org)

Phillips Industries, Inc. (www.phillipsind.com)

Rigidized Steel Corporation (www.rigidized.com)

Steel Deck Institute (SDI) (www.sdi.org)

Steel Joist Institute (SJI) (www.sji.org)

Steel Recycling Institute (SRI) (www.recycle-steel.org)

TRW Nelson Company

Underwriters Laboratories, Inc. (UL) (www.ul.com)

United States Steel Corporation (www.ussteel.com)

Youngstown Sheet and Tube Company

### REFERENCES

We would also like to thank the authors and publishers of the publications in the following list for their contribution to our research for this chapter.

Abate, Kenneth. 1989. "Metal Coatings, Fighting the Elements with Superior Paint Systems, Part I." *Metal Architecture*, 5(6)(June): 10,69 (www.moderntrade.com).

———. 1989. "Metal Coatings, Fighting the Elements with Superior Paint Systems, Part II." *Metal Architecture*, 5(7)(July): 40–41 (www.moderntrade.com).

American Architectural Manufacturers Association (AAMA). Schaumburg, IL (www.aamanet.org). Specifications:

611-98, "Voluntary Specification for Anodized Architectural Aluminum."

612-02, "Voluntary Specification, Performance Requirements, and Test Procedures for Combined Coatings of Anodic Oxide and Transparent Organic Coatings on Architectural Aluminum."

2603-02, "Voluntary Specification, Performance Requirements and Test Procedures for Pigmented Organic Coatings on Aluminum Extrusions and Panels.

609 and 610-02, "Cleaning and Maintenance Guide for Architectural Finished Aluminum (Combined Document)."

2604-02, "Voluntary Specification, Performance Requirements and Test Procedures for High Performance Organic Coatings on Aluminum Extrusions and Panels."

2605-02, "Voluntary Specification, Performance Requirements and Test Procedures for Superior Performing Organic Coatings on Aluminum Extrusions and Panels."

CW-10-97, "Care and Handling of Architectural Aluminum from Shop to Site."

American Institute of Steel Construction (AISC). *LRFD Manual of Steel Construction*, 3rd ed. Chicago: AISC (www.aisc.org).

American Iron and Steel Institute (AISI). Washington DC (www.steel.org). Standards:

———. 2002, "CF02-1 Cold-Formed Steel Framing Design Guide."

———. 2003. CF03-1, "Steel Stud Brick Veneer Design Guide."

———. 2001. SG02-1 and SG02-2, "North American Specification for the Design of Cold-Formed Steel Structural Members and the Commentary."

———. 2004. SG02-3, "Standard for Cold-Formed Steel Framing—General Provisions."

———. 2004. SG02-4, "Standard for Cold-Formed Steel Framing—Prescriptive Method for One and Two Family Dwellings."

———. 2004. SG02-5, "Standard for Cold-Formed Steel Framing—Truss Design."

———. 2002. SG03, *Cold-Formed Steel Design Manual*.

American Welding Society (AWS). 2003. AWS D1.2, *Structural Welding Code—Aluminum*. Miami, FL: AWS (www.aws.org).

———. 2004. AWS D1.1, *Structural Welding Code—Steel*. Miami, FL: AWS. (www.aws.org).

*Architectural Graphic Standards*. See Ramsey et al. (2000).

Arcom Master Systems. *Masterspec®*. Salt Lake City, UT (www.arcomnet.com). Basic Section:

05120, "Structural Steel."

05210, "Steel Deck."

05400, "Cold Formed Metal Framing."

05500, "Metal Fabrications."

05511, "Metal Stairs."

05521, "Pipe and Tube Railings."

05700, "Ornamental Metal."

ASM International Materials Park, OH (www.techstreet.com).

———. 1990. *ASM Handbook*, Vol. 01, *Properties and Selection: Irons, Steels, and High-Performance Alloys*.

———. 1991. *ASM Handbook*, Vol. 02, *Properties and Selection Nonferrous Alloys and Special-Purpose Materials*.

———. 1994. *ASM Handbook*, Vol. 05, *Surface Engineering*.

———. 1989. *ASM Handbook*, Vol. 17. *Nondestructive Evaluation and Quality Control*.

ASTM International (ASTM). West Conshohocken, PA (www.astm.org). Standards:

A6/A6M, "Standard Specification for General Requirements for Rolled Structural Steel Bars, Plates, Shapes, and Sheet Piling."

A27/A27M, "Standard Specification for Steel Castings, Carbon, for General Application.

A36/A36M, "Standard Specification for Carbon Structural Steel."

A47/A47M, "Standard Specification for Ferritic Malleable Iron Castings."

A48/A48M, "Standard Specification for Gray Iron Castings."

A53/A53M, "Standard Specification for Pipe, Steel, Black and Hot-Dipped, Zinc-Coated, Welded and Seamless."

A82, "Standard Specification for Steel Wire, Plain, for Concrete Reinforcement."

A108, "Standard Specification for Steel Bar, Carbon and Alloy, Cold-Finished."

A121, "Standard Specification for Metallic-Coated Carbon Steel Barbed Wire."

A123/A123M, "Standard Specification for Zinc (Hot-Dip Galvanized) Coatings on Iron and Steel Products."

A153/A153M, "Standard Specification for Zinc Coating (Hot-Dip) on Iron and Steel Hardware."

A167, "Standard Specification for Stainless and Heat-Resisting Chromium-Nickel Steel Plate, Sheet, and Strip."

A185, "Standard Specification for Steel Welded Wire Reinforcement, Plain, for Concrete."

A242/A242M, "Standard Specification for High-Strength Low-Alloy Structural Steel."

A269, "Standard Specification for Seamless and Welded Austenitic Stainless Steel Tubing for General Service."

A276, "Standard Specification for Stainless Steel Bars and Shapes."

A283/A283M, "Standard Specification for Low and Intermediate Tensile Strength Carbon Steel Plates."

A307, "Standard Specification for Carbon Steel Bolts and Studs, 60 000 PSI Tensile Strength."

A312/A312M, "Standard Specification for Seamless, Welded, and Heavily Cold Worked Austenitic Stainless Steel Pipes."

A325, "Standard Specification for Structural Bolts, Steel, Heat Treated, 120/105 ksi Minimum Tensile Strength."

A424, "Standard Specification for Steel Sheet for Porcelain Enameling."

A463/A463M, "Standard Specification for Steel Sheet, Aluminum-Coated, by the Hot-Dip Process."

A490, "Standard Specification for Heat-Treated Steel Structural Bolts, 150 ksi Minimum Tensile Strength."

A497, "Standard Specification for Steel Welded Wire Reinforcement, Deformed, for Concrete."

A500, "Standard Specification for Cold-Formed Welded and Seamless Carbon Steel Structural Tubing in Rounds and Shapes."

A501, "Standard Specification for Hot-Formed Welded and Seamless Carbon Steel Structural Tubing."

A510, "Standard Specification for General Requirements for Wire Rods and Coarse Round Wire, Carbon Steel."

A536, "Standard Specification for Ductile Iron Castings.

A554, "Standard Specification for Welded Stainless Steel Mechanical Tubing."

A568/A568M, "Standard Specification for Steel, Sheet, Carbon, and High-Strength, Low-Alloy, Hot-Rolled and Cold-Rolled, General Requirements for."

A572/A572M, "Standard Specification for High-Strength Low-Alloy Columbium-Vanadium Structural Steel."

A575, "Standard Specifications for Steel Bars, Carbon, Merchant Quality, M-Grades."

A588/A588M, "Standard Specification for High-Strength Low-Alloy Structural Steel with 50 ksi [345 MPa] Minimum Yield Point to 4 in. [100 mm] Thick."

A591/A591M, "Standard Specification for Steel Sheet, Electrolytic Zinc-Coated, for Light Coating Weight (Mass) Applications."

A606, "Standard Specification for Steel, Sheet and Strip, High-Strength, Low-Alloy, Hot-Rolled and Cold-Rolled, with Improved Atmospheric Corrosion Resistance."

A615/A615M, "Standard Specification for Deformed and Plain Billet-Steel Bars for Concrete Reinforcement."

A641, "Standard Specification for Zinc-Coated (Galvanized) Carbon Steel Wire."

A653/A653M, "Standard Specification for Steel Sheet, Zinc-Coated (Galvanized) or Zinc-Iron Alloy-Coated (Galvannealed) by the Hot-Dip Process."

A663/A663M, "Standard Specification for Steel Bars, Carbon, Merchant Quality, Mechanical Properties."

A675/A675M, "Standard Specification for Steel Bars, Carbon, Hot-Wrought, Special Quality, Mechanical Properties."

A743/A743M, "Standard Specification for Castings, Iron-Chromium-Nickel, and Nickel-Base, Corrosion-Resistant, for General Application."

A744/A744M, "Standard Specification for Castings, Iron-Chromium-Nickel, Corrosion Resistant, for Severe Service."

A767/A767M, "Standard Specification for Zinc-Coated (Galvanized) Steel Bars for Concrete Reinforcement."

A780, "Standard Practice for Repair of Damaged and Uncoated Areas of Hot-Dip Galvanized Coatings."

A786/786M, "Standard Specification for Hot-Rolled Carbon, Low-Alloy, High-Strength Low-Alloy, and Alloy Steel Floor Plates."

A792/A792M, "Standard Specification for Steel Sheet, 55% Aluminum-Zinc Alloy-Coated by the Hot-Dip Process."

A875/A875M, "Standard Specification for Steel Sheet, Zinc-5% Aluminum Alloy-Coated by the Hot-Dip Process."

A996/A996M, "Standard Specification for Rail-Steel and Axle-Steel Deformed Bars for Concrete Reinforcement."

A1003/A1003M-02a, "Standard Specification for Steel Sheet, Carbon, Metallic- and Nonmetallic-Coated for Cold-Formed Framing Members."

A1011/A1011M, "Standard Specification for Steel, Sheet and Strip, Hot-Rolled, Carbon, Structural, High-Strength Low-Alloy and High-Strength Low-Alloy with Improved Formability."

B26/B26M, "Standard Specification for Aluminum-Alloy Sand Castings."

B32, "Standard Specification for Solder Metal."

B101, "Standard Specification for Lead-Coated Copper Sheet and Strip for Building Construction."

B108, "Standard Specification for Aluminum-Alloy Permanent Mold Castings."

B135, "Standard Specification for Seamless Brass Tube. West Conshohocken, PA: ASTM."

B177, "Standard Guide for Engineering Chromium Electroplating."

B209, "Standard Specification for Aluminum and Aluminum Alloy Sheet and Plate."

B210, "Standard Specification for Aluminum-Alloy Drawn Seamless Tubes."

B211, "Standard Specification for Aluminum and Aluminum-Alloy Bar, Rod, and Wire."

B221, "Standard Specification for Aluminum and Aluminum-Alloy Extruded Bars, Rods, Wire, Profiles, and Tubes."

B247, "Standard Specification for Aluminum and Aluminum-Alloy Die Forgings, Hand Forgings, and Rolled Ring Forgings."

B253, "Standard Guide for Preparation of Aluminum Alloys for Electroplating."

B254, "Standard Practice for Preparation of and Electroplating on Stainless Steel."

B271, "Standard Specification for Copper-Base Alloy Centrifugal Castings."

B320, "Standard Practice for Preparation of Iron Castings for Electroplating."

B370, "Standard Specification for Copper Sheet and Strip for Building Construction."

B429, "Standard Specification for Aluminum-Alloy Extruded Structural Pipe and Tube."

B455, "Standard Specification for Copper-Zinc-Lead Alloy (Leaded-Brass) Extruded Shapes."

B483, "Standard Specification for Aluminum and Aluminum-Alloy Extruded Drawn Tubes for General Purpose Applications."

B584, "Standard Specification for Copper Alloy Sand Castings for General Applications."

B650, "Standard Specification for Electrodeposited Engineering Chromium Coatings on Ferrous Substrates."

B766, "Standard Specification for Electrodeposited Coatings of Cadmium."

C282, "Standard Test Method for Acid Resistance of Porcelain Enamels (Citric Acid Spot Test)."

C283, "Standard Test Method for Resistance of Porcelain Enamel Utensils to Boiling Acid."

C286, "Standard Terminology Relating to Porcelain Enamel and Ceramic-Metal Systems."

C346, "Standard Test Method for 45-deg Specular Gloss of Ceramic Materials."

C448, "Standard Test Method for Abrasion Resistance of Porcelain Enamel."

C538, "Standard Test Method for Color Retention of Red, Orange, and Yellow Porcelain Enamels."

C645, "Standard Specification for Nonstructural Steel Framing Members."

C703, "Standard Test Method for Spalling Resistance of Porcelain Enameled Aluminum."

C955, "Standard Specification for Load-Bearing (Transverse and Axial) Steel Studs, Runners (Track), and Bracing or Bridging for Screw Application of Gypsum Board and Metal Plaster Bases."

C1007, "Standard Specification for Installation of Load-Bearing (Transverse and Axial) Steel Studs and Accessories."

D1730, "Standard Practices for Preparation of Aluminum and Aluminum-Alloy Surfaces for Painting."

D1731, "Standard Practices for Preparation of Hot-Dip Aluminum Surfaces for Painting."

D2092, "Standard Guide for Preparation of Zinc-Coated (Galvanized) Steel Surfaces for Painting."

D3794, "Standard Guide for Testing Coil Coatings."

E94, "Standard Guide for Radiographic Examination."

E165, "Standard Test Method for Liquid Penetrant Examination."

E527, "Standard Practice for Numbering Metals and Alloys (UNS)."

E709, "Standard Guide for Magnetic Particle Examination."

Bocchi, Greg. 1986. "Powder Coatings, A New Technology Takes Off." *The Construction Specifier*, 39(9)(September): 102–05 (www.csinet.org).

———. 1988. "Powder Coatings Making Inroads in Metal Construction Market." *Metal Architecture*, April: 8–9 (www.moderntrade.com).

Brick Industry Association (BIA). *Technical Notes on Brick Construction.* No. 28B Rev. II. Feb. 1999. "Brick Veneer, Steel Stud Panel Walls." Reston, VA: BIA (www.brickinfo.org).

Copper Development Association (CDA). 4050, *Copper in Architecture Design Handbook.* New York: CDA (www.copper.org).

International Code Council. 2003: Falls Church, VA (www.iccsafe.org). International Codes:

———. *International Building Code.*®

———. *International Residential Code.*®

National Association of Architectural Metal Manufacturers (NAAMM). Chicago (www,naamm.org.)

———. *Metal Finishes Manual.*

———. *Code of Standard Practice for the Architectural Metal Industry.*

———. *Metal Stairs Manual.*

———. *Pipe Railing Manual.*

Porcelain Enamel Institute (PEI). Norcross, GA. (www.porcelain.com)

———. 2006. PEI-101, "Design & Fabrication of Metal of Porcelain Enamel."

———. 2006. PEI-1001 (ALS-100), "Specification for Architectural Porcelain Enamel."

———. 2006. "Color Guide for Architectural Porcelain Enamel."

———. 2006. "Weatherability of Porcelain Enamel."

Ramsey, Charles George, Harold Reeve Sleeper, and John Ray Hoke, Jr. 2000. *Architectural Graphic Standards,* 10th ed. Hoboken, NJ: John Wiley & Sons, Inc. (www.wiley.com).

Sagan, Vincent E. 1998. "Steel Stud Wall System Essentials." *The Construction Specifier* (May): 70 (www.csinet.org).

Sheet Metal and Air Conditioning Contractors National Association (SMACCNA). 2003. *Architectural Sheet Metal Manual.* Chantilly, VA: SMACNA.

Society for Protective Coatings. *Good Paint Practice: SSPC Painting Manual Vol. 1,* 4th ed. Pittsburgh: SSPC (www.sspc.org).

———. 2005. *Systems and Specifications: SSPC Painting Manual, Vol. 2* Pittsburgh: SSPC (www.sspc.org).

Society of Automotive Engineers (SAE). 2004. Document Number J1086, *Numbering Metals and Alloy.* Warrendale, PA: SAE (www.sae.org).

Specialty Steel Industry of North America (SSINA). Washington, DC (ssina.com).

———. *Standard Practices for Stainless Steel Roofing, Flashing, Copings.*

———. "Finishes for Stainless Steel."

———. 1995. "Specifications for Stainless Steel."

———. "Stainless Steel Fasteners."

———. "Design Guidelines for the Selection and Use of Stainless Steel."

Steel Deck Institute (SDI). *Design Manual for Composite Decks, Form Decks and Roof Decks—No. 30.* Fox River Grove, IL: SDI (www.sdi.org).

Steel Joists Institute (SJI). 1987. "Handling and Erection of Steel Joists and Joists Girders." *SJI Technical Digest,* No. 9, Myrtle Beach, SC: SJI (www.sji.org).

———. 2002. *Catalog of Standard Specifications and Load Tables for Steel Joists and Joist Girders,* 41st ed. Myrtle Beach, SC: SJI (www.sji.org).

Steel Stud Manufacturers Association (SSMA). *Product Technical Information.* Chicago: SSMA (www.ssma.com).

———. *Cold Formed Steel Details.* Chicago: SSMA (www.ssma.com).

# 6

# Wood, Plastics, and Composites

# Introduction

Wood is a renewable resource. Wise forest management assures a permanent supply of timber to meet foreseeable future needs. Foresters can grow trees faster than nature can without their assistance. As a result of reforestation and other conservation practices, the yearly growth of timber nationwide exceeds that harvested or lost to fire or disease.

Wood is highly prized for its decorative character, which is derived from a variety of grain, figure, texture, hue, and other natural markings. These properties are important in lumber, plywood, and laminated timber, as well as in other wood products, such as flooring, windows, doors, and millwork.

# Sustainability Considerations

There is some controversy in the industry concerning the use of wood in the current movement toward sustainability. Some wood products producers and the associations that represent them have expressed dismay at the implication that the use of wood is less desirable than the use of other materials, naming LEED as the source of their problems in this regard. They say that the LEED guidelines are prejudiced against the use of wood and penalize designers and constructors who use wood. They point out that wood, unlike some competing materials, is a renewable resource. There is no intent in this book to enter this dispute in any way. The authors and publishers take no position on either side of this issue.

The Architectural Woodwork Institute (AWI) claims that of all the common building materials, wood has the least impact on total energy use, greenhouse gases, air and water pollution, solid waste, and ecological resource use.

AWI says that 70% less energy is required in the use of wood compared with any other building materials. Recent studies in Canada seem to support this point of view, even though the harvesting of wood, transportation of wood materials, production of wood products, and construction with wood consume large quantities of energy. It is true that removal of forests increases the amount of carbon dioxide, a greenhouse gas, in the atmosphere, but more new forest area is added in the United States and Canada than is cut every year.

It is true that wood is the only major construction product in current use in this country that is renewable. Unfortunately, this renewability requires forest management, which is not always forthcoming.

Some major wood products producers do a good job of managing the forests they own, using a process called *sustainable forestry*, but much of the wood used today comes from uncontrolled forest by producers who just cut the forest and move on. Some foreign producers make no attempt to manage the forest at all, denuding entire countries in some cases.

*Clearcutting* is a process in which a forest is cut without regard to sustainability. All trees are cut down or pushed over. Undesirable trees are left to decay, as are stumps and debris. Even so, such a forest will renew itself in time.

Much wood is wasted in the forestry and mill processes and during construction. Bark is removed from trees. Some is shredded into mulch, but much is either burned or buried. Scraps and cutoffs can be cycled into usable materials as components of pressed wood products such as oriented strand board, but much of these materials is burned to produce steam to run the mills or simply buried. Construction debris is burned or carried to landfills and buried. There is no question that a much better job could be done to use these waste materials in a constructive way.

One way a wood specifier or buyer can help in this process is by requiring that wood come from certified forests. Unfortunately, such a requirement may be costly or impossible to implement in many areas where certified forests do not exist.

Some wood products can contribute to indoor air pollution. Formaldehyde, a colorless, strong-smelling gas, is one of the large family of chemical compounds called *volatile* (they vaporize at room temperatures) *organic compounds* (VOCs).

Formaldehyde is contained in the urea-formaldehyde glues used in the adhesive in pressed wood products, such as particleboard, hardwood plywood, and medium density fiberboard (MDF). Some people are very sensitive to formaldehyde, while others may have no noticeable reaction. Reactions range from watery eyes and burning sensations in the eyes, nose, and throat to nausea, coughing, chest tightness, wheezing, skin rashes, and, rarely, severe allergic reactions (allergic skin disease and hives). The risk of cancer to humans is believed to be small.

In recent years, manufacturers have reduced formaldehyde emissions from pressed wood products by as much as 90% of the levels of the early 1980s.

Consumer Product Safety Commission (CPSC) Document 725 recommends using pressed wood products such as particleboard and hardwood plywood that are labeled or stamped to be in conformance with American National Standards Institute (ANSI) criteria that require lower formaldehyde emission levels. Particleboard should be in conformance with ANSI A208.1, MDF with ANSI A208.2, and hardwood plywood with ANSI/HPVA. Furniture and cabinets should contain a high percentage of panel surface and edges that are laminated or coated. Unlaminated or uncoated (raw) panels of pressed wood products will generally emit more formaldehyde than those that are laminated or coated.

The ultimate way to reduce the formaldehyde levels in the indoor atmosphere introduced by pressed wood products is by specifying or buying alternative products such as wood panel products that are not made with urea-formaldehyde glues, lumber, or metal.

# 6.1  Properties of Wood

The wide range of properties in wood makes it suitable for use in many structural, nonstructural, and decorative products and in hundreds of end uses. It has high strength relative to its weight in compression, tension, bending, and resistance to impact. It can be worked easily to desired shapes with simple tools, has high durability when simple precautions are observed, and has excellent insulating qualities. The general properties of wood that set it apart from concrete, metals, plastics, and other materials should be understood so that wood and wood products are used properly to perform satisfactorily.

## 6.1.1  IMPORTANT TREE SPECIES

Figure 6.1-1 shows important commercial species of wood and the areas in which they grow. The division of species into two classes, *softwood* and *hardwood*, is a botanical difference and is not always descriptive of the softness or hardness of the wood. Because of fundamental differences in microstructure and properties, softwoods and hardwoods differ in use, size standards, and method of grading.

### 6.1.1.1  Softwoods

About 75% of the lumber produced in this country comes from softwood trees. This lumber is used as structural and framing lumber, sheathing, roofing, subflooring, exterior siding, flooring, trim, and interior paneling. Softwoods are called *conifers* or *coniferous* because most species bear cones. With few exceptions, softwoods have scale-like or needle-like leaves and are evergreen (Fig. 6.1-2).

### 6.1.1.2  Hardwoods

Lumber produced from hardwood trees is used as flooring, interior paneling, and for building cabinets and furniture. Hardwood trees, which are broad-leaved, are called *deciduous* because they shed their leaves at the end of each growing season, except in warmer climates (see Fig. 6.1-2).

## 6.1.2  WOOD GROWTH

A tree grows by the production of new cells in the *cambium* layer, a three- to five-cell-thick zone of active cell growth just beneath the inner bark (Fig. 6.1-3).

As the cells mature, the inner layer of the cambium becomes sapwood and the outer layer becomes inner bark. The inner bark changes to outer bark as the cells become inactive. A wood cell does not develop further after it is formed and matures. All tree growth, consisting of increases in diameter and height, results entirely from new cell formation.

Each year, completely new layers of wood cells envelop the preceding year's growth beneath the bark over the entire tree. Each year's growth can be seen visually as an annual growth ring consisting of a light-colored layer and a darker

| | WEST COAST REGION<br>Douglas fir  west coast hemlock<br>western red cedar  Sitka spruce | | NORTHERN FORESTS<br>northern white pine  eastern spruce<br>jack pine  aspen  northern hemlock | | NORTHERN FORESTS<br>maple  birch  beech  ash<br>black cherry |
| --- | --- | --- | --- | --- | --- |
| | REDWOOD REGION<br>redwood  Douglas fir<br>WESTERN PINE REGION<br>ponderosa pine  Idaho white pine<br>Douglas fir  white fir  sugar pine<br>inland red cedar  western larch<br>Engelmann spruce  lodgepole pine<br>incense cedar | | SOUTHERN PINE REGION<br>lobolly pine  slash pine  shortleaf<br>pine  longleaf pine  cypress | | CENTRAL AND SOUTHERN<br>HARDWOOD FORESTS<br>oaks  yellow poplar  gum  hickory<br>black walnut  basswood |

**FIGURE 6.1-1** Areas of timber growth in the United States.

**a**     **b**

**FIGURE 6.1-2** (a) Typical growth from a coniferous (softwood) tree. (b) A representative leaf from a deciduous (hardwood) tree.

layer. In many species, growth is more active during the spring, and cells formed then (*springwood*) tend to be large and thin-walled compared to cells formed later in the year, when growth is slower (*summerwood*). The smaller, thickwalled cells of summerwood are denser and darker than those of springwood (Fig. 6.1-4 [also see Fig. 6.1-3]). Because of its density, summerwood is both stronger and harder than springwood. In species such as Douglas fir and southern yellow pine, the amount of summerwood can be used to estimate the wood's strength and density.

The wood between the *pith* at the center of a tree and the *cambium* is divided into light-colored *sapwood*, whose cells are active in storing food and carrying sap from the roots up to the leaves, and generally darker *heartwood*, whose cells are inactive in the growth process but provide structural support for the tree. Generally speaking, sapwood and heartwood are of compa-

rable strength. The gradual change from sapwood to heartwood involves the depositing of chemicals in the cellular structure of the heartwood, sometimes filling the cell cavities. In some species these chemicals are toxic, giving heartwood greater resistance to wood-destroying fungi and insects.

Most wood cells run parallel to the height of the tree and are called *fibers* because of their needle-like shape. However, running transversely from the *pith* to the *bark* are cells called *rays* which convey food across the grain (see Fig. 6.1-4). It is these rays that produce the decorative grain effects seen in lumber and plywood cut from species such as oak, where the rays are conspicuous.

### 6.1.2.1 Growth Characteristics

Certain growth characteristics affect the grade and use of wood. These include *knots*, *shakes*, and *pitch pockets* (Fig. 6.1-5).

#### KNOTS

*Knots* are common growth characteristics that are formed when branches of a tree, which originate in the center at the pith, become enclosed within the wood during subsequent growth. If a branch dies and later falls off, the dead stubs are incorporated in the wood, become overgrown with new wood, and appear as knots when the wood is cut into lumber. Knots

are detrimental to the structural properties of wood.

#### SHAKES

A *shake* is a grain separation occurring between annual growth rings, running along the grain parallel to the height of the tree (see Fig. 6.1-5). Shakes seldom develop in lumber unless they are present in the tree before it is felled.

#### PITCH POCKETS

A *pitch pocket* is a small, well-defined grain separation that may contain solid or liquid resin (see Fig. 6.1-5). Pitch pockets may be found in Douglas fir, western larch, pine, and spruce.

### 6.1.3 PHYSICAL PROPERTIES

The properties of wood are determined by its physical and chemical composition. Most characteristics of wood are related to its cellular structure.

#### 6.1.3.1 Chemical Composition

Wood cells, or fibers, are primarily *cellulose* cemented together with *lignin*. The wood structure is about 70% cellulose, between 12% and 28% lignin, and up to 1% ash-forming materials. These constituents give wood its hygroscopic properties, its susceptibility to decay, and its strength. The bond between individual fibers is so strong that when fibers are tested in tension, they commonly tear apart rather than separate. The remainder of the wood, although not part of its structure, consists of extractives that give different species distinctive characteristics such as color, odor, and natural resistance to decay.

It is possible to dissolve the lignin in wood chips using chemicals, thus freeing the cellulose fibers. By further processing, these fibers can then be turned into

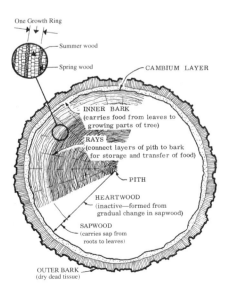

**FIGURE 6.1-3** Cross section of a living tree showing active and inactive cell layers.

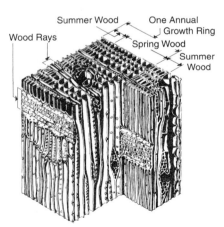

**FIGURE 6.1-4** The cell structure of wood. This minute cube represents a block of wood about $1/32$ in. (0.794 mm) on a side. Its top surface is parallel with the end surface of a log. (Reprinted from the Forest Products Laboratory, U.S. Forest Service, USDA publication. Not subject to copyright in the United States.)

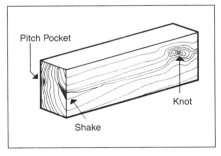

**FIGURE 6.1-5** Growth characteristics of wood.

a *pulp* from which paper and paperboard products are made. It is also possible to chemically convert cellulose so that it may be used to make textiles (such as rayon), plastics, and other products that depend on cellulose derivatives.

### 6.1.3.2 Hygroscopic Properties

Wood is *hygroscopic*, meaning that it expands when it absorbs moisture and shrinks when it dries or loses moisture. This property affects the end use of wood. Although the wet (green) condition is normal for wood throughout its life as a tree, most products made of wood require that it be used in a dry condition; therefore, *seasoning* by drying to an acceptable moisture content is necessary.

#### MOISTURE CONTENT

The *moisture content* of wood is the weight of water it contains, expressed as a percentage of the weight of the wood when oven-dry. The weight of the water in wet wood can be twice that in wood that is oven-dry. The oven-drying method, used to determine moisture content in the laboratory, is shown in Figure 6.1-6. For field use, a moisture meter is more practical: it gives an instantaneous reading by measuring the resistance to current flow between two pins driven into the wood.

In living trees, the amount of moisture varies widely between different species, among individual trees of the same species, among different parts of a tree, and between sapwood and heartwood. Many softwoods have a large proportion of moisture in the sapwood and far less in the heartwood, while most hardwoods have about the same moisture content in both sapwood and heartwood. The extreme limits of moisture content in green softwoods can be shown by comparing the moisture content of the heartwood of Douglas fir and southern pine, which may be as low as 30%, to the moisture content of the sapwood of cedars and redwoods, which may be as high as 200%.

#### FIBER SATURATION POINT

Moisture in green wood is present in two forms: in the cell cavities as free water and within the cell fibers as absorbed water (Fig. 6.1-7). When wood dries, its cell fibers give off their absorbed water only after all the free water is gone and the adjacent cell cavities are empty. The point at which the fibers are still fully saturated but the cell cavities are empty is called the *fiber saturation point*. In most species, this occurs at about 30% moisture content. The significance of this condition is that it represents the point at which shrinkage begins. Even lumber cut with a green mois-

ture content as high as 200% can dry to the fiber saturation point (30% moisture content) with no shrinkage of the wood. Only when the cell fibers begin to give off their absorbed water and start to constrict does the wood shrink.

Therefore, all of the shrinkage wood can experience takes place between its fiber saturation point and a theoretical moisture content of 0% (oven-dry condition). Within this range, shrinkage is proportional to moisture loss. Once wood has reached a 30% moisture content and below that level, for every 1% loss or gain in moisture content, it shrinks or swells, respectively, about $1/30$ of the total expansion or contraction. For example, at 15% moisture content, wood will have experienced half of its total possible shrinkage. However, wood in service almost never reaches a 0% moisture content because of the influence of water vapor in the surrounding atmosphere. Therefore, the total possible shrinkage is far less important than the probable shrinkage under ordinary conditions.

#### EQUILIBRIUM MOISTURE CONTENT

After a tree is felled and cut into lumber, its moisture content begins to drop as moisture in the wood is lost to the surrounding air. Wood will then continue to give off or take on moisture until the moisture within the wood has come to a point of equilibrium with the moisture in the air. The moisture content of the wood at this point is called the *equilibrium moisture content*. The relation of atmospheric temperature and humidity to the equilibrium moisture content. of wood is given in Figure 6.1-8, which shows, for

**FIGURE 6.1-6** Oven-drying method of measuring the moisture content of wood.

**FIGURE 6.1-7** Moisture content and wood-cell shrinkage.

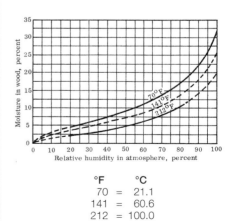

**FIGURE 6.1-8** Equilibrium moisture content for various temperatures and relative humidities.

example, that wood kept constantly in air at 70°F (21.1°C) with a 60% relative humidity eventually will balance at a moisture content of about 11%. Since temperature and humidity are not constant in service, wood is subject to variations in moisture content as it seeks an equilibrium with the surrounding air. These variations tend to be gradual and seasonal. In addition, the equilibrium moisture content of wood is usually less on the interior of heated buildings than in unheated buildings or in an exterior exposure.

Knowing the equilibrium moisture content in a particular location is significant because it permits one to predict the moisture content a wood will attain in service there. To ensure that wood will experience only minor dimensional changes, it should be fabricated and installed at a moisture content as close as possible to the equilibrium moisture content it will attain in use (Fig. 6.1-9). This is especially critical when dealing with strip flooring, wood doors, and items of millwork, where a proper fit between adjoining wood elements is important. Since the equilibrium moisture content of interior wood items in most areas of the country varies between 6% and 12%, these products are usually fabricated with a moisture content in this range. However, since wood may lose or absorb moisture during shipment, it may end up at the site with a moisture content substantially higher or lower than the local equilibrium moisture content. Therefore, it is advisable to condition such products

by storing them in the space in which they will be installed for several days before installation.

Exterior trim, siding, and board sheathing are subject to greater variations in atmospheric humidity and temperature, and therefore also in equilibrium moisture content. In addition, a given wood's equilibrium moisture content is usually higher on the exterior than it is indoors, and may differ from season to season within the same area, as well as from region to region (see Fig. 6.1-9). However, the conditions of installation and service are not as demanding for exterior elements as they are for interior ones, so exterior elements are normally fabricated with a moisture content between 12% and 15%, which is considered adequate for exterior use in all locations and for all purposes.

Framing lumber is surfaced either with a moisture content higher than 19% (grademarked S-GRN, meaning surfaced green) or at a maximum moisture content of 19% (grademarked S-DRY). Framing lumber may be surfaced at a moisture content of 15% or lower (grademarked MC 15). These grademarks should be those of an agency approved by the Board of Review of the American Lumber Standards Committee.

To increase the chances that no individual piece exceeds the maximum moisture content required by a particular grade stamp, lumber is seasoned to an average moisture content several percentage points lower than the maximum.

Therefore, S-DRY lumber is dried to an average moisture content of 15%, and MC 15 lumber is dried to an average moisture content of 12%. Since the equilibrium moisture content of wood in most areas of the country is less than 15%, lumber that has been milled while green may, after air drying, actually have a much lower moisture content. In arid regions, framing lumber should have a maximum moisture content of 15% when it is installed on the interior of a building. In every other location, including damp regions, lumber for interior use should have a moisture content of 19% or lower (see Fig. 6.1-9). In arid regions, either S-GRN or S-DRY lumber may achieve the desired maximum 15% moisture content if it is permitted to air-dry at the yard or site, because the lumber loses moisture quickly under dry atmospheric conditions. Specifying MC 15 merely provides the visual assurance of a grademark that this moisture content has been achieved at the mill. In other parts of the country, where air drying is not so rapid, it is best to specify dry lumber (19% maximum moisture content) for all 2-in. (50-mm)-thick or thicker framing lumber.

Large quantities of lumber less than 2 in. (50 mm) thick are used in residential construction. Boards and dimension lumber 2 in. (50 mm) thick or thinner should be purchased either as S-DRY or MC 15 and should be so identified by an appropriate grademark.

In most species, boards and 2-in. (50-mm)-thick dimension lumber are graded and sold either dry or green. Southern pine lumber that has been graded according to Southern Pine Inspection Bureau (SPIB) rules is usually surfaced at 15% maximum moisture content and carries the KD or KD 15 grademark. Lumber 3 in. (75 mm) and 4 in. (100 mm) thick is not readily available in dry grades in many locations, because most mills fabricate these sizes from green wood. However such lumber, even when bearing an S-GRN grademark, often has an acceptable moisture content, because it has been air dried at the yard. Larger pieces, 3 in. (75 mm) and 4 in. (100 mm) thick, should have a 19% maximum moisture content and, in the absence of a grademark, should be checked to ensure that the moisture content is at that level.

Moisture contents should be checked in accordance with the procedures outlined in ASTM International (ASTM) Standards D4442 and D4444.

Damp Southern and Coastal Areas

Remainder of U.S. All Other Areas

Dry Western Areas

**FIGURE 6.1-9** Atmospheric humidity regions of the United States.

## EFFECTS OF MOISTURE CONTENT ON PROPERTIES

As stated above, size variations in wood due to shrinkage occur only with changes in moisture content below the fiber saturation point. Shrinkage or swelling parallel to the grain (longitudinally with the height of the tree) is practically negligible and has little significance in construction applications. Across the grain, however, wood will shrink appreciably in both width and thickness. Shrinkage is greatest in the direction tangent to the annual growth rings and is about one-half to two-thirds as much across these rings (radial). Figure 6.1-10 shows the combined effects of tangential and radial shrinkage for various shapes due to drying from the green condition.

In general, hardwoods shrink more than softwoods and heavier species shrink more than lighter ones. Members with a large cross section do not shrink as much proportionately, because drying is not simultaneous in the inner and outer parts of the piece. Since the inner part dries more slowly than the outer portion, the wood near the surface is prevented from shrinking normally. The outer layers become set and tend to keep the inner portions from shrinking as they dry out. A 6 × 6 (150 × 150) or larger softwood structural lumber member will shrink about $1/32$ in. (0.79 mm) per 1 in. (25.4 mm) of face width as is dries from the green state to ordinary conditions in use.

Softwood structural lumber 2 in. (50 mm) and 3 in. (75 mm) thick is usually partially or fully seasoned when marketed. The average amount of shrinkage in service for material of these sizes usually does not exceed $1/32$ in. (0.79 mm) per 1 in. (25.4 mm) of face width (Fig. 6.1-11).

The strength of wood increases as its moisture content decreases. This increase in strength is caused by (1) strengthening and stiffening of the cell fibers as they dry out and (2) increase in the compactness of the wood in a given volume. Wood dried from green to a 5% moisture content may add 2.5% to 20% to its density (weight of wood per unit of volume) and, in small pieces, end-crushing strength and bending strength may easily be doubled or, in some woods, even tripled. The increase in strength in small, clear specimens is much greater than in large timbers, because the increase is offset by the influence of defects that develop in large members during seasoning.

Not all strength properties are equally affected by decreases in moisture content. For example, crushing and bending strengths increase greatly, but some properties, such as stiffness, increase only moderately. Still others, like shock resistance, may show a very slight decrease. This is because shock resistance and toughness depend on pliability as well as strength, and while drier wood is stronger, it does not bend as far before failure as does green wood. The increase in various strength properties due to a change from

**FIGURE 6.1-11** Approximate shrinkage during seasoning of a 2 × 10 (50 × 250) from green to a theoretical 0% moisture content.

the fiber saturation point at 30% moisture content down to a moisture content of 19% is shown in Figure 6.1-12.

The decay resistance of wood is assured at a moisture content below 20%. In fact, the most effective and practical method of preventing decay is to control the moisture content of wood. Optimum conditions for decay occur when the moisture content is about 25%. Wood that is installed and maintained below a moisture content of 20% will not decay. Most wood in use under protected conditions has a moisture content below this level. If the moisture content will exceed 20% due to use or exposure conditions, the wood should be treated with toxic chemicals to prevent decay (see Section 6.4) or the naturally durable heartwood of certain species may be used (see Section 6.4).

**FIGURE 6.1-10** Characteristic shrinkage and distortion of members as affected by the direction of the annual rings.

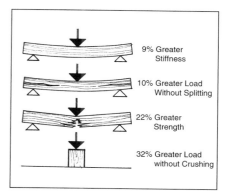

**FIGURE 6.1-12** Approximate increase in wood structural properties with a decrease in moisture content from 30% to 19%.

Paintability varies slightly with moisture content. The caveat that wet wood should not be painted refers to free water present on the surface and not to the moisture content of the wood. Wood painted at 16% to 20% moisture content holds paint slightly longer than wood painted at 10% moisture content

Although the time required for paint to dry (harden) depends greatly on the paint, intensity of sunlight, temperature, and relative humidity, drying is retarded when the moisture content of a wood substrate is excessive. Therefore, wood to be painted should have a moisture content of 20% or less.

Many wood products depend extensively on gluing. The moisture content of the wood at the time of gluing affects both the strength of the bond and, later, the checking and warping of glued members. In general, glue bond improves at reduced moisture contents. Wood glues adhere well to wood with moisture contents of up to 15% and even higher, but when large changes in moisture content occur after gluing, shrinking or swelling stresses may weaken both the wood and the glued joints. The moisture content of wood products at the time of gluing should be increased by the moisture in the glue, and the total should approximate the average moisture content that the glued member will attain in service.

Lumber used for interior millwork can be satisfactorily glued at a 5% to 6% moisture content and exterior millwork at a 10% to 12% moisture content. Veneers used to make plywood must be dried to a 2% to 6% moisture content at the time of gluing.

### 6.1.3.3 Specific Gravity

*Specific gravity* is a ratio comparison of the weight of a material to the weight of water. It is obtained by dividing the weight of the material by the weight of an equal volume of water. Water has a specific gravity of 1. Wood fiber, with a specific gravity of about 1.5, is heavier than water. However, dry wood of most species floats in water because a proportion of the volume is occupied by air-filled cell cavities. The specific gravity of most species when oven-dry ranges from about 0.36 to 0.70 (Fig. 6.1-13). Variation among species in the size of cells and in the thickness of cell walls affects the amount of solid wood substance present and therefore the specific gravity. Specific gravity is an approximate measure of solid wood substance and a general index of strength properties, although specific gravity values may be affected somewhat by gums, resins, and extractives which contribute little to strength. The relationship of specific gravity to wood strength is evident in the practice of assigning higher basic stress values to lumber such as Douglas fir and southern yellow pine designated as *dense*.

### POROSITY OF WOOD

Void space in wood occurs as (1) coarse microscopic capillaries within the basic cell system and (2) submicroscopic capillaries within the cell walls which are at a maximum when the wood is saturated with moisture, as in the green or freshly cut condition. A single cubic foot of green wood has about 15 sq ft (1.394 m²) of internal cell wall surface area and about 22,000 sq ft (2043.87 m²) (about half the surface area of a football field) of the internal submicroscopic capillary surface area in the cell walls (see Fig. 6.1-4).

This enormous internal surface area gives wood many of its unique properties, including:

- Ease of impregnation with toxic chemicals for decay and insect protection.
- Ease of impregnation with moisture repellents for slowing down moisture exchange to minimize shrinkage.
- Excellent insulating properties.
- Shrinking and swelling in response to variations in moisture content.
- Excellent adherence of paint films and adhesives. Some synthetic resins adhere with such tenacity that they resist many exposure conditions.

### 6.1.3.4 Structural Properties

Unlike crystalline materials such as steel and concrete, wood is fibrous. The length of fibers varies from about $1/25$ in. (1.016 mm) in hardwoods to between $1/8$ in. (3.2 mm) and $1/3$ in. (8.47 mm) in softwoods. The strength of wood, however, does not depend on the length of the fibers, but rather on the thickness of their

**Figure 6.1-13** Design Values for Visually Graded Douglas Fir-Larch Dimensional Lumber*

| Grade | Size | Bending $F_b$ | Tension Parallel to Grain $F_t$ | Shear Parallel to Grain $F_v$ | Compression Perpendicular to Grain $F_{c\perp}$ | Compression Parallel to Grain $F_c$ | Modulus of Elasticity $E$ |
|---|---|---|---|---|---|---|---|
| Select Structural | | 1500 | 1000 | 95 | 625 | 1700 | 1,900,000 |
| No. 1 & Btr | 2 to 4 in. | 1200 | 800 | 95 | 625 | 1550 | 1,800,000 |
| No. 1 | (50 to 100 mm) thick | 1000 | 675 | 95 | 625 | 1500 | 1,700,000 |
| No. 2 | 2 in. (50 mm) | 900 | 575 | 95 | 625 | 1350 | 1,600,000 |
| No. 3 | and wider | 525 | 325 | 95 | 625 | 775 | 1,400.000 |
| Stud | | 700 | 450 | 95 | 625 | 850 | 1,400,000 |
| Construction | 2 to 4 in. | 1000 | 650 | 95 | 625 | 1650 | 1,500,000 |
| Standard | (50 to 100 mm) thick | 575 | 375 | 95 | 625 | 1400 | 1,400,000 |
| Utility | 2 to 4 in. (50 to 100 mm) thick | 275 | 175 | 95 | 625 | 900 | 1,300,000 |

*Douglas Fir-Larch is graded under the rules of the WCLIB and the WWPA.

walls and on the direction of the fibers relative to applied loads. The fibers in wood are oriented with their length essentially parallel to the vertical axis of the tree, and the strength of wood parallel to these fibers (parallel to the grain) is quite different from its strength perpendicular to the fibers (perpendicular to the grain).

External forces, called *loads* (such as wind, gravity, and the forces caused by such natural phenomena as earthquakes), produce internal resisting forces in structural members known as *stresses*. *Tensile stresses* result from stretching forces, *compressive stresses* from squeezing forces. When a member is loaded so that the force is acting more or less perpendicular to its length, as is the case in a beam, the resulting stress is called *bending*. A simple span member in bending develops compressive stresses in the upper part and tensile stresses in the lower part of its cross section.

The strength values of wood are determined by testing small, clear specimens in bending, compression, shear, and so forth, and assigning basic stresses (basic strength values) to different species. Since small, clear pieces are seldom used in construction, lower working stresses (also called *allowable unit stresses*) must be derived to take into account the effect of knots, slope of grain, checks, and other characteristics that reduce the strength of a member to a value less than that for clear wood.

Accordingly, the maximum allowable unit stress (*working stress*) that should be permitted to exist on a grade of lumber is determined by the effect of the maximum size and most unfavorable position of any strength-reducing characteristics permitted in the grade. The working stresses of various species of dimension lumber are shown in tables in the *National Design Specifications (NDS) for Wood Construction®* developed by the American Forest & Paper Association (AF&PA) and published by the AF&PA affiliate, the American Wood Council (AWC). The *NDS* tables list the various species of structural lumber, subdivided into their grades as established by the various grading agencies. For each grade and species, the tables list the design characteristics in pounds per square inch. There are separate tables for visually graded lumber and mechanically graded lumber.

Figure 6.1-13 shows some of the data that might be included in a typical table for visually graded lumber. Similar design data are given in the *NDS* tables for mechanically graded lumber. The species used in Figure 6.1-13 is Douglas fir-larch. The data in Figure 6.1-13 were compiled from data in a table in the *NDS* and may not be accurate at the time you want to use it. Always check the source grading agency or the *NDS* for the particular species you will specify. Also note that the *NDS* includes requirements for increasing some of the values in the *NDS* tables based on a repetitive member factor when the listed members are used as joists, truss chords, rafters, studs, planks, decking, or similar members; and are in contact or spaced not more than 24 in. on center; and are not fewer than three in number and are joined by floor, roof, or other load-distributing elements adequate to support the design load.

Conversely, the listed data must be reduced by a wet service factor when the wood is used with a moisture content larger than 19%.

The design bending stress must also be adjusted by a member size factor.

There are also other multipliers for stud grade, construction and standard grade, and utility grade members, and additional factors for use of a member in the flat position, and when the length of shakes and splits is less than that contemplated when the tables were developed.

The allowable unit stresses shown in the *NDS* tables are adjusted using size, repetitive member, flat use, wet service, and shear stress factors, based on what *NDS* calls *normal load duration*. The *NDS* defines this condition as fully stressing a member to its allowable design value by the application of the full design load for a cumulative duration of approximately 10 years, or applying 90% of the full design load continuously throughout the remainder of the life of the structure. Values for normal loading generally are used for designing members to carry dead loads (such as floor or roof construction) and live loads due to occupants, equipment, and furnishings. For loads of brief duration, such as those produced by snow, earthquake, or wind, working stresses are higher than those required for normal loads (Fig. 6.1-14).

Detailed descriptions of the various adjustments and the required values for them are included in the *NDS*.

### BENDING

Wood develops high fiber strengths in bending. When a wood beam, loaded as in Figure 6.1–15a, deflects, bending stresses are produced in the fibers of the member. If the load is great enough to produce stresses that exceed the fiber strength of the wood, the beam will fail by breaking (Fig. 6.1-15b). The stresses produced in the fibers are greater in magnitude the farther they are from the cen-

**FIGURE 6.1-14** Load duration factors. The graph plots duration of load on the horizontal axis and load duration factors on the vertical axis. The tabular allowable unit stresses in the *NDS* are multiplied by the load duration factor determined using this graph. The duration factors plotted are for load duration other than the normal duration of 10 years, which has a load duration factor of 1. (Courtesy American Forest & Paper Association.)

**FIGURE 6.1-15** Stresses produced in a beam subjected to bending.

tral axis. If one fiber is located twice as far from the central axis as another fiber, it will have twice the stress. Therefore, in a beam, bending stresses are maximum in the outermost fibers (called the *extreme fibers*) at the top and bottom of the beam. In beams subjected to bending, the fiber stresses in bending should be held below the working stress values ($F_b$), for the grade of wood used (see Fig. 6.1-13).

### HORIZONTAL SHEAR

A beam also tends to fail by dropping down between supports (see Fig. 6.1-15c) as a result of vertical shear, the tendency for one part of the beam to move vertically with respect to an adjacent part. Wood beams are more likely to fail due to the tendency of the top fibers in the beam to move horizontally with respect to the bottom fibers. This produces stresses in fibers called *horizontal shear* (see Fig. 6.1-15d). Horizontal shearing stresses are largest at each end of a beam along the central (neutral) axis. In beams, the fiber stress in horizontal shear should be held below the working stress values ($F_v$) for the grade of wood used (see Fig. 6.1-13).

### MODULUS OF ELASTICITY

The stiffness of a material is measured by its *modulus of elasticity*, ($E$) (see Fig. 6.1-13). This measure of wood's stiffness determines a beam's resistance to deflection. Modulus of elasticity is also used in computing the load on long columns, because such columns depend on stiffness to resist buckling.

Mechanical grading techniques based on modulus of elasticity permit greater utilization of wood's strength properties (see Section 6.2.4.3).

### TENSION

Theoretically, tensile strength parallel to the grain is the strongest property of wood, but in practice, this is greatly affected by knots, splits, and checks. Since the allowable unit stress in *tension parallel to grain* ($F_t$), is a value applied to the entire cross section (less holes) of a piece of wood, it must reflect an average internal resistance to loading which is much more affected by wood characteristics than a value applied to a narrow band of the cross section, such as the extreme fiber in bending stress ($F_b$). For most species, the allowable $F_t$ values are approximately two-thirds of $F_b$ values. Wood is weak in *tension perpendicular to the grain*, which permits the fibers to be pulled apart under heavy loading. Therefore, wood should not be loaded so as to create this type of stress. Allowable values for tension perpendicular to the grain usually are not tabulated.

### COMPRESSION

The allowable unit stress in compression parallel to the grain is two to five times greater than it is perpendicular to the grain (Fig. 6.1-16). There are only minor differences in the strength properties in the two directions perpendicular to the grain (radially and tangentially). The values for compressive strength perpendicular to the grain ($F_c$) and parallel to the grain ($F_c$) are shown for Douglas fir-south in Figure 6.1-13.

### 6.1.3.5 Gluing Properties

The gluing properties of the woods most widely used for glued products are shown in Figure 6.1-17. These classifications are based on the average quality of typical joints of wood when glued with animal, casein, starch, urea resin, and resorcinol

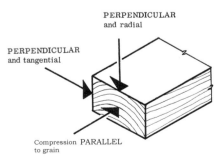

**FIGURE 6.1-16** Compression parallel and perpendicular to the grain, both radially and tangentially.

resin glues. A joint is considered to be glued satisfactorily when its strength is approximately equal to the strength of the wood.

Whether it will be easy or difficult to obtain a satisfactory joint depends on the density of the wood used, its structure, the presence of extractives or infiltrated materials in the wood, and the kind of glue. In general, hardwoods are more difficult to glue than softwoods, and heartwood is more difficult than sapwood. Several species vary considerably in their gluing characteristics with different glues.

### 6.1.3.6 Nonstructural Properties

Many unique characteristics of wood result in its selection as a material for many nonstructural uses. These nonstructural properties include decorative characteristics, weathering qualities, thermal expansion and conductivity, and natural decay resistance (see Section 6.4.3.3).

**FIGURE 6.1-17** Gluing Properties of Some Wood Species

| Glueability* | Hardwoods | Softwoods |
|---|---|---|
| **Excellent** | Aspen Chestnut Cottonwood Black willow Yellow poplar | Bald cypress Western red cedar White fir Western larch Redwood Sitka spruce |
| **Good** | Alder Basswood Butternut Rock elm Hackberry | Eastern red cedar Douglas fir West Coast hemlock Northern white pine Southern yellow pine Ponderosa pine |
| **Satisfactory** | White ash Black cherry Soft maple Oak Pecan Sycamore Tupelo Black walnut | Alaska cedar |
| **Poor** | Beech Birch Hickory Hard maple | |

*Ratings based on the ability of the wood to be glued under a wide range of conditions and with a variety of glues.

## DECORATIVE CHARACTERISTICS

Factors influencing the decorative value of wood include color, figure patterns, texture, and luster; ability to take filler, stains, bleaches, and other finishes; and the method of cutting or sawing. The terms *figure*, *grain*, and *texture* are often incorrectly used interchangeably. *Figure* is the pattern or design created more by the abnormal than by the normal growth of a tree (wavy or swirly) and is incorrectly used synonymously with *grain*.

*Grain* is often used in referring to the annual rings (fine grain and coarse grain), but it is also used to indicate the direction of the fibers (straight grain, vertical grain). *Finishes* refer to the relative size of the pores (open-grained, closed-grained), which determines the need for a filler.

The term *texture*, though often used synonymously with *grain*, usually refers more to the finer structure and the visual dimension of depth in the wood than to the annual growth rings. In plain-sawn boards and rotary-cut veneers, the annual rings often form ellipses and parabolas making striking figures. In quarter-sawn boards and sliced veneers, the growth rings form stripes, and in species such as oak, the rays add conspicuous figures. Examples of plain- and quarter-sawn boards are shown in Figure 6.1-18. In open-grained hardwoods (such as oak, mahogany, and walnut) and sliced veneers, the appearance can be varied greatly by using fillers of different colors, and in softwoods the annual growth layers can be accentuated by applying stain.

## WEATHERING

Weathering of exposed boards with no protective coating may cause:

1. Change in color (cedar and cypress weather to a light gray with a silvery sheen, Douglas fir and redwood to dark gray with no sheen).
2. Roughening and checking of the surface except in cedar, cypress, and redwood, on which weather checks are inconspicuous.
3. Warping tendencies (a slight tendency in cedar, cypress, and redwood; a distinct tendency in most of the other softwoods; and a pronounced tendency in most hardwoods). Some resistance to warping can be obtained by keeping the width of an exposed, unpainted board to not more than eight times its thickness.

Lumber cut tangent to the annual rings is:
Plain-sawn for hardwoods
Flat-sawn for softwoods

Lumber cut radially to the annual rings and parallel to the ray is:
Quarter-sawn for hardwoods
Edge-grained or vertically grained for softwoods

**FIGURE 6.1-18** The two basic ways to cut lumber from a log. (Drawing by HLS.)

## THERMAL EXPANSION

Like most materials, wood expands when heated and contracts when cooled. Except in special applications, such as buildings kept dry but subject to wide temperature ranges, the thermal expansion and contraction of wood are ignored, because they are much smaller than the swelling and shrinkage due to moisture content variations occurring under normal exposure conditions.

## THERMAL CONDUCTIVITY

The thermal conductivity of wood is influenced by several of its properties, the most important of which are specific gravity and moisture content. Lighter-weight woods are better insulators, and the insulation value increases as the moisture content decreases. Because of its good insulating characteristics, wood fiber is a principal constituent of insulating fiberboards.

# 6.2 Lumber

*Lumber* is the designation given to a large number of products produced in sawmills

and planing mills. It is manufactured by sawing, edging, and, trimming logs into square or rectangular pieces of wood. Veneers, barrel staves, and hammer handles, however, are not lumber. *Yard lumber* includes (1) *boards* used for sheathing, siding, flooring, paneling, trim, and patterned millwork, (2) *dimension* stock used for light framing, joists, planks, and roof and floor decking, and (3) *timbers* used for posts, beams, and stringers.

Lumber varies widely in quality and therefore must be divided into commercial lumber grades. Each grade covers a narrow range of quality to provide a basis for selecting lumber for each of its many uses.

## 6.2.1 MANUFACTURING PROCESS

The manufacture of lumber begins in the forest. After trees are felled, the logs are transported to a lumber mill for processing. Following manufacture by a sawmill, a wholesaler locates lumber and resells it to wood-using plants and retail lumber yards that distribute about 75% of all lumber to the ultimate user. Many retail yards and some wholesalers maintain a planing mill and other equipment to satisfy individual user requirements.

### 6.2.1.1 Logging Practice

Modern logging methods make it possible to transport most of the trunk of a tree from a forest to a mill. Portable power saws are used for felling trees and cutting them into logs. Depending on the terrain and other variable factors, logs are removed from the forests by trucks, tractors, or trains or by fast, powerful cable logging systems. The logs are grouped at a landing where mobile or other types of loaders place them on trucks for the haul to the mill (Figs. 6.2-1 and 6.2-2).

### 6.2.1.2 Sawmill Practice

Operations vary from mill to mill, depending on the rate of output, the type of machinery, and the end product desired, but in the main, the pattern of operations is as follows. Automatic machinery handles much of the sawmill process (Fig. 6.2-3), but plants still depend heavily on the judgment of skilled workers at each step during manufacture.

In some mills, logs are cut to a shorter length. In some cases, low-quality parts are removed. In other cases, logs are cut to fit the mill's machinery or to produce

**FIGURE 6.2-1** Logs are loaded aboard trailer trucks for transport to a sawmill. (Courtesy Weyerhaeuser.)

**FIGURE 6.2-2** At a sawmill, a log unloader can remove an entire load from a truck in one bite. (Courtesy Weyerhaeuser.)

**FIGURE 6.2-3** A sawmill operator. (Courtesy Weyerhaeuser.)

lumber of certain lengths. Usually, logs are sent first to a debarker, where their bark is removed by peeling, scraping, or blasting using high-pressure water jets. This reveals the shape of each log and its knots and irregularities. The log is then placed on a carriage and fed into a *headsaw*, which may be either a circular saw or a band saw. The main saw mechanism, called a *headrig*, is the point where a log is broken down into rough cants, timbers, planks, and boards. Some mills employ as the head saw a gang saw with as many as 40 straight saws. Others use a gang saw in combination with a band or circular head saw to break large cants rapidly into boards and dimension lumber.

The next step involves the *edger*, a series of small saws that rip the lumber into the desired width and remove the rounded edge. The edger also rips wide pieces into narrow widths. Edged lumber moves onto the trimmer, where the saws trim off the rough ends and defects to produce the most desirable lengths.

The final step in the sawmill is called a *green chain*, where the edged and trimmed pieces are sorted visually according to grade, species, and size (Figs. 6.2-4 and 6.2-5). From the green chain, lumber may be sent to a seasoning yard or dry kiln (Fig. 6.2-6) or to remanufacturing plants for resawing, dressing, ripping, planing, or other treatment.

### 6.2.1.3 Seasoning Practice

Lumber *seasoning* is the process of reducing the moisture content of the wood. Kiln drying and air drying are the methods used. In either case, the advantages to be gained over unseasoned, or green, lumber include increased strength, reduced shrinkage, checking, and warping in service, less susceptibility to fungus attack in storage, and improvement in the capacity to receive pressure preservative treatment. Kiln drying can further reduce the moisture content to almost any desired value, can result in even greater reduction in weight, hastens the drying time over that required in air seasoning, and kills stains, decay fungi, and insects in the wood. In the case of hardwoods, air drying in outside yards is often combined with kiln drying. Following seasoning, lumber is sent to a planing mill for sizing and surfacing, after which it is graded, tallied, and prepared for shipment.

Under the *American Softwood Lumber Standard* (Product Standard PS 20), mills can fabricate dimension lumber in either the green or seasoned state, provided that green lumber is surfaced to slightly larger sizes to compensate for eventual shrinkage (Fig. 6.2-7). PS 20 also requires that lumber up to, but not including, 5 in. (127 mm) thick be grade-marked S-DRY if it was surfaced at a

FIGURE 6.2-4 Operator making sure that boards do not pile up. He is also tossing culls. (Courtesy Weyerhaeuser.)

moisture content of 19% or less; if it was surfaced at a higher moisture content it must be marked S-GRN. Although not required by the standard, lumber milled at a 15% moisture content is identified in most species as MC 15; in grading Southern pine, the term KD 15 (kiln-dried) is used in place of MC 15.

### KILN-DRIED LUMBER

In a dry kiln, heat, humidity, and air circulation are controlled to reduce the moisture content to any practical percentage. With few exceptions, the appearance grades of lumber intended for exterior and interior finish are kiln dried, and some mills also kiln dry dimension and other yard grades. Specifying kiln-dried lumber does not necessarily ensure that it is dry, since drying may be terminated in the kiln at any time. The moisture content desired should always accompany the words *kiln dried* unless the moisture content established by the grading rules is suitable for the intended use. This is usually 6% to 12% in the finish grades of softwoods and hardwoods and 15% to 19% in the common yard grades of softwoods. Kiln drying takes several days.

### AIR-DRIED LUMBER

In the air-dried method, lumber dries simply by evaporation when exposed to the atmosphere, either outdoors or in unheated storage over a period of several months. Ordinarily, dimension and lower grades of softwood lumber used for framing are sent to a yard for air drying, but sometimes they are shipped unseasoned (green). Structural timbers are not usually held long enough to be seasoned, though some drying may take place between their sawing and the time of shipment.

The moisture content of thoroughly air-dried lumber averages between 12% and 15% in the United States, but varies widely from as low as 6% in summer in the arid Southwest to 24% in winter in the Pacific Northwest. The various grading rules governing softwood lumber establish the maximum moisture content for each species of seasoned lumber, whether it is air or kiln dried.

Hardwoods are commonly classified according to grade and size at the time of sawing, with all stock then sent to the air-drying yard. After air drying, it may be kiln dried at the mill or shipped to a remanufacturing plant where it is kiln dried before being made into flooring, interior finish, cabinet work, or other finished products.

FIGURE 6.2-5 Lumber grading. (Courtesy Weyerhaeuser.)

FIGURE 6.2-6 Stacks of lumber being loaded into a restraint dryer. (Courtesy Weyerhaeuser.)

**FIGURE 6.2-7** Nominal and Minimum Dressed Sizes of Boards, Dimension, and Timbers

Thicknesses section — Minimum Dressed (Dry: mm, inch; Green: mm, inch):

| Item | Nominal inch | Dry mm | Dry inch | Green mm | Green inch |
|---|---|---|---|---|---|
| Boards* | 3/4 | 16 | 5/8 | 17 | 11/16 |
| | 1 | 19 | 3/4 | 20 | 25/32 |
| | 1¼ | 25 | 1 | 26 | 1 1/32 |
| | 1½ | 32 | 1¼ | 33 | 1 9/32 |
| Dimension | 2 | 38 | 1½ | 40 | 1 9/16 |
| | 2½ | 51 | 2 | 52 | 2 1/16 |
| | 3 | 63 | 2½ | 65 | 2 9/16 |
| | 3½ | 76 | 3 | 78 | 3 1/16 |
| | 4 | 89 | 3½ | 90 | 3 9/16 |
| | 4½ | 102 | 4 | 103 | 4 1/16 |
| Timbers* | 5 and thicker | | 13 off | ½ off | |

Widths section — Minimum Dressed (Dry: mm, inch; Green: mm, inch):

| Item | Nominal inch | Dry mm | Dry inch | Green mm | Green inch |
|---|---|---|---|---|---|
| Boards* | 2 | 38 | 1½ | 40 | 1 9/16 |
| | 3 | 64 | 2½ | 65 | 2 9/16 |
| | 4 | 89 | 3½ | 90 | 3 9/16 |
| | 5 | 114 | 4½ | 117 | 4 5/8 |
| | 6 | 140 | 5½ | 143 | 5 5/8 |
| | 7 | 165 | 6½ | 168 | 6 5/8 |
| | 8 | 184 | 7¼ | 190 | 7½ |
| | 9 | 210 | 8¼ | 216 | 8½ |
| | 10 | 235 | 9¼ | 241 | 9½ |
| | 11 | 260 | 10¼ | 267 | 10½ |
| | 12 | 286 | 11¼ | 292 | 11½ |
| | 14 | 337 | 13¼ | 343 | 13½ |
| | 16 | 387 | 15¼ | 394 | 15½ |
| Dimension | 2 | 38 | 1½ | 40 | 1 9/16 |
| | 2½ | 51 | 2 | 52 | 2 1/16 |
| | 3 | 64 | 2½ | 65 | 2 9/16 |
| | 3½ | 76 | 3 | 78 | 3 1/16 |
| | 4 | 89 | 3½ | 90 | 3 9/16 |
| | 4½ | 102 | 4 | 103 | 4 1/16 |
| | 5 | 114 | 4½ | 117 | 4 5/8 |
| | 6 | 140 | 5½ | 143 | 5 5/8 |
| | 8 | 184 | 7¼ | 190 | 7½ |
| | 10 | 235 | 9¼ | 241 | 9½ |
| | 12 | 286 | 11¼ | 292 | 11½ |
| | 14 | 337 | 13¼ | 343 | 13½ |
| | 16 | 387 | 15¼ | 394 | 15½ |
| Timbers* | 5 and wider | | 13 off | ½ off | |

*Boards less than the minimum thickness for nominal 1 in. but 5/8 in. (16 mm) or greater thickness dry [11/16 in. (17 mm) green] shall be regarded as American Standard Lumber, but such boards shall be marked to show the size and condition of seasoning at the time of dressing. They shall also be distinguished from nominal 1-in. boards on invoices and certificates.

(Source: Table 3, Voluntary Product Standard PS 20, "American Softwood Lumber Standard," U.S. Department of Commerce. Not subject to copyright in the United States.) (Refer to Figure 25.7-1 for metric conversions.)

## UNSEASONED LUMBER

Lumber more that 2 in. (50 mm) thick is rarely seasoned at the mill. Lumber 3 in. (75 mm) and 4 in. (100 mm) thick requires longer periods of air drying to achieve a 19% moisture content. This thicker lumber is usually fabricated green and is not often available in a retail lumber yard with an S-DRY grademark. However, such lumber frequently achieves an acceptable moisture content of 19% or less after air drying at the lumber yard. In the absence of a grademark, the wood can simply be measured. If a representative sample of pieces is close to the thickness and width requirements of PS 20 for dry lumber of that size (see Fig. 6.2-7), the lumber is likely to have an acceptable moisture content. An electric moisture meter, of course, provides a more reliable estimate of moisture content.

### 6.2.1.4  The Planing Mill

As previously stated, lumber can be surfaced (planed) either unseasoned or seasoned. Except for some framing lumber, most lumber is surfaced after seasoning, because the planing removes some of the imperfections caused by shrinkage during seasoning. Surfacing is done at a planing mill, where the lumber is planed and worked to patterns. Equipment varies from a few machines for surfacing and resawing in some mills to complete fabricating operations in others. Dried, rough-sawed products are converted into boards, or dimension, flooring, ceiling, partition, siding, casing, and other finished stock.

After going through a planing mill, unseasoned lumber is returned to be seasoned; seasoned lumber may be removed to storage sheds to await shipment, but most of it usually is moved directly to cars and loaded for immediate shipment to a wholesale or retail yard or to a woodworking or industrial plant (Fig. 6.2-8).

### 6.2.1.5  Reforestation

The lumber industry has had to adapt production to its source of supply, trees. Raw material requirements must be geared to the volume of timber which the forests can grow. This adaptation has been accomplished by greater wood research and utilization, coupled with the application of forest conservation principles such as tree farming.

**FIGURE 6.2-8** Lumber waiting to be loaded aboard ship. (Courtesy Weyerhaeuser.)

tree cutting do not afford this same protection in any significant way.

### 6.2.2 MARKETING AND MEASURE

PS 20 states that lumber should be "tallied board measure," which is defined as using the board foot as the unit of measure, and most lumber is marketed and sold that way. A board foot is the equivalent of a piece 1 in. thick and 1 ft square. In metric measure, 100 board feet are equivalent to 0.236 m³. Some products, such as moldings, which have been machined further, are sold by the lineal foot (running foot), which is equivalent to 0.2048 m. Wood shingles are packaged in bundles and sold according to the surface area they are intended to cover. Posts, poles, and pilings are sold by the piece or the lineal foot (meter).

#### 6.2.2.1 Sizes

Lumber is bought and sold by its *nominal* dimensions. The nominal size of lumber is its rough unfinished size, sold commercially as 1 × 6 (50 × 150), 2 × 4 (50 × 100), 4 by 8 (100 × 200), and so on (Fig. 6.2-9). The dressed size is smaller than the nominal dimensions as a result of seasoning and surfacing. Thus, a 2 × 4 (50 × 100) is actually $1^1/_2$ in. (38.1 mm) × $3^1/_2$ in. (88.9 mm). Members up to 6 in. (150 mm) wide lose $^1/_2$ in. (12.7 mm); wider members lose $^3/_4$ in. (19.1 mm). Wood in buildings constructed before 1973 is larger. A 2 × 4 (50 × 100), for example, will be $1^5/_8$ in. (41.275 mm) by $3^5/_8$ in. (92.08 mm). Even older buildings may have full-sized members. A 2 × 4 (50 × 100) may actually be 2 in. (50.8 mm) by 4 in. (101.6 mm). In worked lumber (Fig. 6.2-10), the face is that portion of the piece that will be exposed to view after it has been installed.

Four approaches are generally used for timber harvesting and management: *clearcutting*, *selective cutting*, *shelterwood cutting*, and *seed-tree cutting*. Clearcutting involves harvesting all the trees in a given area of forest at one time. The lumber industry prefers clearcutting for many species because it is more efficient and results in quicker regeneration. Conservationists, on the other hand, charge that this method is unsightly, causes erosion, and destroys wildlife habitats.

*Clearcutting* is advisable only when there is a great deal of rainfall. Without relatively high rainfall, seedlings will burn up, as they have no relief from the hot sun. Many West Coast forests have the amount of rainfall needed for clearcutting. Clearcutting is also done regardless of rainfall amounts when a disease is discovered in the forest or is moving toward the forest from adjacent forests. The wood is then said to have been *saved* from the disease, which would soon destroy the living trees. Clearcutting is also done when part of a forest has become very ragged from selective cutting. Foresters are now able to grow good trees faster than can nature unassisted by man. As soon as a stand of trees is clearcut, forestry crews start planting new trees on the same areas as soon as possible. Unfortunately, this replanting must be done by hand, which is expensive.

Entire forests are not usually clearcut at the same time. Lumber companies instead lease a limited number of acres of a government-owned forest to cut or cut

a limited number of acres of their own land at a time. Some people call this *patch clearcutting* because it sometimes resembles a checkerboard or patch quilt pattern.

*Selective cutting* involves taking only the older, slower-growing, and defective trees, leaving behind enough trees for natural reseeding, wildlife shelter, and erosion control. However, regardless of the method of cutting and reseeding, seedlings require 60 years or more before they produce harvestable trees. Unfortunately, selective cutting is often done too frequently and is sometimes too extensive for forest self-perpetuation. Therefore, selective-cut forests tend to eventually run out of harvestable wood and trees that would soon produce it.

*Shelter-wood cutting* and *seed-tree cutting* are processes in which some trees are left standing while others are harvested. The difference between the two is that in shelter-wood cutting, 10 to 20 trees are left per acre, while in seed-tree cutting, only about 4 or 5 trees are left per acre. Shelter-wood cutting is used when there is insufficient rainfall for clearcutting, when there are trees that cannot tolerate continuous direct sun, and when the subject tree species has a shallow root system, as does spruce, and therefore cannot withstand significant wind. The trees left standing reseed the cut area, eliminating the need for reseeding by hand. The 10 to 20 trees left per acre in shelter-wood cutting also provide some shade and wind protection for the seedlings. The fewer trees left by seed-

**FIGURE 6.2-9** Nominal, dressed, and face sizes of worked lumber.

DRESSED AND MATCHED (D & M)
Used for subflooring, flooring,
sheathing, or where tight joints are
required between boards

SHIPLAP
Used for subflooring, wall and roof
sheathing or low-cost siding

Vertical siding          Drop siding          Moldings
PATTERNED

**FIGURE 6.2-10** Typical cross sections of worked lumber.

Under the provisions of PS 20, the dressed size of dimension lumber less than 5 in. (125 mm) thick is related to the moisture content of the wood at the time of manufacture. Therefore, lumber fabricated at a moisture content of more than 19% is considered green and must be dressed to slightly larger dimensions (see Fig. 6.2-9) than lumber fabricated at a moisture content of 19% or less, which is considered dry. The objective of this requirement is to ensure that both green and dry lumber achieve approximately the same actual dimensions (and strength) after air drying in service.

Softwood lumber is usually sold in multiples of 2 ft (610 mm) lengths, ranging from 6 ft (1.829 m) to 24 ft (7.32 m), but many mills and lumber yards will precut members to exact dimensions to meet the demands of preassembled component construction and other special customer requirements. Hardwood lumber usually is sold by odd- and even-foot lengths or in random lengths for remanufacture into other products.

#### 6.2.2.2 Board Measure

As stated earlier, *board measure* is used to indicate quantities of lumber. A board foot is the amount of lumber in a piece with nominal dimensions of 1 in. thick, 1 ft wide, and 1 ft long. This is the equivalent of 0.002 36 m³. Therefore, a 1-in. (25-mm)-thick board that is 1 ft (0.3048 m) wide and 10 ft (3.048 m) long has 10 board feet (0.0236 m³) of lumber in it. The actual dimensions of this board are about ³/₄ in. (19.1 mm) by 11¹/₄ in. (285.75 mm) by 10 ft (3.048 m). To calculate the board feet in a piece of lumber, multiply the thickness in inches by the width in feet by the length in feet, us-

ing all nominal dimensions. Since the width of a board is usually expressed in inches, it is necessary to divide by 12 to convert inches to feet (see the example in Fig. 6.2-11). If the nominal thickness is less than 1 in., the number of board feet equals the width in feet multiplied by the length in feet. To convert board feet to cubic meters, multiple by 0.002 359 74.

### 6.2.3 LUMBER QUALITY

As cut from a log, lumber varies widely in quality, because of either natural growth characteristics or the manufacturing process. The product of a log is divided into commercial lumber grades, each having a relatively narrow range of quality. The visual grading of lumber should be done in accordance with ASTM D245 and the provisions of PS 20.

In terms of quality, or number of undesirable growth characteristics, lumber can be classed as common or select. *Common grades* are suitable for general

construction purposes. *Select grades*, as the name implies, are selected to provide a surface suitable for painting or natural finishes. The bulk of yard lumber is graded as common lumber, with only boards (lumber up to 1¹/₂ in. [38.10 mm] thick and 2 in. [50 mm] wide or wider) available in select grades. Lumber over 1¹/₂ in. (38.10 mm) thick also may be graded as select but is used for fabrication into millwork and generally is not stocked as yard lumber. The word *select* in *Select Structural* and *Select Decking* does not identify a select grade. Timbers and dimension lumber grades are classed only as common grades.

Quality-reducing characteristics are faults which detract from the appearance, strength, or general utility of a piece in the use for which it is intended. The definition of such faults is often difficult in grading because a slight or small defect in common lumber is often a medium or large defect in select (finish) lumber. In select lumber the defects are light, small, or serious, depending on the size of the piece or as they come into combination with other defects. Most of the natural and manufacturing characteristics having an effect on the quality of lumber as it is graded are briefly described below.

#### 6.2.3.1 Natural Characteristics

Some characteristics that affect the appearance, grading, or use of wood are knots, shakes, pitch and pitch pockets, decay, molds and stains, and insect damage. Reference should be made to the grading rules covering each species for the specific limitations of these characteristics in a given grade.

ONE BOARD FOOT

L = 1'-0"     W = 12"

T x W x L
1" x 12" x 1' = 1 Board Foot
     (12)
2" x 4" x 12' = 8 Board Feet
     (12)

2×4

12'

2 × 4 = (50 × 100)
1" = 25.4 mm
1'-0" = 304.8 mm
12" = 304.8 mm
25.4 × 304.8 × 304.8 = 2 359 737.2
2 359 737.2 × 0.000 000 001 (mm³/m) = 0.002 36 m³ = 1 board ft.
50.8 × 101.6 × 304.8 × 12 = 1 88 77 897
1 88 77 897 × 000 000 001 = 0.018 88 m³ = 8 board ft.

**FIGURE 6.2-11** Board foot measurement.

## KNOTS

Contrary to the common notion, lumber containing knots is not seriously defective for ordinary use. Knots may be undesirable because of their appearance in select grades, but they are characteristic of all common grades (Fig. 6.2-12). Knots are restricted by size and location in grading because they do reduce the strength value of wood used for structural purposes. Strength reduction is a result of the combined effects of the local cross grain, produced by knots interrupting the direction of the grain, plus checking that may develop in and around knots during drying. Strength reduction is not caused by an inherent inferiority of the material comprising the knots. Therefore, knots in lumber are more accurately defined as strength-reducing characteristics than as defects.

## SHAKES

The effect of shakes (Fig. 6.2-13) on wood is largely in appearance, since they only slightly affect the compressive and tensile strength of members. Shakes do reduce the shear resistance of wood used for beams, however, so limitations are imposed during grading.

## PITCH POCKETS

Whether pitch pockets affect strength depends on their number, size, and location in the piece. Their effect on strength has often been overestimated. Certain restrictions are imposed in the lumber grading rules.

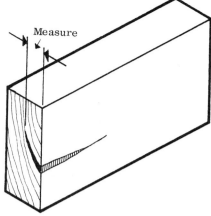

**FIGURE 6.2-13** A shake is a lengthwise separation between, not across, the annual rings.

## DECAY

The presence of decomposed or destroyed wood as a result of decay is restricted by grading rules in structural grades of lumber to specific proportions of the cross section.

## MOLDS AND STAINS

Molds and stains that cannot be removed by surfacing are quality-reducing factors only for appearance grades.

## INSECT DAMAGE

The effect on strength of pinholes, grub holes, and other insect damage is taken into account during grading. The strength of such lumber is acceptable when properly graded. Such insect damage is of primary concern in the appearance grades.

### 6.2.3.2 Manufacturing Characteristics

Some characteristics that affect the appearance, grading, or use of lumber are the result of the manufacturing process. Manufacturing faults may be present due to sawing practice or finishing operations at either the saw or planing mill, or they may be caused during seasoning.

## EFFECTS OF SAWING

Theoretically, lumber is cut from a log in one of two distinct ways (see Fig. 6.1-18): (1) tangential to the annual growth rings, producing lumber called *plain-sawn* in hardwoods and *flat-grained* or *slash-grained* in softwoods, or (2) on a radial to the rings (parallel to the rays), producing lumber called *quarter-sawn* in hardwoods and *edge-grained* or *vertical-grained* in softwoods. Not all lumber can be cut to fit these definitions exactly. In commercial practice, a piece sawn so that the growth rings when viewed from the end of the piece form an angle of 45 degrees or more with the wide face is classified as quarter-sawn, and when the rings form an angle less than 45 degrees it is classified as plain sawn.

For many uses, including structural applications, plain-sawn and quarter-sawn lumber are equally satisfactory, but each has some advantages. When compared with *quarter-sawn* lumber, *plain-sawn* lumber:

- Is less expensive to produce because less labor and waste are involved
- Shows some figure patterns involving growth rings and other figure sources more conspicuously
- Is subject to less shrinking and swelling in thickness
- Has round or oval knots that usually have a lesser weakening effect and affect surface appearance less than the spike knots that are more common in quarter-sawn lumber
- Has shakes and pitch pockets that extend through fewer pieces

Conversely, *quarter-sawn* lumber:

- Is more costly to produce
- Displays some figure patterns peculiar to the radial surface more conspicuously
- Shrinks and swells less in width
- Tends to wear more evenly, because the radial surface is more uniform
- Holds paint better on radial surfaces in some species

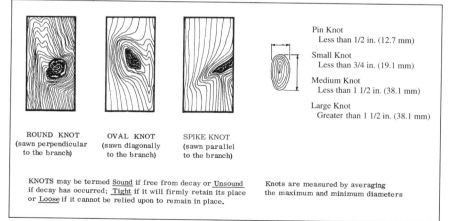

ROUND KNOT (sawn perpendicular to the branch)

OVAL KNOT (sawn diagonally to the branch)

SPIKE KNOT (sawn parallel to the branch)

Pin Knot Less than 1/2 in. (12.7 mm)

Small Knot Less than 3/4 in. (19.1 mm)

Medium Knot Less than 1 1/2 in. (38.1 mm)

Large Knot Greater than 1 1/2 in. (38.1 mm)

KNOTS may be termed Sound if free from decay or Unsound if decay has occurred; Tight if it will firmly retain its place or Loose if it cannot be relied upon to remain in place.

Knots are measured by averaging the maximum and minimum diameters

**FIGURE 6.2-12** Knots are limbs embedded in a tree that have been cut through during lumber manufacture.

- Works smoother for items run to patterns or profiles
- Twists and cups less and has less tendency to develop surface checks and splits in seasoning and in use

*Cross-grain* results when the fibers in a piece of wood are not parallel to the edges. One type of cross-grain, called *diagonal grain*, occurs when a log is sawn at an angle not parallel with the bark. If the center of the log is lined up with the center of the headrig carriage, the first few cuts do not cut the full length of the log because most logs are tapered, and cross-grain may result. Cross-grain may cause warpage, end shrinkage, rough faces, torn or lifted grain, and weak lumber.

Like knots, cross-grain has a depreciating influence on the strength of lumber due to the deviation in grain alignment. But unlike knots, which are highly localized, cross-grain may involve an entire piece. The various grading rules control cross-grain in structural grades by a measurement of the slope of the grain. Figure 6.2-14 shows how this is measured.

### EFFECTS OF SURFACING

Occasionally, imperfections in lumber are created during surfacing or finishing. Depending on their size and location in the piece, the imperfections or blemishes resulting from manufacture defined below are limited in the grading rules so that they will have no more effect on the utility of the piece than will other characteristics permitted in the same grade.

*Torn grain* is a roughened area caused by a machine tearing out bits of the wood during dressing.

*Skip* is an area on a piece that is not surfaced smoothly as it passes through a planer.

*Mismatched lumber* results when adjoining pieces in worked or patterned lumber do not fit tightly or will not align in the same plane.

*Wane* is the presence of bark or the absence of wood on the edge or corner of a piece. It is restricted in grading rules because of such limitations as bearing area, nailing edge, and appearance rather than direct effect on strength (Fig. 6.2-15).

*Machine burn* is the darkening of the wood due to overheating by machine knives or rolls when pieces are stopped in a machine.

### EFFECTS OF SEASONING

The lowering in quality of lumber due to seasoning, which is called *de-grade*, is usually a result of unequal shrinkage, which includes surface and end checking, loosening of knots, and various types of warp. Seasoning de-grade can be largely eliminated by proper practice in either kiln drying or air drying at the point of manufacture, in retail yard storage, and at the construction site. Lumber dried too rapidly will check and split. If it is dried too slowly and the temperature is favorable for fungal growth, stain or decay may develop.

Grading rules specify the amount of seasoning de-grade permitted in the various grades, some of which are described and illustrated below.

*Warp* is a deviation from a true or plane surface and includes bow, crook, twist, cup, or combinations of these (Fig. 6.2-16). Warp is measured at the point of greatest deviation from a straight line.

*Checking* is separation across the growth rings, occurring as a surface check if present on only one surface or as a through check if it extends throughout the entire piece. Checking is measured by depth and length (Fig. 6.2-17). Checks seldom affect strength but do affect appearance, especially in select grades of softwoods where large grain separations are objectionable. In structural lumber, checks have little, if any, effect on tension or compression members. Checks may reduce shear resistance, and limitations in the grading rules are based on this reduction.

*Splits* are lengthwise grain separations extending through the piece from one surface to the other. They are measured as the penetration of the split from the end of the piece and parallel to the edge of the piece (Fig. 6.2-18).

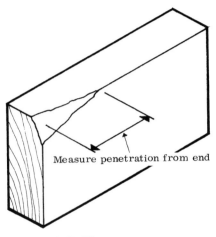

Measure penetration from end

**FIGURE 6.2-15** Wane.

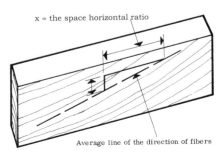

x = the space horizontal ratio

Average line of the direction of fibers

**FIGURE 6.2-14** The slope of the grain affects strength.

Point of Greatest Deflection

CUP

DIAMOND

BOW

Point of Greatest Deflection

CROOK

Rise of Fourth Corner

TWIST

**FIGURE 6.2-16** Warp.

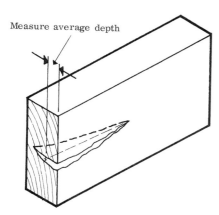

Measure average depth

**FIGURE 6.2-17** Check.

Measure penetration from end

**FIGURE 6.2-18** Split.

## 6.2.4 SOFTWOOD LUMBER

In the early days, sawmilling was entirely a local industry without need for definite standards of size or quality; manufacturers produced lumber to local customer requirements. As the lumber industry and transportation systems expanded, however, markets were no longer local. Lumber of various species originating in widely separated localities produced by different mills could not be sold and used successfully without common standards of quality.

Around 1900, several lumber manufacturers' associations were formed, largely to formulate grading rules for the various species manufactured by the producers in their region. While these grading rules had much in common, they differed in detail. Industry standardization conferences were held, beginning in 1919, and by 1924 countrywide softwood lumber standards had emerged. The consensus document, known as the *Ameri-*

*can Lumber Standard*, was revised several times and stayed in force for many years as *Simplified Practice Recommendation SPR 16–53.* This standard underwent a major revision in 1965, and was adopted and published in 1970 as a National Bureau of Standards voluntary Product Standard PS 20–70, *American Softwood Lumber Standard.* Under this standard, for the first time, dressed lumber sizes were related to moisture content in order to ensure reliability of actual in-service dimensions and strengths. PS 20 was amended again in 1986 and in 1994 and stands today as the industry standard for softwood lumber.

A major advance of PS 20–70 was the establishment of an American Lumber Standards Committee and under it a National Grading Rules Committee. The function of the latter is to establish and maintain the National Grading Rule for *dimension* lumber. Its members include representatives from all the major grading associations and model code groups, and from many other organizations representative of the industry and designers and consumers.

*Dimension* lumber, as defined in PS 20, is "lumber from 2 in. (50 mm) to, but not including 5 in. (125 mm) in nominal thickness, and 2 in. (50 mm) or more in nominal width. Dimension may be classified as framing, joists, planks, rafters, studs, small timbers, etc."

Under the National Grading Rule, all grading authorities apply the same grade names and descriptions to dimension lumber of similar quality, regardless of species. However, the Rule does not restrict grading authorities from including additional grades of special products in their own grading rules. For example, the Southern Pine Inspection Bureau grades, and its members market, many grades not covered by the National Grading Rule. In addition, the National Grading Rule does not cover decking lumber, so grade names for it are not the same for all associations. Nevertheless, PS 20 and the National Grading Rule considerably simplify grade recognition and the assignment of working stresses for softwood lumber.

### 6.2.4.1 Major Producing Regions

The majority of softwood used in construction is produced in three major regions. The lumber species, approximate yearly volume of production, association

that represents producers, and the grading authorities that establish and supervise the lumber grading practices for the different species are shown for each region in Figure 6.2-19.

### WESTERN WOODS REGION

The western woods region is the largest of the producing areas. It includes Oregon, Washington, California, Idaho, Arizona, New Mexico, Texas, Colorado, Montana, South Dakota, Nevada, Utah, and Wyoming. This region produces lumber cut from ponderosa pine, Douglas fir, Engelmann spruce, Idaho white pine, incense cedar, lodgepole pine, sugar pine, Sitka spruce, western hemlock, western larch, western red cedar, and white fir.

### REDWOOD REGION

The redwood region is the smallest in area, encompassing but 12 northern California counties. About half of the total production there consists of lumber cut from redwood; the remainder comes from other softwood species.

### SOUTHERN PINE REGION

The southern pine region covers the entire Southeast, from the eastern part of Texas to Virginia, with some production in Maryland and Missouri. This region produces lumber cut from shortleaf, loblolly, longleaf, slash, Virginia, and pond pine.

### 6.2.4.2 Standards of Quality

PS 20 does not contain detailed requirements for selecting or specifying lumber. It does, however, establish the basic principles, provisions, and procedures for developing commercial grading rules for various species.

The commercial grading rules for each region not only conform to PS 20 and the National Grading Rule, but also reflect the properties of individual species and recognize the specific qualities of lumber desired by users. The grading rules are detailed requirements used to facilitate the manufacture, specifying, sale, and use of lumber. Their basic aim is to separate into marketable and usable classifications the wide range of lumber qualities obtained from a log. Grading is based mainly upon quality-reducing characteristics that affect the appearance, strength, durability, or utility of wood.

Commercial grading rules are formulated and maintained by grading authorities, each of which covers the species

FIGURE 6.2-19 Major Softwood-Producing Regions of the United States

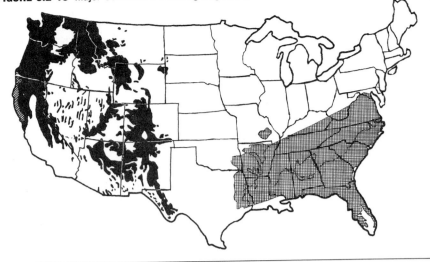

| Key | Region | Species | Approximate Yearly Production | Lumber Association | Grading Authority |
|---|---|---|---|---|---|
| ■ | Western Wood Region | Douglas fir<br>Ponderosa pine<br>Western red cedar<br>Incense cedar<br>Western hemlock<br>White fir<br>Engelmann spruce<br>Sitka spruce<br>Western larch<br>Lodgepole pine<br>Idaho white pine<br>Sugar pine | 16.9 billion board feet | Western Wood Products Association | Western Wood Products Association<br>West Coast Lumber Inspection Bureau |
| ▓ | Redwood | Redwood<br>Douglas fir | 2.7 billion board feet | California Redwood Association | Redwood Inspection Service |
| ▦ | Southern Pine Region | Longleaf pine<br>Shortleaf pine | 6.7 billion board feet | Southern Forest Products Association | Southern Pine Inspection Bureau |

and individual products manufactured in a given producing region. These authorities publish grading rules for the species produced in their regions and maintain staffs of qualified lumber inspectors who supervise grading at the mills.

Mills that manufacture lumber according to PS 20 usually stamp each piece of dimension lumber with an appropriate grademark. Examples of official grademarks of the four chief grading authorities are shown in Figure 6.2-20. American grading authorities are approved by the Board of Review of the American Lumber Standards Committee operating under procedures established by PS 20. The Board of Review reviews the adequacy, competency, and reliability of U.S. agencies it has approved. The Board of Review has approved agencies in Canada on the same basis. Therefore, Canadian lumber that has been graded and stamped by an authorized agency is compatible with PS 20 and U.S. standard engineering practice.

### 6.2.4.3 Classifications

Softwood lumber is classified by use, manufacture, size, and grade.

#### USE CLASSIFICATIONS

The use classifications of softwood lumber are yard lumber, structural lumber, and factory and shop lumber.

*Yard lumber* consists of grades, sizes, and patterns generally intended for ordinary construction and general building purposes.

*Structural lumber* is lumber that is 2 in. (50 mm) or more in nominal thickness, and is intended for use where working stresses are required. Structural lumber is sometimes called *stress grade* lumber. Working stresses (allowable unit stresses) for commercially available species and grades are published by the AF&PA's *NDS*. *NDS* is the industry-recognized standard of proper engineering practice for structural wood design.

*Factory and shop lumber* is intended primarily for further manufacture into windows, doors, millwork, and similar items. Therefore, the grade of a piece is based on the amount of usable clear wood that can be cut from it.

#### MANUFACTURING CLASSIFICATIONS

The extent to which lumber is processed is covered in this classification, which includes *rough* lumber, *dressed* (surfaced) lumber, and *worked* lumber.

*Rough lumber* is lumber that has not been dressed (surfaced) but is sawed, edged, and trimmed at least to the extent of showing saw marks in the wood on four longitudinal surfaces of each piece for its overall length.

*Dressed (surfaced) lumber* is dressed by a surfacing machine to attain a smooth surface and uniform size. It is designated as follows:

S1S—surfaced on one side
S2S—surfaced on two sides
S1E—surfaced on one edge
S2E—surfaced on two edges
S4S—surfaced on both sides and both edges

These designations are linked to describe pieces finished in other ways. For example, a piece finished on one side and two edges would be designated as S1S2E.

*Worked lumber*, in addition to being dressed, is matched, shiplapped, or patterned. *Matched lumber* has been worked with a tongue on one edge and a groove on the opposite edge to provide a close tongue-and-groove joint by fitting two pieces together. For end matching, the tongue and groove are worked in the ends (see Fig. 6.2-10).

*Shiplapped lumber* has been rabbeted on both edges of each piece to provide a close-lapped joint by fitting two pieces together (see Fig. 6.2-10).

**FIGURE 6.2-20** Typical Grademarks for Dimension Lumber

| | |
|---|---|
| MILL 10 WCLB CONST D. FIR S-DRY | **West Coast Lumber Inspection Bureau** Construction-grade light framing Douglas fir maximun 19% m.c. Mill No. 10 |
| 12 SEL DECK WWP MC 15 INC CDR | **Western Wood Products Association** Select-grade decking Incense cedar maximum 15% m.c. Mill No. 12 |
| SPIB No. 1 KD19 (7) | **Southern Pine Inspection Bureau** No. 1-grade light framing Kiln-dried less than 19% m.c. Mill No. 7 |
| (50) STUD S GRN REDWOOD RIS. | **Redwood Inspection Service** Stud-grade light framing Redwood over 19% m.c. Mill No. 50 |

(Courtesy West Coast Lumber Inspection Bureau, Western Wood Products Association, Southern Pine Inspection Bureau, and Redwood Inspection Service, a Division of the California Redwood Association.)

*Patterned lumber* is shaped to a pattern or to a molded form in addition to being dressed, matched, or shiplapped and includes siding and other millwork.

## SIZE CLASSIFICATION

Softwood lumber is subdivided into three major categories by size: nominal size, rough size, and dressed size. Refer to Section 6.2.2.1 for additional discussion.

*Nominal Size* The nominal size classification is further subdivided into three categories: boards, dimension lumber, and timbers. Sizes here are given in nominal dimensions.

*Boards* are pieces less than 2 in. (50 mm) thick and 2 in. (50 mm) or more wide. They include sheathing, subflooring, roofing, finish, trim, siding, and paneling. Boards less than 6 in. (150 mm) wide may be classified as strips.

*Dimension lumber*, as defined earlier, is used principally for the structural elements of wood-framed buildings. It makes up most common lumber.

*Timbers* are lumber pieces 5 in. (125 mm) or more in the smallest dimension. These may be further classified as beams, stringers, posts, caps, sills, girders, purlins, and other members. Lumber thicker than 5 in. (125 mm) in either face is seldom used in ordinary construction and, when stocked by a retail yard, is fre-

quently in a green, rough (undressed) condition. When heavy timbers are used, shrinkage of the assembly should be expected, and the design of the structure should be planned to minimize the effects of this shrinkage. Laminated timber, made by gluing up smaller pieces that can readily be kiln dried, should be used when seasoned timbers are mandatory.

Most dimension and timber lumber is stress graded, with working stresses assigned to each grade. Such lumber, intended for load-bearing conditions, also is called *structural*.

## GRADE CLASSIFICATIONS

Another method of classifying softwood lumber is into *Select*, *Finish*, or *Common* grades. *Select* and *Finish* grades have good appearance and are used where a natural or high-quality paint finish is required. They are used for the highest quality of interior and exterior finish, trim, moldings, paneling, flooring, and siding. In most species the two highest grades, A and B, are combined as *B and Better* (B&B or B&Btr). Grades C&Btr, C Select, and D Select are more economical and serve many finishing purposes where cost savings are more important than a perfect appearance. *Finish* grades include *Superior*, *Prime*, and *E* (Fig. 6.2-21). *Common* grades are suitable for general construction and utility purposes.

The various grades of softwood framing lumber are grouped according to size and intended use into several categories as follows:

*Light framing* is lumber 2 in. (50 mm) to 4 in. (100 mm) thick and 2 in. (50 mm) to 4 in. (100 mm) wide and consists of several categories of products. Ordinary light framing grades (Construction, Standard, and Utility) are intended for uses not requiring high strength, such as general framing, plates, and blocking.

*Structural light framing* grades (Select Structural, No. 1, No. 2, and No. 3) are designed to fit engineering applications where higher bending strength ratios are needed, such as in trusses and box beams.

*Structural joists and planks* are lumber of rectangular cross section 2 in. (50 mm) to 4 in. (100 mm) thick and 5 in. (125 mm) wide or wider, graded with respect to strength in bending when loaded either on the narrow face of a joist or the wide face of a plank. The grades are the same as in Structural Light Framing: Select Structural, No. 1, No. 2, and No. 3.

*Studs* grade members are suitable for stud use in both non-load-bearing and load-bearing walls. Studs (2 in. [50 mm] to 4 in. [100 mm] thick, 2 in. [50 mm] to 4 in. [100 mm] wide) are made in lengths of 10 ft (3048 mm) and shorter primarily for use in walls. Their strength in bending is comparable to that of No. 3 Structural Light Framing lumber.

*Decking* is lumber 2 in. (50 mm) to 4 in. (100 mm) thick, 4 in. (100 mm) wide and wider intended for exposed view as planking in ceilings and roofs. The moisture content usually is limited to 15%,

**FIGURE 6.2-21** Grademark stamps for finish and select grades. (Courtesy Western Wood Products Association.)

and the lumber is worked to a tongue-and-groove pattern. It may be ordered with V-shaped or rounded edges and with striated or grooved patterns on the face. Typical grades are Select and Commercial.

*Timbers* are classified as either stress rated or non-stress rated. The grades of non-stress-rated timbers are Square Edge and Sound (SE&S), No.1, No. 2, and No. 3. Stress-rated grades are Select Structural SR, No. 1SR, and No. 2SR. Timbers that are nearly square (that is, their width does not exceed their thickness by more than 2 in. [50 mm]) are generally intended for use as posts and have slightly lower allowable bending strengths than rectangular timbers, which are intended for use mainly as beams.

Most lumber is graded visually by trained inspectors. In stress-grade lumber, limitations are imposed on slope-of-grain and the size, number, and placement of knots. Some structural grades in some species (for example, Douglas fir and southern pine) have additional grade requirements based on density and growth-ring count. Although most dimension lumber is graded visually, some lumber 2 in. (50 mm) or less in thickness is now being mechanically stress rated at the sawmill, allowing manufacturers to standardize the load-bearing capacity of each piece. This more refined method of selection permits assignment of higher stresses for much of the lumber in a grade which had been subjected to visual inspection.

In mechanical rating, lumber is fed through a machine that subjects each piece to bending and measures its modulus of elasticity, *E*. The machine then electronically computes the stress grade, taking into account the interacting effect of knots, slope-of-grain, growth rate, density, and moisture content. As the piece leaves the machine it is automatically stamped MACHINE RATED, with an assigned fiber stress in bending ($F_b$ on the grademark) and a corresponding modulus of elasticity ($E$) rating (Fig. 6.2-22).

The close relationship between $E$ and $F$ has led to the establishment of categories of stress grades for lumber 2 in. (50 mm) or less in thickness (Fig. 6.2-23) by the Southern Pine Inspection Bureau (see also Fig. 6.1-15). Machine stress-rated lumber is also visually graded to limit certain characteristics (checks, splits, wane, warp).

### 6.2.4.4 Size and Grade Standardization

Historically, lumber grade names and grade requirements have been developed independently on a regional basis and, even within regions, have varied among some species. Methods of establishing stress values for stress-grade lumber also have differed among lumber-producing regions, and changes in structural grading are continuing as the newer methods of machine rating are refined.

Under PS 20, a National Grading Rule Committee (NGRC) was established for the purpose of developing uniform countrywide grade requirements for dimension lumber. The functions of the NGRC, according to PS 20, were to "establish, maintain, and make fully and fairly avail-

**FIGURE 6.2-22** Machine stress-rated (MSR) lumber is automatically stamped with an identifying grademark. (Courtesy Western Wood Products Association.)

**FIGURE 6.2-23** Machine Stress Rated Lumber, Allowable Design Values (in psi)

| Extreme Fiber Stress in Bending, $F_b$[a] | f-E Classification | Modulus of Elasticity, E | Tension Parallel to Grain, $F_t$ | Compression Parallel to Grain, $F_{c\perp}$ |
|---|---|---|---|---|
| 1200 | 1200f-1.2E | 1.2 | 600 | |
| 1350 | 1350f-1.3E | 1.3 | 750 | 1600 |
| 1400 | 1400f-1.2E | 1.2 | 800 | 1600 |
| 1500 | 1500f-1.3E | 1.3 | 900 | 1650 |
| | 1500f-1.4E | 1.4 | | |
| 1600 | 1600f-1.4E | 1.4 | 950 | 1675 |
| 1650 | 1650f-1.4E | 1.4 | 1020 | 1700 |
| | 1650f-1.5E | 1.5 | | |
| 1800[b] | 1800f-1.5E | 1.5 | 1300 | 1750 |
| 1800 | 1800f-1.6E | 1.6 | 1175 | 1750 |
| 1950 | 1950f-1.5E | 1.5 | 1375 | 1800 |
| | 1950f-1.7E | 1.7 | | |
| 2000 | 2000f-1.6E | 1.6 | 1300 | 1825 |
| 2100 | 2100f-1.8E | 1.8 | 1575 | 1875 |
| 2250 | 2250f-1.6E | 1.6 | 1750 | 1925 |
| | 2250f-1.7E | 1.7 | | |
| | 2250f-1.9E | 1.9 | | |
| 2400 | 2400f-1.7E | 1.7 | 1925 | 1975 |
| | 2400f-1.8E | 1.8 | | |
| | 2400f-2.0E | 2.0 | | |
| 2550 | 2550f-2.1E | 2.1 | 2050 | 2025 |
| 2700 | 2700f-2.2E | 2.2 | 2150 | 2100 |
| 2850 | 2850f-2.3E | 2.3 | 2300 | 2150 |
| 3000 | 3000f-2.4E | 2.4 | 2400 | 2200 |
| 3150 | 3150f-2.5E | 2.5 | 2500 | 2250 |
| 3300 | 3300f-2.6E | 2.6 | 2650 | 2325 |

Horizontal shear $F_v$ – 90 psi
Compression perpendicular to grain $F_{c\perp}$ – 565 psi

[a]The tabulated Extreme Fiber in Bending values, $F_b$, are applicable to lumber loaded on edge. When loaded flatwise, these values may be increased by multiplying by the following factors:

| Nominal Width (in.) | 4 | 6 | 8 | 10 | 12 | 14 |
|---|---|---|---|---|---|---|
| Factor | 1.1 | 1.15 | 1.15 | 1.2 | 1.2 | 1.2 |

[b]Grade requires $F_t$ certification and quality control.

able grade strength ratios, nomenclature, and descriptions of grades for dimension lumber conforming to American Lumber Standard." The requirements developed by the NGRC are known as the National Grading Rule for dimension lumber and constitute a part of all grading authority grading rules.

Individual grading rules are much more extensive and detailed than the National Grading Rule. Bending strength ratios are standard relationships between the basic stresses of an ideal, clearwood, piece of lumber free of strength-reducing characteristics (knots, sloping grain, splits) and the stresses developed in a particular grade of lumber in that species.

Since lumber shrinks in direct proportion to moisture loss in drying from a green condition of about 30% moisture content to the moisture content it ultimately attains in service, it is possible to establish, with reasonable tolerances, separate sizes for green and dry surfaced lumber so that both types achieve essentially the same size in service and provide equal strength and stiffness for the same spans and loading conditions.

The *NDS* takes into account the slight oversizing of green lumber and allows the same values for green and dry surfaced lumber when both are used at a moisture content of 19% or less. Lower values are assigned to lumber which will be used under exposed conditions where the moisture content is likely to exceed 19% for extended periods.

## 6.2.5 HARDWOOD

Hardwood accounts for about 25% of the total board foot production of lumber. It comes from many species and has uses ranging from small specialty items to the heaviest structural work. The predominant hardwood species are the oaks and walnut, which constitute more than 50% of total hardwood lumber production.

### 6.2.5.1 Areas of Growth

Most of the commercial hardwood species in the United States grow east of the Great Plains (see Fig. 6.1-1). The hardwoods of the West, which grow principally in California, Oregon, and Washington, amount to about 2% of the total hardwood produced.

### 6.2.5.2 Standards of Quality

The National Hardwood Lumber Association (NHLA) publishes size and grading rules for sawn hardwood lumber. These rules, however, are used within the lumber industry to describe hardwood lumber used to manufacture other products and are not normally used to specify hardwood used in building construction. In the construction industry, hardwood lumber is made to special order for the specific requirements of the purchaser, and the grade descriptions and characteristics are those which are agreed upon by both buyer and seller.

The grading rules set down by the Architectural Woodwork Institute in their *Architectural Woodwork Quality Standards, Guide Specifications, and Quality Certification Program* can be used to control hardwood lumber for cabinets, paneling, shelving, miscellaneous ornamental items, stairs and handrails, millwork, doors, windows, and casework.

### 6.2.5.3 Hardwood Dimension

A large proportion of the hardwood lumber produced at sawmills is used in paneling, furniture, flooring, and other finished or semifinished products. *Dimension stock*, or *hardwood dimension*, produced by hardwood sawmills is shipped to furniture, tool, piano, cabinet, and similar factories.

Hardwood *dimension* lumber is usually kiln dried and processed to the point where it will deliver maximum utility to the user. It is manufactured from rough boards and flitches to the specific requirements of a particular plant or industry in specified thicknesses, widths, and lengths. It may be solid or glued up and is classified as rough dimension, surfaced dimension, semifabricated dimension, or completely fabricated dimension.

*Rough hardwood dimension* consists of planks sawn and ripped to certain sizes.

*Surfaced* and *semifabricated* are rough dimension carried one or more steps further, which may include one or more of several operations such as surfacing, molding, tenoning, drum-sanding, equalizing, trimming, or mitering. However, such operations do not make the product a completely fabricated one ready for assembly.

*Completely fabricated* is lumber ready for assembly into finished products.

Hardwood parts are often fabricated in hardwood dimension plants adjoining the sawmill site, and many hardwood flooring plants are operated in conjunction with hardwood sawmills.

The manufacture of hardwood lumber and its by-products is one of the more complex branches of the lumber industry because of the variety of uses for hardwood and the many species involved.

# 6.3 Engineered Wood Products

Engineered wood products include veneered products such as softwood and hardwood plywood, nonveneered products such as orientated strand board, and wood or cane fiber panel products.

## 6.3.1 PLYWOOD

Plywood is an engineered panel composed either of several wood veneer sheets placed at right angles to one another or veneer sheets placed across a lumber, mineral, or oriented strand board core. These components are then bonded together under high pressure using either a waterproof or water-resistant adhesive. Placing thin veneer sheets in this manner tends to balance the positive and negative aspects of wood's strength properties, improves its dimensional stability, and increases its resistance to both checking and splitting.

Both decorative and construction grades of plywood are manufactured. Construction plywood is used wherever strength is an important factor but occasionally also for paneling; decorative plywood (chiefly hardwood) is used for wall paneling, cabinet work, and furniture.

### 6.3.1.1 Construction Plywood

Construction plywood is manufactured primarily of softwoods, but hardwood species are also employed. However, a plywood panel need not be composed entirely of the same species. Strength and stiffness requirements for inner plies are less demanding, and these frequently are of a different species than the outer plies. Structural strength or stiffness is not significantly changed, because the outer faces of a plywood panel provide most of its strength and stiffness lengthwise, which is the normal direction to apply a plywood panel when it is used structurally.

## MANUFACTURE

Manufacturing begins by debarking a tree, cutting it into peeler logs, and cutting these into shorter log-shaped *blocks*. These blocks are placed in a giant lathe and rotated against a long knife that peels the wood into long, continuous veneer sheets, mainly $^1/_{10}$ in. (2.54 mm) to $^3/_{16}$ in. (4.76 mm) thick (Fig. 6.3-1). Veneer cut in this manner is called *rotary cut*. Veneer primarily intended for decorative plywood is sometimes *sliced* rather than peeled from the block (Fig. 6.3-2), but an insignificant amount of softwood plywood is manufactured of sliced veneer.

The veneer is next conveyed to clippers, cut into desired widths, then run through long drying ovens that reduce its moisture content to between 2% and 6% to provide the best possible glue bond and panel stability (Fig. 6.3-3).

The 4 ft (1200 mm) × 8 ft (2400 mm) veneers are then graded and placed on a long 4-ft (1200-mm)-wide conveyer, where they pass under a device that drops glue in lines or sprays glue downward onto the veneers as they move along. A parallel conveyer above this lower system delivers other veneers to the top of the glued veneers. The interior veneers, which have their grain running perpendicular to the face veneers, are not 4 ft (1200 mm) × 8 ft (2400 mm), as are the face veneers. Because they are made from poorer-quality veneers and because of their grain orientation, they are 4 ft (1200 mm) measured parallel to their grain and some dimension less than 4 ft (1200 mm) in the other direction, which is the direction of their movement. Consequently, it takes several of these smaller veneers side by side to match the length of the 8 ft (2400 mm) veneers moving down the conveyer. Workers standing beside the conveyer quickly position the inner veneers dropping on the line from the upper conveyer into a 4 ft (1200 mm) × 8 ft (2400 mm) format. Once these inner veneers have been positioned, they pass under a glue dropping machine. The last veneer to drop from the upper conveyer onto the stack is another 4 ft (1200 mm) × 8 ft (2400 mm) face veneer.

These laid-up panels are stacked in contact with each other and inserted into

**FIGURE 6.3-1** At a lath, a sharp steel blade peels a block into thin layers. (Courtesy APA—The Engineered Wood Association, formerly American Plywood Association.)

**FIGURE 6.3-2** A flitch is held in clamps and moved against a knife.

a hydraulic press in this stack, where pressure of about 150 psi (105.46 Mg/m²) is exerted to compress the veneer and glue together to make plywood panels. The panels are then separated, and each is placed into a separate shelf of a final pressing machine (Fig. 6.3-4). The panels are not in contact with each other. The machine applies heat at about 300°F (148.9°C) to each panel and presses up from the bottom, closing the shelves against each other. This step takes 10 to

**FIGURE 6.3-3** A dryer oven reduces the moisture content of veneer. (Courtesy APA—The Engineered Wood Association, formerly American Plywood Association.)

**FIGURE 6.3-4** Laid-up panels are placed in a press that presses veneers into plywood panels. (Courtesy APA—The Engineered Wood Association, formerly American Plywood Association.)

15 seconds. The bottom of the machine then lowers and tilts the shelves. The panels slide out, one at a time, onto an elevated moving table that collects the panels with no contact by human hands. After removal from the press, these panels are trimmed to exact size, run through large drum sanders if they are of a sanded grade, and, finally, checked for grade and appearance, emerging as finished plywood ready for use.

## STANDARDS OF QUALITY

Construction plywood, except for a few specialty panels, should conform with the requirements of the latest version of U.S. Product Standard PS 1, "Construction and Industrial Plywood." PS 1 contains the recognized grading rule for construction plywood and the accepted bond, panel construction, dimensional tolerance, and workmanship requirements for maintaining a high level of quality. It also includes procedures for testing, grading, labeling, and identifying plywood. Conformance with the requirements of PS 1 is indicated by the grade trademark or other suitable designation of a qualified testing and certifying agency performing the functions outlined in the Product Standard.

About 85% of all construction plywood is produced by mills whose owners belong to the American Plywood Association (APA). This organization maintains a quality control program to ensure that plywood carrying an official grade trademark conforms with PS 1. Typical APA grade-trademarks are shown in Figure 6.3-5.

Figure 6.3-6, which is excerpted from a more extensive APA table, can be used to help select plywood for general use. The grade of plywood needed can be determined by examining the data in the "Description and Use" column. It is then necessary to select either a thickness or a span rating and, for appearance grades, the desired species group. When engineering design is needed, such as for stress skin panels or unusual loading conditions, a designer should refer to the more extensive design data produced by APA.

Plywood is graded and marked according to species, exposure durability, veneer grade, panel grade, and application.

## WOOD SPECIES

PS 1 classifies about 70 species of wood for use in construction and industrial ply-

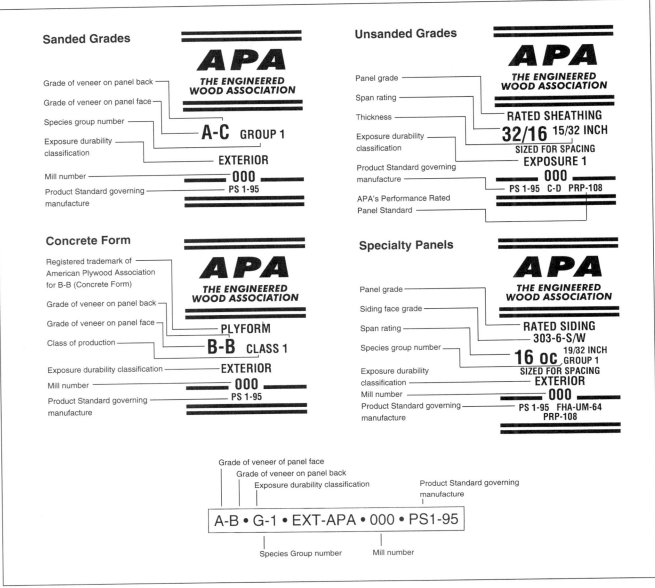

**FIGURE 6.3-5** Typical grade trademarks of APA—The Engineered Wood Association. (Courtesy APA—The Engineered Wood Association.) (Refer to Figure 25.7-1 for metric conversions.)

wood. These are divided into five groups according to their strength, stiffness, and other properties (Fig. 6.3-7), with Group 1 being the strongest and stiffest. The group number shown in the grade stamp (Fig. 6.3-5) refers to the species used in the face and back veneer plies.

## EXPOSURE DURABILITY

Two basic types of construction plywood are produced to comply with the requirements of PS 1. These are *exterior type* and *interior type*, which refer to the exposure durability of the panel. Within each type there are several classifications

based on the types of veneers and adhesives used.

***Exterior Type (Waterproof)*** The veneers used in exterior panels must be C grade or better. Exterior plywood panels are designed to retain their glue bond under one of the three levels of exposure designated by the exposure durability classifications:

■ *Exterior* panels maintain their integrity when repeatedly wetted and dried or otherwise subjected in a permanent exterior exposure to weather or moisture.

■ *Exposure 1* panels may exposed for a long time during construction but finally protected. They have the same adhesives used in exterior type panels but are not designed for permanent exposure.

■ *Exposure 2* panels are used where exposure time will be limited before the panel is protected.

***Interior Type (Moisture-Resistant)*** Interior type plywood is made with D grade veneer or better. There are three levels of adhesive durability within the interior type classification.

**FIGURE 6.3-6** Guide to Use of Plywood Panels

## Interior or Protected Applications

| Plywood Grade | Description and Use | Typical Trademarks | Veneer Grade | | | Common Thicknesses |
|---|---|---|---|---|---|---|
| | | | Face | Back | Inner | |
| APA-Rated Sheathing Exp 1 or 2[a] | Unsanded sheathing grade for wall, roof, subflooring, and industrial applications such as pallets and for engineering design, with proper stresses. Manufactured with intermediate and exterior glue.[b] For permanent exposure to weather or moisture only. Exterior type plywood is suitable. | *APA* THE ENGINEERED WOOD ASSOCIATION RATED SHEATHING 32/16 15/32 INCH SIZED FOR SPACING EXPOSURE 1 000 PS 1-95 C-D PRP-108 | C | D | D | $5/16$, $3/8$, $15/32$, $1/2$, $19/32$, $5/8$, $23/32$, $3/4$ |
| APA Structural I Rated Sheathing Exp 1[a] or APA Structural II[c] Rated Sheathing Exp 1 | Plywood grades to use where strength properties are of maximum importance, such as plywood-lumber components. Made with exterior glue only. STRUCTURAL I is made from all Group 1 woods. STRUCTURAL II allows Group 3 woods. | *APA* THE ENGINEERED WOOD ASSOCIATION RATED SHEATHING STRUCTURAL I 24/0 3/8 INCH SIZED FOR SPACING EXPOSURE 1 000 PS 1-95 C-D PRP-108 | C | D | D | $5/16$, $3/8$, $15/32$, $1/2$, $19/32$, $5/8$, $23/32$, $3/4$ |
| APA Rated Sturd-I-Floor Exp 1 or 2[a] | For combination subfloor-underlayment. Provides smooth surface for application of carpet. Possesses high concentrated and impact load resistance during construction and occupancy. Manufactured with intermediate and exterior glue. Touch-sanded.[d] Available with tongue and groove.[e] | *APA* THE ENGINEERED WOOD ASSOCIATION RATED STURD-I-FLOOR 20 OC 19/32 INCH SIZED FOR SPACING T&G NET WIDTH 47-1/2 EXPOSURE 1 000 PS 1-95 UNDERLAYMENT PRP-108 | C plugged | D | C and D | $19/32$, $5/8$, $23/32$, $3/4$, $1\frac{1}{8}$ (2-4-1) |
| APA Underlayment Exp 1, 2, or INT | For underlayment under carpet. Available with exterior glue. Touch-sanded. Available with tongue and groove.[e] | *APA* THE ENGINEERED WOOD ASSOCIATION UNDERLAYMENT GROUP 1 EXPOSURE 1 000 PS 1-95 | C plugged | D | C and D | $1/2$, $19/32$, $5/8$, $23/32$, $3/4$ |
| APA C-D Plugged Exp 1, 2, or INT | For built-ins, wall and ceiling tile backing, NOT for underlayment. Available with exterior glue. Touch-sanded.[e] | *APA* THE ENGINEERED WOOD ASSOCIATION C-D PLUGGED GROUP 2 EXPOSURE 1 000 PS 1-95 | C plugged | D | D | $1/2$, $19/32$, $5/8$, $23/32$, $3/4$ |
| APA Appearance Grades Exp 1, 2, or INT | Generally applied where a high-quality surface is required, Includes APA N-N, N-A, N-B, N-D, A-A, A-B, A-D, B-B, and B-D INT grades.[e] | *APA* THE ENGINEERED WOOD ASSOCIATION A-D GROUP 1 EXPOSURE 1 000 PS 1-95 | B or better | D or better | C and D | $1/4$, $11/32$, $3/8$, $15/32$, $1/2$, $19/32$, $5/8$, $23/32$, $3/4$ |

*(Continues)*

FIGURE 6.3-6 (Continued)

## Exterior Applications

| Plywood Grade | Description and Use | Typical Trademarks | Veneer Grade | | | Common Thicknesses |
|---|---|---|---|---|---|---|
| | | | Face | Back | Inner | |
| APA-Rated Sheathing Ext[a] | Unsanded sheathing grade with waterproof glue bond for wall, roof, subfloor, and industrial applications such as pallet bins. | *APA* THE ENGINEERED WOOD ASSOCIATION RATED SHEATHING 48/24 23/32 INCH SIZED FOR SPACING EXTERIOR 000 PS 1-95 C-C PRP-108 | C | C | C | $5/16$, $3/8$, $15/32$, $1/2$, $19/32$, $5/8$, $23/32$, $3/4$ |
| APA Structural I Rated Sheathing Ext[a] or APA Structural II[c] Rated Sheathing Ext | "Structural" is a modifier for this unsanded sheathing grade. For engineered applications in construction and industry where full exterior-type panels are required. STRUCTURAL I is made from Group 1 woods only. | *APA* THE ENGINEERED WOOD ASSOCIATION RATED SHEATHING STRUCTURAL I 24/0 3/8 INCH SIZED FOR SPACING EXTERIOR 000 PS 1-95 C-C DDD-108 | C | C | C | $5/16$, $3/8$, $15/32$, $1/2$, $19/32$, $5/8$, $23/32$, $3/4$ |
| APA Rated Sturd-I-Floor Ext[a] | For combination subfloor-underlayment where severe moisture conditions may be present, as in balcony decks. Possesses high concentrated and impact load resistance during construction and occupancy. Touch-sanded.[d] Available with tongue and groove.[e] | *APA* THE ENGINEERED WOOD ASSOCIATION RATED STURD-I-FLOOR 20 oc 19/32 INCH SIZED FOR SPACING EXTERIOR 000 PS 1-95 C-C PLUGGED PRP-108 | C plugged | C | C | $19/32$, $5/8$, $23/32$, $3/4$ |
| APA Underlayment Ext and APA C-C Plugged Ext | Underlayment for floor where severe moisture conditions may exist. Also for controlled atmosphere rooms and many industrial applications. Touch-sanded. Available with tongue and groove.[e] | *APA* THE ENGINEERED WOOD ASSOCIATION C-C PLUGGED GROUP 2 EXTERIOR 000 PS 1-95 | C plugged | C | C | $1/2$, $19/32$, $5/8$, $23/32$, $3/4$ |
| APA B-B Plyform Class I or II[c] | Concrete-form grade with high reuse factor. Sanded both sides, mill-oiled unless otherwise specified. Available in HDO. For refined design information on this special-use panel see *APA Design/Construction Guide: Concrete Forming*, Form No. V345. Design using values from this specification will result in a conservative design.[e] | *APA* THE ENGINEERED WOOD ASSOCIATION PLYFORM B-B CLASS 1 EXTERIOR 000 PS 1-95 | B | B | C | $19/32$, $5/8$, $23/32$, $3/4$ |
| APA Marine Ext | Superior Exterior-type plywood made only with Douglas fir or western larch. Special solid-core construction. Available with MDO or HDO face. Ideal for boat hull construction. | MARINE • A-A • EXT- APA • 000 • PS 1-95 | A or B | A or B | B | $1/4$, $3/8$, $1/2$, $5/8$, $3/4$ |
| APA Appearance Grades Ext | Generally applied where a high-quality surface is required. Includes APA A-A, A-B, A-C, B-B, B-C, HDO, and MDO Ext.[e] | *APA* THE ENGINEERED WOOD ASSOCIATION A-C GROUP 1 EXTERIOR 000 PS 1-95 | B or better | C or better | C | $1/4$, $11/32$, $3/8$, $15/32$, $1/2$, $19/32$, $5/8$, $23/32$, $3/4$ |

[a]Properties and stresses apply only to APA-rated Sturd-I-Floor and APA-rated Sheathing manufactured entirely with veneers.

[b]When exterior glue is specified, i.e., Exposure 1, stress level 2 (S-2) should be used.

[c]Check local suppliers for availability before specifying Structural II and Plyform Class II grades, as they are rarely manufactured.

[d]APA rated Sturd-I-Floor 2-4-1 may be produced unsanded.

[e]May be available as Structural I.

(Courtesy APA—The Engineered Wood Association.) (Refer to Figure 25.7-1 for metric conversions.)

**FIGURE 6.3-7** Classification of Species

| Group 1 | Group 2 | | Group 3 | Group 4 | Group 5 |
|---|---|---|---|---|---|
| Apitong[a,b] | Cedar, Port Orford | Mengkulang[a] | Alder, red | Aspen | Basswood |
| Beech, American | Cypress | Meranti, red[a,d] | Birch, paper | Bigtooth | Poplar, balsam |
| Birch | Douglas fir 2[c] | Mersawa[a] | Cedar, Alaska | Quaking | |
| Sweet | Fir | Pine | Fir, subalpine | Cativo | |
| Yellow | Balsam | Pond | Hemlock, eastern | Cedar | |
| Douglas fir 1[c] | California red | Red | Maple, bigleaf | Incense | |
| Kapur[a] | Grand | Virginia | Pine | Western red | |
| Keruing[a,b] | Noble | Western white | Jack | Cottonwood | |
| Larch, western | Pacific silver | Spruce | Lodgepole | Eastern | |
| Maple, sugar | White | Black | Ponderosa | Black (western popular) | |
| Pine | Hemlock, western | Red | Spruce | Pine | |
| Caribbean | Lauan | Sitka | Redwood | Eastern white | |
| Ocote | Almon | Sweetgum | Spruce | Sugar | |
| Pine, southern | Bagtikan | Tamarack | Engelmann | | |
| Loblolly | Mayapis | Yellow poplar | White | | |
| Longleaf | Red lauan | | | | |
| Shortleaf | Tangile | | | | |
| Slash | White lauan | | | | |
| Tanoak | Maple, black | | | | |

[a]Each of these names represents a trade group of woods consisting of several closely related species.

[b]Species from the genus *Dipterocarpus* are marked collectively: Apitong if originating in the Philippines; Keruing if originating in Malaysia or Indonesia.

[c]Douglas fir from trees grown in the states of Washington, Oregon, California, Idaho, Montana, and Wyoming and the Canadian provinces of Alberta and British Columbia shall be classed as Douglas fir No. 1. Douglas fir from trees grown in the states of Nevada, Utah, Colorado, Arizona, and New Mexico shall be classed as Douglas fir No. 2.

[d]Red Meranti shall be limited to species having a specific gravity of 0.41 or more based on green volume and oven-dry weight.

(Table 1, Voluntary Product Standard PS 1, "Construction and Industrial Plywood," U.S. Department of Commerce. Not subject to copyright in the United States.)

- *Moisture-resistant* plywood is *bonded with interior glue*, and is intended for interior applications where it may be temporarily exposed to the elements.
- *Plywood bonded with intermediate glue* has lower moisture resistance than that bonded with exterior glue but greater than that bonded with interior glue. Intermediate glues have a high level of resistance to bacteria, mold, and moisture. They are suitable for use in protected construction where moderate delays in providing protection may be expected or conditions of high humidity and water leakage may exist.
- *Plywood bonded with exterior glue* is intended for protected construction where maximum performance is required for protection against moisture exposure from long construction delays or other conditions of similar severity.

## GRADES

The term *grade* refers to both veneer grade and panel grade. The veneer grade is a definition of the quality of the face veneer. The panel grade is identified in terms of the veneers used, by a name identifying the panel's intended use, or both.

*Veneer Grades* Each type of plywood panel is manufactured with several grades of face veneers. Veneer grade designation is based on the species group and the quality of the veneer used on the face and back. Panels are built up of veneers glued together at right angles to each other, using one or more of five veneer grades ranging from N (highest) through A, B, C, and D (lowest). Veneer may be repaired to raise its quality to a higher grade. Repairs consisting of patches or plugs, usually oval or circular, along with long, narrow shims, fill voids in the veneer sheet or replace characteristics such as knots and pitch pockets. There are also free-form synthetic (plastic) patches. These are put into a panel after the veneers have been bonded together. A flaw is routed out, and the synthetic patch is squirted into the void from a hand-held device resembling a glue gun. The synthetic material is then cured and sanded. In the grade stamp designation for Rated Siding in Figure 6.3-5, the 303-6-S/W designation means the following: 303 is the number code for a particular line of siding, S stands for synthetic patches, W stands for wood patches, and 6 is the number of patches permitted in the face. In this case, six synthetic or wood patches are permissible.

Veneer grades are illustrated in Figure 6.3-8 and described as follows:

*N grade*: A specialized veneer available on special order used for a natural finish. It is select, all heartwood or all sapwood, and free of open defects, with only a few small, well-matched repairs (not illustrated). N grade veneer is of such high quality and so expensive that it is rarely specified.

*A grade*: A smooth paint grade permitting a number of neatly made repairs, plugs, or patches with certain restrictions. This is the grade usually specified for high-quality work.

*B grade*: A solid surface veneer, except for specified minor characteristics. It may have a considerable number of neatly made repairs and is suitable for concrete formwork where a smooth surface is desired.

*C plugged grade*: A special improved C grade intended for underlayment for floors and combination subfloor underlayment (not illustrated).

**"A" Grade**

**"C" Grade**

**"B" Grade**

**"D" Grade**

**FIGURE 6.3-8** Plywood veneer grades. (Courtesy APA—The Engineered Wood Association, formerly American Plywood Association.)

*C grade*: The lowest quality of veneer permitted in exterior type panels. It is commonly used as the face of sheathing-grade panels. Permitted defects include tight knots up to $1^1/_2$ in. (38.1 mm), 1 in. (25.4 mm) knotholes; occasional $1^1/_2$ in. (38.1 mm) knotholes under certain conditions; splits up to $1/_2$ in. (12.7 mm) tapering to a point; and other minor defects.

*D grade*: The lowest quality of veneer permitted in interior type panel. This grade permits $2^1/_2$ in. (63.5 mm) knotholes; occasional 3 in. (76.2 mm) knotholes under certain conditions; pitch pockets up to $2^1/_2$ in. (63.5 mm) wide; and tapering splits up to 1 in. (25.4 mm).

**Panel Grades** The grade of a panel is indicated by two letters, such as A-D, describing a panel with an A grade front face veneer and a D grade back face veneer. The inner plies for an interior type panel are D or better, and those for an exterior type panel are C or better. Unless a decorative or other surface is specified, panels in most grades are run through large drum sanders, where they are sanded smooth on both sides for appearance and balanced construction.

### PERFORMANCE-RATED PANELS

*Performance-rated panels* are also sometimes called *structural panels* or *structural-use panels*. These are panels designed for use where design properties such as tension, compression, shear, cross-panel flexural properties, and nail bearing may be significant. Included in this group are panels designed for use as floor, wall, or roof sheathing; for subflooring, underlayment, or both; for siding; for construction of stress-skin panels, box beams, and other structural elements; and for other structural uses. This category includes most of the plywood used in building construction. It also includes oriented strand board (see Section 6.3.2) and Com-ply, a product made with alternating plies of wood veneer and reconstituted wood.

Performance-rated panels possess good stiffness, strength, nail-holding ability, durability, dimensional stability, lateral bracing ability, and resistance to puncture and impact. Some of them must also provide a suitable base for a covering material, such as shingles, built-up roofing, siding, or flooring.

In small buildings with conventional framing, sheathing, and subflooring, no special engineering design is required if recognized standards and construction details are followed. Correct use of plywood involves selecting the appropriate panel type, grade, and span rating for the spans and loads to be carried, in addition to using proper nailing.

Performance-rated panels are identified according to their application, the quality of the face and back veneer, or both (see Fig. 6.3-6). When necessary, they are also identified by their structural grade (Structural I or Structural II). They are classified into one of the three exterior exposure durability classifications discussed earlier (Exterior, Exposure 1, or Exposure 2).

Performance-rated panels for use as sheathing, subflooring, underlayment, or siding are assigned a *span rating* to set the support spacing required for each. This rating is included as part of the grade trademark (see Fig. 6.3-5) as a single number or as two numbers separated by a slash. When a single number appears, that is the required maximum spacing of the supports in inches when used as a flooring material. When two numbers separated by a slash appear, the two numbers denote different spacings, depending on the use of the panel. For example, a panel may be used as either wall or roof sheathing, but the support spacing is not the same in the two different uses. The first number denotes the maximum spacing in inches when the panel is used as a roof decking. The second number is the maximum spacing in inches when the panel is used for flooring. Wall sheathing has its own single-number system which denotes the maximum span in inches of the supporting studs when sheathing spans in its weak direction

Some performance-rated panels are left unsanded for greater stiffness and strength, as well as economy. However, some panels, such as those used for underlayment, are either sanded or touch sanded to bring adjacent panels to a uniform thickness, and some, such as siding, are sanded for appearance.

### SPECIALTY PLYWOODS

Special manufacturing processes and treatments produce a number of variations from standard panels, each designed to meet specific use requirements.

**Overlaid Plywood** Overlaid plywood panels are high-grade panels of *exterior* plywood to which cellulose-fiber sheets containing prescribed amounts of resin have been bonded to one or both faces. Most overlaid plywood has an overlay on both sides. The resin is usually phenolic, but other resins may also be used if they meet the requirements stated in PS 1. The overlay sheets are bonded to the veneers with adhesives under pressure. Overlaid plywood is classified as either *high density* (HDO), *medium density* (MDO), or *special overlay* panels as follows:

*High density overlay* (HDO): The face veneer quality of HDO plywood may be either A-A or B-B. The inner plies are C Plugged veneer. Concrete form overlay HDO has B-B face and back veneers. The overlay sheets on HDO are impregnated with at least 45% resin solids based on the volatile-free weight of fiber and resin exclusive of glueline. The total thickness of a resin-impregnated overlay sheet is 0.012 in. (0.305 mm) before pressing, and the total weight is at least 60 lb (27.22 kg) per 1000 sq ft (92.903 m²) or 0.293 kg/m².

HDO has a hard, smooth, chemically resistant, durable surface that requires no further finishing by paint or varnish. It is available in several basic colors, but also provides an excellent paint base and can be painted if color variation is preferred. HDO is especially suitable for and widely used in concrete formwork for architectural finishes, highway signs, and other severe exposures.

*Medium density overlay* (MDO): The face veneer quality of MDO plywood is B-B. The inner plies are C veneer. The overlay sheets on MDO must contain at least 17% resin solids for a beater-loaded sheet and 22% resin solids for an impregnated sheet, both based on the volatile-free weight of fiber and resin exclusive of glueline. The total thickness of a resin-impregnated overlay sheet is 0.012 in. (0.305 mm) before pressing, and the total weight is at least 58 lb (26.31 kg) per 1000 sq ft (92.903 m²) or 0.28 kg/m².

MDO is intended to receive high-quality paint finishes and provides an opaque, smooth, nonglossy surface that generally blanks out the grain, although some evidence of the underlying grain is permitted. A number of mills manufacture special MDO precut siding, frequently mill-primed, in widths of up to 16 in. (406.4 mm).

*Special overlays:* Overlaid plywood surfacing panels that do not fit the definitions of either HDO or MDO are called *special overlays*. These must meet the glue bond requirements for overlaid plywood, and the panels must be of the exterior type.

*Textured Siding Panels*  Textured plywood siding panels are used for exterior siding, gable ends, and fencing, and as forms for special effects in concrete. Refer to Section 7.7 for additional information.

*Paneling*  Most paneling plywoods are manufactured from hardwood species (see Section 6.3.1.2). Some softwood species, such as California redwood, knotty pine, and western red cedar, are used for interior paneling. Other manufacturers assemble softwood veneers into paneling plywood with surface textures such as embossing, striations, or relief grain. Some manufacturers produce softwood panels having printed overlays simulating hardwood grains.

*Tongue-and-Grooved Panels*  Plywood in ¹⁄₂ in. (12.7 mm), ⁵⁄₈ in. (15.9 mm), and ³⁄₄ in. (19.1 mm) thicknesses, with tongue-and-groove joints, is produced by a number of mills in underlayment grades and occasionally other grades. Tongue-and-groove panels are used to eliminate the need for blocking at joints. One performance-rated panel is available in thicknesses from ¹⁹⁄₃₂ in. (15.08 mm) up to 1³⁄₃₂ in. (27.78 mm) for use as a combination subfloor and underlayment, as well as for use over joists spacings varying from 16 in. (410 mm) to as wide as 48 in. (1220 mm), based on the panels being continuous over at least two spans. These types of panels are tongue-and-grooved (Fig. 6.3-9) to eliminate the need to block the joints.

*Concrete Form Panels*  The face veneers for plywood panels designed for concrete formwork are not less than B grade. The inner plies are not less than C grade. These panels are usually sanded on both sides and oiled at the mill. They are classified as either Class I or Class II according to the combination of wood species groups used in faces, crossbands, and center veneer plies. Class I panels require the use of stronger and stiffer species groups in faces and crossbands than do Class II panels. Concrete form panels are designed for extensive reuse. HDO panels are also sometimes used as concrete forms.

*Marine Plywood*  Marine plywood panels are exterior grade plywood of either grade A-A, A-B, B-B, HDO, or MDO, but with certain additional requirements.

**FIGURE 6.3-9** Tongue-and-grooved panels. (Photograph courtesy APA—The Engineered Wood Association, formerly American Plywood Association.)

For example, the only wood species permitted are Douglas fir 1 and western larch; the number of repairs in the face veneers is more limited than normal; inner plies must be B grade or better; wood repairs must be glued with an exterior adhesive; and there are additional restrictions as well.

### SIZES

Plywood is most readily available in panels 4 ft (1220 mm) wide by 8 ft (2400 mm) long. Some mills also produce 5-ft (1500-mm)-wide panels and lengths of up to 12 ft (3600 mm), but these are not always stocked. Beyond these limits, widths of up to 10 ft (3000 mm) and any reasonable length (which has been more than 50 ft [15.24 m]) can be made by scarf-jointing or finger-jointing panels together. Scarf- or finger-jointed construction consists of gluing panels either end-to-end or side-to-side continuously along a tapered or fingered cut to form a single panel (Fig. 6.3-10).

Plywood panels are permitted by PS 1 to be up to ¹⁄₁₆ in. (1.6 mm) smaller but may be no larger in the horizontal dimension than the specified size. Nominal available thicknesses and thickness tolerances for plywood panels are shown in Figure 6.3-11. Plywood panels always have an odd number of *layers* for balanced construction, the minimum number being three. A layer may consist of a

**FIGURE 6.3-10** Panels can be scarf-joined (top) or finger-jointed (bottom).

single thickness of veneer, or it may be a lamination of two veneers with their grain running in the same direction. Each single thickness of veneer is called a *ply*. The minimum number of plies for various thicknesses of plywood is given in Figure 6.3-12. A diagram of how the layers are typically arranged is given in Figure 6.3-13.

## PHYSICAL AND MECHANICAL PROPERTIES

Aside from the characteristics of the veneer used, the most important properties of plywood derive from its cross-laminated construction, Plywood utilizes

the strength and dimensional stability of wood along the grain by placing alternate layers at right angles. This cross-lamination provides resistance to splitting, improved dimensional stability, warp resistance, and other properties, as discussed in the following sections.

***Durability*** In order to comply with the PS 1 requirements for exterior type plywood, the glueline used to bond the veneers must be as durable as the wood itself. Theoretically, exterior panels may be exposed outside without paint or other treatment and without affecting the durability of the glue bond or the life of the panels. In practice, exposing the edges of any plywood to weather is not a good idea. The glueline at the edge of a plywood panel is the most vulnerable part of the panel to damage by water and weather. Usually, therefore, plywood, like other wood products, is painted or stained when exposed outdoors to protect it against weathering and checking. Preservative treatments are available for plywood that is to be used in areas of high decay or termite hazard.

***Bending Strength and Stiffness*** The cross-laminating of layers in plywood panels capitalizes on the more desirable physical and mechanical properties of wood. The strength of wood parallel to the grain is many times greater than that perpendicular to the grain. Cross-lam-

inated construction provides a longitudinal grain crosswise and improves significantly solid wood's crosswise strength and stiffness. However, plywood panels are both stiffer and stronger along the face grain than across the face grain and should be applied with the face grain perpendicular to its supports for maximum resistance to bending.

***Racking Resistance*** Plywood functions in its own plane as a structural bracing material when used on a wall, roof, or floor. Although $5/16$ in. (7.94 mm) is the minimum thickness for sheathing, even $1/4$-in. (6.4-mm)-thick plywood on a standard wood stud-framed wall provides twice the stiffness and more than twice the strength furnished by 1-in. (25-mm)-thick diagonal boards, which has long been recognized as adequate. The racking strength of plywood and other sheathing materials is discussed further later in this chapter.

Two properties give plywood superior bracing qualities: (1) high shear strength in all directions for loads applied perpendicular to panel faces and (2) relatively high nail-bearing strength, allowing nails to be driven as close as $1/4$ in. (6.4 mm) from the edge of a panel. Splitting is virtually impossible lengthwise or crosswise since no cleavage plane exists.

***Impact and Concentrated Load Resistance*** Stresses resulting from impact or concentrated loading are distributed over a wide area by the cross veneers in a plywood panel. This is an important property in sheathing, subflooring, siding, and wall paneling.

***Dimensional Stability*** Like solid wood, plywood is subjected to dimensional changes due to variations in moisture content, temperature, and other causes.

***Hygroscopic Expansion and Contraction*** Expansion due to absorption and contraction due to loss of water in solid wood are generally insignificant parallel to the grain, even with large moisture content changes. Perpendicular to the grain, dimensional changes can be considerable (see Section 6.1.3.2). In plywood, any tendency for veneers to expand or contract crosswise is greatly minimized by one or more adjacent veneers running lengthwise. The total expansion both across the width and along the length of a 4 × 8 ft (1200 × 2400

**FIGURE 6.3-11** Plywood Panel Thicknesses and Tolerances According to U.S. Product Standard PS 1

| Plywood Type | Normally Available Nominal Thicknesses, in. (mm) | Tolerances (plus or minus), in. (mm) |
|---|---|---|
| **Unsanded panels** | $5/16$ (7.9), $3/8$ (9.5), $15/32$ (11.9), $1/2$ (12.7), $19/32$ (15.1), $5/8$ (15.9), $23/32$ (18.3), $3/4$ (19.1) | $1/32$ (0.79) |
| | $7/8$ (22.2), 1 (25.4), $1 1/8$ (28.6) | 5% of specified thickness |
| **Sanded panels** | $1/4$ (6.4), $9/32$ (7.14), $11/32$ (8.73), $3/8$ (9.5), $15/32$ (11.9), $1/2$ (12.7), $19/32$ (15.1), $5/8$ (15.9), $23/32$ (18.3), $3/4$ (19.1) | $1/64$ (0.38) |
| | $7/8$ (22.2), 1 (25.4), $1 1/8$ (28.6) | 3% of specified thickness |
| **Touch-sanded panels** | $1/2$ (12.7), $19/32$ (15.1), $5/8$ (15.9), $23/32$ (18.3), $3/4$ (19.1) | $1/32$ (0.79) |
| | $1 1/8$ (28.6) | 5% of specified thickness |
| **Marine plywood** | $1/4$ (6.4), $3/8$ (9.5), $1/2$ (12.7), $5/8$ (15.9), $3/4$ (19.1) | $1/64$ (0.38) |
| **Overlaid plywood** | $11/32$ (8.73), $3/8$ (9.5), $15/32$ (11.9), $1/2$ (12.7), $19/32$ (15.1), $5/8$ (15.9), $23/32$ (18.3), $3/4$ (19.1) | $1/32$ (0.79) |

**FIGURE 6.3-12** Panel Constructions

| Panel Grades | Finished Panel Nominal Thickness Range, in. (mm) | Minimum Number of Plies | Minimum Number of Layers |
|---|---|---|---|
| **Exterior**<br>Marine<br>Special exterior<br>B-B concrete form<br>High-density overlay<br>High-density concrete form overlay | Through ³⁄₈ (9.5)<br>Over ³⁄₈ (12.7), through ³⁄₄ (19.1)<br>Over ³⁄₄ (19.1) | 3<br>5<br>7 | 3<br>5<br>7 |
| **Interior**<br>N-N, N-A, N-B, N-D, A-A, A-B, A-D, B-B, B-D<br>Structural I (C-D, C-D plugged, and underlayment)<br>Structural II (C-D, C-D plugged, and underlayment)<br><br>**Exterior**<br>A-A, A-B, A-C, B-B, B-C<br>Structural I and Structural II (C-C and C-C plugged)<br>Medium density and special overlays | Through ³⁄₈ (9.5)<br>Over ³⁄₈ (12.7), through ¹⁄₂ (15.9)<br>Over ¹⁄₂ (15.9), through ⁷⁄₈ (22.2)<br>Over ⁷⁄₈ (22.2) | 3<br>4<br>5<br>6 | 3<br>3<br>5<br>5 |
| **Interior** (including grades with exterior glue)<br>Underlayment<br><br>**Exterior**<br>C-C plugged | Through ¹⁄₂ (12.7)<br>Over ¹⁄₂ (12.7), through ³⁄₄ (19.1)<br>Over ³⁄₄ (19.1) | 3<br>4<br>5 | 3<br>3<br>5 |
| **Interior** (including grades with exterior glue)<br>C-D<br>C-D plugged<br><br>**Exterior**<br>C-C | Through ⁵⁄₈ (15.9)<br>Over ⁵⁄₈ (15.9), through ³⁄₄ (19.1)<br>Over ³⁄₄ (19.1) | 3<br>4<br>5 | 3<br>3<br>5 |

(Table 4, Voluntary Product Standard PS 1, "Construction and Industrial Plywood," U.S. Department of Commerce. Not subject to copyright in the United States.)

mm) panel in service under normal variations in moisture content (8% to 14%) should average between 0.03 in. (¹⁄₃₂ inch [0.79 mm]) and 0.05 inch (¹⁄₂₀ in. [1.27 mm]).

***Thermal Expansion*** Expansion and contraction due to thermal changes in solid wood are about 10 times as great across the grain as they are parallel to the grain, where changes are quite small. In plywood, thermal expansion along the face is somewhat greater than in plain wood because of the influence of the crossbands. In an 8-ft (2400-mm)-long Douglas fir panel with five equal piles, the expansion caused by a 120°F (48.89°C) rise in temperature should average approximately 0.021 in. (¹⁄₅₀ in. [0.53 mm]). This is about 40% more than for plain Douglas fir wood but still insignificant for most uses. Across the grain, expansion due to a 120°F (48.89°C) rise is approximately 0.029 in. (0.74 mm), or only about one-sixth that for plain wood.

***Warp*** Bow, cup, or twist may occur in a plywood panel if it is subjected to unequal absorption of moisture along its edges, different exposure to moisture on opposite faces, or unbalanced painting or coating of its faces. However, plywood that is flat at the time of manufacture will remain so unless conditions of storage or use subject it to uneven moisture content changes.

***Checking*** Checking may appear on any wood surface, plywood included, in the form of hairline cracks or even slightly open splits. Under severe moisture and dry conditions, such as an unpainted, unprotected plywood panel exposed to the weather, checks may in time become open cracks penetrating virtually the full thickness of the face. With moisture pickup, the exposed veneer surface expands while the undersurface glued to the crossband is held in a relatively fixed position. The expanding outer fibers push against one another and acquire a *compression set*. As the fibers dry out and shrink they pull away from each other, creating checks of different intensity.

Remedies to control checking within acceptable limits include protection from dampness or moisture during transit and storage, protection during use at the job site, and the use of protective coatings, such as conventional paint, and water-repellent dips to minimize moisture absorption and loss.

**FIGURE 6.3-13** Each layer is perpendicular to adjacent layers for strength and stiffness.

*Workability* Plywood is easily worked, cut to various sizes and shapes, and may be bent to fairly sharp radii (Fig. 6.3-14).

*Fire and Decay Resistance* A high degree of resistance to either fire or decay can be imparted to plywood through pressure-impregnation treatment similar to that used for lumber (see Section 6.4). Only plywood bonded with exterior glue is treated, because the adhesives used in it are not affected by pressure impregnation of fire-retardant chemicals. Fire-retardant treatments have little effect on the rate of heat transfer through a panel, but virtually eliminate fuel contribution to a fire and prevent the wood from burning by causing it to char.

Plywood used indoors may be protected by the application of a fire-retardant paint to its surface. When exposed to high temperatures, these paints blister or bubble, forming air pockets which insulate the plywood, retard combustion, and greatly reduce flame spread along the surface.

### 6.3.1.2 Decorative Plywood

The division of plywood products into two categories, (1) *construction and industrial* and (2) *decorative*, is based on the two governing standards: the previously discussed PS 1 and the Hardwood Plywood and Veneer Association (HPVA) standard HP-1.

Construction and industrial plywood covered by PS 1 is made mostly of softwood species, although a few hardwood species are permitted. Decorative plywood covered by HP-1 consists entirely of hardwood-faced panels. The interior layers of these panels are often of softwood or particleboard, however, to help lower their cost and to conserve the more rare hardwood species.

Decorative plywood panels are most often used as wall paneling, in cabinet and furniture work, as thin face skins in door manufacture, and as face veneer for block flooring.

When panels are exposed on one surface only, back and face plies are of species and thicknesses to balance the face veneers. The species of wood used for the face veneers of panels identifies the name of hardwood plywood panels. For example, birch plywood has a face veneer and probably a back veneer, of birch.

### MANUFACTURE

Differences between the manufacture and appearance of hardwood and softwood plywood, as well as differences among hardwood plywoods, may result from the (1) tree species, (2) part of the tree used, (3) cutting methods used to produce the veneer, (4) methods used to match the veneers, and (5) type of core construction.

*Tree Species* More than 99,000 species of hardwoods have been classified. About 240 of these are available in commercial quantities in the United States, and about 50 are widely used for hardwood plywood. These include lauan, walnut, birch, gum, oak, maple, ash, cherry, elm, beech, African mahogany, Honduras mahogany, butternut, pecan, cottonwood, sycamore, hackberry, teak, and rosewood. The most popular at a given time varies with the public's changing preferences. Some typical imported and domestic hardwood veneer species and their characteristics are listed in Figure 6.3-15.

*Parts of the Tree* The portion of the tree used greatly affects hardwood plywood figure patterns. Most plywood comes from relatively straight portions of the tree, but some of the more interesting and valuable patterns are cut from a crotch, a burl, or a stump. Veneers cut from a crotch have a plume-like design. A burl produces veneers with a pattern of swirls. A stump yields rippling patterns with rich contrasts (Fig. 6.3-16).

FIGURE 6.3-14 Minimum Plywood Bending Radii

| Panel Thickness, in. | Bending Radii for Panel Bend in Direction, ft | |
|---|---|---|
| | Across Grain | Parallel to Grain |
| 1/4 | 2 | 5 |
| 5/16 | 2 | 6 |
| 11/32 and 3/8 | 3 | 8 |
| 15/32 and 1/2 | 6 | 12 |
| 19/32 and 5/8 | 8 | 16 |
| 23/32 and 3/4 | 12 | 20 |

(Courtesy APA—The Engineered Wood Association, formerly American Plywood Association.) (Refer to Figure 25.7-1 for metric conversions.)

FIGURE 6.3-15 Visual Characteristics of Representative Hardwood Species

| Commercial Name | Color | Type of Figure |
|---|---|---|
| Ash, American | White to light brown | Medium open grain |
| Basswood | White to pale brown | |
| Beech | White to reddish brown | |
| Birch | White to light reddish brown | Curly grained, figured flat cut, plain rotary |
| Butternut | Pale brown | Leafy grain |
| Cherry, American | Light to dark reddish brown | Plain to rich mottle |
| Cottonwood | White to light grayish brown | |
| Elm | Light brownish red | Strong |
| Gum, black and tupelo | White, grayish white Greenish to grayish black | |
| Gum, red | Pink to reddish brown | Medium to highly figured |
| Hackberry | Pale yellow, yellowish gray to light brown | |
| Lauan | Pale grayish, yellowish brown to reddish brown | |
| Mahogany, African | Pink to reddish brown | Plain stripe to highly figured |
| Mahogany, tropical American | Pink to gold brown | Straight to rich mottle |
| Mahogany, crotch and swirls | Pink to reddish brown | Moon and feather crotch, plain and figured swirl |
| Maple | White to tan | Plain, curly burls |
| Oak, red | Pink tan to ochre | Plain to flake |
| Oak, white | Gray tan to ochre | Plain to flake |
| Rosewood, Brazilian | Pink brown and violet | Wide range figure |
| Rosewood, East Indian | Purple to straw | Striped and figured |
| Teak | Light tan, dark brown | Plain, ripple, mottle, stripe |
| Walnut, American | Soft gray brown | Typical figure or stripe |

(Used with permission of the Hardwood Plywood & Veneer Association.)

**FIGURE 6.3-16** Portions of a tree from which various types of figure patterns are obtained. **Some species produce several figure types.** (Used with permission of the Hardwood Plywood & Veneer Association and the Fine Hardwood Veneer Association.)

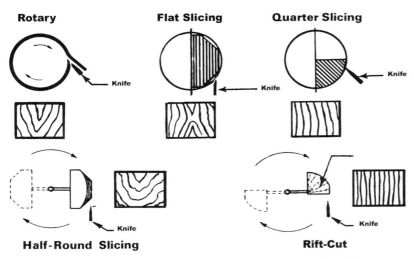

**FIGURE 6.3-17** Visual characteristics of veneers depend on the method of manufacture.

*Veneer Types* The manner in which veneers are cut is important in producing various visual effects. Two logs of the same species that are cut differently will have entirely different visual characteristics even though their veneer colors are similar. Five principal methods of cutting veneers are used (Fig. 6.3-17).

*Rotary cut*: To produce a rotary cut, a log is mounted centrally in a lathe and turned against a blade, producing exceptionally wide veneer compared to that cut by other methods. Since this cut follows the log's annular growth rings, the resulting pattern has a bold, variegated ripple figure. This is the type of cut used for most softwood veneers.

*Flat slicing*: Flat slicing results in a variegated wavy figure. To produce it, a half-log, or flitch, is mounted with the heart side flat against the guide plate of a slicer, and the slicing is done parallel to a line through the center of the log (see Fig. 6.3-2).

*Quarter slicing*: Quarter-sliced veneer has a series of stripes, straight in some woods, varied in others. To produce this type of cut, a quarter log (flitch) is mounted on a guide plate so that the growth rings of the log strike the knife at approximately right angles.

*Half-round slicing*: Half-round sliced veneer shows modified characteristics of both rotary and plain-sliced veneers. It is a variation of rotary cutting in which segments of a log (flitches) are mounted off-center in a lathe. This results in a cut slightly across the annual growth rings.

*Rift-cut veneer*: Various species of oak and some other species are sometimes cut using a rift cut. In these woods, the ray cells which radiate from the center of the log are conspicuous, and a rift or comb grain effect can be obtained by cutting perpendicularly to these rays either on a lathe or on a slicer.

*Veneer Matching* The methods used to match face veneers also produce different visual characteristics. Basic and special matching methods are shown in Figure 6.3-18 and are described as follows:

*Book matching*: Any type of veneer can be book matched. In book matching, every other sheet is turned over like the leaves of a book. Thus, the back of one veneer meets the front of the

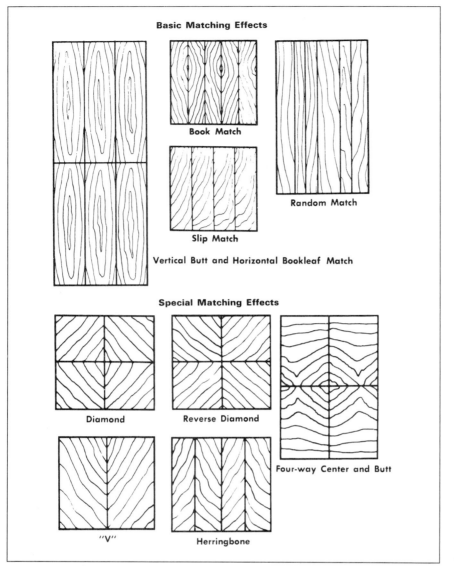

**Basic Matching Effects**

Book Match

Random Match

Slip Match

Vertical Butt and Horizontal Bookleaf Match

**Special Matching Effects**

Diamond

Reverse Diamond

Four-way Center and Butt

"V"

Herringbone

**FIGURE 6.3-18** Basic and special matching effects for wood veneers.

adjacent veneer, producing a matching joint design. Book matching is also called *edge matching*.

*Slip matching*: Quarter-sliced veneers are often slip matched, but this method may also be used for other types of veneer cuts. In this type of match, adjacent sheets are joined side by side, without turning, to repeat the flitch figure. Slip-matched veneers are common in *premium*, *good*, and *speciality* hardwood plywood grades and are often used in doors and matched, ungrooved wall panels.

*Random matching*: The joining of veneers with the intention of creating a casual unmatched effect is called random matching. In this type of matching, veneers from several logs may be used in the manufacture of a single panel. Random matching is generally used in prefinished V-grooved wall panels.

*Vertical butt and horizontal bookleaf matching*: Flitches whose length does not permit its fabrication in the desired height of panel can be matched using the vertical butt and horizontal bookleaf method. In this type of matching, the veneer may be matched horizontally as well as vertically. This type of matching is often used in custom wall paneling and in furniture.

*Four-way center and butt matching*: Several *special* matching effects are used in furniture manufacture and in cus-

tom wall paneling. One of these, four-way center and butt matching, is ordinarily applied to crotch or stump veneers to reveal the figure. Occasionally, flat-sliced veneers are matched in this manner when panel length requirements exceed the length of available veneers.

***Core Construction*** The innermost ply in a hardwood plywood panel is called the *core*. The plies between the core and the face piles are called *crossbands*. Four basic types of cores are used in hardwood plywood: *veneer*, *lumber*, *particleboard*, and *mineral*.

*Veneer core*: The veneer core method is the same as that for softwood plywood. In hardwood plywood, three, five, or more plies are laid with the grain direction of adjacent plies at right angles to each other. The total number of plies is always odd so that the panels are balanced.

*Lumber core*: Some panels have a core of sawed lumber to which crossband plies and face veneers are glued (Fig. 6.3-19). Lumber core construction is often used in cabinet work and furniture construction in which exposed edges or edge treatments such as doweled, splined, or dovetailed joints are desired or where butt hinges are to be used. A lumber core is made of narrow wood strips edge-glued together. Panels with hardwood bands that match the face applied on the edges may be specially ordered.

*Particleboard core*: Plywood for use as table, desk, and cabinet tops often has a particleboard core. This is an aggregate of wood particles (also called *chipboard*, *chipcore*, *particle board*, etc.) fused under heat and pressure. The face veneers are usually glued directly to this core, although crossbanding is sometimes used.

*Mineral core*: Fire-resistant panels are made with mineral cores. In this construction, the face veneers are bonded to a core of hard, noncombustible material.

***Sizes*** Hardwood plywood is most commonly sold in 4 × 8 ft (1200 × 2400 mm) panels, although panels are custom-made up to 40 ft (12.19 m) long and up to 7 ft (2.134 m) wide. Stock panels may also be purchased in lengths of 10 ft (3048 mm) and 12 ft (3658 mm). Hardwood plywood is manufactured with an uneven number of plies in thicknesses ranging

Face veneer (of fine hardwood)

Crossband (usually of poplar or gum)

Lumber core (gumwood, poplar, basswood, maple, etc.)

Crossband (poplar, etc.)

Back veneer (fine hardwood if to be exposed)

**FIGURE 6.3-19** Typical five-ply lumber core panel. Each ply (indicated by arrows) runs at right angles to adjacent plies. (Used with permission of the Hardwood Plywood & Veneer Association and the Fine Hardwood Veneer Association.)

**FIGURE 6.3-20** Standard Thicknesses of Veneer-Core Hardwood Plywood*

| Plies | Thickness, in. |
|---|---|
| 3 | $\frac{1}{8}$, $\frac{3}{16}$, $\frac{1}{4}$ |
| 5 | $\frac{5}{16}$, $\frac{3}{8}$, $\frac{1}{2}$ |
| 5 and 7 | $\frac{5}{8}$ |
| 5, 7, and 9 | $\frac{3}{4}$ |

*Hardwood plywood is square within $\frac{1}{16}$ in. measured on short dimension. Particleboard and MDF-core hardwood plywood panels are typically 3- and sometimes 5-ply in thicknesses up to $\frac{3}{4}$ in. and thicker.

(Used with permission of the Hardwood Plywood & Veneer Association.) (Refer to Figure 25.7-1 for metric conversions.)

from $\frac{1}{8}$ in. (3.2 mm) to as much as 3 in. (76.2 mm) in custom panels. Common thicknesses include $\frac{1}{4}$ in. (6.4 mm), $\frac{3}{8}$ in. (9.5 mm), $\frac{1}{2}$ in. 12.7 mm), and $\frac{3}{4}$ in. (19.1 mm) veneer core and $\frac{3}{4}$ in. (19.1 mm) in lumber and particleboard core. Figure 6.3-20 shows common thicknesses for varying numbers of plies.

*Pattern and Figure Characteristics of Veneer* The term *figure* refers to the highlights of so-called crossfire running at right angles to the grain direction. The grain character and direction are described by using the word *pattern*. Those in the industry may describe a figure by saying that it "has a great deal of cross-

fire" or "has a straight or broken stripe" or "is highly figured." Figure 6.3-21 shows some common veneer figures and the terms often used to describe them.

### GENERAL CLASSIFICATIONS

Hardwood plywood falls generally into two categories: (1) *specialty* plywoods, which are not typically or necessarily manufactured to meet industry standards, and (2) *standard* plywoods, which are produced to meet HPVA HP-1 (Fig. 6.3-22).

*Specialty Plywood* There are no industry-recognized standards covering *specialty plywoods*. Species, type, match, core construction, and other requirements are as agreed upon by the seller and the buyer. Generally, panels containing the fewest number of blemishes are sold at a higher price than those with large knots and filled knotholes, splits, and other surface defects. Often, the intention is that color and grain vary.

Each manufacturer develops his own grade terminology. However, such terminology is rarely by number or letter but instead by an arbitrary descriptive system. For example, *aristocratic* and *homestead* are names selected to appeal to different tastes rather than to establish degrees of quality. The aristocratic panels might contain few blemishes and very few filled places, while the homestead panels might contain more knots, large filled knotholes, and more character. There may be little, if any, price difference between the two types of panels.

*V-grooved wall panels*, either natural or prefinished, account for much of the

Cross-fire     Ribbon stripe     Swirl     Flaked

Mottle     Blistered     Fiddle-back     Pencil stripe

**FIGURE 6.3-21** Commonly used figure descriptions. (Used with permission of the Hardwood Plywood & Veneer Association and the Fine Hardwood Veneer Association.)

**FIGURE 6.3-22** Classification of hardwood plywood.

hardwood plywood used in residential construction. There are two types of V-grooved panels: *random matched* and *gang-grooved*.

In random-matched panels the individual veneer sheets, of which there may be four to nine in a 4 × 8 ft (1200 × 2400 mm) panel, are selected to simulate lumber planking. The V-grooving at the joints of the veneer sheets produces a mismatched panel with veneer sheets dissimilar in grain and color.

Gang-grooved panels are produced by passing one-piece or multipiece faces under a machine with grooving knives set at spaced intervals. When multipiece face veneers are used, no attempt is made to groove on the joints of the veneer. In gang-grooved panels, the spacing of grooves is identical from one panel to the next.

*Architectural custom panels* are made to special order. The method is to select the desired hardwood face veneer from flitch samples before the plywood is manufactured. The selected flitch is then fabricated at a plywood plant into panels to meet the custom specification. The only limitations of architectural custom panels are those imposed by the patterns which are available in the length and width of the selected veneer.

Increasing amounts of hardwood plywood are produced for uses too numerous to discuss individually here. Among them are wall panels of varying widths less than 4 ft (1200 mm) and flooring products such as laminated block.

***Standard Plywood*** Plywood that conforms with the requirements of HPVA HP-1 is called *standard* hardwood plywood. It is manufactured primarily for use as stock panels, as door skins, and in furniture manufacture. The HPVA maintains a quality control program which ensures that plywood carrying its official grade stamp conforms with HP-1 (Fig. 6.3-23).

*Stock* panels, as the name implies, are fabricated and stocked by a plywood manufacturer and sold through local distribution warehouses to many different consumers. The character and quality of face veneers are dependent on market conditions, and the buyer selects from the available stock at the time of purchase. Panels are typically 4 × 8 ft (1200 × 2400 mm) of varying thicknesses, such as $1/4$ in. (6.4 mm), $1/2$ in. (12.7 mm), and $3/4$ in. (19.1 mm), and may be put to a variety of uses.

*Doorskin* plywood comprises a large percentage of hardwood veneer and is used as door skins to face solid and hollow core doors (see Section 8.2).

The furniture industry and other wood product manufacturers make extensive use of hardwood plywood. Outside the construction industry, hardwood plywood finds almost endless uses in automobile, ship, boat, aircraft, and railroad car construction, as well as in sporting goods, musical instruments, caskets, and many other items.

***Standard Types*** The word *type* is used here to indicate a panel's durability, that is, the ability of the plies to stay together under various conditions of exposure. Three basic types of hardwood plywood are produced to conform with the requirements of HPVA HP-1.

*Type I* (exterior) hardwood plywood is made with waterproof adhesive and is used where it might come in contact with water or must withstand full weather exposure. The adhesives used are usually resins of phenol, resorcinol, phenolresorcinol, melamine, or melamine urea. Veneer core construction is preferred for Type I plywood.

The *technical type* is made with the same adhesive requirements as Type I but varies in the thickness and arrangement of plies to provide approximately equal tensile and compressive strengths in length and width.

*Type II* (interior) plywood is made with water-resistant adhesive (urea resin) and is used where it would not be subjected to contact with water, although it should retain practically all of its strength if occasionally subjected to thorough wetting and drying. The majority of hardwood plywood is Type II.

***Standard Grades*** A plywood panel's *grade* designates the quality of the face, back, and inner plies. Hardwood veneer grades used under HP-1 are *face, speciality, back,* and *inner ply*.

*Face grades*: Face veneers may be made from more than one piece. With most species, multipiece faces must be

| HARDWOOD PLYWOOD & VENEER ASSOCIATION | | |
|---|---|---|
| FORMALDEHYDE EMISSION 0.2 PPM CONFORMS TO HUD REQUIREMENTS | SIMULATED DECORATIVE FINISH ON PLYWOOD | FLAME SPREAD 200 OR LESS ASTM E84 |
| LAY UP 16 3.6 mm THICK HP-SG-96 | **hpva** MILL # | GLUE BOND TYPE II ANSI/HPVA HP-1-1994 |

**FIGURE 6.3-23** Grade stamp of the Hardwood Plywood & Veneer Association. (Used with permission of the Hardwood Plywood & Veneer Association.)

book matched or slip matched. The quality of these veneers is high: only a few small burls, occasional pin knots, slight color streaks, and inconspicuous small patches are permitted. No other defects, such as splits, shakes, worm holes, or decay, are allowed. Face grades include AA, A, B, C, D, and E, in descending order of quality.

*Speciality grade*: Speciality grade includes face veneers that are unlike those in the face grades. Examples include wormy chestnut, birdseye maple, and English brown oak, all of which have unusual decorative features.

*Back grades*: Back grade veneers are labeled 1, 2, 3, and 4, with 1 being the most restrictive. These grades are used on surfaces that will not be exposed in the finished work.

*Inner ply grades*: Four grades are used to identify the inner plies of hardwood plywood panels: J, K, L, and M. J is the most restrictive. M is reserved for plies not directly adjacent to facing plies.

## 6.3.2 NONVENEER PANELS

Particleboard panels are manufactured from small fragments of wood bound together with an adhesive. These types of products have been variously called *flakeboard*, *waferboard*, and *particleboard*. Today, most of them are called *oriented strand board* (OSB). Most structural nonveneer wood panels produced today are OSB, in which panels are made using strands or wafers of wood mostly oriented in the long direction of the panel. Several individual panels are laid with the direction of their strands or wafers perpendicular to each adjacent sheet and bonded together to form a final panel. This cross-lamination makes such panels strong and stable for the same reasons that plywood has those characteristics.

In general, particleboard should comply with the requirements of ANSI A208.1, but some such products are actually extruded. Particleboard may be either fire-retardant treated or untreated.

In some panel products, particleboard panels are used in combination with wood panels to produce a board. The particleboard sheets are used as crossbands with a wood core and wood face veneers. The panels are bonded in the same manner as is standard plywood. Some such products are made in a one-step pressing operation in which the voids in the veneers are filled by the wood particles in the particleboard layers.

## 6.3.3 FIBERBOARD

Wood or cane fiber panels are used primarily for sheathing wood frame walls. Panels are coated or impregnated with asphalt or given some other treatment to provide water resistance. Occasional wetting and drying during construction will not damage this material. Refer to Section 6.10 for additional data on fiberboard sheathing panels.

# 6.4 Wood Treatment

The term *wood treatment* as used in this text refers to all methods used to protect wood from decay, fungus, and insect damage and from premature destruction due to fires. These methods include selecting decay- and insect-resistant species, controlling moisture, separating wood from the ground, blocking paths that may be used by insects to achieve access to the wood, preservative treatment, and fire-retardant treatment.

The hazards associated with decay and insect damage can be divided into fungus damage and insect damage.

## 6.4.1 FUNGUS DAMAGE

*Fungi* (microscopic plants) cause decay, molds, and stains. Fungus growth can develop in wood only under the following conditions:

1. There must be an adequate supply of organic material, which for some fungi is the wood itself.
2. Temperatures must range from 41°F (5°C) to 104°F (40°C).
3. There must be sufficient oxygen, which is always present unless the wood is completely below the ground water line or completely submerged in water.
4. The wood's moisture content must exceed 20%.

### 6.4.1.1 Dry Rot

The term *dry rot* is an often-used misnomer, because the fungi that cause decay must have access to water in order to function. Decay will not occur in wood that is maintained below a 20% moisture content, which is typical of air-dried wood. Therefore, even wood that is dry when decay is detected must have been wet earlier. Most decay in wood occurs when its moisture content is above the fiber saturation point (30% moisture content). But wood that is continuously water soaked will not decay either, because insufficient oxygen will be present.

In the later stages of decay, a wood-destroying fungus seriously reduces the strength of wood. Therefore, most grading rules limit the amount of decay that may be present in structural lumber.

### 6.4.1.2 White Pocket

Also known as *white speck*, *white pocket* is caused in living softwood trees by a relatively harmless fungus that ceases to develop when a log is cut into lumber and put into normal use. This fungus is as harmless as other natural characteristics normally found in lumber, such as knots, shake, or wane. It will not spread from one piece to another or even within an individual piece. Lumber with white pocket is serviceable, and certain amounts of this fungus are accepted by grading authorities in all but the higher grades of lumber (Fig. 6.4-1). However, white pocket does reduce the strength of lumber; more importantly, nail-holding values are greatly reduced in lumber having white pocket.

### 6.4.1.3 Molds and Stains

Molds and some stains are caused by fungi which do not destroy the infected wood. Stock with such stains and molds is practically unimpaired except in appearance.

Powdery surface molds vary from white or light colors to black. These discolorations are largely superficial and often are easily brushed off or removed by surfacing the wood.

Stains appear as speck, spots, streaks, or patches that vary most commonly from bluish to bluish-black or brown. These discolorations usually cannot be removed by brushing or surfacing because they extend into the sapwood.

## 6.4.2 INSECT DAMAGE

Holes caused by insects may occur in standing trees and in either unseasoned

**FIGURE 6.4-1** White pocket (white speck) is permitted in certain lumber grades and is serviceable and acceptable for many uses. (Courtesy West Coast Lumber Inspection Bureau.)

or seasoned lumber. These are generally classified as *pinholes* or *grub holes*. Such damage is taken into account when lumber is graded, but the strength of a piece of wood damaged by insects after it has been installed cannot be estimated visually. Therefore, when strength is an important factor, every installed piece found containing insect holes should be removed and replaced by a new member.

Many precautions taken to prevent decay are also effective against various insects that attack and destroy wood.

Subterranean termites account for about 95% of all termite damage in the United States and for a significant portion of the total insect damage. Nonsubterranean termites (called *dry-wood* termites) occur only in the southern part of the country and the lower east and west costal areas, but they are also a hazard where they exist. Refer to Section 21.6 for a discussion about the location and control of termites.

### 6.4.3 SOLUTIONS

Suitable protection of wood structural elements from fungal and insect damage can be provided by appropriate methods of design and construction, controlling the moisture content of the wood, select-

ing naturally resistant wood species, treatment of the soil with pesticides (see Section 21.6), and preservative treatment of the wood.

#### 6.4.3.1 Design and Construction

When lumber is used under continuously dry conditions and is adequately protected from attack by insects and fungi, preservative treatment is unnecessary. Preventive measures against these hazards during the manufacturing and merchandising stages are well established and widely practiced by reputable lumber producers. Serious decay or insect problems in buildings result most often from poor construction techniques.

The design and construction of foundations which support buildings using wood structural floor, wall, and roof systems must include precautions for the protection of the wood from the effects of decay and attack by subterranean termites.

Some methods used to protect wood from damage are:

- Positive site and building drainage
- Separation of wood elements from the ground
- Termite barriers
- Monolithic concrete slab construction

- Ventilation and condensation control in enclosed spaces.

### SEPARATION OF WOOD FROM THE GROUND

Figure 6.4-2 shows the distances by which wood should be separated from the ground in order to (1) maintain wood at a safe moisture content for decay protection, (2) provide a termite barrier, and (3) facilitate periodic visual inspection. When it is impracticable to comply with the clearances recommended, all heartwood of a naturally durable species or pressure-treated lumber should be used. However, when termites are known to exist, access space always should be provided for periodic visual inspection. Greater clearances may be required where mechanical equipment is located in a crawl space.

### TERMITE BARRIERS

A *termite barrier* is any material that cannot be penetrated by termites and which drives the insects into the open, where they can be detected and eliminated. Where local experience indicates that additional protection against termites is required, one or more of the following barriers is recommended, depending on the degree of hazard: (1) chemical soil treatment (see Section 21.6), (2) preservative-treated lumber for all floor framing up to and including the subfloor, or (3) properly installed termite shields.

Additional precautions include using poured concrete foundations in which no cracks greater than $1/64$ in. (0.381 mm) are present or poured reinforced concrete caps at least 4 in. (100 mm) thick on unit masonry foundations. Such caps

**FIGURE 6.4-2** Minimum separation of wood from the ground.

should have no cracks greater than $^1/_{64}$ in. (0.381 mm).

## 6.4.3.2 Controlling Moisture Content

The decay resistance of wood is assured at a moisture content below 20%. In fact, the most effective and practical method of preventing decay is to control the moisture content of wood. Optimum conditions for decay occur when the moisture content is about 25%. Wood that is installed and maintained below a moisture content of 20% will not decay. Most wood in use under protected situations is below this level. If the moisture content will exceed 20% due to use or exposure conditions, the wood should be treated with toxic chemicals to prevent decay or the naturally durable heartwood of certain species may be used.

## 6.4.3.3 Natural Decay Resistance

Natural toxic substances deposited in the cell structure of some species make their heartwood resistant to decay and termite attack. In all conditions of exposure, except for structural members embedded in the ground and used as part of the foundation of permanent buildings, and except in other locations where the applicable code requires the use of pressure-treated wood, heartwood grades of naturally resistant species can be used in place of pressure-treated lumber. Some acceptable grades and species of naturally resistant dimension lumber are: California redwood—Select Heart, Construction Heart, or Foundation 2 Grade; tidewater red cypress—Heart Common; and Eastern red cedar—Foundation Grade. Other decay-resistant species are black locust and black walnut.

For maximum durability, 100% heartwood of resistant species selected during grading should be used. For such species as redwood and cedar, several grading authorities, including the Redwood Inspection Service, the West Coast Lumber Inspection Bureau, and the Western Wood Products Association, have foundation grades selected from 100% heartwood with a high toxic extractive content. Lumber with natural decay and termite resistance should be marked with the grademark of an inspection agency authorized by the American Lumber Standards Committee. Where wood is embedded in the ground for the support of permanent structures, it should be pressure-treated.

## 6.4.3.4 Preservative Treatment Types

Preservative treatments protect wood against decay and insect attack.

In all areas of the country, lumber and wood products used in contact with the ground or embedded in the ground should be preservative pressure-treated.

Lumber and plywood used in exterior foundation walls below an upper top plate and wood footings or bottom plates placed in contact with concrete or the ground should be pressure-treated. Treatment should be in accordance with the American Wood Preservers Association (AWPA) Standard C22, "Lumber and Plywood for Permanent Wood Foundations, Preservative Treatment by Pressure Process."

Treated lumber and plywood should bear the appropriate American Wood Preservers Bureau (AWPB) Quality Mark or that of an acceptable independent inspection agency that maintains continuing control, testing, and inspection of the quality of the product. Treated lumber should be dried to a moisture content not exceeding 19% after treatment. To the extent practicable, fabricating, cutting, boring, and trimming of lumber and plywood should be done before treatment. Lumber or plywood cut after treatment should have material exposed by the cutting brush-coated with a 3% solution of the same preservative used in the original treatment.

In areas of severe termite infestation (see Section 21.6) or with a high decay hazard, such as damp coastal areas (see Fig. 6.1-9), lumber and wood products should be preservative pressure-treated in the following locations:

1. Members in contact with masonry or concrete
2. Members at grade, below grade, or less than 8 in. (200 mm) above grade
3. Members in a crawl space (floor joists less than 18 in. [460 mm] above interior grade and wood girders less than 12 in. [305 mm] above interior grade)
4. Members used in conjunction with roofing or flashing
5. Wood siding closer than 6 in. (150 mm) to exterior finish grade

Other wood members and products may require preservative treatment where dictated by local experience.

Wood preservatives may be divided into three major classes: (1) waterborne preservatives, (2) oil-borne preservatives, and (3) creosote and solutions containing creosote.

### WATERBORNE PRESERVATIVES

Waterborne preservatives are most often used to pressure-treat lumber and plywood for construction because they leave the wood clean, odorless, and easy to paint, and because they present less of an environmental hazard than do other preservatives. Three types have proven to be durable and will protect wood to be used in ground contact supporting permanent building structures. These are ammoniacal copper arsenate (ACA), ammoniacal copper zinc arsenate (ACZA), and chromated copper arsenate (CCA). The last is available in three different formulations, but all are considered to be equivalent in performance. Waterborne preservatives are relatively odorless and paintable, provided that the wood has been dried to the moisture content required for untreated wood. These preservatives will not cause discoloration in interior finishes unless a wet-process finish, such as plaster, is applied to treated wood without a waterproof backing paper.

There has been a movement recently to reduce or eliminate the use of preservative that contain arsenic. Two new materials that may replace them are alkaline copper quat (ACQ) and copper boron azole (CBA).

### OIL-BORNE PRESERVATIVES

Oil-borne preservatives are available today, but are much less used in the construction industry than their waterborne equivalents.

Lumber treated with oil-borne preservatives can be used in any installation that does not place it in contact with salt water, except where its use is not permitted by code, law, or other restriction. Less oil-borne preservative-treated lumber is used each year, and its use may eventually be banned altogether due to environmental concerns. Such lumber is often prohibited for use where it will come into contact with people, especially children.

The oil-born preservative Penta (pentachlorophenol), which has been the most widely used oil-borne preservative for several years, is highly toxic to both fungi and insects, and is insoluble in water and thus permanent. It is this permanence that has caused such great concern about its use. Because it is almost impossible to break down and because of its high toxicity, Penta has been designated a toxic environmental hazard and is no longer used in the building construction industry.

Other oil-borne preservatives used today include copper naptheanate, tributyl tinoxide, and copper 8-quinolinolate.

## CREOSOTE

Creosote is a widely used preservative for some applications, but it is also being prohibited in some jurisdictions for use where it is likely to come into contact with people. Creosote and solutions containing it are used where protection against wood-destroying organisms is of first importance, where painting is not required, and where its odor is not objectionable. Wood may be protected in severe exposures with high retentions and in less severe exposures with lower retentions.

*Creosote-coal tar mixtures* are the most widely accepted preservatives for marine or saltwater installations and are ideally suited for treating pilings used for shore dwellings.

*Creosote-petroleum mixtures* are used where economy is of first importance, but they should not be used for marine installations.

### 6.4.3.5 Treatment Methods

There are two methods of preservative treatment: (1) pressure processes and (2) nonpressure processes.

## PRESSURE PROCESSES

Pressure treatments, in which preservative chemicals are applied under pressure to obtain maximum penetration, afford the greatest protection for lumber and plywood against decay or termite attack.

There are two pressure preserving processes in general use: the *full cell process* and the *empty cell process*. Both are dependable means of ensuring uniform penetration and distribution of preservative. Timber to be treated by either of these pressure processes is loaded on tram cars and rolled into a long steel cylinder called a *retort*, which is then sealed.

The *full-cell process* is used for treating with creosote, creosote mixtures, waterborne preservatives, and fire-retarding chemicals (see Section 6.4.3). It provides high retention of chemicals, as is required for wood to be installed in a marine environment, for example.

In the full cell process, much of the air is removed from the cylinder and con-sequently from the wood. Then the cylinder is filled with preservative under various combinations of temperature and pressure to force the preservative deep into the wood. In this process, preservative is not only absorbed by the cell walls, but the cell cavities also are left full. After the required penetration and retention are achieved, the preservative charge is removed. This process leaves the maximum amount of preservative in the wood but, when used with creosote or its solutions, will not leave as clean a surface as do the other processes. In addition, especially when waterborne preservatives are used, the wood is left wet and must be air or kiln dried before shipment.

The *empty cell process* is primarily used for treatment with oil-borne preservatives. In this process, the wood is subjected to high air pressure before the preservative is introduced. When this pressure is released, the expanding trapped air expels the excess preservative from the wood cell cavities, thus reducing the net retention of preservative. This method provides treatment of the cell walls but leaves the cell cavities empty. Thus it yields a drier product, gives deeper and more uniform penetration, and permits maximum penetration for a given weight. Job-cut ends of preservative-treated lumber should be brush-coated with not less than a 3% solution of an approved preservative, usually the same one used in the factory process.

## NONPRESSURE PROCESSES

Nonpressure treatment can be effective for exterior millwork such as precut door, window, and trim components, where the wood is generally dry but may be exposed to moisture or wetting intermittently. Nonpressure treatment processes should not be used on wood that will be in contact with the ground or exposed to severe conditions of decay or termite attack, because the preservative is not sufficiently impregnated in the wood to afford adequate protection. Nonpressure-treated lumber should be seasoned to a maximum moisture content of 19%, or lower if necessary for the intended use. Surfaces which are cut or milled after treatment should be treated again with a brush application of preservative before final assembly or installation. *Vacuum* and *brief dip* nonpressure treatments usually provide clean, paintable surfaces.

In the *vacuum process*, a vacuum or partial vacuum is introduced to exhaust air from the cells and pores of the wood. Atmospheric pressure is then reintroduced to force the preservative into the wood.

The *brief dip* or 3-minute immersion process consists of completely submerging the wood in an open tank of the preservative solution for at least 3 minutes.

## STANDARDS AND CERTIFICATION OF QUALITY

Standards have been established for the wood preserving industry, and certification seals are available for treated products.

*Pressure-Treated Wood* The American Wood Preservers Association (AWPA) is a technical and research organization serving the wood preserving industry. AWPA develops and promulgates requirements for preservative formulations, methods of treatment, and minimum chemical retentions, which are the recognized industry standards for the pressure preservative treatment of wood. The most commonly used AWPA standards are shown in Figure 6.4-3. A related organization, the American Wood Preservers Institute (AWPI), performs the promotional and educational functions for the industry.

*Non-Pressure-Treated Wood* The Window and Door Manufacturers Associa-

**FIGURE 6.4-3** Frequently Used Preservative Treatment Standards

| AWPA Standards | |
|---|---|
| C1 | Preservative treatment by pressure process |
| C2 | Lumber, timbers, bridge ties, and mine ties, pressure treatment |
| C9 | Plywood, pressure treatment |
| C15 | Wood for commercial-residential construction, pressure treatment |
| C22 | Lumber and plywood for permanent wood foundations, pressure process |
| C23 | Round poles and posts used in building construction, preservative treatment by pressure processes |

tion (WDMA) administers a testing and plant inspection program, resulting in certification and issuance of its seal. The WDMA seal assures that a product conforms to WDMA IS-4. WDMA issues a list of products that comply with IS-4.

### 6.4.4 FIRE-RETARDANT TREATMENT

The natural fire resistance of wood can be greatly enhanced by its impregnation with inorganic salts which react chemically at temperatures below the ignition point of wood. This reaction causes the combustible vapors normally generated in the wood to break down into non-flammable water and carbon dioxide.

Fire-retardant-treated wood has been marketed since World War II. Today's treatment methods and chemicals provide a uniformity that was unattainable earlier. Products tested and approved by the Underwriters' Laboratories, Inc. (UL) which bear its label are recognized by model building code authorities and federal agencies as suitable for structural use in fire-resistive and noncombustible construction.

After treatment, plywood and lumber 2 in. (50 mm) nominal or less in thickness should be dried to a moisture content of 19% or less. Designers may require a lower moisture content for fire-retardant-treated wood to be used for millwork, cabinets, office paneling, and other special uses. Fire-retardant-treated wood is generally not suitable for transparent finishes. Most of it is painted when exposed to view.

Except where exterior-grade treatment complying with rain-testing requirements has been employed, fire-retardant-treated wood should not be used where it will be exposed directly to the weather. The presence of fire-retardant chemicals in treated wood may prohibit the use of some wood preservatives, such as CCA.

## 6.5 Foundations

Foundations used for light wood framed buildings include concrete walls and footings (Chapter 3), masonry walls (Chapter 4), and permanent wood foundations, which are discussed in this section.

Permanent wood foundation system principles were originally developed cooperatively by the AWPI, the U.S. Forest Service, and the National Forest Products Association (now the American Forest & Paper Association [AF&PA]). Permanent wood foundation systems literature is currently published by the Southern Pine Council, a promotional body coordinated and supported by the members of the Southern Forest Products Association (SFPA) and the Southeastern Lumber Manufacturers Association (SLMA).

This section is based on the SPC publication *Permanent Wood Foundations: Design and Construction Guide*, which was compiled by SFPA.

Most local building officials and all major model building code authorities, including the International Code Council's *International Residential Code*, the U.S. Department of Housing and Urban Development/Federal Housing Administration (HUD/FHA), and the Farmers Home Administration, accept permanent wood foundation systems.

Specific design and construction recommendations for a permanent wood foundation system are based on studies conducted by the National Association of Home Builders (NAHB) Research Foundation and by APA—the Engineered Wood Association.

The system provides a practical foundation for light-frame buildings with the following advantages over more conventional foundation systems:

- A reduction in construction delays due to inclement weather
- Savings in cost
- Less erection time
- Creation of a more comfortable below-grade living area

The system requires good quality control to prevent seemingly minor errors from resulting in serious problems later. For example, without such quality control, a worker may substitute a piece of untreated lumber where treated lumber is required or careless backfilling may damage the protective film layer and expose the wood to below-grade earth.

### 6.5.1 GENERAL DESCRIPTION

A permanent wood foundation system is fabricated from preservative pressure-treated lumber and pressure-treated plywood. The structural elements of the substructure are wood. Foundation wall sections of nominal 2 in. (50 mm) lumber and plywood cladding may be factory fabricated or constructed at the job site.

Either conventional concrete footings or wood footing plates may be used. Wood footing plates are nominal 2 in. (50 mm) or thicker pressure-treated wood planks resting on a 4 in. (100 mm) or thicker bed of fine gravel or crushed stone.

The walls are of wood stud construction with a plywood facing on the earth side. Where a basement occurs, a finish on the interior of the wall is optional. When habitable space is enclosed, the wall is insulated.

The system can be used for both basement and crawl space construction. The below-grade portions of exterior basement walls are covered with a polyethylene film bonded to the plywood at the top. Joints are lapped and sealed with adhesive. Concrete basement floors are poured over a gravel base.

Typical details of wood foundation systems for basement and crawl space construction are shown in Figures 6.5-1 and 6.5-2. There are several permitted variations, however, so details different from those shown may be required.

### 6.5.2 SITE PREPARATION

For basement construction, a house site should be excavated and leveled to the required elevation. Utility lines located below the basement floor should then be installed, and provision should be made for foundation drainage in accordance with local requirements.

When habitable space is planned below grade, unless the soil is very porous, a sump is required to ensure adequate drainage (Fig. 6.5-3). A sump can be drained by gravity or by pump to a storm sewer or drainage swale.

After grading and installation of plumbing lines and drainage system, the basement site should be covered with a layer of gravel. For crawl space construction, trenches should be dug to the required depth and gravel placed as a base for the footing plates.

### 6.5.3 FOOTINGS

The width of footing plates is determined by loads on the foundation wall and the bearing capacities of the gravel base and soil. When brick veneer is used, larger footing plates are required to provide support for the veneer. Brick height should not exceed 8 ft (2400 mm). To reduce the amount of brick required, such veneer is typically supported below grade on a wood stud knee-wall which rests on and is nailed to the treated wood footing.

Field applied 2x top plate

Floor joist

Framing anchors for deep backfill

2x stud wall

Required insulation

Vapor retarder

Interior finish

Concrete slab

Polyethylene sheeting

1x screed board (optional)

Sheathing or exterior siding may overlap field applied top plate for shear transfer (Flashing not required if siding overlaps). See Note

8" min. between grade and 2x top plate

2x top plate*

Caulk

1x or plywood strip protecting top of polyethylene sheeting (12" nom.)

Finish grade slope 1/2" per foot min. 6' from wall

Backfill**

Asphalt or polyethylene strips

Polyethylene sheeting

2x bottom plate

2x footing plate

3/4 d

Below frost line

Gravel, coarse sand, or crushed stone fill (4" for Group I and II soils, 6" for Group III)

d

2 d

\* Not required to be treated if backfill is more than 8″ below bottom of plate. Typical for all details.
\*\* Backfill with crushed stone or gravel 12″ for Group I soils, and half the backfill height for Group II and Group III soils.
*Note:* For daylight basement foundations, use double header joists (stagger end joints) or splice header joist for continuity on uphill and daylight sides of building.

**FIGURE 6.5-1 Basement foundation.** (Courtesy Southern Forest Products Association.)

Where basement construction is involved, footings under exterior foundations should be placed on a strip of polyethylene film laid directly on the gravel base.

A load-bearing wall may be used as center support under the first floor (Fig. 6.5-4). If a girder support is used instead, the footings under the foundation ending walls and under the posts should be designed for concentrated loads.

Footing plates must be located below the frost line.

### 6.5.4 FOUNDATION WALLS

Detailed structural requirements, including member spacings and sizes, for the foundation wall of typical one and two-story wood frame houses are shown in

*Permanent Wood Foundations: Design and Construction Guide.*

#### 6.5.4.1 Design Method

Lumber framing members in exterior walls enclosing a basement are designed to resist the lateral pressure of fill and the vertical forces resulting from live and dead loads on the structure. The plywood sheathing is designed to resist the maximum inward soil pressures at the bottom of the wall.

Foundation walls for crawl space construction are designed to resist only vertical loads when the difference between the outside grade and ground level in the crawl space is 12 in. (300 mm) or less. Framing is covered with 1/2 in. (12.7 mm) plywood to keep out vermin and provide additional stability.

#### 6.5.4.2 Construction

Foundation walls are built over a previously laid wood footing. Common practice is to prefabricate a foundation wall in modular sections. Components are fastened together using corrosion-resistant nails. Treated foundation wall studs are end-nailed to treated top and bottom plates. Treated plywood panels, 1/2 in. (12.7 mm) thick or thicker, are attached to studs. Usually, panels are applied with their face grain vertical and parallel to the studs. Vertical joints between panels within wall sections should be located over studs.

Foundation wall sections should be nailed to the wood footing and to adjacent sections. Wall panel joints should not occur over joints in the footing. An upper top plate should be face-nailed to the lower top plate of the foundation wall to tie the wall sections together.

An alternate method of attaching the upper top plate is to fabricate the foundation wall sections with the edge of the plywood wall panels extending 3/4 in. (19.1 mm) above the top of the treated top plate. The field-applied top plate is then face-nailed to the top plate of the wall section and is secured to the overlapping edge of the plywood.

Joints in the field-applied top plate should be staggered with respect to joints in the treated wood plate of the wall panels. At corners and wall intersections, the field-applied upper plate of one wall should be extended across the treated plate of the intersecting wall to tie the building together. The first floor is then installed following standard platform frame construction practices.

Plywood panel joints in the foundation wall should provide the minimum space recommended by the manufacturer of the sealant compound selected. In both basement and crawl space construction, the exterior side of foundation walls should be covered below grade with a 6 mil (0.1524 mm) polyethylene sheet to act as a vapor retarder. This film should be bonded to the plywood, and should have its joints lapped and sealed (see Figs. 6.5-1 and 6.5-2).

The top edge of the polyethylene sheeting should be adhered to the plywood wall with adhesive. Film areas near grade level can be protected from mechanical damage and exposure using treated plywood, brick, stucco, or other covering appropriate to the architectural design. Where plywood is used, the top edge of a 12-in. (305-mm)-wide panel of treated material is attached

| $h_i$ (in.) | Max. $h_o$ (in.) |
|---|---|
| 6 | 17 |
| 8 | 23 |
| 10 | 28 |
| 12 | 33 |
| 14 | 37 |

**FIGURE 6.5-2** Crawl space foundation. (Courtesy Southern Forest Products Association.)

to the wall several inches above the finish grade level. The top inside edge of the panel should be caulked full length before the panel is fastened to the wall.

Where appropriate, suitable insulation should be installed within the wall stud space. When habitable space occurs, an interior vapor retarder and wall

**FIGURE 6.5-3** Sump for poorly drained soils. (Courtesy Southern Forest Products Association.)

finish are usually applied to the stud framing.

### 6.5.5 FLOORS

For basement construction, a 6 mil (0.1524 mm) polyethylene film is applied over the gravel bed, and a concrete slab at least 3 in. (75 mm) thick is placed over the film. Placement of the film and the slab can be delayed until after the building is under roof. Backfill should not be placed against the foundation walls, however, until after the concrete slab is in place and set and the top of the wall is adequately braced. Where the height of backfill exceeds 4 ft (1200 mm), gravel should be used for the lower portion.

For crawl space construction, a 6 mil (0.1524 mm) polyethylene film ground cover should be used. Film edges should be lapped and sealed.

The first floor framing on a wood foundation is installed according to conventional platform framing methods. Floor joists should be suspended at least 18 in. (457 mm) above the ground.

### 6.5.6 CENTER SUPPORT

Framing members in load-bearing walls acting as center support for a first floor joist system are similar to exterior foundation walls (see Fig. 6.5-4), except that only vertical loads need be considered during design.

#### 6.5.6.1 Load-Bearing Wall

In basement construction, load-bearing interior walls should be made with pressure-treated bottom plates and, in areas of low termite hazard, with untreated studs and top plates. A double bottom plate is used to provide clearance between the top of the concrete slab floor and the untreated vertical framing in the wall (see Fig. 6.4-4). In areas of high termite hazard, pressure-treated studs should be used. The interior wall framing can be covered on both sides with gypsum wallboard or another interior finish as desired.

Interior load-bearing walls in crawl space construction should be made of treated framing and covered on one side with 1/2-in. (12.7-mm)-thick treated plywood.

#### 6.5.6.2 Post-and-Girder Support

Where a post-and-girder center support system is used, the posts should be sup-

**Subfloor**

**Floor joist (lapped)**

**Field applied 2x top plate**

**Treated 2x top plate**

**Treated 2x stud wall per local code. Structural panel sheathing or wall board on at least one side**

**Moisture barrier**

**Optional interior finish**

**Double 2x bottom plates**

**2x footing plate**

**Concrete slab**

3/4d

**Polyethylene sheeting**

d

**4" or 6" gravel or crushed stone fill**

2d

**FIGURE 6.5-4** Basement bearing partition. (Courtesy Southern Forest Products Association.)

ported on blocks of sufficient size to ensure that the vertical concentrated loads will not exceed the bearing capacity of the gravel base and soil. Foundation walls supporting a center girder should also be designed to carry and distribute the concentrated load from the girder to a wood footing and gravel base so that the allowable gravel and soil-bearing capacities are not exceeded. Design of a suitable post-and-girder support system requires engineering calculations.

### 6.5.7 PRESERVATIVE TREATMENT

Lumber and plywood used in permanent wood foundation systems should be pressure treated in accordance with AWPA Standard C22, "Lumber and Plywood for Permanent Wood Foundations." Refer to paragraph 6.4.2.6 for additional requirements related to pressure treatment in permanent wood foundations.

## 6.6 General Framing Requirements

A variety of methods and systems are used in the construction of wood-framed buildings across the country. Local climatic conditions, economic considera-

tions, and esthetic preferences may influence the selection of the system that will be used. This section includes some of the requirements common to the various systems used. Sections 6.7 through 6.11 address specific aspects of various framing systems.

### 6.6.1 ANCHORAGE REQUIREMENTS

The superstructure of a wood-framed building should be anchored to the foundation to provide resistance to lateral forces, primarily from wind. Wind pressures produce compressive, tensile (uplift), shearing, and racking loads on foundations (Fig. 6.6-1).

#### 6.6.1.1 Overturning, Translation, and Rotation

*Overturning* forces may produce critical stresses in foundations that support light superstructures subjected to high wind pressures. Connections anchoring a building's superstructure to its foundation and the foundation itself must have adequate strength in tension to resist these forces.

When the ratio of the exposed building surface to the width of the building is large, high wind forces tend to cause *translation* (lateral movement) of the superstructure. When the wind forces are not symmetrical, building *rotation* can occur. Forces tending to cause translation and rotation are transferred to a foundation by the first floor acting as a diaphragm. Connections anchoring the superstructure must develop enough shearing strength to transmit these forces to the foundation. When a first floor does not have enough rigidity to provide diaphragm action, high shear and bending stresses may be produced in the foundation walls. In this case, consideration

**FIGURE 6.6-1** Wind forces tend to lift or overturn light superstructure, move them laterally, or rotate them on their foundations.

should be given to the use of shear walls, pilasters, or bond beams to stiffen the walls.

### 6.6.1.2 Anchorage Recommendations

Superstructures should be anchored to both masonry and concrete foundations as applicable. These anchoring requirements are covered in the International Code Council's *International Building Code* and *International Residential Code.* Anchorage of light-gauge steel framing is also addressed in Section 5.7. In general, the following requirements apply to wood-framed systems, but each should be verified with the requirements in the applicable code.

The sill plates of a wood joist floor system should be anchored to masonry walls with $1/2$ in. (12.7 mm) bolts, extending at least 15 in. (381 mm) into the filled cores of masonry units. Anchor bolts should be spaced not more than 6 ft (1830 mm) apart, with one bolt not more than 12 in. (305 mm) from each end of the sill plate.

For concrete walls, sill plates should be anchored with $1/2$ in. (12.7 mm) bolts embedded 7 in. (178 mm), with the same maximum spacing as for masonry walls. Hardened steel studs driven by powder-actuated tools are not specifically covered in the international codes, but may be allowed as an alternate, except when earthquake design is required. Their use is subject to the requirements of the ICC evaluation reports. Their maximum spacing should not exceed that required by the reports.

### 6.6.2 NAILING PRACTICES

Fastening of light-gauge steel framing components is addressed in Section 5.7. Of primary consideration in the construction of a light wood-framed building is the method used to fasten the various wood members together. These connections are most commonly made with nails, but in certain cases metal straps, lag screws, bolts, staples, or adhesives may be used. Proper fastening of frame members and covering materials provides the rigidity and strength to resist windstorms, earthquakes, and other destructive hazards. Adequate nailing also is important for suitable performance of wood parts. In general, nails provide stronger joints when driven into the side grain (perpendicular to wood fibers) than into the end grain of wood. For instance, the allowable lateral (shear) loads for nails driven into the side grain are 50% higher than those allowed when the nails are driven into the end grain. Whenever possible, connections should be designed in such a way that nails are loaded in shear, exerting a sideways pull on the nail, rather than in direct withdrawal, pulling straight out. Connections should never be designed in a way which would load an end-nailed joint in direct withdrawal (Fig. 6.6-2).

The international codes have tables that show the required nailing practices for the framing of a conventional wood-framed building. Penny designations and actual sizes of common wire nails are shown in Figure 6.6-3.

**FIGURE 6.6-2** Common nailing methods: (a) slant nailing at a 30-degree angle to the face or edge of the attached piece (toenailing); (b) nailing into the end grain of the supporting piece (end nailing); (c) nailing into side grain through the wide dimension of the attached piece (face nailing); and (d) toenailing through the edge of a plank to conceal the nail head (blind nailing).

**FIGURE 6.6-3** Standard sizes of common wire nails. Thicker nails in sizes 10d to 60d are known as spikes. (Refer to Figure 25.7-1 for metric conversions.)

# 6.7 Conventional Framing and Furring

Modern light wood framing is an outgrowth of the *balloon frame* which originated around the middle of the nineteenth century in Chicago. It was a revolutionary development which drastically changed the direction of the housing industry and contributed substantially to the colonization of the West. Historians have claimed that if not for the balloon frame, Chicago and San Francisco could not have grown from villages to great cities in a single year. Houses and other wood frame structures were built with a rapidity made possible only with this revolutionary type of construction.

Before the balloon frame evolved, frame buildings were built of heavy timbers, mortised-and-tenoned together and pinned with hardwood dowels (see Section 6.11). The principle of balloon framing involves the repetitive use of slender structural members called *studs* for walls and *joists* for roofs and floors, all tied together with simple nailed connections. It changed house construction from a highly skilled craft to a relatively simple mechanical procedure that could be learned quickly by most handymen familiar with hand tools. For that reason, perhaps, it was regarded with suspicion and acquired its derisive name on the assumption that such a house would blow away like a balloon the first time a strong wind hit it.

The development of the balloon frame is closely associated with the level of industrialization reached in the United States in the middle of the nineteenth century. The evolution of sawmill technology and the ability to mass produce cut nails cheaply are credited with the impetus to this revolutionary idea. When the manufacture of cut nails began at the turn of the century, wrought nails cost 25 cents a pound. With the introduction of more sophisticated machinery, the price was reduced to 8 cents in 1820, to 5 cents in 1833, and to a mere 3 cents a pound by 1842.

Over the years, balloon framing was modified to permit the use of shorter (and cheaper) wall studs spanning from floor to floor rather than from foundation to roof, as in the original system. This new approach made it possible to erect a floor platform first and then to nail together a full-length wall on a horizontal surface, where it could be done more conveniently and faster than in an upright position. This modified framing system evolved as wood framing moved west with the early settlers and came to be known as *Western* or *platform framing*. Most new wood-framed buildings today are platform framed.

The search for economies in labor and materials is a continuing process. As the cost of labor became a more significant factor in the total housing cost, plank-and-beam construction (see Section 6.11) and other floor and roof assemblies evolved, utilizing greater spacing of fewer and larger structural members. Such systems were more compatible with post-and-beam wall framing (see Section 6.11) and, as with the skeleton structure of steel and concrete buildings, enabled designers to treat the space between posts as a curtain wall, intended mainly to keep out the weather rather than to resist loads. The use of prefabricated trusses (see section 6.11) for roof support permitted fewer, and sometimes no, interior walls.

In spite of innovations in light wood framing, the essential construction principles inherent in the balloon frame persist. The nail still remains an essential connective element in light wood framing systems, and the exterior covering between framing members, the sheathing, is relied upon to provide structural continuity to an assembly of light members.

Much of the material in this and related sections on wood framing is based on data in the U.S. Department of Agriculture's Agriculture Handbook No. 73, *Wood-Frame House Construction*; the publications of the National Forest Products Association; the publications of the American Plywood Association; and the other publications listed in Section 6.16, or noted in the text, or both.

The superstructure of a wood-framed building is made up of floor and ceiling assemblies, including their subflooring (see Section 6.10), walls, and roof and ceiling assemblies, including their sheathing (see Section 6.10). Some components of the superstructure may be preassembled (see Section 6.11). The superstructure may be conventionally framed or may be constructed using other framing systems (see Section 6.11). The superstructure may be supported directly on foundation walls (see Chapters 3 and 4; Section 6.5) or may rest on posts, beams, and girders that in turn are supported by the foundations.

In the design, selection, and construction of a floor and ceiling assembly, a wall assembly, or a roof and ceiling assembly, it is necessary to consider (1) the required level of functional performance, (2) the costs, and (3) the desired finished appearance. These considerations often are influenced by the systems and finishes used in other basic elements of the construction.

The ultimate costs of a floor and ceiling assembly, wall assembly, or roof and ceiling assembly include in-place costs, maintenance costs, and operating costs. In-place costs are the sum of the costs of materials, labor, and overhead required to assemble and install the assembly. The type of assembly selected may affect other costs, such as those for installing insulation and a vapor retarder, applying finishes, providing for openings (windows, doors, stairs, skylights, etc.), constructing and supporting partitions and other components of the building, and providing space to accommodate heating, air conditioning, electrical, and plumbing components and equipment.

The necessary time and costs to enclose the building and the costs of installing heating, air conditioning, electrical, and plumbing equipment may also be influenced by the type of assembly.

Maintenance and operating costs, such as those attributable to heat loss or gain through floors and floor, wall, and ceiling assemblies, have an important influence on the ultimate cost of a building and should not be overlooked if long-term economy is to be achieved.

## 6.7.1 FLOORS AND FLOOR AND CEILING ASSEMBLIES

The floor framing in a conventional wood-framed building consists of posts, beams (or girders), sill plates, and joists. Floor sheathing (subflooring) ties the various framing elements together and stabilizes the joists to prevent twisting and buckling. When these elements are assembled properly on a foundation, they form a level anchored platform for the erection of partitions, additional floors, and the roof. Posts and center beams, which support the inside ends of joists, sometimes are replaced with a wood frame or masonry wall when the basement area is divided into rooms. Wood-framed buildings also may be constructed over a crawl space with floor framing similar to that used over a basement. In a slab-on-ground construction, wood floor framing is limited to the upper floors.

In multilevel buildings, the combination of each intermediate floor and the ceiling systems below it is usually recognized as a single basic element of the superstructure. A combined floor and ceiling system is called a *floor and ceiling assembly*.

A floor and ceiling assembly should have:

1. Adequate strength to support the materials used in the floor and ceiling assembly and other elements in the building which it supports (dead load), as well as loads imposed by the occupants and furnishings (live loads)
2. Provisions for lateral support of walls
3. Satisfactory resistance to the transmission of airborne and structure-borne sound
4. Suitable fire resistance
5. Suitability for the application of finish materials
6. Adaptability to economical methods of assembly and erection
7. Space to accommodate heating, air conditioning, electrical, and plumbing equipment
8. Control of heat loss and the flow of water vapor

Floor and ceiling assemblies must support superimposed loads safely and without excessive deflection or vibration. Design dead and live loads are usually established by building codes which require structural members to be sized so that an allowable fiber stress in bending is not exceeded when the assembly is fully loaded. Codes do not always limit deflection, because it seldom has a bearing on structural failure or safety. Deflection limitations should be established in floor and ceiling assembly design, however, to reduce deflection to acceptable limits, to ensure the integrity of applied finish materials, and to limit vibration to acceptable levels of human perception. Deflection, rather than bending strength, is generally the controlling factor in the structural design of wood floor and ceiling assemblies.

Depending on the design of the foundation system, a first-floor system may be required to provide lateral support to foundation walls. Floor and ceiling assemblies may also be required to provide lateral support for other elements of the superstructure against loads caused by wind or earthquake.

The makeup of a floor and ceiling assembly is influenced by the desired finish applications. The finish materials selected may:

1. Influence the spacing and span of structural members
2. Determine the need for floor underlayment or wall or ceiling furring when varying structural component spacing is impractical, as is often the case, or when such variation would violate code requirements
3. Dictate methods of construction

Prefinished floor or ceiling surfaces may not require the application of an additional finish after the system has been placed. However, the floor should provide a working platform for finishing operations during the construction period. Therefore, a floor and ceiling assembly is often selected that has a subfloor or sheathing application that is later provided with finished surfaces.

Wood-framed floor and ceiling assemblies include wood floor joists or floor trusses, plank-and-beam construction (see Section 6.11), heavy timber construction (see Section 6.11), and panelized construction (see Section 6.9). Concrete slab-on-grade floor construction is discussed in Section 3.12.

### 6.7.1.1 Lumber Materials

Framing lumber should be of the proper grade for the intended use as graded by a recognized grading authority. The moisture content of framing lumber is important in the performance of finishes. It should be as close as possible to the equilibrium moisture content (see Section 6.1) for the locality and should not exceed 19% when the finish material is installed.

However, a 15% or lower moisture content is better in all regions of the country and essential in arid regions, especially when a finish, such as gypsum board, will be applied directly to the framing. Dimension lumber generally can be obtained at a moisture content of 19% or less when S-DRY is specified or at 15% or less when MC 15 is specified. In arid parts of the country, lumber of higher moisture content can be specified if the lumber is allowed to approach its equilibrium moisture content before interior finishes and trim such as baseboard, shoe mold, and door casings are applied.

In light wood-framed buildings, wood studs are typical framing members for walls and partitions; wood joists and trusses are the most common ceiling and roof framing members. These framing members generally are spaced 16 in. (406 mm) or 24 in. (610 mm) on centers, which permits direct application of most finishes, such as gypsum board and plaster. However, furring (see Section 6.7.4) may be required when member spacing exceeds the maximum recommended for the particular finish material or when special performance, such as increased sound isolation, is desired.

The National Grading Rule for dimension lumber, established in accordance with PS 20 (see Section 6.2), provides a series of uniform grade designations applicable to all species. Joists and planks (2 in. [50 mm] to 4 in. [100 mm] thick and 5 in. [130 mm] wide or wider) are graded as Select Structural, No. 1, No. 2, and No. 3 grades. All grades are suitable for floor joists if they have adequate strength characteristics for the intended spacing, span, and loading conditions. Light framing (2 in. [50 mm] to 4 in. [100 mm] thick and 2 in. [50 mm] to 4 in. [100 mm] wide) is sorted into as many as seven grades. These grades, in order of decreasing strength, are Select Structural, No. 1, No. 2, Construction, No. 3, Standard, and Utility. An eighth, Stud grade, is similar to No. 3 grade but is cut to 10 ft (3050 mm) or shorter lengths, is straighter, and has a better nailing surface. The lower light framing grades should be used for sills and plates rather than for studs and posts.

### 6.7.1.2 Posts, Beams, and Girders

Wood or steel posts generally are used in a basement to support wood girders or steel beams. Masonry piers also can be used in basements but are most commonly employed in buildings with crawl spaces.

Round steel posts, known as *pipe columns* or *lally columns* (steel pipes filled with concrete), can be used to support both wood girders and steel beams. Columns should be provided with a steel bearing plate at each end. Secure anchoring of a girder or beam to a post or column is essential (Fig. 6.7-1).

Wood posts for freestanding use should be of solid or built-up lumber and not less than the equivalent cross-sectional area of a solid 6 × 6 (150 ×

**FIGURE 6.7-1** (a) Steel top plate provides a means of anchoring a beam; (b) base plate distributes the load and anchors the steel post to the footing. (Figure 29, *Wood Frame House Construction*, U.S. Department of Agriculture, p. 41. Not subject to copyright in the United States.)

150). When combined with a frame wall, posts may be 4 × 6s (100 × 150s) to conform to the stud depth. Loads may dictate other sizes. Spaced columns may also be used. Refer to *Wood Structural*

*Design Data* by AF&PA for design requirements for wood columns. Wood posts should be squared at both ends and securely fastened to the girder (Fig. 6.7-2a). The bottom of a post should rest on and be pinned to a concrete pedestal 2 to 3 in. (50 to 75 mm) above the finish floor (Fig. 6.7-2b). It is good practice to treat the bottom of a post with a decay preservative or use a vapor-resistant covering such as heavy roll roofing over the concrete pedestal. A sheet metal covering is preferred in areas subject to termite infestation.

Both wood girders and steel beams are used in present-day wood-framed construction. Standard steel S- and W-shaped (wide-flange) beams are the most commonly used steel beam shapes. Wood girders may be either solid or built up (Fig. 6.7-3). A built-up beam is preferred because it can be made up from drier dimension lumber and is more dimensionally stable. Solid lumber or commercially available glue-laminated girders may be desirable when a beam will be exposed in a finished room.

The use of the term *beam* in connection with steel members and *girder* with wood members is based on common usage rather than a technical difference in meaning. In engineering terminology, a beam is any member with loads applied perpendicular to its direction of span and a girder is a principal beam, frequently supporting other beams. In this broader sense, joists, girders, headers, and trimmers are types of beams. Wood joists are seldom called beams, but both built-up and solid wood members carrying joists frequently are referred to as beams or girders.

A built-up girder (see Fig. 6.7-3) usually is made up of two or more pieces of 2 in. (50 mm) lumber nailed together, with the ends of the pieces joined over a supporting post. Building codes dictate the required type, location, size, and number of fasteners. For example, a code might require that a two-piece girder be face-nailed from one side with 10d nails, two nails at the end of each piece and others driven near the top and bottom in two rows 16 in. (406 mm) on centers, with the nails staggered. A three-piece girder might be required to be face-nailed from each side with 20d nails, two nails near each end of each piece and others in two rows driven 32 in. (810 mm) on centers staggered.

Examples of member sizes and the number of members in built-up wood girders supporting floor loads for various spans are shown in Figure 6.7-4. The

**FIGURE 6.7-2** (a) A wood post should be connected to a girder with metal angles of framing anchors; (b) the base of a wood post should be pinned and protected against decay and termites. (Figure 30, *Wood Frame House Construction*, U.S. Department of Agriculture, p. 41. Not subject to copyright in the United States.)

**FIGURE 6.7-3** At least one member of a built-up wood girder should be continuous across a supporting post. In termite-infested areas, a beam pocket should be lined with metal to form a termite shield. (Figure 31, *Wood Frame House Construction*, U.S. Department of Agriculture, p. 42. Not subject to copyright in the United States.)

**FIGURE 6.7-4** Maximum Allowable Spans for Built-Up Girders Supporting One or Two Floors

Load diagram annotations (supporting one floor): LL + DL = 50 psf; "L/2 ± 1"; Girder; "L"; "S"; "S"; Load diagram applies also to 2-st house below.

Load diagram annotations (supporting two floors): LL + DL = 40 psf; Wall load = 50 plf; LL + DL = 50 psf; "L/2 ± 1"; Girder; "L".

### Maximum Allowable Spans ("S") for Girders Supporting One Floor[a]

| Fb, psi | Lumber Size | 20 ft | 22 ft | 24 ft | 26 ft | 28 ft | 30 ft | 32 ft | 34 ft | 36 ft |
|---|---|---|---|---|---|---|---|---|---|---|
| 1000 | (2) 2 in. × 6 in. | 4 ft 5 in. | 4 ft 0 in. | — | — | — | — | — | — | — |
| | (3) 2 in. × 6 in. | 5 ft 6 in. | 5 ft 3 in. | 5 ft 0 in. | 4 ft 10 in. | 4 ft 8 in. | 4 ft 5 in. | 4 ft 2 in. | — | — |
| | (2) 2 in. × 8 in. | 5 ft 10 in. | 5 ft 3 in. | 4 ft 10 in. | 4 ft 5 in. | 4 ft 2 in. | — | — | — | — |
| | (3) 2 in. × 8 in. | 7 ft 3 in. | 6 ft 11 in. | 6 ft 7 in. | 6 ft 4 in. | 6 ft 2 in. | 5 ft 10 in. | 5 ft 5 in. | 5 ft 1 in. | 4 ft 10 in. |
| | (2) 2 in. × 10 in. | 7 ft 5 in. | 6 ft 9 in. | 6 ft 2 in. | 5 ft 8 in. | 5 ft 3 in. | 4 ft 11 in. | 4 ft 8 in. | 4 ft 4 in. | 4 ft 1 in. |
| | (3) 2 in. × 10 in. | 9 ft 3 in. | 8 ft 10 in. | 8 ft 5 in. | 8 ft 1 in. | 7 ft 10 in. | 7 ft 5 in. | 6 ft 11 in. | 6 ft 6 in. | 6 ft 2 in. |
| | (2) 2 in. × 12 in. | 9 ft 0 in. | 8 ft 2 in. | 7 ft 6 in. | 6 ft 11 in. | 6 ft 5 in. | 6 ft 0 in. | 5 ft 8 in. | 5 ft 4 in. | 5 ft 0 in. |
| | (3) 2 in. × 12 in. | 11 ft 3 in. | 10 ft 9 in. | 10 ft 3 in. | 9 ft 10 in. | 9 ft 6 in. | 9 ft 0 in. | 8 ft 5 in. | 7 ft 11 in. | 7 ft 6 in. |
| 1500 | (2) 2 in. × 6 in. | 5 ft 3 in. | 4 ft 10 in. | 4 ft 5 in. | 4 ft 1 in. | — | — | — | — | — |
| | (3) 2 in. × 6 in. | 6 ft 9 in. | 6 ft 5 in. | 6 ft 2 in. | 5 ft 11 in. | 5 ft 8 in. | 5 ft 3 in. | 4 ft 11 in. | 4 ft 8 in. | 4 ft 5 in. |
| | (2) 2 in. × 8 in. | 7 ft 0 in. | 6 ft 4 in. | 5 ft 10 in. | 5 ft 4 in. | 5 ft 0 in. | 4 ft 8 in. | 4 ft 4 in. | 4 ft 1 in. | — |
| | (3) 2 in. × 8 in. | 8 ft 11 in. | 8 ft 6 in. | 8 ft 1 in. | 7 ft 9 in. | 7 ft 5 in. | 7 ft 0 in. | 6 ft 6 in. | 6 ft 2 in. | 5 ft 10 in. |
| | (2) 2 in. × 10 in. | 8 ft 11 in. | 8 ft 1 in. | 7 ft 5 in. | 6 ft 10 in. | 6 ft 4 in. | 5 ft 11 in. | 5 ft 7 in. | 5 ft 3 in. | 4 ft 11 in. |
| | (3) 2 in. × 10 in. | 11 ft 4 in. | 10 ft 10 in. | 10 ft 4 in. | 9 ft 11 in. | 9 ft 6 in. | 8 ft 11 in. | 8 ft 4 in. | 7 ft 10 in. | 7 ft 5 in. |
| | (2) 2 in. × 12 in. | 10 ft 10 in. | 9 ft 10 in. | 9 ft 0 in. | 8 ft 4 in. | 7 ft 9 in. | 7 ft 2 in. | 6 ft 9 in. | 6 ft 4 in. | 6 ft 0 in. |
| | (3) 2 in. × 12 in. | 13 ft 9 in. | 13 ft 2 in. | 12 ft 7 in. | 12 ft 1 in. | 11 ft 7 in. | 10 ft 10 in. | 10 ft 2 in. | 9 ft 6 in. | 9 ft 0 in. |

### Maximum Allowable Spans ("S") for Girders Supporting Two Floors[a]

| Fb, psi | Lumber Size | 20 ft | 22 ft | 24 ft | 26 ft | 28 ft | 30 ft | 32 ft | 34 ft | 36 ft |
|---|---|---|---|---|---|---|---|---|---|---|
| 1000 | (2) 2 in. × 8 in. | 4 ft 7 in. | 4 ft 2 in. | 4 ft 0 in. | — | — | — | — | — | — |
| | (3) 2 in. × 8 in. | 4 ft 9 in. | 4 ft 4 in. | — | — | — | — | — | — | — |
| | (2) 2 in. × 12 in. | 5 ft 10 in. | 5 ft 4 in. | 4 ft 11 in. | 4 ft 7 in. | 4 ft 3 in. | 4 ft 0 in. | — | — | — |
| | (3) 2 in. × 12 in. | 7 ft 1 in. | 6 ft 6 in. | 6 ft 0 in. | 5 ft 6 in. | 5 ft 2 in. | 4 ft 10 in. | 4 ft 6 in. | 4 ft 3 in. | 4 ft 0 in. |
| 1500 | (3) 2 in. × 6 in. | 4 ft 2 in. | — | — | — | — | — | — | — | — |
| | (2) 2 in. × 8 in. | 4 ft 8 in. | 4 ft 3 in. | — | — | 4 ft 1 in. | — | — | — | — |
| | (2) 2 in. × 10 in. | 5 ft 6 in. | 5 ft 0 in. | 4 ft 7 in. | 4 ft 3 in. | 4 ft 1 in. | — | — | — | — |
| | (2) 2 in. × 12 in. | 5 ft 8 in. | 5 ft 2 in. | 4 ft 9 in. | 4 ft 5 in. | — | — | — | — | — |
| | (3) 2 in. × 10 in. | 7 ft 0 in. | 6 ft 5 in. | — | 5 ft 6 in. | 5 ft 1 in. | 4 ft 9 in. | 4 ft 6 in. | 4 ft 3 in. | 4 ft 0 in. |
| | (3) 2 in. × 12 in. | 8 ft 6 in. | 7 ft 9 in. | 7 ft 2 in. | 6 ft 8 in. | 6 ft 2 in. | 5 ft 9 in. | 5 ft 5 in. | 5 ft 1 in. | 4 ft 10 in. |

[a] Maximum spans are based on 30 psf live load for second floor and 40 psf live load for first floor: 10 psf dead load for either floor: 50 psf dead load for partition above girder.

[b] Spans given may be interpolated on a straight line basis for house widths intermediate between those shown.

table indicates required lumber sizes for various conditions of loading and framing, assuming a minimum fiber stress in bending ($F_b$ of 1000 psi [703.07 Mg/m$^2$] and 1500 psi [1054.6 Mg/m$^2$]). Species and grades meeting this requirement can be selected from tables in the *NDS*. These tables should be verified with the requirements of the applicable building code, which may vary from the table data. Solid beams with an equivalent cross section are usually permitted in lieu of built-up beams.

The ends of wood girders are generally required to bear at least 4 in. (100 mm) on foundation walls and pilasters. When the lumber is not preservative-treated, a $1/2$ in. (12.7 mm) air space is often required at each end and at each side of wood girders framing into a masonry or concrete wall (see Fig. 6.7-3). In termite-infested areas, these pockets should be lined with sheet metal, which should extend on the inside about $3^{1}/_2$ in. (90 mm) to form a termite shield. A girder should be accurately aligned (true) and level. Wood shims should not be used to level girders or beams. Where adjustment is required, slate, metal, or another durable, noncompressible material should be used to support the girder ends. Shim space under a beam or girder should be grouted solid.

### 6.7.1.3 Girder-Joist Connection

The simplest method of floor framing is one in which joists bear directly on a wood girder or steel beam. Then the top of the beam coincides with the top of the anchored sill plate. This method is used when adequate headroom exists and a dropped girder is not objectionable. It is simple to frame and has the advantage of providing ample joist cavities for ductwork and electrical and mechanical equipment.

### WOOD GIRDER OR BEAM

For more uniform shrinkage and to provide greater headroom, joists can be framed into a wood girder with joist hangers or ledger strips (Fig. 6.7-5). A ledger strip should be at least 2 × 2 (50 × 50) in size and should be face-nailed to the girder with three 16d nails at each point where a joist frames into the girder.

Depending on the sizes of the joists and girders, joists may be supported on a ledger strip. In the rare case where indi-

**FIGURE 6.7-5** Typical joist-girder connection showing how joists can be tied together when a wood board subfloor is used.

vidual boards rather than plywood are used for subflooring, continuity across the supporting beam or girder must be provided by elements other than the subfloor. A continuous horizontal tie between exterior walls can be obtained by nailing notched joists together (Fig. 6.7-5a), or a connecting scab at each pair of joists can provide this continuity and also serve as a nailing area for the subfloor (Fig. 6.7-5b). A steel strap can be used to tie joists together when the tops of beams and joists are level (Fig. 6.7-5c). When notched joists or scabs ride over the top of a girder, it is important that a small space be allowed above the beam to provide for shrinkage of the joists. Sometimes a ledger strip is positioned higher on a girder, and the joists are notched so that the joist shoulders bear on the ledger strip and the joist bottoms are level with the girder. In such situations, the notch depth should not exceed one-quarter of the depth of the joist.

When a space is required for heating ducts in a partition supported on a girder, a spaced wood girder sometimes is necessary (Fig. 6.7-6). Solid blocking is used at intervals between the two members. A single post support for a spaced girder usually requires a bolster (wide capital) of sufficient width to support the spaced girders.

### STEEL BEAM

Joists may be supported by or framed into a steel beam (Fig. 6.7-7). In general, connections between joists and beam should provide (1) lateral stability for the beam, (2) resistance against twisting and overturning of the joists, (3) structural continuity and diaphragm action to resist lateral pressures against the foundation walls, and (4) anchorage for the entire floor assembly. This is accomplished by bolting a 2 × 4 (50 × 100) or 2 × 6 (50 × 150) plate to the upper or lower beam flange, on which the joists are sup-

**FIGURE 6.7-6** Spaced girder detail showing connections to steel post and floor joists.

**FIGURE 6.7-7** Wood joists can be supported (a) on a wood ledger or (b) directly on lower steel beam flange.

ported (Fig. 6.7-7a). Each joist should be toenailed to the wood plate and should be bolted at 24 in. (610 mm) on centers, either through the beam's web or through its flange.

When it is not feasible to use a nailing plate, special clips can be nailed to the sides of the joists to develop the kind of connection described above. In Figure 6.7-7b, the nailing plate has been omitted, and a metal strap tie provides continuity at each joist and anchors the floor assembly, while 2 × 4 (50 × 100) blocking fitted between the joists keeps them

from twisting or overturning. Since strap ties only prevent the joists from pulling apart, the joists should be installed so that the ends are in contact with the beam web to provide compressive resistance as well.

### 6.7.1.4 Sill Plate Construction

The sill plate over a foundation wall should be securely anchored (see Section 6.6.1.2) and should be separated from the finish grade by at least 8 in. (200 mm). In the Gulf States and Florida, the sill plate and all wood members within 18 in. (450 mm) of grade should be preservative-treated against decay and termites. In areas of high humidity but where termite infestation is not a hazard, inherently decay-resistant grades of redwood, cypress, or red cedar can be used.

Because of its adaptability to the use of tilt-up and component construction methods, and because it provides a stable working platform during construction, platform framing (Fig. 6.7-8) provides great flexibility for the wide variety of possible design requirements in modern light wood-framed buildings.

A *box sill assembly* is commonly used in platform framing (Fig. 6.7-9). It consists generally of a 2 × (50 ×) (nominal 2 in. [50 mm] thick of some width) *band joist* and a 2 × 4 (50 × 100) or 2 × 6 (50 × 150) *sill plate*. The plate should be wide enough to provide a full base for the band joist plus at least a $1\frac{1}{2}$-in. (38-mm)-wide bearing ledge for the floor joists. When the band joist is perpendicular to the floor joists, it is also called a *header*; when parallel, a *stringer*. To differentiate between the two conditions, this terminology has been used in this book.

Each floor joist should be toenailed to the sill plate with three 8d or two 10d nails; a header should be end-nailed to each joist with two 16d nails. If suitable rigid wall sheathing is not nailed to the sill plate, the header should be toenailed to the sill plate with 8d nails 16 in. (406 mm) on centers. A stringer similarly should be toenailed to the sill plate with 8d nails 16 in. (406 mm) on centers. A sill plate should be anchored to the foundation wall over a resilient sill sealer or bed of grout. A termite shield should be used in areas subject to termite infestation.

### 6.7.1.5 Floor Joists

A wood joist floor is part of the floor and ceiling assembly portion of a building's

FIRESTOPPING

SHEET ROOF SHEATHING

TRUSSES

JOIST

SHEET SHEATHING

STEEL GIRDER

STUD WALL

SHEET SUBFLOORING

FOUNDATION

**FIGURE 6.7-8** Customary platform framing, characterized by floor joists and story-height studs spaced 16 in. (406 mm) or 24 in. (610 mm) on center; roof trusses spaced up to 4 ft (1220 mm) on center; panel sheathing, such as plywood, gypsum, or fiberboard; and elimination of unnecessary floor bridging and corner bracing. Floor construction forms a work platform, permitting efficient tilt-up methods. It adapts easily to panelized construction.

superstructure, which also includes the walls and a roof and ceiling assembly. A joist floor is essentially the same whether the walls are wood-framed, solid masonry, or concrete. It also makes no difference whether wood-framed walls have brick veneer or some other exterior finish.

Platform-frame constructions typically include wood floor joists of nominal 2 × (50 ×) lumber spaced 12 in. (305 mm), 16 in. (406 mm), or 24 in. (610 mm) on centers. Contemporary floor joists, however, are frequently composite truss-type units made up of plywood webs and lumber top and bottom chords (Fig. 6.7-10a). Longer spans are sometimes achieved using floor trusses made with top and bottom chords and diagonal wood framing members arranged in a framework of triangles in much the same way as in flat roof trusses (see Section 6.11). Subflooring of wood boards or plywood is applied over floor joists and floor trusses to form a structural floor and a working platform (Figs. 6.7-10b and 6.7-11).

Generally, joists span from the exterior walls to center load-bearing walls or beams. Interior load-bearing walls may restrict the flexibility of room planning, however. For relatively narrow buildings (about 16 ft [4.88 m] or less), joists may span from exterior wall to exterior wall without a center support.

Joist floor systems provide space for heating, air conditioning, electrical, and plumbing equipment and devices, and are easily insulated to minimize heat loss and improve sound transmission characteristics.

Some types of floor finishes (such as hardwood flooring) may be installed directly over wood or plywood subflooring; other finishes (such as resilient flooring) may require an underlayment. A variety of ceiling finishes may be applied, usually directly to the underside of the joists.

### MATERIALS

Floor joist lumber is selected primarily to meet strength and stiffness requirements, which depend upon the dead and live loads to be carried. For average spans encountered in residential construction, stiffness requirements are more critical than strength requirements. Therefore, typical joist sizes are selected on the basis of their stiffness (resistance to deflection) rather than strength (ability to carry loads without failure). Stiffness require-

FIGURE 6.7-9 Box sill assembly. The studs rest on a sole plate. The subflooring may be boards, as shown, but is more likely to be a sheet material.

FIGURE 6.7-11 Wood joist floor/ceiling construction. Subflooring may be boards, as shown, but is more likely to be a sheet material.

ments mandate an arbitrary maximum deflection under loads: $1/360$ of the joist span in inches. For instance, the maximum allowable deflection for a 15 ft joist span would be 180 in. $(15 \times 12)$ divided by 360 (180/360), or $1/2$ in. (12.7 mm).

Other desirable qualities for joist lumber are good nail-holding ability and freedom from warp. Wood floor joists are usually 2-in. (50-mm)-thick nominal lumber 6 in. (150 mm) to 12 in. (305 mm) deep. The size depends upon loading, length of span, joist spacing, and the species and grade of lumber used.

Floor joists for simple frame buildings and ordinary construction can be selected by referring to *Span Tables for Joists and Rafters*, published by AF&PA. The table there for floor joists for 40 psf (195.3 kg/m²) live load and 10 psf (48.8 kg/m²) dead load is reproduced in Figure 6.7-12. These tables indicate the minimum modulus of elasticity, $E$, and extreme fiber strength in bending, $F_b$, required for various spans, lumber sizes, and conditions of loading. *Span Tables for Joists and Rafters* also includes tables for 30 psf (146.47 kg/m²), 50 psf (244.12 kg/m²),

and 60 psf (292.95 kg/m²) live load and for 20 psf (97.65 kg/m²) dead load. Suitable species and grades meeting the indicated strength requirements can be selected from the tables of allowable unit stresses in the *NDS*.

In these tables, the allowable span is the clear distance between supports. Tabulated spans are applicable to green or dry lumber, provided that it is used in covered structures, where it will normally

FIGURE 6.7-10 (a) Truss-type wood joists. (b) A plywood subfloor over wood joists. (Courtesy American Plywood Association.)

**FIGURE 6.7-12** Maximum Allowable Spans for Floor Joists with L/360 Deflection Limits[a,b]

| Joist Size, in. | Spacing, in. | Modulus of Elasticity, $E$, in 1,000,000 psi | | | | | | | | | | | | | | | | |
|---|---|---|---|---|---|---|---|---|---|---|---|---|---|---|---|---|---|---|
| | | 0.8 | 0.9 | 1.0 | 1.1 | 1.2 | 1.3 | 1.4 | 1.5 | 1.6 | 1.7 | 1.8 | 1.9 | 2.0 | 2.1 | 2.2 | 2.3 | 2.4 |
| 2 × 6 | 12.0 | 8-6 | 8-10 | 9-2 | 9-6 | 9-9 | 10-0 | 10-3 | 10-6 | 10-9 | 10-11 | 11-2 | 11-4 | 11-7 | 11-9 | 11-11 | 12-1 | 12-3 |
| | 16.0 | 7-9 | 8-0 | 8-4 | 8-7 | 8-10 | 9-1 | 9-4 | 9-6 | 9-9 | 9-11 | 10-2 | 10-4 | 10-6 | 10-8 | 10-10 | 11-0 | 11-2 |
| | 19.2 | 7-3 | 7-7 | 7-10 | 8-1 | 8-4 | 8-7 | 8-9 | 9-0 | 9-2 | 9-4 | 9-6 | 9-8 | 9-10 | 10-0 | 10-2 | 10-4 | 10-6 |
| | 24.0 | 6-9 | 7-0 | 7-3 | 7-6 | 7-9 | 7-11 | 8-2 | 8-4 | 8-6 | 8-8 | 8-10 | 9-0 | 9-2 | 9-4 | 9-6 | 9-7 | 9-9 |
| 2 × 8 | 12.0 | 11-3 | 11-8 | 12-1 | 12-6 | 12-10 | 13-2 | 13-6 | 13-10 | 14-2 | 14-5 | 14-8 | 15-0 | 15-3 | 15-6 | 15-9 | 15-11 | 16-2 |
| | 16.0 | 10-2 | 10-7 | 11-0 | 11-4 | 11-8 | 12-0 | 12-3 | 12-7 | 12-10 | 13-1 | 13-4 | 13-7 | 13-10 | 14-1 | 14-3 | 14-6 | 14-8 |
| | 19.2 | 9-7 | 10-0 | 10-4 | 10-8 | 11-0 | 11-3 | 11-7 | 11-10 | 12-1 | 12-4 | 12-7 | 12-10 | 13-0 | 13-3 | 13-5 | 13-8 | 13-10 |
| | 24.0 | 8-11 | 9-3 | 9-7 | 9-11 | 10-2 | 10-6 | 10-9 | 11-0 | 11-3 | 11-5 | 11-8 | 11-11 | 12-1 | 12-3 | 12-6 | 12-8 | 12-10 |
| 2 × 10 | 12.0 | 14-4 | 14-11 | 15-5 | 15-11 | 16-5 | 16-10 | 17-3 | 17-8 | 18-0 | 18-5 | 18-9 | 19-1 | 19-5 | 19-9 | 20-1 | 20-4 | 20-8 |
| | 16.0 | 13-0 | 13-6 | 14-0 | 14-6 | 14-11 | 15-3 | 15-8 | 16-0 | 16-5 | 16-9 | 17-0 | 17-4 | 17-8 | 17-11 | 18-3 | 18-6 | 18-9 |
| | 19.2 | 12-3 | 12-9 | 13-2 | 13-7 | 14-0 | 14-5 | 14-9 | 15-1 | 15-5 | 15-9 | 16-0 | 16-4 | 16-7 | 16-11 | 17-2 | 17-5 | 17-8 |
| | 24.0 | 11-4 | 11-10 | 12-3 | 12-8 | 13-0 | 13-4 | 13-8 | 14-0 | 14-4 | 14-7 | 14-11 | 15-2 | 15-5 | 15-8 | 15-11 | 16-2 | 16-5 |
| 2 × 12 | 12.0 | 17-5 | 18-1 | 18-9 | 19-4 | 19-11 | 20-6 | 21-0 | 21-6 | 21-11 | 22-5 | 22-10 | 23-3 | 23-7 | 24-0 | 24-5 | 24-9 | 25-1 |
| | 16.0 | 15-10 | 16-5 | 17-0 | 17-7 | 18-1 | 18-7 | 19-1 | 19-6 | 19-11 | 20-4 | 20-9 | 21-1 | 21-6 | 21-10 | 22-2 | 22-6 | 22-10 |
| | 19.2 | 14-11 | 15-6 | 16-0 | 16-7 | 17-0 | 17-6 | 17-11 | 18-4 | 18-9 | 19-2 | 19-6 | 19-10 | 20-2 | 20-6 | 20-10 | 21-2 | 21-6 |
| | 24.0 | 13-10 | 14-4 | 14-11 | 15-4 | 15-10 | 16-3 | 16-8 | 17-0 | 17-5 | 17-9 | 18-1 | 18-5 | 18-9 | 19-1 | 19-4 | 19-8 | 19-11 |
| $F_b$ | 12.0 | 718 | 777 | 833 | 888 | 941 | 993 | 1043 | 1092 | 1140 | 1187 | 1233 | 1278 | 1323 | 1367 | 1410 | 1452 | 1494 |
| $F_b$ | 16.0 | 790 | 855 | 917 | 977 | 1036 | 1093 | 1148 | 1202 | 1255 | 1306 | 1357 | 1407 | 1456 | 1504 | 1551 | 1598 | 1644 |
| $F_b$ | 19.2 | 840 | 909 | 975 | 1039 | 1101 | 1161 | 1220 | 1277 | 1333 | 1388 | 1442 | 1495 | 1547 | 1598 | 1649 | 1698 | 1747 |
| $F_b$ | 24.0 | 905 | 979 | 1050 | 1119 | 1186 | 1251 | 1314 | 1376 | 1436 | 1496 | 1554 | 1611 | 1667 | 1722 | 1776 | 1829 | 1882 |

[a]The required bending design value, $F_b$, in pounds per square inch (psi) is shown at the bottom of the table and is applicable to all lumber sizes shown. Spans are shown in feet-inches and are limited to 26 ft. and less. Check sources of supply for availability of lumber in lengths greater than 20 ft.

[b]Design Criteria: Deflections—for 40 pounds per square foot (psf) live load. Limited to span in inches divided by 360. Strength—live load of 40 psf plus dead load of 10 psf determines the required bending design value.

(Courtesy American Forest & Paper Association.)

achieve a moisture content of 19% or less.

For an example of floor joist selection, assume a required span of 12 ft 9 in. (3886 mm), a live load of 40 psf (195.3 kg/m²), and joists spaced 16 in. (406 mm) on centers. Figure 6.7-12 shows that a 2 × 8 (50 × 200) having an $E$ value of 1,600,000 psi (1 124 912 Mg/m²) (1.6 in the table) and an $F_b$ value of 1255 psi (882.35 Mg/m²) would have an allowable span of 12 ft 10 in. (3911.6 mm) when spaced at 12 in. (304.8 mm) on centers. Suitable species and grades can be selected from the *NDS*. In our example, a review of the data in Figure 6.1-13, which were compiled from an *NDS* table, shows that a visually graded Douglas Fir-Larch No.1 and Btr grade has an $F_b$ of 1200 psi (843.68 Mg/m²) and an $E$ of 1,800,000 psi (1 265 526 Mg/m²). The *NDS* permits an increase in the allowable stress shown in the table by a repetitive member factor ($C_r$), a wet service factor ($C_M$), a flat use factor ($C_{fu}$), a size factor ($C_F$), and a shear stress factor ($C_H$). These allowable increases are shown in the *NDS*. Of them, only the repetitive member and size factors are applicable. The $F_b$ can be increased by a factor of 1.15 because of the repetitive member factor ($C_r$), producing an allowable $F_b$ of 1380 psi (970.24 Mg/m²). A further increase permitted by the size factor is 1.2. Therefore, the $F_b$ can be increased to 1656 psi (1164.28 Mg/m²). Similarly, data in the *NDS* tables for machine-stress-graded lumber (not reproduced here) show a lumber, 1700f–1.6E, of WCLIB species that has an $F_b$ of 1700 psi (1195.22 Mg/m²) and an $E$ of 1,600,000 psi (1 124 912 Mg/m²).

AF&PA advises that the tabulated size factor increases apply only to Douglas Fir-Larch, Douglas Fir-Larch (North), and Douglas Fir South. The tables already take this factor into account for Southern Pine.

Joist grades should be selected for the specific design conditions. Where several combinations of joist size and spacing satisfy the given requirements, a deeper joist at greater spacing generally is preferred, provided, of course, that the subfloor and the ceiling finish are selected to span the greater spacing. Using deeper joists usually requires less lumber and labor and results in a stiffer floor and ceiling assembly.

### JOIST INSTALLATION

Floor and ceiling assemblies intended for the direct application of gypsum board or any other board finish should have all joists evenly spaced, with bottom faces aligned in a level plane. After the sill plates have been anchored to the foundation walls, joists are laid out over the sill plates and center supports. The floor joists are spaced uniformly at 12 in. (305 mm), 16 in. (406 mm), or 24 in. (610 mm) on centers, as determined by the design requirements. Extra joists are added where required to support bearing partitions (Fig. 6.7-13) or to frame floor openings (Fig. 6.7-14). Joists are braced temporarily with wood strips nailed across the top until the subflooring is applied.

When designing a wood floor system, it is important to equalize shrinkage of the framing at the exterior walls and at the center support. By using approximately the same total depth of cross-grain wood at the center girder as at the exterior wall, differential shrinkage can be minimized. This reduces plaster cracks, prevents sticking doors, and avoids other inconveniences caused by uneven shrinkage. For instance, if there is a total of 12 in. (305 mm) of cross-grain wood at the foundation wall (including joists and sill plate), this should be balanced with an equal amount of cross-grain wood at the center girder. To avoid uneven shrinkage, a steel beam center support generally is preferred. If a wood girder is used, MC 15 lumber should be specified, or the joists should be supported on ledger strips rather than over the top of the girder.

Excessively bowed or crooked joists should not be used, but joists with a slight crown may be used if they are installed with the crown up. After the subfloor and normal floor loads are applied, the joist will tend to straighten out. The largest edge knots also should be placed on top, since the upper side of the joist is in compression and the knots will have a lesser effect on strength and deflection.

Header joists should be fastened by nailing into the end of each joist with two 16d nails. In addition, the header joists and the stringer joists (parallel to the exterior walls) should be toenailed to the sill with 8d nails spaced 16 in. (406 mm) on centers, unless rigid wall sheathing covers the sill plate and is nailed to it with 8d nails 8 in. (200 mm) on centers.

Each floor joist should be toenailed to the sill and center beam or girder with two 10d or three 8d nails; joists should also be nailed to each other with at least three 16d nails when they lap over the center beam. The lap should be not less than 4 in. (100 mm), and the overhang of either joist beyond the edge of the girder or beam should not exceed 12 in. (305 mm). The limit on overhangs is imposed to reduce the possibility of lifting the subfloor resting on the overhang when a heavy, concentrated load is placed on the joist near its midspan. If a nominal 2 in. (50 mm) scab is used across butt-ended joists, it should be nailed to each joist with at least three 16d nails at each side of the joint.

For greater lumber economy, an in-line joist system sometimes is used in floor and roof framing when center supports are present. This system normally

**FIGURE 6.7-13 Spaced joist construction under a parallel bearing partition.** (Figure 43, *Wood Frame House Construction*, U.S. Department of Agriculture, p. 52. Not subject to copyright in the United States.)

**FIGURE 6.7-14** Framing for a large floor opening (a) with the long dimension of the opening parallel with the joists and (b) with the long dimension of the opening perpendicular to the joists. (Figure 48, *Wood Frame House Construction*, U.S. Department of Agriculture, p. 57. Not subject to copyright in the United States.)

Floor joists under parallel bearing partitions should be doubled; if access to the partition is required for air ducts, the double joists should be spaced and solid blocking should be used between them, as in spaced wood girder construction (Fig. 6.7-16 [see also Figs. 6.7-6 and 6.7-13]).

### 6.7.1.6 Bridging

Cross-bridging between joists, which has been used in house construction for many years and which is still required by some building codes, has been demonstrated by the National Association of Home Builders Research Foundation and several other independent laboratories to be virtually useless under most circumstances. In fact, even when bridging is tight-fitting and well installed, there is no significant improvement in a floor's ability to transfer vertical loads after the subfloor and finish flooring have been installed. Therefore, in most wood-framed buildings, bridging between floor joists is not necessary. Properly nailed subflooring and the nailing of the ends of joists to band joists or headers is sufficient to keep the joists aligned and to prevent them from twisting.

However, when the depth-to-thickness ratio of joists exceeds 6 or when joists twisting has occurred, bridging is often added at midspan or in intervals of not more than 8 ft (2440 mm). This bridging may consist of solid wood blocks, or of either metal struts or 1 × 3 (25 × 75) or larger wood members installed in an X fashion, with one end at the top of a joist

allows the use of joists one size smaller than in conventional floor framing. Briefly, the system consists of uneven-length joists with the long, overhanging joist cantilevered over the center support and connected to the supported joist with a metal connector or plywood splice plate (Fig. 6.7-15). In typical single-family construction, the overhang varies between about 1 ft 10 in. (558.8 mm) and 2 ft 10 in. (863.6 mm). For Douglas fir and southern pine No. 1 lumber, total spans (between foundation walls) of up to 22 ft (6.71 m) are possible with 2 × 6s (50 × 150s); spans of up to 30 ft (9.14 m) with 2 × 8s (50 × 200s); and spans of up to 34 ft (10.36 m) with 2 × 10 (50 × 250) joists. Other innovative floor systems which depend on special attachment of a plywood subfloor are discussed in Section 6.10.

**FIGURE 6.7-15** In-line joist system framed with pairs of overhanging joists and supported joists. (Figure 35, *Wood Frame House Construction*, U.S. Department of Agriculture, p. 46. Not subject to copyright in the United States.)

and the opposite end at the bottom of the adjacent joist.

In addition, either cross-bridging or solid blocking using 2 × 4 (50 × 100) or larger lumber installed vertically between joists may improve the diaphragm action of a floor and increase its resistance to lateral (horizontal) loads imparted to it by foundation walls. Therefore, bridging may be required in situations where the engineering design so indicates.

Solid bridging between joists should be used to provide a more rigid base for nonbearing partitions which happen to fall between joists (see Fig. 6.7-16). It also provides a firestop against lateral flame spread. Solid blocking should be well fitted and securely nailed to the joists. Load-bearing partitions running parallel to the joist direction should be supported by doubled joists, either by placing the partition over a normally spaced joist which has been doubled or by providing an additional pair of joists under the partition.

### 6.7.1.7 Floor Openings

Large openings are often necessary in floors for ducts, chimneys, and stairs (see Fig. 6.7-14). Where joists have been cut away, the remaining *tail joists* are framed into single or double cross-joists called *headers*. The headers, in turn, are framed into abutting full-length joists known as *trimmers*.

A single header can be used if the dimension of the opening along the header is less than 4 ft (1220 mm); a single trimmer can be used to support a single header when the opening is near the end of the joist span and less than 4 ft (1220 mm) long. Tail joists shorter than 6 ft (1800 mm) should be connected to headers with three 16d end nails and two 10d toenails. Longer tail joists should be connected with framing anchors or supported on a ledger strip, as in joist-girder connections. Headers should be connected to trimmers with three 16d end nails and two 10d toenails.

When framing openings are larger than 4 ft (1220 mm) by 4 ft (1220 mm), both the trimmers and headers should be doubled. Connections between headers and trimmers should be made with framing anchors, joist hangers, or ledger strips. Tail joists more than 6 ft (1800 mm) long should be similarly attached to headers; shorter tail joists can be attached as recommended above for smaller openings.

**FIGURE 6.7-16** Typical platform-framed floor construction. (Figure 34, *Wood Frame House Construction*, U.S. Department of Agriculture, p. 46. Not subject to copyright in the United States.)

The nailing sequence for double headers is shown in Figure 6.7-17; the opening is first framed out using single headers and trimmers; then the second joists are applied and face-nailed to the first.

The largest floor openings are usually for stairs; for straight runs and typical floor heights, they are usually about 10 ft (3000 mm) long and 3 ft (900 mm) to 4 ft (1220 mm) wide. The long dimension of stairway openings may be either parallel or at right angles to the joists. However, it is much easier to frame a stairway opening when its length is parallel to the joists (see Fig. 6.7-14b).

When the length of a stair opening is perpendicular to the joists, a long doubled header is required (see Fig. 6.7-14a). A header under these conditions without a supporting wall beneath it is usually limited to a 10 ft (3000 mm) length and should be designed as a beam. A load-bearing wall under all or part of this opening substantially simplifies the framing, as the joists then bear on the top plate of a wall rather than on a header.

### 6.7.1.8 Floor Projections

Framing for floor projections beyond the supporting wall below, such as for

**FIGURE 6.7-17** Nailing sequence for floor openings: (1) nailing trimmer to first header; (2) nailing header to tail beams; (3) nailing header together; (4) nailing trimmer to second header; and (5) nailing trimmers together.

---

a bay window or balcony, generally consists of extended floor joists if they run perpendicular to the supporting wall (Fig. 6.7-18a). This extension normally should not exceed 24 in. (610 mm) unless it is specifically structurally designed for a greater projection. The joists forming each side of such a bay should be doubled. Nailing procedures should conform to those used in framing stair openings. The subflooring should be extended flush with the outer framing member.

In framing the roof of such a projection, rafters often are supported on a header over the bay area in the plane of the supporting wall below. The header thus supports the roof load and transmits it to the foundation wall below so that the bay wall carries less load.

Projections parallel to the direction of the floor joists should generally be limited to small areas and extensions of not more than 24 in. (610 mm) (Fig. 6.7-18b). In this construction, the stringers should frame into doubled joists. Joist hangers or a ledger strip can be used to frame

joists into the double header, as in a joist-girder connection.

### 6.7.1.9 Framing for Mechanical and Electrical Equipment

It is frequently necessary to notch, drill, or cut framing members to accommodate heating, plumbing, and electrical equipment. Preconstruction planning can minimize the amount of cutting required. If mechanical and electrical distribution plans are worked out in advance, many time- and cost-saving adjustments in framing can be made.

Floor joists should be notched or drilled in a way that will minimize the effect on their structural integrity. The top or bottom of a joist should be notched only in the end one-third of the span and to a depth of not more than one-sixth of the joist depth. Holes may be bored in joists, provided that the hole is not more than 2 in. (50 mm) in diameter and not less than $2\frac{1}{2}$ in. (63.5 mm) from either edge (Fig. 6.7-19). Where notching or drilling to a greater extent

**FIGURE 6.7-19** Recommended limits for boring of holes in load-bearing floor joists. (Figure 43, *Wood Frame House Construction*, U.S. Department of Agriculture, p. 52. Not subject to copyright in the United States.)

is necessary, the weakened joist should be reinforced with scabs on both sides or by adding another joist. If reinforcing is not possible because of conflict with the mechanical equipment, the joist should be cut away and the area treated as a floor opening, with tail beams sup-

**a** Continuation of floor joists

**b** Projection perpendicular to floor joists

**FIGURE 6.7-18** Framing for wall projections: (a) when joists are perpendicular to wall; (b) when joists are parallel with wall. (Figure 39, *Wood Frame House Construction*, U.S. Department of Agriculture, p. 51. Not subject to copyright in the United States.)

ported on headers. When a plumbing wall adjacent to a bathroom also serves as a load-bearing partition, the supporting double joists should be spaced and blocked (Fig. 6.7-20), as recommended for a spaced girder. The joist nearest the front apron of a tub should also be doubled to prevent excessive deflection when the tub is full of water.

## 6.7.2 WALL ASSEMBLIES

Wall assemblies are a basic element of a framed building's superstructure. The functional performance requirements of a wall assembly include:

1. Adequate strength to support the wall and other building elements it supports (dead loads) and live loads, such as those created by wind, earthquakes, and those transmitted into the wall from supported construction, such as floors
2. Weathertightness (control of the flow of heat, moisture, air, and water vapor)
3. Satisfactory levels of sound and visual privacy
4. Suitable fire resistance
5. The ability to accommodate heating, air conditioning, electrical, and plumbing equipment
6. Suitability for the application of various finish materials
7. Adaptability to economical methods of assembly and erection
8. Provision for the installation of doors, windows, and other openings

When other requirements exist, the conventional constructions described in this section may not be adequate. Structural walls must safely support superim-

**FIGURE 6.7-20 Recommended framing to minimize floor deflection under tub.** (Figure 41, *Wood Frame House Construction*, U.S. Department of Agriculture, p. 52. Not subject to copyright in the United States.)

posed vertical loads and resist horizontal and racking loads due to lateral forces, such as wind and earthquake, and transmit all loads to the foundation. Rarely are walls specifically engineered. Ordinarily, building codes or rules of thumb based on established practices determine materials and methods of construction. Occasionally, some of the structural elements in a wood-framed building are left exposed, but wood-framed wall assemblies are usually covered with applied materials to provide the desired finished appearance and performance. The selection of finish materials may:

1. Influence the spacing and unsupported span of primary structural members
2. Determine the need for wall sheathing and other finish bases
3. Dictate the kind, type, and physical properties of the subsurface material
4. Dictate methods of construction

A building's appearance is influenced by the location, size, and type of doors, windows, and other openings that form a part of its wall assemblies. In addition, wall assemblies must be compatible with the roof and floor assemblies used in the construction.

Typical wall assemblies include conventional stud wall construction, panelized construction (see Section 6.9), masonry construction (see Chapter 4), and post-and-beam construction (see Section 6.11).

### 6.7.2.1 Materials

Wall framing lumber should be of the proper grade for the intended use as graded by a recognized grading authority. Its moisture content is important in the performance of applied finishes, and should be as near as possible to the equilibrium moisture content for the particular area. In no case should it be more than 15% when finishes are applied.

### 6.7.2.2 Stud Walls and Partitions

In light wood-framed construction, wood studs are the typical framing members for walls and partitions; wood joists and trusses are the most common ceiling and roof framing members. These framing members generally are spaced 16 in. (407 mm) or 24 in. (610 mm) on centers, depending on the unsupported height of the member and the number of floor and roof structures supported. Member sizes and spacings are often dictated by the build-

ing code. The usual spacings permit direct application of most finishes, such as gypsum board and plaster. However, furring may be required when member spacing exceeds the maximum recommended for the particular finish material or when special performance, such as increased sound isolation, is desired.

The following discussion of conventional stud wall construction refers to stick-built erection methods, where carpenters on the job handle individual studs rather than prefabricated panels (see Section 6.9).

Buildings built with platform-frame construction typically uses wood stud wall assemblies (Fig. 6.7-21). Walls may be sheathed with wood boards, plywood, fiber, or gypsum boards (see Section 6.10) which provide rigidity, form a weather barrier, and may be necessary as a nailing base to receive exterior finishes. Some types of exterior and interior finishes can be applied directly to studs; other finishes must be applied over sheathing or other bases. In load-bearing partitions, studs should be of 2 × 4s (50 × 100s) or 2 × 6s (50 × 150s) usually spaced at 16 in. (406 mm) or 24 in. (610 mm) on centers, depending on the loads to be encountered. In non-load-bearing or staggered-stud partitions, 2 × 3 (50 × 75) or 2 × 4 (50 × 100) studs may be used, spaced 16 in. (406 mm) on centers when openings are present or cabinets and other elements are supported on them, and 24 in. (610 mm) on centers when no openings are included and no additional loads are imposed on them. Backup members should be provided at interior corners for support and as a nailing base for gypsum board and other finish materials. Warped or crooked studs should not be used.

Because of the existing body of building codes and other regulations and practices, stud spacing rarely is specifically engineered. The conventional 16 in. (406 mm) and 24 in. (610 mm) spacing of studs has evolved from years of established practice and is based more on accommodating finish materials applied to the framework than on actual engineering for imposed loads. For example, in one-story light wood frame construction, 2 × 4 (50 × 100) studs could be placed at 6 ft (1800 mm) intervals and adequately support the generally imposed roof loads. But few exterior or interior finish materials can span such intervals economically. The 16 in. (406 mm) and

16" = 406 mm
24" = 610 mm

**FIGURE 6.7-21** Typical wood stud construction showing framing members. Structural wall sheathing or diagonal bracing, which is required to resist wind and other stresses, is not shown here.

24 in. (610 mm) spacings permit the use of standard widths and lengths of panel sheathing and siding, and therefore remain the convention in use. Stud spacings of 12 in. (305 mm), though rare in low-rise buildings, are sometimes used when walls are tall or lateral loads are to be superimposed. The use of 12 in. (305 mm) spacings is a structural decision based on engineering. There is currently a trend toward using 2 × 6 (50 × 150) studs at 24 in. (610 mm) on center for exterior walls to take advantage of the increased depth to install more thermal insulation.

The suitable performance of most types of siding depends in part on the wall structure that supports it. Wall studs should be adequately sized and spaced to carry superimposed loads. Conservatively, structural load requirements normally are met using:

1. 2 × 4 (50 × 100) studs at a maximum spacing of 24 in. (610 mm) on center when a wall supports only a ceiling and roof
2. 2 × 4 (50 × 100) studs at a maximum spacing of 16 in. (406 mm) on centers when the wall supports a floor in addition to the roof
3. 2 × 6 (50 × 150) studs at a maximum spacing of 16 in. (406 mm) on center when the wall supports two floors in addition to the roof

Codes permit the use of smaller members and wider spacings under certain conditions. Often, the spacing of exterior wall studs is dependent on the type and thickness of sheathing rather than on structural load requirements. Stud spacings for sheathing support are discussed in Section 6.10.

In stud systems, roofs and other vertically imposed loads are carried by the studs. Sheathing and applied finishes resist lateral forces but are not assumed to share in carrying vertical loads.

Wood stud assemblies lend themselves to on-site or off-site fabrication. Preframed panel wall assemblies are wood stud assemblies fabricated in sections away from the job site (see Section 6.9). Stud wall assemblies provide flexibility in design and erection and require only the use of simple hand tools and fastenings. Changes can be made easily during construction, particularly with on-site fabrication. Space between studs provides for insulation and heating, air conditioning, electrical, and plumbing equipment. Studs may be erected individually in the traditional manner, but most stud walls today are framed on the subfloor with sheathing and siding applied and windows and doors installed. The assembly is then tilted into place. Wood stud walls may be framed, erected, and sheathed quickly, and the building can be enclosed quickly. Interior finishing operations and mechanical installations then may take place under cover with maximum flexibility of job scheduling.

Wood stud wall systems are economical and widely used. They provide excellent structural performance, make use of relatively inexpensive basic materials, permit the application of many finishes, and are adaptable to practically any methods of on- or off-site fabrication (see Section 6.9).

In masonry veneer construction (see Chapter 4), wood framing assumes all vertical loads, and the masonry performs essentially as an exterior finish surface and provides a weather barrier for the assembly. The space between the studs typically is used for the installation of insulation and heating, air conditioning, electrical, and plumbing lines.

### 6.7.3 ROOF AND CEILING ASSEMBLIES

In many cases, especially in wood-framed buildings, the ceiling construction is an integral part of the roof construction. When it is, the entire system is called a *roof and ceiling assembly*. Such assemblies are a basic element of a building's superstructure.

Light wood-framed buildings' roof and ceiling assemblies include joist-and-rafter construction, plank-and-beam construction (see Section 6.11), and panelized construction (see Section 6.8).

Roof and ceiling assemblies are required to:

1. Have adequate strength to support the roof and ceiling assembly (dead loads) and live loads (such as from snow, ice, and wind)
2. Control the flow of heat, water, air, and water vapor
3. Provide satisfactory levels of sound privacy
4. Have suitable fire resistance
5. Be suitable for the application of finish materials
6. Be adaptable to economical methods of assembly and erection
7. Accommodate heating, air conditioning, electrical, and plumbing equipment

The form of a roof is determined by the roof and ceiling assembly selected, but usually, the assembly is covered with applied materials that provide the desired finished appearance and performance. The finish materials selected may:

1. Affect the spacing, span, and slope of structural members
2. Determine the need for roof sheathing or underlayment
3. Dictate methods of construction

A roof and ceiling assembly should be compatible with the wall assembly and the floor and ceiling assemblies used in the construction.

In light wood-framed building construction, wood joists and rafter construction (this section) and trusses (Section 6.11) are the most common ceiling and roof framing members.

These framing members generally are spaced 16 in. (406 mm) or 24 in. (610 mm) on centers, depending on the loads to be carried, which permits direct application of most finishes, such as gypsum board and plaster. However, furring (Section 6.7.4) may be required when member spacing exceeds the maximum recommended for the particular finish material or when special performance, such as increased sound isolation, is desired.

Ceiling assemblies intended for the direct application of gypsum board or any other board finish should have all joists evenly spaced, with bottom faces aligned in a level plane. Joists with a slight crown should be installed with the crown up. Slightly crooked or bowed joists can be straightened and leveled by nailing stringers approximately at midspan (Fig. 6.7-22) (see also Section 6.11). Excessively bowed or crooked joists should not be used.

### 6.7.3.1 Materials

Framing lumber should be of the proper grade for the intended use, graded by a recognized grading authority. The moisture content of framing lumber should be as near as possible to the equilibrium moisture content for the particular area and should not exceed 15% at the time the finish material is installed.

### 6.7.3.2 Joist-and-Rafter Roofs

Joist-and-rafter assemblies consist of nominal 2 in. (50 mm) lumber ceiling joists and roof rafters usually spaced 16 in. (406 mm) or 24 in. (610 mm) on centers. Typical construction for joist-and-rafter roofs is shown in Figure 6.7-23.

Joists span roughly one-half the width of the building from the exterior walls to a center load-bearing wall or beam. The rafters are supported on exterior walls and bear against each other or a ridge board at the high point of the roof. When loaded, the ridge tends to lower and the rafters exert a thrusting force which tries to spread the outside walls. This thrust is resisted by the ceiling joists acting as ties from wall to wall. Roof and ceiling framing should be of the proper grade for the intended use, graded by a recognized grading authority. Its moisture content should not be more than 15% at the time finishes are applied.

Joists and rafters for simple frame buildings and ordinary construction can be selected by referring to *Span Tables for Joists and Rafters*, published by AF&PA. These tables indicate the minimum modulus of elasticity, $E$, and extreme fiber strength in bending, $F_b$, required for various spans, lumber sizes, and conditions of loading. Suitable species and grades meeting the indicated strength requirements can be selected from the tables of Allowable Unit Stresses in the *NDS*.

A table of maximum allowable spans for ceiling joists carrying a drywall ceiling and a 20 lb attic live load is presented in Figure 6.7-24. *Span Tables for Joists and Rafters* also contains a table for ceiling joists with a ceiling but not attic storage. A table for rafters carrying a drywall ceiling, a 40 psf (195.3 kg/m²) live load, and a 10 psf (48.8 kg/m²) dead load is presented in Figure 6.7-25. *Span Tables for Joists and Rafters* also includes tables for rafters carrying live loads of 20 psf (97.65 kg/m²), 30 psf (146.47 kg/m²), and 50 psf (244.12 kg/m²) and dead loads of 15 psf (73.24 kg/m²) and 20 psf (97.65 kg/m²) for rafters carrying a drywall ceiling and for rafters carrying no ceiling.

In each case, the allowable span is the clear distance between supports. For rafters, the span is measured along the horizontal projection (Fig. 6.7-26). Tabulated spans are applicable to green or dry lumber, provided that it is used in covered structures, where it will normally achieve a moisture content of 19% or less.

As an example of rafter selection, assume a horizontal projection span of 15 ft (4.57 m) in a rafter carrying a drywall ceiling, a live load of 40 psf (195.3 kg/m²), and a 10 psf (48.8 kg/m²) dead load, with rafters spaced 16 in. (406 mm)

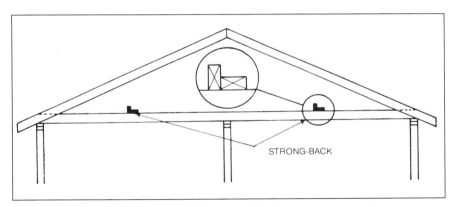

**FIGURE 6.7-22** Joist spacing can be regulated with 2 × 4 (50 × 100) stringers at midpoints, and can be leveled with 2 × 6 (50 × 150) stringers set on edge. (Courtesy Gypsum Association.)

**FIGURE 6.7-23** Wood joist-and-rafter roof and ceiling assembly.

**FIIGURE 6.7-24** Maximum Allowable Spans for Ceiling Joists with L/240 Deflection Limits[a,b]

| Joist Size, in. | Spacing, in. | Modulus of Elasticity, $E$, in 1,000,000 psi | | | | | | | | | | | | | | | | |
|---|---|---|---|---|---|---|---|---|---|---|---|---|---|---|---|---|---|---|
| | | 0.8 | 0.9 | 1.0 | 1.1 | 1.2 | 1.3 | 1.4 | 1.5 | 1.6 | 1.7 | 1.8 | 1.9 | 2.0 | 2.1 | 2.2 | 2.3 | 2.4 |
| 2 × 4 | 12.0 | 7-10 | 8-1 | 8-5 | 8-8 | 8-11 | 9-2 | 9-5 | 9-8 | 9-10 | 10-0 | 10-3 | 10-5 | 10-7 | 10-9 | 10-11 | 11-1 | 11-3 |
| | 16.0 | 7-1 | 7-5 | 7-8 | 7-11 | 8-1 | 8-4 | 8-7 | 8-9 | 8-11 | 9-1 | 9-4 | 9-6 | 9-8 | 9-9 | 9-11 | 10-1 | 10-3 |
| | 19.2 | 6-8 | 6-11 | 7-2 | 7-5 | 7-8 | 7-10 | 8-1 | 8-3 | 8-5 | 8-7 | 8-9 | 8-11 | 9-1 | 9-3 | 9-4 | 9-6 | 9-8 |
| | 24.0 | 6-2 | 6-5 | 6-8 | 6-11 | 7-1 | 7-3 | 7-6 | 7-8 | 7-10 | 8-0 | 8-1 | 8-3 | 8-5 | 8-7 | 8-8 | 8-10 | 8-11 |
| 2 × 6 | 12.0 | 12-3 | 12-9 | 13-3 | 13-8 | 14-1 | 14-5 | 14-9 | 15-2 | 15-6 | 15-9 | 16-1 | 16-4 | 16-8 | 16-11 | 17-2 | 17-5 | 17-8 |
| | 16.0 | 11-2 | 11-7 | 12-0 | 12-5 | 12-9 | 13-1 | 13-5 | 13-9 | 14-1 | 14-4 | 14-7 | 14-11 | 15-2 | 15-5 | 15-7 | 15-10 | 16-1 |
| | 19.2 | 10-6 | 10-11 | 11-4 | 11-8 | 12-0 | 12-4 | 12-8 | 12-11 | 13-3 | 13-6 | 13-9 | 14-0 | 14-3 | 14-6 | 14-8 | 14-11 | 15-2 |
| | 24.0 | 9-9 | 10-2 | 10-6 | 10-10 | 11-2 | 11-5 | 11-9 | 12-0 | 12-3 | 12-6 | 12-9 | 13-0 | 13-3 | 13-5 | 13-8 | 13-10 | 14-1 |
| 2 × 8 | 12.0 | 16-2 | 16-10 | 17-5 | 18-0 | 18-6 | 19-0 | 19-6 | 19-11 | 20-5 | 20-10 | 21-2 | 21-7 | 21-11 | 22-4 | 22-8 | 23-0 | 23-4 |
| | 16.0 | 14-8 | 15-3 | 15-10 | 16-4 | 16-10 | 17-3 | 17-9 | 18-1 | 18-6 | 18-11 | 19-3 | 19-7 | 19-11 | 20-3 | 20-7 | 20-11 | 21-2 |
| | 19.2 | 13-10 | 14-5 | 14-11 | 15-5 | 15-10 | 16-3 | 16-8 | 17-1 | 17-5 | 17-9 | 18-1 | 18-5 | 18-9 | 19-1 | 19-5 | 19-8 | 19-11 |
| | 24.0 | 12-10 | 13-4 | 13-10 | 14-3 | 14-8 | 15-1 | 15-6 | 15-10 | 16-2 | 16-6 | 16-10 | 17-2 | 17-5 | 17-9 | 18-0 | 18-3 | 18-6 |
| 2 × 10 | 12.0 | 20-8 | 21-6 | 22-3 | 22-11 | 23-8 | 24-3 | 24-10 | 25-5 | 26-0 | | | | | | | | |
| | 16.0 | 18-9 | 19-6 | 20-2 | 20-10 | 21-6 | 22-1 | 22-7 | 23-1 | 23-8 | 24-1 | 24-7 | 25-0 | 25-5 | 25-10 | | | |
| | 19.2 | 17-8 | 18-4 | 19-0 | 19-7 | 20-2 | 20-9 | 21-3 | 21-9 | 22-3 | 22-8 | 23-1 | 23-7 | 23-11 | 24-4 | 24-9 | 25-1 | 25-5 |
| | 24.0 | 16-5 | 17-0 | 17-8 | 18-3 | 18-9 | 19-3 | 19-9 | 20-2 | 20-8 | 21-1 | 21-6 | 21-10 | 22-3 | 22-7 | 22-11 | 23-4 | 23-8 |
| $F_b$ | 12.0 | 896 | 969 | 1040 | 1108 | 1174 | 1239 | 1302 | 1363 | 1423 | 1481 | 1539 | 1595 | 1651 | 1706 | 1759 | 1812 | 1864 |
| $F_b$ | 16.0 | 986 | 1067 | 1145 | 1220 | 1293 | 1364 | 1433 | 1500 | 1566 | 1631 | 1694 | 1756 | 1817 | 1877 | 1936 | 1995 | 2052 |
| $F_b$ | 19.2 | 1048 | 1134 | 1216 | 1296 | 1374 | 1449 | 1522 | 1594 | 1664 | 1733 | 1800 | 1866 | 1931 | 1995 | 2058 | 2120 | 2181 |
| $F_b$ | 24.0 | 1129 | 1221 | 1310 | 1396 | 1480 | 1561 | 1640 | 1717 | 1793 | 1866 | 1939 | 2010 | 2080 | 2149 | 2217 | 2283 | 2349 |

[a]The required bending design value, $F_b'$, in pounds per square inch (psi) is shown at the bottom of the table and is applicable to all lumber sizes shown. Spans are shown in feet-inches and are limited to 26 ft. and less. Check sources of supply for availability of lumber in lengths greater than 20 ft.

[b]Design Criteria: Deflections—for 20 pounds per square foot (psf) live load. Limited to span in inches divided by 240. Strength—live load of 40 psf plus dead load of 10 psf determines the required bending design value.

(Courtesy American Forest & Paper Association.)

**FIIGURE 6.7-25** Maximum Allowable Spans for Rafters with L/240 Deflection Limits

| Rafter Size, in. | Spacing, in. | Bending Design Value, $F_b$, psi | | | | | | | | | | | | | | | | | | | | | |
|---|---|---|---|---|---|---|---|---|---|---|---|---|---|---|---|---|---|---|---|---|---|---|---|
| | | 300 | 400 | 500 | 600 | 700 | 800 | 900 | 1000 | 1100 | 1200 | 1300 | 1400 | 1500 | 1600 | 1700 | 1800 | 1900 | 2000 | 2100 | 2200 | 2300 | 2400 |
| 2 × 6 | 12.0 | 5-6 | 6-4 | 7-1 | 7-9 | 8-5 | 9-0 | 9-6 | 10-0 | 10-6 | 11-0 | 11-5 | 11-11 | 12-4 | 12-8 | 13-1 | 13-6 | 13-10 | 14-2 | | | | |
| | 16.0 | 4-9 | 5-6 | 6-2 | 6-9 | 7-3 | 7-9 | 8-3 | 8-8 | 9-1 | 9-6 | 9-11 | 10-3 | 10-8 | 11-0 | 11-4 | 11-8 | 12-0 | 12-4 | 12-7 | 12-11 | | |
| | 19.2 | 4-4 | 5-0 | 5-7 | 6-2 | 6-8 | 7-1 | 7-6 | 7-11 | 8-4 | 8-8 | 9-1 | 9-5 | 9-9 | 10-0 | 10-4 | 10-8 | 10-11 | 11-3 | 11-6 | 11-9 | 12-0 | 12-4 |
| | 24.0 | 3-11 | 4-6 | 5-0 | 5-6 | 5-11 | 6-4 | 6-9 | 7-1 | 7-5 | 7-9 | 8-1 | 8-5 | 8-8 | 9-0 | 9-3 | 9-6 | 9-9 | 10-0 | 10-3 | 10-6 | 10-9 | 11-0 |
| 2 × 8 | 12.0 | 7-3 | 8-4 | 9-4 | 10-3 | 11-1 | 11-10 | 12-7 | 13-3 | 13-11 | 14-6 | 15-1 | 15-8 | 16-3 | 16-9 | 17-3 | 17-9 | 18-3 | 18-9 | | | | |
| | 16.0 | 6-3 | 7-3 | 8-1 | 8-11 | 9-7 | 10-3 | 10-10 | 11-6 | 12-0 | 12-7 | 13-1 | 13-7 | 14-0 | 14-6 | 14-11 | 15-5 | 15-10 | 16-3 | 16-7 | 17-0 | | |
| | 19.2 | 5-9 | 6-7 | 7-5 | 8-1 | 8-9 | 9-4 | 9-11 | 10-6 | 11-0 | 11-6 | 11-11 | 12-5 | 12-10 | 13-3 | 13-8 | 14-0 | 14-5 | 14-10 | 15-2 | 15-6 | 15-10 | 16-3 |
| | 24.0 | 5-2 | 5-11 | 6-7 | 7-3 | 7-10 | 8-4 | 8-11 | 9-4 | 9-10 | 10-3 | 10-8 | 11-1 | 11-6 | 11-10 | 12-2 | 12-7 | 12-11 | 13-3 | 13-7 | 13-11 | 14-2 | 14-6 |
| 2 × 10 | 12.0 | 9-3 | 10-8 | 11-11 | 13-1 | 14-2 | 15-1 | 16-0 | 16-11 | 17-9 | 18-6 | 19-3 | 20-0 | 20-8 | 21-4 | 22-0 | 22-8 | 23-3 | 23-11 | | | | |
| | 16.0 | 8-0 | 9-3 | 10-4 | 11-4 | 12-3 | 13-1 | 13-10 | 14-8 | 15-4 | 16-0 | 16-8 | 17-4 | 17-11 | 18-6 | 19-1 | 19-7 | 20-2 | 20-8 | 21-2 | 21-8 | | |
| | 19.2 | 7-4 | 8-5 | 9-5 | 10-4 | 11-2 | 11-11 | 12-8 | 13-4 | 14-0 | 14-8 | 15-3 | 15-10 | 16-4 | 16-11 | 17-5 | 17-11 | 18-5 | 18-11 | 19-4 | 19-10 | 20-3 | 20-8 |
| | 24.0 | 6-6 | 7-7 | 8-5 | 9-3 | 10-0 | 10-8 | 11-4 | 11-11 | 12-6 | 13-1 | 13-7 | 14-2 | 14-8 | 15-1 | 15-7 | 16-0 | 16-6 | 16-11 | 17-4 | 17-9 | 18-1 | 18-6 |
| 2 × 12 | 12.0 | 11-3 | 13-0 | 14-6 | 15-11 | 17-2 | 18-4 | 19-6 | 20-6 | 21-7 | 22-6 | 23-5 | 24-4 | 25-2 | 26-0 | | | | | | | | |
| | 16.0 | 9-9 | 11-3 | 12-7 | 13-9 | 14-11 | 15-11 | 16-10 | 17-9 | 18-8 | 19-6 | 20-3 | 21-1 | 21-9 | 22-6 | 23-2 | 23-10 | 24-6 | 25-2 | 25-9 | | | |
| | 19.2 | 8-11 | 10-3 | 11-6 | 12-7 | 13-7 | 14-6 | 15-5 | 16-3 | 17-0 | 17-9 | 18-6 | 19-3 | 19-11 | 20-6 | 21-2 | 21-9 | 22-5 | 23-0 | 23-6 | 24-1 | 24-8 | 25-2 |
| | 24.0 | 7-11 | 9-2 | 10-3 | 11-3 | 12-2 | 13-0 | 13-9 | 14-6 | 15-3 | 15-11 | 16-7 | 17-2 | 17-9 | 18-4 | 18-11 | 19-6 | 20-0 | 20-6 | 21-1 | 21-7 | 22-0 | 22-6 |
| E | 12.0 | 0.14 | 0.22 | 0.31 | 0.41 | 0.51 | 0.63 | 0.75 | 0.88 | 1.01 | 1.15 | 1.30 | 1.45 | 1.61 | 1.77 | 1.94 | 2.12 | 2.30 | 2.48 | | | | |
| E | 16.0 | 0.12 | 0.19 | 0.27 | 0.35 | 0.44 | 0.54 | 0.65 | 0.76 | 0.88 | 1.00 | 1.12 | 1.26 | 1.39 | 1.54 | 1.68 | 1.83 | 1.99 | 2.15 | 2.31 | 2.48 | | |
| E | 19.2 | 0.11 | 0.18 | 0.24 | 0.32 | 0.41 | 0.50 | 0.59 | 0.69 | 0.80 | 0.91 | 1.03 | 1.15 | 1.27 | 1.40 | 1.54 | 1.67 | 1.81 | 1.96 | 2.11 | 2.26 | 2.42 | 2.58 |
| E | 24.0 | 0.10 | 0.16 | 0.22 | 0.29 | 0.36 | 0.44 | 0.53 | 0.62 | 0.71 | 0.81 | 0.92 | 1.03 | 1.14 | 1.25 | 1.37 | 1.50 | 1.62 | 1.75 | 1.89 | 2.02 | 2.16 | 2.30 |

The required elasticity, E, in 1,000,000 pounds per square inch (psi) is shown at the bottom of the table, is limited to 2.6 million psi and less, and is applicable to all lumber sizes shown. Spans are shown in feet-inches and are limited to 26 ft. and less. Check sources of supply for availability of lumber in lengths greater than 20 ft.

Design Criteria: Strength—live load of 40 pounds per square foot (psf) plus dead load of 10 psf determines the required bending design value. Deflection—For 40 psf live load. Limited to span in inches divided by 240.

(Courtesy American Forest & Paper Association.)

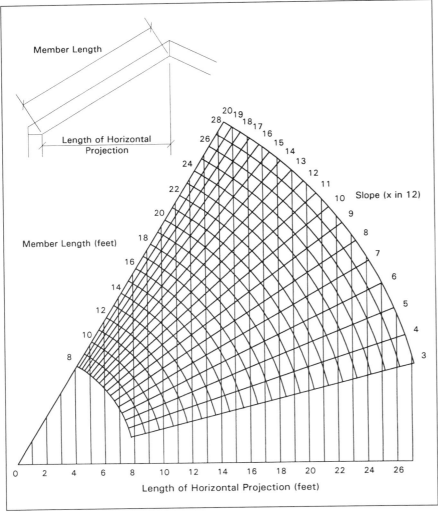

**FIGURE 6.7-26** Conversion diagram for rafters. (Courtesy American Forest & Paper Association.)

ber (not reproduced here) show that a lumber, 1250f-1.4E, of WCLIB species has an $F_b$ of 1250 psi (878.84 Mg/m$^2$) and an $E$ value of 1,400,000 psi (984 298 Mg/m$^2$).

Joist-and-rafter construction generally requires longer erection time than truss construction and may expose materials longer to possible weather hazards.

Interior load-bearing walls somewhat restrict the flexibility of room planning. However, for relatively short spans of 16 ft (4880 mm) or less, load-bearing interior walls or beams generally are not required. If a building's plan includes a number of wings of different spans, joist-and-rafter framing may be more economical than truss construction.

Joist-and-rafter assemblies provide overhead space for the installation of insulation and mechanical equipment; in addition, with adequate rafter slopes, economical attic storage or expansion space can result.

### 6.7.3.3 Joist Roofs

In joist roof and ceiling assemblies, roof joists alone provide the function of both ceiling joists and roof rafters (Fig. 6.7-27). These joist members generally are of nominal 2 in. (50 mm) dimension lumber spaced 12 in. (305 mm), 16 in. (406 mm), or 24 in. (610 mm) on centers, depending on the spans and the loads to be carried. Joists may have a clear span between exterior walls or they may be centrally supported on a load-bearing partitions or beams. When the roof is

on centers. Figure 6.7-25 shows that a 2 × 10 (50 × 250) having an $F_b$ value of 1100 psi (773.4 kg/m$^2$) and an $E$ value of 880,000 psi (618 702 kg/m$^2$) would have an allowable horizontal projection of 15 ft 4 in. (4673.5 mm). Suitable species and grades can be selected from the tables in the *NDS*. For our example, Figure 6.1-13, which was compiled from data in an *NDS* table, shows that a visually graded No. 2 Douglas Fir-Larch has an $F_b$ of 900 psi (632.76 kg/m$^2$) and an $E$ value of 1,600,000 psi (1 124 912 Mg/m$^2$). The $F_b$ can be increased by a factor of 1.15 because of the repetitive member factor ($C_r$), producing an allowable $F_b$ of 1035 psi (727.68 Mg/m$^2$). A further increase permitted by the size factor is 1.2. Therefore, the $F_b$ can be increased to 1242 psi (873.21 Mg/m$^2$). Similarly, data in the *NDS* tables for machine-stress-graded lum-

**FIGURE 6.7-27** Wood joist roof and ceiling assembly.

**FIGURE 6.7-28** Wood furring on masonry is shimmed to provide a plumbed, true, firm base for gypsum wallboard attachment. (Courtesy American Plywood Association.)

FURRING STRIPS ATTACHED TO STUDS

**FIGURE 6.7-29** Wood cross furring supporting gypsum board. (Courtesy American Plywood Association.)

sloped rather than flat, the construction also may be called a *ridge-supported rafter assembly*, which often is used to achieve an interior sloped ceiling effect. Roof sheathing materials may be attached directly to the top of the joists, and ceiling finishes may be applied directly to the underside. Joists and rafters may be se-

lected from the tables in *Span Tables for Joists and Rafters* by AF&PA.

Ductwork can be accommodated in the joist space if it is run parallel to the joists. When it is necessary to run ductwork perpendicular to the joists, it is common to use dropped-ceiling areas over hallways. Wiring can be concealed in the joist space, and the system is relatively easy to insulate. Interior and exterior finishes are applied in the same manner as for truss and joist-and-rafter roofs.

### 6.7.4 FURRING

Wood furring strips are used with wood-framed and masonry or concrete construction (Fig. 6.7-28) where incombustibility, decay resistance, or termites are not a consideration. When the framing is of irregular spacing or the spacing is too great for the intended wall finish material, cross furring should be applied perpendicular to framing members (Fig. 6.7-29). Recommendations for wood cross furring are given in Figure 6.7-30.

Cross furring of wood smaller than 2 × 2s (50 × 50s) is not sufficiently stiff to permit the nailing of gypsum wallboard or gypsum plaster lath without excessive hammer rebound, which may result in loose board attachment. Thinner cross furring is acceptable for screw-attached board because little or no impact is involved when screws are driven. Thinner furring also may be acceptable if the board is applied with stud adhesive and nailing is limited to locations where the furring is fully supported by underlying framing.

When the primary purpose of furring is to increase the assembly thickness to conceal piping or to accept additional thermal insulation, 1 × 2 (25 × 50) or larger parallel furring can be applied directly over the primary framing, nailed 16 in. (406 mm) on centers, with nails

providing penetration recommended for cross furring (see Fig. 6.7-30).

Furring strips over masonry or concrete can be secured with cut nails, helically threaded concrete nails, or powder-actuated fasteners of appropriate size. Maximum fastener spacing should be 16 in. (406 mm) on centers for 1 × 2 (25 × 50) strips and 24 in. (610 mm) on centers for 2 × 2 (50 × 50) or larger strips.

## 6.8 Structural Panels

Structural panels including stressed-skin panels and sandwich panels are discussed in this section. Floor, wall, and ceiling assemblies constructed with structural panels are covered in Section 6.9.

Off-site, shop-fabricated structural panels consist of plywood panels either nailed to precut lumber or fastened with glue, using either pressure- or nailed-glued techniques under controlled shop conditions. Full pressure gluing, which requires rather expensive equipment and careful control, produces excellent results (Fig. 6.8-1). Nail gluing, where nails furnish the pressure required for an adequate glue bond, also requires careful and precise techniques but less equipment. Nail gluing leads to satisfactory, predictable results in the laboratory but has not always been successful in the field. This is due to poor gluing techniques, unfavorable temperatures, use of improperly dried or surfaced lumber, insufficient nailing for proper pressure, and dried-out gluelines. With proper supervision and control, field gluing can be successful.

Aside from the advantages in less erection time and reduced in-place costs that may result from the use of almost any preassembled components (see Section 6.9), plywood components derive

**FIGURE 6.7-30** Minimum Requirements for Wood Cross Furring*

| Location | Support Spacing | | Furring | Nails | |
| | in. | mm | | Spacing | Penetration |
|---|---|---|---|---|---|
| Ceilings | 16 o.c. | 406 o.c. | 2 in. × 2 in. | 2 at each support | $\frac{7}{8}$ in. (22 mm) (threaded) |
| | 24 o.c. | 610 o.c. | 2 in. × 2 in. | | $1\frac{1}{4}$ in. (32 mm) (smooth shank) |
| Walls | 16 o.c. | 406 o.c. | 1 in. × 2 in. | 1 at each support | $\frac{3}{4}$ in. (19 mm) (threaded or smooth shank) |
| | 24 o.c. | 610 o.c. | 1 in. × 3 in. | | |

*Wallboard screw attached; if nailed, 2 in. × 2 in. furring should be used in all cases.

(Courtesy American Plywood Association.)

**FIGURE 6.8-1** Inserting insulation and foil vapor retarder in a stressed-skin panel. (Courtesy American Plywood Association.)

sheets glued to the top and, in most cases, also to the bottom faces of longitudinal framing members so that the assembly will act integrally in resisting bending stresses (Fig. 6.8-2). The plywood faces carry the compressive and tensile stresses induced by bending, while the framing members principally carry shear stresses. Stressing the skin allows a reduction in framing member size. For example, wall and partition studs, which normally are 2 × 4s (50 × 100s), can be reduced to 1 × 3s (25 × 75s) or 2 × 3s (50 × 75s) with stressed-skin design. The plywood must be bonded firmly with glue to the frame in order to transfer applied shearing stresses (Fig. 6.8-3). Stressed-skin panels may be used as walls, floors, and roofs (Fig. 6.8-4) (see also Section 6.9).

## 6.8.2 SANDWICH PANELS

*Sandwich panels* follow the general principles of stressed-skin panel design, but the faces are glued to and separated by weaker, lightweight core material such as resin-impregnated paper honeycomb or a plastic foam core designed for this use (Fig. 6.8-5). Sandwich panels possess many of the advantages of stressed-skin panels but are lighter in weight, needing no inner panel framing except, perhaps, around the perimeter. The cores provide a uniform, stabilizing support for the strong, relatively thin faces. The cores usually have little strength in flexure but have adequate compressive strength for normal floor panels. Some cores, particularly plastic foams, also provide excellent insulation.

Sandwich panels can be used as entire walls and, when combined, can comprise entire structures (Fig. 6.8-6).

added benefits from the plywood itself. Plywood's tensile and compressive strength, panel shear strength, split resistance, and dimensional stability enable it to function structurally as part of the component instead of serving just as a base for covering materials.

Although the use of structural panels provides many of the advantages of component fabrication, the incorporation of heating, air conditioning, electrical, and plumbing equipment requires careful design of the system. Designs have been made to incorporate mechanical and electrical equipment into stressed-skin and sandwich panels, but this incorporation requires special handling and erection procedures in the field and may not be acceptable to local code authorities.

Therefore, such panels have had limited success to date.

In addition, many local building codes specifically require 2 × 4 (50 × 100) studs spaced at 16 in. (406 mm) or 24 in. (610 mm) on centers, which would negate the advantage of using stressed-skin or sandwich panels unless an exception can be obtained. Some codes, however, permit the use of any system that performs as well as the system specified in the code. In this case, a properly engineered and constructed panelized system might be accepted.

### 6.8.1 STRESSED-SKIN PANELS

*Stressed-skin* panels are panels with stressed covers consisting of plywood

**FIGURE 6.8-2** Typical two-faced stressed-skin panel. (Courtesy American Plywood Association.)

**a**

**b**

**FIGURE 6.8-3** (a) Stressed-skin panels with plywood surfaces glued to a 2 in. (50 mm) lumber frame. After the glue has been placed and the plywood positioned, the entire panel is treated with heat and pressure in a press; (b) stressed-skin panel being lowered into place by crane. (Courtesy American Plywood Association.)

**FIGURE 6.8-4** Typical uses for stressed-skin panels.

**FIGURE 6.8-5** Sandwich panel with (left) honeycomb core and (right) foam core. (Courtesy American Plywood Association.)

a

b

c

**FIGURE 6.8-6** (a) Sandwich panel with foamed plastic core; (b) an assembly panel made of a foamed plastic core sandwich panel; (c) an entire house structure made of sandwich panels. (Courtesy American Plywood Association.)

# 6.9 Shop-Fabricated Structural Wood

Shop-fabricated structural wood consists of preassembled components that are parts of a building's framing, floor and ceiling assemblies, wall assemblies, or roof and ceiling assemblies and that are manufactured in a location other than that in which they will be finally used. They may range from small wall panels to complete structures.

Elements used in preassembled components include the types of structural panels discussed in Section 6.8 and other units made from plywood and lumber, including preframed panels, rigid frames, box beams, various curved units, roof trusses, and truss joists.

Panelized assemblies for floors, walls, and roofs consist of either normal framing systems that are built as a unit and then erected, or one or more of the components listed in the preceding paragraph, or both.

## 6.9.1 PANELIZED ASSEMBLIES

Floor and ceiling assemblies, wall assemblies, and roof and ceiling assemblies can be manufactured and brought to the erection site as panelized units.

### 6.9.1.1 Panelized Floor and Ceiling Assemblies

A common nonpanelized floor and ceiling assembly consists of $1\frac{1}{8}$-in. (28.58-mm)-thick APA Rated Sturd-I-Floor combination subfloor and underlayment plywood panels spanning between beams spaced 48 in. (1220 mm) on centers (see Fig. 6.10-6). A common *panelized* floor and ceiling assembly consists of preframed panels in which the members are precut and to which subflooring is applied to form a preassembled component (Figs. 6.9-1 and 6.9-2). Frame members of nominal 2 in. (50 mm) dimension lumber function as floor joists. Panels may be fabricated on- or off-site in practically any size, with insulation and vapor retarder installed.

Panelized floor and ceiling assemblies may also utilize stressed-skin panels or sandwich panels (see Section 6.8). In panelized floor assemblies, panels may span between transverse or longitudinal beams or be centrally supported by bearing walls or beams, as with conventional floor and ceiling assemblies.

**FIGURE 6.9-1** This preframed floor panel consists of plywood sheathing attached to 2 × 4 (50 × 100) stiffeners that span 4 ft (1220 mm) from beam to beam. (Courtesy American Plywood Association.)

**FIGURE 6.9-2** Preframed floor panel with top and bottom plywood surfaces applied to 2 × 8 (50 × 200) frame members. (Courtesy American Plywood Association.)

**FIGURE 6.9-3** Panels may be built off-site in sizes that can be conveniently handled by two workers. (Courtesy American Plywood Association.)

**FIGURE 6.9-4** Preframed panels 4 ft (1220 mm) wide being erected. Studs are spaced 24 in. (610 mm) on centers. The space at the top of the panels is to receive a continuous header. (Courtesy American Plywood Association.)

a

b

**FIGURE 6.9-6** Off-site fabrication may also include (a) panels with exterior finish and windows installed; (b) panels with siding, windows, and insulation installed. (Courtesy American Plywood Association.)

**FIGURE 6.9-7** This preframed roof panel has been assembled on the ground, partially sheathed, and lifted into position. (Courtesy American Plywood Association.)

### 6.9.1.2 Panelized Wall Assemblies

Panelized wall assemblies may consist of preframed panels using precut wood studs to which sheathing has been applied (Figs. 6.9-3 and 6.9-4). Fabrication may include the installation of insulation, air infiltration barriers, interior and exterior finishes, windows, and doors (Figs. 6.9-5 and 6.9-6), and panels may be fabricated off-site in any size that is conveniently handled.

Panelized wall assemblies may also include stressed-skin and sandwich panels (see Section 6.8).

### 6.9.1.3 Panelized Roof and Ceiling Assemblies

Panelized roof and ceiling assemblies may consist of preframed, precut, and sheathed panels (Fig. 6.9-7). Nominal 2 in. (50 mm) dimension frame members of the panels function as rafters or ceiling joists. Panels may be fabricated on- or off-site in any size that can be conveniently handled.

**FIGURE 6.9-5** Large wall panel with sheathing applied and windows installed can be fabricated off-site. (Courtesy American Plywood Association.)

Panelized roof and ceiling assemblies may also include stressed-skin or sandwich panel construction (see Section 6.8). Panelized roof and ceiling assemblies may be used in a variety of ways to form sloped or flat roofs and ceilings (Fig. 6.9-8). Panels may span between transverse or longitudinal beams or can be centrally supported by bearing walls or beams, as in joists roofs. Membrane or shingle roofing finishes may be applied to panelized systems. Interior finishes may be field-applied or may be an integral, prefinished part of the panel.

### 6.9.1.4 Finishes

Finishes that can be applied to a conventional assembly can be applied to a panelized floor and ceiling, wall, or roof and ceiling assembly. Finishes can be field-applied or may be an integral, prefinished part of the panelized assembly.

**FIGURE 6.9-8** Vaulted stressed-skin panels can create unusual roof and ceiling forms. (Courtesy American Plywood Association.)

and rafters with the same span. Trusses require stress-rated lumber and must be built from engineered designs. They commonly are spaced 24 in. (610 mm) to 48 in. (1220 mm) on centers, and like other preassembled components, they can be erected easily, permitting the building to be enclosed quickly (Fig. 6.9-13).

Truss roof systems may be designed for flat (see Fig. 6.9-13) or sloped roofs to receive membrane or shingle roofing finishes (Fig. 6.9-14). Plywood or board sheathing is fastened to trusses to receive roofing. Ceilings can be either flat, sloped (see Fig. 6.9-13), or curved (Fig. 6.9-15).

Plaster bases and gypsum board finishes commonly are applied directly to the lower chord member of a truss, but almost any finish may be used. When trusses are spaced more than 24 in. (610 mm) on centers and when other than board finishes are

### 6.9.2 WOOD ROOF TRUSSES

Wood roof trusses for light wood-framed buildings are composed of a number of wood framing members (chords) arranged in a framework of triangles to form a supporting structural element for both a roof and a ceiling (Fig. 6.9-9). In modern trusses, members, generally 2 × 4s (50 × 100s) or 2 × 6s (50 × 150s), are connected by split ring connectors or metal gang-nail plates (Figs. 6.9-10 and 6.9-11) that permit members to be assembled without having to be overlapped at the joints. Older trusses were often connected by wood or plywood gusset plates.

A truss acts as a unit to support roof and ceiling loads and to span the entire width of a building between exterior walls. This clear span allows interior walls to be non-load-bearing and permits great flexibility in placing walls (Fig. 6.9-12) while also providing space for insulation and heating, air conditioning, electrical, and plumbing equipment.

In a light truss, individual diagonal members resist compression and tension forces. The top and bottom chords also resist bending because roof and ceiling loads are applied along their entire length. However, since the chords span only between panel points, their spans are much shorter than those of normal rafters and ceiling joists. Therefore, the loads they carry are lighter and the bending stresses they must resist are smaller than those of conventional framing members. As a result, truss top and bottom chords are much smaller than conventional joists

**FIGURE 6.9-9** A wood truss roof and ceiling assembly.

**FIGURE 6.9-10** The types of metal plates used to connect wood truss components. (Courtesy Wood Truss Council of America.)

**FIGURE 6.9-11** Spliced joints may have metal plates on one or both sides. (Courtesy Wood Truss Council of America.)

**FIGURE 6.9-14** Trusses with sloped roof and ceiling. Ceilings can also be flat. (Courtesy Wood Truss Council of America.)

**FIGURE 6.9-12** Trusses provide clear spans and eliminate the need for interior load-bearing walls. (Courtesy Wood Truss Council of America.)

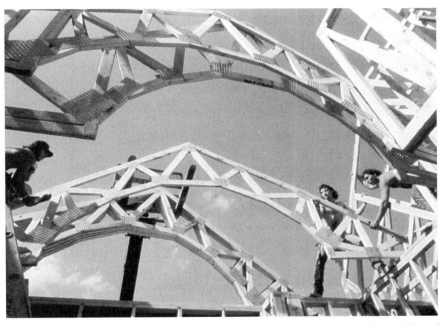

**FIGURE 6.9-15** Sloped roof trusses with segmented bottom chords to permit installation of a curved ceiling. (Courtesy Wood Truss Council of America.)

**FIGURE 6.9-13** Individual trusses are positioned easily by hand. Trusses shown will produce a flat roof and ceiling. (Courtesy Wood Truss Council of America.)

used, it may be necessary to apply nailing or furring strips across the trusses to receive the interior finish.

Trusses are economical and widely used, even for large structures (Fig. 6.9-16). They provide excellent structural performance, make efficient use of materials, are easy to assemble and erect, and provide maximum interior flexibility.

Irregularities in spacing and leveling often are magnified in wood truss construction because relatively small members are used for bottom chords and larger spans are involved than in joist construction. Cross furring generally is recommended to solve these problems.

However, when gypsum board is nailed directly to the bottom chords, the chords can be aligned and leveled by using stringers at approximately one-third points continuously across the entire ceiling. Built-in camber to compensate for future deflection of bottom chords also is recommended (Fig. 6.9-17).

With trussed roof construction, exterior walls and the ceiling sometimes are finished before interior partitions are erected and finished. When this method is used, the roofing and other construction elements which would increase roof loads should be installed before interior partitions are erected. If substantial roof loads

**FIGURE 6.9-16** Large open spaces can be created economically using wood trusses. (Courtesy Wood Truss Council of America.)

**FIGURE 6.9-17** When cross furring is not used; 1 × 6 (25 × 150) or 2 × 4 (50 × 100) stringers at one-third points may be installed to regulate truss spacing. Additional 2 × 6 (50 × 150) stringers may also be used to level the bottom chords when necessary.

**FIGURE 6.9-18** In truss construction, ceiling distortion may result if substantial loads are introduced after interior partitions have been erected. (Courtesy Gypsum Association.)

are introduced after partitions are installed, ceiling distortion may result near interior partitions as trusses deflect (Fig. 6.9-18).

### 6.9.3 MISCELLANEOUS ASSEMBLIES

Other prefabricated assemblies are also often used. These include, among others, preframed panels, rigid frames, box beams, and curved and other roof units.

*Preframed panels* consist of fabricated, precut lumber and plywood that are nailed or mechanically fastened together. They may be used for walls, floors, or roofs (Fig. 6.9-19).

a

b

**FIGURE 6.9-19** Preframed panels being placed (a) by crane and (b) by hand. (Courtesy American Plywood Association.)

Scarf Joint in Plywood Web
No Stiffener or Splice
Plate Req'd as in Butt Joint

Scarf Joint in
Lumber Flange

Pressure Laminated
Lumber Flanges
Top Flange

Butt Joint Between
Plywood Webs

Glue Joint

Lumber
Intermediate Stiffener

Glue Joint

Lumber Stiffener

Plywood Splice Plate

Bottom Flange

Plywood Web

Lumber
Bearing Stiffener

**typical cross sections showing beam types**

A          B          C1          C2

**FIGURE 6.9-20** Typical Plywood box beams. (Courtesy American Plywood Association.)

tribute concentrated loads and resist web buckling.

Box beams have been used successfully for many purposes, including rather spectacular 48-in. (1220-mm)-deep girders for factory roof systems and other members spanning up to 120 ft (36.58 m), as well as in many conventional applications (Fig. 6.9-21).

Either nail gluing or nailing alone may be satisfactory, but pressure gluing with accurate control is the preferred method to give maximum stiffness and strength to box beams.

*Curved and other roof units* permit construction of unique design shapes. Curved panels consist of built-up panels with plywood faces glued at the top and bottom to various core materials. These usually form an arc of a circle, as shown in Figure 6.9-22. Other design shapes, some of which are shown in Figure 6.9-23, also can be produced.

## 6.10 Sheathing

Sheathing includes subflooring and underlayment and wall and roof sheathing. Plank flooring is addressed in Section 6.11, wood finish flooring in Chapter 9, and air infiltration barriers in Chapter 7.

*Rigid frames* are specialized construction elements consisting of arch-shaped frames, usually comprising four pieces of straight lumber (two for posts and two for rafters). Plywood gussets, rigidly nailed or glued, connect adjacent members, transmit stresses across the joints, and consolidate the frames into integral units.

*Box beams* are hollow structural units consisting of two or more vertical plywood webs which are attached, usually by gluing or nailing, to lumber flanges. Typical box beam cross sections are shown in Figure 6.9-20. As in a steel beam, the flanges carry the bending forces, while the plywood webs transmit the shear. In this manner, each of the materials is employed more effectively. At intervals along the beam, stiffeners may be inserted between flanges and attached to the webs to dis-

**FIGURE 6.9-21** A plywood box beam. (Courtesy American Plywood Association.)

## Typical Panel Using
## Curved Plywood Ribs

Preferred method: Scarf joint in plywood skin

Alternate: Spliced butt joint where permitted by design

Glue joint

Plywood upper skin

Set out edge member to form tongue of tongue & groove joint

Insulation (optional)

Set in edge member to form groove

Depth

Headers

Plywood laminated ribs (laminated & curved prior to panel assembly)

Lumber ribs may be used if desired

Width

Length

Lower plywood skin

## Typical Panel Using
## Solid Plywood Core

Scarf joint in plywood

Plywood upper skin

Glue joint

Plywood core set in to form groove

Tight butt joint in plywood core unless scarf joint required

Plywood core set out to form tongue of tongue & groove joint

Width

Length

Plywood core

Lower plywood skin

## Typical Panel Using
## Lightweight Sandwich Core

Upper plywood skin

Edge members & headers may be plywood laminated sections

Scarf joint in plywood

Lumber edge member set out to form tongue

Depth

Lumber edge member set in to form groove of tongue & groove joint

Glue

Resin-impregnated paper honeycomb core

Length

Lower plywood skin

Width

Lumber headers

Scarf joint in plywood

**FIGURE 6.9-22** Curved stressed-skin panels. (Courtesy American Plywood Association.)

**FIGURE 6.9-23** Plywood stressed-skin construction permits use of many geometric shapes. (Courtesy American Plywood Association.)

## 6.10.1 SUBFLOORING AND PLYWOOD UNDERLAYMENT

Subflooring is required beneath most kinds of finish flooring. An additional underlayment is also needed for some types of flooring.

### 6.10.1.1 Subflooring

Subflooring serves as a working platform and as a base for floor finishes. The two main types of subflooring used as a base for flooring in light wood-framed construction are wood and concrete. Concrete floors are discussed in Chapter 3. Additional requirements for subfloors are discussed in the sections where the various floor finishes are covered.

Tile, terrazzo, resilient flooring, and carpet may be installed either over a double-layer base composed of a separate subfloor and an underlayment or a single-layer construction consisting of a combination subfloor/underlayment material specifically designed for the purpose. Wood flooring is usually installed over a single-layer subfloor.

Whether it is covered with underlayment or not, wood subflooring should be firmly fastened and structurally adequate to carry the intended loads without excessive deflection. The maximum deflection of such subflooring under full dead and live loads after the finished flooring has been installed should not exceed $^1/_{360}$ of any span.

Figure 6.10-1 shows wood joist floor construction suitable for the application of resilient flooring, carpet, and ceramic tile. The only acceptable wood material base for terrazzo is plywood. Wood subflooring for wood floors may be boards or plywood.

Suspended wood floors above grade usually are sufficiently removed from excessive external moisture sources, but floors over crawl spaces at or near grade may require the crawl space to be ventilated and the ground inside the crawl space to be covered with a suitable vapor retarder (see Section 21.4).

## BOARD SUBFLOORING

Board subflooring is permissible without underlayment beneath structural flooring such as wood strip flooring, beneath tile set in a mortar bed, and beneath terrazzo with an underbed. However, underlayment is recommended over board subflooring beneath thin-set tile, thin-set terrazzo, resilient flooring, and carpet.

Conventional board subflooring consists of nominal 1-in. (25-mm)-thick boards 4 in. (100 mm) to 8 in. (200 mm) wide. These boards may be applied either diagonally or at right angles to the joists. When boards are placed at right angles to the joists, wood strip or plank flooring must be laid at right angles to the boards. Diagonal board installation is preferred with wood finish flooring because it permits the finish flooring to be laid either parallel or at right angles to the joists. Boards may have square, tongue-and-groove, or shiplap edges. Except for tongue-and-grooved boards, board ends should rest over the joists. Subfloor boards less than 8 in. (200 mm) wide should be nailed to each joist with two 8d nails; those 8 in. (200 mm) wide or wider should be nailed with three 8d nails. Where desired, 7d annularly threaded nails can be substituted for the 8d smooth shank nails.

The moisture content of lumber boards used for subflooring at the time of installation should be not more than 15%.

## PLYWOOD SUBFLOORING

Plywood is a suitable subflooring material for tile, terrazzo, wood flooring, resilient flooring, and carpet. Additional details are covered in sections where the various types of flooring are addressed. A single-layer subfloor without underlayment may be used beneath structural flooring such as wood strip flooring, beneath tile with a thick-set mortar bed, beneath terrazzo with an underbed, and beneath carpet with a pad. Plywood subflooring requires an underlayment for receipt of thin-set tile, thin-set terrazzo, and resilient flooring.

A conventional plywood subfloor should be APA Rated Sheathing, either Exposure 1 or 2, Exterior, Structural I, Exposure 1, or Structural I, Exterior, depending on its exposure to moisture. The more moisture-resistant grades should be selected when a hazard of prolonged exposure to moisture exists, such as in bathrooms, in floors adjacent to plumbing fixtures, or when the subfloor may be exposed to the weather for long periods during construction. The thickness varies from $^1/_2$ in. (12.7 mm) to $^3/_4$ in. (19.1 mm) thick, depending on the span rating, type of finish floor, and spacing of the supporting joists. Plywood subfloor panels to be covered with underlayment may be unsanded.

Plywood subflooring should be installed with the long dimension perpendicular to the joists, with end joints centered over the joists (Fig. 6.10-2). Panels should be staggered so that the ends of adjacent panels do not fall on the same support. The longitudinal edges should be supported on 2 × 3 (50 × 75) or heavier wood blocking. The blocking may be eliminated, however, when tongue-and-groove plywood is used. Supports should be spaced and panels should be nailed as shown in Figure 6.10-3.

Subfloor panels should be spaced approximately $^1/_8$ in. (3.2 mm) apart at end and edge joints, and placed about $^1/_8$ in. (3.2 mm) from adjacent vertical surfaces, such as walls and pipes, to allow for expansion caused by changes in humidity. Nailing should start in the middle of the panel and proceed outward to the edges, with nail heads set flush or slightly countersunk. Perimeter nails should be located not more than $^3/_8$ in. (9.5 mm) from the edges. Panel joints should be sanded smooth and flush before finish flooring is installed. It is important to protect single-layer floors from damage during construction. However, the difficulty of furnishing adequate protection during construction may warrant the use of unsanded or touch-sanded plywood and power sanding of the entire floor area just before the flooring is installed.

**FIGURE 6.10-1** Typical subflooring construction for the application of carpet, tile, or resilient flooring.

Carpet, Tile, or Resilient Flooring

Plywood Subfloor Perpendicular to Joists

Lumber Blocking at Edge Joints Except for T&G Plywood

**SINGLE-LAYER WOOD FLOOR CONSTRUCTION**

Carpet, Tile, or Resilient Flooring

Panel Underlayment (Plywood, Hardboard or Particleboard)

Lumber or Plywood Subfloor

Wood Joists

**DOUBLE-LAYER WOOD FLOOR CONSTRUCTION**

FACE GRAIN

**FIGURE 6.10-2** Tongue-and-groove-edged APA Rated Sturd-I-Floor with the proper span rating can be installed on spacings of up to 48 in. (1220 mm) and tongue-and-groove-edged APA Rated Sheathing with the proper span rating can be installed on spans of up to 24 in. (610 mm), both without edge blocking, but should be protected during construction. (Courtesy APA—The Engineered Wood Association, formerly American Plywood Association.)

### 6.10.1.2 Plywood Underlayment

An underlayment is required over both wood board and plywood subflooring beneath thin-set tile, thin-set terrazzo, and resilient flooring. Using an underlayment is often desirable even when not absolutely required for the type of finish flooring, because the subfloor, which provides a convenient working platform during most of the construction period, can be easily damaged, and installation of an underlayment can be scheduled closer to the installation of the flooring.

Underlayment beneath resilient flooring is often either Underlayment or C-C Plugged panel grades marked *sanded face*. Other grades may also be used, however, when acceptable to the flooring manufacturer. Thin underlayment for other purposes may be APA Underlayment Interior, Exposure 1, C-C Plugged Exterior, or APA C-C Plugged Exterior. Underlayment thicker than $^{19}/_{32}$ in. (15.08 mm) may be APA Rated Sturd-I-Floor Exposure 1 or 2 or APA Rated Sturd-I-Floor, Exterior. The more moisture-resistant grades should be used in unusual moisture conditions, such as those encountered in bathrooms and laundry areas and beneath tile and terrazzo.

Underlayment panels should be installed with the finished face up and, over plywood subflooring, with the long edge across the supports. When laid over lumber board subflooring, the face grain of plywood underlayment should be laid perpendicular to the boards (see Fig. 6.10-1). The ends and edges should be staggered and should not fall within 2 in. (50 mm) of the joints in the subflooring. However, it is not necessary to stagger the joints in underlayment beneath carpet with a pad. In general, joints between panels should be left open $^{1}/_{32}$ in. (0.79 mm) to permit some movement. However, wider spacings between panels may be required by some flooring manufacturers for certain applications, and their recommendations should be followed. Joints between underlayment and other material, such as adjacent vertical surfaces and pipes, should be left open $^{1}/_{8}$ in. (3.2 mm) beneath resilient flooring and carpet and $^{1}/_{4}$ in. (6.4 mm) beneath tile and terrazzo unless the manufacturers of these products recommend different spacings. Surfaces of adjacent panels should be level within $^{1}/_{32}$ in. (0.79 mm).

Underlayment should be of sufficient strength to withstand loads that will be encountered. In general, panels beneath tile should not be less than $^{5}/_{8}$ in. (15.9 mm), except that $^{1}/_{2}$ in. (12.7 mm) is

**FIGURE 6.10-3** APA Panel Subflooring (APA Rated Sheathing)

| Panel Span Rating | Minimum Panel Thickness, in. | Maximum Span, in. | Nail Size[a] and Type | Maximum Nail Spacing, in. | |
|---|---|---|---|---|---|
| | | | | Supported Panel Edges | Intermediate Supports |
| 24/16 | $^{7}/_{16}$ | 16 | 6d common | 6 | 12 |
| 32/16 | $^{15}/_{32}$ | 16[b] | 8d common[c] | 6 | 12 |
| 40/20 | $^{19}/_{32}$ | 20[b,d] | 8d common | 6 | 12 |
| 48/24 | $^{23}/_{32}$ | 24 | 8d common | 6 | 12 |
| 60/32 | $^{7}/_{8}$ | 32 | 3d common | 6 | 12 |

[a]Other code-approved fasteners may be used.

[b]Span may be 24 in. if $^{3}/_{4}$-in. wood strip flooring is installed at right angles to joists.

[c]6d common nail permitted if panel is $^{1}/_{2}$ in. or thinner.

[d]Span may be 24 in. if a minimum $1^{1}/_{2}$ in. of lightweight concrete is applied over panels.

(Courtesy APA—The Engineered Wood Association, formerly American Plywood Association.) (Refer to Figure 25.7-1 for metric conversions.)

sometimes acceptable in small residential applications beneath terrazzo not less than $^3/_8$ in. (9.5 mm).

Panels $^1/_2$ in. (12.7 mm) thick or thinner should be nailed with 3d ring-shank nails at 6 in. (150 mm) on centers along the edges and at 8 in. (200 mm) on centers each way throughout the remainder of each panel. Thicker panels should be nailed using 4d nails at 6 in. (150 mm) on center along the edges and at 12 in. (305 mm) on center throughout the remainder of each panel. Fastener length should be the same as the total thickness of the subflooring and the underlayment.

### 6.10.1.3 Combination Subfloor and Underlayment

APA Rated Sturd-I-Floor is a proprietary *combination subfloor and underlayment* plywood panel. It is designed specifically for use as a single-layer subfloor beneath carpet with pad, but is also usable under other flooring types and may be used even when a separate underlayment is required. It is available in thicknesses from $^{19}/_{32}$ in. (15.08 mm) to $1^1/_8$ in. (28.58 mm) and with span ratings from 16 to 48. It is available in Exposure Durability Classifications: Exterior, Exposure 1, and Exposure 2.

The support system for APA Rated Sturd-I-Floor may be a conventional floor framing system with the framing members (joists or floor trusses) at 16 in. (406 mm), 20 in. (508 mm), or 24 in. (610 mm) on centers (Fig. 6.10-4), a joist system with the joists spaced 32 in. (760 mm) on centers (Fig. 6.10-5), or a girder system with the girders at 48 in. (1220 mm) on centers (Fig. 6.10-6).

APA Rated Sturd-I-Floor panels are manufactured either as conventional plywood, as a composite, or as a nonveneer panel. Panels are available both square-edged and tongue-and-grooved. The face ply can be either sanded or touch-sanded; the latter is used when field sanding is intended at the end of construction, just before the floor finish is installed.

The installation of APA Rated Sturd-I-Floor panels is similar to that for other plywood subfloors: face plies perpendicular to supports; end joints staggered; panels continuous over at least two supports; panel ends resting on the supports; end and edge joints spaced $^1/_8$ in. (3.2 mm); and solid 2 in. (50 mm) nominal blocking beneath unsupported panel edges of square-edged panels (blocking

**FIGURE 6.10-4** APA Rated Sturd-I-Floor, 16 in. (406 mm), 20 in. (560 mm), and 24 in. (610 mm) on center. (Courtesy APA—The Engineered Wood Association, formerly American Plywood Association.) (Refer to Figure 25.7-1 for metric conversions.)

**FIGURE 6.10-5** APA Rated Sturd-I-Floor 32 in. (813 mm) and 48 in. 1220 mm) on center (over supports 32 in. [813 mm] on center). (Courtesy APA. The Engineered Wood Association, formerly American Plywood Association.) (Refer to Figure 25.7-1 for metric conversions.)

**FIGURE 6.10-6** APA Rated Sturd-I-Floor 48 in. (1220 mm) on center (over supports 48 in. [1220 mm] on center). (Courtesy APA—The Engineered Wood Association, formerly American Plywood Association.) (Refer to Figure 25.7-1 for metric conversions.)

is not required at tongue-and-groove joints).

A glued installation is recommended for APA Rated Sturd-I-Floor (see Section 6.10.1.4), but this material can also be installed with nailing only. Unless the code requires a different spacing, the spacings should be in accordance with Figure 6.10-7.

### 6.10.1.4 Glued Floor System

In the APA glued floor system, the flooring and the joists are glued together with such strength and stiffness that they act as a series of integral T-beams. In addition, the gluing virtually eliminates the squeaks which can result from shrinking lumber (Fig. 6.10-8). Under some circumstances, using the APA glued floor system will permit the use of smaller joists or longer spans for the same size joists than will a conventional nailed system. The glued system also has better creep resistance. The APA glued subfloor system can be installed quickly using readily available materials and techniques.

**FIGURE 6.10-7** APA Rated Sturd-I-Floor[a]

| Span Rating (Maximum Joist Spacing), in. | Minimum Panel Thickness,[b] in. | Fastening: Glue-Nailed[c] | | | Fastening: Nailed Only | | |
|---|---|---|---|---|---|---|---|
| | | | Maximum Spacing, in. | | | Maximum Spacing, in. | |
| | | Nail Size and Type | Supported Panel Edges | Intermediate Supports | Nail Size and Type | Supported Panel Edges | Intermediate Supports |
| 16 | $^{19}/_{32}$ | 6d ring- or screw-shank[d] | 12 | 12 | 6d ring- or screw-shank | 6 | 12 |
| 20 | $^{19}/_{32}$ | 6d ring- or screw-shank[d] | 12 | 12 | 6d ring- or screw-shank | 6 | 12 |
| 24 | $^{23}/_{32}$ | 6d ring- or screw-shank[d] | 12 | 12 | 6d ring- or screw-shank | 6 | 12 |
| | $^7/_8$ | 8d ring- or screw-shank[d] | 6 | 12 | 8d ring- or screw-shank | 6 | 12 |
| 32 | $^7/_8$ | 8d ring- or screw-shank[d] | 6 | 12 | 8d ring- or screw-shank | 6 | 12 |
| 48 | $1^3/_{32}$ | 8d ring- or screw-shank[e] | 6 | f | 8d ring- or screw-shank[e] | 6 | f |

[a]Special conditions may impose heavy traffic and concentrated loads that require construction in excess of the minimums shown.

[b]Panels in a given thickness may be manufactured in more than one Span Rating. Panels with a Span Rating greater than the actual joist spacing may be substituted for panels of the same thickness with a Span Rating matching the actual joist spacing. For example, $^{19}/_{32}$-in.-thick Sturd-I-Floor 20 on centers (oc) may be substituted for $^{19}/_{32}$-in.-thick Sturd-I-Floor 16 oc over joists 16 in. oc.

[c]Use only adhesives conforming to APA Specification AFG-01, applied in accordance with the manufacturer's recommendations. If OSB panels with sealed surfaces and edges are to be used, use only solvent-based glues; check with panel manufacturer.

[d]8d common nails may be substituted if ring- or screw-shank nails are not available.

[e]10d common nails may be substituted with $1^1/_8$-in. panels if supports are well seasoned.

[f]Space nails maximum 6 in. for 48-in. spans and 12 in. for 32-in. spans.

(Courtesy APA—The Engineered Wood Association, formerly American Plywood Association.) (Refer to Figure 25.7-1 for metric conversions.)

Note:
Provide adequate ventilation and use ground cover vapor barrier in crawl space. Panels must be dry before applying finish floor.

Carpet and pad

1/8" spacing is recommended at all edge and end joints unless otherwise indicated by panel manufacturer

Stagger end joints

Tongue-and-groove edges (or 2" lumber blocking between supports)

Strength axis

Site-applied glue, both joints and tongue-and-groove joint (or between panels and edge blocking)

2x joists, "I" joists or floor trusses—16", 19.2", 24", or 32" oc (4x supports for 48" oc spacing)

APA RATED STURD-I-FLOOR 16, 20, 24, 32 or 48 oc

a                                                                 b

**FIGURE 6.10-8** (a) APA glued floor system; (b) applying adhesive. (Courtesy APA—The Engineered Wood Association, formerly American Plywood Association.) (Refer to Figure 25.7-1 for metric conversions.)

The subflooring in the APA glued floor system should be either APA Rated Sheathing with a plywood overlay or structural flooring, such as a wood strip floor, or a one-piece combination subfloor and underlayment of APA Rated Sturd-I-Floor. The subflooring will vary in thickness from $1/2$ in. (12.7 mm) to $1 1/8$ in. (28.58 mm), depending on the joist spacing and the finish flooring material to be supported. Tongue-and-groove material is recommended. If square-edge material is used, it is necessary to provide solid blocking under the panel edges between joists. The plywood is glued to this blocking. The basic joist framing is the same as for a conventional nailed subfloor.

When the finish floor is a resilient nontextile material, a separate underlayment is required even over APA Rated Sturd-I-Floor. This underlayment may be either veneer-faced APA Rated Sturd-I-Floor with a sanded face or another underlayment (see Section 6.10.1.3). In a glued system, the plywood subflooring panels should be placed with the long dimension perpendicular to the joists, with end joints staggered and spaced $1/8$ in. (3.2 mm) at both ends and edges. Before each panel is set in place, a bead of glue is applied to the supports with a caulking gun; for extra stiffness, the edge groove should also be glued. The panel then is secured with ring- or screw-shank nails. These nails should be 6d for $3/4$ in. (19.1 mm) and thinner panels and 8d for thicker panels. Nail spacing varies from 6 in. (150 mm) to 12 in. (305 mm) along panel ends and edges and in the field of the panels at each support, depending on the joist spacing and the panel thickness (see Fig. 6.10-7).

More detailed information on the APA glued floor system, including joist span tables, nailing sizes and spacings, application sequence, and recommended adhesives, is contained in the APA booklet *Engineered Wood Construction Guide.*

## 6.10.2 WALL AND ROOF SHEATHING

Most exterior wall finishes and roofing materials applied over wood-framed con-

struction require sheathing. Some require sheathing paper. Wall sheathing may be lumber, plywood, oriented strand board, or a nonwood material such as fiberboard or gypsum board. Roof sheathing is usually lumber, plywood, or oriented strand board.

Wood shakes and shingles can be attached to wood nailing strips fastened to the wall studs through a sheathing material. These nailing strips are discussed in the sections where the finish materials are covered.

### 6.10.2.1 Wall Sheathing

Wall sheathing is applied to framing members for one or more of the following reasons:

1. When properly fastened, it is believed to help a structural frame resist racking (being forced out of shape or plumb). However, tests by the Division of Building Research, National Research Council of Canada, indicate that interior wall finishes, like plaster and gypsum board, provide about four

times as much resistance to racking as does lumber sheathing applied horizontally with corner bracing. Therefore, the National Building Code of Canada does not require sheathing except when the exterior finish materials require it as a nailing base.

2. It contributes insulation, which reduces heat transmission and resists infiltration of wind and moisture.
3. It functions in many cases as a nailing base for exterior finish materials.

When wall sheathing is used as a nailing base, it should be smooth and securely attached to its supports. Shingles may be applied over nailable solid sheathing consisting of wood boards, plywood, or nailable fiberboard. They may also be applied to wood nailing strips over almost any type of sheathing. Wood shakes and shingles can also be applied over spaced sheathing. Plywood siding can be installed either directly over the framing or over plywood sheathing. The types of sheathing required and additional details concerning it are discussed in the sections covering each exterior wall finishing material.

## LUMBER WALL SHEATHING

Lumber sheathing may be either solid, spaced, or nailing strips, depending on the type of finish siding to be applied.

Lumber sheathing boards beneath wall shingles should be seasoned to a moisture content between 12% and 18%. They should be free from loose knots and large knotholes. Warped boards that would prevent the siding from lying flat should not be used.

Sheathing boards up to 8 in. (200 mm) wide are face-nailed at each framing member with two 7d annularly threaded nails or two 8d common wire nails. Boards 10 in. (250 mm) and 12 in. (305 mm) wide require three nails at each support. Supports for nominal 1 in. (25 mm) lumber sheathing should not be more than 24 in. (610 mm) on centers.

***Solid Sheathing*** Boards for *solid lumber sheathing* should be nominal 1 in. (25 mm) thick. Boards used for solid sheathing beneath mineral fiber cement shingles should be not more than 8 in. (200 mm) nominal wide and should have shiplapped or tongue-and-groove edges. Boards used for sheathing beneath wood shakes and shingles should be not more than 10 in. (250 mm) nominal wide. These may be

square-edged, shiplapped, or tongue-and-groove.

When lumber sheathing boards are placed horizontally, structural corner bracing is required. Therefore, solid sheathing boards are usually placed diagonally, except where the finish requires horizontal placement (see Chapter 7).

***Spaced Sheathing*** For the application of wood shingles and shakes, wood spaced sheathing should be at least 1 × 3 (25 × 75), 1 × 4 (25 × 100), or 1 × 6 (25 × 150) boards, as required for the finish material (see Chapter 7).

## PLYWOOD WALL SHEATHING

Plywood sheathing should be APA Rated Sheathing of either Exposure 1 or 2; Exterior; Structural I, Exposure 1; or Structural I, Exterior. It should have the appropriate span rating according to the stud spacing. The span rating may be either 16 or 24 for a 16 in. (406 mm) stud spacing but must be 24 for a 24 in. (610 mm) stud spacing (Fig. 6.10-9). These materials may include particleboard, as discussed in Section 6.3.

*Nailable plywood sheathing* is sheathing to which a finish material, such as siding, will be directly nailed. However, sheathing on brick-veneered and stuccoed walls should be installed in the same manner. Such sheathing may be placed with its long edge either horizontal or vertical, except that three-ply panels at studs spaced 24 in. (610 mm) on centers must be placed with the long edge across the studs, and panels $\frac{3}{8}$ in. (9.5 mm) thick or less at studs spaced 16 in. (406 mm) on centers must be placed with the long edge across the studs. Some codes require the application of nailers (back

blocking) at joints in plywood sheathing that do not fall over supports (Fig. 6.10-10). Because greater than normal stiffness is required in sheathing beneath stucco, panels there are often applied with the long side across the studs, and blocking is usually installed at all panel edges, whether or not required by code.

Plywood wall sheathing $\frac{1}{2}$ in. (12.7 mm) thick or less should be applied with 6d common, smooth shank, annular, spiral-thread, or galvanized box nails. Thicker panels should be installed with similar 8d nails. Nails should be spaced at 6 in. (150 mm) centers along supported panel edges and at 12 in. (305 mm) centers at intermediate supports.

## PLYWOOD CORNER BRACING

When nonwood sheathing is used that does not provide corner bracing, such as fiberboard and gypsum board, the corners of a building must be braced. This bracing may be plywood sheathing (Fig. 6.10-11). Such sheathing should be one of the materials mentioned above for plywood wall sheathing. The thickness should match that of the adjacent sheathing material.

## NONLUMBER SHEATHING

Nonlumber sheathing materials in common use include fiberboard, gypsum board, and insulating board sheathing.

When fiberboard, gypsum board, insulating board, or sheathing materials other than wood or plywood are recommended by the manufacturer for use as wall sheathing under a type of siding, the sheathing should be installed and attached to its supports according to its manufacturer's recommendations. The recommendations of the manufacturer supersede the

**FIGURE 6.10-9** APA Panel Wall Sheathing Material

| Panel Span Rating | Maximum Stud Spacing, in. | Nail Size[a,b] | Nail Spacing, in. | |
| --- | --- | --- | --- | --- |
| | | | Supported Panel Edges | Intermediate Supports |
| 12/0, 16/0, 20/0, or Wall-16 oc | 16 | 6d for panels $\frac{1}{2}$ in. thick or less; 8d for thicker panels | 6 | 12 |
| 24/0, 24/16, 32/16, or Wall-24 oc | 24 | | | |

[a]Common, smooth, annular spiral thread, or galvanized box.

[b]Other code-approved fasteners may be used.

(Courtesy APA—The Engineered Wood Association, formerly American Plywood Association.) (Refer to Figure 25.7-1 for metric conversions.)

**FIGURE 6.10-10** APA Rated Sheathing installation on a wall. (Courtesy APA—The Engineered Wood Association, formerly American Plywood Association.) (Refer to Figure 25.7-1 for metric conversions.)

Labels on figure:
- 1/8" spacing is recommended at all edge and end joints unless otherwise indicated by panel manufacturer
- "Block" horizontal joints in panels used for bracing
- Filter strip if required
- APA RATED SHEATHING applied with long dimension parallel to studs
- APA RATED SHEATHING applied with long dimension across studs
- Siding
- 6" minimum clearance, siding to grade

**FIGURE 6.10-11** APA panel corner bracing. (Courtesy APA—The Engineered Wood Association, formerly American Plywood Association.) (Refer to Figure 25.7-1 for metric conversions.)

Labels on figure:
- Supports 16" or 24" o.c.
- 1/2" gypsum or fiberboard
- 15/32" or 1/2" APA RATED SHEATHING nailed to studs with 6d common nails spaced 6" o.c. along panel edges and 12" o.c. at intermediate supports*
- *When corner panels are plywood, 1-1/2" roofing nails spaced 4" o.c. along panel edges and 8" o.c. at intermediate supports may be used.

suggestions in this section for installing and fastening such products.

Fiberboard, gypsum board, and insulating sheathing are not adequate nailing bases. When these sheathing materials are used, siding should be applied by nailing through the sheathing and into framing members or furring strips. These materials also do not provide adequate frame bracing. When they are used, another form of bracing must be provided. One way to brace such a frame is to use $^{15}/_{32}$ in. (11.91 mm) or $^{1}/_{2}$ in. (12.7 mm) APA Rated

Sheathing plywood at the corners (see Fig. 6.10-11).

***Fiberboard Sheathing*** Fiberboard sheathing is a homogeneous cellulosic material with square edges. It is available in $^{1}/_{2}$ in. (12.7 mm) and $^{25}/_{32}$ in. (19.84 mm) thicknesses and in 4 ft (1220 mm) widths and 8 ft (2440 mm) and 9 ft (2700 mm) lengths. It should comply with the requirements of American National Standards Institute (ANSI) A 194.1, Type IV. Two densities are available: Class 1, Regular Density and Class 2, Intermediate Density. Class 3, High Density or nail-base fiberboard is no longer made. Class 1 requires the addition of frame corner bracing. Class 2 may or may not require corner bracing, depending on the building's structural requirements.

Studs for fiberboard sheathing should be placed 16 in. (406 mm) on centers. Nailers must be provided, and panels must be nailed along all edges.

Unless otherwise recommended by its manufacturer, fiberboard sheathing should be applied with its long dimension parallel to the framing members (vertically), with sufficient bearing for satisfactory nailing along all edges. About $^{1}/_{8}$ in. (3.2 mm) should be left open between the edges and ends of adjoining boards.

Fiberboard sheathing should be installed according to its manufacturer's instructions. Usually, these boards are nailed or stapled at 3 in. (75 mm) on centers along the edges and at 6 in. (150 mm) on centers at intermediate supports. Fasteners for $^{1}/_{2}$ in. (12.7 mm) sheathing should be 11 gauge (3.9 mm) galvanized roofing nails, 6d common galvanized nails, or 16 gauge (1.6 mm) galvanized staples. Fasteners for $^{25}/_{32}$ in. (19.84 mm) material should be correspondingly larger.

***Gypsum Board Sheathing*** Gypsum sheathing may be used beneath wood, vinyl, composition, metal, and plywood siding and beneath masonry veneer, stucco, and exterior insulation and finish systems. It is available in $^{1}/_{2}$ in. (12.7 mm) regular and $^{5}/_{8}$ in. (15.9 mm) Type X (fire-resistant) boards. The $^{1}/_{2}$ in. (12.7 mm) boards come in 2 ft (610 mm) by 8 ft (2440 mm) and 4 ft (1220 mm) by 8 ft (2440 mm) sizes. The $^{5}/_{8}$-in. (15.9-mm)-thick boards are 4 ft (1220 mm) by 8 ft (2440 mm). The 4 ft (1220 mm) by 8 ft (2440 mm) panels have square edges; the 2-ft (610-mm)-wide ones have tongue-and-groove edges. All sizes may have either a plain or water-resistant core. Their faces, backs, and edges are covered with either paper or a glass fiber mat.

The maximum framing spacing for gypsum sheathing is 24 in. (610 mm) on centers. Sheathing of the 2 ft (610 mm) by 4 ft (1220 mm) size should be installed with the long edge across the framing (horizontal) and with the tongue edge up. The ends should occur over framing members. The 4 ft (1220 mm) by 8 ft (2440 mm) sheathing boards should be installed with the long edges vertical, and with each edge occurring over a framing member. A continuous nailer should be installed when the ends of vertically applied boards do not fall over a sill or plate.

Gypsum sheathing may be fastened in place using nails for the $1/2$ in. (12.7 mm) and $5/8$ in. (15.9 mm) thicknesses. Staples may also be used for the $1/2$ in. (12.7 mm) thickness. Nails should be 11 gauge (3.9 mm) with a $7/16$ in. (11.11 mm) head, and $1^1/2$ in. (38.1 mm) long for $1/2$-in. (12.7-mm)-thick boards and $1^3/4$ in. (44.45 mm) long for $5/8$ in. (15.9-mm)-thick boards. Staples should be 15 gauge (1.75 mm) or 16 gauge (1.6 mm), $7/16$ in. (11.11 mm) or $1/2$ in. (12.7 mm) wide, and $1^1/2$ in. (63.5 mm) long. Fasteners should be placed not closer than $3/8$ in. (9.5 mm) from the edge or end of a board. Boards may also be applied using screws and often are, especially when the wall framing consists of metal studs. The size and spacing of drywall screws should be in accordance with the board manufacturer's recommendations.

Fasteners should be located in 4 ft (1220 mm) by 8 ft (2440 mm) panels at not more than 4 in. (100 mm) on centers along the edges and ends and not less than 8 in. (200 mm) on centers along each intermediate support. Fasteners in 2 ft (610 mm) by 4 ft (1220 mm) tongue-and-groove boards should be located where recommended by the manufacturer of the board, but no fewer than four fasteners shall be used in each 2 ft (610 mm) width at each framing member when diagonal bracing is used and no fewer than seven fasteners in each 2 ft (610 mm) width when bracing is not used.

Codes vary on the use of diagonal bracing with gypsum board sheathing. Some do not require bracing under some circumstances. Some permit the gypsum sheathing to be counted as bracing under some conditions. Others require bracing but do not permit gypsum sheathing to serve in that capacity. In these cases, ply-

wood panels or let-in wood or metal corner bracing may be used.

***Insulated Sheathing*** Several types of insulated sheathing are available. Common insulating materials in them include polystyrene and isocyanurate. Some are plain boards without coverings. Others are covered with reflective foil or other materials that increase their thermal effect and, in some cases, themselves provide air and moisture barriers, or can be made to so function by taping the joints, thus eliminating the need for a separate layer of material to provide these functions. It is necessary, however, to take special precautions, such as installing vapor relief strips, to ensure that moisture does not become trapped in the stud cavities behind insulated sheathing.

Most insulated sheathing products are combustible and cannot be left exposed. Therefore, their use is limited by codes to wall assemblies that protect them. They are often used, for example, in exterior insulation and finish systems (see Chapter 7). Another limitation of these products is that they are not usable or do not perform well under some kinds of finishing materials. The manufacturer of the siding or other finishing material to be used should be consulted and its advice followed regarding the specific use of insulated sheathing products.

The installation and fastening of insulated sheathing boards vary from product to product, and they should be installed according to their manufacturers' instructions.

### 6.10.2.2 Roof Sheathing

Suitable performance of roofing is dependent on the entire roof structure. Roof joists, rafters, and other supports should be adequately sized and spaced to carry the necessary superimposed loads over the required spans.

Many types of roofing require a bituminous sheet underlayment for proper performance. These underlayments are discussed in the section where each roofing type is covered.

Roof sheathing should be smooth, securely attached to its supports, and provide a base to receive roofing nails and fasteners. Some roofing types are applied over solid sheathing composed of wood boards or plywood. Other types are applied over spaced sheathing. Refer to Chapter 7 for a discussion of the sheathing type required for each roofing type.

## LUMBER SHEATHING

Solid sheathing boards should be seasoned to a maximum moisture content of 15%. Boards should be tongue-and-groove edged, in nominal 1 in. (25 mm) thickness, and not more than 6 in. (150 mm) in nominal width to minimize shrinkage. Boards should be tightly matched and securely face- and edge-nailed with two 7d annularly threaded or 8d common wire nails at each framing member. Maximum spacing of supports for nominal 1 in. (25 mm) lumber boards is 24 in. (610 mm) on centers.

Spaced sheathing of $1 \times 3$ ($25 \times 75$), $1 \times 4$ ($25 \times 100$), or $1 \times 6$ ($25 \times 150$) boards may be used in blizzard-free areas (generally where the outside design temperature is above 0°F [−17.8°C]) to receive the application of wood shingles or wood shakes (see Chapter 7). The center-to-center spacing of these boards should be equal to the exposure of the shingle or shake, but not more than 10 in. (250 mm), and the openings between boards should not be wider than the individual boards. Seasoning and face nailing of spaced boards should be the same as described for solid wood board sheathing.

## PLYWOOD SHEATHING

Plywood sheathing panels should be APA Rated Sheathing of one of the following grades: Exposure 1 or 2; Exterior; Structural I, Exposure 1; or Structural I, Exterior. Panels exposed to the exterior should be Exterior grade. Plywood sheathing should have a span rating, panel thickness, support spacing, and edge support appropriate to the load supported (Fig. 6.10-12). Panel edges may be square or tongue-and-groove.

Plywood sheathing should be applied with its long edges perpendicular to the framing and with a $1/8$ in. (3.2 mm) gap between panels to permit thermal movement. Square-edged boards should be held level and supported with back blocking or panel clips. The number and spacing of panel clips depend on the support spacing; one panel clip between each pair of supports that are 24 in. (610 mm) apart and two clips in 48-in. (1220-mm)-wide spaces is normal.

Plywood sheathing up to $1/2$ in. (12.7 mm) thick should be fastened in place with 6d nails; plywood sheathing from $1/2$ in. (12.7 mm) to 1 in. (25.4 mm) thick should be fastened with 8d nails. These nails may be either common smooth

| Sturd-I-Floor Span Rating | Sheathing Span Rating | Maximum Span, in. | Allowable Live Loads, psf* | | | | | | |
|---|---|---|---|---|---|---|---|---|---|
| | | | Joist Spacing, in. | | | | | | |
| | | | 12 | 16 | 20 | 24 | 32 | 40 | 48 |
| 16 oc | 24/16, 32/16 | 16 | | 100 | | | | | |
| 20 oc | 40/20 | 20 | | 150 | 100 | | | | |
| 24 oc | 48/24 | 24 | | | | 100 | | | |
| 32 oc | 60/32 | 32 | | | | | 100 | | |
| 48 oc | | 48 | | | | | | 100 | 55 |

*10 psf dead load assumed. Live load deflection limit is f/360.

(Courtesy APA—The Engineered Wood Association, formerly American Plywood Association.) (Refer to Figure 25.7-1 for metric conversions.)

shank or deformed shank nails. Panels $1^1/_8$ in. (28.58 mm) thick should be nailed with 8d ring- or screw-shank nails or 10d common smooth-shank nails. Nails should be spaced at 6 in. (150 mm) on centers around panel edges and 12 in. (305 mm) on centers at intermediate supports, except that in panels over support spacings of 48 in. (1220 mm) or more, all nails should be spaced at 6 in. (150 mm) on centers.

### NONWOOD SHEATHING

When nonwood sheathing materials are recommended by their manufacturers for use as roof sheathing, the sheathing manufacturer should specify the type of fastening required to attach sheathing to supports. Such materials should not be used without the concurrence of the roofing manufacturer.

### 6.10.3 PREASSEMBLED COMPONENTS

Preassembled components are parts of a building's framing, floor and ceiling assemblies, wall assemblies, or roof and ceiling assemblies that are manufactured in a location other than that in which they will be finally used. Preassembled components are discussed in Section 6.9.

## 6.11 Other Framing Systems

Other wood framing systems include post-and-beam, plank-and-beam, heavy timber, laminated timber, and wood pole construction.

### 6.11.1 POST-AND-BEAM

Post-and-beam systems (Fig 6.11-1) consist of:

- Posts that support a series of beams
- The supported beams, which carry the loads of floors, roofs, and ceilings
- Non-load-bearing walls located between the posts (Fig. 6.11-2)

In a post-and-beam system the posts typically are 4 × 4s (100 × 100s) or are built up of several 2 by 4s (50 × 100s) and are spaced 4 ft (1220 mm) to 8 ft (2400 mm) apart. Wall areas between posts are framed, sheathed, and finished similar to other wood wall systems to serve as weather barriers, provide re-sistance to racking, and give lateral support to the posts. Resistance to lateral forces is achieved with sheathing materials or diagonal bracing. Since these wall areas do not carry vertical loads, framing members may be placed horizontally to receive vertical exterior siding, and areas between posts can easily accommodate large windows or doors. Glass panels often are set (stopped) into the posts. No headers are required, but the placement of windows and doors is determined by the spacing and location of the posts.

Post-and-beam systems are often integrated with plank-and-beam roof and ceiling assemblies. Planks and beams are often left exposed to form a finished ceil-

**FIGURE 6.11-1** Post-and-beam framing.

---

**WOOD, PLASTICS, AND COMPOSITES** ■ Other Framing Systems  **399**

**FIGURE 6.11-2** Wall construction for use with post-and-beam framing.

ing, and posts may be partially exposed on the interior and exterior surfaces as part of the finished wall. In such designs, the expression of the basic structural system largely determines the interior and exterior appearance and character.

### 6.11.2 PLANK-AND-BEAM

Plank-and-beam construction is similar to wood joist construction except that the subfloor is 2 in. (50 mm) nominal or thicker and structural members are larger and spaced farther apart. Underlayment is recommended over plank subfloors.

When heavier structural members are used for framing, support spacing can be increased, and heavier 2-in. (50-mm)-thick boards called *planks* generally are used. A 2-in. (50-mm)-thick member laid flat and stressed in bending across its 2 in. (50 mm) dimension is referred to as a *plank*, and hence this type of subfloor also is called *planking*. Planking is usually tongue-and-grooved to facilitate the transfer of concentrated loads from one plank to another. In lumber grading, such tongue-and-groove material is called *decking*.

The meanings of the terms *beam*, *plank*, and *decking* used in lumber grading do not completely agree with those in general construction terminology. In light wood-framed construction, 4-in. (100-mm)-thick framing members are called *beams*; yet, in lumber grading, they are classed as *joists* and *planks*, and the term *beams and stringers* is restricted to rectangular members 5 in. (127 mm) thick and thicker. In lumber grading, *decking* is lumber 2 in. (50 mm) to 4 in. (100 mm) thick, 4 in. (100 mm) wide and wider, with a maximum moisture content of 15% 19% and with tongue-and-groove edges. However, in common usage, both installations with this special product and ordinary surfaced four-sided lumber

joists and planks are referred to as *planking*, or as *decking* when used in a roof assembly, which is called a *roof deck*. For instance, a traditional plank-and-beam framing system uses planking which is really decking and beams which are really joists, according to lumber grading terminology.

#### 6.11.2.1 Plank-and-Beam Floors

Plank-and-beam floor assemblies are characterized by the use of nominal 2 × 6 (50 × 150) or 2 × 8 (50 × 200) tongue-and-groove or splined *planks* spanning between beams generally spaced 4 ft (1220 mm) to 8 ft (2440 mm) apart (Fig. 6.11-3). The planks serve as subflooring, as a working platform, and to transmit floor loads to fewer but larger members than occur in conventional wood joist floor systems.

Plank-and-beam floor and ceiling assemblies often are integrated with plank-and-beam roof and ceiling assemblies and post-and-beam wall systems. In such designs, the basic structural system largely determines the interior appearance and character.

Conventional joist framing requires a maximum support spacing of 24 in. (610 mm) on centers. Usin APA Rated Sturd-I-Floor combination subfloor and underlayment can push this spacing up to 48 in. (1220 mm) on centers. But plank-and-beam framing permits support spacings of 6 ft (1800 mm) or more.

Tongue-and-groove or splined plank subfloors have a finished surface on the underside, which can be left exposed to provide a finished ceiling. Using such a system, however, does not provide the usual cavities for mechanical and electrical distribution equipment, and in the absence of a separate ceiling finish there is relatively poor sound isolation, particularly from impact noise. Therefore, plank

floors are most suitable over unfinished spaces or where impact sound isolation from above is not an important consideration. In light wood-framed construction, plank-and-beam construction is generally more adapted to roof assemblies or to floor systems in one-story buildings. When deep girders are not needed beneath them, plank floors permit a lower floor-to-grade elevation, resulting in a lower building height.

Floor finishes such as strip wood flooring may be applied directly over such planks, but resilient flooring and adhesively applied ceramic tile or thin-set terrazzo require an underlayment. Planks can be used as a base for carpeting with suitable cushioning (padding) if the subfloor surface is reasonably level and smooth. Heavy cushioning and deep-pile carpeting, with or without underlayment, also considerably improves impact sound isolation and is recommended where sound transmission is a factor.

Plank subfloors can be built with tongue-and-groove or surfaced-four-sides (S4S) lumber. Tongue-and-groove lumber is graded in most species as Select Decking and Commercial Decking, both of which are suitable for subfloors.

**FIGURE 6.11-3** Plank-and-beam construction.

Decking is manufactured 2 in. (50 mm) to 4 in. (100 mm) nominal thick and in widths ranging from 4 in. (100 mm) to 12 in. (305 mm). The maximum width for subfloors should not exceed 8 in. (200 mm) at a 19% moisture content, but wider boards can be used when their moisture content is limited to 15%. Thinner (2 in. [50 mm]) planks generally are made with a single tongue and groove, while 3-in. (75-mm)- and 4-in. (100-mm)-thick planks have two. Planks with grooves on both edges, for use with splines, also are available. One face usually is of better quality and is intended for an exposed view, such as in a paneled ceiling. This face also may be worked to a special pattern with V-joints, rabbets, or rounded edges.

S4S lumber suitable for planking is the same as for joists and is graded as Select Structural No. 1, No. 2, and No. 3.

A plank-and-beam system is essentially a skeleton framework. Planks are intended to support moderate, uniformly distributed loads. These loads are carried to the beams, which in turn transmit the loads to posts or walls supported on foundations. Foundations for plank-and-beam framing are similar to those used for other wood framing systems (Fig. 6.11-4). With beams (and sometimes posts) spaced 6 ft (1800 mm) or more on centers, this system is well adapted to pier and crawl space foundations.

Posts should be of adequate size to carry the load and large enough to provide full bearing for the ends of beams. In general, posts should be at least 4 × 4 (100 × 100), but 4 × 6 (100 × 150) posts are preferred, particularly where beams join or where spaced beams are used. Where the ends of beams abut over a post, the 6 in. (150 mm) dimension should be parallel to the beams (Fig. 6.11-5).

Suitable beam sizes vary with grades, species, span, loading, and spacing. Beams may be solid lumber, glue-laminated members, or built up of several joists securely nailed to each other or to spacer blocks between them. Wood beams supporting plank floors or roofs should not exceed the maximum allowable spans tabulated in the AF&PA publication *Plank and Beam Framing for Residential Buildings*. Figure 6.11-6 is an example of one of these tables for beams supporting a 20 psf (97.65 kg/m²) live load with an allowable deflection of L/240. *Plank and Beam Framing for Residential Buildings* also includes tables for

**FIGURE 6.11-4** Spaced beam (or girder) for plank floors can be (a) supported on a foundation sill or (b) framed into a wall pocket. (Courtesy American Forest & Paper Association.)

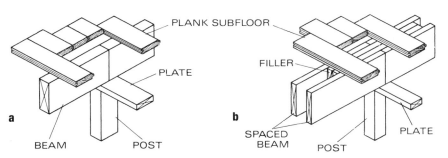

**FIGURE 6.11-5** Intermediate support for plank floors can be provided by a wood post under (a) a solid beam or (b) a spaced beam. (Courtesy American Forest & Paper Association.)

live loads of 30 psf (146.47 kg/m²) and 40 psf (195.3 kg/m²) and allowable deflections of L/300 and L/360. To use these tables, first determine the span, the live loads to be supported, and the deflection limitation. Then select from the tables the proper size of beam with the corresponding required values for *f* and *E*. The beam used should be of a grade and species that meet or exceed the selected values.

Beams should be securely fastened to posts with framing anchors or clip angles. Details of upper-floor beams framing into an exterior frame wall are illustrated in Figure 6.11-7. Greater advantage can be taken of the strength and stiffness of plank subflooring material by making the planks continuous over more than one span. For example, using the same span and uniform load in each case, a plank which is continuous over two spans is nearly two and one-half times as stiff as a plank which extends over a single span (Fig. 6.11-8).

A nominal 2-in. (50-mm)-thick plank should not exceed the maximum allowable spans shown in a table in *Plank and Beam Framing for Residential Buildings* (Fig. 6.11-9). The types

of plank arrangements shown in the table are defined as:

Type A—extending over a single span
Type B—continuous over two equal spans
Type C—continuous over three equal spans
Type D—a combination of types A and B

To use this table, determine the plank arrangement type, the span, the supported live load, and the deflection limit, and select from the table the corresponding values for *f* and *E*. The plank used should be of a grade and species that meets or exceeds the minimum values listed. The maximum span for a given thickness, grade, and species of plank can be obtained by reversing this procedure. The maximum allowable *f* and *E* for most commercial grades and species are tabulated in the *NDS* (see Section 6.1.3.4).

Regardless of their width, planks should be blind nailed to each support with one 10d nail; in addition, planks 4 in. (100 mm) or 6 in. (150 mm) wide should be face-nailed with one 16d nail at each support; 8 in. (200 mm) planks should be face-nailed with two 16d nails; and 10 in. (250 mm) and 12 in. (305 mm)

**FIGURE 6.11-6** Required *f* and *E* Values for Floor and Roof Beams

| Span of Beam, ft. | Nominal Size of Beam | 6 ft.-0 in. *f* | 6 ft.-0 in. *E* | 7 ft.-0 in. *f* | 7 ft.-0 in. *E* | 8 ft.-0 in. *f* | 8 ft.-0 in. *E* |
|---|---|---|---|---|---|---|---|
| 10 | 2-3 × 6 | 1070 | 780000 | 1250 | 910000 | 1430 | 1040000 |
|  | 1-3 × 8 | 1235 | 680000 | 1440 | 794000 | 1645 | 906000 |
|  | 2-2 × 8 | 1030 | 570000 | 1200 | 665000 | 1370 | 760000 |
|  | 1-4 × 8 | 880 | 485000 | 1030 | 566000 | 1175 | 646000 |
|  | 3-2 × 8 | 685 | 380000 | 800 | 443000 | 915 | 506000 |
|  | 2-3 × 8 | 615 | 340000 | 720 | 397000 | 820 | 453000 |
|  | 2-2 × 10 | 630 | 273000 | 735 | 219000 | 840 | 364000 |
| 11 | 2-3 × 6 | 1295 | 1037000 | 1510 | 1210000 | 1730 | 1382000 |
|  | 1-3 × 8 | 1490 | 905000 | 1740 | 1056000 | 1990 | 1206000 |
|  | 2-2 × 8 | 1245 | 754000 | 1450 | 880000 | 1660 | 1005000 |
|  | 1-4 × 8 | 1065 | 647000 | 1245 | 755000 | 1420 | 862000 |
|  | 3-2 × 8 | 830 | 503000 | 970 | 587000 | 1105 | 670000 |
|  | 2-3 × 8 | 745 | 453000 | 870 | 529000 | 995 | 604000 |
|  | 2-2 × 10 | 765 | 363000 | 890 | 424000 | 1020 | 484000 |
| 12 | 2-3 × 6 | 1545 | 1346000 | 1800 | 1571000 | 2060 | 1794000 |
|  | 1-3 × 8 | 1775 | 1175000 | 2070 | 1371000 | 2370 | 1566000 |
|  | 2-2 × 8 | 1480 | 980000 | 1725 | 1144000 | 1970 | 1306000 |
|  | 1-4 × 8 | 1270 | 840000 | 1480 | 980000 | 1690 | 1120000 |
|  | 3-2 × 8 | 985 | 653000 | 1150 | 762000 | 1315 | 870000 |
|  | 2-3 × 8 | 890 | 588000 | 1035 | 686000 | 1185 | 784000 |
|  | 1-6 × 8 | 755 | 483000 | 880 | 564000 | 1005 | 644000 |
|  | 2-2 × 10 | 910 | 472000 | 1060 | 551000 | 1210 | 629000 |
|  | 1-3 × 10 | 1090 | 566000 | 1275 | 660000 | 1455 | 754000 |
| 13 | 2-3 × 6 | 1815 | 1711000 | 2110 | 1997000 | 2415 | 2281000 |
|  | 1-3 × 8 | 2085 | 1494000 | 2430 | 1743000 | 2780 | 1991000 |
|  | 2-2 × 8 | 1740 | 1245000 | 2025 | 1453000 | 2315 | 1660000 |
|  | 1-4 × 8 | 1490 | 1067000 | 1735 | 1245000 | 1985 | 1422000 |
|  | 3-2 × 8 | 1160 | 830000 | 1350 | 969000 | 1545 | 1106000 |
|  | 2-3 × 8 | 1045 | 747000 | 1215 | 872000 | 1390 | 996000 |
|  | 1-6 × 8 | 885 | 614000 | 1040 | 716000 | 1185 | 818000 |
|  | 2-2 × 10 | 1070 | 600000 | 1245 | 700000 | 1420 | 800000 |
|  | 1-3 × 10 | 1280 | 719000 | 1495 | 839000 | 1710 | 958000 |
| 14 | 2-2 × 8 | 2015 | 1555000 | 2350 | 1815000 | 2685 | 2073000 |
|  | 3-2 × 8 | 1340 | 1037000 | 1570 | 1210000 | 1790 | 1382000 |
|  | 2-3 × 8 | 1210 | 933000 | 1410 | 1089000 | 1610 | 1244000 |
|  | 1-6 × 8 | 1025 | 766000 | 1200 | 894000 | 1370 | 1021000 |
|  | 1-3 × 10 | 1485 | 899000 | 1730 | 1049000 | 1980 | 1198000 |
|  | 2-2 × 10 | 1235 | 749000 | 1445 | 874000 | 1650 | 998000 |
|  | 1-4 × 10 | 1060 | 642000 | 1240 | 749000 | 1415 | 856000 |
|  | 3-2 × 10 | 825 | 499000 | 965 | 582000 | 1100 | 665000 |
|  | 2-3 × 10 | 740 | 449000 | 865 | 524000 | 990 | 598000 |
| 15 | 3-2 × 8 | 1540 | 1275000 | 1800 | 1488000 | 2055 | 1699000 |
|  | 2-3 × 8 | 1390 | 1148000 | 1620 | 1340000 | 1850 | 1530000 |
|  | 1-6 × 8 | 1180 | 943000 | 1375 | 1100000 | 1570 | 1257000 |
|  | 1-3 × 10 | 1705 | 1105000 | 1990 | 1289000 | 2270 | 1473000 |
|  | 2-2 × 10 | 1420 | 921000 | 1660 | 1075000 | 1895 | 1228000 |
|  | 1-4 × 10 | 1220 | 789000 | 1420 | 921000 | 1625 | 1052000 |
|  | 3-2 × 10 | 950 | 614000 | 1105 | 717000 | 1265 | 818000 |
|  | 2-3 × 10 | 850 | 553000 | 995 | 645000 | 1135 | 737000 |
|  | 1-6 × 10 | 735 | 464000 | 855 | 541000 | 980 | 618000 |
|  | 4-2 × 10 | 710 | 461000 | 830 | 538000 | 945 | 614000 |
|  | 2-2 × 10 | 960 | 512000 | 1120 | 597000 | 1280 | 682000 |
| 16 | 3-2 × 8 | 1755 | 1548000 | 2045 | 1806000 | 2340 | 2063000 |
|  | 2-3 × 8 | 1580 | 1393000 | 1840 | 1626000 | 2105 | 1857000 |
|  | 2-2 × 10 | 1615 | 1118000 | 1890 | 1305000 | 2155 | 1490000 |
|  | 1-4 × 10 | 1385 | 958000 | 1615 | 1118000 | 1845 | 1277000 |
|  | 3-2 × 10 | 1075 | 745000 | 1260 | 869000 | 1435 | 993000 |
|  | 2-3 × 10 | 970 | 671000 | 1130 | 783000 | 1290 | 894000 |
|  | 1-6 × 10 | 835 | 563000 | 975 | 657000 | 1130 | 750000 |
|  | 4-2 × 10 | 810 | 559000 | 945 | 652000 | 1080 | 745000 |
|  | 1-8 × 10 | 615 | 413000 | 715 | 482000 | 815 | 550000 |
|  | 1-3 × 12 | 1310 | 746000 | 1530 | 871000 | 1750 | 994000 |
|  | 2-2 × 12 | 1090 | 621000 | 1275 | 725000 | 1455 | 828000 |

*(Continues)*

planks should be face-nailed with three 16d nails.

A wood strip floor should be laid at right angles to a plank subfloor using the same procedure followed in conventional construction. Where the underside of the plank is to serve as a ceiling, flooring nails should not penetrate through the plank.

In a plank-and-beam system, interior partitions are usually nonbearing. Where bearing partitions occur, they should be placed over beams and the beams enlarged to carry the added load. If this is not possible, supplementary beams must be placed in the floor framing. Nonbearing partitions, which are parallel to the planks, should be supported on a sole plate consisting of double 2 × 4s (50 × 100s) set on edge and spiked together (Fig. 6.11-10b). Where door openings oc-

cur, double 2 × 4s (50 × 100s) may be placed under the plank floor and framed into the beams with framing anchors (Fig. 6.11-10a). Nonbearing partitions located at right angles to the planks do not require supplementary framing, since the partition load is distributed across a number of planks.

### 6.11.2.2 Plank-and-Beam Roofs

Plank-and-beam roofs are typically used on one-story buildings and for low-slope and flat roofs. These systems are characterized by the use of 2 × 6 (50 × 150) and 2 × 8 (50 × 200) tongue-and-groove or splined planks spanning between longitudinal or transverse beams (Fig. 6.11-11). Beam spacing and planking thickness are determined by imposed loads, spans, and material strength. Nom-

inal 2 in. (50 mm) plank may be used on spans of up to about 8 ft (2440 mm). The AF&PA publication *Plank and Beam Framing for Residential Buildings* includes tables for the minimum required stress in bending and modulus of elasticity for beams and planks used in plank-and-beam roofs. Figures 6.11-6 and 6.11-7 reproduce some of these tables. The use of these tables is explained in Section 6.11.2.1.

Longitudinal beams span parallel to the long dimension of a building and are supported on end walls or columns. In buildings up to about 36 ft (10.97 m) wide, beams are typically placed at quarter points. This divides the width of the building into two equal spans for planking on each side of the center ridge beam. Planks may be supported along the exterior sides by walls or beams. Interior

**FIGURE 6.11-6** *(Continued)*

| Span of Beam, ft. | Nominal Size of Beam | 6 ft.-0 in. f | 6 ft.-0 in. E | 7 ft.-0 in. f | 7 ft.-0 in. E | 8 ft.-0 in. f | 8 ft.-0 in. E |
|---|---|---|---|---|---|---|---|
| 17 | 2-2 × 10 | 1825 | 1341000 | 2125 | 1565000 | 2430 | 1781000 |
| | 1-4 × 10 | 1565 | 1149000 | 1825 | 1341000 | 2085 | 1532000 |
| | 3-2 × 10 | 1215 | 894000 | 1420 | 1043000 | 1620 | 1192000 |
| | 2-3 × 10 | 1095 | 804000 | 1280 | 938000 | 1460 | 1072000 |
| | 1-6 × 10 | 945 | 675000 | 1100 | 788000 | 1260 | 900000 |
| | 4-2 × 10 | 910 | 670000 | 1065 | 782000 | 1215 | 894000 |
| | 1-8 × 10 | 690 | 495000 | 805 | 578000 | 910 | 660000 |
| | 1-3 × 12 | 1480 | 894000 | 1725 | 1043000 | 1975 | 1192000 |
| | 2-2 × 12 | 1235 | 745000 | 1440 | 869000 | 1645 | 993000 |
| | 1-4 × 12 | 1060 | 639000 | 1230 | 746000 | 1410 | 852000 |
| | 3-2 × 12 | 820 | 497000 | 960 | 580000 | 1095 | 663000 |
| 18 | 2-2 × 10 | 2045 | 1592000 | 2385 | 1858000 | 2725 | 2123000 |
| | 1-4 × 10 | 1755 | 1364000 | 2045 | 1592000 | 2340 | 1819000 |
| | 3-2 × 10 | 1365 | 1061000 | 1590 | 1238000 | 1815 | 1415000 |
| | 2-3 × 10 | 1270 | 955000 | 1480 | 1114000 | 1695 | 1273000 |
| | 1-6 × 10 | 1060 | 801000 | 1235 | 935000 | 1415 | 1068000 |
| | 4-2 × 10 | 1020 | 796000 | 1195 | 929000 | 1365 | 1062000 |
| | 1-8 × 10 | 780 | 588000 | 910 | 686000 | 1040 | 784000 |
| | 1-3 × 12 | 1660 | 1062000 | 1935 | 1239000 | 2210 | 1416000 |
| | 2-2 × 12 | 1380 | 885000 | 1615 | 1033000 | 1845 | 1180000 |
| | 1-4 × 12 | 1185 | 758000 | 1385 | 885000 | 1580 | 1011000 |
| | 3-2 × 12 | 920 | 590000 | 1075 | 688000 | 1230 | 786000 |
| | 2-3 × 12 | 830 | 531000 | 970 | 620000 | 1105 | 708000 |
| 19 | 3-2 × 10 | 1520 | 1248000 | 1775 | 1456000 | 2025 | 1664000 |
| | 2-3 × 10 | 1365 | 1123000 | 1595 | 1310000 | 1825 | 1497000 |
| | 1-6 × 10 | 1170 | 943000 | 1365 | 1100000 | 1560 | 1257000 |
| | 4-2 × 10 | 1140 | 936000 | 1330 | 1092000 | 1520 | 1248000 |
| | 2-4 × 10 | 975 | 802000 | 1140 | 936000 | 1300 | 1070000 |
| | 1-8 × 10 | 860 | 691000 | 1005 | 806000 | 1145 | 921000 |
| | 1-3 × 12 | 1850 | 1249000 | 2155 | 1457000 | 2465 | 1665000 |
| | 2-2 × 12 | 1540 | 1041000 | 1800 | 1215000 | 2055 | 1388000 |
| | 1-4 × 12 | 1320 | 892000 | 1540 | 1041000 | 1760 | 1190000 |
| | 3-2 × 12 | 1025 | 694000 | 1200 | 810000 | 1370 | 926000 |
| | 2-3 × 12 | 925 | 624000 | 1080 | 728000 | 1230 | 832000 |
| | 1-6 × 12 | 805 | 531000 | 940 | 620000 | 1070 | 708000 |

| Span of Beam, ft. | Nominal Size of Beam | 6 ft.-0 in. f | 6 ft.-0 in. E | 7 ft.-0 in. f | 7 ft.-0 in. E | 8 ft.-0 in. f | 8 ft.-0 in. E |
|---|---|---|---|---|---|---|---|
| 20 | 3-2 × 10 | 1685 | 1456000 | 1965 | 1699000 | 2245 | 1942000 |
| | 2-3 × 10 | 1515 | 1310000 | 1770 | 1529000 | 2020 | 1747000 |
| | 1-6 × 10 | 1300 | 1099000 | 1515 | 1282000 | 1735 | 1465000 |
| | 4-2 × 10 | 1260 | 1092000 | 1475 | 1274000 | 1685 | 1456000 |
| | 2-4 × 10 | 1080 | 936000 | 1265 | 1092000 | 1445 | 1248000 |
| | 1-8 × 10 | 960 | 806000 | 1120 | 941000 | 1280 | 1075000 |
| | 2-2 × 12 | 1705 | 1214000 | 1990 | 1417000 | 2275 | 1619000 |
| | 1-4 × 12 | 1465 | 1040000 | 1710 | 1214000 | 1950 | 1387000 |
| | 3-2 × 12 | 1140 | 809000 | 1330 | 944000 | 1520 | 1079000 |
| | 2-3 × 12 | 1025 | 728000 | 1195 | 850000 | 1365 | 971000 |
| | 1-6 × 12 | 970 | 620000 | 1130 | 723000 | 1295 | 826000 |
| | 2-4 × 12 | 730 | 520000 | 855 | 607000 | 975 | 694000 |
| 21 | 3-2 × 10 | 1855 | 1685000 | 2165 | 1966000 | 2475 | 2247000 |
| | 2-3 × 10 | 1670 | 1516000 | 1950 | 1827000 | 2225 | 2088000 |
| | 1-6 × 10 | 1430 | 1273000 | 1670 | 1485000 | 1905 | 1697000 |
| | 4-2 × 10 | 1390 | 1264000 | 1625 | 1475000 | 1855 | 1686000 |
| | 2-4 × 10 | 1195 | 1083000 | 1390 | 1264000 | 1590 | 1444000 |
| | 1-8 × 10 | 1050 | 933000 | 1225 | 1089000 | 1400 | 1244000 |
| | 2-2 × 12 | 1880 | 1405000 | 2195 | 1640000 | 2510 | 1874000 |
| | 1-4 × 12 | 1615 | 1204000 | 1880 | 1405000 | 2150 | 1606000 |
| | 3-2 × 12 | 1255 | 937000 | 1465 | 1093000 | 1670 | 1249000 |
| | 2-3 × 12 | 1130 | 843000 | 1320 | 984000 | 1505 | 1124000 |
| | 1-6 × 12 | 970 | 717000 | 1130 | 837000 | 1295 | 956000 |
| | 2-4 × 12 | 805 | 602000 | 940 | 702000 | 1075 | 802000 |
| 22 | 1-6 × 10 | 1580 | 1463000 | 1845 | 1707000 | 2105 | 1951000 |
| | 4-2 × 10 | 1525 | 1453000 | 1780 | 1696000 | 2035 | 1938000 |
| | 2-4 × 10 | 1310 | 1245000 | 1530 | 1455000 | 1745 | 1660000 |
| | 1-8 × 10 | 1160 | 1073000 | 1355 | 1252000 | 1545 | 1431000 |
| | 1-4 × 12 | 1770 | 1384000 | 2065 | 1615000 | 2360 | 1846000 |
| | 3-2 × 12 | 1375 | 1077000 | 1605 | 1257000 | 1835 | 1436000 |
| | 2-3 × 12 | 1240 | 969000 | 1445 | 1130000 | 1655 | 1291000 |
| | 1-6 × 12 | 1080 | 825000 | 1260 | 963000 | 1440 | 1100000 |
| | 2-4 × 12 | 885 | 692000 | 1035 | 807000 | 1180 | 922000 |
| | 4-2 × 12 | 1035 | 808000 | 1205 | 943000 | 1375 | 1078000 |
| | 5-2 × 12 | 825 | 646000 | 965 | 754000 | 1105 | 862000 |
| | 3-3 × 12 | 825 | 639000 | 965 | 746000 | 1105 | 852000 |

(Courtesy American Forest & Paper Association.)

**FIGURE 6.11-7** A plank floor at an upper floor can be supported by (a) a built-up beam or (b) a spaced beam framed into an exterior wall over a post or top plate. (Courtesy American Forest & Paper Association.)

planning flexibility is slightly more restricted than when transverse beams are used because of the greater number of beam supports.

Transverse beams, generally spaced 6 ft (1800 mm) to 8 ft (2440 mm) apart, span perpendicular to the long dimension of a building. They may be supported by load-bearing walls or beams along the ridge or center of the building.

The system may be used with wood stud exterior walls and is often used with post-and-beam exterior walls.

A plank ceiling and roof deck does not provide space for concealment of ductwork and wiring. If the system requires insulation, it is necessary to use rigid insulation applied to the top of the deck,

## PLANK CONTINUOUS ACROSS ONE SPAN

## PLANK CONTINUOUS ACROSS TWO SPANS

**FIGURE 6.11-8** The resultant deflection under similar loads is less when individual planks are long enough to reach across two spans than it is in single-span floors. (Courtesy American Forest & Paper Association.)

which in turn may limit the selection of roofing finishes.

Plank-and-beam roof assemblies, in contrast to truss and joist-and-rafter roof and ceiling assemblies, require different design and construction methods and more precision in erection, since the finished appearance generally is an integral part of the system. Planks, beams and columns generally are left exposed as the finished interior surface.

### 6.11.3 HEAVY TIMBER

Before the development of balloon frame construction (see Section 6.7), heavy timber framing was the normal method used to build most wood-framed buildings. Timbers for early timber-framed buildings were hewed from trees by hand using axes and similar cutting tools and two-man pit saws. Development of saws powered by moving water made production of square and rectangular timbers easier, and led to the use of heavy timber framing in most of the industrial buildings of the nineteenth century and many of the homes and farm buildings of that time.

**FIGURE 6.11-9** Required *f* and *E* Values for 2-In.-Thick Plank

| Plank Span, ft. | Live Load, psf | Deflection Limitation | Type A f psi | Type A E psi | Type B f psi | Type B E psi | Type C f psi | Type C E psi | Type D f psi | Type D E psi |
|---|---|---|---|---|---|---|---|---|---|---|
| 6 | 20 | $\ell/240$ | 360 | 576000 | 360 | 239000 | 288 | 305000 | 360 | 408000 |
|   |    | $\ell/300$ | 360 | 720000 | 360 | 299000 | 288 | 318000 | 360 | 509000 |
|   |    | $\ell/360$ | 360 | 864000 | 360 | 359000 | 288 | 457000 | 360 | 611000 |
|   | 30 | $\ell/240$ | 480 | 864000 | 480 | 359000 | 384 | 457000 | 480 | 611000 |
|   |    | $\ell/300$ | 480 | 1080000 | 480 | 448000 | 384 | 571000 | 480 | 764000 |
|   |    | $\ell/360$ | 480 | 1296000 | 480 | 538000 | 384 | 685000 | 480 | 917000 |
|   | 40 | $\ell/240$ | 600 | 1152000 | 600 | 478000 | 480 | 609000 | 600 | 815000 |
|   |    | $\ell/300$ | 600 | 1440000 | 600 | 598000 | 480 | 762000 | 600 | 1019000 |
|   |    | $\ell/360$ | 600 | 1728000 | 600 | 717000 | 480 | 914000 | 600 | 1223000 |
| 7 | 20 | $\ell/240$ | 490 | 915000 | 490 | 380000 | 392 | 484000 | 490 | 647000 |
|   |    | $\ell/300$ | 490 | 1143000 | 490 | 475000 | 392 | 605000 | 490 | 809000 |
|   |    | $\ell/360$ | 490 | 1372000 | 490 | 570000 | 392 | 726000 | 490 | 971000 |
|   | 30 | $\ell/240$ | 653 | 1372000 | 653 | 570000 | 522 | 726000 | 653 | 971000 |
|   |    | $\ell/300$ | 653 | 1715000 | 653 | 712000 | 522 | 907000 | 653 | 1213000 |
|   |    | $\ell/360$ | 653 | 2058000 | 653 | 854000 | 522 | 1088000 | 653 | 1456000 |
|   | 40 | $\ell/240$ | 817 | 1829000 | 817 | 759000 | 653 | 968000 | 817 | 1294000 |
|   |    | $\ell/300$ | 817 | 2287000 | 817 | 949000 | 653 | 1209000 | 817 | 1618000 |
|   |    | $\ell/360$ | 817 | 2744000 | 817 | 1139000 | 653 | 1451000 | 817 | 1941000 |
| 8 | 20 | $\ell/240$ | 640 | 1365000 | 640 | 567000 | 512 | 722000 | 640 | 966000 |
|   |    | $\ell/300$ | 640 | 1707000 | 640 | 708000 | 512 | 903000 | 640 | 1208000 |
|   |    | $\ell/360$ | 640 | 2048000 | 640 | 850000 | 512 | 1083000 | 640 | 1449000 |
|   | 30 | $\ell/240$ | 853 | 2048000 | 853 | 850000 | 682 | 1083000 | 853 | 1449000 |
|   |    | $\ell/300$ | 853 | 2560000 | 853 | 1063000 | 682 | 1354000 | 853 | 1811000 |
|   |    | $\ell/360$ | 853 | 3072000 | 853 | 1275000 | 682 | 1625000 | 853 | 2174000 |
|   | 40 | $\ell/240$ | 1067 | 2731000 | 1067 | 1134000 | 853 | 1444000 | 1067 | 1932000 |
|   |    | $\ell/300$ | 1067 | 3410000 | 1067 | 1417000 | 853 | 1805000 | 1067 | 2415000 |
|   |    | $\ell/360$ | 1067 | 4096000 | 1067 | 1700000 | 853 | 2166000 | 1067 | 2898000 |

(Courtesy American Forest & Paper Association.)

FIGURE 6.11-10 Interior nonbearing partitions running parallel to a plank floor can be supported on (a) a suitable underfloor beam or (b) a double vertical sole plate. (Courtesy American Forest & Paper Association.)

FIGURE 6.11-11 Wood plank-and-beam roof and ceiling assemblies.

Heavy timber construction using large timbers is rarely used today but can be seen in many older buildings. One reason for the demise of this type of construction is the development of balloon framing and, later, platform framing, accompanied by the ease of producing lumber for them by modern mechanical means. Another reason is the lack of old-growth timber of the size necessary to produce large timbers. Typical heavy timber columns, for example, are 8 × 8 (200 × 200), and 12 × 12 (305 × 305) or even 16 × 16 (406 × 406) columns are often found in early warehouses and other storage-type buildings and in buildings used in manufacturing. Heavy timber beams supporting floor loads are 6 × 10s (150 × 250s) or larger. Trusses might be made using 8 × 8 (200 × 200) members. Decking is often 3 in. (75 mm) thick with a 1 in. (25 mm) wood finish floor.

Even where old-growth timber of the appropriate size exists, laws often prohibit harvesting it for construction purposes. Where such wood is available, it is usually expensive when compared to other available materials, such as steel, that will do the same job. However, heavy timber construction is acceptable in most codes and is still used in expensive residential construction projects.

### 6.11.4 LAMINATED TIMBER

Even though heavy timber buildings using large wood members are seldom built today, people still enjoy the look and feel of wood, and many buildings are made using heavy timber principles and laminated timber members. *Laminated timbers*, which are also called *structural glued laminated members* and *gluelams*, make possible sizes and shapes that are difficult and sometimes impossible to obtain in solid timber. Arches, beams, trusses, and other laminated members are custom-fabricated from pieces of stress-graded lumber bonded together with structural glues stronger than the wood itself.

Laminated timbers are used in every type of building today. In residences, they appear in utilitarian uses, such as beams to span openings and to produce large open areas. In commercial structures, laminated timbers are used to produce large beams for utilitarian and decorative purposes. In churches, athletic arenas, and residences, they are used to produce arches and domed structures that are at once structural and decorative.

Laminated timbers are used to produce large, rigid space frames and trusses with spans of more than 100 ft (30 m).

Laminated timbers and their installation should conform with the standards included in the publications of the American Institute of Timber Construction (AITC) that are listed in Section 6.15 and other applicable AITC publications

### 6.11.5 WOOD POLE CONSTRUCTION

Pole construction has been used for millennia. From prehistoric lake dwellers to twenty-first-century farmers, people have been embedding poles in the ground and building simple, elevated structures to protect themselves from unfriendly humans, ravaging animals, fierce winds, and engulfing floods.

Over time, people have learned to recognize naturally decay-resistant species such as redwood, cypress, and cedar. However, it was not until after World War II, when pressure-treated poles became widely available, that pole construction of farm buildings and beach cottages became commonplace.

The use of chemically treated poles gained impetus from the favorable experience of West Coast builders in the 1950s. They erected these houses on steep slopes, on periodically flooded lowlands, and in areas subject to earthquakes, and yet they were able to build them at reasonable cost with a high degree of permanence.

The information in this section is primarily based on a study by the American Wood-Preservers Institute (AWPI) done for the Department of Housing and Urban Development (HUD). Some recommendations and illustrations also were obtained from publications of the Forest Products Laboratory of the U.S. Department of Agriculture and from *Pole Building Design* by Donald Patterson for the Architectural Woodwork Institute (AWI).

### 6.11.5.1 General Description

A pole building is a structure whose contact with the ground is through a series of decay- and termite-resistant round tree trunks (poles) or squared wood members (timbers). Poles are cheaper and longer-lasting, are more frequently used, and, therefore, give their name to the system (Fig. 6.11-12).

Although cedar, redwood, and cypress have inherent decay resistance, their availability in long, thick poles is limited and their decay resistance varies with local growing conditions. Chemical treatment of Douglas fir, southern pine, ponderosa pine, and red pine in accordance with the American Wood Preservers' Association's (AWPA) standards ensures long-term resistance to infestation and in-ground durability.

In typical pole construction, poles ranging in diameter from 6 in. (150 mm) to 12 in. (305 mm) are spaced 6 ft (1800 mm) to 12 ft (3600 mm) apart to support floor and roof areas of up to 144 sq ft (13.38 m²) (Fig. 6.11-13). Most such poles are dropped into holes made either with a power auger or by hand digging. Occasionally, poles are "jetted" with a stream of water or driven with a pile driver. Proper pole embedment and back-filling of the hole are critical to the structural soundness of a pole building. A concrete pad under the pole or a concrete necklace around the pole may be required to increase the contact area with the soil. The pad and necklace help to distribute the load over a larger area.

Poles may extend substantially above ground to form the load-bearing frame of a superstructure or may terminate at the first floor level, thus permitting conventional stud-and-joist construction above the pole-framed platform (Fig. 6.11-14).

Embedded poles can support a superstructure either by spreading the loads with a concrete footing or by bearing on rock or very stiff soils. The following rec-

**FIGURE 6.11-12** Treated poles are frequently combined with timbers to form a structural frame and enhance a design. (Courtesy American Wood Preservers Institute.)

7′ = 2.134 m
10′ = 3.048 m
28′ = 8.534 m

**FIGURE 6.11-13** A plan for a small house with all poles inside the house. Short poles could be eliminated and poles placed outside the walls if floor girders were used. (Courtesy American Wood Preservers Institute.)

a

b

**FIGURE 6.11-14** In a pole-frame building (a), the poles continue up past the main floor. In a pole-platform building (b), the poles stop at the main floor. Lag bolts and joists are used for relatively light loads. Timbers and through-bolted shear connectors usually are required for efficient pole loading. (Courtesy American Wood Preservers Institute.)

ommendations are applicable only where a concrete footing pad, necklace, or backfill will be used to spread the building loads.

All wood pole structures should be designed by a licensed professional engineer.

### 6.11.5.2 Advantages and Disadvantages

Pole construction is most advantageous when conventional methods are either uneconomical or do not meet the desired objectives of minimum disturbance to the terrain, freedom from flooding, or resistance to strong winds and earthquakes. Therefore, pole buildings are generally more economical on steep slopes because they use relatively light wood members. Because pole construction does not require heavy equipment for excavation, it retains existing drainage patterns and preserves ground cover (Fig. 6.11-15).

On level ground subject to periodic flooding (as in coastal areas from high winds or in river plains from spring floods), pole buildings permit raising of the lower floor safely above flood levels. When there is adequate embedment and proper connections, flood waters can flow under the building with minimum

damage to the structure (Fig. 6.11-16). However, pole buildings in flood-prone areas should be carefully engineered for specific local conditions rather than the generalized guidelines of this section.

In hurricane-plagued areas, whether on flat or sloping ground, pole construction permits secure anchorage of a structure from roof to footings. Proper connections and adequate embedment are especially important in high-wind areas subject to flooding.

In most parts of the country, it is necessary to insulate the first floor and to provide a finish to protect the insulation.

**FIGURE 6.11-15** A pole house "floats" above the terrain. (Courtesy American Wood Preservers Institute.)

In heavily wooded areas, vegetation and combustible materials should be removed from the area that will be beneath the building when it is raised above grade only enough to create a crawl space. A skirt may be desirable to keep out animals and prevent accumulation of debris under the building.

The underfloor cavity should remain accessible to permit periodic inspection of the poles and the underside of the floor.

### 6.11.5.3 Site Preparation

Using pole construction provides a great opportunity to preserve the natural features of a site. Buildings can be fitted to the terrain while observing varying setbacks and sideyards, as dictated by codes and good design principles.

Drainage can be crucial on hillside sites. Pavements, roofs, road cuts, and utility trenches may concentrate surface waters into erosive streams. Heavy machinery should be used as little as possible, and necessary roads should be terraced along contour lines. It is generally preferable to hand-dig holes and to remove as little natural vegetation as possible. Hand digging theoretically is possible for holes up to 6 ft (1800 mm) deep, but may be uneconomical except for shallow poles, as described in Section 6.11.5.6.

**FIGURE 6.11-16** Pole substructure elevates a home above the natural terrain, permitting unobstructed runoff of storm and flood waters. (Courtesy American Wood Preservers Institute.)

On relatively flat sites, machine digging is more practical, but most truck-mounted augers are limited to 18 in. (460 mm) hole diameters. While these hole sizes are adequate for close pole spacing in good soils, larger diameters are required for most soil conditions.

The number of poles used should be the minimum number that is consistent with design and structural requirements. The most efficient use of poles usually occurs when the vertical load per pole is about 8000 lb (3630 kg). Heavier loads require special engineering; one-half of the indicated loading is generally inefficient.

Dug holes should be at least 8 in. (200 mm) larger than the diameter of the pole to permit accurate aligning and plumbing (Fig. 6.11-17). In most cases, the hole is wider than this minimum in order to develop a bearing area of sufficient size for the loads.

### 6.11.5.4 Pole Selection

The main considerations in pole selection are (1) type of treatment and (2) size and length.

Poles should be pressure preservative-treated in accordance with AWPA Standard C-23 (see Section 6.4), which specifies how much preservative should be retained in each pole and how deep it should penetrate into the wood. Treated poles are frequently branded with the manufacturer's name and location, the year of treatment, pole size, length, wood species, and type of preservative.

### 6.11.5.5 Construction Methods

There are two principal methods of pole construction: pole frame and pole platform. Footing size and depth of embedment are partly dependent on the method of construction.

In *pole-frame* buildings, the main vertical members extend from below the ground surface to the roof. The poles not only support the vertical weight of the building, but give it considerable lateral resistance to the forces of wind, flood, and earthquake.

In *pole-platform* buildings, treated poles form the substructure, permitting conventional construction methods to be used above. Pole-platform structures generally require greater pole embedment than pole-frame types.

### GENERAL RECOMMENDATIONS

Adhering to a few general rules will avoid problems in matching conventional framing methods with pole construction. Poles should not be placed within stud walls, because it is costly to apply finishes around poles which vary in diameter and shape. Beams, girders, and posts should be spaced so that standard, readily available sizes of lumber and panel materials can be used with a minimum of waste.

On level ground and on steep slopes with good soils (6000 psf [29.29 Mg/m$^2$] and better), noncementitious, granular backfill can be used, provided that a concrete pad or necklace is used (Fig. 6.11-18). Common backfill materials include sand, gravel, and crushed rock. Clean sand is often the least expensive of these and can achieve 100% compaction when properly tamped. Excavated soil from the site also can be used, provided that it is free of organic matter and is compacted by tamping.

Under these conditions, backfilling with concrete, or with soil-cement (excavated soil mixed with portland cement in a 5:1 ratio), will generally reduce the required hole depth by 1.5 ft (460 mm) to 2 ft (610 mm) (see Fig. 6.11-18).

On steep slopes with average or below-average soils (3000 psf [14.65 Mg/m$^2$] and lower), only concrete or soil-cement backfills are suitable. When concrete or soil-cement is used, a separate pad or necklace is not necessary.

### DESIGN ASSUMPTIONS

Pole embedment depends on many factors. To reduce the variables to a smaller number which can be tabulated readily, certain design assumptions are made. These are included in the following list. Note that other design assumptions would produce different results from those outlined in the following paragraphs.

Soil bearing values fall into three general categories, as shown in Figure 6.11-19.

The building height of pole-frame structure is 9 ft (2743 mm) from floor to top plate and 4 ft (1220 mm) from top plate to ridge (Fig. 6.11-20).

Embedment conditions:

*Condition A:* each pole either rests on a concrete pad or has a concrete necklace and is backfilled with granular material.

*Condition B:* each pole is completely backfilled to grade with either concrete or soil-cement.

### POLE-FRAME BUILDINGS

Embedment depths vary according to the slope of the site, as well as the soil, pole spacing, and unsupported pole height.

**FIGURE 6.11-17** Poles are aligned and plumbed, then braced temporarily before holes are backfilled. Heavier horizontal members are likely to be used in typical home construction.
(Courtesy American Wood Preservers Institute.)

**FIGURE 6.11-18** Recommended pole footings: (a) concrete pad; (b) concrete necklace; (c) concrete or soil-cement backfill. Dimensions A, B, and D can be obtained from Figures 6.11-23 through 6.11-25. Pad thickness C should be determined by engineering calculations.

***Flat Slope*** On sites with a slope of less than 1 in 10, poles should be embedded and supported, as recommended in Figure 6.11-21.

***Steep Slope*** On sites with a slope of 1 in 10 and up to 1 in 1, poles should be embedded and supported as recommended in Figures 6.11-22 and 6.11-23.

Embedment recommendations differ for uphill and downhill poles. Uphill poles, which have a short unsupported height, provide the necessary rigidity for the structure. In order to transfer this rigidity to the longer downhill poles, the floor should be designed as a diaphragm. Properly nailed conventional joists and

$1/2$ in. (12.7 mm) or thicker plywood floors provide such diaphragm action.

### POLE-PLATFORM BUILDINGS

Embedment depths of poles vary according to the slope of the site, as well as the other factors tabulated for pole-frame buildings.

***Flat Slopes*** On sites with a slope of less than 1 in 10, poles should be embedded and supported as recommended in Figure 6.11-24.

***Steep Slopes*** On sites with a slope of 1 in 10 and up to 1 in 1, uphill poles should be embedded and supported as recom-

mended in Fig. 6.11-25. Recommendations for downhill poles are given in Figure 6.11-22. The floor should be designed as a diaphragm to transfer the rigidity of the uphill poles to the rest of the structure.

### CONNECTIONS

It is generally better to frame poles with pairs of beams by lag bolting, through

**FIGURE 6.11-19** Soil Classification for Pole Construction

| Rating | General Description | Soil Groups | Maximum Bearing Value, psf | Lateral Pressure per ft. of Embedment, psf |
|---|---|---|---|---|
| Good | Compact, well-graded sand and gravel; hard clay or graded fine and coarse sand | GW, GP, SW, and SP | 6,000 | 400 |
| Average | Loose gravel, medium clay, or more compact compositions | GM, GC, SM, SC, and SL | 3,000 | 200 |
| Below average | Sand or clay containing large amounts of silt | CL, OL, MH, CH, and OH | 150 | 100 |

(Courtesy American Wood Preservers Institute.) (Refer to Figure 25.7-1 for metric conversions.)

**FIGURE 6.11-20** Building height design assumptions and embedment conditions: (a) concrete pad or necklace and granular backfill; (b) concrete or soil cement backfill. (Courtesy American Wood Preservers Institute.) (Refer to Figure 25.7-1 for metric conversions.)

**FIGURE 6.11-21** Embedment Depths for Pole-Frame Buildings on Sites with Slopes Less Than 1 in 10

| H, ft. | Pole Spacing, ft. | Good Soil | | | | Average Soil | | | | Below-Average Soil | | | |
|---|---|---|---|---|---|---|---|---|---|---|---|---|---|
| | | Embedment Depth, ft. | | D, in. | Tip Size, in. | Embedment Depth, ft. | | D, in. | Tip Size, in. | Embedment Depth, ft. | | D, in. | Tip Size, in. |
| | | A | B | | | A | B | | | A | B | | |
| 1½ to 3 | 8 | 5.0 | 4.0 | 18 | 6 | 6.5 | 5.0 | 24 | 6 | * | 6.0 | 36 | 6 |
| | 10 | 5.5 | 4.0 | 21 | 7 | 7.0 | 5.0 | 30 | 7 | * | 6.5 | 42 | 7 |
| | 12 | 6.0 | 4.5 | 24 | 7 | 7.5 | 5.5 | 36 | 7 | * | 7.0 | 48 | 7 |
| 3 to 8 | 8 | 6.0 | 4.0 | 18 | 7 | 7.5 | 5.5 | 24 | 7 | * | 7.0 | 36 | 7 |
| | 10 | 6.0 | 4.5 | 21 | 8 | 8.0 | 6.0 | 30 | 8 | * | 7.5 | 42 | 8 |
| | 12 | 6.5 | 5.0 | 24 | 8 | * | 6.0 | 36 | 8 | * | 8.0 | 48 | 8 |

*Embedment depth is greater than 8 ft., considered excessively expensive.

**Note:** Where a concrete floor slab is used at grade, embedment depths may be reduced to 70% of those shown.

(Courtesy American Wood Preservers Institute.) (Refer to Figure 25.7-1 for metric conversions.)

**FIGURE 6.11-22** Embedment Depths for Downhill Poles of Pole-Frame Buildings on Slopes Up to 1 in 1

| Soil Strength | Embedment Depth, ft. | | |
|---|---|---|---|
| | Slope of Grade | | |
| | Up to 1:3 | Up to 1:2 | Up to 1:1 |
| Below average | 4.5 | 6.0 | * |
| Average | 4.0 | 5.0 | 7.0 |
| Good | 4.0 | 4.0 | 6.0 |

*Embedment depth is greater than 8 ft., considered excessively expensive.

(Courtesy American Wood Preservers Institute.) (Refer to Figure 25.7-1 for metric conversions.)

bolting, or nailing to the supporting pole (Fig. 6.11-26). To make a stronger connection, poles can be faced or notched (Fig. 6.11-27). Use of a spike-grid (Fig. 6.11-28) is preferred to squaring or notching because it minimizes the damage to the preservative-treated outside layer and increases the load-bearing value of the connection.

Accepted design practice is outlined in AWPI's *Pole Building Design* and in the *Timber Construction Manual* published by the AITC. Strength values used in designing bolted and nailed connections should conform to the *National Design Specifications for Wood Construction* by the American Forest and Paper Association (AFPA).

### 6.11.5.6 Special Construction on Rocky Slopes

Pole bracing as an alternative to embedment, discussed below, should not be attempted without competent engineering advice.

On rocky slopes, where deep embedment cannot be developed without expensive machine digging or rock drilling, it is possible to use shallow hand-dug holes, provided that the pole structure is adequately braced and anchored.

Alternates to deep embedment include (1) pole seating and crossbracing and (2) shear walls and floor diaphragms.

### POLE SEATING AND CROSSBRACING

If the understructure is adequately braced, the poles need to be embedded only far enough to prevent slipping downhill. The embedment should be at least 4 ft (1220 mm), and the poles should be placed in specially dug seats in the rock. Since a rock surface is likely to be irregular, tamped earth or wetted sand should be used to make a flat bearing surface under each seated pole. If a pole seat cannot be dug into the rock, a concrete pad may be used over the rock, and the pole should be anchored to the pad with lag bolts or standard post anchors. The

**FIGURE 6.11-23** Embedment Depths for Shorter Uphill Poles in Pole-Frame Buildings on Slopes Up to 1 in 1

| H, ft. | Pole Spacing, ft. | Good Soil | | | | Average Soil | | | | Below-Average Soil | | | |
|---|---|---|---|---|---|---|---|---|---|---|---|---|---|
| | | Embedment Depth, ft. | | D, in. | Tip Size, in. | Embedment Depth, ft. | | D, in. | Tip Size, in. | Embedment Depth, ft. | | D, in. | Tip Size, in. |
| | | A | B | | | A | B | | | A | B | | |
| 1½ to 3 | 6 | 7.0 | 5.0 | 18 | 8 | a | 6.5 | 18 | 8 | a | a | a | a |
| | 8 | 7.5 | 5.5 | 18 | 9 | a | 7.0 | 24 | 9 | a | a | a | a |
| | 10 | a | 6.0 | 21 | 9 | a | 8.0 | 30 | 9 | a | a | a | a |
| | 12 | a | 6.5 | 24 | 10[b] | a | a | a | a | a | a | a | a |
| 3 to 8 | 6 | 7.5 | 5.5 | 18 | 8 | a | 7.0 | 18 | 8 | a | a | a | a |
| | 8 | 8.0 | 6.0 | 18 | 9 | a | 8.0 | 24 | 9 | a | a | a | a |
| | 10 | a | 7.0 | 21 | 10[b] | a | a | a | a | a | a | a | a |
| | 12 | a | 7.0 | 24 | 11[b] | a | a | a | a | a | a | a | a |

[a]Embedment depth is greater than 8 ft., considered excessively expensive.

[b]These tip diameters may be decreased 1 in., provided embedment is increased by ½ ft.

(Courtesy American Wood Preservers Institute.) (Refer to Figure 25.7-1 for metric conversions.)

**FIGURE 6.11-24** Embedment Depths for Poles in Platform Buildings on Slopes Less Than 1 in 10

| H, ft. | Pole Spacing, ft. | Good Soil | | | | Average Soil | | | | Below-Average Soil | | | |
|---|---|---|---|---|---|---|---|---|---|---|---|---|---|
| | | Embedment Depth, ft. | | D, in. | Tip Size, in. | Embedment Depth, ft. | | D, in. | Tip Size, in. | Embedment Depth, ft. | | D, in. | Tip Size, in. |
| | | A | B | | | A | B | | | A | B | | |
| 1½ to 3 | 8 | 4.0 | 4.0 | 18 | 5 | 5.5 | 4.0 | 24 | 5 | 7.0 | 5.0 | 36 | 5 |
| | 10 | 4.5 | 4.0 | 21 | 5 | 6.0 | 4.0 | 30 | 5 | 8.0 | 5.5 | 42 | 5 |
| | 12 | 5.0 | 4.0 | 24 | 5 | 6.5 | 4.5 | 36 | 5 | * | 5.5 | 48 | 5 |
| 3 to 8 | 8 | 5.0 | 4.0 | 18 | 6 | 6.5 | 4.5 | 24 | 6 | * | 6.0 | 36 | 6 |
| | 10 | 5.5 | 4.0 | 21 | 7 | 7.0 | 5.0 | 30 | 7 | * | 6.5 | 42 | 7 |
| | 12 | 6.0 | 4.5 | 24 | 7 | 7.5 | 5.5 | 36 | 7 | * | 7.0 | 48 | 7 |

*Embedment depth is greater than 8 ft., considered excessively expensive.

(Courtesy American Wood Preservers Institute.) (Refer to Figure 25.7-1 for metric conversions.)

**FIGURE 6.11-25** Embedment Depths for Uphill Poles in Platform Buildings on Slopes Up to 1 in 1

| H, ft. | Pole Spacing, ft. | Good Soil | | | | Average Soil | | | | Below-Average Soil | | | |
|---|---|---|---|---|---|---|---|---|---|---|---|---|---|
| | | Embedment Depth, ft. | | D, in. | Tip Size, in. | Embedment Depth, ft. | | D, in. | Tip Size, in. | Embedment Depth, ft. | | D, in. | Tip Size, in. |
| | | A | B | | | A | B | | | A | B | | |
| 1½ to 3 | 6 | 5.5 | 4.0 | 18 | 6 | 7.5 | 5.0 | 18 | 6 | a | 7.0 | 24 | 6 |
| | 8 | 6.5 | 4.5 | 18 | 7 | a | 6.0 | 24 | 7 | a | 8.0 | 36 | 7 |
| | 10 | 7.0 | 5.0 | 21 | 7 | a | 6.5 | 30 | 7 | a | a | a | a |
| | 12 | 7.5 | 5.5 | 24 | 8 | a | 7.0 | 36 | 8 | a | a | a | a |
| 3 to 8 | 6 | 7.0 | 5.0 | 18 | 8 | a | 6.5 | 18 | 8 | a | 8.0 | 24 | 8 |
| | 8 | 7.5 | 5.5 | 18 | 9 | a | 7.0 | 24 | 9 | a | a | a | a |
| | 10 | a | 6.0 | 21 | 10[b] | a | 7.5 | 30 | 10[b] | a | a | a | a |
| | 12 | a | 6.5 | 24 | 10[b] | a | a | a | a | a | a | a | a |

[a]Embedment depth is greater than 8 ft., considered excessively expensive.

[b]These tip diameters may be decreased 1 in., provided embedment is increased by ½ ft.

(Courtesy American Wood Preservers Institute.) (Refer to Figure 25.7-1 for metric conversions.)

**FIGURE 6.11-26** Double headers at exterior end walls frame a ridge-high pole and are bolted to it. (Courtesy American Wood Preservers Institute.)

**FIGURE 6.11-27** Poles are faced or notched to provide better contact with beams and make suitable connections with shear connectors such as the spike grid shown in Figure 6.11-28. (Courtesy American Wood Preservers Institute.)

**FIGURE 6.11-28** A single curved spike grid with a ³/₄ in. (19.1 mm) bolt has a carrying capacity of 3800 lb (1723.6 kg); with a 1 in. (25.4 mm) bolt 4100 lb (1859.7 kg). (Courtesy American Wood Preservers Institute.)

pad should be 8 in. (200 mm) thick and of the same diameter as the hole, which can be 8 in. (200 mm) larger than the pole.

In one crossbracing method, $^5/_8$ in. (15.9 mm) or $^3/_4$ in. (19.1 mm) steel rods are fitted through drilled holes, flooded with preservative, and then fastened with nuts, using beveled washers to avoid crushing the wood. Steel rod crossbracing is not usually needed between every pole; several diagonal rods often are adequate.

A pole structure also can be knee-braced with $2 \times 6$ ($50 \times 150$) wood members, nailed, or bolted to form a diagonal connection from the poles to the floor beams. The bracing can be left exposed or covered with skirt boards around the perimeter of the house.

### SHEAR WALLS AND FLOOR DIAPHRAGMS

A masonry or concrete fireplace block, which is supported on concrete footings at least 4 ft (1220 mm) below grade, or a masonry or concrete wall can act as a shear wall to resist horizontal forces. Similarly, a masonry or concrete utility room below the first floor can act to brace the structure and anchor it.

Key walls also can be used to resist the horizontal forces pushing a structure down a slope (Fig. 6.11-29). For both shear walls and key walls, floors should be designed as diaphragms to transfer the lateral rigidity of the walls to the poles. Such floors can be built of conventional wood joists and $^1/_2$ in. (12.7 mm) or thicker plywood, with nailing as recommended by the APA.

## 6.12 Finish Carpentry

Finish carpentry, which is sometimes called *architectural woodwork*, traditionally includes exterior wood siding and shingles (see Chapter 7), casework and cabinets (see Chapter 12), and interior and exterior standing and running trim, interior paneling, and other wood elements addressed in this section. Finish carpentry also traditionally includes wood doors and windows, which are addressed in Chapter 8.

Finish carpentry is generally required to comply with the recommendations of the *Manual of Millwork* of the Woodwork Institute (WI) (formerly the Woodwork Institute of California [WIC]) or *Architectural Woodwork Quality Standards Illustrated*, which has been developed jointly by AWI and the Architectural Woodwork Manufacturers Association of Canada (AWMAC). The WI standard is applicable in some western states, and the *Architectural Woodwork Quality Standards Illustrated* is applicable throughout most of the United States and Canada. These two standards are similar in many respects but differ in others. For example, they use the same

names for material and construction grades (Premium, Custom, and Economy), but the WIC manual adds an additional Laboratory grade for casework and countertops in laboratories where chemical resistance is required. They also use different names for some finish carpentry components. A designer should obtain the standard applicable in the area of the project and use it as a guide. This section refers to the AWI/AWMAC standards in most cases.

### 6.12.1 MATERIALS AND GRADES

The materials used in finish carpentry are those discussed in Section 6.1. Both hardwoods and softwoods are used. Plywood is also used for casework and in paneling (see Section 6.3).

The grades in the AWI/AWMAC system are Premium, Custom, and Economy, the Premium grade being of the highest quality. The grade refers to both materials and workmanship. Sometimes a different quality is required of the workmanship than of the materials, but most of the time the same grade is applied to both.

### 6.12.2 STANDING AND RUNNING TRIM AND RAILS

*Standing trim* is a term used to identify items such as door and window casings, stops, stools, sills, aprons, and the like. *Running trim* includes bases, shoe molding, soffits, fascias, chair rails, cornices, crown molding, and similar items. *Rails* include hospital corridor rails and guard rails.

Figure 6.12-1 shows a few of the many molding shapes that are available as stock items from wood suppliers and lumber mills and that are used to make standing and running trim and rails. In addition, almost any desired shape can be milled at the shop or on the job. Numerous cutter blade shapes are available, and any shape can be ground in a metal shop. Trim shapes and moldings can be combined to form larger units (Figs. 6.12-2 and 6.12-3).

Right angle intersections of most standing and running trim items are mitered (Fig. 6.12-4a) to make neat corners. Corner joints are either doweled, splined, or held together with plate joiners (biscuits) (Fig. 6.12-4a). Running joints are sometimes back cut and sometimes half-lapped (Fig. 6.12-4b). Inter-

**FIGURE 6.11-29 Key wall bracing.** (Courtesy American Wood Preservers Institute.) (Refer to Figure 25.7-1 for metric conversions.)

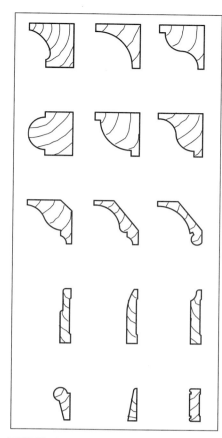

**FIGURE 6.12-1** Various molding shapes. (Drawing by HLS.)

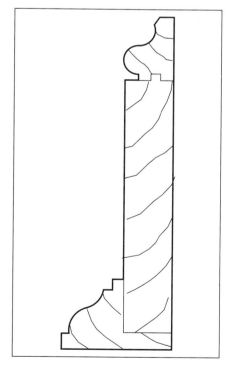

**FIGURE 6.12-2** A combination of molding shapes joined to make a wall base. (Drawing by HLS.)

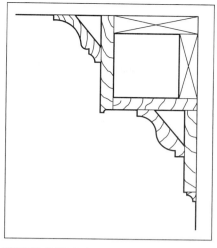

**FIGURE 6.12-3** A combination of molding shapes joined to make a complex cornice. (Drawing by HLS.)

secting joints are sometimes made with mortise-and-tenon jointing (Fig. 6.12-4c).

### 6.12.3 PANELING

Wood paneling is generally available in either $^7/_{16}$-in. (11-mm)- or $^3/_4$-in. (19.1-mm)-thick panels. Joints in flat paneling may be installed using applied molding (Fig. 6.12-5a) or may be butted

and splined to hold them in alignment (Figs. 6.12-5b and 6.12-5c). Articulated joints (Fig 6.12-6) are also used. These figures show a few of the many possible variations. Raised panels are also often used.

Pattern matching in wood paneling is described in Section 6.3 (see Fig. 6.3-18).

### 6.12.4 CLOSET AND STORAGE SHELVING

Closet and utility shelving is constructed of either wood boards (Section 6.2) or panel products (Section 6.3). The edges of shelving made from products other than solid wood are usually edge banded with wood or plastic edging strips.

Shelves are generally required to be $^3/_4$ in. (19.1 mm) thick, and those more than 48 in. (1220 mm) long must have intermediate supports. Unsupported shelves more than 48 in.(1220 mm) long must be thicker or be supported along the front edge by a 2 × (50 mm) wood apron.

### 6.12.5 STAIRWORK AND HANDRAILS

*Architectural Woodwork Quality Standards Illustrated* contains requirements for stairs and handrails. In addition, stair riser heights and tread widths and their

**FIGURE 6.12-4** Jointing methods for standing and running trim: (a) mitered corner with a plate joiner; (b) a half-lap joint; (c) a mortise-and-tenon joint. (Drawing by HLS.)

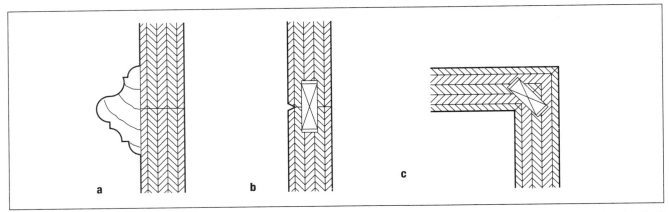

**FIGURE 6.12-5** Butt jointing methods for paneling: (a) an applied molding; (b) a butted and splined running joint; (c) a mitered and splined corner joint. (Drawing by HLS.)

ratio to one another, as well as heights and types of handrails and safety guards, are dictated by the applicable code. Free standing and cantilevered stairs should be designed by a licensed engineer.

Decorative stairs, especially curved and cantilevered stairs, are often constructed in a woodwork shop and delivered to the project in sections for final field assembly. This process offers better control of the process and often results in a higher-quality finished product than can be achieved when the stair is built completely at the job site.

### 6.12.6 OTHER FINISH CARPENTRY ITEMS

Many other wood elements are used in modern buildings. Among them are mantles, ornamental grills, pediment heads, pilasters, columns, tracery, and the like. *Architectural Woodwork Quality Standards Illustrated* addresses these as well.

## 6.13 Plastics

Plastics are relative newcomers to the construction field compared with traditional materials such as wood, glass, and concrete. The first commercial plastic in the United States, cellulose nitrate, was developed a little over a century ago. It was created by John Wesley Hyatt to help overcome a shortage of ivory, from which billiard balls were made.

Progress in plastics technology was slow at first. The next commercial plastic to be developed was phenol-formaldehyde, in 1909. Gradually, a few others were introduced, such as casein (1919), alkyds (1926), and polyvinyl chloride (1927). It was not until the 1930s, however, that the modern commercial plastics industry began; plastics were not widely used in building construction until after World War II.

Despite their relative newness, plastics have become important materials in

**FIGURE 6.12-6** Articulated jointing methods for paneling: (a) a running joint; (b) an inside corner joint; (c) an outside corner joint. (Drawing by HLS.)

the construction field. As a group, they offer a wide range of properties possessed by few other building materials: they are moisture- and corrosion-resistant, lightweight, tough, and easily molded into complex shapes. Furthermore, the great variety of plastic materials available, each of which offers slightly different advantages, makes plastics well suited for a multitude of purposes.

Sheet and tile flooring is a common use for plastics in construction. Vinyl flooring is noted for excellent abrasion and stain resistance and good resilience. Plastic siding is becoming increasingly popular because of its good durability and freedom from maintenance; plastic-covered wood windows and doors offer similar advantages. Plastic laminates are standard materials for counter- and table-tops; they are produced in many decorative designs and bright colors, and have good heat and chemical resistance. Plastics also are used for break-resistant glazing, skylights, and roof domes. Plastic films make excellent vapor retarders; plastic foams can be used for a variety of insulation purposes.

Plastics do have limitations, however, and they can create serious problems if used incorrectly. For satisfactory service, a suitable plastic for the intended use should be chosen. The plastic product should be correctly designed and properly integrated with other products. This section discusses the composition and properties of plastics and describes the chief applications of the most important types. For more detailed information on molecular structure and how it affects the specific properties of materials, see Chapter 24.

## 6.13.1 COMPOSITION

Twenty to thirty families of materials are classified as plastics. Within each of these families, a multitude of variations and combinations are possible. The complex details of the chemistry involved in the production of these compounds and the terminology used to classify them are beyond the scope of this book. A brief explanation of their composition, molecular structure, and basic types is included, however, to provide a basic understanding of what plastics are, how they behave, and how they can be used in construction.

Despite having widely varying properties, all plastics have several characteristics in common, which warrants classifying them as a single materials group:

1. At one stage in their formation they exhibit plastic behavior; that is, they soften and can be formed into desired solid shapes, with or without the use of heat and pressure.
2. They are made up of molecules built around a carbon atom and, therefore, are classed as organic materials. There are some exceptions, however. Some plastics are built around a silicon atom rather than a carbon atom, yet are still considered plastics.
3. They are *synthetic* materials, that is, they are man-made and do not occur in nature.
4. They are high polymers, which are giant molecules made of numerous relatively simple repeating units linked together into large aggregations.

Not all materials which possess the listed characteristics are necessarily plastic. For example, synthetic rubbers have all of these characteristics but are classed as rubbers.

### 6.13.1.1 Molecular Structure

Some plastics are soft and flexible; others are hard and brittle. Some will become soft and may melt or even evaporate when exposed to intense heat; others are virtually unaffected by extremes of temperature. An examination of the molecular structure of high polymers can help to explain many of these differences in the properties and behavior of plastics.

### HIGH POLYMERS

Most plastics are based on carbon chemistry. This means that they are dependent on the peculiarities and properties of the carbon atom (Fig. 6.13-1). Carbon atoms form bonds with other atoms through covalent bonding (see Chapter 24). They have a valence of 4, meaning that there are four points to which other atoms can attach themselves to form covalent bonds.

Atoms of other elements have different valences. For example, hydrogen atoms have a valence of 1, and oxygen atoms have a valence of 2. Therefore, four hydrogen atoms can attach themselves to a single carbon atom (one at each valence point) to produce $CH_4$, the gas known as *methane*. Two oxygen atoms can attach themselves to a single carbon atom (each oxygen atom taking up two valence points) to produce $CO_2$, *carbon dioxide* (see Fig. 6.13-1).

Two adjacent carbon atoms can attach to each other. If this attachment involves only a single valence point, thus creating a single bond between them, and the other valence points are taken up by other atoms, the molecule is said to be *saturated*. In other words, a saturated molecule contains only single bonds between the carbon atoms and has no free (unattached) valence point.

Two carbon atoms may also attach to each other at two valence points, creating a double bond, or at three valence points, creating a triple bond (Fig. 6.13-2). In these cases, the molecules are said to be *unsaturated*. Under certain conditions, the double or triple bonds of an unsaturated molecule can be opened up (activated), creating free valence points. Other atoms, or groups of atoms, can then attach themselves to these free valence points. For example, if the double bonds in a large number of ethylene molecules are activated, they can join together to form a chain (Fig. 6.13-3), which is then known as *polyethylene*. Such chain formation, called *polymerization*, is essential in the production of plastics. Because only unsaturated molecules will react together to form chains, these are the only kinds of molecules that are useful in producing plastics.

**FIGURE 6.13-1** Two molecules based on the carbon atom.

**FIGURE 6.13-2** Molecules can have double bonds (ethylene) or triple bonds (acetylene) between the carbon atoms. Such molecules are called *unsaturated*.

**FIGURE 6.13-3** Several ethylene molecules can be linked to form polyethylene, a polymer.

A single molecule such as ethylene is known as a *monomer.* Several of these molecules linked together become a *polymer.* When the number of units in a chain is very large (between several hundred and several thousand), it is called a *high polymer.* The number of monomeric units in different chains can vary widely. When chain growth stops, free valence points are left at each end of the chain. These may be taken up by single atoms, such as hydrogen, or by groups of atoms with a single net valence. In some cases, the two ends of a chain loop around and connect, forming a ring.

Plastics are composed of these giant chain-like molecules and are therefore classed as high polymers. This is evidenced by the fact that many of the chemical names for plastics start with the prefix *poly,* which means "many." Examples are polyethylene, polystyrene, and polyvinyl chloride.

### BASIC CATEGORIES

Plastics can be divided into two basic categories based on their molecular structure: (1) *thermoplastic* materials, which can be repeatedly softened and hardened by heating and cooling, and (2) *thermosetting* materials, which are set into permanent shapes during forming and cannot be softened again by reheating.

***Thermoplastics*** Thermoplastic materials are made up of long chain-like molecules such as the polyethylene molecule described earlier. These molecules are characterized by their attachment to each other. The chemical attraction along the chain is very strong, but the attraction between adjacent chains is relatively weak. Furthermore, these chains are generally saturated, having no double or triple bonds that can be activated. This means that no free valence points can be created

to cause a strong attraction to adjacent chains.

At any temperature above absolute zero, molecules are in constant random motion. As a result of the unconnected molecular structure of thermoplastics, the molecules can slide past each other and change shape, as in a liquid. At room temperature, this random motion is slight and the plastic holds its shape. However, as the temperature rises, molecular motion increases and the attractive forces between molecules weaken, causing the material to expand and become soft and flexible.

When a plastic reaches a high enough temperature, it will flow and can then be molded into new shapes. The temperature at which this occurs is called the *softening point,* which will vary according to the type of plastic involved. When the material is cooled again, it assumes its former stiffness and holds its new shape at normal-use temperatures. With thermoplastic materials, the process of heating to soften and cooling to harden can be repeated again and again without affecting the properties of the plastic.

The configuration of the molecules in thermoplastics also affects the final properties. Some molecules are arranged in neat, continuous chains (linear molecules); others are highly branched. Highly branched chains do not pack together well, whereas linear molecules can be stacked together closely. As a result, plastics composed of branched molecules tend to be softer, less strong, and lower in density than those composed of linear molecules.

The molecular configuration in a given plastic can be controlled to a certain extent by the method used in its formation. For example, when ethylene is polymerized by high temperature and pressure, the molecules are often highly

branched. However, ethylene can also be polymerized through the use of catalysts, which are substances added to the basic chemicals to speed up the chemical reaction. When polyethylene is made by this process, the molecules are generally linear. This means that variations in strength and stiffness can be produced in different batches of the same type of plastic simply by changing the method of polymerization. There are hundreds of different thermoplastics. A brief description of the most important types is given in Section 6.13.3. However, a few words about the way these different types are developed may be helpful here.

Starting with the ethylene molecule, which is the basic building block of many thermoplastics, substitutions can be made for one or more of the hydrogen atoms involved. Single atoms such as chlorine or fluorine, or whole groups of atoms such as the styrene monomer, can be substituted for the hydrogen atoms (Fig. 6.13-4). For example, if one hydrogen

**FIGURE 6.13-4** Starting with an ethylene molecule, substitutions can be made for a hydrogen atom, producing (a) polyvinyl chloride or (b) polystyrene.

atom is replaced by a chlorine atom, a plastic known as *polyvinyl chloride* results. If one hydrogen atom is replaced by a phenyl group, *polystyrene* results. If one hydrogen atom is replaced by a methyl group ($CH_3$) and a second hydrogen atom is replaced by a methoxycarbonyl group ($COOCH_3$), *polymethyl methacrylate* (acrylic) results.

Hundreds of such variations are possible, and each variation produces a plastic with slightly different properties. A large group of plastics called *vinyls* is based on the ethylene molecule. Other types are based on different molecules. For example, *cellulosics* are based on modifications of the basic celluloid molecule.

Still other variations are possible by combining different monomeric units into the same molecular chain. These substances then are called *copolymers*. Therefore, instead of having a single repeating monomer along a polymer chain, two different monomers may alternate along the same chain. Numerous variations in the properties of plastics are made possible through such copolymerization. For example, vinyl chloride may be copolymerized with vinyl acetate, producing a substance with some of the properties of each of the original polymers (Fig. 6.13-5). Acrylonitrile-butadiene-styrene is another copolymer which has proven to be very useful because of its unique combination of properties.

*Thermosets* Thermosets differ from thermoplastics in that once they are set (cured), they cannot be softened again.

This is because, during forming, they go through an irreversible chemical process which develops strong chemical bonds between adjacent molecules. Therefore, the molecules cannot slide past each other, as they can in thermoplastic material, but instead form a strongly interconnected molecular structure.

An example of this process is the production of thermosetting plastics from unsaturated polyesters. Being unsaturated, the chain-like polyester molecules contain at least occasional double carbon bonds. These double bonds can be activated, creating free valence points. If other activated molecules are present in the mixture, strong cross-links can be developed between them. For example, activated styrene molecules will react with activated polyester molecules, forming a continuous, cross-linked network. The degree of cross-linking that occurs determines the final properties of the material, that is, whether it will be resilient and rubbery, hard and glassy, or somewhere in between.

Another type of thermosetting reaction, called *condensation polymerization*, produces a by-product such as steam. An example of this is the reaction between phenol and formaldehyde, producing the thermoset known as *phenol-formaldehyde*. During this reaction, the phenol molecules each give up one or more hydrogen atoms and the formaldehyde molecules give up an oxygen atom. This leaves open valence points on both molecules, allowing them to join together to form a strongly interlinked structure. The leftover hydrogen and oxygen atoms

join together to form water, which is given off as steam.

Thermosetting reactions are irreversible. Therefore, these plastics go through a pliable stage only once. After they have set, they remain rigid and hard unless actual chemical decomposition (break down) occurs due to extreme heat, pressure, or environmental effects.

### 6.13.1.2 Modifiers

Many plastics in their pure form are difficult to work with, expensive to produce, or lack some property needed for a particular application. These difficulties can sometimes be overcome by the addition of one or more of several types of modifiers.

These modifiers are added to the basic plastic ingredient (the resin) before fabrication in a process known as *compounding*. It is important that the different ingredients be thoroughly compounded into a homogeneous mass, because the uniformity of the mix contributes greatly to the final properties of the plastic.

The type and amount of modifiers used with different plastics vary greatly. Some mixes consist of 90% to 95% basic resin, with only small amounts of modifiers. Other mixes contain only 20% to 30% resin, the remainder being comprised of various plasticizers, fillers, stabilizers, and other additives.

### PLASTICIZERS

At ordinary-use temperatures, many thermoplastics are inherently hard and rigid. In order to produce sheets, tubes, films, and other flexible plastic products, plasticizers must be added to these materials (Fig. 6.13-6).

**FIGURE 6.13-5** Two different monomers may alternate along a polymer chain, producing a copolymer.

**FIGURE 6.13-6** The addition of plasticizers imparts flexibility to vinyl sheet flooring, making it easier to handle and install.

Plasticizers are liquid or solid materials that are blended with plastic resin to impart increased flexibility, impact resistance, resiliency, and moldability. Basically, a plasticizer produces the same effect in a thermoplastic as does high temperature, that is, it lessens the forces of attraction between the long chain-like molecules, allowing them to slide more freely past each other. Unfortunately, this greater flexibility also often reduces the strength of the plastic and lowers its heat resistance and dimensional stability.

To obtain the optimum effects from a plasticizer, it is important that it be compatible with the plastic resin being used. This means that the plasticizer should mix well with the resin and should not separate out during the life of the object. A poor plasticizer may migrate to the surface of the object, creating a greasy film and causing the original plastic to become hard and brittle again. Furthermore, a plasticizer should not deteriorate with age or cause the plastic itself to deteriorate.

### FILLERS

Fillers are used most commonly with thermosets, although there are some filled thermoplastics as well. Fillers are used to improve a particular property of a plastic, such as moldability, heat resistance, or toughness. However, some fillers are used primarily to increase the bulk and therefore decrease the cost of a plastic. These latter ones are called *extenders*. A wide variety of materials may be used as fillers, depending on the properties desired in the finished product. The amounts used generally vary between 10% and 50% of the total plastic mix. Fillers are used to increase the following properties:

- *Mobility and bulk.* Wood flour, made of finely ground hardwood or nut shells, is commonly added to thermosets to make a plastic that is more easily molded and less expensive to produce.
- *Hardness.* Mineral powders and metallic oxides may be added to a plastic to increase its hardness. However, if too much of these materials are added, the plastic may become brittle.
- *Toughness.* Fibrous fillers are often added to plastics to decrease their brittleness and increase their impact resistance. Natural fibers such as cotton,

sisal, or hemp may be used, as may synthetic fibers such as rayon, nylon, and polyester.

- *Heat resistance.* Certain kinds of mineral fiber fillers, and other inorganic fillers such as silica, clay, and finely ground limestone, are used to increase the heat resistance of some plastics.
- *Electrical resistance.* Finely ground mica or quartz may be added to increase the electrical resistance of a plastic.

### REINFORCING AGENTS

Reinforced plastics are a large and important group in the construction industry. Their strengths and stiffnesses have been significantly increased through the addition of high-strength fillers that act as reinforcing agents. Generally, thermosetting resins such as polyester and epoxy are used in reinforced plastics. Fibrous fillers such as sisal and mineral fibers can be used as reinforcing agents, but the strength of glass fiber is so superior that it is almost always the material selected for this purpose.

Sheet glass is quite brittle, but when drawn into fibers, its tensile strength rises enormously. In fact, pound for pound, glass fiber is the strongest commercially available construction material. Therefore, adding glass fibers to a plastic greatly increases its strength and stiffness. Glass fiber can be used either as chopped strands, yarns, pressed matting, or woven fabrics, depending on the design and properties desired for the finished product.

### STABILIZERS

The effects of weathering on construction materials is an important consideration. Many plastics degrade when exposed to the ultraviolet rays of the sun and to oxygen and other gases in the air. As a result, small amounts of *stabilizers* are added to these plastics to act as antioxidants and ultraviolet light absorbers.

Several different substances may be used as stabilizers. For example, carbon black, which is commonly used, changes polyethylene from a quickly degrading material to one that stands up well under extended outdoor exposure.

### COLORANTS

The addition of colorants to plastics is a complex subject. Some plastics, such as acrylic and polystyrene, are inherently

clear and can be produced in a wide range of colors. Others, such as the phenolics, are dark and opaque and therefore have a more limited range of possible colors.

Both pigments and dyes can be used to color plastics. Pigments produce opaque colors; dyes, transparent ones. The pigment or dye chosen must be compatible with the other ingredients in the plastic mix. It should not bleed out of the plastic to discolor or stain adjacent materials. Furthermore, it should have good light and heat stability so that it will not fade or change in appearance with age.

### 6.13.2 PROPERTIES

The range of properties of various plastics is quite broad. They can be flexible or stiff, tough or brittle, transparent or opaque. However, plastics also have certain limitations, which must be considered when choosing them for applications in buildings. These limits can be seen most clearly when the general properties of plastics are compared with those of other construction materials such as steel, concrete, wood, and glass.

#### 6.13.2.1 Strength and Stiffness

Strength can be measured in terms of tension, compression, and bending, tensile strength being the most important. The tensile strength of most plastics is near that of wood, which is much lower than that of steel. However, some plastic laminates and reinforced plastics approach steel in strength. The highest tensile strengths are found in glass-fiber-reinforced plastics. Stiffness is more important than strength in most plastics applications. Stiffness is measured by the modulus of elasticity. The higher the stiffness, the greater the modulus of elasticity. Generally, plastics are less stiff than steel, glass, and other structural materials, and, therefore, they deflect more than these other materials under a given applied load. Thermosets are somewhat stiffer than thermoplastics. Reinforced plastics have the highest modulus of elasticity (about the same as those of wood and concrete), but even they are only about 10% as stiff as steel. Plastics weigh considerably less than other structural materials. Unmodified plastic resins range from a specific gravity of about 0.9 for polypropylene to about 1.5 for polyvinyl chloride. Both filled and reinforced plastics are more dense. In contrast, steel has a specific gravity of about 7.5. Therefore, when strength comparisons are made between plastics and metals on an equal-weight basis, plastics appear to have the advantage.

For this reason, plastics are employed in applications where strength combined with lightness is of prime importance, such as in spacecraft. In other construction, they can provide very strong, lightweight components (such as in sandwich panels), which can appreciably decrease the overall weight of a structure (Fig. 6.13-7).

### 6.13.2.2 Toughness and Hardness

*Toughness*, which is strength combined with resilience, is an outstanding characteristic of many plastics. This property is difficult to measure precisely, but it is generally expressed in terms of a material's *impact strength*, or ability to resist impact, such as that from a falling ball or a striking pendulum.

Plastics differ greatly in impact strength, and different formulations of the same plastic also show considerable variation. For example, regular polystyrene is quite brittle, whereas copolymers of styrene, such as acrylonitrile-butadiene-styrene, are noted for toughness.

Compared with other materials such as glass, concrete, and wood, plastics are quite resistant to impact. For example, a $1/8$-in. (3.2-mm)-thick sheet of acrylic can withstand 25 times as much impact as can a $1/8$-in. (3.2-mm)-thick sheet of double strength window glass. This is illustrated by the widespread use of plastics for break-resistant glazing and lighting fixtures. Another property of importance is *hardness* (abrasion resistance). In this respect, plastics do not per-

**FIGURE 6.13-7** Low-density polyurethane foam is injected between two strong, rigid facings to produce a lightweight sandwich panel. (Courtesy Dow USA.)

form as well as glass or steel. Plastic glazing scratches much more readily than glass, for example. Conversely, many plastics are more resistant to abrasion than wood.

### 6.13.2.3 Creep

*Creep* (cold flow) is a permanent change in the dimensions of a material due to prolonged stress. All materials are subject to creep, but in steel the rate of deformation at room temperature is so slight that it is negligible.

Because of their molecular structure, many plastics exhibit creep at room temperature, and the amount of deformation which results can be large. This is particularly true of thermoplastics, whose unconnected molecules begin to slide past each other and assume new forms when an external force is applied. If the force is applied only for a short time, the plastic may recover, but if the force is long term, deformation can be permanent and may eventually result in failure of the material.

Thermosets, laminates, and reinforced plastics have greater resistance to creep, but even they cannot resist prolonged high stresses, as can steel and concrete. Nevertheless, plastics can be used in self-supporting and even some load-bearing applications if the proper formulation is chosen.

### 6.13.2.4 Thermal Properties

The important thermal properties of plastics are expansion and contraction, conductivity, and maximum service temperatures.

#### EXPANSION AND CONTRACTION

All building materials expand and contract with changing temperatures. Thermal expansion is much higher in plastics than in most building materials. Thermosets exhibit a lower rate of expansion than thermoplastics, while reinforced plastics and laminates generally show the lowest rates among plastics.

Changes in dimension caused by expansion and contraction can cause buckling or cracking of plastic components used in assembles. These difficulties can be avoided through proper design of the component and through the use of expansion joints to allow for movement. Particular care must be taken when plastics are joined to metals because of the wide difference in rates of expansion of

the two materials. Generally, thermosets such as phenolics or reinforced plastics work best in such applications.

#### CONDUCTIVITY

Plastics are generally good heat insulators. They have a much lower thermal conductivity than metals. Pure, unmodified plastics also compare favorably with concrete, glass, and brick. Wood, however, has a lower conductivity than most plastics.

The exception is the foam plastics, which outperform even wood in this respect and are among the best heat insulators available. Polyurethane foam can have a conductivity as low as 0.15, which is lower even than that of mineral wool, with 0.27. Values for other foams vary according to density, type of foaming agent used, and whether the foam is of the open-cell (interconnecting) or closed-cell variety.

Because of these low conductivity values, plastic foams make excellent insulation materials. However, some problems have been encountered with their use in construction due to their flammability, the toxic fumes they can produce when burned, and the tendency of some of them to reduce their insulating value over time. Usually, these problems can be overcome with proper material selection and good design. Codes require that some plastic insulations not be left exposed, but rather covered with more fire-resistive material.

#### SERVICE TEMPERATURES

The maximum *continuous service temperature* of a material is the highest prolonged temperature at which it can be used without softening. Service temperatures of plastics are low compared with those of other building materials. Many thermoplastics begin to soften at temperatures near the boiling point of water. Thermosets have better heat resistance, and some can be used at temperatures up to the wood char point (380°F [193.3°C] to 400°F [204.4°C]). Fluorocarbons have the highest continuous service temperature of unmodified plastics and can be used in non-load-bearing applications at 500°F (260°C).

Because additives can substantially improve the heat resistance of plastics, many filled and reinforced plastics can withstand higher temperatures than the pure, unmodified resins. Service temperatures are important in determining the suitability of a plastic for a particular application. Most plastics can withstand the temperatures encountered in ordinary building structures; for more demanding applications, such as piping for hot water supply lines, special heat-resistant formulations may be required.

### 6.13.2.5 Corrosion Resistance

Plastics have excellent corrosion resistance. They resist attack by most chemicals, with the exception of certain organic solvents. As with other properties, performance varies according to composition. Fluorocarbons are extremely inert and can be attacked only by the most powerful chemical reagents. Other plastics are attacked selectively by certain groups of solvents. Therefore, care must be taken to choose a suitable plastic for the corrosive conditions involved in the application.

### 6.13.2.6 Fire Resistance

Flammability of materials is an important consideration in construction. Being organic in nature, plastics will burn. Some plastics, such as cellulose nitrate, are extremely flammable and are generally not permitted as construction materials by most building codes. Others are slow-burning, and some will stop burning after the source of flame has been removed. Flammability can be somewhat controlled through the use of additives. For example, compounds of chlorine, fluorine, bromine, and phosphorus are commonly added to reinforced plastics to retard flame spread.

The production of smoke and toxic fumes is another element of fire hazard. The amount of smoke and the types of fumes produced depend on the composition of the plastic; those plastics rendered flame retardant through the use of additives generally produce the heaviest smoke. Furthermore, if a plastic contains compounds of chlorine, fluorine, nitrogen, or sulfur, these compounds will also be present in the gases given off during combustion. These gases, in combination with carbon monoxide and carbon dioxide, which are by-products of all organic combustion, may prove highly toxic.

Plastics are no more flammable than wood, fabrics, or other organic materials. However, because plastics often substitute for traditional nonflammable materials like glass and metals, they should be used with discretion and only after care-ful consideration of their potential fire hazard.

To deal with these and other problems associated with the use of plastics in construction, most of the major building codes, including the *International Building Code*, have chapters devoted specifically to plastics applications. Tests developed or accepted by ASTM International (ASTM) and published as ASTM standards can be used to rate plastics in terms of their burning rate, fire endurance, and flame spread. The *International Building Code* and other building codes require that plastics suitable for installation in buildings be able to pass the tests in one or more of these ASTM standards. In addition, the *International Building Code* and other codes restrict the use of plastics in certain areas by special requirements, such as requiring that foamed plastics be covered with noncombustible materials, requiring that plastic glazing be spaced with intermittent noncombustible surfaces, and limiting the amount of plastic materials that may be used as interior finish.

### 6.13.3 MAJOR TYPES AND USES

The multitude of commercially available plastics can be broken down into several major groups. This section includes a brief description of the chief characteristics and application of plastics commonly used in construction. A listing of plastics and their abbreviations compiled by ASTM is given in Figure 6.13-8.

### 6.13.3.1 Thermoplastics

Thermoplastics are by far the largest group of plastics. Although thermo-

**FIGURE 6.13-8** ASTM Abbreviations for Plastics

| Term | Abbreviation | Thermoplastic | Thermosetting |
|---|---|---|---|
| Acrylonitrile-butadiene-styrene | ABS | x | |
| Acrylic | | | |
|   Poly(methyl methacrylate) | PMMA | x | |
| Cellulosics | | | |
|   Cellulose acetate | CA | x | |
|   Cellulose acetate-butyrate | CAB | x | |
|   Cellulose acetate-propionate | CAP | x | |
|   Cellulose nitrate | CN | x | |
|   Ethyl cellulose | EC | x | |
| Epoxy, epoxite | EP | | x |
| Fiberglass-reinforced plastics | FRP | | x |
| Fluorocarbons | | | |
|   Polytetrafluoroethylene | PTFE | x | |
|   Fluorinated ethylene propylene | FEP | x | |
| Melamine-formaldehyde | MF | | x |
| Nylon | | | |
|   Polyamide | PA | x | |
| Phenolic | | | |
|   Phenol formaldehyde | PF | | x |
| Polycarbonate | PC | x | |
| Polyester | — | | x |
| Polyethylene | PE | x | |
| Polypropylene | PP | x | |
| Polyurethane | | | |
|   Urethane plastics | UP | | x |
| Styrene | | | |
|   Polystyrene | PS | x | |
|   Styrene-acrylonitrile | SAN | x | |
|   Styrene-butadiene plastics | SBP | x | |
| Silicone plastics | SI | | x |
| Urea-formaldehyde | UF | | x |
| Vinyls | | | |
|   Poly(vinyl acetate) | PVAc | x | |
|   Poly(vinyl alcohol) | PVAL | x | |
|   Poly(vinyl butyral) | PVB | x | |
|   Poly(vinyl chloride) | PVC | x | |
|   Poly(vinyl fluoride) | PVF | x | |

(Copyright ASTM, reprinted with permission.)

plastics can be softened by heating, they are dimensionally stable at normal-use temperatures.

## ACRYLONITRILE-BUTADIENE-STYRENE (ABS)

ABS copolymers are noted for their unique combination of properties. They are an outgrowth of the high-impact polystyrenes, which are a combination of styrene and butadiene-rubber compounds. Polystyrene plastics alone are very brittle, but their toughness can be greatly improved by adding a butadiene-rubber compound to them. The further addition of acrylonitrile to the mixture improves other properties, such as chemical resistance, rigidity, and tensile strength.

ABS plastics are widely used in the construction industry for piping and pipe fittings. They are used for water and gas supply lines and in drain, waste, and vent systems. They are tough and will withstand rough usage and fairly high temperatures for extended periods. ABS plastics are also used to make hardware items such as door knobs, handles, and latches, and to manufacture handrails, furniture, appliances, and machine parts.

## ACRYLICS

Polymethylmethacrylate is the best known of the acrylic plastics. These transparent plastics are noted for their superior optical properties and can be used readily for glazing. Unlike glass, they also have the ability to *pipe* light. This means that a beam of light can be bent around a corner due to the internal reflections of the light by the outer layer of the plastic.

Acrylics have good resistance to breakage and outstanding weathering properties. Their transparency and physical stability are not adversely affected by long outdoor exposure. However, they cannot withstand high temperatures and begin to soften at 200°F (93.3°C). In addition, their low abrasion resistance makes them more susceptible to scratching than glass. Acrylics are used for skylights and roof domes and for glazing in public buildings where the breakage hazard is high. They also are used in lighting fixtures as translucent ceiling panels, lamp enclosures, and light diffusers. Their ability to pipe light is used in surgical instruments and in lighted outdoor signs. Other applications include clear or translucent corrugated sheets for roofing,

films or sheets bonded to wood or metal for exterior finishes, and various molded pieces of hardware.

In addition, acrylics can be dispersed as finely divided particles in a liquid or mastic medium, producing a *latex*. In this form they are used widely for making paints (see Chapter 9) and building sealants (see Chapter 7).

## CELLULOSICS

Cellulosics are based on modifications of the cellulose molecule. Cellulose acetate (CA) and cellulose-acetate-butyrate (CAB) are the most common cellulosics.

Cellulosics are exceedingly tough and withstand a great deal of rough usage. They can be produced clear or translucent, although their optical properties are not as good as those of the acrylics. CA is not suitable for extended outdoor exposure, but CAB can be formulated so as to be quite resistant to weathering.

CA is known primarily for its use in photographic film and recording tape. CAB is used for piping and pipe fittings (primarily for gas and chemicals), as well as for outdoor lighting fixtures. Because of their toughness, both types of cellulosics are also used for a variety of hardware items, for handrails, and for tool handles. Cellulosics also find use as adhesives and coating compounds.

## FLUOROCARBONS

Polytetrafluoroethylene (PTFE), which is commonly known by the trade name Teflon, is the primary fluorocarbon used in the construction industry. It is remarkably inert to chemical attack. In fact, strong chemicals such as boiling aqua regia, chlorine, and bromine do not affect it at at. Furthermore, it remains serviceable over an extensive range of temperatures (from −450°F [−267.8°C] to +500°F [260°C]), which is unmatched by most other plastics. It also has one of the lowest coefficients of friction of any known solid, which means that it has superior antistick properties. However, it is quite difficult to form and is consequently expensive to produce.

PTFE is used in particularly demanding applications such as piping for extremely corrosive chemicals at high temperatures, for low-friction slider pads to permit movement in steam lines, on the surface of bearing pads between beams and columns of steel structures to permit movement, and as nonstick linings for cooking utensils.

## NYLONS

Nylon, which was originally a trade name, is now the common name for a group of plastics known as polyamides. These plastics have good all-around toughness, high strength, and good chemical resistance. They are also resistant to abrasion, but they are not recommended for continuous outdoor exposure.

Nylons are perhaps best known for their use in synthetic fibers and filaments. They produce high-strength fabrics which are durable and tear resistant. These fabrics are used for sails, parachutes, and many articles of clothing. In the construction industry, nylon fabrics, which may be coated or reinforced as necessary, have been used in air-supported structures similar to that shown in Figure 6.13-9. Nylon fibers are also used to make carpets.

Because of its toughness, nylon is also used to produce molded parts such as locks, latches, rollers for sliding windows and drawers, gears, and cams.

## POLYCARBONATES

Polycarbonates are transparent plastics with exceptionally high impact strength, good heat resistance, and good dimensional stability. They are used in break-resistant safety glazing, lighting fixtures, lighted signs, and various pieces of molded hardware.

## POLYETHYLENE

Polyethylene is a well-known plastic made by polymerization of the ethylene molecule. It is strong, light, and flexible, and it retains its flexibility even at low temperatures. It is noted for its good water resistance and low moisture-vapor transmission. It is produced in several different formulations, including linear and branched molecular configurations, which may vary considerably in strength, density, and stiffness.

Polyethylene is used widely for vapor retarders in floors and walls (Fig. 6.13-10), for dampproofing of basement walls, including permanent wood foundation walls, and for sealant backer rods. It is also used for some types of piping (cold water, gas, and chemicals), but it is not suitable for use under severe conditions involving high temperatures or extremely corrosive chemicals. Another common application of polyethylene is for wire and cable insulation.

FIGURE 6.13-9 Air-supported structures are lightweight and portable, making them ideal for such uses as temporary construction enclosures. (Courtesy Thermo-Flex, Inc.)

FIGURE 6.13-11 Polystyrene foam insulation is available in blocks or boards, which can be bonded directly to masonry walls. (Courtesy Dow USA.)

FIGURE 6.13-10 A polyethylene vapor retarder on the interior of a wall assembly reduces vapor condensation inside the wall. (Photograph furnished by Owens-Corning Fiberglas Corporation.)

## POLYPROPYLENE

Polypropylene is similar to polyethylene in many respects—both are classed as polyolefines, for example—but it is much stiffer and more heat-resistant than polyethylene. This makes it suitable for piping and pipe fittings used under more severe conditions, such as for hot-water supply lines and drainage systems. Although polypropylene lacks the flexibility of polyethylene and is more expensive to produce, it is used for some films and sheeting, primarily in the packing industry.

Another common use for polypropylene is in the production of high-strength fibers for carpeting.

## POLYSTYRENE

Polystyrene, made from the styrene monomer, is a transparent, water-resistant, and dimensionally stable plastic. It is not affected by extended low temperatures, but it begins to soften at temperatures near that of boiling water (212°F [100°C]). The main drawbacks of polystyrene are that it is quite brittle and not very weather-resistant, becoming even more brittle and turning yellow with outdoor exposure.

To improve its properties, polystyrene is often copolymerized with other substances. One such copolymer is high-impact polystyrene produced by adding small amounts of butadiene to the styrene. This results in a much tougher plastic with greater impact resistance and greater flexibility. Another well-known copolymer of styrene is ABS, discussed earlier.

Polystyrene is used for lighting fixtures and lighted signs, and also for various molded pieces of hardware. Expanded (foamed) polystyrene, of which Styrofoam® is a well-known example, is used widely in the construction industry for duct and pipe insulation; in insulation for freezers and walk-in refrigerators; and in insulation for walls, floors, and ceilings of residences and commercial buildings (Fig. 6.13-11). Polystyrene foam is also used as the core material in the manufacture of doors and sandwich panels.

## VINYLS

Vinyls are a large group of plastics based on modifications of the ethylene molecule. Included in this group are such plastics as polyvinyl acetate, polyvinyl alcohol, polyvinyl butyral, polyvinyl chloride, polyvinylidene chloride, and many others. These plastics exhibit a wide range of properties. They generally are characterized by good strength and toughness, fair chemical resistance, and a slow rate of water absorption. They perform well at normal-use temperatures as well as at relatively low (freezer) temperatures, but they cannot withstand prolonged exposure to high temperatures. In fact, some vinyls begin to soften at 130°F (54.4°C). Most types are recommended for indoor use, but some are suitable for outdoor exposure.

The most widely used vinyl in the construction industry is polyvinyl chloride (PVC). This is a very versatile plastic with high impact resistance, good dimensional stability, abrasion resistance, and good aging characteristics. It is used to produce a large variety of products such as sheet and tile flooring, gutters and downspouts, moldings, siding, window frames, piping, and drainage systems. In sheet form it is used for facings in sandwich construction, and it is bonded to wood or metal for exterior building finishes, doors, and window frames.

PVC is also available as a foam (either rigid or flexible), and is used for the core material in sandwich construction and for cushions in seating. PVC is copolymerized with many other plastics to produce a variety of adhesives, binders for terrazzo, and other floor toppings.

Other vinyls encountered in construction are polyvinyl butyral, which is used for the interlayer in safety glass and also as a coating for upholstery fabrics; polyvinyl acetate, which is used in adhesives, mortars, and paints; and vinyl coate polyester, which is used in demountable air-supported structures.

### 6.13.3.2 Thermosets

Thermoset plastics are permanently set during forming and cannot be softened again by reheating. In general, they have better heat resistance and greater stiffness than thermoplastics, but are more difficult to form and more brittle. They have a more limited range of applications than do thermoplastics.

### EPOXIES

Epoxies were introduced in the 1950s and were quickly adopted for a wide variety of uses because of their excellent adhesive strength, combined with good chemical and moisture resistance. They will adhere to, and can be used to bond metals, glass, masonry, other plastics, and most other building materials.

Epoxies are used primarily in coating compounds and adhesives. They are often produced as two-component systems, consisting of a resin and a hardener, which must be mixed together just before application. Because they bond well to metals, they are used as protective coatings for appliances, automobiles, and piping. As adhesives, they are used in the construction of sandwich panels; for bonding facings to walls, doors, and other assemblies; and for applying tile finishes. They make excellent high-strength mortars for concrete masonry units and for patching cracked monolithic concrete. Compounds are available for patching deteriorated wood members in existing construction. Epoxy compounds also can be mixed with mineral aggregate or plastic chips to produce terrazzo. Most glass-fiber-reinforced plastic products are made with polyesters, but sometimes epoxies are used instead. Cured, pure, unmodified epoxy is hard and brittle, but when reinforced with glass fibers, it becomes a strong, tough material that can be used in structural and semistructural applications.

### MELAMINES

Melamines are produced by the reaction between melamine and formaldehyde. They are classified as amino plastics, as are ureas. Melamines are hard and scratch-resistant, and can be used at temperatures approaching the boiling point of water. They also stand up well to chemical attack.

The most important application for melamine plastics is in high-pressure laminates, which are used widely for countertops and cabinet finishes. They are also used as adhesives (notably for plywood) and as a protective treatment for fabrics and paper.

As molding compounds, unmodified melamines are somewhat expensive and difficult to work with. Therefore, they are usually mixed with mineral or wood flour fillers. In this form, they are used for molded hardware and for electrical fittings.

### PHENOLICS

Phenolics are produced by the reaction of phenol and formaldehyde. Like many of the thermosets, pure phenolics are quite brittle and hard, and are generally mixed with fillers to improve their strength and impact resistance. Many different formulations with varying properties are available. Phenolics have excellent heat resistance, and glass-fiber-reinforced types will perform well at temperatures of up to 400°F (204.4°C). They also have excellent thermal and electrical insulating properties.

Phenolics are among the more widely used thermosetting plastics. As molding compounds, they are used for electrical parts (sockets, switch boxes, circuit breakers), for a great variety of hardware items, and for appliance cabinets and parts.

Phenolic resins are important in the production of high-pressure laminates. They are used to impregnate the sheets of kraft paper that form the base of most laminates. These resins are also used to stiffen paper honeycomb, which is then bonded to a variety of facings to produce lightweight, strong sandwich panels for walls, doors, and room dividers. Adhesives and protective coatings are also made from phenolic resins.

Another major use for phenolics is as foamed insulation. In this form, they are used as the core for sandwich panels and for insulation around piping and ducts.

### POLYESTERS

Polyester is the generic name for a large group of plastics that are used for a wide variety of purposes. Some examples of different types of polyesters are oil-modified polyesters (sometimes referred to as *alkyds*), used in paints and protective coatings; saturated polyesters, used as plasticizers in other plastic compositions; and saturated polyesters of high molecular weight, used to form fibers and film. The most important polyesters for building purposes are the unsaturated polyesters, which form the basic constituent of reinforced plastics.

Unsaturated polyesters can be cross-linked with a compatible monomer (such as styrene) to form a hard, strong thermosetting material. They have become the most widely used resin for the production of reinforced plastics because of an important combination of properties: they are easy to work with and combine readily with a variety of fibers and reinforcing agents; they cure rapidly at ordinary room temperatures without giving off volatile by-products; and they have excellent dimensional stability combined with high strength and toughness. Furthermore, they can be specially formulated to provide good weathering properties and good heat and flame resistance.

For maximum strength and stiffness, glass fiber is used as the reinforcing agent in polyester compounds. This produces an exceptionally strong, lightweight material that can be used in many structural building applications. Some of the common uses for glass-fiber-reinforced polyesters are prefabricated sections for roof structures; translucent sheets for roofing and interior partitions; window frames and sash; facings for sandwich panels; and many large, molded articles such as bathtubs, shower stalls, sinks, and cabinets. Products such as bathtubs and countertops may be made by molding reinforced polyester within an acrylic shell to give a combination of rigidity, strength, and attractive appearance.

### POLYURETHANES

Polyurethanes, including polyisocyanurate, are a class of plastics known primarily for their use as low-density foams, which can be produced with widely varying properties. These range from soft, flexible open-cell types to tough, rigid closed-cell types. Urethane foams have excellent heat and chemical resistance and good tensile strength. They can be formulated to be fire-resistant.

Rigid urethane foams make excellent insulation material. They have the lowest thermal conductivity of any building material, including mineral wool. They are used widely in construction for wall, floor, and ceiling insulation, for insulation around pipes and ducts, and as the core material in sandwich panels. Flexible urethane foams are used for cushions, upholstery, and padding.

Urethane resins also are used in the production of elastomers (synthetic rubbers), adhesives, sealing and glazing compounds, and no-seam flooring.

### SILICONES

Silicone plastics are based on the silicon atom rather than the carbon atom. They are therefore an exception to the general rule that plastics are organic compounds.

Silicones are very stable compounds, characterized by excellent corrosion resistance and good electrical insulating properties. They can be used over a wide range of temperatures (from $-80°F$ [$-62.2°C$] to $+500°F$ [$+260°C$]). They also stand up extremely well to most weathering conditions.

Silicones, noted for being extremely water repellent, are used as water-resistant coatings on concrete masonry and brick (Fig. 6.13-12), although this use is somewhat controversial. They also make excellent sealants, and are used in that capacity in metal curtain walls, as joint sealers around windows and doors, and in other joints in exterior walls and in wet locations, such as kitchens, bathrooms, showers, and toilets. Small amounts of silicone are frequently added to reinforced plastics to improve the adhesion of the polyester resin to the glass fibers

**FIGURE 6.13-12** Silicone water repellent being applied to a masonry wall. (Courtesy SIKA Corporation.)

and to increase the strength and moisture resistance of the product.

### UREAS

The most common urea plastic, urea-formaldehyde, is produced by the reaction between urea and formaldehyde. Ureas are chemically related to melamines and are classed with them as amino plastics.

Ureas have lower heat and chemical resistance than melamines, but they are more scratch-resistant and have good electrical resistance. They are used for molded hardware items and electrical fittings. Urea resins provide adhesives for use in sandwich panels, and they are also available as insulating foams.

## 6.14 Additional Reading

More information about the subjects discussed in this chapter can be found in the references listed in Section 6.15 and in the following publications.

Allen, Edward. 2004. *Fundamentals of Building Construction: Materials and Methods*, 4th ed. Hoboken NJ: John Wiley & Sons, Inc. (www.wiley.com).

Ambrose, James E. 1997. *Simplified Design of Wood Structures*. Hoboken NJ: John Wiley & Sons, Inc. (www.wiley.com).

Ballash, Evon. 2002. "Developing a Continuous Lateral Load Path for Wood Construction." *The Construction Specifier* (September): 34 (www.csinet.org).

Faherty, Keith F. and Thomas G. Williamson. 1999. *Wood Engineering and Construction Handbook*. New York: McGraw-Hill (www.books.mcgraw-hill.com).

Halperin, Don A. and G. Thomas Bible. 1994. *Principles of Timber Design for Architects and Builders*. Hoboken NJ: John Wiley & Sons, Inc. (www.wiley.com).

Newman, Morton. 1995. *Design and Construction of Wood Framed Buildings*. New York: McGraw-Hill (www.books.mcgraw-hill.com).

Parker, Harry and James Ambrose. 1997 *Simplified Design of Wood Structures*,

5th ed. Hoboken NJ: John Wiley & Sons, Inc. (www.wiley.com).

Williamson, Thomas G. 2002. *APA Engineered Wood Handbook*. New York: McGraw-Hill (www.books.mcgraw-hill.com).

## 6.15 Acknowledgments and References

### ACKNOWLEDGMENTS

We gratefully acknowledge the assistance of the following organizations and individuals in preparing this chapter. We are also indebted to them for permission to use their illustrations when requested and for the use of their publications as references.

Addison Wesley Professional (formerly Addison Wesley Publishing Company (www.awprofessional.com)

AIA and ACSA Council on Architectural Research (www.acsa-arch.org)

American Forest & Paper Association (AF&PA) (www.afandpa.org)

American National Standards Institute (ANSI) (www.ansi.org)

APA—The Engineered Wood Association (APA) (www.apawood.org)

American Wood Council (www.awc.org)

American Wood Preservers Association (AWPA) (www.awpa.com)

American Wood Preservers Institute (AWPI) (www.preservedwood.com)

Architectural Woodwork Institute (AWI) (www.awinet.org)

Armstrong World Industries (www.armstrong.com)

ASTM International (ASTM) (www.astm.org)

California Redwood Association (CRA) (www.calredwood.org)

Dow USA (www.dow.com)

Gypsum Association (GA) (www.gypsum.org)

Hardwood Plywood and Veneer Association (HPVA) (www.hpva.org)

Kawneer Company (www.kawneer.com)

National Association of Home Builders (NAHB), Research Foundation (www.nahbrc.org)

National Hardwood Lumber Association (NHLA) (www.natlhardwood.org)

National Institute of Building Sciences (NIBS) (www.nibs.org)

National Institute of Standards and Technology (NIST) (www.nist.gov)

National Terrazzo and Mosaic Association (NTMA) (www.ntma.com)

Owens Corning Fiberglas Corporation (www.owenscorning.com)

Redwood Inspection Service (RIS), California Redwood Association (CRA) (www.calredwood.org)

Rohm and Haas Company (www.rohmhass.com)

Sika Corporation (www.sikaconstruction.com)

Society of the Plastics Industry (www.socplas.org)

Southeastern Lumber Manufacturers Association (SLMA) (www.slma.org)

Southern Forest Products Association (SFPA) (www.sfpa.org)

Southern Pine Council (SPC) (www.southernpine.com)

Southern Pine Inspection Bureau (SPIB) (www.spib.org)

Thermo-flex, Inc.

United States Gypsum Company (USG) (www.usg.com)

U.S. Department of Agriculture (USDA) (www.usda.gov)

U.S. Department of Housing and Urban Development (HUD) (www.hud.gov)

U.S. Forest Products Laboratory (FPL) (www.fpl.fs.fed.us)

West Coast Lumber Inspection Bureau (WCLIB) (www.wclib.org)

Western Wood Products Association (WWPA) (www.wwpa.org)

Weyerhaeuser Company (www.weyerhaeuser.com)

Window & Door Manufacturers Association (WDMA) (www.nwwda.org)

Wood Truss Council of America (WTCA) (www.woodtruss.com)

Woodwork Institute (WI) (formerly Woodwork Institute of California [WIC]) (www.wicnet.org)

## REFERENCES

We would also like to thank the authors and publishers of the publications in the following list for their contribution to our research for this chapter.

American Institute of Timber Construction (AITC) *Timber Construction Manual*. Englewood, CO: AITC (www.aitc-glulam.org).

———. AITC 104-2003, *Typical Construction Details*.

———. AITC 108-1993, *Standard for Heavy Timber Construction*.

———. AITC 112-1993, *Standard for Tongue-and-Groove Heavy Timber Roof Decking*.

———. AITC 117-2004, *Standard Specifications for Structural Glued Laminated Timber of Softwood Species, Design Requirements*.

———. AITC 119-1996, *Standard Specifications for Structural Glued Laminated Timber of Hardwood Species*.

American National Standards Institute (ANSI). New York (www.ansi.org). Standards:

———. A58.1-1982, "Minimum Design Loads in Buildings and Other Structures."

———. ICC/ANSI A117.1-2003, "Accessible and Usable Buildings and Facilities."

———. A190.1-2002, "Wood Products-Structural Glued Laminated Timber."

———. A194.1-1985, "Cellulosic Fiber Board."

———. A208.1-1999, "Particleboard."

American Wood Council (AWC). *Design Values for Joists and Rafters*. 1992. Washington, DC: AWC (www.awc.org).

———. *National Design Specifications for Wood Construction*. 2005.

———. *Design Values for Wood Construction (NDS Supplement)*. 2005.

———. *Span Tables for Joists and Rafters*. 1993.

———. *Wood Structural Design Data*. 1986.

———. WCD No. 1—*Details for Conventional Wood Frame Construction*. 2001.

———. WCD No. 4—*Plank and Beam Framing for Residential Buildings*. 2003.

———. WCD No. 5.—*Heavy Timber Construction Details*. 1989.

———. WCD No. 6—*Design of Wood Frame Structures for Permanence*. 1988.

———. *Wood Frame Construction Manual for One- and Two-Family Dwellings*. 2001.

———. *Load and Resistance Factor Design Manual for Engineered Wood Construction*. 1996.

———. *Engineered Wood Construction Guide*.

American Wood Preservers Association (AWPA). 2003. *Book of Standards*. Selma, AL: AWPA (www.awpa.com).

American Wood Preservers Institute (AWPI). *Pole Building Design Manual*. Fairfax, VA: AWPI (www.preservedwood.com).

———. *How to Design Pole-Type Buildings Manual*.

APA—The Engineered Wood Association (APA). Tacoma, WA (www.apawood.org).

A310, "Nonresidential Roof Systems."

———. B360, "HDO/MDO Plywood."

———. E30, "Engineered Wood Construction Guide." 2005.

———. E445, "Performance Standards and Qualification Policy for Structural-Use Panels."

———. F405, "Performance Rated Panels."

———. K435, "Sanded Plywood."

———. L335, "Underlayment for Resilient Floor Covering."

———. N335, "Proper Installation of APA Rated Sheathing for Roof Applications." 2004.

———. Q220, "Preservative-Treated Plywood."

———. Q300, "Builder Tips: Steps to Construct a Solid, Squeak-Free Floor System." 2005.

———. S811, "Plywood Design Specifications, Supplement #1."

———. S812, "Plywood Design Specifications, Supplement #2."

———. U813, "Plywood Design Specifications, Supplement #3."

———. U814, "Plywood Design Specifications, Supplement #4."

———. X505, "Panel Handbook and Grade Glossary."

Architectural Woodwork Institute (AWI). *Architectural Woodwork Quality Standards Illustrated*. Reston, VA: AWI (www.awinet.org).

*Architectural Graphic Standards. See* Ramsey et al. (2000).

Arcom Master Systems Salt Lake City, UT. (www.arcomnet.com). *Masterspec®*. Basic Sections:

06100, "Rough Carpentry."

06105, "Miscellaneous Carpentry."

06130, "Heavy Timber Construction."

06150, "Wood Decking."

06160, "Sheathing."

06176, "Metal-plate-connected Wood Trusses."

06185, "Structural Glue Laminated Timber."

06200, "Finish Carpentry."

06401, "Exterior Architectural Woodwork."

06402, "Interior Architectural Woodwork."

06420, "Paneling."

ASTM International (ASTM). West Conshohocken, PA (www.astm.org). Standards:

D245, "Standard Practice for Establishing Structural Grades and Related Allowable Properties for Visually Graded Lumber."

D1600, "Standard Terminology for Abbreviated Terms Relating to Plastics."

D1784, "Standard Specification for Rigid Poly(Vinyl Chloride) (PVC) Compounds and Chlorinated Poly (Vinyl Chloride) (CPVC) Compounds."

D2000, "Standard Classification System for Rubber Products in Automotive Applications."

D2287, "Standard Specification for Nonrigid Vinyl Chloride Polymer and Copolymer Molding and Extrusion Compounds."

D4442, "Standard Test Methods for Direct Moisture Content Measurement of Wood and Wood-Base Materials."

D4444, "Standard Test Methods for Use and Calibration of Hand-Held Moisture Meters."

E84, "Standard Test Method for Surface Burning Characteristics of Building Materials."

E96, "Standard Test Method for Water Vapor Transmission of Materials."

E119, "Standard Test Methods for Fire Tests of Building Construction and Materials."

E136, "Standard Test Method for Behavior of Materials in a Vertical Tube Furnace at 750 deg. C."

Forest Products Laboratory. 1999. *Wood Handbook—Wood as an Engineering Material*. Gen. Tech. Rep. FPL-GTR-113. Madison, WI: U.S. Department of Agriculture, Forest Service, Forest Products Laboratory (www.fpl.fs.fed.us).

———. 1998. *Wood-frame House Construction*. Madison, WI: U.S. Department of Agriculture, Forest Service, Forest Products Laboratory (www.fpl.fs.fed.us).

Gypsum Association. GA-253, "Application of Gypsum Sheathing." Washington, DC: Gypsum Association (www.gypsum.org).

GA-254, "Fire Resistant Gypsum Sheathing."

Hardwood Plywood & Veneer Association (HPVA). 2004. ANSI/HPVA HP-1, *American National Standard for Hardwood and Decorative Plywood*. Reston, VA: HPVA (www.hpva.org).

International Code Council. 2003. Falls Church, VA (www.iccsafe.org). International Codes:

*International Building Code®*.
*International Residential Code®*.

National Institute of Building Sciences (NIBS). 1993. Catalog No. 4010-1, *Wood Protection Guidelines, Protecting Wood From Decay Fungi and Termites*. Washington DC: NIBS (www.nibs.org).

Ramsey, Charles George, Harold Reeve Sleeper, and John Ray Hoke, Jr. 2000. *Architectural Graphic Standards*, 10th ed. Hoboken, NJ: John Wiley & Sons, Inc. (www.wiley.com).

Southern Pine Council (SPC). (www.southernpine.com). 2000. *Permanent Wood Foundations, Design and Construction Guide*. Kenner, LA: SPC.

Southern Pine Inspection Bureau (SPIB). 2002. *Standard Grading Rules for Southern Pine Lumber*. Pensacola, FL: SPIB (www.spib.org).

Truss Plate Institute (TPI). 2002. *National Design Standard for Metal Plate-Connected Wood Truss Construction*. Madison, WI: TPI (www.tpinst.org).

Underground Space Center. 1988. *Building Foundation Design Handbook*. Minneapolis: Underground Space Center.

U.S. Consumer Products Safety Commission. (CSPC). 1997. Document 725, "An Update on Formaldehyde: 1997 Revision." Washington, DC: CSPS (www.csps.gov).

U.S. Department of Agriculture. 1994. HGB064—*Subterranean Termites: Their Prevention and Control in Buildings*. Washington, DC: U.S. Department of Agriculture (www.usda.gov).

U.S. Department of Commerce. Washington, DC (www.commerce.gov).

Voluntary Product Standard PS 1, "Construction and Industrial Plywood." 1995.

———. PS 20, "American Softwood Lumber Standard." 1999.

West Coast Lumber Inspection Bureau (WCLIB). *Standard Grading Rules*. Portland, OR: WCLIB (www.wclib.org).

Western Wood Products Association (WWPA). Portland, OR. (www.wwpa.org).

*Western Lumber Grading Rules*, No. 17.

———. "WWPA grade stamps."

———. A-2, "Specifying Lumber."

———. *Western Woods Use Book*.

Window and Door Manufacturers Association (WDMA). I.S.4-2000, "Industry Standard for Water-Repellent Preservative Non-Pressure Treatment for Millwork." Des Plaines, IL: WDMA (www.nwwda.org).

Woodwork Institute (WI) (formerly Woodwork Institute of California [WIC]). 2003. *Manual of Millwork*. West Sacramento, CA: WI (www.wicnet.org).

# 7

# Thermal and Moisture Protection

# Introduction

A building's shell is required to prohibit weather penetration into the building and to maintain a comfortable environment within. Assault from the exterior comes from temperature variations, wind, precipitation, and groundwater. This chapter discusses methods and materials used in modern construction to help maintain a livable environment within buildings, including the exclusion of water and control of heat migration through building walls and roofs.

# Sustainability Considerations

For a discussion of sustainability, the systems addressed in this chapter can be separated into the materials used and the design of the systems in which those materials are used.

## Materials

The materials covered in this chapter are varied in their effects on the environment and the sustainability of the construction in which they are employed. Considerations regarding the sustainability of the materials addressed in this chapter are much the same as those described in the various chapters related to the various materials. For example, sustainability considerations for concrete shingles and roofing tile are much the same as those addressed in Chapter 3; for metal roofing, flashing, and siding, are similar to those addressed in Chapter 5; for wood shingles, are similar to those addressed in Chapter 6; and so on.

Products made from asphaltic compounds, which are made from coal or petroleum, such as some dampproofing, waterproofing, and roofing materials, help to deplete a nonrenewable resource and are often subject to outgassing that introduces harmful products into the atmosphere. Hot asphalt roofing materials give off harmful fumes during their installation. Some single-ply membranes are also made from petroleum.

Asbestos is no longer permitted in roofing products but is still to be found in older roofing. Extreme environmental and health problems can result when these materials must be removed unless extensive precautions are observed in their removal.

Welding and adhesive bonding of single-ply roofing materials can give off volatile organic compounds (VOCs), and some single-ply membranes themselves emit VOCs.

Sustainability considerations related to thermal insulation materials vary, depending on the material. Foamed insulation products were transitionally made using ozone-depleting blowing agents. Fortunately, most such agents are no longer used. Newer agents have been developed that do not harm the ozone layer. Polystyrene foam insulation products are made from petroleum. Foam insulation products have not been shown to be harmful to installers or building occupants. However, when burned, some plastic foam products emit carbon monoxide gas and other harmful pollutants (see Section 7.3.1.1).

While glass and mineral fiber insulations are not carcinogenic, installers must wear protective masks to protect their lungs. Some fiber products emit small amounts of formaldehyde, which is harmful to humans, but designers can specify one of the available fiber products that do not outgas formaldehyde.

Installers of cellulose thermal insulation also should wear protective masks, but the borate compound used to make cellulose insulation and the insulation itself are not carcinogenic and not harmful to building occupants. In addition, some cellulose products contain as much as 85% recycled materials, and the energy needed to produce a pound of cellulose insulation is only a small fraction of the energy required to produce a pound of glass wool, mineral wool, or foam insulation products.

## Design

The design of a roof can greatly influence the comfort of a building's occupants and reduce the amount of energy needed to heat and cool the building. For example, roof overhangs can be used to reduce the amount of sunlight falling on walls and thereby reduce the cooling load. The color of the roofing can affect heating and cooling loads. Lighter shades reflect sunlight and reduce cooling loads. Darker roofing increases absorption and reduces heating loads. Selection of roof coloring is influenced by location. In southern areas where cooling is the major problem, reducing cooling loads may be paramount, while in northern climates, heating is the major problem. Some roofing manufacturers make dark-colored shingles containing a product that reduces the heat absorption of the shingles.

A properly oriented roof can act as a platform for solar collectors. Properly designed low-slope roofs can be covered with soil and developed as green areas where gardens, grass, shrubs, and even trees can produce a pleasant environment and decrease both the cooling and heating loads on the building.

A roof can be designed to capture and hold rainwater for eventual distribution to a holding pond on the site or into a storm drainage system to control the quantity of water that is introduced into the drainage system at one time.

# Applicable *MasterFormat* Sections

The following *MasterFormat* 2004 Level 2 sections are applicable to this chapter:

07 10 00  Dampproofing and Waterproofing

07 20 00  Thermal Protection
07 25 00  Weather Barriers

# 7.1 Moisture Control

Moisture in its visible liquid state—water—is relatively simple to understand and to cope with in construction. Roofs shed it, earth is graded to divert it, pipes conduct it, and flashing leads it back to the exterior when it penetrates the outer layer of a building. The technology needed to prevent liquid water penetration is available. Unfortunately, many designers do not give adequate consideration to preventing this penetration. No competent architect will say that leaks are acceptable, but many will say privately that they do not permit potential maintenance problems to interfere with their design goals. As a result, more than half of the lawsuits involving architects are related to water leaks.

Water is also present as an invisible gas—water vapor—in the air all around us and in construction assembly cavities such as stud and joist spaces, masonry cores, furred spaces, and chases containing plumbing or heating, ventilating, and air conditioning ducts. As absorbed water, moisture is present in most construction materials and affects their properties, especially dimensional stability. Moisture may also be present as a solid—ice—within the range of temperatures encountered in most parts of the United States.

Moisture is not damaging to construction in its vapor form. However, it becomes dangerous as it condenses (liquefies) or freezes (solidifies). Over time, hidden condensation can cause organic materials to decay, metal products to corrode, and paint coatings to blister. When moisture freezes, it can cause concrete and masonry to crack or spall.

Condensation and freezing may occur as a result of either a temperature drop or the migration of water vapor to areas of lower temperature. Therefore, moisture problems occur (1) in cold climates in heated occupied buildings, which generate substantial amounts of water vapor and (2) in warm, humid climates in cooled buildings.

Buildings built 50 or more years ago were not as well built or sealed against the weather as they are today. This permitted considerable air leakage, dissipating moisture harmlessly through the structure. Economic and comfort considerations require that modern buildings be built tighter and kept warmer in winter and cooler in summer.

In addition, just as dehumidification is an important aspect of summer comfort conditioning, deliberate humidification is an integral part of typical winter heating systems. All these factors create conditions conducive to moisture problems unless provisions to cope with them are included in design and construction.

The properties and behavior of water in its various forms are important considerations in many aspects of building construction, ranging from soil mechanics to roofing. This section addresses the principles of water vapor movement, problems associated with its condensation to visible water, and methods of controlling condensation in buildings. Refer to Section 17.1 for detailed discussion of water vapor properties, relative humidity and its effect on human comfort, and the principles of vapor movement within a building and of water vapor transmission through building materials.

## 7.1.1 WATER VAPOR MOVEMENT

Water vapor becomes a hazard when it condenses (liquefies) inside the structural cavity of a floor, wall, or roof assembly. Here it can reduce the effectiveness of insulating materials and cause organic materials to decay, metals to corrode, and protective coatings to deteriorate. Wall, floor, and ceiling assemblies should be designed and constructed so as to minimize vapor migration into structural cavities.

Vapor moves largely independently of air movement, from areas of higher vapor concentration and temperatures to areas of lower concentration and temperatures. Water vapor establishes a pressure

proportional to the amount of water vapor present within the air mix. Air with more vapor has a higher vapor pressure. Vapor moves through air by *diffusion* from regions of high vapor pressure to regions of lower pressure without relying on air circulation to carry it.

Vapor pressure is commonly expressed in inches of mercury (Hg). At 70°F (21.1°C) and 100% relative humidity, vapor pressure is 0.739 in. Hg (0.091 Pa), and at 0°F ($-17.8$°C) and 100% relative humidity, vapor pressure is only 0.0377 in. Hg (0.128 Pa), or about $1/20$ of the saturated condition.

With a winter condition of 0°F ($-17.8$°C) and 75% relative humidity, an outside vapor pressure of 0.027 in. Hg (0.091 Pa) would exist, while inside a building heated to 70°F (21.1°C) and with 35% relative humidity, vapor pressure would equal 0.259 in. Hg (0.877 Pa). The vapor pressure inside would be nearly 10 times the pressure outside. Like other gases, water vapor moves from an area of high pressure to an area of low pressure until equilibrium is established. During cold weather, the difference in pressure between inside and outside environments causes vapor to move out through every available crack and directly through permeable materials.

## 7.1.2 ADSORPTION AND ABSORPTION

The surfaces of most materials have an affinity for water, and a film of water molecules often is held to the surface by molecular forces of attraction. This phenomenon is known as *adsorption*. Surface films at low humidities may be only one molecule thick; at moderate humidities, multimolecular films may be established. The film thickness, and therefore the amount of water held in equilibrium with the surrounding atmosphere, are roughly proportional to the relative humidity. At humidities close to 100%, the films become so relatively thick that small surface pores become completely filled and larger capillaries may be partially filled. At saturation conditions, all voids in the material are completely filled.

Fibrous materials such as wood, plywood, and fiberboards present very large effective surface areas to water molecules, so that the amount of water held on the surface, and subsequently absorbed into these materials through surface pores, is relatively large even at moderate humidities. These materials are said to be *hygroscopic*. Materials which present relatively small surfaces, such as most metals, are not penetrated by water molecules and take on only small amounts of water as surface film.

### 7.1.2.1 Dimensional Instability

Changes in moisture content can cause significant dimensional changes in construction materials. For example, when wood is dried from an air-dry condition of 12% or 25% moisture content to an oven-dry condition (0% moisture content), it shrinks about 0.1% longitudinally, 2% radially, and 4% tangentially.

The continuous movement of dimensionally unstable materials must be carefully considered in building design and detailing. For example, millwork elements lap one another to permit movement without developing unsightly cracks; air gaps around wood strip floors allow expansion without buckling in humid weather.

### 7.1.2.2 Deterioration of Materials

Moisture can cause or contribute to the breakdown of materials by (1) chemical changes such as the rusting of steel, (2) physical changes such as masonry spalling by frost action, or (3) biological processes such as wood decay. Measures to control liquid water (other than water vapor) include dampproofing and waterproofing (Section 7.2), flashing (Section 7.9), and storm drainage (Sections 21.4 and 23.3).

Since air saturation is more likely to take place at low temperatures when the risk of freezing is also enhanced, condensation and freezing are both major design considerations. Moreover, condensation may occur over a period of time in concealed locations within wall, floor and ceiling, or roof assemblies and not become apparent until a conspicuous failure occurs.

### 7.1.2.3 Condensation Tolerance of Materials

Some materials are able to tolerate condensation with little damage; others are

not. Each material should be assessed individually. The metal panels in refrigerated storage boxes, for example, almost always contain a frost coating with no resulting damage. Some materials can store condensation with little or no effect. Concrete masonry units, for example, can easily have their saturation level raised from 40% to 50% by absorbing condensation without being noticeable. Metal panels absorb little moisture, but can rapidly release collected moisture or frost through evaporation when weather conditions change.

### 7.1.2.4 Effect on Heat Flow

Moisture can have a marked effect on heat transmission through building materials. When present in liquid form, it increases the conductivity of materials by increasing available heat flow paths.

Vegetable fiber products and other hygroscopic insulations, which readily absorb and hold moisture, do not dry out as quickly as nonabsorbent fiber insulations and lose their thermal effectiveness for longer periods. Nonabsorptive products, however, have limited moisture storage capacity and begin "dripping" sooner, possibly damaging interior finishes.

### 7.1.3 TRANSMISSION OF WATER VAPOR

The water vapor permeance of component materials should be considered when designing exterior walls, floors, and roofs of habitable spaces. The amount and rate of water vapor transmitted through building materials are dependent on (1) the *gradient* (differential) in vapor pressure from one side of the material to the other, (2) the material's area, and (3) its *permeance* (ability to permit vapor passage).

### 7.1.3.1 Permeance and Permeability

Water vapor *permeance* is a measure of water vapor flow through a material of specific thickness or an assembly of several materials. The unit of permeance, a *perm*, states the amount of vapor flow in grains per hour, per square foot of surface, per inch of Hg (mercury) (3.386 Pa) vapor pressure gradient. One perm is 57.452 5 nanograms per pascal second square meter (ng/(Pa.s.m$^2$)).

Water vapor *permeability* is the permeance of a 1 in. (25.4 mm) thickness of a homogeneous substance. Permeability is designated in perm-inches and states

the amount of vapor flow through 1 in. (25.4 mm) of material, in grains per hour (ng/(Pa.s.m) per square foot (meter) of surface, per inch of Hg (3.386 Pa) vapor pressure gradient. One perm-inch is 1.459 29 nanograms per pascal second meter (ng/(Pa.s.m)).

Most building materials are permeable to a degree. Some, such as metals and glass, are almost completely impermeable, permitting almost no passage of water vapor. Such materials have a perm rating (permeance) approaching zero. Materials that are either impermeable or have a permeance of 1 perm (57.45 ng/(Pa.s.m$^2$)) or less were formerly referred to as *vapor barriers*. However, in recognition that even the best of these materials permit some vapor transmission, ASHRAE now calls them *vapor retarders*. This book uses this more accurate term. In addition, modern construction methods demand better vapor retarders than those approaching 1 perm (57.45 ng/(Pa.s.m$^2$)). Therefore, vapor retarders are frequently required to have a rating of 0.5 perm (28.73 ng/(Pa.s.m$^2$)) or less. Permeances for representative construction materials are given in Figure 7.1-1.

Many of the permeance numbers in Figure 7.1-1 were obtained by different test methods, which can produce somewhat varying ratings. Therefore, it is not valid to compare the relative performance of those materials with small variations in their permeances. For instance, it is not meaningful to compare hot melt asphalt, 2 oz/sq ft (56.68 g/0.093 m$^2$) (perm: 0.5 [28.73 ng/(Pa.s.m$^2$)]), with asphalt-saturated paper (perm: 0.2–0.3 [11.49–17.24 (ng/(Pa.s.m$^2$)]). Vapor retarder materials rated at under 1 perm (57.45 [ng/(Pa.s.m$^2$)]) can be more readily compared to products with permeances substantially higher than 1 perm (57.45 [ng/(Pa.s.m$^2$)]).

The overall permeance ($M_t$) of an assembly of materials can be calculated from the permeances of the individual components. The calculation is similar to that used in determining the overall coefficient of heat transmittance (U) from individual material conductances (C):

$$M_t = \cfrac{1}{\cfrac{1}{M_1} + \cfrac{1}{M_2} + \cfrac{1}{M_3} + \ldots + \cfrac{1}{M_n}}$$

where $M_t$ = overall total permeance and $M_1, M_2, M_3, \ldots M_n$ = the individual permeances of materials in an assembly.

**FIGURE 7.1-1** Water Vapor Permeance of Building Materials

| Material | Permeance (Perm) |
|---|---|
| **Materials used in construction** | |
| 1 in. concrete (1:2:4 mix) | 3.2 |
| Brick masonry (4 in. thick) | 0.8 |
| Concrete block (8 in. cored, limestone aggregate) | 2.4 |
| Plaster on metal lath ($^3/_4$ in.) | 15.0 |
| Plaster on plain gypsum lath (with studs) | 20.0 |
| Gypsum wallboard ($^3/_8$ in. plain) | 50.0 |
| Gypsum sheathing ($^1/_2$ in. asphalt impregnated) | 10.0 |
| 1-in. structural insulating board (sheathing quality) | 20–50 |
| Structural insulating board (interior, uncoated, $^1/_2$ in.) | 50–90 |
| Hardboard ($^1/_8$ in. standard) | 11.0 |
| Hardboard ($^1/_8$ in. tempered) | 5.0 |
| Built-up roofing (hot-mopped) | 0.0 |
| 1-in. wood sugar pine | 0.4–5.4 |
| Plywood (douglas-fir, interior, blue, $^1/_4$-in. thick) | 1.9 |
| **Thermal insulations, 1 in. thick** | |
| Air (still) | 120.0 |
| Cellular glass | 0.0 |
| Corkboard | 2.1–9.5 |
| Mineral wool (unprotected) | 116.0 |
| Expanded polyurethane (R-11 blown) | 0.4–1.6 |
| Expanded polystyrene, extruded | 1.2 |
| Expanded polystyrene, bead | 2.0–5.8 |
| Unicellular synthetic flexible rubber foam | 0.02–0.15 |
| **Plastic and metal foils and films** | |
| Aluminum foil (1 mil) | 0.0 |
| Aluminum foil (0.35 mil) | 0.05 |
| Polyethylene (2 mil) | 0.16 |
| Polyethylene (4 mil) | 0.08 |
| Polyethylene (6 mil) | 0.06 |
| **Building papers, felts, roofing papers*** | |
| Duplex sheet, asphalt laminated, aluminum foil one side (43) | 0.002 |
| Saturated and coated roll roofing (326) | 0.05 |
| Kraft paper and asphalt laminated, reinforced 30-120-30 (34) | 0.3 |
| Blanket thermal insulation backup paper, asphalt coated (31) | 0.4 |
| Asphalt saturated and coated vapor-barrier paper (43) | 0.2–0.3 |
| Asphalt saturated but not coated sheathing paper (22) | 3.3 |
| 15-lb. asphalt felt (70) | 1.0 |
| 15-lb. tar felt (70) | 4.0 |
| Single-kraft, double infused (16) | 31.0 |
| **Protective coatings** | |
| Paints, 2 coats | |
|     Aluminum varnish on wood | 0.3–0.5 |
|     Enamel on smooth plaster | 0.5–1.5 |
|     Various primers plus 1 coat flat oil paint on plaster | 1.6–3.0 |
|     Water emulsion on interior insulating board | 30.0–85.0 |
| Styrene-butadiene latex coating (2 oz./sq. ft.) | 11.0 |
| Polyvinyl acetate latex coating (4 oz./sq. ft.) | 5.5 |
| Asphalt cut-back mastic ($^1/_{16}$ in. dry) | 0.14 |
| Hot melt asphalt (2 oz./sq. ft.) | 0.5 |

*Numbers in parentheses are weights in lb. per 500 sq. ft.

(Reprinted with permission of the American Society of Heating, Refrigerating and Air-Conditioning Engineers.)

## 7.1.3.2 Water Vapor Transmission

Sometimes it is more significant to report the vapor resistance of materials at a specific pressure difference, rather than at 1 in. Hg (3.386 Pa), as in permeance and permeability. Water vapor transmission (WVT) states the weight of vapor (grains [kg]) transmitted per unit of time (typically 1 hour) per unit of area (one sq ft

[m²]); values may be high or low, depending on the vapor pressure gradient chosen for the test. WVT data can be converted to perms by the following formula:

$$Permeance = WVT\ rating/\,\Delta_p$$

where *WVT rating* equals the weight of vapor transmitted in grains per hour per square foot and $\Delta p$ equals the vapor pressure gradient in the test, in inches (Pa) of mercury.

## 7.1.4 CONDENSATION CAUSES AND PREVENTION

Condensation occurs when vapor-laden air at a high enough humidity contacts cool surfaces. Condensation problems may occur in cooler climates in occupied, heated buildings and in humid, warm climates in air-conditioned buildings. Production of condensation-causing moisture within buildings depends on the types of activities there. Commercial kitchens and shower rooms, for example, can generate enormous quantities of vapor. Four people in a normal home environment generate between 22.5 and 50 lb (10.21 and 22.68 kg) of vapor in 24 hours. In winter, condensation may collect as water or frost on cold surfaces such as glass, metal wall panels, metal window frames, and metal doors and frames.

Although condensation forming on a surface may enter surface pores as fast as it forms and thus become invisible, any condensation on a visible surface is referred to as *surface condensation* to distinguish it from concealed condensation, which occurs within a material or assembly.

### 7.1.4.1 Surface Condensation

Condensation occurs when a surface is colder than the dew point of the nearby air. The temperature of interior surfaces such as walls, ceilings, and floors is dependent on (1) air temperatures both inside and outside the building and (2) heat transmittance of the wall, ceiling, or floor assembly. Ratings of overall heat transmittance for an assembly can be derived by a formula from individual transmittances of the materials in the assembly.

Surface condensation occurs when a surface loses heat fast enough to reach the dew point (condensation temperature). Adequate insulation reduces heat loss and increases surface temperature, thereby reducing the condensation hazard. Therefore, when enough insulation is provided in exterior walls, ceiling and floor assemblies, and roof and ceiling assemblies to keep surfaces that are in contact with moisture at a temperature below the dew point, there will be no surface condensation. Where installing sufficient insulation to accomplish this is not feasible, warm air currents can be induced to increase surface temperature and dissipate moisture. Keeping interior relative humidity within the range recommended in Figure 17.1-13 also will minimize surface condensation.

Figure 7.1-2 shows the relative humidities at which condensation will appear on interior surfaces with various U-values in a room with a 70°F (21.1°C) air temperature. Since condensation can appear at any sufficiently cold spot, the values given in Figure 7.1-2 should be used with caution. U-values constitute an average for large surface areas, within which there may be spots where heat transmittance is higher, such as near structural supports or metal outlet boxes. Also, the inside surface temperature of a stud wall will often be lower at the base due to air stratification inside the stud cavity. As a result, the maximum limit of relative humidity for a nonhomogeneous wall is lower than might be inferred from its average U-value.

Glass areas are normally the most likely surfaces in heated buildings on which condensation might appear; however, an uninsulated wall surface shielded from radiation and convection by furniture may also be quite cold. Over time, surfaces with temperatures near the dew point temperature of the adjacent air may experience swelling, mold, staining, or decay.

Condensation on interior room surfaces can be controlled both by suitable construction and by operating precautions such as (1) reducing the interior dew point temperature or (2) raising the temperatures of interior surfaces above the dew point.

Interior air dew point temperatures can be lowered by removing moisture from the air through either ventilation or dehumidification. Surface temperatures in winter can be maintained sufficiently high by incorporating adequate thermal insulation, by using double glazing, by circulating warm air over the surfaces, or by directly heating the surface.

### BELOW-GRADE CONSTRUCTION

Visible surface condensation may occur in summer on concrete basement walls and floors and on exposed cold-water pipes. Because of their large mass and the insulating effect of the adjacent soil, basement floors and walls tend to maintain a relatively cool and constant temperature. If the dew point temperature within the space rises above the temper-

ature of walls, floors, and pipes, condensation occurs. Finish surfaces in below-ground spaces can be seriously damaged by condensation.

Condensation on interior surfaces can be controlled by (1) reducing the air intake from the outside when outdoor dew point temperatures are high, (2) warming the surfaces, or (3) most successfully, by deliberate dehumidification. For instance, pipes to water closets can be warmed by using lukewarm water for flushing; in climates having low temperatures at night, it may be feasible to ventilate only at night, when the air contains less moisture, and thus reduce the moisture absorption of hygroscopic materials.

### SLAB-ON-GRADE CONSTRUCTION

Like basement walls, slabs on grade tend to maintain a relatively constant temperature. Surface condensation can develop when such a slab is sufficiently cold in relation to a warm interior atmosphere that has a high dew point. In climates where high dew point temperatures frequently occur in summer, condensation may damage finish flooring laid on a slab on grade. Thick rugs or carpets aggravate the problem because they insulate the slab from interior warmer air and retain the slab at a low temperature. Because woven, knitted, and tufted floor coverings are often vapor permeable, moisture may condense under the covering and may promote decay, mold, and a musty odor.

Methods of controlling surface condensation are similar to those for above-grade construction: deliberate dehumidification and increasing surface temperature. A slab on grade can be kept warm by insulating the slab from cooler ground temperatures with rigid insulation at the perimeter and with a granular, well-drained base course under the entire slab.

### 7.1.4.2 Concealed Condensation

In winter, in most parts of the country, the continual production of vapor inside an occupied building normally raises the vapor pressure above that outside the building. This pressure difference provides the force that causes vapor diffusion into exterior walls and ceilings. The vapor pressure rise in the building is directly proportional to the amount of vapor produced and inversely proportional to its ability to escape. In other words, the pressure rises as more vapor is pro-

**FIGURE 7.1-2 Relative humidity and condensation.** (Reprinted with permission of the American Society of Heating, Refrigerating and Air-Conditioning Engineers.)

duced, but the pressure is relieved if it is permitted to diffuse freely to areas of lower concentration.

Exterior walls, floors over unheated spaces, and roof and ceiling assemblies should be protected with vapor retarders and provisions for ventilation. In most areas, because moisture sources are usually within a building, a vapor retarder should be placed near the interior finish to prevent moisture migration into concealed structural spaces. However, a suspended floor above a heated crawl space need not be insulated. Since the ground constitutes the chief source of moisture (Fig. 7.1-3), a vapor retarder should be used as a ground cover.

As water vapor moves through a portion of a structure, such as a wall, floor, or roof and ceiling assembly, it may reach a point within the assembly at which it is cooled to the point of condensation. In older frame buildings with little or no insulation, so much heat was lost through the building's shell that the entire structural cavity often was warm enough to be above the dew point (Fig. 7.1-4a). The dew point temperature was reached somewhere in the sheathing or siding, but condensation was rarely a problem because moisture was carried off by air leakage through joints in the sheathing and siding, which were usually of the board type.

Although the theoretical dew point line in an insulated frame wall generally lies well within the cavity or even within the insulation (Fig. 7.1-4b), the vapor does not condense at this plane but moves on through the vapor-permeable fibrous insulation until it reaches the sheathing. Modern sheathing and siding products frequently are made in broad panels rather than narrow boards, thus reducing the opportunity for moisture diffusion by air leakage through joints. In addition, many siding and sheathing materials, such as plywood and hardboard, are inherently vapor resistant so that only small amounts of vapor can move directly through the material itself. These factors, coupled with the generally higher interior humidity levels maintained by modern heating systems, account for the greater incidence of condensation problems in modern construction if an adequate vapor retarder is not provided.

## WALL CONSTRUCTION

When water vapor is allowed to enter a wall assembly and condensation occurs within its outer cold elements, frost or water may develop, depending on the outside temperature (Fig. 7.1-5). In weather that is continuously cold for a long period, the frost may build back into the wall cavity or fibrous insulation, forming an ice mass. When the outside temperature rises, this frost or ice may melt and be absorbed by hygroscopic materials such as wood, or it may run down nonabsorbing or already-saturated surfaces (Fig. 7.1-6).

FIGURE 7.1-3 Daily evaporation from a crawl space floor. (U.S. Forest Products Laboratory. Not subject to copyright in the United States.)

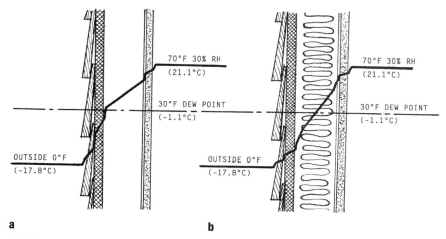

FIGURE 7.1-4 When there is no insulation in a wall cavity (a), so much heat is lost that the temperature in the cavity often remains above the dew point. When insulation is added (b), the dew point will usually occur within the cavity.

FIGURE 7.1-5 Sheathing reversed to show accumulation of frost on an inside surface. This cavity was insulated but did not have a vapor retarder. (U.S. Forest Products Laboratory. Not subject to copyright in the United States.)

**FIGURE 7.1-6** Condensation formed frost and ice behind this siding during cold weather, then weeped through laps in the siding as the weather moderated, staining the siding. (U.S. Forest Products Laboratory. Not subject to copyright in the United States.)

In masonry walls, water seepage to the outside may occur harmlessly when the temperature is above freezing, but water seepage into the building must be prevented. The inclusion of weep holes and base flashings encourages runoff to the outside and bars seepage to the interior (see Sections 4.6 and 4.7).

An accumulation of moisture within insulation lowers its thermal efficiency, causing greater heat loss and condensation problems. Trapped moisture in siding and sheathing is a principal cause of paint failure. In seeking to escape from behind vapor-resistant paint films, moisture can cause paint to blister and peel, especially when it is warmed by solar radiation (Fig. 7.1-7). In extreme cases, moisture trapped in wall cavities will eventually cause decay (Fig. 7.1-8).

The installation of an adequate vapor retarder is the most important moisture control device in exterior wall construction.

## ROOF AND CEILING ASSEMBLIES

The principles of vapor movement and control within a roof and ceiling assembly are much the same as for walls. Water vapor moves from the side with the higher vapor pressure toward the side with the lower vapor pressure. When the side with the lower pressure is also colder, the water vapor may condense in concealed spaces. Some roofing materials, such as wood shingles, can "breathe" and are affected little by trapped vapor; others, such as built-up roofs, are likely to be ruptured by excessive moisture. When a built-up roof blisters and cracks, permitting water to enter the roofing, rapid deterioration is imminent.

In building spaces with high relative humidities, such as laundry, sauna, and pool enclosures, special consideration must be given to properly placed vapor retarders and sufficient insulation.

In unventilated roof and ceiling assemblies, condensation may collect in the form of ice, frost, or water on the framing members, roof deck, and projecting fasteners. Condensation problems seldom exist in roof and ceiling assemblies with insulation above the ceiling, adequate ventilation, and a properly installed vapor retarder on the warm side of the insulation. But roof constructions with rigid insulation above a wood or metal deck may require a vapor retarder or details that permit venting of the insulation. On sunny days, even at low temperatures, frost or ice may melt and seep to the ceiling below, causing staining and possible damage to the ceiling (Fig. 7.1-9). In extreme cases, rapid deterioration and decay of wood framing and rusting of steel roof structure members are also possible.

Adequate vapor retarders on the warm side of insulation and ventilation of roof and ceiling assemblies above the insulation are essential (see "Roof and Ceiling Assemblies" in Section 7.1.5.2).

**FIGURE 7.1-7** Blistering and deterioration of paint surfaces caused by vapor pressure behind a paint film. (U.S. Forest Products Laboratory. Not subject to copyright in the United States.)

**FIGURE 7.1-8** Decay of studs and sheathing due to frequent condensation in the wall cavity. (U.S. Forest Products Laboratory. Not subject to copyright in the United States.)

**FIGURE 7.1-9** Repeated wetting caused this plaster ceiling to crack and produced a visible stain. (U.S. Forest Products Laboratory. Not subject to copyright in the United States.)

## CRAWL SPACE CONSTRUCTION

The presence of water in a crawl space may produce considerable amounts of water vapor. This accumulation of water vapor may increase the possibility of decay and termite attack on wood members and rusting of steel members and result in unpleasant odors within the building. Therefore, it is important to ventilate enclosed foundation spaces adequately to provide for the dissipation of water vapor. The amount and methods of ventilation depend on the type of space being ventilated, the degree of water vapor present, and the type of vapor retarder installed.

Water vapor rising from the ground below a crawl space is usually a greater problem than vapor coming from inhabited spaces above. Moisture in the soil can rise as much as 11 ft (3352.8 mm) above the water table. Depending on the level of the groundwater table, the character of the soil, and drainage around the foundation, large amounts of water may be introduced into the crawl space by evaporation from the ground (Fig. 7.1-10).

Excessive vapor can cause condensation on structural members that have been cooled by heat loss to the outside. Water vapor from a crawl space may also enter walls, move upward within them, and even reach the roof construction when the wall construction permits.

These problems can be prevented by reducing evaporation from the ground with a vapor retarder and providing adequate ventilation.

### 7.1.5 DESIGN AND CONSTRUCTION RECOMMENDATIONS

Moisture condensation in walls, floors, and roof and ceiling assemblies can be controlled (1) by providing a vapor retarder on the interior side of the assembly to limit vapor entrance into the construction and (2) by ventilating structural spaces such as attic areas and crawl spaces to dissipate vapor to the outside. This double line of defense is desirable in all areas where a severe or even moderate condensation hazard exists. However, in wall assemblies, reliance is placed on a retarder, and air movement through the cavity is discouraged to minimize heat transfer. The continental United States can be divided into three condensation zones, based mainly on the isotherms of the −20, 0, and 20°F (−28.9, −17.8, and −6.7°C) winter design temperatures (Fig. 7.1-11).

Humid climates present slightly different condensation problems, because in air-conditioned buildings located in such climates, the flow of water vapor is from outdoors to indoors. This means that vapor retarders should be located on the ex-

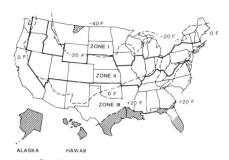

**FIGURE 7.1-11 Condensation zones in the United States.** (Reprinted with permission of the American Society of Heating, Refrigerating and Air-Conditioning Engineers.)

terior side of insulation rather than on the inside. Figure 7.1-12 is a map produced by ASHRAE that shows the location of such areas in the United States.

When construction limitations make it impractical to provide a vapor retarder, as would be the case, for example, in a masonry wall with interior insulation and no applied finish, reliance must be placed on ventilation. Conversely, when adequate ventilation is hard to achieve, vapor-resistant interior construction is essential.

In this discussion, the term *ventilation* is used both in its broad, all-inclusive sense—as any attempt to dissipate moisture by using air as the vehicle—and in a more specific sense. In this narrower meaning, *ventilation* describes typical methods of dissipating moisture from structural spaces by positive air movement through continuous air passages linked to the general atmosphere. In this use, the term *ventilation* is often accom-

**FIGURE 7.1-10** Beads of condensation formed on joists in a crawl space not provided with foundation vents or ground cover. (U.S. Forest Products Laboratory. Not subject to copyright in the United States.)

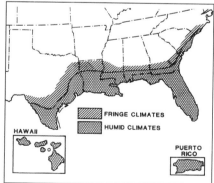

**FIGURE 7.1-12** Humid climates in the continental United States. (Reprinted with permission of the American Society of Heating, Refrigerating and Air-Conditioning Engineers.)

panied by a quantitative recommendation, such as a ratio of vent area to total area. This narrower meaning can be contrasted with another type of ventilation, *venting*, which describes the inclusion of air leaks or gaps in an otherwise vapor-resistant material, finish, or membrane, such as vinyl siding, as a means of vapor diffusion. In venting, air spaces are not necessarily interconnected and air movement is nonexistent or sluggish; ventilation infers much more positive air movement and more rapid moisture dissipation.

### 7.1.5.1 Vapor Retarder Selection

Although a vapor retarder by definition is any material with a permeance of 1 perm (57.45 ng/Pa.s.m$^2$) or less, not all materials are equally suitable for use in a particular assembly. Some materials are inherently better vapor retarders than others; however, the effectiveness of a retarder depends on the method of installation, as well as its physical properties. For instance, 1-mil-(0.0254-mm)-thick aluminum foil has a permeance of 0.00; the permeance of a typical 0.35 mil (0.00889 mm) paper-backed product is likely to be about 0.05 (28.73) when installed in large sheets across the face of studs. The practice of stapling aluminum-faced batts to the sides of wood studs may increase the installed permeance about 20-fold, to as high as 1 perm (57.45 ng/Pa.s.m$^2$).

Vapor retarder selection depends on a variety of factors, such as prevailing interior and exterior humidities, length and severity of the winter season, and materials and methods of construction. In general, the severity of the winter season, as

expressed by the winter design temperature, is a prime factor in determining the degree of vapor protection required (see Fig. 7.1-11).

Figure 7.1-13 summarizes vapor retarder and ventilation recommendations for roof, wall, and floor assemblies in the three U.S. condensation zones. Structural space ventilation is recommended for attic spaces and unheated or cooled crawl spaces in all zones. Vapor retarders are recommended for most assemblies in Zones I and II.

In Zone III, the warm southern part of the United States, average winter temperatures are high enough so that large temperature and vapor pressure differentials between the inside and outside are not developed. Therefore, vapor retarders are often omitted from walls and roof and ceiling assemblies in Zone III; however, a retarder should be installed beneath slabs on grade and in crawl spaces as a general precaution against moisture in this zone, as in Zones I and II. Figure 7.1-13 shows three levels of retarder efficiency: (I) permeance of 0.1 perm (5.75 ng/Pa.s.m$^2$) or less, usually aluminum foils or plastic films; (II) 0.5 perm (28.73 ng/Pa.s.m$^2$) or less, asphalt-coated felts or asphalt-laminated papers; (III) 1.0 perm (57.45 ng/Pa.s.m$^2$) or less, same as above, but with less asphalt per square foot (square meter).

Depending on the type, vapor retarders may be incorporated with, or installed separately from, other materials in the construction. Typical vapor retarders in frame construction include (1) a separate plastic or metallic membrane, job-installed over framing members and under

the interior finish material; (2) a metallic foil, factory-applied on the back of an interior finish material such as gypsum wallboard; (3) aluminum-faced paper or asphalt-laminated paper, factory-applied over batt or blanket insulation; and (4) insulations made of inherently vapor-resistant materials, such as layers of reflective metallic foil or closed-cell plastic foam.

### RETARDER-FACED INSULATION

Retarder-faced insulation is discussed in Section 7.3.1. Where the vapor retarder material on such insulation is integral with the batts or blankets and the flanges are nailed to stud sides, as recommended by most manufacturers, it is almost impossible to avoid air and vapor leaks. Therefore, this installation method should be restricted to areas of moderate condensation hazard, as in Zone II. In Zone I, the vapor retarder flanges should be secured across the stud faces or a separate, continuous vapor retarder should be used.

### FOIL-BACKED BOARDS

Foil-backed gypsum boards are discussed further in Section 9.3. Foil-backed gypsum lath and gypsum wallboard have an aluminum foil retarder laminated to the back of each board (Fig. 7.1-14). However, breaks exist at board joints, which are typically 16 in. (406.4 mm) apart vertically and 48 in. (1219 mm) apart horizontally for lath and 48 in. (1219 mm) apart horizontally and 96 in. (2438 mm) apart vertically for wallboard. These joints permit some vapor leakage (especially where the joint is not backed up by

**FIGURE 7.1-13** Vapor Retarders and Structural Space Ventilation in Different Condensation Zones

| Condensation Zone | Atticless Joist Roof/Ceiling | | Attic-Type Joist Roof/Ceiling | | Crawl Space | | Floor Over Open Unheated Area, V.R., Perms | Exterior Walls, V.R., Perms | Slab on Ground V.R., Perms |
|---|---|---|---|---|---|---|---|---|---|
| | Ceiling V.R., Perms | Structural Ventilation[a] | Ceiling V.R., Perms | Structural Ventilation[a] | Ground V.R., Perms | Structural Ventilation[a,b] | | | |
| Zone I | 0.1 | $1/300$ | 0.5 | $1/300$ | 1.0 | $1/1500$ | 1.0 | 1.0[c] | 3.0 |
| Zone II | 0.5 | $1/300$[d] | 1.0 | $1/300$[d] | 1.0 | $1/1500$ | 1.0 | 1.0[c] | 3.0 |
| Zone III | 1.0 | $1/300$[d] | None | $1/300$[d] | 1.0[e] | $1/1500$[e] | None | None | 3.0 |

[a]Ratio of net free ventilating area (NFA) to total floor or ceiling area.

[b]Ventilation to atmosphere not required when crawl space is heated or cooled and foundation wall is insulated.

[c]Venting of walls may be desirable where highly impermeable exterior finish is used.

[d]Ratio of $1/150$ preferred to relieve high summer solar heat gain.

[e]When crawl space is not heated or cooled, vapor retarder may be omitted provided vent area is increased to $1/150$.

(Reprinted with permission of the American Society of Heating, Refrigerating and Air-Conditioning Engineers.)

**FIGURE 7.1-14** Vapor control can be provided by foil-backed gypsum board. (Courtesy USG Corporation.)

a structural member), and the efficiency of the retarder is reduced.

Foil-backed gypsum lath boards are generally installed horizontally with the long edges perpendicular to the studs. This method permits a great deal of vapor leakage and is not recommended as the sole means of vapor control in either Zone I or II.

Using foil-backed wallboard results in effective vapor isolation if the boards are installed vertically, with all joints over studs. It is suitable for vapor control in exterior walls in all zones.

### SEPARATE SHEETS

The most efficient vapor retarder installation is one in which the vapor retarder is applied separately from the insulation and the interior finish. It is installed in the largest possible sheets and with joints backed up by framing members. Insulation can be provided as pressure-fit blankets or rigid boards. When combined with a suitable insulation, a separate sheet vapor retarder is particularly suitable for severe condensation conditions, as in Zone I.

Polyethylene film and kraft-paper-supported aluminum foil are typical prod-

ucts available in roll widths from 3 to 20 ft (0.914 to 6.1 m). Best results are obtained when joints are made over a structural member. This vaporproofing method is highly recommended for Zone I.

### INSULATING BOARDS

A number of synthetic insulating boards, such as foamed glass, polystyrene, and polyurethane, are being marketed; many of these materials have a closed-cell construction and are inherently good vapor retarders. However, their permeability varies with the manufacturing process and should be checked in each case to determine that the required permeance can be achieved in the thickness contemplated. Boards should also be butted tightly together, and joints should be sealed with a vapor-resistant mastic.

### PAINTS AND OTHER COATINGS

Although many types of paints have proven to be good vapor retarders under laboratory conditions, the uncertainties of field application make reliance on paint coatings somewhat questionable. In new construction, the more reliable factory-produced vapor retarders are rec-

ommended. Paints and other surface coatings should not be used as the sole means of vapor control in Zones I and II.

However, paint coatings can sometimes be used correctively in older structures and as an adjunct to other vapor control methods. Several manufacturers market products for specific use as vapor retarders. Alkyd, aluminum, and bituminous coatings generally have low permeability (high resistance) to vapor. Because the vapor properties of specific products depend on the formulation, the manufacturer should be consulted. Most coatings are more vapor-resistant when their finish appearance is shiny or glossy.

### 7.1.5.2 Vapor Retarder and Ventilation Design

Ventilating openings in exposed locations often are fitted with louvers to keep out wind and rain; screening is usually included to bar insects and vermin. Ventilating requirements are generally given as net free area, which is the space actually available to the moving air after deducting obstructions such as the screening and louvers. Figure 7.1-15 shows how to calculate the required gross area of various ventilating opening types.

### WALLS

Ventilating exterior wall cavities may provide adequate vapor control in a moderate climate. However, wall openings designed to move substantial amounts of air and vapor may waste considerable heat in colder climates. Therefore, in

**FIGURE 7.1-15** Required Gross Area of Vent Openings

| Vent Covering | RGA[a] / NFA[b] |
|---|---|
| Hardware cloth, $\frac{1}{4}$-in. mesh | 1 |
| Screening, $\frac{1}{8}$-in. mesh | $1\frac{1}{4}$ |
| Insect screen, $\frac{1}{16}$-in. mesh | 2 |
| Louvers and hardware cloth $\frac{1}{4}$-in. mesh | 2 |
| Louvers and screening, $\frac{1}{8}$-in. mesh | $2\frac{1}{4}$ |
| Louvers and insect screen, $\frac{1}{16}$-in. mesh | 3 |

[a]Required gross area.

[b]Net free area (which can be determined from Fig. 7.1-13).

(Reprinted with permission of the American Society of Heating, Refrigerating and Air-Conditioning Engineers.)

walls, as in most other assemblies, vapor retarders are the primary means of vapor control; wall ventilation (more typically venting), when provided, is an adjunct to vapor control.

Venting of an exterior surface should be considered when the siding is a vapor-resistant material in sheet or panel form. Recognizing the high vapor resistance of plastic and metal sidings, manufacturers of these products usually include vent holes. No special venting efforts are required with wood and plywood horizontal board sidings, which are self-venting at the overlap. However, plywood and hardboard panel sidings require more effective interior vapor retarders and may require special venting details.

In older buildings that have no vapor retarders, painted siding may resist vapor flow sufficiently to trap moisture within stud spaces. In Zones I and II, the resulting vapor pressure under the paint film may cause blistering and peeling. In these cases, ventilating stud spaces with individual vents at the top and bottom (Fig. 7.1-16) will help dry out the stud cavities. However, if vents become clogged with cobwebs and dust, it will be necessary to provide an interior vapor retarder.

It is essential to form a continuous retarder envelope on the inside of walls that contain any material that would be damaged by moisture or freezing. This includes most thermal insulation, wood and metal studs, and sheathing. This vapor retarder material should have a permeance of 1.0 perm (57.45 ng/(Pa.s.m$^2$)) or less. Relatively permeable materials should be used on the outside of a wall. Sheathing paper, for example, should be a "breath-ing" type, with 6 perms (344.72 ng/(Pa.s.m$^2$)) or more.

Special emphasis is placed on sheathing paper because the Number 15 asphalt-saturated felt typically used for this purpose also is used for built-up roofing and often is too vapor resistant for use in a wall. Figure 7.1-1 shows Number 15 asphalt felt and tar felt as having a permeance of 1.0 and 4.0 perms (57.45 and 229.81 ng/(Pa.s.m$^2$), respectively. These are average figures. The permeance of a particular product can be higher or lower, depending on the degree of saturation with bitumens.

An often used rule of thumb to ensure safe outward vapor diffusion is that the combination of exterior materials should have a total installed permeance of at least five times the total permeance of the interior vapor retarder plus the interior finish materials. Therefore, if the interior retarder-and-finish permeance is 1 perm (57.45 ng/(Pa.s.m$^2$)), the total exterior permeance should be 5 perms (287.26 ng/(Pa.s.m$^2$)) or more. Conversely, when vapor-resistant finishes such as plywood or hardboard panels will be used on the exterior, a more effective retarder and installation method should be used.

In air-conditioned buildings located in the humid areas shown in Figure 7.1-12, water vapor diffuses from the exterior to the interior. Therefore, in these areas, the exterior portions of a wall should have a higher vapor resistance than the interior, and when vapor retarders are used, they should be on the outside. Water vapor is then directed toward the interior rather than the exterior, as is normal in other climate zones, where it can be removed by the air conditioning system. The air conditioning system must be designed with sufficient dehumidification potential to accomplish this removal. Normal systems employed in other areas of the country may not be appropriate. Reheating of circulating air may be necessary. Ventilating air must be dehumidified as well to prevent introducing high outside humidity levels into a building.

## CRAWL SPACES

When a crawl space underlies a heated or air-conditioned area, it is generally more economical to limit heat loss (or gain) by insulating the foundation wall rather than the floor above. Under these conditions, there is no need to ventilate the crawl space to the outside, but the crawl space should be heated or cooled. Introducing nominal amounts of air from the supply side (or conducting air to the return side) of an air-handling unit will usually ensure adequate crawl space ventilation.

In closed crawl spaces not equipped with a means of ventilation, moisture rising by capillary action from the ground can cause condensation problems. To stop the moisture at its source, buildings with either heated or cooled crawl spaces in all zones should be provided with a vapor retarder laid directly over the soil as a ground cover.

A vapor retarder for this purpose should be a material not subject to decay or insect attack. Typical materials are 4 or 6 mil (0.102 or 0.152 mm) polyethylene film, a coated 30 psf (146.5 kg/m$^2$) or heavier roofing felt, or roll roofing. This retarder should be installed in the largest possible widths, with edges lapped 6 in. (152.4 mm) and taped or sealed (unless ballast is used), and the sides turned up approximately 6 in. (152.4 mm) against the foundation walls. A pea gravel or sand ballast over the retarder is desirable to keep it in place and to prevent mechanical damage (see Fig. 21.4-3).

Ventilation to the atmosphere is recommended when a foundation wall is not insulated, heated, or cooled. At least four vents (see Fig. 7.1-16) should be provided, with one not more than 3 ft (9.14 mm) from each corner. They should have a total net free area not less than 1/1500 of the crawl space area. Vents should be placed as high as possible. Vents are not required if the crawl space is completely open on one side to a basement; however, if the crawl space area is greater than the basement, it should be separately ventilated by supplying or returning air.

In older buildings or where the crawl space is not cooled or heated, the ground cover can be omitted, provided that the net free area of vents is increased 10 times, to $^{1}/_{150}$ of the crawl space area. In the absence of a ground cover, more positive ventilation is required and at least four vents should be provided, with at least one in each wall.

In Zone I, crawl spaces are usually heated and ventilation is limited. When a crawl space will be ventilated in the winter, it is necessary to insulate the floor above it to reduce heat loss and improve occupant comfort. During cold weather, the air in the crawl space will have a low temperature and low moisture content, so there will be vapor pressure outward from the building. To prevent condensa-

**FIGURE 7.1-16** Prefabricated circular vent. (Courtesy Easy-up Home Ventilation from The Solar Group.)

tion on the framing, a vapor retarder should be installed on the upper, warm side of the insulation. Normal retarder-faced blankets have nailing flanges on the same side as the retarder. If the insulation is installed from below the joists, special reverse flange blankets or retarder-wrapped blankets should be used (see Section 7.3). Wide polyethylene sheets laid over the subfloor or joists are most effective, but retarder-faced or retarder-wrapped blankets are acceptable.

## ROOF AND CEILING ASSEMBLIES

For purposes of this discussion, roof and ceiling assemblies can be divided into three categories: *insulated framed systems*, *compact conventional systems*, and *protected membrane systems*. Unfortunately, the industry is not in complete agreement about the use of ventilation and vapor retarders in each of these systems.

*Insulated Framed Systems*  In insulated framed systems, an air space separates the ceiling and insulation from the roof deck. Such assemblies may have the roof deck in contact with the framing (Fig. 7.1-17) or may have an attic space above the ceiling framing. Of the three roof and ceiling assembly categories, this one is the most likely to experience condensation problems. Not only are such roofs more susceptible to condensation, but because many are framed in wood, the result is more damaging.

In most areas of the United States, the ceiling in an insulated framed system should have a continuous vapor retarder membrane on the warm side of the insulation. In the humid areas shown in Figure 7.1-12, however, a vapor retarder should be installed only if local experience so dictates. In these areas, condensation may form on the outside of a vapor retarder whose temperature is below the dew point of the outside air.

Ventilating insulated framed systems is particularly important because the area above their ceilings is a natural depository of moisture from inside the building. Convection currents can carry moisture-laden air up wall cavities and into the roof space through openings inadvertently left at the top of partitions or even through the voids in concrete masonry units. Light fixtures and other equipment installed in ceilings can also create air and vapor leaks.

Ideally, such vapor leaks should be eliminated; in practice this is not possible, and a fair amount of moisture will penetrate past the ceiling. Fortunately, suitable ventilation details can dissipate this moisture safely to the atmosphere. In fact, ventilation alone can take care of all such moisture, except in the coldest part of Zone I, with winter design temperature of $-10°F$ ($-12.2°C$) or lower.

In such systems having attic spaces, wind direction and velocity, as well as the vent type, greatly affect air movement distribution and volume within the attic spaces. "Dead" spots of minimum air flow create potential condensation pockets. Therefore, ventilation based only on the ratio of vent opening to ceiling area may be inadequate. Vents should be selected and located for maximum air flow and uniform air distribution; generally, at least one-half of the required vent area should be located at the ridge or upper half of the roof, the rest at the eaves.

The amount of ventilation required is $1/300$ of the ceiling area below when gable end ventilators can be used on both ends of the area. When gable end vents cannot be used, a vent area equal to $1/600$ of the attic area should be uniformly distributed at the eaves, and an additional $1/600$ of the attic area should be provided at the ridge. All attic spaces should be interconnected.

In older buildings that do not have vapor retarders, vent openings should have a net free area not less than $1/150$ of the ceiling area. This larger ventilation volume may also be desirable in moderate and southern zones, regardless of the presence of retarders. The effectiveness of a ventilation system in removing both heat and moisture depends on the volume of air moved and the extent to which the air flow reaches into all parts of the void space.

*Roof louvers* (Fig. 7.1-18a), when used alone, contribute little to effective air flow through an attic space. However, when they are placed in the upper half of the attic space and combined with soffit

FIGURE 7.1-18 Typical roof ventilators: (a) roof louvers, (b) soffit vent, (c) gable end louver, and (d) continuous ridge vent. (a and c courtesy Easy-up Home Ventilation from The Solar Group; b and d courtesy Alcoa Building Products.)

FIGURE 7.1-17 Insulated framed roofing system. (Courtesy H. Leslie Simmons.)

Framing
Deck
Air space
Roofing membrane
Ceiling
Insulation
Vapor retarder

vents (Fig. 7.1-18b) in equal amounts, the air distribution is improved and air flow is almost doubled. The combined system also becomes more responsive to wind, increasing air volume in direct proportion to wind velocity.

*Continuous soffit vents* (see Fig. 7.1-18b) produce substantial air flow regardless of wind direction, and the flow is directly proportionate to wind velocity. However, most air flow remains near the attic floor, and additional vent openings in the upper roof or ridge are needed for good circulation.

*Gable end louvers* (Fig. 7.1-18c) are most effective for wind perpendicular or diagonal to the louver face. However, air flow is one-half to one-third of this amount when the wind direction is parallel to the louver face, perpendicular to the ridge. Although the addition of soffit vents does not improve air flow for the ideal condition, under adverse wind conditions the addition of continuous soffit vents increases air flow by 50%.

*Continuous ridge vents* (Fig. 7.1-18d) are unique in that they provide air flow due to temperature differences with roof slopes as low as 4/12. They provide about the same air flow at zero wind velocity as at 10 mph (16.09 km/hr); however, air flow volume is several times larger with wind parallel to the ridge as it is with the wind perpendicular. Continuous ridge vents combined with continuous soffit vents produce the greatest volume and most uniform air flow under most wind conditions.

Ventilation of insulated framed systems with no attic space (see Fig. 7.1-17) should include outlets from every joist space, either with individual vents (see Fig. 7.1-15) or with continuous soffit ventilation on the sides perpendicular to the joists. Soffit ventilation can be provided either with prefabricated vents (see Fig. 7.1-18b) or job-built vent details (Fig. 7.1-19). A continuous ³/₄-in. (19.1-mm)-wide screened opening is usually adequate for buildings of average width, but the vent free area should be verified as discussed earlier.

***Compact Conventional Systems*** Compact conventional systems may have either metal decks, as shown in Figure 7.1-20, or wood plank decks.

Because built-up roofing and singleply roofing membranes, which are typically used with compact conventional roofing systems, are nearly perfect vapor retarders (0.00 perms), it is impossible for vapor to be diffused through them. A vapor retarder on the underside of insulation beneath such roofing actually can cause more problems than it solves by trapping moisture within the insulation. If a vapor retarder is used in such systems, it should be located so that the calculated dew point will fall on the low vapor pressure side of the retarder. At any rate, the circumstances under which a retarder should be used are controversial.

In the past, it was believed that a vapor retarder was not needed in a conventional compact system when the interior winter relative humidity averaged less than 40%. It was thought that a retarder might be needed when the interior relative humidity exceeded 40% and the average January temperature was 35°F (7.2°C) or lower. Currently, however, vapor drive maps are a better way to determine the need for a vapor retarder.

**FIGURE 7.1-20** Compact conventional roofing system. (Drawing by HLS.)

The 35°F (7.2°C) January isotherm coincides approximately with the 0°F (−17.8°C) isotherm shown in Figure 7.1-11; humidities of over 40% are encountered in public shower rooms, kitchens, laundries, and pool areas. In single-family structures, only pool areas merit special consideration; other high-moisture areas, such as baths and kitchens, are usually equipped with exhaust fans to draw off the moisture.

Unless a vapor retarder is used, it is not necessary to dissipate moisture with structural ventilation or venting. The insulation generally is absorptive enough to hold the vapor for short periods without condensation. The roof deck is generally permeable enough to permit dissipation of vapor back into the interior when the vapor pressure is reversed.

However, if a vapor retarder is provided in cold climates and with high interior relative humidities, the possibility

**a**                    **b**

**FIGURE 7.1-19** Continuous soffit vents can be used to ventilate both attic-type and atticless insulated framed roofs.

**FIGURE 7.1-21** Insulated wood plank deck showing (a) edge venting and (b) stack venting. Insulated steel deck systems are similar. (U.S. Forest Products Laboratory. Not subject to copyright in the United States.)

of trapping moisture between the retarder and the roofing is greatly increased. The moisture may be present during construction, or may enter the insulation later through leaks from above or below. In any case, moisture cannot readily escape through the roofing or the vapor retarder by diffusion, and peripheral air leaks (venting) must be provided to permit dissipation of moisture.

*Edge venting* is a method of connecting roof insulation with the exterior atmosphere through built-in vent openings at a roof's perimeter (Fig. 7.1-21a). Edge venting is adequate for roofs not exceeding 40 ft (12.19 m) in width; for wider roofs, stack venting as well as eave venting is required.

*Stack venting* consists of simple sheet metal stacks about 3 in. (914 mm) in diameter reaching through the roofing membrane into the insulation (Fig. 7.1-21b). One stack should be provided for each 900 sq ft (83.61 m²) of roof area.

Unfortunately, vents can produce more problems than they solve. They can admit water vapor and even free water from drifting snow into the roofing system. In addition, water leaks can occur where the vents are flashed into the roof. Vents may also act as chimneys, carrying moisture-laden air into the roof system from faults in the vapor retarder.

*Protected Membrane Systems* A protected membrane roofing system is a compact system in which the roofing membrane is below at least a portion of the insulation (Fig. 7.1-22). The roofing lies on the warm side and acts itself as a vapor retarder. In most areas of the coun-

try, no condensation is associated with this type of assembly when it has been properly designed and installed. However, in the humid areas shown in Figure 7.1-12, where the indoor vapor pressure may be lower than that outdoors, condensation may occur in such roofing assemblies over air-conditioned buildings. Calculations are required to determine their suitability.

## SLABS ON GRADE

Whether at grade or below grade (basements), slabs on grade are in direct contact with the soil, which often contains large amounts of vapor or free water. Although 4 in. (101.6 mm) of concrete has good resistance to vapor (perm rating of less than 1.0 [57.45 ng/(Pa.s.m²)]), the vapor pressure from the moist slab underside to the interior may be sufficiently high and prolonged enough to cause large

**FIGURE 7.1-22** Protected membrane roofing system. (Courtesy H. Leslie Simmons.)

amounts of vapor to be diffused through the slab. Additional moisture, in vapor or liquid form, can seep through cracks and structural joints in the slab. Therefore, a separate vapor retarder should be placed under every building slab on grade. It may seem that a vapor retarder is unnecessary when a slab on grade is underlain by well-drained granular fill, and is in a region where irrigation and heavy sprinkling are not practiced. Nevertheless, because a retarder can be installed at nominal cost during construction and because it is costly to correct moisture problems later, a retarder should always be placed under slabs of habitable rooms. The vapor retarder should have a permeance of less than 0.30 perms (17.24 ng/Pa.s.m²), should be installed with 6 in. (152.4 mm) laps, and should be fitted carefully around utility and other service openings.

Vapor retarder materials suitable for slab-on-ground construction include single-layer membranes such as polyethylene film and multiple-layer membranes such as reinforced waterproof paper with a polyethylene film extrusion coated onto both sides. Polyethylene film used over gravel or crushed stone, or under a reinforced slab, should be at least 6 mil (0.15 mm) nominal thickness; thinner 4 mil (0.10 mm) film can be used over sand or firmly compacted soil.

### 7.1.5.3 Vapor Retarder Installation

Three fundamental principles should be observed when applying vapor retarders: (1) the retarder should be as near as possible to the warm (heated in winter) side of the insulation or warm face of the assembly; (2) the retarder should be as continuous as possible, with a minimum number of joints, breaks, openings, and air leaks; and (3) the retarder should be installed by a method suitable for the specific condensation hazard.

Because a vapor retarder is frequently an integral part of insulation, effective vapor control generally depends on the proper selection and attachment of insulation. Insulation and retarder should be fitted neatly behind piping and equipment. The flanges of retarder-faced insulating blankets should be snugly stapled about 6 in. (152 mm) on centers to framing members to avoid air leaks.

### WALLS

Gaps in insulation near wall cavity bottoms or tops encourage convection currents, which carry heat and vapor from the inside to the outside face of the cavity. Therefore, insulation should be pushed tightly against the back of sheathing and should fill the cavity from top to bottom.

Batts and blankets are made to fit a standard $14^1/_2$-in. (368.3-mm)-wide cavity resulting from 16 in. (406.4 mm) on-center wood stud or joist spacings. Nonstandard spaces less than $14^1/_2$ in. (368.3 mm) wide should be fitted with blankets cut 1 in. (25.4 mm) oversize to form a stapling flange on the cut size. Nonstandard spaces wider than $14^1/_2$ in. (368.3 mm) can best be protected by a separate retarder (Fig. 7.1-23) or by foil-faced gypsum board.

If the standard spacing of wood members in a structure is 12 or 24 in. (305 or 610 mm), pressure-fit insulation with a separate retarder should be used. Cracks around window and door frames should be stuffed with loose insulation and covered with a separate retarder.

Metal studs do not lend themselves to the use of retarder-faced blankets due to the difficulty of maintaining a continuous retarder. When metal studs are used, retarders should be formed either with foil-faced gypsum board or with separate sheet retarders.

### FLOORS

The perimeter of a floor assembly just behind the band joist in wood framing or at the ends of steel joists is frequently overlooked when insulation and a vapor retarder are installed. This is particularly true when the floor is between two habitable spaces and no insulation is used in the joist cavities.

When a floor is over an open-air area or an unheated crawl space, insulation is likely to be required in the floor. Here, too, care should be taken to provide thermal and vapor control near the perimeter. The vapor retarder should be on the upper (heated in winter) side of the insulation. When the blankets are installed from above, they can be readily stapled to the joists. When installed from below, either reverse flange insulation should be used or the blankets should be placed with both retarder and flanges facing up, using wire supports below.

Retarder-faced blankets are usually not the best way to provide a vapor retarder in a metal-framed (joist) floor system. A continuous sheet retarder is likely to be more successful. When retarder-faced blankets are used in metal-framed floors, they can be supported with wire supports or mesh, but providing a continuous retarder using this type of installation is difficult.

### ROOF AND CEILING ASSEMBLIES

Blankets in wood ceiling joists should be extended partway over the top plate but not enough to block eave ventilation. When blankets are installed from below with recessed stapled flanges, the gap at the end should be stuffed with loose insulation. Insulation and vapor retarder should extend over dropped soffits in bathrooms and kitchens to provide thermal and vapor control. Tightly wedged glass fiber or rock wool insulation above the soffit also isolates the attic area from the soffit space and prevents the spread of fire. Holes for mechanical equipment through top plates should be filled with loose insulation to minimize air and vapor leaks into the roof space.

A separate vapor retarder should be provided when poured insulation is used between ceiling joists.

## 7.2 Dampproofing and Waterproofing

The below-grade walls of buildings, tunnels, and pits, as well as horizontal decks below paving or earth, are subject to ground water penetration. *Dampproofing* and *waterproofing* are used to prevent this penetration.

### 7.2.1 DAMPPROOFING

*Dampproofing* is usually a bituminous coating applied to prevent materials from absorbing adjacent moisture or to prevent absorbed moisture from penetrating further into an invaded system. Bituminous dampproofing is not appropriate when the adjacent water is under hydrostatic pressure. Bituminous dampproofing may be used either on the interior or exterior but is usually concealed in either case. Interior uses include interior surfaces of exterior above-grade walls; interior surfaces of the exterior wythe of exterior cavity walls; and for coating structural steel subject to high-moisture conditions. Bituminous dampproofing is available in either hot- or cold-applied forms.

#### 7.2.1.1 Hot-Applied Bituminous Dampproofing

Hot-applied bituminous dampproofing consists of either an asphalt primer conforming with ASTM Standard D41 and dampproofing asphalt conforming with ASTM Standard D449, Type I, or a coal-tar-based primer conforming with ASTM Standard D43 and coal-tar bitumen conforming with ASTM Standard D450, Type II or Type III.

Hot-applied bituminous dampproofing is used on the exterior of buildings, usually underground, and especially on large projects when the use of a hot system with solvents is permitted. It is seldom used on the interior because of the problem of transporting and using hot

**FIGURE 7.1-23** Polyethylene vapor retarder being installed across an entire wall surface. (Photograph furnished by Owens-Corning Fiberglas Corporation.)

products indoors. It cannot be used when the dampproofing must be able to transmit water vapor (be able to breath). Coal-tar systems are almost never used on the interior of buildings because they have a distinctively unpleasant odor that never completely disappears.

Hot-applied dampproofing may be applied directly on smooth concrete walls or on a coat of parging over masonry.

### 7.2.1.2 Cold-Applied Dampproofing

Cold-applied dampproofing may be either a coal-tar- or asphalt-based system. It may be used either on the interior or the exterior, and is especially useful on small projects where hot products would be difficult or too costly to use. As mentioned earlier, however, coal-tar products are seldom used on the interior of buildings.

Asphalt-based cold-applied bituminous dampproofing systems consist of either cut-back asphalt or asphalt emulsion materials. Both types are available as either liquid, semifibrated, or heavy fibrated types. Fibrated types are reinforced with mineral fibers. Most of them also contain various fillers. Dampproofing that must breathe should be an asphalt emulsion type.

Coal-tar-based cold-applied bituminous dampproofing systems consist of a coal-tar bitumen compound including ASTM Standard D450, Type II coal tar pitch and solvent compound, an inorganic fiber reinforcement, and various fillers.

Cold-applied dampproofing should be applied on exterior wall surfaces over smooth concrete or masonry that has been given a parging coat. Coal-tar products or fibrated asphalt products may be applied over a rougher concrete surface, but it is better to first parge rough concrete surfaces.

Cold-applied dampproofing is sometimes applied over the interior face of exterior above-grade walls or on the interior of the exterior wythe of masonry cavity walls. However, the Brick Industry Association says that mortar coatings are more effective in such locations and cautions against the use of dampproofing there. The problem is that dampproofing applied on an interior surface allows water to enter brick but prohibits its escape, thus contributing to efflorescence and spalling of the masonry when the water in it freezes.

The National Concrete Masonry Association recommends that an impermeable barrier, such as a cut-back asphalt dampproofing, be used on the interior face of concrete masonry unit walls. Under these circumstances, the exterior finish must be completely impermeable. Even oil- or alkyd-based paints may not be used there.

Liquid cold-applied dampproofing materials may be applied by brush or spray; semifibrated materials by brush or spray; and heavy fibrated ones by trowel. In every case, they should be concealed from view.

### 7.2.2 WATERPROOFING

Waterproofing is a membrane, coating, or sealer used in concealed locations to prevent water from entering a building. Waterproofing will work even when the adjacent water is under hydrostatic head. Vertical waterproofing is usually used in conjunction with a foundation drainage system (see Section 21.4). Horizontal waterproofing systems also require a positive means of draining away water that penetrates the wearing surface.

There are three basic types of waterproofing materials: *membrane*, *clay*, and *cementitious*.

### 7.2.2.1 Membrane Waterproofing

On vertical surfaces, a membrane waterproofing system usually consists of a membrane, a protection course, a drainage medium, and a layer of filter fabric (Fig. 7.2-1). Parging may also be required over a masonry substrate.

Several arrangements are possible on horizontal surfaces, but they all have characteristics similar to those of the system shown in Figure 7.2-2. They have a membrane, a protection course, a drainage medium (or space), a layer of filter fabric (when a drainage membrane is used), and a wearing surface. In some systems, a layer of insulation may be required. Sometimes this layer of insulation also serves as the protection course. Some systems require a working slab beneath the wearing surface.

Membranes may be either a fluid-applied, sheet, or bituminous, which is a combination product. Fluid-applied membranes are usually thicker than sheet membranes and more foolproof when properly installed. In addition, it is easier to locate leaks in them. The materials used in fluid-applied membranes tend to

**FIGURE 7.2-1 Vertical membrane waterproofing system.** (Courtesy H. Leslie Simmons.)

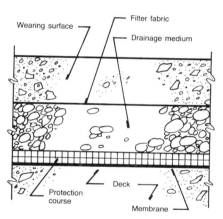

**FIGURE 7.2-2 Horizontal membrane waterproofing system.** (Courtesy H. Leslie Simmons.)

be less expensive than sheet membranes, but they cost more to install because their installation is more difficult. A major advantages of fluid-applied membranes is that they conform easily to irregular surfaces.

Connections between two portions of fluid-applied membrane is often easy using the same material as in the membrane. Separate sheets of the same or compatible material may be bonded at intersections in sheet membranes. Flashing of the edges of fluid-applied and sheet membranes is essentially the same process. Sheet materials are sometimes finished in a reglet, but usually the edges of both should finish beneath a compatible flashing material. Metal flashing is not appropriate when the two materials must be bonded because of the differential rates of thermal expansion between the waterproofing materials and the metal. Metal can be used, however, as a cap flashing where appropriate.

A major disadvantage of fluid-applied membranes is that they require considerable skill to obtain a uniform coating thickness. Another is that these materials are incompatible with some substrate materials.

Most bituminous membranes are built-up courses of felts and bitumen. Some bituminous membranes, however, are totally fluid-applied systems consisting of bitumens reinforced with nylon or glass fiber filaments.

## FLUID-APPLIED MEMBRANES

Most concealed fluid-applied horizontal waterproofing membranes and many concealed vertical waterproofing membranes used today are either polyurethane or hot rubberized asphalt. Fluid-applied silicone and neoprene systems are also available. These have wearing and weathering abilities beyond those of polyurethane or rubberized asphalt systems but are also more expensive. They are, therefore, most often used where they will be exposed, such as on roofs and decks. These systems may, however, also be used as concealed waterproofing where conditions warrant their use.

*Polyurethane* Fluid-applied polyurethane waterproofing is available either as one- or two-part materials. One-part materials are convenient to use and offer more quality control than two-part ones. The proportions of two-part materials can be adjusted by skilled applicators to suit the peculiarities of a particular project, and they tend to cure faster. The in-place performance of one- and two-part systems is comparable.

*Hot Rubberized Asphalt* Rubberized asphalt waterproofing is not as elastic as polyurethane and requires a thicker application. This is desirable over some substrates but not over others. The manufacturer's advice should be sought before selecting this system.

## SHEET MEMBRANES

Sheet membrane materials include butyl synthetic rubber sheet, ethylene propylene diene monomers (EPDM) sheet, premolded bituminous sheet, rubberized asphalt sheet, and self-adhesive butyl sheet.

*Butyl* synthetic rubber sheet is uniform, flexible sheet material about 60 mils (1.52 mm) thick. EPDM sheet is a flexible sheet material that is available in several thicknesses, the minimum being $1/16$ in. (1.6 mm). *Bituminous* sheet waterproofing is a proprietary seven-ply material with a water vapor transmission rating of zero. *Rubberized asphalt* sheet waterproofing is a self-adhering membrane bonded to a polyurethane sheet, with a total thickness of about 56 mils (1.42 mm). *Self-adhering butyl* waterproofing sheet consists of a butyl rubber core bonded to polyethylene plastic to form a sheet about 60 mils (1.52 mm) thick. It requires an adhesive primer.

## BITUMINOUS MEMBRANES

Bituminous waterproofing is usually applied on vertical below-grade surfaces, but in some instances it may be appropriate for horizontal surfaces as well. It is usually either hot coal-tar pitch or hot asphalt. Its number of plies, overall thickness, and content will vary, depending on the hydrostatic head (Fig. 7.2-3). Because it has a proven record of successful use and is highly resistant to deterioration, coal-tar pitch should be used when the melting point can be kept low and there is no incompatibility between that material and adjacent materials. Hot asphalt is necessary when a higher melting point is needed.

Cold asphalt materials are less costly for small applications, and they reduce the danger of air pollution present with hot-applied materials, eliminate the danger of injury due to fire, and produce less objectionable odor. They are not, however, as reliable or foolproof as the hot-applied materials, and they are suitable only when the hydrostatic head is of short duration and never exceeds 3 ft (914 mm). Cold asphalt waterproofing is usually composed of two plies of glass fiber mat

and three coats of emulsion, each applied at a rate of 3 gal/100 sq ft (1.147 L/m$^2$).

Materials used in bituminous waterproofing include:

- Coal-tar-based priming oil conforming with ASTM International (ASTM) D43
- Coal-tar bitumen conforming with ASTM D450
- Number 15 coal-tar felt conforming with ASTM D227
- Cut-back type asphalt primer conforming with ASTM D41
- Waterproofing asphalt conforming with ASTM D449 (generally Type I for vertical surfaces, Type II for unexposed horizontal surfaces, and Type C for exposed surfaces)
- Number 15 asphalt-saturated organic felt conforming with ASTM D226
- Bituminous-saturated organic fabric conforming with ASTM D173
- Glass fiber fabric conforming with ASTM D1668
- Glass fiber mat conforming with ASTM D2178
- Base sheet conforming with ASTM D2626
- Number 5 rosin-sized building paper as a slip sheet
- Either $1/8$-in. (3.2-mm)-thick semirigid asphalt saturated and coated laminations or felt, $1/2$-in. (12.7 mm)-thick fiberboard protection board

The manufacture of bituminous sheets is discussed in Section 7.6-2.

## INSTALLATION

Substrates should be installed and prepared, and fluid-applied, sheet, and bituminous membrane waterproofing materials should be applied in accordance with their manufacturers' recommendations.

Decks to receive waterproofing membranes should be sloped at least $1/2$ in./ft

**FIGURE 7.2-3** Composition of Reinforced Hot Membrane*

| Hydrostatic Head, ft (m) | Number of Fabric Plies | Number of Felt Plies | Bitumen, lb (kg) | Number of Moppings |
|---|---|---|---|---|
| 3 (0.914) | 1 | 1 | 75 (34.02) | 3 |
| 10 (3.05) | 1 | 2 | 100 (45.36) | 4 |
| 25 (7.62) | 2 | 2 | 125 (56.7) | 5 |
| 50 (15.24) | 2 | 3 | 150 (68.04) | 6 |
| 100 (30.48) | 3 | 3 | 175 (79.38) | 7 |

*Possible to use one ply fewer when a coated base sheet is used as an underlayment.

(9.75 mm/m) and preferably $1/4$ in./ft (19.5 mm/m) or more. This slope should be present even after the substrate has deflected. The best surface to receive horizontal waterproofing is a structural concrete slab. Toppings are more likely to move and crack. Precast concrete decks should be covered with a $1^1/2$-in. (38.1-mm)-thick or thicker reinforced topping.

Concrete substrates should be cured for the number of days required by the waterproofing manufacturer, usually at least 14 days. Substrates should be prepared by chipping or grinding away projections that would interfere with or puncture the waterproofing. Substrates should be cleaned of curing compounds, dirt, dust, and other materials that would interfere with membrane placement or bond.

A primer should be applied to substrates when so recommended by the waterproofing material manufacturer. Primed areas should be covered with waterproofing within 24 hours.

Waterproofing materials should be thoroughly bonded to vertical substrates and, where recommended by the manufacturer, to horizontal substrates as well. Seams and edges should be sealed and flashed to prevent water penetration. Expansion joints should be provided at each expansion joint in the substrate, and where the substrate material changes, and in other locations where recommended by the manufacturer.

Except over slabs on grade, horizontal waterproofing should be tested for leaks by covering it with 1 in. (25.4 mm) of water for 24 hours and observing the drop in water level, if any. Repairs should be made before a protection course is installed.

### PROTECTION COURSE

Most applications of membrane waterproofing should be covered with a protecting material to prevent damage to the waterproofing during subsequent construction operations, including backfilling. This *protection course* is usually a layer of insulating material or other board materials. Each waterproofing material requires a different type of protection course. The waterproofing manufacturer should be consulted to determine the correct type.

### 7.2.2.2 Clay Waterproofing

Clay waterproofing is a layer of an expanding material called *bentonite clay*.

When wetted, bentonite expands to between 10 and 15 times its dry volume. It is available loose for mixing with water and spraying in place. It is also available in dry sheets between cardboard, adhered to a drainage board, and adhered to a plastic sheet.

Bentonite is applied below grade to the exterior surfaces of building walls and tunnels. It should not be installed beneath a wearing surface or under earth or any other material that is less than 18 in. (457 mm) thick. A masonry substrate requires a layer of parging before bentonite is installed over it. Cracks wider than $1/8$ in. (3.2 mm) must be filled in all substrates because bentonite cannot bridge wider gaps.

Bentonite is especially useful in tight locations where access is not available to install membrane waterproofing. This is often the case where existing construction occurs.

Bentonite sheets are usually nailed in place or adhered with bitumen. But installation methods vary, depending on hydrostatic pressures, substrate materials, edge conditions, and other factors. Therefore, the manufacturer should be consulted in each instance.

### 7.2.2.3 Cementitious Waterproofing

Some forms of cementitious materials are primarily decorative, even though they may have some waterproofing characteristics. One type that is specifically used as waterproofing is composed of portland cement, aggregate, and an acrylic or plastic admix. These may or may not also contain pulverized iron fillers. The type containing iron is know as *iron oxide waterproofing*. An iron-oxidizing agent in the mix causes the iron to rust quickly and expand to fill the pores in the plaster. These mixes are used either on the exterior or interior of below-grade walls, pits, sumps, and the like.

A second form of cementitious waterproofing is *hydraulic cement*. It is a compound of cement and rapid-setting, nonshrinking hydraulic material. It is used to seal holes, cracks, and open joints, and will set even in free-flowing water.

Nondecorative cementitious waterproofing tends to be brittle and unable to span even hairline cracks. Therefore, its substrate must be smooth and crack-free.

## 7.3 Thermal Insulation

In some climates, using structural or finish materials with high thermal capacities is sufficient to adequately control heat gain or loss through a building's envelope and thereby maintain a healthy and comfortable indoor environment. In most cases, though, a building's structural and finish materials permit excessive heat flow, and insulating products should be added in the roof, walls, and floors to make them more energy efficient.

### 7.3.1 MATERIALS

The most common insulation product types are rigid board, mineral fiber, loose fill, and foamed-in-place. Other materials that provide an insulating function are also sometimes used, although they are not all intended specifically for that function. These include insulated sheathing, fiberboard, reflective insulation, and structural insulating roof deck.

Today manufacturers tend to list their thermal insulation products according to their ability to resist heat flow, which is called *thermal resistance* or *R-value*. Thermal resistance is the reciprocal of *thermal conductivity* (*k-value*). Some manufacturers still list their products according to their k-values, and some even list U-factors. R-values are the preferable listing, however, because they can be added directly to calculate an assembly's overall thermal resistance. The overall R-value is the reciprocal of the assembly's *thermal coefficient* (*U-factor*), which is finally used to calculate the heat loss through the assembly.

The R-values for some insulation products are given in Figure 7.3-1. Note that in most cases, the values listed are averages taken from several competitive products. Values for specific products may vary somewhat from those shown.

Acceptable R-values have increased significantly since the large jump in energy costs that occurred in the 1970s. Figure 7.3-2 shows some typical values. The actual values used for a particular project should be calculated, taking into account such economic factors as insulation costs, increased construction costs that may be necessary to provide more insulation, cost of heating equipment, and building operation cost. In addition, some codes dictate minimum required R-values.

## FIGURE 7.3-1 Resistance (R) Values of Some Insulating Products[a]

| Product | R-Value per 1 in. (25.4 mm) Thickness |
|---|---|
| **Loose fill** | |
| Mineral fiber | 2.20–3.00 |
| Cellulose | 2.80–3.70 |
| Perlite | 2.70 |
| Vermiculite | 2.13 |
| **Batts and blankets** | |
| Mineral fiber batts | 3.00–3.70 |
| Glass fiber batts | 3.00–4.00 |
| **Rigid** | |
| Cellular glass | 2.78 |
| Expanded polystyrene, extruded | 5.00 |
| Expanded polystyrene, molded | 3.60–4.40 |
| Expanded polyurethane | 6.25 |
| Polyisocyanurate | 5.60–7.70 |
| Phenolic | 8.30 |
| Organic fiberboard | 2.50–2.80 |
| Mineral fiberboard | 3.45 |
| Glass fiberboard | 3.80–4.80 |
| **Reflective** | |
| Aluminum foil[b] | 3.48 |
| **Foamed-in-place** | |
| Polyurethane | 6.25 |
| Polyisocyanurate | 7.00–8.00 |
| Phenolic based | 4.50 |

[a]Values are for 75°F (23.9°C) mean temperature.

[b]Thickness of foil not a factor; $3/4$ in. (19.1 mm) air space on room side.

### 7.3.1.1 Rigid Insulation

Rigid insulations are usable in all parts of a building. They are made in board form for use in walls, foundations, and roofs, and in special shapes for use in the cores of concrete masonry units. Most rigid insulations are either polystyrene, polyurethane, polyisocyanurate, perlite, phenolic, cellular glass block, organic fiber, or glass fiber. Various composites made from more than one of the listed materials and other materials are also available.

### POLYSTYRENE

Polystyrene is a preformed nonstructural plastic cellular material. It is made in board form for use over structural roof decks, beneath promenade walking surfaces, at foundation perimeters, and in masonry cavities (Fig. 7.3-3). These boards are also used as wall sheathing when wind bracing is provided by other means.

Polystyrene board products are made in two forms: extruded and molded. *Extruded* polystyrene is a closed-cell material made in five densities ranging from 1.35 to 3 lb/cu ft (21.62 to 48.06 kg/m³). *Molded* polystyrene is an open-cell material made in four densities ranging from 0.9 to 1.8 lb/cu ft (14.42 to 28.83 kg/m³). Therefore, the extruded material is denser and stronger. Its closed-cell structure also means that it is more resistant to mois-

ture absorption and less permeable to water vapor than the molded form. The molded type is commonly called "beadboard" because it is made up of compressed pellets or beads.

Polystyrene insulations have properties similar to those of some other plastic foam insulations but lower R-values. They are sensitive to daylight and deteriorate after prolonged exposure. Like other cellular plastics, they give off carbon monoxide gas while burning. They do not, however, smolder or add significant fuel to a fire.

Polystyrene is also produced in tapered form. When used under a waterproof membrane, the tapered form permits positive drainage of a roof or promenade.

### POLYURETHANE

Urethane foam is a cellular plastic product noted for its excellent resistance to heat transmission and deterioration from chemicals. When formed into rigid boards, its low-density, closed-cell structure produces a lightweight and thermally efficient product. Urethane foam has a resistance per inch (25.4 mm) of about 6.25. Polyurethane boards are made with asphalt-impregnated felt bonded to both side for use in roofing systems.

Urethane will burn and must be protected by some type of finish material if

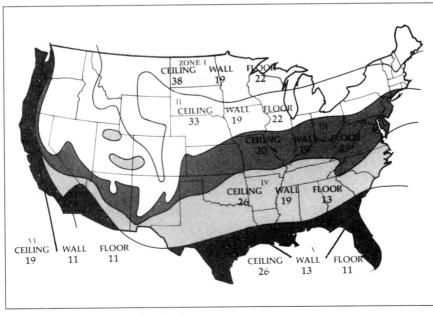

**FIGURE 7.3-2** Target insulation R-values.

**FIGURE 7.3-3** Insulation is inserted in a cavity as a wall is built. (Courtesy Dow USA.)

used for interior applications or on roofs. It may also expand after prolonged exposure to moisture.

## POLYISOCYANURATE

Polyisocyanurate foam (a form of urethane) is manufactured in rigid boards faced with reflective aluminum foil, foilcraft material, or glass fiber felt. It is used for wall sheathing and in roofing systems. It has a high R-value per inch (7.2). Like other urethanes, it must be protected from flame in interior applications.

## PERLITE

Expanded perlite, mineral binders, and waterproofing agents are combined to make boards with weight characteristics similar to those of the plastic insulations. Unlike the cellular plastics, however, perlite boards are inorganic and will not burn or smolder. However, the R-value per inch is only about 2.78, which is considerably lower than that of the plastics. Perlite should be kept dry because of its tendency to absorb water.

## PHENOLIC

Phenolic insulation is a closed-cell foam core material with a density of about 2.5 lb/cu ft (40.05 kg/m$^3$). It is made in board form with a foil-kraft liner or a glass fiber scrim. It can be used in walls and in roofing systems. It has a very high R-value (8.3) and tends to hold that value better than the other foam plastic insulations, but it must be handled carefully because it tends to break easily. It is also reported to corrode metal.

## CELLULAR GLASS BLOCK

Cellular glass is also used to produce a noncombustible, lightweight board insulation. Its closed-cell structure makes it impervious to water vapor and moisture absorption. Its R-value per inch (25.4 mm) is about 2.5, which is comparable to that of expanded perlite insulation.

This product is used on roof decks, exterior decks and promenades, and as perimeter insulation for slabs and foundations. It is available in thicknesses ranging from 1$^1$/$_2$ to 4 in. (38.1 to 101.6 mm) and can be produced in tapered form on special order. When used under a waterproof membrane, the tapered form permits positive drainage of a roof or promenade. Cellular glass must be handled carefully because it breaks easily even though it has high compressive strength. It will deterio-rate severely if exposed to freezing and thawing and should be protected from this by a coating of bituminous compound. In severe conditions, it should be coated on all faces and set with adhesive-filled joints. The manufacturer's advice on this issue should be sought and followed.

## ORGANIC FIBER

Organic fiber rigid insulation is composed of wood fibers or other organic fibers and water-resistant binders. These materials are asphalt impregnated and chemically treated to prevent their deterioration and treated on the top surface to reduce asphalt absorption from roofing. They are used primarily in roofing systems and are available tapered to provide a positive roof slope. They are available in both regular and high-density versions.

## GLASS FIBER

Rigid glass fiber boards are made from inorganic glass fibers bonded together by a resinous binder. They are used primarily in roofing systems. For this use, they are given an integrally bonded, asphalt-impregnated felt or kraft paper top covering. They are available both flat and tapered, and in thicknesses varying from $^1$/$_2$ to 3 in. (12.7 to 76.2 mm). Semirigid glass fiber boards and glass fiber batts and blankets are covered in Section 7.3.1.2.

## COMPOSITES

Composites of urethane or polyisocyanurate and perlite, waferboard, or fiberboard take advantage of the high R-values of the plastic foams and the stability and higher heat resistance of the other materials. Composites are also available with gypsum board and glass fiber boards.

### 7.3.1.2 Mineral Fiber Insulation

Mineral fiber insulation includes products made from fibers manufactured from rock, slag, or glass. Mineral fiber insulations produced in the United States are made from either slag or glass. Products produced from slag are sometimes called *rock wool* or *mineral wool* but are more properly termed *semirefactory fiber insulations* because of their natural resistance to high temperatures. Because glass fiber and semirefractory fiber insulations are inorganic, they are naturally resistant to fire, moisture, and vermin.

Mineral fiber insulation is made in semirigid and rigid boards, as well as in batts and blankets.

## SEMIRIGID BOARDS

Semirigid mineral fiber insulation boards are available 2 by 4 ft (609.6 by 1219 mm) and 1 to 3 in. (25.4 to 76.2 mm) thick. They can be cut to size in the field. They are used in curtain walls, between exterior wall furring members, as horizontal support panels for other insulation, and in other locations where a semirigid product is needed.

## RIGID BOARDS

Rigid mineral fiber insulation boards are used primarily in roofing applications (see Section 7.5).

## BATTS AND BLANKETS

Batts and blankets are similar in appearance and composition but differ in length. Batts are 48 in. (1219 mm) long; blankets are longer and are packaged in large rolls (Fig. 7.3-4). These are used in areas where it is not practical to install loose fill or where the attached foil or kraft paper facing is desired as a vapor retarder. Typical uses include thermal insulation in walls and partitions, below floors in crawl spaces, above ceilings in attics and the like, and as sound insulation.

Batts and blankets are produced in several types: (1) wrapped with kraft paper on the edges and a vapor retarder on one or both sides; (2) faced with a vapor retarder on one side only; and (3) friction-fit without any covering, in which the interlaced fibers have sufficient resilience to remain upright in a cavity. Insulation with kraft paper or foil paper facing should be protected from flame.

Batts and blankets are manufactured to fit between wood studs placed at standard spacings of 16 and 24 in. (400 and

**a**          **b**

**FIGURE 7.3-4** Batt insulation (a) is produced in panels. Blanket insulation (b) is made in rolls that can be cut to size.

600 mm) on centers. Faced and wrapped products are provided with a stapling flange for secure installation. Product thickness ranges from $3\frac{1}{3}$ to 12 in. (86.6 to 304.8 mm). Thickness, however, is not a reliable measure of insulating effectiveness, and should be accompanied by an R-number stamped on the wrapper or facing.

### 7.3.1.3 Loose Fill Insulation

Fibers, granules, or chips are the usual components of loose fill insulations. They are either poured or blown into attics or wall cavities. Most loose fill insulation used in attics and walls is manufactured from mineral wool, glass wool, or cellulosic fibers. Loose perlite and vermiculite are used to fill the voids in concrete masonry units and the spaces between wythes in masonry walls.

Loose fill insulation that is other than free-flowing should not be used in walls or other closed cavities where the large fibers may catch on nails or other obstructions that could leave uninsulated pockets.

### POURING AND BLOWING WOOL

This category includes slag and glass wool products. Both are moisture and vermin resistant. They are less often used in walls than semirigid boards, batts, or blankets because the large clumps of fiber tend to catch on nails or other obstructions, leaving unfilled voids. The most frequent application is between ceiling joists in attics where the wool can be poured from bags or blown by machine (Fig. 7.3-5).

### VERMICULITE AND PERLITE

Vermiculite and perlite are used in areas where they can be readily poured or blown, such as in masonry cavities (Fig. 7.3-6). Vermiculite is manufactured by expanding aluminum magnesium silicate, a form of mica. Perlite is an expanded volcanic glass. They are both noncombustible, vermin and moisture resistant, and do not compress in a cavity.

### CELLULOSIC FIBER

Cellulosic fiber insulations are composed of recycled newsprint, wood chips, and other organic fibers. They can be installed by being either blown or poured into cavities, and can be combined with adhesives for sprayed application.

The fibers are chemically treated to resist fire and to minimize smoke contribution and vermin. However, certain

**FIGURE 7.3-5** Large areas can be quickly insulated with a pressurized hose.

**FIGURE 7.3-6** Vermiculite and perlite can be poured directly from the sack into a masonry cavity.

clearances adjacent to heat-producing fixtures must be maintained. They have a neutral pH to prevent reaction with metal and will not decay after extended exposure to moderate dampness. Cellulosic insulation should be UL listed and labeled to ensure that it is fire resistant.

### 7.3.1.4 Foamed-in-Place Insulation

Most foamed-in-place insulations are either polyurethane-, polyisocyanurate-, or phenolic-based materials. They are created by a chemical reaction that expands a mixture of components by as much as 30 times. When cured, which takes 24 to 72 hours, they solidify into a cellular plastic.

### URETHANE FOAM

Urethane foam has a high resistance per inch (R-6.25 or higher), is resistant to moisture and most chemicals, and has fire-retardant capabilities if chemically treated. It has many applications, including floor and wall cavities, roofs, and structural sandwich panels. The R-value per inch may vary with field conditions at the time of placement.

Urethane foam is particularly effective where a high R-value must be achieved in limited space or on irregularly shaped construction. Care must be taken not to overfill a cavity, because it expands to its ultimate volume after injection.

This type of insulating foam should be used in open cavities or where it is possible to accurately compute the amount of foam needed. Even when fire-retardant treated, it will burn and should be protected from flame by a fire-resistant finish.

## POLYISOCYANURATE FOAM

Polyisocyanurate foam can be used in the same locations as polyurethane foams. It has a higher R-value than polyurethane (7 to 8) and a comparable density.

## PHENOLIC-BASED FOAM

Phenolic-based foams may be used in most of the same locations as polyurethane foams. They have a lower R-value (about 4.5) and a lower density than polyurethane.

### 7.3.1.5 Other Insulation Materials

Some other materials used as insulation are fiberboard, reflective insulation, insulated sheathing (see Section 6.6.5.1), and structural insulating roof decking (see Section 3.4).

## FIBERBOARD

Fiberboard is a general-purpose, low-cost insulation board sometimes used as an interior wall finish in cottages and temporary buildings. However, because of their softness, fiberboard wall panels are seldom used in permanent buildings.

Fiberboard for wall use is available with a flame-resistant, predecorated factory finish. It is manufactured $1/2$ and $3/4$ in. (12.7 and 19.1 mm) thick in tongue-and-grooved plank or panel form. The $1/2$ in. (12.7 mm) thickness can be used when the supporting members are spaced 16 in. (400 mm) on centers; $3/4$ in. (19.1 mm) thickness is recommended when 24 in. (600 mm) spacing is used.

Fiberboard ceiling tile is stapled to furring strips or applied with mastic to a surface that is plane and smooth. It can also be installed with wood or metal hangers to form a suspended ceiling.

## REFLECTIVE INSULATION

Reflective insulation is composed of one or more layers of metallic foil. It is different from other insulations in that its insulating effectiveness depends on how shiny the reflective surface remains over time, whether a $3/4$ in. (19.1 mm) air space is provided on the warm side, and whether it is intended to reduce heat flow into or out of the structure. Therefore, reflective insulation should face an air space of at least $3/4$ in. (19.1 mm) and should remain free of dust or other materials that could reduce its reflective qualities.

### 7.3.1.6 Standards of Quality

Insulating products are manufactured to various standards, including those established by the ASTM (Fig. 7.3-7).

## 7.3.2 DESIGN AND INSTALLATION

Heat flow through ceilings, walls, and floors can be controlled by one or more of the following techniques: (1) by constructing the assembly with materials that are inherently good insulators; (2) by using concrete, stone, solid masonry, or other dense materials that have the capacity to store heat and slow down heat flow; or (3) by adding thermal insulation to the assembly. The first and second methods are discussed in Section 17.2; the third is considered in this section. Whichever method is employed, the material must be correctly installed to achieve the full benefit.

### 7.3.2.1 General Recommendations

The envelopes of most conventionally heated and cooled buildings should be designed to minimize heat flow through floors, walls, and ceilings that surround habitable areas. Except in the high-humidity areas discussed in Section 7.1.3.2, to most effectively block heat flow, insulation is placed near the inside of the building envelope. Insulated sheathing, of course, is an exception.

However, in some passive solar heated-buildings, the insulation is placed on the outside face of the wall, rather than on the inside, as in conventional systems. Such systems are designed to absorb solar heat in masonry or concrete walls during sunshine hours and to release (radiate) this heat slowly into the interior during sunless periods. To prevent rapid cooling of warm masonry during sunless periods, insulation is placed on its outside face.

In exterior insulation and finish systems (see Section 7.4), the insulation is also placed on the exterior of the building.

Some insulation products are made from flammable materials or materials that produce toxic combustion discharges when they are heated or burned. These should be protected by gypsum wallboard or another fire-resistant finish, as most building codes require.

Polyurethane, polyisocyanurate, and polystyrene should not be exposed to daylight to prevent ultraviolet deterioration. Molded polystyrene and expanded perlite boards must also be protected from contact with water to prevent absorption. As mentioned earlier, cellular glass boards must be protected from freezing.

### 7.3.2.2 Slab on Grade

Heat loss related to floor slabs built on ground is more nearly proportional to

**Figure 7.3-7** Summary of Insulating Products

| Material | Standard | Material Covered by the Standard |
|---|---|---|
| **Polystyrene** | ASTM C578 | Rigid cellular polystyrene thermal insulation |
| **Polyisocyanurate** | | |
| Unfaced | ASTM C591 | Unfaced polyisocyanurate thermal insulation board |
| Faced | ASTM C1289 | Faced polyisocyanurate thermal insulation board |
| **Phenolic** | ASTM C1126 | Faced or unfaced rigid cellular thermal insulation |
| **Perlite** | | |
| Loose fill | ASTM C549 | Perlite loose fill thermal insulation |
| Board | ASTM C728 | Perlite thermal insulation board |
| **Vermiculite** | ASTM C516 | Vermiculite loose fill thermal insulation |
| **Cellular glass** | ASTM C552 | Cellular glass thermal insulation board |
| **Organic fiber** | ASTM C208 | Cellulosic fiber thermal insulation board |
| **Mineral fiber** | ASTM C553 | Mineral fiber blanket thermal insulation for commercial and industrial applications |
| | ASTM C612 | Mineral fiber block and board thermal insulation |
| | ASTM C665 | Mineral fiber blanket thermal insulation for light frame construction |
| | ASTM C726 | Mineral fiber roof insulation boards |
| | ASTM C764 | Mineral fiber loose fill thermal insulation |

**FIGURE 7.3-8** Interior locations for perimeter insulation at slabs.

slab perimeter than to slab area. Heat loss to the ground below is minor. Therefore, in areas where heating systems are required, insulation should be installed around the entire slab perimeter to prohibit excessive heat loss directly to areas of low outside air or ground temperature. Uninsulated slab perimeters may be uncomfortably cold, may produce condensation, and may result in uneconomical heating efficiency.

This *perimeter* insulation should be (1) virtually nonabsorptive of moisture (noncapillary), (2) not harmed permanently by wetting or contact with wet concrete, (3) sufficiently rigid to resist damage and compression during placing of concrete, and (4) resistant to decay and insect damage.

Perimeter insulation should be at least 1 in. (25.4 mm) thick and preferably 2 in. (50.8 mm) thick. Even greater thicknesses are common in very cold climates. Insulation should be provided around the entire perimeter of the slab, and should extend either vertically 24 in. (600 mm) below grade level, or below the frost line if it is deeper, or horizontally for a distance of 24 in. (600 mm) under the slab. Alternate methods of perimeter slab insulation are shown in Figure 7.3-8. Insulation may also be placed on the exterior of the foundation wall (Fig. 7.3-9).

Glass fiber insulation board and other plastic foam boards are sometimes used for perimeter insulation, but most perimeter insulation is polystyrene.

### 7.3.2.3 Crawl Spaces and Floors

Heat loss through floors over heated crawl spaces can be reduced by installing insulating material against the foundation

**FIGURE 7.3-9** Exterior location for perimeter insulation.

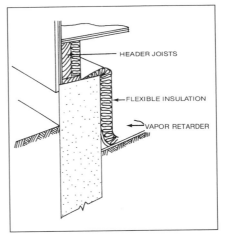

**FIGURE 7.3-10** Insulation location on the perimeter walls of a heated crawl space.

walls (Fig. 7.3-10). If the space is unheated, the insulation should be placed between the floor joists (Fig. 7.3-11). In either case, the vapor retarder should face the living space.

**FIGURE 7.3-11** Insulation location over an unheated crawl space. Insulation may be stapled in place if installed from above or if reverse flange blankets are used. It may be held up by wires, as shown here, when installed from below. Wires can also be used when the framing is metal.

**FIGURE 7.3-12** Sill sealer insulation is placed on top of a foundation wall. Insulation should be fitted snugly against the band joist.

In basements and heated crawl spaces, sill sealer insulation should be used to prevent air infiltration between sill and foundation or between header joists and foundation (Fig. 7.3-12). In addition, the insulating sheathing should extend to the base of the plate.

The space behind the header joist should be covered with short pieces of flexible insulation. If the wall is insulated, as in Figure 7.3-10, the end of the blanket can be extended upward and pushed against the header. Cantilevered floors should be insulated as shown in Figure 7.3-13. In every case, the outside edge of a floor cavity should be insulated with the same insulation value and vapor retarder as the adjacent walls (Fig 7.3-14).

vapor retarder side

**FIGURE 7.3-13** Small cantilevered areas should not be overlooked, but should be carefully insulated comparable with the remaining building.

THERMAL INSULATION

VAPOR RETARDER

JOIST

AIR SPACE

**FIGURE 7.3-14** Detail showing insulation and a vapor retarder at the outside edge of a floor cavity. (U.S. Forest Products Laboratory. Not subject to copyright in the United States.)

### 7.3.2.4 Walls

Heat transmission through wall assemblies can be reduced by selecting materials that are inherently good insulators and by adding insulation. The amount of thermal insulation installed in a wall depends on human comfort and the effect of total heat loss and heat gain and, therefore, on the cost of heating and cooling the building. Estimated savings in both heating and cooling costs may be used as a guide to determine the amount of insulation that can be installed economically. In some mild climates, insulation may be necessary only to raise or lower the inside surface temperature of a wall to increase human comfort. Cold wall surfaces absorb radiant heat from the body and may produce a feeling of chill even though the inside air temperature is 70°F (21.1°C) or above. In most climates, insulation is added to every exterior wall.

Hourly heat loss or gain through an exterior wall is measured by its U-value. The U-value is the amount of heat in British thermal units (Btus) (W) transmitted in 1 hour per square foot (0.93 m²) of wall area, for a difference in temperature of 1°F (17.2°C) between the inside and outside air. Figure 7.3-15 gives U-values of 8 in. (203 mm) masonry walls with various combinations of finish and insulation. The R-value is the reciprocal of the U-value (R = I/U). R-values are useful for estimating the effect of the individual components of a building section on the total heat flow because R-values can be directly added. Figure 7.3-16 lists R-values for single-wythe concrete masonry walls. Figure 7.3-17 gives R- and U-values for some typical concrete masonry walls.

The insulation values recommended in most locations today are often difficult to obtain using a single layer of insulation. For example, the cavity in a 2 × 4 (50 × 100) wood stud wall will only permit $3^1/_2$ in. (88.9 mm) of blanket insulation with an R-value ranging from about 11 to about 17, depending on the insulation material and its density. A similar situation exists when exterior walls are framed with metal studs. In most areas of the country, total recommended wall R-values are 19 or higher. This leaves a gap between the R-value of the insulation alone and the required value for the entire wall. If insulation in the higher R-value range is selected, the other portions of the wall, such as interior and exterior

finishes and sheathing, may produce a sufficient total wall R-value. When such is not the case, one way to increase the insulation thickness in wood- or metal-stud-framed structures is to use 6 in. (150 mm) wall studs or staggered studs, but even this is not always enough. Insulations with higher R-values and other construction methods must be used when high R-values are needed. Batt and blanket insulation will most likely not be the proper selection in such cases.

### RIGID INSULATIONS

When high R-values are required, polyurethane foam insulations may appear to be the proper choice because of their relatively high R-values. However, all plastic foam insulations are good vapor retarders and will prevent moisture that finds its way into a wall from escaping unless certain precautions are taken. In most climates, a continuous polyethylene vapor barrier on the room side of the assembly and venting of the wall cavity are highly recommended. Specific venting methods vary with the manufacturer, but its recommendations should be followed. Refer to Section 7.1.3.2 for a discussion about the placement of vapor retarders and venting in warm, humid climates.

### BATT AND BLANKET INSULATIONS

When doing so will not interfere with installation of finish materials, insulation flanges in wood stud framing may be attached to the stud face to make a continuous vapor retarder. For stud spaces narrower than $14^1/_2$ in. (114 mm), standard barrier-faced blankets may be cut down to form a stapling flange. These flanges should then be stapled to the inside faces of the studs. However, facing the studs with an insulation flange will interfere with the installation of board products such as gypsum board, and this type of application should not be used when such products are to be installed. Batts and blankets in stud spaces should touch the sheathing or siding and fit tightly against the top and bottom plates.

When a stud face must be kept clear, or friction fit insulation or other insulation without a vapor retarder is used, or the stud spacing is less than 16 in. (400 mm), the inside face of the wall should be completely covered with a vapor retarder such as polyethylene film (see Fig. 7.1-23) or aluminum foil. Necessary openings should be cut out later. The re-

**FIGURE 7.3-15** Estimated U-Values for 8 in. (200 mm) Hollow Concrete Masonry Walls

| Wall Details | U-Value Based on Density of Concrete Used in Block | | | | |
| --- | --- | --- | --- | --- | --- |
| | 60 (292.9) pcf (kg/m²) | 80 (390.6) pcf (kg/m²) | 100 (488.2) pcf (kg/m²) | 120 (585.9) pcf (kg/m²) | 140 (683.5) pcf (kg/m²) |
| No insulation | 0.32 | 0.34 | 0.38 | 0.43 | 0.55 |
| No insulation, ½ in. (12.7 mm) gypsum board on furring strips | 0.21 | 0.23 | 0.25 | 0.27 | 0.31 |
| No insulation, ½ in. (12.7 mm) foil-backed gypsum board on furring strips | 0.15 | 0.15 | 0.16 | 0.17 | 0.19 |
| Loose-fill insulation in cores | 0.12 | 0.14 | 0.18 | 0.21 | 0.35 |
| Loose fill in cores, ½ in. (12.7 mm) gypsum board on furring | 0.10 | 0.12 | 0.14 | 0.17 | 0.24 |
| Loose fill in cores, ½ in. (12.7 mm) foil-backed gypsum board, furring | 0.08 | 0.10 | 0.11 | 0.12 | 0.16 |
| 1 in. rigid glass fiber, ½ in. (12.7 mm) gypsum board | 0.14 | 0.14 | 0.15 | 0.15 | 0.17 |
| 1 in. polystyrene, ½ in. (12.7 mm) gypsum board | 0.12 | 0.12 | 0.12 | 0.13 | 0.14 |
| 1 in. polyurethane, ½ in. (12.7 mm) gypsum board | 0.10 | 0.10 | 0.11 | 0.11 | 0.12 |
| Loose fill in cores plus 1 in. (25.4 mm) rigid glass, ½ in. (12.7 mm) gypsum board | 0.08 | 0.09 | 0.10 | 0.11 | 0.14 |
| Loose fill in cores plus 1 in. (25.4 mm) polystyrene, ½ in. (12.7 mm) gypsum board | 0.07 | 0.08 | 0.09 | 0.10 | 0.12 |
| Loose fill in cores plus 1 in. (25.4 mm) polyurethane, ½ in. (12.7 mm) gypsum board | 0.07 | 0.07 | 0.08 | 0.09 | 0.11 |
| R-7 blanket insulation, ½ in. (12.7 mm) gypsum board, furring | 0.09 | 0.10 | 0.10 | 0.10 | 0.11 |

tarder material should be installed in the largest possible sheets to minimize vapor leakage. Retarder joints should be made over framing member.

Cracks around window and door openings should be packed with loose insulation and covered with polyethyl-ene to reduce air infiltration (Fig. 7.3-18). Insulation should also be packed around pipes, ducts, and electrical boxes to prevent "cold spots" in the wall (Fig. 7.3-19). Gaps in the insulation at the top or bottom of the cavity may allow con-vection currents which carry heat and vapor from the inside to the outside of the cavity.

### INSULATING MASONRY WALLS

Masonry walls can be insulated by in-stalling semirigid, batt, or blanket insu-lation between wood or metal furring strips (Fig. 7.3-20) or by applying min-eral fiber or plastic foam board insulation directly to the wall (Fig. 7.3-21). Board insulation that is not inherently fire re-sistant should be protected with gypsum board.

Gypsum board may be adhesively at-tached to a mineral fiber semirigid board but should not be glued directly to plas-tic foam materials. If it is, heat will pass through the gypsum board in a fire and melt the plastic, permitting the gypsum board to fall away, leaving the foam board exposed. Instead, such gypsum board should be attached to furring.

Closed-cell plastic boards are highly resistant to moisture and vermin and can be installed on the outer faces of masonry and frame walls. However, polystyrene and polyisocyanurate board insulation must be painted or otherwise protected from exposure to daylight. All plastic boards should also have some type of

**FIGURE 7.3-16** R-Values of Single-Wythe Concrete Masonry Walls[a]

| Nominal Wall Thickness, in. (mm) | Insulation in Cells[b] | R-Value Based on Concrete Unit Weight | | | | |
| --- | --- | --- | --- | --- | --- | --- |
| | | 60 (292.9) pcf (kg/m²) | 80 (390.6) pcf (kg/m²) | 100 (488.2) pcf (kg/m²) | 120 (585.9) pcf (kg/m²) | 140 (683.5) pcf (kg/m²) |
| 4 (100) | Filled | 3.36 | 2.79 | 2.33 | 1.92 | 1.14 |
| | Empty | 2.07 | 1.68 | 1.40 | 1.17 | 0.77 |
| 6 (150) | Filled | 5.59 | 4.59 | 3.72 | 2.95 | 1.59 |
| | Empty | 2.25 | 1.83 | 1.53 | 1.29 | 0.86 |
| 8 (200) | Filled | 7.46 | 6.06 | 4.85 | 3.79 | 1.98 |
| | Empty | 2.30 | 2.12 | 1.75 | 1.46 | 0.98 |
| 10 (250) | Filled | 9.35 | 7.45 | 5.92 | 4.59 | 2.35 |
| | Empty | 3.00 | 2.40 | 1.97 | 1.63 | 1.08 |
| 12 (300) | Filled | 10.98 | 8.70 | 6.80 | 5.18 | 2.59 |
| | Empty | 3.29 | 2.62 | 2.14 | 1.81 | 1.16 |

[a]R-values do not include the sums of the effect of air film or surface conductance on the inside of the walls ($1/f_i = 0.68$) and on the outside ($1/f_o = 0.17$).
[b]Loose-fill insulation such as perlite, vermiculite, or others of similar density.

## FIGURE 7.3-17 R- and U-Values of Some Typical Concrete Masonry Walls

| Components | Heat-Resistance Values, R | | | | | | | |
|---|---|---|---|---|---|---|---|---|
| | A | B | C | D1 | D2 | E | F1 | F2 |
| Surface film (outside) | 0.17 | 0.17 | 0.17 | 0.17 | 0.17 | 0.17 | 0.17 | 0.17 |
| 8 in. (200 mm) hollow concrete masonry at 100 pcf (1601.9 kg/m³) density (cores open) | 1.75 | | | 1.75 | 1.75 | | | |
| Cores filled with bulk insulation | | 4.85 | | | | | 4.85 | 4.85 |
| Concrete brick at 140 pcf (2242.6 kg/m³) density | | 0.44 | | | | 0.44 | | |
| 2 in. (50 mm) air cavity | | 0.97 | | | | | | |
| 4 in. (100 mm) hollow concrete masonry at 100 pcf (1601.9 kg/m³) density | | 1.40 | | | | 1.40 | | |
| 2 in. (50 mm) bulk insulation in cavity | | | | | | 4.00 | | |
| Batt insulation between furring strips: 1 in. | | | | 3.70 | | | 3.70 | |
| 2 or 2¼ in. | | | | | 7.00 | | | 7.00 |
| ½ in. (12.7 mm) gypsum board interior finish | | | | 0.45 | 0.45 | | 0.45 | 0.45 |
| ⅝ in. (15.9 mm) plaster, lightweight aggregate | | 0.39 | | | | 0.39 | | |
| Surface film (inside) | 0.68 | 0.68 | 0.68 | 0.68 | 0.68 | 0.68 | 0.68 | 0.68 |
| TOTAL resistance, R | 2.60 | 4.05 | 5.70 | 6.75 | 10.05 | 7.08 | 9.85 | 13.15 |
| U-value, 1/R | 0.384 | 0.247 | 0.175 | 0.148 | 0.100 | 0.141 | 0.102 | 0.076 |

A
8 in. (200 mm) hollow concrete block

B
Concrete brick, 4 in. (100 mm) hollow concrete block, ⅝ in. (15.9 mm) plaster

C
8 in. (200 m) concrete block, insulation in hollow cells

D1
8 in. (200 mm) concrete block, gyp. board with 1 in. (25.4 mm) insulation in furring space

D2
2 in. (50 mm) insulation used

E
2 in. (50 mm) bulk insulation in cavity

F1
8 in. (200 mm) concrete block, gyp. & 1 in. (25 mm) insulation and filled cores

F2
2 in. (50 mm) insulation used

protection to prevent dents and chips if used in an exposed area.

In general, the methods used to add insulation to masonry walls include (1) batt or blanket insulation placed in a furring space on the interior face of a wall (Fig. 7.3-22; also see Fig. 7.3-20); (2) rigid or semirigid insulation placed in or applied to the face of a wall (Figs. 7.3-23, 7.3-24, and 7.3-25; also see Fig. 7.3-21); and (3) fill insulation poured or blown in the masonry cavity or core space (Fig. 7.3-26; also see Fig. 7.3-6).

Loose fill insulation used in cavity walls and the cores of hollow masonry should support their own weight without settlement. Granular fills should be pourable in lifts of at least 4 ft (220 mm). In addition, loose fill insulation should be inorganic or have comparable resistance

**FIGURE 7.3-18** Detail showing stuffing of cracks around a window with insulation and covering it with a vapor retarder.

**FIGURE 7.3-19** Insulation should be packed behind ducts, pipes, and electrical boxes, filling the space between them and the exterior finish or sheathing.

**FIGURE 7.3-20** Batts, blankets, or semirigid insulation can be applied between furring strips behind a wall finish. This illustration shows blankets or batts with a nailing flange attached to the face of the furring.

a

b

**FIGURE 7.3-21** (a) Insulation board is bonded to the interior face of masonry using adhesives. (b) Gypsum board is bonded to glass-fiber insulation using adhesives. (Courtesy Dow USA.)

**FIGURE 7.3-22** Friction-fit glass-fiber blanket insulation being installed between wood furring at a masonry wall. (Photograph furnished by Owens-Corning Fiberglas Corporation.)

to decay, fire, and vermin; it should not transmit moisture from exterior to interior wythes; and its insulating efficiency should not be impaired by retained moisture.

Types of loose fill meeting the above criteria are water-repellent, expanded vermiculite and silicone-treated, expanded perlite.

**FIGURE 7.3-23** Rigid foam insulation board installed in a cavity wall. (Courtesy Brick Institute of America.)

**FIGURE 7.3-24** Glass-fiber insulation being installed in a cavity wall. Brick ties hold insulation against the inner masonry wythe, leaving an air space between the outer wythe and the insulation. (Courtesy Brick Institute of America.)

a

b

**FIGURE 7.3-25** (a) Insulation inserted in the face shell of specially formed blocks. (b) Specially formed insulation inserted in cores of concrete masonry units. (Courtesy Concrete Block Insulating Systems, Inc.)

**FIGURE 7.3-26** Granular insulation fills a brick cavity wall. It can also be used to fill the cores in concrete masonry units. (Courtesy Brick Institute of America.)

Rigid boards having the required properties include expanded or molded polystyrene, polyurethane, polyisocyanurate, cellular glass, preformed glass fiber, and perlite board.

### MASONRY WALL CAPACITY INSULATION

In warm climates and in the summer months in cooler climates, the heat gain characteristics of walls are important. Heat gain in buildings with masonry walls depends on the *capacity insulation*, which is the wall's heat storage capacity. The greater the capacity insulation, the lower the heat gain and the cooler the inside temperatures. In general, heavier walls have a greater capacity insulation. They absorb and retain heat during the day and dissipate it to the outside during the cooler night hours. However, the wall must be thick enough so that its entire thickness does not reach a temperature greater than that of the interior space, in which case the wall will radiate heat into the cooler space. The more difference there is between the daytime and nighttime outdoor temperatures, the more efficiently a wall will radiate the heat built up in it in the daytime.

### 7.3.2.5 Ceilings and Roofs

#### CEILINGS

Except in the humid areas discussed in Section 7.1.3.2, flexible insulation with vapor retarders should be installed in ceilings so that the retarder faces the living space. Loose fill or batt or blanket insulation without a vapor retarder should have separate polyethylene film placed below the insulation facing the living space.

For maximum effectiveness against

vapor transmission, the kraft vapor retarder facing on the insulation should overlap the faces of the framing members. If such a procedure is not appropriate because it interferes with tight installation of gypsum board or other finish materials, a separate vapor retarder should be installed.

The insulation should extend entirely across the top plate to prevent heat loss at the joint. If necessary, the junction between the insulation and the plate should be stuffed with loose insulation. When eave vents are used, the insulation should not block the air movement between the vent and the space above the insulation.

Eave baffles or special cardboard ducts should be installed to prevent loose fill insulation from covering the vents and ensure air movement from the eaves to the general attic. Cellulosic insulation should also be kept from contact with potential sources of heat such as electrical fixtures by providing at least a 3 in. (76.2 mm) clearance along each face of the fixture.

### ROOFS

Heat transmission through a roof can be reduced by (1) using optimum amounts of insulation, (2) using shiny or light-colored roof materials to reflect heat, (3) ventilating attic spaces, (4) increasing the thermal mass to achieve a time lag in the heat flow, (5) using louvered rather than solid overhangs to free heated air trapped beneath eaves, and (6) using steeply pitched roofs on the windward side of the building to deflect cold winds and reduce the roof exposure.

All of the listed rigid insulations are used over structural roof decks as a base for roofing, as well as to provide thermal resistance.

Section 7.5.5 contains additional information about roof insulation.

### 7.3.3 REINSULATION OF EXISTING BUILDINGS

Although the heat control concepts discussed here apply to existing construction, it is not always possible to insulate an older structure in the same way and to the same level as a new building. Often it is not economical to add insulation to less accessible parts of a completed building such as walls and floors.

In general, attics with no insulation should be insulated to the same levels as new construction. If the attic already has

6 in. (152 mm) of batt insulation or 8 in. (203 mm) of loose fill insulation, the benefits of additional insulation should be examined on a case-by-case basis. Other parts of the building envelope, such as floors and walls, should be reinsulated to those levels when it is economically practical.

Whenever insulation is added to an existing structure, proper vapor control is essential. This involves installation of a vapor retarder facing the living space, provision for adequate cavity ventilation, or both. Section 7.1 provides a more detailed description of proper techniques.

#### 7.3.3.1 Foundations

Existing foundations of concrete or masonry can be insulated from either inside or outside the wall. Outside application is most frequently used when the interior already has a wall finish that would be covered by the new insulation.

#### INSIDE THE WALLS

Mineral wool batts or blankets or rigid plastic foam boards are the typical materials used on the inside of exterior walls.

Batts or blankets are installed by furring out the wall with wood strips and stapling the insulation to the strips (Fig. 7.3-27). The added wall thickness may not be acceptable for inhabited above-ground rooms, but can easily be accommodated in basements that are not used as primary living space.

Mineral wool has the advantage of being fireproof if a paper backing is not attached. If the batts are paper-faced, they should be protected from flame by wall covering (Fig. 7.3-28).

Rigid plastic boards can be installed on the interior of an exterior wall between traditional $1 \times 2$ ($25 \times 50$) and $2 \times 2$ ($50 \times 50$) wood furring strips with the wall finish attached to the strips. There are also various proprietary systems that use mastics, adhesives, or metal channels in place of furring. Adhesives or mastics are thermally more efficient than the furring system and can be installed at a lower cost by a contractor familiar with their use. However, these systems can only be used if the wall is straight and true, which basement walls often are not. In addition, mastics and adhesives can only be used on clean, dry, unpainted surfaces.

A major advantage of plastic boards is that they often require less space to

**FIGURE 7.3-27** If $2 \times 4$ ($50 \times 100$) studs are used, a wall can be insulated to R-11; $2 \times 3$ ($50 \times 75$) studs will accommodate only R-7 insulation.

**FIGURE 7.3-28** A wall covering should be applied over paper-faced insulation.

achieve a particular R-value. Extruded polystyrene has an aged R-value per inch of 5 or more. Mineral wool and glass fiber boards have an R-value per inch of about 4, batts of about 3. Thus, a required R-value of 19 would demand 5 in. (127 mm) of mineral wool or glass fiber boards and $6^{1}/_{2}$ in. (165 mm) of mineral wool or glass fiber batts, but only 4 in. (101.6 mm) of polystyrene. Some densities of glass fiber boards have R-values per inch of 4.4, which is close to, but still not as large as, that of extruded polystyrene. The real difference occurs when the foam is polyurethane, polyisocyanurate, or closed-cell phenolic. With their R-values of 7 to 8, only $2^{1}/_{2}$ to 3 in. (63.5 to 76.2 mm) are required to reach the R-value of 19 or more in our example. A major disadvantage of foamed plastic board is that it must be covered with gypsum board or another fire-resistant material to be safe from fire.

## OUTSIDE THE WALLS

Only plastic foam boards, such as ure-thane, polyisocyanurate, and extruded polystyrene, are suitable for installation on the outside of walls because of their resistance to moisture. The mastic or adhesive normally used to secure the board to the masonry requires a sound, unpainted masonry, stucco, or parged surface for a good bond. To be most effective, the insulation should extend below the frost line, but not less than 2 ft (600 mm) below ground.

Plastic foamboards are soft materials that can be dented and chipped if not protected by some harder material (Fig. 7.3-29). Several commercial systems available involve bonding a layer of glass fiber reinforcement to the insulation and applying a stucco-like finish. Polystyrene and polyisocyanurate are also sensitive to daylight and must be protected to prevent deterioration.

Installation of insulation on the exterior side of an existing wall is often less economical than installation on the indoor side of the same wall because of higher installation and material costs. However, it has the advantage that installation can proceed without disrupting interior spaces or covering up interior finishes.

### 7.3.3.2 Crawl Spaces

Accessible crawl spaces can be insulated in the same way as new construction. Special attention should be given to the area between the floor joists and behind the header joist—an often ignored or underinsulated spot (Fig. 7.3-30). If the crawl space is not accessible, it can be in-

FIGURE 7.3-29 Plastic foamboard should be protected at the top edge with metal flashing and on its face with a hard waterproof covering.

FIGURE 7.3-30 Strips of insulation should be stuffed behind header joists.

sulated on the outside of the wall, as described for basements.

### 7.3.3.3 Walls

Three basic types of walls—conventional stud, brick veneer, and masonry cavity—are most economically insulated at the time of construction. They can be effectively insulated later only if no insulation was installed initially and if there are no major mechanical or electrical obstructions in the wall.

### STUD WALLS

The thermal efficiency of stud walls can be improved by adding insulation to the stud cavity or to the exterior wall face before adding new siding.

Stud cavities are typically reinsulated with either blown mineral wool, cellulose, or foamed-in-place plastic foam.

Before insulation is added, a wall should be checked for existing insulation by removing the faceplate of a wall receptacle. If no insulation is present, the cavity should be probed with a plumb bob or other weight to detect obstructions that might cause blockages. The insulation should be blown or squirted into the wall through holes drilled between the studs (Fig. 7.3-31). Exterior installation is usually preferable, because most often siding can more readily be removed and replaced than can parts of the interior finish.

Exterior residing can provide an opportunity for adding plastic foam sheathing board under the new finish. The simple procedure requires only that the insulation be attached to the existing siding before new siding is applied (Fig. 7.3-32). Special metal channels accommodate the extra wall thickness around

a

b

c

FIGURE 7.3-31 Installing insulation in an existing wall is a three-step process: (a) open the wall cavity, (b) add insulation, and (c) close the wall cavity.

door and window frames. Although this procedure is not as thermally effective as fully insulating the stud cavity, the addition of 1 in. (25.4 mm) rigid polystyrene board, for example, can add 3.57 to the total R-value of the wall.

### BRICK VENEER

Insulation can be added to a brick veneer wall by blowing or foaming. It is often

**FIGURE 7.3-32** Foam insulation board completely covers the existing walls, thus reducing air infiltration as well as thermal transmission.

difficult, however, to gain access to a stud cavity unless the fascia board or a few bricks near the eaves can be removed to expose an opening into the cavity.

### MASONRY CAVITY

Masonry cavity walls are rarely reinsulated because of the difficulty of gaining access to the cavity. However, it may be possible to remove several bricks or a fascia board and squirt the insulation into the cavity. In addition to plastic foams and blown mineral wool, vermiculite and perlite can be used for such applications.

### 7.3.3.4 Attics

Attics are often the most economical area to reinsulate because they are normally accessible and account for a substantial part of heat loss or gain. Where the insulation is placed usually depends on whether the attic has been left unfinished or whether finish materials have been added to floors, walls, and ceilings.

### UNFINISHED ATTICS

When the attic is unfloored, loose fill insulation can be blown into the joist cavities or mineral wool batts can be placed between the joists in the same way that new attics are insulated. If the attic already has some insulation, additional loose fill can be blown or poured over the old, or another layer of unfaced batts can be added. If the old insulation was wrapped or if the new insulation is faced, the kraft paper or aluminum foil facing on the top face should be punctured at 12 in. (305 mm) intervals to prevent moisture condensation within the insulation. Condensed moisture could reduce insulating effectiveness and cause structural member decay.

If the attic is floored, it is often possible to remove several boards and blow insulation under the remaining floor before replacing the boards (Fig. 7.3-33). If the space is not used for storage, unfaced batts or blankets can be laid directly on the floor (Fig. 7.3-34). If there is a possibility that the space will be finished and heated in the future, it may be preferable to attach the insulation to the sloping rafters.

### FINISHED ATTICS

A finished attic usually has three parts that may need insulation: the upper ceiling, sloping ceiling, and outer attic (Fig. 7.3-35).

Upper ceilings can be insulated like an unfloored attic with either blown loose fill insulation or flexible batts or blankets if the space is accessible.

Proprietary systems are available that provide insulation and ceiling finish in one integral panel. In such systems, mineral wool batts or blankets are attached to the back side of fiberboard ceiling panels and are fastened to the ceiling with

**FIGURE 7.3-33** Cavities should not be overfilled to the extent that the insulation is compacted when the floor boards are reinstalled.

**FIGURE 7.3-34** Insulation over an existing floor should be fitted snugly together to prevent thermal leaks.

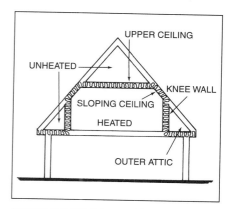

**FIGURE 7.3-35** A layer of insulation should be provided between heated or air-conditioned spaces and unheated or unconditioned spaces.

metal grids. They typically require 3 in. (76.2 mm) of ceiling clearance and provide an R-value of about 12.

Sloping ceilings can be insulated if either the outer attic or upper attic is accessible to permit stuffing the rafter cavities with loose fill or batts or blankets. It may also be possible to install the thermal ceiling panels described above. Such panels can be applied to the face of existing sloping or cathedral ceilings if the panel thickness does not reduce room dimensions below acceptable levels.

Outer attics can be insulated either at the knee wall and outer attic floor or in the sloping rafters. It is generally preferable to insulate the knee wall and outer attic floor when there are eave vents, or eave vents can be installed, and if the outer attic is sealed from the living space and is not to be used for storage.

## 7.4 Exterior Insulation and Finish Systems

Exterior insulation and finish systems (EIFS) are often called *synthetic stucco* because they look like stucco. Most of them consist of a layer of a reinforced finish (lamina) over a layer of insulation, which is fastened to a layer of sheathing applied to furring or framing. Other configurations are possible, however, and many have been used since EIFS were introduced into this country from Europe in the 1960s. EIFS are available either as field-applied systems or prefabricated into panels.

National standards applicable to EIFS include "Exterior Insulation and Finish Systems (EIFS)" published by the American National Standards Institute (ANSI) and the Exterior Insulation Manufacturers Association (EIMA), and several ASTM standards including C1382, C1397, C1481, C1535, E2098, EW2110, E2134, E2273, and E2321, as well as ANSI/EIMA 99-A.

## 7.4.1 SYSTEM CLASSES

EIFS are classed as either Class PB (polymer based) or Class PM (polymer modified). Both classes consist of a textured protective finish, a reinforced base coat, a layer of insulation, a layer of sheathing, and necessary accessories. The configuration and materials are slightly different, however (Figs. 7.4-1 and 7.4-2).

### 7.4.1.1 Finish Coats

Both classes use either a polymer-based or polymer-modified finish coat. Many colors and texture combinations are available. Class PB finishes are, however, softer and more flexible than Class PM finishes. For this reason, and because of the difference in their insulation material, PB systems are sometimes called *softcoat* systems and Class PM systems are called *hardcoat* systems. The composition of the finish coat varies from manufacturer to manufacturer. Some coating mixes contain random fibers; others do not. Some contain sand. Some finish coat materials may reemulsify if allowed to become saturated or to stand in free water. Therefore, horizontal surfaces should be sloped to provide adequate drainage.

### 7.4.1.2 Base Coats

In Class PB systems, base coats vary from $1/16$ in. (1.6 mm) to $1/4$ in. (6.4 mm), depending on the manufacturer. They are typically reinforced with a glass fiber mesh embedded during installation.

In Class PM systems, base coats vary from $1/4$ in. (6.4 mm) to $3/8$ in. (9.5 mm), depending on the number of layers of reinforcement. A typical method of reinforcing a base coat for Class PM systems is to fasten a layer of wire lath or glass fiber mesh to the insulation. The base coat is then applied over the reinforcement. In some but not all systems, another layer of reinforcement (usually glass fiber mesh) is embedded in the base coat as it is applied. Different weights of glass fiber mesh are selected according to the abuse expected for different segments of an installation.

Additional reinforcement should be provided at corners of openings, such as windows and doors, in both classes.

The composition of a base coat varies with the manufacturer.

### 7.4.1.3 Insulation

The type of insulation varies with the manufacturer. Most Class PB systems use molded rigid cellular polystyrene. Most Class PM systems contain extruded rigid cellular polystyrene, which is more expensive, less vapor permeable, and more dense than the molded product. However, some Class PM systems have molded polystyrene

The insulation of some Class PB systems is mechanically attached, and some is attached both mechanically and with adhesives. Conversely, the insulation in Class PM systems is always mechanically fastened.

### 7.4.1.4 Sheathing

Failures and fear of failures have created some controversy about the type of sheathing that should be used in EIFS. The traditional sheathing for both Class PB and Class PM systems has been gypsum, because it is relatively inexpensive, fire resistant, light, and thin. Unfortunately, failures have occurred in adhesively applied EIFS related to the adhesive interface with the sheathing. EIFS manufacturers have claimed that such failures are small in number and are due to failures in the adhesives themselves.

Gypsum sheathing represents a small portion of the cost of an EIFS wall sys-

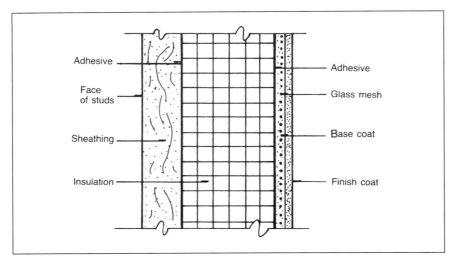

**FIGURE 7.4-1** Typical Class PB EIFS. (Courtesy H. Leslie Simmons.)

Adhesive — Adhesive
Face of studs — Glass mesh
Sheathing — Base coat
Insulation — Finish coat

**FIGURE 7.4-2** Typical Class PM EIFS. (Courtesy H. Leslie Simmons.)

Face of studs
Sheathing
Insulation

Glass mesh or wire lath reinforcement
Synthetic stucco base coat
Finish coat
Second layer of glass mesh reinforcement (optional)

tem, but some gypsum products manufacturers and the Gypsum Association, apparently fearing that such systems will fail because the paper facing delaminates from the core of their sheathing when used in these systems, have stated that they will not be responsible for the use of their products in adhesively attached EIFS systems. One sheathing manufacturer has specifically stated that its gypsum sheathing is not endorsed for use in EIFS other than with insulation that is mechanically fastened through the sheathing into the support system. Some local codes require mechanical fasteners for locations higher than the lower two stories.

Other types of sheathing can also be used in EIFS, but they are more expensive than gypsum sheathing and are not likely to replace the gypsum product in spite of the controversy.

The sheathing used should be selected, installed, and protected as recommended by its manufacturer and the manufacturer of the EIFS. Conflicts between the two should be resolved before a system is installed.

### 7.4.2 CHARACTERISTICS OF EIFS

The two basic EIFS (PB and PM) vary in their appearance and performance characteristics. Failures may occur for several reasons related to failure to recognize the characteristics of the system being used and to accommodate them in the design and installation of the system.

#### 7.4.2.1 Vapor Permeability and Condensation

Both classes of EIFS have protective coatings that are vapor permeable. However, this permeability is reduced considerably in systems with a second layer of reinforcement added to increase its impact resistance. EIFS systems should always have a complete vapor retarder on the interior. Vapor that enters the system in spite of the retarder can cause the coating to separate from the insulation, the insulation to separate from the sheathing, or the sheathing to deteriorate. Condensation can also cause loss of thermal efficiency if it saturates the insulation either in the EIFS or in the wall behind it. Both types of polystyrene insulation act as a vapor retarder. However, extruded polystyrene is a much better vapor retarder than the molded product. In addition, the extruded material has a much lower rate of absorption, which may help reduce loss of insulation effectiveness in the event that condensation does occur.

To help prevent condensation from forming, the dew point should be calculated to fall within the cellular plastic board insulation in an EIFS.

#### 7.4.2.2 Cracking

The finish coats on Class PM systems are comparatively brittle, and the insulation usually employed (extruded polystyrene) is more rigid than the materials used in a typical Class PB system. Therefore, Class PM systems must be divided into panels by control joints, much like stucco. Class PB systems do not require such division and need joints only where expansion or control joints are needed in the substrates.

In addition, EIFS should have built-in drainage system to allow water that enters through cracks to escape.

### 7.4.3 SELECTION OF A SYSTEM

The building, including its construction, use, environment, and construction budget, should be analyzed to determine the most effective EIFS to use. EIFS are proprietary products, and each is different from the next, even when in the same class. It is necessary to either name a specific product and design around it or to specify the required characteristics. The physical properties used include a system's accelerated weathering characteristics, wind-driven rain resistance, salt-spray resistance, absorption-freeze resistance, mildew resistance, abrasion resistance, water vapor transmission, impact resistance, and negative wind load performance.

# 7.5 Weather Barriers

According to *MasterFormat* 2004, weather barriers include vapor retarders that are not part of another product and air infiltration barriers. Vapor retarders are covered in Section 7.1. Air infiltration barriers are discussed in this section.

Air infiltration barriers reduce the number of air changes in a building and help create a more energy-efficient interior environment. They also protect sheathing against deterioration and resist the further penetration of moisture that finds its way through the siding or other finish material. Air infiltration barriers in common use over wood-framed walls today include asphalt saturate organic felt (building paper) and more modern products such as polyethylene sheets and woven polyolefin sheet.

These materials may be applied directly to studs to resist the infiltration of air and moisture when sheathing is not used and when spaced sheathing is used. They may also be applied to the studs behind water-resistant, nonnailable sheathing to make locating the studs easier for nailing finishing materials. They may also be installed over the sheathing. While not strictly required with plywood sheathing or other water-resistant sheathing materials, polyethylene and polyolefin materials may be installed anyway as an air infiltration barrier only. Local codes may require a vapor retarder beneath masonry veneer, and may or may not permit the more modern plastic materials to be used in this application. Building paper is required beneath stucco. An air infiltration barrier is not appropriate in exterior insulation and finish systems.

Sheathing paper should be a material such as No. 15 or No. 30 asphalt-saturated felt, which has low vapor resistance. Materials such as coated felts or laminated waterproof papers which have high vapor resistance and act as vapor retarders should not be used. Such materials may permit moisture or frost to accumulate as a result of moisture condensation between the sheathing paper and the surface of the wall sheathing. Coal-tar-saturated felts should not be used because they may stain siding.

When sheathing paper is used, it should be applied over the entire wall as soon as possible with at least a 4 in. (100 mm) top lap at horizontal joints and a 4 in. (100 mm) side lap at end joints. It should be lapped 6 in. (150 mm) from both sides around corners. Only sufficient fasteners need be used to hold the sheathing paper securely in place until the finish materials are applied. Wall finish materials should not be applied over wet sheathing paper.

Installation, fastening, and sealing of air infiltration barriers should be done in accordance with the recommendations of their manufacturer. Fastening is done using staples, nails, or adhesives. Overlaps should be sealed. These products degrade in ultraviolet light and should, therefore, not be left exposed for more than 1 month.

# 7.6 Steep-Slope Roofing

*Steep-slope roofs* are those whose slope exceeds $1^1/_2$ in./ft (125 mm/m). They are generally covered with the materials and systems that function by shedding water. This section discusses the materials and methods needed to provide steep-slope roof finishes using composition shingles and wood shakes and shingles, and the hard materials, slate, clay and concrete tile, and mineral fiber-cement shingles. Sheet metal roofing, which also may be used on steep-slope roofs, is discussed in Section 7.10. By contrast, the types of roofing used on roofs with slopes less than $1^1/_2$ in./ft (125 mm/m) provide a waterproof membrane (see Section 7.8). Mineral-surfaced roll roofing, which functions like steep-slope roofing and can be applied on roofs with slopes of 1 in./ft (83.33 mm/m) or more, is also discussed in Section 7.8.

Roofing for steep-slope roofs comes in many forms, including ancient traditional shapes and newer configurations, and in materials ranging from the stone and clay materials that have been used for centuries to modern equivalents made from concrete, mineral-fiber-cement, or metal.

Stone and clay tiles and slate shingles were early hard roofing materials. Concrete and mineral-fiber-cement tiles were originally produced, and are still used, primarily to mimic clay tile, slate roofing, or wood shingles or shakes. Those newer materials share several general advantages when compared with the earlier forms. They are often lighter, more readily available, and less susceptible to damage. In many cases, the newer materials cost less than the roofing they mimic. Wood shingles are an exception, usually costing less than clay, concrete, or mineral-fiber-cement substitutes, but wood shingles do not have the fire resistance characteristics shared by the others.

The newer forms of hard roofing materials share the disadvantage of being machine-made. The hand-made forms have somewhat irregular surfaces, which appear more rustic and have greater charm than roofs made with their modern, mass-produced counterparts. Another disadvantage is that, in most cases, the modern materials will not stand up to close visual comparison with the products they are imitating; the difference is immediately apparent when they are installed alongside the real thing.

When used alone, though, some modern materials are often mistaken for the materials they mimic when viewed from the ground by people who are not experts in recognizing the materials imitated. By contrast, composition shingles that are supposed to look like wood shingles do not fool anybody.

Stone tile, perhaps the earliest heavy roofing, is generally used today only when repairing historic buildings. Stone tile was never used much in the United States and is not found often enough to warrant detailed discussion here.

In addition to cost, other criteria to be considered in selecting steep-slope roofing and its application methods include roof slope, expected service life, wind resistance, fire resistance, and local climate. Depending on the color and kind of material selected, the total heat loss or gain of a roof may also be affected.

## 7.6.1 GENERAL RECOMMENDATIONS

Individual shingle and tile units are applied to steep-slope roofs in an overlapping fashion to shed water and resist weather penetration. They can be applied effectively on most roofs that have sufficient slope to ensure good drainage.

Depending on the material type and application method selected, esthetic effects range from surfaces having relatively smooth, uniform coursing to those with random coursing, which give a heavily textured appearance. Color can be achieved by a variety of factory- or field-applied coatings or by leaving the material in its natural color.

In addition to the shingle or tile units themselves, many accessory materials are required to prepare a roof surface for their application and to apply them. Depending on the shingles or tile type and the application method, accessory materials may include sheathing, underlayment, sheathing paper, backer board, undercoursing, flashing, drip edges, roofing cement, caulking, and nails and other fasteners. Depending on the kind of shingle or tile, other accessories, such as starter shingles and hip and ridge units, may also be required. Typical shingle and tile materials and application methods for roofs are discussed in Sections 7.6.2 and 7.6.3 but the shingle or tile manufacturer's instructions and recommendations should always be consulted to ensure the most suitable performance.

In this section, the underlying roof is assumed to be correctly and adequately ventilated, a practice that is essential to the prevention of condensation. Condensation, the actual cause of many roof leakage complaints, can be eliminated as a hazard by providing sufficient ventilation in attic areas or enclosed air spaces beneath the roof. Refer to Section 7.1 for a discussion of vapor retarders and ventilation.

### 7.6.1.1 Nomenclature

Some definitions of roof and roofing terminology are given here to assist in understanding the text and illustrations.

#### ROOF TERMINOLOGY

Figure 7.6-1 shows several common steep-slope roof types. These simple shapes are often complicated by intersecting walls or roofs, as well as projections through the roof such as dormers, chimneys, plumbing vents, and roof ventilators.

The composite drawing in Figure 7.6-2 names the elements of a steep-slope roof.

***Slope and Pitch*** The terms *slope* and *pitch* are often incorrectly used synonymously when referring to the incline of a steep-slope roof. Both are defined here and in Figure 7.6-3, which also compares some common roof slopes to corresponding roof pitches. *Roof slope* is the term used throughout this text.

*Slope* indicates the roof incline as a ratio of vertical rise to horizontal run. It is expressed sometimes as a fraction but typically as $\times$ in 12. For example, a roof that rises at the rate of 4 in. (101.6 mm) for each foot (12 in. [304.8 mm]) of run is designated as having a 4 in 12 slope. The triangular symbol above the roof in Figure 7.6-3 conveys this information.

*Pitch* indicates the incline of a roof as a ratio of the vertical rise to twice the horizontal run. It is expressed as a fraction. For example, if the rise of a roof is 4 in. (101.6 mm) and the run is 12 in. (304.8 mm), the roof is designated as having a pitch of $^1/_6$.

**FIGURE 7.6-1** Types of sloped roofs.

$$\text{Assume:}$$
$$\text{Rise} = 4 \text{ ft.}; \; \text{Run} = 12 \text{ ft.}$$
$$\text{Slope: } 4/12, \text{ or } 4 \text{ in } 12$$
$$\text{Pitch: } \frac{4}{2 \times 12} = \frac{4}{24} = \frac{1}{6}$$

$$\text{Slope} = \frac{\text{Rise}}{\text{Run}}$$

$$\text{Pitch} = \frac{\text{Rise}}{2 \times \text{Run}}$$

| Slope | Pitch |
|---|---|
| 2 in 12 | 1/12 |
| 3 in 12 | 1/8 |
| 4 in 12 | 1/6 |
| 5 in 12 | 5/24 |
| 6 in 12 | 1/4 |
| 7 in 12 | 7/24 |
| 8 in 12 | 1/3 |
| 10 in 12 | 5/12 |
| 12 in 12 | 1/2 |

**FIGURE 7.6-3** Slope and pitch.

## SHINGLE AND TILE ROOFING TERMINOLOGY

The following terms are peculiar to shingle and tile roofing. Other terms are defined when they first appear in the text or are given in the Glossary.

*Coverage* The amount of weather protection provided by the overlapping of shingles or tile. Depending on the kind of shingle and method of application, shingles may furnish one (single coverage), two (double coverage), or three (triple coverage) thicknesses of material over the surface of a roof.

*Endlap* The shortest distance in inches (millimeters) which adjacent shingles, sheets, or tiles horizontally overlap each other (Fig. 7.6-4).

*Exposure* The shortest distance in inches (millimeters) between exposed edges of overlapping shingles or tiles (see Fig. 7.6-4).

*Headlap* The shortest distance in inches (millimeters) from the lower edges of overlapping shingles to the upper edge of the unit in the second course below (see Fig. 7.6-4).

*Shingle Butt* The lower, exposed edge of a shingle.

*Shingles* Relatively small individual roofing units designed to overlap other similar units to provide weather protection. Typically, shingles are applied to a nailing base, such as sheathing or horizontal nailing strips, which supports the shingles between structural framing members.

*Sidelap* See Endlap.

*Square* A square of roofing is the amount required to cover 100 sq ft (9.29 m²) of roof surface. Roofing is estimated and sold by the *square*, although packages of one-third square are usually available so that individuals or contractors do not have to buy a full square of roofing when the amount needed is not an exact multiple of squares.

**FIGURE 7.6-2** Composite diagram of sloped roof types showing terminology.

$E$ = *Exposure*
$TL$ = *Toplap*
$HL$ = *Headlap*
$SL$ = *Sidelap*
$W$ = *Width for Strip Shingles*
*or Length for Individual Shingles*

AMERICAN METHOD

DUTCH LAP METHOD

**FIGURE 7.6-4** Shingle roofing terminology.

*Toplap* The shortest distance in inches (millimeters) from the lower edge of an overlapping shingle or sheet to the upper edge of the lapped unit in the first course below (that is, the width of the shingle minus the exposure) (see Fig. 7.6-4).

### 7.6.1.2 Roof Slope Limitations

Free and effective drainage from sloped roof surfaces is essential for long shingle life. As the roof slope decreases, runoff is slower and the susceptibility to leakage from wind-blown rain or snow being driven underneath the roofing increases. Accordingly, the roof slope governs, or may impose limitations on, the following: (1) roofing material selection, (2) exposure, (3) underlayment requirements, (4) eaves flashing requirements, and (5) methods used in roofing application.

In this text, when a reduction in roof slope requires precautions other than normal practice, the roof is referred to as being *of lower slope than normal*; otherwise, the term *normal slope* is used.

The minimum slope recommended for each shingle and tile type is shown in Figure 7.6-5. Slope limitations for underlayment and eaves flashing requirements vary with each shingle and tile type and are discussed later in this section. Application methods are discussed in the following subsections for each shingle type.

### 7.6.1.3 Roof Framing and Sheathing

Steep-slope roofing performance is dependent on the supporting structure. Roof joists, rafters, headers, and other supports should be adequately sized and spaced to carry the necessary superimposed loads over the required spans.

Sheathing should be smooth, securely attached to supports, and provide an adequate base to receive nails and other fasteners. All shingle types are applied over solid sheathing composed of wood boards or plywood. Wood shingles and shakes are also applied over spaced sheathing. Roof framing and sheathing are discussed in Chapter 6.

### 7.6.1.4 Roofing Cements

Roofing cements include plastic asphalt cements, lap cements, quick-setting asphalt adhesives, roof coatings, and primers. They are used for installing eaves flashing, for flashing assemblies, for cementing asphalt shingles tabs and laps in sheet materials, and for roof repairing. They are discussed in Section 7.8.

Materials and application methods for steep-slope roofing should be as recommended by the roofing material's manufacturer.

### 7.6.1.5 Roofing Underlayment

Roof underlayment performs several functions:

- Protecting sheathing from moisture absorption until roofing is applied
- Providing important additional weather protection by preventing the entrance of wind-driven rain below the roofing onto the sheathing or into the structure
- Preventing direct contact between asphalt shingles and resinous areas in wood sheathing which, because of chemical incompatibility, may damage the shingles

**FIGURE 7.6-5** Summary of Recommended Minimum Roof Slope

| Type of Roofing | Minimum Slope[a] |
|---|---|
| Mineral fiber-cement shingles | 3 in 12[b] |
| Asphalt shingles | 2 in 12[c] |
| Wood shakes | 3 in 12[d] |
| Wood shingles | 3 in 12[e] |
| Slate shingles | 4 in 12[b] |
| Clay tile | |
| Flat | 5 in 12[b] |
| Other types | 3 in 12[b] |
| Concrete tile | |
| Flat | 5 in 12[b] |
| Other shapes | 4 in 12[b] |
| Mud set | 4 in 12 |

[a]Where eave flashing is required, it should be cemented.

[b]Requires double underlayment.

[c]Strip shingles only; requires double underlayment and wind-resistant shingles or cemented tabs.

[d]Requires solid sheathing, underlayment in addition to interlayment, and reduced weather exposure.

[e]Requires reduced weather exposure.

Underlayment should be asphalt-saturated felt or another material having a low vapor resistance. Laminated waterproof papers, coated felts, and other materials that act as vapor retarders should not be used. Such materials may permit moisture or frost to accumulate between the underlayment and the surface of the sheathing.

Underlayment requirements for different steep-slope roofing types for various roof slopes are summarized in Figure 7.6-6.

Underlayment should be applied over the entire roof immediately after application of the roof sheathing. It should lap at least 2 in. (50.8 mm) at horizontal joints and 4 in. (101.6 mm) at end joints (Fig. 7.6-7), and should be lapped 6 in. (152.4 mm) from both sides over hips and ridges. Only a sufficient number of fasteners need be used to hold the underlayment securely in place until roofing is applied. Roofing should not be applied over wet underlayment.

### 7.6.1.6 Flashing and Caulking

Roofs are often intersected by other roofs, walls, and projections, creating opportunities for leakage. Flashing must be installed at these locations to weatherseal the roof. Flashing must be watertight and water-shedding.

**FIGURE 7.6-6** Summary of Underlayment for Shingle Roofs

| Material | No. 15 | No. 30 | 1 Layer | 2 Layers | Remarks |
|---|---|---|---|---|---|
| Roll roofing | X | | X | X | a, b, c |
| Composition shingles | | | | | |
| 4 in./ft (333.3 mm/m) and steeper | X | | X | | b |
| 3 in./ft (250 mm/m) to 4 in./ft (333.3 mm/m) | X | | | X | c |
| 2½ in./ft (208.3 mm/m) to 3 in./ft (250 mm/m) | X | | | | d |
| Slate | | | | | |
| 4 in./ft (333.3 mm/m) and steeper | | | | | |
| ¼ in. (6.4 mm) thick and thinner | X | X | | | b |
| More than ¼ in. (6.4 mm) thick | X | | | X | c, e |
| Less than 4 in./ft (333.3 mm/m) | X | | | | i |
| Graduated, all slopes | X | | | X | c |
| Wood shingles | X | X | | | f |
| Wood shakes | X | X | | | f, g |
| Clay tile | X | X | | | h |
| Concrete tile | X | X | | | h |
| Mineral fiber-cement tile | | | | | |
| 4 in./ft (333.3 mm/m) and steeper | X | X | | | f, g |
| 3 in./ft (250 mm/m) to 4 in./ft (333.3 mm/m) | X | X | | | g |
| Class A roofs | | | | | j |
| Metal tile | X | X | | | b |

<sup></sup>aUse one layer on slopes 4 in./ft (333.3 mm/m) and more. Use two layers on flatter slopes.

bWhere one layer of underlayment is required, install two layers of No. 15 felt or one No. 50 felt from eave line to 24 in. (610 mm) inside wall line as an ice shield when mean January temperature is 30°F (−1.1°C) or lower. At slate roofing and metal tile, use two layers of No. 30 felt. Set ice-shield felt in hot asphalt or mastic. As an alternative, use adhered bituminous membrane manufactured for the purpose as an ice shield.

cWhere two underlayment layers are required, apply underlayment from eave to 24 in. (610 mm) inside wall using hot asphalt or mastic when mean January temperature is 30°F (−1.1°C) or less.

dEntire underlayment applied using hot asphalt or mastic.

eAlternative: One layer of No. 55 or No. 65 roll roofing over entire roof, with portion from eave to 24 in. (610 mm) inside wall applied using hot asphalt or mastic when mean January temperature is 30°F (−1.1°C) or less.

fUnderlayment scheduled is optional except in snow areas (see text).

gFelt eave strips and felt interlayers between shakes are essential (see text).

hUnderlayment scheduled is often used for roof slopes of 4 in./ft (333.3 mm/m) and higher, but there is disagreement in the industry about underlayment requirements for clay and concrete tile (see text). In applicable areas, the requirements in notes b and c apply.

iSlate on slopes less than 4 in./ft (333.3 mm/m) should be installed over membrane roofing or waterproofing (see text).

jSee text.

(Courtesy H. Leslie Simmons.)

## FLASHING MATERIALS

Corrosion-resistant metal flashing should be at least 26-gauge (0.55 mm) galvanized steel, 0.019 in. (0.4826 mm) aluminum, or 16 oz (453.6 g) copper. Refer to Section 7.9 for a discussion of flashing materials.

Additional requirements and typical flashing methods are described in the following subsections for each steep-slope roofing type.

## ROOF FLASHINGS

**Valley Flashing** The joint formed by two sloping roofs meeting at an angle is called a *valley*. Because drainage concentrates in a valley, causing runoff toward and along the joint, valleys are especially vulnerable to leakage. An unobstructed drainage way must be provided with enough capacity to carry away the water rapidly.

**FIGURE 7.6-7** Underlayment requirements for steep roofing.

**Chimney Flashing** A chimney is usually built on a foundation separate from the foundation that supports the structure and is normally subject to some differential settling. To permit this movement between a chimney and the associated roof structure without damage to the waterseal, it is necessary to secure base flashings to the roof deck and secure counter flashing, which is bent down over the base flashing, to the masonry.

A saddle (see Fig. 7.6-2) should be built between a roof and the back face of a chimney that falls below the ridge. A saddle will prevent snow and ice accumulation behind the chimney and deflect water down around the chimney.

**Eaves Flashing** Eaves flashing is recommended in areas where the outside design temperature is 30°F (−1.1°C) or colder (Fig. 7.6-8) or wherever there is a possibility of ice forming along the eaves. This protection is important to avoid leaks and moisture penetration caused by water backing up under shingles behind ice dams and snow that collects at the eaves.

Different flashing methods are used to prevent leakage from this cause, depending on roof slope, icing conditions severity, and roofing type.

For normal slopes, eaves flashing is usually formed by a smooth or mineral-surfaced coated roll roofing course applied over the underlayment. It extends up the roof to cover a line at least 12 in. (304.8 mm) inside the interior wall line of the building (Fig. 7.6-9a).

On roofs with lower slope than normal, a double layer of underlayment should be applied around the entire perimeter of the roof. The two layers should be cemented together with asphalt cement up the roof to a line at least 24 in. (609.6 mm) inside the interior wall line of the building (Figs. 7.6-9b and 7.6-10).

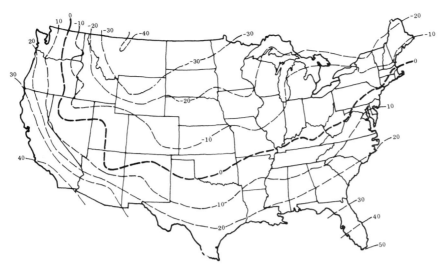

**FIGURE 7.6-8** Outside design temperatures. (Reprint courtesy National Weather Service.)

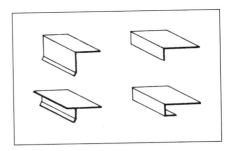

**FIGURE 7.6-11** Typical drip edge shapes.

tion. Typical preformed drip edge shapes are shown in Figure 7.6-11.

A drip edge is applied to the sheathing and under the underlayment at the eaves, but over the underlayment up the rake (Fig. 7.6-12).

***Flashing at Wall and Roof Intersections***
Sheet metal flashing should be installed at horizontal intersections and at vertical intersections when the exterior finish material does not provide a self-flashing joint.

### 7.6.1.7 Nails and Fasteners

No single step in applying steep-slope roofing is more important than proper nailing. Suitability is dependent on (1) selecting the correct nail for the roofing and sheathing type (Fig. 7.6-13), (2) using the correct number of nails, (3) locating them correctly, and (4) choosing nail metal compatible with the metal used for flashings.

Specific recommendations for the type, size, number, and spacing of roofing nails are given in the following subsections for each roofing material. Some general recommendations follow.

**a**

**b**

**FIGURE 7.6-9** Eave flashing for normal slopes (a) extends up the roof at least 12 in. (304.8 mm) inside the exterior wall; for a lower than normal slope (b), an additional course of underlayment should be cemented down and extend up the roof at least 24 in. (609.6 mm) inside the interior face of the wall.

In areas where roofs are subject to severe icing, a double layer of underlayment should be installed over the entire roof, regardless of the roof slope.

Specific recommendations for eaves flashing are described in the following subsections for each kind of steep slope roofing.

***Drip Edge*** A continuous corrosion-resistant, nonstaining material such as 26-gauge (0.55 mm) galvanized steel, formed to provide a drip, and nailed along the eaves and rakes is recommended for most shingle roofs. Drip edges should be designed and installed to protect deck edges and prevent leaks at this point by allowing water to drip free of underlying eave and cornice construc-

**FIGURE 7.6-10** Application of a double layer of felt underlayment and eave flashing. (Courtesy Asphalt Roofing Manufacturers Association.)

**FIGURE 7.6-12** Application of drip edge at the rake and eave. (Courtesy Asphalt Roofing Manufacturers Association.)

### LUMBER BOARDS, PLANK DECKING, AND PLYWOOD SHEATHING

Roofing nails should be long enough to penetrate through the shingle and through lumber boards or plywood sheathing. They should penetrate at least 1 in. (25.4 mm) into plank roof decking. Nails for applying shingles over plywood sheathing should have threaded shanks.

### NONWOOD SHEATHING

When gypsum, concrete, fiberboard, or other nonwood sheathing materials are used, special fasteners and special details for fastening are often necessary to provide adequate anchorage. The nonwood material manufacturer's recommendations should be followed to ensure acceptable performance.

**FIGURE 7.6-13** Types of smooth and threaded shank nails used for application of shingle roofing other than for application of asphalt shingles (see Fig. 7.6-38).

### 7.6.2 BITUMINOUS ROOFING PRODUCTS

Bituminous roofing includes shingles, roll roofing, and built-up membrane roofing. Bituminous shingles are discussed in this section. Roll roofing and built-up membrane roofing are covered in Section 7.8

Products used in bituminous roofing include (1) shingles, mats, and felts and (2) the materials used to install and coat them. These products are made using four basic raw materials: (1) a reinforcing base which is either organic dry felt or glass fiber mat; (2) bitumen (asphalt or coal tar); (3) mineral stabilizers; and (4) either fine or coarse surfacings.

#### 7.6.2.1 Shingle, Mat, and Felt Raw Materials

The raw materials and some of the processing steps required to manufacture bituminous shingles, mats, and felts are shown diagrammatically in Figure 7.6-14.

### DRY ORGANIC FELT

Dry organic felt, used as a base for making reinforcing felts for built-up roofing and waterproofing, underlayment, smooth and mineral-surfaced roll roofing, and asphalt shingles, is made from combinations of cellulose fibers such as those derived from rags, paper, and wood on a machine similar to the type used for making paper (Fig. 7.6-15). The fibers are prepared by various pulping methods, then blended and proportioned to produce felt with the necessary weight, tensile

strength, absorptive capacity, and flexibility required to make a suitable roofing product. The felt must be able to absorb one and one-half to two times its weight in asphalt saturants and must be strong and flexible enough to withstand strains placed on it during the manufacturing process.

As the felt comes off the end of the machine in a continuous sheet, it is cut into specified widths and wound into individual rolls from 4 to 6 ft (1219 to 1829 mm) in diameter weighing up to a ton or more.

### GLASS FIBER MAT

Dry glass fiber mats are used as a base for reinforcing felts for built-up roofing and waterproofing, coated smooth rolls and mineral-surfaced roll roofing, and asphalt shingles. To make glass fiber mats, sand, soda ash, and limestone are first combined to make chopped strand glass fibers. Then these fibers are mixed with a binding agent and cured to produce a mat with the thickness, tensile strength, tear strength, weight, and flexibility required to produce roofing and waterproofing products.

### BITUMENS

The bitumens used in bituminous roofing products are asphalt and coal tar. They are used to saturate and coat dry organic felts and coat glass fiber mats in the manufacture of roofing felt and to weld or fuse roofing or waterproofing felts together to form a built-up roof or waterproofing membrane. Both asphalt and coal tar bitumens are thermoplastic, which means that they become more fluid when heated and return to their former solid state when they cool.

A bitumen's equiviscous temperature (EVT) is the optimum temperature at which it should be used. Each asphalt and coal tar bitumen has its own EVT. The lowest type number has the lowest EVT.

*Asphalt* Asphalt used to make roofing and waterproofing products, known as *asphalt flux*, is a petroleum product obtained from the fractional distillation of crude oil. Asphalt flux is processed to produce roofing grades of asphalt called *saturants* or, when combined with mineral stabilizers, *coating asphalts*. Both forms combine with dry felt to make asphalt roofing or waterproofing membranes.

**FIGURE 7.6-14** Raw materials and processing steps used to make asphalt roofing products. (Courtesy Asphalt Roofing Manufacturers Association.)

can be further protected by a covering of opaque mineral granules. A primary difference between saturants and coating asphalts is the temperature at which they soften. The softening point of saturants varies from 100 to 160°F (37.8 to 71.1°C); that of coatings is as high as 260°F (126.7°C). Saturant asphalt is not used with glass fiber mat. Rather, coating mixtures are used to completely surround the glass fibers and to provide a layer of coating on both sides of the mat. Asphalt flux is also the base material used to make asphalt plastic cement, quick-setting roof adhesives, asphalt primers, other roof coatings, and adhesives.

*Coal Tar*   Coal tar used to make roofing products is a distillation of bituminous coal. It is considered basically superior to asphalt for use in membrane roofing. It is most effective at a slope of $1/4$ in./ft (20.83 mm/m) and can be used on slopes of up to $1/2$ in./ft (41.67 mm/m), but cannot be used on slopes higher than $1/2$ in./ft (41.67 mm/m).

The two types of coal tar recommended for use in built-up roofing are classified in ASTM D450 as Type I, *coal-tar pitch*, and Type III, *coal-tar bitumen*. Type III has a higher softening point than does Type I.

## MINERAL STABILIZERS

Finely ground minerals, such as silica, slate dust, talc, micaceous materials, dolomite, and trap rock, which are called *stabilizers*, when combined with coating bitumens, control hardness, elasticity, adhesion, and weathering. Coating bitumens containing stabilizers resist weather better, are more shatter- and shockproof in cold weather, and significantly increase roofing product life.

The four types of asphalt used in built-up roofing are classified in ASTM D312 as Type I, *dead level asphalt*; Type II, *flat asphalt*; Type III, *steep asphalt*; and Type IV, *special steep asphalt*. Type I is limited to slopes of $1/2$ in./ft (41.67 mm/m) and lower; Type II may be used on slopes of up to $1^1/2$ in./ft (125 mm/m); Type III up to 3 in./ft (250 mm/m); and Type IV up to 6 in./ft (500 mm/m). However, steep asphalts are not as durable or as easy to apply as the lower-slope types and should be used only when the slope is steeper.

The preservative and waterproofing characteristics of asphalt reside largely in certain oily constituents. In making roofing products, the body of the highly absorbent felt sheet is first impregnated (saturated) to the greatest possible extent with saturants which are oil-rich asphalts. The saturant is then sealed in with an application of a harder, more viscous coating asphalt which, in turn,

**FIGURE 7.6-15** Flow diagram of a typical felt mill. (Courtesy Asphalt Roofing Manufacturers Association.)

## SURFACINGS

Bituminous roofing product surfaces may be either smooth or surfaced with coarse mineral granules.

*Finely Ground Minerals* Powders, usually talc and mica, are dusted on the surface of smooth roll roofings and the backs of mineral-surfaced roll roofings and shingles to prevent layers in packages from sticking together. They are not a permanent part of the finished product and gradually disappear from exposed surfaces after application.

*Coarse Mineral Granules* Coarse granules are usually colored slate or rock granules (either natural or colored by a ceramic process) that are used on shingles and on certain roll products. These mineral-surfaced products have increased weather and fire resistance and provide a wide range of colors and color blends. To protect the underlying bitumen from light ray impact, the granules should be opaque, dense, and well graded for maximum coverage.

### 7.6.2.2 Shingle, Mat, and Felt Manufacture

In general, organic felt bituminous roofing product manufacture includes processing cellulose fibers into dry felt, saturating and coating the felt with asphalt or coal tar, and then, depending on the finished product, surfacing the coated felt with selected mineral aggregates. When glass fiber mat is used as a reinforcing base instead of organic felt, the mat goes directly from a dry looper into a coater, bypassing the saturating process.

Raw materials are processed into finished roofing products in a continuous sequence (Fig. 7.6-16). The principal steps are described in the following subsections.

### DRY LOOPER

A roll of dry felt or glass fiber mat installed on the felt reel is unwound onto a dry looper. This looper provides a felt material reservoir that can be drawn on by the machine as circumstances demand, eliminating stoppages when a new roll must be put on the felt reel or when imperfections in the felt must be cut out.

### SATURATION OF FELT

The felt is then subjected to a hot saturating process which eliminates moisture and fills the felt fibers and intervening spaces as completely as possible with saturant. Glass fiber mat does not undergo this saturating process. Rather, coating bitumen is used to fill the spaces between the glass fibers.

### WET LOOPER

Following the saturating process, excess saturant usually remains on the surface of the sheet. Therefore, the sheet is held for a time on a wet looper so that natural bitumen shrinkage upon cooling will draw the excess into the felt, resulting in a high degree of saturation. At this point, uncoated products, such as saturated underlayment felt, move directly to the cooling looper and onto the winding mandrel for final packaging.

### COATER

Material to be coated continues to the coater, where coating bitumen is applied to both the top and bottom surfaces. Coating thickness is regulated by rollers, which can be brought together to reduce the amount of coating or separated to increase its thickness.

The final weight of the product is controlled by the roofing machine operator. Often machines are equipped with automatic scales that weigh the sheets during manufacture to warn the operator when the material is running over or under weight specifications. When smooth roll roofing is being made, talc or mica is applied to both sides of the sheet by spreading and pressing it through a pressure roll.

### MINERAL SURFACING

When shingles and other mineral-surfaced products are made, granules of specified color or color combinations are applied from a hopper and spread thickly on the hot coating bitumen; talc or mica is then applied to the back (smooth) side. The sheet is then run through a series of press and cooling rollers, which subject the sheet to a controlled pressure that embeds the granules in the coating to the desired depth.

### TEXTURE

At this stage, some products are textured by the pressure of an embossing roller, which forms a pattern in the sheet's surface.

### COOLING LOOPER

The sheet then travels into a cooling or finish looper, where it is cooled so that it can be cut to length or into shingles and packed. The sheet is cut from the back (smooth) side as it passes between a cutting cylinder and an anvil roll. Shingles are then separated into stacks of the proper number for packaging. The stacks

**FIGURE 7.6-16** Asphalt roofing products are produced in a continuous process, as shown in this diagram. (Courtesy Asphalt Roofing Manufacturers Association.)

are moved to packaging equipment, where the bundles are prepared for warehousing and shipment.

## ROLL ROOFING WINDER

When roll roofing is made, a sheet is drawn directly from the finish looper to a winding mandrel. There it is wound, and the material is measured to length as it turns. After sufficient material accumulates, it is cut off, removed from the mandrel, and wrapped for warehousing and shipment.

## CARE OF MATERIALS

The bituminous products discussed here provide a waterproof membrane after application but can be harmed if stored outside without protection, particularly if they are applied wet. The following simple rules should be observed in storing finished products in the warehouse and on the job to ensure suitable performance:

- Store roll goods on end and upright; if they are stored in tiers, place boards between the tiers to prevent damage to the ends of rolls.
- Do not store shingle bundles to a height of more than 4 ft (1219 mm) to eliminate the possibility of sticking or discoloration.
- Rotate warehouse stock, particularly white and pastel roofing, by moving out of storage first the material held longest.
- Cover bituminous roofing and waterproofing materials stored outside for weather protection.

### 7.6.2.3 Products Made on an Organic Felt or Glass Fiber Felt Base

Bituminous roofing and siding products made on an organic or glass fiber felt base may be classified broadly into two groups: (1) shingles and (2) rolls. Sizes, exposures, and other information covering shingles are discussed in Section 7.6.3.

Sizes, exposures, and other information covering roll products are given in Figure 7.6-17. Selection, installation, and other requirements for roll products are addressed in Sections 7.6.3, 7.8.7, and 7.8.8. Bituminous roll products are usually referred to by a number (No. 15, No. 30, No. 40, etc.). These numbers represent the approximate weight in pounds as shipped of the amount of that material needed to cover a square (100 sq ft [9.29 m²]) with one ply of the material.

### 7.6.2.4 Products Used for Installation and Coating of Bituminous Roofing

Bituminous roofing installation and coating products include (1) bitumens, (2) asphalt coatings and cements, (3) sheet products made on an organic felt base, and (4) sheet products made on a glass fiber mat base.

## COATINGS AND CEMENTS

Coatings and cements are bituminous materials combined with special ingredients and either (1) mixed with cut-back type solvents or (2) emulsified in water. These materials are processed to various consistencies, depending on their use. They are flammable and should never be warmed over an open fire or placed in direct contact with a hot surface. They may be softened by placing the container, unopened, in hot water or by storage in a warm place.

Cut-back asphalt coatings and cements should be applied to a dry, clean surface and troweled or brushed vigorously to force the material into cracks and openings and to eliminate air bubbles. Emulsions may be applied to damp or wet surfaces, but should never be applied in an exposed location if rain is anticipated within 24 hours.

Coatings and cements may be classified into several groups, including plastic roofing cements, flashing cements, lap cements, quick-setting adhesives, roof coatings, asphalt water emulsions, primers, and roofing tape. These may be used in accordance with information provided in Figure 7.6-18.

## PLASTIC ROOFING CEMENT (ROOFING MASTIC) AND FLASHING CEMENT

These materials are processed with mineral fibers and cut-back asphalt or coal tar, which are especially plastic and elastic, combined with nonoxidizing minerals, oils, and solvents to produce black plastic repair and flashing cements. They will not flow at high temperatures or become brittle at low temperatures. They are sufficiently elastic after setting to compensate for normal roof deck expansion and for contraction and movement between the deck and other structural elements. They have excellent acid and alkali resistance, do not absorb water, and provide good adhesion to all surfaces. These materials are similar but not interchangeable. Flashing cement is the stiffer

of the two because it has more and longer fibers. It should be used for vertical applications and on cants. Plastic roofing cement is suitable for horizontal applications in flashing assemblies, patching, and roof repairing.

## LAP CEMENTS

Lap cements are of various consistencies but are not as thick and heavy as plastic cements. They should be used as recommended by the manufacturer. Lap cements are intended to make a watertight bond between lapping elements of roll roofing. They should be spread over the entire lapped area. Nails used to secure the roofing should pass through the cement so that the shank of the nail will be sealed where it penetrates the deck material.

## QUICK-SETTING ASPHALT ADHESIVE CEMENT

This material, available in brush, trowel, or gun consistency, is used for cementing down the free tabs of strip shingles and for laps of roll roofing.

Quick-setting asphalt adhesive cement has about the same consistency as plastic asphalt cement but is more adhesive. It contains a solvent that evaporates rapidly on exposure, causing the cement to set quickly.

## ROOF COATINGS

Roof coatings are of various types and consistencies but are usually thin enough to be applied with a spray or brush. Coatings are used to resurface existing roll roofing and built-up asphalt roofs.

## ASPHALT WATER EMULSIONS

Asphalt water emulsions are a special type of roof coating made with asphalt, sometimes combined with other ingredients and emulsified with water. Emulsions must be stored where they will not freeze and should be used at times when they will not be rained on for at least 24 hours after being applied.

## PRIMERS

Primers are very fluid bituminous materials that are applied with a brush or spray. They must be thin enough to penetrate rapidly into masonry or concrete surface pores without leaving a continuous surface film. Their purpose is to prepare masonry or concrete to receive other

**Figure 7.6-17** Summary of Typical Asphalt Roll Roofing Products

| | | | | Typical Asphalt Rolls | | | | | |
|---|---|---|---|---|---|---|---|---|---|
| Product | Appx. Shipping Weight per Roll (lb) | Appx. Shipping Weight per Square (lb) | Squares per pkg. | Width (in.) | Length (in.) | Selvage (in.) | Exposure (in.) | ASTM Fire and Wind Ratings | |
| Mineral surface roll | 75–79 | 75–90 | 1 | 36–39³/₄ | 32.7–38 | 2–4 | 32–34 | Some Class C | |
| Mineral surface roll (double coverage) | 55–70 | 110–140 | ¹/₂ | 36–39³/₄ | 32.7–36 | 19 | 17 | Some Class C | |
| Smooth surface roll | 50–86 | 40–65 | 1–2 | 36–39³/₄ | 32.7–72 | 2–4 | 34–37³/₄ | None | |
| Non-perforated felt underlayment | 24–60 | 6–30 | 2–8 | 36 | 72–288 | 2–19 | 17–34 | May be a component in a complete fire-rated system. Check with manufacturer for details. | |
| Self-adhered eave and flashing membrane | 35–82 | 33–40 | 1–2¹/₄ | 36 | 36–75 | 2–6 | 34 | May be a component in a complete fire-rated system. Check with manufacturer for details. | |

(Courtesy Asphalt Roofing Manufacturers Association.)

**FIGURE 7.6-18** Uses and Properties of Asphalt Coatings and Cement Products

| Product | Uses | Properties |
|---------|------|------------|
| Plastic roofing cement | Flashing assemblies; patching; roof repairing; caulking | Remains elastic after setting; will not flow at summer temperatures or become brittle at low temperatures; excellent resistance to acids and alkalies; will not absorb water; good adhesion to all surfaces |
| Lap cement | Watertight bond between lapping elements of roll roofing | Not as thick and heavy as plastic cements |
| Quick-setting adhesive | Cementing down the free tabs of strip shingles and laps of roll roofing | Available in brush, trowel, or gun consistency; same consistency as plastic asphalt cement, but more adhesive; sets quickly |
| Roof coating | Resurfacing old roll roofing and built-up asphalt roofs | Thin enough to be applied with spray or brush |
| Asphalt water emulsion | Roof coating | Must be kept from freezing and applied when no rain is forecast for 24 hours |
| Primer | Preparing masonry surfaces to receive other bituminous products; ensures good bond between the masonry and built-up roofing, plastic cement, or roof coatings | Very fluid; applied with brush or spray |

(Courtesy Asphalt Roofing Manufacturers Association.)

bituminous products and ensure a good bond between the masonry or concrete and built-up roofing, plastic cement, or bituminous coatings.

### ROOFING TAPE

Some manufacturers make a roofing tape of cotton fabric or other porous material saturated with asphalt or coal tar. It is available in widths varying from 4 to 36 in. (101.6 to 914.4 mm) and is provided in rolls up to 50 yd. (45.72 m) long. It is used with bituminous coatings and plastic flashing cements, as well as in patching seams, breaks, and holes in bituminous and metal roofing.

### 7.6.2.5 Standards of Quality

Standards of quality vary among shingles and roll products used in bituminous roofing.

### GENERAL

Asphalt roofing should conform to Underwriters Laboratories, Inc. (UL) standards. The UL specifications constitute the recognized standard for the asphalt roofing industry and include provisions for (1) fire resistance (see Fig. 7.6-19), (2) wind resistance, (3) felt saturation tests to determine saturant quantity and saturation efficiency, (4) coating asphalt thickness and distribution, (5) granule adhesion and distribution, (6) finished product weight, count, size, coloration, and other characteristics before and after packaging, and (7) application instructions.

UL maintains a check on labeled products by making periodic factory inspections. UL inspectors are afforded the privilege of using the manufacturers' control laboratories and visit, inspect, and analyze product samples. When necessary, they recommend to the manufacturer measures that are required to maintain the product's standards to remain eligible to use the UL label.

Products that conform to UL standards are identified by Underwriters Laboratories labels.

Application of asphalt roofing systems should follow the specifications of the manufacturer of the system and the recommendations of the National Roofing Contractors Association (NRCA). In addition, metal flashing and accessories should be fabricated and installed in accordance with the recommendations of the Sheet Metal and Air Conditioning Contractors National Association, Inc. (SMACNA). Conflicts among the recommendations of these various parties should be reconciled before installation of a roofing system.

### SHINGLES

Shingles should conform to the UL 790 with a minimum of a Class C rating. Many shingles today have a Class A rat-

**FIGURE 7.6-19** Classifications Established by Underwriters Laboratories Inc. Standard UL 790, "Test Methods for Fire Resistance of Roof Covering Materials"

| | |
|---|---|
| Class A | Includes roof coverings that are effective against severe fire exposures. Under such exposures, roof coverings of this class are not readily flammable and do not carry or communicate fire; afford a fairly high degree of fire protection to the roof deck; do not slip from position; possess no flying brand hazard; and do not require frequent repairs to maintain their fire-resisting properties. |
| Class B | Includes roof coverings that are effective against moderate fire exposures. Under such exposures, roof coverings of this class are not readily flammable and do not readily carry or communicate fire; afford a moderate degree of fire protection to the roof deck; do not slip from position; possess no flying brand hazard. May require infrequent repairs to maintain their fire-resisting properties. |
| Class C | Includes roof coverings that are effective against light fire exposure. Under such exposures, roof coverings of this class are not readily flammable and do not readily carry or communicate fire; afford at least a slight degree of fire protection to the roof deck; do not slip from position; possess no flying brand hazard. May require occasional repairs or renewals to maintain their fire-resisting properties. |

(Courtesy Asphalt Roofing Manufacturers Association.)

ing. Each shingle bundle should be identified with the UL class label. In areas subject to winds of hurricane force, which the U.S. Weather Service defines as those of 75 mph (120.7 k/hr) or more, *wind-resistant* shingles in accordance with UL 997 Type I are recommended. Such shingles are provided with factory-applied adhesive or integral locking tabs, and the bundles carry the UL class label and include the words "Wind Resistant."

## MEMBRANE ROOFING PRODUCTS

Temperature affects bitumen flow rate. Bitumens that have been heated at too high a temperature flow too readily, which affects the amount of bitumen in a roofing system. They may also separate and leave voids between sheets of membrane, which are potential leak points. Not raising the bitumen temperature high enough can also cause problems. Bitumen that is not heated enough will not flow as it should, and may not fuse with and thereby adhere properly to the membranes.

NRCA recommends that each bitumen package be labeled with its *equiviscous temperature* (EVT) and its *flash point* (FP). A bitumen's EVT is its optimum temperature. A bitumen should never be heated above its FP, and should be applied at its EVT. In addition, bituminous materials should not be kept at high temperatures or allowed to stand in luggers for prolonged periods.

## 7.6.3 ASPHALT SHINGLES

Unless specific exceptions are made in the following text, the provisions of Section 7.6.1 apply to asphalt shingles and their application.

### 7.6.3.1 Materials

Refer to Section 7.6.2 for a general discussion of the materials and products used in asphalt roofing and their manufacture. Raw materials are processed into finished shingle roofing products in a continuous sequence, as described in Section 7.6.2. The data in the next subsection expand the material in Section 7.6.2 to more fully cover asphalt shingles.

### SHINGLE TYPES

Many weights and patterns of both individual and strip shingles are available for both new construction and reroofing projects. Sizes, exposures, and other information covering some shingles are given

in Figure 7.6-20. Available products may differ in weight, dimension, and design from those shown.

Asphalt roof shingles are of four basic types: (1) *strip* shingles (2) *laminated* (multithickness shingles), (3) *interlocking* shingles, and (4) *giant individual* shingles for application by either the *American* or *Dutch Lap* methods. Most strip shingles have cutouts to separate shingle *tabs*, but some are made without cutouts. Three-tab shingles are the most popular type in use today.

Wind-resistant strip shingles have tabs with factory-applied adhesive or integral locking tabs. Lightweight individual shingles applied by the Dutch Lap method are intended primarily for reroofing existing roofs.

Each kind of shingle is applied in a distinct way, giving a different appearance to a roof. Starter units, hip and ridge shingles, and valley roll are also manufactured to match asphalt shingles.

### 7.6.3.2 Preparation for Asphalt Shingle Installation

Limitations imposed by the slope of the roof are shown in Figure 7.6-3. Asphalt shingles may safely be applied using normal application methods on roofs with slopes of 4 in 12 or greater. Asphalt shingles may be used on slopes as low as 2 in 12 when applied using the lower than normal slope methods described in this section.

### ROOF SHEATHING

Sheathing types, thicknesses, spans, and nailing requirements are discussed in Chapter 6. If lumber boards are used, they should not be over 6 in. (152.4 mm) in nominal width. Wider boards may swell or shrink in width enough to buckle or distort the shingles. Badly warped boards and those containing excessively resinous areas or loose knots should be rejected. If any of these defects appear after the sheathing has been applied, they should be covered with 26-gauge (0.55 mm) galvanized steel patches before the underlayment is placed.

### UNDERLAYMENT

Underlayment should be No. 15 asphalt saturated felt. No. 30 asphalt felt should not be used as general underlayment beneath asphalt shingles, because it is too stiff and may cause wrinkling of shingles applied over it. However, on lower than

normal slope applications (2 in 12 to 4 in 12), two layers of No. 15 felt should be installed over the entire roof.

### EAVE FLASHING

Eave flashing is recommended in areas where the outside design temperature is 30°F (−1.1°C) or colder (see Fig. 7.6-8) or where there is a possibility of ice forming along the eaves and causing water backup.

*Normal Slope (4 in 12 or Greater)* A course of No. 90 mineral-surfaced roll roofing or not less than No. 55 smooth roll roofing should be installed at the eaves over the underlayment. It should overhang the underlayment and drip edge from $^1/_4$ to $^3/_8$ in. (6.4 to 9.5 mm) and should extend up the roof to cover a line at least 12 in. (304.8 mm) inside the interior wall line of the building (see Fig. 7.6-9a). When the eave overhang requires the flashing to be wider than 36 in. (914.4 mm), the necessary horizontal lap joint should be cemented and located outside the exterior wall line of the building.

*Lower Slope Than Normal (as Low as 2 in 12)* Asphalt shingles may be applied on slopes as low as 2 in 12. For slopes less than 4 in 12 down to 2 in 12, or in areas subject to severe icing, a double layer of underlayment should be used over the entire roof. Eaves flashing may be formed by cementing the two layers of underlayment together with asphalt cement up the roof to a line that is at least 24 in. (609.6 mm) inside the interior wall line of the building (see Figs. 7.6-9b and 7.6-10).

This eave flashing should be formed by applying a continuous plastic asphalt cement layer, at the rate of 2 gal/100 sq ft (.815 L/m²), to the surface of the underlayment starter course before the first full underlayment course is applied. Cement should also be applied to the 19 in. (482.6 mm) underlying portion of each succeeding course which lies within the eaves flashing area before the next course is placed. This cement should be applied uniformly with a comb trowel so that underlayment does not touch underlayment at any point when the application has been completed. The overlying sheet should be pressed firmly into the entire cemented area (see Fig. 7.6-10).

**Figure 7.6-20** Typical Asphalt Shingles

**Typical Asphalt Shingles**

| Product | Configuration | Appx. Shipping Weight per Square (lb) | Shingles per Square | Bundles per Square | Width (in.) | Length (in.) | Exposure (in.) | ASTM Fire and Wind Ratings |
|---|---|---|---|---|---|---|---|---|
| Laminated self-sealing random tab shingle | Various edge, surface texture, and application treatments | 240–360 | 64–90 | 3–5 | $11\frac{1}{2}$–$14\frac{1}{4}$ | 36–40 | $4$–$6\frac{1}{8}$ | Class A or C fire rating. Many wind resistant. |
| Multi-tab self-sealing square tab strip shingle | 55–70 Various edge, surface texture and application treatments | 240–300 | 65–80 | 3–4 | 12–17 | 36–40 | 4–8 | Class A or C fire rating. Many wind resistant. |
| Multi-tab self-sealing square tab strip shingle | Three-tab or four-tab | 200–300 | 48–80 | 3–4 | 12–$13\frac{1}{4}$ | 36–40 | $5$–$5\frac{5}{8}$ | Class A or C fire rating. All wind resistant. |
| No-cutout self-sealing square tab strip shingle | Various edge and surface texture treatments | 200–300 | 65–81 | 3–4 | 12–$13\frac{1}{4}$ | 36–40 | $5$–$5\frac{5}{8}$ | Class A or C fire rating. All wind resistant. |
| Individual interlocking shingle (basic design) | Several design variations | 180–250 | 72–120 | 3–4 | 18–$22\frac{1}{4}$ | 20–$22\frac{1}{2}$ | n/a | Class A or C fire rating. Many wind resistant. |

(Courtesy Asphalt Roofing Manufacturers Association.)

## DRIP EDGES

A corrosion-resistant, nonstaining drip edge should be installed along the eaves and rakes (see Fig. 7.6-12).

## FLASHING

See Section 7.6.1.6.

***Valley Flashing*** Either the open, closed-cut, or woven method may be used to construct valley flashing. In constructing *open* valley flashing, a valley underlayment strip of No. 15 asphalt-saturated felt, 36 in. (914.4 mm) wide, is applied first, as shown in Figure 7.6-21, centered in the valley, and secured with enough nails to hold it in place. When laps are necessary, they are formed so that the upper strip overlaps the lower by at least 12 in. (304.8 mm) and is bonded with asphaltic plastic cement. The lower edge is trimmed flush with the eave drip strip. In areas experiencing heavy rainfalls, a layer of self-adhered eave and flashing membrane should be installed beneath the valley underlayment.

The horizontal roof underlayment courses are cut to overlap this valley strip a minimum of 6 in. (152.4 mm). Where eaves flashing is required, it is applied over the valley underlayment strip.

A 26-gauge (0.55 mm) galvanized steel or other corrosion-resistant metal strip is placed over the underlayment strip from ridge to eave and trimmed

flush with the eave drip edge (Fig. 7.6-22). The metal is secured without puncturing it by means of metal cleats spaced 8 to 12 in. (203.2 to 304.8 mm) apart. Overlaps should be 12 in. (304.8 mm) and cemented.

A chalkline should be snapped on each side of the valley for its full length before shingles are applied. The chalkline should start 6 in. (152.4 mm) apart at the ridge and spread wider apart at the rate of $1/8$ in./ft (10.42 mm/m) down to the eave. These chalklines should serve as a guide in trimming the last shingle units to fit the valley, ensuring a clean, sharp edge. The upper corner of each end shingle should be clipped as shown in Figure 7.6-23 to direct water into the valley and prevent water penetration between courses. Each shingle should be cemented to the valley lining with asphalt cement to ensure a tight seal. No exposed nails should appear along the valley flashing.

Closed (woven) valleys can be used only with strip shingles. These methods double the shingle coverage throughout the length of the valley, increasing weather resistance at this vulnerable point. Individual shingles cannot be used because nails would be required at or near the center of the valley lining.

To provide either a closed-cut or woven valley flashing, first a valley lining made from a 36-in. (914.4-mm)-wide strip of No. 50 or heavier roll roofing

should be placed over the valley underlayment and centered in the valley (see Fig. 7.6-18).

In a closed-cut valley flashing (Fig. 7.6-24), the shingles on one side of the valley are laid and extended at least 12 in. (304.8 mm) onto the roof surface on the opposite side of the valley (see Fig. 7.6-18). A joint is never made in the valley. Normal shingle nailing is used, except that no nail falls within 6 in. (152.4 mm) of the valley center and each shingle crossing the valley is nailed at its end with two nails. A chalkline is placed about 2 in. (50.8 mm) from the valley centerline on the unshingled side. The shingles are then placed on the unshingled side of the valley and trimmed to the chalkline. The upper corner of each shingle on the chalkline is trimmed 1 in. (25.4 mm) to help direct water into the valley. Then each shingle end is embedded in a 3-in. (76.2-mm)-wide strip of asphalt plastic cement.

In a woven valley flashing installation (Fig. 7.6-25), valley shingles should be laid over the lining either by (1) applying them on both roof surfaces at the same time, weaving each course in turn over the valley, or (2) covering each surface first to the point approximately 36 in. (914.4 mm) from the center of the valley and weaving the valley shingles in place later.

In either case, the first course at the valley should be laid along the eaves of

**FIGURE 7.6-21** Flashing underlayment in a valley. (Courtesy Asphalt Roofing Manufacturers Association.)

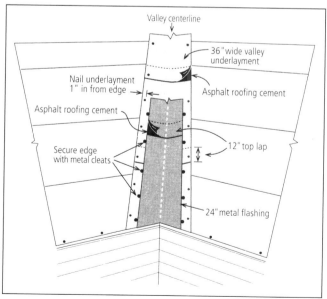

**FIGURE 7.6-22** Open valley flashing using corrosion-resistant metal. (Courtesy Asphalt Roofing Manufacturers Association.)

**FIGURE 7.6-23** Shingle application in an open valley. (Courtesy Asphalt Roofing Manufacturers Association.)

**FIGURE 7.6-25** Shingle application in a woven valley. (Courtesy Asphalt Roofing Manufacturers Association.)

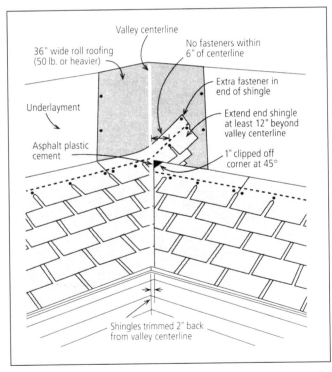

**FIGURE 7.6-24** Shingle application in a closed-cut valley. (Courtesy Asphalt Roofing Manufacturers Association.)

**FIGURE 7.6-26** Application of step flashing. (Courtesy Asphalt Roofing Manufacturers Association.)

one roof surface over the valley lining, extending it along the adjoining roof surface for a distance of at least 12 in. (304.8 mm). The first course of the adjoining roof surface should then be carried over the valley on top of the previously applied shingle. Succeeding courses should then be laid alternately, weaving the valley shingles over each other.

The shingles should be pressed tightly into the valley and nailed normally, except that no nail should be located closer than 6 in. (152.4 mm) from the valley center line and two nails should be used at the end of each terminal strip.

***Vertical Sidewall Flashing*** When the rake of a roof abuts a vertical wall, the most satisfactory method of protecting the joint is to use stepped metal base flashing applied over the end of each shingle course (Fig. 7.6-26).

Individual pieces of this metal flashing are rectangular in shape, approximately 7 in. (152.4 mm) by 10 in. (254 mm). With strip shingles laid 5 in. (127 mm) to the weather, each flashing strip will lap the shingle by 2 in. (50.8 mm). This flashing should be bent with one end extending 5 in. (127 mm) onto the roof underlayment, the remainder extending up the wall surface. Each flashing piece should be placed up-roof from the exposed edge of the shingle that overlaps it and secured to the wall sheathing with one nail in the top corner. When the shingles are laid 5 in. (127 mm) to the weather,

each 7 in. (177.8 mm) width of flashing will lap the next flashing piece by 2 in. (50.8 mm). The wall siding should be finished over the flashing to serve as counter (cap) flashing, but should be held far enough away from the shingles so that the ends of the siding can be painted for weather protection (Fig. 7.6-27).

***Chimney Flashing*** Before flashings are placed, shingles should be applied over the underlayment up to the front face of a chimney and a coat of asphalt primer applied to the masonry to seal the surface at all points where plastic cement will be applied.

Base flashing may be No. 90 mineral-surfaced roll roofing. The chimney front (Fig. 7.6-28) should be applied first by securing the lower section over the shingles in an asphalt plastic cement bed. The vertical upper section should be secured against the masonry with plastic cement and nails driven into the mortar joints; the triangular ends should be bent around the corners of the chimney and cemented into place.

The base flashing for chimney sides should be applied as shown in Figure 7.6-29. It should be secured to the deck with plastic cement and to the brick with plastic cement and nails. The triangular ends of the upper section should be turned around the chimney corners and cemented in place over the front base flashing.

The base flashing covering a saddle should be secured in plastic cement over

the underlayment. The corners should be installed as shown in Figure 7.6-30. The flashing over the cricket should be installed as shown in Figure 7.6-31, with the standing portion extending up the masonry 6 to 12 in. (152.4 to 304.8 mm). This should be capped as shown in Figure 7.6-32. The capping should be bedded tightly in plastic cement, centered over the saddle flashing, and extend up the roof. This provides added protection at the point where the ridge of the saddle joins the roof.

Metal counter flashing should be applied over the base flashing to complete the flashing installation (Figs. 7.6-33 and 7.6-34). One method of securing counter flashing to masonry is to rake out the mortar joint to a depth of $1\frac{1}{2}$ in. (38.1 mm) and insert the bent back edge of the flashing in the raked joint between bricks. The joint should then be filled with portland cement mortar or plastic cement and the flashing bent down over the base flashing.

The counter flashing on the chimney front should be one continuous piece. On the sides and rear, the sections should be cut to conform to brick joint locations and roof pitch and should lap each other at least 3 in. (76.2 mm).

***Stack Flashing*** Pipes projecting through a pitched roof require special flashing methods. Either corrosion-resistant metal sleeves which slip over stacks and have adjustable flanges to fit any roof slope or

**FIGURE 7.6-27** Vertical wall flashing using step flashing. (Courtesy Asphalt Roofing Manufacturers Association.)

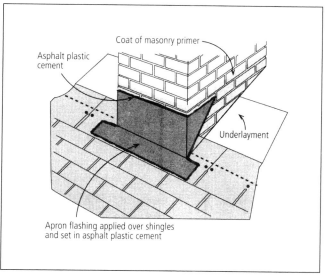

**FIGURE 7.6-28** Chimney front apron flashing. (Courtesy Asphalt Roofing Manufacturers Association.)

**FIGURE 7.6-29** Step flashing at the side of a chimney. (Courtesy Asphalt Roofing Manufacturers Association.)

**FIGURE 7.6-30** Corner flashing at the rear of a chimney. (Courtesy Asphalt Roofing Manufacturers Association.)

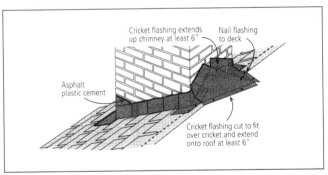

**FIGURE 7.6-31** Flashing over a cricket. (Courtesy Asphalt Roofing Manufacturers Association.)

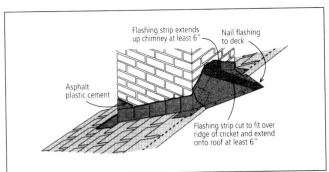

**FIGURE 7.6-32** Flashing over the ridge of a cricket. (Courtesy Asphalt Roofing Manufacturers Association.)

**FIGURE 7.6-33** Counter flashing at the front and side of a chimney. (Courtesy Asphalt Roofing Manufacturers Association.)

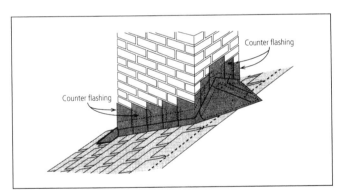

**FIGURE 7.6-34** Counter flashing at the side and rear of a chimney. (Courtesy Asphalt Roofing Manufacturers Association.)

asphalt products may be used with metal stacks. Plastic flanges are available for use with plastic plumbing stacks.

Roofing should be applied up to the point where the stack projects and fitted around it (Fig. 7.6-35). A shingle is cut to fit over the pipe and set in asphalt roofing cement. A preformed flange is then fitted snugly over the pipe and sealed to the pipe and to the shingles with asphalt plastic cement (Fig. 7.6-36). Plastic flanges are welded to plastic pipes using adhesives recommended by the pipe manufacturer. The flange should lay flat on the roof. Laying of shingles continues with shingles being cut to fit around the pipe and embedded in plastic roofing cement. The completed installation should appear as it does in Figure 7.6-37.

### 7.6.3.3 Shingle Application

Shingles should be laid so that they overlap and cover each other to shed water.

**FIGURE 7.6-35 Application of shingle over a vent pipe.** (Courtesy Asphalt Roofing Manufacturers Association.)

**FIGURE 7.6-36 Flashing at a vent pipe.** (Courtesy Asphalt Roofing Manufacturers Association.)

**FIGURE 7.6-37 Application of shingle around stack flashing.** (Courtesy Asphalt Roofing Manufacturers Association.)

The extent of overlapping depends on their shape and method of application and furnishes single, double, and sometimes triple coverage over different roof areas. Strip shingles, some interlocking shingles, and giant individual shingles applied by the American method, providing either double or triple coverage, are suitable for new construction, but as was stated earlier, strip shingles are the only ones in general use today.

Strip and individual hexagonal shingles and giant individual shingles, applied by the Dutch lap method, provide only single coverage and are used only to reroof existing construction. Refer to the manufacturer's literature for details of their application.

Shingles should be applied to a tight roof deck providing a suitable nailing base, with underlayment, drip edge, and flashings in place. Chimneys should be completed and counter flashing installed. Stacks and other equipment requiring openings through the roof should be in place, with provision for counter flashing where necessary; gutters should be hung.

## NAILS AND FASTENER

Nails (Fig. 7.6-38) for applying asphalt roofing should be made of hot-dipped galvanized steel, aluminum, or another corrosion-resistant metal, and should have sharp points and large, flat heads $3/8$ to $7/16$ in. (9.5 to 11 mm) in diameter. Shanks should not exceed 0.135 in. (3.429 mm) or be less than 0.105 in. (2.67 mm) in overall outside diameter (10- to 12-gauge (3.5–2.75 mm) wire). Shanks may be smooth or threaded, but threaded nails are preferred for increased holding power in lumber board sheathing and should definitely be used in plywood sheathing. Aluminum nails should have screw threads with approximately a $12^{1}/_{2}$ degree thread angle. Galvanized steel nails, if threaded, should have annular threads. Nail lengths typically required are given in Figure 7.6-39.

## STRIP SHINGLES (NORMAL SLOPE)

On small roofs, strip shingles may be laid from either rake. On roofs 30 ft (9.15 m) and longer, they should be started at the center and worked both ways from a vertical line to ensure more accurate vertical alignment and to provide for meeting and matching above a projection such as a dormer or chimney.

**FIGURE 7.6-38 Asphalt shingle nails: (a) smooth, (b) annular-threaded, and (c) screw-threaded.**

**FIGURE 7.6-39 Asphalt Shingle Nail Lengths**

| Application | 1 in. (25.4 mm) Sheathing | $3/8$ in. (9.5 mm) Plywood |
|---|---|---|
| Strip or individual shingle (new construction) | $1^{1}/_{4}$ in. (31.8 mm) | $7/8$ in. (22.2 mm) |
| Over asphalt roofing (reroofing) | $1^{1}/_{2}$ in. (38.1 mm) | 1 in. (25.4 mm) |
| Over wood shingles (reroofing) | $1^{3}/_{4}$ in. (44.45 mm) | — |

When a roof surface is broken, shingles should be started from a rake and laid toward a dormer or into a valley. Where the roof surface is unbroken, they should be started at the rake that is most visible. If both rakes are equally visible, shingles should be started in the center and worked both ways.

Minor variations in the dimensions of asphalt shingles are unavoidable. Variations seldom exceed plus or minus $1/4$ in. (6.4 mm) in either length or width in a 12 by 36 in. (304.8 by 914.4 mm) strip shingle and usually are no more than plus or minus $1/8$ in. (3.2 mm). However, to control the placement of shingles so that cutouts will be accurately aligned horizontally, vertically, and diagonally, horizontal and vertical chalklines should be used to control alignment.

## STARTER COURSE

The purpose of a starter course is to back up the first regular course of shingles and fill in the spaces between tabs. The starter course may be either a 7-in. (177.8-mm)-wide (or wider) mineral-surfaced roll roofing *starter strip* of a color to match the shingles or a row of inverted *starter course* shingles or strips made from the top lap portion of the shingles with the tabs cut off.

A *starter strip* should be applied over an eave flashing strip, even with its lower edge along the eave. The starter strip should be fastened with roofing nails 3 or 4 in. (76.2 or 101.6 mm) above the eave edge and spaced so that the nail heads will not be exposed at the cutouts between tabs on the first course (Fig. 7.6-40).

*Starter course shingles* should be applied with tabs facing up the roof over the eaves flashing strip. Nails should be spaced so that they will not be exposed at the cutouts of the first course (see Fig. 7.6-40).

***First and Succeeding Courses*** The first course should be started with a full shingle. Succeeding courses start with portions removed, depending on the shingle style and the pattern desired. Shingles with no cutouts and those with variable butt lines should be installed according to their manufacturer's directions. There are three methods of applying three-tab shingles: the 6 in. (152.4 mm) method, the 5 in. (127 mm) method, and the 4 in. (101.6 mm) method.

In the 6 in. method (Fig. 7.6-41), the second course is installed with the first 6 in. (152.4 mm) removed. Each succeeding course through the sixth is started with a shingle having an additional 6 in. (152.4 mm) removed. Adjacent shingles are installed full width. The seventh course then starts with a full shingle, and the pattern described is followed again.

In the 5 in. method (Fig. 7.6-42), the second course is installed with 5 in. (127 mm) removed. Each succeeding course through the seventh is installed with an additional 5 in. (127 mm) removed. Adjacent shingles are installed full width. Additional courses continue the 5 in. (127 mm) offset pattern. One method is to start the eighth course with 11 in. (279.4 mm) removed from the first shingle.

In the 4 in. method (Fig. 7.6-43), the second course is started with a shingle having 4 in. (101.6 mm) removed. Each succeeding course through the ninth is installed with an additional 4 in. (101.6 mm) removed. Adjacent shingles are installed full width. The 10th course then starts with a full shingle, and the pattern described is followed again.

## NAILING AND FASTENING STRIP SHINGLES

The correct number of nails per strip and correct nail placement are vital for suitable roof application. To prevent buckling, each shingle should be in perfect alignment before nails are driven. Nailing should be started at the end nearest the shingle last applied and proceed to the opposite end, with nails driven straight to avoid cutting the fabric of a shingle with the edge of a nail head. Nail heads should be driven flush, not sunk below the surface of the shingles. When a nail fails to

**FIGURE 7.6-40** Application of starter strip. (Courtesy Asphalt Roofing Manufacturers Association.)

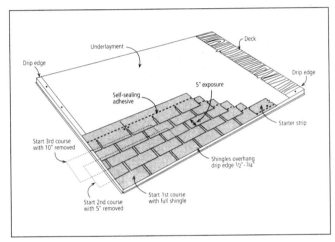

**FIGURE 7.6-42** Application of shingles using the 5 in. (127 mm) **method.** (Courtesy Asphalt Roofing Manufacturers Association.)

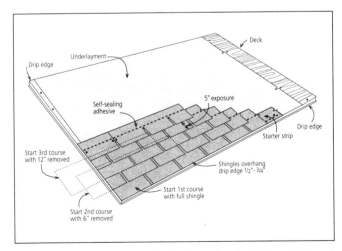

**FIGURE 7.6-41** Application of shingles using the 6 in. (152.4 mm) **method.** (Courtesy Asphalt Roofing Manufacturers Association.)

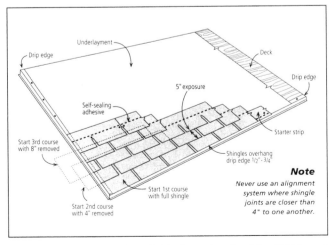

**FIGURE 7.6-43** Application of shingles using the 4 in. (101.6 mm) **method.** (Courtesy Asphalt Roofing Manufacturers Association.)

penetrate the sheathing, an additional nail should be driven in a new location.

***Nailing Three-Tab Shingles*** Three-tab shingles installed in locations having normal weather conditions require four nails for each three-tab strip. When such shingles are applied with a 5 in. (127 mm) exposure, the nails should be placed $^5/_8$ in. (15.9 mm) above the top of the cutouts and located horizontally with one nail 1 in. (25.4 mm) back from each end and one nail on the centerline of each cutout.

Three-tab shingles installed in high-wind regions require six nails for each three-tab strip. When such shingles are applied with a 5 in. (127 mm) exposure, the nails should be placed $^5/_8$ in. (15.9 mm) above the top of the cutouts and located horizontally with one nail 1 in. (25.4 mm) back from each end and two nails at each cutout, located 1 in. on each side of the cutout.

Windy locations require additional protection. In areas where wind velocities are 75 mph (120.7 km/hr) or greater, shingles manufactured with factory-applied adhesives or integral locking tabs that conform to the UL standard for Class C Wind Resistant Shingles should be used.

If UL Class C Wind Resistant Shingles are not used, an effective way to obtain protection against high winds is to cement the free tabs in place with an asphaltic plastic roofing cement. A spot of cement about the size of a quarter should be applied on each tab with a putty knife or caulking gun. The free tab should then be pressed against the cement. In applying the cement, shingles should not be bent back farther than necessary.

## HIPS AND RIDGES

Hips and ridges may be finished by using hip and ridge shingles furnished by the manufacturer or by cutting pieces at least 9 by 12 in. (228.6 by 304.8 mm) either from 12 by 36 in. (304.8 by 914.4 mm) shingle strips or from mineral-surfaced roll roofing of a color to match the shingles. They should be applied by bending each shingle lengthwise down the center with equal exposure on each side of the hip or ridge. In cold weather, the shingles should be warmed before bending. They should be applied beginning at the bottom of a hip or at one end of a ridge with a 5 in. (127 mm) exposure. Each shingle should be secured with one nail at each side, $5^1/_2$ in. (139.7 mm)

back from the exposed end and 1 in. (25.4 mm) up from the edge.

Metal ridge rolls subject to corrosion may discolor the roof and should never be used with asphalt roofing products.

## STRIP SHINGLES (LOWER SLOPE THAN NORMAL)

Strip shingles may be used on roof slopes less than 4 in 12 but not less than 2 in 12. Lower slope than normal application methods require (1) double underlayment, (2) a cemented eaves flashing strip, and (3) shingles provided with factory-applied adhesives conforming to the UL standard for Class C Wind Resistant Shingles, or each free tab of strip shingles should be cemented in place. Any shingle arrangement described under normal slope application methods may be used.

### 7.6.3.4 Reroofing Existing Construction

When a roof is to be reroofed with asphalt shingles, existing wood shingles, asphalt shingles, roll roofing, built-up asphalt roofing, or slate need not be removed before applying the new shingles, provided that (1) the strength of the existing roof framing is adequate to support the new dead and live loads, (2) the existing roof sheathing is sound and will provide adequate anchorage for nails used to apply new roofing, and (3) not more than two roofs are currently in place (the original and one reroof). Tile and cedar shake roofing must be removed. When either the framing or sheathing is inadequate, the old roofing, regardless of type, must be removed to correct the defects.

## EXISTING ROOFING TO BE REMOVED

When old roofing must be removed, a roof should be prepared for new roofing by (1) stripping the old roofing down to the roof deck, (2) repairing and reinforcing existing roof framing where required to provide adequate strength, (3) removing existing damaged sheathing and installing new sheathing in its place, (4) filling spaces between boards with securely nailed wood strips of the same thickness, or removing existing sheathing altogether and resheathing, (5) removing protruding nails, and (6) covering large cracks, slivers, knotholes, loose pitchy knots, and excessively resinous areas with sheet metal securely nailed to the sheathing.

## EXISTING ROOFING TO REMAIN

When inspection discloses that existing roofing need not be removed, the existing roofing's surface should be prepared to receive new roofing, as described below for various roofing types.

***Existing Wood Shingles*** Preparation includes:

- Removing loose or protruding nails and nailing the shingles in a new location
- Nailing down loose shingles
- Splitting badly curled or warped shingles and nailing down the segments
- Replacing missing shingles with new ones

When shingles and trim at eaves and rakes are badly weathered or the location is subject to high winds, the shingles there should be cut back and 1 × 4 (25 × 100) or 1 × 6 (25 × 150) boards nailed with their outside edges projecting beyond the edges of the deck the same distance as did the wood shingles (Fig. 7.6-44). Beveled wood "feathering strips" should be used along the butts of each course of old shingles to provide a relatively smooth surface to receive new asphalt roofing.

***Existing Asphalt Shingles*** Preparation includes:

- Nailing loose and protruding nails
- Removing and replacing badly weathered edging strips
- Removing or renailing loose, curled, lifted, or broken shingles
- Replace missing shingles with new ones

***Existing Roll Roofing*** Preparation includes:

- Splitting buckles and nailing segments down smoothly
- Removing loose and protruding nails, and placing new fasteners
- Trimming damaged and torn roofing to provide square or rectangular sides
- Patching tears in the roofing with new roofing

***Existing Built-Up Roofs*** If the slope of the roof is not less than 2 in 12 and if the roof sheathing is sound and can be expected to provide good nail-holding power, new shingles can be applied. Old slag, gravel, and other coarse surfacing materials should first be removed, leaving the surface of the underlying felts smooth

a

b

**FIGURE 7.6-44** Preparation of an existing roof to receive new shingles. (Reprint courtesy National Roofing Contractors Association.)

**FIGURE 7.6-45** Application of drip edge at a rake (a) and at an eave (b). (Courtesy Asphalt Roofing Manufacturers Association.)

and clean. New asphalt shingles should be applied directly over the felts as described under "Strip Shingles (Lower Slope Than Normal)" in Section 7.6.3.

## UNDERLAYMENT (REROOFING)

When existing roofing must be removed before applying new material, underlayment is required over the entire roof, as it would be for new construction.

When new shingles can be applied directly over old roofing, no underlayment is required, as the old roofing adequately serves the same purpose. Underlayment may be required by code over old wood roofing.

## DRIP EDGE AND EAVES FLASHING STRIP (REROOFING)

When installing new asphalt shingles over an existing wood-shingled roof, drip edges and eaves flashing strip should be installed as for new construction (Fig. 7.6-45).

## FLASHING (REROOFING)

Existing metal flashing, unless perforated or deteriorated, may be left in place and reused.

*Valley Flashing*   When the existing roof has an open valley, the valley is built up with mineral-surfaced roll roofing and valley flashing is installed in the same way as for new roofing.

*Flashing Against a Vertical Wall*   When new asphalt shingles are being applied over old asphalt shingles, the joint between a vertical wall and the rake of an abutting roof may be treated as shown in Figure 7.6-46. When asphalt shingles are being installed over old wood shingles, a 6 or 8 in. (152.4 or 203.2 mm) strip of smooth No. 50 roll roofing or a mineral-surfaced roll roofing placed upside down should be applied over shingles abutting a wall surface, using a row of nails placed about 4 in. (101.6 mm) on centers along each edge. As the work proceeds, the strip should be covered with asphalt plastic cement and the end of each course of new shingles should be firmly secured by bedding it in the cement. An improved appearance and a tight joint can be achieved by using a caulking gun to apply a final cement bead between the ends of the shingles and the siding.

*Chimney Flashing*   When existing flashing is damaged, it should be removed and new flashing installed as described earlier in this section. When existing flashing is in good condition, it may be lifted and shingles installed beneath it. The area of shingle beneath the flashing should be coated with asphaltic plastic cement. At the sides and at metal crickets, the shingles should be trimmed to within 1/4 in. (6.4 mm) of the flashing and the last 3 in. (76.2 mm) of the shingles embedded in

**FIGURE 7.6-46** Wall flashing (reroofing). (Courtesy Asphalt Roofing Manufacturers Association.)

asphaltic plastic cement. Existing shingled crickets are finished the same way as new shingled crickets.

## SHINGLE APPLICATION (REROOFING)

After the existing roof has been prepared, the methods of applying the various shingle types are similar to those described in this section for new construction. The major difference is a nesting procedure used to minimize unevenness that might result from new shingles overlapping the butts of existing shingles (Fig. 7.6-47). This nesting is achieved, in addition to removing the tabs from the starter strip, by removing an additional 2 in. (50.8 mm) from the top so that when installed,

**FIGURE 7.6-47** Application of new asphalt shingles over existing asphalt shingles using the nesting method. (Courtesy Asphalt Roofing Manufacturers Association.)

the starter course does not overlap the existing course above. In addition, 3 in. (76.2 mm) are removed from the rake end of the starter course.

A 2-in. (50.8-mm)-wide strip is also removed from the top of the first course so that the strips fit between the butts of the existing third course and the eaves edge of the new starter strip without overlapping the existing third course.

Full-width shingles are installed for the second and succeeding courses. This leaves a 3 in. (76.2 mm) exposure on the first course instead of the normal 5 in. (127 mm). Succeeding courses, however, have the full 5 in. (127 mm) exposure and do not overlap the butts of existing shingles.

### 7.6.4 SLATE ROOFING

Slate shingles are not widely used today, especially on commercial projects, but so much slate roofing exists that some discussion is appropriate.

#### 7.6.4.1 Materials

##### SLATE

Slate is a hard, practically nonabsorbent rock characterized by its natural proclivity to split along a single plane. This single-direction cleavage characteristic permits slate to be split into thin sheets. Roofing slate is usually split in the direction of its natural grain.

Roofing slate is classified as either *commercial standard*, *textural*, or *graduated*. Commercial standard roofing slate is smooth and about $^3/_{16}$ in. (4.8 mm) thick. Textural roofing slate is $^3/_8$ in. (9.5 mm) thick or thinner. Its color is more variable than that of commercial standard slate, and its surface texture is rougher. Graduated roofing slate varies in thickness from $^3/_{16}$ in. (4.8 mm) to $1^1/_2$ in. (38.1 mm) or thicker. It also varies more

in color and texture than either commercial standard or textural slate. Some slate used in roofing (*ribbon stock*) has strips of other rock embedded in it. Slate without ribbons is called *clear slate*.

Individual slate roofing pieces (*slates*) vary in size from 10 by 6 in. (254 by 152.4 mm) to 24 by 16 in. (609.6 by 406.4 mm). Most slates are square, but corners are sometimes clipped to produce special effects. Most slate is split so that the grain runs with the length of the slate.

Slate is available in several colors and shades, including those classified in the industry as black, blue-black, gray, mottled gray, purple, variegated purple (mottled purple and green), green, and red. Other colors (*specials*) may also be used. Existing slate will almost certainly have changed color over time and may be difficult to match with new material. New slate will, of course, also change color, but there may be a time when new slates are apparent. Slate is classified as *permanent* (or *unfading*) or as *weathering*, denoting the degree of color fading expected.

Before selecting material to match existing slate, it is important to determine the source of the existing slate and try to find the same or similar slate from the same source as a match. Slate is currently mined in Virginia, New York, Maine, Vermont, and Pennsylvania, but existing slate may have come from another location. Slate from different states, and even slate from different quarries in the same state, or from a single quarry, may not match. Fortunately, some variation is normal, especially in some colors, so that an acceptable match is often possible.

Every slate should have at least two nail holes machine punched at the quarry. Slates $^3/_4$ in. (19.1 mm) or more in thickness when 20 in. (508 mm) or more in length should have four nail holes. Hand punching should be done only where absolutely necessary, such as for fitting hips.

#### FASTENERS

Zinc nails, and copper, brass, yellow metal, and other nails with copper content, have all been used to install slate with success. For repairing slate work, however, the best choice is large-head copper slaters' nails. Under no circumstances should nails intended for any other purpose, such as roofer's nails or galvanized steel nails, be used with slate roofing.

Nails should be at least 1 in. (25.4 mm) longer than the slate thickness, but never less than long enough to penetrate the roof sheathing. Nails should not, however, project on the other side of the sheathing unless they will be concealed in the finished work.

Nail gauge depends on the weight of the slates and locations within the roof. For example, 3d nails can be used for commercial standard slates up to 18 in. (457.2 mm) long. Longer commercial standard slates require 4d nails. Hips and ridges of roofs covered with commercial standard slate should be installed using 6d nails. Heavier slates require heavier nails.

#### 7.6.4.2 Preparation for Slate Roofing

##### SHEATHING

In the past, slate was often installed using wood pegs hooked over wood lath. More recently, some slate was hung from wood lath or battens, or directly from steel structural elements by heavy wire hooked to the wood or structural elements. Sometimes slate was held to steel sections using long, bent copper nails. Today, most slate roofing is installed over continuous nailable substrates, often wood, plywood, or a wood fiber product, sometimes over a nailable portland-cement-based material, and fastened in place using nails.

##### UNDERLAYMENT

Surfaces to receive slate roofing $^1/_4$ in. (6.4 mm) or less in thickness, where the roof slope is 4 in./ft (333.3 mm/m) or greater, should be covered with an underlayment consisting of at least one layer of No. 30 asphalt-impregnated, unperforated roofing felt.

Surfaces to receive slate roofing with a thickness greater than $^1/_4$ in. (6.4 mm), where the roof slope is 4 in. (101.6 mm) or more per foot (304.8 mm), should receive an underlayment consisting of two layers of No. 30 asphalt-impregnated, unperforated roofing felt or a single layer of No. 55 or 65 asphalt-impregnated, unperforated roofing felt.

Surfaces to receive slate roofing where the roof slope is less than 4 in./ft (333.3 mm/m) cannot be rightly called steep-slope roofs. In such conditions, slate should be installed as tile (without laps) as a ballast over membrane roofing or as a promenade surface over membrane waterproofing.

Under the conditions outlined earlier in this section, slate roofs should have ice shields. The ice shield should consist of at least two layers of No. 30 felt.

### 7.6.4.3 Slate Installation

Slates installed using nails should each receive no fewer than two nails. Four nails should be used in larger slates. Nails should be driven so that the head just touches the slate. The slate should not be clenched by the nail, but should hang on the nail.

Slates should be set so that joints fall as close as possible to the midpoint of the length of the slate immediately above or below, but never less than 3 in. (76.2 mm) from the end of those adjacent slates (see "Joint Break" in Fig. 7.6-48). Joints should never be open clear through to the underlayment below.

Slate weather exposure depends on the size of the slate and the headlap (see Fig. 7.6-48). Exposure is the shingle length minus the headlap divided by 2. The exposure for a 24 in. (609.6 mm) shingle with a 3 in. (76.2 mm) headlap is 10½ in. (266.7 mm). The amount of headlap varies with the roof slope and region. On slopes from 0 to 4 in./ft (333.3 mm/m), slate should be laid as tile, with no lap. On slopes from 4 to 8 in./ft (333.3 to 666.7 mm/m), headlap is often the standard 3 in. (76.2 mm), but roofers in some parts of the country consider 4 in. (101.6 mm) necessary. On slopes from 8 to 20 in./ft (666.7 to 1666.7 mm/m), the standard 3 in. (76.2 mm) headlap is generally accepted as sufficient. In some southern areas, a headlap of 2 in. (50.8 mm) is considered adequate, but in most of the country, 2 in. (50.8 mm) headlap is restricted to mansards and other roofs with slopes of more than 20 in./ft (1666.7 mm/m).

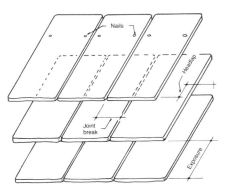

**FIGURE 7.6-48 Slate jointing.** (Courtesy H. Leslie Simmons.)

Slate is sold in squares. A square of slate is the amount needed to cover a square of roof (100 sq ft [9.29 m²] of roof surface) when the headlap is 3 in. (76.2 mm). The actual headlap to be used must be considered when calculating the amount of slate necessary to cover a given area of roof. More headlap will result in less coverage for a square of slate.

Random-appearing slate roofs using standard and textural slate are achieved by varying the slate width while keeping the length equal or by varying both width and length. Equal-width slates are seldom used.

Variation in graduated slate roofing is achieved by varying lengths, widths, and thicknesses. Larger and thicker slates are used in lower courses and graduate to thinner and smaller slates at the ridge. All courses contain various slate thicknesses.

Repaired hips and ridges should match the existing ones. The National Roofing Contractors Association's *The NRCA Steep Roofing Manual* and the National Slate Association's *Slate Roofs* each contain detailed descriptions and drawings showing how several common types of hips and saddles should be formed.

Ridges may be *saddle* type or *comb* type. A saddle ridge is one in which the top slates abut evenly. In a regular saddle ridge, the top slates overlap the adjacent top slate (Fig. 7.6-49a).

Two nails, concealed by the next higher slate, are used in each top slate. In a strip saddle ridge, the top slates do not overlap; they abut, and each slate is fastened down using four nails. Figure 7.6-49b is a section through either a regular or strip saddle ridge.

Comb ridges are formed similarly to strip saddle ridges, except that the top of the ridge slates do not abut, but overlap by ¹⁄₁₆ in. (1.6 mm) to ⅛ in. (Fig. 7.6-50). The projection may occur all on one side, or the projecting slates may alternate from one side to the other in a configuration called a *coxcomb ridge*.

There are several ways to form slate roofing hips. Four common methods are known as *saddle*, *mitered*, *fantail*, and *Boston* hips. *Saddle hips* and *strip saddle hips* are similar to saddle and strip saddle ridges, except that each slate is fastened in place using four nails.

In *mitered hips*, the slates forming the hip are in the same plane as the regular roofing slates and are cut to the slope of the hip (Fig. 7.6-51).

**a**

**b**

**FIGURE 7.6-49** (a) Regular saddle ridge. (b) Section through a saddle ridge. (Courtesy H. Leslie Simmons.)

**FIGURE 7.6-50** Detail at a comb ridge. (Courtesy H. Leslie Simmons.)

**FIGURE 7.6-51** Mitered hip. (Courtesy H. Leslie Simmons.)

*Fantail hips* are a variation of mitered hips in which the bottom edge of the hip shingle is cut at an angle to form a fantail (Fig. 7.6-52).

*Boston hips* are similar in appearance to saddle hips, but are constructed by

**FIGURE 7.6-52 Fantail hip.** (Courtesy H. Leslie Simmons.)

weaving three shapes of cut shingles together (see Figure 7.6-53).

Flashing beneath the hip types mentioned here is usually not necessary. Some existing conditions, however, may have been flashed with metal strips woven in with each course.

Valleys may be either *open, closed,* or *round.* Open valleys are protected using metal flashing. They should start with the slate 4 in. (101.6 mm) apart at the top and widen uniformly from top to bottom at a rate of 1 in. (25.4 mm) in every 8 ft (2400 mm).

In closed valleys, the slate is cut to abut in the valley, and metal flashing is woven in as the slate is installed.

Flashing is not required in properly installed round valleys. Round valleys require considerable underlying support and special slate sizes. Considerable expertise is needed to properly install round valleys. Round valley slates are sometimes bedded in elastic cement. Round valleys built by inexperienced or insufficiently skill personnel should be flashed. Round valleys susceptible to ice damage should always be flashed. Round valley flashing may be metal or bituminous materials.

In all types of ridge and hip conditions, the top regular slate courses should

**FIGURE 7.6-53 Boston hip.** (Courtesy H. Leslie Simmons.)

have their edges set in elastic cement to resist wind uplift. The joint between ridge slates and the overlap in comb ridges should be sealed with elastic cement. Some slate roofing is installed with elastic cement in other locations, such as along flashing lines and in the field of the roofing. Some slate roofing is installed with no elastic cement whatever, and experiences no leaks as a result. The absence of elastic cement does not, in itself, mean that there is a problem or that the installation in inferior.

In every case, however, exposed nails should be coated with elastic cement. Where slates are nailed near flashing, the nails should not penetrate the flashing.

### 7.6.5 TILE ROOFING

Industry standards for tile roofing are scarce. Sheathing, underlayments, fasteners, and tile should be installed in accordance with the tile producers' recommendations, applicable building code requirements, and the highest standards of the trade. The requirements presented in this section are a conglomerate of those generally recommended by the producers. Recommendations of industry-recognized sources, such as the National Roofing Contractors Association, have also been taken into account where they were found. Individual producers and their associations' recommendations and building codes may differ, however.

#### 7.6.5.1 Materials

##### CLAY TILE

Clay tile is dense, hard, nonabsorbent material produced by baking molded clay units in a kiln. It may be either glazed or unglazed.

Clay tile was used before the Golden Age of Greece. Since at least the early twelfth century, some clay tile shapes have remained virtually the same. Clay tile for roofing comes in four general styles: "S" (Spanish), half-round (Mission), pan and cover (Roman), and flat tile. Figure 7.6-54 shows some general clay tile shapes, including three in the flat category. Available shapes may vary from those shown, especially in the flat and pan tile types. It is also possible that an existing roof will have a shape not shown.

Flat shingles may have special butt shapes. These are sometimes called *Persian* tile. Special butt shapes include those

**FIGURE 7.6-54 Clay tile shapes.** (Courtesy H. Leslie Simmons.)

with rounded corners, round ends, half-hexagonal ends, half-octagonal ends, single pointed ends, and many others.

Some flat interlocking tile comes in $1/4$-, $1/2$-, and $1^1/2$-sized units to permit finishing at rakes.

Special shapes are used for finishing hips and ridges, and for trim and closures at ridges, hips, rakes, eaves, and valleys. Special shapes include regular tiles with the ends closed and specially shaped pieces to fit within or between tiles. Shapes vary to work with each tile type.

Round building elements, such as towers and turrets, may be roofed using *graduated* tiles. Graduated tiles are units made to match the regular types but specially shaped to fit a rounded surface. Tile manufacturers furnish such tiles routinely, but matching an existing graduated tile may be a problem. If the particular existing shape is no longer available, it may be necessary to have a match manufactured.

Some clay tile is finished to mimic wood shingles or slate. Broken or cut tiles will show clay's characteristic red color, however, and are easily identified as clay.

## CONCRETE TILE

Concrete tile is made from a mixture of portland cement, aggregate, and water and is cured under pressure at a controlled temperature and humidity. Some tile is cast; other tile is extruded. Concrete mixes vary from manufacturer to manufacturer. Some manufacturers use sand as the aggregate; others use lightweight materials such as perlite. Some concrete tile is finished on the exposed surface while still wet, using a pigmented cement slurry. Some is colored throughout by pigment added to the concrete mix. Some is the natural concrete color with no added coloration. Concrete tile may be glazed or unglazed.

Some concrete tile is made to mimic wood shingles, clay tile, or slate in shape, color, and texture. Except under close scrutiny, the match with clay tile is fairly successful when the tile is new, differing only in individual tile shape. The attempt to match other materials, however, often fails. For example, some concrete tile is too large to successfully mimic wood shingles. Broken edges identify concrete tile made to look like slate or wood. Some concrete tile matches the color of new wood shingles effectively but, of course, does not age as wood does. In addition, the fungus that grows on wet concrete does not look like the fungus that grows on aging wood. Concrete tile does, however, enjoy the major advantage over wood of being fire resistant. And some concrete tile is colored during manufacture to match closely the appearance of mossy, aged wood. A great advantage of using this concrete tile instead of old wood is that the concrete tile will have that same appearance 50 to 75 years later, while wood that far gone will soon need replacing. Such colored concrete tile can be made to look more natural by using different colors placed randomly.

All concrete tile absorbs some moisture. Tile made with some lightweight aggregates absorbs more moisture than tile made with sand aggregate. The moisture absorption does not seem to harm the tile or its reinforcement, but is sometimes unsightly and can contribute to rusting metal or rotting wood in the underlying structure. Some concrete tile producers include water-absorption inhibitors in their mixtures to prevent such damage from occurring.

Most concrete tile used for roofing is either the standard concrete barrel tile shape (Fig. 7.6-55) or a flat interlocking shape similar to the unit called "Flat with interlocking feature" in Figure 7.6-54. Shapes used on an existing building may vary slightly from those shown. A common variation, for example, is in the width of the valley in the barrel tile shape. Concrete tile on an existing roof may also be similar to one of the other styles shown in Figure 7.6-54 for clay tiles or another shape not shown here.

Special shapes are used for finishing hips and ridges, and for trim and closures at ridges, hips, rakes, eaves, and valleys. Special shapes include regular tiles with the ends closed and specially shaped pieces to fit within or between tiles. Shapes vary to work with each tile type. Round building elements, such as towers and turrets, may be roofed using *graduated* tiles, as described for clay tile in Section 7.6.5.1.

## FASTENERS

Tile may be nailed in position or hung from wire hangers, depending on the tile type, substrates, and code requirements.

Fasteners should be noncorrosive, such as stainless steel, aluminum, or a copper-bearing metal such as hard copper, silicon copper, or yellow metal. Nails should be at least 11-gauge (3.1 mm), large-headed (minimum $^5/_{16}$ in. [7.9 mm] diameter), and long enough to penetrate completely through plywood sheathing, not less than $^3/_4$ in. (19.1 mm) into other wood nailers, and the distance into other nailable materials recommended by the tile manufacturer. Nails should not, however, penetrate completely through board plank decks. Nails for use in plywood should be ring-shank. Nails for use in boards should be smooth shank. Nails for use in gypsum plank or nailable concrete should be stainless steel or silicon bronze screw shank nails long enough to penetrate one-half and three-quarters their length, respectively, into the deck. They should never penetrate fully through the deck. For very hard decks, it may be necessary to use smooth shank nails. Some so-called nailable decks may not have sufficient nail-holding power to support concrete tile.

Wire hangers are sometimes used to hang tile from wood battens but are more often used on concrete and steel substrates. Hangers are also used, even where the substrate is wood, in locations where driven nails would penetrate flashings and in locations where driving nails is difficult. Wire hangers and associated ties are available in stainless steel, brass, copper, and galvanized steel. One system employs a $1^1/_2$ in. (38.1 mm) by $^1/_2$ in. (12.7 mm) wire strip hanger and 14-gauge (1.9 mm) tie wire, but a variety of hanger and tie types are available which may be acceptable to a particular tile manufacturer.

*Hurricane clips*, also called *storm clips*, are manufactured from brass or galvanized steel to shapes necessary for use with each tile type. Clips should not be visible in the completed roof. Hurricane clips are fastened positively to the roof deck and clip over tile edges to hold tiles in place during high winds.

## MORTAR AND CEMENT

Most mortar used in clay tile roofing is composed of one part portland cement, four parts sand, and coloring to produce a mortar that matches the color of the tile.

Concrete tile manufacturers do not agree on the mix for mortar for use with concrete tile or even on the materials to be used. It may be difficult or impossible to determine the actual mortar mix that was used in an existing application. Mortar color usually matches tile color.

In *mud-set* tile applications, concrete tile manufacturers usually recommend using Type M mortar as defined in ASTM Standard C270.

Plastic cement should be a heavy-bodied asphalt roofing cement. Some tile manufacturers prefer that a silicone sealant be used in lieu of plastic cement. Plastic cement and sealant should be colored to match the tile or clear.

**FIGURE 7.6-55** Concrete barrel tile type.
(Courtesy H. Leslie Simmons.)

## 7.6.5.2 Preparation

### SUBSTRATES, BATTENS, AND SUPPORT STRIPS

Clay and concrete tile may be installed over wood, metal, or concrete decks. Fastener requirements differ, depending on the tile configuration and deck type.

Wood sheathing for clay or concrete tile should be either exterior-grade plywood or boards. Plywood thickness should be selected to comply with structural and fastener requirements but should not be less than $1/2$ in. (12.7 mm). Plywood joints should be left open $1/16$ in. (1.6 mm) to permit expansion.

Board sheathing should be at least 1 × 6s (25 × 150s), spanning not more than 24 in. (609.6 mm). Supporting structure spans wider than 24 in. (609.6 mm) require heavier boards. Spaced sheathing is not generally recommended for clay tile but may have been used on some existing roofs. Some sources say that spaced sheathing is not recommended in concrete tile installations, but some manufacturers recommend installation over spaced board sheathing and some concrete tile roofs have been installed in that manner. Concrete tile applied over spaced sheathing should be installed in accordance with the tile manufacturer's recommendations.

Some manufacturers and many tile workers prefer that concrete tile be installed over wood battens on every roof, regardless of slope, because using battens makes replacing damaged tile much easier than if battens are not used. Another reason is the general feeling that nails do not hold in plywood well enough to support heavy concrete tile over an extended period. Many low-slope concrete tile roofs have been installed without battens, of course, but battens should be used on slopes of 7 in./ft (583.3 mm/m) and steeper, regardless of tile type or manufacturer's recommendations. Battens are usually 1 × 2s (25 × 50s), fastened to the substrate at spacings recommended by the tile manufacturer. Spacing is dependent on tile type and size. Clay and concrete tile are seldom installed over cast-in-place concrete decks, because few cast-in-place concrete decks slope as much as 3 in./ft (250 mm/m), which is the minimum at which the tile can be nailed to a roof. Such tile applied over cast-in-place concrete is usually fastened in place using wood battens in a config-

uration known as the *counter batten* system. In a counter batten system, wood battens, usually 1 × 2s (25 × 50s), are nailed to wood nailers cast into the concrete deck from eave to ridge at about 20 in. (508 mm) on center. Underlayment is installed over the deck and cast-in nailers before battens are applied. Batten spacing depends on tile size.

Clay and concrete tile installed over precast concrete decks (and sometimes over cast-in-place concrete) is usually fastened to a counter batten system, in which the wood members running up the slope are through-bolted to the concrete. Sometimes the batten running up the slope is omitted and the horizontal battens are themselves fastened directly to the deck by through-bolting.

A system of nailers and wood battens may also be used over metal decks. Often, however, fire ratings or code requirements will dictate that a fire-retardant sheathing be applied over the metal decking. Nailable fire-retardant sheathing, when the sheathing manufacturer so recommends and the tile manufacturer concurs, can be used the same way as a plywood deck.

Battens can also be fastened over underlayment to nailers set flush with nonnailable insulation over any deck. Insulating products that are also nailable can be used like plywood when their manufacturer so recommends and the tile manufacturer concurs.

Manufacturers of some clay tile types recommend that the tile be applied over battens under all, or sometimes just certain selected, conditions. Such recommendations should be followed.

A means should be provided for water that finds its way through the tile to drain out of the roofing system. One way to do this is to separate the end joints between battens about $1/2$ in. (12.7 mm), with end joints about 40 in. (1016 mm) on center. The problem with this method is that since joints must fall over the nailers below, the battens must be in many relatively short pieces. A better way is to raise the battens at least $1/4$ in. (6.4 mm) above the underlayment using cut shingles or wood lath strips laid beneath the battens along the line of the nailers and fastened through the underlayment to the nailers.

Tile may also be attached directly to concrete or steel decks using wire hangers. "S" and half-round clay tile require

nailers at ridges, hips, rakes, and projections. Flat clay and concrete tile requires nailers at ridges and hips. Nailer size varies with tile type and rise, which may be different for different manufacturers and is definitely different for different tile types. Nailers are either 1 in. (25 mm) or 2 in. (50 mm) nominal lumber, depending on the tile configuration and the tile manufacturer's recommendations. Hip and ridge nailers are usually set in cement mortar.

Eave strips are required except for some types of flat clay tile and for concrete tile where starter course elevation is achieved using a specially made tile strip and other tile where special tile accessories are used to elevate the first course. Eave strips are usually 1 × 2 (25 × 50) wood members. Another way to raise the starter course is to elevate the fascia board. In any case, a tapered cant just above the nailer strip or tile starter course, or some other device, is necessary to prevent water from ponding at the eave line.

### UNDERLAYMENT

Surfaces to receive clay tile should be completely covered with at least one layer No. 30 asphalt-impregnated roofing felt. Additional layers of underlayment are required in some circumstances, but industry sources do not all agree on which circumstances dictate additional layers. For example, some sources say that roof slopes of 3 in./ft (250 mm/m) or steeper can receive any type clay tile when completely covered with two layers of No. 30 asphalt-impregnated felt set in hot asphalt or mastic. These same sources imply that flat clay roofing tile needs a double layer of No. 30 felt underlayment on all slopes less than 5 in./ft (416.7 mm/m), but that it can be installed on a single layer of felt for steeper slopes.

Other sources, including many tile manufacturers, however, do not agree on these requirements. Some manufacturers agree that any clay tile shape can be used on any slope exceeding 3 in./ft (250 mm/m) when there is a double layer of No. 30 felt underlayment. Other manufacturers, however, recommend a double layer of No. 30 felt beneath flat tile on all slopes. Some see no need for double layers of felt beneath clay tile except for flat clay tile, regardless of slope. Some say that one layer of No. 40 felt can be substituted for the two No. 30 layers.

Some say that underlayment on steep slopes should consist of a single No. 40 felt layer, and that underlayment on low slopes should consist of No. 60 felt or two layers of No. 30 felt.

The level of disagreement related to underlayment for concrete tile is similar. Some sources say that, to be waterproof, concrete tile should be installed on slopes of 4 in./ft (333.3 mm/m) or steeper, that flat shingle tile requires a slope of 5 in./ft (416.7 mm/m) or more, and that any slope below 4 in./ft (333.3 mm/m) should have two layers of roofing felt underlayment set in hot asphalt or mastic.

Other sources, however, do not agree with these requirements, nor do they agree with each other. Some say that concrete tile can be nail-applied on slopes of 3 in./ft (250 mm/m) or steeper. Some manufacturers recommend that a single-ply No. 40 asphalt-impregnated, nonperforated roofing felt underlayment be applied as a minimum. Some recommend that surfaces with slopes between 3 in./ft (250 mm/m) and 4 in./ft (333.3 mm/m) to receive nailed-on concrete tile should be completely covered with an underlayment consisting of two layers of No. 15 felt with hot asphalt between the plies, or a waterproofing membrane, or a self-adhering bituminous membrane. Some say that slopes of 4 in./ft (333.3 mm/m) or steeper should receive a single layer of No. 30 asphalt-impregnated roofing felt.

Consequently, an existing clay or concrete tile roof may have been installed using any one of those underlayment systems or some other system entirely.

There is general agreement that either clay or concrete tile on slopes below 3 in./ft (250 mm/m) must be installed only over built-up roofing or proper membrane waterproofing. Some manufacturers say that concrete tile on slopes less than 4 in./ft (333.3 mm/m) should be set in mortar beds over membrane roofing or membrane waterproofing. The method is called *mud-set*.

Some manufacturers recommend that mud-set tile applications be made over mineral-surfaced roll roofing weighing not less than 83 lb (37.65 kg) per square. Others have different requirements.

All sources also agree that valleys should have an additional 36-in. (914.4-mm)-wide layer of No. 30 roofing felt beneath the flashing, and that felts should be carried 6 in. (152.4 mm) up vertical surfaces.

Under the conditions outlined earlier, clay and concrete tile roofs should have ice shields. The ice shield should consist of at least two layers of No. 30 felt.

### 7.6.5.3 Installation

The following discussion assumes that clay and concrete tile, other than in mud-set applications of concrete tile, will be installed using nails. Similar requirements also apply to tile installed using wire hangers. Wire hanger installation should be done in strict accordance with code requirements and the tile and wire hanger manufacturers' recommendations.

In some areas of the country, on roof slopes below 4 in./ft (333.3 mm/m), concrete tile is sometimes installed without nails in mortar beds laid directly over roll roofing. These mud-set applications should conform to the tile manufacturer's recommendations.

Clay tile should be installed with one, two, or three nails. Nail-applied concrete tile should be hooked over battens (where battens are used) and nailed in place with one, two, or three nails. In each case, the number of nails depends on the type of tile and the tile manufacturer's recommendations. Most field tile will require only one nail. Hip and ridge covers require one or more nails. Tile overlapping flashing should be supported with wire hangers and set in plastic cement. Nails heads should just clear the tile and should never be driven down to the tile. Nails must be driven into nailable material, never into joints in the substrates. Some codes require that every concrete tile be nailed. Other codes permit nailing alternate tile rows or even every third tile row. Sometimes requirements vary, depending on roof slope.

The first course of tile should be elevated to the proper height by special tile units manufactured for the purpose, by wood strips, or by a raised fascia. Provisions should be made to permit water to drain from the roof beneath the tile. The first tile course should extend beyond the edge of the sheathing or fascia board by $1^1/_2$ to 3 in. (38.1 to 76.2 mm), according to the tile manufacturer's recommendations.

Tile is usually installed with butts parallel to eaves and with tile perpendicular to eaves. Tile should be cut to follow the line of hips and valleys or special tile may be used. Valleys may be open, closed, or rounded. Most valleys in tile roofs, however, are open. Open ends of tile at val-

leys should be sealed with special units or mortar, but tile at valleys should not be sealed in a way that would prevent water that has penetrated through the tile to the underlayment surface from draining into the valley. Fan-shaped valley tile should not be sealed at the laps. Battens along valleys should be broken to permit trapped water to escape onto the valley flashing. Tile should lap valley flashing by at least 4 in. (101.6 mm).

Open valleys should be left open 6 in. (152.4 mm) at the top and should taper wider at the rate of 1 in. (25.4 mm) for every 8 ft (2438 mm) of length.

Roofing tile should be cut and fitted close to and sealed against hip and ridge boards using plastic cement or silicone sealant. Similarly, beads of cement should be placed between tiles at hips and ridges. Hip and ridge covers should be nailed in place and lapped 3 in. (76.2 mm). Laps should be sealed with plastic cement. Open space within hip and ridge covers should not, however, be sealed against air circulation. The space between roofing tiles at ridges should be sealed using special units made for the purpose or mortar. Additional nailers should be applied if special ridge filler tile units are used. Hip and ridge cover ends should be closed using special units designed for the purpose or by cement mortar. Where cement mortar is used, care should be exercised to prevent water intrusion beneath the tile, where it can lead to efflorescence on the face of the tile.

On flat-tile roofs, one-and-one-half-width tiles rather than half-tile units are usually used at gables. The smaller tiles are too light and tend to crack or blow off in windstorms.

Most tile is set with uniform exposure, but some is set with varying exposure. During tile installation, spacing of tiles may need adjustment to achieve uniform exposure. Exposure is determined by tile size and headlap (the distance a tile laps over the preceding tile course). Recommended headlap may vary from manufacturer to manufacturer, but most recommend that headlap not be less than 3 in. (76.3 mm).

Hurricane clips are required on the nose end of each eave-course tile in areas classified by code as wind hazard areas and, whether so classified or not, where winds frequently exceed 70 mph (112.65 km/hr). On steep-slope roofs, hurricane clips should also be used periodically throughout the roof, or the butt of each tile may be embedded in a plastic cement or silicon sealant bead. Some codes may require additional provisions.

Tile on steep or vertical roofs should be protected against lifting by wind, even when not in wind hazard areas, by means of wind lock clips.

Other precautions are necessary in high-wind areas. For example, tile head-lap should be increased to 4 in. (101.6 mm), and a bead of mastic should be applied over nail heads at gable, rake, and ridge tiles.

Open-end tile should be sealed at the eaves using special units made for the purpose or sealed with mortar. Either method must permit trapped water to drain out.

## 7.6.6 MINERAL-FIBER-CEMENT SHINGLES

Because of possible health hazards, roofing and siding shingles that were earlier made using asbestos fibers are now manufactured using nonasbestos mineral fibers. Therefore, the term used here for such shingles is *mineral-fiber-cement shingles*. Similar materials found in older buildings, however, may contain asbestos. Readers are, therefore, strongly advised to consult current OSHA, EPA, state, and local regulations and suggestions with regard to safety precautions, and to obtain professional assistance when dealing with such existing materials.

Refer to Section 7.6.1 for (1) definitions of terminology, (2) limitations imposed by the slope of the roof, (3) sheathing guidelines (see also Chapter 6), and (4) general guidelines for typical accessory materials, such as underlayment, flashing, roofing cements, eaves flashings, drip edge, nails, and caulking for preparation of a roof for application of shingles. Unless specific exceptions are made in the following text, the provisions of Section 7.6.1 apply to mineral-fiber-cement shingles and the methods used to apply them.

### 7.6.6.1 Materials

Mineral-fiber-cement shingles can be discolored by moisture and dampness while still in packages and until applied. Therefore, they should be kept clean, dry, and completely protected from the weather until they have been applied. Shingles should not be applied over wet sheathing, underlayment, or sheathing paper.

## ROOF SHINGLES

Mineral-fiber-cement roof shingles are manufactured in four basic types: (1 and 2) *individual units* and *multiple units* for application by the *American method* and (3 and 4) *Dutch lap* units and *ranch-style* units for application by the *Dutch* (or *Scotch*) *method*. Data covering the four types are summarized in Figure 7.6-56. Each type of shingle should be applied in a distinct way, giving a different appearance. Starter units, hip and ridge shingles, and ridge roll are also available.

Roof shingles are made with prepunched holes correctly located to receive concealed roofing nails. Additional holes are provided in Dutch lap and ranch-style units to receive storm anchors.

Mineral-fiber-cement roof shingles should conform to the requirements of ASTM Standards C1225 and C1530.

## NAILS AND FASTENERS

Roofing nails for applying mineral-fiber-cement shingles should be hot-dip galvanized steel, aluminum, stainless steel, or another corrosion-resistant metal, and should be needle or diamond pointed, with large flat heads approximately $3/8$ in. (9.5 mm) in diameter. Shanks should not exceed 0.140 in. (3.57 mm) in overall outside diameter so as not to exceed the diameter of prepunched holes in the shingles. Shanks may be smooth or threaded, but threaded nails are preferred to increase holding power. Aluminum nails should have screw threads with approximately a $12\frac{1}{2}$ degree thread angle; galvanized steel nails, if threaded, should have annular threads.

Nails for applying shingles over solid lumber sheathing should be approximately $1\frac{1}{4}$ in. (31.8 mm) by 0.120 in. (3.048 mm) and should have threaded shanks.

Nails for application of shingles over plywood sheathing should have threaded shanks and should be approximately $1\frac{1}{4}$ in. (31.8 mm) by 0.120 in. (3.048 mm) or long enough to penetrate through the sheathing.

In addition to nails, *storm anchors* should be used to apply ranch-style and Dutch lap units. Storm anchors should be copper, aluminum, or lead, approximately $3/4$ in. (19.1 mm) long with a flat base, and have sufficient ductility to permit repeated clinching and bending upward without cracking the shank (Fig, 7.6-57).

### 7.6.6.2 Preparation for Mineral-Fiber-Cement Shingle Roofing

#### SLOPE LIMITATIONS

Limitations imposed by the roof slopes are shown in Figure 7.6-5. Mineral-fiber-cement shingles are applied by normal slope methods on roofs with slopes of 5 in 12 or greater. Using lower slope than

**FIGURE 7.6-57** Storm anchors.

**FIGURE 7.6-56** Mineral-Fiber-Cement Roof Shingles*

| Type | Width, in. (mm) | Height, in. (mm) | Thickness, in. (mm) | Exposure, in. (mm) | Weight per sq lb (Mg) |
|------|------|------|------|------|------|
| Individual unit | 8 (203) | 16 (406) | $5/32$ (3.99) | 7 (177.8) | 350 (1587.5) |
| | | | $1/4$ (6.4) | 7 (177.8) | 585 (2653.4) |
| | 9 (229) | 18 (457) | $1/4$ (6.4) | 8 (203.2) | 570 (2585.4) |
| Multiple unit | 30 (762) | 14 (356) | $5/32$ (3.99) | 6 (152.4) | 440 (1995.7) |
| Dutch lap | 16 (406) | 16 (406) | $5/32$ (3.99) | $12 \times 13$ ($305 \times 330$) | 265 (1202) |
| Ranch style | 24 (610) | 12 (305) | $5/32$ (3.99) | $20 \times 9$ ($508 \times 229$) | 255 (1156.6) |

*Products of individual manufacturers may vary slightly from the dimensions listed.

normal methods, they may be used on slopes as low as 3 in 12. Mineral-fiber-cement shingles are not recommended for use on roof slopes less than 3 in 12.

## ROOF SHEATHING

Sheathing types, thicknesses, spans, and nailing requirements are discussed in Chapter 6.

## UNDERLAYMENT

Underlayment material should be unperforated No. 15 asphalt-saturated inorganic roofing felt, or No. 30 asphalt-saturated organic felt. Coal-tar-saturated felt, which may stain the shingles, should not be used. Underlayment should be applied over the entire roof.

## CANT STRIP

When the American method is used with individual-unit and multiple-unit shingles, a $^1/_4$ in. (6.4 mm) by 2 in. (50.8 mm) wood strip should be applied along the eaves to pitch (cant) the starter strip at the proper angle (Figs. 7.6-58 and 7.6-59).

## DRIP EDGE

A corrosion-resistant, nonstaining drip edge should be installed along the eaves and rakes. When cant strips occur, the drip edge should be applied over the cant strips.

## EAVE FLASHING

In areas where the outside design temperature is 30°F (−1.1°C) or less (see Fig. 7.6-8), or where the possibility of ice forming along the eaves could cause a backup of water, eaves flashing should be installed as described below.

**Normal Slopes (5 in 12 or Greater)** Eave flashing is formed by cementing an additional strip of underlayment over the first underlayment course. A continuous plastic asphalt cement layer should then be applied, at the rate of 2 gal/square (0.815 L/m$^2$), from the eave to cover a line at least 12 in. (304.8 mm) inside the interior wall line of the building (see Fig. 7.6-10), and this should be covered with the additional ply of underlayment.

**Lower Slopes Than Normal (3 in 12 to 5 in 12)** A cemented eaves flashing for lower slopes than normal should be formed just as described above for normal slopes. However, this flashing should extend from the eave to cover a line at least 24 in. (609.6 mm) inside the interior wall line (see Fig. 7.6-10).

## FLASHING

The following requirements are in addition to those presented in Section 7.6.1.6.

**Valley Flashing** The open or closed method may be used to construct valley flashing, as shown in Figure 7.6-60.

**Chimney and Side Wall Flashing** When shingles butt against vertical surfaces such as side walls, dormers, or chimneys, individual flashing units of corrosion-resistant metal should be inserted between each course of shingles. These metal flashing units should be folded from a width of 19 in. (482.6 mm) material and have a standing leg 6 in. (152.4 mm) high and a flat leg 4 in. (101.6 mm) wide on the shingle. The length of this flashing should be 3 in. (76.2 mm) greater than the exposure of the shingles. Side wall exterior finish materials should cover the standing leg. At masonry chimneys, a stepped counter flashing of corrosion-resistant metal should cover the standing leg of the flashing units and be trimmed off just above the surface of the shingles.

### 7.6.6.3 Shingle Roofing Application (New Construction)

The *American method* should be used to apply *individual* and *multiple* shingles without a sidelap. The *Dutch method* with sidelap and toplap should be used to apply Dutch lap and ranch-style shingles. In both methods, the underlayment felt should be rolled back and starter shingles applied directly on top of the metal drip edges previously nailed. Starter shingles should overhang the rakes and the eaves about $^3/_4$ in. (19.1 mm). When starters match the field shingle color and are used without a gutter, they are usually applied with the colored side down for a better appearance from below.

### AMERICAN METHOD (NORMAL SLOPE)

**Individual Shingles (Straight Butt Line)** After the underlayment is replaced over the starter shingles, the first shingle

**FIGURE 7.6-58** Application of individual shingles with straight butt lines.

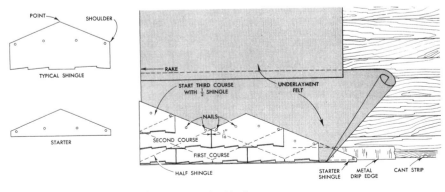

**FIGURE 7.6-59** Application of multiple-unit shingles.

**FIGURE 7.6-60** (a) Open valley flashing. (b) Closed valley flashing.

course should be applied starting with a half-shingle overhanging the rake about $^3/_4$ in. (19.1 mm) and in line at the butt with the starter shingles. Each shingle should be applied with nails through two of the prepunched holes. The second course should be started with a full shingle, and the third course, like the first course, should start with a half-shingle. Chalklines should be marked on the underlayment for each shingle course for accurate alignment (see Fig. 7.6-58).

***Individual Shingles (Staggered Butt Line)*** Individual shingles have two sets of prepunched holes to permit a staggered application. The first course should be applied as described above for a straight butt line. In starting the second course, the exposure of every second shingle should be reduced by dropping it down and nailing through the upper set of nail holes. Staggering can be varied, depending on the appearance desired, but is limited to the distances between the upper and lower holes. The third course should be started with a half-shingle, and alternating shingles should be staggered as in the previous course.

Individual shingles with either a straight or staggered butt line should have a minimum headlap and sidelap of 2 in. (50.8 mm) (Fig. 7.6-61; see also Fig. 7.6-58).

***Multiple-Unit Shingles*** After the underlayment is replaced over the starter shingles, the first shingle course should be applied starting with a half-shingle overhanging the rake about $^3/_4$ in. (19.1 mm) and in line at the butt with the starter shingles. The second course should be started using a full shingle, and the third course, like the first course, should be started with a half-shingle. Chalklines are not necessary if the visual alignment design of the shingles is properly utilized (see Fig. 7.6-59). Overdriving of nails causes the butts of the shingles to cock up and should be avoided to prevent damage to the shingles.

Because the design of multiple-unit shingles may vary among manufacturers, application instructions of the manufac-

**FIGURE 7.6-61** A staggered butt line can be achieved with individual shingles by varying the exposure.

turer for nailing, positioning, and alignment should be followed.

### DUTCH METHOD (NORMAL SLOPE)
The Dutch method is used to apply Dutch lap shingles (Fig. 7.6-62a) and ranch-style shingles (Fig. 7.6-62b). The following application methods refer to shingles having a 4 in. (101.6 mm) sidelap.

Starter shingles should be applied as shown in Figures 7.6-63a and 7.6-63b. After the felt has been replaced over the starter shingles, a 4-in. (101.6-mm)-wide piece cut from a full shingle should be applied to form a rake starter. As shown in Figure 7.6-63c, storm anchor A should be inserted from below and the rake starter secured in place using two nails. A full-size shingle should be applied over the rake starter by first inserting storm anchor B and positioning the shingles so that storm anchor A appears through the free corner (Fig. 7.6-63d). Storm anchor A should be clinched and the shingle secured with two nails (Fig. 7.6-63e). Each succeeding shingle should be applied in the same manner, first inserting a storm anchor in one corner and then engaging the other corner over the previous storm anchor. The second shingle course should be started with a shingle 4 in. (101.6 mm) less than full width, punched with a new hole as required to hold the shingle along the rake, and should be completed using full shingles. The third course should be

**FIGURE 7.6-62** (a) Dutch lap shingles and (b) ranch-style shingles. Holes A and B are for nails; holes D and E are for storm anchors.

started with a shingle 8 in. (203.2 mm) less than full width, the fourth course with a shingle 12 in. (304.8 mm) less than full width, and so on up the rake. Automatic alignment of the shingles is usually provided by the location of the prepunched holes.

### HIPS AND RIDGES

Before hip and ridge shingles (Fig. 7.6-64) are applied, wood nailing strips equal in thickness to the roof shingle assembly should be applied at the hips and ridges to provide a flat surface. Hips or ridges should then be covered with two thicknesses of underlayment (Fig. 7.6-65) or a double thickness of 8-in. (203.2-mm)-wide felt can be adhered with plastic cement over the nailing strips. Application of hip and ridge shingles should be started with a half-length shingle covered with a full-length shingle. Thereafter, full-length shingles should be applied with a maximum of 8 in. (203.2 mm) of exposure (depending on the design of the

shingle used) by nailing each shingle through two nail holes. Alternating shingles should be lapped on either side of the hip or ridge. Finally, the ridge joint should be pointed with plastic asphalt roofing cement.

### RIDGE ROLL

A tapered half-round ridge roll is used in some localities (Fig. 7.6-66). Before a ridge roll is applied, a wood nailing strip should be installed along the apex of hips

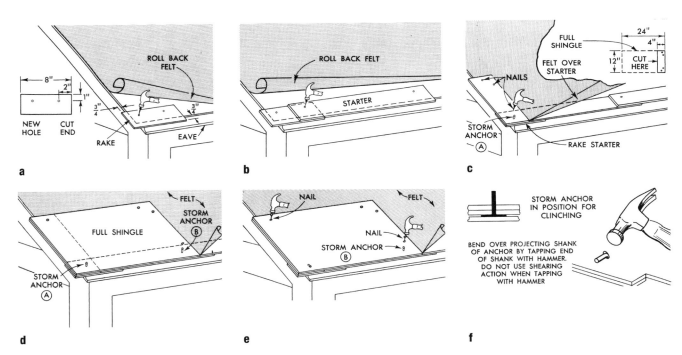

1″ = 25.4 mm;  2″ = 50.8 mm;  8″ = 203.2 mm;  ¾″ = 19.1 mm;  4″ = 101.6 mm;  12″ = 304.8 mm;  24″ = 609.6 mm

**FIGURE 7.6-63** Application of Dutch lap and ranch-style shingles. (a, b) Apply starter shingles. (c) Insert storm anchor A from below and secure the rake starter in place using nails. (d) Insert storm anchor B and position shingles so that storm anchor A appears through the free corner. (e) Clinch storm anchor A and secure the shingle with nails.

**FIGURE 7.6-64** Hip and ridge shingles.

HOLES – ¾″ ON EACH
SIDE OF CENTER LINE

12″

4¾″     5⅜″

4¼″     6½″

| 12″ = 304.8 mm | 4¾″ = 120.65 mm |
| 5⅜″ = 136.5 mm | 6½″ = 165.1 mm |
| 4¼″ = 107.95 mm | ¾″ = 19.1 mm |

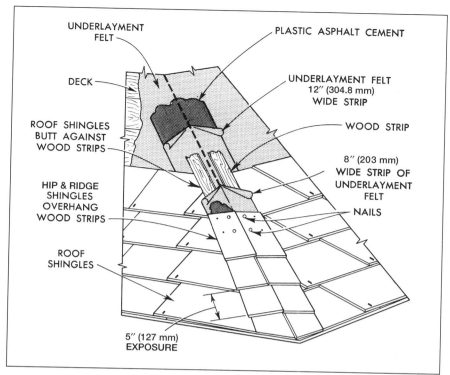

**FIGURE 7.6-65** Application method for finishing hips and ridges.

and ridges. The nailing strip should then be covered with a double thickness of underlayment. Whenever possible, the application of a ridge roll should start at the end of the ridge and be laid in a direction opposite to that of the prevailing winds; at hips, a ridge roll should be started at the eaves.

The first ridge roll section should be placed with the large end at the starting point overhanging the roof shingles ¹/₄ in. (6.4 mm). Each end should be fastened to the nailing strip by nailing at the head end through a corrosion-resistant clip designed for this purpose. The next ridge section should overlay the starter at least 2 in. (50.8 mm). These clips should be bent back to fasten the exposed end of the next ridge roll section. This method should be continued until the entire hip or ridge has been covered. Ridge pieces should be embedded in asphalt plastic cement, and exposed nails should be pointed with plastic cement. The openings at end units may be closed with colored cement mortar or with a closure consisting of a half-round piece cut from a shingle that is nailed or screwed into the end of the nailing strip.

## LOWER SLOPE THAN NORMAL METHOD (3 IN 12 TO 5 IN 12)

Methods for applying shingles on roof slopes less than 5 in 12 but not less than 3 in 12 are the same as described above for normal slopes, 5 in 12 and over, with certain exceptions with regard to underlayment.

For slopes 3 in 12 to 4 in 12, two underlayment layers should be applied over the entire roof. For slopes 4 in 12 to 5 in 12, one underlayment layer is

**FIGURE 7.6-66** Ridge roll may be used on a ridge as an alternate to ridge shingles.

sufficient, except that two layers are required in areas where the outside winter design temperature is 30°F (−1.1°C) or lower.

When eaves flashing is required on a roof with a slope that is lower than normal, the additional layer should be cemented at the eaves as described in Section 7.6.1.6.

### 7.6.6.4 Reroofing Existing Roofs

When mineral-fiber-cement shingles are applied to an existing roof, the surface must be made as smooth as possible. Mineral-fiber-cement shingles may be applied directly over old asphalt shingles, provided that the sheathing and roof framing members are sound and strong enough to support the increased loads.

Loose or curled wood shingles should be split and nailed down and missing shingles replaced. Beveled wood strips should be installed below the butts of each course of shingles (Fig. 7.6-67). When existing wood shingles fail to retain nails with at least the force of 40 lb (18.14 kg) per fastener, the shingles should be removed and the existing sheathing prepared as it would be in new construction. If old wood shingles were applied over spaced sheathing, new sheathing should be fitted between the strips to fill the spaces and provide a solid deck. In this case, it is often simpler to remove the old sheathing strips and construct a new deck.

Nails for applying shingles to existing roofs should be long enough to penetrate through the existing sheathing. Threaded shank roofing nails provide improved holding power.

### REPLACING DAMAGED SHINGLES

Damaged shingles may be removed and replaced as shown in Figure 7.6-68 and as described below.

First, the damaged shingle should be shattered and removed. The nails may then be withdrawn or cut off by hooking a slater's ripper over the nails and striking the offset handle of the ripper with a hammer (Fig. 7.6-68a). New shingle should then be positioned.

With the American method of application, a hole should be drilled through the joint between the overlying shingles and the head of the new shingle. One nail should be driven through this hole, and a 3-in. (76.2-mm)-wide corrosion-resistant metal flashing strip should be placed over the nail head and under the overlying shingle (Fig. 7.6-68b).

a    b

c    d

**FIGURE 7.6-68** Replace damaged shingles as follows: (a) Remove the damaged shingle and nails. (b) Nail new shingle through the side joint and cover the nail with corrosion-resistant metal strip inserted in lap. (c) In the Dutch method, remove the damaged shingle and shear off the nails and storm anchor by engaging them with the hook end of a slater's ripper and striking the offset on the handle with a hammer. Loosen the nail in the exposed shingle, remove the sheared-off storm anchor, and insert the new storm anchor. Drive the loosened nail home and apply plastic cement. (d) Place storm anchor B in the new shingle. Raise the overlapping shingle to allow the storm anchor to align with the holes. Embed the shingle in cement and bend the storm anchor down.

**FIGURE 7.6-67** Preparation of an existing roof for application of new shingles.

For the Dutch method of application, the new shingles should be anchored in place with two spots of roofing cement placed on the head of the underlying shingle and by engaging the storm anchors. Nailing is not required (Figs. 7.6-68c and 7.6-68d).

### 7.6.7 WOOD SHINGLES AND SHAKES

Refer to Section 7.6.1 for (1) definitions of terminology, (2) limitations imposed by the slope of the roof, (3) sheathing guidelines (see also Chapter 6), and (4) general guidelines for typical accessory materials, such as underlayment, flashing, roofing cements, eaves flashing, drip edge, nails, and caulking for preparation of a roof or wall for application of shingles. Unless specific exceptions are made in the following text, the provisions of Section 7.6.1 apply to wood shingles and shakes and methods of their application.

#### 7.6.7.1 Materials

Most wood shingles and shakes used for roofing are produced from western red cedar trees, giant, slow-growing coniferous trees of the Pacific Northwest, but northern white cedar, bald cypress, and redwood are also used. This section discusses only western red cedar shingles and shakes. Those made with other woods are similar and are installed in a similar manner.

Western red cedar has extremely fine and even grain, exceptional strength in proportion to weight, low expansion and contraction with changes in moisture content, and high impermeability to water. Its outstanding characteristics are great durability and resistance to decay.

Cedar shingles are sawn, have smooth faces and backs, and are produced in four grades and a variety of types. Tapersawn shakes are sawn on both sides but have a thickness more like that of shakes. They are produced in two grades. Other shakes have rough split faces and either sawn or split backs, and are produced in one grade and in three types.

Both shingles and shakes are made from selected cedar logs cut to proper lengths—16, 18, and 24 in. (406, 482.6, and 610 mm) for shingles and 18 and 24 in. (482.6 and 609.6 mm) for shakes. These lengths are again cut or split into the proper size blocks for the shingle cutting machine or a convenient size for splitting of shakes. In reducing the large log sections to blocks, each section is quartered, split, and requartered to produce blocks having a true edge grain face.

Each bundle of wood shingles and shakes should carry the official grade-marked label of the Cedar Shake and Shingle Bureau (CSSB).

## CEDAR SHINGLES

Roof shingles are cut from cedar blocks on an upright shingle machine. This cutting machine carries a block past a thin, razor-sharp circular saw which slices a shingle smoothly at each stroke. Shingles cut from one block are jointed or edged and graded by a sawyer. This sawyer can adjust the block in the machine from time to time so that the highest possible grade of shingles will be produced.

A sawyer picks up each shingle as it falls from the saw and places its butt firmly against a guide. Using a circular saw set at right angles to the butt, the sawyer clips one edge of the shingle off smoothly. Flipping the shingle over, the sawyer repeats the process, making another smooth edge parallel to the one first cut. If defects are present, the sawyer trims the shingle in such a way that a narrower shingle or two shingles without defects can be obtained. Depending on the grade, the sawyer then drops the shingle into a chute which leads to the packing bins.

Shingles are graded again as they are packed into bundles in a packing frame. Bundles of shingles are held together by two bandsticks connected by metal straps. Packaged in this manner, bundled shingles are seasoned and remain perfectly flat in the process.

*Grades, Types, and Sizes* Shingles are marketed according to grade, thickness, and type. Nominal shingle lengths are 16, 18, and 24 in. (406.4, 457.2, and 609.6 mm) within $1/4$ in. (6.4 mm). A 1 in. (25.4 mm) tolerance over or under the specified length is permitted in 10% of the shingles. All grades are produced in these lengths, except that Undercoursing Grade is not produced in the 24 in. (609.6 mm) length. A 16 in. (406.4 mm) shingle is known as XXXXX or Five Xs; an 18 in. (457.2 mm) shingle as Perfection; a 24 in. (609.6 mm) shingle as Royal.

Most cedar shingles conform to one of three grades: No. 1 (Blue Label), No. 2 (Red Label), and No. 3 (Black Label). Other available units include No. 4, Undercoursing; No. 1 and No. 2 Rebutted-rejointed; No. 1, Machine Grooved; No. 1 and No. 2, Dimension; No. 1 and No. 2 Preformed; and factory-built Hip and Ridge units.

Except for dimension shingles, all shingles are produced in random widths from 3 in. (76.2 mm) up to 14 in. (355.6 mm). Shingle thickness varies, depending on length. A 16-in. (406.4-mm)-long shingle must be thick enough so that the combined butt thicknesses of five shingles will measure at least 2 in. (50.8 mm). Thus, a shingle thickness is designated as 5 in. 2 in. (50.8 mm) or $5/2$ in. An 18 in. (457.2 mm) shingle has a thickness so that the combined butt thicknesses of five shingles will measure at least $2 1/4$ in. ($5/2$–$1/4$ in.) (57.2 mm), and 24 in. (609.6 mm) shingles have a thickness so that four shingles will measure at least 2 in. (50.8 mm) ($4/2$ in.). These dimensions are for unseasoned shingles measured green. If the shingles are measured after seasoning has occurred, a 3% allowance is made for shrinkage.

Shingles with extreme cross-grain are not permitted in any grade.

*Grade No. 1 (Blue Label)* is the best or premium grade of shingle for roofs. These shingles are 100% edge grain, consist entirely of heartwood, and contain no knots. On 8 in 12 sloped roofs or steeper, Grade No. 1 shingles are estimated to have a service life of more than 35 years under normal climatic conditions; on roof slopes of 6 in 12, a service life of at least 25 years can be anticipated; and on 4 in 12 slopes, 20 years.

*Grade No. 2 (Red Label)* is a good grade for roofs on secondary buildings and may be used on major building roofs when exposures are reduced about 20%.

Shingles of this grade must be clear, or free from blemishes for 10, 11, and 15 in. (254, 279.4, and 381 mm) for 16, 18, and 24 in. (406.4, 457.2, and 609.6 mm) shingles, respectively. A maximum 1 in. (25.4 mm) width of sapwood is permissible in the first 10 in. (254 mm). Mixed edge and flat grain are allowed.

*Grade No. 3 (Black Label)* is a utility grade to be used where economy is the prime consideration. It may be used for roofs of secondary buildings. These shingles may contain flat grain and sapwood, but must be 6 in. (152.4 mm) clear from the butt for 16 and 18 in. (406.4 and 457.2 mm) shingles and 10 in. (254 mm) clear for 24 in. (609.6 mm) shingles.

*Hip and Ridge units* are factory cut, mitered, and assembled to fit hips and ridges of various roof slopes. These units, which reduce costs by eliminating on-the-job trimming and fitting (Fig. 7.6-69), are produced in either No. 1 or No. 2 grades in 16 and 18 in. (406.4 and 457.2 mm) lengths for a 6 in 12 roof slope, and are adjustable to fit any slope between 4 in 12 and 8 in 12.

A summary of cedar shingle types, grades, and coverage is given in Figures 7.6-70 and 7.6-71. The labels shown in Figure 7.6-70 provide assurance that the shingles meet the quality standards of the CSSB. The maximum exposure for shingles is shown in Figure 7.6-72.

## CEDAR SHAKES

Wood shakes are manufactured in four types of units: (1) tapersawn, (2) handsplit and resawn, (2) tapersplit, and (3) straightsplit. Tapersplit shakes are produced in 24 in. (609.6 mm) length. The others are produced in both 18 and 24 in. (457.2 and 609.6 mm) lengths. Each is 100% clear wood and 100% heartwood. Both tapersplit and straightsplit shakes, produced largely by hand, are of 100% edge grain.

**FIGURE 7.6-69** Preassembled hip and ridge units. (Courtesy Cedar Shake and Shingle Bureau.)

**Figure 7.6-70** Types of Red Cedar Shingles

| Grade | Size | Bundles or Cartons* per Square | | | Description | Labels |
|-------|------|------|------|---|-------------|--------|
| | | No. | Weight | | | |
| No. 1 BLUE LABEL | 24" (Royals) 18" (Perfections) 16" (XXXXX) | 4 bdls. 4 bdls. 4 bdls. | 192 lbs. 158 lbs. 144 lbs. | | The premium grade of shingles for roofs and sidewalls. These shingles are 100% heartwood, 100% clear and 100% edge-grain. | CERTIGRADE Red Cedar SHINGLES BLUE 1 LABEL |
| No. 2 RED LABEL | 24" (Royals) 18" (Perfections) 16" (XXXXX) | 4 bdls. 4 bdls. 4 bdls. | 192 lbs. 158 lbs. 144 lbs. | | A good grade for most applications. Not less than 10" clear on 16" shingles, 11" clear on 18" shingles and 16" clear on 24" shingles. Flat grain and limited sapwood are permitted. | CERTIGRADE Red Cedar SHINGLES RED LABEL |
| No. 3 BLACK LABEL | 24" (Royals) 18" (Perfections) 16" (XXXXX) | 4 bdls. 4 bdls. 4 bdls. | 192 lbs. 158 lbs. 144 lbs. | | A utility grade for economy applications and secondary buildings. Guaranteed 6" clear on 16" and 18" shingles, 10" clear on 24" shingles. | CERTIGRADE Red Cedar SHINGLES BLACK 3 LABEL |
| No. 4 UNDER-COURSING | 18" (Perfections) 16" (XXXXX) | 2 bdls. 2 bdls. | 60 lbs. 60 lbs. | | A low grade for undercoursing on double-coursed sidewall applications. | RED CEDAR SHINGLE BUREAU UNDERCOURSING Red Cedar SHINGLES |
| No. 1 or No. 2 REBUTTED-REJOINTED | 18" (Perfections) 16" (XXXXX) | 1 carton 1 carton | 60 lbs. 60 lbs. | | Same specifications as No. 1 and No. 2 Grades above but machine trimmed for exactly parallel edges with butts sawn at precise right angles. Used for side-wall application where tightly fitting joints between shingles are desired. Also available with smooth sanded face. | No. 1 & No. 2 labels (above) |
| No. 1 MACHINE GROOVED | 18" (Perfections) 16" (XXXXX) | 1 carton 1 carton | 60 lbs. 60 lbs. | | Same specifications as No. 1 and No. 2 Grades above; these shingles are used at maximum weather exposures and are always applied as the outer course of double-coursed sidewalls. | CERTIGROOVE CEDAR SHAKES NUMBER 1 GRADE RED CEDAR SHINGLE BUREAU |
| No. 1 or No. 2 DIMENSION | 24" (Royals) 18" (Perfections) 16" (XXXXX) | 4 bdls. 4 bdls. 4 bdls. | 192 lbs. 158 lbs. 144 lbs. | | Same specifications as No. 1 and No. 2 Grades above; except they are cut to specific uniform widths and may have butts trimmed to special shapes. | No. 1 & No. 2 labels (above) |
| No. 1 or No. 2 HIP AND RIDGE | 18" (Perfections) 16" (XXXXX) | — | — | | Same specifications as No. 1 and No. 2 Grades above; factory-cut, mitered and assembled units produced for a 6 in 12 slope and adjust to fit slopes between 4 in 12 and 8 in 12. | No. 1 & No. 2 labels (above) |

*Nearly all manufacturers pack 4 bundles to cover 100 sq. ft. when used at maximum exposures for roof construction; Undercoursing, rebutted and rejointed and machine grooved shingles typically are packed one carton to cover 100 sq. ft. when used at maximum exposure for double-coursed sidewalls.

(Courtesy Cedar Shake and Shingle Bureau.)

Factory-assembled hip and ridge units are also available.

Each bundle of shakes should carry the official grademarked label of the CSSB.

Both faces of tapersawn shakes are sawn like shingles, but they are thicker and heavier than shingles.

Most shakes are split by machine, although some are still split by hand using a hardwood mallet and steel froe. As a result of being split, no two shakes are exactly alike. The three types of shakes produced are tapersplit, straightsplit, and handsplit-and-resawn (Fig. 7.6-73).

Tapersplit and straightsplit shakes (Figs. 7.6-73a and 7.6-73b) are made by machine, or by hand-riving individual shakes from the block. To obtain the butt-to-tip taper for tapersplit shakes, blocks are turned end-for-end after each split. Straightsplit shakes are split from the same end of the block and therefore are of uniform thickness.

Handsplit-and-resawn shakes (see Fig. 7.6-73c) are produced by splitting blocks into boards of the desired thickness. The boards are then passed diagonally through a thin bandsaw to form two shakes, each with a split face and a sawn

**FIGURE 7.6-71** Coverage of Wood Shingles at Varying Exposures

| | Approximate Coverage of 4 Bundles or 1 Carton, sq. ft.[b] | | | | | | | | | | | | | | | | | | |
| | Weather Exposures, in. | | | | | | | | | | | | | | | | | | |
| Length and Thickness[a] | 3½ | 4 | 5 | 5½ | 6 | 7 | 7½ | 8 | 8½ | 9 | 10 | 11 | 11½ | 12 | 13 | 14 | 15 | 15½ | 16 |
|---|---|---|---|---|---|---|---|---|---|---|---|---|---|---|---|---|---|---|---|
| **Random width and dimension** | | | | | | | | | | | | | | | | | | | |
| 16 in. × 5/2 in. | 70 | 80 | 100[c] | 110 | 120 | 140 | 150[d] | 160 | 170 | 180 | 200 | 220 | 230 | 240[e] | — | — | — | — | — |
| 18 in. × 5(2¼) in. | — | 72½ | 90½ | 100[c] | 109 | 127 | 136 | 145½ | 154½[d] | 163½ | 181½ | 200 | 209 | 218 | 236 | 254½[e] | — | — | — |
| 24 in. × 4/2 in. | — | — | — | — | 80 | 93 | 100[c] | 106½ | 113 | 120 | 133 | 146½ | 153[d] | 160 | 173 | 186½ | 200 | 206½ | 213[e] |
| **Rebutted and rejointed** | | | | | | | | | | | | | | | | | | | |
| 16 in. | — | — | — | — | 50 | 59 | 63[d] | 67 | 72 | 76 | 84 | 93 | 100[e] | 100[e] | — | — | — | — | — |
| 18 in. | — | — | — | — | 43 | 50 | 54 | 57 | 61[d] | 64 | 72 | 79 | — | 100[e] | — | — | — | — | — |
| **Machine-grooved** | | | | | | | | | | | | | | | | | | | |
| 16 in. | | | | | (Normally applied at maximum exposure) | | | | | | | | | 86 | 93 | 100[e] | | | |
| 18 in. | | | | | (Normally applied at maximum exposure) | | | | | | | | | | | 100[e] | | | |

[a]Sum of the thickness; e.g., 5/2 in. means 5 butts equals 2 in.

[b]Nearly all manufacturers pack 4 bundles to cover 100 sq. ft. when used at maximum exposures for roof construction; rebutted and rejointed and machine-grooved shingles typically are packed one carton to cover 100 sq. ft. when used at maximum exposure for double-coursed sidewalls.

[c]Maximum exposure recommended for roofs.

[d]Maximum exposure recommended for single-coursed sidewalls.

[e]Maximum exposure recommended for double-coursed sidewalls.

(Courtesy Cedar Shake and Shingle Bureau.)

## FIGURE 7.6-72 Maximum Exposure for Wood Roof Shingles

| | Shingle Type No. 1 | | | Shingle Type No. 2 | | | Shingle Type No. 3 | | |
|---|---|---|---|---|---|---|---|---|---|
| **Shingle Length** | 16 | 18 | 24 | 16 | 18 | 24 | 16 | 18 | 24 |
| **Roof Slope** | Exposure in in. (mm) | | | | | | | | |
| 3½ to 4½ | 3¾ (95.3) | 4¼ (107.9) | 5¾ (146.1) | 3½ (88.9) | 4 (101.6) | 5½ (139.7) | 3 (76.2) | 3½ (88.9) | 5 (127) |
| 4½ and Steeper | 5 (127) | 5½ (139.7) | 7½ (190.5) | 4 (101.6) | 4½ (114.3) | 6½ (165.1) | 3½ (88.9) | 4 (101.6) | 5½ (139.7) |

Taper split

a

Straight split

b

Hand split and resawn

c

**FIGURE 7.6-73** Three types of shakes: (a) tapersplit, (b) straightsplit, (c) handsplit and resawn. (Courtesy Cedar Shake and Shingle Bureau.)

back having thin tips and thick butts and up to 10% flat grain. These shakes are available either as heavy handsplit-and-resawn (also called *heavy resaw*), or medium handsplit-and-resawn (also called *medium resaw*). The latter are thinner and lighter.

Packers assemble finished shakes into bundles in a standard-size frame, compressing the bundles slightly and binding them with wooden bandticks and steel strapping (Fig. 7.6-74).

Each type of shake may be used for roof shingles. Handsplit-and-resawn shakes have the heaviest butt lines and give the most rugged textured appearance to a roof; tapersplit and straightsplit shakes give a more uniform texture.

**FIGURE 7.6-74** Grade labels beneath the bandsticks of both shingles and shakes ensure that they meet or exceed Cedar Shake and Shingle Bureau quality standards. (Courtesy Cedar Shake and Shingle Bureau.)

On 8 in 12 sloped roofs or steeper, shakes are estimated to have a service life of more than 60 years. On lesser slopes, to a minimum of 4 in 12, a life of about 40 years may be expected.

***Grades, Types, and Sizes*** Tapersawn shakes are available in Grades 1 and 2. In Grade 1, no defects are permitted. In Grade 2, flat and cross-grains are allowed and the top half is permitted to have tight knots and other defects.

There is only one grade, No. 1, for other cedar shakes. Shakes of this grade are 100% clear wood, graded from the split face, in the case of handsplit and resawn shakes, and from the best face in the case of tapersplit and straightsplit shakes. All are 100% heartwood, free of bark and sapwood. Tapersplit and straightsplit shakes are 100% edge grain; handsplit and resawn shakes may have up to 10% flat grain.

Shake thickness is determined by measuring the area within ½ in. (12.7 mm) from each edge. If corrugations or valleys exceed ½ in. (12.7 mm) in depth, a minus tolerance of ⅛ in. (3.2 mm) is permitted in the minimum specified thickness.

Shakes of random width, none narrower than 4 in. (101.6 mm), are packed in straight courses in 18- and 20-in. (457.2- and 508-mm)-wide frames. Curvatures in the sawn face of handsplit-and-resawn shakes should not exceed 1 in. (25.4 mm) from the level plane in the

**FIGURE 7.6-75** Starter-finish shakes are more economical than regular shakes for use in the initial undercourse at the eave and in the final course at the ridge. (Courtesy Cedar Shake and Shingle Bureau.)

**FIGURE 7.6-76** Hip and ridge shake units.
(Courtesy Cedar Shake and Shingle Bureau.)

length of the shake. Excessive grain sweeps on the split face are not permitted.

Starter-finish shakes, produced in a 15 in. (381 mm) length, are used as a begin-ning or underlay course at the eave and as the final course at the ridge (Fig. 7.6-75).

Factory-cut, mitered, and assembled hip and ridge units are available to elim-inate on-the-job trimming and fitting (Fig. 7.6-76). They are produced in 18 in. and 24 in. (457.2 and 609.6 mm) lengths for a 6 in 12 roof slope and are adjustable to fit any slope between 4 in 12 and 8 in 12.

A summary of red cedar shake types, grades, and coverage is given in Figures 7.6-77 and 7.6-78. The label shown in Figure 7.6-74 provides assurance that the shakes meet the quality standards of the CSSB. Labels change from time to time and may look slightly different when you read this book. The maximum exposure for shakes is shown in Figure 7.6-79.

## STANDARDS OF QUALITY

The manufacture of cedar shingles and shakes should conform to the specifica-tions of the CSSB. These specifications constitute the recognized standards of the industry.

In addition to formulating specifica-tions, a primary function of the CSSB is to establish and enforce rigid quality con-trols covering the manufacture and grad-ing of red cedar shingles and shakes. To ensure compliance with these standards, products made by members of the CSSB are inspected at frequent unannounced in-tervals by CSSB inspectors. Mills which conform to these standards are privileged to use the *Certigrade* shingle or *Certi-Split* shake labels of the CSSB on their products.

## NAILS

Nails for applying shakes and shingles should be corrosion resistant, such as hot-dipped galvanized steel or bright or colored aluminum alloy. Common wire nails or even electrogalvanized nails

**Figure 7.6-77** Types of Cedar Shakes

| Grade | Length and Thickness | Bundles per Square* | Weight (lbs. per square) | | Description | Label |
|---|---|---|---|---|---|---|
| No. 1 HANDSPLIT & RESAWN | 18" × 1/2" to 3/4" | 4 | 220 | | These shakes have split faces and sawn backs. Cedar blanks or boards are split from logs and then run diagonally through a bandsaw to produce two tapered shakes from each. | |
| | 18" × 3/4" to 1-1/4" | 5 | 250 | | | |
| | 24" × 3/8" | 4 | 260 | | | |
| | 24" × 1/2" to 3/4" | 4 | 280 | | | |
| | 24" × 3/4" to 1-1/4" | 5 | 350 | | | |
| | 32" × 3/4" to 1-1/4" | 6 | 450 | | | |
| No. 1 TAPERSPLIT | 24" × 1/2" to 5/8" | 4 | 260 | | Produced largely by hand, us-ing a sharp-bladed steel froe and a wooden mallet. The nat-ural shingle-like taper is achieved by reversing the block, end-for-end, with each split. | |
| No. 1 STRAIGHT-SPLIT (BARN) | 18" × 3/8" (True-Edge) | 4 | 200 | | Produced in the same manner as taper-split shakes except that by splitting from the same end of the block, the shapes acquire the same thickness throughout. | |
| | 18" × 3/8" | 5 | 200 | | | |
| | 24" × 3/8" | 5 | 260 | | | |
| No. 1 STARTER-FINISH | 15" × 1/2" to 1-1/4" | — | — | | These shakes are used as the starting or underlay course at the eaves and as the final course at the ridge. | |
| No. 1 HIP & RIDGE | 18" × 1/2" to 1-1/4" | — | — | | Factory cut, mitered and as-sembled units are produced for a 6 in 12 slope; they ad-just to fit slopes between 4 in 12 and 8 in 12. | |
| | 24" × 1/2" to 1-1/4" | | | | | |

*Generally, the number of bundles required to cover 100 sq. ft. when used for roof construction at the maximum recommended weather exposure.
(Courtesy Cedar Shake and Shingle Bureau.)

FIGURE 7.6-78 Coverage of Shakes at Varying Weather Exposures

| Length and Thickness | Number of Bundles | Approximate Coverage, sq. ft. Weather Exposures, in. | | | | | | | | | | | |
|---|---|---|---|---|---|---|---|---|---|---|---|---|---|
| | | 5½ | 6½ | 7 | 7½ | 8 | 8½ | 10 | 11½ | 13 | 14 | 15 | 16 |
| Hand-split-and-resawn 18 in. × ½ in. to ¾ in. | 4 | 55[a] | 65 | 70 | 75[b] | 80 | 85[c] | — | — | — | — | — | — |
| 18 in. × ¾ in. to 1¼ in. | 5 | 55[a] | 65 | 70 | 75[b] | 80 | 85[c] | — | — | — | — | — | — |
| 24 in. × ⅜ in. | 4 | — | 65 | 70 | 75 | 80 | 85 | 100 | 115[c] | — | — | — | — |
| 24 in. × ½ in. to ¾ in. | 4 | — | 65 | 70 | 75[a] | 80 | 85 | 100[b] | 115[c] | — | — | — | — |
| 24 in. × ¾ in. to 1¼ in. | 5 | — | 65 | 70 | 75[a] | 80 | 85 | 100[b] | 115[c] | — | — | — | — |
| 32 in. × ¾ in. to 1¼ in. | 6 | — | — | — | — | — | — | 100[a] | 115 | 130[b] | 140 | 150[c] | — |
| Taper-split 24 in. × ½ in. to ⅝ in. | 4 | — | 65 | 70 | 75[a] | 80 | 85 | 100[b] | 115[c] | — | — | — | — |
| Straight-split 18 in. × ⅜ in. (true-edge) | 4 | — | — | — | — | — | — | — | — | — | 100 | 106 | 112[d] |
| 18 in. × ⅜ in. | 5 | 65[a] | 75 | 80 | 90 | 95 | 100[c] | — | — | — | — | — | — |
| 24 in. × ⅜ in. | 5 | — | 65 | 70 | 75[c] | 80 | 85 | 100 | 115[c] | — | — | — | — |
| 15-in. Starter-finish course | | Use supplementary with shakes applied not over 10-in. exposure | | | | | | | | | | | |

[a]Recommended maximum weather exposure for three-ply roof construction.

[b]Recommended maximum weather exposure for two-ply roof construction.

[c]Recommended maximum weather exposure for single-coursed sidewalls.

[d]Recommended maximum weather exposure for double-coursed sidewalls.

(Courtesy Cedar Shake and Shingle Bureau.)

may shorten the life of a shake or shingle roof or wall and should not be used. Nails may have smooth or threaded shanks and medium or blunt diamond points. Threaded nails provide increased holding power, although smooth shank nails are used in most applications over wood sheathing and give suitable service. Nails for applying wood shingles or shakes to plywood sheathing should be threaded. If threaded, galvanized steel nails should have annular threads; aluminum nails

should have screw threads with approximately a 12½ degree thread angle.

Box nails, having smaller diameters and longer lengths than common nails, are also suitable. For new roofs, nail lengths typically required are 6d (2 in. [50.8 mm]) for shakes with a ¾ in. (19.1 mm) to 1¼ in. (31.8 mm) butt thickness and 5d (1¾ in. [19.1 mm]) for shakes with a ½ to ¾ in. (12.7 to 19.1 mm) butt thickness.

Nail dimensions typically required for roof installations are given in Figure

7.6-80. Nails 2d sizes larger should be used to apply hip and ridge units. For new roofing and overroofing, nails should be long enough to penetrate through the sheathing; the 5d size (1¾ in. [44.45 mm]) is normally adequate.

### 7.6.7.2 Preparation for Roof Shingles and Shakes

#### SLOPE LIMITATIONS AND EXPOSURE

Roof slope limitations are discussed in Section 7.6.1.2. The exposure of wood

**FIGURE 7.6-79** Maximum Exposure of Wood Shakes

| Type | Length, in. | Maximum Exposure Double Coverage, in. | Triple Coverage, in. |
|---|---|---|---|
| Hand-split-and-resawn shakes | 18 | 7½ | 5½ |
| | 24 | 10 | 7½ |
| | 32 | 13 | 10 |
| Taper-split shakes | 24 | 10 | 7½ |
| Straight-split shakes | 18 | — | 5½ |
| | 24 | — | 7½ |

*Note:* In areas not subject to wind-driven snow, sheathing such as nominal 1 × 4 or 1 × 6 boards may be spaced equal to the exposure of the shake, but not more than 10 in.

(Courtesy Cedar Shake and Shingle Bureau.)

**FIIGURE 7.6-80** Nail Sizes for Application of Wood Shingles

| Size | Nail Size Length, in. | Gauge | Head, in. | Shingle Length, in. |
|---|---|---|---|---|
| 3d[a] | 1¼ | 14½ | 7/32 | 16 and 18 |
| 4d[a] | 1½ | 14 | 7/32 | 24 |
| 5d[b] | 1¾ | 14 | 7/32 | 16 and 18 |
| 6d[b] | 2 | 13 | 7/32 | 24 |

[a]3d and 4d nails are used for new construction.

[b]5d and 6d nails are used for overroofing.

(Courtesy Cedar Shake and Shingle Bureau.)

shingles and shakes on a roof is dependent on the slope of the roof.

***Shingles*** Standard exposures are shown in Figure 7.6-72. Four bundles of shingles will cover 100 sq ft (9.29 m²) (one square) of roof area when the shingles are applied at standard exposures. Shingles are not recommended on slopes less than 3 in 12.

Shingle exposure should not be larger than the length of the shingle minus 1 in. (25.4 mm) divided by 3. This will provide at least triple coverage (three layers of wood at every point) to ensure complete freedom from leakage in heavy wind-driven rain or snow storms. Approximate coverage of four bundles of shingles at varying exposures is shown in Figure 7.6-71.

***Shakes*** The maximum recommended exposure for double-coverage roofs is 10 in. (254 mm) for 24 in. (609.6 mm) shakes and 7¹/₂ in. (190.5 mm) for 18 in. (457.2 mm) shakes. A triple-coverage roof is achieved by reducing these exposures to 7¹/₂ in. (190.5 mm) for 24 in. (609.6 mm) shakes and 5¹/₂ in. (139.7 mm) for 18 in. (457.2 mm) shakes. Approximate roof coverage for shakes using different exposures is shown in Figure 7.6-78. Maximum exposure to provide double and triple coverage is summarized in Figure 7.6-79.

Shakes are recommended only on slopes 4 in 12 or steeper. However, suitable installations have been achieved on slopes as low as 3 in 12 by taking the following extra precautions: (1) reducing exposures to provide triple coverage and (2) using solid sheathing with an underlayment of No. 30 asphalt-saturated felt applied over the entire roof with No. 30 asphalt-saturated felt interlayment between each course.

### ROOF SHEATHING

Sheathing types, thicknesses, spans, and nailing are discussed in Chapter 6.

***Shingles*** Either spaced or solid sheathing may be used with wood shingles. When solid sheathing is preferred, either nominal 1 in. (25.4 mm) boards or plywood may be used. When spaced sheathing such as nominal 1 × 3 (25 × 75) or 1 × 4 (25 × 100) boards is used, it should be applied so that the space between boards is not greater than the width of the boards themselves.

***Shakes*** Either spaced or solid sheathing may be used with wood shakes, depending on the climatic conditions of the region.

In areas not subject to wind-driven snow, spaced sheathing, such as nominal 1 × 4 (25 × 100) or 1 × 6 (25 × 150) boards, is suitable (Fig. 7.6-81). The center-to-center spacing of boards should be equal to the exposure of the shake but not more than 10 in. (254 mm).

In areas subject to wind-driven snow (generally where the outside design temperature is 30°F [−1.1°C] or lower), solid sheathing of nominal 1 in. (25.4 mm) boards or plywood is recommended.

### UNDERLAYMENT AND INTERLAYMENT

***Shingles*** Underlayment typically is not required between shingles and spaced or solid sheathing, but may be desirable for the protection of sheathing and to ensure against air infiltration.

When underlayment is desired, No. 15 asphalt-saturated felt may be used.

***Shakes*** Interlayment strips and the underlayment course used at eaves should be No. 30 asphalt-saturated felt. The installation of interlayment between each shake course, essential to provide a baffle against infiltration of wind-driven rain or snow, is described in Section 7.6.7.4.

### EAVE FLASHING

In areas where the outside design temperature is 30°F (−1.1°C) or less (see Fig. 7.6-8), or where there is a possibility of ice forming along the eaves and causing a backup of water, eaves flashing is recommended. Sheathing should be applied solidly above the eave line to cover a line at least 24 in. (609.6 mm) inside the interior wall line of the building.

***Shingles*** For 4 in 12 and steeper slopes, eaves flashing should be formed by applying a double layer of No. 15 asphalt-saturated felt to cover the solid sheathing. When the eave overhang requires flashing wider than 36 in. (914.4 mm), the horizontal joint between felts should be cemented and located outside the exterior wall line.

For slopes 3 in 12 up to 4 in 12, or in areas subject to severe icing, eaves flashing may be formed as described above for 4 in 12 slopes, except that the double layer of No. 15 asphalt-saturated underlayment should be cemented. The eaves flashing should be formed by applying a continuous plastic asphalt cement layer, at the rate of 2 gal/100 sq ft (0.815 L/m²), to the surface of the underlayment starter course before the second underlayment layer is applied. Cement should also be applied to the 19 in. (482.6 mm) underlying portion of each succeeding course which lies within the eaves flashing area before placing the next course. The cement should be applied uniformly with a comb trowel, so that at no point does underlayment touch underlayment when the application has been completed. The overlying sheet should be pressed firmly into the entire cemented area (see Fig. 7.6-10).

**FIGURE 7.6-81** (a) Spaced sheathing. (b) Maximum recommended exposure produces a two-ply roof; reducing exposure produces a three-ply roof. (Courtesy Cedar Shake and Shingle Bureau.)

*Shakes* On slopes 4 in 12 or steeper, eaves flashing should be formed by applying an additional course of No. 30 asphalt-saturated felt over the underlayment starting course at the eaves. The eaves flashing should extend up the roof to cover a line at least 24 in. (609.6 mm) inside the interior wall line of the building. When the eave overhang requires the flashing to be wider than 36 in. (914.4 mm), the necessary horizontal joint should be cemented and located outside the exterior wall line of the building (see Fig. 7.6-10).

For slopes 3 in 12 up to 4 in 12, or in areas subject to severe icing, eaves flashing may be formed as described above for 4 in 12 slopes, except that the double layer of No. 30 asphalt-saturated underlayment should be cemented. The eaves flashing should be formed by applying a continuous plastic asphalt cement layer, at the rate of 2 gal/100 sq ft (0.815 L/m²), to the surface of the underlayment starter course before the second underlayment layer is applied. Cement should also be applied to the 19 in. (482.6 mm) underlying portion of each succeeding course which lies within the eave flashing area before next course is placed. It is important to apply the cement uniformly with a comb trowel, so that at no point does underlayment touch underlayment when the application has been completed. The overlying sheet should be pressed firmly into the entire cemented area.

### DRIP EDGE

Wood shingles and shakes should extend out over the eave and rake from 1 in. (25.4 mm) to $1\frac{1}{2}$ in. (38.1 mm) to form a drip. Accordingly, a corrosion-resistant drip edge is not necessary at these points since the shingle starter course and rake shingles serve the same purpose.

### FLASHING

Refer to Section 7.6.1.6 for general comments about flashing.

Unless special precautions are taken (Fig. 7.6-82), copper flashing materials are not recommended for use with red cedar shingles or shakes. Premature deterioration of the copper may occur when the metal and wood are in intimate contact in the presence of moisture.

*Valley Flashing for Wood Shingles* Only the open method should be used to construct valley flashing for wood shin-

**FIGURE 7.6-82** Surface contact of copper with red cedar must be minimized. (Courtesy Cedar Shake and Shingle Bureau.)

gles. Butting shingles together in the valley at the centerline to form a closed valley is not recommended.

On slopes of up to 12 in 12, metal valley sheets should be wide enough to extend at least 10 in. (254 mm) on each side of the valley centerline (Fig. 7.6-83) On roofs of steeper slope, narrower sheets may be used, extending on each side of the valley centerline for at least 7 in. (177.8 mm).

To avoid restricting the delivery of a valley and to prevent water splashing from a steep slope onto the shingles of a lower slope, the edges of the shingles on the lower slope should be lined up at least 1 in. (25.4 mm) farther back from the valley centerline than those on the steeper slope. As an alternative, a vertical ridge, or waterstop, can be made by crimping the metal up in the center of the valley.

Valley sheets should be edge-crimped by turning the edges up and back approximately $\frac{1}{2}$ in. (12.7 mm) toward the valley centerline. This provides an additional waterstop, directing water down the valley. The open portion of the valley should be at least 4 in. (101.6 mm) wide, but valleys may taper from a width of 2 in. (50.8 mm) where they start and increase at the rate of $\frac{1}{2}$ in. (12.7 mm) per 8 ft (2438 mm) of length, becoming wider as they descend.

In areas where the outside design temperature is 30°F (−1.1°C) or less, underlayment should be installed under metal valley sheets.

*Valley Flashing for Wood Shakes* The open method, using roofing felt and sheet metal, or the closed method, using hand-fitted shakes, may be used to construct valley flashing for wood shakes. For longest service life, the open method is highly recommended.

Open valleys are first covered with a No. 30 asphalt-saturated felt underlayment strip, at least 20 in. (508 mm) wide, centered in the valley and secured with enough nails to hold it in place (Fig. 7.6-84). Metal flashing strips, 20 in. (508 mm) wide, are then nailed over the valley underlayment. If galvanized steel flashing is used, it should be 18-gauge (1.3 mm), and never less than 26-gauge (0.55 mm), preferably center-crimped, and painted on both surfaces. Valley sheets should be edge-crimped

**FIGURE 7.6-83** Method for constructing open valley flashing. (Courtesy Cedar Shake and Shingle Bureau.)

**FIGURE 7.6-84 Open valley flashing for shakes.** (Courtesy Cedar Shake and Shingle Bureau.)

by turning the edges up and back approximately $1/2$ in. (12.7 mm) toward the valley centerline. This provides an additional waterstop, directing water down the valley. Shakes laid to finish at the valley should be trimmed parallel with the valley to form a 6-in. (152.4-mm)-wide gutter.

Closed valleys should first be covered with a $1 \times 6$ ($25 \times 150$) wood strip, nailed flat into the saddle and covered with roofing felt, as discussed above for open valleys. Shakes in each course should be edge-trimmed to fit into the valley, then laid across the valley with a metal flashing undercourse having a 2 in. (50.8 mm) headlap and extending 10 in.

(254 mm) under the shakes on each side of the saddle.

### CHIMNEY AND WALL FLASHING

Individual corrosion-resistant metal flashing units should be inserted between each course of shingles or shakes that butt against vertical surfaces such as walls, dormers, or chimneys. One leg of this flashing should extend at least 6 in. (152.4 mm) under the shingles or shakes with the upward leg covered by the vertical wall finish or by metal counter flashing, as shown in Figure 7.6-85.

On new construction, chimney flashing should be built in by the mason (Fig. 7.6-86). If masonry work has been completed, flashing should be inserted between bricks to a depth of $3/4$ in. (19.1 mm) by removing mortar and filling over the flashing with bituminous mastic.

Galvanized steel flashing should be painted on both surfaces before it is placed. Flashing strips which must be bent to sharp angles should be painted after bending. If they are crimped to form a waterstop in the centers of valleys, the crimped surface should be given a double coat of paint.

### 7.6.7.3 Application of Roof Shingles (New Construction)

The first course of shingles at the eaves should be doubled or tripled and should project 1 in. to $1^{1}/_{2}$ in. (25.4 to 38.1 mm) beyond the eaves to provide a drip. The second layer of shingles in the first course should be nailed over the first layer to provide a minimum sidelap of at least $1^{1}/_{2}$ in. (38.1 mm) between joints (Fig.

7.6-87). If possible, joints should be "broken" by a greater margin. A triple layer of shingles in the first course provides additional insurance against leaks at the cornice. No. 3 grade shingles frequently are used for the starter course.

Shingles should be spaced at least $1/4$ in. (6.4 mm) apart to provide for expansion. Joints between shingles should be separated not less than $1^{1}/_{2}$ in. (38.1 mm) from joints in adjacent courses above and below, and joints in alternate courses should not be in direct alignment. When shingles are laid with the recommended exposure, triple coverage of the roof results (see Fig. 7.6-87).

Shingles at a valley should be carefully cut to the proper miter at the exposed butts and should be nailed in place first so that the direction of shingle application is away from the valley. This ensures that shingle joints will not break over the valley flashing.

The application of wood shingles can be speeded by the use of a special shingler's or lather's hatchet (Fig. 7.6-88). The hatchet is used for nailing, carries an exposure gauge, and has a sharpened heel that expedites trimming.

### NAILING

To ensure that shingles will lie flat and for maximum service, only two nails should be used to secure each shingle on roofs. Nails should be placed not more than $3/4$ in. (19.1 mm) from the side edge of shingles, at a distance of not more than 1 in. (25.4 mm) above the exposure line. Nails should be driven flush but not so that the nail head crushes the wood.

**FIGURE 7.6-85 Flashing at a wall.** (Courtesy Cedar Shake and Shingle Bureau.)

**FIGURE 7.6-86 Flashing at a chimney.** (Courtesy Cedar Shake and Shingle Bureau.)

FIGURE 7.6-87 Application of wood shingles over spaced or solid sheathing. (Courtesy Cedar Shake and Shingle Bureau.)

FIGURE 7.6-88 A shingler's hatchet. (Courtesy Cedar Shake and Shingle Bureau.)

FIGURE 7.6-89 Application of hips and ridges. (Courtesy Cedar Shake and Shingle Bureau.)

## HIPS AND RIDGES

Hips and ridges should be of the modified Boston type with protected nailing. Either site-applied or preformed, factory-constructed hip and ridge units may be used. Nails should be increased at least two sizes larger than the nails used to apply the shingles.

Construction of hips and ridges is shown in Figure 7.6-89. Hips and ridges should begin with a double starter course.

## GABLE RAKES

Shingles should project 1 to $1^{1}/_{2}$ in. (25.4 to 38.1 mm) over the rake. End shingles can be canted to eliminate drips, as shown in Figure 7.6-90.

## ALTERNATE APPLICATION METHODS

Several alternate methods of applying shingles, giving a different appearance to the roof, are shown in Figure 7.6-91.

## REROOFING EXISTING CONSTRUCTION

It is not necessary to remove existing wood or asphalt shingles in order to install new wood shingles. However, existing slate, tile, asbestos-cement, or mineral-fiber-cement shingles, which do not provide an adequate nailing base, must be removed.

However, the first course of existing shingles at the eaves should be removed just below the butts of the shingles in the second course. Old shingles should also be cut back and removed along the gable rakes (Fig. 7.6-92). The removed shingles along the eaves and gable rake edges should be replaced with boards having a thickness approximately equal to the thickness of the old roof section. These boards provide a strong base at the perime-

FIGURE 7.6-90 Beveled strip also provides decorative termination at gables. (Courtesy Cedar Shake and Shingle Bureau.)

ter of the new roof, concealing the old roof from view. Ridge shingles should be replaced with strips of bevel siding, laid with the thin edges down the roof.

If necessary, new flashing should be installed over existing flashing.

New shingles should be applied by the same methods used in new construction. If the existing roof sheathing is spaced, it is not necessary for all new nails to strike the sheathing.

### 7.6.7.4 Application of Roof Shakes (New Construction)

First, a 36-in. (914.4-mm)-wide No. 30 asphalt-saturated felt starter strip should be laid along the eave line over the sheathing. The starter course of shakes at the eave line should be doubled, using an undercourse of 24, 18, or 15 in. (609.6, 457.2, or 381 mm) shakes, the last being made expressly for this purpose. Striking visual effects can be achieved by using a double undercourse at the eave line. After each course has been applied, an 18-in. (457.2-mm)-wide strip of No. 30 asphalt-saturated felt interlayment should be applied over the top portion of the shakes extending onto the sheathing (Fig. 7.6-93). The bottom edge of the interlayment should be positioned at a distance above the butt equal to twice the exposure. For example, if 24 in. (609.6 mm) shakes are being laid with a 10 in. (254 mm) exposure, the bottom edge of the felt should be positioned at a distance 20 in. (508 mm) above the shake butts; the strip will then cover the top 4 in. (101.6 mm) of the shakes and extend 14 in. (355.6 mm) onto the sheathing. Individual shakes should be spaced approximately $^{1}/_{4}$ to $^{3}/_{8}$ in. (6.4 to 9.5 mm) apart to allow for possible expansion due to moisture absorption. The joints between shakes should be offset at least $1^{1}/_{2}$ in.

**a** Thatch. Shingles are positioned above and below a hypothetical course line, with deviation from the line not to exceed 1″.

**b** Serrated. Courses are doubled every third, fourth, fifth, or sixth course. Doubled courses can be laid butt edge flush or with a slight overhang.

**c** Dutch Weave. Shingles are doubled or superimposed at random throughout the roof area.

**d** Pyramid. Two extra shingles, narrow shingle over a wide one, are superimposed at random.

**FIGURE 7.6-91** Alternate shingle applications. (Courtesy Cedar Shake and Shingle Bureau.)

(38.1 mm) in adjacent courses. The joints in alternate courses also should be kept out of direct alignment when a three-ply roof is being built.

When straightsplit shakes, which are of equal thickness throughout, are applied, the froe end of the shake (the smoother end from which it has been split) should be laid undermost.

The application of shakes can be speeded by the use of a special shingler's or lather's hatchet (see Fig. 7.6-88). This hatchet is used for nailing, carries an exposure gauge, and has a sharpened heel that expedites trimming.

## NAILING

Only two nails should be used to apply each shake on roofs, regardless of the width of the shake. Nails should be placed about 1 in. (25.4 mm) in from each edge, and from 1 in. to 2 in. (25.4 to 50.8 mm) above the buttline of the succeeding course. Nails should be driven until the heads to meet the shake surface but no farther (Fig. 7.6-94).

## HIPS AND RIDGES

The final course at the ridge line, as well as shakes that terminate at hips, should be secured with additional nails and should be composed of smoother-textured shakes. A strip of No. 30 asphalt-saturated felt, at least 12 in. (304.8 mm) wide, should be applied over the crown of hips and ridges, with an equal exposure of 6 in. (152.4 mm) on each side.

**FIGURE 7.6-92** Preparation of an existing roof to receive new wood shingles or shakes. (Courtesy Cedar Shake and Shingle Bureau.)

**FIGURE 7.6-93** Application of shakes over spaced sheathing. (Courtesy Cedar Shake and Shingle Bureau.)

**FIGURE 7.6-94** Nails should be correctly placed, and heads should not break the shake surface. (Courtesy Cedar Shake and Shingle Bureau.)

# 7.7 Siding

This section discusses the materials and methods needed to provide wall finishes using mineral-fiber-cement shingles, wood shingles and shakes, and aluminum, vinyl, wood, plywood, and hardboard siding.

In addition to cost, other criteria to be considered in selecting exterior siding materials, and the method of their application, include the expected service life of the siding, wind resistance, fire resistance, and local climate. Depending on the siding color and kind selected, the total wall heat loss or gain may also be affected.

Prefabricated hip and ridge units can be used, or the hips and ridges may be cut and applied on the site.

In site construction of hips, shakes about 6 in. (152.4 mm) wide should be sorted out. Two wooden straightedges should be tacked on the roof, 6 in. (152.4 mm) from the centerline of the hip, one on each side. The starting course of shakes should be doubled. The first shake on the hip should be nailed in place with one edge resting against the guide strip. The edge of the shake projecting over the center of the hip should be cut back on a bevel. The shake on the opposite side should then be applied and the projecting edge cut back to fit. Shakes in the following courses should be applied alternately in reverse order (Fig. 7.6-95). Ridges should be constructed in a similar manner. Exposure of hip and ridge shakes is normally the same as for the shakes on the roof.

Ridge shakes should be laid along an unbroken ridge that terminates in a gable

at each end. They should be started at each gable end and terminated in the middle of the ridge. At that point, a small saddle should be face-nailed to splice the two lines. The first shake course should always be doubled at each end of a ridge.

### TILTED STRIP AT GABLES

Dripping water may be eliminated by inserting a single strip of bevel siding the full length of each gable rake with the thick edge flush with the sheathing edge (see Fig. 7.6-90). The resulting inward pitch of the roof surface diverts water away from the gable edge.

### REROOFING EXISTING CONSTRUCTION

Structures that require new roof coverings, either for appearance or serviceability, may be covered with any of the four types of cedar shakes. Except over slate, tile, asbestos-cement, or mineral-fiber-cement shingles, shakes may be applied directly over the existing roof. When the existing roof is to be overroofed with shakes, a 6-in. (152.4-mm)-wide strip of the old roofing should be removed along the eaves and gable rakes, and then filled in with boards having a thickness approximately equal to the thickness of the old roof section (see Fig. 7.6-92). These boards provide a strong base at the perimeter of the new roof, concealing the old roof from view. The old ridge covering should be removed and replaced with bevel siding, overlapping the butt edges at the peak. If necessary, new flashing should be installed over existing flashing. Nails for applying a new roof should be long enough to penetrate the underlying existing sheathing.

## 7.7.1 GENERAL

Shingles, tiles, and aluminum, vinyl, wood, plywood, and hardboard siding units are applied to exterior walls in an overlapping or interlocking fashion to shed water and resist weather penetration. They can be applied effectively on most walls. These products also are used on the interior for esthetic purposes. In either case, they may contribute color and texture to a wall surface, and when properly applied and maintained, they provide a suitable finish offering long service.

Depending on the siding type and application method selected, esthetic effects range from wall surfaces having relatively smooth, uniform coursing to those with random coursing which give a heavily textured appearance. Color can be achieved by a variety of factory- or field-applied coatings or by leaving the material in its natural color.

In addition to the wall finishing units themselves, many accessory materials are required to prepare a wall surface for their application and to apply them. Depending on the type of finishing materials and the method of application, accessory materials may include sheathing, underlayment, sheathing paper, backer board, undercoursing, channel and corner moldings, flashing, caulking, and nails and other fasteners. While some materials and typical methods of applying the described finishes are discussed in the following sections, the siding manufacturers' instructions and recommendations should always be consulted to ensure the most suitable performance.

In this section, the wall construction is assumed to be correctly installed and

**FIGURE 7.6-95** Hips and ridges. The starter course should be doubled as shown. (Courtesy Cedar Shake and Shingle Bureau.)

adequate structurally, and walls are assumed to have been insulated correctly and adequately where climatic conditions require it. It is further assumed that precautions have been taken to prevent water vapor condensation within walls. Condensation within walls can normally be eliminated by providing a vapor retarder on the warm side of the insulation. Refer to Section 7.1 for a discussion of vapor retarders and ventilation.

### 7.7.1.1 Nomenclature

The definitions of the terms *coverage*, *endlap*, *exposure*, *headlap*, *shingles*, *single butt*, *sidelap*, and *toplap* are the same as those given in Section 7.6.1 for shingle and tile roofing. Some additional definitions of siding terminology are provided here to assist in understanding the text and illustrations. Other terms are defined when they first appear in the text or are found in the Glossary. Figure 7.7-2 shows graphic representation of the terms exposure, endlap, headlap, sidelap, and toplap as they apply to siding.

*Backer Board* An undercoursing material, typically fiberboard, used beneath siding materials such as mineral-fiber-cement shingles (Fig. 7.7-1a). It adds insulation value, increases resistance to impact, and provides a heavier shadowline at the butt than shingle siding applied directly without the use of a backing material.

*Double Coursing* A method of applying shingles over an undercourse of lower-grade shingles or other suitable material (Fig. 7.7-1b). Nails are exposed as a result of face (butt) nailing, and greater shingle exposures are possible than with single coursing. Double coursing usually results in greater wall coverage at lower cost.

*Single Coursing* A method of applying shingles without an undercourse (Fig. 7.7-3). In wood shingles, this method results in concealed nails and a smaller exposure than is found in double-coursed walls.

### 7.7.1.2 Wall Framing and Sheathing

The performance of siding is dependent on the supporting structure. Studs, headers, and other supports should be adequately sized and spaced to carry the necessary superimposed loads over the required spans.

Sheathing should be smooth, securely attached to supports, and provide an adequate base to receive nails and other fasteners. Both mineral-fiber-cement and wood shingles are applied over solid sheathing composed of wood boards or plywood. Wood and hardboard siding are applied over wood board, plywood, or nonwood sheathing, such as gypsum board, rigid foam, and fiberboard.

Wall framing and sheathing are discussed in Chapter 6.

### 7.7.1.3 Flashing and Caulking

Walls may contain window or door openings, intersect with sloped roof surfaces, or terminate in exposed locations against other materials, creating opportunities for

E  - Exposure

TL - Toplap

HL - Headlap

W  - Width for Strip Shingles or Length for Individal Shingles

**FIGURE 7.7-2** Shingle siding terminology.

leakage. Flashing must be installed at these locations and be made watertight and water-shedding.

### FLASHING MATERIALS

Corrosion-resistant metal flashing should be at least 26-gauge (0.55 mm) galvanized steel, 0.019 in. (0.4826 mm) aluminum, or 16 oz (453.6 g) copper. Refer to Section 7.9 for a discussion of flashing materials.

Additional requirements and typical methods of flashing are described in the following subsections for each siding type.

### WALL FLASHING

Flashing is discussed in Section 7.9.

*Flashing at Openings* Heads of openings should receive drip caps. Bottoms of openings should receive sills. Drip caps and sills (see Fig 7.9-4) should be flashed. At heads of openings in wood and metal-stud-framed walls, metal flashing should be installed extending from at least 2 in. (50.8 mm) above the trim and turned down over the outside edge of drip caps, forming a drip. However, in areas not subject to wind-driven rain, head flashing may be omitted when the vertical height between the top of finish trim of openings and the bottom of eave soffits is equal to or less than one-quarter of the eave overhang.

Flashing should be installed beneath sills and turned out down to form a drip beyond the face of the siding.

**a**          **b**

**FIGURE 7.7-1** Double-coursed walls with shingles applied over an undercourse: (a) mineral-fiber-cement shingles; (b) wood shingles or shakes.

**a**  **b**

**FIGURE 7.7-3** Single-coursed walls have shingles applied without an underlayment: (a) mineral-fiber-cement shingles; (b) wood shingles and shakes.

Where sheathing paper is omitted, flashing should extend up behind the sheathing.

In unsheathed walls, jambs should be flashed with a 6-in. (152.4-mm)-wide strip of metal, 3 oz (85 g) copper-coated paper, or 6 mil (0.1524 mm) polyethylene film.

***Flashing At Wall and Roof Intersections***
Sheet metal flashing should be installed at horizontal intersections and at vertical intersections when the exterior finish material does not provide a self-flashing joint.

***Flashing at Base of Walls and at Horizontal Material Changes*** Bottoms of walls and horizontal changes in materials should receive watertables. On these watertables, metal flashing should be installed to a line at least 2 in. (50.8 mm) above the bottom of the siding and turned down over the watertable to form a drip (see Fig 7.9-4).

### CAULKING

A nonshrinking caulking compound, either in white or matching color, should be used to weatherseal joints where siding abuts wooden trim, masonry, or other projection and in unprotected vertical joints in panel siding.

### 7.7.1.4 Nails and Fasteners

No single step in applying siding materials is more important than proper nailing, which includes:

■ Selecting the correct nail for the kind of siding and sheathing

■ Using the correct number of nails
■ Locating nails correctly
■ Choosing nail metal compatible with the metal used for flashings

Specific recommendations for the type, size, number, and spacing of nails are given in the following subsections for each wall finish material. Some general recommendations are given here.

### SHINGLES OVER WOOD AND PLYWOOD SHEATHING

Nails for attaching shingles should be long enough to penetrate through the shingles and through wood and plywood sheathing. Nails for applying shingles over wood and plywood sheathing should have threaded shanks.

### SHINGLES OVER NONWOOD SHEATHING

When gypsum, rigid foam, fiberboard, or other nonwood sheathing materials are used, special fasteners and special details for fastening shingles are often necessary to provide adequate anchorage. The nonwood material manufacturer's recommendations should be followed.

### WOOD, PLYWOOD, AND HARDBOARD SIDING

Nails for attaching wood, plywood, and hardboard board and panel siding should penetrate through the sheathing and at least $1\frac{1}{2}$ in. (38 mm) into the underlying studs.

### 7.7.2 MINERAL-FIBER-CEMENT SHINGLES

Unless specific exceptions are made in the following text, the provisions of Section 7.6.1 apply to mineral-fiber-cement shingles and the methods used to apply them.

### 7.7.2.1 Materials

Mineral-fiber-cement wall shingles are usually supplied in rectangular units having straight or wavy butt edges. These units are available in the sizes shown in Figure 7.7-4, and other sizes may be available. They are designed to be applied with a minimum nominal toplap of 1 in. (25.4 mm). Depending on their size, these units range up to approximately $\frac{3}{16}$ in. (4.8 mm) in thickness.

Wall shingles are manufactured with prepunched nail holes correctly sized and located to receive exposed face nails. These holes are aligned the proper distance above the lower edge of the units to establish the recommended amount of toplap (Fig. 7.7-5) and to minimize damage to shingles during their installation.

**FIGURE 7.7-4** Mineral-Fiber-Cement Wall Shingles

| Sizes,[a] in. (mm) | Exposure, in. (mm) | Pieces per Bundle | Pieces per Square[b] |
|---|---|---|---|
| $12 \times 24$ ($305 \times 610$) | 11 (279.4) | 18 | 54 |
| $9 \times 32$ ($229 \times 813$) | 8 (203.2) | 19 | 57 |
| $9\frac{5}{8} \times 32$ ($244.5 \times 813$) | $8\frac{5}{8}$ (219) | 17–18 | 52 |
| $11\frac{3}{4} \times 32$ ($298.5 \times 813$) | $10\frac{3}{4}$ (273) | 14 | 42 |
| $14\frac{5}{8} \times 32$ ($371.5 \times 813$) | $13\frac{5}{8}$ (346) | 11 | 33 |
| $14\frac{3}{4} \times 32$ ($374.7 \times 813$) | $13\frac{3}{4}$ (349) | 11 | 33 |
| $8\frac{3}{4} \times 48$ ($222 \times 1219$) | $7\frac{3}{4}$ (197) | 13 | 39 |
| $12 \times 48$ ($292 \times 1219$) | 11 (279) | 9 | 27 |

[a]Three nail holes per shingle.
[b]Weight per square ranges from 153 to 206 lb (693.97 to 934.37 Mg).

**FIGURE 7.7-5** Prepunched nail holes in mineral-fiber-cement siding shingles.

For standard application, face nails driven through these holes and resting on the top edge of the next lower course serve as guides for the correct amount of toplap, exposure, and unit alignment.

### NAILS AND FASTENERS

Face nails should be long enough to enter into and hold securely to the nailing base. Face nails are usually furnished by the siding manufacturer to ensure the use of the proper kind and type. Nails and fasteners should be permanently corrosion-resistant and stain-resistant. Only those furnished with the siding or recommended by the siding manufacturer should be used.

Figure 7.7-6 is a summary of face nail types recommended for use in applying mineral-fiber-cement wall shingles.

### 7.7.2.2 Preparation for Wall Shingles

### WALL SHEATHING

Mineral-fiber-cement shingle siding can be applied over lumber, plywood, and other suitable types of rigid sheathing that are smooth and dry, and that provide adequate support and an adequate nailing base. Sheathing material thicknesses, spans, and nailing requirements are discussed in Chapter 6.

### SHEATHING PAPER

Sheathing paper for use over lumber and plywood sheathing should be water repellent yet vapor porous (permeable to water vapor), such as No. 15 asphalt-saturated felt. Coal-tar-saturated felt, which may stain the shingles, should not be used. Vapor retarder membranes should not be used for sheathing paper between siding and sheathing.

Sheathing paper for walls is further discussed in Section 7.6.3.2. Sheathing paper used as an air infiltration barrier is discussed in Section 7.5.

### JOINT FLASHING STRIPS

Asphalt (not coal-tar)-saturated and coated, water-repellent joint flashing strips are supplied by the manufacturer with all types of mineral-fiber-cement shingle siding and should be applied behind every vertical joint between wall shingles (Fig. 7.7-7).

### CANT STRIPS

A $^1/_4$ in. (6.4 mm) by $1^1/_2$ in. (38.1 mm) wood strip should be nailed level along the bottom edge of the sheathing so that it overhangs the top of the foundation wall a sufficient amount to seal the joint between the top of the foundation, the wood sill, and the bottom of the sheathing (see Fig. 7.7-7). The bottom edge of the shingles should align with the bottom edge of the cant strip. This strip gives the necessary cant (pitch) to the first shingle course and provides solid bearing to prevent breakage.

### FLASHING

Flashing should be done in accordance with Section 7.6.1.6.

### 7.7.2.3 Wall Shingles Application (New Construction)

Depending on the sheathing type, mineral-fiber-cement shingles may be applied by several methods: (1) *direct application* to nailable sheathing; (2) the *shingle backer method*; (3) the *channel method* over nailable sheathing or directly over framing members without sheathing; and (4) the *wood nailing strip method* for use over any type of sheathing or directly over framing members without sheathing.

The methods described are based on applying 12 by 24 in. (304.8 by 609.6

**FIGURE 7.7-6** Face Nails for Use with Mineral-Fiber-Cement Wall Shingles

| | **New Construction** | |
| --- | --- | --- |
| **Sheathing Type** | **Nailed Directly to Sheathing or Wood Nailing Strips** | **Nailed Through Fiberboard Shingle Undercoursing** |
| Board lumber | $1^1/_8$ in. (28.5 mm) screw thread<br>$1^7/_{16}$ in. (36.5 mm) screw thread<br>$1^1/_2$ in. (38.1 mm) annular thread | 2 in. (50.8 mm) annular thread |
| Plywood | $1^1/_8$ in. (28.5 mm) screw thread | $1^3/_4$ in. (44.5 mm) screw thread |
| Wood nailing strip | $1^1/_8$ in. (28.5 mm) screw thread | $1^3/_4$ in. (44.5 mm) screw thread |
| Nailbase fiberboard | $1^3/_8$ in. (34.9 mm) special annular thread | 2 in. (50.8 mm) special annular thread |

| | **Residing**<br>**(over existing wood siding or shingles)** |
| --- | --- |
| **Nailing Base** | **Nail Types** |
| Sound board lumber | $1^3/_4$ in. (44.5 mm) or longer annular thread, screw thread, or helical thread |
| Wood nailing strip | $1^1/_8$ in. (28.5 mm) screw thread |

| Screw Thread | Helical Thread | Special Annular Thread | Annular Thread |

SHEATHING PAPER WHEN
LUMBER BOARDS ARE USED

CORNER FLASHING  JOINT STRIP  CHALK LINES  11"
12"
CANT STRIP MUST OVERHANG TOP OF FOUNDATION TO SEAL JOINT AT SILL
¼" × 1½"
CANT STRIP

11" = 279.4 mm
12" = 304.8 mm
¼" × 1½" = 6.4 × 38.1 mm

**FIGURE 7.7-7** Direct application of mineral-fiber-cement shingles showing application of joint flashing strips.

mm) wall shingles from left to right on new construction. The manufacturers' instructions should be followed when applying other shingle sizes.

Careful layout should establish the location of the first horizontal course and ensure that all courses will meet and match at corners and are level all around the building. The total number of horizontal courses to complete one wall should be laid out, and adjustments should be made so that the top courses at the eaves will not be too narrow. This may require uniform adjustments in reducing the exposure of each shingle course. A level chalkline should be marked all around the building to fix the location of the top edge of the first course of shingles. Additional horizontal chalklines should mark the top of each succeeding course. The lines should be spaced to provide the desired exposure (see Fig. 7.7-7).

### DIRECT APPLICATION METHOD

The direct application method is used for sheathing types that provide a nailing base, such as wood boards or plywood.

*First and Odd-Numbered Courses* A full-size shingle should be started at the left corner of the wall. This unit should be carefully applied plumb, level, and in alignment with a chalkline because it guides the laying of all other units.

The correct face nails for the sheathing type and application method being used are supplied by the manufacturer with the shingles, as described in Figure 7.7-6. Nails should be driven snug but not too tight.

Before the last nail is driven at the right end of a unit, a felt joint flashing strip should be inserted in place and secured with the last nail. Joint flashing strips should always be used, centered behind the butted joint between wall shingles, and with the lower end overlapping the cant strip or head of the next lower course.

Application of full-size shingles should be continued in the first course with the top edges aligned to the chalkline. The last unit in a course or at an opening should not be less than 6 in. (152.4 mm) wide, and if smaller spaces will remain to be filled, a few inches should be cut from the units applied earlier in the course. Necessary holes for face nails in less than full-size pieces of siding should be punched. Unit ends should be butted tightly end to end, with no space in between. Wedge-shaped spaces between unit ends indicate incorrect application.

*Second and Succeeding Even-Numbered Courses* In succeeding courses of 12 in. by 24 in. (304.8 by 609.6 mm) (305 by 610 mm) shingles, vertical joints should break on *halves*; therefore, the second course should be started with a half-unit. When 32-in. (812.8-mm)-long units are used, vertical joints should also break on halves. When 48-in. (1219.2 mm)-long units are used, they should break on *thirds*, and the second course should be started with a two-third unit. Some bun-

dles of shingles contain the necessary shorter pieces for starting courses.

A partial unit should be started at the left corner with its head edge aligned to the chalkline and its lower (butt) edge overlapping the head of the next lower course the correct distance to provide the necessary toplap between courses. A nail should be inserted in a face nail hole and the siding positioned so that the shank of the nail is resting on the top edge of the next lower course, thus establishing the proper amount of toplap. The unit should then be nailed in place with a felt joint flashing strip installed as described above. The course should be continued with full-size units.

The face nails described in Figure 7.7-6 should always be used. The nails should penetrate completely through the sheathing. A standard threaded face nail should be used when the siding is applied directly to nominal 1 in. (25.4 mm) wood board sheathing or $^5/_{16}$ in. (7.9 mm) and thicker plywood without wood nailing strips or shingle backer.

### SHINGLE BACKER METHOD

The shingle backer method uses rigid insulating fiberboard as a backer for the shingles and may be used over wood boards or plywood, since they provide a nailing base for both the shingle backer and the shingles (Figs. 7.7-8, 7.7-9, and 7.7-10). Special methods may be used to apply shingles directly over framing members.

**a**

WOOD SHEATHING
℄
NO. 15 ASPHALT SATURATED FELT
SHINGLE BACKER
JOINT STRIP
SHINGLE
EDGE OF SHINGLE MUST OVERHANG BOTTOM EDGE OF BACKER BOARD APPROX. ¼"
¼" × ½" (6.4 mm × 38.1 mm) CANT STRIP

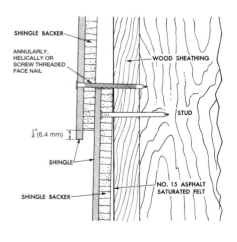

**b**

SHINGLE BACKER
ANNULARLY, HELICALLY OR SCREW THREADED FACE NAIL
WOOD SHEATHING
STUD
¼" (6.4 mm)
SHINGLE
SHINGLE BACKER
NO. 15 ASPHALT SATURATED FELT

**FIGURE 7.7-8** Application of mineral-fiber-cement wall shingles using the shingle backer method.

**FIGURE 7.7-9** Application of mineral-fiber-cement wall shingles using rabbeted nailing strips.

**FIGURE 7.7-10** Application of mineral-fiber-cement wall shingles using grooved shingle backer.

A rigid, water-resistant, insulating fiberboard shingle backer should be at least $^5/_{16}$ in. (7.9 mm) thick and should underlay the full width and length of each course of shingles completely. The shingle backer should be approximately $^1/_4$ in. (6.4 mm) narrower than the shingles. The shingles should be applied with their head edges flush with the head edges of the shingle backer and their butt edges extended down and beyond the butt edges of the underlying shingle backer to provide a drip edge. Shingle backer should be at least 4 ft (1220 mm) long and should be applied with staggered joints in succeeding courses. Vertical joints between

shingles should not occur over the ends of the shingle backer. Shingle backer should be applied with its butt edges overlapping the head of the lower course of shingles about $^1/_4$ in. (6.4 mm) less than the headlap used between shingle courses.

The shingle backer and the shingles should be secured to the wall with face nails driven through holes in the shingles and into the underlying sheathing. Shingle backer should be held in position before shingles are applied by nailing each panel about 3 in. (76.2 mm) from the top edge with two or three $1^1/_4$ in. (31.8 mm) galvanized roofing nails. The type and length of the face nails should be as described in Figure 7.7-6.

Succeeding shingle courses should be applied in a similar manner. Felt flashing joint strips are not required if the shingle backer board is water-repellent and stain-resistant.

Direct attachment to framing members is possible using a rabbeted nailing strip with a special shingle backer that also serves as sheathing (see Fig. 7.7-9). This method provides firm backing and a heavy shadowline. A similar method using a grooved shingle backer is shown in Figure 7.7-10.

### CHANNEL METHOD

The channel method uses one of several styles of metal channel molding designed for securing mineral fiber siding to nailable sheathing or directly to framing members. Metal channel should be corrosion-resistant and stain-resistant metal of sufficient thickness to provide adequate support and rigidity. Moldings usually are prepainted to match the shingle color and also provide either a standard or heavy butt shadow (Fig. 7.7-11).

A $1^3/_{16}$ in. (30.16 mm) channel permits standard prepunched shingles to be used when units are reversed head to butt (Fig. 7.7-12). Channels should be secured at 16 in. (406.4 mm) intervals either to the framing member or to nailable

**FIGURE 7.7-11** Typical profiles of metal channel moldings for installing mineral-fiber-cement shingles by the channel method.

**FIGURE 7.7-12** Application of mineral-fiber-cement shingles using the channel method. See also Figure 7.7-13.

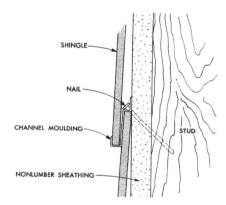

**FIGURE 7.7-13** Channels are nailed directly to framing members when sheathing does not provide a nailing base.

sheathing with 10-gauge (0.135 in. [3.43 mm]) to 12-gauge (0.106 in., 2.7 mm), hot-dip galvanized steel, large head roofing nails long enough to penetrate and hold securely in the nailing base. Nails should be driven at an angle and set well to provide sufficient pressure on the nailing flange to ensure a tight, vibration-free assembly (Fig. 7.7-13).

Channel molding should have a vertical nailing flange at least $^3/_8$ in. (9.5 mm) high that can bear against the face of the sheathing, and should be strong and rigid enough to secure the siding to the wall sheathing.

Direct attachment to framing members requires shingles wide enough to span from center to center of studs. Face nail holes should be punched within $^3/_8$ to $^1/_2$ in. (9.5 to 12.7 mm) from the vertical edge of the shingles to ensure nailing into the stud when two shingles are butted together. Intermediate face nail holes should be provided for nailing each stud, and face nails should be long enough to penetrate at least 1 in. (25.4 mm) into the stud.

When this method of nailing is used, studs must be located accurately on 16 in. (406.4 mm) centers wherever possible. Warped studs should not be used.

When it is not possible to nail directly into a stud, either a *twist nail* or a *self-clinching nail* should be used to attach the corner of the siding to nonlumber sheathing. A twist nail is a slender copper or aluminum nail with flat head and medium needle point designed for twist clinching, that is, manual twisting of the exposed shank. A self-clinching nail is designed so that its shank clinches automatically when the nail has been fully driven.

The inside of exterior walls should be left exposed for final inspection to ensure that shingle nails have been properly driven into the studs and that twist nails and self-clinching nails are properly clinched.

### WOOD NAILING STRIP METHOD

The wood nailing strip method may be used over lumber boards, plywood, and any type or thickness of nonlumber sheathing that does not have adequate nail-holding power, because the strips rather than the sheathing serve as the nail base. No special nails, fasteners, or channels are required, and the standard face nails shipped with the siding can be used. In addition to providing maximum security for the siding and economy in application, this method produces a heavy shadowline along the butt edge of each course of shingles.

Wood nailing strips for the application of mineral-fiber-cement shingles should be at least nominal $3/8$ in. (9.5 mm) by $3 1/2$ in. (88.9 mm) boards that are straight, sound, and free from knots or knotholes that might cause splitting or hinder nailing. Strips should be not less than 4 ft (1220 mm) long, and preferably longer, so that they can span continuously over several framing members. Seasoning conditions and face nailing of spaced boards should be the same as described for solid sheathing.

Wood nailing strips should be placed to overlay the head of the lower course of shingles in an amount equal to the desired toplap less $1/4$ in. (6.4 mm) (Fig. 7.7-14). This firmly clamps the head of the course of shingles to the wall. At each framing member, the wood strip should be nailed through the sheathing and to each underlying stud with two 2 in. (50.8 mm) by 0.113 in. (2.87 mm) (6d) or longer annular threaded nails, penetrating the stud at least 1 in. (25.4 mm). The wood strips should end over framing members; joints in the strips should be staggered in succeeding courses. Vertical joints between shingle units should not occur over the end joints of wood strips.

The next course of shingles should then be applied with its butt edge placed over the wood strip with the lower edge of the shingles extending down over the lower edge of the strip not more than $1/4$ in. (6.4 mm) to provide a drip. The siding should then be nailed through the face nail holes to the wood nailing strips with the threaded face nails supplied with the shingles.

### CORNER FINISHING

When wall surfaces join at inside corners, the shingle courses are usually butted, although metal molding or wood corner strips can be used. Outside corners can be finished by any one of three methods.

***Metal Corner Finish*** Several styles of individual stain-resistant metal corners and finish moldings are available (Fig. 7.7-15). They should be applied as recommended by the manufacturer.

***Wood Corner Board Finish*** Corners finished with wood strips can be made of previously primed or painted nominal 1 in. (25.4 mm) lumber applied along one corner. A 1 × 4 (25 × 100) should be applied along one corner so that its side edge projects a distance beyond the corner equal to the thickness of the board applied on the adjoining wall. A 1 × 3 (25 × 75) should then be applied with its edge butting against the back of the 1 × 4 (25 × 100) first applied. Therefore, the wood corner strips will appear to be the same width on each wall (Fig. 7.7-16). When wood nailing strips, shingle backer, or both, are used, wood corner boards thicker than nominal lumber or blocking may be required.

***Lapped or Woven Corner Finish*** This method is appropriated for white or gray shingles only. For lapped adjoining courses at outside corners, a whole shingle should be placed temporarily against each adjoining wall so that shingles meet and join at the corner. The adjoining vertical edges of the shingles are scribed and cut at the proper angle to form a lapped or woven corner joint, and the edges are dressed with a file. Corner joints should be lapped to the left and right in alternate courses. Courses on adjoining walls should meet and match at corners.

### FLASHING AT OPENINGS AND CORNERS

Metal flashing should be applied over exposed door and window openings (Fig. 7.7-17). On old structures, the flashing must be renewed if it is not in good condition. Drip edges should be provided at heads of openings, under sills, and at projections to keep water away from the finished wall surface below.

16″ = 400 mm
$1/4$″ = 6.4 mm

$3/8$″ × $3 5/8$″ = 9.5 mm × 92.1 mm
$1 1/8$″ = 28.6 mm

**FIGURE 7.7-14** Application details of mineral-fiber-cement shingles using the wood nailing strip method.

**FIGURE 7.7-15** Metal corner for mineral-fiber-cement shingles.

$1 \times 3'' = 1 \times 3 \ (25 \times 75)$
$1 \times 4'' = 1 \times 4 \ (25 \times 100)$

**FIGURE 7.7-16** Wood board corner for mineral-fiber-cement shingles.

**FIGURE 7.7-17** Head flashing over an opening in a mineral-fiber-cement shingle wall.

Outside and inside corners should be flashed with a 12-in. (304.8-mm)-wide strip of underlay material centered in or around the corner when corner boards are used (see Fig. 7.7-16). In other corner treatments, the underlayment should be carried around the corner of each sidewall so that a double thickness of felt results over the corner (see Fig. 7.7-15).

#### 7.7.2.4 Reshingling Existing Walls

An old wall that is to be covered with new mineral-fiber-cement shingles should be repaired so that it has a smooth, substantially true surface with adequate nail-holding capacity and the ability to support the new shingles. When the existing exterior finish consists of relatively thin beveled siding applied directly to widely spaced framing members and does not provide an adequate nailing base for new shingles, the walls should be stripped at nailing locations with beveled wood strips. The face nails described in Figure 7.7-6 should be used for application of new shingles.

#### RESHINGLING OVER NONLUMBER SHEATHING

When mineral-fiber-cement shingles are to be applied to a structure that originally was sheathed with a nonnailable sheathing, and the existing siding to be covered will not provide an adequate nailing base, the only recommended attachment method is the wood nailing strip method discussed in Section 7.7.2.3. Using any other method will require both the existing siding and the existing sheathing to be removed down to the existing framing members and replaced with new sheathing to receive new shingles.

#### RESHINGLING OVER STUCCO

When new shingles are to be applied over old stucco, the best practice is to remove the old stucco, nails, lath, and protruding elements and apply the new shingles directly over the existing sheathing, if any exists. However, if the stucco is in reasonably good condition or is removed to bare studs, new shingles can be applied by the wood nail strip or metal channel methods described in Section 7.7.2.3, modifying the steps as necessary to meet special conditions.

### 7.7.3 WOOD SHINGLES AND SHAKES

Unless specific exceptions are made in the following section, the provisions of Section 7.7.1 apply to wood shingles and shakes and methods of their application.

#### 7.7.3.1 Materials

Most wood shingles and shakes used for side walls are made from the same material and produced in the same way as described for wood roof shingles and shakes in Section 7.6.7.

#### CEDAR SHINGLES

Wall shingles are produced from cedar blocks in the same manner as described in Section 7.6.7.1 for roof shingles.

*Grades, Types, and Sizes*  Shingles are marketed according to grade, thickness, type, and size, as described in Section 7.6.7.1, and the requirements in Section 7.6.7.1 also apply to wood wall shingles and shakes.

The following additional terms also apply to wood wall shingles:

1. *Undercoursing shingles,* a lower grade than No. 3, are available in 16 and 18 in. (406.4 and 457.2 mm) lengths. They are produced expressly for use in the inner and completely concealed course of double-coursed walls (Fig. 7.7-18a). They should not be used for any other purpose. Some mills mix Grade No. 3 and Undercoursing shingles and call them *special undercoursing* shingles.

2. *Rebutted and rejointed shingles* are machine trimmed for exact parallel edges, with butts cut to exact right angles. They are also available with smooth-sanded faces (Fig. 7.7-18b). Produced in No. 1 and No. 2 grades in 16 and 18 in. (406.4 and 457.2 mm) lengths, they can be used on either single- or double-coursed walls where tight joints between shingles are desired.

3. *Machine-grooved shingles* (also known as *machine-grooved shakes*) are similar to rebutted and rejointed shingles, except that the face is machine striated to give a corrugated appearance (Fig. 7.7-18c). They are produced in No. 1 grade only, in 16 and 18 in. (406.4 and 457.2 mm) lengths, and either natural, prime-coated, or finished coated. These units can be used for the surface course in double-coursed wall applications over low-grade undercoursing.

4. *Dimension shingles* are cut to specific uniform widths where special effects are desired (Fig. 7.7-18d). They are available in 5 or 6 in. (127 or 152.4 mm) widths in 16 in. (406.4 mm) lengths in No. 1 and No. 2 grades; in 5 or 6 in. (127 or 152.4 mm) widths in 18 in. (457.2 mm) lengths in No. 1 grade; and in 6 in. (152.4 mm) width in 24 in. (609.6 mm) length in No. 1

**a**

**b**

**c**

**d**

**e**

**FIGURE 7.7-18** (a) A double-coursed wall showing completely concealed undercover shingles; (b) a wall made using rebutted and rejointed shingles; (c) machine-grooved shingles; (d) dimension shingles; (e) fancy-butt shingles (many other shapes are available). (Courtesy Cedar Shake and Shingle Bureau.)

the recognized standards of the industry. Refer to Section 7.6.7.1 for additional discussion.

### NAILS

Nails for applying wall shakes and shingles should be of the same type and material as those required for wood roof shingles and shakes (see Section 7.6.7.1).

Box nails, having smaller diameters and longer lengths than common nails, are also suitable. Nails should be long enough to penetrate completely through the sheathing.

Nail lengths typically required for wall installations are given in Figure 7.7-20.

### 7.7.3.2 Preparation for Wall Shingles and Shakes

### EXPOSURE

The exposure of shakes and shingles is dependent on their length and on whether the walls are to be single- or double-coursed. For single coursing, shake and shingle weather exposure should never be greater than half of the shingle or shake length minus $1/2$ in. (12.7 mm) so that two layers of wood occur over the entire wall.

The maximum recommended weather exposures for single- and double-coursed shingle and shake wall construction are shown in Figure 7.7-20. Approximate wall coverage for shingles using different exposures is shown in Figure 7.6-71.

---

grade. *Fancy-butt* shingles (Fig. 7.7-18e) are produced from dimension shingles. The butt designs are produced by trimming the shingles on a band saw.

A summary of cedar shingle types and grades is given in Figure 7.6-70. The coverage of wood shingles is shown in Figure 7.6-71. The maximum exposure for wall shingles is shown in Figure 7.7-19.

### CEDAR SHAKES

Sizes, types, grades, method of producing, grade trademarks, and packaging are the same for wall shakes, as for roofing shakes as described in Section 7.6.7.1.

### STANDARDS OF QUALITY

The manufacture of cedar shingles and shakes should conform to CSSB specifications. These specifications constitute

**FIGURE 7.7-19** Maximum Exposure for Wood Shingles and Shakes

| Type | Length, in. | Single Coursing, in. | Double Coursing, in. |
|------|-------------|----------------------|----------------------|
| **Hand-split-and-resawn shakes** | 18 | $8\frac{1}{2}$ | 14 |
| | 24 | $11\frac{1}{2}$ | 20 |
| | 32 | 15 | — |
| **Taper-split shakes** | 24 | $11\frac{1}{2}$ | 20 |
| **Straight-split shakes** | 18 | $8\frac{1}{2}$ | 16 |
| | 24 | $11\frac{1}{2}$ | 22 |
| **Random-width and dimension shingles** | 16 | $7\frac{1}{2}$ | 12 |
| | 18 | $8\frac{1}{2}$ | 14 |
| | 24 | $11\frac{1}{2}$ | 16 |
| **Rebutted and rejointed shingles** | 16 | $7\frac{1}{2}$ | 12 |
| | 18 | $8\frac{1}{2}$ | 14 |
| **Machine-grooved shingles** | 16 | — | 12 |
| | 18 | — | 14 |

(Courtesy Cedar Shake and Shingle Bureau.)

**FIGURE 7.7-20** Minimum Nail Lengths for Wood Shingles and Shakes[a]

| Type | Nail Lengths[b] | |
| | Double-Coursed Sidewalls, in. (mm) | Single-Coursed Sidewalls, in. (mm) |
| --- | --- | --- |
| Shingles | 1¾ (44.5) (5d) | 1¼ (28.6) (3d) |
| Shakes[c] | 2½ (63.5) (8d) | 1½ (38.1) (4d) |

[a]For application over plywood sheathing, nails should be threaded.

[b]Nails should be long enough to penetrate through sheathing.

[c]Because thickness and exposure of shakes vary, larger nails may be necessary to obtain suitable penetration in the sheathing.

(Courtesy Cedar Shake and Shingle Bureau.)

## WALL SHEATHING

Sheathing types, thicknesses, spans, and nailing recommendations are discussed in Chapter 6.

## SHEATHING PAPER

To provide moisture protection when lumber boards are used for sheathing, No. 15 or No. 30 asphalt-saturated felt should be applied over the entire wall. When spaced (open) sheathing is used, sheathing paper should be applied first to the framing members (Fig. 7.7-21).

### 7.7.3.3 Application of Walls Shingles and Shakes

The two basic methods used to apply shakes and shingles are (1) single coursing and (2) double coursing. When single-coursed (Fig. 7.7-22), units can be applied in the same manner as wood roof shakes and shingles, except that greater weather exposures can be used. When double-coursed (Fig. 7.7-23), units can

**FIGURE 7.7-21** Spaced sheathing being applied over sheathing paper. (Courtesy Cedar Shake and Shingle Bureau.)

**FIGURE 7.7-22** Single coursing of a wood shingle or shake wall. (Courtesy Cedar Shake and Shingle Bureau.)

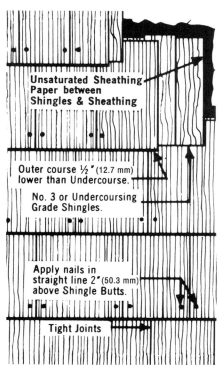

Unsaturated Sheathing Paper between Shingles & Sheathing

Outer course ½″ (12.7 mm) lower than Undercourse.

No. 3 or Undercoursing Grade Shingles.

Apply nails in straight line 2″ (50.3 mm) above Shingle Butts.

Tight Joints

**FIGURE 7.7-23** Double coursing of a wood shingle or shake wall. (Courtesy Cedar Shake and Shingle Bureau.)

be applied at much greater weather exposures over a course of either No. 3 Grade or Undercoursing Grade shingles.

Using either method, at least two layers of shingles (double coverage) occur at every point on the wall surface. Except where specifically noted, the application methods described and illustrated in the following subsections are applicable to both shakes and shingles.

## LAYING OUT

The exposure and number of courses should be determined by measuring the height of the dominant wall from a point 1 in. (25.4 mm) below the top of the foundation to the top of the windows, underside of soffits, or other vertical reference points.

The height is then divided into equal parts, as near as possible to the maximum recommended shingle or shake weather exposure. This measurement and the number of courses can be transferred to a story pole or be established with chalklines all around the building (Fig. 7.7-24). Course exposure can vary slightly to line up with the bottom and top of openings and at the eave. Story pole marks or chalklines should be made after sheathing paper has been applied.

a

b

**FIGURE 7.7-24** Courses are laid out around the building using (a) a story pole or (b) a chalkline. (Courtesy Cedar Shake and Shingle Bureau.)

**FIGURE 7.7-25** Two layers of undercoursing under the first course. (Courtesy Cedar Shake and Shingle Bureau.)

**FIGURE 7.7-26** The straightedge forms a guide for nailing undercoursing and exposure courses. (Courtesy Cedar Shake and Shingle Bureau.)

## SINGLE COURSING

The first course at the foundation should be doubled. Each shake or shingle should be secured with the correct nails (see Fig. 7.7-19), driven about $^3/_4$ in. (19.1 mm) from each edge and not to exceed 1 in. (25.4 mm) above the buttline of the succeeding course. Joints between shingles in one course should be separated at least $1^1/_2$ in. (38.1 mm) from joints in adjacent courses, and joints in alternate courses should not be in direct alignment.

When fiberboard or gypsum sheathing occurs, shingles and shakes should be applied over 1 × 3 (25 × 75) nailing strips (see Fig. 7.7-21) spaced according to shingle exposure, or special fasteners designed for this purpose should be used.

## DOUBLE COURSING

The bottom course should be laid with two under courses and one outer course. This triple coursing eliminates the need for a costly drip cap and makes the bottom course as thick as the succeeding courses (Fig. 7.7-25).

A straightedge, such as a piece of wood siding or 1 × 4 (25 × 100) lumber, should be tacked tightly along a chalkline snapped between the bottom story pole marks. Undercourse shingles, spaced about $^1/_8$ in. (3.2 mm) apart, should be laid on the upper lip of the straightedge and fastened with a single staple near the top (Fig. 7.7-26).

The outer course should then be applied, laid tight without spacing. As each double course is laid, a chalked string should be stretched from opposite corner marks to indicate the level of the next course and relocate the position of the straightedge.

Each shingle or shake on walls should be nailed with a minimum of two nails, applied 2 in. (50.8 mm) above the buttline and $^3/_4$ in. (19.1 mm) from the edge of each shingle or shake. On shingles wider than 8 in. (203.2 mm), two additional nails should be driven near the center and about 1 in. (25.4 mm) apart.

## CORNERS

Outside corners may be lapped alternately. One outer course should protrude slightly past the corner, butting the shake on the other side of it. The protruding corner should then be trimmed with a knife and plane and touched up with stain (Fig. 7.7-27). Metal corner molding designed for this purpose may also be used.

Inside corners should be fitted with a lumber strip about $1^1/_2$ in. (38.1 mm) square to receive the undercourse and outer shakes or shingles with a minimum

**FIGURE 7.7-27** Corner treatments for shakes: (a) jointed inside corner, (b) woven outside corner, (c) mitered inside corner, (d) mitered outside corner. Corner treatments for shingles: (e) overlapping outside corner, (f) mitered inside corner using metal flashing in the joint. (Courtesy Cedar Shake and Shingle Bureau.)

of fitting. The strip should be stained after application.

### 7.7.3.4 Residing Existing Walls

Shingles or shakes can be applied either directly to existing siding or to new sheathing. Building paper should be applied over old siding; other steps are the same as described above (Fig. 7.7-28). Before applying courses around windows and doors, molding strips, to which shakes and shingles should be joined, should be nailed on the face of old casings, flush with the outer edges. When residing over stucco, nailing strips first should be nailed over the wall to provide a good base for nailing double coursing (Fig. 7.7-29). These horizontal strips should be spaced with centers 2 in. (50.8 mm) above the buttlines of each course to receive nails.

The lath strip application shown in Figure 7.7-30 should be used to apply shingles and shakes over insulating sheathing. Lath nailed at the studs provides a nailing base for double-coursing nails. As an alternate for attaching to non-wood sheathing, $^3/_8$ by $1^1/_2$ in. (9.5 by 38.1 mm) treated or decay-resistant lath may be used. Lath should be spaced ac-

**FIGURE 7.7-28** Reshingling is similar to new wall construction. (Courtesy Cedar Shake and Shingle Bureau.)

**FIGURE 7.7-29** Horizontal nailing strips must be used over nonnailable sheathing or stucco. (Courtesy Cedar Shake and Shingle Bureau.)

**FIGURE 7.7-30** Lath strips are used with nonnailable sheathing for double-coursed walls. (Courtesy Cedar Shake and Shingle Bureau.)

cording to shingle exposure, with the butts of the undercoursing shingles resting on the lath. Lath should be nailed to the studs through the sheathing with $2^1/_8$ by 0.109 in. (7d) (53.98 by 2.78 mm) annular threaded or $2^1/_2$ in. by 0.131 in. (8d) (63.5 by 3.4 mm) common wire nails. The outer course of shakes or shingles should be nailed to the lath with small-headed, corrosion-resistant annular threaded nails penetrating the lath. The butts of the outer course should project about $^1/_2$ in. (12.7 mm) below the lath (see Fig. 7.7-30).

Joints should not be closer than $1^1/_2$ in. (38.1 mm) to joints of the undercoursing in the same course. Joints in the following courses should be staggered at least $1^1/_2$ in. (38.1 mm).

### 7.7.4 ALUMINUM AND VINYL SIDING

Aluminum siding was first introduced to the American public following World War II. Vinyl siding came along a little later. Early aluminum siding was made in the natural aluminum finish, but soon afterward an enamel finish was applied, making it available in a choice of colors.

The prominent features of both aluminum and vinyl siding include minimal maintenance, good durability, light weight, and high resistance to corrosion.

For multiple design possibilities, these sidings are available in a variety of colors, textures, sizes, and shapes. They are used in both new and existing construction (Figs. 7.7-31 and 7.7-32).

### 7.7.4.1 Aluminum Siding

The recommendations in this subsection are based largely on the American Architectural Manufacturers Association (AAMA) Publication 1402, "Standard Specifications for Aluminum Siding, Soffit, and Fascia," which is a recognized quality standard for aluminum siding.

#### MATERIAL

Aluminum sheet is fabricated into horizontal and vertical panels by cold rolling. The typical alloy used in aluminum siding is known as 3003H14, a magnesium-

**FIGURE 7.7-31** Vinyl siding on a new building. (Courtesy Wolverine Technologies.)

**FIGURE 7.7-32** Vinyl siding on an existing building. (Courtesy Alcoa Building Products.)

containing alloy distinguished by moderate strength and good workability. See Section 5.2 for additional information about aluminum materials and fabrication. The finishes used on aluminum siding are generally organic enamels such as vinyls, acrylics, or alkyds. They are baked on at the factory after proper surface pretreatment.

***Siding Types*** Horizontal aluminum siding is available in several typical widths. The most commonly used siding type is a horizontal panel with 8 in. (203.2 mm) exposure (face). Other types include double 4 in. (101.6 mm) and double 5 in. (127 mm) exposure effects which are created by grooves bisecting a single 8 or 10 in. (203.2 and 254 mm) horizontal panel.

Aluminum siding is plain or insulated. Some insulated boards have a polystyrene or fiber backer board installed at the factory. Others are supplied with separate backer boards for field installation. Thicknesses vary from $3/8$ to $1/2$ in. (9.5 to 12.7 mm). Insulated siding is available in 8 in. (203.2 mm), double 4 in. (101.6 mm) and double 5 in. (127 mm) styles.

Vertical siding is available in both 12 and 16 in. (304.8 and 406.4 mm) exposures. Both exposures lend a board-and-batten effect, while the 16 in. (406.4 mm) exposure is often bisected by a V-groove for additional interest.

Like other metals, aluminum is an effective vapor barrier. Therefore, aluminum siding should be equipped with vents or weep holes to permit the water vapor to escape and prevent moisture accumulation in the wall.

***Accessories*** Accessories for use with aluminum siding include starter strips, J-channels, window head flashing, inside corner posts, outside corner posts, and individual corner caps. Other accessories include aluminum nails, caulking, touch-up paint, aluminum breather (perforated) foil, and adhesive. Adhesives used to bond backer boards to siding should demonstrate sufficient moisture resistance, strength, and flexibility to maintain a permanent bond under service conditions. AAMA specifications outline specific performance tests for the adhesives as well as for the finish coating and require compliance with other industry standards for accessory products employed in siding manufacture and installation.

## INSTALLATION

In refinishing projects, the quality of the finished job depends heavily on the initial surface preparation. Loose boards should be nailed down and irregular surfaces leveled. Unbacked siding cannot bridge low spots as readily as the backed type, and furring of low spots may be required to provide adequate support. In some cases, the whole wall must be furred. Good initial preparation permits siding joints to close up and avoids the wavy look of poor siding installations.

In new construction, backed siding can be applied directly to stud framing over sheathing paper. Unbacked siding should be installed only over a solid sheathing surface. When wood board sheathing is used, sheathing paper should be provided; when panel sheathing materials are applied in a way that minimizes air leakage, the sheathing paper can be omitted.

Aluminum siding is often applied over an insulated sheathing to increase a wall's R-value, especially in existing construction. This insulated sheathing may be a faced panel with a core of cellular kraft paper, or of a foamed plastic such as polystyrene or polyisocyanurate. Care must be taken to ensure that using an impervious sheathing will not trap moisture in the wall cavity. If it will, a pervious board should be used.

Aluminum foil laminated to kraft paper is sometimes used as sheathing paper under aluminum siding to increase a wall's R-value. It may be installed over an insulated sheathing, or directly on an existing wall or on furring where there is no insulated sheathing. This foil should be applied with its shiny side facing an adjacent air space. However, aluminum foil is an excellent vapor retarder and will trap moisture unless it is deliberately perforated to allow vapor to escape from the wall cavity. Foil products intended for use on an existing wall should be factory perforated and have a perm rating of at least 10 (632 ng/Pa.s.m$^2$). An efficient vapor retarder, with a permeance of 0.5 perm (287.26 ng/Pa.s.m$^2$) or less, should be used on the room side of the wall assembly. Some manufacturers also recommend installing vent cups at the top and bottom of each stud space. In new construction, the foil should be perforated or unperforated according to the wall design and vapor retarder location.

Careful application of accessories such as starter strips and corner trim is essential to provide a level and square installation. In most cases, starter strips are installed first to ensure proper alignment and fastening of siding.

Continuous inside and outside corner posts are used to receive siding at corners, insets, projections, and wall junctions. Since aluminum expands and contracts with temperature variations, a space should be allowed where siding ends abut a post or any rigidly supported surface. Outside corners can also be completed with the use of outside corner caps installed after siding is in place.

In installing horizontal siding, extra care should also be used in the application of the first course, making certain that it is level and securely locked into the starter strip. Each length of siding should be lapped over the preceding one to create the effect of a single panel extended across the building. Factory-finished ends should be exposed and cut ends lapped under the edges of adjoining panels.

To fit siding around areas such as windows, eaves, and doors, panels can be cut to the desired size. Joints should be planned so that, wherever possible, no piece is less than 20 in. (508 mm) long.

Proper nailing is essential to successful installation. Panels should be secured with aluminum nails driven through the slotted holes in the siding at intervals of about 16 in. (406.4 mm) along its length. Siding should be suspended on the nails, not nailed rigidly to the wall, because locked siding will not be able to expand or contract with temperature changes.

Matching gutters, downspouts, fascia boards, and soffit panels are also available. Because of the natural corrosion resistance of aluminum and the durability of the baked-on enamel finish, these products are particularly well adapted to their severe exposures.

To prevent water entry and improve appearance, caulking should be used at junctions where siding abuts wood, brick, stone, or other metals. Caulking compounds should be durable synthetic materials compatible with the aluminum and paint finish.

## GROUNDING

Although the possibility is remote, some building codes require that aluminum siding be grounded as a precaution against electrical shock from faulty wiring. This can be done by connecting a No. 8 copper wire to a cold water pipe or a steel rod embedded in the earth. Grounding allows stray electrical fault currents to be dissipated safely into the ground and eliminates the shock hazard to persons touching the siding.

## MAINTENANCE

Baked-on siding finishes are generally resistant to the effects of intense sunlight, snow, sleet, hail, ice, and salt spray. In addition to the natural cleaning by normal rainfall, siding surfaces should be kept clean by periodic rinsings with a garden hose. Persistent stains resulting from industrial fallout, tree sap, or insecticides can be removed with a nonabrasive household cleaner or special siding cleaner.

Panels damaged by storm, fire, and other hazards can be replaced individually without disturbing adjacent panels. Aluminum siding may also be repainted, if desired, or when necessary due to damage. A special paint intended for aluminum surfaces is recommended, although ordinary house paint can also be used.

### 7.7.4.2 Vinyl Siding

The recommendations in this subsection are based largely on the Vinyl Siding Institute (VSI) publication "Vinyl Siding Installation Manual" and on ASTM Standard D4756, "Standard Practice for Installation of Rigid Poly (VinylChloride) (PVC) Siding and Soffit," which are the recognized standards for installation of vinyl siding. In addition, vinyl siding should comply with ASTM Standard D3679.

## MATERIAL

Vinyl siding and accessories are made from a polyvinyl chloride (PVC) compound and additives such as lubricants, stabilizers, impact modifiers, ultraviolet inhibitors, and colors. PVC is a thermoplastic material, meaning that it can be reformed upon heating. It is flexible and has high impact resistance. The finishes on vinyl siding are integral.

Siding finishes range from almost smooth, through matte, to deeply embossed wood grain surfaces that simulate wood clapboard. Many colors are available.

*Siding Types* Horizontal and vertical vinyl siding is available in several widths and configurations. Figure 7.7-33 shows some more popular sizes and shapes. Special shapes and even vinyl lattice work are also available (Fig. 7.7-34) (see also Fig. 7.7-32).

Drop-in backer board is supplied with some vinyl siding, but VSI does not recommend its use, preferring instead a continuous sheathing backup. The manufacturer should be consulted for advice about drop-in backer board use.

*Accessories* Accessories for use with vinyl siding include starter strips, J-channels, window head flashing, inside corner posts, outside corner posts, and others. Their sizes and shapes vary with the manufacturer. Fasteners for installing vinyl siding should be aluminum, galvanized steel, or other corrosion-resistant metal nails or staples.

Nails should be long enough to penetrate at least $3/4$ in. (19.1 mm) into a solid nailable backing surface. Nails heads should be $5/16$ in. (7.9 mm) in diameter and $1^1/2$ in. (38.1 mm) long for gen-

**FIGURE 7.7-33** Vinyl siding: (1) 4 in. (101.6 mm) double; (2) 6, 8, 9, or 10 in. (152.4, 203.2, 228.6, or 254 mm) single; (3) 5 in. (127 mm) double; (4) 3 in. (76.2 mm) triple; (5) Dutch lap; (6) vertical siding is available in 5 in. (127 mm) double, 3 in. (76.2 mm) triple, and 4 in. (101.6 mm) quad. (Courtesy Vinyl Siding Institute.)

**FIGURE 7.7-34** Shingle shapes over the entranceway are only one of several shapes available in vinyl siding material. (Courtesy Alcoa Building Products.)

eral use. Nails used in residing should be longer.

Power-driven staples may be used to install vinyl siding. A special fitting must be used on power equipment, however. The manufacturer's instructions should be followed.

### INSTALLATION

Vinyl siding does not provide a weather-proof barrier. Therefore, a weather-resistant barrier is required beneath vinyl siding, and flashing and other means are required to lead water that enters the siding back to the exterior.

In refinishing projects, the quality of the finished job depends heavily on initial surface preparation. Loose boards should be nailed down and irregular surfaces leveled. Loose caulking should be removed and joints recaulked to make them watertight.

In new construction, vinyl siding should be applied over solid sheathing. Installing a new layer of sheathing over an existing wall surface will also produce an excellent vinyl siding job. It is also possible to produce an acceptable finished project by installing furring or strapping over an existing wall to even out the face. This furring or strapping should be shimmed where necessary to produce an even face.

Vinyl siding can be applied over insulated sheathing to increase a wall's R-value. The insulation and siding manufacturers' advice should be sought and followed in making this installation.

Careful application of accessories such as starter strips and corner trim is essential to provide a level and square installation. In most cases, starter strips are installed first to ensure proper siding alignment and fastening.

Continuous inside and outside corner posts are used to receive siding at corners, insets, projections, and wall junctions (Figs. 7.7-35a and 7.7-35b). Since vinyl expands and contracts with temperature variations, a space should be allowed where siding ends abut a post or any rigidly supported surface.

Extra care should also be used in applying the first course of horizontal siding, making certain that it is level and securely locked into the starter strip. Each length of siding should be lapped over the preceding one to create the effect of a single panel extended across the building. Factory-finished ends should be exposed

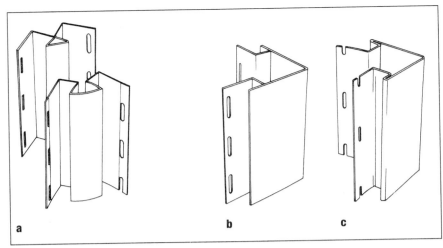

**FIGURE 7.7-35** (a) Inside corner posts; (b) and (c) outside corner posts. (Courtesy Vinyl Siding Institute.)

and cut ends lapped under the edge of adjoining panels (Fig. 7.7-36).

To fit siding around areas such as windows, eaves, and doors, panels can be cut to the desired size using saws or snips. Joints should be planned so that, wherever possible, no piece is less than 20 in. (508 mm) long.

Proper nailing is essential to successful installation. Panels should be secured with nails or staples driven through the slotted holes in the siding at intervals of about 16 in. (406.4 mm) along its length. Siding should be suspended on the nails, not nailed rigidly to the wall, because locked siding will not be able to expand or contract with temperature changes (Figs. 7.7-37 and 7.7-38).

Matching gutters, downspouts, fascia boards, and soffit panels are also available. Because of the natural corrosion resistance of vinyl and the durability of vinyl siding, these products are particularly well adapted to their severe exposures.

Siding overlaps should not be caulked, and siding panels should not be caulked where they meet receivers, corners, or J-trim, but should rather be per-

**FIGURE 7.7-37** Nails should be driven into the center of the slot. It should be possible to move the siding $1/2$ in. (12.7 mm) each way after fastening to allow for normal expansion and contraction. (Courtesy Alcoa Building Products.)

**FIGURE 7.7-36** Lapped end joints. (Courtesy Vinyl Siding Institute.)

**FIGURE 7.7-38** Staples should be driven so that the panels remain loose and movable. (Courtesy Alcoa Building Products.)

mitted to float free to account for expansion and contraction.

## MAINTENANCE

Vinyl siding is generally resistant to the effects of intense sunlight, snow, sleet, hail, ice, and salt spray. In addition to the natural cleaning by normal rainfall, siding surfaces should be kept clean by periodic washings with a long-handled car-washing brush mounted on the end of a water hose. Hard-to-remove soil may be removed with a solution of $1/3$ cup (80 mL) of detergent and $2/3$ cup (160 mL) of household cleaner in 1 gal (3.79 L) of water. Mildew may be removed using that same solution but substituting 1 qt (0.95 L) of liquid laundry bleach for 1 qt (0.95 L) of the water. More aggressive means and chemicals such as solvents, cleaning fluids, oxalic acid, and lacquer thinner can be used to remove more stubborn stains, but the advice of the siding manufacturer or VSI should be sought first.

## 7.7.5 WOOD AND HARDBOARD BOARD SIDING

The following section describes wood and hardboard siding intended for exterior use on building walls. Interior paneling is discussed in Chapter 6. Miscellaneous boards used for decking and other purposes are not included.

Unless specific exceptions are made in the following text, the provisions of Section 7.7.1 apply to wood and hardboard board siding and methods of their application.

### 7.7.5.1 Materials

#### WOOD BOARD SIDING

Wood board siding is made from softwoods such as cedar, Douglas fir, various species of pine, cypress, redwood, western hemlock, spruce, yellow poplar, western larch, and others. It should be of high grade and free from knots, pitch pockets, and waney edges. Moisture content at the time of application should be between 10% and 12%, except that in very dry areas, moisture content should be between 8% and 9%.

There are four basic types: *bevel, tongue-and-groove, lap,* and *board and batten* siding. Figure 7.7-39 shows a few of the many wood board siding profiles available.

*Bevel Siding* Bevel siding has both a smooth side and a rough-sawn side. When staining is desired, the rough-sawn side may be exposed. For painting, the smooth side is exposed. Widths vary from $3^{1}/_{2}$ in. (88.9 mm) to 12 in. (304.8 mm). The taper is usually $3/_{16}$ in. (4.8 mm) at the thin edge and varies from $5/_{8}$ to $1^{1}/_{4}$ in. (15.9 to 31.8 mm).

*Tongue-and-Groove Siding* Tongue-and-groove siding may be used either vertically or horizontally. Nominal board widths are generally 4, 6, or 8 in. (101.6, 152.4, or 203.2 mm). Thickness is usually $3/_{4}$ in. (19.1 mm).

*Lap Siding* Lap siding boards include channel siding and other lap siding configurations. Lap siding may be used vertically, horizontally, or diagonally. Various profiles are available in widths of 6, 8, and 10 in. (152.4, 203.2, and 254 mm), and other widths may be available. Overall thickness is usually $3/_{4}$ in. (19.1 mm).

*Boards for Board and Batten Siding* Flat boards are available in various nominal widths ranging from 2 in. (50.8 mm) to 12 in. (304.8 mm) and in thicknesses nominally 1, $1^{1}/_{4}$, and 2 in. (25, 32, and 50 mm) They may be rough, surfaced on one side and two edges, or surfaced on all four sides.

#### HARDBOARD SIDING

Hardboard board sidings are proprietary products made from thermoset hardboard, usually finished with a fused resin-oil finish. They should comply with all code requirements, should have a flame spread rating of Class III, <200, should be certified by the American Hardboard Association (AHA), and should comply with ANSI A135.6, "Hardboard Siding."

*Types and Size* Hardboard board siding is available in various widths, depending on the profile of the siding. Flat lap siding comes in 6, 8, $9^{1}/_{2}$, and 12 in. (152.4, 203.2, 228.6, and 304.8 mm) widths. Most interlocking flat-faced lapped siding comes in nominal 8 in. (203.2 mm) widths. Beveled lap siding usually comes in $11^{1}/_{2}$ in. (292.1 mm) widths. Other widths are also available. Most are available in $7/_{16}$ in. (11 mm) thickness, but some are $1/_{2}$ in. (12.7 mm) thick. Figure 7.7-40 shows some of the available profiles.

Trim boards made from the same material are available in 4, 5, 6, 8, 10, and 12 in. (101.6, 127, 152.4, 203.2, 254, and 304.8 mm) widths.

### 7.7.5.2 Preparation for Wood and Other Board Siding

#### EXPOSURE

Overlap for bevel siding should be not less than 1 in. (25.4 mm). Lap for drop siding and diagonal and vertical siding is dictated by the profile of the boards.

#### WALL SHEATHING AND SHEATHING PAPER

Sheathing paper is recommended over lumber and plywood sheathing, and is recommended by some sources over gypsum board sheathing, but is often omitted from nonwood sheathing.

**a**    **b**    **c**

Bevel    Rabbeted Bevel    Tongue and Groove    Channel Lap    Other Lap

**FIGURE 7.7-39** Wood siding: (a) beveled, (b) tongue-and-groove, (c) lap. Other shapes are available in each category. (Drawing by HLS.)

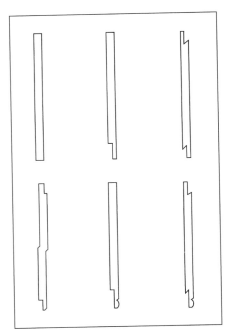

**FIGURE 7.7-40** Typical shapes of hardboard siding. Many others are available. (Drawing by HLS.)

## NAILS

Nails may be either aluminum, stainless steel, or hot-dip galvanized. The preferred material is No. 304 stainless steel for most locations, with No. 316 being used in seacoast locations. Other nail materials should not be used.

Nails should be special siding nails with thin shanks and blunt points. Nails should be ring shank or spiral type. Nails with textured heads present an appearance closer to that of the siding when finished.

### 7.7.5.3 Application of Wood and Hardboard Board Siding

#### INSTALLATION OF SIDING

***Wood Bevel Siding*** Installation begins at the bottom course with a starter strip (Fig. 7.7-41). Each succeeding course laps the lower course. One nail should occur in each board at each stud. Nail sizes should be 7d, 8d, or 10d, depending on the thickness of the siding and the sheathing. For 1-in. (25.4-mm)-thick foam, 10d nails should be used. If ring shank nails are used, sizes may be smaller, but penetration must be at least $1^{1}/_{2}$ in. (38.1 mm) into the studs, regardless of the nail size. Nails should be located high enough to miss the course be-

low by at least $^{1}/_{8}$ in. (3.2 mm) to permit movement in the boards.

***Tongue-and-Grove Siding*** Tongue-and-groove siding is installed in a manner similar to that of bevel siding, except that the exposed face is fixed and is always the same from board to board. Boards up to 6 in. (152.4 mm) wide can be blind-nailed through the base of each tongue with one nail into each stud. Wider boards should be face-nailed with two nails into each stud. Nails should be long enough to penetrate at least $1^{1}/_{2}$ in. (38.1 mm) into the studs.

***Lap Siding*** Lap siding is installed in a manner similar to that of tongue-and-groove siding, except that a $^{1}/_{8}$ in. (3.2 mm) gap should be left at the top of each tongue to provide for expansion. Boards should be face-nailed into each stud, with one nail in boards up to 6 in. (152.4 mm) wide and two nails in wider boards. Nails should penetrate at least $1^{1}/_{2}$ in. (38.1 mm) into the studs.

***Board and Batten Siding*** Horizontal blocking or furring strips are required to provide bearing at the proper locations for fastening board and batten siding. A minimum $^{1}/_{2}$ in. (12.7 mm), space should be left between boards for expansion. Battens should overlap boards by at least $^{1}/_{2}$ in. (12.7 mm) with more overlap on wider boards. Boards up to 6 in. (152.4 mm) wide should be fastened with a single nail in the center of each board. Wider boards should be fastened to each support with two nails approximately 3 in. (76.2 mm) apart. Battens should be nailed with one nail into each support. Nails

**FIGURE 7.7-41** Condition at the bottom of beveled siding. (Drawing by HLS.)

should penetrate at least $1^{1}/_{2}$ in. (38.1 mm) into the blocking.

***Hardboard Board Siding*** Installation, including nail types and sizes, and methods of starting and finishing should be in accordance with the manufacturer's directions.

## END JOINTS

End joints in horizontal board siding should be avoided when possible. Unavoidable end joints should be made over a stud and staggered from course to course. Cut ends should be brushed with a preservative.

End joints in vertical board siding may be made using the configuration of the boards (lap or tongue-and-groove), or may be made using battens or board on board designs (Fig. 7.7-42).

## CORNERS

When wall surfaces join at inside corners, the courses of wood and hardboard board siding are either butted into a wood corner strip or trimmed with a wood corner strip (Fig. 7.7-43a). Outside corners of wood siding are either mitered (Fig 7.7-43b) or finished with wood corner boards (Figs. 7.7-43c and 7.7-43d). Hardboard board sidings are usually finished at corners with wood corner boards or corner boards made of the same materials as the siding (see Figs. 7.7-43c and 7.7-43d).

***Mitered Finish*** Mitering is appropriated for wood siding only. Mitered joints are cut with a compound miter saw so that the corners finish neatly.

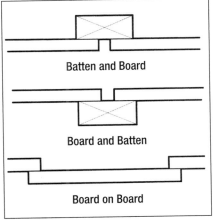

**FIGURE 7.7-42** Joints in panel siding. Similar joints are sometimes used in vertical board siding applications. (Drawing by HLS.)

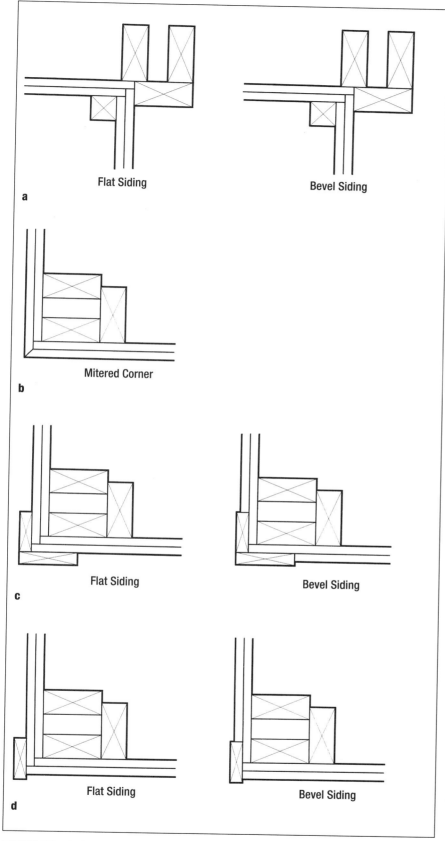

**FIGURE 7.7-43** Corners in board and panel siding applications: (a) inside corners, (b) mitered corner, (c) two-board corner, (d) single-board corner. (Drawing by HLS.)

*Wood Corner Board Finish* Corners finished with wood strips can be made of previously primed or painted nominal 1 in. (25.4 mm) lumber applied along one corner. A 1 × 4 (25 × 100) should be applied along one corner so that its side edge projects a distance beyond the corner equal to the thickness of the board applied on the adjoining wall. A 1 × 3 (25 × 75) should then be applied with its edge butting against the back of the 1 × 4 (25 × 100) first applied. Therefore, the wood corner strips will appear the same width on each wall (see Fig. 7.7-43c). For lower-quality work, a single corner board may be used (see Fig 7.7-43d).

*Hardboard Corner Board Finish* Corners may be made using special hardboard products made of the same material as the siding especially for this purpose. The method is similar to that described above for wood corner boards.

### FLASHING AT OPENINGS AND CORNERS

Metal flashing should be applied over exposed door and window openings. Drip edges should be provided at heads of openings, under sills, and at projections to keep water away from the finished wall surface below.

Outside and inside corners should be flashed with a 12-in. (12.7-mm)-wide strip of underlay material centered in or around the corner or with metal flashing when corner boards are used.

In other corner treatments, the underlayment should be carried around the corner of each side wall so that a double thickness of felt results over the corner (see Fig. 7.7-15).

### 7.7.6 PLYWOOD AND HARDBOARD PANEL SIDING

Unless specific exceptions are made in the following section, the provisions of Section 7.7.1 apply to plywood and hardboard siding and their application.

#### 7.7.6.1 Materials

##### PLYWOOD

Plywood is discussed in Chapter 6. Plywood used for siding includes paper-overlaid and textured panels made under their own trade names. Three available textures are shown in Figure 7.7-44.

**FIGURE 7.7-44** Three textures of plywood that serve as combination siding and sheathing. (Courtesy APA—The Engineered Wood Association, formerly American Plywood Association.)

### HARDBOARD

Hardboard panel siding is made in 4 ft by 8 ft (1219.2 by 2438.4 mm) and 4 ft by 10 ft (1219.2 by 3048 mm) sizes and is usually $^7/_{16}$ or $^5/_8$ in. (11 or 15.9 mm) thick, but other sizes may also be available. It is available with flat surfaces and with vertical grooves of varying widths and spacings. Wood grain surface textures are available.

#### 7.7.6.2 Preparation for Plywood and Hardboard Siding

##### WALL SHEATHING

Plywood paneling can serve as a combination sheathing and siding or can be applied over another sheathing, such as insulating sheathing. Hardboard panel siding can be applied over wood, plywood, or nonwood sheathing.

##### SHEATHING PAPER

Sheathing paper should be used over wood, plywood, and gypsum sheathing but is often omitted over nonwood sheathing.

#### 7.7.6.3 Application of Plywood and Hardboard Siding

##### PLYWOOD SIDING

Some plywood siding can be installed horizontally, but most is installed vertically. Sheets should be nailed to each stud at 6 in. (152.4 mm) intervals around the perimeter and at 12 in. (304.8 mm) intervals at intermediate members. Nails should penetrate supports by at least $1^1/_2$ in. (38.1 mm). Joints may butt and be caulked, or may have battens or board on board design (see Fig 7.7-42).

##### HARDBOARD SIDING

Hardboard panel siding should be installed in strict accordance with the recommendations of its manufacturer.

##### CORNERS

Corners in panel siding can be finished as described above for wood and hardboard board siding.

# 7.8 Membrane Roofing

This section discusses three types of membrane roofing: bituminous roll roofing, built-up bituminous roofing, and single-ply roofing. In general, membrane roofing is used on roofs having slopes less than $1^1/_2$ in./ft (125 mm/m). However, mineral-surfaced roll roofing can be used on higher slopes than that.

Roll roofing sheds water in a manner analogous to that of shingles, but employs sheet materials instead of small individual units and incorporates bituminous materials between the sheets. Both built-up roofing and single-ply roofing form a waterproof barrier. Sheet materials are discussed in Section 7.6.

Membrane roofing is a system of compatible components selected to work together to form a waterproof membrane. Each component has a unique function. Possible components include a vapor retarder, underlayment, insulation, waterproof membrane, and ballast.

## 7.8.1 ROOF CONFIGURATIONS

The three typical roof configurations in which membrane roofing is installed are called *insulated framed*, *compact conventional*, and *protected membrane*.

Most membrane roofing systems are insulated. In a *framed* system, the insulation is separated from the roofing by the roof deck or sheathing. A system in which a ventilated air space is provided between the insulation and the deck or sheathing is called a *ventilated* system (see Fig. 7.1-17).

In *compact systems* (see Fig. 7.1-20) the insulation, membrane, and vapor retarder, if there is one (see Section 7.1), are all in contact. The conventional order of components in a compact system starting from the exterior is ballast, membrane, insulation, vapor retarder (if there is one), insulation, and roof deck. In some systems, the ballast is replaced by another surfacing material.

In a *protected membrane* system, the order is ballast, pervious fabric, insulation, membrane, underlayment, and deck (see Fig. 7.1-22). The membrane serves as a vapor retarder. Ballast must be used to hold the insulation in place.

A system in which there is insulation both above and below the roof deck is called a *combination framed and compact* system.

## 7.8.2 WARRANTIES AND MAINTENANCE AGREEMENTS

Two types of warranties are issued covering membrane roofing: a manufacturer's warranty, or guarantee, and a roofing contractor's warranty and maintenance agreement. There is considerable controversy about the value of both.

Some experts believe that some warranties issued by roofing manufacturers limit the rights an owner would otherwise have under current law rather than expanding the owner's protection.

The effectiveness of contractors' warranties and maintenance agreements is only as good as the organization that signs them. In addition, some standard warranty agreements are biased toward the roofing contractor, offering it more protection than is granted to the building owner. A normal period of such warranties is 2 years, because about 65% of roofing problems occur during that period.

The purpose of a maintenance agreement is to ensure that the roof is properly maintained and repaired if a problem appears. Such agreements usually cover about a 5-year period and are usually renewable.

To be more sure of proper protection, a building owner should seek legal counsel before agreeing to any roofing warranty or maintenance agreement.

## 7.8.3 SUBSTRATES

Proper functioning of membrane roofing is dependent not only on the roofing material and its application, but on the roof structure as well. This structure must be sound and smooth, have sufficient slope to prevent ponding of water on the roof, and be designed to support the loads that will be imposed. There must be enough roof drains of the proper size to remove water from the roof.

Essentially three types of substrates are used beneath built-up bituminous and single-ply roofs: *nailable*, *insulated*, and *concrete*.

### 7.8.3.1 Nailable Substrates

Nailable decks include wood planks, plywood, cement-wood fiber planks, lightweight insulating concrete, poured-in-place gypsum, and precast gypsum planks.

### 7.8.3.2 Insulated Substrates

Insulated decks are covered with one or more layers of insulating material. The underlying deck may be nailable, metal, or concrete. Insulation may be boards, sprayed-in-place plastic foams, or a thermosetting insulating fill. Steel decks are covered with a layer of insulation (usually boards), even when the insulation is not necessary for thermal purposes, be-cause membrane roofing materials are not designed to span the flutes in steel decking.

### 7.8.3.3 Concrete Substrates

Concrete decks may be either poured-in-place or precast.

## 7.8.4 SELECTING SYSTEMS

Important factors in selecting a membrane roofing system include the configuration of the system and the components used in it. Two overriding selection criteria are (1) compliance with the applicable codes and (2) compliance with the applicable insurance requirements.

The decision about whether to use a built-up roofing system or a single-ply one is influenced by the advantages of each in the particular circumstances. For example, single-ply roofing can be installed in a much wider range of temperatures than can a built-up bituminous system. The availability of competent applicators is another factor. The success of a single-ply roofing system in preventing leaks and other problems depends to a much larger extent on the ability of the installer than is true for conventional built-up roofing. The sealing of single-ply roofing seams must be perfect. Even a slight error will usually result in a leak. In addition, roof decks that will receive single-ply membranes should be sloped to provide positive drainage, because they are susceptible to damage from standing water and ice. Some single-ply systems can be installed directly over existing roofing.

Figure 7.8-1 is a summary of criteria for selecting slope roofing types.

### 7.8.4.1 Configuration

Which components a roofing system should contain depends on the building's design, construction, and location. Other factors influencing this decision include energy considerations, budget considerations, esthetics, and codes. For example, the applicable code may dictate which type of insulation can be used in contact with a steel or wood deck.

Whether a vapor retarder is required is dependent on the local weather conditions and the roof configuration selected. Review Section 7.1 for a discussion about vapor retarders in roofing assemblies.

## 7.8.5 ROOF INSULATION

### 7.8.5.1 Materials

The type of roof insulation selected depends on applicable code requirements, the thermal characteristics desired, the selected roofing system configuration, economic considerations, fire resistance, traffic indentation resistance, weight, and available insulation panel dimensions. Different insulation materials may be selected, depending on whether the material should be vapor permeable or relatively impermeable. This may depend on whether the selected system contains a vapor retarder. If a vapor retarder is included, the insulation should not be impermeable, because there would then be two vapor retarders, which could trap moisture between them.

Applying hot-applied, built-up roofing directly over polyisocyanurate, polyurethane, and phenolic foam insulations sometimes leads to delamination and blistering problems. Therefore, the NRCA recommends that a layer of wood fiber,

**FIGURE 7.8-1** Selection Criteria for Membrane Roofing Type

| Selection Item | Criteria |
| --- | --- |
| Overriding criteria | Code compliance; insurance requirements |
| Built-up vs. single-ply | Climate; availability of competent installers; presence of existing roofing; condition of the substrates |
| Configuration | Building's design, construction, and location; energy use; construction budget; esthetics; code compliance; insurance requirements |
| Vapor retarder | Climate; roofing configuration; insulation material; roofing membrane material |
| Insulation type | Code compliance; climate; desired thermal characteristics; economy; roofing system configuration; fire resistance; traffic indentation resistance; presence of a vapor retarder; use of hot-applied or solvent-applied roofing membrane; roofing materials used |

perlite, or glass fiber board insulation be placed over these plastic foam insulations when a hot-applied, built-up roofing system will be applied.

Insulation beneath loose-laid, single-ply roofing can be installed without adhesives, and the roofing membrane is not attached using adhesives. Therefore, compatibility problems are reduced and the selection of usable insulation types is much wider.

Some roofing materials are not compatible with certain insulation products. Therefore, the roofing membrane manufacturer's recommendations must be followed when selecting roof insulation that will be in contact with roofing membranes. This is especially true for roofing membranes that are installed using the adhered method discussed in Section 7.8.9.

### 7.8.5.2 Application of Insulation

Most roof insulation should be applied in at least two layers, with the joints staggered in the two layers to reduce thermal bridges through the joints, to reduce ridging that sometimes occurs in single-layer applications, and to make mechanical attachment easier (Fig. 7.8-2).

Beneath built-up and adhered single-ply systems, insulation should be mechanically fastened to nailable decks, except that it may be laid in strips of hot bitumen when a vapor retarder has been installed over the deck.

Over a steel deck, single-layer insulation and the first layer of two-layer board insulation beneath built-up or adhered single-ply roofing should be mechanically fastened in place (Figs. 7.8-3 and 7.8-4). When fasteners are not used, the boards should be bedded in strips of hot bitumen laid along the deck flutes (Fig.

7.8-5). Boards should be placed with the long dimension parallel to the deck flutes and perpendicular to the roof slope (see Fig. 7.8-6).

Board insulation beneath built-up or adhered single-ply roofing should be applied over nonnailable decks in hot bitumen or adhesive.

Board insulation may be loose-laid beneath loose-laid single-ply systems.

The second layer of board insulation should be laid in a full mopping of bitumen (see Fig. 7.8-4).

### 7.8.6 FLASHING

Flashing is necessary in roofing systems to seal the gaps between the roofing and adjoining materials, such as walls, pipes, and other penetrations. Flashing may be one of several products, including metal, fabric, and plastics.

Flashing associated with membrane roofing systems should be a part of the roofing. The materials selected should be

**FIGURE 7.8-3** Single-layer insulation over steel deck. (Reprint courtesy National Roofing Contractors Association.)

**FIGURE 7.8-4** Two-layer insulation over steel deck. (Reprint courtesy National Roofing Contractors Association.)

**FIGURE 7.8-5** Placing strips of bitumen on flutes of steel deck. (Reprint courtesy National Roofing Contractors Association)

**FIGURE 7.8-6** Laying insulation boards over steel deck in bitumen strips. (Reprint courtesy National Roofing Contractors Association.)

**FIGURE 7.8-2** Typical insulation over steel deck. A vapor retarder is not always present. (Reprint courtesy National Roofing Contractors Association.)

compatible with the roofing both chemically and in thermal expansion characteristics. This is especially true when there is a large differential in the coefficient of expansion of the roofing material and the flashing. Therefore, metal flashing is not appropriate as a base flashing with most roofing materials. Conditions where metal flashings, such as gravel stops, are built into roofing membranes are particularly susceptible to failure.

Refer to Section 7.9 for a detailed discussion about flashings.

### 7.8.7 ROLL ROOFING

#### 7.8.7.1 Materials

Roll roofing is made by adding a more viscous, weather-resistant asphalt coating to a felt that has first been impregnated with a saturant asphalt or to a glass fiber mat. Some roll roofing is surfaced with mineral granules providing a wide color range, while some styles are furnished in split rolls designed to give an edge pattern when applied to the roof. Details of some roll roofing products are shown in Figure 7.6-17.

Roll roofing may be either mineral-surfaced or smooth, but organic felts are not appropriate, because roll roofing is designed to be water-shedding, rather than waterproof, and organic felts tend to absorb water.

One entire side of *standard mineral-surfaced roll roofing* is coated with granules. Most of it weighs between 75 and 100 lb per square (100 sq ft [9.29 m²]). This is between 34.02 and 45.36 kg per square. Double-coverage roll roofing is coated with granules for only the first 17 in. (431.8 mm) of a 36-in. (914.4-mm)-wide sheet. The remainder is smooth. These rolls weigh between 55 and 70 lb per square (24.95 and 31.75 kg per square).

#### 7.8.7.2 Application

##### SHEATHING

Most roll roofing is applied over plywood or wood board sheathing (see Chapter 6), but it can also be applied over most nailable decks.

##### APPLICATION OF UNDERLAYMENT

Roll roofing requires the application of a continuous underlayment. This is usually No. 15 asphalt-impregnated, unperforated roofing felt applied with the long edge parallel with the eaves. On slopes of 4 in./ft (333.3 mm/m) and steeper, one layer is sufficient. Two layers should be used on lower slopes.

Layers of underlayment should be laid shingle fashion so that they shed water, and they should be nailed using enough nails to hold them in place. Ends should be lapped at least 6 in. (152.4 mm), and edge joints should be lapped at least 2 in. (50.8 mm). The top layer in double-layer applications should lap 19 in. (482.6 mm) over the lower layer. Underlayment should lap over hips and ridges so that at least two layers cover them.

When the mean January temperature is 30°F (−1.1°C) or lower, an ice shield should be applied. When one layer of underlayment is used, this ice shield should consist of an additional layer of No. 15 felt laid and set in mastic from the eave to a line 24 in. (609.6 mm) inside the inner face of the exterior wall. When two layers of underlayment are used, the ice shield should be formed by applying the underlayment layers in hot asphalt from the eave to a line 24 in. (609.6 mm) inside the inner face of the exterior wall.

##### APPLICATION OF DRIP EDGE

A metal drip edge should be placed along rakes and eaves (Fig. 7.8-7). The underlayment should be placed beneath the drip edge at the rakes and above it at the eaves.

##### APPLICATION OF FLASHING

Flashings at penetrations are usually metal. Base flashings at walls, chimneys, curbs, and the like are usually bituminous but may also be metal. Valley flashings may be either bituminous or metal. Metal flashings are discussed in Section 7.9.

##### USE OF FASTENERS

Roll roofing is usually applied using 11- or 12-gauge (3.1 or 2.7 mm) annular or screw-threaded, galvanized steel, or aluminum roofing nails having 3/8 in. (9.5 mm) diameter heads and lengths between 7/8 in. (22.2 mm) and 1 1/4 in. (31.8 mm). Nails should penetrate at least 3/4 in. (19.1 mm) into the underlying nailable deck or sheathing.

##### APPLICATION OF ROOFING

Roll roofing sheets may be placed either parallel or perpendicular to the eaves. When perpendicular, the sheets should be tilted toward the rake at a rate of 1/8 in./ft (10.42 mm/m) of rake.

*Standard Mineral-Surfaced Roll Roofing* Sheets may be applied either with concealed or exposed nails. The concealed nailing method (Fig. 7.8-8) should be used (1) when the sheets are applied parallel with the eaves and (2) when the slope is 3 in./ft (250 mm/m) or more, even when the sheets are laid perpendicular to the eaves.

The exposed nailing system is limited to (1) slopes exceeding 2 in./ft (166.7 mm/m) when the roofing is applied parallel with the eaves (Fig. 7.8-9) and (2) slopes 4 in./ft. (333.3 mm/m), or more

**FIGURE 7.8-7** Metal drip edge at eaves and rake beneath roll roofing. (Reprint courtesy National Roofing Contractors Association.)

**FIGURE 7.8-8** Standard roll roofing installed by the concealed nailing method, with sheets laid parallel with eaves. (Reprint courtesy National Roofing Contractors Association.)

**FIGURE 7.8-9** Standard roll roofing installed by the exposed nailing method, with sheets laid parallel with eaves. (Reprint courtesy National Roofing Contractors Association.)

when the roofing is applied perpendicular to the eaves (Fig. 7.8-10). End laps are usually 6 in. (152.4 mm); edge laps are 2 in. (50.8 mm).

In the concealed nailing system, a 12-in. (304.8-mm)-wide starter strip is applied completely around the perimeter of the roof and subsequently covered with roofing. Roofing cement is applied over the entire area of each lap of roofing material, whether over the starter strip or over other roofing material.

In the exposed nailing system, edges at rakes and eaves and laps of roofing material should be sealed with lap cement before nailing.

Ridges and hips should be covered with 12-in. (304.8-mm)-wide strips of roofing. In concealed nailing systems,

these strips are nailed, then cemented, and the next section lapped over the succeeding one to conceal the nails. In exposed nailing systems, the sheets are laid with a 2-in. (50.8-mm)-wide band of roofing cement along each edge and nailed in place. In either case, laps should be fully bedded in roofing cement.

***Double-Coverage Mineral-Surfaced Roll Roofing*** Sheets may be applied with the long dimension either parallel or perpendicular to the eave (Figs. 7.8-11 and 7.8-12). Nailing should be done in the selvage area. Laps should be fully bedded in roofing cement. Sheets should be laid so that only the area with a granular finish is exposed.

### 7.8.8 BUILT-UP MEMBRANE ROOFING

NRCA has developed a four-letter system of generic designations to describe membrane roofing systems, depending on their makeup. The four letters stand for the deck type, the bitumen used, the felt used, and the surface type (Fig. 7.8-13). For example, a NAGA system has a nailable deck, asphalt bitumen, glass fiber felt, and aggregate surfacing.

#### 7.8.8.1 Built-up Roofing Felts

Built-up roofing felts are broadly called *reinforcing felts*. They consist of base sheets, ply sheets, and cap and flashing sheets. They may be saturated, saturated and coated, impregnated, or prepared in some special way for a specific purpose.

**FIGURE 7.8-10** Standard roll roofing installed by the exposed nailing method with sheets laid perpendicular to eaves. (Reprint courtesy National Roofing Contractors Association.)

**FIGURE 7.8-11** Double-coverage roll roofing laid parallel with eaves. (Reprint courtesy National Roofing Contractors Association.)

**FIGURE 7.8-12** Double-coverage roll roofing laid perpendicular to eaves. (Reprint courtesy National Roofing Contractors Association.)

**FIGURE 7.8-13** NRCA Four-Letter Designations for Built-Up Roofing Systems

**First Letter (Deck Type)**
N = nailable
I = insulated
C = concrete

**Second Letter (Bitumen Type)**
A = asphalt
C = coal tar

**Third Letter (Surfacing)**
A = aggregate surface
E = emulsion coating
M = mineral surface

**Fourth Letter (Ply Felt Type)**
O = organic
G = glass fiber

**Example**
NAAO = nailable deck type (N), asphalt bitumen (A), aggregate surfaced (A), organic felt (O)

## FELT SELECTION

Dry felts impregnated with asphalt or coal tar saturants, but otherwise untreated, are called *saturated felts*. They are also used as underlayment for shingles and as sheathing paper. Saturated felts are made in several weights; the most common are No. 15 and No. 30, weighing about 15 and 30 lb per square (6.8 and 13.6 kg per square), respectively.

Most felts used in built-up roofing today are glass fiber base felts, because they are stronger and more resistant to slippage than the organic felts that were favored in the past. Although glass fiber felts, which are impregnated with bitumen but otherwise untreated, may be classified under the broad category of saturated felts, they are not actually saturated, since glass fibers do not absorb bitumen. In reality, the mat is lightly coated so that all fibers are thoroughly surrounded by bitumen.

The polymer-modified bitumen sheets discussed in Section 7.8.10.1 may also be used in built-up roofing.

*Saturated and coated* sheets are saturated felts that are then surfaced on one or both sides with a fine mineral material to prevent adhesion before application. These are generally used as base sheets in built-up roofing systems.

*Impregnated* sheets have less coating material than coated sheets, which makes them lighter and able to ventilate vapors during their application.

*Prepared roofing materials* are saturated and coated or impregnated with fine decorative mineral granules. Some are combination products made of layers of glass fiber felt, organic fiber felt, or both.

## BASE SHEETS

Base sheets are used in some, but not all, built-up roofing systems. Usual materials are No. 30 asphalt-saturated organic felt, No. 40 asphalt-saturated and coated organic felt, asphalt glass fiber mat roll roofing, and venting asphalt-saturated and coated inorganic sheet.

The supposed function of a base sheet is to protect the roofing system against moisture in the substrates and from condensation on the bottom of the membrane. Their effectiveness in this regard is not universally accepted, however. Following local practice is best when deciding whether to include a base sheet in a roofing system. When the base sheet is omitted, an additional ply sheet should be added.

## PLY SHEETS

Ply sheets are usually No. 15 asphalt-saturated organic felt, No. 15 coal-tar-saturated organic felts, asphalt-impregnated glass fiber felts Type III, or asphalt-impregnated glass fiber felts Type IV. Type IV glass-fiber-impregnated sheets are heavier and have a breaking strength roughly twice that of Type III sheets.

## CAP AND FLASHING SHEETS

Cap and flashing sheets are usually wide-selvage asphalt roll roofing cap sheet surfaced with granules, No. 90 mineral-surfaced organic roll roofing, No. 50 or No. 60 smooth-surfaced asphalt roll roofing, inorganic cap sheet with mineral granules, or reinforced bituminous flashing material.

## BITUMEN SELECTION

Coal tar is considered superior to asphalt bitumen, but is limited to use on slopes of $\frac{1}{2}$ in./ft (41.67 mm/m) or less in the coldest areas of the United States and to even lower slopes in more southern climates. In the hottest areas, the slope should be less than $\frac{1}{4}$ in./ft (20.8 mm/m).

Asphalt bitumen can be used over a wide range of slopes and climatic conditions. In general, the lowest-numbered type that will work should be selected for a given slope.

Soft asphalt and coal tar bitumens are incompatible and should not be used in the same roof system. Some roof systems, however, have successfully combined hard asphalt in the felts with coal tar bitumen between the felts, but such systems must be applied more carefully than fully compatible systems.

## PLY COUNT

A built-up roofing system's ply count is the number of felts in the system. Bitumen coats between plies and topping flood coats and ballast above the top ply do not count as plies. Normal built-up roof systems for commercial projects have either three or four plies.

## SURFACING

In most areas of the country, built-up roofing is usually made up of smooth-surfaced felts covered with stone or unit-paver-type ballast. Stone ballast, however, is limited to slopes of 3 in./ft (250 mm/m) or less. In some areas, however, mineral-surfaced felt is often used as the

topping sheet over built-up roofing, and no ballast is applied.

When roof slope prohibits ballast use, the top ply is often coated with either bitumen or another coating material such as a fibered or nonfibered aluminum coating.

### 7.8.8.2 Application

Built-up roofing application should be scheduled and carried out so that neither insulation nor roofing is left directly exposed to precipitation or condensation at any time. Roofing plies should be laid shingle fashion with laps sufficient to provide the selected number of plies throughout (Fig. 7.8-14). The shingles should be placed in the proper direction to shed water. On slopes in excess of 2 in./ft (166.7 mm/m), the sheets should run parallel with the slope.

Built-up roofing plies should be nailed to nailable substrates as directed by the system's manufacturer and the standards noted earlier, and to built-in nailers where appropriate. Nailers should occur at flashing lo-

**a**

**b**

**FIGURE 7.8-14** Built-up membrane roofing (a) without a base sheet and (b) with a base sheet. (Reprint courtesy National Roofing Contractors Association.)

cations and in other locations as needed. Nails should be concealed in the finished roofing.

Forty-five-degree cant strips and tapered edge strips should be installed where the roofing abuts vertical surfaces and at gravel stops.

In systems where a base sheet will be used, it should be applied first. The base sheet should be mopped and fastened in place.

The selected number of interply sheets should be applied in full hot bitumen beds having the bitumen weight per square specified for the selected system. A glaze-coat top surface of bitumen should then be applied over the top roofing ply. Bitumen should be heated to a temperature that will permit its application at its EVT, but never to its flash point.

Plies should be installed over cants and tapered edge strips and extend at least 2 in. (50.8 mm) above the top of the cant or tapered edge (Fig. 7.8-15). A folded-back envelope should be built in at the edges of built-up roofing that does not terminate at a cant or tapered edge strip to prevent bitumen from flowing over the edge of the roofing onto adjacent surfaces (Fig. 7.8-16). This envelope should be turned back over the plies and bedded in hot bitumen.

Expansion joints (Fig. 7.8-17), area dividers (Fig. 7.8-18), accessories, and flashings (Fig. 7.8-19) should be built into the roofing according to the standards listed earlier. Counter flashings is discussed in Section 7.9.

As soon as possible after the roofing has been installed, the top should be flood-coated with hot bitumen, and the selected aggregate ballast should be laid in it (Figs. 7.8-20 and 7.8-21). Where aggregate ballast will not be used, the selected aggregate-surfaced asphalt felt, cap sheet, emulsion top coating, or aluminum reflective coating should be promptly applied.

### 7.8.9 SINGLE-PLY MEMBRANES

Single-ply sheet roofing technology was first developed in Europe, and many of the currently available systems first appeared there.

### 7.8.9.1 Membrane Materials

This section covers single-ply membranes that fall into one of the following

**FIGURE 7.8-15** Base flashing for a wall-supported deck. When the deck is not supported by a wall, built-up roofing should be applied against an expansion joint to prevent wall movement from damaging the roofing membrane. (Reprint courtesy National Roofing Contractors Association.)

**NOTES:**

THIS DETAIL SHOULD BE USED ONLY WHERE THE DECK IS SUPPORTED BY THE OUT-SIDE WALL.

METALS OF 22-GAUGE STEEL, 0.050" ALUMINUM, 24-GAUGE STAINLESS STEEL OR HEAVIER ARE APPROPRIATE FOR THIS DETAIL. METALS OF THIS WEIGHT ARE VERY RIGID WHEN FORMED, AND FASTENING AT THE CENTER-LINE AND JOINT COVER WILL ALLOW EXPANSION AND CONTRACTION WITHOUT DAMAGING THE BASE FLASHING MATERIAL.

ATTACH NAILER TO MASONRY WALL. REFER TO FACTORY MUTUAL DATA SHEET 1-49.

WOOD-BLOCKING MAY BE SLOTTED FOR VENTING WHERE REQUIRED.

**FIGURE 7.8-16** Built-up roof edge at a heavy metal gravel stop. (Reprint courtesy National Roofing Contractors Association.)

categories: vulcanized elastomers, non-vulcanized elastomers, thermoplastics, and polymer-modified bitumen (Fig. 7.8-22). The materials covered are probably the most popular in use today, but they are not the only ones. Single-ply roofing technology has developed rapidly and continues to develop. Some materials that were available a few years ago have disappeared from the market, and new ones appear from time to time. In addition, some materials discussed here may also disappear in time and new systems will surely be developed.

The thicknesses given here are standard today, but the trend is toward heavier membranes.

### VULCANIZED ELASTOMERS

Factory-cured (vulcanized) elastomers include EPDM and vulcanized Neoprene. EPDM is made from ethylene ($E$), propylene ($P$), a diene monomer ($DM$), and plasticizers, accelerators, and fillers. Neoprene (polychloroprene) is a synthetic rubber. Because vulcanized elastomers cannot be molded, reshaped, or welded by heat, they can only be bonded to themselves using adhesives.

Most EPDM is 45 mils (1.143 mm) thick, but 60 mil (1.524 mm) material is also available. Neoprene is usually 60 mils (1.524 mm) thick.

Some formulations of Neoprene will accept a chlorosulfonated polyethylene (CSPE) color coating.

### NONVULCANIZED ELASTOMERS

Nonvulcanized (uncured) elastomer sheet materials include those made from chlorosulfonated polyethylene (CSPE), chlorinated polyethylene (CPE), poly-isobutylene (PIB), and nonvulcanized neoprene. Uncured elastomers can be molded, shaped, and welded by heat. Therefore, seams in them can be heat-welded or chemically welded. Most of these materials cure in place after they have been installed and are then said to be "thermoset." Thermoset materials can be installed using adhesives in the seams.

*CSPE (DuPont Hypalon)* CSPE is a synthetic rubber produced in many colors. Sheets are made of CSPE laminated to a mineral fiber backing or a polyester scrim. Sheets are usually 60 mils (1.524 mm) thick. CSPE is also available in liquid form for use as a coating. It is weather, chemical, and pollutant resistant,

FLEXIBLE VAPOR RETARDER TO
SERVE AS INSULATION RETAINER—
ATTACHED TO TOP OF CURB

CHAMFER TOP
OF BOTH WOOD
CURBS TO DRAIN
TO ONE SIDE

DRIVE CLEAT OR
STANDING SEAM

DRAINAGE SLOPE

8" MIN.

FASTENERS APPROX.
8" O.C. BOTH SIDES

FASTENERS APPROX.
12" O.C.

BASE FLASHING—
COVER TOP OF BASE
FLASHING WITH VAPOR
RETARDER

COMPRESSIBLE
INSULATION

WOOD CANT TO
PROVIDE STRUCTURAL
STRENGTH

NAIL TOP AND BOTTOM
APPROX. 16" O.C.

WOOD NAILER EACH SIDE
SECURED TO DECK, WITH
APPROPRIATE FASTENERS
APPROX. 24" O.C.

**FIGURE 7.8-17** Expansion joint with a high curb in built-up roofing. Expansion joints should be carried through the roofing whenever they occur in the structure. Expansion joints may also be covered with preformed elastomeric or combination metal and elastomeric expansion caps. Expansion joints bonded into built-up roofing without high curbs are not recommended. (Reprint courtesy National Roofing Contractors Association.)

8" MIN

FASTENERS APPROX. 24" O.C

FASTENERS APPROX. 8" O.C.

BASE FLASHING

FIBER CANT STRIP —
SET IN BITUMEN

FASTEN WOOD BLOCKING
TO METAL DECK WITH
MECHANICAL FASTENER

**FIGURE 7.8-18** Area dividers should be located in built-up roofing between expansion joints at 150 to 200 ft (45.72 to 60.96 m) intervals to provide expansion relief in the membrane. They should not restrict water flow to drains. (Reprint courtesy National Roofing Contractors Association.)

**FIGURE 7.8-19** Installation of bituminous base flashing at a roof penetration. (Reprint courtesy National Roofing Contractors Association.)

**FIGURE 7.8-20** Spreading aggregate ballast into hot bitumen adjacent to roof penetrations. (Reprint courtesy National Roofing Contractors Association.)

**FIGURE 7.8-21** Placing aggregate ballast over built-up roofing. (Reprint courtesy National Roofing Contractors Association.)

**FIGURE 7.8-22** Summary of Single-Ply Roofing Types

| Type | Materials | Thickness, mils (mm) |
|---|---|---|
| Vulcanized (cured) elastomers | EPDM | 45–60 (1.14–1.52) |
| | Neoprene[a] | 60 (1.52) |
| Nonvulcanized (uncured) elastomers | CSPE[b] | 60 (1.52) |
| | CPE | 40–48 (1.02–1.22) |
| | PIB | 100 (2.54) |
| | Neoprene[a] | 60 (1.52) |
| Thermoplastics | PVC | 40–60 (1.02–1.52) |
| Polymer-modified bitumen | APP | 40–60 (1.02–1.52) |
| | SBS | 40–60 (1.02–1.52) |
| | SBR | 40–60 (1.02–1.52) |

[a]Neoprene is available cured and uncured.

[b]CSPE (Hypalon) is also available in liquid form for use as a coating.

which makes it usable as an exposed finish.

***CPE*** CPE is an unreinforced or polyester-reinforced sheet material usually 40 to 48 mils (1.016 and 1.219 mm) thick. It can be pigmented with many colors. The raw material used to make CPE is manufactured by the Dow Chemical Company. This material is compatible with bitumen and can, therefore, be installed directly over an existing asphalt or coal tar pitch roof. It is also resistant to weathering, chemicals, and pollutants and is suitable for exposed locations.

***PIB*** PIB is formulated from isobutylene and other polymers, carbon black, and other additives. A sheet consists of a 60 mil (1.524 mm) PIB membrane laminated to a 40 mil (1.016 mm) polyester backing. Sealing edges are left unbacked. It can be used in exposed locations, and in contact with asphalt, but must be protected from petroleum distillates, organic oils and fats, and tar.

## THERMOPLASTICS

Thermoplastic materials can be softened and reformed throughout their life. PVC, when combined with plasticizers, forms a thermoplastic roofing membrane that is usually between 40 and 60 mils (1.016 and 1.524 mm) thick. It is available unreinforced and reinforced with polyester or glass fiber. It can be used as exposed roofing but is not compatible with bitumens. PVC comes without a finish for field finishing with coatings or with a factory-applied acrylic coating for weather exposure.

## POLYMER-MODIFIED BITUMEN

Polymer-modified bitumen sheets are made from asphalt that has been modified. Typical modifiers include atactic polypropylene (APP), styrene-butadiene-styrene (SBS), and styrene-butadiene-rubber (SBR). They are reinforced with glass fiber mats or scrim, polyester mats or scrim, polyethylene sheets, or copper or aluminum foils. Most are between 40 and 60 mils (1.016 and 1.524 mm) thick.

Some polymer-modified bitumen sheets are surfaced at the factory with mineral granules to protect them from the elements and to improve their appearance. APP-modified bitumen sheets must have a coating of granules. Some sheets are not factory-coated, but are suitable for field application of granules.

APP-modified bitumen sheets are not compatible with asphalt. SBS and SBR sheets are compatible with asphalt and can be used in built-up roofing applications. All three can be made suitable for torch application. Some ABS and SBR sheets are self-adhering. In general, APP sheets work better in warm climates. The other two are better in cold climates.

## POLYURETHANE FOAM

Sprayed-in-place, closed-cell polyurethane rigid foam systems provide excellent waterproof roofing for both new buildings and reroofing applications. They have no joints to open up and will not shrink. They are flexible, so they move with the structure. They are ultraviolet, mildew, and fungus resistant. They are lightweight, which reduces the dead load on the structure. These systems are self-flashing by forming a tight bond between the roofing and vertical surfaces. Sprayed-in-place polyurethane foam is coated with a silicone or acrylic formulation and may be finished with a broadcast application of roofing granuals. It may also be finished with a fleeceback membrane layer applied with a polyurethane adhesive.

### 7.8.9.2 Coatings and Surfacings

Liquid Hypalon (CSPE) can be used to coat EPDM and some formulations of neoprene to provide color. Silica sand coatings are also sometimes used on EPDM.

PVC and some modified bituminous membranes are sometimes coated with acrylic emulsions to improve their appearance.

Some modified bitumen membranes that do not have a factory-applied finish require a field-applied protective layer of mineral granules or aggregate to make them fire-resistant. Aluminum coatings are also sometimes used over modified bitumen membrane roofing.

### 7.8.9.3 Application of Single-Ply Membranes

Single-ply membrane roofing application should follow the recommendations of the roofing system manufacturer, NRCA, and the Single-Ply Roofing Institute (SPRI). Associated metal flashing and accessories should comply with the recommendations of the Sheet Metal and Air Conditioning Contractors National Association, Inc. (SMACNA). Conflicts in these recommendations should be reconciled before roofing system installation.

The methods available for installing single-ply roofing include *loose laid and ballasted*, *mechanically fastened*, and *adhered*. Each manufacturer issues specific instructions concerning application of its product which must be followed if a warranty is to be issued.

Regardless of the system selected, plies should be laid shingle fashion, starting at low points in the deck so that water drains across the roof without interruption. In all three methods, the membrane should be laid over a rosin-sized building paper separator unless the manufacturer requires another type of separator.

## LOOSE LAID AND BALLASTED SYSTEMS

EPDM, neoprene, hypalon, PIB, PVC, and some modified bitumen sheets may be laid using the loose laid and ballasted system. In this system, the roofing is independent of the roof deck, which allows the structure to move without affecting the roofing. This same feature, however, means that such roofing will blow off quite easily due to wind uplift. This tendency can be reduced, but not necessarily eliminated, by a layer of ballast. Most installations include washed river gravel as the ballast, but concrete pads are also sometimes used. Ballast adds weight to the roof and often requires additional structure. Loose laid roofing with too little ballast or ballast that is too light may suffer damage due to wind uplift.

Loose laid roofing is placed over the substrates with only minimal fastening around the edges and at penetrations (Fig. 7.8-23). Adjoining sheets should be lapped and bonded together using the roofing manufacturer's recommended sealant.

Aggregate ballast should be applied uniformly at a rate recommended by the roofing manufacturer. This weight should be determined based on the location and height of the building, but should not be less than 10 to 12 lb sq/ft (48.82 to 58.59 kg/m²). Precast concrete paver ballast should be applied in accordance with the roofing manufacturer's recommendations. Ballast weight must be taken into account when designing the roof structure. Care should be taken to ensure that the membrane is not damaged during ballast installation.

## MECHANICALLY FASTENED SYSTEMS

When reducing the dead load on a roof is desirable, or the roof is more steeply sloped than can be accommodated by ballast, or the building is in a location that is subject to high winds, single-ply roofing can be installed using a mechanically fastened system. Mechanically fastened systems should not be used when a vapor retarder is required, because the fasteners penetrate into the decking and, therefore, through the installed vapor retarder.

Single-ply membranes may be applied using either penetrating or nonpenetrating fasteners. Penetrating fasteners pass through the membrane (Fig. 7.8-24). Nonpenetrating fasteners are anchored to the structural deck, and the membrane is fastened to them using clamps or snap-on caps (Fig. 7.8-25).

The spacing, location, and type of fasteners should be in accordance with the roofing manufacturer's recommendations (Figs. 7.8-26 and 7.8-27).

**FIGURE 7.8-24** Penetrating-type fastener for single-ply roofing. Several different types are available, but their principle is similar to that of the device shown. (Drawing by HLS.)

**FIGURE 7.8-25** Nonpenetrating fasteners do not pass through the membrane. These are proprietary systems. (Drawing by HLS.)

## ADHERED SYSTEMS

Adhering a single-ply membrane to the roof deck or to insulation permits applications over complex shapes, curves, and steep slopes. Membranes may be either fully or partially adhered.

**FIGURE 7.8-23** The home court of the NBA's Minnesota Timberwolves is protected by a ballasted, 45 mil (1.143 mm) EPDM membrane system. (Courtesy Firestone Building Products Company.)

**FIGURE 7.8-26** A polymer batten bar is used to mechanically attach sheets of single-ply EPDM roofing membrane to the substrate atop JVC Company of America's distribution center in Aurora, Illinois. (Courtesy Firestone Building Products Company.)

FIGURE 7.8-27 Fasteners are installed along the leading edge of this white thermoplastic alloy roofing membrane to securely attach it to a $2^3/_4$-in. (69.9-mm)-thick layer of poly-isocyanurate insulation. (Courtesy Firestone Building Products Company.)

FIGURE 7.8-28 Fully adhered single-ply roofing systems are particularly useful on complex roof shapes, such as domes, barrels, and parabolas, and, of course, on flat roofs like the one shown here. (Courtesy the Goodyear Tire and Rubber Company.)

FIGURE 7.8-29 Installing a fully adhered single-ply roofing system. (Courtesy The Goodyear Tire and Rubber Company.

In a *fully adhered* system, the membrane is completely attached to the substrates using hot- or cold-applied bitumen, cold-applied adhesives, solvents by heating (torching) the back of the membrane, or by pressing self-adhering membrane in place (Figs. 7.8-28 and 7.8-29).

In a *partially adhered* system, the roofing membrane is laid into beads or strips of bitumen, adhesive, or solvent and rolled, or is adhered by similar materials placed on the top plates of the fasteners that hold down the insulation.

## JOINTS

Seams should be made by lapping adjacent sheets and sealing the joint. Different joining methods are required for each roofing type. The manufacturer's recommendations should be followed. Before laps can be sealed, separation coatings, such as talc or mica, must be removed from the area to be joined. This can be done using a solvent.

Joints in nonvulcanized elastomers can be sealed using either heat or chemical welding.

Vulcanized and nonvulcanized materials that have been cured must be sealed using adhesives (Fig. 7.8-30). The type of adhesive should be the one recommended by the roofing manufacturer. Some sheets require contact adhesives, some require adhesives made from the same material as the membrane, and oth-

FIGURE 7.8-30 Applicator ensures that adjoining EPDM sheets are lapped a minimum of 3 in. (76.2 mm), cleaned, and spliced with splice adhesive to form a continuous, watertight membrane atop the Target Center in downtown Minneapolis. (Courtesy Firestone Building Products Company.)

ers will work with generic adhesives. Some sheets are naturally self-adhering. Some are made self-adhering by the addition of an adhesive coating or self-adhering strips bonded to the sheets.

Joints in thermoplastic materials can be either heat- or solvent-welded (Fig. 7.8-31).

Joints in self-adhering modified bitumen sheets should be made by pressing the sheets together and rolling. Joints in non-self-adhering APP sheets should be torch welded (Fig. 7.8-32). Joints in non-

FIGURE 7.8-31 This white thermoplastic alloy roofing system uses heat-weldable seams to form a continuous watertight membrane. (Courtesy Firestone Building Products Company.)

FIGURE 7.8-32 Torch-down application of APP-modified bitumen systems. (Courtesy Firestone Building Products Company.)

self-adhering SBS and SBR sheets can be either torch-welded or sealed with hot asphalt.

## FLASHING

Base and penetration flashings are usually the same material as the sheet membrane or a nonvulcanized compatible elastomeric sheet material (Fig. 7.8-33). Metal base flashings are inappropriate, and joints between elastomeric base flashings and metal cap (counter) flashings should be avoided when possible. Gravel stops may be metal, or may be made from compatible plastic materials or coated with the same elastomer used in the roofing membrane to permit proper bonding with the roofing.

Joints between roofing membrane and elastomeric flashings should be sealed using the same material used in the seams in the roofing membrane (Fig. 7.8-34). Expansion joint covers should be made using compatible materials (Fig. 7.8-35).

When the roofing manufacturer so requires, nailers should be provided and elastomeric flashings should be securely fastened in place.

Metal cap flashings are addressed in Section 7.9.

# 7.9 Flashing and Sheet Metal

Flashing is installed at vulnerable locations to help prevent water from entering a building and to collect water that has entered the exterior shell and divert it back to the exterior. Flashings are formed from sheet metal, bituminous fabric membranes, plastic sheet materials, or a combination of these materials.

## 7.9.1 MATERIALS

Because replacement costs can exceed the initial installed costs, it is prudent to select a durable flashing material for the original installation.

Flashing materials must possess at least four important properties: (1) imperviousness to moisture penetration, (2) resistance to corrosion caused by exposure either to the atmosphere or to caustic materials in adjacent materials, such as the alkalis present in mortars, (3)

SMOOTH CONCRETE—
EXPOSED SURFACES MUST BE WATERPROOFED

CAULK WITH
ELASTOMERIC SEALANT

ANGLE CLAMPING BAR
WITH SLOTTED
ANCHOR HOLES

FASTENERS IN
EXPANSION SHIELDS

COMPRESSIBLE
ELASTOMERIC TAPE TO
SPAN IRREGULARITIES

CURED OR UNCURED FLASHING
MEMBRANE ADHERED TO WALL AND
MEMBRANE WITH APPROPRIATE
ADHESIVE

EPDM MEMBRANE

PRIME CONCRETE, IF REQUIRED

SEAL EDGE OF FLASHING WITH
LAP EDGE SEALANT

FASTENING STRIP APPLIED TO
DECK OR WALL—FASTENED
12" O.C.

**FIGURE 7.8-33 Wall flashing for a single-ply membrane.** (Reprint courtesy National Roofing Contractors Association.)

SEALANT

3" MINIMUM

JOINT COVER 4" TO 6" WIDE—
SET IN COMPATIBLE MASTIC

METAL GRAVEL STOP

SEAL EDGE OF FLASHING MEMBRANE WITH
LAP EDGE SEALANT

CURED OR UNCURED FLASHING MEMBRANE
ADHERED TO METAL AND
MEMBRANE WITH APPROPRIATE
ADHESIVE

FASTENERS APPROX. 4" O.C.—
DO NOT STAGGER

EPDM MEMBRANE

12" TO 18" TAPERED
EDGE STRIP

1½"

CONTINUOUS CLEAT

NOTES:
ATTACH NAILER TO MASONRY WALL. REFER TO FACTORY MUTUAL DATA SHEET 1-49.

THIS DETAIL SHOULD BE USED ONLY WHERE DECK IS SUPPORTED BY THE OUTSIDE WALL.

THIS DETAIL SHOULD BE USED WITH LIGHT-GAUGE METALS, SUCH AS 16-OZ. COPPER, 24-GAUGE GALVANIZED METAL OR 0.040" ALUMINUM. A TAPERED EDGE STRIP IS USED TO RAISE THE GRAVEL STOP. FREQUENT NAILING IS NECESSARY TO CONTROL THERMAL MOVEMENT.

WOOD BLOCKING MAY BE SLOTTED FOR VENTING WHERE REQUIRED.

**FIGURE 7.8-34 A light-metal gravel stop at the edge of a single-ply membrane.** Some roofing manufacturers recommend that a heavy gravel stop coated with an elastomeric material compatible with the roofing be used in such a condition. When this is not acceptable for esthetic or other reasons, the detail shown can be used. (Reprint courtesy National Roofing Contractors Association.)

FLEXIBLE VAPOR RETARDER TO
SERVE AS INSULATION RETAINER—
ATTACHED TO TOP OF CURB

CHAMFER EACH SIDE
OF WOOD CURB
TO DRAIN

COMPRESSIBLE
INSULATION

WOOD NAILER EACH SIDE
SECURED TO DECK WITH
APPROPRIATE FASTENERS
APPROX. 24" O.C.

SEALANT

3" MINIMUM

CURED OR UNCURED
FLASHING MEMBRANE
ADHERED TO BELLOWS
AND MEMBRANE WITH
APPROPRIATE ADHESIVE

SEAL EDGE OF FLASHING
WITH LAP EDGE SEALANT

EPDM MEMBRANE

FASTENERS APPROX.
4" O.C.—DO NOT STAGGER

**FIGURE 7.8-35** An expansion joint in an elastomeric single-ply roofing membrane. Some manufacturers recommend other types of expansion joints, but some type of expansion provision should be made in roofing over every expansion joint in the structure. However, separate expansion joints may not be needed in loose-laid systems if a metal plate is placed over the joint beneath the insulation and fastened to the deck at one side of the joint only. The roofing manufacturer's advice should be followed. (Reprint courtesy National Roofing Contractors Association.)

toughness to resist puncture, abrasion, and other damage during installation, and (4) formability so that the flashing can be worked to the correct shape and retain this shape after installation.

### 7.9.1.1 Sheet Metals

The metal materials generally used in flashing for walls and roofs include aluminum, copper, lead-coated copper, galvanized steel, lead, and stainless steel.

#### ALUMINUM

Aluminum is a light, bluish silver-white, ductile, malleable metal. It is resistant to oxidization but is subject to attack by the caustic alkalis present in fresh, unhardened mortar. Although dry, seasoned mortar will not affect aluminum, corrosion can occur if the adjacent mortar becomes wet. Therefore, aluminum should not be used as a masonry or stone wall flashing. If it must be so used, it should be coated to completely protect it from contact with both fresh and cured mortar.

Aluminum sheet flashing exposed to weather should be at least 0.019 in. (0.48 mm) thick and should have a 16,000 psi (11249.12 mg/m$^2$) tensile strength. Aluminum sheet that is completely concealed

should be not less than 0.015 in. (0.38 mm) thick and should have a 14,000 psi (9842.98 mg/m$^2$) tensile strength.

#### COPPER, PLAIN AND LEAD-COATED

Copper is a reddish, ductile, malleable metal. The three types of sheet copper available are cold-rolled copper, soft copper, and lead-coated copper. Cold-rolled copper sheet is less malleable than soft copper but much stronger. Accordingly, it is used much more often. The copper in lead-coated copper is similar to cold-rolled copper.

Though more costly than most other flashing materials, copper is durable, available in special preformed shapes, and an excellent moisture barrier. Because it is not materially affected by the caustic alkalis present in masonry and stone mortars, copper can safely be embedded in fresh mortar and will not deteriorate in continuously saturated, hardened mortar unless excessive chlorides are present. When using copper flashing, adding calcium chloride or other chloride-based additives to the mortar should be prohibited.

Copper sheet used for flashing should be cold-rolled copper conforming with

ASTM Standard B370 or lead-coated copper conforming with ASTM Standard B101, Type I or Type II, Class A. Type I has the lead coating applied by hot dipping, Type II by electro-depositing. Class A denotes a lead coating on both sides of the sheet with the total lead weight on both sides at between 12 and 15 oz (340 and 425 g) of lead on each 100 sq ft (9.29 m$^2$) of sheet.

Copper thicknesses for various uses should be in accordance with the recommendations of the Copper Development Association's (CDA) *Copper in Architecture Design Handbook.* In general, copper flashing that will be completely concealed should weigh not less than 10 oz/sq ft (3051.52 g/m$^2$); when exposed to the weather, it should weigh not less than 16 oz/sq ft (4882.4 g/m$^2$). Base, valley, gable end, and some other flashings and gravel stops, gutters, and downspouts should weigh between 20 and 24 oz/sq ft (7323.7 g/m$^2$), depending on their size and use.

When copper is exposed to weather, its normal oxidization will wash across adjacent surfaces. This may stain or discolor those surfaces, which is particularly objectionable when copper flashing is installed in light-colored masonry or stone, or when the wash-off from copper will pass over other metals, such as aluminum, or flow down walls. Where such copper staining would be objectionable, lead-coated copper can be used.

## GALVANIZED STEEL

Galvanized steel is steel that has been coated with a layer of zinc, either electrolytically or by the hot-dip process. It will corrode in fresh mortar and may corrode even when placed in contact with cured mortar. Therefore, it should not be used in masonry or stone where such use is avoidable, and when it is so used it should be protected with a bituminous coating. It will also corrode eventually in the air unless protected with a coating such as a bituminous compound or paint. However, because of its low cost, galvanized steel is widely used for flashing, especially in small commercial and residential buildings. While it can perform adequately in such a use when it is well protected, exposed galvanized steel requires periodic maintenance and should be accessible for inspection and painting.

Galvanized steel for wall flashings should conform with the requirements of ASTM A653. It should be 28-gauge (0.37 mm) where completely concealed and 26-gauge (0.55 mm) where exposed.

## LEAD

Lead is a bluish, soft, heavy, malleable, ductile, plastic (but not elastic) metal. Some lead roofs have been in constant use for hundreds of years. Lead will corrode in fresh mortar and may corrode even when placed in contact with cured mortar. Therefore, it should not be used in masonry or stone where such use is avoidable, and when so used it should be protected with a bituminous coating. For this reason, and because elastomeric flashings are less expensive and easier to work with, lead is seldom used today for wall flashing.

## STAINLESS STEEL

Stainless steel is an alloy of steel containing more than 12% chromium and sometimes also nickel, magnesium, or molybdenum. Most stainless steel will not rust and corrodes only under very severe conditions. It is available in either hard or soft temper.

Stainless steel flashing materials are available in several gauges and finishes. They are durable, highly resistant to corrosion, excellent moisture barriers, and surprisingly workable. A number of types are available, the varieties depending basically on the proportions of chromium, nickel, and steel. These varieties have differing degrees of resistance to corrosion and are priced accordingly. Stainless steel for masonry wall flashing should be ASI Type 304 with a No. 2B finish, and should be specified by number and not by the generic term alone. Stainless steel flashing that will be exposed to the weather should be 28-gauge (0.47 mm) and 30-gauge (0.39 mm) when completely concealed. Associated fascias should be formed from 22-gauge (0.85 mm) material, but base, cap, and counter flashings may be 28-gauge (0.47 mm).

In addition to its other favorable properties, stainless steel will not stain adjacent materials, resists rough handling, and can be soldered, welded, or brazed.

## SHEET METAL FINISHES

Sheet metal materials associated with prefinished metal roofing are usually finished to match the roofing. Those not associated with prefinished roofing are usually painted when exposed to view or when a protective coating is required. Copings, gravel stops, gutters, fascias, and downspouts are often factory finished. Those used with single-ply roofing are often coated with the same material used in the roofing.

Copper, lead, aluminum, stainless steel, and galvanized steel do not require painting other than for esthetic purposes. However, if left unpainted, galvanized steel will eventually rust.

Terneplate made with stainless steel does not need painting. That made with normal steel does, because the terne coating contains pin holes that will permit the underlying steel to rust.

Painted sheet metals are usually painted with an oil- or alkyd-based paint. Some are factory-primed; others must be primed in the field.

Factory-finished sheet metals may be finished with porcelain enamel, fluoropolymer, acrylic enamel, siliconized polyester, or the same materials used in single-ply roofing membranes. Aluminum may be clear or color-anodized.

### 7.9.1.2 Flexible Sheet Flashings

Flexible sheet flashings include elastic sheet materials made from PVC or modified polymers and membranes composed of glass fiber or organic felt fabrics saturated with bitumen. These are used as damp checks and as low-cost substitutes for metal flashing at wall bases, opening heads, window sills, floor lines, and in other locations where wall flashings are needed. Elastic sheet materials are also used in conjunction with single-ply roofing and as base flashings and penetration flashings in some types of built-up bituminous roofing (Fig. 7.9-1). Elastomeric sheets, usually plain or reinforced chlorinated polyethylene, are used as roof expansion joint covers. Often this joint covering is bonded to a layer of foam insulation.

Some elastic flashings are tough, resilient, and highly resistant to corrosion and to acid and salt attack. This is particularly true of flashing materials, such as EPDM synthetic rubber, recommended by the manufacturers of single-ply roofing membranes for use with their products. Elastomeric products are also able to withstand greater movement without failure in walls and roofs than are metal flashings, and can be more easily formed into loops or bellows to accommodate movement. Since the chemical composition of plastics varies so widely, they cannot all be used in every location. Some plastic flashings, for example, cannot withstand the corrosive effects of masonry mortars, and others, such as PVC materials, cannot be exposed to sunlight.

When properly installed, both bituminous and plastic flashings can be effective. However, flashing must permanently exclude the passage of water. Unfortunately, bituminous and plastic fabric flashings are often inadvertently cut during construction by trowels or abraded by masonry units so that they do not provide a continuous membrane. In addition, thin fabrics may sag and not re-

**FIGURE 7.9-1** Elastic flashing is installed easily around a pipe vent.

tain the desired shape as a result of changes in temperature or poor workmanship. This is a particular problem when such flashings are concealed, because the changes will not be apparent until the wall begins to leak or the water held in pockets in the deformed flashing freezes and damages the wall.

Since recognized standards have not yet been established for either bituminous or plastic wall flashing materials, it is necessary to rely on material performance and test records, and the manufacturer's reputation, to ensure suitable quality and performance.

### 7.9.1.3 Combination Sheet Flashing

Combination flashings consist of materials combined to utilize the best properties of each. Figure 7.9-2 shows two examples. Other examples are a copper sheet bonded with asphalt between two layers of glass fiber cloth and a rubberized asphalt compound sheet bonded to a polyethylene film. There are many oth-

**a**

**b**

**FIGURE 7.9-2** Two types of combination wall flashing: (a) a copper sheet bonded to asphalt-saturated kraft paper (aluminum may be used in lieu of copper); (b) a five-layer assembly composed of a copper or aluminum sheet sandwiched between layers of asphalt-saturated cotton fabric, which is then faced with outer layers of ductile mastic.

ers. The combination of materials can provide a lower-cost flashing by reducing the metal thickness required, or it may permit the use of corrodible metals which might otherwise prove unsatisfactory unless coated.

Some combination flashing materials are made by coating sheet metal with another material. Coating materials commonly used include bitumen, kraft paper, bituminous-saturated cotton fabrics, glass fiber fabrics, and combinations of them. The sheet metals commonly used for combination flashing include 3, 4, and 5 oz/sq ft (915.5, 1220.6, and 1525.8 g/m$^2$) copper and 0.004- or 0.005-in. (0.102- or 0.127-mm)-thick aluminum.

Composites consisting of a reinforced, insulated, chlorinated polyethylene joint cover bonded on each edge to a galvanized steel, stainless steel, or copper flange are used as roof expansion joints.

While no recognized standards exist for combination flashing, the weights of the copper and the thicknesses of the aluminum sheet metal given above, when coated on both sides with any of the coatings listed above and when installed with unbroken skins, provide suitable concealed flashings.

### 7.9.1.4 Fasteners

Aluminum sheets should be fastened in place using screws. Other metals should be fastened in place using either screws or twisted or threaded nails. Screws are often used in roof specialties, nails in flashings. Fasteners should extend not less than 1$^1/_2$ in. (38.1 mm) into solid, pressure-preservative-treated wood.

Cleats in lead roofing should be installed using hard copper wire nails. Most other fasteners in sheet metal flashings,

roof accessories, and roofing should be of the same material as the metal being fastened. When that is impracticable, the fasteners should be noncorrosive and compatible with the metal being fastened so that their contact will not produce galvanic corrosion.

Lead sheets, regardless of their size, and other metal sheets more than 18 in. (457.2 mm) wide should not be nailed. These should instead be held in place using cleats of the same metal and thickness as the sheets, spaced about 12 in. (304.8 mm) on centers (Fig. 7.9-3). However, cleats should be continuous (1) on slopes less than 3 in. (76.2 mm) per foot, (2) where the sheets are subject to high winds, and (3) in other locations where the metal manufacturer so recommends. Cleats should not be less than 2 in. (50.8 mm) wide and 3 in. (76.2 mm) long. They should be nailed in place using at least two nails and lock-seamed into the sheet being held.

Narrow sheets should be nailed on one edge only. In general, nails should be staggered at about 3 in. (76.2 mm) apart and $^1/_2$ in. (12.7 mm) from the edge of the sheet.

Metal flashing flanges should be nailed to wood members (nailers) at a maximum spacing of 6 in. (152.4 mm).

### 7.9.2 WALL FLASHING

Refer to Section 4.7 for information about wall flashing in masonry walls.

Except in areas of very slight exposure to precipitation where local experience supports omitting it, flashing should be installed in other types of walls at the following locations (Fig. 7.9-4): (1) at the bottom of walls, (2) under wall opening

**FIGURE 7.9-3** A typical cleat. (Courtesy American Iron and Steel Institute.)

Metal reinforcement must extend down over flashing

Counter flashing

Flashing

ROOF

Flashing over drip cap

HEAD

SILL — Flashing under sill

Where wood contacts stucco, cover with waterproof paper

Metal lath must extend down over flashing

Flashing

WATERTABLE

Grade

**FIGURE 7.9-4** Wall flashing locations in wood stud construction. (Drawing by HLS.)

sills, (3) over wall opening heads, (4) at intermediate floor lines, depending on construction type, (5) where a wall penetrates a roof surface, (6) beneath copings, and (7) at expansion joints.

### 7.9.2.1 Base of Walls

Flashing and damp checks are used at the base of walls. Flashing diverts to the exterior moisture that enters the wall above the flashing location. A damp check stops the upward capillary travel of ground moisture. If installed properly, metal shields used for termite protection may serve these purposes.

### 7.9.2.2 Wall Openings

#### SILLS

Sill flashing should be placed under and behind every sill. The ends of sill flashing should extend beyond the sides of an opening and turn up at least 1 in. (25.4 mm) into the wall. Sills should slope and project from the wall to drain water away from the building and to prevent staining. When the undersides of sills do not slope, a drip notch should be provided or the flashing may be extended and bent down to form a drip.

#### HEADS

Head flashing should be placed over all openings except those completely protected by overhanging projections. At steel and wood lintels, the flashing should be placed under and behind the facing material with its outer edge bent down over the lintel to form a drip.

### 7.9.2.3 Floor Lines

Flashing should be used at floor lines where water entering a wall might be able to find its way into the building through floor edge construction. It should extend from the exterior face of the wall below the floor construction, up the exterior face of the floor construction, pass beneath the wall above, and turn up. This flashing may be either of the materials described in Section 7.9.1.

### 7.9.2.4 Copings

Flashing should be carried beneath copings where the construction will not otherwise prevent water penetration. Coping flashing may consist of any of the materials listed in Section 7.9.1.

### 7.9.2.5 Walls Penetrating Roofs

At chimneys and other walls passing through a roof, the base flashing described in Section 7.9.3 should be capped with metal counter (cap) flashing that turns down over the base flashing and is anchored securely in the mortar joints.

### 7.9.2.6 Expansion Joints

Flashings are required at expansion joints in walls where water might enter the building. These flashings may be preformed of metal or plastic or may be formed in the field, usually from metal fashioned into a bellows.

### 7.9.3 ROOF FLASHING

Refer to Section 7.8 for information about flashing associated with built-up and single-ply roofing, and to Section 7.6 for similar information about flashing associated with shingle and tile roofs.

Flashings should be installed in roofs:

- At eaves, rakes, and valleys of pitched roofs
- Where walls, pipes, curbs, and other elements penetrate
- At roof drains

Except for counter (cap) flashings, flashings associated with field-formed

metal roofing are often locked into the roofing using a lock seam joint.

Flashing for performed roofing should be made specifically to fit the configuration of the roofing or should be flexible enough to fit the particular roofing's shape. Flashings should be manufactured or supplied by the manufacturer of the preformed roofing.

### 7.9.3.1 Eave, Rake, and Valley Flashings

Metal eave, rake, and valley flashings are described and illustrated in Section 7.6 for each type of pitched roofing covered there.

### 7.9.3.2 Roof Penetration Flashings

Roof penetrations include major penetrations, such as walls and chimneys, and minor penetrations, such as pipes, curbs, guy wires, roof drains, and other elements that penetrate a roof. Essentially, flashings at penetrations consist of a base flashing turned up against the penetration and a counter (cap), flashing turned down over the base flashing. However, several variations of such flashings are used, depending on the roofing and the element to be flashed.

#### BASE AND CAP FLASHINGS

Separate base and cap flashings are used with walls, chimneys, and curbs. Metal base flashings of the same material and thickness as the roofing are used with metal roofs. Both metal base flashings and bituminous base flashings are used with shingles (see Section 7.6). Sheet metal base flashings are not usually used with built-up or single-ply roofing, because metals have coefficients of expansion that are not compatible with those of most roofing membranes. Base flashings for built-up roofing are usually bituminous or an elastomeric material compatible with the roofing. Base flashings for single-ply roofing are usually the same materials as the roofing membranes or a compatible elastomeric material. Refer to Section 7.8 for additional discussion about base flashings for use with built-up and single-ply roofing.

Metal base flashings in steep roofing may be either continuous or pieced. Pieced base flashing is woven into the roofing as it is placed, with the pieces overlapped by at least 3 in. (76.2 mm) in the direction of water flow. Sloped-roof base flashing should turn out onto the roofing and up the wall not less than 4 in. (101.6 mm) when there is a cap flashing, 6 in. (152.4 mm) when a wall finish forms the cap flashing, and 8 in. (203.2 mm) where there is no cap flashing. When the slope is away from the wall, the base flashing should extend up the wall at least 5 in. (127 mm).

Every base flashing requires some kind of counter (cap) flashing. A wall finish may act as a cap flashing, but most cap flashings are metal. The metal material used in cap flashings varies, depending on the base flashing material and the type of penetration being flashed. Figure 7.9-5 shows a typical type of cap flashing, but many other configurations are possible. A cap flashing may extend beneath a window sill, for example. Cap flashings at sloped roofs should be in sections, with no section more than 8 in. (203.2 mm) high. Cap flashings should overlap base flashings by at least 4 in. (101.6 mm).

## FLASHINGS FOR SMALL PENETRATIONS

Small penetrations such as pipes, conduits, ducts, and columns are sometimes flashed by building a curb around them and providing base and cap flashings. More often, though, such elements are flashed using sleeves, formed metal housings, or collars.

Flashings for some penetrations consist of a factory-formed collar of an elastomeric material shaped to fit the penetrating element. This is especially true in single-ply roofs, where such a collar can be sealed to the roofing membrane using the same materials used to seal seams in the roofing membrane.

**FIGURE 7.9-5** A typical cap flashing in a reglet. (Courtesy H. Leslie Simmons.)

Pipes are flashed with metal in several ways. One common way consists of a lead-sheet flange that extends out onto the roofing about 4 in. (101.6 mm) to be flashed in with roofing membranes. A sleeve is soldered or welded to this flange, extends up the pipe for its full height, and is then turned down into the pipe at least 2 in. (50.8 mm).

Penetrations such as shaped steel columns, conduit, and ducts are sometimes flashed using a two-piece metal housing. The lower portion of the housing is a sleeve that extends 8 in. (203.2 mm) up the penetrating element. It may be fitted to the penetration's shape or may be square or rectangular. This sleeve has a 4-in. (101.6-mm)-wide flange that is flashed into the roofing using roofing membrane. The upper portion of the housing is a hood that is attached firmly to the penetrating element and extends down over the sleeve. The joint between the hood and the penetrating element should be filled with elastomeric sealant.

Difficult-to-flash penetrating items, such as guy wires and sloping-rooftop-mounted sign supports, are sometimes flashed using a pitch pan. This is a two-piece soldered metal pan set over the roofing membrane and flashed into the roofing by the roofer. The pan is then filled with a layer of mortar or a mixture of roofing cement and sand and topped off with a layer of a polyurethane sealant, coal tar pitch, or another self-leveling topping. All materials used must be compatible with the adjacent roofing. The use of pitch pans is discouraged by most authoritative sources but persists nevertheless, because there is sometimes no other way to provide flashing for a penetrating element.

## ROOF DRAIN FLASHING

Roof drains in membrane roofs are usually flashed by extending the roofing membrane into the drain's flange ring. Then a 30 in. (762 mm) square of 4 lb (1.8 kg) lead or 16 oz (4882.4 g/m²) copper is placed on the roofing membrane, extending into the drain's flashing ring, and bedded in roofing cement. This metal pan should be stripped into the roofing by the roofer. In some single-ply roofing systems, the roofing membrane is extended into the drain's flashing flange and the metal sheet is omitted.

Sometimes the roofing insulation is tapered down to the drain. When the flashing material can be exposed to sunlight, a gravel stop is sometimes installed to surround the drain.

### 7.9.4 SHEET METAL INSTALLATION

Additional information about installing flashing in masonry walls is presented in Section 4.7; in built-up and single-ply roofing in Section 7.8; in roof shingles and tile in Section 7.6; in wall shingles and siding in Section 7.7; and in sheet metal roofing in Section 7.10.

#### 7.9.4.1 Preparation

Surfaces to receive flashing should be smooth and free from projections that would puncture the metal material or destroy its effectiveness.

Metal flashings over a wood or cementitious substrate should be installed over a red rosin paper slip sheet and an underlayment of polyethylene sheet. The flanges of metal flashing should be bedded in roofing cement.

#### 7.9.4.2 Installation

Sheet metal flashings and roofing should be installed in accordance with the SMACNA's, *Architectural Sheet Metal Manual*. Elastic flashings should be installed as recommended by the manufacturer.

#### JOINTS AND EDGES

Seams in metal flashing must be overlapped and thoroughly bonded or sealed to prevent water penetration. There are many variations of such joints. Most sheet metal flashing materials other than aluminum can be soldered, but lockslip joints are required at intervals to permit thermal expansion and contraction. Joints in lead sheets are sometimes welded. Most joints in sheet metals are soldered with a tin and lead solder. The exact composition of the solder depends on the metal being soldered. Aluminum sheets are usually joined by lapped, lock-seamed, or butted and back-flashed. Aluminum gutters and downspouts are often lapped and riveted. Joints in aluminum should be filled with sealant.

Many plastics can be permanently and effectively joined by heat or an appropriate adhesive. The elastic pliability of plastic flashings eliminates the need for most expansion and contraction seams.

Metal flashings to be sealed with sealants should be overlapped at least 3 in. (76.2 mm). Others should be connected with seams to form a continuous watertight system.

Sheet metal edges that are not attached or locked to adjacent metal should be hemmed for strength.

### EXPANSION AND CONTRACTION

There is usually a large differential in the expansion coefficient of most sheet metals and that of their supports. Therefore, it is necessary to place expansion joints in sheet metal. The normal intervals are 32 ft (9.75 m) for aluminum and 40 ft (12.2 m) for most other metals. Additional joints are also required when the distance from the last joint to the end of the sheet metal is more than half the normal interval.

Extruded aluminum gravel stops, copings, and fascias should have expansion joints about 12 ft (3657.6 mm) apart.

### FLASHING RECEIVERS

Most metal counter flashings should be set into mortar joints as the masonry is built or later inserted into flashing receivers. Reglets should be used at concrete walls. Flashing receivers are made from the same metal as the cap flashing. Reglets are made from various metals and plastics. The material used should be compatible with the flashing.

Flashing should be snapped or inserted into metal flashing receivers and held in place by the receiver's snaplock feature or by fasteners.

Depending on reglet design, flashing may be snapped into it or inserted and wedged in place with lead wedges or lead wool (see Fig. 7.9-5). After the flashing has been inserted, a reglet should be filled with an elastomeric sealant. Lead flashing, however, is an exception. It should be locked to a separate lead strip, which can then be caulked directly into a reglet.

Cap flashings are sometimes inserted into a *raggle* (cut slot) left by the mason or cut into a finished masonry joint and held in place by lead wedges. The raggle should then be filled with sealant.

## 7.10 Metal Roofing

Metal roofs have protected buildings for at least 2000 years. The Pantheon in Rome had a lead roof. The 1400-year-old lead roof on the Church of Hagia Sophia in Istanbul is still in usable condition. Notre Dame Cathedral in Paris and St. Paul's Cathedral in London both have lead roofs. The balcony of the Castle of Frustenburg, built in the fifteenth century, has a gilded copper roof. The Potala Palace in Lhasa, Tibet, built between 1645 and 1693, also had a gilt copper roof. Many buildings erected during the Revolutionary War period had copper roofs. Copper sheets have been rolled specifically for roofing since about 1740.

Today, galvanized steel, stainless steel, aluminum, lead-coated copper, zinc, and terne-coated stainless steel have been added to the old metal roofing mainstays, copper and lead. Composite panels are also available that combine insulation with a metal skin.

The requirements related to sheet metals discussed in Section 7.9 apply to metal roofing except where they are specifically modified as noted in the following discussion.

Sheet metal roofs are used on every building type. They are made from a variety of sheet metals (Fig. 7.10-1) and in many configurations.

Metal roofing must (1) shed water and prevent its penetration into the building, (2) resist corrosion caused by exposure to the atmosphere and pollutants, (3) resist puncture, abrasion, and other damage during installation and use, and (4) be made from a formable metal that can be worked to the correct shape and retain this shape after installation.

Because replacement costs can exceed the initial installed costs, it is prudent to select a durable metal roofing material for the original installation.

### 7.10.1 SYSTEMS

#### 7.10.1.1 Preformed Metal Roofing

Preformed metal roofing consists of either composite panels, called *metal roof and wall panels*, or sheet metal panels. Preformed roofing is ready to install with joint configurations made on the panels when they are shipped to the project site.

Many preformed panels are also used on vertical surfaces.

**FIGURE 7.10-1** A sheet copper roof. (Courtesy Copper Development Association.)

## COMPOSITE PANELS

Composite panels have a sheet metal facing over an insulating material, such as polyurethane, polyisocyanurate, or another rigid foam. Some have uninsulated honeycomb cores. The metal facing may be any of the metals described in this section but is usually either stainless steel, galvanized steel, aluminum, galvalume, copper, or zinc. Panels are standing seam, corrugated, or flat with interlocking joints (Fig. 7.10-2). Some flat panels have ridges or grooves in them. Some have embossed finishes.

Composite panels are available in a variety of widths up to 42 in. (1066.8 mm) and generally in thickness from $1^{1}/_{4}$ in. (45 mm) to 6 in. (152 mm). Wider panels and thicker panels may be available from some manufacturer. R-values run from 4 to 50.

## INTERLOCKING INSULATED PANELS

Uninsulated interlocking panels are formed into various configurations at the factory (Fig. 7.10-3). They are made from stainless steel, galvanized steel, aluminum, and galvalume sheets.

Interlocking panels come in a variety of widths, including 10, 12, 16, and 18 in. (254, 304.8, 406.4, and 457.2 mm), and in depths of usually $^{3}/_{4}$ in. (19.1 mm) to $1^{1}/_{2}$ in. (38.1 mm). Other widths and depths are available from some manufacturers.

## PANELS FOR SEAMED ROOFING

Preformed panels for standing seam, batten seam, and step seam roofing are formed into U-shaped configurations at the factory (Fig. 7.10-4). They are made from stainless steel, galvanized steel, aluminum, and galvalume sheets.

Panels for seamed roofing are available in 8, 12, 16, 18, and 24 in. (203.2, 304.8, 406.4, 457.2, and 609.6 mm) nominal widths and with seams finishing $1^{3}/_{4}$, $2^{1}/_{2}$, 3, and 4 in. (44.5, 63.5, 76.2, and 101.6 mm) high. Actual widths and seam heights vary from manufacturer to manufacturer and from system to system. Structural panels tend to be thicker and have higher ribs than architectural panels. The 16- and 18-in. (406.4- and 457.2-mm)-wide panels are the most popular. Most are made from 24-gauge (0.70 mm) metal or the equivalent, but some are made from thinner material.

Long panels are more economical than short ones, but expansion and contraction limit their length to 48 ft (14.63 m) for steel pans and 32 ft (10.5 m) for aluminum pans. Materials producers and their associations may recommend even shorter panels.

### 7.10.1.2 Field-Formed Systems

Field-formed systems are formed to their final U-shaped configurations by roll forming them over anchor clips or cleats at the construction site. They may be made from any of the materials outlined later in this section.

Panel widths for field-formed metal roofing are usually 16 or 18 in. (406.4 or 457.2 mm) nominal. Ribs should be at least 1 in. (25.4 mm) high.

## 7.10.2 TYPES

Sheet metal roofs are either structural or architectural.

### 7.10.2.1 Structural Sheet Metal Roofing

Structural metal roofing is a structural roof that also forms a weather barrier. It is designed to span between structural roof supports. No roof deck is required. It can be used on slopes as low as $^{1}/_{4}$ in./ft (20.83 mm/m). Most such roofs consist of steel or aluminum panels joined by raised seams that provide stiffness. Clips are used to attach such roofing to structural supports, usually purlins. Each manufactured product has its own specific installation method. The manufacturer's instructions should be followed.

### 7.10.2.2 Architectural Sheet Metal Roofing

Architectural metal roofing is solely a weather barrier. It has no structural component and must be installed over a supporting roof deck. It is designed to shed water rather than act as a total barrier to it and is, therefore, usually installed on slopes of 3 in./ft (150 mm/m) or steeper. Some systems can be installed on slopes as low as 1 in./ft (83.3 mm/m), but these require special precautions to prevent water intrusion. Among these are welding every joint in lead roofing, filling every joint in aluminum roofing with a nonhardening sealant, and soldering every joint in other types of metal.

Architectural sheet metal roofing may be either preformed or formed in place. Installation methods are basically the same for both.

## 7.10.3 MATERIALS

Metals used in roofing are the same as those used in flashings, and the descriptions of them in Section 7.9 apply. Additional description of these materials can be found in Chapter 5. The descriptions in this section related to materials supplement those presented in Section 7.9.

**FIGURE 7.10-2** A typical insulated composite panel. Many configurations are available. (Drawing by HLS.)

**FIGURE 7.10-3** Some typical interlocking or overlapping panels. Many configurations are available. (Drawing by HLS.)

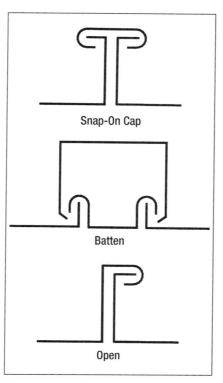

Snap-On Cap

Batten

Open

**FIGURE 7.10-4** Typical standing seams made with shop-formed panels. Other configurations are available. (Drawing by HLS.)

### 7.10.3.1 Aluminum

Sheet aluminum is seldom used to make field-formed roofing. Preformed sheet aluminum roofing, aluminum-coated sheet steel roofing, and aluminum-covered composite panels, however, are quite common. The usual thickness of aluminum roofing is usually about 0.032 in. (0.813 mm). Roof accessories such as copings, gravel stops, gutters, and downspouts are usually thicker, varying with the manufacturer.

### 7.10.3.2 Copper

Copper thicknesses for various uses should be in accordance with the recommendations of the CDA's *Copper in Architecture Design Handbook*.

Most batten seam, flat seam, and standing seam copper roofings should weigh 20 oz/sq ft (6103 g/m$^2$). Some types of roofing may weigh less (12 to 16 oz [3662 to 4882 g/m$^2$]) when the pans are 16$^3$/$_4$ in. (425.5 mm) wide or smaller.

### 7.10.3.3 Galvanized Steel

Galvanized steel is steel that has been coated with a layer of zinc, either electrolytically or by the hot-dip process. Because of its low cost, galvanized steel is widely used for roofing, especially in small commercial and residential buildings. While it can perform adequately in such a use when it is well protected, exposed galvanized steel requires periodic maintenance and should be accessible for inspection and painting.

Galvanized steel roofing should be 26-gauge (0.55 mm) for panels up to 24 in. (609.6 mm) wide and 24-gauge (0.70 mm) for wider panels.

### 7.10.3.4 Lead

Most sources recommend that sheet lead roofing and associated flashings and roof specialties be formed of not lighter than 3 lb (14.65 kg/m$^2$) hard lead. Small pieces may be formed from 2$^1$/$_2$ lb (12.21 kg/m$^2$) hard lead. Batten seam roofing may also be formed of 2$^1$/$_2$ lb (12.21 kg/m$^2$) lead when the battens are kept not more than 18 in. (457.2 mm) apart.

Lead roofing should follow the recommendations of the Lead Industries Association publication *Lead Roofing and Flashing*.

### 7.10.3.5 Stainless Steel

Stainless steel roofing should be 28-gauge (0.47 mm) when the roofing panels do not exceed 18 in. (457.2 mm) and 24-gauge (0.70 mm) for wider panels.

Stainless steel roofing should be selected and installed in accordance with the Specialty Steel Industry of North America publication *Standard Practice for Stainless Steel Roofing, Flashing, Copings*.

### 7.10.3.6 Terneplate

Terne is an alloy of lead and tin, usually in a ratio of 4 parts lead to 1 part tin. Terne-coated steel or stainless steel (terneplate), which is the terne used in roofing, is made by coating steel or stainless steel sheet with a layer of terne metal.

Terneplate used for roofing should be 0.015 in. (0.381 mm) thick for panels up to 24 in. (609.6 mm) wide and 0.178 in. (4.5 mm) thick for wider panels.

### 7.10.3.7 Zinc

Zinc is a bluish-white crystalline metal. It is brittle at ordinary temperatures but ductile when heated. The zinc sheets used in roofing are an alloy of zinc, copper, and titanium.

Preformed zinc roofing panels are usually 0.027 in. (0.69 mm) thick, but thicker sheets are sometimes available. They carry a guarantee of up to 20 years.

### 7.10.3.8 Aluminum-Zinc Alloy on Steel

The most durable coated steel products available today have an aluminum-zinc alloy coating applied to them. This is referred to as *galvalume* or *zincalume*. They carry warranties exceeding 20 years.

### 7.10.4 FINISHES

Factory-finished sheet metals may be finished with porcelain enamel, fluoropolymer, acrylic enamel, siliconized polyester, or the same materials used in single-ply roofing membranes. Aluminum may be clear or color-anodized.

Steel and aluminum sheets are also available with flexible and colorful factory-applied coatings that can be shaped on a brake without damaging the coatings.

Sheet metal materials associated with prefinished metal roofing are usually finished to match the roofing. Those not associated with prefinished roofing are usually painted when exposed to view or when a protective coating is required. Copings, gravel stops, gutters, fascias, and downspouts are often factory-finished. Those used with single-ply roofing are often coated with the same material used in the roofing.

Copper, lead, aluminum, stainless steel, and galvanized steel do not require painting other than for esthetic purposes. However, if left unpainted, galvanized steel will eventually rust.

Terneplate made with stainless steel does not need painting. That made with normal steel does, because the terne coating contains pin holes that will permit the underlying steel to rust.

Painted sheet metals are usually painted with an oil- or alkyd-based paint. Some are factory-primed; others must be primed in the field.

### 7.10.5 INSTALLATION

Surfaces to receive roofing should be smooth and free from projections that would puncture the metal material or destroy its effectiveness.

### 7.10.5.1 Underlayment

An underlayment of not less than No. 15 asphalt-impregnated, unperforated roofing felt should be installed beneath metal roofing. Edges should be lapped 2 in. (50.8 mm), ends 6 in. (152.4 mm). Felt should be nailed in place using large-headed roofing nails. In addition to the underlayment, an ice shield should be provided where the January mean temperature is 30°F ($-1.1$°C) or less. This ice shield should consists of two layers of at least No. 15 asphalt-impregnated roofing felt, installed using hot asphalt, laid to cover an area from the eave to a line 24 in. (610 mm) inside the interior face of the exterior wall. As an alternative, a single-layer membrane produced specifically for this purpose may be used as an ice shield.

The underlayment beneath metal roofing should be covered by a rosin-sized building paper slip sheet to prevent the roofing from sticking to the felts. However, this paper is sometimes omitted from beneath roofing installed using flat seams, since some designers do not demand that such roofing be able to move.

### 7.10.5.2 Fasteners

#### FASTENERS IN COPPER ROOFS

Each cleat in a copper roof should be fastened in place using two nails. Nails should be flathead, barbed, wire slating nails at least 1 in. (25.4 mm) long of 12-gauge (2.7 mm) hard copper or brass. The required length and gauge may vary with the condition of the substrate and should be verified with the roofing producer.

Screws used in conjunction with a copper roof should be hard copper or brass. They should have round heads and flat seats. A lead washer should be installed between the screw head and the copper. Where screws are exposed and subject to water penetration, copper caps should be installed over the screw heads.

### FASTENERS IN ALUMINUM ROOFS

Aluminum sheets should be fastened in place using screws. Other metals should be fastened in place using either screws or twisted or threaded nails. Screws are often used in roof specialties, nails in flashings. Fasteners should extend not less than $1^{1}/_{2}$ in. (38.1 mm) into solid, pressure-preservative-treated wood.

### FASTENERS IN LEAD ROOFS

Cleats in lead roofing should be installed using hard copper wire nails. Most other fasteners in sheet metal flashings, roof accessories, and roofing should be of the same material as the metal being fastened. When that is impracticable, the fasteners should be noncorrosive and compatible with the metal being fastened so that their contact will not produce galvanic corrosion.

Lead sheets, regardless of their size, and other metal sheets more than 18 in. (457.2 mm) wide should not be nailed. These should instead be held in place using cleats of the same metal and thickness as the sheets, spaced about 12 in. (304.8 mm) on centers (see Fig. 7.8-4). However, cleats should be continuous (1) on slopes less than 3 in./ft (250 mm/m), (2) where the sheets are subject to high winds, and (4) in other locations where the metal manufacturer so recommends. Cleats should not be less than 2 in. (50.8 mm) wide and 3 in. (76.2 mm) long. They should be nailed in place using at least two nails and lock-seamed into the sheet being held.

### FASTENERS IN STAINLESS STEEL ROOFS

Nails for fastening stainless steel clips to secure stainless steel roofing should be flat head, annular thread, diamond point stainless steel nails long enough to penetrate the backing by at least 1 in. (25.4 mm).

### FASTENERS IN OTHER METAL ROOFING MATERIALS

Fasteners should be compatible with the metal being fastened. They should penetrate the supporting material by at least 1 in. (25.4 mm).

### 7.10.5.3 Cleats and Clips

Sheet metal roofing should be fastened in place using cleats or clips generally of the same material as the roofing sheets. Lead roofing, however, should be installed using 16 oz ($4882.4$ g/m$^2$) hard copper cleats, except that exposed cleats in lead roofing may be of 3 lb (14.65 kg/m$^2$) lead.

In general, cleats should be placed about 8 in. (203.2 mm) apart. On very steep slopes, however, cleats should be continuous. Some materials and field conditions may require other spacings. The recommendations of the association representing the roofing material manufacturer should be followed.

Cleats and clips should be fastened to the roof deck and locked into the roofing sheets at the joints. Fasteners should be concealed from view and protected from the weather by the roofing sheets.

On nailable decks, cleats and clips may be nailed or screwed directly to the deck. On insulated decks, cleats should be fastened to wood nailing strips embedded in the insulation to a layer of plywood or oriented strand board attached to the insulation or fastened to Z-clips attached to the decking. Attaching cleats or clips by driving them through the insulation may void some warranties. Metal decking directly over concrete slabs requires the installation of nailers. On uninsulated steel or concrete decks, cleats or clips can be fastened directly to the decking, but on large surfaces it is better to install Z-shaped purlins over the deck.

### 7.10.5.4 General Installation Requirements

Sheet metal roofing should be installed in accordance with the SMACNA's *Architectural Sheet Metal Manual*. Elastic flashings should be installed as recommended by the manufacturer.

### DESIGN CRITERIA

*Air Leakage* Manufacturers of roofing products vary somewhat in their product's performance relative to air infiltration. A typical system might resist air leakage at not more than 0.06 cfm/ft$^2$ (0.3 L/s/m$^2$) of roof area at a static pressure of 1.56 psf (0.07 kPa) when tested in accordance with ASTM E283. Another might calculate leakage on the basis of seam length and say that the system has a maximum leakage of 0.06 cfm per linear foot (0.33 m$^3$/h per linear meter) of seam at a static pressure of 6.24 psf (3 kPa) when tested per ASTM E283. Both criteria are acceptable minimums, depending on the type of system. Slightly more leakage might be tolerable in some systems.

*Water Penetration* The system should have no water penetration at a positive pressure differential of 6.24 psf (0.3 kPa) or 20% of the design wind pressure, whichever is larger at a test pressure between 6.24 psf (0.3 kPa) and 12 psf (0.57 kPa) in accordance with ASTM E1646 or ASTM E331.

*Wind Uplift* The system should comply with specific wind uplift resistance criteria, such as UL90, which is 90 psf (300 Pa) when measured in accordance with UL Standard 580.

### JOINTS AND EDGES

Seams in metal roofing must be overlapped and thoroughly bonded or sealed to prevent water penetration. There are many variations of such joints. Although most sheet metal roofing materials can be soldered, lockslip joints are required at intervals to permit thermal expansion and contraction. Joints in lead sheets are sometimes welded. Most joints in sheet metals are soldered with a tin and lead solder. The exact composition of the solder depends on the metal being soldered. Aluminum sheets are usually joined by lapped, lock-seamed, or butted and back-flashed. Aluminum gutter and downspouts are often lapped and riveted. Joints in aluminum should be filled with sealant.

The edges of sheet metal that are not attached or locked to adjacent metal should be hemmed for strength.

### EXPANSION AND CONTRACTION

There is usually a large differential in the expansion coefficient of most sheet metals and that of their supports. Therefore, it is necessary to place expansion joints in metal roofing. The spacing of expansion joints in metal roofing depends on the type of roofing and the seaming method. In general, when the ends of roofing are fastened, joints should be at about 15 ft (4.57 m) on centers; where ends are free to move, joints may occur at 30 ft (9.15 m) intervals. Locations where roofing abuts other materials or penetrating walls should have provisions for expansion. Some metals may require closer or wider spacings. Stainless steel, for example, requires expansion joints between 24 ft (7.32 m) and 40 ft (12.19 m) on continuous runs.

### 7.10.5.5 Installation of Factory-Formed Metal Roofing

Factory-formed roofing, including both composite and formed panel roofing, should be installed in accordance with its manufacturer's instructions. Methods may differ from manufacturer to manufacturer, depending on the design and materials used in a particular roofing system. The general procedures, however, including the use of cleats, are similar to those used to install formed-in-place metal roofing. Different profiles are made by different manufacturers (Fig. 7.10-5). Many profiles are available. The building in Figure 7.10-6 was roofed with factory-formed standing-seam metal roofing.

### 7.10.5.6 Installation of Formed-in-Place Metal Roofing

The seaming methods most often used for formed-in-place metal roofing are stand-

**FIGURE 7.10-6** A standing-seam metal roof. (Courtesy Metal Sales Manufacturing Corporation.)

ing seams, flat seams, and batten seams. All three may be used when the roof slope exceeds 3 in./ft (250 mm/m). Most seams today are made using mechanical seaming machines.

Standing seams (Fig. 7.10-7) should be at least 1 in. (25.4 mm) high.

Flat seams (Fig. 7.10-8) are made by interlocking adjoining sheets. Flat seams are used in both directions in flat seam roofing and as cross seams in standing seam and batten seam roofing. Flat seam roofing may be laid with the seams parallel and perpendicular to the eaves or in a diamond-shaped pattern with the seams at a 45-degree angle with the eaves.

In batten seam roofing (Fig. 7.10-9), the battens may run up and down the roof from eaves to ridge or form ornamental patterns.

### 7.10.5.7 Oil Canning

Sheet metal roofing characteristically displays a waviness called *oil canning*. This condition is normal, is to be expected, and is almost impossible to eliminate

**FIGURE 7.10-8** Flat lock seam. (Courtesy H. Leslie Simmons.)

**FIGURE 7.10-5** A portion of a roof using one manufacturer's standing seam roof profile. These seams are finished using a mechanical seamer. (Courtesy Metal Sales Manufacturing Corporation.)

**FIGURE 7.10-7** Standing seam. (Courtesy H. Leslie Simmons.)

**FIGURE 7.10-9** Batten seam. (Courtesy H. Leslie Simmons.)

completely. Dark colors increase thermal movement and increase the amount of oil canning. Low-gloss paints reduce the visual effect of oil canning. Bad workmanship can significantly increase this waviness. Using tensioned, leveled, and resquared metal will reduce this condition. Using wedge-shaped panels will make it worse.

### 7.10.5.8 Flashing and Accessories

Roof flashings, specialties, and accessories are discussed in Sections 7.6, 7.8, 7.9, and 7.11, respectively.

# 7.11 Roof Specialties and Accessories

## 7.11.1 ROOF SPECIALTIES

Roof specialties include copings, expansion joint covers, gravel stops, fascias, scuppers, gutters, and downspouts.

### 7.11.1.1 Copings

Most metal wall coping covers are either formed sheet metal or extruded aluminum. They may be factory-finished or primed for field painting. Most are installed with open joints backed by flashing or with joint covers. Sections are usually 8 or 10 ft (2438.4 or 3048 mm) long. Most slope down toward the roof side so that water falling on them will not run down the face of the building.

Installation methods vary with the coping, but some characteristics apply to most. Some come complete with their own compatible support system. Others are supported on pressure-preservative-treated wood nailers. Most are installed over a layer of building paper. Outer edges are snapped down in place over continuous cleats. Inner edges are screwed to wood framing or snapped over metal cleats.

### 7.11.1.2 Expansion Joint Covers

Roof expansion joints may be either preformed metal, field-formed metal, preformed elastomeric, or a combination preformed unit of metal and an elastomeric material. See Figures 7.8-17 and 7.8-35.

Except in some single-ply roofing systems, roof expansion joints should be installed over pressure-preservative-treated wood curbs. These wood curbs should be flashed into the roofing with base flashing. The expansion joint cover acts as a cap flashing.

Elastic expansion joints are sometimes installed flush with the roofing, but this is discouraged by most roofing experts. This type of installation should be used only in single-ply roofing where specifically recommended by and warranted by the roofing membrane manufacturer and where an independent competent roofing expert concurs.

### 7.11.1.3 Gravel Stops and Fascias

Gravel stops and fascias are usually prefabricated from either formed sheet metal (see Fig. 7.8-34) or extrusions. The length of sections is usually 8 ft (2438.4 mm). Roof flanges should extend 4 in. (101.6 mm) out onto the roof, be securely nailed in place, and be stripped into the roofing by the roofer. Light fascias and gravel stops should have hook strips at the bottom to hold them in place. Heavy extruded items may not require a hook strip.

### 7.11.1.4 Scuppers

Scuppers are outlets through a parapet or gravel stop that permit water to flow from a roof without traveling through a pipe inside the building. Water flowing through a scupper may fall to the ground with no connection to a drain system or fall directly into a conductor head, which is short section of gutter with the ends sealed. Conductor heads are usually connected to a downspout.

### 7.11.1.5 Gutters

Gutters collect water from a roof and distribute it to a downspout. They are built in many shapes of formed sheet metal or plastic (see Fig. 5.3-6). Some are built in, but most are hung from a fascia. Most hung gutters are made from aluminum, copper, galvanized steel, or stainless steel in rectangular, half-round, or ogee shapes. Built-in gutters are usually made from one of the metals used for flashing and sheet metal roofing. Built-in gutters are seldom made from aluminum, however, because it cannot be soldered and making joints watertight is difficult.

Hung gutters are supported by straps, brackets, or spikes and ferrules of the same metal as the gutter. The size, thickness, and spacing of these hangers should be based on the size and material of the gutter supported. Spacings are usually 30 in. (762 mm) maximum on centers.

Gutters should be set with a slight slope toward the downspouts to permit them to drain. Screens should be used over gutters to prevent them from becoming clogged with leaves.

Expansion should be accounted for in gutter design, especially in built-in gutters. In built-in gutters, the method depends on the design of the gutter. At the least, watertight expansion joints must be installed. Expansion in hung gutters can be accommodated by lapping sections of the gutter with a water barrier between the sections. Expansion can also be accomplished by building separate sections of gutter with ends and closing the gap between them with a cover plate.

### 7.11.1.6 Downspouts

Downspouts draining built-in gutters are plumbing items. Downspouts for hung gutters are usually built of the same material as the associated gutters (see Fig. 5.3-6). Their size depends on their spacing and the area of roof drained.

Downspouts should be supported at intervals, depending on their size, but never more than 5 ft (1524 mm) apart.

Many downspouts terminate with elbow-type fittings draining out onto splash blocks or pans. However, on many commercial buildings and some residences, downspouts terminate in boots connected to the site storm sewer.

## 7.11.2 ROOF ACCESSORIES

Roof accessories include ridge vents, curbs, gravity ventilators, skylights, heat and smoke vents, and roof hatches.

Roof accessories often come with built-in 4-in. (101.6-mm)-wide roof flanges and counter flashings. Base flashing should be extended up the accessory's curb and beneath the built-in cap flashing. Roofing accessory flanges for use with metal roofing should conform with the roofing's profile.

Roof accessories for formed-in-place metal roofing usually have the portions that contact the roofing made from the same material as the roofing.

Specialties for use with preformed roofing should be specifically designed to fit the roofing profile. They should be made from the same materials as the roofing and be finished to match it. Using roofing specialties with standard curbs not specifically designed for use with the particular preformed roofing product will usually lead to leaks.

## 7.12 Fireproofing

Fires cause loss of property and lives. The primary consideration in a fire is to evacuate all occupants from the building as quickly as possible.

The National Fire Protection Association's NFPA 550, *Guide to the Fire Safety Concepts Tree*, identifies the various components of fire safety and shows their relationship to each other. The major components are prevention, detection, containment, control, and extinguishment. This section discusses prevention, containment, and control. Fire detection and alarm are covered in Section 20.2, manual fire extinguishers in Section 10.1, fire sprinkler systems in Section 15.1.

The tree approach of NFPA 550 demonstrates that no one of the components of fire safety is sufficient by itself. Rather, a balanced approach is necessary, employing several or preferably all of them. The NFPA *Life Safety Code* also recognizes the need for a balanced approach to fire safety. This section discusses some of the measures that can be taken to ensure fire safety. For a more detailed review of the tree concept, refer to NFPA 550.

### 7.12.1 FIRE PREVENTION

The only way to prevent damaging fires from occurring is to either eliminate or control potential sources of fire, or to separate them from combustible materials, or to eliminate the combustible materials.

#### 7.12.1.1 Fire Sources

A faulty electrical system is a potential source of fire and a potential source of injury to individuals through electrocution. Methods of controlling this source of fire are discussed in Chapter 18.

A lightning strike can also be a source of fire. Systems to eliminate that danger are covered in Section 18.4.

Kitchens are among the most fire-prone spaces in both dwellings and other building types. Fire-producing devices, such as gas ranges or stoves, should not be located near flammable materials. In

a residence, for example, such devices should be kept away from windows because of the risk of curtains catching fire. Flammable wall coverings and other flammable finishes should not be used adjacent to such devices.

Careless smoking is a major cause of fires in both residential and other buildings. Many commercial, institutional, and industrial buildings prohibit smoking indoors.

#### 7.12.1.2 Fire-Retardant Products

Consumer protection laws require that mattresses and carpeting be treated for flame resistance. Some states have adopted restrictions on upholstered furniture to ensure that the room a fire starts

in will remain at temperatures from which occupants can escape for 5 minutes after a fire starts and remain safe for 30 minutes from toxic levels of carbon monoxide. Older mattresses, carpeting, curtains, and drapes can be treated with homemade fireproofing chemicals (Fig. 7.12-1).

#### 7.12.1.3 Fire-Resistant Plants

In some parts of the country, forest and brush fires can be hazardous to buildings, especially dwellings. In those areas where such plants will grow, fire-resistant ground cover should be planted within 100 ft (30.48 m) of the building, and shrubs and trees should not be closer than 30 ft (9.14 m) from the building and not less than 18 ft (5.49 m) apart

**FIGURE 7.12-1** Flameproofing Recipes

| Material to Be Treated | Recipe* |
| --- | --- |
| Permanent-press fabrics (blends of vegetable and synthetic fibers) | 12 oz (354.9 mL) diammonium phosphate<br>2 qt (1.89 L) hot water |
| Untreated vegetable-fiber fabrics | 7 oz (207 mL) borax<br>3 oz (88.7 mL) boric acid<br>2 qt (1.89 L) hot water |
| Paper and cardboard | 7 oz (207 mL) borax<br>3 oz (88.7 mL) boric acid<br>5 oz (147.9 mL) diammonium phosphate<br>1 tsp (5 mL) liquid dishwashing detergent<br>$3\frac{1}{2}$ qt (3.3 L) hot water |
| Christmas tree | $4\frac{1}{2}$ gal (17 L) sodium silicate (water glass)<br>2 qt (1.89 L) hot water<br>2 tsp (10 mL) liquid dishwashing detergent |

*To treat fabrics and paper products: (1) dissolve the recipe ingredients in hot water, (2) spray or dip the product in the appropriate solution, (3) allow to dry. Treat Christmas trees twice.

**FIGURE 7.12-2** Fire-Resistant Plants

| Type of Plant | Plant Name |
| --- | --- |
| Ground covers | Rosea ice plant (*Drosanthemum hispidum*)<br>Jelly beans (*Sedum rubrotinctum*)<br>Trailing gazania (*Gazania uniflora*)<br>Trailing African daisy (*Osteospermum fruticosum*)<br>Kentucky bluegrass (*Poa pratensis*)<br>Rye grass (*Lolium*, several species)<br>White clover (*Trifolium repens*) |
| Shrubs | Oleander (*Nerium oleander*)<br>Elephant bush (*Portulacaria afra "variegata"*)<br>Bluechalksticks (*Senecio serpens*)<br>Toyan (*Heteromeles arbutifolia*)<br>Common lilac (*Syringa vulgaris*) |
| Trees | Gum trees (*Eucalyptus*, several species)<br>Carob (*Ceratonia siliqua*)<br>California pepper (*Schinus molle*)<br>California laurel (*Umbellularia californica*)<br>Cottonwood (*Populus deltoides*) |

(Fig. 7.12-2). Under these conditions, a fire cannot easily jump from one to the next and then to the building.

## 7.12.2 FIRE CONTAINMENT AND CONTROL

The UL publication *Fire Resistance Directory* contains a list and description of fire protection systems and fire ratings for various floor and ceiling assemblies, roof and ceiling assemblies, and structural members.

The UL ratings for assemblies specify the type, size, and spacing of the floor framing (including bar joists); the type and size (material thickness, depth, and configuration) of floor and roof decking; and the methods of fastening framing and decking in place.

The UL ratings for columns, beams, and girders spell out the materials and thicknesses of fireproofing that may be used to achieve the various hourly ratings. Among the many materials that may be used to fireproof steel are concrete, masonry, gypsum board, plaster, mineral fiber, cementitious materials, and intumescent mastics.

### 7.12.2.1 Encasement

Before the advent of modern lightweight fireproofing materials, structural steel requiring a fire resistance rating was encased in poured concrete or masonry (Fig. 7.12-3). These methods were effective but considerably increased building weight and therefore increased framing member sizes, which also increased the construction cost.

### PLASTER

Attempts to decrease fireproofing weight led first to the use of plaster as an encasing material (Fig. 7.12-4). The weight was further reduced by a change from sand aggregate to lightweight aggregates such as vermiculite.

### GYPSUM BOARD

Plaster encasement fireproofing is seldom used today, having been largely replaced by gypsum board (Fig. 7.12-5). Plaster and gypsum board encasements also serve as a finish.

### MINERAL FIBERBOARD

When a structure to be fireproofed is concealed from view, other fireproofing types are more often used today. These

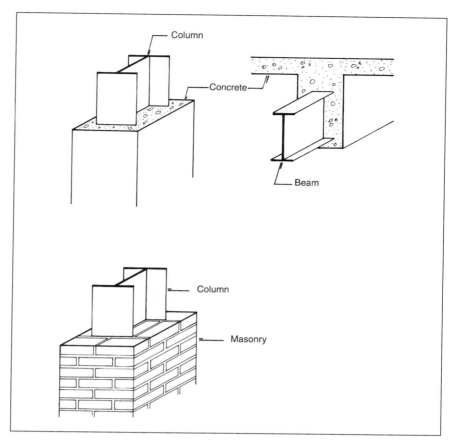

**FIGURE 7.12-3** Concrete and masonry fireproofing. (Drawing by HLS.)

include mineral fiber and gypsum slabs which are held against the steel by welded-on attachments (Fig. 7.12-6) and sprayed-on contour materials.

### 7.12.2.2 Spray-Applied Fireproofing

Sprayed-on contour fireproofing products (Fig. 7.12-7) are applied following the contour of the structural element being protected. They are also used to protect steel decking and, sometimes, bar joists and trusses. These uses should be approached with caution, however, because thermal and other substrate movement, as might happen in a steel floor or roof deck, can damage sprayed-on fireproofing. Both sprayed-fiber and cemen-

**FIGURE 7.12-4** Plaster fireproofing. (Drawing by HLS.)

**FIGURE 7.12-5** Gypsum board fireproofing.
(Drawing by HLS.)

**FIGURE 7.12-6** Fireproofing slab enclosure.
(Drawing by HLS.)

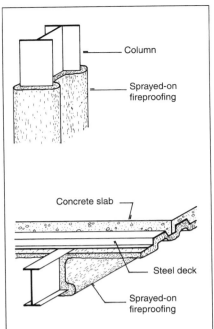

**FIGURE 7.12-7** Sprayed-on fireproofing.
(Drawing by HLS.)

titious products are available. Sprayed-fiber products are made from cement and mineral fiber. Cementitious fireproofing materials include:

- Cement-mineral aggregate
- Mineral fiber
- Foamed magnesium oxychloride
- Magnesium cement

### 7.12.2.3 Intumescent Coatings

Intumescent coatings are used primarily on structural steel that is exposed to view, but can also be used as general surface coatings in areas subject to high fire hazards and in portions of exitways, such as corridors. These coatings include both paints and mastics, which may be either clear or pigmented. They work by expanding when heated to form a thick carbonaceous foam that thermally insulates the surface on which they are placed and resists further damage from a fire. They also slow the progress of a fire, reduce smoke, and delay the buildup of toxic gases. These coatings are available in colors, and some can serve as a base coat for other paints.

### 7.12.2.4 Ceilings

Other methods are also used to protect steel from fire. One is to hang a fire-resistant ceiling, such as one of plaster,

beneath the structure. Refer to Chapter 9 for a discussion of plaster, gypsum board, and acoustical ceilings.

### 7.12.2.5 Flame Shields

Exterior steel members may be partly protected from contact with a fire by steel flame shields (Fig. 7.12-8).

### 7.12.2.6 Through-Penetration Fireproofing

Fire must be prevented from passing from one fire zone to another through openings in fire-rated floor, wall, and other assemblies. Such openings occur where pipes, ducts, and other elements penetrate walls and floors and where exterior curtain walls pass by floor slab edges. This through-penetration fireproofing, which is also called *firestopping* or *fire-safing*, is accomplished by filling open spaces around the penetrations and at the edges of floors with firestopping. The following are among the many firestopping materials available today:

- Mineral wool felt
- Mineral fiber blankets
- Nonexpanding cementitious compounds
- Fiber-reinforced foamed cement
- Fireproof sealants

**FIGURE 7.12-8** Flame shields. (Drawing by HLS.)

- Intumescent putty, wrap strips, pillows, and spray-applied products
- Endothermic spray that expands when heated
- Fire-rated premanufactured collars

# 7.13 Joint Protection

*Sealants* are materials that are placed in joints in buildings and pavings where a weathertight seal is needed and in other locations to close building and paving joints for esthetic reasons or as a base for painting. Before the advent of curtain wall construction in the 1950s, most buildings were finished in stone, brick, or terra cotta. These materials generate relatively small movements across joints in them. In addition, most openings in walls were small. Most windows and doors were wood-framed. Oil-based caulking compounds were the only joint sealants around, and they were adequate in most cases.

Curtain walls, however, were made of metals, mostly aluminum, and glass, with their high thermal expansion coefficients. Movement across joints in and around them was larger than oil-based caulking compounds could accommodate. These materials would harden, crack, and fall out of the moving joints, leaving openings through which wind and water could easily pass. The solution was synthetic rubber (elastomeric) sealants.

The first elastomeric sealants were based on polysulfide polymers. These were quickly followed by elastomeric sealants based on acrylics, silicone, and polyurethane.

Each sealant type has its own characteristics and preparation requirements. A designer must understand these differences and select the proper sealant for the job at hand. The following steps are necessary to ensure proper functioning of a sealant:

- Proper joint design
- Proper joint surface preparation
- Selection of the proper sealant
- Proper sealant installation and finishing

The sealants discussed in this section are *building sealants*, which includes those designed for use on horizontal and vertical surfaces in or on buildings, and *paving joint sealants*, which are designed for use in joints in concrete-paved roads,

curbing, and walkways, and between asphalt and concrete pavements. Glazing sealants are addressed in Section 8.8.

## 7.13.1 BUILDING SEALANTS

Building sealants include:

- Elastomeric (rubber-like) sealants based on polysulfides, silicone, or polyurethane
- Solvent-release-curing sealants
- Latex emulsion sealants
- Tape sealants
- Acoustical sealants
- Preformed foam sealants
- Oil-based caulking compounds

Until recently, federal specifications were the standards for most sealants. Today, these have been largely replaced by ASTM standards, but some manufacturers still list the old federal specifications as the standards by which their products are manufactured. Fortunately, most manufacturers also list the current ASTM standard. The following descriptions list the applicable standards, which are summarized in Figure 7.13-1.

### 7.13.1.1 Elastomeric Sealants

Elastomeric sealants are resin-based, rubber-like compounds intended prima-

rily for use in expansion joints and other moving joints, such as those in curtain walls, and those between stone panels or precast concrete panels. The polymers used to make most of them are polysulfide, silicone, and polyurethane. A few other polymers also exist, but they appear mostly in proprietary sealants and are not widely used.

### STANDARD

The industry standard for elastomeric joint sealants is ASTM C920. It classifies elastomeric sealants by type, grade, class, and use.

*Type* Elastomeric sealants are available as single-component (one-part) and multicomponent types. ASTM C920 classifies one-part sealants as Type S and multicomponent sealants as Type M. Type S sealants are more expensive than those of Type M, because they are prepackaged in small cartridges or other units for direct field application. Type M sealants can be packaged in bulk form. They must be mixed in the field, which requires more skill than is necessary for Type S sealants, but they cure faster than those of Type S and are often made with better components, which results in a superior finished sealant. In addition, the

**Figure 7.13-1** Standards Governing Sealants

| Sealant | Standard |
|---|---|
| Elastomeric | |
|     Polysulfide | ASTM C920 |
|     Silicone | ASTM C920 |
|     Urethane | ASTM C920 |
| Solvent-release curing | |
|     Acrylic | FS A-A-1556A; AAMA 800 Series |
|     Butyl | ASTM C1805; FS A-A-272A |
|     Small joint | AAMA 800 Series |
| Latex-emulsion | ASTM C834 |
| Tape | AAMA 800 Series |
| Acoustical | None |
| Oil-based caulking compound | ASTM C570; FS A-A-272A |
| Cold-applied paving joint | |
|     Elastomeric | ASTM C920 |
|     Jet-fuel-resistant, coal-tar-modified polysulfide | FS SS-S-200E(2) |
|     Jet-fuel-resistant urethane | FS SS-S-200E(2) |
| Hot-applied paving joint | |
|     Concrete to concrete | ASTM D3406 |
|     Concrete to asphalt | ASTM D3405 |
|     Jet-fuel resistant | |
|         Concrete to concrete | ASTM D3569 |
|         Concrete to tar-concrete | ASTM D3581 |

color of three-or-more-part Type M sealants can be more closely controlled because their coloring is a separate component that is added in the field.

*Grade* The flow characteristics of a sealant are defined by Grades P and NS. Grade P sealants are also called *self-leveling*, because they seek a level and remain level and smooth when applied at temperatures as low as 40°F (4.4°C). They are used on flat horizontal surfaces.

Grade NS sealants are used on vertical surfaces and in sloping horizontal surfaces where a Grade P sealant would flow down the slope. They are also called *non-sag* because they will stay in place even when applied at temperatures as high as 122°F (50°C).

*Class* There are two classes in ASTM C920: Class $12^1/_2$ and Class 25. The numbers relate to the amount of joint movement a sealant can withstand without cohesive failure. The numbers are the percentage of the joint at the time of sealant application and are plus or minus measurements. A Class 25 sealant, then, is able to withstand a 25% expansion and a 25% contraction without failure—in other words, a total of 50% of the original joint width.

Some elastomeric sealants withstand greater joint movement than 25% in laboratory tests; as a result, their manufacturers claim greater capability. However, most authoritative sources believe that joints should be designed and spaced so that the movement across them does not exceed 25% in each direction, because actual sealant performance can be affected by several imponderables. These factors include the character of joint preparation and sealant application, varying joint widths in the field due to allowable construction tolerances and other factors, and the effect of temperature at the time of application on later joint movement. A sealant may perform exactly within the rated range when the application temperature is about 70°F (21.1°C) but may not perform the same way if the application temperature is far below or above that midrange temperature.

*Use* There are two use classifications in ASTM C920. The first is related to exposure. The classes are Use T, which covers sealants intended for use on traffic surfaces, and Use NT, which are designed for nontraffic areas.

The second use classification has to do with the ability of a sealant to adhere to various substrates. The classifications are Use M (mortar), Use G (glass), Use A (aluminum), and Use O (everything not in the first three use classifications). Because the designations M, G, and A refer to specific laboratory material samples, it is advisable to select materials that fall into the Use O category. Even then, it is necessary to obtain verification that a selected sealant is satisfactory for use with the specific substrates to be found on a given project. This should be demonstrated by actual tests of the selected sealant on the joint substrates that will be encountered.

## POLYSULFIDE SEALANTS

Polysulfide sealants were the first elastomeric sealants and have a proven good performance record. Unfortunately, they tend to be more expensive than sealants made with silicone or urethane polymers, and some manufacturers have made them with less polymer to remain competitive. In addition, silicone and urethane sealants are available that have recovery and resistance to ultraviolet light and ozone properties that are equal or superior to those of polysulfide products. As a result, the percentage of the market is lower today for polysulfides than it once was. Polysulfide sealants are still made that will give the excellent performance found in earlier polysulfide products, so they should not be disregarded. However, when selecting a polysulfide sealant today, it is prudent to compare its polymer content and other properties with those of earlier polysulfide products that have a proven record of dependability.

## SILICONE SEALANTS

Silicone sealants perform well in terms of weather resistance, color fading, and durability. They also display low shrinkage and no increase in hardness over time, regardless of the temperatures encountered. Some silicone sealants are marketed for use in traffic-bearing joints, but most manufacturers limit the use of their products to nontraffic joints.

Silicone sealants are available in *high-*, *medium-*, and *low-modulus* formulations. Unfortunately, these terms are useful only in relative terms, since there are no universally recognized standards defining them. In general terms, *high-modulus* silicone sealants are to be used

where joint movement is 25% (each way) or less. They are intended for use in stopless glazing, curtain wall sealing, and structural glazing (see Section 8.8).

*Medium-modulus* silicone sealants can be used in the same applications as high-modulus silicone sealants, but they have a higher recovery capability and greater tear resistance. Some of them have demonstrated movement resistance capabilities far above ASTM C920's Class 25 category. Some can withstand movement as much as 50% each way.

Some *low-modulus* silicones appear to be able to withstand a joint movement equal to 100% each way without failure.

Both acid-curing and neutral-curing formulations are available. A major difference is that the acid-curing formulations have shown a tendency to not adhere well to certain substrates, such as marble, cement-based materials, copper, galvanized steel, and materials that corrode under the influence of the acids in the sealant.

## URETHANE SEALANTS

Urethane sealants rate highly in terms of weather resistance and durability. Their high performance level will persist for a considerable time, but they tend to harden with age, which eventually reduces their ability to withstand movement in the joints without failure. However, urethanes are available that show resistance to movement without failure of up to 50% each way. Formulations are available for use in vertical joints and others for use in horizontal ones. Some urethane sealants may be used underwater.

### 7.13.1.2 Solvent-Release-Curing Sealants

Solvent-release-curing sealants include acrylic, butyl, and pigmented small-joint sealants. These sealants are used primarily in joints that are fastened together (nonworking joints), such as those in building cladding.

## ACRYLIC

Solvent-release-curing acrylic sealants can be used in exterior joints, but are limited to use where the joints will not move more than $12^1/_2\%$ either way, and their performance is inferior to that of the elastomeric sealants discussed in Section 7.13.1.1. However, they are able to adhere to a great many surfaces, mostly without application of a primer and with little joint cleaning necessary. The applicable standards are ASTM C920 and the AAMA 800 Series, "Voluntary Specification and Test Methods for Sealants."

## BUTYL

Solvent-release-curing butyl sealants are somewhat more limited than solvent-release-curing acrylic sealants in that they are limited to a joint movement of $7^1/_2\%$ each way. Butyl sealants are covered by ASTM Standard C1311 and Federal Specification A-A-272A.

## SMALL-JOINT SEALANTS

Solvent-release-curing small-joint sealants are more limited still. They cannot be used in joints wider than $^3/_{16}$ in. (4.8 mm). Movement must be almost nil. Some of these sealants come in spray cans with small nozzles. Others can be applied to very small joints using needles.

### 7.13.1.3 Latex-Emulsion Sealants

Latex-emulsion sealants are usually used to seal small interior joints having little or no movement. They should comply with ASTM C834. These sealants are used to fill joints and crevices around doors and windows in wood, concrete, brick, and other masonry surfaces.

### 7.13.1.4 Tape Sealants

Tape sealants consist of a bead of butyl- or butyl-polyisobutylene-based sealant with a release paper on one side. Tapes may be reinforced with string or unreinforced. They are used for a variety of purposes, the majority being associated with glazing and concealed lap joints. Tape sealants should conform with the applicable portions of the AAMA 800 Series.

### 7.13.1.5 Acoustical Sealants

Acoustical sealants are used mostly in concealed locations to seal off sound transmission paths around partitions, electrical outlets, and other openings. They should be nondrying synthetic rubber formulations. No ASTM standard is available for acoustical sealant materials. Acoustical sealant application should be in accordance with ASTM C919.

### 7.13.1.6 Preformed Foam Sealants

Preformed polyurethane foam sealants are installed in joints in a compressed state and expand to fill the joint while remaining compressed to a specified degree.

There are no current standards covering preformed foam sealants. The manufacturer's instructions should be followed.

### 7.13.1.7 Oil-Based Caulking Compound

Oil-based caulking compounds have an oil or oleoresinous base and are used in relatively fixed joints subject to very limited movement. There is no ASTM standard for oil-based caulking compounds. The federal specification for them is FS A-A-272A.

Oil-based caulking compounds can be used to fill joints and crevices around doors and windows in wood, concrete, brick, and other masonry surfaces that are subject to little or no movement. They are supplied in a gun grade suitable for use with a caulking gun and a knife grade which can be spread with a putty knife. Oil-based caulking compounds tend to dry on the surface but remain soft and tacky inside. The exposed surface of the caulking bead should be painted each time the surrounding area is painted to help extend its life.

Other sealant types are preferred in most instances to oil-based caulking compounds because of their better adhesion and greater elasticity even at low temperatures. Although most are more expensive than oil-based caulking compounds, their greater reliability and durability are often worth the difference in cost.

### 7.13.2 PAVING JOINT SEALANTS

Because paving joints are exposed to sun, weather, ice-melting chemicals, and vehicular traffic, they are subject to more severe expansion and contraction conditions and wear than most building joints. For this reason, most sealants that will work quite well in building joints will not perform adequately in paving joints. Special sealants have been developed to handle such conditions.

These sealants fall into two categories: *cold-applied* and *hot-applied*. The one selected often depends on the amount of work to be done. Hot-applied sealants require the use of kettles and special installation equipment and, therefore, are usually restricted to those projects where there are many joints to be sealed, such as highway paving.

Several paving sealants are discussed in the following two subsections, but they are not the only ones currently available.

In addition, the list will surely grow longer in time and the products available will become better. There is a trend, for example, toward products with higher resistance to joint movement and abrasion, and the ability to withstand both higher and lower temperatures without loss of properties.

### 7.13.2.1 Cold-Applied Sealants

Cold-applied paving joint sealants include pourable elastomeric sealants complying with ASTM C920, which may be a jet-fuel-resistant sealant for concrete made with either urethane, coal-tar-modified polymer, or bitumen-modified urethane.

Jet-fuel-resistant, coal-tar-modified polysulfide and jet-fuel-resistant urethane sealants for concrete joints are also available that comply with FS SS-S-200E(2).

Cold-applied silicone paving joint sealants are also available, one for concrete-only joints and another for joints between concrete and asphalt. Both should comply with ASTM C920.

### 7.13.2.2 Hot-Applied Sealants

Hot-applied elastomeric paving joint sealants are available for concrete-to-concrete and concrete-to-asphalt joints. Jet-fuel-resistant formulations are also available for concrete-to-concrete and concrete-to-tar-concrete. Those for concrete-to-concrete joints should comply with ASTM D3406; for concrete-to-asphalt joints with ASTM D6690; for jet-fuel-resistant concrete-to-concrete joints to ASTM D3569; and for jet-fuel-resistant concrete-to-tar-concrete joints with ASTM D3581.

### 7.13.3 ACCESSORY MATERIALS

Accessories for sealed joints include joint backings and bond breakers for building sealants, and joint fillers and bond breakers for paving joints.

Joint backings and joint fillers serve as bond breakers, limit sealant depth to the desired one, and support the sealant during installation and tooling. In traffic joints, they also tend to support the sealant under applied loads.

### 7.13.3.1 Backings and Joint Fillers

Joint backings, also called *backer rods*, include plastic foams and elastomeric tubing. Foams may be made of either open-cell polyurethane, closed-cell polyethylene, or another plastic material that is acceptable by the sealant manufacturer. Tubing may be either neoprene, butyl, EPDM, or silicone complying with ASTM D1056. Closed-cell foams and elastomeric tubing tend to support a sealant better during installation and tooling. Open-cell foams are better able to accommodate variations in joint width and may help sealants applied over them cure faster. Elastomeric tubing can provide a temporary weather seal until the sealant has been applied and can act as a secondary seal after sealant application.

Joint fillers for use in joints in traffic surfaces include expanding cork, cork sponge rubber, bituminous-saturated fiberboard, and proprietary products that are composites of cork or fiberboard encased in organic or glass fiber felt.

Joint backing and joint filler materials must be compatible with the sealant and the substrate and should be acceptable to the sealant manufacturer. Backer rods and fillers for use with hot-applied sealants should be those specifically selected by the sealant manufacturer.

### 7.13.3.2 Bond Breakers

Bond breakers are polyethylene or other plastic tapes designed for installation in joints that have no backer rod or filler to prevent the sealant from bonding to the substrate at the back of the joint. The material used should be acceptable to the sealant manufacturer.

### 7.13.4 JOINT DESIGN

Proper joint design is essential in ensuring that sealants will perform satisfactorily.

### 7.13.4.1 Joint Types

Two joint types occur in buildings and pavements: *construction joints* and *separation joints*.

### CONSTRUCTION JOINTS

Construction joints that require sealants are planned joints that divide a building into buildable segments and prevent damage due to building movement after erection. They include *expansion joints* and *control joints*. Another construction joint type is used to interrupt pours of concrete and break slabs into panels for ease of construction (see Section 3.12), but these joints are seldom left open and do not usually require a sealant. When they do

require a sealant, the principles discussed here apply.

*Expansion joints* are relatively large joints intended to permit a structure to expand and contract without damage to the structure. These are sometimes called *isolation joints*. They usually pass completely through the structure and its finishes.

*Control joints* are relatively small joints used to divide thin materials such as pavements, floors, walls, and finish materials such as ceramic tile into panels of a size that will not be damaged by thermal or other movement.

### SEPARATION JOINTS

*Separation joints* are joints between materials, such as between windows or door frames and adjacent materials, or between two portions of the same material, such as in sheet metals. These joints should be tightly sealed in all climates, whether or not the need for cooling or heating is predominant. To achieve a tight seal around windows and doors, a sealant should be placed at the joint between their frames and the adjoining walls.

### 7.13.4.2 Joint Movement

Joints can be classified as either dynamic or static. *Dynamic joints*, which are also called *movement joints* or *working joints*, are joints that move with changes in temperature or other building movements. This includes all construction joints and some, but not all, separation joints. Dynamic joints in exterior walls are sealed to prevent air, water, and sound passage through them. Dynamic joints in both interior and exterior horizontal and vertical surfaces are sealed to prevent light, air, water, sound, and vermin passage through them and to make them more esthetically pleasing.

*Static joints* have little or no movement. They are also called *nonmovement joints* or *nonworking joints*. They include such joints as those between interior door frames and adjacent masonry, plaster, or gypsum board. Static joints are usually sealed to prevent light, air, and sound passage through them and to provide a base for painting.

### 7.13.4.3 Joint Design Procedure

Poor joint design is a major cause of joint sealant failures. The first step in designing dynamic joints is to estimate the amount of movement that will occur

across them. To do this, it is necessary to determine the expected temperature extremes and the coefficient of expansion of the materials associated with the joint. Using these data, it is possible to calculate the amount of theoretical movement. Actual movement, however, is very difficult to determine accurately because of many imponderables, such as the exact thermal movement that will occur when more than one material is involved and construction tolerances that result in variable joint widths. Therefore, it is necessary to determine an approximate anticipated movement range and select a joint width that will accommodate this movement. The temperature at the time a sealant will be applied in a joint must be taken into account in joint width selection to ensure that the joint will be of the necessary size at that time. It is usual to assume a temperature range of between 40 and 100°F (4.4 and 37.8°C) at the time the sealant will be installed and select the joint width so that it will work over that entire range.

Two basic joint types are used in buildings: *butt joints* and *shear joints* (Fig. 7.13-2). Shear joints protect the sealant and place less strain on the sealant. However, they are difficult to design, and their appearance may not always be desirable. In addition, it is difficult to prepare them to receive sealant and to install the sealant. Shear joints are sometimes used in such applications as sheet metal joints, certain joints in curtain walls, and articulated stone walls.

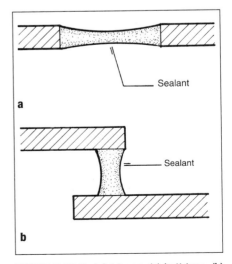

**FIGURE 7.13-2** Joint types: (a) butt type, (b) shear type. (Drawing by HLS.)

For both dynamic and static joints, the minimum joint width at any time during or after sealant application must be greater than the selected sealant's minimum performance width. For most sealants, this is $1/4$ in. (6.4 mm), but each sealant's minimum performance width should be determined from that sealant's manufacturer and not assumed. Small joint sealants, for example, cannot be used in joints wider than $3/16$ in. (4.8 mm). Their minimum joint widths are sometimes very small, but are still significant in design and sealant selection.

Joints that are wider than the limitation of a sealant should be avoided. It is better to reduce joint width by providing more joints, but if this is impracticable or undesirable, the joints should be protected by metal plates or preformed elastomeric and metal expansion joints. This is applicable both for joints in traffic surfaces and for vertical joints.

The depth of the space allotted for a sealant must be such that the sealant has the proper shape. If a sealant is too thin, it may tear when it expands to fill a joint. If a sealant is too thick, it may fail to compress properly and may protrude from the joint or even be squeezed out of it. Under normal conditions, the thickness should be one-half the width, but not less than $1/4$ in. (6.4 mm) or more than $1/2$ in. (12.7 mm). This rule of thumb limits the desirable width of joints to be closed with a sealant to 1 in. (25.4 mm).

## 7.13.5 SEALANT SELECTION

A sealant should be selected that will provide the necessary esthetic and functional performance. Some sealants are available in many colors, others only in white or black, and still others are clear. Joint type and size are prime considerations when selecting a sealant. The sealant must be capable of performing within the confines of the joint in which it is installed. For example, high-modulus silicone sealants should not be used in joints subject to movement of more than $+25\%$ or $-25\%$. Narrow joints require sealants specifically formulated for them.

Sealants must also be selected to suit the joint location. Traffic joints demand self-leveling sealants; vertical joints, nonsag ones. Traffic joints also require sealants with a higher hardness number. Traffic sealants should have a hardness of at least 25 but not more than 50 when tested in accordance with ASTM C920.

Special use conditions must also be considered when selecting a sealant. Special conditions include the presence of high temperatures, the need for a sealant to be submersed, or the presence of vehicular traffic.

Selecting elastomeric sealants using only the ASTM C920 classification system is possible but often impracticable. A perfectly good sealant product may have slightly different characteristics than those required by the ASTM C920 system. In addition, the ASTM C920 system is not a completely adequate gauge of sealant performance. Products that exceed its limits should not be rejected on that basis alone. A product that has the desired features for a particular project may be selected even if it does not comply with another aspect of the ASTM C920 system. The usual practice is to select specific proprietary sealant products. When competition is required, several comparable products may be specified.

## 7.13.6 INSTALLATION

The sealant manufacturer's instructions should be followed when preparing for and installing sealants.

### 7.13.6.1 Joint Preparation

Cleaning of surfaces to receive sealants is essential. There must be no contaminants of any kind in joints to receive sealants. This includes, but is not limited to, incompatible paint and other coatings, rust and other corrosion, mill scale, water, frost, grease, oil, laitance, and form release agents. Joints to receive sealant must also be dry and free from dust and grit. These should be removed using dry compressed air or brushes. The sealant manufacturer's instructions should be followed in determining if a particular substrate is compatible with its sealant, and in cleaning and treating such substrates as rubber, plastics, wood, bituminous materials, water-repellent coatings, ceramic materials, glass, and paints and other decorative coatings, chemical cleaners and solvents must not leave a residue. A corrosion-resistant coating may be used on corrodible joint surfaces, but it must be compatible with the sealant and dry when the sealant is applied.

Joints should have no protrusions into them, such as aggregate projecting from concrete, and should be the same width throughout. Projections should be ground

down. Recesses should be filled solidly and permanently.

Joints that are too narrow should be cleaned out or cut wider, using power tools to produce the proper width. Under no circumstances should sealant be placed in a joint that is more narrow than the minimum performance width of the sealant.

Deep joints should be filled with a joint filler, backer rod, or both to create a sealant pocket of the correct depth (Fig. 7.13-3).

When a backer rod is not installed, a bond breaker tape should be placed over the joint filler or in the back of a joint, as applicable, to prevent the sealant from adhering there (Fig. 7.13-4). If a sealant adheres to the back of a joint, it will not be able to contract properly when it stretches with an expanding joint, and will tear apart internally or separate from the sides of the joint.

When the sealant manufacturer recommends doing so, joints to receive a sealant

**FIGURE 7.13-3** A backer rod should be used to control joint depth. Sealant depth should not be more than one-half the joint width. (Drawing by HLS.)

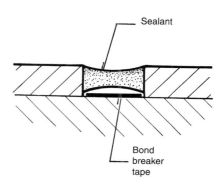

**FIGURE 7.13-4** Bond breaker tape is necessary when the joint depth is shallow. (Drawing by HLS.)

should be primed using a primer recommended by the sealant manufacturer.

Before sealants are applied, adjacent materials should be protected with tape or other means to ensure that sealant does not contaminate them and that the lines of an unconfined sealant run straight and true.

### 7.13.6.2 Sealant Application

In general, sealants should be applied in accordance with their manufacturer's recommendations in such a manner that they completely fill the joint and completely *wet* the sides of the joint. Hot-pour sealants should be melted within the manufacturer's recommended temperature limits.

Sealants should be confined within a joint when possible. If they finish below the face of the adjoining materials (Fig. 7.13-5), they will be less subject to damage from contact and deterioration by being somewhat shielded from the sun and weathering.

When a sealant must be applied as a triangular fillet seal (Fig. 7.13-6), it should be applied so that the fillet is not less than $^3/_8$ in. (9.5 mm) across the face. The face of a fillet seal should be convex to ensure an adequate volume of sealant.

Mixing procedures for multicomponent sealants should be those recommended by the sealant manufacturer. Care should be taken to prevent raising the temperature of cold-applied sealants during mixing and to prevent air from becoming trapped in the sealant.

Application devices should be those recommended by the sealant manufacturer. Nonsag sealants should be extruded into a joint so that they are forced against the sides. Caulking guns are used to do this with most sealants. Figure 7.13-7 shows a hand-operated caulking gun used with one-part sealants. Similar hand-operated guns (Fig. 7.13-8) and pneumatic guns (Fig. 7.13-9) are used to apply multicomponent bulk sealants.

**FIGURE 7.13-6** A triangular fillet seal. If the joint is more than $^3/_{16}$ in. (4.8 mm) wide, a bond breaker tape should be used to prevent sealant from adhering to the sides of the joint. (Drawing by HLS.)

**FIGURE 7.13-7** A hand-operated caulking gun for premixed sealants. (Courtesy National Institute of Standards and Technology.)

Hot-pour sealants are placed through a pouring nozzle of a size that just fits in the joint. Cold-pour sealants can be poured directly from the mixing vessel or placed using a special pouring device.

Tape sealants are placed in open joints by first removing the paper slip sheet and laying the tape in the joint. The second layer of substrate is then installed and pressed into the tape sealant.

Preformed foam sealants are forced into a joint and allowed to expand to fill it. Some preformed strips are heated after they have been inserted in a joint to cause them to achieve continuous contact with the joint sides.

Care must be taken with both tape sealants and preformed tape sealants to ensure that they make positive contact where sections join.

### 7.13.6.3 Tooling

Nonsag sealants should be tooled before *skinning* (curing of the top layers of sealant) occurs. Tooling should be done using the devices recommended by the sealant manufacturer. Tooling should compact the sealant and force it into contact with the sides of the joint. The surface of tooled joints should be left slightly concave.

### 7.13.6.4 Protection

Sealants must be protected from contact with people and traffic until they completely cure.

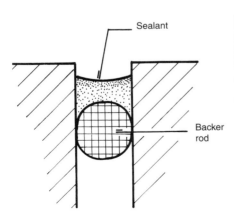

**FIGURE 7.13-5** Sealant depressed below the surface. (Drawing by HLS.)

**FIGURE 7.13-8** A hand-operated caulking gun for bulk sealants. (Courtesy Sealant Waterproofing and Restoration Institute.)

**FIGURE 7.13-9** A pneumatic caulking gun for bulk sealants. (Courtesy Sealant Waterproofing and Restoration Institute.)

# 7.14 Additional Reading

More information about the subjects discussed in this chapter can be found in the references listed in Section 7.15 and in the following publications.

Adler, David L.1994. "Sloped Roof Design Principles." *The Construction Specifier* (July): 101 (www.csinet.org).

Adler, David L. and Carl G. Cash. 1995. "Asphalt Shingles: A Primer." *The Construction Specifier* (November): 24 (www.csinet.org).

Ahoph, Arnold J. 1988. "Traditional Building Methods; Moisture Control." *Custom Builder* (February): 9–15 (www.housingzone.wm).

Amstock, Joseph S. 2001. *Handbook of Adhesives and Sealants in Construction.* New York: McGraw-Hill (www.mcgraw-hill.com).

Anis, Wagdy. 2002. "Insulation Strategies for Exterior Walls." *The Construction Specifier* (August): 40 (www.csinet.org).

Aulisio, Lorraine M. 1996. "Foam Plastic Insulation: Building Codes and Fire Testing." *The Construction Specifier* (October): 68 (www.csinet.org).

Bakus, Jason. 2005. "Pre-Cured Silicon Sealant Advantages." *The Construction Specifier* (April): 80 (www.csinet.org).

Buchanan, Robert, Jason Pitcole, and Steve Tyler. 1997. "Specifying Firestopping to Save Lives." *The Construction Specifier* (June): 40 (www.csinet.org).

Byrd, Stacy. 2004. "Specifying Bentonite Waterproofing." *The Construction Specifier* (June): 36 (www.csinet.org).

Cash, Carl G. 1997. "Asphalt Roof Shingle Performance." *The Construction Specifier* (November): 43 (www.csinet.org).

———. 2000. "Selecting the Appropriate Roofing System." *The Construction Specifier* (January): 25 (www.csinet.org).

Ceruti, Jeffry J. 1997. "Metal Roofing Systems: Dos and Don'ts." *The Construction Specifier* (September): 64 (www.csinet.org).

———. 2001. "Metal Roofing Systems." *The Construction Specifier* (October): 39 (www.csinet.org).

Council on Tall Buildings and Urban Habitat. 1992. *Fire Safety in Tall Buildings.* New York: McGraw Hill (www.mcgraw-hill.com).

*Custom Builder.* 1988. "Slate Roofs; State of the Industry." *Custom Builder* (September): 29–31 (www.housingzone.com).

Crosbie, Michael J. 1997. "A Chip Off the Old Slate." *The Construction Specifier* (November): 49 (www.csinet.org).

D'Annunzio, John A. 2004. "Proper Roof System Design Standards." *The Construction Specifier* (January): 34 (www.csinet.org).

Davis, Richard. 2001. "Preserving the Protective Shell." *The Construction Specifier* (September): 26 (www.csinet.org).

Doelp, Gregory R. 1998. "Rooftop Penetrations: Round Penetrations Are Easier to Flash." *The Construction Specifier* (May): 9 (www.csinet.org).

———. 1999. "Wall Leakage Disguised as Roof Problems." *The Construction Specifier* (June): 59 (www.csinet.org).

Egan, William F. 1993. "Specifying EIFS." *The Construction Specifier* (December): 66 (www.csinet.org).

———. 1995. "Specifying Premium Class PB EIFS." *The Construction Specifier* (February): 21 (www.csinet.org).

Gansen, Neal E. 1988. "Avoiding Roof Ice Dams." *Custom Builder* (November): 35–36 (www.housingzone.com).

Gill, Cris, and Chad Stroike. 2001. "Specifying Firestop Systems for Building Perimeter Joints." *The Construction Specifier* (May): 23 (www.csinet.org).

Haddock, Rob. 1994. "Metal Roofing Tips." *The Construction Specifier* (April): 50 (www.csinet.org).

Hardy, Steve. 1996. "Metal Roofing Fundamentals." *The Construction Specifier* (November): 26 (www.csinet.org).

Haws, Richard B. 1993. "Metal Roofing: The Environmental Choice." *The Construction Specifier* (November): 107 (www.csinet.org).

Henshell, Justin. 1995. "Watertight Details: A Checklist." *The Construction Specifier* (May): 125 (www.csinet.org).

Hering, Achim. 1995. "The Problem with Intumescents." *The Construction Specifier* (September): 80 (www.csinet.org).

Ingalls, Kelly McArthur. 1997. "In Search of Green Building Insulation."

*The Construction Specifier* (August): 26 (www.csinet.org).

———. 2000. "Sustainable Aspects of Thermal and Moisture Protection Products: New Options." *The Construction Specifier* (July): 59 (www.csinet.org).

Kalwara, Joseph J. and Chester T. Chmiel. 1996. "Seaming Technology for EPDM Roofing Membranes." *The Construction Specifier* (November): 40 (www.csinet.org).

Klamke, Stephan E. 1997. "Exterior Insulation and Finish Systems: Proven Performance." *The Construction Specifier* (December): 11 (www.csinet.org).

Klein, Kenneth A. and Daniel G. Gibbons. 2001. "Don't Get Blindsided: Techniques for Blind-Side Waterproofing." *The Construction Specifier* (April): 35 (www.csinet.org).

Kluttz, Robert Q. 2001. "Polymer Modified Asphalt in Roofing." *The Construction Specifier* (December: 41 (www.csinet.org).

Kroll, Richard E. and Steven J. Collins. 1993. "Pressure-Equalized EIFS." *The Construction Specifier* (December): 69 (www.csinet.org).

Kubal, Michael. 1993. *Waterproofing the Building Envelope.* New York: McGraw-Hill (www.mcgraw-hill.com).

———. 2000. *Construction Waterproofing Handbook.* New York: McGraw-Hill (www.mcgraw-hill.com).

Levine, Jeffrey S. 1994. "Historic Slate Roofs." *The Construction Specifier* (July): 86 (www.csinet.org).

———. 1995. "What Kind of Slate Is That?" *The Construction Specifier* (November): 38 (www.csinet.org).

Lstiburek, Joseph and John Carmody. 1999. *Moisture Control Handbook: Principles and Practices for Residential and Small Commercial Buildings.* New York: John Wiley & Sons, Inc. (www.wiley.com).

Luft, Rene W. and Richard C. Schroter. 1996. "Wall Leaks: Lessons Learned from Sealant Failure Investigations." *The Construction Specifier* (July): 84 (www.csinet.org).

Mackinlay, Ian, and Richard S. Flood. 2000. "Roof Slopes, Ice Dams, and Cold Roofs." *The Construction Specifier* (January): 31 (www.csinet.org).

Nicastro, David H. 1991. "Water Vapor Transmission." *The Construction Specifier* (June): 76 (www.csinet.org).

———. 1994. "12,000 Points of Water

Entry." *The Construction Specifier* (July): 160 (www.csinet.org).

O'Brien, Sean M., and Stephen J. Condren. 2002. "Design Guidelines for Vapor Retarders in Warm Climates." *The Construction Specifier* (December): 42 (www.csinet.org).

O'Connor, Daniel J. 1993. "Fire Stops and Fire Safety." *The Construction Specifier* (September): 136 (www.csinet.org).

O'Connor, Jerome P. 1997. "Exterior Insulation and Finish Systems: Obvious Weaknesses." *The Construction Specifier* (December): 11 (www.csinet.org).

Peterson, Wayne and Ron Deibel. 2000. "Understanding Polystyrene Insulation." *The Construction Specifier* (March): 47 (www.csinet.org).

Petrie, Edward. 2000. *Handbook of Adhesives and Sealants.* New York: McGraw-Hill (www.mcgraw-hill.com).

Quirouette, Rick. 1994. "The Rainscreen Roof System." *The Construction Specifier* (October): 51 (www.csinet.org).

Raeber, John A. 1990. "Deciding When to Use Negative-Side Waterproofing." *The Construction Specifier* (December): 35 (www.csinet.org).

———. 1991. "Steep Roof Systems." *The Construction Specifier* (November): 19 (www.csinet.org).

———. 1992. "Preformed Metal Roofing Systems." *The Construction Specifier* (April): 23 (www.csinet.org).

———. 1992. "Roofing and Waterproofing Old or New?" *The Construction Specifier* (June): 29 (www.csinet.org).

———. 1992. "Asphalt Shingle Roofing." *The Construction Specifier* (October): 25 (www.csinet.org).

———. 1995. "Clay Tile Roofing." *The Construction Specifier* (March): 86 (www.csinet.org).

Reid, Robert N. 2000. "Penetrations Through the Roofing System Membrane." *The Construction Specifier* (October): 27 (www.csinet.org).

Remmele, Thomas E. 1993. "Detailing Class PB EIFS." *The Construction Specifier* (December): 78 (www.csinet.org).

Roodvoets, David L. 2000. "Specifying Wind Resistance in Shingles." *The Construction Specifier* (October): 33 (www.csinet.org).

Ruggiero, Stephen S. and Dean A. Rutila. 1991. "Plaza Waterproofing Design Fundamentals." *The Construction Specifier* (January): 74 (www.csinet.org).

Rush, Richard D. 2000. "The Future of Fire: Performance-Based Codes." *The Construction Specifier* (January): 39 (www.csinet.org).

Sather, Lloyd. 1992. "Firestopping." *The Construction Specifier* (April): 111 (www.csinet.org).

Schade, Anne L. 2004. "Specifying and Designing Copper Roofs." *The Construction Specifier* (December): 28 (www.csinet.org).

Scharff, Robert. 1995. *Roofing Handbook.* New York: McGraw-Hill (www.mcgraw-hill.com).

Schroeder, Edward K. 1993. "Regarding Roof Edges." *The Construction Specifier* (November): 40 (www.csinet.org).

Schroter, Richard C. 1994. "Metal Roof Specifications." *The Construction Specifier* (August): 76 (www.csinet.org).

Seeley, James I. 1997. "The Protocols of White Roofing." *The Construction Specifier* (November): 35 (www.csinet.org).

Serino, Roland and Tony Stall. "EIFS and Pressure-Equalized Rainscreens." *The Construction Specifier* (March): 28 (www.csinet.org).

Sievert, Norman. 1988. "Wood Siding Basics." *Custom Builder* (February): 16–20 (www.housingzone.com).

Simmons, H. Leslie. 1999. "Architectural Standing Seam Metal Roofing." *The Construction Specifier* (February): 49 (www.csinet.org).

Sobol, John. 2001. "Understanding Fire Barriers for Expansion Joints." *The Construction Specifier* (September): 33 (www.csinet.org).

Spencer, William. 1996. "Achieving Lasting Impermeability with a Roof." *The Construction Specifier* (November): 32 (www.csinet.org).

Sterling, Walter P. 1996. "Preserving Firestopping Systems." *The Construction Specifier* (July): 26 (www.csinet.org).

Sternthal, Daniel. 2000. "Copper Flashings in Contemporary Construction." *The Construction Specifier* (October): 21 (www.csinet.org).

Subasic, Christine A., Brian E. Trimble, and Janet Peddycord. 1996. "Water Repellents for Brick Walls." *The Construction Specifier* (February): 23 (www.csinet.org).

Thomas, Robert, Jr. 1995. "Getting the Best from EIFS." *The Construction Specifier* (February): 18 (www.csinet.org).

Trial, Tim. 2005. "EPDM Roofing Membranes and Long-Term Performance." *The Construction Specifier* (February): 24 (www.csinet.org).

Truesdell, David E. 1998. "Rooftop Penetrations: Odd-Shape Penetrations Can Be Specified with Confidence." *The Construction Specifier* (May): 9 (www.csinet.org).

Tuluca, Adrian. 1996. "Thermal Bridges in Buildings." *The Construction Specifier* (October): 38 (www.csinet.org).

Verrastro, Salvatore B. and Roy Schauffele. 2001. "Foamed Plastic Insulation Basics." *The Construction Specifier* (April): 43 (www.csinet.org).

Weaver, Martin E. 1989. "The Basics of Slate Roof Restoration." *The Construction Specifier* (November): 49 (www.csinet.org).

———. 1991. "Copper Top." *The Construction Specifier* (November): 40 (www.csinet.org).

Wimberly, David. 1995. "Thermoplastic Roofing Systems." *The Construction Specifier* (November): 32 (www.csinet.org).

## 7.15 Acknowledgments and References

### ACKNOWLEDGMENTS

We gratefully acknowledge the assistance of the following organizations and individuals in preparing this chapter. We are also indebted to them for permission to use their illustrations when requested and for the use of their publications as references.

ABTCO Vinyl Siding (www.abtco.com)
Airline Products Company (www.nystrom.com)
Alcoa Building Products (www.alcoa-homes.com)
American Architectural Manufacturers Association (AAMA) (www.aamanet.org)
American Board Products Association
American National Standards Institute (ANSI) (www.ansi.org)
American Society of Heating, Refrigerating and Air Conditioning Engineers (ASHRAE) (www.ashrae.org)

Arcom Master Systems (www.arcomnet. com)

Asphalt Institute (www.asphaltinstitute. org)

Asphalt Roofing Manufacturers Association (www.asphaltroofing.org)

ASTM International (www.astm.org)

Brick Industry Association (BIA) (www. bia.org)

Carlisle SynTec (www.carlisle-syntec. com)

Cedar Shake and Shingle Bureau (www. cedarbureau.org)

Celotex Corporation (www.calotex.com)

CertainTeed (www.certainteed.com)

Coastal Atlantic Associates, Inc.

Concrete Block Insulating Systems, Inc. (www.cbisinc.com)

Copper Development Association (CDA), Inc. (www.copper.org)

Dow Chemical Company (www.dow. com)

E/S Products, Inc.

Firestone Building Products Company (www.firestonebpco.com)

Georgia Pacific Corporation (www.gp. com)

Goodyear Tire and Rubber Company (www.goodyear.com)

House and Home Magazine (www. canadianhouseandhome.com).

Independent Nail Corporation

John Manville Corporation (www.jm.com)

Koppers, Inc. (www.koppers.com)

Lead Industries Association, Inc. (www. leadinfo.com)

W. H. Maze Company (www.mazenails. com)

Metal Sales Manufacturing Company (www.metalsales.com).

Mineral Fiber Products Bureau

Mineral Insulation Manufacturers Association

National Association of Home Builders (NAHB) Research Center (www.nahbrc.org)

National Institute of Building Sciences (NIBS) (www.nibs.org)

National Institute of Standards and Technology (NIST) of the U.S. Department of Commerce (www.nist.gov)

National Mineral Wool Insulation Association

National Roofing Contractors Association (NRCA) (www.nrca.net)

National Slate Association (NSA) (slate-association.org)

National Tile Roofing Manufacturers Association (NTRMA) (www.ntrma.org)

Nickel Institute (www.nickelinstitute. org)

Owens-Corning Fiberglas Corporation (www.owenscorning.com)

Perlite Institute (www.perlite.org)

Portland Cement Association (PCA) (www.portcement.org)

Sealant Engineering and Associated Lines (SEAL), Inc. (www.sealgroup. org)

Sealant, Waterproofing, and Restoration Institute (www.swrionline.org)

Sheet Metal and Air Conditioning Contractors National Association (SMACNA) (www.smacna.org)

Single Ply Roofing Institute (SPRI) (www.spri.org)

Solar Group, Inc. (www.thesolargroup. com)

Specialty Steel Industry of North America (www.ssina.com)

Steel Recycling Institute (SRI) (www. recycle-steel.org)

Underwriters Laboratories, Inc. (www. ul.com)

U.S. Department of Agriculture, Forest Service, Forest Products Laboratory (www.usda.gov)

U.S. Department of the Army, Corps of Engineers (www.usace.army.mil)

U.S. Department of Housing and Urban Development (HUD) (www.hud. gov)

U.S. Weather Service (www.nws.noaa. gov)

Vinyl Siding Institute (VSI) (www.vinyl-siding.org)

Wolverine Vinyl Siding (www.vinylsiding. com)

## REFERENCES

We would also like to thank the authors and publishers of the publications in the following list for their contribution to our research for this chapter.

Air-Conditioning and Refrigeration Institute (ARI) (www.ari.org), Sheet Metal and Air Conditioning Contractors National Association (SMACNA) (www. smacna.org), and National Roofing Contractors Association (NRCA) (www.nrca.net), *Roof Mounted Outdoor Air-Conditioner Installations*.

Air Movement and Control Association International, Inc. (AMCA). Arlington Heights, IL (www.amca.org). 2003. *Application Manual for Air Louvers.*

———. 1999. *Test Methods for Louvers, Dampers, and Shutters.*

———. 1999. *Laboratory Methods of Testing Louvers for Rating.*

Alcoa Building Products. 1990. *Vinyl Siding, Soffit, and Designer Accessories Installation Manual."* Sidney, OH: Alcoa Building Products (www. alcoahomes.com).

Aluminum Association (AA). AA DAF-45, "Designations from Start to Finish." Washington, DC: AA (www. aluminum.org).

American Architectural Manufacturers Association (AAMA). Des Plaines, IL (www.aamanet.org).

1986. AAMA 1402. "Standard Specifications for Aluminum Siding, Soffit, and Fascia."

———. 1992. AAMA 800. "Voluntary Specifications and Test Methods for Sealants."

American National Standards Institute/ Exterior Insulation Manufacturers Association (ANSI/EIMA). ANSI/ EIMA 99-A, "Exterior Insulation and Finish Systems (EIFS). Washington, DC: ANSI (www.ansi.org) and EIMA (www.eima.com).

American Society of Heating, Refrigerating and Air-Conditioning Engineers (ASHRAE). 2001. *ASHRAE Handbook—Fundamentals.* Atlanta, GA: ASHRAE (www.ashrae.org).

*Architectural Graphic Standards. See* Ramsey et al. (2000).

Arcom Master Systems. Salt Lake City, UT (www.arcomnet.com). *Masterspec®.* Basic Sections:

07115, "Bituminous Dampproofing," 2/00 ed.

07131, "Self-Adhering Sheet Waterproofing," 5/04 ed.

07132, "Elastomeric Sheet Waterproofing," 5/04 ed.

07133, "Thermoplastic Sheet Waterproofing," 5/04 ed.

07141, "Cold Fluid-Applied Waterproofing," 2/99 ed.

07142, "Hot Fluid-Applied Waterproofing," 2/99 ed.

07163, "Metal-Oxide Waterproofing," 5/02 ed.

07170, "Bentonite Waterproofing," 8/02 ed.

07190, "Water Repellents," 11/02 ed.

07210, "Building Insulation," 8/03 ed.

07240, "Exterior Insulation and Finish Systems," 11/00 ed.

07241, "Exterior Insulation and Finish Systems—Class PB," 11/00 ed.

07242, "Exterior Insulation and Finish Systems—Class PM," 11/00 ed.

07311, "Asphalt Shingles," 11/00 ed.

07313, "Metal Shingles," 2/02 ed.

07315, "Slate Shingles," 5/01 ed.

07317, "Wood Shingles and Shakes," 2/01 ed.

07320, "Roof Tiles," 5/01 ed.

07321, "Clay Roof Tiles," 5/01 ed.

07322, "Concrete Roof Tiles," 5/01 ed.

07411, "Metal Roof Panels," 8/00 ed.

07412, "Metal Wall Panels," 8/00 ed.

07460, "Siding," 5/01 ed.

07510, "Built-Up Bituminous Roofing," 2/00 ed.

07511, "Built-Up Asphalt Roofing," 2/00 ed.

07512, "Built-Up Coal-Tar Roofing," 2/00 ed.

07530, "Elastomeric Membrane Roofing," 11/99 ed.

07531, "Epdm Membrane Roofing," 11/99 ed.

07532, "CSPE Membrane Roofing," 11/99 ed.

07540, "Thermoplastic Membrane Roofing," 2/00 ed.

07550, "Modified Bituminous Membrane Roofing," 5/00 ed.

07553, "Self-Adhering Modified Bituminous Membrane Roofing," 8/99 ed.

07561, "Hot Fluid-Applied Roofing," 8/99 ed.

07610, "Sheet Metal Roofing," 8/00 ed.

07620, "Sheet Metal Flashing and Trim," 11/00 ed.

07710, "Manufactured Roof Specialties," 11/01 ed.

07716, "Roof Expansion Assemblies," 11/03 ed.

07720, "Roof Accessories," 11/02 ed.

07811, "Sprayed Fire-Resistive Materials," 5/04 ed.

07821, "Board Fire Protection," 2/02 ed.

07841, "Through-Penetration Firestop Systems," 11/02 ed.

Asphalt Roofing Manufacturers Association (ARMA). 1990. *Built-Up Roofing Systems Design Guide*. Washington, DC: ARMA (www.asphaltroofing.org).

ASTM International (ASTM). West Conshohocken, PA (www.astm.org). Standards:

A167, "Standard Specifications for Stainless and Heat-Resisting Chromium-Nickel Steel Plate, Sheet, and Strip."

A463, "Standard Specification for Sheet Steel, Cold-Rolled Aluminum-Coated by the Hot-Dip Process."

A591, "Standard Specifications for Steel Sheet, Electrolytic Zinc-Coated for Light Coating Mass Applications."

A606, "Standard Specifications for Steel Sheet and Strip, High Strength, Low-Alloy, Hot-Rolled and Cold-Rolled, with Improved Atmospheric Corrosion Resistance."

A653/A653M, "Standard Specification for Steel Sheet, Zinc-Coated (Galvanized) or Zinc-Iron Alloy-Coated (Galvannealed) by the Hot-Dip Process."

A755/A755M, "Standard Specification for Steel Sheet, Metallic Coated by the Hot-Dip Process and Prepainted by the Coil-Coating Process for Exterior Exposed Building Products."

A792/A792M, "Standard Specification for Steel Sheet, 55% Aluminum-Zinc Alloy-Coated by the Hot-Dip Process."

A875, "Standard Specification for Steel Sheet, Zinc 5%, Aluminum Alloy, Coated by the Hot Dip Process."

B32, "Standard Specification for Solder Metal."

B101, "Standard Specifications for Lead-Coated Copper Sheet and Strip for Buildings."

B209, "Standard Specification for Aluminum and Aluminum-Alloy Sheet and Plate."

B210, "Standard Specification for Aluminum and Aluminum-Alloy Drawn Seamless Tubes."

B211, "Standard Specification for Aluminum and Aluminum-Alloy Bar, Rod, and Wire."

B221, "Standard Specification for Aluminum and Aluminum-Alloy Extruded Bars, Rods, "Wire, Profiles, and Tubes."

B370, "Standard Specifications for Copper Sheet and Strip for Building Construction."

B429, "Standard Specification for Aluminum-Alloy Extruded Structural Pipe and Tube."

B749, "Standard Specifications for Lead Alloy Strip, Sheet, and Plate products."

C150, "Standard Specification for Portland Cement."

C208, "Standard Specification for Cellulosic Fiber Insulating Board."

C209, "Standard Test Methods for Cellulosic Fiber Insulating Board."

C270, "Standard Specification for Mortar for Unit Masonry."

C481, "Standard Test Method for Laboratory Aging of Sandwich Construction."

C516, "Standard Specification for Vermiculite Loose Fill Thermal Insulation."

C549, "Standard Specification for Perlite Loose Fill Insulation."

C552, "Standard Specification for Cellular Glass Thermal Insulation."

C553, "Standard Specification for Mineral Fiber Blanket Thermal Insulation for Commercial and Industrial Applications."

C578, "Standard Specifications for Rigid Cellular Polystyrene Thermal Insulation."

C591, "Standard Specification for Unfaced Preformed Rigid Cellular Polisocyanurate Thermal Insulation."

C612, "Standard Specification for Mineral Fiber Block and Board Thermal Insulation."

C665, "Standard Specification for Mineral Fiber Blanket Thermal Insulation for Light Frame Construction and Manufactured Housing."

C687, "Standard Practice for Determination of Thermal Resistance of Loose-fill Building Insulation."

C726, "Standard Specification for Mineral Fiber Roof Insulation Board."

C728, "Standard Specification for Perlite Thermal Insulation Board."

C764, "Standard Specification for Mineral Fiber Loose-fill Thermal Insulation."

C834, "Standard Specification for Latex Sealants."

C919, "Standard Practice for Use of Sealants in Acoustical Applications."

C920, "Standard Specification for Elastomeric Joint Sealants."

C1126, "Standard Specification for Faced or Unfaced Rigid Cellular Phenolic Thermal Insulation."

C1193, "Standard Guide for use of Joint Sealants."

C1225, "Standard Specification for Non-Asbestos Fiber-Cement Roofing Shingles, Shakes, and Slates."

C1289. "Standard Specification for Faced Preformed Rigid Cellular Polysocyanurate Thermal Insulation Board."

C1311, "Standard Specification for Solvent Release Sealants."

C1382, "Test Method for Determining Tensile Adhesion Properties of Sealants When Used in Exterior Insulation and Finish Systems (EIFS) Joints."

C1397, "Standard Practice for Application of Class PB Exterior Insulation and Finish Systems."

C1481, "Standard Guide for Use of Joint Sealants with Exterior Insulation and Finish Systems (EIFS)."

C1530, "Standard Specification for Non-Asbestos Fiber-Cement Roofing Shakes, Shingles, and Slates with Designed Varying Profiles and Thicknesses."

C1535, "Standard Practice for Application of Exterior Insulation and Finish Systems Class PI."

D41, "Standard Specification for Asphalt Primer Used in Roofing, Dampproofing, and Waterproofing."

D43, "Standard Specification for Coal Tar Primer Used in Roofing, Dampproofing, and Waterproofing."

D173, "Standard Specification for Bitumen-Saturated Cotton Fabrics Used in Roofing and Waterproofing."

D225, "Standard Specification for Asphalt Shingles (Organic Felt) Surfaced with Mineral Granules."

D226, "Standard Specification for Asphalt-Saturated Organic Felt Used in Roofing and Waterproofing."

D227, "Standard Specification for Coal-Tar Saturated Organic Felt Used in Roofing and Waterproofing."

D312, "Standard Specification for Asphalt Used in Roofing."

D449, "Standard Specification for Asphalt Used in Dampproofing and Waterproofing."

D450, "Standard Specification for Coal-Tar Pitch Used in Roofing, Dampproofing, and Waterproofing."

D1056, "Standard Specification for Flexible Cellular Material—Sponge or Expanded Rubber."

D1079, "Standard terminology Relating to Roofing, Waterproofing, and Bituminous Materials."

D1227, "Standard Specification for Emulsified Asphalt Used as a Protective Coating for Roofing."

D1668, "Standard Specification for Glass Fabrics (Woven and Treated) for Roofing and Waterproofing."

D1863, "Standard Specification for Mineral Aggregate Used on Built-up Roofs."

D970, " Standard Specification for Self-Adhering Polymer Modified Bituminous Sheet Materials Used as Steep Roofing Underlayment for Ice Dam Protection."

D2178, "Standard Specification for Asphalt Glass Felt Used in Roofing and Waterproofing."

D2626, "Standard specification for Asphalt-saturated and Coated Organic Felt Base Sheet Used in Roofing."

D2822, "Standard Specification for Asphalt Roof Cement."

D2824, "Standard Specification for Aluminum-Pigmented Asphalt Roof Coatings, Nonfibered, Asbestos Fibered, and Fibered Without Asbestos."

D3018, "Standard Specification for Class A Asphalt Shingles Surfaced with Mineral Granules."

D3161, "Standard Test Method for Wind Resistance of Asphalt Shingles (Fan-Induced Method)."

D3406, "Standard Specification for Joint Sealant, Hot-applied, Elastomeric-type, for Portland Cement Concrete Pavements."

D3462, "Standard Specification for Asphalt Shingles Made from Glass Felt and Surfaced with Mineral Granules."

D3569, "Standard Specification for Joint Sealant, Hot-applied, Elastomeric, Jet-Fuel-Resistant-Type for Portland Cement Concrete Pavements."

D3581, "Standard Specification for Joint Sealant, Hot-Poured, Jet-Fuel-Resistant-Type for Portland Cement Concrete and Tar-Concrete Pavements."

D3679, "Standard Specification for Rigid Poly (Vinyl Chloride) (PVC) Siding."

D3909, "Standard Specification for Asphalt Roll Roofing (Glass Mat) Surfaced with Mineral Granules."

D4434, "Standard Specification for Poly (Vinyl Chloride) Sheet Roofing."

D4586, "Standard Specification for Asphalt Roof Cement, Asbestos Free."

D4637, "Standard Specifications for EPDM Sheet Used in Single-Ply Roofing Membranes."

D4756, "Standard Practice for Installation of Rigid Poly (VinylChloride) (PVC) Siding and Soffit."

D4869, "Standard Specification for Asphalt-Saturated Organic Felt Underlayment Used in Steep Slope Roofing."

D6380. "Standard Specification for Asphalt Roll Roofing (Organic Felt)."

D6690. "Standard Specification for Joint and Crack Sealants, Hot Applied, for Concrete and Asphalt Pavements."

E84, "Standard Method for Surface Burning Characteristics of Building Materials."

E96, "Standard Test Method for Water Vapor Transmission of Materials."

E108, "Standard Test Methods for Fire Tests of Roof Coverings."

E119, "Standard Test Method for Fire Tests of Building Construction and Materials."

E136, "Standard Test Method for Behavior of Materials in a Vertical Tube Furnace at 750°C."

E283, "Standard Test Method for Determining the Rate of Air Leakage Through Exterior Windows, Curtain Walls, and Doors Under Specified Pressure Differences Across the Specimen."

E331, "Standard Test Method for Water Penetration of Exterior Windows, Curtain Walls, and Doors by Uniform Static Air Pressure Difference."

E662, "Standard Test Method for Specific Optical Density of Smoke Generated by Solid Materials."

E699, "Practice for Criteria for Evaluation of Agencies Involved in Testing, Quality Assurance, and Evaluating Building Components."

E1592, "Standard Test Method for Structural Performance of Sheet Metal Roof and Siding Systems by Uniform Static Air Pressure Difference."

E1646, "Standard Test Method for Water Penetration of Exterior

Metal Roof Panel Systems by Uniform Static Air Pressure Difference."

E2110, "Standard Terminology for Exterior Insulation and Finish Systems (EIFS)."

E2134, "Standard Test Method for Evaluating the Tensile-Adhesion Performance of an Exterior Insulation and Finish System (EIFS)."

E2273, "Standard Test Method for Determining the Drainage Efficiency of Exterior Insulation and Finish Systems (EIFS) Clad Wall Assemblies."

E2321, "Standard Practice for Use of Test Methods E 96 for Determining the Water Vapor Transmission (WVT) of Exterior Insulation and Finish Systems (EIFS)."

Brick Industry Association (BIA). Reston, VA (www.bia.org). *Technical, Notes on Brick Construction.*

———. 2004. No. 7C, "Moisture Control in Brick and Tile Walls: Condensation

———. 2004. No. 7D, "Moisture Resistance of Brick Masonry Walls: Condensation Analysis.

Cash, Carl G. 2000. "Selecting the Appropriate Roofing System." *The Construction Specifier* (January): 25 (www.csinet.org).

Cedar Shake and Shingle Bureau (CSSB). Sumas, WA: (www.cedarbureau.org). CSSB. 2004. *New Roof Construction Manual.*

———. *Exterior and Interior Walls Construction Manual.*

Copper Development Association (CDA). *Copper in Architecture Design Handbook.* New York: CDA (www.copper.org).

Federal Specifications, all: *see* U.S. General Services Administration Specifications Unit.

Lead Industries Association, Inc. (LIA). New York: LIA (www.leadinfo.com). *Lead Roofing and Flashing.* "Sheet Lead: the Protective Metal."

Litvin, Albert. 1968. "Portland Cement Association Research and Development Laboratories, Development Department Bulletin D137: Clear Coatings for Exposed Architectural Concrete. Portland Cement Association." (Reprinted from the May 1968 *Journal of the PCA Research and Development Laboratory,* 10[2]: 49–57.)

Mackinlay, Ian and Richard S. Flood. 2000. "Roof Slopes, Ice Dams, and Cold Roofs." *The Construction Specifier* (January): 31 (www.csinet.org).

Mack, Robert C. 1975. "Preservation Brief No. 1: The Cleaning and Waterproof Coating of Masonry Buildings." Washington, DC: Technical Preservation Services, U.S. Department of the Interior.

———. 1978. "The Cleaning and Waterproofing of Masonry Buildings." *The Construction Specifier* (November): 65–70 (www.csinet.org).

National Concrete Masonry Association (NCMA). *TEK Manual—Concrete Masonry Design and Construction.* Herndon, VA: NCMA (www.ncma.org).

National Fire Protection Association (NFPA). Quincy, MA (www.nfpa.org).

———. 2002. "Guide to the Fire Safety Concepts Tree."

———. 2003. NFPA 101®. *Life Safety® Code.*

National Roofing Contractors Association (NRCA). Rosemont, IL: NRCA (www.nrca.net).

———. 2005. *The NRCA Roofing and Waterproofing Manual.* Rosemont, IL: NRCA.

———. 2005. *The NRCA Waterproofing Manual.*

———. 2003. *The NRCA Steep-Slope Roofing Manual,* 5th ed.

———. 2006. *The NRCA Architectural Sheet Metal and Metal Roofing Manual,* 5th ed.

National Slate Association (NSA). (www.slateassociation.org). 1926. *Slate Roofs.* Poultney, VT: NSA. (Reprinted in 1977 by Vermont Structural Slate Co., Inc., Fair Haven, VT.)

Poore, Patricia. 1983. "What Most Roofers Don't Tell You About Traditional and Historic Roofs." *The Old-House Journal* (April): 56–58 (www.oldhousejournal.com).

Portland Cement Association (PCA). 1965. *Moisture Migration—Concrete Slab-on-Ground Construction.* Skokie, IL: PCA (www.portcement.org).

Ramsey, Charles George, Harold Reeve Sleeper, and John Ray Hoke, Jr. 2000. *Architectural Graphic Standards,* 10th ed. Hoboken, NJ: John Wiley & Sons, Inc. (www.wiley.com).

Sealant Engineering and Associated Lines (SEAL). "SEAL Guide Specification for Sealants and Caulking." Tucker, GA: SEAL (www.sealgroup.org).

Sealant, Waterproofing, and Restoration Institute (SWR Institute). 1995. "Sealants: The Professional's Guide." Kansas City, MO: SWR Institute (www.swrionline.org).

Sheet Metal and Air Conditioning Contractors National Association (SMACNA). 2003. *Architectural Sheet Metal Manual.* Chantilly, VA: SMACNA (www.smacna.org).

Shmedin, G. 1987. "Bibliography for Slate Roofing." *Association for Preservation Technology Newsletter* 4(2): 89 (www.apti.org).

Simmons, H. Leslie. 1989. *Repairing and Extending Weather Barriers.* New York: Van Nostrand Reinhold.

Single Ply Roofing Institute (SPRI). *Flexible Membrane Roofing: A Professional's Guide to Specifications,* 7e. Waltham, MA: SPRI (www.spri.org).

———. 2001. *Wind Load Design Guide for Low Sloped Flexible Membrane Roofing Systems.* Waltham, MA: SPRI (www.spri.org).

Specialty Steel Industry of North America (SSINA). *Standard Practices for Stainless Steel Roofing, Flashing, Copings.* Washington, DC: SSIAN (ssina.com).

———. "Finishes for Stainless Steel."

———. "Specifications for Stainless Steel."

———. "Stainless Steel Fasteners."

———. "Design Guidelines for the Selection and Use of Stainless Steel."

Sweetser, Sarah M. 1978. "Roofing for Historic Buildings." Preservation Brief No. 4. Washington, DC: Technical Preservation Services, U.S. Department of the Interior. (www.cr.nps.gov).

Tile Roofing Institute (TRI). Chicago (www.tileroofing.org). *Concrete and Clay Tile Installation Manual for Moderate Climate Regions.*

———. *Concrete and Clay Tile Installation Manual (Florida).*

———. *Concrete and Clay Tile Design Criteria for Cold and Snow Regions.*

———. "Concrete Tile Roofs."

Tobiasson, Wayne, and Marcus Harrington. *Vapor Drive Maps of the U.S.A.* Cold Regions Research and Engineering Laboratory (CRREL), Corps of Engineers. Hanover, NH (www.crrel.usace.army.mil).

Underground Space Center. 1988. *Building Foundation Design Handbook.* Minneapolis: Underground Space Center.

Underwriters Laboratories, Inc. (UL). Northbrook, IL. (www.ul.com).

2005. *Building Materials, Roofing Materials and Systems, Fire Protection Equipment, and Fire Resistance Directories.*

———. 2004. 55A, "Standard for Materials for Built-Up Roof Coverings."

———. 1994. 580, "Standard Tests for Uplift Resistance of Roof Assemblies."

———. 2004. 790, "Standards for Standard Tests for Fire Tests of Roof Covering Materials."

———. 1995. 997, "Standard for Wind Resistance of Prepared. Roof Covering Materials."

U.S. Department of Agriculture, Forest Service, Forest Products Laboratory. 1989. *Wood-Frame House Construction.* Washington, DC: USDA. For technical information: Forest Products Laboratory, One Gifford Pinchot Drive, Madison WI 53705 (tel: 608-231-9223). For copies: U.S. Department of Commerce, National Technical Information Service, 5285 Port Royal Road, Springfield, VA 22161 (Order No. PB90-180662). Tel: 703-605-6000.

U.S. General Services Administration Specifications Unit. (GSA). Washington, DC. Federal Specifications:

———. A-A-272A, "Caulking Compounds."

———. SS-S-200E(2), "Sealants, Joint, Two-Component, Jet-Blast Resistant, Cold-Applied., for Portland Cement Concrete Pavement."

Vinyl Siding Institute (VSI). *Vinyl Siding Installation Manual.* Washington, DC: VSI (www.vinylsiding.org).

# Doors and Windows

# Introduction

Exterior doors and aluminum entrances provide access openings into a building. They minimize heat gain and loss through these access openings and keep the weather outside, and they provide privacy for the building occupants and security for persons and property. They also help control the passage of sound through access openings.

Interior doors control passage between interior spaces. They also provide privacy and security, reduce sound trans-mission between spaces, and screen storage and utility areas from view.

Windows, storefronts, and curtain walls keep out the weather while admitting light and providing view. Windows also facilitate ventilation.

# Sustainability Considerations

For a discussion of sustainability, the systems addressed in this chapter can be separated into the materials used and the design of the systems in which those materials are used.

## Materials

The materials covered in this chapter are varied in their effects on the environment and the sustainability of the construction in which they are employed. Considerations regarding the sustainability of the materials addressed in this chapter are much the same as those described in the chapters that cover the various materials. For example, sustainability considerations for aluminum and other metals used in the products discussed in this chapter are addressed in Chapter 5; for plastics in Chapter 6; and so on.

## Design

The design of a building to minimize the amount of glazing can serve to re-duce the quantity of energy needed to heat and cool the structure. It is particularly important to reduce the amount of glazing in orientations that will contribute to heat gain in summer or heat loss in winter, and to shade glazed surfaces when excessive heat gain is expected, such as from south-facing windows. Refer to Chapter 17 for additional discussion.

It is essential that unglazed areas of a building's skin be well insulated.

Doors can be a source of significant heat leakage. Such loss can be greatly reduced by the use of foam-core doors. Storm doors can also reduce heat loss through doors, and adequate weatherstripping is essential. In commercial installations, the use of revolving doors or the addition of vestibules can reduce heat loss as doors are opened.

The metals used in windows, doors, and frames conduct heat readily. The introduction of a nonconducting material within such assemblies will prohibit such thermal bridging.

Glass is a good conductor of heat. Heat passage through glass in openings can be greatly reduced by the use of insulating glass.

Both wood and polyvinyl chloride (PVC) are good insulators. Their use can significantly reduce the heating and cooling loads on a building.

Air infiltration should be eliminated where undesirable by the installation of air barriers and the adequate sealing of openings.

Refer to Chapter 17 for a discussion of the current trend toward permitting occupants to open and close windows as an energy-saving strategy.

Chapter 17 also includes a discussion of the use of passive solar design to conserve energy. A building's glazed surfaces can be a major component of such systems.

# Applicable *MasterFormat* Sections

The following *MasterFormat* 2004 Level 2 sections are applicable to this chapter:

08 14 00  Wood Doors
08 15 00  Plastic Doors
08 16 00  Composite Doors
08 32 00  Sliding Glass Doors
08 40 00  Entrances, Storefronts, and
          Curtain Walls
08 51 00  Metal Windows
08 52 00  Wood Windows
08 53 00  Plastic Windows
08 54 00  Composite Windows
08 71 00  Door Hardware
08 80 00  Glazing

# 8.1 Metal Doors and Frames

This section discusses interior and exterior doors, frames, and associated sidelights formed from carbon steel, stainless steel, aluminum, and copper alloy sheets. Extruded aluminum and other types of doors and frames are discussed in other sections of this chapter.

## 8.1.1 GENERAL REQUIREMENTS

The two types of formed doors and frame units are *standard* and *custom*. Standard units are mass-produced and sold to retail outlets such as lumber yards and home centers. Standard units are also made for a particular project where they have been specified. Custom units are manufactured especially for a particular project to suit an owner's or architect's design. Both types should meet the same standards of quality. Formed carbon steel, stainless steel, and aluminum doors may be either standard or custom. Copper-alloy doors and frames are custom.

### 8.1.1.1 Standards

Both standard and custom carbon steel doors and frames should comply with the American National Standards Institute/Steel Door Institute's publication ANSI A250.8, "Recommended Specifications for Standard Steel Doors and Frames." In addition, a good guide for custom steel door and frame work is the *Hollow Metal Manual*, a collection of National Association of Architectural Metal Manufacturers (NAAMM) publications related to steel, doors, and frames. It includes the publications of NAAMM's Hollow Metal Manufacturers Association (HMMA) division and the NAAMM *Metal Finishes Manual* series.

In most cases, formed aluminum, stainless steel, and copper-alloy doors and frames should also comply with these same standards. In addition, they should conform to the applicable standards of the appropriate organization, which include:

Aluminum Association (AA)
American Architectural Manufacturers Association (AAMA)
American Iron and Steel Institute (AISI)
American National Standards Institute (ANSI)
American Society for Testing and Materials (ASTM)
Copper Development Association (CDA)
National Association of Architectural Metal Manufacturers (NAAMM)

Formed doors and frames should conform with the following standards:

- Fire doors and frames: National Fire Protection Association (NFPA) 80; tested and labeled by Underwriters Laboratories, Inc. (UL), Factory Mutual (FM), or another independent testing laboratory
- Sound-rated Assemblies: ASTM Standard E90
- Insulated doors: ASTM Standard C1199
- Door hardware: ANSI Standards A156.1 through A156.31

### 8.1.1.2 Materials

Refer to Chapter 5 for a general discussion about the metals addressed in this section.

Carbon steel for *interior* door facings and frames should be commercial-quality steel sheet and strip conforming with ASTM Standards A568 and A1011. It should have a *stretcher-level* standard of flatness.

Carbon steel for *exterior* door facings and frames should be hot-dip galvanized commercial quality carbon steel sheets complying with ASTM Standard A653. The zinc coating should be not less than Class G60. Sheets should be *mill phosphatized* and have a *stretcher-level* standard of flatness.

Sheet aluminum is often used for door faces. Sheet aluminum and bent plate components should comply with ASTM Standard B209.

Stainless steel is used to make formed doors and frames, sidelights, and storefronts. Door facings are usually flat sheets, but patterned sheets are also used.

Copper-alloy materials are used to make formed doors and frames for both exterior and interior locations. Most door facings on such doors are also copper-alloy sheets.

Concealed components, such as subframes, stiffeners, edge channels, reinforcements, and the like, should be fabricated from the same materials as the exposed portions of the door or frame.

### 8.1.1.3 Finishes

Carbon steel doors are either given a factory finish, such as a baked enamel or fluorocarbon finish, or are painted in the field (see Section 9.15).

The mill finish is usually left on concealed surfaces and sometimes even on the exposed surfaces of aluminum doors. However, most exposed portions of aluminum entrance doors are anodized (see Section 5.9).

Most exposed stainless steel is polished, but organic coatings are also applied to some assemblies.

Most exposed copper-alloys are given a patina or architectural bronze finish.

Exposed fasteners are usually finished to match the adjacent metal.

### 8.1.1.4 Supports, Anchors, Fasteners, and Flashing

Supports, anchors, fasteners, and flashing materials should be compatible with the material being anchored or fastened and with the substrate so that galvanic and chemical reactions do not occur between different materials. Carbon steel should be anchored and fastened with galvanized steel, stainless steel, or cadmium-plated steel devices; aluminum and stainless steel with nonmagnetic stainless steel, cadmium-plated steel, or galvanized steel; copper-alloy with copper-alloy.

Inserts for use in concrete may be cast iron, malleable iron, or hot-dip galvanized steel.

### 8.1.1.5 Weatherstripping

Weatherstripping controls air infiltration, prevents rain penetration, minimizes dust infiltration, and withstands atmospheric conditions. It must be durable under extended usage, resist corrosion, and be readily replaceable. Refer to Section 8.8 for an additional discussion of weatherstripping.

### 8.1.2 CARBON STEEL DOORS

Carbon steel doors and frames are often called *hollow metal*. Many members of the industry that makes them are represented by the HMMA, a division of NAAMM. Its manual of practice is the *Hollow Metal Manual*.

### 8.1.2.1 Construction

Carbon steel doors consist of a core, faces, tops, and bottoms. There are two types of core construction. In one, cores are hollow and braced with steel stiffeners. Most are filled with the manufacturer's standard fireproof, rotproof, sound-deadening material. Insulated doors are filled with glass fiber or plastic foam insulation.

In the second form of core construction, a door's face sheets are laminated to a core of impregnated kraft paper honeycomb, plastic foam insulation, or mineral fiber blocks (Fig. 8.1-1).

The tops and bottoms of standard doors are finished with a recessed, welded-in closure. Exterior doors have tops and bottoms formed flush as an integral part of the door construction or are finished with a welded-in inverted channel. Bottoms may also be fitted with a recess for weatherstripping (Fig. 8.1-2).

Doors should be sized so that the *maximum* clearances between the door and frame, between doors in a pair, and between door and sill are in accordance with the recommendations of the Steel

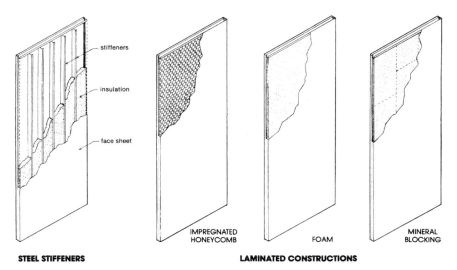

**FIGURE 8.1-1** Steel door construction. (Reprinted with permission of NAAMM, the National Association of Architectural Metal Manufacturers.)

**FIGURE 8.1-2** Bottom and top details of a typical steel door. (Reprinted with permission of NAAMM, the National Association of Architectural Metal Manufacturers.)

**FIGURE 8.1-4** Plastic foam insulation fills the inner core of insulated doors.

250.8 and SDI 108. SDI classifies doors in three grades:

- Grade I: Standard
- Grade II: Heavy Duty
- Grade III: Extra Heavy Duty

In Grades I and II there are two models:

- Model 1: Full Flush Design
- Model 2: Seamless Design

In Type III, there are five models:

- Models 1 and 1A: Full Flush Design
- Models 2 and 2A: Seamless Design
- Model 3: Stile-and-Rail Flush Panel

Figure 8.1-5, which is derived from SDI 108, shows the uses of the three grades recommended by SDI. It also shows eight basic door designs, their recommended uses, and the abbreviations and acronyms used to identify them. Steel doors that imitate the look of wood stile-and-rail panel doors are also available.

### 8.1.3 CARBON STEEL FRAMES

Carbon steel frames are brake-formed of steel sheets into one of two basic shapes (Fig. 8.1-6). However, many variations of these basic shapes are made, including cased opening frames that have no stops. Some special frames have tapered or molded shapes. Regardless of their shape or size, frames should be rigid, neat in appearance, and free from defects, warp, and buckle. They should be fitted and assembled in the manufacturer's plant when possible.

#### 8.1.3.1 General Requirements

The metal thickness of frames is dependent on the door grade and model as follows:

Grade I: 18- or 16-gauge (1.3 or 1.6 mm)
Grade II: 16-gauge (1.6 mm)
Grade III, Models 1, 2, and 3: 16- or 14-gauge (1.6 or 1.9 mm)
Grade III, Models 1A and 2A: 14- or 12-gauge (1.9 or 2.7 mm)

Door Institute's (SDI) publication ANSI/SDI A250.8. These clearances, which are subject to a tolerance of plus or minus $^1/_{32}$ in. (0.8 mm), are shown in Figure 8.1-3. As discussed in Section 20.1, security requirements may dictate less space between door and frame than

**FIGURE 8.1-3** Door Clearances

| Joint | Clearance, in. (mm) |
|---|---|
| Door head to frame | $^1/_8$ (3.2) |
| Door jambs to frame | $^1/_8$ (3.2) |
| Between meeting edges of pairs (unrated doors) | $^1/_8$–$^1/_4$ (3.2–6.4) |
| Between meeting edges of pairs (fire rated doors) | $^1/_8$ (3.2) |
| Door bottom | $^3/_4$ (19.1) |
| Face of door to stop | $^1/_{16}$ (1.6) |

that shown in Figure 8.1-3. Codes and NFPA 80, which is an applicable requirement in most codes for fire-rated doors and frames, may require closer spacings and less, or even no, tolerances.

#### 8.1.2.2 Insulated Doors

Insulated doors have an inner core of plastic foam insulation, have high R-values (R-8 to R-15), and are even more effective than the combination of prime and storm doors in reducing heat flow (Fig. 8.1-4). They are often installed with magnetic weatherstrips that are relatively airtight and minimize heat loss by infiltration.

#### 8.1.2.3 Door Types, Styles, and Designs

Most of the steel door types, styles, and designs available today fall into the classification system described in ANSI/SDI

# FIGURE 8.1-5 Basic Door Grades and Applications

| Building Types | Standard Steel Door Grades | | | Door Thickness | | Door Design Nomenclature | | | | | | Basic Door Design |
|---|---|---|---|---|---|---|---|---|---|---|---|---|
| | Grade I Standard Duty, 1¾ in. (44.5 mm) or 1⅜ in. (35 mm) | Grade II Heavy Duty, 1¾ in. (44.5 mm) only | Grade III Extra Heavy Duty, 1¾ in. (44.5 mm) only | 1¾ in. (44.5 mm) | 1¾ in. (44.5 mm) or 1⅜ in. (35 mm) | F | G | V | FG | NL | L | |
| **Apartment** | | | | | | | | | | | | |
| Main entrance | | • | | • | | | • | | • | • | | |
| Unit entrance | • | • | | • | | • | | | | | | |
| Bedroom | • | | | | • | • | | | | | | |
| Bathroom | • | | | | • | • | | | | | | |
| Closet | • | | | | • | | | | | | • | |
| Stairwell | | • | • | • | | | | | • | | | |
| **Dormitory** | | | | | | | | | | | | |
| Main entrance | | • | | • | | | • | | • | • | | |
| Unit entrance | • | • | | • | | | • | | | | | |
| Bedroom | • | | | • | | | • | | | | | |
| Bathroom | • | | | • | | | • | | | | | |
| Closet | • | | | • | | | | | | | • | |
| Stairwell | | • | • | • | | | | | • | | | |
| **Hotel, Motel** | | | | | | | | | | | | |
| Unit entrance | • | • | | • | | | • | | | | | |
| Bathroom | • | | | | • | • | | | | | | |
| Closet | • | | | | • | | | | | | • | |
| Stairwell | | • | • | • | | | | | • | | | |
| Storage and utility | • | | • | | • | | | | | | | |
| **Hospital, Nursing Home** | | | | | | | | | | | | |
| Main entrance | | • | | • | | | • | | • | • | | |
| Patient room | | • | | • | | | • | | | | | |
| Stairwell | | • | • | • | | | | | • | | | |
| Operating and exam room | | • | • | • | | • | | | | | | |
| Bathroom | • | | | | • | • | | | | | | |
| Closet | • | | | | • | • | | | | | • | |
| Recreation | | • | | • | | • | | | | | • | |
| Kitchen | | • | • | • | | • | | • | | | | |
| **Industrial** | | | | | | | | | | | | |
| Entrance and exit | | • | | • | | | • | | • | • | | |
| Office | • | • | | • | | | • | | • | • | | |
| Production | | • | | • | | • | | | | | | |
| Toilet | | • | • | • | | • | | | | | | |
| Tool | | • | | • | | • | | | | | • | |
| Trucking | | • | | • | | | • | | | | | |
| Monorail | | • | | • | | | • | | • | | | |
| **Office** | | | | | | | | | | | | |
| Entrance | | • | | • | | | • | | • | • | | |
| Individual office | • | | | • | | • | • | | | | | |
| Closet | • | | | | • | • | | | | | • | |
| Toilet | | • | • | • | | • | | | | | | |
| Stairwell | | • | • | • | | | | • | | | | |
| Equipment | | • | • | • | | • | | | | | | |
| Boiler | | • | • | • | | • | | | | | • | |
| **School** | | | | | | | | | | | | |
| Entrance and exit | | • | | • | | | • | | • | • | | |
| Classroom | | • | | • | | | • | | | • | | |
| Toilet | | • | • | • | | • | | | | • | | |
| Gymnasium | | • | • | • | | • | | | | | | |
| Cafeteria | | • | • | • | | | • | | | | | |
| Stairwell | | • | • | • | | | | • | | | | |
| Closet | • | | | | • | | | • | | | | |

Basic Door Design:

Flush (F)

Half glass (G)

Narrow light (NL)

Vision light (V)

Bottom louver (L)

Dutch door (D)

Full glass (FG)

Full louver (FL)

(Courtesy Steel Door Institute.)

**FIGURE 8.1-6** Profiles of basic steel door frames and their parts. (Reprinted with permission of NAAMM, the National Association of Architectural Metal Manufacturers.)

Except for double-acting, smoke, and fire doors, frames that are not weather-stripped should be drilled and have three rubber or plastic *silencers* (small, resilient bumpers) installed on the strike jambs. Frames for double doors should have silencers on the head.

Some codes and regulations require that interior doors in hospitals, nursing homes, and some other medical buildings have *terminated stops* (also called *cutoff* or *sanitary stops*) (Fig. 8.1-7). These must be omitted, of course, from fire, smoke, lead-lined, and sound-rated doors. The bottom of such a terminated stop is closed with a filler plate.

**FIGURE 8.1-7** Terminated stops. (Reprinted with permission of NAAMM, the National Association of Architectural Metal Manufacturers.)

Plaster guards or mortar boxes are required on frames where plaster or mortar might intrude into the operating hardware. These are sheet metal boxes welded to the frame.

Glazing beads used to hold glass panels in steel frames should be removable. They may be either mitered or butted in the corners but should be factory fitted. Removable beads should occur on the security side of openings so that they cannot be removed for access.

Frames wider than 49 in. (1245 mm) should be reinforced with a 14-gauge (1.9 mm) steel channel. Lead-lined frames and frames for power-operated doors usually also require reinforcement to support the loads.

#### 8.1.3.2 Frame Types

There are two basic frame types, *welded* and *knockdown*, and several special frame types, including *drywall slip-on*, *storefront*, and *lead-lined* frames.

In *welded* frames, corners and intersections are fully welded or attached together using concealed welded splice plates. Corners should be mitered.

*Knockdown* frames are made with mechanical joints. Most, but not all, have mitered corners. They are similar in design and appearance to welded frames, except that they are delivered to a project site disassembled for field assembly.

*Drywall slip-on* frames are designed for installation in drywall partitions after the partitions are in place. Several designs are available.

*Storefront type* frames are essentially welded frames. Frame sections that are too large to be shipped in one piece are shipped in several pieces and assembled in the field.

*Lead-lined* frames are fabricated with lead linings or are of a modified design that will receive lead sheeting in the field. They are used in x-ray rooms and wherever else radiation is a hazard.

### 8.1.4 FORMED STAINLESS STEEL AND COPPER-ALLOY DOORS AND FRAMES

Most stainless steel and copper-alloy doors and frames are custom-made formed shapes. Therefore, most of the door and frame designs available for carbon steel doors are also available in stainless steel or copper-alloy.

### 8.1.5 INSTALLATION

The finest door and frame design and the highest quality of manufacture will not compensate for poor installation. Weathertightness, although simple to achieve, can only be ensured if the installation is made by experienced workers following the guidelines and instructions supplied by the manufacturer and the appropriate industry standards. A rough opening should be prepared to receive each unit in such a manner that it can be installed and will finish out square, plumb, level, straight, and true. The design of most assemblies provides for minor adjustments at the job site, but no unit will operate or weather properly if it is twisted and misaligned during installation.

A properly selected and manufactured door and frame will be weather-resistant as a unit. Other things must be done, however, to achieve a completely weathertight installation. For example, if the construction is insulated, the space between the rough opening and the unit should be filled with insulation. In addition, the joints between units and adjacent materials must be closed with a sealant, trim members, or both to prevent the passage of air, dust, and water around the frame.

Door frames are seldom designed to support loads other than themselves. Therefore, unless the frame is specially reinforced to support other loads, a separate lintel or other support is usually required.

Door frames should be supported around the entire perimeter and anchored securely to the supporting construction in a firm and rigid position. They should also be flashed at their heads and sills to provide a path back to the exterior for water that finds its way into the wall.

Some metals, such as aluminum and stainless steel, are compatible and can be used in contact with each other. Other metals, however, react with one another chemically or electrolytically, and should be insulated from direct contact with each other by waterproof, nonconductive materials such as neoprene, waxed paper, or coated felt. Such dissimilar metals that are located where water passing over them may contact another surface should be painted to prevent staining. Refer to Section 5.2 for a discussion about the effect of dissimilar metal materials.

Concealed aluminum in contact with concrete, masonry, or an absorbent material such as wood, paper, or insulation should be protected by coating either the aluminum or the adjacent material with a bituminous or aluminum paint, or by coating the aluminum with a zinc chromate primer to minimize the chance of chemical corrosion of the aluminum by acids, alkalis, and salts leached out of the adjacent materials. Creosote and tar coatings, which might damage the aluminum, should not be used.

Sealants should be used in locations where frames adjoin other materials to make the joints there airtight and watertight.

### 8.1.5.1 Formed Frames and Doors

Formed door frames should be installed in accordance with SDI 105. Fire-rated frames should be installed in accordance with NFPA Standard No. 80.

When practicable, frames should be placed before the surrounding construction is built. Each frame should be anchored with jamb anchors near the top and bottom and not more than 24 in. (610 mm) apart in between (Fig. 8.1-8). Jambs should be anchored at the floor with adjustable floor anchors (Fig. 8.1-9).

Anchors in masonry should extend at least 8 in. (203 mm) into the masonry. Anchors for stud walls should be clips designed to attach to the types of studs used. Several different methods are used to anchor frames in previously placed masonry or concrete (Fig. 8.1-10).

**FIGURE 8.1-8** Common types of jamb anchors for conditions in which the frames are installed before the wall. (Reprinted with permission of NAAMM, the National Association of Architectural Metal Manufacturers.)

**FIGURE 8.1-9** Adjustable floor anchor. (Reprinted with permission of NAAMM, the National Association of Architectural Metal Manufacturers.)

Carbon steel and stainless steel frames in masonry should be filled solid with mortar. Copper-alloy and aluminum frames, however, should not be filled with mortar, because the mortar will corrode the metal.

## 8.2 Wood and Plastic Doors

Because of its ready availability, easy workability, and insulating properties, wood was a very early door material. Before the advent of plywood, doors were assembled from solid wood members. The methods used to conceal joints resulted in the familiar panel treatment of surfaces. Later designs substituted thinner wood or plywood panels in a supporting framework to make interior panel doors and glass to make sash doors.

Later, the mass production of plywood brought about the widespread use of flush doors, which now account for most wood doors used in all types of construction.

The following discussion outlines the variety of types and uses, methods of manufacture, standards, certification of quality, and installation of wood doors.

**FIGURE 8.1-10 Anchorage of frames in previously placed openings.** (Reprinted with permission of NAAMM, the National Association of Architectural Metal Manufacturers.)

Plastic doors are a relatively new innovation in door manufacture. They are used mostly in residential applications but sometimes appear in other types of buildings.

### 8.2.1 TYPES AND USES

Wood and plastic doors may be classified according to their construction as:

- *Flush*: Plywood, hardboard, or plastic laminate face panels bonded to solid or hollow cores; plastic doors with flat faces.
- *Stile-and-rail*: Vertical and horizontal members enclosing wood, glass, or louver inserts, creating panel, sash, storm, screen, or louver designs.
- *Accordion folding* wood and plastic doors are discussed in Section 8.3.2.

#### 8.2.1.1 Flush Doors

In addition to providing the desired surface appearance, face panels of flush wood doors act as stressed skins to impart strength, rigidity, and stiffness. Face panels may be hardwood veneer, plywood, high-pressure plastic laminate, or hardboard. Wood face panels are usually of two or three plies with a total thickness of at least $1/12$ in. (2.13 mm). Door construction is described as three-, five-, or seven-ply by counting the total number of plies in both face panels and counting the core as one ply.

Flush doors are available with glazed lights and louver inserts. The U.S. Consumer Product Safety Commission's (CPSC) "Safety Standard for Architectural Glazing Materials" requires that all glazing materials in doors be safety glazing material. In addition, industry standards limit the size of the cutout to a maximum of 40% of the door area (Fig. 8.2-1).

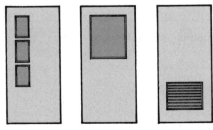

**FIGURE 8.2-1 Flush doors with glass and louver inserts.** (Courtesy Window and Door Manufacturers Association [WDMA] [formerly National Wood Window and Door Association].)

## SOLID CORE DOORS

Solid core construction increases the dimensional stability of a door, provides thermal insulation, and may add sound isolation and fire resistance. Solid core doors are commonly used as entrance and other exterior doors, but may also be used where sound or fire ratings are required. Doors for which specific fire ratings are required must carry a fire door label from an approved testing agency, such as UL. Fire ratings for wood doors range from 20 minutes to $1^{1}/_{2}$ hours, depending on materials and constructions. Many regular solid core doors are capable of a 20-minute fire rating. Applicable codes should be consulted for specific requirements and applications.

Solid core doors provide sufficient sound isolation for most uses. Where sound isolation is more critical, properly gasketed flush doors with a Sound Transmission Class (STC) rating of up to 42 may be required (see Section 13.1).

Solid cores may be of wood, particleboard, mineral composition, or a combination of these materials. Wood cores are made of strips or blocks of wood forming a glued-block core, framed-block core, or stile-and-rail core (Fig. 8.2-2).

Solid cores for plastic doors may be wood or plastic foam.

***Glued Block Core*** Most solid core doors have a glued-block core construction. They consist of wood blocks bonded together with the end joints staggered. The stiles and rails are bonded to the core, and the entire assembly is sanded to a smooth, uniform thickness. The face panels are bonded to this assembly and must be a minimum of $1/12$ in. (2.13 mm) thick.

***Framed-Block Core*** Framed-block core doors consist of a stile-and-rail frame surrounding an interior core of wood blocks, which may or may not be bonded to each other but which are bonded to the face panels. Face panels must be a minimum of $1/12$ in. (2.13 mm) thick.

***Stile-and-Rail Core*** Stile-and-rail core doors consist of wood blocks bonded into stile-and-rail panel units. These units are assembled to comprise the core of the door and are bonded to the face panels. Again, face panels must be a minimum of $1/12$ in. (2.13 mm) thick.

**FIGURE 8.2-2** Typical solid core constructions for flush doors: (a) glued block core; (b) framed-block glued core; (c) framed-block nonglued core; (d) stile-and-rail core; (e) mat-formed composition-board core; (f) wood-block, lined core. (Courtesy Window and Door Manufacturers Association [WDMA] [formerly National Wood Window and Door Association].)

*Particleboard Core* Particleboard core doors are constructed of one or more pieces of particleboard, each the full thickness of the core, which may or may not be edge- or end-glued together. The stiles and rails of the door may or may not be bonded to the core, depending on the type of construction. Open spaces between core pieces or between the core and frame are controlled by industry standards. Face panels must be a minimum of $1/12$ in. (2.13 mm) thick.

*Mineral-Composition Core* Doors with a mineral-composition core are normally

not available in other than fire-rated doors. These doors are constructed similarly to particleboard core doors, except that the core is comprised of inorganic, noncombustible materials formed into a rigid slab. Special blocking and framing are normally required in doors of this type to accommodate hardware, glazed openings, and other cutouts.

*Wood-Block, Lined Core* Wood-block, lined core doors are a combination of the constructions described previously. This type of door consists of a central wood-

block core with a $3/16$ in. (4.8 mm) or thinner liner of particleboard or other material bonded to each face to comprise the full core thickness. Face panels are then bonded to this assembly and to the stile-and-rail frame. These doors may provide added sound isolation, fire resistance, or other performance features, depending on the properties of the liner material.

### HOLLOW CORE DOORS

Flush doors with hollow cores are used for interior locations in residential and some light-commercial buildings. They are occasionally used as exterior doors in residential buildings and can serve satisfactorily in that use if bonded with waterproof adhesives and protected from the weather. They generally do not provide as much heat or sound insulation as solid core doors, but they can be used where these factors are not critical.

Face panels like those used on solid core doors are available for use on hollow core doors. Wood face panels must be at least two plies and must have a total thickness of at least $1/10$ in. (2.54 mm). The basic hollow core constructions are mesh (cellular) core and ladder core (Fig. 8.2-3).

*Mesh (Cellular) Core* Mesh core doors consist of wood or wood derivative material joined, interlocked, or woven to form a mesh or grid throughout the core area. Honeycomb cores of expanded paper or other material are the most common types of hollow cores. Face panels are bonded to the core and to the stiles, rails, and lock blocks of the door.

*Ladder Core* Ladder core doors consist of wood or wood derivative strips placed horizontally or vertically throughout the core area. Face panels are bonded to these strips and to the stiles, rails, and lock blocks of the door.

Lock blocks are provided in hollow core doors for lock mortising and backset up to $2 5/8$ in. (66.7 mm). Doors with special blocking may be obtained for other hardware mounting.

### INSULATED DOORS

Insulated doors have an inner core of plastic foam insulation, have high R-values (R-8 to R-15), and are even more effective than the combination of prime and storm doors in reducing heat flow. Wood-framed insulated doors often

**a**　　　　　　　　　　　　　　**b**

**FIGURE 8.2-3** Typical hollow core construction for seven-ply flush doors: (a) mesh or cellular core; (b) ladder core. (Courtesy Window and Door Manufacturers Association [WDMA] [formerly National Wood Window and Door Association].)

have metal skins. Insulated doors are often installed with magnetic weatherstrips that are relatively airtight and minimize heat loss by infiltration.

### 8.2.1.2 Stile-and-Rail Doors

Stile-and-rail doors derive their strength mainly from a supporting framework of vertical members (stiles) and horizontal members (rails) interlocked and glued together. Stiles and rails may be solid or veneered and are usually made of softwoods.

The chief categories of stile-and-rail doors are panel, sash, louver, storm, screen, and combination.

#### PANEL AND SASH

*Panel* doors consist of stile-and-rail frames enclosing flat plywood or raised wood panel fillers. They are available in a variety of designs (Fig. 8.2-4) and may be used for either interior or exterior application.

*Sash* doors are similar to panel doors in appearance and construction but have one or more glass lights replacing the wood panels. They are used often for main entrance and exterior service doors (Fig. 8.2-5). Casement (French) doors are fully glazed sash doors, and are referred to as *rim* doors when they have a single large glass panel and as *divided light* doors when the glass is divided into smaller lights (Fig. 8.2-6).

Stiles and rails are assembled with glued dowel or blind mortise-and-tenon construction (Figs. 8.2-7a and 8.2-7b). Together with mullions, muntins, and bars, they frame the glass lights or wood panels (Fig. 8.2-7c). The contoured profiles, grooves, or rabbets (sticking) milled on the edges of these members hold the glass or wood panels (Fig. 8.2-7d).

#### LOUVERED DOORS

Wood louvered doors are made of horizontal slats contained within stile-and-rail frames. The slats usually are blind-mortised. They are available in several common designs and two basic

**FIGURE 8.2-4** Panel door designs. (Courtesy Window and Door Manufacturers Association [WDMA] [formerly National Wood Window and Door Association].)

slat styles, chevron and standard (Fig. 8.2-8).

Wood louvered doors are used in interior locations primarily as visual screens and where sound privacy is not required. Appropriate uses include closet doors, space dividers, and closures around residential laundry areas, where ventilation and dissipation of moisture are desirable.

### WOOD SCREEN DOORS

Wood screen doors are lightly constructed in configurations similar to the louvered doors shown in Figure 8.2-8, but substituting insect screening material for the louvers. Such doors provide ventilation while excluding insects. Today this type of door is used mostly in utilitarian structures and porches. Screened doors used on residences and other

**FIGURE 8.2-5** Sash door designs. (Courtesy Window and Door Manufacturers Association [WDMA] [formerly National Wood Window and Door Association].)

**FIGURE 8.2-6** Casement (French) doors. (Courtesy Window and Door Manufacturers Association [WDMA] [formerly National Wood Window and Door Association].)

FIGURE 8.2-7 Typical construction of stile-and-rail doors: (a) glued dowel; (b) blind mortise-and-tenon; (c) shown together with mullions, muntins, and bars; (d) edges milled with contoured profiles, grooves, or rabbets.

FIGURE 8.2-8 Typical louver doors: (a) door designs, (b) slat styles.

structures are usually combination wood storm and screen doors of the type discussed in Section 8.8.

### 8.2.2 DOOR OPERATION

The operation of wood doors can be classified as either swinging, sliding, or folding (Fig. 8.2-9). The use and location of a door determine the most suitable manner of operation. Fully glazed wood sliding glass doors are addressed in Section 8.4.

#### 8.2.2.1 Swinging Doors

Swinging doors typically operate on hinges secured to the side jambs, although they may also operate on pivots supported by the door frame head and the floor. The majority of wood swinging doors are hinged (Fig. 8.2-9a). Either pivot hardware or double-acting hinges are used for double-acting swinging doors (doors that swing in both directions).

Hinged operation provides the greatest degree of security, weather resistance, heat insulation, and sound isolation. Where these features are particularly important, doors may be weatherstripped or gasketed for increased effectiveness.

Swinging door operation is most convenient for frequently used entry and passage doors, but may be used in any location unless door swing into usable space is objectionable.

*Door swing* is either right hand, left hand, right hand reverse, or left hand reverse. Assume that the viewer is standing outside a door (on the exterior of an entrance door, in the corridor at a room door, or on the side opposite the hinges on a connecting door). A door that swings away from the viewer with the hinges on the right side is a right-hand door; the same door swinging toward the viewer is a right-hand reverse door. When the hinges are on the left and the door swings away from the viewer, it is a left-hand door; if it swings toward the viewer, it is a left-hand reverse door.

#### 8.2.2.2 Interior Sliding Doors

Interior sliding wood doors are suspended on overhead tracks and operate on rollers attached to the door. These rollers may be nylon or another plastic or metal with plastic bearings. Floor guides usually are provided to prevent sliding doors from swinging laterally. Rarely, sliding doors are mounted on rollers operating on floor tracks.

Sliding wood doors are used on closets and as room dividers in all kinds of buildings and in storage areas in residential construction, where they are suitable for visual screens. Sliding doors used on closets and other storage areas are usually of the bypass type (Fig. 8.2-9b), but pocket doors are also sometimes used in these locations. Closet sliding doors are often louvered but may also be either flush or panel doors.

Surface sliding and pocket sliding doors are more likely to be found as passage doors (Figs. 8.2-9c and 8.2-9d). Bi-parting doors (two doors sliding in opposite directions) are used for openings 3 ft (914 mm) or wider and may be either surface or pocket type. For extremely large openings, multitrack installations permit stacking of several sliding doors at one or both ends. Surface and pocket sliding doors are usually either flush or panel doors but may also be louvered.

Sliding operation does not give a high degree of privacy, sound isolation, or weather resistance unless custom hardware or preengineered door units are used. The chief advantage of sliding operation is that it eliminates door swings which might interfere with the use of interior space.

#### 8.2.2.3 Folding Doors

Interior folding wood doors are hung on overhead tracks with rollers similar to those used on sliding doors. Some types also require a floor track. Folding operation, like sliding, is generally appropriate for locations requiring primarily visual screening.

Flush, louver, and panel doors are often used for closets in single or paired bi-fold arrangements (Fig. 8.2-9e) or as space dividers and movable partitions with special hardware for multifolding action and stacking (Fig. 8.2-9f). Accor-

**a Swinging**  **b Bypass Sliding**  **c Surface Sliding**

**d Pocket Sliding**  **e Bifolding**  **f Multifolding**

**FIGURE 8.2-9** Door operation: (a) swinging, (b–d) sliding, and (e, f) folding. Surface sliding doors (c) are used mostly on industrial and agricultural (usually metal prefabricated) buildings. They are seldom found in modern commercial, institutional, or residential buildings.

dion folding doors (see Section 8.3.1) are used as closet doors or as space dividers and may be pocketed or end-stacked within the opening.

### 8.2.3 MANUFACTURE

The following discussion of door manufacture presumes conformance with industry standards.

#### 8.2.3.1 Materials

##### WOOD DOORS

Some common species of hardwoods for door manufacture are oak, ash, elm, birch, butternut, cherry, maple, mahogany, walnut, basswood, and lauan. Softwood species are ponderosa (western) pine, Douglas fir, Sitka spruce, and western hemlock.

Most wood-faced flush doors are faced with hardwood veneer. Most stile-and-rail doors are made from ponderosa pine, but some are produced from other softwoods. Hardwood veneered stile-and-rail doors and softwood veneered flush doors, though uncommon, are available.

Lumber used in door manufacture is kiln-dried to a moisture content of 12% or less before being assembled into a door.

The solid wood core for a flush door should contain a single species of wood which has a specific gravity of .42 or less. Composition core materials range in density up to about 37 lb/cu ft (592.7 kg/m$^3$) at the same moisture content.

Doors for exterior use are made with waterproof (Type I) adhesives such as thermosetting melamine urea or phenolic resins. Interior doors are made with water-resistant (Type II) adhesives such as casein (sometimes fortified with urea resins and formaldehyde), capable of resisting a limited number of wetting cycles.

Some wood doors are made from a wood composition material that is pressed into a form resembling a stile-and-rail door. Some of these doors are given a wood-grained texture for painting. Others receive a face veneer of real wood.

##### PLASTIC DOORS

Plastic doors are manufactured from PVC and other plastics into flush panel or stile-and-rail configurations. They may be hollow core or have a solid core of wood or plastic foam.

Some doors are now being made from glass-fiber-reinforced plastic (fiberglass) in both flush and stile-and-rail configurations.

Most of the requirements included in this section for wood door types and uses, door operation, standards of quality, and installation apply as well to plastic doors.

Plastic doors perform as well as wood doors without the inherent maintenance problems associated with wood doors. Some plastic doors come prefinished; others must be painted.

### 8.2.3.2 Fabrication and Assembly of Wood Doors

Fabrication varies somewhat, depending on the type of door.

#### FLUSH DOORS

Stiles, end rails, lock rails, and lock blocks for hollow core flush wood doors are cut from kiln-dried, surfaced lumber. Stiles and rails are joined with corrugated metal fasteners or with machined dovetail joints. The stile-and-rail frame is laid up in adhesive with one face panel. Cores, lock blocks, and sometimes lock rails are then set and glued in place. Cores may be ladder, mesh, or implanted blanks (Figs. 8.2-10 and 8.2-11). Assembly is completed by bonding the second skin to the frame and the core.

Most hollow core flush wood doors are bonded with water-resistant adhesives which do not require heat to set.

**FIGURE 8.2-10** Implanted-blanks core for hollow core door is made by feeding wood spirals into an assembled frame.

**FIGURE 8.2-11** Cellular core construction is made by gluing paper honeycomb into an assembled frame.

These doors are cold pressed (Fig. 8.2-12) and are allowed to cure overnight, prior to trimming and further working.

Most solid core wood doors are made with *thermosetting* adhesives. Continuous block cores are assembled from wood blocks, glued together, and cured. After

**FIGURE 8.2-12** Hollow core flush doors are cold pressed to ensure proper bonding of face panels to core.

additional air drying for 24 hours, face panels are adhesive-bonded to the core, and the entire assembly is *hot-pressed* to set the glueline (Fig. 8.2-13).

Further operations on both solid and hollow core wood doors may include routing for louver and glass openings and in installing muntins and bars prior to glazing (Figs. 8.2-14 and 8.2-15).

Glass and louver openings in flush doors should not be more than 40% of the door area, and openings should not be closer than 5 in. (127 mm) from the door edge. The height of openings in a hollow core door should not exceed half of the door, and face skins should be supported with blocking, glued in place, to frame the opening.

#### STILE-AND-RAIL DOORS

Kiln-dried lumber is surfaced to exact thickness, and boards are then cut to proper widths and lengths. Pieces for stiles are bored for dowels on a boring machine (Fig. 8.2-16); a molding machine shapes the sticking on stile and rail edges (Fig. 8.2-17); and a double-end tenoner cuts rails to exact lengths, coping both ends to fit the stacking on the stiles. Another machine bores holes in rails, inserts glue, and drives hardwood dowels into the holes (Fig. 8.2-18).

For panel doors, wood or plywood panels are cut to size and may be shaped

**FIGURE 8.2-13** Hot pressing sets the glue line of solid core doors bonded with thermosetting adhesives.

**FIGURE 8.2-14** Flush doors are routed for glass lights and louvers.

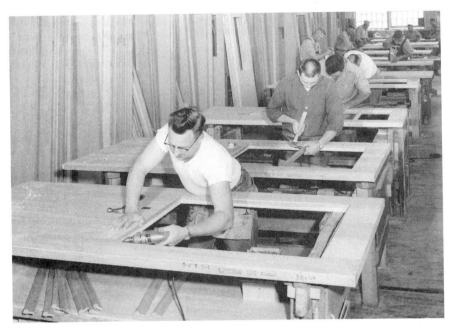

**FIGURE 8.2-15** Muntins and bars are fitted and installed before glazing.

**FIGURE 8.2-16** For stile-and-rail doors, holes for dowels are bored in stiles automatically fed from a hopper.

to form raised panel designs (Fig. 8.2-19). Stiles and rails are assembled with water-resistant adhesives around the panels and clamped in a jig. When the glue has set, the doors are sanded on both sides to a smooth finish (Fig. 8.2-20).

Stile-and-rail frames, and sometimes panels, muntins, and bars for sash doors, are fabricated and assembled in a similar manner. Glass is cut to size and bedded in glazing compound, and stops are accurately mitered and nailed in place (Fig. 8.2-21).

### 8.2.3.3 Prefabrication

Wood doors may be further worked and finished in the plant to reduce job-site preparation and installation. They may be seal-coated to prevent moisture absorption and soiling or may be completely surface-finished. Standard doors and prefinished custom doors are usually prefit to exact opening dimensions so that the finished edges will not be removed by job-fitting. Doors for prefabricated frames sometimes are machined for locks and hinges (Fig. 8.2-22), and bifold doors may have all operating hardware installed.

Plastic laminate-surfaced doors are usually prefabricated and prefit.

Prehung door units are assembled complete with frames, trim, and hardware. The stop may be part of the frame or left loose to be applied during installation. Frames are braced at jamb bottoms and carefully packed to prevent racking and distortion in shipment.

### 8.2.3.4 Sizes

Door sizes vary, depending on the door type. Figure 8.2-23 shows the more commonly used sizes of both wood and plastic doors.

### 8.2.3.5 Finishes for Wood Doors

Finishes for wood doors may be either opaque or transparent. *Opaque* finishes are essentially paint. Several types of *transparent* finish systems are used on doors. These include lacquers, conversion varnishes, catalyzed vinyls, polyesters, polyurethanes, and oils.

Factory finishes are graded as standard, select, or super.

*Standard* finishes are simply the manufacturer's standard. They usually include lacquers, conversion varnishes, and catalyzed vinyls. *Select* finishes are similar but permit the designer to select colors and

**FIGURE 8.2-17** An endless feeder chain carries rails or stiles through a molding machine to form sticking on edges.

**FIGURE 8.2-18** A single machine bores holes, inserts glue, and drives dowels.

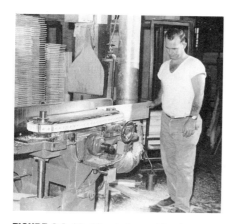

**FIGURE 8.2-19** After being cut to size, panels are shaped to make raised designs.

sheens. Select finishes are usually more expensive than standard finishes.

*Super* finishes are generally superior to both standard and select finishes but are usually the most expensive of the three. They include polyester, polyurethane, and oil finishes. A designer may select the color and sheen.

### 8.2.4 STANDARDS OF QUALITY

Standards of quality for various types of wood doors are based on construction grades, face material selection, core construction, and vertical and horizontal edge and core interface. The Window and Door Manufacturers Association

(WDMA), the Architectural Woodwork Institute (AWI), and the Woodwork Institute (WI) have developed standards applicable to wood doors. The applicable standards are:

Flush Doors:
- WDMA, I.S.1-A, "Architectural Wood Flush Doors"
- AWI, *Architectural Woodwork Quality Standards Illustrated*, Section 1300, "Architectural Flush Doors"
- WI, *Manual of Millwork*

Stile-and-Rail Doors:
- WDMA, I.S.6, "Architectural Wood Stile and Rail Doors"
- AWI, *Architectural Woodwork Quality Standards Illustrated*, Section 1400, "Stile and Rail Doors"
- WI, *Manual of Millwork*

These publications are the recognized standards of quality for the wood door industry. They are similar but vary slightly in detail. The AWI and WI standards act as supplements to the WDMA standards. Before making choices related to door selection and design, a designer should examine the requirement of the selected standard and not rely solely on citing the name of the standard. It may be necessary to modify these standards to suit a particular project. For example, one standard may require a particular face veneer thickness for one facing grade, while another standard may require a different minimum thickness or may be silent on the issue.

In addition to establishing minimum requirements for material, design, construction, and waterproof performance under test conditions, these standards provide a basis for quality grading of wood doors.

#### 8.2.4.1 Grades

Wood doors are graded primarily according to their exposed face veneers. However, the standards also include in the grades requirements for vertical edge bands, pair matching, transom matching, beads for light and louver openings, and quality of workmanship. The grades limit the materials that may be used in each grade based on the type of finish to be used (opaque, transparent, or laminate). Flush doors are also limited by core construction.

#### WOOD FLUSH DOORS

Solid and hollow core flush doors are available in *premium*, *custom*, and *economy*

**FIGURE 8.2-20** Stile-and-rail doors are sanded to a smooth finish.

**FIGURE 8.2-21** After glazing, accurately mitered stops are nailed in place.

**FIGURE 8.2-22** Doors are mortised for lock-sets on an automatic router.

**Figure 8.2-23** Commonly Used Door Sizes

| Door Type | Thickness | Width | Height |
|---|---|---|---|
| **Flush (hollow)** | $1^3/_8$ in. (35 mm) and $1^3/_4$ in. (44.5 mm) | 1 ft 6 in. (457 mm) to 3 ft 0 in. (914 mm) | 6 ft 8 in. (2032 mm) and 7 ft 0 in. (2134 mm) |
| **Flush (solid)** | $1^3/_8$ in. (35 mm), $1^3/_4$ in. (44.5 mm), and $2^1/_4$ in. (57.2 mm) | 2 ft 4 in. (812.8 mm) to 3 ft 6 in. (1067 mm) | 6 ft 8 in. (2032 mm) and 7 ft 0 in. (2134 mm) |
| **Panel** | $1^3/_8$ in. (35 mm) and $1^3/_4$ in. (44.5 mm) | 1 ft 6 in. (457 mm) to 3 ft 4 in. (1016 mm) | 6 ft 8 in. (2032 mm) and 7 ft 0 in. (2134 mm) |
| **Sash** | $1^3/_8$ in. (35 mm) and $1^3/_4$ in. (44.5 mm) | 2 ft 0 in. (610 mm) to 3 ft 6 in. (1067 mm) | 6 ft 8 in. (2032 mm) and 7 ft 0 in. (2134 mm) |
| **Screen** | $1^1/_8$ in. (28.6 mm) | 2 ft 6 in. (762 mm) to 3 ft 0 in. (914 mm) | 6 ft 7 in. (2007 mm) and 6 ft 9 in. (2057 mm) and 7 ft 1 in. (2159 mm) |
| **Louver** | $1^1/_8$ in. (28.6 mm) and $1^3/_8$ in. (35 mm) | 1 ft 3 in. (381 mm) to 3 ft 0 in. (914 mm) | 6 ft 6 in. (1981 mm), 6 ft 8 in. (2032 mm), and 7 ft 0 in. (2134 mm) |

grades, depending on the quality of face panels. Flush doors are not recommended for exterior use. Interior doors may be made using either *waterproof* (Type I) or *water-resistant* (Type II) adhesives depending on their location. Doors in wet or humid areas should be made using Type I adhesives.

***Premium Grade*** Premium-grade face panels are available for either transparent, opaque, or laminate finishes. This is the highest grade available from commercial stock.

For *transparent* finishes, the face veneers are tight, smoothly cut, and free from knots, worm holes, splits, shakes, and torn grain. When the face consists of more than one piece, the pieces are of approximately equal width and are matched at the joints for color and grain. Several different matching patterns are used. Many species are used for transparent finishes (see Section 8.2.3.1).

For *opaque* finishes, custom-grade doors have *medium density overlay* plywood faces.

*Laminate* faces and edges are nominal 0.050-in. (1.27-mm)-thick high-pressure decorative laminate. The laminate may be either solid or have a grain or texture.

***Custom Grade*** Flush doors with custom-grade face veneers are also intended for either a transparent or opaque finish. They are intended for use in high-quality projects.

For *transparent* finishes, the face panels for custom-grade flush doors are made of veneers essentially similar to those used in premium-grade doors, but need not be matched at joints for color and grain if sharp contrasts between adjacent pieces are avoided.

For *opaque* finishes, the face veneer may be any close-grained hardwood. The grain pattern may show through. Split veneer joints and other wood characteristics may also show through the finish unless special precautions such as filling are taken to prevent this.

*Laminate* faces are nominal 0.050-in. (1.27-mm)-thick high-pressure decorative laminate. The laminate may be either solid or have a grain or texture. Edges are close-grained hardwood.

***Economy Grade*** The face veneers in economy-grade flush doors are intended for painting and may contain a variety of color, grain, and surface defects which will not be visible after two coats of paint have been applied. Their selection is generally left up to the mill producing the doors. Some manufacturers make economy-grade doors with hardboard face veneers.

There is no economy grade for laminate finish doors.

### WOOD STILE-AND-RAIL DOORS

Grades for stile-and-rail doors are *premium*, also sometimes called *select*, and *standard*, also called *custom*. Both grades are made with water-resistant adhesives and are often treated with water repellents.

***Premium Grade*** Premium-grade stile-and-rail doors are made from stock that is practically free from defects. They are intended to receive a transparent finish.

***Standard Grade*** Standard-grade stile-and-rail doors are made from lumber stock that may contain stains and various other minor defects. They are intended to receive a paint finish.

### 8.2.4.2 Certification of Quality

Wood doors that conform to the industry standards may be grademarked or labeled by the manufacturer when requested by the buyers. *Grademarks* are usually stamped or branded on the edge of a door; *labels* are written statements attached to the door and containing information similar to that of grademarks.

The WDMA administers an independent certification program for hardwood veneered doors based on continuous, impartial inspection and testing of doors. A WDMA seal ensures that the materials, construction, and manufacture of doors conform to applicable industry standards. A red plastic plug in the top edge of a door indicates a door bonded with waterproof exterior Type I adhesives.

Exterior doors should be treated with a water-repellent preservative for increased dimensional stability and resistance to decay, fungus, and insect attack. WDMA administers a certification program for preservative-treated millwork products which includes softwood doors. The WDMA seal is assurance of adequate preservative treatment in accordance with WDMA I.S.4, "Water-Repellent Preservative Non-Pressure Treatment for Millwork."

### 8.2.5 INSTALLATION

The satisfactory performance of doors depends on proper handling, storage, hanging, and finishing.

### 8.2.5.1 Handling and Storage

Unfinished doors should be handled with clean gloves to prevent soiling. They should be stored flat (not on edge) on a level surface in a clean, dry, well-ventilated space. They should not be stored in excessively hot, dry, or humid areas, but should be conditioned for several days to the average prevailing local humidity before hanging or finishing. If doors appear warped, they should be stacked flat under uniformly distributed weights to restore flatness.

### 8.2.5.2 Hanging

Finished door frames should be square and plumb, and doors should be fitted in the openings with a total clearance of approximately $3/16$ in. (4.8 mm) in both horizontal and vertical dimensions to prevent binding under humid conditions. Doors may be trimmed for fitting but should not be cut to smaller nominal sizes. If a substantial amount of the perimeter is removed, the balance of the door could be destroyed and warping could result.

Installed doors should conform to NFPA 80.

### 8.2.5.3 Finishing

Unfinished wood doors should receive a seal coat of the finish to be used on them or a suitable sealer as soon as they are delivered to the job site to prevent undue moisture absorption and soiling. Finishing should be delayed until doors have been trimmed and fitted in the openings. Door surfaces, including both faces and top, bottom, and side edges, should be sealed against moisture with two coats of paint, varnish, or sealer before final hanging.

Door surfaces should be clean and dry before a finish is applied. Soil marks and other surface defects should be removed by light sanding. Doors should not be finished in excessively humid weather. The proposed finish should be tested on a sample of the same species of wood to determine its appearance and suitability.

# 8.3 Specialty Doors

There are many types of specialty doors in general use. They include sliding glass doors, revolving doors, access doors and panels, folding doors and grills, coiling doors and grills, vertical lift doors, cold storage doors, hanger doors, industrial doors, lightproof doors, darkroom doors, radiofrequency interference shielding doors, radiation shielding doors, security doors, vault doors, detention doors, sound control doors, traffic doors, airtight doors, watertight doors, blast-resistant doors, and cellar doors.

Sliding glass doors are addressed in Sections 8.6 and 8.7; revolving doors in Section 8.4; and accordion folding doors, coiling doors, and vertical lift doors in this section. The other specialty doors mentioned are beyond the scope of this book.

### 8.3.1 ACCORDION FOLDING DOORS

Accordion folding doors are assembled from narrow slats which give a long, drape-like appearance (Fig. 8.3-1). There are currently no industry standards covering accordion folding doors.

The folding doors discussed here should not be confused with *folding partitions*, which are similar to accordion folding doors but tend to be much sturdier and larger. Accordion folding doors, for example, are usually limited to heights of about 14 ft (4267 mm), but most range from 6 ft 8 in. (2032 mm) to 10 ft 0 in. (3048 mm). Folding partitions can be much taller. Accordion folding doors are generally designed to fold to a width less than the thickness of a normal interior partition. Folding partitions usually have much wider panels.

Accordion folding doors are primarily visual screens for locations such as closets, space dividers, residential laundry enclosures, and areas where air circulation is desirable. They are not suitable where sound isolation or privacy are prime requirements. The folding operation also makes them useful where door swings are objectionable. Conversely, many folding partitions are designed to provide both privacy and sound transmission control.

Folding partitions are used where visual and sound privacy are needed, such as to separate meeting rooms or classrooms.

Folding partitions are usually constructed of wood or metal frames covered with plywood or hardboard and a wood, fabric, or plastic laminate panel. They often have a plastic foam or other solid core.

Accordion folding doors are made with a framework or core of plastic, metal, or wood and a covering of solid wood, wood veneer, vinyl, fabric, plastic laminate, or other materials. The wood

| Solid | Woven | Corded |

**FIGURE 8.3-1** Typical accordion folding doors.

used may be any of a variety of hardwoods and softwoods

Such doors are made in several styles (see Fig. 8.3-1). In a solid accordion folding door, wood slats are connected to each other with continuous metal, vinyl, or nylon fabric hinges. The wood slats vary in width from $3^1/_2$ in. (89 mm) to 5 in. (127 mm).

Woven doors are commonly of the basketweave type in which vertical strips $3/_8$ to 1 in. (9.5 to 25.4 mm) wide are interwoven with nylon-reinforced vinyl tape.

Corded doors are made with similar wood slats connected with cotton cord.

Accordion folding doors are usually suspended from ceiling-mounted tracks and operate on nylon rollers or glides.

### 8.3.2 ACCESS DOORS AND PANELS

As their name implies, access doors and panels are installed to permit entry into concealed portions of a building to allow access to the building's mechanical, plumbing, and electrical components. They are installed in ceilings, walls, and floors. Floor access doors and panels are called *floor hatches*.

Access doors are small spring-operated, swing-down or swing-up doors. Figure 8.3-2 shows a typical swing-down ceiling access door. Wall access doors are similar. Operable floor hatches usually have a hold-open bar or device for the safety of a user. Access panels, including inoperable floor hatches, are generally not hinged and must be lifted out of place to provide access.

Access doors and panels and their frames are constructed of galvanized or stainless steel. Floor hatches are usually made of aluminum or steel.

Access doors and panels may have flat metal panels or may be recessed to receive a finish like that of adjacent surfaces. For example, a ceiling panel may be recessed to receive acoustical tile or gypsum board; an inoperable floor hatch cover may be recessed to receive stone, tile, wood, carpet, or another floor finish.

Access doors and panels may be thermally insulated for use in locations where the space beyond the door or panel is not heated or cooled. Fire-rated doors and panels are available.

### 8.3.3 VERTICAL LIFT DOORS

*Vertical lift doors* (sometimes called *overhead doors*) are used for vehicular access

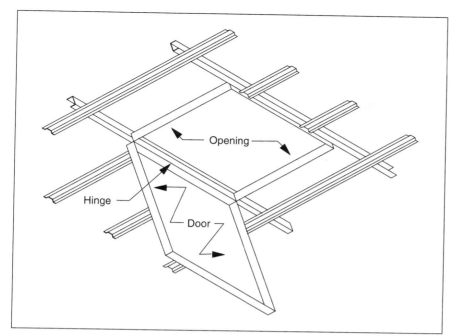

**FIGURE 8.3-2** A ceiling access door. Such doors may be frameless, as shown, or have an exposed frame. Door panels may be flush, as shown, or recessed to accept the same material as in the adjacent ceiling. In either case, the door finish is flush with the ceiling. The ceiling framing shown is typical for a gypsum board ceiling with channel runners and hat channels. Depending on the weight of the door, additional channel framing may be required. The finished ceiling has been omitted for clarity. (Drawing by HLS.)

in residential garages and commercial applications. They are generally of two types, sectional and one-piece. One-piece doors consist of a single panel that can be raised either manually or with a motorized door operator. They are available with and without tracks. This type of door tilts as it is lifted and slides back into the interior space to provide an opening the full size of the door (Fig. 8.3-3).

*Sectional doors* are by far the more common type of vertical lift doors. They are assembled from panels of aluminum, steel, or fiberglass that are hinged together. Rollers attached to each panel

ride in a track that rises vertically, then turns and finishes in a horizontal position (Fig. 8.3-4).

Sectional door panels are nominally 24 in. (609.6 mm) wide. Door widths vary from 8 ft (2438 mm) to 24 ft (7315 mm). Maximum heights are 14 ft (4267 mm) for aluminum panels, 16 ft (48.77 mm) for fiberglass panels, and 18 ft (5486 mm) for steel panels. Panels may be flat, horizontally ribbed, or glazed, and may have an embossed texture similar to wood grain or another texture.

Doors are lifted by a cable. Small doors may be either manually lifted or mechanically lifted by a motor drive. Larger doors are mechanically lifted. Remote operators are often used.

### 8.3.4 COILING DOORS

Coiling doors are assembled from interlocking steel or aluminum slats (Fig. 8.3-5) or by rods and links to form an open grid. Operation may be horizontal (rolling door) or vertical (shutter). A common use is as a counter shutter, but coiling doors are also used extensively as security screens for commercial spaces in malls, airports, and the like. In certain sizes and configurations, coiling doors may be fire-rated.

Operation may be mechanical or manual (push-up, hand crank, or pull chain). Mechanical operators are driven by electric motors. Torsion springs assist in the operation.

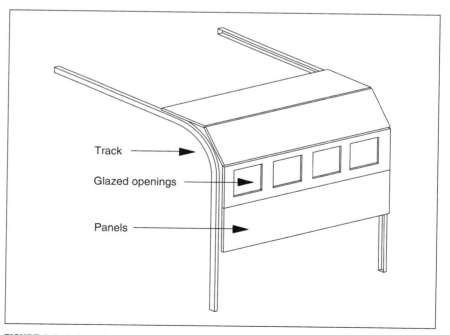

**FIGURE 8.3-3** A one-piece vertical lift door. (Drawing by HLS.)

**FIGURE 8.3-4** A sectional vertical lift door. (Drawing by HLS.)

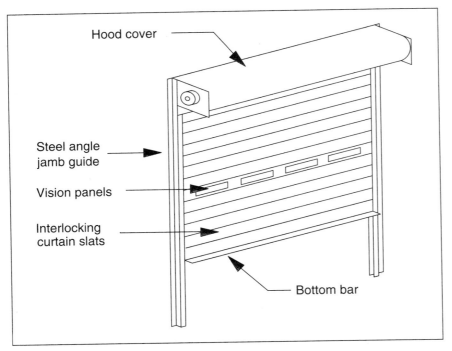

**FIGURE 8.3-5** A coiling door. (Drawing by HLS.)

Panels may be flat, textured, or articulated. Glazed vision panels are available.

Coiling doors may be fitted for use either on the interior or on the exterior. Weatherstripping must be added for exterior use.

# 8.4 Entrances and Storefronts

Entrances and storefronts include both exterior and interior metal doors, frames, and storefront systems, as well as revolving doors. Frames for entrance doors and associated glazed sidelights and other openings that are constructed of wood are considered to be finish carpentry items (see Section 6.12) and are not addressed in this section. Because most entrances and storefronts installed today are made from extruded aluminum components, this section primarily addresses extruded aluminum systems. However, entrances and storefronts are also constructed from formed sheet aluminum (as opposed to being extruded) and from formed steel, stainless steel, or copper-alloy sheets. The principles associated with the extruded aluminum systems described in

this section also apply to entrances and storefronts constructed with formed components. For example, formed systems should comply with the stated performance requirements for strength, water resistance, air infiltration, and condensation resistance; components should also have thermal breaks; and finish hardware should be comparable. Additional requirements related to formed frames are addressed in Section 8.1.

## 8.4.1 STANDARDS

Extruded aluminum entrances and storefronts should conform to the applicable standards of the appropriate organization, including:

- Aluminum Association (AA)
- American Architectural Manufacturers Association (AAMA)
- ASTM International (ASTM)
- National Association of Architectural Metal Manufacturers (NAAMM)
- American National Standards Institute (ANSI)

Extruded aluminum components should comply with AAMA publication SFM-1, *Aluminum Store Front and Entrance Manual*.

Door hardware should conform to ANSI Standards A156.1 through A156.21.

## 8.4.2 MATERIALS

Refer to Chapter 5 for a general discussion about the metals addressed in this section.

A variety of aluminum alloys and tempers are used to make extruded door and frame components. Alloys should conform with ASTM Standard B221. Common alloys used are 6061, 6063, and 6463. The temper should be suited to the finish to be applied. A common temper is T5.

Sheet aluminum is often used for door faces. Sheet aluminum and bent plate components should comply with ASTM Standard B209.

Subframes, stiffeners, edge channels, reinforcements, and other concealed components are fabricated from the same materials as the exposed portions of the door or frame.

## 8.4.3 FINISHES

The mill finish is usually left on concealed surfaces and sometimes even on the exposed surfaces of aluminum doors, but most exposed portions of aluminum doors and storefronts are anodized. Fluoropolymer finishes are also used but are not recommended for surfaces subject to excessive wear or damage, such as on doors and adjacent frames.

Refer to Section 8.1.1.3 for additional requirements. Finishes for metals are addressed in Section 5.9.

Exposed fasteners are usually finished to match the adjacent metal.

## 8.4.4 SUPPORTS, ANCHORS, FASTENERS, AND FLASHING

Supports, anchors, fasteners, and flashings should be compatible with the material being anchored or fastened and with the substrate so that galvanic and chemical reactions do not occur between different materials. Aluminum should be anchored and fastened with aluminum or with nonmagnetic stainless, cadmium-plated, or galvanized steel.

Inserts for use in concrete may be cast iron, malleable iron, or hot-dip galvanized steel.

Refer also to Section 8.1.1.4.

## 8.4.5 WEATHERSTRIPPING

Weatherstripping controls air infiltration, prevents rain penetration, minimizes dust infiltration, and withstands atmospheric conditions. It must be durable under extended usage, resist corrosion, and be readily replaceable. Refer to Section 8.9 for additional discussion of weatherstripping.

## 8.4.6 EXTRUDED ALUMINUM DOORS, FRAMES, AND STOREFRONTS

Most main entrance doors and their frames, as well as packaged entrances and storefronts in commercial and institutional buildings, are made from extruded aluminum components.

### 8.4.6.1 Design Requirements

Extruded aluminum doors, frames, and storefronts should be tested by the manufacturer or by an independent testing agency to ascertain their resistance to deflection due to wind loads, air infiltration, and water penetration. These characteristics are determined by the effect on a door and frame, packaged entrance, or storefront system when tested at specific levels of wind pressure. Since these elements are located on the ground floor, they are not subjected to the high levels of wind pressure that may affect windows and glazed curtain walls. The wind pressures used in the tests, therefore, are lower.

To select a door, frame, packaged entrance, or storefront, first determine the type desired (e.g., stick framing or package, type of door, frame profile). Then determine the required design pressure.
*Design pressure* is determined using the fastest wind speed likely to be encountered at a building's location. Data found in ANSI/ASCE 7-02 can be used to estimate the wind loads likely to be encountered.

The assembly selected should have been tested using the expected wind loading and the procedures listed in ASTM Standard E330. Permanent deflection normal to the plane of the unit should be no more than 0.2% of the clear span, and there should be no glass breakage or permanent damage to fasteners, anchors, or hardware. Deflection during the test should not exceed $1/175$ of any span. This limit may be decreased to $1/360$ of any span if a nonflexible finish, such as plaster, is to be used.

The water resistance of an assembly should be determined by testing it in accordance with ASTM Standards E331 and E547. The test pressure is usually 6.24 lb/sq ft (30.47 kg/m²). For a unit to pass this test, no water may pass through the unit during the test.

Air infiltration should be tested in accordance with ASTM E283. The usual pressure is 1.57 lb/sq ft (7.67 kg/m²); the permitted infiltration is 0.06 cu ft (0.002 m³) per minute per square foot (0.09 m²) of fixed area.

Condensation resistance should be tested for frames having a thermal break feature. Condensation resistance should be determined by procedures outlined in AAMA 1503. An assembly's condensation resistance gives an indication of that assembly's ability to resist condensation. The higher the condensation resistance factor, the less likely it is that condensation will occur.

As discussed in Chapter 17, heat will flow quite readily through entrances and storefronts unless some means are taken to reduce this flow. The methods for determining conductive heat transmission through these assemblies is a part of AAMA 1503.

### 8.4.6.2 Manufacture

The manufacture of extruded aluminum doors and packaged entrances consists of fabricating aluminum extrusions of the correct sizes and shapes and assembling them together with glass, glazing materials, hardware, weatherstripping, and screws and other fasteners into finished, preassembled components.

### 8.4.6.3 Doors

Each manufacturer has its own standards for its doors, even within the limits of the standards listed earlier. However, there are only two basic types of swinging doors: (1) stile-and-rail doors and (2) flush-panel doors (Fig 8.4-1).

#### STILE-AND-RAIL DOORS

Aluminum stile-and-rail doors consist of a tubular extruded aluminum frame usually infilled with glass (Fig. 8.4-1a). Stiles may be *thin* (less than $1^3/_4$ in. [44.5 mm] wide), *narrow* (usually 2 in. [50.8 mm] wide), *medium* (4 in. [102 mm] wide), or *wide* (5 in. [127 mm] or more wide). The thickness of most stiles is $1^3/_4$ in. (44.5 mm).

Thermal break doors are available, but most aluminum stile-and-rail doors are standard. The usual glazing is $1/_4$-in. (6.4-mm)-thick glass, but most thermal break doors are glazed with 1-in. (25.4-mm)-thick insulating glass.

#### FLUSH-PANEL DOORS

Aluminum flush-panel doors consist of an extruded channel or angle frame and bracing members and sheet aluminum facing panels (Fig. 8.4-1b). Most such doors are filled with a core made of a kraft paper honeycomb, rigid plastic foam, or mineral insulation. Faces may be either flat or embossed. Flush-panel doors may have glazed openings or louvers.

### 8.4.6.4 Frames and Storefronts

Extruded aluminum frames and storefront sections come complete with accessories such as glazing gaskets, glazing stops, clips, trim, fins, flashings, anchors, and fasteners. Incidental components, such as closers, channels, flats,

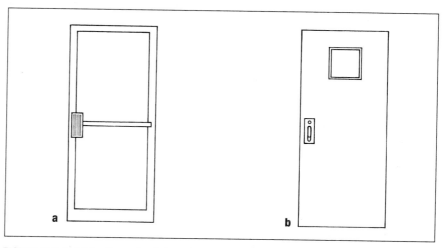

**FIGURE 8.4-1** (a) A stile-and-rail door; (b) a flush-panel door. (Drawings by HLS)

Recessed glazing　　　　　Applied glazing stops

**FIGURE 8.4-2** Typical extruded aluminum entrance and storefront frames. Many other types and shapes are available. (Courtesy H. Leslie Simmons.)

trim, sleeves, and the like, are also usually delivered with the frames.

Frames and storefront sections should be accurately cut and fitted to make hairline joints. Where required, joints should be designed and erected to be watertight. Joints between two pieces of metal should be filled with a nonhardening sealant. Welds should be made on the inside so that they will not be visible in the finished installation.

Frames and storefronts are usually glazed with $1/4$ in. (6.4 mm) glass, except where insulating glass is desired. Glaz-

ing stops and gaskets should be of the correct size to receive the glass to be used.

Frames and storefront sections are usually either tubes or channels (Fig. 8.4-2). They are typically $1^3/4$ or 2 in. (44.5 or 50.8 mm) thick and vary in depth from 4 to 7 in. (102 to 178 mm). However, other sizes are available.

Frames may be either standard (Fig. 8.4-2) or thermal break. They may be either flush glazed or with applied stops (see Figs. 8.4-3 and Fig 8.4-4).

### THERMAL BREAK

Where design calculations indicate the need to reduce the heat transmittance of storefronts, significant reduction can be accomplished by incorporating double or triple glazing or insulating glass. Where frame design permits, thermal breaks in the frames will also help. At

any rate, these assemblies, when located where the winter design temperature is 10°F ($-12.2$°C) or less, should incorporate such devices, regardless of calculations.

A metal frame that holds insulating glass should have a *thermal break* to prevent a thermal "short circuit," that is, heat flow being resisted by the insulating glass but flowing freely through the aluminum frame. A thermal break separates the metal parts on the warm side of a frame from the metal on the cold side using an insulating material such as polyurethane, vinyl, or another material of low conductivity to keep heat from being conducted through the frame or sash. The insulating separator is also sometimes called a *thermal barrier*.

An aluminum frame with a good thermal break has an R-value of 1.66, which is about equal to that of double-pane insulating glass. A comparable frame with no thermal break would have an R-value of 1.18.

In some areas, it may be economically practical to use three layers of glass separated by air spaces. Three layers of sealed glass with two $1/4$ in. (6.4 mm) air spaces, for example, have an R-value of 2.13 compared to 1.72 for typical insulating glass.

### INTERIOR EXTRUDED FRAMES

Extruded aluminum entrance and storefront components are also used for interior doors and frames. They are available in a number of sizes and shapes to suit various conditions. The requirements for interior frames are similar to those for exterior frames, except that watertightness is not required.

**FIGURE 8.4-3** The shaded thermal break reduces conductive heat transmission through the metal frame.

Thermal break is formed by a glazing gasket　　　　　Thermal break is formed by an imposed barrier

**FIGURE 8.4-4** Two typical frames with a thermal break feature. Many other configurations are also available. (Courtesy H. Leslie Simmons.)

## ALL-GLASS ENTRANCES

All-glass entrances feature doors made of glass without metal frames. They are supported on special pivot-type hardware. Such doors are often fitted with automatic operators.

### 8.4.7 FINISH HARDWARE

Extruded aluminum doors and packaged entrances usually come equipped with all operating hardware except cylinder locks.

Hardware for these entrances may be either standard, handicapped-access, automatic, or security type. Standard hardware includes all the devices necessary to operate, lock, and weatherstrip a door, including thresholds.

Handicapped-access hardware includes counterbalanced closers that permit easy operation of a door and specially designed door pulls with extended or oversized grips to permit easy operation by a handicapped person.

Automatic operators open doors when a electromechanical, hydraulic, or pneumatic operator is triggered by an actuating device, such as a floor treadle, motion sensor, sonic sensor, push button, photoelectric cell, or remote switch. Automatic hardware also includes detectors that automatically close doors when smoke or fire is present.

Security hardware consists of high-security cylinders and special control devices such as electromechanical, electromagnetic, or electronic locks and three-way locks in which a single key operates bolts in the head, jamb, and sill of a door. Security locks may operate by key, push buttons, or card readers.

Refer to Section 8.9 for additional discussion about hardware.

### 8.4.8 INSTALLATION

The finest entrance or storefront design and the highest quality of manufacture will not compensate for poor installation. Weathertightness, although simple to achieve, can only be ensured if the installation is made by experienced workers following the guidelines and instructions supplied by the manufacturer and the appropriate industry standards. A rough opening should be prepared to receive each unit so that it can be installed and will finish out square, plumb, level, straight, and true. The design of most assemblies provides for minor adjustments at the job site, but no unit will operate or weather properly if it is twisted and misaligned during installation.

A properly selected, manufactured, and installed entrance or storefront will be weather resistant within itself. Other things must be done, however, to provide a weathertight installation. For example, spaces between rough openings and units should be filled with insulation, and joints between units and adjacent materials must be closed with sealant, trim members, or both to prevent the passage of air, dust, and water.

Entrance frames and storefront sections are seldom designed to support loads other than themselves. Therefore, unless the frame is specially reinforced to support other loads, a separate lintel or other support is required.

Entrance and storefront frames should be supported around their entire perimeter and firmly and rigidly anchored to supporting construction. They should also be flashed at their heads and sills to provide a path back to the exterior for water that finds its way into the wall.

Some metals, such as aluminum and stainless steel, are compatible and can be used in contact with each other. Other metals, however, react with one another chemically or electrolytically, and should be insulated from direct contact by waterproof, nonconductive materials such as neoprene, waxed paper, or coated felt. Such dissimilar metals that are located where water passing over them may contact another surface should be painted to prevent staining. Refer to Section 5.2.2.4 for a discussion about the effect of dissimilar metal materials.

Concealed aluminum in contact with concrete, masonry, or an absorbent material such as wood, paper, or insulation should be protected by coating either the aluminum or the adjacent material with a bituminous or aluminum paint, or by coating the aluminum with a zinc chromate primer, to minimize the chance of chemical corrosion of the aluminum by acids, alkalis, and salts leached out of the adjacent materials. Creosote and tar coatings, which might damage the aluminum, should not be used.

Sealants should be used in locations where frames, packaged entrances, and storefront sections adjoin other materials to make the joints there airtight and watertight.

Both interior and exterior extruded aluminum entrances and storefronts should be installed in accordance with AAMA's *Aluminum Store Front and Entrance Manual* and the manufacturer's recommendations.

Each interior frame should be anchored with jamb anchors near the top and bottom and not more than 24 in. (610 mm) apart in between. Jambs should be anchored at the floor with adjustable floor anchors.

Anchors in masonry should be long enough to extend at least 8 in. (203 mm) into the masonry. Anchors for stud walls should be clips designed to attach to the types of studs used. Several different methods are used to anchor frames in previously placed masonry or concrete.

### 8.4.9 REVOLVING DOORS

Revolving doors are both always open and always closed. They prevent passage of cold and warm air and noise while permitting people to pass through easily. They are proprietary packaged units that come complete with hardware (Fig. 8.4-5). They are made from stainless steel, aluminum, and copper alloys. Most of them are between 6 and 8 ft (1800 and 2400 mm) in diameter, but some are larger. Both three- and four-wing designs are available. Many options are available. Custom design is frequent.

Revolving doors operate either manually or with a power assist. They have speed controls for the safety of users and protection of the door itself. Their optimal speed is 10 revolutions per minute, but slower operation is often desirable in such settings as hospitals, airports, and

**FIGURE 8.4-5** Revolving doors usually include a canopy, which can be glazed or solid, as shown here. (Courtesy Boonedam Entrance Technology.)

**FIGURE 8.4-6** Large doors are available to permit the passage of wheelchairs or of several people at once. (Courtesy Boonedam Entrance Technology.)

train stations. Slow-down switches are available to ensure safe passage of persons in wheelchairs (Fig. 8.4-6).

Revolving doors can be fitted with manual or automatic locks. Collapsible units are available so that doors provide open access in the event of fire or other emergencies.

### 8.4.10 ADJUSTMENT, CLEANING, AND PROTECTION

Because aluminum entrances and storefronts are prefinished building components, their installation should be delayed as long as the construction schedule will permit. Once installed, they must be protected to prevent damage to their finish, appearance, and ability to operate properly. Units should be protected from splashing of plaster, mortar, and cleaning solutions.

## 8.5 Glazed Aluminum Curtain Walls

A building wall that carries no superimposed loads is called a *curtain wall*. A glazed aluminum curtain wall is an exterior curtain wall that consists essentially of an aluminum (usually extruded) framework infilled with vision panels (fixed or windows) or opaque panels such as spandrels and column covers. Curtain walls are defined as different from *storefronts* in that they generally do not contain entrances and usually occur above the first floor (Fig. 8.5-1). Curtain wall manufacturers call their products *curtain walls*, and storefront manufacturers call theirs *storefronts*. Aluminum doors, frames, entrances, and storefronts are discussed in Section 8.4; windows used in aluminum-glazed curtain walls are discussed in Section 8.6; and glazing in metal and wood curtain walls and structural glass curtain walls are covered in Section 8.10.

This section discusses only aluminum-framed glazed curtain walls, because they are the only type specifically manufactured and sold for use as glazed curtain walls. Their components are manufactured to fit together. Other metals and wood are also used in glazed curtain walls, and some glazed curtain walls are made up of unframed glass with or without metal supporting angles or clips. The principles discussed here apply to curtain walls made from other materials, even though they may contain formed rather than extruded members.

### 8.5.1 SYSTEMS

Most aluminum-glazed curtain walls consist of a manufactured system composed of standard extruded components with metal fill-in panels where glass is not used. Some designers, however, choose to create *custom* systems. Such custom systems may vary in relatively simple ways such as substituting another material, such as stone, for the metal panels in standard systems. Other custom systems, however, have mullions or other components of configurations and sizes different from those of any standard system. It is important for a designer choosing to introduce a custom aluminum curtain wall system to recognize that such systems are inherently more costly than standard systems, that costly testing of such systems is advisable, and that regardless of the results of such tests, the designer will ultimately be held accountable for the performance of the system.

AAMA classifies glazed curtain walls by their installation methods in five systems: *stick, unit, unit-and-mullion, panel,* and *column-cover-and-spandrel.*

#### 8.5.1.1 Stick Systems

Most aluminum-glazed curtain walls are stick systems. They consist of mullions, the extruded vertical elements that provide the major support for the system; vision panels that may be either fixed glass or plas-

**FIGURE 8.5-1** One Skyline Tower, Baileys Cross Roads, Virginia; Architect The Wiehe Partnership. (Photograph courtesy Bruce Wall Systems Corporation.)

tic sheets or windows; and opaque panels, which are usually insulated (Fig. 8.5-2). They may also include horizontal rails, sills, fascias, copings, and trim. The components in most stick systems are stock off-the-shelf pieces used without modification. Some stick systems, however, use custom-made components. Others are a mixture of stock and custom-made pieces.

Stick systems are usually glazed in the field from the interior using gaskets or a glazing compound. Systems with lock strip glazing from the exterior are also available. In these systems, a glazing gasket is inserted into the frame and the glazing is installed in it using a special zip-in strip gasket. Special tools are needed to install this type of glazing system.

*Sloped glazing* is a special type of stick system in which portions are either sloped or curved (Fig. 8.5-3).

### 8.5.1.2 Unit Systems

Unit curtain wall systems are modular panels that are preassembled in the factory and shipped to a construction site in a single piece (Fig. 8.5-4). These panels are usually glazed at the factory and come with spandrel panels already in place. In some designs, field-installed joint covers are required.

### 8.5.1.3 Unit-and-Mullion Systems

Unit-and-mullion systems are a combination of stick and unit systems (Fig. 8.5-5). The framing (vertical) mullions are shipped loose and installed in the field. The infill panels are factory-made units that are inserted between the mullions in the field. In some unit-and-mullion systems, the infill panels contain spandrels and vision panels in a single unit. In others, the spandrels and vision panels are separate units, shipped and installed separately in the space between the mullions.

### 8.5.1.4 Panel Systems

A panel curtain wall system is similar to a unit system, except that a panel system has homogeneous sheet or cast panels with few joints and may not have separate mullions. Unit systems are made up of smaller components. There are two types of panel systems: (1) architectural and (2) industrial.

*Architectural* panels are custom-designed systems that tend to be expensive (Fig. 8.5-6). They are used mostly on very large projects. *Industrial* systems are made from sheet materials and in large quanti-

**FIGURE 8.5-2** Details of a stick-type glazed curtain wall system. (Drawing by HLS.)

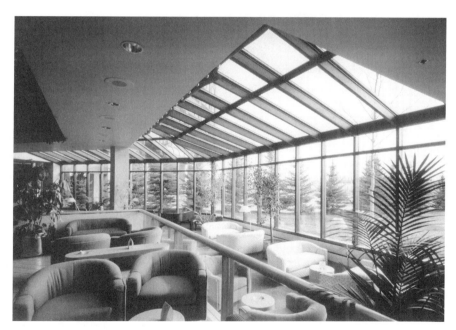

**FIGURE 8.5-3** A sloped glazing system. (Courtesy Tubelit Indal Architectural Systems.)

ties. They are, therefore, relatively inexpensive. These are the types of panels often used on metal buildings (Fig. 8.5-7).

### 8.5.1.5 Column-Cover-and-Spandrel Systems

Column-cover-and-spandrel systems consist of aluminum column covers and spandrels that fit between them. They are similar to a unit-and-mullion system, except that the building's structure is expressed by the column covers. The space between columns is filled with glazing panels, with or without separate mullions, and spandrel panels (Fig. 8.5-8).

**FIGURE 8.5-4** A panel-unit glazed curtain wall system being installed. (Courtesy Kalwall Corporation, Manchester, NH, USA.)

## 8.5.2 STANDARDS

Aluminum glazed curtain walls should conform to the applicable standards of the organizations listed in Section 8.4.1.

In general, aluminum-glazed curtain walls should comply with AAMA publications 501, 1503, CW-10, CW-11, CW-12, CW-13, CW-DG-1, TIR-A1, and TIR-A7. The names of these publications are listed in Section 8.12. Other standards are also mentioned elsewhere in this chapter and listed in Section 8.12.

## 8.5.3 MATERIALS AND FINISHES

Refer to Sections 5.2 and 5.9 for a general discussion about the metals and finishes addressed in this section.

### 8.5.3.1 Materials

A variety of aluminum alloys and tempers are used to make extruded curtain wall components. Alloys should conform with ASTM Standard B221. A common alloy used is 6063. Common tempers are

T5 and T6. Other tempers and alloys may be used, however, depending on the design and finish.

Sheet aluminum is often used for closures and trim, bent plate for structural support and trim. Sheet aluminum and bent plate components should comply with ASTM Standard B209.

Concealed components, such as subframes, stiffeners, edge channels, reinforcements, and the like, should be fabricated from the same materials as the exposed portions of the curtain wall.

Many types of opaque panels are used in glazed aluminum curtain walls. Spandrel glass is a common one. Others include metal-faced, ceramic tile-faced, stone, brick, and fiber-reinforced cementitious, color-coated materials. Many stock systems come with aluminum-faced panels. Some systems will accommodate only certain types of panels. Therefore, panel types must often be selected before the major framing components can be chosen. For example, many stock sections will not accept stone pan-

els. They must either be adapted or a different section selected.

Louvers and grills are also common. These may be part of a system's opaque panels or be themselves infill between the mullions. They may be glazed into a system using glazing beads or gaskets or may be mounted mechanically into the frames.

### 8.5.3.2 Finishes

The mill finish is usually left on concealed surfaces and sometimes even on the exposed surfaces of glazed aluminum curtain wall components. However, most exposed portions are either anodized or finished with a fluoropolymer (see Section 5.9).

Exposed fasteners should be finished to match the adjacent metal.

### 8.5.3.3 Supports, Anchors, Fasteners, and Flashing

Supports, anchors, fasteners, window cleaners' bolts, and flashing materials

FIGURE 8.5-5 A unit-and-mullion glazed curtain wall system (Drawing by HLS.)

Labels in figure: Mullions (usually one or two stories long); Floor construction; Glazing; Panel; Opaque panel; Floor construction; Previously placed panels

FIGURE 8.5-6 Architectural panels may take almost any form that can be made in a metal shop, cast in concrete, or molded in plastic. (Drawing by HLS.)

FIGURE 8.5-7 Industrial panels take many shapes, including corrugated ones. (Drawing by HLS.)

should be of materials that are compatible with the material being anchored or fastened and with the substrate. Aluminum should be anchored, fastened, and flashed with aluminum, nonmagnetic stainless steel, cadmium-plated steel, or galvanized steel. Occasionally a manufacturer's design will require the use of aluminum castings as structural elements. This is acceptable when the design takes the strength characteristics of the casting into account. However, aluminum castings are not generally used this way, because they sometimes lack sufficient shear strength.

Inserts for use in concrete may be cast iron, malleable iron, or hot-dip galvanized steel.

## 8.5.4 PERFORMANCE REQUIREMENTS AND TESTING

Standard glazed extruded aluminum curtain wall systems should have been tested by their manufacturer or by an independent testing agency to ascertain their resistance to deflection due to wind loads, water penetration, air infiltration, condensation resistance, thermal transmittance, and, where applicable, sound transmission. The results of these tests should be ob-

tained from the manufacturer and taken into consideration when selecting a system for a particular project. The characteristics of the selected system should be compatible with conditions that are expected

**FIGURE 8.5-8** A column-cover-and-spandrel system. The glazed panels may contain windows or may be made up of strip windows. (Drawing by HLS.)

to exist at the project site, applicable code requirements, and the demands of the occupancy.

Testing the installed system is advisable for custom-designed systems and often also for standard systems with special requirements, such as those in very tall structures. Such additional testing is expensive, however, and is seldom justified for standard systems or for relatively small projects. This testing should be done by an independent testing laboratory (one not affiliated with the manufacturer or the contractor). It may take place on a mockup of the actual system constructed in the testing laboratory's testing facilities before work on the curtain wall at the building begins. This mockup should be built exactly like the actual work on the building. As an alternative, these tests may take place on a mockup of the curtain wall constructed at the site. This mockup is sometimes separate from the building and sometimes is a portion of the actual curtain wall. When the designer approves, a satisfactory field-constructed mockup is sometimes incorporated into the building.

The performance characteristics of a system are determined by the effect on the system when tested at specific levels of wind pressure. The test pressure to use in these tests is determined using the fastest wind speed likely to be encountered at a building's location. Data in the American Society of Civil Engineers' (ASCE) "Minimum Design Loads for Buildings and Other Structures" can be used to estimate this maximum wind speed. After the expected wind velocity has been determined, the resulting wind loads to be used are determined from load tables. Such load tables are based on wind velocities ranging from 70 mph (112.65 km/hr) to 110 mph (177 km/hr) and the height of the curtain wall above ground from zero (at grade) to 500 ft (152.4 m).

The test pressures given in this section represent those recommended by the listed standards. Higher pressures may be required for a given project.

Sometimes it is necessary to reinforce aluminum mullions and other components to make them comply with performance requirements.

### 8.5.4.1 Structural Performance

A glazed aluminum curtain wall system should be tested using the expected wind loading and the procedures listed in ASTM Standard E330. The test method in ASTM Standard E1233 may also be used.

Vertical systems should be tested both perpendicular and parallel to the plane of the wall and in both the inward-acting and outward-acting directions. Sloped glazing should also be tested for upward- and downward-acting loads, and in addition should be tested for a uniform load, usually of 200 lb (90.72 kg) acting on the system at any point. This is done to ensure that a person walking on the system will not damage it. This load may be higher if a larger concentrated load will be superimposed on the system.

The test pressures for perpendicular-to-the-wall tests should be 150% of the wind pressures to be encountered. Pressures may vary with the system's elevation above ground and proximity (within 15 ft [4572 mm] or so) to corners, where pressures are higher. The actual loads to be used should be established by a registered structural engineer. Some building codes dictate the minimum pressures to be used.

The maximum allowable deflection under these loads is usually $1/175$ of a span, but project conditions and the system manufacturer's requirements may require limits of as much a 1/1000 of a span. In any case, the maximum deflection should not exceed $3/4$ in. (19.1 mm) over the total length of any member. In addition, under the selected loads:

- No panels should be dislodged.
- No permanent deflection should occur.
- Sealant joints should not be reduced to less than half of the designed width.

The allowable movement parallel to the wall is dictated by panel edge conditions. The bite on glazing and opaque panels should not be reduced to less than 75% of the design value. In addition, the edge clearance between panels and other elements should not be reduced to less than $1/8$ in. (3.2 mm). Greater restrictions may be imposed by the method in which the system is assembled.

### 8.5.4.2 Water Penetration

The water penetration resistance of a glazed aluminum curtain wall system is determined by testing it in accordance with ASTM Standard E331. The test pressure suggested by AAMA 501 is 20% of the inward design wind load pressure but not less than 6.24 psf (30.47 kg/m²) or more than 12 psf (58.59 kg/m²). For a system to pass this test, no water may pass through it during the test.

### 8.5.4.3 Air Infiltration

Air infiltration is tested in accordance with ASTM E283. The usual pressure is 1.57 psf (7.67 kg/m²); the permitted infiltration is 0.06 cfm (0.002 m³/min) per square foot (0.09 m²) of fixed area.

### 8.5.4.4 Condensation Resistance

A metal frame that holds insulating glass should have a *thermal break* as described in Section 8.4.6.4.

An aluminum frame with a good thermal break (Fig. 8.5-9) has an R-value of 1.66, which is approximately equal to that of double-pane insulating glass. A comparable frame with no thermal break would have an R-value of 1.18.

Condensation resistance is tested for frames having a thermal break feature. A system's condensation resistance is an indication of that system's ability to resist condensation. It is determined by procedures outlined in AAMA 1503. The higher the value, the less likely it is that condensation will occur.

### 8.5.4.5 Thermal Transmittance

As discussed in Section 17.1, heat will flow quite readily through aluminum curtain walls unless some means are taken to reduce this flow. The methods for determining conductive heat transmission through these assemblies are included in AAMA 1503.

In addition, it is usual to require that an aluminum-glazed curtain wall be able to withstand the highest ambient temperature differential that it will encounter without damage of any kind. In most parts of the country the ambient temperature difference is about 120°F (48.89°C), which can result in a metal surface temperature differential of 180°F (82.2°C). Some extreme southern and northern areas may experience even larger ambient temperature differentials.

### 8.5.4.6 Sound Transmission

When a specific sound transmission level is required, an aluminum-glazed curtain wall should be tested in accordance with ASTM Standard E90 to establish its Sound Transmission Class (STC). The required STC depends on the project and its location. Buildings adjacent to high noise levels, such as one adjacent to the flight pattern of an airport or near a major highway, will require much higher STCs than buildings in other settings. A manufacturing facility will not require as high an STC as a school, office building, or apartment unit.

### 8.5.5 DESIGN AND FABRICATION

Aluminum-glazed curtain wall systems designed to exclude water simply by having tight or sealant-filled joints are not always successful. This is because the pressure within the system is often lower than the pressure on its exterior, causing water to be sucked into every fault faster than it can run out through weep holes. Water can even be sucked in through weep holes unless they are placed where water cannot reach them.

The answer to this dilemma is to design the system so that minor amounts of water are permitted to infiltrate the system, but are then conducted back to the exterior through a series of water channels and weep holes without entering the building. All sealant-filled joints are located behind the water channels where water cannot reach them. This mechanism works on what is called the *rain screen principle*. A portion of the frame screens the sealants so that water does not come into contact with them. Sealants in curtain walls are called *differential seals*, meaning that the difference in pressure on the two sides of the sealant is almost always more than 1 psf (4.88 kg/m²). Water will penetrate almost all sealed joints under those circumstances. However, the water channels are open to the air, so that pressure within them is the same as that outside the screen; thus, there will be no pressure differential to suck water into the channels. Water that does enter the channels will be quickly led back to the exterior without reaching the sealant joints. Therefore, systems with this type of design are less likely to leak. The concealed sealants should be installed in the shop when possible.

In sloped glazing systems, it is also important to prevent water that runs across the system from building up against horizontal members, where it can cause leaks. This is usually done by a series of gutters and downspouts designed into the system to collect water and lead it quickly to the ground.

Even with these safeguards, glazed curtain walls will sometimes leak through imperfections in glazing seals, incompletely sealed joints, or other inadvertent openings. In addition, even carefully designed systems will sometimes collect condensation. Therefore, a water-collection and removal system on the interior of such systems is advisable. This is particularly important on sloped glazing systems, where interior gutters are often advisable.

Glazed aluminum curtain wall systems come complete with accessories such as glazing gaskets, glazing stops, clips, trim, fins, flashings, anchors, and fasteners. Normal glazing gaskets are made from one of several common materials, including, but not limited to, extruded PVC, extruded or molded neoprene, and extruded or molded ethylene propylene diene monomer (EPDM). Lockstrip-type gaskets are usually structural rubber. Incidental components, such as closers, channels, flats, trim, sleeves, and the like, are also usually delivered with the systems.

Components should be cut and fitted to make hairline joints, except where wider joints are needed to accept sealants or to permit thermal and other movement in the system. Joints in exterior frames should be designed and erected to be watertight. Joints between two pieces of metal should be filled with a nonhardening sealant. Welds should be made on the inside so that they will not be visible in the finished installation.

Mullions are usually either tubes or structural shapes, such as T or I shapes. Structural-shape mullions are often, but not always, encased in a rectangular cover. Mullions vary considerably in size and shape. They may be either standard (Fig. 8.5-10) or thermal break (see Fig. 8.5-9). Most systems are flush glazed (see Fig. 8.5-9), but face (butt)-glazed systems are also used (Fig. 8.5-11).

A system and its anchorages should be designed to permit both horizontal and vertical movement due to thermal

**FIGURE 8.5-9** A typical glazed curtain wall mullion containing a thermal break feature. (Drawing by HLS.)

Plastic thermal break material

Insulating glass

**FIGURE 8.5-10** A typical standard glazed curtain wall mullion. (Drawing by HLS.)

**FIGURE 8.5-11** A typical glazed curtain mullion with butt glazing. (Drawing by HLS.)

changes in the system and building and deflection in the structure.

### 8.5.6 INSTALLATION

The finest glazed curtain wall design and the highest quality of manufacture will not compensate for poor installation. Proper installation can only be achieved if the work is done by experienced workers following the guidelines and instructions supplied by the manufacturer and the appropriate industry standards. Aluminum glazed curtain walls should be installed in accordance with AAMA's *Aluminum Curtain Wall Design Guide Manual* and the manufacturers recommendations.

Glazed curtain wall installations include many components and can become very complicated. For example, in addition to the components necessary for a basic system, curtain walls include other elements that would ordinarily be assigned to different trades. These include such items as louvers, copings, fascias,

flashings, parapet caps, operable windows, venetian blind tracks, drapery rods, insulation, firestopping, heating, ventilating, and air conditioning unit enclosures, solar shades, masonry infill, and joint sealing. It is, therefore, better that all work directly associated with glazed aluminum curtain walls be done by a single subcontractor to ensure proper coordination of the different elements. This often means that the curtain wall manufacturer is also made responsible for the design and installation of the curtain wall system and all related elements. Unfortunately, not all manufacturers are qualified to offer such a service, and union jurisdictions sometimes prevent it as well. Therefore, single-subcontractor responsibility is often not possible, except with stock stick systems with few of the additional items mentioned.

Curtain wall components should be protected during shipment and handling, and should be installed as soon as possible after they reach the project site to prevent them from being damaged. The substrates should be installed and prepared to receive curtain wall components so that they can be properly installed and will finish out square, plumb, level, straight, and true. The design of most systems provides for minor adjustments at the job site, but no system will function properly if its framing members are twisted or misaligned during installation.

A properly designed, manufactured, and installed aluminum glazed curtain wall system should be weather-resistant as a unit. Other things must be done, however, to achieve a completely weathertight installation. For example, joints between system components and adjacent materials must be closed with a sealant, trim members, or both to prevent the passage of air, dust, and water around the system.

Glazed aluminum curtain walls should be supported around their perimeter and anchored securely to the supporting construction. They should also be flashed where appropriate to provide a path back to the exterior for water that finds its way into the wall.

The installed system should be plumb, level, and in line within specified tolerances. Permissible tolerances vary from project to project and from system to system. However, systems are usually required to be plumb within $^1/_8$ in. (3.2 mm) in any 10-ft (3048-mm)-long segment with a maximum of $^1/_4$ in. (6.4 mm) in

40 ft; level to within $^1/_8$ in. (3.2 mm) in any 20-ft (6096-mm)-long segment, with a maximum of $^1/_4$ in. (6.4 mm) in 40 ft; and in line within $^1/_8$ in. (3.2 mm), except that the limit should be $^1/_{16}$ in. (1.6 mm) when the elements are supposed to be flush with each other. Systems should also be within $^3/_8$ in. (9.5 mm) of their planned location.

Curtain wall installation should allow for changes due to thermal movement in the system and in the building and for building structure deflection.

Metals with dissimilar electrolytic or chemical characteristics should be insulated from direct contact with each other by waterproof, nonconductive materials such as neoprene, waxed paper, or coated felt. Such dissimilar metals located where water passing over them may contact another surface should be painted to prevent staining. Refer to Section 5.2 for a discussion about the effect of dissimilar metal materials.

Concealed aluminum in contact with concrete, masonry, or an absorbent material, such as wood, paper, or insulation, should be permanently protected by coating either the aluminum or the adjacent material with a bituminous or aluminum paint, or by coating the aluminum with a zinc chromate primer, to minimize the chance of chemical corrosion of the aluminum by acids, alkalis, and salts leached out of the adjacent materials. Creosote and tar coatings might damage the aluminum and should not be used.

Sealants should be used within the system, in concealed locations where needed, to make the system airtight and watertight.

Firestopping should be installed in glazed aluminum curtain wall systems where required by code or other governing regulations. It should always be installed at the edges of floor construction where a glazed curtain wall passes. Firestopping in such locations is usually accomplished by filling the space between the floor construction and the curtain wall with a glass fiber or mineral wool insulation specifically made for this purpose.

### 8.5.7 GLAZING

Glazing material selection depends on wind load requirements, thermal requirements, aesthetics, and safety. The type and thickness are often established by the applicable building code. For example, some codes state that glazing materials close to floors must meet the requirements for "safety type" and be in compliance with ANSI Z-97.1 or another similar standard. Refer to Section 8.9 for a discussion of glass and glazing.

Aluminum curtain wall systems are usually glazed with $1/4$ in. (6.4 mm) glass, except where insulating glass is desired. Glazing stops and gaskets should be of the correct size to receive the glass to be used.

Where design calculations indicate the need to reduce heat transmittance, significant reduction can be accomplished by incorporating insulating glass. Thermal breaks in the frames will also help. At any rate, curtain walls located where the winter design temperature is $10°F$ ($-12.2°C$) or less should incorporate such devices regardless of calculations.

Some curtain wall components can be glazed at the factory, which is the most satisfactory method. Stick systems, of course, must be glazed in the field. When glazing is done at the job site, the curtain wall manufacturer's recommendations should be followed to ensure a satisfactory result.

### 8.5.8 ADJUSTMENT, CLEANING, AND PROTECTION

Immediately after aluminum-glazed curtain walls are installed, they should be cleaned. Thereafter, curtain walls should be protected to prevent damage to their finish, appearance, and ability to function properly. Units should be protected from splashing of plaster, mortar, paint, cleaning solutions, and other foreign substances.

# 8.6 Metal Windows and Sliding Glass Doors

Metal windows are made from aluminum, steel, and other metals. Metal sliding glass doors are made primarily from aluminum. This section addresses aluminum windows and sliding glass doors and steel windows.

### 8.6.1 GENERAL REQUIREMENTS FOR METAL WINDOWS AND SLIDING GLASS DOORS

Aluminum and steel windows are similar in many ways. Types, for example, are very much the same. Installation is basically the same. There are differences, of course. Most steel windows are face glazed, while most aluminum windows are glazed with gaskets or aluminum stops. Many aluminum windows are available with thermal-break construction; few steel windows are so equipped.

#### 8.6.1.1 Glazing

Glass and glazing materials are discussed in Section 8.9. The standards discussed there apply to glazing in metal windows and sliding glass doors. Most windows are glazed with glass, but plastic glazing is also used occasionally, and some windows are even glazed with opaque panels. The glass used may be single panel, multiple pane, or insulating glass.

Many aluminum and steel windows and most sliding glass doors have the glass installed at the factory, which eliminates glazing on the job site and allows controlled methods of glazing not feasible with on-site glazing.

In one method of factory glazing, a glazing compound is applied to the glazing leg of a window as a back-bedding material; the glass is then dropped in and held in place by aluminum or rigid plastic face stops. The stops snap into place to serve as trim molding and hold tension on the assembly until the glazing compound has set. Another method widely used in the factory to glaze units is to apply a channel-shaped gasket over the edge of the glass and then force the channel shape into the window or door frame around it. Flexible vinyl plastic is most often used for this application.

When windows or sliding glass doors are glazed at the job site, the manufacturer's recommendations should be followed to ensure a satisfactory result.

Glass selection by size and grade for windows depends primarily on wind load requirements and is often established by the applicable building code. In the absence of code requirements, recommendations of AAMA 101/1.S.2 can be used. For sliding glass doors, in addition to wind load requirements, the glass must be of "safety type" and be in compliance with ANSI Z-97.1 or another similar standard as regulated by codes or laws.

Other functional or esthetic requirements often dictate the selection of different types and higher grades of glass. These include double glazing to minimize heat loss and heat gain; tinted glass for privacy and decorative purposes; tem-

pered glass for additional strength; and tempered, wire, or laminated glass for additional safety.

Glazing materials for both aluminum windows and sliding glass doors include gaskets made from neoprene or PVC and mastic-type compounds such as butyl- or rubber-base materials that are compatible with aluminum and do not require painting.

#### 8.6.1.2 Weatherstripping

Some steel windows have no weatherstripping, but others do. Aluminum windows and sliding glass doors are weatherstripped. The function of weatherstripping is to control air infiltration, prevent rain penetration, minimize dust infiltration, and withstand atmospheric conditions. It typically has a resilient part that prevents air penetration when it is pressed between the window and its sash.

Weatherstripping must be durable under extended usage, resist corrosion, and be readily replaceable. Many aluminum windows and sliding glass doors are equipped in the factory with fabric pile weatherstripping (similar to that used in automobiles) or with flexible vinyl plastic neoprene gaskets or fins. In some cases, these materials are combined. Thin stainless steel, which compresses to achieve a seal, is sometimes used.

#### 8.6.1.3 Accessories

Window accessories are assembled and attached during all stages of fabrication and assembly. For example, weatherstripping is frequently pulled into retaining grooves in long lengths of extruded frame members before cutting and punching take place. Operating hardware such as sash balances, pivot arms, guides, and guide followers are usually assembled after the main frames are together. Accessories such as locking hardware and handles are attached by riveting, bolting, screwing, staking, or interlocking with frame members. Manually rotating operators on casements, awnings, and jalousie windows are usually the last accessories to be assembled, sometimes being left until after installation of the unit at the job site.

### 8.6.2 ADDITIONAL REQUIREMENTS FOR ALUMINUM WINDOWS AND SLIDING GLASS DOORS

The use of aluminum for window construction has increased significantly since aluminum windows were introduced for industrial application in the United States about 60 years ago. In contemporary buildings, glass areas have increased until the size of aluminum windows permits them to provide entire walls of glass and, in the case of aluminum sliding glass doors, movable walls of glass.

### 8.6.2.1 Manufacture

The manufacture of aluminum windows and sliding glass doors consists of fabricating wrought aluminum alloys into the correct size and shape, using various techniques of die cutting, punching, notching, piercing, and sawing. The aluminum members thus fabricated are assembled with glass, hardware, weatherstripping, and other components into finished preassembled units.

The workability of aluminum, coupled with the ease and simplicity with which aluminum windows and sliding glass doors can be manufactured, has contributed immeasurably to their widespread availability and use. Modern machinery and production techniques produce a finished product whose parts can be fabricated to extremely close tolerances, resulting in weathertight assemblies having precision operation at relatively low cost.

### MATERIALS

Pure aluminum has few commercial applications, but when it is combined with other metals, such as manganese, silicon, iron, and magnesium, alloys are formed that are useful in a variety of applications. Refer to Section 5.2 for additional information about aluminum.

The principal materials used in aluminum windows and sliding glass doors are wrought (worked into shape by mechanical means) alloys. The alloys used for windows and sliding glass doors must possess strength, be resistant to corrosion, be readily worked and machined for efficient fabrication and simple assembly, and provide surfaces suitable to receive protective and decorative finishes. Although other alloys are also used, the wrought aluminum alloy known as 6063 is a common one for making aluminum extrusions. It contains silicon and magnesium, and possesses good formability and corrosion resistance with medium strength. To make aluminum products, aluminum alloy is formed by the extrusion process and artificially aged by heat

treatment. After heat treatment, usually in temper designation T5 (or T6), this alloy, known commercially as 6063-T5 (or 6063-T6), is commonly used for the main frame, sash, and ventilator members in both windows and sliding glass doors. Extruded shapes permit infinite possibilities of cross-sectional configuration to satisfy the requirements of glazing, weatherstripping, condensation control, and assembly.

Window and door locks, hinges, handles, and corners are often made from castings (formed by pouring or forcing molten metal into a mold) of aluminum alloy.

Other aluminum wrought alloys made by roll forming (passing strips of flat aluminum sheet through a series of forming rolls that bend it into a tube or other shape) also have a variety of uses in aluminum windows and sliding glass doors. These include such shapes as screen frames and glazing strips. Aluminum sheet is also cut and formed by stamping in a punch press where accessories such as corner locks and screen clips can be made.

### FINISHES

The silvery sheen of the natural mill finish surface of aluminum is a common finish used on low-cost aluminum windows and doors. This natural oxide coating, which forms on aluminum, provides an adequate degree of protection against corrosion in normal atmospheres and can be readily restored by rubbing with stainless steel wool if damaged or discolored. Normal oxidation over time will convert the silvery sheet to a soft gray. Other finishes often used may be either protective or decorative.

***Protective Finishes*** Protective finishes are necessary when an abnormal atmosphere is present, such as in heavily industrial and seacoast areas. Protective finishes are also necessary over certain types of decorative finishes, particularly where the surface is etched or polished.

Anodizing, an electrolytic process that provides a heavy oxide film on the surface, is a permanent protective finish against moderate abrasion and atmospheric attack. Clear anodizing, which has a frosty appearance when an etching material is used during the cleaning cycle, is the most economical finish when corrosion protection is a concern.

Certain permanent color tones on aluminum are possible using the anodizing

process. These include several shades of gray, bronze, gold, and black. They are produced by varying the alloy constituents, the electric current, the electrolyte, and the time cycle. Practically any color is possible by impregnating the porous surface of an anodic film with dye prior to sealing. But since dye colors are not always permanent, they are not recommended for exterior use. (See Section 5.9 for a discussion about anodizing and other finishing process.)

Organic coatings, such as paint, that are sometimes applied to aluminum for decorative purposes, may also provide some protection (see Section 9.15).

***Decorative Finishes*** Decorative finishes, such as a backed synthetic resin enamel, are colored or opaque coatings. They can be expected to have a life of up to 20 years when the coating is applied and baked on at the factory under controlled conditions. Organic coatings should comply with AAMA 2603.

### FABRICATION

Fabrication of aluminum alloy sections into window or sliding glass door components is accomplished by sawing, drilling, punching, and other traditional methods of metal working. Small punch presses are often used. Aluminum's machinability permits the use of a variety of intricate punches and dies in a press to cut, form, shape, notch, and pierce the metal. Thus, components can be joined accurately and easily to make weathertight assemblies. A multiple-punch press can cut a long extrusion to the proper length and punch the required notches, holes, and slots, producing a complete framing member in a single operation.

### ASSEMBLY

Materials other than aluminum, such as glass, plastics, and steel, are required to complete the assembly of windows and sliding glass doors.

***Fasteners and Hardware*** A variety of fasteners and hardware are required to hold aluminum window and sliding glass door parts together and to make them operable. Although aluminum is used extensively for this purpose, other metals are also used. These other metals must be either compatible with aluminum or insulated from direct metal-to-metal contact with it to prevent electrolytic corrosion (galvanic action). Therefore, mild steel fasteners must be plated with zinc, cadmium, or another suitable material. Nonmagnetic stainless steel with a chromium content of not less than 16% need not be plated and can be used in exposed locations. Nylon, polypropylene, and other plastic materials are used for slides, bushings, bearings, and similar contact surfaces.

Anchoring materials and flashing should be of aluminum or material compatible with aluminum, such as stainless steel, cadmium-coated steel, or galvanized steel. Dissimilar materials, such as copper or bronze, should be insulated from direct contact with aluminum by waterproof, nonconductive materials, such as neoprene, waxed papers, or coated felts. Dissimilar metals located where water passing over them may contact a window or door should be painted to prevent staining the aluminum. Refer to Section 5.2 for a discussion about the effect dissimilar materials can have on aluminum.

Concealed aluminum in contact with concrete, masonry, or absorbent material, such as wood, paper, or insulation, should be permanently protected by coating either the aluminum or the adjacent material with a bituminous or aluminum paint, or by coating the aluminum with a zinc chromate primer, to minimize the chance of chemical corrosion of the aluminum by acids, alkalis, and salts leached out of the adjacent materials. Creosote and tar coatings, which might damage the aluminum, should not be used.

***Frame Assembly*** The main frame, sash, and ventilators of aluminum windows and sliding glass doors are assembled by screwing or welding.

*Screwed* assemblies are shipped from the factory KD (knocked down) and later assembled by a distributor, dealer, or contractor. A slot to receive the screw threads is often extruded as an integral part of the aluminum member, and holes are prepunched to ensure perfect alignment of the corner joints (see Fig. 5.3-4). Lower corners of the main frame are usually sealed with a mastic to make them watertight. Units that are completely assembled at the factory are either staked (see Fig. 5.3-7) or welded together.

Several welding methods are used. *Flash* (butt) welding produces a continuous weld across the entire cross section of a shape. It is made by forcing two members together at their intersection under pressure and applying a voltage which melts the parent metal. Flash welding uses no filler material or flux and forms a rigid, strong, neat, watertight joint.

*Inert gas* welding uses filler material and is similar to conventional gas welding used on other metals. It is seldom continuous and is generally used to join members that are of dissimilar shape. Inert-gas-welded joints can be made reasonably watertight, but a mastic sealing material may be required at the lower corners to prevent water leakage.

***Thermal Break*** A metal frame that holds insulating glass should have a thermal break to prevent a thermal "short circuit," in which heat flow is resisted by the insulating glass but flowing freely through the aluminum frame. A thermal break is polyurethane, vinyl, or another low-conductive material that provides a separation between the outside and inside portions of a frame.

An aluminum frame with a good thermal break has an R-value of 1.66, which is approximately equal to that of double-pane insulating glass. A comparable frame with no thermal break would have an R-value of 1.18.

In some areas, it may be economically practical to use three layers of glass separated by air spaces. Three layers of sealed glass with two $1/4$ in. (6.4 mm) air spaces, for example, have an R-value of 2.13 compared to 1.72 for typical insulating glass. This technique of triple glazing can also be accomplished by installing storm sash over double-glazed windows.

## 8.6.2.2 Standards of Quality

The design and construction of aluminum windows and sliding glass doors should conform to AAMA 101/1.S.2, which is the recognized standard for the aluminum window and sliding glass door industry. It was developed by AAMA and adopted by ANSI. It includes minimum provisions for frame strength and thickness, corrosion resistance, air infiltration, water resistance, wind load capability, and other requirements necessary to ensure adequate performance.

### CLASSES

For rating purposes, AAMA lists aluminum windows in five performance classes as follows:

R: Residential
C: Commercial
LC: Light Commercial
HC: Heavy Commercial
AW: Architectural

The first four (R, C, LC, and HC) are also applicable to sliding glass doors.

### DESIGNATION SYSTEM

Aluminum windows and sliding glass doors are given designations that describe the type of window or door, its performance class, its design pressure, and other information. For example, DH-DW-R-15 conveys the following information: The first letters describe the window or door type (DH means double-hung window, C means casement window, SGD means sliding glass door, etc.); DW signifies a dual window; R means Residential grade (C means Commercial grade, HC means Heavy Commercial grade); and the number designates the design pressure (DP).

### R, C, LC, and HC WINDOWS AND SLIDING GLASS DOORS

Most windows used in low-rise buildings fall into either the R, C, LC, or HC class.

*Types* Various types of aluminum windows and doors covered in the R, C, LC, and HC classes are illustrated and described in Figure 8.6-1.

*Performance Requirements* AAMA-certified windows and sliding glass doors are required to have certain performance characteristics. These characteristics are determined by the effect on the window or sliding glass door when tested at specific levels of design pressure (DP), structural test pressure (STP), water resistance test pressure (WTP), and air infiltration test pressure and rate of infiltration.

Condensation resistance and heat transmission are two other factors that should be taken into account when selecting windows, depending on a building's location, size, and design.

Selection of a window or sliding glass door begins by determining the type desired (awning, double hung, projected, sliding glass door, etc.). Then the required DP is calculated. R windows and sliding glass doors are rated at a minimum DP of 15 lb/sq ft (psf) (73.24 kg/m²); C windows and sliding glass doors at a minimum of 20 psf (97.65 kg/m²); and HC units at a minimum of 40 psf (195.3 kg/m²). Windows in high-rise structures

**FIGURE 8.6-1** Summary of Aluminum Windows and Sliding Glass Doors

| Type | Designation and Application | Description |
|---|---|---|
| **Double Hung** | DH-R15 Residential<br>DH-DW-R15 Dual Window Residential<br>DH-C20 Commercial<br>DH-DW-C20 Dual Window Commercial<br>DH-HC40 Heavy Commercial<br>DH-DW-HC40 Duel Window Heavy Commercial | *Double-hung windows* have two operable sash. The sash move vertically within the main frame with the assistance of mechanical balancing mechanisms that minimize the effort required to raise the sash. 50% of the window area is available for ventilation. |
| **Casement** | C-R15 Residential<br>C-C20 Commercial<br>C-HC40 Heavy Commercial | *Casement windows* contain side hinged ventilators that operate individually and swing outward. They frequently have nonoperating fixed lights (shown) which the ventilators close to. Casements may be open 100% for ventilation. |
| **Projected** | P-R15 Residential<br>P-C20 Commercial<br>P-HC40 Heavy Commercial | *Projected windows* consist of horizontally mounted ventilators that may project out or in. They differ from awning and hopper windows in that the hinged side of the ventilator moves in a track up or down when the ventilator is operated. Used commonly where large glass areas are desired and where it is desired to deflect incoming air in an upward direction. |
| **Awning** | A-R15 Residential<br>A-C20 Commercial | *Awning windows* consist of a number of top-hinged ventilators operated by a single control device that swings the lower edges outward giving an "awning" effect when open. Operation is through a roto operator which also serves as the lock. Ventilating area is 100% of the window area. |
| **Horizontal Sliding** | HS-R15 Residential<br>HS-DW-R15 Dual Window Residential<br>HS-C20 Commercial<br>HS-DW-C20 Dual Window Commercial<br>HS-HC40 Heavy Commercial<br>HS-DW-HC40 Duel Window Heavy Commercial | *Horizontal sliding windows* have one or more operable sash arranged to move horizontally within a main frame. Ventilating area is 50% of the window area. This type of window is sometimes combined with fixed lights and constructed in large sizes reaching the proportion of a "window wall." |

*(Continues)*

**FIGURE 8.6-1** *(Continued)*

| Type | Designation and Application | Description |
|---|---|---|
| **Jalousie** | J-R15 Residential | *Jalousie windows* (louver windows) consist of a series of overlapping horizontal glass louvers that pivot together in a common frame. Ventilating area is 100% of the window area. Jalousies are not as resistant to air and dust infiltration as other types, but are used in moderate climates and as porch enclosures in all climates. They should be used only where large heat losses or heat gains can be tolerated. Insulating effectiveness may be improved by inside storm panels, designed to be interchangeable with screens. |
| **Jal-Awning** | JA-R15 Residential | *Jal-awning windows* resemble an awning window in appearance but use different operating and locking mechanisms. These mechanisms are usually separate and require individual operation. Ventilating area is 100% of the window area. |
| **Vertical Sliding** | VS-R15 Residential<br>VS-DW-R15 Dual Window Residential | *Vertical sliding windows* resemble double-hung windows in appearance but use no sash balancing devices. They are operated manually by lifting and are held in various open positions by mechanical catches that engage in the jamb or hold by friction. Ventilating area is 50% of the window area. |
| **Vertical Pivot** | VP-C20 Commercial<br>VP-HC40 Heavy Commercial | *Vertically pivoted windows* consist of a ventilator sash mounted on pivots located in the center of the top and bottom main frame members. This allows the ventilator to be reversed by rotating and permits the glass to be cleaned from the interior. This window is not used for normal ventilation and is operated solely for cleaning purposes or emergency ventilation. It is used generally in completely air-conditioned high rise buildings. |
| **Top-Hinged** | TH-C20 Commercial<br>TH-HC40 Heavy Commercial | *Top-hinged, inswinging windows* consist of a ventilator hinged to the main frame at the top that swings into the room. This permits the glass to be cleaned from the interior. This window is not used for normal ventilation and is operated solely for cleaning purposes and emergency ventilation. Its use is generally in completely air-conditioned high-rise buildings. |

*(Continues)*

**FIGURE 8.6-1** *(Continued)*

| Type | Designation and Application | Description |
|------|------------------------------|-------------|
| **Sliding-Glass Door**  | SGD-R15 Residential<br>SGD-C20 Commercial<br>SGD-HC40 Heavy Commercial | *Sliding glass doors* consist of two or more framed glass panels contained in a main frame designed so that one or more of the panels move horizontally in the main frame. The most common arrangement is of two panels. "Single-Slide" has one operable panel, "Double-Slide" has two. The ventilation area of a single- and double-slide door is 50% of the main frame area. Doors are also made in three- and four-panel arrangements: three-panel arrangement can provide one-third or two-thirds of its area for ventilation, and four-panel arrangement can provide one-half or in some cases three-fourths of its area for ventilation. |
| **Fixed**  | F-R15 Residential<br>F-DW-R15 Dual Window Residential<br>F-C20 Commercial<br>F-DW-C20 Dual Window Commercial<br>F-HC40 Heavy Commercial<br>F-DW-H40 Dual Heavy Commercial | *Fixed windows* may utilize any of the other types. They consist of a frame with no operating sash. Fixed windows may have mullions or muntins. |

may require much higher design pressures than these minimums, varying up to as high as 100 psf (488.24 kg/m²).

DP is determined using the fastest wind speed likely to be encountered at a building's location. A map in AAMA 101/1.S.2 can be used to estimate this maximum wind speed. After the expected wind velocity has been determined, the resulting VP, DP, STP, and WTP are determined from load tables in AAMA 101/1.S.2. These load tables are based on wind velocities from 70 mph (112.65 km/hr) to 110 mph (177 km/hr) and the height of a window above the ground from zero (at grade) to 500 ft (152.4 m). The *velocity pressure* (VP) taken from these load tables varies with building height above grade and wind speed. For example, a 70 mph (112.65 km/hr) wind will produce a VP of 10 at grade and 17.3 at 100 ft (30.48 m) up. VP is determined by the formula

$$\text{VP} = 0.00256 \, K_Z (IV)^2$$

where

$V$ = the wind velocity in miles per hour taken from the map in AAMA 101/1.S.2.

$I$ = the importance factor as described in AAMA 101/1.S.2.

$K_Z$ = the velocity pressure exposure coefficient as described in AAMA 101/1.S.2. This depends on the height above ground and the exposure category.

DP is defined as 1.25VP, and the *uniform load* STP for aluminum windows and sliding glass doors is 1.25DP.

The *water resistance* of AAMA-rated windows and sliding glass doors is determined by testing them in accordance with ASTM Standards E331 and E547. WTP is established at 15% of DP but never more than 2.86 psf (13.96 kg/m²). This is equal to a wind velocity of 33.4 mph (53.75 km/hr). At this test pressure, water is applied to the test sample at a rate of 5 gal (18.93 L) per hour, which is equivalent to 8 in. (203 mm) of rain per hour.

The *air infiltration* of AAMA-certified windows and sliding glass doors is tested in accordance with ASTM E283. The usual pressure is 1.57 psf (7.67 kg/m²) for residential and commercial sliding glass doors and windows of all types;

1.57 psf (7.67 kg/m²) for double-hung and sliding heavy commercial windows; and 6.24 psf (30.47 kg/m²) for other heavy commercial units. The normal maximum rate of air infiltration is 0.37 cubic feet per minute (cfm) (0.01 m³/min) per foot (305 mm) of operating ventilator or sash crack for most windows. The rate is 1.50 cfm (0.042 m³/min) for jalousie windows, 0.15 cfm (0.004 m³/min) for fixed windows, and 0.37 cfm (0.01 m³/min) for sliding glass doors, all based on the total area of the unit in square feet (square meters).

The *condensation resistance factor* (CRF) is determined by procedures outlined in AAMA 1503. It indicates a window's ability to resist condensation. The higher the CRF, the less likely it is that condensation will occur.

As discussed in Section 17.1, heat will flow readily through windows and sliding glass doors unless some means are taken to reduce this flow. The methods for determining conductive heat transmission through windows and doors is a part of AAMA 1503.

Where design calculations indicate the need to reduce the heat transmittance of aluminum windows and sliding glass doors, significant reduction can be accomplished by incorporating double or triple glazing and by thermalizing the frames and sash. At any rate, windows and sliding glass doors in areas where the winter design temperature is 10°F (−12.22°C) or less should incorporate such devices regardless of calculations. In fact, double or triple glazing is used in most new windows installed today, especially in commercial applications.

Thermalizing involves separating the aluminum parts on the warm inside portion of a window from those on the cold outside portion through the use of a thermal break, as discussed earlier.

## ARCHITECTURAL (AW) WINDOWS

Windows in the AW class include high-quality windows that were previously known as *monumental windows*. Much of the discussion in previous paragraphs related to windows in the other four classes also applies to architectural windows. There are some differences, however, in types, grades, and performance criteria.

*Types*  Architectural window types are similar to those in the other four classes, with some exceptions. For example, there are no awning, jal-awning, jalousie, or greenhouse windows in the AW class, and sliding glass doors are not included. There are, however, several window types that not included in the other classes, including horizontally pivoted windows, which are similar to vertically pivoted windows turned on their side, and side-hinged inswinging windows, which are similar to casement windows but swing inward. Some of the features of AW windows with the same names as those covered by the other classes are slightly different.

*Grades*  Windows in the AW class are divided into three categories: standard, modified, and custom. *Standard Architectural windows* are made entirely from the manufacturer's stock components. In *modified Architectural windows* the manufacturer's stock components may be altered slightly. *Custom Architectural windows* are entirely custom-designed units.

*Performance Requirements*  The major difference between the requirement of windows in the AW class and those in the other four classes is in performance requirements. For example, the minimum

test unit size required for windows in the AW class is larger than most of the sizes required for the other classes. In addition, the static air pressure required in the air infiltration test is 6.24 psf (30.47 kg/m²) for all units, which is larger than that required for all except a few heavy commercial units. The permitted infiltration rates are also smaller.

The air pressure difference for the water resistance test for windows in the AW class is 8 psf (39.06 kg/m²), which is much higher than required for the other classes.

Most of the requirements for windows in the AW class are higher than those for the other classes, but not all. For example, the minimum STP for windows in the AW class is 30 psf (146.47 kg/m²); for the other classes, it is 15 psf (73.24 kg/m²). However, this test pressure for HC windows is 40 psf (195.3 kg/m²), which is higher than the minimum required for AW windows.

## CERTIFICATION OF QUALITY

In addition to developing standard specifications, AAMA sponsors a program of independent testing, evaluating, inspecting, and labeling aluminum products to certify compliance with AAMA standards. Products that conform with the requirements of the applicable specifications are identified by the AAMA label.

The manufacturer's code number, appropriate standard, and maximum size tested are indicated on the label. The label must be visible on the metal after installation.

## 8.6.3 ADDITIONAL REQUIREMENTS FOR STEEL WINDOWS

### 8.6.3.1 Materials

The most common steel products used to produce steel windows are hot-rolled or cold-rolled carbon-steel shapes. Formed sheet steel may be used for interior trim.

### FINISHES

Steel windows may be finished in the shop or shop-primed and painted in the field. Shop-applied finishes are usually acrylic or polyester baking-finishing enamel. Galvanized steel windows are sometimes shop-primed for field painting and are sometimes left unpainted.

Some steel windows have components that are made from aluminum or another material that is not steel. These

are usually finished to match the steel components.

### FASTENERS

Fasteners for steel windows should be steel, galvanized steel, or stainless steel.

### HARDWARE

Hardware used to operate and lock operating window sashes may be manually locking or key locking. Operators may operate a single window or may be a bank of devices that operates several windows simultaneously. Each type of window has a specific operating device designed for that type. Awning windows, for example, are supported on pivots and balance-support arms and operate by worm gears and a crank handle, but can be operated by pole operators or remote devices. Casement windows hang on hinges and are operated by gear-type rotary operators or manually with a finger pull.

Hardware for steel windows is often a die-cast or solid-copper alloy with steel or copper-alloy operating bars and rods.

### STANDARDS OF QUALITY

The Steel Window Institute (SWI) produces standards for steel windows. Unfortunately, the SWI standards are often not adequate for high-quality installations and must be supplemented with ASTM or other standards. For very large projects, it is sometimes necessary to require specific testing of the units selected. For most projects, the standard is a specific proprietary product. Even then, it may be necessary to specifically select air infiltration and water penetration requirements and require testing in accordance with ASTM E283 and ASTM E331, respectively, to ensure compliance.

### GRADES

SWI places steel windows in 11 categories: Standard Intermediate, Heavy Intermediate, Heavy Custom, Architectural Projected, Commercial Projected, Manufacturer's Custom Grade, Residential, Guard, Psychiatric, Security, Awning, and Curtain Wall. The first three are of the highest quality established by SWI. The next two are of lower quality. The Manufacturer's Custom Grade is applicable only when a proprietary product is used as the standard. The Residential Grade category applies only to residential windows. The last five are related to window types and are not well defined in terms of quality.

## WINDOW TYPES

Steel window types and their names are essentially the same as the aluminum window types shown in Figure 8.6-1 with the addition of several types. For example, side-hinged emergency egress and top-hinged outswinging types are available.

## 8.6.4 INSTALLATION

The finest window or door design and the highest quality of manufacture will not compensate for poor installation. Weathertight installation, although simple, can only be ensured if a window or sliding glass door is installed by an experienced worker following the guidelines and instructions supplied by the manufacturer. A rough opening should be prepared to receive the unit so that it can be installed, and will finish out square, plumb, level, straight, and true. The design of most windows and doors provides for minor adjustments at the job site, but no unit will operate or weather properly if it is twisted and misaligned during installation.

A properly selected and manufactured window or sliding glass door will be weather-resistant as a unit, but if the building is insulated, the space between the rough opening and the windows or sliding glass doors should be filled with insulation. In addition, the spaces between the unit and adjacent materials must be closed with a sealant, trim members, or both to prevent the passage of air, dust, and water.

Windows and sliding glass doors should be rigidly supported and nailed, screwed, or otherwise anchored securely around the entire frame to the supporting construction. They should also be flashed at their heads and sills to provide a path back to the exterior for water that finds its way into the wall.

Because aluminum windows and sliding glass doors, and often steel windows as well, are prefinished building components, they should be handled and treated to prevent damage. Once installed, they should be protected to prevent damage to their finish, appearance, and ability to operate properly. Units should be protected from splashing of plaster, mortar, and cleaning solutions.

# 8.7 Wood and Plastic Windows and Sliding Glass Doors

For maximum comfort at minimum heating and air conditioning costs, windows and doors must be weathertight, insulated, and condensation free. Also, they must be designed and built to permit ready adaptation to varying installation conditions.

This section outlines the manufacture, nomenclature, standards of quality, and installation of wood and plastic windows and sliding glass doors. These windows and doors are manufactured at mills or factories in a variety of types and sizes.

## 8.7.1 STANDARDS

Most wood and plastic windows and sliding glass doors are manufactured in accordance with standards developed by AAMA, WDMA, and CSA. The publication containing these standards is AAMA/WDMA/CSA 101/I.S.2/A440-05, *Standard/Specification for Windows, Doors and Unit Skylights.* Sliding door manufacture employs materials and methods generally similar to those of window manufacture. The discussion in this section presumes a level of quality conforming with this standard.

### 8.7.1.1 Performance Criteria

The referenced standard establishes minimum performance criteria, and rates the products while broadly regulating material, design, and manufacture. For example, this standard recognizes six performance grades: 15, 20, 25, 30, 35, and 40. The numbers represent the minimum structural performance test pressure in pounds per square foot (psf) (4.88 kg/m$^2$) that a unit must be able to sustain without damage. These grades also include requirements for water resistance and air infiltration. The acceptable levels are different in each grade, becoming more stringent as the grades get higher.

Other items covered by this standard are moisture content of the lumber, preservative treatment, screening, and glass requirements.

### 8.7.1.2 Certification of Quality

The referenced standard does not provide for inspection, testing, or certification of products. These functions are performed by certification programs sponsored by the WDMA or by independent testing laboratories. Issuance of a label ensures that the design, manufacture, and tested performance of the product meet or exceed the appropriate standard. This type of certification program is based on periodic inspection of the manufacturing facilities, quality control procedures, and laboratory testing.

Wood and plastic windows and sliding glass doors are certified for conformance to AAMA/WDMA/CSA 101/I.S.2/A440, and water-repellent preservative treating is certified for conformance to WDMA I.S.4, "Water-Repellent Preservative Non-Pressure Treatment for Millwork." Each program consists of initial qualification testing, in-plant inspection, and periodic follow-up testing and inspection to verify that products are manufactured in conformance with the applicable standards. Conformance to the standards results in certification and issuance of a seal. This seal is rubber-stamped on various parts of the sash and frame or imprinted on labels affixed to the product. The seal indicates the fabricator of the product and the industry standard under which the product was produced and is evidence of compliance with the standard.

## 8.7.2 BURGLAR PROTECTION

The construction of windows generally makes them more vulnerable to unauthorized entry than doors. Nine out of ten times, a potential intruder will find an open window.

Even when windows are closed and locked, the glass can be broken if an intruder does not mind making noise. When the panes are too small to crawl through, an intruder will be forced to spend more time and make more noise in the process of enlarging the opening. Windows can also be forced open with a small crowbar.

In lieu of regular glass, one of the following security glasses may be used: (1) laminated, consisting of a vinyl panel sandwiched between two glass panels, (2) wired, or (3) tempered. Acrylic or polycarbonate glazing can also be used. Unfortunately, plastics are subject to surface scratches, which can detract from their appearance over time.

Exterior ornamental grillwork and interior window gates can be used to protect windows. Gates or grillwork should be installed with nonretractable screws. Grillwork should never be installed on windows needed as fire exits. Interior window grates should have an integral locking knob system, rather than a padlock, so that they do not become an obstacle in case of fire.

Refer to Section 8.9 for additional discussion about burglar protection.

### 8.7.3 TYPES AND NOMENCLATURE

Wood and plastic windows may be classified according to design and manner of operation into *fixed, double- (or single-) hung, awning, hopper, casement, horizontal sliding,* and *basement* types.

Sliding glass doors are essentially large horizontal sliding windows. However, their increased size, weight, and use result in greater demands on frame and sash construction and on sliding hardware. Sliding panels usually are roller supported and operate on metal tracks built into the threshold. Metal parts are covered with wood on the interior to prevent condensation and present a uniform appearance. Door panels are sometimes reinforced with steel to provide necessary strength while retaining a slim profile.

A summary of the most common window and sliding glass door types is given in Figure 8.7-1.

### 8.7.4 MATERIALS

#### 8.7.4.1 Wood Windows and Sliding Glass Doors

Wood was one of the earliest materials used in the manufacture of windows and sliding glass doors, and over the years has proven its suitability for this purpose. Wood's insulating properties, ready availability, ability to take either natural or painted finishes, and ease of fabrication and repair with simple tools all contribute to its widespread use in window and door manufacture.

Recent technological innovations have further improved wood as a material for window and door construction. Kiln drying of lumber reduces shrinkage, resulting in better operation and weathertightness.

Water-repellent preservative chemical treatments reduce swelling and warping, improve paint retention and increase wood's resistance to decay and insect attack under all climatic conditions. Windows that cover entire walls

consist of combinations of fixed glass for light and view and an operating sash for light and ventilation. In some window walls, sliding glass doors also provide access to the outdoors in addition to light and ventilation.

Most stock windows and sliding glass doors are manufactured from ponderosa (western) pine. This species has excellent workability, gluability, nail-holding capacity, and a uniform light color suitable for natural or painted finishes. Other species commonly used are southern yellow pine, Idaho white pine, sugar pine, and Douglas fir. Lumber used is sound, free from defects such as loose knots and excessive checking, and kiln dried to a moisture content between 6% and 12%.

Wood windows and sliding glass doors clad with PVC (vinyl) or factory-finished aluminum have been used extensively in recent years due to their ease of maintenance. Combination windows made of wood interior portions and plastic exterior portions are also being used today.

#### 8.7.4.2 Plastic Windows and Sliding Glass Doors

Plastic windows and sliding glass doors are relatively new products that have replaced aluminum, wood, and clad wood units in many new residential and small commercial construction projects. Plastic units are less expensive than their aluminum or wood competitors. They require less maintenance than the unclad, and thus lowest-cost, wood units in that they need no painting.

Most plastic window and sliding glass door frame and sash components are made by extruding PVC (vinyl) into tubular or other shaped sections. Glass-fiber reinforced plastics (GFRPs) (composites) are also being used. These are made by coating glass fiber strands with plastic resin and drawing the combination through a die.

Unlike aluminum, PVC and GFRP are good thermal insulators, and windows made from them do not require thermal breaks. They do, however, have relatively high coefficients of thermal expansion when compared with aluminum, though GFRP's coefficient is lower than PVC's. Neither plastic is as stiff as aluminum, so components made from them must be much thicker than aluminum serving the same purpose.

### 8.7.5 FABRICATION AND ASSEMBLY

For wood units, kiln-dried lumber is delivered to a millwork shop for fabrication into window and sliding glass door components. There it is cut to specified lengths and widths on rip saws and crosscut saws, leaving cut stock substantially clear and without defects. Cut lengths are then sorted and marked to indicate the window or door component for which they are intended. Like pieces are conveyed to molders or lineal milling machines which shape the pieces to the desired profile. The next step is cross milling, in which necessary grooves and channels (dados, mortises, and tenons) (Fig. 8.7-2) are cut across each piece to facilitate weathertight joining of units. Milling of components is completed by routing machines which cut notches to receive recessed hardware.

When shaping has been completed, milled components and assembled sash are preservative-treated with a solution of toxic, water-repellent agents. Nonpressure treatments, such as the vacuum or immersion processes, are commonly used. Besides imparting water-repellent properties and increasing resistance to decay fungus and insect attack, preservative treatment gives the wood increased dimensional stability, minimizing intermittent swelling and shrinking due to varying atmospheric conditions or contact with water. Preservative treatment should be in accordance with WDMA I.S.4, "Water-Repellent Preservative Non-Pressure Treatment for Millwork." Exterior components may also be paint-primed or factory-finished for additional protection.

Following preservative treatment, weatherstripping is pressure-fit into grooves or surface-tacked, and some hardware is installed. Sash and frames are then assembled from components on automatic nailing machines. A continuous sander gives the assembled sash three separate sandings—coarse, medium, and fine—in one trip through the machine.

The final steps consist of glazing, fitting sash to the frames, and assembly with all operating hardware. Completed window units are usually provided with angle and spacer braces to prevent racking and distortion in shipment.

| Type and Industry Standard | Description |
|---|---|
| **Fixed** | *Fixed windows* usually consist of a frame and glazed stationary sash. They are often flanked with double hung and casement windows, or stacked with awning and hopper units, to make up windows of custom designs. To keep the sight lines (width or height of view area) consistent, fixed sash members are made to same appropriate cross sectional dimensions as adjacent operating sash. |
| **Double Hung**<br><br>**Single Hung** | *Double hung windows* have two operating sash; single hung have only the lower sash operative. The sash move vertically within the window frame and are maintained in the desired position by friction fit against the frame or with balancing devices. Balancing devices also assist in raising the sash. 50% of the window area is available for ventilation. |
| **Casement** | *Casement windows* have side-hinged sash, generally mounted to swing outward. They may contain one or two operating sash and sometimes a fixed light between the pair of sash. When fixed lights are used, a pair of casements may close on a mullion, or against themselves, providing an unobstructed view when open. Operating sash may be opened 100% for ventilation. |
| **Horizontal Sliding** | *Horizontal sliding windows* have two or more sash of which at least one moves horizontally within the window frame. In three-sash design, the middle sash is usually fixed; in two-sash units, one or both sash may be ventilating. This type of window sometimes is increased in size to door proportions. Ventilating area is 50% of the window area in most designs. |
| **Awning** | *Awning windows* have one or more top-hinged, outswinging sash. Single awning sash often is combined with fixed and other types of sash into larger window units. Several sash may be stacked vertically and may close on themselves or on meeting rails which separate individual sash. When awning windows have sliding friction hinges, which move the top rail down as the sash swings out, they are said to have *projected action*. Ventilating area is considered to be 100% of operating sash area. |
| **Hopper** | *Hopper windows* have one or more bottom-hinged, inswinging sash. Hopper sash is similar in design and operation to the awning type, and may actually be an inverted awning sash with minor hardware and weatherstripping modifications. For this reason, windows with hopper sash sometimes are referred to as awning windows. Operating sash provides 100% ventilating area. |
| **Basement** | *Basement windows* generally are single sash units of simplified design intended for less demanding installations, particularly in masonry and concrete foundations. The sash may be of awning type, hopper or top-hinged inswinging (shown) type. 100% of the window is available for ventilation. |

*(Continues)*

**FIGURE 8.7-1** *(Continued)*

| Type and Industry Standard | Description |
|---|---|
|  **Sliding Glass Doors** | *Sliding glass doors* contain at least one fixed and one operating panel in a main frame. Door types are commonly designated by the number and arrangement of panels (as viewed from the outside) using the letter "X" for operating panel, and "O" for fixed panel. The two panel door is most common, but three and four panel designs are also available; two and four panel doors (OX & OXXO) have 50% ventilating area, the three panel unit (OXO) has 33% ventilating area. |

## 8.7.6 WINDOW AND DOOR COMPONENTS

A window unit consists of a frame, one or more sashes, and hardware and weatherstripping necessary to make a complete operating unit. Storm sash and screens may be included as integrated elements of the main frame or contained in a separate combination subframe (Fig. 8.7-3).

### 8.7.6.1 Sash Components

A wood window sash is made up of horizontal rails, vertical stiles, and sometimes muntins and bars (see Fig. 8.7-2). A wood sliding glass door is composed of a main frame containing one or more

operating or fixed door panels. Like window sash, door panels are composed of stiles and rails and may hold either single glass or insulating glass. Most doors provide self-storing sliding screens in a separate track of the door frame.

**FIGURE 8.7-3** The principal parts of window units, as shown here for a double-hung window.

### 8.7.6.2 Frame Components

Figures 8.7-2 and 8.7-4 show the parts of a double-hung window sash. The components of other window types vary somewhat from those shown, of course, but the other basic elements—*sill, side jambs, head jamb, stops,* and *exterior casing*—are present in most window and door designs. *Blind stops* are included in some window frames between jambs and outside casings, and may vary in thickness for proper window positioning in wood or metal stud walls with $1/2$ or $3/4$ in. (12.7 or 19.1 mm) sheathing. *Jamb extenders* adapt windows to varying interior finish thicknesses. They are not usually a part of the window frame assembly but may be ordered loose as optional items. *Interior trim* is not usually supplied with stock windows. Typically it is selected separately and in some installa-

**FIGURE 8.7-2** (a) Sash component terminology; (b) common millwork methods of joining sash corners.

**FIGURE 8.7-4** Terminology for typical double-hung windows. Jamb extenders are also called *jamb liners*.

tions may be replaced by a gypsum board or plaster reveal.

### 8.7.6.3 Weatherstripping

Weatherstripping provides a seal against infiltration (leakage of air) and penetration of dust and wind-blown rain. It should be incorporated on all ventilating sash to reduce infiltration substantially below the maximum established by referenced standards. The maximum is 0.34 cfm (0.0096 m$^3$/min) for Grade 20 (see Section 8.7.7.1) windows and doors; 0.25 cfm (0.007 m$^3$/min) for Grade 40; and 0.10 cfm (0.003 m$^3$/min) for Grade 60. The measurement is per linear foot (305 mm) of sash crack length for operating sash and per square foot (0.093 m$^2$) of surface for fixed sash. The test pressure is a uniform 1.57 psf (0.16 m$^2$).

Weatherstripping typically has a resilient part that prevents air penetration when it is pressed between the door or window and its sash.

#### TYPES

The most common weatherstripping types for wood and plastic windows are foam rubber, felt stripping, rolled vinyl, casement stripping, spring metal, interlocking metal, and compressible bulb.

***Foam Rubber*** Foam rubber weatherstripping has either an adhesive or wood backing. It is easy to install but lacks durability. The adhesive type can be used wherever there is no friction, such as at the top and bottom of double-hung windows.

***Felt Stripping*** Felt weatherstripping has the same advantages and disadvantages as foam rubber, but it also collects dirt more readily. Metal-backed felt, which provides more durability, is also available.

***Rolled Vinyl*** Rolled vinyl weatherstripping is available with or without a metal backing. Some types are manufactured with tubular gaskets that can either be hollow or filled with foam (see Fig. 8.9-10c). Rolled vinyl is very durable and can be installed on the exterior of double-hung, sliding, and casement windows.

***Casement*** Casement window weatherstripping can be one of several kinds, depending upon whether the window is wood or metal. Wood casements can be sealed with felt, adhesive-backed foam, or spring metal.

***Spring Metal*** A common weatherstripping used in wood windows is *spring tension weatherstripping* of bronze, aluminum, rigid vinyl, stainless, or galvanized steel (Fig. 8.7-5a). Extruded aluminum, cold-formed galvanized steel, or rigid plastic are used as sash guides and integral weatherstripping in many double-hung windows. These may be either

**a Spring Tension**

**b Woven Felt**

**c Compression Sash Guide**

**d Compressible Bulb**

**FIGURE 8.7-5** Common types of weatherstripping: (a) spring tension; (b) woven felt; (c) compression sash guide; (d) compressible bulb.

fixed to the window frame or cushioned with plastic foam or springs to provide constant pressure against the sash stiles. In the fixed type, *woven felt* in the stiles sometimes engages a projecting fin in the sash guide for improved weathertightness (Fig. 8.7-5b). The cushioned *compression sash guide*, in addition to weather protection, often provides for easy removal of the sash (Fig. 8.7-5c). Metal or plastic sash guides eliminate the need for finishing sash runs and the possibility of becoming clogged with paint or varnish.

Spring metal stripping is available in two types: flat stripping, which is most commonly used for double-hung windows, and V-strip, which is used for doors. Flat stripping is inserted into the narrow side channels of a window sash on the bottom rails of the upper and lower sashes and on the top rail of the upper sash.

***Interlocking Metal*** Interlocking metal stripping is produced as interlocking channels and interlocking J-strips. These are difficult to install because alignments must be exact for a secure interlock. Both provide excellent, durable seals. The channels are subject to damage, however, because they are exposed. The J-strips are not visible when installed; they are not exposed and, therefore, not subject to damage.

***Compressible Bulb*** Compressible bulb weatherstripping of vinyl or neoprene is found most often in storm sash, in screens, and in sliding glass doors. It is employed by some manufacturers in main sash (Fig. 8.7-5d). In addition to flexible vinyl, woven fabric weatherstripping also is common in sliding glass doors.

### 8.7.6.4 Hardware

#### OPERATING HARDWARE

Hardware provides effective closure, operates a sash, and holds it stationary at the desired degree of openness. Hardware types range from simple hinges and locks on basement windows to complex arrangements on windows which require interior operation without removing screens. Underscreen rotary gear, push bar, and lever-type operators are commonly used for awning windows, in combination with sliding friction hinges which provide projected action (Fig. 8.7-6a). Operators are usually capable of weathertight closure,

crank operator     push bar

sash lock     snap lock

HOPPER

**a Awning**

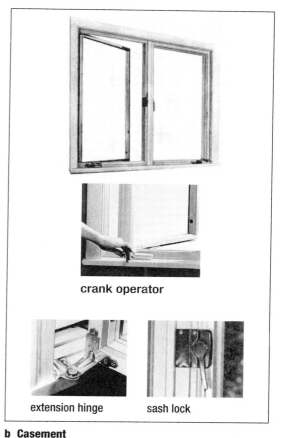

crank operator

extension hinge     sash lock

**b Casement**

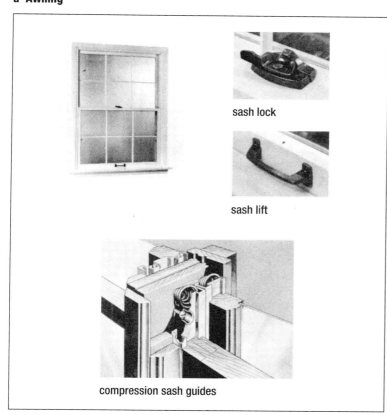

sash lock

sash lift

compression sash guides

**c Vertical Sliding**

Compression sash guide

Sill track

**d Horizontal Sliding**

**FIGURE 8.7-6** Commonly used hardware for various window types. Note in (a) and (b) that sliding friction hinges provide projected action; for example, when the awning sash is opened, the top rail moves downward.

but on larger sash, additional awning locks may be provided. Outswinging casement windows often utilize similar rotary gear operators, with either sliding hinges or extension hinges (Fig. 8.7-6b); one or more casement fasteners usually are necessary to draw up relatively long sash stiles against the frame.

Double-hung and horizontal sliding windows generally employ cam-action sash locks which draw the sash together at the meeting stiles (or rails) while pushing the sash outward against the frames to compress the weatherstripping. Sliding windows often are equipped with a spring-cushioned metal or plastic track at the head lamb, similar to the sash guide of the double-hung window; fixed metal or plastic tracks are commonly used at the sill (Figs. 8.7-6c and 8.7-6d).

Double-hung windows are usually equipped with balancing hardware that assists in raising the sash and holding it in position. Spiral spring, spring, and coiled tape are the most common types of balancing devices (Fig. 8.7-7). Smaller and lighter sash may not require mechanical assistance for raising. In these types, the friction fit provided by a compression sash guide may be sufficient to hold sash in place (see Fig. 8.7-6c).

Exposed hardware such as sash locks, sash lifts, and various operators are available in corrosion-resistant standard (zinc, bronze, aluminum, chrome) and proprietary finishes.

### LOCKS

Most window locks are vulnerable to prying, and therefore are equipped with various locking devices (Fig. 8.7-8).

**a  Spiral Spring**　　**b  Spring**

**c  Coiled Tape**　　**d  Counterweight**

**FIGURE 8.7-7** Balancing devices for double-hung windows.

**FIGURE 8.7-8** Window locks include (a) friction type, (b) pin type, (c) wedge type, (d) stop type, and (e) key-operated cam latches.

A friction-type device has a pin and a receptacle into which the pin is screwed. Friction is supposed to keep the window in place, but it requires very little effort to nudge the pin from the receptacle.

A pin-type lock is a keyed device. Its bolt slips into a hole drilled into the upper window frame. Though stronger than the friction type, this lock is vulnerable to prying. Because of the keyed lock, it may represent a safety hazard in case of fire.

A wedge-type ventilating lock is self-defeating. It holds the window shut by spreading the upper and lower sashes apart. A potential burglar can easily slip a knife blade between the sashes and cut off the wedge.

A stop-type ventilating lock uses a knob or other protrusion to limit the distance that a window can be opened. Though it does not spread the window sashes apart, it too is vulnerable to prying.

A key-operated cam latch replaces an ordinary double-hung window cam latch.

Like a pin-type lock, it is vulnerable to prying and may represent a safety hazard in case of fire.

An eyebolt, installed according to the recommendations of Consumer's Union, is the most effective device to prevent the forcing of a window.

To accommodate the eyebolts, holes are drilled on sides of the interior of the window. They should go all of the way through the inside sash and three-quarters of the way through the outside sash at a slight downward angle, as shown in Figure 8.7-9. Two sets of holes can be drilled in the outside sash so that the window can be partially opened for ventilation. Eyebolts, $5/_{16}$ in. (7.9 mm) in diameter, inserted into the holes act as locking pins.

### CLEANING AND MAINTENANCE

To make exterior surfaces accessible for cleaning, double-hung and horizontal sliding windows usually have special

**FIGURE 8.7-9** A simple pry-proof window lock: a ⁵/₁₆ in. (23.8 mm) eyebolt is inserted through holes drilled through the inside and outside sashes.

hardware which permits easy removal of sash (see Figs. 8.7-6c and 8.7-6d). Awning window sash may be removable, but more often, awning windows have sliding friction hinges which create an access space of several inches between sash and frame when the top rail is moved down. Casement windows similarly provide accessibility for cleaning by sliding or extension hinges. The inswinging operation of hopper sash makes both surfaces automatically accessible from the inside. Most sash not ordinarily considered removable may be removed with a few simple tools.

### 8.7.6.5 Glass and Glazing

Single-strength (³/₃₂ in. [2.38 mm]) glass is limited to sizes less than 76 *united in.* (1930 united mm) (width plus height) and double-strength (¹/₈ in. [3.2 mm]) glass to sizes less than 100 united in. (2540 united mm). Additional information on grades and sizes of glass is given in Section 8.10. Most stock windows and sliding glass doors are available with either insulating glass or single glazing; tinted and patterned glass is sometimes available on special order. The *International Building Code* requires compliance with the Consumer Product Safety Commission's (CPSC) "Safety Standard for Architectural Glazing Materials," which requires safety glazing material in sliding doors.

Much wood window sash is face glazed by bedding (back puttying) the glass in a rabbet, securing it with glazing points, and sealing it with a bevel face putty bead (Fig. 8.7-10a). Sometimes

**a  Face**

**b  Wood Stop**

**c  Groove**

**FIGURE 8.7-10** With face glazing (a), glazing compound holds glass in a wood seat; with wood stop glazing (b), glass is bedded in compound and secured by a wood stop; with groove glazing (c), glass is bedded in compound and set in a continuous groove.

pointing is eliminated by using adhesive bedding materials which cushion the glass, seal it, and secure it in place.

Wood stop glazing replaces both pointing and face puttying (but not bedding) and provides a glass seal of greater strength, greater durability, and better appearance (Fig. 8.7-10b). Many wood sliding glass doors and large, fixed windows are glazed in this manner.

Groove glazing, employed by some manufacturers, has the same advantages as wood stop glazing but eliminates the need for handling a separate stop (Fig. 8.7-10c). In this method, grooves in the sash are filled with glazing compound, and sash members are assembled around the glass, clamped in a press, and nailed or pinned together. To facilitate reglazing, one of the sash members can be removed by drawing out the pins that join

it at the corners. The sash is reassembled after the new glass has been placed in the groove.

### 8.7.6.6 Screening

Outswinging hinged sash (casements and awnings) typically have interior removable screens; double-hung and hopper sash have exterior screens. Basement windows may have either interior or exterior screens, depending on whether sash swings in or out.

Aluminum, galvanized steel, bronze, or vinyl-coated glass fiber insect screening may be used. Mesh size is limited to 18 by 14 or 18 by 16 (number of openings per inch [24.4 mm]).

Screening is secured to wood sash by rolling it into a groove and splining, by tacking, or by stapling. A sash is usually "trimmed out" with a screen mold to conceal tacks, staples, and splines. In metal-rimmed screens, the screening is rolled into a groove in the rim and held in place by a vinyl or metal spline.

Screens may be a part of a combination storm sash and screen (see Section 8.8).

### 8.7.6.7 Storm Sash

For greater heat control, any window sash may be glazed with insulating glass (Fig. 8.7-11) or provided with *storm sash* (see Section 8.8).

### 8.7.7 SELECTION

The selection of windows and sliding glass doors depends on the use and loca-

**FIGURE 8.7-11** Groove-glazed insulating glass.

tion of the units. The first step is to decide on the type of unit desired (see Section 8.7.4). Then the unit's grade and other required performance criteria should be selected.

### 8.7.8 INSTALLATION

Even the best manufactured windows and sliding glass doors will not perform satisfactorily if they are not installed properly. Although these units are commonly squared and braced before shipping, they may be distorted in transit and should always be checked for squareness and equal spacing of jambs before they are installed. Braces should be renailed if necessary and left in place until the unit has been installed and is ready for trimming. The frame should be plumbed and fastened securely in the rough opening. Adequate shimming should be provided so as not to distort the frame during nailing. The spaces between finished frame and rough frame should be completely filled with insulation. Drip caps and flashing over the cap or exterior casing are strongly recommended to minimize the possibility of leakage. The entire unit should be well caulked at the juncture of the exterior casing and the wall finish after both elements have been prime-coated but before finish painting.

Where windows and doors are shipped with loose interior stops, it is essential that these be installed carefully for a snug, weathertight fit against the sash. Weatherstripping and hardware should be adjusted for proper operation and closure according to the manufacturer's instructions. Interior trim should be applied to seal the wall cavity between the rough frame and the finish frame.

Wood and plastic windows and sliding glass doors should be handled carefully to prevent damage. Wood units should be protected with a prime coat as soon after arrival at the building site as possible. They should be stored out of the weather in a well-ventilated space. When finishing, it is important to keep paint or varnish off weatherstripping and interior finish hardware.

Exterior hinges and storm sash hardware may be painted for improved corrosion resistance, but the metal or plastic sash guides of double-hung and siding windows should not be painted. Oil-based glazing compound should be painted, with the paint lapped slightly over the glass on both face putty and back putty runs. Other sealants may or may not be painted, depending on the sealant type and use.

# 8.8 Storm and Screen Doors and Windows

Storm doors and windows are glazed panels or inserts which are sometimes added to doors and windows as an extra line of defense against the elements. Their major use is in residential and small commercial buildings.

Properly installed storm doors and windows provide an effective barrier to the transfer of heat. They are recommended for use in the northern part of the country, above the 4500-winter-degree-day line, and are desirable down to the 2500-degree-day line. (Fig. 8.8-1), where they can significantly reduce heating costs and minimize condensation on windows. By preventing cold air from leaking around windows and doors, they also reduce drafts and improve interior comfort. They can also increase comfort in summer, especially in southern areas, by keeping cooled air inside.

Above the 4500-degree-day line, storm doors or an entrance vestibule should be provided as part of the initial construction. Storm windows can be omitted in original construction if the prime windows are equipped with insulating glass and a thermal break in the sash and frames.

Although the term *storm doors and windows* generally refers to units intended for weather protection, it is often used collectively to include screen units as well. Accordingly, the following discussion covers both combination units, containing interchangeable glass and screen inserts, and single-purpose units with fixed glass or screening.

### 8.8.1 STANDARDS

Storm doors and windows should conform to standards developed by AAMA and adopted as American National Standards by the ANSI, AAMA 1002.10 and AAMA 1102.7. These constitute the recognized industry standards and establish minimum performance requirements for air infiltration, water leakage, wind load, and structural strength.

Screens should conform to requirements of the Screen Manufacturers Association (SMA) listed in Section 8.12.

AAMA also sponsors an independent certification program of testing, random sampling, and product evaluation to ensure that aluminum storm doors and windows meet ANSI standards. Products which meet all requirements of the standards are identified with an AAMA certification label.

### 8.8.2 MATERIALS AND FINISHES

Most storm doors and windows sold today are made of aluminum. Their manufacture is similar to that of aluminum

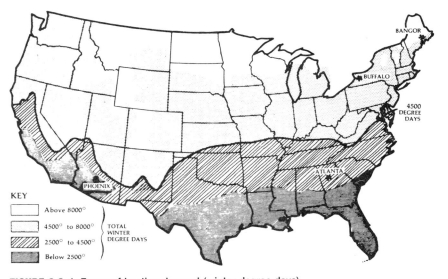

**FIGURE 8.8-1** Zones of heating demand (winter degree days).

prime windows and sliding glass doors. They are made in a variety of finishes and colors. A natural mill finish is adequate for many moderate-cost installations, where appearance is not a prime consideration. For more demanding projects, where both corrosion resistance and appearance are important, either an anodized or paint finish is appropriate. Anodized finishes can be of natural aluminum color or deeper shades of bronze or brown-black. Paint finishes are available in a variety of colors.

Many storm panels and screens are framed with aluminum or stainless steel rims of thinner cross section which provide greater flexibility in positioning within the limited thickness of a typical frame wall. They can be manufactured with wood, metal, or rigid plastic frames. Storm doors are most often made of aluminum, although wood is still sometimes used.

### 8.8.3 TYPES

Storm doors and windows can be of the *self-storing* type or may depend on a glazed *storm panel* that is applied and removed seasonally as needed. Self-storing storm doors and windows generally have both glass and screen inserts sliding vertically or horizontally in a frame. In self-storing *combination* (storm and screen) doors, glass and screen inserts slide vertically.

Combination frames eliminate the need to remove external screens for winter storage and provide a year-round choice of storm panel or screen. However, combination frames provide for screening only half of the door or window glazed area. When it is desirable to open both the top and bottom sashes of a double-hung window (or right and left sashes of a horizontal sliding window), full screens may be used to cover the entire window.

#### 8.8.3.1 Storm and Screen Doors

Entry doors are a great source of heat loss. The two most effective structures to reduce such heat loss are storm doors and insulated doors. Insulated doors are discussed in Sections 8.1 and 8.2.

Storm doors reduce heat loss by reducing both conduction and infiltration. For example, a typical $1^3/_4$ in. (44.5 mm) solid wood exterior door has an R-value of about 2. When a wood storm door is added, the R-value is increased to approximately 3.70 (Fig. 8.8-2).

**FIGURE 8.8-2** Door R-Values

| Door Type | R-Value |
|---|---|
| $1^3/_4$ in. (44.5 mm) exterior door | 2.04 |
| $1^3/_4$ in. (44.5 mm) exterior door and aluminum storm door (50% glass) | 3.03 |
| $1^3/_4$ in. (44.5 mm) exterior door and wood storm door (50% glass) | 3.70 |
| Metal-clad insulated door | 8.00–15.00 |

Storm, screen, and combination storm and screen doors are lightly constructed glazed doors used with exterior doors to improve weather resistance. Screen doors are of similar construction with screening, which provides ventilation while excluding insects.

Storm doors are made either (1) with fixed glass lights, exclusively for weather protection, or (2) with both screen and glass inserts to permit ventilation with insect control as well. In parts of the country with moderate year-round temperatures, screen doors (without glass inserts) are frequently used. Screen door construction is similar to that of storm doors. Both must be constructed rigidly to resist sagging, bowing, or rattling from repeated impact.

Combination doors combine the functions of both storm and screen doors with interchangeable screen and glass panels (Fig. 8.8-3).

Some self-storing combination doors are equipped with a screen and a pair of glass panels that operate in a vertical track similarly to a double-hung window. Either ventilation or weather protection is possible without removing the panel inserts, but the ventilating area available

Screen          Combination

**FIGURE 8.8-3** Storm and screen doors give added weather protection and ventilation. (Courtesy Window and Door Manufacturers Association [WDMA] [formerly National Wood Window and Door Association].)

is limited to one-half that of a full screen door. Glass jalousie inserts in combination doors have similar self-storing advantages with practically no loss of ventilating area. However, even when closed, jalousies permit a high rate of air infiltration and leak water in high-wind storm conditions.

Many storm doors are of the combination type, providing both weather protection and insect control. Such doors are stile-and-rail assemblies, framing a storm window with glass and screen inserts which can be interchanged seasonally. Doors with openings limited to the upper half of the door generally accommodate only one insert at a time.

Self-storing storm doors contain the equivalent of a two-track window, accommodating two inserts in one track and another in the adjacent track. The glass and screen inserts slide up and down just as they do in a vertical storm window.

Both prime doors and storm doors can be of the jalousie type. To provide ventilation and storm protection, interchangeable screen and glass panels can be clipped to the inside of the door. Clip-on glass storm panels not only provide added insulation, but also reduce air leakage between the louvers, a common problem with this type of door.

#### DOOR CONSTRUCTION

Storm and screen doors are opened and closed many times a day and are subject to wind buffeting and other stresses. It is therefore important that they be durably constructed and properly installed. Rigid corner connections are particularly important to resist racking forces.

Storm doors are available in sizes to fit the openings for exterior doors. Standard door thickness is $1^1/_8$ in. (28.6 mm). Available sizes include 6 ft 7 in. (2007 mm), 6 ft 9 in. (2057 mm), and 7 ft 0 in. (2134 mm) high and 2 ft 6 in. to 3 ft 0 in. (762 to 914 mm) wide. Standard door frames are 1 or $1^1/_4$ in. (25.4 or 31.8 mm) thick, with top rail and stiles 2 to $2^1/_2$ in. (50.8 to 63.5 mm) wide. The bottom rail (kickplate) varies in width, depending on the size of the opening. To avoid the hazards of broken glass, storm doors should be safety glazed with transparent plastic or with tempered, laminated, or wired glass. In some locations, laws stipulate this requirement.

Doors currently are manufactured in hundreds of types and styles, utilizing many different glass sizes. To minimize waste, storm windows and doors should be required to accommodate standard glass sizes adopted by the industry. Limiting the number of glass sizes not only reduces production costs but also facilitates replacement, particularly of tempered glass.

Construction of glass and screen inserts for doors is similar to that of windows.

### 8.8.3.2 Storm Windows and Panels

Window storm sash are commonly called *storm windows*. They control heat flow by making windows more airtight to reduce infiltration, and by providing an insulating layer of air between the window sash and the exterior to reduce conductive transfer.

In the past, inexpensive *storm panels*, made of wood or metal frames with single glazing and hung in place with special clips, were commonly used with hinged sash such as casements, awnings, and hoppers (Fig. 8.8-4). Sliding and double-hung sash were often equipped with either clip-on storm panels or an external *combination frame* which held both storm sash and screens (Fig. 8.8-5). Such storm panels were independent of the screens, so they were installed in the fall and taken down in the spring. If the building was air-conditioned and not properly oriented for natural ventilation, they could be left up year-round to reduce cooling loads. These types of windows are not often used in new construction.

Contemporary storm windows are usually permanently installed combination types that include both storm sash and screens (Fig. 8.8-6). Frames are usually aluminum and have an anodized or baked enamel finish. They should move smoothly in the frame tracks and seal tightly when closed. They should have tight joints and weatherstripping to ensure a tight seal.

Fixed-in-place interior storm windows are inexpensive temporary installations used almost exclusively in existing residential buildings. They consist of a frameless sheet of rigid plastic held in place with tape or snap-on moldings (Fig. 8.8-7). It is also possible to use polyethylene film taped to the frame. Temporary plastic windows provide an insulating value comparable to that of ordinary storm windows because it is the air films along the glazing surfaces and the air space between them that reduce the heat loss, not the material thickness.

**FIGURE 8.8-5** A combination storm and screen sash.

**FIGURE 8.8-4** (a) Clip-on metal-rimmed storm panels are typically used with awning or casement windows; (b) storm panels for double-hung windows are attached to the frame.

### STORM WINDOW TYPES

Clip-on storm panels and single-track storm windows are available, but the most widely used storm windows are of the triple-track and two-track types. When practicable, storm windows should be installed on the outside of the prime sash. Where the prime sash swings out, as in casement and awning windows, storm windows may have to be on the room side.

To prevent rapid heat transfer and condensation on the interior window frame, an insulating thermal break should be provided between the prime window and storm window frames. A thermal break is recommended regardless of

**FIGURE 8.8-6** These windows are also called *triple-track* because of the three panels that move in separate tracks.

whether the storm window is installed on the room side or outside. Some windows are provided with a plastic or wood liner which serves as the thermal break. An air gap or a small wood molding can be used instead.

***Clip-on Storm Panels*** Clip-on storm panels are available for outswinging prime windows such as casements, awnings, and jalousies. These consist of single glass panels which are attached either to the individual sash or across the

**FIGURE 8.8-7** These temporary storm panels attach directly to the window frame.

entire prime window, including both fixed and vent sash. When the panel is clipped to the sash only, heat transfer through the glass is reduced, but heat lost through the frame and by air leakage is not affected.

Larger panels covering the entire window control heat loss more effectively but cannot be left in place for the summer.

***Single-Track Storm Windows*** Single-track windows have two glass inserts and one screen insert, but only two inserts can be housed in the track at any one time. This makes changing from weather protection to ventilating somewhat less convenient, because one insert must always be stored elsewhere.

***Two-Track Storm Windows*** Two-track windows are generally used with vertically sliding single-hung windows and with two-light horizontal sliders. These units have three inserts which are stored in the frame. In vertical windows, the bottom light is usually screened; in horizontal sliders, the screen is placed right or left, depending on the prime vent sash location.

***Triple-Track Storm Windows*** Triple-track windows are intended for use mainly with vertically sliding double-hung and horizontal sliding triple-light windows. A typical unit consists of a frame, with each insert riding in its own track. This self-storing feature eliminates the need for switching storms and screens every spring and fall. Instead, selection of glass or screen can be made easily from inside the house.

## STORM WINDOW CONSTRUCTION

Extrusions used for the main frame and insert frames should be of wrought aluminum alloy with a minimum tensile strength of 22,000 psi (15 468 Mg/m²) and a yield strength of 16,000 psi (11 249 Mg/m²). Reinforcing members which are not in plain view can be of cadmium, zinc-plated steel, or other noncorrosive materials compatible with aluminum.

Storm windows should be provided with weatherstripping to minimize air infiltration and reduce rattling. Glass inserts generally consist of a light-gauge aluminum frame into which the glass is set with bedding compound, plastic channel gaskets, or glazing tape. These materials should cover the glass edge continuously so that there will be no glass-to-metal contact.

Glass used for storm inserts in windows should not be less than B quality and of adequate thickness. Glass inserts in vertically sliding windows should have self-activating latches to hold the insert level in different ventilating positions. Insert frames should be designed to permit reglazing without special tools.

Screen inserts consist of insect screening stretched on an extruded or roll-formed frame. Screening can be of aluminum, vinyl-coated glass fiber, or other synthetic fibers in 18 by 14 or finer mesh.

### 8.8.4 INSTALLATION

Most storm windows are sold as complete units, with necessary accessories and hardware. This ensures smooth, convenient operation and adequate security. Installation consists simply of attaching the window frame in a previously prepared opening.

Storm doors are marketed both prehung on an aluminum frame, with hardware attached, and unassembled, with frame parts and hardware supplied loose. The unassembled type provides somewhat more flexibility in adapting to special situations but requires field assembly and attachment of suitable hardware.

A spring-loaded safety chain or other device should be provided to limit door swing and cushion the impact of a sudden opening, as by a gust of wind. A door closer is recommended. A durable latch to maintain the door in a closed position and to permit locking the door from the inside is also desirable.

## 8.9 Door Hardware

Finish hardware for formed carbon steel, stainless steel, and copper-alloy doors and interior extruded aluminum doors is usually supplied by a hardware supplier. Extruded aluminum exterior doors and packaged entrances usually come equipped with all operating hardware except cylinder locks (see Section 8.4).

### 8.9.1 FINISH HARDWARE

Hardware may be either standard, handicapped access, automatic, or security type. Standard hardware includes all the devices necessary to operate, lock, and weatherstrip a door, including thresholds.

Door hardware should comply with the following standards:

- Material: American National Standards Institute/Builders Hardware Manufacturers Association (ANSI/BHMA) A156 series
- Installation: American National Standards Institute/Door and Hardware Institute (ANSI/DHI) A115 series

#### 8.9.1.1 Locks

Local fire codes should be checked before installing a new lock or upgrading an old lock. Fire codes sometimes restrict the use of locks which might slow down escape in case of fire. Lock types include key-in-knob, key-in-lever, mortise, interconnected, auxiliary, security locks, and padlocks.

#### KEY-IN-KNOB LOCKS

One of the most common and least secure locks is the key-in-knob type (Fig. 8.9-1). Such locks are relatively inexpensive but have several disadvantages:

- The locking mechanism is in the exterior knob and can be easily stolen.

**FIGURE 8.9-1 Key-in-knob lock.** (Courtesy Falcon Lock, Brea, California.)

- The latch is often short and tapered, and can easily be pried out of position with a credit card or another similar object (Fig. 8.9-2).

However, a trigger bolt attached to the latch bolt can make it impervious to credit card attack. Some models of these locks feature longer latches to make prying less effective, but none deal with the problem of the primary locking mechanism in the exterior knob.

Key-in-knob locks do not provide adequate burglar protection and therefore should be replaced or supplemented with one of the auxiliary locks discussed below.

### KEY-IN-LEVER LOCKS

Key-in-lever locks (Fig. 8.9-3) have the same advantages and disadvantages as key-in-knob locks.

### MORTISE LOCKS

A mortise lock (Fig. 8.9-4) offers more security than a key-in-knob lock because the locking mechanism is inside a metal enclosure mounted within the door. It not only has a latch, like the key-in-knob type, but also a deadbolt which cannot be violated by a credit card. Push buttons located under the latch allow the exterior knob to lock in a stationary position or unlock and turn freely.

However, mortise locks have a number of disadvantages:

- A user may fail to engage the deadbolt.

**FIGURE 8.9-2** Some door locks are easy to pry open.

**FIGURE 8.9-3** Key-in-lever lock. (Courtesy Falcon Lock, Brea, California.)

**FIGURE 8.9-4** Mortise lock. (Courtesy Falcon Lock, Brea, California.)

- The large cavity within a door needed to house this lock may impair the structural integrity of the door.
- The keyed cylinder is often poorly anchored and vulnerable to forceful removal.
- It may be defeated by loosening the set screw that holds the cylinder in place.

Some locks have faceplates which protect the cylinder. Because of the large cavity occupied by the lock case, the mortise lock is difficult to replace with any other type.

There are also locks that combine elements of both the key-in-knob and mortise lock types.

### INTERCONNECTED LOCKS

An interconnected lock has a key-in-knob cylinder, which makes it look like a mortise lock from the outside. However, it lacks the push buttons under the latch that lock and unlock the knob and differs in its method of installation. An interconnected lock requires the boring of holes. It can replace a key-in-knob lock because the old lock hole can be used and the other necessary holes added.

As with a mortise lock, the user must remember to engage the bolt. Also, if the cylinder is unprotected, it is vulnerable to pulling. Since an interconnected lock removes less of the door structure than a mortise lock, the door may stand up to more abuse before breaking.

### AUXILIARY LOCKS

A second lock is often the easiest and cheapest method of adding security to a door. A cylinder deadbolt fits within a door (Fig. 8.9-5). As with other internal locks, this may weaken the door. Another disadvantage is that two locks may keep out an intruder but may also slow escape in case of an emergency.

A vertical deadbolt lock is also called a *rim lock* because it is surface-mounted on the inside rim of the door. This is an obvious but effective type of lock. If installed with long enough screws, it resists kicking and picking. Some locks have special shutters that freeze the bolt in place if someone tries to pry it open.

### SECURITY LOCKS

Security hardware consists of high-security cylinders, special control devices such as electromechanical, electromagnetic, or electronic locks, and three-way locks in which a single key operates bolts in the head, jamb, and sill of a door. Security locks may operate by a key, push buttons, or card readers.

### PADLOCKS

A good padlock consists of (1) a hardened steel shackle, (2) a brass or laminated steel lock case, and (3) a five-pin cylinder. A padlock can add protection to any door, but it is most often used on garage or shed doors. Only the side of the door to which the padlock is attached is accessible.

### REINFORCEMENT

Many adequate locks come with inadequate strikes or screws or are installed on inadequate doors.

A *strike* is a jamb fastening that receives the bolt in the locked position. Several kicks can tear away a conventional strike along with its screws and part of the jamb, leaving the lock intact but the door open. A high-security strike consists of a small metal box which fits into the frame's

**FIGURE 8.9-5** A cylinder deadbolt is an easy and inexpensive means of adding extra security to a door. (Courtesy Falcon Lock, Brea, California.)

bolt hole and fastens securely to the building framing (Fig. 8.9-6).

The screws that attach the various kinds of locks to the door and the strikes to the jamb are often too short. In general, the longer the screw, the more securely the lock or strike will be fastened. The screws for the average rim lock should be 2 in. (50.8 mm) long. The screws on a strike should be at least 2 in. (50.8 mm) long, and preferably 3 in. (76.2 mm) long, so that they will secure the strike to the framing studs.

On soft-wood doors, a wraparound channel made of metal fits on the door and reinforces the area around the lock, preventing the wood from splitting (Fig. 8.9-7).

### SPECIAL LOCKS

Doors such as sliding glass doors and garage doors need special locks.

**FIGURE 8.9-6** This type of strike fastens securely to the building studs and helps prevent a locked door from being kicked open.

**FIGURE 8.9-7** A wraparound channel helps prevent a door from splitting and breaking in a kicking assault.

Sliding glass doors can be protected by a keyed deadbolt at the door base (Fig. 8.9-8). The disadvantage of this method is that it can slow escape in case of fire.

Some manufacturers make special locks that replace and upgrade the normal locks provided on siding glass doors. These include double deadbolts operated by both a handle and a key, which permits opening from the outside with a key.

Garage doors can be fitted with a padlock in a hole drilled through the end of a garage door bolt (Fig. 8.9-9). The padlock will secure the door even if the door's regular lock mechanism is destroyed.

### 8.9.1.2 Hinges

Several types of door hinges are available. Butt hinges are mounted on the door and frame. They may be full mortise, half mortice, full surface, and half surface types. They may be standard, heavy-duty, security, or industrial types. They are available in three-knuckle and five-knuckle types, and in standard bearing and ball bearing types. Special hinges are available for sliding and folding doors. Decorative hinges are also available. "Invisible" and piano hinges are also available but are used mostly on cabinet work.

Offset hinges allow doors to swing wider than normal. Spring hinges cause doors to close automatically.

**FIGURE 8.9-8** A keyed deadbolt prevents a door from sliding open.

**FIGURE 8.9-9** A supplementary padlock can be installed on a garage door.

Pivot hinges are a special type that are mounted in the tops and bottoms of doors and into head frames and floors. They are frequently used in aluminum entrances.

Hinges are made from prime-coated steel, polished plated steel, chromium-plated metals, brass, bronze, aluminum, and stainless steel.

Hinges are sized according to the size and type of door. For example, all exterior, interior solid core, and interior mineral core doors, and other types as well when they are 7 ft (2134 mm) or more in height, should be hung with a minimum of three hinges per door, aligned so that the door will not twist out of shape. Interior hollow core doors less than 7 ft (2134 mm) in height may be hung with two hinges.

### 8.9.1.3 Closers

*Closers* are devices that close a door using spring or hydraulic action. Types include overhead surface-mounted closers, overhead units that are concealed within the door or frame, and floor-concealed closers.

### 8.9.1.4 Automatic Operators

Automatic operators open doors when an electromechanical, hydraulic, or pneumatic operator is triggered by an actuating device. These include floor treadles, motion sensors, sonic sensors, push buttons, photoelectric cells, and remote switches. Automatic hardware also includes smoke or fire detectors that automatically close doors when smoke or fire is present.

### 8.9.1.5 Handicapped-Access Operators

Handicapped-access hardware includes counterbalanced closers that permit easy operation of a door, as well as specially designed doorknobs with extended or oversized grips to permit easy operation by a handicapped person.

### 8.9.1.6 Exit Devices

Exit devices consist of a bar across a door which, when depressed, unlatches the door to permit easy exit, even when the door is locked. Many styles and materials are available. Doors with exit devices are equipped with closers and automatically latch upon closing. Exit devices are required by code on exit doors from most public buildings.

### 8.9.1.7 Other Hardware Devices

Other door hardware devices include:

- Pushes and pulls, which are generally used on entrance doors
- Doors stops
- Hold-open devices, which are often, but not necessarily, part of a closer
- Coordinators, which cause closers on two doors to operate in sequence or simultaneously
- Rollers, tracks, and guides for pocket and other sliding doors
- Electronic and key-pad locks
- Finger guards
- Door edges

### 8.9.2 WEATHER PROTECTION

#### 8.9.2.1 Weatherstripping

Weatherstripping is applied to door frames to prevent air leakage. It typically has a resilient part that prevents air penetration when it is pressed between the door and its frame.

The most common weatherstripping types are:

- Foam rubber
- Felt
- Rolled vinyl
- Spring metal
- Interlocking metal

*Foam rubber* weatherstripping has either an adhesive or wood backing. It is easy to install but lacks durability. The adhesive type can be used wherever there is no friction, such as at the top and sides of frames (Fig. 8.9-10a). Wood-backed foam rubber is more conspicuous than the adhesive type (Fig. 8.9-10b).

**FIGURE 8.9-10** Common types of weatherstripping: (a) adhesive-backed foam rubber, (b) wood-backed foam rubber, (c) rolled vinyl, (d) V-strip.

*Felt* stripping has the same advantages and disadvantages as foam rubber, but it also collects dirt more readily. Metal-backed felt, which provides more durability, is also available.

*Rolled vinyl* is available with or without a metal backing. Some types are manufactured with tubular gaskets that can either be hollow or filled with foam (Fig. 8.9-10c). Rolled vinyl is very durable.

*Spring metal* stripping is available in two types: flat stripping, which is most commonly used for double-hung windows, and V-strip, which is used for doors. A V-strip is a double-over strip of springy metal that fits between the door edge and jamb (Fig. 8.9-10d). When the door is open, the metal springs apart; when the door is closed, it is forced together to fit tightly between door and jamb. It is durable and inconspicuous after installation, provides an excellent seal, and will last indefinitely.

*Interlocking metal* stripping is produced as interlocking channels and interlocking J-strips. Both are difficult to install because alignments must be exact for a secure interlock. Both provide excellent, durable seals. The channels are subject to damage, however, because they are exposed. The J-strips are not visible when installed; they are not exposed and, therefore, not subject to damage.

#### 8.9.2.2 Door Bottoms

Air leakage through cracks at door bottoms can be stopped with (1) sweeps attached to the door bottom, (2) weatherproof thresholds, or (3) door shoes.

The most common types of *sweeps* consist of a felt or rubber flap mounted in a metal channel (Fig. 8.9-11a). They seal most effectively on flat thresholds. They drag on carpets, however, and will eventually wear down from friction. To avoid carpet drag, automatic door sweeps can be used. These are spring-loaded and flip up when the door is opened, down when it is closed.

An *interlocking threshold* consists of a metal sill that is shaped to interlock tightly with a metal channel attached to the door bottom (Fig. 8.9-11b). It is durable and provides an excellent weather seal but is difficult to install because proper alignment is critical.

*Weatherproof thresholds* can be either vinyl-bulb or interlocking type. These thresholds are difficult to install because a good fit requires that the door bottom be trimmed at or manufactured with a $1/8$

**FIGURE 8.9-11** Door bottoms: (a) sweeps, (b) interlocking threshold, (c) vinyl-bulb threshold, (d) door shoe.

in. (3.2 mm) bevel. A vinyl-bulb threshold contains a flexible vinyl strip which is compressed when the door is closed (Fig. 8.9-11c). The threshold itself is metal and very durable, and although the vinyl bulb will wear out, it can be replaced.

*Door shoes* are metal frames with attached vinyl bulbs that press against the threshold to form a tight seal (Fig. 8.9-11d). They are very durable because the bulbs can be replaced when worn. Door shoes are best suited for flat thresholds.

### 8.9.3 BURGLARY PROTECTION

In most localities today, it is necessary to make a special effort to reduce the vulnerability of buildings to unauthorized entry by thieves, vandals, and pranksters, which is called *burglary*. Burglars do not like challenges. They prefer to work quickly, quietly, and in the dark. If more than 4 minutes are required to gain entry, they will often give up the attempt. No lock will stop a determined professional burglar, but good burglary protection hardware on doors and windows and the design of doors and windows to resist intrusion can slow a burglar down. Care should be taken, however, to ensure that devices intended to deter intrusion do not endanger occupants by preventing their exit in an emergency.

To be effective, the devices discussed in this section need to be combined with proper site planning, warning devices, and adequate lighting. Site planning measures to deter unwanted intruders, and devices intended to make noise when triggered so that a burglar will not be able to work quietly and to warn occupants and neighbors of an attempted intrusion, are discussed in Section 20.1. Outdoor lighting designed to deny a burglar a dark place to work is discussed in Section 18.4.

### 8.9.3.1 Unauthorized Entry

Before effective measures can be designed to prevent intrusion, it is necessary to know how burglars gain entry. Even locked doors and windows offer only a slight challenge to experienced burglars.

Simple lightweight tools can be used to pry open doors equipped with automatic latching devices of the type commonly used in many cylinder locks. Deadbolts are somewhat harder to pry open, and require more force and heavier tools.

A credit card or a similar thin piece of plastic can be slipped between a door and its frame to push back the latch and open the door (see Fig. 8.9-2).

A pair of pliers can be used to grip a lock cylinder and pull it out of the lock, making the mechanism accessible. The cylinder puller sometimes uses a screwdriver to twist and wrench the core of the locking pins away from the cylinder.

A crowbar may be used to jimmy doors. By inserting the crowbar between the sash and frame, a lock forcer can pop the bolt out of place. Even if the lock holds, the frame may break.

### 8.9.3.2 Doors and Frames

Adequate locks are frequently installed on inadequate doors. A lock is only as good as the door on which it is installed. In general, exterior entry doors should be carefully selected and properly installed. Selection of door and frame types and materials, and the types of locks and other security devices used, depends on the project budget and the level of security required. In some cases, the company insuring a building's contents may dictate the level of security required.

### HINGED DOORS

Flush wood solid core or metal doors, $1^3/_4$ to 2 in. (44.5 to 50.8 mm) thick, are the least vulnerable to attack. Wood hollow core doors and stile-and-rail doors with wood or glass panels are more vulnerable. A burglar can easily break through a hollow door or panels and reach the lock from inside.

Exterior metal doors consist of a core and a metal frame surrounded by steel skins. Most security experts prefer the use of metal doors over any type of wood door. Doors should fit in their frames snugly, with only enough clearance to ensure easy operation.

### SLIDING DOORS

Security problems with siding glass doors include (1) inadequate locks, (2) flimsy frames, (3) door panels that can be lifted out of their frames, and (4) vulnerability of the glass.

A series of screws along the upper track of a sliding glass door can prevent the door from being lifted out of its track (Fig. 8.9-12). A piece of wood or metal slightly shorter than the door width can be laid in the lower track (Fig. 8.9-13) to supplement the lock and prevent the door from being forced open. To keep the wood or metal bar from being lost, it can be hinged at the frame and leaned against the side jamb when not in use. Commercial devices are available to serve this purpose.

### VIEWING DEVICES

One-way viewing devices (peepholes) in doors allow occupants inside a building to see outside without having to open the door.

### DOOR FRAMES

Even a strong door with a good lock and proper hinges can be opened easily by a

**FIGURE 8.9-13** A metal or wood bar in the track prevents a sliding glass door from being forced open from outside. The bar can be hinged to prevent its loss.

burglar if the door's frame is inadequate. If the rough opening is oversized and the door frame is inadequately shimmed, a burglar can often pry the frame from the lock to release the door. A space as small as $^1/_{16}$ in. (1.6 mm) may be large enough to insert a crowbar, screwdriver, or hacksaw. The space between the door and its frame should be $^1/_{64}$ in. (1.6 mm) or smaller. The same tolerance should be observed at the side jambs and at the top of the door because an automobile jack can be used to spring the frame at a point just above the lock.

**FIGURE 8.9-12** Screws at periodic intervals along the upper track prevent a would-be intruder from lifting a sliding glass door off its floor track.

### 8.9.4 HANDICAPPED PROVISIONS

Requirements for compliance with the Americans with Disabilities Act (ADA) are discussed in Section 1.4. The requirements here are supplementary to those in Section 1.4 specifically related to door hardware.

The general principle is that all places of public accommodation must be accessible to persons with disabilities. The major concerns are the ability of a person with a disability to:

■ Grasp and operate the pull side of a door exit device
■ Avoid contact with obstructions on the surface of a door where a wheelchair foot rest is used to push open the door
■ Avoid obstructions that protrude into an exitway
■ Open and close a door

A solution to the grasping requirement is lever handle devices. Unfortunately, lever devices are much more easily damaged by vandals than are standard knob devices. Free-wheeling and break-away lever designs are available, which are more resistant to deliberate damage than rigid lever devices.

Contact at the bottom of a door can be avoided by not specifying surface-mounted vertical operating rods for closers and instead using concealed rods. Several proprietary devices are also available to solve this problem.

The clear area required at doors is specifically delineated in ANSI A117.1 (see Section 1.4).

Door closers must have sufficient force to close doors completely and to permit latching hardware to operate properly, especially on fire-rated doors. The pressure needed is often more than a person with a disability can exert. Automatic operators and handicapped-access hardware, as described earlier, can solve these problems.

## 8.10 Glazing

This section discusses glass and plastic glazing materials and their application (glazing). The term *glazing* refers to the installation of vision panels of glass or plastic in a door, formed frame, window, packaged entrance, storefront, or glazed curtain wall. Spandrel glass in curtain walls is also glazing. The other types of opaque panels in a curtain wall are also sometimes said to be *glazed* into the system but are not generally regarded as glazing materials.

### 8.10.1 GLASS

Glass results from the fusion of certain common minerals at high temperatures and subsequent controlled cooling. This process causes solidification without crystallization. The chief ingredients of glass are silica, sodium oxide, and calcium oxide (sand, soda, and lime), all of which are abundant minerals throughout the United States.

Glass-like mineral formations resulting from the accidental fusion of silica sand and metallic oxides by the heat of a volcano or lightning (obsidian and lightning stone) can be readily found in nature and may have served as the impetus for the early invention of glass. Glass has been made artificially for more than 4500 years.

The first window glass we know of was made by casting and grinding during Roman times. In eleventh-century Germany, glass was hand blown in spheres, elongated into cylinders by a swinging process, then cut apart and flattened to make a product called *broad glass.* *Crown glass* was produced in France during the fourteenth century by a blowing and twirling process. Thomas Jefferson's Monticello contains Bohemian broad glass, and there is crown glass in Williamsburg. Plate glass was produced in the seventeenth century mainly to provide the smooth surfaces needed to make mirrors. It was cast and then ground and polished. In the nineteenth century, a product called *cylinder glass* was made by a blowing and swinging process. It could achieve a maximum size of 10 by 49 in. (254 by 1245 mm). It was this size restriction that governed the building module (48 in. [1219 mm]) of the Crystal Palace built in London in 1851, which contained 900,000 sq ft (83613 m²) of cylinder glass. Later innovations led to the drawing of molten glass cylinders vertically to a length of up to 50 ft (15.24 m).

The real revolution in glassmaking occurred in the early twentieth century, when the Fourcault and Colburn processes for mass producing large sheets of flat glass were developed.

Although they were all known earlier, glass ranks with concrete, steel, aluminum, and plaster as a true material of the twentieth century. Glass has exerted a tremendous influence on our technology, our modern way of life, and our architectural expression. From the divided-light window of colonial days to modern floor-to-ceiling glazing, glass technology has progressed with building design needs and has in turn influenced them.

Just as colonial builders limited the size of windows because of high glass cost and shelter needs, today's designers continually increase glass areas as the cost of glass declines and our ability to control the interior environment improves (Fig. 8.10-1). Largely glass or all-glass buildings are commonplace and create psychological impacts on their inhabitants (Fig. 8.10-2). The ever-changing panorama of the natural outdoor environment is psychologically satisfying to most people. Possible adverse effects from overexposure to the elements can be minimized by proper ori-

**FIGURE 8.10-1** University Tower in Durham, North Carolina, is an example of the kind of glass use that influences architecture today. (Architect Pickle/Thomas; Photograph courtesy Guardian Industries.)

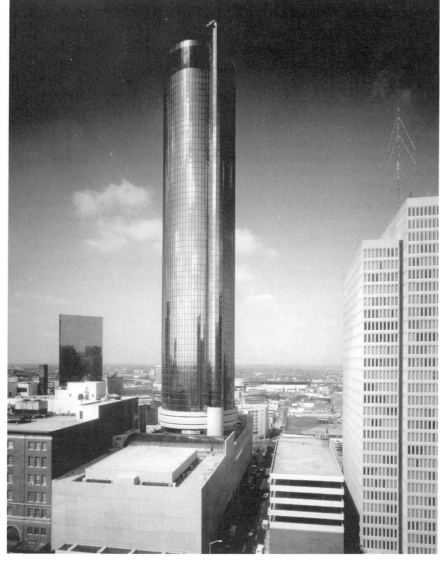

**FIGURE 8.10-2 Peachtree Hotel, Atlanta, Georgia.** (Architects: John Portman and Associates; photograph courtesy Bruce Wall Systems Corporation.)

entation and judicious selection of special glass types such as glare-reducing, heat-absorbing, reflective coated, or insulating glass.

### 8.10.1.1 Properties of Glass

Glass is unusual in terms of its internal structure. It is mechanically rigid and thus behaves like a solid. However, most mineral solids have a crystalline structure, with the atoms arranged in a definite geometric pattern. In glass, the atoms are arranged in a random or disordered fashion, as is characteristic of a liquid. But the atoms, though arranged at random, are "frozen" in position by the rapid cooling process used to make glass.

Therefore, glass combines some of the aspects of a solid and a liquid. For this reason, it is sometimes called a *super-cooled* liquid.

The physical properties of glass can be varied over an extensive range by modifying either the composition of the glass, the production techniques used, or both. In selecting an individual glass for a particular product, a combination of properties must be taken into consideration, because one property usually cannot be changed without causing a change in other properties. The following is a brief discussion of some of the properties of glass which are important in architectural applications.

## MECHANICAL PROPERTIES

Mechanical properties deal with the action of forces on a material and the effects these forces produce within the material.

An important mechanical property of glass is that it is *fatigue-resistant*. This means that if pressure is applied to glass, causing it to bend or stretch, the glass will return exactly to its original shape when the bending or stretching force is removed. Of course, if increasing force is applied, the glass will eventually break, but at any point short of breakage, it will not be permanently deformed.

The mechanical strength of glass is determined by its ability to withstand forces that cause breakage. Glass generally breaks from stretching or bending (tensile forces). Thus, tensile strength is the chief factor used in determining its mechanical strength.

Unlike most metals, glass does not have a clearly defined tensile strength. The *theoretical* tensile strength of glass has been estimated to be as high as 4,000,000 psi (2 812 280 Mg/m$^2$). The actual tensile strength of annealed glass ranges between 3000 and 6000 psi (2109.2 and 4218.42 Mg/m$^2$).

The main reason for the wide variation between theoretical and actual strength in glass is that actual strength is dependent on the surface condition of the glass. No matter how carefully it is handled, glass will acquire some small nicks and scratches in the course of manufacture and later in use. These surface defects then create weak planes in the material. When a force is applied to glass, stresses tend to concentrate at the weak planes rather than distribute themselves uniformly, as is assumed in theoretical strength calculations. Because of variations in the surface condition of different lights of glass, a group of seemingly identical glass specimens will exhibit a wide range of breaking strengths when tested. Thus, the mechanical strength of glass cannot be stated as a clearly defined figure; it must always be calculated on a probability/stress curve.

The mechanical strength of glass can be improved by both heat and chemical treatments. These processes are discussed in Section 8.10.1.2.

## OPTICAL PROPERTIES

An obvious and important property of glass is its transparency. Despite appearances, however, no glass is completely transparent. When light falls on a piece of glass, some of that light is reflected and some is absorbed. The amount of light that does pass through the glass is called *transmittance*. Commonly used clear commercial glass has a transmittance of about 85% to 90% of visible light.

Glass may be treated in many different ways to reflect or absorb varying degrees of light. Glass that has been tinted absorbs a higher percentage of light than clear glass and, depending upon the color and thickness of the glass, may transmit anywhere from 21 to 75% of visible light. Glass that has been treated with a thin reflective coating reflects a high proportion of light and may transmit 8 to 50% of visible light, depending on the nature of the coating. The metallic coating on the backs of mirrors produces the maximum amount of reflectance possible, allowing no visible transmittance.

Another important optical property of glass is its *fidelity* (lack of distortion). When the two surfaces of a sheet of glass are perfectly parallel, the image that is seen through, or is reflected from, the glass will be free of waves or distortion (Fig. 8.10-3). However, small variations in the thickness of a sheet of glass can cause considerable distortion in the image seen. For example, in a large area of glass, a variation in thickness of only a few hundred thousandths of an inch (millimeter) can be quite apparent because the variation acts as a lens to magnify or diminish portions of the reflection.

### THERMAL PROPERTIES

The amount of heat that passes through the glass areas of a building is an impor-

**FIGURE 8.10-3** The uniform thickness and flatness of float glass ensure distortion-free views.

tant consideration in determining heating and air conditioning loads. Heat can be transferred by conduction, convection, and radiation. Glass has lower thermal conductivity than most metals, but a great deal of solar energy can be transferred directly through glass by radiation, and air currents that come into contact with glass surfaces can transfer heat by convection. Therefore, overall heat loss in winter and heat gain in summer are generally greater through the glass areas of a building than through ordinary masonry, metal, or wood-framed walls.

However, heat gain (or loss) through windows can be controlled by insulation, reflection, absorption, or shading. Special glass products such as insulating glass, reflective coated glass, and tinted glass make use of these principles and can substantially reduce heat gain. In addition, shading devices such as blinds, draperies, and awnings can be used with glass to further reduce heat gain (see Section 12.2).

The thermal expansion of glass is another important consideration in the design of buildings. This factor is particularly important when using tinted or reflective coated glasses, as these glasses absorb a higher percentage of solar energy than ordinary clear glass, and therefore will expand and contract more.

When different portions of a single pane of glass heat up at different rates, they expand at different rates, causing thermal stresses to develop within the pane. What often happens under ordinary glazing conditions is that the center portion of the pane heats up faster than the edges (which are shaded by the edge framing), causing the center portion to expand more than the cooler edge regions. This produces stress differentials within the glass, which tend to concentrate at weak points in the glass and may eventually cause breakage. Most thermal failures can be traced to minor edge damage or imperfections, as this is where weak points in a sheet of glass are most likely to occur. Therefore, the best way to ensure against such thermal breakage is to be sure that glass edges are strong and clean-cut.

The thermal expansion of glass also is important when two glasses or a glass and a metal are to be sealed together, as in insulating glass and some types of coated glass. The rates of expansion of the two materials must be closely matched to avoid damaging differential stresses in the finished product.

### CHEMICAL PROPERTIES

Glass is much more resistant to corrosion than most metals and other common materials. It is nonporous and nonabsorptive, and thus is impervious to many common elements that affect other materials. For this reason, glass windows, after many years of exposure to the elements, remain clear and apparently unchanged.

Nevertheless, glass will corrode and even dissolve under some conditions. For example, hydrofluoric acid will attack soda-lime glass, causing extreme corrosion; similarly, hot, concentrated alkali solutions and superheated water will cause soda-lime glass to dissolve. However, these chemicals are usually encountered only in laboratory situations. Under ordinary environmental conditions, glass is an extremely inert and durable substance.

### ELECTRICAL PROPERTIES

Because it offers high resistance to the passage of electricity, glass is a good electrical insulator. Therefore, glass is used in making many electrical products.

#### 8.10.1.2 Manufacture of Glass

The progress in glass manufacture and the development of glass technology have been outstanding since the turn of the twentieth century. Today glass is manufactured in a wide variety of compositions and can be fabricated by several different techniques into a tremendous array of products.

#### RAW MATERIALS

Glass is manufactured in several generic types which vary according to chemical composition (Fig. 8.10-4). The most common and least expensive type is *soda-lime glass*, made from the oxides of silicon, sodium, and calcium. This type of glass is used for virtually all glazing purposes in both commercial and residential buildings and in automobiles.

Soda-lime glass is composed of approximately 72% silica, 15% sodium, 9% calcium, and 4% various minor ingredients. Other glass types, which have been derived from this basic glass, have slightly different ingredients. These will be mentioned only briefly here, as they are of minor importance to the construction industry.

*Silica* Silica, the principal ingredient of glass, is obtained from sand. It is one of

**FIGURE 8.10-4** Representative Glass Types, Their Composition, and Their Uses

| | | | Composition[a] | | | | | | | |
|---|---|---|---|---|---|---|---|---|---|---|
| | Sand | Soda Salt Ash or Cake | Lime-stone | Dolomite | Potash | Boric Acid | Litharge | Feldspar | | |
| Type of Glass | $SiO_2$ | $Na_2O$ | CaO | MgO | $K_2O$ | $B_2O_3$ | PbO | $Al_2O_3$ | Applications or Uses |
| 100% silica | 100%[b] | — | — | — | — | — | — | — | Glass laboratory ware |
| Soda lime silica[c] | 68–74 | 11–16 | 8–12 | 1–4 | — | — | — | 0–4 | Float glass |
| Borosilicate | 65–80 | 2–4 | 3 | — | 0–4 | 13–28 | — | 1–3 | Heat-resistant glasses |
| Lead alkali-silicate | 35–63 | 0–10 | 0–1 | — | 10–13 | 0–1 | 15–58 | 0–1 | Radiation-shielding products |
| Optical (flint)[d] | 35 | — | — | — | 7 | — | 58 | — | Ophthalmic lenses |
| Aluminosilicate | 55–57 | 0–1 | 5 | 8–12 | 4 | 4–8 | — | 20–23 | High-thermal-impact glasses |

[a]In addition to the raw materials shown, small amounts of other elements may be added to help oxidizing, coloring, reducing, etc. Cullet (crushed glass) is also used to help melting.

[b]All figures are percentages of the ultimate oxides in glass compositions.

[c]Most important type based on volume and value.

[d]Optical glass often contains oxides of zinc, titanium, zirconium, and boron in addition to oxides shown.

the most common elements of the earth's crust. Unlike the sand found at the seashore, sand for making glass must be more than 99% silica, perfectly white and not too fine. Sand of this quality usually is obtained from sandstone deposits, which are abundant in the United States, particularly in Pennsylvania, West Virginia, Illinois, and Missouri.

Silica is the glass former in the glassmaking process. It can be used alone to produce glass, but it requires a very high melting temperature (3100°F [1704°C]) and is very difficult to form. Thus, it is usually combined with other ingredients to make a more workable glass.

*Sodium* Sodium for glassmaking is usually obtained from soda ash made from salt, but sodium nitrate and sodium sulfate may also be used. Sodium is added in glassmaking as a flux, to lower the melting point and to make a more workable mix. When sodium is added to silica, it forms a mixture that will melt at 1460°F (793°C), a considerably lower temperature than for silica alone.

However, silica and sodium alone form a glass which is not very durable. In fact, it slowly dissolves when placed in contact with water. It also has low resistance to chemical attack. Therefore, an ingredient that will act as a stabilizer must be added to every glass batch.

*Calcium* The stabilizer usually added to glass is calcium. It is obtained either from limestone or dolomite. When added

to silica and sodium, it produces a glass that is both durable and easily worked.

*Additives* Special elements may be added to, or substituted for, the basic ingredients in glassmaking in order to produce glass with specific properties or characteristics.

*Potash*, a form of potassium, was used instead of soda ash in the early days of glass manufacture in the United States. At that time, wood was the principal fuel for glassmaking and potash was a byproduct of wood ash. Potassium is used infrequently today because it is much more costly than sodium. Glass made with potassium is finer, harder, and more brilliant than that made with sodium.

In soda-lime glass, lead is sometimes used to replace the calcium and sometimes to replace part of the silica. It produces a soft, easy-to-melt glass and, when used in quantity, imparts a canary yellow color to the glass. *Lead glass* is relatively expensive, but it has excellent electrical insulating and radiation shielding properties and is more brilliant in appearance than soda-lime glass. It is used primarily in making electrical products, objects of art, and fine tableware.

Boron and aluminum are used to produce types of glass which are both heat- and shock-resistant. *Borosilicate* glass and *aluminosilicate* glass, as they are called, are used primarily for laboratory ware and oven-cooking ware.

Several additives are used specifically to produce color in glass (Fig. 8.10-5).

For example, nickel oxide produces a tint ranging from yellow to purple, depending on the glass base. Cobalt oxide gives an intense blue, and cobalt in combination with nickel produces a gray tone. Red glass is made with gold, copper, or selenium oxides, and many other colors may be produced with other chemicals. Also, arsenic is sometimes added to glass in small quantities to counteract the greenish color caused by iron impurities in the glass. Some of these colored and tinted glasses are used widely in architectural applications.

## MIXING AND MELTING

The first major step in glassmaking is mixing and melting the raw materials. They are granulated to nearly uniform particle size, then carefully weighed and mixed in the proper proportions. Some *cullet* (scrap glass) from a previous melt of the same kind of glass is added to the mixture to assist in the melting process and to conserve materials. The mixed batch is then loaded into a melting unit. The type of unit used depends upon the type and quantity of the glass needed.

Some glasses that are required in small quantities are melted in *refractory pots*, which are clay containers that can hold up to 3000 lb (1361 kg) of glass. Up to 20 of these pots can be heated in a single furnace, thus allowing several different types of glass to be melted at the same time.

For somewhat larger quantities of glass, *day tanks* are used. Day tanks hold

**FIGURE 8.10-5** Additives for Coloring Glass

| Additive | % in Glass | Resulting Color |
|---|---|---|
| Cadmium sulfide | 0.03–0.1 | Yellow |
| Cadmium sulfoselenide | 0.03–0.1 | Ruby and orange |
| Carbon and sulfur compounds | | Amber |
| Chromium oxide | 0.05–0.2 | Green to yellow-green |
| Cobalt oxide | 0.001–0.1 | Blue |
| Copper | 0.03–0.1 | Ruby |
| Copper oxide | 0.2–2.0 | Blue-green |
| Ferric oxide | Up to 4.0 | Yellow-green |
| Ferrous oxide | | Blue-green |
| Gold | 0.01–0.03 | Ruby |
| Iron oxide | (1.0–2.0%) with manganese oxide (2.0–4.0%) | Amber |
| Manganese oxide | 0.5–3.0 | Pink-purple |
| Neodymium oxide | Up to 2.0 | Pink |
| Nickel oxide | 0.05–0.5 | Brown and purple |
| Selenides | | Amber |
| Selenium | | Bronze |
| Uranium oxide | 0.1–1.0 | Yellow with green fluorescence |

enough glass for a single day's operation—up to 20,000 lb (9072 kg). The tank is filled with raw materials at the beginning of the operation, the glass is melted, and then all of the glass is used before the furnace is refilled.

The units most commonly used in commercial glassmaking operations are continuous tanks, holding from 50 to 1000 tons (45 359 to 907 184 kg) of molten glass. These operate continuously, with raw materials being fed into the loading end as rapidly as the molten glass is removed from the working end. This continuous operation can go on for years, depending on how long it takes for the heat-resistant clay walls of the furnace to be worn away by the hot glass.

### GLASSMAKING PROCESSES

After molten glass leaves a melting tank, different manufacturing processes are employed to make various glass products. Processes currently used to make flat glass include (1) *rolling* for making patterned and wired glass and (2) *floating* for making float glass.

Other processes, such as *pressing* and *foaming*, are used to make such products as glass block and cellular glass insulation, respectively. Blowing is, of course, the oldest and best-known glassmaking

process, but today it is primarily employed in the manufacture of containers and art objects rather than for making architectural products.

*Drawing*, an older process for making sheet glass, is no longer used but will be discussed here from a historical perspective.

***Drawing*** In the early days of glass manufacture, sheet glass was made by the blowing process. The molten glass was blown into the shape of a cylinder; the cylinder was then split open, reheated, and flattened to form a small sheet of glass. Glass made by this process could not be produced in large sheets and was of poor quality. It was characterized by uneven surfaces and longitudinal wrinkles that distorted one's vision.

The next evolution in sheet glassmaking was the drawing process (Fig. 8.10-6). Three different methods were used: the *Fourcault*, the *Colburn*, and the *Pittsburgh* methods. All three methods used the same basic procedure.

The raw materials were first melted in a continuous melting tank. The drawing operation was then started by inserting a *bait* (steel rod) into the molten glass. Molten glass has very high viscosity, making it sticky and highly resistant to

flow. Therefore, the glass adhered to the bait and was drawn up with the bait as it was raised. The molten glass was drawn directly upward between jets of flame which fire-polished the surfaces and made them transparent.

Pulling rollers then kept the drawing process going. A continuous sheet was drawn from the molten glass and was pulled slowly through an *annealing lehr*, a continuous oven. Inside the lehr, tem-

AUTOMATIC CUTTER
CUT-OFF FLOOR
2nd BALCONY
DRAWING ROLLS
1st BALCONY
CONTINUOUS SHEET
DRAWBAR
MOLTEN GLASS

**FIGURE 8.10-6** The Pittsburgh process for drawing sheet glass vertically.

peratures were regulated so that the glass was cooled gradually, thus avoiding excessive amounts of strain in the finished product.

After sheet glass emerged from the annealing lehr, no further finishing operations were required other than cutting to size and packing.

There were some minor differences in technique between the three methods of drawing sheet glass.

The *Fourcault* process was the first process to be developed, and it was the simplest one. The glass sheet was drawn vertically from the tank to a height of about 40 ft (12.19 m).

The *Colburn process* was similar to the Fourcault process, except that the glass sheet was drawn vertically for only about 3 ft (914.4 mm) and was then bent over horizontal rollers. The chief advantage of this process was that the glass sheet could be drawn through a horizontal rather than a vertical annealing lehr. The length of a vertical lehr was limited by the height to which it was practical to draw glass vertically—that is, about 30 to 40 ft (9.15 to 12.19 m). The horizontal lehr used in the Colburn process, however, could be up to 200 ft (60.96 m) long, which allowed the glass to be cooled much more slowly, producing a glass that was less brittle and easier to cut.

In the *Pittsburgh* process, the glass was drawn vertically, as in the Fourcault process. It differed from the other two processes mainly in that the drawing machine was constructed so that the sheet of glass was not contacted by center rolls until the glass was hard enough to resist being marred by the rolls.

Several problems were encountered in the drawing process, no matter which method was used. One problem was that the drawing process stretched the sheet, causing "waisting-in" (variations in the width) of the glass. This problem was overcome by water-cooled rollers at each edge which maintained sufficient side pull to keep the width constant.

In addition, the drawing process caused waves and ripples to form in the rising sheet of glass before it had cooled enough to become solid, which left an inherent wave or distortion, usually running in one direction. For most purposes, this wave was not noticeable enough to cause a problem; however, when large pieces of glass were required or when an oblique angle of view was likely, these imperfections could be quite apparent.

This problem largely excluded sheet glass from use for large commercial mirrors, automobile windshields, and display windows, for which the superior qualities of plate or float glass were required.

*Rolling* Plate glass was originally made by *casting*, a very slow, laborious process. The glass was poured onto a casting table, then rolled flat, cooled, and finally ground and polished.

Between 1922 and 1924, however, the major glass manufacturers began using a continuous rolling process for making both plate and patterned glass. In this process, the glass first was mixed and melted in a continuous melting tank. The molten glass then flowed through an opening in the tank into the forming rolls. These rolls were water-cooled, electrically driven, and had precision-ground steel surfaces which could not be harmed by the extremely high temperatures of the glass. They rolled the molten glass to form a semisolid continuous ribbon of glass. The rolls could be adjusted for various thicknesses of glass, in widths of up to 11 ft (3.35 m).

From the forming rolls, the ribbon of glass moved over a series of other steel rolls until it became solidified enough to enter the annealing lehr. There it was cooled gradually to remove internal strains. After leaving the lehr, the glass was cooled still further until it reached room temperature. At this point, the glass was called *rough-rolled* glass.

If a translucent effect was desired, rough-rolled glass could be used without further processing. When patterned glass was desired, the forming rolls were embossed with various decorative designs; this glass, like rough-rolled glass, did not have to be ground and polished. However, if plate glass was to be made, both surfaces still needed to be ground and polished.

Most plate glass was made by the twin-grinding method, in which the glass was ground on both sides simultaneously. This method ensured smooth, parallel surfaces and, thus, distortion-free glass. A mixture of silica sand and water was used in the grinding process.

The sand was graded into several different sizes; the largest particles were used in the first grinding units, and then the size of the particles gradually was reduced. A considerable quantity of glass was removed in the grinding process. Thus, the rough-rolled glass that was

used to make plate glass was produced in a thickness greater than that desired in the finished product.

After grinding, the glass was moved to polishing tables, where it was passed along on a conveyor belt between a series of polishing wheels. The wheels were covered with buffs of felt, leather, or a soft metal, and the polishing agent most often used was rouge (ferric oxide) mixed in a solution of copperas water (ferrous sulfate and water). After polishing, the glass was cut to size and packed for shipment.

Rolling is no longer used to make polished plate glass by any manufacturer in this country. Rolling is still used, however, to make patterned and wired glass.

*Floating* Ground and polished plate glass has been replaced by *float glass* because the latter is substantially cheaper to produce. The float process was introduced in 1959 by Pilkington Brothers, Ltd., of England in an attempt to combine the fire-polish and low cost of sheet glass with the flatness and low-distortion properties of plate glass.

The first stage in the production of float glass is the same as that once used to produce sheet and plate glass. The batch is melted in a huge continuous tank. From there on, however, the process is quite different.

The molten glass leaves the tank and is floated on a bed of molten metal composed mostly of tin. When a liquid floats on another liquid of higher density, both surfaces of the upper liquid must become flat and parallel. Therefore, when the liquid glass moves across a float bath, it forms a thin layer with flat, parallel surfaces.

The float bath is divided into three sections (Fig. 8.10-7). In the first section, heat from the molten tin and a top heater causes the glass to float uniformly over the flat surface of the tin. In the second section, the glass receives a brilliant fire-polished surface. In the third section, the glass is cooled until it becomes hard enough to be conveyed to rollers in the lehr. Very precise control of heat all along the length of the float bath is required.

After the glass leaves the float bath, it enters an annealing lehr, where it is cooled gradually to ensure that the glass will not contain excessive amounts of strain. It is then washed and moved to a warehouse for inspection, cutting, and shipping.

**FIGURE 8.10-7** The float process for making glass.

## HEAT AND CHEMICAL TREATMENTS

Several heat and chemical treatments are used to increase the strength and usefulness of flat glass products. *Annealing* relieves internal strain and facilitates fabrication. *Tempering* and *heat strengthening* cause physical characteristics to be developed in the glass so that it cannot be fabricated afterward. Fabrication, such as cutting, grinding, sandblasting, drilling, and notching, must be performed before these heat treatments.

***Annealing*** All flat glass, after forming, is subjected to a controlled cooling process called *annealing* to relieve internal stress. If a piece of glass is not annealed, parts of it will cool and contract at different rates, and when the glass approaches room temperature, it may fracture due to differential stresses throughout the sheet.

Annealing is performed by moving the glass on a belt through a continuous oven called a *lehr*, in which temperatures are carefully regulated. The glass is first raised to a high enough temperature to relieve strain developed in the forming process. Then the temperature is brought down slowly so that all parts of the sheet cool at the same rate. Therefore, when the glass reaches room temperature, it is free of internal stress and can be easily cut and processed.

***Tempering*** Tempering is a heat treatment used to increase the strength of flat glass. The process involves heating the glass to just below the softening point and then chilling it suddenly by subjecting both surfaces to jets of cool air. This causes the surfaces to shrink and harden quickly while the interior is still fluid. The interior begins to cool and shrink next, but since the surfaces are already hardened, they cannot flow and adjust to the shrinkage of the interior. This causes the surfaces and edges to be placed in a state of compression while the interior of the glass is in a state of tension. In properly tempered glass, the opposing stresses of the surface and the interior balance each other out. The resultant glass is much stronger than ordinary flat glass because the built-in compression of the surface must be overcome before breakage can occur.

This type of glass usually is referred to as *fully tempered* glass, to distinguish it from *heat strengthened* glass, which is only partially tempered. Fully tempered glass is three to five times more resistant to wind loads, impact, and thermal stresses than ordinary annealed glass of the same thickness.

***Heat Strengthening*** The heat treatment known as *heat strengthening* is also used to increase the strength of flat glass. It involves heating the glass and then cooling it in a manner similar to that used in tempering. However, heat strengthening produces lower surface compression stresses in the glass than the tempering process. Therefore, heat-strengthened glass is only about twice as strong as annealed glass of the same thickness.

***Chemical Strengthening*** Glass can be strengthened by chemical treatment as well as by heat treatment. The most commonly used chemical treatment involves an exchange of ions in the surface of the glass. The glass is immersed in a molten salt bath and the large potassium ions in the salt migrate into the glass surface, replacing the smaller sodium ions there. This action crowds the surface, resulting in compression stresses similar to those produced in glass by thermal tempering.

Full temper in soda-lime glass is difficult to achieve by chemical tempering methods. A high degree of temper by chemical means generally requires accurate formulation of ingredients selected to ensure that glass ions interchange readily with ions in the molten salt bath. Furthermore, in chemically strengthened glass, the compressed surface layer is appreciably thinner than that in thermally tempered glass. Therefore, it is more readily weakened by surface abrasions such as cuts and scratches.

## SURFACE FINISHES

The surface finish of glass depends on both the method of manufacture and the subsequent finishing processes to which it is subjected.

Float glass emerges from the annealing lehr with a smooth, glossy, transparent finish known as a *fire finish*. This type of finish is characteristic of the glass-making process, and no further finishing operations are required. One reason plate glass was more expensive to produce than float glass was that it emerged from the annealing lehr with a rough, translucent surface that required grinding and polishing to produce the smooth, transparent finish desired for most flat glass products.

There are special cases, however, where a smooth, transparent glass is not essential. A translucent glass may be desired to provide privacy or to limit the amount of light transmitted into a room; a specially coated or tinted glass may be needed to reduce glare or to control the amount of solar heat transmitted into a room; a patterned or specially treated glass may be preferred just to produce a decorative effect. Some of the surface finishes associated with these special types of glass are discussed in the following subsections.

*Etching*   Many different degrees of transparency can be produced in glass by etching. These range from a frosted, almost opaque quality to a semipolished, translucent appearance.

Hydrofluoric acid or one of its compounds is used to etch glass, as it is the only chemical that interacts actively with glass. Either the glass is dipped in the chemical or the chemical is sprayed on; the type of surface produced depends on the composition of the glass, the concentration of fluorides, and the time involved. If an etched design is desired, the glass is first painted with an acid-resistant chemical (the *resist*) to protect the parts of the glass outside the desired pattern. The acid eats away the unpainted surface of the glass, leaving the design.

*Sandblasting*   The translucent finish obtained by sandblasting usually is rougher than that obtained by etching. Sandblasting is done by using compressed air to blow coarse, rough-grained sand against the glass surface. Often this is done through a rubber stencil to produce a decorative design.

Both etching and sandblasting reduce the strength of glass because these processes introduce surface flaws. Sandblasting, for example, reduces the strength of glass by about 50%.

*Enameling*   Translucent and solid colors can be produced on glass surfaces with vitreous enamels fired on at high temperatures. The firing process also imparts a partial or full temper to the glass. This type of glass is used primarily for spandrel areas and other parts of curtain wall construction.

*Silvering*   In the early days of mirror manufacture, silvering was done by hand, the silvering solutions being poured from a pitcher onto the flat glass surface. Some fine mirrors are still produced this way today.

Most mirrors, however, are manufactured by an automatic process in which the silver is deposited directly on the glass as it passes on a conveyor belt under sprays of silver nitrate and tin chloride. The silvering can then be protected with a coating of shellac, varnish, or paint. For almost permanent protection, an electroplated layer of copper can be applied on the silver.

*Reflective Coatings*   With the advent of the glass-enclosed high-rise, an entire new family of architectural glasses was developed to control the transmission of solar energy into buildings. These products have a thin reflective coating applied either (1) to one surface of a single sheet of glass or (2) to an inside surface of a laminated or insulating glass unit.

Two processes are used to apply reflective coatings to glass: (1) the pyrolytic process, in which the coating material is sprayed onto heated glass and becomes fused to the glass surface, and (2) vacuum deposition, in which the glass is bombarded with metallic ions, which form an integral molecular bond with the glass.

Under appropriate lighting conditions, reflective coated glass can become a type of one-way mirror, permitting vision only to the side with the greater amount of light. During the daytime, it creates a mirror-like effect from the outside of a building, while from the inside the glass is transparent. At night, when the interior lights are on, the effect is reversed, although to a lesser degree. Reflective coatings can be made to reflect varying degrees of light and heat, providing control over the solar energy transmitted into a building.

*Tinting*   Tinting is not really a type of surface finish, because it involves adding special ingredients to the basic glass batch. However, it is used widely to produce glass with heat-absorbing and glare-reducing qualities.

Many different ingredients can be added to glass to produce various tints. Ferrous iron, for example, imparts a bluish-green tint to glass, cobalt oxide in combination with nickel imparts a grayish tint, and selenium is used to make bronze-tinted glass. These different ingredients are added to the glass mixture at the very beginning of the melting process. Float glass can be tinted in this manner.

Tinted glass expands more than untinted glass due to its greater absorption of heat, which must be taken into account when designing the surrounds of the glass.

### 8.10.1.3 Flat Glass Products

The multitude of available glazing products makes possible a high degree of specialization in meeting specific design requirements. The method of glass manufacture determines the basic type of glass. By modifying the ingredients, any of these types of glass can be made either clear, to provide true vision, or tinted to reduce glare and absorb solar energy. Clear or tinted glass may be coated with a reflective surface to further reduce light and heat transmission.

Where resistance to impact and thermal stresses is important, these glasses can be strengthened by heat treatments, such as tempering or heat strengthening. Glass, either clear or tinted, can be rolled with a wire mesh reinforcement to make wired glass. Finally, almost any two or more of these glasses can be combined to form a composite multilayer glass. Different types of composite glass are designed to meet specific construction requirements, such as laminated glass for safety and security and insulating glass for thermal and sound insulation.

#### STANDARDS OF QUALITY

In the United States, primary float glass, wired glass, and patterned glass are manufactured to meet the requirements of ASTM C1036; heat-treated glass to meet the requirements of ASTM C1048; and laminated glass to meet the requirements of ASTM C1172. Figure 8.10-8 summarizes the standards governing some typical glass types. Both ASTM C1048 and C1172 reference C1036 for basic glass

**FIGURE 8.10-8** Summary of Glass Standards

| Glass Type | Standard |
| --- | --- |
| **Primary float** | ASTM C1036 |
| **Tinted float** | |
| Heat-absorbing and light-reducing | ASTM C1036 |
| Light-reducing only | ASTM C1036 |
| **Patterned** | |
| Heat-strengthened | ASTM C1048 |
| Fully tempered | ASTM C1048 |
| Not heat-treated | ASTM C1036 |
| **Wired** | ASTM C1036 |
| **Heat-strengthened** | |
| Coated and uncoated | ASTM C1048 |
| **Fully tempered** | |
| Coated and uncoated | ASTM C1048 |
| **Laminated glass** | ASTM C1172 |
| **Insulating glass** | ASTM E774 |

used in the glass types covered by them. The ASTM standards establish the dimensional, optical, strength, allowable bow (warpage), and other requirements for these glazing products. Figure 8.10-9 gives the typical glass classifications found in ASTM C1036 and C1048.

Sealed insulating glass products should comply with ASTM Standard E774.

Another important standard for glass is ANSI Z97.1, which covers the safety requirements for fully tempered, laminated, and wired glass as well as rigid plastics. It is directed primarily towards ensuring that safety glazings have fail-safe characteristics when broken by human impact.

The use of safety glazing in hazardous locations such as door sidelights, sliding glass doors, storm doors, and tub-shower enclosures is required by the U.S. Department of Housing and Urban Development's (HUD) *Minimum Property Standards for Housing*, the *International Building Code*, the *International Residential Code*, many state and municipal codes, and the Consumer Product Safety Commission, Code of Federal Regulations (CFR) 16.

### Certification of Sealed Insulating Glass

The Insulating Glass Certification Council (IGCC), founded in 1977, provides independent laboratory testing and periodic unannounced on-site plant inspections of participating manufacturers. The Associ-

**FIGURE 8.10-9** Summary of ASTM C1036 and ASTM C1048 Glass Classifications

| Types | I | Transparent glass |
| --- | --- | --- |
| | II | Patterned or wired glass |
| **Classes** | 1 | Clear |
| | 2 | Tinted, heat-absorbing, light-reducing |
| | 3 | Tinted, light-reducing |
| **Quality** | $q^1$ | Mirror select |
| | $q^2$ | Mirror |
| | $q^3$ | Glazing select |
| | $q^4$ | Glazing A |
| | $q^5$ | Glazing B |
| | $q^6$ | Greenhouse |
| | $q^7$ | Decorative |
| | $q^8$ | Glazing |
| **Form** | 1 | Wired |
| | 2 | Patterned and wired |
| | 3 | Patterned |
| **Finish** | $f^1$ | Patterned one side |
| | $f^2$ | Patterned both sides |
| **Mesh** | $m^1$ | Diamond |
| | $m^2$ | Square |
| | $m^3$ | Parallel strand |
| | $m^4$ | Special |
| **Pattern** | $p^1$ | Linear |
| | $p^2$ | Geometric |
| | $p^3$ | Random |
| | $p^4$ | Special |
| **Kind** | HS | Heat-strengthened |
| | FT | Fully tempered |
| **Condition** | A | Uncoated surfaces |
| | B | Spandrel glass with one surface ceramic-coated |
| | C | Other coated glass |

ated Laboratories, Inc. (ALI) and the National Certified Testing Laboratories also provide certification and testing programs. These certification programs are conducted in compliance with ASTM E774.

### Certification of Quality-Safety Glazing

The Safety Glazing Certification Council (SGCC) administers a program of testing, random sampling, product evaluation, and certification to ensure that safety glazing materials comply with ANSI Standard Z97.1. Products meeting the requirements of Z97.1 are given a certification label. There is no one universal SGCC label. Instead, each manufacturer submits his own label to SGCC for approval. This label must include reference to ANSI Z97.1, plus an SGCC identification number which indicates the plant in which the product was made. The label must be permanently affixed to the

certified product at the time and place of manufacture.

*Manufacturers' Warranties* Specialty products such as laminated, reflective, and insulating glass are warranted by their manufacturers against manufacturing defects. Laminated glass typically is covered for 5 years against edge separation or defects which obstruct vision. Reflective glass is warranted for 10 years against peeling or deterioration of the reflective coating. Insulating glass usually carries a warranty against any material obstruction inside the glass unit which impairs vision, such as condensation, dust, or foreign particles. The rainbow "oil slick" effect observed occasionally at certain angles is not a manufacturing defect, but the result of light refraction between accurately machined parallel glass surfaces.

Insulating glass carries either a 5-year or 10-year warranty. Stock windows and sliding doors containing insulating glass are normally preglazed with 5-year quality units.

In small window units enclosed by an accurately manufactured sash, the likelihood of the seal breaking is remote, and the optical quality is not critical if the windows are used primarily for daylighting rather than viewing. For these uses, insulating units with a 5-year warranty generally are acceptable.

However, larger insulating units (often used in sliding glass doors and in field-glazed fixed sash) which are intended for viewing should carry long-term assurance against annoying visual obstructions such as internal fogging. These units generally are subject to more mechanical stresses, and are larger and more expensive to replace. They should be covered, therefore, by a standard 10-year warranty.

Most glass warranties provide protection against manufacturing defects but specifically exclude damage due to faulty installation or use in other than ordinary buildings. For instance, the warranty does not apply for applications in ships, vehicles, or commercial refrigeration. The manufacturer usually agrees to deliver a replacement unit without charge to the shipping point nearest the installation, but the cost of forwarding the unit and the actual installation must be borne by others.

### FLOAT GLASS

Until recently, there were three basic types of flat glass in general use in the

United States: (1) sheet glass, (2) plate glass, and (3) float glass. Today, float glass has completely replaced the other two and will be the only one discussed here.

According to the Flat Glass Marketing Association (FGMA), float glass comprises as much as 98% of the glass manufactured in the United States. It is used today for many of the purposes which were once served only by plate glass (high-quality mirrors, display windows, and automobile glazing) and in thinner sheets for the purposes formerly served by sheet glass.

Most of the other types of glass discussed in the following subsections are variations of float glass.

According to ASTM Standard C1036, clear float glass (Type I, Class 1) is available in six qualities, as shown in Figure 8.10-10. It is usually available in thicknesses from $3/32$ to $7/8$ in. (5.56 to 22.2 mm), but some qualities may only be available in certain thicknesses. For example, $q^5$ (see Fig. 8.10-9) is usually available only in $3/32$ in. (2.38 mm) thickness and $q^4$ only in $1/8$ in. (3.2 mm).

## PATTERNED GLASS

Patterned glass is translucent, with a linear or geometric pattern embossed on one or both sides of the glass. It is produced by the same method as was plate glass (rolling) but is not ground and polished. Patterned glass is classified in ASTM Standard C1048 when it is heat-strengthened or fully tempered and in ASTM Standard C1036 when it is neither of these. Its normal classifications in all three cases are shown in Figure 8.10-10.

Patterned glass is available in a wide range of textures and patterns, which vary according to the manufacturer. These patterns offer different degrees of light transmission and obscurity, depending upon the nature of the design and whether it is rolled into one or both surfaces of the glass. Additional obscurity may be obtained by etching or sandblasting the patterned glass. These processes also reduce the strength of the glass, which should be considered when determining suitable sizes and uses.

Patterned glass is available mainly in thicknesses of $1/8$ and $7/32$ in. (3.2 to 5.56 mm), although some patterns are produced in other thicknesses on a limited basis (Fig. 8.10-11). Some patterns also are available heat treated.

Patterned glass can provide visual privacy while at the same time allowing diffused light transmission. It has a decorative appearance and is often used for this reason alone. It is used for interior partitions and exterior windows, such as in toilet and shower rooms, where obscured vision is desired. It is also used in tub and shower enclosures; glass in these locations should be tempered for safety.

## WIRED GLASS

Wired glass has a wire mesh or parallel wires rolled into the center of the glass thickness. It is available either polished or patterned. Most of it is $1/4$ in. (6.4 mm) thick, but some manufacturers provide other thicknesses (see Fig. 8.10-11). It should comply with ASTM Standard C1036 (see Fig. 8.10-10).

The wire in this type of glass holds it together under low levels of impact, thereby reducing injuries from broken glass. It is considered a safety glazing material and, when properly made, meets the requirements of ANSI Standard Z97.1. It is used widely for corridor and entrance doors, partitions, skylights, and other potentially hazardous locations. It is widely used in fire doors and windows because of its ability to remain in place even though cracked by excessive heat. When used as a fire-resistant material, it should conform to UL standards.

## TINTED GLASS

To make tinted glass, an admixture is included in the glass batch to impart glare-reducing and heat-absorbing qualities. Float glass is available tinted. Tinted glass is covered by ASTM C1036. It may be either simply light-reducing (Class 3) or light-reducing and heat-absorbing (Class 2) (see Fig. 8.10-10).

Most gray- and bronze-tinted glasses are produced in thicknesses of up to $1/2$ in. (12.7 mm) and in select glazing quality ($q^3$) only. Green-tinted glass is generally available only in thicknesses of up to $1/4$ in. (6.4 mm). Equal thicknesses of gray-, bronze-, and green-tinted glass all absorb the same percentage of solar energy. Tinted glasses offer a wide range of visible light transmittance, varying from 20% to 78%, depending on the thickness and color of the glass (Fig. 8.10-12).

This type of glass generally is used where reduced light or heat transmission is desired for indoor visual comfort or where the color of the glass can contribute to the design of the building. Because of its heat-absorbing characteristics, special precautions should be taken when using tinted glass to avoid damage

**FIGURE 8.10-10** Usual Classifications for Some Types of Flat Glass

| Product | Classifications* |
|---------|------------------|
| Clear float | Type I; Class 1; Quality $q^1$, $q^2$, $q^3$, $q^4$, $q^5$, or $q^6$ |
| Tinted heat-absorbing, light-reducing float | Type II; Class 2; Quality $q^3$, $q^4$, $q^5$, or $q^6$ |
| Tinted light-reducing float | Type I; Class 3; Quality $q^3$, $q^4$, $q^5$, or $q^6$ |
| Patterned | Type II; Class 1, 2, or 3; Quality $q^7$ or $q^8$; Form 3; Finish $f^1$ or $f^2$; Pattern $p^1$, $p^2$, $p^3$, or $p^4$ |
| Wired | Type II; Class 1, 2, or 3; Quality $q^7$ or $q^8$; Form 1; Mesh $m^1$, $m^2$, $m^3$, or $m^4$ |
| Wired/patterned | Type II; Class 1, 2, or 3; Quality $q^7$ or $q^8$; Form 2; Finish $f^1$ or $f^2$; Mesh $m^1$, $m^2$, $m^3$, or $m^4$; Pattern $p^1$, $p^2$, $p^3$, or $p^4$ |
| Heat-strengthened | Type I; Class 1, 2, or 3; Quality $q^3$, $q^4$, or $q^5$; Kind HS; Condition A, B, or C |
| Heat-strengthened patterned | Type II; Class 1, 2, or 3; Quality $q^7$ or $q^8$; Form 3; Finish $f^1$ or $f^2$; Pattern $p^1$, $p^2$, $p^3$, or $p^4$; Kind HS; Condition A, B, or C |
| Fully tempered | Type I; Class 1, 2, or 3; Quality $q^3$, $q^4$, or $q^5$; Kind FT; Condition A, B, or C |
| Fully tempered patterned | Type II; Class 1, 2, or 3; Quality $q^7$ or $q^8$; Form 3; Finish $f^1$ or $f^2$; Pattern $p^1$, $p^2$, $p^3$, or $p^4$; Kind FT; Condition A, B, or C |

*Refer to Figure 8.10-9 for meaning of classification symbols.

**FIGURE 8.10-11** Summary of Patterned and Wired Glass Types and Sizes

| Product | Type | Nominal Thickness in. | Nominal Thickness mm | Maximum Area, in. (mm)[a] | Weight, lb./sq. ft. (kg/m$^2$) | Visible Light Transmission, % |
|---|---|---|---|---|---|---|
| Patterned Glass | Floral Hammered Stippled | 1/8 | 3 | 60 × 132 (1524 × 3353) | 1.60–2.10 (7.81–10.25) | 80–90 |
| | Ribbed Fluted[a] Striped[b] | 7/32 | 5.5 | 60 × 132 (1524 × 3353) | 2.40–3.00 (11.72–14.65) | 80–90 |
| Wired Glass | Polished (square or diamond mesh) | 1/4 | 6 | 60 × 144 (1524 × 3658) | 3.50 (17.09) | 80–85 |
| | Patterned (square or diamond mesh) | 1/4 | 6 | 60 × 144 (1524 × 3658) | 3.50 (17.09) | 80–85 |
| | Parallel wired[c] | 7/32 | 5.5 | 54 × 120 (1372 × 3048) | 2.82 (13.77) | 80–85 |
| | | 1/4 | 6 | 60 × 144 (1524 × 3658) | 3.50 (17.09) | 80–85 |
| | | 3/8 | 10 | 60 × 144 (1524 × 3658) | 4.45 (21.73) | 80–85 |

[a]Maximum area varies according to producer; larger sizes may be available from some producers.

[b]These are just a few of the most common patterns available; many patterns are patented and made by one producer only.

[c]This type of wired glass does not carry a UL fire-retardant rating; it is used mainly for decorative partitions.

**FIGURE 8.10-12** Summary of Tinted Glass Types and Sizes

| Product | Quality | Nominal Thickness in. | Nominal Thickness mm | Maximum Area, in. (mm)[a] | Weight, lb/sq ft (kg/m$^2$) | Visible Light Transmission, % | Solar Energy Transmission, % | Principal Uses |
|---|---|---|---|---|---|---|---|---|
| Bronze | Glazing | 1/8 | 3 | 35 sq ft (3.25 m$^2$) | 1.64 (8.01) | 68 | 65 | — |
| | | 3/16 | 5 | 120 × 144 (3048 × 3658) | 2.45 (11.96) | 58 | 55 | |
| | | 1/4 | 6 | 128 × 204 (3251 × 5182) | 3.27 (15.97) | 50 | 46 | |
| | | 3/8 | 10 | 124 × 200 (3150 × 5080) | 4.90 (23.9) | 37 | 33 | |
| | | 1/2 | 12 | 120 × 200 (3048 × 5080) | 6.54 (31.93) | 28 | 24 | |
| Gray | Glazing | 1/8 | 3 | 35 sq ft (3.25 m$^2$) | 1.64 (8.01) | 62 | 63 | Where reduced heat and light transmission are desired |
| | | 3/16 | 5 | 120 × 144 (3048 × 3658) | 2.45 (11.96) | 51 | 53 | |
| | | 1/4 | 6 | 128 × 204 (3251 × 5182) | 3.27 (15.97) | 42 | 44 | |
| | | 3/8 | 10 | 124 × 200 (3150 × 5080) | 4.90 (23.9) | 28 | 31 | |
| | | 1/2 | 12 | 120 × 200 (3048 × 5080) | 6.54 (31.93) | 20 | 24 | |
| Green | Glazing | 3/16 | 5 | 96 × 128 (2438 × 3251) | 2.45 (11.96) | 78 | 55 | — |
| | | 1/4 | 6 | 96 × 128 (2438 × 3251) | 3.27 (15.97) | 74 | 48 | |

[a]Maximum area varies according to producer; larger sizes may be available.

Transmission values vary slightly from one producer to another.

to the edges and consequent breakage from thermal stresses.

## HEAT-ABSORBING GLASS

If a metallic oxide is added to float glass during the manufacturing process, the glass can absorb significantly more sunlight, especially ultraviolet light. This characteristic distinguishes heat-absorbing glass from tinted glass that merely excludes the visible light useful for illumination (light-reducing).

The sunlight absorbed by glass is reradiated and convected as heat to the inside or outside, depending upon which side is cooler. The result is that in summer, more heat is dissipated to an air-conditioned interior because of its cooler temperature. In winter, more heat is dissipated to the outside because it is then cooler.

However, the net effect is still beneficial because 17% less heat-producing sunlight is admitted through heat-absorbing glass than through clear glass.

The performance of heat-absorbing glass is substantially improved when it is used as the outer sheet of an insulating unit. In an insulating unit, the absorbed heat must first bridge the air space to the inner sheet of glass before it can be radiated and convected into the building interior. Because heat transfer is slowed by the insulating air space, more heat is dissipated to the outside air, which is in direct contact with the heat-absorbing glass. Furthermore, the outward rate of heat dissipation is greatly accelerated if there is a wind.

## COATED GLASS

When sunlight strikes a pane of glass, it is either transmitted, reflected, or absorbed and radiated as heat. By increasing the amount of reflected sunlight, the amount that can be absorbed and transmitted or radiated is reduced.

The category of coated glass called *high-performance glass* has on one side a transparent film of metal or metal oxide. This coating is applied either by *pyrolytic deposition* or *vacuum sputtering*. Those applied by pyrolytic deposition are called *hard-coat* coatings. Some producers claim that they can be exposed to wear. Others do not agree. Coatings applied by vacuum sputtering are called *soft-coat* coatings. Again, some producers recommend them for use where they will be exposed to contact. Others claim that they cannot be used that way and must only be applied on the protected surfaces of panes, such as those in insulated glass or spandrel glass.

Because such coated glass appears mirror-like when viewed from the exterior, it is called *reflective glass*. Some designers select it as much for this esthetic property as for its ability to reduce the solar energy transmitted into a building. In insulating units, it also minimizes ambient heat flow into and out of the building. Several of the more durable coatings for this glass, such as cobalt oxide, chromium, stainless steel, and titanium, are available on single glass as well as on insulating and laminated glasses. Other available coatings, such as copper, aluminum, nickel, and gold, are less durable and are mostly furnished on an inside surface of insulating or laminated glasses, where the coating is protected from weather and accidental abrasion.

A wide range of metallic colors is available with differing visual and thermal properties. Daylight reflectances range from about 10% for the more subdued coatings to about 45% and visible light transmittance from 8% to 55%. The more common colors are neutral, gray, gold, bronze, blue, silver, green, and copper. Many coatings are also available on tinted, heat-absorbing glass. When a coating is applied to such glass, it must often be heat-strengthened or fully tempered.

Coated glass products vary widely from one manufacturer to another, and certain colors and tones are offered by only one company. Consequently, individual producers should be consulted for specific product information.

Reflective glass is more effective as the outer sheet of insulating glass than as single glazing. This is because, despite its name, reflective glass absorbs more sunlight than clear glass, which permits most of the sunlight that strikes it to pass through. There is, therefore, more heat buildup in reflective glass. When that heat is concentrated in the outer sheet of insulating glass, it is dissipated to the outside air because the air space between glass layers impedes its flow inward.

There are some rather important disadvantages of reflective glass:

- In winter, beneficial heat gain is reduced by it.
- Transmission of visible light for general illumination is severely reduced year round, making occupants feel as though they are looking through dark sunglasses.
- In spite of its large reduction in light transmission, it does not significantly increase the insulating value of the glass.
- Its high reflectivity can create problems for its neighbors and even for traffic on the streets below. In fact, some cities now restrict the amount of reflectivity on buildings.

Most of these disadvantages have been overcome, however, by advances in coating technology. New metallic coatings are of a different type than those used on earlier versions. Some of these coatings produce a more subdued mirror effect, reducing the amount of reflected light. These are called *low-reflectance* glasses. Others are almost transparent. Much light passes through them, but their emissivity is reduced sharply. *Emissivity* refers to the amount of heat absorbed by a material and then reradiated by it. This type of high-performance glass is called *low-emissivity* or simply *low-E* glass. It can reflect as much as 90% of the radiant energy that strikes it while admitting most of the light. Low-E glass is available in many colors and with many levels of performance characteristics.

*Spandrel glass* is used most often to cover the spandrel beams and floor construction in curtain walls. It is also sometimes used in other locations on the exterior of a building. It is often coated using ceramic materials. It may also be coated with a high-performance coating resembling that used on the vision panels in the same building. Sometimes the metal oxide coating on spandrel glass is protected with a thin sheet of plastic called an *opacifier*.

## HEAT-TREATED GLASS

There are two types of heat-treated glass: fully tempered and heat-strengthened. When either of these types of glass is to be used, the exact sizes desired must be specified, as the glass cannot be cut, drilled, or notched after the heat treatment. These types of glass are governed by ASTM Standard C1048.

*Fully Tempered Glass* Except for wired glass and patterned glass with deep patterns, any glass $1/8$ in. (3.2 mm) or thicker may be tempered. Tempered glass may be incorporated into insulating glass or laminated glass units; heat-absorbing and heat-reflective glasses may also be tempered.

An inherent characteristic of all tempered glass is a slight deviation from flatness, particularly near the edges. The degree of deviation depends on thickness, width, length, and other factors. Usually, greater thicknesses yield flatter products.

Fully tempered glass has three to five times the resistance of annealed glass to uniform loading, thermal stresses, and most impact loads. When fractured, it breaks into a safe pattern of relatively small, harmless particles (Fig. 8.10-13). It is considered a safety glazing material and, when properly made, meets the requirements of ANSI Standard Z97.1. It is frequently required by state or local codes in areas where human contact with glass is probable, such as entrances and sidelights, tub and shower enclosures, storm doors, and sliding glass doors.

*Heat-strengthened Glass* Heat-strengthened glass is similar to fully tempered glass, except that it has only a partial heat temper. It is twice as strong as ordinary annealed glass of the same thickness. When it fails, it breaks into fragments much larger than those of fully tempered glass, and therefore it is not considered a safety glazing material. Heat-strengthened glass generally is used in applications that require a stronger product than regular annealed glass but not the full strength of tempered glass.

Any type of $^1/_4$-in. (6.4-mm)- or $^5/_{16}$-in. (7.9-mm)-thick glass that can be tempered can be heat-strengthened. The most widely used form of heat-strengthened glass is spandrel glass. This glass has a fired-on ceramic frit on the interior

**FIGURE 8.10-14** Laminated glass was used in this sloped glazing application at the IBM facility in Charlotte, North Carolina. (Courtesy Laminated Glass Corporation.)

surface and is used commonly in spandrel areas in curtain wall construction.

## DECORATIVE GLASS

Patterned, articulated, and other forms of decorative glass are available. These may be etched glass or layered glass with different colors in different layers or raised portions of clear or colored glass. Leaded or *stained* glass is another form of decorative glass. Most decorative glass is more expensive than other forms of glass and is, therefore, used primarily in special locations, such as chapels, where the esthetic effect outweighs the additional expense.

## COMPOSITE GLASS

The two main types of fabricated glass, *laminated* and *insulating* glass, are products made by combining two or more layers of glass into a single unit.

*Laminated Glass* Laminated glass consists of two or more layers of glass with interlayers of *polyvinyl butyral plastic* sandwiched between them, bonded together under heat and pressure to form a single unit. The thickness of the glass layers may range from $^3/_{16}$ in. to $^7/_8$ in. (4.8 to 22.2 mm). The plastic interlayer is usually 0.015 or 0.030 in. (0.381 or 0.762 mm), although it may be as thick as 0.090 in. (0.29 mm). The face of the glass that

**FIGURE 8.10-13** Broken fully tempered glass.

**FIGURE 8.10-15** Laminated security glass protects the users of the CN Tower in Toronto. (Courtesy Laminated Glass Corporation.)

is to be bonded to the plastic cannot be patterned or otherwise irregular. Heat-absorbing, heat-reflecting, fully tempered, heat-strengthened, and wired glasses may all be laminated with themselves and in combination with other types.

When laminated glass is broken, the fragments of glass generally adhere safely to the plastic interlayer and do not evacuate the opening. This reduces the potential hazards of flying glass and minimizes cutting injuries. All varieties of laminated glass are considered safety glazing mate-rials and, when properly made, meet the requirements of ANSI Standard Z97.1 (Figs. 8.10-14 and 8.10-15).

Most laminated glass is made of two lights of glass of varying thicknesses with a single plastic interlayer. It is used commonly for interior partitions, doors, and automotive glazing. There are also sev-eral special types of laminated glass which are made for particularly demand-ing applications (Fig. 8.10-16).

*Burglar-resistant glass* is a laminated glass usually consisting of two layers of $\frac{1}{8}$ in. (3.2 mm) glass with a 0.060 to 0.090 in. (1.52 to 2.29 mm) plastic inter-layer. It is made primarily for use in the protection of show windows displaying valuable merchandise. It has superior re-sistance and serves as a major deterrent to the "smash-and-run" burglar.

*Bullet-resisting glass* is a laminated glass having four or more layers totaling $\frac{3}{4}$ in. (19.1 mm) to 3 in. (76.2 mm) in thickness. Most manufacturers have $1\frac{3}{16}$, $1\frac{9}{16}$, $1\frac{3}{4}$, and 2 in. (30.16, 49.69, 44.5, and 50.8 mm) thicknesses listed by

**FIGURE 8.10-16** Summary of Laminated Glass Products

| Product | Glass Type | Plastic Interlayer, in. (mm) | Nominal Thickness in. | Nominal Thickness mm | Maximum Area, in. (mm)[a] | Weight, lb/sq ft (kg/m²) | Principal Uses[b] |
|---|---|---|---|---|---|---|---|
| Laminated glass | SS/SS[c] | .015–.030 (0.381–0.762) | $\frac{13}{64}$ | 5 | 48 × 80 (1219 × 2032) | 2.45 (11.96) | Partitions, doors, and automotive glazing |
| | Lam/Lam[c] | .015–.030 (0.381–0.762) | $\frac{15}{64}$ | 6 | 48 × 100 (1219 × 2540) | 2.90 (14.16) | |
| | DS/DS[c] | .015–.030 (0.381–0.762) | $\frac{1}{4}$ | 6 | 48 × 100 (1219 × 2540) | 3.30 (16.1) | |
| | 2 lights $\frac{1}{8}$ in. | .015–.030 (0.381–0.762) | $\frac{1}{4}$ | 6 | 72 × 120 (1829 × 3048) | 3.30 (16.1) | |
| | 2 lights | .015–.030 (0.381–0.762) | $\frac{3}{8}$ | 10 | 72 × 120 (1829 × 3048) | 4.80 (23.44) | |
| | | | $\frac{1}{2}$ | 12 | 72 × 120 (1829 × 3048) | 6.35 (31) | |
| | | | $\frac{5}{8}$ | 16 | 72 × 120 (1829 × 3048) | 8.00 (39.1) | |
| | | | $\frac{3}{4}$ | 19 | 72 × 120 (1829 × 3048) | 9.70 (47.36) | |
| | | | $\frac{7}{8}$ | 22 | 72 × 120 (1829 × 3048) | 11.40 (55.66) | |
| | | | 1 | 25 | 72 × 120 (1829 × 3048) | 12.95 (63.73) | |
| Burglar resistant | 2 lights $\frac{1}{8}$ in. | .060–.090 (1.524–2.286) | $\frac{5}{16}$ | 8 | 60 × 120 (1524 × 3048) | 3.50 (17.1) | Display windows |
| Bullet resistant | At least 4 lights | At least 3 plies of .015 (0.381) | $1\frac{3}{16}$ | 30 | 40 sq ft (3.72 m²) | 15.50 (61) | Bank teller windows and payroll booths |
| | | | $1\frac{9}{16}$ | 40 | 30 sq ft (2.79 m²) | 20.00 (97.65) | |
| | | | $1\frac{3}{4}$ | 45 | 25 sq ft (2.3 m²) | 23.00 (112.3) | |
| | | | 2 | 50 | 22 sq ft (2.04 m²) | 26.20 (127.9) | |
| Acoustical[d] STC-36 STC-40 | 2 lights | One or more plies of 0.45 (1.143) | $\frac{9}{32}$ | 7 | 60 × 120 (1524 × 3048) | 3.40 (16.6) | Office partitions, radio and TV studios, airport glazing |
| | | | $\frac{1}{2}$ | 12 | 60 × 120 (1524 × 3048) | 6.36 (31.05) | |
| STC-43 | | | $\frac{3}{4}$ | 19 | 60 × 120 (1524 × 3048) | 9.74 (47.55) | |

[a]Maximum area varies according to producer. Larger sizes may be available from some producers.

[b]Double-layer products using thicker glass and interlayers and multilayered laminated glass products are also used in other demanding installations, such as zoological animal enclosures and observation platforms (see Figs. 8.8-14 and 8.8-15) and incarceration facilities.

[c]Lam indicates $\frac{7}{64}$ in. (2.78 mm) float glass; SS, $\frac{3}{32}$ in. (2.38 mm) float glass; DS, $\frac{1}{8}$-in. (3.175 mm).

[d]Refers to sound transmission class ratings.

**a**            **b**

**FIGURE 8.10-17** Insulating glass types: (a) spacer bar edge and (b) fused glass edge.

UL for resistance to medium-, high-, and superpower small arms and high-power rifle arms, respectively. Thicknesses of up to 7 in. (177.8 mm) are available on special order. This type of glass is used for bank teller windows, payroll booths, and other security applications.

*Acoustical glass* is a laminated glass with one or more plastic interlayers of 0.045 in. (1.14 mm) thickness. It is particularly effective in reducing sound transmission in the frequency ranges of speech, radio, and TV (250 to 4000 cycles per second range). It is widely used for office partitions and in radio and TV studios and similar applications.

***Insulating Glass*** Insulating glass is a factory-produced, airtight unit consisting of two layers of glass with a dehydrated air space between them. Insulating units are produced with two different edge types: spacer bar edge and fused glass edge (Fig. 8.10-17). Insulating glass is governed by ASTM Standard E774.

Any glass type with at least one smooth surface may be used in an insulating glass unit. When heat-absorbing or reflective coated glass is used in an insulating glass unit, and it is known that the glass will be subjected to severe thermal stresses or high wind loads, one or both lights of the unit should be heat-strengthened or fully tempered.

Insulating glass is designed primarily to protect glazed areas from excessive heat loss in winter and heat gain in summer. Use of insulating glass can substantially reduce both heating and air conditioning costs.

Insulating glass has the added advantage of allowing a much higher interior relative humidity than single glass be-

cause it minimizes condensation problems during cold weather (Fig. 8.10-18).

*Spacer-edged insulating units* consist of two lights of glass separated by a hollow metal or rubber spacer around the edges and sealed airtight with an organic sealant such as polysulfide or butyl. The trapped air is kept dehydrated by a desiccant (chemical dehumidifier) located in

the spacer. Until recently, the edge construction and glass in many units were held together by a metal channel that wrapped around the edge of the unit, encasing both panes of glass. This was designed to provide edge protection and, in some cases, to hold the assembly together. This type of "metal-edged" insulating glass unit may be found in some existing buildings, but modern insulating glass units do not have this feature.

Spacer-edged insulating units are most commonly used in commercial construction. Sealed insulating units are manufactured and stocked in a limited number of standard sizes and thicknesses. For example, units with $1/4$ and $1/2$ in. (6.4 and 12.7 mm) air spaces and glass thicknesses ranging from $1/8$ to $3/8$ in. (3.2 to 9.5 mm) are available on that basis (Fig. 8.10-19). Special sizes can be provided by some manufacturers at a slight additional cost and longer delivery time.

*Fused glass-edge insulating units* consist of two lights of SS ($3/32$ in. [2.38 mm]) or DS ($1/8$ in. [3.2 mm]) float

**FIGURE 8.10-18** Maximum relative humidity causing condensation on insulating and other glass.

**FIGURE 8.10-19** Summary of Insulating Glass Products

| Product | Glass Type and Thickness, in. | Nominal Air Space in. | mm | Nominal Unit Thickness in. | mm | Maximum Area, sq ft (m²)* | Weight, lb/sq ft (kg/m²) | Principal Uses |
|---|---|---|---|---|---|---|---|---|
| **Spacer edge units** | 2 lights DS, ⅛ | ¼ | 6 | ½ | 12 | 22.2 (2.06) | 3.27 (15.97) | Residential and commercial glazing |
| | | ½ | 12 | ¾ | 19 | | | |
| | 2 lights, ³⁄₁₆ | ¼ | 6 | ⅝ | 16 | 34 (3.16) | 4.90 (23.9) | |
| | | ½ | 12 | ⅞ | 22 | 41.7 (3.87) | | |
| | 2 lights, ¼ | ¼ | 6 | ¾ | 19 | 50–70 (4.6–6.5) | 6.54 (31.9) | |
| | | ½ | 12 | 1 | 25 | | | |
| **Glass edge units** | 2 lights SS, ³⁄₃₂ | ³⁄₁₆ | 5 | ⅜ | 10 | 10 (0.93) | 2.40 (11.72) | Smaller lights in residential and commercial glazing |
| | 2 lights DS, ⅛ | ³⁄₁₆ | 5 | ⁷⁄₁₆ | 11 | 24 (2.23) | 3.20 (15.6) | |

*Maximum area given is for units constructed with clear glass; when tinted or reflective coated glass is used, either the maximum area must be reduced or special fabrication is required.

glass with a ³⁄₁₆ in. (4.8 mm) air space between them. The two sheets are fused together to form a glass-to-glass sealed edge. The air space in glass edge units may be filled with dehydrated air or inert gas.

Fused glass-edge units are available in a more limited range of sizes and glass combinations than spacer-edged units (see Fig. 8.10-19). They were developed mainly to provide economical insulating glass for smaller lights in residential and commercial buildings.

## 8.10.2 PLASTIC GLAZING MATERIALS

Plastic glazing materials are either polycarbonate or acrylic plastic. Both are tough, breakproof, shatterproof, and crack-resistant thermoplastics. They have a higher coefficient of expansion than glass, are lighter, and resist impact better. They are, however, much more susceptible to scratching and are flammable, which glass is not.

Acrylics are harder and more impact resistant than polycarbonates, and they wear better. Both can be cold-formed into curves. Polycarbonates are the more heat-resistant of the two. Both can be coated with an abrasion-resistant silicone-based material to increase their resistance to weather and chemicals. Polycarbonate sheets can also be coated with polymethyl methacrylate (acrylic) material to reduce damage normally caused to it by sunlight.

Refer to Chapter 6 for additional discussion of acrylics and polycarbonates.

## 8.10.3 GLAZING ACCESSORY MATERIALS

### 8.10.3.1 Glazing Sealants

Glazing sealants should be selected for their intended use and for their compatibility with the glazing materials and the other surfaces they will contact. For example, silicones may become discolored and even fail to function as sealants if they come into contact with neoprene gaskets or neoprene or EPDM setting blocks or spacers. A particular sealant should not be selected unless its compatibility with adjacent materials has been tested. The results of such tests are generally available from sealant manufacturers.

### FACE GLAZING COMPOUNDS

Oil-based glazing putty is still occasionally used to install small lights in wood and metal sash, but most such glazing today is done using elastic synthetic materials specifically formulated for the purpose. The materials should be suitable for painting.

### CHANNEL-GLAZING COMPOUNDS

Channel-glazing compounds may be either elastomeric or elastic oil- and resin-based materials. They are used both for bedding and face glazing into wood or metal. Glazing compounds set firmly but retain limited flexibility which is generally greater than that of putty. Glazing compounds have a life expectancy of 5 to 20 years when protected by periodic painting of the exposed surface.

### TAPES

Glazing tape is polybutene or polyisobutylene-butyl tape. It should be ¹⁄₁₆ in. (1.6 mm) thicker than the finished seal and be compressed during application.

### SILICONE SEALANTS

Silicone formulations are used as elastic glazing compounds, especially, but not exclusively, in butt-glazing applications. Special formulations of chemically curing, one-part silicone are used in structural glazing applications. These must be specifically formulated for this purpose.

### 8.10.3.2 Gaskets

Glazing gaskets should be the standard types and materials of the manufacturer of the unit being glazed. Gaskets may be vinyl, neoprene, or another elastomeric material. Lockstrip gaskets are rubber.

### 8.10.3.3 Miscellaneous Glazing Materials

Setting blocks and spacers should be neoprene, EPDM, silicone, or rubber. These should be of the size, shape, and hardness recommended by the glass and sealant manufacturers.

Compressible filler rods should be closed-cell or waterproof-jacked rods of synthetic rubber or plastic foam.

Glazing clips should be as required by the conditions.

### 8.10.4 GLAZING

Glass selection depends on wind load requirements, thermal requirements, esthetics, and safety. The type and thickness are often established by the applicable building code. For example, the *International Building Code* requires glass in entrance doors and the lower portions of sidelights to meet the requirements for "safety type" and be in compliance with ANSI Z-97.1 or another similar standard.

Watertight and airtight installation of glazing materials is the goal of glazing. Installed glazing materials must be able to withstand the effects of temperature changes, wind, and impact with no breakage of glazing material, failure of sealants or gaskets, deterioration of glazing materials, or other defects.

The installation should comply with the requirements of the appropriate standards and codes. It should also comply with the recommendations of the manufacturers of the glazing material; of the sealant, gaskets, and other glazing materials manufacturers; and of the window, door, frame, entrance, storefront, or curtain wall manufacturers.

The locations and conditions where glazing materials will be installed should be inspected and unsatisfactory conditions corrected before glazing is started. Surfaces to be glazed should be cleaned and removable coatings removed. Ventilating sash should be properly adjusted before they are glazed.

### 8.10.4.1 General Requirements for Glazing

Glazing materials vary, depending on the component being glazed. Formed door and frame units are usually glazed using stops and channel glazing compound, tape sealant, or gaskets. Most extruded aluminum storefront sections and doors are glazed using compressible gaskets and tape.

When windows or sliding glass doors are glazed at the job site, the manufacturer's recommendations should be followed to ensure a satisfactory result.

Glazing includes the installation of glazing beads and gaskets that are furnished by the manufacturer of the unit being glazed. Wood and metal doors and some portions of curtain walls may be either factor preglazed or glazed in the field. Most windows are glazed in the factory. Frames, sidelights, storefronts, and most curtain walls are glazed in the field.

The design of a glazing channel (Fig. 8.10-20) must provide the necessary minimum bite and edge clearance and allow for the proper glazing sealant thickness. The person who installs glazing materials is called a *glazier*. The glazier is responsible for providing the correct glass size for each opening, but realistically, many of today's glazing materials leave little room for adjustment. For example, heat-treated and preassembled insulating glass cannot be cut in the field. Therefore, the sizes of openings and glazing materials must be coordinated in the shop. Panes of glass and plastic should be fabricated in the shop with the proper edge clearances. Glass clearances should be in accordance with the Glass Association of North America's (GANA) *Glazing Manual* or the requirements of the glass manufacturer.

Glazing systems must have provisions to prevent moisture from accumulating in them. This requires them to have weep holes open to the exterior (see Fig. 8.10-20).

There are many glazing systems in use today. Small openings in wood and steel are sometimes glazed by placing the glass in a bed of glazing compound (Fig. 8.10-21). Stop glazing is also used (Fig. 8.10-22). Glazing in metal is often done with stops as well. There, the exterior glazing material may be a liquid or tape sealant but is sometimes a gasket. The interior glazing material may be a tape or liquid sealant but more often is a gasket (Fig. 8.10-23). In high-performance sys-

**FIGURE 8.10-20** Typical glazing channel terminology. (Drawing by HLS.)

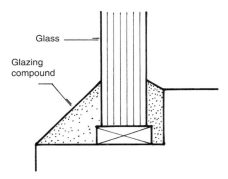

**FIGURE 8.10-21** Face glazing for small lights in wood and metal. (Drawing by HLS.)

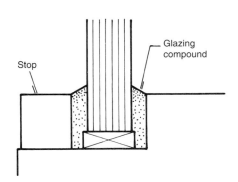

**FIGURE 8.10-22** Stop glazing. (Drawing by HLS.)

**FIGURE 8.10-23** A typical glazing into metal system. Many others are also used. (Drawing by HLS.)

tems, the interior and exterior systems may consist of more than one sealant. For example, there may be a tape topped off with a liquid silicone sealant or a gasket topped off with an acrylic sealant. In lock-seam gasket glazing, there are no liquid glazing materials (Fig. 8.10-24). The glass is set into a groove on the gasket, and the gasket is squeezed against the glass by insertion of the lockstrip using a special tool.

Insulating glass should be installed with the tinted glass toward the exterior, with coated glass, where used, as the outer light. Opaque glass should have its coated side toward the interior.

Insulating glass, laminated glass, and transparent mirrors should not be allowed to come into contact with chlorinated solvents or benzine-related compounds, such as toluene, nor should glazing compounds used with these glass types be thinned with these compounds. They will damage the interlayer in such glasses.

The use of setting blocks, spacers, and edge blocking should be in accordance with the GANA's *Glazing Manual.*

Wire glass should be installed in accordance with the requirements of NFPA 80.

Two pieces of glass that abut end to end or at a corner should be sealed using beads of silicone sealant following the manufacturer's recommendations.

### 8.10.4.2 Structural Silicone-Sealant Glazing

In a structural silicone-sealant glazing system, the glass is bonded to metal or

FIGURE 8.10-25 Detail of a typical structural glass. The other two sides of the glass may be similarly supported or may be supported by a metal frame in the conventional manner (see Figs. 8.5-9 and 8.5-10). (Drawing by HLS.)

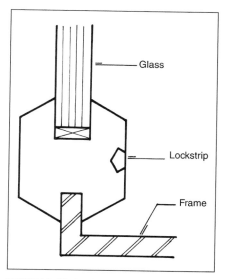

FIGURE 8.10-24 A lockstrip glazing gasket. (Drawing by HLS.)

glass supports using a structural-grade silicone sealant.

There are three basic types of structural silicone-sealant glazing systems: (1) those with silicone sealant supporting two sides of the glass and with the other two sides of the glass supported by a metal frame (Fig. 8.10-25); (2) those with structural silicone sealant supporting four sides of the glass; and (3) those with glass mullions called *fins* (Fig. 8.10-26).

The butt joints of glass in these systems are sealed using a silicone sealant, but not necessarily a structural sealant.

### 8.10.4.3 Multiple Glazing

Because of its thinness, a single sheet of glass is a good conductor of heat. One square foot (0.093 m$^2$) of $^1/_4$ in. (6.4 mm) clear glass can conduct 6 to 10 times more Btus per hour (watts) than a square foot (0.093 m$^2$) of a typical frame wall.

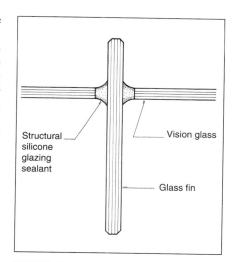

FIGURE 8.10-26 A typical glass mullion. Other types place the glass mullion behind the vision glass and attach the vision glass to the mullions using metal connectors. (Drawing by HLS.)

The rate of heat transfer through the glass is so rapid that added thickness is of almost no value. However, if layers of glass are separated by an air space, the path of conduction is interrupted and the rate of heat flow is reduced. Heat can then pass through the air space primarily by radiation and convection and only minimally by conduction.

Storm windows can be effective in reducing conduction and infiltration with an air separation between panes of glass of as much as 6 in. (152 mm). However, an air space is most effective in improving thermal performance when it is between $3/16$ in. (4.8 mm) and $5/8$ in. (15.9 mm) wide. Within this range, heat flow is reduced markedly as the width is increased from $3/16$ in. (4.8 mm) toward $5/8$ in. (15.9 mm). Increasing the air space beyond $5/8$ in. (15.9 mm) does not reduce heat flow to the same extent, because a wider space allows the air in it to circulate more freely and develop convection currents. These currents transport heat from the warmer glass to the colder glass. This increased convection heat loss more than offsets the slight reduction in conductive heat loss through the air. Conversely, a space of less than $3/16$ in. (4.8 mm) is of little value because heat is readily conducted by the air across such a short distance.

### 8.10.4.4 Glazing with Plastic

Openings for plastic glazing should be accurately measured and the plastic sheets cut to fit, leaving the appropriate room for expansion.

The size of the sheets should be determined in accordance with the manufacturer's recommendations for edge clearances.

Plastic sheets should be set using the sealants recommended by the plastics manufacturer.

### 8.10.4.5 Cleaning and Protection

Labels should be removed from insulating glass to prevent thermobreakage or staining due to the labels. Labels may be left on other types of glass until final cleaning unless the manufacturer of the glass recommends removal.

Glass should be protected from breakage immediately after installation by barriers and warning tapes. Warning tapes should be attached to the frames, and not to the glass, to prevent damage to the glass by the warning devices. Glazing materials should also be protected from contaminants, such as masonry cleaning material, that would harm the glass.

Glazing materials should be thoroughly cleaned of foreign materials and polished on both sides. Cleaning procedures should be as recommended by the glazing materials' manufacturers.

# 8.11 Additional Reading

More information about the subjects discussed in this chapter can be found in the references listed in Section 8.12, and in the following publications.

Altenhofen, David W. 2003. "Fenestration Beyond Standards—Successfully Integrating Windows, Storefronts, and Curtain Walls." *The Construction Specifier* (July): 39 (www.csinet.org).

Beers, Paul E. 1997. "Security Glazing." *The Construction Specifier* (October): 66 (www.csinet.org).

————. 2000. "Hurricane Protection Requirements for Glazing: Codes and Standards Update." *The Construction Specifier* (August): 43 (www.csinet.org).

Behr, Richard A. and Scott A. Warner. 2003. "Earthquake-Resistant Architectural Glass." *The Construction Specifier* (May): 29 (www.csinet.org).

Chambers, Michael D. 2002. "Knowing a Stile from a Rail—Specifying Flush Wood Doors." *The Construction Specifier* (February): 35 (www.csinet.org).

Clelland, Thomas S. 2000. "Architectural Hardware: A Security Primer." *The Construction Specifier* (April): 33 (www.csinet.org).

Curkeet, Rick. 2003. "User's Guide to Fire-Rated Glazing." *The Construction Specifier* (January): 50 (www.csinet.org).

————. 2003. "Test Methods and Common Sense Solve Safety Glazing Requirements." *The Construction Specifier* (May): 34 (www.csinet.org).

Davidson, Allen. 2000. "Specifying Revolving Doors." *The Construction Specifier* (August): 25 (www.csinet.org).

Johnson, Christopher. 2005. "Open and Shut Cases—Automatic Doors and Physical Requirements." *The Construction Specifier* (January): 34 (www.csinet.org).

Harpole, Tom. 1995. "Is Tempered Glass Safe?" The Construction *Specifier* (December): 36 (www.csinet.org).

————. 1996. "Protective Glazing: Making Credible Choices." *The Construction Specifier* (April): 60 (www.csinet.org).

————. 1997. "Security Glazing Options." *The Construction Specifier* (July): 52 (www.csinet.org).

Kirby, Don. 2002. "Cam-and-Roller and Rack-and-Pinion Designs—Influencing Door Closer Performance." *The Construction Specifier* (July): 38 (www.csinet.org).

Makepeace, Chris B. and John O'Connor. 1996. "The Glass Roof: Sloped Glazing Need Not Leak." *The Construction Specifier* (November): 50 (www.csinet.org).

Myers, Donald L. 1996. "Security Locking Systems." *The Construction Specifier* (April): 68 (www.csinet.org).

————. 1996. "Security Standards." *The Construction Specifier* (April): 70 (www.csinet.org).

Nashed, Fred. 1996. *Time-Saver Details for Exterior Wall Design*. New York: McGraw-Hill (www.mcgraw-hill.com).

Nicastro, David H. 1995. "Tempered Glass." *The Construction Specifier* (July): 96 (www.csinet.org).

————. 1996. "Glazing System Leaks." *The Construction Specifier* (January): 80 (www.csinet.org).

————. 1996. "Falling Snap Covers." *The Construction Specifier* (July): 96 (www.csinet.org).

————. 1996. "Structural Sealant Glazing." *The Construction Specifier* (August): 112 (www.csinet.org).

————. 1996. "Zipper Gaskets." *The Construction Specifier* (September): 96 (www.csinet.org).

————. 1996. "Sloped Glazing Slips." *The Construction Specifier* (December): 88 (www.csinet.org).

Niehaus, Paul. 2003. "Don't Be Negative about Positive Pressure Fire Doors." *The Construction Specifier* (May): 52 (www.csinet.org).

Razxick, Jerry. 2002. "Specifying Fire-Rated Glass and Framing." *The Construction Specifier* (July): 48 (www.csinet.org).

Rose, Helen. 1997. "Choosing Gasketing Products." *The Construction Specifier* (April): 73 (www.csinet.org).

Schechter, John. 1996. "Stainless Steel Doors." *The Construction Specifier* (March): 66 (www.csinet.org).

West, Heather. 2002. "When Is a Window as Good as a Wall?—Performance Considerations for Windows." *The Construction Specifier* (February): 50 (www.csinet.org).

———. 2003. "Understanding Environmentally Certified Architectural Wood Doors." *The Construction Specifier* (January): 57 (www.csinet.org).

Wiemer, David. 2004. "Do the Research Before Selecting the Hardware." *The Construction Specifier* (July): 48 (www.csinet.org).

Winandy, Joseph and John O'Neil. 1996. "Doors, Frames, and Hardware." *The Construction Specifier* (October): 54 (www.csinet.org).

Zarghamee, Mehdi S. and Rasko P. Ojdrovic. 1996. "Seismic Design for Curtain Walls." *The Construction Specifier* (September): 74 (www.csinet.org).

Zeimetz, Terry. 2004. "Specifying the Appropriate Glazing System." *The Construction Specifier* (October): 32 (www.csinet.org).

# 8.12 Acknowledgments and References

## ACKNOWLEDGMENTS

We gratefully acknowledge the assistance of the following organizations and individuals in preparing this chapter. We are also indebted to them for permission to use their illustrations when requested and for the use of their publications as references.

The Adams Rite Manufacturing Company (www.adamsrite.com)

American Architectural Manufacturers Association (AAMA) (www.aamanet.org)

Aluminum Association (AA) (www.aluminum.org)

American National Standards Institute (ANSI) (www.ansi.org)

Architectural Woodwork Institute (AWI) (www.awinet.org)

ASTM International (ASTM) (www.astm.org)

Boon Edam, Inc. (www.boonedam.nl)

Bruce Wall Systems Corporation (www.brucewall.com)

Builders Hardware Manufacturers Association (BHMA) (www.builershardware.com)

Cardinal Glass Industries (www.cardinalcorp.com)

Consumer Product Safety Commission (CPSC) (www.cpsc.gov)

Copper Development Association (CDA) (www.copper.org)

Door and Hardware Institute (DHI) (www.dhi.org)

Falcon Lock (www.falconlock.com)

Glass Association of North America (GANA) (www.glasswebsite.com)

Hollow Metal Manufacturers Association (HMMA), Division of the National Association of Architectural Metal Manufacturers (NAAMM) (www.naamm.org)

Insulating Glass Certification Council (www.igcc.org)

International Code Council (ICC) (www.iccsafe.org)

Kalwall Corporation (www.kalwall.com)

Laminated Glass Corporation

National Association of Architectural Metal Manufacturers (NAAMM) (www.naamm.org)

National Association of Home Builders (NAHB) Research Foundation (www.nahbrc.org)

National Fire Protection Association (NFPA) (www.nfpa.org)

National Institute of Building Sciences (NIBS) (www.nibs.org)

National Institute of Standards and Technology of the U.S. Department of Commerce (NIST) (www.nist.gov)

Pease Company (www.peasecompany.com)

Pella Corporation (www.pella.com)

Screen Manufacturers Association (www.smacentral.org)

Steel Door Institute (SDI) (www.steeldoor.org)

Tubelite Company Inc. (www.tubelite.com)

U.S. Department of Housing and Urban Development (HUD) (www.hud.gov)

Window and Door Manufacturers Association (WDMA) (www.wdma.com)

Woodwork Institute (WI) (formerly Woodwork Institute of California [WIC]) (www.wicnet.org)

## REFERENCES

We would also like to thank the authors and publishers of the publications in the following list for their contribution to our research for this chapter.

American Architectural Manufacturers Association (AAMA). Schaumburg, IL (www.aamanet.org):

AAMA 501-94, "Methods of Test for Exterior Walls."

AAMA 609 and 610-02, "Cleaning and Maintenance Guide for Architecturally Finished Aluminum."

AAMA 701 and 702-00, "Voluntary Specifications for Pile Weather Stripping and Replaceable Fenestration Weatherseals."

AAMA 800-92, "Voluntary Specifications and Test Methods for Sealants."

AAMA 850-91, "Fenestration Sealants Guide Manual."

AAMA 901-96, "Voluntary Specifications for Rotary Operators in Window Applications."

AAMA 902-98, "Voluntary Specifications for Sash Balances."

AAMA 904-01, "Voluntary Specifications for Multi-Bar Hinges in Window Applications."

AAMA 906-05, "Voluntary Specifications for Sliding Glass Door Roller Assemblies."

AAMA 1002.10-93, "Voluntary Specifications for Insulating Storm Products for Windows and Sliding Glass Doors."

AAMA 1102.7-89, "Voluntary Specifications for Aluminum Storm Doors."

AAMA 1302-76, "Voluntary Specifications for Forced-Entry Resistant Aluminum Prime Windows."

AAMA 1303.5-1976, "Voluntary Specifications for Forced-Entry Resistant Aluminum Sliding Glass Doors."

AAMA 1402-86, "Standard Specifications for Aluminum Siding, Soffits and Fascia."

AAMA 1503-98, "Voluntary Test Method for Thermal Transmittance and Condensation Resistance of Windows, Doors and Glazed Wall Sections."

AAMA 1504-97, "Voluntary Standard for Thermal Performance of Windows, Doors and Glazed Wall Sections."

AAMA 2603-02, "Voluntary Specification, Performance Requirements and Test Procedures for Pigmented Organic Coatings on Aluminum Extrusions and Panels."

AAMA 2604-02, "Voluntary Specifications, Performance Requirements, and Test Procedures for High Performance Organic Coatings on Aluminum Extrusions and Panels."

AAMA CW-DG-1, *Aluminum Curtain Wall Design Guide Manual.*

AAMA CWG-1-89, "Installation of Aluminum Curtain Walls."

AAMA CW-10, "Care and Handling of Architectural Aluminum from Shop to Site."

AAMA CW-11, *Curtain Wall Manual, Volume 11, Design Windloads for Buildings and Boundary Layer Wind Tunnel Testing.*

AAMA CW-12-84, "Structural Properties of Glass."

AAMA CW-13-85, "Structural Sealant Glazing Systems."

AAMA CW-RS-1-96, "The Rain Screen Principle and Pressure-Equalized Wall Design."

AAMA GAG-1-97, "Glass and Glazing."

AAMA MCWM-1-89, *Metal Curtain Wall Manual.*

AAMA SFM-1, *Aluminum Store Front and Entrance Manual.*

AAMA SHDG-1, "Skylight Handbook Design Guidelines."

AAMA TIR-A4-97, "Recommended Glazing Guidelines for Reflective Insulating Glass."

AAMA TIR-A7-83, "Sloped Glazing Guidelines."

AAMA Vol-1, *Windows and Doors.*

AAMA Vol-2, *Metal Curtain Walls.*

AAMA Vol-3, *Storefront and Entrance Manual.*

AAMA Vol-4, *Skylights and Space Enclosures.*

AAMA WSG-1-95, "Window Selection Guide."

American Architectural Manufacturers Association (AAMA) (www.aamanet.org), Window and Door Manufacturers Association (WDMA) (www.wdma.com), Canadian Standards Association (CSA) (www.csa.ca). AAMA/WDMA/CSA 101/I.S.2/A440-05, *Standard/Specification for Windows, Doors and Unit Skylights.* Schaumburg, IL: AMA/WDM/CSA.

American National Standards Institute (ANSI). New York (www.ansi.org). Standards:

A117.1-2003, "Accessible and Usable Buildings and Facilities."

Z97.1-2004, "Safety Glazing Materials Used in Buildings, Safety Performance Specifications and Methods of Test."

American National Standards Institute/Builders Hardware Manufacturers Association (ANSI/BHMA) (www.ansi.org)/(www.buildershardware.com). New York: ANSI. Standards:

A156.1, "Butts and Hinges."

A156.2, "Bored and Preassembled Locks and Latches."

A156.3, "Exit Devices."

A156.4, "Door Controls—Closers."

A156.5, "Auxiliary Locks and Associated Products."

A156.6, "Architectural Door Trim."

A156.7, "Templet Hinge Dimensions."

A156.8, "Door Controls—Overhead Stops and Holders."

A156.10, "Power Operated Pedestrian Doors."

A156.12, "Interconnected Locks and Latches."

A156.13, "Mortise Locks and Latches."

A156.14, "Sliding and Folding Door Hardware."

A156.15, "Closer Holder Release Devices."

A156.16, "Auxiliary Hardware."

A156.17, "Self-Closing Hinges and Pivots."

A156.18, "Materials and Finishes."

A156.19-1997, "Power Assist and Low Energy Power Operated Doors."

A156.21, "Thresholds."

A156.27, "Power and Manual Operated Revolving Pedestrian Doors."

American National Standards Institute/Door and Hardware Institute (ANSI/DHI) (www.ansi.org)/(www.dhi.org). New York: ANSI. Standards:

ANSI A115-00, Set Wood, *Wood Door Hardware Standards.*

A115.1-1990, "Preparation for Mortise Locks for $1^3/_8''$ and $1^3/_4''$ Doors."

A115.2-1996, "Preparation for Bored Locks for $1^3/_8''$ and $1^3/_4''$ Doors."

A115.4-1994, " Preparation for Lever Extension Flush Bolts."

A115.5-1992, "Preparation for Mortise Auxiliary Deadbolt Strikes."

A115.6-1993, "Preparation for Preassembled Locks."

A115.12-1994, "Preparation for Offset Intermediate Pivots."

A115.13-1991, "Preparation for Auxiliary Bored Deadlocks and Deadlatches."

A115.14-1994, "Preparation for Open Back Strikes."

A115.15-1994, "Preparation for Bored or Cylindrical Locks in Prehung Insulated Steel Doors and Steel Frames."

A115.16-1990, "Preparation for Double-Type Locks in Prehung Insulated Steel Doors and Steel Frames."

A115.17-1994, "Preparation for Double Type Locks in Standard $1^3/_8''$ and $1^3/_4''$ Steel Doors and Steel Frames."

A115.18-1994, "Preparation for Bored Locks and Latches with Lever Handles for $1^3/_8''$ and $1^3/_4''$ Steel Door and Steel Frames."

A115.W6-1993, "Specification for Preparation of $1^3/_4''$ in Flush Wood Doors."

A115.W7-1997, "Specification for Preparation of $1^3/_4''$ in Flush Wood Doors."

A115.W8-1993, "Specification for Preparation of $1^3/_4''$ in Flush Wood Doors for Bored Auxiliary Deadlocks and Deadlatches."

American National Standards Institute/Steel Door Institute (ANSI/SDI). Cleveland (www.steeldoor.org). Standards:

A250.4, "Test Procedure and Acceptance Criteria for Physical Endurance for Steel Doors and Hardware Reinforcings."

A250.6, "Hardware on Standard Steel Doors."

A250.7, "Nomenclature: Standard Steel Doors Frames."

A250.8, "Recommended Specifications for Standard Steel Doors and Frames."

A250.10, "Test Procedure and Acceptance Criteria for Prime Painted Steel Surfaces for Steel Doors and Frames."

American Society of Heating, Refrigerating and Air-Conditioning Engineers (ASHRAE). 1997. *ASHRAE Handbook; 1997 Fundamentals.* Atlanta: ASHRAE (www.ashrae.org).

———. 2004. *Handbook—Fundamentals.* Atlanta: ASHRAE (www.ashrae.org).

American Welding Society (AWS). 2004. AWS D1.1 *Structural Welding Code—Steel.* Miami, FL: AWS (www.aws.org).

———. 2003. AWS D1.2 *Structural Welding Code—Aluminum.* Miami, FL: AWS (www.aws.org).

*Architectural Graphic Standards. See* Ramsey et al. (2000).

Architectural Woodwork Institute (AWI). *Architectural Woodwork Quality Standards Illustrated.* 8e, Version 2.0. Reston, VA: AWI (www.awinet. org).

ASTM International (ASTM). West Conshohocken, PA (www.astm.org). Standards:

A123, "Standard Specification for Zinc (Hot-Dip Galvanized) Coatings on Iron and Steel Products."

A153, "Standard Specification for Zinc Coating (Hot-Dip) on Iron and Steel Hardware."

A167, "Standard Specification for Stainless and Heat-Resisting Chromium-Nickel Steel Plate, Sheet, and Strip."

A568, "Standard Specification for Steel, Sheet, Carbon, and High-Strength, Low-Alloy, Hot-Rolled, and Cold-Rolled, General Requirements for."

A653, "Standard Specification for Steel Sheet, Zinc-Coated (Galvanized) or Zinc-Iron Alloy-Coated (Galvannealed) by the Hot-Dip Process."

A1011/A1011M, "Standard Specification for Steel, Sheet and Strip, Hot-Rolled, Carbon, Structural, High-Strength Low-Alloy and High-Strength Low-Alloy with Improved Formability."

B209, "Standard Specification for Aluminum and Aluminum-Alloy Sheet and Plate."

B221, "Standard Specification for Aluminum and Aluminum-Alloy Extruded Bars, Rods, Wire, Profiles, and Tubes."

C236, "Standard Test Method for Steady-State Thermal Performance of Building Assemblies by Means of a Guarded Hot Box."

C481, "Standard Test Method for Laboratory Aging of Sandwich Construction."

C509, "Specification for Elastomeric Cellular Preformed Gasket and Sealing Material."

C542, "Standard Specification for Lock-Strip Gaskets."

C716, "Standard Specification for Installing Lock-Strip Gaskets and Infill Glazing Materials."

C719, "Standard Test Method for Adhesion and Cohesion of Elastomeric Joint Sealants Under Cyclic Movement."

C864, "Standard Specification for Dense Elastomeric Compression Seal Gaskets, Setting Blocks, and Spacers."

C920, "Standard Specification for Elastomeric Joint Sealants."

C964, "Standard Guide for Lock-Strip Gasket Glazing."

C1036, "Standard Specification for Flat Glass."

C1048, "Standard Specification for Heat Treated Flat Glass—Kind HS, Kind Coated and Uncoated Glass."

C1115, "Standard Specification for Dense Elastomeric Silicone Rubber Gaskets and Accessories."

C1172, "Standard Specification for Laminated Architectural Flat Glass."

C1199, "Standard Test Method for Measuring the Steady-State Thermal Transmittance of Fenestration Systems Using Hot Box Methods."

D2000, "Standard Classification System for Rubber Products in Automotive Applications."

D2287, "Standard Specification for Nonrigid Vinyl Chloride Polymer and Copolymer Molding and Extrusion Compounds."

D4726. "Standard Specification for Rigid Poly(Vinyl Chloride) (PVC) Exterior-Profile Extrusions Used for Assembled Windows and Doors."

E84, "Standard Test Method for Surface Burning Characteristics of Building Materials."

E90, "Standard Method for Laboratory Measurement of Airborne-Sound Transmission Loss of Building Partitions and Elements."

E96, "Standard Test Method for Water Vapor Transmission of Materials."

E283, "Standard Test Method for Determining the Rate of Air Leakage Through Exterior Windows, Curtain Walls, and Doors Under Specified Pressure Differences Across the Specimen."

E330, "Standard Test Method for Structural Performance of Exterior Windows, Curtain Walls, and Doors by Uniform Static Air Pressure Differential."

E331, "Standard Test Method for Water Penetration of Exterior Windows, Skylights, Doors, and Curtain Walls by Uniform Static Air Pressure Differential."

E413, "Classification for Rating Sound Insulation."

E547, "Standard Test Method for Water Penetration of Exterior Windows, Skylights, Doors, and Curtain Walls by Cyclic Air Pressure Differential."

E662, "Standard Test Method for Specific Optical Density of Smoke Generated by Solid Materials."

E773, "Standard Test Method for Accelerated Weathering of Sealed Insulating Glass Units."

E774, "Standard Specification for the Classification of the Durability of Sealed Insulating Glass Units."

E783, "Standard Test Method for Field Measurement of Air Leakage Through Installed Exterior Windows and Doors."

E987, "Standard Tests Methods for Deglazing Force of Fenestration Products."

E1105, "Standard Test Method for Field Determination of Water Penetration of Installed Exterior Windows, Skylights. Doors, and Curtain Walls by Uniform or Cyclic Static Air Pressure Difference."

E1233, "Standard Test Method for Structural Performance of Exterior Windows, Curtain Walls, and Doors by Cyclic Static Air Pressure Differential.

F588, "Standard Test Methods for Measuring the Forced Entry Resistance of Window Assemblies, Excluding Glazing Impact."

F842, "Standard Test Methods for Measuring the Forced Entry Resistance of Sliding Door Assemblies, Excluding Glazing Impact."

Consumer Product Safety Commission (CPSC). Part 1201, "Safety Standards for Architectural Glazing Materials." Bethesda, MD: CPSC (www.cpsc. gov).

Door and Hardware Institute (DHI) Chantilly, VA (www.dhi.org). 1976. "Hardware Reinforcements on Steel Doors and Frames."

———. 1985. "Basic Architectural Hardware."

———. 1990. "Processing Hardware for Custom Aluminum Doors."

———. 1990. "Recommended Locations for Architectural Hardware for Standard Steel Doors and Frames."

———. 1994. "Installation Guide for Doors and Hardware (ANSI/DHI A115.IG)."

———. 1996. "Recommended Locations for Builder's Hardware for Custom Steel Doors/Frames."

———. 1996. DHI-WDHS-1, "Templet Book Criteria for Wood Doors."

———. 1996. DHI-WDHS-2, "Recommended Fasteners for Wood Doors."

———. 1993. DHI-WDHS-3, "Recommended Locations for Architectural Hardware for Flush Wood Doors."

Glass Association of North America (GANA). Topeka, KS (www.glass-website.com).

———. 1990. *FGMA Sealant Manual.*

———. 2004. *GANA Glazing Manual.*

Glass Tempering Association (GTA). 1989. 89-1-6, "Specification for Environmental Durability of Fully Tempered or Heat-Strengthened Spandrel Glass with Applied Opacifiers." Topeka, KS: GTA (www.glasswebsite.com).

Hollow Metal Manufacturers Association. *Hollow Metal Manual.* Chicago: National Association of Architectural Metal Manufacturers (www.naamm.org).

Insulating Glass Certification Council (IGCC). *Certified Products Directory—Sealed Insulating Glass.* Henderson Harbor, NY: IGCC (www.igcc.org).

International Code Council (ICC). Falls Church, VA: (www.iccsafe.org). Codes:
2006. *International Building Code®.*
2006. *International Residential Code®.*

Laminated Safety Glass Association (LSGA). 1987. *LSGA Design Guide.* Topeka, KS: LSGA (www.glasswebsite.com).

National Association of Architectural Metal Manufacturers (NAAMM). 1988. *Metal Finishes Manual.* Chicago: NAAMM (www.naamm.org).

National Fire Protection Association (NFPA). Quincy, MA (www.nfpa.org).
1999. *NFPA 80: Standard for Fire Doors and Fire Windows.*

———. 2003. *NFPA 101: Life Safety Code.*

———. 2003. *NFPA 105: Recommended Practice for the Installation of Smoke-Control Door Assemblies.*

National Glass Association (NGA). McLean, VA (www.glass.org).
2003. "Guide to the Glass and Glazing Requirements of the Model Building Codes."

———. "Guide to the Design and Construction Considerations of a Glass and Metal Framing System."

———. "New Expanded Sealant Manual: Basic Instructions on the Essentials of Sealant Installation."

National Institute of Standards and Technology. *See* U.S. Department of Commerce.

Ramsey, Charles George, Harold Reeve Sleeper, and John Ray Hoke, Jr. 2000. *Architectural Graphic Standards,* 10th ed. Hoboken, NJ: John Wiley & Sons, Inc. (www.wiley.com).

Safety Glazing Certification Council (SGCC). *Certified Products Directory—Safety Glazing Materials Used in Buildings.* Henderson Harbor, NY: SGCC (www.scgc.org).

Screen Manufacturers Association (SMA). Palm Beach, FL (www.smacentral.org). Standards:
SMA 1004-1987-R1998, "Specifications for Aluminum Tubular Frame Screens for Windows."
SMA 1201-2002, "Specifications for Insect Screens for Windows, Sliding Doors, and Swinging Doors."
SMA 2006-1987-R1998, "Specifications for Aluminum Sliding Screen Doors."
SMA 3001-1987-R1998, "Specifications for Aluminum Swinging Screen Doors."

Sealed Insulating Glass Manufacturers Association (SIGMA). Chicago. (www.sigmaonline.org/sigma). 1990. "TB-3001: Voluntary Guidelines for Sloped Glazing."
1997. "TM-3000: Glazing Guidelines for Sealed Insulating Glass Units."

Steel Door Institute (SDI). Cleveland (www.steeldoor.org). Technical Documents:
SDI-106, "Recommended Standard Door Type Nomenclature."
SDI SDI-108, "Recommended Selection and Usage Guide for Standard Steel Doors."
SDI-109, "Hardware for Standard Steel Doors and Frames."
SDI-110, "Standard Steel Doors and Frames for Modular Masonry Construction."
SDI-111, "Recommended Standard Details Steel Doors and Frames."
SDI-112, "Zinc-Coated (Galvanized/Gavannealed) Standard Steel Doors and Frames."
SDI-113, "Test Procedure and Acceptance Criteria for Apparent Thermal Performance of Steel Door and Frame Assemblies."
SDI-117, "Manufacturing Tolerances for Standard Steel Doors and Frames."
SDI-118, "Basic Fire Door Requirements."
SDI SDI-122, "Installation and Troubleshooting Guide for Standard Steel Doors and Frames."
SDI-128, "Guidelines for Acoustical Performance of Standard Steel Doors and Frames."

Society for Protective Coatings (SSPC). Pittsburgh (www.sspc.org). *SSPC Painting Manual, Vol. 1, Good Painting Practice,* 4th ed.
2005. *SSPC Painting Manual, Vol. 2, Systems and Specifications.*

U.S. Department of the Army. 1995. Corps of Engineers Guide Specifications, Military Construction, CEGS-08110, "Steel Doors and Frames." Office of the Chief of Engineers, Department of the Army (www.army.mil).

U.S. Department of Commerce, National Institute of Standards and Technology (NIST). 1977. *Window Design Strategies to Conserve Energy.* Washington, DC: NIST (www.nist.gov).

U.S. Department of Housing and Urban Development (HUD). 2003. *Minimum Property Standards for Housing.* Washington, DC: HUD (www.hud.gov).

U.S. Department of the Navy, Naval Facilities Engineering Command. 1995 (July). Guide Specifications Section 08110, "Steel Doors and Frames." Washington, DC. (https://portal.navfac.navy.mil)

———. 1995 (July). Guide Specifications Section 08120, "Aluminum Doors and Frames."

———. 1995 (July). Guide Specifications Section 08210, "Wood Doors."

———. 1995 (July). Guide Specifications Section 08520, "Aluminum Windows."

———. 1995 (July). Guide Specifications Section 08610, "Wood Windows."

———. 1995 (July). Guide Specifications Section 08710, "Finish Hardware."

———. 1995 (July). Guide Specifications Section 08800, "Glazing."

Window and Door Manufacturers Association (WDMA). Des Plaines, IL (www.wdma.com). *Specifiers Guide to Windows and Doors.*

———. 2004, "How to Store, Handle, Finish and Maintain Wood Doors."

———. Care and Finishing of Wood Windows."

———. ANSI/WDMA I.S.6-01, "Architectural Wood Stile and Rail Doors."

———. WDMA I.S.1-A-04, "Architectural Wood Flush Doors."

———. WDMA I.S.4-00, "Water-Repellent Preservative Non-Pressure Treatment for Millwork."

Woodwork Institute (WI) (formerly Woodwork Institute of California [WIC]). 2003. *Manual of Millwork.* West Sacramento, CA: WI (www.wicnet.org).

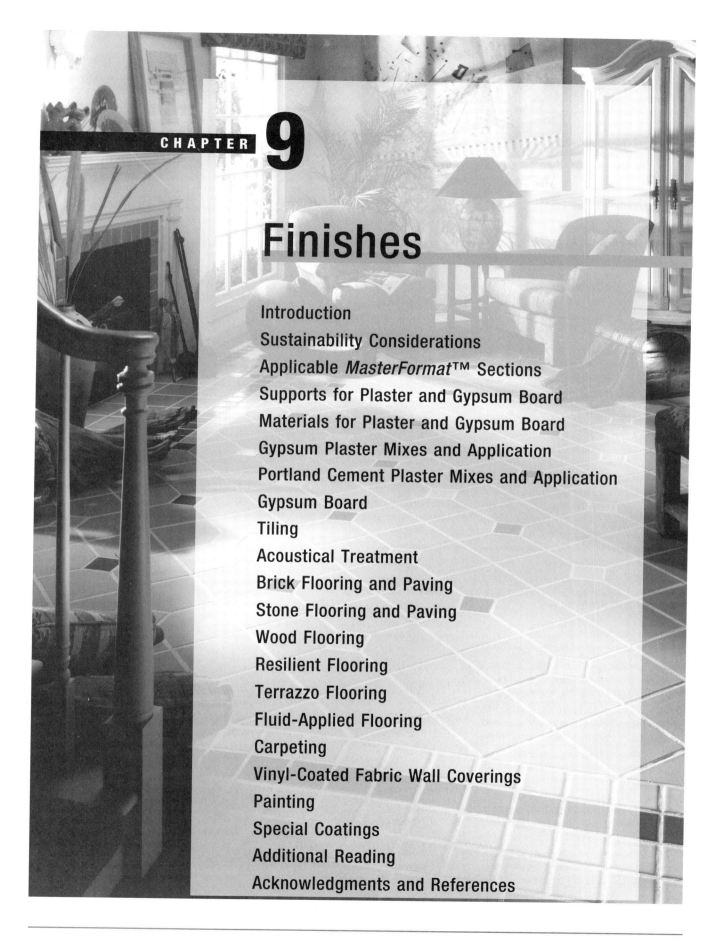

# Introduction

Finishes include materials and systems used to make the visible surfaces of the interior floors, walls, and ceilings of a building. Some of the same products are also used as exterior finishes. There are many products and systems in use today as interior finishes, and there are usually several variations within a particular product or system. Selection of these materials and systems for interior application is influenced by the use of the spaces in which they will be applied and the wear to which they will be subjected. This chapter covers some of the many finishes available.

# Sustainability Considerations

For a discussion of sustainability, the systems addressed in this chapter can be separated into the materials used and the products and systems in which those materials are used.

## Materials

The materials used in the products and systems discussed in this chapter are varied in their effects on the environment and the sustainability of the construction in which they are employed. Considerations regarding the sustainability of the materials addressed in this chapter are much the same as those described in the chapters that cover the various materials. For example, sustainability considerations for the masonry and stone used in the flooring products discussed in this chapter are addressed in Chapter 4; aluminum and other metals used in the products discussed in this chapter are addressed in Chapter 5; wood and plastics in Chapter 6; and so on.

## Products and Systems

### Plaster and Gypsum Board

Sustainability considerations related to portland cement are discussed in Chapter 3.

Gypsum is an abundant mineral but is naturally occurring and is, therefore, non-renewable. Most gypsum is acquired by surface mining, which disturbs the natural habitat. The paper used in gypsum board is mostly recycled material.

Large quantities of gypsum dust are raised during cutting and installation of gypsum board products, which can cause discomfort to workers and is a general nuisance, but this dust has not been demonstrated to be otherwise harmful either to people or the environment.

During demolition, most gypsum ends up in landfills, although some is recycled into new gypsum board products and some is used in agricultural applications to help make clayey soils looser.

### Acoustical Treatment

Most acoustical ceiling and wall treatment products are made from glass or mineral fiber with a painted surface. While the glass and mineral fiber materials used have not been demonstrated to be carcinogenic, installers must wear masks to protect their lungs. Some fiber products emit small amounts of formaldehyde, which is harmful to humans, but designers can specify one of the available fiber products that do not outgas formaldehyde.

Sustainability considerations for metal, ceramics, plastics, and other materials used in acoustical treatment products are discussed in the chapters covering those materials. Paint is discussed later in this section.

Most acoustical treatment products end up in landfills when demolition occurs.

### Brick and Stone Flooring and Paving

Sustainability considerations related to brick and stone flooring and paving are essentially the same as those discussed for brick and stone in Chapter 4.

### Wood Flooring

Sustainability considerations related to wood products are discussed in Chapter 6. Sustainability considerations related to finishes for wood flooring are discussed later in this section.

### Resilient Flooring

The primary ingredient in many resilient flooring products is polyvinyl chloride (PVC, vinyl).

The manufacture of vinyl produces large quantities of toxic air pollutants. Plasticizers and some of the other materials used in flooring products contain toxic materials that may outgas over time. Burning vinyl emits dioxins and hydrochloric acid.

Adhesives used in installing resilient flooring may outgas toxic chemicals.

Most resilient flooring products are not recycled upon demolition and end up in landfills.

### Terrazzo Flooring

Sustainability considerations related to the basic ingredients in terrazzo, portland cement and sand, and stone are addressed in Chapters 3 and 4, respectively.

### Fluid-Applied Flooring

A particular concern with fluid-applied flooring materials is the propensity of some of them to exude volatile organic compounds (VOCs). Avoiding this problem is usually a matter of selecting products that do not contain VOC.

Some fluid-applied flooring products contain other potentially harmful chemicals. Most of these are present before and during application but are not a hazard after the product cures. Appliers must exercise care and wear protective clothing and masks when the product warrants doing so.

In addition, many fluid-applied flooring products contain highly flammable resin components. Applicators must maintain suitable precautions to prevent catastrophic fires from these materials.

Some of the components of fluid-applied flooring can produce severe allergic reactions in susceptible people.

Cured fluid-applied floor products are generally nontoxic and safe.

### Carpeting

The finishing and dyeing of carpet traditionally uses a large quantity of water. Recent innovations have led to less use of water in these processes, but copious quantities of water are still involved. It is possible to reclaim much of this water and reuse it in the carpet manufacturing and dyeing process. Carpet manufacturers have been working to conserve water by reclaiming water for reuse, returning wastewater to its source, and increasing the treatment of wastewater so that it returns to the environment cleaner.

Carpet manufacture introduces into the atmosphere pollutants from outgassing of solvent-based compounds used in the process, as well as large quantities of dust and smoke.

The carpet manufacturing process results in large quantities of solid waste. In the past, as much as 90% of this waste went into landfills. Recent innovations in recycling have reduced this waste disposal to less than 10% of the waste produced. For example, waste from nonwoven fabric is being used as raw material for geotextile fabrics. Carpet waste can be recycled into carpet padding.

Carpet waste can also be used in the manufacture of fan housings and transmission parts for the automotive industry, fast-food trays, chair casters, flower pots, roll-out recycle containers, garbage bags, truck seat covers, parking lot bumpers, and marine lumber. There are many other examples.

Carpet is being manufactured from recycled products other than carpet. For example, postconsumer plastic bottles are being used as the fiber source for some polyester carpets.

### Vinyl-Coated Fabric Wall Covering

The manufacture of PVC (vinyl) produces large quantities of toxic air pollutants. Plasticizers and some of the other materials used in vinyl contain toxic materials that may out gasover time. Burning vinyl emits dioxins and hydrochloric acid. However, most vinyl-fabric wall coverings are Class A fire-rated and self extinguishing.

The growth of mold, mildew, and fungus is always a problem with any fabrics, but most vinyl-coated fabric wall covering materials contain antimicrobial additives to combat these hazards.

### Painting

Some paint and coating materials are toxic in the can, but most are relatively nontoxic when dry. A notable exception is products containing lead. Such products are prohibited from use in most jurisdictions in locations where they may be accessible to children, but they are still used in concealed locations.

Some paints and many pretreatment products are volatile, and others are highly toxic. Acid and alkali cleaning agents can cause severe harm to those exposed to them. Appliers must handle these products with care and protect themselves from the harmful effects of these substances with goggles and rubber gloves, and with respirators where appropriate.

Some paints and many cleaning and pretreatment materials are highly flammable. Caution must be exercised when using such materials.

Some paints and associated products contain VOCs. Many jurisdictions now limit the allowable VOC content of such materials, and many more restrictions will apply in the future. Selection of products that contain no VOCs is prudent, since resurfacing will require recoating such products with compatible materials, which may not be available in the future.

### Special Coatings

Many of the comments in this section related to sustainability issues for fluid-applied flooring apply as well to special coatings. These include comments related to VOC content, flammability, and toxic materials content.

## Applicable *MasterFormat* Sections

# 9.1 Supports for Plaster and Gypsum Board

Before applying a plaster or gypsum board finish, it is necessary to select, install, and prepare supporting construc-

tion, select and apply a plaster base (Fig. 9.1-1), and select and install proper accessories. Nonstructural framing, ceiling suspension systems, and furring systems for support of plaster and gypsum board, as well as lathing, accessories, and trim for use with plaster, are addressed in this section. In accordance with normal industry practices, trim and accessories for

use with gypsum board and gypsum board joint taping and finishing are covered in Section 9.3.

### 9.1.1 SUPPORT SYSTEMS

Support systems for plaster and gypsum board include concrete, masonry, and wood and metal framing and furring. Wood fram-

**FIGURE 9.1-1** Applying plaster over a gypsum plaster base. (Courtesy United States Gypsum Company.)

ing and furring are addressed in Chapter 6; the others are addressed in this section.

This supporting construction must support design loads without excessive deflection and without introducing stresses that might cause the finish to crack. The calculated deflection of ceiling members for plaster should not exceed $1/360$ of the span after the finish has been applied. Generally, the deflection of ceiling members supporting gypsum board finishes should not exceed $1/240$ of the span under the full design load. Headers or lintels of adequate size should be provided over openings, special construction should be provided to support wall-hung equipment and fixtures, and all construction should be true and plumb or level.

### 9.1.1.1 Concrete and Masonry

Both gypsum and portland cement plasters and gypsum board can be applied directly to some concrete and masonry surfaces under certain conditions.

Supporting construction should be strong and stable, but also capable of accommodating structural movement. Concrete and masonry construction should have control and expansion joints as dictated by good design practice. Control joints in the finish should be placed directly over joints in the supporting construction and in other locations where required to relieve stresses and minimize cracking (see Section 9.4.4).

### PLASTER

When plaster is applied directly to concrete or masonry, the concrete or ma-

sonry serves as both support system and plaster base (see Section 9.1.2).

### GYPSUM BOARD

Gypsum board can be applied directly to above-grade interior masonry or concrete when the surface is dry, smooth, plumb, and true. Above-grade exterior cavity walls also may be suitable for direct lamination if the cavity is properly insulated. However, predecorated wallboard with a decorative film that is impermeable to vapor should not be laminated directly to concrete or masonry, because moisture may become trapped within the gypsum core of the board.

Before gypsum board is directly applied, rough or protruding edges in the substrate should be removed and depressions greater than 4 in. (101.6 mm) in diameter or $1/8$ in. (3.2 mm) deep should be filled with grout or a similar filler. Mortar joints should be cut flush with the face of the wall. Surfaces should be free from form oils, curing compounds, loose particles, dust, grease, and efflorescence to ensure an adequate bond. Concrete must be allowed to cure for at least 28 days before gypsum board is applied to it.

Exterior and below-grade walls and horizontal surfaces subject to moisture or excessive heat loss, and surfaces that cannot be prepared readily for adhesive lamination, are not suitable for direct application of gypsum board. These should be furred to separate the gypsum board from possible moisture in the wall and to provide a suitable base for attaching gypsum board.

### 9.1.1.2 Wood and Metal Framing and Furring

Wood and metal supporting construction for plaster and gypsum board usually consists of:

1. *Self-supporting framing*: studs, joists, and trusses
2. *Furring members*: wood strips or metal channels supported by underlying construction

Framing and furring should be accurately aligned in a single plane and spaced not to exceed the maximum recommended for the plaster or gypsum board being installed. Furring is necessary when the spacing of framing exceeds the maximum allowable span for the plaster lath or gypsum board.

### PLASTER

*Wood Framing and Furring* Gypsum plaster can be installed over wood framing and furring with either a gypsum lath base or metal reinforcement. Portland cement plaster can be installed over wood framing and furring only using metal reinforcement.

Exterior wood frame construction is either unsheathed (open-frame) or sheathed. Exterior wall studs will usually be as follows:

- Supporting a ceiling and roof only: $2 \times 4s$ ($50 \times 100s$) at 24 in. (610 mm) on centers
- Supporting one or two floors: $2 \times 4s$ ($50 \times 100s$) at 16 in. (406 mm) on centers
- Supporting more than two floors: $2 \times 6s$ ($50 \times 150s$) at 16 in. (406 mm) on centers

*Unsheathed Construction* The structural frame should be properly braced by let-in $1 \times 4s$ ($25 \times 100s$) or larger diagonal bracing or by other means to resist racking forces. Paper-backed reinforcement can be applied directly. If backing paper is applied separately, a soft annealed 18-gauge (1.3 mm) or heavier steel line wire should be stretched across the faces of the studs in horizontal strands about 6 in. (139.7 mm) apart. The wire should be stretched taut by first nailing or stapling it at every fifth stud, then securing it to the center of intermediate studs and stretching it tightly by raising and lowering the attachments (Fig. 9.1.2a). Backing paper then is nailed over the line wire with edges lapped at least 3 in. (76.2 mm) to prevent the paper from sagging and to produce a more uniform thickness of portland cement plaster that is less likely to crack.

Application of flexible reinforcement should start at least one full stud away from corners and should be bent around them to avoid a joint at the corner. Prefabricated corner reinforcement should be used when reinforcement is too rigid to be bent easily. Horizontal laps should be wired at least once between each stud space but not more than 9 in. (228.6 mm) on centers. Vertical laps should be made over supports. The bottom of a plaster panel should be terminated with a metal drip screed to help support the plaster and permit drainage of water that might otherwise penetrate the plaster (Fig. 9.1-3).

**FIGURE 9.1-2** (a) Unsheathed (open frame) construction; (b) sheathed construction. Line wire is not required with paper-backed reinforcement. (Courtesy Portland Cement Association.)

**FIGURE 9.1-3** (a) Properly installed drip screed allows drainage and reduces the possibility of water penetration into a wall; (b) no drip screed and bond between concrete and portland cement plaster (stucco) may result in moisture backup. (Courtesy Portland Cement Association.)

As plaster is applied, it should be forced through the openings in the reinforcement against the paper backing to ensure at least 90% coverage of the reinforcement. An alternate method sometimes used to ensure complete embedment of the reinforcement is to eliminate the backing paper and backplaster the reinforcement between supports. This backplaster coat should be at least $^3/_8$ in. (9.5 mm) thick, and should be applied only after the scratch coat has hardened sufficiently so that pressure will not break the plaster keys. The surface should be uniformly dampened before application and moist-cured for at least 24 hours after application.

**Sheathed Construction** Wood boards, plywood, or other suitable panel sheathing should be applied horizontally and fastened securely to each stud (see Fig. 9.1-2b). Lumber boards used for sheathing should not be more than 8 in. (203 mm) wide, and should be fastened with at least two nails at each support to minimize the possibility of damage from warping. The sheathing should be covered with backing paper, lapped at least 3 in. (76.2 mm), followed by metal reinforcement fastened through the sheathing into the wood supports. Paper-backed reinforcement can be attached directly over the sheathing. The reinforcement should be returned at least 6 in. (139.7 mm)

around corners. Vertical laps over supports and horizontal laps should be nailed or stapled at 6 in. (139.7 mm) on centers into the sheathing. Unless the reinforcement is self-furring, nails should be of the furring type; staples can be used for self-furring reinforcement.

***Metal Framing and Furring*** The thermal insulating and fire-resistant properties of lightweight aggregates, such as perlite and vermiculite, have encouraged the use of portland cement plaster in commercial and institutional structures, especially where fire ratings are required. Lightweight aggregates are not recommended for exterior plaster, however, so they are used primarily for backplaster coats on exterior walls and for fireproofing.

Both gypsum and portland cement plaster are supported by metal framing and furring in interior locations where high-quality construction or fire resistance ratings are required.

Assemblies consisting of steel studs, furring channels (C-shaped and hat-shaped), and reinforcing rods should be spaced and securely fastened according to the specialty manufacturer's recommendations and the recommendations of the Metal Lath/Steel Framing Association Division of the National Association of Architectural Metal Manufacturers (NAAMM) publication ML/SFA 920, "Specifications for Metal Lathing and Furring." The recommended spacing of metal supports for various types of reinforcement closely parallels that of wood (Fig. 9.1-4). Metal stud framing is as described in Section 9.5 for support of gypsum board.

### GYPSUM BOARD

Gypsum board can be applied directly to wood or metal supporting construction that is capable of supporting the applicable design loads and that will provide a firm, level, plumb, and true base.

***Wood Framing*** Wood framing for gypsum board is usually 2 × 4s (50 × 100s) at 24 in. (610 mm) on centers or 2 × 4s (50 × 100s) at 16 in. (406 mm) on centers. Larger or smaller framing may also be used, depending on structural or other requirements (see Chapter 6).

***Metal Framing*** Metal framing and furring are common in commercial, institutional, and high-rise residential buildings where noncombustible construction is

**FIGURE 9.1-4** Support Spacing for Metal Reinforcement[a]

| Reinforcement | Weight, lb./sq. yd. (kg/m²) | Support Spacing, in. (mm) on centers |
|---|---|---|
| Diamond mesh | 3.4 (1.8) | 16 (406) |
| Stucco mesh[b] | 1.8 and 3.6 (.98 and 1.95) | 16 (406) |
| Flat rib lath | 3.4 (1.8) | 19 (483)[c] |
| ³⁄₈-in. rib lath | 3.4 and 4.0 (1.8 and 2.17) | 24 (610) |
| Woven wire | 1.7 (.9) (1 in. hex.) (25.4 mm hex.) | 16 (406) |
| | 1.4 (.96) (1½ in. hex.) (38 mm hex.) | 16 (406) |
| Welded wire | 1.4 (.96) (1 in. × 1 in.) (25.4 × 25.4 mm) | 16 (406) |
| | 1.4 (.76) (2 in. × 2 in.) (50.8 × 50.8 mm) | 16 (406) |
| | 1.9 (1) (1½ in. × 2 in.) (38 × 50.8 mm) | 16 (406)[d] |

[a]Support spacing applicable to both wood and metal framing for both vertical and horizontal surfaces unless otherwise noted.

[b]Not recommended for vertical surfaces over metal framing or for horizontal surfaces.

[c]Spacing of metal framing for horizontal surfaces should not exceed 13½ in. (343 mm) on centers.

[d]Spacing may be increased to 24 in. (610 mm) on centers when stiffener wires are provided 6 in. (152 mm) on centers.

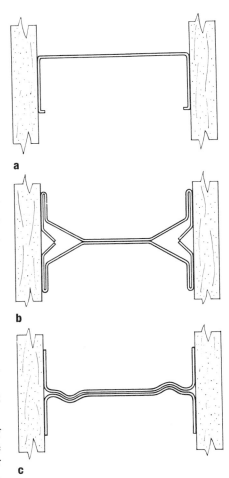

**FIGURE 9.1-5** Metal studs: (a) C-shaped stud designed to accept drywall screws (the type most often used today); (b) nailable stud for special ring-shank nails; (c) nailable stud for smooth-shank nails. Types (b) and (c) are seldom used today and may not be available in many areas.

required. In such instances, non-load-bearing partitions typically are framed with metal studs; ceilings usually are framed with open-web joists or steel beams.

Metal studs are usually C-shaped, 25-gauge (0.6 mm), cold-formed, electro-galvanized steel units conforming with ASTM C645. They are available in 20- and 25-gauge (1.0 and 0.6 mm) thicknesses, which are compatible with most drywall sheet metal screws. Proprietary nailable studs (Fig. 9.1-5) were readily available in the past and may be found in some existing construction. They are difficult or impossible to obtain today.

Floor and roof framing often consists of steel beams or open-web joists. These members are not designed to receive gypsum board directly and often are spaced more than 24 in. (609 mm) apart. Therefore, it is necessary to provide furring or suspension systems for applying gypsum board on the ceilings of metal framed buildings (Fig. 9.1-6).

*Furring* Wood or metal furring will:

- Provide plumb, true, and properly spaced supports
- Eliminate capillary transfer of moisture in exterior and below-grade walls
- Minimize moisture condensation on interior wall surfaces
- Improve the thermal insulating properties of exterior walls

- Increase assembly thickness to accommodate mechanical equipment
- Improve sound isolation by resilient mounting

Metal furring can be used with all types of supporting construction, and is preferable to wood furring when noncombustible or sound isolating assemblies are required. Metal furring members may be either cold-rolled channels or drywall channels (Fig. 9.1-7). *Cold-rolled channels* are multipurpose members used for both plaster and gypsum board wall and suspended ceiling assemblies. They usually are of 16-gauge (1.6 mm) steel with either a galvanized or black asphaltum finish. They are ³⁄₄ to 2 in. (19.1 to 50.8 mm) wide and 16 to 20 ft (4880 to 6100 mm) long. Because they are too thick to receive drywall sheet metal screws, these channels usually are wire-tied in place. In gypsum board assemblies, cold-rolled channels are used primarily as a supporting grid for lighter drywall channels to which wallboard can be screw-attached (see Fig. 9.1-6).

*Drywall channels* are roll-formed of 25-gauge (0.6 mm) electro-galvanized steel, designed for gypsum board attachment with self-drilling sheet metal screws. They are installed perpendicular to wood framing with a 1¹⁄₄ in. (28.6 mm) screw or a 5d nail at each support and are clipped to cold-rolled furring channels. Over masonry or concrete surfaces, they may be installed horizontally or vertically, whichever is more convenient.

Proprietary clips, runners, and adjustable brackets are available with engineered metal furring systems to facilitate rapid erection over irregular masonry walls (Fig. 9.1-8).

Drywall channels can be either plain (hat-shaped) or resilient. Plain channels are used primarily with unit masonry and concrete wall assemblies that possess inherent sound isolation because of their mass and are not significantly improved by resilient mounting. In suspended ceiling assemblies, a grillage composed of cold-rolled channels and plain drywall channels is used to support a gypsum board finish (Fig. 9.1-9). If sound or vibration isolation is desired, resilient hangers can be wired into the suspension system to isolate the ceiling from the framing.

FIGURE 9.1-6 Open-web bar joists usually require a grid of cold-rolled channels and hat-shaped furring channels to support gypsum board.

FURRING CHANNEL DETAILS

| | |
|---|---|
| 4'-0" | = 1200 mm |
| 1½" | = 38 mm |
| 24" | = 300 mm |
| ½" | = 12.7 mm |
| ⅝" | = 15.9 mm |

FIGURE 9.1-9 In suspended ceilings, wire hangers are used to support and level the channel grid. (Courtesy Gypsum Association.)

Resilient channels are used over both metal and wood framing (Fig. 9.1-10). In addition to improving sound isolation, they help to isolate the gypsum board from structural movement, reducing joint cracking. Resilient channels are nailed or screwed perpendicular to wood framing members through predrilled holes, using a 1¼ in. (28.6 mm) screw or 5d nail at each support.

### 9.1.2 PLASTER BASES

Plaster bases are constructions or surfaces which develop an adequate bond for the direct application of plaster. Bond can be provided by mechanical keying or by suction. Mechanical keying results from the interlocking of a plaster mix with openings or projections in the base

FIGURE 9.1-10 Resilient channel mounting substantially improves sound isolation in wood-framed construction. (Courtesy Gypsum Association.)

Plain Channel

Resilient Channel

Cold-Rolled Channel

FIGURE 9.1-7 Metal furring channels. Plain channels are also called *hat-shaped* channels. (Courtesy Gypsum Association.)

FIGURE 9.1-8 Adjustable brackets have notched surfaces to permit plumb and true installation of a channel grid. (Courtesy Gypsum Association.)

surface; suction develops bond by drawing part of the plaster paste into minute pores in the base surface by capillary action. Over metal reinforcement, bond is developed by mechanical keying; over concrete and masonry, bond is developed by either mechanical keying, suction, or both.

Both gypsum and portland cement plaster can be applied over metal reinforcement and directly over some concrete, concrete masonry units (CMUs), and brick. Gypsum plaster can also be applied directly to gypsum products such as other gypsum plaster, gypsum masonry, or gypsum lath, but these bases do not provide an adequate bond for portland cement plaster. Metal lath should be installed over such materials before portland cement plaster is applied; backing paper (see Section 9.2.2.3) should also be used in exterior locations. Wood lath, fiber insulating lath, and sheet metal lath should not be used as a base for portland cement plaster.

Gypsum veneer (thin coat) plaster is applied over *veneer plaster base* (see Section 9.3).

### 9.1.2.1 Concrete and Masonry

Both gypsum and portland cement plaster can be applied directly over concrete and over CMU and brick that have sufficient bonding capabilities. However, gypsum plaster should not be applied directly to concrete or masonry surfaces that are the interior surfaces of exterior walls or the opposite surfaces of walls or partitions that are exposed to wet conditions, such as would be found in a sauna, shower, or tub and shower enclosure.

The following discussion references several methods that may be used when it is doubtful that portland cement plaster can obtain a good bond with concrete or masonry. Among these are the application of a dash bond coat or a bonding agent, which may be used with both concrete and masonry.

A *dash bond coat* consists of a mixture of 1 part by volume of portland cement and 1 to 2 parts sand. A small straw broom or long-fibered stiff brush is dipped in the mix, and the material is splattered on the wall with a quick throwing motion (Fig. 9.1-11). A dash coat should not be troweled and should be allowed to set before the plaster base coat is applied. If there is a delay in applying the base coat, the dash coat

**FIGURE 9.1-11** A dash bond coat being applied. (Courtesy Portland Cement Association.)

should be kept moist until the next coat is applied.

A *bonding agent* is a proprietary waterproof chemical compound recommended by its manufacturer for exterior use to assist the bonding of plaster to concrete and masonry. Before a bonding agent is applied, the supporting surface should be structurally sound and rigid, as well as free from laitance, dust, grease, and materials that are soluble in water, such as water-based paints or glue.

### CONCRETE

Before plaster is applied directly to concrete, sandblasting should be used to clean the surface of laitance, form oil, and other substances that might impair the bond. Smooth or slick spots on fresh concrete caused by form oil also can be removed by scrubbing with strong soap and water and allowing the concrete to dry to restore suction. However, form oil should not be used for concrete forms when direct application of plaster is intended.

Rough lumber can be used for the forms of concrete that are to receive plaster, or the freshly placed concrete can be lightly sandblasted to roughen the surface immediately after the forms have been stripped. Hardened, smooth concrete may require more extensive preparation such as chipping, acid etching, or roughening with a bush hammer. Metal reinforcement should be used when gypsum plaster is to be applied over smooth concrete where roughing is not practical. A bonding agent or a dash coat also may be used where portland cement plaster is to be applied and roughening of the surface is impractical.

A solution of 1 part muriatic acid and 6 parts water can be used for *acid etching* of slick concrete where a dash coat or bonding agent might not adhere well. Before an acid solution is applied, the surface should be wetted so that the acid

will only act on the surface. Several applications may be necessary. After treatment, concrete should be thoroughly washed with water to remove all traces of the acid and allowed to dry to restore suction.

### MASONRY

When plaster is to be applied directly to masonry, mortar joints should be struck flush or slightly raked. The masonry surface should be clean, sound, and firm and should not contain efflorescence, grease, waterproofing compounds, paint, or other substances detrimental to a good bond. Efflorescence, oil, and grease can be removed by washing with a 10% solution of muriatic acid and water. Other coatings or substances that will adversely affect the bond should be removed with appropriate solvents and rinsed clean.

Medium-hard common and face brick, standard-weight concrete block, and many types of stone generally provide adequate suction and often a mechanical bond. The surface can be tested for suction by spraying it with clean water to see how quickly the moisture is absorbed. If the water is readily absorbed, good suction is likely to result; if water droplets form, the surface does not provide adequate suction and a bond will depend on mechanical keying. It can be determined by observation whether the surface is sufficiently rough to ensure mechanical keying with the plaster. Glazed brick will probably be too slick, but standard-weight CMU will usually be rough enough.

On surfaces that absorb moisture too quickly, such as soft common brick and lightweight CMU, portland cement plaster will quickly stiffen and become difficult to apply properly. Excessive suction can be controlled by fine-spraying (not soaking) the masonry with several applications of water immediately before applying the first plaster coat. The surface should appear slightly damp, but no free water should be present when the plaster is applied. When stucco is machine-applied, it may not be necessary to dampen the surface because the mix usually contains more water than those for hand application.

Hard-burnt brick, glazed CMUs, and some types of stone may provide a poor or nonuniform bond. If suction is not uniform over the entire surface, parts of the wall may draw more moisture than others, and the final finish will be spotty in color. Moreover, nonuniform bonding is likely to cause cracking, because the plaster will move with the supporting base in the areas of good bonding and independently of the base in areas of poor bonding.

If an adequate or uniform bond cannot be obtained, the surface should be prepared with metal reinforcement, a dash coat, or a bonding agent. When metal reinforcement is used with portland cement plaster, backing paper should also be used to isolate the plaster membrane from the base and prevent dissimilar suction. When plaster is to be applied as a continuous finish over dissimilar rigid bases, a control joint or a strip of metal reinforcement should be placed over the juncture of the dissimilar materials.

Plaster should not be applied directly to a masonry chimney, but should be applied over metal reinforcement. However, backing paper should not be used, because heat might cause deterioration. Flashings should be installed under the chimney cap and at junctures with the roof to prevent water penetration (see Chapter 4).

### 9.1.2.2 Metal Reinforcement and Accessories

Metal reinforcement may be used in lieu of gypsum lath for gypsum plaster installed over wood or metal framing or furring. It should also be used in gypsum plaster over concrete or masonry that does not provide a sufficient bond for direct plaster application.

Metal reinforcement should also be used when portland cement plaster is applied over either sheathed or unsheathed wood frame construction, steel frame construction, concrete and masonry providing an unsatisfactory bond, and at flashings and chimneys.

Plaster applied over metal reinforcement acts as a thin, reinforced slab which, though supported by the underlying construction, is not in sufficiently rigid contact to permit direct transfer of stresses. The reinforcement should be attached securely to supporting construction and stretched tightly between supports to eliminate slack. Loosely attached reinforcement generally results in an uneven thickness of plaster, with the thinner and weaker sections being more likely to crack. At least a $1/4$ in. (6.4 mm) space should be maintained between the reinforcement and the supporting construction by use of either furring fasteners or self-furring reinforcement (Fig. 9.1-12).

a   b

**FIGURE 9.1-12** (a) Self-furring wire lath is crimped every 3 in. (75 mm); (b) fiber spacers on furring nails keep plain wire lath $1/4$ in. (6.4 mm) away from the backing paper.

The metal reinforcement should be of a suitable type and weight for the support spacing (see Fig. 9.1-4) and should be attached with fasteners recommended for the particular supporting construction (Fig. 9.1-13).

#### TYPES OF REINFORCEMENT

The most common commercially available types of metal reinforcement are *expanded metal lath* and *wire lath* (Figs. 9.1-14 and 9.1-15). Openings in the lath should be large enough to permit the scratch coat to be forced through the openings and to embed the reinforcement, thereby preventing corrosion. Metal reinforcement should conform to the standards listed in Figure 9.2-4 and the weight requirements given in Figure 9.1-15.

*Expanded Metal Lath*  Expanded metal lath is formed from light-gauge steel sheets that have been cut in a regular pattern and expanded (stretched) to form diamond-

**FIGURE 9.1-13** Fastening of Reinforcement Over Wood Supports[a]

| Reinforce-ment | Nails | | Staples | |
| | Type[b] | Maximum Spacing, in. (mm) on Centers | Type | Maximum Spacing, in. (mm) on Centers |
|---|---|---|---|---|
| Diamond mesh | | 6 (150) | | 6 (150) |
| Stucco mesh[c] | | 6 (150) | $7/8$ in. (22.7) leg, 16-gauge (1.6), $3/4$ in. (19.1) crown | 6 (150) |
| Flat rib lath | $1^{1}/_{2}$ in. (38 mm), 11-gauge (3.9), barbed, $7/_{16}$ in. (11) head | 6 (150) | | 6 (150) |
| $3/8$ in. (9.5 mm) rib lath | | 6 (150)[d] | $1^{1}/_{4}$ in. (28.6) leg, 16-gauge (1.6), $3/4$ in. (19.1) crown | $4^{1}/_{2}$ (114) |
| Woven and welded wire | | 6 (150) | $7/8$ in. (22.2) leg, 16-gauge (1.6), $7/_{16}$ in. (11) crown | 6 (150) |

[a]Applicable to both vertical and horizontal surfaces unless otherwise noted.

[b]One-inch roofing nails suitable on vertical surfaces for all but $3/8$ in. (9.5 mm) ribbed lath, $1^{1}/_{2}$ in. (38 mm), 12-gauge (2.7 mm), $3/8$ in. (9.5 mm) head furring nails suitable to attach welded and woven wire on vertical surfaces.

[c]Not recommended for horizontal surfaces.

[d]Maximum spacing $4^{1}/_{2}$ in. (114 mm) on centers for horizontal surfaces using 2 in. (50.8 mm) nails.

**EXPANDED METAL LATH**

STUCCO MESH  DIAMOND MESH  FLAT RIB LATH  3/8" RIB LATH

SELF FURRING

**WIRE LATH**

WELDED WIRE  WELDED WIRE (BACKED)  WOVEN WIRE

**FIGURE 9.1-14** Metal reinforcement suitable for portland cement plaster (stucco) finishes: all but stucco mesh and 3/8 in. (9.5 mm) rib lath are available in plain, self-furring, and paper-backed types.

**FIGURE 9.1-15** Metal Reinforcement for Portland Cement Plaster (Stucco) Finishes

| Type | Weight, lb./sq. yd. (kg/m²) | Opening Size, in. (mm) | Dimensions |
|---|---|---|---|
| **Expanded Metal Lath** | | | |
| Diamond mesh[a] | 3.4 (1.8) | 5/16 × 3/8 (7.9 × 9.5) | 27 in. × 96 in. (686 mm × 2438 mm) |
| Stucco mesh | 1.8 and 3.6 (0.98 and 1.95) | 1 3/8 × 3 1/8 (34.9 × 79.4) | 48 in. × 99 in. (1219 mm × 2515 mm) |
| Flat rib lath | 3.4 (1.8) | 5/16 × 3/8 (7.9 × 9.5) | 24 in. × 96 in. (610 mm × 2438 mm) and 27 in. × 96 in. (686 mm × 2438 mm) |
| 3/8-in. (9.5 mm) rib lath | 3.4 and 4.0 (1.8 and 2.17) | 5/16 × 3/8 (7.9 × 9.5) | 27 in. × 96 in. (686 mm × 2438 mm) and 27 in. × 99 in. (686 mm × 2515 mm) |
| **Wire Lath** | | | |
| Woven wire[a] | 1.7 (0.92) (18 gauge) (1.3 mm) | 1 (25.4) hex. | 3 ft. × 150 ft. (.91 m × 45.7 m) (rolls)[b] |
| | 1.4 (0.76) (17 gauge) (1.45 mm) | 1 1/2 (38) hex. | |
| Welded wire[a] | 1.4 (0.76) (16 gauge) (1.6 mm) | 2 × 2 (50.8 × 50.8) | 3 ft. × 150 ft. (.91 m × 45.7 m) (rolls)[b] |
| | 1.4 (0.76) (18 gauge) (1.3 mm) | 1 × 1 (25.4 × 25.4) | |
| | 1.9 (1) (16 gauge) (1.6 mm) | 1 1/2 × 2 (38 × 50.8) | |

[a]Available in plain, self-furring, and paper-backed types.

[b]Paper-backed and self-furring types also available in sheet form.

(Courtesy Portland Cement Association.)

shaped openings (see Fig. 9.1-14). Both expanded zinc-coated (galvanized) steel sheet and cold-rolled steel sheet coated with corrosion-resistant paint after fabrication are used.

*Plain, self-furring,* and *paper-backed* types of diamond mesh are produced in sheet form. The plain type generally should be installed with furring fasteners. Diamond mesh is produced in weights of 2.5 and 3.4 lb/sq yd; (1.36 and 1.84 kg/m²); only the heavier 3.4 lb (1.84 kg/m²) weight is suitable for exterior portland cement plaster use.

The *self-furring* type is dimpled or contains other devices that hold it away from the surface to which it is attached. These devices should keep the back of the lath at least 1/4 in. (6.35 mm) away from the supporting construction. Self-furring lath can be nailed directly to smooth reinforced concrete, masonry, and other solid surfaces to receive plaster.

*Paper-backed diamond mesh,* with paper strips attached to the back surface, is produced primarily for machine application of exterior portland cement plaster. Spaces are provided between the paper strips to permit fastening to studs or other supporting construction.

*Stucco mesh* is a metal lath similar in pattern to diamond mesh, but with larger openings. It is designed primarily as a base for plaster over solid surfaces. It should be installed with backing paper and furring fasteners.

*Rib lath* is a type of expanded metal lath that consists of integrally formed V-shaped ribs which provide greater stiffness and permit wider spacing of supports (see Fig. 9.1-4). Laths with 1/8, 3/8, and 3/4 in. (3.2, 9.5, and 19.1 mm) ribs are produced (see Fig. 9.1-15). The 1/8 in. (3.2 mm) rib lath referred to as *flat rib lath* generally requires furring fasteners and can be used for most plaster applications. It is available in weights of 2.75 and 3.4 lb/sq yd (1.49 and 1.84 kg/m²), but only the heavier weight is acceptable as a base for exterior portland cement plaster.

The 3/8 in. (9.5 mm) rib lath does not require furring but is not recommended for severe weather exposure because it is difficult to encase the lath in plaster. It can be used, however, in milder climates and in locations protected from direct wetting, such as soffits. Its chief advantage is its ability to provide a base for support spacings up to 24 in. (610 mm) on centers (see Fig. 9.1-4).

**Wire Lath** Wire lath is produced as woven or welded wire fabric. Wire lath should be fabricated from copper-bearing galvanized cold-drawn steel wire of the appropriate gauge (see Fig. 9.1-15). It is used extensively as a base for exterior portland cement plaster.

*Woven wire fabric* (stucco netting) is a hexagonal woven wire fabric made by interweaving wire into hexagonal openings 1 in. (25.4 mm) to $2^1/_4$ in. (57.2 mm) wide (see Fig. 9.1-14). The plain type is nailed to supporting construction with furring nails over backing paper. Woven wire fabric also is available in the self-furring type, with or without backing paper, which eliminates the need for furring fasteners or separate installation of backing paper, depending on the type of lath used. The minimum wire size for 1 in. (25.4 mm) openings should be 18-gauge (1.3 mm); 17-gauge (1.45 mm) for $1^1/_2$ in. (38 mm) openings (see Fig. 9.1-15).

*Welded wire fabric* lath consists of a grid formed by welding steel wire at intersections into 1 in. (25.4 mm) by 1 in. (25.4 mm) and 2 in. (50.8 mm) by 2 in. (50.8 mm) squares. It is available with backing paper and stiffener wires and in a self-furring type. This fabric should be galvanized, coated with corrosion-resistant paint after welding, or fabricated from galvanized wire. Wire should be at least 16-gauge (1.6 mm) for fabric with 2 in. (50.8 mm) by 2 in. (50.8 mm) openings and at least 18-gauge (1.3 mm) for fabric with 1 in. (25.4 mm) by 1 in. (25.4 mm) openings.

## METAL PLASTER ACCESSORIES

In most residential construction, and in many other buildings as well, wood trim often covers the edges of a plaster finish. When wood trim is not used, exposed edges requiring protection from damage should be trimmed with metal casing beads (Fig. 9.1-16a). Control joints (Fig. 9.1-16b) are incorporated to break large areas into self-contained smaller panels, and drip screeds are used as terminating lines at the bottom of walls (Fig. 9.1-16c). Other accessories, such as *strip lath* and *cornerite* (Figs. 9.1-16d and 9.1-16e), are used for reinforcing flat surfaces and internal corners, respectively. External corners often are provided with corner reinforcement (Fig. 9.1-16f) when the metal lath is not flexible enough to be bent around corners.

Corner reinforcement without a solid nosing is designed to permit complete embedment of the metal within the plaster. However, corner beads made with a solid nosing which remains exposed can be used, provided that they are made of white alloy zinc (99% pure zinc) where they will be exposed to wetting. Casing beads should be formed from metal not less than 24-gauge (0.70 mm); other accessories should be at least 26-gauge (0.55 mm). For plaster exposed to wetting, exposed metal accessories (not completely embedded in the plaster) should not be less corrosion-resistant than white alloy zinc.

## INSTALLATION

Reinforcement should be attached to concrete and masonry with suitable fasteners spaced 6 in. (152 mm) on centers vertically and 16 in. (406 mm) on centers horizontally. Horizontal and vertical laps should be secured with fasteners or wire ties 6 in. (152 mm) on centers. Self-furring reinforcement can be attached with powder-actuated fasteners or concrete stub nails; plain reinforcement should be attached with concrete furring nails. These nails should be long enough to provide at least $^3/_8$ in. (9.5 mm) penetration into the concrete or masonry.

Generally, metal reinforcement should be installed over framing or furring with the long dimension perpendicular to the supports and with end laps staggered, forming a continuous network of metal over the entire surface. Expanded metal lath should be lapped $^1/_2$ in. (12.7 mm) at the sides and 1 in. (15.4 mm) at the ends. Edge ribs should be nested. Wire lath should be lapped at the ends and sides at least one full mesh but not less than 1 in. (25.4 mm). Generally, lower sheets should be lapped over upper sheets at horizontal laps; however, paper-backed reinforcement should be applied shingle fashion, with upper sheets sandwich-lapped over the lower sheets with metal on metal.

Metal reinforcement should be attached to wood framing and furring with corrosion-resistant nails or staples, sized and spaced as shown in Figure 9.1-13. Nails generally should provide at least $^7/_8$ in. (22.2 mm) penetration into supporting construction on walls and $1^3/_8$ in. (35 mm) penetration on ceilings. Aluminum nails react chemically with fresh portland

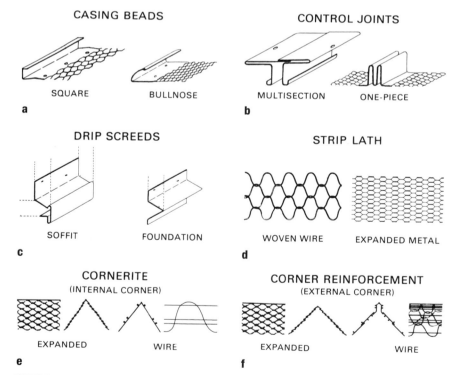

**FIGURE 9.1-16** Typical accessories for portland cement plaster (stucco): (a) Casing beads are available with or without expanded flanges. (b) External corner reinforcement is also available with solid metal nosing (cornerbead) in white alloy zinc. Other accessories are produced to meet special conditions.

cement and should not be used with it. For most types of reinforcement, 1 in. (15.4 mm) or $1^1/_2$ in. (38 mm) nails are generally suitable. However, $^3/_8$ in. (9.5 mm) riblath usually is nailed at the rib, and $1^1/_2$ in. (38 mm) or 2 in. (50.8 mm) nails are recommended to ensure the minimum penetration for walls and ceilings.

Metal reinforcement should be attached to metal framing and furring using screws or clips supplemented by screws. At suspended ceilings, metal reinforcement may be wire-tied to furring members.

When sound ratings are required, metal reinforcement may be attached using resilient clips.

External corners can be formed with *cornerbead*, with prefabricated corner reinforcement (see Fig. 9.1-16e), or by bending the metal reinforcement around the corner (Fig. 9.1-17). Internal corners can be reinforced with metal corner reinforcement (Fig. 9.1-16f) wired to the reinforcement or the metal reinforcement itself can be bent to conform to the angle (Fig. 9.1-18).

### 9.1.2.3 Backing Paper

Metal reinforcement for portland cement plaster should have a backing paper in exterior locations, in wet interior areas and when the plaster will be sprayed on. Backing paper is used to (1) isolate the plaster from the supporting construction, (2) prevent loss of plaster mix through lath openings, (3) facilitate complete embedment of the lath, and (4) resist water penetration through breaks in the plaster membrane. Backing paper is not required for hand-applied plaster on horizontal surfaces protected from direct wetting.

Backing paper should have a vapor permeance of at least 5 perms to permit the plaster to dry and to prevent moisture from becoming trapped within the wall. Waterproof and vapor-permeable products used for sheathing paper, such as 15 lb (8.14 kg/m²) asphalt-saturated roofing felt, are suitable. Backing paper may be factory-attached to metal reinforcement or can be installed separately. Absorptive paper attached to some types of metal reinforcement is not water resistant and is not suitable for exterior applications where weather protection is required.

### 9.1.2.4 Gypsum Lath

Gypsum lath consists of a gypsum core enclosed by a multilayered, fibrous paper

a b

**FIGURE 9.1-17** (a) External corner reinforced by bending lath around the corner; (b) external corner being formed with zinc cornerbead.

**FIGURE 9.1-18** Woven wire lath can be easily bent to form internal corners.

covering designed to ensure a good bond with gypsum plaster. A suction bond is created by tiny needle-like crystals of plaster forming in the porous paper covering. Thicknesses are shown in Figure 9.1-19. Edges are of the rounded type (see Fig. 9.5-3). The governing standard of quality for gypsum lath is ASTM C1396.

### TYPES

The main types of gypsum lath are plain, perforated, foil-backed, and fire-resistant (Type X).

***Plain Lath*** Plain lath is ordinary gypsum lath consisting of a gypsum core and paper covering, with none of the special features incorporated in the other available types.

***Perforated Lath*** Before the development of Type X lath, perforated lath, with $^3/_4$ in. (19.1 mm) holes drilled 4 in. on centers in each direction, was used to improve the lath-to-plaster bond under fire exposure. The covering and core material are the same as for plain lath.

**Figure 9.1-19** Typical Dimensions of Gypsum Lath Products

| Gypsum Lath Type | Thickness, in. (mm) | Width, in. (mm) | Length, ft (mm) |
|---|---|---|---|
| Plain | $^3/_8$ (9.5) | 16, $16^1/_5$, 24 (406, 411, 610) | 4, 8, 12 (1200, 2400, 3600) |
| | $^1/_2$ (12.7) | 16, $16^1/_5$, 24 (406, 411, 610) | 4 (1200) |
| Perforated | $^3/_8$ (9.5) | 16 (406) | 4 (1200) |
| | $^1/_2$ (12.7) | 16 (406) | 4 (1200) |
| Foil-backed | $^3/_8$ (9.5) | 16 (406) | 4, 8 (1200, 2400) |
| | $^1/_2$ (12.7) | 16 (406) | 4 (1200) |
| Type X | $^3/_8$ (9.5) | 16 (406) | 4, 8 (1200, 2400) |

**FIGURE 9.1-20** Application of gypsum lath over framing.

*Foil-backed Lath* To make foil-backed lath, a bright-finish aluminum foil is laminated to the back of plain lath. The foil acts as a vapor retarder and can act as a reflective insulation when the reflective surface faces a $^3/_4$-in. (19.1-mm)-wide dead air space.

*Fire-Resistant (Type X) Lath* Special additives and glass fibers in the gypsum core of Type X lath improve its fire-resistant qualities. Covering materials are the same as for plain lath.

### INSTALLATION

Gypsum lath may be stapled, screwed, nailed, or clipped to supporting construction (Fig. 9.1-20). Gypsum lath should be fastened to wood wall and ceiling framing and furring, using screws or nails, and to metal framing and furring using screws and clips (Fig. 9.1-21). Where sound rat-

ings are required, gypsum lath may be attached using resilient clips.

### 9.1.2.5 Veneer Plaster Base

*Veneer plaster base* is a type of gypsum board with a surface especially formulated to provide a stable base for gypsum veneer plaster.

In general, veneer plaster base should be installed in the same manner as other types of gypsum board (see Section 9.3), and the joints should be taped. Some manufacturers recommend using paper tape; others prefer open mesh glass-fiber tape. Both base coat veneer plaster and joint compounds are used to embed paper tape, depending on the manufacturer of the plaster.

The manner of board placement should produce the least possible number of end joints, which are difficult to conceal.

### 9.1.3 FLASHING

Properly installed flashing to prevent water from entering the construction is essential to durable and satisfactory service of plaster surfaces. Flashing (see Section 7.9) should be provided in all locations where the hazard of water penetration exists. Water entering behind a cement plaster finish could cause corrosion of metal reinforcement and accessories and damage to interior finishes. Only corrosion-resistant materials, such as galvanized steel, copper, stainless steel, and suitable plastics, should be used for flashing. Metal reinforcement should be installed over flashing because these materials do not provide adequate bond for plaster.

**FIGURE 9.1-21** Gypsum plaster lath applied to metal studs using screws, with clips at corners and end joints. (Courtesy United States Gypsum Company.)

# 9.2 Materials for Plaster and Gypsum Board

Both gypsum and portland cement plaster were used extensively in the past in all types of buildings. Recently, however, use of these wet materials has been largely supplanted by the application of dry materials such as gypsum board, mineral-fiber cement products, glass-fiber reinforced concrete, exterior insulation and finish systems, wood, and acoustical tile and panels. Portland cement (stucco) is still used in new work, mostly in exterior applications. Most gypsum plaster used today is in the form of veneer (thin coat) plaster. Conventional plaster applications are addressed here because they are still occasionally used today in new work and in repairing and extending existing plaster.

This section discusses the manufacture and use of materials used in gypsum plaster and board products, materials used in portland cement plaster, and plaster products. Support systems and accessories for plaster are covered in Section 9.1; gypsum plaster mixes and application in Section 9.3; and portland cement plaster mixes and application in Section 9.4. Additional information about the manufacture of gypsum board products and other gypsum building products, as well as gypsum board installation, is provided in Section 9.5.

### 9.2.1 GYPSUM PLASTER AND BOARD MATERIALS

#### 9.2.1.1 Gypsum

**MATERIAL**

Gypsum, a rock-like mineral known chemically as *hydrous calcium sulfate*, is found in many parts of the world, usually combined with impurities such as clay, limestone, and iron oxides. In its pure form gypsum is white, but in combination with impurities it may be gray, brown, or pinkish. Relatively pure calcined gypsum is known as *plaster of paris*, after the huge beds of it that underlie the city of Paris, where it was mined in abundance at the turn of the twentieth century. Gypsum is quarried or mined throughout the United States where deposits are accessible to transportation (Fig. 9.2-1).

**a**                    **b**

**FIGURE 9.2-1** Gypsum is (a) quarried or (b) mined throughout the United States. (Courtesy Gypsum Association.)

Two properties of gypsum make it useful as a plaster and board material: (1) it gives up some of its chemically combined water and becomes a fine powder when intensely heated (calcined), and (2) when water is added, it returns chemically to its original rock-like state by forming interlocking crystals (crystallizing). A mixture of calcined gypsum powder and water remains plastic for a short time and can then be shaped or molded. After it hardens by crystallizing (sets), it forms an effective fire barrier.

Gypsum was used in construction as early as 2000 B.C. Early gypsum plasters, such as plaster of paris, had short, uncontrolled setting times and were difficult to use. Late in the nineteenth century, retarders and accelerators capable of controlling the setting time were added to processed gypsum, expanding the use of gypsum plasters for decorative and fire-resistant purposes.

The first large-scale use of gypsum plaster was in the "white palaces" (exhibition buildings) of Chicago's World Columbian Exposition in 1893. Soon afterward, gypsum board was invented for use as a plaster base, similar to contemporary gypsum lath. It consisted of layers of gypsum sandwiched between plies of felt paper. Further refinements in ingredients, designs, and manufacturing processes resulted in modern varieties of plaster, lath, and other gypsum products.

Gypsum provides fire resistance in the following manner. When the temperature of crystalline gypsum is raised to 212°F

(100°C) water in the form of steam is released. In addition to absorbing some of the heat and dissipating it, the conversion of water to steam prevents the temperature of the gypsum from rising above 212°F (100°C), as long as steam is present. With continued exposure to heat this action progresses inward, releasing more water, and the surface becomes calcined, turning to a white, chalky powder. The calcined layers form an insulator, protecting the underlying uncalcined areas and retarding further calcination.

## MANUFACTURE

Most gypsum building products can be classed as plaster products or board products. Plaster products consist principally of dry, calcined gypsum that requires the addition of water to develop the chemical reaction necessary for setting. Board products are prefabricated at the factory from crystallized gypsum enclosed by paper or vinyl and are delivered to a job site in immediately usable form (see Section 9.5). Figure 9.2-2 shows a manufacturing sequence for plaster and board products.

About two-thirds of all the crude gypsum mined goes into products for building construction. These products are used extensively where fire resistance or sound control is required and where hard and dense wall surfaces are desired. Many new, engineered gypsum products and assemblies combine the advantages of board and plaster products to meet contemporary construction needs.

After gypsum is mined, it is crushed to a 2 in. (50.8 mm) size and passed through a hammer mill, where it is further reduced to a $^1/_2$ in. (12.7 mm) maximum particle size. It is suitable in this form for treating in a rotary calciner, but additional grinding to a fine powder is necessary for treating it in a kettle calciner.

A *rotary calciner* is an inclined steel cylinder lined with fire brick, typically 150 ft (45.72 m) long and 12 to 15 ft (3.66 to 4.57 m) in diameter. As a calciner rotates, the rock inside gravitates to the lower end, while heat is injected to produce a temperature of about 350°F (176.67°C). This heat drives off about three-quarters of the combined water, changing the gypsum into a dehydrated form with an affinity for water.

A *kettle calciner* is a vertical cylinder 10 to 12 ft (3.05 to 3.66 m) in diameter with internal agitators that rotate around a central shaft. As in the rotary calciner, heat is introduced near the bottom, driving off the combined water in the powdered gypsum (see Fig. 9.2-2).

Calcined gypsum from the kettle is further treated in a *tube mill*, a rotary cylinder containing thousands of steel balls. The action of the steel balls produces a finer grind, which contributes to better plasticity and workability.

Calcined gypsum at this point sets too quickly to be suitable for plastering. The type of aggregate used (see Section 9.1.3), local impurities in aggregates and in water, and seasonal variations all affect the setting time. Therefore, the amounts of retarder added vary and are formulated by manufacturers for specific locations and uses to give the desired job set, which is usually a maximum of 4 hours.

Organic fibers are added to fibered gypsum plasters at the mill. Ready-mixed plasters usually are combined with lightweight aggregates before they are packaged and shipped.

### 9.2.1.2 Gypsum Plaster Aggregates

The aggregates commonly used in connection with gypsum products are sand, vermiculite, and perlite.

### SAND

Most sands used in gypsum plaster are primarily quartz and silica, with small amounts of mica, feldspar, clay, and other minerals. Natural deposits from all over

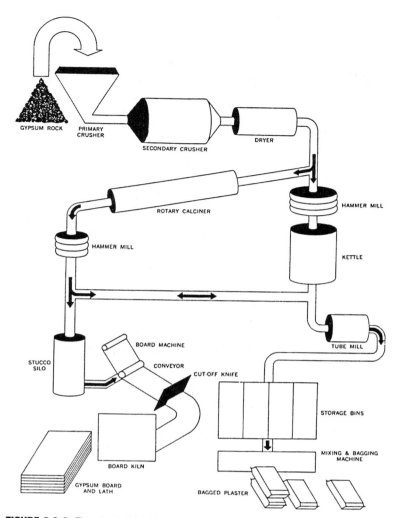

FIGURE 9.2-2 Flowchart of gypsum board and plaster manufacture. (Courtesy Gypsum Association.)

FIGURE 9.2-3 Magnified expanded vermiculite.

the United States are used as the source of plastering sands. The sand is graded according to particle size and is often washed to remove impurities. Artificially produced sand (stone screenings) of the proper size and type is also suitable.

The natural characteristics of sands vary according to location and must be controlled to ensure their suitability as plaster aggregates. Standards limiting size, gradation, and percentage of impurities, and testing procedures to measure these factors, have been established by ASTM International and ANSI. In general, rough, sharp, angular sand particles produce stronger plasters than do smooth, round ones.

### LIGHTWEIGHT AGGREGATES

Lightweight aggregates used in gypsum plaster include vermiculite and perlite.

***Vermiculite***  Vermiculite consists of silica, magnesium oxide, aluminum oxides, and other minerals combined with 5% to 9% water by weight. In the United States, vermiculite ore is mined primarily in Montana and South Carolina. This micaceous mineral is composed of a series of parallel layers which expand under the pressure of water trapped between them when subjected to intense heat (Fig. 9.2-3). The name vermiculite originates from the vermicular (worm-like) movement of the layers during expansion.

Processing consists of crushing the raw ore, grading it according to particle size, and removing impurities. Ore particles then are heated at 1600 to 2000°F (871 to 1093°C) in a furnace for 4 to 8 seconds to produce expansion, and are cooled rapidly to impart pliability and toughness to the expanded particles. As the material cools, it is air-lifted to remove unexpanded rock particles and to classify particle sizes. The expanded material is soft and pliable, silvery or gold

in color, and contains less than 1% water by weight. Expanded vermiculite is classified into five types according to particle size. Type II, the lightest of the standard aggregates, is used in plaster products.

***Perlite***  Raw perlite is a volcanic ore consisting primarily of silica and alumina, with 4% to 6% combined water by weight. It is quarried mostly in the western part of the United States.

Perlite ore is processed by crushing, screening, and grading according to size. It then is expanded in a furnace at 1400 to 2000°F (760 to 1093°C). At this temperature range the glassy ore particles approach their melting point and begin to soften. A small amount of the combined water is converted to steam, which pops the particles into frothy glass-like bubbles up to 20 times the original volume (Fig. 9.2-4). After expansion, an air blast separates and grades the particles according to size.

### 9.2.2 GYPSUM PLASTER PRODUCTS

Powdered, calcined gypsum is the prime raw material in gypsum plaster products and in the cores of board products. It acts as a binder. Other materials used are aggregates, mineral or organic fibers, and lime. Aggregates are fillers that provide dimensional stability, and may also provide increased hardness and fire resistance in some plasters. Wood or other organic fibers are sometimes included in plasters to control their working quality and their keying action with metal lath. Lime putty is added in finish coat plasters to increase bulk and improve workability. Proprietary chemical retarders and accelerators usually are introduced at the plant to control setting time.

This discussion is limited to a general description of the properties and uses of

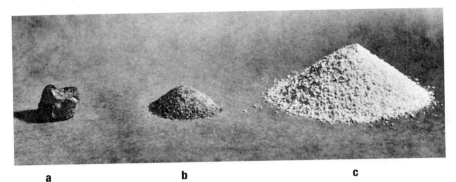

**FIGURE 9.2-4** Perlite: (a) crude ore, (b) crushed, (c) expanded.

packaged plaster products (*bagged goods*) and does not describe the many plasters which may be produced on the job for specific applications. Packaged plaster products and job-prepared plaster mixes are both referred to generally as *plasters*.

*Packaged gypsum plaster products* consist primarily of powdered gypsum and additives to control setting time and working quality. These products also may contain aggregates and fibers. *Unfibered neat plasters* generally require the addition of aggregates on the job. Ready-mixed basecoat and prepared gypsum finish plasters contain all of the necessary ingredients, including aggregates. Adding water to properly proportioned ingredients produces plasters suitable for application over appropriate plaster bases.

Gypsum plasters are suitable for interior plastering in locations not subject to severe moisture conditions. They can also be used in exterior locations like the ceilings of open porches and carports or for eave soffits, where they are not directly exposed to water. They are applied over plaster bases such as gypsum lath, metal lath (sheet, expanded, welded, or woven wire), and masonry or concrete bases.

Plasters used for scratch and brown coats are called *basecoat* plasters; those used for the final coat are called *finish coat* plasters. The governing ASTM standards, and the physical and working properties of basecoat and finish coat plasters, are summarized in Figure 9.2-5.

### 9.2.2.1 Gypsum Plaster Types

#### BASECOAT PLASTERS

*Basecoat* plasters are classed according to ingredients as *neat, wood-fibered, ready-mixed,* and *bond* plasters. Basecoat plasters are formulated for hand-trowel or machine applications.

*Neat Plaster*   *Neat* gypsum basecoat plaster does not contain aggregates. It is used where adding aggregates on the job is desired, such as where a specially formulated plaster is required to match an existing plaster or where the typical aggregate in ready-mix plaster is too fine for the desired texture. It may be obtained lightly fibered with cattle hair, sisal, or other fine organic fibers, or specially formulated for use with lightweight aggregates such as vermiculite or perlite.

*Wood-Fibered Plaster*   *Wood-fibered plaster* consists of calcined gypsum mixed with selected coarse cellulose fibers which provide bulk and greater coverage. It is formulated to produce high-strength basecoats for fire-resistant ceiling assemblies.

*Bond Plaster*   In addition to gypsum, bond plaster contains from 2% to 5% lime by weight and chemical additives that improve its bond with dense nonporous surfaces, such as concrete. Bond plaster is seldom used today because of a number of proprietary bonding agents that permit the application of standard plasters to nonporous surfaces.

*Ready-Mixed Plaster*   *Ready-mixed plaster* is a mill-prepared plaster consisting of calcined gypsum and an aggregate (usually perlite, but sometimes sand or vermiculite) and requires only the addition of water. This plaster possesses all the features and advantages of neat plaster mixed on the job, but offers mill control of the quality, gradation, and proportioning of aggregates.

### FINISH COAT PLASTERS

The finish coat is the final plastering operation. It may serve as a base for further decorating or as a final decorative surface. *Finish coat plasters* usually consist of calcined gypsum, lime, and sometimes an aggregate. Some finish coat plasters may require the addition of lime or sand on the job.

*Gauging Plaster*   *Gauging plaster* consists of coarsely ground gypsum of low consistency which soaks up mixing water readily and blends well with lime putty. This plaster controls (gauges) the setting time of plasters when mixed with lime putty (which by itself does not set). Gauging plaster is available as white plaster, which produces a brilliant white finish when gauged with lime putty, and local plaster, which may be somewhat darker in color.

*Gypsum Keene's Cement Plaster*   *Gypsum Keene's cement* plaster is similar to gypsum gauging plaster, but unlike ordinary gypsum plasters, gypsum Keene's cement is burned in a kiln instead of a calciner or kettle. In this burning process, most of the combined water is removed, producing a *dead-burned* gypsum. The resulting material is denser than ordinary gypsums and has greater impact, abrasion, and moisture resistance. Select white gypsum rock is used to produce the hardest white gypsum available.

Gypsum Keene's cement plaster has an initial set of about $1^{1}/_{2}$ hours and a final set of 4 to 6 hours. A quick-setting gypsum Keene's cement which provides a final set of about 2 hours is also available.

*Gypsum Keene's cement-lime putty-trowel plaster* produces a dense, hard finish.

*Gypsum Keene's cement-lime-sand float plaster* is a highly crack- and spall-resistant plaster.

*Prepared Gypsum Finish Plaster*   *Prepared gypsum finish plaster* is a ready-mixed plaster that includes all necessary ingredients except water. It consists of finely ground, calcined gypsum with or without fine aggregates. It is available as a white or gray *trowel* finish and as *sand float, colored float,* or *textured* finishes. These finishes are comparable in hardness and appearance to those obtained with job-mixed Keene's cement, but factory blending of ingredients ensures somewhat better results.

**FIGURE 9.2-5** General Properties of Gypsum Plasters

| Basecoat Plasters | | | | | |
|---|---|---|---|---|---|
| **Plaster Mix** | **Standard** | **Compressive Strength—Dry, psi (g/m²)** | | **Average Tensile Strength—Dry, psi (g/m²)** | **Setting Time** | **Characteristics** |
| | | **Average** | **Minimum** | | | |
| Neat (with 2 parts sand) | ASTM C28 | 900 (633) | 750 (527) | 175 (123) | 2 to 15 hrs.[a] | Greater versatility because aggregates and/or fibers are job-mixed; less control of plaster quality |
| Ready mixed (with perlite) | ASTM C28 | 675 (475)[b] | 400 (281) | 155 (109) | 1½ to 8 hrs.[a] | 3½ times better thermal insulator than sanded plaster; closely controlled quality |
| Wood fibered | ASTM C28 | 2100 to 2500 (1476 to 1758) | 1200 (844) | 400 to 500 (281 to 352) | 1½ to 9 hrs.[a] | May be used without aggregate; about 50% greater surface hardness than sanded plasters; generally higher strength and fire resistance |
| Bond | ASTM C28 | 2300 (1617) | — | 400 (281) | 2 to 9 hrs. | Bonds with nonporous surfaces such as concrete |

| Finish Coat Plasters | | | | | |
|---|---|---|---|---|---|
| **Plaster Mix** | **Standard** | **Compressive Strength—Dry, psi (g/m²)** | | **Setting Time** | **Characteristics** |
| | | **Average** | **Minimum** | | |
| Gauging (with lime putty) Regular | ASTM C28 | 2000 to 3000 (1406 to 2109) | 1200 (844) | 20 to 40 min. | Hard surface; good working properties |
| Special high strength | ASTM C28 | Variable | 5000 (3515) | 20 to 40 min. | |
| Gypsum Keene's cement (with lime putty) | ASTM C61 | 4000 to 5000 (2812 to 3515) | 2500 (1758) | 20 min. to 6 hrs. | Very hard surface; becomes harder with decreased proportions of lime; less susceptible to moisture; good working properties; can be retempered; can be polished |
| Prepared | None at present | Variable[c] | Variable[c] | Variable[c] | Moderately hard surface; no alkali reaction with paint; not as workable as gypsum–lime putty plasters |
| Acoustical | None at present | Variable[c] | Variable[c] | Variable[c] | Moderately soft; better sound absorption than conventional plasters |

[a]Four-hour maximum for ASTM C842 Standard Specification for Gypsum Plastering.

[b]Not applicable to ready mix plasters for masonry bases.

[c]Proprietary plaster mixes containing different ingredients, which have variable effects on strength and setting time of plasters.

(Courtesy Gypsum Association.)

**Veneer (Thin-Coat) Plaster** Veneer plaster is a high-strength plaster for application as a $\frac{1}{16}$-in. (1.6-mm)- to $\frac{1}{8}$-in. (3.2-mm)-thick one- or two-coat finish over a special plaster base produced specifically for this purpose. Veneer plaster is a dominant type of finish in many parts of the country because it is economical and speeds up job completion.

**Molding Plaster** Molding plaster is a finely ground gypsum plaster used for casting in molds and for ornamental plaster work. Its fineness is necessary for the smooth, intricate surfaces and close tolerances common in ornamental work.

**Acoustical Plasters** Acoustical plasters are proprietary plasters intended to increase sound absorption to reduce sound reverberation. They consist usually of ground gypsum with chemical ingredients which encourage the formation of air bubbles that increase the available surface area so that more sound is absorbed.

Some products touted as acoustical plasters today are actually other plaster products with finishes made to simulate the appearance of acoustical plaster. These may provide some sound absorption due to their rough surface, but they are not true acoustical plasters.

### 9.2.3 PORTLAND CEMENT PLASTER MATERIALS

Portland cement plaster is used in both interior and exterior locations where it will be exposed to direct wetting or high moisture. In much of the country, exterior portland cement plaster is called *stucco*, while interior portland cement plaster is called simply *plaster*. In some regions, however, especially along the West Coast, both interior and exterior portland cement plaster is referred to as *stucco*. Portland cement plaster is also used as a parging coat over masonry. *Parging* is often applied below ground to increase a wall's resistance to moisture penetration and to serve as a base and leveling coat to receive dampproofing and waterproofing.

The ingredients of portland cement plaster are essentially the same as those of concrete and of portland cement mortar. These include cementitious materials, which act as a binder; aggregate, which acts as an inert filler and provides strength; water, which, together with the cementitious materials, forms a paste and initiates the reaction that binds the aggregates together into a solid mass; and admixtures, which improve the plasticity or cohesiveness of the mix or retard or accelerate the setting time.

Portland cement plaster requires the use of metal reinforcement and accessories for edge protection and crack control. Exterior portland cement plaster (stucco) also employs backing paper (or felt) and flashings.

#### 9.2.3.1 Cementitious Materials

The cementitious materials used in portland cement plaster are portland cement, masonry cement, plastic cement, and lime. These products should conform to the requirements of the standards listed in Figure 9.2-6. Lime is discussed in Section 9.2.4.

Finishes made with plastic cement or masonry cement, or with portland cement mixed with less than 50% by volume of lime, are called *portland cement plaster*. When lime is added to portland cement in amounts greater than 50% by volume, the finish is referred to as *lime-cement* (or *portland cement-lime plaster*). The principles and methods described here apply to both, whether used on the interior or the exterior.

### PORTLAND CEMENT

The types of portland cement suitable for plaster (see Fig. 9.2-6) are also used in concrete and are discussed in detail in Chapter 3.

Type I (normal) portland cement is used for most portland cement plaster work, except where the special properties of Types II and III are required. Types IA, IIA, and IIIA (air-entraining) portland cement provide improved workability, as well as resistance to bleeding (concentration of water and fines at the surface) and to damage from freezing.

Type I is available in natural gray or white. Truer colors can be obtained in finish coats containing integral coloring or colored aggregate by using a white portland cement. *Waterproof* white portland cement is also available. It contains moisture-resistant additives to increase its inherent moisture resistance. These products must contain lesser amounts of iron oxides than standard portland cement, but in all other respects should conform to the requirements of ASTM C150 for Type I portland cement.

Type II (moderate heat) portland cement has greater resistance to decomposition from the action of sulfates and lower heat of hydration. Resistance to sulfate attack may be a useful property when portland cement plaster is applied to brick or CMU containing soluble sulfate materials.

Type III (high early strength) portland cement develops strength faster in the first few days after application. Therefore, Type III may be desirable in cold weather to minimize the hazard of freezing damage and to speed application of successive coats.

***Masonry Cement and Plastic Cements***
Masonry cement and plastic cements consist primarily of portland cement with varying amounts of such materials as natural cements, finely ground limestone, air-entraining agents, plasticizers, which make the field addition of lime unnecessary, and sometimes lime. They are used extensively for portland cement plaster finishes in some parts of the country.

Plastic cement is produced primarily in the southwestern United States. It is similar in properties and formulations to Types I and II portland cement, but may contain larger amounts of air-entraining agents, insoluble residues, and plasticizing agents not in excess of 12% of the volume. With the exception of these variations, plastic cement should conform to the requirements of ASTM C150 for Types I and II portland cement.

***Prepared Finish Coat*** Factory-prepared mixes (stucco finishes) are available for application over base coats. These prepared mixes contain all the necessary finish-coat ingredients except water. They are available in white or a variety of colors and, in general, provide more satisfactory results than those obtained by the addition of pigments in the field.

#### 9.2.3.2 Aggregates

Aggregates constitute most of the volume of a portland cement plaster mix. One cubic foot of a typical portland cement plaster mix contains about 0.97 cu ft (0.007 m$^3$) of aggregate and 0.25 to 0.33 cu ft (0.007 to 0.009 m$^3$) of cementitious material. Therefore, the aggregate must be clean, sound, and well graded, with

**Figure 9.2-6** Summary of Portland Cement Plaster Materials

| Material | Standard |
|---|---|
| **Cementitious Material** | |
| Portland cement | |
|    Types I, II, III | ASTM C150 |
|    Types IA, IIA, IIIA | ASTM C150 |
| Masonry cement | ASTM C91 |
| Plastic cement | ASTM C150 |
| Lime, hydrated* | |
|    Special finishing | ASTM C206 |
|    Normal finishing | |
| Mason's, Types S and N | ASTM C207 |
| Quicklime | ASTM C5 |
| **Aggregates** | |
| Sand | ASTM C144 |
| Lightweight | ASTM C332 |
| **Metal Reinforcement** | ASTM C847 |
| **Backing Paper** | |
| Roofing felt | ASTM D226 |
| Building paper | FS-UU-B-790 |

\* Both special and normal limes are available as air-entraining types (SA and NA).

(Courtesy Gypsum Association.)

particles ranging from fine up to a maximum size of $1/8$ in. (3.18 mm).

The best and most practical mix results when the aggregate size is proportioned so that a minimum of water-cement paste is required to coat the aggregate and fill the voids between particles. An excess of either fine or coarse particles can result in a mix containing too much cement or too much water, either of which can be detrimental to a portland cement plaster finish. Portland cement plaster mixes are discussed in Section 9.4.1.

### SAND

Sand aggregate for portland cement plaster should consist of natural or manufactured sand which is clean, granular, and free from deleterious amounts of loam, clay, soluble salts, or vegetable matter. It should be well graded within the limits given in Figure 9.2-7 and, with the exception of grading, should conform to the requirements of ASTM C144.

### LIGHTWEIGHT AGGREGATE

Expanded shale, clay, slate, and slag sometimes are used where sand is not available locally. Vermiculite and perlite have been used in portland cement plaster where light weight, insulating properties, or fire resistance are prime factors. Proportioning of these materials is determined by laboratory testing or by making test panels. Thermal insulating properties, gradation, and unit weight for lightweight aggregate should conform to the requirements of ASTM C332.

### 9.2.4 LIME

Lime is used in both gypsum and portland cement plasters. It is obtained principally

**FIGURE 9.2-7** Grading for Sand Aggregate

| Sieve Size | Percent of Aggregate by Weight Retained on Sieve | |
|---|---|---|
| | Minimum | Maximum |
| No. 4 | — | 0 |
| No. 8 | — | 10 |
| No. 16 | 10 | 40 |
| No. 30 | 30 | 65 |
| No. 50 | 70 | 90 |
| No. 100 | 95 | 100 |
| No. 200 | 97 | 100 |

from limestone, a common mineral consisting of calcium carbonate (calcite) and magnesium carbonate (dolomite). To produce building lime, crushed limestone is calcined in a kiln at 2000°F (1093°C). Limestone used to make lime for construction purposes comes from dolomitic deposits in Ohio or calcitic deposits in Arizona, California, Missouri, Pennsylvania, Texas, and Virginia.

The physical and chemical properties of lime depend on the limestone from which it was derived. Building limes may be manufactured from *dolomitic* limestone, which normally contains substantial amounts of magnesium carbonate, or from *calcitic* limestone, which is naturally high in calcium carbonate. When dolomitic limestone is calcined during manufacture, magnesium oxides result. These oxides produce a plastic putty with better working properties than those of the putty made from the calcium oxides derived from calcitic limestone, but it takes longer to hydrate than the putty made with calcium oxides.

Building limes can be classed as (1) *quicklime* or (2) *hydrated lime*.

### 9.2.4.1 Quicklime

During the calcining process, limestone gives off carbon dioxide and changes from the carbonate form to calcium oxide or magnesium oxide. *Quicklime* was used as the chief binder in lime plasters for several hundred years before the introduction of controlled-setting gypsum plasters.

Before quicklime can be used in plaster, it must be *slaked* (combined with water) and soaked for 1 to 2 weeks to hydrate its oxides and make a plastic putty usable in mortar or plaster. Slaking initiates a chemical reaction which transforms oxide to a more stable hydroxide state. The type of raw material affects the degree and speed of hydration possible during this soaking period.

Quicklime identified as calcitic or high-calcium can be expected to have greater hydration of oxides, and is acceptable for use in plaster only if it is identified as *calcitic* or *high-calcium* and if it is properly slaked as recommended by its manufacturer. For this reason, some quicklime is sold in the natural pulverized form, but to eliminate the slaking operation, most quicklime is partially hydrated (slaked) at the mill.

### 9.2.4.2 Hydrated Lime

The material produced by partially hydrating quicklime at the mill is called *hydrated lime*. The addition of water to hydrated lime results in *lime putty*, a plastic form which is combined with gypsum and water to make some finish coat plasters.

If oxides are not completely hydrated during soaking or during the manufacture of hydrated lime, they may be present in the plaster. Under conditions of high humidity they could hydrate there, expand, and crack the plaster containing them. Therefore, dolomitic limes often are autoclaved at the mill to ensure that a minimum of unhydrated oxides remain in the manufactured hydrated lime. To be suitable for use in finish coat plasters, lime must have no more than 8% unhydrated oxides.

Available hydrated limes include *mason's* hydrated lime, intended primarily for masonry mortars, and *finishing* hydrated lime intended for use in plasters. Both are available in Type S (special), Type SA (special air-entraining), Type N (normal), and Type NA (normal air-entraining).

### TYPE N (NORMAL) HYDRATED LIME

Type N hydrated lime is produced in accordance with the requirements of ASTM C206, except that it has no limitation on unhydrated oxides. However, when calcitic limestone is used in its manufacture, the resulting product (identified as calcitic or high-calcium finishing hydrated lime) consists of at least 95% calcium oxide, which hydrates readily during manufacture. Some oxides may remain unhydrated, but these can be expected to hydrate substantially during the required 12- to 16-hour preuse soaking period for Type N limes. Often, high-calcium finishing lime will meet ASTM limitations for unhydrated oxides for Type S lime but may not develop the necessary degree of plasticity within the time limit established by ASTM.

Type N hydrated finishing lime produced in accordance with ASTM C206 is acceptable for use in plaster if it is identified as calcitic or high-calcium and if it has been properly soaked as recommended by its manufacturer.

### TYPE S (SPECIAL) HYDRATED LIME

Type S (special) finishing hydrated lime conforming with the requirements of

ASTM C206 requires no preliminary soaking and can be mixed with other ingredients before water is added. Because it is easy to use, and because unhydrated oxides are limited during manufacture, Type S lime most often is used and is specifically recommended for plaster mixes.

Type S hydrated lime conforming to ASTM C206 must contain not more than 8% unhydrated oxides, regardless of the raw materials used.

## 9.2.5 WATER

Water for mixing and curing should be clean and free from deleterious amounts of oils, acids, alkalis, salts, or organic substances that could adversely affect proper hydration or cause corrosion of metal reinforcement and accessories.

The presence of water promotes hydration of portland cement and carbonation of lime, the chemical processes by which these materials harden. However, more water generally is added to a mix than is necessary to provide for hydration and carbonation in order to bring the mix to a workable consistency. The amount of water added for workability depends on the characteristics of the cementitious material, gradation of aggregate, suction of the supporting construction, and drying conditions.

## 9.2.6 ADMIXTURES

Admixtures called *plasticizers*, such as mineral, animal, vegetable, or glass fibers, sometimes are added to a plaster mix to improve its plasticity. These plasticizers can be added to the first (*scratch*) coat where greater cohesiveness of the mix is required, particularly for ceilings and soffits, and to reduce the amount of mix lost through openings in metal reinforcement when backing paper is not used. Mineral fibers also may be used for machine-applied base coats to prevent segregation of the aggregate and to provide lubrication in the hose.

Fibers should be clean and free from harmful amounts of any substance, such as unhydrated oxides, that may be injurious to the plaster ingredients. Animal hair and vegetable fibers should be $1/2$ to 2 in. (12.7 to 50.8 mm) in length and free from grease, oil, dirt, and other impurities.

Other admixtures intended to accelerate or to retard the setting time generally are not recommended, but they may be used where local practice indicates satisfactory results. When used, admixtures should not reduce the compressive strength of the mix more than 15% below the strength of a comparable mix without admixtures.

# 9.3 Gypsum Plaster Mixes and Application

Gypsum plaster mixes and application methods vary with different gypsum plaster products and the bases to which they will be applied.

## 9.3.1 MIXES

The materials used in gypsum plaster include (1) powdered, calcined gypsum, (2) aggregates, (3) mineral or organic fibers, and (4) lime. The last three may be included in a packaged plaster product or added in the field to neat plaster (see Section 9.2.2)

### 9.3.1.1 Basecoat Plaster Mixes

Basecoat plasters include *neat plaster*, *wood-fibered plaster*, and *ready-mixed plaster* (see Section 9.2.2).

#### NEAT PLASTER

Neat gypsum basecoat plaster usually is combined on the job with aggregates in proportions of 1 part plaster to 2 or 3 parts aggregate by weight.

#### WOOD-FIBERED PLASTER

Although wood-fibered plaster is often used without aggregates, 1 cu ft (0.03 m³) of sand is recommended for each 100 lb (45.4 kg) of wood-fibered plaster to obtain a base coat of superior hardness, strength, and added crack resistance or for use over masonry.

#### READY-MIXED PLASTER

Ready-mixed plaster is used as it comes from the bag with only the addition of water.

### 9.3.1.2 Finish Coat Plaster Mixes

Finish coat plasters usually consist of calcined gypsum, lime, and sometimes an aggregate. Some finish coat plasters require the addition of lime or sand on the job.

#### GAUGING PLASTER

Different *gauging plasters* provide either a quick set or a slow set. This eliminates the need for adding accelerators or retarders in the field.

#### GYPSUM-LIME PUTTY-TROWEL PLASTER

*Gypsum-lime putty-trowel plaster* is a low-cost finish plaster that leaves a smooth surface suitable for most uses. The field mix consists of 100 lb (45.4 kg) of gauging plaster to 200 lb (90.7 kg) of dry hydrated lime. Although it is generally used without aggregates, adding fine silica sand or perlite fines in an amount of at least $1/2$ cu ft (0.03 m³) per 100 lb (45.4 kg) of dry gauging plaster will increase the crack and crazing resistance of the finish when it is used over basecoat plasters containing lightweight aggregates.

#### GYPSUM KEENE'S CEMENT-LIME PUTTY-TROWEL PLASTER

This plaster does not require aggregates except when it is applied over a lightweight aggregate base coat. In that case, at least $1/2$ cu ft (0.03 m³) of fine sand or perlite is added to each 100 lb. (45.4 kg) of gypsum Keene's cement and 50 lb (22.7 kg) of hydrated lime; an extra-hard finish can be obtained by reducing the lime amount to 25 lb (11.3 kg).

#### GYPSUM KEENE'S CEMENT-LIME-SAND FLOAT PLASTER

This type of plaster is usually mixed with sand. The resulting textured surface may be painted or an integral coloring agent may be introduced in the plaster. For a float finish, 150 lb (68 kg) of gypsum Keene's cement is mixed with 100 lb (45.4 kg) of hydrated lime and 400 to 600 lb (181 to 272 kg) of silica sand, depending on the texture desired.

#### PREPARED GYPSUM FINISH PLASTER

Prepared gypsum finish plaster requires only the addition of water at the job site.

#### MOLDING PLASTER

Most molding plaster mixes consist of 100 lb (45.4 kg) of molding plaster to 50 lb (22.7 kg) of dry hydrated lime.

## 9.3.2 APPLICATION

The minimum thickness and number of coats required for gypsum plaster depend on the plaster base. Three coats are required over metal lath and some gypsum lath installations. Two coats are adequate over most gypsum lath and over masonry and concrete bases.

### 9.3.2.1 Three-Coat Work

Three-coat work consists of a *scratch coat*, a *brown coat*, and a *finish coat*, each applied separately and allowed to partially set before the next coat is applied. The scratch coat derives its name from the cross-raking of the wet surface to improve the bond with the following brown coat (Fig. 9.3-1).

### 9.3.2.2 Two-Coat Work

In two-coat work, the scratch coat is applied as in three-coat work, but is not cross-raked, and the brown coat is applied immediately, before the scratch coat has set. The finish coat is applied after the basecoat has set and is partially dry (Fig. 9.3-2).

**FIGURE 9.3-1** A scratch coat is raked for three-coat work. (Courtesy Portland Cement Association.)

### 9.3.2.3 Application Methods

Basecoat plasters are formulated for hand-trowel or machine applications. There are three basic methods of applying finish coat gypsum plasters: trowel, float, and spray. The method used determines the surface appearance (Fig. 9.3-3).

Trowel finishes are flat, smooth, and nonporous. They often form the base for further decorating (Fig. 9.3-4a). When an aggregate is used, it is fine silica sand or perlite. Float finishes produce surface textures (Fig. 9.3-4b). The coarseness of the texture depends on the aggregate, mix proportions, and type of float used.

### 9.3.2.4 Composition of Coats

The composition of base and finish coats varies considerably. The term *basecoat* includes scratch coats and brown coats.

#### BASECOATS

Gypsum neat plaster is suitable for use as a basecoat over metal or gypsum lath, gypsum or clay tile, CMU, and brick of moderate suction (see Section 4.4). Fibered neat plaster often is used as a scratch coat over metal lath because less plaster is lost as mechanical keys are formed behind the lath. When neat plaster is used as a brown coat, sand usually is added to prevent excessive shrinkage.

Wood-fibered plaster may be used over the same bases as neat plaster, but it has greater fire resistance than neat plaster and other sanded basecoat plasters.

Ready-mixed plaster with a composition similar to that of gypsum neat plaster or wood-fibered plaster may be applied where those two plaster types are acceptable.

a

b

c

**FIGURE 9.3-3** Finish coat plasters may be smooth trowel finish or one of the following: (a) float finish, (b) spray finish, (c) texture finish. (Courtesy United States Gypsum Company.)

#### FINISH COATS

Either trowel or float finishes may be obtained with *gauging plaster*, *gypsum Keene's cement plaster*, prepared *gypsum finish plaster*, or *veneer (thin-coat) plaster*. *Acoustical plaster* may be hand-applied but is usually machine-sprayed (Fig. 9.3-4d).

Gypsum Keene's cement is used for gauging lime finish coat plaster, much the

a

b

**FIGURE 9.3-2** A Brown coat (a) is applied over a scratch coat, followed by (b) a finish coat in separate operations for three-coat work.

**a**      **b**

**c**      **d**

**FIGURE 9.3-4** Methods of applying finish coats: (a) hand trowel, (b) power trowel, (c) hand float, (d) machine spraying.

from 0.40 to 0.60 may be obtained with these products (see Section 13.1). Surface textures vary from fine to rough, and crack resistance is high. Indentation resistance, however, is much lower than with other plaster products, which limits the use of acoustical plasters to ceiling and upper wall areas not subject to abrasion or impact. Acoustical plasters may be left in their natural off-white or gray color or may be spray painted.

### 9.3.2.7 Application Over Framing and Furring

Gypsum plaster may be applied to either gypsum or metal lath over wood or metal framing or furring.

**a**      **b**

**c**

**FIGURE 9.3-5** Hollow plaster partitions: (a) With channel studs. (b) With trussed studs. Trussed studs are seldom used in new work today, but may be found in many existing buildings. (c) With braced double studs. Double-stud construction accommodates mechanical equipment and plumbing.

same as gauging plaster, but produces a harder, more moisture-resistant surface. Hardness is proportionate to the amount of gypsum Keene's cement. Both trowel and float finishes are possible with gypsum Keene's cement.

Gypsum Keene's cement-lime-sand float plaster may be applied over all types of basecoat plasters but is particularly recommended over lightweight aggregate basecoat plasters. It provides greater crack resistance than a trowel finish.

#### 9.3.2.5 Veneer (Thin-Coat) Plaster

Veneer plaster may be applied over conventional framing and furring (Sections 9.1 and 9.3.2.7), as part of specialized partition systems and assemblies (Section 9.2.3.8), and in most locations where gypsum board assemblies are applicable (Section 9.5). In assemblies where gypsum board is the outer layer, veneer plaster base is substituted. The rest of the assembly remains as described. Care must be exercised to ensure that fire or acoustical rate assemblies still comply with these requirements after veneer plaster

has been substituted for a layer of gypsum board.

Before veneer plaster is applied, joints in the veneer plaster base are treated with a glass fiber tape or reinforced with drywall paper tape and a compatible joint compound, and corner beads are installed. Because of the composition and thinness of the plaster coat, it dries rapidly and may be decorated 24 hours after installation. Refer to Section 9.5 for additional information.

Veneer plaster can be applied as a smooth trowel finish or as a textured, sprayed finish. Both are extremely hard and abrasion-resistant. Veneer plaster also may be used as a finish coat over bond plaster or as a base coat for gypsum-lime and gypsum Keene's cement lime finishes with exceptional strength and crack resistance.

#### 9.3.2.6 Acoustical Plasters

Depending on their formulation, acoustical plasters may be hand- or machine-applied, but most are applied by machine. Noise reduction coefficients ranging

## PARTITIONS

Gypsum plaster finished partitions may be framed with either wood or metal studs 16 to 24 in. (406.4 to 609.6 mm) on centers and supported by wood or metal floor and ceiling runners. Wood studs normally are 2 × 4s (50 × 100s) or 2 × 6s (50 × 150s). Metal channel studs are available in $1^5/_8$ in. (41.3 mm), $2^1/_2$ in. (63.5 mm), $3^5/_8$ in. (92.1 mm), 4 in. (101.6 mm), and 6 in. (152.4 mm) sizes (Fig. 9.3-5a). Greater wall cavities can be developed by using two sets of studs braced with allboard (Fig. 9.3-5c).

Hollow lath and plaster partitions have gypsum lath screwed, clipped, or nailed to studs, or metal lath wired, nailed, clipped, or stapled to studs. Two or three coats of plaster are applied to this plaster base.

## CEILINGS

Gypsum plaster is sometimes applied to ceilings over gypsum lath, but most installations are over metal lath (see Section 9.1).

### 9.3.2.8 Specialized Partition Systems and Other Assemblies

In addition to application over gypsum or metal lath applied to standard framing or furring (Section 9.1), gypsum plaster is used in proprietary assemblies of framing, lath, board, and plaster with metal accessories designed for each system. A variety of such systems have been developed by gypsum product manufacturers to minimize the number of operations and speed up partition erection. The chief types using gypsum plaster are (1) solid and (2) semisolid partitions. A third type, hollow and demountable (movable) systems, are designed for gypsum board products and do not include gypsum plaster.

## SOLID PARTITIONS

Solid partitions are field-erected from plaster on metal or gypsum lath. These include wood or proprietary metal ceiling and floor runners to stabilize the partitions. In lath and plaster partitions, an integral flush metal base acts as a screed. Partitions also may be trimmed out with a resilient or wood base.

Lath and plaster partitions use self-supporting ribbed lath (Fig, 9.3-6a) or ordinary expanded mesh wired to steel bars or channels (Fig. 9.3-6b). To this base, three coats of plaster are applied to each side in thicknesses varying from 1 in. to $1^1/_4$ in. (31.75 mm). Solid plaster partitions also may be made of $^1/_2$-in. (12.7-mm)-thick gypsum lath, with two $^3/_4$-in. (19.05-mm)-thick plaster faces built up of two or more coats (Fig. 9.3-6c).

Other solid partitions consist of a backing board, a layer of veneer plaster base board, and a veneer plaster finish coat. These are similar to the single and double solid wallboard partitions shown in Figure 9.5-43.

## SEMISOLID PARTITIONS

Semisolid partitions similar to the gypsum board semisolid partitions shown in Figure 9.5-42 are not usually constructed with conventional plaster, but veneer plaster base may be substituted for the outer layer of gypsum board and finished with veneer plaster.

## SPECIAL CEILING ASSEMBLIES

Gypsum plaster may be used in some proprietary electric-radiant heating systems. In this case, the heating cables are attached to or through standard gypsum lath and are then embedded in the plaster. Such installations should be made strictly in accordance with the instructions of the manufacturer of the heating panels (see Section 9.5).

# 9.4 Portland Cement Plaster Mixes and Application

Portland cement plaster is used on both the interior and the exterior. It can be used where it will be exposed to wetting.

## 9.4.1 GENERAL REQUIREMENTS

Portland cement plaster on exteriors, and in such interior spaces as shower rooms, steam baths, saunas, laundries, tub-shower enclosures, commercial kitchens, and similar locations, is exposed to hazards not normally present in other interior applications. These hazards include

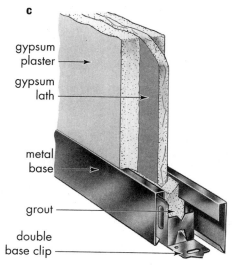

a — gypsum plaster, rib lath, metal base, double base clip

b — gypsum plaster, metal lath, $^3/_4$" cold-rolled channel, metal base, double base clip

c — gypsum plaster, gypsum lath, metal base, grout, double base clip

**FIGURE 9.3-6** Solid plaster partitions: (a) on self-supporting ribbed lath, (b) on stud-supported woven wire lath, (c) on gypsum lath.

water penetration, which may corrode reinforcement and accessories, and stresses due to variations of temperature and humidity. The presence of such hazards requires different practices and precautions than those needed in less hazardous applications.

A portland cement plaster membrane is, in effect, a thin concrete slab, built up with plastering tools and methods and having many of the desirable properties of concrete. Therefore, the discussion of portland cement in Chapter 3 will be helpful in understanding the data in this section.

Properly applied portland cement plaster membranes are hard, strong, and decorative. They can be formed to almost any shape, and can be finished with a wide variety of surface textures and colors. They are resistant to fire, weather, and fungus attack. They are used predominantly in the southern, southwestern, and western parts of the country but have proven to be durable and economical in both warm and cold climates.

### 9.4.2 MIXES

About 0.97 cu ft (0.028 m³) of each cubic foot (cubic meter) of a typical portland cement plaster mix is aggregate, and 0.25 to 0.33 cu ft (0.007 to 0.009 m³) is cementitious material. Therefore, the aggregate must be clean, sound, and well graded, with particles ranging from fine up to a maximum size of $1/8$ in. (3.3 mm).

In the best mixes, aggregate size is proportioned, so that a minimum of water-cement paste is required to coat the aggregate and fill the voids between particles. An excess of fine or coarse particles can result in a mix containing too much cement or too much water, either of which is detrimental to the plaster.

Figure 9.4-1 lists several possible mixes, classified according to plaster groups (C, L, F, and P). The letter designations broadly indicate the type of mix. The selection of a suitable mix for a specific coat depends in part on the bonding properties of the plaster base (Fig. 9.4-2). Mixes containing substantially more portland cement than those shown should not be used.

In general, proportions should be 1 part cement (including lime, when used) to at least 3 and not more than 5 parts by volume of damp, loose sand. Since wide variations in sand quality and gradation exist in different locations, the correct proportion of sand to cement should be determined from trial mixes or previous experience with aggregate from local sources.

Durable finishes depend on proper gradation of aggregate from fine to coarse (see Fig. 9.2-7). When there are too many fine particles, more water-cement paste than is desirable is needed to coat the larger surface area of the particles. When an excess of coarse particles is present, more paste is required to fill the spaces between particles.

Having too little water-cement paste in a mix causes poor workability, but adding either too much water or too much cement to improve workability results in a plaster that is subject to high drying shrinkage and cracking. Sand fines can contain a high percentage of clay particles which shrink excessively upon drying. An excess of coarse particles produces a harsh and difficult-to-apply mix. Therefore, aggregate should be selected carefully and mix proportions should not be relied on to overcome deficiencies in aggregate quality or gradation.

A good mix is easily workable and easy to trowel. It exhibits good adhesion to the base and sufficient cohesiveness so that the mix will not sag on vertical surfaces. Once the mix has been determined, uniformity in proportioning of materials from batch to batch will help to ensure uniform suction and color.

Dry plaster ingredients should be mixed to a uniform consistency before water is added. Then only enough water should be added to ensure workability. A power mixer is recommended for uniform blending of materials. It should be operated for at least 5 minutes after all ingredients are in the mixer. Materials should be mixed only in quantities which can be used within $3^1/2$ hours. Remixing to restore plasticity is permissible; retempering with additional water is not.

In portland cement plaster to be applied by machine, the amount of water should be adjusted to produce a mix of proper consistency for the length of hose, temperature and humidity conditions, and type of plaster base. When measured with a standard 2 by 4 by 6 in. (50.8 by 102 by 152 mm) cone, the slump should be from $2^1/2$ to 4 in. (63.5 to 101.6 mm) at the mixer and from 2 to $3^1/4$ in. (50.8 to 82.6 mm) at the nozzle. The hose length from the machine to the working surface should be as straight and short as possible (not longer than 200 ft [60.96 m]).

#### 9.4.2.1 Masonry and Plastic Cement Mixes

Masonry cement and plastic cement come from the factory with plasticizing agents in them. Adding lime or plasticizers to them is not necessary. This simplifies field mixing because only sand and water need to be added. Either masonry cement or plastic cement can be combined with portland cement in equal volumes when greater strength is required (see Figs. 9.4-1 and 9.4-2). Plastic and masonry cements, alone or in combination with portland cement, can be used for any of the three plaster coats.

#### 9.4.2.2 Portland Cement Mixes

Lime, plastic cement, or masonry cement is added to portland cement to impart suitable workability and plasticity to portland cement plaster mixes. For base coats, 1 part portland cement can be mixed with up to $1^1/4$ parts lime or 1 part masonry or plastic cement. For finish coats, 1 part portland cement can be mixed with 1 part plastic cement or $1/2$ to 2 parts lime (see Figs. 9.4-1 and 9.4-2). Truer colors will occur if white portland cement and light-colored sand are used for field-mixed finish coats.

Plasticizers such as mineral, animal, vegetable, or glass fibers can be added to mixes for the first coat, but only in quantities sufficient to obtain the required plasticity or cohesiveness. The kind and exact amount of plasticizing agent will vary with the type of aggregate, plaster base, and manner of application. The amount of a mineral fiber used should not exceed 3% of the cementitious material by weight. The best combination of ingredients should be determined by trial mix.

### 9.4.3 APPLICATION

Portland cement plaster should not be applied in freezing weather or to surfaces containing frost. The temperature of the plaster should be maintained at 50°F (10°C) or higher during application and for at least 48 hours after the finish coat has been applied. When temperatures are falling, portland cement plaster should be protected against freezing by heated enclosures (Fig. 9.4-3) (see also Section 9.4.4).

Hand application of portland cement plaster involves traditional tools and practices that have proven to be successful over centuries. However, machine application enjoys some advantages. Wet

**FIGURE 9.4-1** Stucco Mixes

| Group* | Portland Cement | Lime | Masonry Cement Type II | Plastic Cement | Sand |
|---|---|---|---|---|---|
| | | | **Proportions, Parts by Volume** | | |
| C | 1 | 0 to ¼ | — | — | 3 to 4 |
| C | 1 | — | 1 | — | 6 to 7½ |
| C | 1 | ¼ to ½ | — | — | 4 to 6 |
| L | 1 | 2 to 1¼ | — | — | 4½ to 9 |
| L | — | — | 1 | — | 3 to 4 |
| F | 1 | 1¼ to 2 | — | — | 5 to 10 |
| P | 1 | — | — | 1 | 6 to 10 |
| P | — | — | — | 1 | 3 to 4 |

*C = (portland) cement; L = lime; F = finish coat; P = plastic cement.

(Courtesy Portland Cement Association.)

**FIGURE 9.4-2** Stucco Mix Selection Guide

| Stucco Base | Scratch | Brown[b] | Finish[c] |
|---|---|---|---|
| | **Recommended Groups[a] for Stucco Coats** | | |
| Low absorption (placed concrete, dense clay brick) | C, P | C, L, P | L, F, P |
| High absorption (concrete unit masonry, clay brick, structural tile) | L, P | L, P | L, F, P |
| Metal reinforcement[d] | C, P | C, L, P | L, F, P |

[a]C = (portland) cement; L = lime; F = finish coat; P = plastic cement.

[b]Use as base coat in two-coat work over concrete and masonry bases.

[c]Finish coat may be factory prepared "stucco finish."

[d]Over any type of supporting construction.

(Courtesy Portland Cement Association.)

**FIGURE 9.4-3** Protective enclosures in mild climates prevent rapid loss of moisture and facilitate curing. (Courtesy Portland Cement Association.)

mix can be delivered under pressure to scaffolding in multistory buildings; application is speeded, and obvious lap and joint marks, sometimes difficult to avoid with hand application, are eliminated. Machine application results in a more uniform texture and deeper and darker colors than are possible with hand application. Also, some textures, such as a uniform rough texture, are obtained best by spraying. Consequently, finish coats are sometimes applied by machine over a hand-applied base coat.

Generally, portland cement plaster is applied either in two or three coats, depending on the supporting construction.

### 9.4.3.1 Three-Coat Work

Three-coat applications are required over metal reinforcement and are sometimes also used over a rigid base. They consist of a *scratch coat*, a *brown coat*, and a *finish coat*, totaling ⁷⁄₈ in. (22.2 mm) in thickness. Each coat is applied separately and allowed to set partially. The scratch coat and brown coat are sometimes referred to as *base coats*.

The first (scratch) coat provides mechanical keying with the reinforcement and a rigid surface for the following coat. Over a rigid base, a *dash bond* coat is sometimes used instead of a scratch coat. The second (brown) coat provides additional thickness and strength and ensures a level surface for the application of the final decorative *finish coat*. Each coat should be moist-cured before the following coat is applied (see Section 9.4.3).

### DASH BOND COAT

A dash bond coat's surface is sufficiently rough so that cross-raking is not required (Fig. 9.4-4). A dash bond coat should be moist-cured for 12 to 24 hours before the brown coat is applied.

### SCRATCH COAT

A scratch coat is the first coat over metal reinforcement and can be the first coat over a rigid base that has sufficient bonding capability to make a dash bond coat unnecessary. A scratch coat should be at least ³⁄₈ in. (9.5 mm) thick when measured from a rigid base or backing paper. It should be applied evenly, with sufficient pressure to squeeze the mix through the reinforcement to form mechanical keys. The space between the metal and the backing should be filled thoroughly so that the metal will be sufficiently embedded. The coverage on the surface should be such that no more than 10% of the surface of the reinforcement remains exposed. As soon as the scratch coat has become firm, its entire surface should be cross-raked (*scratched*) to provide good mechanical bond for the brown coat (Fig. 9.4-5). On walls, horizontal cross-raking is recommended. Regardless of whether application is by hand or by machine, cross-raking is done manually.

### BROWN COAT

A brown coat is applied after the scratch coat has been moist-cured and is sufficiently firm to carry the weight of the next coat (Fig. 9.4-6). If a scratch coat becomes dry, it should be dampened again to control suction before a brown coat is applied. A brown coat should be

**FIGURE 9.4-4** A dash bond coat applied to a wall surface to enhance the bond between a solid base and a portland cement plaster. (Courtesy Portland Cement Association.)

tire surface to ensure uniformity of texture. Deep scratches or surface defects should be smoothed out, because such irregularities may cause unequal suction or may show through the finish coat. External corners should be formed with a straightedge when corner reinforcement is not used.

After moist curing, the surface should be allowed to dry uniformly before a finish coat is applied (see Section 9.4.4).

### FINISH COAT

A brown coat should be uniformly dampened when a finish coat is to be applied by hand. Dampening is not required for machine-applied finishes. A finish coat should be at least $^1/_8$ in. (3.2 mm) thick, bringing the total thickness to at least $^7/_8$ in. (22.2 mm) (Fig. 9.4-8a). Machine-applied, sprayed finish coats are often made in two or more consecutive passes (Fig. 9.4-8b). The first pass is made with a thinner mix applied primarily to ensure complete coverage of the brown coat. Successive applications are thicker to develop the desired finished texture.

Factory-prepared stucco finishes ensure greater uniformity of color in the finish coat. A variety of decorative surface finishes are possible for both machine and hand application (Fig. 9.4-9) (see Section 9.4.6). A plain finish coat that is going to

approximately $^3/_8$ in. (9.5 mm) thick, bringing the thickness of both coats to at least $^3/_4$ in. (19.1 mm).

When possible, a brown coat should be applied over an entire wall panel without stopping. Interruptions in the work should be made at screeds, control joints, corners, pilasters, belt courses, doorways, openings, or other lines where changes in

appearance, which might show through the finish coat, will not be obvious.

A brown coat should be *rodded* (struck off) (Fig. 9.4-7) with a straightedge and then darbied to impart a level, uniformly rough texture. Even with machine application, these operations are performed by hand. The same type of darby or float should be used for the en-

**FIGURE 9.4-5** A stucco scratch coat showing horizontal scoring. (Courtesy Portland Cement Association.)

**FIGURE 9.4-6** A second coat of stucco is applied over the scratch coat. (Courtesy Portland Cement Association.)

**FIGURE 9.4-7** Worker rodding a stucco coat to the desired plane. (Courtesy Portland Cement Association.)

**a**  **b**

**FIGURE 9.4-8** A finish coat can be (a) hand-floated or (b) sprayed by machine.

**a**  **b**

**FIGURE 9.4-9** Rock dash (marblecrete) finish: (a) small aggregate can be machine-sprayed; (b) a larger aggregate is usually placed by hand. (Courtesy Portland Cement Association.)

Finish coat approx. $\frac{1}{4}$" thick

Rough-floated base coat approx. $\frac{3}{4}$" thick

Joints struck flush

**FIGURE 9.4-10** Two coats are adequate over a suitably rough concrete or masonry surface. (Courtesy Portland Cement Association.)

### 9.4.4 CURING

Portland cement plaster requires sufficient water and favorable temperatures to facilitate the process of hydration by which the cement gains strength and hardens to a solid mass. The large exposed area of plaster relative to its volume encourages rapid loss of moisture by evaporation. Moist curing of each coat replaces this moisture and maintains an adequate supply for the chemical reaction to continue. An ordinary hose nozzle can be adjusted to provide a fine spray for curing; a steady stream that could erode the surface should not be used.

Moist curing does not require spraying at set intervals, but rather periodic observation and spraying when the surface appears to be drying out. Since evaporation increases with increased air circulation and higher temperatures, more frequent spraying is generally required in breezy, warm weather. Also, wind breaks or hangings are needed to protect fresh portland cement plaster from hot, dry winds.

In addition to adequate moisture, proper hydration requires that a portland cement plaster mix and each plaster coat be maintained at a temperature of at least 50°F (10°C). A finish coat should be maintained at this temperature for at least 48 hours after it has been applied. In cold weather, it may be necessary to heat the mixing water and provide a heated enclosure around the work area.

While it is generally accepted that adequate water to ensure hydration is essential in producing quality portland cement plaster, there is no general agreement on exact curing procedures. The curing recommendations given in Figure 9.4-11 are based on the premise that deliberate moist curing ensures a greater measure of success than does dependence

be painted should be moist-cured. Prepared stucco finishes often contain curing agents, and moist curing is not necessary; in fact, it is specifically discouraged by some manufacturers because it may cause streaking of the color surface. The finish coat should be maintained at the proper temperature for at least 48 hours after application (see Section 9.4.4).

### 9.4.3.2 Two-Coat Work

Sometimes three-coat applications are used over concrete and masonry bases, especially where increased strength or thickness is required. Usually, however, two-coat applications give sufficient strength and durability over such solid bases. In two-coat applications, a single base coat takes the place of the scratch and brown coats and is followed by a finish coat. The total thickness should be at least $^5/_8$ in. (15.9 mm) thick.

### BASE COAT

Except when a chemical bonding agent is used, a surface to be plastered should be dampened to ensure uniform suction, A $^3/_8$-in. (9.5-mm)-thick base coat is then applied. It is not scratched, as in a three-coat application. Mechanical bond for the finish coat is provided by rough floating of the base coat in the same manner as the brown coat in three-coat work (Fig. 9.4-10).

A machine-applied base coat is built up with several applications, with the final application being rodded and darbied manually.

### FINISH COAT

A finish coat should be about $^1/_4$ in. (6.4 mm) thick. It can be applied in a variety of textures by hand or machine, as in a three-coat application.

**FIGURE 9.4-11** Curing Schedule for Portland Cement Plaster

| Application[a] | Moist Curing[b] | Total Setting Time[c] |
|---|---|---|
| Scratch coat | 12 to 24 hr. | At least 48 hr. (between coats) |
| Brown coat | 12 to 24 hr. | At least 7 days (between coats) |
| Finish coat | 12 to 24 hr.[d] | At least 48 hr. (after application) |

[a]In two-coat work, requirements for base and finish coats are the same as for brown and finish coats, respectively.

[b]Moist curing should be delayed until scratch and brown coats are sufficiently set to prevent erosion and for 12 to 24 hours for finish coat.

[c]At least 50°F (10°C) temperature should be maintained during and after application.

[d]Not recommended for prepared colored "stucco finishes."

(Courtesy Portland Cement Association.)

on favorable climatic conditions or on water contained in the mix.

In some parts of the country, no deliberate moist curing is performed, and each coat is applied as soon as the previous coat is sufficiently hard to support it. This method, sometimes called *doubling back*, relies on the protective nature of each covering coat to retain the moisture and facilitate curing of the underlying coat.

Another practice involves application of coats at not less than 24-hour intervals, with continuous moist curing between coats and after the finish coat is applied.

The curing method chosen should be based on accepted local practice which has a demonstrated history of satisfactory performance. All curing procedures attempt either to maintain the previous coat uniformly damp or to permit it to become uniformly dry. This is particularly true of the base (or brown) coat, because uniformity of color in the finish coat is dependent on equal suction in the base coat.

The long waiting period between base and finish coats shown in Figure 9.4-11 ensures that the brown coat will be uniformly dry. When shorter intervals are allowed, it is often necessary to keep the base coat damp by continuous moist curing until the finish coat is applied. Unless a rigid schedule for rapid successive applications can be maintained, the need for continuous moist curing may be more demanding than the recommended short curing period and subsequent drying delay.

### 9.4.5 CRACK CONTROL

Proper design of supporting construction, selection of materials, and application of portland cement plaster will minimize cracking. Cracks develop in plaster from (1) shrinkage stresses, (2) structural movement, (3) weak sections in the finish due to thin spots, openings, or changes in materials in the supporting construction, or (4) restraints due to penetration of mechanical or electrical equipment or intersecting walls or ceilings.

#### 9.4.5.1 Shrinkage Stresses

As a wet portland cement plaster mix hydrates and dries, it tends to shrink, setting up stresses in the plaster membrane. Cracking occurs when shrinkage stresses exceed the tensile strength of the membrane. It is desirable, therefore, to minimize the degree of shrinkage. Proper moist curing and protection from temperature extremes allow the plaster to dry slowly and uniformly, thus increasing strength and reducing shrinkage. Shrinkage stresses also are minimized by proper proportioning of mixes and by proper placement of control joints.

#### PROPORTIONING MIXES

The chief factors in proportioning mixes which affect the balance between tensile strength and shrinkage stresses in the fin-

ished plaster are the cement content, the water content, and the aggregate quality and gradation.

Mixes that are high in cement content (rich mixes) generally result in plaster with higher initial shrinkage, balanced to some extent by increased strength. Excessive water in a mix increases the volume of minute voids in the plaster as water evaporates or is used up in hydration, which results in higher shrinkage and lower strength.

Poorly graded sand requires the addition of more water or more cement, or both, to obtain suitable workability, either of which can be detrimental and cause shrinkage cracking (see Section 9.4.1).

#### CONTROL JOINTS

When portland cement plaster is applied directly to masonry or concrete and no reinforcement is used, control joints in the plaster should coincide with joints in the base. Additional control joints to resist shrinkage cracking in the plaster finish are not usually required, because the underlying base will resist stresses due to shrinkage.

In portland cement plaster applied to metal reinforcement, control joints should be installed in the plaster not more than 18 ft (5486 mm) apart, forming panels of not more than 150 sq ft (13.94 m²).

Either one-piece or multisection control joints (see Fig. 9.1-16b) can be used for vertical joints in walls and for joints in protected soffits. A multisection control joint is particularly effective in relieving stresses caused by shrinkage and differential structural movement in the supporting construction (Fig. 9.4-12). For horizontal joints, a watertight control joint designed for the purpose should be used (Fig. 9.4-13).

#### 9.4.5.2 Structural Movement

Structural movement may be caused by foundation settlement, deflection of structural members, racking due to wind

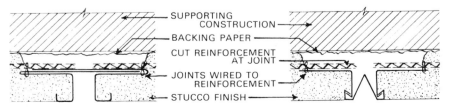

**FIGURE 9.4-12** Multisection control joints effectively relieve stress caused by drying shrinkage, structural movement, and temperature variations. (Courtesy Portland Cement Association.)

**FIGURE 9.4-13** Specially designed horizontal control joints provide weathertightness as well as crack control. (Courtesy Portland Cement Association.)

forces, and expansion and contraction of materials due to temperature and humidity variations. Good structural design will minimize structural movements. Specially designed joints can then be installed to localize movements which cannot be reduced to negligible proportions.

Relief joints and control joints are used to localize movement and minimize stresses within a portland cement plaster finish. Relief joints at ceilings and walls should be used where the plaster is intersected (Fig. 9.4-14) or penetrated by structural elements such as beams, columns, or load-bearing walls (Fig. 9.4-15). Location of control joints is dependent on the type of supporting base.

### RIGID BASES

Where portland cement plaster is applied directly to a concrete or masonry base, firm and continuous support of the plaster membrane will tend to resist shrinkage, temperature, and impact stresses in the plaster. However, the finish will be subjected to structural movements in the

base and is likely to crack wherever excessive stress concentrations develop. It is particularly important, therefore, in this type of installation to provide proper control joints in the base. Metal control joints should be placed in a plaster finish directly over each underlying joint. Control joints should also be placed in the plaster finish over construction, isolation, or expansion joints that occur in the base and at the juncture of dissimilar materials.

### METAL REINFORCEMENT

Portland cement plaster applied over metal reinforcement, with or without backing paper, acts as a thin, reinforced concrete slab. Because the plaster is somewhat isolated from the supporting construction, it is not subject to structural stresses to the same degree as plaster applied directly to rigid bases. However, it is more subject to shrinkage, impact, and temperature stresses because it is not supported and restrained by an underlying rigid base. Of these, shrinkage stresses are the most critical and involve the greatest movement. Therefore, control joints located as recommended for shrinkage purposes also will relieve stresses occasioned by other forces acting on the plaster membrane.

### 9.4.5.3 Weakened Sections

Stresses caused by shrinkage, humidity, or temperature variations are evenly distributed in plaster of uniform thickness but tend to be higher at the thin spots. Plaster applied over an irregular, rigid base, or over sagging metal reinforcement, will be of unequal thickness and will contain weakened sections. Concentration of stresses can result in cracking at these weakened points. To minimize

this hazard, plaster should be applied over a level base in a uniform thickness.

Weakened sections are also created at corners of cutouts and openings in a plaster finish. Metal reinforcement will resist the concentration of stresses at the corners and generally prevent diagonal cracking. When reinforcement is not used and plaster is applied directly to a rigid base, strip reinforcement (*strip lath*) can be installed diagonally at each corner to prevent cracking.

Weakened sections can be introduced deliberately in locations where cracking would not be objectionable. A trowel cut adjacent to a door or window frame will result in a straight, inconspicuous crack rather than random cracking. Often, the trowel cut can be concealed by trim; where trim does not occur, neatly formed cuts can be left exposed and caulked.

### 9.4.5.4 Restraints

Free movement of a plaster membrane is generally restrained at corners and wherever the membrane is penetrated by fixtures or structural elements. Cracking due to restraint is not common at external corners, particularly if control joints are provided in each plane for shrinkage control.

Cracking is more frequent at internal corners formed by the intersection of load-bearing elements (columns, walls, pilasters) and non-load-bearing walls, and at the juncture of walls and suspended or furred ceilings. At such corners, perimeter relief should be provided by omitting cornerite, stopping the metal reinforcement in each plane, and including relief joints which will permit independent movement in each plane.

Relief joints can be formed by a trowel cut (Fig. 9.4-14a) or a pair of casing beads (Fig. 9.4-14b). The casing

**FIGURE 9.4-14** Perimeter relief at an internal corner between the wall and soffit: (a) using trowel cut; (b) formed with casing beads. (Courtesy Portland Cement Association.)

**FIGURE 9.4-15** Perimeter relief joint for use where structural elements or fixtures penetrate the surface. (Courtesy Portland Cement Association.)

beads should be wired to the reinforcement, not rigidly fastened to supporting construction. The corner should be caulked with a nonhardening resilient compound after initial shrinkage has taken place.

A relief joint should be provided where plaster abuts or is penetrated by a structural element or a fixture. The plaster should terminate with a casing bead, and the space between the bead and the structural element or fixture should be caulked (see Fig. 9.4-15).

### 9.4.6 SURFACE FINISHES

In addition to providing the desired decorative surface, the finish selected affects the weather resistance and maintenance requirements of portland cement plaster. Smooth-troweled plaster finishes tend to make surface irregularities more conspicuous. Rougher textures hide slight color variations, lap joints, uneven dirt accumulation, and streaking.

The range of portland cement plaster textures is practically limitless. Brick, stone, and travertine marble can be simulated, and many other finishes can be produced by hand or machine, depending on the desired texture. The finish may be natural or integrally colored so that painting is not required. Colored finishes can be produced either by using a prepared stucco finish, by adding pigment to the mix, or by embedding colored aggregate in the surface.

Figure 9.4-16 shows several common textures. Textures can vary substantially with each applicator. Therefore, a standard sample should be produced at the beginning of the work, and as the work progresses, it should be checked with the sample for conformity.

## 9.5 Gypsum Board

This section covers the manufacture, properties, and uses of gypsum board products, and the materials and methods necessary to provide suitable single- and multi-ply gypsum board installations. Gypsum board panels and their supports and accessories are covered, but there are only general recommendations for sound isolation, fire resistance, thermal insulation, and moisture resistance. Additional information about these subjects is given in Chapters 7 and 13. The material *gypsum* is discussed in Section 9.2.1.

The use of gypsum board as a finish material started in about 1915, when small $^3/_8$-in. (9.5-mm)-thick sheets were developed primarily for remodeling. In the following decades, erection and joint treatment techniques gradually evolved, resulting in an improved surface appearance and the use of gypsum board in other types of construction. After World War II, greater emphasis on total system and assembly performance brought gypsum board widespread use (Fig. 9.5-1).

Gypsum board finishes are also called *drywall* finishes because they have initial low moisture content and require little or no water during application. Plywood, hardboard, and fiberboard are also sometimes called *drywall*, but the term usually means gypsum board.

Gypsum board finishes consist of one or more plies of gypsum board having noncombustible gypsum cores with surfaces of paper or other sheet material. Face (exposed) plies either have plain calendered manila paper intended for decorating with paint, wall covering, or wall tile, or are predecorated with colored, textured, or simulated wood-grained paper or vinyl coverings.

Most gypsum board installations are made with plain boards, which require fastener and joint treatment and decorating. Predecorated board is applied directly to studs or furring with color-matched fasteners or is laminated to a

a  b  c

d  e  f

**FIGURE 9.4-16** Some typical portland cement plaster (stucco) finishes: (a) *float finish*, made with carpet- or rubber-faced float; (b) *wet-dash*: coarse aggregate mix is dashed on with a brush; (c) *dash-troweled*: high spots or dashed surface are troweled smooth; (d) *stippled-troweled*; the surface is stippled with a broom; then high spots are troweled smooth; (e) *rock-dash*: decorative pebbles are thrown by hand or machine against wet surface; (f) *combed*: surface is formed with a notched templet. (Courtesy Portland Cement Association.)

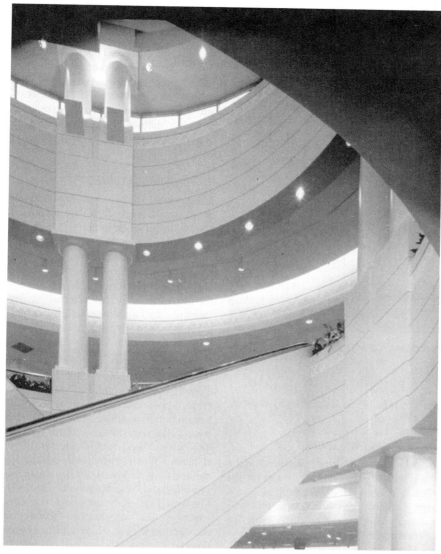

**FIGURE 9.5-1** We usually think of gypsum board being used for flat walls and ceilings. As this photograph shows, gypsum board can also be used in highly imaginative ways. (Courtesy Gypsum Association.)

sum Board," is the currently recognized industry standard and the basis for much of the data in this section.

### 9.5.1 BOARD MANUFACTURE

As shown in Figure 9.2-2, calcined gypsum without tube milling is used to make gypsum board products. This calcined gypsum is mixed with water to form a slurry. Additives and glass fibers are added for highly fire-resistant gypsum boards. This slurry is fed onto a conveyor belt between two layers of paper. After traveling down the belt for 3 or 4 minutes, the core material has set sufficiently to be cut to length (see Fig. 9.5-2), and then boards are conveyed to kilns for drying. Boards not requiring additional finishes may be bundled and stacked for shipping. Others may receive special backings such as aluminum foil, decorative vinyl paper finishes, or an organic texture coating.

### 9.5.2 BOARD TYPES

Gypsum board products are panels or slabs consisting of a noncombustible gypsum core, surfaced and edged with covering material specifically designed for various uses with respect to performance, application, location, and appearance.

Gypsum board products are produced for specific uses. They include:

- *Gypsum wallboard:* board product with either a prefinished surface or a surface suitable for paint finishing after appropriate treatment of fasteners and joints
- *Gypsum sheathing:* boards used in exterior frame walls as a base for finish materials and for increased fire resistance
- *Gypsum backing:* boards that provide a base for adhesive application of gypsum wallboards in multi-ply construction and of acoustical tiles in suspended ceilings
- *Gypsum formboards:* boards that serve as both permanent forms and finished ceilings in poured gypsum concrete roof decks (see Section 3.17)
- *Gypsum lath:* see Section 9.1.2.4

Board products are made in conformance with ASTM standards in different sizes and thicknesses, with several types of edges (Fig. 9.5-3). The dimensions and edge types of generally used board products are summarized in Figure 9.5-4. The

base layer of gypsum backing board. Joints are either left exposed, treated with special moldings or battens, or concealed with a loose flap of the decorative material. This section refers to plain wallboard finishes except when predecorated panels are specifically mentioned.

In addition to their other functions, walls and ceilings are often required to provide sound isolation, fire resistance, thermal insulation, and moisture control. A variety of gypsum boards and engineered systems are available to help meet such specific needs. In general, however, performance is closely related to the properties, arrangement, and installation of the wall or ceiling assembly.

For gypsum board finishes to have suitable quality, consideration should be given to:

- Preparation of supporting construction
- Selection of board and accessory materials
- Job conditions
- Board application, including joint and fastener treatment
- Special constructions techniques where sound isolation, fire resistance, thermal insulation, or moisture resistance is required

The Gypsum Association publication GA-216, "Recommended Specifications for the Application and Finishing of Gyp-

**FIGURE 9.5-2** Gypsum board on a continuous belt during manufacture. (Courtesy United States Gypsum Company.)

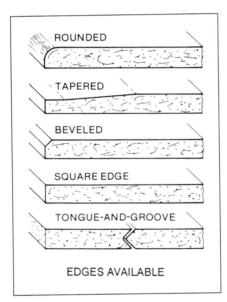

ROUNDED

TAPERED

BEVELED

SQUARE EDGE

TONGUE-AND-GROOVE

EDGES AVAILABLE

**FIGURE 9.5-3** Typical edge designs for gypsum board products. (Courtesy Gypsum Association.)

governing standard of quality for all the board types shown in Figure 9.5-4 is ASTM C1396. Figure 9.5-5 shows some of the terms used in gypsum board work.

### 9.5.2.1 Gypsum Wallboard

The main types of gypsum wallboard are regular core, predecorated, foil-backed, fire-resistant (Type X), and water-resistant (WR).

### REGULAR CORE WALLBOARD

Regular core wallboard is surfaced on the back with a gray liner paper and on the face and along the longitudinal edges with a calendered manila paper that provides a smooth, even finish suitable for decorating.

Most regular core wallboards are produced with tapered edges which permit taping and finishing of joints to provide a homogeneous, smooth surface with inconspicuous joints.

### PREDECORATED WALLBOARD

Predecorated wallboard is regular wallboard with a factory-applied decorative finish on one face. Common finish materials are paper and vinyl in a variety of colors, patterns, and simulated wood finishes (Fig. 9.5-6). These boards usually have accurately formed square or beveled longitudinal edges which do not require further treatment in the field. Beveled-edged and some square-edged boards simply are butted together to create a shadow line repeated at regular intervals. Square edges are trimmed with matching moldings or decorative battens to conceal joint irregularities and fasteners. Some

manufacturers provide a loose flap of the decorative covering material which is cemented down after installation to conceal the joint. Predecorated boards may be secured with matching prefinished nails or by adhesive bonding. A lighter-weight, predecorated product, MH Board, is often used in manufactured construction, primarily mobile homes. It is available factory painted or faced with paper or vinyl.

### FOIL-BACKED WALLBOARD

Foil-backed wallboard is regular wallboard with a bright-finish aluminum foil bonded to the back to act as a vapor retarder, eliminating the need for a separate vapor retarder in an exterior wall assembly. When the reflective surface faces a $3/4$-in. (19.1-mm)-wide dead air space, it can also provide increased insulating performance. Installation and edge treatment of foil-backed gypsum board are the same as for regular wallboard.

### FIRE-RESISTANT (TYPE X) WALLBOARD

Type X board is similar in exterior covering, appearance, edge treatment, and installation to regular core wallboard, but it has a core specially formulated with additives and glass fibers for greater fire resistance. When prescribed installation methods are followed, fire-resistant assemblies of 45 and 60 minutes are possible with Type X board and frame construction. Fire-resistant wallboard also is available with an aluminum foil vapor retarder.

### WATER-RESISTANT WALLBOARD

Water-resistant (WR) boards are used in toilet rooms, bathrooms, kitchens, utility rooms, and other humid areas. They are not recommended for high-moisture areas such as shower rooms, saunas, or steam rooms. They can be painted or used as a base for adhesively applied ceramic, metal, and plastic wall tile. Water resistance is achieved by a multilayered covering of chemically treated paper and a gypsum core formulated with asphaltic additives. Edges usually are tapered. Installation is the same as for regular wallboard.

Glass mesh mortar units, a highly water-resistant backing board used as a base for tile in areas subject to direct wetting such as shower stalls, is described in Section 9.6.

**Figure 9.5-4** Summary of Gypsum Board Products

| Type | Thickness, in. (mm) | Width, in. (mm) | Length, ft (mm) | Edge Detail |
|---|---|---|---|---|
| **Wallboard** | | | | |
| Regular | 1/4 (6.4) | 48 (1200) | 8, 10, 12 (2400, 3000, 3600) | Tapered |
| | 5/16 (7.9) | 48 (1200) | 8, 10, 12, 14 (2400, 3000, 3600, 4000) | |
| | 3/8 (9.5) | 48 (1200) | 7, 8, 9, 10, 12 (2000, 2400, 2700, 3000, 3600) | |
| | 1/2 (12.7) | 48 (1200) | 7, 8, 9, 10, 12, 14 (2000, 2400, 2700, 3000, 3600, 4000) | |
| | 5/8 (15.9) | 48 (1200) | 8, 10, 12 (2400, 3000, 3600) | |
| Foil backed | 3/8 (9.5) | 48 (1200) | 8, 10, 12 (2400, 3000, 3600) | |
| | 1/2 (12.7) | 48 (1200) | 8, 9, 10, 12 (2400, 2700, 3000, 3600) | |
| Type X | 1/2 (12.7) | 48 (1200) | 8, 9, 10, 12 (2400, 2700, 3000, 3600) | |
| | 5/8 (15.9) | 48 (1200) | 8, 9, 10, 12 (2400, 2700, 3000, 3600) | |
| **Backing Board** | | | | |
| Regular | 1/4 (6.4) | 48 (1200) | 8 (2400) | Square |
| | 3/8 (9.5) | 48 (1200) | 8 (2400) | Square |
| | 1/2 (12.7) | 24 (610) | 8 (2400) | Tongue-and-groove |
| Foil-backed | 3/8 (9.5) | 48 (1200) | 8 (2400) | Square |
| Type X | 5/8 (15.9) | 24 (610) | 7 (2000) | Tongue-and-groove |
| | 1 (25.4) | 24 (610) | 7 (2000) | Tongue-and-groove |
| **Sheathing** | | | | |
| Regular | 3/8 (9.5) | 24 (610) / 48 (1200) | 8 (2400) / 8, 9 (2400, 2700) | Tongue-and-groove / Square |
| | 4/10 (10) | 24 (610) / 48 (1200) | 8, 9 (2400, 2700) / 8, 9 (2400, 2700) | Tongue-and-groove / Square |
| | 1/2 (12.7) | 24 (610) / 48 (1200) | 8, 9 (2400, 2700) / 8, 9 (2400, 2700) | Tongue-and-groove / Square |
| | 5/8 (15.9) | 24 (610) / 48 (1200) | 8, 9 (2400, 2700) / 8, 9 (2400, 2700) | Tongue-and-groove / Square |
| Type X | 5/8 (15.9) | 24 (610) / 48 (1200) | 8 (2400) / 8, 9 (2400, 2700) | Tongue-and-groove / Square |

**FIGURE 9.5-5** Terms used in gypsum board work. See Glossary for definitions. (Courtesy Gypsum Association.)

**FIGURE 9.5-6** Predecorated wallboard.

### 9.5.2.2 Gypsum Backing Board

Gypsum backing boards serve as a base to which wallboards or acoustical ceiling tiles are laminated. Figure 9.5-7 shows two-ply construction of wallboard bonded to backing board. Backing boards differ from regular wallboards mainly in that the paper covering on both faces and edges is a gray liner paper not suitable for decorating. Backing boards may be secured to framing with staples, screws, or nails. The main types are regular, fire-resistant (Type X), and foil-backed. *Coreboard*, a 3/4- or 1-in. (19.1- or 25.4-mm)-thick, factory-laminated board, which

FIGURE 9.5-7 Double-layer (laminated) gypsum board applied to wood construction. Application over metal framing is similar.

consists of two layers of regular backing board, is used to build solid gypsum partitions and shaft walls.

A highly water-resistant backing board with treated paper and core is produced as a base for tile in areas subject to direct wetting, such as shower stalls.

### 9.5.2.3 Gypsum Sheathing

Gypsum sheathing provides fire resistance, wind bracing, and a base for exterior finishes. It is made of a water-resistant gypsum core completely enclosed with a firmly bonded, water-repellent paper, which eliminates the need for sheathing paper. Standard 2 by 8 ft (609 by 2438 mm) boards have V-shaped tongue-and-groove edges to shed water. These are nailed or screwed horizontally across the framing members (Fig. 9.5-8). The 4 by 8 ft (1219 by 2438 mm) and 4 by 9 ft (1219 by 2743 mm) boards have square edges and are installed vertically where improved wind bracing and infiltration resistance are desired.

### 9.5.2.4 Gypsum Formboard

Gypsum formboards are used as permanent forms and as finished ceilings for gypsum concrete roof decking (see Section 3.17). Formboards are surfaced on the exposed face and longitudinal edges with calendered manila paper specially treated to resist fungus growth. The back face is surfaced with gray liner paper, as in regular wallboard. A vinyl-faced board, which provides a white, highly reflective, and durable surface, is also available.

FIGURE 9.5-8 Gypsum sheathing applied horizontally to wood construction.

### 9.5.3 ACCESSORY MATERIALS

Accessory materials for the application of gypsum board include mechanical fasteners, adhesive, joint compounds, tape, and metal edge and corner trim.

#### 9.5.3.1 Fasteners

Nails and screws are used to apply both single- and multi-ply finishes. Clips and staples are limited to attaching the base ply in multi-ply construction.

#### NAILS

Nails for applying gypsum board can be bright, coated, or chemically treated. Shanks may be either smooth or annularly threaded, with medium or long diamond points. Nail heads are flat or slightly concave, thin at the rim and not more than $5/16$ in. (7.9 mm) in diameter. Nail heads of about $1/4$ in. (6.4 mm) diameter provide adequate holding power without cutting the face paper.

Nails may be either annularly threaded nails developed for wallboard (GWB-54) or smooth- or deformed-shank nails suitable for the application of gypsum board. Nails should conform to ASTM C514. Casing nails and common nails have heads that are too small in relation to the shank or are too thick and should not be used.

Nails should be of the proper length for the wallboard thickness (Fig. 9.5-9). Generally recommended penetration into supporting construction for smooth shank nails is $7/8$ in. (22.2 mm). Annularly threaded nails provide more withdrawal resistance, require less penetration, and generally minimize nail popping. How-

FIGURE 9.5-9 Nails for Gypsum Board Application

| Drywall Nails | Recommended Penetration | | Board Thickness | | Nail Length | |
|---|---|---|---|---|---|---|
| | in. | mm | in. | mm | in. | mm |
| Annularly threaded[a] ASTM C 514 | $3/4$ | 19 | $3/8$ | 9.5 | $1\frac{1}{8}$ | 29 |
| | | | $1/2$ | 12.7 | $1\frac{1}{4}$ | 32 |
| | | | $5/8$ | 15.9 | $1\frac{3}{8}$ | 35 |
| Smooth shank[b] ASTM C 514 | $7/8$ | 22 | $3/8$ | 9.5 | $1\frac{1}{4}$ | 32 |
| | | | $1/2$ | 12.7 | $1\frac{3}{8}$ | 35 |
| | | | $5/8$ | 15.9 | $1\frac{1}{2}$ | 38 |

*Note:* For fire-rated constructions, building code and/or manufacturer's specifications should be followed.
[a]Typically, GWB 54: $1/4$-in. (6.4 mm)-diameter head, 0.098-in. (2.49 mm)-diameter shank.
[b]Typically, $1/4$-in. (6.4 mm)-diameter head, 0.099-in. (2.51 mm)-diameter shank, with dished head.
(Courtesy Gypsum Association.)

ever, for fire-rated construction, 1 in. (25.4 mm) or more penetration usually is required and longer, smooth-shank nails are used.

## SCREWS

Both regular and fire-rated gypsum board can be fastened to both wood and metal supporting construction with *drywall screws*. The usual finish for drywall screws is a zinc phosphate coating with baked-on linseed oil. These screws typically are self-drilling, and have self-tapping threads and flat Phillips heads for use with a power screwdriver. Their contour head design makes a uniform depression free of ragged edges and fuzz.

Screws pull gypsum board tightly to the framing without damaging the board, thus minimizing fastener surface defects due to loose board attachment. The three

basic types of drywall screws for wood, sheet metal, and gypsum construction, with recommended penetration for various uses, are shown in Figure 9.5-10. Other specially designed screws are available for attaching wood or metal trim and for wallboard attachment to heavier-gauge load-bearing studs.

***Drywall Wood Screws*** Type W and similar screws are designed for fastening to wood framing or furring. Type W screws are diamond-pointed to provide efficient drilling action through both gypsum and wood and have a specially designed thread for quick penetration and increased holding power. Recommended minimum penetration into supporting construction is $5/8$ in. (15.9 mm), but in two-ply application when the face layer is being screw-attached, the additional

holding power developed in the base ply permits the penetration into supports to be reduced to $1/2$ in. (12.7 mm). Type W screws are available in $1^1/4$ in. (28.6 mm) length. Drywall sheet metal screws, more readily available in longer sizes, may be substituted in two-ply construction.

***Drywall Sheet Metal Screws*** Type S and similar screws are designed for fastening gypsum board to 25-gauge (0.6 mm) metal studs or furring. Type S screws have a self-tapping thread and a mill-slot drill point designed to penetrate sheet metal with little pressure. This is important because steel studs are flexible. The threads should be of adequate depth and should be turned within $1/4$ in. (6.4 mm) of the head to eliminate stripping. They are available in several lengths, from 1 to $2^1/4$ in. (25.4 to 57.2 mm). Other lengths and head profiles are available for attaching wood trim, metal trim, and metal framing components. Recommended minimum penetration through sheet metal for drywall sheet metal screws is $3/8$ in. (9.5 mm).

***Drywall Gypsum Screws*** Type G and similar screws are used for fastening gypsum board to gypsum board. Type G screws are similar to Type W screws but have a deeper special thread design. They generally are available in a $1^1/2$ in. (38 mm) length only. Drywall gypsum screws require penetration of at least $1/2$ in. (12.7 mm) of the threaded portion into the supporting gypsum board. Allowing approximately $1/4$ in. (6.4 mm) for the point, this results in a minimum penetration of $3/4$ in. (19.1 mm). For this reason, drywall gypsum screws should not be used to attach wallboard to $3/8$-in (9.5-mm)-thick backing board. In two-ply construction with a $3/8$-in. (9.5-mm)-thick base ply, nails or longer screws should be used to provide the necessary penetration into supporting wood or metal construction.

## STAPLES

Staples are recommended only for attaching base ply to wood members in multi-ply construction. They should be of 16-gauge (1.6 mm) flattened, galvanized wire with a minimum $7/16$-in. (11-mm)-wide crown and with divergent sheared bevel points. Staples should be long enough to provide a minimum penetration of $5/8$ in. (15.9 mm) into supporting construction.

**FIGURE 9.5-10** Typical Screws for Gypsum Board Application

| Drywall Screws* | Recommended Minimum Penetration | Screw Length | | Application |
|---|---|---|---|---|
| | | in. | mm | |
| For fastening to wood ASTM C1002 | $5/8$ in. (15.9 mm) | $1^1/4$ | 32 | Single-ply or base-ply application |
| | $5/8$ in. (15.9 mm) | $1^5/8$ | 41 | Face-ply application in two-ply construction |
| For fastening to sheet metal ASTM C1002 | $3/8$ in. (9.5 mm) | 1 | 25.4 | Single-ply or base-ply application |
| | $3/8$ in. (9.5 mm) | $1^5/8$ | 41 | Face-ply application in two-ply construction |
| For fastening to gypsum board ASTM C1002 | $1/2$ in. (12.7 mm) $3/4$ in. (19 mm) | $1^1/2$ | 38 | Gypsum board to gypsum board in multi-ply construction |
| | $3/4$ in. (19 mm) $3/4$ in. (19 mm) | | | |

*Note:* Applications are generally acceptable for fire-rated construction.

*Screws shown are Types W, S, and G, respectively; other drywall screws are available.

(Courtesy Gypsum Association.)

### 9.5.3.2 Adhesives

Adhesives can be used to attach singly-ply wallboard directly to wood framing, masonry, or concrete, or to laminate gypsum board to gypsum board, sound-deadening board, or rigid foam insulation. Adhesives are generally used in combination with nails or screws which provide temporary or permanent supplemental support.

Adhesive formulations vary, and manufacturers' recommendations should be consulted in each case. Adhesives for applying wallboard finishes may be classed as *stud adhesives*, *laminating adhesives*, and *contact adhesives*.

#### STUD ADHESIVES

Stud adhesives are formulated for attaching single-ply gypsum board to wood supporting construction. They generally have a rubber or asphaltic base, are of caulking consistency that will bridge framing irregularities, and are applied with a gun in a continuous bead. Structural stud adhesives should conform to ASTM C557, which establishes rigid standards for strength, bridging ability, aging, and other qualities necessary for long-term structural performance. Fewer permanent supplemental fasteners are required for structural stud adhesives than are required for adhesive-nail attachment with nonstructural stud adhesives.

#### LAMINATING ADHESIVES

Most laminating adhesives are casein-base adhesives in powder form to which water is added. They are used to laminate gypsum boards to each other and to suitable masonry or concrete surfaces. They are not suitable for adhesive-bonding to wood. In addition to the several adhesives specifically formulated for laminating purposes, some powder embedding compounds (see Section 9.5.3.3) can be used when mixed at the recommended consistency. When specifically recommended by their manufacturers, some stud adhesives and contact adhesives can also be used for laminating purposes.

Only as much adhesive should be mixed as can be used within the working time specified by the manufacturer. Laminating adhesive is applied over the entire board area with a spreader. Installed boards require temporary support or supplemental fasteners until the adhesive develops sufficient bond strength.

#### CONTACT ADHESIVES

Contact adhesives have a synthetic rubber base, sometimes with resins added, in either solvent or emulsion formulations. They are used to laminate gypsum boards to each other and to bond wallboard to metal studs in demountable partitions. The adhesive is applied by roller, spray gun, or brush in a thin, uniform coating to both surfaces to be bonded. For most contact adhesives, some open time is required before surfaces can be joined and a bond can be developed. To ensure maximum adhesion between mating surfaces, the face board should be impacted over its entire surface with a suitable tool such as a rubber mallet.

Contact adhesives provide an immediate bond, excellent long-term strength, resistance to fatigue, and no temporary or permanent fasteners required in the field of the board. However, they are unable to bridge irregularities in the mating surfaces, thus creating discontinuity in the bond film. In addition, most contact adhesives do not permit repositioning of the boards once contact between surfaces has been made. A slip sheet of polyethylene film or building paper can be used to facilitate gradual bonding of surfaces as the slip sheet is withdrawn.

#### MODIFIED CONTACT ADHESIVES

Modified contact adhesives combine good long-term strength and bridging ability with sufficient immediate bonding to permit erection with a minimum of temporary fasteners. In addition, they have an open assembly time of about $1/2$ hour before a bond develops, during which time boards can be repositioned. Modified contact adhesive is intended for attaching wallboard to all types of supporting construction, including framing, furring, gypsum board, and rigid insulating board.

### 9.5.3.3 Joint Tape and Joint Compounds

Joint tape and joint compounds are used to produce smooth, monolithic wallboard finishes with inconspicuous joints and no visible fasteners. Joint compounds and tape should conform to ASTM C475.

#### JOINT TAPE

Tape used for joint reinforcement is a strong fibered tape with chamfered edges. It resists tensile stresses across the joint as well as longitudinally.

#### JOINT COMPOUNDS

Joint compounds include:

- *Embedding compounds*: for embedding and bonding joint tape
- *Topping compounds:* for final smoothing and leveling over joints and fasteners
- *All-purpose compounds:* combine the features of the other two

Joint compounds are either premixed or packaged in powder form to be mixed with water on the job. Except for quick-setting embedding compounds, these compounds harden by evaporation of the water and can be kept in a wet-mix state for several days if covered with a moist cloth or a thin layer of water. Both premixed and job-mixed compounds should be protected from freezing.

***Embedding Compounds*** Embedding compounds are casein-based adhesives similar to laminating adhesives. They are available either premixed or as a powder which is mixed with water. They are used for embedding tape and as a first coat for concealing fasteners and edge and corner trim. Some embedding compounds also can be used for laminating gypsum board in multi-ply construction when mixed at the proper consistency. Usually, embedding compounds require at least overnight drying before following coats can be applied. However, quick-setting compounds which dry in 2 to 4 hours also are available, which permits three-coat application of embedding and topping compounds in 1 day.

***Topping Compounds*** Topping compounds are surface fillers used to conceal and smooth over embedded tape, fasteners, and trim. They are casein or casein-vinyl formulations available either in premixed or powder form. They bond well with joint tape and compound, gypsum board, and fasteners. They sand easily and provide a surface with sufficient "tooth" and suction for painting.

***All-Purpose Compounds*** All-purpose compounds work as both adhesive and filler. They are available in either powder or premixed form and in either a machine or hand tool consistency. All-purpose compounds can be used for embedding tape, topping over tape, finishing over metal trim, and concealing fasteners. However, they should not be used for laminating gypsum boards unless specifically recommended by their manufacturer.

### 9.5.3.4 Edge and Corner Trim

Exposed wallboard edges and ends usually are protected by trim. Wood casing trim is often used at window and door frames, and wood or resilient wall base is used to protect wallboard edges at the floor. Metal trims, such as corner beads and casing beads, are used in similar locations when inconspicuous edge protection is desired and applied trim is not to be used. Plastic trim may be used when exposure of the trim is acceptable or desired.

Metal trim is available in standard shapes for different conditions and varying board thicknesses (Fig. 9.5-11). On outside corners, metal corner beads are used to protect gypsum board from damage and provide neat, slightly rounded corners, which are difficult to obtain with tape and joint compounds alone. Casing beads and corner beads with metal flanges generally are nailed or screwed to supporting construction at about 6 in. (150 mm) on centers. Beads with paper flanges are held in place by embedding the flanges in joint compound.

### 9.5.4 JOB CONDITIONS

Temperature and humidity affect the performance of joint treatment materials, the scheduling of joint finishing operations, the quality of a joint, and sometimes the bonding properties of adhesives.

Interior finishes should not be installed when the outside temperature is less than 55°F (12.8°C) unless the building is completely enclosed and has a controlled temperature of 55°F (12.8°C) to 70°F. (21°C). This temperature should be maintained for 24 hours before installation, during installation, and until a permanent heating system is in operation. A building need not be completely glazed during summer months, but all materials and installation should be protected from the weather.

Ventilation should be provided to eliminate excessive humidity. If the building has been glazed, windows can be kept partially open to provide air circulation. Temporary air circulators should be used in enclosed areas that have no natural ventilation. Additional drying time between coats of joint compound should be allowed when drying conditions are poor. During hot, dry

| USASI Designation | Trim Profile | Application |
|---|---|---|
| **CORNERBEAD**[1]<br><br>CB—100 x 100<br>CB—118 x 118<br>CB—114 x 114<br>CB—100 x 114<br>CB—PF (paper flange steel corner combination bead) | | |
| **U BEAD**[2]<br><br>U-38<br>U-12<br>U-58<br>U-34 | | |
| **L BEAD**[2]<br><br>LB-38<br>LB-12<br>LB-58<br>LB-34 | | |
| **LK BEAD**[3,4]<br><br>LK-S<br>LK-B | | |
| **LC BEAD**[2]<br><br>LC-38<br>LC-12<br>LC-58<br>LC-34 | | |

[1]Numbers indicate width of flanges: e.g., 114 × 114 has two 1¼″ (28.6 mm) flanges.
[2]Numbers indicate thickness of board intended: e.g., U-12 is U Bead intended for ½″ (12 mm) wallboard.
[3]Letter (S) indicates square nose; letter (B), bull or round nose.
[4]For use with kerfed jamb.

**FIGURE 9.5-11** Standard Drywall Metal-Trim Shapes. (Courtesy Gypsum Association.)

weather, drafts should be avoided to prevent joint compounds from drying too rapidly.

The delivery of gypsum board should coincide as closely as possible with the installation schedule. Boards should be stored flat, indoors, and under cover. Stacking long lengths on short lengths should be avoided to prevent the longer boards from breaking. For short periods of time, boards may be placed vertically against the framing with the long edges of the boards horizontal. Materials should remain in their wrappings or containers until ready for use.

### 9.5.5 BOARD APPLICATION

Gypsum board finishes may be either single-ply, using wallboard alone, or multiply using wallboard over wallboard, backing board, or sound-deadening board.

Installing wallboard perpendicular to supporting members results in a stiffer assembly and is preferred. Whenever possible, boards should be applied to span ceilings and walls in one piece to avoid butt-end joints, which are difficult to finish inconspicuously. Unavoidable butt-end joints should be staggered and located as far from the center of walls and ceilings as possible so that they will be less conspicuous. With the exception of face layers in two-ply construction, board ends and edges that are parallel to supporting members should fall on these members.

Successful application of gypsum board finishes depends on proper measuring, cutting, and fitting, board attachment, and joint and fastener treatment.

#### 9.5.5.1 Measuring, Cutting, and Fitting

Measurements should be taken accurately at the intended location for each edge or end of a board. Accurate measuring usually will indicate irregularities in framing and trimming so that allowances can be made in cutting.

Straight-line cuts across the full width or length of a board are made by scoring through the face paper into the gypsum core (Fig. 9.5-12a). The core of the board then is snapped, and the back paper is cut (Fig. 9.5-12b). Cut edges should be smoothed with a rasp or trimmed with a sharp knife (Fig. 9.5-12c). Panels should be cut so that they fit easily into place. Ceiling panels should be installed first. Joints should be loosely butted without forcing the boards into position. Tapered edges should be placed next to tapered edges, and square-cut ends should be butted to square ends. Square ends should not be placed next to tapered edges because excessive buildup of joint compound over the tapered edge results in a conspicuous ridge caused by differential shrinkage.

**FIGURE 9.5-12** (a) Face paper is scored; (b) gypsum core is snapped along the score line, and back paper is cut; (c) rough edges are smoothed with coarse sandpaper. A rasp or sharp knife could also be used. (Courtesy Gypsum Association.)

#### 9.5.5.2 Board Attachment

Wallboard and backing board can be attached to supporting construction with mechanical fasteners, with adhesive, or with a combination of adhesive and mechanical fasteners.

##### MECHANICAL FASTENERS

Nails and screws are used to attach gypsum board in both single- and multi-ply installations. In two-ply construction, however, staples and clips can be used to attach the base ply. Adhesives, when used, generally are supplemented with mechanical fasteners.

Fasteners should be of a type recommended for the intended method of application. Ordinary wood screws, sheet metal screws, and common nails are not designed to hold gypsum board tightly or to countersink neatly and should not be used. Fasteners should be placed at least $3/8$ in. (9.5 mm) from board edges and ends. Application should start in the middle of a board and proceed outward toward the perimeter. Fasteners should be driven as nearly perpendicular as possible while the board is held in firm contact with supporting construction. Nails should be driven with a crown-headed hammer forming a uniform depression around the nail head. Fastener heads should not be seated deeper than about $1/32$ in. (0.8 mm) below the board surface, and particular care should be taken not to break the face paper (Fig. 9.5-13).

##### ADHESIVE-BONDING

Adhesive can be used to secure single-ply wallboard to wood framing and furring, masonry, and concrete and to laminate a face ply to a base layer of gypsum board, sound-deadening board, or rigid foam insulation. The surface to be bonded should be free from dirt, grease, oil, and other foreign material.

#### 9.5.5.3 Single-Ply Construction

In single-ply construction, a single layer of gypsum board is permanently nailed, screwed, or adhesive-bonded to supporting wood or metal studs or furring (Fig. 9.5-14). After appropriate fastener and joint treatment, the surface may be painted or finished with wall covering or wall tile.

**FIGURE 9.5-13** A properly driven drywall nail. (Courtesy Gypsum Association.)

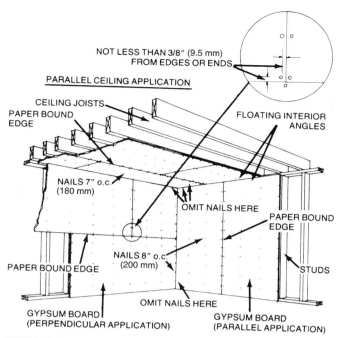

NOT LESS THAN 3/8" (9.5 mm) FROM EDGES OR ENDS

PARALLEL CEILING APPLICATION

CEILING JOISTS

PAPER BOUND EDGE

FLOATING INTERIOR ANGLES

NAILS 7" o.c. (180 mm)

OMIT NAILS HERE

PAPER BOUND EDGE

NAILS 8" o.c. (200 mm)

PAPER BOUND EDGE

STUDS

GYPSUM BOARD (PERPENDICULAR APPLICATION)

OMIT NAILS HERE

GYPSUM BOARD (PARALLEL APPLICATION)

**FIGURE 9.5-14** Gypsum board application over wood-framed construction. (Courtesy Gypsum Association.)

This subsection discusses assemblies, methods, and precautions needed to obtain suitable finishes with single-ply construction. Refer to Section 9.5.5.6 for a more detailed discussion about provisions for sound isolation, fire resistance, thermal resistance, moisture resistance, and other special construction. Also refer to Chapter 13 for additional information about sound isolation.

Sound ratings ranging from 35 to 50 Sound Transmission Class (STC) (see Section 13.1) can be achieved with single-ply construction. The higher values require special construction systems such as resilient mounting, separated construction, or sound-absorbing materials. Fire ratings for single-ply framed partitions generally are limited to 1 hour.

Because of the large number of fasteners in the surface of the board and the greater dependence on accurately aligned supporting construction, single-ply finishes are more susceptible to surface problems such as joint ridging, corner cracking, and nail popping. However, with the selection of proper supporting construction and methods of attachment; attention to ridging, cracking, and nail popping precautions; and care in joint treatment, it is possible to produce visually attractive, durable finishes.

Framing or furring members should be spaced as recommended in Figure

9.5-15. Except for $^3/_8$-in. (9.5-mm)-thick wallboard on ceilings, which should be applied perpendicular to the supports, boards can be applied either perpendicular or parallel to supporting members (Fig. 9.5-16). Perpendicular application provides greater strength and shorter joint lengths.

Except in end-floated construction, gypsum board edges and ends should occur over supporting members. Butt-end joints should be staggered with respect to each other on the same side of a partition

and with respect to joints on the opposite side of the partition.

## WOOD FRAMING AND FURRING

Gypsum board is attached to wood furring and framing with nails or screws or by a combination of adhesives and nails. Resilient metal furring can be used with wood framing where increased sound isolation is desired.

**Screw Attachment**  Fewer fasteners are required when screws are used, and the number of fasteners requiring treatment and possible surface defects is reduced. Drywall wood screws, such as Type W, providing at least $^5/_8$ in. (15.9 mm) penetration into supports should be used, spaced as shown in Figure 9.5-17.

**Nail Attachment**  Nailed gypsum board panels must be drawn tightly against framing and furring members so that the board will not move on the nail shank, which will result in nail popping. Wallboard can be attached by either single or double nailing, both in accordance with GA-216. In general, nail spacing is based on requirements for fire resistance and on intimate board-to-stud contact to reduce possible surface defects rather than on structural support. Depth of nail penetration should be as shown in Figure 9.5-9.

*Single nailing* can provide satisfactory results if the methods in GA-216 are carefully followed. Nails should be spaced 7 in. (180 mm) on centers on ceilings and 8 in. 200 mm) on centers on walls along supports (Fig. 9.5-18a).

**FIGURE 9.5-15** Support Spacing for Single-Ply Construction

| Wallboard Thickness | | | | Maximum Support Spacing, o.c. | |
| in. | mm | Location | Application[a] | in. | mm |
|---|---|---|---|---|---|
| $^3/_8$ | 9.5 | Ceilings[b,c] | Perpendicular | 16 | 406 |
| | | Walls | Perpendicular or parallel | 16 | 406 |
| $^1/_2$ and $^5/_8$ | 12.7 and 15.9 | Ceilings[c] | Perpendicular | 24 | 610 |
| | | | Parallel | 16 | 406 |
| | | Walls | Perpendicular or parallel | 24 | 610 |

[a]Direction of wallboard edges relative to supports.

[b]$^3/_8$-in. (9.5 mm) ceilings should not support insulation.

[c]For ceilings to receive a water-based spray texture finish, use $^5/_8$-in. (15.9 mm) board applied perpendicular to 24-in. (610 mm) supports or $^1/_2$ in. (12.7 mm) perpendicular to 16-in. (406 mm) supports. (Courtesy Gypsum Association.)

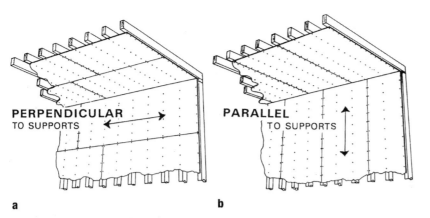

**PERPENDICULAR**
TO SUPPORTS

**PARALLEL**
TO SUPPORTS

a                                                        b

**FIGURE 9.5-16** (a) Horizontal and (b) vertical application of gypsum board to wood-framed construction using the single nailing method. (Courtesy Gypsum Association.)

**FIGURE 9.5-17** Single-Ply Screw Attachment to Wood Supports

| Support Spacing, o.c. | | Screw Spacing, o.c.* | | | |
|---|---|---|---|---|---|
| | | Walls | | Ceilings | |
| in. | mm | in. | mm | in. | mm |
| 16 | 406 | 16 | 406 | 12 | 305 |
| 24 | 610 | 12 | 305 | 12 | 305 |

*Spacing applicable to both field and perimeter of board.
(Courtesy Gypsum Association.)

*Double nailing* ensures consistent board-to-stud contact and fewer defects than does single nailing. A set of nails is driven into the field of the panel at 12 in. (300 mm) on centers, followed by a second set of nails 2 in. (50.8 mm) from each of the first nails. Edges and ends falling over supports are nailed at 7 in. (180 mm) on centers on ceilings and 8 in. (200 mm) on centers on walls (Fig. 9.5-18b). The total number of nails per panel is about 10% higher than for single nailing, but the number of nail spots that must be concealed is 20% fewer.

*Adhesive-Nail Attachment* Adhesive-nail attachment reduces the number of nails in the middle of a panel by at least 50% over the number required for conventional nailing, thereby reducing the amount of fastener treatment. The adhesive between board and supporting members produces stronger, stiffer assemblies and fewer fastener defects. Screws can be used in lieu of nails, but seldom are, because good board-to-stud contact is ensured by the adhesive and the number of face fasteners has already been reduced to the minimum.

A straight bead of *stud adhesive* is applied to the face of members supporting the field of a panel with a caulking gun (Fig. 9.5-19a). Where adjacent panels join over a supporting member, two beads are applied (Fig. 9.5-19b). When a predecorated wallboard joint is to be left untreated, two parallel beads of adhesive should be applied, one near each edge of the member. Generally, the volume of the bead should be such that when the board is applied, the adhesive will spread to an average width of 1 in. (25.4 mm) at least $1/16$ in. (1.6 mm) thick.

Care should be taken in applying adhesive to prevent it from being squeezed out through the joints. Adhesive should not be applied to members that are not required for board support, such as diagonal bracing, blocking, and plates.

The number and spacing of supplemental nails (or screws) vary, depending on the adhesive properties.

*Nonstructural stud adhesives* require fasteners in the field as well as at the perimeter of panels for wall and ceiling installation. Fastener spacing varies with the type of fastener, support spacing, and load-bearing conditions (Fig. 9.5-20).

*Structural stud adhesives* conforming to ASTM C557 require only perimeter fasteners for wall application. The fasteners should be spaced 16 in. (400 mm) on centers for edges or ends that fall parallel to supports and at each support for edges or ends that fall perpendicular to supports. For ceiling application, the perimeter is nailed as for walls, but additional fasteners are required in the field spaced 24 in. (600 mm) on centers (Fig. 9.5-21).

When installing *predecorated wallboard* with beveled or finished edges not requiring joint treatment, it often is desirable to avoid fasteners at the vertical joints. For these and plain wallboard applied vertically on walls, fasteners may be omitted along panel edges if appropriate means such as prebowing or temporary bracing are used to ensure contact with the adhesive until it develops full bond strength. For plain wallboard, temporary bracing or nailing should be allowed to remain in place and joint or nail treatment should not be started for at least 24 hours after installation.

### METAL FRAMING AND FURRING

Gypsum board should be attached to metal framing and furring with drywall sheet metal screws spaced not more than 12 in. (300 mm) on centers along supports for both walls and ceilings. For fire-rated construction, gypsum board is applied parallel with supports using screws spaced 8 in. (203.2 mm) on centers along edges and 12 in. (300 mm) on centers in the field.

### CONCRETE AND MASONRY

Concrete and masonry are usually furred before gypsum board is applied. However, gypsum board can be laminated directly to interior and some above-grade exterior masonry (see Section 9.1) if the surfaces are sufficiently dry, smooth, plumb, and true (Fig. 9.5-22). Some laminating adhesives and stud adhesives are suitable for this purpose, but manufacturers' recommendations vary and should be consulted to ensure proper selection and application. While the adhesive is setting, temporary support is provided using threaded stub nails or cut nails.

### 9.5.5.4 Multi-Ply Construction

Multi-ply finishes consist of two or more layers of gypsum board and include (1) supported ceiling and wall assemblies

FIELD

3/8" to 1/2" (9.5 to 12.7 mm)

3/8" to 1" (9.5 to 25.4 mm)

7" (180 mm) on ceilings
8" (200 mm) on walls

**a**

3/8" to 1/2" (9.5 to 12.7 mm)

3/8" to 1" (9.5 to 25.4 mm)

Approx 12" (300 mm)

2" to 2-1/2" (50.8 to 63.5 mm)

8" (200 mm) side walls

7" (180 mm) ceilings

Row 4    Row 3    Row 2    Row 1    Row 2A    Row 3A    Row 4A

**b**

**FIGURE 9.5-18** (a) Single nailing method. (b) Double nailing method. First nails (shown by black dots) are installed starting with Row 1, then Rows 2 and 2A, and so on, working outward. Second nails (shown by circles) are installed in the same row sequence. (Courtesy Gypsum Association.)

and (2) self-supporting all-gypsum partitions. Face layers are laminated to a base layer of wallboard, backing board, or sound-deadening board with a few fasteners to ensure adhesive bond (see Fig. 9.5-7). Multi-ply construction, therefore, allows fastener treatment and minimizes surface defects such as shadowing and nail popping. It further provides inherent joint-reinforced construction and reduces the incidence of joint ridging. After appropriate fastener and joint treatment, the surface may be painted or finished with wall covering or tile.

Multi-ply finishes are particularly suitable where increased sound isolation or fire resistance is desired and generally provide better surface quality than single-ply construction. Sound ratings of up to 60 STC and 2-hour fire ratings are possible with some two-ply framed assemblies and multi-ply all-gypsum partition assemblies. Framed load-bearing wall assemblies with two-ply finishes receive somewhat lower fire ratings than non-load-bearing assemblies.

Two-ply construction on framed ceilings does not provide substantially higher fire or impact sound ratings than single-ply construction. Therefore, the decision to use two-ply finishes on ceilings is usually based on improved surface quality rather than on fire or sound ratings. Unfortunately, two-ply construction is inherently more expensive than single-ply construction.

The gypsum board finishes discussed in this subsection are of two-ply construction supported by either wood or metal framing and furring, or by concrete or masonry. The spacing for framing and furring depends mainly on the base-ply thickness, whether the base ply is attached perpendicular or parallel to supporting construction, and whether the face ply is adhesively bonded to the base ply (Fig. 9.5-23). Regardless of these recommendations, support spacing in most commercial installations is 16 in. (400 mm) on centers.

**BASE-PLY ATTACHMENT**

The base ply of supported wall and ceiling finishes may be regular core wall-

**FIGURE 9.5-19** Beads of stud adhesive are applied: (a) single under field of board or (b) parallel under joints. (Courtesy Gypsum Association.)

**FIGURE 9.5-20** Fastener Spacing for Nonstructural Adhesive

| Location | Support Spacing, o.c. | | Fastener Spacing, o.c.* | | | |
|---|---|---|---|---|---|---|
| | | | Nails | | Screws | |
| | in. | mm | in. | mm | in. | mm |
| **Ceilings** | 16 | 406 | 16 | 406 | 16 | 406 |
| | 24 | 610 | 12 | 305 | | |
| **Load-bearing walls** | 16 | 406 | 16 | 406 | 24 | 610 |
| | 24 | 610 | 12 | 305 | 16 | 406 |
| **Non-load-bearing walls** | 16 | 406 | 24 | 610 | 24 | 610 |
| | 24 | 610 | 16 | 406 | | |

*Applies to both field and perimeter of board.
(Courtesy Gypsum Association.)

**FIGURE 9.5-21** Supplemental fastener spacing for structural adhesives. (Courtesy Gypsum Association.)

board, but backing board is used more often because it costs less. Sound-deadening board may be used where increased sound isolation is required. Fastener size and spacing for applying sound-deadening board vary for different fire- and sound-rated constructions.

A base ply can be attached to supporting framing or furring with nails, staples, resilient clips, or screws, but resilient clips are limited to walls. Base-ply attachment for self-supporting partitions is discussed in Section 9.5.5.6.

***Wood Framing and Furring*** In wood framing and furring, the fastener spacing for nails, screws, and staples varies primarily with the method of face-ply attachment (Fig. 9.5-24).

*Nail attachment* of base ply should be done with drywall nails long enough to provide the recommended penetration into supporting construction. Where the face ply is to be laminated, the base ply should be nailed as recommended for single-ply construction. Double nailing is generally not used because nail heads will be covered by the face ply and fastener treatment is not a consideration.

When the face ply is to be attached with mechanical fasteners, fewer nails are required in the base ply because the

a

b

**FIGURE 9.5-22** Laminating adhesive is applied with a notched spreader. Stud adhesive is applied in beads with a caulking gun. (Courtesy Gypsum Association.)

**FIGURE 9.5-23** Support Spacing for Two-Ply Construction

| Thickness of Base Ply | | Location | Application of Base Ply[a] | Maximum Support Spacing, o.c. | |
|---|---|---|---|---|---|
| in. | mm | | | in. | mm |
| 3/8 | 9.5 | Ceilings[b,c] | Perpendicular | 16 | 406 |
| | | Walls | Perpendicular or parallel | 24[d] | 610 |
| 1/2 and 5/8 | 12.7 and 15.9 | Ceilings[c] | Perpendicular | 24 | 610 |
| | | Walls | Perpendicular or parallel | 24 | 610 |

*Note:* Installation with mechanical fasteners with or without adhesive.

[a]Direction of board edges relative to supports.

[b]Two-ply 3/8-in. (9.5 mm) ceilings should not support insulation.

[c]For two-ply ceilings to receive a water-based spray texture finish, use 5/8-in. (15.9 mm) board applied perpendicular to 24-in. (610 mm) supports or 1/2 in. (12.7 mm) perpendicular to 16-in. (406 mm) supports.

[d]16 in. (406 mm) o.c. if no adhesive used between plies.

(Courtesy Gypsum Association.)

**FIGURE 9.5-24** Base Ply Fastener Spacing for Two-Ply Application in Inches (mm) O.C.

| Location | Nail Spacing | | | Screw Spacing | | Staple Spacing | |
|---|---|---|---|---|---|---|---|
| | Framing Spacing in. (mm) | Laminated Face Ply in. (mm) | Nailed Face Ply* in. (mm) | Laminated Face Ply in. (mm) | Screwed Face Ply* in. (mm) | Laminated Face Ply in. (mm) | Screwed Face Ply* in. (mm) |
| **Walls** | 16 (406) | 8 (200) | 24 (610) | 16 (406) | 24 (610) | 7 (178) | 16 (406) |
| | 24 (610) | 8 (200) | 24 (610) | 12 (305) | 24 (610) | 7 (178) | 16 (406) |
| **Ceilings** | 16 (406) | 7 (178) | 16 (406) | 12 (305) | 24 (610) | 7 (178) | 16 (406) |
| | 24 (610) | 7 (178) | 16 (406) | 12 (305) | 24 (610) | 7 (178) | 16 (406) |

*Fastener spacing for face ply should be the same as for single-layer application.
(Courtesy Gypsum Association.)

nails in the face ply will support the base ply as well.

Except for predecorated wallboard, base plies on walls should be installed parallel to framing members. On ceilings, base plies should be installed perpendicular to framing members. At inside corners, only the overlapping base ply should be nailed and fasteners should be omitted from the face ply to create a *floating corner* more capable of resisting structural stresses (Fig. 9.5-25).

*Screw attachment* of the base ply to wood framing or furring should be done with drywall wood screws providing at least $5/8$ in. (15.9 mm) penetration into supporting construction. Spacing of screws depends on the method of face-ply attachment and supporting construction (see Fig. 9.5-24). Inside corners should be floated as described for nailing.

*Staple attachment* is applicable to wood supporting construction only. Staples should provide at least $5/8$ in. (15.9 mm) penetration into supporting mem-

bers. Recommended spacing for staples is given in Figure 9.5-24. Staples should be driven with the crown perpendicular to the edges of the board, except that where edges fall over supports, they should be driven parallel to the edges (Fig. 9.5-26). Staple crowns should bear tightly against the board without breaking the face paper.

***Metal Framing and Furring*** Base plies should be attached to metal framing or furring with drywall sheet metal screws. Screw spacing on walls should be 12 in. (300 mm) on centers in the field, 8 in. (200 mm) on centers along the edges when the face ply is laminated, and 16 in. (400 mm) on centers throughout the board when the face ply is attached with screws. Screws should be long enough to penetrate metal studs or furring by at least $3/8$ in. (9.5 mm).

Two-ply ceiling finishes are not commonly used on metal framing or furring because they do not substantially im-

prove fire resistance or impact sound isolation. They do provide increased airborne sound isolation and may be used where this is an important consideration.

***Concrete and Masonry*** Concrete and masonry have inherently good fire resistance and sound isolation, which two-ply finishes do not substantially improve. However, when single-ply gypsum board cannot be laminated directly to masonry or concrete and the walls must be furred, two-ply finishes may be desired for superior surface finish. Base plies are applied the same way as they are on other furring.

### FACE-PLY ATTACHMENT

Joints in the face ply should be offset from joints in the base ply by at least 10 in. (250 mm). Gypsum board can be applied either horizontally or vertically, whichever results in less waste of materials and shorter joint lengths. Wallboard applied horizontally on walls usually re-

**FIGURE 9.5-25** At floating corners, base ply is nailed on only one side of the interior corner. (Courtesy Gypsum Association.)

VERTICAL APPLICATION

STAPLES SPACED 8 IN. (203.2 mm) O.C. CROWNS PERPENDICULAR TO EDGES EXCEPT AT EDGES OVER SUPPORTS

SUPPORT SPACING 16 in. (406 mm) OR 24 in. (610 mm) O.C.

HORIZONTAL APPLICATION

**FIGURE 9.5-26** Staple attachment of base ply in wood supporting construction. (Courtesy Gypsum Association.)

sults in shorter and less conspicuous joints and therefore is preferred. However, predecorated wallboard which is to be left with joints untreated and plain wallboard which is to be trimmed with battens are installed vertically.

***Nail and Screw Attachment*** When the face ply is attached with mechanical fasteners with no adhesive, the maximum spacing and minimum penetration recommended for nails and screws in singleply construction should be used in accordance with GA-216. However, the specifications for some proprietary sound- and fire-rated partition assemblies may differ.

***Adhesive Attachment*** The face ply is typically attached to gypsum board and sound-deadening board by *sheet lamination* or *strip lamination*. Generally, strip lamination is specified for systems with sound-deadening board.

*Sheet lamination* involves covering the entire back of the face ply with laminating adhesive using a notched spreader or another suitable tool (Fig. 9.5-27). The size and spacing of the notches vary, depending on the adhesive. The wallboard is then erected using moderate pressure. Adhesive squeezed out at joints should be promptly removed. In *strip lamination* the face ply is applied vertically, and adhesive is spread in vertical ribbons spaced 16 to 24 in. (400 to 600 mm) on centers, depending on the adhesive and assembly (Fig. 9.5-28).

Supplemental mechanical fasteners or temporary wood bracing are required to ensure a close adhesive bond between base and face plies. When structural ceiling members are used to anchor temporary wood bracing, ceiling installation should be delayed until after the wall finish has been installed.

Temporary fasteners may be used when the face ply is laminated to gypsum board in wood construction and a fire rating is not required. These may be double-headed nails, stucco furring nails, or drywall nails, generally spaced to provide face support approximately 24 in. (600 mm) on centers in each direction. Nails should be long enough to ensure proper penetration into supporting construction. Nails should be driven through gypsum board washers or other rigid material to facilitate removal without damaging the face ply. After nails are removed, nail holes should be *dimpled*

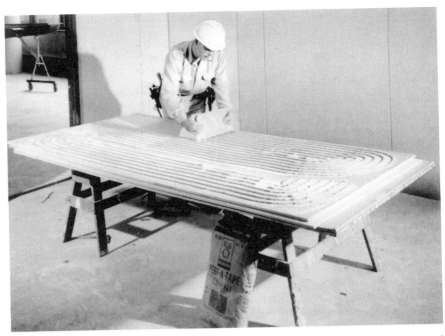

**FIGURE 9.5-27** Sheet lamination. A notched spreader or laminating box (shown) can be used to apply adhesive. (Courtesy United States Gypsum Company.)

(slightly depressed with a curved-face hammer) and finished with compound, as for permanent nails.

Some permanent mechanical fasteners are recommended in each laminated gypsum board panel. Usually, these can be spaced at the perimeter of the board, where they will be concealed by joint treatment or trim. In the face ply of fire-rated assemblies, permanent fasteners are usually required in the field as well as at the perimeter. Fastener spacing depends on the tested assembly construction and is not related to whether adhesive is used between plies.

In sound-rated partitions where fire resistance is not a consideration, the face ply is generally applied vertically over sound-deadening board with permanent mechanical fasteners at the tops and bottoms of boards. Intermediate fasteners may be omitted along vertical joints and in the field if temporary bracing is provided until adhesive has developed sufficient bond strength,

#### 9.5.5.5 Cracking and Ridging Precautions

Improper installation and adverse job conditions may cause surface imperfections such as cracking at corners and ridging at joints. *Corner cracking* may result from differential structural stresses at inside corners where adjacent walls or

walls and ceilings meet. *Joint ridging* often occurs when joints are treated during cold or humid weather; it occurs more frequently when joint compound is applied too thickly and is feathered over a wide area. These finish defects are not commonly encountered in multi-ply construction.

In single-ply construction, cracking at interior angles can be minimized by *corner floating*; ridging can be substantially reduced by *back blocking* or *strip rein-*

**FIGURE 9.5-28** Strip lamination. The laminating head applies adhesive beads on the base play in parallel strips. (Courtesy United States Gypsum Company.)

*forcing*. However, two-ply finishes also are effective for both of these purposes and often may be more practical. A proprietary bullnose-edge wallboard has been developed specifically to overcome the problem of ridging in single-ply construction. It utilizes a special refill joint compound with nonstriking and fast-setting properties which develops high initial bond strength Fig. 9.5-29).

## CORNER FLOATING

*Corner floating*, which can be used with either screw or nail attachment, consists of omitting some fasteners at interior corners (Fig. 9.5-30). When corners are *floated*, the gypsum board should be applied to the ceiling first. It should be fitted snugly at both wall-to-wall and ceiling-to-wall intersections. Framing members or blocking should be used at corners to provide solid backup even though they may not receive fasteners.

At ceiling-to-wall intersections, the last row of fasteners should be omitted on the wall so that nails start 8 in. (200 mm) from the intersection. On ceilings, where ceiling joists are parallel to the intersection, nailing should start at the intersection; where the joists are perpendicular, nails should start 7 in. (180 mm)

FIGURE 9.5-29 Bullnose-edged wallboard requires prefilling joint channels before standard joint treatment is applied. (Courtesy Gypsum Association.)

FIGURE 9.5-30 Corners are floated by omitting fasteners at the intersection of walls and of walls and ceilings, as shown by the shaded areas. (Courtesy Gypsum Association.)

from the intersection (see Fig. 9.5-30). For interior corners at the meeting of walls, the last row of fasteners should be omitted from the board first applied only, and the overlapped board should be nailed conventionally. When double nailing, the same unrestrained clearances at corners should be maintained as for single nailing, described in Section 9.5.5.3.

## BACK BLOCKING

*Back blocking* is the reinforcing of joints against stresses that cause surface defects, particularly ridging. Simple back blocking consists of laminating wallboard blocks between framing members. To reinforce the wallboard ends by back blocking, it is necessary to float the ends by locating end joints midway between framing members. However, because ends are not tapered, the need for relatively high crowns and wide feathering at such joints still may cause ridging when joint treatment is done under adverse weather conditions. End-joint tapering can be used in conjunction with floating to reduce the need for high crowns and wide feathering at joints (Figs. 9.5-31 and 9.5-32).

Back blocking and associated constructions most often are performed on ceilings because imperfections are more conspicuous on such large, unobstructed, obliquely lighted surfaces.

## STRIP REINFORCING

*Strip reinforcing* is a method of reinforcing joints against ridging that offers many of the advantages of two-ply construction

FIGURE 9.5-31 Cross section shows how a floated end joint is tapered and back-blocked. A brace is temporarily nailed over a wood strip, which depresses the end of the panels. (Courtesy United States Gypsum Company.)

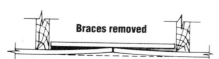

FIGURE 9.5-32 When the strips shown in Figure 9.5-31 are removed, the tapered formation remains. (Courtesy United States Gypsum Company.)

at reduced cost. By combining adhesive lamination with mechanical fastening, the number of nails in the face ply is reduced. Although it results in thicker assemblies than back blocking, strip reinforcing can achieve the same results in possibly less time.

Strips of $^3/_8$-in. (9.5-mm)-thick backing board or scrap wallboard 8 in. (200 mm) wide are installed 24 in. (600 mm) on centers perpendicular to framing members, similar to cross furring. Strips may be 4 in. (100 mm) wide at the base of a partition and at a partition-ceiling intersection. This cross stripping supports the edges of the board and provides intermediate support in the field of the board. To provide support at the butt ends of a board, additional short strips are secured directly over the supporting member between the cross stripping. Strips are nailed about 1 in. (25.4 mm) from the edge. Arrangement of strips, fasteners, and boards is illustrated in Figure 9.5-33.

Laminating adhesive is applied to the strips with a strip laminating head in parallel beads. Wallboard should be $^1/_2$ in. (12.7 mm) thick or thicker and should be installed perpendicular to the framing members so that long edges fall on the cross stripping and the ends are supported on the short intermediate strips. When supporting members are spaced 16 in. (400 mm) on centers, the nail spacing should be 24 in. (600 mm) on centers at

CEILINGS: PERIMETER, FIELD NAILS (X) AT EACH SUPPORT

WALLS: PERIMETER NAILS (X) ONLY AT EACH SUPPORT

FIGURE 9.5-33 Strip reinforcing. When support spacing exceeds 16 in. (406 mm) on centers, additional screws (dots) are required midway between supports. (Courtesy Gypsum Association.)

the ends and over each support at the edges for walls, with temporary double-headed nails at every other stud in the field until the adhesive sets. On ceilings, nails should be spaced 24 in. (600 mm) on centers along each support.

When supporting members are spaced more than 16 in. (400 mm) on centers, $1\frac{1}{2}$ in. (38 mm) drywall screws are required in the gypsum board strips midway between framing members for both walls and ceilings, as shown in Figure 9.5-33. Appropriate drywall nails should be used to attach both stripping and gypsum board panels.

### 9.5.5.6 Special Constructions

Most special constructions designed to improve sound, fire, moisture, and thermal control assume the use of gypsum board finishes with complete joint treatment and panel caulking. Materials used for caulking should be nonhardening, nonstriking, resilient sealants. In each case, the manufacturer's specific installation recommendations should be followed for individual products and systems.

### SOUND-ISOLATING CONSTRUCTIONS

The properties of sound and the principles of room acoustics and sound isolation are discussed in Section 13.1. Section 13.1 also includes detailed information on specific assemblies and additional information related to construction of assemblies to resist sound transmission.

Transmission of airborne and structure-borne sound can be reduced in wall and ceiling assemblies by:

- Eliminating flanking paths
- Separating the supporting structural construction of each finish surface
- Resilient mounting of finishes with channels, clips, or sound-deadening board
- Including sound-absorbing materials in the assembly cavity
- Increasing the weight of the surface construction

*Airborne Sound Through Flanking Paths* Failing to eliminate airborne sound transmission through flanking paths will seriously impair the effectiveness of other sound control methods and the performance of the entire assembly. Reduction of flanking paths includes constructing an airtight assembly and avoiding back-to-back location of recessed fixtures such as medicine cabinets

**FIGURE 9.5-34** Caulking with acoustical sealant to reduce sound transmission (a) around electrical outlet boxes and (b) beneath partition runners. (Courtesy Gypsum Association.)

and electrical, telephone, television, intercom, and other devices that perforate a gypsum board surface. In addition, openings for such fixtures and related piping outlets should be caulked (Fig. 9.5-34). The entire perimeter of a sound-isolating partition should be caulked at wallboard edges and under runners to make it airtight.

Assemblies intended for sound isolation should not contain heating or air conditioning ducts or ventilating outlets, and should be designed to eliminate flanking paths such as those through ceilings or around windows and doors adjacent to the walls, or through floors into crawl spaces below.

*Separated Construction* Separated construction in wood framing can be obtained by using separate sets of studs, each supporting a surface of one or more plies of gypsum board. This is ac-

complished by staggering the studs on the same top and bottom plates or by placing the studs on separate plates (Fig. 9.5-35). When separated wall-framing systems are used, cabinets, ceramic tile, lavatories, and other wall-hung fixtures can be installed in the conventional manner, which is not possible on walls with resilient mounting. Separated walls also can be developed with gypsum double-solid and semi-solid partitions.

*Resilient Mounting* Resilient-mounting systems and separated structural framing systems are the most effective and practical sound control methods capable of bringing lightweight framed assemblies to acceptable STC and Impact Insulation Class (IIC) ratings. Gypsum board finishes can be mounted on walls using resilient channels or resilient clips over sound-deadening board. Wall cabinets, lavatories, ceramic tile, medicine cabinets, and other fixtures should not be supported on resiliently mounted walls, as their weights may "short out" the construction acoustically.

*Resilient channels* are used in both wall and ceiling assemblies to isolate gypsum board from structural framing. Channels are screw-attached perpendicular to framing members (see Fig. 9.1-10). In ceiling assemblies, nails used to attach channels to wood joists may work loose as the framing lumber dries. Therefore, screw attachment is preferred.

*Resilient clips* can be used to attach the base ply of two-ply finishes to wood studs, but they are not recommended for ceiling application.

**FIGURE 9.5-35** Separated construction: (a) staggered studs on a single plate, (b) a double row of studs using separate plates. (Courtesy Gypsum Association.)

**Sound-Deadening Board** Sound-deadening board includes (1) wood- or cane-based fiberboard, (2) mineral (including glass) fiberboard, (3) plastic foamboard, and (4) regular core gypsum board. Sound-deadening board is manufactured $^1/_4$ to $^5/_8$ in. (3.2 to 15.9 mm) thick and 4 ft (1219 mm) wide by 8 to 9 ft (2438 to 2743 mm) long. It is used as a base layer to which wallboard is attached. It is installed parallel to framing members with end joints staggered on alternate studs (Fig. 9.5-36). Gypsum wallboard can then be laminated to the sound-deadening board either parallel or perpendicular to supporting construction. Nailing or screwing through the sound-deadening board into the framing is required to satisfy fire ratings, but it reduces sound-isolating effectiveness.

Wood fiberboard and plastic foamboard are used over wood construction or where noncombustible materials are not required. Mineral fiberboard is used with metal framing where noncombustible materials are required.

**Sound-Absorbing Materials** Mineral fiber insulating batts and blankets may be used in the cavity space of assemblies to improve resistance to sound transmission. The full height of the cavity should be filled with batts or blankets carefully fitted behind electrical outlets, blocking, and fixtures, and around cutouts necessary for plumbing lines.

Insulating batts and blankets may be paper- or vapor retarder-faced with flanges and are installed by stapling within the stud space. Stapling flanges across the stud face is not recommended because wrinkled flanges may prevent intimate board-to-stud contact. In metal stud and laminated all-gypsum-board partitions, the blankets can be attached to the back of the gypsum board. Batts and blankets without facings required in noncombustible construction are friction-fit within the stud space.

## FIRE-RESISTANT CONSTRUCTION

Fire-resistant construction in gypsum board assemblies requires the use of Type X gypsum board.

**Fire-Resistant Ceiling Panels** Fire-resistant gypsum board panels are cut into 2 by 2 ft (609 by 609 mm) and 2 by 4 ft (609 by 1219 mm) panels for use in some fire-rated ceilings. Common thicknesses are $^3/_8$, $^1/_2$, and $^5/_8$ in. (9.5, 12.7, and 15.9 mm). Fire ratings of up to 2 hours are possible with an approved suspension system (Fig. 9.5-37). Such ceilings also reduce sound transmission but have few sound-absorptive properties. Standard 4 by 8 ft (1219 by 2438 mm) fire-resistant gypsum board panels also may be used to provide fire-rated floor and ceiling assemblies for steel (see Fig. 9.1-6) and wood (Fig. 9.5-38) construction.

**Fire-Resistant Wall Panels** Fire resistance ratings of gypsum board assemblies can be improved by:

- Using fire-resistant Type X board
- Increasing gypsum board thickness
- Filling the stud cavity with mineral wool

**FIGURE 9.5-37** Fire-resistant $^5/_8$ in. (15.9 mm) ceiling panels in a fire-rated T-grid suspension system provide 2-hour fire protection for steel and concrete construction. (Courtesy Gypsum Association.)

**FIGURE 9.5-36** Resilient mounting with sound-deadening board improves sound isolation in metal frame construction. (Courtesy Gypsum Association.)

**FIGURE 9.5-38** Properly constructed wood joist floor and ceiling assemblies with $^5/_8$ in. (15.9 mm) fire-resistant gypsum board receives a 1-hour fire rating. (Courtesy Gypsum Association.)

Unless all of the materials used in the tested assembly are classified as noncombustible, the term *combustible* often is used after the assigned hourly classification. The following ratings are possible with gypsum board construction:

- *1 hour:* ordinary 2 by 4 (50 by 100) wood stud partitions, when finished with ⁵/₈-in. (15.9-mm)-thick Type X wallboard on both sides and properly installed
- *1 hour:* wood joist floor and ceiling assemblies with single-layer wallboard
- *2 hours:* wood and metal stud partitions with two layers of ⁵/₈-in. (15.9-mm)-thick Type X board on each side
- *2 hours:* metal studs with ³/₄-in. (19.1-mm)-thick gypsum board panels on both sides
- *2 hours and more:* steel-and-concrete assemblies protected by a suspended gypsum board ceiling

Another criterion for the fire hazard of exposed surface materials is the rate of surface flame spread. The flame spread rating of gypsum board is under 20, based on a flame spread of 0 for mineral fiber-cement board and 100 for red oak. Gypsum board's low flame spread rating more than satisfies most interior finish requirements.

### THERMAL INSULATING CONSTRUCTION

In some parts of the country, exterior wall and ceiling assemblies require thermal insulating construction to minimize heat gain or loss through the assembly. In such areas, a vapor retarder usually is required to retard the migration of moisture and condensation within the assembly. Refer to Chapter 17 for a discussion of moisture and heat control.

Typical materials used to increase the thermal resistance of assemblies with gypsum board finishes include (1) flexible fiber insulation (batts and blankets), (2) loose fill insulation, (3) rigid plastic insulation, (4) reflective foil-faced air space, and (5) air spaces provided by furring over solid materials such as masonry and concrete. For a discussion of thermal insulations, see Section 7.3.

Vapor control can be provided as an integral part of the gypsum board or as part of the insulation. Foil-backed gypsum board will usually provide some insulating value when combined with a dead air space, is an effective vapor retarder, and can be used with or without

other insulating materials. Foil-backed gypsum board should not be used:

- As backing material for ceramic tile
- As the face ply in laminated two-ply construction
- For laminating directly to masonry or concrete
- In connection with electric heating cables in electric panel heating systems

Rigid plastic foam and reflective foil insulation are materials with inherently low vapor permeability and generally do not require a separate vapor retarder. Flexible insulations such as batts and blankets are available with or without vapor retarders. Loose insulation requires either a separate vapor retarder, such as foil or polyethylene film, or an integral retarder provided by foil-backed gypsum board. Thermal insulation batts and blankets should be attached as recommended for sound-absorbing batts and blankets.

Two important considerations should not be overlooked in insulating design: (1) if reflective foil is to provide effective insulating value, a dead air space of at least ³/₄ in. (19.1 mm) should be provided, and (2) when a vapor retarder is used, it should be located on the warm side of the insulation. The installation of insulating and vapor control materials that are independent of a gypsum board finish is essentially the same as for other finishes. Only wood and metal furring and rigid plastic insulation which provide support for a gypsum board finish require special consideration in this section.

Rigid plastic insulation is intended primarily for use on exterior and below-grade concrete or masonry walls. Rigid insulating foamboards are made from expanded polystyrene or polyurethane in panels ¹/₂ to 3 in. (12.7 to 76.2 mm) thick, up to 4 ft (1219 mm) wide and 12 ft (3658 mm) long. These foamboards may be adhesive-bonded to masonry or concrete surfaces. They should not be laminated to masonry or concrete having a surface temperature of less than 45°F (7.2°C). They should be fitted together so that there will be no gaps between the boards which could create cold spots or allow moisture migration. Gypsum board may be bonded to rigid foam insulation with an approved adhesive or screw-attached to special channels provided with the foamboard. At least 24 hours of drying time should be allowed before gypsum board is laminated to rigid plastic insulation using an adhesive. In addition, the

gypsum board should be held in place by temporary bracing or double-headed nails which can be removed after sufficient bond strength develops. Supplemental fasteners are required.

### MOISTURE-RESISTANT CONSTRUCTION

Most gypsum board is sensitive to moisture and should not be used in locations subject to direct wetting or continuous high humidity. However, it can be used in limited exterior locations such as eave soffits, and porch and carport ceilings when properly finished and painted, and in interior locations such as tub, shower, and laundry areas when protected with wall tile. Such locations require special construction details and precautions, however.

This subsection covers the use of regular core and moisture-resistant gypsum board. Glass mesh mortar units that can serve as a base for ceramic tile in lieu of moisture-resistant gypsum board are discussed in Section 9.6.

*Exterior Locations* Regular or water-resistant wallboard used on horizontal surfaces in exterior locations should be at least ¹/₂ in. (12.7 mm) thick. Fascia boards at gypsum board soffits should be designed to form a drip edge to prevent water runoff across the board surface and to protect the board edge (Fig. 9.5-39). In addition, edges and ends at the perimeter of the board should be protected by metal

**FIGURE 9.5-39** Gypsum board can be protected from damp exterior surfaces by (a) air spaces and wood moldings or (b) metal trim.
(Courtesy Gypsum Association.)

edge trim. Unless protected by metal trim or another water stop, edges should be spaced at least $^1/_4$ in. (3.2 mm) from abutting masonry.

Exposed wallboard surfaces and metal trim should be painted with two coats of exterior paint. Adequate ventilation of eaves, ceilings, and attic spaces above gypsum board soffits should be provided.

***General Interior Location*** Water-resistant gypsum board is recommended for all interior walls subject to high humidity, except for high-moisture spaces like shower rooms, saunas, and steam rooms. Moisture-resistant gypsum board should not be used on ceilings. Foil-backed gypsum board should not be used in high-humidity interior areas because the foil backing may prevent moisture from being dissipated, thus trapping it within the board core. But where a high-humidity area is located on an exterior wall, adequate insulation and a vapor retarder on the warm side of the wall are necessary to prevent condensation in the wall cavity.

***Tub-Shower Areas*** Water-resistant gypsum board and proprietary highly water-resistant backing board are recommended as a base for wall tile in tub-shower areas.

Gypsum board in tub-shower areas should be at least $^1/_2$ in. (12.7 mm) thick. Stud spacing should not exceed 24 in. (609 mm) on centers. When the stud spacing exceeds 16 in. (406 mm) on centers, blocking should be installed not more than 4 in. (102 mm) on centers between studs, with the first row of blocking about 1 in. (25.4 mm) above the top of the tub or shower receptor. When a wall is to be finished with tile (see Section 9.6) or handicapped accessories are to be installed (see Section 1.4), closer stud spacings may be required.

Gypsum board should be applied perpendicular to framing members, with paperbound edges parallel to the top of the tub or shower receptor and spaced at least $^1/_4$ in. (3.2 mm) above it (Fig. 9.5-40). This space should be filled with a continuous bead of waterproof sealant. Generally, gypsum board should be fastened with nails spaced not more than 8 in. (203 mm) on centers or screws spaced not more than 12 in. (305 mm) on centers. However, when ceramic tile more than $^5/_{16}$ in. (7.9 mm) thick will be installed, nails should be spaced 4 in. (101 mm) on centers and screws 8 in. (203 mm) on centers.

**FIGURE 9.5-40** Gypsum board and wall tile installation details for tub-shower areas. (Courtesy Gypsum Association.)

Special water-resistant backing board or regular core gypsum board should be installed as recommended by the manufacturer or the Gypsum Association. When regular core gypsum board is used, joints between adjacent boards, including those at corners, should be taped and treated as for painting but should not receive finishing coats. Before installing tile, a water-resistant sealer should be applied to the entire wallboard surface, including treated joints and corners. When tile adhesive is used as a sealer, it should be spread separately from the bonding coat in a continuous, uniform coating approximately $^1/_{16}$ in. (1.6 mm) thick, covering all cut ends and all cutouts for fixtures. A Type I, highly water-resistant organic adhesive should be used to apply ceramic wall tiles.

A tile finish should cover all surfaces, as illustrated in Figure 9.5-41. The space between tile and tub rim, shower receptor, or subpan top should be caulked with a continuous bead of sealant (see Fig. 9.5-40). To prevent this caulked joint from opening due to settlement, the tub should be supported at the walls on metal hangers or vertical blocking fastened to the studs. Waterproof shower receptor pans or subpans should have an upstanding leg or flange at least 1 in. (25.4) higher than the water dam or threshold in the entry to the shower.

### ELECTRIC PANEL HEATING

Electric resistance elements may be incorporated in the gypsum board itself for one-ply application, or in the backing board for two-ply construction. Another type of radiant electric heating ceiling system is a built-up one incorporating a gypsum backing board, heating cables, a filler material, and a gypsum wallboard

FIGURE 9.5-41 Shaded areas show tile coverage of gypsum board in tub-shower areas. (Courtesy Gypsum Association.)

finish. Standard gypsum lath and plaster products also can be assembled into a radiant heating ceiling system by embedding heating cables in the plaster coats over gypsum lath. In each case, the manufacturer's detailed recommendations should be followed.

Heat loss calculations indicate the number of square feet of heating panels required in each space, which is usually less than the total ceiling area. The areas adjacent to heating panels in each space are covered with backing board or wallboard of matching thickness. Wallboard face panels may be finished in the normal manner, but installation of acoustical tile over such a system is not recommended.

Proper insulation design is important to the efficient operation of any electric panel heating installation. Thermal insulating batts used in the ceiling above a panel heating system should be unfaced friction-fit mineral or glass wool.

### SELF-SUPPORTING GYPSUM PARTITIONS

Self-supporting gypsum partitions are non-load-bearing, noncombustible, multiply partitions fabricated on the job almost entirely from gypsum boards. Inner plies consist of backing board in rib or sheet form to which wallboard face plies are laminated. Unlike the ceiling and wall assemblies described elsewhere in this section, which depend on wood, metal, concrete, or masonry construction to support the surface finish, gypsum partitions are entirely self-supporting and require only floor and ceiling runners to stabilize them. Runners should be attached securely to ceiling and floor construction with fasteners spaced 24 in. (609 mm) on centers, and should be carefully caulked

at partition perimeters and under runners to close air leaks. The maximum height of partitions is limited by the design and dimensions of the components. Therefore, the manufacturer's recommendations should be followed.

A variety of proprietary partition systems have been developed by manufacturers of gypsum products to minimize the number of operations and speed field erection of partitions. The chief types are:

- Semisolid, utilizing backing board ribs
- Solid, with coreboard (thicker backing board) inner plies
- Hollow
- Demountable

All gypsum board components such as ribs, coreboard, backing board, and wallboard are typically installed in the vertical position.

Gypsum partition products are used in self-supporting partition systems, engineered assemblies of board, and plaster products with special metal accessories designed for use with each system.

***Semisolid Partitions*** In semisolid partitions, the backing board core is replaced by 6- to 8-in. (150- to 203-mm)-wide gypsum board ribs laminated at the factory in 1 in. (25.4 mm) and $1^5/_8$ in. (41.3 mm) thicknesses (Fig. 9.5-42a). Ribs are adhesive-bonded in the field to the wallboard faces for single-ply construction or to backing boards for two-ply construction (Fig. 9.5-42b). In two-ply construction, regular or predecorated wallboard is bonded to form the finished surface. Veneer plaster finish may be used over an appropriate plaster base laminated to the ribs. Semisolid partitions generally are not made with standard plaster finishes. Partition thicknesses vary between 2 and 3 in. (50.8 and 76 mm), depending on rib and face panel thicknesses.

The ribs are cut 6 in. (150 mm) shorter than the gypsum board and are laminated to the face ply on one side of the partition either before or after it is erected. Perimeters of door and window openings are reinforced with additional ribs. Adhesive is spread on the erected face ply or on the ribs, keeping it 1 in. (25.4 mm) from the rib edges. These rib-reinforced panels are secured to wood or metal runners at the top and bottom with appro-

FIGURE 9.5-42 Semisolid gypsum partitions: (a) single-faced with both faces laminated to each rib and (b) double-faced with each two-ply face laminated to alternate ribs, forming separated rib construction.

priate drywall screws spaced 12 in. (305 mm) on centers.

The ribs on the erected panels then are spread with adhesive, and facing panels are installed and screw-attached to runners with screws spaced 12 in. (305 mm) on centers. To ensure an adequate bond at ribs, $1\frac{1}{2}$ in. (38 mm) drywall screws are provided at each rib 12 in. (305 mm) from the top and bottom and spaced not more than 36 in. (914 mm) on centers.

For partitions with two-ply faces, the ribs and backing board base plies are assembled and installed as for single-ply faces. Face plies then are sheet-laminated as for two-ply supported partitions using supplemental drywall screws.

Separated rib construction to improve sound-isolating effectiveness can be provided in either single- or two-ply ribbed partitions. Installation procedure is essentially the same as for ordinary ribbed partitions, but twice as many ribs are required, spaced 12 in. (305 mm) on centers and staggered on opposite sides of the partition (see Fig. 9.5-42b).

*Solid Partitions* Solid gypsum partitions are an extension of the concept of semisolid partitions. They consist of a solid coreboard to which wallboard faces are laminated on each side (Fig. 9.5-43a). The vertical joints in the face ply are offset at least 3 in. (76 mm) from joints in the coreboard. The face ply is attached by sheet lamination with supplemental drywall gypsum screws. Double- or triple-solid partitions separated by air spaces provide increased fire resistance and sound isolation (Fig. 9.5-43b).

Solid gypsum partitions may be field-erected from combinations of lath and plaster or backing boards and wallboards. Both generally employ wood or proprietary metal ceiling and floor runners to stabilize the partitions. In lath and plaster partitions, an integral flush metal base acts as a screed. Partitions also may be trimmed out with resilient or wood base.

*Backing board and wallboard partitions* consist of a 1-in. (25.4-mm)-thick backing board core to which $\frac{1}{2}$ or $\frac{5}{8}$ in. (12.7 or 15.9 mm) wallboard faces are laminated with adhesives. The core may consist of two job-laminated $\frac{1}{2}$-in. (12.7-mm)-thick backing boards or a 1 in. (25.4 mm) factory-laminated product referred to as *coreboard* (see Fig. 9.5-43a). Regular wallboards usually are used, requiring joint treatment and decorating, but in some instances, predecorated wallboards

with V-edges may be installed in a visually attractive repeat pattern. When veneer plaster finish is desired, veneer plaster base is used in place of wallboard. Double-solid partitions are constructed of two separate layers of lath or backing board separated by an air space, with plaster or wallboard faces on the exterior (Fig. 9.5-43b).

*Hollow Partitions* Hollow partitions are made with wood or metal studs supporting faces of plaster or wallboard (see Fig. 9.3-5). Studs usually are spaced 16 to 24 in. (406 to 609 mm) on centers and

are supported by wood or metal floor and ceiling runners. Wood studs normally are 2 × 4s (50 × 100s) or 2 × 6s (50 × 150s). Metal channel studs are available in $1\frac{5}{8}$ in. (41.4 mm), $2\frac{1}{2}$ in. (63.5 mm), $3\frac{5}{8}$ in. (92.1 mm), 4 in. (101.6 mm), and 6 in. (177.8 mm) sizes. Greater wall cavities may be produced using parallel sets of studs braced with wallboard.

Hollow partitions may be either of single-ply or two-ply construction. For single-ply construction, wallboard is nailed or screwed directly to the studs (Fig. 9.5-44a). For two-ply construction, backing board is clipped, nailed, stapled, or

**FIGURE 9.5-43** Solid gypsum partitions: (a) single solid partition, (b) double solid partition with air space between the two faces.

**FIGURE 9.5-44** Hollow gypsum board partitions: (a) screwed single ply, (b) two-ply construction with resilient mounting for improved sound control.

screwed to the studs, and wallboard is adhesive-bonded to this backing with a minimum of supplemental fastening. Where soundproofing is important, resilient channels may be used in either one- or two-ply construction to mount the board (Fig. 9.5-44b). When regular core wallboard is used as the exterior face, fasteners and joint treatment are required; predecorated wallboard normally is adhesive-bonded to backing board and has V-edge joints which can be left exposed.

***Demountable (Movable) Systems*** In situations where it may be desirable to rearrange partitions, demountable panel systems are frequently used. They are designed for speedy erection and for minimum loss of components when being demounted and rearranged.

Demountable partition systems use wallboard face panels attached to interior metal studs for stability. Various pressure-fit, clip, or screw systems are used to secure the wallboard panels to the studs. One system uses an H-section stud 24 in. (609 mm) on centers, the flanges of which are pressure-fit into grooves of factory-laminated, predecorated panels. In other systems, predecorated square-edge wallboard is adhesive-bonded directly to studs 24 in. (609 mm) on centers, and the edges are secured with nails or screws. Decorative batten trim then is used to conceal fastenings at 24 or 48 in. (609 or 1219) on centers (Fig. 9.5-45).

***Shaft Walls*** Shaft walls are lightweight gypsum board assemblies used for ele-

**FIGURE 9.5-45** One type of demountable partition.

vator shafts, electrical and mechanical enclosures, and stairwells. Shaft wall assemblies with metal stud framing weigh 10 to 13 psf (48.8 to 63.5 kg/m²), whereas masonry or masonry-and-plaster assemblies with similar fire ratings weigh between 35 and 65 psf (161 and 317.4 kg/m²).

Shaft walls built with gypsum board are not only light in weight, but also easier and less expensive to install than conventional masonry construction. They can be erected floor by floor from outside the mechanical or elevator shaft; they do not delay installation of the mechanical equipment inside the shaft due to masonry scaffolding; and they do not require extensive cleanup inside the shaft upon completion.

### RESURFACING EXISTING CONSTRUCTION

Gypsum board may be used to provide a new finish on existing walls and ceilings. If the existing surface is sufficiently smooth and true after appropriate preparation, and if it provides solid backup for the new wallboard without shimming, $1/4$-in. (3.2-mm)-thick gypsum board can be applied using adhesive, nails, or screws. Nails and screws should be long enough to provide recommended penetration into supporting members under the existing finish (see Figs. 9.5-9 and 9.5-10).

Existing surfaces that are not sufficiently smooth to receive gypsum board should be furred with wood or metal furring. Furring members should be shimmed as required to provide a suitably level base. The minimum gypsum board thickness for various support spacings and installation methods should be as recommended for new construction over furring.

Surface trim for mechanical and electrical equipment such as switch plates, outlet covers, and ventilating grilles should be removed and reinstalled over the new gypsum board finish.

### 9.5.6 JOINT AND FASTENER TREATMENT

After gypsum board has been erected, specific areas require treatment to achieve a smooth, homogeneous appearance. Panel joints in the same plane and interior corners require taping. Exterior corners and exposed edges require metal corner and edge trim. Both taped and trimmed areas, as well as fasteners, require finishing with compound (Fig. 9.5-46).

**FIGURE 9.5-46** Gypsum board joints are finished by embedding a perforated tape in bedding compound and covering it with several coats of topping compound. (Courtesy Gypsum Association.)

#### 9.5.6.1 Joint Taping and Finishing

A minimum of three coats of joint compound is recommended for taped joints. These include an embedding coat to bond the tape and two finishing coats over the tape. Each coat should be allowed to dry thoroughly so that the surface can be readily sanded and so that the compound will obtain maximum shrinkage before the next coat is applied.

Depending on atmospheric conditions, 12 to 24 hours may be required for each successive coat to dry unless quick-setting compound is used. The second and third coats should be lightly sanded when dry to remove surface irregularities and to blend them evenly with the adjacent board. Care should be taken not to raise the fibers on the board's face paper, as this may result in uneven surface texture and a spotty appearance.

#### FLUSH JOINTS

Embedding compound should be spread into the depression formed by the tapered edges of adjacent boards (Fig. 9.5-47a) and over all butt-end joints. Then tape is centered over the joint, smoothed to avoid wrinkling, and pressed into the compound by drawing a knife along the joint with enough pressure to squeeze out excess compound (Fig. 9.5-47b). This excess compound may be removed or redistributed as a skim coat over the tape (Fig. 9.5-47c). When the embedding coat is dry, a second coat is applied with the edges feathered 2 to 4 in. (50.8 to 102 mm) beyond the tape edges (Fig. 9.5-47d). Over the dry second coat, a third coat is spread with edges feathered

The labels in Figure 9.5-45: metal ceiling trim, metal ceiling runner, metal stud, predecorated wallboard, batten trim, metal floor runner, metal base.

a b c

d e f

**FIGURE 9.5-47** Finishing knives of varying widths are used to spread embedding compound, embed joint tape, and apply successively wider coats of topping compound. (Courtesy United States Gypsum Company.)

2 to 4 in. (50.8 to 102 mm) beyond the second coat to blend smoothly with the board surface (Fig. 9.5-47e). Joints can be treated in a similar manner by machine application of tape and joint compound (Fig. 9.5-48).

### INTERIOR CORNER JOINTS

Compound should be applied to both sides of interior corners. Then tape should be folded along the center crease and embedded snugly in the corner to form a right angle (Fig. 9.5-49). Excess compound is removed, and the surfaces are subsequently finished with additional coats of compound, as described for flush joints. Both hand and mechanical tools are available for corner finishing.

#### 9.5.6.2 Edge and Corner Trimming

All trim shapes except metal U-bead are installed after the gypsum board has been fastened in place (see Fig. 9.5-11). Metal cornerbeads and casing beads such as L,

LC, and LK are installed with the flange outward and are nailed, crimped, or screwed 5 in. (127 mm) on centers through the gypsum board into the supporting construction. Metal cornerbeads with paper flanges are secured by embedding the flanges in joint compound. All edge and corner trim except U-bead require finishing with joint compound to obtain a smooth surface and to conceal fasteners (Fig. 9.5-47f). The methods and sequence of three-coat application are the same as for fastener treatment.

Unlike other trims, U-bead is fastened through the back flange first; then the gypsum board edges are inserted into the trim. Finishing with compound is not required because the fasteners are concealed.

#### 9.5.6.3 Fastener Treatment

The first coat of embedding compound is applied over fastener heads at the same time that joints are taped. Fasteners may be treated (*spotted*) using an automatic

spotter or a finishing knife (Fig. 9.5-50). This knife should be pressed against the surface to finish, level, and remove excess compound. The second and third coats of compound are applied over the fasteners when similar coats are applied to taped joints, edges, and corners. Each successive coat is feathered slightly beyond the one previously applied. Compound is sanded between coats as required. Care should be taken not to raise the fibers in the face paper when sanding. Each coat should be thoroughly dry before the next coat is applied.

### 9.5.7 FINISH DEFECTS

Improper installation or adverse job conditions may result in surface imperfections such as nail popping, shadowing, cracking at corners, or ridging at joints. Such finish defects can be minimized or completely eliminated if their causes are understood and appropriate precautions are taken.

a

b

a

c

d

b

c

**FIGURE 9.5-48** An automatic taping machine applies embedding compound and joint tape on (a) ceiling and (b) wall; (c) a broad knife removes excess compound; and (d) a finishing tool with an interchangeable head applies successively wider coats of topping compound. Occupational Safety and Health Administration (OSHA) standards now require workers to wear hard hats (as in b) when using this type of equipment. Eye protection should also be worn. (Courtesy United States Gypsum Company.)

**FIGURE 9.5-49** Tape can be embedded in the corner (a) by hand or (b) with a tool. Finishing can also be done by hand or (c) with a corner tool. (Courtesy United States Gypsum Company.)

### 9.5.7.1 Nail Popping

Nail popping is a surface defect in which compound over a nail head is pushed out beyond the surface of the board. It is due primarily to loose boards caused by wood shrinkage, the use of excessively long nails, or improper nailing.

#### WOOD SHRINKAGE

Wood shrinks across the grain as its moisture content is reduced (see Section 6.1). As a piece of lumber shrinks, it tends to expose the shank of a nail driven into its edge (Fig. 9.5-51). If shrinkage is substantial and nails are too long, separation between board and framing member can result. A separation as small as $^1/_{32}$ in. (0.8 mm) can result in an objectionable nail pop. It is essential that framing and furring lumber have as low an initial moisture content as possible, that lumber be kept dry during storage and installation, and that nail lengths do not

substantially exceed those recommended for specific applications. Framing and furring lumber should be at the equilibrium moisture content for the area and should not exceed 15% moisture content. Nail penetration should be as near as possible to those recommended in

**FIGURE 9.5-50** Slightly recessed screw or nail heads are treated with (a) an automatic spotter or (b) a finishing knife. A hard hat and eye protection should be worn. (Courtesy United States Gypsum Company.)

**FIGURE 9.5-51** As a wood stud shrinks, separation may be (a) ¹/₁₆ in. (1.6 mm) when the nail penetrates 1 in. (25.4 mm) or (b) ¹/₈ in. (3.2 mm) when the nail penetrates 2 in. (50.8 mm). (Courtesy United States Gypsum Company.)

Figure 9.5-9. Annularly threaded nails require less penetration than smooth shank nails and tend to minimize nail popping.

## IMPROPER NAILING

When gypsum board is improperly nailed, loose boards may result due to loose nails, loose attachment, or punctured face paper.

*Loose Nails* Nails that either miss the underlying construction or enter it at an angle near the edge (Fig. 9.5-52a) can

**FIGURE 9.5-52** Nail popping can result from (a) loose nails that miss framing members, (b) loose boards, or (c) obliquely driven nails that puncture the face paper.

eventually be loosened by vibration until they protrude beyond the finish surface. To guard against such defects, the location of supporting members should be marked carefully on the gypsum board prior to nailing, nails should be driven perpendicular to the surface into the member, and nails that have missed the member should be removed promptly. After nails have been removed, nail holes should be dimpled and treated with compound.

***Loose Attachment*** Loose attachment can occur when too little pressure is applied to a board while it is being nailed (Fig. 9.5-52b). Even when substantial pressure is applied properly, it may be impossible to produce snug contact with supporting members if the board is cut too large and must be forced into place or if nailing is begun at the perimeter and proceeds inward. To prevent loose board attachment, panels should be cut to fit easily in place, pressure should be applied while nailing, and the proper nailing sequence from the middle of the panel outward should be observed. Use of screws and double nailing also tends to minimize loose boards.

***Punctured Face Paper*** Improper nail head design, obliquely driven nails (Fig.

9.5-52c), excessive and repeated impact during nailing, and using boards with extremely dry face paper can all result in the face paper being punctured during board installation. When the paper is punctured, the gypsum core is easily crushed and offers less resistance to loosening of the board from normal impact and vibration. In hot, dry periods, the face paper may become more brittle, increasing the hazard of puncturing. This hazard can be minimized by increasing the air humidity during storage and installation of gypsum board.

Other common causes of punctured face paper are misaligned or twisted members and projections caused by improperly installed blocking or bracing (Fig. 9.5-53). Intimate contact between gypsum board and supporting members is impossible in such cases, and hammer impact causes the board to rebound, which often ruptures the paper. Such defects should be corrected before the board is installed. Use of screws, adhesive, or two-ply construction will minimize problems resulting from such defects.

### 9.5.7.2 Shadowing

Shadowing is actually a surface discoloration, but it may appear similar to nail popping when viewed at a distance.

Shadowing is the result of an accumulation of dust or dirt caused by increased thermal conductivity and moisture condensation over a nail head. It occurs commonly on roof and ceiling assemblies and on exterior wall assemblies that are inadequately insulated, and in regions with great indoor-outdoor temperature variations. Shadowing can be remedied by periodic washing or decorating. It can be avoided by proper insulation design and use of laminated two-ply construction. Where a permanent ventilating system is employed, adequate air filtration and humidity control will be helpful.

### 9.5.7.3 Cracking and Ridging

Corner cracking may result from differential structural stresses at interior corners where adjacent walls or walls and ceilings meet. Joint ridging often results when joints are treated during cold or highly humid weather and occurs more frequently when joint compound is applied too thickly and is feathered over a wide area. These finish defects are not commonly encountered in multi-ply construction.

In single-ply construction, cracking at interior angles can be minimized by corner floating (see Section 9.5.5.3 and Fig. 9.5-14); ridging can be substantially re-

**FIGURE 9.5-53** Punctured face paper and loose boards can be caused by (a) misaligned members, (b) twisted members, or (c) projecting blocking. (Courtesy United States Gypsum Company.)

duced by back blocking (see Fig. 9.5-32) or strip reinforcing (see Fig. 9.5-33). A bullnose-edged wallboard (see Fig. 9.5-29) may also be used to help overcome ridging problems in single-ply construction.

## 9.5.8 DECORATING AND REDECORATING

A gypsum board installation should be inspected before it is decorated or redecorated. Special attention should be paid to joints and fasteners. Imperfections should be sanded or repaired.

It is essential that joint compound be thoroughly dry before any form of decoration is applied. Depending on temperature and humidity conditions, 24 hours or more of good drying time is required.

Gypsum board surfaces can be decorated with textured coatings, emulsion, or oil-base paints, or with paper, fabric, or vinyl wall coverings. It is necessary to prepare the board surface with a sealer or primer. A good sealer will seal the pores and lay down the fibers in the paper surface so that suction and texture differences between paper and joint compound will be equalized; it also will seal in surface impurities sometimes present in the wallboard surface. A good primer will provide uniform texture and will conceal color or surface variations. Some sealers incorporate the properties of primers as well.

A sealer or primer should be used on gypsum board beneath oil-base paints, lacquers, and enamels, because they tend to raise the fibers in the face paper. Sealers are used under wall coverings primarily to facilitate later removal without marring the board surface and to provide an adequate base for redecorating.

Most emulsion paints such as latex, resin, and casein do not raise the fibers in the face paper. Good latex paints have inherently good sealing and priming properties and require no special preparation of gypsum board surfaces. However, applying a pigmented emulsion sealer as a first coat or using at least two coats of emulsion paint generally will provide a durable, attractive finish if the gypsum board has been properly installed and if joints, fasteners, and trim have been properly treated.

Glue size, shellac, and varnish are not suitable as sealers or primers under paints or wall coverings.

If nail popping and other surface defects occur after decorating, it is advisable to wait until one full heating cycle has been completed before repairing and redecorating.

The information here is general in nature. Refer to Section 9.16 for further discussion about painting gypsum board.

# 9.6 Tiling

The term *ceramic* derives from the Greek word, *keramos*, which means "potter's clay." Ceramic tile is made from nonmetallic minerals fired at high temperatures and manufactured in modular unit sizes (tiles) which facilitate installation. The term *ceramic tile* includes several products of varying dimensions, properties, and appearance:

- *Glazed wall tile*: 4 by 4 in. (101.6 by 101.6 mm) or $4^1/_4$ by $4^1/_4$ in. (108 by 108 mm) units used mainly on walls, but some are sufficiently abrasion-resistant to make them suitable for floors.
- *Ceramic mosaic tile*: 1 by 1 in. (25.4 by 25.4 mm), 1 in. by 2 in. (25.4 by 50.8 mm), or 2 by 2 in. (50.8 by 50.8 mm) units used mostly for floors but used also on walls.
- *Paver tile*: Floor units similar to mosaic tile but generally 4 by 4 in. (101.6 by 101.6 mm) or larger.
- *Quarry tile*: 6 by 6 in. (177.8 by 177.8 mm) or larger natural clay units for floors. Other sizes and shapes are also available (Fig. 9.6-1).

**FIGURE 9.6-1** Pentagon-shaped tiles lend an elegant appearance to this setting. (Courtesy American Olean. [Ceramic] Tile Co. [now Dal-Tile Corporation].)

- *Decorative or commemorative tile*: ceramic materials made to serve purely decorative or commemorative uses (Fig. 9.6-2).

Ceramic tiles possess good or superior abrasion resistance, a variety of patterns and colors, and a high degree of resistance to moisture. These properties make them uniquely suited for areas exposed to severe foot traffic and in intermittent or continuous contact with water or corrosive chemicals. Ceramic tile is used for most interior areas of residential, commercial, institutional, and industrial buildings. Many types are also suited for exterior use.

The origins of ceramic tile date back 5000 years, as evidenced by burnt natural clay tiles found by archaeologists in early Egyptian and Babylonian excavations. Decorative and functional clay tiles were also an essential part of later Greek and Roman architecture. The Italian town of Faenza is credited with developing, in the fourteenth century, what is now known as *faience tile*, a highly decorative tile with a textured, irregular handmade appearance. A century later, Holland started production of *delft* tile, characterized by a figure or landscape in blue or violet.

In the United States, production of ceramic tile was spurred by the introduction of the dust-press method at the end of the nineteenth century. Samuel Keys is credited with first producing floor and wall tile in Pittsburgh in 1867. By 1937 there were 52 manufacturers of ceramic tile in the United States. It is generally agreed that more improvements in tile manufacturing and installation techniques have been made by the U.S. industry during the past 70 years than in the preceding 70 centuries.

Much of the progress is due to the research and educational activities of the Tile Council of America (TCA). In the early 1950s, the TCA sponsored research that resulted in the development of several adhesives and mortars that eliminated the need for thick setting beds, laborious presoaking of tile, and extended curing. To ensure satisfactory tile installations, the TCA:

- Licenses manufacturers of mortars and grout according to TCA formulations.
- Promulgates standards for the manufacture and installation of ceramic tile, which has resulted in a series of American National Standards Institute (ANSI) standards.
- Publishes and annually updates the *Handbook for Ceramic Tile Installation*. Much of the discussion and most of the installation details in this section have been abstracted from TCA's *Handbook*.

## 9.6.1 GENERAL RECOMMENDATIONS

Satisfactory ceramic tile installations depend on a sound, rigid, and dimensionally stable backing surface. Ceramic tile is installed on floors (Fig. 9.6-3) either in an unbonded, isolated, setting bed (thickbed method) or by direct bonding to the backing (thinset method). Direct bonding is made possible by synthetic resins possessing greater bonding power than conventional portland cement. The following subsections generally employ terms used in the tile industry. Refer to the Glossary for unfamiliar terms.

### 9.6.1.1 Materials

Materials involved in ceramic tile installation include:

- Tile
- Mortars and adhesives for setting tile
- Grouts for filling joints
- Related materials to complete the installation, such as portland cement, reinforcing metal lath or mesh, and cleavage membrane

#### CERAMIC TILE

Ceramic tile is made from clay or from clay mixed with other ceramic materials. Oxides may be included for glaze color-

**FIGURE 9.6-2** Artist's mosaic of cut tiles. Mosaic murals are also made commercially. (Courtesy American Olean. [Ceramic] Tile Co. [now Dal-Tile Corporation].)

**FIGURE 9.6-3** Basic tile installation methods: (a) thin-set method (tile is bonded to the supporting substrate); (b) conventional mortar (thickbed) method (tile and mortar setting bed are isolated from the backing by a membrane).

ing. Ceramic tile products are available in a broad range of sizes (Fig. 9.6-4), appearance characteristics, and functional properties (Fig. 9.6-5), which include perviousness to moisture, abrasion resistance, and other properties. These factors dictate the suitability of a particular tile for a specific purpose.

In discussing the appearance or properties of tile, it is convenient to separate the face from the body of the tile. Some tile faces are glazed; others are not. Glazed tile may have an opaque decorative glaze over a light-colored body or a transparent glaze which reveals the natural color of the body. Unglazed tiles, such as some mosaic tiles and all quarry tiles, are hard, dense, units of homogeneous composition throughout, including the face.

Tile bodies have varying degrees of density and porosity, resulting in greater or lesser perviousness to moisture. Tile with a body which has an absorption of less than 0.5% is *impervious*; tile with an

absorption of 0.5% to 3% is *vitreous*. The low absorption of impervious and vitreous tiles makes them suitable for exterior installations subject to freezing and to interior installations, such as swimming pools, that are subject to continuous immersion in water. *Semivitreous* tile, with an absorption of 3% to 7%, and *nonvitreous* tile, with an absorption between 7% and 18%, are not suitable for exposure to freezing or immersion in water.

Only two grades of tile are recognized by the TCA and the ANSI: *standard grade* and *seconds*. Standard-grade tile is as perfect as it is commercially practicable to manufacture. Individual tiles are harmonious in color, but may vary in shade, and are free from spots and defects visible from a distance of more than 3 ft (900 mm). Seconds may have minor blemishes and defects not permissible in standard grades, but are free from structural defects or cracks.

ANSI Standard 137.1 is the recognized industry standard for the manufacture, testing, and labeling of ceramic tile. According to this standard, tile should be shipped in sealed cartons with the grade of contents indicated by grade seals, except for quarry tile, which has the grade printed on the carton in lieu of grade seals. A grade seal is a strip of paper, not less than 2 in. (50.8 mm) by 4 in. (101.6 mm), applied over the container in such a way that it cannot be opened without breaking the seal.

Ceramic tile used where appearance is an important factor should be standard grade as defined in ANSI 137.1. Additional evidence of compliance with ANSI 137.1 is recommended in the form of a manufacturer's written certification, which is called a Master Grade Certificate.

Ceramic tile products are grouped by TCA 137.1 into four major types:

■ Glazed wall tile
■ Mosaic tile
■ Quarry tile
■ Paver tile

*Special-purpose* tile products combining the properties of the basic type also are manufactured, and those produced by reputable manufacturers have a proven record of success. Figure 9.6-5 summarizes the generally available types, sizes, and uses of ceramic tile. A variety of trim shapes are produced, such as caps, corners, and bases, in matching colors, both for thin-set and conventional installation (Fig. 9.6-6).

*Glazed Wall Tile*  Most glazed wall tile is $1/4$ in. (3.2 mm) to $5/16$ in. (7.9 mm) thick and 4 by 4 in. (101.6 by 101.6 mm) or $4^1/4$ by $4^1/4$ in. (108 by 108 mm), although larger units are also made. It is available in plain and scored surfaces, in bright, matte, and crystalline finishes, and in hand-decorated and sculptured designs (see Fig. 9.6-5). Glazed wall tile will not withstand impact or freezing and thawing conditions. Some glazed tile is suitable for flooring in areas subject to moderate foot traffic, as in most residential and some commercial areas. However, ANSI 137.1 does not cover this type of tile, and the manufacturer's specifications must be relied on to determine suitability for the intended exposure. Glazed tile suitable for flooring usually has a heavier glaze, with a matte or crystalline finish.

Most wall tile is made with an impervious glaze over a *nonvitreous* (absorbent) body and is intended for interior

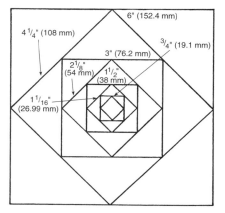

**FIGURE 9.6-4** Derivation of normal tile sizes of less than 36 in. (914 mm).

**FIGURE 9.6-5** Summary of Ceramic Tile Products

| Tile and Description | Suitable Uses | Limitations | Nominal Sizes, in.* |
|---|---|---|---|
| **Glazed Wall Tile**<br>Nonvitreous body, matte or bright glaze | All interior walls in residential, commercial, and institutional buildings | Not suitable for floors or areas subject to freezing | $\frac{5}{16}$ in. (7.9 mm) thick: $4\frac{1}{4}$ in. × $4\frac{1}{4}$ in. (107.95 × 107.95 mm), 6 in. × $4\frac{1}{4}$ in. (152.4 × 107.95 mm), 6 in. × 6 in. (152.4 × 152.4 mm), scored, octagon, hexagon, and other special shapes |
| Nonvitreous body crystalline glaze | All interior walls, as well as light-duty residential floors and countertops | Not suitable for areas subject to freezing | |
| Vitreous body, matte or bright glaze | Interior walls, countertops, refrigerator and chemical-tank linings, and exterior areas subject to freezing | Not suitable for floors | |
| **Ceramic Mosaic Tile**<br>Porcelain body, unglazed | All uses, interior and exterior, floors and walls; especially suitable for bathrooms, kitchens, and swimming pools | None except limitations imposed by installation method and grout | $\frac{1}{4}$ in. (6.4 mm) thick: 1 in. × 1 in. (25.4 × 25.4 mm), 2 in. × 1 in. (50.8 × 25.4 mm), 2 in. × 2 in. (50.8 × 50.8 mm) |
| Porcelain body, glazed | Interior and exterior walls in residential, commercial, and institutional buildings; decorative inserts with above tile | Not suitable for floors | |
| Natural clay body, unglazed | Swimming pool runways; floors of porches, entrances, and game rooms in the home; floors and walls of commercial, institutional, and industrial buildings | Exterior use requires special frostproof body | |
| Natural clay body, glazed | Walls in residential, commercial, and institutional buildings; decorative inserts with above tile | Exterior use requires special frostproof body | |
| Conductive | Hospital operating rooms and adjoining areas; where required to safely dissipate dangerous charges of static electricity | Requires special installation procedures according to NFPA No. 56A | |
| **Quarry Tile** | Exterior and interior; from moderate to extra-heavy-duty floors in all types of construction | Recommended for interior use and exteriors not subject to freezing | $\frac{1}{2}$ in. (12.7 mm) thick: 3 in. × 3 in. (76.2 × 76.2 mm), 4 in. × 4 in. (101.6 × 101.6 mm), 6 in. × 3 in. (152.4 × 76.2 mm), 6 in. × 6 in. (152.4 × 152.4), 8 in. × 4 in. (203.2 × 101.6 mm), and special shapes |
| **Paver Tile** | Same as quarry tile | None | $\frac{3}{8}$ in. (9.5 mm) thick: 4 in. × 4 in. (101.6 × 101.6 mm), $\frac{1}{2}$ in. (12.7 mm) thick: 4 in. × 4 in. (101.6 × 101.6 mm), 6 in. × 6 in. (152.4 × 152.4 mm), 8 in. × 4 in. (152.4 × 101.6 mm) |

*Sizes shown are those considered typical by the governing industry standard, ANSI A137.1; many more sizes and shapes are produced by tile manufacturers. Special purpose tiles that combine properties of several types also are produced.

use. Some wall tile, however, is made with an *impervious* or *vitreous* body suitable for exposure to moisture or freezing. All wall tile is glazed, the majority with a *bright* (shiny) or *matte* (dull) finish. Some tile is made with other special glazes which are more appropriate for light residential traffic than are bright or matte glazes.

Most wall tile is made with a cushioned-edge profile, but square-edge tile is also available. Soap dishes, paper holders, towel bar posts, and other accessories for toilet room and bathroom uses are available in matching colors.

***Ceramic Mosaic Tile*** Ceramic mosaic tile is made in two body types, *porcelain*

and *natural clay*, in relatively small sizes (Figs. 9.6-7 and 9.6-8). It is generally about $\frac{1}{4}$ in. (6.4 mm) thick, with a surface area of less than 6 sq in. (3871 mm²). Typical sizes are 1 by 1 in. (25.4 by 25.4 mm), 1 by 2 in. (25.4 by 50.8 mm), and 2 by 2 in. (50.8 by 50.8 mm). This tile may be either glazed or unglazed. Unglazed tiles have a wide range of nat-

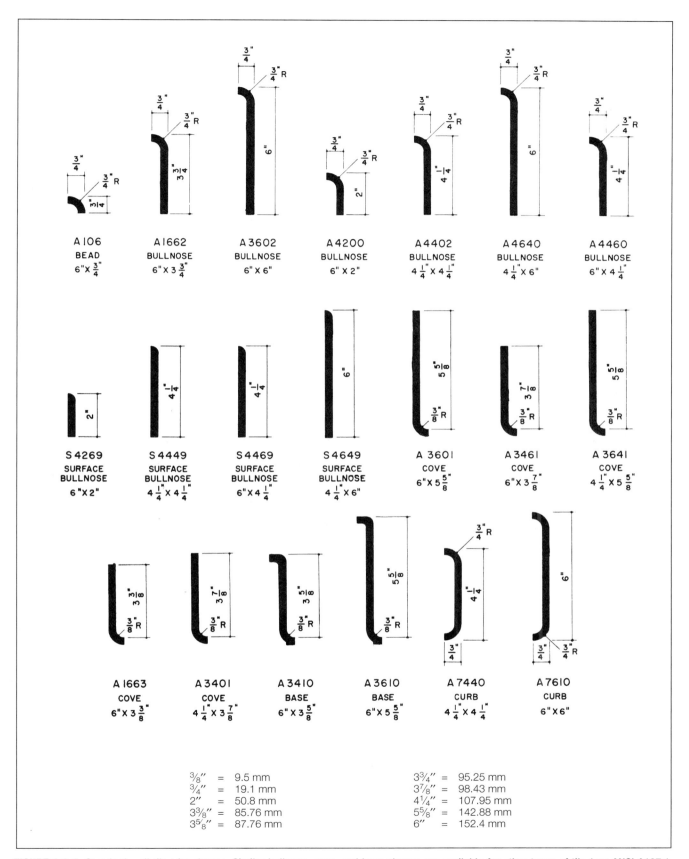

| | | |
|---|---|---|
| A 106 BEAD 6" X 3/4" | A 1662 BULLNOSE 6" X 3 3/4" | A 3602 BULLNOSE 6" X 6" |

A 106
BEAD
6" X 3/4"

A 1662
BULLNOSE
6" X 3 3/4"

A 3602
BULLNOSE
6" X 6"

A 4200
BULLNOSE
6" X 2"

A 4402
BULLNOSE
4 1/4" X 4 1/4"

A 4640
BULLNOSE
4 1/4" X 6"

A 4460
BULLNOSE
6" X 4 1/4"

S 4269
SURFACE
BULLNOSE
6" X 2"

S 4449
SURFACE
BULLNOSE
4 1/4" X 4 1/4"

S 4469
SURFACE
BULLNOSE
6" X 4 1/4"

S 4649
SURFACE
BULLNOSE
4 1/4" X 6"

A 3601
COVE
6" X 5 5/8"

A 3461
COVE
6" X 3 7/8"

A 3641
COVE
4 1/4" X 5 5/8"

A 1663
COVE
6" X 3 3/8"

A 3401
COVE
4 1/4" X 3 7/8"

A 3410
BASE
6" X 3 5/8"

A 3610
BASE
6" X 5 5/8"

A 7440
CURB
4 1/4" X 4 1/4"

A 7610
CURB
6" X 6"

| 3/8" | = | 9.5 mm | 3 3/4" | = | 95.25 mm |
|---|---|---|---|---|---|
| 3/4" | = | 19.1 mm | 3 7/8" | = | 98.43 mm |
| 2" | = | 50.8 mm | 4 1/4" | = | 107.95 mm |
| 3 3/8" | = | 85.76 mm | 5 5/8" | = | 142.88 mm |
| 3 5/8" | = | 87.76 mm | 6" | = | 152.4 mm |

**FIGURE 9.6-6** Standard wall tile trim shapes. Similar bullnose, cove, and base shapes are available for other types of tile (see ANSI A137.1 or the manufacturer's literature). (Courtesy American Olean. [Ceramic] Tile Co. [now Dal-Tile Corporation].)

**FIGURE 9.6-7** The range of ceramic mosaic tile uses is virtually limitless, including decorative walls and floors indoors and outdoors. The large, flat wall on this hotel is finished in ceramic mosaic tile. (Courtesy American Olean. [Ceramic] Tile Co. [now Dal-Tile Corporation].)

**FIGURE 9.6-8** This decorative pattern was made with ceramic mosaic tile. (Courtesy American Olean [Ceramic] Tile Co. [now Dal-Tile Corporation].)

soluble adhesive, so that it can be easily removed before the joints are grouted.

Available types include:

- *Unglazed porcelain (frostproof) tile:* Made by firing refined ceramic materials; dense, impervious, smooth, and highly stain- and wear-resistant; clear, luminous colors or granular blends; available with an abrasive surface in a few sizes and colors.
- *Glazed porcelain (frostproof) tile:* Made with the same body as unglazed porcelain tile; glazed with clear (transparent) glaze, ceramic glaze, textured glaze, metallic glaze, or special decorator glaze.
- *Unglazed natural clay (frostproof or nonfrostproof) tile:* Dense and abrasion-resistant; made from unwashed clays; rugged and slightly textured surface tough enough to minimize slipping; easily cleaned; also available with a slip-resistant face in 1 by 1 in. (25.4 by 25.4 mm) squares only, in limited colors.
- *Glazed natural clay (frostproof or nonfrostproof) tile:* Same body as unglazed natural clay tile with a clear, ceramic, metallic, or decorator glaze.
- *Conductive tile:* Impervious; with additives that minimize electrostatic buildup and explosion hazard in rooms where explosive gas mixtures are used, such as hospital rooms and laboratories; tile and installation must comply with National Fire Protection Association (NFPA) Bulletin No. 56A.

***Quarry Tile*** Quarry tile is an unglazed tile with a surface area of 9 sq in. (58.06 mm$^2$) or more, $^1/_2$ in. (12.7 mm) thick or thicker, made by the extrusion process from natural clay or shale (Fig. 9.6-9). Typical tile sizes are 3 by 6 in. (76.2 by 152.4 mm), 6 by 6 in. (152.4 by 152.4 mm), 8 by 4 in. (203.2 by 101.6 mm), 9 by 6 in. (228.6 by 152.4 mm), and 9 by 9 in. (241.3 by 241.3 mm) in square, hexagonal, and other special shapes. Tile thicker than $^3/_4$ in. (19.1 mm), made of the same materials by the same process, is called *packing house tile.*

***Paver Tile*** As the name implies, paver tiles are used primarily for floors. They are similar in appearance to quarry tile but more similar to mosaic tile in composition and physical properties, though relatively thicker ($^1/_2$ in. [12.7 mm] is typical) and larger (6 sq in. [3871 mm$^2$] or more). Typical tile sizes are 4 by 6 in.

ural earth colors that extend throughout the body of each tile. Additives can be mixed in with the other components for special effects: abrasives for *slip-resistant* tile and carbon black for *conductive* tile.

Individual tiles generally are face- or back-mounted to form larger units, to facilitate handling, and to speed installation. Back-mounted tile has perforated paper, fiber, resin adhesive, or another suitable bonding material permanently attached to the back and edges of the tile so that a portion of each tile is exposed to the mortar or adhesive. Face-mounted tile has a layer of kraft paper applied to the face of the tile, usually with a water-

(101.6 by 152.4 mm), 6 by 6 in. (152.4 by 152.4 mm), and 4 by 8 in. (101.6 by 203.2 mm) (see Fig. 9.6-9).

## TILE-SETTING PRODUCTS

Commercially available tile-setting products possess a variety of special properties, each intended to make them suitable for setting tile over certain backings or conditions. Figure 9.6-10 summarizes several common tile-setting products and the governing industry standards. Tile-setting products not covered by industry standards may also be used if they are produced by reputable manufacturers and have a proven record of success.

Portland cement mortar, dry-set mortar, latex-portland cement mortar, epoxy mortar, and organic adhesives and their installation should conform to the appropriate industry standard (see Fig. 9.6-10). Products not covered by these standards should conform with the recommendations of their manufacturers and good industry practice.

***Neat Portland Cement (Cement-Water Paste)*** Neat portland cement can be used as a bond coat for tile over a still plastic portland cement mortar setting bed. Neat cement bond coats are considered to be mortars. Dry-set and latex-portland cement mortar can be used to bond ceramic tile to a mortar bed that has been cured for at least 24 hours. Dry-set mortar can also be used as a $1/16$-in. (1.6-mm)-thick bond coat over a still plastic mortar bed. Mortar setting beds to receive organic adhesives must be fully cured and dry. To ensure satisfactory installations in the proper plane, mortar beds should be installed by the tile-setting contractor.

Resinous tile-setting materials, such as epoxies and furans, offer greater bonding strength and chemical resistance than cementitious mortars. However, special tile-setting skills are required, and both labor and materials are more costly than those for cementitious mortars.

***Portland Cement Mortar*** Portland cement mortar is a mixture of portland cement, sand, water, and sometimes hydrated lime. It is the traditional material for installing ceramic tile. The proportions vary, depending on whether the mortar is to be used as a leveling bed or a setting bed and whether it is to be used on a vertical or a horizontal surface.

**FIGURE 9.6-9** Quarry and paver tiles provide attractive, low-maintenance floor finishes in heavy-duty traffic areas. (Courtesy American Olean. [Ceramic] Tile Co. [now Dal-Tile Corporation].)

**FIGURE 9.6-10** Summary of Tile-Setting Products

| Type | Description | Industry Standards | | Limitations |
|------|-------------|---------|--------------|-------------|
| | | **Product** | **Installation** | **Limitations** |
| Portland cement mortar | Field mix of portland cement, sand, and water for setting bed; mix portland cement and water for bond coat. Interior or exterior. | ASTM C 150 (Portland cement) | ANSI A108.1 | Requires $1\frac{1}{4}$ in. setting bed, resulting in additional thickness and weight. Requires presoaking of most glazed wall tile. |
| Dry-set mortar | Prepared mix[a] of portland cement, sand, and certain additives that impart water retentivity and eliminate moist curing. High resistance to moisture, impact, and freezing. Interior and exterior. | ANSI A118.1[b] | ANSI A108.5[c] | Three days of moist curing before grouting recommended for heavy-duty floors. |
| Latex portland cement mortar | Factory mix of portland cement and sand to which liquid latex resins are added in the field to impart greater bonding strength and a resilience. Interior or exterior. | ANSI A118.4 | ANSI A108.5 | Not suitable for wood, metal, hardboard, or particleboard backings. |
| Epoxy mortar | Two-part system of (1) epoxy resin and finely graded sand and (2) hardener. High bond strength and chemical resistance. Interior and exterior. | ANSI A118.3 | ANSI A108.6 | Requires special installation skills. Material and labor costs relatively high. Not suitable for walls. |
| Epoxy adhesive | Two-part system of (1) epoxy resin and (2) hardener. Good bond strength and chemical resistance. Interior or exterior. A three-part system containing portland cement powder is also available. | ANSI A118.3 | ANSI A108.6 | Requires special skills. Installation cost high, but lower than epoxy mortar. |
| Furan mortar | Two-part system of (1) furan resin and (2) hardener. Highest chemical resistance and installation cost. Interior or exterior. | ANSI A118.5 | ANSI A108.8 | Same as epoxy mortar. |
| Organic (mastic adhesive) | Prepared organic material, ready to use without addition of liquid or powder. Adequate bond strength and resilience for light duty on floors and most walls. Interior use only. | ANSI A136.1 | ANSI A108.4 | Not suitable for exterior use or interior use in continuous contact with water. Applied $\frac{1}{16}$ in. thick; requires smooth backing. |

[a]Requires addition of water in the field.
[b]For conductive dry-set mortar, ANSI A118.2.
[c]For conductive dry-set mortar, ANSI A108.7.

Portland cement mortar is suitable for most surfaces and ordinary types of installation over masonry, monolithic concrete, concrete/glass-fiber reinforced backerboard, wood boards, plywood, foam insulation board, gypsum wallboard, and gypsum plaster. Their relatively thick beds, $3/4$ to 1 in. (19.1 to 25.4 mm) on walls and $3/4$ to $1\frac{1}{4}$ in. (19.1 to 28.6 mm) on floors, facilitate accurate slopes and true planes in the tile finish.

***Dry-Set Mortar*** Dry-set mortar is a mixture of portland cement, sand, and resinous additives that impart water retentivity. Improved retentivity prevents rapid loss of water to the backing surface and improves the bond.

Dry-set mortar is suitable for use over masonry, monolithic concrete, concrete/glass-fiber reinforced backerboard, cut-cell expanded polystyrene and rigid closed-cell urethane insulation board, gypsum wallboard, cured portland cement mortar, ceramic tile, and marble. It has excellent impact resistance, is nonflammable, and does not require presoaking of absorptive tile, as is required with most cementitious mortars. It is the preferred mortar for absorptive glazed tile.

Dry-set mortar is available either unsanded or factory-sanded. Presanded mortars, to which only water needs to be added, offer greater control over the amount of sand in the mix. Dry-set mortar is not affected by prolonged contact with water and does not form a water barrier. It is limited in truing or leveling the work of other trades to about $1/4$ in. (6.4 mm) thickness but can be used in a layer as thin as $1/16$ in. (1.6 mm) over a plastic mortar bed.

***Latex-Portland Cement Mortar*** Latex-portland cement mortar is a mixture of portland cement, sand, and a latex additive, which imparts greater flexibility to the mortar. The uses of latex-portland cement mortar are similar to those of dry-set mortar. It should be applied over a cured mortar bed in a coat $1/8$ to $1/4$ in. (3.2 to 6.4 mm) thick. Since latex formulations vary; the manufacturer's directions should be followed.

*Epoxy Mortar* Epoxy mortar is available in a two-part mortar, consisting of an epoxy resin and a hardener, or a less expensive three-part system consisting of an epoxy resin, a hardener, and portland cement powder. When the parts are mixed, a chemical reaction occurs, and the mortar must be used within a limited time (*working time*). Epoxy mortar is suitable for use where chemical resistance or high bond strength is an important consideration. It can be used over concrete, plywood, steel plate, and ceramic tile. It is applied in a layer $1/16$ to $1/8$ in. (1.6 to 3.2 mm) thick. Working time, bond strength, water cleanability before cure, and chemical resistance vary with the manufacturer. Epoxy mortar is frequently used in combination with epoxy grout, resulting in a highly durable water-resistant and chemically resistant installation.

*Epoxy Adhesive* Epoxy adhesive is a two-part system consisting of an epoxy resin and a hardener. It is less costly than epoxy mortar but also less chemically resistant. It is intended for thin-set application of tile on walls, floors, and counters, where high bond strength and ease of applications are overriding considerations. A typical use is in the application of a tile base over a metal refrigeration enclosure in a commercial kitchen. It has a better chemical resistance than do organic adhesives.

*Furan Mortar* Furan mortar is a two-part mortar consisting of furan resin and a hardener. It is suitable where chemical resistance is the most critical consideration and is, therefore, used primarily on floors in laboratories and industrial plants. Acceptable subfloors include concrete, steel plate, and ceramic tile. Furan mortar is used in combination with furan grout.

*Organic Adhesive (Mastic)* Organic adhesives are prepared organic materials that are ready to use with no addition of liquid or powder. They harden by evaporation. Organic adhesives are used in a layer $1/16$ in. (1.6 mm) thick on floors, walls, and counters. Suitable backings include monolithic concrete, gypsum wallboard, portland cement mortar beds, gypsum plaster, mineral-fiber cement board, brick, ceramic tile, marble, and plywood.

Organic adhesives eliminate presoaking of tile and are suitable for most residential and commercial uses. They can be used for the walls of residential tub and shower areas, but are not suitable for high-moisture areas such as shower rooms and swimming pools or for exterior installations. Their chief advantages are low cost and a flexible bond. Bond strength varies greatly among the numerous available brands. The solvents in some adhesives may be irritating to the skin, and some are flammable.

## GROUTS

Grouting materials are available to meet the requirements of the different kinds of ceramic tile and types of exposure. Cementitious grouts consist of portland cement modified to provide specific qualities such as whiteness, uniformity, hardness, flexibility, and water retentivity. Resinous grouts (epoxies and furans) possess special properties such as high bond strength and chemical resistance but are harder and costlier to apply. Figure 9.6-11 summarizes commonly available tile grouts and the governing standards.

*Commercial Cement Grout* Commercial cement grout is a water-resistant, uniformly colored mixture of portland cement and other ingredients. It is the most commonly specified grout for tile walls subject to ordinary use. Soaking of nonvitreous wall tile is required to prevent moisture loss by the grout to the tile. Moisture in the grout is essential to a good bond and full grout strength. Damp curing is needed to ensure an adequate supply of moisture for the hardening process (hydration). Commercial cement grout is usually white, but colored grouts are also available.

*Sand-Portland Cement Grout* Sand-portland cement grout is an on-the-job mixture of normal portland cement, fine-graded sand, and sometimes hydrated lime. It is used with ceramic mosaic tile, quarry tile, and paver tile. Mixes should be 1 part portland cement to 1 part sand for joints up to $1/8$ in. (3.2 mm) wide; 1 part to 2 parts for joints up to $1/2$ in. (12.7 mm) wide; and 1 part to 3 parts for joints over $1/2$ in. (12.7 mm) wide. Up to $1/5$ part lime may be added for improved workability. Presoaking of nonvitreous wall tile is necessary, Damp curing of the grouted surface is required

*Dry-Set Grout* Dry-set grout is a prepared mixture of portland cement and sand, with additives to improve water retentivity. Dry-set grout has the same properties as dry-set mortar and is suitable for grouting walls subject to ordinary use. It eliminates soaking of tile, although dampening sometimes is desirable under very dry conditions.

*Latex Grout* Latex grout is a mixture of one of the three preceding products, consisting mainly of portland cement with a latex additive to increase stain resistance and resilience of the joints. Latex grout is suitable for installations subject to ordinary use. It is less rigid and less water-permeable than sand-portland cement grout, allowing for more movement and greater exposure to moisture. This grout therefore is particularly suitable for tub and shower areas.

*Silicone Rubber Grout* Silicone rubber grout is an engineered elastomeric grout system for interior use employing a single-component nonslumping silicone rubber which, upon curing, is resistant to staining, moisture, mildew, cracking, crazing, and shrinking. It adheres tenaciously to ceramic tile, cures rapidly, and withstands exposure to moisture as well as subfreezing temperatures and hot, humid conditions.

*Epoxy Grout* Epoxy grout is a two-part system consisting of an epoxy resin and a hardener. It is made in several formulations, each intended for a specific ceramic tile. It is highly stain-resistant and impervious, and is used with epoxy mortar or epoxy adhesive. Its use requires special skills and is usually expensive.

Epoxy grout is formulated without coarse fillers that might scratch glazed wall tile or mosaic tile. It can be used on walls, floors, and counters subject to food staining.

It is also formulated with a coarse silica filler for use with quarry tile and pavers. It is formulated for industrial and commercial installations where chemical resistance is of paramount importance. It has excellent bonding characteristics.

**FIGURE 9.6-11** Summary of Tile-Grouting Products

| Type | Description | Limitations |
|------|-------------|-------------|
| Commercial portland cement grout | Prepared mix* of portland cement and other ingredients resulting in a water-resistant, uniformly colored material; usually white, but available in colors. Commonly used for tile walls and floors subject to ordinary use. Suitable for interior and exterior. | Requires presoaking of most glazed wall tile and damp curing of grouted surface. |
| Sand–portland cement grout | Field mixture of portland cement, fine-graded sand (sometimes lime), and water. Most commonly used grout for ceramic mosaic tile. Suitable for floors and walls, interior and exterior. | Same as for commercial cement grout. |
| Dry-set grout | Prepared mix* of portland cement, sand, and additives that provide water retentivity and eliminate need for presoaking of glazed wall tile. Damp curing may develop greater strength in portland cement grouts. Generally used on walls, interior and exterior. | Dampening of the tile surface prior to grouting may be required under extremely dry conditions. |
| Latex–portland cement grout | Prepared mix* of any of the above formulations plus latex additives. It is less rigid and less permeable than regular cement grout. The addition of latex to portland cement grouts eliminates the necessity of damp curing for 72 hours. Can be used on walls and floors, interior and exterior. | Same as for dry-set grout. |
| Silicon rubber grout | Ready-to-use one-part caulking-type material that cures from moisture vapor in the air. Used mainly for glazed tile walls. Flexible, mildew resistant, stain resistant. | Bonds only to clean, dry surfaces. Must be used with proper adhesive to prevent staining. |
| Epoxy grout | Ready-to-use two-part system consisting of epoxy resin and hardener. Formulation for wall tile omits coarse fillers, which might scratch glaze; coarse silica fillers generally are included in grout intended for quarry tile and pavers. Highly resistant to chemical staining, hence used for industrial and kitchen floors, countertops. | Requires special skills for proper application. Material cost and labor higher than with other grout system. |
| Furan resin grout | Ready-to-use two-part system consisting of furan resin and hardener. Highest chemical resistance; used primarily in industrial installations with quarry tile and pavers. | Same as for epoxy grout. |

*Note:* Epoxy grout should comply with ANSI A118.3; other grouts with ANSI A108.10 and A118.6.
*Requires addition of water in the field.

**Figure 9.6-12** Summary of Related Materials and Products

| Material or Product | Standard |
|---------------------|----------|
| Portland cement, gray or white, Type I | ASTM C150 |
| Sand for mortar | ASTM C144 |
| Hydrated lime, Type S | ASTM C206 |
| Sealants | ASTM C920 |
| Metal reinforcement | |
|   Welded wire fabric | ASTM A82 and ASTM A185 |
|   Expanded metal lath | ASTM C847 |
| Cleavage membrane | |
|   Polyethylene film | ASTM D4397 |
|   Asphalt roofing felt | ASTM D226 |

**Furan Resin Grout** Furan resin grout is a two-part grout consisting of a furan resin and a hardener. It is intended for use with quarry tile and pavers, mainly in industrial areas requiring a maximum of chemical resistance. It is generally used in conjunction with furan mortar.

## RELATED MATERIALS AND PRODUCTS

Installation of ceramic tile involves a variety of related materials and products, described below and summarized in Figure 9.6-12.

**Portland Cement** Normal (Type I) portland cement is used for mortar setting beds, for leveling beds, for neat cement bond coats, and for field-mixed cement grout. White grout requires the use of white portland cement in the mortar bed and in the grout.

**Sand** Sand should be clean, free from impurities, and well graded. When fine sand is specified, it should pass through a 16-mesh (1.18 mm) screen.

**Hydrated Lime** Hydrated lime is used in portland cement-sand mortar to make it more plastic for application on walls and sometimes on floors. Type S (special) lime should be used to ensure a dimensionally stable mortar bed.

**Sealants** Sealants for use in expansion joints and control joints should be suitable for the exposure expected. Conditions that cause sealants to deteriorate include temperature variations, direct sunlight, repeated stretching and compression, frequent wetting and drying, and immersion in water. These sealants can be single-component or two-component types. The sealant color can be selected to match or blend with the tile or adjacent materials. A sealant used on floors must be hard enough to withstand occasional or repeated foot traffic. It should have a Shore A hardness index of not less than 25 for joints subject to occasional foot traffic and not less than 35 in high-traffic areas.

**Metal Reinforcement** Isolated (unbonded) mortar setting beds should contain a suitable metal reinforcement. Reinforcement for a mortar bed consists of metal lath on walls and welded wire fabric on floors. This welded wire fabric is usually 2 by 2 in. (50.8 by 50.8 mm) mesh, WO.3 by WO.3 (16 AWS gauge or 0.0625 in. [1.6 mm] diameter). Metal lath should be expanded lath weighing not less than 3.4 lb/sq yd (1.84 kg/m$^2$) or sheet lath weighing not less than 4.5 lb/sq yd (2.44 kg/m$^2$).

***Cleavage Membrane*** A cleavage membrane should be installed beneath every unbonded ceramic tile setting bed. It should be one of the following:

- 15 lb (8.14 kg/m$^2$) asphalt (or 13 lb [7.05 kg/m$^2$] coal tar)-saturated roofing felt
- Reinforced waterproof duplex asphalt paper
- 4 mil (0.10 mm) or thicker polyethylene film
- Membrane waterproofing

### 9.6.1.2 Job Conditions

Materials should be delivered in the manufacturer's original sealed containers and stored under cover to prevent damage or contamination. Tile should be set and grouted when the temperature is at least 50°F (10°C) and rising, unless the mortar manufacturer requires a higher temperature or permits a lower temperature.

### 9.6.1.3 Supporting Construction

A tile installation is only as good as the underlying assembly of structural members and backing (subfloor or wall surface) that supports it. If the supporting construction deflects, expands, or contracts, these movements will be passed on to the tile finish. This is particularly true in bonded installations. Except in small areas, a bonded tile finish cannot absorb these movements, and internal stresses are likely to result in cracks. It is essential, therefore, to provide supporting construction with a minimum amount of movement. Where movement is anticipated, expansion or control joints should be provided in the supporting construction and extended into the tile.

Movement is less critical when the mortar bed is isolated from the backing by a cleavage membrane. Although the membrane permits some independent movement in the supporting construction, the movement which can be tolerated still is limited. Therefore, dimensionally stable, rigid construction is desirable.

The *backing* is that part of the supporting construction to which the tile is bonded. It can be a subfloor, wall finish, or mortar leveling coat. In thin-set installations, the backing must be dimensionally stable, sound, true to line, plumb, and sloped or level, as designed. Also, the backing must be free of foreign substances or loose materials that might weaken the bond with the tile finish. Trueness of the backing is less critical when a mortar bed is used, but where slopes are required, they should be built into the backing, not developed by varying the mortar bed thickness.

### FLOORS

Floors should be designed so that the maximum deflection under full load, including live loads and the weight of the tile, does not exceed $1/360$ of the span. For instance, on a span of 15 ft (180 in.) (4572 mm), the deflection should be not more than 180/360 (4572/360) or $1/2$ in. (12.7 mm). The most common floor assemblies encountered in small buildings involve a concrete or wood subfloor; in large buildings, almost always a concrete subfloor.

***Concrete Subfloor*** Tile can be installed on a concrete subfloor either by the thin-set or the portland cement mortar bed method. For thin-set bonding directly to a floor slab, the surface should be steel-troweled and fine-broomed, with no variations in the plane or slope exceeding $1/8$ in. (3.2 mm) in 10 ft (3000 mm). For portland cement installations, a variation of $1/4$ in. (6.4 mm) in 10 ft (3000 mm) is acceptable. Concrete slabs should be thoroughly moist-cured before tile application. Liquid curing compounds or other coatings which may reduce the bond should not be used. Slabs should not be saturated with water at the time of installation. To prevent cracking, a slab should be properly reinforced, concrete quality should be controlled, and suitable joints should be provided in the slab.

***Wood Subfloor*** Supporting joists should be spaced not more than 16 in. (400 mm) on centers. For thickbed installations, either a single-layer subfloor of 1 in. (25.4 mm) nominal boards or $5/8$-in. (15.9-mm)-thick plywood are acceptable. For thin-set installations with most tile-setting products (except epoxy mortar), a double floor is required, consisting of a subfloor as described above and

an underlayment. The underlayment should be exterior type CC-plugged plywood, $1/2$ in. (12.7 mm) thick for residential use and $5/8$-in. (15.9-mm) thick for commercial buildings. When epoxy mortar is used in residential construction, the underlayment can be omitted if a $5/8$-in. (15.9-mm)-thick plywood subfloor is used and the longitudinal joints between panels are supported with solid wood blocking $2 \times 4$ $(50 \times 100)$ or larger members.

A $1/4$ in. (3.2 mm) space should be left between the underlayment and vertical surfaces, such as walls, and projections such as pipes or posts. For organic adhesives, $1/16$ in. (1.6 mm) should be left between underlayment ends and edges. For epoxy mortar, $1/4$ in. (3.2 mm) should be left at the ends and edges of panels, and these joints should be filled with epoxy. Adjacent underlayment edges should not be more than $1/32$ in. (0.8 mm) above or below each other.

*Concrete/glass fiber-reinforced backer board* may be used in lieu of a portland cement mortar bed over a $1/2$-in. (12.7-mm)-thick plywood subfloor. This backer board is designed for use with ceramic tile floors, is available in various sizes, and is approximately $1/2$ in. (12.7 mm), usually $7/16$ in. (11 mm) thick. It can be nailed or screwed to wood joists through $1/2$-in. (12.7-mm)-thick plywood subflooring. Ceramic tile can be bonded to it with either dry-set or latex-portland cement mortars. Adhesives are not recommended for use with this product.

### WALLS

Wall surfaces should be sound, plumb, and true, with square corners. For thin-set application, solid backing is usually installed by workers other than the tile installer. The backing surface should not vary from a true plane by more than $1/8$ in. (3.2 mm) in 8 ft (2400 mm). For portland cement mortar bed installations, either a solid backing or open framing is suitable (Fig. 9.6-13). In either case, the tile contractor must attach metal lath (over a membrane) to the framing or backing, then apply a scratch coat and build up a mortar bed $1/4$ to $3/4$ in. (6.4 to 19.1 mm) thick to which the tile is bonded. A similar thickbed installation can be used over irregular masonry or concrete surfaces.

Usually, tile is installed on walls by the thin-set method over gypsum wallboard, gypsum plaster, plywood, concrete/

**FIGURE 9.6-13** Typical supporting construction for ceramic tile. In areas subjected to wetting, a portland cement mortar bed should be used instead of gypsum plaster. Concrete/glass-fiber-reinforced backer board can also be used where gypsum board is indicated here.

glass fiber-reinforced backing board, or other dimensionally stable solid backing. However, gypsum plaster and plywood should not be used as backing in tub enclosures and shower stalls unless waterproof grout such as silicone rubber is used.

***Gypsum Wallboard*** Wallboard should be installed in the maximum practical lengths to minimize the number of end joints. Joints should be taped, and both joints and fasteners should be treated with at least two coats of joint compound. In tub and shower areas, water-resistant backing board should be used. Its joints should be taped and filled with one coat (maximum 4 in. [400 mm] wide) of joint compound.

***Gypsum Plaster*** Plaster walls that will receive tile should be steel-troweled, using a brown coat mix (1 part gypsum to 3 parts sand) or a prepared finish coat plaster. When the tile finish will be flush with an adjacent plastered surface, the finish coat should be omitted and the tile set on the brown coat. Otherwise, a full-thickness three-coat plaster surface is employed (see Section 9.3). The surface to receive tile should be troweled smooth, with no trowel marks or ridges more than $1/_{32}$ in. (0.8 mm) high.

***Plywood*** A plywood backing should be Exterior Type CC-plugged, $3/_8$ in. (3.2 mm) thick or thicker. Panels should be applied with the face grain perpendicular to the studs or furring. Solid 2 × 4 (50 × 100) wood blocking should be used under panel edges between studs. Nails should be driven 6 in. (150 mm) on centers at panel perimeters and 12 in. (300 mm) on centers at intermediate supports.

Panels should be spaced $1/_8$ in. (3.2 mm) apart, and there should be not more than a $1/_{32}$ in. (0.8 mm) variation between the faces of adjacent panels. When $1/_2$-in. (12.7-mm)-thick plywood is used, horizontal blocking between studs can be eliminated and panels can be applied parallel to studs. Before its installation, plywood edges should be sealed with an exterior primer or aluminum paint.

***Concrete/Glass Fiber-Reinforced Backing Board*** Concrete/glass fiber-reinforced backing board is a backer board designed for use with ceramic tile in wet areas around tubs and showers and on floors. Available in various sizes and about $1/_2$ in. (12.7 mm) thick, this material can be nailed or screwed in place over wood or metal studs or over a $1/_2$-in. (12.7-mm)-thick plywood subfloor. Ceramic tile can be bonded to it with either dry-set or latex-portland cement mortar. It can be used in place of metal lath, portland cement scratch coat, and a mortar bed. It is not suitable as a base for tile installed using adhesives.

***Masonry and Cement*** Masonry and cement backings include monolithic concrete, concrete/glass fiber-reinforced backing board, brick, and CMU. For thin-set installations, the wall surface should be clean, sound, and free of paint, efflorescence, and cracked or chipped areas. When latex-portland cement or dry-set mortars are to be used, smooth concrete or brick surfaces should be roughened by sandblasting or bush hammering to ensure a good bond.

For mortar bed installations, the surface qualities of the backing are not as critical. If a leveling bed is to be used, the wall surface can be more irregular

than if the tile is bonded directly to the wall, but it should be free of surface coatings and loose material which might affect the bond. When a cleavage membrane is used, surface contamination or irregularity is not important, but the wall should be structurally sound and dimensionally stable.

### 9.6.1.4 Tile Layout and Setting

An area to be covered with tile should be checked for squareness, and midlines should be established in both directions. The tile pattern should be laid out so that a tile is either butted to the midline or bisected by it. The most desirable method is the one which produces larger than half-tiles around the perimeter of the area.

Tiles should be cut with a suitable tool rather than split. Wall tile is cut with a tile cutter; mosaics are cut with "nippers"; quarry tile may be cut with a water-cooled masonry saw or heavy-duty cutter. Tiles should be closely fitted to projections, smoothing all ragged edges with a carborundum stone. Joint lines should be straight, plumb, level, and of even width. Tilework accessories should be evenly spaced, properly aligned with tile joints, and set in the proper plane (Fig. 9.6-14). When the tile is not of the self-spacing type, joint widths should be as shown in Figure 9.6-15, and setting

**FIGURE 9.6-14** Projections make tile layout more difficult and more important. Tile should be symmetrical with respect to projections and corners. Tiles should be larger than half size. (Courtesy American Olean. [Ceramic] Tile Co. [now Dal-Tile Corporation].)

**FIGURE 9.6-15** Joint Widths for Ceramic Tile

| Tile Description | Joint Width |
| --- | --- |
| Ceramic mosaic tile*<br>2 in. (50.8 mm) square or smaller | $\frac{1}{32}$ in. to $\frac{1}{8}$ in. (0.8 mm to 3.2 mm) |
| Paver and special tile<br>$2\frac{3}{16}$ in. to $4\frac{1}{4}$ in. (71.44 mm to 107.95 mm) | $\frac{1}{16}$ in. to $\frac{1}{4}$ in. (1.6 mm to 6.4 mm) |
| 6 in. (152.4 mm) square and larger | $\frac{1}{4}$ in. to $\frac{1}{2}$ in. (6.4 mm to 12.7 mm) |
| Quarry tile<br>3 in. to 6 in. (76.2 mm to 152.4 mm) | $\frac{1}{8}$ in. to $\frac{3}{8}$ in. (3.2 mm to 9.5 mm) |
| 6 in. (152.4 mm) square and larger | $\frac{1}{4}$ in. to $\frac{1}{2}$ in. (6.4 mm to 12.7 mm) |
| Glazed tile<br>3 in. (76.2 mm) square and larger | $\frac{1}{8}$ in. to $\frac{1}{2}$ in. (3.2 mm to 12.7 mm) |

*Ceramic mosaic tile is mounted, so joint widths are predetermined by manufacturer.

be filled flush with the tile surface; joints of cushion-edged tile should be tooled to the depth of the cushion. The finished grout surface should be uniform in color, smooth, and free of pinholes or low spots. The tile surface should be washed and sponged thoroughly, then polished with a dry, clean cloth or burlap (Fig. 9.6-20).

Glazed wall set using portland cement mortar usually requires soaking before it is set. When a portland cement grout will be used, glazed tile set with a material other than portland cement mortar and not presoaked before it was laid should be soaked after setting. Soaking is not required for other types of grout.

frames or grids are used for setting tile more quickly and accurately.

Tile is usually set into a previously spread layer of mortar, adhesive, or neat cement (Figs. 9.6-16 and 9.6-17), but sometimes it is more convenient to brush a coat of the tile-setting material on the back of the tile. Alignment of joints should be checked frequently for trueness and squareness (Fig. 9.6-18). After face-mounted tile is set, the mesh or paper is soaked off and removed. Mortar or adhesive is removed from tile faces with a solvent. Strings or ropes used to space tile should be removed as soon as the mortar or adhesive has set sufficiently to hold the tile firmly in place.

### 9.6.1.5 Grouting and Curing

Grout should be applied by troweling diagonally across the joints (Fig. 9.6-19). Enough grout should be forced into each joint so that the joint is filled down to the mortar. Joints of square-edged tile should

**FIGURE 9.6-17** Spreading a setting bed for wall tile. (Courtesy American Olean. [Ceramic] Tile Co. [now Dal-Tile Corporation].)

**FIGURE 9.6-16** Quarry tile is laid in a bond coat of dry-set mortar and tapped in place with a trowel handle. (Courtesy Tile Council of America.)

**FIGURE 9.6-18** Quarry tile is spaced to provide $\frac{1}{4}$- to $\frac{1}{2}$-in. (6.4- to 12.7-mm)-wide joints. A straightedge and a metal square are used periodically to check joint alignment and squareness. (Courtesy Tile Council of America.)

**FIGURE 9.6-19** Joints are filled with portland cement grout by troweling diagonally across the joints. Kneeling boards prevent damage to newly laid tile. (Courtesy Tile Council of America.)

**FIGURE 9.6-20** Excess grout is wiped from the tile surface with burlap or rough-textured cloth. (Courtesy Tile Council of America.)

Commercial cement and portland cement grout should be damp-cured for at least 72 hours. Dry-set grout need not be damp-cured, except when used (1) on floors, (2) on walls exposed to periodic wetting, or (3) in exterior locations. Damp curing can be accomplished by periodically applying a spray mist to the tile surface or by covering it with polyethylene film after initial and daily spraying.

### 9.6.1.6 Control and Expansion Joints

To ensure an attractive crack-free installation, control and expansion joints should be provided in the backing and in the tile finish.

#### JOINTS IN BACKING

Control joints are provided in cementitious materials such as concrete and stucco to localize shrinkage cracks due to loss of moisture by hydration and evaporation. When control joints are provided, expansion joints usually are not necessary because shrinkage exceeds the likely thermal expansion.

Because shrinkage is minimal in brick masonry, thermal movement is critical and expansion joints must be provided (see Chapter 4). Where differential settlement or movement of adjacent elements is expected, isolation joints should be used. Expansion joints frequently act also as isolation joints or control joints. To conform to usage in the tile industry, such joints in a tile finish will be referred to here as *expansion joints*.

#### JOINTS IN TILE FINISH

Expansion joints are not required in areas less than 12 ft (3660 mm) wide or over old, well-cured concrete. However, expansion joints should always be provided in a tile finish (1) directly over expansion, isolation, construction, or control joints in the backing, (2) where the tile finish abuts projecting surfaces, and (3) at proper intervals in floor areas measuring over 24 ft (7300 mm) in either direction. Expansion joints in a tile finish should never be narrower than the underlying joint in the backing (Fig. 9.6-21).

Tile edges to which sealant will bond should be clean and dry. Sanding or grinding of these edges is recommended to obtain a maximum sealant bond. Some sealant manufacturers recommend priming edges. Primers must be kept off the tile face. Sealant depth should be approximately one-half the joint width. Compressible backup strips should be used in a joint to prevent bonding of the sealant to the joint filler and to ensure a proper sealant profile (Fig. 9.6-22).

***Exterior Locations*** Exterior installations should be provided with expansion joints at least $^3/_8$ in. (9.5 mm) wide for joints 12 ft (3660 mm) on centers and at least $^1/_2$ in. (12.7 mm) wide for joints 16 ft (4880 mm) on centers. These minimum joint widths should be increased $^1/_{16}$ in. (1.6 mm) for each 15°F (−9.4°C) of temperature differential (range between summer high and winter low) over 100°F. (37.8°C). Decks exposed to the sky in the northern United States should have $^3/_4$-in. (19.1-mm)-wide joints 12 ft (3660 mm) on centers.

***Interior Locations*** Thickbed tile installations in interior areas should have ex-

**FIGURE 9.6-21** Expansion joints in tile surfaces. (Courtesy Tile Council of America, *Handbook for Ceramic Tile Installation.*)

**FIGURE 9.6-22** Proper sealant depth. Good bond at the sides and bond-free contact with backup strip are essential to long-term effective performance. (Courtesy Tile Council of America, *Handbook for Ceramic Tile Installation.*)

pansion joints in the tilework spaced 24 ft (7320 mm) to 36 ft (10 973 mm) each way. Joints for interior quarry tile and paver tile should be the same width as the grout joints but not less than $^1/_4$ in. (6.4 mm) wide. Joints for ceramic mosaic tile and glazed wall tile should be $^1/_4$ in. (6.4 mm) wide but never less than $^1/_8$ in. (3.2 mm) wide. Large thin-set installations may benefit from appropriately located expansion joints, but no generalizations can be drawn. Joints always are desirable at offsets and changes in backing construction.

### 9.6.1.7 Thresholds

When it is not practical to depress the subfloor for a conventional mortar bed installation and adjacent finishes are subsequently thinner, a transition is required. A resilient flooring contractor may use rubber or vinyl thresholds; carpenters are likely to use hardwood thresholds in connection with wood flooring; carpeting might be butted to tile with a metal or vinyl edge strip. However, the preferable material for this purpose is a marble or slate threshold located in a doorway, or in a cased opening, and set with the same bonding agents as ceramic tile (Fig. 9.6-23).

### 9.6.1.8 Shower Receptors

Shower receptors can be precast terrazzo, plastic, field-installed ceramic tile, or a synthetic material. Tile can be installed over wood or concrete subfloors; a mortar bed must be used over wood sub-

**FIGURE 9.6-23** The high end of the threshold should be located under the door. (Courtesy Tile Council of America, *Handbook for Ceramic Tile Installation.*)

**FIGURE 9.6-24** In a shower stall, including a membrane, crushed stone, and weep holes ensures drainage of water that might seep under the tile finish. (Courtesy Tile Council of America, *Handbook for Ceramic Tile Installation.*)

**FIGURE 9.6-25** Applying sealant in the space between tile and tub apron ensures a watertight joint. (Courtesy Tile Council of America, *Handbook for Ceramic Tile Installation.*)

floors. In either case, a waterproof membrane or pan must underlie the tile finish, and it must be sloped $^1/_4$ in. per foot (6.4 mm per 300 mm) to the drain. The sides of the membrane or pan should be turned up the wall 5 in. (127 mm) (Fig. 9.6-24).

Although only the mortar bed method is recommended for shower receptors, the slope should be developed in the backing rather than in the mortar bed to ensure positive runoff of water that penetrates the tile finish. Crushed stone or tile around the shower drain and weep holes in the drain permit this occasional water to seep into the drainage system (see Fig. 9.6-24).

Concrete subfloors can themselves be sloped to the drain. With a wood subfloor, a separate portland cement mortar fill should be used to create the slope. The fill can vary from $^1/_2$ in. (12.7 mm) to $1^1/_4$ in. (28.6 mm) in thickness. It should include a latex additive to improve its bond strength and make it possible to apply it in thin layers.

A waterproof membrane or pan should be installed over the sloped fill, followed by the tile finish set in a mortar bed. Any commercial grout is suitable for grouting the tile joints. Figure 9.6-25 shows proper detailing of tiled tub-shower areas; Figure 9.5-41 shows recommended tile coverage of gypsum board in bathing areas.

### 9.6.1.9 Tile Over Other Finishes

Ideally, existing finishes or coatings should be completely removed and new tile placed on the original backing. However, if this is impractical, new tile can be installed over existing tile, resilient flooring, hardwood flooring, painted concrete, gypsum plaster, and gypsum wallboard by either installing a new backing

surface for thin-set application or installing the new tile using the conventional mortar bed method. Figure 9.6-26 shows several tile-over-tile installation possibilities.

Thickbed installations present no special problems over existing finishes. For thin-set installations, the backing surface must be free of dirt, dust, efflorescence, oil, grease, and other contaminants. The maximum variation in a backing surface should not exceed those indicated in Section 9.6.1.3. Glossy or scaly painted surfaces should be sandblasted or treated with a chemical paint remover. The adhesive or mortar manufacturer should be consulted to determine whether a sealer

or primer is needed. When the old floor surface is questionable and the structural floor assembly is capable of carrying the weight, the more reliable isolated thickbed installation method should be used.

### 9.6.1.10 Countertops

Either mosaic, quarry, or glazed wall tile can be used as a durable finish for countertops. There, tile is installed typically over a wood, plywood, or cementitious board backing either by the mortar bed (thickbed) or the thin-set method (Fig. 9.6-27).

The backing for a mortar bed installation should be either $1 \times 6$ ($25 \times 150$) boards spaced $^1/_4$ in. (6.4 mm) apart (Fig. 9.6-27a) or $^3/_4$ in. (19.1 mm) exterior plywood. Plywood should be scored with saw cuts 6 to 8 in. (150 to 203.2 mm) on centers to prevent warping. Thin-set installation require $^3/_4$ in. (19.1 mm) exterior plywood, with a $^1/_4$ in. (6.4 mm) gap panels. This gap should be closed from below with a batten strip (Fig. 9.6-27b).

Either an epoxy grout or a stain-resistant grout recommended for that purpose by the manufacturer should be used on countertops installed using the thin-set method.

### 9.6.2 INSTALLATION ON FLOORS

Selection of the most suitable installation method depends on the desired performance level, the supporting construction, and whether the installation is interior or exterior.

### 9.6.2.1 Installation Performance Levels

The type of traffic (wheeled or foot), impact loads, and contaminants (water, grease, alkalis, chemicals) to which a floor will be subjected are prime considerations in the selection of a suitable performance level (Fig. 9.6-28). Once the performance level has been determined, an appropriate installation method can be selected (Fig. 9.6-29).

In addition to the installation method, consideration should be given to:

■ Wear properties of the selected tile
■ Fire-resistance properties of the wall or floor assembly
■ Acoustical properties (especially the Impact Noise Rating) of the floor assembly
■ Slip resistance of the floor tile

**FIGURE 9.6-26** New tile may be bonded directly to existing tile or a new backing, such as gypsum board, may be imposed between the new and existing tile. (Courtesy Tile Council of America, *Handbook for Ceramic Tile Installation.*)

Unglazed tile (mosaic, quarry, and paver tile) used for floor surfacing will give satisfactory wear for all the performance levels listed. Decorative glazed tile and soft-body unglazed tile should not be used on floors unless this use is specifically recommended by the tile manufacturer. When determining the suitability of glazed tile for a specific use, color pattern, surface texture, and glaze hardness should be considered. All glazed tile will show traffic wear over time.

#### 9.6.2.2 Interior Concrete

An interior concrete slab on grade is not subjected to the drastic temperature and humidity fluctuations encountered outdoors. In addition, if such a slab is made up of a properly engineered concrete mix and has suitable reinforcement, adequate finishing, proper curing, a well-drained granular base, and proper jointing, the likelihood of excessive movement due to settlement and temperature changes is slight. Therefore,

such slabs will accept the complete range of tile installation methods, including a bonded mortar bed, an isolated mortar bed, or direct application with a thin-set mortar or adhesive.

A supported slab, on the other hand, will often deflect enough under loads to cause tile set using other than an isolated mortar bed to crack or even delaminate. Supported slabs also may encounter enough thermal movement to damage tile set using a bonded system.

**FIGURE 9.6-27** Ceramic tile countertops may be installed by (a) the cement mortar method or (b and c) the thin-bed method. (Courtesy Tile Council of America, *Handbook for Ceramic Tile Installation.*)

**FIGURE 9.6-28** Installation Performance Level Ratings

| Area Description | Type of Use or Traffic | Performance Level |
|---|---|---|
| Office space, commercial reception areas, public spaces in restaurants and stores, corridors, shopping malls | (a) Light high-heel traffic<br>(b) Heavy high-heel traffic | (a) Moderate<br>(b) Heavy |
| Kitchens | (a) Residential<br>(b) Commercial<br>(c) Institutional | (a) Light or moderate<br>(b) Heavy<br>(c) Extra heavy |
| Toilets, bathrooms | (a) Residential<br>(b) Commercial<br>(c) Institutional | (a) Light or moderate<br>(b) Moderate or heavy<br>(c) Heavy |
| Hospitals | (a) General<br>(b) Kitchens<br>(c) Operating rooms | (a) Heavy<br>(b) Extra heavy<br>(c) Heavy |
| Exterior decks | (a) Roof decks, large on-grade patios<br>(b) Walkways, small patios | (a) Extra heavy<br>(b) Heavy or extra heavy |

*Note:* Performance level ratings are defined as follows: Light: Normal residential foot traffic; occasional 300 lb (136 kg) loads on soft rubber wheels. Moderate: Better residential, normal commercial, and light institutional use; 300 lb (136 kg) loads on rubber wheels and occasional 100 lb (45.4 kg) loads on steel wheels. Heavy: Heavy high-heel traffic; 200 lb (90.7 kg) loads on steel wheels, 300 lb (136 kg) loads on rubber wheels. Extra heavy: High-impact service, as in institutional kitchens and industrial work areas; 300 lb (136 kg) loads on steel wheels.

**FIGURE 9.6-29** Suitable Methods for Various Performance Levels

| Performance Level | Installation | | | | |
|---|---|---|---|---|---|
| | Subfloor | Tile | Mortar or Adhesive | Grout | Remarks |
| **Light** | Wood | Ceramic mosaic or quarry | Organic | Latex–portland cement | Low cost; OK for home bath, entry |
| | Wood | Ceramic mosaic | Epoxy | Sand–portland cement[a] | Costlier; OK for kitchen, dining room |
| | Existing tile finish | Ceramic mosaic or quarry | Epoxy | Mastic or latex–portland cement | Residential renovation |
| **Moderate** | Wood | Ceramic mosaic | Portland cement mortar bed | Sand–portland cement[a] | OK for depressed wood subfloor |
| | Wood | Ceramic mosaic or quarry | Epoxy | Epoxy | Very costly; very durable even over wood subfloor |
| | Concrete | Ceramic mosaic | Dry-set | Sand–portland cement[a] | Moderate cost; smooth surface |
| | Concrete | Ceramic mosaic | Dry-set | Latex–portland cement | Moderate cost; eliminates damp curing of grout |
| **Heavy**[b] | Concrete | Ceramic mosaic or quarry | Dry-set[c] | Epoxy | Costlier, good chemical resistance |
| | Concrete | Ceramic mosaic or quarry | Portland cement mortar bed | Sand–portland cement[a] | Moderate cost; bed ensures smooth surface |
| | Concrete | Ceramic mosaic or quarry | Dry-set | Sand–portland cement[a] | Low cost; good all-purpose floor |
| | Concrete | Ceramic mosaic or quarry | Epoxy | Epoxy | Very costly; most durable and easy to maintain |

[a]Damp-cured for 3 days after grouting.

[b]Damp-cured for 3 days before grouting.

[c]All installations suitable for extra-heavy performance when packing house tile is substituted for mosaic tile.

The specific procedure for installing tile depends mostly on the tile-setting material selected.

## PORTLAND CEMENT MORTAR

Neat portland cement can be used as a bond coat only when the concrete surface is still plastic and suitably rough to bond with the neat cement. This condition is difficult to achieve on a structural slab. In addition, an isolated, reinforced mortar setting bed with a cleavage membrane should always be used over a structural (supported) slab so that stresses induced in the slab are not transmitted to the tile finish.

In a proper concrete slab on grade, as described above, the likelihood of excessive movement is minimized and tile can be applied either by the thin-set method directly to the slab or over a bonded setting bed. However, even on slabs on grade, when a mortar bed is used, the bed is often reinforced with wire mesh, and sometimes a cleavage membrane is also installed.

A portland cement mortar bed should not exceed $1^{1}/_{4}$ in. (28.6 mm) in thickness. After the bed has been installed, a layer of neat cement paste, $^{1}/_{32}$ in. (0.8 mm) to $^{1}/_{16}$ in. (1.6 mm) thick, should be troweled over it or onto the back of the tile. Alternatively, a layer of dry portland cement, $^{1}/_{32}$ in. (0.8 mm) to $^{1}/_{16}$ in. (1.6 mm) thick, can be dusted over the setting bed as a bond coat and then, immediately before the tile is set, worked lightly with a trowel or brush until damp.

Glazed semivitreous tile should be dampened by placing it on a wet cloth or in a shallow pan. However, no free moisture should remain on the backs of the tile when set. Vitreous and impervious

tile, such as mosaic tile and quarry tile, do not require soaking.

To obtain maximum contact and a strong bond, tile should be set in position and tamped firmly into the mortar (Fig. 9.6-30a). Surfaces should be brought to a true level at proper elevation. Tamping and leveling should be completed within 1 hour after the tiles have been placed.

When installing face-mounted ceramic mosaic tile, the paper should be wetted within 1 hour of setting, using no more water than necessary for removing the paper and glue (Fig. 9.6-30b). Tiles that are out of line or not level should be adjusted before the mortar assumes its initial set.

Sand-portland cement grout is used with quarry tile, pavers, and ceramic mosaic tile. For quarry tile and pavers, the grout consists of 1 part portland cement to 2 parts sand and $^{1}/_{5}$ part hydrated lime by volume. For ceramic mosaic tile, a mix of 1 part portland cement to 1 part finely screened sand is used (Fig. 9.6-30c). Before the joints are grouted, they are filled and pointed with neat cement paste. Grout in paver, quarry, and cushion-edged glazed tile joints should be tooled (compressed) with a pointing tool to firmly bond the grout to the pointing mortar and to the tile.

Excess grout should be removed from the surface with a clean cloth or burlap. Voids and gaps should be filled before the grout sets. When portland-cement-based grouts are used, the entire installation should be damp-cured for 72 hours.

The tile surface should be thoroughly cleaned while the grout is still plastic. Muriatic acid may be used to clean glazed tile; sulfamic acid can be used to clean unglazed tile 10 days after setting. Joints should be soaked before acid is applied

and flushed thoroughly with water after cleaning. Metal and enameled surfaces should be protected with grease or a strippable coating.

***Isolated Setting Bed***   To create an isolated mortar setting bed, a cleavage membrane is spread over the slab and folded at the edges and ends to form a locked joint. Metal reinforcement should be lapped at least one full mesh, and should be supported approximately in the middle of the setting bed. The mesh should be cut to fit accurately and kept free of vertical surfaces. Figure 9.6-31 illustrates typical isolated bed construction.

The setting bed mix should consist of one 1 part portland cement, 5 parts dry sand (or 6 parts damp sand), and $^{1}/_{10}$th part hydrated lime by volume. Only enough water should be added to the mix to obtain the desired consistency. The setting bed should be applied $^{3}/_{4}$ in. to $1^{1}/_{4}$ in. (19.1 to 28.6 mm) thick, tamped firmly, and screeded to a level surface.

**FIGURE 9.6-31** Metal reinforcement should be installed in every isolated thick-bed installation. (Courtesy Tile Council of America, *Handbook for Ceramic Tile Installation.*)

a

b

c

**FIGURE 9.6-30** Mounted ceramic mosaic tile is (a) laid in a neat cement bond coat; (b) the tile is tamped into place, and the paper is sponged off; (c) grout is troweled into joints, and the excess is removed with a clean cloth. (Courtesy American Olean [Ceramic] Tile Company [now Dal-Tile Corporation].)

***Bonded Setting Bed*** The bonded bed method should be used infrequently because it does not possess the chief advantage of unbonded installations, namely, separation from stresses in the concrete subfloor. Its chief advantages over thinset methods are the ability to produce a smooth finish over a rougher subfloor and reliance on familiar, readily available materials such as portland cement and sand. Bonded setting bed construction is similar to isolated construction except that the cleavage membrane and reinforcement are omitted (Fig. 9.6-32).

The mortar bed can be bonded with a synthetic bonding compound or neat portland cement. For the latter, a fairly rough, wood-floated surface is required. The slab should be thoroughly saturated with water before neat cement and mortar are spread. The mortar mix, grout mix, and installation procedures for a bonded setting bed are the same as those described above for an isolated setting bed.

## LATEX-PORTLAND CEMENT AND DRY-SET MORTARS

These mortars generally are used to bond tile directly to a concrete slab (Fig. 9.6-33) that has a variation of not more than $1/4$ in. (6.4 mm) in 10 ft (3048 mm). On slabs with more variation, tile should be bonded to a cured mortar leveling bed. In addition, the slab should be designed to ensure a minimum of movement. It should be well cured and free of cracks, waxy or oily films, and curing compounds. The slab surface should be steel-troweled and fine-broomed. Dry-set or latex-portland cement mortar should be mixed as recommended by the manufacturer. Mortar should be troweled in a layer about $1/4$ in. (6.4 mm) thick for mo-saic tile and $3/8$ to $1/4$ in. (9.5 to 6.4 mm) thick for quarry and paver tile. The tile should be tamped into the mortar to ensure a level surface and intimate contact.

Paper should be removed from face-mounted tile within 1 hour of setting using a minimum of water. Grouting should be delayed for at least 24 hours or until the tile is firmly set. Suitable grouts include dry-set grout, latex grout, sand-portland cement grout, and commercial cement grout. Dry-set and portland cement grouts should be damp-cured for 72 hours.

## ORGANIC ADHESIVES

Since these products are recommended for use on floors only in light traffic areas, they are used infrequently over a concrete slab, except in small areas and over relatively smooth surfaces. For organic adhesives, which are applied only $1/16$ in. (1.6 mm) thick, a slab surface should be steel-troweled and fine-broomed, and the surface should not vary by more than $1/16$ in. (1.6 mm) in 3 ft (900 mm).

Tile installation with organic adhesives requires the same general precautions as with other thin-setting products (see Fig. 9.6-33). The tile should be tamped into place to ensure maximum contact with the adhesive. Latex-portland cement and epoxy grouts are usually used with organic adhesives, but any grout can be used.

## EPOXY MORTAR AND ADHESIVE

The performance and working properties of epoxy adhesives are similar to those of epoxy mortars. Most of the following discussion concerning epoxy mortars applies to adhesives as well.

Epoxy mortar is likely to be used in conjunction with an epoxy grout to provide chemical resistance in laboratories, industrial plants, food preparation areas, and similar clean environments. However, the high bond strength of the mortar, combined with the general stain resistance of the grout, make this also a preferred system for many hard service areas (such as entries, public toilet rooms, lobbies, and restaurants), particularly when combined with quarry tile. Since the mortar is installed in a layer $1/4$ to $3/8$ in. (6.4 to 9.5 mm) thick, it can absorb minor variation in the subfloor surface of not more than $1/4$ in. (6.4 mm) total in 10 ft (3048 mm). A wood-floated concrete surface is suitable (Fig. 9.6-34).

Epoxy mortar should be of the same color as the grout. The tile should be tamped into the setting bed, taking care not to damage the face of the tile or smear mortar on it. Mortar should be immediately removed from the tile face.

Grouting should be delayed at least 16 hours; then grout should be forced into the joints with a squeegee or rubber trowel to completely fill them. The installation should be shaded from direct sun and kept at a relatively even temperature (more than 60°F [33.3°C]) for at least 8 hours after grouting to facilitate curing. Traffic should be kept off the floor for at least 40 hours. Light traffic can be permitted after this period; heavy traffic, after 7 days.

Epoxy grout also can be used with other tile-setting mortars. Where chemical resistance at the tile surface is desir-

**FIGURE 9.6-34** Epoxy mortar and grout possess high bond strength and chemical resistance. Epoxy adhesives can be substituted in less demanding situations. (Courtesy Tile Council of America, *Handbook for Ceramic Tile Installation*.)

**FIGURE 9.6-32** Reinforcement can be omitted when the mortar bed is bonded to the backing. (Courtesy Tile Council of America, *Handbook for Ceramic Tile Installation*.)

**FIGURE 9.6-33** Dry-set and latex-portland cement mortar can be used for thin-set placement of tile on a concrete slab. (Courtesy Tile Council of America, *Handbook for Ceramic Tile Installation*.)

able but a strong bond with the backing is not essential, epoxy grout can be used with latex or dry-set mortars. These cementitious mortars generally produce an adequate bond and are less expensive than epoxy mortar.

### 9.6.2.3 Exterior Concrete

Except for installations over built-up roofing or waterproofing membranes, exterior tile installations are likely to be bonded directly to a concrete slab (Fig. 9.6-35). Over a suitably finished concrete surface, dry-set mortar can be used in a coat $1/4$ to $3/8$ in. (6.4 to 9.5 mm) thick to bond tile directly to a slab.

When a structural slab is not suitable for direct application of tile, a mortar setting bed will ensure a suitable substrate (Fig. 9.6-36). Since a mortar bed is installed by the tile contractor, he can follow up with a neat portland cement bond coat while the bed is still plastic.

**FIGURE 9.6-35** Dry-set mortar can be used for thin-set application of tile in both interior and exterior locations. (Courtesy Tile Council of America, *Handbook for Ceramic Tile Installation.*)

**FIGURE 9.6-36** Expansion joint in an exterior mortar bed tile installation.

Expansion joints extending through tile and mortar bed should be provided at intervals of 12 to 16 ft (3658 to 4877 mm) each way (see Fig. 9.6-36). Expansion joints also should be provided where the tile finish abuts restraining vertical surfaces and where expansion, isolation, or control joints exist in the slab. Both the structural slab and the tile surface should be sloped to drain. See Section 9.6.1.6 for additional discussion of expansion joints.

### 9.6.2.4 Wood Subfloors

Wood subfloors usually consist of wood joists and plywood or board sheathing. Unless stiffness is carefully considered in the design, joists are likely to deflect and vibrate more than massive structural systems such as those made of concrete or steel. The boards and plywood, being hygroscopic, also are likely to expand and contract with humidity variations.

Therefore, tile should be either completely isolated from the subfloor to permit differential movement or bonded to the subfloor with a tile-setting material that can resist the movement without breaking the bond. When the installation depends on a very strong bond and no setting bed is used, the grout also must be strong enough to resist the internal stresses transmitted to the tile joints.

A typical isolated conventional installation consists of a reinforced mortar setting bed and a latex, dry-set, or neat cement bond coat over a cleavage membrane. A typical bonded thin-set installation utilizes epoxy mortar or epoxy adhesive. In relatively small areas, such as residential bathrooms, tile also can be bonded to a plywood subfloor with organic adhesives. Although this material is not as strong as epoxy, it is adequate to resist the smaller stresses developed in a small, bathroom-sized area.

#### PORTLAND CEMENT MORTAR

The procedure for installing a mortar setting bed is the same as for concrete subfloors. The grout can be any of the several commercially available formulations or sand-portland cement grout. The complete tile installation, including the mortar bed, should be about 1 in. (25.4 mm) to $1\frac{1}{2}$ in. (38 mm) thick (Fig. 9.6-37).

Since many interior floor finishes are much thinner than $1\frac{1}{2}$ in. (38 mm), a transition problem develops where a tile finish abuts a thinner finish. To minimize this problem, the subfloor can be de-

**FIGURE 9.6-37** A reinforced isolated mortar bed and tile over a single-layer wood subfloor. (Courtesy Tile Council of America, *Handbook for Ceramic Tile Installation.*)

pressed between the joists, thus gaining $5/8$ to $3/4$ in. (15.9 to 19.1 mm) in height (Fig. 9.6-38). The subfloor, usually plywood, is supported on cleats level with the tops of the joists. This method of tile installation is somewhat more durable in high-moisture areas, such as bathrooms, than the thin-set method. It is also more resistant to impact due to the mass of the mortar bed underlying the tile finish. Depressing the subfloor below the tops of the joists and leveling the joists is not recommended.

#### ORGANIC ADHESIVE

The organic adhesive method is not suitable for tile installation on the floors of commercial or institutional buildings. It is, however, suitable for all areas in a home, except those subject to excessive moisture, such as in a tiled sunken tub or

**FIGURE 9.6-38** A portland cement mortar bed and tile installation over a depressed subfloor.

a swimming pool. It will not withstand heavy traffic or impact loads. Double floor construction generally is recommended to achieve a stiffer subfloor (Fig. 9.6-39).

The subfloor can be 1 in. (25 mm) nominal boards or $^5/_8$ in. (15.9 mm) plywood. The underlayment should be $^3/_8$ in. (9.5 mm) or thicker plywood. For bathroom floors and other areas subject to occasional moisture, Type I adhesive should be used. This type is capable of withstanding more prolonged exposure to moisture than Type II adhesive. The adhesive should be spread with a notched trowel recommended by the manufacturer and the tile firmly pressed into the adhesive. Grout can be latex-portland cement or epoxy.

## EPOXY MORTAR AND ADHESIVE

Epoxy mortar and adhesive is a very strong adhesive suitable for bonded thin-set installations over a double wood floor consisting of subfloor and underlayment (Fig. 9.6-40). The subfloor should be $^5/_8$ in. (15.9 mm) plywood or 1 in. (25-mm) nominal boards; the underlayment should be $^1/_2$ in. (12.7 mm) or thicker plywood ($^5/_8$ in. [15.9 mm] in nonresidential uses). The underlayment should be installed with $^1/_4$ in. (6.4 mm) gaps between panels, and these gaps should be completely filled with epoxy mortar. In residential uses, the underlayment may be omitted if the subfloor is $^5/_8$ in. (15.9 mm) or thicker plywood installed perpendicular to the joists and with the edges blocked.

The mortar should be spread with a trowel recommended by the manufacturer. The tile should be tamped into the mortar to ensure a smooth surface and a proper bond.

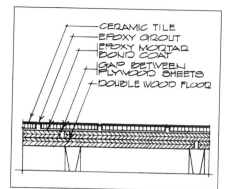

**FIGURE 9.6-40** An epoxy mortar-set tile over a double wood subfloor. (Courtesy Tile Council of America, *Handbook for Ceramic Tile Installation.*)

Grouting should be delayed for 16 hours. The installation should be shaded from direct sunlight and kept at a stable temperature of more than 60°F (15.6°C) for 8 hours or more after grouting. Grout and mortar should be cleaned off the face of the tile as soon as possible and before it hardens.

## 9.6.3 INSTALLATION ON WALLS

The ceramic products most often used on walls are glazed wall tile and ceramic mosaic tile. Quarry tile and paver tile are seldom used as wall finishes. Where the wall surface is sufficiently sound, true, clean, and free of coatings, tile can be bonded to the wall by the thin-set method. The surface should not vary more than $^1/_8$ in. (3.2 mm) in 8 ft (2438 mm) if thin-set materials are to be used and not more than $^1/_4$ in. (6.4 mm) in 8 ft (2438 mm) for portland cement mortar beds.

A mortar setting bed should be used on wall surfaces that are irregular, chipped, cracked, or coated with materials that would impede a bond. A mortar setting bed is installed over a base coat of mortar (scratch coat) applied to metal lath over a cleavage membrane.

A *scratch coat* consists of 1 part portland cement, 5 parts damp sand, and $^1/_5$ part hydrated lime. It is normally applied by a plastering contractor but can also be installed by the tile contractor. A *float coat* generally is applied by the tile contractor to ensure a plastic surface for setting the tile with neat cement.

A solid backing is not essential when tile is to be installed by the mortar bed method. Over wood or metal studs, the

backing can be omitted and the membrane and metal lath can be attached directly to the studs. A lath-reinforced mortar bed generally is rigid enough by itself to support tile without excessive deflection between the studs; to ensure adequate vertical rigidity, the metal studs should be at least $2^1/_2$ in. (63.5 mm) deep at 16 in. (400 mm) on centers or $3^1/_2$ in. (88.9 mm) deep when spaced 24 in. (600 mm) on centers.

The materials, installation methods, and suitable backings for tile finishes vary, depending on whether the installation is interior or exterior. Exterior installations are much more critical because of exposure to moisture and temperature variations.

### 9.6.3.1 Interior Surfaces

Suitable backings for tile finishes include masonry and concrete walls, as well as plywood, gypsum board, and plaster over furring or stud framing. In tub and shower areas, the thin-set method is acceptable, but it is essential that the backing be inherently water-resistant or properly prepared to render it so. Dry-set mortar can be used as a bond coat for thin-set application in sunken tubs and pools over a portland cement mortar bed.

## PORTLAND CEMENT MORTAR

Traditional installations on walls require a preliminary *scratch coat* followed by a *mortar bed* (*two-coat method*). The scratch coat can be omitted when the wall provides a solid backing for the application of metal lath, mortar, and tile (*one-coat method*). The tile actually is set in a separate bond coat, but it is not counted in describing the installation.

The *one-coat method* can be used over existing wall finishes that are not suitable for thin-set application because of surface conditions (Fig. 9.6-41). It is also a preferred method of tiling tub and shower areas in new work. The backing can be masonry, concrete, plaster, wallboard, plywood, or any other solid wall surface. The single bed coat is $^3/_8$ to $^3/_4$ in. (9.5 to 19.1 mm) thick and consists of 1 part portland cement, $^1/_2$ half part hydrated lime, and 5 parts damp sand.

The *two-coat method*, involving both a *scratch coat* and a *bed coat*, is used most frequently over open framing of furring or studs (Fig. 9.6-42). On metal studs, the overall thickness of the scratch

**FIGURE 9.6-39** A thin-set tile application over a double wood subfloor. (Courtesy Tile Council of America, *Handbook for Ceramic Tile Installation.*)

**FIGURE 9.6-41** The one-coat method can be used over a solid backing that is too uneven for thin-set tile. (Courtesy Tile Council of America, *Handbook for Ceramic Tile Installation.*)

**FIGURE 9.6-42** Two-coat method over open framing or solid backing. (Courtesy Tile Council of America, *Handbook for Ceramic Tile Installation.*)

**FIGURE 9.6-44** For thin-set tile applied directly on masonry, the backing surface should be smooth and the joints struck flush. Since exterior thin-set applications directly over masonry are expected to move with the masonry, expansion joints are not as important as in other types of installations. (Courtesy Tile Council of America, *Handbook for Ceramic Tile Installation.*)

coat and mortar bed should not exceed 1 in. (25.4 mm). On wood studs, the overall mortar thickness should not be more than $1^{1}/_{2}$ in. (38 mm). The mortar mix is the same as for one-coat installations. The scratch coat should be allowed to cure for at least 24 hours and should be dampened thoroughly before the mortar bed is applied. The mortar bed should be applied in areas no greater than may be covered with tile while the bed is still plastic. Figure 9.6-43 illustrates the steps in applying tile over a portland cement setting bed.

A neat cement *bond coat* is used for setting tile with both the one- and two-coat methods. The bond coat should be troweled $^{1}/_{32}$ to $^{1}/_{16}$ in. (0.08 to 1.6 mm) thick over the float coat, and the tile should be tamped into it. The portland cement for the bond coat should be ordinary gray or white cement, depending on whether the grout is to be gray or white. The grout can be neat portland cement or portland cement mixed with equal parts sand, latex, or epoxy. Where commercial portland cement or portland cement grout is used, the finish should be cured for at least 72 hours.

### DRY-SET AND LATEX-PORTLAND CEMENT MORTARS

Dry-set and latex-portland cement are thin-set mortars suitable for sound, clean, and true surfaces such as gypsum plaster, gypsum board, monolithic concrete, or concrete and brick masonry (Fig. 9.6-44). They also can be used over a bonded mortar bed used mainly for leveling an irregular solid surface (Fig. 9.6-45) or

**FIGURE 9.6-43** In conventional mortar bed installations, (a) a scratch coat is applied to metal lath, and temporary screeds are inserted; (b) the float coat is leveled, and screeds are removed; (c) a neat cement coat is troweled on; and (d) tile is applied and tamped in place. The glazed tile shown here must be soaked before it is set. (Courtesy American Olean. [Ceramic] Tile Company [now Dal-Tile Corporation].)

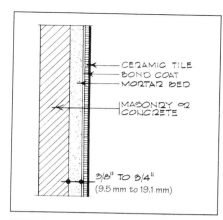

**FIGURE 9.6-45** Before tile is applied over uneven masonry, a leveling bed is placed and cured. Then tile is set using either neat cement or thin-set mortar. (Courtesy Tile Council of America, *Handbook for Ceramic Tile Installation.*)

when a full-thickness mortar backing on open framing is already in place (Fig. 9.6-46). When such a mortar backing is used, the backing should be allowed to cure for several days until firm.

The backing should be dampened just before mortar is spread over it. Mortar should be troweled on in the appropriate thickness ($3/32$ to $1/4$ in. [2.4 to 6.4 mm] maximum) in an area that can be readily covered with tile before the mortar starts skinning over (Fig. 9.6-47a). The tile generally should be tamped into place to obtain at least 80% coverage; 100% coverage in tubs, showers, and other high-moisture areas (Fig. 9.6-47b). Any grout can be used, but in selecting a suitable type, consideration should be given to the type of exposure (Fig. 9.6-47c).

**FIGURE 9.6-46** Open framing requires a two-coat mortar bed over metal lath. Except with dry-set mortar, curing the bed is required for thin-set applications. (Courtesy Tile Council of America, *Handbook for Ceramic Tile Installation.*)

a

b

c

**FIGURE 9.6-47** Thin-set application of glazed wall tile directly to masonry: (a) dry-set mortar is applied with a notched trowel; (b) edge-glued tiles are set and tamped in place; (c) grout is forced into joints and the excess wiped off. (Courtesy American Olean [Ceramic] Tile Company [now Dal-Tile Corporation].)

## ORGANIC ADHESIVE

The use of organic (mastic) adhesive is limited to light-duty interior areas and over smooth surfaces such as plaster, gypsum board, mineral fiber-cement board, or plywood (Fig. 9.6-48). It also can be used over concrete or masonry if the surface is relatively smooth (maximum variation $1/8$ in. [3.2 mm] in 8 ft [2438 mm]). Type I adhesive should be used for installations requiring prolonged water resistance, such as in tub and shower areas, and on bathroom floors. Type II adhesive can be used in areas subject to occasional or intermittent wetting. However, neither type is suitable for installations requiring continuous contact

**FIGURE 9.6-48** Organic adhesives can tolerate limited moisture exposure (such as in residential shower stalls) if a water-resistant adhesive and backing are used and the board edges are sealed. (Courtesy Tile Council of America, *Handbook for Ceramic Tile Installation.*)

with water, such as in sunken tubs or pools. The manufacturer's recommendations should be followed to determine which type is best for the specific use intended or whether a primer is required.

The adhesive should be spread with a notched trowel recommended by the manufacturer, leaving no bare spots (Fig. 9.6-49a). The tile should be tamped into the adhesive to ensure maximum contact (Fig. 9.6-49b). The average contact area of tile removed for inspection should be not less than 75%, and no individual tile or mounted assembly should have less than 40% coverage of adhesive. Adjustments should be made while the adhesive is still plastic. Grouting should be delayed for at least 24 hours to permit solvent evaporation from the adhesive. Joints should be grouted with a product suitable for the conditions. Drywall and latex grouts commonly are used with organic adhesives (Fig. 9.6-49c).

### 9.6.3.2 Exterior Surfaces

Tile installations on exterior walls require the same precautions regarding movement as exterior floors. The supporting construction, including the backing, must be designed to accommodate movement due to temperature and moisture variations. This means having:

■ Suitable control, isolation, and expansion joints
■ Proper metal reinforcement in concrete and masonry

a

b

c

**FIGURE 9.6-49** Over previously applied organic adhesive, (a) edge-mounted tile is applied; (b) tile is tamped in place with a "rubber float" trowel; and (c) the same trowel is used to force grout into joints and remove excess grout. (Courtesy American Olean [Ceramic] Tile Company [now Dal-Tile Corporation].)

■ Careful control of mortar and concrete mixes
■ Sound engineering design
■ Good workmanship

Exterior tile can be applied either by the conventional mortar bed or the thin-set method, depending on:

■ Whether the backing is suitable for direct bonding
■ Dimensional stability of the backing
■ Provisions for crack control

On large areas exposed repeatedly to direct wetting, the conventional isolated method usually gives better results. The thin-set bonded method generally is more reliable in covered outdoor areas and when broken into smaller panels by decorative trim, battens, or expansion joints. However, tile can be thin-set over an isolated mortar bed with dry-set mortar.

Regardless of installation methods, expansion joints should be provided in the tile finish over all underlying structural joints, at the tile perimeter, and against any projecting surface. In isolated installations, expansion joints in the tile finish should be provided at 12 to 16 ft (3660 to 4880 mm) on centers. Where these expansion joints do not coincide with joints in the backing, they should be provided in addition to those over the underlying joints. In a thin-set installation over a suitable stable backing, the additional joints in the tile finish are not as essential because the tile finish is intended to move with the backing.

In exterior tile installations, it is essential to provide suitable flashings, overhangs, or copings to prevent entry of water into or behind the tilework. Exterior walls of habitable spaces should include a suitable vapor retarder on the warm side of the wall to prevent condensation in the wall cavity. Additional information on the design and construction of expansion joints is given in Section 9.6.1.6.

### PORTLAND CEMENT MORTAR BED

Procedures for installing tile over a mortar bed outdoors are essentially the same as those described earlier for interior locations. The entire installation should be backed by cleavage membrane. Metal lath should be firmly attached to the backing and cut at expansion joints (Fig. 9.6-50).

A scratch coat consisting of 1 part portland cement, 5 parts damp sand, and $^1/_5$ part lime usually is installed by a plastering contractor. The mortar bed, consisting of 1 part portland cement, 5 parts damp sand, and $^1/_2$ part hydrated lime, is normally installed by a tile contractor. The overall thickness of both coats should be $^3/_4$ to 1 in. (19.1 to 25.4 mm) thick. The mortar bed mix can be varied up to 1 part portland cement, 1 part lime, and 7 parts damp sand. However, extremely rich mixes (too much portland cement) will result in greater drying and shrinkage cracking. Similarly, a total

**FIGURE 9.6-50** Expansion joints 12 to 16 ft (3658 to 4877 mm) on centers are essential in exterior isolated thickbed tile installations. (Courtesy Tile Council of America, *Handbook for Ceramic Tile Installation.*)

mortar bed thickness in excess of 1 in. (19.1 mm) should be avoided.

Neat cement can be used for setting tile over a still plastic mortar bed. When dry-set or latex-portland cement mortar will be used, the mortar bed should be cured for at least 24 hours, but longer curing periods of up to 10 days are preferred. Suitable grouts include commercial cement, dry-cure.

### THIN-SET INSTALLATION

Waterproof backings suitable for bonding of tile with thin-set mortars include monolithic concrete, concrete and brick masonry, structural clay tile, and a cured portland cement mortar bed (see Fig. 9.6-44). Latex-portland cement and dry-set mortar typically are used for setting tile. Either latex or dry-set grout can be used for filling tile joints. Damp curing for at least 72 hours is essential.

# 9.7 Acoustical Treatment

Chapter 13 includes general requirements for and the theory behind sound control in buildings. This section discusses in more detail that part of sound control generally called *acoustical treatment*. This includes acoustical ceilings, integrated ceilings, and acoustical wall treatment and baffles. More than one of these are often used together (Fig. 9.7-1). Acoustical treatment also includes the furring and suspension systems needed to mount acoustical materials.

Before proceeding, it is helpful to define some of the terms used in this section.

- *Accoustical Baffle*: An acoustical panel that is suspended from an edge or floats cloud-fashion in a space.
- *Acoustical Ceiling*: The entire ceiling system, including acoustical material and associated insulation, furring, and suspension systems. It does not include auxiliary thermal or acoustical insulation in addition to that normally associated with the acoustical ceiling being discussed. A layer of acoustical insulation placed above a ceiling and over or adjacent to a partition to achieve increased acoustical performance is not part of the acoustical ceiling. An acoustical blanket or pad which is necessary to achieve the rated acoustical performance of a metal ceiling is a part of the acoustical ceiling. Metal ceilings, as defined below, are also acoustical ceilings.

- *Acoustical Material*: A general term that covers both composition and metal acoustical materials as defined in this chapter.
- *Acoustical Wall Panels*: Acoustical panels that are mounted on the surface of a wall or partition, regardless of their mounting method or size. Acoustical wall panels may completely cover a wall surface or may be mounted as plaques with space between them.
- *Composition*: When referring to acoustical materials, includes cellulose, mineral fiber, glass fiber, asbestos cement, ceramic-faced, plastic-faced, gypsum, and wood fiber acoustical materials. Plastic- and metal-faced acoustical ceiling materials in which the plastic or metal is adhered or bonded to a composition material as defined here are also called composition materials.
- *Integrated Ceiling*: A ceiling system consisting of acoustical material, a suspension system, air distribution outlets, and lighting fixtures.
- *Metal Acoustical Ceiling Material*: Linear metal strips, as well as metal

panels and pans, are all called metal acoustical materials, even though some of them may be of questionable acoustic value. The term *metal acoustical material* means the ceiling material only. It does not include the associated suspension system or acoustical insulation blankets or pads.
- *Metal Ceiling*: The entire ceiling construction, including metal acoustical materials, acoustical pads or blankets, and the suspension system.

### 9.7.1 ACOUSTICAL CEILINGS

Acoustical ceilings include composition tiles and panels, metal acoustical materials, and associated furring and suspension systems. Integrated ceilings are discussed in this chapter.

#### 9.7.1.1 Acoustical Materials

The characteristics of acoustical materials are identified according to ASTM E1264. It classifies acoustical materials according to type, pattern, fire class, acoustical performance, and light reflectance coefficient.

#### MATERIAL TYPES AND PATTERNS

New acoustical materials will usually be one of the types shown in Figure 9.7-2, with one of the patterns shown in Figure 9.7-3. Other materials are also sometimes available, including ceramic-faced units for use in wet and high-impact areas; units with other abuse-resistant or sanitary plastic faces; gypsum; and wood fiber composition. Acoustical units in existing buildings may contain asbestos. Metal ceiling panels are made from aluminum, steel, and stainless steel. Suspended ceilings may also be of wood, plastic, or sheet metal, but most of those types are more properly classified as decorative rather than acoustical ceiling materials.

ASTM E1264 lists three fire classes with the following flame spread ratings:

- *Class A:* 0 to 25
- *Class B:* 26 to 75
- *Class C:* 76 to 200

#### FINISHES

Most composition acoustical units are painted on the exposed face. Most such painted finishes are considered washable, although gentleness is advised. Some manufacturers list some products as having a *scrubbable* finish, and these are

**FIGURE 9.7-1** Acoustical ceiling and wall panels are both used in this auditorium. (Photograph provided by Metal Building Interiors Products Company.)

**FIGURE 9.7-2** Acoustical Material Types

| Type I | Cellulose base with painted finish |
|---|---|
| Type II | Cellulose composition with membrane-faced overlay |
| Type III | Mineral base with painted finish; type includes:<br>    Form 1 Nodulated, cast, or molded<br>    Form 2 Water felted<br>    Form 3 Dry felted |
| Type IV | Mineral base with membrane-faced overlay; type includes:<br>    Form 1 Nodulated, cast, or molded<br>    Form 2 Water felted<br>    Form 3 Dry felted |
| Type V | Perforated steel facing (pan) with glass fiber base backing |
| Type VI | Perforated stainless steel facing (pan) with mineral or glass fiber base backing |
| Type VII | Perforated aluminum facing (pan) with mineral or glass fiber base backing |
| Type VIII | Cellulose base with scrubbable pigmented or clear finish |
| Type IX | Mineral base with scrubbable pigmented or clear finish; type includes:<br>    Form 1 Nodulated, cast, or molded<br>    Form 2 Water felted<br>    Form 3 Dry felted |
| Type X | Mineral base with plastic or aluminum membrane-faced overlay or both |
| Type XI | Mineral base with fabric-faced overlay; type includes:<br>    Form 1 Nodulated, cast, or molded<br>    Form 2 Water felted<br>    Form 3 Dry felted |
| Type XII | Glass fiber base with membrane-faced overlay; type includes:<br>    Form 1 Plastic<br>    Form 2 Cloth<br>    Form 3 Other |
| Type XIII | Aluminum or steel strip with mineral or glass fiber base backing; type includes:<br>    Form 1 Perforated<br>    Form 2 Nonperforated |
| Type XX | Other types (this category is left for additional types) |

(Copyright ASTM, reprinted with permission.)

**FIGURE 9.7-3** Acoustical Material Patterns

A. Perforated, regularly spaced large holes
B. Perforated, randomly spaced large holes
C. Perforated, small holes
D. Fissured
E. Lightly textured
F. Heavily textured
G. Smooth
H. Printed
I. Embossed
J. Embossed-in-register
K. Surface scoured
Z. Other patterns (describe)

(Copyright ASTM, reprinted with permission.)

listed in ASTM E1264, but there is no universally accepted definition of the term.

A variety of finishes are available for metal ceiling panels, including flouro-polymer, baked enamel, anodized aluminum, acrylic lacquer, and even wood veneer.

Light reflectance is rated as one of four coefficients with the following light reflectances:

- *LR 1:* 75% or more
- *LR 2:* 70% to 74%
- *LR 3:* 65% to 69%
- *LR4:* 60% to 64%

### ACOUSTICAL PERFORMANCE

Acoustical products are classified by Sound Transmission Class (STC) and Noise Reduction Coefficient (NRC) or by Articulation Class (AC) or Speech Privacy Isolation Class (NIC′). An acoustical ceiling is rated by Ceiling Sound Transmission Class (CSTC). STC, NRC, and CSTC ratings are used for closed-room situations. AC and NIC′ are used for large spaces that contain several activities, such as occur in an open-plan office space. AC- and NIC′-rated products have been tested for their ability to both control sound reflections and permit white-noise transmission from plenum speakers into a space. White noise is used in open-space designs to mask background noises and help provide speech privacy.

The sound absorption quality of an acoustical material is indicated by the NRC, which shows a material's ability to reduce sound levels within a space. An acoustical material's noise isolation control ability is indicated by the STC (CSTC for ceilings), which shows a material's ability to prevent sound from passing through it into another space. Usually there is an inverse relationship between the two properties. Materials that reduce sound reverberations are usually light and porous. To prevent sound transmission, materials must be heavy and airtight. STC and CSTC are measured by test procedure AMA 1-II, "Ceiling Sound Transmission Test by the Two Room Method," which gets its name from the no longer existing Acoustical Materials Association. The standard is available from the Ceilings and Interior Systems Construction Association (CISCA).

Where sound control is an essential element in an existing building, the advice of an acoustical specialist should be sought before changes or repairs are made.

### COMPOSITION ACOUSTICAL TILE VERSUS PANELS

Composition acoustical ceiling materials may be either tile or panels, and all the types, patterns, and other characteristics listed in previous subsections are found in both of them. In the past, much tile was used. In recent years, the advent of better acoustical panel materials, more attractive and functional suspension systems, an increased need for access to the space above ceilings, changes in construction methods, their lower cost, and other factors has led to a much greater use of panels. Some manufactures have stopped making tile products and concealed grid suspension systems. An advantage of panels over tile is that panels are loosely laid into a grid, which makes them easily removable and makes the space above them readily accessible. A major disadvantage is that they are loosely laid into a grid, which makes them subject to dislocation by wind, pressure applied during cleaning, and air passing through a plenum space above them. Tile and panel edges are square, beveled, or have a special edge such as the reveal edge shown in Figure 9.7-4.

*Tile* Most modern tile is either 12 or 24 in. (305 by 610 mm) square or 12 by 24 in. (305 b 610 mm). Tile in older buildings will probably be 12, 16, or 20 in. (305, 406, or 508 mm) square. Tile joints are flanged, rabbeted, kerfed (splined), or tongue-and-groove. Figure 9.7-5 shows some types, but there are many others in use today and still others have been used in the past. Sometimes a tile's edges are not all cut the same way. The tongue-and-

**FIGURE 9.7-4** Acoustical tile and panel edges. Many variations of beveled and reveal edges are available. For example, reveal edges may have two, three, or more steps. Bevels may be combined with steps. (Courtesy H. Leslie Simmons.)

**FIGURE 9.7-5** Typical acoustical tile joint conditions. (Courtesy H. Leslie Simmons.)

groove joints on two sides of a tile may be quite different from those on the other two sides so that the tile will nest properly.

*Panels* Most panels are either 24 in. (610 mm) square or 24 by 48 in. (610 by 1220 mm). Some are scored so that they appear from the floor to be smaller units. Scoring may produce squares or linear strips and may be parallel with panel sides or diagonal to them.

Panel joints are usually square or reveal type (see Fig. 9.7-4). In systems where cross suspension members are concealed (semiconcealed systems), the ends of the panels are kerfed and rabbeted to receive the concealed suspension member.

### METAL ACOUSTICAL CEILING MATERIALS

In addition to metal-faced composition products, there are many metal acousti-

cal ceiling material styles (Fig. 9.7-6). They include, but are not limited to, linear shapes (strips), flat panels, shaped (corrugated and other shapes) panels, and flat or textured pans. Surfaces may be continuous or perforated and either flat or embossed. Metal grids and louvers are also available, but whether they are acoustical is debatable. Metal ceiling members may be lay-in or snap-in types. Snap-in types require special support systems. Some lay-in types also require special support systems; others do not. Some panels come with glass fiber or mineral wool sound-absorbing pads; others do not.

### 9.7.1.2 Support Systems

Composition acoustical materials may be directly attached to solid or supported substrates or supported on a suspension system. Metal ceilings usually include a suspension system. The term *suspension*

*system* usually refers to a metal grid that is hung by wires or bars from the structure, but it is also used to refer to a system of wood or metal furring to which acoustical materials, mainly tiles, are screwed, stapled, or nailed. Sheet metal decorative ceilings are often held in place using fasteners, but metal acoustical ceiling materials are seldom so installed.

### WOOD FURRING

Except in single-family residences and other small buildings, little acoustical tile is installed directly on wood furring today (Fig. 9.7-7), but in the past, such installations were common. The normal installation when wood furring is used today is to install a layer of gypsum board, called *backing board*, over the furring and fasten the acoustical tile to the gypsum board.

### METAL FURRING

In the past, acoustical materials were sometimes screwed directly to metal furring channels (Fig. 9.7-8), but this is seldom done today. Today, when composition acoustical tile is installed over metal furring bars, the method usually used is that described in ASTM Standard C635 as the "Furring Bar Suspension System," in which a layer of gypsum board, backing board, is applied to suspended nailing or furring bars. The acoustical tile is then fastened to the gypsum board.

Metal acoustical materials are seldom installed over metal furring bars today. They are occasionally installed on snap-in or other grid members that are directly clipped to structural elements or inserts in concrete slabs, but that installation is so similar to the one described later in this chapter for a metal suspension system that the requirements addressed there apply.

### HUNG METAL SUSPENSION SYSTEMS

Metal suspension systems for acoustical ceilings should comply with the applicable requirements of ASTM C635 and C636. Acoustical treatment manufacturers may, however, have more stringent requirements.

There are two categories of hung suspension systems for acoustical materials. In *direct-hung* systems, the main runners are suspended from the structure (Fig. 9.7-9). In *indirect-hung* systems, the main runners are suspended from carrying channels (Fig. 9.7-10). The main run-

**FIGURE 9.7-6** An angled ceiling made from flat, perforated metal acoustical ceiling panels. (Manufactured by Hunter Douglas Architectural Products.)

**FIGURE 9.7-7** Acoustical tile on wood furring. (Courtesy H. Leslie Simmons.)

**FIGURE 9.7-8** Acoustical tile directly applied to metal furring. (Courtesy H. Leslie Simmons.)

**FIGURE 9.7-9** One type of direct-hung acoustical ceiling suspension system. (Courtesy H. Leslie Simmons.)

**FIGURE 9.7-10** One type of indirect-hung acoustical ceiling suspension system. (Courtesy H. Leslie Simmons.)

ners may be any type that is appropriate for the acoustical material being supported. Some common types are shown in Figure 9.7-11, but there are many others.

*Lay-in* suspension systems may be either direct- or indirect-hung, depending on the system members and the necessary spacing of hangers. *Concealed grid* systems may also be either direct- or indirect-hung, but systems having splines or other less rigid cross members are usually indirect-hung. Except for the carrying channels, most members in both types of systems are proprietary, meaning that they are manufactured or furnished by a single supplier or producer for installation as a system.

Hung suspension systems for composition materials include concealed grid types for tile and exposed or semiexposed grid types for lay-in panels.

***Concealed Grid Systems for Composition Tile*** Most hung suspension systems designed to be concealed in the finished ceiling are H and T systems, which include H-shaped main runners and tee splines similar to those shown in Figure 9.7-11; concealed single- or double-web systems using tee-shaped main runners similar to the standard tee shown in Figure 9.7-11 and spacer bars similar to those shown in the same figure at right angles to the main runners for stiffening;

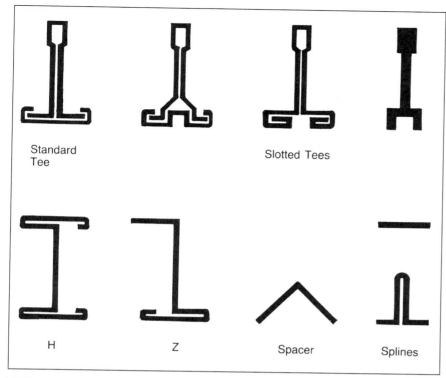

**FIGURE 9.7-11** Typical acoustical ceiling suspension system components. (Courtesy H. Leslie Simmons.)

Standard Tee

Slotted Tees

H

Z

Spacer

Splines

*Grids for Metal Ceilings* Hung suspension systems for metal acoustical ceiling materials include snap-in types and lay-in types. Many metal pans and some panels are installed into snap-in grids. The T-bar snap-in member shown in Figure 9.7-12 is only one of several snap-in-device types in general use. Snap-in grid members are manufactured specifically to fit the edge configuration of the pans or panels being supported. A grid member made by one manufacturer may not correctly fit or support a panel made by a different manufacturer.

Some types of metal ceiling grids are direct-hung but do not have cross furring members because the metal ceiling material itself is stiff enough to stabilize the system.

Some metal pans and panels lay into the same type of grid used for composition material. Others hang from similar grids (Fig. 9.7-13).

Support grids for linear metal ceilings, while similar to that shown in Figure 9.7-14, vary from manufacturer to manufacturer to suit their particular ceiling material.

*Structural Classification* ASTM C635 divides hung suspension systems into three structural strength categories: *light-duty*, *intermediate-duty*, and *heavy-duty*. Light-duty systems, which are used in residential and light commercial installations, will support only the acoustical ceiling materials themselves. Intermediate-duty systems, which are used in most commercial structures, will support light fixtures and air diffusers in addition to the acoustical materials themselves. Heavy-duty systems will support heavier loads than the other two classifications.

*Edge Moldings and Trim* Except for some linear metal ceilings, most acoustical ceiling suspension systems have matching edge moldings at all vertical surfaces to which the acoustical ceiling abuts, including penetrations such as columns. Acoustical tile ceilings applied using fasteners or adhesives should also have edge moldings and trim.

The edge molding and trim type and configuration should be suitable for the suspension system used. Some ceiling types have specialized trim members used only for that type.

*Metals and Finishes* In exterior installations and in interior installations where

or zee systems (see Fig. 9.7-10) using zees and splines similar to those shown in Figure 9.7-11. Other systems are also used today and still others have been used in the past, so an existing installation may be different from those noted here.

Concealed suspension systems for composition tile are designed to lock the tile in place so that it cannot be easily removed. To overcome this drawback, manufacturers have devised ways to make sections of tile hinged or removable so that access can be gained where necessary. In direct-hung systems, hinged tile sections, after being released by a special tool, may be moved upward into the plenum or downward into the room. In both types, more than one tile is usually movable. Strips of tile between cross framing members either lift up or can be pulled down to provide access. The movable panels of tile may be hinged along the cross tees or along the main tees. Once the movable tile has been moved out of the way, the remainder of the tile can be removed in successive pieces. Downward access systems are not recommended in some seismic zones.

In indirect-hung systems, access is sometimes gained with special cross runners or split splines which permit acousti-cal unit removal. Such special runners or splines are identified with unobtrusive permanent markers. Some indirect-hung systems are made accessible with special tools that work only with those systems.

Not all systems, however, have a removable feature. Some older tile may be removable only by breaking out a tile and then removing the adjacent tile until the necessary opening has been achieved. Where the space above the ceiling is not accessible through the tiles, access doors must be provided.

*Exposed and Semiexposed Grids for Composition Panels* Suspension systems designed for lay-in application of acoustical panels usually consist of T-shaped main runners and interlocking cross tees (see Fig. 9.7-9). The cross members of semiconcealed systems are usually T-splines. The main runners are usually directly hung from the structure, with no carrying channels present. Main runner and cross-runner shapes are proprietary and vary with each system manufacturer. They are available with wide or narrow flat-bottom surfaces, roll-formed caps, slotted, with reveals, and in other configurations. A few of the many configurations available are shown in Figure 9.7-11.

**FIGURE 9.7-12** One type of T-bar snap-in metal ceiling system. (Courtesy H. Leslie Simmons.)

Clip

Carrying channel

Snap-in T-bar

Metal pan

Main runner

Metal acoustical panel

Acoustical pad

Main runner

Metal acoustical panel

Acoustical pad

Main runners

Acoustical pad

Metal acoustical panel

**FIGURE 9.7-13** Some typical hung metal ceiling systems. (Courtesy H. Leslie Simmons.)

**FIGURE 9.7-14** Typical support grid member for a linear metal ceiling. (Courtesy H. Leslie Simmons.)

high humidity will be present, such as swimming pool areas or commercial kitchens, concealed metals should be either aluminum, stainless steel, zinc alloy, or hot-dip galvanized steel.

In these same types of locations, exposed grids and edge members should be either aluminum or stainless steel. Aluminum is usually protected by a baked-on paint, flouropolymer, or anodized finish but may also be left natural and protected by a coat of lacquer. Stainless

steel may be protected by a coating or left exposed.

In such locations, the finishes on both concealed and exposed metals should comply with ASTM C635 requirements.

Concealed metals in grids supporting composition tile and panels in locations other than those listed in the preceding paragraphs may be the manufacturer's standard steel products with the manufacturer's standard galvanized or painted finish.

Exposed grids and edge members in locations other than those mentioned in the preceding paragraph are usually steel, finished with the manufacturer's standard paint finish, but may be aluminum or stainless steel where those decorative finishes are selected. Where the acoustical ceiling material has a decorative finish, such as anodized aluminum or stainless steel, exposed suspension system members often match the ceiling material and finish.

Colors of exposed finished metal vary, though white and off-white are the most common.

### 9.7.1.3 Accessories and Miscellaneous Materials

The types of accessories needed for acoustical ceilings include the following:

*Acoustical pads for use in metal pan and panel ceilings.* Mineral or glass fiber. Some are wrapped or faced with materials such as polyvinyl chloride. Others are bare.

*Acoustical sealant.* Permanently elastic, water-based, nondrying, nonbleeding, and nonstaining.

*Adhesives for installing acoustical tile.* Underwriters Laboratories, Inc. (UL) labeled with a Class 0 to 25 flame spread rating.

*Angle shaped hangers.* Formed from nominal 15- or 16-gauge (1.75 or 1.6 mm) galvanized sheet steel. Legs should not be less than $^7/_8$ in. (22.2 mm) wide.

*Carrying channels.* $1^1/_2$ in. (38 mm) cold-rolled steel channels weighing a minimum of 0.475 lb/lin ft (0.71 kg/m), painted. Where spans require, 2 in. (50.8 mm) cold-rolled steel channels weighing a minimum of 0.590 lb/lin ft (0.88 kg/m) are used.

*Clips.* Commercial types acceptable under the building code for support.

*Extension hangers for metal ceilings and hanger rods and flat bars.* Zinc- or cadmium-coated mild steel. Alternatively, such hangers may be finished with rust-inhibitive paint.

*Hanger anchorage devices.* Rust-inhibitive screws, clips, bolts, concrete inserts, expansion anchors, or other devices sized to support five times the calculated hanger load.

*Hanger wire.* For typical installation, galvanized carbon steel wire of a size necessary to properly support three times the hanger design load, but in no case less than 12-gauge (2.75 mm) wire. Some codes may require heavier wire, regardless of the load. For support of aluminum or stainless steel grid systems, and in exterior locations and air plenums, stainless steel or aluminum wire of a gauge recommended by the suspension system manufacturer is used.

*Hold-down clips.* Suspension system manufacturer's standard types. Hold-down clips are necessary where uplift is a problem from air pressure or to permit cleaning. Clips are also required in some fire-rated assemblies. Security clips are sometimes used to make metal pan ceilings nonremovable.

*Screws and other fasteners.* Rust-inhibitive and of a type standard with the suspension system manufacturer. Tile fasteners should be stainless steel or cadmium-plated of the type recommended by the tile manufacturer and should penetrate the substrate by at least $1/2$ in. (12.7 mm). Exposed fasteners should match the material fastened in appearance.

*Sound attenuation blankets.* Semirigid mineral fiber blankets with no membrane facing; with a Class 25 flamespread; $1^1/2$ in. (38 mm) thick or more.

*Tie wire.* Galvanized wire at least 16-gauge (1.6 mm).

### 9.7.1.4 Fire Performance

The ASTM E1264 Fire Classes refer only to the acoustical material itself. The UL *Building Materials Directory* contains a listing showing the ratings of acoustical ceiling materials by manufacturer and product. Acoustical materials manufacturers usually indicate the flame spread rating of their products.

Hung suspension system are classified as fire-resistance-rated or non-fire-resistance-rated, but sometimes it is hard to tell which is which from the manufacturers' product literature.

Unfortunately, using a rated acoustical material, suspension system, or both together does not alone result in compliance with code or other requirements for fire-rated construction. Fire-rated construction is a function of the entire assembly, including the floor or roof construction, flooring or roofing, insulation, and the installed ceiling. UL's *Fire Resistance Directory* includes a list of construction systems, which includes the structure (floor, roof, etc.) and finishes, including acoustical ceilings. Ceiling materials are also listed there by manufacturer and product.

Fire resistance ratings are established according to ASTM Standard E119.

### 9.7.1.5 Installation

Layout plans should be prepared for spaces where acoustical ceilings will be installed to ensure coordination with lighting fixtures and other ceiling-mounted equipment and devices.

Manufacturers' brochures and installation details and ASTM and CISCA standards should be obtained and followed. Metal suspension systems should be installed in accord with ASTM Standard C636 unless the acoustical treatment manufacturer has more stringent requirements.

The manufacturer's recommendations for cleaning and refinishing acoustical units and suspension system components should be obtained and stored in a safe place so that they will be available when maintenance and repair are necessary in the future.

Extra material, totaling 2% or 3% of the amount installed, is often left at the site for the owner to use in making repairs.

Acoustical materials and suspension system components should be kept dry before and after use. Acoustical materials should be allowed to acclimate to the temperature and humidity of the space before they are installed to prevent dimensional changes which might affect the appearance of the installed material.

Watermarked, rusted, or otherwise damaged units should not be installed.

Acoustical ceiling components should not be installed until the building is weathertight, glazing has been completed, exterior openings have been closed, and mechanical, electrical, fire-fighting, and other work above ceilings has been completed.

Wet work, such as masonry, concrete, and plaster, should be allowed to dry to

the moisture content recommended by the acoustical materials manufacturer before acoustical materials are installed. A temperature of at least 60°F (15.6°C) should be maintained in the space during and after installation of acoustical materials.

Ceilings where cleanliness is paramount, such as in hospital operating rooms, often require sealant around the perimeter of the ceiling grid and sometimes even around the perimeter of each panel at its supports. Health department rules and regulations and building codes should be examined to determine the applicable requirements.

Suspension systems for use where fire-resistance ratings are required should comply with governing regulations, and their materials and installations should be identical with assemblies which have been tested and listed by recognized authorities such as UL.

Suspension system members that abut vertical or overhead building structural elements should be isolated from structural movement sufficiently to prevent transfer of loads into the suspension system. Provisions should also be made in the suspension system to accommodate movement that occurs across control and expansion joints in the substrates.

Suspended acoustical ceilings should have a seismic restraint system in accordance with ASTM Standard E580 where required by code or usual practice.

Flat surfaces of suspension system members should not be warped, bowed, or out of plumb or level by more than $1/8$ in. (3.2 mm) in 12 ft (3658 mm) in both directions or by more than $1/16$ in. (1.6 mm) across any joint or $1/8$ in. (3.2 mm) total along any single member.

Before acoustical ceiling installation begins, mechanical, electrical, and other work above or within the acoustical ceilings should have been completed.

Acoustical units should be handled carefully to prevent damage to edges and faces.

### FURRING INSTALLATION

Furring should be installed as is appropriate to support the acoustical ceiling. It should be placed on lines and levels necessary to cause the acoustical material to fall into the proper location. It should correct unevenness in the supporting structure or substrate.

The surface of wood furring to which acoustical material will be fastened should be not less than $1^1/2$ in. (38 mm) wide. Generally, 2 × 2 (50 × 50) lumber is used for furring that is attached to supporting structure and 1 × 2s (25 × 50s) or 1 × 3s (25 × 75s) where furring is laid directly over solid substrates. Nominal 1-in. (25.4-mm)-thick lumber is too flexible to use over framing. Sometimes acoustical ceilings are furred using 2 × 4s (50 × 100s) hung from structural floor or roof framing by hangers spaced not more than 48 in. (1200 mm) on center. Frequently in older buildings, the hangers were wood members instead of wire.

Furring should form a complete system adequate to properly support the acoustical tile. Closers should be installed at edges and openings.

Bolts, lag screws, and other anchors are used to anchor wood furring in place. Fasteners are placed near the ends of furring members and not more than 36 in. (900 mm) on center between them. Shorter members, however, should be anchored at 30 in. (750 mm) on center.

Screws, bolts, clips, and other anchors are used to anchor metal furring in place. Fasteners are placed near the ends of furring members and not more than 24 in. (600 mm) on center between them, but other support spacings may be recommended by the furring bar manufacturer. Some types of furring bars may require closer spacings. The correct spacing should be verified with the furring bar manufacturer.

Furring spacings may vary with the substrate and the acoustical material being supported. Minimum spacings are dictated by acoustical material thickness and size, and may vary somewhat from manufacturer to manufacturer and system to system. Spacing must be such as to adequately support the acoustical ceiling.

## HUNG SUSPENSION SYSTEM INSTALLATION

Suspension system hangers should be secured to wood, concrete, or steel structural supports by connecting directly to the structure where possible. Where direct connection is not possible, the hangers should be connected to inserts, clips, anchorage devices or fasteners. Except for hanger attachments installed specifically for the purpose by the steel deck installer, hangers should not be attached to steel decks. Since failures have occurred even when deck tabs acceptable to the deck manufacturer have been used, some sources recommend that ceilings not be supported from steel deck under any cir-

cumstances. Hangers should never be attached to pipes, conduit, ducts, or mechanical or electrical devices.

Hangers should be located plumb in relation to primary suspension members. Contact with insulation covering ducts and pipes and other items within the plenum space should be avoided. Hangers should be splayed only where obstructions or conditions prevent plumb, vertical installation. Horizontal forces generated by splayed hangers should be offset by counter-splaying, bracing, or another suitable method.

Frames for grilles, registers, and other items that occur within acoustical ceilings should be properly installed in the correct locations. Additional hangers should be provided for support of the grid systems at lights, grilles, registers, and other items to prevent excessive deflection. At grid systems with a light-duty structural classification, lights, grilles, registers, and other items should be supported separately from the ceiling grid.

Main runner and cross furring members should be placed in proper positions, leveled, and fastened together in the manner required by the system. T-shaped members are usually locked together without accessories. Some systems, however, require clips.

Except in self-edging systems, edging strips should be provided for all acoustical ceilings, including those with concealed grids, along the entire perimeter and where they abut vertical surfaces. Short lengths should be avoided. Standard moldings should be attached through the web 3 in. (76.2 mm) from each end and not more than 16 in. (406 mm) on center between. Special moldings may require other spacings. Edging strips are usually of the same material and finish as exposed portions of grid systems.

Light fixtures and other penetrations are often boxed, using mineral wool boards to comply with UL rating requirements where ceilings are rated.

Suspension systems should achieve a tight, level ceiling at the spacings and directions recommended by the manufacturer and to comply with UL label requirements.

### Installation of Concealed Grid Systems for Composition Acoustical Tile   Maximum hanger spacing is usually 48 or 60 in. (1220 or 1524 mm), depending on the system and the components selected. The first hanger on each end of each runner

or carrying channel should be no more than 6 in. (150 mm) from the ends. Some systems require closer spacings. Where the hangers must be placed at wider spacings than those recommended for the system, the main runners can be supported by carrying channels. In indirect-hung suspension systems, main runners are usually supported by carrying channels.

### Installation of Exposed Grid Systems for Composition Acoustical Panels   Exposed members should be installed in line, square, level, in one plane, and within the tolerances permitted by ASTM Standard C636. Tees should be of adequate size to span between supporting structures. It may be necessary to support main tees from carrying channels.

Main tees are usually located 48 in. (1220 mm) on center and all in the same direction. Cross tees are then installed to span between the main runners with cross tees at 24 in. (610 mm) on center. The tees are then locked together using whatever method is applicable to the system being used. The tees should rest on wall moldings at vertical surfaces. Where 24 in. (610 mm) square panels are used, additional cross tees are placed from cross tee to cross tee and securely fastened (or locked) to the cross tees.

Main tees should be supported by hangers placed not more than 48 in. (1220 mm) on center. Where the structure will not permit hangers at that spacing, additional carrying channels should be provided at 48 in. (1220 mm) on center, and the main runners should then be supported from the carrying channels. The first hanger on each end of each runner or carrying channel should be no more than 6 in. (150 mm) from the ends.

## COMPOSITION ACOUSTICAL TILE INSTALLATION

Tiles should be placed over the entire grid area and cut to fit around fixtures, penetrating hangers, and accessories. Border units should be cut and fit neatly and snugly into the grid, leaving no voids in the finished ceiling.

Splines and other grid members are placed into the kerfed edges of tiles or tile tongues are inserted into tile grooves so that every joint is tightly fitted and properly supported. The tile field should be held in compression with leaf-type spring steel spacers between the tiles and the edge moldings. Spacers should be placed 12 in. (305 mm) apart.

When a grid system is not used, tiles are held in place with fasteners, adhesive, or both. Before adhesive application, loose dust should be removed from the backs of the tiles, and they should be primed with a thin coat of adhesive. Surfaces to receive tiles should be sound, free of foreign materials, clean, and dry. New plaster and concrete should be cured thoroughly. Bare plaster, gypsum board, and concrete should be primed. The alkalinity of surfaces to receive adhesive-applied tiles should be tested with litmus or Hydrian pH paper. Surfaces with a pH in excess of 9 should have a neutralizer applied to lower the pH to 8 or less.

Tiles are installed with adhesives by applying spots of adhesive to each tile and pressing the tiles in place using a sliding motion. The thickness of the adhesive after application should be about $1/8$ in. (3.2 mm).

Splines are installed in the joints between tiles to keep them in proper alignment and flush with each other. Tiles are scribed to fit at penetrations and edges.

When tiles are fastened in place using fasteners or adhesives, it is necessary to provide access doors in the ceiling where mechanical, electrical, or other equipment must be accessible. Such panels are usually recessed so that they can be faced with tiles like those installed in the ceiling.

### COMPOSITION ACOUSTICAL PANEL INSTALLATION

Lay-in panels should be placed into each opening of the exposed grid system. Panels should be cut neatly around fixtures, diffusers, penetrating hangers, accessories, and other items. Border units should be cut to fit snugly into the grid, with no cracks or voids.

### METAL ACOUSTICAL MATERIALS INSTALLATION

Linear metal ceiling strips are snapped into supporting grid members to cover the entire ceiling.

Lay-in metal panels and pans are laid into each opening in the grid.

Snap-in panels and pans are placed in each opening in the grid and shoved into place so that the holding feature fully engages and the panels or pans are held securely in place. Where a ceiling is to be nonaccessible, security clips are installed as the tile or pans are placed.

### SOUND INSULATION

Sound insulation blankets may be installed over the full area of an acoustical ceiling or only in certain locations, such as close to sound-rated partitions. Unless the ceiling is specifically designed to support it, sound insulation should be separately supported so that its weight does not rest on the acoustical materials or any part of the suspension system.

In metal ceilings, sound insulation pads should fit snugly into each pan, panel, or strip over the entire ceiling area.

Insulation should be tightly butted with joints in solid contact and without voids. It should be fitted neatly to adjacent surfaces.

### ACOUSTICAL SEALANT

Where light leaks, sound leaks, and air movement at the edge of acoustical ceilings are not acceptable, a continuous bead of acoustical sealant is applied on the concealed leg of the ceiling edge molding before it is installed.

## 9.7.2 INTEGRATED CEILINGS

Integrated ceilings consist of acoustical materials, a suspension system, outlets (grills, vents, or vanes) for air distribution, and electrical devices, such as lights, combined into a single system the components of which are compatible visually and mechanically. For the most part, statements made earlier in this section about acoustical ceilings materials and suspension systems and their installation apply as well to materials used in integrated ceilings.

### 9.7.2.1 System Components

#### ACOUSTICAL MATERIALS

The acoustical materials used in integrated ceilings include the composition and metal acoustical materials listed in ASTM E1264 and discussed in Section 9.7.1.1. Besides the types of acoustical materials specifically listed in ASTM E1264, the other types of materials mentioned earlier that may be used in acoustical ceilings may also be used in integrated ceilings.

#### SUPPORT SYSTEMS

Unlike suspension systems for normal acoustical ceilings, those for integrated ceilings are usually designed to support all components of the ceiling, including

such devices as light fixtures, and the grid members of an exposed grid integrated system are often designed to provide an air supply or return, or both.

Grids for integrated ceilings may be either concealed or exposed. In linear metal integrated ceilings and in metal snap-in systems, the supporting members are usually not exposed.

### AIR DIFFUSION

Using grid members that are designed to contain air supply and return outlets of the heating, ventilating, and air conditioning system can result in effective concealment of diffusers and grills. Some integrated systems employ air supply and return slots that cannot be discerned as such from the floor level.

Integrated ceilings employ boots or manifolds to connect the integral air supply and return elements to the building's ductwork. Connections from the ceiling to the ductwork are usually flexible.

### LIGHTING

Spaces with acoustical ceilings are often lit by standard fixtures not specifically designed for the purpose, which are mounted on the surface of the ceiling or recessed into the ceiling. Often recessed fixtures in acoustical ceilings are standard units laid into the suspension grid. Some integrated ceilings also use standard recessed or surface-mounted fixtures, but many of them include light fixtures that have been designed specifically for the particular integrated ceiling system.

Lighting design is usually an integral part of the design of an integrated ceiling. Fixture locations and types are sometimes limited, but many integrated ceiling designs result in effective and comfortable lighting. Care must be exercised in some cases, especially where standard fixtures are used, to prevent glare.

### ACOUSTICAL PERFORMANCE

Integrated ceilings are often used in open-plan spaces. Such installations are sometimes accompanied by so-called *white noise* generators to mask background noises.

### FIRE PERFORMANCE

The fire-resistance classification of an integrated ceiling system must take into account the electrical and mechanical components of the system. When an integrated ceiling must be time-rated for fire resistance, the limitations may be severe. Light fixture and mechanical opening sizes and locations may be restricted. The method of enclosing such devices will probably be dictated by the rating designation. The appearance of the installed system may be affected drastically by such limitations. A fire-rated integrated ceiling may not look much like a nonrated system using the same basic components.

### 9.7.2.2 Installation

Unlike some acoustical ceilings, integrated ceilings are seldom fastened directly to wood or metal framing or furring. Most integrated ceiling installations, almost by definition, include some sort of metal suspension system.

Suspension systems for integrated ceiling systems may be either direct-hung or indirect-hung. Grids may be either concealed or exposed, depending on the particular ceiling system. In every case, the installation should follow the integrated ceiling manufacturer's recommendations.

Light fixtures and air distribution components should be correctly installed so that their functions are properly carried out. Their installation must be coordinated with the mechanical and electrical system installations.

### 9.7.3 ACOUSTICAL WALL PANELS AND BAFFLES

Acoustical wall panels include standard and custom *spline-mounted* and *back-mounted* panels. Baffles are either double or single thickness. Wall panels and baffles are faced with either standard, custom, or owner-furnished facing materials.

### 9.7.3.1 Panel and Baffle Materials and Fabrication

Most acoustical wall panels and baffles consist of a fabric or vinyl facing over mineral, glass, or wood fiberboard. Materials and fabrication may vary, however, between spline-mounted and back-mounted panels.

Metal acoustical wall panels and baffles are also available.

#### SPLINE-MOUNTED PANELS

The core of most spline-mounted panels is faced on one side only, because the panels are mounted with butted joints. The finish is wrapped around the long edges back to the line of the kerf, however. The short panel ends are unfinished.

*Panel Cores* The cores of spline-mounted panels are:

- Perforated water-felted mineral fiberboard with a fairly high density, such as 17 to 20 lb/cu ft (272 to 320 kg/m$^3$)
- Low-density mineral fiberboard
- Glass fiberboard with a nominal density between 6 and 8 lb/cu ft (96.1 and 128.1 kg/m$^3$)
- High-density wood fiberboard

The long edges of spline-mounted panels are kerfed and rabbeted to receive mounting splines (Fig. 9.7-15).

*Accessories* Mounting and trim accessories for spline-mounted panels include H- and J-shaped splines and J-, H-, and angle-shaped moldings, as shown in Figure 9.7-16. Some manufacturers' accessories may vary from those shown. Accessories may be either extruded aluminum or plastic.

Wood battens are sometimes used to cover the joints and sometimes to support panels. Wood battens are usually decorative hardwood but may be painted softwood.

*Panel Size and Thickness* Spline-mounted panels are either 9 or 10 ft (2743 or 3048 mm) long and either 24, 30, or 48 in. (610, 762, or 1220 mm) wide. High-density fiberboard panels vary in thickness from $1/2$ in. (12.7 mm) to $1^1/8$ in. (28.6 mm). Glass fiberboard and low-density fiberboard panels vary in nominal thickness from 1 in. (25.4 mm) to $1^1/8$ in. (28.6 mm).

**FIGURE 9.7-15** Typical edge detail of a spline-mounted acoustical wall panel. (Courtesy H. Leslie Simmons.)

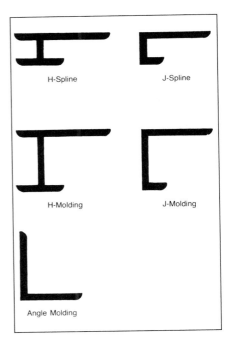

**FIGURE 9.7-16** Some typical accessories for spline-mounted acoustical wall panels. (Courtesy H. Leslie Simmons.)

#### BACK-MOUNTED PANELS

The core face that will be exposed after the panel is installed and the edges of back-mounted panels are usually finished.

*Panel Cores* The core of back-mounted panels may be glass fiberboard with a nominal density between 4 and 7 lb/cu ft (64 and 112 kg/m$^3$) or high-density wood fiberboard. Glass fiber panels may have impact resistant faces which usually consist of a $1/8$-in. (3.2-mm)-thick layer of 18 lb/cu ft (288 kg/m$^3$) density glass fiberboard laminate to the face of the core board.

The edges of the cores of back-mounted panels are either chemically hardened to reinforce the edge and help prevent warpage and damage or framed with aluminum, zinc-coated steel, or wood.

Panel edges that are not framed are usually wrapped with the facing material. Edge shapes of panels with wrapped edges may be square, beveled, radiused, notched, mitered, or a combination of those types (Fig. 9.7-17).

Panel corners may be either square or rounded.

Some panels are faced on the back side with a thin aluminum foil to prevent air passage.

**FIGURE 9.7-17** Some typical edge profiles for back-mounted acoustical wall panels. (Courtesy H. Leslie Simmons.)

**FIGURE 9.7-18** Typical mounting accessories for back-mounted acoustical wall panels. (Courtesy H. Leslie Simmons.)

*Accessories* Mounting accessories for back-mounted panels include:

- Conventional Z-clips
- Strip-clips (Z-clips mounted on a continuous strip)
- Magnetic pads (Fig. 9.7-18)
- Other proprietary clip assemblies
- Hook and loop panels (©Velcro)
- Adhesives
- Screws (sometimes used to mount high-density wood fiber panels)
- Wood battens (sometimes used to cover joints between panels and sometimes to hold panels in place)

*Panel Size and Thickness* Back-mounted panels vary in thickness from $1/2$ in. (12.7 mm) to 3 in. (76.2 mm). Panels are available in a large variety of sizes from 15 in. (381 mm) square up to 60 by 144 in. (13.24 by 3657 mm).

## METAL WALL PANELS

Some of the metal ceiling products discussed earlier in this section are sometimes used on walls, but here we will discuss only products that are specifically designed for use on walls. Such panels are available in both aluminum and galvanized steel.

Metal wall panels are available flat or shaped (corrugated, for example), perforated or not perforated, and either plain, embossed, or textured. Most panels come with glass fiber or mineral wool sound-absorbing pads. Some pads are wrapped or faced with materials such as polyvinyl chloride. Others are bare.

Special brackets and other accessories are required to install some metal acoustical wall panels.

Most metal wall panel sizes vary from 18 to 48 in. (457 to 1220 mm) wide and from 18 to 144 in. (457 to 3658 mm) long, but some may be smaller or larger.

## BAFFLES

A baffle is often nothing more than a standard acoustical wall panel of the back-mounted type that is finished on both faces and all edges. Some baffles are the equivalent of two acoustical wall panels laminated back to back. Some double-thickness baffles are completely encased with finish material. Others have a joint between the two sides of the baffle. Most baffles are flat panels, but cylindrical, octagonal, triangular, and other shaped baffles are also available. Their basic panel construction is similar to that of flat baffles.

Baffle edges may be finished in the same fabric or vinyl as the face, or framed in wood or metal. Edge configurations are similar to those for back-mounted acoustical wall panels.

Metal baffles may be flat panels similar to the metal acoustical ceiling materials discussed earlier or more like the metal wall panels discussed in the preceding subsection.

Baffles are made to be mounted or suspended by special hangers either built into or attached to the baffle. Some baffle hangers are nothing more than eye bolts mounted in the narrow edge of the panel or grommets through the panel face. Others use more elaborate hanger systems, which vary from manufacturer to manufacturer. Hangers may be cable, wire, decorative chain, or almost invisible nylon fishing line. Sometimes hangers are solid metal bars. Some baffles are designed to be clipped to overhead framing.

### WALL PANEL AND BAFFLE FINISHES

*Fabric*  Standard panels and baffles are available with both woven and nonwoven synthetic fabrics, the majority of which are largely or entirely polyester. Other fabrics are also used. Some manufacturers state that almost any upholstery-weight fabric may be used, but fire-resistance requirements may not permit the use of just any fabric. Some sources state that some wool, nylon, metallic, rayon, and acetate fabrics are not suitable.

*Vinyl*  Perforated vinyl is often used as a facing for acoustical wall panels, especially in spaces, such as classrooms, corridors, and the like, where the material is likely to become soiled quickly. It is also sometimes used for facing on baffles. In older installations, the vinyl may have had large holes punched in it to provide the proper acoustical performance. Vinyl in newer installations probably has been punched using a process called *microperforating*, which eliminates large holes.

*Metal*  Metal panels designed for wall panel or baffle use are often painted with either acrylic or polyester baked-on paint. Other finishes may be available.

*Other Finishes*  More utilitarian baffles may be faced or wrapped with polyethylene or glass fiber cloth.

### ACOUSTICAL PERFORMANCE

Acoustical wall panels and baffles are generally classified by their NRC. Panels and baffles fall into two general categories. The first includes panels that will absorb between 60% and 70% (NRC 0.60 to 0.70) of the sound that strikes them. They are used in normal-sized rooms, such as conference rooms and offices. The second category includes panels that absorb between 80% and 95% (NRC 0.80 to 0.95) of the sound that strikes them. They are used in large spaces, such as open-plan offices and spaces where high noise levels are generated, such as computer rooms, swimming pools, and gymnasiums (Fig. 9.7-19).

#### 9.7.3.2 Support Systems

Spline-mounted acoustical wall panels may be directly attached to supported substrates or supported by wood or metal furring applied over wood or metal studs or solid substrates. Back-mounted acoustical wall panels are usually directly attached to supported substrates using one of the accessories discussed later in this section.

Metal acoustical wall panels may be directly attached to solid or supported substrates, or supported by metal furring applied over studs or solid substrates.

### WOOD FURRING

Spline-mounted and metal acoustical wall panels are sometimes installed over horizontal wood furring (Fig. 9.7-20). Such furring is most often used when the substrates are solid ones such as concrete or masonry, and the substrates are too irregular to directly apply the splines and provide a finished surface within acceptable tolerances. Wood furring may also be used over existing wood framing that is not in proper alignment to receive acoustical wall panels, but such use is rare in new buildings. Wood furring may also be used over wood framing for support of acoustical baffles when the framing is not at the proper spacings or locations for the baffle hangers.

Back-mounted acoustical wall panels are usually mounted directly on a supported substrate, such as plaster or gypsum board. Where wood furring is re-

**FIGURE 9.7-19** Acoustical baffles in an arched ceiling in a large armory. (Photograph provided by Metal Building Interiors Products Company.)

**FIGURE 9.7-20** Mounting detail for spline-mounted acoustical wall panels over a solid substrate. (Courtesy H. Leslie Simmons.)

**FIGURE 9.7-21** Mounting detail for spline-mounted acoustical wall panels over metal studs. Similar details may be used when the wall panels are installed over solid substrates that are irregular. (Courtesy H. Leslie Simmons.)

quired in such installations, it is used to furr the supported substrate and not the back-mounted acoustical panels.

### METAL FURRING

Spline-mounted and metal acoustical wall panels are sometimes installed over horizontal hat-shaped galvanized steel furring (Fig. 9.7-21). Such metal furring is most often used when the substrates are solid ones such as concrete or masonry, and the substrates are too irregular to directly apply the splines and provide a finished surface within acceptable tolerances. Metal furring may also be used over existing metal framing, such as studs, that is not in proper alignment to receive acoustical wall panels, but such use is rare in new buildings. Galvanized or painted steel channel furring may also be used over wood or metal framing for support of acoustical baffles when the framing is not at the proper spacings or locations for the baffle hangers.

Back-mounted acoustical wall panels are usually mounted directly on a supported substrate, such as plaster or gypsum board. Where metal furring is required in such installations, it is used to furr the supported substrate and not the back-mounted acoustical panels.

Metal furring is usually screwed, clipped, or wired to the supporting structure.

### 9.7.3.3 Fire Performance

Acoustical wall panels and baffles should be tested for flame spread and smoke-developed ratings as complete panels containing the same components they will have when installed. They should rate as Class A according to ASTM Standard E84. The flame spread rating of a Class A material ranges from 0 to 25. When the individual components are tested separately, it is difficult to determine a composite value for the entire panel.

Since not all manufacturers have had their acoustical wall panels or baffles tested by independent testing laboratories such as UL, it is often necessary to obtain specific approval of a particular panel or baffle by local authorities before using it.

### 9.7.3.4 Installation

Acoustical wall panels should be coordinated with electrical outlets and switches, thermostats, and other wall-mounted equipment and devices to ensure that such items are properly accommodated.

Manufacturers' brochures and installation details should be obtained and followed. The manufacturer's recommendations for cleaning and refinishing wall panels and baffles should be obtained and stored in a safe place so that they will be available when maintenance and repair are necessary in the future. They should include precautions against materials and methods which may be detrimental to finishes and acoustical efficiency.

Extra material is occasionally left at the site for the owner to use in making repairs.

Acoustical wall panels and baffles should be kept dry before and after use. They should be allowed to acclimate to the temperature and humidity of the space before they are installed to prevent dimensional changes which might affect the appearance of the installed panels or baffles.

Watermarked, stained, or otherwise damaged units should not be installed.

Acoustical wall panels and baffles should not be installed until the building is weathertight, glazing has been completed, and exterior openings have been closed.

Wet work, such as masonry, concrete, and plaster, should be allowed to dry to the moisture content recommended by the acoustical wall panel or baffle manufacturer before wall panels or baffles are installed. A temperature of at least 60°F (15.6°C) should be maintained in the space during and after installation.

Furring members should be isolated from building structural elements to prevent transfer of loads into the furring. Provisions should also be made to accommodate movement that occurs across control and expansion joints in the substrates. Furring or panels should not span such joints.

Flat surfaces of furring members and supported substrates should not be warped, bowed, or out of plumb or level by more than $1/8$ in. (3.2 mm) in 12 ft. (3658 mm) in both directions or by more than $1/16$ in. (1.6 mm) across any joint or $1/8$ in. (3.2 mm) total along any single member.

Acoustical wall units should be handled carefully to prevent damage to edges and faces.

Hangers should be properly designed and sufficiently strong to support the baffles. Where splaying of hangers is necessary to avoid ducts, pipes, conduit, or other items, the forces so generated should be compensated for by counter-splaying or another suitable means.

### FURRING INSTALLATION

Furring should be placed on lines and levels necessary to cause the acoustical panels to fall into the proper location and to properly support baffle hangers at the correct spacings. Where the furring will support directly applied panels or baffles, it should correct unevenness in the supporting structure or substrate. The surface of wood furring to which acoustical panel splines will be fastened should be not less than $1^{1}/_{2}$ in. (38 mm) wide. Generally, $2 \times 2$s (50 × 50s) are used for furring that is attached to the supporting structure, and $1 \times 2$s (25 × 50s) or $1 \times 3$s (25 × 75s) are used where furring is laid directly over solid substrates. Nominal 1-in. (25.4-mm)-thick lumber is too flexible to be used over framing. Furring to support baffles should be the proper size to span between the framing members without undue deflection in the furring.

Over a solid substrate, furring should be installed over a layer of polyethylene film or other vapor-retarding material as an air flow barrier and vapor retarder.

Furring should form a complete system adequate to properly support the acoustical wall panels or baffles. Closers should be installed at edges and openings.

Bolts, lag screws, and other anchors are used to anchor wood furring in place. Generally, fasteners should be placed near the ends of furring members and not more than 36 in. (900 mm) on center between. Shorter members, however, should be anchored at 30 in. (750 mm) on center. Bolts should have nuts and washers.

Screws, bolts, clips, wires, and other anchors should be used to anchor metal furring in place. Generally, fasteners should be placed near the ends of furring members and not more than 24 in. (600 mm) on center between, but some types of metal furring members may require closer spacings. The correct spacing should be verified with the furring member manufacturer.

Furring spacings may vary with the substrate and the acoustical panels or baffles being supported. Minimum spacings for wall panels are dictated by the panel thickness and size, and may vary somewhat from manufacturer to manufacturer and system to system. Spacing must adequately support the panels. Spacing of furring for baffle hangers should be such that the baffle hangers fall in the correct locations.

### SPLINE-MOUNTED PANELS

The usual method of mounting spline-mounted acoustical wall panels is to place splines vertically and fasten them to a supported substrate such as plaster or gypsum board (Fig. 9.7-22). Then the panels are slipped in place so as to engage the splines in the kerfed and rabbeted panel edges. Spline-mounted pan-

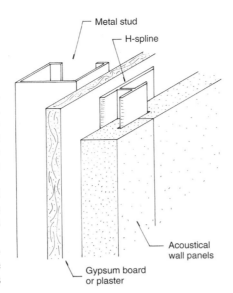

**FIGURE 9.7-22** Mounting detail for spline-mounted acoustical wall panels over a supported substrate. (Courtesy H. Leslie Simmons.)

els can also, however, be mounted by fastening the splines to wood or metal furring over masonry, concrete, or another solid substrate (see Figs. 9.7-20 and 9.7-21).

Spline-mounted acoustical wall panels may rest directly on the floor or be supported at the bottom by a base retainer such as a J or angle molding (see Fig. 9.7-16).

Panels should not extend into the ceiling plenum. Sometimes the panels are stopped short of the ceiling, leaving a decorative reveal.

When panels must be cut for fitting or other reasons, the panel backs should be resealed using heavy mastic tape or foil to prevent air flow through the panel. Cut edges should be covered by a molding or wrapped with fabric.

Exposed panel edges should be covered with fabric or a molding.

### BACK-MOUNTED PANELS

Back-mounted panels are usually mounted directly on a supported substrate such as plaster or gypsum board. Most older panels and many newer panels have been mounted using a conventional Z-clip or continuous Z-clip or some similar proprietary device. Some have been mounted using adhesives. Others are supported by magnetic pads. Some have been screw-attached or supported by battens. Many modern installations are supported using hook and loop (Velcro) strips.

### METAL PANELS

Many metal panels are installed using clips or screws. Some are installed using accessories similar to the moldings shown in Figure 9.7-16 or the clips and magnetic pads shown in Figure 9.7-18.

### BAFFLES

Most baffles are suspended vertically by wire, cables, chains, bars, or fishing line from framing or furring. They should be installed in line and vertically or at the proper angle according to the design. Baffles that are suspended in other configurations should be installed according to the design. In every case, hangers should be securely fastened to the supports and to the baffles.

### SOUND INSULATION

Where they are used, sound insulation blankets should be installed over the full area occupied by the acoustical wall panels.

Sound insulation pads in metal wall panels should be fitted snugly into each pan, panel, or strip over the entire area of the wall panel installation.

Insulation should be tightly butted with joints in solid contact and without voids. It should be fitted neatly to adjacent surfaces.

## 9.8 Brick Flooring and Paving

Brick *paving* is installed on the exterior, where it is generally exposed to weather. Brick *flooring* is installed on the interior, where it is not exposed to weather. However, even on the interior, a brick traffic surface is exposed to severe wear and must therefore be highly durable. Because a unit's resistance to freeze/thaw cycles is a good measure of its durability, units with this characteristic are often chosen even for indoor use and for use outdoors even where severe weather is not a problem.

Brick flooring and paving are laid in one of many possible patterns. Figure 9.8-1 shows some of the more popular patterns.

In addition to the requirements in this section, brick paving and flooring should be handled, protected, and cleaned in accordance with the requirements for brick

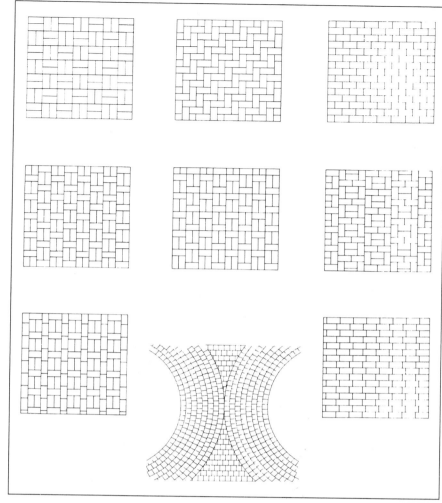

**FIGURE 9.8-1** Brick flooring patterns. Patterns may be laid parallel with the borders or on a diagonal. (Courtesy Brick Institute of America [now renamed Brick Industry Association].)

in general that are discussed in Chapter 4. Chapter 4 also includes additional information about mortar and grout materials and mixing that are applicable to brick paving and flooring, as well as to brick used in walls.

## 9.8.1 MATERIALS

### 9.8.1.1 Paving Brick

Brick units used in flooring and paving are called *pedestrian and light-traffic paving brick.* ASTM C902 covers paving units designed to support pedestrian and light vehicular traffic in patios, walkways, floors, plazas, and driveways. It lists three classes, SX, MX, and NX, based on the severity of weather exposure, and three types, I, II, and III, based on the severity of traffic end exposure. The physical requirements for the classes are given in Figure 9.8-2 and its uses in Figure 9.8-3. Paving brick is also classified in ASTM C902 by use in the following applications:

1. PS: general-use classification for brick to be used with mortar- and grout-filled joints or in patterns, such as running bond, laid without mortar where a close tolerance in the units is necessary
2. PX: for units to be laid without mortar in patterns that require a high degree of dimensional accuracy
3. PA: for units with special appearance characteristics where dimensional tolerances are not restricted

### 9.8.1.2 Mortar and Grout

Thick-set portland cement mortar beds for brick flooring and paving should be a mixture of portland cement, hydrated lime, and fine aggregate. This material is mixed with water, but a latex additive is often added in place of some or all of the water to provide better adhesion between the brick and the setting bed and to produce a more flexible setting bed than can be accomplished using water alone. Setting bed mortar should comply with ASTM C270, either Type S or Type N, with Type S being preferable in most instances for interior use and Type M for exterior use.

Thin-set brick flooring is set in a bed of either dry-set portland cement mortar conforming with ANSI A118.1 or latex-

**FIGURE 9.8-2** Physical Requirements of Pedestrian and Light Traffic Paving Brick[a] (ASTM C902)

| Designation Class | Minimum Compressive Strength (brick flatwise), Gross Area, psi | | Maximum Cold Water Absorption | | Maximum Saturation Coefficient[b] | |
|---|---|---|---|---|---|---|
| | Average of 5 Bricks | Individual | Average of 5 Bricks | Individual | Average of 5 Bricks | Individual |
| SX | 8000 | 7000 | 8% | 11% | 0.78 | 0.80 |
| MX | 3000 | 2500 | 14% | 17% | No limit | No limit |
| NX | 3000 | 2500 | No limit | No limit | No limit | No limit |

[a]Minimum modulus of rupture values should be considered by the purchaser for uses of brick where support or loading may be severe.

[b]The saturation coefficient is the ratio of absorption by 24-hour submersion in room temperature water to that after 5-hour submersion in boiling water.

(Copyright ASTM; reprinted with permission.)

**FIGURE 9.8-3** Uses of Pedestrian and Light-Traffic Paving Brick (ASTM C902) Based on Class and Type

| Class (based on weathering index) | Uses |
|---|---|
| SX | Where brick may be frozen while saturated with water |
| MX | Where resistance to freezing is not a factor |
| NX | Interior use when an effective sealer, wax, or other suitable surface coating will be applied |

| Type (based on traffic) | Uses |
|---|---|
| I | Where brick will be exposed to extensive abrasion, such as in driveways and entranceways to public or commercial buildings |
| II | Where brick will be exposed to intermediate traffic, such as floors in restaurants, stores, and exterior walkways |
| III | Where brick will be exposed to low traffic, such as floors or patios in single-family houses |

portland cement mortar conforming with ANSI A118.4.

Grout for brick pavers should be either a mixture of portland cement and aggregate mixed with water or with a latex additive, or a dry-set grout complying with ANSI A118.6, or a portland cement grout complying with ANSI A118.6.

### 9.8.2 INSTALLATION

Brick flooring and paving are classified as either *mortared* or *mortarless*. Most brick paving is installed in a mortarless system. Conversely, most brick flooring is set in mortar because of the difficulties of sealing unfilled joints or joints filled with sand to prevent liquids from entering them when the floor is cleaned.

#### 9.8.2.1 Mortared Flooring and Paving

Mortared brick flooring and paving have mortar or grout in the joints and are laid in a mortar setting bed, either over a concrete slab on grade or on a suspended diaphragm, such as a structural concrete, steel, or wood floor or roof assembly.

Most mortared brick flooring and paving are laid in a thick-set, latex-modified portland cement mortar setting bed with its joints grouted using a portland cement grout mixed with a latex additive (Fig. 9.8-4). Thick-set beds may be bonded to a slab on grade, but should be placed over a *cleavage membrane* on slabs subject to bending or deflection so that these movements do not disrupt the flooring. Such installations typically weigh about 29 lb/ft$^2$ (141.59 kg/m$^2$).

In residential construction, and in some small buildings of other types, brick floor-ing is sometimes laid in a thin-set mortar bed either over a concrete slab or on a properly supported wood subfloor. A thin-set bed is lighter and requires less floor-to-floor height than a thick-set bed, but the substrate must be sufficiently level and properly sloped to permit the thinner application. In addition, curing compounds cannot be used on a slab that will receive brick flooring set in a thin-set mortar bed.

#### 9.8.2.2 Mortarless Flooring and Paving

Mortarless brick flooring and paving can be laid on a rigid concrete slab, on an asphaltic concrete paving base, or over a suspended diaphragm structure. A separation membrane or thin layer of sand is often used over concrete. A thin asphalt setting bed is needed over an asphalt base. On a suspended diaphragm, a roofing or waterproofing membrane, with or without insulation, may occur beneath the pavers when on a roof or plaza deck. Most mortarless paving and flooring, however, is set in a leveling bed of sand, gravel, crushed stone, or a mixture of sand and cement. This bed can be supported directly on earth or on a rigid base, such as a concrete slab or asphalt concrete paving.

Mortarless flooring and paving laid on a sand or gravel cushion (Fig. 9.8-5) typically weigh about 27 lb/ft$^2$ (131.82 kg/m$^2$).

#### 9.8.2.3 Joints

Joints in mortared brick flooring and pavers are usually left open and filled with mortar or latex-modified grout.

**MORTAR SET—BED AND ALL JOINTS** (Type M Mortar)

Weight:  29 lb/sq ft$^2$ (141.59 kg/m$^2$)

**FIGURE 9.8-4** Mortar-set brick flooring or paving with mortar-setting bed and joints.

**DRY-SET AND BUTTED** (Sand Swept into Joints)

Weight:  27 lb/sq ft$^2$ (131.82 kg/m$^2$)

**FIGURE 9.8-5** Morterless brick flooring or paving.

Joints in mortarless brick paving or flooring may have open joints or the bricks may be placed with hand-tight closed joints. Open joints are filled with latex-modified cement grout. Hand-tight joints are filled by sweeping a dry mixture of portland cement and sand into them and fogging the surface with water.

## 9.9 Stone Flooring and Paving

Dimension stone is used for interior flooring and as exterior paving. Most paving and flooring stone is either slate, marble, granite, limestone, or soapstone, but some other stones are also used. Granite is particularly well suited for ex-terior paving. Basic stone materials and exterior use of dimension stone are described in Section 4.8. It is suggested that a reader review Section 4.8 in conjunction with studying this section.

Stone flooring and paving units are available in standard sizes and shapes, depending on the stone material, but stone can be cut to almost any shape using modern cutting tools including saws and powerful waterjets. Spectacular detail can be achieved in stone projects large and small. Natural stone can be a flexible graphic medium for architectural and artistic expression.

Stone flooring and paving are used today in almost every building type. Stone flooring occurs in entryways, mudrooms, kitchens, bathrooms, foyers, lounges, dens, offices, retail stores, building lobbies, restaurants, and almost every other type of space (Figs. 9.9-1 and 9.9-2). Stone paving is used in patios, walkways, approaches to entrances, and many other locations.

All stone types can be cut economically into unique designs. Special inlays are frequently used in lobbies, retail stores, restaurant foyers, and other buildings for thematic representations. Designs can be either contemporary or traditional, copies or unique creations. Medallions, insets, border designs, emblems, and logos are easily fabricated with computerized waterjet technology. Also possible are murals, ecclesiastical icons, animal motifs, and representational art. It has been said that "anything that can be drawn and computerized can be realized" (Fig. 9.9-3). Figure 9.9-4 shows a few of the almost unlimited number of possible patterns.

Using waterjet technology, metals from brass to stainless steel can be combined with marble, granite, and all other stones to achieve spectacular results.

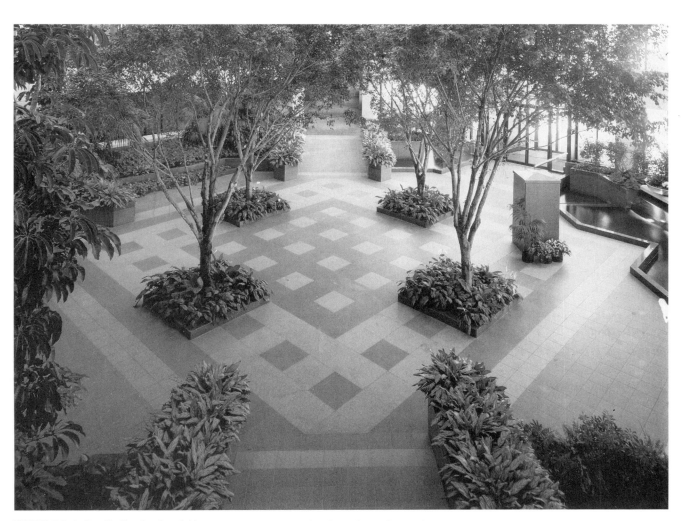

**FIGURE 9.9-1 Granite flooring in a lobby.** (Photograph courtesy of Cold Spring Granite Company.)

**FIGURE 9.9-2** A granite foyer extending into the next room. (Photograph courtesy of Cold Spring Granite Company.)

## 9.9.1 STONE FLOORING AND PAVING MATERIALS

### 9.9.1.1 Stone

#### GENERAL CHARACTERISTICS

All stone flooring and paving should meet certain standards. Their surface abrasion resistance should fall within certain limits. The standard for hardness value, $H_a$ is ASTM C241. The Marble Institute of America (MIA) recommends that all floor tile have a hardness value of at least 10 in residential locations and at least 12 for commercial applications. Different stones in the same area should have a hardness value within 5 of each other.

MIA does not recommend the use of stone flooring less than $^3/_4$ in. (19.1 mm) thick. This advice is widely ignored, however, and there are many successful applications with thinner tile. There can be a problem with thinner tile over substrates that are likely to move or where a

**FIGURE 9.9-3** A patterned granite floor showing one of the many designs possible with stone flooring. (Photograph courtesy of Cold Spring Granite Company.)

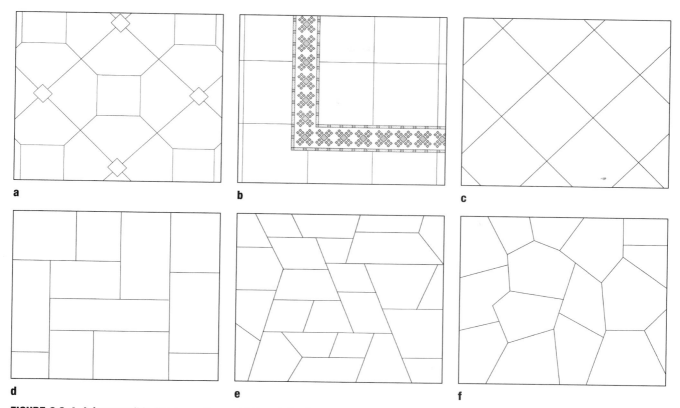

**FIGURE 9.9-4** A few possible flooring patterns: (a) French border pattern, (b) decorative border pattern, (c) square pattern, (d) random rectangular pattern, (e) trapezoidal pattern, (f) random shapes pattern. (Drawing by HLS.)

solid mortar bed cannot be achieved, which is sometimes difficult with thin tile. Without a solid bed, tile tends to crack.

*ADA Accessibility Guidelines for Buildings and Facilities*, which forms a part of the Americans with Disabilities Act (ADA), encourages the use of flooring materials having a static coefficient of friction of not less than 0.6 for flat surfaces and 0.8 for sloped surfaces. This is also the value required by ASTM D2047.

### SLATE

Slate has a characteristic cleavage that allows it to be split easily along one plane but not another. Slate is chiefly comprised of quartz and illite, with mica, calcite, and minor quantities of various other minerals. Slate has long life and durability. It has high strength and resists chipping, cracking, and abrasion. A few types of slate are capable of withstanding rapid freeze–thaw cycles, but most are not. Slate absorbs only 0.25% to 1% of water when immersed for 24 hours. That, together with its dense, compact structure and resistance to chemical acids, gives it

a high sanitary value and makes it easy to clean.

Slates come from China, India, Africa and Brazil, as well as, of course, the United States.

The standard for slate used in flooring and paving is ASTM C629.

*Color* Slate varies in color, but is most commonly black, gray, red, purple, green, and multicolored variations of these. A red color is due to the presence of hematite, while green colors are due to chlorite. Gray and black are due to carbon and/or graphite. Shades of red (purple) and tan are related to varying amount of iron oxides. Slates from a particular area may be available in only one color. Pennsylvania slate, for example, comes only in light or dark gray.

*Sizes* Slate tiles, sometimes called *gauge rubbed tile*, are available in slabs and large-format units, but most are stocked only in 12 × 12 in. (304.8 × 304.8 mm) or 24 × 24 in. (609.6 × 609.6 mm) sizes. In some slates, irregular random shapes are available up to 4 ft (1.2 m) to a side. Some slates come in sizes from 6 × 6 in.

(152.4 × 152.4 mm) up to 12 × 12 in. (304.8 × 304.8 mm) in 3 in. (76.2 mm) increments. Custom sizes are also available. Size range varies with thickness. In some slates, the pieces furnished are roughly rectangular, with about 85% of their edges sawn. Trimming is required to achieve uniform joints and a "broken ice" appearance.

Sawn slate flooring for mortar bed applications usually consists of rectangles or squares clean sawn on all four sides. Their nominal sizes range from 6 × 6 in. (152.4 × 152.4 mm) up to 12 × 12 in. (304.8 × 304.8 mm) in 3 in. (76.2 mm) increments in 1/2 in. (12.7 mm) thickness.

The length of stones for stair treads and risers is up to 7 ft (2.13 m).

*Thickness* Slate tiles to be set in a thinset bed are 1/4 in. (6.4 mm), 3/8 in. (9.5 mm), or 1/2 in. (12.7 mm) in thickness. For gauged tile, the thickness is usually 1/4 in. (6.4 mm). In heavy-use areas, however, the thickness is usually increased to 3/8 in. (9.5 mm) for tiles up to 12 × 12 in. (304.8 × 304.8 mm) and up to 1/2 in. (12.7 mm) for larger tile.

Sawn slate flooring for mortar bed applications is available in $1/2$ in. (12.7 mm), $5/8$ in. (15.9 mm), $3/4$ in. (19.1 mm), and 1 in. (25.4 mm) thicknesses.

The thickness for stair treads and rises is usually 1 in. (25.4 mm), but $1^1/4$ in. (28.6 mm), $1^1/2$ in. (38.1 mm), and 2 in. (50.8 mm) thicknesses are also available.

**Weight** The nominal weight of slate flooring is $3^3/4$ psf (18.31 kg/m$^2$) for $1/4$-in. (6.4-mm)-thick tile, $4^3/4$ psi (23.19 kg/m$^2$) for $3/8$-in. (9.5-mm)-thick tile, $7^1/2$ psf (36.62 kg/m$^2$) for $1/2$-in. (12.7-mm)-thick stone, $11^1/4$ psi (54.93 kg/m$^2$) for $3/4$-in. (19.1-mm)-thick stone, and 15 psi (73.24 kg/m$^2$) for 1-in. (25.4-mm)-thick stone.

**Finishes** Natural cleft face and bottom are usually specified for interior tile surfaces having random shapes arranged in an irregular pattern and for paving. This finish provides a long-wearing, slip-resistant surface. Bottoms should be gauge-rubbed for thin-set application.

The finish for sawn slate flooring is usually a natural cleft surface, but some floorings are sand-rubbed and others have a machine face or are honed.

The finish for stair treads and risers is usually a sand-rubbed one, but a honed finish is sometimes used.

### GRANITE

Granite, used for floor and paving tile, is one of the hardest such materials available. It is particularly suited for exterior paving. It offers durability and easy maintenance. It is available in 28 or more colors and several finishes.

The physical characteristics of granite are defined in ASTM C615.

**Colors** Granite is available in a variety of colors, including white, black, pearl, and pink.

**Sizes** Tile is available in 6 × 6 in. (152.4 × 152.4 mm) 6 in. × 12 in. (152.4 × 304.8 mm), 12 in. × 12 in. (304.8 × 304.8 mm), and 18 × 18 in. (457.2 × 457.2 mm) sizes. Special sizes are also available.

Tiles in metric sizes such as 300 mm × 300 mm × 10 mm and 400 mm × 400 mm × 12 mm can be ordered subject to minimum quantities. Custom orders are also available.

**Thickness** Granite tiles to be set in a thin-set bed are $1/4$ in. (6.4 mm), $3/8$ in.

(9.5 mm), $1/2$ in. (12.7 mm), or $5/8$ in. (15.9 mm) thick. Standard 12 × 12 in. (304.8 × 304.8 mm) tiles are usually $3/8$ in. (9.5 mm) thick; 18 × 18 in. (457.2 × 457.2 mm) tiles are usually $1/2$ in. (12.7 mm) thick. Tolerances are usually less than $1/32$ in. (0.8 mm).

**Weight** A $3/8$ in. (9.5 mm) tile weighs approximately 5.6 psf (27.3 kg/m$^2$).

**Finishes** Granite finishes include polished, honed, fine-rubbed, thermal, rubbed, sawn, and other special finishes. Tile backs are usually diamond ground for thickness uniformity and enhanced bonding.

Granite floors in heavy-traffic areas of shopping malls, airports, and other commercial facilities and granite pavers are typically given a honed finish. Residential floor applications are usually give a polished finish.

### MARBLE

On interior and exterior walking surfaces, the quality of marble has been proven many times over. The physical properties of marble give it extraordinary strength and impart a water- and dirtproof surface. Unusually large crystals interlock in such a way as to block moisture and grime that can destroy marble. Usually, the more colorful marbles are not as strong as the white varieties.

The physical requirements for marble are defined in ASTM C503.

**Colors** Marble comes in well over 60 colors, including white, gray, black, white to gray with charcoal veining, light gray, dark gray, gray with charcoal veins, pink, pink with charcoal veins, beige, gold, green, brown, pink, and red. Black marble may be streaked with veins of white or gold. Other veining may be red. Veining may be fine and web-like or strong and multicolored. Some marbles have no veining at all but rather a fine crystalline texture, large breccia, or small fossils. As a natural product, the colors and textures of marble vary frequently.

**Sizes** The standard size of marble tile is 12 × 12 in. (304.8 × 304.8 mm), but 6 × 6 in. (152.4 × 152.4 mm), 6 × 12 in. (152.4 × 304.8 mm), 8 × 8 in. (203.2 × 203.2 mm), and 16 × 16 in. (406.4 × 406.4 mm) tiles are also available, as are other custom sizes.

**Thickness** The usual marble tile thickness is $3/8$ in. (9.5 mm).

**Finish** Standard finishes are *honed* and *polished*. Other finishes include *fine abrasive* and *textured*.

**Tumbled Marble** In spite of its appearance, tumbled marble is real marble. It is ideal for use in countertops and fireplace surrounds, but it also can be used as accent pieces in general flooring and as the flooring in small spaces, such as showers.

Tumbled marble is available in a variety of sizes, usually relatively small, such as 2 × 2 in. (50.8 × 50.8 mm), 4 × 4 in. (101.6 × 101.6 mm), or 8 × 8 in. (203.2 × 203.2 mm), but larger sizes, such as 12 × 12 in. (304.8 × 304.8 mm) are also available. The thickness of tumbled marble, regardless of panel size, is $3/8$ in. (9.5 mm). It is ideal for mixing with ceramic tile and all types of stone flooring.

Tumbled marble is also available in a variety of precut shapes for use in borders and the like.

### LIMESTONE

All Indiana limestone meets or exceeds the strength requirements set forth in ASTM C568 for Type II Dimension Limestone.

When used for flooring, paving, or steps, the abrasion resistance of limestone should be 6 minimum to 17 maximum when measured according to ASTM C241

**Color** Limestone comes in many colors. Among these are beige, gray, and green.

**Thickness** The nominal thickness of most limestone flooring tiles is $3/8$ in. (9.5 mm).

**Finish** Limestone flooring tile finishes include polished, honed, and special finishes.

#### 9.9.1.2 Mortar and Grout

Stone flooring may be installed using either thin-set or thick-set beds in accordance with ANSI standards, but most thin stone tiles are usually set using thin-set methods. Thicker stones are usually set in thick-set mortar beds.

Portland cement mortar, dry-set mortar, latex-portland cement mortar, and epoxy are all acceptable setting materials for stone flooring. However, the one used should be verified for compatibility by the producer of the stone.

Usually, the setting beds for stone flooring and paving are made of portland cement mortar that complies with ANSI A108.1. Grout may be either sand-portland cement, commercial portland cement, dryset, or latex-portland cement type.

Mortar for interior setting beds is usually Type N in accordance with ASTM C270.

Stone paving bed mortar can be either Type M portland cement-lime mortar conforming with ASTM Standard C270 or a mixture of 1 part portland cement and 3 parts sand mixed with water. Part or all of the water is sometimes replaced with a latex admixture to make the mortar bond more readily to the stone, to make the setting bed somewhat more flexible, and to reduce water penetration through the mortar. Adding a latex admixture increases the setting time of a bed and may not be advisable in all cases, especially if the additive also contains a retarder.

Pointing mortar should be a mixture of 1 part portland cement and 2 parts aggregate. The aggregate may be either natural or ground white sand or ground marble, granite, or another stone to produce the desired color.

## 9.9.2 STONE FLOORING AND PAVING INSTALLATION

### 9.9.2.1 Preparatory Work

Stone flooring and paving should be handled and stored in accordance with the material producer's instructions. Installed stone should not be chipped, broken, or otherwise damaged. It should be stored in the producer's packaging as long as possible to avoid damage.

### 9.9.2.2 Installation Methods

Dimension stone tiles can be installed thin-set as ceramic tiles on interior surfaces or in accordance with guidelines supplied by the MIA, the Terrazzo Tile and Marble Association of Canada, and the Tile Council of America (TCA). Complete installation recommendations are available from the manufacturers of the products to be used.

### 9.9.2.3 Applicable Standards

The following standards are applicable to stone tile installation, and provide recog-

nized and accepted technical and installation guidelines.

- American National Standards Institute (ANSI)
  a. ANSI A108.1—Thick Bed Mortar Method
  b. ANSI A108.4—Thin Bed Mortar Method
- ASTM International (ASTM)—ASTM C615,
- International Organization for Standardization (ISO)—ISO 9002, *Quality Systems—Model for Quality Assurance in Production, Installation, and Servicing*
- Tile Council of America, *Handbook for Ceramic Tile Installation*.
- MIA, *Dimension Stone Design Manual IV*

### 9.9.2.4 Building Codes

Installation must comply with the requirements of all applicable local, state, and national code jurisdictions.

### 9.9.2.5 Support Systems

Substrates must be prepared according to the stone supplier's recommendations. Supporting surfaces must be structurally sound, solid, stable, level, plumb, and true. Floors must be free from dust, oil, grease, paint, tar, wax, curing agents, primers, sealers, form release agents, or other deleterious substances and debris. Concrete must be at least 28 days old; 6 weeks is preferable.

Concrete to receive stone flooring should be steel-troweled smooth and then fine-broomed. If deflection will be greater than $1/720$th of the spans, the concrete should be finished with a steel trowel and a cleavage membrane should be installed.

Wood, ceramic tiles, and terrazzo, if clean and firmly adhered, can also provide a suitable base for a dimension stone floor.

### 9.9.2.6 Installation

The usual method of placing stone paving is to set the stone in a thick portland cement mortar bed having a latex additive. Joints are pointed with grout or filled with sealant.

Stone flooring may also be set in a thick setting bed, but thin stone tile flooring is often set using a thin-set method similar to that used to install ceramic tile.

In general, the recommendations of the TCA for tile setting can be followed

when laying thin stone tile flooring. There is one caveat, however. Always obtain and follow the recommendations of the stone provider for materials and methods to be used for each particular type of stone flooring.

Stone flooring should be installed only when the ambient temperature is 50°F (10°C) and remains at that level or above for at least 7 days after installation.

Honed-finish tiles can be laid with narrow joints of $1/8$ in. (3.2 mm). Textured tiles should be laid with joints of $1/4$ in. (6.4 mm) minimum.

### INSTALLATION METHODS

TCA-recommended setting methods are shown in figures in Section 9.6 as follows:

- Figure 9.6-31: concrete slab subject to deflection; portland cement bed; cleavage membrane
- Figure 9.6-32: concrete slab on grade; portland cement bed
- Figure 9.6-33: concrete; dry-set and latex-portland cement bed
- Figure 9.6-34: concrete; epoxy mortar
- Figure 9.6-37: wood support system; portland cement mortar; cleavage membrane
- Figure 9.6-39: wood support system; thin-set
- Figure 9.6-40: wood support system; epoxy mortar

For other possibilities, refer to the publications mentioned in Section 9.9.2.3.

### GROUTING

Beds must be allowed to set before grouting can begin. Unless recommended otherwise by their producers, grout honed tiles with neat cement and textured tiles with a mixture of 1 part cement to 3 parts sand. Proprietary resin-based and cement-based grouts are also available. Stones set in an epoxy bed are often grouted with an epoxy grout.

### SEALING

Stone sealers are manufactured specifically to seal and protect stone. Unless their use is specifically not recommended by their producer, stone flooring should be given a sealer, applied in accordance with the sealer manufacturer's recommendation.

Some slate producers recommend that their stone be sealed. Others say that the use of stone sealer is not required to ensure permanence or durability, but sealers are recommended where a darkening effect or a high luster is desired.

The application of impregnating coatings to thermal and sandblasted granite tiles can aid maintenance, but may alter their color and texture as well as their slip-resistance characteristics. Sample testing on a small, inconspicuous area should always be done before application of coatings to the entire surface.

Despite its appearance, tumbled marble is still marble and generally needs to have a penetrating sealer applied to it. Any penetrating sealer recommended for marble will work and leave the stone's appearance unchanged. If a dark look is desired, a penetrating sealer that contains a color enhancer can be used

The Indiana Limestone Institute of America states that sealers may change the color of limestone or make it shiny. They recommend that sealers be tested on samples of the stone before they are used.

### 9.9.3 OTHER STONE TYPES

Quartz-containing stone and other stone materials are also used in flooring. Some are available from a limited number of quarries, sometimes only a single quarry. Some such stones are selected because of their low cost. Conversely, their uniqueness makes some of these stones very expensive. Some have distinctive colors or graining characteristics. Some are selected because they are local native stones.

In the absence of more detailed data, the weight of such stones can be assumed to be at least 7.4 psf (36.13 kg/m$^2$).

These materials are generally installed and finished in the same manner as the other stone flooring materials discussed in this section. In addition, the same precautions related to substrate and structural support apply.

# 9.10 Wood Flooring

The widespread use of wood flooring can be attributed to its distinctive natural appearance, excellent wearing qualities, moderate cost, underfoot comfort, and ease of installation and maintenance. It can be used in almost all rooms above grade. Improved methods of protecting wood from dampness have also encour-

aged its use in rooms below grade as well, and even in residential kitchens.

Because of its ready availability and the relative ease of working it with hand tools, early flooring was made mostly of softwood. This flooring often was in the form of thick planks 3$^1$/$_2$ to 10 in. (89 to 254 mm) wide, just as they came from the sawmill. Subsequently, improvements in manufacturing and seasoning methods brought about the development of carefully machined, narrower hardwood strips (3$^1$/$_4$ in. [82.6 mm] and narrower), which were easier to install.

Today, hardwood strip flooring is the most popular type (Fig. 9.10-1a), but demand for floors simulating early colonial random-width plank installations has encouraged the production of plank products (Fig. 9.10.1b). Various parquet (pattern) types, such as blocks and uniform-length strips, are also available (Fig. 9.10-2).

Factory-finished (*prefinished*) flooring suitable for service immediately after installation is available in strip, plank, and block and in a variety of woods.

A characteristic property of wood that must be given special attention in flooring installations is its tendency to shrink and swell as its moisture content changes. Through modern research and manufacturing methods, products with improved dimensional stability have been developed, but products vary in their sensitivity to moisture and should be selected with a view to the conditions of use.

### 9.10.1 MATERIALS

The terms *hardwood* and *softwood*, popularly applied to the two major groups of trees cut for lumber, actually have no bearing on the degree of hardness of the wood. In fact, many softwoods are harder than some hardwoods. These terms are used primarily to distinguish the botanical characteristics of the trees. Arbitrarily, trees having broad leaves are known as hardwoods, while coniferous trees—those bearing needles and cones—are known as softwoods.

About 12 types of wood are regularly manufactured into flooring. Of these, the

a      b

**FIGURE 9.10-1** Contemporary and traditional interiors are complemented by (a) strip and (b) plank wood floors. (Courtesy Bruce Hardwood Floors.)

a      b

**FIGURE 9.10-2** Patterned (parquet) floors may be created with (a) laminated plywood blocks or (b) solid wood slats either preassembled into blocks or individually laid in mastic. (Courtesy Bruce Hardwood Floors.)

hardwoods account for about 80%. Their greater popularity can be attributed to their appearance and, in the species used, their substantially greater hardness and wear resistance.

### 9.10.1.1 Hardwoods

Various commercial species of oak comprise more than 90% of all hardwood flooring produced in a typical year. Maple accounts for about 6%. The rest is beech, birch, pecan, and several other hardwoods in limited quantities.

### OAK

There are about 20 species of oak in the United States that are considered commercially important in lumber production. Of these, about half are classified as *red oak* and half as *white oak*. As growing trees, the several species within each group are readily distinguishable, but in lumber form, the differences are fairly inconspicuous. Therefore, precise separation of the various species within each group of oaks is impractical and unnecessary in flooring manufacture.

Oak is lumbered throughout the southern, eastern, and central states in forests of the Atlantic Plain and the Appalachian Mountains. In these regions, species of oak grow under a wide range of climatic conditions in many different kinds of soil. There is, accordingly, much variation in the color of the wood, especially the heartwood. The sapwood shades from white to cream color in all species of oak. In the standard grading rules for oak flooring, color is entirely disregarded except in the amount of light-colored sapwood allowed. Sapwood is limited only in *clear grade*, the top grade of flooring.

Red oak and white oak are about equal in mechanical properties. Both make satisfactory floors when properly finished. A special feature of white oak is the prominence of large rays that make an interesting flake pattern in quarter-sawn flooring.

Although grading rules do not differentiate between red oak and white oak, practically all manufacturers supply either all-red or all-white oak flooring except in the lowest grades. Red oak flooring generally is higher in price and more uniform in color than white oak.

### MAPLE

Maple flooring is made from sugar maple, logged largely in the Northeast,

the Appalachians, and the lake states (Minnesota, Michigan, and Wisconsin). Sugar maple is known as *hard maple* or *rock maple*. It is strong, hard, and abrasion-resistant, making it particularly suitable for hard-use locations, such as factories and gymnasiums, as well as residences.

The so-called soft maples (silver maple, red maple, and bigleaf maple) are not as hard, heavy, or strong as hard maple and therefore are not used commonly for flooring.

The heartwood of both sugar maple and black maple is light reddish-brown, and the sapwood, which in mature trees is several inches thick, is creamy white, slightly tinged with brown. The contrast in color between heartwood and sapwood in maple is much less pronounced than in oak, and the standard grading rules permit natural color variation in the wood.

### BEECH AND BIRCH

Beech and birch are lumbered in the northeastern part of the country and around the Great Lakes. They are used sparingly in the manufacture of flooring. Only two of the almost 20 species of birch that grow in the United States are manufactured into flooring. Of these, *yellow birch* is by far the most abundant and the most important commercially; the other is *sweet birch*. Only one species of beech is native to the United States.

The heartwood of these species is reddish-brown, with a slight variation in color among them. Similar slight variations also exist in the color of their sapwood, which is of a lighter shade than the heartwood. The varying color of the natural wood is an accepted characteristic in grading beech and birch flooring.

### 9.10.1.2 Softwoods

In a typical year, more than 50% of the softwood flooring produced is southern pine, more than 40% is Douglas fir, and the rest is western hemlock, eastern white pine, ponderosa pine, western larch, eastern hemlock, redwood, spruce, cypress, or the true firs. Western larch is similar to Douglas fir in strength and often is sold in a mixture with Douglas fir of the northern interior region of the western states. Ponderosa pine, eastern white pine, and redwood are softer than is desirable where wear is a prime factor. However, the formidable decay resistance of redwood in the all-heartwood grade has

prompted its use for porch and deck flooring.

### SOUTHERN PINE

*Southern pine* is a commercial name applied to a group of yellow pines that grow principally in the southeastern states. They include longleaf, shortleaf, loblolly, slash pines, and several other pines of minor importance. Except in dimension lumber and structural timbers, no differentiation between the species is made in marketing the products of this group.

The wood of all southern pines is much alike in appearance. The sapwood and heartwood often are different in color, the former being yellowish-white and the latter reddish-brown. However, the contrast in color between sapwood and heartwood in southern pine generally is not conspicuous in a finished floor, and the standard grading rules permit sapwood in all grades of southern pine flooring.

When uniform color is desired, the standard flooring specifications can be amended to require all-sap-face stock (for a light color) or all-heart-face material (for a reddish-brown color). However, special selection of stock for color increases the cost somewhat over the established grade.

### DOUGLAS FIR

*Red fir, yellow fir, coast Douglas fir,* and *Oregon pine* are other names by which Douglas fir is known in the western part of the United States and Canada, where it grows. Douglas fir occupies the same important position in the western and Pacific coast states that southern pine does in the southeastern states.

The sapwood of Douglas fir is creamy white. The heartwood is reddish-brown, and as in southern pine, the contrast in color between the two is not so pronounced as to be objectionable in a finished floor. Pieces containing both heartwood and sapwood are permitted in all grades.

### WEST COAST HEMLOCK

Western hemlock grows along the Pacific Coast from northern California to Alaska and as far inland as northern Idaho and northwestern Montana. The bulk of hemlock lumber (West Coast hemlock) comes from Oregon and Washington. Both the heartwood and sapwood of western hemlock are almost white with a pinkish tinge and with very little contrast, although the sapwood is sometimes lighter in color.

Western hemlock has clear color and good finishing qualities, which account for its use in moderate-wear areas such as bedrooms, where good appearance is the principal requirement. Western hemlock flooring is relatively free from warping and is easy to cut and nail, but is not as hard and wear-resistant as are Douglas fir or larch.

### 9.10.1.3 Supports

Generally, the joist spacing to support board subflooring should not exceed:

- For $^3/_4$ in. (19.1 mm) strip or plank flooring laid parallel to joists: 16 in. (400 mm) on centers
- For parquet flooring: 16 in. (400 mm) on centers
- For $^3/_4$ in. finish flooring laid at right angles to joists: 24 in. (600 mm) on centers

### 9.10.2 MANUFACTURE

Wood flooring is made in three basic styles: *strip*, *plank*, and *block* (Fig. 9.10-3). The following discussion of its manufacture is limited to strips and planks, as blocks may be made in many different ways. Methods of manufacture of hardwood flooring differ from those of softwood flooring.

### 9.10.2.1 Hardwoods

Hardwood flooring production begins at a sawmill, where logs are rough-sawn into boards, planks, and timbers. Each log is cut to produce the best grade of lumber in each thickness. Thicker pieces containing the highest percentage of clear lumber are reserved for furniture and millwork.

The common nominal thickness of lumber for flooring is 1 in. (25.4 mm), from which the popular $^3/_4$ in. strips and planks are milled. After air-drying in a yard for several months, rough-sawn lumber is kiln-dried to reduce the moisture content to 6% to 9%. Sawmill operations are described and illustrated in Section 6.2.1.2.

### SIZING AND DRESSING

Rough-sawn lumber often is air-dried at a sawmill, but kiln drying is normally performed at a flooring plant. After cooling for at least 24 hours, the lumber is ready to enter a flooring mill. The first step consists of cutting the boards into the desired widths with a ripsaw (Fig. 9.10-4). Lumber which is to be tongue-and-grooved is cut wider than the face width of the flooring to compensate for the tongue edge and for the material taken off in planing the edges. In some flooring mills, the pieces are then run through a planing machine, which dresses (smooths) the top and bottom surfaces. Then the pieces proceed on conveyor belts to cut-off saws (Fig. 9.10-5), where

**FIGURE 9.10-5** Natural defects are eliminated on a cut-off saw. (Courtesy Bruce Hardwood Floors.)

they are cut to eliminate natural defects that are undesirable in finish flooring. The pieces emerge in random lengths. Then these strips are conveyed to a machine known as a *side matcher*.

**FIGURE 9.10-3** Basic wood flooring types. (Courtesy Bruce Hardwood Floors.)

**FIGURE 9.10-4** Rough-sawn lumber is cut lengthwise in a ripsaw. (Courtesy Bruce Hardwood Floors.)

## SIDE AND END MATCHING

A side-matching machine smooths the top surface accurately, and cuts one or two channels in the bottom side, a tongue on one edge, and a groove on the other (Fig. 9.10-6). The pieces then continue on a belt conveyor to end-matching machines, where one end of each strip is grooved and the other is tongued.

## GRADING AND BUNDLING

Next, the pieces go to grading tables, where they are separated by grade, color, and length (Fig. 9.10-7). Each piece is trade- and grademarked, and the pieces of each grade are separated and bundled according to length in multiples of 1 ft (304.8 mm). An allowance of 6 in. (152.4 mm) under and 6 in. (152.4 mm) over the nominal length is permitted in each bundle, so that a 4 ft (1219 mm) bundle might contain pieces 3 ft 6 in. (1067 mm) long to 4 ft 6 in. (1372 mm) long. However, the most popular flooring size, $^3/_4$ by $2^1/_4$ in. (19.1 by 57.2 mm), is also bundled in an assortment called *nested flooring*, which includes various lengths ranging from 9 to 102 in. (228.6 to 2591 mm).

Prefinished flooring pieces undergo further treatment, including fine sanding (Fig. 9.10-8), finishing, and final waxing and polishing. Bundles are then shipped to distributors in sealed boxcars or trucks which protect them from the weather and moisture.

### 9.10.2.2 Softwoods

Softwood flooring production begins at a sawmill, where green (unseasoned) logs are rough-sawn. The resulting lumber is then seasoned to meet the moisture requirements of the grading rules. Kiln-dried flooring requires a maximum of 12% moisture content. The actual moisture content at the time of installation usually averages about 9% or 10%.

After flooring has been seasoned, it goes directly to a planing mill, where it is manufactured to the desired profile in one operation. Square-end flooring is then trimmed and packaged in equal lengths, six pieces to a bundle. End-matched flooring bundles may contain pieces ranging from 1 ft (25.4 mm) to 8 ft (2438 mm) in length laid end to end and stacked in one or two tiers of six courses each.

### 9.10.2.3 Standards of Quality

Of the three basic types of flooring produced, strip flooring and block flooring are regulated by industry-wide specifications. There are no generally accepted standards for plank flooring, but most manufacturers market products in appearance grades similar to those established for strip flooring. The standard for block flooring is the American Parquet Association's, APA 1.

Strip flooring is manufactured to comply with the grading rules of the particular trade association governing the type of wood involved. The Wood Flooring Manufacturers Association (NOFMA) promulgates and regulates the production of oak flooring through its *Official Flooring Grading Rules*. Since some NOFMA members also produce maple, beech, birch, and pecan flooring, this association also promulgates grading rules for these woods. However, the majority of maple, beech, and birch flooring is produced by members of the Maple Flooring Manufacturers Association (MFMA), which promulgates its own grading rules for these species.

### 9.10.2.4 Certification of Quality

Strip flooring must be kiln-dried, grademarked, and trademarked in order to

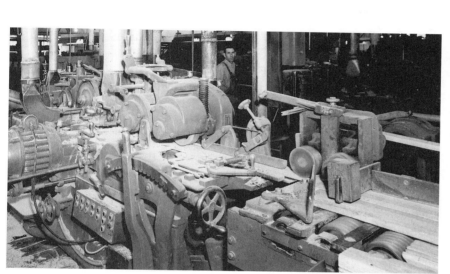

**FIGURE 9.10-6** The top face is smoothed, and the sides are tongue-and-grooved on a side-matching machine. (Courtesy Bruce Hardwood Floors.)

**FIGURE 9.10-7** Each piece is carefully inspected and graded before it is packaged. (Courtesy Bruce Hardwood Floors.)

**FIGURE 9.10-8** Disc sanders impart a satin-smooth finish to pieces that will be factory-finished. (Courtesy Bruce Hardwood Floors.)

a

b

**FIGURE 9.10-10** Strip flooring application: (a) the first and last strips are face-nailed; (b) others are blind-nailed. (Courtesy Bruce Hardwood Floors.)

comply with NOFMA and MFMA grading rules. The two trade associations operate certification programs based on inspections of their members' manufacturing facilities and products, and only qualifying members are permitted to imprint the trademark of the association. This mark is found on the back of each piece of wood and provides visual assurance of conformance with the grading rules. In addition to the grade- and trademarks, the imprint often contains the mill number and other identifying information.

Softwood strip flooring is manufactured to conform to the grading rules of the Southern Pine Inspection Bureau (SPIB), the Western Wood Products Association (WWPA), and the West Coast Lumber Inspection Bureau (WCLIB). These associations do not require trademarking or grademarking under the grading rules.

### 9.10.3 STRIP FLOORING

Hardwood flooring is manufactured commonly to a standard pattern (Fig. 9.10-9), producing flooring which is side-matched and end-matched. The top face is generally slightly wider than the bottom, so that when the strips are driven tightly together, the upper edges make contact but the lower edges are slightly separated. Standard tongue-and-groove strip flooring usually is installed over

wood subfloors or wood sleepers by *blind* (concealed) nailing at the intersection of the tongue and shoulder (Fig. 9.10-10).

Most strip flooring is either hollow-backed (Figs. 9.10-11a and 9.10-11b) or scratch-backed (Fig. 9.10-11d). The most widely used standard pattern is $^3/_4$ in. (19.1 mm) thick and has a face width of $2^1/_4$ in. (57.2 mm), but other widths and thicknesses are also available. The strips are random in length, and the proportion of short pieces depends on the grade.

Other strip flooring, which is used to a limited degree, has square edges and a

**FIGURE 9.10-9** Detail dimensions of typical (*standard pattern*) $^3/_4$ in. (19.1 mm) tongue-and-groove flooring. (Courtesy Bruce Hardwood Floors.)

**FIGURE 9.10-11** Flooring patterns: (a–d) for installation with nails or screws; (e,f) for mastic installation. (Courtesy Bruce Hardwood Floors.)

flat back (Fig. 9.10-11c). It is generally thinner than standard flooring and is usually installed by face nailing. The nails are driven and set so that the matching filler used in finishing will fill and conceal the holes.

While standard strip flooring generally is installed by blind nailing to wood subfloors or sleepers (Fig. 9.10-12), uniform short lengths of tongue-and-groove

**FIGURE 9.10-12** Typical installation methods: (a) wood subfloor and (b) concrete slab; 1 × 2 (25 × 50 mm) sleepers can also be used over a wood subfloor. Polyethylene film and ³/₄ in. (19.1 mm) plywood can also be installed over a slab as a subfloor. (Courtesy Bruce Hardwood Floors.)

flooring (*single-slat* flooring) also can be installed with *mastic* (paste-like adhesive). Products intended for mastic installation generally have flat backs for greater contact with the adhesive and have a recess milled into the lower surface below the tongue to reduce the possibility of mastic being forced up through the joint (Figs. 9.10-11e and 9.10-11f). Flooring intended for mastic installation may have tongue-and-groove edges or may be grooved for insertion of a hardwood or metal spline.

Softwood flooring is either end-matched (see Fig. 9.10-11a) or has plain ends (see Fig. 9.10-11e). It may have a wide, hollow back like ³/₄ in. (19.1 mm) standard pattern hardwood flooring (see Fig. 9.10-11a), a double groove like the thinner standard pattern flooring (see Fig. 9.10-11b), or a single V-shaped groove in the back (see Fig. 9.10-11d).

Standard flooring grades are based almost wholly on appearance. They exclude or severely limit such defects as knots, wormholes, and the like in the higher grades and permit increasing sizes and numbers of these characteristics in the lower grades. Natural variations in color generally are not limited, except that in the clear grade of oak the amount of light-colored sapwood is restricted.

Traditionally, hardwood flooring has been packaged and sold in bundles containing pieces of more or less the same length, 6 in. (150 mm) shorter or longer than the nominal length. The shortest

pieces permitted per bundle and the percentage of short lengths permitted in each grade are regulated by established grading rules of NOFMA and MFMA.

NOFMA grading rules include pieces ³/₄ in. (19.1 mm) by 1¹/₂ in. (38 mm), by 2 in. (50.8 mm), by 2¹/₄ in. (57.2 mm), and by 3¹/₄ in. (82.6 mm) *nested flooring*, which is laid end-to-end continuously to make either 7-ft (2134-mm)-long or shorter or 8-ft (2438-mm)-long bundles. A nested 8 ft (2438 mm) bundle is four tiers high and three courses wide or six tiers high and two courses wide, each course being 7¹/₂ ft (2286 mm) to 8¹/₂ ft (2591 mm) long. Individual pieces may range from 9 in. (228.6 mm) to 102 in. (2591 mm), but both the proportion of short pieces (9 to 18 in. [228.6 to 457.2 mm]) and the minimum average length are regulated by the grading rules and depend on the grade and species.

The short pieces included in nested flooring are not detrimental to an installation. Experiments have demonstrated that even if the subflooring were omitted and a 9 in. (228.6 mm) piece of strip flooring were placed so that no part of it rested on a support (Fig. 9.10-13), it could support almost 2¹/₂ times the concentrated load under the leg of a concert grand piano. This represents an adequate safety factor for what may be assumed to be the most critical loading condition in most uses. Installations over subfloors, or those involving longer pieces, produce lower stresses and certainly are adequate.

#### 9.10.3.1 Oak

Oak is regularly manufactured into plain-sawn and quarter-sawn flooring (Fig. 9.10-14). Most oak flooring is plain-sawn, the lower-priced of the two. Quarter-sawn oak is characterized by a striking grain and flake pattern and by greater dimensional stability, which minimizes shrinking and swelling in width. Oak flooring is graded under the rules of NOFMA.

The most common oak flooring thickness is ³/₄ in. (19.1 mm). Thinner flooring is available for use over existing floors and for light service conditions. Flooring thicker than ³/₄ in. (19.1 mm) is intended for heavy service conditions.

Although most oak strip flooring is end- and side-matched, some square-edged strip flooring is made. This is intended primarily for industrial, institutional, and commercial use to facilitate replacement of damaged areas, which would be more difficult with tongue-and-grooved flooring. Prefinished oak flooring is made in four grades: Prime, Standard, Standard & Better, and Tavern. It is available in all-red and all-white oak selections. Prime grade is available on special order only.

#### 9.10.3.2 Maple, Beech, and Birch

Maple, birch, and beech flooring is graded both under the rules of MFMA and NOFMA. Although the grade names are sometimes slightly different, the re-

$$11\frac{1}{4}'' = 285.8 \text{ mm}$$
$$9'' = 228.6 \text{ mm}$$
$$3\frac{1}{2}'' = 88.9 \text{ mm}$$
$$2\frac{1}{2}'' = 63.5 \text{ mm}$$

$$1\frac{1}{8}'' = 28.6 \text{ mm}$$
$$4\frac{5}{8}'' = 92.1 \text{ mm}$$
$$16'' = 406 \text{ mm}$$
$$48'' = 1220 \text{ mm}$$

**FIGURE 9.10-13** Short pieces of nested flooring (shaded) resting entirely on tongue-and-groove of adjacent strips (a) supported a concentrated load of 845 lb (383.28 kg), while pieces partly resting on a joist (b) supported more than 915 lb (415 kg).

**FIGURE 9.10-14** The angle the annual rings make with the surface determines whether wood is (a) quarter-sawn, which shows edge grain, or (b) plain-sawn, which shows flat grain.

quirements under both sets of rules are similar. Both generally disregard normal color variations.

Because these woods are very dense and hard in plain-sawn as well as quarter-sawn stock, the grading rules make no special requirements in this respect. However, quarter-sawn material can be obtained when specified. MFMA defines edge-grain (quarter-sawn) stock as pieces whose annual rings form an angle of over 30 degrees with the face. This definition is slightly more permissive than the standard definition requiring an angle of 45 degrees or more (see Fig. 9.10-14)

### 9.10.3.3 Southern Pine

Southern pine is graded under the rules of SPIB. It may be specified in side- and end-matched flooring and also in square-end flooring that is side-matched only. Regardless of the pattern to which it is worked, all southern pine flooring is available in both *flat-grain* and *edge-grain* stock. However, edge-grain material is in limited supply, available through special order only, and may be hard to obtain. Edge-grain material in this species must have at least six annual rings per inch across the face, in addition to the

standard requirement of grain direction of more than 45 degrees.

A class of southern pine flooring intermediate between edge-grain and flat-grain is known as *near-rift* flooring. This material has annual rings that form an angle of less than 45 degrees with the face, but the rings are so close to this position that each piece shows at least 6 annual rings anywhere across the face.

### 9.10.3.4 Douglas Fir and Other Western Softwood Species

With the exception of the B & Better Grade, which is manufactured only from *vertical-grain* material, Douglas fir flooring in the upper grades is regularly manufactured in both vertical-grain (VG) and flat-grain (FG) stock (see Fig. 9.10-14).

The term *vertical grain* used in these grading rules is almost synonymous with *edge grain* used for southern pine. However, Douglas fir Grade C requires four rings per inch (25.4 mm), and only B & Better Grade requires at least six rings per inch (25.4 mm) of face width, as does southern pine. Vertical-grain Douglas fir flooring is harder and possesses better wearing qualities than flat-grain material. Douglas fir flooring generally is tongue-and-grooved but not end-matched. Flooring of this wood is graded under the rules of the WCLIB and the WWPA.

### 9.10.4 PLANK FLOORING

Flooring boards wider than $3\frac{1}{4}$ in. (82.6 mm) are referred to as *planks*. The term *strips* is usually reserved for narrower pieces. However, many random plank installations include both plank- and strip-sized boards, and manufacturers often refer to such assortments as *plank flooring*. Strip and plank floors are installed in end-to-end random pattern, with end joints staggered, using nails or screws.

Plank flooring is available end- and side-matched in the same thicknesses as strip flooring. Most plank flooring is white or red oak $^3/_4$ in. (19.1 mm) thick and $3^1/_2$ to 8 in. (88.9 to 203.2 mm) wide. It is available prefinished in dark or light shades and graded according to NOFMA rules for prefinished flooring.

Some plank flooring is designed to simulate the appearance of the random-width planks commonly used in colonial-era buildings. These rustic floors had a distinctive appearance resulting from the hardwood pegs used to secure the planks to the subfloor.

Contemporary plank flooring is installed with wood screws in the face of each plank. These screws are then covered with wood plugs to simulate pegs. Narrower, strip-sized boards which can be secured by blind nailing sometimes have decorative pegs inserted during their manufacture and are sold as plank flooring.

Because the cross-grain dimension in plank flooring is greater than in strip flooring, it is affected more by variations in atmospheric humidity. To avoid unsightly loose joints, plank edges are often eased or beveled to accentuate rather than conceal the joints.

Plank flooring is sometimes made in three-ply laminated form with a center cross ply. This form is more dimensionally stable in width than the solid form, making it possible to install such plank floors with tight joints. Plank flooring generally is sold by the square foot rather than by the board foot, as strip flooring is sold.

## 9.10.5 PARQUET FLOORING

Parquet (pattern) floors consist of individual strips of wood (*single slats*) or larger units (*blocks*) installed to form a decorative geometric pattern (Fig. 9.10-15). Blocks may be made by laminating several hardwood veneers or by gluing a number of solid hardwood pieces into a unit to facilitate installation. Unlike strip and plank flooring, parquet flooring usually is installed with mastic.

Oak is by far the most predominant species used in all types of pattern flooring, but other species, such as maple, walnut, cherry, and East Indian teak, also

9" × 9" by 3/4" (typ.)

pieces of strip flooring

UNIT BLOCK

a

9" × 9" by 15/32" (typ.)

3 or more plies of wood

LAMINATED BLOCK

b

variable, 9" to 30" square, by 5/16" (typ.)

thin slats of hardwood

SLAT BLOCK

c

9" × 9" by 3/4" = 228.6 × 228.6 × 19.1 mm
9" × 9" by 5/32" = 228.6 × 228.6 × 3.99 mm
9" = 228.6 mm
30" = 762 mm
5/16" = 7.9 mm

**FIGURE 9.10-15** Typical construction and dimensions of basic block flooring forms: (a) unit block, (b) laminated block, (c) slat block. Slat blocks often consist of 16 or more basic squares.

are available. Sometimes a mixture of hardwoods, such as hickory, ash, elm, pecan, sycamore, beech, and hackberry, is used at random in a single block.

### 9.10.5.1 Block Flooring

Block flooring is manufactured in several basic types. In *unit block* (also known as *solid unit block*), short lengths of strip flooring are joined together edgewise to form square units (Fig. 9.10-15a). In lam-

inated (plywood) block, three or more plies of veneer are bonded with adhesive to obtain the desired thickness (Fig. 9.10-15b). *Slat block* (sometimes called *mosaic parquet hardwood slat*) flooring utilizes narrow slats or "fingers" of wood preassembled into larger units to facilitate installation (Fig. 9.10-15b).

Most unit and laminated block flooring is tongued on two adjoining or opposing edges and grooved on the other two to ensure alignment between adjoining blocks. Some manufacturers produce square-edged blocks, while others include grooves on all four block edges and furnish splines for insertion between adjoining blocks. Both types are designed to be installed with mastic over a wood subfloor or concrete slab. Prefinished blocks usually have eased or beveled edges.

Unit blocks typically consist of several $^3/_4$ in. (19.1 mm) tongue-and-groove strips, all laid parallel or alternating in each quarter of the block checkerboard fashion (Fig. 9.10-16). Consequently, typical block sizes are multiples of the strip width used (Fig. 9.10-17). Common strip widths are $1^1/_2$, 2, and $2^1/_4$ in. (38, 50.8, and 57.2 mm), resulting in typical block dimensions of 6, $6^3/_4$, $7^1/_2$, 8, 9, 10, and $11^3/_4$ in. (152.4, 171.5, 190.5, 203.2, 228.6, 245, and 298.5 mm). The individual strips are held together with wood or metal splines embedded in the lower surface.

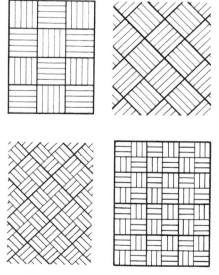

**FIGURE 9.10-16** Typical unit block patterns; heavy lines show individual blocks laid in square or diagonal patterns.

**FIGURE 9.10-17** Typical Unit Block Dimensions*

| Strip Width, in. (mm) | Strip Thickness, in. (mm) | | |
|---|---|---|---|
| | ½ in. (12.7 mm) | ¾ in. (19.1 mm) | 33/32 in. (26.19 |
| 1½ (38) | 7½ (190.5) 9 (228.6) | 7½ (190.5) 9 (228.6) | 9 (228.6) |
| 2 (50.8) | 8 (203.2) 10 (254) | 8 (203.2) 10 (254) | |
| 2¼ (57.2) | | 6¾ (171.5) 9 (228.6) 11¼ (285.8) | 9 (228.6) 11¼ (285.8) |

*All dimensions are for square blocks.

## LAMINATED BLOCK

This type of block is typically 15/32 in. (11.9 mm) thick in 9 in. (228.6 mm) squares, but other sizes, such as 8 in. (203.2 mm) and 8½ in. (215.9 mm), also are available from some manufacturers. Appearance grades are Prime and Standard.

Because of its cross-laminated construction in three to five plies, shrinking and swelling of individual blocks is minimized. This type of wood flooring has good dimensional stability and often is recommended for damp locations such as slabs on grade.

Adhesives used in the manufacture of laminated block flooring should be capable of resisting the temperature and humidity variations to which the flooring may be subjected. Melamine-urea resin adhesive gives good results at moderate cost and is the typical adhesive used for most laminated block flooring.

Highly water-resistant adhesives such as phenols, resorcinols, and melamines may also be used if the application warrants it; however, the use of these glues adds to the cost of the product. Adhesives which create high-strength dry bonds but are adversely affected by moisture are not recommended for the manufacture of laminated block flooring.

## SLAT BLOCK

Appearance grades are based on grading rules developed by the American Parquet Association's (APA) ANSI/APA 1, "Mosaic-Parquet Hardwood Slat Flooring." These products are suitable for installation in mastic over concrete surfaces, both above and on grade. The basic components of these products are solid, 5/16-in. (7.9-mm)-thick, generally square-edged slats of hardwood, ¾ in. (19.1

mm) to 1¼ in. (28.6 mm) wide and 4 to 7 in. (101.6 to 177.8 mm) long, assembled into basic squares. These squares, in turn, are factory-assembled checkerboard fashion, with the grain in each adjoining square reversed, into larger flooring blocks up to 30 in. (76.2 mm) long and wide (Fig. 9.10-18).

Several types of slat block flooring are available. Some are assembled into panels held together with a face paper. These are 9½ in. (241.3 mm), 18 in. (457.2 mm), or 19 in. (482.6 mm) squares and are marketed unfinished. Others are made up of single 6 in. (152.4 mm) tongue-and-groove squares. These are held together by mechanical attachments and are generally factory finished. Still others are assembled into panels held together with

a backing material such as asphalt-saturated felt, textile webbing, or another type of felt or nonwoven interfacing. These products generally are 9½, 11, or 12 in. (241.3, 279.4, or 304.8 mm) square. They may be square-edged or grooved and splined, and may be either unfinished or prefinished.

### 9.10.5.2 Single-Slat Parquet

Many decorative motifs can be created with single-slat parquet flooring (Fig. 9.10-19), herringbone being the most common. This type of flooring consists of side- and end-matched strips machined to uniform lengths ranging from 6 to 18 in. (152.4 to 457.2 mm). Typical thicknesses are ¾ and 33/32 in. (19.1 and 26.2 mm); common face widths are 1½, 2, and 2¼ in. (38, 50.8, and 57.2 mm). Since single-slat parquet flooring is essentially short lengths of strip flooring, it is available in all types of woods, grades, and sizes available in strip flooring.

### 9.10.6 MATERIALS HANDLING

To maintain proper moisture content, flooring products should not be transported or unloaded in rain, snow, or excessively humid weather. Flooring should not be delivered to a construction site until the building is enclosed, concrete and plaster work has been completed, and all building materials are dry.

**FIGURE 9.10-18** Typical slat block patterns. Pickets forming borders may be part of the block or separate.

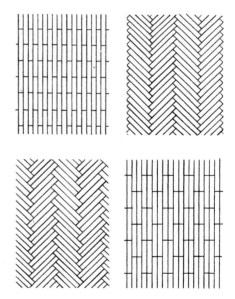

**FIGURE 9.10-19** Popular single-slat patterns achieved with uniform-length precut wood strips.

An interior temperature of 65°F (18.3°C) to 70°F (21°C) should be maintained for at least 5 days before flooring is delivered. Flooring should be stored for several days in the rooms where it will be installed to allow it to become acclimated to local conditions.

In crawl space construction, adequate cross-ventilation should be provided under the floor. The total area of vent openings should equal 1.5% of the first-floor area. A ground cover of 4 to 6 mil (0.1 to 0.15 mm) polyethylene film is essential as a moisture retarder. Inadequate moisture control can harm any floor installation by contributing to warping or discoloration of the flooring.

## 9.10.7 MAINTENANCE

Properly finished wood floors are durable and easy to maintain. It is best to vacuum or dust-mop floors once a week. Wood floors should never be washed or wet mopped because water seepage between boards can cause staining and warping. Regardless of the floor finish type, flooring manufacturers recommend that the finish be protected with a solvent-based liquid buffing wax-cleaner or a paste wax. Spills should be wiped up immediately to prevent damage.

# 9.11 Resilient Flooring

Resilient flooring is a class of flooring products distinguished by resilience and dense, nonabsorbent surfaces. Resilience contributes to indentation resistance and quietness, while the dense surface ensures durability and relative ease of maintenance. A variety of decorative colors, patterns, and textures are available in sheet or tile form.

Availability of self-adhesive tile has made installation of some products easier. No-wax surfaces on both tile and sheet goods have reduced maintenance.

The first resilient flooring material, linoleum, was invented more than 100 years ago in England. After the advent of modern vinyl and vinyl composition materials, linoleum fell into disfavor. Today, linoleum is enjoying a comeback.

Asphalt tile evolved after World War I from a troweled-on mastic flooring which consisted of a mixture of asphalt, asbestos fibers, and naphtha solvents. Today, asphalt tile has been largely replaced by sheet vinyl and vinyl composition tile products.

Early composition tile products contained asbestos, but since it was discovered to have harmful effects on people, it has been restricted by law and replaced with other, less harmful fibers. The introduction of vinyl and vinyl composition products, with their excellent wearing properties and decorative potential, substantially increased the range of resilient flooring uses. Today, a variety of resilient flooring products is available for residential and nonresidential construction.

Resilient flooring is used where economical, dense, nonabsorbent wear surfaces are desired. Depending on the product, resilient floor finishes offer a wide range of decorative effects and may provide underfoot comfort, good durability, and relative ease of maintenance. Specific products with enhanced properties are available to suit most functional requirements.

Suitability of a flooring product is related to the product's physical properties and intended use. In addition to reasonable cost, ease of maintenance and durability are important considerations in determining suitability. These factors, in turn, depend on subfloor conditions, proper installation, and maintenance.

## 9.11.1 FLOORING MATERIALS

Resilient flooring is manufactured in tile and sheet forms, in different thicknesses and from a variety of ingredients. Tile sizes range from 9 by 9 in. (228.6 by 228.6 mm) up to 20 × 20 in. (508 by 508 mm), depending on the material; sheets are available in widths from 4 ft 6 in. (1350 mm) to 12 ft (3600 mm), also depending on the material (Fig. 9.11-1). Matching resilient accessories such as wall base, thresholds, stair treads, and feature strips are available to complete flooring installations.

Resilient flooring products may be classified according to basic ingredients as (1) *vinyl*, (2) *vinyl composition*, (3) *rubber*, (4) *cork*, and (5) *linoleum*. Many resilient flooring products are manufactured to meet the requirements of ASTM standards and Federal Specifications, which together constitute the industry-recognized standards of quality. In some types, the basic ingredients run the full depth of the flooring; in others, backing material adds desired physical characteristics or reduces costs. Figure 9.11-1 summarizes the ingredients and backing materials, sheet widths, tile sizes, gauges, and applicable standards.

### 9.11.1.1 Vinyl Composition Tile

Vinyl composition tile is composed of mineral fibers, ground limestone plasticizers, pigments, and polyvinyl chloride (PVC) resin binders. These resins permit light colors and a great variety of designs. Some tile products are available with self-adhesive backs and "no-wax" surfaces. Older vinyl composition tile contained asbestos fibers, which have been found to cause cancer and other physical problems in humans, especially diseases of the lungs, and are no longer used in resilient flooring products.

ASTM Standard F1066 classifies vinyl composition tile as either Class 1, which is solid color tiles; Class 2, in which a pattern extends completely through the tile; or Class 3, in which a pattern is only on the surface of the tile.

Ingredients are mixed under heat and pressure and rolled into blankets to which decorative chips may be added. Additional rolling and calendering produce smooth sheets of desired thickness. After waxing, sheets are cut into tiles.

### 9.11.1.2 Vinyl Sheet and Tile

The chief ingredient of vinyl products is PVC resin. Other ingredients include mineral fillers, pigments, plasticizers, and stabilizers. Plasticizers provide flexibility; stabilizers fix the mixture to ensure color stability and uniformity. Vinyl products sometimes are referred to as *flexible* vinyl to distinguish them from vinyl composition products, which also are made with PVC resins and are termed *semiflexible* vinyl.

#### VINYL SHEET

Vinyl sheet products are made with a vinyl wear surface bonded to a backing. The backing may be vinyl, polymer-impregnated mineral fibers, asphalt- or resin-saturated felt, nonfoam plastic, or foamed plastic. Some vinyl products have a layer of vinyl foam bonded either to the backing or between the wear surface and the backing. Backings are classified in ASTM Standard F1303 as Class A for fibrous formulations, B for nonfoamed plastics, and C for foamed plastics.

**FIGURE 9.11-1** Summary of Resilient Flooring Products

| Type | Standards | Backing Materials | Gauges, in. (mm)[a] | Sizes, in. (mm) | Basic Ingredients |
|---|---|---|---|---|---|
| **Asphalt Tile** Plain | ASTM F1066 | — | ⅛ (3.2), 3⁄16 (4.8) | 9 × 9 (228.6 × 228.6), 12 × 12 (304.8 × 304.8)[b] | Asphaltic and/or resin binders, mineral fibers, and limestone fillers |
| Greaseproof | ASTM F1066 | — | | | |
| **Vinyl-Composition Tile** | ASTM F1066 | — | 1⁄16 (1.6), 0.080 (2), 3⁄32 (2.4), ⅛ (3.2) | 9 × 9 (228.6 × 228.6), 12 × 12 (304.8 × 304.8)[b] | PVC resin binders, limestone, mineral fibers |
| **Vinyl Tile** Solid | ASTM F1066 | Scrap vinyl, mineral or rag felts | 1⁄16 (1.6), 0.080 (2), 3⁄32 (2.4), ⅛ (3.2), 0.050 (1.27) to 0.095 (2.4) | 9 × 9 (228.6 × 228.6), 12 × 12 (304.8 × 304.8) | PVC resin binders and mineral fillers |
| Backed | ASTM F1066 | | | 4 × 36 (101.6 × 914.4),[b] 9 × 9 (228.6 × 228.6), 12 × 12 (304.8 × 304.8), and sizes up to 36 × 36 (914.4 × 914.4)[b] | |
| **Vinyl Sheet**[c] | ASTM F1303 | Scrap vinyl, mineral or rag felts | 0.065 (1.7) to 0.160 (4.1) (0.010 [.25] to 0.050 [1.3] wear surface) | 72 (1829), 144 (3658), and 54 (1372) wide | PVC resin binders and mineral fillers; clear PVC film for clear vinyl surface |
| | | | 0.065 (1.7) to 0.160 (4.1) (0.010 [.25] to 0.050 [1.3] wear surface) | | |
| **Rubber Tile** | ASTM F1344 | — | 0.080 (2), 3⁄32 (2.4), ⅛ (3.2), 3⁄16 (4.8) | 9 × 9 (228.6 × 228.6), 12 × 12 (304.8 × 304.8), 18 × 36 (457 × 914.4), 36 × 36 (914.4 × 914.4)[b] | Synthetic or natural rubber, mineral fillers |
| **Cork Tile** Protected surface Wood surface | None[d] | — | ⅛ (3.2), 3⁄16 (4.8), ¼ (6.4), 5⁄16 (7.9),[b] ½ (12.7)[b] | 9 × 9 (228.6 × 228.6), 6 × 6 (152.4 × 152.4),[b] 6 × 12 (152.4 × 304.8), 12 × 12 (304.8 × 304.8)[b] | Cork particles and resin binders, wax or resin finish |
| Vinyl surface | | | ⅛ (3.2), 3⁄16 (4.8) | 9 × 9 (228.6 × 228.6), 12 × 12 (304.8 × 304.8) | Cork particles and resin binders, clear vinyl finish |
| **Linoleum Sheet** | None[e] | Burlap | (0.050 [1.27] wear surface), ⅛ (3.2) | 72 (1829) wide | Cork and/or wood flour with linseed oil binders |
| Plain and marbelized Inlaid and molded | | Rag felt | 0.090 (2.3) | | |
| | | Burlap | ⅛ (3.2) | | |
| | | Rag felt | 0.090 (2.3) | | |
| Battleship | | Burlap | ⅛ (3.2) | | |
| **Linoleum Tile** | None[e] | | (0.050 (1.27) wear surface), ⅛ (3.2) | 9 × 9 (228.6 × 228.6), 12 × 12 (304.8 × 304.8)[b] | Cork and/or wood flour with linseed oil binders |
| | | Rag felt | 0.090 (2.3) | | |

[a]Thinnest gauges and wear surfaces indicated are minimums suitable for adhesive installation.

[b]Not common, available on special order from some manufacturers.

[c]Also produced with foam backing, self-adhesive back, and no-wax surface; not included in data shown.

[d]Previously covered by Federal Specification LLL-T-00431, now cancelled.

[e]Previously covered by Federal Specification LLL-F-1238A, now cancelled.

In addition to PVC resins, the wear surface of vinyl-surfaced sheets may contain decorative vinyl chips, filler, pigments, and other ingredients. Powdered vinyl resins, fillers, plasticizers, stabilizers, and pigments are mixed, rolled into sheets, and chopped to form these vinyl chips. Vinyl chips of various colors are mixed with additional resins, spread evenly over the backing, and bonded to it under high heat and pressure (Fig. 9.11-2).

The vinyl resins and plasticizers together are called the *binder*.

ASTM F1303 classifies vinyl sheet flooring as either Type I or Type II. In Type I flooring, the binder constitutes at least 90% of the wear layer; in Type II, at least 34%.

ASTM F1303 designates three grades within each type to define wear layer thickness:

- *Grade 1:* For commercial, light commercial, and residential projects; requires a 0.020 in. (0.508 mm) wear layer in Type I and a 0.050 in. (1.27 mm) wear layer in Type II
- *Grade 2:* For light commercial and residential projects; requires a 0.014 in. (0.36 mm) wear layer in Type I and a 0.030 in. (0.76 mm) wear layer in Type II
- *Grade 3:* For residential projects only; requires a 0.010 in. (0.254 mm) wear layer in Type I and a 0.020 in. (0.51 mm) wear layer in Type II

Greater thicknesses are required in Type II than in Type I, because the abrasion resistance and durability per unit of thickness of a wear layer is greater in the binder than in the fillers. The more binder, the thinner can be the wear layer for the same service.

Sheet products with vinyl backings may have a design imprinted on top of the backing or on the underside of the wear surface. They are usually printed with vinyl inks. A clear PVC wear surface is then calendered to the desired thickness and laminated to the backing with heat and pressure (Fig. 9.11-3).

## VINYL TILE

Vinyl tile may be of homogeneous solid composition or may be backed with other materials such as organic felts, mineral fibers, or scrap vinyl.

Ingredients for solid vinyl tiles are mixed at high temperature and hydraulically pressed or calendered into homogeneous sheets of the required thickness. These sheets are then cut into tile sizes. Some tiles are made with a self-adhesive back.

### 9.11.1.3 Rubber Tile

Natural or synthetic rubber is the basic ingredient of rubber flooring. Clay and fibrous talc or mineral fillers provide the desired degree of reinforcement; oils and resins are added as plasticizers and stiffening agents. Color is achieved by nonfading organic pigments. Chemicals are added to accelerate the curing process.

The ingredients are mixed thoroughly and rolled into sheets. The sheets are calendered to uniform thickness and vulcanized in hydraulic presses under heat and pressure into compact, flexible sheets with a smooth, glossy surface. The backs then are sanded to gauge, ensuring uniform thickness, and the sheets are cut into tiles.

### 9.11.1.4 Cork Tile

Cork tile is composed chiefly of the granulated bark of the cork oak tree, native to Spain, Portugal, and North Africa. Synthetic resins are added to the granulated cork, and the mixture is pressed into sheets or blocks and baked. Surfaces are finished with a protective coat of wax, lacquer, or resin applied under heat and pressure. Sheets then are cut to tile sizes. Vinyl cork tile has a film of clear PVC vinyl fused to the top surface to improve durability, water resistance, and ease of maintenance.

### 9.11.1.5 Linoleum

Linoleum consists of a blend of oxidized linseed oil binders, rosins, wood flour, and cork flour fillers bonded to a burlap, felt, jute, or polyester backing. Linoleum is available in sheet and tile form.

### 9.11.2 SELECTION CRITERIA

The suitability of a resilient flooring material for a particular location or condition depends on its physical properties. *Moisture resistance* is important in determining whether a material can be used in areas subject to surface or ground moisture. *Grease and alkali resistance* have a bearing on suitability for use in kitchens or over concrete, respectively. *Resilience* is related to quietness, indentation resistance, and underfoot comfort.

These and other properties may be combined into ratings for *durability* and *ease of maintenance*. Moisture and alkali resistance must be considered together in determining a suitable location with respect to whether the flooring is *suspended*, *on grade*, or *below grade* (Fig. 9.11-4). Installed cost is another factor in flooring selection, but should be considered along with long-term costs which are affected by physical properties, durability, and ease of maintenance. Selection criteria for resilient flooring products are summarized in Figure 9.11-5.

**FIGURE 9.11-2** Ingredients are keyed in a series of presses to make vinyl sheet flooring.

**FIGURE 9.11-3** Felt backing is coated to form a light, smooth background for printed designs for vinyl sheet.

**FIGURE 9.11-4** Subfloor locations. Subfloors are considered suspended when built at least 18 in. (457 mm) over a ventilated crawl space.

**FIGURE 9.11-5** Selection Criteria for Resilient Flooring Products[a]

| Type of Flooring | Location[b] | Resistance to: | | | | | Resil-ience[d] | Quietness | Ease of Mainte-nance | Durability |
|---|---|---|---|---|---|---|---|---|---|---|
| | | Grease | Alkalies | Stain[c] | Cigarette Burns | Indenta-tion | | | | |
| **Vinyl-Composition Tile** | BOS | 2 | 2 | 4 | 2 | 3 | 6 | 6 | 2 | 2 |
| **Vinyl Tile** Solid | BOS | 1 | 1 | 1–2 | 1–5 | 1–4 | 2–5 | 2–5 | 3 | 1 |
| Asphalt felt-backed[e] | BOS | 1 | 1 | 1–3 | 3–5 | 2–5 | 2–4 | 2–5 | 3 | 1 |
| Rag felt-backed | S | 1 | 2 | 1–3 | 3–5 | 2–5 | 2–4 | 2–5 | 3 | 1 |
| **Vinyl Sheet** Asphalt felt-backed[e] | BOS | 1 | 1 | 1–3 | 3–5 | 2–5 | 2–4 | 2–4 | 2 | 1 |
| Rag felt-backed[e] | S | 1 | 2 | 1–3 | 3–5 | 2–5 | 2–4 | 2–4 | 2 | 1 |
| **Rubber Tile** | BOS | 3 | 3 | 2 | 1 | 4 | 2 | 2 | 3 | 2 |
| **Cork Tile** | OS | 4 | 4 | 5 | 4 | 4 | 1 | 1 | 5 | 5 |
| **Vinyl Cork Tile** | OS | 1 | 3 | 1 | 5 | 3 | 3 | 3 | 1 | 4 |

[a]Numerical rating indicates rank of each product compared with other products. Highest rating is 1 (excellent); 2 = very good; 3 = good; 4 = fair; 5 = poor.

[b]B = below grade, O = on grade, S = suspended.

[c]Varies with staining agent.

[d]Also indicates potential underfoot comfort.

[e]Foam-cushioned products are rated highly for resilience, indentation resistance, ease of maintenance, and quietness within the ranges indicated.

### 9.11.2.1 Moisture Resistance and Location

Linoleum contains linseed oil binders and needs sufficiently dry conditions so that it will not be softened from exposure to moisture. Products with organic ingredients or backings are subject to mold growth and decay. Asphalt, vinyl composition, solid vinyl, and sheet vinyl with mineral fiber backings are least affected by moisture and alkalis which may be leached by moisture from concrete floors.

The suitability of some flooring products is shown in the "Location" column in Figure 9.11-5. Consideration has been given to the alkali resistance of the backing, which may not be the same as that of the surface material. In sheet vinyl with asphalt felt backing, for instance, the vinyl surface has excellent alkali resistance, while the backing does not.

In addition to the effect on flooring materials directly, moisture may prevent proper bonding to the subfloor or underlayment. Water-based adhesive will not set up in the presence of water, and asphaltic emulsion adhesives eventually may be displaced under sustained exposure to moisture. Excessive moisture also may deteriorate organic underlayments such as hardboards or plywood.

The main types of moisture affecting resilient floors are (1) *surface moisture*, (2) *subfloor moisture*, and (3) *ground moisture*.

#### SURFACE MOISTURE

Spilled water, floor moppings, and tracked-in moisture are the primary sources of surface moisture. Spilled water may be common in laundry areas, bathrooms, toilet rooms, and kitchens. Entries and corridors may be subjected to tracked-in moisture and excessive wet mopping.

Surface moisture affects flooring materials and adhesives by entering through the seams. Sheet flooring products minimize the number of seams, and when provided with moisture-resistant backings are suitable for use in such areas. The seams of many sheet products also may be sealed to prevent water seepage through them.

#### SUBFLOOR MOISTURE

Concrete subfloors and mastic underlayments mixed with water release large amounts of moisture as they cure. With lightweight-aggregate concrete (weighing less than 90 lb/cu ft [1442 kg/m³]) or under conditions which retard the curing process, moisture may be released over long periods of time. Field tests described under "Concrete Subfloors" in Section 9.11.5.2 should be used to establish that the subfloor is sufficiently dry to receive the intended product. Since it originates in the original mix of the concrete, subfloor moisture may be present in below-grade, on-grade, and suspended subfloors.

#### GROUND MOISTURE

Moisture from the ground generally is limited to below-grade and on-grade locations. Concrete floor slabs in direct contact with the ground, unless protected by a vapor retarder, may transmit moisture by capillary action. A well-drained fill of sand or gravel will retard the migration of moisture.

When resilient flooring is installed on wood subfloors over crawl spaces, a vapor retarder and adequate ventilation of the subfloor space are required. A vapor retarder placed over the ground will prevent transmission of moisture from there. Ventilation will encourage dissipation of moisture resulting from breaks in the vapor retarder and other sources. Both crawl spaces and concrete slabs on ground require proper surface drainage of exterior finished grade to prevent groundwater from becoming trapped under the floor.

### 9.11.2.2 Alkali and Grease Resistance

Major sources of grease are oils spilled in kitchens, laboratories, and shops; alkalis may be present from various cleaner residues. If flooring is properly protected with a floor wax or finish and spilled oils or residues are removed promptly before they seep into the seams, such temporary exposure is not a hazard, except for materials with the lowest resistance ratings (see Fig. 9.11-5). Vinyl composition and solid vinyl tile are not affected substantially by either grease or alkali.

### 9.11.2.3 Resilience

*Resilience* is a measure of the instantaneous yielding and recovery of a surface from impact. Indentation resistance, quietness, and underfoot comfort are closely related to resilience.

#### INDENTATION RESISTANCE

In assessing indentation resistance, the momentary indentations produced by foot traffic and dropped objects are of primary importance. These impact pressures can be quite high and demanding. A 105 lb (47.63 kg) woman in spike heels, for example, exerts a pressure on the floor of approximately 2000 psi (1 406 140 kg/m$^2$), while a 225 lb (102 kg) man with his weight spread over 3 by 3 in. (76.2 by 76.2 mm) heels exerts only 25 psi (17 577 kg/m$^2$).

Permanent indentation from heavy stationary objects, such as a piano or desk, may be minimized by using floor protectors to distribute the load. Indentation resistance of thinner flooring materials is greatly affected by the subfloor or underlayment and may be increased by selecting harder subsurface materials. Homogeneous vinyl tile and foam-cushioned vinyls have the highest indentation resistance.

Permanent indentation in some flooring types and under certain conditions cannot be entirely prevented. However, these indentations may be less conspicuous in patterned, textured, and low-luster floors.

#### QUIETNESS

The quietness rating indicates the effectiveness of a floor in reducing sound from foot traffic and other impact noises. Resilient floors soften impact sounds, but will not reduce reverberated noises originating from other sources because they have little sound absorption ability. Cork, linoleum, and foam-cushioned vinyl flooring have high resilience and quietness ratings. Asphalt tile is at the low end of the range of resilience and quietness.

### 9.11.2.4 Resistance to Sunlight

The actinic rays in sunlight may cause fading, shrinking, or brittleness in some resilient floors. Linoleum and vinyl are resistant to such deterioration. Color pigments are the critical factor in fade-resistant properties. Neutral colors show the best light resistance; pastel tones, especially yellows, blues, and pinks, are least effective in retaining colors under prolonged exposure to sunlight. Cork tile has the same tendency as natural wood to fade under strong sunlight.

### 9.11.2.5 Ease of Maintenance

The term *ease of maintenance* represents an appraisal of the relative expense (labor and materials) in keeping flooring at an acceptable level of cleanliness and attractiveness. Relative ratings are based on a record of experience with various flooring types as well as laboratory testing. Textured surfaces and darker colors show less scuffing and soiling and generally receive higher ease-of-maintenance ratings. No-wax vinyl products and vinyl cork tile are the easiest to maintain, followed by other vinyl products, vinyl composition, and linoleum.

### 9.11.2.6 Durability

Its durability is the ability of a flooring product to retain serviceability and attrac-

tiveness over time. Durability ratings are based on laboratory tests and are related to the physical properties of the material. Homogeneous vinyl tile is the most durable. Vinyl sheet flooring, rubber tile, and vinyl composition also are rated highly. Foam-cushioned products have good durability because the surfaces of these products absorb impact and resist abrasion.

### 9.11.2.7 Slip Resistance

The Americans with Disabilities Act of 1990 (see Section 1.4) requires flooring materials to have a static coefficient of friction of not less than 0.60 for level surfaces and 0.80 for ramps. Some manufacturers publish these data for their flooring products, even though there is no consensus in the industry concerning the test methods needed to ensure compliance. It would be prudent to verify the current status of slip-resistance requirements before selecting a resilient flooring for any project.

### 9.11.3 FLOORING ACCESSORY MATERIALS

The most common resilient accessories used with resilient flooring materials are (1) *wall bases*, (2) *stair treads, risers, and nosings*, and (3) and *thresholds, feature strips*, and *reducing strips*. These accessories are available in more limited color and design variations than the flooring materials.

### 9.11.3.1 Wall Bases

Rubber or vinyl wall bases are available in 2$^1/_2$ in. (57.2 mm), 4 in. (101.6 mm), and 6 in. (152.4 mm) heights, in straight type, and in either set-on or butt types with coved-bottom designs (Fig. 9.11-6). Vinyl base is available in 50 ft (15.2 m) coils and in 4 ft (1200 mm) lengths. Rubber base normally is available in 4 ft (1200 mm) lengths only. Exterior and interior corners of rubber base commonly are premolded (Fig. 9.11-7). End

**FIGURE 9.11-6** Resilient wall base types.

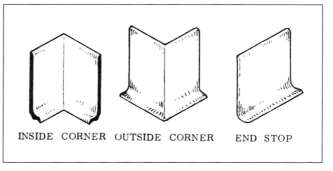

**FIGURE 9.11-7** Premolded base accessories.

stops with a finished edge for terminating wall base runs are available in both rubber and vinyl. The greater flexibility of vinyl base permits field-molding around exterior corners (Fig. 9.11-8); inside corners usually are scribed and fitted on the job.

### 9.11.3.2 Stair Treads, Risers, and Nosings

Vinyl and rubber stair treads and nosings are made with smooth or textured surfaces and may have inlaid abrasive grit strips to improve slip resistance (Fig. 9.11-9). They may be combined with standard 6-in. (152.4-mm)-high wall base

**FIGURE 9.11-8 Vinyl floor and rubber wall base.** (VPI Solid Vinyl Tile and VPI Rubber Wall Base courtesy Vinyl Plastics Incorporated.)

or higher stair risers to form complete stair coverings. Lengths vary from 3 to 12 ft (900 to 3600 mm).

### 9.11.3.3 Thresholds, Caps, Feature Strips, and Reducing Strips

Thresholds, caps, and feature and reducing strips are illustrated in Figure 9.11-10.

Resilient *thresholds* are either vinyl or rubber, $2^3/_4$ in. (69.9 mm) and $5^1/_2$ in. (139.7 mm) wide, about $^1/_2$ in. (12.7 mm) high, and 3 or 4 ft (900 or 1200 mm) long. *Feature strips* are made of asphalt, vinyl composition, linoleum, and vinyl in widths of $^1/_8$, $^1/_4$, $^1/_2$, and 1 in. (3.2, 6.4, 12.7, and 25.4 mm) widths and in $^1/_2$ in. (12.7 mm) increments up to 4 in. (101.6 mm). Thicknesses usually are $^1/_{16}$ to $^1/_8$ in. (1.6 to 3.2 mm), and lengths usually are 3 and 4 ft (914 and 1219 mm). Rubber- and vinyl-beveled *reducing strips*, 1 to $1^1/_2$ in. (25.4 to 38 mm) wide, 3 to 12 ft (914 to 3658 mm) long, and $^1/_{16}$ to $^3/_8$ in. (1.6 to 9.5 mm) thick, permit transition in finished floor heights.

### 9.11.4 FLOORING INSTALLATION MATERIALS

Resilient flooring installation materials include underlayments, lining felt, and adhesives.

### 9.11.4.1 Underlayments

Underlayments are used over subfloors to provide a smooth, level, hard, clean surface to receive flooring products. In general, panel underlayments are used over wood subfloors, mastic underlayments over concrete.

#### PANEL

Panel underlayments may be used over any structurally adequate suspended subfloor such as plywood or lumber boards. The three common types of panel underlayment are plywood, hardboard, and particleboard.

Stiffer floors result when joints in the underlayment are staggered with respect to joints in a plywood subfloor and blocking can be omitted. Over a board subfloor, underlayment joints should not fall over joints in the floor boards (Fig. 9.11-11). To allow for expansion from humidity changes, underlayment panels should be spaced approximately $^1/_{32}$ in. (0.8 mm) at joints and about $^1/_8$ in. (3.2 mm) apart at intersections with vertical surfaces. End joints in underlayment panels should be staggered ashlar fashion (see Fig. 9.11-11).

Nailing should start in the middle of the panel and proceed outward to the edges, with nail heads set flush or slightly countersunk. Panel joints should be

a

b

c

d

**FIGURE 9.11-9 Stair treads and nosings: (a) smooth, flat tread; (b) textured (diamond) tread; (c) abrasive strip tread; (d) smooth or ribbed nosings, which may be used with carpet or sheet or tile treads.** (a–c courtesy The R.C. Musson Rubber Company; d courtesy Mercer Products Company, Incorporated.)

**FIGURE 9.11-10** Resilient accessories. From left to right, cap for coved sheet goods, tile reducer, vinyl reducer, feature strip, and cove stick (supports sheet flooring for coving at a wall). (Courtesy Mercer Products Company, Incorporated.)

**FIGURE 9.11-11** Underlayment layout over a board subfloor.

sanded smooth and flush before the flooring is installed.

Minimum underlayment thicknesses are given in the following subsections for each material; however, greater thicknesses may be warranted where increased stiffness is desired.

*Plywood* A Sanded Interior Underlayment grade, at least $1/4$ in. (6.4 mm) thick, is suitable for most plywood underlayment. However, in areas subject to excessive moisture, such as bathrooms and laundry areas, an Exterior C-C (Plugged) or Underlayment grade with Exterior adhesive should be used. Un-

derlayment may be installed over a plywood subfloor with the face grain either parallel or perpendicular to the floor joists. However, if the underlayment is installed with the face grain perpendicular to the floor joists, a stiffer floor assembly results. Plywood should be installed over a board or plank subfloor with the face grain perpendicular to the direction of the floor boards (see Fig. 9.11-11), and preferably with long edges over joists. Plywood underlayment should be installed with the fasteners and spacing shown in Figure 9.11-12.

*Hardboard* Hardboard underlayment will bridge small irregularities and provide a smooth surface over an otherwise structurally sound wood floor. Underlayment-grade hardboard is an untempered service type surfaced on one side. It is manufactured in a nominal $1/4$ in. (6.4 mm) (0.215 or 0.200 in. [5.46 or 5.08 mm] actual) thickness, in 3 by 4 ft (914 by 1219 mm), 4 by 4 ft (1219 by 1219 mm), and 4 by 8 ft (1219 by 2438 mm) panels, and should conform to the ANSI standard ANSI A135.4, Class 4 (service), surface S1S. Installation of panels is the same as for other panel underlayment using the fasteners and spacing recommended in Figure 9.11-12.

**FIGURE 9.11-12** Fastening for Panel Underlayment

| Underlayment, in. (mm) | Fastener Length, in. (mm)[a] | | Fastener Spacing |
|---|---|---|---|
| | Nails[b] | Staples[c] | |
| **Plywood** | | | |
| $1/4$ (6.4) | 1 (25.4) (2d) | $7/8$ (22.2) | Nails: 8 in. (203 mm) o.c. each direction in middle of panel, 6 in. (152 mm) o.c. around panel perimeter. |
| $3/8$ (9.5) | $1 1/4$ (31.8) (3d) | $1 1/8$ (28.6) | |
| $1/2$ (12.7) | $1 1/4$ (31.8) (3d) | $1 1/4$ (31.8) | |
| $5/8$ (15.9) | $1 1/2$ (38) (4d) | $1 1/2$ (38) | Staples: 6 in. (152 mm) o.c. each direction in middle of panel, 3 in. (76 mm) o.c. around panel perimeter. |
| $3/4$ (19.1) | $1 1/2$ (38) (4d) | $1 5/8$ (41.3) | |
| **Hardboard[d]** | | | |
| $1/4$ (6.4) | 1 (25.4) (2d) | $7/8$ (22.2) | Nails and staples: 6 in. (152 mm) o.c. each direction in middle of panel, 3 in. (76 mm) o.c. around panel perimeter. |
| **Particleboard[e]** | | | |
| $3/8$ (9.5) | $1 1/4$ (31.8) (3d) | $1 1/8$ (28.6) | Nails: 10 in. (254 mm) o.c. each direction in middle of panel, 6 in. (152 mm) o.c. around panel perimeter. |
| $5/8$ (15.9) | $1 1/2$ (38) (4d) | $1 1/2$ (38) | |
| $3/4$ (19.1) | $1 1/2$ (38) (4d) | $1 5/8$ (41.3) | Staples: 6 in. (152 mm) o.c. each direction in middle of panel, 3 in. (76 mm) o.c. around panel perimeter. |

[a]Fastener lengths shown are minimum to provide $3/4$ in. (19.1 mm) penetration for nails and $5/8$ in. (15.9 mm) for staples. If longer fasteners are used, penetration should not exceed 1 in.

[b]Annularly threaded; common nails may be used by increasing the sizes $1/4$ in. (6.4 mm) (1d).

[c]Galvanized divergent chisel, 16-gauge (1.6 mm), $3/8$ in. (9.5 mm) crown width.

[d]Nominal thickness, actual thickness 0.215 in. (5.5 mm).

[e]Moisture content as shipped from mill should not exceed 7% average.

*Particleboard* Particleboard underlayment should not be used in areas subject to excessive moisture, and its moisture content as shipped from the mill should not exceed an average of 7%. Particleboard underlayment should be at least $^3/_8$ in. (9.5 mm) thick, should conform to ANSI A208.1, and should bear the American Forest and Paper Association grademark "Floor Underlayment," Grade 1-M-1. Particleboard underlayment is manufactured in 4 ft (914 mm) by 8 ft (2438 mm) panels and in thicknesses ranging from $^3/_8$ to $^3/_4$ in. (9.5 to 19.1 mm). Nailing recommendations for various panel thicknesses are given in Figure 9.11-12.

Some resilient flooring manufacturers do not recommend using particleboard as an underlayment for their materials. However, some particleboard manufacturers issue a warranty covering the entire resilient flooring installation if the flooring and underlayment are installed to their specifications.

## MASTIC

Mastic underlayments contain a chemical binder such as latex, asphalt, or polyvinyl acetate resins and portland, gypsum, or aluminous cement. Mixtures consisting of powdered cement and sand to which only water has been added function only as crack fillers and, when applied in thin coats, break down under traffic.

Latex underlayments are most suitable for applications requiring a thin layer ($^3/_8$ in. [9.5 mm] or less), for skim coating, and for patching where the fill must be feather-edged (Fig. 9.11-13).

Skim coating smooths surface irregularities. It does not raise the elevation of the floor.

Asphaltic and polyvinyl acetate underlayments are used where thicker

**FIGURE 9.11-13** Latex underlayment can be feathered to a thin edge when patching subfloor depressions. (Courtesy Kentile Floors.)

(more than $^3/_8$ in. [9.5 mm]) underlayments are needed.

Mastic underlayments must be troweled smooth and true, with not more than a $^1/_8$ in. (3.2 mm) variation from a straight line in 10 ft (3000 mm). Subfloors should be free of wax, oil, and surface coatings, such as concrete curing compounds, before mastic underlayment is applied.

*Latex* Latex underlayments are mixtures containing a synthetic latex emulsion, a cementitious material such as portland cement, fine sand, and water. They are troweled on in layers not more than $^1/_8$ in. (3.2 mm) thick and may be feathered to a thin edge (see Fig. 9.11-13). They are built up in several applications totaling not more than $^3/_8$ in. (9.5 mm). Rough surfaces are machine-sanded after the mastic has dried completely. Because latex underlayment is a quick-setting compound, splashed material should be cleaned off promptly.

Asphaltic solvent-base primers and adhesives used over latex underlayment for installing asphalt or vinyl composition tile should be allowed to set for the time recommended by their manufacturer before the tile is installed.

*Asphaltic* Asphaltic underlayment is a mixture of an asphaltic emulsion, a cementitious material such as portland cement, fine sand, and water. Proper proportioning varies, depending on the ingredients and the thickness of the underlayment, and the manufacturer. Asphaltic underlayment is not recommended for installations less than $^3/_8$ in. (9.5 mm) thick.

Before placing an asphaltic underlayment, it is necessary to prime the concrete subfloor with a thin solution of the same asphaltic emulsion used in the underlayment mixture. The mastic is then poured uniformly over the entire surface using wood screeds to establish the desired thickness. A straightedge is used to level the mastic over the screeds, after which the surface is troweled to a smooth finish. The setting time of the fill should not be accelerated with excessive heat, as this may cause cracking or crazing. Where it is necessary to trowel the mastic to a feather edge, the floor should be given a brush coat of undiluted asphaltic emulsion to prevent brittleness after the mastic has dried.

Solvent-base asphaltic primers and adhesives should not be applied over as-

phaltic underlayments; asphaltic emulsion or latex adhesives should be used.

*Polyvinyl Acetate* Polyvinyl acetate underlayment is formed by mixing portland cement and sand with a specially compounded polyvinyl acetate resin emulsion. The polyvinyl acetate liquid also should be applied to the concrete subfloor to improve the bond with the underlayment. When a fill of $^1/_2$ in. (12.7 mm) or thicker is required, a portland cement concrete "topping" may be used without polyvinyl acetate resins, but the polyvinyl acetate liquid still should be used as a bonding agent on the concrete subfloor.

This type of underlayment may be used with any adhesive and flooring type. Its installation is similar to that described for asphaltic underlayment.

### 9.11.4.2 Lining Felt

Some manufacturers recommend using lining felt over panel underlayment to minimize the possibility of joints opening and tile splitting due to underlayment movement. However, this is not a significant hazard if the underlayment or subfloor is sufficiently thick and is properly installed. Of greater significance is that the indentation resistance of flooring materials is substantially reduced by lining felt, which imposes limitations on its use.

Lining felt is used in conjunction with some vinyl sheet products over nonporous subfloors, such as old resilient floors, terrazzo, ceramic tile, and concrete that has been treated with curing agents. The liner's purpose is to provide a means for the water in the adhesive to be absorbed, which allows the adhesive to cure.

### 9.11.4.3 Adhesives

Most adhesives used for the installation of resilient flooring are of a troweling grade for application with a notched trowel (Fig. 9.11-14). Some types, however, are of a more fluid consistency for application with a brush. These often are used with thin-gauge products to eliminate the possibility of trowel marks "telegraphing" to the surface of the finished floor. The major types of adhesive are listed in Figure 9.11-15 and discussed below. Figures 9.11-15 and 9.11-16 summarize the properties and applications of the various adhesives. Adhesives with slightly varying properties are produced

**FIGURE 9.11-14** Latex adhesive is gradually spread to a uniform thickness by a swirling motion with a notched trowel. *(Courtesy Kentile Floors.)*

in each type, so the recommendations of the flooring and adhesive manufacturers should be followed.

## LINOLEUM PASTE

Linoleum paste is a low-cost adhesive used for installing linoleum, vinyl sheet flooring, and lining felt. It is a water-based adhesive which remains water-soluble even after hardening, and should not be used on concrete slabs on or below grade or in other areas subject to moisture.

Flooring should be installed immediately after spreading this adhesive, then rolled with a 100 lb (45.36 kg) roller to ensure a proper bond. The paste dries to a firm film within 24 hours and becomes hard within a few days. The rigidity of the hardened adhesive restrains sheet flooring dimensionally and prevents curling.

## ASPHALTIC ADHESIVE

Asphaltic adhesives are low-cost, water-resistant adhesives used exclusively for

**FIGURE 9.11-15** Summary of Adhesives by Type

| Adhesive | Type | Flooring | Installation | Precautions |
|---|---|---|---|---|
| **Linoleum Paste** | Clay: Water-base Troweling grade | Lining felt All sheet flooring Vinyl tile with backing Rubber tile Cork and vinyl cork tile | Over suspended wood or concrete subfloors, panelboard, and latex underlayments; install flooring immediately. | Do not use on or below grade, over suspended subfloors subject to moisture, or for solid vinyl, asphalt, and vinyl-asbestos tile. Remove with damp cloth. |
| **Asphalt**[a] Cutback asphalt | Asphalt: Solvent-base Troweling grade | Vinyl-composition tile | Over all concrete subfloors and panelboard underlayment primed with asphaltic primer. Allow to set 30 min; install tile within 4 to 18 hr, as recommended. | Do not use under or over lining felt or asphaltic underlayments. Combustible. Remove with fine steel wood and soapy water. |
| Asphalt emulsion | Clay and asphalt: Water-base Troweling grade | Vinyl-composition tile $\frac{1}{16}$ in. and thicker | Over all subfloors and over lining felt. Allow to set 30 to 60 min; install tile within 8 to 24 hr, as recommended. | Do not use to install lining felt. Keep from freezing. Remove with fine steel wool and soapy water. |
| **Asphalt-Rubber** | Asphalt and rubber: Water-base Brush and troweling grades | Vinyl-composition tile | Over all subfloors and over lining felt. Allow to set 30 min before installing tile or as recommended. | Use brushing grade with $\frac{1}{16}$ in. vinyl-asbestos tile; keep from freezing. Remove with fine steel wool and proprietary cleaner. |
| **Waterproof Resin** | Resin: Solvent-base Troweling grade | Vinyl with rag felt, vinyl, or rubber backing Rubber tile Cork tile Solid vinyl | Over all suspended subfloors; install flooring within 15 min | Do not use on or below grade, for asphalt or vinyl-asbestos tile, or to install lining felt. Remove with fine steel wool and proprietary cleaner. Combustible. |
| **Latex**[b] | Rubber: Water-base Brush and troweling grades | Vinyl with rubber backing Rubber tile Solid vinyl tile Lining felt All sheet flooring | Over all subfloors; install flooring immediately. | Keep from freezing. Remove with fine steel wool and soapy water. |
| **Epoxy** | Resin and catalyst: Troweling grade | Solid vinyl and rubber tile on or below grade | Over all types of subfloors. Mix only as much as can be spread within 30 min and covered within $2\frac{1}{2}$ hr or as recommended. | Guard against tile slipping while adhesive sets. |
| **Wall Base** Synthetic rubber cement | Synthetic rubber resin: Solvent-base Brush grade | Vinyl cove base Metal nosings and edges All resilient tile except asphalt and vinyl-asbestos to metal surfaces | Apply to both wall and material; install material within 20 min. | Combustible. Remove with proprietary remover. |
| Standard cove base cement | Solvent-base Troweling grade | Vinyl and rubber cove base | Over dry walls above grade; install base within 15 min | Combustible. Remove with fine steel wool and soapy water. |

*Note:* Formulations and working properties of adhesives vary and manufacturers' specific recommendations should be followed.
[a]May be used over latex underlayment if sufficient setting time is allowed.
[b]May be used to install asphalt and vinyl-asbestos tile; however, asphaltic adhesives are used more commonly because of lower cost.

**FIGURE 9.11-16** Summary of Adhesives by Flooring Types and Locations

| Flooring | Subfloor and Location | | |
| --- | --- | --- | --- |
| | Wood (Suspended) | Concrete (Suspended) | Concrete (On and Below Grade) |
| **SHEET FLOORING** | | | |
| **Vinyl** | | | |
| Rag felt-backed | Waterproof resin | Waterproof resin | Not recommended |
| Vinyl- or rubber-backed | Latex Waterproof resin | Latex Waterproof resin | Latex |
| Mineral fiber backed | Latex | Latex | Latex |
| **TILE FLOORING** | | | |
| **Vinyl** | | | |
| Solid | Waterproof resin Latex | Waterproof resin Latex | Latex Epoxy |
| Rag felt-backed | Linoleum paste Waterproof resin | Linoleum paste Waterproof resin | Not recommended |
| Mineral fiber-backed | Latex Linoleum paste | Latex Linoleum paste | Latex |
| **Rubber** | Linoleum paste Waterproof resin | Waterproof resin Linoleum paste | Latex Epoxy |
| **Vinyl-Composition** | Asphalt emulsion Asphalt cutback Asphalt rubber | Asphalt emulsion Asphalt cutback Asphalt rubber | Asphalt emulsion Asphalt cutback Asphalt rubber |
| **Cork** | Linoleum paste Waterproof resin | Linoleum paste Waterproof resin | Not recommended |

*Foam-cushioned products are installed by specially formulated adhesives recommended by flooring manufacturer.

installing vinyl composition tile. They may be useful on all types of below-grade, on-grade, and suspended subfloors. Asphaltic adhesives work with most underlayments, except that cutback asphalt adhesive and solvent-base primer should not be used over emulsion-based underlayment. Asphaltic adhesives retain their bonding qualities while the tile gradually conforms to minor surface irregularities. They eventually harden to provide a durable water- and alkali-resistant bond. Asphaltic adhesives are either solvent-base or water-emulsion.

*Solvent-Base* Solvent-base asphaltic adhesives consist of a solution of asphalt in hydrocarbon solvents. They are available in brushable and thicker formulations. Cutback asphalt adhesive, the most common of the solvent-base asphaltic adhesives, is of a troweling grade consistency. Because the volatile solvents are flammable and mildly toxic, these adhesives should not be used near an open flame, and the area should be well ventilated. They should be allowed to set approxi-

mately 30 minutes before applying tile. The working time ranges from 4 to 18 hours.

Solvent-base adhesives may cause or aid atmospheric damage and are not permitted in some jurisdictions. In time, they may be prohibited altogether.

*Water-Emulsion* Water-emulsion asphaltic adhesives consist essentially of tiny asphalt particles suspended in water. They become water-resistant upon drying. They are of troweling grade consistency and are spread with a notched trowel. The adhesive should be allowed to set for 30 to 60 minutes before installing tile, but the working time ranges from 8 to 24 hours. This permits application to relatively large areas, overnight setting of adhesive, and installation of tile the following day.

### ASPHALT-RUBBER ADHESIVE

An asphalt-rubber adhesive is an asphalt and rubber water-emulsion used primarily for the installation of vinyl composition tile. The working characteristics of

this adhesive are similar to those of asphaltic adhesive, and it can be used over any type of subfloor or underlayment.

### WATERPROOF-RESIN ADHESIVE

Most adhesives of the waterproof-resin type are resins in an alcohol solvent. They have approximately the same application and working characteristics as linoleum paste but are not soluble in water. However, they are adversely affected by alkaline moisture and are unsuitable for use over concrete in contact with the ground on or below grade. With the exception of asphalt and vinyl composition tile, waterproof-resin cements may be used for applying most resilient flooring in suspended locations requiring a waterproof bond. The volatile alcohol content makes these adhesives flammable, and they should not be used near an open flame or in poorly ventilated areas.

### LATEX ADHESIVE

Latex cements consist of rubber resins and water. They are extremely moisture- and alkali-resistant after they set. On drying, the adhesive turns to a rubbery film which eventually hardens and becomes insoluble in water. Latex cements generally provide good adhesion and may be used for installing all types of resilient flooring on any type of subfloor or underlayment. They are relatively higher in cost and normally are not used for installing vinyl composition and asphalt tile where cheaper asphaltic adhesives also are suitable. Latex cements more commonly are used for installing rubber tile, vinyl tile, and fiber-backed sheet products on or below grade. Flooring may be installed immediately or as recommended by the manufacturer.

### EPOXY ADHESIVE

Epoxy adhesives are extremely moisture-resistant formulations with high bonding strength and excellent durability. However, their relative high cost, difficulty in spreading, and excessive slipping of tiles immediately after installation moderate the use of these adhesives.

Epoxy adhesives consist of two parts, an epoxy resin and a catalyst which activates the mixture. The parts are mixed at the job site. A chemical reaction begins as soon as the components are mixed, so the adhesive must be used within 2 to 3 hours. The actual setting time varies, however, due to variations in the formulations. Therefore, the manufacturer's recommendations should be followed closely.

## WALL BASE CEMENT

The two basic adhesives used to secure resilient wall base are standard cove base cement and synthetic rubber cement. Both are solvent-base formulations which set as the volatile solvents escape from the mix; both are flammable and should not be used near an open flame or in poorly ventilated areas.

***Standard Cove Base Cement*** Low-cost standard cove base cement is used for installing resilient wall base and premolded corners. It develops bonding power about 10 minutes after application to either the wall surface or the wall base (Fig. 9.11-17).

***Synthetic Rubber Cement*** Synthetic rubber cement is a contact adhesive which must be applied to both the wall and the resilient base. Bond is developed as soon as the two surfaces touch. Synthetic rubber cement also is used for installing metal and resilient nosings, edges, and binding strips. It is particularly useful in bonding resilient accessories in the field where movement after fitting would be undesirable.

## 9.11.5 GENERAL INSTALLATION RECOMMENDATIONS

Successful installation of resilient flooring depends on (1) job conditions, (2) adequate subfloor construction, and (3) proper selection of underlayment, adhesives, and flooring material.

Because of the variety of proprietary flooring materials, underlayments, and adhesives, the individual manufacturers' recommendations always should be consulted to ensure proper use in each case.

**FIGURE 9.11-17** Standard cove base cement may be applied with a notched spreader or by a cartridge type gun, as shown here.

### 9.11.5.1 Job Conditions

Installation of finish flooring should be delayed until all other interior work, with the possible exception of door trim and touch-up painting, has been completed. If it is necessary to install flooring sooner, the finished floor must be protected with plain, undyed, unsaturated building paper. Door trim may be installed prior to flooring if it is shimmed up with scraps of flooring to permit later installation of the flooring under the trim.

Care should be taken to prevent freezing and damage to materials during delivery, handling, and storage. Materials should be stored on a smooth, level surface in a warm, dry place. Sheet flooring should not be stored for long periods of time, and the rolls should be stored vertically, on end. Tile flooring should be stacked not more than five cartons high.

A temperature of at least 70°F (21°C) and not more than 90°F (32.2°C) should be maintained for 48 hours before installation, during installation, and for 48 hours after installation. A minimum temperature of 55°F (12.8°C) should be maintained thereafter, and adequate ventilation should be provided at all times. Ventilation is particularly important when using solvent-based adhesives and primers which are highly volatile and flammable.

Figure 9.11-18 shows some of the terms used to describe the locations to be covered in a resilient flooring application. The terms used are defined in the body of the text when they occur or in the Glossary.

### 9.11.5.2 Supporting Construction

The two main types of subfloor that occur below resilient flooring are wood and concrete. All subfloors, whether covered with underlayment or not, should be

**FIGURE 9.11-18** Flooring terms.

firmly fastened and structurally adequate to carry the intended loads without excessive deflection.

Both wood joist and concrete slab floor constructions are suitable for the application of resilient flooring. Plank-and-beam construction is similar to wood joist construction, but the subfloor is 2 in. (50.8 mm) thick or thicker, and structural members are larger and spaced farther apart. Underlayment is recommended over plank subfloors.

Concrete slab floors may be supported on ground or suspended above grade. Basement floors are concrete slabs supported on ground below grade.

## WOOD SUBFLOORS

Suspended wood floors above grade usually are sufficiently removed from excessive external moisture sources, but floors over crawl spaces at or near grade may require that the crawl space be ventilated and the ground inside the crawl space be covered with a suitable vapor retarder.

Wood floors may be of single-layer construction, consisting of a combination subfloor and underlayment, or of double-layer construction composed of a separate subfloor and underlayment. Subfloor and combination subfloor and underlayment materials, minimum subfloor thickness for various joist spacings, and nailing requirements are discussed in Section 6.10.1. Thicker subfloors will result in stiffer floors. Single lumber board construction is not acceptable for the direct application of resilient flooring.

## CONCRETE SUBFLOORS

Concrete subfloors should have a density of at least 90 psf (439.3 kg/m$^2$). If concrete of lower density has been used, it should be refinished with a 1-in. (25.4-mm-)-thick topping of a regular concrete mix, bonded to the existing concrete with a bonding agent such as polyvinyl acetate compound.

Subfloors on or below grade should be separated from the ground with a suitable vapor retarder such as polyethylene film (see Section 21.4).

Concrete subfloors generally provide a suitable surface for resilient flooring but require precaution against excessive moisture. The moisture content of concrete floors should be checked, particularly when lightweight concrete is used, because it cures slowly. In some instances, dampness can be observed under stationary objects on a floor on or below grade, but most often moisture cannot be detected readily and tests should be performed to determine if the subfloor is sufficiently dry to receive flooring.

Concrete slabs that have been treated with curing, hardening, or separating compounds also should be tested for adequate bond.

*Suitability Test* Several tests may be performed to determine acceptable moisture levels and bonding properties of concrete subfloors. These are the relative humidity test, mat moisture and bonding test, and primer test.

A *relative humidity test* is intended for testing new suspended concrete floors for the installation of moisture-sensitive products such as linoleum, cork, and rag-felt-backed vinyl. To conduct this test, a relative humidity meter is placed on the concrete next to an interior wall or column. Depending on atmospheric conditions, the meter should read 40% to 65%. The meter is then covered with an 18 in. (450 mm) square of polyethylene film, and the edges are sealed to the floor with adhesive or tape. When substantial temperature fluctuations are expected, the meter should be protected with several layers of burlap or insulating batts. If there are no leaks in the polyethylene film, the reading should stabilize after 24 hours on a normal slab and after 72 hours on thicker than normal slabs. A sufficiently dry slab will register a reading of 80% or less. If the reading is more than 80%, the slab should be allowed to cure further and the test should be repeated.

A *mat moisture and bonding test* is designed for checking the moisture or surface bonding properties of concrete floors to establish their suitability for installation of (1) rubber tile, solid vinyl tile, and vinyl sheet with mineral fiber backing on or below grade and (2) any resilient flooring on slabs from which paint, oil, concrete hardeners, or curing compounds have been removed.

Linoleum and vinyl sheet mats approximately 36 in. (900 mm) square are applied with two bands of adhesive in several locations on the subfloor. One band is a water-soluble adhesive, the other a water-resistant latex adhesive. The mats are then sealed to the floor with adhesive or industrial tape (Fig. 9.11-19). After 72 hours, they are removed and the adhesive is examined. If too much mois-

**FIGURE 9.11-19** The mat moisture and bonding test for suspended concrete subfloors: (a) water-resistant latex adhesive; (b) water-soluble adhesive.

ture is present, the water-soluble adhesive will be partly or completely dissolved and the water-resistant adhesive will be stringy and provide little bond. If excessive moisture is still indicated after further curing of the concrete and repeated testing, the above-mentioned flooring types should not be used and consideration given to asphalt or vinyl composition tile, which have greater moisture resistance. However, for these products, the primer test also should be performed.

A *primer test* determines suitability for the installation of asphalt and vinyl composition tile over concrete. Patches of asphaltic primer are applied in several places on the floor and are allowed to dry for 24 hours (Fig. 9.11-20). If the primer does not readily peel and sticks to the floor when scraped with a putty knife, the subfloor is dry enough to receive asphalt or vinyl composition tile.

*Surface Preparation* Expansion joints, scored joints, cracks, and depressions must be filled with latex floor-patching compound or mastic underlayment to obtain a smooth surface. Concrete with an

**FIGURE 9.11-20** The primer test.

irregular surface may require grinding or smoothing with mastic underlayment. Curing or hardening compounds may leave surface films which will adversely affect adhesive bonding. If a mat bonding test reveals insufficient bonding properties, the film should be removed by power wire brushing or grinding with a terrazzo grinder. An asphaltic primer is required on dusty or porous concrete floors for the installation of asphalt and vinyl composition tile.

### 9.11.6 SHEET FLOORING INSTALLATION

The following subsections describe the installation of sheet flooring in new construction. Reflooring of existing construction is discussed in Section 9.11.9.

#### 9.11.6.1 Preparation for Flooring

The effectiveness of a flooring installation greatly depends on the proper selection and preparation of the elements which make up the installation. Subfloors that are to receive flooring directly must be firm, smooth, and dense and must possess good bonding properties. When these properties are lacking, it is necessary to prepare the subfloor by grinding or sanding or by installing underlayment. Adhesives must be compatible with all elements, including the flooring itself, and all elements must be suitable for the intended location.

#### SUBFLOOR

Vinyl sheet products with mineral fiber backing may be installed on below-grade, on-grade, and suspended subfloors. Products with rag felt backs are subject to decay and alkali attack in the presence of moisture and are limited to suspended locations. Linoleum should be installed on suspended floors only.

Concrete subfloors in all locations should be tested for moisture and bonding properties to ensure suitability for the proposed flooring and adhesive (see Section 9.11.5.2).

#### UNDERLAYMENTS

The need for underlayment is determined by the condition of the subfloor (see Section 9.11.4.1).

#### ADHESIVES

The adhesives generally recommended for use with sheet products in various locations are summarized in Figures 9.11-15 and 9.11-16.

## LINING FELT

Lining felt should be used in the installation of certain sheet flooring products on nonporous floors, as described in Section 9.11.4.2.

### 9.11.6.2 Application of Flooring

The installation of sheet materials should be planned to minimize the number and total length of seams. Necessary seams should be placed in inconspicuous locations, out of the path of heavy foot traffic. In a rectangular room, running the flooring strips parallel to the side walls generally results in an economical installation with a minimum number of seams. However, sheet flooring installed directly over wood strip floors should run perpendicular to the floor joints and may result in seams running in the short dimension, parallel to end walls (Fig. 9.11-21).

Installation of sheet flooring consists generally of *fitting and cutting*, *adhesive bonding*, and *seam treating*.

### FITTING AND CUTTING

Many sheet flooring products must be kept rolled face out until the time of installation. Some manufacturers warn against rolling sheet flooring face in because it may set up internal stresses that can result in residual growth of the sheet flooring long after it is installed. When fitting adjacent strips which have a definite geometric repeat pattern, the pattern should be aligned and matched. Sheet flooring with random overall patterns should be reversed (turned end for end) to minimize differences in graining and shading.

Methods of cutting and fitting sometimes are related to the intended method

of bonding the sheet goods. Adhesive application is facilitated by folding the strip back on itself, thus working only half of the strip at a time. The strip is folded back along its length in the *tubing method* and back along its width in the *lapping method* (Fig. 9.11-22).

Sheet flooring is cut and fitted to walls and to itself by *knifing*, *scribing*, and *seam cutting*.

After the end of a roll is squared (Fig. 9.11-23), strips are cut 3 in. (76.2 mm) longer than the distance between end walls to provide excess material fitting to wall recesses or irregularities. The first strip is placed in position with the side edges butted against the side wall, and the ends of the material are flashed (bent up) against the end walls. The material is scribed along the length of the side walls with dividers or a scribing tool (see "Scribing" below) and cut to fit snugly to the wall surface. Ends which have been flashed against the end walls also may be scribed or may be trimmed by cutting away small pieces at a time until the desired fit is achieved.

*Knifing* Knifing is a method of end fitting that involves some waste. It is used with less expensive felt-backed products. After a strip of flooring is fitted in place against a side wall or adjacent strip, excess material is cut away gradually to fit the end walls and other vertical surfaces.

*Safety cuts* are made at various points to allow the material to be fitted into difficult locations without cracking. One or more safety cuts should be made first wherever the material tends to buckle as

**FIGURE 9.11-23** The end of a roll is squared with a straightedge and square.

it is flashed against the wall. When the flooring has to be fitted into a corner before it can be knifed at the walls, the *inside corner safety cut* is the simplest (Fig. 9.11-24a). This allows the material to be dropped down snugly into a corner and then be flashed up the walls without cracking. An *outside corner safety cut* is used to fit flooring around a curved or cornered projection (Fig. 9.11-24b). A *crossover safety cut* permits bending of the flooring under an obstacle such as a radiator.

*Scribing* When flooring must be fitted against a vertical surface such as a wall, column, or stair riser with irregular corners, the exact contour of the vertical surface must be transferred to the flooring material. Scribing generally is used in fitting vinyl flooring to side walls and end walls and is the preferred method for all sheet flooring.

In scribing with dividers (Fig. 9.11-25), the flooring is pushed close to the wall

**FIGURE 9.11-21** Sheet flooring direction over floor joints.

**FIGURE 9.11-22** Methods of handling sheet flooring for adhesive bonding.

a

b

**FIGURE 9.11-24** Safety cuts facilitate fitting (a) the inside corner, (b) the outside corner.

**FIGURE 9.11-25** When scribing with dividers, the dividers should be kept perpendicular to the wall surface.

or the projection, and the dividers are opened so that one leg touches the wall at the farthest point and the other rests about 1 in. (25.4 mm) inside the edge of the flooring. Keeping the dividers perpendicular to the wall and one point always in contact with it, the profile is transferred to the flooring. The material

**FIGURE 9.11-26** An 18 in. (457 mm) scriber permits scribing under obstacles which would be impossible with dividers. (Photograph courtesy the makers of Armstrong vinyl flooring.)

then may be cut with a linoleum knife, or for vinyl sheet with a notched-blade knife, and further fitted in place. Cutouts for isolated obstacles such as pipes also can be scribed with a scribing tool in a similar manner (Fig. 9.11-26).

*Pattern scribing* is used when the size of the room or wall projections make it difficult to scribe the flooring itself. Complex profiles are first scribed to a piece of felt or heavy paper, checked for fit, trimmed, and transferred to the flooring (Fig. 9.11-27). This is particularly desirable when wall base is not used or when the base is a straight type that will not cover minor irregularities in scribing.

*End scribing*, when the flooring is tubed, is similar to that for side walls. However, when the lapping method is used, each strip must be partly cemented before scribing. The unfitted end of the sheet is lapped back and, working toward the end wall from the center of the room, adhesive is applied to within 3 ft (900 mm) of the end wall. The sheet is laid into the adhesive and rolled. The loose, uncemented end of the material then is scribed to the end wall by drawing a mark perpendicular to the edge of the material 9 to 12 in. (230 to 300 mm) from the wall and extending it from the material onto the subfloor. The material then is buckled and butted against the wall line (Fig. 9.11-28). The perpendicular distance between the line on the subfloor and the line on the material is measured with the scriber. With the scriber set at this distance, the material is scribed to the wall line and cut.

**Seam Cutting**   To make a neat, inconspicuous seam between adjacent strips of flooring, it is necessary to cut away the selvage (factory-cut edge). Sheet widths are generally $1/2$ in. (12.7 mm) greater than the nominal width to permit this excess material to be cut away. On goods

a

b

c

d

**FIGURE 9.11-27** Pattern scribing: (a) a felt pattern is scribed, (b) the pattern is transferred to the flooring, (c) flooring is cut to the pattern, and (d) flooring is checked for fit.

with a geometric repeat pattern, the amount of selvage cut away cannot be varied if the pattern at the seam is to match and be aligned properly. In goods with a random overall pattern, the selvage to be trimmed is less critical and may vary (Fig. 9.11-29).

Seams between adjacent strips are made chiefly by *underscribing*, *single*

FIGURE 9.11-28 When lapping, each strip is partly cemented before end scribing.

FIGURE 9.11-30 Underscribed seam.

FIGURE 9.11-31 Single-cut seam.

FIGURE 9.11-32 Double-cut seam.

*cutting*, or *double cutting*. To prevent misalignment due to shifting, strips are partially or completely bonded to the floor before seams are cut. When using the single- or double-cutting methods, adhesive is kept 4 in. (100 mm) or more from the seam until both edges are cut. This prevents adhesive from being pulled up when the lower edge (selvage) is removed. In both straight edging and seam cutting, a piece of scrap flooring is inserted under the cut line to prevent the backing from fraying. The knife is held as low as possible, with the blade perpendicular to the floor surface.

In *underscribing*, a recess scriber is used to transfer the exact edge profile from the previously straight-edged strip to the adjacent strip. As it is guided along the finished edge, the tool leaves a score mark on the overlapped adjacent strip directly above. The scribe mark then is used as a guide for cutting (Fig. 9.11-30).

FIGURE 9.11-29 The selvage is trimmed with a straightedge and either a linoleum knife or a straight-blade utility knife.

Underscribing is the preferred way of making seams for flooring with filled vinyl wear surfaces. The trimmed edge can be cemented down completely before underscribing, which often minimizes the number of operations involved in fitting and cementing the flooring.

*Single cutting* (Fig. 9.11-31) is used for making seams in patterned flooring where the pattern must be carefully aligned. The straight-edged strip is laid over the untrimmed strip, overlapping to fit the pattern. The lower sheet then is scored with the linoleum knife guided along the upper trimmed edge, and is then cut along the score line.

*Double cutting* (Fig. 9.11-32) is similar to single cutting except that neither edge is pretrimmed and the cut is made through both strips at the same time. Using a straightedge as a guide, the upper sheet is cut in the middle of the overlapped edges, scoring the sheet below. The lower sheet is then cut along the score line. Double cutting is an older method and is not used as commonly as the underscribing and single-cutting methods, but it may still be used with some clear (unfilled) vinyl-surfaced sheet which are not as easily scored with an underscriber.

### ADHESIVE BONDING

To facilitate spreading adhesive prior to bonding, the flooring is folded back by either tubing or lapping. The sequence of cutting and fitting, adhesive spreading, and cementing operations depends on which folding method is used. Lapping is the preferred method, but tubing is used where wall projections or the shape of the room make lapping difficult or impossible. Tubing also is recommended for heavy-gauge vinyl sheet and for burlap-backed linoleum.

Adhesive is spread with a notched trowel, spreader, or brush, depending on the type of adhesive. As each strip is pasted down, the seam is smoothed with a hand roller and the entire strip is rolled with a 100 lb (45.36 kg) or heavier roller (Fig. 9.11-33).

**b**

**FIGURE 9.11-33** Adhesive bonding is completed by (a) rolling the flooring and (b) hand rolling the seam.

**FIGURE 9.11-34** Flashed cove base with (a) an outside metal corner, (b) an inside metal corner, (c) a metal binding strip, and (d) a wax cove fillet.

## SEAM TREATING

Linoleum is softer than vinyl sheet and tends to "flow" around the knife blade without creating ragged edges. In cutting vinyl sheet, however, burrs often are created on the cut edges and must be removed to make the seams inconspicuous. After hand rolling of the seams, these burred edges may be removed with the back of a linoleum knife moved in a "chisel" motion along the seam, or a hammer head or scrap of flooring doubled in half (face out) may be rubbed along the burrs while the seam is kept damp to avoid scuffing.

An *electric iron* also may be used to treat seams: a 1-in. (25.4-mm)-wide strip of heavy-duty aluminum foil is laid over the seam (dull side down) and rubbed to make the burrs show through. The foil then is pressed with an electric iron held at an angle and moved in repeated strokes toward the mechanic. When the foil has cooled it is removed, and the seam is cleaned with a damp cloth and a light abrasive cleaner.

## TENSION FLOORS

Tension floor installation is a unique method of perimeter bonding recommended for certain sheet flooring products. Flooring may be installed by this method over well-bonded existing resilient floors over concrete and wood subfloors on all grade levels. Special adhesives allow the floor to be bonded only at the room perimeter, at seam lines, and at columns or fixtures.

The two types of adhesive used are a two-part epoxy and a special solvent-based mastic. The installation method varies slightly with the adhesive type, flooring type, and manufacturer's instructions. As in all resilient floor installations, the subfloor surface must be free of moisture, dust, paint, wax, oil, and grease, and must be structurally sound and smooth.

A tension floor installation should be allowed to dry for 4 or 5 days before it is cleaned. Floor maintenance generally is the same as for other resilient floors. However, when moving heavy objects such as appliances, a piece of plywood or particleboard should be placed under the object to prevent tearing or buckling the flooring material.

### 9.11.6.3 Cove Base Flashing

Sheet flooring may be trimmed out with standard set-on cove base (see Section 9.11.3), or the material may be extended partly up the wall to make an integral flashed cove base (Fig. 9.11-34). To make the cove profile at the meeting of the wall and floor, a $^7/_8$ in. (22.2 mm) wax fillet strip is cemented or a wood cove molding is nailed in place. A binding strip then is nailed or cemented to the wall at the desired height to protect the top of the base. The flooring may be softened with a torch to facilitate molding it against the fillet strip. In heavy-traffic areas, inside and outside metal corners also may be desirable (see Fig. 9.11-34). Similar metal endstops are available for closing base ends at door openings without casing trim.

In rooms requiring a decorative sheet border, the full-width strips may be pasted down in the field first; then the narrower border strips are flashed up the wall, scribed against binding strips, and cemented in place. After the border strip has been cut to the proper width, inside and outside corners without metal corner pieces may be scribed and installed.

Where no decorative borders are used, it is necessary to flash the ends and edges of the flooring strips up the wall. When working with such large pieces, it is difficult to cut flooring accurately by fitting and cutting in place. Therefore, it is necessary to make a felt pattern of the entire floor and base, carefully scribing and cutting around projections to the binding strips (Fig. 9.11-35). The pattern is transferred to the sheet flooring, which then can be readily cut, fitted, and cemented in place by either the tubing or lapping method. Inside corners can be made by

**FIGURE 9.11-35** A felt pattern is scribed to the binding strip, and a 45-degree V cutout is formed for the outside corners of a flashed base.

scribing, as shown in Figure 9.11-36 for resilient base. Figure 9.11-37 illustrates the preferred method of making an outside corner. A less durable corner can be made simply by mitering the flashed material at the corner.

A 100 lb (45.36 kg) roller should be used to roll the floor, but a flashed cove base requires a small hand roller. Care should be taken not to allow the edge of the heavy roller to ride over the fillet, as it may damage the installation.

## 9.11.7 TILE FLOORING INSTALLATION

The following subsections describe the installation of resilient tile flooring in new construction.

Resilient floor tile is available in vinyl, vinyl composition, rubber, asphalt, cork, and linoleum. Tile products range in size from 9 by 9 in. (28.6 by 28.6 mm) up to 36 by 36 in. (914 by 914 mm) in various gauges and in varying colors and patterns. Some tiles are made in no-wax finishes and with self-adhesive backs. Ingredients, gauges, and standards of quality are summarized in Figure 9.11-1, selection criteria in Figure 9.11-5.

**FIGURE 9.11-36** Inside base corner. The right side is butted to the corner and bonded; the left side is scribed to the right side and bonded.

### 9.11.7.1 Preparation for Flooring

The effectiveness of a resilient tile flooring installation greatly depends on the proper selection and preparation of the elements which make up the installation. Subfloors which are to receive the flooring directly must be firm, smooth, and dense and possess good bonding properties. When the subfloor does not possess these qualities, it will be necessary to prepare the surface by grinding or sanding or with the use of underlayment. Adhesives should be selected that are compatible with all elements, including the flooring itself. All elements must be suitable for the intended locations.

### SUBFLOOR

Solid vinyl, mineral fiber-backed vinyl, vinyl composition, and rubber tile may be installed on subfloors in suspended, on-grade, and below-grade locations. Cork, vinyl cork, rag felt-backed vinyl, and linoleum tile should be installed on suspended subfloors only.

Concrete subfloors in all locations should be tested for moisture and bonding properties to ensure their suitability for the proposed flooring and adhesive (see Section 9.11.5).

### UNDERLAYMENT

The condition of the subfloor will determine the need for underlayment. (See Section 9.11.4.1 for a discussion of underlayments and recommended uses.) Mastic underlayments should be compatible with intended adhesives and primers, as discussed below.

### ADHESIVES

The adhesives recommended for tile products in various locations are summarized in Figure 9.11-16. For a discussion of adhesive types and their properties, see Section 9.11.4.3 and Figure 9.11-15. Solvent-base asphaltic adhesives and primers used for the installation of asphalt and vinyl composition tile should not be used over asphaltic underlayments. For such installations, asphaltic emulsions should be used. When cutback asphalt, which is a solvent-base adhesive, is used over latex underlayment, the amount of time recommended by its manufacturer must be allowed before tile is installed. Porous or dusty concrete floors and panel underlayments intended for the installation of asphalt or vinyl composition tile should be primed with an asphaltic primer before adhesive is applied. Some tiles are made with self-adhesive backs, but these are not generally used in commercial or institutional projects.

### 9.11.7.2 Installation of Flooring

The first step in the installation of resilient tile flooring is to make a trial design layout of the tiles. The tiles then are cemented down according to the best layout, and walls are trimmed out with resilient or other base.

### DESIGN LAYOUT

In determining the design layout, it is important that tile joints do not fall over joints in the subfloor or in the underlayment below. Because few walls are seldom exactly parallel, it is necessary to center and square off the room. Tiles may be laid in either *square* or *diagonal designs*. Over existing wood strip floors, a diagonal installation is preferred.

*Square Designs*   Tile joints for square designs run parallel to the walls. The trial layout is started by snapping chalklines to divide the room into four equal quadrants (Fig. 9.11-38). If the floor area is

a          b          c

**FIGURE 9.11-37** Outside one-piece corner: (a) a V cutout is transferred from the pattern to the flooring before installation; (b) a corner piece is cut out separately with a square; (c) before the matching corner piece is fitted, a narrow strip of backing is removed at the fold.

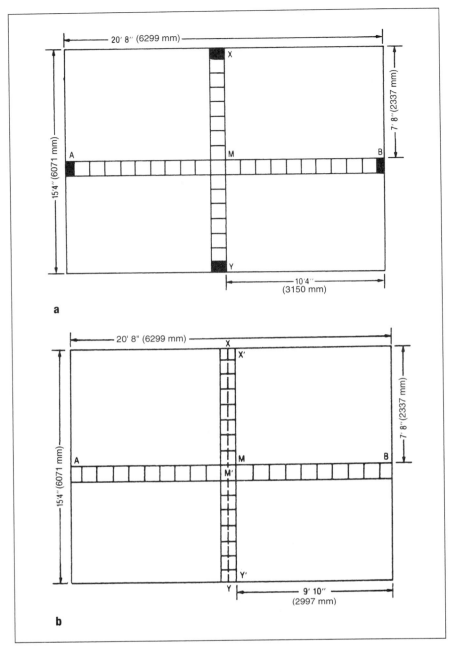

**a**

**b**

**FIGURE 9.11-38** Trial layout, square design: (a) unacceptable layout (sizes of border tiles differ); (b) acceptable layout (border tiles are equal and wider than one-half of a tile). (Illustration by the makers of Armstrong vinyl flooring.)

out the trial design so that a full repeat of the pattern will occur at all walls (Fig. 9.11-39). Such patterns commonly are used with one-color borders to emphasize the design in the field.

***Diagonal Designs*** Tile joints are run at 45-degree angles to the walls for diagonal designs. A trial row of tiles is laid out along each centerline and diagonally to the corner to determine the resulting condition at the border (Fig. 9.11-40a).

Two-color designs, requiring a border laid on the square, also must be checked to ensure that the half-tiles along each of the walls will be of the same color. To achieve this, it may be necessary to shift tiles from the trial position shown in Figure 9.11-40a so that tiles will abut rather than be bisected by diagonal lines (Fig. 9.11-40b). The proper starting position of tiles at the center of the room is determined by the dimensions of the room and must be established in each case by a trial layout.

If it is determined that the design should start with single tiles at the intersection of diagonals, new diagonals should be struck one-half a tile over from the original location to serve as guidelines for the first row of tiles.

One-color designs without a border present no special problem, and the starting point need only be adjusted to ensure that the layout will not result in excessively narrow tiles at the walls. In one-color installations with tiles having a directional grain, it is important that the half-tiles near the border fit properly into a basketweave pattern. If the tiles are cut properly with respect to grain, tiles at both the right and left sides will fit the pattern (Fig. 9.11-41).

extremely large or job conditions prevent working the entire floor area, each quadrant may be subdivided further. Full tiles should be laid starting from the center and working toward the walls. The first trial may result in unequal border tiles at side walls and end walls (see Fig. 9.11-38a).

To equalize border tiles at walls, the horizontal row of tiles may be shifted so that it is bisected by the chalklines, as shown in Figure 9.11-38b. In general,

layout lines should be shifted so that approximately equal border tiles of at least one-half of a tile or wider will result.

*Checkerboard* patterns may be made by alternating tiles of two different colors or alternating the grain direction in directional tiles. Either a tile which matches the field or a contrasting one-color sheet or tile can be used for the border.

For a *hollow square* design, another two-color pattern, it is important to lay

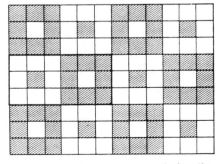

**FIGURE 9.11-39** Hollow square design; the repeated unit is indicated by the dark outline.

(Illustration by the makers of Armstrong vinyl flooring.)

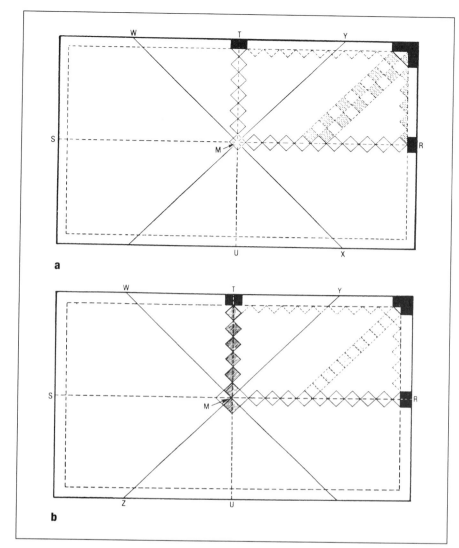

**FIGURE 9.11-40** Diagonal design, trial layout: (a) unacceptable layout; (b) acceptable layout—all half-tiles at the border are of the same color. (Illustration by the makers of Armstrong vinyl flooring.)

## ADHESIVE BONDING

To allow sufficient working space to handle materials, only part of the floor area, depending on the working time of the adhesive, should be covered with adhesive at one time. Care should be taken in applying the adhesive so that chalklines will not be completely concealed. This may be done by leaving ends and intermediate points on the chalkline uncovered.

Adhesive such as cutback asphalt develops holding power by setting before tiles are applied. However, with some adhesives, such as waterproof resin and epoxy, the possibility of tile slippage may be a problem. Temporary lattice strips nailed at the chalklines will prevent the tiles from sliding out of place during installation.

Tiles should be installed with tight, straight, carefully aligned joints in both directions. Each tile should be butted to the preceding tiles and lowered in place without sliding.

Installation on the square with a two-color checkerboard pattern can be speeded by starting with a row of 10 tiles along each centerline and laying tiles of one color at a time in diagonal rows (Fig. 9.11-42).

For diagonal designs, tiles may be cemented down in one quadrant at a time, starting at the center, laying the tiles along the diagonal, and proceeding clockwise around the room.

### FITTING AND CUTTING

Border tiles may require scribing and fitting against walls and obstructions. They are scribed essentially as described for sheet flooring, using dividers or a scrib-

**FIGURE 9.11-42** Checkerboard square designs can be installed more quickly by laying tiles of one color in diagonal rows as numbered. (Illustration by the makers of Armstrong vinyl flooring.)

**FIGURE 9.11-41** Tiles must be cut according to the grain to obtain half-tiles that will fit into a basketweave pattern. (Illustration by the makers of Armstrong vinyl flooring.)

*Right Half-Tile*

*Left Half-Tile*

*Right Half-Tile* — *Base*

*Left Half-Tile* — *Base*

ing tool to transfer the contour of the vertical surface (see Fig. 9.11-25).

Most resilient tiles can be cut at room temperature with a linoleum or straight-blade knife. However, it is easier to cut vinyl composition and asphalt tiles by heating the tile and the knife with a blowtorch or a heat gun.

## 9.11.8 INSTALLATION OF RESILIENT FLOORING ACCESSORIES

### 9.11.8.1 Wall Base Installation

The two main types of wall base used with resilient tile are set-on and butt-cove base (see Fig. 9.11-6). Most tile installations are made with set-on base because it is easier and more economical to install. This type of base is set on top of the finished floor and requires less accurate fitting of both flooring and base. Butt-cove base provides a more sanitary condition and may be required in spaces where cleanliness is mandated, such as in laboratories and some hospital spaces. Both set-on and butt-cove base are available in rubber and vinyl.

### RUBBER BASE

Rubber base is manufactured in straight runs and in premolded outside and inside corners. These facilitate turning corners with a minimum of field labor, the only fitting required being at butted joints. Rubber base is used to match rubber flooring and in cases where premolded corners are desired.

### VINYL BASE

The greater pliability of vinyl base permits it to be molded around corners, and therefore it is manufactured mainly in straight runs. Inside corners are made by notching a V in the back of the base at the fold and heating it to increase pliability. The base may then be formed and pressed into the corner. Inside corners also can be made by butting the first piece squarely in the corner and scribing the second piece around it (see Fig. 9.11-36).

Outside corners are formed by first cutting away about half of the thickness from the back of the material at the point of bend. At the back of the cove heel, additional material should be cut away approximately 1 in. (25.4 mm) to each side of the bend. The base then can be molded readily around the corner and set in place with standard base cement or synthetic rubber cement. Synthetic rubber cement

is preferred because it provides immediate holding power and prevents movement from contraction as a bond is developed.

### 9.11.8.2 Stair Treads, Risers, and Nosings

Resilient sheet materials can be used to provide a covering over wood and concrete stairs. The sheet material can be flashed from the tread up the riser (Fig. 9.11-43a) or used with a set-on coved rubber or vinyl riser (Fig. 9.11-43b). The nosing of a wood tread may be wrapped with the sheet flooring or protected by a separate metal or rubber nosing. Treads on riserless wood stairs may be completely wrapped with sheet materials (Fig. 9.11-43c). Formed concrete stairs can be covered with continuous runs of sheet flooring on both treads and risers (Fig. 9.11-43d). Methods of installing sheet stair covering are similar to those described for sheet flooring. Accurate layout, scribing, fitting, and proper selection of adhesive are essential.

Resilient tiles can also provide an attractive wearing surface on wood and concrete stairs. Tile treads usually are installed with rubber, vinyl, or metal nosings (Fig. 9.11-44). Risers for tile treads may be of matching tile or a one-piece

set-on coved riser similar to wall base. Matching one-piece treads with premolded nosings also are available from some manufacturers (see Fig. 9.11-44). Methods of installation are similar to those described for tile floors. Accurate layout, scribing, fitting, and proper selection of adhesives are essential.

### 9.11.9 REFLOORING EXISTING CONSTRUCTION

Both sheet and tile products may be installed over existing finished floors that have been suitably prepared to provide a firm, smooth, stable base for the new flooring. The main types of floor finishes encountered in reflooring are (1) wood strip, block, or plank, (2) concrete, (3) ceramic tile, (4) terrazzo, and (5) marble. New flooring may be installed over old resilient flooring, provided that the old flooring is not cushioned or deeply embossed. For specific recommendations, the manufacturer of the new flooring should be consulted.

Existing floors intended for reflooring should be inspected to ensure a firm base free of loose particles. Existing wax, oil, grease films, and most paints should be removed. Surfaces should be smooth and reasonably level or leveled with an ap-

**FIGURE 9.11-43** Stair coverings with sheet flooring: (a) one-piece tread and riser; (b) sheet tread and set-on cove riser; (c) wrapped riserless wood tread; (d) continuous risers and treads on concrete.

**FIGURE 9.11-44** Stair coverings with tile flooring and resilient accessories.

propriate underlayment when necessary. Recommendations for flooring and adhesive selections with respect to grade locations should be adhered to as in new construction.

#### 9.11.9.1 Wood Floors

A suitable panel underlayment is recommended over existing wood floors. However, sheet products may be applied directly to double-layer floor construction consisting of a lumber board or plywood subfloor and a finish wood strip floor of tongue-and-grooved boards at least $^3/_4$ in. (19.1 mm) thick and not wider than 3 in. (76.2 mm) if the existing floor is sufficiently smooth, level, firm, and well seasoned.

When panel underlayment is not used, a protective film on wood floors may be removed with a power sander or a blowtorch. Mild chemical solutions in water should not be used because the large quantity of water required for scrubbing and rinsing may cause the boards to warp. After a wood floor has been sanded, one brushed coat of floor sizing consisting of equal parts of a wax-free shellac and denatured alcohol should be applied to seal the wood. This prevents excessive moisture absorption from adhesive and other sources and minimizes the possibility of boards warping in the future.

Other double wood floors of block and plank and single-layer plank floors should be covered with underlayment consisting of at least $^1/_4$ in. (6.4 mm) plywood, $^1/_4$ in. (6.4 mm) hardboard, or $^3/_8$ in. (9.5 mm) particleboard.

#### 9.11.9.2 Concrete Floors

Most paint films can be removed from concrete floors by power grinding, burning with a blowtorch, or scrubbing with a solution of trisodium phosphate and water. After the paint has been removed, the floor should be scrubbed with a strong soapy solution, rinsed with clear water, and allowed to dry thoroughly. Paint removers should not be used because they react adversely with many adhesives. Because the use of trisodium phosphate and some other paint removers is not permitted in some jurisdictions, it is necessary to check with local authorities before using them.

Paints with a chlorinated rubber or resin base are often tightly bonded to the concrete surface and, if the mat bonding test reveals good adhesion, may be left in place.

#### 9.11.9.3 Ceramic Tile, Terrazzo, and Marble

Ceramic tile, terrazzo, and marble floors should be cleaned with a scrubbing ma-

chine using soapy water and clean, sharp sand. A floor grinder used dry with $4^1/_2$ grit open-coated paper may be preferred to avoid delay due to drying. Badly fitted joints and cracks should be leveled with latex underlayment or a patching compound.

### 9.11.10 CLEANING

A resilient floor can be damp-mopped immediately and should be polished within 48 hours. However, it should not be given a thorough cleaning until 4 to 5 days after installation. The floor then should be washed with a mild solution of a nonabrasive, nonalkaline cleaner or soap and thoroughly rinsed with cold water. It should be allowed to dry and then waxed with a floor finish recommended by the flooring manufacturer. A soft cotton mop or wax applicator should be used to apply the floor finish as thinly as possible.

No-wax tile and sheet floors should be cleaned the same way as other resilient floors; however, waxing is not necessary. Because they have no renewable surface finish, it is important to protect no-wax floors from severe abrasion, such as from gritty dirt and furniture without floor protectors.

## 9.12 Terrazzo Flooring

The term *terrazzo* is derived from the Italian word *terrassa* ("terrace on the roof") because many early installations were made in upper-level outdoor living areas. The origins of terrazzo date back to the beginning of the Christian era, having evolved from *mosaic* work, the art of placing colorful pieces of stone, vitreous enamel, or marble (tesserae) in decorative designs on walls and floors. The earliest terrazzo installations in Roman times were made of a mixture of crushed brick in lime mortar, which was ground and polished after it hardened. Later, the decorative chips salvaged from mosaic work were introduced in tiled outdoor terraces to create colorful and durable floor finishes.

Terrazzo has been used for centuries as an economical and durable indoor-outdoor finish. Traditionally, it has been defined as a form of mosaic flooring made by embedding small pieces of marble in mortar and polishing. In modern times, the mortar (matrix) for true terrazzo finishes has been portland cement. Historically, the decorative chips were limited to marble. More recently, synthetic resinous binders have been developed, and other hard rocks capable of being polished are also used.

Because of the expense involved in hand assembly of decorative chips in complex designs for large areas, mosaic work survives today primarily as custom inserts in terrazzo installations. The tesserae usually are assembled in the shop and mounted on paper; then the assembly is installed in a mortar setting bed in the desired location. After the paper is removed, joints are grouted and the entire installation is ground and polished.

Grinding and polishing are distinctive operations which differentiate terrazzo from seamless floorings which also utilize resinous binders. A notable exception to this general rule is *rustic* portland cement terrazzo, which is not ground but is washed to reveal the decorative chips. In rustic terrazzo, selected colorful gravels sometimes are used instead of marble chips.

Terrazzo is a durable, low-maintenance floor finish suitable for both exterior and interior locations. It can be cast in place or precast in a variety of shapes and colors. In precast form, its chief use has been for shower receptors, window sills, wall bases, and stair treads. More decorative products intended for counters, lavatory tops, and tables are marketed under the names *art marble* and *cultured marble*. These precast terrazzo products involve installation methods different from those associated with cast-in-place terrazzo floor finishes and are not included in this discussion.

Since many terrazzo floors use portland cement mortar and are installed over concrete surfaces, it is recommended that the requirements discussed in Chapter 3 be reviewed.

## 9.12.1 MATERIALS

The decorative wear layer of terrazzo finishes is called a *topping*. In some installations, the topping is placed directly on a wood, metal, or concrete subfloor; in others, an intervening portland cement underbed is required. When an underbed is used, it may be rigidly *bonded* to a tile subfloor, or it may be *unbonded* by sep-

**FIGURE 9.12-1** Basic construction elements for unbonded and bonded terrazzo installations.

arating it from the subfloor with an *isolation membrane* (Fig. 9.12-1). When no underbed is used, the topping always is bonded to the subfloor. Other terms relating to terrazzo installations are illustrated in Figure 9.12-1 and defined in the Glossary.

The basic ingredients of terrazzo toppings are:

- Binders
- Decorative chips
- Fine aggregate
- Pigments
- Divider strips, *expansion strips*, and other accessories
- Metal reinforcement
- Isolation membranes
- Curing compounds
- Sealing materials

Materials for terrazzo should conform to the industry standards listed in Figure 9.12-2.

### 9.12.1.1 Binders

*Binders* in terrazzo are either *cementitious* or *resinous*. Cementitious binders are the older, traditional terrazzo materials; resinous binders are the result of developments in organic chemistry since World War II.

#### CEMENTITIOUS BINDERS

The primary cementitious binder used today is either white or natural (gray) Type I portland cement. Natural portland cement is less expensive and can be used if a vivid or light-colored matrix is not essential to the color scheme. The National Terrazzo and Mosaic Association (NTMA), however, cautions that using gray cement may result in an uneven color. White portland cement is more expensive but offers a lighter background for displaying decorative chips and produces clearer, truer colors with mineral pigments. Type IA air-entraining portland cement is recom-

mended for exterior applications subject to freezing and thawing.

#### RESINOUS BINDERS

Resinous binders possess most of the desirable properties of portland cement binders, but have greater strength and chemical and abrasion resistance. In addition, the ability of resinous binders to bond with most subfloor materials and to be applied in $1/4$- to $1/2$-in. (6.4- to 12.7-mm)-thick toppings makes them suitable for many installations where a portland cement binder would be too thick. However, their high material costs limit resinous terrazzo to uses where their special properties are specifically required.

Synthetic binders for terrazzo finishes include epoxy, polyacrylate, and poly-

**FIGURE 9.12-2** Summary of Terrazzo Materials

| Material | Standard |
|---|---|
| **Cementitious binders** | |
| Portland cement, Type I | ASTM C150 |
| **Resinous matrixes** | |
| Epoxy | NTMA* |
| Polyester | NTMA* |
| Polyacrylate | NTMA* |
| **Decorative chips** | NTMA* |
| **Aggregate** | |
| Sand | ASTM C33 |
| **Pigments** | NTMA* |
| **Metal reinforcement** | |
| Welded wire fabric | ASTM A185 |
| **Isolation membranes** | |
| Roofing felt | ASTM D226 |
| Polyethylene film | ASTM D2103 |
| **Curing compounds** | ASTM C309 |

*Specifications of the National Terrazzo and Mosaic Association.

ester. Because of their high cost and the desire to impart special properties, these resins are combined with various additives and fillers which together form the matrix. Epoxy and polyacrylate resins are also combined with portland cement to form *epoxy-modified cement* terrazzo and *polyacrylate-modified cement* terrazzo.

Materials and application for epoxy, polyacrylate, and polyester terrazzo should conform to the requirements of the NTMA specifications, details, product data, and other requirements, as well as the binder manufacturer's recommendations.

Epoxy and polyester matrixes are two-part formulations (a resin and a catalyst) to which a dry mix (fillers, pigment, additives, and decorative chips) is added. These resins are distinguished by good chemical resistance and adhesion, good abrasion resistance, and high impact strength. Epoxy matrixes also exhibit negligible curing shrinkage. In addition to their high cost, they are toxic to the skin until cured.

Polyester matrixes are particularly resistant to acids of high concentration. They are also resistant to water and weather, and are strong and durable. However, they have a somewhat lower resistance to alkali than epoxy and a stronger odor during the installation and curing period.

Polyacrylate-modified cement terrazzos have a high bond strength and are resistant to snow-melting salts, foodstuffs, and urine. They are also nontoxic and relatively free from objectionable odors.

## 9.12.1.2 Decorative Chips

The decorative chips in terrazzo are mainly marble, which is defined to include all calcareous (limestone) rocks such as onyx and travertine, as well as attractive serpentine rocks capable of being ground and polished. Granite, which is not a marble, is extremely hard and weather-resistant, and is often used for heavy-duty industrial and rustic terrazzo.

Domestic and imported chips, available in a wide range of colors, are graded by number according to size, as shown in Figure 9.12-3. Most producers supply #O, #1 and #2 as separate sizes; some produce #3 and #4 as separate sizes. Larger sizes are frequently grouped as #3-4 mixed, #5-6 mixed, #7-8 mixed, and #4-7 mixed.

The appearance and minimum thickness of toppings are determined largely

**FIGURE 9.12-3** Grading of Decorative Chips for Terrazzo Toppings

| Type of Topping and Minimum Thickness | Chip Number | Passes Screen, in. (mm) | Retained on Screen, in. (mm) |
|---|---|---|---|
| $^{1}/_{4}$ in. (6.4 mm) Thin-set | 0 | $^{1}/_{8}$ (3.2) | $^{1}/_{16}$ (1.6) |
| | 1 | $^{1}/_{4}$ (6.4) | $^{1}/_{8}$ (3.2) |
| $^{5}/_{8}$ in. (15.9 mm) Standard* $^{1}/_{2}$ in. (12.7 mm) Thin-set | 2 | $^{3}/_{8}$ (9.5) | $^{1}/_{4}$ (6.4) |
| $^{5}/_{8}$ in. (15.9 mm) Intermediate | 3 | $^{1}/_{2}$ (12.7) | $^{3}/_{8}$ (9.5) |
| | 4 | $^{5}/_{8}$ (15.9) | $^{1}/_{2}$ (12.7) |
| $^{3}/_{4}$ in. (19.1 mm) Venetian | 5 | $^{3}/_{4}$ (19.1) | $^{5}/_{8}$ (15.9) |
| | 6 | $^{7}/_{8}$ (22.2) | $^{3}/_{4}$ (19.1) |
| | 7 | 1 (25.4) | $^{7}/_{8}$ (22.2) |
| | 8 | $1^{1}/_{8}$ (28.6) | 1 (25.4) |

*Palladiana toppings consist of broken marble slabs set in a standard chip mix and should not be less than $^{1}/_{2}$ in. (12.7 mm) thick.

by the size and type of chips used. Toppings are identified as *standard*, *intermediate*, *Venetian*, and *Palladiana* (see Fig. 9.12-3). The first three use crushed chips mixed with resinous or portland cement matrix and spread uniformly in the topping. The fourth, Palladiana toppings, are custom finishes similar to mosaics. They contain cut slabs of marble about $^{1}/_{2}$ in. (12.7 mm) thick that are set by hand in the desired pattern. Spaces between these slabs are filled with a standard topping mix.

The NTMA and some producers of chips publish standard color plates as an aid in selecting and specifying decorative chips and color pigments. These marble chips and color combinations are designated by plate number. However, finished terrazzo should not be expected to duplicate the color plate exactly because of slight variations in color and veining between different shipments of chips. Physical samples should be prepared as a standard of comparison at the beginning of each project.

## 9.12.1.3 Fine Aggregate

The term *aggregate* refers to sand and fillers. It does not include decorative chips. Fine silica sand and fillers usually are included in the proprietary formulations of resinous matrixes. Sand aggregate is not used in a portland cement topping but is a basic ingredient of a mortar underbed. It should be natural or manufactured and well graded, clean, granular, and free of deleterious amounts of clay, salts, and organic matter.

## 9.12.1.4 Pigments

Terrazzo toppings may be the natural color of the binder or can be treated with pigments to produce a wide variety of colors. Pigments should be stable, nonfading mineral or synthetic formulations compatible with the binder.

## 9.12.1.5 Accessories

Accessories for terrazzo installations include divider strips, expansion strips, and base beads. Metal accessories are made from half-hard brass or white alloy zinc (about 99% zinc). Plastic-top strips and beads in various colors with brass, zinc, or galvanized steel supports also are available. All-brass, stainless steel, and zinc-supported plastic-top accessories are suitable for exterior use. Aluminum strips are suitable for use in resinous terrazzo, but should not be used with portland cement terrazzo because aluminum is subject to attack by alkalis present in the topping mix. Only plastic-top strips and beads should be used in conductive floors to prevent the accumulation of dangerous electrostatic charges. White alloy zinc accessories are not suitable in exterior locations because a chemical reaction often causes blooming (a white, chalky deposit on the strips). The standard length of strips and beads is 6 ft (177.8 mm), but longer lengths are available.

### DIVIDER STRIPS

Divider strips are installed to:

- Localize shrinkage cracking
- Ensure uniformity in topping thickness

- Act as decorative elements
- Permit accurate placement of the different colors forming a floor pattern
- Act as construction joints, subdividing a floor into convenient working areas

A variety of divider strips are available to suit particular installations (Fig. 9.12-4). The thickness of the top of the strip (exposed face) can range from 20-gauge (1.0 mm) up to $^1/_2$-in. (12.7-mm)-thick bars; 18-gauge (1.3 mm) is the recommended minimum. Divider strips with a $^1/_8$ in. (3.2 mm) or wider face usually have a heavy plastic or metal top section and a thin metal bottom section. For a topping thickness of $^5/_8$ in. (15.9 mm), a $1^1/_4$-in. (28.6-mm)-deep divider is recommended; $1^1/_2$-in. (38-mm)-deep strips are recommended for Venetian toppings $^3/_4$ in. (6.4 mm) and thicker. For thin-set installations, special angle or T-shaped strips $^1/_4$ to $^1/_2$ in. (6.4 to 12.7 mm) deep are used. Other special strips are available for wall base and for edgings at the juncture with other flooring materials.

## EXPANSION STRIPS

Provision for crack control is necessary where substantial structural movement can be expected, particularly over beams and bearing walls and over expansion or isolation joints in the subfloor. Expansion strips for this purpose consist of a pair of divider strips separated by a resilient material such as neoprene (see Fig. 9.12-4). Expansion strips are manufactured in types and sizes similar to those of divider strips.

## BASE BEADS

All-metal or plastic-top base beads are used to terminate terrazzo wall base. They are made in several profiles and to accommodate vertical terrazzo thicknesses varying from $^1/_2$ in. (12.7 mm) to $1^1/_2$ in. (38 mm).

### 9.12.1.6 Metal Reinforcement

Metal reinforcement is required in the underbed, except where the underbed is bonded to a concrete subfloor. Underbed reinforcement should be corrosion-resistant welded wire fabric of at least 16-gauge (1.6 mm) with wires spaced not more than 2 in. (50.8 mm) on centers. Thin underlayments for resinous toppings may use welded or woven wire weighing at least 1.4 lb/sq yd (0.76 kg/m$^2$) or expanded metal lath weighing at least 2.5 lb/sq yd (1.36 kg/m$^2$).

### 9.12.1.7 Isolation Membrane

An isolation membrane is required in all unbonded installations to separate the underbed from the subfloor. The membrane may be 15 lb (8.14 kg/m$^2$) asphalt-saturated roofing felt or 4 mil (0.1 mm) polyethylene film.

### 9.12.1.8 Curing Materials

Curing is necessary for portland cement toppings to develop optimum wear properties. Materials suitable for curing include polyethylene film, nonstaining (nonasphaltic) water-resistant building paper, and clean water. Spray-on curing compounds are sometimes used for curing terrazzo surfaces, but may prevent bonding if used to cure concrete slabs that are to receive a thin-set, chemically bonded, or monolithic topping.

### 9.12.1.9 Sealing Materials

Sealing of terrazzo is discussed in Section 9.12.2.5.

### 9.12.2 APPLICATION

Suitable terrazzo finishes require attention to (1) general recommendations common to most installation methods, (2) specific application procedures for particular methods, and (3) various finishing procedures for special surface finishes.

Terrazzo systems can be classed as *bonded* or *unbonded*. Bonded systems include thin-set monolithic, chemically bonded, and bonded-underbed methods. *Thin-set* installations consist of a $^1/_4$- to $^1/_2$-in. (6.4 to 12.7-mm)-thick resinous topping placed directly over a sound wood, metal, or concrete subfloor (Fig. 9.12-5a). In *monolithic* installations, a somewhat thicker ($^5/_8$ in. [15.88 mm] thick or thicker) portland cement topping is applied directly over a rough-finished concrete subfloor (Fig. 9.12-5b). If the concrete subfloor is an existing, smooth-troweled concrete slab, a portland cement topping can be *chemically bonded* with a bonding agent (Fig. 9.12-5c). Monolithic and chemically bonded installations depend on the subfloor to resist shrinkage stresses and con-

**FIGURE 9.12-4** Accessories: shallow T- and L-shaped (angle) strips for installation without underbed are available in thin-top, heavy-top, and edging profiles.

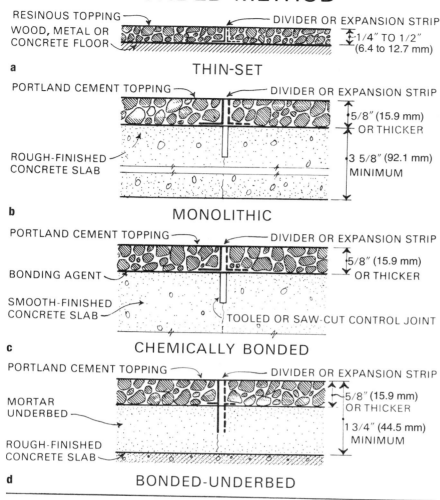

# BONDED METHOD

**a** RESINOUS TOPPING
WOOD, METAL OR
CONCRETE FLOOR — DIVIDER OR EXPANSION STRIP
1/4″ TO 1/2″ (6.4 to 12.7 mm)

### THIN-SET

**b** PORTLAND CEMENT TOPPING — DIVIDER OR EXPANSION STRIP
5/8″ (15.9 mm) OR THICKER
ROUGH-FINISHED CONCRETE SLAB
3 5/8″ (92.1 mm) MINIMUM

### MONOLITHIC

**c** PORTLAND CEMENT TOPPING — DIVIDER OR EXPANSION STRIP
5/8″ (15.9 mm) OR THICKER
BONDING AGENT
SMOOTH-FINISHED CONCRETE SLAB
TOOLED OR SAW-CUT CONTROL JOINT

### CHEMICALLY BONDED

**d** PORTLAND CEMENT TOPPING — DIVIDER OR EXPANSION STRIP
5/8″ (15.9 mm) OR THICKER
MORTAR UNDERBED
ROUGH-FINISHED CONCRETE SLAB
1 3/4″ (44.5 mm) MINIMUM

### BONDED-UNDERBED

# UNBONDED METHOD

**e** PORTLAND CEMENT TOPPING — DIVIDER OR EXPANSION STRIP
5/8″ (15.9 mm) OR THICKER
MORTAR UNDERBED
METAL REINFORCEMENT
ISOLATION MEMBRANE
ANY SUBFLOOR
2 1/2″ (63.5 mm) MINIMUM
UP TO 1/4″ (6.4 mm) SAND

### FLOATING (SAND CUSHION)

**FIGURE 9.12-5** Details of basic terrazzo installation methods.

spicuous cracking in portland cement toppings. However, *bonded-underbed* installations provide more positive crack control by creating weakened planes at each divider strip (Fig. 9.12-5d).

Unbonded installations provide the most crack control and are preferred wherever structural movement is expected. Separation between subfloor and underbed can be accomplished by an isolation membrane alone or with a membrane and sand cushion (Fig. 9.12-5e). When a sand cushion is used, it is called a *floating* installation.

### 9.12.2.1 General Recommendations

The installation of terrazzo involves (1) preparation of the subfloor and, depending on the method of installation, (2) placement of the topping, strips, and underbed when used.

### SUBFLOOR PREPARATION

Regardless of the installation method, the subfloor must be sound, free of loose material, and structurally adequate to support the expected live loads without excessive deflection, as well as the additional dead load imposed by the terrazzo. This dead load will be about 11 lb/sq ft for each inch of thickness (53.71 kg/m² for each 25.4 mm of thickness) for both resinous and cementitious terrazzo.

For unbonded installations, the surface quality of the subfloor is not critical. In bonded installations, the quality of the installation depends on an adequate bond with the subfloor. Therefore, the subfloor must be free of coatings such as paint, wax, and sealing or curing compounds that could prevent a proper bond.

For monolithic installations, the surface of the subfloor should not be troweled but should be wood-floated and broomed or otherwise roughened. A chemical bonding agent should be used over a smooth-troweled concrete surface.

Concrete intended for a thin-set resinous installation should be smooth-troweled and be at least 21 days old. Wood floors should consist of a suitable single plywood subfloor or double floor including at least a 3/8-in. (9.5-mm)-thick plywood underlayment (Fig. 9.12-6). Plywood should be Exterior Type CC-plugged grade or better. The recommendations of the manufacturer of the resinous matrix should be followed for subfloor preparation. Over most subfloor materials, the topping can be adhesively bonded, and priming of the subfloor is required to achieve a bond. Over plywood, some manufacturers recommend direct adhesive bonding after plywood edges have been bound with glass fiber tape.

The subfloor for monolithic, chemically bonded, and thin-set installations should conform to the profile of the finished surface so that the topping will be of uniform thickness. There should be no more than a 1/8 in. (3.2 mm) variation in the subfloor when tested with a 10 ft (3000 mm) straightedge. A topping should not be used to fill voids or develop pitch, because excessive variations in topping

**FIGURE 9.12-6** Wood Floor Construction for Thin-Set Terrazzo[a]

### Single-Floor Construction

| Combination Subfloor/ Underlayment[b,c] | Maximum Support Spacing, in. (mm) o.c. | Nails[d] | Nail Spacing |
|---|---|---|---|
| ⅝ in. (15.9 mm) plywood | 16 (400) | 2 in. (50.8 mm) (6d) threaded | 6 in. (150 mm) o.c. around panel edges; 10 in. (250 mm) o.c. at intermediate supports |
| ¾ in. (19.1 mm) plywood | 20 (500) | | |
| ⅞ in. (22.2 mm) plywood | 24 (600) | | |
| 1⅛ in. (28.6 mm) plywood (2-4-1) | 48 (1200) | 2½ in. (63.5 mm) (8d) threaded | 6 in. (150 mm) o.c. at panel edges and intermediate supports |

### Double-Floor Construction

| Subfloor[b] | Maximum Support Spacing, in. (mm) o.c. | Nails | Nail Spacing |
|---|---|---|---|
| ½ in. (12.7 mm) plywood | 16 (400) | 2 in. (50.8 mm) (6d) common | 6 in. (150 mm) o.c. around panel edges; 10 in. (250 mm) o.c. at intermediate supports |
| ⅝ in. (15.9 mm) plywood | 20 (500) | 2½ in. (63.5 mm) (8d) common | |
| ¾ in. (19.1 mm) plywood | 24 (600) | | |
| ¾ in. (19.1 mm) lumber boards less than 6 in. (152.4 mm) wide | 16 (400) | 2¼ in. (53.98 mm) (7d) threaded | 2 per board at each support |
| ¾ in. (19.1 mm) lumber boards wider than 6 in. (152.4 mm) and less than 8 in. (203.2 mm) wide | | | 3 per board at each support |

| | Fasteners[f] | | |
|---|---|---|---|
| Underlayment[b,e] | Nails[d] | Staples[g] | Fastener Spacing |
| ⅜ in. (9.5 mm) plywood | 1¼ in. (28.6 mm) (3d) threaded | 1⅛ in. (28.6 mm) | Nails: 8 in. (200 mm) o.c. each direction in middle of panel, 6 in. (150 mm) o.c. around panel perimeter |
| ½ in. (12.7 mm) plywood | | 1¼ in. (31.8 mm) | |
| ⅝ in. (15.9 mm) plywood | 1½ in. (38 mm) (4d) threaded | 1½ in. (38 mm) | Staples: 6 in. (150 mm) o.c. each direction in middle of panel, 3 in. (75 mm) o.c. around panel perimeter |
| ¾ in. (19.1 mm) plywood | | 1⅝ in. (41.3 mm) | |

[a]Based on 40 psf (195.3 kg/m$^2$) live load.

[b]Plywood thicknesses given are for Group 1 softwoods in accordance with U.S. Product Standard PS1. For Groups 2 and 3, increase thickness ⅛ in. (3.2 mm); for Group 4, increase thickness ¼ in. (6.4 mm).

[c]Blocking at longitudinal edges is required unless plywood with tongue-and-groove edges is used.

[d]Common nails may be substituted for annularly threaded nails by increasing their size ¼ in. (6.4 mm) (1d).

[e]Blocking not required if underlayment joints are staggered with respect to subfloor joints.

[f]Fastener lengths shown are minimum to provide ¾ in. (19.1 mm) penetration for nails and ⅝ in. (15.9 mm) for staples. If longer fasteners are used, penetration should not exceed 1 in. (25.4 mm).

[g]Galvanized divergent chisel, 16-gauge (1.6 mm), ⅜ in. (9.5 mm) crown width.

thickness cause cracking. Where an existing subfloor is not sufficiently smooth or does not have the desired profile, it should be leveled or pitched with a suitable mastic underlayment such as latex. Over wood strip flooring, these underlayments should be installed at least ⅜ in. (9.5 mm) thick using a 15 lb (8.14 kg/m$^2$) felt and expanded metal or wire lath secured to the subfloor at 6 in. (152.4 mm) on centers along the supports.

## UNDERBED

An underbed is necessary in unbonded installations and may be required in some bonded ones. An underbed can be relatively thin, because it is not a structural slab but functions as a surface preparation for a topping. The thickness of an underbed varies, depending on the installation method (see Figs. 9.12-5d and 9.12-5e).

The underbed should be reinforced with welded wire fabric in unbonded installations and in bonded installations over wood or metal subfloors. Except where the underbed is bonded, an isolation membrane should be spread over the subfloor, and metal reinforcement should be placed over the membrane. Except at expansion strips, the adjacent sheets of this reinforcing should overlap by at least two mesh spaces. No conduit or pipe should be located within the minimum recommended underbed thickness.

Underbed mortar should consist of 1 part portland cement, 4 to 5 parts sand, and sufficient water to make a workable mortar. The surface over which the underbed is placed should be sufficiently wet to prevent water from being drawn out of the mix. Isolation membranes need only be uniformly dampened; concrete slabs should be thoroughly saturated by overnight soaking.

Underbed mortar should be placed and screeded to a height that will permit installation of the selected topping thickness at the desired finish floor elevation.

## STRIP PLACEMENT

When an underbed is used, conventional divider strips (see Fig. 9.12-5) should be installed while the underbed is still semiplastic. Standard $1\frac{1}{4}$ or $1\frac{1}{2}$ in. (28.6 or 38 mm) strips should be set in the desired pattern, with the top of the strip at the finished floor elevation. The strips then should be troweled firmly along the edge to ensure adequate anchorage in the underbed.

In bonded installations without an underbed, where the terrazzo is applied directly to the subfloor, and in monolithic and chemically bonded installations, usually it is not possible to install strips while the concrete subfloor is still plastic because the subfloor often is installed by a different contractor. For such installations, angle or T-shaped strips (see Fig. 9.12-4) are secured to the concrete with adhesive or mechanical fasteners.

Divider strips should be provided wherever a discontinuity exists in the subfloor, such as at construction, isolation, and control joints and at changes in materials. Expansion strips may also be used in such locations if desired, and are specifically required at structural expansion joints where a greater degree of movement is anticipated. Divider or expansion strips should be used over beams and girders supporting the subfloor, because tensile stresses are likely to develop due to negative bending moments. The spacing of divider and expansion strips for various methods of installation is given in Figure 9.12-7.

***Divider Strips*** Divider strips are not required in resinous terrazzo installations for control of shrinkage cracking, but often are desirable for decorative purposes and for convenience in placing and fin-

**FIGURE 9.12-7** Strip Placement

| Matrix | Installation Method | Divider Strip Spacing[a] | Expansion Strip Spacing[a] |
|---|---|---|---|
| Portland cement | Unbonded | 6 ft (1800 mm) o.c. maximum | As needed[b] |
| | Bonded-underbed | | |
| | Chemically bonded | 15 ft to 20 ft (4500 to 6000 mm) o.c.— panels not over 300 sq ft[c] | 30 ft (9000 mm) o.c. maximum[c] |
| | Monolithic | | |
| Resinous | Thin-set | As desired for decorative or convenience purposes | 30 ft (9000 mm) o.c. maximum[d] |

[a]Locate divider strips or expansion strips on beams and girder lines wherever possible.

[b]Provide expansion strips over all expansion joints in the subfloor or at other joints where movement is expected.

[c]Divider strips and expansion strips can be alternated, or expansion strips can be used at 15 ft (4500 mm) o.c.

[d]Recommended for large areas or long corridors; otherwise, not required.

ishing the topping. In thin-set installations, strips ensure a uniform topping thickness and conserve material. They also minimize the appearance of waviness in large areas.

In portland cement terrazzo installations using an underbed, strips should be spaced not more than 6 ft (177.8 mm) on centers. When a portland cement topping is placed directly over a concrete slab, the slab should be provided with cut or tooled control joints 15 to 20 ft (4.57 to 6.1 m) on centers in each direction, forming panels of not more than 300 sq ft (27.87 m²).

The length of these panels should not exceed their width by more than 50%. Divider strips should be placed at all such control joints, forming similar panels in the topping (Fig. 9.12-8). In slabs on grade where a minimum of stripping is desired, control joints and divider strips can be placed in doorways and adjacent to partitions where they will be concealed by base trim (Fig. 9.12-9).

Divider strips should be continuous from wall to wall and should form right-angled intersections rather than acute angles. Where a terrazzo wall base is

**FIGURE 9.12-8** Control joints in monolithic and chemically bonded installations: (a–c) inconspicuous controlled cracking is encouraged adjacent to a strip; (d) uncontrolled cracking results from improper installation.

INTERIOR WALL FINISH
RESILIENT (OR OTHER) WALL BASE
DIVIDER STRIP OVER CONTROL JOINT
TERRAZZO
CONTROL JOINT

**WALL SECTION**

INTERIOR WALL FINISH
TYPICAL DOOR FRAME & CASING TRIM
DIVIDER STRIP UNDER DOOR

**PLAN AT DOOR FRAME**

**FIGURE 9.12-9** Strip placement in monolithic and chemically bonded terrazzo over a slab on grade when strips are not a design feature.

used, strips should be carried up to the base screed. When terrazzo forms a wainscot, strips should be continued from the base in a straight line up the wall. The recommended strip spacing for portland cement terrazzo wainscoting is 3 ft (914 mm) on centers maximum. Strip spacing for resinous wainscoting is not critical.

*Expansion Strips* Greater consideration should be given to the need for and location of divider and expansion strips in bonded systems because the terrazzo finish will tend to move with the subfloor. In monolithic, chemically bonded, and thin-set installations, expansion strips should be provided about 30 ft (914 mm) on centers where unobstructed large areas or long corridors are involved. In monolithic and chemically bonded installations, expansion and divider strips can be alternated 15 ft (4.57 m) on centers. Expansion strips are not required in such installations if the subfloor is continuous, homogeneous, and 1000 sq ft (92.9 m²) or smaller in area.

Unbonded and bonded-underbed installations will usually have enough di-

vider strips distributed over the floor so that expansion strips are not required except over expansion joints in the subfloor.

## TOPPING

This subsection covers portland cement terrazzo with a regular finish. See "Thin-Set Method" in Section 9.12.2.2 for resinous terrazzo and Section 9.12.2.4 for rustic and conductive terrazzo.

*Mixing* A power mortar mixer should be used to mix a carefully measured and consistently proportioned portland cement topping matrix. A conventional portland cement mix consists of 200 lb (90.72 kg) of marble chips for each 94 lb (42.64 kg) bag of cement. Pigment, when used, is measured by weight (not volume) and should be added to the dry mix. After thorough mixing of dry ingredients, water should be added at the ratio of 5 to 5½ gal (18.93 to 20.82 L) per bag of cement.

Resinous binder formulations vary among manufacturers. Therefore, their proprietary mixing and installation recommendations should be carefully followed.

*Placing* Installation of a topping over an underbed should be delayed for 12 to 24 hours to permit the underbed to harden. Immediately before a topping is placed, a slush coat of neat cement, pigmented to match the topping mix, is broomed into the underbed (Fig. 9.12-10). The various colored mixes with decorative chips are then placed according to the desired pattern. The topping is

**FIGURE 9.12-10** A slush coat of neat cement is broomed into the surface just before the topping is placed.

then troweled even with the tops of divider strips, which act as screeds for leveling the surface (Fig. 9.12-11).

To ensure even distribution and adequate chip content, the surface is seeded with additional chips of the type contained in the mix (Fig. 9.12-12). These chips are wetted, sprinkled over the surface, and embedded by additional troweling. Then the entire area is rolled with a heavy roller until all excess water has been brought to the surface (Fig. 9.12-13). Then the surface is again troweled smooth

**FIGURE 9.12-11** Topping is spread and troweled even with the divider strips.

**FIGURE 9.12-12** Additional chips are seeded and troweled into the surface where necessary for a uniform appearance.

**FIGURE 9.12-13** Rolling compacts the topping mix, brings up excess water, and creates a harder, more durable surface.

and uniform, revealing the tops of the divider strips.

Before finishing operations are started, a portland cement topping should be wet-cured for 3 to 7 days either by ponding, covering it with watertight covers, or spraying it with a curing compound.

***Preliminary Finishing*** After proper curing, the preliminary finishing operations of grinding, grouting, and curing are performed.

The surface is first ground using a terrazzo grinding machine with number 24 or finer abrasive stones, followed by a second grinding with number 80 or finer stones (Fig. 9.12-14). These grinding operations are performed while the surface is covered with water.

After grinding, the surface is thoroughly rinsed with water to remove dust and loose, fine particles. Then a skim coat of neat cement grout of the type used in the topping mix (including pigment to match the topping) is troweled over the entire surface, filling all voids.

The installation is then wet-cured again for at least 3 days using a watertight cover or curing compound. The compound or cover should be allowed to remain in place as protection against physical damage or staining until final finishing is performed.

***Final Finishing*** Final finishing includes polishing, cleaning, and sealing. A floor machine with number 80 or finer stones is used for polishing. The polishing operation will continue until the grout skim coat is removed (except in voids and depressions) to reveal the decorative chips in between 70% and 75% of the area.

A solution of neutral liquid cleaner is used to clean the polished floor. It is then

**FIGURE 9.12-14** After proper curing, the surface is flooded and wet-ground to reveal the decorative chips.

rinsed with clean water and allowed to dry. The cleaned, polished terrazzo surface is then treated with a sealing compound. These compounds are proprietary formulations and should be applied in accordance with their manufacturer's recommendations.

Until all work on the premises is completed, finished terrazzo should be protected with a covering, such as heavy building paper.

#### 9.12.2.2 Installation Methods

The following recommendations are in addition to those in Section 9.12.2.1.

##### UNBONDED METHOD

Unbonded installations require the use of a reinforced underbed to support the topping. The combined thickness of underbed and topping should be at least $2^1/2$ in. (38 mm) (see Fig. 9.12-5e). Separation between subfloor and underbed can be provided by an isolation membrane alone or by a membrane spread over a $1/16$- to $1/4$-in. (1.6- to 6.4-mm)-deep layer of sand. Over a smooth subfloor, a double layer of polyethylene film will provide isolation between surfaces.

Metal reinforcement should be placed over the membrane, lapped at joints at least two mesh spaces, and wired together. After some mortar has been poured in place, the reinforcement should be lifted to about 1 in. (25.4 mm) above the membrane. Underbed, strips, and topping should be installed as indicated in Section 9.12.2.1.

##### BONDED METHOD

Bonded installations have a smaller dead load and less finish thickness. Most bonded methods allow the topping to be placed directly over the subfloor without an intervening underbed. However, these methods do not generally develop the degree of crack control obtained when an underbed is used with conventional divider strips. Therefore, the bonded-underbed method (see Fig. 9.12-5d) often is used if the greater dead load and thickness can be accommodated. Since the difference in thickness between bonded-underbed and unbonded methods is only $3/4$ in. (19.1 mm) (9 lb/sq ft [43.94 kg/m$^2$]), the unbonded method should always be considered because it is most effective in reducing cracking.

In the monolithic, chemically bonded, and thin-set bonded methods, the rela-

tively thin topping by itself cannot resist the various stresses acting on it. An adequate bond between topping and subfloor is necessary so that stresses are resisted across the full depth of the floor construction. Therefore, special care should be taken to remove foreign substances, which might impair the bond.

Moreover, a topping should be of uniform thickness to prevent concentration of stresses and consequent cracking at thin spots. The subfloor must be level or pitched to conform with the desired finished floor profile.

***Bonded-Underbed Method*** The bonded-underbed method, which can be used over a metal or concrete subfloor, provides a measure of crack control for portland cement terrazzo not available in other bonded installations. The total recommended thickness of the underbed and topping is $1^3/4$ in. (44.5 mm) for bonded-underbed as compared to $2^1/2$ in. (63.5 mm) for unbonded installations.

The crack control advantage of this method is derived from the relatively frequent spacing of divider strips. The maximum spacing of 6 ft (1800 mm) on centers for conventional divider strips induces unnoticeable stress-relieving cracks along the strips. The minimum overall terrazzo thickness results from the normal strip depth ($1^1/4$ or $1^1/2$ in. [28.6 or 38 mm]) plus a $1/4$ in. (6.4 mm) tolerance for subfloor irregularity.

The underbed must be bonded to the subfloor. A natural bond will occur between a portland cement underbed and a suitably roughened concrete slab. A chemical bonding agent can be used over a smooth-troweled concrete slab to bond the underbed. Over a metal deck, an adequate bond can be expected without special treatment if the surface is free of corrosion, pickling acids, oil, and other foreign substances. An additional bond is gained by the keying action of the underbed with depressions in a cellular metal deck. Corrugated and ribbed metal decks generally depend on composite action of the deck and a reinforced structural concrete slab. In such floor assemblies, the top of the concrete slab constitutes the subfloor and topping should be installed by the monolithic or chemically bonded method.

Underbed and topping generally are installed as described in Section 9.12.2.1.

***Monolithic Method*** A monolithic bonded terrazzo finish is a portland cement topping installed over a concrete subfloor where a natural bond is developed between the portland cement surfaces. Monolithic installations can be used in interior as well as exterior locations.

The concrete subfloor should be reinforced properly and be at least $3^5/_8$ in. (92.1 mm) thick. The surface should be brushed with a wire broom or otherwise roughened to remove laitance and ensure a good mechanical bond with the topping. As soon as the concrete is sufficiently hard to resist tearing with a jointing tool or power saw, control joints should be installed in the slab about 15 to 20 ft (4.57 to 6.1 m) apart. Divider or expansion strips should be installed in the topping over these control joints in the subfloor. The surface should be protected from damage and dirt with suitable coverings until the topping is placed.

A subfloor that has not been adequately protected from dirt, paint, plaster, or other construction residue should be thoroughly scrubbed with detergents and rinsed clean. Then the subfloor should be saturated with water by ponding overnight. Excess water should be removed, and a slush coat of neat portland cement, pigmented to match the topping, should be broomed into the surface immediately before the topping mix is applied. The topping should be placed, finished, and cured as described in Section 9.12.2.1.

*Chemically Bonded Method* This method of installing portland cement topping is intended for smooth-troweled concrete subfloors, and is often suitable for older concrete floors not originally prepared for a terrazzo topping. Where a terrazzo topping is planned in the initial design, a roughened subfloor surface is preferred.

The subfloor should be inspected, and where random cracks exist or control joints are not provided 15 to 20 ft (4.57 to 6.1 m) on centers, such joints should be installed with a power saw. Divider strips or expansion strips should be placed over all joints in the subfloor.

Chemically bonded installations depend on a bonding agent such as epoxy, epoxy polysulfide, or PVC resin. The subfloor surface should be free of coatings such as paint, curing compounds, and wax, and the bonding agent should be applied in accordance with its manufacturer's recommendations.

The topping should be placed, finished, and cured as described in Section 9.12.2.1.

*Thin-Set Method* Using a resinous binder, it is possible to install very thin toppings over a sound, firm subfloor. Epoxy and polyester toppings should be at least $^1/_4$ in. (6.4 mm) thick. When they are installed over concrete, the subfloor surface should be smooth-troweled and level to within $^1/_8$ in. (3.2 mm) in every 10 ft (3000 mm).

Although some manufacturers allow resinous toppings $^1/_8$ in. (3.2 mm) thinner than those recommended above, the industry consensus favors the slightly heavier installations. This is particularly true of epoxy and polyester, where a $^1/_8$ in. (3.2 mm) thickness would require decorative aggregate the size of sand and an acceptable minimum thickness over subfloor irregularities would be difficult to achieve.

Thin-set terrazzo finishes should not be installed over slab-on-grade construction subject to excessive moisture or with a high alkaline content. To test for moisture, a piece of polyethylene film about 2 ft (1290 mm) square should be taped down in several locations and allowed to remain for 8 hours. Visible moisture accumulation under the mat is an indication of excessive moisture in or under the slab. Under such conditions, the alkalinity of the moisture also should be tested and the thin-set matrix manufacturer should be consulted.

Divider strips are not required with resinous binders for shrinkage control, but should be provided where there is a discontinuity in the subfloor, such as at isolation joints, control joints, and changes in materials. Divider strips may be provided for decorative purposes or as construction joints at the termination of a pour. Since wall base often is installed as a separate operation, a strip should be provided at the toe of the base. In large areas or long corridors, expansion strips should be installed at 30 ft (9.5 m) intervals. Closer spacing of expansion or divider strips may be preferred for decorative purposes, to minimize the illusion of waviness, or to ensure uniform topping thickness.

The thin-set matrix manufacturer's recommendations should be followed with respect to the use of primers, mixing of ingredients, and application of the topping mix, as well as for grinding and sealing of the surface. Over wood sub-

floors, the terrazzo installer's specific recommendations should be consulted for details of taping plywood joints and for use of mastic underlayment and wire or metal reinforcement.

Generally, the dry components such as fillers and chips are mixed first, then added to the liquid resin. The mix is poured onto the subfloor, then spread and leveled with a trowel. The setting time before grinding depends on the specific resin formulations. Epoxy and polyester toppings can be ground the day after application.

In residences, sealing of the surface is often not required, because the resinous matrixes are quite dense and resistant to most staining agents. Marble chips, however, vary in their stain resistance and, depending on anticipated traffic conditions, a sealer may be desirable. The matrix manufacturer's recommendations should be consulted in each case.

## VIBRATION

Traditional terrazzo installations can develop surface imperfections due to portland cement shrinkage, and the appearance of the finished floor sometimes suffers from uneven chip distribution. To overcome these problems, the Terrazzo Industry Research Institute, a branch of NTMA, has developed a special vibrating machine which improves the performance and appearance of most terrazzo floor installations, including rustic and polyacrylate thin-set ones.

### 9.12.2.3 Special Surface Finishes

In addition to other characteristics discussed earlier, terrazzo floors can be described according to their surface finish (regular, rustic, or conductive) (Fig. 9.12-15).

Terrazzo in homes, apartments, offices, stores, building lobbies, and other locations of normal foot traffic requires no special additives or surface treatment. The usual binder formulations, after normal grinding and polishing, provide adequate resistance to abrasion, staining, discoloration, and slipping hazard.

Special surface finishes have been developed for situations where unusual hazards exist. In exterior locations, a greater degree of protection from slipping hazard is afforded by *rustic terrazzo*. In hospitals and laboratories where oxygen is used and potentially explosive atmospheres exist, *conductive terrazzo* will prevent electrostatic sparks. For locations

**FIGURE 9.12-15** Surface Finishes for Various Installation Methods

| Installation Method | Binder | Chip Size | Surface Finish |
|---|---|---|---|
| Thin-set | Polyester and epoxy | #0 and #1 | Regular |
| | Latex | Standard | Conductive |
| Monolithic | Portland cement | Standard | Regular |
| Chemically bonded | | Intermediate Venetian | Rustic |
| Bonded-underbed | Portland cement | Standard | Regular |
| Unbonded | | Intermediate Venetian | Rustic Conductive |

subject to damage from acids, such as hospitals, laboratories, and industrial buildings, many resinous matrixes are suitable. Where greater slip resistance than that provided by regular terrazzo is required for interior locations, as on ramps and stairs, regularly spaced abrasive strips are recommended (Fig. 9.12-16). Some resinous matrixes can be formulated to provide inherently good slip-resistant properties. The regular and special surface finishes can be used with most installation methods (see Fig. 9.12-15).

### RUSTIC TERRAZZO

Rustic (washed) terrazzo is suitable for nonslip and decorative exterior surfaces. The decorative chips are either a weather-resistant marble, quartz, or granite chips, or decorative gravel. Although resinous

**FIGURE 9.12-16** Abrasive inserts for nonslip surfaces: (a) prefabricated strips and nosing, (b) abrasive compound poured into a channel.

binders can be used, they generally are limited to wall installations. Portland cement is the common material for rustic terrazzo floors, and in areas subject to freezing, an air-entraining agent is added. Rustic terrazzo should not be installed where the surface will be subjected to de-icing salts or chemicals.

After the topping has been placed, rolled, and leveled as described in Section 9.12.2.1, the surface is washed with sufficient water pressure to remove the paste between chips or gravel to a depth of about $\frac{1}{16}$ in. (1.6 mm). The surface is then scrubbed with a solution of 1 part muriatic acid to 10 parts water to remove particles of paste clinging to the aggregate. A final rinsing with clean water removes the acid solution and neutralizes the surface. Light grinding may be desirable for exposed chips in pool areas, and a floor may be sealed, but usually no further operations are necessary.

### CONDUCTIVE FLOORS

Conductive floors prevent electrostatic sparks in potentially explosive atmospheres. They consist of resinous matrixes that have a conductive ingredient, usually carbon black, added during manufacture. Chip sizes in conductive terrazzo should be no larger than #1 size.

#### 9.12.2.4 Special Elements

Cast-in-place terrazzo can be carried up a wall 4 to 12 in. (100 to 300 mm) to form a coved wall base (Fig. 9.12-17) or can be continued to form a protective wainscot (Fig. 9.12-18). Full-height terrazzo finishes are not practical because of the labor involved in grinding vertical surfaces with relatively small, hand-held wheels (Fig. 9.12-19). Cast-in-place terrazzo stairs often are practical and economical (Fig. 9.12-20). Precast treads, risers, and wall base (Fig. 9.12-21) are available.

#### 9.12.2.5 Protection

Portland cement matrixes are porous and require protection from deterioration and staining from various chemical agents. Therefore, portland cement terrazzo requires protection with a proprietary pen-

**FIGURE 9.12-17** Coved wall base: (a) straight type and (b) splayed type. The underbed thickness may be varied, creating a recessed, flush (shown), or projecting base.

**FIGURE 9.12-18** Wainscoting can be installed over studs and metal lath (shown) or over any masonry base that will provide an adequate bond for portland cement mortar.

**FIGURE 9.12-19** Base grinding: small hand-held wheels are used to grind and polish terrazzo base and wainscoting.

**FIGURE 9.12-20** Cast-in-place stairs: (a) continuous treads and risers over concrete substair, (b) treads cast in steel pan forms.

**FIGURE 9.12-21** Precast stair treads and risers can be used with (a) concrete or (b and c) metal stair construction.

etrating sealer capable of internally sealing the pores. Resinous matrixes are denser, more stain-resistant, and chemically inert. However, all terrazzo finishes consist of at least 70% stone or marble, which possess varying degrees of porosity and stain resistance. Resinous terrazzo may require similar protection, depending on the chip type and conditions of use. Sealers should be of a type specifically recommended by the manufacturer for terrazzo surfaces and should be applied in accordance with the manufacturer's recommendations.

Surface coatings such as wax, which provide abrasion protection for softer floor finishes, are not required. Foot traffic on unwaxed terrazzo tends to polish and impart further luster to the surface.

In addition, surface coatings tend to make the floor slippery when wet and to discolor the surface in low-traffic areas.

# 9.13 Fluid-Applied Flooring

There are many fluid-applied seamless resinous flooring products on the market for use in laboratories, food processing facilities, pharmaceutical plants, chemical processing plants, manufacturing and other industrial facilities, high-tech manufacturing plants, chemical processing facilities, electronics facilities, trans-

portation facilities, food and beverage production facilities, light manufacturing and textile plants, warehouses, clean rooms, commercial kitchens, food stores, medical facilities, parking and bridge decks, stadiums, wastewater treatment plants, automobile facilities, and other locations where sanitary, dustless, high-performance flooring is needed (Fig. 9.13-1). Some of these products can be applied when the temperature is less than 24°F (−4°C), which makes them ideal for coolers, freezers, and cold weather applications. Resinous flooring systems may have an integral cove base.

## 9.13.1 SYSTEMS

Fluid-applied flooring systems range from thin-film dustproofing sealers to heavy-duty, trowel-applied systems that provide flexible installation options, easy maintenance, and long service life. Antimicrobial, thermal shock-resistant formulations, and low-temperature curing acrylics enhance their performance. Thin-film products range in thickness from 4 to 25 mils (0.12 to 0.63 mm); thick-film products are generally 60 mils (1.52 mm), but some epoxy thick-film coatings are thinner. Many fluid-applied flooring products are USDA approved with 100% solids and zero volatile organic compounds (VOCs).

The acrylic reactive resins in fluid-applied flooring chemically bond to each other. Fresh resin dissolves the surface of cured material. The redissolved material cures with the fresh material. The need for surface preparation between layers is eliminated. After years of service, resurfacing or additions without cold joints are simple and fast. This welded bond allows different resinous materials to be combined in monolithic systems for specific performance requirements.

Fluid-applied resinous flooring is usually pigmented in beige, red, green, medium gray, concrete gray, or charcoal gray, but many other colors are possible. Small quantities of vinyl chips in contrasting colors can be randomly broadcast onto fresh resinous flooring for a decorative appearance. A wide variety of premixed quartz color blends are available.

Fluid-applied flooring systems are available in many different forms. These include:

- General service systems having excellent chemical, impact, and abrasion resistance.

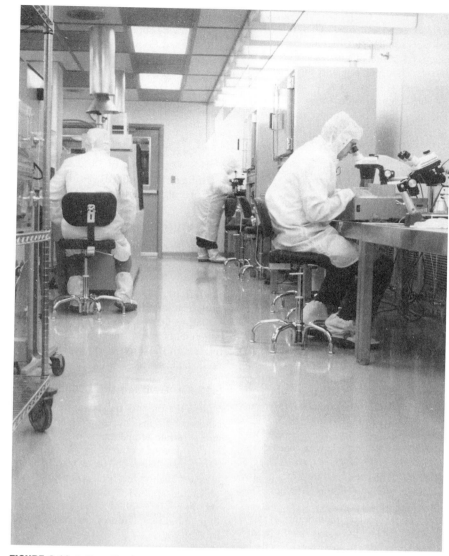

**FIGURE 9.13-1** Free-flowing static dissipative polymer floor system in a clean room. (Photo courtesy of Stonhard, Inc.)

cast and troweled formulations. Their nonporous, monolithic surface makes them ideal for clean rooms, medical, pharmaceutical, commercial, and institutional facilities, restaurants, and kitchens.

- Self-leveling and colored quartz systems for fast-cure, high-strength, monolithic overlays and toppings for industrial and commercial flooring. These systems are suited for heavy-traffic industrial environments where superior toughness, wear, and flexibility are required. Industrial flooring is well suited for heavy metal-working and machining, paper and pulp, high-volume food processing, and other general manufacturing operations that need quick flooring replacement.
- Colored, very smooth, self-leveling overlay products that combine a primer, a matrix layer of filler-enhanced resin, and a standard or chemical-resistant topcoat. Some such systems have an antimicrobial additive that does not support bacterial growth. These products offer exceptional chemical, stain and impact resistance. Rapid curing ensures short application down times.

### 9.13.2 MATERIALS

Fluid-applied flooring products include latex mastic, polyester, urethane, vinyl ester, methylmethacrylate, polyurethane, and epoxy. The last three are discussed in this section. The others are beyond the scope of this book.

### 9.13.2.1 Methylmethyacrylate

Methylmethacrylate is a transparent resin to which pigment, aggregate, and fillers are added on site. Methylmethacrylate floors exhibit excellent resistance to moisture and chemicals, impact abrasion, and thermal shock. They can be placed on concrete, asphalt, and wood. Colored quartz crystals are added to them as an aggregate to produce colored quartz products. A solvent-free, 100% reactive acrylic resin may also be based on methylmethacrylate. Polymerization (cure) at the job site is effected by adding a hardener and a peroxide initiator. These products feature skid resistance, easy cleaning, and USDA/FDA approval for federally inspected food and beverage facilities.

Self-leveling, pigmented products having 100% solid methylmethacrylates provide easy maintenance and high dura-

- Power-trowelable systems for reduced installation time in larger open areas.
- Static-dissipative and conductive formulations made to meet exacting electrical specifications. These are designed for electrical manufacturing and assembly areas, computer/control rooms, and clean environments.
- High-build, protective resurfacing systems. These have epoxy or silica aggregate mortar, high-build grout, and seal coats.
- Self-leveling formulations in smooth or skid-inhibiting systems, in an easily applied and durable self-leveling slurry designed to withstand moderate to heavy-traffic, high-impact, and thermal conditions.

- Waterproofing products formulated to eliminate water migration through concrete slabs on parking decks and ramps and mechanical equipment rooms.
- Resurfacers and slurry floor systems designed to restore old concrete or to provide a protective surface for new concrete in heavy-traffic areas. These are highly abrasion, impact and chemical resistant. Their excellent adhesion properties make them suitable for use on concrete, brick, quarry tile, and metal surfaces.
- Blended decorative quartz systems that are compacted tightly with a nonyellowing binder and top coat. These systems are available in broad-

bility for heavy-traffic floors. Their satin finish maintains its appearance for years and features superior stain, chemical, and slip resistance.

#### 9.1.3.2.2 Polyurethane

Polyurethane floors exhibit excellent wear, ultraviolet light resistance, and stain and chemical resistance. Colored quartz can be added to make colored quartz products. Vinyl chips can also be added for color. Formulations are available that offer enhanced chemical resistance. Polyurethane formulations are available in both solvent- and water-based systems.

Trowel-applied, textured polyurethane mortar system are designed to resist thermal shock/cycling conditions. These are formulated especially for food processing applications.

Electrostatic discharge polyurethane systems are available.

#### 9.13.2.3 Epoxy

Epoxy flooring is available in both solvent- and water-based systems. Its appearance ranges from smooth, high-gloss finishes to orange-peel finishes. Decorative quartz aggregates can be added for color, as can vinyl chips.

Epoxy flooring offers excellent resistance against many chemicals and is easy to clean. These coatings are recommended for assembly and light manufacturing areas, aircraft hangars, food processing areas, laboratories, clean rooms, pedestrian walkways, warehouses, and similar industrial/commercial areas.

Epoxy flooring comes in both thin-film and high-build systems. High-build systems fill in small imperfections in a floor's surface.

Electrostatic discharge control epoxy systems are available in both thin-film and thick-film formulations.

Available epoxy systems include high-chemical-resistant formulations with high organic/inorganic acid resistance.

Solvent-free, flexible epoxy waterproofing membranes are used as a crack isolation system under decorative flooring or as a stand-alone coating system for mechanical rooms and high-traffic areas.

There are polymer epoxy industrial flooring systems that combine resins with hard aggregate to provide a tough traffic-bearing flooring. In these systems, flexibility is achieved without plasticizers or other additives, which can separate or migrate as the system ages. Therefore, these products remain flexible and continue to function for many years. Fiberglass scrim may be incorporated into these systems for added tensile strength. Hard aggregate provides abrasion, impact resistance, and skid inhibition properties. Various topcoats can be specified to provide protection against water, oils, chemicals, and ultraviolet light.

In addition, the following systems are available:

- High-gloss epoxy coating systems with 100% solids.
- High-build, low-odor epoxy coatings having 100% solids, either clear or pigmented.
- High-build, chemical-resistant, low-odor epoxy coatings having 100% solids, either clear or pigmented. These coatings offer excellent overall resistance to a wide spectrum of chemicals, including aromatic and aliphatic hydrocarbons, acids, alkalis, alcohols, detergents, salts, cutting oils, gasoline, jet fuels, and many solvents.
- High-build, quick-release, low-odor epoxy coatings having 100% solids, which allow for easy removal of buildups of rubber, wax, mastic, and so on.
- High-build, reduced-yellowing, low-odor epoxy coatings having 100% solids.
- Flexible epoxy membrane systems. These specially formulated, flexible epoxy coating systems are ideal for applications requiring protection against thermal shock and areas that use wet processes.
- Two-component, solvent-borne polymer epoxy primer sealers.
- Two-component, waterborne polymer epoxy primer sealers.
- Two-component, 100% solids polymer epoxy coatings.
- Multiple-component, seamless polymer epoxy flooring systems.
- Trowel-applied polymer epoxy repair mortar.
- High-performance polymer epoxy topcoat products

#### 9.13.2.4 Limitations

Most resin systems have very low vapor permeability and act as vapor barriers. Therefore, substrates must be protected against rising moisture.

#### 9.13.2.5 Applicable Standards

The following standards are applicable to fluid-applied flooring:

- American Concrete Institute (ACI): 503.1, 503.2, 503.3, 503.4, 503.5R, 503.6R
- ASTM International (ASTM): C109, C307, C501, C579, C580, C696, D570, D635, D638, D695, D790, D2047, D2240, D4541
- U.S. Department of Defense (DOD): Military Standard MIL-D-3134, "Deck Covering Materials"

### 9.13.3 INSTALLATION

Each fluid-applied flooring and ancillary products require a different installation procedure. The following should be considered as generic, and generally applicable, but the flooring manufacturer's installation instructions should be strictly adhered to for environmental restrictions, mixing, installation methods, and cure times and methods.

#### 9.13.3.1 Environmental Conditions

Unless the product manufacturer requires other conditions, the minimum ambient and substrate temperatures must be 50°F (10°C) before commencing installation, during installation, and for at least 72 hours after installation has been completed.

Liquid-applied flooring systems should not be applied when the substrate or material temperatures are above 104°F (40°C).

#### 3.13.3.2 Precautions

Liquid-applied flooring resin components are flammable, with a 50°F (10°C) flashpoint, which is about the same as that of alcohol. Current regulations controlling flammable liquids must be observed—for example, posting No Smoking signs and taking other precautions. Minutes after placement, the flooring material will no longer support combustion.

Resins can present a minimal dermatological hazard to allergic individuals. Workers should wear protective clothing such as, but not limited to, rubber boots, gloves, and splashproof goggles. Cured materials are completely nontoxic and safe. For further information, consult the manufacturer's Material Safety Data Sheets (MSDS).

Adequate ventilation must be provided, especially in closed areas. For application in food processing plants and slaughterhouses, the odor of uncured resin requires special ventilation or removal of foodstuffs during application.

### 9.13.3.3 Installer's Qualifications

Some, but not all, manufacturers of resinous flooring products permit their products to be installed only by their authorized representative or by a contractor acceptable to and certified in writing by the manufacturers.

### 9.13.3.4 Preparatory Work

Bond tests, as directed by the manufacturer of the flooring, should be conducted to determine the adequacy of surface preparation and bonding.

### SURFACE PREPARATION

The substrate must be clean, durable, level, dry, and free of grease, wax, solvents, curing membranes, oil, and other contaminants. Proper surface preparation requires the actions described in the following subsections.

*Decontamination* Decontamination is the removal of oils, grease, wax, and other hydrocarbon-based materials. It may require the use of chemical cleaning, microbial decontaminants, high-pressure water, steam cleaning, scrubbing with an industrial-grade degreasing compound or nonionic detergent solution, and mechanical cleaning.

Surfaces to which industrial cleaners or acids have been applied must be scrubbed and flushed with copious amounts of potable water. A moist litmus paper reading should indicate a pH in excess of 10. Material that cannot be chemically removed must be removed mechanically.

*Creation of Surface Profile* The surface profile of horizontal concrete subjected to traffic should vary less than 5 mils (0.254 mm).

*Repair of Surface Irregularities* Surface irregularities, including spalls, cracks, deteriorated joints, and the like, must be repaired prior to placement of the resinous flooring, or the system must be designed to offset the thickness irregularities.

*Removal and Replacement of Nondurable Concrete* Unsound concrete and laitance must be removed by acid

etching or appropriate mechanical means. Nondurable, weak, or deteriorated concrete must be removed and new concrete installed before a resinous flooring is installed. Depressions, cracks, holes, and enlarged joints should be filled with a material recommended by the flooring manufacturer, usually epoxy.

*Methods of Surface Preparation* The preferred methods of removing dirt, dust, laitance, and curing compounds from concrete slabs are vacuum-grit blasting, sand or grit blasting, or mechanical scarification. Other surfaces, such as cove base (floor and wall intersection) and wall surfaces, should be grit-blasted or prepared with other mechanical abrasive techniques.

Fluid-applied flooring systems can be applied to a variety of substrates, if the substrate is properly prepared according to the flooring manufacturer's directions. Preparation of surfaces other than concrete will depend on the type of substrate.

Fluid-applied flooring is usually installed directly over nonmoving control joints and cracks that have been treated with the manufacturer's standard crack treatment materials, such as 100% solids epoxy. It should not be installed across isolation or expansion joints.

### PRIMING

Most fluid-applied flooring systems require priming of the substrate with a primer supplied by the flooring manufacturer. Low-viscosity primers penetrate, seal, and reinforce the substrate, providing a uniform acrylic surface to ensure a deep, strong bond of the topping to the substrate.

### 9.13.3.5 Installation

#### MIXING

Flooring resins are mixed at the job site with the appropriate quantities of colored quartz sands or pigment and fillers. A hardener is added just before application of the flooring. Mixing is done in small batches with plastic buckets and electric drills with special mixing blades. The pot life of mixed materials is usually 10 to 15 minutes at 68°F (20°C) but may vary with different products.

#### APPLICATION

After mixing, the flooring material is spread over a properly leveled and

primed area, then lightly troweled or gauge-raked to an even thickness dictated by the manufacturer. After troweling, a special roller is used to release trapped air from the topping. For increased skid resistance, a #16 to #25 mesh, clean, dried, angular, quartz sand can be broadcast into the fresh flooring mix. When random scattering of colored vinyl chips is desired, they are also broadcast into the fresh flooring. The chips should rain upon, not be thrown into, the surface.

### 9.13.3.6 Protection

Fluid-applied floor systems should be protected from damage and wear during other phases of the construction operation, using temporary coverings as recommended by the manufacturer if required.

### 9.13.3.7 Maintenance

#### CLEANING

Cleaning materials and procedures should be acceptable to the system manufacturer. In general, resinous flooring can be steam-cleaned or cleaned using conventional industrial cleaners and wax, but solvent-containing materials should not be used. Mineral spirits may be used to remove rubber marks. Some cleaners will affect the color, gloss, or texture of resinous floor surfaces. Cleaners should be tested in a small, inconspicuous area before they are used.

#### REPAIR

Chemical bonding allows fast, easy resurfacing when this is necessary.

## 9.14 Carpeting

The original definition of *carpet* was a rug with a repeating pattern. Wall-to-wall broadloom carpet was developed in the United States, but the art of carpet making has been passed down for many centuries from many cultures. The immediate roots of the American carpet industry are in Western Europe.

The Moorish invasion of Spain in A.D. 711 introduced the carpet craft to Europe. The patterns of Spanish carpet intertwined the living forms of humans, animals, or plants and the symbols of church and state with rich arabesque ornamentation.

When Queen Eleanor of Castille married Edward I of England in 1251, she took many carpets with her to England and thereby introduced the art of carpet making to the English.

Carpet making excelled in France, particularly during the reign of Louis XIV. The patterns of French carpets were not borrowed from other cultures. They usually featured floral designs complemented with elaborate scrolls and noble insignias. The secrets of French carpet making were kept primarily in France until 1685, when Louis XIV revoked the Edict of Nantes. This caused many Protestant carpet masters to flee to England and build the first hand looms there.

The English changed carpet making from a craft to an industry. Previously, carpets, like many other works of art, were primarily commissioned by nobles.

As the wealth of Great Britain increased during the eighteenth century, so did the carpet industry. For the first time, classes other than the nobility were able to purchase floor coverings.

English inventions during the Industrial Revolution formed the basis for present-day methods of carpet manufacture. Like their needlework, the English incorporated their passion for naturally arranged flowers into their carpets.

The carpet industry in North America began to grow after the American Revolution. The first power loom was invented for weaving carpets in Lowell, Massachusetts, in 1839. For the first time, carpets could be mass-produced. To simplify the process of mass production, carpet makers made carpet without patterns. Carpet was sold primarily for function. However, this monotone carpet was merely a step in the evolution of carpeting.

Today, carpet making has advanced to the point where carpet makers can reincorporate patterns into affordable carpet. Just as European nobles once commissioned carpets to display their insignias, many businesses today order pattern carpet to display their logos. Other businesses order patterned carpet to enhance the esthetic appeal of their workplaces.

Carpet is often selected for its comfort and decorative values, but it is also effective in reducing impact sound transmission through floor and ceiling assemblies. Therefore, its acoustical performance should be a prime consideration in flooring selection. Many common installations will provide a performance considerably higher than the range of $-5$ to $+10$ on the Impact Noise Reduction (INR) scale (see Section 13.1). Specially selected combinations of carpet and cushioning may be rated as high as $+17$ over wood floors and $+29$ over concrete slabs.

Carpet is also capable of absorbing sound and reducing sound reflection within a room, much like acoustical ceiling tile. The Noise Reduction Coefficient (NRC) of most carpet ranges between 0.35 and 0.55, which compares favorably with a range of 0.55 to 0.75 for acoustical ceiling tile. Properly selected carpet may achieve absorption coefficients equal to those of acoustical tile. However, carpet is not particularly effective in controlling the transmission of airborne sound, and where this is an important consideration, special sound-isolating construction should be used.

This section describes carpet suitable for long-term wall-to-wall installation, as distinguished from rugs, which usually have bound edges and are laid loose over finished flooring (Fig. 9.14-1).

Carpet can be identified by its type, which defines how it is made, by the fibers used to make it, and by its face construction.

## 9.14.1 CARPET CONSTRUCTION

Most carpet consists of pile yarns, which form the wearing surface, and backing yarns, which interlock the pile yarns and hold them in place. Therefore, carpet can be identified according to its construction, which describes the method of interlocking the backing and pile yarns. Comparative factors related to carpet construction, such as pile yarn weight and pile thickness, number of tufts per square inch, and other factors, may be useful in comparing carpet of similar construction.

Its face construction is another factor by which carpet can be identified. The following terms describing face style are defined in the Glossary: *level-loop pile, cut pile, level tip shear, multilevel loop, random shear, frieze* or *twist,* and *sculptured* or *carved* (Fig. 9.14-2). Some face styles are dictated by the method of manufacture. For example, fusion-bonded goods can only be cut pile. Similar restrictions are identified in the following discussion.

### 9.14.1.1 Carpet Types

The term *broadloom* is often mistakenly thought to define a type of carpet construction, but actually it refers merely to carpet wider than 6 ft (1.83 m). With the exception of rubber-backed loomed carpet and carpet tiles, the carpet types discussed in this section are manufactured in widths of 9, 12, 15 and sometimes 18 ft (2.74, 3.66, 4.57, and sometimes 5.49 m). Loomed carpet generally is made in 4 ft 6 in. (1.37 m) widths, with $^3/_{16}$ in. (4.88 mm) attached rubber cushioning.

**FIGURE 9.14-1** Carpet goes from wall to wall; rugs do not. (Courtesy Carpet and Rug Institute.)

a

b

c

d

*Weft Yarn*

a

*Stuffer Yarn*

b

*Chain Yarn*

c

**FIGURE 9.14-2** Some carpet texture variations: (a) multilevel loop, (b) twist, (c) single-level looped pile, (d) random shear, combining cut and loop pile. (Courtesy Carpet and Rug Institute.)

**FIGURE 9.14-3** Backing yarns of woven carpet identified by heavy lines: (a) also called *weft shots* or *filling*; (b) also called *stuffer warp*; (c) also called *chain warp* or *binder* or *binder warp*. (Courtesy Carpet and Rug Institute.)

Carpet can be classified according to the method of manufacture as woven, tufted, knitted, fusion-bonded, needle-punched, and flocked. Of these, tufted and woven are the most commonly used methods for making yard goods and fusion-bonded is the most prevalent method for making carpet tile. In general, weaving and knitting interlock pile and backing yarns in a single operation, while a tufting loom stitches pile yarns onto a premade backing. Tufted carpets can be produced much faster than woven or knitted types, are less expensive to produce, and currently account for about 95% of all carpet manufactured in this country.

## WOVEN CARPET

Woven carpet is made on looms that interweave pile yarns and backing yarns in one operation. Backing yarns consist of *weft* (crosswise) yarns and *warp* (lengthwise) yarns. Because the tuft-forming yarns in woven carpet run lengthwise, they too are classified as either *face warp* or *pile warp*. Though more properly applied to woven carpet, the terms *warp* and *weft* are loosely applied to knitted carpet as well.

Weft yarns run crosswise, passing over the pile yarns (Fig. 9.14-3a). *Stuffer yarns*, a type of warp yarn, give strength and dimensional stability to carpet (Fig.

9.14-3b). Typically, there are two or three stuffer yarns per tuft running the length of the carpet. *Chain yarns*, another type of warp, run the length of the carpet, alternately passing over and under the weft yarns, forming a chain which locks the weft in place and pulls the pile yarns down tightly (Fig. 9.14-3c).

Four weaving processes are used from which woven carpets derive their names: *velvet*, *Wilton*, *Axminster*, and *loomed*.

*Velvet Weave* A velvet weave is the simplest of all weaves (Fig. 9.14-4). Chain, stuffer, and pile yarns are alternately raised and bound together with the weft yarns. The pile is formed by *wires* inserted between pile and warp yarns. The height of the wire determines the depth of the pile. After a row of loops has been formed over the wire and secured by the weft, the wire is withdrawn, leaving a row of loops. Cut pile is created by a razor edge on the wire which cuts the loops as the wire is withdrawn.

Because a velvet loom feeds pile yarn from a single spool, it is best suited for solid colors. However, strands of different color can be twisted together into a single yarn (moresque), producing tweeds, salt-and-pepper effects, and stripes. A wide range of textural effects is possible by using uneven wires forming high and low loops.

**FIGURE 9.14-4** Velvet weave with double chain warp and typically two or three stuffer yarns per pitch. (Courtesy Carpet and Rug Institute.)

Velvet weave can be recognized at a cut edge or open selvage where the chain warp is visible, binding all the construction yarns together in a zigzag pattern (see Fig. 9.14-4).

*Wilton Weave* A Wilton loom (Fig. 9.14-5) is basically a velvet loom fitted with a special *Jacquard* mechanism that uses punched cards to select the various colored yarns. These cards, locked together to match the entire surface design of the carpet, are suspended in the loom. Colored pile yarns are fed from spools

**FIGURE 9.14-5** A Wilton loom. The punched cards are at the upper left. (Courtesy Carpet and Rug Institute.)

in velvet weaves, the chain warp also is visible at a cut edge.

***Axminster Weave*** Axminster carpet is woven on a loom which draws pile yarns from small spools wound with yarn of various colors. The spools are locked together in a frame equal to the width of the carpet. Since each frame provides one row of tufts across the carpet and each strand of yarn provides a single pile tuft, the sequence of colors determines the finished surface pattern.

An Axminster loom simulates hand weaving because each tuft of yarn is individually inserted, and theoretically each tuft could be of a different color. This flexibility offers unlimited design possibilities, and intricate patterns are produced even in lower-priced carpet. Considerable delicacy and subtlety of design are possible in the higher-priced carpets, where the rows of tufts are closer together.

The Axminster weave is almost always a cut pile with an even pile height and elaborate patterning with many colors. The backing is so heavily ridged and *sized* (coated) that it usually can be rolled easily only lengthwise. The characteristic chain warp can be seen along a cut edge, as in other woven carpets (Fig. 9.14-7).

**FIGURE 9.14-6** Wilton weave with four frames (colors), double chain warp, two weft shots per wire, and stuffer yarns. (Courtesy Carpet and Rug Institute.)

**FIGURE 9.14-7** Axminster weave with single chain warp, three double shots per row, and stuffer yarns. (Courtesy Carpet and Rug Institute.)

assembled on frames (racks) in the back of the loom, one frame for each color. As the cards pass, the Jacquard mechanism selects and lifts the appropriate pile yarns into position to be looped over the wires, while other yarns remain buried in the body of the carpet. Depending on the number of colors used, for each tuft of yarn showing on the surface there may be as many as four strands of yarn buried in the backing, adding to the stiffness of the carpet (Fig. 9.14-6).

A carpet woven on a Wilton loom with just one colored pile yarn would have the same construction as a velvet weave and would be designated as a velvet carpet. Wilton carpets are designated as two-, three-, four-, or five-frame Wiltons, depending on the number of colored pile yarns used in their manufacture. Generally speaking, for carpets of the same density and yarn size, the more frames used, the greater the pile yarn weight and hence the better the quality.

A Wilton loom can weave sharply delineated sculptured and embossed textures or patterns, which are created by varying the pile height, using high and low loops or combinations of cut and uncut piles. Wilton carpets with uncut loop pile are called *roundwire* Wilton.

In multicolor Wilton weaves, the different colors of pile fibers are usually visible on the back, and the body of the carpet is thicker than that of other types, As

**Loomed Weave** Loomed carpet is intended specifically for use with a bonded rubber cushioning. It is characteristically a high-density nylon with single-level, low-loop pile about $1/8$ in. (3.2 mm) high woven on a modified upholstery loom (Fig. 9.14-8). The pile back is treated with a waterproof adhesive which serves as a bond coat for the rubber backing. Sponge or foam rubber is applied to the pile fabric, and the backed carpet is heated to cure the rubber. Loomed carpet generally is manufactured in a 4 ft 6 in. (1371 mm) width with a rubber backing about $3/16$ in. (4.8 mm) thick.

## TUFTED CARPET

Tufted carpet is manufactured by inserting tufts of pile into a premade backing by a machine which resembles a huge sewing machine with thousands of needles operating simultaneously (Fig. 9.14-9). The prime backing is made from jute, kraftcord, cotton, or polypropylene olefin and is coated with a heavy layer of latex to hold the tufts permanently in place (9.14-10). On most tufted carpets, a loosely woven, coarse jute fabric (*scrim*) is added as secondary backing to lend additional strength and dimensional stability (Fig. 9.14-11).

Tufted pile can be level-loop, plush, multilevel, cut or uncut, or a combination of cut and uncut. Striated and carved effects can be achieved. Tufted carpets are made in solids, moresques, and multicolor patterns made possible by new dyeing methods.

Tufted carpet generally can be identified on the reverse side, where the even rows of tufts punched through the prewoven backing with punched-in tufts can be seen.

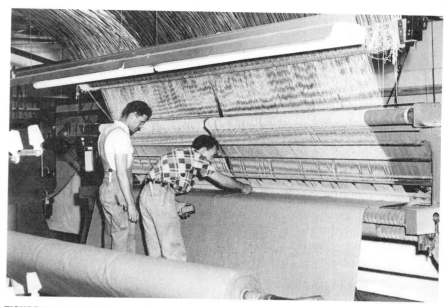

**FIGURE 9.14-9** A tufting machine. (Courtesy Carpet and Rug Institute.)

**FIGURE 9.14-10** A roll of carpet with a latex coating applied to the back by a special roller machine. Latex is used as a backing on all tufted and knitted carpet and on many woven carpets. A second backing is also laminated to good-quality tufted carpets. (Courtesy Carpet and Rug Institute.)

**FIGURE 9.14-8** Loomed weave with double chain warp, single weft shot, and low-level loop pile bonded to rubber cushioning. (Courtesy Carpet and Rug Institute.)

## KNITTED CARPET

Knitted carpet is somewhat similar to woven carpet in that the pile and backing yarns are fabricated in one operation. Unlike weaving, the knitting process loops together the backing, stitching, and pile yarns with three sets of needles in much the same way as hand knitting. A coat of latex is applied to the back as a fixative and to give additional body to the fabric (Fig. 9.14-12).

Because a single pile yarn is used in the knitting operation, as in velvet weaving, knitted carpets are usually solid colors or tweeds, though patterns are also possible. Knitted carpets usually have either single-level or multilevel uncut pile, but cut pile can be achieved with modification of the knitting machine.

**FIGURE 9.14-11** Tufted carpet construction: secondary backing usually is bonded to a prime backing with latex compound. (Courtesy Carpet and Rug Institute.)

**FIGURE 9.14-12** Knitted carpet with typical continuously looped pile, single chain warp, and stuffer yarns, all backed with latex. (Courtesy Carpet and Rug Institute.)

If a knitted carpet is *grinned* (folded back) to expose the backing, it will show the continuous looping of the pile yarn from row to row, held in place by the stitching yarns. The rows of pile run in an irregularly diagonal direction with a random, homespun look, not in straight rows, as in woven and tufted carpets.

### FUSION-BONDED CARPET

Fusion-bonded materials are those most often used for carpet tile. Fusion-bonded carpet is made by sandwiching a yarn bundle between substrates, implanting it there, and heat fusing the two. A blade is then run between the substrates, producing two pieces of carpet. The method of manufacture limits this type of carpet to cut pile face construction.

### NEEDLEPUNCHED CARPET

Needlepunched carpets are made by having hundreds of barbed needles punch loose, unspun fibers (usually polypropylene or nylon) through woven batts or fleeces of fibers to mesh them together, which results in a homogeneous pressed layer of fibers much like an army blanket or heavy felt. Needlepunched carpet is used mostly in the low-cost field.

Since needlepunched carpet is made entirely of synthetic fibers which are not subject to decay, it can be used on both exterior and interior surfaces. Needlepunched carpet made from solution-dyed polypropylene is often used as outdoor carpet. Needlepunched nylon carpet is often foam-backed for use indoors.

### FLOCKED CARPET

Flocked carpet is produced by spraying or electronically depositing short, loose nylon fibers on an adhesive-coated backing, similar to flocked wall coverings. In them, polypropylene might be substituted for jute, making them suitable for exterior and high-moisture locations.

### 9.14.1.2 Comparative Factors

Comparative factors related to carpet construction, such as pile yarn weight and pile thickness, number of tufts per square inch, and other factors, may be useful in comparing carpets of similar construction.

The traditional rule of thumb, "the deeper and denser (the pile), the better," is often used in making visual comparisons between carpets of similar construction and fiber type. However, such a generalization is superficial, and more detailed measures of carpet construction are generally necessary to judge carpet quality properly. Pile density is actually a combination of pile thickness and pile weight.

Pile thickness (in inches [millimeters]) and pile weight (in ounces per square yard [grams per square meter]) can be established accurately only by laboratory analysis. Pile density (in ounces per cubic yard [grams per cubic meter]) can be derived by a simple formula from established information about pile thickness and weight. All of these criteria are used in determining standards of quality for carpet (see Section 9.14.2.4).

Other comparative factors have been developed as a substitute for, or supplement to, the basic criteria of pile density, weight, and thickness. The number of *tufts per square inch*; *pitch*, *gauge*, and *needles*; and *rows*, *wires*, and *stitches* are terms which describe the spacing of individual tufts in various carpets. When considered together with yarn size (yarn weight) and number of plies (strands) twisted together to form the pile yarn, these factors provide measures of pile density and other important basic criteria relating to quality.

### PITCH, GAUGE, AND NEEDLES

The spacing of tufts across the width of a woven carpet can be described in terms of pitch, gauge, or needles. *Pitch* dictates the number of pile warp yarns (tufts) in a 27 in. (685.8 mm) width of carpet and is a measure of the tightness of crosswise construction (Fig. 9.14-13). Pitch may range between 90 and 256 for velvet and Wilton carpets and between 189 and 216 for Axminster carpet. The higher the pitch, the tighter the weave and the greater the pile density.

In knitted and tufted carpets where no continuous lines are present, *gauge* describes the actual spacing of tufts across the carpet width, expressed as a fraction such as $1/8$ in. (3.2 mm), $3/16$ in. (4.8 mm), or $5/32$ in. (3.99 mm). For example, $1/8$ in. (3.2 mm) gauge describes carpet with 8 tufts per inch (25.4 mm); $3/16$ in. (4.8 mm) gauge indicates 16 tufts in 3 in. (76.2 mm) (a little more than 5 tufts per inch [25.4 mm]).

In tufted carpets, the term *needles* often is used instead of *gauge* to describe the number of tufts per inch across the width. Tufted construction may vary from 4 to 13 needles (tufts) per inch (25.4 mm).

### ROWS, WIRES, AND STITCHES

*Rows*, *wires*, and *stitches* are terms describing the spacing of tufts lengthwise. The number of pile tufts per inch in Axminsters is described in *rows* per inch; for Wiltons and velvets, in *wires* per inch (Fig. 9.14-14). Single or double weft shots are usually visible on the back of the carpet and correspond to the rows of tufts in the length of the carpet. To give an accurate indication of rows or wires per inch, it is customary to count the vis-

**FIGURE 9.14-13** Pitch. (Courtesy Carpet and Rug Institute.)

<weft shots→|←row→|

**FIGURE 9.14-14 Wires or rows per inch (25.4 mm).** (Courtesy Carpet and Rug Institute.)

ible single or double weft shots in a 3 in. (76.2 mm) length of carpet and express this value in rows per inch (25.4 mm), as a fraction where necessary. Thus, 17 rows in 3 in. (76.2 mm) would be expressed as $5^2/_3$ rows or wires per inch (25.4 mm). Woven carpets range from 4 to 11 rows or wires per inch (25.4 mm).

In tufted carpets, there are no weft yarns which can be related to the spacing of pile tufts. Therefore, lengthwise construction is expressed as the number of tufts per inch. To distinguish this count from tuft count across the carpet (needles), the term *stitches* is used. Tufted carpets may have 6 to 10 stitches per inch (25.4 mm).

## TUFTS PER SQUARE INCH (MM²)

The overall density of carpets can be computed in tufts per square inch from information describing both lengthwise and crosswise construction (Fig. 9.14-15) or can be determined by actually counting the tufts in a 1-in. (25.4-mm)-square area. This comparative method is preferred for tufted and knitted carpets but also can be applied to woven carpet. It is helpful in comparing the overall density of dissimilar carpets when different terms are used to describe lengthwise and crosswise construction.

## YARN SIZE AND PLIES

Wool fibers and synthetic staple fibers are spun into single-strand yarns. Continuous-filament synthetic fibers are not spun, but emerge from the extrusion machine in bundles which are formed into a single-ply yarn. The strands of yarn may then be further twisted together into thicker yarns, each forming a single-pile tuft. Depending on the number of single strands, the pile yarn is identified as *single-*, *two-*, *three-*, or *four-ply* yarn. The size of both single-ply and multi-ply

yarns can be measured by *woolen count* (yards [meters] of yarn per ounce [gram]) or denier (grams per 9000 meters of yarn) and other methods, depending on the spinning method employed. Synthetic fibers and yarns are most often measured according to the denier system.

Along with spacing of tufts, the yarn size and number of plies affect carpet density, resilience, compression resistance, and other important performance criteria.

## BACKING

In woven and knitted carpets, the backing yarns and pile yarns are interlocked simultaneously during manufacture. The backing yarns typically are jute (sisal fiber), kraftcord (wood pulp fiber), polyester/cotton, nylon, rayon, or polypropylene. Backing yarns hold the pile tufts in place and provide the necessary stiffness so that the carpet will lie flat without wrinkling. In addition, they provide sufficient strength to prevent the carpet from tearing when heavy objects are moved across the surface.

Woven carpets such as velvet, Wilton, and Axminster may have single- or double-chain yarns and several stuffer

yarns per tuft running lengthwise; the number of *weft shots per wire* running across the carpet also may vary, resulting in such designations as *one-*, *two-*, or *three-shot* carpet. Increasing the number of backing yarns results in a stiffer, stronger backing and therefore in improved carpet quality. The better Wilton and velvet carpets may be woven through the back by hooking the pile yarns around the bottom shot yarns, thus providing better anchorage of the pile tufts.

In tufted carpets, pile yarns are stitched into a separately prepared *primary backing*. Most such primary backing today is polypropylene, although jute fabric is still sometimes used. An adhesive such as latex is used to secure the pile yarns in place, and a *secondary backing* of polypropylene or jute is bonded to the primary backing for additional strength, stiffness, and dimensional stability.

Similar *secondary backings* sometimes are used in knitted carpets as well. There is usually no secondary backing on woven carpet.

### 9.14.1.3 Pile Fibers

Carpet can also be selected according to the material of the pile fibers from which

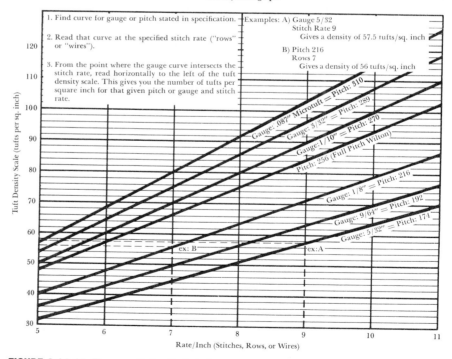

**FIGURE 9.14-15 The overall density in tufts per square inch is useful in comparing dissimilar carpets, such as tufted (example A) and woven (example B).** (Courtesy Carpet and Rug Institute.)

the pile yarns are made. These currently include wool, nylon, acrylics, polyester, and polypropylene.

Although pile fibers differ in their physical properties, all fiber materials included in this discussion provide adequate service when used in a carpet construction suitable for the traffic exposure.

Wool, the traditional fiber used in carpet manufacture, has proven its performance over the years and remains the classic fiber to which synthetic fibers often are compared. However, synthetic fibers, marketed under a variety of proprietary names, now account for about 90% of all carpet fiber used. The commonly used synthetic fibers can be classified generically as nylon, acrylic, and polypropylene (olefin). Cotton and two other man-made fibers, acetate and rayon, have not proven suitable for long-term wall-to-wall installations and therefore are not considered here. Polyester, one of the newest synthetic fibers, shows considerable promise but has not been in use long enough to demonstrate its properties fully.

The type of fiber used does not in itself guarantee carpet quality or performance. The various fibers have unique characteristics which may affect carpet performance and styling. Selected physical properties of pile fibers are given in Figure 9.14-16; other characteristics of fiber related to carpet performance are given in Figure 9.14-17.

In blended carpets, the higher the percentage of a particular fiber, the more the carpet will reflect the characteristics of that fiber. For example, wool may be reinforced by tough nylon in a 70/30 blend, but the carpet will look and feel more like wool. Generally, at least 20% of a fiber must be used for its characteristics to be apparent.

## WOOL

The outstanding characteristic of wool is resilience, which, in combination with moderate fiber strength (see Breaking Tenacity in Fig. 9.14-16) and good resistance to abrasion, produces excellent appearance retention. Wool's relatively high specific gravity contributes to greater pile density.

Carpet wool is obtained from Syria, Iraq, Argentina, Pakistan, New Zealand, Australia, Scotland, and other countries. Wools from different regions have unique qualities, as various breeds of sheep yield different wools. Some wools are fine and lustrous, while others are coarse and springy; also, fibers can vary in length from $3^1/_2$ in. (88.9 mm) to 7 in. (177.8 mm). *Woolen yarn* is made up of interlocked long and short fibers; *worsted yarn* uses only long fibers. Woolen yarns are more irregular, and hence are softer and bulkier than worsted yarns. The pile of most wool carpet is made of woolen yarns spun from fibers $3^1/_2$ in. (88.9 mm) to $6^1/_2$ in. (165 mm) long (see Fig. 9.14-16).

## NYLON

Nylon comprises about 85% of the commercial carpet market and a large portion of the housing market as well. The extensive use of nylon can be attributed to its lower cost, availability in many bright colors, exceptional resistance to abrasion, and high fiber strength (see Breaking Tenacity in Fig. 9.14-16).

Chemically, there are two types of nylon: Type 6, which represents a long-chain polymer, a chemical containing six carbon atoms; and Type 6,6, which results from the polymerization of two chemicals, each containing six carbon atoms. Although the two types perform similarly as yarn, they differ slightly in their ability to be processed and dyed.

Nylon fiber is produced in two forms. *Staple* nylon is composed of $1^1/_2$-in. (38-mm)- to 6-in. (152.4-mm)-long fibers, which are spun into yarn much like wool or other short fibers. *Continuous filament* nylon consists of *bundles* of continuous fibers which are formed into yarn without spinning. Both staple nylon and continuous filament nylon yarns are made in round and modified cross sections (Fig. 9.14-18). Staple fibers are mechanically crimped; continuous filament yarns are texturized (Fig. 9.14-19) to impart bulk and covering power to the yarns. Texturizing and crimping give the yarns or fibers an irregular alignment which further increases their bulk and covering power.

## ACRYLICS AND MODACRYLICS

Acrylics and modacrylics (modified acrylics) are synthetic fibers that closely resemble wool in abrasion resistance and texture.

Chemically, acrylic fibers are composed of 85% or more by weight acrylonitrile. Modacrylics contain at least 35% of acrylonitrile in chemical combination with modifying materials. Acrylic fiber is produced in round, bean-, and mushroom-shaped cross sections (see Fig. 9.14-18b). The last is a bicomponent fiber of two filaments welded parallel to each other. Because the shrinking ratios of the two filaments are different, the fiber assumes a coiled shape in its dry

**FIGURE 9.14-16** Selected Properties of Pile Fibers

| Type | Length | Specific Gravity | Breaking Tenacity, grams/denier | Degrading Temperature[a] | Moisture Absorption, % |
|---|---|---|---|---|---|
| Wool | Staple | 1.32 | 1.17 | 420°–570°F (233–317°C) | 15–16 |
| Nylon | Staple and CF[b] | 1.14 | 4–6 | 428°–482°F (238–268°C)[c] | 4–5 |
| Acrylic | Staple | 1.18 | 2–4 | 400°–500°F (222–277°C) | 1.3–2.5 |
| Modacrylic | Staple | 1.3–1.4 | 2.5–3.0 | 300°F (167°C) | 0.4–4.0 |
| Polypropylene | CF[b] | 0.91 | 3–6 | 333°F (185°C) | up to 0.10 |

[a]For wool, scorching to charring range; for synthetics, melting temperature.
[b]Continuous filament.
[c]428°F (238°C) for Type 6; 482°F (268°C) for Type 6.6.
(Courtesy Carpet and Rug Institute.)

**FIGURE 9.14-17** Selection Criteria for Carpet Pile Fibers[a]

| Criteria | Wool | Nylon | Acrylic | Modacrylic | Polypropylene |
|---|---|---|---|---|---|
| Resistance to: | | | | | |
| Abrasion | Good | Excellent | Good | Good | Excellent |
| Alkalies | Fair | Good to Excellent | Good to Excellent | Good to Excellent | Good to Excellent |
| Acids | Fair | Good to Excellent | Good | Good | Good to Excellent |
| Insects and fungi | Excellent[b] | Excellent | Excellent | Excellent | Excellent |
| Burns | Good | Fair | Fair | Fair | Fair |
| Compression | Good | Good | Good | Good | Fair |
| Crushing | Excellent | Good | Good to Excellent | Good | Fair to Good |
| Staining | Good | Good to Excellent | Good to Excellent | Good to Excellent | Excellent |
| Soiling | Good | Fair to Good | Good to Excellent | Good to Excellent | Excellent |
| Static buildup | Fair to Good | Fair | Good | Good | Excellent |
| Texture retention | Excellent | Good to Excellent | Good to Excellent | Good | Fair to Good |
| Wet cleanability | Fair to Good | Good | Good to Excellent | Good to Excellent | Excellent |
| Durability | Good to Excellent | Excellent | Good to Excellent | Good to Excellent | Good |
| Appearance retention | Excellent | Good to Excellent | Good to Excellent | Good to Excellent | Good |
| Ease of maintenance | Good to Excellent | Good | Good to Excellent | Good to Excellent | Excellent |

[a]In each case, criteria vary depending on the properties of the specific fiber type used, carpet construction, and installation procedures.
[b]When chemically treated.

(Courtesy Carpet and Rug Institute.)

equilibrium state, a property which improves texture retention.

Fibers of the acrylic family are now available only in staple form, and are characterized by an appearance and high durability comparable to those of wool. Acrylics often are blended with modacrylics in commercial carpet to reduce potential flammability. However, with substantial improvement in acrylic fibers, the trend is toward 100% acrylic fiber.

### POLYESTER

Polyester is a fiber-forming thermoplastic polymer. It is made from terephthalic acid and ethylene glycol. Polyester fibers are staple fibers, and polyester yarns are spun yarns.

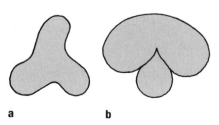

**FIGURE 9.14-18** Modified-fiber cross sections: (a) trilobal nylon and (b) bicomponent acrylic. (Courtesy Carpet and Rug Institute.)

### POLYPROPYLENE OLEFIN

There are two classes of olefin, polyethylene and polypropylene, but only polypropylene has been produced in a fiber suitable for carpet construction. Polypropylene fiber consists of at least 85% propylene by weight and is generally produced in bulked continuous filament yarn, although staple yarn also is available.

Polypropylene has the lowest moisture absorption rate of any carpet fiber, which gives it superior stain resistance as well as excellent wet cleanability (see

a

b

**FIGURE 9.14-19** Continuous filament nylon: (a) before texturizing; (b) after texturizing. (Courtesy Carpet and Rug Institute.)

Fig. 9.14-16). Its low specific gravity results in superior covering power, and its high fiber strength contributes to good wear resistance.

#### 9.14.1.4 Dyeing Methods

Color can be introduced at various stages of manufacture. For both synthetic and wool pile carpets, dyes can be added at either the fiber, yarn, or carpet manufacturing stage. The dyeing method used may vary with the general properties of the fibers, the intended carpet construction, and the color effect desired or may depend on special methods developed for proprietary fibers.

#### SOLUTION DYEING

Many synthetic fibers are colored by a dye or pigment introduced into the liquid chemical from which fibers are extruded. In this method, the dye becomes chemically a part of the fiber; therefore, solution-dyed yarns are generally colorfast and more resistant to discoloration from wet cleaning.

#### STOCK DYEING

Wool and synthetic staple fibers can be dyed before being spun into yarn by dipping them in kettles of dye. This method ensures a uniform distribution of dye and is suitable where large quantities of a single-colored yarn are desired.

### SKEIN AND PACKAGE DYEING

When smaller quantities of colored yarn are required, *skein dyeing* is often used. In this method, skeins of natural-colored spun yarn or continuous filament fibers are immersed in vats of hot dye. In *package dyeing*, yarn is wound on a perforated tube and dye is forced through the perforations to soak the yarn.

### SPACE DYEING

Both staple and continuous filament yarns can be space dyed in alternating bands of color, generally for use in multicolored tufted carpet. In the *knit sleeve* method of space dyeing, the yarn is knitted into a fabric, print rollers apply the desired colors, and then the fabric is unraveled again into yarn. In the *warp print* method, parallel yarns are fed continuously into a machine where color is applied by rollers.

### PIECE DYEING AND RESIST PRINTING

Some woven and many tufted carpets are piece-dyed by immersing the entire carpet in a hot dye bath. However, the carpet need not be dyed only one color; some synthetic fibers can be chemically formulated to absorb certain compatible dyes and repulse others. Thus, it is possible to produce complex color patterns in the same dye bath. A similar technique of selective dyeing can be applied in skein dyeing. *Resist printing*, another method of selective dyeing, involves printing tufted carpet with a dye-resistant agent before piece dyeing.

### PRINT DYEING

*Print dyeing* provides good dye penetration and pattern definition in tufted carpet. With possible color variation of up to six colors, print-dyed tufted carpets compare favorably in appearance with Wiltons and Axminsters but are cheaper to produce. The process involves silk-screening plain tufted carpet with a premetalized dye and applying an electromagnetic charge under the carpet which drives the dye deeply into the pile. The final pattern is built up by continuously processing the carpet through a series of troughs, each applying a different color.

## 9.14.2 SELECTION CRITERIA

Carpet performance cannot be attributed to any single element of carpet construction or physical property of its pile fiber. The performance of installed carpet is the combined effect of its surface texture, construction, backing, and cushioning, and its pile thickness, weight, density, and other variables. Of these, pile weight and pile density are the most uniformly significant, regardless of fiber type or carpet construction.

Although pile fibers differ in their physical properties (see Fig. 9.9-16), all the fibers discussed here will provide adequate service when used in a carpet construction suitable for the traffic exposure (see Section 9.14.2.5). The variables of carpet construction are so great that it is difficult to assess their individual effect on performance and selection criteria. Therefore, in the following discussion, selection criteria are outlined for fibers only (see Fig. 9.14-17); the effects of carpet construction are described only where appropriate.

Individual selection criteria may be useful in carpet selection for special conditions of traffic or exposure. For example, *alkali*, *acid*, and *stain resistance* may be important in food preparation and serving areas. *Soiling resistance* and *wet cleanability* may be factors in entry areas. *Insect and fungus resistance* may require consideration in potentially damp locations. Possible *static buildup* should be considered in low-humidity areas and where such buildup might affect equipment or activities, such as in computer rooms. These and other criteria have been combined into composites such as *durability*, *appearance retention*, and *ease of maintenance* for average conditions (see Fig. 9.14-17).

### 9.14.2.1 Durability

Pile weight and density, backing and cushioning firmness, quality of installation, and maintenance significantly affect carpet durability. Durability is also affected by hazards that tend to destroy pile fibers. The resistance of fibers to such hazards is shown in ratings for resistance to abrasion, alkalis, acids, insects, fungi, and burns. The rating for durability is a composite reflecting overall resistance to destruction or loss of pile fibers and is a measure of the service life of a carpet.

### ABRASION RESISTANCE

Abrasion resistance is the resistance of the fiber to wearing away due to foot traffic or other moving loads. It is measured by the actual loss of fiber when exposed to a machine that simulates abrasive forces due to foot traffic. Since loss of fiber occurs as a result of breaking, fiber strength is one of the physical properties that influences abrasion resistance. Nylon and polypropylene have superior fiber strength and hence excellent abrasion resistance.

### ALKALI AND ACID RESISTANCE

Foods and beverages are generally mildly alkaline or acid in nature; many soaps and detergents are strongly alkaline. If alkaline or acid solutions are not promptly removed when spilled, they may attack the pile fibers. Wool fibers are more vulnerable to such attacks because they have a higher moisture absorption rate and are less inert chemically than the synthetic fibers.

### INSECT AND FUNGUS RESISTANCE

The synthetic fibers in modern carpets contain no organic nutritive value and are, therefore, immune to attack by insects, such as carpet beetles and moths, and to fungi, such as mold and mildew. To render wool carpet resistant to these hazards, most manufacturers treat wool yarn in the dye bath with a chemical preservative that provides long-term protection from insect attack, even after wet and dry cleaning. Since many carpets intended for interior use employ at least some organic fibers in the backing, some fungus hazard exists with most carpets, even those with synthetic fiber piles.

In general, the hazards of insect and fungus attack can be reduced by regular vacuuming and exposing the carpet to light and air. This is particularly important where large or heavy pieces of furniture make part of the carpet inaccessible to regular cleaning and ventilation. In such areas, greater insect protection can be provided for wool carpets by periodic spraying with mothproofing agents.

### BURN RESISTANCE

Most synthetic carpet fibers are flame-resistant to some degree, but they will melt and fuse when exposed to the heat of a burning cigarette or glowing ash which produce heat of more than 500°F (260°C). Prolonged exposure to such concentrated heat may result in complete local loss of fiber or a slightly fused spot. The fused spot is more resistant to wear and abrasion than the surrounding area, and this causes a visible discoloration in the carpet surface.

Although wool fiber will actually burn by charring, the damage from a lit cigarette is usually less severe and the charred spot will wear off more readily or can be removed with fine sandpaper. The fused spot on synthetic carpets similarly can be snipped off if it is not too severe, and if followed by brushing, an acceptable appearance usually will result. If the burn is severe, replacement of the burned area may be necessary.

The behavior of a carpet in a fully developed fire is controlled by building codes and laws. All carpet manufactured in the United States must pass a methenamine "pill" test in accordance with ASTM Standard D2859. In this test, a carpet is exposed to the burning of a methenamine tablet. The medium *flame spread index* or *critical radiant flux* must be 0.04 watt/sq cm.

Often, carpet will be required by code to have a higher rating than the pill test minimum. The National Institute of Standards and Technology (NIST) recommends that carpet used in corridors and exit ways have a radiant flux of at least 0.22 watt/sq cm in commercial occupancies and 0.45 watt/sq cm in institutional occupancies.

A carpet's flame spread during a fully developed fire is measured by the Flooring Radiant Panel Test of ASTM E648. This flame spread index is called the *critical radiant flux*. It represents the minimum energy necessary to sustain flame. The higher the number, the more resistant the carpet to flame propagation.

### 9.14.2.2 Appearance Retention

Appearance retention depends on factors that change the original carpet color, texture, or pattern, such as fading, soiling, and staining, compression, and crush resistance. Other factors common to many types of fiber and carpet construction that affect appearance without impairing serviceability are pilling, shedding, sprouting, and shading (see the Glossary). Ratings for appearance retention are composites of these factors.

#### COMPRESSION RESISTANCE

The extent to which the pile will be compressed under heavy loads or by exten-

sive foot traffic is called *compression resistance*. For instance, of two samples tested under the same conditions, the one that compresses less is said to have better compression resistance. With the exception of polypropylene, which is rated somewhat lower, all fibers are approximately equal in their ability to resist compression loads. Of greater importance in resisting such forces are the cushioning, pile height, and pile density. Looped pile construction with closely spaced tufts and tightly twisted yarns is more resistant to compression.

#### CRUSH RESISTANCE

Crush resistance is dependent on the ability of the fibers to recover from short- and long-term compression loads. This property is measured by the extent to which the fiber springs back after a load has been removed as a percentage of the original height. Wool and acrylic have the highest crush resistance, nylon and modacrylic moderate resistance, and polypropylene somewhat lower resistance. Pile density, spacing of tufts, and tightness of yarns affect crush resistance more than the relative rating of the fibers.

#### TEXTURE RETENTION

The ability of carpets to retain the surface texture imparted during manufacture is related to the compression and crush resistance of the fibers, as well as the density of the pile and tightness of the yarns. Generally, all fibers have adequate texture retention, with wool rated highest and polypropylene somewhat lower than the other synthetic fibers. Within this range, special fiber types such as heat-set nylon and bicomponent acrylic (mushroom cross section) have improved texture retention.

### 9.14.2.3 Ease of Maintenance

The ease of maintaining an installed carpet depends mainly on its resistance to soiling and staining, as well as its wet cleanability. Although dry powder cleaning products are useful for moderately soiled areas, wet cleaning (shampooing) remains the most satisfactory method for thorough in-place carpet cleaning under heavier soil conditions.

#### STAINING RESISTANCE

*Staining* is a local discoloration of a carpet as compared with *soiling*, which is more or less uniform over the carpet sur-

face. Many foodstuffs (oil, grease, eggs, chocolate), beverages, and beauty aids may cause staining. If promptly and properly treated, such stains need not be permanent. Many carpet manufacturers publish detailed recommendations for stain removal and indicate the type of solvent or cleaning agent that should be used with each type of stain. Since stains are often moisture-borne, the rate of moisture absorption is a factor in staining resistance. Polypropylene and acrylics have low moisture absorption and are therefore rated highly for staining resistance.

#### SOILING RESISTANCE

Soiling is the result of dirt or soil particles deposited on the carpet surface by foot traffic or dust. Wool and nylon generate higher electrostatic charges in low humidity and, therefore, tend to attract airborne dust and soil faster than some other fibers. However, the scaly surfaces of wool fibers make soiling less apparent. Polypropylene and acrylics have higher soiling resistance because of low static buildup and low moisture absorption. Bulked fibers with modified cross sections and texturized yarns spun more loosely tend to retain dirt more tenaciously than round, tight, high-density yarns. Also, tightly woven, uncut-loop piles will keep soil near the surface, where it is easier to remove. Multicolored and textured piles of medium shades will conceal soiling better than light or dark, smooth plush surfaces.

#### WET CLEANABILITY

Since it is impractical to remove wall-to-wall carpet for cleaning, the ability of carpet to respond successfully to in-place wet cleaning is important. Polypropylene, with its low moisture absorption rate, is considered the best in this respect. Generally, synthetic fibers, with their lower moisture absorption rates and inert chemical compositions, clean more readily, dry more rapidly, and are less subject to mildew odors while drying than is wool. However, a variety of effective cleaning formulations specifically suited to wool carpet are available.

### 9.14.2.4 Static Generation

Static electricity is an annoyance in every carpeted space but can become actually harmful in areas housing computers or other electrically sensitive equipment. Static occurs whenever two different materials come into contact with each other. People usually do not become aware of static until it reaches about 3.5 kilovolts (kv). The normal carpet requirements for solely human occupancies require a level of 3 kv or less. Computers require 2 kv or less.

Of the three factors involving static electricity—static generation, static dissipation (conductivity), and static decay time—only the first is usually limited in the selection of carpets in nonsensitive locations. Static generation in carpet is measured by the American Association of Textile Colorist and Chemists' "Step Method," which simulates conditions of actual use and measure the charge generated by someone walking across the carpet.

Static dissipation may be also limited in spaces that will house computers. It is measured by a method recommended by the IBM Corporation.

Nylon is inherently high in static buildup. To combat this problem, some nylon carpet manufacturers include carbon-loaded nylon fibers that carry a lifetime antistatic guarantee. Some carpet is laced with metal wires that are in contact with a grounded backing. In very sensitive areas, it may be impossible to reduce static sufficiently using antistatic fibers or wires alone. Then, the antistatic properties of such carpet must be supplemented with conductive shoes, furniture, mats, or other devices.

### 9.14.2.5 Standards of Quality

The Department of Housing and Urban Development (HUD) has issued "Use of Materials Bulletin UM-44d," which covers residential carpet, but it is also referenced in specifying commercial and institutional carpet.

HUD does not permit the use of carpet in baths, kitchens, and service areas such as laundry, utility, and furnace rooms. Carpet is also not appropriate in public toilets or shower rooms. Living areas within a residential unit, such as living, dining, recreation, sleeping, and dressing rooms, are designated as moderate-traffic areas. Public areas such as corridors, entrances, lobbies, waiting rooms, stairways, and elevators are considered heavy-traffic areas. Offices and institutional spaces may be either moderate- or heavy-traffic areas, depending on their actual use.

HUD requirements for carpet in heavy-traffic areas are summarized in Figure 9.14-20. Carpet construction requirements for moderate-traffic areas are discussed below and illustrated in Figures 9.14-21 and 9.14-22. For both types of traffic exposure, carpet should conform to the following requirements for pile fibers, backing, and cushioning.

### CARPET CONSTRUCTION

"Use of Materials Bulletin UM-44d" requirements are based on average values for pile yarn weight, pile thickness, and pile density.

*Pile density* is expressed as either average pile density (*AD*) or weight density (*WD*). *Pile yarn weight* (*W*) is the aver-

**FIGURE 9.14-20** Physical Requirements for Carpet in Heavy-Traffic Areas

| | Wilton, Wool or Acrylic | Velvet | | | Tufted, Wool or Acrylic | Knitted, Wool or Acrylic |
|---|---|---|---|---|---|---|
| | | Wool or Acrylic | Nylon* | Wool or Acrylic | | |
| Pile Description | Single-level loop woven through back | Single-level loop pile woven through back | Single-level loop pile woven through back | Multilevel loop woven through back | Single-level loop | Single-level loop |
| Tufts/sq. in. (mm²) | 52 (0.081) | 60 (0.093) | 32 (0.049) | 34 (0.053) | 60 (0.093) | 26 (0.040) |
| Frames | 3 | — | — | — | — | — |
| Shots/wire | 2 | 2 | 2 | 2 | — | — |
| Weight oz./sq. yd. (g/m²) Pile | 48 (1627.5) | 42 (1424) | 29 (983.3) | 44 (1491.9) | 42 (1424) | 37 (1254.5) |
| Total | 72 (2441) | 60 (2034) | 50 (1695.3) | 64 (2170) | — | 58 (1966.5) |
| Pile height in. (mm) | Min. 0.250 (6.4) Max. 0.320 (8.1) | Min. 0.210 (5.3) Max. 0.310 (7.9) | Min. 0.200 (5.1) Max. 0.290 (7.4) | Min. 0.190 (4.8) Max. 0.370 (9.4) | Min. 0.250 (6.4) Max. 0.320 (8.1) | Min. 0.230 (5.8) Max. 0.290 (7.4) |
| Chain | Cotton and/or rayon | Cotton and/or rayon | Cotton and/or rayon | Cotton and/or rayon | — | Cotton, rayon, or nylon |
| Filling | Cotton and/or rayon, or jute | Cotton and/or rayon, or jute | Cotton and/or rayon, or jute | Cotton and/or rayon, or jute | — | Jute or kraftcord |
| Stuffer | Cotton, jute, or kraftcord | Cotton, jute, or kraftcord | Cotton, jute, or kraftcord | Cotton, jute, or kraftcord | — | — |
| Back coating, oz./sq. yd. (g/m²) | — | 8 (271.2) | 6 (203.4) | 8 (271.2) | — | — |
| Ply twist turns/in. (turns/mm) | Min. 1.5 (0.06) Max. 3.5 (0.14) | Min. 1.5 (0.06) Max. 3.5 (0.14) | Min. 1.0 (0.04) Max. 3.0 (0.12) | Min. 1.5 (0.06) Max. 3.5 (0.14) | Min. 2.5 (0.09) Max. 4.5 (0.18) | Min. 1.5 (0.06) Max. 3.5 (0.14) |

*High bulk or textured continuous filament.

(Courtesy Carpet and Rug Institute.)

## FIGURE 9.14-21 Physical Requirements for Carpet in Moderate-Traffic Areas[a]

| Pile Fiber | Minimum Pile Yarn Weight (W), oz./sq. yd. (g/m$^2$)[b] | Minimum Weight-Density Factor (WD)[c] |
|---|---|---|
| Wool | 25 (847.6) | 70,000 (25 × 2800) |
| Acrylic | | |
| Modacrylic | | |
| Nylon (staple and filament) | 20 (867) | 56,000 (20 × 2800) |
| Polypropylene olefin | | |

[a]Based on HUD Use of Materials Bulletin UM-44A.

[b]Computed for blended yarns by multiplying percentage of each fiber by minimum pile weight and adding the two amounts.

[c]Pile density (D) = 2800, the minimum density that historically has given satisfactory service for wall-to-wall installations.

(Courtesy Carpet and Rug Institute.)

age weight of the face pile yarn in ounces per square yard (grams per square meter). *Pile thickness* (t) is the height of the tufts above the backing in inches. *AD* is a calculated quantity in ounces per cubic yard (grams per cubic yard) obtained by the formula $AD = (36) (W)/t$. *AD* is a four-digit number that should be above 3000 for commercial carpet and above 2800 for other carpet. *WD* is also a calculated quantity in ounces per cubic yard (grams per cubic meter) obtained by the formula $WD = (AD)(W)$.

UM-44d establishes minimums for pile yarn weight *W* and *WD*, depending on the type of fiber (see Fig. 9.14-22). The minimum *WD* values given are always the minimum *W* values multiplied by a pile density (*D*) of 2800. The HUD requirements are based on minimum weight-density factors rather than a con-

stant pile density of 2800. If only the pile density were specified, it would establish a simple proportionate relationship between actual pile thickness and pile yarn weight which would require increasing the minimum pile weight to the same extent as the pile thickness. This requirement would be excessive for moderate-traffic areas. Figure 9.14-22 illustrates calculation of the weight-density factor.

### PILE FIBERS

Pile yarn should be made of 100% wool, nylon, acrylic, modacrylic, or polypropylene olefin fibers or blends of these fibers in yarn (exclusive of ornamentation). Not less than 20% of any of the above fibers should be used when blended with other fibers, and such blends should not consist of more than two types of fiber. All fibers should be entirely new, not processed or reclaimed from previously knitted, woven, tufted, or felted products. Wool yarn should contain at least 97% wool fibers by dry weight; synthetic fibers should be 15 denier or thicker and should contain not more than 2% chloroform-soluble material.

Carpet pile should be resistant to fading from wet cleaning and exposure to light. Light colors should be rated as satisfactory after laboratory exposure to 20 standard fading hours and dark colors after 40 fading hours.

### BACKING

Backing materials should be those customarily used and accepted in the industry for each type of carpet. For tufted carpet, there should be a secondary backing reinforcement of at least 4 oz/sq yd (135.6 g/m$^2$), exclusive of the adhesive coating.

### FIGURE 9.14-22 Weight-Density Calculations for Nylon or Polypropylene Carpet

Weight density (*WD*) calculated for nylon or polypropylene carpet from pile thickness (t) and pile yarn weight (W).

Pile thickness (t) = 0.25 in.

Pile yarn weight (W) = 20 oz./sq. yd.

Calculated pile density (D) = 36W/t

$$D = \frac{36 \times 20}{0.25} = 2880 \text{ oz./cu. yd.}$$

Calculated weight-density factor (*WD*)

$WD = 20 \times 2880 = 57,600*$

*Referring to Figure 9.12-21, pile yarn weight (20 oz./sq. yd.) equals minimum requirements; weight density factor (57,600) exceeds minimum requirement (56,000).

(Courtesy Carpet and Rug Institute.)

### 9.14.2.6 Certification of Quality

There is no industry-wide program for certification of carpet quality at the present time. However, many manufacturers will certify that their product meets or exceeds HUD requirements for moderate- or heavy-traffic areas.

Federal law requires that all textiles be identified with an appropriate label specifying the fiber content for each fiber type that constitutes 5% or more of the total. In the case of carpets consisting of backing yarns and pile yarns, only the fibers in the pile yarn need be identified. In addition, carpet samples shown to consumers must be similarly identified as to fiber content, name of manufacturer, and country of origin, if imported. Several producers of synthetic fibers have agreements with carpet manufacturers which stipulate that only carpet meeting certain minimum construction requirements can display their fiber trademark.

### 9.14.2.7 Cushioning

All broadloom carpet should be installed over cushioning to increase resilience and durability. Some carpets, especially those used in direct glue-down installations (Section 9.14.3.2), have an integral secondary backing that serves this function. In loomed carpets, cellular (foam or sponge) rubber cushioning usually is bonded directly to the carpet at the mill. For most carpets, however, cushioning is provided in the form of a separate padding. Currently available paddings include felted hair, rubberized fibers, sponge rubber, and urethane foam.

### FELTED HAIR

The most economical traditional type of conventional padding is made of felted animal hair. This padding has a waffle design to provide a skidproof surface and improve resiliency. It is sometimes reinforced with a jute backing or a burlap centerliner. When reinforced with burlap, the hair is punched through the burlap fabric and compressed to a uniform thickness. Sizing (adhesive) sometimes is used to strengthen the bond between the fibers and the burlap core. Many manufacturers also sterilize and mothproof hair padding.

Felted padding of hair or hair and jute may mat down in time or may develop mildew, especially if the fibers become wet, as during cleaning. However, when properly cleaned, sterilized, and treated, hair padding is suitable for reasonably dry floors at all grade levels and on conventional radiant-heated floors.

## RUBBERIZED FIBERS

Some padding made of jute or hair is coated with rubber on one or both sides to hold the fibers together and provide additional resilience. Sometimes this padding has an animal hair waffle top and a jute back reinforced with a patterned, rubberized application.

## SPONGE RUBBER

In addition to cushioning bonded directly to the carpet, foam or sponge rubber is produced in sheet form with waffle, ripple, grid, or V-shaped rib designs. A scrim of burlap fabric usually is bonded to the rubber sheet to facilitate installation of the carpet. When laid with the fabric side up, this permits a taut and even stretch of the carpet.

Rubber padding is more expensive, but it retains its resilience longer than hair padding. It is highly resistant to decay and mildew and is nonallergenic. It can be used at all grade levels, but the denser cushionings are not recommended for radiant-heated floors.

## URETHANE FOAM

Urethane foam padding is a high-density polymeric foam available in two major types: prime and bonded. *Prime* urethane foam padding is available as either prime or densified prime. The latter is modified chemically to make it wear better.

*Bonded* urethane foam padding is made from heat-fused scraps of urethane rather than being a homogeneous product, as are prime and modified prime urethane padding.

The quality of urethane foam padding is defined by its density or weight per cubic foot (0.028 m³) of foam.

## SELECTION

As a minimum, the characteristics of separate carpet padding should meet or exceed the recommendations of the Carpet Cushion Council (CCC).

*Residential Applications* The CCC publication "Minimum Recommendations for Residential Applications" recommends the density ($D$), weight ($Oz$), and thickness ($t$) of carpet cushions for residential use based on two classes, as follows:

- *Class 1,* Light and Moderate Traffic: Living rooms, dining rooms, bedrooms, and recreational rooms

- *Class 2,* Heavy Duty Traffic: Lobbies and corridors in multifamily facilities and all stair applications

Class 2 cushion may be used in Class 1 applications.

$D$ is measured in lb/ft³ (kg/m³); $Oz$ in oz/yd² (kg/m²); $t$ in decimals of inches (millimeters).

The recommended density, weight, and thicknesses of padding vary within each class, depending on the padding material, and from class to class.

*Commercial Applications* The CCC publication "Minimum Recommended Criteria for Satisfactory Carpet Cushion Performance in Contract Installations" recommends the specific density ($D$), weight ($W$), compression resistance ($CR$), and compression force deflection ($CFD$) of each available carpet padding type for each type of commercial use as follows:

- *Class I*, Moderate Traffic: Executive and administrative spaces in such buildings as schools, office buildings, airports, retail establishments, health care facilities, libraries, and museums; sleeping rooms in hotels and motels
- *Class II*, Heavy Traffic: The more public areas in the same types of buildings as listed for Class I, but in clerical areas, moderate-use corridors, patients' rooms, lounges, dormitories, classrooms, small retail stores, bank lobbies and corridors, and moderate-use public areas
- *Class III*, Extra Heavy Traffic: Heavy-use public corridors, lobbies, locker rooms, dining areas, nurses' stations, cafeterias, major retail stores, restaurant dining rooms, and convention centers

As in residential applications, the recommended criteria for padding vary within each class, depending on the padding material, and from class to class.

$D$ is measured in lb/ft³ (kg/m³); $Oz$ in oz/yd² (kg/m²); $t$ in decimals of inches (millimeters).

The CR of carpet padding is measured in accordance with ASTM D3676. The $CR$ is the force in pounds per square inch (kg/m²) needed to compress padding to a specified percentage of its original thickness.

The CFD of carpet padding is measured in accordance with ASTM D3574. The $CFD$ is the number of pounds (kilograms) needed to depress a piece of padding to a specified percentage of its original thickness over an 8-in. (200-mm)-diameter test area.

### 9.14.2.8 Carpet Tile

Carpet tiles are used in commercial, institutional, and residential buildings. They meet the demands of business environments, including functionality, flexibility, style, value, and performance. Many carpet tiles can be installed with almost invisible seams so that the installation looks like broadloom. Most are available in 36 × 36 in. (914 × 914 mm) and 18 × 18 in. (457 × 457 mm) sizes. Other sizes may be available from some manufacturers.

Carpet tiles are more adaptable, flexible, and easily installed and removed than broadloom carpet. They also provide unlimited underfloor access to electrical, data, and communications systems. Their backings are designed for easy handling and installation. Most are PVC-free, which means that they can be installed over many existing finishes without complete removal of old adhesives and sealers.

Carpet tiles should be required to comply with the Carpet and Rug Institute's (CRI) Indoor Air Quality Carpet Testing Program. They should have a Class I flammability rating in accordance with ASTM D648.

### 9.14.3 INSTALLATION OF CARPET

Carpet can be installed over finished flooring, subflooring of concrete, plywood, particleboard, or any other suitable, smooth surface. The carpet manufacturer's recommendations should be followed with regard to seam locations and direction of carpet. Carpet edge guards should be located where carpet edges would otherwise be exposed. Edge guards should be securely fastened to the substrate.

#### 9.14.3.1 Tacked and Tackless Installation

Carpet that is not rubber-backed can be readily installed over nailable surfaces by tacking the edges or by using *tackless strips* around the perimeter. Tackless strips consist of pieces of plywood of the thickness of the padding with two or three rows of metal pins which pass through the plywood at an angle and are designed to grip carpet when it is stretched over the pins (Fig. 9.14-23). Installations on concrete can be made by tacking if wooden plugs are provided 6 to 9 in. (150 to 230 mm) on centers in the concrete slab to receive the tacks. Also, tackless strips can be installed on concrete with powder-actuated fasteners or epoxy adhesive. Two-row strips can be used when the carpet is 20 ft (6.1 m) or less in width, but three-row strips should be used with wider carpet. Strips should be located at the entire perimeter of the carpet.

Padding is placed between the tackless strips and bonded to the substrate with an adhesive. The joints are placed at right angles to the carpet joints. Seams in padding should be taped.

Regardless of the installation method, tacked or tackless carpet should be located, trimmed, stretched sufficiently to eliminate wrinkles and buckles, and attached firmly with tacks or to strips. Seams should be butted tightly and taped or sewn to make them strong enough to resist damage during stretching and to keep them closed without gaps.

### 9.14.3.2 Direct Glue-Down Installation

In *direct glue-down* applications, carpet is glued to the substrates with no padding between them. Such carpet is fitted in place before adhesive is applied. The adhesive is then applied uniformly to the substrate. Carpet edges are butted tightly so that there are no gaps. The carpet is then rolled lightly to ensure a uniform bond to the substrates with no air pockets.

# 9.15 Vinyl-Coated Fabric Wall Coverings

Vinyl-coated fabric wall coverings are decorative wall finishes for interior wall surfaces.

### 9.15.1 APPLICABLE STANDARDS

Industry standards applicable to vinyl-coated fabric wall coverings include:

- ASTM International (ASTM) Standards: D751, D1308, E84, G21
- Chemical Fabrics and Film Association (CFFA) CFFA-W-101-D
- Federal Specifications: CCC-W-408D

**FIGURE 9.14-23** Using a power stretcher, a skilled carpet mechanic can achieve a smooth, taut installation of carpet over tackless strips

### 9.15.2 MATERIALS

Vinyl-coated fabric wall coverings are fabric-backed PVC sheets. They are used in both commercial and residential applications. They are easily cleanable, damage-resistant, and decorative finishes. They are durable, scrubbable, and stain- and fade-resistant. They hide cracks, gaps, and uneven surfaces. They are available in hundreds of colors and many patterns. They are available with antimicrobial additives that inhibit mold, mildew, and fungus growth.

Most vinyl-coated fabric wall coverings are Class A fire-rated. In addition, when heated to 300°F (149°C) or higher, some vinyl wall coverings emit an odorless, harmless vapor that triggers ionization fire detectors. Most are also self-extinguishing; they will not continue to burn when the fire source is removed.

Vinyl-coated fabric wall coverings hide imperfections better and are easier to clean than paint. Their overall lifetime cost is lower than that of paint due to their long life. Warranties of up to 5 years are available.

These wall coverings are available in several thicknesses, measured by their weight, such as the following:

- 12 oz/lin yd (407 g/m$^2$)
- 15 oz/lin yd (509 g/m$^2$)
- 20 oz/lin yd (678 g/m$^2$)
- 24 oz/lin yd (814 g/m$^2$)

The heavier materials generally have higher resistance to tearing than do their thinner counterparts.

Vinyl-coated fabric wall coverings may be applied on new or existing plaster, gypsum wallboard, or concrete masonry units (CMU). With proper substrate preparation, they may also be applied over existing painted walls, ceramic tile, marble, glass, and other materials. New surfaces must be cured for the length of time recommended by the wall covering manufacturer.

### 9.15.3 INSTALLATION

Vinyl-coated fabric wall coverings are installed using adhesives of a type recommended by the wall covering manufacturer. Wheat paste should not be used.

Primers are usually required. The types used should be those recommended by the wall covering manufacture.

Installation methods and cleanup should also follow the manufacturers' recommendations.

Most vinyl-coated wall covering manufacturers recommend that the substrate moisture content not exceed 5% at the time of application.

Material should be wrapped around inside and outside corners so that joints do not occur there. Vinyl-coated fabric wall covering should be smoothed with a bristle brush or flexible broad knife to ensure positive adhesion and eliminate bubbles. Excessive adhesives must be cleaned from surfaces immediately to prevent staining.

# 9.16 Painting

There are no generally accepted definitions in the industry that clearly state the difference between *paints* and *coatings*. The two terms are often used interchangeably. Because of the possible confusion created by this double usage, and to clarify some other terminology used in the painting industry, this book will define these terms as follows:

*Paints*: Air-drying materials; applied as a liquid in the field, except that primers are applied either in the field or in a shop or factory. The complete system of materials needed to produce a paint film is called a *paint system*. The term *paint* includes primers, emulsions, enamels, opaque stains, sealers, fillers, and other applied materials used as prime, intermediate, or finish coats in a paint system.

*Coatings*: Applied only in the shop or factory; similar in materials and formula to paint; either air-drying or baked-on; include such materials as Kynar coatings on metal, anodizing on aluminum, and baked enamel on lockers or furniture.

*Special Coatings*: Thick, high-performance, field-applied architectural coatings, such as high-build glaze coatings, fire-retardant coatings, industrial coatings such as those used in sewage disposal plants, and cementitious coatings.

*Transparent Finish*: A system of materials used to form a protective film on wood that lets the wood color, grain, or both remain visible; applied either in the shop or factory or in the field; sometimes called *clear* wood finishes.

Paints, transparent finishes, and a few special coatings are discussed in this section. Some coatings used on metal are covered in Chapter 5. Other special coatings are discussed in Section 9.17.

The comments in this section related to materials have been drawn from data developed by the Scientific Committee of the National Paint and Coatings Association and from recommendations included in the product and use literature of paint manufacturers. The paints and coatings discussed here are not the only ones available. Some manufacturers have other formulations, designed for specific purposes, that have advantages in their intended applications over the more commonly used paint and transparent finishes discussed here.

The information in this section related to methods of applying paint and transparent finishes has been developed from data contained in earlier editions of this book. These data were abstracted from the *Paints and Protective Coatings Manual of the Army, Navy, and Air Force* compiled by the National Bureau of Standards (now the National Institute of Standards and Technology) and from data contained in the preparation and application recommendations of paint manufacturers.

Paints and coatings are used to coat, beautify, and protect an underlying surface. They form a protective shield between construction materials and the elements that attack and deteriorate them. Typical causes of paint and transparent finishes failure are sunlight, temperature variations, saltwater, water vapor, decay, mildew, chemicals, and abrasion. When regularly renewed, paint and transparent finishes offer long-term protection that extends the useful life of the materials and the structure. However, as will be discussed later, the time between renewals varies considerably, being usually shorter for transparent finishes than for opaque ones.

Paints may serve other functions as well. For example, paint is an essential part of sanitary programs for hospitals, as well as for commercial, institutional, and home kitchens and washrooms. Gloss and semigloss paints provide smooth, nonabsorptive surfaces that can be washed easily and kept free of dirt and germs.

White and light-tinted paints reflect light, brighten rooms, and increase visibility. Flat paints soften and evenly distribute interior illumination. The functional use of color creates comfortable living and working conditions in the home, shop, and office.

Color-coded painting frequently is used to enhance public comfort, safety, and efficiency. Certain colors are stimulating, others are relaxing, and still others are universally associated with potentially dangerous situations.

## 9.16.1 MATERIALS

There is rarely one best paint or transparent finish for any surface. Among the factors that must be considered are the condition of the surface, the method of application, the curing conditions, the service expected, and the relative weight of initial cost against appearance and durability. Because the entire paint or transparent finish system must be considered as a unit, it is unwise and often impossible to choose a paint or transparent finish without regard to the surface over which it is to be applied. While it is generally impossible to single out the best paint or transparent finish, it is possible to list commonly encountered surfaces and to recommend the materials most suitable for each surface.

Paint is a mixture of minute solid particles known as *pigment* suspended in a liquid medium called the *vehicle*. The pigment provides hiding power and color. The vehicle combines (1) the *solvent* (thinner), which ensures the desired consistency for application by brush, roller, or spray, and (2) the *binder*, which bonds the pigment particles into a cohesive paint film during the drying process. Some paints dry and harden simply by evaporation of the solvent. Others involve chemical reactions as well as solvent evaporation. Most paints are described according to the type of binder used in them (alkyd, latex, oil, oleoresinous, rubber, urethane).

Materials in any of the types described here may be available in varying qualities. The reputation of the manufacturer is a significant factor when selecting paint and transparent finishes. The reputation of most major paint manufacturers is well known within the construction industry.

Many states, the District of Columbia, and many cities and towns have enacted air pollution regulations that restrict the amounts and types of solvents that may be emitted into the atmosphere by paints, transparent finishes, and coatings, and by the materials used to clean surfaces to receive them and remove them after they have been applied. In addition, many jurisdictions have regulations concerning the use and disposal of chemicals they do not prohibit. Therefore, paint removers or chemical cleaning materials are used, and before a paint, transparent finish, or coating is applied, it is necessary to verify its acceptability by the authorities having jurisdiction.

### 9.16.1.1 Interior Primers

Primers are paint or transparent finish materials intended for application over bare surfaces as a base for succeeding coats. Some topcoat materials also serve as a primer on some surfaces.

#### WALL PRIMERS AND PRIMER-SEALERS

Interior wall primers and primer-sealers may have a latex, alkyd, oil, or alkali-resistant binder, depending on the material to be covered and the topcoat that will be applied.

*Latex*  Latex primers and primer-sealers are emulsion paints intended for use on gypsum board surfaces, but they are also used on masonry. They do not raise the fibers on gypsum board, as do solvent-thinned primers. They have excellent alkali resistance. Equipment cleaning is easy using water. They dry quickly, which allows recoating within hours.

*Alkyd*  Alkyd primers and primer-sealers are usually made of odorless alkyd, meaning that the binder is an alkyd resin dissolved in odorless mineral spirits. They require overnight drying before recoating. They are suitable for all interior surfaces except paper-surfaced gypsum board and surfaces containing active alkali, such as partially cured or damp plaster or masonry.

*Oil*  The binder for oil primers and primer-sealers is a processed drying oil, often containing some resin. The characteristics of these materials are similar to those of alkyd primers, except that they are softer and dry more slowly. Even so, next-day recoating is possible. They may be thinned with odorless or ordinary mineral spirits.

*Alkali-Resistant*  Alkali-resistant primers and primer-sealers usually are based on a butadiene-styrene copolymer or on chlorinated rubber, with one or more softer resins, such as chlorinated paraffin, as plasticizers. Sometimes small amounts of

drying oil are added. Aromatic solvents often are required. Application is likely to be more difficult than for other primers. They are particularly suitable for alkaline surfaces, such as damp or partly cured plaster or masonry. They will not resist hydrostatic pressure, such as may occur below grade, but may be used on above-grade masonry.

#### WOOD PRIMERS

Interior wood primers include enamel undercoaters, clear wood sealers, and wood fillers.

*Enamel Undercoaters*  Enamel undercoaters may be alkyd, latex, or oil-based, but most are alkyds. Undercoaters are characterized by good hiding ability and low gloss. They are hard, tight films that prevent the penetration of enamel coats applied over them, a property known as *enamel holdout*. They may be used under any pigmented interior finish but are definitely used under gloss and semigloss enamel, where a smooth finish is important. Enamel undercoaters are usually white but may be tinted for use under colored finish coats. They contain no rust-inhibitive pigment, but they may be used as primers on metals where no rusting is expected. They dry sufficiently overnight to be painted the following day. Latex undercoats may be painted in 2 to 4 hours; oil-based undercoats should dry for at least 16 hours before recoating.

*Clear Wood Sealers*  Clear wood sealers seal the pores in wood surfaces without impairing their natural appearance. Many sealers contain a transparent pigment to improve sealing and reduce penetration. This reduces the prominence of the grain to some extent, which usually is desirable. These products generally are used under transparent finishes, although they can be used under pigmented materials as well.

*Wood Fillers*  Wood fillers include paste and liquid wood fillers. They are used beneath transparent finishes.

Paste wood filler is used to fill the pores of open-grain woods, such as oak or mahogany. A transparent and relatively coarse pigment is combined with sufficient binder to make a stiff paste, which requires thinning with a volatile solvent before application. Stain may be added. When the solvent has evaporated,

the filler is rubbed across the grain to force the filler into the pores and to remove excess material. This is followed by a cleanup wipe in the direction of the grain using a coarse cloth such as burlap.

Liquid wood fillers are designed to eliminate thinning. While they generally do not perform as well as paste fillers on many woods, they are adequate when extensive filling is not required. All fillers impart some color to the wood. Fillers containing stearates should not be used under urethane finishes.

#### MASONRY AND CONCRETE WALL PRIMERS AND FILLERS

*Primers*  Interior masonry and concrete wall primers may be either latex or alkyd. The first coat of an epoxy finish often serves as a primer, but conditions where a heavy-duty finish is needed may require a different epoxy formulation for the primer. Primers are used as sealers on normally smooth masonry surfaces.

*Block Fillers*  When a smoother finish is desired on masonry and concrete, a block filler can be applied beneath the paint primer. Block fillers are relatively thick materials that can be applied by brush, roller, or spray. When applied by spray, they must then be brushed into the surface to ensure that voids are filled. The binder may be latex or a solvent-thinned material such as an epoxy ester. Latex materials have excellent alkali resistance and may be used on damp surfaces. Solvent-thinned materials may have better chemical resistance.

*Cement Grout*  Instead of a block filler, a thin portland cement mortar may be applied with a brush or trowel to give a smooth surface to rough masonry or concrete. These materials are available premixed with latex or in powder form, which is mixed with a latex emulsion in the field. Mortar made without latex is permeable and is not a good substrate for paints.

### 9.16.1.2 Exterior Primers

The following primers are used on exterior surfaces.

#### WOOD PRIMERS

Exterior wood primers include both oil and alkyd primers, but those used most often today are alkyd.

**Oil Primers** Oil primers contain oil to control penetration. They generally do not contain zinc oxide because it can cause blistering and peeling, particularly under severe moisture conditions. However, paint systems without zinc oxide are more difficult to render mold-resistant.

**Alkyd Primers** Alkyd primers are similar to oil primers, except that their binder is alkyd. This promotes faster drying; shorter recoating time; resistance to bleeding, mold, and moisture; and controlled penetration.

### METAL PRIMERS

Metal primers may have oil, oleoresinous, alkyd, or latex binders. Most metal primers used today have an alkyd binder.

Some of the primers listed in this subsection contain lead. Lead pigment is toxic and coatings containing more than 0.06% lead should not be used in residential buildings or on structures with which children may come in contact. Many jurisdictions limit the lead content of paint by law, and some prohibit its use altogether.

**Red Lead (Red Lead in Oil)** Red lead in oil is probably the oldest type of metal primer. It is still in use today because it is very reliable. The oil binder wets the surface well and seeps into cracks and crevices to ensure protection of the metal, even on poorly prepared surfaces. However, drying is comparatively slow, and the film may be too soft for some finish coats. The red lead pigment will carbonate and turn white if left exposed too long. This primer should be covered with a finish coat within 6 months.

**Red Lead/Mixed Pigment** Red lead/mixed pigment primers are either oil- or alkyd-based primers that depend on the red lead for their corrosion-inhibitive properties. However, they also contain iron oxide and selected extenders which contribute to the general performance of the coating, adding hiding power, hardness, and general durability. In other respects, their properties are similar to those of the red lead in oil primers.

Alkyd-based red lead/mixed pigment primers are similar to their oil-based counterparts but will dry faster and harder, making them suitable for use under harder finish coats. The surface must be cleaned more completely before application than is required for oil-based primers.

**Zinc Chromate (Zinc Yellow)** Zinc chromate primers are made with zinc potassium chromate and often contain iron oxide as part of the pigment. The binder most often used today is alkyd. These primers depend on the zinc chromate in them for their anticorrosive effect. They generally dry harder and faster than red lead primers but require more thorough cleaning of the surface before application. When lead-free, they often are used for priming aluminum and magnesium surfaces.

**Zinc-Dust/Zinc-Oxide** Zinc-dust/zinc-oxide primers are available with oil, alkyd, or phenolic resin as the binder. They are especially effective on galvanized steel but may be used on other metallic surfaces as well. The oil type usually is the best for old, rusty, or weathered surfaces. The alkyd type is for use on clean surfaces and as a heat-resistant coating.

Phenolic primers generally are recommended only when the coated surface is to be immersed in water, but they are not recommended for exterior use. When these primers are used, the surface should be given a phosphate treatment before painting.

**Latex Inhibitive Primers** Latex inhibitive primers are water-soluble before they cure, which makes them easy to apply and easy to clean up. They are available in white, permitting coverage with one finish coat, usually within hours. These primers will give good corrosion protection but may show rust staining if not protected by a moisture-impervious finish coat.

**Wash Primers** Because wash primers usually contain phosphoric acid, they clean as well as prime a surface. They are useful in promoting adhesion of subsequent coats of paint and frequently are used to protect freshly sandblasted surfaces from further corrosion. Wash primers are furnished as either one- or two-part materials. One-part wash primers have the advantage of longer pot life after thinning but generally sacrifice some corrosion resistance when compared with the two-part type. Two-part formulations must be used the day they are mixed.

**Portland Cement Paints** Portland cement paints are either alkyd- or oil-based paints containing portland cement. They are not to be confused with cement grout or dry-powder portland cement paints which are designed for use on masonry. They are effective primers for galvanized steel. If white portland cement is used, a white primer results, permitting tinting to shade and contributing to the ease with which the primer can be covered by finish coats.

### MASONRY PRIMERS

Exterior masonry and concrete wall primers include latex and alkyd primers. The first coat of an epoxy finish often serves as a primer, but conditions in which a heavy-duty finish is needed may require a different epoxy formulation for the primer. Masonry primers are used as sealers on normally smooth masonry surfaces. A clear sealer sometimes is recommended to stabilize weathered surfaces. If latex finish coats are to be used, the sealer should be thinned so that it will dry to a low gloss.

Silicone water repellents are also used on exterior masonry surfaces when a water-repellent surface is required without a change in surface appearance. They consist of a solution of silicone resin in a solvent, usually an alkyd. Since they are transparent solutions, it is difficult to determine visually when an adequate coating has been obtained. To ensure that the desired water repellence is achieved, the spreading rate should be as recommended by the manufacturer. These materials should not be finished with other coatings until they have weathered for several years.

When a smoother finish is desired on masonry and concrete, block fillers or cement grouts are applied beneath the paint primer as indicated under "Masonry and Concrete Wall Primers" in Section 9.16.1.1.

#### 9.16.1.3 Interior Topcoat Paints

A wide range of interior topcoat paints are available with various surface characteristics.

### GLOSS FINISHES

Gloss finishes can be achieved using either paint (opaque) or transparent materials.

**Gloss Enamels** Most gloss enamels used today are alkyd-based. Modern formulations offer easier application than the older oleoresinous gloss paints. They also have much better gloss retention and resistance to yellowing and alkaline cleaners.

Some latex paints are also available in gloss formulations. They do not produce as high a sheen as alkyd enamels, but their ease of application and cleanup, continued advances in their quality, and more restrictive VOC requirements will eventually (if they haven't already) make them more popular than alkyd gloss enamels.

**Floor Enamels** Alkyd and oleoresinous enamels are formulated for relatively fast drying and for resistance to abrasion and impact. They are impermeable to water but are not notably resistant to alkalis, such as may arise from fresh concrete. These enamels will blister and peel if moisture accumulates behind the film. Most floor enamels used today are alkyd-based.

**Alkali-Resistant Enamels** Alkali-resistant enamels usually are based on a solvent-thinned styrene-butadiene resin or chlorinated rubber. They have high resistance to alkalis but relatively poor resistance to solvents. They flow and level well but may sag on vertical surfaces. Bubbling sometimes will occur when they are applied by roller on concrete floors.

**Epoxy and Urethane Enamels** Epoxy and opaque urethane enamels are available in one- or two-part formulations. The one-part materials are similar to alkyd and oleoresinous enamels but have somewhat improved properties, particularly abrasion resistance. The two-part materials have extremely high adhesion, abrasion resistance, and resistance to water and solvents, but they require great care in surface preparation to develop these properties to the maximum extent.

**Dry-Fallout Spray Gloss** Dry-fallout spray gloss is similar to alkyd gloss enamels but has a much faster-acting solvent and usually a faster-curing resin. It is designed for spray application only. An advantage is that much of the overspray dries quickly before reaching a surface and therefore can be removed easily.

**SEMIGLOSS FINISHES**

Semigloss paints are available with a range of glosses. Some manufacturers call their semigloss paints *eggshell* or *satin* finishes, and some call different amounts of gloss by different names. Here we will refer to all nonflat and non-high-gloss finishes as semigloss.

**Semigloss Wall Paint** Semigloss wall paint is an oleoresinous enamel with an initial intermediate gloss, which it may lose over time. It is likely to have low resistance to grease and alkaline cleaners, and is often too soft for use on horizontal surfaces. Its use is decreasing in favor of other types of semigloss enamels and paints.

**Semigloss Enamel** Semigloss enamel is an alkyd-based product with better gloss retention and grease and alkali resistance than semigloss oleoresinous wall paint. Application may be somewhat more difficult over large areas. The gloss may be reduced rapidly the first few days after application but will then stabilize.

**Semigloss Latex Paint and Enamel** Semigloss latex paints and enamels have the usual advantages of latex paints: ease of application and cleanup, rapid drying, little odor, nonflamability, and good film leveling. The hiding power may not be as good as that of alkyd products, but this disadvantage is rapidly disappearing as improved latex paints are developed.

**Dry-Fallout Spray Semigloss** This material, intended for spray application only, has the properties of its gloss counterpart but at a lower gloss level.

**FLAT FINISHES**

Flat paint finishes include oil and oleoresinous flats, alkyd flat wall paints, latex flat wall paints, and dry-fallout spray flat paints.

**Oil and Oleoresinous Flat Paints** Oil and oleoresinous flat paints are seldom used today. They have a predominantly limed oil vehicle, which reduces penetration and permits easy application and excellent leveling. The finished appearance is excellent, but washability and resistance to yellowing are likely to be poor. Because of good hiding power in thin films, these flat paints are sometimes used in repainting acoustical surfaces.

**Alkyd Flat Wall Paints** Alkyd flat wall paints usually are superior to latex paints in washability and hiding power, although their advantage in hiding power is dimin-ishing as better latex paints are developed. Odorless alkyd flat paints may have very little odor during application, but an after-odor usually occurs as the oil portion of the resin oxidizes. Whether odorless or not, these products should be used only with good ventilation.

**Latex Flat Wall Paint** Latex flat wall paint is the most popular kind of interior flat wall paint. High hiding power and good washability may be built into the product, but often one is achieved at the expense of the other. These paints are quick-drying and easy to touch up, cleaning of tools is easy, and they have no flammable solvents or unpleasant odors during or after application. However, ease of application may lead to excessive spreading rates and poor hiding ability. Even quality products may fail to form a coherent film and, hence, show poor washability if applied over a porous surface or under conditions of low humidity or temperature extremes.

**Dry-fallout Spray Flat Paints** Dry-fallout flat paints are made with fast-drying solvents and may be either of the alkyd or oleoresinous type. They are used on ceiling areas of industrial or commercial buildings where application is by spray, and where it is not practicable to protect all adjacent areas from overspray.

**Latex Floor Paints** Latex floor paints have most of the application advantages of latex wall paints, as well as fairly good abrasion resistance. The absence of toxic or flammable solvents makes them useful in basements and other areas where ventilation is poor. They may be applied to damp (but not wet) surfaces even if alkalis are present. Continuous exposure to standing water is likely to soften the paint film.

### 9.16.1.4 Exterior Topcoat Paints

Exterior finish coats include many formulations with oil, alkyd, latex, and other synthetic resins.

A wide range of alkyd- and latex-based house paints are available in flat or glossy sheens. Because of excellent color retention, durability, and ease of cleanup, latex exterior paints are gaining favor over the solvent-thinned alkyd types.

**OIL-BASE PAINTS**

Oil-based paints include *gloss house paint* and *barn paint*.

**Gloss House Paints** Traditional house paints were white lead in linseed oil. Later, various blends with zinc oxide, talc, and other extenders were developed. Finally, blends with various titanium dioxide pigments were introduced. These paints have been the foundation of the paint industry. Linseed oil is still the binder used in many quality oil-type house paints, although modifications with other materials may provide improved properties. A number of long-oil alkyds (resins with 55% or more oil or fatty acid) are available which are similar to oleoresinous trim enamels but are softer and more durable.

Lead has been found to be harmful to health and is no longer allowed in residential paints or paints in other locations where children may come into contact with it. Paints containing more than 0.06% lead (nonvolatile content) must carry a warning label and may be banned in some jurisdictions.

Zinc oxide is often included to control mold and to aid in the control of chalking; it is also a factor in regulating the hardness of the paint film. Extender pigments are used to control consistency and to develop an optimum blend of other properties.

Titanium dioxide is the most important pigment for providing hiding power in white and tinted paints, and contributes to the control of chalking and fading. Some of these factors are less important with dark-colored paints, which contain little or no white pigment.

**Barn Paints** Barn paints possess most of the characteristics of oil-based house paints. However, they are more likely to contain vehicle modifications, such as rosin, and they usually employ red iron oxide as the chief hiding pigment. Green, black, and several other colors are also available. Barn paints are formulated to give a uniform appearance over poorly prepared and nonuniform surfaces but lack the long-term protective qualities of house paints. They also may be used on roofs, particularly those of sheet metal.

## OLEORESINOUS PAINTS

Oleoresinous paints are rapidly disappearing from the market as alkyd-based and latex paints improve in quality. However, some paints for exterior metal are still made with oleoresinous vehicles, notably those for structural steel. They give good service and protection but usually

lose their gloss earlier than do similar products made with alkyds.

## ALKYD ENAMELS

Alkyd enamels for exterior use include trim enamels, equipment enamels, and masonry paints.

**Trim Enamels** Trim enamels are designed for use on exterior wood. They usually are made with long oil resins. For this reason and because they usually are designed for easy brush application, they are not fast-drying. Drying usually ranges from a few hours to overnight. Typically, they are made in high-gloss and bright colors, and show good retention of color and gloss. They are not suitable where high resistance to acid, alkali, or other chemicals is required. Silicone alkyds are substantially more durable than conventional alkyd enamels.

**Equipment Enamels** Equipment enamels are similar to trim enamels, but because they are intended chiefly for metal, a hard-drying resin may be employed and some rust-inhibitive pigment may be added. If they are designed for spray application, a faster solvent also will be used and drying may be quite rapid. These products often are called *automotive enamels*.

**Masonry Enamels** Masonry enamels are similar to alkyd trim enamels, except that they contain a higher proportion of pigment. Because very little flexibility is required of a masonry finish, these alkyd-based products perform very well, except when excessive moisture and alkali are present.

## EXTERIOR LATEX PAINTS

Exterior latex paints include latex house and trim paints, semigloss maintenance finish, masonry paint, and special latex paints.

**Latex House and Trim Paints** Latex house and trim paints are intended mainly for wood, but some of them can also be used over masonry and metal. They are rapidly gaining in popularity over solvent-thinned alkyd paints because of their excellent color retention, durability, and ease of cleanup. Primers will almost always be required over bare wood, badly weathered surfaces, and bare metal. Where primers are needed, it is best to employ one specifically recommended

for use with the intended finish-coat material. These products are similar to other latex paints in their general properties, and they are formulated for good flexibility so that they will accommodate dimensional changes of wood.

Most latex house and trim paints provide a matte finish and offer better color retention than oil and alkyd flats, even in light colors. They are also available with semigloss and gloss finishes.

**Semigloss Maintenance Finish** Semigloss latex maintenance paints applied over suitably primed metal have shown good performance. However, moisture vapor permeability is fairly high. They have good resistance to pitting and rusting, but some rust staining may appear unless an impermeable primer is used.

**Masonry Paint** Paints for masonry were the first large-scale exterior use of latex systems. Because masonry surfaces show less dimensional change than wood and metal, and also are rougher, adhesion is less of a problem. For surfaces with efflorescence, a clear or lightly pigmented, solvent-thinned sealer is often recommended. Like latex house paints, these products usually give a low-sheen finish.

**Special Latex Paints** Special latex exterior paints may be used on mineral-fiber cement and tile roofs. Special latex paints are available for painting roll roofing. They may be used on asphalt shingles but may curl them.

## ALUMINUM PAINTS

Aluminum paints include general-purpose and heat-resistant paints.

**General-Purpose Paints** General-purpose aluminum paints may be made with aluminum powder or paste and a vehicle of drying oil, alkyd resin, or one of a number of varnishes. The quality may vary widely, depending upon the amount of aluminum, the ratio of aluminum to binder, and the quality of the binder.

**Heat-Resistant Paints** The most heat-resistant aluminum paints are made with pure silicone resin. When other materials replace part of the silicone vehicle, there will be a reduction in the temperature the paint will withstand and in the paint's durability. Another type, intended for high-temperature surfaces, uses a small amount of vehicle (usually an oil) which burns off, leaving the aluminum melted and fused to the surface.

## ALKALI-RESISTANT PAINTS

Alkali-resistant paints are formulated with styrene-butadiene resin or chlorinated rubber. They dry flat, and ultimate deterioration of the film is caused by gradual chalking. These materials are used on masonry walls or decks, where higher resistance to moisture penetration is required than can be obtained with latex paints. They may be used as primers under less resistant finish coats. Their resistance to water, alkali, and acid is excellent, but their resistance to solvents may be poor.

### 9.16.1.5 Miscellaneous Opaque Coatings

Miscellaneous opaque coatings include fire-retardant paints, fire-retardant intumescent coatings, portland-cement powder paints, zone-marking paints, and roof coatings.

## FIRE-RETARDANT PAINTS

Fire-retardant paints are made in a number of types. They can be formulated as strictly functional or both functional and decorative coatings. Products are available with good fire retardance after repeated washing and with all the properties of conventional interior coatings. Since some of these paints are sensitive to water and high humidity, their suitability for the intended purpose should be determined.

Fire-retardant capability is usually expressed as a flame spread ratings determined and listed by UL. Flame spread is not a property of paint but depends on the number of coats and their thickness, the surface to which the paint is applied, and other factors. The flame spread rating of a paint may vary under different conditions. In particular, a paint may be non-combustible and have a low flame spread on noncombustible surfaces, but may have little fire retardance on combustible surfaces. However, a paint with a low rating on combustible surfaces will always have a low flame spread on noncombustible surfaces. If coatings are not applied at the listed spreading rates, the flame spread rating will not apply.

## FIRE-RETARDANT INTUMESCENT COATINGS

Fire-retardant intumescent coatings slow the progress of a fire by forming a pro-tective foam. In many cases, the charred foam may be scraped off and the sound substance repainted.

This material works when heat-sensitive ingredients in it begin to intumesce (swell) at 300°F (149°C). The resulting foam, which is hundreds of times thicker than the original paint film, slows the fire, reduces smoke, and delays the buildup of toxic gases.

Intumescent coatings may be used either in strategic places or throughout a building. These coatings are especially recommended for corridors or areas leading to fire exits. They are available pigmented or clear and are scrubbable.

## PORTLAND-CEMENT POWDER PAINTS

Portland-cement powder paints are made from white portland cement, suitable pigments, and usually small amounts of a water-repellent agent. They are mixed with water just before application. Since hydration of the portland cement requires moisture, painted surfaces should be kept damp as recommended by the manufacturer. These paints are useful, low-cost finishes for rough masonry. Their principal drawback is that they are not good bases for other paints or coatings.

## ZONE-MARKING PAINTS

Zone-marking paints, which are sometimes called *traffic paints*, are fast-drying and have little flow, good hiding ability, resistance to bleeding over asphalt, and normally low gloss. Zone-marking paints for highway use may be reflectorized or nonreflectorized. Nonreflectorized paints are used where there is overhead lighting, as, for example, on many parking lots. Reflectorized paints may be of the drop-in type, where glass beads or similar material are added after application of the paint, or premixed with beads incorporated in the paint. The drop-in type usually is more durable and has better visibility early in the life of the paint. The premixed type is easier to apply. A small number of beads may be dropped into the premixed paint after application for early visibility.

Normal zone-marking paints are flat and therefore are not satisfactory for such interior uses as factory floors or school corridors where dirt collection is a problem. Marking paints for this purpose must have a gloss or semigloss finish.

## ROOF COATINGS

Bituminous roof coatings are made of special weather-resistant asphalt dis-solved in a solvent. They are intended for use on asphalt roofing surfaces. Mineral fiber and other fillers may be added to prevent sagging on sloping roofs and to permit the application of relatively thick coatings. The addition of aluminum powder enables colors other than black to be obtained. For some roof coatings the asphalt is emulsified in water, permitting application to damp surfaces.

### 9.16.1.6 Transparent Finishes

Transparent finishes are applied on wood. Transparent finishing materials include wood stains, linseed oil, varnish, shellac, lacquer, and polyurethane.

Transparent finishes are used mostly in interior locations, but some are available for exterior use. In exterior locations, they do not usually perform as well as paints do, because they cannot approach the durability of the same vehicle protected by pigment.

## STAINS

Wood stains are made from color pigments suspended in linseed oil or another drying oil. They are then thinned to make them easy to apply. Types of stains include pigmented, dye-type, and shingle stains.

*Pigmented Stains* Pigmented stains contain insoluble pigments, generally permanent in color. If applied too heavily, they will obscure the texture, giving a painted effect to the wood. They are particularly suitable for subduing the conspicuous irregular grain effect of softwood plywood. These stains will not bleed into varnish or enamel coats applied over them.

*Dye-Type Stains* Dye-type stains employ an organic coloring material that is soluble in the vehicle. They have better clarity and transparency than pigmented stains, and they produce brighter colors. However, they are more likely to bleed into succeeding coats, and the color is less permanent.

*Shingle Stains* Shingle stains are pigmented products designed to provide color, only partial hiding ability, maximum penetration, and some mold-resistant protection. They may contain creosote or another preservative; linseed oil and alkyd resin are the usual binders.

## LINSEED OIL

Boiled linseed oil is the classic wood finish. It is inexpensive and easy to apply, maintain, and repair. It is normally reduced with terpentine in ratios ranging from equal parts to twice as much oil as terpentine.

## VARNISH

Varnish is a homogeneous mixture of resin, drying oil, dryer, and solvent. It can produce a transparent film that is either flat, satin, or high gloss. Varnish is available in colors ranging from clear to dark brown and in several qualities. Most varnishes for interior use include alkyd resins, but epoxy-ester varnishes are also available. The old oleoresinous varnishes have been almost completely replaced today by alkyd varnishes, which often are paler and have better color retention. Epoxy varnishes have good color quality and alkali resistance.

Alkyd varnishes have good initial color and color retention but may crack and peel over time. Some synthetics have good durability but may darken on exposure. A long oil phenolic, marine spar varnish, has the best history of satisfactory performance, although it also will darken and yellow with time. Penetrating coats of oil or thinned varnish should be used on the bare wood, followed by unthinned varnish. Varnishes containing ultraviolet screening agents exhibit less discoloration and greater durability.

*Flat and satin varnishes* have an added flatting agent such as silica. They do not have the undesirable highlights associated with gloss varnishes, but they are less abrasion resistant.

*Floor varnishes* are relatively fast-drying, with good resistance to yellowing and excellent abrasion resistance.

*Counter varnishes* are hard-drying, printproof films. *Bar top* varnishes should also be resistant to alcohol. Polyurethanes often are used for this purpose.

*Paneling and trim varnishes* need not be as hard as varnish for floors or counters, but should be resistant to checking and should dry tack-free. Alkyds usually exhibit the best color quality and color retention.

*Spar varnish* is a varnish in which additives have been included to give the material resistance to salt water.

*Clear sealers* are solvent-thinned varnishes that penetrate wood to prevent grain rising and seal porous wood. They

are applied liberally, and the excess is removed by wiping while still wet or by buffing when dry (after about 15 hours). Buffing provides better protection but involves more time and labor. These products give a satin finish with fair abrasion resistance and color retention.

## SHELLAC

Shellac produces a finish similar to that of varnish. It is fast-drying, light-colored, and has excellent color retention. Although its abrasion resistance is poor, the finish is easily repaired. Resistance to water and alcohol is poor, but resistance to petroleum solvents is good. Shellac is often used as a wash coat on wood before final sanding or as a clear primer or stain sealer.

A major disadvantage of shellac is its short shelf life; it is unusable within 4 to 6 months after manufacture.

## LACQUER

Lacquer is available in either a flat or glossy finish but is seldom used for field-applied finishing of building surfaces. It is more likely to appear as a finish for furniture or shop-finished casework.

A major disadvantage of lacquers is that they are not usable over previously finished surfaces. In fact, lacquer can be used as a paint remover.

## POLYURETHANE

Some manufacturers classify polyurethanes as varnishes. Others call them lacquers. In fact, although they have some characteristics of each, they are not truly either. They are available either in oil-modified or moisture-curing formulations. Oil-modified polyurethanes are transparent materials that are mostly used on interior surfaces. They are water-resistant and highly durable. They can produce either matte or gloss finishes but tend to lose their gloss quickly when used on the exterior. They cannot be used over shellac, paste wood fillers, and some other materials. The manufacturer's advice should be sought about where they can be used.

Moisture-curing polyurethanes can be either clear or pigmented to produce an opaque finish.

Polyurethanes are often used on bar tops and countertops and on floors.

Urethanes provide better abrasion resistance than varnishes or lacquers but generally require more thorough surface preparation prior to application.

## 9.16.2 SURFACES

This subsection outlines the broad categories of surfaces on which paints, transparent finishes, and coatings are used, and describes products suitable for each surface under varying exposure conditions. Unusual surfaces and surfaces that are rarely painted (glass, underground metal, underwater concrete, etc.) require specialized coatings designed for the special conditions of use and are not included in the following discussion. Products recommended in the following discussion are identified by generic name (see Section 9.16.1).

### 9.16.2.1 Interior Walls and Ceilings

#### PRIMING

Most surfaces require a primer as a base for topcoats. The importance of proper selection and application of primers cannot be overemphasized. Regardless of the quality of subsequent finishing, failure of the primer will mean failure of the system.

*Plaster* New plaster should be primed with a latex, alkyd, or oil-type primer-sealer. A latex flat finish coat may be used as a primer, but if semigloss or gloss paints are to be used as a finish coat, the primer should be tested for enamel *holdout* (resistance to penetration by the finish coat), which results in uneven gloss. Highly alkaline or fresh, damp plaster requires an alkali-resistant primer or latex system. Wet plaster and plaster having a continuing flow of moisture from the back cannot be painted successfully. Textured or swirl plaster and soft, porous, or powdery plaster should be given an acid wash with a solution of 1 pt (473.2 ml) of white household vinegar in 1 gal (3.79L) of water. This procedure should be repeated until the surface is hard; it should then be rinsed well and allowed to dry before painting.

*Gypsum Board* Latex primer-sealers are recommended for gypsum board and other paper-surfaced materials because solvent-thinned primers raise the fibers of the paper, giving it an unsightly appearance. A latex finish coat can be used as a primer, but semigloss or gloss paints used as primers should be tested for enamel holdout.

*Brick* Brick interior walls are usually somewhat porous, but if the mortar is well cured, they will not be highly alkaline. They may be sealed with a latex primer-sealer or with an enamel undercoater containing oil or varnish. Latex or solvent-thinned exterior masonry paints may also be used for priming. In every case, efflorescence should be removed by vigorous scrubbing, and the brick should be treated with a clear resin sealer. It should be noted that the Brick Industry Association (BIA) does not recommend painting brick.

*Concrete Masonry*  Most concrete masonry (CMU), especially the lightweight type, has a very rough, porous surface. Paints and coatings usually cannot bridge the surface voids in these materials. One or two coats of block filler can be used to smooth such surfaces, thus reducing dirt accumulation and improving surface color uniformity. Fillers may be omitted when these considerations are not important, when low-cost finishing is required, and when the sound-absorbing effect of a rough surface is desirable.

*Concrete*  The problems associated with painting concrete are similar in many respects to those of concrete masonry. However, the concrete surface usually is denser, less regularly formed, and more likely to be contaminated with oils and other bond-breaking materials. Foreign substances should be removed unless it can be ensured that they will not affect paint adhesion. Rough or honeycombed surfaces can be smoothed with cement grout. This grout is usually thinned with a latex solution, especially when it is to be applied in thin layers. Grout is primed as described earlier for priming plaster. One or two coats of block filler may be used instead of grout, thus eliminating the need for an alkali-resistant primer.

*Ferrous Metals*  A rust-inhibitive primer is always desirable and is essential for quality work, for damp locations, when water-thinned paints are used for finishing, and when the finish coats do not produce an impervious film. A primer may be omitted in dry locations and if the finish coats are to be of a solvent-thinned type producing a tight film. Slow-drying oil-base primers wet the surface better and give more positive results, but quick-drying primers are more resistant to moisture penetration.

*Zinc-Coated (Galvanized) Metals*  When applied over zinc-coated surfaces such as galvanized steel, many paints and rust-inhibitive primers develop a layer of white, powdery material which eventually separates the paint film from the metal. This condition can be prevented by using (1) a nonoxidizing coating such as a chemical-resistant coating or, more commonly, (2) a primer containing zinc dust or portland cement. There are proprietary primers for galvanized steel (often latex) which are very effective for interior use.

*Wood*  In many interior locations, wood may be primed at the same time and with the same sealer used for plaster walls. One or two coats of paint may give a sufficiently smooth base for enamels, but such water-thinned paints will raise the grain and roughen the surface of many woods. Good enamel holdout is necessary under gloss or semigloss finish coats. Solvent-thinned wall primer-sealers provide better adhesion and flexibility. For the best enamel work, an enamel undercoater should be used either over a wall primer-sealer or directly on the wood. A clear wood sealer is recommended for natural finishes, preceded by a paste wood filler if necessary.

*Acoustical Surfaces*  Acoustical tile and acoustical plaster are rarely painted at the time of their installation but may require finishing later. Since acoustical properties depend on a suitably porous or rough surface, a thin paint should be used so that the pores are not bridged or filled. Paints used on plaster surfaces may be thinned and used on acoustical surfaces. Metal acoustical surfaces should be treated as light-duty areas using one coat to avoid excessive paint buildup.

## FINISHING

Finishing differs slightly, depending on the use of the location.

*Light-Duty Areas*  Some building areas have low concentrations of moisture, grease, or dirt in the air and require relatively little maintenance. These include small private offices and other similar spaces in commercial and institutional buildings, as well as living rooms, adult bedrooms, and similar areas in residential facilities. In such spaces, a matte finish is both desirable and practical on the ceilings and most walls.

Latex flat wall paints are usually preferred because of their easy application, quick drying, easy cleanup, and lack of odor. Alkyd flats can produce thicker films, have better hiding power, and give a more uniform appearance. These properties are particularly desirable on rough surfaces and when a considerable change is made in color. Both latex and alkyd paints have good resistance to soiling and marring and are relatively easy to clean.

Odorless alkyd flats exhibit little solvent odor during application, but they must be used with adequate ventilation because the solvent fumes, while odorfree, may cause headaches and other physiological symptoms. The curing of these air-drying paints results in odors which can persist for several days, requiring continued ventilation.

*Moderate-Duty Areas*  Moderate-duty areas are similar to light-duty areas in most respects, except that the amount of traffic or airborne dirt may dictate a more intensive cleaning schedule. Typical areas are private corridors and large office spaces in commercial and institutional buildings, as well as children's rooms, entries, halls, and closets in residential occupancy. Finish coats should be chosen from among the more washable flat and semigloss paints available. Glossier materials usually have better wear resistance and washability, but the accompanying higher surface shine makes surface irregularities more conspicuous, especially when natural or artificial light strikes the surface at a sharp angle. This consideration becomes less important in smaller spaces or where wall areas are broken up by interior furnishings.

*Heavy-Duty Areas*  Heavy-duty areas are those in which there may be large amounts of moisture and grease in the air or where the amount and type of traffic requires a more durable paint film to withstand greater wear or more intensive cleaning. These areas include kitchens, recreation rooms, toilets, public corridors, entrance lobbies, and service areas in institutional, commercial, and apartment buildings, and bathrooms in homes and apartments.

The usual choice is a gloss or semigloss enamel. Where the exposure is particularly severe, hard-gloss coatings can be used. The light reflectivity factors discussed above for moderate-duty areas always should be considered when selecting gloss finishes.

### 9.16.2.2 Interior Trim and Paneling

The selection of primers for interior use on trim and paneling depends primarily on the type of surface material being coated and the desired finish characteristics.

#### PRIMING

*Ferrous Metals*  Although most interior exposures are not particularly corrosive, a rust-inhibitive primer is usually desirable for ferrous metal surfaces. Any metal primer (oil, alkyd, or latex) may be used. The choice is dictated by the speed of drying, the amount of surface preparation, and the desired color. For very mild exposure, enamel undercoaters may be used, but these products have no corrosion-resistant effect and rust will spread from breaks in the film.

*Zinc-Coated Metals*  Interior galvanized surfaces which are dry and do not receive heavy wear may be coated with interior latex paints. However, if moisture, abrasion, or repeated cleaning is expected, an inhibitive primer will provide better service.

*Wood, Clear Finish*  Priming will depend on both the wood and the type of finish desired. Trim should be primed on all surfaces before installation. Most softwoods will require a clear sealer, especially those, such as pine, that show marked contrast between springwood and summerwood. A sealer will provide a more uniform surface over hard and soft grains; thus, it is especially recommended if the surface is to be stained.

Dense, close-grained hardwoods (maple, birch, etc.) have less need for a sealer, and the first coat may be a thinned version of the finish coat if so recommended by the manufacturer. In stain systems, the products used and the order in which they are applied will depend on the effect desired. For example, applying the sealer before the stain will minimize both the depth of color and the amount of grain contrast.

Open-grained hardwoods (oak, walnut, mahogany) require a wood filler, which also will serve as a sealer. When the grain is to be subdued, stain may be applied after the filler.

*Wood, Opaque Finish*  Interior wall primers and enamel undercoaters may be used as a primer on wood to receive an opaque finish, but water-thinned primers are apt to raise the grain of the wood. An enamel undercoat will give a smoother surface and is preferred under gloss and semigloss enamels, although many semigloss enamels are designed to be self-priming.

#### FINISHING

Finish coat selection is determined largely by the type of exposure expected.

*Light- and Moderate-Duty Areas*  The selection of clear finishes for wood depends on decorative and exposure requirements. For light- and moderate-duty areas, satin or low-luster finishes often are used.

*Opaque finishes* for wood and metal usually are semigloss or latex enamels. In light- and moderate-duty areas, the predominant interior finish used on the adjacent walls may be used.

*Heavy-Duty Areas*  A clear finish for wood in heavy-duty areas may be floor varnish or counter varnish. Floor varnish is most resistant to abrasion, while counter varnish has good resistance to water and solvents. The selection should be made on the basis of the exposure anticipated.

Opaque finishes for wood and metal are usually hard-drying enamels. For special exposures, such as in laundries, a special type may be indicated.

### 9.16.2.3 Interior Floors

The selection of finishes for interior floors depends primarily on the type of surface to be coated.

#### WOOD FLOORS

*Clear Finishes*  Open-grained woods such as oak may use a filler before coating. Clear finishes include:

- *Shellac:* For light-duty floors where color quality and retention are important; good gloss and color; not water-resistant; scratches easily but is easily patched
- *Floor varnish:* More resistant to wear and water but somewhat darker in color; may be slippery when wet
- *Wood sealer:* Lower gloss; usually requires waxing
- *Specially designed finishes:* For heavy-duty areas such as gymnasium floors
- *Polyurethane:* For heavy-duty floors

*Opaque Finishes*  Alkyd enamels and latex floor paints generally are used for opaque floor finishes. Other fast-drying gloss enamels and paints may be used if recommended by the manufacturer. However, products not specifically recommended for floors may be too soft and slow-drying or too hard and brittle.

#### CONCRETE FLOORS

Enamels suitable for wood floors are generally suitable for concrete floors as well. However, alkali-resistant enamels should be used when a concrete slab is not completely cured or is placed below grade or on the ground without a vapor retarder. Even with alkali-resistant enamels, adhesion problems may be encountered if hydrostatic pressure forces water up through the floor.

Latex floor paints usually dry to a comparatively low gloss. Because they are nonflammable and have little odor, latex paints are preferred for basements and enclosed areas that are difficult to ventilate adequately. Since they are water emulsions and have excellent alkali resistance, the floor need not be completely dry before application. However, latex paints are softened by continued exposure to moisture or solvents and are not suitable for laundry and service areas where solvents are handled. In such exposures, specialized highly resistant finishes should be used.

#### STEEL FLOORS

A fast-drying, tough enamel applied over a corrosion-resistant primer is recommended for most steel floors. Floors subject to chemical agents as in industrial installations should be finished with specialized industrial coatings.

#### CERAMIC FLOORS

Ceramic floors are seldom painted, but they may be coated with an abrasion-resistant clear coating to seal the surface and facilitate cleaning.

#### RESILIENT FLOORS

Resilient flooring is rarely coated when new, but coating may be required to restore its appearance or change its color. The coating chosen will depend on the flooring material, its condition, the adhesive used in the installation, and the effect desired. Specific recommendations should be obtained from the paint manufacturer for each case.

### 9.16.2.4 Exterior Priming

As with the finishing of interior surfaces, the selection and application of primers are critical to the performance of an exterior coating system.

#### WOOD

*Siding* Smooth siding should be primed with a wood primer that is compatible with the intended finish coat. The manufacturer's recommendations for the finish coat should govern the choice of a primer. Nail heads, cracks, and deep surface imperfections should be filled; caulking should be provided at the perimeters of doors and windows and wherever required for weather resistance. All filling and caulking should be done after the siding is primed.

For rough siding, shakes, and shingles, a wood primer may be used, or the finish coat may be thinned and used as a primer if it is zinc-free and is recommended by the manufacturer. Flat house paint and shingle stain do not require special primers. Factory-primed siding does not require a field-applied primer unless the surface has been heavily damaged during shipping or application. Minor scratches, dents, and raw edges should be spot-primed.

*Plywood* After prolonged exterior exposure, plywood surfaces will check and the coating will crack unless an overlaid plywood is used. Medium-density overlaid plywood with a resin-impregnated paper covering is recommended for an opaque paint finish. Painting recommendations for wood siding also are applicable to plywood. Textured plywood, such as grooved, rough-sawn, and striated surfaces, is not intended for painting and only pigmented stains should be used on it.

*Transparent Finishes* Wood surfaces intended for a transparent finish usually are primed with a coat of the same material to be used as the finish coat. To improve penetration, some manufacturers recommend that the finish coat be thinned when used for priming. However, some products may be damaged by thinning; therefore, it should be done only when specifically recommended by the manufacturer.

#### MASONRY

The cost of painting a clay masonry wall in most cases must be charged to appearance alone, not to increased performance or resistance to weather. The only exception is cement-water paint applied to correct a leaky wall condition. In the case of concrete masonry, paint not only provides an attractive finish but may also be necessary to make the masonry surface watertight.

Masonry paints are of four general types:

- *Cement-based paints:* Powdered form, mixed with water before application
- *Water-thinned emulsion paints:* Butadiene-styrene, vinyl acrylic, alkyd, and multicolored lacquers
- *Fill coats:* Similar to cement-based paints but contain an emulsion paint in place of some water, giving improved adhesion and a tougher film than unmodified cement paints
- *Solvent-thinned paints:* Oil-base, alkyd, synthetic rubber, chlorinated rubber, and epoxy; except for special-purpose paints and special applications, should be used only on interior masonry walls that are not susceptible to moisture penetration

Masonry surfaces generally do not require a special primer, and the first coat can be the same material as that used for the finish coat. The finish coat material may require thinning, but the manufacturer's recommendations should be followed. Common brick should be sealed with a penetrating type of exterior varnish to minimize spalling and efflorescence, which may cause peeling.

Cement grout thinned with latex may be applied to masonry surfaces that have large voids. Sound but rough surfaces, especially in concrete block, may be surfaced with block filler. Cast-in-place concrete should be cleaned of form oils, and other bond-breaking agents should be removed unless they are guaranteed not to affect paint adhesion. Old surfaces which have become excessively chalky may be bonded or stabilized with a clear or lightly pigmented phenolic or alkyd coating, but if the existing surface is soft or crumbly, it should be removed to achieve a sound layer, usually by sandblasting. Where only water repellence is required, a silicone solution may be used above grade; hot melt or cutback asphalt can be used below grade.

#### METAL

*Ferrous Metal* Coating materials cannot compensate for inadequate surface preparation. Rust, mill scale, dirt, oils, and old, loose paint must be removed. The standards for surface preparation recommended by the Society for Protective Coatings should be followed.

An appropriate primer should be used, depending on conditions and the finish materials intended. Oil and oleoresinous materials generally have better bonding properties and are more suitable for hand-cleaned surfaces. Alkyd and other synthetic primers dry more rapidly but require more careful surface preparation. Chemical-resistant and industrial coatings usually are marketed as complete systems of primer and finish coat, and the manufacturer's recommendations concerning the appropriate primer should be followed.

*Galvanized Steel* Zinc-dust/zinc-oxide, latex, and portland cement paints give excellent results, as do certain proprietary paints designed for use on galvanized steel. Conventional exterior paints usually lose adhesion rapidly unless the surface has weathered for a considerable length of time. If the galvanized surface has weathered to the point where substantial amounts of rust are showing, the recommendations for painting ferrous metals should be followed. Chemical pre-treatments, usually based on phosphoric acid, improve the adhesion of paints to galvanized metal. However, other acids or chemical solutions sometimes recommended for the preparation of galvanized surfaces rarely contribute to the performance of paints; actually, they may be harmful, and their use generally should be avoided.

*Aluminum and Magnesium* Zinc chromate primers are usually recommended for aluminum and magnesium.

*Copper, Brass, and Bronze* Copper and copper-bearing metals oxidize rapidly when exposed to air and moisture. To preserve the bright metallic luster of new polished copper and brass, a clear coating should be applied promptly. Old copper and brass should be buffed or polished to a bright color, cleaned with a mild phosphoric acid cleaner, and coated before the surface becomes discolored by oxidation. Specially formulated clear coatings are available which will preserve the bright luster and protect the metal from discoloration by the elements.

Copper gutters and downspouts generally are not painted or coated. The rapidly forming oxide (patina) darkens the surface but protects the underlying metal from progressive deterioration. However, copper may be cleaned and coated either to match the color of adjacent surfaces or to protect them from staining by runoff from the copper. Coating is especially recommended for copper surfaces located above aluminum products to prevent electrolytic deterioration of the aluminum by water wash from above.

### 9.16.2.5 Exterior Finishing

The choice of finish should be influenced by the type of exposure expected and the performance desired. High temperatures and humidity, particularly in the absence of direct sunlight, encourage the growth of mold (mildew). Therefore, it is important to choose fairly hard-drying finish coats, avoiding those types which may dry to a rough, soft surface, as these finishes will tend to trap and hold mold spores. Finish coats containing substantial amounts of zinc oxide are effective in controlling fungus growth. Various other additives are used to provide further mold resistance.

If a high degree of mold resistance is required, the primer as well as the finish coat should be designed for this purpose. However, the use of coatings containing zinc oxide is not recommended on bare wood. The degree of mold resistance is determined by the total formulation rather than by a single component, and the specific product should have adequate resistance under conditions similar to those to be encountered.

Industrial fumes may darken painted surfaces by the reaction of hydrogen sulfide with lead or occasionally mercury compounds. Paints used in areas subject to hydrogen sulfide should be free from lead and mercury.

Soot and other airborne materials may settle on and discolor a paint film. Hard-drying films will not retain dirt as tenaciously as softer films. Paints may be formulated to encourage moderate chalking of the surface, which has a self-cleaning effect, as the soot and dirt are carried away with the surface layer.

However, for severe weather exposures, resistance to chalking is a desirable property. Repeated exposure to rain and high winds generally will keep the surface clean but will erode the film too quickly unless a comparatively non-chalking paint is chosen.

### WOOD SIDING

***Smooth Siding*** Oil paints or exterior latex paints are preferred for whites and light tints on smooth siding. Oil paints have better hiding ability and will show less staining from wood resins. Latex paints are easier to apply, may be used on slightly damp surfaces, and often have better color retention, particularly in light tints. Where factory-primed siding is used, the recommendations of the siding manufacturer should be followed. Transparent finishes may be used where the natural appearance of the wood is desired, but their durability is much lower than that of pigmented products.

***Rough Siding, Shingles, and Shakes*** Flat house paint, shingle stains, or exterior latex paints are usually used on rough siding, shingles, and shakes.

Latex paints generally give good service if rust-resistant nails have been used. Wood discoloration can be avoided by sealing with a solvent primer or using a latex primer specially designed for resistance to staining. Rough surfaces have a substantially greater surface area than smooth surfaces, and more paint must be applied to obtain a suitable film thickness.

Shingle stains are often the choice for wood shingles. These stains penetrate the wood but crack rather easily. Impermeable coatings usually are not desirable because they seal in moisture and are likely to peel. Staining shingles before installation is preferred because many areas are inaccessible after installation.

### WOOD TRIM

Alkyd trim enamels should be used where high gloss and good abrasion resistance are desired, as around doors and windows. Any of the finish coats suggested above for smooth siding also may be used.

### METAL

Oleoresinous and alkyd paints are suitable finishes over compatible primers. Alkyd paints tend to be harder and more abrasion resistant. Paints recommended for wood, such as alkyd enamels, can be used if the surface is properly primed.

### CONCRETE AND CONCRETE MASONRY

An alkali-resistant coating is required over concrete and concrete masonry because, even after some aging, they still contain alkali, which will attack oils and oil-based materials. Latex paints or resin-based finishes may also be used. Resin-based finishes are preferred on flat areas such as ledges and window sills and on areas subject to foot traffic and standing water, such as patios and roofs. Other types of paint may be used on well-cured concrete, on concrete that is sealed with an impervious primer, and on previously painted surfaces.

### BRICK AND STONE MASONRY

Because brick and stone usually do not contain alkali, almost any exterior paint may be used on them with satisfactory results if the mortar is well cured. However, since fresh mortar is usually strongly alkaline, only paints recommended for concrete should be used on new walls. Where the natural appearance of brick or stone is desired, a silicone-resin solution may be used to retard the penetration of water. The solution and spreading rate should be as recommended by the manufacturer. Silicone finishes should not be painted until they have weathered for at least 2 years.

### MINERAL-FIBER-CEMENT

Mineral-fiber-cement compositions such as shingles, siding, corrugated boards, and insulating sandwich panels often are supplied with a transparent water-repellent coating which should be weathered or removed before painting. For most products, trapped moisture is seldom a problem, but alkalinity may be too high for oleoresinous or alkyd paints. However, the insulating core of sandwich panels may absorb moisture, releasing it later to damage the coating. It is recommended that the edges of insulating panels be sealed after cutting and before installation. Exterior latex or solvent-thinned resin paints are generally preferred.

### ASPHALT ROOFING

Asphalt shingles, roll roofing, and built-up roofing are rarely coated when new, unless for decorative or heat-reflecting purposes. Most solvent-thinned paints are too brittle to be used over asphalt. The solvent will dissolve the asphalt, and the paint will become stained.

Bituminous roof coatings are most commonly used, and if heat reflection is required, these coatings may be obtained pigmented with aluminum or in a limited range of colors. Where white is required, certain latex paints are sufficiently flexible to give satisfactory performance. However, latex paints are likely to curl shingles and are not recommended for these products. Because severe exposure conditions are encountered on roofs, thick coatings are required to give satisfactory durability.

## METAL ROOFING

In general, paints recommended for metal roofs follow the recommendations for other exterior metals. However, because foot traffic is rarely a consideration, softer coatings such as house paints or barn paints may be used for finish coats. As on other roofs, a relatively heavy coating is desirable for resistance to weathering. Terneplate roofs are usually painted with a special iron oxide primer and then finished as suggested here.

### 9.16.3 APPLICATION

Satisfactory painting and transparent finishing depend on (1) proper selection of paint and transparent finishing materials, (2) adequate preparation of materials and surfaces and (3) suitable application methods and procedures.

Paints, transparent finishes, and coatings should be standard commercial brands with a history of satisfactory use under conditions equal to or similar to the conditions present in the area concerned. They should be recommended by the manufacturer for the specific use proposed. The printed application instructions should clearly identify the suitability of the material for the type of exposure (exterior or interior), the type of surface to be covered (wood, metal, masonry, concrete, gypsum board, plaster), and the type of service to which the paint will be subjected, such as exposure to moisture, frequent washing, or heavy traffic.

#### 9.16.3.1 Materials Handling

Proper storage conditions and stocking procedures are essential to avoid material damage, explosion hazards, and application problems.

#### STORAGE

Paint materials should be stored in dry, and well-ventilated areas maintained at a temperature between 50°F (10°C) and 90°F (32.2°C). Lower temperatures generally increase material viscosity and make it necessary to temperature-condition materials before use. Freezing temperatures may permanently damage water-based paints.

If a paint material is sensitive to heat, temperatures higher than 100°F (37.78°C) may bring about a chemical reaction within the container. This generally results in increased viscosity to the point of gelation, and the paint becomes unusable. In addition, overheating may cause covers to blow off, creating a serious fire hazard.

Other hazards to be considered are high humidity, which may cause containers to corrode and labels to deteriorate, and poor ventilation, which may allow excessive concentration of solvent vapors that are both toxic and combustible.

#### STOCKING

Paint materials should be rotated so that the oldest stocks are used first. It is desirable to make paints ready for use by conditioning them at the proper temperature and mixing them thoroughly. Using leftover paint should be avoided. When using old paint is necessary, such material should be strained and temperature conditioned again, as recommended below for unopened containers.

Paint material should be stored in full, tightly sealed containers. Required quantities should be carefully estimated so that little or no material is left at the end of a job. It is generally safer to discard small quantities than to use paint that has skinned. When saved, leftover paint should be placed in smaller containers, filled full, and tightly sealed. All information from the original container should be transferred to the new container.

#### 9.16.3.2 Conditioning and Mixing

Unless stored at the recommended temperature, paint materials should be temperature conditioned at 65°F (18.3°C) to 85°F (29.4°C) for at least 24 hours before use. After this time, paints should be mixed, thinned, or tinted as required and strained if necessary, all according to the manufacturer's recommendations.

Paints consist of two principal components, solid pigment and liquid vehicle. Suspension agents are included in most paints to minimize settling of solids. However, when paints are stored for long periods, separation of the pigment and vehicle may take place. The purposes of mixing are to reblend settled pigment with the vehicle and to eliminate lumps, skins, and other hindrances to proper application.

#### MIXING

Paint materials should always be mixed prior to issuance or delivery to the job site. Mixing can be done by either manual or mechanical methods, but the latter is preferred to ensure maximum uniformity. Some hand stirring should be done just prior to use and periodically during application. Regardless of the procedure used, caution is recommended to avoid the incorporation of excessive air through overmixing. Figure 9.16-1 outlines the recommended mixing procedures for different coating types.

It is important to mix ready-mixed paints before as well as after introducing thinners or other additives. It is best to use the same conditioning procedures for all components of multicomponent paint materials before mixing. Manufacturers' label directions regarding proper mixing should be followed.

*Mechanical Mixers* The two most common mechanical mixers either vibrate the full sealed container or utilize propellers that are inserted into the paint. Vibrating shakers are used for full containers with capacities of up to 5 gal (18.93 L). Propeller mixers are simple hand drill attachments and generally are used on the job for containers holding 1 qt (946.53 mL) or more.

*Manual Mixing* Manual mixing is more time-consuming than mechanical mixing, but can be used when mechanical equipment is not available and for mixing materials that have been recently well mixed. Paint should be stirred from the bottom up in order to dislodge any settled matter on the bottom or lower sides of the container. Properly mixed paint should have a completely blended appearance with no evidence of varicolored swirls or lumps at the top, indicating unmixed pigment solids or foreign matter.

When two or more cans of the same color are to be used in the same area, the paint may be "boxed" by pouring it back and forth from one container to the other until it is uniform. This procedure mini-

**FIGURE 9.16-1** Mixing Procedures for Various Paint Types

| Coating Type | Mixing Equipment | Remarks |
|---|---|---|
| Enamel, semigloss, or flat oil paints | Manual, propeller, or shaker | Mix until homogeneous. |
| Water-based latex paints | Manual, propeller, or shaker | Use extreme care to avoid air entrapment. |
| Clear finishes, such as varnishes and lacquers | Manual | Generally require little or no mixing. |
| Extremely viscous finishes, such as coal tar paints | Propeller | Use extreme care to avoid air entrapment. |
| Two-package metallic paints, such as aluminum paint* | Propeller | Add small amount of liquid to paste and mix well; slowly add remainder of vehicle, while stirring, until coating is homogeneous. |
| Two-component systems | Manual, propeller, or shaker | Mix until homogeneous; check label for special instructions. |

*With metallic powder, first make into a paste with solvent, then proceed as recommended above.

mizes the visual difference when switching from one container to the next.

## TINTING

Tinting should be done by experienced personnel in a paint shop or at a paint store. On-the-job tinting should be avoided to minimize errors and to eliminate color-matching problems. An exception is the practice of tinting intermediate coats for identification and to ensure complete coverage under the finish coat. Tinting should be done with care, using only colors which are known to be compatible, and preferably adding not more than 4 oz (118.3 mL) per gallon (3.79 L) of paint, unless otherwise approved by the manufacturer. Tinting of chalking-type exterior paints is not recommended except for identification of intermediate coats.

Many paint manufacturers employ tinting systems for the addition of colorants to their base paints. The colorants are formulated to precise standards of tinting strength and are packaged in tubes or in bulk for use with a colorant-dispensing machine at the paint shop. Each color in the system has an identifying number and mixing formula indicating the proportions of base paint and colorants to be combined. Thus, any color can be reproduced accurately at a later time without guesswork.

***Tinting Colors*** There are two types of tinting colors in general use:

*Colors-in-oil* are limited to use with standard oil-base paints, alkyd resins, chlorinated rubbers, and other solvent systems. They should not be used with other synthetics or with water-thinned paints.

*Universal tinting colors* are more compatible with a wide variety of paint materials. Many can be used with both solvent-thinned and water-thinned paints.

***Tinting Procedure*** Paint should be at application consistency before tinting. Ingredients should be compatible and fluid so as to blend readily. Mechanical agitation is important to ensure uniform color distribution, but overmixing should be avoided. When trying to achieve a paint color not included in a manufacturer's tinting system, a paint sample should be allowed to dry before it is compared with the color chip used for reference. When the desired color has been achieved, it is important to maintain a written record of the tinting formula. The container should be identified by name or number, and a spot of paint should be applied to the cover for visual reference.

## STRAINING

Paint from newly opened containers normally does not require straining. However, a paint showing evidence of skins, lumps, color flecks, or foreign materials should be strained through a fine sieve or commercial paint strainer. Paint that is to be applied by spray gun should be strained to prevent clogging the spray nozzle.

## THINNING

Paints are formulated and packaged at the proper viscosity for application by brush or roller without thinning. However, thinning frequently is required and recommended by manufacturers for spray application. The arbitrary addition of thinners should be avoided; unnecessary or excessive thinning may result in an inadequate film thickness and may reduce the durability of the applied coating.

Thinning should be performed by competent personnel using only compatible thinning agents recommended in label instructions or manufacturers' specifications. Thinning should not be used as a means of improving the brushing or rolling properties of cold paint materials; instead, materials should be conditioned at 65°F (18.3°C) to 85°F (29.4°C) before use.

### 9.16.3.3 Preparation of Surfaces

Proper surface preparation is essential to achieve maximum coating life. Paint will not perform effectively if applied on a poorly prepared surface. The initial cost of adequate surface preparation is rewarded by longer coating life and improved appearance. Selection of the surface preparation method is dependent on (1) the nature and condition of the surface to be painted, (2) the type of exposure, (3) the surface texture and appearance desired, and (4) material and labor costs.

Surfaces to be painted or to receive a transparent finish or coating often contain contaminants such as dirt, grease, rust, and moisture which reduce adhesion and cause blistering, peeling, flaking, or cracking of the finished surface. Surface defects that also adversely affect adhesion include metal burrs and irregular weld areas, wood knots and torn grain, powdery concrete and masonry, and old paints in various stages of deterioration. In general, sharp edges, irregular areas, cracks, and holes create conditions conducive to early finish failure.

Because iron and steel corrode readily unless properly prepared and painted, special attention is given in the following discussion to suitable preparation methods for these and other metals. Painting of ferrous metals should be started as soon as possible after the surface has been prepared. The suitability of treatments for various material surfaces is covered in Figure 9.16-2.

## MECHANICAL TREATMENTS IN THE FIELD

Mechanical treatments in the field are intended to remove corrosion, mill scale, and old paint. Solvent cleaning to remove surface oil or grease is recommended prior to other treatments.

Metals can be prepared for painting by cleaning with hand tools, power tools, flame, and abrasive blast. The use of goggles and gloves is essential for blast and flame cleaning and is recommended for the other methods as well. Good ventilation is desirable. Dust respirators also may be required.

*Hand Tool Cleaning* Hand tool cleaning will remove only superficial contaminants such as rust, mill scale, and loosely adhering paint. It is primarily recommended for spot cleaning where corrosion is not a serious factor. Since manual cleaning removes only surface contamination, it is essential to use primers which will thoroughly wet the surface and penetrate to the base metal.

Before cleaning is started, the surface should be treated with solvent cleaners to remove oil, grease, dirt, and chemicals.

Rust scale and heavy buildup of old coatings should then be removed with impact tools such as chipping hammers, chisels, and scalers. Loose mill scale and nonadhering paint can be removed with wire brushes and scrapers. Caution should be exercised to avoid deep marking or scratching of the surface by the tools used.

*Power Tool Cleaning* Power tool cleaning provides faster and more adequate surface preparation than hand tool methods but is less economical than blasting for effective cleaning of large areas. Power tools can be used for removing small amounts of tightly adhering contaminants that are difficult to dislodge with hand tools.

Power tools are driven either electrically or pneumatically and include a variety of attachments for the basic units. Chipping hammers are used for removing tight rust, mill scale, and heavy paint coats. Rotary and needle scalers are handy for removing rust, mill scale, and old paint from large metallic and masonry areas. Wire brushes can be used for removing loose mill scale, old paint, weld flux, slag, and dirt deposits. Grinders and sanders are useful in removing old paint, rust, or mill scale on small surfaces and for smoothing rough surfaces.

As with hand tools, care is recommended to avoid cutting too deeply into the surface, since this may result in burrs that are difficult to protect satisfactorily. When using wire brushes, care should be taken to avoid polishing the metal surface, which may prevent adequate coating adhesion. Power tool cleaning should be preceded by solvent or chemical treatment; painting should be started and completed as soon after cleaning as possible.

*Flame Cleaning* Flame cleaning is satisfactory for both new and maintenance work. Oil and grease should be removed prior to flame cleaning, both for safety and for adequacy of preparation. Wire brushing normally follows flame cleaning to remove loose matter. A high-velocity oxyacetylene flame is used in this treatment. Extreme caution and adequate ventilation are necessary to prevent fires.

New paint is generally applied while the surface is still warm, thereby speeding up drying time and making it possible to paint when air temperatures are somewhat below 50°F (10°C). However, when solvent-thinned paints are used, the painting operation should be sufficiently separated from the cleaning operation to avoid a fire hazard.

*Blast Cleaning* Blast cleaning involves the high-velocity impact of abrasive particles such as sand or glass beads on a surface. The abrasive is discharged, either wet or dry, under pressure. In the wet system the water, and sometimes a rust inhibitor, is either mixed with the abrasive in the pressure tank or introduced into the blast stream near the nozzle. Protective clothing, dust masks, and eye protection are needed when blast cleaning to prevent injury.

Metal surfaces that have been blast cleaned should be prime coated the same day, because rust forms rapidly on newly cleaned surfaces. Metal or synthetic shot, grit, glass beads, corn cobs, or similar abrasives are used where recovery of the abrasive is possible; otherwise, sand is generally used. The abrasive should be of a size sufficient to remove surface contamination without pitting the surface or creating extreme peaks and valleys. Pitted surfaces require protection, and high peaks represent areas of possible paint failure.

As in other mechanical cleaning, surface grease or oil should be removed by solvent cleaning prior to blasting. Dry

**FIGURE 9.16-2** Preparation of Various Surfaces

| Preparation Method | Wood | Metals | | Concrete and Masonry |
| | | Steel | Other | |
| --- | --- | --- | --- | --- |
| **Mechanical** | | | | |
| Hand tool cleaning[a] | OK | OK | OK | OK |
| Power tool cleaning | OK[b] | OK | — | OK |
| Flame cleaning | — | OK | — | — |
| Brush-off blast cleaning | — | OK | OK | OK |
| Other blast cleaning | — | OK | — | — |
| **Chemical and Solvent** | | | | |
| Solvent cleaning | OK | OK | OK | — |
| Alkali cleaning | — | OK | — | OK |
| Steam cleaning | — | OK | — | OK |
| Acid cleaning | — | OK | — | OK |
| Pickling | — | OK | — | — |
| **Pretreatments** | | | | |
| Hot phosphate | — | OK | — | — |
| Cold phosphate | — | OK | — | — |
| Wash primers | — | OK | OK | — |
| **Primers** | — | — | — | OK |
| **Sealers** | OK | — | — | — |
| **Fillers** | OK | — | — | OK |

[a]Also suitable for gypsum wallboard and plaster.
[b]Sanding only recommended.

blasting of steel should be avoided if the temperature of the metal is less than 50°F (10°C) above the dew point to prevent condensation and subsequent rusting.

Blast cleaning is the most effective mechanical treatment and is often used on metal structures but also may be used, with caution, on masonry surfaces. The depth to which the surface is abraded can be varied, depending on the intended paint system and the severity of exposure. There are four methods of blast cleaning; the most effective methods, reaching down to or near the untarnished "white metal," require more labor and are therefore costlier.

*White metal blast* completely removes all rust, mill scale, and other contaminants from the surface and is the preferred technique in blast cleaning. It is used for coatings that must withstand exposure to very corrosive atmospheres and where a high initial cost of surface preparation is warranted.

*Near-white metal blast* surfaces are nearly free of all contaminants but will show streaks and discolorations distributed across the white metal surface. This technique is 10% to 35% cheaper than white metal blasting and has proven adequate for many protective coatings in moderately severe exposures.

*Commercial blast* is the removal of superficial contaminants such as loose scale and rust. This method results in a surface quality generally adequate for the majority of paint systems under normal exposure conditions.

*Brush-off blasting* is a relatively low-cost method used to remove old finishes, loose rust, and mill scale. It is not suitable where severe corrosion is prevalent, but is intended as a more efficient substitute for manual and power tool cleaning. Brush-off blasting also is used for the removal of loose or deteriorated paint from masonry.

## CHEMICAL AND SOLVENT TREATMENT

Chemical cleaning methods are more suited to shop application, while mechanical methods, other than blast cleaning, are more practical for use in the field. Chemical cleaning is both more efficient and more effective than mechanical methods other than blast cleaning, but chemical cleaning in the field is more difficult than even blast cleaning there because of the potential environmental hazards associated with chemical cleaning in the field.

The use of goggles and rubber gloves is recommended, and additional protective clothing is required where acid or alkaline solutions are used. Adequate ventilation is essential during chemical and solvent cleaning. Respirators should be used when operating in confined areas; proper fire precautions should be taken when flammable cleaners are used.

Some chemicals, such as trisodium phosphate, that have been traditionally used in cleaning and the removal of existing paint and coatings are no longer permitted in some areas because of their potential harm to people and the environment. Some jurisdictions permit the use of such chemicals but severely limit the means of their disposal after use. Therefore, before using any chemical for removal of existing paints or coatings, or for cleaning surfaces before applying paint, transparent finishes, or coatings, it is advisable to check with local authorities to ascertain the restrictions on their use and disposal.

*Solvent Wiping and Degreasing* Solvent cleaning is a procedure used for dissolving and removing surface contaminants such as oil, grease, dirt, and paint remover residues before painting or mechanical treatment. It is best to first remove soil, cement spatter, and other dry materials with a wire brush. The surface then should be scrubbed with brushes or rags saturated with solvent. Other effective methods include immersing the work in a solvent or spraying solvent over the surface.

The solvent quickly becomes contaminated, so it is essential that several clean solvent rinses be applied to the surface. Clean rags should be used for rinsing and wiping dry. Mineral spirits is an effective solvent for cleaning under normal conditions. Toxic solvents and solvents with low flash points constitute health and safety hazards. Use of solvents with flash points below 100°F (37.78°C) or solvents with a maximum safe concentration of less than 100 parts per million is not recommended. Rags should be placed in fireproof containers after use.

*Alkali Cleaning* Alkali cleaning is more efficient, less costly, and less hazardous than solvent cleaning but is more difficult to carry out. It is suitable for most metals, except stainless steel and aluminum. Alkali cleaner solutions are applied at relatively high temperatures

(150 to 200°F) (65.6 to 93.3°C) since cleaning efficiency increases with temperature. Alkalis attack oil and grease, converting them into soapy residues that can be washed away with water.

Commercial cleaners contain other active ingredients that aid in removing surface dirt and other contaminants such as mildew. These cleaners also are effective in removing old paint by breaking up the dried vehicle. The most commonly used alkaline cleaners are trisodium phosphate, caustic soda, and silicated alkalis. For steel, these cleaners should contain 0.1% chromic acid or potassium dichromate to prevent corrosion. Cleaners can be applied by brushing, scrubbing, or spraying, or the item to be cleaned can be immersed in a soak tank.

After cleaning, thorough rinsing is necessary to remove the soapy residue as well as all traces of alkali. Rinsing also avoids the possibility of a chemical reaction with the applied paint. The rinse water should be hot (175°F [79.4°C]) and applied under pressure. Universal pH test paper should be used to ensure that the surface is neutral and that no free alkali is present on the surface after rinsing.

*Steam Cleaning* Steam cleaning involves the use of steam or hot water under pressure, sometimes with a detergent included for added effectiveness. The steam or hot water removes oil and grease by emulsifying or diluting them, then flushing them away with water. When steam is applied to old paint, the paint's vehicle swells, so that it loses its adhesiveness and is easily removed. Under ordinary conditions, steam or hot water alone may be used to remove dirt deposits, soot, and grime. In severe cases, wire brushing or brush-off blast cleaning may be necessary to augment the steam cleaning.

*Acid Cleaning* Acid cleaning is used to remove oil, grease, dirt, and other surface contaminants and is suitable for cleaning iron, steel, concrete, and masonry. It should not be used on stainless steel or aluminum.

Iron and steel surfaces are treated with solutions of phosphoric acid containing small amounts of solvent, detergent, and wetting agent. Unlike alkali cleaners, acid cleaners also remove light rust and etch the surface to ensure better adhesion of applied coatings. There are many types of metal acid cleaners and rust removers, each formulated to perform a specific cleaning job. The most common formulations are based on phosphoric acid solutions. The procedures usually involve applying the cleaner by brush or rag, allowing time for it to act, thorough rinsing, wiping, and air drying. Smaller objects, particularly in the shop, can be immersed in hot cleaner and followed by several rinses.

Concrete and masonry surfaces can be washed with a solution of muriatic (hydrochloric) acid to remove efflorescence and laitance, to clean the surface, and to improve its bonding properties. Efflorescence is a white, powdery deposit often found on masonry surfaces. Laitance is a fine gray powder which sometimes develops on the surface of improperly finished concrete. Paints applied over efflorescence or laitance may loosen prematurely and result in early coating failure. It is best to remove as much as possible of these deposits using a stiff fiber or wire brush; putty knives or scrapers also may be used.

Oil and grease should be removed prior to acid cleaning, either by solvent wiping or by steam or alkali cleaning. To start acid cleaning, the surface is first thoroughly wetted with clean water, then scrubbed with a 5% solution (by weight) of muriatic acid using a stiff fiber brush. In extreme cases, up to a 10% muriatic acid solution may be used, and it may be allowed to remain on the surface for up to 5 minutes before scrubbing. Work should be done on small areas, not greater than 4 sq ft (0.37 m²) in size. Immediately after the surface has been scrubbed, the acid solution should be rinsed completely from the surface by thoroughly sponging or flushing with clean water. The surface should be rinsed immediately to avoid formation of salts on the surface which are difficult to remove.

This procedure also may be used to etch hard and dense concrete surfaces with poor surface bonding properties. The etching procedure breaks up the glaze so that paint will adhere to the surface. To determine whether etching is required, a few drops of water are poured on the surface. If the water is quickly absorbed, the surface can be readily painted; if the drops remain on the surface, etching is recommended.

Surface glaze also may be removed by rubbing with an abrasive stone, light sandblasting, or allowing the surface to weather for 6 to 12 months. The surface also may be etched with a solution of 3% zinc chloride and 2% phosphoric acid; this solution should not be washed off but allowed to dry to produce a paintable surface.

Concrete and masonry surfaces are generally alkaline in nature, and it may be necessary to use acid cleaning methods to neutralize these surfaces before applying alkali-sensitive coatings. The presence of free alkali may be detected by testing with pH paper.

*Pickling*   Pickling is used only in the mill or shop to remove mill scale, rust, and rust scale from iron and steel products. Sulfuric, hydrochloric, nitric, hydrofluoric, and phosphoric acids are used individually or in combination. Pickling is usually accomplished by immersing the piece in a tank, followed by several intermediate rinses to remove acids and salts. The final rinse usually is performed with a weak alkali solution to retard rusting and neutralize lingering acid solution.

*Paint Removers*   Paint and varnish removers are used on small areas. Solvent-type removers or solvent mixtures are selected according to the type and condition of the old finish, as well as the nature of the surface. Removers are available as flammable or nonflammable types and in liquid or semipaste consistency. While most paint removers require scraping to remove the softened paint, some removers loosen the paint sufficiently to permit flushing off the debris with steam or hot water.

To retard evaporation, some removers contain paraffin wax, which often remains on the surface after the paint has been removed. It is essential that this residue be removed from the surface prior to painting to ensure proper bonding of the paint coat. The wax can be removed with mineral spirits or a solvent specifically recommended by the manufacturer. Most paint removers are toxic to a degree, and some are flammable. Consideration should be given to adequate ventilation and fire control whenever they are used.

## METAL PRETREATMENTS

Pretreatment of metal surfaces improves coating adhesion, reduces underfilm corrosion, and extends coating life.

*Hot Phosphate Treatments*   Hot phosphate treatments use zinc or iron phosphate solutions to form crystalline deposits on the surface of a metal. They increase the bond and adhesion of paints while reducing underfilm corrosion. Zinc phosphate usually produces the best results and is most widely used. Its use is limited to the shop or plant.

When a gloss finish is desired, it is often necessary to apply thicker paint coats over a zinc phosphate coating, because phosphate coatings absorb considerably more paint. If a higher-gloss finish is desired, iron phosphate is preferred to zinc phosphate, because it produces a finer crystalline structure and therefore a thinner film. Hot phosphate treatments are very effective in ensuring good coating adhesion and tight bonds.

*Cold Phosphate Treatments*   Cold phosphate treatments are produced with a mixture of phosphoric acid, wetting agent, water-miscible solvent, and water. An acid concentration of about 5% to 7% by weight may be adequate when the area is not exposed during application to high temperatures, direct sunlight, or high wind velocities. Such adverse environmental conditions cause rapid evaporation and consequent high acid concentration.

When a dry, grayish-white powdery surface develops within a few minutes after application, acid has reacted properly and has the proper dilution. If a dark color develops and the surface is somewhat sticky, the acid probably is too concentrated. Then, if the area is small, wiping with damp rags may bring about the desired appearance; otherwise, the surface should be rinsed with water and retreated with a more dilute solution.

Although cold phosphate treatments may produce a crystalline deposit on the metal surface, the density of the deposit is not as great as that achieved with the hot phosphate treatment; therefore, paint adhesion may be less effective. The procedures used for cold phosphating are adaptable to field use on large or small structures.

*Wash Primers*   Wash primers are a form of cold phosphatizing. They are said to perform more efficiently than standard cold phosphating treatments and have generally replaced them in field use. Wash primers derive their name from the method of applying them in thin "wash" coats. They usually contain polyvinyl butyral resin, phosphoric acid, and a rust-inhibitive pigment such as basic zinc chromate or lead chromate. Wash primers develop extremely good adhesion to blast-cleaned or pickled steel and provide a sound base for topcoating. They also may be used to improve adhesion of coatings to hard-to-paint surfaces, such as galvanized steel or stainless steel and aluminum. Wash primers can be applied in the shop or in the field.

## PRIMERS, SEALERS, AND FILLERS

Application of a finish coat should always be preceded by a suitable primer, filler, or sealer, or by a base coat of the same paint as the finish coat which performs the same functions.

Primers are often applied on masonry to seal chalky surfaces and to improve adhesion of water-based topcoats. Sealers are used on wood to prevent resin exudation or bleeding. Fillers help to produce a smooth finish on open-grain wood and rough masonry.

*Primers* Ordinary latex paints do not adhere well to chalky masonry surfaces. To overcome this problem, modified latex paints can be used or the surface can be coated with an oil-based primer before latex paint is applied. The entire surface is vigorously wire-brushed by hand or power tools, then dusted to remove loose particles and chalk residue. The primer is then brushed on to ensure effective penetration and allowed to dry. Primers are not as weather-resistant as finish coat paints and should not be used as a topcoating.

*Knot Sealers* Sealers are used on bare wood to prevent resin from bleeding through newly applied paint. Freshly exuded resin, while still soft, may be scraped off with a putty knife, and the affected area may be cleaned with alcohol. Hardened resin may be removed by scraping or sanding. Since a sealer is not intended as a priming coat, it should be used only when necessary and applied only over the affected area. When a previously painted softwood lumber surface becomes discolored over knots, the knots should be covered with sealer before new paint is applied.

*Fillers* Fillers are used on porous wood, concrete, and masonry to fill the pores and to provide a smoother finish coat.

*Wood fillers* are used on open-grained hardwoods. In general, hardwoods with pores larger than those in birch (open

grain) should be filled (Fig. 9.16-3). When filling is necessary, it should be done 24 hours after staining. If staining is not warranted, natural (uncolored) filler should be applied directly to the bare wood. When desired, the filler may be colored with stain in order to accentuate the grain pattern of the wood.

Wood filler should be thinned with mineral spirits to a creamy consistency, then liberally brushed across the grain, followed by a light brushing along the grain. It should be allowed to stand for 5 to 10 minutes until most of the thinner has evaporated and the finish has lost its glossy appearance. Before it has a chance to set and harden, the filler should be wiped off across the grain, using burlap or other coarse cloth, rubbing the filler into the pores of the wood while removing the excess. The operation is finished by stroking along the grain with clean rags, making sure that all excess filler is removed.

Wiping too soon may pull a filler out of the pores. Allowing a filler to set too long may make it difficult to wipe off. A simple test for dryness consists of rubbing a finger across the surface. If the filler forms into a ball, it is time to wipe; if the filler slips under the pressure of the

finger, it is still too wet for wiping. A filler should be allowed to dry for 24 hours before a finish coats is applied.

*Masonry fillers* can be applied by brush or spray to both old and new bare concrete, stucco, concrete block, and other masonry surfaces. These surfaces should be rough, sound, and free of loose and powdery materials. If surface voids are relatively large, it is preferable to apply two coats of filler rather than one heavy coat. In large areas, filler can be sprayed on, then troweled smooth. One to 2 hours' drying time should be allowed between filler coats, and painting should be delayed 24 hours after the final filler coat.

### 9.16.3.4 Repair of Surfaces

Surfaces that will be painted must be clean, sound, and uniform in texture. To achieve this condition:

- Repair or replace damaged materials and surfaces.
- Replace broken windows.
- Remove loose putty and glazing compounds and provide new materials.
- Fill cracks, crevices, and joints with caulking compound or sealants.
- Patch cracks or holes in wood, masonry, and plaster.

**FIGURE 9.16-3** Finishing Characteristics of Major Wood Species

| Wood Species | Grain Type | Remarks |
|---|---|---|
| **Hardwoods** | | |
| Ash | Open | Requires filler |
| Aspen | Closed | Paints well |
| Basswood | Closed | Paints well |
| Beech | Closed | Paints poorly; varnishes well |
| Birch | Closed | Paints and varnishes well |
| Cherry | Closed | Varnishes well |
| Chestnut | Open | Requires filler; paints poorly |
| Cottonwood | Closed | Paints well |
| Cypress | Closed | Paints and varnishes well |
| Elm | Open | Requires filler; paints poorly |
| Gum | Closed | Varnishes well |
| Hickory | Open | Requires filler |
| Mahogany | Open | Requires filler |
| Maple | Closed | Varnishes well |
| Oak | Open | Requires filler |
| Teak | Open | Requires filler |
| Walnut | Open | Requires filler |
| **Softwoods** | | |
| Alder | Closed | Stains well |
| Cedar | Closed | Paints and varnishes well |
| Fir | Closed | Paints poorly |
| Hemlock | Closed | Paints fairly well |
| Pine | Closed | Variable depending on grain |
| Redwood | Closed | Paints well |

(Courtesy Architectural Woodwork Institute.)

**FIGURE 9.16-4** Materials for Repair of Surfaces

| Material | Where Used | Characteristics |
|---|---|---|
| Caulking compound | Filling joints and crevices around doors and windows in concrete, wood, brick, block, and other masonry surfaces | Tends to dry on the surface but remain soft and tacky inside. Exposed surface of caulking; bead should be painted each time surrounding area is painted to help extend its life (5–15 years). |
| Sealants | Filling joints in locations where movement occurs and a weathertight joint is essential at all times | Better adhesion and greater elasticity than caulking compound, even at low temperatures. Life expectancy 15–30 years. |
| Putty | Filling nail holes, cracks, and imperfections in wood surfaces; installing small lights in wood and metal sash; not used for joints or crevices | Dries to a harder surface than caulking compound. |
| Glazing compound | Bedding and face glazing of windows | Sets firmly but retains limited flexibility, generally greater than putty. Life expectancy of 5–15 years when protected by periodic painting over exposed surface. |

Products and materials suitable for these purposes are described in Figure 9.16-4. Refer to Section 7.13 for information concerning caulking and sealants.

In general, patching of cracks and holes should be done by the installer and finisher of the underlying material. Rarely, the painter may patch minor cracks and fill minor holes. In addition to the materials described in other sections, including Section 7.13, the materials described here may be used for patching holes and cracks.

Cracks, holes, and crevices in masonry, plaster, gypsum board, and wood may be filled with various patching materials. These are supplied either ready for use or as a dry powder to which water is added.

When using patching materials on masonry, plaster, or gypsum board, the crack should first be opened with a putty knife or wall scraper so that weak material is removed and the patching compound can be forced in completely. The compound should be applied with a putty knife or trowel, depending on the size of the hole. Then the surface should be smoothed off slightly convex to allow for shrinkage.

One of the oldest painters patching materials is *putty*, an oil-base material used to fill nail holes, cracks, and imperfections in wood surfaces. Putty generally is supplied in bulk form and is applied with a putty knife. It should be applied after priming to prevent the wood from drawing the oils out. It dries to a harder surface than caulking compound and should not be used for joints or crevices. Special glazing putty is used to install small lights in wood and metal sash.

*Plastic wood* is suitable for filling gouges and nail holes in wood. It is applied in a manner similar to that of putty. After the plastic wood has completely dried, it should be sanded until smooth.

A soupy mixture of portland cement, sand, and water called *grout* can be used to repair cracks in concrete and masonry. Synthetic resins such as latex or epoxy may be added to improve bonding qualities and to permit application in thin layers. Hydrated lime sometimes is added to lengthen the working time. This grout should be moist-cured for several days or longer.

*Patching plaster* is used to repair plaster. It is similar to ordinary plaster, except that it hardens more quickly. It is supplied as a powder to be mixed on the job with water.

*Spackle* is used to fill cracks and small holes in plaster and gypsum board. It is easy to apply and to sand smooth after it hardens. It is supplied both as a paste and as a powder.

*Joint compound* is intended primarily for sealing joints between gypsum board panels, but it can also be used to repair large cracks. It is supplied as a powder and is used in conjunction with joint tape, which gives it added strength.

### 9.16.3.5 Paint Application

The most common methods of applying paint are by brush, roller, and spray. Dip-and-flow coat methods are generally limited to shop work. Of the three methods designed for field use, brushing is the slowest and spraying is the fastest. One painter can cover up to 1000 sq ft (92.9 m²) a day by brushing, 2000 to 4000 sq ft (186 to 372 m²) by rolling, and up to 12,000 sq ft (1115 m²) by spraying. Coatings applied by brush may leave brush marks in the dried film; rolling leaves a stippled effect; spraying yields the smoothest finish if done properly. The choice of method is based on (1) environmental conditions, (2) the type of surface and coating, and (3) the appearance of the finish desired.

General surroundings may prohibit the use of spray application because of possible fire hazards or potential damage from overspray. Adjacent areas not intended to be coated should be masked when spraying is performed. If the time spent in masking preparations is extensive, it may offset the advantage of the speed of spraying operations.

Roller coating is efficient on large, flat surfaces. Corners, edges, and odd shapes, however, generally require brushing. Spraying is especially suitable for large surfaces, but it can also be used for round or irregular shapes. Brushing is ideal for small surfaces or for cutting-in corners and edges. Dip-and-flow coat methods are appropriate for volume production painting of small items in the shop.

Rapid-drying lacquer-type products such as vinyls are usually sprayed. Application of such products by brush or roller is difficult, especially in warm weather or outdoors on breezy days.

### JOB CONDITIONS

To obtain optimum performance from a coating, certain basic application procedures should be followed, regardless of the type of equipment used. Cleaned, pretreated surfaces should be prime-coated within the specific time limits recommended by the paint manufacturer.

Paint manufacturers generally recommend the following:

- Surface and air temperatures should be between 50°F (10°C) and 90°F (32.2°C) for water-thinned coatings, and between 45°F (7.2°C) and 95°F (35°C) for other coatings.
- Paint should be maintained at a temperature between 65°F (18.3°C) and 85°F (29.4°C) at all times.
- Paint should not be applied when the air temperature is expected to drop to freezing before the paint has dried.
- Wind velocity should be below 15 mph (24 km/h).
- Relative humidity should be below 80%.

However, some manufacturers may recommend other limits for their products under some circumstances.

Masonry surfaces that are damp (not wet) may be painted with latex or cementitious paints. Other surfaces should be completely dry before painting. Paints should be applied at the spreading rates recommended by their manufacturers. When successive coats of the same paint are used, each coat should be tinted a different color to ensure complete coverage with each coat. Sufficient time should be allowed for each coat to dry thoroughly before the next coat is applied. A finish coat (topcoat) should be allowed to dry as long as is practicable before the space is occupied or put into service.

The line between different colors or finishes should be straight and neat. Previously finished surfaces should be protected by suitable masking or covering. Stain or paint splatter from adjacent surfaces should be removed.

Unless otherwise recommended by the manufacturer, a minimum of two coats should be used. Painted exterior wood surfaces should have a total thickness of 4.0 mils (0.1.16 mm) or more (Fig. 9.16-5).

**FIGURE 9.16-5** Dry Film Thickness of Applied Paints

| Coverage Rate, sq ft/gal (m²/L) | Approximate Dry-Film Thickness of Finish Coat, mils (mm) |
|---|---|
| 450 (11.05) | 2.25 (0.057) |
| 500 (12.27) | 2.00 (0.051) |
| 550 (13.5) | 1.75 (0.044) |
| 600 (14.73) | 1.50 (0.038) |
| 650 (15.95) | 1.30 (0.033) |
| 700 (17.18) | 1.10 (0.028) |
| 750 (18.41) | 1.00 (0.025) |

# 9.17 Special Coatings

Special coatings are used on both the interior and exterior of buildings and other structures to waterproof, protect, and beautify. Several types of coating systems and many coating products are available tody. This section addresses some of them.

## 9.17.1 GENERAL CHARACTERISTICS

Comparing special coatings based on their chemical content is usually difficult and often impossible. Some manufacturers do not disclose the components of their systems. Others use terms so generic as to be useless in determining the real contents. In the end, an architect should base coating selection decisions on achieving a desired result. If it is necessary to waterproof a deck or coat a concrete wall, an architect should look for a product that does this, and make selections on a performance basis, based on the reputation of the coating system's manufacturer and the proven performance of the product.

When comparing special coatings, it is necessary to take into account their general characteristics. These properties vary considerably from product class to product class and from manufacturer to manufacturer even within the same product class. The characteristics of similar products of different manufacturers may vary slightly. If the variation is great, however, it can be concluded that the products are significantly dissimilar and may not be equal in performance.

Most manufacturers list many of the characteristics of their products in their basic catalogs. Other data can usually be obtained directly from the manufacturer or from the manufacturer's MSDS. Some of these important characteristics are:

- *Recommended use:* This is the first selection criterion and a critical one.
- *Coating manufacturer:* The manufacturer's reputation is paramount.
- *Product history:* Successful installations are a very good guide.
- *Water vapor transmission:* Perm ratings vary considerably, ranging from 25 to 200, often depending on the coating thickness but also due to the texture of the coating. Additives that impart texture may affect the perm rating.

- *Coating thickness:* Thickness will vary widely with different types of coatings, but the real significance is when it varies within the same coating type.
- *Solids content:* Solids content is measured by weight and by volume. These vary considerably from product to product and within a product depending on the texture.
- *Pigment volume concentration:* This varies considerably from product to product and within a product due to texture.
- *Abrasion resistance.*
- *Chemical resistance:* Some coatings are more resistant to chemicals than others. This is important when a coating will come into contact with chemicals, such as ice-melting materials, or harsh cleaning agents. Coatings used in high chemical exterior or interior environments must be selected with increased care.
- *Resistance to salt spray.*
- *Resistance to scaling.*
- *Water penetration resistance:* Waterproofing coatings must resist water penetration measured by the static head of water pressure. For example, a coating may be required to resist penetration of water at up to 4 ft (1200 mm) of water pressure.
- *Scrub resistance:* Scrub resistance varies somewhat within a class of coatings, depending on the thickness.
- *Impact resistance.*
- *Tensile properties.*
- *Surface-burning properties.*
- *Chalking resistance:* Most special coatings should have no chalking after 1000 hours of testing.
- *Freeze-thaw properties.*
- *Radon gas penetration:* This characteristic can be important for some below-grade waterproofing installations.
- *Viscosity.*
- *VOC content:* This is usually less than 400 gm/L per ASTM D5090.
- *Compressive strength:* This is an important characteristic for cementitious coatings.

## 9.17.2 APPLICABLE STANDARDS

Industry standards are used to test special coatings and identify their characteristics. The following is a partial list of applicable standards. Their titles are listed in Section 9.19.

- American Concrete Institute: ACI 201.1R
- ASTM International (ASTM): Many ASTM standards apply. The following is a partial list: B117, C67, C91, C109, C666, C672, C926, D522, D870, D968, D1308, D2243, D2370, D2486, D4141, D4258, D4259, D4260, D4261, D5090, E84, E96, E514

## 9.17.3 SPECIAL COATING SYSTEMS

The following classifications were determined by examining the constituents of a wide array of coating products. This section does not mention trade names or manufacturers' names, nor does it suggest which product types are best for a particular situation. It merely displays the variety of coatings available and offers some guidance as to their properties and where they are usually used.

Where the components of a system were identifiable, the system is listed in this section under a heading containing the names of the components (acrylic latex terpolymer coatings, for example). Products for which the components were not identifiable are listed under the generic names (not trade names) the manufacturers use to identify those systems (polymer-modified waterproof cement coatings, for example). Some of the products listed here under different names may in fact be very similar. Other manufacturers may use different nomenclature from that shown here for similar products.

### 9.17.3.1 Acrylic Emulsion Coatings

Acrylic emulsion coatings are premixed coatings available in either smooth, fine, medium, or coarse textures. They are abrasion, mildew, and weather resistant while permitting vapor transmission. They bind friable surfaces together and increase the bond strengths of coatings. They are available in tint bases and 50 or more standard colors.

These products are used in the following ways:

- In both vertical and overhead applications as above-grade decorative finishes
- As a primer for water-based coatings, especially on friable, dusty concrete substrates or prepainted chalky surfaces

- As a finish coat over properly prepared concrete, cementitious backer board, CMU, stucco, portland cement plaster, brick, drywall, stone, and painted surfaces
- As deck coatings (some textured acrylic emulsion coatings)

### 9.17.3.2 Acrylic Emulsion Block Fillers

Block fillers are a special class of acrylic emulsion coatings. They are premixed, smooth, water resistant, ready-to-use, flat-finish, white coatings. They are mildew and weather resistant. They may be applied by brush, roll, or spray.

Block fillers are used:

- To fill concrete masonry surface pores and voids prior to coating and as a primer for specialized coatings
- In both vertical and overhead applications on the interior or exterior
- As a fill coat over concrete, cementitious backer board, block, stucco, plaster, brick, drywall, and stone, and as a primer for latex, enamel, epoxy, urethane, and elastomeric coatings.

### 9.17.3.3 Acrylic Polymer Coatings

Acrylic polymer coatings are premixed high-impact coatings available in both elastomeric and synthetic stucco form. They are flexible and crack resistant. They are also alkali, water, mold, and mildew resistant while permitting vapor transmission. They are available in tint bases and 50 or more standard colors.

They are used as follows:

- On both interiors and exteriors
- As a coating over concrete, cementitious backer board with a brown coat, CMU, exterior stucco, brick, EIFS, and gypsum wallboard

### 9.17.3.4 Silicone Emulsion, Acrylic Polymer Coatings

Silicone emulsion, acrylic polymers are premixed coatings available in both elastomeric and synthetic stucco form. They are flexible, repel water, are recoatable, and are mildew, dirt, algae, and alkali resistant. They are available in tint bases and 50 standard colors.

These coatings are used over concrete, cementitious backer board with a brown coat, CMU, exterior stucco, brick, and EIFS.

### 9.17.3.5 Acrylic Terpolymer Coatings

Acrylic terpolymer coatings are textured, high-build, 100% acrylic terpolymer, water-base coatings with elastomeric properties. They are both decorative and protective. They are available in 16 or more standard colors, and special colors are also available. When roller applied, they produce a smooth, fine texture. Spray application gives them a smooth, fine, coarse, or extra-coarse texture. An embossed texture is also available to help minimize uneven block and mortar joints.

These coatings are used above grade over concrete, stucco, masonry, metal, and other manufacturer-approved surfaces.

### 9.17.3.6 Acrylic Latex Coatings

Acrylic latex color coating products are 100% acrylic latex exterior finishes. They are weather, moisture, and mildew resistant. They are available in a smooth finish only.

These products are used as a topcoat over textured primers and over water-base and oil-base textured coatings.

### 9.17.3.7 Acrylic Latex Terpolymer Coatings

Acrylic latex terpolymer coatings are high-performance coatings with a 100% base of elastomeric resin providing up to 400% elongation. These coatings are flexible in below-freezing temperatures. They resist alkali. They also resist hairline cracking and have excellent adhesion characteristics.

These products are used above grade over concrete, cement plaster, CMU, brick, galvanized metal, cement fiberboard, wood siding, and other manufacturer-approved surfaces.

### 9.17.3.8 High-Build Acrylic Resin Coatings

High-build acrylic resin coatings are water-based, 100% acrylic resin, textured, waterproof coatings. They are VOC compliant, durable, breathable and ultraviolet, blister, abrasion, wind-driven rain, and mildew resistant. They are available in smooth, fine, and course textures, and either with or without algae-resistant additives. They are available for airless spray application, in heavy-textured formulations to emulate stucco, and with a fine texture for use for foot traffic. They can be highly reflective. They are available in 48 or more standard colors. When properly applied, they hide minor construction imperfections and provide a uniformly textured, weather-resistant surface.

They are used:

- Over vertical concrete, masonry, and stucco
- On parking garages, stairways, traffic barriers, and in other locations

### 9.17.3.9 High-Build Latex Coatings

High-build latex coatings are water-based, textured coatings designed for either spray or roll application. They dry rapidly, adhere well, and are flexible.

They are used above grade only over concrete, cement plaster, block, brick, cement fiberboard, metal, and other manufacturer-approved surfaces.

### 9.17.3.10 Vinyl Toluene Resin Coatings

Some textured primers and waterproofers are high-build, plasticized, pigmented, solvent systems based on a vinyl toluene resin binder. The cured primer membrane provides fill, texture, and weatherproofing properties over cured or green concrete masonry surfaces. These materials can be used over damp or highly alkaline concrete. They are available in both smooth and coarse textures. Some of these products can be applied at temperatures as low as 20°F (−6°C). Others require that the temperature be between 40°F (4.4°C) and 100°F (37.8°C).

These coatings are used:

- Above grade over concrete, cement plaster, CMU, brick, previously painted surfaces, and other manufacturer-approved surfaces
- For one-coat application on concrete highway abutments and barriers and on parking garages

These coatings are not recommended for wood or metal surfaces.

### 9.17.3.11 Vinyl Toluene Acrylic Resin Systems

Some industrial maintenance coatings are high-build, plasticized, vinyl toluene acrylic resin systems formulated to provide a weatherproof coating over cured or uncured concrete masonry surfaces. They can be placed over damp or high-alkaline surfaces. They are available in both standard and special colors. They may be roller-applied for a smooth, fine texture or spray-applied to give a smooth, fine, coarse, or extra-coarse texture.

These coatings are used above grade over concrete, cement plaster, CMU, brick, and other manufacturer-approved surfaces.

### 9.17.3.12 Acrylic Elastomeric Polymer Coatings

Acrylic elastomeric polymer coatings are smooth and premixed.

### 9.17.3.13 Siliconized Acrylic Elastomeric Polymer Coating

Siliconized acrylic elastomeric polymer coatings are smooth and premixed.

### 9.17.3.14 Acrylic Conditioners, Primers, and Sealers

#### CLEAR CONDITIONERS, PRIMERS, AND SEALERS

Acrylic conditioners, primers, and sealers are ready-to-use, clear, water-based, alkali-resistant coatings. They reduce potential for efflorescence while allowing moisture vapor transmission. They consolidate light surface chalk and concrete dust. They are water repellent, stain, abrasion, and mildew resistant, VOC compliant, and nonflammable. They can be cleaned up with water.

These coatings are used:

- As above-grade surface conditioners and primers over concrete, stucco, stone, CMU, brick, plaster, and drywall
- As surface conditioners and primers for elastomeric coatings, synthetic stuccos, water-based textured coatings, acrylic latex paints, and oil-based paints
- As two-coat surface sealers over concrete, stucco, brick, CMU, stone, and plaster

#### PIGMENTED SEALERS

Pigmented sealers are high-build, VOC-compliant, 100% acrylic waterproofing sealers. They stop wind-driven rain of up to 100 mph (161 km/hr). They allow the surface to breath. They span hairline cracks and hide minor surface imperfections. They can be applied at temperatures as low as 20°F (−6°C).

These sealers are used on above-grade masonry, including slick precast, glazed brick, and marginally prepared surfaces without the use of primers and conditioners.

### TINTABLE PRIMERS

Tintable primers are 100% acrylic, ready-mix, opaque, water-based, VOC-compliant primers that dry to a gritty finish to help reduce suction in the substrate, providing easier application and better adhesion of various acrylic coatings. They will not degrade in the presence of intermittent moisture. They are breathable. They may be applied by brush, roller, or spray. They retard efflorescence. They provide uniform background color prior to topcoat applications, and they improve workability and ease of acrylic topcoat applications.

These primers are used on masonry.

### 9.17.3.15 Polymer-Modified Waterproofing Cement Coatings

Polymer-modified waterproofing cement coatings are used to waterproof masonry.

### 9.17.3.16 Acrylic Polymer Crack and Surface Repair Materials

Acrylic polymer crack and surface repair materials are premixed, flexible, brush- or knife-grade stucco repair products. They are available either smooth or textured. They are mildew and weather resistant. Brush-grade products are usually white; knife-grade products are usually off-white.

These materials are used:

- Above grade in both interior and exterior vertical and overhead applications
- To fill and repair static cracks in concrete, stucco, cementitious backer board, drywall, plaster, block, and brick

### 9.17.3.17 Penetrating Sealers and Masonry Stains

Penetrating sealers and masonry stains are water-repellent sealers suitable for a wide range of surfaces. They feature long-term water repellence and resistance to discoloration, efflorescence, dirt, and mildew. They are classed as water repellent and are not intended to serve as waterproofing materials.

These products are used:

- Over concrete, brick, stucco, block, wood, limestone, and aggregate panels
- To protect and maintain bridges, parking garages, airport pavements, industrial plants, buildings, and other concrete and masonry structures

Sealers and stains protect and preserve architectural substrates. Treated surfaces are resistant to staining, spalling, weathering, efflorescence, water intrusion, salt intrusion, fungi and mildews, deterioration, freeze-thaw scaling, and reinforcing steel corrosion. There are several kinds of these sealers.

## REGULAR CLEAR SEALERS

Regular clear sealers are penetrating water-repellent sealers. Some contain a hydrocarbon plasticizer which is white, non-glossy, and stain and oxidation resistant.

They are used on both interior and exterior concrete, cement plaster, painted surfaces, natural stone, brick, and wood and other porous surfaces.

## HEAVY-DUTY CLEAR SEALERS

Heavy-duty clear sealers are specially formulated for more porous surfaces. Their hydrocarbon plasticizer content is 20%. There are two-coat water-repellent systems which require a 24-hour curing period under proper weather conditions prior to second-coat application. Upon drying, the surface is enriched and slightly darkened, with a low sheen.

These sealers are used over concrete masonry, especially split-face block, and cinder block, slump stone, adobe block, lightweight concrete, and exposed aggregate systems.

## OPAQUE STAIN SEALERS

Opaque stain sealers are solvent-based resin products that penetrate masonry for water repellence and greater depth of color while allowing moisture vapor to escape. They are available in standard and special colors.

These sealers are used on concrete, CMU, brick, natural stone, and stucco.

## SILOXANE AND SILANE SEALERS

Siloxane and silane sealers protect and preserve architectural substrates without altering their surface texture or appearance. Treated surfaces retain their vapor permeability and are resistant to staining, spalling, weathering, efflorescence, water intrusion, salt intrusion, fungi and mildews, deterioration, freeze-thaw scaling, and reinforcing steel corrosion.

These sealers are used on flagstones and other pavers, concrete, synthetic stone, CMU, brick, limestone, marble, sandstone, stucco, and synthetic stucco.

*Siloxane Sealers* Siloxane sealers are one-component penetrating sealers. They react chemically with the substrate to repel salt and water while allowing the substrate to retain its permeability. They do not change the substrate's texture, appearance, or skid/slip resistance. They are compatible with most solvent-base surface finishes.

These sealers are used in both traffic-bearing and non-traffic-bearing applications and on both horizontal and vertical surfaces.

*Silane Clear Sealers* Silane clear sealers are a chemical treatment which renders concrete, masonry, limestone, and wood repellent to salt and water. They are available in two grades for concrete, masonry, and limestone and one grade for wood. They are claimed to be superior to oligomeric polysiloxanes, stearates, siliconates, silicates, urethanes, butadienes, chlorinated rubber, epoxies, and oils.

Premium-grade silane weatherproofing products contain 40% alkyltrialkoxy silane in mineral spirits. Some will penetrate up to $1/2$ in. (12.7 mm) of the substrate with each application. They are used to protect and maintain bridges, parking garages, airport pavements, industrial plants, buildings, and other concrete and masonry structures in the most severe environments.

There are also silane weatherproofing products containing 20% alkyltrialkoxy silane in mineral spirits. They penetrate up to $1/4$ in. (6.4 mm) with each application. They are used for bridges, parking garages, pavements, plants, and other brick and concrete structures in severe environments

Silane sealers are useable for traffic conditions. They do not change the substrate's skid/slip resistance, vapor permeability, texture, or appearance. They are usable for both horizontal and vertical applications. They are compatible with most solvent-base surface finishes. They are fast-acting. Treated substrates, such as bridge decks, are usable in 20 to 45 minutes. They flow into microcracks and penetrate substrates. They contain no hydrocarbons and are not susceptible to ultraviolet or chemical degradation.

*Microemulsions* Microemulsions are VOC-compliant, microemulsifiable concentrates. When silanes and alkoxysilanes are mixed with water, a microemulsion is produced. This diluted form will impart water repellence. These are water-repellent products. They are not intended to serve as waterproofing materials. Topcoating of these products with other paints or coatings can proceed following a minimum 72-hour curing time in acceptable environmental conditions.

These materials are used above grade over absorbent mineral surfaces.

### 9.17.3.18 Cementitious Coatings

Cementitious coatings are decorative, water-resistant, portland-cement-based finish coats. They are durable and water-resistant. They eliminate the need to rub concrete to finish it. They are available in 50 or more factory-blended colors. Building codes may classify cement-based coatings as dampproof rather than waterproof. The architect needs to verify the performance requirements against product technical data.

These materials are used as a decorative finish coat over concrete, brick, and stone.

## POLYMER-MODIFIED, WATERPROOF CEMENT COATINGS

Some polymer-modified, waterproof cement coatings are *block fillers*. They are usually one-component systems. They are nontoxic, reduce radon gas penetration by more than 99%, and are waterproof under a maximum of 4 ft (1200 mm) of water pressure. They are abrasion and moisture resistant and are breathable. They have a white reflective surface.

These coatings are used:

- Both above and below grade and on either interiors or exteriors
- To fill pores and voids in concrete or masonry substrates
- For waterproofing of basements, parapets, cisterns, reservoirs, grain elevators, silos, foundations, and glass-block mortar joints
- For decorating stone, brick, concrete, CMU, concrete, planters, fountains, and glass-block mortar joints

## POLYMER-MODIFIED, WATER-RESISTANT CEMENT COATINGS

Polymer-modified, water-resistant, lightweight portland-cement-based base coats and finish coats are crack resistant, durable, water resistant, and lightweight.

They are used as a base coat and finish over autoclaved, aerated concrete.

## CEMENT-BASED WATERPROOF COATINGS

Portland-cement-based waterproofing/dampproofing coatings, when used in conjunction with an acrylic admixture, become an integral part of the substrate when cured. They resist hydrostatic pressure. They are breathable. They are National Sanitation Foundation Approved (N.S.F. 61.1).

These coatings are used:

- On both interior and exterior surfaces
- To fill and seal pores and voids in concrete and masonry surfaces
- To correct surface irregularities, eliminating the need for concrete rubbing

### READY-TO-USE CEMENT-BASED WATERPROOF COATINGS

These are high-build, one-component, water-based, ready-mixed, acrylic waterproofing coatings. They provide a dense surface, which reduces radon infiltration. They should pass a 2000 hour weatherometer test. They clean up with water. They are VOC compliant and environmentally friendly.

These coatings are used:

- On interiors in below-grade locations
- For protection of new or existing concrete and masonry surfaces

### EXTERIOR BELOW-GRADE WATERPROOF COATINGS FOR FOUNDATIONS

These are portland-cement-based waterproofing/dampproofing coatings. When used in conjunction with an acrylic admixture, they become an integral part of the substrate.

They are used on below-grade exterior concrete and masonry surfaces.

### 9.17.3.19 Flecked, Multicolored Interior Finishes

Flecked, multicolored interior finishes are spray-applied, textured, decorative coatings that offer a decorative pattern different from that of most coatings, with virtually unlimited color combinations. They hide minor surface irregularities and give surfaces added dimensions in color and texture. Some are finished after application with a clear urethane that offers VOC compliance and features outstanding hardness, durability, abrasion resistance, and gloss retention. They are applied using conventional, airless, or high-velocity, low-pressure (HVLP) spray equipment. They are stain and mildew resistant and do not support bacterial growth. Their variegated coloration helps conceal dirt. They are scrubbable with liquid detergents. Touchups are nearly invisible. Their fire resistance is Class A. They are odorless and meet EPA Clean Air (VOC/VOS) standards. They dry quickly (2–4 hours to the touch). Recoating can be done in 24 hours.

These finishes are used:

- On walls, ceilings, and trim
- Over a variety of substrates, including drywall, plaster, masonry, wood, and metals
- Over existing paint or vinyl

### 9.17.3.20 Antigraffiti Coatings

Antigraffiti coatings are water-based, VOC-compliant coatings that facilitate easy removal of most kinds of graffiti, such as spray paints and marking pens. They are available in clear or pigmented formulations. The clear formulations do not alter the appearance of the substrates. The pigmented formulations leave a glossy surface.

These coatings are used on metal, concrete, brick, vinyl, stone, stucco, concrete masonry, wood, and EIFS. Some of them are not recommended for glass, polyurethane, or polypropylene surfaces.

### 9.17.4 INSTALLATION OF SPECIAL COATINGS

The following requirements are applicable to most coatings. However, preparation for and installation of coatings should be done only in accordance with the coating manufacturer's specific recommendations, which may vary with the following recommendations.

### 9.17.4.1 Preparation

Surfaces should be cleaned and repaired in accordance with the requirements of ASTM D4258, ASTM D4259, ASTM D4260, ASTM D4261, and ACI 201.1R, as applicable.

Surfaces to receive most coatings must be clean, sound, and dry. Some coatings, however, are designed to be applied over damp surfaces. In no case should precipitation be falling or imminent.

Loose, flaking, or oxidized paint should be removed by sandblasting, water blasting, wire brushing, or scraping. Old paint in sound condition can usually be left in place, washed to remove excess chalk, and allowed to dry before coating application.

Most of the time, surface contaminants such as dust, dirt, chalk, mildew, form oils, and other foreign matter must be removed, but for some coatings, some of these may simply be stabilized with an appropriate primer.

Decayed cement plaster must be removed and new stucco installed using a bonding agent.

Cracks, holes, and voids must be filled with cement patching compounds and bonding agents recommended by the coating system manufacturer. The texture of patches should match that of the existing surface.

Joint sealants used must be compatible with the coating and must cure at least 48 hours prior to priming.

New concrete, masonry, and cement plaster must be allowed to cure for the minimum number of days recommended by the coating manufacturer. Some coatings require a 7-day cure. Others require 28 days.

### 9.17.4.2 Application

The ambient temperature during application of most coatings must be between 45°F and 100°F (7°C and 38°C).

Coating must be applied at the rate recommended by the manufacturer. A uniform film of the recommended thickness should be provided over the entire surface. Material coverage rates can vary due to surface porosity and texture.

To prevent lap marks, a wet edge must be maintained during spraying, brushing, or rolling. Application should not be started or stopped in the middle of walls or large areas. Application should continue to a natural break such as a panel edge, seam, or corner. Back rolling while spraying will help minimize spray patterns.

Coatings should be cured in accordance with their manufacturers' recommendations. For most coatings, primers must be cured a minimum of 4 days prior to the application of the color coat. However, for most coatings, primers must not have been exposed to ultraviolet rays for more than 4 weeks.

Subsequent coats should be allowed to dry for the time recommended by the manufacturer. Under ideal conditions, most, but not all, coatings will dry to the touch in about 2 hours and dry hard in 24 hours. Residual matter in the film will then continue to cure for some additional time. These times will vary if the weather is not favorable and may be different from those indicated here for some coating systems.

Coatings can damage materials for which they are not intended and may be difficult to remove. It is, therefore, important to protect adjacent work and surrounding areas from contact with coatings. Adjacent surfaces that become inadvertently contaminated with coatings should be cleaned immediately to help prevent damage.

In their liquid state, some coatings contain chemicals, such as ethylene glycol, that will cause serious illness or death if swallowed, or even if large quantities are absorbed by the body. Therefore, locations where these types of coatings are being applied must be well ventilated, and appliers must be protected using face masks, eye shields, and protective clothing. Some liquid constituents may also be flammable and must be kept away from heat sources.

# 9.18 Additional Reading

More information about the subjects discussed in this chapter can be found in the references listed in Section 9.19 and in the following publications:

Anis, Wagdy A.Y. and Donald P. Gatley. 2003. "Danger and Opportunity with Vinyl Wall Covering." *The Construction Specifier* (March): 28 (www.csinet.org).

Blackburn, Lane. 1997. "The Waterborne Coatings Revolution." *The Construction Specifier* (August): 58 (www. csinet.org).

Bukowski, Steven. 2003. *Flooring Instant Answers*. New York: McGraw-Hill (www.mcgraw-hill.com).

Eich, Sandra. 2005. "Cracking Considerations—Protecting Tile and Stone Flooring with Crack-Isolation Systems." *The Construction Specifier* (June): 53 (www.csinet.org).

Fogg, Brian. 2003. "Reduced Carpet Underlayment VOC Emissions." *The Construction Specifier* (August): 70 (www.csinet.org).

Gillespie, William A. 2003. "Floor Finish Tolerances for Suspended Slabs." *The Construction Specifier* (August): 42 (www.csinet.org).

Jacob, John. 2003. "The 411 on Wall coverings." *The Construction Specifier* (November): 64 (www.csinet.org).

McIlvain, Jess. 1994. "Ceramic Tile." *The Construction Specifier* (May): 126 (www.csinet.org).

Mercolino, Nancy and Michael Chusid. 1998. "Advanced Technology in Curved Metal Ceilings." *The Construction Specifier* (May): 75 (www.csinet.org).

Murphy, Tom. 2004. "Works of Art Underfoot—Thin-Set Epoxy Terrazzo Flooring." *The Construction Specifier* (February): 64 (www.csinet.org).

Olson, Eric K. 2005. "Avoiding the Perils of Paper-Faced Exterior Gypsum Sheathing." *The Construction Specifier* (February): 65 (www.csinet.org).

Pompo, Donato. 2003. "Substrate Prep and Quality Controls for Ceramic Tile and Stone." *The Construction Specifier* (August): 83 (www.csinet.org).

Pruter, Walter F. 2005. "Crack Control in Portland Cement Plaster." *The Construction Specifier* (April): 101 (www.csinet.org).

Remmele, Thomas E. 2003. "Specifying High-Performance Coatings for Concrete." *The Construction Specifier* (September): 49 (www.csinet.org).

Schroeder, Steve. 2003. "Controlling Moisture Underfoot for Successful Finish Flooring." *The Construction Specifier* (August): 54 (www.csinet.org).

Simmons, H. Leslie. 1996. "Field-Applied Organic Finish Failures." *The Construction Specifier* (July): 54 (www.csinet.org).

Sovinski, Rob W. 1999. *Brick in the Landscape: A Practical Guide to Specification and Design*. Hoboken NJ: John Wiley & Sons, Inc. (www.wiley.com).

Turner, R. Carroll. 2005. "Appearance Retention Ratings—New Help for Carpet Specifications." *The Construction Specifier* (June): 66 (www.csinet.org).

Weaver, Martin E. 1991. "Covering the Past: Historic Wallpapers." *The Construction Specifier* (March): 51 (www.csinet.org).

———. 1992. "Building and Monumental Stones of the United States." *The Construction Specifier* (March): 29 (www.csinet.org).

———. 1997. "A Systematic Approach to Graffiti Removal." *The Construction Specifier* (June): 24 (www.csinet.org).

# 9.19 Acknowledgments and References

## ACKNOWLEDGMENTS

We gratefully acknowledge the assistance of the following organizations and individuals in preparing this chapter. We are also indebted to them for permission to use their illustrations when requested and for the use of their publications as references.

American National Standards Institute (ANSI) (www.ansi.org)

American Olean Tile Company, Inc. (part of Dal-Tile Corporation, Dallas, TX) (www.daltile.com)

American Slate Company (www.americanslate.com)

Ancor Granite Tile Inc. (www.ancor.ca)

APA—The Engineered Wood Association (APA) (www.apawood.org)

Architectural Woodwork Institute (AWI) (www.awinet.org)

Armstrong World Industries, Inc. (www.armstrong.com)

ASTM International (ASTM) (www.astm.org)

Athenian Marble (www.athenianmarble.net)

Binks Manufacturing Company (www.binks.com)

Bonsal American (www.bonsal.com)

Brick Industry Association (BIA) (www.bia.org)

Bruce Hardwood Floors (www.bruce.com)

Buckingham-Virginia Slate Corporation (www.bvslate.com)

Burke/Mercer Division of Burke Industries, Inc. (www.mercerproducts.com)

Burlington Natstone Inc. (www.burningtonstone.com.uk)

Camara Slate Products, Inc. (www.camaraslate.com)

Carpet and Rug Institute (CRI) (www.carpet-rug.org)

Carpet Cushion Council (CCC) (www.carpetcushion.org)

Ceiling and Interior Systems Construction Association (CISCA) (www.cisca.org)

Celotex Corporation (www.celotex.com)

Clemco Industries Corporation (www.clemcoindustries.com)

Cold Spring Granite Co. (www.coldspringgranite.com)

Congoleum Corporation (www.congoleum.com)

Coronado Paint Co. (www.coronadopaint. com)

Creative Edge Corporation (www.cec-waterjet.com)

Crossfield Dex-o-Tex Products (www.dexotex.com)

Dur-A-Flex, Inc. (www.dur-a-flex.com)

E-Z Painter Corporation (www.ezpaintr.com)

Fluid Management, Inc. (www.fluidman.com)

Garland Floor Company (www.garlandfloor.com)

General Polymers (www.generalpolymers.com)

Georgia Marble Company (www.emporis.com)

Georgia Pacific Corporation (www.gp.com)

Greek Trade Commission (www.hepogreektrade.com/marble)

Guard Contract Wall coverings

Gypsum Association (GA) (www.gypsum.org)

Harris Specialty Chemicals, Inc.

Hercules, Incorporate

Hunter Douglas Architectural Products (www.luxalon.com)

Indiana Limestone Co., Inc. (www.indianalimestonecompany.com)

Indiana Limestone Institute of America (www.iliai.com)

International Institute for Lath and Plaster (www.iilp.org)

Johns Manville Corporation (www.jm.com)

Johnsonite Division of Duramax, Inc. (www.johnsonite.com)

K-Lath Divison, Georgetown Wire Co., Inc.

Kentile Floors

Kurfees Paint Co. (progresspaint.com)

Mab Paints (www.mabpaints.com)

MAC Publishers Incorporated

Maple Flooring Manufacturers Association (www.maplefloor.org)

Marble Institute of America (www.marble-institute.com)

Mercer Products Company, Inc. (see Burke/Mercer)

Metal Building Interior Products Company (www.mbiproducts.com)

Metal Lath/Steel Framing Association Division of the National Association of Architectural Metal Manufacturers (www.naamm.org.)

Mid-States Terrazzo and Tile Company

Milliken Carpet (www.millikencarpet.com)

Modac (see Mab Paints)

Mohawk Industries, Inc. (www.mohawkind.com)

Monsanto Company (www.monsanto.com)

National Association of Home Builders (NAHB) Research Foundation (www.nahbrc.org)

National Concrete Masonry Association (www.ncma.org)

National Gypsum Company (www.national-gypsum.com)

National Institute of Building Sciences (NIBS) (www.nibs.org)

National Institute of Standards and Technology (NIST) (www.nist.gov)

National Lime Association (www.lime.org)

National Paint and Coatings Association (NPCA) (www.paint.org)

National Terrazzo and Mosaic Association, Inc. (NTMA) (www.ntma.com)

Nora Flooring Systems, Inc. (www.nora.com)

Painting and Decorating Contractors of America (PDCA) (www.pdca.org)

Panel Solutions, Inc. (www.panelsolutions.com)

Perlite Institute (www.perlite.org)

Portland Cement Association (PCA) (www.cement.org)

Resilient Floor Covering Institute (RFCI) (www.rfci.com)

Ruberoid Company (www.ruberoid.com.uk)

Sherwin-Williams Company (www.sherwin.com)

Sivento Inc., a Hüls Group Company

Society for Protective Coatings (SSPC) (www.sspc.org)

Solnhofen Natural Stone, Inc. (www.solnhofen.com)

Stonhard, Inc., an RPM Company (www.stonehard.com)

Southern Pine Inspection Bureau (SPIB) (www.spib.org)

Structural Slate Company (structuralslate.com)

Terrazzo Marble and Supply Company Textured Coatings of America (www.texcote.com)

Thoro Consumer Products (www.thoro-products.com)

Tile Council of America, Inc. (TCA) (www.tileusa.com)

Trend Mills, Incorporated

United States Gypsum Company (USG) (www.usg.com)

U.S. Department of Housing and Urban Development (HUD) (www.hud.gov)

Vermiculite Association, Inc. (www.vermiculite.org)

Walter E. Selck and Company

West Coast Lumber Inspection Bureau (WCLIB) (www.wclib.org)

Western Wood Products Association (WWPA) (www.wwpa.org)

Wood Flooring Manufacturers Association (NOFMA) (www.nofma.org)

Wool Bureau, Inc.

Wunda Weve Carpet Company, Inc. (see Mohawk Industries)

## REFERENCES

We would also like to thank the authors and publishers of the publications in the following list for their contribution to our research for this chapter.

Acoustical Materials Association. AMA 1-II, "Ceiling Sound Transmission Test by the Two Room Method." Glenview, IL: Ceilings and Interior Systems Construction Association (CISCA) (cisca.org).

American Association of Textile Chemists & Colorists (AATCC). Research Triangle Park, NC: AATCC. (www.aatcc.org). Standards:

30-2004, "Antifungal Activity, Assessment on Textile Materials: Mildew, and Rot Resistance of Textile Materials."

100-2004, "Antibacterial Finishes on Textile Materials: Assessment of."

134-2001, "Electrostatic Propensity of Carpets."

American Concrete Institute (ACI). Farmington Hills, MI. (www.concrete.com).

117–90 (Reapproved 2002), "Standard Specifications for Tolerances for Concrete Construction and Materials."

201.1R-92, "Guide for Making a Condition Survey of Concrete in Service."

302.1R-04, "Guide for Concrete Floor and Slab Construction."

503.1-92 (Reapproved 2003), "Standard Specifications for Bonding Hardened Concrete Steel, Wood, Brick and Other Materials to Hardened Concrete with a Multi-Component Epoxy Adhesive."

503.2-92 (Reapproved 2003), "Standard Specifications for Bonding Plastic Concrete to Hardened Concrete with a Multi-Component Epoxy Adhesive."

503.3-92 (Reapproved 2003), "Standard Specifications for Producing a Skid-Resistant Surface on Concrete by the Use of a Mulit-Component Epoxy System."

503.4-92 (Reapproved 2003), "Standard Specifications for Repairing Concrete with Epoxy Mortars."

ACI 503.5R-92 (Reapproved 2003), "Guide for the Selection of Polymer Adhesives with Concrete."

ACI 503.6R-97 (Reapproved 2003), "Guide for the Application of Epoxy and Latex Adhesives for Bonding Freshly Mixed and Hardened Concrete."

American Iron and Steel Institute (AISI). Washington, DC (www.steel.org). 2002. CF02-1, "Cold-Formed Steel Framing Design Guide."

———. 2003. CF03-1, "Steel Stud Brick Veneer Design Guide."

———. 2001. SG02-1 and SG02-2, "North American Specification for the Design of Cold-Formed Steel Structural Members and the Commentary."

———. 2004. SG02-3, "Standard for Cold-Formed Steel Framing—General Provisions."

———. 2004. SG02-4, "Standard for Cold-Formed Steel Framing—Prescriptive Method for One and Two Family Dwellings."

———. 2002. SG03, *Cold-Formed Steel Design Manual.*

American National Standards Institute (ANSI). New York (www.ansi.org). Standards:

A10.20-1988, "Safety Requirements, Ceramic Tile, Terrazzo, Marble."

A108.1/A118/A136-1999, "Specifications for the Installation of Ceramic Tile."

A135.4-2004, "Basic Hardboard."

A137.1-1988, "Specifications for Ceramic Tile."

A208.1-1999, "Standard for Particleboard."

American Parquet Association (APA). APA 1, "For Floors and Flooring-Mosaic-Parquet Hardwood Slat." APA.

*Architectural Graphic Standards. See* Ramsey et al. (2000).

Association of the Wall and Ceiling Industry (AWCI (www.awci.org). 1994. *Veneer Plaster Manual.* Falls Church, VA.: AWCI.

ASTM International (ASTM). West Conshohocken, PA (www.astm.org). Standards:

A82, "Standard Specification for Steel Wire, Plain, for Concrete Reinforcement."

A185, "Standard Specification for Steel Welded Wire, Fabric, Plain, for Concrete Reinforcement."

B117, "Standard Practice for Operating Salt Spray (Fog) Apparatus."

C5, "Standard Specification for Quicklime for Structural Purposes."

C11, "Standard Terminology Relating to Gypsum and Related Building Materials and Systems."

C28/C28M, "Specification for Gypsum Plasters."

C33, "Standard Specification for Concrete Aggregate."

C35, "Standard Specification for Inorganic Aggregates for Use in Gypsum Plaster."

C59, "Standard Specification for Gypsum Plaster Casting and Gypsum Molding Plaster."

C61/C61M, "Standard Specification for Gypsum Keene's Cement.

C67, "Standard Test Methods for Sampling and Testing Brick and Structural Clay Tile."

C91, "Standard Specification for Masonry Cement."

C97, "Standard Test Method for Absorption and Bulk Specific Gravity of Dimension Stone."

C99, "Standard Test Method for Modulas of Rupture of Dimension Stone."

C109/C109M, "Standard Test Method for Compressive Strength of Hydraulic Cement Mortars (Using 2-in. or [50-mm] Cube Specimens)."

C119, "Standard Terminology Relating to Dimension Stone."

C144, "Standard Specification for Aggregate for Masonry Mortar."

C150, "Standard Specification for Portland Cement."

C170, "Standard Test Method for Compressive Strength of Dimension Stone."

C206, "Standard Specification for Finishing Hydrated Lime."

C207, "Standard Specification for Hydrated Lime for Masonry Purposes."

C241, "Standard Test Method for Abrasion Resistance of Stone Subject to Foot Traffic."

C242, Standard Terminology of Ceramic Whitewares and Related Products."

C270, "Standard Specification for Mortar for Unit Masonry."

C307, "Standard Test Method for Tensile Strength of Chemical-Resistant Mortar, Grouts, and Monolithic Surfacings."

C309, "Standard Specification for Liquid Membrane-Forming Compounds for Curing Concrete."

C332, "Standard Specification for Lightweight Aggregate for Insulating Concrete."

C346, "Standard Test Method for 45-Deg Specular Gloss of Ceramic Materials."

C423, "Standard Test Method for Sound Absorption and Sound Absorption Coefficients by the Reverberation Room Method."

C472, "Standard Methods for Physical Testing of Gypsum Plasters and Gypsum Concrete."

C475/C475M, "Standard Specifications for Joint Compound and Joint Tape for Finishing Gypsum Board."

C481, "Standard Test Method for Laboratory Aging of Sandwich Construction."

C501, "Standard Test Method for Relative Resistance to Wear of Unglazed Ceramic Tile by the Taber Abraser."

C503, "Standard Specifications for Marble Dimension Stone (Exterior)."

C514, "Standard Specifications for Nails for the Application of Gypsum Board."

C557, "Standard Specifications for Adhesive for Fastening Gypsum Wallboard to Wood Framing."

C568, "Standard Specification for Limestone Dimension Stone."

C579, "Standard Test Method for Compressive Strength of Chemical-Resistant Mortars, Grouts, Monolithic Surfacers and Polymer Concretes."

C580, "Standard Test Method for Flexural Strength and Modulus of Elasticity of Chemical-Resistant Mortars, Grouts, Monolithic Surfaces and Polymer Concretes."

C587, "Standard Specifications for Gypsum Veneer Plaster."

C588/C588M, "Specifications for Gypsum Base for Veneer Plaster."

C615, "Standard Specification for Granite Dimension Stone."

C627, "Standard Test Method for Evaluating Ceramic Floor Tile

Installation Systems Using the Robinson-Type Floor Tester."

C629, "Standard Specification for Slate Dimension Stone."

C631, "Standard Specifications for Bonding Compounds for Interior Gypsum Plastering."

C635, "Standard Specification for the Manufacture, Performance, and Testing of Metal Suspension Systems for Acoustical Tile and Lay-in Panel Ceilings."

C636, "Standard Practice for Installation of Metal Ceiling Suspension Systems for Acoustical Tile and Lay-in Panels."

C641, "Standard Test Method for Staining Materials in Lightweight Concrete Aggregates."

C645, "Standard Specifications for Nonstructural Steel Framing Members."

C666, "Standard Test Method for Resistance of Concrete to Rapid Freezing and Thawing."

C672, "Standard Test Method for Scaling Resistance of Concrete Surfaces Exposed to Deicing Chemicals."

C696, "Standard Test Methods for Chemical, Mass Spectrometer, and Spectrochemical Analysis of Nuclear-Grade Uranium Dioxide Powders and Pellets."

C754, "Standard Specifications for Installation of Steel Framing Members to Receive Screw-Attached Gypsum Panel Products."

C840, "Standard Specifications for Application and Finishing of Gypsum Board."

C841, "Standard Specifications for Installation of Interior Lathing and Furring."

C842, "Standard Specifications for Application of Interior Gypsum Plaster."

C843, "Standard Specifications for Application of Gypsum Veneer Plaster."

C844, "Standard Specifications for Application of Gypsum Base to Receive Gypsum Veneer Plaster."

C847, "Standard Specifications for Metal Lath."

C880, "Standard Test Method for Flexural Strength of Dimension Stone."

C897, "Standard Specifications for Aggregate for Job-Mixed Portland Cement-Based Plasters."

C902, "Standard Specification for Pedestrian and Light Traffic Paving Brick."

C919, "Standard Practices for Use of Sealants in Acoustical Applications."

C920, "Standard Specifications for Elastomeric Joint Sealants."

C926, "Standard Specifications for Application of Portland Cement-Based Plaster."

C932, "Standard Specifications for Surface-Applied Bonding Agents for Exterior Plastering."

C955, "Standard Specifications for Load-Bearing (Transverse and Axial) Steel Studs, Runners (Track), and Bracing or Bridging, for Screw Application of Gypsum Board and Metal Plaster Bases."

C1002, "Standard Specifications for Steel Self-Piercing Tapping Screws for the Application of Gypsum Panel Products or Metal Plaster Bases to Wood Studs or Steel Studs."

C1007, "Standard Specifications for Installation of Load-Bearing (Transverse and Axial) Steel Studs and Related Accessories."

C1036, "Standard Specification for Flat Glass."

C1047, "Standard Specifications for Accessories for Gypsum Wallboard and Gypsum Veneer Base."

C1063, "Standard Specifications for Installation of Lathing and Furring to Receive Interior and Exterior Portland Cement-Based Plaster."

C1363, "Standard Test Method for Thermal Performance of Building Materials and Envelope Assemblies by Means of a Hot Box Apparatus."

C1396/C1396M, "Standard Specification for Gypsum Wallboard."

D16, "Standard Terminology for Paint, Related Coatings, Materials and Applications."

D226, "Standard Specification for Asphalt-Saturated Organic Felt Used in Roofing and Waterproofing."

D522, "Standard Test Methods for Mandrel Bend Test of Attached Organic Coatings."

D570, "Standard Test Method for Water Absorption of Plastics."

D578, "Standard Specification for Glass Fiber Strands."

D635, "Standard Test Method for Rate of Burning and/or Extent and Time of Burning of Self-Supporting Plastics in a Horizontal Position."

D638, "Standard Test Method for Tensile Properties of Plastics."

D648, "Standard Test Method for Deflection Temperature of Plastics Under Flexural Load in the Edgewise Position."

D660, "Standard Test Method for Evaluating Degree of Checking of Exterior Paints."

D661, "Standard Test Method for Evaluating Degree of Cracking of Exterior Paints."

D662, "Standard Test Method for Evaluating Degree of Erosion of Exterior Paints."

D695, "Standard Test Method for Compressive Properties of Rigid Plastic."

D751, "Standard Test Method for Coated Fabrics."

D790, "Standard Test Methods for Flexural Properties of Unreinforced and Reinforced Plastics and Electrical Insulating Materials."

D870, "Standard Practice for Testing Water Resistance of Coatings Using Water Immersion."

D968, "Standard Test Method for Abrasion Resistance of Organic Coatings by Falling Abrasive."

D1187, "Standard Specification for Asphalt-Base Emulsions for Use as Protective Coatings for Metal."

D1308, "Standard Test Method for Effect of Household Chemicals on Clear and Pigmented Organic Finishes."

D1653, "Standard Test Method for Water Vapor Transmission of Organic Coating Films."

D1730. "Standard Practices for Preparation of Aluminum and Aluminum-Alloy Surfaces for Painting."

D1731. "Standard Practices for Preparation of Hot-Dip Aluminum Surfaces for Painting."

D1784, "Standard Specification for Rigid Poly(Vinyl Chloride)(PVC) Compounds and Chlorinated Poly(Vinyl Chloride)(CPVC) Compounds."

D2000, "Standard Classification System for Rubber Products in Automotive Applications."

D2047, "Standard Test Method for Static Coefficient of Friction of Polish-Coated Floor Surfaces as Measured by the James Machine."

D2092, "Standard Guide for Preparation of Zinc-Coated (Galvanized) Steel Surfaces for Painting."

D2103, "Standard Specification for Polyethylene Film and Sheeting."

D2240, "Standard Test Method for Rubber Property-Durometer Hardness."

D2243, "Standard Test Method for Freeze-Thaw Resistance of Water-Borne Coatings."

D2287, "Standard Specification for Nonrigid Vinyl Chloride Polymer and Copolymer Molding and Extrusion Compounds."

D2370, "Standard Test Method for Tensile Properties of Organic Coatings."

D2486, "Standard Test Method for Scrub Resistance of Wall Paints."

D2859, "Standard Test Method for Ignition Characteristics of Finished Textile Floor Covering Materials."

D3276, "Standard Guide for Painting Inspectors (Metal Substrates)."

D3574, "Standard Test Method for Flexible Cellular Materials—Slab, Bonded, and Molded Urethane Foams."

D3676, "Standard Specification for Rubber Cellular Cushion Used for Carpet or Rug Underlay."

D3794, "Standard Guide for Testing Coil Coatings."

D4141, "Standard Practice for Conducting Black Box and Solar Concentrating Exposures of Coatings."

D4214, "Standard Test Method for Evaluating Degree of Chalking of Exterior Paint Films."

D4258, "Standard Practice for Surface Cleaning Concrete for Coating."

D4259, "Standard Practice for Abrading Concrete."

D4260, "Standard Practice for Liquid and Gelled Acid Etching of Concrete."

D4261, "Standard Practice for Surface Cleaning Concrete Unit Masonry for Coating."

D4397, "Standard Specification for Polyethylene Sheeting for Construction, Industrial, and Agricultural Applications."

D4541, "Standard Test Method for Pull-out Strength of Coatings Using Portable Adhesion Testers."

D5090, "Standard Practice for Standardizing Ultrafiltration Permeate Flow Performance Data."

D5324, "Standard Guide for Testing Water-borne Architectural Coatings."

E84, "Standard Test Method for Surface Burning Characteristics of Building Materials."

E90, "Standard Test Method for Laboratory Measurement of Airborne Sound Transmission Loss of Building Partitions and Elements."

E96, "Standard Test Methods for Water Vapor Transmission of Materials."

E119, "Standard Test Methods for Fire Tests for Building Construction and Materials."

E136, "Standard Test Method for Behavior of Materials in a Vertical Tube Furnace at 750 deg C."

E413, "Classification for Rating Sound Insulation."

E488, "Standard Test Method for Strength of Anchors in Concrete or Masonry Elements."

E514, "Standard Test Method for Water Penetration and Leakage Through Masonry."

E580, "Standard Practice for Application of Ceiling Suspension Systems for Acoustical Tile and Lay-in Panels in Areas Requiring Seismic Restraint."

E648, "Standard Test Method for Critical Radiant Flux of Floor-Covering Systems Using a Radiant Heat Energy Source."

E662, "Standard Test Method for Specific Optical Density of Smoke Generated by Solid Materials."

E1264, "Standard Classification for Acoustical Ceiling Products."

E1347, "Standard Test Method for Color and Color-Difference Measurement by Trist Mulus (Filter) Colorimeter."

F141, "Standard Terminology Relating to Resilient Floor Coverings."

F142, "Standard Test Method for Indentation of Resilient Floor Tiles (McBurney Test)."

F150, "Standard Test Method for Electrical Resistance of Conductive and Static Dissipative Resilient Flooring."

F373, "Standard Test Method for Embossed Depth of Resilient Floor Coverings."

F386, "Standard Test Method for Thickness of Resilient Flooring Materials Having Flat Surfaces."

F510, "Standard Test Method for Resistance to Abrasion of Resilient Floor Coverings Using an Abrader with a Grit Feed Method."

F693, "Standard Practice for Sealing Seams of Resilient Sheet Flooring Products by Use of Liquid Seam Sealers."

F710, "Standard Practice for Preparing Concrete Floors to Receive Resilient Flooring."

F924. "Standard Test Method for Resistance to Puncture of Cushioned Resilient Floor Coverings."

F925, "Standard Test Method for Resistance of Chemicals of Resilient Flooring."

F1066, "Standard Specification for Vinyl Composition Floor Tile."

F1265, "Standard Test Method for Resistance to Impact for Resilient Floor Tile."

F1303, "Standard Specification for Sheet Vinyl Floor Covering with Backing."

F1344, "Standard Specification for Rubber Floor Tile."

F1482, "Standard Practice for Installation and Preparation of Panel Type Underlayments to Receive Resilient Flooring."

F1516, "Standard Practice for Sealing Seams of Resilient Flooring Products by the Heat Weld Method."

F1700, "Standard Specification for Solid Vinyl Floor Tile."

F1859, "Standard Specification for Rubber Sheet Floor Covering without Backing."

F1860, "Standard Specification for Rubber Sheet Floor Covering with Backing."

F1861, "Standard Specification for Resilient Wall Base."

F1869, "Standard Test Method for Measuring Moisture Vapor Emission Rate of Concrete Subfloor Using Anhydrous Calcium Chloride."

F1913, "Standard Specification for Vinyl Sheet Floor Covering without Backing."

F2034, "Standard Specification for Sheet Linoleum Floor Covering."

F2169, "Standard Specification for Resilient Stair Treads."

F2170, "Standard Test Method for Determining Relative Humidity in Concrete Floor Slabs Using in Situ Probes."

F2195, "Standard Specification for Linoleum Floor Tile."

G21, "Standard Practice for Determining Resistance of Synthetic Polymeric Materials to Fungi."

G152, "Standard Practice for Operating Open Flame Carbon Arc Light Apparatus for Exposure of Nonmetallic Materials."

G153, "Standard Practice for Operating Enclosed Carbon Arc Light Apparatus for Exposure of Nonmetallic Materials."

Bakhalov, G.T. and A.V. Turkovakaya. 1965. *Corrosion and Protection of Metals.* Elmsford, NY: Pergamon Press.

Banov, Able. 1973. *Paints and Coatings Handbook for Contractors Architects and Builders.* Farmington, MI: Structures Publishing Co.

Brick Industry Association (BIA). Reston, VA (www.bia.org). Technical Notes on Brick Construction:

No. 14. Rev. September 1992. "Brick Floors and Pavements—Design and Detailing, Part I."

No. 14A. Rev. January 1993. "Brick Floors and Pavements—Materials and Installation, Part II."

No. 14B. November/December 1975 (Reissued September 1998). "Brick Floors and Pavements: Part III."

Carpet and Rug Institute (CRI). Dalton, GA (www.carpet-rug.org). *The Carpet Primer.*

———. 1999. "Carpet and Rug Flammability Tests."

Ceilings and Interior Systems Construction Association (CISCA). *Acoustical Ceilings Use and Practice.* St. Charles, IL: CISCA (cisca.org).

Chemical Fabrics and Film Association (CFFA). Cleveland (www.chemicalfabricsandfilm.com). 2002. CFFA-W-101-D, "Quality Standard for Vinyl Coated Fabric Wall covering." Cleveland, OG. CFFA.

Dean, Sheldon W. and T.S. Lee, eds. 1988. *Degradation of Metals in the Atmosphere.* West Conshohocken, PA: ASTM (www.astm.org).

Diehl, John R. 1965. *Manual of Lathing and Plastering, Gypsum Association Edition.* New York: MAC Publishers Association.

Factory Mutual Insurance Company (FM Global). Johnston, RI: (www.fm-global.com).
*Approval Guide.*
*Loss Prevention Data Sheets.*

Federal Specifications, all: *See* U.S. General Services Administration Specifications Unit.

Gypsum Association (GA). Washington, DC (www.gypsum.org).

GA-214, "Recommended Levels of Gypsum Board Finish."

GA-216, "Application and Finishing of Gypsum Board."

GA-223, "Gypsum Panel Products Types, Uses, Sizes and Standards."

GA232, "Painting New Gypsum Board."

GA-253, "Application of Gypsum Sheathing."

GA-254-86, "Fire Resistant Gypsum Sheathing."

GA 530, "Design Data: Gypsum Board."

GA-600, "Fire Resistance Design Manual."

Maple Flooring Manufacturers Association (MFMA). Northbrook, IL (www.maplefloor.org).

MFMA-PQ (Parquet), "Grading Rules for Hard Maple."

MFMA-RL (Random Length), "Northern Hard Maple Flooring Grading Rules."

MFMA-FJ (Finger Jointed), "Northern Hard Maple Flooring Grading Rules."

"Sanding, Sealing, Court Lining and Finishing Maple Gym Floors."

MFMA "Source Book"

Marble Institute of America, Inc. 2003. *Dimension Stone—Design Manual VI.* Cleveland, OH: Marble Institute of America (www.marble-institute.com).

McIlvain, Jess. 1982. "The Problem with Dry-Set Mortars." *Tile and Decorative Surfaces* (September): 17–20 (www.telemagazine.com).

———. 1982. "Ceramic Tile Maintenance." *Tile and Decorative Surfaces* (October/November): 48–57 (www.telemagazine.com).

———. 1982. The Preconstruction Conference." *Tile and Decorative Surfaces* (December): 102–3.

———. 1983. "Caution: Latex-Portland Cement Mortar Over Plywood." *Tile and Decorative Surfaces* (January/February) (www.telemagazine.com).

———. 1983. "Plaster vs. Mortar Beds." *Tile and Decorative Surfaces* (April): 41–56.

———. 1983. "Never Again." *Tile and Decorative Surfaces*, July: 138–40 (www.telemagazine.com).

———. 1983. "Dodging the Dry-Set Mortar Gremlins." *The Construction Specifier* (June): 77–82 (www.csinet.org).

———. 1984. "Prefabricated Exterior Ceramic Tile Cladding." *The Construction Specifier* (June): 76–91 (www.esinet.org).

———. 1984. "Prefabricated Exterior Ceramic Tile Cladding, Part 1." *Tile and Decorative Surfaces* (January/February): 80–85 (www.telemagazine.com).

———. 1984. "Prefabricated Exterior Ceramic Tile Cladding, Part 2." *Tile and Decorative Surfaces* (March): 41–53 (www.telemagazine.com).

———. 1984. "Prefabricated Exterior Ceramic Tile Cladding, Part 3." *Tile and Decorative Surfaces* (April): 50–58 (www.telemagazine.com).

———. 1984. "Let's Talk Tile." *Tile and Decorative Surfaces* (June): 69–74 (www.telemagazine.com).

———. 1984. "Mortar Beds." *Tile and Decorative Surfaces* (September): 118–20 (www.telemagazine.com).

———. 1984. "The Preconstruction Conference." *Tile Letter* (October): 32–33.

———. 1985. "Floor Renovation." *Tile and Decorative Surfaces* (April): 48–49 (www.telemagazine.com).

———. 1985. "Moisture Expansion." *Tile and Decorative Surfaces* (June): 72–86 (www.telemagazine.com).

———. 1985. "The Goose That Laid the Rotten Egg." *Tile Letter* (August): 42–46 (www.tele-assn.com).

———. 1986. "Writing Ceramic Tile Specifications." *Tile and Decorative Surfaces* (October/November): 81–82. (www.telemagazine.com).

———. 1986. "Lights, Action." *Tile Letter* (July): 36–40 (www.tele-assn.com).

———. 1987. "Selecting Ceramic Tile Floor Installation Methods." *Tile and Decorative Surfaces* (August): 38–42 (www.telemagazine.com).

———. 1988. "Grout: It Doesn't Have to Be a Problem." *Tile and Decorative Surfaces* (June): 42–45 (www.telemagazine.com).

Metal Lath/Steel Framing Association (ML/SFA). Chicago.

1987. *Lightweight Steel Framing Systems Manual.*

1991. *Specifications for Metal Lathing and Furring.*

National Association of Architectural Metal Manufacturers (NAAMM).

ML/SFA 920, "Specifications for Metal Lathing and Furring." Chicago: NAAMM (www.naamm. org).

National Concrete Masonry Association (NCMA). TEK 09-03, "Plaster and Stucco for Concrete Masonry." Herndon, VA: NCMA (secure.ncma.org).

National Fire Protection Association (NFPA). Quincy, MA (www.nfpa.org). NFPA 253-2005, "Standard Method of Test for Critical Radiant Flux of Floor Covering Systems Using a Radiant Heat Energy Source."

NFPA 255-00, "Standard Method of Test of Surface Burning Characteristics of Building Materials."

National Terrazzo & Mosaic Association, Inc. (NTMA). Purcellville, VA (www.ntma.com). "Basic Terrazzo Systems Information."

"Guide Specifications."

National Tile Contractors Association (NTCA). *Reference Manual.* Jackson, MS: National Association of Tile Contractors (www.tile-assn.com).

Paint & Decorating Retailers Association (PDRA). *Paint Problem Solver.* Fenton, MO: PDRA (www.pdra.org).

Painting and Decorating Contractors of America (PDCA). Falls Church, VA (www.pdca.org).

1982. *Painting and Decorating Encyclopedia.*

1984. *Painting and Wall Covering: A Century of Excellence.*

*Architectural Painting Specification Manual.**

1988. *Hazardous Waste Handbook.*

1988. *The Master Painters Glossary.*

1995. *Painting and Decorating Craftsman's Manual and Textbook,* 5th ed.

*Washington State Council of PDCA.

2004. *PDCA Estimating Guide.*

Portland Cement Association (PCA). Skokie, IL (www.cement.org). 2003. *Portland Cement Plaster (Stucco) Manual.*

———. 1992. "Painting Concrete, IS134."

———. 1988. "Removing Stains and Cleaning Concrete Surfaces."

———. 2001. "Effects of Substances on Concrete and Guide to Protective Treatments."

Ramsey, Charles George, Harold Reeve Sleeper, and John Ray Hoke, Jr. 2000. *Architectural Graphic Standards,* 10th ed. Hoboken, NJ: John Wiley & Sons, Inc. (www.wiley.com).

Resilient Floor Covering Institute (RFCI). Rockville, MD (www.rfci. com). 1998. "Recommended Work Practices for the Removal of Resilient Floor Coverings."

———. "Recommended Installation Practice for Homogeneous Sheet Flooring, Fully-Adhered."

———. "Static Load Limit Testing of Resilient Flooring Products."

Society for Protective Coatings (SSPC). Pittsburgh (www.sspc.org). *Good Paint Practice: SSPC Painting Manual, Vol. 1,* 4th ed.

———. *Systems and Specifications: SSPC Painting Manual Vol. 2,* 2005 ed.

Southern Pine Inspection Bureau (SPIB). 2002. *Standard Grading Rules for Southern Pine Lumber.* Pensacola, FL: SPIB (www.spib.org).

Steel Stud Manufacturers Association (SSMA). Chicago (www.ssma.com). 2001. *Product Technical Information.*

———. *Cold Formed Steel Details.*

Tile Council of America, Inc. (TCA) (www.tileusa.com). 2005. *Handbook for Ceramic Tile Installation.* Anderson, SC: TCA.

Underwriters Laboratories, Inc. (UL).

Northbrook, IL (www.ul.com). 2006. *Building Materials Directory.*

———. 2006. *Fire Resistance Directory.*

United States Gypsum Company (USG). *Gypsum Construction Handbook.* Chicago: USG (www.usg.com).

U.S. General Services Administration Specifications Unit. Washington, DC (apps.fss.gsa.gov). Federal Specifications:

A-A-50574, "Enamel, Odorless, Alkyd, Interior, Semigloss, White and Tints."

RR-T-650E, "Treads, Metallic and Nonmetallic Skid Resistant."

TT-L-201A, "Linseed Oil, Heat Polymerized."

TT-P-19D, "Paint (Latex Acrylic Emulsion): Exterior."

TT-P-645B, "Primer, Paint, Zinc-Molybdate, Alkyd Type."

TT-P-650D, "Primer Coating, Latex Base, Interior, White (For Gypsum Wallboard or Plaster)."

UU-B-790A, "Building Paper, Vegetable Fiber (Kraft, Waterproofed, Water-Resistant, and Fire-Resistant)."

CCC-W-408D, "Wall covering, Vinyl Coated."

West Coast Lumber Inspection Bureau (WCLIB). *Standard Grading Rules.* Portland, OR: WCLIB (www.wclib. org).

Western Wood Products Association (WWPA). *Western Lumber Grading Rules, '05."* Portland, OR: WWPA (www.wwpa.org).

Wood Flooring Manufacturers Association (NOFMA). Memphis, TN (www.nofma.org). *Installing Hardwood Floors.*

———. *Finishing Hardwood Floors*

———. *Flooring Grading Rules.*

Zulandt, David F. 1985. "Consider an Alternative: Veneer Plaster." *The Construction Specifier* (July): 84–85.

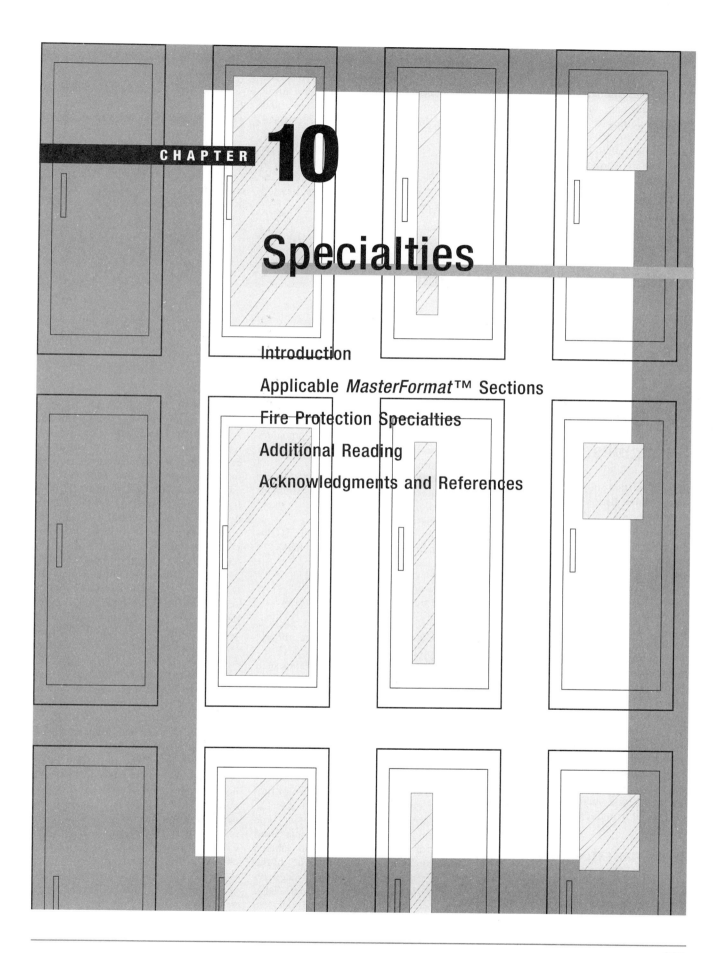

CHAPTER **10**

# Specialties

# Introduction

This chapter discusses portable fire extinguishers and their housings. The other specialties listed in *MasterFormat* 2004 Division 10 include, but are not necessarily limited to, the following: visual display surfaces such as chalkboards and tackboards; display cases; directories; signage; telephone enclosures; compartments and cubicles; wire mesh and demountable partitions; toilet and bath accessories; fireplaces and stoves; lockers; postal specialties; protective covers such as carports and awnings; flagpoles; and many others. Collectively, these would require a great deal of space to discuss adequately, and they are, therefore, beyond the scope of this book.

Subjects related to portable fire extinguishers include fire sprinkler systems, which are addressed in Chapter 15, and fire detection and alarm systems, which are covered in Chapter 20. The fixed chemical fire extinguishing equipment and systems used in such locations as commercial kitchens and industrial buildings are beyond the scope of this book.

# Applicable *MasterFormat* Sections

The following *MasterFormat* 2004 Level 2 section is applicable to this chapter:

10 44 00  Fire Protection Specialties

# 10.1 Fire Protection Specialties

Most fires start out small. Often a handheld extinguisher or sprinkler system can keep them small or put them out entirely.

Every residential kitchen should be equipped with a fire extinguisher. Others should be put in a garage, in a basement, and near bedrooms and any other place a fire hazard exists. Many insurance companies offer homeowners insurance discounts when fire extinguishers are present.

Most codes require portable fire extinguishers in other types of buildings and stipulate their locations. The *International Building Code*, for example, requires compliance with the *International Fire Code*, which requires both automatic fire control systems, such as sprinklers, and portable fire extinguishers.

Portable fire extinguisher units are usually required in corridors, storage areas, mechanical and electrical equipment rooms, and other areas where a fire hazard might exist, such as kitchens, laboratories, and manufacturing facilities.

## 10.1.1 FIRE EXTINGUISHERS

Fire extinguishers smother a fire by depriving it of oxygen. They are made with a variety of fire-fighting agents, including:

- A dry chemical (primarily fluidized and siliconized monoammonium phosphate and siliconized sodium bicarbonate)
- Pressurized water
- Liquid carbon dioxide under pressure
- Sodium chloride
- Potassium acetate
- Halon replacement chemicals (Halotron 1, DuPont FE-36, and others)

Each of these extinguisher types is effective against one or more classes of fires. Fires can be classed as follows:

- A—fire fed by a solid fuel, such as wood, paper, textiles, plastic, rubber, or other nonliquid combustibles
- B—fire fed by flammable liquids such as grease, cooking oils, gasoline, and paint solvents
- C—fire in a live electrical circuit
- D—fire in combustible metals such as magnesium
- K—fire in liquid cooking media

Figure 10.1-1 shows the types of extinguishers required for each class of fire.

For complete protection, it may be necessary to have more than one type of extinguisher. An extinguisher used against a fire for which it is not labeled can actually increase the intensity of the fire. Extinguishers should be rated by Underwriters Laboratories, Inc. (UL), and marked with its classification. The marking for extinguishers rated for Class A and B fires includes a letter designating the class of fire for which the extinguisher is effective and a number indicating its relative effectiveness. Extinguishers for use on Class C or D fires are not required to have the number designation. Codes usually specify the class of extinguisher required. Selection is then a matter of comparing the code requirements with the manufacturer's listing. In NFPA 10, "Portable Fire Extinguishers," the National Fire Protection Association (NFPA) spells out detailed requirements for extinguisher selection. Conformance with NFPA 10 is listed as a requirement in the *International Fire Code*.

Fire extinguishers should be located where required by code. There should be one near each doorway and as far as permitted by code from the likely source of a fire. NFPA guidelines require the top of fire extinguisher be no more than 5 ft (1524 mm) above the floor, but within that limit, a fire extinguisher's mounting height is dictated by Americans with Disabilities Act (ADA) requirements and depends on the reaching distance to the extinguisher. For example, if the fire extinguisher handle is 48 in. (1219.2 mm) above the floor, the maximum side reach is 54 in. (1371.6 mm).

Extinguishers should be checked once a month to ensure that they are fully

**FIGURE 10.1-1** Type of Extinguisher Used for Each Class of Fire

| Extinguisher Type | Fire Class | | | | |
|---|---|---|---|---|---|
| | A | B | C | D | K |
| Dry Chemical | X | X | X | | |
| Carbon Dioxide | X | | | | |
| Sodium Chloride | | | | X | |
| Potassium Acetate | | | | | X |
| Halon Replacement | X | X | X | | |

**FIGURE 10.1-2** Typical cabinet front designs.

charged. Once an extinguisher has been used, no matter how briefly, it must be recharged. Otherwise, the material left within the discharge valve will gradually allow the pressure to bleed off.

## 10.1.2 FIRE EXTINGUISHER CABINETS

Fire extinguishers may be mounted on wall brackets, but in most visible locations, esthetic considerations dictate enclosing them in cabinets. Fire extinguisher cabinets are made from steel, stainless steel, aluminum, and brass in several types and styles. Basic door styles are shown in Figure 10.1-2. Combination units are also available that hold both extinguishers and fire hoses or both extinguishers and valves.

Most fire extinguisher cabinets are made of a single wall thickness of steel or aluminum. Insulated fire-rated units are also available, however, for those locations where they are required.

Cabinet sizes are dictated by the fire extinguishers they will hold.

Cabinets may be surface mounted, fully recessed, or partially recessed. The amount they may project from a wall, including their handles, in a space controlled by ADA Accessibility Guidelines is 4 in. (101.6 mm) when their leading edge is between 27 and 80 in. (685.8 and 2032 mm) above the floor.

The ADA guidelines do not require fire extinguisher handles or signage to be in compliance with the guidelines for other doors and signs.

## 10.2 Additional Reading

More information about the subjects discussed in this chapter can be found in the references listed in Section 10.3 and in the following publications:

Council on Tall Buildings and Urban Habitat. 1992. *Fire Safety in Tall Buildings*. New York: McGraw-Hill Book Companies (www.mcgraw-hill.com).

McGowan, Maryrose. 1993. "Specifying Portable Fire Extinguishers." *The Construction Specifier* (January): 25 (www.csinet.org).

## 10.3 Acknowledgments and References

### ACKNOWLEDGMENTS

We gratefully acknowledge the assistance of the following organizations and individuals in preparing this chapter. We are also indebted to them for permission to use their illustrations when requested and for the use of their publications as references.

American National Standards Institute (ANSI) (www.ansi.org)

Consumer Product Safety Commission (CPSC) (www.cpsc.gov)

National Fire Protection Association (NFPA) (www.nfpa.org)

U.S. Department of Housing and Urban Development (HUD) (www.hud.gov)

### REFERENCES

We would also like to thank the authors and publishers of the publications in the following list for their contribution to our research for this chapter.

National Fire Protection Association (NFPA). Quincy, MA (www.nfpa.org). 2002. NFPA 550, "Guide to the Fire Safety Concepts Tree."

———. 2003. NFPA 101.® *Life Safety Code*.

Raeber, John A. 1991. "Redundancy in Fire Protection." *The Construction Specifier* (February): 17–19 (www.csinet.org).

Reid, Ed, and Bob Schifiliti. 1991. "A Balanced Approach to Fire Protection Detection and Suppression Systems." *The Construction Specifier* (February): 30–41 (www.csinet.org).

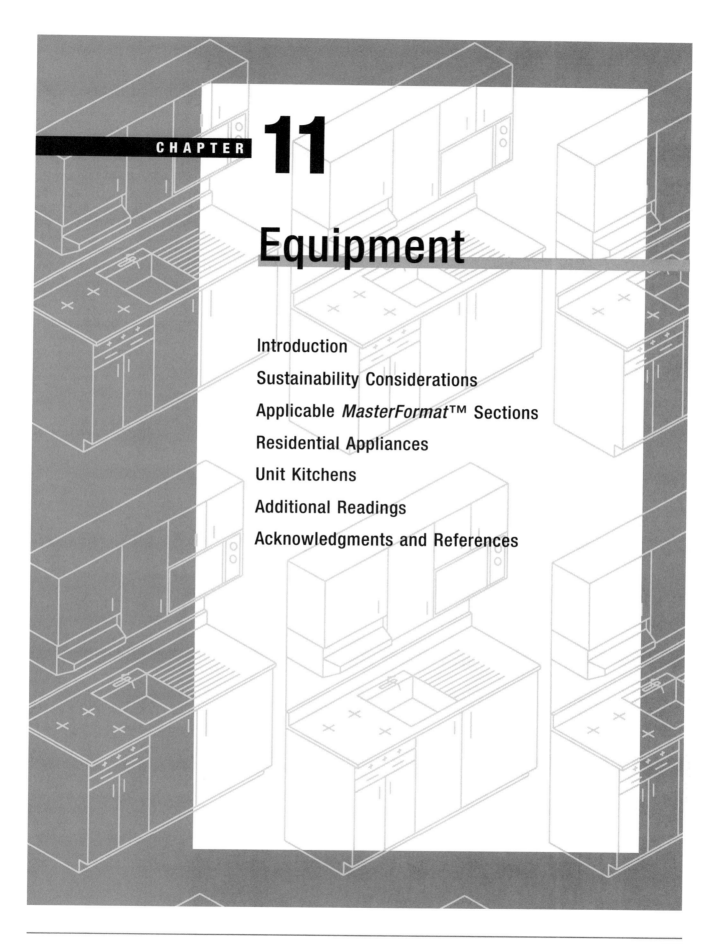

CHAPTER **11**

# Equipment

# Introduction

*Equipment*, as covered by *MasterFormat* 2004 Division 11 and discussed in this chapter, includes devices and materials used for many purposes:

- Vehicle and pedestrian equipment, including vehicle service equipment; parking control equipment; loading dock equipment, including dock levelers, lifts, ramps, bridges, and bumpers; and pedestrian control equipment
- Security detention and banking equipment, including vault doors and gates, security and emergency equipment, safes, safe deposit boxes, teller windows and associated equipment, package transfer units, and automatic banking systems; and detention equipment
- Commercial equipment, including mercantile equipment, including barber and beauty shop equipment, cash registers, and vending equipment; refrigerated display equipment; commercial laundry and dry cleaning equipment, including washers, dry cleaning equipment, extractors, and drying and finishing equipment; building maintenance equipment, including vacuum cleaning systems, window washing equipment and floor and wall cleaning equipment; hotel registration equipment; checkroom equipment; unit kitchens; darkroom equipment, including transfer cabinets and film processing equipment; office equipment, including computers, printers, and copiers; and postal, packaging, and shipping equipment
- Residential equipment, including kitchen and laundry appliances, and retractable stairs
- Food service equipment, including food storage, preparation, and dispensing equipment; cleaning and disposal equipment; and bar and soda fountain equipment
- Educational and scientific equipment, including library equipment, including book stacks, carrels, theft protection equipment, and book depositories equipment; audiovisual equipment, including projectors and screens; laboratory equipment for scientific purposes, including cabinets, fume hoods, incubators, sterilizers, refrigerators, and safety devices; planetarium equipment; observatory equipment, including telescopes; vocational shop equipment; and exhibit equipment
- Entertainment equipment, including theater and stage equipment, including curtains, rigging, and acoustical shells; and musical equipment such as organs, carillons, and bells
- Athletic and recreational equipment, including such athletic equipment as exercise equipment, gymnasium equipment, and interior scoreboards
- Recreational equipment, including bowling alley, shooting range, table game, and game room equipment
- Play field equipment and structures, including playground equipment, exterior court athletic equipment, athletic field equipment, and exterior scoreboards
- Health care equipment, including medical sterilizing equipment, examination and treatment equipment, patient care equipment, dental equipment, optical equipment, operating room equipment, radiology equipment, mortuary equipment, mortuary refrigerators, crematorium equipment, mortuary lifts, and therapy equipment
- Collection and disposal equipment, including such solid waste handling equipment as incinerators, compactors, recycling equipment, and composting equipment
- Other equipment, including religious equipment including baptistries and chancel fixtures; agricultural equipment; and horticultural equipment

Many of these types of equipment would require a book of their own to demonstrate them adequately. They are, therefore, beyond the scope of this book. Two of them, residential appliances and unit kitchens—which architects encounter frequently—are covered here.

# Sustainability Considerations

Home appliances are a major user of energy and may contain materials that are harmful to people and the environment. Many older refrigerators and freezers, for example, contained chlorofluorocarbon (CFC), a class of synthetic compounds of carbon, hydrogen, chlorine, and fluorine, which are harmful to the ozone layer, in both the insulation used in them and in their coolant system. Refrigerators and freezers are available today that have no CFCs, and these should be used where possible.

Home appliances are made from metals or plastics. Some of them contain insulation products. The effects of these products on the sustainability of buildings is discussed in the chapters covering the various materials.

Major home appliance standards are included in the National Appliance and Conservation Act of 1987 (NAECA), which authorizes the Department of Energy (DOE) to set energy efficiency standards for major home appliances according to a statutory time schedule. DOE regulations issued under NAECA apply to refrigerator/freezers, clothes dryers, clothes washers, dishwashers, range/ovens, microwave ovens, and room air conditioners.

NAECA amends the federal Energy Policy and Conservation Act. In turn, NAECA has been amended by the National Appliance Energy Conservation Amendments of 1988 and by the Energy Policy Act of 1992.

Federal law requires that appliances have attached to them an Energy Guide card. For room air conditioning units, the card should show the unit's Energy Efficiency Rating (EER) and an indication of the ratings for the least efficient and most efficient models of similar units. The law further requires minimum ratios, ranging from 8 to 10, depending on the type of unit. An EER of 11 is generally considered the lowest a consumer should accept, and some sources say that 12 should be the minimum. The higher the EER, the more energy efficient the unit. For other appliances, the Energy Guide card should show the estimated yearly operating cost for the particular unit and the same data for the highest-cost and lowest-cost similar units. These cards should also show the estimated yearly operating cost for the particular unit for various energy costs.

Every manufacturer must provide units that at least meet the minimum federal standards. Some ultraefficient refrigerators and freezers that use up to 90% less energy than conventional energy-efficient units are available today. As the national move toward conservation grows stronger, more such units are sure to become available.

Electric ranges are available that heat using induction coils beneath glass tops. Such units are more energy efficient than conventional electric range elements.

## Applicable *MasterFormat* Sections

The following *MasterFormat* 2004 Level 2 sections are applicable to this chapter:

11 31 00  Residential Equipment
11 26 00  Unit Kitchens

# 11.1 Residential Appliances

This section contains data related to *residential appliances*, which is the industry term for kitchen and laundry appliances. The appliances discussed in this section include ranges, cooktops, ovens, refrigerator/freezers, freezers, dishwashers, clothes washers, clothes dryers, and room air conditioners. Plumbing for this equipment is covered in Chapter 16; electrical supply and connections in Chapter 18. Central heating and air conditioning systems are addressed in Chapter 17. The general requirements in this section are also applicable to such other residential equipment as food waste disposal units, trash compactors, and instant hot water dispensers.

## 11.1.1  USE BY HANDICAPPED PERSONS

The Americans with Disabilities Act (ADA) and ANSI A117.1 of the American National Standards Institute (ANSI) impose certain restrictions on residential appliances. The requirements in this section (11.1) are supplementary to those addressed in Section 1.4. Many wheelchair users have problems using appliances designed for standing rather than seated users. An ideal cooking arrangement for a wheelchair user consists of the following:

- A lowered, front-controlled cooktop
- A lowered sink
- A lowered work counter at least 30 in. (760 mm) wide
- An accessible oven installed as described in the following section

### 11.1.1.1  Cooktops and Ranges

More accidents are related to *cooktops* and *ranges* than to any other kitchen appliance. To prevent such injury, cooktop and range controls should be in the front of the appliance so that a cook does not have to reach over hot burners to adjust them. People with limited arm coordination must drag rather than lift pots and pans. A cooktop, if separate from the oven, should be adjustable with the same knee space and height dimensions as sinks and work counters. The knee space should be finished like the rest of the floor. To protect a user from injury, the underside of a cooktop should be smooth, insulated, and free of sharp projections. Electric cooktops should be equipped with a light to indicate that a burner is on.

When residents include small children, controls should be out of sight on a horizontal surface and should either have a protective lid or require an extra movement, such as pushing, before a control can be turned to activate a burner.

Ovens should have side-hinged doors with a pull-out shelf beneath and should be self-cleaning. Ovens without a self-cleaning feature, and the latch side of those that are side-hinged, should be located next to a lowered counter area with accessible knee space. Side-hinged ovens should have a heatproof pull-out shelf below the oven running the full width of the oven; the depth of the shelf should not be less than 10 in. (255 mm).

### 11.1.1.2  Refrigerators and Freezers

Many refrigerators with freezers (side-by-side and some over-under styles) are accessible for wheelchair users. *Over-under* refrigerators should have at least 50% of the freezer and most of the refrigerator space and the controls at a height lower than 54 in. (1370 mm). Freezers that have any of their storage volume above the 54 in. (1370 mm) reach limit should be self-defrosting.

### 11.1.1.3  Dishwashers

Dishwashers should open in the front, and the racks should be within easy reach. Pull-out racks are the easiest to use.

### 11.1.1.4  Washers and Dryers

Washers and dryers should be the front-loading type, and should be hinged on the side rather than on the bottom.

## 11.1.2  STANDARDS

Residential equipment should be listed by the Underwriters Laboratories, Inc. (UL), and should bear UL labels.

Gas-burning residential equipment should bear the American Gas Association's (AGA) Seal of Approval, and should comply with the ANSI Series Z21 standards.

Appliances that bear the applicable seal of the Association of Home Appliance Manufacturers (AHAM) comply with AHAM standards. AHAM standards cover such things as shelf area ratings for refrigerators and freezers, refrigeration capacity, and amperage ratings. Refer to Section 11.4 for a listing of some AHAM standards. Note, though, that some manufacturers do not belong to AHAM. This does not mean that their products are inferior or would not meet AHAM standards were they eligible for that distinction.

Each appliance should bear the manufacturer's standard warranty. This warranty should cover all factory defects for not less than 1 year. Some states require longer or more complete warranties.

### 11.1.3 SPECIFIC APPLIANCES

Appliances are available with such a wide range of features and options that it is impossible to list them all here. In addition, the available features and options change regularly as competition increases and technological advances occur. Therefore, the features included here should be considered basic and not fully representative of the many product features available. Appliances should be selected from up-to-date manufacturers' catalogs.

In selecting appliances, great care should be taken to determine the features available and the energy efficiency of the particular unit under consideration, and to compare the units of several reputable manufacturers. Although there are few generally accepted industrywide standards for residential appliances, units produced by most major manufacturers

are usually, but not always, comparable. In addition, some lesser-known manufacturers produce products that compare favorably with those of the major companies, but greater care is advisable in considering their products. While not a guarantee of quality, a manufacturer's reputation can become very important when something goes wrong. Those with good reputations keep their reputations by servicing the products they sell and by upholding warranty agreements.

When practicable, all the appliances in a particular residence or other unit should be obtained from the same manufacturer. This will ensure compatibility of color and design and make service contracts less complicated and often less costly. Many manufacturers and vendors discount service contracts when more than one appliance is insured.

#### 11.1.3.1 Ranges

Ranges may be either gas or electric units and either a freestanding (Fig. 11.1-1a) or a drop-in type (Fig. 11.1-1b). Typical units have four plug-in burners, a self-cleaning oven, chrome oven racks, a

timer, a clock, and a storage drawer below the oven. Some have glass panels in the oven door, automatic rotisseries, and other features. Some units have built-in microwave ovens. Some gas units have pilot-free electronic ignition systems.

#### 11.1.3.2 Cooktops

Cooktops may be either gas or electric units. Typical units have four plug-in heating elements (Fig. 11.1-2). Some units are available with built-in griddles or grilles; others have flush heating elements or either glass or ceramic tops. Some gas units have pilot-free electronic ignition systems.

#### 11.1.3.3 Wall Ovens

Wall ovens (see Fig. 11.1-2) may be either gas or electric units. They have the same features generally as the oven portion of freestanding and drop-in ranges.

#### 11.1.3.4 Exhaust Hoods

Exhaust hoods are available in a variety of types. Figure 11.1-2 shows a standard undercounter unit. These come in 24 in. (610 mm), 30 in. (762 mm), and 36 in.

**FIGURE 11.1-1** Ranges. (a) freestanding type, (b) drop-in type. (Drawing by HLS.)

**FIGURE 11.1-2** Cooktop, double wall oven, and exhaust hood. Single ovens are also available. Some manufacturers make unit with a conventional oven and a microwave oven as a unit. Exhaust hoods may be either ductless or with a duct connection to the outdoors. (Drawing by HLS.)

FIGURE 11.1-3 Microwave ovens may be either freestanding or built in. Some can be hung beneath cabinets; others may be wall-mounted. Some built-ins contain an exhaust system, which may be either ducted or ductless. (Drawing by HLS.)

Built-in microwave oven

Freestanding microwave oven

door for water, cubed ice, and crushed ice. Some have separate small doors to hold high-use items, such as milk and soft drinks. Some manual-defrost units have a single door. Automatic-defrost types typically have separate doors for the freezer and refrigerator compartments.

### 11.1.3.7 Freezers

Freezers are available as either an upright type (Fig. 11.1-5a) or a chest type (11.1-5b). Capacities range from 4 cu ft (0.11 m$^3$) to 26 cu ft (0.74 m$^3$). Both manual and automatic-defrost models are available.

### 11.1.3.8 Dishwashers

Dishwashers are either built-in (Fig. 11.1-6a), freestanding (Fig. 11.1-6b), or convertible (can be used either as an undercounter or a freestanding unit). Standard features include at least two wash cycles, two wash levels, and energy-saving drying cycles. Standard units have porcelain enamel interior liners and insulated cabinets. Racks should extend fully outside the cabinet for loading. Optional features include a forced air drying cycle, a built-in auxiliary hot water heater, additional wash cycles, a built-in soft-food disposer, a plastic liner, delay

(889 mm) widths. Special units include 36-in. (889-mm)-wide by 30-in. (762-mm)-deep, 24-in. (610-mm)-tall island hoods and 30-in. (762-mm)-wide, 24-in. (610-mm)-high wall-mounted hoods.

### 11.1.3.5 Microwave Ovens

Microwave ovens (Fig. 11.1-3) are available either as freestanding countertop units or undercounter-mounted units. Most have electronic digital controls, but a few have manual dial controls. Wattages vary from 500 to 800 or even more. Units are available with rotisseries, convection cooking, convection broiling, turntables, and recirculating charcoal filters.

### 11.1.3.6 Refrigerator/Freezers

Refrigerator/freezers are combination units housing a freezer compartment and a refrigerator compartment. They are available as a side-by-side type (Fig. 11.1-4a) and a top-freezer type (Fig. 11.1-4b).

Units may be either a manual-defrost type or an automatic-defrost type. Units are also available with bottom-mounted freezers. Some have dispensers on the

a                                    b

FIGURE 11.1-4 Refrigerators with freezers. The side-by-side type (a) may be as much as 40 in. (1016 mm) wide and slightly taller than the top-freezer type. Some top-freezer types (b) fit into a nominal 36 in. (914 mm) counter opening, but others are narrower. Many combination refrigerator/freezer units are self-defrosting. Upright, single-door refrigerators are also available, with a smaller freezer inside the refrigerator compartment. These are usually not self-defrosting. Single-door, undercounter, combination refrigerator/freezer units are also available. (Drawing by HLS.)

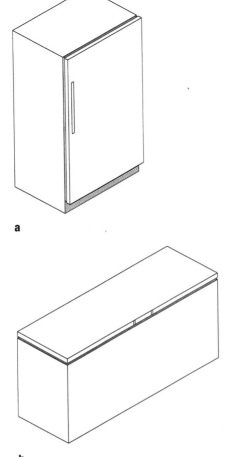

**a**

**b**

**FIGURE 11.1-5** Freezers. The upright type (a) may be 36 in. (914 mm) wide and as much as 72 in. (1830 mm) tall. Some are self-defrosting; others are not. Chest-type freezers (b) are a nominal 36 in. (914 mm) high and range from 20 in. (508 mm) to 75 in. (1905 mm) wide. Most chest-type freezers must be manually defrosted. (Drawing by HLS.)

start operation, and an automatic rinsing aid dispenser.

### 11.1.3.9 Clothes Washers

Most clothes washers today are of the automatic top-loading type (Fig. 11.1-7). However, units that must be ADA accessible must be front-loading, which are available in undercounter and space-saving stacked units. Most clothes washers have metered fill-level controls, presettable wash and rinse water temperatures, variable speed and cycle fabric selectors, removable lint filters, and bleach and fabric softener dispensers. Some units have a removable

**a**

**b**

**FIGURE 11.1-6** Dishwashers. (a) Undercounter (built-in) type; (b) freestanding type. Most built-in units are designed to fit into a 24 in. (610 mm) counter opening. Freestanding units are generally the same size as the built-in type. Some manufacturers make dishwashers that can be either freestanding or rolled into an opening beneath a counter. (Drawing by HLS.)

tub for washing small and delicate fabric loads.

### 11.1.3.10 Clothes Dryers

Clothes dryers (Fig. 11.1-8) may be either gas or electric. Standard features include timed cycle selection, a fabrics

**FIGURE 11.1-8** A typical clothes dryer. Dryers are generally 27 in. (686 mm) wide, 27 in. (686 mm) deep, and 43 in. (1092 mm) to the top of the control panel. Undercounter units are also available. (Drawing by HLS.)

**FIGURE 11.1-7** A typical clothes washer. Washers are generally 27 in. (686 mm) wide, 25 in. (635 mm) deep, and 43 in. (1092 mm) to the top of the control panel. Front-loading, undercounter units are also available. (Drawing by HLS.)

Dryer

Washer

**FIGURE 11.1-9** Space-saving washer/dryer combination units combine both functions into a single stacked unit. The washer portion is about the same size as a freestanding washer. The overall height is about 66 in. (1676 mm). Units with front-loading washers and larger dryers stacked directly on top of the washer are also available. (Drawing by HLS.)

selector, a cycle-end signal, a safety starting control, a removable lint filter, and a no-heat drying cycle. Some gas units have an automatic electric ignition (no pilot light).

### 11.1.3.11 Space-Saving Washers and Dryers

Washers and dryers are available as sets designed to save space for use in apartments and other small home laundry areas (Fig. 11.1-9). Features are essentially as described for separate washers and dryers.

### 11.1.3.12 Air Conditioning Units

Room air conditioners are available in sizes ranging from 4000 Btu (4 220.24 kJ) to 23,000 Btu (24 266.4 kJ) or even more. Some have heating coils, and some are actually heat pumps. Most units may be mounted either in a window or through a wall. Large units, units with heating coils, and heat pumps usually require 240 volt service. Most smaller units work on 120 volt standard house power.

## 11.2 Unit Kitchens

Unit kitchens, sometimes called *kitchenettes* are self-contained units including a countertop with cabinets under it, a sink, and wall cabinets (Figs. 11.2-1, 11.2-2, and 11.2-3). Some unit kitchens also contain a range or burner top and a range hood. Some have an undercounter refrigerator built in. Some are designed for use by people in wheelchairs. Special units provide refrigeration and water along with storage cabinets for use as wet bars. Other appliances sometimes available are ice makers, dishwashers, garbage disposals, and microwave ovens. Unit kitchens are used in offices, elderly housing facilities, and as nourishment centers in many types of facilities.

Unit kitchens are made of metal or wood with many finishes available, in-

**FIGURE 11.2-1** Large unit kitchen containing a sink in a countertop with a drain board, a range containing a conventional oven and surface cooking units, a ductless exhaust hood, a refrigerator, a dishwasher, wall and base cabinets, a continuous backsplash, and a microwave oven. (Drawing by HLS.)

**FIGURE 11.2-2** A smaller unit kitchen of a type generally used in offices, conference rooms, and similar spaces. It has a sink in a counter-top with a drain board, surface cooking units, a ductless exhaust hood, a refrigerator, wall and base cabinets, a continuous backsplash, and a microwave oven. (Drawing by HLS.)

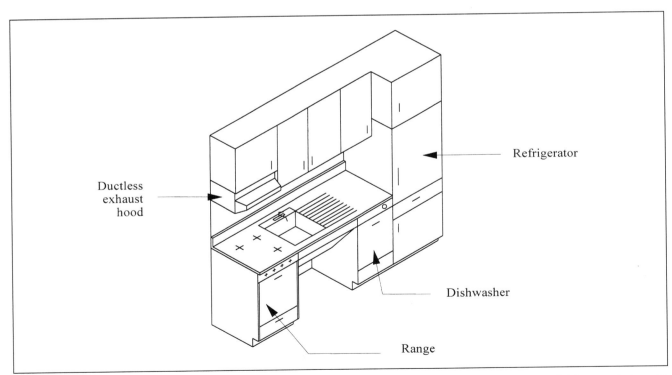

**FIGURE 11.2-3** A unit kitchen designed specifically for use by someone in a wheelchair. The unit shown contains a sink in a countertop with a drain board, a range containing a conventional oven and surface cooking units, a ductless exhaust hood, a refrigerator, a dishwasher, wall and base cabinets, and a continuous backsplash. A microwave oven could be easily substituted for the conventional one. Some manufacturers make units of this type that disassemble easily so that they can be converted for conventional use. On such units, the countertop can be raised or lowered, and doors can be easily installed to provide an enclosed under-sink storage space. (Drawing by HLS.)

cluding natural wood, stainless steel, baked enamel on steel, and plastic laminate.

Unit kitchens that comply with the ADA are available. They are 34 in. (863.6 mm) high (even above undercounter refrigerators and ovens), handles and controls are designed for easy visual and mechanical operation, and they meet or exceed all "reach" restrictions. Many models can be specified to meet additional state or local accessibility/adaptability requirements as well.

# 11.3 Additional Reading

More information about the subjects discussed in this chapter can be found in the references listed in Section 11.4 and in the following publications.

Association of Home Appliance Manufacturers (AHAM). 2003. *Home Appliance Industry Fact Book*. Washington, DC: AHAM (www.aham.org).

National Kitchen and Bath Association (NKBA). *Kitchen Planning Guidelines*. Hackettstown, NJ: NKBA (www.nkba.org)

# 11.4 Acknowledgments and References

## ACKNOWLEDGMENTS

We gratefully acknowledge the assistance of the following organizations and individuals in preparing this chapter. We are also indebted to them for permission to use their illustrations when requested and for the use of their publications as references.

American National Standards Institute (ANSI) (www.ansi.org)

Association of Home Appliance Manufacturers (AHAM) (www.aham.org)

Dwyer Products Corporation (www.dwyerkitchens.com)

Hotpoint Division of the General Electric Company (www.hotpoint.com)

Whirlpool Corporation (www.whirlpoolcorp.com)

## REFERENCES

We would also like to thank the authors and publishers of the publications in the following list for their contribution to our research for this chapter.

American National Standards Institute (ANSI). *Z21 Series Documents*. New York: ANSI (www.ansi.org).

Arcom Master Systems (www.arcom-net.com. *Masterspec*. Salt Lake City, UT.

Basic Section: 11400, "Food Service Equipment," 8/02 ed.

————. Section 11451, "Residential Appliances," 2/03 ed.

————. Section 11460, "Unit Kitchens," 8/01 ed.

Association of Home Appliance Manufacturers (AHAM) Washington, DC (www.aham.org). 2003. ANSI/AHAM DH-1, "Dehumidifiers."

————. 2004. AHAM DW-1, "Dishwashers."

————. 1992. ANSI/AHAM ER-1, "Electric Ranges."

————. 1992. ANSI/AHAM FWD-1, "Food Waste Disposers."

————. 1992. ANSI/AHAM HLD-1, "Clothes Dryers."

————. 2003. ANSI/AHAM RAC-1, "Room Air Conditioners."

————. 2004. ANSI/AHAM HRF1. "Energy, Performance and Capacity of Household Refrigerators, Refrigerator-Freezers and Freezers."

————. 1992. ANSI/AHAM TC-1, "Trash Compactors."

————. 2002. HLW-1, "Clothes Washers."

Ramsey, Charles George, Harold Reeve Sleeper, and John Ray Hoke, Jr. 2000. *Architectural Graphic Standards*, 10th ed. Hoboken, NJ: John Wiley & Sons, Inc. (www.wiley.com).

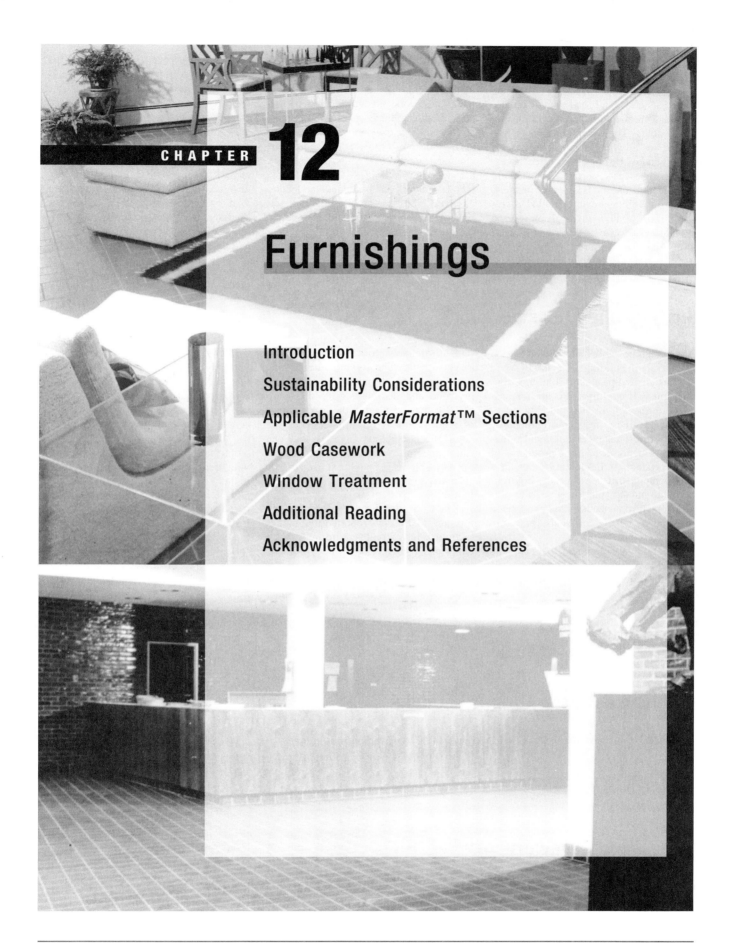

CHAPTER **12**

# Furnishings

Introduction

Sustainability Considerations

Applicable *MasterFormat*™ Sections

Wood Casework

Window Treatment

Additional Reading

Acknowledgments and References

# Introduction

MasterFormat 2004 Division 12, "Furnishings," covers a broad range of materials and products used in the furnishing of buildings. These include, but are not limited to, artwork, manufactured casework, window treatment, furniture, accessories, rugs, mats, theater and stadium seating, and interior plants and planters. Space limitations prevent detailed coverage of most of the many furnishings used in buildings. However, two types of these furnishings are sufficiently important to

include here. These are manufactured wood casework, including both wood-finished and laminated-plastic-faced units, and some types of window treatment.

In addition to those addressed in this chapter, many other kinds of casework are used in buildings. These include metal casework and both metal and wood casework units that are used in banks, libraries, restaurants, cafeterias, medical facilities, veterinary facilities, ecclesiastical facilities, and as display units in re-

tail establishments. A reader requiring specific data on one of these should first contact the various manufacturers of these products and obtain the necessary catalog data.

The types of window treatment discussed in this chapter are those usually designed and specified by architects. Window treatments used solely for decorative purposes are generally selected by interior design professionals and are not addressed in this book.

# Sustainability Considerations

Wood is a renewable material that is used extensively in buildings. The use of wood as a sustainable product in building construction will continue into the foreseeable future, so long as replanting of replacement trees and environmentally friendly forest management occurs. It is therefore incumbent on the producers of forest products to implement procedures and methods that preserve the environment and manage the forest so as to provide for current needs while preserving the supply of wood products for future generations.

Among the methods that can be adopted to reduce the use of wood is the use of the otherwise unusable lumber left by harvesting and milling in wood derivative products, such as oriented strand board, which includes flakeboard, waferboard, and particleboard.

Unfortunately, formaldehyde is used as part of the adhesive in some pressed wood products (particleboard, hardwood plywood, and medium density fiberboard [MDF]) that are used in the manufacture of wood cabinetry. Formaldehyde is a colorless, strong-smelling gas that can cause uncomfortable and sometimes serious symptoms in people exposed to levels in the air that are above 0.1 ppm (parts in a million parts of air).

The Consumer Products Safety Commission (CSPC) has reported that manufacturers have reduced formaldehyde emissions from pressed wood products by 80% to 90% from the levels of the early 1980s. Unfortunately, some products used in cabinetry still contain formaldehyde.

CSPC Document #725 recommends that products be purchased that are la-

beled or stamped to be in conformance with American National Standards Institute (ANSI) criteria (see Chapter 6) or that contain a high percentage of panel surface and edges that are laminated or coated. Document #725, also suggests the use of alternative products such as wood panel products that are not made with urea-formaldehyde glues, lumber, or metal.

Refer to Chapter 6 for additional comments about the use of wood as it relates to the sustainability of buildings and to the dangers of using products containing formaldehyde.

The types of window treatments addressed in this chapter are by nature conducive to energy savings and thereby contribute to energy conservation. These advantages are addressed throughout Section 12.2 as they are applicable.

# Applicable *MasterFormat* Sections

The following *MasterFormat* 2004 Level 2 sections are applicable to this chapter:

23 00 00  Heating, Ventilating, and Air-Conditioning (HVAC)

23 20 00  HVAC Piping and Pumps
23 30 00  HVAC Air Distribution
23 40 00  HVAC Air Cleaning Devices
23 50 00  Central Heating Equipment

23 60 00  Central Cooling Equipment
23 70 00  Central HVAC Equipment
23 80 00  Decentralized HVAC Equipment

# 12.1  Wood Casework

Wood casework has been used in commercial, institutional, and residential

buildings for centuries. The origins of residential bath and kitchen cabinets can be traced back to the sixteenth century, when *ambries* (cupboards) were used to keep rats and other vermin away from

food. Ambries were closed food storage areas above open shelving for utensils. More extensive built-in cabinets emerged in the seventeenth century, when the space between the chimney and the sidewalls was filled with open or closed shelves, giving the effect of built-in kitchen cabinets.

Wood is still used extensively for cabinets. However, increased cost has led to substitution of plywood and particleboard in cabinets and more use of metal cabinets, especially in commercial and institutional buildings.

Because of the high cost of manufacturing custom cabinets, they are seldom used in residential buildings today, but they still appear in commercial and institutional buildings. Many wood cabinets are mass-produced in factories that manufacture thousands of units a day, using sophisticated milling machines, presses, sanders, laminating devices, conveyors, and finishing equipment.

This section addresses applicable quality standards and certification programs for, and the typical construction, assembly, and installation of, some of the many cabinet and countertop types available today. The methods described here are not the only ones in use, and other methods are not necessarily inferior. Any construction that produces cabinets that comply with the indicated quality standards is satisfactory.

### 12.1.1 QUALITY STANDARDS

Wood cabinets in commercial and institutional buildings and in higher-priced residential buildings are usually required to comply with *Architectural Woodwork Quality Standards Illustrated*, published by the Architectural Woodwork Institute (AWI) and the Architectural Woodwork Manufacturers Association of Canada (AWMAC). This standard requires that the designer select the desired grade based on appearance and service requirements from three grade options, *Premium*, *Custom*, and *Economy*. The highest grade is the strongest and most durable, but is usually the most expensive.

In parts of the western United States, designers sometimes require compliance with the standard *Manual of Millwork*, published by the Woodwork Institute (WI), in lieu of compliance with the AWI standard. The WI standard has the same grade listings as the AWI/AWMAC standards, but adds a fourth *Laboratory* grade.

The AWI/AWMAC and WI standards differ in some aspects and use different names for the same cabinet components. A designer should obtain the one of these two standards that will be applied and ascertain its requirements for each aspect of cabinets.

Often, cabinets for laboratories, hospitals, and schools are selected by specifying one or more acceptable manufacturers and product lines. Occasionally, manufacturers are required to modify their stock units to comply with either the AWI/AWMAC or WI standard, or with even more stringent requirements, such as thicker materials.

The standard usually applied to residential kitchen and bath vanity cabinets is ANSI/KCMA A116.1, "Performance and Construction Standard for Kitchen and Vanity Cabinets." It is a publication of the Kitchen Cabinet Manufacturers Association (KCMA) that has been adopted by the ANSI. It applies to manufactured, factory-finished bath and kitchen cabinets of metal, wood, or other nonmetallic materials, such as plastic laminate, plastics, hardboard, and particleboard.

To ensure high quality, bath and kitchen cabinets in homes should also comply with the requirements of the U.S. Department of Housing and Urban Development (HUD) 4910.1 *Minimum Property Standards for Housing*, whether or not the housing will be financed with a Federal Housing Administration (FHA) guaranteed loan. HUD 4910.1 requires compliance with ANSI/KCMA A161.1.

ANSI/KCMA A161.1 is a performance standard. It does not prescribe specific requirements for cabinet design or materials. This allows considerable freedom in the selection of materials and methods of fabrication so long as the final product meets stringent testing procedures for endurance and quality. A161.1 does, however, require that:

- Lumber and plywood parts be kiln dried to a moisture content of 10% or less at the time of fabrication.
- Metal cabinets be rust resistant.
- All materials be suitable for use in kitchen and bath environments.
- Cabinets be enclosed with sides, backs, and bottoms, except that backs are not required on vanity units, kitchen sink fronts, sink bases, oven cabinets, or refrigerator cabinets. Bottoms are not required on vanity bowl or drawer fronts, or on kitchen drawer units, oven cabinets, refrigerator cabinets, or sink fronts.

- Countertops enclose the tops of base cabinets.
- Wall cabinets have individual tops.
- Swinging doors have a catch or other device to hold them closed.
- Adjacent cabinets and doors be in proper alignment with each other.
- Doors be of balanced (warp-resistant) construction and operate freely.
- Drawer slides, shelf standards, brackets, rotating shelf hardware, and other miscellaneous hardware support the design loads and operational functions described in ANSI/KCMA A161.1.
- Hardware complies with the Builders Hardware Manufacturers Association ANSI/BHMA A156.9, "Cabinet Hardware."

A KCMA Certification Seal placed on the inside of a cabinet door or drawer offers assurance that cabinets bearing it have been manufactured in conformance with KCMA standards. Any manufacturer of prefinished, factory-engineered kitchen cabinets may apply for a license to participate in the KCMA's Certification Program. KCMA membership is not required. To become a participant, a manufacturer must enter into an agreement with the KCMA spelling out the conditions under which it may affix a KCMA Certification Seal to its cabinets. However, some manufacturers of high-quality cabinets, even ones who are KCMA members, do not participate, because a KCMA Certification Seal represents a *minimum* standard and they do not want to have their products identified with it. Cabinets bearing a Certification Seal must meet performance tests that simulate many years of normal use. These include tests for structural integrity, door operation, drawer operation, and exterior cabinet finish.

### 12.1.2 CABINETS

Figure 12.1-1 shows normal work heights, mounting heights, and clearances for cabinets. These may vary, especially in commercial and institutional uses, and where accommodations for the handicapped are required (see Section 1.4).

Figure 12.1-2 shows some of the available types and sizes of stock cabinets for residential kitchen use. Cabinets for laboratories, hospitals, and schools are often specifically designed for their applications. These include laboratory fume hoods, specimen cases, work benches with racks, balance tables,

COUNTER HEIGHT FOR STOOLS    COUNTER HEIGHT FOR CHAIRS    HIGHEST SHELF ADVISABLE    ABOVE SINK    ABOVE RANGE    ABOVE CABINETS

**FIGURE 12.1-1** Typical work heights and cabinet clearances. Refer to Figure 25.7-1 for metric conversions.

## WALL CABINETS

BLANK CORNER CABINET    CORNER CABINET    WALL COMBINATION CABINET

9" - 12" - 15"
18" - 21" - 24"
DEPTHS 12"-13" VARIES ACCORDING TO MANUFACTURER.

24" - 27" - 30'
33" - 36" - 39'
42" - 45" - 48"

24" - 27" - 30"
33" - 36" - 39'
40" - 42" - 45"
48"

24" - 30" - 36"
42" - 48" - 60"
66" - 72" - 84"
MULTIPLE DOORS ON CABINETS 42" AND OVER.

21"- 24"- 27"
REQUIRED ALONG EACH WALL.

54" - 60" - 66" - 72" - 84" - 90" - 96"

## BASE CABINETS

SINGLE DOOR DRAWER   DOUBLE DOOR DRAWER    SINK OR RANGE BASE    BLANK CORNER CABINET       SINK OR RANGE BASE COMBINATION

9" (NO DRAWER)
12" - 15" - 18"
21"- 24"
DEPTHS 23"- 23-1/2'- 24"- 24-1/2' VARIES ACCORDING TO MANUFACTURER.

27" - 30" - 33" - 36'
39" - 42" - 45" - 48'

24" - 30" - 33" - 36"
39" - 42" - 45" - 48"

24" - 36" - 39"
42" - 48" - 60"
66" - 72" - 84"
MULTIPLE DOORS & DRAWERS ON 60" AND OVER.

33"- 36"- 42"
REQUIRED ALONG EACH WALL.

54" - 60" - 66"- 72" - 84" - 90" - 96"

## MISCELLANEOUS CABINETS

ALL DRAWER CABINET    SINK OR RANGE FRONT    STACK·ON RANGE BASE     UTILITY OR BROOM CABINET    OVEN CABINET    REFRIGERATOR CABINET

12' - 15" - 18'
21" - 24"-
BASE UNITS ALSO AVAILABLE WITH THREE OR FIVE DRAWERS.

24" - 30" - 36"
42" - 48"

24" - 30" - 39"- 40"
VANITIES SIMILAR

18" SINGLE DOORS
24"- 30'- 36'
AVAILABLE IN, 12"- 13"
18" & STANDARD BASE DEPTHS.

24" - 27" - 30'- 33"
66" HIGH (NO TOP DOORS)
CUT OUT SIZE VARIES FOR DIFFERENT OVEN MODELS.

36" - 39"- 42"
OPENING SIZE VARIES FOR DIFFERENT REFRIGERATOR MODELS.

## VANITY CABINETS

24" - 30" - 39"- 40"

54" - 60" - 66"- 72" - 84"- 90"- 96"

**FIGURE 12.1-2** Summary of kitchen and bath cabinet types and sizes. Refer to Figure 25.7-1 for metric conversions.

Reveal Overlay          Flush Overlay          Flush

**FIGURE 12.1-3** Cabinet face styles. (Drawing by HLS.)

chemistry and physics tables, speciality storage cabinets, reagent racks, glass-front cabinets, and many more. Base and wall cabinets, in general appearance not unlike those shown in Figure 12.1-2, are also used in such applications, but these are usually quite different in construction. For example, they are likely to have heavier and thicker sides, backs, fronts, and tops and stronger joints.

## 12.1.3 TYPES AND STYLES

Cabinets are classed generally according to function as *base cabinets*, *wall cabinets*, and *miscellaneous cabinets*. The third category includes laboratory cabinets, school science tables, medicine cabinets, broom cabinets, oven cabinets, and other special pieces.

Cabinets are also classified by style, which refers to either *design style* or cabinet *face style*. Cabinets are manufactured in many decorative design styles, some with glass doors, others with raised panels, applied panel molds, and other articulation.

In standard manufactured kitchen and vanity cabinets there are three basic face styles: *reveal overlay*, *flush overlay*, and *flush* (Fig. 12.1-3). Other styles may be used in custom cabinets.

## 12.1.4 MATERIALS AND FINISHES

Cabinets may be made either from wood and plywood with a wood finish, from various materials with a plastic laminate finish, or from metal. Hardboard and particleboard are used for some components in some cabinets. Common species of hardwood used are ash, oak, birch, maple, and walnut; common softwood species are ponderosa (western) pine, Douglas fir, and western hemlock. Other wood species are also used, however.

The AWI standard lists both transparent and opaque finish systems for the wood portions of custom cabinets. These same finishes can be specified for standard fabricated units, but not all manufacturers will be able to supply them. ANSI/KCMA A161.1 lists no specific finishes. Rather, it requires that the finish meet certain performance requirements, leaving the finish up to the manufacturer. The designer must select from among the manufacturer's standard finishes. These include various wood grains, stains, and opaque colors, varying with the popularity of a particular color or other finish at the time.

### 12.1.4.1 Construction—General

Performance standards for metal cabinets are basically the same as those discussed here; the construction details, of course, differ. There is no nationally recognized organization representing metal cabinet manufacturers and no coordinated consensus on construction criteria. Therefore, when specifying metal cabinets, it is necessary to rely on the manufacturer's reputation.

Wood cabinets may have either *face frame* or *frameless* construction. Most cabinets have *face frame* construction, such as that shown in Figure 12.1-4. In

**FIGURE 12.1-4** Typical base cabinet construction showing component parts. Cabinets may have two end panels as shown, no end panels for use within a series of cabinets, or one end panel to be placed at the end of a series.

**FIGURE 12.1-5** Wood-Finished Cabinet Criteria

| Component | KCMA Standard | AWI Standard |
|---|---|---|
| Face Frames | ¹⁄₂- to ³⁄₄-in. (12.7- to 19-mm)-thick hardwood | Same |
| Rails: Back, Top, and Bottom | 3-in. (76.2-mm)-wide, ³⁄₄-in. (19-mm)-thick solid wood or panel product | Same |
| Backs | ¹⁄₈-in. (3.2-mm)-thick hardboard or ³⁄₈-in. (9.5-mm)-thick hardwood plywood | ¹⁄₄-in. (6.4-mm)-thick panel product |
| Tops and Bottoms of Base Cabinets; Tops of Wall and Special Cabinets | ¹⁄₂-in. (12.7-mm)-thick particleboard, or ¹⁄₄-in. (6.4-mm)-thick hardboard, or ³⁄₈-in. (9.5-mm)-thick hardwood plywood | ³⁄₄-in. (19-mm)-thick lumber or panel product |
| End Panels | ¹⁄₂-in. (12.7-mm)-thick hardwood plywood | ³⁄₄-in. (19-mm)-thick panel product |
| Flush Overlay Doors | ¹⁄₂-in. (12.7-mm)-thick plywood with hardwood face veneers | From ³⁄₄-in. (19 mm) to 1¹⁄₄ in. (31.8 mm) thick for particleboard core doors, and ³⁄₄ in. (19 mm) thick for veneer core doors |
| Stile and Rail Doors | ⁵⁄₈-in. (15.9-mm)-thick with ³⁄₄-in. (19-mm)-thick solid wood center panels, or ³⁄₄-in. (19-mm)-thick with applied ¹⁄₄-in. (6.4-mm)-thick center panels | From ³⁄₄-in. (19 mm) to 1¹⁄₄ in. (31.8 mm) thick for particleboard core doors, and ³⁄₄ in. (19 mm) thick for veneer core doors |
| Shelves | From ¹⁄₂- to ³⁄₄-in. (12.7- to 19-mm)-thick, lumber, plywood, or particleboard | ³⁄₄ in. (19 mm) thick for short spans (32 to 36 in. [812.8 to 900 mm] and 1 in. to 1¹⁄₁₆ in. [25.4 to 27 mm]) for larger spans (42 to 48 in. [1066.8 to 1219.2 mm]) |

frameless construction, the face frame is omitted, and the cabinet ends are finished on their exposed edges. Other components are similar to those in cabinets that have face frames.

Wood cabinets are further classified as either *wood-finished* or *plastic-laminate-clad*.

### 12.1.4.2 Wood-Finished Cabinets

The main components of *wood-finished* cabinets are *face frames; back, top, and bottom rails; backs; tops and bottoms; end panels; doors; drawers; shelves;* and *hardware* (see Fig. 12.1-4).

Some manufacturers use panel components similar to those in plastic-laminate-clad cabinets in concealed or semiexposed (visible only with doors or drawers open) locations in wood-finished cabinets. Typical locations are end panels, back panels, and bottoms. Sometimes plastic laminate components are used for shelves in wood-finished cabinets. With that exception, the materials generally used and their nominal thicknesses are shown in Figure 12.1-5. Note that the AWI/AWMAC standard requires that some components be a *panel product*. This is defined as plywood, particleboard, fiberboard, or hardboard faced with a hardwood plywood veneer.

*Face frame rails, stiles, and mullions* are doweled (or mortised and tenoned), as well as glued and stapled or nailed for

rigidity (Fig. 12.1-6). Side frames are often tongued and grooved into the face frames. Rails are rabbeted to receive top and bottom panels. Back rails are screwed and glued to secure them in place.

*Backs* are generally fastened to end panels and top and bottom rails.

The *bottoms* of base cabinets and the tops of wall and special cabinets are usually *dadoed* (let) into the sides and interlocked into the hanging rails of wall cabinets.

*End panels* are typically glued and mechanically fastened to side frames.

*Door* construction varies with the face style of the cabinet. Doors may be of flush overlay style or may have solid

wood stiles and rails. In some cabinets, the doors are hollow; in others, they have particleboard, wood, or high-density fiber cores with hardwood or plastic veneer. Sometimes the core is plastic with hardwood veneer overlays. Other doors have particleboard or plywood cores with plastic faces, and in still others the entire door is plastic. The required thickness for cabinet doors complying with the AWI/AWMAC standard varies with the composition of the core of the door and its grade, width, and height. Face veneers are hardwood plywood.

*Drawers* often have hardwood lumber sides and backs and plywood bottoms,

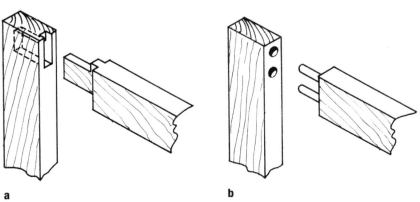

**FIGURE 12.1-6** Rails and stiles of frames can be joined by (a) mortises and tenons or (b) dowel joints.

but the sides may also be plywood or particleboard. Bottoms are often particleboard. Drawer construction varies with cabinet quality. In high-quality cabinets, the sides are connected to the front and back with multiple *dovetail* joints; in others, with *lock-shouldered* or *square-shouldered* joints (Fig. 12.1-7). Drawer bottoms are *dadoed* into the sides, front, and back for more rigid construction. Some drawer units are molded plastic (except for the front), with rounded inside corners for ease of cleaning.

*Drawer fronts* usually match the *door fronts* and are of the same or similar construction. They should be positively fastened to a wood *subfront* with screws. The subfront should be *rabbeted* into the sides and fastened to them with mechanical fasteners and glue. There is as wide

**a**

**b**

**c**

**FIGURE 12.1-7** Drawer construction: (a) multiple-dovetail, (b) lock-shouldered, and (c) square-shouldered joints.

a variation in drawer front styling as there is in door design.

Most *shelves* are lumber, plywood, or particleboard, with rounded front edges. Plywood or particleboard shelves may be banded with hardwood front edges.

### 12.1.4.3 Plastic-Laminate-Clad Cabinets

Many *plastic-laminate-clad* cabinets are constructed in the same way as *wood-finished cabinets,* except that the finish veneers are made from any suitable grade of sound hardwood, or in some cases from softwood, and are covered with plastic laminate.

A second type of plastic-laminate-clad cabinet is made from panels prefinished with plastic laminate and assembled into a finished cabinet. The main components of this type of plastic-laminate-clad cabinet are similar to those of wood-finished cabinets, but there are some differences.

*Frameless* construction is more likely to be used in plastic-laminate-clad cabinets. In *reveal overlay style* plastic-laminate-clad cabinets, face frames are usually not required.

The *concealed* (never visible) surfaces of plastic-laminate-clad cabinets are usually plywood or particleboard. The *exposed* and *semiexposed* faces of components are finished in *high-pressure decorative laminate* (HPDL) or *thermoset decorative panels.* The HPDL faces should comply with the requirements of the National Electrical Manufacturers Association (NEMA) standard, "High Pressure Decorative Laminates," which has been adopted by ANSI as ANSI/NEMA LD-3. LD-3 designates laminate types according to the International Organization for Standardization (ISO) nomenclature and thicknesses, such as:

HGS—Horizontal General purpose, Standard—0.048 in. (1.2 mm)
HGP—Horizontal General purpose, Postformable—0.038 in. (1.0 mm)
VGP—Vertical General purpose, Postformable—0.028 in. (0.7 mm)
HGF—Horizontal General purpose, Fire-rated—0.048 in. (1.2 mm)
VGF—Vertical General purpose, Fire-rated—0.028 in. (0.7 mm)
CLS—Cabinet Liner, Standard—0.020 in. (0.5 mm)

The uses of the various types are mostly self-explanatory. For example,

materials designated HG are used in horizontal applications; VG in vertical applications; CL for the insides of cabinets where subjected to little or no wear. Postformable sheets can be bent to form curved surfaces, such as a rounded front edge or a coved backsplash.

The panel thicknesses listed are those most often used in manufactured cabinets, but thicker materials are readily available, and are often used in commercial and institutional casework.

Edges in semiexposed locations are often finished with *high-impact plastic edging.* Exposed edges are usually finished with the same plastic laminate used on the adjacent surface. *Thermoset,* or *low-pressure,* decorative laminate panels are used in semiexposed locations where they are not subjected to wear. They should comply with the American Laminators Association (ALA) standard 1985, "Performance Standard for Thermoset Decorative Panels," which is roughly equivalent to CLS as defined by ISO.

The core materials generally used in plastic laminate cabinets and their nominal thicknesses are shown in Figure 12.1-8.

*Face frames* are made with mortise and tenon joints.

*Stretchers* are needed at the top of base cabinets to support the countertop. Sides, dividers, tops, bottoms, shelves, and stretchers are covered on both sides with CLS HDLP.

*Backs* generally have a *thermoset* decorative laminate on the surface that will be inside the cabinet. Backs are fastened to end panels and to top and bottom rails.

*Doors* usually have a VGP plastic laminate finish on both sides. The thickness of doors for cabinets complying with the AWI standard varies, depending on the core type and the height and width of the door. Permissible core types are particleboard and plywood veneer.

*Drawer fronts, backs, and bottoms* should be *rabbeted* into the sides and fastened with screws and glue. Some drawer units are molded plastic (except for the front), with rounded inside corners for ease of cleaning.

*Drawer fronts* usually match the door fronts and are of the same construction. They should be positively fastened to a particleboard subfront with glue and fasteners. The subfront should be *rabbeted* into the sides and fastened to them with mechanical fasteners and glue.

**FIGURE 12.1-8** Plastic-Laminate-Clad Cabinet Criteria

| Component | KCMA Standard | AWI Standard |
|---|---|---|
| Face Frames | 1 in. by ⅝ in. (25.4 mm by 15.9 mm) solid wood | |
| Sides, Dividers, Bottoms, Tops, Shelves, and Stretchers | ½-in. (12.7 mm)-thick particleboard | ¾ in. (19 mm) thick with shelves being 1 in. thick for spans of more than 36 in. (914 mm) |
| Backs | ⅛-in. (3.2-mm)-thick hardboard | ¼-in. (6.4-mm)-thick panel product |
| Doors | ⅝-in. (15.9-mm)-thick particleboard | ¾ to 1¼ in. (19 to 31 mm) thick |
| Drawers | ⅜-in. (9.5-mm)-thick particleboard sides and backs | ½-in. (12.7-mm)-thick lumber, plywood, or particleboard; bottoms from ¼-in. (6.4-mm)-thick panel product |
| Drawer Fronts | Match door fronts with a ⅜-in. (9.5-mm)-thick particleboard subfront | Match door fronts with ½-in. (12.7-mm)-thick lumber, plywood, or particleboard subfront |

#### 12.1.4.4 Hardware

Cabinet hardware includes, *pulls, catches, drawer suspension devices,* and *adjustable shelf supports.* Many types of each are available. Hardware should comply with ANSI/BHMA A156.9.

*Drawer suspension devices* should provide for a minimum capacity of 50 lb (22.7 kg) and should have nylon rollers.

*Shelves* may be either fixed or adjustable, with metal hardware of various types to permit adjustment.

#### 12.1.4.5 Manufacture

There are two basic kinds of cabinet manufacturing operations: *integrated plants* and *assembly plants.* In an *integrated wood-finished-cabinet plant,* wood is processed from raw material (logs) to the finished products (cabinets) with minor purchases of plywood, particleboard, plastic components, and hardware. Such facilities have vast milling operations for the many wood-processing steps needed to produce the final products.

In an *integrated plastic-laminate-clad cabinet plant,* plastic laminates are manufactured and processed on site. Wood may either be milled on site or purchased already milled. Plastic components, plywood, particleboard, and hardware are usually purchased.

In an *assembly plant,* raw wood materials have already been either completely or partially milled, and plastic laminate and other materials are purchased from their manufacturing source. Production is mainly the assembly of components produced elsewhere.

### 12.1.5 COUNTERTOPS

Many materials are used as *countertops.* These include, but are not necessarily limited to, plastic laminate, filled polymer, cultured marble, ceramic tile, wood, natural stone—such as granite and marble—stainless steel, and the special tops used in laboratories and the like, such as soapstone, synthetic stone, and epoxy resin.

#### 12.1.5.1 Plastic Laminate

The most common type of countertop in residential and general commercial and institutional buildings is finished with horizontal-grade plastic laminate. In a plastic laminate countertop, the backsplash, top, and front piece are assembled first on a wood or particleboard blank. Then adhesive is applied to the wood or particleboard surface, and a plastic laminate sheet is laid over the top and pressed into intimate contact with the surface. If a countertop blank has curved surfaces, such as a rounded front edge and a coved backsplash, the plastic sheet is first heated and then *postformed* (molded) to the blank surface by a special press. Openings for sinks are cut and holes for pipes are drilled at the factory or shop.

#### 12.1.5.2 Filled Polymer

*Filled polymer,* or *acrylic-modified polyester,* tops are made from rigid-cast acrylic plastic that is homogeneous, solid, and nonporous. It can be cut and shaped with ordinary woodworking tools and sanded to remove burns and scratches. Probably the best-known example of this material is *Corian* by the DuPont Company, but similar materials are made by other manufacturers.

#### 12.1.5.3 Other Materials

Countertops made from synthetic materials called *cultured marble, synthetic stone,* and various other names are manufactured by proprietary methods and shipped to a project site ready for final cutting and assembly. These materials may or may not have marble dust in their mix, and may either be homogeneous or just have a decorative surface applied.

Marble, granite, and other real stone countertops are cut to fit in the shop and shipped to the project site ready for installation. In both real stone and synthetic tops, cutouts for plumbing fixtures and pipes are usually made at a factory or shop.

Ceramic tile countertops are normally built at the construction site from standard tile components. Tops supporting the ceramic tile are shop cut and field fitted.

Wood tops are usually manufactured to fit, but some cutting may be done in the field.

Special tops for use in laboratories are manufactured or cut to fit at the factory.

### 12.1.6 INSTALLATION

Cabinets, countertops, and devices built into them should be installed level, plumb, and true, in accordance with the manufacturers' instructions and with the requirements of the standards referenced earlier.

#### 12.1.6.1 Cabinets

Attach wall and base cabinets with screws (not nails) to studs or other framing members.

To ensure plumb and level installation, remove high spots and shim low spots in walls and floors. Remove the wall base and chair rail from behind cabinets to ensure a flush fit.

Install wall cabinets first, beginning with a corner unit. Insert screws through *hanging rails* built into the backs of cabinets at both the top and bottom. After the cabinets have been leveled and plumbed and doors perfectly aligned, tighten the screws.

Start installing base cabinets in a corner. After installing the base corner unit, use C-clamps to hold adjacent cabinets in alignment. Then use T-nuts to fasten them together. Check each cabinet from front to back and across the front edge with a level as it is attached to the wall. The front frame should be plumb and the top should be level. Attach base cabinets with screws through the back frame into wall studs. After base cabinets have been secured in place, install the countertop and finish the toestrip with a tile, resilient, or wood base.

### 12.1.6.2 Countertops

Install *plastic laminate* and *filled polymer* countertops using screws through *corner blocks* in the base unit into the top. Spline and glue joints in plastic laminate countertops and then pull them tight with mechanical fasteners. Finish seams in *filled polymer* tops with the manufacturer's joint adhesive.

Install and finish other types of tops in accordance with their manufacturers' instructions. Refer to Chapter 9 for information about installing ceramic tile.

# 12.2 Window Treatment

Heat flow through windows can be reduced by means of *window treatment*, including interior and exterior shading devices and interior blinds, shades, and draperies. This section addresses window treatment intended to control heat loss or gain. Shutters, blinds, and draperies used primarily for decoration are beyond the scope of this book, even though some of them—lined draperies, for example—may incorporate features that reduce heat transfer.

Measures often imposed to reduce heat gain and loss through glass include tightly sealed openings, multiple glazing, insulating glass, and storm windows. These measures do result in less heat loss and gain. Unfortunately, even double-glazed windows, lose heat 6 to10 times

faster than typical insulated wood-framed walls. Therefore, additional thermal protection is often desirable.

The most effective form of additional protection is provided by *exterior* shading devices, such as *shutters, blinds, shades, awnings,* and *architectural projections.* Though less effective, *interior* devices such as *blinds, roll shades, thermal shutters,* and *draperies* are also worthwhile.

### 12.2.1 EXTERIOR SHADING DEVICES

Shading devices are most effective when installed on the exterior of a building where their shade can prevent sunlight from entering. This is critical because only a portion of the solar radiation that enters a building can be reflected back to the exterior. The remainder is absorbed and then radiated and conducted to the interior (see Section 17.3).

### 12.2.1.1 Shutters

A shutter that blocks out all direct sunlight can reduce solar heat gain through a window by 80%. Shading performance, however, depends on how well heat absorbed into the shutter is dissipated to the outside air. Shutters that absorb large amounts of heat radiate a significant percentage of that heat into the building. Therefore, light-colored shutters, which reflect most sunlight before it can be absorbed, are superior to dark shutters.

The most effective shutters have operable-tilt slats that can be adjusted to block the sun and also let air circulate through them. Nonoperable louvered shutters will also permit natural air flow but cannot be adjusted to efficiently track the sun's angle. Air should be allowed to circulate between windows and shutters to dissipate heat absorbed by the shutters. Shutters used for reduction of heat gain or loss are usually manufactured of wood or aluminum. Decorative shutters are often vinyl. Shutters having a finished surface on one side only are not suitable for use as operable shutters. Several common shutter types are shown in Figure 12.2-1.

Because they reduce air flow, tight-fitting shutters provide more resistance to heat loss than do loose-fitting ones. However, even when the fit is not tight or the louvers cannot be completely closed, shutters will reduce heat loss by reducing infiltration and preventing the scouring action of the wind from disturbing the

a

b

c

**FIGURE 12.2-1** Common shutter types: (a) rolling shutters move in horizontal tracks; (b) Bahama shutters cover the entire window opening; (c) Sarasota shutters extend over the top half of the window opening and are typically combined with side-hinged shutters.

insulating film of air at the outer surface of the glass.

### 12.2.1.2 Roll Blinds and Shades

Roll blinds and shades range from simple bamboo or canvas shades to motorized roll blinds that can be operated from the interior. Such devices are highly effective in stopping sunlight from entering a building, even low-angle sun from west elevations. Light colors are most effective because they reflect heat-producing sunlight.

Motorized roll blinds are usually made of aluminum, vinyl, or wood, producing the added benefit of year-round use and durability (Fig. 12.2-2). Acting

a

b

**FIGURE 12.2-2** (a) Soffits above the windows house the motorized roller that stores the blinds. (b) Even larger windows can be protected, such as on this screened porch. (Courtesy Levelor.)

like tight-fitting storm windows, they can also reduce heat loss in winter by as much as 50%. Independent shade control of each window and ease of operation are other advantages. Their prime disadvantage is that exterior view and natural ventilation are reduced. Blinds are available, however, that can be projected away from the window like an awning to provide view and ventilation.

### 12.2.1.3 Awnings

How well an *awning* shades a window depends on how opaque the awning material is. The more opaque it is, the more complete is the shade provided.

The exposed surfaces of an awning should be a light color to minimize the amount of sunlight absorbed. When sunlight is absorbed, an awning's temperature is raised. This heat is then radiated through the shaded window.

Clean white canvas or slatted, light aluminum awnings reflect between 80% and 91% of the sunlight that strikes their surfaces. A dark green canvas awning reflects only 21%, and a dark green plastic awning reflects 27% percent of the sunlight.

A light-colored awning can reduce heat gain by 64% on south-facing win-

dows and 77% on west-facing windows. Figure 12.2-3 lists the net effectiveness of several types of awnings. Heat from sunlight absorbed by an awning will build up under the awning and be transferred to the window unless air is permitted to escape. Therefore, fabric awnings are often installed with a continuous gap between the top of the awning and the wall. Slatted aluminum awnings provide air circulation between the horizontal slats.

Awnings should provide adequate coverage of a window for its orientation. South-facing windows require only a minimal horizontal projection to be completely shaded throughout the day, but the sides of such awnings should be closed to prevent sunlight from angling in behind them. Awnings on east- or west-facing windows should extend down a substantial distance to block out early morning or late afternoon sun, as applicable.

### 12.2.1.4 Architectural Projections

Windows can be shaded by horizontal or vertical planes that extend beyond the face of the glass to intercept the summer sun. If properly designed, these projections will not interfere with the penetration of the beneficial winter sun or block the view.

Roof overhangs are one of the most common ways to shield the windows in small buildings, especially houses, from sunlight. Figure 12.2-4 lists the horizontal projection necessary to provide complete summer window shading with a roof overhang.

A deeply recessed window, or one with closely spaced horizontal or vertical fins, can also be effective. In general, east- and west-facing windows are more effectively shaded by vertical projecting planes; south-facing windows, by horizontal planes.

The underside of an overhang should be dark so that it will transmit less ground-reflected sunlight through the glass. The net result of such shading is that heat gain through a completely protected window can be reduced by 80%.

### 12.2.2 INTERIOR HEAT GAIN/LOSS DEVICES

Although not as effective as exterior shading devices, interior heat gain/loss devices have the advantage of greater accessibility and ease of operation. Interior devices used to reduce heat gain in the summer and heat loss in the winter include *blinds*, *roll shades*, *thermal shutters*, and *thermal draperies*.

**FIGURE 12.2-3** Heat Gain Through Single-Glazed Windows with Awnings

| Window Orientation | Awning Type | Heat Gain/100 sq ft Glass Surface, Btu/day (W/m²) | Heat Excluded Total, Btu/day (W/m²) | % Reduction |
|---|---|---|---|---|
| South | No awning | 62,200 (82) | — | — |
| | White canvas | 22,500 (30) | 39,700 (52) | 64 |
| | Dark green canvas | 27,700 (36) | 34,500 (46) | 55 |
| | Dark green plastic | 35,600 (47) | 26,600 (35) | 43 |
| West | No awning | 84,200 (111) | — | — |
| | White canvas | 19,500 (26) | 64,700 (85) | 77 |
| | Dark green canvas | 23,900 (32) | 60,300 (79) | 72 |
| | Dark green plastic | 34,800 (46) | 49,400 (65) | 59 |

**FIGURE 12.2-4** Roof Overhang for Complete Summer Window Shading *

| | Overhang Width | |
|---|---|---|
| Latitude • North, Degrees | South Exposure | Southeast/Southwest Exposure |
| 28 | 1 ft 5 in. (432 mm) | 3 ft 10 in. (1168 mm) |
| 32 | 1 ft 11 in. (584 mm) | 4 ft 2 in. (1270 mm) |
| 36 | 2 ft 5 in. (737 mm) | 4 ft 8 in. (1422 mm) |
| 40 | 3 ft 0 in. (914 mm) | 5 ft 4 in. (1626 mm) |
| 44 | 3 ft 6 in. (1067 mm) | 6 ft 0 in. (1829 mm) |

*Assumes window sill is 2 ft from floor.

### 12.2.2.1 Blinds

The most common such devices are *venetian blinds* (Fig. 12.2-5). These are slatted horizontally and can be tilted to reflect a maximum amount of sunlight back outdoors or direct daylight toward the ceiling for glare-free, uniform lighting. Blinds, however, provide only minimal reduction of heat loss in winter because they do not effectively trap a layer of air. Vertical blinds offer similar protection against heat gain. They reduce heat loss more efficiently in winter, however, because they trap an insulating layer of air between themselves and the glass (Fig. 12.2-6).

**FIGURE 12.2-5** Venetian blinds can be made to cover large glazed areas such as sliding glass doors. (Courtesy Levelor.)

**FIGURE 12.2-6** Vertical blinds extend to the floor, thus reducing heat-robbing convection currents. (Courtesy Levelor.)

Some proprietary systems encase venetian blinds between layers of glass (Fig. 12.2-7). When fully closed, this system can provide an R-value of 2.50 compared to an R-value of 2.04 for a similar window with no blinds.

### 12.2.2.2 Roll Shades

The most effective type of interior protection against window heat gain is an opaque or translucent *roll shade*. Such shades can be easily installed with simple brackets on the window frame and, when not in use, rolled up for unobtrusive storage. A completely opaque roll shade with a white outer face can reflect 80% of the sunlight back through the glass.

A roll shade can have a dark color on one side which effectively absorbs sunlight and a white surface on the reverse side which effectively reflects it. Reversing such a shade, from dark side facing out in winter to reflective side facing out in summer, will permit it to perform both as a solar collector and as a shading device.

**FIGURE 12.2-7** With blinds fully closed, solar heat gain is reduced by about 82%. (Courtesy Pella Corporation.)

### 12.2.2.3 Thermal Shutters

Covering windows with a rigid insulating material such as polyurethane or polystyrene is an effective way to control heat loss. Figure 12.2-8 lists several common insulating materials and the resulting R-values when they are combined with $1/4$-in. (6.35-mm)-thick plate glass. They are usually installed on the interior, but they can be placed on the exterior if the insulation is protected from the weather.

A common way to achieve protection from the elements is to face the insulation with aluminum, galvanized steel, or exterior grade plywood. Foam plastic must also be protected from fire. Plywood or metal cladding can serve this function.

Thermal shutters can be hinged, tracked, or mounted with clips. Removable shutters can be taken down and stored during warm weather. Hinged or tracked shutters can be folded or pushed back out of the way when not needed.

When an insulating material is placed in direct contact with glass, heat loss is reduced only in proportion to the R-value of that material. A space left between a thermal shutter and a sash creates an insulating dead air space, which more effectively reduces conductive heat loss. Tightly sealing cracks around a frame will reduce infiltration. Loosely fitting shutters permit air to circulate between the glass and the shutter, carrying away heat when the air passes over the glass and dramatically reducing overall insulating effectiveness.

### 12.2.2.4 Thermal Drapes

Drapes offer a more traditional way of controlling window heat flow. Because the surface temperature of a drape is much higher than that of window glass, drapes reduce the radiant heat loss from occupants and therefore improve thermal comfort.

**FIGURE 12.2-8** Thermal Shutter Materials

| Material | R-Value/ 1 in. (25.4 mm) | Window R-Value | |
|---|---|---|---|
| | | With 1 in. (25.4 mm) Insulation | With 2 in. (50.8 mm) Insulation |
| Expanded polystyrene (extruded) | 4.00 | 5.00 | 9.00 |
| Expanded polystyrene (molded) | 3.57 | 4.54 | 8.11 |
| Polyurethane | 6.25 | 7.14 | 13.39 |

For maximum benefit from window coverings, room air must not pass between the glass and the covering material. When air is allowed to pass over the glass, the improvement from a loose-fitting drape amounts to an increase from the R-0.85 of an undraped window to only R-0.94. If the heated air from a wall or floor register passes between the covering and the glass, heat loss is actually more rapid than if no covering were provided.

By contrast, when warm room air is prevented from circulating behind the drapes, the R-value of the window and drapes can increase to 1.13. To achieve this, drapes must be tightly sealed at the bottom and sides. Thermal drapes can be fastened to a window frame with hooks, snaps, or Velcro fasteners. Long drapes should touch the floor.

Thermally efficient drapes are usually made of tightly woven fabric. Drapes can be further improved by adding a liner or reflective material, foam backing, or additional layers of fabric.

## 12.3 Additional Reading

More information about the subjects discussed in this chapter can be found in the references listed in Section 12.4 and in the following publications.

Bianco, John. 1995. "Commercial Window Treatments." *The Construction Specifier* (July): 78 (www.csinet. org).

McGowan, Maryrose. 1992. "Horizontal Louver Blinds." *The Construction Specifier* (June): 47 (www.csinet. org).

Pearce, John. 1984. "Sunlight Solutions: Operable Woven Sunscreens." *The Construction Specifier* (March): 44 (www.csinet.org).

## 12.4 Acknowledgments and References

We gratefully acknowledge the assistance of the following organizations and individuals in preparing this chapter. We are also indebted to them for permission to use their illustrations when requested and for the use of their publications as references.

### ACKNOWLEDGEMENTS

American National Standards Institute (ANSI) (www.ansi.org)
Formica Corporation (www.formica. com)
Kitchen Cabinet Manufacturers Association (www.kcma.org)
Levelor (www.levolor.com)
Nevamar Company (www.nevamar.com)
Pella Corporation (www.pella.com)
U.S. Department of Housing and Urban Development (HUD) (www.hud.gov)
Wilsonart International (www.wilsonart. com)
Woodwork Institute (www.woodworkinstitute.com)

### REFERENCES

We would also like to thank the authors and publishers of the publications in the following list of references for their contribution to our research for this chapter.

American National Standards Institute (ANSI). New York: ANSI (www. ansi.org). Standards:
A117.1-2003. "Accessible and Usable Buildings and Facilities."
A208.1-1979, "Particleboard, Mat-Formed Wood."
*Architectural Graphic Standards. See* Ramsey et al. (2000).
Architectural Woodwork Institute (AWI). *Architectural Woodwork Quality Standards Illustrated.* Reston, VA: AWI (www.awinet.com).
Arcom Master Systems (www.arcom-net.com). *Masterspec®.* Salt Lake City, UT Basic Sections:
12356, "Residential Casework," 2/03 ed.
12491, "Horizontal Louver Blinds Manually Operated and Motorized Horizontal Louver Blinds," 11/03 ed.
12492, "Vertical Louver Blinds Manually Operated and Motorized Vertical Louver Blinds," 11/03 ed.
12494, "Roller Shades Manually Operated and Motorized Roller Shades," 2/04 ed.
12497, " Draperies and Tracks," 8/03 ed.
Building Hardware Manufacturers Association (BHMA). A 156.9, "Cabinet Hardware." New York: BHMA (www. buildershardware.com).
Kitchen Cabinet Manufacturers Association. ANSI/KCMA A161.1, "Performance Standard for Kitchen and Vanity Cabinets." Reston, VA: ANSI/ KCMA (www.kcma.org).
National Electrical Manufacturers Association. 2000. NEMA LD 3, "High Pressure Decorative Laminates." Washington, DC: NEMA (www.nema.org).
Ramsey, Charles George, Harold Reeve Sleeper, and John Ray Hoke, Jr. 2000. *Architectural Graphic Standards,* 10th ed. Hoboken, NJ: John Wiley & Sons, Inc. (www.wiley.com).
U.S. Consumer Products Safety Commission. Document 725, "An Update on Formaldehyde: 1997 Revision." Washington, DC: CSPS (www.csps. gov).
U.S. Department of Housing and Urban Development (HUD). Washington, DC: HUD (www.hud.gov).
1988. *Uniform Federal Accessibility Standards (UFAS).*
1994. HUD-4910.1, *Minimum Property Standards for Housing.*
Woodwork Institute (WI). *Manual of Millwork.* West Sacramento, CA: WI (www.wicnet.org).

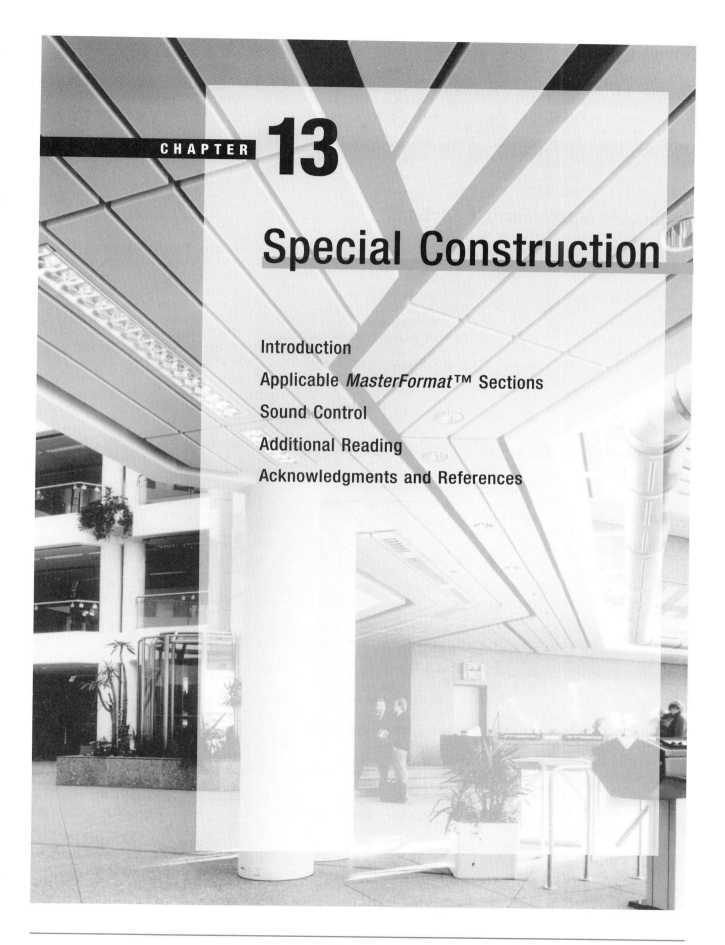

CHAPTER **13**

# Special Construction

# Introduction

The type of special construction addressed in this chapter is sound control. *MasterFormat* Division 13 includes other types of special construction that are beyond the scope of this book.

These include, but are not limited to, swimming pools, aquariums, amusement park structures and equipment, hot tubs and therapeutic pools, ice rinks, air-supported structures, radiation protection, kennels and animal shelters, preengineered structures, incinerators, and seismic and meteorological instrumentation.

# Applicable *MasterFormat* Sections

The following *MasterFormat* 2004 Level 2 sections are applicable to this chapter:

| | |
|---|---|
| 06 10 00 | Rough Carpentry |
| 07 20 00 | Thermal Protection |
| 08 10 00 | Doors and Frames |
| 08 50 00 | Windows |
| 08 80 00 | Glazing |
| 09 20 00 | Plaster and Gypsum Board |
| 09 30 00 | Tiling |
| 09 50 00 | Ceilings |
| 09 60 00 | Flooring |
| 09 70 00 | Wall Finishes |
| 09 80 00 | Acoustic Treatment |
| 12 00 00 | Furnishings |
| 13 48 00 | Sound, Vibration, and Seismic Control |
| 22 40 00 | Plumbing Fixtures |
| 23 30 00 | HVAC Air Distribution |
| 23 50 00 | Central Heating Equipment |
| 23 60 00 | Central Cooling Equipment |
| 23 70 00 | Central HVAC Equipment |
| 26 30 00 | Facility Electrical Power Generating and Storing Equipment |
| 27 30 00 | Voice Communications |
| 27 40 00 | Audio-Video Communications |

# 13.1 Sound Control

Computer printers, typewriters, calculators, intercom systems, background music systems, clothes washers and dryers, dishwashers, garbage disposers, refrigerators, vacuum cleaners, radios, television sets, and other electrical devices make life easier and more enjoyable, but at the same time they create noise. Heating, ventilating, and air conditioning equipment produce noise and carry it into every interior space. Outdoor noise levels have also increased. People work and live close to their neighbors. Apartments and townhouses continue to increase in number. Sounds created by high-speed roads, fast trains, and jet airliners add to the problem. Open offices allow workplace noise to travel into adjacent spaces.

Many construction materials exacerbate our noise problems. Glass, plaster, gypsum board, wood paneling, masonry, tile, metal panels, and other hard, relatively flat surfaces reflect up to 98% of the sound that strikes them. Thin, lightweight materials, including some preassembled components, in walls and in floor and ceiling assemblies permit noise to travel easily from room to room, from floor to floor, and from one living or working unit to another.

The first major use of the science of sound control occurred in the twentieth century, applied to churches, theaters, and auditoriums to make the living environment more pleasant. Its use quickly spread to schools, offices, hospitals, restaurants, retail stores, and other commercial and institutional buildings. Today, sound control is considered an essential element in the design of all buildings. When incorporated into a building's design, sound control can usually be accomplished inexpensively. Conversely, alterations to control unwanted sound transmission after a building has been completed are almost always costly.

This section addresses the properties and sources of sound, identifies common acoustical problems, and discusses methods of measuring and controlling noise. It also covers design and construction techniques and the criteria essential to providing and maintaining the acoustical performance of walls, floor and ceiling assemblies, doors, and windows and to resist transmission of airborne and structure-borne sound through them. It contains sound performance ratings, including the Sound Transmission Class (STC), the Impact Noise Rating (INR), and the Impact Insulation Class (IIC) for some commonly used walls and floor and ceiling assemblies. It also discusses *room acoustics*, in which the goal is to provide good hearing conditions within a space, and *sound transmission,* in which the goal is to control sound transmission, by airborne sound or structural vibrations, between spaces.

Chapter 16 addresses control of the generation and transmission of unwanted sound by and through plumbing systems and fixtures, and describes methods that will reduce them. Chapter 17 discusses the role of heating, ventilating, and air conditioning systems in noise generation and transmission, and describes methods that will reduce them. Chapter 18 discusses methods that will reduce sound transmission through and around electrical system devices, and noise generation by appliances.

## 13.1.1 PROPERTIES OF SOUND

Sound is the vibration of an elastic medium. In air, it is the movement of air molecules in a wave motion. When the frequency and intensity of this motion are within certain ranges, they produce the sensation of hearing.

A vibrating body causes the layer of air particles next to it to vibrate. These air particles, in turn, transmit their motion to the next layer of particles. Because air is compressible and has weight, a small time interval is required for the first air layer to set the second layer in motion. Succeeding air layers are similarly set in motion after time intervals that are directly proportional to their distance from the source of the sound. This action constitutes a sound wave. The waves produced when a stone is dropped into water are analogous (Fig. 13.1-1).

### 13.1.1.1 Wave Motion

Water waves travel along the surface and are circular in shape. Sound waves spread out in three dimensions spherically. Each air particle vibrates only in the direction in which the wave is traveling, that is,

along a straight line drawn between the particle and the source of the sound wave. Sound travels in air at a speed of about 1130 ft/sec (344.42 m/s) or about 5 seconds per mile (1.608 344 km) (Fig. 13.1-2). In water, sound travels at about 4500 ft/sec (1371.6 m/s); and in steel, at about 15,000 ft/sec (4572 m/s) The speed in a given medium is always the same, regardless of the pitch or loudness of the sound. We hear echoes because we hear the sound from the source and then a reflected sound, which has to travel farther, a few moments later. This same property of sound also causes the unpleasant reverberations we sometimes experience indoors.

### 13.1.1.2 Frequency

One complete round trip of a vibrating body, starting from its neutral position, moving to one side, then to the other side, and back to neutral, is called a *cycle* (Fig. 13.1-3). The number of cycles performed in 1 second is called *frequency*. Frequency, then, is measured in cycles per second (cps) and may be expressed as *hertz* (*Hz*) (1 cps). Each air particle in a sound wave has the same frequency as the source. For example, if the sound from a loudspeaker has a frequency of 1000 cps, then the loudspeaker diaphragm and every air particle through which the sound wave passes are vibrating 1000 times each second. The performance and efficiency of acoustical materials in absorbing sound and the ability of materials to transmit sound vary, depending on the frequency of the sound.

The audible frequency range is from about 20 to 20,000 cps, but the range narrows as people grow older. Music covers most of this range, and speech extends from about 100 to 5000 cps. The most important range of frequencies for speech is shown in Figure 13.1-4. However, the human ear is more sensitive to sounds with frequencies between 1000 and 4000 cps, shown by the low dip on the Threshold of Hearing curve in Figure 13.1-4. Sounds between 250 and 1000 cps or between 4000 and 8000 cps must be more intense and those below 250 or above 8000 cps must be considerably more intense in order to be heard. Most sounds consist of a mixture of numerous frequencies at different relative intensities.

### WAVELENGTH

The length of a sound wave is calculated by dividing the velocity of sound by the frequency:

$$\text{wavelength} = \frac{\text{velocity}}{\text{frequency}}$$

*Wavelength* is the distance a sound travels during a single vibration. Since the velocity of sound is constant, each frequency has a different wavelength. A sound with a frequency of 1000 cps has a wavelength equal to about 1 ft (0.304 8 m). To reflect a sound effectively, a reflecting surface must be equal to or greater than the wave length in size. A 1 sq ft (0.092 903 m²) surface will reflect sound of 1000 cps and higher, but sound below 1000 cps will bend around such a surface and pass beyond it. A sound with

**FIGURE 13.1-1** Sound is a wave motion in air caused by a vibrating source. Wave motion in water is analogous.

**FIGURE 13.1-2** Light travels almost instantly, but sound requires about 5 seconds for each mile (1.609 km) it travels.

**FIGURE 13.1-3** Wavelength is the distance a sound wave travels during one vibration (cycle).

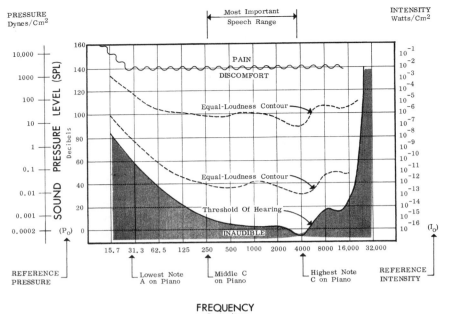

Most Important
Speech Range

PAIN

DISCOMFORT

Equal-Loudness Contour

Equal-Loudness Contour

Threshold Of Hearing

INAUDIBLE

SOUND PRESSURE LEVEL (SPL)
Decibels

15.7  31.3  62.5  125  250  500  1000  2000  4000  8000  16,000  32,000

REFERENCE
PRESSURE

Lowest Note
A on Piano

Middle C
on Piano

Highest Note
C on Piano

REFERENCE
INTENSITY

FREQUENCY
Cycles/Sec

**FIGURE 13.1-4** Audible frequency range. Two sounds of the same intensity but of different frequencies do not necessarily sound equally loud. Equal-loudness contours represent sounds of different frequencies that seem to the average human ear to be equally loud.

a frequency of 100 cps has a wavelength of 11.3 ft (3.45 m). This dimension usually exceeds the ceiling height and often the wall length of residential rooms; thus the sound wave does not fit within the room, and the sound is not reproduced within the room with fidelity. This accounts for the poor reproduction in small rooms of low-frequency sounds from radios, tape and disk players, televisions, and similar sources.

***Octaves and Tones*** The frequency scale is laid out in octaves in the same manner as a piano keyboard. The term *octave*, taken from musical terminology, is the interval between two sounds having a frequency ratio of 2:1. For example, frequencies between 125 and 250 cps or between 250 and 500 cps constitute an octave. A pure tone is a sound having only one frequency, whereas a complex tone is a sound composed of a number of different frequencies.

***Pitch*** The hearing sensation produced by the frequency of a sound is called *pitch*. The greater number of vibrations per second, the higher the pitch. High-frequency sounds such as squeaks and whistles are heard as high pitches, and low-frequency sounds such as rumbles and roars are heard as low pitches. Pitch is expressed in tones of the musical scale.

***Quality*** The *quality* of a tone is determined by the distribution of its numerous frequencies. Higher-frequency components are called *harmonics* or *overtones* when they are exact multiples of the fundamental frequency. The distribution and relative intensity of the overtones of a musical sound determine the quality (*timbre*) of the sound. Each of the two sounds illustrated in Figure 13.1-5 is a complete tone composed of a combination of related frequencies. The vertical distances of the graphs are proportional to the loudness.

## SOUND LEVELS

The intensity of the vibration of a sound source depends on the amount of force that sets it into vibration. Sound levels are expressed in *decibels* (dB). The deci-

**FIGURE 13.1-5** Musical instruments produce sounds of many related frequencies.

bel scale extends from 0 dB (threshold of hearing) to more than 120 dB (threshold of feeling). Zero decibel represents a fixed value of sound intensity that is slightly less than that of the faintest audible sound. Sound at 120 dB will cause the sensation of tickling in the ear. Sound at 180 dB, generated by some rocket engines, can cause structural damage to buildings and can be fatal to humans.

A decibel, which cannot be measured directly and has no fixed, absolute value, is the smallest change in sound intensity that can be detected by the average human ear. It is a logarithmic unit that expresses the ratio between a sound level being measured and a reference point. A decibel tells by what proportion one sound level is greater or less than another. The greater the sound level, the higher the rating on the decibel scale.

Sound levels in decibels are calculated from either a ratio of *sound pressures* ($P$) or *sound intensities* ($I$) using a reference point that corresponds to the faintest audible sound.

*Pressure* is a force per unit of area. Sound pressure at a given distance from a source is the force of moving air molecules spread over a spherical area and is expressed in *dynes* (units of force) per square centimeter.

*Power* is expended when a force is applied over a certain distance in a given unit of time. Sound power in *watts* (units of power) is a basic expression of the flow of sound energy. Sound power is spread over a sphere of increasing area as the sound wave radiates farther from the source. Sound power per unit of area is called *sound intensity* ($I$) and is expressed in watts per square centimeter ($W/cm^2$).

The reference pressure, marked ($P_o$) in Figure 13.1-4, is 0.0002 dyne/cm² and is roughly equivalent to the smallest amount of pressure that will cause an eardrum to vibrate. The corresponding reference intensity, marked ($I_o$) in Figure 13.1-4, is $10^{-16}$ W/cm².

The ratio in decibels between any measured pressure ($P$) and the reference pressure ($P_o$) is called the *Sound Pressure Level* (SPL). It is found by formula (a) in Figure 13.1-6. The ratio in decibels between any intensity ($I$) and the reference intensity ($I_o$) is called the *Sound Intensity Level* (SIL). It is given by formula (b) in Figure 13.1-6.

The decibel scale makes values computed for either pressure or intensity di-

**FIGURE 13.1-6** Formulas for computing the ratio in decibels between (a) the pressure of two sounds and (b) the density of two sounds.

rectly comparable. Although intensity levels are convenient for discussion, they cannot be measured directly. Therefore, in practice, sound levels are determined by measuring pressures as read on a sound level meter and expressing the measurement as SPL on a decibel scale.

The easiest way to understand sound levels on the decibel scale is to compare them to common, easily recognized sounds (Fig. 13.1-7).

An increment in sound level of 1 dB (say, the difference between 50 and 51 dB) is barely perceptible. A 2 to 3 dB increment is insignificant, a 5 dB increment is clearly noticeable, and a 10 dB increment is quite perceptible. It corresponds to a subjective reaction of doubling (or halving) the loudness.

*Loudness* The *loudness* of a sound depends on the intensity and frequency of the sound and on the characteristics of the human ear. People hear sound over a wide range of intensity, extending from a low whisper to a deafening roar. Beyond a certain point, pain in the ears becomes more evident than the sensation of increasing loudness.

There are limitations in the frequency response of the human ear as well as a definable range of intensities to which it will respond. The psychological sensation of loudness is closely related to the intensity of a sound. The greater the intensity, the louder a sound appears to a listener, but a given sound may appear to be much louder when it is in a middle frequency range than when it is in a high or low range. In fact, Figure 13.1-4 shows that a 15 dB tone, heard easily at 1000 or 2000 cps, is inaudible at 200 and 10,000 cps. The *Equal Loudness Contours* shown in Figure 13.1-4 connect the tones of different frequencies that sound equally loud. The loudness of sound also depends on the characteristics of the listener's ear. Thus, when the intensity of a given sound striking the ear of a normal-hearing person and of a hard-of-hearing person is exactly the same, the sensation of loudness is quite different.

*Noise* Unwanted sound is called *noise*. Usually a sound becomes noise when it is loud enough to be a nuisance. Certain sounds, such as scratching, are noise because they are nonuniform vibrations. Even pleasant or otherwise acceptable sound becomes noise when it interferes with concentration, conversation, or other tasks. How loud a sound must be before it becomes objectionable depends on the kind of sound and the person who hears it.

## 13.1.2 ROOM ACOUSTICS

Room acoustics is a branch of architectural acoustics that deals with both *acoustical correction* and *noise reduction*.

### 13.1.2.1 Acoustical Correction

*Acoustical correction* is the physical shaping of a space to produce the best possible hearing conditions. It is usually

**FIGURE 13.1-7** Levels of Typical Sounds in Decibels

| Loudness | Sound Level in Decibels (dB) | |
|---|---|---|
| | 160 | Near jet engine |
| | 130 | Threshold of painful sounds; limit of ear's endurance |
| Deafening | 120 | Threshold of feeling (varies with frequency) 18 ft (5.48 m) from airplane propeller |
| | 110 | |
| | | Express train passing at high speed |
| | 100 | Loud automobile horn 23 ft (7 m) away |
| | | Noisy factory |
| Very loud | 90 | |
| | | Loud hi-fi |
| | 80 | Subway train |
| | | Motor trucks 15 to 50 ft (4.572 to 15.24 m) away |
| Loud | 70 | Stenographic room, average TV, loud conversation |
| | 60 | Average busy street |
| | | Noisy office or department store |
| Moderate | 50 | Moderate restaurant clatter |
| | | Average office, noisy room, average conversation |
| | 40 | |
| | | Soft radio music in apartment, average residence |
| Faint | 30 | |
| | 20 | Average whisper 4 ft (1.22 m) away |
| Very faint | 10 | Rustle of leaves in gentle breeze |
| | 0 | Threshold of audibility |

done only for public spaces, such as auditoriums, concert halls, and churches, but is sometimes done for other types of spaces as well. The objective is to shape an environment where sounds are heard distinctly, in ample loudness, and with as faithful a replication of the source as possible everywhere in the space. Acoustical correction requires a knowledge of the principles of acoustics, engineering calculations, design of the space to take best advantage of the properties of sound, and the selection and location of acoustical and other finish materials. Sound amplification systems are often involved. Such design requires the skill and judgment of an acoustical specialist and is therefore beyond the scope of this book. In this section, room acoustics deals only with noise reduction.

### 13.1.2.2 Noise Reduction

*Noise reduction* is the treatment of interior surfaces with acoustical materials to alleviate the discomfort and distraction caused by the reflection of unwanted sound.

### REFLECTION OF SOUND

When a sound source is operating, sound waves travel outward in all directions radially from the source. When the sound waves encounter an obstacle or surface, such as a wall, their direction of travel is changed, or reflected.

The reflection of sound follows the same laws as the reflection of light from a mirror: (1) the direction of travel of the reflected sound always makes the same angle with the surface as that of the incident sound, that is, the angle of incidence equals the angle of reflection; and (2) the reflected sound waves travel in exactly the same manner as they would if they had originated at an "image" of the sound source, located the same distance behind the wall as the real source is in front of the wall (Fig. 13.1-8).

***Multiple Reflection*** When a sound is produced in a room having reflective surfaces, sound waves are reflected back and forth by these surfaces, and almost immediately the room is filled with sound waves traveling in every direction. This is called *multiple reflection* (Fig. 13.1-9).

Multiple reflection causes a buildup of sound intensity to a level greater than there would be if no reflection occurred. It makes noises unnaturally and unnecessarily loud and makes noise last longer.

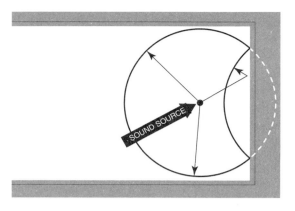

**FIGURE 13.1-8** Sound travels radially in all directions and is reflected by room surfaces.

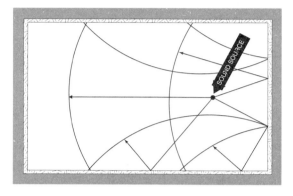

**FIGURE 13.1-9** Multiple reflection of sound waves causes sound levels to build up and sounds to reverberate longer.

**FIGURE 13.1-10** Relation of the distance from the source to the resulting sound level in decibels.

Aside from causing annoyance and fatigue, excessive reflection interferes with telephoning, conversation, detection of alarms and sound signals, and other types of work or play that depend on accurate hearing.

***Reverberation*** When a sound source is stopped, the reflected waves do not cease to exist at that instant but continue to travel back and forth, *reverberating* (bouncing) between room surfaces. The total sound energy in the room gradually diminishes due to the partial absorption of the sound at each contact with a sound-absorbing surface (Fig. 13.1-10).

Reflected sound waves strike the ear of a listener in rapid succession, so the listener does not hear them as distinct repetitions of the original sound. Instead, the listener hears the original sound being drawn out or prolonged after the source is stopped and steadily dying out until it becomes inaudible.

***Reverberation Time*** The *reverberation time* of a room is the time in seconds required for a sound to die out to an intensity one millionth of the intensity it had at the moment the source was stopped. On the decibel scale this is a drop of 60 dB (Fig. 13.1-11).

The reverberation time ($T$) of a room depends only on its size (volume, $V$) and sound-absorbing properties (absorption units, $a$), not on the power of the source (Fig. 13.1-12). Values of reverberation time have been established which give the best hearing conditions for speech and music (Fig. 13.1-13). Speech intelligibility

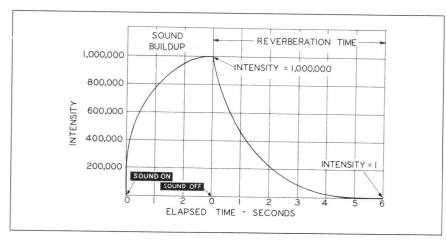

**FIGURE 13.1-11** Curve showing building up and dying out of sound in a room.

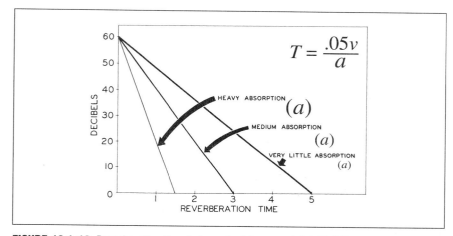

$$T = \frac{.05v}{a}$$

**FIGURE 13.1-12** Reverberation time is important for speech intelligibility and quality reproduction of music. $T$ = reverberation time in seconds; $V$ = room volume in cubic feet, and $a$ = room absorption in sabins.

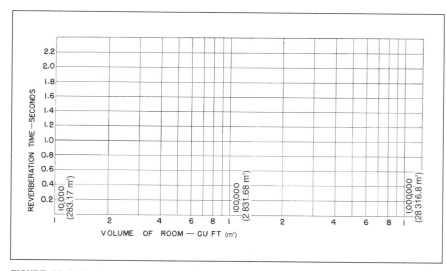

**FIGURE 13.1-13** Range of acceptable reverberation times at a frequency of 500 cycles per second (cps) for rooms of various sizes.

becomes poorer as loudness decreases or as reverberation time increases.

***Echoes*** An *echo* is a reflection that can be heard as a distinct repetition of the original sound. Echoes are likely to occur in large rooms where there is an appreciable time lag between the original sound and the reflected sound. To be heard as an echo, the reflected sound must be heard at least $1/17$th of a second later than the direct sound. Therefore, reflected sound must travel a path from source to listener at least 65 ft (19.81 m) longer than the path of the direct sound. Large rooms make possible the time lag for hearing an echo, and when reverberation time is short, the echo is heard more easily above the general reverberation.

Echoes produced by concave surfaces are louder than those produced by flat surfaces because a curved surface converges and focuses the reflected sound, increasing its intensity in the same manner that the curved reflector in a flashlight focuses light (Fig. 13.1-14). A ceiling having a center of curvature near the floor line is the most objectionable type of curved ceiling from the standpoint of focused echoes, since the focal point is brought closer to the audience than when the center of curvature is higher or lower.

Focusing is best corrected by breaking up the curve or changing its shape.

a

b

**FIGURE 13.1-14** (a) Reflections from a flat surface are divergent. (b) Concave surfaces focus sound at point (F) that may be louder than the source (S).

Treatment with acoustical materials is rarely enough.

## ABSORPTION OF SOUND

Smooth, hard building surfaces reflect up to 98% of the sound that strikes them. This reflection causes the sound to build up to a high level and to die out slowly.

Acoustical materials are designed to absorb most of the sound striking them and reflect less than 50% back into the room. They owe their absorption efficiency to their high porosity.

The amount of absorption of a porous material depends on its thickness, the size and number of its pores, and the frequency of the sound. Air particles moving in and out of the pores cause friction, and sound energy is dissipated as heat. Various materials designed for this purpose are discussed in Chapter 9.

When sound-absorbing materials are placed over the hard surfaces in a room, sound does not build up as much and dies out faster. Although the sound level goes down, no matter how much sound-absorbing material is introduced into the room, the level cannot drop below the level of the direct sound from the source to the listener. Sound-absorbing materials cannot "suck up" the sound from the source; they can only absorb what reaches them. This is analogous to light in a room. When the walls are white (reflective), the whole room is bright, but if the walls are painted black (absorptive), the whole room becomes dark, even though the same source at the same brightness level continues to emit the same amount of light. Acoustical materials act essentially the same way with sound as black paint acts with light.

In practice, it is impractical and uneconomical to obtain sound-level reductions of much more than 10 dB by using sound-absorbing materials. For example, 5 to 7 dB is the reduction usually accomplished when the entire ceiling of a normal office is covered with acoustical tile. However, such a reduction is obvious to the user of the space, and the improvement in communications and comfort is usually conspicuous.

***Measuring Sound Absorption*** One unit of absorption (called a *Sabin*) is equal to 1 sq ft (0.092 903 m²) of totally absorbent surface. An open window 1 sq ft (0.092 903 m²) in area is equivalent to 1 sound-absorbing unit, because, theoretically, all sound that strikes it passes through; none

is reflected. Therefore, an open window has 100% absorption.

An *Absorption Coefficient* (*a*) is assigned to materials to measure the percentage of incident sound energy absorbed by them. These vary theoretically from 0 to 1. A perfectly reflective material has an $a = 0$, and a perfectly absorptive material has an $a = 1$. A material with an absorption coefficient of $a = 0.50$ will absorb 50% of the sound that strikes it and reflect the other 50%. Because absorption coefficients vary with the frequency of the sound, it is necessary to test materials at more than one frequency in order to cover the complete range of sound frequencies. Materials are normally tested at six frequencies: 125, 250, 500, 1000, 2000, and 4000 cps. The test results are plotted as a curve (Fig. 13.1-15). The absorption coefficients for typical construction surfacing materials are given in Figure 13.1-16.

The numerical average of the absorption coefficients at the four middle frequencies, 250, 500, 1000, and 2000 cps rounded off to the closest 5%, is called the *Noise Reduction Coefficient* (NRC). This is a single-number rating used to calculate the required amount of sound-absorbing materials needed. It permits comparison of such materials according to their acoustical performance.

Because small differences in acoustical efficiency cannot be detected by the human ear, and since tests that determine this efficiency cannot be carried out with pinpoint accuracy, the selection of an acoustical material is usually based on a 10-point range (.10) of NRCs (Fig. 13.1-17). A difference of less than 10 points is insignificant, because it is seldom detectable in a completed installation.

**FIGURE 13.1-15** Absorption coefficients and other information is published by the manufacturers of acoustical materials.

In addition to the sound-absorption rating of a material, such other properties as light reflectivity, flame resistance, and appearance are also generally considered. Only in special cases where high absorption is required does NRC alone determine the selection. The sound absorption and noise reduction coefficients of acoustical materials are published by their manufacturers.

***Controlling Noise with Sound-Absorbing Materials*** Acoustical treatment controls noise by reducing multiple reflection of sound waves, thereby reducing the loudness level of reflected sound and shortening the reverberation time, so that the noise is not prolonged unnecessarily.

There are two basic noise reduction principles that serve as rough guides in determining the amount of treatment within a given room for satisfactory noise control: (1) the total number of absorption units in spaces with low ceilings should be between 20% and 30% of the spaces' total area of interior surface in square feet, and (2) to produce a satisfactory improvement in an existing space, the total absorption after treatment must be 3 to 10 times the absorption before treatment.

The following outlines one procedure for calculating the percentage of reduction in loudness of reflected sound produced by acoustical treatment.

1. Determine the number of absorption units in the space before treatment ($a_1$). To do this, multiply the absorption coefficient of each material in the space (see Fig. 13.1-16) by the area occupied by that material in square feet. This is the number of absorption units contributed by that material. It is usually satisfactory to use the material's absorption coefficient at 500 Hz. Add all the resultant units together. To these add an estimated number of units for furniture at about 0.03 times the area covered by furniture. Then add about 4 units per occupant.

2. Determine the number of units provided by the treatment. To do this, obtain the manufacturer's stated NRC of the selected treatment. This might be .70, for example. Multiple this by the area covered by the treatment in square feet.

3. From $a_1$ subtract the number of units originally calculated for the material to be covered by the treatment. Then add the number of units provided by

**FIGURE 13.1-16** Absorption Coefficients for Typical Construction Materials

| Material | Absorption Coefficients | | | | | |
|---|---|---|---|---|---|---|
| | 125 Hz | 250 Hz | 500 Hz | 1000 Hz | 2000 Hz | 4000 Hz |
| **Brick** | | | | | | |
| Unglazed | .03 | .03 | .03 | .04 | .05 | .07 |
| Unglazed, painted | .01 | .01 | .02 | .02 | .02 | .03 |
| **Carpet (heavy) on concrete slab** | .02 | .06 | .14 | .37 | .60 | .65 |
| With 40-oz (1134 g) hair-felt or foam-rubber pad | .08 | .24 | .57 | .69 | .71 | .73 |
| With impermeable latex backing and 40-oz (1134 g) hair-felt or foam pad | .08 | .27 | .39 | .34 | .48 | .63 |
| **Concrete Block** | | | | | | |
| Porous | .36 | .44 | .31 | .29 | .39 | .25 |
| Painted | .10 | .05 | .06 | .07 | .09 | .08 |
| **Fabrics** | | | | | | |
| Light velour (10 oz/sq yd) (339.06 g/m²) hung straight in contact with wall | .03 | .04 | .11 | .17 | .24 | .35 |
| Medium velour (14 oz/sq yd) (474.68 g/m²) draped to half the area | .07 | .31 | .49 | .75 | .70 | .60 |
| **Floors** | | | | | | |
| Concrete or terrazzo | .01 | .01 | .015 | .02 | .02 | .02 |
| Resilient flooring on concrete | .02 | .03 | .03 | .03 | .03 | .02 |
| Wood | .15 | .11 | .10 | .07 | .06 | .07 |
| Parquet in asphalt on concrete | .04 | .04 | .07 | .06 | .06 | .07 |
| **Glass** | | | | | | |
| Large panes heavy plate glass | .18 | .06 | .04 | .03 | .02 | .02 |
| Standard window glass | .35 | .25 | .18 | .12 | .07 | .04 |
| **Gypsum board, ½ in. (12.7 mm) nailed to 2-in. (50.8 mm) × 4-in. (101.6 mm) studs 16 in. (406.4 mm) o.c.** | .29 | .10 | .05 | .04 | .07 | .09 |
| **Marble or glazed tile** | .01 | .01 | .01 | .01 | .02 | .02 |
| **Plaster** | | | | | | |
| Smooth finish on brick or tile | .013 | .015 | .02 | .03 | .04 | .05 |
| Rough finish on lath | .02 | .03 | .04 | .05 | .04 | .03 |
| Smooth finish on lath | .02 | .02 | .03 | .04 | .04 | .03 |
| **Plywood paneling, ⅜ in. (9.53 mm) thick** | .28 | .22 | .17 | .09 | .10 | .11 |

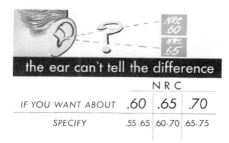

**FIGURE 13.1-17** NRC rates the efficiency of acoustical materials.

**FIGURE 13.1-18** Relation of reduction in loudness (in percent) to the absorption ratio.

carpeted, additional absorption in the form of acoustical ceilings will probably be unnecessary.

***Transmission Through Sound-Absorbing Materials*** Acoustical materials are used to absorb sound. Although some materials may be effective in preventing sound transmission, most acoustical surfacing materials are not. Acoustical tile and panels, carpets and pads, and other sound-absorbing surfacing material placed in a room where noise originates act indirectly in reducing sound transmission to adjoining rooms by lowering the noise level in the room where the sound originates. When used in an adjoining room, they will similarly lower the level of the transmitted noise. These reductions, however, usually amount to only a few decibels. Acoustical surface treatment will supplement but will not take the place of sound isolation. *Sound isolation* is the control of sound transmission through walls and floor and ceiling assemblies.

### 13.1.3 SOUND TRANSMISSION

Sound is transmitted through buildings when wall, floor, or ceiling surfaces vibrate. This vibration may be caused by (1) the alternating air pressure of incident sound waves, called *airborne sound transmission*, or (2) direct mechanical contact or impact such as might be caused

the treatment. This is the total absorption in the room after treatment ($a_2$).

4. Divide $a_2$ by $a_1$ to obtain the absorption ratio $a_2/a_1$.
5. Determine the percentage of loudness reduction from Figure 13.1-18.

It is desirable to reduce the reverberation time in small spaces to about 0.5 second and, in large spaces, especially those used for music as well as speech, to not more than 1 second. In most spaces, the problem is usually one of reducing reverberation enough to lower noise levels about 5 to 8 dB. Unless the room shape presents special problems, covering the ceiling with a good sound-absorbing material is usually adequate treatment.

Spaces such as corridors generally need sound-absorbing materials on the floor as well as the ceiling to reduce the *piping* of sound along the corridor to other spaces.

In small rooms, especially those containing upholstered furniture and heavily by vibrating equipment, a footstep, an object dropped on a floor, or a slamming door. Transmission as a result of direct mechanical contact or impact is called *structure-borne sound transmission*. In either case, vibrating surfaces generate new sound waves of reduced intensity in the space where the sound originates as well as in spaces adjacent to or above or below the origin (Fig. 13.1-19).

### 13.1.3.1 Airborne Sound Transmission

Airborne sound is produced by sources that radiate sound directly into the air, such as people talking, a TV on or a CD playing, mechanical equipment humming, or traffic rumbling. Airborne sound may be transmitted to adjoining rooms or spaces (1) directly through the intervening constructions or (2) by way of *flanking paths* such as open doors or along interconnecting structures (Fig. 13.1-20). Airborne sound can be isolated by walls and by floor and ceiling assemblies that reduce the level of transmission to the

PLAN

ELEVATION

**FIGURE 13.1-20** Sound may be transmitted (1) through the intervening construction (2) through open air paths, or (3) along the interconnecting structure.

**a** Airborne

**b** Direct Impact or Mechanical Contact

**FIGURE 13.1-19** Sound is transmitted by (a) alternating air pressures of airborne sound and (b) impact or direct mechanical contact.

point where it is either inaudible or not annoying or interfering.

#### MEASURING AIRBORNE SOUND TRANSMISSION

To determine the effectiveness of walls and of floor and ceiling assemblies in isolating airborne sound transmission, the *transmission loss* (TL) of the wall or assembly must be measured. TL is the number of decibels a sound loses while passing through a wall or floor and ceiling assembly when tested in a laboratory in accordance with ASTM International's ASTM E90. In both laboratory and field tests of walls for airborne transmission, the procedure is to generate a steady sound on one side of the wall and measure the sound level in decibels on both sides. The TL of the wall is the measured difference in sound level on the two sides. Therefore, a sound of a given frequency with an intensity level of 70 dB on one side of a wall will have a reduction in intensity to 40 dB on the other side when the wall has a TL of 30 dB at this frequency (Fig. 13.1-21). The higher the TL, the more efficient the wall or assembly.

TL is a physical property of a wall or floor and ceiling assembly. It depends on the materials and methods used in the

construction, not on the loudness of sound striking the surface. However, because TL varies considerably, depending on sound frequency, assemblies are tested at numerous frequencies and plotted as a curve (Fig. 13.1-22). Most assemblies are more efficient at reducing high-frequency sounds than they are at blocking low-frequency ones.

It is seldom necessary or economically feasible to build walls or floor and ceiling assemblies that are capable of completely blocking sound transmission. Sound transmitted through a wall or floor will be inaudible or "masked" if its in-

**FIGURE 13.1-21** TL is the difference in sound level on opposite sides of a construction. The wall here has a TL of 30 dB.

**FIGURE 13.1-22** The TL of a wall consisting of ½-in. (12.7-mm)-thick gypsum board nailed directly to both sides of wood studs placed at 16 in. (400 mm) on centers.

tensity level is below the level of the background noise in the receiving room (Fig. 13.1-23).

***Single-Number Rating Systems*** It is generally impractical to use the TL values at each frequency tested because it is difficult to identify the frequency of a noise source in sufficient detail. For convenience in selecting or comparing walls or floor and ceiling assemblies, two single-number rating systems have been developed, the *Average TL* system and the *Sound Transmission Class* (*STC*) system. Either will indicate whether a

wall or floor and ceiling assembly will offer sufficient resistance to sound transmission, but the STC system is the one generally used today.

The Average TL is the numerical average of the transmission loss values measured at nine or more frequencies, including 125, 175, 250, 350, 500, 700, 1000, 2000, and 4000 cps. Average TL was used in the past by the National Bureau of Standards (National Institute of Standards and Technology), Riverbank Acoustical Laboratories, and other laboratories for rating airborne sound transmission through walls and floor and ceiling assemblies. It was also once the most commonly used basis for comparison of published data.

To estimate the minimum Average TL required of a wall or floor and ceiling assembly to obtain speech privacy, subtract the level of the background noise on one side of the construction from the level of sound that strikes the other side. Hearing conditions on the listening side of walls with varying average TLs are shown in Figure 13.1-24. Based on the curve in Figure 13.1-22, the Average TL of a commonly used wood stud partition is 31 dB, which is not much resistance to sound transmission.

The Average TL system gives a fair indication of the sound isolation value if there are no sharp dips on the test curve and if the values being compared are based on identical tests performed by the same laboratory under identical conditions. Sharp dips on the curve in a narrow frequency band, which are normal for partitions, are harmful, but the numerical averaging in this system misses

**FIGURE 13.1-24** Sound-Insulating Properties of Construction According to Transmission Loss (TL)*

| Transmission Loss (TL) | Hearing Conditions |
|---|---|
| 30 dB or less | Normal speech can be understood quite easily and distinctly through the wall. |
| 30 to 35 dB | Loud speech can be understood fairly well. Normal speech can be heard but not easily understood. |
| 35 to 45 dB | Loud speech can be heard but is not easily intelligible. Normal speech can be heard only faintly, if at all. |
| 40 to 45 dB | Loud speech can be faintly heard but not understood. Normal speech is inaudible. |
| 45 dB or greater | Very loud sounds, such as loud singing, brass musical instruments, or a radio at full volume, can be heard only faintly or not at all. |

*Background noise level on the listening side assumed at 30 dB.

them, making it inaccurate for use with partitions.

The STC system of rating airborne-sound isolating performance is preferable to the Average TL system because it compensates for dips in the test curve by lowering the overall evaluation when they are present. This gives a more reliable indication of speech or noise transmission, especially by partitions.

The STC is a single-number rating for airborne sound transmission which represents the TL of a wall or floor and ceiling assembly at all test frequencies. The higher the STC rating, the more efficient the construction.

An STC rating is derived by plotting the laboratory TL test curve for a given assembly, tested at 16 frequencies in accordance with ASTM E90, and comparing this curve to a standard frequency curve (*standard contour*) (Fig. 13.1-25). The standard frequency curve is drawn on a transparent overlay which is placed over the test curve and moved up and down until the test curve remains above the standard curve at every frequency within the following tolerances: (1) no single deviation where the sound transmission loss value falls below the standard curve may exceed 8 dB, and (2) the sum of these unfavorable deviations may

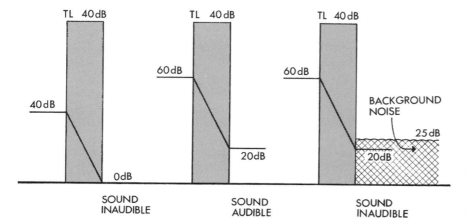

**FIGURE 13.1-23** Normal background noise on the listening side of a wall has the effect of masking transmitted sound that may otherwise be objectionable.

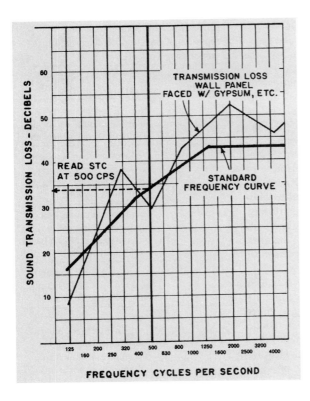

FIGURE 13.1-25 Method for obtaining an STC rating.

FIGURE 13.1-26 Sound Isolation of Walls in Terms of STC Values

| Sound Transmission Class (STC) | | | | |
|---|---|---|---|---|
| | Background Noise[a] | | | |
| None | 20 dB | 30 dB | 40 dB | Speech Audibility[b] |
| 40 to 50 | 25 to 35 | 15 to 25 | | Normal speech easily understood. |
| 50 to 60 | 35 to 45 | 25 to 35 | 15 to 25 | Loud speech easily understood. Normal speech 50% understood. |
| 60 to 70 | 45 to 55 | 35 to 45 | 25 to 35 | Loud speech 50% understood. Normal speech faintly heard but not understood. |
| 70 to 80 | 55 to 65 | 45 to 55 | 35 to 45 | Loud speech faintly heard but not understood. Normal speech usually inaudible. |
| | 65 to 75 | 55 to 65 | 45 to 55 | Loud speech usually inaudible. |

[a]Background noise levels on *A* scale of sound level meter.

[b]Loud speech taken as 10 dB higher than normal speech; intelligibility taken as percent of sentences understood; normal source and receiving room absorption considered.

not exceed 32 dB. The STC rating is the numerical value which corresponds to the sound transmission loss value at 500 Hz (see Fig. 13.1-25).

The ASTM test procedure was first issued in 1950 and has been extensively revised since that time. Some currently published STC ratings for assemblies are based on tests performed before the current revision of ASTM E90. STC values derived from earlier tests may vary somewhat from the values of tests on similar assemblies performed in accordance with

ASTM E90. Very old references may vary considerably. Variables that might influence test results include the number of test frequencies used, the size of the panel tested, and the application of the standard frequency curve.

In general, a wall or floor and ceiling assembly provides acceptable privacy when the conversational speech in one room is reduced to a level at which it cannot be understood in another room. To determine the minimum STC required to obtain speech privacy, subtract the level

of the background noise on one side of the wall or assembly from the level of sound that strikes the other side. The examples in Figure 13.1-26 illustrate the approximate effectiveness of walls with varying STC values as influenced by the background noise. Values of STC for some assemblies are given in Section 13.1.5.

## CONTROLLING AIRBORNE SOUND TRANSMISSION

Variables that affect the control of airborne sound transmission to provide a suitable acoustical environment include (1) the frequency and sound level of the sources, (2) the sound level that will be acceptable within the space, and (3) the sound-reducing characteristics of the intervening construction.

Methods to control airborne sound transmission include (1) selecting walls and floor and ceiling assemblies that reduce transmission by vibration, (2) closing paths of direct air transmission, and (3) installing sound-absorbing materials within the free air space of frame construction. Usually, a combination of these control measures is necessary to achieve suitable sound privacy. Methods of controlling airborne sound transmitted through walls and through floor and ceiling assemblies are discussed in Section 13.1.4 "Design and Construction."

The sound reducing characteristics of a wall or floor and ceiling assembly directly influence its cost. Therefore, it is important to minimize or eliminate the need for special construction techniques by proper planning using some basic sound control principles and common sense. These are discussed in Section 13.1.4.

### 13.1.3.2 Structure-Borne Sound Transmission

Structure-borne sound transmission occurs when walls or floor and ceiling assemblies are set into vibration by direct impact or direct mechanical contact. The vibrating surface generates new sound waves on both sides of the construction. The new sound waves may even occur in different parts of the building. Floors are especially critical in this type of noise transmission. Some structure-borne noise is caused by objects striking or sliding on a floor such as footsteps, furniture, and dropped objects; by objects that bump against a wall, such as furniture; and by the slamming of doors. Other such noise is generated by mechanical equipment, fixtures, and appliances that transmit their own vibrations to adjacent construction. These include motors and fans on heating, cooling, and ventilating equipment; vacuum cleaners; water closets; bathtubs; and showers.

Isolating structure-borne sound requires providing wall and floor and ceiling assemblies that will reduce the level of transmission to the point where the sound is either inaudible or not annoying or interfering.

## MEASURING IMPACT SOUND TRANSMISSION

Resistance to impact noise transmission is determined using a standard test method described in ASTM E492, which is based on the International Organization for Standardization (ISO) Recommendation 140, "Field and Laboratory Measurements of Airborne and Impact Sound Transmission." This test is performed by placing on the floor to be tested a standard tapping machine which produces a series of uniform impacts at a uniform rate. The *Impact Sound Pressure Level* (ISPL) (measured in decibels) is then recorded in the receiving room below and electronically separated into different frequency bands. A curve can then be plotted showing how the sound energy in the receiving room is distributed over the audible frequency range (Curve 1 in Fig. 13.1-27).

To determine airborne sound transmission resistance, the difference between the sound levels in the sending and receiving rooms is measured and is expressed as a sound transmission loss. Conversely, ISPL curves measure the actual sound levels from a standard noise source that are transmitted through the construction. The lower the transmitted level, the better the acoustical performance of the construction.

ISPL determination assumes that an assembly that will transmit little sound generated by a standard tapping machine will perform similarly with other types of impact sound. Unfortunately, the noise produced by a standard tapping machine does not duplicate the sound produced by many types of impact. For example, much footfall noise is in a frequency below the lowest of the standard test frequencies. Because of the problems with duplicating actual conditions (and other limitations of the test used), model build-

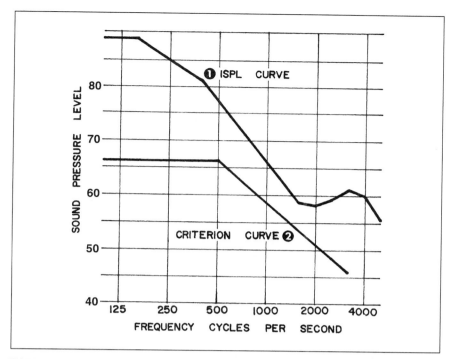

**FIGURE 13.1-27** Impact Noise Rating (INR) is obtained by comparing the ISPL test curve (1) with a standard criterion curve (2).

ing codes often require that specific construction assembly types be used rather than relying on test performance criteria.

***Impact Noise Rating*** An *Impact Noise Rating* (INR) is a single-number rating in decibels for evaluating the impact sound isolation performance of a floor and ceiling system. It is determined by analyzing a test performed in accordance with ISO R140. This method has been superseded by the *Impact Insulation Class* (IIC) system discussed later in this section. The INR system is discussed here because some still available published material uses it to compare the performance of floor and ceiling assemblies.

The INR for an assembly is determined by first measuring the assembly's ISPL, plotting it on a transparent overlay, and placing the overlay over a standard criterion curve (Curve 2 in Fig. 13.1-27). The standard curve is then moved up or down until the ISPL test curve falls over the standard criterion curve within the following tolerances: (1) the mean amount by which the test curve exceeds the standard curve must be 2 dB or less, as averaged over the 16 one-third octave frequency bands between 100 and 3200 Hz; (2) the test curve must not exceed the criterion curve by more than 8 dB at any one frequency.

The INR value for a given assembly is the distance the ISPL curve falls below or above the standard criterion curve. Plus (+) INR values indicate better than standard criterion performance; minus (−) INR values indicate less than standard criterion performance. When the test curve meets the standard curve within the allowable tolerance, the INR is 0. If the standard curve must be moved down 5 dB for the test curve to conform, the construction is given a +5 rating, indicating that the assembly is 5 dB better than the standard; if the standard curve must be moved up 5 dB for the test curve to conform, the assembly is given a −5 rating, indicating that it is 5 dB poorer than the standard.

The criterion curve is based on average background noises that are typical of apartments in moderately quiet suburban neighborhoods. For quieter areas, floor constructions with an INR of +5 to +10 should be used. For noisy urban districts, a construction with a −5 INR may suffice. INR values for various floor and ceiling constructions are given later in this section.

***Impact Insulation Class*** In an effort to reconcile and correlate the various rating systems for airborne sound transmission and impact sound transmission, the Federal Housing Administration (FHA) devised a single-number rating system for impact sound transmission called the *Impact Insulation Class (IIC)*. The IIC system supersedes the INR system described earlier.

On the average, an IIC rating for a given construction is approximately 51 dB higher than an INR rating. For example, an INR rating of +4 would be comparable to an IIC rating of about 55. However, exact values for IIC ratings should be determined on the basis of individual test data.

To develop the IIC of an assembly, the ISPL is measured and a curve plotted as discussed earlier (see Fig. 13.1-27). A standardized reference contour is plotted on a transparent overlay, placed over the ISPL curve, and moved up and down until only the following allowable tolerances exist: (1) a single unfavorable deviation in which an ISPL value lies above the reference contour may not exceed 8 dB, and (2) the sum of the unfavorable deviations may not be greater than 32 dB. The IIC rating is the value on the right scale corresponding to the ISPL value at 500 Hz (Fig. 13.1-28).

The HUD publication "A Guide to Airborne, Impact, and Structure-Borne Noise Control in Multifamily Dwellings" presents a three-segment standard contour that offers three options, depending on the outdoor nighttime background noise level expected. The contours presented are labeled Grade I (IIC 55), Grade II (IIC 52), and Grade III (IIC 48). The higher the outdoor level, the greater the masking will be and the less the in-

sulation needed. The most often used of the three is the Grade II (IIC 52) contour.

The IIC rating system parallels the STC rating system in that performance of a floor and ceiling assembly is expressed with positive values in ascending degrees of impact sound isolation. The larger the IIC rating, the better the assembly.

## CONTROLLING STRUCTURE-BORNE SOUND TRANSMISSION

Most building materials, especially hard or dense ones, transmit vibrations readily. A building's steel frame, for example, can telegraph a vibration throughout a large structure. Clicking of high heels on a concrete floor can be heard distinctly in the room below if no other construction intervenes. The vibration of a structure sets up sound waves which can be heard distinctly at great distances from the source. Impact noise caused by a floor or wall being vibrated by direct mechanical contact is radiated from both sides. This vibration may also be transmitted throughout the structure to walls and reradiated as sound to adjoining spaces. Since footsteps and moving furniture on floors constitute a major impact problem, control of impact noise is treated here as primarily a floor problem.

Impact isolation methods in general use are more effective in reducing the high-frequency components of impact noise than the low-frequency "thumping" noises caused by heavy impacts. Design requirements for low-frequency isolation are more stringent for wood joist floors than for concrete floors because of the lighter weight and greater flexibility of wood floor joist construction. Methods to control structure-borne sound transmission are discussed in Section 13.1.4.

## 13.1.4 DESIGN AND CONSTRUCTION

This section deals with (1) criteria and planning principles to aid in the design and selection of walls and floor and ceiling assemblies to control noise within buildings; (2) construction recommendations, including precautions essential to obtaining and maintaining suitable sound isolation between rooms and between living units; and (3) selection criteria for obtaining suitable sound isolation. It also includes the design of doors and windows and the selection of appropriate hardware to control sound transmission. Criteria and planning principles that will reduce sound generation by and transmission

through plumbing systems are covered in Section 16.5; heating, ventilating, and air conditioning systems in Section 17.5.2; and electrical systems and appliances in Section 18.2.11.

### 13.1.4.1 Design and Planning

Sound control involves (1) reducing loud noises at their source; (2) orienting quiet spaces away from noise sources, thus separating the source from the listener by distance; (3) careful interior design and building layout, especially with respect to building shape and the location of door and window openings and other sound leaks; and (4) reducing the amount of noise that reaches a listener by selecting walls and floor and ceiling assemblies that have improved resistance to sound transmission.

### BACKGROUND NOISE SOURCES AND LEVELS

The most recent tool developed to measure and describe background noise levels is the *Day Night Average Sound Level* system (DNL). DNL is symbolized mathematically as $L_{DN}$. The DNL system has been adopted by the Environmental Protection Agency (EPA), the Department of Defense (DOD), the Department of Housing and Urban Development (HUD), and the Federal Aviation Administration (FAA) and will most likely obtain more widespread use in the future. DNL is the 24-hour average sound level, after the addition of a 10 dB penalty for sound levels that occur between 7 P.M. and 8 A.M. DNL is expressed in decibels.

EPA recommends that the DNL in residential zones not exceed $L_{DN}$ 55. EPA's recommendation is not a standard, does not take cost or feasibility into account, and cannot be currently enforced. This level is also a HUD goal, but HUD, using a more realistic approach, currently divides high-noise locations into three zones: the *acceptable* zone, where the $L_{DN}$ is 65 or less, in which all types of housing projects could be built; the *normally unacceptable* zone, where the $L_{DN}$ exceeds 65 but does not exceed 75, in which mitigation measures are required and every project would be required to have approval by HUD; and the *unacceptable* zone, where the $L_{DN}$ exceeds 75, in which almost all housing projects would be prohibited. Attenuation measures in the normally unacceptable zone depend on the $L_{BN}$ levels. Of course,

**FIGURE 13.1-28** The IIC is obtained by comparing the ISPL curve with a standard reference contour and is read as a positive rating.

these restrictions apply only to housing projects. Other types of projects must be evaluated based on their use and possible noise mitigation procedures that are employed.

The source and intensity of outdoor noise at the site affect the noise-reducing effectiveness of exterior walls and roofs and also have an important effect on the selection of interior wall construction and floor and ceiling assemblies. The noise characteristics of the site should be studied to determine existing and potential noise levels (Fig. 13.1-29).

Theoretically, airborne sound travels from a *point noise source* in a spherical pattern (see Fig. 13.1-1). The sound level of a point noise source is reduced 6 dB each time the distance from the source is doubled (see Fig. 13.1-10). But fast-moving traffic is a *line noise source* rather than a point source. Sound from a line noise source travels in a cylindrical rather than a spherical pattern. In order to reduce the noise level from a line source 6 dB, the distance between noise source and listener must be increased four times.

The direction and speed of air movement, air temperature, and obstructions such as buildings, trees, and terrain all affect sound level. For example, sites upwind of a major noise source will be quieter than sites an equal distance away but downwind of the noise.

Hard exterior surfaces of buildings reflect a large proportion of the sound energy that strikes them (Fig. 13.1-30). Multiple reflection of sound striking a large building can increase the noise problem at a site. Curved buildings, or buildings designed to form courtyards, having the concave side facing a noise

source also will result in an especially difficult noise problem. Future development of nearby vacant land should be anticipated, and the type and level of probable or possible noise sources should be investigated.

When a proposed site is near high-level noise sources, such as a business district, highway, expressway, railroad line, or airport, both daytime and nighttime outdoor noise levels should be surveyed. When the average background noise level is unusually high or if there are infrequent but intermittent unusually loud or annoying noises, expert acoustical advice should be sought.

*Ground Vehicular Traffic* Ground vehicular traffic (by automobiles and trucks) is a primary source of outdoor background noise. Background noise levels are affected by topography, landscaping, and the masking or magnifying effect of other buildings. Typical outdoor noise levels are shown in Figure 13.1-31. High-speed highways and dense city traffic generally produce higher noise levels than do moderate-speed suburban roads carrying an equivalent number of cars. Trucks increase highway noise about 15 dB, and trucks going uphill produce the

highest highway noise levels. Unless the highway is already saturated with traffic, a future increase in traffic generally can be expected, with a resultant increase in noise.

*Aircraft Traffic* The proliferation of jet passenger and transport operations and the rapid expansion of small commercial and business jet aircraft create noise problems even in small communities. Methods for minimizing aircraft noise include techniques for quieting engines, reducing noise exposure by takeoff and landing operational procedures, and reducing the number of people affected by controlling land use around airports through zoning regulations.

Two types of aircraft noise are of concern: (1) noise generated by revving engines on the ground and (2) flyover noise. Figure 13.1-32a illustrates that the noise level during flyover changes very rapidly with time, with the maximum noise sustained for perhaps a second or less, and that sound waves during this short time interval strike a building from all directions.

Figure 13.1-32b illustrates the noise produced by revving engines while the aircraft is stationary on the ground. Here the relationship between aircraft and building is fixed; only one or two exterior surfaces of the building are directly exposed to aircraft noise, and other surfaces are partially shielded. Intervening obstacles, such as hills, houses, or other terrain features, will partially block the sound.

Figure 13.1-33a illustrates typical noise level contours of jet aircraft during takeoffs; Figure 13.1-33b is an example of noise level contours for the revving of one engine of a jet aircraft.

Of all exterior wall and roof surfaces, windows offer the least resistance to noise transmission. The difference in perceived noise level (PNdB, Fig. 13.1-33) between exterior and interior noise levels from aircraft is illustrated in Figure 13.1-34.

The lower values in each category will be for buildings having window areas comprising more than 20% of the total exposed wall area, with relatively lightweight wood frame walls and roofs (weighing less than 8 psf). The higher values will occur in buildings with relatively small, airtight windows and airtight storm windows of good quality. An example of aircraft noise reductions through various exterior wall and roof

**FIGURE 13.1-30** Hard surfaces reflect sound; soft, porous ones, such as are found in plants, absorb some of it.

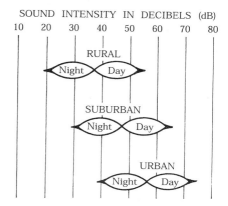

SOUND INTENSITY IN DECIBELS (dB)

10   20   30   40   50   60   70   80

RURAL

Night  Day

SUBURBAN

Night  Day

URBAN

Night  Day

**FIGURE 13.1-29** Typical outdoor background noise levels.

**FIGURE 13.1-31** Approximate Sound Levels for Automobile Traffic at Ground Level

| Cars/ Minute* | Distance from Traffic Lane | | |
| | 40 ft (12.2 m) | 100 ft (30.5 m) | 500 ft (152.4 m) |
|---|---|---|---|
| 1 | 50 dB | 43 dB | 29 dB |
| 10 | 59 dB | 52 dB | 38 dB |
| 100 | 63 dB | 60 dB | 47 dB |

*The maximum capacity of one lane of cars moving freely at approximately 30 mph (48.28 km/h) is 2000 cars per hour, or 34 cars per lane per minute.

a

b

**FIGURE 13.1-32** (a) Flyover noise strikes buildings from all angles, but its maximum intensity lasts only a second or less; (b) noise from revving engines on the ground strikes fewer building surfaces but lasts longer.

**FIGURE 13.1-34** Perceived Noise Levels from Aircraft

| Window Condition | Difference Between Interior and Exterior Noise Levels |
|---|---|
| Windows open | 10 to 20 PNdB |
| Windows closed* | 20 to 30 PNdB |
| Storm sash installed* | 25 to 35 PNdB |

*For maximum performance, glass should be sealed airtight within frames; storm sash should be separated by at least 2 in. (50.8 mm) air space between panes.

construction assemblies is summarized in Figure 13.1-35.

### SITE PLANNING

Site planning can reduce problems caused by exterior noise by (1) increasing the distance between the noise source and buildings, (2) siting buildings to minimize the effect of direct sound transmission, reflection, and reverberation from existing or possible future buildings, and (3) using natural topography, other buildings, barrier walls, and landscaping to shield buildings from major noise sources.

***Building Location*** Buildings should be located as far as possible from noise sources. Buildings reflect noise readily. An orientation which places rows of buildings parallel to each other results in a buildup or focusing of the noise by multiple reflection of the sound. Acoustically, the best orientation for a building occurs when its long axis is perpendicular to noise sources (Fig. 13.1-36a).

U-shaped courtyards are especially noisy when the open end of the U faces a major noise source (Fig. 13.1-36b). It is acoustically better to have the end of such a courtyard facing away from noise sources. Landscaping in such a courtyard will help to reduce reverberation of sound within the courtyard.

***Noise Barriers*** A building's walls, landscaping, and topography can be used to help shield against an exterior noise source. A barrier's effectiveness depends on the frequency of the sound, the angle of the sound shadow, and the height of the barrier (Fig. 13.1-37). Higher-frequency sounds are controlled more effectively by barriers than are low-

a

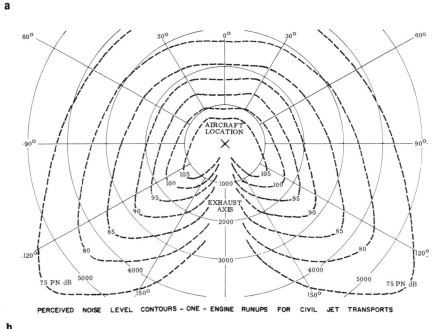

b

**FIGURE 13.1-33** Aircraft noise level contours expressed as perceived noise level (PNdB).

**FIGURE 13.1-35** Summary of Noise Reduction Through Exterior Materials and Assemblies

| Typical Constructions | Difference Between Outside and Inside Noise Levels (PNdB) | | | | | |
|---|---|---|---|---|---|---|
| | 0 | 10 | 20 | 30 | 40 | 50 |
| **Windows** | | | | | | |
| ⅛ in. (3.18 mm) glass in double-hung wood frame | | ▬ | | | | |
| ¼ in. (6.35 mm) plate glass sealed in place | | | ▬ | | | |
| Aluminum frame window, glass panels set in neoprene gaskets, ¼ in. (6.35 mm) and ⁷⁄₃₂ in. (5.56 mm) panes, 3¾ in. (95.25 mm) air space | | | | | ▬ | |
| **Exterior walls** | | | | | | |
| Wood sheathing (or stucco), 2 × 4 (50 × 100) studs, ½ in. (12.7 mm) plaster board interior surface | | | ▬ | | | |
| Wood sheathing (or stucco), 2 × 4 (50 × 100) studs, ⅞ in. (22.2 mm) plaster interior surface | | | | ▬ | | |
| 4½ in. (114.3 mm) brick or 6 in. to 8 in. (140 mm to 200 mm) lightweight concrete block | | | | ▬ | | |
| 9 in. (230 mm) brick wall | | | | | ▬ | |
| **Roofs** | | | | | | |
| Built-up roofing on 1 in. (25.4 mm) wood decking, ½ in. (12.7 mm) gypsum board on 2 × 8 (50 × 200) joists | | | ▬ | | | |
| Built-up roofing on 1 in. (25.4 mm) wood decking, ⅞ in. (22.2 mm) plaster on 2 × 8 (50 × 200) joists | | | | ▬ | | |
| Built-up roofing or shingles on wood siding, ventilated attic space, ½ in. (12.7 mm) gypsum board ceiling | | | | | ▬ | |
| Built-up roofing or shingles on wood sheathing, ventilated attic space, ⅞ in. (22.2 mm) plaster ceiling | | | | | ▬ | |

**FIGURE 13.1-36** (a) Perpendicular orientation exposes less wall area to a noise source; (b) sound reverberates in a U-shaped courtyard.

**FIGURE 13.1-37** Buildings and walls serving as barriers should be as high as possible and as close as possible to the noise source.

**FIGURE 13.1-38** Plants reduce noise levels only slightly, but may provide beneficial visual and psychological barriers.

frequency sounds. For greatest effectiveness, barriers should be as high as possible, located as close as possible to the noise source, and impervious to air flow. Barriers should also have some mass, but great weight will not appreciably increase their effectiveness.

Plants absorb some sound and can help to reduce reverberation noise between buildings, but trees and shrubs reduce noise levels only slightly. A 2-ft-thick, dense cypress hedge will reduce high-frequency noise about 4 dB (Fig. 13.1-38). A single row of trees does little to reduce noise levels. Even a dense tropical jungle will reduce levels only about 6 or 7 dB per 100 ft (30.48 m) of depth, which is only 1 or 2 dB more than

can be achieved by distance alone. Since noise has a psychological effect, plants that provide a visual barrier to noise sources such as traffic may help to reduce sensitivity to the noise.

Nonresidential buildings parallel and adjacent to a highway can serve as a noise barrier between the highway and residential buildings farther away (Fig. 13.1-39a). Hills and highway cuts can provide natural barriers to major noise sources if the buildings are located to take advantage of these sound shadow barriers (Fig. 13.1-39b).

## INTERIOR PLANNING

It is difficult and expensive to correct sound transmission problems caused by

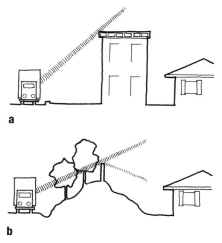

**FIGURE 13.1-39** (a) Nonresidential buildings and (b) natural terrain and man-made road cuts and fills can serve as noise barriers.

faulty interior design after a building has been built. Therefore, it is essential that acoustical principles be recognized in the initial design stages. In addition, design of interior spaces with sound privacy in mind can minimize or eliminate the need for costly construction techniques to control noise. Sound transmission through walls and floor and ceiling assemblies can be reduced by selecting suitable sound-insulating systems and controlling the methods used in the field, as discussed in Section 13.1.4.2.

Sound control should create an environment that is acoustically comfortable, not render rooms soundproof. The basic principles of such planning include (1) locating occupied spaces as far as possible from noise sources, (2) separating noisy areas from quiet areas by isolating the noise source or using buffer spaces, (3) eliminating air paths that permit sound to travel between spaces, especially through the design, sizing, and placement of doors and windows, and (4) reducing the noise level at the source by selecting mechanical equipment and appliances that operate quietly.

In planning for sound control in homes and apartments, most of the emphasis is usually placed on sound originating in adjacent homes or apartments or from outdoors. This is because people can control the intensity of noise sources within their own home and tend to be more tolerant of noises created by their own family than they are of those that come from other sources. The following discussion uses apartment living units to illustrate design principles, but most of these principles are equally applicable to commercial and institutional buildings.

***Space Layout*** Spaces can be zoned acoustically into three general areas: (1) areas that are relatively noisy, (2) areas that should be quiet, and (3) areas that fall between the two extremes.

People and their activities establish acceptable noise levels for different areas. Noisy areas should be grouped together. Buffer spaces such as corridors, stairways, and other spaces can be used to provide acoustical buffers against airborne sound transmission between spaces. Closets, storage rooms, and bookcases backed up to common walls help to reduce airborne sound transmission between rooms, and in addition force the placing of furniture away from common walls, thereby increasing the distance between a noise source and the occupants of adjacent rooms (Fig. 13.1-40).

In apartment design, it is acoustically preferable that a plan be mirrored horizontally and matched vertically (Fig. 13.1-41). This places noisy areas adjacent to other noisy areas and quiet areas adjacent to other quiet areas so that the sound control requirements can be as low as possible, thereby reducing the cost. For example, kitchens and bathrooms which have similar noise levels are located adjacent to and above and below kitchens and bathrooms in adjoining apartments. Similarly, quieter areas such as bedrooms and living rooms should be located adjacent to and above and below quiet areas in adjoining apartments. The noise level requirements on both sides of walls and floor and ceiling assemblies will be the same, and the full effect and cost of improved sound control measures will benefit all apartments. Some additional precautions that should be observed in mirror planning are illustrated in Figure 13.1-42.

Row plans where apartments are arranged in lines provide better acoustical performance than block plans where each apartment may have up to three common walls (Fig. 13.1-43b). In row planning, with halls between every other apartment, each apartment has only one common wall (Fig. 13.1-43a).

Room proportions can be planned to reduce sound transmission. The higher the ratio of common wall to floor area, the more sound the common wall will transmit (Fig. 13.1-44).

Adding large, absorptive surfaces, such as acoustical tile, carpeting, or heavy drapes, to rooms will slightly reduce the amount of sound available that can be transmitted. In a room with large, hard-surfaced areas, a large proportion of the sound is reflected, whereas large, soft-surfaced areas will absorb sound.

## DOORS AND WINDOWS

Doors and windows are acoustically weak. Maximum acoustical performance requires that they be properly placed and that their edges and frames be sealed airtight. Packing and caulking around their frames is necessary to reduce sound transmission through cracks.

***Door Construction and Installation*** Solid core and mineral-filled doors are generally better sound isolators than hol-

**FIGURE 13.1-40** Closets and bookshelves can serve as buffers to separate listeners from a noise source.

**FIGURE 13.1-41** Mirroring the plan results in about equal noise level requirements on both sides of common walls and floor and ceiling assemblies.

**FIGURE 13.1-42** Plumbing should be located in walls within each living unit rather than in common walls between units.

**FIGURE 13.1-43** (a) Row planning reduces the number of common walls between units. (b) Block planning can result in as many as three common walls in each unit.

**FIGURE 13.1-44** Room proportions can reduce the ratio of common wall to floor area. Hard-surfaced areas should be minimized and soft, absorbent areas increased.

low core doors (Fig. 13.1-45). However, the amount of air space around the edges of a door is usually the controlling acoustical factor. Doors should be gasketed if sound privacy is required (see Fig. 13.1-45). The relative performance of solid and hollow core doors, gasketed and ungasketed, is given in Figure 13.1-46.

Soft weatherstripping provides a good seal at the top and sides of a door, but automatic threshold closers are needed to seal the bottom. While it may be costly

to use gasketed doors, the provision of a cushion such as sponge rubber between the door and its jamb will reduce the impact energy of a door slam and make door closure quieter. Unfortunately, properly sealed doors, because they are tight-fitting, are harder to close and more prone to warping.

Undercutting doors to allow for rugs or return air movement in forced air ventilation systems essentially destroys their acoustical effectiveness.

Sliding doors provide little sound privacy, because a tight air seal cannot be maintained.

Heavy-duty doors made of metal or metal clad over very heavy core materials, such as fire-rated metal doors, will not necessarily provide good acoustical performance in spite of their mass, even when two doors are installed, unless they are sealed with gaskets and doorsills (Fig. 13.1-47a). When such doors are installed in locations where loud noises occur, such as in mechanical equipment rooms, a second door with better acoustical performance can be installed (Fig. 13.1-47b).

### Window Construction and Installation
About 80% of the acoustical problems involving windows results from the quality

of airtightness of construction and about 20% from the selection of the glass. Sturdy, well-weatherstripped, and tightly closing windows are better sound insulators than loosely fitting ones; a fixed sash performs better than an operating sash. Refer to Figure 13.1-48 for the relative acoustical performance of various types of windows.

Operating single-glazed wood windows in grooved wooden tracks with weatherstripping, which are pressure sealed when closed, perform well. The performance of wood windows in metallic or plastic slides is fair against air infiltration but poor against resonant frequencies. Metal windows are available in many grades and quality levels. The better grades will perform acceptably if tight-fitting, properly sealed, correctly installed, and sufficiently rigid to reduce air infiltration and resonant vibration.

In operating double-glazed (storm) windows, the greater the separation between two panes of glass, the more efficient the sound isolation, but to be significant, the separation must be relatively large. For example, a separation of 2 to 4 in. (50.8 to 101.6 mm) increases the STC from 33 to only 34. The minimum useful distance is 4 in. (101.6 mm); 8 in. (203.2 mm) spacing is effective but not

**FIGURE 13.1-45** Solid core doors should be used in sound-conditioned walls, with the door sealed at the jambs, head, and sill.

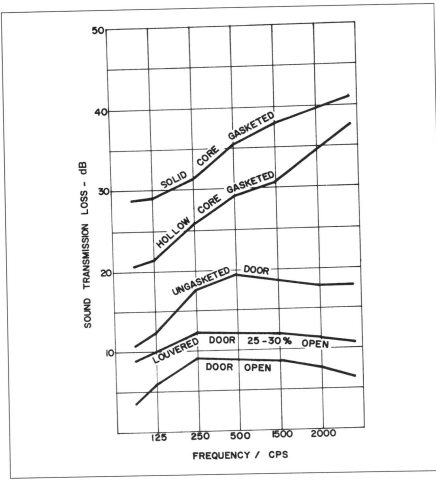

**FIGURE 13.1-46** Relative acoustical performance of hollow and solid core wood doors.

absorbing some of the sound energy in the air space.

*Location* Doors, windows, and other openings are direct air-to-air paths which transmit sound readily. Exterior doors, windows, and other openings should be placed on the quiet side of a building when possible. Where they are necessary on the noisy side, their number and size should be minimized when sound control has a higher priority than light, view, or ventilation. Exterior doors and windows should be located as far as possible from, or in positions that do not interfere with, doors and windows in adjacent apartments (Fig. 13.1-49).

When possible, interior doors should be staggered on opposite sides of corridors so that the least amount of sound is radiated directly from one door across to another. Offsetting doors allows the greatest amount of diffusion and dissipation of sound before it reaches adjoining units (Fig. 13.1-49b).

Noise is transmitted in both directions from open windows. To minimize noise from one apartment to another, the windows should be placed as far as possible from the common wall. Casement windows should be arranged to operate in the same direction so that the sound from one room is not reflected into another (Fig. 13.1-50).

always practical, and 12 in. (304.8 mm) is required to isolate severe noise. If only a 2 in. (50.8 mm) space is possible, it is more economical to double the weight of the single glass. Storm windows attached to casements are credited for added weight of glass only.

Each glass pane should be mounted in a heavy frame and securely fastened in place. Using flexible or preextruded sealant improves the damping characteristics. Each glass pane should be of a different thickness or weight per square foot (square meter) so that one can offset the deficiency of the other and they will not vibrate together or resonate.

A fixed window having a single pane of plate glass is a fair sound barrier, compared with an operating window, but a window is still the weakest part of a wall construction. Year-round air conditioning allows glass to be sealed in less expensive frames and reduces air infiltration.

Glass thickness should meet a project's decibel loss requirements. Two layers of insulating glass with a $1/2$ in. (12.7 mm) air space should be considered for fuel savings and prevention of condensation. Acoustically, two layers of closely spaced $1/4$ in. (6.4 mm) glass are approximately equal to one layer of $1/2$ in. (12.7 mm) glass.

Windows should be installed with flexible mounting, making certain that the mounting is installed in a firm frame and is airtight to prevent flanking sound from bypassing the glass.

Fixed, double-glazed sash should be used for the most demanding noise problems. This consists of two panes of glass separated by at least 8 in. (203 mm) of enclosed air space. The addition of a sound absorber, such as a 1-in. (25.4-mm)-thick resilient material around the perimeter between the two glass panes will generally improve performance by

DOUBLE DOORS WITH GASKETS

a

ACOUSTICAL LOCK TO MECHANICAL ROOM

b

**FIGURE 13.1-47** Acoustical traps can be used to reduce sound transmission from high-noise sources, such as service rooms and mechanical equipment rooms.

**FIGURE 13.1-48** Relative Acoustical Performance of Various Types of Windows

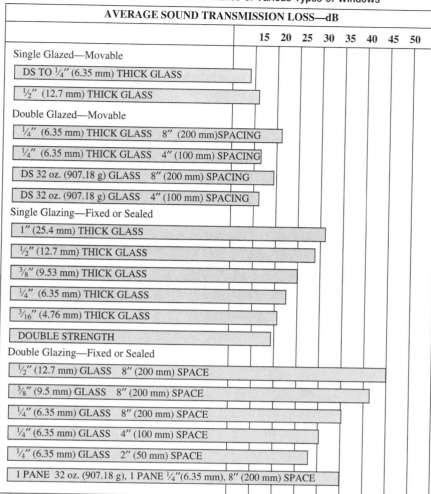

| AVERAGE SOUND TRANSMISSION LOSS—dB | | | | | | | | |
|---|---|---|---|---|---|---|---|---|
| | 15 | 20 | 25 | 30 | 35 | 40 | 45 | 50 |

Single Glazed—Movable
- DS TO ¼″ (6.35 mm) THICK GLASS
- ½″ (12.7 mm) THICK GLASS

Double Glazed—Movable
- ¼″ (6.35 mm) THICK GLASS  8″ (200 mm) SPACING
- ¼″ (6.35 mm) THICK GLASS  4″ (100 mm) SPACING
- DS 32 oz. (907.18 g) GLASS  8″ (200 mm) SPACING
- DS 32 oz. (907.18 g) GLASS  4″ (100 mm) SPACING

Single Glazing—Fixed or Sealed
- 1″ (25.4 mm) THICK GLASS
- ½″ (12.7 mm) THICK GLASS
- ⅜″ (9.53 mm) THICK GLASS
- ¼″ (6.35 mm) THICK GLASS
- 3/16″ (4.76 mm) THICK GLASS
- DOUBLE STRENGTH

Double Glazing—Fixed or Sealed
- ½″ (12.7 mm) GLASS  8″ (200 mm) SPACE
- ⅜″ (9.5 mm) GLASS  8″ (200 mm) SPACE
- ¼″ (6.35 mm) GLASS  8″ (200 mm) SPACE
- ¼″ (6.35 mm) GLASS  4″ (100 mm) SPACE
- ¼″ (6.35 mm) GLASS  2″ (50 mm) SPACE
- 1 PANE 32 oz. (907.18 g), 1 PANE ¼″ (6.35 mm), 8″ (200 mm) SPACE

**FIGURE 13.1-49** (a,b) Doors and windows should be placed so that sound will not travel through them from one unit to another. (b) Interior doors should be staggered on opposite sides of a hallway.

**FIGURE 13.1-50** Casement windows should not reflect sound from one unit to another; (a) improper and (b) proper arrangement.

## PLUMBING, MECHANICAL, AND ELECTRICAL EQUIPMENT AND SYSTEMS

Plumbing and heating, ventilating, and air conditioning equipment and systems produce airborne as well as structure-borne noise. It may be that the noise at the source cannot be eliminated, but it can be minimized by selection of equipment, design layout, and construction precautions.

### 13.1.4.2 Construction Recommendations

The ability of materials and assemblies to reduce noise transmission varies with the frequency of the sound. Some assemblies are efficient at high frequencies, others at low frequencies. A wall or floor and ceiling assembly that will adequately reduce the transmission of conversational speech may not be able to block the low-frequency thump of an air compressor.

Acceptable sound conditioning requires proper selection and construction of sound-isolating walls and floor and ceiling assemblies. It also requires proper selection and installation of plumbing, heating, ventilating, air conditioning, and electrical systems, equipment, and appliances of types that will reduce noise generation and transmission to acceptable levels.

The sound-isolating efficiency of a wall or floor and ceiling assembly depends on (1) its weight or mass, (2) effective separation of the surfaces of the assembly, and (3) the addition of sound-absorbing materials withing the assembly.

To reduce sound transmission through an assembly, do one or more of the following: (1) increase the weight of the assembly; (2) devise discontinuous construction methods that will permit the surfaces of an assembly to be independent of each another; (3) add sound-absorbing materials in the cavities within the assembly.

The ability of a wall or floor and ceiling assembly to transmit sound is inversely dependent on its mass. The heavier or more massive it is, the greater will be its resistance to sound transmission. A wall or floor and ceiling assembly with an infinite mass cannot be set into vibration by airborne sound pressures, regardless of its intensity. If the mass is extremely small, even faint sound pressures will move it. However, massiveness soon reaches a point of diminishing returns, since doubling the mass of a wall or floor and ceiling assembly will improve its resistance to sound transmission by only about 5 dB. For example, to achieve a 20 dB reduction in sound transmission in a wall having a single layer of $^1/_2$-in. (12.7-mm)-thick gypsum board would require adding 16 layers of the same board.

Separating opposite surfaces by discontinuous construction in such a manner that the coupling between the surfaces is very ineffective reduces transmission, because the surface opposite the source is not set into as vigorous vibration as the surface on the source side.

One method is to build the opposite faces of a wall independently of each other, with no structural connections between them (Figs. 13.1-51 and 13.1-52). In wood frame or metal stud constructions such as those with staggered studs or with two rows of studs, the framing provides vibration isolation between the two wall surfaces. The surfaces can also be isolated with resilient underlayment materials between the outer wall surfaces and the structural framing (Fig. 13.1-53), thus reducing transmission by isolating the surfaces, just as an automobile spring isolates the body of the auto from the road. Sound striking one side of a wall causes it to vibrate but, because of mechanical separation, the vibration of the other side is greatly reduced. Double walls are much more effective than single walls of the same overall weight.

Another method is to attach each face to the structural support, yet still have them mechanically separated using flexible connections or resilient mountings such that the opposite faces do not vibrate sympathetically (Fig. 13.1-54). Then much of the sound energy is dissipated by vibrating one surface and its mountings instead of being transmitted to the opposite surface. The performance improvement due to discontinuous construction or resilient mountings can be increased significantly by placing sound-absorbing insulation blankets within the wall or floor and ceiling assembly to absorb sound energy in the cavity, as shown in Figure 13.1-51.

**FIGURE 13.1-51** Staggered studs provide discontinuous construction by effectively separating opposite surfaces. Adding insulation improves performance.

**FIGURE 13.1-53** Sound-deadening board reduces transmission by resilient mounting of surfaces and can be used with single-, double-, or staggered-stud walls.

**FIGURE 13.1-52** A double row of studs, each row on separate plates, provides discontinuous construction. Adding insulation improves performance, and performance increases as insulation thickness is increased.

**FIGURE 13.1-54** Resilient channels (on one side of the wall) flexibly mounted on the surface. The performance increases when insulation is added and continues to increase as the thickness of the insulation is increased.

Discontinuous construction can be obtained in floor and ceiling assemblies by floating the finish flooring and separating the ceiling construction from the floor construction. Attempting to isolate airborne sound transmission by separating opposite surfaces alone has limitations. Doubling the free air space between the opposite surfaces of a wall or floor and

ceiling assembly improves its resistance to sound transmission by only about 5 dB, assuming a constant mass. Within practical limits of conventional building materials, both mass and discontinuous construction are required in order for a wall to reach a 50 to 55 dB STC. STC values greater than that are not practical, regardless of mass and isolation, without construction methods that will control transmission through flanking paths.

Presuming that walls and floor and ceiling assemblies that are selected on the basis of test data will actually perform according to their ratings, the greatest single problem in achieving satisfactory sound conditioning is improper installation. Only when every detail has been correctly taken care of will the actual walls and floor and ceiling assemblies live up to the ratings of the tested assemblies. When possible, workers and other key personnel should be taught the reasons for care in building and applying sound-control measures. Supervision to ensure that the proper methods and techniques are used, and coordination between the various trades, are essential. Seemingly insignificant variations from the correct technique can nullify the effectiveness of costly sound control measures. For example, nails that are too long can short-circuit resilient mountings and make them useless. Failing to seal small cracks and holes will create a flanking path and negate the effectiveness of a wall or floor and ceiling assembly. The manufacturer's instructions for installing sound-control devices, such as resilient clips and sound insulation, must be followed precisely. The actual walls and assemblies must be built exactly as were the tested ones.

It is normal to construct walls and floor and ceiling assemblies as continuous surfaces to divide living units into rooms. Windows, doors, and other openings in such surfaces are required. These horizontal and vertical surfaces and their openings, as well as construction imperfections such as cracks and holes, create flanking paths that transmit sound. Just as snapping a rope will cause it to snake the entire length, sound impulses in a wall or floor and ceiling assembly create vibrations that may transmit sound around, over, or under a partition even though the partition itself has high resistance to transmission (Fig. 13.1-55).

Open-air flanking paths are openings through which sound travels virtually undiminished. A doorway or window is an enormous hole through a wall. It is an ideal sound leak and, unless properly sealed, can render a wall almost acoustically ineffective.

To reduce the number of flanking paths, doors and windows should be airtight, be heavy enough to minimize transmission through them, and be installed to maintain their airtightness.

CONCRETE FLOOR

a

WOOD FLOOR

b

**FIGURE 13.1-55** Sound can be transmitted around, over, and under partitions through (a) structure-borne flanking paths and (b) airborne flanking paths.

Flanking paths through construction imperfections are illustrated in Figure 13.1-20. Walls, ceilings, and floors are pierced by pipes, ducts, conduits, lighting fixtures, electrical boxes, and similar equipment. Because airborne sound will travel wherever there is an air passage, they all constitute potential sound leaks. Often they are inadvertently arranged to become ideal "speaking tubes." Air conditioning and ventilating ducts are frequently offenders of this type. Medicine cabinets or outlet boxes placed back-to-back are common sites of sound leaks. Such air leaks must be closed, openings

caulked or sealed, fixtures staggered, and ducts lined with absorbent materials or equipped with baffles that interrupt the direct path of the sound and assist in absorbing it.

Airborne sound transmission cannot be controlled satisfactorily solely by acoustical surfacing materials. Covering the surfaces of walls or ceilings with acoustical materials, or a floor with carpeting, adds little to the STC of the walls or floor and ceiling assemblies. It is true that the amount of sound available to be transmitted will be reduced somewhat by being absorbed by acoustically treated surfaces, but that reduction is almost never enough.

However, flexible insulation placed in the free air space of frame walls and floor and ceiling assemblies improves performance by absorbing sound within the cavity, and can result in as much as a 10 dB improvement when the surfaces are effectively separated.

## WALL SYSTEMS

Sound-isolating wall systems may be constructed of wood or metal framing, concrete, or unit masonry. Regardless of the system selected, it is essential that its surfacing materials be installed over the entire wall area, including areas behind soffits, cabinets, and duct enclosures and above suspended ceilings. This precaution is particularly critical for common walls between units to ensure that sound will not be transmitted through an unfinished space (Fig. 13.1-56). Resilient, nonhardening caulking must be used at the floor and ceiling in partitions, where partitions abut other walls and partitions,

CARRY WALL & CEILING FINISH BEHIND DUCTS & SOFFITS

SOFFIT

DUCT

KITCHEN CABINET

COMMON PARTY WALL

**FIGURE 13.1-56** To prevent sound transmission through unfinished spaces, wall and ceiling finishes should cover all common wall and ceiling areas, including those inside furred areas and above ceilings and those concealed by stairs or other construction.

and around electrical outlets and switches, plumbing pipes, and other penetrations to make the wall airtight (Fig. 13.1-57).

Most sound transmission test data for wood and metal frame construction have been obtained with gypsum board as the surfacing material because of its extensive use and ease of installation in the laboratory, without the necessity of drying and curing. The following discussion is based primarily on the evaluation of such testing, but the principles and conclusions are applicable to lath and plaster and other surfacing materials of comparable weight applied to similar construction. Representative STC ratings for lath and plaster, as well as other typical constructions, are given in Section 13.1.5.

***Wood Framing Systems*** Conventional single-stud walls finished with one layer of gypsum board on each side (Fig. 13.1-58) have an STC of about 34, which represents poor resistance to sound transmission.

Sound-absorbing blankets placed in the stud cavity will not appreciably increase the sound isolation performance of this kind of wall. Either the weight of the covering material must be doubled using multiple layers of gypsum board (Fig. 13.1-59) or the two wall surfaces must be mechanically separated to achieve an STC rating of more than 40. A summary of the acoustical performance of wood-framed walls is given in Figure 13.1-60.

Resilient mounting of the surface finish of a single-stud wall with resilient channels on one side only (see Fig. 13.1-54) will increase the performance of

the wall about 6 dB. Adding channels to the other side will not further improve performance because of the resiliency of the wall through sympathetic vibration of both surfaces.

Installing insulation in a resiliently mounted wall increases its acoustical performance in proportion to the thickness of the insulation and can result in as much as a 10 dB improvement. Noise isolation performance depends on the surfacing materials and how they are installed. Backer strips for attaching baseboards or horizontal blocking at the top and middle height for firestopping decreases the performance. Resilient channels must be handled and installed carefully to avoid permanent deflection. The wall should contain no openings, and heavy objects such as fixtures and cabinets (which would destroy resiliency) should not be mounted on the resilient channel side of the wall.

To obtain an STC of 50 or more, without regard for fire ratings, channels must be used throughout, instead of gypsum backing strips, to totally float the surfacing materials. Resilient sealants must also be installed around the entire perimeter and around every penetration.

A wall with resilient channels on one side generally results in the thinnest possible sound-isolating system in wood frame construction and can be constructed with proper blocking to give a 1-hour fire rating.

Sound-deadening board applied to both sides of a single-stud wall provide resilient mounting if the surfacing materials are correctly laminated to the sound-deadening board (see Fig. 13.1-53) and

**FIGURE 13.1-58** Conventional single-stud walls permit loud speech and other noises on one side to be heard clearly on the other.

**FIGURE 13.1-59** An additional layer of gypsum board on each side slightly improves the performance of walls like the one shown in Figure 13.1-58.

can result in a wall with an STC of about 46. However, if the surfacing materials are nailed in order to meet fire rating requirements, the STC will be reduced about 10 dB when nails are used at spacings of 8 in. (203.2 mm) on center, 8 dB with nails spaced 16 in. (406.4 mm) on center, and 4 dB with nails spaced 24 in. (609.6 mm) on center.

Staggered-stud walls, consisting of 2 × 4 (50 × 100) studs staggered on a single 2 × 6 (50 × 150) plate (see Fig. 13.1-51), with one layer of gypsum board

**FIGURE 13.1-57** To prevent sound transmission through them, openings in sound-insulated walls, floors, and ceilings must be sealed with nonhardening, resilient sealants or a resilient material.

**FIGURE 13.1-60** Acoustical Performance of Typical Wood-Framed Walls[a]

| Wall Construction | Approximate STC Rating | | |
| --- | --- | --- | --- |
| | Without Insulation | With Insulation[b] | Sound-Deadening Board Both Sides[c] |
| Single-stud wall | | | |
|    Single-layer surface | 34 | 36 | 46 |
|    Double-layer surfaces | 40 | — | — |
| Single-stud wall with resilient channel mounting one side | 40 | 50[d] | — |
| Staggered-stud wall | 41 | 49 | 50 |
| Double-stud wall | 43 | 53 | 50 |

[a]As illustrated in Figures 13.1-51, 13.1-52, 13.1-53, 13.1-54, 13.1-58, and 13.1-59.

[b]Minimum thickness of insulation (2 or 3 in.) (50 or 75 mm).

[c]Gypsum board surfaces laminated, with exception of double-stud wall, which was laminated and nailed 8 in. (200 mm) o.c. into studs and floor and ceiling plates.

[d]Resilient channels installed throughout; fire-rated wall with firestopping top and middle heights or backer strips for attaching wall base, STC 46.

on each side, have an STC of about 42. The addition of insulation in the stud cavity or sound-deadening board to both sides increases performance of the wall to a range of 45 to 49 STC. The STC rating does not increase appreciably as the thickness of the insulation is increased, because sound will travel through the single plate. Therefore, 2 or 3 in. (50.8 or 76.2 mm) of mineral fiber blanket insulation is all that is necessary.

One advantage of using staggered-stud construction is that cabinets can be surface-mounted on the wall framing without impairing the acoustical effectiveness of the wall, because they are mounted on framing members that do not contact both sides of the wall.

For maximum performance, a wall base should not be nailed to the bottom plate of a staggered-stud wall.

A double-stud wall consists of two separate 2 × 4 (50 × 100) stud walls on two separate 2 × 4 (50 × 100) plates spaced 1/4 to 1 in. (19 to 25.4 mm) apart (Fig. 13.1-52). This kind of wall has an STC of about 43, which is 9 dB higher than a conventional single-stud wall. Adding sound-deadening board to both sides will increase the rating to about 50. The isolation performance of a double-stud wall improves when the thickness of insulation in the cavity is increased. Ratings as high as 59 STC can be achieved.

A double-stud wall can accommodate surface-mounted cabinets without reduced effectiveness. It also has fewer performance variations due to bad materials or workmanship. However, to perform acoustically, the separate walls should not be tied together in any way when plumbing or wiring is installed.

***Metal Framing Systems*** Because metal studs are themselves resilient, the two surfaces of a wall built with them vibrate independently of each other and may be considered resiliently mounted. Therefore, the acoustical performance of non-load-bearing metal stud walls differs from that of wood stud walls. For example, the addition of resilient channels, which can be effective with wood stud walls, does not appreciably improve the performance of metal stud walls.

When resistance to sound transmission is required in metal-framed walls, the measures to be used should be selected on the basis of individual test data for the system and should be installed in accordance with specified procedures.

The precautions recommended for resiliently mounted wood-framed walls also apply to metal-framed walls. For maximum performance, the wall should contain no openings, and heavy objects such as cabinets or fixtures should not be mounted on either side of the wall.

The difference between the acoustical performance of walls constructed with 1 5/8- and 2 1/2-in. (41.3- and 63.5-mm)-deep metal studs, or using 1/2- or 5/8-in. (12.7- or 15.9-mm)-thick gypsum board, is negligible. A summary of the acoustical performance of metal-channel-stud walls is given in Figure 13.1-61. The effects of adding gypsum board, insulation, and sound-deadening board in metal-channel-stud construction are cumulative.

Single metal-channel-stud walls with single-layer surfaces on both sides have an STC of about 37. Adding 2 or 3 in. (50.8 or 76.2 mm) of insulation in the cavity (Fig. 13.1-62) will improve the acoustical performance to an STC of about 46. Adding one layer of mineral fiber sound-deadening board to one side of the wall will improve the STC to 43 without insulation within the cavity and to 49 with insulation. Adding sound-deadening board to both sides, and insu-

**FIGURE 13.1-61** Acoustical Performance of Typical Metal-Channel-Stud Walls[a]

| Gypsum Board Surfacing | Approximate STC Rating | | | | | |
| --- | --- | --- | --- | --- | --- | --- |
| | | | Sound-Deadening Board One Side | | Sound-Deadening Board Two Sides | |
| | Without Insulation | With Insulation[b] | Without Insulation | With Insulation[b] | Without Insulation | With Insulation[b] |
| Single layer each side | 38 | 45 | 43 | 49 | 47 | 54 |
| Double layer one side, single layer one side | 44 | 50 | 46 | 51 | 52 | 54 |
| Double layer each side | 49 | 57 | 53 | 57 | 54 | 57 |

[a]As illustrated in Figures 13.1-62, 13.1-63, and 13.1-64.

[b]Minimum thickness of insulation (2 or 3 in.) (50 or 75 mm).

**FIGURE 13.1-62** A metal-channel-stud wall acts as a resilient system without the need for resilient mountings.

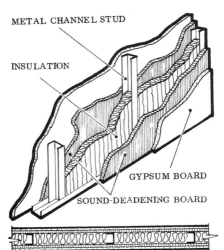

**FIGURE 13.1-63** The sound-isolating properties of metal-channel-stud walls are measurably improved when insulation, sound-deadening board, or additional layers of gypsum board are added.

**FIGURE 13.1-64** When the cavity depth or insulation thickness is increased and sound-deadening board is used, a metal-channel-stud wall can achieve excellent resistance to sound transmission.

lation in the cavity (see Fig. 13.1-59), can produce an STC of 52 to 54.

Adding layers of surfacing material to either one or both sides will increase the STC rating of a wall by 5 or 6 dB for each layer added. Adding insulation to a wall with two layers of gypsum board on one side will improve performance an additional 6 to 8 dB. A layer of sound-deadening board will increase the STC by about 5 dB, which is about the same as adding a single layer of gypsum board.

The effects of adding gypsum board, insulation, and sound-deadening board in metal-stud construction are cumulative (Fig. 13.1-63). However, the improvement is small when sound-deadening board is added to both sides of a wall that already has two layers of gypsum board on both sides. This is probably because the added mass of the various layers of finish begins to overshadow the resiliency of the studs.

Double-channel-stud walls consisting of $1^5/_8$-in. (41.3-mm)-deep studs tied together with bracing ties (Fig. 13.1-64) to accommodate plumbing, wiring, and other systems will achieve a rating of 42 STC when surfaced with one layer of gypsum board on each side. This rating can be improved to 50 by adding a layer of sound-deadening board to both sides and to 52 by adding 2 to 3 in. (50.8 to 76.2 mm) of insulation in the cavity. Double-layer gypsum board and sound-deadening board added to both sides results in an STC of 53. Increasing the cavity to 12 in. (304.8 mm) and adding three layers of insulation (9 in.) (228.6 mm) to

a wall with single-layer gypsum board on each side can result in an STC of 55; another layer of gypsum board added to one or both sides of this construction can result in a 59 or 60 STC.

***Concrete and Unit Masonry Walls*** The sound-isolating performance of walls with sealed surface pores is closely correlated with the weight of the wall, and

the performance of such walls parallels what is known as the *limp mass law relationship*. The mass law relationship holds true only for limp homogeneous partitions that are nonporous and have uniform physical properties throughout the entire wall panel. This relationship between transmission loss and mass can be expressed by the formula

$$TL = 20 \log m + 20 \log f - 33$$

where *m* is the mass of the wall stated in lb/sq ft and *f* is the frequency in cps

Heavy walls tend to reduce the transmission of sound more than light walls, but doubling the mass of a wall increases the transmission loss only 6 dB. As can be seen in Figure 13.1-65, to achieve a transmission loss of greater magnitude, the increase in mass must be comparatively large.

Nonporous concrete and masonry walls do not permit sound to pass through them, but rather transmit sound by acting as a diaphragm. In a porous wall, the pore structure is the principal transmission path, particularly at low frequencies. When a porous wall's surface is completely sealed with paint or plaster, transmission loss no longer depends on pore structure but instead is determined by the weight and stiffness of the wall. Therefore, the acoustical performance of a painted or plastered porous wall will be essentially the same as that of a nonporous wall of the same weight and stiffness.

It is possible to predict the STC of concrete and masonry walls that are nonporous to the extent that sound transmission through them is due to vibration rather than the passage of airborne sound. This prediction, which is based on the weight of the wall, can be reasonably accurate when there are no openings in the wall and all joints and cracks are sealed. Figure 13.1-65 shows that STC test values are consistently 5 or 6 dB below STC values calculated according to the mass law formula.

Concrete masonry units are produced in hollow and solid units manufactured from either lightweight or normal-weight aggregates such as sand and gravel (see Section 4.5). Normal-weight, or dense, aggregate units have a relatively tight pore structure, are considerably heavier than lightweight aggregate units, and will perform acoustically according to their mass.

Lightweight concrete blocks typically contain a larger number of pores (air

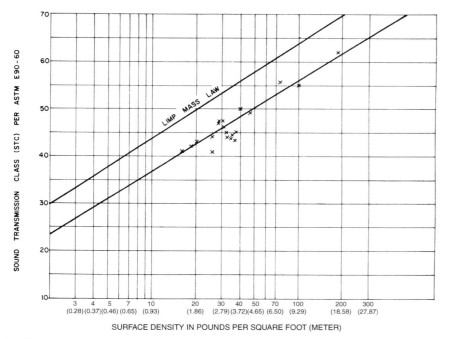

FIGURE 13.1-65 The performance of well-built, nonporous masonry walls can be predicted from their weight according to a mass law formula.

There is evidence that hollow units may be better sound insulators than solid units of the same thickness and made with the same aggregate type. A suspected reason is that because the core surfaces are denser than the remainder of the block due to the manufacturing process, a *skin effect* occurs, producing a greater than expected resistance to sound transmission.

Because clay masonry units are relatively dense and nonporous, the acoustical performance of tightly constructed walls made with them can be estimated according to their mass. Cavity wall construction will provide improved sound isolation with brick masonry, just as it does with concrete masonry.

Regardless of the material used to build it, the sound-isolating performance of a masonry wall will depend largely on the extent to which it is airtight. Even relatively fine cracks and open joints, such as may occur when a wall is poorly built or as occur at a wall's perimeter, will destroy the acoustical effectiveness of a masonry wall.

Because of the great number of mortar joints (and therefore possible acoustical leaks), it is recommended that brick masonry walls be sealed with plaster when acoustical privacy is desired.

spaces between aggregate particles and cement matrix) than dense units. For this reason, they excel in sound absorption but are not effective in resisting sound transmission. Lightweight blocks are subject to a wide variation in transmission loss when not sealed because of large differences in the pore structure of blocks made by different manufacturers. Completely sealing their surface pores increases the performance of lightweight units about 7 dB when painted and about 10 dB when plastered (Fig. 13.1-66). Conversely, the increase from painting or plastering dense units is only about 2 dB. But because of their greater mass, dense units perform considerably better than lightweight units in the first place.

Cavity wall construction and combinations of dense and porous units improve acoustical performance and can provide STCs of 50 or better (see Fig. 13.1-66). Using a combination of two 4 in. (101.6 mm) lightweight block wythes with a $^3/_8$ in. (9.5 mm) or wider air space between them, and parging one of the wythes on the cavity side, eliminates the need to seal the faces of the units, thereby retaining intact the sound-absorbing porosity of lightweight units.

Filling the cores of hollow masonry units with sand or another aggregate will increase the acoustical performance of a well-built wall by only 1 or 2 dB. Filling

the cores in a poorly built wall will improve its performance more, but it will never be as good as a properly constructed wall.

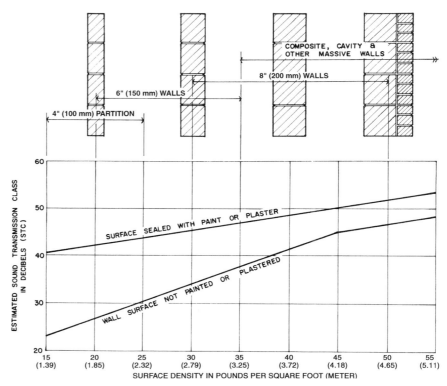

FIGURE 13.1-66 Estimated STC values for sealed and porous concrete masonry walls according to their weight in pounds per square foot.

Concrete and masonry wall surfaces that need to be sealed should be completely sealed over their entire surfaces, including areas behind cabinets and surfacing materials, in closets, in furred spaces, and above ceilings, to eliminate flanking paths through these areas. Porous masonry walls that are relied on to act as a sound barrier should be plastered or painted even when they are to be furred and surfaced with gypsum board or paneling as the interior finish. A sheet of polyethylene film over a masonry wall does not seal the surface acoustically.

There are often openings at the top and ends of a masonry wall that allow sound to pass to the other side of the wall. Sometimes such openings will occur even at the bottom of a masonry wall. Such openings must be closed with acoustical sealant, resilient sound barriers, or lead sheeting.

## FLOOR AND CEILING SYSTEMS

Airborne sound transmission is a consideration in selecting floor and ceiling assemblies, but structure-borne (impact) sound transmission is the major problem. Most conventional wood, concrete, and metal floor and ceiling assemblies provide negligible resistance to impact sound transmission. Therefore, some sound-isolating provisions must be added when acoustical privacy is required. Most measures that will control impact sound will also provide a good degree of resistance to airborne sound, provided that essential construction practices are followed.

Measures used to control structure-borne sound include (1) cushioning impact with soft floor-surfacing materials, (2) providing discontinuous construction by floating the finish flooring over the structural framing or subflooring, (3) providing discontinuous construction between the structural floor and the finished ceiling by mounting the ceiling on resilient devices or on separate ceiling joists, and (4) installing sound-absorbing insulation in the cavity of frame construction. Singly and in combination, these construction measures reduce im-

pact sound transmission by breaking the direct structural connections that could carry vibration into other rooms and by absorbing some of the sound within cavities.

Impact cushioning and floated floors are effective with any type of structural floor system. Discontinuous ceilings generally are used only with wood- or metal-framed construction or for critical areas of reinforced concrete, such as ceilings under equipment rooms and service areas. Installing sound-absorbing materials in the cavity of frame construction is significantly effective only when a resiliently mounted ceiling is also installed.

There is little difference in the acoustical performance of the various hard floor surfacing materials such as wood strip, wood block, and resilient tile and sheet flooring. IICs for representative floor and ceiling assemblies are given in Section 13.1.5.

***Cushioning the Impact*** The most effective method for controlling impact

**FIGURE 13.1-67** Effect of Carpet in Controlling Impact Sound[a]

| Carpet[b] | Pad | Floor/Ceiling Construction | |
| | | Concrete Slab[c] IIC | Wood Joist[d] IIC[e] |
| --- | --- | --- | --- |
| 20 oz (678.11 g/m²) wool | (none) | 53 | — |
| 40 oz (1356.23 g/m²) wool | (none) | 57 | — |
| 60 oz (2034.34 g/m²) wool | (none) | 65 | — |
| 44 oz (1491.85 g/m²) wool | 44 oz (1491.85 g/m²) sponge rubber bonded to carpet | 68 | 54 |
| 20 oz (678.11 g/m²) wool | 40 oz (1356.23 g/m²) felt | 69 | — |
| 44 oz (1491.85 g/m²) wool | 40 oz (1356.23 g/m²) hair and jute covered with foam-rubber face and back | 70 | 59 |
| 40 oz (1356.23 g/m²) wool | 40 oz (1356.23 g/m²) | 72 | — |
| 44 oz (1491.85 g/m²) wool | 40 oz (1356.23 g/m²) all hair | 72 | 59 |
| 44 oz (1491.85 g/m²) wool | 66 oz (2237.78 g/m²) | 72 | 63 |
| 60 oz (2034.34 g/m²) wool | 40 oz (1356.23 g/m²) felt | 73 | — |
| 44 oz (1491.85 g/m²) wool | 40 oz (1356.23 g/m²) hair and jute | 73 | 61 |
| 44 oz (1491.85 g/m²) wool | 20 oz (678.11 g/m²) combination hair and jute, foam-rubber face, latex back | 73 | 61 |
| 44 oz (1491.85 g/m²) wool | Urethane foam | 75 | — |
| 44 oz (1491.85 g/m²) wool | 40 oz (1356.23 g/m²) sponge rubber | 76 | 65 |
| 44 oz (1491.85 g/m²) wool | 31 oz (1051.08 g/m²) foam rubber on burlap (⅜ in. [9.53 mm] thick) | 79 | 67 |
| 44 oz (1491.85 g/m²) wool | 80 oz (2712.46 g/m²) sponge rubber | 80 | 68 |

[a]Based on tests performed by Kodaras Acoustical Laboratories, Inc. for the American Carpet Institute. IIC values approximated by adding 51 to original INR test values.

[b]Carpet pile weight given in oz/sq yd (g/m²).

[c]5 in. (127 mm) concrete slab (sand-gravel aggregate), no ceiling finish.

[d]2 × 8 (50 × 200) wood joists, 16 in. (400 mm) o.c.; ⅝ in. (15.88 mm) tongue-and-grroove plywood nailed to joists as subfloor. ⅝ in. (15.88 mm) gypsum board nailed to joists as ceiling.

[e]Values given only where test data are comparable to tests performed on concrete slab construction.

FIGURE 13.1-68 Conventional wood-joist floor construction provides poor resistance to impact sound transmission.

FIGURE 13.1-69 Resilient channels separate the finished ceiling from the structural framing and reduce both airborne and structure-borne sound transmission.

sound transmission is to cushion the impact with soft, resilient floor finish materials, such as carpet and padding. By dissipating a substantial amount of impact energy as it strikes the surface, the amount of sound transmitted to the supporting structural system is substantially reduced. The IIC of the carpet and pad combination varies with the thickness of the carpet, the degree of matting of the pile, and the weight of both carpet and pad in ounces per square yard (grams per square meter) (Fig. 13.1-67). Unfortunately, there is no appreciable gain in airborne sound isolation by carpet and padding or by acoustical surfacing materials adhered to the ceiling below. Therefore, when cushioning is the only measure to be used to control impact sound transmission, the basic floor and ceiling assembly should be designed to control airborne sound transmission.

Cushion-backed vinyl sheet flooring can be effective in controlling impact sound transmission when carpet is not desired, as might be the case in kitchens, laboratories, toilets, and corridors, for example. Test data for a representative cushion-backed vinyl sheet product show that an IIC of 54 was achieved when this material was installed on a 6-in. (152.4-mm)-thick precast hollow concrete slab. When the ceiling beneath the slab was resiliently mounted, an IIC of 59 was recorded. Adding a 2-in. (50.8-mm)-thick concrete topping to the slab increased these values by 4 dB.

*Wood Frame Construction* A conventional wood joist floor and ceiling assembly, finished with hard-surfaced flooring and with a gypsum board ceiling attached directly to the joists (Fig. 13.1-68), will have an IIC of approximately 26. Adding plywood or other subflooring will not appreciably affect this rating. Where an IIC of 51 or better is required, one of the following methods should be employed, depending on the degree of sound isolation required: (1) apply carpet and padding, either alone or in combination with a floated floor or separated ceiling; (2) use separated ceiling construction, by resiliently mounting either the ceiling material or using separate ceiling joists, in conjunction with a floated floor.

The previously mentioned methods for reducing impact sound transmission can also be used in wood-framed construction. As with other floor and ceiling assemblies, cushioning the impact with carpet and padding of sufficient thickness and density is the most effective method. Carpet and padding used alone or with any of the following tested sound-isolating systems will result in IICs ranging from 42 to 85.

Separating a finished ceiling from an associated structural floor system can be done either by applying the finished ceiling with resilient channels (Fig. 13.1-69) or by attaching the finished ceiling to separate ceiling joists (Fig. 13.1-70). Either of these discontinuous construction methods can effectively reduce both impact sound from above and airborne sound from either direction. As is true of every sound control measure, separating a ceiling from a structural system will not produce an effective acoustical barrier unless the entire floor and ceiling assembly is built correctly, with careful attention to the blocking of flanking paths, and measures are taken to prevent sound from bypassing the ceiling through walls and other structural connections. As with re-

FIGURE 13.1-70 The impact-sound-isolating performance of floor and ceiling assemblies like that shown in Figure 13.1-68 can be improved by either (a) resilient mounting of the finished ceiling or (b) carrying the finished ceiling on independent ceiling joists.

siliently mounted walls, heavy objects, such as lighting fixtures, should not be mounted on or recessed in the ceiling, and airtight construction should be maintained.

Resiliently mounting a finished ceiling or attaching it to separate joists can improve the IIC of an assembly as much as 7 dB; sound-absorbing material in the cavity can provide an additional 7 dB. But separated ceiling construction alone will not produce a floor and ceiling assembly with much more than 45 IIC when the floor above is hard-surfaced, even though an acceptable STC of 50 is obtained. However, if a floated floor is used instead, even when the flooring has a hard surface, the IIC can jump to 57 and the STC will still be 50 or more.

Floated-floor construction over wood joists and subflooring can be accomplished with a resilient underlayment separating the finished flooring and underlayment from the structural floor system (Fig. 13.1-71a) or by placing lightweight (cellular) concrete over the resilient underlayment (Fig. 13.1-71b). Section 13.1.5 shows other possible methods. When resilient underlayment is used (Fig. 13.1-72), the resilient material must be capable of resisting major deformation from superimposed loading. Most such material is designed to carry only the relatively light loading of the finished floor, people, and furniture.

In general, the greater the resilience of the underlayment material, the greater the isolation. Like carpeting, resilient underlayment works primarily as a cushion and only secondarily to resist airborne sound transmission through the assembly.

The performance of a floated floor depends on effectively separating the finished floor from the structural elements. The floating part should not be rigidly connected to other elements. Adhesive generally is used to attach resilient underlayment. Nailing or stapling the subflooring or finished flooring through the resilient underlayment into the supporting construction will nullify the sound-isolating properties of the floated floor.

A basic wood-joist construction can be improved approximately 7 dB with a floated-floor system using resilient underlayment. To achieve a rating of more than 51 IIC, the floated floor can be supplemented by a separated finished ceiling. With the addition of sound-absorbing material in the joist space (Fig. 13.1-73), an IIC of 57 can be achieved. The addition of a mineral-fiber insulating blanket in the space between the floor joists absorbs approximately 7 dB of the transmitted noise. Unfortunately, the addition of sound-absorbing materials within the free air space of a framed floor and ceiling assembly is significantly effective in controlling impact sound transmission only when used with a resiliently

**FIGURE 13.1-72** Resilient underlayment (shaded) reduces impact sound transmission by floating the finished flooring above the subfloor or structural floor.

**FIGURE 13.1-73** A resiliently mounted ceiling combined with a floated floor effectively resists impact sound transmission.

mounted ceiling. However, it is effective in reducing airborne sound with all types of discontinuous construction.

Lightweight concrete placed directly over a plywood subfloor does not appreciably improve the impact sound resistance of a basic wood-joist floor and ceiling assembly. When the concrete is resiliently floated, however, the improvement is about 10 dB, but the IIC will still be less than 51 unless carpet and padding are added or the ceiling is resiliently mounted.

**a**   **b**

**FIGURE 13.1-71** Two methods of making a floated floor. In (a) a resilient underlayment separates the finished flooring from the structural floor. In (b) lightweight cellular concrete has been placed over a resilient underlayment to form a floating floor system.

*Concrete Slab Construction* A plain concrete slab with resilient tile flooring and a plaster ceiling applied directly to it (Fig. 13.1-74) will have an IIC of about 36. Increasing the weight or thickness of the slab will have little effect on the impact sound transmission loss produced by it. Sound-control measures that will improve the acoustical performance of a concrete slab are limited to floating the flooring and cushioning impacts with carpet and padding. Because doing so is impracticable and costly, the ceiling under a concrete slab is seldom resiliently mounted and test data for that construction are limited.

Adding carpet and padding to a concrete slab provides somewhat better performance than does adding carpet and padding on a wood frame construction, with IICs ranging from 53 to 88 or even more, depending on the weight and density of the carpet and padding (see Fig. 13.1-67).

Floating to isolate a finished floor from a concrete slab is accomplished in much the same manner as in a wood frame (Fig. 13.1-75). Provisions should be made to keep the floated floor from rigid contact with structural elements. When possible, floating floors should not touch partition edges. One way to prevent such contact is to install a cushion between the edge of the partition and the floor.

A hard-surfaced floor floated on a resilient underlayment over a 6-in. (150-mm)-thick concrete slab with a directly applied plaster ceiling will have an IIC of about 55.

The performance of precast concrete floor and ceiling assemblies is similar to

**FIGURE 13.1-75** Floating the finished flooring over resilient underlayment considerably improves the acoustical performance of a concrete slab assembly like that shown in Figure 13.1-74.

that of cast-in-place concrete assemblies, but available test data on such assemblies are limited. Data for representative precast floor systems are given in Section 13.1.5.

Steel bar joist and other metal-framed floor systems can be improved by measures similar to those used to improve wood-framed assemblies. Methods for floating finished flooring and resiliently mounting finished ceilings closely parallel those used in wood-framed assemblies. The acoustical effectiveness of steel bar joist construction, as well as that of metal decks, concrete floor joists, and similar systems, depends primarily on airtight sealing of wall and floor intersections. Walls must be extended to the bottom of the slab or deck above, and the space at their tops must be made airtight.

When walls run perpendicular to joists, proper sealing is difficult and expensive.

Concrete slabs over steel bar joists are normally about 2 in. (50.8 mm) thick, which gives them sufficient mass to produce an STC only in the high 40s. Therefore, when a higher STC is required for either walls or floor and ceilings assemblies, this type of construction is not effective. Even when a wall itself has a high STC, sound will bypass it through the flanking path provided by a relatively lightweight floor system.

Current test data for evaluating metal frame floor and ceiling assemblies are not as extensive as data for other construction systems. Available test data for representative steel bar joist systems are given in Section 13.1.5.

## WALL AND FLOOR INTERSECTIONS

Noise problems from flanking paths are encountered where walls and floors intersect.

Unless precautions are taken, both airborne and structure-borne sound can penetrate around and through the walls and floors, even though the wall and floor constructions are structurally adequate.

Walls and associated floor and ceiling assemblies should have comparable sound-isolating abilities so that a weakness in one will not diminish the effectiveness of the other.

The spaces between floor joists should be blocked where walls and floor and ceiling assemblies join to reduce airborne sound transmission through such flanking paths (Fig. 13.1-76). Joints at the top and bottom of partitions should be sealed airtight with a resilient, nonhardening

**FIGURE 13.1-74** A typical concrete slab with hard-surfaced flooring and a directly applied plaster ceiling offers poor resistance to impact sound transmission.

**FIGURE 13.1-76** Openings must be closed to eliminate airborne sound flanking paths. The spaces between joists must be blocked, especially when perpendicular to walls.

acoustical sealant or a resilient material (Fig. 13.1-77).

Loads applied to a floated floor by a structural wall will cause the resilient underlayment to collapse, damaging the floor and maybe even causing the structure to settle. An underlayment material that is sufficiently rigid to support structural loading is too rigid to provide resilient isolation in a floated floor system. Therefore, the bottom plate of a bearing partition should not be supported on a floating floor, but rather should sit directly on and be fastened to the subfloor or supporting structure.

A break in the subflooring under partitions will help to reduce impact noise transmission between adjacent rooms by breaking direct contact of the structural elements (Fig. 13.1-78a). When possible, a cushion should be provided between a partition and the edge of a floating floor so that they do not touch (Fig. 13.1-78b).

Floated concrete floors on wood joist construction should be confined to individual rooms to prevent transmission of impact noise along the floor (see Figs. 13.1-78b and 13.1-79). Isolation joints around the perimeter of rooms and across doorways are essential. Before the concrete is placed, walls should be supported on a floor plate, polyethylene film should be installed over the resilient underlayment, and isolation joints should be installed around the perimeter of each room. The concrete should not be directly in contact with the walls. Moldings and quarter round trim can be used to cover isolation joints.

Wall surfacing material should not be tightly forced against resiliently mounted ceilings. Doing so may cause resilient hangers, channels, or clips to collapse and short-circuit the resilient mounting.

### STAIRS

Stair stringers should be fastened only at the top and bottom and should not be nailed to adjacent studs. Header joists should be used at landings to support floors instead of nailing the floor joists to the studs (Fig. 13.1-80).

When continuous wall surfacing material is installed on the entire wall adjacent to a stairwell before the stairs are installed, sound transmission between the stair and adjacent rooms will be reduced.

When stairs are on both sides of a common wall, such as in townhouses, duplexes, or other attached housing units, staggered- or double-stud walls with in-

**FIGURE 13.1-77** Acoustical sealant should be used to seal small, open air paths. This figure shows the sealing of paths at the bottom of a wall. Paths at the top and ends of sound-isolating walls should be similarly sealed.

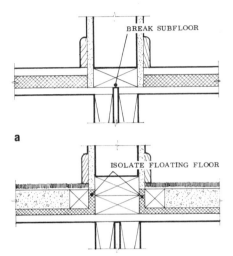

**FIGURE 13.1-78** Flanking structure-borne noise can be reduced by breaking direct contact between structural elements, such as the subflooring shown in (a), and by isolating floated floors from walls (b).

**FIGURE 13.1-79** A floated floor should be confined to a single room by separating it from adjacent construction at doors and walls.

**FIGURE 13.1-80** Stair stringers should not be nailed or otherwise fastened to adjacent walls. If they are, footfalls will be transmitted into the walls.

sulation in the cavity should be used. Stair runners on the wall side should be held away from the studs to accommodate later installation of a finishing material behind the runners. If the walls are to be plastered, temporary construction ladders, such as those used in fire-resistant construction, should be used to permit complete plastering and to eliminate breaks in the wall.

Finishing them with carpet or a resilient material, such as a rubber mat, will substantially reduce impact noise transmission from stairs to adjacent areas.

### CORRIDORS

Air ducts often are installed before ceiling and wall finishes that will be behind them are in place, making it difficult or impossible to install these finishes later. This permits impact noise from a corridor to pass through the ceiling and down the open stud spaces in the walls of adjoining spaces. This problem can be solved by installing only register outlets and the first duct section in a corridor, thus deferring the balance of the duct installation until the interior finishes of the walls and ceilings have been completed. Ducts then can be installed and a furred ceiling hung below them (Fig. 13.1-81).

### 13.1.4.3 Selection Criteria

The following criteria are suggested for selecting walls and floor and ceiling assemblies to provide resistance to airborne and structure-borne sound transmission. Generally, acceptable overall airborne sound privacy is said to exist when the level of conversational speech is reduced by a wall or floor and ceiling assembly

**FIGURE 13.1-81** Unless finishes are placed above them, furred spaces become sound transmission paths. Ducts should be installed after the finishes have been placed.

to the point where it cannot be understood in an adjoining room. To estimate the minimum STC required to obtain privacy, the level of background noise on one side of a construction is subtracted from the level of sound that strikes the other side. Figure 13.1-26 illustrates the approximate effectiveness of walls and floor and ceilings assemblies with varying STC values at varying background noise levels.

General recommendations for common walls and floor and ceiling assemblies separating spaces in terms of STC ratings are given in Figure 13.1-82. Exact values for STC ratings should be determined on the basis of individual test data in accordance with prescribed procedures (see Section 13.1.3).

The criteria recommended by the FHA of HUD for noise control were often used in the past as a basis for the design and construction of multifamily housing units. HUD previously issued detailed standards for indoor noise levels, but HUD regulations no longer include such standards. HUD now sets a goal of 45 dB for all interior spaces, whether the source of the sound is outside or within the building. In terms of exterior sounds, HUD's assumption is that if the exterior $L_{DN}$ is restricted to 65 dB, a reduction to 45 dB is fairly easily achievable using normal construction materials and configurations.

The criteria used in this section are based on (1) the level of ambient background noise represented by urban, suburban, and rural locations and (2) the specific function of the wall or floor and ceiling assembly within the building.

Walls and floor and ceiling assemblies should have STC ratings equal to or greater than those shown in Figure 13.1-82. A floor and ceiling assembly that may provide adequate impact sound isolation may not provide suitable airborne sound isolation. The criteria for both airborne and impact sound isolation of floor and ceiling assemblies must be met.

Recommended basic criteria for airborne and impact sound isolation of common wall and floor and ceiling

**Figure 13.1-82** General Criteria for Airbourne and Impact Sound Isolation Between Building Spaces

| Sound Source Location | Receiving Location | Sound Transmission Class (STC) of Wall or Floor/Ceiling Assembly |
|---|---|---|
| **All Building Types** | | |
| Building exterior | Building interior | 35 to 60 depending on exterior sound level |
| Mechanical equipment space, elevator shaft, trash chute, garage, and similar areas | Occupied space | 54 to 65 in multifamily residences Obtain advice of acoustical consultant for other building types |
| **Multifamily Dwelling Units** | | |
| Dwelling unit | Another dwelling unit | 48 to 55 |
| Within a dwelling unit: | Within same dwelling unit | |
| Bedroom | Bedroom | 48 to 55 |
| Living room | Bedroom | 50 to 57 |
| Kitchen, dining, family, or recreation room | Bedroom | 52 to 58 |
| Bathroom | Bedroom | 52 to 59 |
| Corridor | Bedroom | 48 to 55 |
| Kitchen, dining, family, or recreation room | Living room | 48 to 55 |
| Bathroom | Living room | 50 to 57 |
| Corridor | Living room | 48 to 55 |
| Bathroom | Kitchen | 48 to 55 |
| Corridor | Kitchen | 48 to 55 |
| Bathroom | Bathroom | 46 to 52 |
| Corridor | Bathroom | 46 to 50 |
| **Educational Buildings** | | |
| Classroom or laboratory | Classroom or laboratory | 50 |
| Corridor | Classroom or laboratory | 50 |
| Multiple-purpose space | Classroom or laboratory | 60 |
| Multiple-purpose space | Music room | 60 |
| Music room | Music room or drama area | 55 |
| **Office Buildings** | | |
| Executive offices, conference rooms, and other areas requiring privacy | Other similar spaces and general office areas | 50 to 55 |
| General office spaces | Other similar spaces | 45 to 50 |
| General office spaces | Corridors and other public spaces | 40 to 45 |
| **Entertainment Facilities** | | |
| Theaters, lecture halls, recreation buildings, electronic media studios | Similar facilities | Obtain advice of acoustical consultant |

assemblies which separate dwelling units of equivalent function (see Fig. 13.1-82) are based on STC ratings derived from laboratory measurements. Standard test methods for field measurements have not yet been formally adopted.

Double walls are usually necessary to achieve adequate acoustical privacy. When sensitive areas are above or adjacent to noisy areas, airborne sound isolation is important, but impact sound isolation is not significant when structure-borne vibration is minimal. However, when mechanical equipment rooms are above sensitive areas, airborne sound isolation must be maintained, impact sound isolation becomes critical and specially designed sound-isolating construction is required to ensure quiet areas.

Placing dwelling units vertically or horizontally adjacent to commercial establishments such as restaurants, bars, community laundries, and similar businesses should be avoided when possible. If such situations cannot be avoided, the recommended airborne isolation criteria between business areas and sensitive living areas are STC 56 to STC 60, depending on the type and level of sound involved. In addition, when living areas are situated above such areas, adequate impact isolation criteria should be between STC 56 and STC 60.

However, if the business areas are above living areas, the impact isolation criteria values should be increased by at least 5 points.

If noise levels in mechanical equipment rooms or business areas exceed 100 dB, as measured using the C scale of a standard sound-level meter, the airborne-sound isolation criteria given above should be raised 5 points.

Townhouses and rowhouses in which the living unit occupies more than one story should be separated by a double-wall construction with an STC rating of 60 or greater. Suggested criteria for ratings between rooms in a given dwelling are the same as those listed in Figure 13.1-82. In addition, the floor and ceiling assemblies should have IIC

ratings that are at least numerically equivalent to or greater than the listed STC criteria.

Rooftop or indoor swimming pools, bowling alleys, ballrooms, tennis courts, and gymnasiums require constructions specifically designed in accordance with acoustical analysis.

## 13.1.5 SOUND PERFORMANCE RATINGS FOR SELECTED ASSEMBLIES

The following data and the accompanying figures include sound performance data for walls and floor and ceiling assemblies. Data on systems that provide poor sound isolation are included to indicate the degree of improvement required to render such systems sufficiently resistant to sound transmission.

Unless otherwise noted, the data presented in the figures are based on tests performed by nationally recognized acoustical laboratories in accordance with the test procedures in ASTM E90, "Method for Laboratory Measurement of Airborne-Sound Transmission Loss of Building Partitions" and International Organization for Standardization (ISO) Recommendation R-140, "Field and Laboratory Measurements of Airborne and Impact Sound Transmission." The data include those from tests performed by Cedar Knolls Acoustical Laboratories, Geiger and Hamme, Kodaras Acoustical Laboratories, the National Bureau of Standards (now the National Institute of Standards and Technology), National Gypsum Company, Ohio Research Corporation, Owens-Corning Fiberglas Sound Laboratory, Riverbank Acoustical Laboratories, and the IIT Research Institute. The STC, INR, and IIC ratings shown have been drawn from a variety of manufacturers' literature and testing laboratory reports. In some instances, only very old test results were available. In others, few correlating test results could be found. Therefore, some ratings have been estimated. Before relying on the data shown for the final design of an

actual project, it would be wise to ask the manufacturers of the materials or systems proposed to furnish the results of their latest sound performance tests.

When possible, a range of values is given to account for (1) variation in construction materials, (2) variation in application procedures, and (3) variation in testing facilities. The reproducibility of laboratory tests on different samples can range up to 5 dB at any one frequency. However, overall STC rating should be within 2 dB.

### 13.1.5.1 Wall Assemblies

Refer to Section 13.1.4 for a discussion of the parameters dictating the use of the various materials and construction types in the assemblies illustrated in Figure 13.1-83.

The STC ratings given for clay masonry walls have been interpolated from Figure 13.1-66.

### 13.1.5.2 Floor and Ceiling Assemblies

The use of plywood, fiberboard, hardboard, or another underlayment is generally determined by the type of finish flooring or other practical construction techniques. Except for resilient underlayment in floated floor systems, underlayments contribute insignificant impact sound isolation. Unless otherwise noted, the ranges given in Figure 13.1-84 for INR and IIC include assemblies both with and without underlayment. Refer to Section 13.1.4 for additional discussion.

Because floor and ceiling assemblies that control impact sound isolation will also provide a good degree of resistance to airborne sound transmission, much of the test data for floor and ceiling assemblies are given in terms of INR and IIC, only. However, the STC values for a particular assembly can be estimated, provided that the essential precautions of construction practices are followed. Where test data for STC ratings were not available, the STC values have been estimated.

**FIGURE 13.1-83** Wall Systems

## Wood Frame Load-Bearing Walls—Gypsum Board Finish[1]

### SINGLE-STUD WALLS

**Single-Layer Finish[2]**

STC 32 to 36

**Multilayer Finish[2]**

STC 38 to 41

### SINGLE-STUD WALLS

**Single-Layer Finish**

STC 40 to 42

With insulation[7]

STC 47 to 51[4]

**Sound-Deadening Board[5]**

STC 47 to 51[4]

### STAGGERED-STUD WALLS

**Resilient Finish[3]**

STC 38 to 42[4]

With insulation[8]

STC 46 to 48[4]

**Sound-Deadening Board[5]**

STC 46 to 50[4]

### DOUBLE-STUD WALLS

**Single-Layer Finish**

STC 42 to 44

With insulation[8]

STC 50 to 53[4]

**Sound-Deadening Board[5]**

STC 50 to 53[9]

With insulation[8]

STC 54 to 58

---

1. Ranges include values for $\frac{3}{8}$ in. (9.53 mm), $\frac{1}{2}$ in. (12.7 mm), and $\frac{5}{8}$ in. (15.33 mm) thicknesses of gypsum board finish.

2. Addition of insulation does not appreciably improve this construction; an increase of approximately 2 to 4 dB can be expected regardless of thickness or density of insulation.

3. Resilient channels or clips may be used one side only to totally float the finish; use of blocking for fire stopping and use of backer strips to attach wall base reduces performance approximately 4 to 5 dB.

4. Range includes values derived from ASTM E90-61T, ASTM E90-66T, and field test procedures.

5. Gypsum board finish laminated to sound-deadening board; if gypsum board is nailed or screwed to meet fire ratings, performance decreases approximately 10 dB for fasteners spaced 8 in. (200 mm) o.c., approximately 8 dB for fastener spacing 16 in. (400 mm) o.c., and approximately 4 dB with fastener spacing 24 in. (600 mm) o.c.

6. Ranges include values for 2 × 3 (50 × 75) studs staggered on single 2 × 4 (50 × 100) plate and 2 × 4 (50 × 100) double-stud walls on separate 2 × 3 (50 × 75) plates.

7. A minimum thickness of 2 to 3 in. (50 to 75 mm) of insulation can be used effectively as performance does not increase proportionately with increased thickness of insulation.

8. Ranges include 1$\frac{1}{2}$ in. (38 mm) to 3 in. (75 mm) thickness of insulation; performance increases with increased thickness. On double-stud walls, an STC of 58 to 60 can result with up to 9 in. (230 mm) of insulation.

9. Based on ASTM E90-61T test procedures.

*(Continues)*

**FIGURE 13.1-83** *(Continued)*

**Wood Frame Load-Bearing Walls—Plaster Finish[11]**

| SINGLE-STUD WALLS | | SINGLE-STUD WALLS | |

**Metal Lath & Plaster[2]**    **Resilient Gypsum Lath & Plaster[3]**    **Plaster Base, Veneer Plaster[2]**    **Resilient Plaster Base, Veneer Plaster[3]**

STC    41[10]

STC    43[9]

STC (est) 36
2 layer base 2 sides
STC (est) 42

With insulation[8]
STC 49 to 50[4]

| SINGLE-STUD WALLS | | SINGLE-STUD WALLS | |

**Gypsum Lath & Plaster[3]**    **Resilient Metal Lath & Plaster[3]**    **Resilient Plaster Base & Multilayer Base, Veneer Plaster[3]**    **Gypsum Lath & Plaster**

STC 41[10]

STC 45 to 47[10]
With insulation[8]
STC 50 to 54[4]

With insulation[8]
STC 53

STC 42 to 43[9]
With insulation[7]
STC 50[10]

10. Based on tests performed prior to ASTM E90-61T.

11. Test data can vary widely, depending on weight and thickness of plaster and water content (due to length of curing) of plaster.

12. With exception of double-stud walls, ranges include values for $\frac{1}{2}$ in. (12.7 mm) or $\frac{5}{8}$ in. (15.88 mm) gypsum board thickness of 3 in. (75 mm) or 4 in. (100 mm) studs.

13. Ranges include values for gypsum board screwed, or laminated and screwed, to sound-deadening board or base layer.

14. Unbalanced Construction: single-layer gypsum board one side, double-layer gypsum board one side.

15. Ranges include values for $\frac{1}{2}$ in. (12.7 mm) and $\frac{5}{8}$ in. (15.88 mm) board thickness on 2 in. (50 mm) studs.

16. Up to 60 STC with 9 in. (230 mm) of insulation; lower values result with lesser thicknesses of insulation.

17. Performance of concrete and unit masonry systems is highly dependent on minimizing the number of openings through the construction (to accommodate piping, wiring, and other equipment) and on all openings being sealed airtight.

18. Sealed both sides unless otherwise indicated. Sealing with plaster is most effective; when surfaces are to be painted, use of block sealer and/or 2 coats of cement-based paint is more effective than 2 coats of resin-emulsion paint to seal against airborne sound transmission.

**FIGURE 13.1-83** *(Continued)*

## Metal Frame Non-Load-Bearing Walls—Gypsum Board Finish[12]

### SINGLE-STUD WALLS

**Single-Layer Finish**

|  |  |
|---|---|
| | **STC 37 to 41**[4] |
| With insulation[8] | |
| | **STC 44 to 48** |

**Single-Layer Finish Sound-Deadening Board**[13]

| | | |
|---|---|---|
| SDB 1 side | STC | **43** |
| SDB 2 sides | STC | **46 to 50**[4] |
| With insulation[8] | | |
| SDB 1 side | STC | **47 to 50**[4] |
| SDB 2 sides | STC | **50 to 53**[4] |

### SINGLE-STUD WALLS

**Unbalanced Finish**[14]

|  |  |
|---|---|
| | **STC 44 to 47** |
| With insulation[8] | |
| | **STC 49 to 52** |

**Unbalanced Finish**[14] **Sound-Deadening Board**[13]

| | | |
|---|---|---|
| SDB 1 side | STC | **48**[9] |
| SDB 2 sides | STC | **52**[9] |
| With insulation[8] | | |
| SDB 1 side | STC | **49**[9] |
| SDB 2 sides | STC | **54**[9] |

### SINGLE-STUD WALLS

**Multilayer Finish**

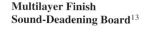

|  |  |
|---|---|
| | **STC 48 to 50**[4] |
| With insulation[8] | |
| | **STC 53 to 58**[4] |

**Multilayer Finish Sound-Deadening Board**[13]

| | | |
|---|---|---|
| SDB 1 side | STC | **52**[9] |
| SDB 2 sides | STC | **54**[9] |
| With insulation[8] | | |
| SDB 2 sides | STC | **57**[9] |

### DOUBLE-STUD (CHASE) WALLS

**Single-Layer Finish**

|  |  |
|---|---|
| | **STC 42** |
| With insulation[8] | |
| | **STC 52 to 55**[16] |

**Single-Layer Finish Sound-Deadening Board**[13]

|  |  |
|---|---|
| | **STC 50**[9] |
| With insulation[8] | |
| | **STC 53** |

1. Ranges include values for $\frac{3}{8}$ in. (9.53 mm), $\frac{1}{2}$ in. (12.7 mm), and $\frac{5}{8}$ in. (15.33 mm) thicknesses of gypsum board finish. *(Continues)*

2. Addition of insulation does not appreciably improve this construction; an increase of approximately 2 to 4 dB can be expected regardless of thickness or density of insulation.

3. Resilient channels or clips may be used one side only to totally float the finish; use of blocking for fire stopping and use of backer strips to attach wall base reduces performance approximately 4 to 5 dB.

4. Range includes values derived from ASTM E90-61T, ASTM E90-66T, and field test procedures.

5. Gypsum board finish laminated to sound-deadening board; if gypsum board is nailed or screwed to meet fire ratings, performance decreases approximately 10 dB for fasteners spaced 8 in. (200 mm) o.c., approximately 8 dB for fastener spacing 16 in. (400 mm) o.c., and approximately 4 dB with fastener spacing 24 in. (600 mm) o.c.

6. Ranges include values for $2 \times 3$ ($50 \times 75$) studs staggered on single $2 \times 4$ ($50 \times 100$) plate and $2 \times 4$ ($50 \times 100$) double-stud walls on separate $2 \times 3$ ($50 \times 75$) plates.

7. A minimum thickness of 2 to 3 in. (50 to 75 mm) of insulation can be used effectively, as performance does not increase proportionately with increased thickness of insulation.

8. Ranges include $1\frac{1}{2}$ in. (38 mm) to 3 in. (75 mm) thickness of insulation; performance increases with increased thickness. On double-stud walls, an STC of 58 to 60 can result with up to 9 in. (230 mm) of insulation.

**FIGURE 13.1-83** *(Continued)*

## Metal Frame Non-Load-Bearing Walls—Gypsum Board Finish[12] *(Cont.)*

### DOUBLE-STUD (CHASE) WALLS[15]

**Multilayer Finish**

**Multilayer Finish Sound-Deadening Board**[13]

METAL CHANNEL STUDS
TIES @ 1/4 POINTS
INSULATION
GYPSUM BOARD
12" (300 mm) CAVITY

METAL CHANNEL STUD
TIE @ 1/4 POINTS
GYPSUM BOARD
SOUND DEADENING BOARD
6" (150 mm) CAVITY

With insulation[8]
Unbalanced Finish[14]
STC **59**[16]
Multilayer finish[13]
STC 55 to 60[4][16]

Unbalanced finish[14]
STC **51**[9]
Multilayer finish[13]
STC **53**[9]

## Metal Frame Non-Load-Bearing Walls—Plaster Finish[11]

### SINGLE CHANNEL STUD WALLS

### SINGLE CHANNEL STUD WALLS

**Plaster Base & Veneer Plaster**

METAL CHANNEL STUD
VENEER PLASTER
GYPSUM BOARD

STC 40 to 45[4]

**Multilayer Base**[13] **& Veneer Plaster**

METAL CHANNEL STUD
INSULATION
VENEER PLASTER
MULTI-LAYER GYPSUM BOARD

STC 49 to 50[4]
With insulation[8]
STC 52 to 53[4]

**Gypsum Lath & Plaster**

METAL CHANNEL STUD
RESILIENT CLIP
INSULATION
GYPSUM PLASTER
GYPSUM LATH

STC 42 to 43[4]
With insulation
STC 46

**Resilient Gypsum Lath**[3] **& Plaster**

METAL CHANNEL STUD
RESILIENT CLIP
INSULATION
GYPSUM PLASTER
GYPSUM BOARD

STC 45
With insulation[8]
STC 49

9. Based on ASTM E90-61T test procedures.

10. Based on tests performed prior to ASTM E90-61T.

11. Test data can vary widely depending on weight and thickness of plaster, and water content (due to length of curing) of plaster.

12. With exception of double stud walls, ranges include values for $\frac{1}{2}$ in. (12.7 mm) or $\frac{5}{8}$ in. (15.88 mm) gypsum board thickness of 3 in. (75 mm) or 4 in. (100 mm) studs.

13. Ranges include values for gypsum board screwed, or laminated and screwed, to sound-deadening board or base layer.

14. Unbalanced Construction: single layer gypsum board one side, double-layer gypsum board one side.

15. Ranges include values for $\frac{1}{2}$ in. (12.7 mm) and $\frac{5}{8}$ in. (15.88 mm) board thickness on 2 in. (50 mm) studs.

16. Up to 60 STC with 9 in. (230 mm) of insulation; lower values result with lesser thicknesses of insulation.

17. Performance of concrete and unit masonry systems is highly dependent on minimizing the number of openings through the construction (to accommodate piping, wiring, and other equipment) and on all openings being sealed airtight.

18. Sealed both sides unless otherwise indicated. Sealing with plaster is most effective; when surfaces are to be painted, use of block sealer and/or 2 coats of cement-based paint is more effective than 2 coats of resin-emulsion paint to seal against airborne sound transmission.

**FIGURE 13.1-83** *(Continued)*

## Metal Frame Non-Load-Bearing Walls—Plaster Finish[11] *(Cont.)*

| SINGLE-TRUSS STUD WALLS | | SINGLE-TRUSS STUD WALLS | |
|---|---|---|---|
| **Metal Lath & Plaster** | **Resilient Metal Lath[3] & Plaster** | **Gypsum Lath & Plaster** | **Resilient Gypsum Lath[3] & Plaster** |

| STC 38 to 41[9] | STC 46 to 50[4] | STC 41 to 45[4] | STC 46 to 50 |
|---|---|---|---|

With insulation[8]

STC 46 to 52[4]

## Concrete and Concrete Masonry Walls[17]

| NONPOROUS (NORMAL-WEIGHT) HOLLOW CONCRETE BLOCK WALLS | | POROUS (LIGHTWEIGHT) HOLLOW CONCRETE BLOCK WALLS | |
|---|---|---|---|
| **Unsealed** | **Sealed**[18] | **Unsealed** | **Sealed**[18] |

| 4″ (100 mm) Block **STC 42 to 45**[4] | 4″ (100 mm) Block **STC 44 to 48**[4] | 4″ (100 mm) Block **STC 26 to 34**[10] | 4″ (100 mm) Block **STC 41 to 44**[4] |
|---|---|---|---|
| 6″ (150 mm) Block **STC 43 to 46**[4] | 6″ (150 mm) Block **STC 47 to 51**[4] | 6″ (150 mm) Block **STC 33 to 37**[10] | 6″ (150 mm) Block **STC 44 to 49**[4] |
| 8″ (200 mm) Block **STC 46 to 49**[9] | 8″ (200 mm) Block **STC    53**[9] | 8″ (200 mm) Block **STC 34 to 38**[10] | 8″ (200 mm) Block **STC 44 to 49**[4] |

1. Ranges include values for $\frac{3}{8}$ in. (9.53 mm), $\frac{1}{2}$ in. (12.7 mm), and $\frac{5}{8}$ in. (15.33 mm) thicknesses of gypsum board finish.     *(Continues)*

2. Addition of insulation does not appreciably improve this construction; an increase of approximately 2 to 4 dB can be expected regardless of thickness or density of insulation.

3. Resilient channels or clips may be used one side only to totally float the finish; use of blocking for fire stopping and use of backer strips to attach wall base reduces performance approximately 4 to 5 dB.

4. Range includes values derived from ASTM E90-61T, ASTM E90-66T, and field test procedures.

5. Gypsum board finish laminated to sound-deadening board; if gypsum board is nailed or screwed to meet fire ratings, performance decreases approximately 10 dB for fasteners spaced 8 in. (200 mm) o.c., approximately 8 dB for fastener spacing 16 in. (400 mm) o.c., and approximately 4 dB with fastener spacing 24 in. (600 mm) o.c.

6. Ranges include values for 2 × 3 (50 × 75) studs staggered on single 2 × 4 (50 × 100) plate and 2 × 4 (50 × 100) double-stud walls on separate 2 × 3 (50 × 75) plates.

7. A minimum thickness of 2 to 3 in. (50 to 75 mm) of insulation can be used effectively, as performance does not increase proportionately with increased thickness of insulation.

8. Ranges include $1\frac{1}{2}$ in. (38 mm) to 3 in. (75 mm) thickness of insulation; performance increases with increased thickness. On double-stud walls, an STC of 58 to 60 can result with up to 9 in. (230 mm) of insulation.

9. Based on ASTM E90-61T test procedures.

**FIGURE 13.1-83** *(Continued)*

Concrete and Concrete Masonry Walls[17] *(Cont.)*

## POROUS (LIGHTWEIGHT) HOLLOW CONCRETE BLOCK WALLS

### Gypsum Board Laminated Both Sides

6″ (150 mm) Block **STC** **49**

### Resilient Gypsum Board 1 Side, Sealed[18] 1 Side

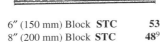

6″ (150 mm) Block **STC** **53**
8″ (200 mm) Block **STC** **48**[9]

## POROUS (LIGHTWEIGHT) HOLLOW CONCRETE BLOCK WALLS

### Resilient Gypsum Board Both Sides

8″ (200 mm) Block **STC** **49**[9]

### Cavity Wall Unsealed

**STC** **38**[10]

## HOLLOW CONCRETE BLOCK WALLS

### Porous Cavity Wall Sealed[18]

**STC 49**[10]

### Porous/Nonporous Cavity Wall Sealed[18]

**STC 51 to 52**[10]

## NONPOROUS (NORMAL-WEIGHT) SOLID CONCRETE MASONRY WALLS

### Unsealed

| | |
|---|---|
| 4″ (100 mm) Block | **STC 30** |
| 6″ (150 mm) Block | **STC 38** |
| 8″ (200 mm) Block | **STC 47** |

### Sealed[18]

| | |
|---|---|
| 4″ (100 mm) Block | **STC 44** |
| 6″ (150 mm) Block | **STC 48** |
| 8″ (200 mm) Block | **STC 53** |

10. Based on tests performed prior to ASTM E90-61T.

11. Test data can vary widely depending on weight and thickness of plaster, and water content (due to length of curing) of plaster.

12. With exception of double-stud walls, ranges include values for $\frac{1}{2}$ in. (12.7 mm) or $\frac{5}{8}$ in. (15.88 mm) gypsum board thickness of 3 in. (75 mm) or 4 in. (100 mm) studs.

13. Ranges include values for gypsum board screwed, or laminated and screwed, to sound-deadening board or base layer.

14. Unbalanced Construction: single layer gypsum board one side, double-layer gypsum board one side.

15. Ranges include values for $\frac{1}{2}$ in. (12.7 mm) and $\frac{5}{8}$ in. (15.88 mm) board thickness on 2 in. (50 mm) studs.

16. Up to 60 STC with 9 in. (230 mm) of insulation; lower values result with lesser thicknesses of insulation.

17. Performance of concrete and unit masonry systems is highly dependent on minimizing the number of openings through the construction (to accommodate piping, wiring, and other equipment) and on all openings being sealed airtight.

18. Sealed both sides unless otherwise indicated. Sealing with plaster is most effective; when surfaces are to be painted, use of block sealer and/or 2 coats of cement-based paint is more effective than 2 coats of resin-emulsion paint to seal against airborne sound transmission.

**FIGURE 13.1-83** *(Continued)*

Concrete and Concrete Masonry Walls[17] *(Cont.)*

| POROUS (LIGHTWEIGHT) SOLID CONCRETE MASONRY WALLS | | POURED-IN-PLACE SOLID CONCRETE WALLS |

**Unsealed**                  **Sealed**[18]

| | | | | | | | | | | |
|---|---|---|---|---|---|---|---|---|---|---|
| 4″ (100 mm) Block | **STC** | **27** | 4″ (100 mm) Block | **STC** | **42** | 4″ (100 mm) Thick | **STC** | **46** |
| 6″ (150 mm) Block | **STC** | **36** | 6″ (150 mm) Block | **STC** | **46** | 6″ (150 mm) Thick | **STC** | **50** |
| 8″ (200 mm) Block | **STC** | **45** | 8″ (200 mm) Block | **STC** | **50** | 8″ (200 mm) Thick | **STC** | **55** |

Clay Masonry Walls[17]

| SOLID WALLS | CAVITY WALL | COMPOSITE WALL |

**Brick Wall Sealed**[18]    **Brick Cavity Wall Sealed**[18]    **Composite Brick, Bonded to Concrete Block**

| | | | | |
|---|---|---|---|---|
| 6″ (150 mm) Brick | **STC 45 to 50** | **STC 45 to 50** | **STC 45 to 50** |
| 8″ (200 mm) Brick | **STC 50 to 52** | | |
| 12″ (300 mm) Brick | **STC 55** | | |

**FIGURE 13.1-84** Floor and Ceiling Assemblies

## Wood Joist Floors—Gypsum Board Ceiling Finish[2]

| HARD SURFACE FLOORING | CARPETING | HARD SURFACE FLOORING[3] RESILIENT CEILING | CARPETING RESILIENT CEILING |
|---|---|---|---|
| STC 38 to 42 (No Ceiling STC 25) | STC 38 to 42 | STC 46 to 51 | STC 46 to 48 |

**INR –19 to –16**   **IIC 32 to 35**
(No Ceiling)
**INR   –28**   **IIC   23**

Carpet Only
**INR –10 to –9**   **IIC 41 to 42**
Lightweight Carpet & Pad[4]
**INR   –5**   **IIC   46**
Standard Carpet & Pad[5]
**INR +8 to +10**   **IIC 59 to 61**

**INR –13 to –9**   **IIC 38 to 42**
With Insulation[6]
Vinyl Flooring
**INR –5 to –3**   **IIC 46 to 48**
Wood Block Flooring
**INR   –1**   **IIC   50**

**INR +15 to +16**   **IIC 66 to 67**
With insulation[6]
Lightweight Carpet & Pad[4]
**INR   +7**   **IIC   58**
Standard Carpet & Pad[5]
**INR +18 to +20**   **IIC 69 to 71**

| HARD SURFACE FLOORING[3] SEPARATE CEILING JOISTS | CARPETING[5] SEPARATE CEILING JOISTS | FLOATED HARD SURFACE FLOORING | FLOATED HARD SURFACE FLOORING, RESILIENT CEILING |
|---|---|---|---|
| | | | STC 50 to 52 With insulation[6] |
| STC 50 to 53 | STC   51 | STC   46 | STC 51 to 53 |

With insulation[5]
**INR –6 to –2**   **IIC 45 to 49**

With insulation[6]
**INR   +29**   **IIC   80**

**INR   –13**   **IIC   38**

**INR –3 to –2**   **IIC 48 to 49**
With insulation[6]
**INR +1 to +2**   **IIC 52 to 53**

1. INR values are based on tests performed in accordance with ISO Recommendation R-140, "Field and Laboratory Measurements of Airborne and Impact Sound Transmission," January, 1960, IIC values approximated by adding 51 to original INR tests values.
2. Ranges include value for ³/₈ in. (9.53 mm), ¹/₂ in. (12.7 mm), and ⁵/₈ in. (15.88 mm) thicknesses of gypsum board finish.
3. Hard surface finish flooring (i.e., stripwood, wood block, vinyl, and vinyl asbestos) generally performs similarly in isolating impact sound. Unless otherwise indicated, ranges given include values for construction with any typical hard surface finish flooring.
4. Lightweight carpet and pad having pile weights of 20 oz per sq yd (678.11 g/m²) and 28 oz per sq yd (949.36 g/m²), respectively.
5. Standard-weight carpet and pad having pile weights of 44 oz per sq yd (1491.85 g/m²) and 40 oz per sq yd (1356.23 g/m²), respectively. Ratings shown are based on standard weight combinations unless otherwise noted.
6. A minimum thickness of 2 to 3 in. (50 to 150 mm) of insulation can be used effectively, as performance does not increase with increased thickness in floor and ceiling assemblies.
7. Ratings vary with thickness of concrete slabs. The low end of the range indicates value for 6 in. (150 mm) thickness; high end, 8 in. (200 mm) thickness.

**FIGURE 13.1-84** *(Continued)*

Wood Joist Floors—Gypsum Board Ceiling Finish[2] *(Cont.)*

| CARPETING ON FLOATED FLOOR RESILIENT CEILING | FLOATED HARD SURFACE FLOORING[3], ON FURRING RESILIENT CEILING | CARPETING[5], ON FLOATED FLOOR, FURRING RESILIENT CEILING | FLOATED HARD SURFACE FLOORING[3] SEPARATE CEILING JOISTS |
|---|---|---|---|
| **STC 50** | **STC 48 to 53** | **STC 52** | **STC 53** |

**CARPETING ON FLOATED FLOOR RESILIENT CEILING**

With insulation[6]
  Lightweight Carpet & Pad[4]

| INR | +8 | IIC | 59 |
|---|---|---|---|

  Standard Carpet & Pad[5]

| INR | +18 | IIC | 69 |
|---|---|---|---|

**FLOATED HARD SURFACE FLOORING[3], ON FURRING RESILIENT CEILING**

| INR | −7 | IIC | 44 |
|---|---|---|---|

With insulation[6]

| INR −2 to +7 | | IIC 49 to 58 |
|---|---|---|

**CARPETING[5], ON FLOATED FLOOR, FURRING RESILIENT CEILING**

With insulation[6]

| INR | +27 | IIC | 78 |
|---|---|---|---|

**FLOATED HARD SURFACE FLOORING[3] SEPARATE CEILING JOISTS**

With insulation[6]

| INR | +3 | IIC | 54 |
|---|---|---|---|

| HARD SURFACE FLOORING ON CONCRETE FILL | CARPETING ON CONCRETE FILL | HARD SURFACE FLOORING ON CONCRETE FILL RESILIENT CEILING | CARPETING ON CONCRETE FILL RESILIENT CEILING |
|---|---|---|---|
| **STC 46 to 48** | **STC 46 to 48** | **STC 50 to 53** | **STC 50 to 53** |

**HARD SURFACE FLOORING ON CONCRETE FILL**

| INR −14 to −10 | | IIC 37 to 41 |
|---|---|---|

**CARPETING ON CONCRETE FILL**

Lightweight Carpet & Pad[4]

| INR | −2 | IIC | 49 |
|---|---|---|---|

Standard Carpet & Pad[5]

| INR | +10 | IIC | 61 |
|---|---|---|---|

**HARD SURFACE FLOORING ON CONCRETE FILL RESILIENT CEILING**

| INR | −5 | IIC | 46 |
|---|---|---|---|

With Insulation[6]
  Wood Block Flooring

| INR | +1 | IIC | 52 |
|---|---|---|---|

  Cushion Backed Vinyl

| INR | +5 | IIC | 56 |
|---|---|---|---|

**CARPETING ON CONCRETE FILL RESILIENT CEILING**

With insulation[6]
  Lightweight Carpet & Pad[4]

| INR | +13 | IIC | 64 |
|---|---|---|---|

  Standard Carpet & Pad[5]

| INR | +23 | IIC | 74 |
|---|---|---|---|

*(Continues)*

**FIGURE 13.1-84** *(Continued)*

**Wood/Steel Truss Floors—Gypsum Board Ceiling Finish[2]**

| HARD SURFACE FLOORING[3] | CARPETING[5] | HARD SURFACE FLOORING[3] RESILIENT CEILING | CARPETING[5] RESILIENT CEILING |
|---|---|---|---|
| STC 45 | STC 45 | STC 47 | STC 47 |
| INR −18 IIC 33 | INR +5 IIC 56 | INR −15 IIC 36 | INR +15 IIC 66 |
| | | | With insulation[6] |
| | | | INR +18 IIC 69 |

| HARD SURFACE FLOORING[3] ON CONCRETE FILL | CARPETING[5] ON CONCRETE FILL |
|---|---|
| STC 46 | STC 46 |
| INR −14 IIC 37 | INR +11 IIC 62 |

1. INR values are based on tests performed in accordance with ISO Recommendation R-140, "Field and Laboratory Measurements of Airborne and Impact Sound Transmission," January, 1960, IIC values approximated by adding 51 to original INR tests values.
2. Ranges include value for $\frac{3}{8}$ in. (9.53 mm), $\frac{1}{2}$ in. (12.7 mm), and $\frac{5}{8}$ in. (15.88 mm) thicknesses of gypsum board finish.
3. Hard surface finish flooring (i.e., stripwood, wood block, vinyl, and vinyl asbestos) generally performs similarly in isolating impact sound. Unless otherwise indicated, ranges given include values for construction with any typical hard surface finish flooring.
4. Lightweight carpet and pad having pile weights of 20 oz per sq yd (678.11 g/m$^2$) and 28 oz per sq yd (949.36 g/m$^2$), respectively.
5. Standard-weight carpet and pad having pile weights of 44 oz per sq yd (1491.85 g/m$^2$) and 40 oz per sq yd (1356.23 g/m$^2$), respectively. Ratings shown are based on standard weight combinations unless otherwise noted.
6. A minimum thickness of 2 to 3 in. (50 to 150 mm) of insulation can be used effectively, as performance does not increase with increased thickness in floor and ceiling assemblies.
7. Ratings vary with thickness of concrete slabs. The low end of the range indicates value for 6 in. (150 mm) thickness; high end, 8 in. (200 mm) thickness.

**FIGURE 13.1-84** *(Continued)*

**Poured-in-Place Reinforced Concrete Slabs**

| HARD SURFACE FLOORING[3] WITH OR WITHOUT PLASTER CEILING | FLOATED HARD SURFACE FLOORING[3], WITH OR WITHOUT PLASTER CEILING | CARPETING WITH OR WITHOUT PLASTER CEILING |
|---|---|---|
| STC 44 to 48 | STC 48 to 50 | STC 44 to 48 |

No Flooring
**INR** −17 **IIC** 34

Vinyl Flooring
**INR** −15 **IIC** 36

Wood Block Flooring
**INR** −9 to −6 **IIC** 42 to 45

**INR** −7 to +4 **IIC** 44 to 55

Lightweight Carpet Only
**INR** +2 **IIC** 53

Standard Carpet Only
**INR** +6 **IIC** 57

Standard Carpet & Pad[5]
**INR** +17 to +22 **IIC** 68 to 73

*(Continues)*

**FIGURE 13.1-84** *(Continued)*

**Precast Concrete Slabs**[7]

| **HARD SURFACE FLOORING**[3] **INTEGRAL CEILING** | **CARPETING**[5] **INTEGRAL CEILING** | **FLOATED HARD SURFACE FLOORING**[3] **INTEGRAL CEILING** |
|---|---|---|
| STC 45 to 47 | STC 45 to 47 | STC 47 to 51 |

Vinyl Flooring
**INR −28 to −23**     **IIC 23 to 28**     **INR +18 to +22**     **IIC 69 to 73**     **INR +2 to +4**     **IIC 53 to 55**

Wood Block Flooring
**INR −4 to −3**     **IIC 47 to 48**

| **HARD SURFACE FLOORING**[3] **ON 1½″ (12.7 mm) CONCRETE INTEGRAL CEILING** | **CARPETING**[5] **ON 1½″ (12.7 mm) CONCRETE INTEGRAL CEILING** |
|---|---|
| STC 49 | STC 45 |

**INR +2 to +4**     **IIC 53 to 55**     **INR     +25**     **IIC     76**

1. INR values are based on tests performed in accordance with ISO Recommendation R-140, "Field and Laboratory Measurements of Airborne and Impact Sound Transmission," January, 1960, IIC values approximated by adding 51 to original INR tests values.
2. Ranges include value for ⅜ in. (9.53 mm), ½ in. (12.7 mm), and ⅝ in. (15.88 mm) thicknesses of gypsum board finish.
3. Hard surface finish flooring (i.e., stripwood, wood block, vinyl, and vinyl asbestos) generally performs similarly in isolating impact sound. Unless otherwise indicated, ranges given include values for construction with any typical hard surface finish flooring.
4. Lightweight carpet and pad having pile weights of 20 oz per sq yd (678.11 g/m²) and 28 oz per sq yd (949.36 g/m²), respectively.
5. Standard-weight carpet and pad having pile weights of 44 oz per sq yd (1491.85 g/m²) and 40 oz per sq yd (1356.23 g/m²), respectively. Ratings shown are based on standard weight combinations unless otherwise noted.
6. A minimum thickness of 2 to 3 in. (50 to 150 mm) of insulation can be used effectively, as performance does not increase with increased thickness in floor and ceiling assemblies.
7. Ratings vary with thickness of concrete slabs. The low end of the range indicates value for 6 in. (150 mm) thickness; high end, 8 in. (200 mm) thickness.

**FIGURE 13.1-84** *(Continued)*

## Steel Bar Joist Floors

| CARPETING[5]<br>ON 2½″ (63.5 mm) CONCRETE<br>LATH & PLASTER CEILING | CARPETING[5]<br>ON 1⅝″ (41.3 mm) CONCRETE<br>GYPSUM BOARD CEILING | CARPETING[5]<br>RESILIENT GYPSUM<br>BOARD CEILING |
|---|---|---|
| **STC 49** | **STC 46** | **STC 47** |

| INR | +11 | IIC | 62 | INR | +11 | IIC | 62 | INR | +18 | IIC | 69 |
|---|---|---|---|---|---|---|---|---|---|---|---|

| CARPETING[5]<br>ON 2½″ (63.5 mm) CONCRETE<br>LATH & PLASTER CEILING | CARPETING[5]<br>ON 2″ (50.8 mm) CONCRETE<br>RESILIENT GYPSUM<br>PLASTER CEILING |
|---|---|
| **STC 47 to 50** | **STC 49** |

| INR | +22 | IIC | 73 | INR | +23 | IIC | 74 |
|---|---|---|---|---|---|---|---|

With insulation[6]

| INR | +36 | IIC | 87 |
|---|---|---|---|

*(Continues)*

1. INR values are based on tests performed in accordance with ISO Recommendation R-140, "Field and Laboratory Measurements of Airborne and Impact Sound Transmission," January, 1960, IIC values approximated by adding 51 to original INR tests values.
2. Ranges include value for ⅜ in. (9.53 mm), ½ in. (12.7 mm), and ⅝ in. (15.88 mm) thicknesses of gypsum board finish.
3. Hard surface finish flooring (i.e., stripwood, wood block, vinyl, and vinyl asbestos) generally performs similarly in isolating impact sound. Unless otherwise indicated, ranges given include values for construction with any typical hard surface finish flooring.
4. Lightweight carpet and pad having pile weights of 20 oz per sq yd (678.11 g/m²) and 28 oz per sq yd (949.36 g/m²), respectively.
5. Standard-weight carpet and pad having pile weights of 44 oz per sq yd (1491.85 g/m²) and 40 oz per sq yd (1356.23 g/m²), respectively. Ratings shown are based on standard weight combinations unless otherwise noted.
6. A minimum thickness of 2 to 3 in. (50 to 150 mm) of insulation can be used effectively, as performance does not increase with increased thickness in floor and ceiling assemblies.
7. Ratings vary with thickness of concrete slabs. The low end of the range indicates value for 6 in. (150 mm) thickness; high end, 8 in. (200 mm) thickness.

**FIGURE 13.1-84** *(Continued)*

**Steel Bar Joist Floors**

**HARD SURFACE FLOORING[3]**
**ON 2½″ (63.5 mm) CONCRETE**
**LATH & PLASTER CEILING**

**STC 49**

| INR | −16 | IIC | 35 |
|-----|-----|-----|-----|

| **HARD SURFACE FLOORING[3]** **ON 2½″ (63.5 mm) CONCRETE** **LATH & PLASTER CEILING** | **HARD SURFACE FLOORING[3]** **ON 2″ (50.8 mm) CONCRETE** **RESILIENT GYPSUM** **BOARD CEILING** |
|---|---|
| **STC**    **47** | **STC**    **49** |
| With insulation[6] | |
| **STC**    **50** | |

| INR | −25 | IIC | 26 | INR | −10 | IIC | 41 |
|-----|-----|-----|-----|-----|-----|-----|-----|

With insulation[6]

| INR | −22 | IIC | 29 |
|-----|-----|-----|-----|

## 13.2 Additional Reading

Additional information about the subjects discussed in this chapter can be found in the references listed in Section 13.3 and in the following publications.

Ambrose, James and Jeffrey Ollswang. 1995. *Simplified Design for Building Sound Control*. Hoboken, NJ: John Wiley & Sons, Inc. (www.wiley.com).

Cavanaugh, William J. and Joseph A Wilkes, eds. 1998. *Architectural Acoustics: Principles and Practice*. Hoboken, NJ: John Wiley & Sons, Inc. (www.wiley.com).

Egan, M. David 1998. "Sound Isolation for Wall Openings: Doors, Windows." *Architecture* (November): 89–90 (www.architecturemag.com).

Haas, Steve, 1998. "Five Ways to Reduce HVAC Duct Noise." *The Construction Specifier* (February): 63 (www.csinet.org).

———— 1998. "Noisy, Rumbling, Hissing." *The Construction Specifier* (February): 64 (www.csinet.org).

Ramsey, Charles George, Harold Reeve Sleeper, and John Ray Hoke, Jr. 2000. *Architectural Graphic Standards*, 10th ed. Hoboken, NJ: John Wiley & Sons, Inc. (www.wiley.com).

## 13.3 Acknowledgments and References

We gratefully acknowledge the assistance of the following organizations and individuals in preparing this chapter. We are also indebted to them for permission to use their illustrations when requested and for the use of their publications as references.

### ACKNOWLEDGMENTS

American National Standards Institute (ANSI) (www.ansi.org)
Architectural Woodwork Institute (AWI) (www.awinet.org)
ASTM International (ASTM) (www.astm.org)
Brick Industry Association (BIA) (www.bia.org)
Carpet and Rug Institute (CRI) (www.carpet-rug.com)
Cedar Knolls Acoustical Laboratories (CKAL) (part of NU Laboratories [NU]) (www.nulabs.com)
Construction Specifications Institute (CSI) (www.csinet.org)
Consumer Product Safety Commission (CPSC) (www.cpsc.gov)
Flexicore Manufacturers Association
Geiger and Hamme Kodaras Acoustical Laboratories
Gypsum Association (GA) (www.gypsum.org)
International Organization for Standardization (ISO) (www.iso.org)
Kodaras Acoustical Laboratories
National Association of Home Builders (NAHB) Research Foundation (www.nahbrc.org)
National Bureau of Standards (now the National Institute of Standards and Technology) (NIST) (www.nist.gov)
National Concrete Masonry Association (NCMA) (www.ncma.org)
National Gypsum Company (www.national-gypsum.com)
National Wood Window and Door Association (NWWDA) (www.nwwda.org)
Ohio Research Corporation
Owens-Corning Fiberglas Corporation (www.owenscorning.com)
Portland Cement Association (PCA) (www.cement.org)
Riverbank Acoustical laboratories (RAL) (www.riverbank.alionscience.com)
Rodale Publishing Company (www.rodale.com)
United States Gypsum Company (USG) (www.usg.com)
U.S. Department of Housing and Urban Development (HUD) (www.hud.gov)

### REFERENCES

We would also like to thank the authors and publishers of the publications in the following list of references for their contribution to our research for this chapter.

American Architectural Manufacturers Association (AAMA). AAMA TIR-A1-04, "Sound Control for Fenestration Products." Des Plaines, IL: AAMA (www.aamanet.org).

American Society of Heating, Refrigerating and Air-Conditioning Engineers (ASHRAE). 2001. *ASHRAE Handbook—Fundamentals*. Atlanta, GA: ASHRAE (www.ashrae.org).

ASTM International (ASTM) (www.astm.org). West Conshohocken, PA Standards:

C919, "Standard Practice for Use of Sealants in Acoustical Applications."

C920, "Standard Specifications for Elastomeric Joint Sealants."

E90, "Standard Test Measurement for Laboratory Measurement of Airborne Sound Transmission Loss of Building Partitions and Elements."

E413, "Classification for Rating Sound Insulation."

E492, "Standard Test Method for Laboratory Measurement of Impact Sound Transmission Through Floor-Ceiling Assemblies Using the Tapping Machine."

International Organization for Standardization (ISO). 1997. *Acoustics—Measurement of Sound Insulation in Buildings and of Building Elements—Part 1: Requirements for Laboratory Test Facilities with Suppressed Flanking Transmission*, #140-1. Falls Church, VA: ISO (www.iso.org).

Ramsey, Charles George, Harold Reeve Sleeper, and John Ray Hoke, Jr. 2000. *Architectural Graphic Standards*, 10th ed. Hoboken, NJ: John Wiley & Sons, Inc (www.wiley.com).

U.S. Department of Housing and Urban Development (HUD). 2004. *The Noise Guidebook*. Washington, DC: HUD (www.hud.gov).

CHAPTER **14**

# Conveying Systems

Introduction

Applicable *MasterFormat*™ Sections

Elevators and Lifts

Escalators and Moving Walks

Additional Reading

References

# Introduction

Many types of conveying systems are used in and around buildings. Among them are dumbwaiters, elevators, escalators, moving walks, lifts, material handling systems, hoists, cranes, turntables, scaffolding, and transportation systems.

*Dumbwaiters* are essentially small elevators intended for moving materials instead of people. They may be packaged or field-assembled and may be motorized or hand-operated. Motorized units may operate either electrically or hydraulically.

*Elevators* are mechanisms designed to move people from one building level to another.

*Escalators* are moving stairs designed to convey passengers.

*Moving walks* consist of moving belts installed either horizontally or on a slant. They are intended to move passengers.

*Lifts* include chair lifts, people lifts, wheelchair lifts, platform lifts (including stage and orchestra lifts), sidewalk lifts, and vehicle lifts. They are designed to lift either people or equipment from one level to another. *MasterFormat* 2004 lists ski lifts, service station lifts, and book and library cart lifts in Division 11, but these are not covered in detail in this book.

*Material handling systems* include chutes, conveyors (baggage, postal, bucket, screw, vertical, roller, belt, etc.), tube systems (pneumatic tube, etc.), and similar equipment.

*Hoists* and *cranes* include fixed and trolley hoists; bridge, gantry, jib, tower, and mobile cranes; and derricks that are part of a completed building.

*Turntables* are devices designed to rotate sections of a building or portions of a building such as displays and vehicles. They are used in restaurants, stages, exhibit halls, and rail and other transportation facilities.

*Scaffolding* classified as vertical transportation includes suspended scaffolding, rope climbers, and telescoping platforms that are part of a completed building.

*Transportation systems* include monorail systems, people movers, passenger loading bridges, and other similar devices used in railroad, subway, monorail, and aircraft facilities. Transportation systems are addressed in *MasterFormat* 2004 Division 34, "Transportation."

Elevators, escalators, wheelchair lifts, stair lifts, and moving walks are addressed in this chapter. The other conveying systems mentioned above are beyond the scope of this book.

# Applicable *MasterFormat* Sections

The following *MasterFormat* 2004 Level 2 sections are applicable to this chapter:

14 11 00 Manual Dumbwaiters
14 12 00 Electric Dumbwaiters

14 14 00 Hydraulic Dumbwaiters
14 21 00 Electric Traction Elevators
14 24 00 Hydraulic Elevators
14 26 00 Limited-Use/Limited-
Application Elevators
14 27 00 Custom Elevator Cabs

14 28 00 Elevator Equipment and
Controls
14 31 00 Escalators
14 32 00 Moving Walks
14 33 00 Moving Ramps
14 42 00 Wheelchair Lifts

# 14.1 Elevators and Lifts

Elevators and lifts are devices for moving people or material vertically from one building level to another.

## 14.1.1 HISTORY

Since ancient times, people have been using hoists powered by water, people, or animals to move loads vertically. Such devices were first powered by steam around 1800, but even the steam-powered devices had no safety features to prevent them from falling when a rope or some other part of the mechanism failed. This limited their use to freight hauling. In addition, early commercial elevators ran only between two floors.

In 1852, Elisha Graves Otis, a master mechanic in a bedstead factory, introduced the first practical safety device for passenger elevators, a system of spring-operated cams that engaged an elevator's guide rails in the event that the cable broke. The first elevator specifically intended to move people was installed in a New York store in 1857.

The first hydraulic elevator was installed in 1878. The hydraulic fluid was water; otherwise, the operation was similar to that of a modern hydraulic elevator. Early hydraulic elevators were of two types. In one, the cylinder was directly beneath the car. In the other, the cylinder pulled ropes that in turn lifted the car.

The first electrically driven elevator was installed in a New York City office building in 1889. Its use made the building of skyscrapers practical.

Early elevator controls were not automatic; a human operator was required. Automatic elevators were available for use in homes as early as 1894, but the first gearless traction elevator was not introduced until 1903, and advanced control systems were not introduced until 1924. Fully automatic elevators for use in tall buildings came into general use in 1949.

Improvements have come rapidly since then. Speeds have increased, acceleration and deceleration functions have become smoother, leveling at floor landings has improved, and the combination of developing technology and stringent laws has increased safety immensely.

## 14.1.2 ELECTRIC TRACTION ELEVATORS

*Electric traction* elevators consist of a car traveling in a shaft (Fig. 14.1-1). The car is lifted by an electric motor housed in an equipment room at the top of the shaft. The motor turns a grooved wheel (*drive sheave*) over which pass multiple steel cables that are connected to a cast iron counterweight on one end and the elevator car on the other end. As the sheave turns, the cable moves, raising or lowering the car. The combined weight of the car and counterweight keeps the cables taunt. There is usually also a second sheave for added traction.

In an *underslung* traction elevator, also called a *basement traction* elevator, the motor is placed in a room adjacent to the bottom of the shaft. The drive wheels on the motor project into the shaft. Cables pass under the drive wheels and extend vertically up the shaft, where they pass over sheaves. One end is connected to the car and the other end is connected to a counterweight. Underslung elevators are used in existing buildings and in low-rise new buildings where a penthouse would be objectionable or illegal and where overhead is limited.

In traction elevator installations, both the car and the counterweights have guide rails along the side of the shaft to keep them in proper alignment.

Traction elevators are classified as either geared or gearless. Geared elevators are used in low-rise buildings where speeds between 200 ft (61 m) and 450 ft (137 m) per minute are sufficient. At speeds of 500 ft (152 m) to 2000 ft (610 m) per minute, which are required for tall buildings, geared machines generate excessive noise and are inefficient. Gearless machines are smoother, quieter, and more efficient than the geared type.

Long, flexible power and control cables run down the shaft to the car to provide the required power for the car. They supply electricity to operate lights, doors, fans, and control systems and connect the control systems in the car with those in the machine room.

### 14.1.2.1 Safety Features

There are several safety systems protecting elevator cars. The car is connected by cable to an electric governor at the top of the shaft. If the car descends faster than a preset limit, the governor slows down the drive sheaves. If the car then fails to stop,

**FIGURE 14.1-1** Electric traction elevator. (Drawing by HLS.)

the governor grips the governor cable, which causes safety clamps that provide a wedge between the car and the guide rails, stopping the car. Elevators also have electrical controls that prevent the car from operating when a door is open.

### 14.1.2.2 Operation and Performance

Until recently, most traction elevators were powered by direct current (dc) power.

Converting alternating current (ac) building power to dc for elevator use was done by a motor generator (MG) set. MG sets consume power when idling, generate copious amounts of heat, and are noisy. Later dc-powered elevators have solid-state motor drives, which use silicon-controlled rectifier (SCR) units instead of MG sets to convert house current to dc. SCR units use power only when the elevator is running. They are also quieter and produce less heat than MG sets. While dc-powered elevators are still available, the trend today is toward elevator drive units powered by variable frequency alternating current (ACVF).

### 14.1.3 OIL HYDRAULIC ELEVATORS

Hydraulic elevators are classified as either *inground*, *holeless*, or *roped*.

### 14.1.3.1 Inground Oil Hydraulic Elevators

Inground hydraulic elevators are supported from below by a steel plunger that moves in a vertical cylinder sunk into a shaft drilled below the elevator shaft (Fig. 14.1-2). The cylinder extends below ground to a distance equal to the rise of the elevator or a little more. For this reason, most hydraulic passenger elevators are not practicable for buildings more than eight stories high. Most are found in shorter buildings. The maximum travel of most hydraulic freight elevators is between 40 and 50 ft (12 and 15 m).

The basic operation of an inground hydraulic elevator is as follows:

- To raise the car, a pump driven by an electric motor forces oil under pressure into the casing, which raises the plunger, lifting the car.
- To lower the car, electrically controlled valves open, permitting the oil to return slowly to a storage tank, thus lowering the plunger and, in turn, the car. The motor and the pump are idle when the car is descending.

**FIGURE 14.1-2** Hydraulic elevator. (Drawing by HLS.)

A hydraulic elevator can fall if oil leaks from the cylinder. When this rare event occurs, oil leaks out very slowly, lowering the car at a safe speed onto a buffer at the bottom of the shaft. Therefore, safety clamping systems, such as those used on traction elevators, are not required to prevent hydraulic elevators from falling.

### 14.1.3.2 Holeless Oil Hydraulic Elevators

Holeless hydraulic elevators are driven by a mechanism similar to that used in an inground hydraulic elevator. The cylinder, however, telescopes to make a deep hole unnecessary. Some telescoping cylinders are in shafts that extend into the ground, but the shaft extends a shorter distance into the ground than the shaft of an inground unit. Holeless units may have either single-stage or two-stage operation. The basic difference between the two is the height of building the elevator can serve. A single-stage holeless installation might be limited to 15 ft (4.57 m) travel distance. A partially inground machine might be able to travel 20 ft (6 m). A two-stage machine might be able to travel about 35 ft (10.7 m), or for four stacked stops. Two-stage machines with additional pit depth and overhead height can travel as much as 41 ft (12.5 m).

### 14.1.3.3 Roped Oil Hydraulic Elevators

A roped hydraulic elevator is driven by an oil hydraulic mechanism. Plungers (which are in turn supported by devices called *jackstands*, sliding in aboveground cylinders) are located in the shaft beside the car. Cables are attached to the top of the jackstands, pass over a sheave, and tie to the bottom of the car frame. As the plunger is forced up by hydraulic pressure, it raises the sheave, shortening the cable length from the sheave to the car and thus raising the car. The other end of the cables, being fixed to the jackstands, cannot move.

Roped systems have a traction governor and other safety features similar to those of an electric traction elevator.

Roped machines work in applications of up to 60 ft (18.3 m).

### 14.1.4 PASSENGER ELEVATORS

Passenger elevators primarily move passengers. Passenger elevators that are also used to move supplies, furniture, and other items in addition to passengers are generally designed and finished the same way as other passenger elevators.

#### 14.1.4.1 Passenger Elevator Cars

##### FEATURES

A passenger car should include the following features:

- An emergency light with battery and charger.
- An emergency exit.
- Finished flooring.
- A suspended ceiling. Various translucent plastic, metal, aluminum eggcrate, and other designs are available. Fluo-

rescent light fixtures are often provided above the suspended ceiling, but incandescent lights are also used.
- A handrail. Many types are available, including wood rails and metal flat bars, tubes, and cylindrical shapes.
- A single-speed or two-speed fan for power ventilation.
- As an option, protective pads and pad hooks for side and rear walls.

See also Sections 14.1.5 and 14.1.6.

#### FINISHES

Many finishes are available for use in passenger elevator cars. Doors, articulation, wall bases, ceilings, and operating panels may be wood, plastic laminate, or metal. Typical metals are stainless steel, aluminum, Muntz metal, and bronze. Metal panels may be finished in baked enamel in almost any color.

Floor finishes in passenger elevators are usually carpet, but some are resilient flooring.

#### CAR AND HOISTWAY DOORS

Passenger type hoistway entrances are usually required to be Underwriters Laboratories, Inc. (UL) "B"-labeled. They are usually hollow metal but may be finished in any noncombustible material, such as stainless steel or bronze, or may receive a baked enamel coating.

Opening sizes vary with use and elevator size and capacity. For example, doors for 2000 lb (907 kg) elevators may be 3 ft (914 mm) wide; for 2500 lb (1134 kg) to 3500 lb (1588 kg) elevators, 3½ ft (1067 mm) wide; for 4000 lb (1814 kg) to 5000 lb (2268 kg) elevators, 4 ft (1220 mm) wide.

#### DOOR OPERATION

Passenger car doors are horizontal sliding and may be *center opening*, *single slide*, or *two-speed* (Figs. 14.1-3, 14.1-4, and 14.1-5). Although other door sizes are sometimes used, in general, elevators with small capacities (2000 lb [907 kg] to 2500 lb [1134 kg]) generally have single-slide doors; those with midrange capacities (3000 lb [1360 kg] to 3500 lb [1588 kg]), center-opening doors; and those with higher capacities (4000 lb [1814 kg] to 5000 lb [2268 kg]), two-speed doors.

#### DOOR OPERATOR EQUIPMENT

A door operator opens and closes car and hoistway doors simultaneously.

**FIGURE 14.1-3** Elevator with center-opening doors. Cab and hoistway doors operate simultaneously. Pairs of doors stack with one on each side of the jamb opening and between it and the hoistway wall. (Drawing by HLS.)

**FIGURE 14.1-4** Elevator with two-speed doors. The doors operate at different speeds so that the two hoistway doors and the two cab doors are fully open at the same time and fully closed at the same time. All four doors stack on the same side of the hoistway opening. (Drawing by HLS.)

**FIGURE 14.1-5** Elevator with single-slide doors. Cab and hoistway doors operate simultaneously. Both car and hoistway doors stack on the same side of the jamb opening and between it and the hoistway wall. (Drawing by HLS.)

Door movement is cushioned at both limits of travel. An electric contact on each car entrance prevents the elevator from operating unless the car door is closed. An electromechanical interlock prevents elevator operation until all of its doors are closed and locked.

Doors are designed so that, if the electric power is interrupted, the doors can be opened from within the car by hand. Emergency devices and keys for opening doors from the landing are required by code.

Doors open automatically when the car has arrived at a landing. They close after a predetermined time interval or can be closed immediately by pressing a *door close* button in the car. Momentary pressing of a *door open* button in the car reopens the doors and resets the time interval.

Each elevator car door has an electronic sensing device which prevents the doors from closing when an obstruction is present and will cause them to stop and reopen if they encounter an obstruction while they are closing.

### 14.1.4.2 Passenger Elevator Dispatch

Overall elevator dispatch is governed by a logic control system. For passenger elevators, two of the logic control systems used are *selective collective* (S/C) and *car group*. Others may also be used.

In an S/C control system, multiple hall and car calls may be registered at the same time. The control system remembers each call and answers the calls in the order in which the landings are reached based on the direction of travel of a particular elevator, unrelated to the order in which the calls were registered.

Modern group logic systems are microprocessor controlled. A processor in each car controls the operations of that car, such as door openings and closings, safety and stop circuits, load weighing, and timing devices. The in-car processors are tied together through a *group controller*, which receives and processes elevator demands from car and hall calls and dispatches individual cars to answer them in the most efficient manner.

### 14.1.5 FREIGHT ELEVATORS

*Freight elevators* are intended primarily to move freight and, as such, are utilitarian in nature. Freight elevators are designed according to their load class, as defined in ASME A17.1:

A.  General Freight Loading
B.  Motor Vehicle Loading
C1. Industrial Truck Loading (this class permits a forklift to enter the elevator)
C2. Industrial Truck Loading (forklifts do not usually enter the elevator)
C3. Other Loading with Heavy Concentrations

#### 14.1.5.1 Freight Elevator Cars

The interior walls and ceilings of freight elevators are usually of 14-gauge (1.9 mm) cold rolled steel with no applied finishes. Perimeter bumper guards are often installed to reduce potential wall damage caused by forklifts and loads contacting the walls. Depending on the capacity and class of load, flooring in freight elevators may be checkered steel plate, galvanized steel, or high-density wood.

#### CAR AND HOISTWAY DOORS

A freight elevator car typically has a gate. In most cases, this is a vertically rising device. Freight elevator hoistway doors are usually biparting units in which the upper panel travels upward and the lower panel travels downward. Some freight elevator doors are one-piece units that travel upward. The car riding gate is usually power-operated. Hoistway doors may be either manual or power-operated.

Power-operated doors are opened and closed by means of push buttons located at each landing and within the elevator. In a normal operation, when the car arrives at a landing, the doors and gate open either automatically or by momentary pressure on a door-open push button. Closing the doors and gate requires constant pressure on the door-close push button. Door closing may also be done automatically.

When passengers are permitted to use a freight elevator, special capacity calculations must be made, special doors are required, and additional door operation features are required. For example, a safety edge that automatically opens the door if it meets an obstruction is required on the lower edge of the car riding gate, and an audible signal must be added to sound 5 seconds before the automatic door-closing sequence begins.

#### 14.1.5.2 Freight Elevator Operation

Overall elevator operation is governed by a logic control system. The two most frequently chosen elevator logic controls for freight elevators are S/C control and *full automatic push button* (FAPB) control.

S/C control systems are described in Section 14.1.4.2. In a FAPB logic control system, only one hall call will be registered at a time, so that the elevator may be used without interruption. When not in use, the elevator will answer a registered hall call, but hall calls will not be registered while the elevator is in use.

### 14.1.6 AUXILIARY ELEVATOR EQUIPMENT

Many items of auxiliary equipment and devices are necessary for proper elevator operation. Some of these pieces of equipment and devices are located in the car, some in the hall or lobby, some on the outside of the car, and others in locations remote from the car.

#### 14.1.6.1 In-Car Equipment and Devices

#### CAR OPERATION STATION

Each elevator is equipped with a main car operating station located integrally in a vertical swing panel. This station contains call registration buttons in accordance with the logic operation selected, an emergency stop switch, an alarm button, tactile plates, and a light switch, as well as any other devices required by the applicable code.

The Americans with Disabilities Act (ADA) requires that devices operable by the general public and mounted in an elevator car operating panel be identified with braille or tactile symbols. In addition, the car operating panel must be located vertically in accordance with ADA requirements.

ADA also requires that an emergency telephone be installed in each elevator cab. It can be a separate unit or part of the main car operating station. Such telephones require specialized controls for handicapped person use.

#### CAR POSITION INDICATOR

A car position indicator must be provided in each car. This indicator must show the car's position by floor designation.

## CAR DIRECTION SIGNS

Each car must contain a car direction sign that indicates the direction in which the car is moving. The direction sign must have an associated audible signal sounding once for up and twice for down.

### 14.1.6.2 Hall Operating Equipment and Devices

#### HALL PUSH-BUTTON STATIONS

Hall push-button stations must be provided at each landing. The push buttons in the station must be illuminated to notify waiting passengers that the call has been registered and must remain illuminated until the call has been answered. ADA requires that these push buttons be identified with braille and tactile symbols. In addition, hall push-button stations must be located vertically in accordance with ADA requirements.

In a FAPB logic control system, a single call button is supplied at each landing. When the elevator is already in use, an illuminated signal is activated indicating that no additional hall call can be registered until the light is extinguished.

#### HALL POSITION INDICATORS

Hall position indicators are sometimes provided. These show the position of the car in the hoistway with floor numbers or letters.

#### CAR DIRECTION SIGNS AND HALL LANTERNS

Car direction signs and hall lanterns include directional indications and an audible signal. The appropriate arrow is illuminated to show the direction in which the car is set to travel. An audible signal alerts passengers on the landing, sounding once for up and twice for down.

#### EMERGENCY PANEL

An emergency status panel may be provided and, especially in large buildings, in hospitals, and in some other building types, may be required by code. These panels contain the following items:

- Digital position indicator for each elevator within the group
- Controller indicator lights
- Controller key switches
- Emergency power indicator lights and key switches
- Fire service indicator lights

- Fire recall key switches
- Emergency telephone jack

### 14.1.6.3 Equipment and Devices Outside of the Car

Each elevator must have an inspector's operating station mounted on top of the car. It must contain the following items:

- Up and down constant pressure buttons
- An emergency stop switch
- A light with a guard and switch
- A duplex, 120 VAC outlet

In addition, a light with a guard and switch and a duplex, 120 VAC outlet must be located under the bottom of each car.

### 14.1.6.4 Remote Devices

An emergency stop switch is required in each elevator pit. It must cut off the power supply to the motor and bring the car to rest independent of other devices.

An alarm bell must be located in or adjacent to the hoistway. It must be connected to an alarm button mounted in the main operating panel in each car.

### 14.1.7 ELEVATOR SELECTION CRITERIA

Elevators are generally package units or are constructed using packaged components. The following criteria should be addressed in determining which type elevator is needed and the types of features it should have:

1. The type of unit (passenger, hospital, service, freight, residential). This influences the types of controls required, the size of the car, the types of doors, and the load-carrying capacity needed. Passenger use loads generally range in 500 lb (226.8 kg) increments from 2000 lb (907 kg) to 3500 lb (1588 kg). For hospital or service use, capacity is usually 4000 lb (1814 kg), 4500 lb (2041 kg), or 5000 lb (2268 kg). Freight elevator capacities range from 2000 lb (907 kg) up to 100,000 lb (45 359 kg). An *observation elevator* is a special type basically made up of one of the other types but arranged to ride up the exterior of a building or in a multistory atrium space in a glass hoistway.

2. For a freight elevator it is necessary to determine:
- The type of freight that is to be carried, which can range from general freight to loaded semitrucks

- How the elevator will be loaded and unloaded (e.g., by hand-carrying freight or by loading material with a forklift)
- The maximum weight of any single item to be moved
- The maximum size of any single item to be moved (width, length, height)
- Whether passengers will use the elevator

3. The number of stops. This influences whether a traction or hydraulic elevator should be selected and which type of machine should be used (geared or gearless for a traction elevator and inground or holeless for a hydraulic elevator).

4. Speed. The speed at which an elevator will ascend and descend influences the type of machine (geared or gearless) and is in turn influenced by the number of stops and the height of the building. Buildings up to about six stories can operate well at speeds between 200 ft (61 m) per minute and 450 ft (137.4 m) per minute. Taller buildings need greater speed. Hydraulic elevators generally operate at speeds of up to 150 ft (45 m) per minute. Geared traction elevators operate up to about 450 ft (137.4 m) per minute. Higher speed requirements dictate the use of gearless traction machines. Very tall buildings require very high speeds. The elevators in the John Hancock Center in Chicago move at speeds of up to 1800 ft (550 m) per minute. Freight elevators, on the other hand, have little use for high speeds. They can operate at 50 ft (15 m) to 350 ft (107 m) per minute, but most operate in the 80 ft (24.4 m) to 100 ft (30.5 m) per minute range.

5. Travel. The total travel distance of an elevator influences the type of drive and whether the elevator should be of the traction or hydraulic type. Hydraulic elevators work well up to about five stories and can be used for up to eight floors. Higher buildings should have traction elevators.

6. Operation. The type of operation is dictated by the height of the building and the type and configuration of the elevators.

7. Number of passengers and wait time. The number of passengers and their distribution within the building control the number and size of the elevators required. Generally, there should be one elevator for every 250 persons in a building of six or more floors.

The number and distribution of elevators are also influenced by the length of time passengers will have to wait for an elevator. People become annoyed when made to wait as little as 20 seconds in an office building but might be willing to wait for about a minute in an apartment building.

When the number of passengers is large, elevators are arranged in groups (*banks*) of up to eight cars.

In buildings exceeding 15 floors, there may be one or more banks of high-speed express elevators that travel directly from the lobby level to an upper floor, bypassing lower floors. The lower floors are served by banks of *local* elevators.

Double-deck elevators are sometimes used to increase capacity without reducing floor space. Similarly, in a *sky-lobby* system, express elevators take passengers to a level high in the building. There they leave the express elevator and take local elevators to their destination floors. These local elevators start at the sky-lobby floor, which leaves space below them open for rental use. Several levels may be used as sky lobbies.

**8.** Platform size. Platform size is dictated by the use of the elevator and the number of passengers to be accommodated or the amount or size of freight to be carried. Cars for hospital elevators are deeper from front to back than passenger cars to accommodate patient beds and stretchers. Platform size can be limited by the load capacity of the elevator.

**9.** Passenger car and hoistway entrance size. The car and hoistway entrance size is dictated by the size and type of elevator.

**10.** Door operation. Passenger car doors are either single-slide, center-opening (may not be available for 2000 lb cars), or two-speed.

**11.** For a freight elevator, the following steps are necessary:

■ Determine the minimum door size (finished opening dimension—width and height).
■ Determine the type of doors (i.e., manually or power-operated, biparting or other).
■ Determine the material, configuration, and construction of the doors.
■ Determine the load class.

**12.** Machine room location. For overhead traction elevators, the machine room is located above the shaft in a penthouse. Machine rooms for underslung traction elevators must be adjacent to the shaft and at a lower level. A machine room for a hydraulic elevator can be in any location, but the preferred location is at the lowest level and as close to the shaft as possible, preferably immediately adjacent to the shaft.

**13.** Weight. The weight of the elevator and associated machinery is a major factor in selecting the type of elevator to use in freight-handling applications. For the same capacity and size, traction elevators tend to be heavier than hydraulic elevators. In addition, the heavy load capacity associated with some freight elevators adds considerable weight and increases the structural requirements of the building substantially. This factor may be an influence when the building under consideration does not exceed the height limit of hydraulic elevators. The shaft depth for heavy-load-carrying inground hydraulic elevators might have to be greater than the travel distance of the elevator. If there are underground water problems or if drilling is difficult because of underground conditions, the cost of drilling might exceed the cost of the additional framing required for the use of a traction elevator. In general, however, within their practical limits, hydraulic elevators are more economical for freight-carrying applications than traction elevators.

## 14.1.8 ELEVATOR INSTALLATION

Elevator installation is done in accordance with the manufacturer's shop drawings and following the manufacturer's instructions. To complete an elevator installation, the following additional work must be performed according to governing codes:

A legal hoistway must be constructed with the required fire rating, with properly sized rough openings, adequate fastening for hoistway entrance assemblies, and clear overhead of adequate height including suitably sized support for necessary hoisting. The front entrance wall at the main landing is not constructed until all elevator material has been located in the hoistway. The other front entrance walls are not constructed until door frames and sills are in place.

Supports for guide rails must be installed. Separator beams must also be installed where required.

There must be a properly located, ventilated, heated, lighted, and sound-isolated room large enough to house the elevator machinery. The temperature must be maintained between 55°F (12.7°C) and 90°F (32.2°C), with relative humidity not to exceed 95%. The entrance door to the machine room must be large enough to allow placing of the elevator equipment.

A dry pit, reinforced to sustain impact loads on guide rails and buffers, must be provided.

Landings must be configured to receive entrance sills.

A fixed vertical metal ladder must be installed in the pit. It must extend at least 42 in. (1066 mm) above the sill of the lowest terminal entrance.

The building structure must be designed to support the loads that will be imposed by the elevator and its equipment.

If sprinklers are installed in the hoistway, machine room, or machinery spaces, the electric power must automatically be disconnected before water is applied. Smoke detectors may not be used to activate these sprinklers or to disconnect the power supply.

The proper type of electric power must be brought to the machine room and connected to controller terminals there.

Electric power with fused disconnect switches must also be installed for car lights and other devices requiring power.

Switched light fixtures and convenience outlets with a ground fault circuit interrupter must be installed in the machine room.

A suitable switched light fixture and a convenience outlet with a ground fault circuit interrupter must be provided in the pit.

Where required, a means of standby power must be provided to operate the elevators in the event of a general or building power failure.

## 14.1.9 LIFTS

Elevators usually transport people and materials more than one floor. Vertical wheelchair lifts transport people only from one level to the next. Inclined wheelchair lifts transport people for up to three floors. Other types of lifts are intended to transport either materials or people only from one level to the next.

### 14.1.9.1 Wheelchair Lifts

Wheelchair lifts are often used in building remodeling projects when the existing building is not barrier-free in accordance with code and legal accessibility requirements (see Section 1.4). ANSI A117.1 is the standard for wheelchair lifts.

## VERTICAL WHEELCHAIR LIFTS

Vertical wheelchair lifts may transport people up to 12 ft (3658 mm) vertically, except that the maximum travel distance in a residence is 10 ft (3048 mm). ANSI A117.1 requirements include operating speeds of lifts not in excess of 19 fpm (0.097 m/s) and a capacity of 500 to 750 lb (226.8 to 340.2 kg).

ANSI A117.1 requires doors not less than 42 in. (1067 mm) wide at the top and bottom of wheelchair lifts, except that doors in residences can be 36 in. (914 mm) wide and guards can be installed at the bottom instead of a door.

## INCLINED WHEELCHAIR LIFTS

Inclined wheelchair lifts may travel up to three floors. They are required to have 42 in. (1067 mm) self-closing doors at the top and bottom, except that in a private residence no door is required at the bottom, and the door at the top may be 36 in. (914 mm) wide. Retractable guards are required to prevent the wheelchair from rolling off the platform.

### 14.1.9.2 Stair Lifts

Stair lifts are installed at existing stairs. They can be used at straight or spiral stairs. Most stair lifts consist of a power-driven seat supported on a rail or tube fastened to the stair or to adjacent walls. The seats usually fold up out of the way when not being used.

The operating speed of a chair lift should not exceed 25 fpm (0.127 m/s), except that on spiral stairs the maximum should be 10 fpm (0.05 m/s). Typical capacity is 500 lb (226.9 kg). Stair lifts are controlled by buttons on the seat arm and at the top and bottom of the lift.

# 14.2 Escalators and Moving Walks

Like an elevator, an escalator is a device for moving people vertically from one building level to another. A mov-

ing walk is a device for moving people along at the same level or at small changes on levels.

## 14.2.1 ESCALATORS

Escalators are very efficient in moving large numbers of people from one building level to another. Their use is usually limited, however, to changes of five or fewer floors. Elevators are generally used when more floors are involved.

Escalators require more floor space than elevators, but this is balanced by the lack of need for separate lobbies and for large equipment rooms. In addition, there is no waiting for the next car, and crowds are more dispersed. For movement from one floor to the next, escalators are generally faster than elevators because there is no waiting time. It takes longer, however, to move multiple floors on escalators than in elevators because elevators move much faster than escalators.

### 14.2.1.1 Operation

Escalator tread movement is generated by a step chain and drive connected to a worm-gear drive machine and brake assembly containing an electric motor. This drive equipment is mounted in the truss space beneath the escalator.

### 14.2.1.2 Configuration and Materials

Most escalators offer a straight run (Fig. 14.2-1), but one manufacturer has introduced a spiral configuration that may one day become as prevalent as the present configuration.

Available escalator finishes include stainless steel, anodized aluminum, and enameled steel.

Steps, risers, and top and bottom floor plates are stainless steel, extruded aluminum, or die-cast aluminum, all with an abrasive finish.

Balustrades may be stainless steel or glass. The glass may be clear or colored.

Moving handrails are synthetic or plastic. Various colors are available.

Nominal escalator widths between rails are 24 in. (606 mm), 32 in. (800 mm), 40 in. (1000 mm), and 48 in. (1200 mm).

An escalator's carrying capacity varies from 2880 to 9000 persons per hour, depending on the escalator's speed, rise, and width.

The usual angle of inclination is 30 degrees, forward speed is 100 fpm (30 mpm), and maximum rise is 21 ft 4 in. (6500 mm).

High-speed, high-rise escalators are available, however, for use when conditions warrant. Consult manufacturers for width, carrying capacity, angle of inclination, speed, and maximum rise.

### 14.2.1.3 Safety Features

Safety features required by ANSI A17.1 include a skirt guard safety switch that will shut off the escalator if an article, such as a shoe, is caught in the tread or riser, a step-chain safety device that will stop the escalator if there is a problem with the step chain, an emergency stop button that will stop the escalator instantly when the button is depressed, a drive chain safety device that will stop the escalator if the drive chain breaks, skirt guards, handrail inlet guards, a reversal stop device that will automatically stop the escalator if the drive direction is inadvertently reversed while the escalator is ascending, a comb step impact de-

**FIGURE 14.2-1** Section cut through an escalator. (Drawing by HLS.)

vice, a missing step device, a handrail speed monitoring device, a step level device, a step upthrust device, and wedge guards between the balustrades and adjacent surfaces.

### 14.2.1.4 Installation

An escalator is supported on a steel truss that is furnished and installed by the escalator supplier. When the escalator rise exceeds 18 ft (6.1 m), additional support may be necessary. The floor opening for an escalator is sized according to the escalator manufacturer's directions. The design of the surrounds for this opening must comply with applicable codes, sprinkler heads, and smoke and fire detectors where required.

An escalator should be installed in strict accordance with the escalator manufacturer's instructions and approved shop drawings.

### 14.2.2 MOVING WALKS

Moving walks are very efficient at moving large numbers of persons. They are used in airport terminals, train stations, subway stations, shopping centers, sports arenas, and other locations where large crowds gather. They operate horizontally or at slopes with a maximum of 12°. Vertical rises of 30 (9.14 m) or more are possible.

Moving walks are generally limited to lengths between 100 and 300 ft (30.48 and 91.44 m), but some permit lengths of 490 ft (149.35 m) or more.

Widths of 26, 32, 40, 48, and 56 in. (660.4, 812.8, 1016, 1219.2, and 1422.4 mm) are available. Widths of 26 in. (660.4 mm) accommodate one person, and widths of 40 in. (1016 mm) permit walking passengers to pass standing passengers.

ANSI A17.1 applies to moving walks. It dictates required dimensions, speeds, and safety features.

A moving walk requires a pit to accommodate its operating mechanism, motor, and drive chains. Some moving walks are supported by trusses in a manner similar to those that support escalators. Others do not require trusses.

Balustrades are available in glass in a variety of colors and as stainless steel, aluminum, and powder-coated aluminum or steel. Handrails are available in a variety of colors. Palates (the walking surface) are stainless steel or ribbed aluminum.

A moving walk should be installed in strict accordance with the manufacturer's instructions and approved shop drawings.

## 14.3 Additional Reading

More information about the subjects discussed in this chapter can be found in the references listed in Section 14.4 and in the following publications.

Barker, Frederick. 1997. "Is 2,000 Feet per Minute Enough?" *Elevator World* (www.elevator-world.com) (3), March. 1995. © Otis Elevator Company, Farmington, Connecticut, U.S.A., All Rights Reserved. Paper originally presented at the International Conference on High Technology Buildings, Council on Tall Buildings and Urban Habitat, São Paulo, Brazil, October 25–26, 1995.

Fortune, James W. 1995. "Elevator Modernizations." *The Construction Specifier* (November): 60 (www.csinet.org).

Gray, Lee E. 2004. "From Elevator Boy to Automatic Operator: A Brief History of Elevators." *The Construction Specifier* (August): 40 (www.csinet.org).

Knorr, Janis. 1989. "The Ups and Downs of Elevator Planning." *The Construction Specifier* (April): 96 (www.csinet.org).

Lacob, Miriam. 1997. "Elevators on the Move." *Scientific American*, 10 (October) (www.sciam.com).

Strakosch, George R., ed. 1998. *The Vertical Transportation Handbook*, 3rd

## 14.4 References

We would like to thank the authors and publishers of the publications in the following list for their contribution to our research for this chapter.

American National Standards Institute (ANSI). New York (www.ansi.org).
2004. ANSI A17.1, *Safety Code for Elevators and Escalators.*
2003, ANSI A117.1, *Accessible and Usable Buildings and Facilities.*

American Society of Mechanical Engineers (ASME). 2000. A17.1, *Handbook on Safety Code for Elevators and Escalators.* New York: ASME (www.asme.org).

Arcom Master Systems. Salt Lake City, UT (www.arcomnet.com). *MasterSpec.® Basic Sections:*
14100, "Dumbwaiters," 5/04 ed.
14210, "Electric Traction Elevators," 5/04 ed.
14240, "Hydraulic Elevators," 5/04 ed.
14310, "Escalators," 5/04 ed.
14320, "Moving Walks," 5/04 ed.
14420, "Wheelchair Lifts," 5/00 ed.
14560, "Chutes," 2/03 ed.

National Fire Protection Association (NFPA). Quincy, MA: NFPA (www.nfpa.org). 2005. NFPA 70, *National Electrical Code.*
———. 1999. NFPA 80, *Standard for Fire Doors and Fire Windows.*
———. 2003. NFPA 101. *Life Safety Code.*®

Ramsey, Charles George, Harold Reeve Sleeper, and John Ray Hoke, Jr. 2000. *Architectural Graphic Standards*, 10th ed. Hoboken, NJ: John Wiley & Sons, Inc. (www.wiley.com).

Underwriters Laboratories, Inc. (UL). 2001. "10B, Fire Tests of Door Assemblies." Northbrook, IL: UL (www.ul.com).

ed. Hoboken, NJ: John Wiley & Sons, Inc. (www.wiley.com).

# Fire Suppression

# Introduction

This chapter addresses water-based fire suppression systems, including standpipe systems and sprinkler systems, and fire extinguishing systems, including those using foams and those using gas. *MasterFormat* 2004 Division 21, "Fire Suppression," includes components of fire suppression systems that are beyond the scope of this book. These include, but are not limited to, fire hydrants, fire department connections, hoses, nozzles, racks, valves, and cabinets, fire pumps, and fire suppression water storage tanks. Fire prevention, containment, and control are discussed in Section 7.12, manual fire extinguishers in Section 10.1, and fire detection and alarm in Section 20.2.

# Sustainability Considerations

Water used in fire suppression systems often destroys or severely damages the contents, finishes, and sometimes even construction elements of a building. This use causes loss of the materials affected. Water can also vaporize into steam and severely harm firefighters or cause a threat to humans if electricity is present.

Other fire suppression systems do not cause these particular problems, but some of them have their own environmental disadvantages. Carbon dioxide suppression systems, for example, may not damage the building's structure or contents, but they work by displacing oxygen, which can be fatal to people. Until 1994, halon 1301 gas systems were used to suppress fires in computer rooms, on commercial aircraft, and in museums, libraries, and the like. Halon was not directly harmful to people and did not damage property, but it became recognized as a significant contributor to damage to the stratospheric ozone layer and is no longer used. Inert gas and clean agent systems have been introduced as substitutes for halon, These agents do not harm people or the ozone layer, but they are more successful at protecting a building's contents than its structure. More successful systems will no doubt become available in the future. These agents must be so constituted that they will suppress fires without harming people or the environment.

# Applicable *MasterFormat* Sections

Subjects addressed in this chapter are included in *MasterFormat* 2004, Division 21, "Fire Suppression." The following *MasterFormat* Level 2 sections are applicable to this chapter:

21 11 00  Facility Fire-Suppression
          Water-Service Piping
21 12 00  Fire-Suppression Standpipes
21 13 00  Fire-Suppression Sprinkler
          Systems
21 21 00  Carbon-Dioxide Fire-
          Extinguishing Systems
21 22 00  Clean-Agent Fire-
          Extinguishing Systems
21 24 00  Wet-Chemical Fire-
          Extinguishing Systems
21 24 00  Dry-Chemical Fire-
          Extinguishing Systems
21 41 00  Storage Tanks for Fire-
          Suppression Water

# 15.1 Fire Suppression Standpipe Systems

Fire suppression standpipe are a series of pipes fed by the public water system or by a separate storage tank on site or within a building. Standpipe systems are required in some building types by the *International Fire Code*. The required provisions are covered in the code.

Standpipe systems are divided into three classes. Class I systems are the ones most used today. They consist of $2^{1}/_{2}$ in. (64 mm) pipe and are used in buildings with sprinkler systems. They are intended for use by trained fire fighters.

Class II systems have a $1^{1}/_{2}$ in. (38 mm) pipe. They are intended for use by building occupants and for initial response by fire fighters.

Class III systems are a combination of Class I and Class II systems, having both a $2^{1}/_{2}$ in. (64 mm) and a $1^{1}/_{2}$ in. (38 mm) pipe.

Standpipe systems are one of the following types:

- *Automatic dry type*, which is filled with compressed air. It has an automatic valve which will admit water when a hose valve is opened.

- *Automatic wet type*, which is filled with water.

- *Manual dry type*, which has no water supply. It requires fire fighters to connect the standpipe system to a water supply when required.

- *Manual wet type*, which has water in the pipes, but in insufficient quantities to completely provide the building's needs in the event of a fire. This type requires fire fighters to connect the standpipe system to a water supply.

- *Semiautomatic dry type*, which has no water but is automatically connected to a water supply when a valve is activated at a hose outlet.

Fire suppression standpipe systems require hoses and nozzles, hose reels, hose racks, hose valves, and valve and hose cabinets.

## 15.2 Fire Suppression Sprinkler Systems

Sprinkler systems are expensive, but many insurance companies offset some of the costs by offering reductions in premiums for professional installations. In addition, most jurisdictions require sprinklers in most or all commercial and institutional buildings, and some require them in multifamily and even in single-family houses. Some codes permit taller buildings or increased floor areas when sprinklers are used. Installing a sprinkler system throughout an entire existing building can be prohibitively expensive and is seldom justified unless required by law. A sprinkler system should be considered, however, in every new building, even when not required by code, except where other extinguishing systems are more appropriate (see Section 15.3). Other extinguishing systems may be appropriate for several reasons. For example, water can damage some equipment beyond repair, even if the fire is extinguished before harming the equipment. In addition, once a sprinkler system is activated, it will continue to deliver water until it is manually turned off, which can result in significant damage to a building and its contents even after the fire has been extinguished.

The *International Fire Code* requires fire suppression sprinkler systems in all Group A (assembly occupancy) buildings (see Section 1.3) where the conditions stipulated in the *Code* exist, except when local officials permit other types of extinguishing systems and in the case where sprinkler use would drastically damage building contents, such as in computer facility spaces, telecommunications complexes, and spaces housing electrical power distribution equipment, provided that those spaces have acceptable fire alarm systems and are separated from the rest of the building by approved fire-resistant construction. Mitigating stipulations include the area within a fire zone, the fire zone's occupancy load, and the location within the building. These stipulations are complex and require that a designer determine from the *Code* in which locations sprinklers might not be required.

The *International Fire Code* also requires sprinklers in Group E (educational) occupancy areas where the area with a fire zone exceeds 20,000 ft$^2$ (1858 m$^2$), and in every area in a level below the exit level, regardless of the area involved, except where each classroom has an exit directly to the exterior.

The *International Fire Code* also requires sprinklers in buildings in all other occupancy groups except two, with some exceptions related to size. Sprinklers are required throughout some Group R (residential) buildings and in specified areas of other Group R buildings, such as storage rooms. A notable exception is granted for Group R3, which consists primarily of single-family and two-family dwellings and certain other incidental uses, such as day care facilities having five or fewer occupants. Sprinklers are not required in buildings classified as Group U (utility and miscellaneous) buildings, such as storage sheds, barns, carports, and agricultural buildings.

The purpose of a sprinkler system is to save property and minimize structural damage. There is controversy regarding the lifesaving benefits of sprinkler systems as compared with smoke detectors (see Section 20.2) in small buildings. The question is whether the benefits outweigh the higher costs of installation and maintenance of sprinkler systems relative to smoke detectors.

The National Fire Protection Association's (NFPA) Standards 13D and 13R contain industry standards for the installation of automatic sprinkler systems in residential buildings. Some municipalities have adopted less stringent standards but require sprinkler systems in all new construction. Other NFPA publications cover sprinkler systems in other building types. Some of these standards are listed in Section 15.5.

Sprinkler system types include *dry-pipe*, *wet-pipe*, *closed loop*, *preaction*, *deluge*, and *mist* systems. They all generally consist of water pipes running at right angles through or under overhead structural framing and sprinkler heads spaced along the pipes to serve as nozzles.

A *dry-pipe* system's pipes contain air. The loss of air pressure from the opening of a sprinkler head or the detection of a fire condition causes the release of water into the piping system and out of the opened nozzle. Dry-pipe systems are used in areas subject to freezing and in other areas where leaks would cause severe damage.

A *wet-pipe* system's pipes contain water. This system type should be used in areas not subject to freezing, in lieu of a dry-pipe system, because it tends to deliver water to the fire a little faster than dry-pipe systems.

A *closed-loop* system is a wet pipe system in which the water filling the pipes is also used as the circulating water for heating, ventilation, and air conditioning water heat pumps.

*Preaction* systems are similar to dry-pipe systems, except that the system is activated by a separate alarm that opens valves to allow water to fill the pipes before the heads open, thus permitting time for possible manual extinguishment before the heads release their water.

*Deluge* systems are dry systems with open heads. Separate detectors cause valves to open for water to enter the pipes. Therefore, all heads emit water at the same time.

*Mist* systems emit water in a fine mist that is less damaging to a building's contents than the streams emitted by other water-based systems.

Sprinkler head types are *upright* (SSU), *pendant* (SSP), and *side wall*. Most sprinkler systems use pendent types, which hang beneath the pipes. Pendent sprinkler heads usually project through suspended ceilings where the pipes are concealed above the ceiling. Upright types mount on top of the pipes. Side wall types are mounted on vertical surfaces.

In case of a fire, a fusible link in a sprinkler head melts or a glass bulb breaks, allowing the struts in the center of the head to fall out and release a plug. Emerging water hits a detector, creating a wheel-shaped spray pattern. Sprinkler systems are effective against Class A fires (those that can be controlled with water).

Sprinkler heads fall into several categories, depending on their use and the type of hazard. One way to measure response time is by use of the *response time* index (RTI), which is a measure of the time required for a sprinkler head to open. The RTI of a sprinkler head is the square root of this time in meters-seconds. *Standard-response* sprinkler heads have a RTI of 80 or more. *Quick-response* sprinkler heads, which are also called *fast-response* heads, are required in all low-hazard occupancies. They have a RTI of 50 or less. *Residential* sprinkler heads are *fast-response* types, designed for maximum protection of occupants.

*Early suppression fast-response* (ESFR) sprinkler heads are used in specialized situations, such as in areas where material is stacked highly. They operate at higher pressures and flow rates than do ordinary quick-response sprinklers.

*Extended coverage* sprinkler heads are designed specifically for use where ceilings are completely unobstructed by hanging lights, diffusers, or other devices.

Sprinkler system design, including the type of system and the type of sprinkler heads and their spacing, is complex and is, therefore, usually done by a professional engineer.

## 15.3 Fire Extinguishing Systems

Other fire extinguishing systems are available for use in areas where water-based fire suppression systems are inappropriate. Use of such systems is acceptable to comply with the requirements of the *International Fire Code*, when they are appropriate for the hazard involved, where the use of water-based systems is not appropriate, and when the local officials who are responsible approve of their use. These other fire extinguishing systems include foam systems, gas-based systems, and chemical-based systems.

*Foams* are gas-filled bubbles generated either mechanically by the delivery heads or chemically produced. Some foams are dispensed in water in a foam-water sprinkler system. Some foams are produced by blowing air through a nylon net that has a coating of soap or chemicals. Foams form an oxygen-blocking barrier on the surface of materials, thus extinguishing fires and preventing their spread. They are particularly effective on burning liquids, because they float on the surface of the liquid and block oxygen contact with it. Foams are generally not harmful to equipment, but fire fighters require specialized respirators and life lines because foams exclude oxygen and block visibility.

*Gas* fire extinguishing systems include carbon dioxide ($CO_2$) systems and clean-agent gases. The earlier use of halon 1301 gas as a fire extinguishing agent has been discontinued. $CO_2$ systems have been used for many years to fight chemical and combustible-liquid fires. They are effective and are generally not harmful to equipment. They are, however, dangerous in areas where there are people, because they exclude oxygen to the extent that unconsciousness results in a short time, perhaps preventing escape.

The proliferation of numerous *clean-agent gases* came about as a result of the discontinuation of halon 1301. One such agent is heptafluoropropane, CF3-CHF-CF3, which bears the American Society of Heating, Refrigerating, and Air-Conditioning, Engineers (ASHRAE) designation HFC-227ea. Unlike halon, it does not affect the ozone layer, which is also the case with most other clean-agent gases. Clean-agent gas systems are appropriate for use where water would harm computer, telecommunication, and other sensitive equipment.

## 15.4 Additional Reading

More information about the subjects discussed in this chapter can be found in the references listed in Section 15.5 and in the following publications.

Dagostino, Frank R. and Joseph Wujek. 2005. *Mechanical and Electrical Systems in Construction and Architecture*, 4th ed. Upper Saddle River, NJ: Prentice Hall (www.prenhall.com).

Kutz, Myer. 1998. *Mechanical Engineers' Handbook*, 2nd ed. Hoboken NJ: John Wiley & Sons, Inc. (www.wiley.com).

Lechner, Norbert. 2000. *Heating, Cooling, Lighting: Design Methods for Architects*, 2nd ed. Hoboken, NJ: John Wiley & Sons, Inc. (www.wiley.com).

Stein, Benjamin, et al. 2006. *Mechanical and Electrical Equipment for Buildings*, 10th ed. Hoboken, NJ: John Wiley & Sons, Inc. (www.wiley.com).

Tao, William K. Y. and Richard R. Janis. 2004. *Mechanical and Electrical Systems in Buildings*, 3rd ed. Upper Saddle River, NJ: Prentice Hall (www.prenhall.com).

## 15.5 References

We would like to thank the authors and publishers of the publications in the following list for their contribution to our research for this chapter.

Dagostino, Frank R. and Joseph Wujek. 2005. *Mechanical and Electrical Systems in Construction and Architecture*, 4th ed. Upper Saddle River, NJ: Prentice Hall (www.prenhall.com).

Mawhinney, Jack R. 2000. "Testing for the Design of Water-mist Fire-suppression Systems." *The Construction Specifier* (December): 45 (www.csinet.org).

National Fire Protection Association (NFPA). Quincy, MA (www.nfpa.org). 1999. *Automatic Sprinkler Systems Handbook*.

———. 2000. NFPA 12, "Standard on Carbon Dioxide Extinguishing Systems."

———. 2002. NFPA 13, "Installation of Sprinkler Systems."

———. 1999. NFPA 13D, "Standard for the Installation of Sprinkler Systems in One- and Two-Family Dwellings and Mobile Homes."

———. 2002. NFPA 13R, "Standard for Installation of Sprinkler Systems in Residential Occupancies Up to and Including Four Stories in Height."

———. 2003. NFPA 20, "Standard for Installation of Stationary Fire Pumps for Fire Protection."

———. 2002. NFPA 25, "*Water-Based Fire Protection Systems Handbook*."

———. 2003. NFPA 101, "Life Safety Code."

———. 1996. NFPA 750, "Standard lon Water Mist Fire Protection Systems."

———. 2000. NFPA 2001, "Standard on Clean Agent Fire Extinguishing Systems."

———. 2004/2005. "National Fire Codes."

Ramsey, Charles George, Harold Reeve Sleeper, and John Ray Hoke, Jr. 2000. *Architectural Graphic Standards*, 10th ed. Hoboken, NJ: John Wiley & Sons, Inc. (www.wiley.com).

Razwick, Jerry. 2001. "Fire Sprinklers and Fire-rated Systems." *The Construction Specifier* (June): 37 (www.csinet.org).

Stein, Benjamin, et al. 2006. *Mechanical and Electrical Equipment for Buildings*, 10th ed. Hoboken, NJ: John Wiley & Sons, Inc. (www.wiley.com).

**CHAPTER 16**

# Plumbing

# Introduction

The earliest discovered plumbing systems were found in the ruins of Minoan palaces on the island of Crete. Four thousand years ago, the Minoans produced plumbing systems that incorporated most of the features of systems used today, including flush toilets and vented drainage systems. It was the Romans, however, who gave plumbing its name, based on *plumbum* the Latin word for lead, which they used for piping. Although famed more for their aqueducts, the Romans also gave their buildings both hot and cold water supply systems.

Nineteenth-century England was the place where the flushing water closet was perfected. Thomas Crapper was knighted for this accomplishment.

Versions of this chapter in earlier editions of this book, and the illustrations in them, were prepared for the United States League of Savings Institutions under a research grant by the staff of the Building Research Council, University of Illinois at Urbana/Champaign. This chapter and many of its illustrations have been updated for this edition, but much of the basic information remains essentially the same as it was in earlier editions.

**FIGURE 16-1** Components of a residential plumbing installation. (Courtesy Small Homes Council, Building Research Council, University of Illinois, Urbana-Champaign.)

Residential and commercial plumbing installations generally consist of a water supply system, a drainage and venting system to conduct away wastes, and plumbing fixtures and equipment. Figure 16-1 shows these basic systems for a residential building. These systems for other building types and larger buildings are much more complicated, but their basic components are the same.

Public and private water supply utilities that serve a building site, public sewage disposal utilities, storm water disposal systems, and subdrainage systems are discussed in Chapter 23. This chapter covers water supply and drain, waste, and vent (DWV) systems within a building, storm water collection and removal within a building, and private disposal of sewage on the building site.

# Sustainability Considerations

Sustainability considerations related to the materials used in plumbing systems are similar to those discussed in the various sections related to the materials used. Considerations for metal piping, for example, are similar to those discussed in Chapter 5. Additional considerations related to water and sewerage systems are discussed in Chapter 23. Consideration regarding the use of roof ponds to control storm water are discussed in Chapter 7.

### Design

The design of water supply and DWV systems, and the selection of plumbing fixtures can greatly reduce the amount of water that must be taken from the ecosystem and the amount of sewage that must be disposed of. One method for accomplishing this kind of reduction consists of using high-efficiency fixtures and dry fixtures such as composting toilets and waterless urinals to reduce wastewater volumes. Another is to use storm water for sewage conveyance. In larger projects, on-site wastewater treatment can be used to eliminate or reduce the amount of sewage that must be sent into the public system. These measures not only reduce the amount of sewage, they also reduce the amount of water required, thus lessening the demand on local water utility supply systems.

Another method that can be used to reduce the demand on public or private water supply systems is the use of collection and storage devices to collect rain water for reuse in irrigation and for such domestic uses as flushing toilets, clothes cleaning, and dishwashing. Wastewater that is not from toilets or urinals (gray water) can also be recycled for some domestic use, but it requires a separate collection system and some filtration or other treatment before use. Properly treated gray water can be used for irrigation, toilet flushing, and other domestic uses where deemed acceptable by the applicable code and local jurisdictional requirements.

### Conservation

Until recently, water was used without regard for future supplies. Today, as groundwater supplies in many areas dwindle, water is regarded as a limited natural resource. Using it wastefully in plumbing systems not only incurs municipal costs in purifying it for consumption, in pumping it to the point of use, and in treating it before discharge into a body of water, but also reduces the amount available in underground aquifers. *Aquifers* are large underground stores of water contained either in large cavities or distributed in smaller voids of porous, granular soils. Although the earth has an enormous reserve of water, 97% of it is salty and more than 2% is trapped in polar ice caps and glaciers. Therefore, less than 1% of all the earth's water is available to us in lakes, rivers, and underground aquifers. Rapid urbanization and industrialization of many areas increases pumping from the aquifer while at the same time reducing the replenishment process from melting snow and rainwater soaking back into the ground. Surface waters are rushed from streets and parking lots to storm sewers and

thence to salty oceans or polluted freshwater lakes. This water is lost as a valuable resource until it comes back in the form of rainwater or is purified for reuse.

Spurred by recurring rainfall shortages, groundwater depletion, and the high cost of water treatment, water conservation has become a regional and national issue. Although the design of facilities and selection of water-saving equipment are essential, conservation depends largely on the habits and decisions of the people using the water.

The *International Building Code* requires the use of water-conserving devices or fixtures. Flow restrictors can be installed on existing shower heads, for instance, and cut the flow rate by about

50%. New energy-saving shower heads are fairly low in cost and can cut water flow from the usual 5 gpm (18.9 L/min) to between 1.8 and 5 gpm (6.8 to 9.5 L/min). Older water closets required 5 to 7 gal (18.93 to 26.5 L) per flush. The U.S. government currently restricts new water closets to those that use no more than 1.6 gal (6 L) of water per flush. Plastic bottles or dams and other devices specifically manufactured for this purpose inserted into older water closet tanks can reduce water consumption, but energy-saving water closets are generally much more effective than modified older fixtures because they are designed to provide equivalent cleansing action while using less water.

## Applicable *MasterFormat* Sections

Subjects addressed in this chapter are included in *MasterFormat* 2004, Division 22, "Plumbing." The following *MasterFormat* Level 2 sections are applicable to this chapter:

22 10 00  Plumbing Piping and Pumps

22 12 00  Facility Potable-Water Storage Tanks
22 13 00  Facility Sanitary Sewerage
22 14 00  Facility Storm Drainage
22 33 00  Electric Domestic Water Heaters
22 34 00  Fuel-Fired Domestic Water Heaters

22 41 00  Residential Plumbing Fixtures
22 42 00  Commercial Plumbing Fixtures
22 43 00  Healthcare Plumbing Fixtures
22 45 00  Emergency Plumbing Fixtures
22 46 00  Security Plumbing Fixtures
22 47 00  Drinking Fountains and Water Coolers

## 16.1  Codes, Laws, and Industry Standards

The design of plumbing systems is generally governed by a local or statewide plumbing code. Many of these codes are based on the *International Plumbing Code* produced by the International Code Council (ICC) (see Section 1.3.3.3), but some are still based on one of the model codes discussed in Section 1.3.3.3. The code requirements discussed in this section are those contained in the *International Plumbing Code*.

Information concerning water supply and wastewater treatment can be obtained from the U.S. National Drinking Water Clearinghouse. The power to enforce water quality standards rests with state governments, which in turn generally adopt U.S. Environmental Protection Agency (EPA) standards. The cur-

rent EPA standard is *National Primary Drinking Water Standards*. These regulations set maximum limits on trace metal and organic chemical content, turbidity, and chlorine residuals, and establish sampling and monitoring procedures. In addition to federal regulations, some local codes impose their own limitation on impurities, and in some cases mandate water-conserving practices or devices.

Many industry standards have evolved to govern the manufacture and installation of plumbing pipe, fixtures, and equipment. Chapter 6 of the *International Plumbing Code* includes tables listing most of the standards related to the various types of piping and fittings materials acceptable for use in plumbing systems. Figure 16.1-1 lists the names of standard-setting organizations and their abbreviations. Figure 16.1-2 lists many generally accepted standards.

Some types of equipment are either *listed* (placed on a list of acceptable products) or *labeled* by a nationally recognized organization to indicate compliance with certain safety standards. Gas water heaters and pressure relief valves, for instance, should be listed and labeled by the American Gas Association (AGA). Electric and oil-fired water heaters should be listed by Underwriters Laboratories, Inc. (UL). These listings mean that the manufacturer has agreed to comply with standards which UL or AGA prescribes and monitors. Lists of approved products and manufacturers are published periodically by AGA, UL, and other testing organizations and are available to the public. Labels are affixed to the equipment to give graphic evidence of compliance, but labeling is not an automatic adjunct of listing. Some products may be listed but not labeled.

FIGURE 16.1-1 Abbreviations for Standards-Writing Organizations

| Abbreviations | Standard Designation |
|---|---|
| AGA | American Gas Association |
| AHAM | Association of Home Appliance Manufacturers |
| ANSI | American National Standards Institute |
| ARI | Air-Conditioning and Refrigeration Institute |
| ASME | American Society of Mechanical Engineers |
| ASSE | American Society of Sanitary Engineering |
| ASTM | ASTM International |
| AWS | American Welding Society |
| AWWA | American Water Works Association |
| CABO | Council of American Building Officials |
| CISPI | Cast Iron Soil Pipe Institute |
| CSA | Canadian Standards Association |
| EEI | Edison Electric Institute |
| FHA | Federal Housing Administration (now HUD) |
| FS | Federal Specifications, Genera; Services Administration |
| HUD | U.S. Department of Housing and Urban Development |
| ICC | International Code Council |
| ISEA | Industry Safety Equipment Association |
| NAHB | National Association of Home Builders |
| NBBPVI | National Board of Boiler and Pressure Vessels Inspectors |
| NFPA | National Fire Protection Association |
| NSF | National Sanitation Foundation |
| PDI | Plumbing and Drainage Institute |
| SSPMA | Sump and Sewage Pump Manufacturers Association |
| UL | Underwriters Laboratories, Inc. |
| WQA | Water Quality Association |

(Based on a table produced by and reprinted by courtesy of Small Homes Council, Building Research Council, University of Illinois, Urbana-Champaign. The table has been modified to show current applicable organization names.)

# 16.2 Water Supply

A *water supply system* is everything used to obtain water from a source, treat it to make it drinkable, and deliver it to its point of use. This section covers water distribution within buildings. Section 23.1 addresses available sources of water and its distribution to building sites.

## 16.2.1 BUILDING WATER SUPPLY SYSTEM DESIGN

A *water supply system* within a building consists of service piping, distribution piping, branches to plumbing fixtures and appliances, and the necessary fittings and control valves. Section 16.4 discusses plumbing fixtures. Chapter 11 discusses appliances that receive water and may or may not discharge liquid or solid waste, including, but not limited to, dishwashers, clothes washers, and some refrigerators.

Designing a building's water supply system includes laying out the system and selecting and sizing pipes, flow control and pressure-regulating devices, and valves.

A water supply system is always filled with water at greater than atmospheric pressure. Once in a building, it is separated into two parts. The first forms a cold-water supply tree. The second goes through a water heater and becomes a hot-water supply tree. Each of these trees then feeds the appropriate plumbing fixtures and appliances. The trees are composed of vertical supply pipes called *risers* and horizontal pipe runs called *branches* that are connected to the risers. Plumbing fixtures may be connected to risers but more frequently are connected through branch runs.

The basic goals in the design of a water supply system are to:

1. Keep the water from becoming contaminated
2. Maintain adequate pressure and flow rate for operation of plumbing fixtures and appliances
3. Prevent excessive flow and consequent waste of valuable water
4. Prevent excessive pressures, velocities, corrosion, and scale buildup
5. Provide water without noise, fixture staining, or odor

The efficient functioning of a water supply system depends on pipe sizing, fixture and materials selection, and installation.

### 16.2.1.1 Consumption

Consumption levels for residential and small commercial buildings are generally estimated from past experience with similar occupancies. The number, frequency, and pattern of use of various plumbing fixtures are estimated based on the number and type of occupants. From this estimate, the total building demand may be determined. Figure 16.2-1 shows some cold water consumption requirements for residential and other small building occupancies.

For larger installations, a probability method is used which takes into account the fact that not all fixtures will be in use at any one time. Consequently, the probability method assigns a weighted mathematical factor known as a *water supply fixture unit* (wsfu) rating to each fixture in proportion to the load it imposes on the water supply when in use (Fig. 16.2-2). The wsfu values may be combined in a residential application into a *bathroom group* for each bathroom. The usual wsfu for a private bathroom with a flush tank water closet is 2.7 cold and 1.5 hot; for a bathroom with a flush valve water closet the wsfu is 6 cold and 3 hot. The wsfu values for all fixtures in the building are summed and converted to a water demand figure by a graph which expresses the probability factor of simultaneous use of various fixtures (Fig. 16.2-3), taking into account volume, duration of flow, and intervals between operations.

### 16.2.1.2 System Layout

The physical configuration of a water supply system should take into consideration efficiency of materials use, simplicity of construction, protection from freezing, and ease of maintenance and repair.

A primary principle in efficient layout of water supply systems is that all runs should be as short, straight, and direct as possible, and should have as few fittings as possible. Following this principle not only conserves materials, but also reduces internal friction in the pipes. This, in turn, keeps to a minimum pressure loss associated with flow from the water main to fixtures.

**FIGURE 16.1-2** Industry Standards for Plumbing Pipe and Equipment

| Subject or Title of Standard | Standard Designation |
|---|---|
| **Water Piping Materials Standards** | |
| ABS pipe | ASTM D2751 |
| Cast-iron soil pipe | ASTM A74, ASTM C564 |
| Cast-iron water pipe | ASTM A377 |
| Clay pipe | ASTM C700, ASTM C425 |
| Copper drainage pipe | ASTM B306 |
| Copper water tube and pipe, K, L, M | ASTM B88 |
| Brass pipe | ASTM B43 |
| CPVC pipe | ASTM F891 |
| Galvanized steel threaded pipe | ASTM A53 |
| No-hub cast-iron pipe | CISPI 302 |
| Plumbing fixtures | FS WW-(-541ANSI A 112 |
| Polybutylene water distribution pipe | ASTM D3309 |
| PVC thin wall | ASTM D2949 |
| Stainless steel—grade TP-409 | ASTM A312 |
| **Conditioning of Domestic Water Supply** | |
| Water softeners | WQA S-100 |
| Water filters | WQA S-200 |
| **Sewage Disposal System Standards** | |
| Minimum design standards for community sewage systems | HUD 4940.3 |
| Tanks, septic, bituminous-coated metal | UL 70 |
| Installing vitrified clay pipe systems | ASTM C12 |
| Sewage treatment plant design | ASCE manuals |
| **Plumbing System Standards** | |
| Joints and connections | ASTM C564 |
| Pipe protection | AWWA C-203 |
| Water softening | WQA S-100 |
| Sump pumps | SSPMA Recommended Standards |
| Plumbing fixtures | |
| Plastic bathtubs, shower receptors and shower stalls | ANSI Z21.1 and Z21.2, NAHAB Research Foundation or US Testing Company Inc. Label |
| **Domestic Water Heater Standards** | |
| Electric water heaters | EEI Standards, UL 174 |
| Gas water heaters | AGA Standards and listed |
| Oil-fired water heaters | UL 732 |
| Water heater controls | ANSI Z21.22 |
| Pressure relief valves—water heaters | NBBPVI and AGA Standards |
| Hot water tank construction | ANSI Z21.10 |
| Indirect heaters | IBR Code by Hydronics Institute |
| Safety devices | |
| Energy shutoff | UL or AGA* listed and labeled |
| Pressure relief valves | AGA* or NBBPVI listed and labeled by AGA or ASME |

*AGA listed shutoff devices and relief valves shall comply with ANSI Z21.22 and shall be labeled.
(Based on a table produced by and reprinted by courtesy of Small Homes Council, Building Research Council, University of Illinois, Urbana-Champaign.)

In the schematic planning stage of a building, an effort should be made to locate rooms that will contain plumbing fixtures as near as possible to the point of entry of the water supply to the building and the point of exit to the sewer line. If possible, rooms that will contain plumbing fixtures should be clustered so that plumbing fixtures in more than one room share the same wall for pipe distribution (Fig. 16.2-4). This can be done by arranging rooms so that fixtures in two or more rooms are placed back-to-back against the same wall or are vertically aligned or stacked on different floors (Fig. 16.2-5).

Jogs and offsets in plumbing walls containing horizontal lines should be avoided to simplify the plumbing lines. Fixtures in a given room should be arranged so that all of their supply, drain, and vent lines fall in one wall. This practice shortens the piping runs and simplifies framing of the building by reducing the number of penetrations through floors and eliminating the need to furr or double frame more than one wall (Fig. 16.2-6).

Walls used in this way for vertical distribution of plumbing are known as *chase* walls, *wet* walls, or *plumbing* walls (Fig. 16.2-7). They are usually thicker than typical interior walls, allowing sufficient space to accommodate the drain pipe plus one supply pipe crossing behind the drain pipe. In most wood-framed construction, chase walls are built using 2 × 6s (50 × 150s) rather than the typical 2 × 4s (50 × 100s).

When more space is needed, a double wall, each built with wood 2 × 4s (50 × 100s) or metal studs, is used. Sometimes it is necessary to separate the two walls of a double-wall chase to provide even more space. The double-wall method is usually used when the partitions are framed with metal studs. The two metal stud walls forming a plumbing chase are required to be tied together and braced, either with sections of stud or with pieces of gypsum board (see Sections 9.1 and 9.3).

Future expansion of a supply system can be facilitated by including the appropriate tees with plugs in the original installation or *roughing in* the plumbing for fixtures to be added at a later time. Roughing in means installing those parts of a plumbing system that can be completed before plumbing fixtures or appliances have been installed. This includes water supply, drainage, and vent piping and the necessary fixture supports.

To reduce the probability of pipes freezing, supply lines should not be located in exterior walls. If such location is unavoidable, the supply pipes should be run on the inside face of the wall with at least $3\frac{1}{2}$ in. (90 mm) of mineral-fiber, or equivalent, insulation separating them from the outside (cold) face of the wall and with no insulation on the warm side. This practice keeps pipes near the heated interior of the building and slows heat loss from the pipes to the exterior. Every effort should be made to keep water-carrying pipes out of areas subject to freezing. In practice, this means providing freezeproof valves for outdoor

**FIGURE 16.2-1** Water Consumption Estimating Guidelines

| Building Use | Daily Usage Per Occupant | |
|---|---|---|
| | Gallons | Liters |
| Single family residence | 50 to 100 | 189 to 568 |
| Multifamily apartment (per resident) | 60 | 227 |
| Theaters (per seat) | 5 | 19 |
| Schools (day students only) | 15 to 25 | 57 to 95 |
| Stores (per toilet room) | 400 | 1514 |
| Restaurants (per patron) | 7 to 10 | 26 to 38 |
| Hotels and motels (per bed) | 40 to 50 | 151 to 189 |

To determine the required daily consumption of one of these buildings, simply multiply the consumption by the number of occupants.

**FIGURE 16.2-2** Fixture Unit Ratings and Water Flow

| Fixtures | Fixture Units | Water Flow, gpm (L/m) | |
|---|---|---|---|
| | | Cold | Hot |
| Water closet, flush valve | 10 | 30–45 (113.56) | 0 |
| Water closet, tank type | 5 | 18 (68.14) | 0 |
| Kitchen sink | 2 | 5 (18.93) | 5 (18.93) |
| Lavatory | 1 | 5 (18.93) | 5 (18.93) |
| Bathtub | 2 | 6 (22.71) | 6 (22.71) |
| Shower stall | 2 | 6 (22.71) | 5 (18.93) |
| Laundry tray | 3 | 10 (37.85) | 10 (37.85) |
| Clothes washer | 3 | 5 (18.93) | 5 (18.93) |
| Dishwasher | 3 | 5 (18.93) | 5 (18.93) |

(Courtesy Small Homes Council, Building Research Council, University of Illinois, Urbana-Champaign.) Metric added by the author.

**FIGURE 16.2-3** Water demand curves. Locate fixture units on the curve and read demand on the left scale. Curve 1 is for buildings with flush valve WCs; curve 2 is for the tank type. (Courtesy Small Homes Council, Building Research Council, University of Illinois, Urbana-Champaign.) Metric added by the author.

sillcocks, providing drain valves for seasonal drainage of outdoor piping, and locating supply pipes away from exterior walls.

When laying out a water supply system and fixture locations, the resulting

drainage system location must also be considered. Care should be taken in locating framing so that a plumbing wall will not fall in the same plane as roof trusses or floor or ceiling joists, which would obstruct the passage of supply, drain, and vent lines in and out of the plumbing wall. Likewise, if major framing elements run perpendicular to the plumbing wall, care should be taken that the fixtures requiring below-floor drains are located so that the drains do not conflict with the floor framing. Wood joists should not be notched in the middle third of the span, and holes should not be larger than $2^{1}/_{2}$ in. (63.5 mm) in diameter.

During the design process, thought should also be given to the future accessibility of plumbing for repairs. In buildings with crawl spaces, sufficient height should be provided for a person to conveniently access the piping. The minimum crawl space depth required by many codes is 18 in. (457 mm); however, this is the depth required for moisture protection rather than access to equipment in the crawl space. For easy access, a crawl space should be at least 2 ft (610 mm) and preferably 3 ft (914 mm) high. Access panels into plumbing walls simplify repairs and are usually required by code for access to valves.

Cold water systems are generally noncirculating, meaning that the water travels to the fixtures and remains stationary in the pipes when no fixture is being used.

**FIGURE 16.2-4** A back-to-back plumbing arrangement is economical and separates noise-producing equipment from living spaces. (Courtesy Small Homes Council, Building Research Council, University of Illinois, Urbana-Champaign.)

**FIGURE 16.2-5** Vertical alignment of fixtures conserves piping and framing. (Courtesy Small Homes Research Council, University of Illinois, Urbana-Champaign.)

**FIGURE 16.2-6** In-line arrangement of plumbing fixtures simplifies piping and reduces costs. (Courtesy Small Homes Council, Building Research Council, University of Illinois, Urbana-Champaign.)

**FIGURE 16.2-7** A 6 in. (150 mm) stud wall is wide enough to permit a $^3/_4$ in. (19.1 mm) water pipe to cross a 3 in. (76.2 mm) plastic vent stack. (Courtesy Small Homes Council, Building Research Council, University of Illinois, Urbana-Champaign.) Metric added by the author.

Hot water systems may be either noncirculating or circulating. In a noncirculating hot-water system, water stays in the pipes and cools off. When a fixture is opened, the cold water in the line must be drained off before hot water arrives. In a circulating system, a complete loop is made from the heater to each fixture and back to the heater. As hot water rises in the system, colder water is forced downward and back to the heater by gravity. Water stays in the branch piping and may cool off, but cold water in the risers flows back to the heater. Therefore, more hot water is available in the piping and much less water is wasted while waiting for the hot water in the system to arrive. Pumps are used in that portion of such a system where cold water cannot flow by gravity back to the heater.

### 16.2.1.3 System Configuration

Both hot and cold water supply systems for small buildings are usually of the upfeed type (see Fig. 16-1). For larger buildings, cold and hot water systems may be either upfeed or downfeed systems or a combination of the two (Figs. 16.2-8 and 16.2-9). Upfeed systems are more simple and less expensive. They are preferred where appropriate.

To determine whether an upfeed system can be used effectively, it is necessary to determine the height of the topmost fixture that can be serviced by the available water service. Upfeed systems are limited in height by service water pressure, static head, friction caused by piping and meters, and the demand of the fixtures. To determine the permissible height of an upfeed system, first make a general estimate as follows:

1. Determine the pressure in the service main. Many municipal systems deliver water at 50 to 60 psi (245 to 413.7 kPa). Private wells usually deliver water at between 40 and 50 psi (275.8 and 245 kPa).

2. Obtain the pressure for the fixture that requires the most pressure at the topmost building floor. A flush-valve water closet, for example, may require 20 psi (137.9 kPa) of water pressure to operate. Subtract this from the street pressure. If the street pressure is 50 psi (245 kPa), the pressure remaining to overcome pressure losses in the system is 30 psi (206.8 kPa); for a 60 psi service, 40 psi (410 kPa) would remain.

**FIGURE 16.2-8** Combination cold water supply system. (Drawing by HLS.)

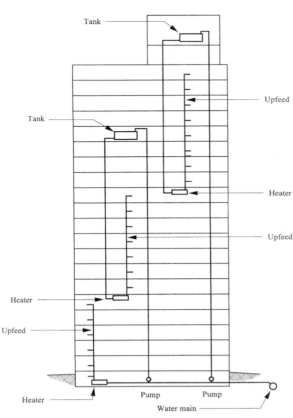

**FIGURE 16.2-9** Combination hot water supply system. (Drawing by HLS.)

3. Determine the pressure loss that will occur when the water flows upward to the topmost fixture (static head). The pressure drop required to raise water in a pipe is 0.434 psi per foot (9.8 kPa/m) of rise. Divide the pressure available by the static head per foot. If there are no fixtures attached and no friction loss to contend with, the maximum rise for water at 30 psi (206.8 kPa) would be 30 divided by 0.434, or about 69 ft (21 m). Water at 40 psi (275.8 kPa) could rise about 92 ft (28 m).

4. Estimate the friction loss for pipe. The loss for a 2 in. (50.8 mm) pipe is about 3.2 psi (22.1 kPa) for each 100 ft (30.48 m) of length. Therefore, the static head for a 69 ft (21 m) pipe riser would be about 2.2 psi (15.2 kPa). If the distance from the street main to the riser is 100 ft (30.48 m), its loss would be 3.2 psi (22.1 kPa). The pressure loss in a meter might be 4 psi (27.6 kPa). The loss for the three would then be 9.4 psi (64.8 kPa). This would lower the permissible height 21.6 ft (6.58 m) to a height of 47.4 ft

(14.4 m). Friction loss in pipe fittings lower this height even further. The practical limit is about 40 ft (12.19 m) for 50 psi (245 kPa) service and 60 ft (18.3 m) for 60 psi (414 kPa) service.

If the building is not higher than the limits determined, an upfeed system may be used. If the building is higher than the limits determined, a downfeed or combination system is required.

## UPFEED SYSTEMS

The steps in designing an upfeed water supply system, and the upfeed portion of a combination system (see Figs. 16.2-8 and 16.2-9), are as follows:

1. Determine the pressure of the incoming service.
2. Determine the pressure needed to raise the water to the height of the topmost fixture (*static head*).
3. Subtract the pressure loss due to the static head from the street main pressure. The result is the pressure available to overcome friction and serve the fixtures.

4. Determine the pressure required by the fixtures on the topmost floor. To do this, find the total number of wsfu's of the fixtures to be served and the amount of water they will require. Figure 16.2-2 shows the number of wsfu's for various fixtures. The water demand curve in Figure 16.2-3 can be used to determine demand, but tables in the *International Plumbing Code* give a more accurate estimation. For example, these tables may show that a riser containing 30 wsfu will demand 23.3 gpm; for 70 wsfu, the demand is 35 gpm; for 100 wsfu, 43.5 gpm. Be sure to use the hot water wsfu when designing the hot water system and the cold water units for a cold water system.
5. Subtract the pressure loss by the fixtures from the pressure left after taking off for the static head. This is the pressure left to overcome friction.
6. Determine the friction loss in the system. This is a combination of losses from pipe, fittings, and a meter. First, determine the size and type of pipe and meter to be used. Experience has

shown that for one- and two-family dwellings of average size, the pipe sizes listed in Figure 16.2-10 will provide adequate service. Selecting pipes for other building types is initially done by trial and error. An educated guess is made, the calculations are run, and then the pipe and meter sizes are refined and the calculations run again. The objective in sizing pipe is to ensure that there is adequate pressure and flow at plumbing fixtures and appliances. Losses due to internal friction in pipes are affected by:

■ The type of pipe and its internal surface characteristics (rougher surfaces generate more friction)
■ The length of the pipe (more length, more total friction)
■ The number of turns, fittings, and valves (these each generate friction loss)
■ The velocity of the water

The friction loss of the selected pipe, its fittings, and the meter is determined from tables in the plumbing code, and this is multiplied by the length of the run. Different pipe sizes may be appropriate for different parts of the system. The horizontal pipe from the street main to the riser may be larger than an individual riser. If the total of all friction losses does not exceed the pressure available after subtracting the loss for static head and fixtures, then the system is satisfactory. If not, different sizes of pipe and meter

**FIGURE 16.2-10** Recommended Water Supply Pipe Sizes[a]

| Fixture | Pipe Size, in. (mm)[b] |
| --- | --- |
| House main | 1 (25.4) |
| House service | ¾ (19.1) |
| Supply riser | ¾ (19.1) |
| Water closet | ½ (12.7) |
| Bathtub | ½ (12.7) |
| Lavatory | ½ (12.7) |
| Kitchen sink | ½ (12.7) |
| Laundry tray | ½ (12.7) |
| Clothes washer | ½ (12.7) |
| Dishwasher | ½ (12.7) |
| Shower stall | ½ (12.7) |

[a]For one- and two-family dwellings of average size.
[b]All pipe sizes except main and service are for either hot or cold.

can be selected, and the calculations re-run until a satisfactory result is obtained. If necessary, fixtures with lower pressure requirements may be selected.

Run through the same process for each floor to determine pipe sizes at each level.

## DOWNFEED SYSTEMS

Design of a downfeed system is similar to that of the upfeed system described earlier, but the configuration differs. The major difference is that the water is pumped up into storage tanks and serves the fixtures by gravity flow (see Fig. 16.2-8). In the usual design for a hot water downfeed system, the water from the tank is allowed to build up sufficient pressure so that it can serve the fixtures in an upfeed system (see Fig. 16.2-9). The steps in the design of a downfeed system are as follows:

1. Determine the number of wsfu's in each zone and the demand load for them.
2. Verify the height of the storage tank in each zone above the topmost fixture served by that tank. The topmost fixtures must be far enough below the tank for the water to build up the pressure required to operate the fixtures. For example, a tank-type water closet might require a pressure of 15 psi (100 kPa) to operate properly. The pressure of falling water increases at 0.434 psi per foot (9.8 kPa/m) of drop. This water will also lose pressure due to friction in the pipe. Therefore, 34.6 ft (10.55 m) of drop is required to serve the fixture, and an additional 5 ft (1.83 m) is required to overcome friction. Flush-valve water closets require a higher pressure (20 psi [137.9 kPa] or so). Therefore, flush-tank fixtures are often selected in tall buildings to reduce the height required above the topmost fixture for the holding tank.
3. Determine the maximum height of each downfeed zone. To do this, first determine the maximum pressure of each fixture from the fixture manufacturer. This pressure may vary from 45 to 60 psi (275 to 415 kPa). Then determine the pressure of the water. Remember that there will be a small loss in pressure due to friction. If the water pressure starts at zero and increases 0.434 psi for each foot (9.8 kPa/m) of fall, the maximum zone height for a fixture requiring a maxi-

mum pressure of 40 psi (275 kPa) will be 103.7 ft (31.61 m). It can be slightly more due to friction loss.

Each riser is then sized in a manner similar to that described for an upfeed system, except that loss due to wsfu is not a factor. If street main pressure is insufficient to fill the tanks, pumps are employed to raise the water to the tanks.

## COMBINATION SYSTEMS

The upfeed portion of a combination water supply system is designed as described earlier for an upfeed cold water system; the downfeed portion is designed as described for a downfeed system.

### 16.2.1.4 Pipe Materials

The water supply pipe material to be used for a particular installation is influenced by several criteria. Plumbing code requirements are, of course, paramount. Within the applicable code parameters, the characteristics of the local water supply—that is, its acidity, and its air, carbon dioxide, and mineral contents—affect the selection. The original cost is also important, as are the ease of installation and repair of a particular piping material. Water supply piping should bear the stamp of approval of the National Sanitation Foundation (NSF). Acceptable piping materials are listed in Figure 16.2-11. Service pipe shown is for exterior underground connection to main or supply tanks. Distribution piping is for indoor use.

### GALVANIZED STEEL

*Galvanizing* is a process that deposits a coating of zinc on the surface of a steel pipe to protect it against corrosion. Galvanized steel (also called *galvanized iron*) pipes are low in cost and high in strength. They are resistant to high pressures and to water hammer (see Section 16.2.5). Their major disadvantage is that joints are typically threaded, although other methods of joining are sometimes used. Threading the initial installation is laborious, and the need to thread new joints in place makes later tapped-in connections to existing systems difficult.

Galvanized pipe has a tendency to corrode or form scale on the pipe interior to a degree that depends on local water chemistry. Galvanized pipe performs best in areas with moderately hard, alkaline water. Naturally soft water should be treated with a silicate inhibitor to produce

**FIGURE 16.2-11** Acceptable and Unacceptable Water Supply Piping Materials[a]

| Material | Service | Distribution Piping | |
| --- | --- | --- | --- |
| | | Under Slab | Other |
| **Cast Iron** | | | |
| Water pipe | A | | |
| No-hub pipe and fittings | NA | NA | NA |
| **Copper Tubing** | | | |
| Type K — Hard temper | A | A[e] | A |
|         Soft temper | A | A[b] | A |
| Type L — Hard temper | A | A[c] | A |
|         Soft temper | A | A[c] | A |
| Type M — Hard temper | A | NA | A |
| DWV — Hard temper | NA | NA | NA |
| **Galvanized Open Hearth or Wrought Iron** | A[c] | A | A |
| **Galvanized Steel Pipe** | A[c] | A | A |
| **Black Steel Pipe** | A[c] | NA | A |
| **Plastic** | | | |
| ABS | A[d] | NA | A[d] |
| CPVC-41 | NA | A[b] | A |
| PVC | A[d] | NA | A[d] |
| PVC thin wall | NA | NA | NA |
| Polybutylene | A | A[b] | A |
| Polyethylene | A[d] | NA | A[d] |
| Styrene rubber | NA | NA | NA |

[a]A = Acceptable; NA = Not acceptable.

[b]Install without joints in or under slab floors.

[c]Protected when installed underground.

[d]For cold water only, outside dwelling or in unfinished basement or crawl space.

[e]Brazed joint only.

a protective film on steel pipe, and with caustic soda or lime to raise its pH (see Section 16.2.3.5). Very hard water should be softened to control scaring on steel pipes. When corrosion or scaling occurs, the interior pipe diameter can become greatly reduced, changing the carrying capacity of the pipe. Because it will react chemically, galvanized steel pipe should not be laid in concrete.

There are three available means of joining galvanized steel pipe:

1. *Threaded connections* are the most common way of joining galvanized pipe (Fig. 16.2-12a). Generally, the outside of the pipe run is threaded to match a female galvanized steel or malleable cast iron fitting with the desired geometry. The threads should be coated with pipe joint compound or wound with plastic joint tape before the fitting is attached.

2. *Unions* (Fig. 16.2-12b) are a means of joining galvanized steel pipe that can be disconnected and reconnected at future times, but they are more costly than conventional fittings. A union consists of two *union nuts* with machined brass mating surfaces, each of which is threaded onto the respective pipe end and a *ring nut* which is tightened, drawing the two union nuts together. The brass components of unions, however, can cause localized corrosion in steel pipes which restricts flow.

3. *Slip couplings* are a third type of connector available for galvanized steel pipes (Fig. 16.2-13). They consist of a sleeve and compression nuts. The sleeve is slipped over the end of one pipe. The two pipes are then aligned, and the sleeve is positioned to cover the joint between them. *Compression nuts* at each end of the sleeve seal it to the pipes by rubber gaskets. Some types of slip couplings also have a screw connection in the center for easier installation.

Where steel pipe is being joined to a dissimilar metal such as brass, and the water supply is corrosive or has a high solids content, a *dielectric* (nonconductor of direct electric current) fitting must be used to prevent galvanic corrosion. Where the water supply is sufficiently hard to produce a protective carbonate or silicate film on the pipe, a dielectric fitting is unnecessary.

## COPPER

Although relatively costly, copper pipe is convenient to work, easy to join, and sufficiently corrosion resistant for use with most water supplies. Water that is very soft, very high in carbon dioxide, or very aggressive and hard should be treated before it enters a copper plumbing system. Red brass (85% copper, 15% zinc) and 90-10 copper-nickel pipes are good choices where the water supply is corrosive. Copper should not be used in systems designed for velocities greater than 4 ft/sec (1.22 m/s). While the piping itself is durable, the joints are weaker than the piping and are subject to damage from water hammer, which necessitates having air chambers at the ends of supply lines (see Section 16.2.5).

internal threads (fitting)    external threads (pipe)

**a**

union nut

matched joint

ring nut

union nut

**b**

**FIGURE 16.2-12** Galvanized pipe joints: (a) threaded fitting receives field-threaded pipe; (b) union fitting to be threaded onto pipe ends.

**FIGURE 16.2-13** A section through a slip coupling.

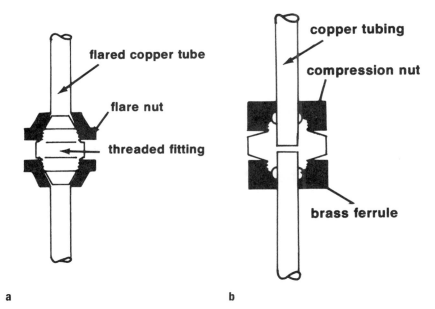

a                           b

**FIGURE 16.2-15** Unsoldered fittings for flexible copper tubing: (a) flared fitting; (b) compression fitting.

Copper supply piping falls into two major categories: rigid pipe and flexible tubing. *Rigid pipe*, which is hard tempered, is generally connected using copper fittings which are sealed with tin-lead solder and a noncorrosive flux (Fig. 16.2-14). Excessive use of self-cleaning flux can lead to later corrosion problems. Rigid copper pipe comes in three wall thicknesses: (1) type K (thick), (2) type L (medium), and (3) type M (thin), and in 10- to 20-ft (3.05- to 6.1-m)-long straight lengths. The three types are equal in exterior diameter and are designated by the same nominal sizes, but the interior diameters vary due to differences in wall thickness.

*Flexible tubing*, which comes in 30, 60, and 100 ft (9.14, 18.3, and 30.5 m) coils, has several advantages over rigid copper pipe. Flexible tubing:

- Is easier to install, especially in cramped areas
- Can be bent into smooth turns, which reduces friction and sometimes permits the use of smaller pipes
- Can withstand several freezing cycles without cracking, although with each freeze the pipe expands and the pipe wall becomes thinner

Flexible tubing also provides more options for connections. Soldered fit-

**FIGURE 16.2-14** Rigid copper pipe being joined with a soldered fitting.

tings, which require the use of a torch, and flared and compression fittings can all be used. A *flared fitting* (Fig. 16.2-15a) is made by slipping a *flare nut* over the end of the copper tube, reforming the end of the tube so that it has a flared tip, then screwing the flare nut onto a threaded fitting. A *compression fitting* (Fig. 16.2-15b) is similar, except that the nut bears against a brass ferrule as it is screwed into the fitting rather than against the flared end of the tube. A push-in connector for copper tubing is also available. It consists of stainless steel and neoprene grabbers mounted in a nylon body.

Flexible copper tubing is also available in type K or type L wall thickness. Type K is typically used for underground installations, and type L is used for indoor plumbing.

## PLASTIC

Not all of the many varieties of plastic pipe are suitable for carrying drinking water. Those which are suitable bear the stamp of the NSF. Plastic pipe is lightweight, being about $1/20$th of the weight of equal-sized steel pipe. It is also easily joined, produces low friction, and does not corrode. Joints are typically made using a solvent and special fittings (Fig. 16.2-16). Although these joints are easy to make, the solvent acts quickly, and the joint becomes permanent. The only way to correct mistakes is to saw the joint out

and start over (Fig. 16.2-17). The corrosion resistance of plastic allows its internal surface to remain smooth, which maintains the carrying capacity of the pipe as the system ages.

Plastic is much less thermally conductive than metal pipe. Therefore, it is a better insulator, which reduces heat loss from hot water lines and condensation on cold water lines. Plastic and metal piping differ greatly in their rates of thermal expansion, amount of softening at the same temperature level, electrical conductivity, flammability, and weathering characteristics. Plastic has a much larger co-

**FIGURE 16.2-16** Plastic pipe joint. First cleaner and then solvent are brushed on the outside of the pipe and the inside of the fittings. Then the two are joined, and the pipe is turned about one-quarter turn. A good joint will show a fused solvent bead on the fitting end.

**FIGURE 16.2-17** Cutting plastic pipe. Courtesy BE&K, Inc.)

efficient of thermal expansion than metal. Because it tends to soften more and often has less strength and stiffness than metal pipe, it is necessary to support plastic pipe more frequently than metal pipe to prevent sagging. Because of its high rate of thermal expansion, plastic pipe is often snaked (placed in gentle, undulating curves rather than straight) when it is laid in underground trenches. The upper limit of operating temperature for plastic pipes ranges from 120° to 300°F (48.9 to 148.9°C), depending on the type of plastic.

Plastic piping should be kept away from high-temperature areas such as heating ducts. Unlike metal piping, it is not a good electrical conductor. This has the advantage of preventing galvanic or electrolytic corrosion, but it also prevents the use of a cold water pipe as a ground for a building's electrical system. Depending on the particular material's formulation, plastic pipe varies in its degree of flammability. Plastics weather differently than metal piping; instead of corroding or oxidizing, they chemically degrade when exposed to ultraviolet radiation and become brittle. This can be controlled by using plastics that are formulated with ultraviolet stabilizers.

Four major types of plastic pipe are used in small buildings: *polyvinyl chloride* (PVC), *polyethylene* (PE), *chlorinated polyvinyl chloride* (CPVC), and *polybutylene* (PB). All four types are suitable for cold water supplies, but only CPVC and PB are suitable for hot water supplies. *Acrylonitrile butadiene styrene*

(ABS) piping is also used, but normally only for underground water service (outdoor) piping.

PVC is most frequently used for cold water supply lines. It can be used for the entire cold water supply system within a building. Joints in PVC pipe can be either solvent-welded or made with threaded fittings which are commonly available in heavier weights of pipe. PVC pipe is manufactured in two types. Type I is unplasticized and has a maximum working temperature of 150°F (65.6°C). Type II is plasticized by the addition of rubber, which makes it less brittle than Type 1. This, however, makes its strength and maximum working temperatures lower than those of Type I. In addition to water lines, PVC pipe is used for natural gas and electrical conduit lines.

PE pipe is used for cold water supply from a municipal water main or well to a building. It has a lower range of working temperature than most other plastics: from −67°F (−55°C) to 122°F (50°C). PE pipe comes in Types I, II, and III, in order of increasing density. Joints in PE pipe are made with tapered, serrated inserts of polystyrene or galvanized steel which slip into the end of the PE pipe (Fig. 16.2-18). The pipe is then compressed against the insert by means of a metal hose clamp.

CPVC is more costly than PVC pipe and is, therefore, generally restricted to hot water supply lines, where it is the most prevalent type of plastic pipe. It can be used with water temperatures up to 212°F (100°C), although the water heater relief valve in a system using CPVC pipe should be set no higher than 180°F (82.22°C). Support clamps should be provided on horizontal runs every 3 ft (914 mm) to prevent sagging. The use of CPVC pipe is usually restricted to hot water lines. Joints are solvent welded.

PB pipe is suitable for either hot or cold water supply pipes. It is so flexible that most turns can be made without fittings. It can be obtained in a high-temperature grade for use up to 221°F

**FIGURE 16.2-18** Clamped connection for PE or PB pipe.

(105°C). PB pipe is similar to Type III PE pipe but is stronger. As with PE pipe, joints are made with tapered inserts and external clamps.

### BRASS

In the past, red brass pipe was frequently used in areas where the water supply was so corrosive as to make galvanized steel piping unsatisfactory. Today, because of its high cost, brass pipe is seldom used in water supply systems. Modern plastic piping can do the same job at a much lower cost.

Brass piping may still be encountered in old buildings. Its connections are threaded, similar to those in galvanized steel pipe. While brass is highly resistant to corrosion generally, it is subject to galvanic corrosion when connected to galvanized steel. There a dielectric coupling or union should be used.

### LEAD

Lead piping was used for water piping in the past and still may be found in some older buildings. It is no longer used for water supply piping inside buildings, however, because minute quantities of lead oxides on the piping dissolve in water, which can result in lead poisoning.

### 16.2.1.5 Flow Control

An unsteady rate of water flow from a supply system into a fixture is due to changes in the flow pressure in the supply lines. This happens because of either inconsistent pressure from the water source (main or well) or the startup or shutdown of other fixtures or appliances on the same plumbing tree. If the problem is severe, the inconsistent flow can be corrected with a flow-control device (Fig. 16.2-19). This is a fitting with a flexible orifice on which the cross-sectional area varies inversely with the

**FIGURE 16.2-19** In a flow controller, increasing the velocity deforms a resilient orifice to reduce the opening. When velocity is normal, the orifice increases to normal. (Courtesy Small Homes Council, Building Research Council, University of Illinois, Urbana-Champaign.)

flow pressure. As pressure in the line increases, a flexible insert in the fitting is pushed forward against a stop, decreasing the cross-sectional area of the pipe and thereby regulating the flow rate. Flow-control devices are not to be confused with (or used as) pressure-reducing valves.

### 16.2.1.6 Pressure Regulation

Most municipal systems deliver water to a building at pressures of 40 to 60 psi (276 to 413.7 kPa). In designing a supply system, it is critical to know the average supply pressure for summer months, when the main pressure is generally lowest due to extensive use of water for lawn watering. The water distribution system in a building is then designed in such a way that its length and friction characteristics will deliver an acceptable flow rate and pressure at the fixtures in the building. If the final pressure at the fixtures is inadequate, which is rarely the case in residential and other small buildings, it can be supplemented by installing a gravity tank system, a hydropneumatic tank system, or a booster pump.

### 16.2.1.7 Valves

Valves used in building water supply systems can be classed according to their operation as *gate valves*, *globe valves*, and *alignment valves*.

Valves of various kinds are located throughout a water supply system. They are, in fact, a system's only moving parts. A *supply main*, *street main*, or *distribution main* is the water line from a treatment plant or well into which individual building supply lines are tapped. The principal, and usually lowest, horizontal artery in a building's water supply system is also sometimes called a *main*, but we will not refer to it as such here in the interest of clarity. Where a building supply line leaves the supply main, there is a shutoff (*corporation*) cock. A main shutoff valve is installed just inside the building if the water meter is remotely located. Shutoff valves are installed on each side of in-line appliances such as a water meter, water heater, or water treatment device. These are typically *gate* or *globe* valves.

Most water valves are made of brass because of its resistance to corrosion. Valves are made with threaded ends for use with galvanized steel and rigid copper pipe and with solderable ends for use with copper tubing. Plastic-bodied valves are available for use with plastic pipe. It is also good practice to install a shutoff valve where a supply pipe makes a turn of more than 45 degrees or enters or leaves a wall. A type of globe valve known as an *angle stop valve* (because it is a shutoff valve through which the supply line makes a right angle turn) should be installed just upstream from each fixture and appliance in a system to facilitate repairs.

Automatic appliances such as dishwashers and clothes washers have internal valves to cycle water flow into them. These are usually electrically controlled *solenoid valves* which open and close very quickly, sometimes causing water hammer (see Section 16.2.5). Finally, there are valves within the faucet assemblies of each plumbing fixture. These are usually globe valves or cocks.

### GATE VALVES

Gate valves are used when full, unobstructed flow is desired when a valve is open and positive closure is needed when it is shut (Fig. 16.2-20a). When the handle on a gate valve is turned, a wedge is lowered into the pipeway, obstructing the flow.

### GLOBE VALVES

Globe valves contain two chambers connected by a small opening (Fig. 16.2-20b). When the handle is turned, a plunger (*stem*) is lowered, blocking the flow of water from one chamber to the other. The seal over the opening is usually made with a washer mounted on the tip of the stem.

### ALIGNMENT VALVES

*Alignment valves* are used primarily in single-lever faucets (Section 26.4.3.2). These valves are washerless and are, therefore, less likely to leak than valves with washers. The lever on a single-lever valve controls the position in the valve of a disk, cylinder, or ball with holes in it. When the holes are aligned with the openings between two chambers in the valve body, water flows. When the holes are not aligned with the openings, flow stops.

## 16.2.2 WATER HEATING

There are two basic types of systems for heating water for use at plumbing fixtures and appliances. They can be designated as *direct* and *indirect*. Solar systems are a form of indirect water heater.

### 16.2.2.1 Direct Water Heaters

Direct water heaters are available in several types.

### AUTOMATIC STORAGE TYPE

*Automatic storage heaters* heat a volume of water and store it until it is needed.

**FIGURE 16.2-20** (a) Gate valves are used in key locations to turn water completely on or off. (b) Globe valves can control flow in a partly open or closed position.

They are the type usually found in residential buildings, where their capacity generally runs from 20 gal (75.7 L) to 100 gal (378.5 L). The automatic storage heaters used in other building types may be identical to those found in a residence, but units with capacities of as much as 4000 gal (15 141.6 L) are also available.

In automatic storage heaters, the water is maintained at a constant temperature by a thermostatically controlled heat source (Fig. 16.2-21). Generally, the thermostat is set between 120° and 140°F (48.9° and 60°C), but different temperatures may be required in some instances, particularly where use by handicapped persons is required and for use in institutional buildings. A higher temperature is required for some automatic dishwashers and other appliances and devices. However, modern energy-saving dishwashers and some other equipment are provided with a booster heater so that 120°F (48.9°C) water can be used.

When hot water is used at a plumbing fixture or appliance, the tank is refilled automatically, because it is connected with an open valve to the pressurized cold water supply line. This refill lowers the temperature of water in the tank, which the thermostat notes and calls for heat from the heat source. The thermostatic control also calls for heat to compensate for heat lost through the tank wall when no hot water is being consumed. For this reason, new water heaters should be internally insulated with at least R-7 insulation. Kits are available for applying

insulating blankets externally to older models.

The life expectancy of a residential-type automatic storage water heater is limited by corrosion of the tank wall to 10 to 15 years. The usual tank material is glass-lined steel. If the local water supply is particularly hard or corrosive, water treatment equipment may be installed to lengthen tank life. An alternative, though expensive, material for tank interiors is monel metal. It forms an oxide coating which resists corrosion. Although aluminum and unlined galvanized steel tanks are available, they fail quickly due to corrosion. Some water heaters are fitted with a sacrificial anode of magnesium, which gives up certain ions to the water more readily than does the tank wall. The anode becomes a substitute material for the water to corrosively attack, thus sparing the tank liner material.

Excessively high temperatures and pressures are dangers associated with heating water in closed tanks. High temperatures can result in scalding or burning of persons using water at fixtures. Pressure buildup in a tank can be an explosion hazard. Under normal operating circumstances, a thermostat cuts off the heat long before either of these two hazards is approached. However, as a safety measure in the event that thermostatic control fails, UL-approved water heaters are fitted with a *pressure and temperature relief valve* (see Fig. 16.2-21).

A *pressure valve* vents a closed water heater tank when a dangerous level of either pressure or temperature is reached. This valve is located on top of the tank and connects to a drip pipe which directs relief flow from the tank to a safe drainage area. Relief valves used on water heaters should be of the reseating type, which vents the system until an acceptable pressure or temperature level is reached.

Relief valves are specified for water heaters based on the British thermal unit (Btu) rating of the heater. The relief valve must always have a larger Btu rating than the heater. Valves are factory-set for the pressure and temperature levels at which they will discharge and should not be altered. The limits set for residential water heaters are approximately 125 psi (862 kPa) of pressure and 200°F (93.3°C) of temperature.

Water heaters are generally chosen on the basis of their volume and their recovery rate. *Volume* simply refers to the

number of gallons the heater storage tank holds. *Recovery rate* is the number of gallons that the heater can raise 100°F (37.8°C) in 1 hour. This is a measure of a heater's ability to reestablish the thermostat-set temperature after rapid drawdown of the stored hot water. In general, oil-fired water heaters have a high recovery rate, gas heaters have a medium recovery rate, and electric heaters have a low recovery rate. Heaters are also rated in terms of energy input and output, usually quoted in Btus. The ratio of the input rating to the output rating is an indicator of the heater's fuel efficiency.

### DEMAND-TYPE WATER HEATERS

*Demand*-type direct water heaters heat water only when more hot water is called for and do not store a reserve of hot water. Therefore, they are much smaller than storage-type heaters. Because hot water is not stored, standby losses at night and when occupants are away are much lower. Demand heaters are not as popular in this country for residential use as are *automatic storage heaters* because of their limited recovery rate, but many demand heaters are used in commercial and institutional facilities.

### 16.2.2.2 Combination Water Heaters

The second type of water heaters is the indirect type, also called *combination heaters*. These use a heat exchanger on a space-heating boiler or furnace and a separate hot water storage tank. The same boiler or furnace that is used to heat the structure is used to heat domestic water. This works well in cold climates, where heat is required all or most of the year. In milder climates, however, the drawback is that the heater must run even when heat is not required. Sometimes an auxiliary water heater is employed to overcome this difficulty, but operating costs can still be higher and the initial cost will certainly be more.

Sometimes it is economical to use an indirect heater and a direct heater in the same installation. The indirect heater raises the temperature of domestic water to a lower level than is required at the discharge point. Then a secondary, direct water heater is used to elevate the temperature to the desired level.

### 16.2.2.3 Fuels

Heat sources for water heaters can be electricity, piped-in fossil fuels (natural

**FIGURE 16.2-21** Gas water heater. The heater should be located near the floor drain to permit periodic draining of sediment and to handle occasional overflow.

gas or heating oil), or sunshine. Electric heaters are limited to small (20 to 100 gal [75.7 to 378.5 L]) direct water heaters. Gas and oil can be used on every water heater type discussed in this chapter. *Electric* water heaters usually have two electrical elements inside a storage tank. They require no supply of combustion air or venting of combustion products (Fig. 16.2-22).

At the base of its storage tank, a *gas* water heater has a burner with a flue to exhaust its combustion products. This flue extends up through the center of the tank, acting as a heat exchanger as it does, and is finally connected to a chimney.

The burners on *gas* heaters can be designed either for natural gas or liquefied propane. Adequate combustion air must be supplied to the burners, and an exhaust flue and chimney must be provided. Requirements similar to those for gas heaters also apply to *oil-fired* heaters, although the burner design is slightly different.

*Solar* water heaters use the sun as a partial source of energy. The intermittent nature of this heat source makes it necessary to provide extra storage capacity so that hot water will be available when needed at night and on cloudy days (Fig. 16.2-23).

In a solar water heater, solar energy is absorbed by a south-facing solar collector mounted outdoors through which a heat-transfer fluid flows. When heat is available from the collector, the transfer fluid carries it to a heat exchanger coil in a storage tank. The domestic water heater

FIGURE 16.2-23 Heated water from a collector is conducted to a heat exchanger, where most of the heat is transferred to a loop conducting heat to a storage tank, from where pre-heated water is drawn as needed to supply a domestic water heater.

uses this preheated water when replacing water in its own tank.

Except in a few locations and during limited periods, the solar energy available is insufficient to heat water to the desired temperature. Therefore, solar heating is usually practical only as an auxiliary heat source that must be backed up with an electric-, gas-, or oil-fired conventional heater. This limitation does not, however, invalidate the potential of solar energy as a means of domestic water heating. In many parts of the country, from one-third to two-thirds of needed heating energy can be obtained from the sun.

## SIZING

*Residential Water Heaters* Water heater size for residences is based on an estimate of hot water use. For a single-family dwelling, this is sometimes estimated as 7 to 10 gal (26.5 to 37.9 L) of hot water per person per hour times the number of persons in the household. The U.S. Department of Housing and Urban Development (HUD) has generated standards based on use and tank recovery rates. For example, in a residence with one to one and one-half baths, the required capacity for a gas heater is 20 gal (76 L) for a one-bedroom residence and 30 gal (114 L) for a two- or three-bedroom residence. For an electric heater, the capacity is 20 gal (76 L) for one bedroom, 30 gal. (114 L) for two bedrooms, and 40 gal (150 L) for three bedrooms. For an oil-fired heater, the

capacity is 30 gal (114 L) for one, two, or three bedrooms.

Capacities for larger residences increase accordingly. For a three-bedroom residence with three to three and one-half baths, the capacity of a gas heater is 40 gal (150 L), and for a four-bedroom residence it is 50 gal (190 L). Required capacities are higher for electric heaters and lower for oil-fired heaters because of the difference in their recovery rates.

HUD also has requirements for tank-type and tankless-type indirect heaters. A table showing recommended capacities for residential water heaters is contained in the American Society of Heating, Refrigerating, and Air-Conditioning Engineers (ASHRAE) 2003 *Handbook-HVAC Applications.*

*Commercial and Institutional Water Heaters* Water heaters for small office buildings and small shops can be determined much like those for residences. The sizing of water heaters for larger buildings of types other than residences is based on a calculation of the maximum hot water demand per hour. The requirement is based on calculations using wsfu's. Refer to Section 6.1.2 for a discussion about the use of wsfu's.

An estimate of hot water usage per hour may be calculated using the number of wsfu's and a water supply demand graph as explained in Section 6.1.2. Fortunately, however, the ASHRAE, *HVAC Handbook Applications* contains tables with rates-of-use recommendations for

FIGURE 16.2-22 Electric water heaters are generally larger than gas- or oil-fired heaters because their recovery rate is slower.

buildings other than residences that make such calculations unnecessary.

Water heater size is a function of the recovery rate of the heater and the capacity of the associated storage tank. The recovery rate for a water heater depends largely on the type of fuel being used. In oil-fired heaters, which have a fast recovery rate, a heater with a small tank can be specified and the maximum demand per hour can still be met. In gas-fired heaters, which have a moderate recovery rate, a medium-sized tank would be necessary. In electric systems, with their slow recovery rates, a relatively large tank would be necessary. The actual recovery rate for water heaters is determined by consulting the manufacturer's specifications data for the unit.

A small heater will operate more steadily than a large heater to maintain the water temperature in the tank, thus consuming less energy than would a larger heater running more sporadically. However, the larger storage tank required will take up more floor space, weigh more, and probably cost more initially than would a tank associated with a larger heater.

The ASHRAE *HVAC Handbook Applications* contains graphs that have the *usable* storage capacity on the horizontal axis and the recovery capacity of the heater on the vertical axis. There is a graph for each building type included in the ASHRAE table (dormitories, nursing homes, office buildings, food service establishments, apartment houses, and schools). A curve in the graph represents maximum hourly consumption. A designer would select a recovery rate, extend a line from the vertical axis to the curve, and read the storage capacity required by dropping down to the horizontal axis.

When a building has more than one use, such as a dormitory in a building also housing a cafeteria, the capacities for each use are calculated separately and added to obtain the design requirements.

Using these data, it is possible to select a large range of recovery rates and corresponding usable storage capacities. For example, the graphs show a minimum recovery rate which use would result in the largest tank permissible. A faster recover rate would result in a smaller tank.

The recovery rate used is often a percentage of the maximum hourly capacity given in the ASHRAE hot-water-use tables. In addition, the ASHRAE tables and graphs do not take into account all the variables in storage tank and heater selection. For example, they do not include adjustments for heat losses from tanks and pipes. The usable capacity of a storage tank is usually considered to be between 60% and 80% of the total tank capacity. The selection, then, becomes one of engineering judgment and should be made by a qualified plumbing engineer.

#### 6.2.2.4 Temperature Controls

Water temperature controls are often advisable, especially where handicapped people, the elderly, or small children will use plumbing fixtures (see Section 1.4). Two types of temperature control devices are available, primarily for tubs and showers. One type maintains constant relative pressures in the hot and cold supply lines to compensate for sudden draws by other nearby fixtures. Consequently, the flow rate, once set, remains reasonably constant. This does not, however, compensate for temperature changes of the water from either supply line, which occur when the hot water heater is drawn down. The second type of temperature control device is thermostatically controlled and changes the flow rate of the hot and cold supplies to maintain a constant temperature of mixed output.

### 16.2.3 WATER TREATMENT

Water from some sources may need to be processed further to make it suitable for domestic use, even though it meets public health standards for safe drinking water. Problems which may require treatment include excessive hardness, resulting in ineffective soaping; materials in the water that cause corrosion or scaling of piping; and unpleasant odor, color, or taste. Treatment to correct such problems usually involves mechanical filtering or chemical treatment of the water.

Filters require periodic cleaning or replacement to prevent bacterial growth in them. Bacterial contamination of a building's water supply can cause problems more severe than those filters are intended to correct. Treatment is often confined to the hot water supply, because higher temperatures accelerate the undesirable chemical reactions and make sensory characteristics more noticeable.

#### 16.2.3.1 Water Hardness

The most common type of domestic water treatment combats what is known as water hardness. *Hardness* refers to the amount of calcium and magnesium carbonate dissolved in the water; *total hardness* refers to the amount of carbonates plus several other compounds dissolved in the water.

Some degree of hardness is desirable, since it causes a thin film of calcium carbonate to deposit on the interior walls of pipes, forming a protective coating against corrosion. Too much hardness, especially in conjunction with high temperatures, has two undesirable effects. It causes excessive carbonate deposition (scaling), and it renders the water ineffective at rinsing soap away. Calcium carbonate scale appears as a cream-colored buildup in piping and fixtures.

Using a *water softener* alleviates both scaling and soaping problems. An *ion-exchange* water softener contains a sodium-zeolite exchange medium which develops sodium ions. These then interact with calcium (or magnesium) ions in the water. When all sodium ions have been consumed, the filter must be recharged. Some systems are recharged in place by backwashing and soaking with a salt solution; others depend on a service company to supply recharged replacements (Fig. 16.2-24).

Softening water too much raises the sodium content of the treated water, which may make it harmful to persons with heart or kidney disorders. Fortunately, the adverse effects of hardness are more pronounced in the hot water than the cold water supply. Humans more often drink water from the cold water supply than from the hot one. Consequently, by installing the water softener on the hot water system only, human consumption of softened water can be limited.

In addition, oversoftened water not only fails to provide the protective calcium carbonate film on pipes, it actually makes the water corrosive. For these reasons, it is not advisable to soften water to less than 60 mg of calcium carbonate per liter. If the output of the softening equipment falls below this level, it should be tempered with untreated water from the cold water supply before it enters the plumbing system.

#### 16.2.3.2 Corrosiveness

A second major problem that can be treated is corrosiveness. A water supply can be rendered corrosive by high levels of dissolved carbon dioxide (yielding car-

**automatic recharge control**

**FIGURE 16.2-24** As hard water passes through resin, calcium and magnesium in the water are retained and sodium is given up. Either the salt or the entire tank must be replaced periodically.

bonic acid) or by low levels of calcium and magnesium carbonate (excessive softness). The former occurs naturally in water supplies in the East and Northwest; the latter can occur naturally or can be the result of overzealous water treatment. Corrosion due to either cause is accelerated by high temperatures and therefore is a larger problem in a hot water system than in a cold water one.

Symptoms of corrosion are staining of fixtures, reduced flow or pressure resulting from clogging, and leaky pipes. The stains are red if the supply piping is steel, green if it is copper. Clogging can occur due to either accumulated rust particles or to localized scaling triggered by the changes in pH that accompany corrosion.

There are several solutions to corrosion problems. Mildly corrosive water from the natural supply can be treated with a neutralizing filter containing marble or limestone (calcite) chips. In severe cases, liquid sodium silicate and caustic soda can be added to raise the pH of the water. While this relieves the corrosion problem in most of a plumbing system, it does not sufficiently protect low-

velocity areas such as storage tanks. Tanks with special protective coatings or sacrificial anodes should be considered.

### 16.2.3.3 Unpleasant Appearance, Odor, or Taste

Unpleasant appearance, odor, or taste in a water supply can also be treated chemically or mechanically. For example, yellow, cloudy water and rust-stained fixtures are frequently caused by suspended iron particles. In mild cases, these can be removed by an ion-exchange water softener; in more severe cases, an oxidizing filter containing manganese-treated sand is effective. Particles of organic matter, silt, or algae can degrade the appearance and taste of water. This problem can also cause pipe clogging and reduced water-heater efficiency. Such particles may be inherent in the water supply or may be introduced by backflow of drainage water into the supply system. If this water quality problem persists, fixtures should be checked for air gaps, and appliances such as clothes washers, laundry tubs, and dishwashers should be checked for vacuum breakers (backflow preventers) (see Section 16.4.7).

If the source is the water supply, a charcoal filter can be installed where the supply piping enters the building. In severe cases, a chemical feeder with a chlorine or alum solution can be used, followed by a sand filter. For small amounts of high-purity potable water, a reverse osmosis unit can be installed on individual fixtures. It features a membrane through which only pure water can pass.

*Rotten egg odor* is caused by high levels of dissolved hydrogen sulfide gas. The source of the hydrogen sulfide can be anaerobic decomposition of organic matter somewhere in the supply chain. This occurs at many locations in the Southwest, Midwest, and Southeast. It can also be produced by magnesium anodes used to protect water heater tanks from corrosion. Water with dissolved hydrogen sulfide also has a black tinge and tarnishes silver. An oxidizing filter with manganese-treated sand can be used to alleviate this problem. Extreme cases call for chlorination followed by sand filtering.

## 16.3 Drainage

A *drainage system* consists of piping that conveys solid and liquid wastes from their source to a point of disposal. It includes the system portions within a building and on its site, as well as the mains of a public sewer system or a private septic system.

There are three kinds of drainage systems. A *sanitary drainage system* carries *sewage*, which is liquid and solid waste matter, including fecal matter and food, and industrial waste, as opposed to the kind of material carried off by storm sewers. The portion of a sanitary drainage system that lies within a building and up to a point 3 ft (914 mm) outside of the walls is called a *building drainage system* or, more frequently, a *drain, waste, and vent* (DWV) system (Fig. 16.3-1).

**FIGURE 16.3-1** Diagram of major components of a DWV system showing a loop vent.

This collection of pipes facilitates the removal of liquid and solid wastes as well as the dissipation of sewer gases within a building. DWV systems are covered in this section. The portion of a drainage system that falls outside a building is discussed in Section 23.2.

The second type of drainage system, a *storm drainage system*, carries rainwater, surface water, condensate, cooling water, and similar liquid wastes, exclusive of sewage and industrial waste. The portion of a storm drainage system that occurs outside of a building and subdrainage systems beneath a building are discussed in Section 23.3. The rest of the facility (in the building) portion of storm drainage systems is covered in this section.

The third type of drainage system is a combined system that carries both sewage and storm water. Refer to Section 23.2.2 for a discussion of combined sewers.

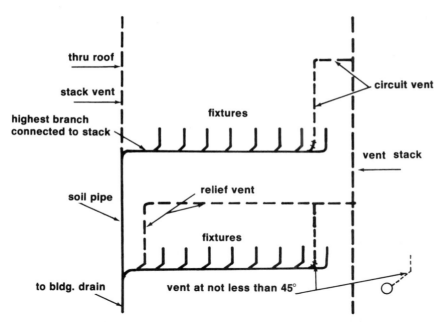

**FIGURE 16.3-2** Drainage diagram for a larger building showing two ways of venting the drain branch: with a circuit vent (above), and a circuit and relief vent (below). (Courtesy Small Homes Council, Building Research Council, University of Illinois, Urbana-Champaign.)

### 16.3.1 SANITARY SEWAGE DISPOSAL

#### 16.3.1.1 DWV System

A DWV system should be designed and installed so that it (1) maintains separation of drain fluids from the water supply, (2) contains and discharges sewer gases to prevent fouling the interior atmosphere, and (3) effectively moves fluids and solids through the pipes.

#### COMPONENTS

DWV systems are larger and often more complex than water supply systems. Their geometry is more critical because the pipes are larger and must be installed at a slope for gravity drainage. They consist of drains (waste and soil pipes), traps, and a vent system. A *drain* is any pipe that carries sewage in a building drainage system. *Waste pipes* carry water-diluted waste products; *soil* pipes carry the discharge of water closets, urinals, and other fixtures with similar functions to the building drain or building sewer (Fig. 16.3-2). Soil pipes may, in addition, carry wastes from other fixtures.

A *trap* is a fitting or device designed to provide, when properly vented, a liquid seal which will prevent the back passage of air without materially affecting the flow of sewage or wastewater through it (Fig. 16.3-3).

A *vent system* is a collection of pipes installed to provide a flow of air to or

**FIGURE 16.3-3** A P-trap showing the water seal. (Courtesy Small Homes Council, Building Research Council, University of Illinois, Urbana-Champaign.) Metric added by the author.

from a drainage system, or to provide circulation of air within such a system to protect trap seals from siphonage and back pressure.

All the components of a DWV system are interconnected. In fact, a given pipe may serve in one portion of its length as in another portion as a vent, or under some circumstances may serve as a drain and a vent simultaneously at the same location.

At the center of a DWV system is one or more vertical combination *drain and vent* pipes that extend from a point above the roof down to the building drain.

A *building drain* is the lowest part of a building's drainage system. It is the principal artery to which branches may be connected, and as such it is sometimes called the system's *main*. It receives the discharge from soil, waste, and other drainage pipes inside the walls of a building and conveys it through the foundation walls to a point 3 ft (914 mm) outside of the building, where it joins with the building sewer (Fig. 16.3-4).

The *building sewer* is the horizontal piping of a *drainage system* that extends from the end of a *building drain* to a public sewer, private sewer, individual sewage disposal system, or other point of disposal (see Fig. 16.3-4). A building sewer begins at a point 3 ft (914 mm) outside of the building wall, where the building drain ends. A building sewer is sometimes called a *house sewer*.

*Soil stacks*, or just *stacks*, are the central vertical pipes that are the collection points for the drain and vent lines in the DWV system.

A typical soil stack will have connections to one or more branch drains serving several fixtures each (see Fig. 16.3-1). Branch vents serving those same fixtures connect to the stack at points higher than the branch drain connections. Thus, the lower portion of this pipe serves as the system's main vertical drain, or soil stack, which leads to the

**FIGURE 16.3-4** Diagram showing the stack, cleanout, building drain, building sewer, and public or private sewer. (Courtesy Small Homes Council, Building Research Council, University of Illinois, Urbana-Champaign.) Metric added by the author.

building drain and ultimately to the public *sewer*. The portion of the soil stack that lies above the highest horizontal waste-carrying connection to it is called a *stack vent*; sometimes it is called a *waste vent* or *soil vent* (see Fig. 16.3-2). A stack vent may collect one or more vent branches before it continues through the roof to the outdoors, providing a source of atmospheric pressure and thus serving, in this case, as the system's *main vent*. A horizontal pipe to which a number of vent stacks are connected before they go through the roof is also sometimes called a *main vent*.

Branch vents and branch drains are fed by fixture drains (Fig. 16.3-5). A *fixture drain* leaves the drainage point of a fixture and runs to the nearest branch drain. It includes the water-filled plumbing trap which serves to prevent sewer gases from rising through the fixture drain into the building. When the fixture drain line reaches a wall, it intersects with the branch drain/branch vent system. Water flowing out of the fixture drain line falls into the lower portion of the intersected pipe.

A stack vent or branch vent provides an air supply to equalize pressures and smooth the drainage process (see Fig. 16.3-1). The most desirable drainage pattern for water going down a vertical pipe is one in which water swirls down the pipe, clinging to the walls by surface tension.

This process draws air down the pipe with the water. The air is supplied through the vent system. If no air is available through a vent system, or if the flow of water is so heavy that it overcomes the surface tension holding it against the pipe

wall, then drainage can be accompanied by gurgling noises, vibration, and loss of the trap seal.

## GRAVITY FLOW

In a water supply system, flow occurs because the fluid is pressurized. Drainage system components exist at atmospheric pressure, and flow is due to gravity. Consequently, sanitary drain lines must be sloped and arranged to assist gravity flow.

Sufficient slope should be provided in sanitary drainpipes to maintain a flow velocity of at least 2 ft/sec (0.61 m/s), which is necessary to properly scour the pipe and keep solids in suspension. For adequate drainage in a horizontal pipe (one with a slope of less than 45 degrees from horizontal), the pipe should (1) slope at least $1/4$ in. per foot (20.83 mm/m) of run if it is less than or equal to 3 in. (76 mm) in diameter or (2) slope at least $1/8$ in. per foot (10.42 mm/m) of run if it is more than 3 in. (76 mm) in diameter.

Sufficient pipe supports should be provided to maintain uniform slope. Drains laid underground should have a bed of undisturbed, compacted earth or concrete to control differential settlement. Drain lines should not be laid beneath driveways or buildings when this is avoidable.

*Turns and fittings* should be properly detailed to minimize resistance to water flow. When possible, flow should move in a straight line. If turns are unavoidable, sweeping bends are preferred to sharp turns. In horizontal runs, 45 degree wyes, sweeps, and $1/2$, $1/6$, $1/8$, and $1/16$ bends should be used; for vertical-to-horizontal turns, $1/4$ and $1/5$ bends should be used.

Fittings for use in drain systems are designed differently from those for supply systems, primarily to eliminate shoulders and other impediments to gravity flow (Fig 16.3-6). Consequently, fitting names are preceded by the designation *drain*, as in *drain bend* and *drain wye*.

Connection of branch drains to main drains is always done in the upper half of the horizontal pipe. This reduces the chance of clogging, cuts down on interference with flow at the point of connection, and, when there is no flow in the branch, leaves the full diameter of the pipe available to act as a vent.

When gravity flow of liquid or solid waste is not possible, it may be necessary to force the material up mechanically to the level of the sewer. This is customarily accomplished by means of a sump and a sump pump. A *sump pump* is a mechanical device used for pumping out sewage or liquid industrial wastes from locations below a building's gravity drain. These pumps are located in a *sump*, which is a tank or pit that receives sewage or other liquid waste, located below the normal grade of a building's gravity sewer system, and which must be emptied by mechanical means (a sump pump).

**FIGURE 16.3-5** Fixture drains connect fixtures to fixture branches or branch drains.

**FIGURE 16.3-6** Drainage fittings showing elimination of lips that would impede sewage flow. (Courtesy Small Homes Council, Building Research Council, University of Illinois, Urbana-Champaign.)

### DRAIN PIPING

The routing of drain pipes may pose serious problems during construction unless it is carefully considered during the design process. Because a sanitary drainage system drains by gravity, horizontal pipe runs must be sloped downward. This restricts the length of horizontal runs which can be tolerated before the sloping pipe encounters structural members or exceeds the depth of the space in which it is to be concealed.

*Sealing and Venting*  A DWV system must be sealed off from the air to prevent sewer gases from escaping into a building's interior. This is usually accomplished by water seals in fixture traps. A DWV system must, however, be open to the outside air to maintain atmospheric pressure within the system, both to aid in the flow of materials through it and to protect the trap seals from being disturbed. This is usually accomplished through a network of vents.

**FIGURE 16.3-7** The residual water in the bowl of a water closet forms a water seal similar to that in a sink trap. A wall-hung water closet is shown.

*Traps*  In addition to waste, a drainage system contains air and foul-smelling sewer gases. For reasons of sanitation and comfort, it is desirable to prevent sewer gases from entering the building through fixture drains. To prevent this from happening, a *trap* (see Fig. 16.3-3) is installed at each fixture. A trap is simply a length of drain pipe shaped in such a way that it does not drain completely. After each use of the fixture, a trap retains enough water to fill the drain pipe at that point. The trapped water serves as a seal which prevents sewer gas from rising through the fixture drain into the building. Each fixture requires its own trap.

Most plumbing fixture traps are shaped like the letter P and are called *P-traps* (see Fig. 16.3-3).

A special trap called a *drum trap* is usually installed on the drain lines of bathtubs. Such traps, which are frequently clogged with dirt and hair, offer easier access for cleaning. A drum trap is constructed so that the drain line from the bathtub enters its lower portion, and the drain line out into the rest of the drainage system leaves it 2 to 4 in. (50.8 to 100 mm) higher. Therefore, a water seal of 2 to 4 in. (50.8 to 100 mm) is maintained within the drum. Drum traps are generally 3 to 4 in. (76 to 100 mm) in diameter and have a removable top for easy access. Water closets have a trap built into their internal structure (Fig 16.3-7).

A *building trap* is a device intended to prevent the circulation of air between the building drainage system and the building sewer (Fig. 16.3-8).

A *grease trap*, unlike other traps, is not designed to prevent gases from rising through the drain pipes, but rather to collect grease particles and prevent them from reaching the sewer or septic system, where they could cause blockages. Grease traps generally are not found in small residential buildings but may be

**FIGURE 16.3-8** A building trap acts both as a seal and as a cleanout for the building drain. (Courtesy Small Homes Council, Building Research Council, University of Illinois, Urbana-Champaign.)

used in restaurants, institutional kitchens, and large residential structures.

Ordinary traps are generally maintenance free, since they are simple devices with no moving parts. Difficulties arise, however, if the water seal in the trap is lost. There are at least five ways in which this can happen. The water seal may be lost because of any of the following:

1. Suction (*back-siphonage*).
2. Back pressure from a slug of sewage passing through the drain system that forces the water seal up into a lower fixture or sucks it out of a higher trap (Fig. 16.3-9).
3. Evaporation from a rarely used drain.
4. Capillary attraction through an object caught in the trap (such as a diaper, from which the water can readily evaporate).
5. Outdoor wind conditions that create positive or negative pressures within the drainage system through the vents. Positive pressure can push water out of the trap; negative pressure can suck it out of the trap. The latter is much more likely to occur.

Another infrequent maintenance problem with traps is freezing. Freezing of plumbing is generally a bigger problem in the supply system than in the drainage system, but it is possible for damage to occur from freezing of the water seal in a trap. This can result in cracking of the

**FIGURE 16.3-9** A slug of sewage in a soil stack creates positive pressure in front of it and suction behind it. If there were no vents or if the vents were clogged, the left-side trap in this figure could be blown out or the right-side one could be sucked out. (Courtesy Small Homes Council, Building Research Council, University of Illinois, Urbana-Champaign.)

pipes or joints and, consequently, leaks. If a building is to be left unheated during freezing weather, traps should be individually drained through the fitting at their lowest point or an antifreeze solution should be added to the water in the trap.

*Vents* Sometimes, the pressure developed in front of a column of water falling through a drain pipe is great enough to blow the water seal (and sewer gases with it) through a fixture trap and into the interior of the building (see Fig. 16.3-9). Under other conditions, the pressure behind a falling column of water can be low enough to suck the water plug out of a fixture trap and into the drainage system. To prevent either of these events from occurring, a second set of openings is added to the drainage system. These openings comprise a venting system, which provides a free-flowing air connection to the outdoors.

A *vent* is a pipe that connects a drainage system to the atmosphere to permit gases to escape. These are usually made of pipe of the same material and size as the lower components, which serve as drains.

The vent piping of a drainage system consists essentially of the upper half of the fixture branch pipes, *circuit vents*, *loop vents*, *relief vents*, *stack vents*, and *vent stacks* (see Fig. 16.3-2).

A *circuit vent* is a branch vent that serves two or more traps and extends from in front of the last fixture connection of a horizontal branch to a *vent stack*.

A *loop vent* is a branch vent that serves two or more traps and extends from in front of the last fixture connection of a horizontal branch to the *stack vent*.

A *relief vent* is a vent whose primary function is to provide circulation of air between *drainage* and *vent systems* (see Fig. 16.3-2).

A *stack vent* is that portion of a *soil stack* above the highest drain line connection.

A *vent stack* is a vertical vent pipe installed primarily for the purpose of providing circulation of air to and from a part of a drainage system.

The maximum distance between a plumbing fixture's trap and its vent is governed by the diameter of the drain pipe (Fig. 16.3-10). The larger the diameter of a drain pipe, the farther the vent can be located away from the trap and still protect its water seal.

**FIGURE 16.3-10** For code compliance, measure the trap-to-vent distance, as shown in the plan view. (Courtesy Small Homes Council, Building Research Council, University of Illinois, Urbana-Champaign.)

In one-story buildings or at the top floor of multistory structures, where fixtures are located close to a soil stack, stack venting may be permitted by the applicable codes (Fig. 16.3-11). In this case, the drain pipe is large enough and short enough so that it is never entirely filled with water. Therefore, there is always an air space above the water flow, which acts as a vent passage connecting into the soil stack and ultimately to the outdoors through the top of the soil stack/stack vent.

Sometimes it is inconvenient to run a vent from a fixture that is remote from a wall. In such situations, wet venting can

**FIGURE 16.3-11** Stack venting permits elimination of fixture vents.

be used. In *wet venting*, the fixture drain is oversized and used as a vent as well as a drain. Wet venting can be used with fixtures such as sinks and lavatories, but not for venting water closets.

## CLEANOUTS

*Cleanouts* are fittings that provide periodic access to the interior of drain pipes to permit removal of blockages of lint, hair, grease, and other solids. Cleanouts usually consist of a 45-degree *Y-fitting* of the same diameter as the drain pipe, with a removable screw-in plug. When the plug is removed, a cleaning tool can be inserted into the main drain line (Fig. 16.3-12). A cleanout fitting should be located at the base of each soil stack and at the point where the building drain passes through the building's foundation. A cleanout should also be provided at each change of direction of more than 45 degrees, and every 50 ft (15.24 m) in 4 in. (100 mm) pipe and every 100 ft (30.5 m) in larger pipes. Cleanouts should be readily accessible, and their height and clearance should be adequate for inserting cleaning tools.

## PIPE

Using the correct pipe sizes and materials is essential to the proper functioning of a drainage system.

*Pipe Material* The most common materials for drainage system piping are cast iron, galvanized wrought iron or steel, and plastic. Copper, brass, and lead can also be used, but their expense is usually prohibitive. See Section 16.2.1.4 for a discussion of these less typical materials. Figure 16.3-13 identifies pipe materials that are acceptable for drain and vent piping.

**FIGURE 16.3-13** Acceptable Drainage and Vent Piping Materials[a]

| Material | Below Grade | Above Grade |
|---|---|---|
| **Cast Iron** | | |
| Soil pipe | A | A |
| No-hub pipe and fittings | A | A |
| **Copper Tubing** | | |
| Type K — Hard temper | A | A |
| Type L — Hard temper | A | A |
| DWV     Hard temper | A | A |
| **Galvanized Open Hearth or Wrought Iron** | A[b] | A |
| **Galvanized Steel Pipe** | — | A |
| **Plastic** | | |
| ABS[c] | A | A |
| PVC[c] | A | A |
| PVC thin wall | NA | A |
| Polybutylene[c] | A | A |
| Polyethylene[c] | NA | NA |
| Styrene rubber | A[c] | NA |
| **VC Pipe** | A[c] | — |

[a]A = Acceptable; NA = Not acceptable.
[b]Protected when installed underground.
[c]Outside dwelling only.
(Courtesy Small Homes Council, Building Research Council, University of Illinois, Urbana-Champaign.)

*Cast iron* is the long-standing traditional material for drain piping. It is heavy, durable, and serviceable underground or in concrete. *Hubless* joints and fittings for cast iron pipe make it possible to run 3 in. (76 mm) cast iron drain lines in 2 × 4 (50 × 100) wood stud walls (Fig. 16.3-14). Hubless joints consist of a neoprene rubber sleeve gasket that covers the joint, a stainless steel band that fits over the gasket, and worm-screw clamps that compress the band against the pipe. Support must be provided at joints and at 4 ft (1219 mm) intervals along a cast iron pipe.

Older installations of cast iron drain pipe have *bell-and-spigot* or *hub-and-spigot* joints (Fig. 16.3-15). Pipe sections are cast so that one end flares out into a *bell* or *hub*, which fits over the straight end, or *spigot*, of the succeeding section.

The gap between the bell and spigot is packed with oakum or tow hemp and filled with molten lead. Such joints have a shoulder on the inner surface and so must be installed directionally to prevent clogging. The direction of flow should always be from the spigot to the bell.

*Galvanized steel pipe* is sometimes called *galvanized iron*. This pipe is used in drain and vent systems essentially the same way as in water supply systems. Threaded joints are typical, but where galvanized pipe connects to cast iron, it is generally fitted into a bell and leaded. Building and plumbing codes vary in their attitude toward plastic drain piping. The applicable codes should always be

**FIGURE 16.3-12** A stack cleanout. (Courtesy Small Homes Council, Building Research Council, University of Illinois, Urbana-Champaign.)

**FIGURE 16.3-14** A section through a hubless cast iron pipe joint.

**FIGURE 16.3-15** Traditional bell-and-spigot joints in cast iron pipe are bulky and time-consuming.

consulted during the early design of a plumbing system.

The two types of *plastic pipe* suitable for use in drainage systems are polyvinyl chloride (PVC) and acrilonitrile-butadiene-styrene (ABS). PVC pipe used in drain systems is of the same formulation as that used in water supply systems, but is thicker.

ABS pipe has an operating temperature range of −40° to 80°F (−40° to 26.7°C). It can be obtained with solid walls or with foam-cored walls, the latter being less expensive because it uses less resin.

ABS pipe can be solvent welded at joints with or without a primer, or it can be obtained with threaded fittings (Fig. 16.3-16). One disadvantage of using welded joints is that they cannot be unwelded. Correction of errors requires cutting the pipe and remaking the joints (see Section 16.1.1.4)

Plastic drain piping has several advantages. Some manufacturers produce a 3 in. (76 mm) drain pipe with joints compact enough to fit within the clearance of a standard 2 × 4 (50 × 100) stud wall (Fig. 16.3-17). This eliminates the need for special framing for plumbing walls. The light weight of plastic pipe means less load on the supporting structure. Since plastic pipe is a better thermal insulator than metal pipe, liquids with sus-

pended fats are less likely to be cooled to the point of depositing the fats, which clog the pipes.

The disadvantages of plastic pipe include its vulnerability to puncture, to damage by rodents, and to some non-household chemicals.

## SIZING

Another factor which should be considered during the design stage is pipe sizing. Expedient choice of the pipe size or slope to conform to the available floor depth may result in an inefficient and noisy installation. The recommended drain and vent sizes for different parts of a residential system are given in Figure 16.3-18.

Pipe sizing for other building types is designed using drainage fixture units (dfu's). The procedure is as follows:

1. Make a sketch of the drainage system similar to that shown in Figure 16.3-19. Show all components and fixtures.
2. Select the minimum trap size for all traps that are not integral to a fixture. In general, every plumbing fixture requires a trap. Select sizes from tables in applicable plumbing codes. The plumbing code referenced here is the *International Plumbing Code*.
3. Size each fixture branch starting with the section that is farthest from the stack. In Figure 16.3-19, that is the part serving the tub. Determine the dfu

for the fixture. Dfu values are shown in Table 709.1 of the *International Plumbing Code*. The dfu for a bathtub is 2. Using the dfu, select the pipe size from tables in the code. In this case, 1½ in. (38 mm) pipe is required. Continue to size each pipe serving a fixture. Then size each successive portion of the branch, making sure to add all the dfu's together in each case. That is, add the dfu of the tub, lavatory, and sink to size the pipe from the intersection of the tub portion going toward the stack.

4. Select the soil stack size. The soil stack is that which carries waste. To do this, add up all the dfu's above the section being selected and select the size from tables in the plumbing code.
5. Determine the size of the main vent stack. Knowing the length of the vent stack and the size of the soil stack, use tables found in the plumbing code and select the vent stack size directly.
6. Size the building drain. Determine the slope of the drain from the position of the street main and soil stacks. Then select the drain size from a table in the plumbing code using the slope and the number of dfu's in the building.

When designing a building's sanitary drainage system, it is necessary to check carefully the requirements of the applicable plumbing code. Codes often require the drainage component sizes to be determined in accordance with the above-mentioned criteria, but with a minimum

**FIGURE 16.3-16** A section through a PVC pipe joint. This type joint is strong, compact, and easy to make.

**FIGURE 16.3-17** A 3 in. (76.2 mm) PVC vent pipe in a 2 × 4 (50 × 100) stud wall. (Courtesy Small Homes Council, Building Research Council, University of Illinois, Urbana-Champaign.)

**FIGURE 16.3-18** Recommended Drain and Vent Pipe Sizes*

| Fixture | Drain/Trap Size, in. (mm) | Vent Size, in. (mm) |
|---|---|---|
| House sewer | 4 (101.6) | — |
| House drain | 3 (76.2) | — |
| Soil stack | 3 (76.2) | — |
| Branch drain/vent | 1½ (38.1) | 1¼ (31.8) |
| Water closet | 3½ (88.9) | 2 (50.8) |
| Bathtub | 1½ (38.1) | 1¼ (31.8) |
| Lavatory | 1¼ (31.8) | 1¼ (31.8) |
| Kitchen sink | 1½ (38.1) | 1¼ (31.8) |
| Laundry | 1½ (38.1) | 1¼ (31.8) |
| Clothes washer | 2 (50.8) | 1¼ (31.8) |
| Dishwasher | 1½ (38.1) | 1¼ (31.8) |
| Shower stall | 2 (50.8) | 1¼ (31.8) |

*For one- and two-family dwellings of average size.

(Courtesy Small Homes Council, Building Research Council, University of Illinois, Urbana-Champaign.)

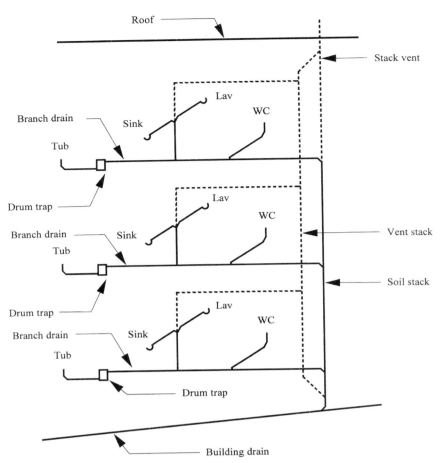

**FIGURE 16.3-19** Drainage riser diagram for an apartment building. A three-story building is shown, but this kind of diagram is essentially the same regardless of the number of stories. It is necessary, however, to draw every floor, because the pipe sizes differ from floor to floor and a convenient way to keep track of them is to write the sizes on the diagram. Diagrams for other building types are similar. (Drawing by HLS.)

Figure labels: Roof, Stack vent, Branch drain, Sink, Lav, WC, Tub, Drum trap, Vent stack, Soil stack, Building drain

that may be larger than would be calculated using the required criteria. For example, the code may say that the main soil stack is to be sized in accordance with the tables in the code, but that it may not be less than 75 mm (3 in.) in diameter.

### 16.3.1.2 Private Sanitary Sewage Disposal Systems

This section covers private sanitary sewage disposal systems. Public sanitary sewage disposal systems are discussed in Chapter 23.

Connection to a public sewage system is the most satisfactory method of sewage disposal. When this is not feasible, and when a large number of residences or other small buildings are to be served, consideration should be given to the construction of a community sewage system

and treatment plant. Specific information on this matter can be obtained generally from the local health authority.

In spite of the immense growth of public sewage disposal systems in this century, many areas are still beyond their reach. Community systems are not always feasible. In locations with no access to either a municipal or community system, sewage disposal systems designed to handle a single residence, a small group of dwellings, or a small commercial project continue to be an important method of sewage disposal.

Not long ago, in rural areas, human and other domestic wastes were disposed of using *cesspools*. These are covered and lined underground pits designed to permit liquids to seep into the ground through the bottom and sides, but retain organic matter and other solids, which are then periodically pumped out. Such sys-

tems not only present an obnoxious nuisance, but are also a major health hazard and are almost universally prohibited in this country. Comparatively, today's *individual sewage (septic) disposal systems* are safe and nuisance free.

These systems consist of a *septic tank*, where the sewage is retained long enough to separate and the solids to decompose, coupled with some type of *soil absorption system* (absorption field, absorption bed, or seepage pit) where the liquids discharged from the septic tank seep into the soil (Fig. 16.3-20).

In these systems, sewage is treated both in the tank and in the soil, as anaerobic (oxygen-free) bacteria decompose the organic matter into simpler, harmless compounds. However, because the anaerobic decomposition of organic matter is accompanied by foul odors and pathogenic (disease-producing) organisms, the partially treated liquid sewage flowing from a septic disposal system (*effluent*) cannot be discharged on the ground surface, or into a lake or stream, without creating a nuisance and a health hazard.

Instead, the effluent from a septic tank is discharged into a properly designed and maintained soil absorption system, which uses the soil for subsurface absorption of treated sewage. Such systems practically eliminate the health and nuisance hazards, but still pose problems in certain areas. A soil absorption system may not work in tightly packed soils or the land area required may be too large.

The removal of solids and the partial biological treatment which takes place in a septic tank is known as *primary treatment*. Most municipal and community sewage plants provide so-called *secondary treatment*, producing relatively odorless but nutrient-rich effluents. Therefore, sewage plants should be equipped to remove or neutralize the nutrients (called *tertiary treatment*) before discharging into lakes or streams.

The discussion of private sewage disposal systems in this section is intended as an aid in understanding the nature of such systems. Much of the discussion and recommendations presented in this section have been abstracted from a publication of the Bureau of Community Environmental Management, U.S. Public Health Service (HEW), entitled *Manual of Septic Tank Practice* (1967), which was developed primarily for single-

ROOF TERMINAL

STACK

PLUMBING FIXTURES TO BE
PROPERLY TRAPPED AND
VENTED

CLEAN OUT

HOUSE SEWER TO BE LAID ON
WELL-COMPACTED EARTH **a**

SEPTIC TANK **b**
COMPACT EARTH AROUND TANK

NONPERFORATED PIPE

ABSORPTION FIELD **c**

PERFORATED PIPE

GRAVEL OR
CRUSHED STONE

PERFORATED PIPE

**FIGURE 16.3-20** Typical septic-tank-type individual sewage disposal system. Solids are separated and retained in the tank, liquids are passed to the absorption field, and gases are vented through the house sewer and vent stack. (Illustration from *Manual of Septic Tank Practices*, Bureau of Community Environmental Management, U.S. Public Health Service [now EPA].)

family residences. There has been some updating here to reflect current practice. These recommendations are made with the full understanding that they may differ in some instances from those of other authoritative sources such as the *International Private Sewage Disposal Code* of the ICC, which will be known in this section as the *Code*. Local, state, and federal laws, codes, and regulations always supersede the recommendations in this section. Therefore, codes and health officials should be contacted and their latest requirements obtained before beginning the design of an individual sewage disposal system. The *Code* contains extensive requirements for private sewage disposal systems for both single-family residences and other types of building occupancies.

## SUITABILITY OF SOIL

A sewage disposal system that uses the soil for subsurface absorption of treated sewage is called a *soil absorption system*. A septic tank system is in this category. The first step in the design of a septic tank system is to determine whether the soil is suitable for the absorption of effluent and, if it is, how much absorption area is required. In general, a site is unsuitable for a soil absorption system unless the following conditions can be satisfied:

1. The percolation rate is within the range required by the applicable plumbing code.
2. The bottom of a trench or seepage pit is at least 4 ft (1219 mm) above the maximum seasonal elevation of the groundwater table.

3. Rock formations and other impervious strata are more than 4 ft (1219 mm) below the bottom of a trench, bed, or pit.
4. The area is large enough to provide the trench and bed bottoms or pit sidewalls required by the applicable code.

*Soil Types* The water absorption of soils can best be determined by percolation tests, but a number of clues help in estimating the likely characteristics of a particular soil. These include soil maps and descriptions of the U.S. Department of Agriculture, on-site visual inspection of soil texture and color, and the record of experience in adjacent areas with similar soils.

For these purposes, soil can be classified as either pervious or impervious. *Pervious soil* is usually granular, such as sand or gravel. It allows water to pass readily through it. *Impervious soil* is a tight, cohesive soil, such as clay, which does not permit the ready passage of water. Most sites contain a mixture of pervious and impervious soils.

Soil particle size governs pore size, which, in turn, influences the rate of water movement. Sandy and gravelly soils generally have large particles and large pores, while clay and silt soils have small particles and small pores (although the total void area may be larger). In some soils, especially clay and silt types, there may be clumping of basic particles into larger particles, giving the soil a well-developed structure. A lump of structured soil will break apart with little force along well-defined cleavage planes. Such structured clay soil is likely to have better water absorption characteristics than one with little or no structure.

Another clue to the absorptiveness of a soil is its color. Bright, uniform, reddish-brown to yellow colors throughout a soil profile indicate favorable oxidation conditions, good air and water movement, and desirable absorption properties.

In some cases, examination of road cuts, stream embankments, or building excavations will give useful information. Well drillers' logs can also be used to obtain data on groundwater and subsurface conditions. In some areas, subsoil strata vary widely within short distances, and borings must be made at the site of the system. If the subsoil appears suitable, as judged by the visual characteristics described above, percolation tests should be made at points and elevations selected as typical of the area in which the soil absorption system will be located.

Subsurface exploration is usually necessary to determine soil conditions: the depth of pervious material, the presence of rock formations, and the level of groundwater.

***Absorption Criteria*** Tables in the *International Private Sewage Disposal Code* show the required percolation rates and area of absorption or seepage area required for soils with various percolation rates. The *percolation rate* is the time required for the level of water in a standard test pit to drop 1 in. (25.4 mm).

The *Code*, for example, divides the required criteria for one- and two-family dwellings into four percolation classes. In class 1, the percolation rate is from 0 to 10. Class 2 includes rates between 10 and 30; class 3, rates between 30 and 45; and class 4, rates between 45 and 60. Soil with a percolation rate higher than 60 is not suitable for any absorption system. Absorption areas increase with the absorption rate from 168 sq ft (15.3 m²) per bedroom for seepage trenches or pits, and 205 sq ft (19.04 m²) per bedroom for seepage beds for class 1, up to 330 sq ft (30.7 m²) and 415 sq ft (38.55 m²), respectively, for class 4.

The procedure for other building types is more complicated. There are four percolation classes with the same percolation rates required for one- and two-family dwellings, but the required absorption areas are smaller, ranging from 110 sq ft (10.22 m²) for seepage trenches or pits, and 140 sq ft (113 m²) for seepage beds for class 1, up to 220 sq ft (20.44 m²) and 280 sq ft (26 m²), respectively, for class 4. These tabulated areas are not related to the building classification and must be adjusted according to it. For example, for a hospital, the tabulated rate must be increased by multiplying the tabulated area by 2 for each bed; for a retail store, the tabulated rate may be reduced by multiplying the tabulated area by 0.03 per customer.

***Percolation Tests*** *Percolation tests* help to determine the acceptability of a site and establish the type and size of the required soil absorption system. The *Code* requires that at least three separate test holes be made but states that others may be required, depending on the conditions at the site. These holes should be spaced uniformly over the proposed soil absorption area and should reach down to the depth of the proposed absorption system. These holes can be 4 to 8 in. (100 to 200 mm) in diameter. Shallow holes can be bored with an auger (Fig. 16.3-21a). Deeper holes may have to be dug by shovel or heavy equipment (Figs. 16.3-21b and 16.3-22).

The *Code* dictates the procedures required for preparing test holes, removing loose materials, protecting the bottom from scouring and sediment, and conducting the tests. *Saturation* means that the void spaces between individual particles of soil are full of water, which can occur in a short period of time. *Swelling* is caused by intrusion of water into the particles of soil, which is a slow process. It is the chief reason for providing an overnight soaking period, especially in clay.

The test procedures required are intended to ensure that the soil is given ample opportunity to swell and to approach the condition that will exist during the wettest season of the year. Thus, the test should give comparable results in the same soil, whether made in a dry or a wet season. In sandy soils containing little or no clay, the swelling procedure is not essential, and the test may be made after the water from a single filling of the hole has completely seeped away.

The length of time required for percolation tests varies, depending on the type of soil.

## SOIL ABSORPTION SYSTEMS

From the results of the percolation tests, it is possible to determine whether a septic tank system is feasible and which type of soil absorption system can be used. When a soil absorption system is usable, three types of design may be considered: absorption bed, absorption field, and seepage pit.

a

b

**FIGURE 16.3-21** Shallow percolation test holes for absorption trench or bed showing (a) a 4 in. (100 mm) auger-bored hole and (b) a larger test hole dug with a shovel. (Illustration from *Manual of Septic Tank Practices*, Bureau of Community Environmental Management, U.S. Public Health Service [now EPA].) Metric added by the author.

**FIGURE 16.3-22** Percolation tests for seepage pits must be made in deep test holes dug down to the level of the proposed bottom of the pit. (Illustration from *Manual of Septic Tank Practices*, Bureau of Community Environmental Management, U.S. Public Health Service [now EPA].) Metric added by the author.

The following definitions apply to soil absorption systems:

- *Absorption Bed*: Similar to an absorption trench, except that it is more than 36 in. (914 mm) wide and contains two or more distribution lines. These are also called *seepage beds*.
- *Absorption Field*: An arrangement of absorption trenches through which treated sewage is absorbed into the soil. These are also called *disposal fields* and *soil absorption fields*.
- *Absorption Trench*: A trench not more than 36 in. (914 mm) in width, containing at least 12 in. (305 mm) of clean, coarse aggregate and a distribution line, through which treated sewage is allowed to seep into the soil.
- *Distribution Line*: An open-jointed or perforated pipe that permits effluent to escape into the soil for absorption. It may consist of 1 ft (305 mm) lengths of 4 in. (100 mm) agricultural drain tile or 2 to 3 ft (610 to 915 mm)

lengths of vitrified clay sewer pipe (both laid with open joints) or perforated plastic pipe. Local experience and applicable codes and ordinances should be reviewed before selecting piping materials.

- *Seepage Pit*: A covered underground pit with a concrete or masonry lining. These pits are designed to permit partially treated *sewage* to seep into the surrounding soil. They are also called *dry wells*. Some jurisdictions do not permit the use of seepage pits.

Only soils with a percolation rate of 30 minutes per inch (25.4 mm) or less are generally considered suitable for seepage pits; if the rate is greater than 60 minutes per inch, the soil is not suitable for any type of soil absorption system. In addition, the selection of a soil absorption system is affected by the location of the system. For example, a safe distance must be maintained between a soil absorption system and the

source of water supply. Since the distance that pollution will travel underground depends on numerous factors, including the characteristics of the subsoil formations and the quantity of sewage discharged, no specified distance is absolutely safe in all localities. In general, the horizontal separation between elements of a soil absorption system and other site improvements should not be less than that outlined in Figure 16.3-23. The *Code* may require greater or permit lesser distances than those shown.

Seepage pits should not be used in areas where water supplies are obtained from shallow wells, or where there are limestone formations and sinkholes through which effluent may easily contaminate the groundwater.

While *absorption fields, absorption beds,* and *seepage pits* should not be intermixed, several of any one of them can be arranged in sequence so that each is forced to utilize its total effective absorption area before liquid flows into the succeeding component. This is called *serial distribution*.

The individual branches of an *absorption field* are called *laterals*. These consist of either:

1. The length of the distribution line between the tee or cross fitting and the farthest point in a closed-loop field (Fig. 16.3-24) or
2. The length of the distribution line between overflow pipes (Fig. 16.3-25)

There should be at least two laterals in every system. No lateral should be more than 100 ft (37.8 m) long. Using more and shorter laterals is preferred because if something happens to disturb one line, most of the field will still be serviceable. The trench bottom and distribution lines should be level. Laterals should be spaced so as to provide at least 6 ft (1829 mm) of undisturbed earth between trenches.

Many different designs may be used in laying out subsurface absorption fields. The choice depends on the size and shape of the available disposal area, the capacity required, and the topography of the area (see Figs. 16.3-24 and 16.3-25).

To provide the minimum required gravel depth and earth cover, the trench depth should be at least 24 in. (610 mm) (Fig. 16.3-26). Additional depth may be needed for surface contour adjustment,

**FIGURE 16.3-23** Minimum Clearances for Components of a Septic Sewage Disposal System[a]

| From | To Well, ft (m)[b] | To Water Line, ft (m)[c] | To Foundation Wall, ft (m) | To Property Line, ft (m) | To Stream, ft (m) |
|---|---|---|---|---|---|
| Building sewer | 50 (15.24) | 10 (3.05) | — | — | 50 (15.24) |
| Septic tank | 50 (15.24) | 10 (3.05) | 5 (1.52) | 10 (3.05) | 50 (15.24) |
| Absorption field | 100 (30.48) | 25 (7.62) | 20 (6.10) | 5 (1.52) | 50 (15.24) |
| Absorption bed | 100 (30.48) | 25 (7.62) | 20 (6.10) | 5 (1.52) | 50 (15.24) |
| Seepage pit | 100 (30.48) | 50 (15.24) | 20 (6.10) | 10 (3.05) | 50 (15.24) |

[a]Based on recommendations of U.S. Public Health Service, HEW.

[b]And to suction line between well and pressure tank.

[c]Pressure type.

(Table excerpted from *Manual of Septic Tank Practices,* Bureau of Community Environmental Management, U.S. Public Health Service [now EPA].) Metric added by the author.

for extra aggregate under the tile, or for other design purposes.

In considering the depth of absorption trenches, the possibility of the lines freezing during prolonged cold periods should be considered. Freezing rarely occurs in distribution lines that are adequately surrounded by gravel at the proper depth when the system is kept in continuous operation. However, pipes under driveways or other surfaces which are usually cleared of snow should be close-jointed and insulated or should be installed at a greater depth.

The total required absorption area is determined by the results of soil percolation tests. As previously discussed, the required area may be obtained from the *Code.* For example, for a three-bedroom dwelling on a lot where the minimum percolation rate was 15 minutes per inch (25.4 mm) (percolation class 2), the necessary absorption area for seepage beds would be three times the tabulated value of 315 sq ft (28.26 m²), or 945 sq ft (87.79 m²). For trenches 2 ft (610 mm) wide, the required total length of the trench would be 945 sq ft (87.79 m²) divided by 2, or 472.5 linear ft (144 m). If three laterals were used on sloping ground (see Fig. 16.3-25), the length of each lateral would be 472.5 ft (144 m) divided by 3, or 157.5 ft (48 m). If a closed-loop system on level ground (see Fig. 16.3-24) were used, each of the half loops would be 165 ft (50.29 m) and the middle lateral 142.5 ft (43.43 m). The spacing of trenches is generally governed by practical construction considerations such as type of equipment, soil, and topography, but the clear space between trenches should not be less than 6 ft (1.83 m) to prevent short-circuiting (flow between trenches).

In the example cited, trenches are 2 ft (0.61 m) wide, there are three trenches, and there are two spaces between trenches at 12.5 ft (3.81 m) each. The total width of trenches and spaces is 6 ft (1.83 m) plus 23 ft (7.01 m), or 29 ft (8.84 m). The required area for the absorption field is 29 ft (8.84 m) times 142.5 ft (43.43 m), which equals 4132.5 sq ft (383.9 m²), plus the additional land required to provide the clearances required by the *Code.*

Ideally, a house site should be large enough to allow room for an additional system if the first one fails or if bedrooms are added later.

Careful construction is important in obtaining a satisfactory soil absorption system. Attention should be given to protecting the natural absorption properties of the soil. Care must be taken to prevent sealing of the surface on the bottom and sides of the trench. Trenches should not be excavated when the soil is wet enough to smear or compact easily. Soil moisture is right for safe working only when a handful will mold with considerable pressure. Open trenches should be protected from surface runoff to prevent the entrance of silt and debris. If it is necessary to walk in the trench, a temporary board laid on the bottom will reduce com-

**FIGURE 16.3-24** Typical absorption field layout for level ground. The length of the lateral should be figured from the start of the absorption trench (dashed lines) to the farthest point.

(Illustration from *Manual of Septic Tank Practices,* Bureau of Community Environmental Management, U.S. Public Health Service [now EPA].) Metric added by the author.

FROM SEPTIC TANK

ALTERNATE DROP BOX METHOD AS SHOWN IN
SECTION BB MAY BE USED.

NOTE:
INVERT OF THE OVERFLOW PIPE
MUST BE AT LEAST 4" (100 mm) LOWER
THAN INVERT OF THE SEPTIC TANK
OUTLET

12"
(305 mm)

UNDISTURBED
EARTH

TIGHT
JOINTS

6' (1.83 m) MINIMUM

SECTION A—A

① TEE*
② ELL*

TIGHT JOINTS

UNDISTURBED EARTH

6' (1.83 m) MINIMUM

SECTION B-B
(ALTERNATE CONSTRUCTION)

* DIFFERING GROUND SLOPES OVER SUBSURFACE DISPOSAL FIELD
MAY REQUIRE USE OF VARIOUS COMBINATIONS OF FITTINGS.

**FIGURE 16.3-25** Typical absorption field layout for sloping ground. (Illustration from *Manual of Septic Tank Practices,* Bureau of Community Environmental Management, U.S. Public Health Service [now EPA].) Metric added by the author.

paction. If smearing and compaction do occur, surfaces should be raked to a depth of 1 in. (25.4 mm), and loose material removed, before the gravel is placed in the trench.

A distribution pipe should be laid in a level trench of sufficient width and depth (see Fig. 16.3-28), and should be surrounded by clean, graded gravel or rock,

broken hard-burned clay brick, or similar aggregate, ranging in size from $1/2$ to $2^1/_2$ in. (12.7 to 63.5 mm). Cinders, broken shell, and similar material are not recommended, because they are usually too fine and may lead to premature clogging. The coarse aggregate should extend from at least 2 in. (50.8 mm) above the top of the pipe to at least 6 in. (152 mm)

below the bottom of the pipe. If open-joint drain tile is used, the upper half of each joint opening should be covered with a strip of Number 15 roofing paper, as shown in Figure 16.3-26.

The top of the stone should be covered with a vapor-pervious barrier, such as untreated building paper, a 2 in. (50.8 mm) layer of hay or straw, or similar pervious material to permit vapor movement and yet prevent the stone from becoming clogged by the earth backfill. An impervious covering should not be used, as this interferes with *evapotranspiration* (absorption and release of moisture by plants) at the surface. Although generally not figured in the calculations, evapotranspiration is often an important factor in the operation of absorption fields and beds.

Open-joint tile is usually laid with the opening covered by strips of roofing paper, but drain tile connectors, collars, clips, or similar accessories are of value in obtaining uniform spacing and protecting tile joints. Most such accessories are made of galvanized iron, copper, or plastics.

Tree and shrub roots sometimes penetrate into transmission lines and block the flow of effluent. Root problems can be prevented by using a liberal amount of gravel or stone over and under the pipe. Roots generally seek ground moisture and, in the few cases where they become troublesome, there is usually some explanation involving changing moisture conditions. For example, at a residence used only during the summer, roots are most likely to penetrate when the house is uninhabited, or when moisture immediately below or around the gravel becomes less plentiful than during the period when the system is in use. In general, trenches constructed within 10 ft (3.05 m) of large trees or dense shrubbery should have at least 12 in. (305 mm) of gravel or crushed stone beneath the pipe.

The backfill over a new absorption trench should be free of large stones and other deleterious materials, and should be hand tamped and overfilled 4 to 6 in. (100 to 152 mm) higher than the adjacent grade to compensate for likely settling. Settlement of the fill in a trench to a lower level is very undesirable because it may cause collection of storm water in the trench, which can lead to premature saturation of the absorption field or even complete washout of the trench.

NOTE: DRAIN TILE LAID WITH JOINTS OPENED FROM 1/8" TO 1/4" (3.2 mm to 6.4 mm). SPECIAL COLLARS MAY BE USED IF DESIRED

SUITABLE PERVIOUS BARRIER

OPENINGS AT 1/8" TO 1/4" (3.2 mm to 6.4 mm)   JOINT COVERING

GRAVEL OR BROKEN STONE

LONGITUDINAL SECTION

12" (305 mm)

2" MIN. (50.8 mm)

5" MIN. (127 mm)

6" MIN. (152.4 mm)

OVERFILL TO ALLOW FOR SETTLEMENT

4" TO 6" (101.6 mm to 152.4 mm)

BACKFILL (EARTH)

12" TO 36" (305 mm to 914.4 mm)

CROSS SECTION

LATERAL OF DRAIN TILE (SHOWN) OPEN JOINTED SEWER PIPE OR PERFORATED PIPE

**FIGURE 16.3-26** Details of a typical absorption trench. (Illustration from *Manual of Septic Tank Practices,* Bureau of Community Environmental Management, U.S. Public Health Service [now EPA].) Metric added by the author.

Heavy vehicles will crush the tile in a shallow absorption field and should not be operated there unless special provisions are made to support the weight. Machine tamping or hydraulic backfilling of absorption trenches should not be permitted. Machine grading should be completed before the field is laid.

Where the ground does not exceed a 6 in. (152 mm) height variation in any direction within the absorption area, effluent may be discharged through a system of laterals connected at their extremes to form a continuous closed loop (see Fig. 16.3-24). Such a system permits the effluent to flow into all the tile lines whenever there is a discharge from the septic tank.

Laterals traditionally were connected to a house sewer through a distribution box, and the ends were not interconnected. A Public Health Service study for the Federal Housing Administration (FHA) on the effectiveness of distribution boxes concluded that distribution boxes generally do not ensure equal distribution as intended and therefore are not needed for single-family systems. However, even when local practice includes distribution boxes, the laterals should be interconnected at their ends, as in Figure 16.3-24. Interconnection of laterals ensures access to each point in the lateral from two sides, so that if one part of the lateral became damaged, effluent would be able to reach the lateral from the other side.

Where sloping ground is used for an absorption field, it is usually necessary to construct a small temporary dike or surface water diversion ditch above the field to prevent the absorption field from being eroded by surface runoff. This dike should be maintained or the ditch kept free of obstructions until the field becomes well covered with vegetation.

The maximum ground slope suitable for soil absorption systems is governed by local factors affecting erosion of the ground used for the absorption field. Excessive slopes which are not protected from surface water runoff or do not have adequate vegetation cover to prevent erosion should be avoided. Generally, ground having a slope greater than 1 vertical to 2 horizontal should be investigated carefully to determine if it is satisfactory from the erosion standpoint. Near banks, the horizontal distance from the side of a trench to the ground surface should be 2 ft (610 mm) or more to prevent lateral flow of effluent and breakout onto the surface.

In sloping ground, each lateral should be connected to the next by a closed overflow line laid on an undisturbed section of ground, as shown in Figure 16.3-25. The arrangement is such that all effluent is discharged to the first trench until it is filled; excess liquid is then carried by means of a closed overflow line to the next succeeding or lower trench. In that manner, each portion of the subsurface system is used in succession.

Design and construction procedures for sloping-ground fields are basically the same as those for level-ground fields. The bottoms of trenches and of distribution lines should be level, and laterals should follow surface contours. Adjacent trenches may be connected with either an overflow line or a drop box arrangement. The *invert* (bottom) of the first overflow line should be at least 4 in. (100 mm) lower than the invert of the septic tank outlet (see Fig. 16.3-25).

Overflow lines should be 4 in. (100 mm) tight-joint pipe with direct connections to the distribution lines in adjacent trenches or to a drop box arrangement. Care must be exercised in constructing overflow lines to ensure at least 6 ft (1.83 m) of undisturbed earth between trenches. The overflow line should rest on undisturbed earth, and backfill should be carefully tamped.

An *absorption bed* is a variation of the typical trench used in absorption fields. The basic design is essentially the same as for an absorption trench, but it is generally much wider than 36 in. (0.914 m) and contains two or more distribution lines 4 to 6 ft (1.22 to 1.83 m) apart. Absorption beds have the advantages of making more efficient use of the land and of modern earth-moving equipment.

The minimum bottom area for absorption beds is the same as for trenches in absorption fields. A bed should have a minimum depth of 24 in. (610 mm) below natural ground level to provide a minimum of 12 in. (305 mm) of gravel and 12 in. (305 mm) of earth backfill. The gravel should extend at least 2 in. (50.8 mm) above and 6 in. (152 mm) below the distribution line. The bottoms of beds and distribution lines should be level. Distribution lines should be spaced not more than 6 ft (1.83 m) apart and not more than 3 ft (0.9 m) from the bed sidewall. When more than one bed is used, there should be a minimum of 6 ft (1.83 m) of undisturbed earth between adjacent beds, and the beds should be connected in series, as illustrated in Figure 16.3-25 for absorption fields in sloping ground.

*Seepage pits* are deeper than absorption trenches or beds and are, therefore, more likely to contaminate groundwater than other absorption methods. In fact, they are not permitted in some jurisdictions. Because pits handle the same amount of effluent as an absorption field or bed in a smaller land area, the likelihood of clogging and saturation is greater, as is the possibility of effluent breaking through to the surface.

Pits should not be used where absorption fields or beds are feasible or when the soil has a percolation rate in excess of 30 minutes per inch (25.4 mm). Where the top 3 or 4 ft (0.914 or 1.29 mm) of soil is not porous enough for an absorption field, but is underlaid with suitable granular material, seepage pits constitute an acceptable substitute. Seepage pits may also be used where the land area is too small, too irregular, or too heavily wooded to accommodate absorption field or bed systems.

The required capacity of deep pits should be computed on the basis of percolation tests made in each vertical stratum penetrated (see Fig. 16.3-22). Soil strata in which the percolation rates are in excess of 30 minutes per inch (25.4 mm) should not be included in computing the absorption area. The weighted average of the results should be computed to obtain an average percolation rate. In shallow pits where the soil is relatively uniform, the required absorption area can be established by soil experts by visual analysis of the soil type (Fig. 16.3-27). In other cases, the absorption area should be established by percolation tests, as previously described for other absorption methods.

Tables in the *Code* give the absorption area requirements for seepage pits for soils with different percolation rates. The absorption area of the seepage pit is the vertical wall area of the actual excavated pit (not the pit lining). Only the area of pervious strata below the inlet should be figured. No allowance should be made for impervious strata or for bottom area. After the required area has been established, Figure 16.3-28 may be used to select the desired dimensions of circular or cylindrical seepage pits.

To demonstrate a typical pit capacity calculation, assume that the absorption system is to serve a three-bedroom house on a lot where the percolation rate is 1 in. (25.4 mm) in 15 minutes. Assume also that the water table does not rise above 17 ft (5.18 m) below the ground surface, that seepage pits with an effective depth of 10 ft (3.05 m) can be provided, and that the house is in a locality where it is common practice to install seepage pits of 5 ft (1.5 m) diameter (i.e., 4 ft [1.2 m] to the outside walls, which are surrounded by about 6 in. [152 mm] of gravel).

**FIGURE 16.3-27** Required Wall Area for Seepage Pits

| Symbol* | Character of Soil | Minimum Wall Area per Bedroom, sq ft (m²) |
|---|---|---|
| GW | Well-graded gravels, gravel-sand mixtures, little or no fines | 50 (4.65) |
| GP | Poorly graded gravels or gravel-sand mixtures, little or no fines | 50 (4.65) |
| SW | Well-graded sands, gravelly sand, little or no fines | 75 (6.97) |
| SP | Poorly graded sands or gravelly sands, little or no fines | 75 (6.97) |
| SM | Silty sand, sand-silt mixtures | 125 (11.6) |
| GC | Clayey gravels, gravel-sand-clay mixtures | 150 (13.94) |
| SC | Clayey sands, sand-clay mixtures | 150 (13.94) |
| ML | Inorganic silts and very fine sands, rock flour, silty or clayey fine sands or clayey silts with slight plasticity | 300 (27.9) |
| CL | Inorganic clays of low to medium plasticity, gravelly clays, sandy clays, silty clays, lean clays | 350 (32.52) |
| OL | Organic silts and organic silty clays of low plasticity | |
| MH | Inorganic silts, micaceous or diatomaceous fine sandy or silty soils, elastic silts | Unsuitable for Seepage Pits |
| CH | Inorganic clays of high plasticity, fat clays | |
| OH | Organic clays of medium to high plasticity, organic silts | |
| Pt | Peat and other highly organic silts | |

*Refer to Section 2.1 for a discussion on soil classification.
(Table excerpted from *Manual of Septic Tank Practices*, Bureau of Community Environmental Management, U.S. Public Health Service [now EPA].) Metric added by the author.

According to the *Code*, the required absorption area for percolation class 2 (10 to less than 30 percolation rate) is 250 sq ft (23.2 m²) per bedroom. The total, then, is 250 × 3, or 750 sq ft (69.6 m²) of absorption area that would be needed. Referring to Figure 16.3-28, it can be determined that a 5 ft (1.52 m) round pit 10 ft (3.05 m) deep would provide 157 sq ft (14.59 m²) of wall area. Dividing 750

**FIGURE 16.3-28** Effective Wall Absorption Areas (sq ft [m²]) of Round Seepage Pits

| Pit Diameter, ft | Depth Below Inlet | | | | | | | | | |
|---|---|---|---|---|---|---|---|---|---|---|
| | 1 ft | 2 ft | 3 ft | 4 ft | 5 ft | 6 ft | 7 ft | 8 ft | 9 ft | 10 ft |
| 3 | 9.4 | 19 | 28 | 38 | 47 | 57 | 66 | 75 | 85 | 94 |
| 4 | 12.6 | 25 | 38 | 50 | 63 | 75 | 88 | 101 | 113 | 126 |
| 5 | 15.7 | 31 | 47 | 63 | 79 | 94 | 110 | 126 | 141 | 157 |
| 6 | 18.8 | 38 | 57 | 75 | 94 | 113 | 132 | 151 | 170 | 188 |
| 7 | 22.0 | 44 | 66 | 88 | 110 | 132 | 154 | 176 | 198 | 220 |
| 8 | 25.1 | 50 | 75 | 101 | 126 | 151 | 176 | 201 | 226 | 251 |
| 9 | 28.3 | 57 | 85 | 113 | 141 | 170 | 198 | 226 | 254 | 283 |
| 10 | 31.4 | 63 | 94 | 126 | 157 | 188 | 220 | 251 | 283 | 314 |
| 11 | 34.6 | 69 | 104 | 138 | 173 | 207 | 242 | 276 | 311 | 346 |
| 12 | 37.7 | 75 | 113 | 151 | 188 | 226 | 264 | 302 | 339 | 377 |

(Table excerpted from *Manual of Septic Tank Practices*, Bureau of Community Environmental Management, U.S. Public Health Service [now EPA].) Refer to Figure 25.7-1 for metric conversions.

(69.6) by 157 (14.59), we get 4.7, that is, four pits 10 ft (3.05 m) deep and one pit 8 ft (2.44 m) deep. If it is desirable to make all pits the same size, six 5 ft (1.52 m) round pits 8 ft (2.44 m) deep can be selected from Figure 16.3-28, providing 6 × 126 or 756 sq ft (70.23 m²) of wall area, which is close enough to the total required. Small-diameter, relatively deep pits require the removal of less soil than larger pits with equivalent wall area but may be harder to excavate. When this is the case, larger pits can be selected.

Where more than one pit is provided, the pits can be connected in series or in parallel. However, they should be separated by a distance equal to three times the diameter of the largest pit. For pits more than 20 ft (6.1 m) deep, the minimum distance between pits should be 20 ft (6.1 m) (Fig. 16.3-29). The area of the lot on which a house is to be built should be large enough to maintain this distance between the pits while still allowing room for additional pits if the first ones fails. If this cannot be done, other sewage facilities should be explored.

The permeability of the soil may be damaged during excavation. Digging in wet soils should be avoided as much as possible. Cutting teeth on mechanical equipment should be kept sharp. Bucket auger pits should be reamed to a larger diameter than the bucket. Loose material should be removed from the excavation.

Pits should be backfilled with clean gravel to a depth of 1 ft (305 mm) above the pit bottom or 1 ft (305 mm) above the reamed ledge to provide a sound foundation for the lining.

Lining materials for seepage pits can be clay or concrete masonry units in brick, block, or ring shapes. Rings should have weep holes or openings to provide for seepage. Concrete brick and block should be laid dry (unmortared) with staggered, slightly open joints. Hollow concrete block is frequently laid on its side with cores horizontally to facilitate soil absorption. Standard clay brick should be laid flat to form a 4-in. (100-mm)-thick wall. The outside diameter of the lining should be at least 1 ft (305 mm) smaller than the excavation diameter to provide at least 6 in. (152 mm), and preferably 1 ft (305 mm), of annular space to be filled with coarse gravel to the top of the lining (Fig. 16.3-30).

Either a corbelled brick dome or a flat concrete cover can be used over a pit. Corbelled bricks should either be laid in cement mortar or have a 2 in. (50.8 mm) covering of concrete. If a flat cover is used, a prefabricated type is preferred. A concrete cover should be reinforced so that it is equivalent in strength to an approved septic tank cover. The cover should rest on undisturbed ground 1 ft (305 mm) beyond the lining. A 9 in. (229 mm) capped opening in the pit cover should be provided for pit inspection.

Connecting lines should be of a sound, durable material, the same type used for the house-to-septic-tank connection. All connecting lines should be laid on a firm bed of undisturbed soil throughout their length. The grade of a connecting line should be at least 2% ($^1/_4$ in. per foot

**FIGURE 16.3-29** Typical septic tank system with two seepage pits connected in series. (Illustration from *Manual of Septic Tank Practices*, Bureau of Community Environmental Management, U.S. Public Health Service [now EPA].) Metric added by the author.

**FIGURE 16.3-30** Details of a typical round seepage pit. (Illustration from *Manual of Septic Tank Practices*, Bureau of Community Environmental Management, U.S. Public Health Service [now EPA].) Metric added by the author.

[30.83 mm/m]). The pit inlet pipe should extend horizontally at least 1 ft (305 mm) into the pit with a tee or ell to divert flow downward to prevent washing and eroding of the sidewalls.

## SEPTIC TANKS

In an *individual sewage disposal system* the building (house) sewer connects the building drain (see Fig. 16.2-8) to a septic tank (see Fig. 16.3-20). A septic tank is a watertight, covered receptacle designed to receive the discharge of sewage from a building sewer; to separate the solids from the liquids, digest the organic matter, and store the digested solids during a period of detention; and to allow the clarified liquids to discharge for final disposal.

Despite their physical position in a system, septic tanks are discussed here—after absorption systems—because it must first be determined whether the soil is suitable for a subsurface absorption system and whether adequate area is available for the type of system intended. Except for choosing the tank's capacity and component materials, there are few options in tank design. Because the interior design of a tank is a matter of code compliance and local practice, designs approved by local health authorities can be specified with confidence.

A septic tank performs three basic functions: (1) removal of solids, (2) storage of scum and sludge, and (3) biological treatment. *Scum* is a mass of sewage matter that floats on the surface of the sewage in a septic tank. *Sludge* is the accumulated solids that settle out of the sewage and form a semiliquid mass on the bottom of a septic tank.

A septic tank's chief function is the separation of solids, but the other functions are also essential if the system is to be more than a holding tank requiring frequent pumping. Biological treatment of the organic matter contained in the scum and sludge ensures that only a small portion of indigestible solid matter remains in the tank permanently between pumpings.

The amount of clogging of the soil by tank effluent varies directly with the amount of suspended solids in the liquid. As sewage from a building sewer enters a septic tank, its rate of flow is reduced so that larger solids sink to the bottom or rise to the surface. These solids are retained in the tank, where most of them

are gradually broken down by bacteria into water-soluble components, which do not clog the soil.

Heavier sewage solids settle to the bottom of the tank, forming a blanket of sludge. Lighter solids, including fats and greases, rise to the surface and form a layer of scum. A considerable portion of the sludge and scum is liquefied through decomposition or digestion. During this process, gas is liberated from the sludge, carrying a portion of the solids to the surface, where they accumulate with the scum. Ordinarily, they undergo further digestion in the scum layer, and a portion settles again to the sludge blanket on the bottom. The settling action is retarded by grease in the scum layer and by gasification in the sludge blanket.

Solids and liquid in the tank are subjected to decomposition by bacterial and natural processes. The bacteria are of a variety called *anaerobic* (oxygen-free) which thrive in the absence of oxygen. This decomposition or treatment of sewage under oxygen-free conditions is termed *septic*, which explains the name of the tank.

Contrary to popular belief, septic tanks do not accomplish a high degree of bacterial removal. Although sewage undergoes treatment while passing through the tank, *pathogenic* (disease-producing) organisms are not removed. Therefore, septic tank effluents cannot be safely discharged onto the surface or into a ditch, lake, or stream. This, however, does not detract from the value of septic tanks. As previously explained, a tank's primary purpose is to condition the sewage to prevent rapid clogging of the subsurface absorption system.

Further treatment of effluent, including the removal of pathogenic organisms, is effected by percolation through the soil. These organisms slowly die out as their nutrients are used up and as they enter the unfavorable environment afforded by the soil.

Although septic tanks are supposed to be watertight when installed, as they age, some leakage may develop through openings in the inlet and outlet pipes or cracks in the tank. Therefore, septic tanks should be located where they are not likely to contaminate a well, spring, or other source of water supply. Underground contamination may travel for considerable distances, usually in the direction of the water table gradient, that is, from a higher to a lower water table area. Be-

cause water tables usually follow ground surface contours, septic tanks should be located downhill from private wells or natural springs.

Unfortunately, however, locating a tank at a lower surface elevation than a well is not a guarantee against contamination. Sewage from soil absorption systems occasionally contaminates wells having higher surface elevations. This happens because elevations of subsurface absorption systems are almost always higher than the level of water in wells which may be located nearby, and contamination may travel downward to the water-bearing stratum (Fig. 16.3-31). It is necessary, therefore, to rely on horizontal as well as vertical distances for protection. Tanks should never be closer than 50 ft (16.34 m) from any source of water supply, and greater distances are preferred where possible (see Fig. 16.3-23).

A septic tank should not be located within 5 ft (1.5 m) of a building foundation, because structural damage may result during construction and seepage may enter the basement. A tank should not be located in a swampy area or in an area subject to flooding. Ideally, tanks should be located where the largest possible area will be available for the absorption system. Consideration should also be given to the location from the standpoint of cleaning and maintenance. A tank should be close enough to a paved area so that it can be readily reached by the hose of

**FIGURE 16.3-31** A seepage pit can contaminate a private water supply if it is inadequately separated from the well or the water-bearing stratum. (Illustration from *Manual of Septic Tank Practices*, Bureau of Community Environmental Management, U.S. Public Health Service [now EPA].)

a septic pumping truck. Where a public sewer may be installed at a future date, provision should be made in the household plumbing system for connection to such a sewer.

Grease interceptors (grease traps) are seldom required in household sewage disposal systems. When one is provided, the discharge from a waste disposer (garbage grinder) should not be passed through it. The septic tank capacities recommended in this discussion are sufficient to receive the grease and solids normally discharged from a typical home.

It is generally advisable to have all sanitary wastes from a household discharge to a single septic disposal system. For household installations, it is also usually more economical to provide a single disposal system than two or more with the same total capacity. Normal household waste, including that from the laundry, bath, and kitchen, should pass into a single system.

Providing adequate capacity is an important consideration in septic tank design. The larger the tank, the longer sewage will be retained in it, the better will be the removal of solids from the effluent, and the more complete will be the biological treatment of the contents. Undersized tanks are more likely to become overloaded, which will result in inadequately treated sewage entering the absorption system and clogging it. The liquid capacities recommended in Figure 16.3-32 allow for the use of all household appliances, including waste disposers, automatic clothes washers, and dishwashers. They also ensure enough excess capacity to retain the sewage a sufficient time to settle out and digest solids. They do not, however, take into account liquids from other sources. For example, water that is free of organic matter should be discharged into a storm sewer or dry well.

A *dry well* is a covered and lined underground pit similar to a *seepage pit* but intended to receive water free of organic matter, such as from roof drains, floor drains, or laundry tubs. Dry wells can be used as an auxiliary to a septic system to avoid overloading the septic system. Dry wells are also called *leaching wells* and *leaching pits*. They should not be confused with *seepage pits* which are intended to receive septic tank effluent, even though these are also sometimes called dry wells.

Septic tanks should be watertight units constructed of corrosion- and decay-resistant materials such as concrete, coated metal, vitrified clay, heavyweight concrete masonry units, or hard-burned bricks. Precast and cast-in-place reinforced concrete tanks are common. Prefabricated steel tanks meeting UL Standard UL70 are acceptable in some states.

Special attention should be given to job-built tanks to ensure watertightness and prevent contamination. Heavyweight concrete blocks should be laid on a solid foundation, and the mortar joints should be well filled. The interior of a tank should be surfaced with two $1/4$-in. (6.4-mm)-thick coats of portland cement sand plaster. Some typical septic tanks are illustrated in Figure 16.3-33.

Precast concrete tanks should have a minimum wall thickness of 3 in. (76 mm) and should be adequately reinforced to facilitate handling. When precast slabs are used as covers, they should be watertight, have a thickness of at least 3 in. (76 mm), and be adequately reinforced, so as to be capable of supporting a dead load of 300 psf (1465 kg/m²).

Backfill around a septic tank should be placed in thin layers and thoroughly tamped in a way that will not produce undue strain on the tank. Backfill may be consolidated using water, provided that the backfill is placed and wetted in layers and the tank is first filled with water to prevent floating.

Adequate access should be provided to each tank compartment for inspection and cleaning. Access to each compartment can be provided by either a removable cover or a manhole at least 20 in. (508 mm) wide. When the top of a tank is located more than 18 in. (457 mm) below the finished grade, manholes and in-

**FIGURE 16.3-33** Septic tanks are precast concrete or welded sheet steel in one of several basic shapes. (Illustration from *Manual of Septic Tank Practices*, Bureau of Community Environmental Management, U.S. Public Health Service [now EPA].)

spection holes should extend to approximately 8 in. (203 mm) below the finished grade (Fig. 16.3-34a), or they can be extended to the finished grade if a *gas seal* is provided to keep odors from escaping. A gas seal is made by filling the top of the manhole with sand or by sealing with a minimum of 1 in. (25.4 mm) of portland cement plaster over an intermediate wood platform (Fig. 16.3-34b). In most instances, an extension can be made using clay or concrete pipe, but attention must be given to the accident hazard involved when manholes are extended close to the ground surface.

Typical single- and double-compartment tanks are illustrated in Figures 16.3-35 and 16.3-36. The inlet invert (bottom) should enter the tank at least 3 in. (76 mm) above the outlet invert and the liquid level in the tank to allow for a momentary rise in the liquid level during discharges into the tank. This free drop prevents water from backing up into the house sewer and minimizes stranding of solid material in the sewer.

A vented inlet tee or baffle should be provided to divert the incoming sewage downward. It should penetrate at least 6 in. (152 mm) below the liquid level, but in no case should the penetration be greater than that allowed for the outlet

**FIGURE 16.3-32** Recommended Septic Tank Capacities*

| Number of Bedrooms | Minimum Tank Capacity (gal) (L) |
|---|---|
| 3 or fewer | 1000 (3785) |
| 4 | 1200 (4543) |
| Additional bedrooms | Add 250 gal (946 L)/bedroom |

*Provides for use of garbage grinder, dishwasher, automatic clothes washer, and other household appliances.

(Table excerpted from *Manual of Septic Tank Practices*, Bureau of Community Environmental Management, U.S. Public Health Service [now EPA].) Metric added by the author.

a                                    b

**FIGURE 16.3-34 Manholes for septic tank access.** (Illustration from *Manual of Septic Tank Practices,* Bureau of Community Environmental Management, U.S. Public Health Service [now EPA].)

| ⅜″ | = 9.5 mm | 8″ | = 203.2 mm |
| 1″ | = 25.4 mm | 10″ | = 254.0 mm |
| 1½″ | = 38.1 mm | 12″ | = 304.8 mm |
| 2″ | = 50.8 mm | | |
| 3″ | = 76.2 mm | 2.5 ft. | = 0.762 m |
| 4″ | = 101.6 mm | 5.0 ft | = 1.5 m |
| 6″ | = 152.4 mm | | |

**FIGURE 16.3-35 Single-compartment septic tank showing recommended proportions, dimensions, and clearances.** (Illustration from *Manual of Septic Tank Practices,* Bureau of Community Environmental Management, U.S. Public Health Service [now EPA].) Metric added by the author.

device. A number of arrangements commonly used for inlet and outlet devices are shown in Figure 16.3-37.

The outlet device should penetrate just far enough below the liquid level of the tank to provide a balance between sludge and scum storage volume; otherwise, part of the advantage of capacity is lost. A vertical section of a properly operating tank would show it divided into three distinct layers: scum at the top, a middle zone free of solids, and a bottom layer of sludge. The outlet device retains scum in the tank while limiting the amount of sludge that can be accommodated without scouring, which results in sludge discharging in the effluent from the tank. The outlet device should generally extend to a distance below the surface equal to 40% of the liquid depth. For horizontal cylindrical tanks, this should be reduced to 35%. For example, in a horizontal cylindrical tank having a liquid depth of 42 in. (1067 mm), the outlet device should penetrate 14.7 in. (42 × 0.35) (373.4 mm) below the liquid level.

The outlet device should extend above the liquid line to about 1 in. (25.4 mm) from the top of the tank. The space between the top of the tank and the baffle allows gas to pass off from the tank into the house vent stack.

The shape of a septic tank does not affect its operation. Shallow tanks function as well as deep ones. However, the smallest plan dimension should be at least 24 in. (610 mm). Liquid depth may range between 30 and 60 in. (762 and 1524 mm). Tank volume more than 72 in. (1829 mm) below the top of the liquid should not be considered in calculating tank capacity. Adequate capacity above the liquid line is required to provide for the approximately 30% of the total scum that floats above the liquid. In addition to provision for scum storage, a 1 in. (25.4 mm) clearance between the tops of inlet and outlet tees (or baffles) and the top of a tank usually should be provided to permit free passage of gas back to the inlet and house vent stack. In tanks with straight vertical sides, the distance between the top of the tank and the liquid line should be about 20% of the liquid depth. In horizontal cylindrical tanks, an area equal to about 15% of the total circle should be provided above the liquid level. This condition is met if the liquid depth (distance from the outlet invert to the bottom of the tank) is equal to 79% of the diameter of the tank.

NOTE: ALL FITTINGS 4" V.C.
OUTLET FITTING SET 3"
BELOW INLET FITTING

PLAN

VENT

9"    8"

FLOW LINE    1'–5"

3'–6"    2'–1"

LONGITUDINAL
SECTION

VENT

LIQUID CAPACITY
750 GALS.

4'–0"

SECTION A–A

| | | |
|---|---|---|
| 3" | = | 76.2 mm |
| 4" | = | 101.6 mm |
| 8" | = | 203.2 mm |
| 9" | = | 228.6 mm |
| 1'–5" | = | 431.8 mm |

| | | |
|---|---|---|
| 2'–1" | = | 635 mm |
| 3'–6" | = | 1066.8 mm |
| 4'–0" | = | 1219.2 mm |
| 4'–10" | = | 1473.2 mm |
| 9'–0" | = | 2743.2 mm |

**FIGURE 16.3-36** Typical design of a double-compartment precast concrete septic tank. Dimensions shown are for a 750 gal (2839 L) tank, which may be found in older homes. The recommended minimum for homes with one to three bedrooms is now 1000 gal (3785 L). (Illustration from *Manual of Septic Tank Practices,* Bureau of Community Environmental Management, U.S. Public Health Service [now EPA].) Metric added by the author.

Although a number of arrangements are possible, the term *compartments*, as used here, refers to a number of units in series. These can be either separate units linked together or sections enclosed in one continuous shell (see Fig. 16.3-36) with watertight partitions separating individual compartments.

A single-compartment tank will give acceptable performance, but a two-compartment tank, with the first compartment equal to one-half to two-thirds of the total volume, generally provides better suspended solids removal. Compartments should be at least 24 in. (610 mm) in the smallest dimension and have a liquid depth ranging from 30 to 72 in. (762 to 1829 mm).

### INSTALLATION INSPECTION

A septic disposal system should be tested and inspected before it is used. Inspection before backfilling is usually required by local regulations, even where drawings and specifications for the disposal system have been checked before issuance of a building permit. A septic tank should be filled with water and allowed to stand overnight to check for leaks.

A soil absorption system should be inspected before it is covered to be sure that the system is installed according to applicable codes and ordinances, contract documents, and good trade practice.

### 16.3.2 STORM WATER DISPOSAL

A *storm drainage system* carries rainwater, surface water, condensate, cooling water, and similar liquid wastes, exclusive of sewage and industrial waste. It consists of underslab drains and footing drains; storm water collection devices on a building, such as gutters, downspouts, and roof drains; overflow pans and other devices for collection of mechanically generated water; and the pipes that convey this collected water from its source to a point of disposal. It includes the system portions within a building and on its site, and the mains of a public storm sewer system or piping that carry water to a disposal site on the building's property.

Section 23.3 discusses underslab and foundation drainage systems and the portions of storm drainage systems that occur outside the building, including on-site and public disposal systems. The rest of the facility (in the building) portion of storm drainage systems is covered in this section. This includes gutters and downspouts, roof drains and leaders, and building storm drains.

Like drain lines in a sanitary sewage disposal system, storm drain lines must be sloped and arranged to assist gravity flow.

Some definitions of terms used in describing storm sewer systems follow:

1. *Building Storm Drain*: A pipe that picks up water from interior sources and interior roof drains and carries it to a point 3 ft (914 mm) outside the building.
2. *Building Storm Sewer*: A pipe that picks up the water at the end of the building storm drain and carries it to a public storm sewer or disposal location on site (see Chapter 23).
3. *Disposal Point:* A location where storm water is returned to the environment, such as a lake, stream, river, or ocean (see Chapter 23).
4. *Drain*: A pipe that carries water in a building's storm drainage system.
5. *Site Storm Drain*: A pipe that picks up the overflow from wells and springs, and storm water that falls on the site and is conveyed to it through curb inlets, site drains, manholes, and other structures, and carries that water to an outfall on site, a public storm sewer, or a disposal point (see Chapter 23).
6. *Storm Sewers*: Pipes that carry collected storm water to a disposal point (see Chapter 23).

Facility (in building) storm drainage systems collect water resulting from groundwater, precipitation, or condensate drainage and carry it outside the building. Storm water is collected from a building's roof, from courtyards, and from the surroundings by a series of inlets and pipes. The roof and site must be sloped or contoured so that water will flow into the system.

Storm drainage systems share many of the characteristics of DWV systems, but

TEE

4" (101.6 mm) CAST IRON SOIL PIPE T BRANCH
4" (101.6 mm) CAST IRON SANITARY T BRANCH
4" (101.6 mm) VITRIFIED CLAY OR CONCRETE
T BRANCHES

PLACE INLET AND OUTLET TEE
IN NOTCH AND FILL WITH MORTAR

PACK MORTAR
AROUND TEE

1" (25.4 mm) MIN. CLEARANCE

STRAIGHT BAFFLE

POURED IN PLACE OR PREFABRICATED
AND DROPPED IN SIDES OF TANK
NOTE: "A" SHOULD BE NO LESS THAN 6" (152.4 mm)
AND NO GREATER THAN B.

"B" PENETRATION OF OUTLET DEVICE
GENERALLY 40% OF LIQUID DEPTH FOR
TANKS WITH VERTICAL SIDES & 35% FOR
HORIZONTAL CYLINDER TANKS.

**FIGURE 16.3-37 Acceptable inlet and outlet devices for septic tanks.** (Illustration from *Manual of Septic Tank Practices*, Bureau of Community Environmental Management, U.S. Public Health Service [now EPA].) Metric added by the author.

there are some differences as well. For example, because sewer gases are not involved, storm drainage systems do not require traps and usually do not require vents. They do require cleanouts. Many of the principles related to DWV systems apply as well to storm water systems.

Footing drains, underslab drains, and other systems for collection of groundwater associated with a building are addressed in Section 23.3.

### 16.3.2.1 Roof Drainage

Drainage from building roofs is a large storm water contributor and, in the case of single-family residences and some other small buildings, perhaps the largest. There are several ways to collect roof water.

In buildings with steeply sloped roofs (see Section 7.6), roof water may be allowed to simply drain from the roof onto the surrounding landscape, or may be collected in a series of gutters and downspouts and deposited either on the ground or in a system of underground pipes leading to a disposal point, as previously described. Buildings with lower-sloped roofs, such as those addressed in Section 7.8, are usually drained by a series of roof drains and pipes, but sometimes these are augmented by overflows called *scuppers*

at the perimeter of the roof. Scuppers may drain into downspouts or simply allow water to flow over the edge of the roof onto the ground below.

The first step in the design of either of these systems is to determine the amount of water to be carried away. Plumbing codes indicate the required rate of rainfall per hour to use in making these calculations. For example, Maine may have a rate of 4 in. (100 mm) per hour throughout the state, while other states may vary internally. Virginia, for instance has a 6 in. (150 mm) per hour rate in the central and eastern parts and a 5 in. (125 mm) per hour rate in the western portions.

#### GUTTERS AND DOWNSPOUTS

To design a gutter and downspout (leader) system for a steeply sloped roof, use the following steps:

1. Lay out the roof plan and determine the location of downspouts.
2. Size the downspouts to verify that their placement is satisfactory. To do this, multiply the rainfall amount from the code by the area of the roof to be served by each downspout. Select the downspout size from tables in the applicable plumbing code. For example,

in a building where four downspouts will serve a roof area of 3200 sq ft (297.68 m²), each downspout will serve 800 sq ft (75 m²) in an area with a 6 in. (150 mm) per hour rainfall, so a 3 in. (75 mm) pipe will be sufficient.

3. Size the gutter based on the roof area to be served by each section of gutter. In most designs, one-half of each gutter section spills into each downspout, so the roof area served by each gutter section will be one-half of that served by each downspout, but this may vary, depending on the layout of the system. Whatever the area served, size the gutter by entering the tables in the applicable code with the area to be served, the rainfall amount, and the slope of the gutter and select the gutter size directly. The slope of a gutter influences the amount of water it can carry in a given time. For example, a 5 in. (125 mm) gutter with a $1/8$ in./ft (10.4 mm/m) slope may serve up to 880 sq ft (81 m²) of roof area where the rainfall per hour is 4 in. (100 mm). Increasing the gutter slope to $1/4$ in./ft (20.9 mm/m) will permit this same pipe to serve 1250 sq ft (116 m²) and a $1/2$ in./ft (41.7 mm/m) slope up to 1770 sq ft (164 m²).

4. If there is a horizontal element, other than the gutters, in a gutter and downspout system, size it by entering the tables in the applicable code with the area to be served, the rainfall amount, and the slope of the pipe and select the pipe size directly. As with a gutter, the slope of a horizontal drain pipe influences the amount of water it can carry in a given time.

#### ROOF DRAINS AND PIPES

To design a roof drainage system for a low-slope roof (see Section 7.8), use the following steps:

1. Lay out the roof plan and determine the location of the roof drains.
2. Determine from the layout the roof area to be served by each drain.
3. Size the leaders connected to each roof drain. To do this, multiply the rainfall amount from the code by the area of the roof to be served by each leader. Select the leader size from tables in the applicable plumbing code. For example, in a building where three roof drains will each serve one-third of a roof area of 3200 sq ft (297.68 m²), each roof drain will serve 1067

sq ft (99 mm/m²). In an area with a 6 in. (150 mm) per hour rainfall, a 3 in. (75 mm) pipe will be sufficient.

4. Size the horizontal portion of a low-slope roof system in the same way as described earlier for a steep-slope roof design.

# 16.4 Plumbing Fixtures

*Plumbing fixtures* fulfill the basic purpose of plumbing systems; that is, they provide clean water and dispose of wastes. They also play a vital role in ensuring that a water supply remains clean by providing a physical separation between supply and drainage piping. If there were no such separation and the positive pressure in a supply line were lost, then dirty water might be sucked back into the supply system (see Section 16.4.7).

When selecting fixtures, one should consider convenience, cost, durability, ease of repair and maintenance, and noise generated. These considerations often are reflected in the type and grade of fixture. In addition to their obvious cosmetic contributions, colors and finishes may also promote ease of maintenance.

Plumbing, health, and handicapped use codes have specific restrictions related to plumbing fixtures, especially, but not limited to, those in some institutional buildings, such as medical care facilities, and in other building types where access and safety are involved. Be sure to check all applicable codes carefully when selecting fixtures and designing rooms to house them. Specific degrees of floor slope may be required in a shower, for example, and curbs may be prohibited. Floor drains may be required in rooms housing fixtures. Specific requirements for the fixtures themselves, their hardware, and their installation will be stated in the applicable codes.

## 16.4.1 LAVATORIES

Lavatories come in three basic types: (1) wall-hung units with a ledge at each side, (2) units installed in or as a part of countertops, and (3) pedestal types. Figure 16.4-1 shows several typical lavatories.

**FIGURE 16.4-1** Lavatory types: (a) drop-in with stainless steel rim, (b) molded lavatory counter, (c) drop-in, self-rimming, (d) undercounter rimless, (e) wall-hung.

*Wall-hung* lavatories often have a raised back ledge that serves as a backsplash. Their back and side ledges vary from 1 to 10 in. (25.4 to 254 mm) in width; at least 4 in. (100 mm) is desirable. Inside bowl dimensions vary from 10 to 16 in. (254 to 406 mm) front to back and from 13 to 17 in. (330 to 432 mm) side to side. Bowl depth varies from about 5¹/₂ to 7¹/₂ in. (140 to 190 mm).

Wall-hung units are mounted on a metal bracket attached to the wall. For a secure attachment to wood studs, this bracket should be bolted to a horizontal wood support mortised into the wall framing. When the walls are framed with metal studs, a lavatory's metal bracket can be bolted to a metal plate that spans several studs and is fastened securely to them. Wood or metal furring should be used as lavatory supports at masonry walls. Additional support can be provided by chrome legs under the front corners of a lavatory.

There are four kinds of *counter lavatories*: (1) a drop-in type with a raised self-rim, (2) a drop-in type with thin-edged metal trim over the joints, (3) a type that is fastened to the bottom of the counter with the counter edge exposed and finished, and (4) a one-piece molded lavatory-counter unit.

Sometimes a single-bowl kitchen sink is installed in a bathroom instead of a lavatory for bathing infants and small children, washing hair, and doing hand laundry. Such a larger basin controls splashing. A hose spray can be added for extra convenience.

Most lavatories are made of vitreous china. Other common materials are porcelain over cast iron or steel, cultured marble (ground marble powder with a plastic binder), thermoformed plastic sheet, and filled polymer (acrylic-modified polyester). Materials differ widely in their resistance to impact and abrasion; to staining or charring by cigarettes; and to the spillage of cosmetics. Vitreous china and enameled metal have the most chemical resistance.

*Molded counter-lavatory units* can be formed of cultured marble, thermoformed plastics, or filled polymer. Their characteristics vary with the types of binders or plastics used. Some lavatories come with faucets and drain fittings; some provide appropriate holes for their insertion. Modern lavatories generally have only one spout but may have separate hot and cold water valves or a single-lever mixing valve. Lavatory faucet-drain assemblies usually include a pop-up drain plug similar to those on bathtubs.

## 16.4.2 SINKS AND LAUNDRY TRAYS

Kitchen sinks are available with one, two, or three bowls to permit simultaneous uses. Most new residences include a two-compartment sink, having the spout high enough and one compartment large enough (usually 15 by 18 in. [381 by 457 mm]) to accommodate large cooking utensils. Typical materials for kitchen sinks are porcelain-enameled cast iron or steel, stainless steel, and plastics. These units are generally designed to drop in to a countertop and are supported by their rims resting on the counter. Some have self-sealing rims; others require a separate stainless steel sealing rim.

A variety of faucet types is available, either with separate hot and cold controls or with a single control and mixing valve. A diverter valve similar to those on tub-shower combinations is often installed to serve a hose-mounted spray accessory. Several appliances are available as attachments to kitchen sinks, including garbage disposal units, instant water heaters with a tap, and small water coolers with a tap.

Laundry sinks (laundry trays) are available in the same materials as kitchen sinks but are larger and deeper. Some are wall-hung; others rest on legs or a trap support.

## 16.4.3 FAUCETS

Faucets are valves (see Section 16.2.1.3) with spouts attached. They are designed to deliver the water flowing through the valve into a fixture. Modern faucets have one spout and either one or two handles.

Faucets should be selected on the basis of compatibility with the fixture served, ease of operation, durability of finish and operating mechanism, and ease of cleaning. When choosing a faucet for a particular fixture, the spout design should be checked to insure that, when installed, an air gap will exist between the tip of the spout and the maximum fill level of the fixture. The height of the air gap should be at least twice the diameter of the spout opening.

A faucet handle that requires a relatively large amount of movement provides easier adjustment of water temperature and flow rate. Round knobs require finger pressure to operate, which is difficult for some persons. For example, persons with arthritis in their hands and wrists prefer handles which can be operated by pressure of the palms or sides of the hands. The simplest faucet to operate is a single-lever type that is moved from side to side for temperature control and up and down for flow rate control. Knobs which must be pulled or pushed are the most difficult to adjust.

Chrome on brass is the most durable finish for faucets and trim. Other finishes may be quite thin and subject to corrosion or wear. This is particularly true for the gold-colored finish. Finely detailed decorative handles are hard to keep clean.

Lavatory faucets can be obtained with a spray head similar to a shower or kitchen sink spray, which is useful for washing hair, pets, and children.

### 16.4.3.1 Stem Faucets

A two-handled faucet has one handle for hot water and the other for cold water and is known as a *stem* or *compression* faucet. Each handle controls a globe valve, or some variation of it, which makes this type of faucet subject to leakage problems. A frostproof outdoor faucet is another type of stem faucet.

### 16.4.3.2 Single-Lever Faucets

Single-lever faucets are often called *mixing faucets*, because both hot and cold water supplies are controlled by the same handle. A single-lever faucet controls the flow of hot and cold water into the spout by the matching of holes in an alignment valve (Section 16.2.1.7). By varying the size of the matched opening on the hot or cold water side, the desired temperature is achieved.

### 16.4.3.3 Antiscald Faucets

Faucets are available with thermostatically controlled valves to maintain the desired temperature regardless of pressure, flow, or temperature variations (Fig. 16.4-2) (see also Section 1.4). These faucets may also have a safety stop so that they cannot be set hotter than 100°F (37.7°C) without first depressing a warning stop. Less expensive temperature control valves compensate for changes in relative pressure between hot and cold supplies, but not for temperature drift in either supply. Antiscald valves are available for lavatory and tub use but are most common in shower installations.

### 16.4.4 WATER CLOSETS

Water closets (WCs) have two basic parts: (1) a bowl that provides a receptacle for wastes and forms the fixture drain trap and (2) a flushing mechanism. They are generally classified according to their flushing mechanism, their bowl type, and their mounting method. Most WCs are made of vitreous china, although some are plastic. WCs can be further classified as: (1) conventional, (2) watersaving, (3) low-consumption, and (4) waterless.

### 16.4.4.1 Conventional Water Closets

#### FLUSHING MECHANISM

Water closets are classified by their flushing mechanisms as either *flush tank* or *pressure valve* units.

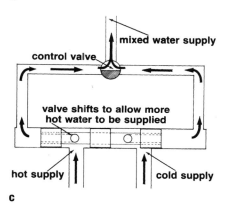

**FIGURE 16.4-2** Antiscald valves maintain selected hot water and cold water flow ratio. (Courtesy Small Homes Council, Building Research Council, University of Illinois, Urbana-Champaign.)

*Flush Tank Units* Flush tank water closets are so called because they contain a tank which holds a reserve of water that drains into the bowl when called for to assist flushing. The most common tank-type water closets are (1) one-piece floor-supported, (2) two-piece-floor-supported, and (3) two-piece wall-hung (Fig. 16.4-3). One-piece wall-hung models are also available.

**a**       **b**       **c**

**FIGURE 16.4-3** WC types: (a) one-piece floor-mounted, (b) two-piece floor-mounted, (c) two-piece wall-hung. (Courtesy Small Homes Council, Building Research Council, University of Illinois, Urbana-Champaign.)

Some one-piece models similar to that shown offer a nonoverflow feature and a cleaner appearance, but they are usually more expensive to purchase and install than standard two-piece units. Standard two-piece units have the tank supported on the bowl. In older WCs, the bowl is generally floor-supported and the tank hangs on the wall.

Submerged in the tank of a flush tank WC are a number of control mechanisms: a flushing valve, a water supply valve, and a combination bowl-refill and tank overflow device (Fig. 16.4-4). The flushing valve is essentially a stopper, or flapper, at the bottom of the tank, which is lifted when the flush lever on the outside of the tank is depressed. This allows the water in the tank to drain into the bowl to promote flushing.

Flush tank supply valves are available in several designs, all of which have a means of sensing when the water reserve is being lost due to action of the flushing valve. The *ballcock valve* is controlled by the position of a float ball riding on the surface of the water in the tank (see Fig. 6.4-4). As the water flows out, the float falls, thereby opening the ballcock supply valve. As the tank refills, the float rises and eventually shuts off the valve.

A *fluid level control valve* works similarly to a ballcock valve, except that it has a float mounted on the supply-inlet post rather than on an arm extending from it (Fig. 16.4-5). It is simpler and quieter than a ballcock valve.

A small tube extending from the high-water level in a flush tank down into the bowl acts as an overflow and refill device. A spur from the water supply line that fills the tank feeds into this tube, refilling the bowl, thereby restoring the trap seal after every flush. Should the water supply valve fail to shut off properly, this tube acts as an overflow drain for the tank.

***Pressure Valve Units***   The flushing of a *pressure valve* WC (also called a *flush valve* unit) is controlled by a *pressure control fill valve* (sometimes called a *flush valve*) which operates on pressure in the serving water line rather than on the pressure of water in a tank. These units are rarely installed in one- and two-family residences due to their noisiness and the lower normal water pressure in residential zones. They are sometimes used, however, in areas with limited water supplies (Fig. 16.4-6). Because of their ease of maintenance and low water consumption, pressure valve WCs are frequently installed in commercial installations where noise is not critical and sufficient water pressure is available. Pressure valve units are available in both floor-supported and wall-hung models.

### TANKS

Traditional flush tank units having a tank attached to the bowl discharged 5 to 7 gal (18.93 to 26.5 L) of water at each flushing action. That flow was reduced in later (shallow trap) models to about 3.5 gal (13.2 L) per flush. Flush tank fixtures that have a separate tank mounted on the wall above the bowl deliver about 2.5 gal (9.5 L). Federal law now requires that all new units function with no more than 1.6 gal (6 L) per flush.

In summer, moisture can condense on the outside of a tank when room humid-

**FIGURE 16.4-4** Flush tank WCs are favored for homes because they are simple and quiet. Flushing of a tank is shown, starting from the top. (Courtesy Small Homes Council, Building Research Council, University of Illinois, Urbana-Champaign.)

**FIGURE 16.4-5** A fluid level control valve is quieter than a ballcock mechanism. (Courtesy Small Homes Council, Building Research Council, University of Illinois, Urbana-Champaign.)

**FIGURE 16.4-6** Pressure valves are noisy but use little water. They are common in commercial buildings. (Courtesy Small Homes Council, Building Research Council, University of Illinois, Urbana-Champaign.)

ity is high and supply water is cold. Tanks are available with factory-installed insulation inside them to minimize condensation. Other remedies for condensation include using a tempering valve to warm the cold water entering the tank; installing interior insulation in the field; providing a decorative fabric jacket around the outside of the tank; and installing a tray at the bottom of the tank to collect the condensate and drain it into the bowl.

## BOWLS

There are several types of conventional bowls in general use today:

1. A *siphon jet bowl* is most frequently used in residential construction (Fig. 16.4-7). In this bowl, a jet of water in the trapway initiates the siphon action that empties the bowl. Its advantages are quietness and effectiveness. Traditional siphon jet bowls, which can be found in many existing residences, require 5 to 7 gal (18.93 to 26.5 L) per flush. Newer *watersaver* siphon jet toilets use between 1.7 and 3.5 gal (6.4 to 13.2 L) per flush.
2. A *siphon vortex bowl* is quieter than a siphon jet bowl and is particularly suitable where the water pressure is low (Fig. 16.4-8). The siphon vortex

**FIGURE 16.4-7** A siphon jet bowl is relatively quiet and has an effective scouring action. (Drawing by HLS)

**FIGURE 16.4-8** Siphon vortex bowls work especially well when the water pressure is low. Therefore, they are more quiet than siphon jet bowls. (Drawing by HLS.)

flushing action requires 6 to 8 gal (22.7 to 30.3 L) of water per flush.
3. *Blowout bowls* flush by the force of a high-pressure jet in the trapway rather than by siphonage (Fig. 16.4-9). They are used primarily in commercial buildings due to their noisiness and requirements for high water pressure. They require 3 to 4 gal (11.4 to 15.1 L) of water per flush.
4. *Washdown bowl* fixtures are not pictured here because they are no longer manufactured in the United States. They are mentioned because they were once the most common type for residential construction and can be found in many older residences. In a washdown-type fixture the trap is on the front, which is evidenced by a bulge on the front of the unit.

## MOUNTING

WCs are either floor-supported or wall-hung. Some flush tank units have floor-supported bowls and wall-hung tanks.

Wall-hung units are supported by a metal carrier concealed in the wall. An advantage of wall-hung WCs is that the floor under the tank is more accessible for cleaning.

The height of floor-supported seats normally ranges from 14 to 18 in. (356 to 457 mm), but a 20 in. (508 mm) seat height may be desirable for the elderly or handicapped. This can be achieved by the use of an extension seat on floor-supported fixtures or adjustment of the carrier height of wall-hung fixtures.

### 16.4.4.2 Watersaver WCs

*Watersaver* WCs are currently available. They require between 1.7 and 3.5 gal. (6.4 and 13.2 L) per flush. This is below the water use of most conventional WCs but still does not comply with current fed-

**FIGURE 16.4-9** Blowout toilets require high water pressure. They use pressure valves (flush valves). Blowout toilets are used mostly in commercial and institutional buildings. (Drawing by HLS.)

eral regulations that require WCs to not exceed 1.6 gal (6 L) of water per flush. These fixtures require some accommodation to achieve the lower flow rate. This can be accomplished by mounting the tank high above the bowl to increase the water pressure or by increasing the pressure from the public source sufficiently using other means to achieve this rate. These fixtures usually change the configuration of the bowl and of the trap to help decrease the amount of water needed for each flush.

### 16.4.4.3 Low-Consumption WCs

*Low-consumption* WCs comply with the latest federal regulations for flow. Early versions of this type of fixture often required more than one flush to completely empty the bowl. Newer versions suffer less from this defect.

Some low-consumption fixtures achieve the required flow by the use of a *flushometer tank*. This type tank employs a siphon jet flush action that is assisted by increasing the water pressure flowing into the bowl by using the pressure of the incoming water to compress the air in the tank. This compressed air then forces the water in the tank to flow into the bowl when flushed at a higher pressure than the pressure in the incoming water service.

Some low-consumption fixtures use two tanks. The first tank holds water at normal atmospheric pressure. On each flush, some water is deposited into a second tank where the air is compressed. When the WC is flushed, water flows under pressure from this second tank into the bowl.

Some low-consumption WCs also employ bowls that hold less water than a conventional or watersaver bowl.

Another innovation in low-consumption WC design employs a sewage outflow tank in which a vacuum has been created. Drain pipes flow from WCs into this tank. A flush opens a valve into the tank, and the sewage is siphoned into the tank under pressure. The tank is then emptied into the sewer system when convenient.

One inherent difficulty with low-consumption WCs that work using compressed air is that an air compressor and the piping necessary to distribute the compressed air to the fixtures are needed. Vacuum systems can use the drain piping that would be needed in any case, but the holding tank and the equipment necessary to achieve the vacuum must be added to a conventional system.

#### 16.4.4.4 Waterless WCs

Some *waterless* WCs employ chemicals or oil instead of water. They are found mostly in airplanes, vehicles, and boats, and occasionally in areas where water is not available. Some waterless urinals are filled with a liquid that permits urine to flow through it and prohibits the passage of sewer gases.

Another innovation in waterless WC design is the *composting WC.* Composting WCs are being used in the same installations where the water supply is inadequate or nonexistent. Composting systems have also been used in isolated cases as a pro-ecological statement. A composting system works essentially the way an old-fashioned outhouse worked, with some additions for safety and esthetics. In a composting system, material from WCs and some organic kitchen waste—but not urinals—is deposited in a composting vault or tank, where it is broken down by aerobic bacteria. Aerobic decomposition is unaccompanied by the odors and methane gas produced by anaerobic bacterial decomposition, such as that in a septic tank system. Composting vaults and tanks require the proper amount of ventilation. The more air supplied, the faster the decomposition. Gases, mostly $CO_2$ and water vapor that accumulates in the vault or tank, must be ventilated through a vent stack. Some composting systems require some type of turning or stirring device to aid the aerobic action.

#### 16.4.5 URINALS AND BIDETS

Urinals are used in the United States mostly in commercial installations. Bidets, which include a washing mechanism in the bowl primarily for women, are not usual in this country, even in public facilities. Urinals and bidets, common in residences in Europe, are seldom installed in residences in the United States. This is due to the expense associated with these units, the floor space they occupy, and cultural differences. In addition, urinals are noisy and look institutional.

Urinals are usually operated by pressure valves, much like flush valve WCs. Both wall-mounted and floor-supported units are available, but most urinals in commercial installations are wall-mounted.

The function of a bidet can also be provided by a spray outlet built into a WC bowl or seat, and several units are commercially available that provide this function. In either case, such mechanisms deliver warm water in a spray with an adjustable intensity.

#### 16.4.6 BATHTUBS

Bathtubs are made in various shapes, sizes, and materials.

#### 16.4.6.1 Shapes and Types

Figure 16.4-10 shows three common shapes of tubs, but many variations are available.

The square and sunken types are more difficult to clean than the rectangular and above-floor types. Typical overall dimensions for rectangular tubs are 14 to 16 in. (356 to 457 mm) high, $28^1/_2$ to 32 in. (724 to 813 mm) front-to-back width, and 60 in. (1524 mm) long. Longer, shorter, and wider tubs are available on special order and at additional cost.

Most residential bathrooms employ a recessed tub with walls on three sides. When walls are available on two sides, a corner tub is used. In custom houses, freestanding tubs are sometimes used; many of these are either surrounded by a raised platform or sunk in the floor. One-piece panalized tub-shower assemblies are available in several types of plastic. They provide low-maintenance shower enclosures with built-in conveniences.

#### 16.4.6.2 Materials and Finishes

The most durable tub material is cast iron with a thick porcelain-enamel finish. Tubs of this type weigh approximately 200 lb (90.72 kg) and require additional floor framing members beneath them when they are placed on a wood-framed floor.

Tubs also are made of steel with a porcelain-enamel finish. These are more prone to enamel chipping and rusting than the cast iron variety. Because they are lighter, they also are noisier when water strikes them. Some models are available with an acoustical treatment on the underside to reduce water-impact noise. Steel and plastic tubs are lighter and less rigid than cast iron ones.

Some tubs are made from synthetic marble, acrylic-modified polyester, or similar homogeneous materials. Special shapes and sizes can be made using these materials, including such unusual shapes as round and octagonal ones. Such tubs frequently have whirlpool or hot-tub devices attached to or built into them.

Plastic tub assembles are made (1) of glass fiber reinforced polyester (FRP) with a smooth gel-coat surface or (2) of a molded acrylic sheet. Built-in ledges are provided in most plastic molded assemblies. Ledges are handy for storing items used during bathing, but the rim to be stepped over should be narrow to make entry into the tub as easy as possible.

Tubs commonly have two openings for plumbing connections: one for the bottom drain and one for an overflow drain. The overflow opening is usually an inconspicuous slot under the edge of a chrome plate mounted high on the tub wall. The overflow drain protects against two hazards associated with overflowing fixtures: water damage to the building and siphonage of the trap seal. Tubs with whirlpool action devices, of course, require additional openings to accommodate those devices and the circulation of water.

#### 16.4.6.3 Tub Fittings

Tub drains are often bothersome because they either do not seal completely or do not drain quickly. Slow drainage is often caused by hair collecting on a drain plug mechanism of the type that is controlled remotely from a lever mounted in the

a            b            c

**FIGURE 16.4-10** Bathtub types: (a) recessed square, with walls on three sides, (b) sunken rectangular, and (c) recessed rectangular, the most common type. (Courtesy Small Homes Council, Building Research Council, University of Illinois, Urbana-Champaign.)

overflow plate. In older versions of this mechanism, the lever raises and lowers a brass cylinder which, when lowered, blocks off the fixture drain (Fig. 16.4-11). In newer versions, the lever raises and lowers a vertical arm which, in turn, controls a horizontal arm with a pop-up drain plug at the end (Fig. 16.4-12).

### 16.4.7 SHOWERS

A shower unit consists of a shower head mounted either in an enclosed bathtub or in a shower stall. Shower heads are sprayers which either have a preset spray pattern or can be adjusted from fine mist to heavy spray. Some models offer built-in flow restrictors, which conserve water. Some units also have a shutoff valve on

**FIGURE 16.4-11** A trip-lever drain uses a cylinder on an adjustable rod to open or close the drain opening. (Courtesy Small Homes Council, Building Research Council, University of Illinois, Urbana-Champaign.)

**FIGURE 16.4-12** A pop-up drain opens when a spring-loaded lever pushes on the end of the pop-up rod. (Courtesy Small Homes Council, Building Research Council, University of Illinois, Urbana-Champaign.)

the head so that the water can be turned off during soaping (Fig. 16.4-13a). This is also an advantage in that the temperature setting on the main supply valve need not be disturbed.

A handheld spray head on a flexible metal-covered hose can be used instead of a fixed shower head (Fig. 16.4-13b). It can be hung on a permanent bracket, or on a sliding bracket, on a support bar to allow adjustment for various heights, or it can be handheld. When such heads are installed over tubs or lavatories, a vacuum breaker should be provided or the shower head should be hung in such a way that at least a 1 in. (25.4 mm) air

**FIGURE 16.4-13** (a) Shutoff valve on a shower head; (b) phone-type adjustable shower head. (Courtesy Small Homes Council, Building Research Council, University of Illinois, Urbana-Champaign.)

gap will be assured when it is hanging fully extended. This prevents back-siphonage of bath water and contamination of the water supply should the spray head fall into the water.

#### 16.4.7.1 Tub-Showers

Showers installed over a bathtub usually share the tub's water supply valve. The *tub-filler* spout includes a *diverter valve* that redirects water coming to the spout to the shower head instead. Adding a shower to a bathtub also requires that a waterproof enclosure be provided up to the height of the shower head so that splatter from the shower is directed into the tub and down the drain. On the three walls surrounding the tub, this can be accomplished with metal or plastic panels specially made for this purpose or with ceramic tile. The risk of leakage is reduced if panel systems without corner joints are used.

In ceramic tile tub enclosures, either cement backer board or water-resistant gypsum board should be used as the backing (see Section 9.4). Care must be taken that grouting between tiles is sound to avoid leaks.

An alternative that eliminates joints and potential leaks from a tub enclosure is a one-piece molded plastic tub-shower unit (Fig. 16.4-14a). The size of such units dictates that they be placed before the bathroom walls are framed. The fourth side of the tub must also be enclosed, either with a shower curtain or with a safety-glazed door.

#### 16.4.7.2 Shower Stalls

Showers not installed in conjunction with a bathtub have their own stalls built around them. Most shower stalls consist of a floor receptor topped by water-resistant walls, either self-supporting or attached to the wall framing. Prefabricated receptors are made of precast concrete, terrazzo, preformed metal, or plastic. Job-built receptors of ceramic tile can also be used if a sloping base and a waterproofing pan are laid beneath them. Complete plastic receptor and wall combinations are available, both as components and as one-piece units (Fig. 16.4-14b). The fourth side of the shower stall can be enclosed with a curtain, but a higher than normal curb is necessary to prevent it from blowing out and allowing water to run onto the floor. Sliding or bifold doors with safety glazing are availa          b

**FIGURE 16.4-14** (a) One-piece molded plastic tub shower; (b) one-piece plastic shower. Construction permits inclusion of conveniences such as soap dishes and seats. (Courtesy Small Homes Council, Building Research Council, University of Illinois, Urbana-Champaign.)

able for wide stalls. Hinged doors are common, but if they open outward, they should have a provision to minimize water dripping onto the floor. A common small shower stall size is 30 in. (762 mm) square. A more convenient stall size is 36 in. (914 mm) square or 36 by 48 in. (914 by 1067 mm). Complete assembles are about 84 in. (2134 mm) high.

A shower stall offers easier entry and exit and more secure footing than does a shower in a tub, especially when provided with grab bars. Portable suction mats or slip-resistant tapes can be used if the floor finish is not slip resistant.

### 16.4.8 DRINKING FOUNTAINS

Drinking fountains may be simply a type of faucet arranged for ease of drinking, but most deliver chilled water. The water may be chilled in a self-contained unit (Fig. 16.4-15) or may be cooled by a remote chiller. Units are available either freestanding or wall-mounted. Recessed and semirecessed units are also available, as are pedestal units and other special designs. Both indoor and outdoor units are available.

### 16.4.9 SEPARATION OF SYSTEMS

Most codes demand that plumbing supply and drainage systems be separated to prevent crossflow of waste fluids into the supply system and vice versa. Contamination of a supply system can have severe effects on the health of building occupants.

Water flows into a fixture from a supply system, is used by humans, and then drains out of the fixture into a drainage system. Fixtures are designed to ensure that the flow of water through them is in one direction only, so that drain water is never allowed back into the supply pipes. The unintentional flow of contaminated water into a supply source is called *back-flow* or *back siphonage*. This can occur when the pressure in a supply pipe drops below that of the fluid, be it air or water, in which the supply pipe terminates. This pressure reversal cannot occur when the supply water pressure is between 40 and 50 psi (276 to 345 kPa), as is usual, and each supply line ends in air, which has a pressure of about 14 to 15 psi (97 to 100 kPa).

Therefore, plumbing codes require that fixtures be designed so that an air gap separates the supply lines from the highest point at which contaminated water can collect (Fig. 16.4-16). An *air gap* is the unobstructed vertical distance between the mouth of a water outlet and the flood level rim of the water receptacle. The water outlet may be a faucet, spout, or other outlet; the receptacle may be a plumbing fixture, tank, or other receptacle. In some cases, the highest point to which contaminated water can reach is permitted to be the level of the overflow drain of a fixture. To ensure that the suction created in the pipe during reversal of flow is not strong enough to draw contaminated water across the air gap between the end of the supply line and the water surface, the required minimum air gap is usually twice the diameter of the faucet or a minimum height of 1 in. (25.4 mm) for lavatories, $1\frac{1}{2}$ in. (38 mm) for sinks and fixtures having $\frac{3}{4}$ in. (19 mm) bathroom faucets, and 2 in. (50.8 mm) for fixtures having 1 in. (25.4 mm) bath faucets.

Where provision of an air gap is not feasible in fixture design, a device called

a          b

**FIGURE 16.4-15** Water coolers: (a) floor model; (b) wall-mounted. (Drawing by HLS.)

FIGURE 16.4-16 The air gap between overflow level and spout prevents contamination of the water supply. (Courtesy Small Homes Council, Building Research Council, University of Illinois, Urbana-Champaign.)

an *atmospheric vacuum breaker* or *siphon breaker* is installed. This is a simple mechanical device consisting essentially of a check valve in a supply line and a valve member (on the discharge side of the check valve) opening to the atmosphere when the pressure in the line drops to atmospheric (Fig. 16.4-17). This device acts as a one-way valve, allowing water to flow through it in one direction, but if flow is reversed it opens, allowing air to enter instead of permitting the reverse flow of water. Vacuum breakers generally are required on dishwashers

and other water-using appliances that otherwise lack an air gap.

## 16.5 Sound Control

High water pressure and velocity and quick-closing valves in supply piping systems combine to produce plumbing noises. If a building does not exceed three stories and the water pressure at the street exceeds 35 to 45 psi (241 to 410 kPa), a pressure-reducing valve can be installed, preferably as far from the building as feasible. In high-rise buildings, combinations of pressure-reducing valves and pressure-regulating systems may be necessary to provide suitable and uniform pressure levels at the various floor levels.

The noises in a plumbing system can be classified as those associated with the piping and those associated with appliances. The first of these include water hammer, creaking, popping, squealing, cavitation, vibration, and other general noises produced by rushing water. Noises associated with fixtures are usually banging and bumping causes by vibration.

### 16.5.1 NOISES ASSOCIATED WITH PIPING

*Water hammer*, which is also called *hydraulic shock*, consists of pounding noises and vibration that sometimes develop in a piping system when its air chambers become filled with water or have been omitted entirely. It also may

be due to worn washers in faucets. This unpleasant condition occurs when there is a sudden change in the velocity of water flowing through a pipe. This causes a shock wave to travel through the piping system, producing a banging as the pipes vibrate against their supports. Such a rapid change in velocity is often caused by the sudden closing of a fast-acting valve, the starting or stopping of certain types of pumps, or improper check valves (Fig. 16.5-1). Continuous water hammer may result in broken pipes, leaky joints, loosened pipe hangers, ruptured tanks, and damage to meters, gauges, regulators, and faucets.

Water hammer can be avoided by using spring-loaded check valves which do not close fully until the movement of water in the pipe has stopped, in lieu of spring or solenoid valves. It can be minimized by installing air-filled extensions of the supply lines (air chambers) at each fixture or near the service entry (Fig. 16.5-2a). These *air chambers* consist of a piece of pipe about 10 in. (254 mm) long installed above the hot and cold valves of fixtures such as sinks, lavatories, and clothes washers, to cushion the rush of water as the valve is closed and prevent water hammer. Another way to reduce water hammer is to use plastic piping which is flexible enough to dissipate shock waves.

In older buildings, water hammer may be due to worn-out washers, which cause faucets to close suddenly, or to waterlogged air chambers. To restore the air in water-logged chambers, faucets should be opened and water drained from the supply system at its lowest point. In both new and old buildings, water hammer problems can be minimized by installing a shock absorber near the service entrance (Fig. 16.5-2b).

*Squealing* noises are sometimes emitted by pipes in which water flowing at a high velocity strikes small protrusions on the inside face of the pipe, causing vibration of the pipe. This problem can usually be solved by reducing the velocity of the flow in the pipe. Squealing can also occur when water passes a loose or worn valve washer, causing it to vibrate. This can be stopped by tightening the screw holding the washer in place or replacing the washer. Proper sizing and anchorage of copper water supply piping will also prevent some noises.

Noise resulting from water rushing through piping can be minimized by

FIGURE 16.4-17 A vacuum breaker uses atmospheric pressure to close the valve when the pressure is reduced on the left side and reopen it when higher than atmospheric pressure is restored on the left side. (Courtesy Small Homes Council, Building Research Council, University of Illinois, Urbana-Champaign.)

**FIGURE 16.5-1** Internal pipe pressures cause water hammer due to a fast-closing valve. (Courtesy Small Homes Council, Building Research Council, University of Illinois, Urbana-Champaign.)

**FIGURE 16.5-2** Water hammer can be minimized by installing (a) an air chamber or (b) a shock absorber near the service entrance. A similar but smaller air chamber should be installed at each valved fixture. (Courtesy Small Homes Council, Building Research Council, University of Illinois, Urbana-Champaign.)

planning the system with a minimum number of bends and by reducing the number of fittings and valves to a minimum. Valves and right-angle fittings produce substantially more noise than straight runs of pipe due to increased water turbulence in them. The noise produced by four elbows is about 10 dB higher than the noise produced in a straight run of equivalent-size pipe under conditions of turbulent flow.

Long runs of hot water supply piping produce creaking noises as the piping expands or contracts from hot or cooled water. Long, straight runs of iron or steel piping should be designed with flexibility, with at least one end having a swing arm to permit movement (Fig. 16.5-3a). Flexible L or M piping bent into an S-shaped curve at one end will allow for movement in long runs of copper piping (Fig. 16.5-3b). Supports should be de-

signed to permit the piping to expand and contract without binding.

Supply piping to units which use large amounts of water, such as dishwashers and WCs, should not be undersized.

When pipes are rigidly connected to structures with hangers, clamps, or straps, sound resulting from pipe vibration can be transmitted through the structure and can be amplified from one apartment unit to another. Resilient pipe insulation between the pipe supports and the pipe should be used to help reduce the transmission of these noises (Fig. 16.5-4a). The number of pipe supports should be kept to the minimum required to structurally support the pipe, and supports should be attached to the most massive structural element available.

The running of drain piping through the walls of bedrooms or other quiet living area should be avoided unless a double sound-isolating wall is provided. Waste dropping down vertical stacks causes piping to vibrate, especially when it hits horizontal runs, unless the pipe is resiliently mounted. When vertical stacks are hung on floor joists, vibrations are transmitted to the structure (Fig. 16.5-4b). Standard pipe clamps and resilient vertical supports or wood brackets should be installed with the end of the support resting on the vertical stack with a 1-in. (25.4-mm)-thick isolation pad resting on the support (Fig. 16.5-4c). A short section of pipe insulation around a pipe where it passes through a wall sill will keep it in horizontal alignment and will isolate it from the structure.

**FIGURE 16.5-3** Flexible connections using (a) swing arms in piping or (b) S curves in tubing minimize noise caused by expansion and contraction in long runs of piping.

FIGURE 16.5-4 Vibrations from a plumbing stack can be isolated from the structure by resiliently mounting vertical supports and resiliently supporting horizontal runs.

It is essential that holes and openings where piping passes through common walls be sealed with a resilient material or tightly packed with insulation, or both, to isolate vibration of the pipe from the structural framing and to seal against air leaks (Fig. 16.5-5).

*Cavitation* is a phenomenon which occurs when water flowing at a high velocity makes a rapid change in direction or passes through a small orifice. Cavitation produces a high-pressure region at the outside of a pipe bend and a low-pressure region at the inside of the bend. If the inside pressure is low enough, boiling occurs in the water, allowing air bubbles to form. Farther down the pipe, these bubbles reencounter pressure and are collapsed by it. This produces a noise similar to gravel rattling in the pipe or balloons popping.

Aside from the noise nuisance, cavitation can cause unnecessary wear on pipes and pumps. A solution is to lower the velocity of the water to less than 10

FIGURE 16.5-5 Where pipes pass through walls, openings must be tightly packed or sealed airtight with resilient materials to isolate vibrations and to close paths for airborne sound transmission.

ft/sec (3.05 m/s). Typical residential installations have water velocities ranging from 4 to 8 ft/sec (1.2 to 2.4 m/s).

## 16.5.2 NOISES ASSOCIATED WITH FIXTURES AND APPLIANCES

Plumbing fixtures and appliances often can be selected and installed to reduce the amount of noise. For example, a siphon jet WC having an adjustable-rate water supply inlet float valve will operate more quietly. A WC can be mounted on a floating floor, and the noise transmission from fixtures can be reduced by using a neoprene or rubber gasket between the base of the WC and the floor (Fig. 16.5-6), or

FIGURE 16.5-6 WC vibrations can be minimized by resilient mounts and floated floor construction.

FIGURE 16.5-7 Flexible piping connections prevent vibrations from being transmitted to the piping.

by a resilient underlayment between two layers of floor sheathing. Flexible caulking or tightly packed insulation should be used to seal the space between a toilet bend and the floor sheathing.

Vibrating appliances such as dishwashers, disposers, and clothes washers and dryers should not be rigidly connected to the structure. Resilient or flexible isolation gaskets should be provided between the appliance and the floor, wall, or counter to isolate vibrations, unless a floated floor system is used. Piping connections to appliances should be flexible enough to prevent vibrations from being transmitted to the piping (Fig. 16.5-7). Electrical connections to appliances should also be flexible whenever possible.

# 16.6 Additional Reading

More information about the subjects discussed in this chapter can be found in the references listed in Section 16.7 and in the following publications.

Dagostino, Frank R. and Joseph B. Wujek. 2005. *Mechanical and Electrical Systems in Construction and Architecture*, 4th ed. Upper Saddle River, NJ: Prentice Hall (www.prenhall.com).

Jahrling, Peter. 2004. "Writing Restroom Specifications." *The Construction Specifier* (March): 58 (www.csinet.org).

Stein, Benjamin and John S. Reynolds. 2000. *Mechanical and Electrical Equipment for Buildings*, 9th ed. New York: John Wiley & Sons, Inc. (www.wiley.com).

Tao, William and Richard Janis. 1997. *Mechanical and Electrical Systems in Buildings*. Upper Saddle River, NJ: Prentice Hall (www.prenhall.com).

# 16.7 Acknowledgments and References

## ACKNOWLEDGMENTS

We gratefully acknowledge the assistance of the following organizations and individuals in preparing this chapter. We are also indebted to them for permission to use their illustrations when requested and for the use of their publications as references.

Delta Faucet Company (www.deltafaucet.com)

National Association of Home Builders (NAHB) Research Foundation (www.nahbrc.org)

Small Homes Council, Building Research Council

U.S. Department of Housing and Urban Development (HUD) (www.hud.gov)

U.S. Environmental Protection Agency (EPA) (www.epa.gov)

## REFERENCES

We would also like to thank the authors and publishers of the publications in the following list for their contribution to our research for this chapter.

American National Standards Institute (ANSI). New York (www.ansi.org). Standards:
A112 Series, "Plumbing."
Z21.10.1 and Z21 10.3, "Gas Water Heaters."
Z21.22, "Relief Valves for Hot Water Supply Systems."
Z124.1, "Plastic Bathtub Units."
Z124.2, "Plastic Shower Units."
Z124.3, "Plastic Lavatories."
Z124.4, "Plastic Water Closet Bowls and Tanks."
Z124.6, "Plastic Sinks."
American Society of Civil Engineers (ASCE). Reston, VA (www.asce.org).
"Sewage Treatment Plant Design." *ASCE Manual of Engineering Practice*, No. 36.
"Design and Construction of Sanitary and Storm Sewers," *ASCE Manual of Engineer Practice* No. 37.

American Society of Heating, Refrigerating and Air-Conditioning Engineers (ASHRAE) 2003. *Handbook—HVAC Applications*. Atlanta, Georgia: ASHRAE (www.ashrae.org).
ASTM International (ASTM). West Conshohocken, PA (www.astm.org). Standards:
A53/A53M, "Standard Specification for Pipe, Steel. Black and Hot-Dipped, Zinc-Coated, Welded and Seamless."
A74, "Standard Specification for Cast Iron Soil Pipe and Fittings."
A312, "Standard Specification for Seamless, Welded, and Heavily Cold Worked Austenitic Stainless Steel Pipe."
A377, "Standard Index of Specifications for Ductile Iron Pressure Pipe."
A888, "Standard Specification for Hubless Cast Iron Soil Pipe and Fittings for Sanitary and Storm Drain, Waste, and Vent Piping Applications."
B43, "Standard Specifications for Seamless Red Brass Pipe, Standard Sizes."
B88, "Standard Specification for Seamless Copper Water Tube."
B108, "Standard Specification for Aluminum-Alloy Permanent Mold Castings."
B134, "Standard Specification for Brass Wire."
B135, "Standard Specification for Seamless Brass Tube."
B210, "Standard Specification for Aluminum and Aluminum-Alloy Drawn Seamless Tubes."
B271, "Standard Specification for Copper-Base Alloy Centrifugal Castings."
B306, "Standard Specification for Copper Drainage Tube (DWV)."
B429, "Standard Specification for Aluminum-Alloy Extruded Structural Pipe and Tube."
B455, "Standard Specification for Copper-Zinc-Lead Alloy (Leaded Brass) Extruded Shapes."
B456, "Standard Specification for Electrodeposited Coatings of Copper Plus Nickel Plus Chromium and Nickel Plus Chromium."
B483, "Standard Specification for Aluminum and Aluminum-Alloy Extruded Drawn Tubes for General Purpose Applications."

B564, "Standard Specification for Nickle Alloy Forgings."
B584, "Standard Specification for Copper Alloy Sand Castings for General Applications."
B587, "Standard Specification for Welded Brass Tube."
B650, "Standard Specification for Electrodeposited Engineering Chromium Coatings on Ferrous Substrates."
C12, "Standard Practice for Installing Vitrified Clay Pipe Lines."
C425, "Standard Specification for Compression Joints for Vitrified Clay Pipe and Fittings."
C564, "Standard Specification for Rubber Gaskets for Cast Iron Soil Pipe and Fittings."
C700, "Standard Specification for Vitrified Clay Pipe, Extra Strength, Standard Strength, and Perforated."
D1785, "Standard Specification for Poly(Vinyl Chloride) (PVC) Plastic Pipe, Schedules 40, 80, and 120."
D2287, "Standard Specification for Nonrigid Vinyl Chloride Polymer and Copolymer Molding and Extrusion Compounds."
D2447, "Standard Specification for Polyethylene (PE) Plastic Pipe, Schedules 40 and 80, Based on Outside Diameter."
D2487, "Standard Classification of Soil for Engineering Purposes (Unified Soil Classification System)."
D2751, "Standard Specification for Acrylonitrile-Butadine-Styrene (ABS) Sewer Pipe and Fittings."
D2949, "Standard Specification for 3.25-in. Outside Diameter Poly(Vinyl Chloride) (PVC) Plastic Drain, Waste, and Vent Pipe and Fittings."
D3309, "Standard Specification for Polybutylene (PB) Plastic Hot- and Cold-Water Distribution Systems."
D3350, "Standard Specification for Polyethylene Plastics Pipe and Fittings Materials."
F891, "Standard Specification for Coextended Poly(Vinyl Chloride) (PVC) Plastic Pipe with a Cellular Core."
American Water Works Association (AWWA). C-203, "Coal-tar Protec-

tive Coatings and Linings for Steel Water Pipelines, Enamel and Tape, Hot-applied." Denver, CO: AWWA.

*Architectural Graphic Standards. See* Ramsey et al. (2000).

Cast Iron Soil Pipe Institute (CISPI). Chattanooga, TN (www.cispi.org). Standards.

1994. *Cast Iron Soil Pipe and Fittings Handbook.*

CISPI 301, "Standard Specification for Hubless Cast Iron Soil Pipe and Fittings for Sanitary and Storm Drain, Waste, and Vent Piping Applications."

CISPI 310, "Specification for Couplings Used to Connect Hubless Cast Iron Soil Pipe and Fittings for Sanitary and Storm Drain, Waste, and Vent Piping Applications."

Dagostino, Frank R. and Joseph B. Wujek. 2005. *Mechanical and Electrical Systems in Construction and Architecture*, 4th ed. Upper Saddle River, NJ: Prentice Hall (www.prenhall.com).

Federal Specifications, all: *see* U.S. General Services Administration Specifications Unit.

Hydronics Institute (HI). 2005. "IBR Ratings for Boilers, Baseboard Radiation, Finned Tube and Indirect-Fired Water Heaters." Berkley Heights, NJ: HI (www.gamanet.org).

International Code Council (ICC). Falls Church, VA (www.iccsafe.org). 2006. International Codes:

*International Energy Conservation Code®.*

*International Plumbing Code®.*

*International Private Sewage Disposal Code®.*

*International Residential Code®.*

Ramsey, Charles George, Harold Reeve Sleeper, and John Ray Hoke, Jr. 2000. *Architectural Graphic Standards*, 10th ed. Hoboken, NJ: John Wiley & Sons, Inc. (www.wiley.com).

Sump and Sewage Pump Manufacturers Association (SSPMA). *Recommended Standards for Sump, Effluent, and Sewage Pumps.* Northbrook, IL: SSPMA (www.sspma.org).

Underwriters Laboratories (UL). Northbrook, IL (www.ul.com). Standards:

70, "Septic Tanks, Bituminous Coated Metal."

174, "Household Electric Storage Tank Water Heaters."

732, "Oil-Fired Storage Tank Water Heaters."

U.S. Department of Housing and Urban Development (HUD). Washington, DC (www.hud.gov).

1994. *Minimum Property Standards for Housing*, HUD-4910.1.

1992. Handbook 4940.3, *Minimum Design Standards for Community Sewerage Systems.*

U.S. Environmental Protection Agency (EPA). Washington, DC (www.epa.gov).

1997. *National Primary Drinking Water Regulations (NPDWRS).*

2002. EPA 625/R-00/008, *Onsite Wastewater Treatment Systems Manual.*

U.S. General Services Administration Specifications Unit. Federal Specification WW-P-541, "Plumbing Fixtures." Washington, DC: GSA Specifications Unit (apps.fss.gsa.gov).

Water Quality Association (WQA). Lisle, IL (www.wqa.org). Standards:

S-100, "Voluntary Industry Standards for Household, Commercial, and Portable Exchange Water Softeners," Vol. 1.

S-200, "Recommended Industry Standards for Household and Commercial Water Filters," Vol. 2.

# CHAPTER 17

# Heating, Ventilating, and Air Conditioning (HVAC)

# Introduction

Early theories defined heat as a material substance called *caloric*, which was a weightless fluid that had the power to penetrate other materials, expand, and transform them into vapor. It was thought that a substance was hot or cold, depending on the proportion of caloric contained in its pores.

The theory of heat was placed on a sound scientific basis in the early nineteenth century by James P. Joule, an English physicist. Joule's experiments led to the principle of *conservation of energy*, which stated that heat energy was not lost as it was used, but was transformed into an equivalent amount of another kind of energy. For instance, heat energy could be converted to kinetic (motion) energy by a steam engine and from kinetic energy to electric energy by a generator.

Animals, including people, sustain life by converting organic food substances into heat energy. Normally, excess heat energy is produced and the body dissipates this excess to the surrounding atmosphere. Each organism has an optimum rate of body heat loss, depending on physical exertion. When the heat loss is at the ideal rate, a person feels comfortable. When heat is lost too rapidly, a person feels cold; when it is lost too slowly, a person feels hot. The objective of interior heat control is to create conditions in which the body loses heat at the optimum rate.

Heat energy generally flows from a warmer region to one that is cooler. Therefore, the flow is to the outside of a building in winter and to the inside in summer. This heat flow affects interior surface and air temperatures.

Cold is the absence of heat. A warm window radiates heat, but a cold window does not radiate cold. A person's sensation of coldness is caused by a rapid loss of body heat to the cold window surface. Likewise, the movement of air over the body produces the sensation of coldness because it carries off heat, not because it brings coldness.

Until recently, buildings were designed to protect occupants from the environment solely by excluding precipitation, wind, and cold air. Heat was provided in isolated locations within a building, and occupants moved to those locations as the day progressed. Fires were often not even stoked at night, and occupants kept warm by increasing the amount of clothing they wore and adding blankets and quilts to their beds. In both residences and other buildings, the world was a cold, drafty place in winter and a hot, humid one in summer.

In the twentieth century, with the increased availability of gas, oil, and electricity, people began to demand a more comfortable interior environment. Designers began to enclose buildings more thoroughly, and central heating and, eventually, central cooling became commonplace. As people demanded even more comfort in their interior environment, designers began to completely isolate the interior from the exterior environment, sealing windows and wrapping buildings with materials that excluded air penetration. The result was an unsafe interior environment, rife with pollutants and foul air, and high energy usage that has become increasingly unacceptable.

People today expect to find a relatively narrow range of uniform temperatures within a building and want the ability to select that range at will. When energy was less costly, it was relatively easy to force an interior environment into a narrow band of comfort. Today, both architects and building owners are examining the benefits of permitting a broader range of interior temperatures and of adjusting the mode of dress to the temperature.

For instance, putting on a sweater in winter might make a 65°F (18.3°C) interior air temperature acceptable, while removing one's tie and turning on an electric fan might make 80°F (26.7°C) tolerable in the summer. The potential gain from accepting a 5°F (−15°C) variation up or down from traditional thermostat settings can produce a 15 to 30% energy savings over the average costs of heating and cooling.

Until the energy crunch of the 1970s, traditional approaches to interior heat control emphasized reliance on energy-intensive mechanical systems. These systems pumped heat in or out of a building and relied on the interior shell only to minimize heat gain in summer and heat loss in winter. Now architects tend to design a building's envelope to permit heat flow through the shell when advantageous. For instance, buildings can be designed to make use of sunshine for heating in the winter and of natural breezes for cooling in the summer. Some of the principles of passive solar energy systems are common in many buildings (see Section 17.7.5).

This chapter discusses the principles of heat flow and human comfort, appropriate building design criteria related to heat control, and the design of heating, ventilating, and air conditioning (HVAC) systems. Section 7.3 discusses the use of insulation to minimize heat flow through a building's envelope. Chapter 8 contains data about strategies for reducing heat loss and gain through doors and windows. Section 12.2 contains information concerning the use of interior and exterior shading devices and of blinds, shades, draperies, and shutters to control the gain or loss of heat through a building's windows. Together, these techniques help to reduce energy consumption and make buildings more comfortable for occupants, with less reliance on mechanical heating and cooling systems.

# Sustainability Considerations

Sustainability requires judicious design of a building and its HVAC system. This includes proper siting and configuration of the building, selection of materials that are recyclable, energy-efficient design of both the building and the HVAC system, and designing both the building and the HVAC system to permit flexibility in use.

Energy-efficient design should reduce or eliminate nonrenewable fossil fuel consumption for heating, cooling, and lighting. Although it is good to create a building with conservative materials, it is critical to ensure that once the building is built, it either continues to conserve energy resources or uses renewable energy resources throughout its life. Consider using durable, low-maintenance materials. Where practical, design full-cycle systems such as solar water heating systems that will capture renewable energy on site and passive solar heating systems that take advantage of naturally occurring conditions with little or no fossil fuel use.

Sustainability considerations related to the materials used in HVAC systems are similar to those discussed in the various sections related to the materials used.

Sustainability considerations are addressed throughout this chapter. For example, the building design recommendations in Section 17.2 focus on methods of obtaining sustainability through design. Section 17.7.1.1 discusses fuel types and the effects of their use on the environment. The use of solar energy is discussed in Section 17.6.3; heat recovery systems, in Section 17.7.5; and heat pumps, in Section 17.8.2.

## Applicable *MasterFormat* Sections

Subjects addressed in this chapter are included in *MasterFormat* 2004, Division 23, "Heating, Ventilating, and Air-Conditioning (HVAC)." The following *MasterFormat* 2004 Level 2 sections are applicable to this chapter:

23 07 00  HVAC Insulation
23 08 00  Commissioning of HVAC
23 09 00  Instrumentation and Control for HVAC
23 10 00  Facility Fuel Systems
23 20 00  HVAC Piping and Pumps
23 31 00  HVAC Ducts and Casings
23 32 00  Air Plenums and Chases
23 33 00  Air Duct Accessories
23 34 00  HVAC Fans
23 36 00  Air Terminal Units

23 37 00  Air Outlets and Inlets
23 40 00  HVAC Air Cleaning Devices
23 51 00  Breechings, Chimneys, and Stacks
23 52 00  Heating Boilers
23 54 00  Furnaces
23 55 00  Fuel-Fired Heaters
23 56 00  Solar Energy Heating Equipment
23 57 00  Heat Exchangers for HVAC
23 61 00  Refrigerant Compressors
23 62 00  Packaged Compressor and Condenser Units
23 63 00  Refrigerant Condensers
23 64 00  Packaged Water Chillers
23 65 00  Cooling Towers
23 71 00  Thermal Storage

23 72 00  Air-to-Air Energy Recovery Equipment
23 73 00  Indoor Central-Station Air-Handling Units
23 74 00  Packaged Outdoor HVAC Equipment
23 75 00  Custom-Packaged Outdoor HVAC Equipment
23 76 00  Evaporative Air-Cooling Equipment
23 81 00  Decentralized Unitary HVAC Equipment
23 82 00  Convection Heating and Cooling Units
23 83 00  Radiant Heating Units
23 84 00  Humidity Control Equipment

# 17.1  Heat and Comfort

Heat is a form of energy. It is produced in a substance by the movement of its molecules—the tiny particles of matter which make up that substance (see Section 24.1). When heat is added, the molecular movement inside that substance is increased. The intensity of movement determines the physical state of the substance, whether solid, liquid, or gas. Temperature, recorded in degrees Fahrenheit (°F) or degrees Celsius (°C) is a measure of that intensity (Fig. 17.1-1). A British thermal unit (Btu) is a measure of heat quantity and is equal to the amount of heat required to raise 1 lb (0.45 kg) of water 1°F ($-17.5$°C). Heat flow rate is measured in Btus per hour. In the Système International (SI) (metric system), heat quantity is measured in joules (J) and heat flow rate is measured by the watt (W). One watt is equal to 1 J/sec, or approximately 0.341 Btu/hr.

## 17.1.1  SENSIBLE AND LATENT HEAT

Energy that causes a rise in temperature but does not alter the physical state of a substance is called *sensible heat*. Heat can be added to a substance without altering its physical state only up to a certain temperature. For example, if a teapot with 1 lb (0.45 kg) of water in it is placed on a burner, each Btu (1.06 kJ) of heat added raises the temperature 1°F ($-17.5$°C). If the water started at 62°F (16.67°C), it would take 150 Btus (158.3 kJ) to bring it to a boil at 212°F (100°C). Heat added beyond the first 150 Btus (158.3 kJ) would change the water to vapor. It would not raise the temperature of the water above 212°F (100°C).

The energy that would be used to convert the water to steam if the pot were left on the burner is the *latent heat of vaporization*. To change the entire 1 lb (0.45 kg) of water to steam would require 970 Btus (1023.4 kJ). To change 1 lb (0.45 kg) of ice at 32°F (0°C) to water would require 144 Btus (151.93 kJ), which is the *latent heat of fusion* of water. Only after all of the first 144 Btus (151.93 kJ) were used and all of the ice had melted would the additional heat be used to raise the water temperature above 32°F (0°C).

## 17.1.2  METHODS OF HEAT TRANSFER

Heat energy moves from a warmer to a colder area by conduction, radiation, or convection. A person gives up heat by all of these methods, as well as by evaporation of perspiration.

### 17.1.2.1  Conduction

The movement of heat through a solid or liquid substance as heated particles transfer their thermal energy to adjacent particles is called *conduction*. Generally,

| | |
|---|---|
| 121 | 250 |
| 100 | 212 — BOIL |
| 75 | 167 |
| 57.2 | 135 |
| 37 | 98.6 — BODY TEMP |
| 20 | 68 |
| 0 | 32 — FREEZE |
| -17.8 | 0 |
| -28.9 | -20 |

$F = \frac{9}{5}C + 32$      $C = (F - 32)\frac{5}{9}$

**FIGURE 17.1-1** Temperature equivalents on the Celsius (metric) and Fahrenheit (English) scales. Thermometers can be calibrated to a temperature of −273.1°C (−459.6°F), which is known as *absolute zero*. This is the point at which all molecular movement stops.

**FIGURE 17.1-2** Reflectivity/Emissivity Chart

| Material | Reflectivity, % | Emissivity, % |
|---|---|---|
| Built-up roof, gravel aggregate | 7 | 93 |
| Slate, dark soil, other dark textures | 15 | 85 |
| Grass, dry | 30 | 70 |
| Copper foil | | |
|    Tarnished | 36 | 64 |
|    New | 75 | 25 |
| Paint | | |
|    Light gray | 25 | 75 |
|    Red | 26 | 74 |
|    Aluminum | 46 | 54 |
|    Light green | 50 | 50 |
|    Light cream | 65 | 35 |
|    White | 75 | 25 |
| Fresh snow | 80 | 20 |
| Aluminum foil | 95 | 5 |

dense materials such as metals can transfer heat faster and are called *conductors*. Less dense substances such as wood and plastic retard heat flow and are called *insulators*.

### 17.1.2.2 Radiation

The transmission of heat by invisible light waves independent of any medium is called *radiation*. The radiant heat from the sun, for example, reaches the earth by traveling through the vacuum of space and the atmosphere. Radiant energy can travel through any nonopaque substance.

An opaque body absorbs radiant energy over the surface struck by the solar radiation and converts it into heat. After absorption, the transfer of heat through the body is by conduction; some heat is also reradiated from the warm body into the air.

### REFLECTIVITY

Materials vary in their acceptance and rejection of the radiant energy that strikes them. Those that reject much of the radiant energy are called *reflectors* because of their high degree of reflectivity. Shiny aluminum foil, for instance, reflects 95% of all radiant energy; white paint reflects about 75%. Dark materials, such as asphalt pavement, which absorb more radiant energy, have a high degree of emissivity (Fig. 17.1-2).

### EMISSIVITY

Emissivity is the converse of reflectivity. Aluminum foil reflects 95% of the radiant heat that strikes it and absorbs the remainder as sensible heat. Therefore, the emissivity of the foil is 5% (see Fig. 17.1-2). The practical significance of such differences is that high-emissivity surfaces absorb and radiate more energy. For example, because asphalt pavements have high emissivity, air immediately above them will be warmed to as much as 20°F (−6.7°C) higher than air in nearby shaded areas.

### 17.1.2.3 Convection

The movement of heat from one region to another through a fluid substance such as air or water is *convection* (Fig. 17.1-3). When air in a room contacts a warm baseboard heater, it is heated by conduction. As it is warmed, the air expands, becomes lighter, and rises, thus permitting colder air to take its place. The warmed air often gives up its heat to a

cold wall or window and settles to the floor to be reheated. Thus, a *convection loop* is established which distributes the heat throughout the space.

### 17.1.2.4 Evaporation

A human body at rest loses heat mainly by convection and radiation. When the person is engaged in strenuous activity or when the air is too warm, those methods are insufficient to carry off all the excess heat. The body then utilizes perspiration, which cools the body by withdrawing from it the *latent heat of vaporization*. Low humidity aids cooling by perspiration; high humidity inhibits it.

### 17.1.3 HUMIDITY

#### 17.1.3.1 Water Vapor Properties

Moisture can exist in several states: as a liquid (water), as a solid (ice), and as an

**FIGURE 17.1-3** Air warmed by a heater rises, gives up its heat to the walls and ceiling, and drops to the floor to be warmed again.

**RELATIVE HUMIDITY IS INCREASED BY:**

decreasing temperature (moisture constant)

or increasing moisture (temperature constant)

**FIGURE 17.1-4** Relative humidity can be increased by decreasing the temperature or by increasing the amount of moisture.

invisible gas (water vapor). When this moisture becomes visible, it is no longer vapor but a liquid. A puff of steam is water in an intermediate stage—vapor in the process of becoming liquid or vice versa. The conversion from a vapor to a liquid is called *condensation*.

The change of liquid water to a gas is *evaporation, Sublimation* is the transition from ice directly to water vapor.

The change of state from water or ice to vapor requires considerable heat input, called *latent heat*, measured in Btus (joules [J]). Therefore, when water evaporates from a surface, that surface is cooled. Perspiring skin, for example, is an effective cooling system for a human body. Like any warm object, the skin gives off heat to the surrounding atmosphere. Heat absorbed by a substance without changing its state is known as *sensible heat* and also is measured in Btus (J).

When water turns to vapor, it mixes with the air and occupies all available space. In many ways, water vapor acts independently of the air because its general properties do not depend on the presence of air. However, when air is moved by wind or a fan, or is heated or cooled, associated water vapor is similarly affected. It is therefore convenient to consider moist air as a mixture of dry air and water vapor.

The effects of water vapor on building materials and means to prevent damage by water vapor on a building are discussed in Section 7.1.

**17.1.3.2 Relative Humidity**

Most air contains water vapor. The measure of the amount of water vapor in air is known as *relative humidity*, which is stated as a percentage. This percentage is a ratio of the actual amount of water vapor in the air to the maximum amount the

same air can contain at a specific temperature and atmospheric pressure. For instance, air with a 50% relative humidity contains half the amount of vapor it could hold at that air temperature and barometric pressure. At 100% relative humidity, air contains the maximum amount of water vapor possible. Such air is said to be *saturated*.

An observable law of physics is that the maximum relative humidity is dependent upon the ambient (general) air temperature. The warmer a mass of air, the more water vapor it can hold (Fig. 17.1-4). Therefore, if the temperature of saturated air is lowered, some vapor must be given up by condensation. Up to the point of condensation, relative humidity increases as air temperature decreases, while the amount of moisture remains the same. When all possible moisture is held by the air (100% relative humidity), its *saturation point* or *dew point* has been reached. As shown in Figure 17.1-5, air at 70°F (21.1°C) can hold twice as much moisture as the same volume of air at 50°F (10°C) when the air is completely saturated. Therefore, at 70°F (21.1°C) the air is said to be *half-saturated* or to have a relative humidity of 50%. If the temperature were lowered below the saturation point, say to 40°F (4.4°C), the relative humidity would remain at 100%, but about 18 grains of moisture per pound (0.00257 kg/kg) of dry air would condense into visible water on cool surfaces. Further cooling to 30°F (−1.1°C) would result in the removal of a total of 30 grains/lb (0.0043 kg/kg) of air (see Fig. 17.1-5).

Relative humidity is measured by a *sling psychrometer* (Fig. 17.1-6). This device consists of two thermometers mounted side by side on a frame. A wetted cloth sack is attached to one of the

**FIGURE 17.1-5** Temperature and Humidity Relationships[a]

| Temperature | Actual Moisture Content at Given Temperature, grains/lb dry air | Maximum Moisture Control at Dew Point, grains/lb dry air | Relative Humidity, %RH | Condensed Moisture,[b] grains/lb dry air |
|---|---|---|---|---|
| 70°F | 54 | 108 | $^{54}/_{108} = 50$ | None |
| 60°F | 54 | 77 | $^{54}/_{77} = 70$ | None |
| 50°F | 54 | 54 | $^{54}/_{54} = 100$ | None |
| 40°F | 36 | 36 | $^{36}/_{36} = 100$ | 54 − 36 = 18 |
| 30°F | 24 | 24 | $^{24}/_{24} = 100$ | 54 − 24 = 30 |

[a]At 29.92 in. Hg pressure.

[b]As air at 50% RH and 70°F drops in temperature (read down in first column).

(Reprinted with Permission of the American Society of Heating, Refrigerating and Air-Conditioning Engineers.)

**FIGURE 17.1-6** A sling psychrometer.

thermometers, and both are whirled through the air. The temperature reading given by the uncovered thermometer is the *dry-bulb* temperature (DBT), which is the common way of measuring and stating temperature. Ambient temperatures are usually stated in DBT. The reading shown by the wetted thermometer is the *wet-bulb* temperature (WBT). This is

the reading more commonly used in the design of mechanical systems.

Rapid movement through the air causes some of the water to evaporate from the wetted thermometer. The process of evaporation uses up latent heat energy, resulting in a lower temperature reading. The greater the difference between WBT and DBT, the lower the relative humidity.

### 17.1.3.3 Psychrometrics

The behavior of moist air (a mixture of dry air and water vapor), involving saturation of air and condensation of water vapor at varying temperatures, generally is referred to as *psychrometrics*. Changes in the properties of moist air with heating and cooling can be followed readily with the aid of a psychrometric chart. Figure 17.1-7 is a somewhat simplified psychrometric chart showing data in the customary system of measure (see Chapter 25). Psychrometric charts are available with much more detail than shown on the chart in Figure 17.1-7. For example, such charts may include data from 20° to 110° of DBT; enthalpy (total heat content) lines may be included, as well

as lines showing the sensible heat factor. Such charts can be used to discern the effects of the following factors used in the design of HVAC systems:

- Humidifying only
- Sensible heating only
- Sensible cooling only
- Evaporative cooling only
- Dehumidifying only
- Chemical dehumidifying
- Cooling and dehumidifying
- Heating and humidifying

Psychrometric charts are also available that show data in SI (metric) units.

A basic use of psychrometric charts can be demonstrated using Figure 17.1-7. The curved saturation line (100% relative humidity) represents the maximum concentrations of water vapor (pounds of water per pound of dry air) (kg of water per kg of dry air) which can exist as vapor, without condensing, at various temperatures.

When the DBT and WBT are known, the relative humidity can be determined from a psychrometric chart. For instance, as shown by the heavy lines on the chart in Figure 17.1-7, if the DBT is 70°F (21.1°C) and the WBT is 58°F (14.4°C), then the relative humidity is 50%.

Figure 17.1-8 is a simplified psychrometric chart for a specific set of conditions. A possible winter condition inside buildings, 70°F (21.1°C) and 40% relative humidity, is represented by point A. This is a partial saturation condition (less than 100% relative humidity). As air is cooled from 70°F (21.1°C) (point A) to 44.6°F (7°C) (point B on the saturation curve), the relative humidity keeps increasing until it becomes 100%.

### 17.1.3.4 Dew Point Temperature

The temperature at which saturation occurs (point B in Fig. 17.1-8) is called the *dew point (DP) temperature (frost point* if below 32°F [0°C]). With further cooling to 35°F (1.67°C), the original amount of vapor (44 grains/lb [0.0063 kg/kg] of dry air) is reduced through condensation to the condition represented by point C (30 grains/lb [0.0043 kg/kg] of dry air). The progression from point A to B to C illustrates what an air-vapor mixture experiences when it comes in contact with a cool window surface. Cooling from point B to point C results in visible condensation on the glass surface. If point C were below 32°F (0°C), the condensation would be frost.

\* With ET lines, for persons at rest, normally clothed, in still air.

**FIGURE 17.1-7 Psychrometric chart.** (Reprinted with permission of the American Society of Heating, Refrigerating, and Air-Conditioning Engineers.) (Refer to Figure 25.7-1 for metric equivalents.)

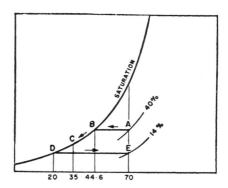

**FIGURE 17.1-8** Use of a psychrometric chart illustrated. (Reprinted with permission of the American Society of Heating, Refrigerating and Air-Conditioning Engineers.)

Once the temperature drops below the dew point (or frost point), vapor pressure is also reduced. This reduction in vapor pressure causes diffusion currents within a room, moving water vapor continuously to window surfaces to be condensed as long as the vapor concentration is maintained in the room.

Moving from point D to point E (see Fig. 17.1-8) illustrates a common winter phenomenon. When outside air at 20°F (−6.7°C) and 100% relative humidity is heated to 70°F (21.1°C), the relative humidity is lowered to 14%. This is the reason that the relative humidity within a building is greatly reduced when cold outside air enters and is heated to room temperature.

High dew point temperatures are common in the summertime. The psychrometric chart shows that for an 80°F (26.7°C) DBT, the dew point temperature is 60°F (15.6°C) when the relative humidity is 50% (see Fig. 17.1-7). However, raising the relative humidity to 90% increases the dew point temperature to 78°F (25.6°C). As the humidity rises, the dew point temperature gets closer to the DBT but cannot exceed it. If the temperature is raised to 90°F (32.2°C) and the relative humidity is maintained at 90% (by adding moisture to the air), the dew point temperature rises to 87°F (30.6°C).

## 17.1.4 HUMAN COMFORT REQUIREMENTS

The human body is a heat-producing mechanism, generating heat energy from oxidized food. In most cases, it produces more heat than is needed for thermal stability. The amount of extra heat that must be dissipated depends on the activity being performed (Fig. 17.1-9) and ranges between 200 Btus (211 kJ) and 4500 Btus (4747.8 kJ) per hour.

### 17.1.4.1 Environmental Control

Control of the interior environment helps the body regulate its temperature and achieve the ideal comfort condition, which is a sensation that is neither too warm nor too cool. Studies conducted by the American Society of Heating, Refrigerating and Air-Conditioning Engineers, Inc. (ASHRAE) and others have shown a rise in the indoor temperatures people require to be comfortable in the winter from between 65° and 70°F (18.3° and 21.1°C) in 1900 to between 72° and 78°F (22.2° and 25.6°C) today. Therefore, the desire for warmer temperatures may be a result of cultural conditioning rather than biological necessity. The high cost of energy today and the growing awareness of the need to conserve energy may not have changed how people feel about the ideal temperature for comfort, but they have somewhat changed the way people act. There is a greater tendency to turn the thermostat down and wear heavier clothes in winter and the reverse in summer.

### 17.1.4.2 Interior Ventilation

In many locations (see Section 7.1.3), summer cooling does not always create serious concealed condensation problems, because cooled interior air seldom produces conditions within a wall that are conducive to condensation. However, during periods of high humidity, condensation may form on cool basement walls and floors. In humid climates, condensation may occur in concealed spaces (see Section 7.1).

Where high water vapor levels are generated indoors, the likelihood of both surface and concealed condensation can be greatly minimized by introducing dry night air or by exhausting excessive water vapor directly at its source, such as in a shower room, kitchen, laundry, or residential bathroom, using exhaust fans. In many buildings, a sufficient amount of fresh air is available through infiltration to make up for the air lost by exhaust. Where this is not the case, it is necessary to introduce a fresh air supply to balance this loss.

During cold weather, interior relative humidity often becomes too low for human comfort, and mechanical humidification of occupied spaces is required. Even then, ventilation in areas where high humidity is present, such as those mentioned earlier, is desirable to reduce excessively high local humidities which could easily result in surface condensation.

The effectiveness of ventilation in reducing humidity is shown in Figure 17.1-10. For simplicity, it uses a 2000 sq ft (185.8 m²) residence as its example. Figure 17.1-10 shows the small amount

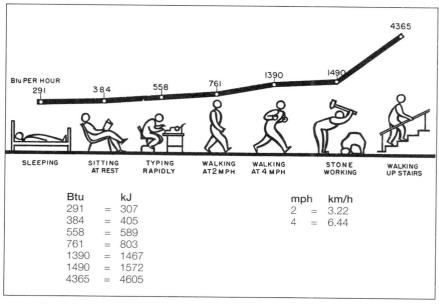

**FIGURE 17.1-9** A body's production of heat depends on how strenuous the activity is.

**FIGURE 17.1-10** Water vapor balance maintained by ventilation. (Reprinted with permission of the American Society of Heating, Refrigerating and Air-Conditioning Engineers.)

of water vapor escaping into well-insulated walls and ceilings of a building provided with vapor retarders (the floor is not considered). With indoor temperature at 70°F (21.1°C) and a relative humidity of 40%, ventilation at the rate of 2000 cu ft (56.6 m³) per hour removes about 21 lb (9.53 kg) of moisture per day. Assuming a production of 22.5 lb (10.21 kg) per day, less than 1.5 lb (0.68 kg) escapes into the walls and ceilings of the structure. As mentioned earlier, 22.5 lb (10.21 kg) in 24 hours is about average for a family of four at home.

In a residence, ventilation at the rate of 2000 cu ft (56.6 m³) per hour is near the minimum for odor control during cooking. Therefore, mechanical ventilation may not be required just to get rid of excess moisture generated in a home. In older buildings of all types, the action of exhaust fans is supplemented by substantial *exfiltration* (air leakage) through walls, windows, and doors to keep moisture in check. However, tight newer buildings will often require more extensive ventilation systems, particularly in periods of cold weather when the relative humidity must be limited to prevent condensation (see Fig. 17.1-11).

Ventilation of attic spaces, crawl spaces, and other locations to prevent condensation is addressed in Section 7.1.

### 17.1.4.3 Comfort Factors

Air movement, relative humidity, and temperature all affect the rates at which perspiration evaporates from the skin and at which a human body loses heat. Either too little or too much humidity can be uncomfortable. Low humidity encourages evaporation and increases the sensation of coolness, which is undesirable in the winter. High humidity reduces skin evaporation, retards body cooling, and makes people feel warmer, which makes them uncomfortable when the air temperature is near the upper limit of comfort. This can happen in any season but is more likely to occur in the summer.

High humidity at higher temperatures can contribute to fatigue and reduced working efficiency. Air that is too dry can irritate nasal and sinus passages, contribute to chapped and dry skin, and encourage the generation of annoying and, in locations like computer rooms, damaging static electricity.

The specific humidity or temperature comfort level is largely a subjective judgment that depends on many factors, such as prevailing climate, seasonal variations, and the individual's age, health, living habits, clothing, and activities.

As mentioned, cold air cannot hold nearly as much water vapor as warm air. Therefore, when cold outside air with a low moisture content (even though its relative humidity may be high) is brought

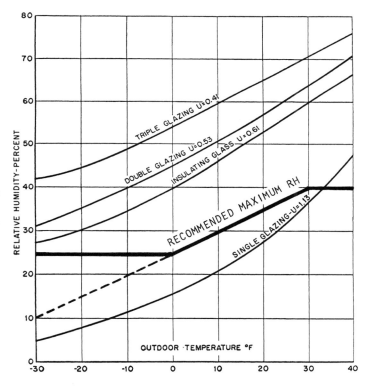

Below 0°F, RH values maintained at 25% to provide acceptable degree of occupant comfort (solid line); however, to minimize condensation on single glazing, use RH values represented by dashed line.

**FIGURE 17.1-11** Recommended maximum interior relative humidity to prevent condensation damage. (Reprinted with permission of the American Society of Heating, Refrigerating and Air-Conditioning Engineers.)

inside and heated, low relative humidity results. For instance, as shown in Figure 17.1-5, saturated air at 30°F (−1.1°C) can hold 24 grains of moisture per pound (0.0034 kg/kg) of air, and half-saturated air at 30°F (−1.1°C) has a moisture content of 12 grains/lb (0.0017 kg/kg) of air. If this air were heated to 70°F (21.1°C), it could hold 108 grains/lb (0.0154 kg/kg), and its relative humidity would be 100% times 12/108 (0.0017/0.0154), or 11.1% relative humidity. The resulting humidity level is too low for comfort. From 25 to 40% relative humidity is usually considered conducive to good health, but such levels are difficult to maintain in cold weather.

Ambient temperature, air movement, relative humidity, and the temperature of a space's surfaces combine to produce the thermal sensations a person experiences within that space. The first three have already been discussed. The fourth is expressed in terms of *mean radiant temperature* (MRT), which represents the collective thermal effect of all the radiant surfaces in the space on an occupant at a given position within the space (Fig. 17.1-12).

## EFFECTIVE TEMPERATURE

The term *effective temperature* (ET) defines an arbitrary index that combines into a single value the effects of temperature, humidity, and air movement on the sensation of warmth or cold felt by a human body. It has been developed by ASHRAE in cooperation with the U.S. Public Health Service and is based on extensive polls of trained individuals subjected to various temperature and humidity conditions.

Effective temperature is the temperature of saturated air (100% relative humidity) that gives the same physical sensation of warmth as various other combinations of air temperature and relative air movement rates ranging from 15 to 25 feet per minute (fpm) (0.076 to 0.127 m/s). Any combination of temperature and humidity with an air movement of 15 to 25 fpm can be stated as a single-number ET. That number is the DBT at 100% relative humidity (or at 50% relative humidity), which produces the same sensation as the given combination of temperature and humidity. This book uses the old ET definition at 100% relative humidity; the newer 50% relative humidity ET is designated here by an asterisk, ET*.

## OPERATIVE TEMPERATURE

*Operative temperature* ($t_{op}$) is an average of the air temperature and the MRT. Operative temperature is sometimes called *adjusted dry-bulb temperature*. It is the temperature at which a person exchanges the same heat by radiation and convection as the temperature of the atmosphere in a space. The *humid operative temperature* is the operative temperature at 100% relative humidity. The *adiabatic equivalent temperature* is the operative temperature at 0% relative humidity.

The ET and the $t_{op}$ are used in ASHRAE Standard 55.

## WINTER COMFORT

Early ASHRAE tests suggested that the majority of people were most comfortable in winter at an ET of 68 (68°F [20°C] [ET* 72.2°F (22.3°C)]). Humidity and temperature combinations that gave this same feeling of warmth are tabulated in Figure 17.1-13. This table shows that in winter, as the relative humidity is reduced, higher temperatures are necessary to maintain the same sensation of warmth. Conversely, a higher relative humidity produces warmth with lower temperatures. However, the thermal and vapor permeance characteristics of exterior construction impose practical limits on interior humidity, which vary with the severity of winters.

From the psychrometric chart in Figure 17.1-7, it is possible to obtain DBT/relative humidity equivalents at more common seasonal humidities. For example, according to Figure 17.1-7, the winter optimum DBT is near 75°F (23.9°C) at 30% relative humidity.

In areas of low winter design temperatures (0°F [−17.8°C] or lower), few buildings can tolerate 40% relative humidity without developing condensation and moisture problems. At this interior humidity and exterior temperature combination, even sealed insulating glass will develop condensation (see Fig. 17.1-11). Single glazing and highly conductive surfaces such as metal wall panels and metal door

$$MRT = \frac{\Sigma t\theta}{360}$$

$t$ = surface temp.
$\theta$ = angle of occupant exposure

Example: An apartment building has masonry and glass exterior walls with interior surface temperatures of 66° F and 20° F. Interior partitions are at 70° F.

Position A: $MRT = \frac{\Sigma t\theta}{360} = \frac{(20 \times 50) + (66 \times 60) + (70 \times 150)}{360} = \frac{18,680}{360} = 51°\,F$

Position B: $MRT = \frac{\Sigma t\theta}{360} = \frac{(20 \times 50) + (66 \times 60) + (70 \times 250)}{360} = \frac{22,460}{360} = 62°\,F$

**FIGURE 17.1-12 Calculation of MRT.** (Reprinted with permission of the American Society of Heating, Refrigerating, and Air-Conditioning Engineers.) (Refer to Figure 25.7-1 for metric equivalents.)

**FIGURE 17.1-13** Temperature-Humidity Equivalents of Optimum Summer and Winter Effective Temperatures[a]

| Winter Optimum, ET 68°F[b] | | Summer Optimum, ET 71°F[b] | |
|---|---|---|---|
| Relative Humidity | Dry Bulb Temperature | Relative Humidity | Dry Bulb Temperature |
| 100% | 68°F | 100% | 71°F |
| 80% | 69°F | 80% | 73.1°F |
| 70% | 70°F | 70% | 74.3°F |
| 60% | 71.1°F | 60% | 75.6°F |
| 50% | 72.2°F | 50% | 76.9°F |
| 40% | 73.5°F | 40% | 78.3°F |
| 30% | 74.8°F | 30% | 79.7°F |
| 20% | 76.1°F | 20% | 81.2°F |
| 10% | 77.5°F | 10% | 83.0°F |

[a]Air movement of 15 to 25 fpm.

[b]Considered comfortable by 98% of people tested.

(Reprinted with permission of the American Society of Heating, Refrigerating and Air-Conditioning Engineers.)

## SUMMER COMFORT

In summer, 98% of those tested in the older test mentioned above were comfortable at an ET of 71 (100% relative humidity and 71°F [21.67°C]) (ET* 76.9°F [24.9°C]). Figure 17.1-12 shows other relative humidity and temperature combinations that give the same physical sensation in summer. The summer optimum DBT is near 77°F (25°C) at 50% relative humidity.

The dashed lines in the ASHRAE comfort chart (see Fig. 17.1-14) represent ETs. Each line represents graphically equivalent temperature and humidity combinations. To select the corresponding DBTs (horizontal coordinate), follow the 68 and 71 ET lines to their intersections with the solid diagonal relative humidity lines. The temperature-humidity equivalents shown in Figure 17.1-13 can thus be derived from this chart.

and window frames without thermal break features will show condensation even at 15% relative humidity. To minimize condensation and yet provide for human comfort, about 25% relative humidity should be maintained when the outside temperature is 0°F (−17.8°C) or lower. The humidity can be maintained at 30% when the outside temperature is 10°F (−12.2°C) or lower, at 35% at 20°F (−6.7°C), and at 40% at 30°F (−1.1°C) outside (see Fig. 17.1-13). Relative humidities above 40% generally are not recommended during the winter heating season.

Although a relative humidity of 25% or higher is considered desirable during the winter season for health reasons, the effect of relative humidity on occupant comfort was overestimated in the past.

Figure 17.1-14 shows the results of both older tests measuring the effect of temperature and humidity on human comfort immediately upon entering a room (curved lines at the top and bottom) and recent tests on occupants after a 3-hour stay (solid vertical lines). Unlike the optimum summer and winter ET lines derived from earlier studies (see Fig. 17.1-12), the newer data suggest (1) that a DBT of 77.5°F (25.3°C) is most comfortable and (2) that humidity has no perceptible effect on the sensation of warmth within a range of 25 to 60% relative humidity. This is discussed more fully in the following subsection on "Summer Comfort."

**FIGURE 17.1-14** ASHRAE comfort chart showing ETs and subjective responses to temperature and humidity combinations. (Reprinted with permission of the American Society of Heating, Refrigerating and Air-Conditioning Engineers.)

CONVECTION **C**
------ RADIATION
......... EVAPORATION

**a**

**b**          **d**

**FIGURE 17.1-15** Both ambient (air) and radiant (wall) temperatures are important for comfort. At rest (a,b), thermal balance can be easily maintained with cooler air if walls are warmer; at work (c,d), cool air moves more body heat, requiring less perspiration to maintain balance. (Refer to Figure 25.7.1 for metric equivalents.)

The optimum summer ET 71 line strikes near the top of the summer curve representing the distribution of responses by the subjects tested. From this curve, it can be established that in summer nearly all subjects were comfortable at an ET of 71, while half were comfortable within an ET range of 68 to 75. Similarly, the winter comfort curve indicates that nearly all subjects were comfortable at an ET of 68, while about 86% were comfortable at ETs between 65 and 70.

### DEFECTS IN USING ET AS A COMFORT INDEX

The ET index is useful only in establishing the human response to temperature-humidity conditions immediately upon entering a room. Humidity played a much smaller role in determining the comfort of subjects tested after a 3-hour stay. These findings are represented by solid lines 3, 4, 5, and 6 in Figure 17.1-14. For both summer and winter, a DBT of 77.5°F (25.3°C) was equally comfortable for relative humidities from 25% to 60%. Only above 60% relative humidity (the curved portion of line 4) must the temperature be reduced to maintain the same degree of comfort. The lines represent av-

eraged responses for both summer and winter conditions; actually, there is a spread of about 1.2°F (−17.1°C) between winter and summer responses. The maximum comfort line, therefore, would be at about 77°F (25°C) in winter and 78°F (25.6°C) in summer.

A second defect in using optimum ETs as a comfort index is that the effect of relative humidity is exaggerated in importance. Later tests concluded that within a temperature range of 60 to 80°F (15.6 to 26.7°C) and a relative humidity range of 25 to 60%, evaporative body cooling is nearly constant. That is, a person at rest loses roughly the same amount of heat by perspiration at 25% relative humidity as he or she does at 60%, so long as the temperature does not exceed about 80°F (26.7°C). However, different temperatures within the 25 to 60% relative humidity range do produce varying sensations of warmth.

Another problem with using optimum ETs as a comfort index is that they ig-

nore the effect of radiant surfaces (Fig. 17.1-15). In summer, cool walls aid body heat dissipation because they absorb radiant body heat which cannot be removed by convection due to high air temperature. A person can therefore lose by radiation heat he or she would otherwise have to eliminate at greater stress, by perspiration. In winter, warmer wall temperatures compensate for low air temperature by preventing radiant body heat loss.

### 17.1.5 ASHRAE THERMAL COMFORT ENVELOPE

ASHRAE Standard 55 specifies generally acceptable comfort conditions for sedentary and slightly active, healthy, normally clothed people at altitudes from sea level to 7000 ft (2133.6 m).

The *comfort envelope* developed in this standard (Fig. 17.1-16) outlines acceptable thermal conditions for 80% of

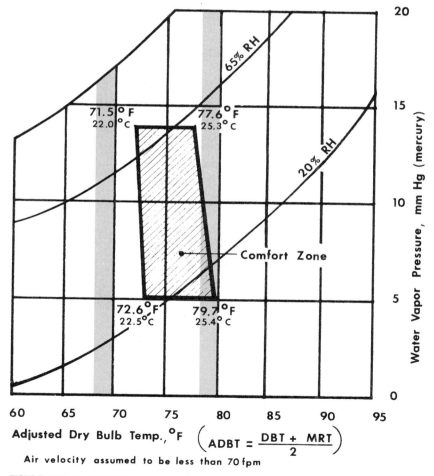

**FIGURE 17.1-16** ASHRAE comfort envelope. (Reprinted with permission of the American Society of Heating, Refrigerating, and Air-Conditioning Engineers.) (Refer to Figure 25.7.1 for metric equivalents.)

tested Americans and Canadians. It integrates the effect of DBT, MRT, relative humidity, and air movement. In this chart, the effects of DBT and MRT are combined in the *adjusted dry-bulb temperature* (ADBT) (operative temperature) scale, and the effect of relative humidity is shown by the water vapor pressure scale. Air movement is assumed to be less than 70 ft/min (21.34 m/min).

Adjusted DBTs and relative humidity levels should be controlled within the ranges shown in Figure 17.1-16. Interior surface temperatures in winter should not be more than 5°F (−15°C) below ambient air temperature to prevent radiant heat loss from bodies and interior spaces. In summer, walls should be kept slightly cooler than air temperature to aid in body heat dissipation.

## 17.1.6 THERMAL DESIGN CRITERIA

For maximum energy conservation, the latest recommendations of the U.S. Department of Energy (DOE) should be used as interior design temperatures. The DOE is working closely with such organizations as ASHRAE to develop better energy standards.

### 17.1.6.1 ASHRAE Energy Conservation Standard

ASHRAE Standard 90.1 provides performance-oriented criteria for the design of building envelopes and of mechanical and electrical systems. These standards refer to the comfort envelope parameters established in ASHRAE Standard 55 for winter and summer design criteria (see Fig. 17.1-16).

Low relative humidities generally accompany the heating season and high humidities are characteristic of the cooling season in most of the United States. Therefore, most designers operate near the bottom of the comfort zone in winter and near the top in summer. If energy conservation is a design objective, it can be further assumed that the design ADBT would be near 73°F (22.78°C) in winter and near 78°F (25.56°C) in summer.

### 17.1.6.2 HUD Standards

HUD's *Minimum Property Standards for Housing* adheres to the historic interior comfort criteria—70°F (21.1°C) DBT for winter and 75°F (23.9°C) DBT for summer ambient air. This standard does not make any specific requirements for relative humidity or radiant temperature.

## 17.1.7 CLIMATE AND WEATHER CONDITIONS

In order to design a building envelope and a mechanical system capable of maintaining interior comfort conditions, it is necessary to predict likely heat gain and loss. To do this, the designer must know not only what indoor temperature is desired, but also the outdoor temperatures that are likely to be encountered.

### 17.1.7.1 Outdoor Design Temperatures

Outdoor design temperatures are determined from weather records over a 25-year period. They are not absolutely the

**FIGURE 17.1-17** Representative Outdoor Design Temperatures, Degree Days, Cooling Hours, and Summer Outdoor Daily Range*

| City | Winter Design Temperature (97½%, °F) (°C) | Degree Days Heating (DDH) | Summer Design Temperature (2½%, °F) (°C) | Annual Cooling Operating Hours | Summer Outdoor Daily Range |
|---|---|---|---|---|---|
| Anchorage, AK | −18 (−27.8) | 10,860 (6016) | 68 (20) | 60 | 15 (−9.4) |
| Phoenix, AZ | 34 (1.1) | 1,680 (916) | 107 (41.7) | 1,930 | 27 (−2.8) |
| Los Angeles, CA | 43 (6.1) | 1,960 (1071) | 80 (26.7) | 460 | 20 (−6.7) |
| San Francisco, CA | 38 (3.3) | 3,040 (1671) | 77 (25) | 170 | 14 (−10) |
| Denver, CO | 1 (−17.2) | 6,150 (3399) | 91 (32.8) | 700 | 28 (−2.2) |
| Washington, DC | 17 (−8.3) | 4,240 (7338) | 91 (32.8) | 960 | 18 (−7.8) |
| Miami, FL | 47 (8.3) | 200 (93) | 90 (32.2) | 2,940 | 15 (−9.4) |
| Chicago, IL | −4 (−20) | 6,640 (3671) | 89 (31.7) | 650 | 20 (−6.7) |
| Wichita, KS | 7 (−13.9) | 4,640 (2560) | 98 (36.7) | 1,030 | 23 (−5) |
| Minneapolis, MN | −12 (−24.4) | 8,250 (4566) | 89 (31.7) | 590 | 22 (−5.6) |
| New York, NY | 15 (−9.4) | 4,880 (2693) | 87 (30.6) | 770 | 16 (−8.9) |
| Cincinnati, OH | 6 (−14.4) | 4,830 (2666) | 90 (32.2) | 870 | 21 (−6.1) |
| Tulsa, OK | 13 (−10.6) | 3,730 (2054) | 98 (36.7) | 1,230 | 24 (−4.4) |
| Rapid City, SD | −7 (−21.7) | 7,370 (4077) | 92 (33.3) | 610 | 28 (−2.2) |
| Houston, TX | 32 (0) | 1,410 (766) | 94 (34.4) | 1,880 | 18 (−7.8) |
| Salt Lake City, UT | 8 (−13.3) | 5,990 (3310) | 95 (35) | 770 | 32 (0) |
| Richmond, VA | 17 (−8.3) | 3,910 (2154) | 92 (33.3) | 1,010 | 21 (−6.1) |
| Spokane, WA | 2 (−16.7) | 6,770 (3743) | 90 (32.2) | 460 | 28 (−2.2) |
| Green Bay, WI | −9 (−22.8) | 8,100 (4482) | 85 (29.4) | 430 | 23 (−5) |

*Data derived from ASHRAE *Handbook of Fundamentals*, ASHRAE *Bin and Degree-hour Weather Data for Simplified Energy Calculations*, USAF AFM88-29, and NCDC *Climatography of U.S.*

highest or lowest temperatures for a given area, but rather sustained extremes. Figure 17.1-17 provides a sampling of design temperatures and other related information for some major U.S. cities.

### 17.1.7.2 Degree Days and Cooling Operating Hours

The severity of winter is expressed in total *degree days heating* (DDH), which is the sum of daily degree days when the temperature drops below 65°F (18.3°C). *Daily degree days* is the numerical difference between 65 and the average recorded temperature of that day. For instance, if there were 10 days in a season when the temperature dropped to 45°F (7.2°C), then the DDH

for each day is 20 and the total DDH for the season is 200.

*Cooling operating hours* measure the intensity of the annual cooling season. These are the hours an air conditioning unit must operate in order to maintain an indoor temperature of 78°F (25.56°C) for 97.5% of the time.

The higher the number of total degree days or annual cooling operating hours for the season, the greater the heating or cooling demand (see Fig. 17.1-17).

### 17.1.8 HEAT TRANSFER THROUGH MATERIALS

All building materials transfer heat to some extent. Heat transfer can be meas-

ured by (1) conductivity (k), (2) conductance (C) (k/t), (3) resistance (R) (1/C), and (4) transmittance (U) (1/R). Figure 17.1-18 shows the heat transfer coefficients of some selected materials for conductivity, conductance, and resistance.

Knowledge of how various materials resist the passage of heat enables designers to calculate heat gain and loss and to select materials that minimize heat flow.

### 17.1.8.1 Conductivity and Conductance

The degree to which a material transfers heat is rated by its conductivity (k), which is the heat flow conductance of 1 in. (25.4 mm) of a substance, or conductance (C), which is the heat flow con-

**FIGURE 17.1-18** Heat Transfer Coefficients of Representative Building Products[a]

| Material or Product | Conductivity (k) (Btu/ft · h · °F) (W/m · K) | Conductance (C)[e] (Btu/ft · h · °F · t) (W/m · K · t) | Resistance (R) (ft² · h · °F/Btu)(m² · K/W) | |
|---|---|---|---|---|
| | | | Per 1-in. (25.4 mm) Thickness (1/k) | Per Thickness Shown (1/C) |
| Concrete | 5.0–10.5 (8.7–18.2) | | 0.20–0.10 (0.12–0.1) | |
| Face brick | 9.00 (15.6) | | 0.11 (0.06) | |
| Hollow concrete block, 8 in. (203.2 mm) | | 0.90 (2.1) | | 1.11 (0.48) |
| Stucco | 5.00 (8.7) | | 0.20 (0.12) | |
| Metal lath and plaster, ¾ in. (19.1 mm) | | 7.70 (17.7) | | 0.13 (0.06) |
| Gypsum board, ½ in. (12.7 mm) | | 2.22 (5.1) | | 0.45 (0.2) |
| Plywood, ½ in. (12.7 mm) | | 1.60 (3.7) | | 0.62 (0.27) |
| Pine, fir, other softwoods | 0.74–1.12 (1.3–1.9) | | 1.35–0.89 (0.78–0.52) | |
| Oak, maple, other hardwoods | 1.06–1.25 (1.8–2.2) | | 0.94–0.80 (0.55–0.46) | |
| Asphalt shingles | | 2.27 (5.2) | | 0.44 (0.19) |
| Built-up roofing, ⅜ in. (9.5 mm) | | 3.00 (6.9) | | 0.33 (0.14) |
| Wood shingles | | 1.06 (3.9) | | 0.94 (0.41) |
| Vegetable board sheathing, ½ in. (12.7 mm) | | 0.76 (1.8) | | 1.32 (0.57) |
| Mineral wool batts, 3–4 in. (76.2–101.6 mm) | | 0.09 (0.2) | | 11.00 (4.8) |
| 5½–6½ in. (139.7 to 165.1 mm) | | 0.05 (0.1) | | 19.00 (8.7) |
| 6–7½ in. (152.4 to 190.5 mm) | | 0.05 (0.1) | | 22.00 (8.7) |
| 9–10 in. (228.6 to 254 mm) | | 0.03 (0.07) | | 30.00 (14.5) |
| Expanded polystyrene, 1 in. (25.4 mm) | | 0.20 (0.35) | | 5.00 (2.2) |
| Inside surface air film[b] | | | | 0.68 (0.29) |
| Outside surface air film[c] | | | | 0.17 (0.07) |
| Airspace ¾ in. (19.1 mm), nonreflective[d] | | 1.00 (2.3) | | 1.01 (0.43) |
| Airspace ¾ in. (19.1 mm), reflective[d] | | | | 3.48 |

[a]At 75°F (23.9°C) mean temperature, from ASHRAE *Fundamentals Handbook*.
[b]Heat flow horizontal, still air.
[c]Heat flow any direction, 15 mph (24.14 km/h) wind.
[d]Heat flow horizontal.
[e]k multiplied by thickness.

ductance of the entire thickness of a material. Therefore, C is k divided by the thickness.

## CONDUCTIVITY

Conductivity (k) is a measure of the number of Btus (kilojoules) per hour which can pass through 1 sq ft (0.09 m²) of a 1 in. (25.4 mm) thickness of a material when its surfaces vary in temperature by 1°F (−17.2°C). For example, a 1 × 1 ft (304.8 × 304.8 mm) by 1 in. (25.4 mm) thick block of polyurethane foam with a k of 0.16 would permit 0.16 Btus (0.168 kJ) to pass in 1 hour if one face of the block were at 74°F (23.3°C) and the opposite face were at 75°F (23.9°C).

## CONDUCTANCE

It is not meaningful to determine heat transfer per 1 in. (25.4 mm) of thickness for products that are not homogeneous or that have air cavities (such as hollow concrete block). Such products are instead given a conductance (C) rating based on the number of Btus transmitted in 1 hour through 1 sq ft (0.09 m²) of a stated thickness when the opposite surfaces vary in temperature by 1°F (−17.2°C). For instance, an 8-in. (203.2-mm)-thick hollow concrete masonry unit has a C-value of 0.90. Therefore, 0.90 Btus (0.95 kJ) is transferred in 1 hour if one face of the unit is at 74°F (23.3°C) and the other is at 75°F (23.9°C).

### 17.1.8.2 Resistance

Materials that are normally considered good insulators have low k- and C-values (see Fig. 17.1-18). However, because k and C for these numbers are stated as multidigit decimals smaller than 1.0, they are difficult to understand and compare. As a result, the use of *resistance* (R) numbers has increased as a method of rating building products and thermal insulations. An R-value is computed by taking the reciprocal of k, (1/k), or the reciprocal of C, (1/C). In the first case, the R-value must be qualified as being for a 1 in. (25.4 mm) thickness; in the second, it is simply the R-value of the product, whatever its thickness.

Resistance (R) represents the ability of a material to retard heat flow. R-values are also convenient because they are directly proportional to the insulating value and can therefore be added or subtracted easily in order to determine the thermal performance of a product or as-

sembly. For example, an 8 in. (203.2 mm) concrete block with a C-value of 0.90 (2.1) would have an R-value of 1.11 (0.48) (see Fig. 17.1-18). If a 1 in. (25.4 mm) slab of polystyrene with an R-value of 5.00 (2.2) were added, the assembly would have a total R-value of 6.11 (2.5), not counting surface films. The molecules of air clinging to the surfaces contribute some insulating value, so R-0.68 (0.29) should be added for still inside air and R-0.17 (0.07) for outside air at 15 mph (24.14 km/h). The final total for the assembly is R-6.96 (2.86).

Many insulating products are labeled with their R-value on the wrapper. The number given is usually the value of the product only, without the air spaces and films, which must be added to obtain the total installed resistance. Some insulations, such as multiple-layer foil, however, depend on air spaces and films for performance and can only be described in terms of installed resistance.

### 17.1.8.3 Transmittance

Transmittance (U) is the number of Btus transferred in 1 hour per 1 sq ft (0.09 m²) of a building assembly for each 1°F

(−17.2°C) temperature difference between inside and outside surfaces. The U-value is calculated by adding the R-values for all elements of the building assembly (R_T) and taking the reciprocal of the total (1/R_T). If the R_T for an assembly equals 20, then the U-value would be 0.05 ([1/20]). Sample U-value calculations are given in Figure 17.1-19.

### 17.1.8.4 Heat Storage Capability of Materials

Traditional heat flow calculations are based on the theory that design conditions are sustained for long periods and that heat flows continually from the warmer to the cooler side at a constant rate. This *steady-state* concept of thermodynamics is useful in designing heating or cooling systems where the building shell is not made of massive materials and the day-to-night outdoor temperature variation is not great.

In actual practice, heavy masonry and concrete buildings behave dynamically and do not generally lose or gain heat as rapidly as would be indicated by the steady-state theory. This is especially true in arid, sunny climates where there

**FIGURE 17.1-19** U-value Computation

| Frame Wall Construction | Insulated U and R_T Values R = (ft²·h·°F/Btu)(m²·K/W) | |
|---|---|---|
| | Construction | R |
| | 1. Outside Air Film | 0.25 | 0.11 |
| | 2. Wood Siding | 0.81 | 0.34 |
| | 3. ½″ Insul. Sheathing | 1.32 | 0.56 |
| | 4. Insulation | 11.00 | 4.68 |
| | 5. ⅝″ Gypsum Board | 0.56 | 0.24 |
| | 6. Inside Air Film | 0.68 | 0.29 |
| | R_T = | 14.62 | 6.22 |
| | 1/R_T = U = | 0.07 | 0.16 |

| Masonry Wall Construction | Insulated U and R_T Values R = (ft²·h·°F/Btu)(m²·K/W) | |
|---|---|---|
| | Construction | R |
| | 1. Outside Air Film | 0.25 | 0.11 |
| | 2. 4″ Common Brick | 0.80 | 0.34 |
| | 3. ½″ Cement Mortar | 0.10 | 0.04 |
| | 4. 4″ Lt. Wt. Conc. Block | 1.50 | 0.64 |
| | 5. ¾″ Air Space | — | — |
| | 6. ⅝″ Gypsum Board | 0.56 | 0.24 |
| | 7. Inside Air Film | 0.68 | 0.29 |
| | R_T = | 3.89 | 1.66 |
| | 1/R_T = U = | 0.07 | 0.16 |

are relatively warm days in winter and cool nights in summer. Under these conditions, massive buildings respond dynamically, acting as storers of heat, rather than as flow resistors, as do insulated lightweight buildings.

In summer, massive structures not only store heat but are able to slow down its movement into the building until late afternoon, when the air and structure start cooling off. In winter, heat movement out of the building during the night is slowed by the mass of the structure until the sun warms the walls, and by late afternoon heat actually moves inward. Designers can use this *thermal lag* to help maintain interior comfort conditions with a minimum of man-made equipment.

The principle of using heavy, dense materials as *thermal mass* was well known to the Indian builders of 1000-year-old Arizona cliff communities, as well as to modern adobe homeowners throughout the sunbelt states. Despite their limited insulating value, earth-sheltered and adobe homes remain surprisingly cool in summer and warm in winter. The effectiveness of a material as a thermal mass, that is, its ability to re-

tard heat flow, depends on its ability to store heat, which is measured by its *specific heat* and *heat capacity* (Fig. 17.1-20). Refer to Section 17.7.5 for a more detailed discussion of this subject.

The dynamic performance of concrete masonry and other massive shells can be accurately calculated by computers programmed with detailed winter and summer weather information. To perform such calculations manually, utilizing the steady-state approach, is difficult, but a reasonable approximation can be made using a dimensionless correction factor called the *M Factor*. The Brick Industry Association can provide more information about the thermal performance of masonry walls.

# 17.2 Building Design Recommendations

Buildings should regulate heat flow in a way which will minimize dependence on mechanical equipment and take advan-

tage of exterior environmental conditions. Energy-saving features can be incorporated economically in buildings if considered in the design phase.

## 17.2.1 SITE PLANNING

In locating a building on a site, the following should be considered: (1) general climate and microclimate, (2) terrain and landscape elements, and (3) orientation to the sun.

### 17.2.1.1 Climate

Climatic analysis can provide valuable data on which to base energy-efficient design. Such an analysis should include data on the amount and angle of sunshine that strikes the site, prevailing winds, temperature ranges, and precipitation.

The U.S. Weather Service and private meteorologists maintain regional data on these factors. However, natural terrain and proximity to bodies of water may substantially modify the weather information recorded for a particular region. To avoid siting a building in an unsuitable area, the climate variations of each site (microclimate) should be examined. For example, sites in warm climates which act as natural dead-air basins should be avoided. In such areas, temperatures may exceed those of surrounding areas by as much as 35°F (1.67°C). In other areas, prevailing winds and natural terrain can combine to produce unusual wind, snow massing, or rain conditions not typical of a particular climate. Air quality also should be studied to determine whether natural ventilation can be used for cooling.

### 17.2.1.2 Terrain and Landscape Elements

Terrain configurations and plant material can be used to control heat flow into and out of buildings. Refer to Section 22.1 for a discussion of landscape elements.

### 17.2.1.3 Orientation

In cold regions, the area of north-facing windows should be limited to reduce heat loss in winter. The area of north-facing walls can be reduced by berming and sloping roofs (Fig. 17.2-1). In hot areas, west wall and glass exposures should be minimized to reduce cooling loads from intense afternoon summer sun.

Buildings in cold regions should be sited so that most windows face south,

**FIGURE 17.1-20** Heat Storage Capabilities of Building Materials

| Material | Specific Heat, Btu/lb./°F (kJ/kg/°C) | Density, lb/cu ft (kg/m³) | Heat Capacity, Btu/cu ft/°F (kJ/m³/°C) |
|---|---|---|---|
| Water (40°F) | 1.000 (4.19) | 62.5 (1001) | 62.5 (4192) |
| Steel | 0.120 (0.5) | 489.0 (7883) | 58.7 (3937) |
| Cast iron | 0.120 (0.5) | 450.0 (7208) | 54.0 (3622) |
| Copper | 0.092 (0.39) | 556.0 (8906) | 51.2 (3434) |
| Aluminum | 0.214 (0.9) | 171.0 (2739) | 36.6 (2455) |
| Basalt | 0.200 (0.84) | 180.0 (2883) | 36.0 (2414) |
| Marble | 0.210 (0.88) | 162.0 (2595) | 34.0 (2280) |
| Concrete | 0.220 (0.92) | 144.0 (2307) | 31.7 (2126) |
| Asphalt | 0.220 (0.92) | 132.0 (2114) | 29.0 (1945) |
| Ice (32°F) | 0.487 (2.04) | 57.5 (921) | 28.0 (1878) |
| Glass | 0.180 (0.75) | 154.0 (2467) | 27.7 (1858) |
| White oak | 0.570 (2.39) | 47.0 (752.9) | 26.8 (1797) |
| Brick | 0.200 (0.84) | 123.0 (1970) | 24.6 (1650) |
| Limestone | 0.217 (0.91) | 103.0 (1650) | 22.4 (1502) |
| Gypsum | 0.260 (1.09) | 78.0 (1249) | 20.3 (1361) |
| Sand | 0.191 (0.8) | 94.6 (1515) | 18.1 (1214) |
| White pine | 0.670 (2.81) | 27.0 (432.5) | 18.1 (1214) |
| White fir | 0.650 (2.72) | 27.0 (432.5) | 17.6 (1180) |

**FIGURE 17.2-1** Sloping roofs act as wind deflectors.

southeast, or southwest to take advantage of the warming sun. Orientation that will buffer against hot breezes is desirable in hot climates. Prevailing summer breezes which can provide ventilation and natural cooling should be considered when siting a building or locating windows.

## 17.2.2 BUILDING DESIGN

Energy use is the greatest impact a building has on the environment. Therefore, buildings should be designed to be energy efficient. Until recently, a major energy-saving strategy in building design was to make buildings as airtight as possible. To this end, high insulation values, thermally efficient windows, and air infiltration barriers were employed. Recently, these strategies have come into question. Current thinking is leaning toward more flexibility, giving occupants control of their individual environment to a much greater extent than was thought reasonable in the recent past. Today, windows are more likely to be operable, and individual access to HVAC controls placement and design often allows individuals to adjust the heating and cooling environments to suit their individual preferences.

In addition, an integrated design approach, in which the interaction between building orientation, windows, lighting, HVAC systems, and selection of energy-efficient building equipment, such as appliances, is being used, takes advantage of the inherent energy savings that result. For example, cooling loads can be minimized through careful building design, glazing selection, lighting design, and landscaping. Energy can be saved through the selection of energy-efficient lighting, HVAC equipment, and appliances. Where practicable, equipment should be selected that obtains its energy from renewable sources.

In addition, thermally efficient buildings can be achieved through room or-ganization and effective building envelope design.

### 17.2.2.1 Room Organization

In areas subject to either extreme heat or cold, compact design reduces weather-exposed areas and reduces heat flow. Garages, closets, corridors, and other service spaces can be located against exterior walls to act as insulating buffers.

Inactive living spaces such as kitchens and laundry areas, which contain heat-producing appliances, should be located on north and east sides. This location permits wintertime use of their excess heat for warming and permits heat dissipation in summer.

Except in extreme southern locations, active living and working spaces should face south or southwest to take advantage of warming afternoon sun in winter. Such spaces should be protected against overheating in summer by external shading devices such as louvers, overhangs, awnings, and trees (see Section 12.2). Interior shading with blinds and drapes is less effective but acceptable. Rooms likely to remain vacant for long periods should be located where they can be closed off readily. Lower ceiling heights can also reduce the volume of air to be heated or cooled. Built-in furnishings should be located away from supply and return air grilles to prevent obstruction of internal air flows.

### 17.2.2.2 Building Envelope

For conventionally heated and cooled buildings, the objective of heat control is to block the flow of heat through the ceilings, walls, and floors that surround habitable areas. This heat flow can be controlled by:

1. Constructing wall, floor, and roof assemblies using materials that are inherently good insulators to minimize heat flow out of the building in winter and into the building in summer.
2. Using solid masonry or another dense material that has the capacity to store heat and slow down heat flow.
3. Installing loose fill, flexible, rigid, reflective, or foamed-in-place insulation. In some climates, using materials with high thermal capacities will sufficiently slow heat gain or loss, but in most locations, adequate insulation of the building envelope is essential to control heat flow. Refer to Section 7.3 for a discussion and recommendations concerning insulation.

4. Earth-sheltered building design, in which all, or a portion, of a building is partially or wholly underground, takes advantage of the stability of underground temperatures.

### FLOORS

Where solar radiation penetrates interior spaces, floors can act as *heat sinks*. Heat sinks are building components consisting of dense, heavy materials such as concrete or masonry. They are desirable in cold climates because such materials absorb and store heat for long periods before releasing it to the interior by radiation and convection.

The floors of areas above unheated crawl spaces or basements and slabs on ground should be insulated as recommended in Section 7.3.

### WALLS

Heat transmission through wall assemblies can be reduced by selecting materials that are inherently good insulators or by adding insulation (Section 7.3).

Wind-deflecting devices such as berms, walls, and fences can also be employed to reduce infiltration along exterior walls exposed to cold winds by steering the wind away from the walls.

### ROOFS

Heat transmission through a roof can be reduced by:

1. Insulation (see Section 7.3)
2. Shiny or light-colored roof materials to reflect heat
3. Ventilation of attic spaces
4. Increasing the thermal mass to achieve a time lag in the heat flow
5. Louvered rather than solid overhangs to free heated air trapped beneath eaves
6. Steeply pitched roofs on the windward side of the building to deflect cold winds away from the rest of the roof and reduce the amount of roof exposed to this wind (see Fig. 17.2-1)

## 17.3 Heat Loss and Heat Gain

In a typical building, windows and doors lose and gain more heat than any other part of the structure. In some climates, this heat loss or gain can account for as much as 70% of the total heating load and 50% of the cooling load, according to a study by the Department of Housing and Urban Development (HUD).

Most heat is lost through *infiltration* (air leakage) through cracks around the frame and *conduction* from warm air coming in contact with the cold window and door surfaces. Most heat is gained through *radiation*, which is transmission of heat by invisible electromagnetic waves through glass. Constructing window and door openings so that unwanted heat loss and heat gain are minimized will conserve energy, create a more comfortable indoor environment, and reduce building operation costs.

Many of the strategies suggested in this section for reducing heat gain and heat loss have been suggested by the National Institute of Standards and Technology publication *Window Design Strategies to Conserve Energy*.

Heat loss through windows can be reduced by tightly sealing openings, using multiple glazing (storm sash or insulating glass), and installing insulating shutters or tightly fitting roll shades.

Besides the other measures discussed here, heat loss through doors is often reduced by vestibules or storm doors.

In most parts of the country, these methods can reduce heating costs, minimize condensation on interior surfaces, and improve interior comfort by reducing drafts and radiant body heat loss. In Zones IV and V of Figure 7.3-2, multiple glazing may not be justified, and its use should therefore be decided on a case-by-case basis, taking into account fuel and material costs.

Heating and air conditioning equipment is sized on the basis of heat loss and heat gain calculations for the particular area in which the building is situated. Such calculations also provide a measure of the thermal efficiency of a building.

## 17.3.1 DEFINITIONS

The following terms are used in performing a heat loss or heat gain analysis:

*Design Temperature Difference*: The difference between the outside design temperature ($t_o$) and the inside design temperature ($t_i$).
*Heat Loss*: The amount of heat flow in Btus (kilojoules [kJ]) per hour or per season, from inside a structure to outside, at given design conditions.
*Indoor Design Temperature*: The ideal comfort temperatures that are to be maintained inside the structure in winter and summer. These should correspond with the ASHRAE comfort envelope unless DOE has developed more stringent guidelines.
*Infiltration*: The leakage of heat into or out of a building through cracks in the building structure, especially around windows and doors (Fig. 17.3-1).
*Latent Heat Gain*: Heat which is introduced into a structure by moisture in the air and which must be removed by the cooling system. The latent heat gain is commonly calculated as 30% of the *sensible heat gain*.
*Outdoor Design Temperatures*: High and low temperatures that can be expected to occur regularly in a particular area (see Fig. 17.1-17).
*R-Value*: A material's resistance to the flow of heat; the larger the R-value, the greater the resistance and the better the insulating value (See Section 7.3).
*Sensible Heat Gain:* The amount of heat gain in Btus (kJ) per hour that can be recorded as a rise in temperature.
*Transmittance (U)*: The number of Btus (kJ) per hour transferred through 1 sq ft (0.09 m$^2$) of a building assembly (such as a wall) for each 1°F (−17.2°C) temperature difference between inside and outside surfaces. It is calculated by adding the R-values for all elements of the building assembly ($R_T$) and taking the reciprocal ($1/R_T$) of the total (Fig. 17.3-2).

## 17.3.2 HEAT LOSS AND HEAT GAIN FORMULAS

Heat flows into or out of a building:

1. Through exterior walls, roofs, and glass areas
2. At the perimeter of a building's ground floor
3. Through floors over unheated spaces
4. By infiltration around windows and doors

The following heat loss and heat gain calculation methods are intended to be used for small building design, but they are applicable to buildings of all types and sizes. Calculating heat gain and heat loss and determining HVAC system design criteria for commercial installations should be done by a licensed mechanical engineer.

Further information on calculating heat loss and design criteria for the sizing and locating of ducts and registers in small residential buildings may be obtained from the National Association of Home Builders (NAHB) Research Foundation and from the publications of the Air Conditioning Contractors of America (ACCA) and the Sheet Metal and Air Conditioning Contractors' National Association (SMACNA).

The following calculation procedures assume use of data and tables in this book, which may not be up-to-date, for determining building heat loss. In an actual case, a designer should obtain and use the latest available data.

**FIGURE 17.3-1** Window and Door Infiltration Values

| Window Type | Infiltration Air Flow, cu ft/ft (m³/m) of Crack* | | |
| --- | --- | --- | --- |
| | 10 mph (16.1 km/h) | 20 mph (32.2 km/h) | 30 mph (48.3 km/h) |
| **Double-hung wood frame** | | | |
| Non-weatherstripped | 21 (2) | 59 (5.5) | 104 (9.7) |
| Weatherstripped | 13 (1.2) | 36 (3.3) | 63 (5.9) |
| **Double-hung metal frame** | | | |
| Non-weatherstripped | 47 (4.4) | 104 (9.7) | 170 (15.8) |
| Weatherstripped | 19 (1.8) | 46 (4.3) | 76 (7.1) |
| **Residential metal casement** | 18 (1.7) | 47 (4.4) | 74 (6.9) |
| **Wood or metal door** | | | |
| Non-weatherstripped | 69 (6.4) | 154 (14.3) | 249 (23.1) |
| Weatherstripped | 19 (1.8) | 51 (4.7) | 92 (8.5) |

*Reduce values by one-third when storm window (or door) is added to weatherstripped construction, by one half when added to nonweatherstripped construction.

**FIGURE 17.3-2** U-value Calculation Procedure

| Building Part | Construction Materials | R-Value $(ft^2 \cdot h \cdot °F/Btu)$ | $(m^2 \cdot K/W)$ |
|---|---|---|---|
| Roof/ceiling | Outside air film | 0.17 | 0.07 |
| | Shingles | 0.44 | 0.19 |
| | Building paper | 0.06 | 0.03 |
| | Plywood, $\frac{1}{2}$ in. | 0.62 | 0.26 |
| | Attic air film | 0.61 | 0.26 |
| | Insulation | 19.00 | 8.1 |
| | Gypsum board, $\frac{1}{2}$ in. | 0.45 | 0.19 |
| | Inside air film | 0.61 | 0.26 |
| | Total R-Value ($R_T$) | 21.96 | 9.35 |
| | **U-Value ($1/R_T$)** | **0.045** | **0.11** |
| Wall | Outside air film | 0.17 | 0.07 |
| | Siding, wood $\frac{1}{2}$ in. × 8 in. lapped | 0.81 | 0.34 |
| | Sheathing, plywood, $\frac{1}{2}$ in. | 0.62 | 0.26 |
| | Insulation | 11.00 | 4.68 |
| | Interior finish gypsum board, $\frac{1}{2}$ in. | 0.45 | 0.19 |
| | Inside air film | 0.68 | 0.29 |
| | Total R-Value ($R_T$) | 13.73 | 5.8 |
| | **U-Value ($1/R_T$)** | **0.073** | **0.17** |
| Header joist | Outside air film | 0.17 | 0.07 |
| | Siding | 0.81 | 0.34 |
| | Sheathing | 0.62 | 0.26 |
| | Header, wood $1\frac{1}{2}$ in. | 1.88 | 0.8 |
| | Insulation | 11.00 | 4.68 |
| | Inside air film | 0.68 | 0.29 |
| | Total R-Value ($R_T$) | 15.16 | 6.45 |
| | **U-Value ($1/R_T$)** | **0.066** | **0.15** |
| Sill | Outside air film | 0.17 | 0.07 |
| | Siding | 0.81 | 0.34 |
| | Sheathing | 0.62 | 0.26 |
| | Sill, wood $5\frac{1}{2}$ in. | 6.88 | 2.93 |
| | Inside air film | 0.68 | 0.29 |
| | Total R-Value ($R_T$) | 9.16 | 3.9 |
| | **U-Value ($1/R_T$)** | **0.109** | **0.26** |
| Foundation | Outside air film | 0.17 | 0.07 |
| | Concrete block, 8 in. | 1.11 | 0.47 |
| | Insulation | 5.00 | 2.13 |
| | Interior finish gypsum board, $\frac{3}{8}$ in. | 0.32 | 0.14 |
| | Inside air film | 0.68 | 0.29 |
| | Total R-Value ($R_T$) | 7.28 | 3.1 |
| | **U-Value ($1/R_T$)** | **0.137** | **0.32** |

### 17.3.2.1 Heat Loss Calculation

The following steps can be followed to calculate heat loss:

1. Determine the winter outdoor design temperature ($t_o$) (see Fig. 17.1-17) and the desired indoor design temperature ($t_i$) from the latest ASHRAE comfort envelope or use the HUD guideline temperature. The design temperature difference is equal to the difference between the established indoor and outdoor design temperatures ($t_i - t_o$).

2. Calculate the U-values for roof wall and glass areas by summing the R-values for all parts of such an assembly, including air films, ($R_T$), and taking the reciprocal ($1/R_T$) of those figures (see Fig. 17.3-2).

3. Calculate the surface area (A) of glass and all similarly constructed roof, wall, and floor areas.

4. Calculate building heat loss ($H_c$) through exterior roof and wall construction and through exterior glass by multiplying the area (A) by its computed U-value, by the design temperature difference ($t_i - t_o$).

5. Calculate the floor heat loss ($H_f$) by the same method used for walls, roof, and glass, with the following qualifications:
   - If the heating unit is in the basement and ducts or pipes are uninsulated, no floor loss calculation is necessary.
   - For floors over unheated basements, use one-third of the design temperature difference.
   - For floors over unvented crawl spaces with insulated crawl space walls, use one-half of the design temperature difference.
   - For floors over vented crawl spaces or other open spaces, use the full design temperature difference.

6. Calculate the building perimeter (P) in feet (meters) if the floor is of slab construction.

7. Calculate the floor slab edge heat loss ($H_e$) by multiplying a factor of 0.55 by the slab perimeter (P) in feet (meters) if the slab edge is insulated, or a factor of 0.8 by (P) in feet (meters) if the slab edge is uninsulated, by the design temperature difference ($t_i - t_o$). Use °F when using feet and °C when using meters.

8. Calculate the length of cracks (L) in feet (meters) around window and door openings.

9. Select the infiltration air flow volume in cubic feet per hour (cfh) ($m^3/h$) per foot (meter) of crack (see Fig. 17.3-1).

10. Calculate the infiltration heat loss ($H_i$) for each opening by multiplying the length of crackage (L) by 0.018 times the per foot (meter) infiltration volume in cfh ($M^3/h$) by the design temperature difference ($t_i - t_o$) in °F for cfh and °C for $M^3/h$.

11. Calculate the total building heat loss ($H_T$) by summing $H_c$, $H_i$, and $H_e$ or $H_f$ if appropriate.

### 17.3.2.2 Heat Gain Calculation

The following steps can be followed to calculate heat gain:

1. Determine the winter outdoor design temperature ($t_i$) (see Fig. 17.2-17) and the desired indoor design tem-

perature ($t_i$) from the latest ASHRAE comfort envelope or the HUD guideline temperature. The design temperature difference is equal to the difference between the established indoor and outdoor design temperatures ($t_i - t_o$).

2. Calculate the U-value for the roof and walls by summing the R-values for all parts of each assembly ($R_T$) and taking the reciprocal ($1/R_T$) of that value (see Fig. 17.3-2).

3. Calculate the surface area (A) of all similarly constructed roofs, walls, and floors.

4. Calculate the heat gain through unglazed (opaque) roof and wall sections ($H_O$) by multiplying the calculated area (A), by the calculated U-value, by the design equivalent temperature difference (Fig. 17.3-3).

5. Calculate the heat gain through glass areas ($H_g$) as follows:
   - Locate the glass according to its building face: north, south, east, or west.
   - Calculate the surface area of each particular type of glass (single, double, or heat-absorbing).
   - Select the appropriate winter glass load factor from tables in the latest ASHRAE handbook.
   - Multiply the glass area (A) by the heat gain factor for the total window heat gain in Btus (kJ) per hour.
   - If the glass is partially or completely shaded by a permanent shading device such as an overhang, the shaded portion is treated as north-facing glass. Unshaded portions are considered separately. The shade line is computed by multiplying the overhang width by the shade line factor found in the ASHRAE *Handbook*. Glass above the line is in shade, and glass below the line is in sun.

6. Heat gain from infiltration ($H_i$) is normally insignificant in summer and can be ignored.

7. Estimate the sensible heat gain from people ($H_p$) at 225 Btus (237.4 kJ) per hour per person. For residences, estimate the number of occupants at twice the number of bedrooms.

8. Estimate appliance loads ($H_a$) at 1200 Btus (1266.1 kJ) per hour.

9. Calculate the total sensible heat gain ($H_{TS}$) by summing $H_o$, $H_g$, $H_p$, and $H_a$.

10. Estimate total latent heat gain ($H_{TL}$) as one-third $H_{TS}$.

11. Calculate total heat gain as the sum of total latent heat gain ($H_{TL}$) and total sensible heat gain ($H_{TS}$).

## 17.4 Commissioning

*Commissioning* is the process of ensuring that a building's HVAC and electrical systems comply with the contract documents and applicable codes and meet the owner's operational requirements. The process is used for large and small, simple and complex commercial, industrial, and institutional projects.

Commissioning is often specified to be accomplished by the general contractor but is also sometimes done by mechanical and electrical engineers. Extensive system tests and reports of the conditions observed are usually required.

SMACNA's *HVAC Systems Commissioning Manual* is a guide to the process and contains sample commissioning specifications, reports, and checklists.

## 17.5 HVAC Controls

### 17.5.1 OPERATING CONTROLS

Operating controls work at several levels, from overall HVAC system controls to individual unit controls, such as those for windows, room thermostats, and boiler and furnace controls.

**FIGURE 17.3-3** Design Equivalent Temperature Differences

| Design Temperature, °F[a] | 85 | | 90 | | | 95 | | | 100 | | 105 | 110 |
|---|---|---|---|---|---|---|---|---|---|---|---|---|
| Daily Temperature Range[b] | L | M | L | M | H | L | M | H | M | H | H | H |
| **Walls and doors** | | | | | | | | | | | | |
| Frame and veneer-on-frame | 17.6 | 13.6 | 22.6 | 18.6 | 13.6 | 27.6 | 23.6 | 18.6 | 28.6 | 23.6 | 28.6 | 33.6 |
| Masonry walls, 8 in. block or brick | 10.3 | 6.3 | 15.3 | 11.3 | 6.3 | 20.3 | 16.3 | 11.3 | 21.3 | 16.3 | 21.3 | 26.3 |
| Wood doors | 17.6 | 13.6 | 22.6 | 18.6 | 13.6 | 27.6 | 23.6 | 18.6 | 28.6 | 23.6 | 28.6 | 33.6 |
| **Ceilings and roofs** | | | | | | | | | | | | |
| Ceilings under naturally vented attic or vented flat roof, dark | 38.0 | 34.0 | 43.0 | 39.0 | 34.0 | 48.0 | 44.0 | 39.0 | 49.0 | 44.0 | 49.0 | 54.0 |
| light | 30.0 | 26.0 | 35.0 | 31.0 | 26.0 | 40.0 | 36.0 | 31.0 | 41.0 | 36.0 | 41.0 | 46.0 |
| Built-up roof, no ceiling, dark | 38.0 | 34.0 | 43.0 | 39.0 | 34.0 | 48.0 | 44.0 | 39.0 | 49.0 | 44.0 | 49.0 | 54.0 |
| light | 30.0 | 26.0 | 35.0 | 31.0 | 26.0 | 40.0 | 36.0 | 31.0 | 41.0 | 36.0 | 41.0 | 46.0 |
| Ceilings under unconditioned rooms | 9.0 | 5.0 | 14.0 | 10.0 | 5.0 | 19.0 | 15.0 | 10.0 | 20.0 | 15.0 | 20.0 | 25.0 |
| **Floors** | | | | | | | | | | | | |
| Over unconditioned rooms | 9.0 | 5.0 | 14.0 | 10.0 | 5.0 | 19.0 | 15.0 | 10.0 | 20.0 | 15.0 | 20.0 | 25.0 |
| Over basement, heated crawl space, or concrete slab on ground | 0 | 0 | 0 | 0 | 0 | 0 | 0 | 0 | 0 | 0 | 0 | 0 |
| Over unheated crawl space | 9.0 | 5.0 | 14.0 | 10.0 | 5.0 | 19.0 | 15.0 | 10.0 | 20.0 | 15.0 | 20.0 | 25.0 |

[a]Interpolate for values between those listed.

[b]Daily Temperature Range (see Fig. 15.2-11):

L (Low) calculation value: 12   M (Medium) calculation value: 20   H (High) calculation value: 30
Range: less than 15°   Range: 15 to 25°   Range: more than 25°

(Refer to Figure 25.7-1 for metric equivalents.)

### 17.5.1.1 Information System Controls

Computer-controlled information systems have revolutionized the control of HVAC systems and produced the innovations variously known as *smart buildings* and *intelligent buildings*. Using information systems, it is possible to control a building's HVAC systems at great distances or on site. Such control enables a building's cooling system to virtually shut down during night hours and restart in time in the morning to make the building comfortable when occupants arrive for the next day's work. Such systems can be used to provide complete environmental control, opening windows when breezes make that reasonable, lowering awnings as the sun strikes the building, and providing fresh air intakes as the need arises. The computer monitors outdoor and indoor temperatures and adjusts the HVAC system to compensate for changes. Delivery of heat can be reduced as the sun falling on building walls raises the heat in the space, and cooling can be provided in warm building areas while heat is provided in cool spaces, all without human interference.

Some more sophisticated control systems can actually learn while operating, making them able to predict such things as usage patterns and allowing them to make adjustment to accommodate such changes without interference from occupants.

### 17.5.1.2 Individual Unit Controls

Traditionally, access by occupants to individual unit controls was limited or prohibited altogether. Temperatures on thermostats, for example, were set by building management personnel, and access was denied to occupants. Windows were sealed in the closed position. It was believed that allowing occupants such control would increase energy use and lead to uncomfortable or unhealthy conditions by admitting pollen, smoke, and air pollutants. Current thinking is somewhat changing that attitude. Under some circumstances, occupants are being allowed to control thermostats in their immediate work area and to open or close windows as they see fit. Of course, such options must be limited where exterior high humidity or extreme cold dictate otherwise, but the option is appearing in more and more new buildings.

Some level of environmental control is transferred to building occupants by introducing *switches* accessible by them. Switches may be either electrically or manually controlled. The hand crank on an operable window, for example, is a switch in this context, as is a room thermostat. Switch settings can be limited or unlimited. For example, an occupant may be able to change a thermostat setting only within certain temperature limits; a window may be opened only to a fixed degree. Switches may be provided for operation of awnings and opaque drapes. Information system controls can be established to limit individual unit control switches. For example, the computer might not permit raising of an awning if the result would be full exposure to sunlight on the shaded wall.

### THERMOSTATS

A *thermostat* is a switch used to measure and control the temperature in a space. It may be located in the space, in the supply system (duct or water pipe), or in the terminal unit or units serving the space. Thermostats in the space or in an air stream are usually operated by a mercury-type electric switch actuated by a temperature-sensitive spring coil. Thermostats in water lines may be either electric or pneumatic.

Room-type thermostats may be simple devices that have a dial to set the temperature desired, a thermometer to indicate the temperature, and a spring coil and switch to open and close the circuit. Other room-type thermostats are complicated time clock- or computer-operated devices that can produce different temperatures at different times of the day and even on different days of the year. ASHRAE Standard 90.2 requires that thermostats be capable of settings between 55° and 85°F (13° and 29°C).

Wall-mounted thermostats should be located on an interior partition. They should not be located near lights, cooking or heating appliances, radiators, convection units, or air supply registers so that the heat and cold air associated with them does not cause a false thermometer reading.

### BOILER AND FURNACE CONTROLS

A HVAC system contains operating controls, limit controls, and interlocks, which are designed to provide a preset controlled sequence of operation and to protect furnaces, boilers, chillers, air handlers, and other equipment from damage by taking corrective action when failure occurs or the equipment is not operating in a prescribed manner. *Operating controls* respond to sensors like thermostats and pressure-actuated switches and send a signal to *limit controllers*, which are preset and respond to operating control signals to open or close valves, dampers, or relays. *Interlocks* link parts of a HVAC system so that a change in one portion affects the operation of other portions as well. In large modern systems, controls are set and operated by computer.

### 17.5.2 SOUND CONTROL

Heating and cooling equipment and the flowing air and liquids they produce and move about are sources of noise vibration—whether they are part of a small individual package unit or large central systems. Heating and cooling equipment can have a wide range of noise levels. It is generally more economical to install a quietly operating (even if more expensive) unit than to attempt to reduce the noise level of a less expensive unit by acoustical construction. In residential occupancy, the greatest air conditioning load occurs between 5:00 P.M. and 7:00 P.M. when exterior and interior conditions are at extremes. Some systems are barely adequate for this peak load time and therefore must operate at full capacity late into the night, when the surrounding noise level is greatly reduced and equipment noise is generally annoying. Therefore, in residences, it is particularly important to select equipment that has greater than adequate capacity, particularly for fans and pumps. Reducing fan speeds only slightly can produce noise level reductions that will more than justify the cost of additional power.

Vibration from mechanical equipment usually is not a significant problem when the equipment is mounted on a concrete slab placed on ground located in a basement, in a crawl space, in a utility room, or in an attached garage. However, on suspended concrete slab floors, structure-borne vibration from equipment can be transmitted great distances. Equipment placed on wood-framed construction also will cause acoustical problems unless it is isolated from the structural framing.

Whenever possible, mechanical equipment should be mounted resiliently, and all related piping, refrigerant lines, condensation lines, electrical connections, and duct systems should be flexibly mounted to the units to prevent transfer of vibrations to walls and floor and ceiling assemblies. Vibration isolation pads should be used between the equipment and the supporting structure. Flues should be isolated from walls by an air gap, especially where they pass through the walls of rooms, such as bedrooms, that are intended to be quiet.

When heating and cooling systems are large or complex, whole machinery rooms can be floated or isolated from the building, or special foundations can be constructed to isolate individual pieces of equipment. Floating floor systems are available to isolate entire floor areas acoustically.

There should be the amount of clear space on each side of a heating unit that is recommended by its manufacturer or required by code, whichever is larger. Locating combustible materials too close to furnaces is a potential fire hazard.

The mechanical equipment room should be buffered from the rest of the building by design. Areas containing furnaces, pumps, compressors, laundries, and similar equipment, and areas enclosing elevator shafts, incinerator chutes, and other potential sources of noise, should be located both horizontally and vertically as far as possible from quiet areas such as bedrooms and living rooms.

In single-family dwellings, low-rise apartments, townhouses, and small commercial buildings, an exterior wall location is usually best acoustically (Fig. 17.5-1a). Unfortunately, in many such buildings, unit forced air heating and cooling equipment is placed in a closet, an alcove, or a utility room near the center of the living unit or working space to produce shorter and therefore more economical duct runs (Fig. 17.5-1b), thus placing major noise sources in the center of the living or working space. In addition, a single air return is located near the center of the floor plan; often a grille in the wall or door of the furnace enclosure is all that is provided. The need for quiet must be weighed against the economics of central location. For air plenum systems, the furnace must be located in the center of the space and other means found to reduce noise levels.

One way to confine mechanical equipment noise in small buildings is to provide three solid walls around the furnace

**a**

**b**

**FIGURE 17.5-1** From an acoustical point of view, a mechanical equipment room should be located on an exterior wall (a) rather than in a central location (b).

room with the entry door to the outside for servicing (see Fig.17.5-1a). In this case, the service door should be a solid core one and should be gasketed around the perimeter to render it airtight.

Mechanical systems are also a source of exterior noise. While a cooling coil is located inside at the furnace plenum, a water- or air-cooled condensing unit must be located outside. Because ordinances sometimes govern the allowable noise level at property lines, condensers often must be placed adjacent to the building in front or rear locations where the distance to the next building is greatest. This creates an exterior noise source at the exterior wall. These units should be as far as possible from windows, porches, patios, and neighboring buildings. An alternate location for the condensing unit would be on a vibration-isolated platform on the roof.

Unit air conditioners in the exterior walls of multifamily residences should be located as far as possible from adjacent apartments so that operation of one unit does not unduly disturb neighbors who do not operate their units at the same time.

Diffusers are a major source of duct noise. Air velocity and the design of diffusers should be considered to produce outlets that do not whistle or cause other

annoying noises. Decreasing the air velocity at the outlet diffuser by 50% will reduce the sound level generated by the grille from 15 to 20 dB, measured 3 to 5 ft (0.9 to 1.5 m) from the diffuser. Wide-angle directional diffusers generate the most noise, while wire mesh or perforated metal diffusers with little air spread capacity are the quietest.

Ducts should have an adequate thickness to minimize vibration, and they should be isolated from mechanical equipment with flexible collar connections to prevent transmission of vibrations (Fig. 17.5-2). The space around ducts should be caulked tightly for isolation where they pass through sound-conditioned construction. Mufflers used in ducts near equipment rooms will absorb transmitted noise (Fig. 17.5-3).

Careful planning and installation of duct systems is critical in controlling sound transmission. Ducts act like speaking tubes in transmitting noises. Noise may result from any of the following:

- Equipment such as fans and motors
- Sound transmitted through the walls into the duct and then conducted through the duct

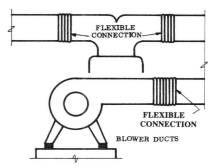

**FIGURE 17.5-2** Flexible connections reduce the transmission of vibrations from equipment to the duct system.

**FIGURE 17.5-3** Duct mufflers are effective in reducing noise transmitted from a furnace.

- Sound created within the duct by air rushing through the system or by expansion and contraction of the duct itself

Sound does not necessarily have to enter or leave a duct through an opening or diffuser. It can be transmitted through walls, particularly where the duct passes vertically through living spaces (Fig. 17.5-4). Planning can minimize the sound transmittance from one room to another through the open air paths of the duct system.

Duct outlets should not be located back-to-back between rooms, because this practice practically eliminates any benefit obtained from sound-conditioned walls (Fig. 17.5-5). A 6 ft (1.8 m) length of fibrous duct material adjacent to the outlet generally will reduce furnace noises and will prevent short-circuiting sound through ductwork between rooms (Fig. 17.5-6).

Duct insulation and duct mufflers will help reduce furnace noises and will help prevent short-circuiting sound through ductwork between rooms.

### 17.5.3 SMOKE AND FIRE CONTROL

Buildings are required by building codes to be divided into fire and smoke zones of a size dictated by the code. The size of the zone is controlled by the building type and the construction class of the structure. Refer to Section 1.3 for a discussion of these classifications. Fire-rated or smoke-rated walls, as applicable,

**FIGURE 17.5-4** A duct system can conduct noise between units. Ducts should not be housed in common walls or floor and ceiling systems between dwelling units.

**FIGURE 17.5-5** Duct outlets placed back-to-back will nullify the effect of sound-isolating walls.

**FIGURE 17.5-6** Fibrous insulation in the duct, especially near outlets, helps reduce transmitted noise.

are required to surround these zones. Dampers are required where ducts cross the rated walls to prevent fire or smoke penetration, as applicable, through the rated walls by means of the ductwork in the event of fire or other smoke generation. Where a wall is both a fire and a smoke barrier, combination smoke/fire dampers can be used.

Underwriters Laboratories, Inc. (UL) dictates the design and type of damper required, and installation requirements for them, for each specific hazard. Codes require compliance with UL assemblies and may have additional requirements as well.

Fire dampers are connected electrically to a fire alarm system containing fire detectors, or may have fusible links to permit attached springs to close them in the event of fire.

Smoke dampers are tied electrically to a smoke alarm system consisting of smoke detectors.

Dampers similar to smoke dampers can also be used to prohibit passage of other gases or radiation in the event of leaks or accidents.

Some dampers must be manually reset after operation. This is an expensive and time-consuming chore. More expensive systems are also available in which the dampers can be opened electrically after an emergency is over. Obviously, dampers containing fusible links cannot be closed electronically. The links must be replaced.

## 17.6 HVAC Delivery Systems

Virtually every habitable building in the United States needs at least a minimal mechanical heating system, including those where much of the heating load is carried by a passive or active solar energy system (see Section 17.6.3). Most buildings also need a mechanical cooling system for occupant comfort and acceptable working and living conditions. A cooling system should definitely be considered in locations where the humidity is often high. In addition, habitable buildings need a method of ventilating toilet rooms, shower rooms, laboratories, bathrooms, and kitchens. Buildings also need a source of fresh air to furnish combustion air for furnaces, boilers, fireplaces, and gas- or oil-burning equipment and devices, such as water heaters, ovens, ranges, and other appliances and devices where open flames are present. Commercial kitchens and laboratories with fume hoods and the like need a source of makeup air for them. These ventilation requirements do not, however, always require mechanical ventilation. In a residence, for example, windows or other openings may satisfy most ventilation requirements.

HVAC systems can be classified as central or decentralized. *Central* systems generally serve an entire building or a major portion of a building. *Decentralized* systems serve small portions of a building, often a single room. The systems discussed in this section are those classified as central systems. Decentralized systems are covered in Section 17.10. Central heating and cooling equipment includes furnaces, boilers, motors and motor-driven compressors, condensing units, cooling towers, heat exchangers, blowers, pumps, fans, and air-handling units (see Sections 17.7 and 17.8).

The heating and cooling equipment for forced air systems in most single-family residential buildings and other small buildings is usually a packaged furnace or heat pump system selected by the architect or the heating and cooling contractor based on calculated heating and cooling loads and experience. Professional engineers are seldom employed to design such systems. Manuals produced by the Air Conditioning Contractors of America may be used to aid in the design of such systems.

More complex systems and those utilizing heavy equipment should be designed by structural and mechanical engineers based on design calculations and acoustical analysis. The noise levels and heating and cooling capacities of equipment should be obtained. Many manufacturers can furnish or develop sound pressure output data based on standard testing for most types of equipment, from which sound pressure levels can be computed for a specific space.

Selecting better-quality heating and cooling equipment with more than adequate capacity may slightly increase its initial cost but will often help to create a quieter installation and will perform more efficiently over the years. Selecting equipment with more than adequate capacity, particularly for fans and pumps, can produce considerable noise level reductions because larger equipment runs more slowly and therefore more quietly.

HVAC systems fall into several categories, depending on the heat medium (air, water, steam, electricity) and the delivery means. For convenience, HVAC systems can be divided into categories, depending on whether the buildings they serve are small or large.

Small-building HVAC systems can be classified by use as heating-only systems, cooling-only systems, and heating-cooling systems. Most large building systems are heating-cooling systems. Both small- and large-building HVAC systems can be classified by delivery mode as all-air systems, all-water systems, steam systems, or air-water systems.

## 17.6.1 HVAC SYSTEM CLASSIFICATION BY USE

### 17.6.1.1 Heating-Only Systems

In some northern locations, the ambient air temperature is rarely high enough to warrant the expenditure of providing a cooling system in small buildings. Systems that provide heat only may be found in some larger older buildings, but in new buildings their use is restricted almost exclusively to small buildings. Even in areas having low ambient temperatures, the interior functions carried out in a building combined with the heat generated by occupants and the need to control pollutants in the air dictates the addition of a cooling system.

Heating-only systems include all-air systems, all-water systems, and air-water systems as discussed in this section, using all the fuel sources discussed in Section 17.7.1.1, as well as the radiators and convectors discussed in Section 17.8.3, the radiant heaters described in Section 17.8.4, and the fireplaces and stoves addressed in Section 17.8.5.

### 17.6.1.2 Cooling-Only Systems

The oldest cooling-only system consisted of fans. The perception that moving air produces a cooler environment than still air has some basis in fact. Studies have shown that people are more comfortable when air is made to move. Large exhaust fans can be used to exhaust hot air to the exterior, which is then replaced by cooler air. This is particularly useful when the exterior temperature drops below that of the interior.

Cooling-only systems include packaged unit air conditioners (see Section 17.8.1) and evaporative air-cooling equipment (see Section 17.7.3.2).

### 17.6.1.3 Heating-Cooling Systems

Most HVAC systems used in modern buildings both heat and cool. They are addressed in the sections covering all-air, all-water, and air-water systems that follow this one. Residences and other small buildings also employ these systems. One such system that is commonly used in residences and other small buildings is an all-air system in which a cooling coil is mounted above and attached to a furnace in a central location in the building. Air is drawn by a fan over the furnace heater or cooling coil as the temperature requirement dictates and passes through a duct system to serve spaces throughout the building. Some larger residences and other small buildings have more than one such system.

Heat pumps (see Section 17.8.2) are also heating-cooling systems.

## 17.6.2 HVAC SYSTEM CLASSIFICATION BY DELIVERY MODE

### 17.6.2.1 All-Air HVAC Systems

In an all-air HVAC system, air is conditioned in a remote location, such as a mechanical equipment room, and delivered by ductwork to the spaces to be conditioned. All heat and all cooling is delivered by warm or cooled air. In most single-family residences and many other small buildings, the air in a forced-air systems is warmed by a gas, oil, or electric furnace and cooled by a unit consisting of cooling coils mounted on the bonnet of the furnace and a condenser located outdoors (see Section 17.2.7.2) or by a heat pump (see Section 17.8.2). In most large installations in buildings other than residences, the air is warmed by hot water or steam generated in a boiler.

#### FORCED-AIR DELIVERY SYSTEMS

Forced-air systems consist of air-handling equipment (see Section 17.7.3.1), ductwork, dampers, and air outlets.

Forced air is delivered through a duct system in one of several possible configurations.

***Single-Duct, Single-Zone Systems*** Most forced-air systems in single-family residences and other small buildings are single-zone, single-duct systems. In these, the plenum at the top of a combined heating and cooling unit is extended through the building. Branch ducts from the extended plenum extend to outlet locations (Fig. 17.6-1). Through these ducts, either heated or cooled air is delivered to each space. The volume of air, and therefore the degree of heating or cooling, delivered to each space are controlled by dampers. In small installations, manual dampers are installed at each outlet. In larger systems,

**FIGURE 17.6-1** Typical plan showing floor register locations.

dampers are controlled by mechanical devices operated by heat sensors.

**Perimeter Systems** Two types of perimeter delivery systems are employed in houses and other small buildings. In a *perimeter radial* (*individual supply*) system, individual ducts extend from the top of the furnace to each outlet. In a *perimeter loop* system, a continuous duct loops around the entire building. This duct is fed by lateral ducts extending from the furnace bonnet. In each case, the amount of air delivered is controlled by dampers.

**Single-Duct Variable Air Volume Systems** Air is supplied at a central source at a constant volume. The delivered volume is adjusted in each space to provide the required amount of air and temperature control. Such systems are used mostly in larger buildings, especially those with extensive interior zones.

**Single-Duct Multizone Systems** Larger residences and some medium-sized commercial buildings are divided into several zones. Each floor, or a portion of each floor, may constitute a zone. Each zone is served by a single-duct system emanating from a single air handler. In larger buildings there may be several air handlers, each serving a single zone or group of zones.

**Single-Duct Reheat** Some single-duct systems in large buildings have hot water or steam reheat coils in the ducts. This permits lower temperatures in the air supply, reducing both duct size and needed fan capacity.

**Double-Duct Systems** Double-duct systems supply air of different temperature in two separate ducts. The air is mixed at the terminal point to be delivered at the proper temperature to each space. Such systems are used in large buildings where interior and exterior spaces and spaces on opposite sides of the building have different heat gain or loss. This difference may be caused by a particular space's proximity to exterior (solar) heating or cooling loads, the number of people in different spaces, or equipment loads. These loads may vary at different times. In a variation of the double-duct system, one duct may carry high-pressure air that varies in temperature as needed, while the second duct may carry air at a constant temperature.

**Ductwork** Forced-air heating and cooling system ductwork includes:

- Diffusers and other outlets through which conditioned air is delivered into habitable spaces
- Supply air ducts that carry air from a heating and cooling unit throughout the building to diffusers
- Return air intakes and ducts that pick up heated or cooled air and return it to the heating or cooling unit for recycling
- Fresh air intakes that provide combustion air and air changes in the building
- Exhaust systems

Supply ducts should be sized to deliver the required amount of air at the desired velocity. Procedures for selecting duct materials and sizes are contained in SMACNA's publication *HVAC Systems—Duct Design.* SMACNA also produces a "Duct System Calculator."

In small buildings, a single-duct system, consisting of a main supply duct with several branch ducts, is usually used. When attic space is available, the main and branch ducts are usually run in that space. When there is no attic, and in intermediate floors in buildings of more than one story, the main supply duct can be dropped below the ceiling. The branches can then also be dropped below the ceiling or run within the framing space. Ducts in individual supply and perimeter loop systems are often run below the floor in a crawl space or within the floor slab.

Combustion-air and fresh-air intakes may be necessary in buildings that are tightly sealed. A combustion-air intake provides outside air for efficient burning of fossil fuel. Outside air may be simply delivered to the room where the furnace is located if the furnace is in a separate room and a blast of cold winter air is not objectionable. To conserve energy and to reduce the discomfort from cold air, the outside-air duct may be connected directly to the combustion chamber. This, in fact, is required by some local utilities and energy conservation codes.

A fresh-air intake helps ventilation and helps reduce the buildup of indoor odors and other pollutants. ASHRAE Standard 62 specifies a ventilation rate of 15 cfm (.74 m³/min) to ensure acceptable air quality. Local codes may require a different amount. Fresh-air intakes may be placed in a vented attic or directly out-

doors and connected to the return-air duct above the furnace's air filter. Air-to-air heat exchangers (see Section 17.7.5) can be used to temper the incoming air and save heating and cooling energy.

Intakes should have insect screens and tight-fitting manual or automatic dampers to control the amount of air intake.

In small buildings, a furnace heats air and the associated blower distributes it through the ducts to supply outlets. The return-air system completes the convection loop by returning air to the furnace, where it can be heated and recycled. There may be a return-air duct system or a single return-air grill directly connected to the furnace.

In larger buildings, air is heated and cooled in air-handling units (see Section 17.7.3.1) and distributed through ducts to supply outlets. Air is returned to the air handler through ducts.

A return-air system should be sized to equal or exceed the supply volume and should induce the return of air from each conditioned space. A properly designed return-air system helps maintain an even temperature level with minimal stratification throughout a building.

In small buildings, nonducted volume air returns allow free passage of return air directly to the furnace, usually through a louvered furnace closet door. In nonducted return-air systems, doors should be undercut to have at least 1 in. (25.4 mm) of clearance from floor tile or carpet to ensure proper air circulation. However, doors that must provide acoustical privacy cannot be used as a means of returning air to the furnace. Separate air return registers can be placed in such rooms and connected to a sound-absorbing duct system. This eliminates blower and furnace noises and reduces crosstalk from room to room. Sound-absorbing air inlets can be used to allow air intake from corridors without admitting noise.

Air-return registers should not be located back to back for the same reasons supply registers should not be (see Section 17.5.2). Similar methods to those suggested for supply air registers can also be used in return-air openings to reduce sound transmission.

Ducted multiple return-air systems are used in larger homes and other buildings with complex room layouts. Return-air grills should be placed near the ceiling. The air is returned from grills in each room through ducts located in the ceiling construction or attic. Smaller rooms such as residential kitchens and baths do not need return-air grills if clear openings, openings fitted with louvers or grills, or undercut doors permit air to move from the small spaces into an adjacent space that does have a return-air grill.

Most codes require that bathrooms without windows be exhausted directly to the exterior. Some codes have similar requirements for bathrooms with windows and for other spaces as well. Most codes require that toilets, shower rooms, and similar spaces in commercial and institutional buildings be directly exhausted. Whether required or not, separate exhaust systems should be installed in laboratories and in rooms housing cooking appliances, laundry equipment, showers, and other equipment or devices that produce excess heat, high humidity, or undesirable fumes.

Exhaust ducts should be sized to fit the volume of the space they are intended to exhaust with the desired number of air changes per hour. Fans should be selected using the same criteria and should usually be oversized to reduce the noise they generate.

## AIR PLENUM SYSTEMS

An air plenum system is a method of distributing heated or cooled air, or both, to the living area of a dwelling or other small building through a tightly sealed underfloor area, usually a crawl space.

The system may be used with poured concrete or concrete masonry unit (block) foundations but is frequently used with wood foundations (see Section 6.5). Although theoretically an air plenum system could be used to heat and cool a two- or three-story building, it is presently recommended only for single-story structures.

An air plenum system is cheaper than conventional forced-air systems because it eliminates ductwork.

In a typical 1400 sq ft (130 m²) air plenum house, energy use can be reduced by 10% compared to a house with a conventional forced-air heating system and crawl space. Uniform air distribution through well-spaced registers prevents stratification of the convected air. Tests conducted at the University of Florida showed an average temperature differential of $2^1/_2°F$ (16.4°C) between air near the floor and 60 in. (1524 mm) above the

floor. Air quantities can be regulated by adjusting dampers.

The floor surface is warm when the heat is on and cool when the air conditioning is on. A warm floor is particularly pleasant underfoot in winter.

Floor framing members, floor sheathing, underlayment, and finished flooring absorb heat from the system and radiate it into the building. Radiant heat makes occupants feel more comfortable in cold weather, thus allowing system operation at a lower temperature.

In studies conducted by Washington State University, a dwelling in Idaho with an air plenum system used less heat than a dwelling with a conventional forced-air heating system and crawl space. The floor surface temperature ranged between 70° and 82°F (21.1° and 27.8°C); 85°F (29.4°C) is the maximum temperature recommended for floor surfaces.

In tests conducted with air conditioning in Florida, the floor surface temperature ranged between 70° and 73°F (22.2° and 22.8°C).

The information and most of the illustrations in this section were abstracted from a manual prepared by the National Association of Home Builders (NAHB) Research Foundation for the American Plywood Association (APA) (now APA—The Engineered Wood Association), the American Wood Preservers Institute (AWPI), the Southern Forest Products Association (SFPA), and the Western Wood Products Association (WWPA). This information has been updated to agree with the pamphlet "Plen-wood System" issued jointly in 1989 by the APA, American Wood Council (AWC), National Forest Products Association (NFPA), SFPA, and WWPA, and some of the illustrations come from that document. An air plenum system uses a downflow air-handling unit (combination furnace and air conditioner) to create enough air pressure under the floor to deliver heated or cooled air to individual rooms through supply registers in the floor (Fig. 17.6-2). Using the crawl space to distribute conditioned air eliminates costly ductwork and provides both convection and radiant heat (Fig. 17.6-3).

Such a system is simple in design and easy to build. It conserves energy and reduces construction costs. With minor variations, an air plenum system meets the requirements of all the major building code groups.

**FIGURE 17.6-2** Diagram of an air plenum system.

**FIGURE 17.6-3** Entering air heats or cools not only interior air but floor surfaces as well.

Air plenum systems are particularly susceptible to the introduction of pollutants, such as radon, into the living environment inside a dwelling. Where radon exists, precautions should be taken to prevent its introduction into the plenum and subsequently into the dwelling.

*Site Preparation* Site clearing and excavation methods should be no different for plenum systems than for conventional foundation systems (see Chapter 21).

No special precautions for drainage are needed when the plenum grade is at or above the exterior grade, the plenum space is well drained, the underlying soils are of types that drain well, and the water table height is normal. When the plenum grade is below the exterior grade, the soil is clayey, or the groundwater level rises close to the plenum floor, foundation drains and sump pumps may be required. Care should also be taken to ensure drainage of rainwater during construction.

When a sloping site results in more of the foundation wall being above grade, thicker insulation should be used on the inside of the wall.

Increased or varying the plenum depth may increase the air volume and operating cost but does not affect satisfactory performance of the system (Fig. 17.6-4).

*Termite Protection* In regions subject to termite infestation, the soil should be

**FIGURE 17.6-4** Neither plenum depth nor slope significantly affects system performance.

chemically treated for a distance of 12 in. (304.8 mm) out from the foundation wall. Any wood in the foundation or wood that is below the level of the first floor should be pressure treated and carry the American Wood Preservers Bureau's AWPB-FDN mark. Refer to Section 6.4 for additional requirements. If a wood foundation system is used, it should follow the preservative treatment methods described in Section 6.5.

A shallow-depth wood plenum (less than 18 in. [457 mm] clearance) is not recommended in areas where a termite hazard exists, because soil-treating chemicals should not be used inside a plenum and because visual inspection is difficult.

No building debris or vegetation should be left in a plenum space.

*Foundation Construction* The basic plenum types are (1) shallow depth, low profile and (2) regular depth, crawl space.

A shallow-plenum house looks like a slab-on-grade house from the outside. It is economical because even on sloping sites, very little site grading is necessary. A shadow-depth plenum has less than 18 in. (457 mm) of clearance below the floor joists and can function with as little as 3 in. (76 mm) of clearance.

A regular-depth plenum is basically the same as a conventional crawl space. A regular-depth crawl space has a clearance of 18 in. (457 mm) or more from the bottom of the floor joists and 12 in. (305 mm) or more from the supporting girders.

A crawl space foundation intended to accommodate an air plenum should be:

- Sealable to prevent conditioned air from escaping and water vapor from entering
- Well drained
- Fire safe
- Free of insects and other vermin

An air plenum system can be constructed with most conventional foundation materials (Fig. 17.6-5). It is also well suited to wood foundations (see Section 6), and this combination is common.

**FIGURE 17.6-5** Air plenum systems may be constructed with foundations of (a) concrete block, (b) cast-in-place concrete, or (c) grade beams on wood posts or concrete piers.

*Thermal Integrity* An air plenum system should be thoroughly caulked and sealed, have a suitable vapor retarder, and be adequately insulated. Foundation vents, typically recommended for crawl spaces, should not be used. Access to an air plenum should be only from inside the dwelling. Materials in a plenum should have a flame spread rating of 200 or less.

An air plenum system should be sealed to prevent leakage of conditioned air and entrance of ground moisture. A sill sealer should be used between foundation walls and sole plates in wood-framed construction. Joints should be caulked.

A polyethylene film or a similar continuous sheet material vapor retarder should be used as a ground cover, and should extend up and across the inside face of foundation walls to reduce air and vapor leakage and to keep the plenum dry.

The underfloor area should be free of sharp stones and debris that could puncture the vapor retarder film. The film should cover the ground, extend up the foundation walls, and be attached to the top of the sill; a few inches should overhang the sill so that the film may be attached to framing members to form a good seal (Fig. 17.6-6). The film should be applied in the largest practicable sheets. Joints should be sealed, except that joints on the ground should be overlapped 12 in. (305 mm) but not sealed. These unsealed joints provide a means of escape if water collects in the plenum area. This film may be covered with 2 in. (50.8 mm) of sand or gravel.

Either rigid or batt insulation should be installed in an air plenum system. This insulation should be foil faced and have an R-value equal to or greater than that recommended for the side walls (R-11 or better). Paper-backed and foamed plastic

**FIGURE 17.6-6** Vapor retarder film in an air plenum.

insulations are fire hazards and are not recommended, unless they are covered with $^{1}/_{2}$-in. (12.7-mm)-thick gypsum board or $^{5}/_{8}$-in. (15.9-mm)-thick plywood. However, covering insulation with such materials increases construction costs.

Insulation should be installed over the vapor retarder on the inside perimeter of the foundation wall and should extend at least 24 in. (610 mm) away from the wall horizontally at the bottom, as shown in Figure 17.6-7. Insulating the entire underfloor area is generally not cost-effective.

***Floor Construction*** The floor over a plenum may be framed conventionally by resting the joists on a wood sill (see Fig. 17.6-7) or in joist hangers. Sometimes joists are supported on a continuous wood ledger mounted on the side of wood foundation walls to lower the dwelling's exterior wall by the depth of the joists. This can reduce the amount of exterior wall sheathing and siding needed by about 9.5%.

Because the moisture content of the air in the plenum is low, the floor system should be built to allow for shrinkage as if it were in a heated basement.

***Underfloor Plumbing*** Seamless copper tubing is recommended for water supply piping to minimize leakage problems. Drain, waste, and vent (DWV) plumbing joints should not be in the plenum, but rather above the floor or below the vapor retarder. DWV cleanouts may be either inside or outside the dwelling but should not be placed in the plenum space. Underground plumbing should be installed before a vapor retarder is laid on the ground. Refer to Section 16.3 for additional information about DWV systems.

In shallow plenums, an access trench is desirable to enable workmen to reach leaky plumbing or faulty equipment.

***Underfloor Wiring*** Electrical wiring within a plenum is restricted by most codes as a potential fire hazard. If permitted, wiring within a plenum should consist of approved types of electrical cable or installed in conduit. Some codes require rigid conduit; others permit thin-wall.

***Mechanical System*** An air plenum system does not require special equipment. The key to optimum efficiency is the proper sizing of system components.

a

b

c

4 in.  = 101.6 mm
18 in. = 457.2 mm

**FIGURE 17.6-7** Vapor retarder and insulation in an air plenum (a) at a wood foundation wall, (b) at a unit masonry wall, and (c) at a concrete foundation wall. (Courtesy American Plywood Association [now APA—The Engineered Wood Association].)

Usually, a counter-flow (down-flow) furnace or heat pump is used. It should be sized using heat load calculations; oversizing should be avoided. Blower controls should be set to produce continuous low-volume air circulation at all times and higher-volume circulation when heating or cooling is supplied.

The furnace selected should have adequate safety mechanisms to prevent excessive temperature buildup. A down-flow furnace should have a safety mechanism to guard against the reversal of air flow should the blower become inoperable.

The heating capacity of the unit and the air volume circulated through the unit by the blower determine the air temperature rise through the furnace. The most efficient underfloor plenum heating units operate at a temperature rise of about 40°F (4.4°C). Figure 17.6-8 shows the recommended furnace output and blower capacities which will result in a 40°F (4.4°C) temperature rise.

Since there are no ducts in the air plenum system, there is little resistance to air flow. Air resistance is mainly due to the air filter and the delivery openings from the plenum into the living or working areas.

The furnace should be located as close to the center of the building as possible. With a central location, air distribution is better and return-air ducting is simpler.

There should be 30 in. (762 mm) of service clearance in front of a furnace or heating unit. The unit should be adequately supported and blocked to prevent vibration. There should be a positive airtight seal around both the return-air duct and the furnace base. For fire safety, there should be a noncombustible base usually made of sheet metal. An air deflector pad made of noncombustible materials such as thin concrete or gypsum board should be placed directly under the furnace (Fig. 17.6-9).

**FIGURE 17.6-8** Recommended Furnace Output and Blower Capacities*

| Heating Unit Output, Btu (kJ) | Blower Capacity, cfm (m³/M) |
|---|---|
| 20,000 (21,101) | 475 (26.19) |
| 30,000 (31,652) | 700 (19.82) |
| 40,000 (42,202) | 925 (26.19) |
| 50,000 (52,753) | 1175 (33.27) |
| 60,000 (63,304) | 1400 (39.64) |
| 70,000 (73,854) | 1625 (46.01) |
| 80,000 (84,405) | 1850 (52.39) |
| 90,000 (94,955) | 2100 (59.47) |
| 100,000 (105,506) | 2325 (65.84) |

*To obtain 40°F (4.4°C) temperature rise through the furnace.

**FIGURE 17.6-9** A noncombustible base and an air deflector prevent excessive heat buildup and fires. (Courtesy American Plywood Association [now APA—The Engineered Wood Association].)

If air conditioning is included in the system, no furnace base may be needed since the cooling coil housing may serve as the base. However, there should be provision for collecting condensation from the cooling coil.

If cooling is not provided, the furnace blower should be turned on occasionally in the summer to evacuate humid air from the plenum. Running the blower in hot weather, especially at night, contributes to occupant comfort.

The blower should be wired to allow continuous operation. A continuous blower sound may be less objectionable than intermittent noise. A two-speed blower can provide a continuous low air volume in addition to a higher air volume when heating or cooling equipment is in operation. Blower operation may also be tied in with a separate thermostat.

The furnace heats air, and the blower distributes it through the plenum and supply outlets. The return-air system completes the convection loop by returning air to the furnace, where it can be heated and recycled (Fig. 17.6-10).

The return-air system should be sized to equal or exceed the supply volume and should induce the return of air from each conditioned space. A properly designed return-air system helps maintain an even temperature level with minimal stratification throughout the dwelling. Nonducted air returns allow free passage of return air directly to a furnace, usually through a louvered furnace room door. Bedroom doors should be undercut to have at least 1 in. (25.4 mm) of clearance from floor tile or carpet to ensure proper air circulation.

Ducted multiple return-air systems are used in larger homes with complex plans. The air is returned from grills in each room, located preferably near the ceiling, through ducts located in the ceiling con-

**FIGURE 17.6-11** A dropped ceiling above the furnace supplies conditioned air through the joist space between floors; the plenum under the first floor returns the air to the furnace.

struction or attic. Smaller rooms, such as a kitchen or bathroom, do not need return-air grills if other means are provided to allow air to return from them to the system.

Combustion-air and fresh-air intakes, as discussed earlier, may also be necessary in buildings with air plenum heating and cooling systems. The same principles apply.

A crawl space cavity can also be used as a return-air plenum rather than as a supply plenum when desired. If it is used as a return-air plenum, an upflow air furnace with overhead ductwork is necessary. Return-air grills near the outside walls can be used as the means of routing air to the plenum.

In single-story construction, an overhead main supply duct can be used; the supply duct can either be dropped below the ceiling or located in the attic. In a two-story dwelling, if the underfloor area of the first floor is used as a return-air plenum, the framing space between stories can serve as a supply plenum (Fig. 17.6-11). A centrally dropped ceiling can provide conditioned air to the space between joists.

Either conventional adjustable floor diffusers, baseboard registers, or a continuous perimeter slot can be used to supply heated and cooled air to living areas (Fig. 17.6-12).

Each *register* should have an adjustable damper to properly balance the air flow. If too much air enters a room,

**FIGURE 17.6-10** Return air grill(s) may be provided in a furnace room wall or ducts may be run from each room through the attic to the furnace.

**a**  **b**

**FIGURE 17.6-12** (a) A conventional floor diffuser. (b) A perimeter slot.

the registers can be partially closed. If there is not enough air, it is easy to replace a small register with a larger one or add a register.

As a general housekeeping measure, sheet metal pans should be installed beneath each register opening to catch condensation that may form on the supply outlet and drip into the plenum (Fig. 17.6-13). These pans may also be required by some codes, especially where sand or gravel is not used over the vapor retarder.

In general, register sizing and location are basically the same as with conventional forced-air systems, with a few additional requirements as outlined here:

1. Heat loss or gain should be calculated for each room and for the entire building. Conventional heat loss calculations can be used for living and working areas; a different method of calculating heat loss is used for a plenum system than for conventional ventilated crawl space construction (Fig. 17.6-14).

2. A heating unit with an output capacity equal to 1.15 of the total calculated heat loss should be selected.

3. Generally, at least one register should be located on each exposed exterior wall, except in bedrooms with less than 7000 Btus (7385 kJ) per hour heat loss. If possible, registers should be located in front of windows (see Fig. 17.6-1). Small rooms near the furnace and interior rooms with no outside wall often need no registers. In cases where kitchen cabinets are on the outside wall, holes can be cut in the floor under the cabinet and conditioned air delivered through a toe-space grill.

4. Register outlet capacity should closely match the air volume of the furnace on a proportional basis. For example, if heat loss for a bedroom is 8400 Btus (8863 kJ) per hour and total dwelling heat loss is 56,000 Btus (59 083 kJ) per hour, then that bedroom should receive about 15% of the air volume. If the blower size is 1600 cfm (45.3 m³/min), 15% of that is 240 cfm (6.8 m³/min). If two registers are used, each should deliver 120 cfm (3.4 m³/min). Because the floor acts as a radiant heat source, rooms nearer the mechanical unit will receive a higher proportion of the radiant heat. Registers in those rooms may be slightly undersized, and registers in the rooms farthest from the heat source can be slightly oversized. This method of determining register size ignores plenum pressurization and will probably result in slightly oversized registers, which can be partially closed if necessary.

Because conditioned air is fed to the registers directly from the plenum, duct length and size considerations can be ignored.

A low plenum pressure (not more than 0.1 in. [2.54 mm] of water) is desirable to maintain more comfortable velocities through the supply registers and to minimize perimeter air loss. Figure 17.6-15 shows a method for determining register sizes if plenum pressure is not more than 0.1 in. (2.54 mm) of water.

Further information on calculating heat loss, and the sizing and locating of registers, may be obtained from the NAHB Research Foundation.

The perimeter slot method of supplying conditioned air is unobtrusive and avoids the problem of obstructing air flow from a grill or register with furniture. Its primary disadvantage is that it requires some unconventional framing and trimming techniques. It may also be a problem with floor-length drapes because the heat will be trapped between the drape and the wall or window.

The perimeter slot method consists of holding floor sheathing about 1/2 in. (12.7 mm) from the inside surface of the exterior walls and then blocking the baseboard out uniformly to create an air delivery gap into the living area (see Fig. 17.6-12b). The width of the slot is determined by heat loss calculations and usually varies between 1/8 in. (3.2 mm) and 3/8 in. (9.5 mm). The amount of air delivered is then fixed by the width of the slot with no further controls.

Some local codes may require short ducts to be stubbed out from the furnace (Fig. 17.6-16), but they are seldom needed for proper air distribution. Some large dwellings with irregular shapes may find their installation helpful in ensuring adequate air distribution in the far corners of the dwelling (Fig. 17.6-17). Stub ducts often eliminate the need for a separate furnace air deflector (see Fig. 17.6-9).

### 17.6.2.2 All-Water Systems

In an *all-water* system, hot and cold water are produced remotely and delivered to terminal units throughout a building. All-water systems are used mostly in large buildings but may also be found in larger, more expensive residences and other medium-sized buildings.

#### HOT WATER SYSTEMS

In hot water systems, the heating medium (water) is generated in a boiler and circulates through the building in pipes. As the pipes pass through heating devices, the medium gives off its heat, which warms the spaces. Three basic configurations are used in hot water heating systems: *series loop*, *one-pipe*, and *two-pipe*.

*Series Loop Systems* In a series loop system, a single pipe carrying hot water from the boiler passes directly through convectors and returns the now cooled water to the boiler for reheating. Series loop systems are used in small buildings such as single-family residences. They are economical to install, but the heat produced is difficult to control because as the water circulates, it cools. Convectors near the end of the loop receive water at a much lower temperature than those at the beginning. Some control can be obtained by dividing the building into

**FIGURE 17.6-13** Sheet metal pans below floor registers. (Courtesy American Plywood Association [now APA—The Engineered Wood Association].)

Every square foot of ground under a dwelling can be accounted for by tracing heat flow through circular arc paths to a point below the surface of the exterior grade (a). Heat loss is based on the average conductivity of the soil. Conductivity varies according to soil type, dampness and looseness; usually, dry loose soil is a better insulator than damped packed soil. The average conductivity of soil is 0.8 BTUs per hour per sq. ft. per foot of path length.

The heat loss factors shown below (b) account for arc path length, insulation, heights of exposed foundation wall, foundation construction, surface films and ground temperatures beneath the exterior grade surface of a 50 ft. by 28 ft. dwelling. Though arc lengths vary according to dwelling size, the values shown below may be used as a reasonable estimate for most dwellings. Under severe climatic conditions or with an extremely large dwelling these values will slightly underestimate plenum heat loss.

## (a) PLENUM HEAT FLOW PATHS

## (b) HEAT LOSS FACTORS, R11/R19 INSULATION

| Dimension B | Dimension A | | | | | | | | | |
| --- | --- | --- | --- | --- | --- | --- | --- | --- | --- | --- |
| | 6" | | 12" | | 18" | | 24" | | 30" | |
| | R11 | R19 | R11 | R19 | R11 | R19 | R11 | R19 | R11 | R19 |
| 6" | .33 | .29 | — | — | — | — | — | — | — | — |
| 12" | .34 | .29 | .37 | .31 | — | — | — | — | — | — |
| 18" | .35 | .29 | .38 | .31 | .41 | .33 | — | — | — | — |
| 24" | .36 | .29 | .39 | .32 | .42 | .34 | .45 | .36 | — | — |
| 30" | .37 | .30 | .40 | .32 | .43 | .34 | .46 | .36 | .49 | .38 |
| 36" | .38 | .30 | .41 | .32 | .44 | .34 | .47 | .36 | .50 | .39 |

### To use (b) Heat Loss Factors:

(1) Select insulation R-value to be used.

(2) Estimate average height of exposed foundation wall and plenum depth from ground to underside of subfloor.

(3) Find appropriate heat loss factor in table by cross-matching exposed foundation in "A" column and plenum depth in "B" column.

(4) From house plan, determine lineal feet of perimeter foundation which encloses the underfloor plenum area.

(5) Determine temperature differential between outside design temperature and plenum perimeter temperature (found to average 85°F).

(6) Multiply heat loss factor times exposed lineal feet of foundation times temperature difference to obtain plenum heat loss: Heat loss factor x exposed lineal feet of foundation wall x temperature difference = Plenum Heat Loss.

(7) Add living area heat loss to plenum heat loss to obtain total estimated heat loss for entire house.

Example: Assume the following characteristics:
• 1400 sq. ft. dwelling.
• Wood foundation, with average 12" exposure to outside.
• Plenum depth (ground to subfloor) about 24".
• R-11 insulation on wall and 24" horizontally around perimeter.

• 156 lineal feet of foundation enclosing the plenum.
• Winter design temperature = 24°F.
• Temperature difference = 85°F minus 24°F = 61°F.

From (b) the heat loss factor is 0.39. Therefore, estimated plenum heat loss is:

0.39 (heat loss factor) x 156 (lineal feet of foundation) x 61°F (temperature difference) = 3,711 BTUs per hour.

Assuming heat loss calculations for the living area of 40,000 BTUs per hour, total heat loss under design conditions for the purpose of sizing equipment will be 40,000 BTUs per hour + 3,711 BTUs per hour. Round off to 44,000 BTUs per hour. It is interesting to note that the plenum heat loss in this example is about the same as would have been added for duct heat loss when insulated ducts are located in a conventional ventilated crawl space. Also, over 5,000 additional BTUs per hour would have been lost in such a dwelling through the floor when insulated with R-11 over its entire area. A net energy saving of over 5,000 BTUs per hour under design conditions results by using the underfloor plenum system in this example.

Cooling load calculations for plenum houses can be made in the conventional manner, neglecting heat gains to the underfloor plenum area, unless there is significant exposed foundation wall. In which case, only heat gain through the exposed portion of the foundation wall should be calculated.

## (c) WOOD FOUNDATION

## (d) CONCRETE OR BLOCK FOUNDATION

**FIGURE 17.6-14** Heat loss calculations for an air plenum system. (Refer to Figure 25.7-1 for metric equivalents.)

## FIGURE 17.6-15 A Second Method of Sizing Registers

Assuming a 56,000 Btu/hr unit with a blower capacity of 1600 cfm and a plenum pressure of 0.1 in. of water, a floor register would deliver about 9 cfm per sq in. of floor register opening. Therefore, the total register free opening for the house would be 1600/9 = 177 sq in.; 10 registers each having 18 sq in. of free area could be distributed uniformly around the house.

**a. Register Air Delivery[a]**

| Free Area of Register | Plenum Pressure, in. of Water | | | | |
|---|---|---|---|---|---|
| | 0.02 | 0.04 | 0.06 | 0.08 | 0.10 |
| 1 sq. in. | 4.4 | 5.8 | 7.3 | 8.4 | 9.4 |

[a]In cfm per sq in. of floor or basement register for different plenum pressures.

The same general procedure applies to the perimeter slot method of air distribution. Assuming a 1600 cfm blower and a plenum pressure of 0.1 in. of water, a $\frac{1}{4}$ in. slot would deliver 19 cfm per lineal foot.

Therefore, total open slot length would be 1600/19 = 84 ft. If a $\frac{3}{16}$ in. slot is used, 13 cfm is delivered, so total slot length would be 1600/13 = 123 ft.

**b. Perimeter Slot Air Delivery[b]**

| Slot Width, in. | Plenum Pressure, in. of Water | | | | |
|---|---|---|---|---|---|
| | 0.02 | 0.04 | 0.06 | 0.08 | 0.10 |
| $\frac{1}{8}$ | 3.5 | 4.7 | 5.8 | 6.8 | 7.8 |
| $\frac{3}{16}$ | 5.8 | 8.4 | 10.0 | 12.0 | 13.0 |
| $\frac{1}{4}$ | 8.5 | 12.0 | 15.0 | 17.0 | 19.0 |
| $\frac{3}{8}$ | 14.0 | 20.0 | 24.0 | 28.0 | 31.0 |

[b]In cfm per sq in. of slot for different plenum pressures.

(Refer to Figure 25.7-1 for metric equivalents.)

zones and installing a separate loop for each zone.

***One-Pipe Systems*** One-pipe systems also have a single pipe running in a loop from the boiler around the building, but each heating devices is served by a tee connection and a pipe to that heating device. Therefore, the amount of water to an individual heating device can be controlled. Heating devices at the beginning of the loop are about the same temperature as those farther along. Therefore, there is more heat left at the end of the loop than in a series loop system, and the heat produced by individual heating devices can be regulated. In addition, larger buildings can be served, because there is less wasted heat in the piping. Even larger buildings can be served by dividing them into zones, each with its own one-pipe system loop.

***Two-Pipe Systems*** In a two-pipe system, one pipe carries heated water from the boiler and a second pipe carries the return water back to the boiler. Heating devices are connected to the supply pipe through a tee. Controls are effected in a manner similar to that used in a one-pipe system but are more effective, since there is no cooled water mixing in the supply pipe. In a *direct return* two-pipe system, there are actually two pipes. The supply pipe connects to the return pipe at the end of a loop. In a *reverse return* system, the supply pipe runs to the end of a loop and then returns to the boiler, passing by each heating device a second time and picking up water from each as it passes. Like a one-pipe system, a two-pipe system can serve a larger building if the building is divided into zones, with a separate two-pipe system serving each zone from a common boiler.

### HOT AND CHILLED WATER SYSTEMS

All-water systems that both heat and cool are usually set up in either two-pipe or *four-pipe* systems

***Two-Pipe Systems*** A two-pipe system for both heating and cooling works in much the same way as a two-pipe system for heating only. The major difference is that the pipes are set up to carry either hot or chilled water. The changeover is controlled by a valve connected into the system's controls (see Section 17.5.1).

***Four-Pipe Systems*** A four-pipe system consists of two separate pipe circuits. One two-pipe circuit is used for circulating hot water, as in a two-pipe system. The second circuit of two pipes is used to circulate chilled water.

### TERMINAL UNITS

Various devices are used to translate the heat generated by a boiler into a usable temperature rise in building spaces. These include fan coil units, unit ventilators, induction units, radiators, convectors, and unit heaters, which are discussed in Section 17.8.2, and radiant heating electric cables, radiant heating hydronic piping, radiant heating electric panels, and electric radiant heaters, which are discussed in Section 17.8.3.

**FIGURE 17.6-16** Stub ducts installed under a furnace help direct air to various parts of the dwelling. (Courtesy American Plywood Association [now APA—The Engineered Wood Association].)

**FIGURE 17.6-17** Directional stub ducts in a nonrectangular design. (Courtesy American Plywood Association [now APA—The Engineered Wood Association].)

### 17.6.2.3 Steam Systems

Steam heating systems, once extensively used in residences, are now used mostly for commercial installations. In steam systems, the heating medium (steam) is generated in a boiler and circulates throughout the building in pipes. As the pipes pass through heating devices, the steam gives off its heat, which warms the spaces, and degenerates back to water, which flows back to the boiler for reuse.

Steam systems use one pipe to supply steam to heating devices and a second pipe to carry condensate water back to the boiler. The system may be either a vapor system or a vacuum system. Piping may be arranged in a downfeed layout, in which steam is pumped to a level above the heating devices and is allowed to fall to them in a series of risers, or in an upfeed system by a series of risers. In both cases, condensate is carried back to the boiler in a separate set of risers.

Terminal units for steam heating systems include radiators and unit heaters, as described for hot water systems.

### 17.6.2.4 Air-Water Systems

In an *air-water* system, heating or cooling of spaces is done partly by conditioned air delivered from a forced-air system but mostly by conditioned water delivered to terminal units in the spaces. Air-water systems are used in both small and large buildings. In small buildings, combination systems are generally used, in which hot water provides heat, while unit air conditioners provide cooling.

In larger buildings, any of the previously described all-water or steam systems and terminal units may be used to provide heating and cooling. Supple-

mentary heat and ventilation are provided by a forced-air system. The forced-air system usually provides only a small amount of the heating and cooling required in spaces where the water or steam system is being used, such as in zones adjacent to exterior walls, but may supply all of the heating and cooling in interior zones.

### 17.6.2.5 Direct Refrigerant HVAC Systems

Direct refrigerant systems supply heating and cooling using devices contained either in the space to be served or nearby. Some direct refrigerant systems are also called *package* or *unitary* systems. These are discussed in Section 17.8.1. Heat

pumps (see Section 17.8.2) are also direct refrigerant units.

### 17.6.2.6 Electric Heating Systems

Electric heating systems are widely used. They are easy to control and have a low installation cost. Unfortunately, they are costly to operate. Decentralized electric heating systems are discussed in Section 17.8. In addition to powering the heaters described in Section 17.8, electricity can be used in centralized systems to power furnaces and boilers (see Section 17.7).

### 17.6.3 SOLAR HEATING AND COOLING SYSTEMS

#### 17.6.3.1 Active Solar Heating Systems

Active solar energy systems consist of collectors and storage devices. Such systems are used to heat buildings, to heat domestic water, and, rarely, to cool buildings. Solar heating systems used solely to heat domestic water are addressed in Section 16.2.2.3.

#### COLLECTORS

The most common type of collector is a flat plate collector (Fig. 17.6-18). There are two types, liquid and air. Both have a metal frame with a metal bottom plate and a glass top panel, and the bottom and sides of the box are lined with insulation. On top of the insulation is an absorber plate. In a water system, the space be-

**FIGURE 17.6-18** A typical flat-plate solar energy collector. Many designs are available. (Drawing by HLS.)

tween the top glazing and the absorber plate contains tubes through which water flows. In the air type, the tubes are omitted and air flows through the space between the glazing and the absorber plate. The sun heats the space below the glazing and subsequently the air or water passing through this space.

Air systems are generally used in locations where freezing is a problem.

## STORAGE DEVICES

Water systems generally use water tanks as their storage devices (Fig. 17.6-19). Air systems generally employ rock beds (Fig. 17.6-20). In a water tank system, the heated water from the collector passes through a heat exchanger in the storage tank, which heats the water in the storage tank. This heated water is then pumped through the heating system to delivery devices, usually fan coil units or space heaters.

In an air system, the heated air flows into a contained bed filled with stone. The air heats the stone and is then pumped back to the collector, where it is again heated. When heat is required, air is made to flow across the stone bed, picking up the heat stored there and delivering it throughout the supply air portion of the heating system. Return air flows through the bottom of the stone bed and is returned to the collector, where it is again heated.

### 17.6.3.2 Passive Solar Heating Systems

#### HISTORY

Primitive peoples intuitively designed their dwellings to take advantage of the local climate. In each region, a style of building evolved that kept buildings warm in winter and cool in summer with a minimum of energy. This was accomplished passively, that is, through the use of the sun, wind, and other natural phenomena.

In the early Sinaqua and Anasazi Indian cliff dwellings of Arizona, known today as Montezuma's Castle, an overhanging rock brow protected southern exposures against the high summer sun and kept interiors cool. In winter, the rock brow did not shut out the low afternoon sun, allowing the rock face to warm up. Since rock is a good conductor and storer of heat, it retained the heat and kept the dwellings warm during the night.

**FIGURE 17.6-19** Diagrammatic illustration of an active solar energy hot-water heating system. (Drawing by HLS.)

**FIGURE 17.6-20** Diagrammatic illustration of an active solar energy air-heating system. (Drawing by HLS.)

The Pueblo Indians of New Mexico also knew how to build energy-efficient dwellings more than 1000 years ago. By facing their structures south and using high, small, deeply cut windows, they were able to delay the entry of heat into the interior of their structures until late in the day, when temperatures started dropping. Exterior surfaces were painted a light color to reflect the sun's rays. The builders crowded their dwellings side by side and on top of one another to reduce temperature fluctuations in both summer and winter and cooked in rooms separated from living spaces to avoid unwanted heat gain in summer.

Descendants of the Pilgrims also evolved a suitable style of building for their northeastern region. In the cold winters of New England, keeping warm meant having efficient heat sources and tight buildings to retain the heat. The typical New England saltbox had warming stoves and fireplaces at the center of the house and a minimum of exterior wall surface. A north-sloping roof provided protection from icy winds and allowed snow to build up for added insulation against the cold. Most windows faced south, and the exterior was painted a dark color to absorb as much solar radiation as possible.

Studies of these and other intuitive responses to local climate have laid the groundwork for a more widespread development of similar naturally adaptive heating and cooling techniques. Some architects and engineers have refined and adapted these concepts to create innovative and energy-efficient buildings. Their work has resulted in buildings which more effectively collect, store, and distribute solar energy for heating, and which better utilize natural climatic factors for cooling. These naturally adaptive *passive* heating and cooling systems still require external energy for their operation, but need less of it than do conventional buildings. Solar heating or cooling installations with sophisticated heat collection, transport, control, and storage methods, known as *active* solar systems, are beyond the scope of this book. This section discusses the basics of passive heating and cooling systems.

### PASSIVE SOLAR SYSTEM PROPERTIES

Buildings are exposed to the warming rays of the sun and the cooling influences of natural breezes. They can be designed and oriented to take advantage of the local climate for heating and cooling, with reduced reliance on mechanical systems. With such an approach, windows, skylights, walls, roofs, floors, and the natural features of the site regulate heat flow into or out of a building. Conduction, convection, radiation, and gravity provide the means to move the heat through a building.

Passive solar systems are distinguished from active systems in that they use no power-consuming pumps and because the collection, storage, and distribution facilities are integrated with the building design. In active systems, the collection and storage elements are manufactured products which can be placed in, on, or near the building and tied to it with pipes or ducts. In general, all effective passive system buildings include the following:

- An envelope that will not leak heat in winter or gain excessive heat in summer
- A solar collection area in the form of windows, skylights, glass wall, water bags, or other solar heat traps
- The capability of storing heat for later use, either in the structural elements or in specially provided facilities

### BUILDING ENVELOPE

A passively heated building can collect and store only a limited amount of heat. In all but a few sunny regions, more heat is required for modern comfort than a passive system can deliver. The extra heat is generally supplied by a supplementary source, such as wood stoves, fireplaces, or a conventional heating system. Refer to Sections 7.3 and Chapter 8 for information about the design and use of insulation, doors, windows, glazing, and window treatments to minimize overall heat loss. Earlier portions of this chapter discuss the basic principles of heat control in buildings.

An energy-efficient building should be insulated to optimum levels; have tightly caulked and sealed windows, door frames, and sill plates; have double or triple glazing; and have as few nonsolar-collecting windows as possible.

### PASSIVE SOLAR COLLECTION

Solar energy reaches the earth as *direct* (parallel) or *diffuse* (nonparallel) rays of the sun. Direct radiation is sunlight that reaches the earth relatively unobstructed. Diffuse radiation is reflected from clouds, dust, and obstructions such as buildings. The relative proportion of direct and diffuse radiation, as well as the total radiation, varies from region to region, making a particular solar energy system more feasible in some areas than in others.

Solar energy that strikes a surface is called *insolation* (incident solar radiation). It is measured in *Langleys* (ly). The amount of energy contained in 1 ly is equal to 1 calorie per square centimeter or 3.69 Btus per square foot of surface.

In designing a building to take advantage of passive solar heating, building shape, orientation, and solar collection area should be given consideration.

A building should be shaped and sited for maximum exposure to sunshine in winter and minimum exposure in summer. In northern latitudes, buildings

should expose as much surface to the south as possible. In Pennsylvania, for example, which is at 40 degrees north latitude, a building receives nearly three times as much solar radiation on its south face in January as it does in June; due south is therefore the optimum orientation for a collector surface. The east and west sides, on the other hand, receive two and one-half times more insolation in summer than in winter, making them inappropriate for heat collection in winter. At latitudes below 35 degrees north, buildings are exposed to even more sun on the south side in winter, while the east and west walls receive two to three times more sun in summer than the south walls.

A square building is never the optimum form for solar collection at any latitude, and should be avoided unless building use and site conditions dictate such a shape. Buildings that are elongated on the north-south axis are even less efficient in winter and summer. The optimum shape is usually a form elongated in the east-west direction.

Solar energy reaches a building in one of two forms: visible light, which is short-wave radiation, and infrared light, which is long-wave radiation. Some short-wave radiation is reflected by glass but much passes through. Long-wave radiation, on the other hand, is either reflected or absorbed and conducted or radiated away. After passing through glass, short-wave radiation strikes an interior surface and is transformed into an equivalent amount of heat energy. Part of this heat energy is conducted and convected away; the rest is radiated in the form of infrared (long-wave) radiation, sometimes called *thermal radiation*. Since thermal radiation cannot penetrate glass, it becomes trapped inside the space. This phenomenon is called the *greenhouse effect*. It accounts for a car parked in the sun on a fall day becoming hot and stuffy inside and for an ordinary window's ability to act as a solar collector.

The type of glass used for solar collection can have a significant effect on the efficiency of the system. Clear, heat-absorbing, and reflecting glass all lose about the same amount of heat by conduction but differ widely in the amount of heat transmitted and trapped under the glass. Clear single glass is the most effective heat gainer but loses the most by conduction. Clear double glazing has a comparable percentage of gain but loses less by conduction.

When preventing summer heat gain is important, heat-absorbing or reflecting glass should be considered for south exposures, but only when external shading or screening is not practical. Because reflective glass reduces heat gain year-round, it is desirable only where cooling energy costs are substantially higher than heating costs.

Although skylights and windows are the primary means of passive heat collection, they also are responsible for the greatest winter heat loss and summer heat gain. Passive systems usually employ an insulating or shading technique to prevent unwanted heat loss or gain through the glass collector. Some more traditional approaches to reducing excessive window heat loss and heat gain are described in Chapters 7 and 8.

## STORAGE

The more glass exposed to the sun, the more heat will be trapped behind it. Windows used as passive solar collectors have the disadvantage that even on very cold days, they can quickly overheat the space they serve. In order to soak up the excess collected heat and prevent overheating, and to retain heat for nighttime use, floor and wall surfaces can be utilized as *heat sinks*; that is, they collect and store heat.

Materials such as water, concrete, brick, tile, and stone are ideal heat sinks because they can absorb and store large quantities of heat in a relatively small volume of material. The collected heat is released by radiation to colder surfaces and to the air to maintain a relatively constant temperature. The greater the temperature differential between the mass and its surroundings, the faster the heat is released. Also, the greater the heat storage capacity (*thermal mass*) of the materials in the building, the longer it will take for room air to overheat and the longer a room can remain comfortable without receiving heat from an auxiliary source.

A heat-storage material should be located in or near the direct sun. When it is, a thermal storage capacity equivalent to 30 lb (13.6 kg) of water or 150 lb (68 kg) of masonry should be used for every 1 sq ft (0.929 m²) of south-facing glass. The thermal mass of heat-storage materials in other locations should be increased four times.

Materials vary in their capacity to store heat. The two most common measures of that ability are specific heat and heat capacity.

A material's *specific heat* is the number of Btus (kilojoules) required to raise 1 lb of it 1°F (−17.2°C). For example, water has a specific heat of 1.0, which means that 1 Btu (1.06 kJ) is required to raise the temperature of 1 lb (0.454 kg) of water 1°F (−17.2°C). Since 1 gal of water weighs 8.4 lb (3.8 kg), 8.4 Btus (8.86 kJ) are required for a 1°F (−17.2°C) rise in temperature.

Various materials absorb different amounts of heat while undergoing the same temperature rise. For example, it takes 100 Btus (105.5 kJ) to heat 100 lb (45.4 kg) of water 1°F (−17.2°C) but only 22.5 Btus (23.74 kJ) to heat the same weight of aluminum 1°F (−17.2°C). Therefore, the specific heat of aluminum is 0.225, while water has a specific heat almost four times as high.

A material's *heat capacity* is the amount of heat needed to raise 1 cu ft (0.028 m³) of it 1°F (−17.2°C). Dense, solid materials are effective as heat storage devices because their high density means that considerable weight is contained in each cubic foot of volume. Such materials are therefore often used as storage devices even though water has a greater specific heat. Concrete, for example, has a specific heat about one-quarter that of water, but because of its density, it has a heat capacity of more than half that of water.

## PASSIVE SYSTEMS

Passive solar heating and cooling systems rely on adaptive design, not mechanical devices. They cannot be grouped according to engineering specifications or performance standards but must be categorized according to the design approach they utilize. This, in turn, makes generic discussion of passive design difficult because most designs are the products of their inventor's ingenuity, not of established standards. What has emerged, however, are three distinct design approaches within which most passive systems can be grouped: direct gain, indirect gain, and isolated gain.

*Direct Gain Systems* A *direct gain* passive solar system is the most common passive solar building solution, having many historic precedents, including the cliff dwellings of the Sinaqua and Anasazi Indians of Arizona. The process can be described as *sun to living space to thermal mass*. Solar radiation is collected in the living space and then stored in a material having a large thermal mass, such as thick masonry or concrete floors or walls. Thus, the living space is heated by the solar radiation penetrating the windows and it serves as a "live-in" collector, storage, and distribution facility.

■ A building with a direct gain passive solar energy system must include the following:

■ A large south-facing glazed collector area with the living space exposed directly behind

■ A large floor and wall thermal mass exposed to solar radiation

■ A method for protecting the collector windows and storage mass from heat loss to the exterior

*Indirect Gain Systems* Indirect gain passive solar systems differ from direct gain systems in that they deposit collected heat in a thermal storage mass before it is introduced into a living space. The storage mass usually is located between the glazed collector and the living space so that when the living area begins to cool, heat radiates from the mass into the living space.

The principal advantage of such an approach is that collected heat does not travel through the living space to reach the storage mass. As a result, more heat can be collected and stored before the space becomes too warm for occupant comfort. In addition, radiant heat flow into living spaces can be controlled more readily in such systems than in direct gain systems. Placing movable insulation on the interior face is one typical regulatory technique. When the space begins to overheat, the insulation can be moved across the wall to reduce heat flow.

The three basic types of indirect heat gain systems are the mass Trombe wall, water Trombe wall, and roof pond.

A *mass Trombe wall*, named after its French inventor, Felix Trombe, consists of a glass or plastic panel over a heat-absorbing storage wall. Commonly used materials for the heat storage wall include concrete, brick, stone, and adobe. When possible, the material is painted a dark color to increase its ability to absorb heat. The glazing prevents immediate reradiation of the collected heat to the exterior.

The system operates by permitting solar radiation to pass through the glazed collector and be transmitted directly into the massive storage wall. When the temperature of the wall becomes higher than that of adjoining interior surfaces, it begins to radiate its heat to the interior.

Heated air in the space between the glazing and the mass Trombe wall can also be introduced into the living space, and controllable vents in the mass itself can regulate the heat supply to the living space and draw cooler air into the mass for heating.

A *water Trombe wall* system uses water as the storage mass. Water is an effective storage medium because it has a higher specific heat than most solid massive materials and can therefore store more heat per cubic foot (cubic meter).

A disadvantage of water Trombe walls is that containing a liquid is more difficult than containing a solid such as concrete. Steel drums are the most common containers, but some designers have experimented with concrete block walls in which the voids are filled with vinyl bags containing water. These containers can be left exposed to the interior or embedded in some massive material, such as adobe or concrete.

Convection currents within the water rapidly transfer heat from the collection surface to the entire volume of water. Heat is therefore delivered to the interior spaces at a faster rate than with mass Trombe walls. In addition, the collector surface operates at a lower temperature, so less heat is lost back through the glazed collector before it is absorbed.

A *roof pond*, a variation of the Trombe wall system, utilizes horizontal water containers on the roof of a building to collect and store the sun's energy. Heat is collected by exposing the ponds to the sun by day and covering them with insulation at night to prevent the collected heat from escaping into the cool night air. The solar heat collected in roof ponds is radiated directly to the rooms through a metal ceiling. Thus a continuous, evenly distributed supply of heat is provided throughout the building. Such a system is capable of delivering heat when the water temperature drops as low as 70°F (21.1°C).

There are several disadvantages to roof pond collectors. They require a stronger than normal structure to support their weight, which increases the cost of the building. The water must be contained in plastic bags or some similar structure. Open ponds have fallen into disfavor for several reasons. They tend to contribute

to roof deterioration and leaks, they make roof maintenance more difficult, and they serve as havens for insects and bacteria.

Even contained ponds have disadvantages. Heat from the low-angle winter sun is difficult to collect on a horizontal surface. On a clear December day in Pennsylvania, for example, the sunlight striking a horizontal roof is less than half that hitting an equal area of a south-facing wall. To compensate for this deficiency, more pond area must be provided. Stratification of the heat within a roof pond is another problem, as the warmer, lighter water remains near the top of the pond at night, losing heat to the outside air. The cooler, denser water falls to the bottom, just above the rooms, where heat is needed the most. An additional problem is that in many climates, the roof must be protected against snow accumulation, which could block the sun from the water bags.

In general, roof pond collectors are better suited to the warmer southern and southwestern states, where the sun climbs higher in the sky on a winter day and snow buildup is not a threat. Roof ponds are also well suited for summer cooling, which is of primary importance in such climates.

Insulated panels should be used to cover a roof pond on summer days to keep the sun from warming the ponds, while the cool water absorbs excess heat from the rooms below. At night, the panels must be removed so that the roof pond can radiate heat to the night air. Heat stratification within a roof pond aids this cooling process because the coldest water lies just above the rooms.

*Isolated Gain Systems* Isolated gain passive systems utilize collectors and storage masses that are thermally isolated from the living space. This concept allows the collector and storage unit to function independently, while the building can draw heat as required.

Isolated gain systems can generally be grouped into two categories: those that use a greenhouse area, or similar structure (often called a *sunspace*), to collect and store heat and those that employ the *thermosiphon* principle to move collected heat.

Using a sunspace allows maximum solar collection in this space (and consequent overheating) without affecting the comfort within the primary living space. The collected heat can be admitted

as needed through doors, windows, or dampers.

Atriums, sun porches, greenhouses, and sunrooms are examples of *sunspaces*. In general, this technique requires a glazed collector space which must be attached to and yet distinct from a living space, it must have the proper southern exposure, and the collector space must contain or be thermally linked to a solar storage mass for heat retention and later distribution.

A *thermosiphon* isolated gain system permits heat to build up within a roof, wall, or room-sized collection area before drawing it off to storage or living areas. Thermosiphon (convection), the natural rise of heated air or fluid, is the usual means of transport.

In a typical installation, a glazed collector is located between the direct sun and the living space and is distinct from the building structure. In this space, a convection heat flow occurs when air heated by the sun rises naturally into an appropriately placed living space or storage mass, causing cooler air or liquid to fall again. Continuous heat-gathering circulation is thereby established.

### SOLAR COOLING SYSTEMS

Within limits, solar energy can be used to cool buildings.

*Active Systems* Active solar collectors can be used to produce chilled water that can subsequently cool air. In such a system, heat from water heated in a solar collector operates a generator that makes a lithium bromide water solution. This fluid then passes into an evaporator, where it produces chilled water. The chilled water is then pumped to terminal devices, such as fan coil units. The heat picked up in the chiller is picked up by a condenser and is then dissipated by a cooling tower.

Active solar water chillers are expensive and complicated, and are therefore suitable only for large buildings. They are not widely used in the United States

Other active solar-powered water chilling systems are being developed and may come into general use in the future.

*Passive Systems* On summer nights, when the outside air is at its lowest temperature, introduction of that air into a house will carry excess heat away from interior floor and wall surfaces. In the process, those surfaces will be cooled and will therefore be able to absorb and store excess heat trapped in the building the following day. This reduces summer heat buildup within the building and aids cooling.

# 17.7 Central HVAC Equipment

## 17.7.1 CENTRAL HEAT GENERATION EQUIPMENT

*Heat generation* is the elevation of a heating medium (air, water, or steam) to a temperature sufficient to heat a building. Devices that perform this function and the fuels used to fire them are discussed in this section.

### 17.7.1.1 Fuel Types

Furnaces in small buildings are powered using gas, oil, or electricity. Boilers in both small and large buildings may be fueled by gas, oil, coal, or electricity.

#### GAS

Furnaces and boilers may be fueled by either natural gas or propane. Most large gas-fired boilers use natural gas.

#### OIL

Oil-fired furnaces generally run on No. 2 fuel oil. Oil-fired boilers run on either No. 2, No. 4, No. 5, or No. 6 fuel oil. In general, the larger the heating load, the heavier the oil. No. 2 oil will produce about 140,000 Btu/gal (39 022.46 kJ/L), while No. 6 will produce about 150,000 Btu/gal (41 809.78 kJ/L).

#### COAL

Anthracite coal was a major furnace and boiler fuel in the past and is still used today, but it has somewhat fallen out of favor because of the difficulty and cost of preventing particulates that result from its burning from reaching the atmosphere.

#### ELECTRICITY

Electricity is used extensively to fuel furnaces, especially in southern regions of the United States. It can also fire boilers, but its use for that purpose is not widespread and is usually limited to small installations.

### 17.7.1.2 Chimneys and Vent Stacks

The purpose of chimneys and vent stacks is to conduct combustion products to a safe point of disposal. This is in the atmosphere outside the building, usually above the roof. The minimum height above a roof is dictated by code, but is usually 3 ft (914 mm) above the roof and 2 ft (610 mm) above any part of a roof within 10 ft (3048 mm) of the chimney or vent. Chimneys are designed to create a draft. Vent stacks work with forced draft designs. Both may be supported on appliances such as a boiler, when the appliance can carry the load, but offset locations are preferable.

### 17.7.1.3 Boilers

Boilers produce heated water or steam by burning fuel adjacent to a tank containing water. Their designed efficiency depends on the number of times the hot gases produced by a burner pass through the water in the boiler (number of passes) and the air-to-fuel ratio in the burner. In large installations, more than one boiler is generally used.

Boilers used in small residences could be, and often are, called *furnaces* but are actually boilers in the sense that they produce hot water or steam by burning fuel. Such boilers are usually made of steel or iron.

There are two basic boiler types in general use in larger buildings: cast iron sectional boilers and tube-type boilers.

#### CAST IRON SECTIONAL BOILERS

A cast iron sectional boiler has a water tank made from hollow cast iron sections bolted together side by side, like donuts stacked in a box. A burner blows hot gases into a combustion chamber either sealed beneath or housed within the water tank. A cast iron sectional boiler is a single-pass boiler, but a large portion of the tank is exposed to the combustion chamber, which increases its efficiency.

#### WATER TUBE BOILERS

A water tube boiler consists of a burner and a combustion chamber, as well as a water tank connected to a series of tubes that contain water. Hot gases produced by the burner pass over the water tubes on their way to the flue. The heated water in the tubes rises into the water tank. A water tube boiler is a single-pass unit.

#### FIRE TUBE BOILERS

Fire tube boilers consist of a burner and its combustion chamber, as well as a water tank containing a series of tubes connected to the combustion chamber at one end and to the flue at the other end. Hot gases produced by the burner pass through the tubes on their way to the flue, heating the water in the tank as they pass through. The water is then available to distribute throughout the heating system.

In a single-pass fire tube boiler, the tubes extend from the combustion chamber directly through the tank to the flue. In a two-pass boiler, some fire tubes extend from the combustion chamber to a secondary chamber on the opposite end of the boiler, and additional tubes extend from the secondary chamber back through the water tank to the flue. In a three-pass boiler, tubes carrying hot gases pass through the tank three times.

#### WET-BACK SCOTCH MARINE BOILER

A wet-back Scotch marine boiler is similar to a fire tube boiler. The major difference is that the combustion chamber is contained within the water tank. These boilers are available as either two-pass or three-pass units.

#### OTHER CONFIGURATIONS

Boilers in other configurations are also used. Often they are a combination of the types mentioned here.

### 17.7.1.4 Furnaces

Most types of furnaces for single-family houses, individual apartment or townhouse units, and small commercial establishments are manufactured for either heating only or for initial or later installation of air conditioning. Those intended to provide cooling have a larger blower for more volume of air output. Such furnaces are quieter because the larger fan can operate more slowly and still move a larger volume of air.

Furnaces for small buildings are available in *upflow*, *downflow*, and *horizontal* configurations. Upflow furnaces may have a vertical design (high boy) or a horizontal design (low boy). In a high boy upflow furnace, conditioned air is discharged at the top of the furnace and return air enters at the bottom. In a low boy upflow furnace, conditioned air leaves from the top of the furnace at one end and return air also enters from the top of the furnace, but at the opposite end.

In a downflow (counter flow) furnace, conditioned air is discharged from the bottom of the furnace and return air enters at the top. Downflow furnaces are used primarily in underfloor air plenum systems (see Section 17.6.2.1).

*Horizontal* furnaces are usually mounted in crawl spaces and attics. In a horizontal furnace, conditioned air is discharged from one end of the furnace and return air enters from the other end.

An air conditioning unit may be added at the discharge end of a furnace in any of the configurations mentioned.

## 17.7.2 CENTRAL COOLING EQUIPMENT

### 17.7.2.1 Refrigeration Cycle

Building cooling systems are made possible by the following physical laws:

1. Compressing a gas releases its *latent* heat and changes it to a liquid.
2. Lowering the pressure of a liquid causes it to absorb latent heat and vaporize.

The cycle of refrigeration consists of compressing a gas to a liquid state and releasing its latent heat, leaving a cooled liquid. The liquid is then caused to absorb heat from the interior air, which vaporizes the liquid and lowers the temperature of the air.

### 17.7.2.2 Fuel

Small cooling units are powered by either gas or electricity. Chillers are powered by electric motors or steam turbine drives, or are gas absorption or gas engine-driven coolers. Pumps, of course, require electricity.

### 17.7.2.3 Cooling Components

Every building cooling system consists of a compressor, a condenser, and an evaporator. The compressor compresses a gas (refrigerant) until it liquefies. In the condenser, which is located close to the compressor, the latent heat in the refrigerant is released to the surrounding medium, which may be either air or water but is usually outside air. The refrigerant flows on to the evaporator, where it absorbs latent heat from the surrounding medium (usually air or water), changing the refrigerant back to a gas and cooling the air or water.

The compressor and the condenser are usually located in a single unit either on the ground or on the roof. The evaporator is usually inside the building.

Where considerable waste heat is available to augment the system, an absorber and a generator sometimes perform the function of the compressor.

Larger systems often employ one or more packaged mechanical refrigeration water-chilling units (*chillers*). Chilled water is circulated from the chillers throughout the building to air-handling units or directly to terminal units, such as fan coil units. There are often one or more cooling towers associated with the condenser in these systems to improve their efficiency. Chilled water systems are complex and vary considerably with the size of the system, type of terminals, and control systems used.

Condenser types include evaporative condensers, water-cooled condensers, dry coolers, evaporative coolers, and cooling towers.

## 17.7.3 CENTRAL HEATING-COOLING EQUIPMENT

### 17.7.3.1 Central Station Air-Handling Units

In single-family residences and other small building systems, the air handler is the furnace. In larger systems, air handlers are located in central locations in mechanical equipment rooms. In very large buildings, more than one air handler will probably be required. Air handlers in large buildings are usually housed in mechanical equipment rooms. Air handlers require access to a supply of fresh air and a means to exhaust excess air to the exterior.

Large air handlers consist of:

- An outside air intake which feeds into an outside-air plenum
- An automatic damper to control the amount of outside air entering the air handler
- A mixing plenum where fresh air and air from the return air system, which is also controlled by an automatic damper, are mixed
- A preheat coil to initially raise the temperature of the mixed fresh and return air
- A cooling coil to cool the air
- A heating coil to produce the desired air temperature near the outflow
- A fan to move the air

All these components are housed in a large metal compartment. A return-air fan is also required, but this is housed in the return-air duct. When excess air is present in the system, a damper on the return air duct opens to permit the excess to vent to the exterior.

### 17.7.3.2 Evaporative Air Cooling Equipment

In evaporative cooling, moisture is added to air, which increases the relative humidity and decreases the dry-bulb temperature. The resulting cooled air is then blown into indoor space. Typically, evaporative cooling equipment is mounted on a building's roof, but it may also be placed inside the building. Evaporative cooling equipment uses water to treat large volumes of hot, humid outside air.

Evaporative cooling is done using the *direct* process or the *direct and indirect process*. In the direct process, outdoor air is drawn into the evaporative cooler, treated, and introduced directly into the treated space. In the daytime portion of the direct and indirect process, outside air is drawn into a rock bed, which functions as a heat exchanger. The air is cooled by contact with the rocks, and is then drawn into an evaporative cooler and discharged into the conditioned space using the direct process. In the nighttime portion of the direct and indirect process, air from the interior is passed over the rock bed, picking up heat stored there during the daytime operation, and is then discharged to the exterior. Cooler exterior air is admitted through dampers to the evaporative cooler, where it is treated and discharged into the treated space.

### 17.7.4 ENERGY RECOVERY SYSTEMS

Heat recovery systems pick up heat that would otherwise be wasted and use it to provide auxiliary space heating, to preheat outside air intakes, to reheat supply air in air conditioning systems, and to preheat domestic water and boiler feed water. Waste heat is recovered from air, liquids, or gas. Sources include heat generated by people, lights, appliances, motors, generators, refrigeration systems, exhaust air from kitchens, toilets, locker rooms, swimming pools, and many other sources.

Heat recovery ventilators extract heat from the air as it is exhausted and transfer it to incoming air (or the reverse in the summer), permitting a tight, energy-efficient building to be ventilated while retaining the heat energy used to maintain the indoor environment. They are sometimes recommended for either very cold or very hot, humid climates. Depending on the climate, heat recovery ventilators can be up to 80% efficient in recovering energy.

## 17.7.5 EMBEDDED PIPES AND DUCTS

When ducts and pipes must pass through slabs, care should be taken to prevent them from breaking by isolating them from the slab.

### 17.7.5.1 Ducts

When galvanized metal ducts are embedded, calcium chloride should not be used in the concrete mixture. Aluminum should not be embedded in concrete.

Heating ducts embedded in a slab on grade may be of metal, fibrous cement, wax-impregnated paper, ceramic tile, concrete, or plastic. Fibrous cement ducts with watertight joints may be set on a sand leveling bed and backfilled with sand to the underside of the slab. Precautions should be taken to ensure that the position of ducts is not disturbed during the placing of concrete.

Metal and paper ducts should be completely enclosed in at least 2 in. (50.8 mm) of concrete to prevent moisture from collecting in them; joints may be taped or cement grouted. These ducts often must be weighted so that they will not rise in freshly poured concrete. When the bottom of ductwork adjacent to a slab perimeter is below exterior finish grade, the duct system should be of ceramic tile, fibrous cement, or concrete pipe. Joints that are completely or partially below exterior grade should be watertight. In areas where termite control is required, these joints should be resistant to termite penetration and soil treatment chemicals.

If the slab is not reinforced (Type I slab) (see Section 3.12), welded wire mesh reinforcement (fabric) should be placed to extend 19 in. (482.6 mm) on each side of the centerline of the duct or to the slab edge, whichever is closer (Fig. 17.7-1a).

### 17.7.5.2 Pipes

Pipes that pass through a slab vertically must be provided with space for expansion and with flexible connections if necessary to accommodate differential movement between the slab and the soil.

**FIGURE 17.7-1** Details for embedding heating sources in a slab.

When galvanized or copper pipe is embedded in a slab (Fig. 17.7-1b), calcium chloride should not be used in the concrete mixture. Aluminum should not be embedded in concrete.

Type II slabs (see Section 3.12) may be heated by circulation of heated liquids through embedded pipes. Usually, ferrous or copper pipe is used with approximately $2\frac{1}{2}$ to 3 in. (63.5 to 76.2 mm) of cover over the pipe. Where pipes pass through control joints or construction joints, provision should be made for possible movement across the joint. The piping should be pressure-tested before the concrete is placed, and air pressure should be maintained in the pipe during concrete placement; water pressure should not be used for testing. After the concrete is placed, the slab should not be heated until curing of the concrete has been completed. Slabs should be warmed gradually using lukewarm liquid in the pipe to prevent the cold concrete from cracking.

Using embedded pipes for summer cooling is not recommended with slab-on-ground construction. Under conditions of high humidity, condensation may form on the cooled slab which may damage flooring and support mold or fungus growth.

## 17.8 Decentralized HVAC Equipment

Decentralized HVAC equipment includes the following:

- Direct refrigeration units
- Terminal units
- Radiant heating units
- Fireplaces
- Stoves
- Humidity control units

### 17.8.1 DIRECT REFRIGERATION UNITS

Direct refrigeration units include packaged units and heat pumps.

#### 17.8.1.1 Packaged Units

*Packaged units*, also called *unitary* systems, are self-contained systems located within or adjacent to the space being conditioned that have all the heating, cooling, or both heating and cooling components in a single unit. Examples include window air conditioning units, rooftop units, and systems that consist of an externally mounted condensing unit and an interior evaporator and air handler. The latter two often provide both heating and cooling.

#### 17.8.1.2 Heat Pumps

Heat pumps draw heat from one medium and transfer it to another medium. The source medium may be exterior air (air exchange), geothermal (ground source), or interior air or water (water loop). The receiving medium may be air or water. Heat pumps have the capacity to either heat or cool.

When cooling, air-exchange and geothermal heat pumps function as direct refrigerant air conditioners, picking up heat from indoor air and depositing it either into the outdoor atmosphere (air exchange) or into the ground (geothermal).

One major difference between a heat pump and an air conditioner is that the heat pump contains a reversing valve, which allows the refrigerant to pick up heat from a warm exterior medium (air, water, or earth) and deposit it indoors. Most residential heat pumps function in the heating cycle by absorbing heat from the exterior air. In geothermal heat pumps, which are used more often in commercial installations, heat is recovered from relatively high earth temperatures below ground. In the summer, the earth acts as a heat sink to help cool the building. Like earth-sheltered building design, ground source heat pumps take advantage of the stability of underground temperatures. Long lengths of copper tubing are buried either horizontally or vertically in the earth and circulated with a heat-exchanging medium.

A form of heat pump called a *water loop* heat pump is used to absorb heat from the air in ducts or water in hot or chilled water piping serving building portions that are too warm and deposit that heat in cooler areas. A water loop heat pump is usually actually a series of heat pumps linked together. Excess heat picked up by them is stored in a water tank until needed. Water loop heat pumps are energy-saving devices that help to balance heat requirements in buildings where heat gain or loss varies considerably from time to time.

Some gas-fired heat pumps are available, but most operate on electricity. Electrically operated heat pumps are not electric heaters. Their efficiency in the heating cycle is directly related to the temperature of the medium from which they pick up heat. As the temperature of the medium drops, the efficiency of a heat pump is reduced, and the pump produces less heat. For this reason, all heat pumps require an auxiliary heater to produce heat when the temperature of the medium is too low. Auxiliary heaters in electrically operated heat pumps are powered by electricity, the least efficient and most costly fuel available. Auxiliary heaters in gas-fired heat pumps, of course, are gas heaters.

Under most circumstances, heat pumps become uneconomical to operate when the temperature of the medium falls below about 40°F (4°C). Therefore, air-exchange heat pumps work well and are economical to operate in southern climates, but in northern climates they may be less economical than other heating/cooling systems. North of a line stretching from Virginia to northern California, air-exchange heat pumps are not very efficient. Geothermal heat pumps are efficient in more northern climates, of course, since the far underground earth temperature is about the same everywhere.

### 17.8.2 TERMINAL UNITS

*Terminal units* are the components of a central heating system that are used to translate the heat generated by a boiler into a usable temperature rise in building spaces. Some terminal units also use water that has been chilled by an HVAC system's cooling components to create a usable temperature fall in building spaces. Terminal units include those discussed in Sections 17.8.2.1 to 17.8.2.7.

### 17.8.2.1 Convection Heating and Cooling Units

Radiators and convectors are capable of heating only, and require that air for ventilation be supplied from another source. Fan coil units are capable of heating and cooling. Air is supplied either through a central system or from air intakes through the wall adjacent to the unit. Induction units are capable of heating and cooling but use air supplied by a central system to distribute the heated or cooled air. Unit ventilators are capable of heating or cooling and obtain fresh air from intakes through an exterior wall connected to the unit.

### 17.8.2.2 Radiators

Cast iron *radiators*, the oldest form of hot water- and steam-heat distributer still in use today, heat the air directly by contact with a water- or steam-filled cast iron device.

### 17.8.2.3 Convectors

*Convectors* are heat transfer units, usually finned-tube radiators, encased in a metal cabinet. Air passes through the cabinet by gravity, rising as it is heated. Heated water for water-based convectors is supplied by pipes from a boiler.

Convection-type electric heaters include *baseboard heaters*, which are electric heating elements encased in a low metal cabinet. A variation, *hydronic baseboard heaters*, contain a liquid that stores heat and increases the efficiency of the unit. A *cabinet convector* is an electric heater in a metal cabinet that can be surface mounted or recessed. Cabinet convectors are higher than baseboard units and therefore can be shorter to deliver the same amount of heat.

When convection heaters are the prime source of heat, ventilation must be provided by another system.

### 17.8.2.4 Fan Coil Units

*Fan coil units* are capable of both heating and cooling. They consist of a finned-tube coil, an air blower and motor, and a return air damper with a filter, all housed in a metal cabinet. When arranged for both heating and cooling, they are supplied either by a four-pipe system and a double or split coil or by a two-pipe system with external mixing valves to supply either hot or chilled water as required. When used for heating only, they are supplied by a two-pipe hot water system.

In the cooling cycle, fan coil units both dehumidify and cool air by sucking it through a filter into the unit from the air in a room. The air then passes over the coil, which contains circulating chilled water, and is blown back into the room. The condensate generated by the process drips into a pan and is carried away from the unit, either into a condensate drain piping system or through an adjacent wall to the exterior.

In the heating cycle, air is drawn into the unit through the filter, is caused to pass over the coil, which now contains hot water, and is then blown back into the room.

Fan coil units can be arranged to take in ventilation air from a louvered opening in the wall immediately behind the unit or can be used with supplementary air supplied by a second system.

Fan coil units are available in several model types. Standard wall-mounted units are about 25 in. (635 mm) tall, 9 to 12 in. (228.6 to 304.8 mm) deep, and vary in length according to design requirements between 30 and 84 in. (762 and 2133.6 mm). High-rise units run from floor to ceiling and vary in size, depending on their capacity. Low-profile units are only about 14 in. (356 mm) high but are somewhat limited in capacity due to their reduced size. Small units that can be furred into soffits are also available

### 17.8.2.5 Induction Units

*Induction units* are capable of both heating and cooling. When supplied by a two-pipe system, they are capable of either heating or cooling, depending on whether hot or chilled liquid is circulated through them. When a four-pipe system is used, induction units can be switched quickly from heating to cooling in response to the thermostat setting within a space. The cooled or heated air circulated by an induction unit is supplied by a central air-supply system. The unit mixes air from that system with room air in a ratio of 20% incoming air to 80% room air to achieve the desired air temperature.

Condensation from induction units is either collected through a central system or passed directly through an exterior wall at the individual unit.

### 17.8.2.6 Unit Heaters

*Unit heaters* consist of a heat transfer unit, often a finned-tube radiator, and a fan encased in a metal cabinet. Air is forced by the fan to flow across the radiator. The heating medium may be either hot water or electricity. Unit heaters are often suspended from the ceiling or a wall, but floor-mounted units are also used. Floor-mounted units may be either surface-mounted or recessed. Electric heaters may be portable.

### 17.8.2.7 Unit Ventilators

*Unit ventilators* are similar to floor-mounted unit heaters but can be designed to take in air through openings in the exterior wall at each unit ventilator.

### 17.8.3 RADIANT HEATING UNITS

*Radiant heating units* are capable of heating only, and require that air for ventilation and cooling be supplied from another source.

### 17.8.3.1 Hot Water Radiant Heaters

RADIANT PANELS

A *radiant panel* is a wall or ceiling surface-mounted or recessed panel containing heating pipes.

RADIANT FLOORS

In a *radiant floor* system, hot water from the boiler passes through pipes embedded in the floor of a space, providing uniform heat throughout the space. It is also possible to embed pipes in a ceiling, but this use is rare because of the possibility of damage due to leaks. In addition, heated air rises, making the placement of a heat source at the ceiling inefficient.

### 17.8.3.2 Electric Radiant Heaters

Electric heating systems are widely used. They are easy to control and have a low installation cost. Unfortunately, they are costly to operate. None of the heaters described here is capable of providing cooling. In addition to the heaters described in this section, electricity can be used in centralized systems to power furnaces and boilers (see Section 17.7).

Electric radiant heaters include *infrared heaters* that heat by means of a heat lamp; *radiant panels*, which can be wall or ceiling mounted; and *resistance cable* systems (see Section 9.3). When radiant heaters are the prime source of heat, ventilation must be provided by another system.

### 17.8.4 FIREPLACES AND STOVES

Open fire is the oldest form of heat used in dwellings. Open wood-burning fireplaces are inefficient and contribute to both indoor and outdoor pollution. Modern fireplaces with closed fronts and heat chambers and circulating fans are much more efficient and produce fewer indoor pollutants. Modern gas logs have an appearance similar to that of wood logs without the polluting side effects. Newer wood- and gas-burning stoves are even more efficient.

### 17.8.5 HUMIDITY CONTROLS

Humidity control equipment includes humidifiers and dehumidifiers. As their names imply, humidifiers add water vapor to the air and dehumidifiers remove moisture from the air. Either adding or removing moisture may be done in the central air-supply system or on a space-by-space basis. The effect of adding moisture to or subtracting it from the air causes a latent heat transfer that can be predicted using psychometrics (see Section 17.1).

# 17.9 Additional Reading

More information about the subjects discussed in this chapter can be found in the references listed in Section 17.10 and in the following publications.

Anastisi, Leonard. 2002. "Air Barrier Systems." *The Construction Specifier* (March): 30 (www.csinet.org).

Dagostino, Frank R. and Joseph B. Wujek. 2005. *Mechanical and Electrical Systems in Construction and Architecture*, 4th ed. Upper Saddle River, NJ: Prentice Hall (www.prenhall.com).

Keeler, Russ. 2002. "Air Conditioning Systems of the Future—Thermal Storage and Low Temperature Air." *The Construction Specifier* (January): 40 (www.csinet.org).

Kutz, Myer. 2005. *Mechanical Engineers' Handbook*, 3rd ed. Hoboken, NJ: John Wiley & Sons, Inc. (www.wiley.com).

Lechner, Norbert. 2000. *Heating, Cooling, Lighting: Design Methods for Architects*, 2nd ed. Hoboken, NJ: John Wiley & Sons, Inc. (www.wiley.com).

McQuiston, Faye C., Jerald D. Parker, and Jeffrey D. Spitler. 2000. *Heating, Ventilating, and Air Conditioning Analysis and Design*, 5th ed. Hoboken, NJ: John Wiley & Sons, Inc. (www.wiley.com).

Stein, Benjamin, et al. 2006. *Mechanical and Electrical Equipment for Buildings*, 10th ed. Hoboken, NJ: John Wiley & Sons, Inc. (www.wiley.com).

Tao, William K. Y. and Richard R. Janis. 2005. *Mechanical and Electrical Systems in Buildings*. Upper Saddle River, NJ: Prentice-Hall (www.prenhall.com).

Vittori, Gail. 2004. "Greening Divisions 15 and 16." *The Construction Specifier* (November): 70 (www.csinet.org).

# 17.10 Acknowledgments and References

ACKNOWLEDGMENTS

We gratefully acknowledge the assistance of the following organizations and individuals in preparing this chapter. We are also indebted to them for permission to use their illustrations when requested and for the use of their publications as references.

American Forest & Paper Association (AF&PA) (www.afandpa.org)

American Society of Heating, Refrigerating, and Air Conditioning Engineers (ASHRAE) (www.ashrae.org)

APA—The Engineered Wood Association(APA) (www.apawood.org)

Brick Industry Association (BIA) (www.bia.org)

National Association of Home Builders (NAHB) Research Foundation (www.nahbrc.org)

Small Homes Council, Building Research Council

Southern Forest Products Association (SFPA) (www.sfpa.org)

U.S. Department of Housing and Urban Development (HUD) (www.hud.gov)

U.S. Environmental Protection Agency (EPA) (www.epa.gov)

Western Wood Products Association (WWPA) (www.wwpa.org)

## REFERENCES

We would also like to thank the authors and publishers of the publications in the following list for their contribution to our research for this chapter.

Air-Conditioning and Refrigeration Institute (ARI). Arlington, VA (www.ari.org).

———. 2003. 210/240, *Unitary Air-Conditioning and Air-Source Heat Pump Equipment.*

———. 1998. 325, *Ground Water-Source Heat Pumps.*

———. 1998. 330, *Water-Source Heat Pumps.*

———. 2000. 340/360, *Commercial and Industrial Unitary Air-Conditioning and Heat Pump Equipment.*

———. 2000. 350, *Sound Rating of Non-Ducted Indoor Air Conditioning Equipment.*

———. 2002. 365, *Commercial and Industrial Unitary Air-Conditioning Condensing Units.*

———. 1986. 370, *Sound Rating of Large Outdoor Refrigerating and Air-Conditioning Equipment.*

———. 1991. 410, *Forced-Circulation Air-Cooling and Air-Heating Coils.*

———. 1999. 430, *Central Station Air Handling Units.*

———. 1998. 440, *Room Fan-Coils and Unit Ventilators.*

———. 1999. ARI 450, *Water Cooled Refrigerant Condensers, Remote Type.*

———. 1997. 520, *Positive Displacement Condensing Units.*

———. 1996. 610, *Central System Humidifiers for Residential Applications.*

———. 1996. 640, *Commercial and Industrial Humidifiers.*

———. 1998. 840, *Unit Ventilators.*

———. 1993. 850, *Commercial and Industrial Air Filter Equipment.*

———. 2001. ARI 870, *Direct Geoexchange Heat Pumps.*

———. 2002. 1010, *Self-Contained, Mechanically-Refrigerated Drinking-Water Coolers.*

———. 1997. Guidelines B, *Roof Mounted Outdoor Air-Conditioner Installations.*

Air Conditioning Contractors of America (ACCA). Arlington, VA (www.acca.org). *Manual C—What Makes a Good Air Conditioning System.*

———. *Manual CS—Commercial Applications, Systems, and Equipment.*

———. *Manual D—Residential Duct Systems.*

———. *Manual G—Selection of Distribution Systems.*

———. *Manual H—Heat Pump Systems: Principles and Applications.*

———. *Manual J—Residential Load Calculations.* 8 ed.

———. 2005. *Manual N—Commercial Load Calculations.*

———. *Manual P—Psychrometrics: Theory and Applications.*

———. *Manual Q—Low Pressure, Low Velocity Duct System Design.*

———. *Manual RS—Comfort Air Quality and Efficiency by Design.*

———. *Manual S—Residential Equipment Selection.*

———. *Manual T—Air Distribution Basics for Residential and Small Commercial Buildings.*

———. *Manual 4—Installation Techniques for Perimeter Heating and Cooling.*

Air Movement & Control Association (AMCA). Arlington Heights, IL (www.amca.org). 2003. 99, *Standards Handbook.*

———. 1986. 500, "Laboratory Methods of Testing Louvers for Rating."

———. 2003. 501, "Application Manual for Air Louvers."

American Plywood Association (now APA—The Engineered Wood Association), American Wood Council, National Forest Products Association, Southern Forest Products Association, Western Wood Products association. *Plen-wood System.* Tacoma, WA: American Plywood Association (www.apawood.org).

American Society of Heating, Refrigerating and Air-Conditioning Engineers (ASHRAE). Atlanta, GA (www.ashrae.org).

———. 2001. *Handbook—Fundamentals.*

———. 2003. *Handbook—HVAC Applications.*

2004. *Handbook—HVAC Systems and Equipment.*

———. Standard 55, "Thermal Environmental Conditions for Human Occupancy."

———. Standard 62.1, "Ventilation for Acceptable Indoor Air Quality."

———. Standard 62.2, "Ventilation for Acceptable Indoor Air Quality in Low-Rise Residential Buildings."

———. Standard 90.1, "Energy Efficient Design of New Buildings Except Low-rise Residential Buildings."

———. Standard 90.2, "Energy Efficient Design of New Low-rise Residential Buildings."

*Architectural Graphic Standards. See* Ramsey et al. (2000).

ASTM International (ASTM). West Conshohocken, VA (www.astm.org). Standards:

C177, "Standard Test Method for Steady-State Heat Flux Measurements and

Thermal Transmission Properties by Means of the Guarded-Hot-Plate Apparatus."

C1199, "Standard Test Method for Measuring the Steady-State Thermal Transmittance of Fenestration Systems Using Hot Box Methods."

Bergoust, Don. 1997. "Preventing HVAC Failures." *The Construction Specifier* (June): 65 (www.csinet.org).

Bobenhausen, William and Catherine Coombs. 1997. "Lean, Clean, and Green HVAC Systems." *The Construction Specifier* (August): 44 (www.csinet.org).

Brick Industry Association (BIA). Reston, VA (www.bia.org). *Technical Notes on Brick Construction:*

No. 4, rev. Jan. 1982 (reissued Sept. 1997). "Heat Transmission Coefficients of Brick Masonry Walls."

No. 4B, rev. Feb. 2002. "Energy Code Compliance of Brick Masonry Walls."

No. 43, rev. Jun. 1981. "Passive Solar Heating with Brick Masonry Walls."

No. 43D Sept./Oct. 1980 (reissued Sept.1988). "Brick Passive Solar Heating Systems, Part 4—Material Properties."

No. 43G Mar./Apr.1981 (reissued Sept. 1986). "Brick Passive Solar Heating Systems, Part 7—Details and Construction."

Dagostino, Frank R. and Joseph B. Wujek. 2005. *Mechanical and Electrical Systems in Construction and Architecture*, 4th ed. Upper Saddle River, NJ: Prentice Hall (vig.prenhall.com).

Eubank, Huston, Gary L. Luepke, Daniel L. Price, and Valery W. Bennett. 2002. "Specifying for Improved Indoor Air Quality." *The Construction Specifier* (August): 30 (www.csinet.org).

Gartland, Lisa. 2000. "Three Overlooked Ways to Save Energy in Your Building." *The Construction Specifier* (September): 17 (www.csinet.org).

Haas, Steve. 1998. "Five Ways to Reduce HVAC Duct Noise." *The Construction Specifier* (February): 63 (www.csinet.org).

International Code Council (ICC). Falls Church, VA (www.iccsafe.org). 2006. International Codes:
*International Building Code*®.
*International Energy Conservation Code*®.
*International Mechanical Code*®.
*International Residential Code*®.

King, Michael. 2001. "Mechanical and Electrical Specifications." *The Construction Specifier* (June): 27 (www.csinet.org).

Long, Jerome A. 1995. "Light Commercial/Residential Heat Pump." *The Construction Specifier* (June): 60 (www.csinet.org).

Mingo, Eric. 1995. "Radiant Floor Heating." *The Construction Specifier* (June): 58 (www.csinet.org).

Nall, Daniel H. 1998. "Underfloor Air Delivery Systems." *The Construction Specifier* (February): 45 (www.csinet.org).

National Climatic Data Center. 1971–2000. *Climatography of U.S. number 81, Monthly Station Normals of Temperature, Precipitation, and Heating and Cooling Degree Days*. Ashville, NC: National Climatic Data Center (www.ncdc.noaa.gov).

National Institutes of Standards and Technology (NIST) of the U.S. Department of Commerce 1977. *Window Design Strategies to Conserve Energy*. Gaithersburg, MD: NIST (www.nist.gov).

Ramsey, Charles George, Harold Reeve Sleeper, and John Ray Hoke, Jr. 2000. *Architectural Graphic Standards*, 10th ed. Hoboken, NJ: John Wiley & Sons, Inc. (www.wiley.com).

Sheet Metal and Air Conditioning Contractors National Association (SMACNA). Chantilly, VA (www.smacna.org).
———. 1987. *HVAC Systems—Applications*.
———. 1990. *HVAC System—Duct Design*.
———. 1994. *HVAC Systems Commissioning Manual*.
———. 1995. *HVAC Duct Construction Standard—Metal and Flexible*.
———. 1995. *Thermoplastic Duct (PVC) Construction Manual*.
———. 1998. *Indoor Air Quality—A Systems Approach*.
———. 1998. *Residential Comfort System Installation Standards Manual*.
———. 2002. *Fire, Smoke and Radiation Damper Installation*.
———. 2002. *HVAC Systems—Testing, Adjusting, and Balancing*.
———. 2003. *Architectural Sheet Metal Manual*.
———. 2003. *Fibrous Glass Duct Construction Standards*.
———. 2003. *Fibrous Glass Duct Construction Standards*.

U.S. Department of Housing and Urban Development (HUD). 1994. *Minimum Property Standards for Housing*. Washington, DC: HUD (www.hud.gov).

# Electrical

# Introduction

Even when an adequate amount of electricity is made available, the extent to which it can be effectively used in a building is determined by the interior wiring. Starting at the service entrance, where electricity enters a building, a wiring system must deliver the correct amount of electricity to every circuit and outlet. Circuits and outlets are either general-purpose ones serving portable devices or are designed to serve a specific purpose. Switch controls are located with both convenience and safety in mind.

Electrical appliances and devices are constantly being improved to provide higher standards of performance and convenience. The resulting increases in wattage are reflected in greater demands on building wiring systems.

The number of appliances and other electrical devices is also increasing. Work-saving and time-saving devices, such as computers, proliferate in offices, shops, factories, institutional buildings, and homes. New homes contain automatic washers, electric clothes dryers, dishwashers, food waste disposers, and freezers. Almost every new commercial, institutional, and residential building is air-conditioned. Even though electric heating systems, a mainstay in some parts of the country for single-family houses when electrical rates were once lower, are not so common today. They have often been replaced by electric heat pumps, which use somewhat less energy than do electrical heating systems but nevertheless still use considerable energy. Such conveniences place large demands on a building's electrical service and wiring system. When loads outstrip the capacities of the wiring system, operating efficiency suffers. When voltage drop becomes excessive in a circuit, poor lighting and inefficient appliance operation result. A 5% voltage loss, for example, produces a 10% loss of heat in an appliance and a 17% loss of light from an incandescent lamp. However, the greatest problems presented by overloading electrical systems are related to safety. Overloaded circuits and undersized circuit breakers may lead to damaged equipment, electric shock, or fires.

This chapter includes a primer on the fundamentals of electricity and discusses facility (within the building) electrical service and distribution; various electrical systems and equipment; the planning of adequate electrical power and lighting systems; and sustainability considerations including conservation of energy. Electrical utilities supplying power to the building are discussed in Chapter 23. Communication systems such as entrance signals, intercommunications systems, television and radio, and telephone are addressed in Chapter 19. Intrusion detection and fire detection and alarm systems are covered in Chapter 20.

# Sustainability Considerations

Materials used in electrical systems can have serious adverse effects on the environment. Wiring and switches are discussed later in this section.

Average consumption of electrical power has increased 10-fold in the past 50 years, but recent energy consciousness and an energy-saving movement, mostly in response to federal government action, have tended to abate the increase somewhat. Despite increased generation of electricity by nuclear plants, hydroelectric stations, and other nonfossil means, most of our electrical production today is generated using coal, gas, or oil as a fuel. Spiraling electrical production reduces our nonrenewable fossil fuel reserves and results in large expenditures to prevent the severe adverse environmental damage caused by burning these fuels. The current national commitment to energy conservation and pollution abatement places responsibility on a designer to make electrical systems not only convenient and efficient, but also frugal users of precious energy. The *International Residential Code* requires compliance with the *International Energy Conservation Code*. The *International Energy Conservation Code* cites the American Society of Heating, Refrigerating and Air-Conditioning Engineers (ASHRAE) Standard 90.1, "Energy Standard for Buildings Except Low-Rise Residential Buildings," as a requirement. Standard 90.1 requires the design of electrical systems that conserve energy.

The proliferating use of electrical equipment in the home causes increased demands not only on home wiring systems, but on national energy resources as well. Residential use of energy accounts for more than 25% of the total consumption of energy in this country.

Conservation of energy is both a worthwhile national goal and a money-saving consumer objective. By careful design of the wiring system and judicious selection of equipment, it is possible to minimize energy waste without sacrificing comfort and convenience.

This section discusses methods that can be used in facility systems, lighting, and appliances to effect the sustainability of a building. Chapter 23 discusses methods that can be used to reduce environmental damage from the generation of electricity.

### Wiring and Switches

Several of the *persistent biochemical toxic substances* (PBTs) identified by the U.S. Environmental Protection Agency (EPA) are associated with products used in electrical systems for buildings. These include mercury used in switches and several toxins associated with the production of vinyl, including lead, cadmium, and dioxins, which are by-products of vinyl chloride monomer (VCM). Lead is also used in solder.

The insulating jackets on wiring are among the worst offenders. *Halogenated* materials, which are used in cable jackets, are those made with chlorine- or fluorine-based compounds. When burned, materials made with such compounds emit hydrogen chloride, hydrogen fluoride, and carbon monoxide. According to one study, between 55% and 60% of all wiring jackets used in the United States are plasticized polyvinyl chloride (PVC). Unfortunately, this plasticizing is accomplished by the addition of harmful heavy metals (lead, cadmium, etc.) to the vinyl. PVC without these harmful materials is available and should be specified when possible. To make the use of lead in plasticized PVC even more unattractive, the lead cannot be electively removed from PVC but does tend to leach out over time, thus polluting landfills.

Fluorinated ethylene propylene (FEP) has been offered as a substitute for plasticized PVC in wiring jackets, but it has as many drawbacks as PVC and is no longer being touted as a substitute. Other plastics have been introduced, but none is used as extensively as PVC. Copper wiring constitutes a major use of nonrenewable materials. Because of impurities in existing copper wiring, recycled copper wiring cannot be used as new wiring. Therefore, new copper wiring must be made from newly mined or rerefined copper. This drawback introduces significant environmental hazards during the mining, smelting, and manufacturing processes. Unfortunately, no alternative is available to mitigate this defect.

Thermostats, silent switches, and mechanical tilt switches contain mercury, which eventually finds its way into the environment when these devices are discarded. Alternatives include electric thermostats, hard-contact switches, and newer float switches that contain no mercury.

### Lightning

In the United States, the annual cost of artificial lighting and the associated cooling energy amount to about $74 billion. Lighting accounts for 30 to 40% of the total electricity used in the United States, ranging from necessary illumination to that used to enhance interior and exterior design. Improving the perception of well-being and increasing the productive capability of workers who are adversely effected by poor lighting have dramatically increased the use of lighting in buildings other than residences. And in residences, lighting accounts for about 80% of the power used.

Fortunately, the allowable energy consumed by lighting is being reduced by codes in many jurisdictions. For example, most states now require that the allowable lighting load in office spaces be limited to 1.5 W/sf (10.7 W/m$^2$). This trend will force designers to find alternative means of providing the necessary illumination. One of these is the more extensive use of task lighting with a corresponding reduction in overall lighting intensity. Another is the increased use of natural light, which is discussed later in this section.

Misplaced or excessive illumination wastes electricity and money. In residence, seldom does an entire room require a high level of illumination, especially when close-up activity is ordinarily restricted to specific areas. For instance, lighting should be concentrated on areas such as a study desk, kitchen counter, and dining table, but can be diffused and dimmer over the rest of the room.

Proper selection and placement of switches in convenient locations in a residence can reduce wasteful consumption by lights left on unintentionally. Rooms with more than one doorway, such as kitchens, dining rooms, and living rooms, should have switches at each doorway.

Larger bulbs are more energy efficient, producing more light per watt of electric energy. However, increased efficiency should be balanced against the increased overall energy consumption that can be caused by indiscriminate use of larger bulbs. In general, light bulbs should not be larger than required for the specific task or design effect.

Another facet of energy conservation is the saving of related materials. For example, long-life luminaires require replacement at longer intervals. This reduces the amount of materials and energy used to produce the fixtures.

For hard-to-reach areas such as roof overhangs, post lights, and high ceilings, long-life bulbs with a life expectancy of 2500 hours can be used. While ordinary incandescent bulbs have a life span of 750 to 2000 hours, fluorescent tubes are rated for between 10,000 and 20,000 hours of life. Moreover, fluorescent tubes produce about 80 lumens per watt, or four times as much light as a standard incandescent bulb. Fluorescent fixtures require more materials than do incandescent ones, but the advantages to outweigh this factor.

Newer fluorescent bulbs are smaller (1 in. [25.4 mm] diameter) than the older ones (1$^1/_2$ in. [38 mm] diameter), which results in savings in their glass, metal, and chemical contents.

A major drawback of fluorescent bulbs is the environmental hazard their disposal presents. Fluorescent bulbs contain mercury, and the older electromagnetic ballasts used for them contain polychlorinated biphenyl (PCB). In newer bulbs, the amount of mercury has been reduced but not entirely eliminated. The newer electronic ballasts do not contain PCB. In addition, they use less material, produce almost no heat—which reduces the air conditioning load—and weigh less. Unfortunately, replacing the old ones with the new ones requires disposing of the old ones.

Newer types of lighting fixtures and components are replacing traditional ones in some specific applications. Light-emitting diodes, high-efficiency ballasts, and high-intensity discharge (HID) lamps are among these innovations.

Another way to reduce energy in lighting systems is by the use of electronic timing devices to turn lights off and to prevent turning them on when they are not needed. Such systems can save as much as 50% of lighting energy use.

Finally, the best way to reduce energy use from lighting is to design buildings to take advantage of the most effective and cheapest source of light, the sun. The intentional use of sunlight to illuminate building interiors is called *daylighting*. This term implies the use of reflected sunlight and diffused light from the exterior, not the use of direct sunlight, which can result in glare and excessive heat gain. In addition to lowering energy costs, daylighting improves health and productivity. The U.S. Department of Energy (DOE) reported recently that in one study of a single building, daylighting reduced energy costs by 66% and annual carbon dioxide emissions by 256,915 lb (116 535 kg).

It might seem that substituting sunlight for artificial light would increase energy costs, but such is not the case. In fact, daylight is more than twice as efficient as artificial light in terms of luminous output verses heat generated.

Daylighting is accomplished by the use of building orientation and space layout. Orientation is affected by geography. For example, in the northern parts of the country, it is easier to control sunlight intrusion into a building on the north and south facades than into those facing either east or west. Therefore, the most efficient orientation is one that places the longest facades toward the north and south. Overhangs and other sun control devices are used to control direct sunlight and its accompanying glare from reaching occupants while permitting access to skylight.

To further take advantage of daylight, occupied spaces are placed on the exterior of the building, and little-used facilities such as storage rooms and bathrooms are placed in the interior. For daylighting to be as effective as possible, careful attention must be paid to selection of interior surface finishes to provide appropriated reflectance factors (see Chapter 9 and Section 18.4), interior window treatments, and exterior shading devices (see Section 12.2). For daylighting, the reflectance levels of interior surfaces should be 50 to 70% for walls, at least 80% for ceilings, and between 20 and 40% for floors, in lieu of the values usually imposed for artificial lighting listed in Section 18.4.3.1.

### Appliances

Appliances that were once luxuries are now commonplace. For example, almost every home has at least one color television set, and many have more than one. Personal computers, unheard of 30 years ago, have proliferated, not just in businesses, but in homes as well. It is diffi-cult to sell a new house today if it does not have all the latest gadgets. There are electric dishwashers, clothes dryers, garbage disposals, can openers, built-in popcorn poppers, microwave ovens, frying pans, hot pots; whirlpool bathtubs, and hot tubs. The list grows every year.

The amount of energy individual appliances use has increased dramatically over the past 50 years. Extra conveniences, such as self-cleaning ovens and frost-free refrigerators, require more energy and cost more to operate than models that do not have these features and are largely responsible for this increased energy consumption per appliance. In selecting appliances, the extra convenience and pleasure of the special features should be balanced by a realistic evaluation of the added initial and long-term operating costs. The local power company can be helpful in making an economic life-cycle cost analysis and in providing other energy-conserving tips.

Energy-efficient appliances can reduce operating costs and minimize demands on a wiring system and on the utility providing power as well. Before the purchase of any appliance, the life-cycle cost (the long-range, lifetime, operating, and maintenance costs) of the product should be determined. Buying an energy-efficient appliance may be more costly initially, but significant savings can be made on subsequent electrical bills and energy demands.

The Energy Efficiency Ratio (EER) established by the National Institute of Standards and Technology can be helpful in determining the relative efficiency of certain household appliances. The EER of air conditioners, for example, is determined by dividing the number of Btus (kilojoules) per hour by the number of watts needed for operation. The EER for air conditioners ranges from 5 to 10 for smaller units and up to 12 for larger units. The higher the number, the more efficient the machine, which means less energy and cost spent on operation. An EER of 7 or more is considered acceptable for household window air conditioners.

Ceiling fans can increase air movement in a space and decrease the need for air conditioning. Except when the humidity is very high, a room with a ceiling fan will remain comfortable at temperatures above 80°F (26.67°C). However, leaving a fan running when no one is present will add heat to the space without producing additional comfort.

## Applicable *MasterFormat* Sections

Subjects addressed in this chapter are included in *MasterFormat* 2004, Division 26, "Electrical." The following *MasterFormat* 2004 Level 2 sections are applicable to this chapter:

26 21 00  Low-Voltage Overhead Electrical Power Systems
26 22 00  Low-Voltage Transformers

26 23 00  Low-Voltage Switchgear
26 24 00  Switchboards and Panelboards
26 25 00  Enclosed Bus Assemblies
26 26 00  Power Distribution Units
26 27 00  Low-Voltage Distribution Equipment
26 28 00  Low-Voltage Circuit Protective Devices

26 29 00  Low-Voltage Controllers
26 41 00  Facility Lightning Protection
26 51 00  Interior Lighting
26 52 00  Emergency Lighting
26 53 00  Exit Signs
26 55 00  Special Purpose Lighting
26 56 00  Exterior Lighting

# 18.1 Fundamentals of Electricity

The generally accepted mental picture of electricity is based on the theoretical structure of an atom. An atom can be visualized as a nucleus orbited by one or more electrons (Fig. 18.1-1). Groups of electrons moving in neighboring orbits can be thought of as belonging to the same orbital shell. Depending on the number of electrons, there can be as many as seven such shells.

The nucleus consists of one or more positively charged *protons* and one or more electrically neutral particles called *neutrons*. Each orbital electron has a neg-ative charge equal in magnitude and opposite to the positive charge of a nuclear proton.

In a neutral atom, the negative and positive charges balance out. But electrons in the outer shell can be dislodged and become free electrons, separated from the parent atom, which is then left with a net positive charge. A free elec-

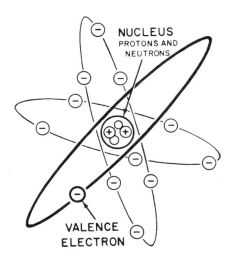

**FIGURE 18.1-1** Valence electrons, moving in orbits that are relatively far from the nucleus of an atom, have the least attraction for it and can be dislodged to create an electron flow, which is electricity.

(ions) which attach themselves to the electrodes, imparting a negative or positive charge. Connecting the two electrodes with a wire allows the electrons collected on the negative electrode to travel to the positive electrode in a balancing effort. The movement of these electrons through a wire is defined as the flow of electricity, or *current*. This current-generating action can be illustrated by a voltaic wet cell which uses a solution of sulfuric acid as the electrolyte (Fig. 18.1-3). In a dry cell, such as a flashlight battery, the zinc casing is used as the negative electrode, while a special paste acts as the electrolyte between the casing and the carbon rod (positive electrode).

Electricity can also be generated directly from solar radiation by means of silicon photovoltaic cells. This technology was developed by the National Aeronautics and Space Administration (NASA) to provide power for satellites and other orbiting and deep space devices. These cells generate low-voltage direct current (dc) power that can then be converted into 120-volt alternating current (ac) using a dc-ac inverter. This type of cell may well be used in the future to generate power in large quantities, but it is an expensive way to generate electricity and so far must be considered experimental, even though a few automobiles and some residential and commercial building projects have made limited use of this technology.

For larger amounts of high-voltage power, an electric generator is the device used today. Unlike batteries, which con-

**FIGURE 18.1-3** The electrolytic solution ($H_2SO_4$) in a wet cell breaks down into positive ($H_2$) and negative ($SO_4$) ions. While the $SO_4$ ions carry electrons to the zinc (Zn) electrode, they also react with it, changing it into $ZnSO_4$.

vert chemical or solar energy into electric power, a generator utilizes kinetic (motion) energy. This kinetic energy usually comes from a steam- or water-driven turbine. Steam is created by burning coal, gas, or oil or from the heat of a nuclear reactor. Water is caused to drop from a reservoir through a turbine. The turbine then turns a generator armature which has loops of wires wound around it. As the armature turns between the poles of a magnet, the loops cut across the invisible lines of magnetic force between the poles. This movement causes current to flow through the wire loops of the armature (Fig. 18.1-4).

### 18.1.1 CURRENT FLOW

An electric current can be visualized as a migration of electrons through a conductor from a negative terminal to a positive one. The process can be visualized as follows: a moving electron lodges on the outer shell of an atom, dislodging a second electron, which then migrates to an adjacent atom with the same result. The action is repeated over and over again in a chain reaction (see Fig. 18.1-2).

To keep electrons flowing, there must be a continuous, uninterrupted circuit (path) from a negative to a positive terminal. If the path is interrupted, electron flow stops, resulting in an open circuit. When a shortcut is inadvertently created,

tron may then attach itself to another neutral atom, lodging on its outer shell and creating a negative charge on the atom (Fig. 18.1-2). Single atoms or clusters of atoms with extra or fewer electrons are called *ions*.

A chemical battery, the simplest source of electricity, is made up of one or more cells. Most chemical batteries deliver small amounts of electricity at a low voltage. A cell consists usually of a carbon rod (positive electrode), a piece of copper or zinc (negative electrode), and a chemical compound (electrolyte). The electrolyte breaks down readily into positively and negatively charged particles

"POSITIVE CHARGE" (SHORTAGE OF ELECTRONS)

+ POSITIVE

"NEGATIVE CHARGE" (SURPLUS OF ELECTRONS)

BATTERY

− NEGATIVE

**FIGURE 18.1-2** Electrons will flow through a conductive material that connects an area with a surplus of electrons to an area that has a shortage of electrons, such as the two terminals of a battery.

**FIGURE 18.1-4** Voltage, or electromotive force (EMF), in a loop of a dc generator armature is highest when that loop moves across the lines of force between the poles of a magnet. Using more coils smooths out the peaks.

permitting a sudden surge of electrons through an easier path, we call that a *short circuit*.

### 18.1.1.1 Voltage

Current is caused to flow by an electromotive force called *voltage*, which is also known as *potential difference*. Voltage is the driving force behind electron flow, somewhat like the pressure in a water pipe or an air duct.

Particles with like charges, such as two electrons, repel each other; unlike particles, such as a positive ion and a negative electron, attract each other. These two factors contribute to the force of voltage. The mutual repulsion of electrons tends to make them rush away from each other; the mutual attraction between unlike charges creates a kind of pull that draws electrons toward positively charged atoms.

Electrons actually flow from a negatively charged to a positively charged body, but the older notion that electricity flows from a positive to a negative terminal still appears in textbooks and electrical diagrams as a convention, a kind of accepted fiction or customary code. If electricity is assumed to flow from a positive to a negative terminal, then current must be composed of positively charged particles. This concept, though contrary to established electron flow theory, is useful in certain calculations.

### 18.1.1.2 Amperage

The rate at which electrons, or current, flows through a conductor is measured in

*amperes* (amps). The speed with which electricity is transmitted is about the same as the speed of light, 186,000 miles (299 338 km) per second. That is, if a current is started at one end of a line 186,000 miles (299 338 km) long, it will be felt at the other end 1 second later. Rate of flow is different from speed because it takes into account *current density*, that is, the number of electrons moving along from atom to atom. Thus, amperage measures the quantity of electricity passing through a circuit per unit of time.

The quantity of electricity without reference to time is measured in coulombs. A *coulomb* is roughly the number of electrons conducted past a point in 1 second. The number is an unimaginable 6,250,000,000,000,000,000 electrons. One ampere equals 1 coulomb of electricity passing a point in the circuit every 1 second.

### 18.1.1.3 Resistance

Some metals become *superconductive* at temperatures near absolute zero (−459.6°F [0°K]). Superconductivity is a complete disappearance of electrical resistance. Recently, researchers have developed materials that become superconductive at much higher temperatures. But at normal-use temperatures, the internal structure of the best conductors available resists the flow of electric current and converts some of it into heat. This internal friction-like effect, which is called *resistance*, is measured in ohms. An *ohm* is

that resistance which allows 1 ampere (1 coulomb per second) to flow when pushed by a pressure of 1 volt.

### 18.1.1.4 Power

The electrical unit of power (energy rate) is the *watt*. In that it measures the rate at which work is done, or energy is used, a watt is similar to British thermal units (Btus) per hour (Btuh) and horsepower (hp) and can be converted directly into either one. For instance, 1 watt = 3.413 Btuh; 1 hp = 746 watts (Fig. 18.1-5).

An electric motor rated at 1 hp is capable of doing mechanical work at a rate of 33,000 foot-lb/min. This means that the motor could lift a 1000 lb (453.6 kg) weight 33 ft (10.1 m) every minute. If it were 100% efficient, the motor would then be converting electric energy into mechanical work at a rate of 746 watts. In electric heating, electricity is converted into heat rather than mechanical work (see Fig. 18.1-5).

Stating the relationship in electrical terms, one gets the basic equation for electrical power, $W = EI$, where wattage ($W$) equals voltage ($E$) times amperage ($I$). The equation says, in effect: the heavier the flow rate (amps) at a given supply pressure (volts), the higher the rate (watts) at which energy is being supplied and used.

**FIGURE 18.1-5** Electric energy can be converted into kinetic (motion) or thermal (heat) energy.

An electric utility, for example, might use this equation to find out the load demand on distribution lines in watts or kilowatts (1000 watts). Or a designer may wish to determine the wattage of an electric heater whose rating is 22 amps. Multiplying 22 amps by 240 volts gives the answer: 5280 watts (about 5.3 kilowatts).

## 18.1.2 OHM'S LAW

In an active circuit, voltage, amperage, and resistance are interrelated. The greater the resistance, the greater the voltage required to produce a given amperage. Current and resistance together, then, imply the required voltage pressure, and that is stated in Ohm's Law: $E = IR$, where voltage ($E$) equals amperage ($I$) times resistance ($R$). Ohm's Law makes it possible to determine any one of these values when the other two are known (Fig. 18.1-6).

By combining the power equation with Ohm's equation, it is possible to derive many equations for the relationship between voltage, wattage, amperage, and resistance (Fig. 18.1-7). The power equation as stated earlier is $W = EI$; substituting $IR$ for $E$ (Ohm's equation), we get $W = (IR)I = I^2R$. In other words, wattage ($W$) equals amperage ($I$) squared times resistance ($R$). This equation helps determine the power used in an electric heater and the power lost in transmitting electricity. In a resistance heater, the power is transformed into useful heat, but in distribution wires it is an undesirable waste of energy.

The usefulness of these equations can be demonstrated by the following example. A 100-ft (30.48-m)-long, No. 14 copper wire has a resistance of about 0.25 ohm. Maintaining a current of 10 amps at 120 volts, what would be the line loss in the conductors? The line loss (see Fig. 18.1-7) is obtained from the power used or loss equation mentioned in the previous paragraph, $W = I^2R$, or $(10)^2$ amps $\times$ 0.25 ohm = 25 watts. The total power would be $EI = W$, or 120 volts times 10 amps equals 1200 watts. The 25-watt line loss would constitute a little over 2% of the total power, which is well within the recommended maximum line loss of 3%.

## 18.1.3 ALTERNATING AND DIRECT CURRENT

So far, it has been convenient to discuss electrical principles in terms of dc rather than the more common commercial form, ac. Direct current always flows in one direction in a circuit, namely, from the negative to the positive terminal of the source. Alternating current also flows from negative to positive terminals, but the terminals change polarity (whether positive or negative) quickly, causing a reciprocating (back-and-forth) surge of electricity along the wire.

Alternating current is produced by an ac generator (Fig. 18.1-8) designed so that the voltage and current reverse direction periodically. In the commonly used 60-cycle power supply, 1 cycle of a complete rise and fall of voltage in each direction takes place in $^1/_{60}$th of a second. The voltage of 120-volt direct current is actually 120 volts at the generator and should not drop below 115 volts at the load. However, because ac consists of rapidly changing values from 0 to 170 volts, the average or useful voltage

appears as 120 volts to the power-consuming device (Fig. 18.1-9). Thus, the effect of both types of current is the same, and the above equations can be used for both ac and dc calculations.

Alternating current is used almost exclusively in residential and commercial wiring because it provides greater flexibility in voltage selection and simplicity of equipment design. When ac is used, the voltage can be stepped up (increased) or stepped down (decreased) by a transformer (Fig. 18.1-10). An ac transformer is a very efficient device, wasting only 1 to 4% of the power in heat. By comparison, an electric generator, which is the only means of changing dc voltage, loses about 25% of its power in waste heat.

Passing current through the primary circuit on one side of the magnet in a transformer generates a current in the secondary circuit. The induced voltage in the secondary circuit has the same relationship to the voltage in the primary circuit as the number of windings in the secondary circuit do to those in the primary circuit. In other words, to double the voltage, it is merely necessary to double the windings in the secondary circuit; to triple the voltage, triple the windings; to reduce the voltage by 50%, use half the number of windings.

# 18.2 Service and Distribution

Electric power enters a building through service entrance equipment and from

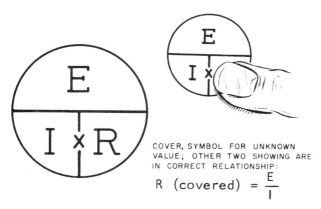

**FIGURE 18.1-6** Ohm's Law gives the relationship between voltage (*E*), amperage (*I*), and resistance (*R*).

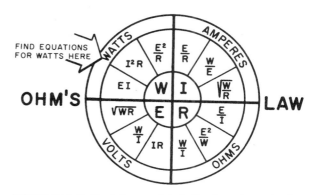

**FIGURE 18.1-7** Ohm's Law and derivative formulas make it possible to find any of the values in the middle of the wheel when any other two are known.

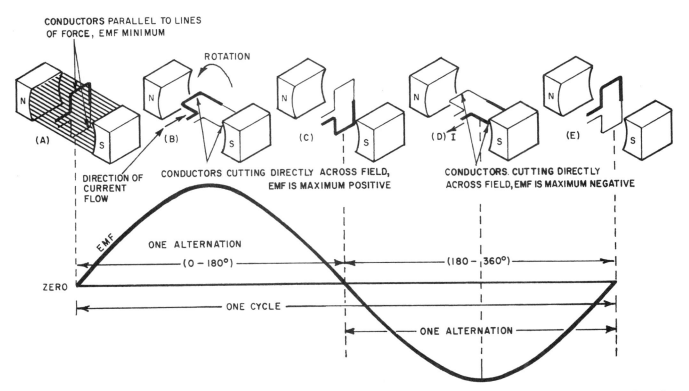

**FIGURE 18.1-8** Just as in a dc generator (Fig. 18.1-4) the voltage (EMF) peaks twice in each cycle in an ac generator, but the alternate peaks are of opposite polarity, as current flow reverses in each alternation.

there is distributed to convenience outlets, lights, appliance outlets, and special-purpose outlets. The design and installation of such systems are subject to compliance with codes and standards. Symbols have been developed to help indicate requirements on drawings. This section discusses some of these aspects of electric service to buildings. Electric power distribution from a public service to a building is discussed in Section 23.3.

## 18.2.1 CODES, STANDARDS, AND SYMBOLS

Familiarity with the applicable codes and standards and the symbols used on drawings is essential to an understanding of this chapter.

Most of the terms used here have different meanings in this context than their standard dictionary meanings and are explained in the body of the text. Those that are not so explained are defined in the Glossary.

### 18.2.1.1 Codes and Standards

The recommendations in this chapter are based generally on the *National Electrical Code* (NEC) and the *International Residential Code* (IRC). The purpose of these recommendations is to suggest design criteria that will produce an efficient and convenient system and minimize physical obsolescence.

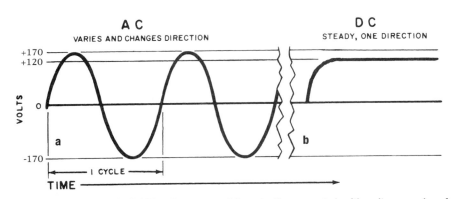

**FIGURE 18.1-9** Nominal 120-volt ac current is actually generated with voltage peaks of 170 volts (a) to approximate the effect of a 120-volt dc current (b).

**FIGURE 18.1-10** A transformer is used to step up or step down voltage. To reduce voltage from 120 to 30 volts would require four times as many turns in the primary as in the secondary winding.

An adequate wiring installation must conform with applicable safety regulations. The standard for electrical safety is the NEC, developed and published by the National Fire Protection Association (NFPA). The requirements in the International Code Council's (ICC) codes, and most local ordinances and state and local building codes, whether they have adopted the ICC codes or not, are based on the NEC or make it a part of their requirements by reference. Therefore, the entire electric system, including its design, layout, equipment, and materials, as well as the methods used to install it, must comply with the requirements of the adopted edition of the NEC, IRC, and local ordinances and laws. It must also be acceptable to the local building authority that has jurisdiction.

Inspection service to determine conformance with safety regulations is available in most communities from the local building official. Where required, a certificate of inspection must be obtained. In the absence of inspection service, an affidavit should be obtained from the electrical contractor attesting to conformity with applicable safety regulations.

### 18.2.1.2 Symbols

The symbols shown in Figure 18.2-1 were adapted from American National Standards Institute (ANSI) Standard Y32.9-1972 (reaffirmed 1989), "Graphic Symbols for Electrical Wiring and Layout Diagrams Used in Architecture and Building Construction." These are the standards recommended by the Institute of Electrical and Electronics Engineers (IEEE). They have been made mandatory by some federal agencies, such as the Department of Defense, and should be used on all drawings for electrical systems, whether they are for public or private buildings.

## 18.2.2 SERVICE POWER REQUIREMENTS

Minimum power requirements are dictated by law, code, or ordinance in many jurisdictions, especially for buildings accessible to the public. Most codes, including the ICC's international codes, either require compliance with the latest edition of the NEC or incorporate the NEC provisions into their electrical chapters.

### 18.2.2.1 Residential Buildings

Regardless of law or code requirements, each separate dwelling unit, including apartments, townhouses, and single-family houses, should have at least a three-wire 120/240-volt electric service. Many older residences were served by two-wire 120-volt services, which is inadequate for today's needs and fails to meet current code requirements.

A 100-amp service will provide a home's needs for lighting, appliances, and fixed equipment totaling not more than 18,000 watts. When the total load is larger than 18,000 watts, as is the case in almost every new home today, the local power company should be consulted. A 150-amp service may be adequate to handle wiring, appliances, and fixed equipment plus 3 to 5 tons of air conditioning. When electric heating, heat pumps, or their equivalent are used, at least 200 amps are called for. In larger homes, especially those having more than one heating system, a 400-amp service is not uncommon.

Inadequate service conductors and service equipment can be serious obstacles to the use of new equipment and appliances in homes. Service entrance conductors should be sized in accordance with the NEC. To provide for probable increases in electrical use, even when electric heating or cooling is not contemplated, service conductors should be oversized to at least No. 1/0 AWG (American Wire Gauge), and service equipment should be of sufficient rating to match the carrying capacity of these wires.

Most portable appliances for residential use are made to work on 110 to 125 volts, but many fixed appliances need 230- to 250-volt circuits. However, some fixed appliances are still manufactured to function on the lower voltages so that they can be used in older residences. When there is a choice, it is more efficient to use the higher voltage.

### 18.2.2.2 Commercial, Institutional, and Industrial Buildings

Design and selection of adequate wiring and service capacity criteria for commercial, institutional, and industrial buildings should be determined on a case-by-case basis by consultation with the local power company. Except for very small commercial buildings, such services should be designed by an electrical engineer.

Mechanical equipment, elevators, heavy industrial machinery, and other major equipment for such buildings will usually operate on 240-volt or even 480-volt service. Ice makers, food preparation equipment, and other lighter equipment may operate on either 120- or 240-volt service. Lighting and convenience circuits, as well as special-purpose circuits for coffee machines, food-dispensing machines, water coolers, and other lighter equipment, will often need only 120-volt outlets. Some lighting fixtures will operate at 240 volts; some equipment and some lights will require low-voltage circuits. Exterior lighting may need low voltage, 120-, 240-, or even 480-volt power, depending on fixture type.

As is true of older residences, many older commercial buildings were built with two-wire 120-volt services. Such service is inadequate for today's needs and violates most codes. When major remodeling or renovation is done in such buildings, codes will often require that the service be brought up to code. This can be a major cost item, possibly requiring complete rewiring of the building and all new outlets, and sometimes new conduits throughout the building.

## 18.2.3 SERVICE ENTRANCE AND DISTRIBUTION EQUIPMENT

A *service entrance* is a system of conductors bringing electricity into a building and the associated service entrance and distribution equipment. *Service entrance and distribution equipment* is defined here as all equipment and devices that occur from the outlet side of the watt-hour meter (see Section 23.3.4) to the outlet side of the distribution panels. It includes the service conductors that bring electricity into a building, the service entrance panel, feeders that connect the service entrance panel to distribution panels, and the distribution panels from which branch circuits distribute electricity to the rest of the building and its grounds. It also includes overcurrent devices mounted in the service entrance and distribution panels. Grounding of the incoming service, also a part of the service entrance equipment, is discussed in Section 18.2.8.5. Figure 18.2-2 gives an overall view of service drop, meter, and service entrance equipment.

### 18.2.3.1 Service Entrance Panel

A *service entrance panel*, which, in small buildings, may also serve as the distribution panel, is an assembly of switches and switch-like devices that permits

## General outlets

○    Ceiling lighting outlet

─○    Wall lighting outlet

Ceiling fluorescent fixture

Wall fluorescent fixture

Ⓕ    Fan outlet

Ⓙ    Junction box

─Ⓒ    Clock outlet

To indicate wall installation of outlets, place symbol near wall and connect to wall with line as shown for clock outlet.

## Convenience outlets

⊖    Duplex convenience outlet

⊖    Triplex convenience outlet

⊖    Duplex convenience outlet—split wired

⊖    Ungrounded duplex convenience outlet[2]
UNG

⊖    Weatherproof convenience outlet
WP

Multi-outlet assembly (extend arrows to limits of installation. Use appropriate symbol to indicate type of outlet. Also indicate spacing of outlets as X in. [mm])
X in.

⊖S    Combination switch and convenience outlet

## Convenience outlets *(Cont.)*

⊖Ⓡ    Combination radio and convenience outlet

⊖    Floor mounted duplex receptacle

⊖R    Range outlet

◭DW    Special purpose outlet (use subscript letters to indicate function. DW = dishwasher, CD = clothes dryer, etc.)

## Auxiliary systems

▣    Push button

▽    Buzzer

◯▭    Bell

◯▭    Combination bell-buzzer

CH    Chime

◇    Annunciator

D    Electric door opener

▭    Interconnection box

BT    Bell-ringing transformer

▶    Outside-line telephone

▷    Interconnecting telephone

R    Radio outlet

TV    Television outlet

## Standard switches

S    Single-Pole switch

$S_3$    Three-way switch

$S_4$    Four-way switch

$S_D$    Automatic door switch

$S_P$    Switch and pilot light

$S_{WP}$    Weatherproof switch

$S_2$    Double-pole switch

Ⓢ    Ceiling pull switch

## Low-voltage systems

$S_L$    Switch for low-voltage relay system

$S_{LM}$    Master switch for low-voltage relay system

Ⓛ    Outlet controlled by low-voltage switching

## Miscellaneous

▨    Service panel

▬    Distribution panel

◯$_{a,b}$
⊖$_{a,b}$
◭$_{a,b}$
▭$_{a,b}$    Special outlets (subscript letters may be added to any standard symbol to designate a variation associated with a particular building. Such variations should be explained in a Key to Symbols, and, if necessary, in the specifications).

**FIGURE 18.2-1 Graphic electrical wiring symbols.** (Drawing by HLS.)

[1]Adapted from USA Standard Y32.9-1972 (reaffirmed 1989) "Graphic Symbols for Electrical Wiring and Layout Diagrams Used in Architecture and Building Construction."

[2]All outlets of every type are grounded, unless marked UNG.

**FIGURE 18.2-2** Service entrance equipment.

disconnecting all power or distributing it to various branch circuits through overcurrent devices such as fuses or circuit breakers.

There are several possible arrangements for a service entrance panel. Sometimes, especially in larger systems, a main disconnect switch is housed in a separate panel from the circuit overcurrent protection devices (fuses or circuit breakers) (Fig. 18.2-3). In houses and other small buildings, a single panel usually includes both the main switch and overcurrent devices.

A main disconnect device provides a means of deliberately deenergizing the entire system and, at the same time, protects the service conductors from unexpected overloads. Depending on the load, the code, and the desires of the designer,

a main disconnect device could be a knife-blade switch, a pull-out fuse, or a circuit breaker. The overcurrent devices may be either fuses or circuit breakers, but today they are usually circuit breakers.

Figure 18.2-4 shows a typical circuit breaker–type service entrance panel. This is the type of panel used in most buildings today. A main knife-blade switch may be inserted between the main disconnect breaker and the incoming service or may replace the main disconnect breaker. The principles shown are applicable to panels with fuses, but wiring is used in lieu of bus bars. In addition, the principles demonstrated are the same, regardless of the amperage rating of the system.

At the upper left in Figure 18.2-4, the service conductors coming from the meter (not shown) enter the panel. The two hot wires carry the current to the main disconnect breaker and through it into two hot bus bars. The hot bus bars carry the current to the individual circuit overcurrent protection devices (circuit breakers). Wires are then extended from each breaker to the outlets or devices on a single circuit. When distribution panels are used, the two hot wires run from a breaker on the service entrance panel to hot bus bars in the distribution panel. Breakers are attached to these bus bars, and wires run from these breakers to

serve the outlets and devices on a single circuit.

The neutral conductor is neither switched nor fused. It connects to the bonded neutral bus bar, to which the service ground wire also is connected. This grounded bus provides terminals for all load neutrals and ground wires in the entire distribution system.

In residences, because of the many major uses of electricity in a kitchen requiring individual equipment circuits, it is usually best to locate the main service equipment or a subpanel near the kitchen. Such a location is also often close to the laundry, which helps to minimize long circuit runs and wiring costs (Fig. 18.2-5).

For similar reasons, the service entrance in commercial, institutional, and industrial buildings is usually close to the heaviest load center. Where practicable, for example, the service entrance is located close to the boiler plant. In facilities within cities, the service entrance is often in a basement room near a vault under the sidewalk or street adjacent to the building, where the building's main transformer is housed. The idea is to reduce the length of the run of the heaviest wiring, thus reducing wiring costs and power loss; the latter is larger in larger wiring.

### 18.2.3.2 Feeders

Conductors connecting a service entrance panel or main disconnect switch to distribution panels are called *feeders*.

### 18.2.3.3 Distribution Panel

A *distribution panel* is connected to the service entrance panel or main disconnect switch by feeders. In small residences and small commercial buildings, the service entrance panel often serves as the only distribution panel. However, in large residences and in most commercial, institutional, and industrial buildings, there are several distribution panels, depending on building size and use. A distribution panel may or may not have a disconnect switch or overload devices like fuses or circuit breakers.

### 18.2.3.4 Branch Circuit

A *branch circuit* is the portion of a wiring system between an overcurrent device and outlets where loads, such as light fixtures or appliances, can be connected. Refer to Section 18.2.8.1 for recommen-

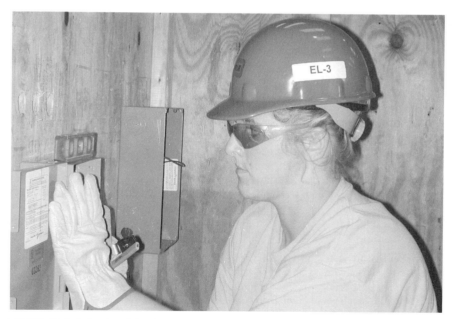

**FIGURE 18.2-3** An electrician installing a service entrance panel. A box for overcurrent protection devices is in the background. (Photo courtesy BE&K, Inc.)

enters the body as well as the point where the current leaves the body. Such electricity interferes with the normal operation of a person's nervous system and, in severe cases, can result in convulsions, severe neurological damage, and even death.

Water is a better conductor of electricity than a human body; it bypasses protection a person might ordinarily have, such as shoes, and it produces multiple contact points for electricity to enter and leave. Therefore, a person in contact with water is likely to sustain more severe injury when coming into contact with an electrical source than when no water is present.

Overcurrent devices are imposed in electrical circuitry to ensure that in case of a short circuit, current will be interrupted before the conductor overheats and possibly burns up the insulation. The first line of defense is at the main circuit control panel. Older systems, and some large special-purpose systems, may employ fuses. Most new systems use circuit breakers. Several other devices and methods are also used in other parts of a typical system.

### 18.2.4.1 Fuses

Fuses have a fusible (meltable) wire designed to melt and break when the current flow exceeds the rated capacity of the circuit. Thus the circuit is opened and current flow is stopped before it becomes a hazard. Circuits of 30 amps or less can use plug fuses; for heavier currents, cartridge fuses are required.

Cartridge fuses are made in all current ratings and voltage ratings from 0 up. Fuses from 0 to 60 amps use a ferrule-type connection; those above 61 amps use a knife-blade connection (see Fig. 18.2-6a).

Plug fuses are obtainable in sizes from 0 to 30 amps (Fig. 18.2-6b). The NEC requires that type S fuses be used for all plug-fused installations. A type S fuse uses a slightly different adapter for each ampere rating; once the adapter is installed in a fuse box, it cannot be removed readily and only a fuse of a certain rating can be inserted. Existing wiring systems which are being modified should be equipped with type S fuses if the fuse box is retained.

### 18.2.4.2 Circuit Breakers

Circuit breakers are simple switch-like devices that automatically open a circuit

**FIGURE 18.2-4** A circuit breaker type service entrance panel. To provide a 120-volt branch circuit, insert a circuit breaker by snapping the clip on its back onto one hot bus bar. Then extend to the location of the circuit, a hot wire (black or red) from the breaker, a neutral wire (gray or white) from the bonded neutral bus bar, and a grounding wire (bare wire) from the bonded neutral bus bar. To create a 120/240-volt circuit, snap a common-trip two-pole breaker with one handle so that one of the two poles contacts each hot bus bar. Then extend to the location of the circuit, one hot wire from each pole of the breaker, and both a neutral wire and a grounding wire from the bonded bus bar. (Drawing by HLS)

dations about the allocation of outlets to circuits.

### 18.2.4 OVERLOAD PROTECTION DEVICES

Inadequate or old wiring, faulty wiring installation, and other flaws in wiring or in appliances are major causes of fires. Electricity may also be a source of personal injury, especially in combination with water. When a person who is in contact with the ground comes into contact with a hot wire, current will flow through the person to the ground, resulting in burns at the point where the electricity

**FIGURE 18.2-5** A second distribution panel is not needed here because the service equipment is centrally located with respect to cooking, laundry, and heating equipment. (Drawing by HLS.)

PLAIN FUSE  KNIFE-BLADE FUSE

a

PLUG FUSE

b

**FIGURE 18.2-6** Fuses are either (a) cartridge type (ferrule type at the left, knife-blade type at the right) or (b) plug type. A type S adapter makes it impossible to insert a 30-amp fuse in a 20-amp socket, thus avoiding dangerous overloads.

when the rated current is exceeded, such as would result from a short circuit. Unlike plug fuses and most cartridge fuses, circuit breakers are not damaged when they trip (open). To reset (close) the breaker, it is simply necessary to push the toggle into the "on" position. Most small circuit breakers are of the thermal-magnetic type. When such a breaker is tripped, a thermal element in it is heated. It may be necessary, therefore, to wait about a minute for the element to cool before the breaker can be reset.

Circuit breakers are available in a wide range of sizes for many applications. Small breakers for 15-, 20-, 30-, 40-, 60-, and even 100-amp branch circuits are used today in service entrance equipment for residential and commercial buildings. All of these can be combined in a single service panel, along with a larger breaker with a rating equal to the combined ratings of all the circuits to serve as a main switch.

### 18.2.4.3 Ground Fault Circuit Interrupter (GFCI)

A GFCI is similar to a fuse or circuit breaker in that it stops the flow of electricity when an abnormal electrical situation develops. Unlike circuit breakers and fuses, which keep large current flows from overloading a circuit, a GFCI will be tripped by even very low current flows. GFCIs are usually set to open the circuit when the ground fault current exceeds 5/1000th of an amp. Refer to Section 18.2.7 for an indication of locations where GFCIs are appropriate.

### 18.2.5 CONDUCTORS AND RACEWAYS

Electric current is carried and distributed by *conductors*. Technically, conductors are the metallic cores which conduct electricity. In practice the entire wire, including the insulating cover, is called a conductor. Conductors are protected from mechanical damage by an outer sheath. When a conductor and its protective sheath are fabricated together, the assembly is called a *cable*. Conductors can also be inserted on the job into *raceways*, which are specially built channels in which wires are placed or pulled.

### 18.2.5.1 Conductors

Conductors consist of an aluminum or copper core covered by an insulating jacket and an outer coating suitable for the anticipated temperature and moisture conditions. Conductors for building wiring are available in AWG sizes ranging from No. 14 to 4/0, and in larger sizes ranging from 250 MCM (thousand circular mils) to 2000 MCM (Fig. 18.2-7). Wiring systems should be designed such that the maximum current carried by a conductor will not exceed the amperages shown in Figure 18.2-7. Copper and aluminum are the typical conductor materials; aluminum is generally limited to larger wire sizes. Due to its lower conductivity, aluminum wire is not made smaller than No. 12 AWG (rated 15 amps). The same types of insulation are used for both copper and aluminum conductors.

Since its insulation determines the suitability of a conductor to certain exposures, such as moisture, heat, or mechanical damage, conductors are identified by the material and properties of the insulating cover and sometimes by use. The following are a few of the many designations in the NEC for conductors:

**FIGURE 18.2-7** Properties of Commonly Used Copper Conductors

| Conductor | | | Maximum Current, ampere[a] | |
|---|---|---|---|---|
| Size: AWG, MCM[b] | One-Half Natural Size | Diameter, in. (mm)[c] | Conductor Type: RUW, T, TW | Conductor Type: RH, RUH, THW, RHW, THWN |
| 14 | • | .0641 (1.63) | 15 | 15 |
| 12 | • | .0808 (2.05) | 20 | 20 |
| 10 | • | .1019 (2.59) | 30 | 30 |
| 8 | • | .1285 (3.26) | 40 | 45 |
| 6 | • | .184 (4.67) | 55 | 65 |
| 4 | • | .232 (5.97) | 70 | 85 |
| 2 | • | .292 (7.42) | 95 | 115 |
| 1 | • | .332 (8.43) | 110 | 130 |
| 0 | • | .372 (9.4) | 125 | 150 |
| 00 | • | .418 (10.62) | 145 | 175 |
| 000 | • | .470 (11.94) | 165 | 200 |
| 0000 | • | .528 (13.41) | 195 | 230 |
| 350 MCM[d] | • | .681 (17.3) | 260 | 310 |

[a]Based on ambient temperature of 86°F (30°C); not more than 3 conductors in raceway, cable, or direct burial. Where number of conductors exceeds 3, maximum current given should be derated as follows: 4–6 cond., 80%; 7–24 cond., 70%; 25–42 cond., 60%; 43 or more, 50%.

[b]Sizes 14 to 8 are solid wires. Sizes 6 to 2 are 7-strand cables. Sizes 1 to 0000 are 19-strand cables. Size 350 MCM is a 37-strand cable.

[c]Exclusive of insulation.

[d]MCM is the designation of wire size in thousands of circular mils (350 MCM = 350,000 circular mils).

- FEP = fluorinated ethylene propylene
- H = heat resistant (for use up to 167°F [75°C] maximum temperature)
- HH = higher heat resistance (up to 194°F [90°C])
- MI = mineral insulation (metal sheathed)
- R = rubber
- T = thermoplastic
- UF = underground feeder
- USE = underground service entrance
- W = moisture resistance

In addition to the basic type letter derived from the material (such as R for rubber), other letters are added to indicate modified or additional characteristics. For example, when an H is added to the basic R designation, the resulting type RH wire may be used in higher-temperature conditions. Adding a W to designate an RHW conductor adds a requirement that the conductor be also moisture resistant.

Insulation thickness varies with wire size and may also vary with the type of material and application and the code limitation. The NEC contains tables showing the requirement for each conductor type by its letter designation (MTW, RH, RHH, etc.). These tables also spell out the maximum temperature range for each conductor type and list its required insulation thickness based on wire size. For example, under circumstances of normal use, an RHW (moisture- and heat-resistant rubber) conductor can be used up to an operating temperature of 167°F (75°C) in dry or wet locations. For voltages up to 600, the following insulation thicknesses apply:

| AWG | Insulation Thickness in mils (mm) |
|---|---|
| 14–10 | 45 (1.14) |
| 8–2 | 60 (1.52) |
| 1–4/0 | 80 (2.03) |
| 213–500 | 90 (2.41) |
| 501–1000 | 110 (2.79) |
| 1001–2000 | 125 (3.18) |

For voltages of about 600, thicker insulation is required.

### 18.2.5.2 Cables

*Cables* are combinations of two or more conductors, factory encased in a protective covering to safeguard them from mechanical damage, They can have metallic coverings, as in armored cable, or plastic or cambric (varnished cotton) jackets, as in nonmetallic sheathed cable.

### ARMORED CABLE

Armored cable, Type AC, which is also sometimes called BX, is a factory assembly of insulated conductors inside a flexible metallic covering intended primarily for use in branch circuits and feeders (Fig. 18.2-8). It may be run in the voids of masonry block or tile walls when such walls are not exposed to excessive moisture. It should not be run below grade; a transition should be made from armored cable to metal conduit or tubing when supplying a basement appliance.

Armored cable should be securely fastened by approved sleeves, staples, or straps, spaced not more than 4 ft 6 in. (1372 mm) apart. The cable should be supported within 12 in. (305 mm) of an outlet box or other termination. When the box is mounted on equipment which needs to be moved for servicing occasionally, the unsupported cable length can be 24 in. (610 mm).

Armored cable should be cut away to allow at least 6 in. (152 mm) of the conductors into the box. The remaining armored portion should be clamped to the box. When the armor is cut back to provide free conductors, care should be taken to prevent damaging the conductor insulation. Also, a fiber insulating bushing should be inserted in the end between the armor and the conductor insulation. This bushing should remain visible within the box.

Armored cable must be handled and bent with care. If it is bent too sharply, the convolutions of the armor can damage the conductor insulation, resulting in possible short circuits or leakage to ground.

### METAL-CLAD CABLE

Metal-clad cable, classified as MC, is similar in construction to armored cable but is intended for use in services, feeders, and branch circuits. It is permitted in systems in excess of 600 volts.

### NONMETALLIC SHEATHED CABLE

This cable, which is also sometimes called Romex, consists of two or more insulated conductors having an outer sheath of moisture-resistant, nonmetallic material (see Fig. 18.2-8). The conductor insulation is rubber, neoprene, thermoplastic, or a moisture-resistant, flame-retardant fibrous material.

Nonmetallic sheathed cable should be securely fastened by approved staples, straps, or other supports in such a manner that the cable will not be damaged. The supports must not be more than 4 ft 6 in. (1372 mm) apart. Generally, this cable must be fastened within 12 in. (305 mm) of each termination. In concealed spaces of existing or prefabricated buildings where such support is impractical, this cable may be fished between points of access. When run at an angle to joists, this cable should either be supported on a running board secured to the joists or run through holes bored through the joists.

Care should be taken not to damage the outer sheath of nonmetallic cable during installation. No bend should have a radius less than five times the diameter of the cable.

There are two types of nonmetallic sheathed cable: type NM and type NMC.

Type NM cable has conductors which satisfy the requirements for type RH, RW, T, and TW wire and a sheath which is flame retardant and moisture resistant. Type NM is limited to use in dry locations such as air voids in masonry or tile walls not exposed to excessive moisture.

Type NMC conductors must satisfy the same requirements as type NM, but the sheath must be not only flame retardant and moisture resistant, but also fungus resistant and corrosion resistant. Type NMC may be used in damp or corrosive locations, as well as in dry areas.

### 18.2.5.3 Raceways

*Raceways* are protective enclosures designed for field insertion of conductors.

**FIGURE 18.2-8** Some common types of electrical cable.

Included in the general term are electrical metallic tubing, rigid and flexible metal conduit, wireways, cable trays, and underfloor ducts.

Raceways should be securely connected and rigidly supported. Cut ends should be reamed (smoothed) to remove rough edges. Combination couplings should be used to make transitions from flexible conduit to electrical metallic tubing or rigid conduit.

### ELECTRICAL METALLIC TUBING (EMT)

This electrical pipe, which is also called *thin-wall conduit*, may be used for both concealed and exposed work (Fig. 18.2-9). It is the most common type of raceway used in single-family, low-rise residential, and commercial construction. With proper joints and connections it can be installed in wet as well as dry locations.

EMT is made in $1/2$ to 4 in. (12.7 to 102 mm) nominal sizes. The actual internal diameter is slightly larger than the nominal size. Connectors and couplings for EMT are secured to the tubing by a setscrew or dimples made on the job in the collar of the fitting and tube after assembly. Watertight compression fittings are also available.

EMT is so often used in residences and commercial buildings because it is light, easy to handle, and requires no special threading tools. However, bending it, which should be done using a special type of bender to prevent flattening (Fig. 18.2-10), requires more skill than does bending of rigid conduit. The radius of the inner edge of a field bend should not be less than that required in the NEC.

FIGURE 18.2-9 EMT, or thinwall conduit, and typical fittings for it.

FIGURE 18.2-10 An EMT bender ensures smooth curves without pinching the interior of a pipe.

### RIGID METAL CONDUIT

This conduit, which is also called *heavy-wall conduit*, resembles plumbing pipe and is shipped in standard lengths of 10 ft (3048 mm) (Fig. 18.2-11). A coupling is furnished for one end of each length of rigid conduit. Both ends are reamed and threaded when shipped, If it is cut during installation, the new ends must be reamed and threaded.

Rigid conduit can be joined by screwing directly into threaded hubs, by locknuts and bushings (see Fig. 18.2-11), or by threadless connectors. When the conduit is connected with a locknut and bushing, the locknut is screwed onto the pipe, the end is inserted through the knockout in a box, and the bushing is screwed on. When the bushing is seated, the outer locknut is screwed tightly against the outer surface of the box. When a conduit bushing is made entirely of insulating material, a locknut must be installed on both sides of the enclosure.

Rigid metal conduit is made in trade sizes of $1/2$ to 6 in. (12.7 to 152 mm). The 6 in. (152 mm) size will accommodate three 2000 MCM (1.63 in. [41.3 mm] diameter) conductors. When greater capacity is required, a bus-type conductor (solid bar) in a bus duct is usually used.

### FLEXIBLE METAL CONDUIT

Flexible metal conduit, which is also called *Greenfield*, is similar in appearance to armored cable but does not contain preinserted conductors (Fig. 18.2-12). It cannot be used where it will be subject to corrosive gases or excessive moisture. For the most part, it must conform to the same regulations as rigid conduit with respect to rewiring existing conduit, reaming cut edges, and other code requirements.

Flexible conduit connectors are fastened to the conduit by setscrews or

FIGURE 18.2-11 Rigid metal (heavywall) conduit and typical fittings for it.

FIGURE 18.2-12 Two types of flexible metal conduit and typical fittings for them.

clamps, and to the box by an expanding mechanism, a locknut, or a threaded hub.

Flexible conduit should be installed taut and then fastened securely to facilitate pulling conductors with a fish tape. It should be supported at intervals not greater than 4 ft 6 in. (1372 mm) and within 12 in. (305 mm) of a box. The radius of bends should not exceed five times the conduit diameter.

### NONMETALLIC CONDUIT

This conduit is made of nonmetallic material that is resistant to moisture and chemical atmospheres. It is flame retardant, resistant to crushing, and resistant to distortion from heat, low temperatures, and sunlight. It can be used underground in floors, walls, and ceilings.

### WIREWAYS

A wireway is a metal trough with hinged or removable covers in which conductors are placed after the wireway has been installed. Because they must be accessible throughout their entire length, wireways can be used only for exposed work.

Splices can be made and laid into a wireway if the total conductor and splice cross-sectional area does not exceed 75%

of the area of the wireway and is accessible. If the wireway is not fully accessible at the splice location, then the percentage of fill should not exceed 20% of the cross-sectional area of the wireway.

## CABLE TRAYS

Cable trays are a variety of wireways that are used above ceilings, in which the wiring is left exposed. They are used extensively in commercial and institutional buildings, such as hospitals, where a variety of systems are required. Fiberoptic and other communications cabling is often laid in cable trays. Essentially, cable trays are wire or pressed metal troughs with open tops.

## UNDERFLOOR RACEWAY

There are a variety of underfloor duct systems in general use. They are used extensively in commercial and institutional builds where the framing is steel with bar joists. Essentially they consist of a metal or fiber header duct that runs parallel to the framing and metal or fiber raceways that cross the steel deck (Fig. 18.2-13). The header duct may be a closed metal or fiber duct or a cellular metal raceway that forms a portion of the steel deck itself (Fig. 18.2-14). Raceways may be a closed duct designed for cover by concrete (see Fig. 18.2-13) or a trench duct with a removable cover (Fig. 18.2-15). The thickness of the concrete cover and the circumstances under which trench duct may be used are dictated by the NEC.

## 18.2.6 PLUGS, OUTLETS AND SWITCHES

### 18.2.6.1 Plugs

Older appliances and portable light fixtures had blades of the same size (Fig. 18.2-16a) that could be inserted in a receptacle in either direction. The problem is that one slot in a receptacle is always connected to a hot wire and the other slot is connected to a neutral wire, and one wire in a lamp cord is connected to the contact that touches the bottom of the bulb; another is connected to the lamp socket. When a plug is inserted in the correct orientation, the wire that is connected to the lamp's socket will be plugged into the neutral slot. If the plug is reversed, the socket will be connected to the hot wire. The lamp will work regardless of which way the cord is plugged in. There-

**FIGURE 18.2-13** A typical underfloor raceway. (Drawing by HLS.)

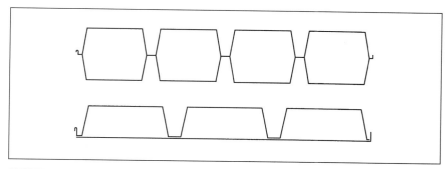

**FIGURE 18.2-14** Cellular metal raceways. These raceways replace and serve instead of standard metal deck. Wiring is run trough the cells. (Drawing by HLS.)

**FIGURE 18.2-15** A typical trench duct. Cover plates are removable for access to wiring. In some ducts of this type the flooring can be run completely across the duct, effectively concealing it from view. (Drawing by HLS.)

**a**

**b**

**c**

**FIGURE 18.2-16** (a) A standard nonpolarized plug. Note that the blades are equal in size. (b) A two-blade polarized plug. (c) A grounded three-prong plug. (Drawing by HLS.)

fore, there is no way to know whether the socket is hot or safe, so, in a lamp with a nonpolarized plug, there is a 50% chance that the threaded portion of the bulb socket is live and could give someone a shock.

Modern portable appliances and light fixtures have polarized plugs with blades of two different sizes so that they can be inserted into an outlet in only one direction (Fig. 18.2-16b). Since a polarized plug can only be inserted with the hot connected to the hot and the neutral connected to the neutral, its use will prevent a shock when the socket is touched. However, polarized plugs can prevent shocks only when they are plugged into outlets wired with the neutral wire on the neutral contact and the hot wire on the hot contact.

Old-fashioned nonpolarized receptacles are made with two slots of a size that

is smaller than the large slot in a polarized receptacle. Therefore, they will not accept a polarized plug. Grinding down a new plug to fit an old receptacle is particularly dangerous because it looks like a modern safe plug while presenting the same chance of shock as an old-fashioned nongrounded plug.

The third prong of a modern three-pronged grounding plug (Fig. 18.2-16c) permits an appliance or piece of equipment to be grounded, thus minimizing the shock hazard from faulty wiring. The third prong also prohibits inserting the plug into a receptacle with the neutral prong in the hot slot.

### 18.2.6.2 Outlets

There are three different types of outlets: convenience, special-purpose, and lighting.

#### CONVENIENCE OUTLETS

Older convenience outlets had two slots of the same size to receive plugs with two blades of the same size (see Section 18.2.6.1). Outlets designed to accept polarized plugs have two slots of different sizes to accept polarized plugs. Modern three-hole grounded receptacles are made with a large and a small slot to accept a polarized plug, and with a hole to accept the third prong of a three-prong plug.

There are two kinds of convenience outlets, standard and multioutlet. Older existing convenience outlets may be either polarized or nonpolarized. For newer installations, the NEC requires that outlets on a 15- or 20-amp general-purpose branch circuit be of the grounding type to minimize the hazard of shock from short circuits.

***Standard Convenience Outlets*** A standard convenience outlet is a plug-in receptacle housed in a protective box with plug-in positions for attachment of portable fixtures or minor appliances. The most common type, *duplex*, has two plug-in positions (Fig. 18.2-17b); *triplex* has three positions; a *single* has only one.

Outlets are sometimes grouped in a single protective box behind a single cover plate. Common groupings are an outlet and a switch, two or three outlets and a switch, and two outlets and two switches.

Regardless of the number of positions, an outlet is considered a single outlet when only one connection has been made

OCTAGONAL BOX FOR FIXTURE

**a**

STANDARD DUPLEX RECEPTACLE

**b**

DRYER OR RANGE RECEPTACLE

**c**

**FIGURE 18.2-17** (a) Box for lighting outlet and junction box use, (b) standard duplex receptacle, (c) special-purpose receptacle. (Drawing by HLS.)

to a branch circuit. Other arrangements are sometimes used, however. For example, a duplex receptacle that has been wired so that each plug-in position is on a separate 120-volt circuit or uses two separate line wires of a single three-wire 240-volt circuit (Figs. 18.2-18 and 18.2-19) is called a *split-receptacle* outlet. Typical uses of such outlets are discussed in Section 18.2.7.

***Multioutlet Assembly*** A multioutlet assembly is a series of plug-in receptacles, usually spaced at regular intervals and contained in a protective raceway (channel). Such assemblies also are called *plug-in strips*. They may be used in most locations where a standard convenience outlet is needed, especially where several outlets are called for.

#### SPECIAL-PURPOSE OUTLETS

A special-purpose outlet is a point of connection to the wiring system for a particular piece of equipment, normally reserved for the exclusive use of that equipment. Such outlets may be plug-in

**FIGURE 18.2-18** Using two hot wires in a duplex split receptacle permits switching off one receptacle while the other remains hot. (Drawing by HLS.)

**FIGURE 18.2-19** Split receptacles minimize overloads and double the capacity of duplex receptacles by connecting each plug-in position to a separate hot wire. When local codes do not permit this type of installation and both plugs on a receptacle must be wired to the same circuit, alternate receptacles can be wired to different circuits. (Drawing by HLS.)

receptacles to which the equipment is connected with a flexible cord, as with clothes dryers or ranges (Fig. 18.2-17c); or they may be protective enclosures, called *junction boxes*, with removable covers containing permanent connections; such junction boxes can be mounted either on the structure or on the equipment.

## LIGHTING OUTLETS

A lighting outlet is a metal protective box to which fixed light fixtures, as opposed to portable lights, can be mounted and in which they may be connected to a branch circuit (see Section 18.2.3). Lighting outlets must be compatible with the lights selected (see Section 18.4). Figure 18.2-17a shows a standard octagonal protective box of a type that is often used as a lighting outlet. Such a box may also serve as a junction box for pulling wires to remote fixtures in coves, valances, and cornices. Circular and square lighting outlet boxes are also used.

### 18.2.6.3 Switches

A switch is a device for interrupting the flow of electricity. Switches are used in service entrance and distribution panels and in branch circuits. The switches in panels are either a part of the panel or are in separate metal boxes. Switches in

branch circuits are connected to lighting, convenience, and special-purpose outlets.

Switches in branch circuits are mounted in the same kinds of protective boxes used for standard convenience outlets. They may be of the push type, toggle type, or slide type. Variable rotary switches are sometimes used to control the power to lights and ceiling fans.

### 18.2.6.4 Copper/Aluminum Devices

Due to the high cost of copper, aluminum wire was used in many residential installations during the mid-1960s. The potential problems associated with using aluminum wire were not realized until the 1970s.

The basic problem is that aluminum expands when warm and contracts when cold at a greater rate than does copper. Over time, repeated expansion and contraction can cause connections in aluminum wire to loosen. In addition, aluminum forms a film of aluminum oxide when exposed to air. This oxide is an insulator rather than a conductor of electricity. The combination of loose wiring and oxide buildup can cause connections to overheat. These problems became apparent when homeowners began to report flickering lights, radio and TV static, unpleasant odors, and occasional fires.

In 1972, copper/aluminum (CO/ALR) devices suitable for use with both copper and aluminum wiring came into use. CO/ALR devices have special brass screws with heads wide enough to prevent aluminum wire from working loose. Proper installation is shown in Figure 18.2-20. Only CO/ALR devices should be used with aluminum wire; other types are incompatible.

According to the U.S. Consumer Product Safety Commission (CSPC), between 1965 and 1973 about 1.5 million homes were wired with incompatible receptacles which CSPC believes to be fire hazards.

Owners of dwellings built or rewired during the 1960s can check for aluminum wiring in unfinished spaces such as the basement or attic; the letters AL or the word *aluminum* are stamped on the insulation covering the wire.

Trouble signs for dwellings wired with aluminum include:

- Cover plates or switches that are warm to the touch
- Smoke or sparks at switches or outlets
- Odors, such as the smell of burning plastic
- Lights that flicker periodically
- Receptacles or entire circuits that do not work

When aluminum wiring is discovered, the old switches and receptacles can be replaced with CO/ALR devices. Unfortunately, this replacement will leave other parts of the circuit, such as fixtures and junction boxes, as potential trouble sources.

### PIGTAILING

A broader solution to the aluminum/copper mismatch involves attaching short lengths of copper wire to existing aluminum wire with a special crimping tool. These copper-wire *pigtails* are then connected to the terminals of the electrical device. While CO/ALR devices might be suitable for new installations using aluminum wiring, pigtailing seems more appropriate for retrofitting existing installations.

**FIGURE 18.2-20** CO/ALR devices have special brass screws with wide heads. Wire should be stripped carefully to avoid nicks and wrapped around a terminal as shown in the drawing of the far left wire. The other wires are incorrectly attached.

### 18.2.7 OUTLET LOCATIONS

#### 18.2.7.1 Commercial, Institutional, and Industrial Buildings

The location of outlets in commercial, institutional, and industrial buildings is dictated by code requirements and the design of the facility. Electrical systems in these kinds of structures must be designed in accordance with sound engineering practices, with outlets and power sufficient to meet the needs of the occupants. The NEC and applicable building codes must also be followed, of course.

#### 18.2.7.2 Residences

This section contains some general recommendations for outlets in both single-family and multifamily residences. These recommendations are necessarily general in nature. Even when unusual conditions make strict compliance impossible, the recommendations given here may still be used as guidelines in developing more appropriate provisions.

#### CONVENIENCE AND SPECIAL-PURPOSE OUTLET NUMBER AND LOCATIONS

This section contains recommendations for the number and location of convenience and special-purpose outlets. Similar information about lighting outlets is presented in Section 16.4. When these recommendations are based on the number of linear or square feet, the number of outlets should be determined by dividing the total linear or square footage by the recommended linear or square footage per outlet. Resulting numbers that contain a fraction larger than one-half should be increased to the next larger number. The floor area should be computed from the outside dimensions of the house or the centerline of the walls surrounding each room. Convenience outlets should be located near the ends of a wall space rather than near the center, thus reducing the likelihood of their being concealed behind large pieces of furniture. In most cases, outlets should be located about 12 in. (300 mm) above the floor line. In every kitchen, living room, dining room, family room, recreation room, den, library, bedroom, and similar habitable space, convenience outlets should be placed so that no point along the floor line of a usable wall space is more than 6 ft (1800 mm) from an outlet in the same wall space. Where floor-to-ceiling windows prevent this requirement from being met with ordinary convenience outlets, other suitable means, such as floor outlets or multioutlet assemblies, can be used.

*Dual-Purpose Rooms* Where a space is intended to serve more than one function, such as a combination living and dining room or a kitchen and laundry, the following outlet recommendations are separately applicable to each specific area. In a living and dining space, for instance, convenience outlets in the dining area should be of the split-receptacle type in which each receptacle is on a different circuit, and ordinary outlets may be used in the rest of the space. This permits use of several portable devices, such as a toaster and a coffee maker, in the dining space without overloading a circuit.

*Exterior Entrances* A weatherproof convenience outlet, located at least 18 in. (457 mm) above the entrance platform near the front entrance, should be provided. This outlet should be controlled by a wall switch inside the building near the entrance for convenient operation of outdoor decorative lighting. Additional waterproof outlets along the exterior of the house are recommended to serve decorative garden treatments and for the use of appliances or electric garden tools, such as lawn mowers and hedge trimmers.

*Living and Recreation Rooms* When general illumination is to be provided by portable lamps in lieu of fixed lighting, at least two wall-switched separate plug-in positions should be planned. These can be provided with two switched regular duplex outlets or one switched plug-in position in each of two split-receptacle outlets (see Fig. 18.2-18).

When regular convenience outlets are wall-switch controlled to accommodate portable lighting, additional unswitched convenience outlets should be provided to prevent limiting the use of portable equipment, such as radios and clocks, in that location. In addition, a single convenience outlet should be installed in combination with the wall switch at one or more of the switch locations for the use of a vacuum cleaner or other portable appliances.

One or more outlets for entertainment equipment, such as a television set, radio, or sound equipment, should be provided at bookcases, shelves, and other suitable locations. In rooms with conventional fireplaces, an outlet may be desirable in or near the mantel shelf.

*Dining Areas* When a dining or breakfast table will be placed against a wall, one of the recommended outlets should be located in that wall, just above table height.

Where open counter space is to be built in, an outlet should be provided above counter height for the use of portable appliances. All convenience outlets in the dining area should be of the split-receptacle type, providing increased capacity at each outlet (see Fig. 18.2-18).

*Bedrooms* Convenience outlets should be placed so that there is one on each side and within 4 ft (1220 mm) of the centerline of each probable individual bed location. The popularity of bedside radios and clocks, bed lamps, and electric blankets makes increased plug-in positions essential at bed locations. Triplex or quadruplex convenience outlets are recommended at these locations (see Fig. 18.2-20c).

Additional outlets should be placed so that no point along the floor line of a wall is more than 6 ft (1800 mm) from an outlet in that space. It is also recommended that a receptacle be provided at at least one switch location in a bedroom to accommodate the use of a vacuum cleaner, floor polisher, or other portable appliance (see Fig. 18.2-20b).

*Bathrooms and Lavatories* A convenience outlet should be provided near the mirror, 3 to 5 ft (900 to 1500 mm) above the floor. An outlet should be installed at each separate mirror or vanity space, and also at any space that might accommodate an electric towel dryer, razor, hair dryer, or other small electrical equipment. A receptacle that is part of a bathroom lighting fixture is not suitable for this purpose unless it is rated at 15 amps, wired with at least 15-amp-rated wires, and easily accessible. A GFCI is required by the NEC in every such condition (see Section 18.2.4).

A wall-switched or timer-operated, built-in ventilating fan capable of providing a minimum of eight air changes per hour should be installed in every bathroom. Such a fan is generally required by code where natural ventilation is not provided through windows or skylights. A fan capable of moving 50 to 55 cu ft (1.42 to 1.56 m$^3$) of air per minute (cu ft/min) (m³/min) will assure eight air changes per hour in a 5 by 10 ft (1500 by 3000 mm) bathroom.

In many parts of the country, a wall-switch-controlled electric space heater is desirable for increased comfort during bathing. It may be part of a ceiling light, fan, heater combination, or a separate unit. Heaters may be of the wall- or ceiling-mounted type and often have a timer switch.

*Kitchens* A kitchen should have one convenience outlet for the refrigerator and additional outlets placed so that no point along the wall line is more than 24 in. (610 mm) from a convenience outlet, with at least one outlet to serve each separate work surface or planning desk. An outlet should be placed in every wall space that may be used for ironing and at every location that may be used for an electric roaster, microwave oven, toaster oven, can opener, popcorn popper, ice-cream maker, wok, or kettle. If a work table or dining table is planned next to a wall, this wall space should have one outlet located just above table level. All kitchen outlets should be protected by a GFCI.

Work-surface outlets should be located about 44 in. (1118 mm) above the finish floor level or about 8 in. (203 mm) above the work surface. Convenience outlets in a kitchen, other than that for a refrigerator, should be of the split-receptacle type (see Fig. 18.2-18).

Possible special-purpose outlets required in a kitchen include those for a range, oven, burner top, ventilating fan, dishwasher, and waste disposer. A special recessed receptacle with a hanger should be installed for an electric clock. A clock should be easily visible from all parts of the kitchen. It is also necessary to provide for a food freezer either in the kitchen or in some other convenient location.

*Laundry Areas* At least one convenience outlet should be provided, preferably near the sorting table if there is one. If properly located, one of the special-purpose outlets (the washer outlet, for example) may also serve as a convenience receptacle for such items as a laundry hot plate or sewing machine. Convenience outlets in the laundry area should be of the split-receptacle type (see Fig. 18.2-18).

A separate special-purpose outlet is generally needed for each of the follow-ing pieces of equipment: automatic clothes washer, hand iron or ironer, and clothes dryer. Outlets for a ventilating fan and clock are desirable. Requirements for an electric water heater should be obtained from the local power company.

*Halls* These recommendations apply to passage halls, reception halls, vestibules, entries, foyers, and similar areas. One convenience outlet should be installed for each 10 linear ft. (3048 mm) of hall measured along the centerline. A hall with more than 25 sq ft (2.32 m²) of floor area should have at least one outlet. In reception halls and foyers, convenience outlets should be placed so that no point along the floor line of a wall is more than 10 ft (3048 mm) from an outlet in that space.

A receptacle at one of the switch outlets is recommended for convenient connection of maintenance equipment such as a vacuum cleaner, floor polisher, or electric broom.

*Stairways* Convenience outlets generally are not required, except at intermediate landings of larger-than-normal size, where a decorative lamp, night light, or portable equipment might be used.

*Utility Rooms* At least one convenience outlet should be provided, preferably near the furnace or near a planned work-bench location.

A special-purpose outlet is needed for each piece of mechanical equipment requiring electrical connection. Such equipment includes a furnace, boiler, water heater, water pump, sump pump, and compressor.

*Basements* At least two convenience outlets should be provided in an open basement area. If a work bench is planned, one of these outlets should be placed at its specific location.

Additional convenience outlets may be needed for a basement laundry, darkroom, or hobby area and for appliances such as a dehumidifier or portable space heater.

A special-purpose outlet must be provided for each piece of mechanical equipment requiring electrical connections such as a furnace, boiler, water heater, water pump, sump pump, or compressor.

An outlet for a food freezer also should be considered.

*Accessible Attics* A convenience outlet is desirable for portable equipment and for trouble lights capable of reaching into distant dark corners. If an open stairway leads to future attic rooms, provide a junction box with direct connection to the distribution panel for future extension to convenience outlets and lights when rooms are finished.

A summer cooling fan, multiple-switch controlled from convenient points throughout the house, often is desirable.

*Porches and Breezeways* At least one GFCI convenience outlet should be provided for each 15 linear ft (4500 mm) of wall bordering a porch or breezeway. These outlets should be weatherproof if exposed to the elements. They should be controlled by a wall switch inside the house. Since such areas are often used for outdoor dining, at least one outlet should be of the split-receptacle type.

*Terraces and Patios* At least one weatherproof GFCI outlet should be provided for each 15 linear ft (4500 mm) of house bordering a terrace or patio. These outlets should be wall-switch controlled from inside the house and located at least 18 in. (457 mm) above the patio level.

*Garages and Carports* At least one GFCI outlet should be provided for one- and two-car garages and carports.

Special-purpose outlets appropriate to the use should be provided when a food freezer, work bench, or automatic door opener is planned. Specific hobbies, such as restoring antique automobiles or making ceramic sculpture, may require special-purpose outlets for compressors, welding rigs, or kilns.

## 18.2.8 WIRING RECOMMENDATIONS

The recommendations given in this section must be supplemented with engineering judgment concerning electrical system adequacy in common-use spaces and for general building services. Particular attention must be paid to limiting voltage drop in feeders to ensure proper operation of appliances and other devices.

### 18.2.8.1 Circuit Recommendations

In addition to branch circuits providing power for lights, appliances, and fixed and portable equipment, a system may include feeder circuits serving secondary distribution panels near heavy-use locations.

## BRANCH CIRCUITS

A branch circuit is that portion of a wiring system between an overcurrent device and outlets where loads such as light fixtures, appliances, or equipment may be connected. Wiring systems usually include three types of branch circuits: general-purpose, small-appliance, and special-purpose. For each circuit, one must consider the voltage, amperage, and number and size of conductor wires.

*General-Purpose Circuits* General-purpose circuits supply lighting and convenience outlets. They should be wired with suitable conductors as required by code, but not smaller than No. 14 AWG if protected by 15-amp overcurrent devices and No. 12 AWG for 20-amp devices. Since No. 14 conductors are more limited in circuit length (30 ft [9144 mm]) and current-carrying capacity (1725 watts), No. 12 conductors with 20-amp devices are recommended for most general-purpose circuits. Some building codes do not permit the smaller No. 14 wires.

The number of general-purpose outlets required is based on the loads to be encountered and the loads permitted per circuit, as determined both in accordance with the NEC and by the branch circuit amperage to be used. In residences, no less than the following number of general-purpose outlets should be provided:

- At least two 20-amp circuits for the first 750 sq ft (69.68 m²) of floor area plus one circuit for each additional 750 sq ft (69.68 m²) or major fraction thereof, or
- At least two 15-amp circuits for the first 500 sq ft (46.45 m²) plus one circuit for each additional 500 sq ft (46.45 m²) or major fraction thereof

These floor areas should be calculated from dimensions taken to the outside of exterior walls, including unfinished spaces which are adaptable to future living use; open porches and garages may be excluded from the calculation.

In commercial, institutional, and industrial buildings, the number of general-purpose outlets is determined in accordance with the NEC and the requirements of the facility. The NEC has requirements for health care facilities, hazardous facilities, mobile homes, outdoor areas, show windows, swimming pools, theaters, and other building use types. It also dictates the amperage ratings required in specific locations. Some building designs may require more general-purpose outlets than the NEC requires.

*Small-Appliance Circuits* The actual number and size of small-appliance circuits should be determined by the loads to be encountered as calculated in accordance with the NEC. The minimum for residences should be one three-wire 120-240-volt, 20-amp circuit (or two two-wire 120-volt 20-amp circuits), equipped with split-receptacle outlets, in areas regularly intended for cooking, dining, or laundering.

Using three-wire circuits to supply convenience outlets in the locations mentioned is an economical means of increasing the capacity and offers practical operating advantages. Such circuits provide greater capacity at individual outlet locations and lessen voltage drop in the circuit. For maximum effectiveness, the upper half of all receptacles should be connected to the same side of the circuit (see Fig. 18.2-19). A separate 20-amp circuit should be provided for each appliance or piece of equipment rated at more than 1400 watts. However, two separate receptacle circuits may have to be used where codes do not permit split receptacles.

*Special-Purpose Circuits* Separate circuits should be provided and conductors should be sized appropriately for the specific equipment. Some residential special-purpose circuits and the required conductor sizes are shown in Figure 18.2-21. Conductor sizes for special-purpose outlets and connection points in other types of buildings are determined in accordance with the NEC based on the load and voltage requirements of the specific equipment being connected.

In residences, spare circuit positions should be made in each distribution panel for at least two future 20-amp, two-wire 120-volt circuits. If the branch circuit or distribution panel is installed in a finished wall, raceways should be extended from the panel to the nearest accessible unfinished space for future use.

In other building types, spare circuits are provided in accordance with the NEC and good design practice for the particular facility. Such requirements may vary in different locations within the same facility. For example, in an office building, where extensive renovation is normal, more spare circuits might be provided than would be required where the use of the facility is expected to remain static.

The majority of portable appliances for residential use are made to operate at 110- to 125-volts. Some more or less fixed appliances, like refrigerators and freezers, also operate in the 110 to 125 volt range. Many fixed appliances, however, operate on 230/250-volt circuits because of improved operating efficiency. When there is a choice, the higher voltage should be used.

In other building types, both 120-volt and 240-volt devices are common, and some heavy equipment may require 480-volt service.

Most new homes today are air-conditioned, and much of the related equipment runs on electricity. Electrical heating systems are not as prevalent today as they were a few years ago because of their high energy use. However, they are being replaced in many instances by electric heat pumps, which use less energy than full electric heating systems but still use considerable electric energy, especially in colder climates. Electric heating and air conditioning systems and heat pumps are usually custom designed to suit the particular house and local climatic conditions. The equipment used is not custom designed, however. It is selected from a range of standard manufactured products to fit the requirements of the individual house. Adequate wiring and service capacity criteria for such devices should be determined on a case-by-case basis by consultation with the local power company or an electrical engineer.

**FIGURE 18.2-21** Interchangeable devices can be installed in the same outlet box in various combinations. (a) Shows a pilot light, switch, and receptacle in the same box, (b) a switch and receptacle, (c) a triplex receptacle, and (d) a switch and a duplex receptacle. (Drawing by HLS.)

## FEEDER CIRCUITS

Consideration should be given to using subpanels served by appropriately sized feeders to serve heavy-use locations throughout a building. In a residence, such areas may be those containing an electric range or clothes dryer.

### 18.2.8.2 Conductor Size

Conductors should be adequately sized for the design loads and length of run. Figure 18.2-22 lists wire sizes for specific loads, but these should be verified with the latest edition of the NEC. The table gives a direct reading of the maximum distance between distribution panel and load for different conductor sizes that will limit the voltage drop to a maximum of 3%. For example, consider a 3000-watt, 120-volt heater that is 75 ft (22.86 m) from the distribution center. First, watts should be converted to amps: $I = W/E = 3000/120 = 25$ amps (see Section 18.1.2). Checking Figure 18.2-23 under the 25-amp column heading, one finds that the maximum allowable distance from the source is 138 ft (42.06 m) for No. 10 AWG wire. This is the maximum distance over which the current will not exceed the permitted 3% voltage drop using that size wire. In selecting a wire size for a known load, the permitted length of run should be greater than the anticipated distance from panel to load.

### 18.2.8.3 Raceway Size

The number and size of conductors which can be placed in a raceway are limited not only by the size of the interior opening, but also by code restrictions.

The major factor limiting the number of conductors that may be put into raceways is *fill*, which is the percentage of the total cross-sectional area of the raceway filled with conductors. The maximum fill is dictated by the NEC. The following percentages are required for non-lead-covered conductors:

One conductor: 53%
Two conductors: 31%
Three or more conductors: 40%

The following percentages are required for lead-covered conductors:

One conductor: 53%
Two conductors: 30%
Three conductors; 40%
Four conductors: 38%
More than four conductors: 35%

The cross-sectional area of the conductors in a raceway should not exceed the maximum percentage of fill listed above. When using several conductors of unequal size in the same raceway, it is necessary to determine the total cross-sectional area of all the conductors combined, including their insulation, and then select the raceway size on the basis of percentage of fill.

The NEC contains extensive tables that list the number of conductors of a given type and size that may be used in each available conduit size so as not to exceed the required maximum fill percentages. For example, it might be possible to install the following number of No. 14 AWG RHW conductors in the listed conduit sizes:

$1/2$ in. (12.7 mm): 6
$3/4$ in. (19.1 mm): 10
1 in. (25.4 mm): 16
$1 1/4$ in. (31.8 mm): 29
$1 1/2$ in. (38 mm): 40
2 in. (50.8): 65
$2 1/2$ in. (63.5 mm): 93
3 in. (76.2 mm): 143

The numbers for No. 12 AWG RHW conductors are:

$1/2$ in. (12.7): 4
$3/4$ in. (19.1 mm): 8
1 in. (25.4 mm): 13
$1 1/4$ in. (31.8 mm): 24
$1 1/2$ in (38 mm): 32
2 in. (50.8 mm): 53
$2 1/2$ in. (63.5 mm): 76
3 in. (76.2 mm): 117

These numbers may not reflect the NEC requirements at the time you read these data. Therefore, it is always best to consult the latest edition of the NEC when sizing wire.

### 18.2.8.4 Fitting and Connections

A wiring installation should be adjusted to fit the design of the building. If compliance with a particular recommendation would require alteration of doors, windows, or structural members, alternate provisions should be made. For example, the extensive use of curtain walls may make it necessary to alter the recommended outlet or switch heights or to resort to surface-mounted raceways.

Conductors should be spliced with approved solderless connectors or by brazing, welding, or soldering. Splices should be made mechanically secure before soldering. Bare ends of conductors should be covered after splicing with an insulation equal to that on the conductors. Connections should be made in an accessible device such as a junction box, switch box, or fixture base.

Splices should never be made outside a raceway and then pulled into it. They should be made in a wireway that has a removable cover or in an accessible box,

**FIGURE 18.2-22** Recommended Conductors for Special-Purpose Circuits

| Item | Typical Wattage | Recommended Conductor Capacity | | |
| --- | --- | --- | --- | --- |
| | | Amperes | Wires | Volts |
| Range or electric oven and drop-in cooktop | 12,000* | 50 | 3 | 120/240 |
| | 5,000 | 30 | 3 | 120/240 |
| | 7,000 | 40 | 3 | 120/240 |
| Automatic washer | 700 | 20 | 2 | 120 |
| Electric clothes dryer | 5,000* | 30 | 3 | 120/240 |
| Fuel-fired heating equipment | 800 | 20 | 2 | 120 |
| Central air conditioning or room air conditioner (1 ton) | 5,000* | 40 | 3 | 120/240 |
| | 1,500 | 20 | 2 | 120 |
| Dishwasher and disposer or dishwasher and waste disposer | 2,700 | 20 | 3 | 120/240 |
| | 1,800 | 20 | 2 | 120 |
| | 900 | 20 | 2 | 120 |
| Food freezer | 400 | 20 | 2 | 120 or 240 |
| Water pump | 500 | 20 | 2 | 120 or 240 |
| Bathroom space heater | 2,000 | 25 | 2 | 240 |
| Electric water heater | 4,500* | 30 | 3 | 240 |

*Typical wattages are shown; verify with manufacturer and/or local power company.

**FIGURE 18.2-23** Maximum Lengths of Two-Conductor Runs

| Conductor Size, AWG | Maximum Length of Run, ft (m) | | | | | | | | | | | | | | | | | | |
|---|---|---|---|---|---|---|---|---|---|---|---|---|---|---|---|---|---|---|---|
| | 5 amp | 10 amp | 15 amp | 20 amp | 25 amp | 30 amp | 35 amp | 40 amp | 45 amp | 50 amp | 55 amp | 70 amp | 80 amp | 95 amp | 110 amp | 125 amp | 145 amp | 165 amp | 185 amp |
| 14 | 274 (85.5) | 137 (41.8) | 91 (27.7) | | | | | | | | | | | | | | | | |
| 12 | | 218 (66.5) | 145 (44.2) | 109 (33.2) | | | | | | | | | | | | | | | |
| 10 | | | 230 (70.1) | 173 (52.7) | 138 (42.1) | 115 (35.1) | | | | | | | | | | | | | |
| 8 | | | | | 220 (67.1) | 182 (55.5) | 156 (47.6) | 138 (42.1) | | | | | | | | | | | |
| 6 | | | | | | | | 219 (66.8) | 193 (58.8) | 175 (53.3) | 159 (48.5) | | | | | | | | |
| 4 | | | | | | | | | 309 (94.2) | 278 (84.7) | 253 (77.1) | 199 (60.7) | | | | | | | |
| 3 | | | | | | | | | | 350 (106.7) | 319 (97.2) | 250 (76.2) | 219 (66.8) | | | | | | |
| 2 | | | | | | | | | | | 402 (122.5) | 316 (96.3) | 276 (84.1) | 232 (70.7) | | | | | |
| 1 | | | | | | | | | | | | 399 (121.6) | 349 (106.4) | 294 (89.6) | 254 (77.4) | | | | |
| 0 | | | | | | | | | | | | 502 (153) | 439 (133.8) | 370 (112.8) | 319 (97.2) | 280 (84.7) | | | |
| 00 | | | | | | | | | | | | | 560 (170.7) | 470 (143.3) | 403 (122.8) | 355 (108.2) | 305 (93) | | |
| 000 | | | | | | | | | | | | | | 588 (179.2) | 508 (154.8) | 447 (136.2) | 385 (117.3) | 339 (103.3) | |
| 0000 | | | | | | | | | | | | | | | 641 (195.4) | 564 (171.9) | 486 (148.1) | 421 (128.3) | 360 (109.8) |

which can be opened for inspection or repairs. Each conductor run should be complete from one opening to the next before conductors are pulled. Conductors inserted before the raceway is completed can be damaged during the process of making up the raceway.

Terminal connections should be made securely. Terminals for securing more than one conductor should be approved for that purpose.

Outlet, junction, and switch boxes used in wiring installations should provide adequate space for the conductors. Conductors should be sized adequately to minimize line losses and voltage drop. Voltage drop between the distribution panel and the load should be limited to less than 3%.

Both raceway joints and conductor connections should be made strong and secure. A mechanically secure raceway connection is one in which the joint is as strong as the raceway itself. An electrically secure connection will carry a full short circuit, that is, the entire return current, if one of the hot wires is accidentally grounded. Such accidental grounding will then open the circuit by blowing the fuse or tripping the circuit breaker. An improperly connected system can result in excessive heating at conductor joints and a possible fire hazard.

An excessive number of bends in a conduit can result in damage to conductors when they are pulled through it. A run of rigid conduit or electrical metallic tubing should not contain more than the equivalent of four quarter bends (360 degrees total), including those bends located at the fitting outlet.

Because of the different characteristics of copper and aluminum, pressure terminals or pressure connectors should be marked as suitable for mixing copper and aluminum conductors when this combination is used.

### 18.2.8.5 Grounding

Two types of grounding are involved in a wiring installation: (1) system grounding and (2) equipment grounding. *System grounding* entails connecting the current-carrying neutral wire to the grounding terminal in the main switch, which in turn is connected to a ground. A system ground is often a pipe or rod driven into the ground or laid in a trench. Sometimes a system ground in a residence is a water pipe A neutral wire is also called a *grounded wire*. *Equipment grounding*

refers to the use of a grounding wire (or the metal raceway) to connect the metal frames of equipment and appliances back to the service cabinet and to ground. The wire connecting a service cabinet to a water pipe or other grounding device is called a *ground wire*.

### SYSTEM GROUNDING

The entire wiring system should be grounded by connecting the neutral wire to the ground, either before it enters the building by connecting it directly to a ground or, in the case of a house or other small building, to a water pipe inside the building. Ground connection to a water pipe should be made on the street side of a water meter; if connected on the building side, a jumper cable should be used around the meter. The two hot wires are generally color-coded with black or red insulating jackets. The neutral wire (also called the *grounded wire*) is usually white but may also be natural gray. The neutral wire is continuous throughout the system, is never interrupted by fuses or switches, and generally is not connected to a hot wire.

For a residential or commercial wiring system, the earth is generally used as the ground. When a metal water supply system terminating in the earth is available, the electrical system is grounded by connecting to a cold water pipe near the service panel in the building (see Fig. 18.2-2). If a metal water pipe cannot be used, a metal rod is driven into the ground at least 2 ft (610 mm) from the building near the service entrance; the neutral wire from the service drop is connected to the rod with a ground wire. The neutral wire of the three-wire service is similarly grounded at the utility pole.

### EQUIPMENT GROUNDING

Unlike the grounded (neutral) wire, the grounding wire and a raceway carry no current, except when a short circuit in the equipment or circuit wiring causes a *fault* current to flow. Under these conditions, equipment grounding minimizes the shock hazard to occupants by providing a low-resistance path for the fault current and by opening the circuit overcurrent device in case of excessive flow.

In an electrical system involving conduit, tubing, or metallic cable, metal boxes are used and equipment or fixtures permanently connected at the box become automatically grounded. Portable

equipment with cords can be grounded through grounding receptacles and three-prong grounding plugs. In either case, a grounding wire, color-coded green, is included in the cord or cable connecting the box or receptacle to the frame of the equipment.

When a nonmetallic sheathed cable system is used, a bare grounding wire is often included to tie the system to the ground. Permanently connected equipment has its green grounding wire attached to the bare copper wire in a junction box. Grounding receptacles housed in nonmetallic boxes have a special terminal for the grounding copper wire. Portable equipment cords include a green grounding wire which contacts the grounding system through the round or U-shaped prong of the plug.

Equipment grounding provides adequate safety from shocks, except in high-hazard areas such as near pools, in bathrooms and shower rooms, and in outdoor locations. Receptacles should not be installed closer than 10 ft (3048 mm) from the edge of a pool. Poolside receptacles 10 or 15 ft (3048 or 4500 mm) from the pool and all outdoor receptacles should be equipped with GFCIs (see Section 18.2.3.2). The use of GFCIs is required by the NEC, because the severity of electrical shock is greatly increased when water is present.

With certain exceptions, the NEC requires GFCIs in new residential construction and in hotels and motels for protection of people. GFCIs are required around swimming pools; in bathrooms, kitchens, garages, crawl spaces, unfinished basements, and in other locations where a damp floor might be expected; and for outdoor receptacles. It may also be desirable to add them in high-risk areas in existing dwellings. Other high-risk areas that need GFCI protection include laundries and workshops. GFCIs should also be considered for any part of a dwelling built on a concrete slab at or below grade.

### 18.2.9 LOW-VOLTAGE CIRCUITS

*Low voltage* refers to ac power supplied by step-down transformers from a 120/240-volt line voltage service. Although the NEC treats anything less than 50 volts as *low voltage*, in practice the term usually refers to circuits of less than 30 volts.

Less insulation is required for this type of system, and the NEC, as well as many local building codes, permit low-voltage wiring without protective raceways. Low-voltage wiring is used for bell or buzzer systems, intercoms, control wiring for heating and cooling systems, certain types of lighting fixtures, and other uses requiring relatively small amounts of power. Because low-voltage wiring is generally used for low-power systems, the conductors are quite small, No. 18 AWG or smaller. In situations requiring multiple and remote switching, low-voltage systems can be used effectively to turn light fixtures or other equipment on and off. The smallness of the wire and elimination of the raceway requirement make this wiring method considerably less costly than 120-volt installations in tubing or conduit. The equipment is fed from the line voltage circuit, as in standard installations, but the switch opening or closing the circuit is activated by a small, low-voltage power relay.

## 18.2.10 PLANNING FOR FUTURE ADDITIONS

When certain facilities or functions are indicated as future additions, the initial wiring should be arranged so that none of it need be replaced or moved when the ultimate plan is realized. Final wiring for a future addition may be left to a later date if desired.

## 18.2.11 SOUND CONTROL

Improper electrical installations can reduce the acoustical performance of walls and floor and ceiling assemblies. Causes for reduced performance include (1) open air transmission paths where wires pass through walls or floor and ceiling assemblies, (2) openings provided for outlets, switches, or fixtures, and (3) noise vibration transmitted by wires or fixtures rigidly connecting two otherwise separated structural members or connecting equipment to the structure.

In multifamily housing, and in every other building type having more than one occupancy, each apartment or separate use should be wired as a unit, keeping wiring within the walls of the same unit (Fig. 18.2-24). Holes should be tightly caulked with a resilient, nonhardening compound.

Convenience outlets, switches, telephones, and such other installations as intercoms should not be placed back-

**FIGURE 18.2-24** Each apartment should be wired independently, and air passages should be sealed airtight.

back. (1) Convenience outlets should be placed at least 36 in. (900 mm) apart on the opposite sides of the wall. In staggered stud walls, installations should be offset horizontally by at least three complete stud spaces (Fig. 18.2-25a). (2) A horizontal distance of 24 in. (610 mm) should be maintained between switches and outlets, and they should not be located in the same stud space (Fig. 18.2-25b). (3) Wall fixtures and equipment should be kept 24 in. (610 mm) apart on opposite sides of the wall (Fig. 18.2-25c).

Where sound control is important, lighting fixtures and other electrical fix-

tures should be surface-mounted (never recessed) and should not be connected to joists above resiliently mounted finished ceilings or connected across separated ceiling joists (Fig. 18.2-25d). If there is no option but to recess a fixture in a sound-controlled wall or ceiling, the fixture should be mounted in an enclosure that provides at least as much sound attenuation as does the penetrated wall or ceiling.

Appliances can often be selected and installed to reduce the amount of noise. For example, a clothes dryer can be mounted on a floating floor, and the noise transmission from the appliance can be

**FIGURE 18.2-25** Electrical outlets, switches, fixtures, and equipment should not be mounted back-to-back, should not be located in the same stud space, and should not be spaced closer than shown in (a), (b), and (c) in this figure. Ceiling fixtures should be surface-mounted, as shown in (d), rather than recessed.

reduced by using a neoprene or rubber gasket or mat between the base of the dryer and the floor, or by using a resilient underlayment between two layers of floor sheathing. Appliances should be selected for quiet operation. Vibrating appliances such as dryers should not be rigidly connected to the structure. Unless a floated floor system is used, resilient or flexible isolation gaskets should be provided between the appliance and the floor, wall, or counter to isolate vibrations,. Electrical connections to appliances should also

**FIGURE 18.2-26** Wiring in rigid conduit can be connected to vibrating equipment with a flexible cable.

be flexible whenever possible; a loop as shown in Figure 18.2-26 will reduce vibration transmission. Vents should also be flexible.

Refer to Chapter 13 for additional discussion of sound control.

# 18.3 Lightning Protection

Lightning strikes cause fires in unprotected buildings. This section discusses lightning protection systems intended to eliminate this hazard.

### 18.3.1 LIGHTNING

Lightning is a discharge of static electricity, a giant spark that jumps between two charged bodies such as two clouds or a cloud and an object on earth. A study by the Lightning Protection Institute shows that dwellings are the most frequent targets of damaging lightning; lightning is also a major cause of rural fires in other types of structures. Lightning often strikes large buildings, especially tall ones, but most remain undamaged because they have lightning protection systems.

Lightning can enter a building by (1) striking the building directly, (2) striking a metal object such as a television antenna or radio tower and then traveling down into the building, (3) striking a nearby tree and leaping over to the building, or (4) following a power line or ungrounded fence. On its way to the ground, lightning follows the path of lowest electrical resistance. Hence it frequently follows wiring or plumbing lines, causing fires and damaging electrical appliances. Television antennas resemble lightning rods in that they are targets for atmospheric discharges, but they cannot safely handle the current of a bolt of lightning even when they are grounded and equipped with a lightning arrester. Lightning can jump from a television antenna to its circuit, to plumbing, or to other electrical wiring and cause damage.

### 18.3.2 LIGHTNING PROTECTION SYSTEMS

Lightning protection should be provided for all buildings and for trees that are taller than an adjacent structure and within 10 ft (610 mm) of the structure.

### 18.3.2.1 Building Protection

Building lightning protection systems are divided by the Underwriter's Laboratory, Inc. (UL) into three classes:

- Class I: ordinary buildings under 75 ft (22.86 m) high
- Class II: buildings more than 75 ft (22.86 m) high
- Class II Modified: structures such as large smokestacks and chimneys

A good lightning protection system provides an easy, direct path for an electrical current to follow to reach the ground. It must prevent injury, damage, or death as a lightning bolt travels that path. Many insurance companies offer a reduction in premiums for buildings with properly installed lightning protection systems.

A lightning protection system includes (1) air terminals (lightning rods), (2) main conductors, (3) bonds to metal roof structures and equipment, (4) branch conductors, (5) lightning arresters, and (6) ground connections (Fig. 18.3-1).

Air terminals, which are pointed copper or aluminum rods (Fig. 18.3-2), are placed on the highest points of roofs and on every projection, such as a chimney or dormer on a house (see Fig. 18.3-1) and along parapet walls on other types of buildings. Terminals should be sized, spaced, and anchored according to the requirements of the National Fire Protection Association's (NFPA) *Lightning Protection Code*. They are usually between 10 in. (254 mm) and 24 in. (610 mm) long but may be longer. Those less than 24 in. (610 mm) long should be spaced not more than 20 ft (6.1 m) apart. Longer ones may be placed up to 25 ft (7.6 m) apart. In any case, there should be a terminal within 2 ft (610 mm) of each roof end. Additional terminals should be placed in large, open, flat and gently sloping roof areas at intervals not exceeding 50 ft (15.24 m).

Arresters should be provided at electric and telephone services and at radio and television lead-ins to protect these services from lightning.

Ground connections can be made by (1) clamping the main conductor cable to a copper-clad or galvanized steel rod which has been driven at least 10 ft (3000 mm) into the ground, (2) burying a stranded copper conducting cable in a trench, (3) clamping copper conducting cable to a sheet metal plate and burying the assembly, or (4) clamping the conductor to a metal water pipe.

The chief requirement for an efficient lightning protection system is a solid connection between the system and the earth. Ground connections of galvanized steel, copper, or copper-clad steel are preferred. Aluminum should not be used for ground

**FIGURE 18.3-2** Air terminals are the only visible parts of a concealed lightning protection system. (From U.S. Department of Agriculture, Farmers Bulletin Number 2136, "Lightning Protection for the Farm.")

connections because it corrodes when in contact with soil.

Every lightning protection system needs at least two ground rods placed as far apart as possible. Usually they are placed at opposite corners of a small building. Additional grounding terminals must be placed to serve larger buildings. These should not be more than 100 ft (30.48 m) apart, and there should be one terminal for every 100 ft (30.48 m), or fraction thereof, of building perimeter. Grounding terminals should reach below and away from the building to avoid lightning damage to exterior walls.

Lightning systems can be added to an existing structure, but if they are installed as part of new construction, (1) protection is effective during the construction process, (2) system parts can be more readily protected and concealed, and (3) installation costs may be lower. Lightning protection system components such as rods, cables, and connectors should be UL labeled.

### 18.3.2.2 Tree Protection

When a building with a lightning protection system is within 25 ft (7.6 m) of a protected tree, the two systems may be interconnected. A common grounding may be used for several large trees, as long as the trees are not more than 80 ft (24.4 m) apart. A tree's lightning protection system (Fig. 18.3-3) consists of:

1. Main terminal (special tree) point placed as high as possible within the tree
2. Miniature terminal points clamped on main branches

**FIGURE 18.3-1** The components of a residential lightning system include (1) air terminals, (2) down-lead conductors, (3) at least two grounds, (4) roof projections tied into the conductor system, (5) tree protection, (6) at least two terminals on chimneys, (7) dormers rodded, (8) arresters on antennas, (9) tie-in to the conductor system of gutters within 6 ft (1800 mm) of the conductor, and (10) arresters on overhead power lines. (From U.S. Department of Agriculture, Farmers Bulletin Number 2136, "Lightning Protection for the Farm.")

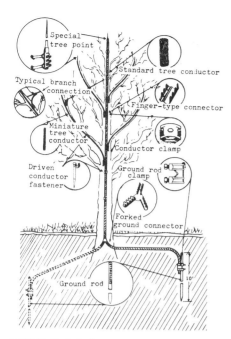

**FIGURE 18.3-3** A tree's lightning protection system consists of (a) a main terminal (special tree point), (b) miniature terminals (branch connection), (c) a main conductor, (d) a branch conductor, and (e) grounds. (From U.S. Department of Agriculture, Farmers Bulletin Number 2136, "Lightning Protection for the Farm.")

3. Main conductor (32-strand, 17-gauge [1.45 mm]) copper cable run from the main terminal down)
4. Branch conductors (14-strand, 17-gauge [1.45 mm]) copper cables run from miniature terminals to the main conductor)
5. Grounds (two or more $^1/_2$- to $^3/_4$-in. [12.7- to 19.1-mm])-diameter rods driven to a 10 ft (3000 mm) depth at points away from the main root system)

Trees having trunks more than 3 in. (76 mm) in diameter should have two main conductors, one on each side of the trunk. Cables should be copper because the color blends unobtrusively with the tree and because other metals may corrode and harm the tree. Cables should be attached with fasteners that hold them away from the trunk at intervals 3 to 4 ft (900 to 1200 mm) apart (see Fig. 18.3-3).

# 18.4 Lighting

Lighting outlets should be located where they will produce the most desirable lighting effects. Their types should conform with the lighting fixtures or equipment to be used. The lighting fixtures addressed in this section are fixed rather than portable.

To understand lighting design and selection, it is necessary to understand some of the special terms used:

*Brightness*: Luminous intensity of a surface.
*Candlepower (cp)*: Luminous intensity of a light source.
*Candle*: Measure of illumination at a surface.
*Coefficient of Utilization*: Measure of the total number of lumens that will reach the working surface.
*Footcandle*: 1 lumen distributed uniformly over 1 sq ft (645 mm$^2$) of surface 1 ft (304.8 mm) from a standard candle.
*Footlambert*: 1 lumen per 1 sq ft (645 mm$^2$).
*Illumination*: Density of light.
*Lamp*: Man-made light source.
*Light*: Radiant energy at a frequency that humans can see.
*Lumen*: The amount of light that falls on a surface that is 1 ft (304.8 mm) from a standard candle; used to indicate the total output of a light source.
*Luminaire*: A complete lighting unit.
*Reflector*: Portion of a luminaire that reflects light.
*Transmission Factor*: The ratio of light that passes through a material to the amount that strikes its surface.

## 18.4.1 FIXTURE TYPES

Light fixtures are either incandescent or of the gaseous discharge type.

### 18.4.1.1 Incandescent Lamps

Incandescent lamps consist of:

■ A base, which screws into a socket; several types are available, including the three most common: *medium* (up to 200 watts), *mogul* (more than 200 watts), and *candelabra* (chandeliers and other decorative fixtures).
■ A tungsten filament, which glows to produce the light.
■ A clear or frosted glass bulb, which encloses the filament.

Incandescent lamps are used in every type of building. In modern commercial, institutional, and industrial buildings, they are used mostly for spot lighting and special-purpose fixtures.

### 18.4.1.2 Gaseous Discharge Lamps

Gaseous discharge lamps include fluorescent and high energy discharge lamps.

#### FLUORESCENT LAMPS

Fluorescent lamps provide more light (lumens) per watt than incandescent bulbs, making them more economical and resulting in fewer power circuits. They also give off less heat, thus reducing air conditioning load. They consist of:

1. One of several configurations, including:
   a. *Standard*, consisting of a sealed glass cylindrical tube. Most are straight, but U-shaped, circular, and right-angle lamps are also available. These are available in standard and slimline shapes, with either single or bipin contacts. They are either preheat, rapid start, or instant start, but most of those used today are of the rapid start type. Available lengths vary from 2 ft (600 mm) to 8 ft (2400 mm), depending on the type of lamp.
   b. *Compact*, which are available in several shapes. Some resemble incandescent lamps. Others are twin-tube cylinders, circular, square, U-shaped, or helical (spiral tubes), reflector types similar to incandescent flood- or spotlights, and several other configurations. Some of these have screw bases to fit into standard lamp holders designed for incandescent lamps; others are plug-in versions with two or four pins.
2. Cathodes at the ends of the tube, which produce electrons; these are classed as either rapid start (most fixtures) or instant start (used mostly in cool areas and outdoors).
3. A low-pressure mercury vapor within the tube, which is caused to glow by the electrons produced by the cathodes.
4. An argon or other inert gas filler.
5. A phosphorus coating on the interior of the tubes, which converts invisible ultraviolet radiation produced by the mercury arc into visible light.

Fluorescent fixtures are used in every type of building. They are available in several colors called variously *white*, *cool white*, *daylight*, *warm white*, and others.

#### HIGH-INTENSITY DISCHARGE LAMPS

High intensity discharge (HID) lamps include mercury vapor, metal halide, high-pressure sodium, and low-pressure sodium lamps. HID lamps make light by passing an electric charge through a mixture of gases. Because they operate under high pressures and temperatures, and because they require startup times of as much as 15 minutes, HID lamps are appropriate only in high-ceilinged interior locations, such as arenas and gymnasiums, in other locations where they are not subject to frequent on-off cycles, and for exterior lighting.

*Mercury vapor lamps* depend on the excitation of mercury in vapor form. They are the least efficient HID type, and their color rendition is only fair. Their startup time ranges from 2 to 5 minutes.

The more efficient *metal halide lamps* contain metal halides and mercury. Because they are subject to explosion when damaged, most metal halide lamps are mounted in a protective enclosure. They have a startup time of about 15 minutes, but because of their efficiency and color qualities, they are generally the choice when HID lamps are appropriate and the color of the light is a consideration.

*Low-pressure sodium* (LPS) and *high-pressure sodium* (HPS) lamps are used in industrial and street security lighting. The startup gas in LPS lamps is neon. At full output the light from a LPS lamp is yellow. The startup gas in HPS lamps is xenon, and the light is produced by passing an electric current through vaporized sodium, which emits a gold-pink light.

## 18.4.2 LIGHT CONTROLS

In most cases, fixed lighting outlets are controlled by either single or multiple wall switches. These are usually located at the latch side of doors or at the traffic side of arches and within the room or area where the outlets are located. Some exceptions to this practice are switches for control of:

- Exterior lights, such as parking lot and other exterior lights, which may be controlled by manual switches or automatically by timers of light sensors.
- Stairway lights.
- Corridor and other exitway lights.
- Lights in infrequently used areas, such as storage areas.
- Lights in large areas, such as office bays and manufacturing plants; these may be controlled directly from lighting panels rather than by separate wall switches.

Wall switches are normally mounted about 48 in. (1220 mm) above the floor but may be mounted lower, if desired, to align them with door knobs or to put them within reach of children. The height of switches in public areas is dictated by handicapped access considerations (see Section 1.4).

Spaces that have more than one entrance for which switched lighting is needed are often equipped with a separate switch at each principal entrance. For instance, a door from a living room to a porch is a principal entrance to the porch. However, this door would not necessarily be considered a principal entrance to the living room unless the front entrance to the house was through the porch. Rooms in commercial, institutional, and industrial buildings often have more than one entry. Whether multiple switches are needed is dictated by the room's use and size. If providing more than one switch in a space would result in placing switches controlling the same light within 10 ft (3000 mm) of each other, one of the switch locations is usually eliminated. Moreover, when a space may be lighted from any one of several sources, as happens when both general and supplementary illumination is provided from fixed sources, multiple switching is required

for one set of controls only, usually for the general illumination source.

## 18.4.3 LIGHTING DESIGN

### 18.4.3.1 Interior Lighting

The amount and type of illumination required should be fitted to the various tasks carried on in a space. In home living areas, for example, there should be suitable lighting for recreation and entertainment as well as for normal day-to-day living.

In many instances, best results are achieved by a blend of lighting from both fixed and portable luminaires. In an office space or laboratory, for example, the best lighting may be achieved by general lighting alone or a blend of general lighting and specific task lighting. The types of tasks being performed will dictate which is best. Special lighting will be required where intensive work is being done, such as in surgical operating rooms or automobile manufacturing assembly lines. Lighting in commercial areas can be a combination of general and display lighting (Fig. 18.4-1). Special lighting is also required for displays of art such as the museum display in Figure 18.4-2. Good lighting, therefore, requires thought-

**FIGURE 18.4-1** A variety of fixture types, including ceiling-mounted recessed fixtures, wall-mounted fixtures, and fixtures in display cases can give both general and display lighting.
(Lytecaster retail 02.tif: image courtesy of Lightolier image archive.)

**FIGURE 18.4-2** Direct down lights illuminate the artworks in the museum shown in this photograph. (Davis Museum 18.tif: photographer, Chuck Pappa, image courtesy of Lightolier image archive.)

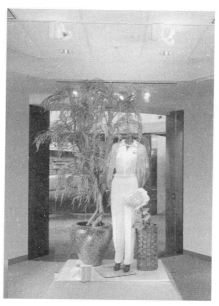

**FIGURE 18.4-3** Spotlights and floodlights from both below and overhead give strong illumination of this display. Down lights also show off the tree in the background. (Tech-Center Mannequin 01.tif: photographer, Jim Kelly, image courtesy of Lightolier image archive.)

ful planning and careful selection of permanent fixtures, portable lamps, and other lighting equipment.

The Illuminating Engineering Society of North America's (IESNA) *Lighting Handbook* contains a chapter that offers lighting design guidance.

## COMMERCIAL, INSTITUTIONAL, AND INDUSTRIAL BUILDINGS

Every habitable space should have adequate lighting for the tasks to be performed there. Adequacy is usually determined by the number of footcandles required. This is often a subjective decision, but there are some generally applicable guidelines. Codes sometimes dictate the minimum requirements. Owners will also sometimes have their own requirements, as might be the case, for example, in a hospital or school. The following are some typical footcandle requirements according to use:

1. *Schools*: corridors, 20; classrooms, 70 with 150 at chalkboard areas; laboratories, 80
2. *Retail stores*: merchandising areas, 150; showcases, 200; display windows, 200 or more (Fig. 18.4-3)
3. *Industrial facilities*: general areas, 50; fine machinery areas, 80 to 400

After determining the number of footcandles required in each space, then calculate the area of the spaces. Next, select the luminaires to be used in each space. From the manufacturer's data for the selected luminaires, determine the coefficient of utilization (cu) for each. This determination depends on wall, ceiling, and floor reflectance factors, the space's cavity ratios, and the lamp maintenance category.

The reflectance factor is the percentage of light that will be reflected by a surface. These factors are supplied by materials manufacturers. They usually range as follows:

1. *Walls*: 10 to 50%, with most being about the 30%
2. *Ceilings*: 50 to 80%
3. *Floors*: Usually about 10%, but may range up to 30%

Cavity ratios are determined by consulting tables in the IESNA *Lighting Handbook*. The data required to enter the tables are the height of the space above the luminaire (ceiling cavity), and the height of the luminaire above the work surface (room cavity), the height of the work surface above the floor (floor cavity). The ratios are designated as:

- CCR: ceiling cavity ratio
- RCR: room cavity ratio
- FCR: floor cavity ratio

Effective cavity reflectances are then determined directly from a table in the IESNA *Lighting Handbook* by entering the cavity ratio and the reflectance factor.

The lamp maintenance category is determined by a table in the IESNA *Lighting Handbook*. It has to do with the configuration of the luminaire.

By entering the data determined, the cu is then taken either from the manufacturer's data or from a table in the IESNA *Lighting Handbook*. If the actual cavity reflectances do not exactly match those used in the manufacturer's data or the IESNA *Lighting Handbook* tables, it will be necessary to interpolate the final cu based on the actual reflectances.

The next step is to determine the luminaire's lamp lumen depreciation factor (LLD), which is taken from the manufacturer's data. Then determine the luminaire dirt depreciation factor (LDD). This is taken from charts in the IESNA *Lighting Handbook* by entering the LLD, the type of dirt expected in the environment, and the number of months the fixtures are expected to be used without replacement.

The maintenance factor (mf) is the multiple of the fixture's LLD and its LDD.

The total lumens required for the space is determined by multiplying the area of the space by the number of footcandles required and dividing the result by a multiple of cu and mf.

To determine the number of luminaries needed, divide the total number of lumens required by the lumens output of each luminaire.

Finally, lay out the correct number of luminaires so that the light falls where needed. The layout is often symmetrical but may be asymmetrical when space shape or lighting needs so dictate.

## RESIDENTIAL BUILDINGS

The following requirements comply with the NEC as a minimum, but some of them exceed the NEC requirements. When two different requirements demand a lighting outlet at the same location, only one need be installed. However, additional wall-switch controls may be necessary.

For example, a lighting outlet in an upstairs hall may be located at the head of the stairway, thus satisfying the need for both hall lighting and stairway lighting with a single outlet. However, multiple-switch control of this lighting outlet will be required at the head and at the foot of the stairway. In addition, the multiple-switch control rule given in Section 18.4.2 may require a third point of control in the upstairs hall because of its length.

***Dual-Purpose Rooms*** When a room is intended to serve more than one function, such as a combination living and dining room, dining room and kitchen (Fig. 18.4-4), or kitchen and laundry, the following outlet recommendations are separately applicable to specific areas. Lighting outlet provisions may be combined in a way that will provide both general overall illumination and local illumination of work surfaces. In determining locations for wall switches, the area is considered as a single room.

***Living and Recreation Rooms*** Some means of general illumination is essential in living rooms, recreation rooms, sunrooms, enclosed porches, television rooms, libraries, dens, and similar areas. This lighting may be provided by ceiling or wall fixtures (Fig. 18.4-5), by lighting in coves, valances, or cornices, or by portable lamps. Wall-switch-controlled outlets in locations

appropriate to the intended lighting method should be provided. Outlets for decorative accent lighting, such as picture illumination and bookcase lighting, should also be considered.

***Dining Areas*** Each dining room, or dining area combined with another room, or breakfast nook should have at least one wall-switch-controlled lighting outlet. This outlet normally is located over the dining or breakfast table to provide direct illumination for it.

***Bedrooms*** Good general illumination is essential in bedrooms. This may be provided from a fixed ceiling fixture or from lighting in valances, coves, or cornices.

Wall-switch-controlled outlets should be placed in locations appropriate to the lighting method selected.

Light sources located over full-length mirrors, or in the bedroom ceiling directly in front of the clothes closet, may serve as general illumination. A master switch in the master bedroom (as well as at other strategic points in the home) controlling selected interior and exterior lights may be desirable.

***Bathrooms and Lavatories*** Illumination of both sides of the face when us-

**FIGURE 18.4-5** Recessed ceiling lights and a floor lamp provide both general and specific lighting to this living room to highlight the art over the mantlepiece and the fireplace wall while also providing task light at the chair. (Lytecaster Living Room 06.tif: photographer, Chet Burak, image courtesy of Lightolier image archive.)

ing a bathroom mirror is essential. This can be accomplished with diffused lighting from a luminous ceiling or with more concentrated light from several sources. A single concentrated light source, either on the ceiling or on the wall, generally is not adequate. A 30- or 40-watt fluorescent tube over the mirror may be adequate in a small room with light-colored walls. All lighting should be controlled by wall switches not readily accessible while standing in the tub or shower stall.

A ceiling outlet located in line with the front edge of the basin will often provide, in addition to improved lighting at the mirror, general room lighting and safety lighting for a combination shower and tub. When more than one mirror location is planned, equal consideration should be given to the lighting in each case.

Recessed ceiling lights can provide excellent lighting for whirlpool tubs as well as general room lighting (Fig. 18.4-6).

A vaporproof fixture should be provided in an enclosed shower stall. It should be controlled by a wall switch which cannot be reached from within the stall.

A switch-controlled night light is also recommended.

**FIGURE 18.4-4** A combination of recessed ceiling lights, hanging fixtures, and undercounter lights gives the needed illumination to this combined dining room and kitchen area. (Lytespan Kitchen 01.tif: image courtesy of Lightolier image archive.)

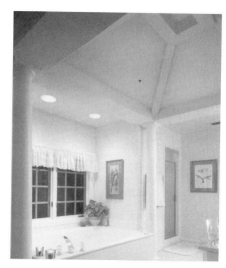

**FIGURE 18.4-6** Recessed ceiling lights at a tub. (Calculite Bath 01.tif image courtesy of Lightolier image archive.)

*Kitchens* Wall-switched outlets for general illumination and for lighting at the sink are essential. Lighting design should provide for illumination of the work areas, sink, range, counters, and tables (see Fig. 18.4-4).

Undercabinet lighting fixtures within easy reach may have integral switches or may be wall-switch controlled. Consideration should also be given to providing lighting inside deep cabinets.

*Laundry Areas* Lighting outlets for fixed lights should be installed to provide illumination of work areas such as laundry tubs, sorting tables, and washing, ironing, and drying centers. At least one outlet in the room should be wall-switch controlled. There should be at least one ceiling lighting outlet centered above laundry trays in unfinished basements or other spaces.

*Closets* Generally, one lighting outlet should be provided in every closet. Where shelving or other conditions make the installation of lights within a closet ineffective, outlets in the adjoining space should be located to provide light within the closet.

*Halls* Wall-switch-controlled outlets should be installed for proper illumination of the entire area of passage halls, reception halls, vestibules, entries, foyers, and similar areas. Particular attention should be paid to irregularly shaped spaces. A switch-controlled night light should be installed in each hall that serves bedrooms.

*Stairways* Fixed wall or ceiling lighting outlets must be installed to provide adequate illumination of each flight of stairs. These outlets should have multiple-switch controls at the head and foot of the stairway. Outlets should be so arranged that the stair is fully illuminated from either floor while also permitting control of bedroom hall lights independently of ground-floor fixtures. Switches should be grouped whenever possible and should never be located so close to steps that reaching for a switch might result in missing a step.

*Utility Rooms* Lighting outlets should be placed to illuminate a furnace area, and a work area if there is one. At least one lighting outlet should be wall-switch controlled.

*Basements* Lighting outlets should be placed to illuminate designated work areas, the foot of the stairway, enclosed spaces, and specific pieces of equipment, such as a furnace, pump, or water heater, and a workbench if one occurs. Additional outlets for open (unpartitioned) basement space should be provided at the rate of one outlet for each 150 sq ft (13.94 m²).

In unfinished basements, the light at the foot of the stairs should be wall-switch controlled near the head of the stairs. Other lights may be pull-switch controlled.

In basements with finished rooms, with garage space, or with other direct access to the outdoors, the stairway lighting provisions mentioned earlier in this section apply.

A pilot light that lights when the basement light is on should be installed at the switch at the head of the stairs when the basement is infrequently visited or when the stair layout makes it difficult to determine visually whether the light is on.

*Accessible Attics* There should be one outlet for general illumination, wall-switch controlled from the foot of the stairs. This switch should have a pilot light to indicate whether it is on. When there are no permanent stairs, the attic light may be pull-switch controlled if located over the access door. When an unfinished attic will later be developed into habitable rooms, the attic-lighting outlet should be multiple-switch controlled at the top and bottom of the stairs. There should be one outlet for each enclosed space.

*Garages and Carports* At least one wall-switched ceiling outlet should be installed in one- and two-car garages.

If the garage has no covered access from the house, one exterior outlet should be provided and multiple-switch controlled from the garage and the residence.

If the garage is also to be used as a work area or laundry, provisions appropriate to these uses should be made. Additional interior outlets often are desirable even if no specific additional use is planned for the garage. Long driveways need additional illumination, such as would be provided by post lighting. These lights should be wall-switch controlled from the house.

### 18.4.3.2 Exterior Lighting

Lights are provided on the exterior of buildings to:

- Illuminate entrances so that occupants and visitors can find their way about
- Make exterior areas usable for social and recreational activities
- Provide for security
- Make parking areas safer and illuminate the way from such areas to the building
- Illuminate driveways and roadways

Lights are needed on the exterior of buildings in the following locations.

### EXTERIOR ENTRANCES

One or more switched lighting outlets should be located at the front (main) and service entrances to every building. Switched lighting outlets should also be installed at other entrances. The principal lighting requirements at entrances are the illumination of walkways and steps leading to the entrance and of faces of people at the door.

In residences, where a single wall outlet is desired, location on the latch side of the door is preferable. Switched outlets in addition to those at the door are often desirable for post lights to illuminate terraced or broken flights of steps or long approach walks.

### PORCHES AND BREEZEWAYS

Porches, breezeways, and similar roofed areas more than 75 sq ft (6.97 m²) in area should have at least one switched lighting outlet. Large or irregularly shaped areas may require more outlets. Multiple-switch control should be installed at entrances when a porch is used as a passage between a house and a garage.

## TERRACES AND PATIOS

One or more fixed outlets on the building wall, or on a post centrally located in the area, are needed for general illumination. These outlets should be switch controlled just inside the door opening onto the area. In commercial and institutional building terraces, much more lighting will be required for safety and to illuminate walkways.

## GENERAL GROUNDS LIGHTING

Floodlights are often used to illuminate the grounds surrounding buildings. Outlets may be located on the building itself, on the exterior of an adjacent building, such as a garage, or on posts. All outlets should be switch controlled from within the building. Multiple and master-switch control from strategic points also is desirable.

## PARKING LOTS

Parking lots for all occupancies must be well lit for safety and security and to light the way from parking spaces to the building. Many different kinds of fixtures are used. In large lots, luminaires are usually mounted on high poles. They may be augmented by ground-mounted walkway lights.

## EXTERIOR SECURITY LIGHTING

Security demands that a facility be made less vulnerable to unauthorized entry by thieves, vandals, and pranksters. This requires siting to eliminate hidden spots, secure door locks, provide intruder detection systems, and provide adequate lighting. Chapter 8 discusses door and window design and hardware that will slow down a burglar. Other security measures are discussed in Section 20.1.

Intruders prefer to work in the dark. Many security experts recommend the installation of all-night security lights to illuminate doorways, yards, public spaces, and other areas subject to intrusion (Fig. 18.4-7). Ordinarily, exterior lights are either manually switched on when they are needed or automatically controlled by

**FIGURE 18.4-7** Good security lighting illuminates doors, windows, and other intrusion points and reduces shadowed areas where intruders might hide.

a programmed timer or light-sensitive switch.

Security lights are sometimes provided in addition to general site lighting, but more frequently the two functions are combined.

***Light Types*** A variety of lamps and light fixtures are available to suit most security needs. An ordinary incandescent bulb can provide adequate illumination at single-family residential entry doors. It does not have to be bright.

An incandescent flood lamp will light a larger area, such as a side yard or driveway.

Flood lamp fixtures hold one to three lamps and allow the lamp to be aimed in whatever direction is necessary.

Mercury vapor lamps cost more than incandescent lamps but emit twice as much light per watt. They also last up to 10 times longer, thus making them the more economical of the two. A standard clear mercury vapor bulb has a blue cast and gives a harsh light; a deluxe white bulb has a pink or orange cast and gives a softer, more balanced light.

Ordinary incandescent fixtures can be adapted for mercury vapor bulbs with components available at an electrical supply store.

***Timing Devices*** Like a deadbolt that is never turned, a light that is never switched on performs no security function. Exterior lights can be automatically switched on by a timer.

A time-of-day device is an electric clock that turns individual fixtures on and off at specified times. In single-family residences, exterior lighting can be wired to a timer through its own circuit. The disadvantage of using a timing device is that nights vary in length throughout the year, which means that the timer will have to be periodically adjusted. Some timers can be set to vary the time at which they go on and off each day, which can give a vacant dwelling a lived-in appearance.

Wired-in timing devices used in commercial, institutional, and industrial facilities are much more sophisticated, overcoming all of the disadvantages listed here for residential units.

A photoelectric device senses light. It will automatically turn a light on at dusk and off at dawn. The device can be an integral part of a fixture or a screw-in device that will adapt a regular fixture.

An automatic sensor (motion detector) can detect anyone walking up to an entrance at night and automatically turn on a light for a few minutes. Such a device is both a convenience for residents and a deterrent to burglars. A photoelectric cell prevents the system from turning on during the day.

# 18.5 Additional Reading

More information about the subjects discussed in this chapter can be found in the references listed in Section 18.6 and in the following publications.

Dagostino, Frank R. and Joseph Wujek. 2005. *Mechanical and Electrical Sys-*

*tems in Construction and Architecture*, 4th ed. Upper Saddle River, NJ: Prentice Hall (www.prenhall.com).

Degrazio, Dawn. 2004. "Designing Lighting for Industrial Spaces." *The Construction Specifier* (April): 60 (www.csinet.org).

Gordon, Gary. 2003. *Interior Lighting for Designers*, 4th ed. Hoboken, NJ: John Wiley & Sons, Inc. (www.wiley.com).

Hay, Mary Ann. 2004. "Lighting Design for Sustainable Buildings." *The Construction Specifier* (April): 52 (www.csinet.org).

Nicolow, Jim. 2004. "Getting the Green Light from the Sun—The Benefits of Daylight Harvesting." *The Construction Specifier* (November): 52 (www.csinet.org).

Ramsey, Charles George, Harold Reeve Sleeper, and John Ray Hoke, Jr. 2000. *Architectural Graphic Standards*, 10th ed. Hoboken, NJ: John Wiley & Sons, Inc. (www.wiley.com).

Schiler, Mark. 1997. *Simplified Design of Building Lighting*. Hoboken, NJ: John Wiley & Sons, Inc. (www.wiley.com).

Stein, Benjamin and John S. Reynolds. 2000. *Mechanical and Electrical Equipment for Buildings*, 9th ed. Hoboken NJ: John Wiley & Sons, Inc. (www.wiley.com).

Vittori, Gail. 2004. "Greening Divisions 15 and 16—Wires, Cables, Pipes, and Environmental Health." *The Construction Specifier* (November): 70 (www.csinet.org).

# 18.6 Acknowledgments and References

## ACKNOWLEDGMENTS

We gratefully acknowledge the assistance of the following organizations and individuals in preparing this chapter We are also indebted to them for permission to use their illustrations when requested and for the use of their publications as references.

American National Standards Institute (www.ansi.org)

Illuminating Engineering Society of North America (www.ieee.org)

Lightning Protection Institute (www.lightning.org)

Lightolier (www.lightolier.com)

National Fire Protection Association (www.nfpa.org)

U.S. Department of Agriculture (www.usda.gov)

U.S. Department of Housing and Urban Development (HUD) (www.hud.gov)

## REFERENCES

We would also like to thank the authors and publishers of the publications in the following list for their contribution to our research for this chapter.

American National Standards Institute (ANSI) (www.ansi.org). 1972 (reaffirmed 1989). *ANSI Y32.9: Graphic Symbols for Electrical Wiring and Layout Diagrams Used in Architecture and Building Construction.* New York: Institute of Electrical and Electronics Engineers (www.ieee.org).

American Society of Heating, Refrigerating and Air-Conditioning Engineers (ASHRAE). 2001. Standard 90.1, "Energy Standard for New Buildings Except Low-Rise Residential Buildings." Atlanta, GA: ASHRAE (www.ashrae.org).

*Architectural Graphic Standards.* See Ramsey et al. (2000).

Arcom Master Systems. *Masterspec®.* Salt Lake City, UT (www.arcomnet.com). Basic Sections:
13100, "Lightning Protection," 12/02 ed.
16050, "Basic Electrical Materials and Methods," 9/02 ed.
16060, "Grounding and Bonding," 9/03 ed.
16130, "Raceways and Boxes Pull, Junction, Outlet, and Cabinets," 6/04 ed.
16139, "Cable Trays Ladder, Trough, Solid-bottom, and Channel Types," 6/04 ed.
16140 Wiring Devices Switches, Receptacles, Lighting-control Devices, and Plates," 6/04 ed.
16145, "Lighting Control Devices," 12/00 ed.
16120, "Conductors and Cables," 3/04 ed.
16410, "Enclosed Switches and Circuit Breakers," 12/01 ed.
16441, "Switchboards," 9/01 ed.
16442, "Panelboards," 6/02 ed.
16443, "Motor-Control Centers," 9/01 ed.
16461, "Dry-Type Transformers (600 V and Less)," 3/00 ed.
16491, "Fuses," 12/01 ed.
16511, "Interior Lighting," 6/03 ed.
16521, "Exterior Lighting," 6/03 ed.

Dagostino, Frank R. and Joseph Wujek. 2005. *Mechanical and Electrical Systems in Construction and Architecture*, 4th ed. Upper Saddle River, NJ: Prentice Hall (www.prenhall.com).

Gordon, Gary. 2003. *Interior Lighting for Designers*, 4th ed. Hoboken, NJ: John Wiley & Sons, Inc. (www.wiley.com).

Illuminating Engineering Society of North America. 1993. *Lighting Handbook*, 8th ed. New York: Illuminating Engineering Society of North America (www.ieee.org).

International Code Council. 2006. Falls Church, VA: ICC (www.iccsafe.org). International Codes:
*International Building Code®.*
*International Residential Code®.*
*International Energy Conservation Code®.*

Lightning Protection Institute (LPI). L.P.I.-175, "Installation Code." Woodstock, IL: LPI (wwwlightning.org).

Loeffler, Mark. 1998. "Environmentally Responsible Lighting." *The Construction Specifier* (April): 57 (www.csinet.org).

National Fire Protection Association (NFPA). Quincy, MA (www.nfpa.org). 2005. NFPA 70. "National Electrical Code."
———. 2002. NFPA 70A, "Electrical Code for One- and Two-family Dwellings."
———. 2003. NFPA 101, "Life Safety Code."
———. 2004 NFPA 780: "Standard for the Installation of Lightning Protection Systems."

Ramsey, Charles George, Harold Reeve Sleeper, and John Ray Hoke, Jr. 2000. *Architectural Graphic Standards*, 10th ed. Hoboken, NJ: John Wiley & Sons, Inc. (www.wiley.com).

Schiler, Mark. 1997. *Simplified Design of Building Lighting*. Hoboken, NJ: John Wiley & Sons, Inc. (www.wiley.com).

Stein, Benjamin. 1997. *Mechanical and Electrical Systems*, 2nd ed. New York: John Wiley & Sons, Inc. (www.wiley.com).

Stein, Benjamin and John S Reynolds. 2000. *Mechanical and Electrical Equipment for Buildings*, 9th ed. New York: John Wiley & Sons, Inc. (www.wiley.com).

Underwriters Laboratories, Inc. (UL). 1982 (revised 1994). UL 96A, *Installation Requirements for Lightning Protection Systems*. Northbrook, IL: UL (www.ul.com).

U.S. Department of Agriculture (USDA) 1978. *Farmers Bulletin Number 2136*. Washington, DC: USDA (www.usda.gov).

**CHAPTER 19**

# Communications

Introduction

Applicable *MasterFormat*™ Sections

Communications Systems

Additional Reading

References

# Introduction

This chapter covers some of the many available communications systems including entrance signals, intercom systems, television and radio, and telephone services. Other communications systems included in *MasterFormat* Division 27, on which this chapter is based, include, but are not necessarily limited to, cabling for communications systems, data (computer) hardware and software, audiovisual communications, health care communications such as patient monitoring and nurse call systems, clock systems, paging systems, and infrared and radio frequency tracking systems.

# Applicable *MasterFormat* Sections

Subjects addressed in this chapter are included in *MasterFormat* 2004, Division 27, "Communications." The following *MasterFormat* 2004 Level 2 sections are applicable to this chapter:

27 21 00  Data Communications
           Network Equipment
27 22 00  Data Communications
           Hardware

27 24 00  Data Communications
           Peripheral Data Equipment
27 25 00  Data Communications
           Software
27 32 00  Voice Communications
           Telephone Sets, Facsimiles
           and Modems

27 41 00  Audio-Video Systems
27 51 00  Distributed Audio-Video
           Communications Systems
27 52 00  Healthcare Communications
           and Monitoring Systems
27 53 00  Distributed Systems

# 19.1 Communications Systems

Entrance signals, radio and TV antenna outlets, telephone outlets, and internal communications (intercom) systems are commonly considered part of a building's wiring system and installed by the electrical trade.

Some other systems using wiring are often installed by other professionals. These include wiring for nurse call systems in hospitals; computer-controlled systems for equipment and devices, such as those in a manufacturing plant; systems for monitoring equipment, such as sensors for patients' vital signs in a hospital or heating, ventilating, and air conditioning equipment sensors in any building; and many other specialized communications and control systems. These specialized systems are beyond the scope of this book.

## 19.1.1 ENTRANCE SIGNALS

In a residence, signal push buttons should be provided at each commonly used entrance. In other types of buildings, similar devices are often installed at normally locked or unattended entrances, such as a delivery entrance or a controlled-entry door. Such devices are sometimes also installed at interior doors in hospitals; in facilities where occupants' movements are restricted, such as jails; and in other facilities where controlled access is desired.

Such devices should be connected to a chime or buzzer that gives a distinctive signal that will not be mistaken as telephone, smoke alarm, or other signals. The electricity supply for entrance signals is usually obtained from a low-voltage transformer installed near the service equipment. In smaller homes, the chime or buzzer is often installed in the kitchen. In larger homes, it should be installed in a central location, where it can be heard throughout the house. In a multilevel house, extension signals may be necessary to ensure being heard in quarters located away from the main entry level. Entrance-signal conductors for residence systems should be no smaller than the equivalent of a No. 18 AWG copper wire and, preferably, should be installed in conduit.

In other types of buildings, sounding devices are installed where they will be heard by the person who will answer the signal. Wiring will be dictated by requirements of the device being used.

## 19.1.2 INTERCOMMUNICATION SYSTEMS

Intercommunication (intercom) systems are quite common even in small houses today and are used extensively in other types of buildings. An intercom system may be a complete sound system with an AM/FM radio, tape deck, or compact disk player, or just a simple push-button two-way talk system. In a residence, an intercom is operated from a master control panel usually located in the master bedroom, recreation room, or kitchen and connected to remote stations throughout the house. In other types of buildings, the control system is usually in a central location within the system.

A larger house, or one with accommodations for a resident servant, may have additional signal and communication devices. These may consist solely of a dining-room-to-kitchen call signal, with a floor treadle or push button connected to an annunciator in the kitchen. More extensive systems may include a push button in each bedroom, living room, and other habitable rooms.

Sometimes an intercom system is tied to the entrance signal or substitutes for it. This permits an occupant to talk with a visitor without opening the door. In apartment buildings, intercom systems are often tied in with electronic locks so that the main entrance door to the building can be operated from each apartment unit. Similar arrangements are also often used in other types of buildings, including at exterior entrances and at interior doors serving restricted areas. Closed-circuit television systems that permit an occupant to see a visitor before opening the door are available. These are expensive and are not commonly used in residences, but are used extensively in hospitals and some other types of buildings.

Intercom systems are sometimes combined with a public address system.

### 19.1.3 TELEVISION AND RADIO

Provisions for television consist mainly of a nonmetallic outlet box at each television set location connected by a suitable transmission line to the available closed-circuit television (CCTV) or master antenna television (MATV) system. MATV sources include off-air antennas, community antenna television (CATV), local television (CCTV), and television receive-only earth stations. A convenience outlet should be provided adjacent to each television antenna outlet. Antenna systems with more than two outlets generally require a booster amplifier, except in very strong signal areas.

In areas with a cable TV service, the cable company should be contacted to ascertain its specific requirements for antenna connections.

In most areas, AM radio receivers used purely for the reception of broadcast communication do not require antenna and ground connections. For FM reception, provisions similar to those for television should be included.

### 19.1.4 TELEPHONE

The number of telephone outlets and the extent of the service in commercial, institutional, and industrial buildings must be determined on a needs basis, depending on the use of the facility. Some types of buildings, such as hospitals, require extensive telephone communications systems. These may be tied into or augment the building's intercom system. Other building types, such as warehouse facilities, may require little more than a single incoming line. Computers require additional telephone lines for Internet access or contact with a companywide network system.

In residential buildings, at least one convenient fixed telephone outlet (jack) should be provided in every living unit. Where a living unit occupies more than one floor, at least one outlet should be provided on each floor. Most homes today have an outlet in every habitable room and one in each work space, such as the laundry room, utility room, basement, and garage. Outlets are also often provided on porches and outdoors on patios and the like.

Advance wiring by the telephone company during construction will generally result in an inconspicuous, more efficient installation. The local telephone company should also be consulted for details about service connection before construction, particularly with regard to the installation of protector cabinets and raceways leading to the service entrance equipment.

## 19.2 Additional Reading

More information about the subjects discussed in this chapter can be found in the references listed in Section 19.3 and in the following publications.

Gill, Michael. 2002. "The Key to Security–Retrofit Systems Enhance Access Control." *The Construction Specifier* (May): 52 (www.csinet.org).

Ingram, Lauren. 2001. "Designing for Tomorrow, Technologically Speaking." *The Construction Specifier* (February): 19 (www.csinet.org).

Rauscher, Thomas C. 2003. "Taking Communications Out of the Closet." *The Construction Specifier* (April): 72 (www.csinet.org).

Stein, Benjamin and John S. Reynolds. 2000. *Mechanical and Electrical Equipment for Buildings*, 9th ed. Hoboken, NJ: John Wiley & Sons, Inc. (www.wiley.com).

## 19.3 References

We would like to thank the authors and publishers of the publications in the following list for their contribution to our research for this chapter.

National Fire Protection Association (NFPA). 2005. NFPA 70, "National Electrical Code." Quincy, MA: NFPA (www.nfpa.org).

Ramsey, Charles George, Harold Reeve Sleeper, and John Ray Hoke, Jr. 2000. *Architectural Graphic Standards*, 10th ed. Hoboken, NJ: John Wiley & Sons, Inc. (www.wiley.com).

Stein, Benjamin and John S. Reynolds. 2000. *Mechanical and Electrical Equipment for Buildings*, 9th ed. Hoboken, NJ: John Wiley & Sons, Inc. (www.wiley.com).

# Electronic Safety and Security

# Introduction

This chapter addresses electronic and other safety and security measures, including intrusion prevention and detection and fire detection and alarm. *MasterFormat* 2004 Division 28, on which this chapter is based, includes other types of devices and systems that are beyond the scope of this book. These include video surveillance systems; electronic personal detection systems; detection and alarm systems for radiation, fuel gas, fuel oil, and refrigerant; and electronic detention monitoring and control systems.

# Applicable *MasterFormat* Sections

Subjects addressed in this chapter are included in *MasterFormat* 2004, Division 28, "Electronic Safety and Security." The following *MasterFormat* 2004 Level 2 sections are applicable to this chapter:

28 13 00 Access Control
28 16 00 Intrusion Detection
28 31 00 Fire Detection and Alarm

# 20.1 Intrusion Prevention and Detection

Security for buildings and occupants has always been important in building design to make a facility less vulnerable to unauthorized entry by thieves, vandals, and pranksters.

Burglary, a frequent problem in both urban and rural areas, is a primary threat to security. Nationwide, a burglary occurs about every 10 seconds. About half of all burglars gain entry through an unlocked window or door.

Would-be intruders prefer to work quickly, quietly, and in the dark. If more than 4 minutes are required to gain entry, most burglars will usually give up the attempt. Good burglary protection devices will either slow down or frighten away a burglar. The intent of security device is to save lives and property by warning residents and neighbors when someone tries to enter a property without authorization.

Building security has become even more important recently in light of international terrorist activities and soaring crime rates in certain jurisdictions.

Building security includes electronic security measures but goes far beyond that to include environmental design measures and bomb resistance. Security requires attention to site layout and building access locations. For example, building entrances should be monitored, the number of stairways that exit to the outside should be limited, roof access should be limited, building parapets should be low to allow surveillance, and it should be possible to block access to the building through the site and prevent access to building entrances when necessary, especially by vehicles.

It is incumbent upon an architect to balance security with cost. It is possible to overreact to possible security threats and introduce measures that are not likely to be needed. The architect must take into account a building's location and use. Some types of buildings are more likely to be targets than others. Financial institutions, chemical manufacturing facilities, and electric power generating facilities are more likely to be terrorist targets than ordinary office buildings. Buildings in high-crime urban areas are more likely ordinary-crime targets than similar buildings in suburbia. Neither being small nor located in an urban area, however, translates into being safe from crime. Gasoline stations and chain food stores, for example, are prime crime targets, even in rural villages, and need to have security measures commensurate with their risk.

*Risk* in this context is defined as the likelihood that a security breach will occur. Risk is measured as high, medium, or low but can also be defined in terms of how frequently a breach is likely to occur. Analyzing the risk leads to a *vulnerability assessment*, which often includes more than one area of vulnerability. To reduce risk to an acceptable level, each vulnerability must be analyzed and a method developed to reduce or eliminate it. These solutions are then considered when designing the building.

Reducing risks is a complex problem. Its solutions can affect every aspect of a building's use and its occupants. Therefore, careful consideration must be given to the effect of one type of risk reduction on other risk factors. Sometimes a possible solution to one vulnerability will introduce another vulnerability. For example, one solution to entrance vulnerability is to reduce the number of entrances. Unfortunately, this can result in reduced fire safety and limit necessary access by police and fire fighters to the building. In this case, making building occupants safe from unauthorized intrusion increases their risk of harm from a fire or insurrection happening inside the building.

Federal government agencies have developed a set of security standards, which are called the General Services Administration (GSA) Security Standards. GSA "Document 8, Security Design," for example, is intended to promote development of a defensible system of space/crime prevention through environmental design (CPTED). These standards are applicable to all federal building construction and are being adopted by state and local governments as well. Security control is defined in layers (levels). The first layer is the site perimeter and site; the second is the exterior building walls; and the third is the interior. This section discusses site planning and intruder detector systems. Locking devices and the design of doors and windows to deter intrusion are discussed in Chapter 8; exterior security lighting is considered in Section 18.4.3.2.

## 20.1.1 SITE PLANNING

The first layer of defense against intruders is the area between a building and its property lines (public space). Building location, terrain, landscaping, lighting, and structural barriers are important considerations in creating physical security. The location of public places, and their convenience and visibility, can encourage the gathering of sociable people and discourage potential criminals.

Barriers that define and protect space may be physical or psychological. Physical barriers include building shape, fences, gates, and doors. The shape of a building can act as a deterrent when all areas are visually accessible and no areas are hidden from view. Psychological barriers do not necessarily bar movement but rather symbolize boundaries. They include building elements such as open gateways, light standards, short runs of steps, and textural changes in walking surfaces.

### 20.1.1.1 Site Choice

Where a dwelling is located on the lot, as well as where it is in comparison to other buildings, affects whether the building discourages criminal activity. For example, placing a building far back from the street can increase its vulnerability to crime. A building should be placed so that its entrances and windows are clearly visible to passersby, neighbors, and police patrols.

However, a building that may be vulnerable to terrorist attack should be at least 100 ft (30.48 m) from the property line to permit employment of measures necessary to protect it from explosive devices in vehicles.

### 20.1.1.2 Fences

Fences of wood, chain links, shrubbery, or masonry serve as both psychological and physical deterrents to burglars and vandals. In many jurisdictions it is difficult to get a trespassing conviction unless property boundaries are clearly marked. A fence should hinder a burglar's attempt to enter the property while not providing cover for a break-in.

A 6-ft (1829-mm)-high brick wall gives a false sense of security. It presents a major barrier but also hides an intruder's preliminary activities. Fences that protect but do not obscure are better security choices.

A wooden picket fence may be a psychological deterrent but is easy to climb over. A thick hedge, particularly a thorny one, is harder to climb through and over. If either one is no more than 40 in. (762 mm) high, it is easy for neighbors or passers-by to see an intruder.

A 6-ft (1829-mm)-high chain link fence is a good physical barrier that does not obscure vision. It should be installed as close to the ground as possible so that a potential intruder cannot crawl under it. A hedge no more than 40 in. (762 mm) high on one or both sides of a chain link fence can soften its appearance.

Access to the protected area should be through a gate. At a residence, a simple bell arrangement can sound a signal whenever the gate is opened or closed. For other types of buildings, a more elaborate signaling device may be necessary.

Unfortunately, with the possible exception of a properly designed masonry wall, none of the types of fences mention here will deter a determined terrorist. Concrete barriers may offer some protection, but even they must be guarded and carefully placed to prohibit a vehicle from circumventing them.

### 20.1.1.3 Landscaping

Plants that hide a building can also hide an intruder. Trees should be far enough away from a building so that a burglar cannot climb them and enter the second story. Bushes should be low enough so that they do not obscure windows or doors. Hedges used as fences should be no more than 40 in. (762 mm) high. Landscaping should not obscure a building from the street. Small plants and floral borders can serve to define private and public spaces.

Landscaping will not block the effects of a car bomb, but judiciously placed landscape elements, such as thick plant beds, ponds, and fountains, may help to keep explosive-carrying vehicles away from a building.

## 20.1.2 ACCESS CONTROLS

Not too many years ago, access control consisted of a key lock. Keying was seldom changed. All it took to gain access to a building was a key, and keys could be easily duplicated in any hardware store. Gradually, key locks became more complex, making their keys more difficult to duplicate.

Push-button lock controls were the next refinement in access control. This system works in essentially the same way as a key lock, except that the function of the lock is controlled by a keyboard with alphanumeric keys. Punching keys in the correct sequence causes the lock to operate.

In time, technological advances made it possible to control access to buildings and sensitive areas within buildings using a *credential and reader system*. A *credential* is a device carried on the person, such as a plastic card. A *reader* is a device that senses the information on a credential and opens a lock. Initially, it was necessary to insert the card into a reader for the card to be read. Later, *proximity readers* were developed that could read a credential placed near the reader. Today, readers are available that can read a credential from a distance and while the credential is moving. It is possible, therefore, to enter a controlled parking lot or garage having such a control device without stopping the vehicle.

Readers today can make a record of the credential use and even prevent its subsequent use for a predetermined time. This makes it difficult for a credential holder to pass the credential to another person, who can then enter using it. Credential readers can act as time clocks, checking a credential in and out and noting the times of entry and departure.

Many credential and reader systems today are controlled by computers that maintain a record of activities and report those activities to persons responsible for security. These systems can be combined with fire detection and alarm systems to change the parameters of the control system when necessary. For example, doorways that might prohibit entry except with the proper credential may be programmed to admit fire fighters in the event of a fire. Doors that prohibit exit, except with a credential, can be opened to permit exit from the building in the event of a natural disaster or fire.

The latest access control systems employ biometric identification to operate locks. Early biometric devices could read fingerprints. Later devices could identify people by reading the configuration of a hand. These *whole-hand geometry* systems examine multiple characteristics, such as finger length, hand width, and hand length, and compare them with files stored in a computer. The reader is a platan on which the hand is placed. A digital camera photographs the hand, and the image is examined and compared with the stored image in a computer. Other similar systems read only parts of a hand, two fingers, for example, in the same way.

Other hand technologies are available, including one that uses infrared light to read veins, arteries, and fatty tissues. Another reads vein patterns on the back of the hand.

Some systems read the patterns in eye retinas or irises. Face recognition systems are available, as are voice recognition systems.

Which of these various biometric systems is most effective in a particular situation must be determined. Some systems are not appropriate is every circumstance. Other issues affecting biometric identification systems include cost (currently higher than that of most other types of entry-control systems), establishing standardized methods of analysis (too many misidentifications), privacy issues (Big Brother is watching you), and hygiene issues (touching surfaces touched by many other persons).

## 20.1.3 INTRUDER DETECTION SYSTEMS

Various alarm systems are available to warn against intruders. Alarm systems predate the telephone. Alexander Graham Bell's assistant, Thomas Watson, acquired much of his electrician's skill by installing burglar alarms during the 1860s. Alexander Pope patented one of the earliest of these systems in 1853.

Intruder detection devices should be used in conjunction with adequate structural protection such as strong doors and windows and secure door and window hardware (see Section 8.7), adequate site planning, and strategically placed lights (see Section 18.4).

### 20.1.3.1 Perimeter Intrusion Detection

Exterior doors, windows, and walls make up the perimeter of a building. Perimeter detectors work by setting up an electronic boundary line around this perimeter; breaching the boundary activates the detector's signaling device.

### DETECTOR ELEMENTS

Most detectors consist of a protective circuit, a control instrument, and a signaling device. The protective circuit may be made up of light beams, sound waves, or wires which detect the presence of an intruder.

A control instrument monitors a protective circuit. When a circuit is breached, the control instrument activates the signaling device.

A control unit may be mounted indoors or outdoors, depending on the system. Many units have a time delay so that occupants may arm or disarm the system without triggering the alarm. A good lock being as important on a control unit as it is on a door, key-operated or keyboard-operated systems are preferable. Some control units will sound an alarm if tampered with.

A signaling device issues the actual warning. It may be a light, bell, buzzer, horn, or siren. The signaling device may turn on exterior floodlights. It may also transmit a warning over the telephone. The telephone warning may go directly to a police station or to the police through a contracted service, which is often the same company that installed the system.

### EFFECTIVENESS

A perimeter detector's alarm or floodlights will frighten a potential burglar away only if it instills a fear of capture. In order for a burglar to be caught, someone else must be able to hear the alarm.

Neighbors or occupants should respond to the warning by calling the police. Trying to capture a burglar is very dangerous and should be left to professionals.

Police response time is the most important factor in capturing a potential burglar. If the response is within 30 seconds of the alarm, the thief will probably be captured (Fig. 20.1-1). Silent alarm devices that alert the police are available. They may make capture more sure, but harbor the inherent danger of permitting an intruder to enter the building thinking he or she has not been detected. This can present a heightened danger to an occupant who knows the alarm has been

**FIGURE 20.1-1** Police Response Time and Probability of Capture of a Burglar

| Response Time after Alarm Sounds | Probability of Capture (%) |
| --- | --- |
| 30 sec | 100 |
| 60 sec | 90 |
| 2 min | 75 |
| 4 min | 50 |
| 10 min | 20 |

sounded and attempts to interfere with the intruder or who simply stumbles on the intruder. It is better to frighten a burglar away than face him or her in the dark.

Unfortunately, both noisy and silent intruder alarm systems are subject to false alarms, which diminish their credibility. If false alarms are common, neighbors and police may not respond when a real alarm is sounded, and even occupants may assume that another false alarm has been sounded when there is a real intruder.

### SIGNALING DEVICES

Signaling devices may be selected to alert occupants and neighbors, a private security company, the police, or all of these at the same time.

A local alarm is usually a loud bell or siren mounted on an exterior wall of the building so that it can be heard by most neighbors. Some alarms continue to ring until they are manually shut off; others sound for a fixed period of time, usually 5 minutes, or are adjustable within a range of 4 to 12 minutes.

A central station alarm is a system in which the protective circuit, or a special dialing apparatus on a regular telephone line, connects to a private company that monitors it. This is often the same company that installed the system.

When an alarm signals, the company will either relay the information to the police or have its own guards check the alarm. The cost includes initial installation and a monthly maintenance fee.

In some jurisdictions, an intruder alarm may be connected directly to a police station. In these locations, extra care should be taken when selecting a detection system, because some of them are very sensitive and prone to signaling false alarms. Some police departments issue fines if they have to respond to too many false alarms. Neighbors may also be annoyed by frequent soundings of a burglar siren.

A telephone dialing system uses an automatic telephone dialing unit with recorded tapes as its signaling device. When activated, the telephone dialer calls a preprogrammed number. Some units can be programmed to dial another number if the first is busy; other units will keep dialing the same number until someone answers.

Many units come with its message on a continuous loop, which is a good idea because of the possibility that an alarm call may be put on hold. These units are called *central station systems* when the programmed number is that of a police department or monitoring company.

### DETECTOR TYPES

Detectors may have their protective circuit, control instrument, and signaling alarm all in one device or each may be in a separate device. They may be either wired or wireless. The normally open (NO) and normally closed (NC) systems discussed in this subsection are called *wired perimeter detection systems*, because their detectors are directly wired to their control and signaling device. The carrier-current and radio-frequency systems, also discussed, are called *wireless*, because their signals are sent to their control and signaling device by a transmitter that is not hardwired to them. While spot detectors are in fact wireless, they actually fit into neither category because they have no separate signaling device.

Some wired systems are versatile and accept both NO and NC intrusion detectors. These combined NO and NC wired systems have the advantage that many forms of protection can be combined into one complete security system.

Both wired and wireless systems have centrally located control and signaling devices, make the same provisions to permit ventilation through windows, and have the same number and type of wired intrusion detectors (Fig. 20.1-2) which attach to doors and windows. Opening the door or window will affect the control device and activate the signaling alarm. These systems may also have window foil or other glass-breakage detectors. Breaking glass that has a glass-breakage detector attached will cause the control instrument to activate the signaling alarm. When these detectors are used with operable windows, contact switches are provided to permit opening the window. Both wired and wireless intrusion detector systems may have photoelectric detectors which send a beam of light between a sending and a receiving station. Stopping the path of the light beam breaks the circuit and causes activation of the signaling alarm.

Installing a wired system can be tedious. Wiring must be run from room to room, connecting each door and window back to the central unit. This wiring may be difficult to conceal and, if left exposed, is unsightly. Wireless systems are easier and more convenient to install because they use transmitters and not extensive wiring to relay a signal to their signaling alarm. Even so, NO and NC wired systems are usually less expensive than other systems.

*NO Wired Systems* An NO wired system has switches in the open position, which is the equivalent of a light switch in the off position. Opening a protected door or window turns the system on. Because current flows through the system and activates the alarm only when electricity completes a loop, removing any switch or breaking any wire may disable the entire system. Accessories such as a smoke detector can be added to an NO system.

*NC Wired Systems* An NC wired system is less susceptible to defeat than an NO system. Electricity flows through an NC system in a constant loop. An interruption in the current, including breaking or cutting a wire, triggers the alarm. This feature can be an advantage, because a broken wire or sensor will set off the alarm. Conversely, a system failure in an NO system will go unnoticed until a routine check is made.

*Carrier-Current Wireless Systems* A carrier-current wireless system sends its signal to the control device via regular 120-volt ac wiring. Its intrusion devices are connected to a transmitter which plugs into a wall outlet. The transmitter is the equivalent of a protective circuit. The receiver is a control device that plugs into an outlet in a central area. It houses the signaling device.

*Radio-Frequency Wireless Systems* A radio-frequency wireless system sends its signal from intrusion device transmitters to a control device receiver via radio frequencies. The transmitters operate on batteries. The receiver usually runs on normal 120-volt current with battery backup. In some radio-frequency systems, the battery may sound an alarm when there is a power failure. In others, the battery operates the system on a standby basis only during a power outage. Batteries usually last a year but should be checked occasionally to make sure that the system is operating properly.

*Spot Detectors* In a spot detector, all of the detector elements are in a single unit designed to be attached to a door or window. The unit buzzes loudly when the door or window is opened. This type of detector is useful in apartments and other areas with few entrance points, but an intruder who takes the time to see where the noise is coming from can often shut it off.

#### 20.1.3.2 Area Controls

Area controls protect the space within a building. They consist of devices to either fool an intruder into thinking a building is occupied or to detect intrusion within the space.

### UNAUTHORIZED ENTRY PREVENTION

Most of the time, a burglar who thinks someone is inside will not break into a building. A burglar can be discouraged

a        b        c

**FIGURE 20.1-2** (a) Separating the housing of a magnetic switch activates the signaling device. Opening a door or window releases the plunger on (b) a surface-mounted and (c) a recessed-mounted detector and activates a signaling device.

by timer-controlled lights and sounds and by decals and labels.

### Timer-Controlled Lights and Sounds
A cycling timer is a 24-hour electric clock that turns on lights or other appliances at designated times and leaves them on for specified intervals. Some models have a cycle override feature which allows both timed and normal use. Others have multiple cycles that switch lights and appliances on and off 1 to 12 times each day. Still others feature variable cycles that switch on at random times (Fig. 20.1-3).

Lights plugged into four timers set in various rooms in a dwelling will give the impression that someone is home. Figure 20.1-4 shows three possible timer use configurations that will suggest various living patterns. If other living patterns are used, it is important to remember to overlap the times that lights are on in at least two rooms.

The sounds produced by radios and television sets can give the illusion that someone is in a building when no one is there. Radios or television sets can be simply left on or can be plugged into a timer.

### Decals and Labels
Some security systems supply decals for doors and windows that indicate that the property is protected. Protective foil tape on windows also serves to label a building as having security protection.

Some police departments and other organizations loan diamond-tipped engraving pens to mark valuable property with the owner's name and identification. Others provide warning decals for the people who have marked their valuables to place in doors and windows to discourage burglary (Fig. 20.1-5).

### INTERIOR AREA DETECTORS
Interior detectors protect the space within a building. In order to activate these detectors, an intruder must have breached the perimeter and entered the building. Interior area detectors can be used instead of, or in addition to, perimeter detectors. Some types of area detectors cannot be used in dwellings with pets or other buildings with guard dogs, because they are unable to distinguish between the movements of animals and those of human intruders. Combination protection systems utilizing various types of perimeter and interior area detector elements are available.

As is true for perimeter detectors, the effectiveness of area control detectors depends on the burglar's fear of capture, people hearing the alarm and calling the police, and police response time.

Area control detectors have a protective circuit, a control instrument, and a signaling device.

Sound waves, radio-frequency waves, and light beams are the equivalents of the protective circuit in a wired system. The control instrument and signaling device are usually relatively small and unobtrusive.

### Photoelectric and Infrared Systems
Photoelectric and infrared systems can be either perimeter or area detectors and can be connected to either NC or NO systems. A transmitter projects either visible light, electromagnetic waves, or an infrared beam to a receiver. The light can be reflected by mirrors to weave a kind of electronic spider web to increase the area protected. If the beam is broken by an intruder, the circuit is broken and the alarm sounds. Some infrared systems detect the presence of body heat and will sound an alarm even if an intruder is wearing heavy clothing and standing 30 ft (9000 mm) away.

### Ultrasonic Motion Detector
An ultrasonic motion detector blankets a room with inaudible sound waves in the 18,000 to 45,000 Hz range. The waves set up an energy field. Some detectors set up an elliptical pattern ranging from 12 × 35 ft (3.66 to 10.67 m) to 5 × 20 ft (1.5 to 6.1 m). A person entering the energy field, also known as the *trap zone*, interrupts the pattern and an alarm sounds. Most of these systems have an adjustable sensitivity control to guard against setting off

**FIGURE 20.1-4** Typical Living Patterns with Four Timers

| Room | A | | B | | C | |
|---|---|---|---|---|---|---|
| | On | Off | On | Off | On | Off |
| Kitchen | 6:00 P.M. | 10:00 P.M. | 7:00 P.M. | 12:00 A.M. | 8:00 P.M. | 1:00 A.M. |
| Living room | 7:30 P.M. | 11:00 P.M. | 8:00 P.M. | 1:00 A.M. | 8:30 P.M. | 12:00 P.M. |
| Bedroom | 10:30 P.M. | 1:00 A.M. | 12:00 A.M. | 3:00 A.M. | 10:00 P.M. | 2:00 A.M. |
| Bathroom | 11:00 P.M. | 6:00 A.M. | 11:30 P.M. | 6:00 A.M. | 9:00 P.M. | 6:00 A.M. |

(Courtesy Intermatic, Inc.)

**FIGURE 20.1-3** This type of variable timer plugs into a wall receptacle. The equipment to be timed is then plugged into the bottom of the timer. (Courtesy Intermatic, Inc.)

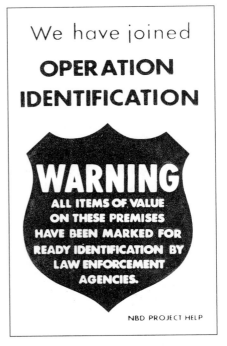

We have joined

**OPERATION IDENTIFICATION**

**WARNING**

ALL ITEMS OF VALUE ON THESE PREMISES HAVE BEEN MARKED FOR READY IDENTIFICATION BY LAW ENFORCEMENT AGENCIES.

NBD PROJECT HELP

**FIGURE 20.1-5** A decal may discourage a burglar because labeled property is hard to dispose of.

the alarm with normal movement or with ordinary noise, such as that from fans.

An ultrasonic detector, which is about the size of a table radio, is sometimes designed to look like a set of books or a small stereo speaker. Some units require no wiring and are portable. If they are used while the building is occupied, people and animals must stay out of the guarded area to avoid tripping the alarm.

*Microwave System*  A microwave detector is a type of radar. Its electromagnetic waves penetrate through wood and plaster but not brick or concrete. This system can protect several rooms using a device that both transmits and receives a signal. This type of transmitter/receiver is placed in a central location in a building. Unfortunately, a microwave system can be a source of false alarms if the energy field is too large. Movement in an adjacent room or apartment, or pedestrian traffic in front of the building, can trigger the alarm. Care in the adjustment and placement of the system should solve these problems.

*Sonic System*  A sonic system resembles an ultrasonic system, except that audible sound waves set off the system. A receiver detects echoes coming from every object in the room. When an intruder enters, the echo changes, triggering the alarm.

*Audio Detector*  An audio detector registers sharp sounds that indicate an intrusion, such as wood splintering or glass breaking.

*Floor Mat Detector*  A floor mat detector is part of a wired perimeter system. It is mentioned here under area controls because such a device cannot be activated until the intruder enters the building.

# 20.2  Fire Detection and Alarm

Fire detection and alarm systems are intended to facilitate the escape of occupants from a building as quickly as possible in the event of a fire.

This section discusses fire detection and alarms. Fire prevention, use of fire-resistant products, and the fireproofing of structures are addressed in Section 7.10; fire protection specialties such as portable extinguishers in Chapter 10; and fire sprinkler systems in Chapter 15.

## 20.2.1  FIRE DETECTION AND ESCAPE

Most residential fires take place between midnight and 6:00 A.M., when people are usually asleep. More than half of the Americans who lose their lives in residential fires yearly could have been saved if they had had smoke and fire alarms and a ready means of escape.

People who die in large buildings usually succumb to smoke and gases rather than the fire itself. These deaths are due to lack of sufficient warning and inadequate means of escape. Current laws have reduced these causes of death, but many older buildings are still inadequately protected. Fire has four distinct stages:

1. Incipient
2. Smoldering
3. Flame
4. High heat

### 20.2.1.1  Smoke and Fire Alarm Systems

Long before flames actually break out, occupants may have been killed or rendered unconscious by smoke, noxious gases, or superheated air. Smoke and heat detection, therefore, are an essential part of any fire alarm system. From the time flames break out, a person may have less than 4 minutes in which to escape.

Smoke signals the presence of fire far sooner than heat; therefore, in most cases, a device sensitive to smoke is more important than one that is sensitive to heat. Smoke detectors typically are of the ionization or photoelectric type. Heat detectors generally have a metal strip which is either melted or distorted by heat to trip the alarm.

In residential installation, cost should not be a problem in smoke detector selection. Battery-powered ionization detectors cost as little as $10; battery-powered photoelectric types may cost a little more. Batteries must be replaced approximately once a year. Many insurance companies provide rate discounts to residents who install the more expensive, permanently wired, combined fire and security protection systems.

Wired-in systems are more expensive, but are required by most codes for buildings other than single-family residences.

### IONIZATION DETECTORS

In an ionization detector, a tiny amount of radioactive material causes the air between two electrodes to conduct an electric current (Fig. 20.2-1). When smoke particles are present, current flow is reduced and the alarm is set off.

The radioactive material in these detectors should not be of concern to building occupants. According to the U.S. Nuclear Regulatory Commission, holding an ionization smoke detector close to the body for 8 hours a day for an entire year would expose a person to $1/10$th as much radiation as would one transcontinental round-trip airline flight.

a                                    b

**FIGURE 20.2-1** In an ionization detector (a) a piece of radioactive material ionizes the air to cause an electric current to flow across an air gap; (b) smoke reduces the current flow, tripping the alarm.

Compared to photoelectric detectors, ionization detectors respond more rapidly to the small, wispy smoke particles of a fast-flaming fire but react more slowly to the large smoke particles of a slow-smoldering fire. The average ionization detector is likely to give 20 to 30 seconds' extra notice of a fast fire.

## PHOTOELECTRIC DETECTORS

A photoelectric detector aims a light beam into a chamber that contains a light-sensitive photocell. The cell detects the light only when smoke enters the chamber and scatters the beam. When enough light reaches the cell, the alarm goes off (Fig. 20.2-2).

A slow-smoldering fire, as would occur when a cigarette is dropped on bedding or upholstered furniture, is the most common type of home fire. A photoelectric detector is likely to give 20 to 25 minutes' quicker notice of a slow-smoldering fire than an ionization detector but is less than a minute slower in detecting a fast fire. For these reasons, many experts prefer the photoelectric type of detector.

## HEAT DETECTOR

Some manufacturers offer a heat-sensing device as an optional part of their smoke detectors or as a separate product. Most heat detectors use a metal strip that is either melted or distorted by excessive heat.

Heat detectors may be of either the fixed-temperature or rate-of-rise type. Heat detectors built into smoke detectors are usually of the fixed-temperature type, which trips the alarm when a certain temperature is exceeded. Similar in appearance, separate detector devices may also be of the fixed-temperature or rate-of-rise type; they sound their own alarms or send electrical signals to a central alarm system.

To be effective, heat detectors must be close to the source of a fire. They are especially useful in environments that could fool or disable a smoke detector, such as a kitchen, where grease particles in the air may cause a smoke sensor to give false alarms. Heat detectors may be useful in other areas that are unsuitable for smoke detectors, such as furnace rooms, attics, and garages.

However, a heat sensor should not be the only fire protection in a building. It is more often smoke than heat that causes injury and death. A heat detector may ignore a smoldering fire that is emitting lethal amounts of smoke, carbon monoxide, and other toxic gases.

## DETECTOR PLACEMENT

The National Fire Protection Association's (NFPA) *Life Safety Code* recommends the installation of at least one smoke detector in each dwelling unit. In some jurisdictions, such a detector is mandated by law. The detector should be powered by the dwelling's electric service and should be placed in the hallway near bedrooms. In multilevel dwellings the NFPA recommends that a detector be placed at the top of the stairs. The *International Fire Code* also requires a hard-wired smoke detection system in new residential buildings. The *International Residential Code* requires a smoke detector in each sleeping room, one outside of and in the vicinity of the sleeping rooms, and one at each level in a multi-level residential building.

Codes dictate detector types and locations in commercial, industrial, institutional, and multifamily residential buildings.

Many electrically powered smoke detectors are part of security systems wired to a central station. These are required by code in most commercial, institutional, industrial, and multifamily residential buildings. Although such systems are expensive to install in single-family residences, some insurance companies offer discounts of 2.5 to 5% or more on the dwelling's fire insurance rate when these detectors have been installed. When central systems are used in a residence, the alarm bell and test button should be installed in the master bedroom or other suitable location in the sleeping area. In other building types, codes dictate the location of alarm bells, and may require that a signal be forwarded electronically to a fire station or alarm-monitoring service. Because of the possibility of failure due to causes originating in another system, the operating power supply should be from a separate and independent transformer rather than from the entrance-signal or another multiple-use transformer.

Some communities require photoelectric detectors. Residents should check local codes before buying and installing smoke detectors.

The Consumer's Union suggests that both photoelectric and ionization detector types may be useful. Although some fire alarm systems combine ionization and photoelectric capacity in one unit, two separate detectors may be less expensive and offer better protection.

For a basic two-detector system in a residence, an ionization detector should be placed in the hallway just outside the bedroom doors, and a photoelectric detector, should be located in the general living area. The photoelectric detector should not be located where cooking might set it off accidentally. In a larger house, a heat detector may be appropriate in the kitchen. In a two-floor house, a detector should also be placed at the top of the stairs.

Ideally, smoke detectors should be mounted in the middle of a ceiling. If this conspicuous location is objectionable, they may be mounted on a wall 6 to 12 in. (150 to 300 mm) below the ceiling or on the ceiling not closer than 6 in. (150 mm) to the wall or the corner. Air currents sometimes do not reach the corners of rooms, thus possibly delaying smoke detection.

A smoke detector should not be mounted in front of an air supply duct or near a furnace cold-air return, because excessive air flow prevents prompt detection of smoke. Such locations, however, are acceptable for heat detectors.

a                    b

**FIGURE 20.2-2** In a photoelectric detector, (a) the sensor does not ordinarily "see" the light; (b) when smoke is present, light rays hit the sensor and sound the alarm.

A smoke detector should not be located on a ceiling or wall that is substantially warmer or colder than the rest of the room. In this case, an invisible thermal barrier near the wall or ceiling surface can keep smoke from reaching the detector. This thermal barrier may be a problem in poorly insulated buildings and in mobile homes.

When a detector must be mounted under any of the adverse conditions discussed above, a heat detector rather than a smoke detector should be used. The heat detector is preferred here because it is activated by temperature change rather than smoke delivered by air flow.

For added protection, a dwelling may be equipped with extra detectors. A photoelectric or heat detector can be placed in the basement near a furnace or boiler. Extra detectors might be placed in storage areas. A heavy smoker may want an extra photoelectric detector in the bedroom. Both an ionization and a photoelectric smoke detector may be placed on each level of a multilevel house.

Detectors operate on either batteries or electric current. The installation of detectors that use batteries does not require professional training. Those that use electricity have to be wired into the building's electrical system, which requires knowledge of such installations. Unless provided with backup batteries or a separate power source, a wired detector system will become inoperable if there is a power failure. Smoke detectors with batteries should be models that beep when the battery becomes weak.

### 20.2.1.2 Escape from Fires

Building codes mandate the number and size of exits required. All public spaces and many other spaces require at least two means of egress. Fire escapes are not usually required in one- and two-family dwellings.

### 20.2.2 FIRE CONTAINMENT AND CONTROL

Fire containment and control measures are dictated by building codes. They include fire walls and fire and smoke doors to limit the spread of fires until inhabitants can escape.

# 20.3 Additional Reading

More information about the subjects discussed in this chapter can be found in the references listed in Section 20.4 and in the following publications.

Chace, Richard. 2002. "Integrating Security into the Design Process." *The Construction Specifier* (May): 79 (www.csinet.org).

Council on Tall Buildings and Urban Habitat. 1992. *Fire Safety in Tall Buildings*. New York: McGraw-Hill (www.mcgraw-hill.com).

Fowler, David W. 2001. "Is It Safe to Go Outside? The Design Process from a Security Perspective." *The Construction Specifier* (April): 23 (www.csinet.org).

General Services Administration. "GSA Document 8, Security Design." Washington, DC: GSA (www.gsa.gov).

Shrive, Charles A. 1993. "Writing Fire Alarm Specifications." *The Construction Specifier* (July): 37 (www.csinet.org).

Tierney, Mike. 2002. "Essential Elements of Security." *The Construction Specifier*, (November): 34 (www.csinet.org).

Wright, Larry J. 2002. "Get in Because of Something You Are." *The Construction Specifier* (March): 53 (www.csinet.org).

# 20.4 Acknowledgments and References

### ACKNOWLEDGMENTS

We gratefully acknowledge the assistance of the following organization in preparing this chapter. We are also indebted to them for permission to use their illustrations when requested and for the use of their publications as references.

Intermatic Incorporated (www.intermatic.com)

### REFERENCES

We would also like to thank the authors and publishers of the publications in the following list for their contribution to our research for this chapter.

Chace, Richard. 2002. "Integrating Security into the Design Process." *The Construction Specifier* (May): 79 (www.csinet.org).

Fowler, David W. 2001. "Is It Safe to Go Outside? The Design Process from a Security Perspective." *The Construction Specifier* (April): 23 (www.csinet.org).

General Services Administration. "GSA Document 8, Security Design." Washington, DC: GSA (www.gsa.gov).

International Code Council. 2006. Falls Church, VA: ICC (www.iccsafe.org). International Codes:
*International Building Code*®
*International Residential Code*®.
*International Existing Building Code*®.
*International Fire Code*®.

National Fire Protection Association (NFPA). Quincy, MA (www.nfpa.org).
2002. NFPA 550: "Guide to the Firesafety Concepts Tree."
2003. *NFPA 101. Life Safety Code*®.

Raeber, John A. 1991. "Redundancy in Fire Protection." *The Construction Specifier* (February): 17 (www.csinet.org).

Ramsey, Charles George, Harold Reeve Sleeper, and John Ray Hoke, Jr. 2000. *Architectural Graphic Standards*, 10th ed. Hoboken, NJ: John Wiley & Sons, Inc. (www.wiley.com).

Reid, Ed and Bob Schifiliti. 1991. "A Balanced Approach to Fire Protection Detection and Suppression Systems." *The Construction Specifier* (February): 30 (www.csinet.org).

Tierney, Mike. 2002. "Essential Elements of Security." *The Construction Specifier* (November): 34 (www.csinet.org).

Wright, Larry J. 2002. "Get in Because of Something You Are." *The Construction Specifier* (March): 53 (www.csinet.org).

# Earthwork

# Introduction

This chapter discusses the below-grade aspects of site construction, including site clearing; dewatering; excavation and backfilling for site improvements and buildings; surface water and groundwater control, including dewatering during construction and prevention of water intrusion into the completed building; and grading. These can have important effects on the construction cost, and the way in which they are done can change the final look of a building site. They can also be crucial to a completed building's structural stability. Improper excavation and backfilling can be dangerous during the work and damaging to the completed structure later, if supporting materials are improperly disturbed or fill is not compacted as it should be. Proper grading will ensure that water drains away from a building rather than toward it, which will prevent potentially disastrous leaks and flooding.

Site characteristics that may affect the design and construction of slabs on grade, building foundations, and site improvements include:

- Underlying soil type and properties
- Moisture conditions
- Thermal conditions
- Geographic factors that may require construction precautions to control termites or unusual condition

The handling of conditions that exist at a construction site before the work under a construction contract begins, including site surveying, demolition of existing construction, and subsurface investigation, is addressed in Chapter 2. Above-grade site construction is addressed in Chapter 22. Water supply sources, sanitary drainage utilities, storm drainage utilities, and electrical distribution to the building are covered in Chapter 23.

# Sustainability Considerations

If care is not exercised, the below-grade sitework associated with a construction project can be very destructive to the environment. Codes and local ordinances mandate measures to protect the environment. Whether or not legally required, care should be taken to protect wetlands and other ecologically important areas from the effects of the sitework and to preserve or restore disturbed local ecosystems damaged by the work.

It is important to protect natural vegetation that can be saved from damage during construction operations and to preserve as much existing topsoil as is practicable.

Pesticides can be very harmful to existing ecosystems, and their use should be avoided when possible.

On-site wastewater should be contained and treated to prevent groundwater pollution. Where the ecosystem has been damaged, it is important to reintroduce native species displaced or destroyed during the work.

# Applicable *MasterFormat* Sections

Subjects addressed in this chapter are included in *MasterFormat* 2004, Division 31, "Earthwork." The following *MasterFormat* 2004 Level 2 sections are applicable to this chapter:

31 11 00  Clearing and Grubbing
31 12 00  Selective Clearing
31 13 00  Selective Tree and Shrub Removal and Trimming
31 14 00  Earth Stripping and Stockpiling
31 22 00  Grading
31 23 00  Excavation and Fill
31 25 00  Erosion and Sedimentation Controls
31 31 00  Soil Treatment
31 32 00  Soil Stabilization
31 37 00  Riprap
31 41 00  Shoring
31 46 00  Needle Beams
31 48 00  Underpinning
31 51 00  Anchor Tiebacks
31 52 00  Cofferdams
31 53 00  Cribbing and Walers
31 62 00  Driven Piles
31 63 00  Bored Piles
31 64 00  Caissons
31 66 00  Special Foundations

# 21.1 Soils

Some buildings rest on a continuous mass of solid crystalline *bedrock*, or a continuous sedimentary rock, but soil underlies most building sites and supports the loads imposed by most low-rise and many high-rise building foundations. Because a building's structural integrity is dependent on the material that supports its foundations, a basic understanding of the nature of soils and how they perform under loads is important.

## 21.1.1 CLASSIFICATION

Proper classification is a critical factor in identifying problem soils. Several soil classification systems are available, but the one in most common use for soils associated with buildings is that developed by the American Society for Testing and Materials (ASTM). The applicable ASTM standards are D2487 and D2488.

In the ASTM system, soils are classed broadly as either *granular* or *cohesive* (Fig. 21.1-1). Granular soils, such as boulders, gravel, and sand, consist of relatively large particles visible to the eye. In cohesive soils the particles are relatively small, and in some soil types, such as fine-grained clays, the particles cannot be seen even through a low-powered microscope. Silts are soils made up of particles finer than sand but coarser than clay (Fig. 21.1-2). Actual soil samples usually consist of mixtures of many soils of different particle sizes. They are described as combinations of the predominant soils, such as silty sand, clayey sand, or gravelly sand.

The soil classification chart in Figure 21.1-3 identifies soils and divides them into coarse-grained and fine-grained soils according to the ASTM system. Each type of soil is given a descriptive name and a letter symbol indicating its principal characteristics. As indicated by the typical names in Figure 21.1-3, field classification of soils includes a variety of different materials. When soil is a combination of two types, it is described as a combination of both. For example, *clayey sand* describes a soil that is predominantly sand but that contains an appreciable amount of clay. A sandy clay has the properties of clay but contains an appreciable amount of sand.

## 21.1.2 PROPERTIES

The stability and structural properties of soil in relation to foundations are determined in large measure by the shearing resistance of the soil. Shearing resistance in most subgrade materials is the combined effect of internal friction and cohesion.

### 21.1.2.1 Internal Friction

The *internal friction* of a soil can be described as the resistance of the soil

S A N D (GRANULAR)     C L A Y (COHESIVE)

**FIGURE 21.1-1** Sand grains are large and angular, and the volume of the voids is relatively small. Clay grains are minute and scaly, and the volume of the voids is quite large.

**FIGURE 21.1-2** Soil Types

| Soil Types | Approximate Size Limits of Soil Particles |
|---|---|
| **Boulders** | Larger than 3 in. (76.2 mm) diameter |
| **Gravel** | Smaller than 3 in. (76.2 mm) diameter but larger than #4 sieve |
| **Sand** | Smaller than #4 sieve[a] but larger than #200 sieve[b] |
| **Silts** | Smaller than 0.02 mm diameter but larger than 0.002 mm diameter |
| **Clay** | Smaller than 0.002 mm diameter |

[a]Approximately $1/4$ in. (6.4 mm) in diameter.
[b]Particles less than #200 sieve are not visible to the naked eye.

grains to their tendency to slide over one another.

### 21.1.2.2 Cohesion

The binding force that holds soil grains together is called *cohesion*. It varies with the moisture content of the soil. Cohesion may be very high in clays, but sand and silt have little if any cohesion.

### 21.1.2.3 Other Properties

Other soil characteristics that influence building foundations and slabs on grade are:

Classification and composition
Grain size
Moisture content
Density
Capillarity
Drainage characteristics
Stratification
Consistency
Compressibility
Climatic variations

#### DENSITY AND CONSISTENCY

The most significant property of cohesionless soils is their *relative density*, which is described in such terms as *very loose, loose, medium, dense,* and *very dense.* Most coarse-grained soils are cohesionless.

The most significant property of cohesive soils is *consistency,* described in such terms as *very soft, soft, medium, stiff, very stiff,* and *hard.* Most fine-grained soils are cohesive. Purposely densifying or increasing the unit weight

of a soil mass by compaction with rollers, tampers, or vibrators can improve its structural properties (see Section 21.3.7).

#### PLASTICITY

If a soil within some range of water content can be rolled into thin threads, it is called *plastic.* All clay minerals are plastic at some range of water content. Most very-fine-grained soils contain clay minerals and therefore are plastic. The degree of plasticity may be defined by the terms *fat* and *lean.* Lean clays are only slightly plastic, because they usually contain a large proportion of silt or sand. The *plasticity index* (PI) is a value expressing the numerical difference between the *liquid limit* and the *plastic limit* of the soil as determined by test, such as ASTM D4318. The PI is equal to the liquid limit (LL) less the plastic limit (PL): (PI = LL − PL). The LL is the amount of moisture present when the soil changes from a plastic to a liquid state. The PL is the amount of moisture present when the soil changes from semisolid to plastic.

A clay soil with a PI greater than 15 may expand when its moisture content increases, as would occur as a result of a rising water table. If the soil beneath a building is not consistent, as is commonly the case, expansion of the part that is of such clay while adjacent, more stable materials remain stable could result in uneven heaving. This will cause cracks in the foundation and the supported structure, which can be severe enough to result in serious structural damage. Structural damage can also occur when the moisture content of such clay beneath exterior foundations is increased by groundwater seeping into the clay while the same clay beneath interior foundations remains stable.

### 21.1.3 BEARING CAPACITY

Soil samples and tests to determine foundation type and size should always be made (see Section 21.1.7) where tall buildings and heavy live loads are present, which is the case for most commercial, institutional, and industrial buildings and high-rise multifamily dwellings. Such testing is also prudent where numerous single-family house foundations are being planned, such as in a subdivision.

Since the loads transmitted to soils by most small buildings are not large, the sizing of their foundations is often based

**FIGURE 21.1-3** ASTM Soil Classification System

| | Major Divisions | Group Symbols | Typical Names | Presumptive Bearing Capacity, tons/sq ft[a] | Modulus of Subgrade reaction, k, lb/sq in./in. |
|---|---|---|---|---|---|
| Coarse-grained soils more than 50% retained on #200 sieve[b] | Gravels 50% or more of coarse fraction retained on #4 sieve — Clean gravels | GW | Well-graded gravels and gravel-sand mixtures, little or no fines | 5 | 300 or more |
| | | GP | Poorly graded gravels and gravel-sand mixtures, little or no fines | 5 | 300 or more |
| | Gravels with fines | GM | Silty gravels, gravel-sand-silt mixtures | 2.5 | 200 to 300 or more |
| | | GC | Clayey gravels, gravel-sand-clay mixtures | 2 | 200 to 300 |
| | Sands more than 50% of coarse fraction passes #4 sieve — Clean sands | SW | Well-graded sands and gravelly sands, little or no fines | 3.75 | 200 to 300 |
| | | SP | Poorly graded sands and gravelly sands, little or no fines | 3 | 200 to 300 |
| | Sands with fines | SM | Silty sands, sand-silt mixtures | 2 | 200 to 300 |
| | | SC | Clayey sands, sand-clay mixtures | 2 | 200 to 300 |
| Fine-grained soils 50% or more passes #200 sieve[b] | Silts and clays, liquid limit 50% or less | ML | Inorganic silts, very fine sands, rock flour, silty or clayey fine sands | 1 | 100 to 200 |
| | | CL | Inorganic clays of low to medium plasticity, gravelly clays, sandy clays, silty clays, lean clays | 1 | 100 to 200 |
| | | OL | Organic silts and organic silty clays of low plasticity | | 100 to 200 |
| | Silts and clays, liquid limit greater than 50% | MH | Inorganic silts, micaceous or diatomaceous fine sands or silts, clastic silts | 1 | 100 to 200 |
| | | CH | Inorganic clays of high plasticity, fat clays | 1 | 50 to 100 |
| | | OH | Organic clays of medium to high plasticity | | 50 to 100 |
| | Highly organic soils | PT | Peat, muck, and other highly organic soils | | |

[a]National Building Code, 1976 Edition, American Insurance Association.

[b]Based on the material passing the 3-in. (75-mm) sieve.

(Courtesy Portland Cement Association.)

on local rules of thumb rather than on soil tests and engineering design. Building codes often list maximum bearing capacities for soils of several general classes that may be used without testing (Fig. 21.1-4). These bearing capacities normally are conservative and rarely take into account compressibility of soils. In addition, the general descriptions of the soils are subject to different interpretations, and the use of generalized tables may lead to erroneous conclusions regarding the true load-carrying capability of the soil at a specific site. As a result, using the loads permitted by a code often results in overdesigned foundations. Soil sampling and testing are, therefore, often valuable even for individual houses and other small buildings, where the reduced costs of smaller foundations may offset the cost of soil sampling and testing.

**FIGURE 21.1-4** Presumptive Bearing Values of Foundation Materials

| Class of Material | Bearing Pressure, lb/sq ft* |
|---|---|
| 1. Crystalline bedrock | 12,000 |
| 2. Sedimentary rock | 6,000 |
| 3. Sandy gravel or gravel | 5,000 |
| 4. Sand, silty sand, clayey sand, silty gravel, and clayey gravel | 3,000 |
| 5. Clay, sandy clay, silty clay, and clayey silt | 2,000 |

*1 psf = 47.88 P.

(BOCA National Building Code/1990, copyright 1989, Building Officials and Code Administrators International, Inc. Published by arrangements with author. All rights reserved.)

## 21.1.4 SETTLEMENT

When the soil being loaded is dense and granular, such as coarse sand or gravel, little settlement is likely and it occurs quickly after the load is applied (Fig. 21.1-5). When the soil is a cohesive clay, settlement is likely to be greater but occurs slowly, perhaps taking several years. The two principal causes of settlement are (1) reduction of the vol-

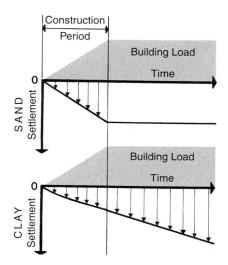

**FIGURE 21.1-5** Building settlement on sand tends to be quick and slight; on clay it tends to be greater and may occur over a long period of time.

ume of soil voids (spaces between soil particles containing air or water) and (2) lateral displacement of the supporting soil.

### 21.1.4.1 Reduction of Voids

The reduction in the volume of the voids within soil when loaded is related to the soil's compressibility. Because sand contains large, angular grains and a relatively small percentage of void spaces (see Fig. 21.1-1), foundation pressures can cause little decrease in the volume of the voids. Clay, however, has a scale-like grain structure and a relatively large percentage of voids. Therefore, decreases in the volume of the voids in clays subjected to foundation loads can be quite large.

When soil voids are filled with water, this water must be pressed out if the volume of the voids is to be decreased. Sand is permeable and offers little resistance to water passage, so total void reduction and settlement occur almost simultaneously with load application. Impermeable clays, on the other hand, require a long time for compression (*consolidation*) to take place, since water cannot pass out of them easily.

Air contained in the voids of sand offers no resistance to compression since it is expelled as soon as a load is applied. A certain amount of compression may occur instantaneously in clays if air is present in the voids, since air is compressible.

### 2.1.4.2 Lateral Displacement

*Lateral displacement* in soil is the horizontal movement of soil under heavy foundation loads. Settlement as a result of lateral displacement is not common in small buildings and generally offers no problem with small building foundation loads, unless the lateral support to the soil under the foundation areas is reduced or disturbed by adjoining excavation. Lateral displacement is more likely when heavy loads are applied and must be accounted for in foundation design for large buildings.

### 21.1.5 NONPROBLEM SOILS

Nonproblem soils (Fig. 21.1-6) are defined as those that are dense, coarse-grained, and properly drained. Most of these cause few problems in light buildings for slab-on-grade construction or foundations. These soils possess excellent bearing capacity, because they are not subject to significant volume change. Gravels, for instance, usually occur in a dense state and are unlikely to cause any settlement problems. Usually, the only consideration is that the natural confinement of the soil not be disturbed by trenching or grading operations that would reduce the necessary lateral support.

Nonproblem soils acceptable for use in the construction of fills, backfills, and embankments for commercial, institutional, and industrial buildings and for large multifamily housing projects are generally limited to those in the GP, GM, GC, SP, and SC categories. General backfill outside of structures is generally limited to soil classified as GP, GM, GC, SP,

or SM. However, in either case, some other classifications may be permitted in some circumstances and subject to additional LL and PL requirements. The decision to permit other soil classifications should be made by a qualified and experienced soils engineer. *Controlled fills* are used to bear loads, usually but not always beneath buildings. Soils permitted in controlled fills are generally limited to those classified as GW, GP, SW, and SP. These are often further limited to an LL not to exceed 35% and a PL not to exceed 12%, with a further limit on the percentage of fines that can be present. The actual requirements and the locations where controlled fills are required or permitted should be determined by a soils engineer.

The load-bearing capabilities of both cohesive and cohesionless soils can be improved by compaction. Soils beneath structures and pavements and soils adjacent to buildings are usually compacted (see Section 21.3.7).

When sand or silt is encountered, it should be determined whether it is dense, medium dense, or loose. Loose sand or silt frequently can be economically compacted to a dense state before slab construction.

*Loess* soils, which are unstratified loamy materials in the ML or MH groups, should be treated in the same manner as other soils in the same groups. The loads in light-building construction generally are not sufficient to cause problems with these soils unless the slab or foundation supports masonry or other heavy, concentrated loads. Heavily loaded slabs and foundations may settle unduly when loess

**FIGURE 21.1-6** Description of Nonproblem Soils

| Soil Classification | Soil Description | Density or Consistency* |
|---|---|---|
| GW, GP | Gravels or sand-gravel mixtures, little or no fines | All densities |
| GM, GC | Gravels or gravel-sand mixtures, with clay or silt | Medium-dense to dense |
| SW, SP | Sands with little or no fines | Medium-dense to dense |
| SM, SC | Silty or clayey sands | Medium-dense to dense |
| ML | Silty or clayey fine sands | Medium-dense to dense |
| ML, MH | Inorganic silts | Medium-dense to dense |
| CL | Inorganic and silty clays of low to medium plasticity | Stiff to hard |

*See Figure 21.1-3 for definition of terms.

soils become saturated; therefore, proper drainage is extremely important when these soils are encountered. Loess soils should not be permitted for bearing or adjacent to larger buildings.

### 21.1.6 PROBLEM SOILS

Problem soils include all soils that are not listed in Figure 21.1-6 as nonproblem soils. They possess less than suitable bearing characteristics for support of slabs, foundations, walks, or pavements, and are usually not satisfactory even for grassy or planted areas adjacent to buildings. Special construction precautions are necessary with soils such as highly compressible or highly expansive clays, highly plastic soils, sands, and silts that are very loose and loess soils that are heavily loaded or not well drained. When a supported slab, foundation, or pavement is not sufficiently stiff or does not possess the necessary strength, detrimental effects to the supported element and even the building's superstructure may result. These effects will occur due to settlement or differential movement caused by lack of density of, or a change in the moisture content of, the supporting soil.

### 21.1.7 EFFECTS OF WATER

Moisture resulting from either surface water from precipitation or groundwater, including the effects of seepage fields, (1) may result in volume change or reduction of the bearing capacity of a soil, (2) may impose additional requirements for site preparation with regard to grading and drainage, and (3) may necessitate special precautions such as a base course or vapor retarder (see Sections 21.3.5 and 21.4.3.2).

The amount and frequency of precipitation have primary influences on problem soils, because they may cause changes in the moisture content of soil supporting a slab or foundation. Exact values for the amount of precipitation, its seasonal variation, the amount and duration of each occurrence, and its effect on supporting soil under a slab or foundation are difficult to measure and almost impossible to predict. The most important considerations are whether precipitation or its absence will change the moisture content of the soil intended for support of slabs or foundations during and after construction, and whether the change will have a detrimental effect on the building.

When variations in moisture content occur on a type of soil that expands as it absorbs moisture and shrinks as it loses moisture, the slab or foundation it supports is subjected in turn to cycles of uplift as the soil expands and settling as the soil shrinks. When a period of high moisture content is followed by drought, soil moisture evaporates more rapidly around the perimeter of a building, and soil moisture under the center of the building, which may be due to capillary action and migration, becomes trapped and sealed from direct exposure. Moisture can be retained under the center of a building long after extended periods of drought.

Trees and shrubbery immediately adjacent to a building's perimeter also can affect the soil moisture content by providing a shield from natural precipitation and by extracting soil moisture during the growing season. Moreover, the watering of gardens and lawns adjacent to a building can increase the underlying soil's moisture content.

When soil changes from a low to a high moisture content, a similar but opposite phenomenon develops. If prolonged periods of alternating drought and moisture occur, considerable differential moisture can develop in different areas under a building. If the soil is a type that undergoes substantial volume changes with changes in moisture, one of two conditions may result. If the building's slabs and foundations are relatively flexible, as in the case of unreinforced or only lightly reinforced slab Types I and II (see Section 3.12), they will follow the uneven contour of the supporting soil that results from uneven volume change, exposing them to distortions and possible damage to both slab or foundation and the supported superstructure (Fig. 21.1-7a).

On the other hand, if a slab or foundation is sufficiently rigid, as is the case with a properly structurally reinforced and stiffened Type III slab or concrete foundation, it will resist deformations even though the contour of the supporting soil is uneven. Higher soil pressures develop under such a slab or foundation over the plateaus with reduced soil pressures at the valleys. The slab or foundation will be subjected to bending between these uneven contours, and the soil at the plateaus may deform due to higher bearing pressures, distributing loads to adjacent areas (Fig. 21.1-7b). Type III slabs are designed to resist bending stresses in order to span between plateaus without

a

b

**FIGURE 21.1-7** Slab distortion and soil movements (highly exaggerated): (a) a flexible slab will distort with volume changes in the supporting soil; (b) a rigid slab is designed to span uneven contours caused by volume changes in the supporting soil.

damage to the slab, as are properly reinforced concrete footings.

### 21.1.8 EFFECTS OF TEMPERATURE

The combined effects of temperature and moisture influence the depth of perimeter grade beams and foundations, as well as the selection of slab types, because some soils are subject to volume change called *frost heave*, which results from the formation of ice layers when water in soil freezes. Unless a Type III slab is used, or if nonheaving soils (GW, GP, SW, SP) are present and the water level is below the frost line, foundations always should be carried below the frost line (Fig. 21.1-8).

The temperature of a building slab can affect the comfort of occupants, especially during cold periods. Insulation is often introduced to help keep slabs warmer. Some buildings are partially heated by radiant heat generated by pipes carrying warm water or ducts carrying heated air built integrally with a slab on grade. This configuration warms the slab as well as the air above it, which can produce a more comfortable environment for occupants than if the air alone is heated.

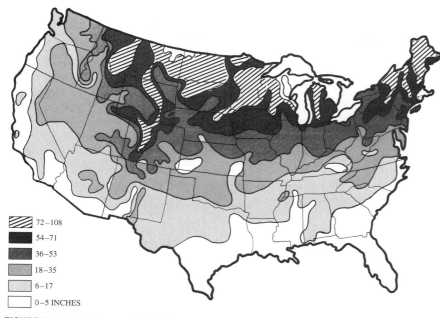

72–108

54–71

36–53

18–35

6–17

0–5 INCHES

**FIGURE 21.1-8** Maximum depth of frost penetration in inches. (Courtesy National Weather Service.)

- Underpinning existing structures to protect them from damage during adjacent excavation
- Slope protection and erosion control
- Chemically treating the soil to protect the building against subterranean termites

### 21.3.1 UNDERPINNING

Underpinning is the time-consuming and expensive process of supporting the foundations of an existing building by placing new footings, pilings, or caissons beneath the existing footings. Underpinning is often needed when the excavation for a new building or an addition to an existing building will undercut existing foundations (Fig. 21.3-1). Underpinning may also be needed when an existing footing is too small to support a vertical addition to a building, when an existing undersized footing has resulted in undue

## 21.2  Site Clearing

Site clearing consists of removing site improvements and vegetation that will not be a part of the new work. This includes removal of trees and other vegetation, clearing and grubbing, topsoil stripping, and removing above- and below-grade site improvements.

Before site clearing is started, adjacent public and private property should be protected from harm by the erection of fences and barricades. Before trees and other vegetation are removed, those to remain should be identified, marked, and protected by temporary fences or barricades. Similarly, water, electric, sewer, and other utility lines should be marked and staked so that they will not be inadvertently damaged.

*Clearing and grubbing* consists of removing trees, shrubs, and other vegetation, including their roots and stumps. Depressions caused by these procedures should be filled with satisfactory backfill material, as discussed in Section 21.3.4.

*Topsoil* should be stripped from areas where new construction is to take place. Topsoil is often stockpiled on the site and later reused in lawn and planting areas. However, topsoil should not be removed from within the farthest extent of the branches (*dripline*) of trees to remain, because doing so will damage the root structure of the trees.

Roadways, walkways, site walls, and other structures that will not be used in the new work should be demolished and the debris removed from the site.

One means of disposing of materials removed from a building site is dumping it in public or private *landfill* or *spoil* areas. Landfill areas often accept organic materials and construction debris, whereas spoil areas usually accept only earth materials, such as excavated soil and unusable topsoil. Burning on the site, which was once the most prevalent means of disposing of combustible waste, is often prohibited today, especially in heavily populated jurisdictions. Even when it is legal, some owners will not permit burning because of environmental and safety concerns.

## 21.3  Earthwork

Earthwork includes:

- Excavation and filling to make way for a building
- Backfilling after the building has been built
- Grading preparatory to installing paving, walkways, lawns, and landscaping

Related site construction components include:

Existing grade

Existing building

Extent of planned excavation for new building

**FIGURE 21.3-1** The existing footing here will most likely slide into the adjacent new excavation unless it is underpinned. (Drawing by HLS.)

settlement and damage, and in other cases when additional support is needed.

Underpinning can be accomplished in many different ways, depending on the nature of the problem. Only two of the many possibilities are covered here:

- Placing a new footing and foundation wall below the existing ones
- Using piles

When new building foundations are lower than those of an existing building, the existing footings can be supported by placing new footings lower than the existing ones (Fig. 21.3-2). One way to accomplished this is to place a series of *needle beams* through the existing building walls to support them until the new foundation wall and footing have been built (Fig. 21.3-3). Another way to place a lower foundation wall and footing is to dig intermittent pits at the existing foundation down to the level of the new footing (Fig. 21.3-4). These pits must be small enough to permit the existing footing and wall to span the pit with support by temporary jacks and shoring. The footing between the pits remains supported by the original soil. After the underpinning foundation wall and footing have been placed (see Fig. 21.3-2), another section of pit is dug until the entire wall has been underpinned.

Underpinning can also be accomplished using piles (see Section 21.3.3).

FIGURE 21.3-3 The exterior ends of needle beams may be supported at the existing grade level or on footings at the bottom of the new excavation. In either case, the soil against the existing wall must not be removed until the beams are in place. (Drawing by HLS.)

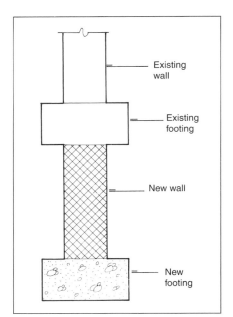

FIGURE 21.3-2 The lower-footing method of underpinning. (Drawing by HLS.)

FIGURE 21.3-4 Pits dug to permit underpinning must themselves be braced and shored to prevent collapse and injury to workers. (Drawing by HLS.)

In general, the method consists of placing new piles adjacent to the existing footings. A section of the foundation wall is then removed at the top of each set of piles, and a concrete cap is poured in place so that it rests on the piles and penetrates the wall (Fig. 21.3-5). A proprietary system is available that substitutes a foundation support and jack assembly for the cap. This assembly rides down on a steel pipe pile, and is rotated toward the existing footing until it extends beneath the footing. It is then raised hydraulically until the loads transmitted by the footing are picked up by the support assembly and transmitted to the pile. This type of assembly is particularly useful when an existing footing has settled.

## 21.3.2 EXCAVATING

Even for single-family residences and other small buildings with floor slabs on grade, excavation is necessary for trenches to contain footings and utilities.

For most buildings, much more excavation than this minimum is required.

Building *excavation* is usually defined as the removal of whatever materials are found down to the level required by the contract documents and disposal of the materials excavated. In some small applications, excavation is done using hand tools, but most excavation requires power equipment (Fig. 21.3-6). This equipment ranges from the bulldozers, bucket loaders, trench excavators, and backhoes common in earth excavation to the rippers, power shovels, pneumatic hammers, and drop balls used to excavate in rock.

Before excavation is permitted to begin, underground utilities that are to remain should be located and staked to prevent damage to them.

The materials that underlie a site are often called *unclassified*, meaning that it is unknown whether they consist of all soil or whether rock is present. This common practice is unfair to contractors, who, when they present their bids, must gamble on the nature of the materials that have to be excavated. Excavation of rock is far more expensive than excavation of earth. It takes more time, requires different equipment, and often requires that blasting be used. Blasting is restricted in many jurisdictions to certain hours and days. In addition, the size of the explosive is sometimes restricted, requiring the use of smaller and more charges. A far fairer method would be to determine whether the materials to be excavated are either earth or rock.

The definitions of the terms *earth* and *rock* vary with the source. One way to define rock is to say that it includes all material that cannot be removed using a specific piece of earth-moving equipment. Earth, then, is any material that is not rock.

Excavation generally includes removal of aboveground site improvements such as pavements, walks, and site walls, as well as underground items like old foundations, piping that will not be used in the new work, wells, and cisterns.

Materials that are suitable for fill and backfill (see Section 21.3.4) should be stockpiled for reuse. Excess and unsuitable materials should be disposed of.

Excavations are classified as *open excavations*, *trenches*, or *pits*. Open excavations are those that cover the large portion of a building excavation. They are usually excavated with power equipment to the level of the lowest floor slab.

Trenches and pits are dug into an open excavation or into soil that has not been otherwise excavated. Trenches for footings and utilities are usually dug by machine and finished by hand. In many cases, pits for isolated footings, sumps, and mechanical devices are dug using hand tools. Larger pits for elevators and large mechanical equipment are usually excavated by machine and finished by hand.

The bottoms of excavations should be protected against erosion, softening due to water saturation, and freezing. Disturbed material at the bottom of excavations for trenches and pits should be removed. Disturbed material beneath slabs that is suitable for that location should be compacted.

## 21.3.3 EXCAVATION SUPPORT SYSTEMS

Except for shallow trenches and pits, the side walls of excavations, including open

**FIGURE 21.3-5** Vertical piles used as underpinning. (Drawing by HLS.)

Existing wall

Existing footing

Pile cap

Pile

FIGURE 21.3-6 Many types and sizes of earth-moving equipment are used in building excavation.

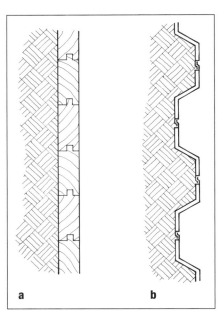

FIGURE 21.3-8 (a) Timber sheet piling and (b) one type of steel sheet piling. Other configurations and other joint types are also available. (Drawings by HLS.)

excavations, trenches, and pits, need to be supported until the permanent construction is in place and able to support them. Support can be obtained by simply sloping the sides of an excavation when the site is large enough to permit it. This slope should be at such an angle that the soil will not slide down it. The angle must be shallow for soils with a large percentage of sand and gravel, but it may be steeper for soil that contains clay. The maximum slope should be determined by testing the actual soil to determine the angle at which it will not slide. Applicable codes and other legal restrictions may dictate the permissible slopes. As excavations become deeper, it is sometimes necessary to step the sides back (Fig. 21.3-7). Excavated materials stockpiled for use in backfill should be stored away from the edges of the excavation to re-

duce the height of the material adjacent to the excavation.

When the space to store excavated soil is small, or when excavations are too close to property lines or other structures to properly slope the sides, other types of support are necessary. Architects are seldom called on to design excavation support systems. This is usually left up to the contractor. Therefore, the type used depends on the conditions at the site and the preferences of the contractor. An architect may, however, be required by law to review a contractor's plans for excavation support systems.

On shallow excavations (12 to 15 ft [3.66 to 4.57 m]), bracing may be done with *sheet piles*. Most of these piles are wood or metal (Fig. 21.3-8), but precast concrete is sometimes used. Sheet pile sections are interlocked and driven into the ground before excavation begins.

Deep excavations are usually braced with *soldier piles* and *lagging* (Fig. 21.3-9). Soldier piles are H-shaped steel beams usually in wide-flange sections. Lagging consists of heavy wood timbers. Soldier piles are driven into the ground before excavation starts. Lagging is placed from the bottom as the excavation is made. Lagging may be placed behind the flanges of the soldier piles, as shown in Figure 21.3-9a, or may be placed on the face of the soldier piles and held in place by battens that are bolted through

the joints in the lagging to the flanges of the soldier piles (Fig. 21.3-9b).

On very shallow excavations, sheet piles driven to sufficient depth to act as cantilevers are sometimes strong enough to support the loads encountered without additional support. When excavations are deeper, however, both sheet and soldier piles and lagging need additional support to withstand the loads imposed by the soil and groundwater behind them. This support is provided by *shoring and bracing*. On very deep excavations, several tiers of bracing may be needed.

Sheet piles and soldier piles and lagging may be supported using *rakers*, *cross bracing*, or *tiebacks*. All three systems are designed to support horizontal, continuous *wales* that extend along the face of the sheet piles or soldier piles. Wales are usually wide-flange steel beams, but other steel shapes may also be used.

The two bracing systems shown in Figure 21.3-10 have a common drawback. They both obstruct activities in the excavation. For example, rakers can be placed only after the excavation has been extended to the level of the concrete footer or heel. But the sheet piles or soldier piles and lagging can be exposed only after the bracing is in place. This means that excavation at the location of

FIGURE 21.3-7 Stepping the sides of an excavation when there is room may preclude the need for shoring. (Drawing by HLS.)

a

b

**FIGURE 21.3-9** (a) Soldier piles and lagging; (b) battens at soldier piles and lagging. (Drawings by HLS.)

the footer or heel must be made to their level, and the rakers must be in place, before the earth can be removed from the face of the sheet piles or soldiers piles and lagging. The dashed line labeled "Limit of excavation" in Figure 21.3-10 indicates the level of the excavation until the rakers have been placed. If more than one line of rakers is needed, the earth beneath them must be removed in stages, extending just below each level. The proper angle of slope must be maintained

so that the remaining soil will stay in place and support the load imposed by the soil and groundwater on the opposite side of the sheet piling or soldier piles and lagging. In addition, excavation of the soil remaining against the sheet piles or soldier piles and lagging is difficult, because the rakers obstruct movement of excavation equipment.

Cross bracing also obstructs the movement of equipment but does permit excavation to continue uninterrupted from

one wale level to the next lower one without slopes. In wide excavations, cross bracing requires additional support, which is provided by driving steel columns at intermediate points across the span and installing additional wales on the columns. As the excavation is lowered, additional levels of braces can be added as needed.

The third method uses earth or rock anchors called *tiebacks*. This method does not obstruct activities in the excavation. Tiebacks are constructed by drilling a hole at each line of wales through the sheet piling or lagging into stable rock or soil. Tiebacks are not appropriate in unstable or very loose soils. An anchor rod or cables are inserted into the drilled hole, and the space around the rod or cables is filled with grout under pressure. When the grout has set, the rod or cables are *post-tensioned* (stretched) using a hydraulic jack and anchored securely to a continuous wale so that the tension is maintained. Figure 21.3-11 shows a wale made from steel channels, but other shapes, such as wide-flange steel beams, are also frequently used. When the soil tends to be somewhat unstable, or a type of soil, such as sand or gravel, is present that would collapse into the hole, a steel pipe may be inserted to keep the hole open until the grout has been placed.

Cross bracing and rakers are removed as soon as permanent construction makes them unnecessary. Tiebacks are left in place but are not relied upon for support after permanent construction is in place. Soldier piles and lagging are often removed as soon as permanent construction negates their need. However, when their removal would disturb new or existing construction, soldier piles and lagging are left in place. Sometimes they are covered with a drainage board or mat and used as the outer form for concrete walls (see Section 21.4.3.3). Sheet piling may be either cut off below ground level and left in place or removed completely, although wood sheet piling is usually removed.

### 21.3.4 FILL AND BACKFILL

Compacted fills are used for building sites in undulating or hilly terrain to create reasonably level ground for slab construction. *Engineered fill*, which is fill whose material selection, installation, and compacting are done according to engineering principles, is also sometimes

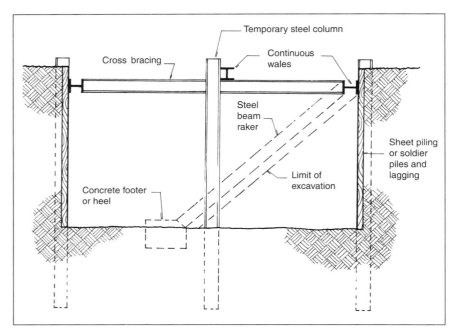

**FIGURE 21.3-10** Two foundation bracing systems. The solid lines show cross-bracing. The dashed lines represent a raker system. The two systems generally are not used together. (Drawing by HLS.)

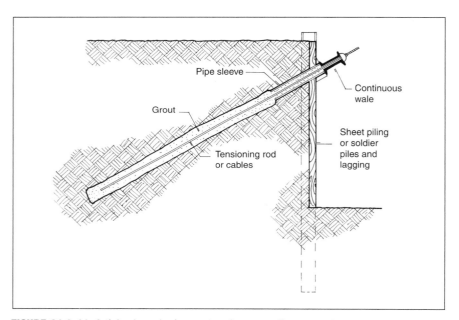

**FIGURE 21.3-11** A tieback anchoring system for excavation supports. (Drawing by HLS.)

used below foundations in small buildings when the existing soil is unsuitable. In any case, the bad soil is removed and replaced with engineered fill. Soils used for fill should contain no vegetation or foreign material that would cause uneven settlement. Beneath slabs, a base course should be provided when necessary.

Topsoil and other unsuitable materials should be removed from areas beneath foundations and slabs. There should be no soil or other materials present that would cause a slab or footing to settle unevenly or excessively.

Fills should be used beneath slabs to establish the desired finished slab eleva-

tion and should be compacted to the maximum practical density. The slab bed for Type I and II ground-supported slabs (see Section 3.12) should not expand or contract due to moisture. Therefore, organic or active clays and silts of the CH, CL, OL, OH, and PT groups, as classified in ASTM D2487, are not suitable for fills under these types of slabs. Soil groups GW, GP, SW, and SP (Fig. 21.3-12) may be used as fill material beneath slabs without a base course. These latter soil groups are also those used in engineered fill and in most fill in other locations. Groups GM and SM are also suitable under some conditions.

Before fill material is placed, *clearing and grubbing* (see Section 21.2) should have been completed where the fill will be placed. Sloped surfaces should be plowed or otherwise broken up so that fill material will mix with existing material and not slip away from it.

Suitable excavated soils may be used to backfill excavations. These include soils classified in ASTM D 2487 as GW, GP, GM, SM, SW, or SP (see Fig. 21.3-12), but some jurisdictions may further limit the acceptable types. In addition, some other qualifications may apply in some projects. For example, a soil that is not in the preceding list may be acceptable if its LL and PL are in a range that is acceptable to the project's soils engineer and its use does not violate the building code or other restrictions. When there are insufficient quantities of suitable excavated materials, acceptable soil should be brought from off site (*borrow material*) to complete the required backfilling and grading.

Filling and backfilling materials should be placed in layers (*lifts*) of about 8 in. (200 mm) loose thickness when compaction will be done by power equipment and about 4 in. (100 mm) loose thickness when hand tampers will be used. When the fill or backfill material has a lower or higher than acceptable moisture content for proper compaction, it should be moistened or aerated, respectively, until it has reached the optimum moisture content. The required moisture content varies with the soil material and should be established by testing.

Backfill should not be placed until underground utilities to be covered are in place and have been tested, concrete formwork has been removed, shoring and bracing have been removed, and sheet

| Major Division | | Group Symbols | Typical Names |
|---|---|---|---|
| Coarse-grained soils (More than half of material is larger than the smallest particle visible to the naked eye) | Gravels (more than half of coarse fraction is larger than ¼ in.) Clean gravels (little or no fines) | GW | Well-graded gravels, gravel-sand mixtures, little or no fines |
| | | GP | Poorly graded gravels or gravel-sand mixtures, little or no fines |
| | Gravels with fines (appreciable amount of fines) | GM | Silty gravels, gravel-sand-silt mixtures |
| | | GC | Clayey gravels, gravel-sand-clay mixtures |
| | Sands (more than half of coarse fraction is smaller than ¼ in.) Clean sands (little or no fines) | SW | Well-graded sands, gravelly sands, little or no fines |
| | | SP | Poorly graded sands or gravelly sands, little or no fines |
| | Sands with fines (appreciable amount of fines) | SM | Silty sands, sand-silt mixtures |
| | | SC | Clayey sands, sand-clay mixtures |
| Fine-grained soils (More than half of material is smaller than the smallest particle visible to the naked eye) | Silts and clays (liquid limit is less than 50) | ML | Inorganic and very fine sands, rock flour, silty or clayey fine sands or clayey silts with slight plasticity |
| | | CL | Inorganic clays of low to medium plasticity, gravelly clays, sandy clays, silty clays, lean clays |
| | | OL | Organic silts and organic silty soils, elastic silts |
| | Silts and clays (liquid limit is greater than 50) | MH | Inorganic silts, micaceous or diatomaceous fine sandy or silty soils, elastic silts |
| | | CH | Inorganic clays of high plasticity, fat clay |
| | | OH | Organic clays of medium high plasticity, organic silts |
| Highly organic soils | | PT | Peat and other highly organic soils |

☐ Generally may be used for fill without a base course

▨ Fill that requires a base course

▨ Not permitted for foundation fill

*Note:* Based on ASTM D 2487.

piling to remain has been cut off below grade level. In addition, backfilling against walls should not be done until building slabs are in place and the walls have been braced either by temporary or permanent construction.

## 21.3.5 BASE COURSE FOR SLABS ON GRADE

For slab-on-grade construction, some preparation must be made to the subgrade, depending on the type of soil and other conditions encountered at the site. At the least, the area beneath the slab must be cleared and grubbed and unsuitable soils, such as topsoil, must be removed. The slab bed for ground-supported slabs should:

- Provide the necessary bearing capacity for slab support
- Control ground moisture
- Establish the proper slab elevation

Fill material and undisturbed soil that is not in a suitably dense state should be properly compacted. Materials that will not support the loads from the slab should be removed and replaced with a compacted fill of suitable materials.

A slab bed may consist of undisturbed soil of the proper type, density, and consistency. Soil groups GW, GP, SW, and SP are in this category (see Fig. 21.3-12). Where other types of soil form the subbase, they should be removed to the proper level and replaced with a base course to control capillary moisture rise and to provide a uniform, level bed for structural support of the slab (Fig. 21.3-13). A base course should be at least a 4-in. (100-mm)-thick compacted layer of a well-graded mixture of gravel, crushed stone, sand, or crushed slag.

## 21.3.6 GRADING

The rough and final grades on a building site are established by *grading* using a combination of power equipment and hand tools. Grading is necessary where lawns, planting, pavement, walks, and building slabs will be placed.

The site around a building should be graded to form protective slopes that will drain surface water away from the foundation so that water will not collect under the slab. Surface runoff should be directed to streets or drainage structures. The finish grade should slope downward in all directions away from a building at

Concrete slab
Undisturbed soil

Concrete slab
Compacted fill
Undisturbed soil

Concrete slab
Base course
Compacted fill
Undisturbed soil

Concrete slab
Base course
Compacted fill
Undisturbed soil
Vapor retarder

**FIGURE 21.3-13** Slab beds for ground-supported slabs. A base coarse or a vapor retarder may be needed to control moisture.

a minimum fall of 6 in. (150 mm) in each 10 ft (3.05 m), which is a slope of about 5%. A percent of slope is the unit of rise (or fall) in each 100 units of run (1 ft [305 mm] in 100 ft [30.48 m], for example). When the area adjacent to a building is paved, a 1% slope, which is about $^1/_8$ in. (3.2 mm) in each foot (305 mm), may be sufficient to direct water away. However, when such a paved area has less than a 5% slope, positive drainage, such as by yard drains or catch basins, should be introduced at sufficient intervals to prevent ponding caused by imperfections in the paving slope and buildup of water during heavy rainfall. Hillside sites should be graded to divert surface water around and away from structures.

Areas beneath lawns and walks should be graded to within 0.10 ft (30.5 mm) above or below the design subgrade elevations, taking into account the thickness of sod and the walkway surface. Areas beneath pavements and building slabs should be graded to within $^1/_2$ in. (121.7 mm) above or below the design subgrade elevations.

## 21.3.7 COMPACTING

Compacting applies energy to soil to consolidate it by compressing air voids to increase the soil's dry density. Proper compacting:

- Minimizes settling
- Increases load-bearing characteristics
- Increases soil stability
- Reduces water penetration

Proper compacting will usually prevent slab and pavement cracking caused by differential settlement of the soil. It will also reduce settling of backfill placed against foundation and retaining walls and in other locations to prevent low spots where water will pond.

Compaction requirements generally are specified as a percentage of the maximum dry density obtainable in the test procedures in ASTM D1557 or D698. The relative density can be specified for fills constructed with free-draining sand and gravel soils (GW, GP, SW, SP) in accordance with ASTM D4253 and D4254. Varying degrees of compaction for the many different types of soil that may be encountered cannot be stated without engineering analysis. Generally, however, clean sand fill (SW, SP) should be compacted to 95% of the optimum dry density and all other fine-grained, nonexpansive fills to 90% of the optimum dry density in accordance with ASTM D1557. Compacting requirements for expansive soils should be determined from engineering analysis and may require somewhat less than 90% of optimum dry density.

Using hand or power equipment (Fig. 21.3-14), compacting of fill material should be performed in 4 to 8 in. (100 to 200 mm) layers (*lifts*) of soil. Regardless of the type of soil being compacted, the thinner the lift, the better the compaction.

Structural slabs (Type IV) which receive their support at the edges and through the center by intermediate foundations require only sufficient compacting beneath the slab portion to provide temporary support of the dead load imposed by the slab until the slab develops the necessary strength to span between its foundations.

Backfill adjacent to foundation walls should be uniformly compacted to at least 90% of optimum dry density unless other requirements dictate a greater degree of compacting. For example, backfill beneath walks, steps, and pavements should

a

b

**FIGURE 21.3-14** (a) Soil compacting at most building sites is done using power equipment. (b) Hand tamping is occasionally used in locations inaccessible to power-driven equipment. Hand tamping may also be used exclusively in small residential buildings and other small buildings when the soil is easily compactible.

be compacted to at least 95% of optimum dry density.

## 21.3.8 SLOPE PROTECTION AND EROSION CONTROL

Exposed earth on a construction site should be protected from erosion; indeed, most jurisdictions have codes, ordinances, or rules that dictate such controls. These controls may be as simple as the staking in place of hay bales to prevent eroded materials from fouling waterways, or they may be more elaborate systems with filter fabrics or riprap. In addition, providing diversion ditches (Fig. 21.3-15) and terracing of slopes (Fig. 21.3-16) may be needed to reduce the speed of water running down a slope and direct it to drainage channels protected with filter fabrics or riprap.

*Filter fabrics* are open woven mats of nylon or other synthetic fibers. They are staked down over a slope to permit water to flow without eroding the underlying soil (Fig. 21.3-17). Some of these products can be left in place and later covered with topsoil to receive lawns or other vegetation.

*Riprap* is a layer of stone or broken concrete designed to form a pathway for the flow of water and to prevent erosion

**FIGURE 21.3-18** Riprap can be either temporary or permanent. It can be used to cover a slope, as shown here, or to line drainage channels or ditches. It is also used to protect roadway embankments and stream beds. (Drawing by HLS.)

WINGED REPRODUCTIVE TERMITE
Thick waist, equal wings

ANT
Thin waist, unequal wings

**FIGURE 21.3-20** Ants are often mistaken for termites. Subterranean termites may be identified by their thick waists and wings of equal length. Ants are thin-wasted and have unequal-length wings.

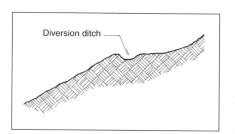

**FIGURE 21.3-15** A diversion ditch at the top of a slope will slow down the flow of groundwater and can be used to direct this flow to a paved outfall or into an interceptor drain that will conduct it to the lower level through pipes. (Drawing by HLS.)

**FIGURE 21.3-16** Terraces cut into a slope will reduce the velocity of groundwater and reduce erosion. (Drawing by HLS.)

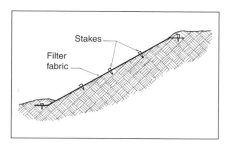

**FIGURE 21.3-17** Filter fabric installed to protect a slope. (Drawing by HLS.)

of underlying soil (Fig. 21.3-18). Riprap consists of pieces large enough to resist their movement down the slope by the force of water. Riprap is sometimes used as a drainage channel and sometimes as protection for an entire slope.

## 21.3.9 TERMITE CONTROL

Two types of termites cause damage to wood in the United States: subterranean and nonsubterranean. As Figure 21.3-19 shows, the hazard is greater in areas having mild temperatures.

Subterranean termites account for about 95% of all termite damage in the United States. These termites (Fig. 21.3-20) live in nests and develop their

colonies underground, but build tunnels of earth through the air to reach wood and other cellulose material such as fiberboard, fabrics, and paper. Termites must have a constant source of moisture or they die. They become most numerous in moist, warm soil containing an abundant supply of food.

Subterranean termites are not carried into buildings in lumber but establish themselves by entering from ground nests

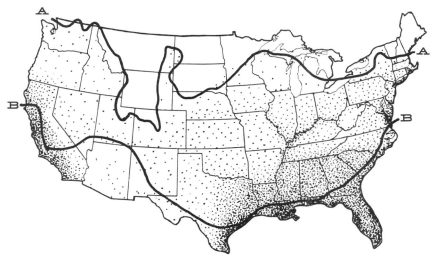

**FIGURE 21.3-19** Relative hazard of termite infestation. Subterranean termites have been found as far north as line A-A. The northern limit of nonsubterranean termites is indicated by line B-B.

after the building has been constructed using three basic routes: (1) attacking wood in direct contact with the ground (porch stairs, trellises, and siding installed too close to grade); (2) entering through cracks and voids as small as $1/32$ in. (0.8 mm) in slabs, piers, or foundation walls; or (3) building mud tubes over materials they cannot go through (Figs. 21.3-21, 21.3-22, and 21.3-23). During certain seasons, winged forms (see Fig. 21.3-20) swarm (fly) from the nest and, if successful, develop new colonies.

The control of subterranean termites may consist of one of the following protective methods or more than one method used in combination: (1) properly installed metal shields that isolate termites from wood or force them to build tubes around the shield; (2) chemically treated soil, which forms a barrier through which termites cannot pass to reach the construc-

FIGURE 21.3-23 The tubes shown here have been built from the wood above down to the ground. Tubes can be built up foundation walls or posts many feet from ground to wood.

FIGURE 21.3-21 Subterranean termite infestation of foundation plates. Improper clearance between wood siding (removed for photo) and ground permitted termite entry.

FIGURE 21.3-22 Termites build mud tubes to permit access from ground to wood. A tube is shown here on a concrete foundation pier. Note that the tube has been built around the metal shield at the top of the pier.

tion; (3) chemically treated, pressure-impregnated wood, which is not subject to termite attack (see Section 6.1); (4) the use of poured concrete foundations, provided that no cracks greater than $1/64$ in. (0.38 mm) are present; or (5) the use of poured reinforced concrete caps at least 4 in. (100 mm) thick on unit masonry foundations, provided that no cracks greater than $1/64$ in. (0.38 mm) occur.

Regardless of the protective method used, its design and construction should permit visual inspection wherever possible.

*Nonsubterranean termites* have been found only south of line "B-B" in Figure 21.3-19. Their total damage is much less than that of subterranean termites, but they may present a problem in the areas where they do occur. Also called *drywood termites* because of their ability to live in damp or dry wood without moisture or contact with the ground, they often are recognized by slightly compressed pellets of partially digested wood pushed through surface openings. Where this type is found, lumber should be inspected to see that it is not infested before arrival at the site. When construction is underway during the swarming season, lumber should be inspected for infestation.

Slab-on-grade construction requires special considerations in areas where the subterranean termite hazard is a significant problem (see Fig. 21.3-19). Slabs are expected to crack in varying degrees, and termites can penetrate through cracks as small as $1/32$ in. (0.8 mm), through joints between the slab and the foundation wall, through control joints, and through openings made for plumbing and conduit.

Some degree of protection of wood structural elements is provided by positive site and foundation drainage and by separation of wood elements from the ground. Foundations and piers of hollow masonry units should have solid masonry caps in which all joints or voids are filled with mortar. Unless the foundation fill is of limited capillarity, at least a 4 in. (100 mm) base course of limited capillarity and a suitable vapor retarder should be provided under the slab.

Where local experience indicates that additional protection against termites is necessary, termite barriers should be provided. Additional protection can be afforded by using pressure-treated wood which will resist decay as well. See Sections 6.1.5 and 6.5.2.

Soil poisoning is the most effective termite barrier. The area under and around a building should be treated, as well as all possible points of entry. The termiticides registered with the U.S. Environmental Protection Agency (EPA) and currently most commonly used are chlorpyrifos, cypermethrin, fenvalerate, isofenphos, and permethrin. The recommended concentrations of these chemicals in treatment solutions range from 0.25 to 1%. The actual concentration, the application of the chemicals, and precautions involved in their use should be in accordance with the manufacturer's recommendations, with U.S. Department of Agriculture Home and Garden Bulletin 64, "Subterranean Termites, Their Prevention and Control in Buildings," and with applicable building codes.

The effective performance of termite shields depends on proper construction and workmanship and by long-term integrity of the slab without cracking, which is unpredictable. Therefore, the effectiveness of termite shields is questionable.

# 21.4 Surface Water and Groundwater Problems

Provisions must be made to prevent surface water and groundwater from entering a building through the walls or floors. If water penetrates basement walls, the habitable space there may become undesirable. Water in a crawl space or other enclosed space may result in decay, musty odors, mold growth, and corrosion of some metals. Water below an improperly constructed slab on grade can migrate through it to cause delamination of floors and can support condensation and frost accumulation at the perimeter of associated walls. In some soils, groundwater can rise high enough to fill heating ducts embedded in or buried beneath a concrete slab.

In addition to the measures discussed here, foundation walls enclosing basement spaces usually require additional protection by applications of *parging* and *dampproofing* on the exterior surface below grade (see Sections 4.7.3 and 7.2.1). When water under pressure (*head*) occurs above a footing, *waterproofing* may be necessary (see Section 7.2.2). Foundation drains and underfloor drains are often used to carry off groundwater that occurs beneath a building. Such subdrainage systems are discussed in Section 23.3.

## 21.4.1 SURFACE WATER

Surface water is water that runs along the surface of the ground as a result of rain, downspout discharge, melting snow, or another source. Most surface water problems come from improper grading of the site adjacent to a building.

## 21.4.2 GROUNDWATER

Groundwater is water that is in the ground most of the time, although its level (*water table*) may fluctuate during dry and rainy seasons. Most groundwater problems result from locating foundations too close to the subsurface water table. Problems then occur when (1) the level of the water in the ground temporarily rises due to a seasonal increase in precipitation or (2) the water in the ground rises due to the capillarity of the soil. When the water table is so high that a foundation would be submerged constantly, either the site should be abandoned or more extensive waterproofing methods must be used to make the foundation watertight (see Section 7.2).

### 21.4.2.1 Water Table Rise

Problems resulting from seasonal fluctuations in precipitation that cause the wa-

ter table to rise temporarily above the level of the foundation may be solved by subsurface drainage. The most effective means of resisting water infiltration through foundations is to collect the water and discharge it away from the site before it reaches the foundation.

### 21.4.2.2 Capillary Rise

Water will rise in certain soils in opposition to gravity due to capillary action. This *capillarity* occurs when the molecular attraction of water molecules to adjacent soil molecules is greater than the molecular attraction of water molecules to each other. This causes water in a small space (capillary tube), such as the voids between soil grains, to spread out upward on the sides of the space (soil grain surfaces), forming a concave water surface (Fig. 21.4-1). The water's surface tension, which always tries to reduce the surface area, pulls the top of the water upward following the edges. The capacity of the water to resist forces tending to pull it apart holds the tube of water together as it creeps upward.

In general, capillarity increases with a decrease in the size of the space between the grains in a soil. In fact, moisture can move upward through the voids within some clays as much as 11 ft (3.35 m) above the water table (Fig. 21.4-2). However, because of the large spaces between soil particles, capillarity does not occur in coarse granular soil types, such as gravel. Therefore, one method of control is to replace soil having high capillarity with granular materials.

Water vapor will also rise through the voids between soil particles, whether or

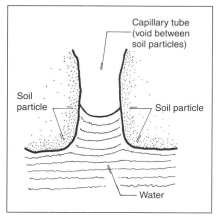

**FIGURE 21.4-1** Water rises in the voids between soil grains in some soils where the grains are tightly packed. (Drawing by HLS.)

| Capillary Rise | Soil Type | Saturation Zone |
|---|---|---|
| Greater Than 8′ (2438 mm) | CLAY | Greater Than 5′ (1524 mm) |
| Greater Than 8′ (2438 mm) | SILT | Greater Than 5′ (1524 mm) |
| 3′ to 8′ (914 mm to 2438 mm) | FINE SAND | 1′ to 5′ (305 mm to 1524 mm) |
| 1′ to 3′ (305 mm to 914 mm) | COARSE SAND | 0 to 1′ (0 to 305 mm) |
| 0 | GRAVEL | 0 |

**FIGURE 21.4-2** Water rises in soils by capillary action. Clays and silts may become fully saturated to almost 6 ft (1800 mm) above a water table, and some water may rise more than 11 ft (3353 mm). Note that even coarse sand may permit a rise of up to 3 ft (914 mm). Coarse gravel prevents the rise of water by capillarity.

not granular soil materials are used. This rise of water vapor can be interrupted and prevented from entering a crawl space or a building by covering the ground surface with a vapor retarder (see Fig. 23.3-1). Refer to Section 7.1.3.2 for a discussion of vapor retarders in crawl spaces.

### 21.4.3 CONTROL OF WATER PROBLEMS

Groundwater and surface water can cause problems both during construction and in a completed building. Means of controlling surface water in the completed project by grading are discussed in Section 21.3.6. This section covers removal of groundwater both during construction and from the completed building.

#### 21.4.3.1 Dewatering

*Dewatering* is the prevention of water from entering an excavation, when possible, and the removal of water that does find its way in. Groundwater can be prevented from entering an excavation by means of berms and drainage channels or ditches.

Water that enters an excavation as rain, snow, or other precipitation can often be removed by means of portable pumps. Such water should be conducted to pump locations by depressions or ditches. Foundation trenches should not be used as drainage ditches, however, because the water will erode or otherwise damage the bearing surface at the bottom of the trench. When the water table is higher than the bottom of an excavation,

it is necessary to continuously lower the level of the water while the work is going on. The most common way of doing this is to sink a series of pipes with screened ends (*well points*) around the perimeter of the excavation to a level below the foundations. These well points may each be fitted with a pump, but more often they are interconnected by *suction lines* (pipes to one or more larger pumps that empty all of them simultaneously). In very deep excavations, two rows of well points may be needed. Well points and pumps lower the water table and keep it lower until pumping stops. The *discharge* from such wells should be conducted to storage ponds or positive outfalls that lead to a stream or storm sewer, depending on which is permitted or required in the jurisdiction of the project.

#### 21.4.3.2 Slabs on Grade

Settlement or differential movement of slabs and foundations caused by moisture fluctuations in problem soils is discussed in Section 21.1.7. Additional provisions for moisture control are necessary for the long-term performance of concrete slab-on-grade construction. Slabs should be protected from water and water vapor to prevent damage to finish flooring materials and pipes or heating ducts embedded in the slab, as well as to guard against reduced effectiveness of thermal insulation. Problems may be caused by either surface water or groundwater.

The methods and extent of protection depend on:

- Slab elevation with respect to the elevation and slope of the finished grade
- Elevation of the groundwater table or man-made water sources, such as seepage fields
- Drainage properties of the soil or fill beneath the slab
- Type of finish flooring

Many of the moisture problems associated with slab-on-grade construction can be minimized or eliminated by proper preliminary grading, correct selection of fill or base course materials, and installation of a vapor retarder.

Slab base courses are discussed in Section 21.3.5; water control by grading in Section 21.3.6.

#### SLAB ELEVATION

The tops of slabs should be located at least 8 in. (200 mm) above the exterior finished grade level. The bottoms of heating ducts placed in or under a slab should be at least 2 in. (50.8 mm) above the adjacent exterior finished grade, unless the ducts are concrete, plastic, mineral-fiber-cement, or ceramic.

#### VAPOR RETARDER

Where no drainage or soil problems exist and in regions where irrigation and heavy sprinkling are not done, a vapor retarder may not be necessary. However, because it is impossible to correct moisture problems after construction and because a vapor retarder usually can be installed at nominal cost during construction, a vapor retarder is recommended for all slab-on-grade construction when impermeable floor coverings will be used and when floor coverings, furnishings, or equipment must be protected from damage by moisture conditions (see Fig. 21.3-13). A vapor retarder membrane should have a permeableness of less than 0.30 perm (17.2 ng/[Pa·s·m²]) and should resist deterioration as well as puncture from heavy traffic during construction. A vapor retarder should be installed with a 6 in. (150 mm) lap and fitted tightly around utility and other service openings.

Vapor retarder materials suitable for slab-on-grade construction include single-layer membranes such as polyethylene sheet or multiple-layer membranes such as glass fiber reinforced waterproof paper with polyethylene film extrusion-coated on both sides. Polyethylene sheet used over sand or firmly compacted soil under unreinforced slabs should be at least 4 mils (0.1 mm) in nominal thickness; when used over gravel or crushed stone or under any structurally reinforced slab, it should be at least 6 mils (0.15 mm) in nominal thickness.

Except when a high-strength concrete with a lower than normal slump is used, a vapor retarder should not be placed directly beneath a slab. Instead, a 2 to 3 in. (50.8 to 76 mm) layer of sand should be placed over the vapor retarder. This permits water that bleeds from a high-slump or low-strength concrete to pass into the sand rather than migrating upward through the slab, where it will cause the slab to crack more than is usual and weaken the surface, resulting in dusting and spalling.

**FIGURE 21.5-1** Point bearing piles reach to strong soils or rock. Friction piles depend on friction between the piles and the soil for their support.

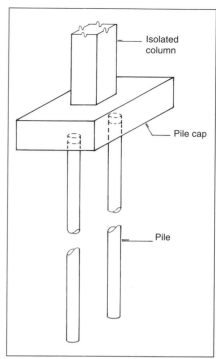

**FIGURE 21.5-2** A cluster of two piles with an isolated pile cap. Similar caps are also used when the supported element is a grade beam instead of a column. Caps may be either square, rectangular, or multisided, depending on the number of piles and the size of the cap. (Drawing by HLS.)

# 21.5 Special Foundations and Load-Bearing Elements

The special foundation systems and load-bearing elements discussed in this section include piles and cassions. Other types of foundation systems, and foundation system components such as grade beams, walls, piers, columns, pilasters, and the turned-down portion of a combination footing and floor system, are covered in Section 3.11. The principles discussed in that section apply to pile and caissons, so far as they are applicable.

## 21.5.1 PILES AND CAISSONS

Piles and caissons are constructed of concrete, masonry, steel, wood, and combinations of these. They are used as foundation support for buildings when the soil is not capable of supporting the loads that will be imposed by a building using spread foundations.

### 21.5.1.1 Driven Piles

*Driven piles* are column-like units that transmit loads through poor soil to rock or lower levels of soil that have adequate bearing capacity (Fig. 21.5-1). In some regions, short piles are called *piers*. Piles serve the same purpose as footings in that they transmit loads to a subbase stratum capable of carrying the load. Piles are usually placed in clusters of two or more spaced 30 to 48 in. (762 to 1 220 mm) on

centers. This arrangement permits them to act together and produces a higher load-carrying capacity than can be achieved with isolated piles. The load capacities mentioned in this section are based on the piles being distributed in clusters. These load capacities are also the maximum permitted. Optimum load capacities are somewhat lower in each instance. Piles receive building loads from isolated columns and from grade beams by means of reinforced concrete pile caps (Fig. 21.5-2). Pile caps are sometimes simply a widened section of a grade beam (Fig. 21.5-3).

Pile foundations are either (1) *point-bearing* types, which transmit loads to lower, stronger soil or rock through their points, or (2) *friction* types, which develop the necessary bearing capacity through surface friction between the pile and the ground (see Fig. 21.5-1). Piles are driven with heavy *hammers* in large machines called *pile drivers*. *Drop hammers* are simply raised and dropped on a pile by the force of gravity. *Differential-acting steam hammers* are rammed into the top of a pile by steam pressure or compressed air. Modern vibratory hammers and diesel-driven hammers are also being used today.

*Friction piles* are driven to a predetermined depth or resistance based on soil boring analysis and field tests. To verify the design analysis, test piles are usually driven and loaded before the remainder of the piles are driven. Point-bearing piles are driven until additional blows of the hammer produce very little movement in the pile (*refusal*).

Piles may be made of wood, concrete, steel, or a combination (composite) of

these. Timber piles have been used for at least 2000 years and probably longer. Some below-water timber piles beneath bridges in Europe have remained in continuous service for more than 1000 years. They are, however, suitable only for relatively light loads (40 tons [390 594 kg/m²] maximum) and must be preservative treated when they extend above the water table. They are also limited in length to the effective height of the tree from which they are cut (45 to 65 ft [13.7 to 19.8 m]), because they cannot be spliced.

Steel piles may be either H-shaped sections or pipes. H-shaped piles are heavy wide-flange sections varying in size from 8 to 14 in. (203 to 356 mm) in both depth and flange width. They can carry loads of 50 to 200 tons (488 243 to 1 952 972 kg/m²) each and may be as much as 150 ft (45.72 m) long. To produce such lengths, sections are welded together as they are being driven.

Steel-pipe piles are later filled with concrete. They are available in several

Grade beam

Pile cap

Pile

**FIGURE 21.5-3** Pile caps formed integrally with a grade beam are sometimes used when the load distributing element is a bearing wall. (Drawing by HLS.)

**FIGURE 21.5-4** Steel pile shapes. (Drawing by HLS.)

**FIGURE 21.5-5** Round and fluted pile shapes. Fluted steel piles are usually tapered. (Drawing by HLS.)

types and shapes (Fig. 21.5-4). They may be heavy-walled types that can be driven directly or thinner-walled types that require a tight-fitting, heavy inner lining (*mandrel*) that is withdrawn before the concrete is placed. They may be smooth-walled or corrugated and round or fluted (Fig. 21.5-5). Sizes range from 8 to 24 in. (203 to 610 mm) in diameter, and maximum load capacity ranges from 75 to 200 tons (732 365 to 1 952 972 kg/m²), depending on the type. Pipes with closed ends may be driven as deep as 120 ft (36.6 m). Other pipe piles are limited to about 80 ft (24.4 m).

Precast concrete piles are either solid concrete or open cylinders that are later filled solid with concrete. They may be square, round, or octagonal in cross section. All precast concrete piles are reinforced. Most are also prestressed. Solid precast concrete piles with simple reinforcement can be up to 80 ft (24.38 m) long. This increases to 150 ft (45.72 m) for

cylinder piles and up to 200 ft (60.96 m) when they are prestressed. Maximum capacity ranges from 100 tons (976 486 kg/m²) for simple reinforced piles to as much as 500 tons (4 882 430 kg/m²) for prestressed cylinder piles. Sizes vary from 10 to 54 in. (254 to 1372 mm) across.

Composite piles are constructed in several configurations. Some have timber or concrete-filled steel pipe lower sections and concrete-filled shell with mandrel upper sections. Others have H-shaped steel lower sections and precast concrete upper sections. Composite piles sometimes permit less expensive applications. For example, the combination of timber and concrete-filled steel shell permits the use of relatively inexpensive timber piles at much lower depths than is ordinarily possible (up to 150 feet [45.72 m]). A disadvantage of using composite piles is that their load-carrying potential is limited to that of the lower of the two elements. For example, a composite pile of timber and concrete-

filled steel shell is limited to 40 tons (39 0594 kg/m²) instead of the 75 to 80 tons (732 365 to 781 189 kg/m²) the concrete-filled steel piles can carry.

### 21.5.1.2 Caissons

*Caisson* are concrete columns placed in auger-drilled or excavated holes. They serve the same purpose as piles. While some caissons are only 6 to 12 in. (152 to 305 mm) in diameter, most are much larger and are capable of carrying much heavier loads. They can reach as much as 6 ft (1.8 m) in diameter and carry 3500 tons (34 177 010 kg/m²) or more. Caissons extend through unsatisfactory soil to a firm bearing soil stratum or to bedrock. As a caisson hole is drilled or dug, a steel casing is lowered into it to keep the hole from caving in. This casing is raised and removed as the concrete is placed. Small, relatively short, lightly loaded caissons, sometimes called *piers*, are frequently poured into cardboard or composition casings that are not removed as the concrete is placed.

Caissons are generally one of four types: (1) *rock caissons* simply rest directly on solid rock; (2) *high-capacity*

**FIGURE 21.5-6** Typical caisson configurations. (Drawing by HLS.)

*rock caissons* rest in a socket cut into rock, thus supporting the load both by resting on the rock and by friction against the sides of the pocket—these are sometimes called *socket caissons*; (3) *hardpan* or *clay caissons* rest on a soil-bearing stratum; and (4) *friction caissons* are supported by the friction acting on the sides of the caisson, much as friction piles are supported. In fact they are sometimes called *cast-in-place piles*.

Hardpan or clay caissons have either straight shafts or an enlarged base called a *bell* (Fig. 21.5-6). Bells are produced by hand excavation or with a special device (*belling bucket*) on the auger.

# 21.6 Additional Reading

More information about the subjects discussed in this chapter can be found in the references listed in Section 21.7 and in the following publications.

Ambrose, James. 1996. *Simplified Design of Concrete Structures*, 7th ed. Hoboken, NJ: John Wiley & Sons, Inc. (www.wiley.com).

Bowles, Joseph E. 1996. *Foundation Analysis and Design*, 4th ed. New York: McGraw-Hill (books.mcgraw-hill.com).

———. 1992. *Engineering Properties of Soils and Their Measurement*, 4th ed. New York: McGraw-Hill (books.mcgraw-hill.com).

Budhu, Muni. 1999. *Soil Mechanics and Foundations*. Hoboken, NJ: John Wiley & Sons, Inc. (www.wiley.com).

Cernica, John N. 1994. *Geotechnical Engineering: Foundation Design*. Hoboken, NJ: John Wiley & Sons, Inc. (www.wiley.com).

Duncan, J. Michael and Stephen G. Wright. 2005. *Soil Strength and Slope Stability*. Hoboken, NJ: John Wiley & Sons, Inc. (www.wiley.com).

Marriott, Betty Bowers. 1997. *Environmental Impact Assessment: A Practical Guide*. New York: McGraw-Hill (books.mcgraw-hill.com).

Monahan, Edward J. 1993. *Construction of Fills*. 2nd ed. Hoboken, NJ: John Wiley & Sons, Inc. (www.wiley.com).

Powers, J. Patrick. 1992. *Construction Dewatering: New Methods and Applications*, 2nd ed. Hoboken, NJ: John Wiley & Sons, Inc. (www.wiley.com).

# 21.7 Acknowledgments and References

## ACKNOWLEDGMENTS

We gratefully acknowledge the assistance of the following organizations and individuals in preparing this chapter. We are also indebted to them for permission to use their illustrations when requested and for the use of their publications as references.

American Concrete Institute (ACI) (www.concrete.org)

American Wood Preservers Association (AWPA) (www.awpa.com)

ASTM International (ASTM) (www.astm.org)

U.S. Department of Commerce, National Weather Service (www.nws.noaa.gov)

## REFERENCES

We would also like to thank the authors and publishers of the publications in the following list for their contribution to our research for this chapter.

American Concrete Institute (ACI). Farmington Hills, MI (www.concrete.org). 2000. ACI 543R-00, "Design, Manufacture, and Installation of Concrete Piles."

———. 2001. ACI 3361R-01, "Reference Specification for the Construction of Drilled Piers."

American Wood Preservers Association (AWPA). 2003. *Book of Standards*. Selma, AL: AWPA (www.awpa.com).

*Architectural Graphic Standards. See* Ramsey et al. (2000).

ASTM International (ASTM). West Conshocken, PA (www.astm.org). Standards:

D427, "Test Method for Shrinkage Factors of Soils by the Mercury Method."

D698, "Standard Test Method for Laboratory Compaction Characteristics of Soil Using Standard Effort (12,400 ft-lbf/ft$^3$ (600kN-m/m$^3$))."

D1143, "Standard Test Method for Piles Under Static Axial Compressive Load."

D1557, "Test Method for Laboratory Compaction Characteristics of Soil Using Modified Effort (56,000 ft-lbf/ft$^3$ (2,700 kN-m/m$^3$))."

D1586, "Standard Test Method for Penetration Test and Split-Barrel Sampling of Soils."

D1883, "Standard Test Method for CBR (California Bearing Ratio) of Laboratory Compacted Soils."

D2487, "Standard Classification of Soils for Engineering Purposes (Unified Soil Classification System)."

D2488, "Standard Practice for Description and Identification of Soils (Visual-Manual Procedure)."

D4253, "Standard Test Methods for Maximum Index Density and Unit Weight of Soils Using a Vibratory Table."

D4254, "Standard Test Method for Minimum Index Density and Unit Weight of Soils and Calculation of Relative Density."

D4318, "Standard Test Method for Liquid Limit, Plastic Limit, and Plasticity Index of Soils."

National Institute of Building Sciences (NIBS). Catalog No. 4010-1, *Wood Protection Guidelines: Protecting Wood from Decay Fungi and Termites*. Washington, DC: NIBS (www.nibs.org).

Ramsey, Charles George, Harold Reeve Sleeper, and John Ray Hoke, Jr. 2000. *Architectural Graphic Standards*, 10th ed. Hoboken, NJ: John Wiley & Sons, Inc. (www.wiley.com).

Underground Space Center. 1988. *Building Foundation Design Handbook*. Minneapolis: Underground Space Center.

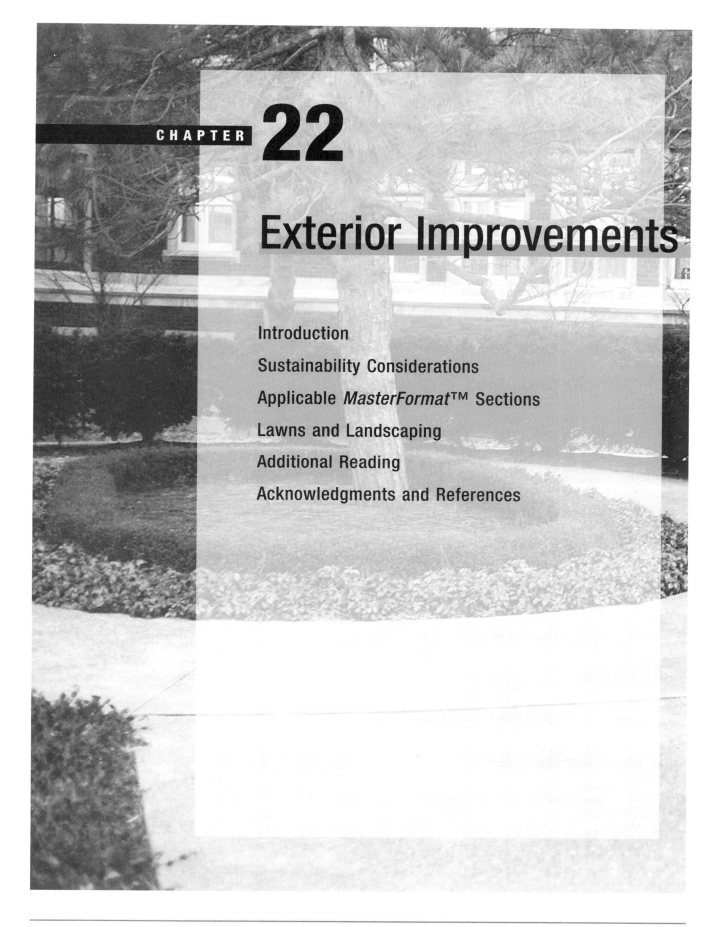

CHAPTER **22**

# Exterior Improvements

# Introduction

Site construction includes work to be done on the grounds, accessways, parking areas, lawns, plantings, and other site features. These have important esthetic and site-use implications, and greatly affect site and building access.

Lawns and plantings are addressed in this section.

The handling of conditions that exist at a construction site before the work under a construction contract begins are addressed in Chapter 2. Below-grade site construction activities that take place after construction begins are covered in Chapter 21. Water supply sources, sanitary drainage utilities, storm drainage utilities, and electrical distribution to the building are covered in Chapter 23.

# Sustainability Considerations

Besides their obvious esthetic implications, pavement and the lawns and plantings around a building can alter the climate immediately at the building and somewhat control the effects of sun and wind on the building. These aspects can increase the comfort of the building occupants and reduce not only the size of its heating and air conditioning systems, but their operating costs as well.

Refer to Section 22.1 for additional sustainability considerations related to lawns and plantings.

# Applicable *MasterFormat* Sections

Subjects addressed in this chapter are included in *MasterFormat* 2004, Division 32, "Exterior Improvements." The following *MasterFormat* Level 2 sections are applicable to this chapter:

32 10 00  Bases, Ballasts, and Paving
32 30 00  Site Improvements
32 70 00  Wetlands
32 80 00  Irrigations
32 90 00  Planting

32 91 00  Planting Preparation
32 92 00  Turf and Grasses
32 93 00  Plants
32 94 00  Planting Accessories

# 22.1 Lawns and Landscaping

Most people think first of the esthetic appeal of lawns and landscaping, but these also serve as buffer zones between adjacent buildings and between buildings and roadways or other property. Properly placed plants can also help keep a building cooler in summer and warmer in winter, thereby reducing dependence on heating and air conditioning equipment.

## 22.1.1 PLANNING REQUIREMENTS

Plantings can reduce reflection and radiation of solar energy from paved areas and adjacent buildings (Fig. 22.1-1). They can also shield exterior walls from the sun, thus reducing heat transmission through them. Earth berms can also shield walls and outdoor living spaces.

Properly located *coniferous* (evergreen) trees can block prevailing winter winds. *Deciduous* (leaf-dropping) trees on southern and western exposures shade walls and windows in summer but drop their leaves to allow desirable winter sun penetration (Fig. 22.1-2).

Ponds and fountains act as natural coolants to reduce the air temperature around building exteriors (Fig. 22.1-3) and should not be blocked by vegetation.

Plants must be selected and placed carefully for several reasons. For example, placing deciduous trees on northern exposures can leave a building unprotected from cold winter winds. In addition, plants that hide a building can also hide an intruder. Trees should be far enough away from a building that a burglar cannot climb them and enter the second story. Bushes should be low enough not to obscure windows or doors. Hedges used as fences should be no more than 40 in. (1016 mm) high. Landscaping should not obscure a building from the street or parking lot. Small plants and floral borders can serve to define the borders between private and public spaces.

Refer to Section 17.2.2 for additional site-planning recommendations related to thermal considerations.

## 22.1.2 TREES AND SHRUBS

Tree and shrub quality should be based on the American Nursery Landscape Association's (ANLA) *American Standard for Nursery Stock*. This publication

**FIGURE 22.1-1** Shading paved areas prevents heat buildup and channels breezes.

WINTER: LEAFLESS TREES LET SUN THROUGH.

SUMMER: LEAVES SHADE HOUSE.

WINTER: EVERGREENS PROTECT HOUSE.

SUMMER: LEAVES DIRECT BREEZE INTO HOUSE.

**FIGURE 22.1-2** Proper trees in the right places will reduce energy needs.

**FIGURE 22.1-3** Air flowing over cool water acts as natural air conditioning. Plants should not block this flow.

contains a standarized system of sizing and describing plants. It covers shade and flowering trees, deciduous shrubs, coniferous evergreens, broadleaf evergreens, rose shrubs, young plants, fruit trees, bulbs, corms, and tubers. It describes and limits the dimensions of container plants, balled and burlapped plants, bareroot plants, balled and potted plants, and processed balled plants. It also contains the names of trees and shrubs that fit into each of its categories.

The U.S. Department of Agriculture (USDA) has identified 11 zones of plant hardiness in the United States. Figure 22.1-4 is an adaptation of the USDA Plant

Hardiness Zone Map in USDA Agricultural Research Service Miscellaneous Publication Number 1475. USDA has also identified many plants as they relate to the Plant Hardiness Zone Map. Figure 22.1-5 contains a list of 10 of the 11 zones and examples of plants that will succeed in each of them. (Zone 11 includes areas where the average annual temperature is above 40°F [4.4°C], which are therefore essentially frost free.) Figure 22.1-6 is a list of additional plants and the zones in which they will survive. The zones listed for each plant are optimum. Listed plants may actually survive in warmer or colder climates, but their performance will probably not be as satisfactory.

Other factors that affect plant growth and hardiness include the amount of rainfall and the soil and its amendments. Plants can be grown in areas that have less than optimum rainfall for them, but they must be watered artificially in those locations.

Before selecting plants for use in a particular area, it is advisable to verify their use with a local horticulturist or

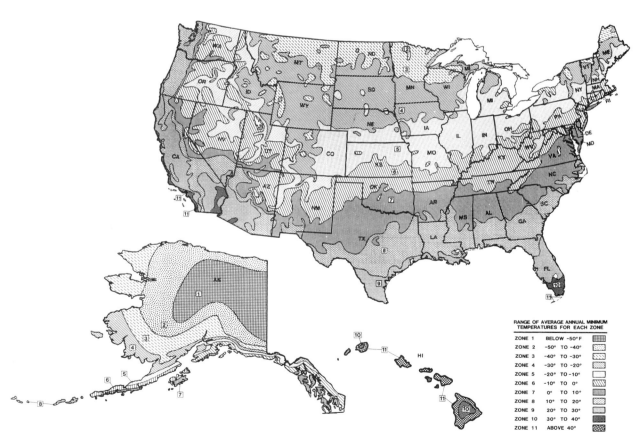

**FIGURE 22.1-4 USDA Plant Hardiness Zone Map.** (From U.S. Department of Agriculture, Agricultural Research Service Miscellaneous Publication 1475, "USDA Plant Hardiness Zone Map.")

**FIGURE 22.1-5** Indicator Plant Examples

| Zone | Botanical Name | Common Name |
|---|---|---|
| **1** <br> °F Below −50 <br> °C Below −45.6 | Betula glandulosa <br> Empetrum nigrum <br> Populus tremuloides <br> Potentilla pensylvanica <br> Rhododendron lapponicum <br> Salix reticulata | Dwarf birch <br> Crowberry <br> Quaking aspen <br> Pennsylvania cinquefoil <br> Lapland rhododendron <br> Netleaf willow |
| **2** <br> °F −50 to −40 <br> °C −45.6 to −40 | Betula papyrifera <br> Cornus canadensis <br> Elaeagnus commutata <br> Larix laricina <br> Potentilla fruticosa <br> Vibrunum trilobum | Paper birch <br> Bunchberry dogwood <br> Silverberry <br> Eastern larch <br> Bush cinquefoil <br> American cranberry bush |
| **3** <br> °F −40 to −30 <br> °C −40 to −34.5 | Berberis thunbergii <br> Elaeagnus angustifolia <br> Junipercus communis <br> Lonicera tatarica <br> Malus baccata <br> Thuja occidentalis | Japanese bayberry <br> Russian olive <br> Common juniper <br> Tatarian honeysuckle <br> Siberian crabapple <br> American arborvitae |
| **4** <br> °F −30 to −20 <br> °C −34.5 to −28.9 | Acer saccharum <br> Hydrangea paniculata <br> Juniperus chinensis <br> Ligustrum amurense <br> Parthenocissus quinquefolia <br> Spiraea x vanhouttei | Sugar maple <br> Panicle hydrangea <br> Chinese juniper <br> Amur River privet <br> Virginia creeper <br> Vanhoutte spirea |
| **5** <br> °F −20 to −10 <br> °C −28.9 to −23.3 | Cornus florida <br> Deutzia gracilis <br> Ligustrum vulgare <br> Parthenocissus tricuspidata <br> Rosa multiflora <br> Taxus cuspidata | Flowering dogwood <br> Slender deutzia <br> Common privet <br> Boston ivy <br> Japanese rose <br> Japanese yew |
| **6** <br> °F −10 to 0 <br> °C −23.3 to −17.8 | Acer palmatum <br> Buxus sempervirens <br> Euonymus fortunei <br> Hedera helix <br> Ilex opaca <br> Ligustrum ovalifolium | Japanese maple <br> Common box <br> Winter creeper <br> English ivy <br> American holly <br> California privet |
| **7** <br> °F 0 to 10 <br> °C −17.8 to −12.3 | Acer macrophyllum <br> Rhododendron Kurume hybrids <br> Cedrus atlantica <br> Cotoneaster microphylla <br> Ilex aquifolium <br> Taxus baccata | Bigleaf maple <br> Kurume azalea <br> Atlas cedar <br> Small-leaf cotoneaster <br> English holly <br> English yew |
| **8** <br> °F 10 to 20 <br> °C −12.3 to −6.6 | Arbutus unedo <br> Choisya ternata <br> Olearia haastii <br> Pittosporum tobira <br> Prunus laurocerasus <br> Viburnum tinus | Strawberry tree <br> Mexican orange <br> New Zealand daisy-bush <br> Japanese pittosporum <br> Cherry-laurel <br> Laurestinus |
| **9** <br> °F 20 to 30 <br> °C −6.6 to −1.1 | Asparagus setaceus <br> Eucalyptus globulus <br> Syzygium paniculatum <br> Fuchsia hybrids <br> Grevillea robusta <br> Schinus molle | Asparagus fern <br> Tasmanian blue gum <br> Australian bush cherry <br> Fuschia <br> Silk-oak <br> California pepper tree |
| **10** <br> °F 30 to 40 <br> °C −1.1 to 4.4 | Bougainvillea spectabilis <br> Cassia fistula <br> Eucalyptus citriodora <br> Ficus elastica <br> Ensete ventricosum <br> Roystonea regia | Bougainvillea <br> Golden shower <br> Lemon eucalyptus <br> Rubber plant <br> Ensete <br> Royal palm |

*Note:* Representative persistent plants listed under the coldest zones in which they normally succeed. Such plants may serve as useful indicators of the cultural possibilities of each zone.

(From U.S. Department of Agriculture, Agricultural Research Service Miscellaneous Publication 1475, "USDA Plant Hardiness Zone Map.")

nurseryman or with a state or local agricultural agent. In some states, agricultural extension services are good sources of such information.

### 22.1.3 LAWNS

There are some 600 grass genera, of which about 40 are suitable for lawns. However, not all of these should be used in every climatic zone. For example, bluegrass (*Poa pratensis*) grows very well in many areas but is often not successful in subtropical zones. St. Augustine grass (*Stenotaphrum secundatum*) needs a warm, moist environment and will not fare well in cold, dry zones.

Some grasses are more suitable for use in shaded areas than others, but none is really suitable for use in deep shade. Those that sometimes work in light shade include Bahia (*Paspalum notatum*), bent (*Agrostis*), carpet (*Axonopus offinis*), centipede (*Eremochloa ophiuroides*), fescue (*Festuca rubra*), rye (*Lolium perenne*), St. Augustine, and zoysia (*Zoysia matrella*). Those suitable for deeper shade include fescue, St. Augustine, and zoysia. Successful growth of a particular grass in shade depends on such other factors as the amount of moisture present or delivered artificially, drainage, soil fertility, and the type of vegetation providing the shade. All shading vegetation will draw some moisture from adjacent grass, but trees and shrubs with shallow roots may present more of a problem than deep-rooted ones. Local sources should be contacted for advice before selecting a grass for use in shaded areas.

Some species, such as bluegrass and fescue, propagate successfully from seed. Many southern species and some northern creeping species, however, should be established from living material such as sod, plugs, sprigs, or stolens. These include zoysia, St. Augustine, centipede, and many northern bent grasses. These methods should also be used, regardless of the grass selected, on slopes of 1:6 and steeper to prevent loss of seed through erosion and when an "instant lawn" is desired.

The name of the proper grass to use in a particular location should be obtained from a local horticulturist or nurseryman, or from a state or local agricultural agent. In general, the following grasses work well in cool, humid zones: bluegrass, both red and alta fescue, and rye. Grasses for dry areas include beach (*Ammophila*), buffalo (*Buchloe*), blue

**FIGURE 22.1-6** Cold Hardiness Ratings for Some Additional Plants

| Botanical and Common Name | Zone | Botanical and Common Name | Zone |
|---|---|---|---|
| *Abeliophyllum distichum* (white forsythia) | 5b | *Ilex crenata* "Convexa" (convexleaf Japanese holly) | 6b |
| *Acer platanoides* (Norway maple) | 4 | *Jacaranda acutifolia* (green ebony) | 10 |
| *Aesculus* x *carnea* (red horsechestnut) | 4 | *Juglans regia* (English or Persian walnut) | 6b |
| *Araucaria araucana* (monkeypuzzle) | 7b | *Juniperus horizontalis* (creeping juniper) | 3 |
| *Arctostaphylos uva-ursi* (bearberry) | 2b | | |
| *Aristolochia durior* (Dutchman's-pipe) | 4b | *Koelreuteria paniculata* (goldenrain-tree) | 6 |
| *Aucuba japonica* (Japanese aucuba) | 7b | | |
| | | *Laburnum* x *watereri* (Waterer laburnum) | 5b |
| *Bauhinia variegata* (purple orchid tree) | 9b | *Lagerstroemia indica* (crapemyrtle) | 7 |
| *Berberis darwinii* (Darwin barberry) | 8 | | |
| *Betula pendula* (European white birch) | 3 | *Mahonia aquifolium* (Oregon hollygrape) | 5b |
| *Bouvardia* "Coral" (Coral bouvardia) | 9 | *Malus* x *arnoldiana* (Arnold crabapple) | 4 |
| *Butia capitata* (Pindo palm) | 8b | *Melia azedarach* (chinaberry) | 7b |
| | | *Metasequoia glyptostroboides* (dawn redwood) | 5b |
| *Camellia reticulata* (*reticulata camellia*) | 9 | *Myrtus communis* (true myrtle) | 8b |
| *Camellia sasonqua* (*sasanqua camellia*) | 7b | | |
| *Carya illinoinensis* "Major" (pecan) | 5 (grows) / 6 (fruits) | *Nandina domestica* (heavenly bamboo) | 7 |
| | | *Nerium oleander* (oleander) | 8b |
| *Casuarina equisetifolia* (Australian pine) | 9b | | |
| *Ceanothus impressus* (Santa Barbara ceanothus) | 8 | *Olea europaea* (common olive) | 9 |
| *Cedrus dodara* (deodar cedar) | 7b | *Osmanthus heterophyllus* (holly osmanthus) | 7 |
| *Cercis chinensis* (Chinese redbug) | 6b | | |
| *Chamaecyparis lawsoniana* (Lawson cypress) | 6b | *Picea abies* (Norway spruce) | 3 |
| *Chamaecyparis pisifera* (Sawara cypress) | 5 | *Pieris japonica* (Japanese andromeda) | 6 |
| *Cinnamomum camphora* (camphor tree) | 9 | *Pinus mugo* var. *mughus* (Mugo pine) | 3 |
| *Cistus laurifolius* (laurel rockrose) | 7 | *Pinus radiata* (Monterey pine) | 7b |
| *Cistus* x *purpureus* (purple rockrose) | 8 | *Pinus strobus* (eastern white pine) | 3b |
| *Cronus alba* (Tatarian dogwood) | 3 | *Prunus yedoensis* (Yoshino cherry) | 6 |
| *Cronus kousa* (Japanese dogwood) | 5b | | |
| *Cunninghamia lanceolata* (cunninghamia) | 7 | *Rhaphiolepis indica* "Rosea" (Indian hawthorn) | 8 |
| *Cytisus* x *praecos* (Warminster broom) | 6 | *Rhododendron* "America" (hybrid rhododendron) | 5 |
| | | *Rhododendron* "Loderi King George" (hybrid rhododendron) | 8 |
| *Elaeagnus multiflora* (cherry elaeagnus) | 5 | *Rhododendron mollis* hybrids (mollis azalea) | 5 |
| *Elaeagnus pungens* (thorny elaeagnus) | 7 | *Rhododendron prinophyllum* (*roseum*) (roseshell azalea) | 4 |
| *Eriobotrya japonica* (loquat) | 8 | *Rhododendron* "Purple Splendor" (hybrid rhododendron) | 7 |
| *Euonymus alatus* (winged euonymus) | 3b | *Rhododendron* southern Indian hybrids (Indian azalea) | 8b |
| *Euphorbia pulcherima* (poinsettia) | 10 | *Rosa rugosa* (rugosa rose) | 3 |
| x *Fatshedera lizei* (botanical-wonder) | 8 | *Schinus terebinthifolius* (Brazilian pepper-tree) | 9b |
| *Forsythia ovata* (early forsythia) | 4b | *Sequoia sempervirens* (redwood) | 8 |
| *Forsythia suspensa* (weeping forsythia) | 5b | *Sequoiadendron giganteum* (giant sequoia) | 7 |
| *Fremontodendron mexicanum* (flannel bush) | 9 | *Stewartia pseudocamellia* (Japanese stewartia) | 6 |
| | | *Syringa vulgaris* (common lilac) | 3b |
| *Ginkgo biloba* (ginkgo, maidenhair-tree) | 5 | | |
| | | *Ulmus americana* (American elm) | 2 |
| *Hibiscus rosa-sinensis* (Chinese hibiscus) | 9b | | |
| *Hibiscus syriacus* (shrub althea) | 5b | *Viburnum burkwoodii* (Burkwood viburnum) | 5b |
| *Hypericum* "Hidcote" (Hidcote St. Johnswort) | 6 | | |
| | | *Zelkova serrata* (Japanese zelkova) | 5b |
| *Iberis sempervirens* (evergreen candytuft) | 5 | | |

(From U.S. Department of Agriculture, Agricultural Research Service Miscellaneous Publication 1475, "USDA Plant Hardiness Zone Map.")

gamma (*Bouteloua*), and wheat (*Agropyron*). In some cases, Bermuda (*Cynodon dactylon*), centipede, fescue, and zoysia grasses may also work in dry zones. Bent grass and bluegrass may be used in dry zones when irrigation or sprinklers are used. Bahia, Bermuda, carpet, centipede, St. Augustine, and zoysia work well in warm, moist zones.

## 22.1.4 PLANTING SOIL

Planting soil for trees, shrubs, and lawns should be topsoil that is fertile, friable, natural loam. It should not contain subsoil, clay lumps, brush, weeds, litter, stumps, roots, rocks, or any other material that would harm the plants or endanger their growth.

In addition, planting soil for trees, shrubs, and lawns must contain the nutrients each particular plant needs. In general, plants need the proper amounts of some 10 to 14 nutrients. Trees and shrubs need a pH value between 6 and 7. The numbers on the pH scale, which runs from 0 to 14, represent the relative level of alkalinity or acidity. Seven is neutral; higher numbers are alkaline; lower ones are acidic. The farther away from 7, the more alkaline or acidic. The required pH value for grasses varies with the species. The actual nutrients needed and their percentage of the soil material vary with the plant species. In addition, some trees and shrubs require a pH lower or higher than 6 to 7. It is necessary to determine the exact requirements for each particular plant from a local horticulturist or nurseryman, or from a state or local agricultural agent. Soil samples taken from the site should be tested in a qualified laboratory to determine their pH level and the amounts of necessary nutrients present. Some state agricultural extension services will perform such tests. Private laboratories are also available in some places. The levels of the proper nutrients can then be adjusted as necessary, using such soil amendments as humus, mulch, sawdust, manure, and fertilizer. The pH of an acidic soil can be lowered by adding lime in the form of natural dolemitic limestone. Alkaline soils can be adjusted by adding aluminum sulfate. Heavy or cohesive soils may be loosened or lightened by adding perlite or vermiculite and, in some cases, sand. Amended soils should be retested to confirm that they have the correct amounts of the proper nutrients and the correct pH value for the plant material to be used.

# 22.2 Additional Reading

More information about the subjects discussed in this chapter can be found in the references listed in Section 22.3 and in the following publications.

Dines, Nicholas T. and Kyle D. Brown. 2002. *Time-Saver Standards for Landscape Architecture.* New York: McGraw-Hill (books.mcgraw-hill.com).

Hightshoe, Gary L. 1987. *Native Trees, Shrubs, and Vines for Urban and Rural America: A Planting Design Manual for Environmental Designers.* Hoboken, NJ: John Wiley & Sons, Inc. (www.wiley.com).

Hopper, Leonard J. 2003. "Between the Building and the Street—Security Design and the Landscape Architect." *The Construction Specifier* (April): 78 (www.csi.net.org).

Johns, Rodney. 2004. *Turfgrass Installation, Management and Maintenance.* New York: McGraw-Hill (books.mcgraw-hill.com).

Rubenstein, Harvey M. 1996. *A Guide to Site Planning and Landscape Construction,* 4th ed. Hoboken, NJ: John Wiley & Sons, Inc. (www.wiley.com).

Stitt, Fred. 1999. *Ecological Design Handbook: Sustainable Strategies for Architecture, Lanscape Architecture, Interior Design, and Planning.* New York: McGraw-Hill (books.mcgraw-hill.com).

# 22.3 Acknowledgments and References

## ACKNOWLEDGMENTS

We gratefully acknowledge the assistance of the following organization in preparing this chapter. We are also indebted to them for permission to use their illustrations when requested and for the use of their publications as references.

U.S. Department of Agriculture (www.usda.gov)

## REFERENCES

We would also like to thank the authors and publishers of the publications in the following list for their contribution to our research for this chapter.

American Nursery and Landscape Association (ANLA). *American Standard for Nursery Stock.* Washington, DC: ANLA (www.anla.org).

Ramsey, Charles George, Harold Reeve Sleeper, and John Ray Hoke, Jr. 2000. *Architectural Graphic Standards,* 10th ed. Hoboken, NJ: John Wiley & Sons, Inc. (www.wiley.com).

U.S. Department of Agriculture, Agricultural Research Service. Miscellaneous Publication Number 1475, USDA Plant Hardiness Zone Map. Washington, DC: USDA (www.usna.usda.gov).

# Utilities

# Introduction

*Utilities* are services distributed to building sites. They include water service, sanitary sewage service, storm drainage systems, ponds and reservoirs, fuel distribution systems, hydronic and steam energy systems, electrical utility transmission and distribution, and communications utilities. Facility (in the building) water distribution, storm drainage systems, and sanitary systems including on-site septic tanks are discussed in Chapter 16. Systems used within a building to heat, ventilate, and cool it are discussed in Chapter 17. Facility electrical system distribution and services are discussed in Chapter 18. This chapter discusses utility water supply, sanitary sewage, storm water, subdrainage systems, and electric power supply.

# Sustainability Considerations

Production and delivery to a building site of water and electricity are by nature destructive to the environment, as are the removal and disposal of storm water and sanitary sewage.

## Storm Water

An aspect of buildings construction that can be severely damaging to the environment is the mismanagement of site storm-water runoff. Adopting an erosion and sediment control plan for the project site for use during construction is required by most codes and, whether required by code or not, is essential. Such a plan will usually include strategies such as temporary and permanent seeding, mulching, earth dikes, silt fencing, sediment traps, and sediment basins. The procedures employed must prevent loss of soil left in place and soil stockpiled at the site for reuse; prevent sedimentation from entering storm sewers and streams; and prevent the polluting of the air with dust and particulate matter.

Equally important is the design of water runoff controls and storage of runoff water after the construction has been completed to prevent overloading streams and subsequent flooding or other destruction of nearby wildlife habitats. One strategy for reducing runoff is the use of pervious paving and the reduction of impervious surfaces when possible. Inserting constructed wetlands and vegetated filler strips to break up water flow will help to reduce the amount of storm water that must be transported off site. Creation of permanent storage ponds on the site also helps reduce the amount of water that leaves the site and allows gradual runoff of the stored water to prevent flooding and destruction of wildlife habitats.

## Sanitary Sewerage

Reduction of sewage amounts generated in a building is an efficient way to reduce the amount of sewage that must be treated. Methods for accomplishing this kind of reduction consists of using high-efficiency fixtures and dry fixtures such as composting toilets and waterless urinals to reduce wastewater volumes (see Chapter 16 for further discussion). Another is to use storm water for sewage conveyance. In larger projects, on-site wastewater treatment can be used to eliminate or reduce the amount of sewage that must be sent into the public system. These measures reduce not only the amount of sewage, but also the amount of water required, thus lessening the demand on local water utility supply systems.

## Electricity

Most electricity generated in the United States is produced using coal-fired generators, which release copious amounts of smoke and pollutants into the atmosphere. Generation of electricity by alternative means, such as nonpolluting and renewable energy from wind, solar, geothermal, low-impact hydro, biomass and biogas systems, or from atomic energy-powered generators, could drastically reduce these pollutants, but these are long-term solutions that are themselves fraught with political and environmental concerns. For the immediate future, the solution seems to be conservation. The Sustainability Considerations section of Chapter 18 contains a discussion of energy conversation.

# Applicable *MasterFormat* Sections

Subjects addressed in this chapter are included in *MasterFormat* 2004, Division 33, "Utilities." The following *MasterFormat* 2004 Level 2 sections are applicable to this chapter:

33 01 00 Operation and Maintenance of Utilities
33 11 00 Water Utility Distribution Piping
33 12 00 Water Utility Distribution Equipment
33 13 00 Disinfecting of Water Utility Distribution
33 16 00 Water Utility Storage Tanks
33 20 00 Wells
33 21 00 Water Supply Wells
33 31 00 Sanitary Utility Sewerage Piping
33 32 00 Wastewater Utility Pumping Stations
33 33 00 Low Pressure Utility Sewerage
33 34 00 Sanitary Utility Sewerage Force Mains
33 36 00 Utility Septic Tanks
33 39 00 Sanitary Utility Sewerage Structures
33 41 00 Storm Utility Drainage Piping

33 42 00  Culverts
33 44 00  Storm Utility Water Drains
33 45 00  Storm Utility Drainage Pumps
33 46 00  Subdrainage
33 47 00  Ponds and Reservoirs
33 49 00  Storm Drainage Structures

33 71 00  Electrical Utility Transmission and Distribution
33 72 00  Utility Substations
33 73 00  Utility Transformers
33 75 00  High-Voltage Switchgear and Protection Devices

33 77 00  Medium-Voltage Utility Switchgear and Protection Devices
33 79 00  Site Grounding

# 23.1 Water Supply Sources

Typically, water originating in lakes, streams, or underground sources is piped to treatment plants, where it is purified. From there it is conducted through a network of water mains to end users such as residences, businesses, and industries (Fig. 23.1-1).

Modern water supply systems provide potable water for drinking, cooking, and bathing and for use in conducting away wastes. *Potable water* is water of sufficient quality to be drinkable. Water may be naturally potable or may be made potable by treatment.

Water entering a building's plumbing system generally comes from either a municipal or community water system or a private well. Water from these sources may differ substantially in quality, pressure, and reliability of supply.

Water for municipal and community systems comes from rivers, reservoirs, wells, or a combination of these. Federal, state, and local health standards are concerned with the safety of water for human consumption, not its usefulness for other purposes or its interaction with common plumbing materials. Many municipalities, however, treat their water supply not only

to make it safe to drink but also to clarify it, soften it, and reduce its corrosiveness. Alum, activated carbon, and coagulant aids are added to remove suspended materials from surface water supplies and to control taste and odor problems. In many localities, fluoride is added to reduce the incidence of tooth decay and chorine is added to control bacterial contamination. The water may also be aerated to remove odors and dissolved gases. Equipment is available to perform similar water treatment procedures on a small scale for private well water.

Private wells are the source of domestic water in locations where community or municipal water is not available. Private wells are also used to supply water for irrigation, even when community or municipal systems are available. Well water must be tested and treated if it is not potable. Treatment for other problems is addressed in Section 16.3.5.

# 23.2 Sanitary Sewage Utilities

Sanitary *sewage* is the name given to liquid and solid waste matter, including fe-

cal matter and food, industrial waste, and other liquid waste that contains organic (animal or vegetable) matter in suspension or solution, as well as liquids containing chemicals in suspension. Without treatment to break them down into nontoxic substances, humans would increase their amounts beyond the ability of the natural environment to absorb them. A *sanitary drainage system* carries sewage from its source to a point of disposal.

Sanitary drainage systems include sewers. A *sewer* is an underground conduit used to carry off liquid or solid waste or both. A system of sewers is called *sewerage*. The term *sewerage* is also used to refer to the removal and disposal of liquid and solid wastes by sewers.

The portion of a sanitary drainage system that lies within a building and up to a point 3 ft (914 mm) outside of the walls is called a *building drainage system* or, more frequently, a *drain*, *waste*, and *vent system* (DWV) (see Section 16.3).

The horizontal piping that extends from a point 3 ft (914 mm) outside a building to the public sewer or private individual sewage disposal system (see Section 16.3) is called a *building sanitary sewer*.

## 23.2.1 PUBLIC SEWAGE DISPOSAL SYSTEM

Large sewage treatment facilities work on the same principle as natural decay and oxidation, except that the action is more concentrated. In a treatment facility, favorable conditions are created for decay bacteria to break down organic solids into chemically simpler water-soluble compounds: phosphates and nitrates. In the natural cycle, limited amounts of phosphates and nitrates are produced, and these become the nutrients for other growing things. Growing plants absorb such nutrients through their roots and, with the help of sunlight, recombine them with carbon dioxide from the air to form new living tissue. In sewage treatment, large amounts of phosphates and

**FIGURE 23.1-1** A water treatment plant must be strategically located between the users and the source, which may be an underground pool (aquifer) or lake.

nitrates are produced and flushed away by the effluent, to be absorbed by the soil or discharged into a body of water.

Phosphates and nitrates are relatively harmless when released into the soil or water in limited quantities. When introduced into a lake or stream in large amounts, they spur the growth of small aquatic plants (algae) which use up valuable dissolved oxygen as they proliferate and die off in large quantities. This overenrichment of a body of water with excessive nutrients, proliferation of algae, and reduction of oxygen content is characteristic of the *eutrophication* process, which gradually destroys aquatic life and renders the water useless for recreation and water supply.

Proper disposal of human excreta (body wastes) is of major importance to the health of individuals. Where public sewers are not available, many diseases, such as dysentery, infectious hepatitis, typhoid, and paratyphoid, and various types of diarrhea are transmitted from one person to another through the fecal contamination of food and water, largely due to the improper disposal of human wastes.

Safe disposal of human wastes is necessary to protect the health of the community and to prevent the occurrence of nuisances. To accomplish satisfactory results, such wastes must be disposed so that they will not:

1. Contaminate a drinking water supply
2. Pollute the waters of any bathing beach, shellfish breeding ground, or stream used for water supply or for recreation
3. Be discharged on or near the surface, where they would be accessible to insects, rodents, or other small animal carriers that may come into contact with food or drinking water
4. Be accessible to children and give rise to a nuisance due to foul odor or unsightly appearance
5. Overburden any body of water with excess nutrients or degrade its quality below officially adopted water quality standards

### 23.2.2 COMBINED SEWERS

A *combined* sewer system carries both storm water and sewage. Such systems have been responsible for polluting many of our waterways when their capacity was insufficient to handle heavy storms. When excessive water overwhelms a storm drainage system, water backs up into streets and even buildings. When this happens to a combined system, what backs up is raw sewage. Combined systems, which may still be found in older installations, are not permitted in many jurisdictions and should be prohibited everywhere.

## 23.3 Subdrainage Systems and Storm Drainage Utilities

A *storm drainage system* carries rainwater, surface water, condensate, cooling water, and similar liquid wastes, exclusive of sewage and industrial waste. It consists of underslab drains and footing drains; storm water collection devices on a building, such as gutters, downspouts, and roof drains; overflow pans and other devices for collection of mechanically generated water; and the pipes that convey this collected water from its source to a point of disposal. It includes the system portions within a building and on its site, and the mains of a public storm sewer system or piping that carry water to a disposal site on the building's property.

This section discusses underslab and foundation drainage systems and the portions of storm drainage systems that occur outside the building, including on-site and public disposal systems. The rest of the facility (in the building) portion of storm drainage systems is covered in Chapter 16.

### 23.3.1 SUBDRAINAGE SYSTEMS

Section 21.4 describes various surface water and groundwater problems that can exist in any building. Part of the solution to some of these problems involves the construction of subdrainage systems. Subdrainage systems include *foundation drains* and *underslab drains*, which are pipes that collect and carry off groundwater and storm water that may seep down into them. Some codes require that all buildings with basements or crawl spaces below grade have foundation drains. This is a good idea in any case, whether or not required by code, to prevent groundwater or surface water from finding its way into crawl spaces or into basements through the walls. Foundation drains are frequently used in conjunction with waterproofing (see Section 7.2). In addition, a grid of drains should be installed to prevent the buildup of hydrostatic head beneath large buildings and buildings where the water table is above the slab on grade or above the grade inside a crawl space.

Many drain pipe materials are available. Among these are nonperforated concrete drain tile, clay drain tile, cast iron soil pipe, and polyvinyl chloride pipe; porous concrete pipe; and perforated concrete, clay, polyvinyl chloride, and bituminous fiber pipe. Foundation and underslab drains are usually made from 4 or 6 in. (100 or 150 mm) pipe.

Nonperforated pipes are laid with open joints. These joints are then covered with a layer of screening to prevent fine particles from the fill above the pipes from entering the pipes through the joints. This screening material is usually heavy mesh burlap, coal tar-saturated felt, copper mesh screen, or synthetic filter fabric.

Perforated pipes are laid with the perforations facing down and the joints closed in accordance with the pipe manufacturer's instructions.

Drainage fill material is gravel or crushed stone evenly graded so that all of it passes a $1\frac{1}{2}$ in. (38.1 mm) sieve and not more than 5% passes a No. 50 sieve.

*Foundation drains* are placed in a bed of compacted drainage fill material around the exterior perimeter of a building's foundations (Fig 23.3-1). Drain pipe should be placed far enough below the floor slab to properly drain and with its lowest *invert* (bottom of the pipe) at least 4 in. (100 mm) above the bottom of the footing to prevent possible washout of the soil beneath the footing. Footings should be lowered if necessary to meet these parameters.

*Underslab drains* are placed in trenches filled with drainage fill material (Fig. 23.3-2). Pipes are usually placed in parallel rows across the building. Depending on the distances involved, interceptor pipes may be placed perpendicular to these rows to reduce the depth of the piping required to establish a positive drainage slope.

In traditional installations, foundation drain pipes are set in a compacted 4-in. (100-mm)-thick bed of drainage fill material laid with its bottom on the undis-

**FIGURE 23.3-1** General moisture control: a sloped, finished grade provides surface water runoff from foundation; a vapor retarder resists the rise of water vapor caused by capillarity; and foundation drains provide subsurface drainage. (Drawing by HLS.)

the next layer is placed. The drainage fill material is then covered with a layer of No. 15 asphalt- or tar-saturated felt or filter fabric to prevent fine materials from infiltrating the drainage fill. An impervious fill material (earth) is then applied to bring the grade to that required, with care taken to slope it away from the building.

Another method is being used today in some installations to overcome the difficulty in a traditional installation of keeping the drainage fill material and adjacent backfill materials separated. The preceding discussion about the traditional method applies to this method as well, except that the drainage fill material is stopped 4 to 6 in. (100 to 150 mm) above the drain pipe and an *in-plane drain* is placed against the foundation wall from the bottom of the drainage fill to a level about 12 in. (305 mm) below grade (Fig. 23.3-3).

In-plane drains (also called *drainage boards*) are thin boards (usually about 1 in. [25.4 mm] thick) consisting of two parts. The drain portion is made of a plastic material configured so that it forms multiple open pathways that permit water to flow freely through them. The side

turbed subgrade at the bottom of the footing. The pipes are then covered with drainage fill material that extends from 4 to 6 in. (100 to 150 mm) on each side of the pipe up to a level about 12 in. (305 mm) below the surface of the ground (see Fig. 23.2-1). The drainage fill should be placed in 3-in. (76-mm)-thick layers, and each layer should be compacted before

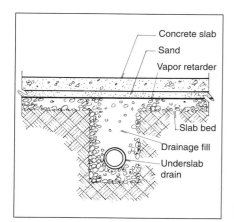

**FIGURE 23.3-2** Underslab drain. (Drawing by HLS.)

**FIGURE 23.3-3** Foundation drain with drainage board showing the connection of underslab drains where they occur. (Drawing by HLS.)

facing the earth is a layer of synthetic filter fabric that permits water to enter the drain but excludes fine soil. These boards can be installed using adhesives or mechanical fasteners as directed by the manufacturer.

Whether a traditional or an in-plane drainage system is used, the drainage system should be tested to be sure that it drains properly before it is covered by drainage fill materials. The test method is usually dictated by code or local practice.

Collected water may drain by gravity to a natural outfall at a lower elevation, or may be discharged by positive machine pumping from one or more collecting sumps located within the foundation area. Some codes dictate the minimum size of pits and pumps, but these may be insufficient for a particular project. Therefore, it is very important that both pit size and pump capacity be based on engineering design.

### 23.3.2 STORM DRAINAGE UTILITIES

Storm drainage systems include sewers. A *storm sewer* is an underground conduit used to carry off liquid waste from a building to a disposal point. A system of storm sewers is called *sewerage*. The term *sewerage* is also used to refer to the removal and disposal of liquid wastes by storm sewers.

A *building storm drain* is the pipe that picks up water from interior sources and interior roof drains and carries it to a point 3 ft (914 mm) outside the building (see Section 16.4).

The horizontal piping that extends from a point 3 ft (914 mm) outside a building to the public storm sewer or private storm water disposal system is called a *building storm sewer*.

A *site storm drain* is a pipe that picks up the overflow from wells and springs, as well as storm water that falls on the site and is conveyed to it through curb inlets, site drains, manholes, and other structures, and carries that water to an outfall on site, a public storm sewer, or a disposal point.

A *disposal point* is a location where storm water is returned to the environment, such as a lake, stream, river, or ocean.

Storm water will be either directed into a community storm drainage system or disposed of in a private system. In a dwindling number of jurisdictions, storm water may be directed into a community combined sewer line (see Section 23.2).

When a community storm sewer is present, storm water may be directed into it by overland flow, as is often the case with single-family residences, or by a system of pipes, as is more usual for larger buildings and extensive sites. The design of parking lots, driveways, and other site conditions to accommodate and collect storm water is beyond the scope of this book. Private storm water system will dispose of water into

- A dry well.
- A low elevation on the property. If the flow is large, riprap or special stone or concrete structures may be required to prevent damage to the surrounding land at the outfall.
- A river, creek, stream, lake, or pond. This may require a permit or be prohibited by the controlling authority of the waterway.

# 23.4 Electric Power Distribution

Electricity from a generating station is transmitted at high voltage to substations and then to transformers near the users. There it is converted to line voltage (120 and 240 volts) for use in small commercial buildings, apartments, and homes, or to 480 volts or higher voltages for use in commercial, institutional, and industrial facilities. This higher voltage, which may be required for heavy equipment or special devices, is then converted again to line voltage for use in power circuits and lighting. Additional transformers drop line voltage down again where needed for low-voltage circuitry.

### 23.4.1 PRIMARY DISTRIBUTION

Most power is generated in power plants at 12,000 volts. For transmission over long distances, the voltage is stepped up by transformers to 115,000 volts or more, depending on the distance. Very high voltage is used because more power can be transmitted with lower power loss in the lines. Utility companies generally use the highest voltage consistent with safety and economy in their transmission lines.

To avoid hazards from high-voltage transmission lines in built-up areas, high voltage is reduced at substations to either 12,000/7200 volts or 4160/2400 volts. Smaller transformers, mounted on poles (Fig. 23.4-1), set on the ground, or placed in underground vaults, further transform current to line voltage (120/240/480 volts).

### 23.4.2 SINGLE-PHASE SERVICE

Single-phase 120/240-volt service is used almost universally for residences in this country and for many small commercial buildings as well. Small electric heating equipment, heat pumps, and most large, fixed appliances such as water heaters, electric ranges, and clothes dryers, use 240-volt service because they operate more efficiently at the higher voltage.

Power at 480 volts is supplied to many larger commercial, institutional, and industrial facilities, where it is often used for heating, ventilating, and air conditioning systems. Step-down transformers then drop the voltage to 240 volts for equipment requiring this voltage and to 120 volts for general power circuits.

General lighting in residences and many commercial, institutional, and industrial buildings, as well as smaller, movable appliances, such as refrigerators, mixers, irons, and toasters, operate adequately on 120 volts, as do dishwashers and some other fixed items. Some types of lighting in commercial, institutional, and industrial buildings, however, operate at 240 volts. To satisfy the needs of most buildings, both 120 and 240 volts must be available. This dual voltage is provided by a 240-volt, center-tapped secondary transformer (Fig. 23.4-2). This transformer makes both 120-volt and 240-volt power available to a three-wire electric service entrance.

**FIGURE 23.4-1** A pole-mounted transformer and service drop. A service drop should be at least 10 ft (3048 mm) aboveground.

A three-wire service drop runs from a utility pole to a building. A three-wire circuit consists of two hot wires, each carrying 120 volts, and a neutral wire having no voltage. Items requiring 240 volts are connected between the two hot wires; loads requiring 120 volts can be connected between either hot wire and the neutral wire (see Fig. 23.4-2).

The incoming service must be grounded. Refer to Section 18.3.4.5 for a discussion about grounding.

### 23.4.3 THREE-PHASE SERVICE

Four-wire, three-phase service is seldom used in residences in this country but is frequently used in commercial, institutional, and industrial applications. This system is particularly suitable for electrical heating and cooling installations and where large motors are involved.

Three-phase service consists of three hot wires and a neutral wire. The voltage in each hot wire is out of phase (out of step) with the others by one-third of a cycle, as if produced by three different generators. Figure 18.1-9 shows single-phase current. In a three-phase diagram there would be three curves, with the peaks in the second and third curves moved one-third of the diagrammed phase length to the right. Connections made between any of the two hot wires produces nominal 208-volt current; connections made between either of the hot wires and the neutral wire result in nominal 120-volt current. Although the service has four wires, branch circuits may have two, three, or four wires, depending on the type of equipment served (Fig. 23.4-3).

A three-phase system should be planned so that the total load is about equally divided among the three phases. It should also be remembered that three wires are hot (rather than two, as in three-wire, one-phase service), and all three require overload protection.

### 23.4.4 SERVICE DROP

When the electrical service comes from overhead, a utility service *drop* is usually connected to the customer's service entrance conductors at the building, forming a drip loop, to prevent water from entering the conductor cable. When the service is underground, the service entrance conductors rise in conduits. From the drip loop or underground cable, conductors go first to a watt-hour meter, which keeps a record of how much power is consumed, and finally to the service equipment. The service equipment is the origin of all the branch circuits which supply a building's electrical needs.

Service equipment provides 120-, 240-, or 480-volt current for general-purpose, appliance, and special-purpose circuits. In addition, line-voltage circuits can be used to power transformers which supply low-voltage circuits.

**FIGURE 23.4-2** A center-tapped transformer can convert two-wire 480-volt (or higher) current to three-wire 120/240-volt service.

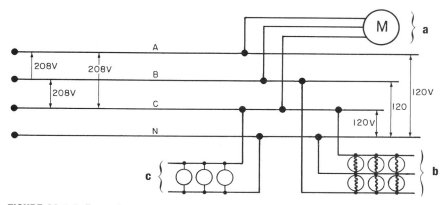

**FIGURE 23.4-3** Four-wire, three-phase service provides flexibility in serving different loads. In this figure, (a) shows a three-wire, three-phase tap for a motor; (b) a three-wire, single-phase circuit for kitchen appliances; and (c) a two-wire, single-phase circuit for lights and receptacles.

## 23.5 Additional Reading

More information about the subjects discussed in this chapter can be found in the references listed in Section 23.6 and in the following publications.

Dagostino, Frank R. and Joseph Wujek. 2005. *Mechanical and Electrical Systems in Construction and Architecture*, 4th ed. Upper Saddle River, NJ: Prentice Hall (www.prenhall.com).

Levy, Sidney M. 2001. *Construction Site Work, Site Utilities and Substructures Databook*. New York: McGraw-Hill (books.mcgraw-hill.com).

Stein, Benjamin and John S. Reynolds. 2000. *Mechanical and Electrical Equipment for Buildings*, 9th ed. Hoboken, NJ: John Wiley & Sons, Inc. (www.wiley.com).

## 23.6 Acknowledgments and References

### ACKNOWLEDGMENTS

We gratefully acknowledge the assistance of the following organization in preparing this chapter. We are also indebted to them for permission to use their illustrations when requested and for the use of their publications as references.

American Society of Civil Engineers (www.asce.org)

### REFERENCES

We would also like to thank the authors and publishers of the publications in the following list for their contribution to our research for this chapter.

American Society of Civil Engineers (ASCE). Reston, VA (www.asce.org). "Sewage Treatment Design Standards." *ASCE Manual of Engineering Practice*, Chapter 5.

American Water Works Association (AWWA). C-203-97, "Coal-Tar Protective Coatings and Linings for Steel Water Pipelines—Enamel and Tape—Hot-Applied." Denver: AWWA (www.awwa.org).

*Architectural Graphic Standards. See* Ramsey et al. (2000).

Arcom Master Systems. *Masterspec®*. Salt Lake City, UT (www.arcom-net.com). Basic Sections:
02080, "Piped Utilities—Basic Materials and Methods," 6/04 ed.
02510, " Water Distribution Utility Service, Domestic, and Fire-protection Water," 3/04 ed.
02530, "Sanitary Sewerage," 12/01 ed.
02620, "Subdrainage Drainage for Foundations, Underslabs, Plaza Decks," 3/03 ed.
02630, "Storm Drainage Gravity and Forced Storm Piping and Underground Structures," 12/01 ed.
02666, "Pond and Reservoir Liners," 11/02 ed.

ASTM International (ASTM). West Conshohocken, PA (www.astm.org). Standards:
C4, "Standard Specification for Clay Drain Tile and Perforated Clay Drain Tile."
C412, "Standard Specification for Concrete Drain Tile."
C444, "Standard Specification for Perforated Concrete Pipe."
C654, "Standard Specification for Porous Concrete Pipe."
C700, "Standard Specification for Vitrified Clay Pipe, Extra Strength, Standard Strength, and Perforated."
D2729, "Specification for Poly (Vinyl Chloride) (PVC) Sewer Pipe and Fittings."
F405, "Standard Specification for Corrugated Polyethylene (PE) Tubing and Fittings."

Dagostino, Frank R. and Joseph Wujek. 2005. *Mechanical and Electrical Systems in Construction and Architecture*, 4th ed. Upper Saddle River, NJ: Prentice Hall (www.prenhall.com).

Ramsey, Charles George, Harold Reeve Sleeper, and John Ray Hoke, Jr. 2000. *Architectural Graphic Standards*, 10th ed. Hoboken, NJ: John Wiley & Sons, Inc. (www.wiley.com).

Stein, Benjamin. 1997. *Mechanical and Electrical Systems*, 2nd ed. New York: John Wiley & Sons, Inc. (www.wiley.com).

Stein, Benjamin and John S. Reynolds. 2000. *Mechanical and Electrical Equipment for Buildings*, 9th ed. New York: John Wiley & Sons, Inc. (www.wiley.com).

U.S. Environmental Protection Agency (EPA). Washington, DC (www.epa.gov).
*National Primary Drinking Water Standards (NPDWRS)*.
AP-42, Vol. I, CH 4.3: "Waste Water Collection, Treatment and Storage."

# CHAPTER 24

# Properties of Materials

# Introduction

In the last 100 years there has been a huge increase in the number of construction products and even in the number of totally new materials used for construction purposes. The entire array of plastic materials we use today is one example. More than 300 new products continue to appear annually. Some of these are made by merely varying the ingredients in or the formulation of existing materials, such as sealants, or by adding new shapes and finishes to a line of masonry units, cladding panels, and other products. A few are entirely new products, often made by combining existing materials. Glass products made by mixing other substances into molten glass are examples of new products with characteristics that did not exist in older products. In addition, many new uses for existing materials are developed annually. The introduction of glass fibers to reinforce poured-in-place or precast concrete is an example. In order to analyze and use these innovations effectively, designers and builders need a basic understanding of the general properties of materials.

A material's suitability for a particular use depends on its ability to perform satisfactorily under the conditions to be encountered. When a material's properties are known, its performance in a particular situation can be predicted. The four general groups of properties of interest to those in the construction industry are mechanical, thermal, electrical, and chemical properties. Each depends on the composition of the material in question.

The millions of materials in our universe are all made from just three basic building blocks: *atoms*, *ions*, and *molecules*. These are bonded together by one of several basic chemical bonds. Materials can be classified according to these basic structural elements and bonds into three major categories with different properties:

1. *Ceramics and glasses*—hard, brittle, poor conductors of heat and electricity
2. *Metals*—more ductile than ceramics, good conductors of heat and electricity
3. *Molecular materials*—low melting temperatures, fair strength, poor conductors of heat and electricity

Since all construction materials fall into one of these three categories, a greater understanding of such materials can be obtained by recognizing the category in which a particular material belongs and remembering the properties of that category. For instance, glass wool and foamed polystyrene may have identical insulating capabilities, but the glass wool falls into the ceramic category and the foam into the molecular category. From that knowledge a designer can surmise, without seeing manufacturers' specifications for the materials, that the foam will have a lower melting temperature than the glass wool, because *the melting temperatures of molecular materials is lower than that of ceramic materials.*

This chapter is intended to expand and sharpen the reader's instinct for categorizing materials. First, the basic structure of matter is discussed; then the four sets of properties that determine a material's performance are considered in some detail. For more information about the properties of a particular material, refer to the chapter in which that specific material is discussed.

# 24.1 Structure of Matter

Matter is formed by the chemical bonding of atoms, ions, and molecules. The particular combination of components and bonding methods determines final material properties.

## 24.1.1 BUILDING BLOCKS OF MATTER

There are 102 chemical substances that cannot be subdivided into other substances. These basic materials are called *chemical elements*. Just two chemical elements, oxygen and silicon, make up more than three-quarters of all the matter on earth; just eight chemical elements constitute more than 97% of all earthly substances (Fig. 24.1-1).

A given quantity of any element can be divided into smaller and smaller quantities until a minute particle is reached.

This entity, the smallest particle which retains the properties of the original material, is called an *atom*.

Ordinarily, atoms are neutral, having no electrical charge. When atoms acquire a positive or negative charge by losing or attracting negatively charged particles, they are called *ions*. Atoms may combine into simple units called *molecules*.

## 24.1.1.1 Atoms

The simplest representation of an atom looks like a miniature solar system (Fig. 24.1-2). The atom has at its core a *nucleus* consisting of positively charged *protons* and electrically neutral *neutrons*. Orbiting about the nucleus are negatively charged *electrons*. Because protons and

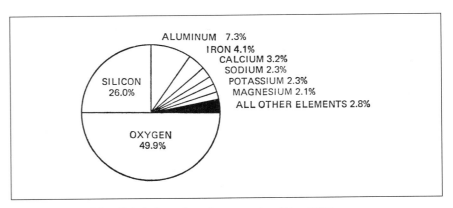

**FIGURE 24.1-1** A handful of elements make up most earthly matter.

**FIGURE 24.1-2** Electrons move in orbits around a nucleus. Helium's two electrons move in the same orbit, but fluorine's nine electrons move in different orbits. Groups of electrons with identical or neighboring orbits constitute a shell; hydrogen and helium have one shell, fluorine two.

neutrons are much denser than electrons, almost all of the weight of an atom resides in the nucleus. The sum of the weights of the protons and neutrons in one of its individual atoms is called the *atomic weight* of an element. In a normal electrically neutral atom, the positive charge of the protons in the nucleus is balanced by an equal negative charge from the electrons. The number of electrons in each of its atoms, which is equal to the number of protons in a single atom, is the *atomic number* of an element.

Groups of electrons moving in neighboring orbits are said to belong to the same *shell* (see Fig. 24.1-2). The shell closest to the nucleus can accommodate only 2 electrons; the second, a maximum of 8; the third, 18; and the fourth, 32. The distribution of electrons around the nucleus gives elements some of their inherent chemical properties. Figure 24.1-3 shows the chemical symbols and common names of all 102 elements, as well as their atomic numbers (electron numbers) and shell distribution.

The elements also can be classified according to ascending atomic numbers and number of electron shells into an arrangement called the *periodic table* (Fig. 24.1-4). The *period number* represents the number of electron shells in each element; thus, all elements with one electron shell (hydrogen [H] and helium [He]) are listed horizontally opposite Period 1; those with two shells, opposite Period 2, and so on up to 7. For each element, the chemical symbol is shown, with its atomic number written above and its atomic weight below.

Each vertical column of the table is labeled with a Roman numeral, and elements within a column possess similar properties. For example, copper (Cu), silver (Ag), and gold (Au) in column IB are dense, soft, and fairly chemically inert metals. The elements on the right side of the table exhibit nonmetallic behavior, while those on the left side exhibit metallic behavior to various degrees.

### 24.1.1.2 Ions

Under certain conditions, electrons may leave an atom, resulting in a net positive charge. The atom then becomes a *positive ion*. This tendency is especially prevalent in elements of groups IA and IIA having just a few electrons in their outermost shells, inasmuch as the attraction of the positive nucleus for these remote electrons is limited. Metallic elements exhibit this tendency for positive ionization most markedly.

Certain other elements show a strong tendency to acquire excess electrons in their outer orbits and to become *negative ions*. These elements, which usually have five or more electrons in their outermost shells to begin with, are almost invariably nonmetallic, appearing at the right side of the periodic table (groups VIA and VIIA).

### 24.1.1.3 Molecule

A material may consist of billions of one or more kinds of atoms (*atomic substances*) or of billions of several kinds of ions (*ionic substances*). Still other substances are formed by the joining together of units, each unit consisting of a number of atoms. These units are called *molecules* and can be broken into their constituent atoms only with difficulty. A

molecule of methane, for example, is composed of one carbon atom and four hydrogen atoms (Fig. 24.1-5).

Substances composed of molecules are called *molecular materials*. Water, for instance, is a molecular material. Each water molecule contains two hydrogen atoms and one oxygen atom and has a triangular shape. When large numbers of molecules are bonded together, the water exists as ice. As the tightness of bonding of the triangular molecule is decreased, water assumes the liquid form. When bonding is very loose, water exists as water vapor. In all three forms of the material the triangular units remain intact, and the individual hydrogen and oxygen atoms are not free to leave their respective molecules.

Molecules can vary greatly in size, as shown in Figure 24.1-6. A molecule of water contains only 3 atoms, whereas a molecule of ethane contains 8 and a molecule of butane 14. Some molecules contain thousands of individual atoms; molecules of this size are called *polymers*, which form the basis of materials such as plastics (see Section 6.9).

### 24.1.2 BONDING OF MATTER

One of the major reasons that atoms, ions, and molecules bond together to form substances is the strong tendency in nature for outermost electron shells to contain as many pairs of electrons as possible. When a shell contains the maximum number of electron pairs, it is said to be *filled*. A filled shell results in a low-energy condition on the atomic scale, meaning that the matter is relatively inert.

Because systems in nature tend to remain in the state of lowest energy, materials with filled outer shells are stable and resist change. Elements that occur naturally with filled outer shells are inert gases, such as helium (He), neon (Ne), and argon (Ar), found in group O at the far right of the periodic table. These elements refuse to combine with other elements, including each other.

Whenever the outer shell of an atom or the ion of a noninert gas element is filled, the atom or ion is said to have achieved the *inert gas configuration* (IGC), in which it is content to remain because of its stability in that situation. Nature moves atoms and ions so strongly toward the IGC that these particles can be thought of as employing many

# FIGURE 24.1-3 The Elements: Their Atomic Numbers, Chemical Symbols, Common Names, and Shell Distributions

| Atomic No. | Symbol | Name | 1 | 2 | 3 | 4 | 5 |
|---|---|---|---|---|---|---|---|
| 1 | H | Hydrogen | 1 | | | | |
| 2 | He | Helium | 2 | | | | |
| 3 | Li | Lithium | 2 | 1 | | | |
| 4 | Be | Beryllium | 2 | 2 | | | |
| 5 | B | Boron | 2 | 3 | | | |
| 6 | C | Carbon | 2 | 4 | | | |
| 7 | N | Nitrogen | 2 | 5 | | | |
| 8 | O | Oxygen | 2 | 6 | | | |
| 9 | F | Fluorine | 2 | 7 | | | |
| 10 | Ne | Neon | 2 | 8 | | | |
| 11 | Na | Sodium | 2 | 8 | 1 | | |
| 12 | Mg | Magnesium | 2 | 8 | 2 | | |
| 13 | Al | Aluminum | 2 | 8 | 3 | | |
| 14 | Si | Silicon | 2 | 8 | 4 | | |
| 15 | P | Phosphorus | 2 | 8 | 5 | | |
| 16 | S | Sulfur | 2 | 8 | 6 | | |
| 17 | Cl | Chlorine | 2 | 8 | 7 | | |
| 18 | A | Argon | 2 | 8 | 8 | | |
| 19 | K | Potassium | 2 | 8 | 8 | 1 | |
| 20 | Ca | Calcium | 2 | 8 | 8 | 2 | |
| 21 | Sc | Scandium | 2 | 8 | 9 | 2 | |
| 22 | Ti | Titanium | 2 | 8 | 10 | 2 | |
| 23 | V | Vanadium | 2 | 8 | 11 | 2 | |
| 24 | Cr | Chromium | 2 | 8 | 13 | 1 | |
| 25 | Mn | Manganese | 2 | 8 | 13 | 2 | |
| 26 | Fe | Iron | 2 | 8 | 14 | 2 | |
| 27 | Co | Cobalt | 2 | 8 | 15 | 2 | |
| 28 | Ni | Nickel | 2 | 8 | 16 | 2 | |
| 29 | Cu | Copper | 2 | 8 | 18 | 1 | |
| 30 | Zn | Zinc | 2 | 8 | 18 | 2 | |
| 31 | Ga | Gallium | 2 | 8 | 18 | 3 | |
| 32 | Ge | Germanium | 2 | 8 | 18 | 4 | |
| 33 | As | Arsenic | 2 | 8 | 18 | 5 | |
| 34 | Se | Selenium | 2 | 8 | 18 | 6 | |
| 35 | Br | Bromine | 2 | 8 | 18 | 7 | |
| 36 | Kr | Krypton | 2 | 8 | 18 | 8 | |
| 37 | Rb | Rubidium | 2 | 8 | 18 | 8 | 1 |
| 38 | Sr | Strontium | 2 | 8 | 18 | 8 | 2 |
| 39 | Y | Yttrium | 2 | 8 | 18 | 9 | 2 |
| 40 | Zr | Zirconium | 2 | 8 | 18 | 10 | 2 |
| 41 | Nb | Niobium | 2 | 8 | 18 | 12 | 1 |
| 42 | Mo | Molybdenum | 2 | 8 | 18 | 13 | 1 |
| 43 | Tc | Technetium | 2 | 8 | 18 | 14 | 1 |
| 44 | Ru | Ruthenium | 2 | 8 | 18 | 15 | 1 |
| 45 | Rh | Rhodium | 2 | 8 | 18 | 16 | 1 |
| 46 | Pd | Palladium | 2 | 8 | 18 | 16 | 1 |
| 47 | Ag | Silver | 2 | 8 | 18 | 18 | 1 |
| 48 | Cd | Cadmium | 2 | 8 | 18 | 18 | 2 |
| 49 | In | Indium | 2 | 8 | 18 | 18 | 3 |
| 50 | Sn | Tin | 2 | 8 | 18 | 18 | 4 |
| 51 | Sb | Antimony | 2 | 8 | 18 | 18 | 5 |

| Atomic No. | Symbol | Name | 1 | 2 | 3 | 4 | 5 | 6 | 7 |
|---|---|---|---|---|---|---|---|---|---|
| 52 | Te | Tellurium | 2 | 8 | 18 | 18 | 6 | | |
| 53 | I | Iodine | 2 | 8 | 18 | 18 | 7 | | |
| 54 | Xe | Xenon | 2 | 8 | 18 | 18 | 8 | | |
| 55 | Cs | Cesium | 2 | 8 | 18 | 18 | 8 | 1 | |
| 56 | Ba | Barium | 2 | 8 | 18 | 18 | 8 | 2 | |
| 57 | La | Lanthanum | 2 | 8 | 18 | 18 | 9 | 2 | |
| 58 | Ce | Cerium | 2 | 8 | 18 | 20 | 8 | 2 | |
| 59 | Pr | Pra'mium | 2 | 8 | 18 | 21 | 8 | 2 | |
| 60 | Nd | Neodymium | 2 | 8 | 18 | 21 | 8 | 2 | |
| 61 | Pm | Promethium | 2 | 8 | 18 | 23 | 8 | 2 | |
| 62 | Sm | Samarium | 2 | 8 | 18 | 24 | 8 | 2 | |
| 63 | Eu | Europium | 2 | 8 | 18 | 25 | 8 | 2 | |
| 64 | Gd | Gadolinium | 2 | 8 | 18 | 25 | 9 | 2 | |
| 65 | Tb | Terbium | 2 | 8 | 18 | 27 | 8 | 2 | |
| 66 | Dy | Dysprosium | 2 | 8 | 18 | 28 | 8 | 2 | |
| 67 | Ho | Holmium | 2 | 8 | 18 | 29 | 8 | 2 | |
| 68 | Er | Erbium | 2 | 8 | 18 | 30 | 8 | 2 | |
| 69 | Tm | Thulium | 2 | 8 | 18 | 31 | 8 | 2 | |
| 70 | Yb | Ytterbium | 2 | 8 | 18 | 32 | 8 | 2 | |
| 71 | Lu | Lutetium | 2 | 8 | 18 | 32 | 9 | 2 | |
| 72 | Hf | Hafnium | 2 | 8 | 18 | 32 | 10 | 2 | |
| 73 | Ta | Tantalum | 2 | 8 | 18 | 32 | 11 | 2 | |
| 74 | W | Tungsten | 2 | 8 | 18 | 32 | 12 | 2 | |
| 75 | Re | Rhenium | 2 | 8 | 18 | 32 | 13 | 2 | |
| 76 | Os | Osmium | 2 | 8 | 18 | 32 | 14 | 2 | |
| 77 | Ir | Iridium | 2 | 8 | 18 | 32 | 17 | 0 | |
| 78 | Pt | Platinum | 2 | 8 | 18 | 32 | 17 | 1 | |
| 79 | Au | Gold | 2 | 8 | 18 | 32 | 18 | 1 | |
| 80 | Hg | Mercury | 2 | 8 | 18 | 32 | 18 | 2 | |
| 81 | Tl | Thallium | 2 | 8 | 18 | 32 | 18 | 3 | |
| 82 | Pb | Lead | 2 | 8 | 18 | 32 | 18 | 4 | |
| 83 | Bi | Bismuth | 2 | 8 | 18 | 32 | 18 | 5 | |
| 84 | Po | Polonium | 2 | 8 | 18 | 32 | 18 | 6 | |
| 85 | At | Astatine | 2 | 8 | 18 | 32 | 18 | 7 | |
| 86 | Rn | Radon | 2 | 8 | 18 | 32 | 18 | 8 | |
| 87 | Fr | Francium | 2 | 8 | 18 | 32 | 18 | 8 | 1 |
| 88 | Ra | Radium | 2 | 8 | 18 | 32 | 18 | 8 | 2 |
| 89 | Ac | Actinium | 2 | 8 | 18 | 32 | 18 | 9 | 2 |
| 90 | Th | Thorium | 2 | 8 | 18 | 32 | 18 | 10 | 2 |
| 91 | Pa | Pr'tinium | 2 | 8 | 18 | 32 | 20 | 9 | 2 |
| 92 | U | Uranium | 2 | 8 | 18 | 32 | 21 | 9 | 2 |
| 93 | Np | Neptunium | 2 | 8 | 18 | 32 | 22 | 9 | 2 |
| 94 | Pu | Plutonium | 2 | 8 | 18 | 32 | 23 | 9 | 2 |
| 95 | Am | Americium | 2 | 8 | 18 | 32 | 24 | 9 | 2 |
| 96 | Cm | Curium | 2 | 8 | 18 | 32 | 25 | 9 | 2 |
| 97 | Bk | Berkelium | 2 | 8 | 18 | 32 | 26 | 9 | 2 |
| 98 | Cf | Californium | 2 | 8 | 18 | 32 | 27 | 9 | 2 |
| 99 | Es | Einsteinium | 2 | 8 | 18 | 32 | 28 | 9 | 2 |
| 100 | Fm | Fermium | 2 | 8 | 18 | 32 | 29 | 9 | 2 |
| 101 | Md | Mendelvium | 2 | 8 | 18 | 32 | 30 | 9 | 2 |
| 102 | No | Nobelium | 2 | 8 | 18 | 32 | 31 | 9 | 2 |

FIGURE 24.1-4 Periodic table of the elements.

| Period | Group IA | Group IIA | | | Transition Elements | | | | | | | | Group IIIA | Group IVA | Group VA | Group VIA | Group VIIA | Group O |
|---|---|---|---|---|---|---|---|---|---|---|---|---|---|---|---|---|---|---|
| 1 | 1 **H** 1.00797 | | Metals | | | | | | | | | | | | | | 1 **H** 1.0079 | 2 **He** 4.0026 |
| 2 | 3 **Li** 6.939 | 4 **Be** 9.012 | | | | | | | | | | | 5 **B** 10.811 | 6 **C** 12.011 | 7 **N** 14.007 | 8 **O** 15.9994 | 9 **F** 18.998 | 10 **Ne** 20.183 |
| 3 | 11 **Na** 22.990 | 12 **Mg** 24.312 | Group IIIB | Group IVB | Group VB | Group VIB | Group VIIB | Group VIII | | | Group IB | Group IIB | 13 **Al** 26.98 | 14 **Si** 28.086 | 15 **P** 30.97 | 16 **S** 32.064 | 17 **Cl** 35.453 | 18 **Ar** 39.95 |
| 4 | 19 **K** 39.102 | 20 **Ca** 40.08 | 21 **Sc** 44.96 | 22 **Ti** 47.90 | 23 **V** 50.94 | 24 **Cr** 52.00 | 25 **Mn** 54.94 | 26 **Fe** 55.85 / 27 **Co** 58.93 / 28 **Ni** 58.71 | | | 29 **Cu** 63.54 | 30 **Zn** 65.37 | 31 **Ga** 69.72 | 32 **Ge** 72.59 | 33 **As** 74.92 | 34 **Se** 78.96 | 35 **Br** 79.91 | 36 **Kr** 83.80 |
| 5 | 37 **Rb** 85.47 | 38 **Sr** 87.62 | 39 **Y** 88.91 | 40 **Zr** 91.22 | 41 **Nb** 92.91 | 42 **Mo** 95.94 | 43 **Tc** 99 | 44 **Ru** 101.07 / 45 **Rh** 102.91 / 46 **Pd** 106.4 | | | 47 **Ag** 107.87 | 48 **Cd** 112.40 | 49 **In** 114.82 | 50 **Sn** 118.69 | 51 **Sb** 121.75 | 52 **Te** 127.60 | 53 **I** 126.90 | 54 **Xe** 131.30 |
| 6 | 55 **Cs** 132.90 | 56 **Ba** 137.34 | 57–71 **La** series* | 72 **Hf** 178.49 | 73 **Ta** 180.95 | 74 **W** 183.85 | 75 **Re** 186.2 | 76 **Os** 190.2 / 77 **Ir** 192.2 / 78 **Pt** 195.1 | | | 79 **Au** 196.97 | 80 **Hg** 200.59 | 81 **Tl** 204.37 | 82 **Pb** 207.19 | 83 **Bi** 208.98 | 84 **Po** 210 | 85 **At** 210 | 86 **Rn** 222 |
| 7 | 87 **Fr** 223 | 88 **Ra** 226 | 89– **Ac** series† | | | | | | | | | | | | | | | |

Non-Metals

*Lanthanide Series
†Actinide Series

| 58 **Ce** 140.12 | 59 **Pr** 140.91 | 60 **Nd** 144.24 | 61 **Pm** 147 | 62 **Sm** 150.35 | 63 **Eu** 151.96 | 64 **Gd** 157.25 | 65 **Tb** 158.92 | 66 **Dy** 162.50 | 67 **Ho** 164.93 | 68 **Er** 167.26 | 69 **Tm** 168.93 | 70 **Yb** 173.04 | 71 **Lu** 174.97 |
|---|---|---|---|---|---|---|---|---|---|---|---|---|---|
| 90 **Th** 232.04 | 91 **Pa** 231 | 92 **U** 238.03 | 93 **Np** 237 | 94 **Pu** 239 | 95 **Am** 241 | 96 **Cm** 242 | 97 **Bk** 249 | 98 **Cf** 252 | 99 **Es** 254 | 100 **Fm** 253 | 101 **Md** | 102 **No** | 103 **Lw** |

tricks—including capturing electrons, surrendering electrons, and sharing electrons with their neighbors—in an effort to achieve chemical stability.

In exchanging or sharing electrons, atoms and ions interact with their neighbors, are attracted to them, and bond into matter. There are four types of bonding: *ionic*, *metallic*, *covalent*, and *secondary*.

### 24.1.2.1 Ionic Bonding

When atoms with almost empty outer electron shells exist in the vicinity of atoms with almost filled shells, the natural tendency of each set of atoms to achieve the IGC can be fulfilled. The atoms with few outer electrons surrender their "free" electrons and become positive ions, whereas atoms with many outer electrons accept these electrons to fill their outer orbit and become negative ions. A situation now exists in which positive ions and negative ions are in close proximity. The oppositely charged particles are attracted and bonded into matter by *ionic bonding*. Metals tend to have almost empty outer orbits, and nonmetals tend to have them almost filled; thus ionic substances usually consist of elements from the left- and right-hand groups of the periodic table bonded together.

Probably the most familiar ionically bonded substance is common table salt, sodium chloride (NaCl). Atoms of sodium (Na) have just a single outer electron and therefore have a strong tendency to form positive ($Na^+$) ions. Atoms of chlorine (Cl) have seven outer electrons and, upon receiving one more electron, fill their outer orbits and achieve the IGC by becoming $Cl^-$ ions. The two sets of ions are then bonded into NaCl (Fig. 24.1-7).

**a**  **b**

FIGURE 24.1-5 Models of a molecule of methane, $CH_4$: (a) a two-dimensional representation; (b) a three-dimensional model.

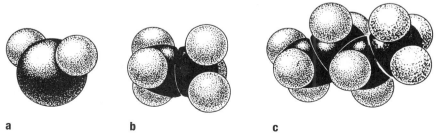

**a**  **b**  **c**

FIGURE 24.1-6 Three molecules of varying sizes: (a) water, (b) ethane, and (c) butane.

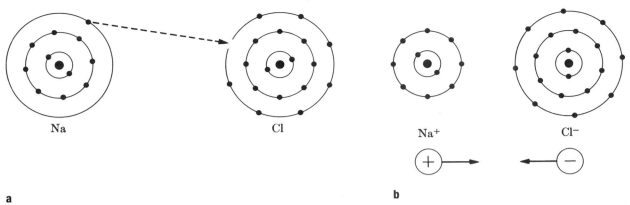

**FIGURE 24.1-7** Ionic bonding in NaCl: (a) Na gives up a single electron, becoming a positive ion, and Cl accepts a single electron, becoming a negative ion; (b) positive and negative ions are then attracted to each other and bond together to become NaCl.

Ceramic materials, being combinations of metallic atoms and nonmetals, are bonded primarily through the ionic mechanism. Examples of this category of materials are brick, tile, portland cement, concrete, and natural stone. Their properties follow from the quality of the ionic bond: they have high melting temperatures and are chemically inert, because both types of ions have achieved the IGC and are tied by a strong ionic bond. Melting involves separating the bonds, and chemical attack requires separation and recombination of ions. Both processes are difficult when the original substance is stable to begin with.

Ceramic materials tend to be brittle, because the regular and extremely rigid arrangement of positive and negative ions resists shape changes. They tend to shatter rather than be forced from one shape into another. Brittleness is generally associated with high strength in compression but low strength in tension. These materials also are poor conductors of electricity and heat, because the electrons are tightly bound into the stable ions and are not free to conduct electrical or thermal energy in response to an applied voltage or temperature.

The typical properties of ceramic materials follow logically from the nature of their constituent ions and of the ionic bond. Hundreds of new ceramic materials with enhanced properties may be developed in the future. It is difficult to predict all of their properties, but in general, each of them will have a high melting temperature, a high compressive strength, chemical inertness, and good thermal and electrical resistance.

### 24.1.2.2 Metallic Bonding

Metallic atoms tend toward the IGC by attempting to surrender their few outer electrons, thus becoming positive ions. When large numbers of atoms are brought together, they each contribute these electrons to a mobile "sea" of electrons which circulates near the original parent atoms. The metal ions are now in a stable situation since they no longer possess their few outer electrons. They exist in a three-dimensional shape, as a metal crystal, with an electron sea dispersed throughout the solid (Fig. 24.1-8).

This peculiar cooperative surrendering of electrons by metal ions results in a *metallic bond*. The characteristic properties of metals follow from the nature of this bond. They are strong and have fairly high melting temperatures because the

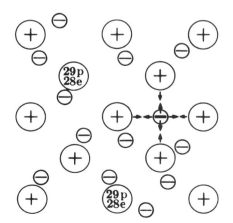

**FIGURE 24.1-8** A metallic bond results from the attraction of the positive cores of the atom (copper) to the negative electrons in the surrounding sea.

metallic bond is strong. Unlike ceramics, metals are good conductors of heat and electricity, because the electrons in the sea are extremely mobile and free to transport thermal and electrical energy.

Metals may or may not be chemically inert. Their ability to transport electrical current makes them susceptible to chemical degradation in the form of corrosion. The tendency toward corrosion sometimes outweighs the intrinsic strength of the metallic bond.

### 24.1.2.3 Covalent Bonding

Many elements found in groups IVA and VA of the periodic table (see Fig. 24.1-4), such as carbon (C) and nitrogen (N), lack the strong tendency to form either positive or negative ions. These elements, with moderate numbers of electrons in their outer shells, reach the IGC by sharing electrons with similar elements. The process of mutual sharing of outer (*valence*) electrons by a cluster of atoms to create a stable entity is known as *covalent bonding*.

Covalent bonding is the most common mechanism whereby small numbers of atoms are bound into molecules. These molecules are then joined together by weak secondary bonds to produce molecular materials. The covalent bond is very strong, as evidenced by the fact that it is very difficult to break up a molecule into its constituent atoms.

A few substances are produced entirely by covalent bonding. Diamond is a familiar example. These substances are extremely hard and have very high melting temperatures due to the strength of the covalent bonds. However, they seldom occur in nature, as suggested by the fact that diamonds are rarely found and then only in small amounts.

### 24.1.2.4 Secondary Bonding

To become molecular materials, atoms are bound into molecules by covalent bonding, and the molecules are then joined by means of weak *secondary bonds*. These bonds arise because positive nuclei or negative electrons in one molecule feel the weak attraction of their opposites in neighboring molecules. The weak attractions bind adjacent molecules into the substance.

Molecular materials such as wood, plastics, and bituminous products derive their properties from the behavior of the secondary bond. They have low strength and low melting temperatures compared to metals and ceramics, because the weak secondary bond can be easily overcome by the application of force or heat. They are poor conductors of heat and electricity, because individual electrons remain tightly constrained within covalently bonded molecules.

Surprisingly, the weak secondary bond is not broken by many of the strong chemical compounds that attack metals and ceramics. Thus molecular materials are chemically inert in a large number of environments. They are attacked by molecular solvents such as acetone, but their resistance to attack by most salts, acids, and industrial atmospheres makes them complementary to metals and ceramics.

# 24.2 Properties of Matter

Anyone who has been alone in a building on a quiet evening knows that materials are "alive." The creaks and groans are material responses to external stimuli such as heat, wind, and gravity loads. These stimuli fall into four categories, which define the four major properties exhibited by materials: *mechanical, thermal, electrical*, and *chemical*.

### 24.2.1 MECHANICAL PROPERTIES

Mechanical properties describe the response of a material to *static* (continuous) or *dynamic* (intermittent) loads.

### 24.2.1.1 Stress and Strain

When a force (load) of $x$ pounds is applied to an object, the object *deflects* (sags) or *deforms* (changes shape) $y$ inches or fractions of an inch. However, knowing the load in pounds or the deflection in inches does not provide enough information to predict the likely response.

For instance, a load of 100 lb (45.36 kg) may induce a negligible reaction when distributed uniformly on the surface of a brick. The same load concentrated at the point of a chisel may shatter the brick. A deformation of 6 in. (152 mm) may be negligible when averaged over the height of an entire building, but the same deformation may have catastrophic consequences if localized. It is therefore necessary to consider forces in relation to the area over which they act and to view deformations with respect to the size of the region being deformed.

Stress ($\sigma$) is equal to the applied load ($L$) divided by the area ($A$) on which the load is acting:

$$\sigma = \frac{L}{A}$$

A 100 lb (45.36 kg) load acting downward on the face of a nominal $4 \times 8$ in. ($100 \times 200$ mm) brick exerts a stress of only 3 psi (20.68 kPa), which is well below the brick's inherent compressive strength. The same load concentrated at the point of a chisel 0.5 in. (12.7 mm) wide and sharpened to 0.02 in. (5.08 mm) thick exerts a stress of 10,000 psi (68.9 MPa), well beyond the strength of the typical brick. In engineering textbooks, stress is usually designated by the Greek symbol $\sigma$ when the applied load is at right angles to the area in question, as in *tension* (stretching) or *compression* (squeezing).

Stress takes into account the fact that an applied load may or may not be shared by a substantial area of a body. Strain ($\epsilon$) correspondingly accounts for the fact that deformation may be distributed over a large area or localized in a small region. If the original dimension of a body along a certain line is designated $d$ and the deformation (change in that dimension) is designated $\Delta d$, then:

$$\varepsilon = \frac{\Delta d}{d}$$

Strain is stated generally in percentage of dimension change; thus, sometimes,

$$\varepsilon = \frac{\Delta d}{d} (100)$$

### ELASTIC DEFORMATION

An applied load, no matter how small, always generates both a stress and a strain in a solid object. A feather floating down and landing on a steel beam will cause the beam to deflect minutely. The resulting deflection cannot be seen, and the resulting strain cannot be measured even with today's sophisticated instruments, but the effect can be theoretically calculated and really exists. When the feather is removed, the strain will disappear and the beam will return to its original dimensions.

In this case, deflection is present only so long as the load is present, and its effect can be reversed by simply removing the load. This reversible type of strain is called *elastic deformation*. Many forces acting on structures cause only elastic deformation. For instance, the swaying of a tall building under the action of wind is an example of elastic deformation. In general, elastic deformation is harmless since the structure is restored to its original configuration when the force is removed.

Most building design criteria contain safety factors to ensure that a structure will not be stressed beyond its elastic range. The most common measure of a material's *stiffness*, or ability to resist elastic deformation, is its modulus of elasticity, $E$.

### PLASTIC DEFORMATION

Permanent or irreversible deformation is called *plastic deformation*. This type of deformation is essential when an aluminum ingot is formed into structural shapes. It would be most inconvenient if the shapes sprang back elastically into the shape of the original ingot. In the final structure, however, load-bearing areas are made large enough so that applied loads will not generate large enough stresses to cause permanent plastic deformation.

### 24.2.1.2 The Stress-Strain Test

Many of a material's mechanical properties can be accurately determined from its performance in a *stress-strain tensile test*, in which a material sample is stretched to fracture (breaking), and a continuous record is made of both the applied stress and the resulting strain. Acceptance of parameters in a stress-strain test as illustrating mechanical behavior requires that such testing be conducted according to rigid standards. Descriptions of test procedures for materials ranging from structural steel to plastic electrical insulation are found in the ASTM standards.

A typical ductile metal, such as aluminum, has a stress-strain behavior as plotted in Figure 24.2-1. The behavior in region OA is elastic and linear. If any stress between O and A is applied and then removed, the strain in the body will return to 0%. The straight line shows that for each unit of increase in stress, the strain will increase by a constant amount; in other words, strain is proportional to stress.

The proportionality constant between stress ($\sigma$) and strain ($\varepsilon$) is the modulus of elasticity ($E$) and is represented by the slope of line OA. The steeper the stress-strain curve, the stiffer the material. It follows, then, that for stresses within the elastic range:

$$E = \frac{\sigma}{\varepsilon}$$

The modulus $E$ is an important property, because it enables calculation of the elastic strain accompanying a particular stress or loading condition, and it enables calculation of deflection. Figure 24.2-2 compares the $E$-values of construction materials belonging to the metals, ceramics, and molecular categories. Metals are generally quite stiff, ceramics slightly less stiff, and plastics still less stiff.

When the stress rises above A into the nonlinear portion of the curve, permanent deformation will appear in the body. If the stress is then reduced to 0, the elastic strain ($\varepsilon_e$) in the body will disappear but some plastic strain ($\varepsilon_p$) will remain. The stress value on the curve marking the transition from elastic to plastic behavior is called the *yield point* or *yield strength* ($\sigma_y$).

The stress-strain curves for most materials do not show clearly defined yield points. For these materials, a theoretical yield strength is calculated from the stress-strain curve. Determining the yield strength of a ductile material is useful, because it is the basis for a common structural design method. The precise definition of $\sigma_y$ and detailed methods for determining it are given in ASTM standards.

If stresses greater than $\sigma_y$ are applied in the test, the material behaves as shown by the AB portion of the curve, with the entire region AB representing plastic behavior. At the stress marked by point B, the material rapidly begins "necking

**FIGURE 24.2-1** Typical stress-strain behavior of a ductile metal. (Courtesy Addison-Wesley Publishing Company.)

**FIGURE 24.2-2** Comparative stiffness of construction materials. (Courtesy Addison-Wesley Publishing Company.)

down," as does a piece of chewing gum stretched to the breaking point. The quantity $\sigma_u$ is the *ultimate* or *tensile strength*.

Beyond point B, deformation continues even with application of increasingly smaller loads for a ductile material and thus leads to the eventual fracture of the material at stress $\sigma_b$, which is the *fracture* or *breaking strength* of the material. The total plastic stretching of the material to failure is given by $\sigma_p$, the *elongation*, which is a measure of the material's ductility.

Figure 24.2-3 shows the stress-strain behavior of four typical construction materials.

Steel has a much higher $E$ than copper and therefore deforms much less in the elastic range. The yield strength of steel is also much higher than that of copper, which indicates that it can withstand greater stresses before it undergoes permanent deformation. On the other hand, steel stretches only about one-half as much as copper, and its greater strength is obtained at the price of higher *brittleness* (less ductility). The behavior of steel and copper is typical of the metallic category of materials and is compared with that of plastics in Figure 24.2-4.

Because of the strength of a metallic bond compared with that of a secondary bond (see Section 24.1.1.2), metals resist permanent deformation more effectively than plastics. Note in Figure 24.2-4 the range of yield strengths in the metals resulting from appropriate processing such as annealing, heat treating, and work hardening.

Ceramics are omitted from Figure 24.2-4 because they do not deform plastically before fracture and are therefore not generally characterized by a specific yield strength. Ceramic materials are typified by the curves for concrete shown in Figure 24.2-3. They have moderately high $E$-values represented by slopes greater than those for plastic but not as great as those for steel. They respond rigidly to elastic stresses, showing small deformations for even fairly large stresses. At the same time they are very brittle, providing little plastic deformation before fracture. The brittleness of ceramic materials is accentuated when they are stressed in tension, which accounts for the common use of concrete, brick, tile, and stone to resist compressive loads but not tensile loads. The absence of an elastic-plastic transition and a maximum in the stress-strain behavior means that

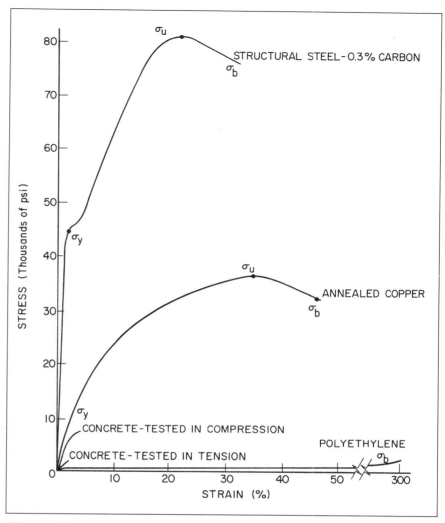

**FIGURE 24.2-3** Stress-strain behavior of four construction materials. (Courtesy Addison-Wesley Publishing Company.)

**FIGURE 24.2-4** Yield strengths of metals and plastics. (Courtesy Addison-Wesley Publishing Company.)

yield strength, ultimate strength, and breaking strength are the same for individual ceramic materials.

The polyethylene curve in Figure 24.2-3 is typical of materials in the molecular category. The linear elastic region is limited, and a small elastic stress produces a large deflection. This behavior is often a disadvantage when molecular materials are used as load-bearing elements. The molecular materials that show the greatest tendency to stretch elastically are the class of rubbers known as *elastomers*, which are used in rubber bands and sealants.

Under the most severe conditions, metals and plastics can be stressed to ultimate strength, $\sigma_u$, before catastrophic fracture occurs, whereas ceramics can be stressed to breaking strength, $\sigma_b$, which for them is identical to yield strength, $\sigma_y$, and ultimate strength, $\sigma_u$. Figure 24.2-5 compares the extreme limits of strength for construction materials falling into each of the three categories.

Figures 24.2-3, 24.2-4, and 24.2-5 substantiate generalizations about material properties derived from their structure. Metals are strong, elastically rigid, and fairly ductile. Ceramics are rigid, strong in compression, and brittle, especially in tension. Molecular materials

have adequate mechanical properties for many construction applications but are limited by other properties.

### 24.2.1.3 Miscellaneous Mechanical Properties

Tabulated stress-strain data provide a quick means for gauging a material's mechanical performance. However, where materials are subject to mechanical loading, knowledge of other properties is necessary to predict performance.

*Hardness* is a measure of a material's ability to resist indentation or penetration. It is determined by various tests, described in ASTM standards, in which indenters are forced into materials under carefully prescribed conditions. In general, the harder a material, the greater its wear and abrasion resistance. Materials with large values of $E$ and $\sigma_y$ tend to be hard and wear-resistant, so in the absence of hardness data, stress-strain data give a measure of wear resistance.

*Fatigue resistance* is a measure of a material's ability to withstand cyclic (repeated) stresses. When repeatedly stressed, even at stresses below yield strength, many materials will fracture without warning. Some airplane acci-

dents have been caused by failures from fatigue. The lifetimes of pumps and other mechanical devices in a building's electrical and mechanical systems depend on the fatigue behavior of the materials employed, and the use of adequate shock-absorbent mountings deserves careful consideration.

*Damping capacity* is a measure of a material's ability to dissipate or deaden mechanical vibration. Because sound is mechanical vibration, a material's ability to absorb sound is directly related to its damping capacity.

*Impact strength* or *toughness* marks a material's capacity to absorb impact without fracturing. It is defined as the total energy, from elastic deformation to fracture, that a material can absorb before breaking under impact. A material's strength and ductility affect this property. Ceramics are strong but do not deform significantly under static or dynamic load, so they lack toughness and shatter under impact. Plastics can stretch up to several hundred percent under load, but their strength is relatively low, so they absorb little energy under impact. Metals, with their good strength and ductility, are the toughest of the common construction materials.

### 24.2.2 THERMAL PROPERTIES

When subjected to temperature changes, a material may change its state (solidify, melt, or vaporize), expand or contract, and conduct or reflect heat.

#### 24.2.2.1 Melting Temperature

As noted earlier, certain mechanical properties, such as tensile strength, give an indirect measure of other properties, such as hardness. A material's melting temperature can be similarly employed. Materials with high melting points usually retain their mechanical properties over a greater temperature range. In addition, they tend to be stronger and more chemically inert than materials with lower melting points.

Materials with strong ionic bonding and high melting temperatures, such as ceramics, tend to perform best at high temperatures. Metals perform moderately well, and molecular materials, with their weak secondary bonding, perform less well. Figure 24.2-6 compares graphically the temperature limits for continuous service of common construction materials. A more quantitative summary of the

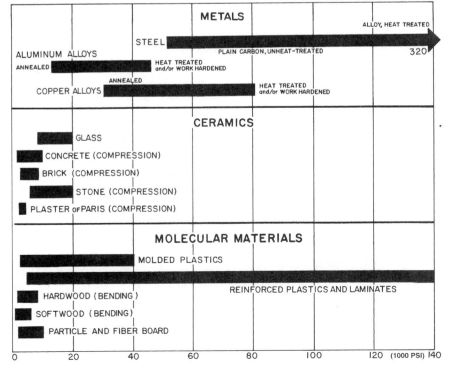

**FIGURE 24.2-5** Ultimate strengths of representative materials. (Courtesy Addison-Wesley Publishing Company.)

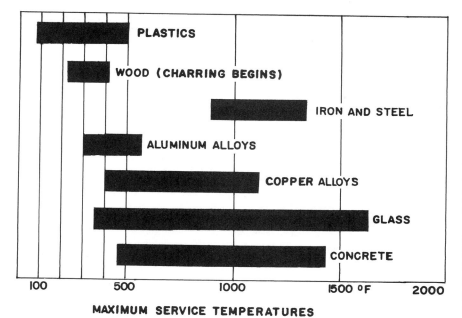

FIGURE **24.2-6** Service temperatures of construction materials. (Courtesy Addison-Wesley Publishing Company.)

thermal properties of typical construction materials is given in Figure 24.2-7.

### 24.2.2.2 Thermal Conductivity

A building designer is concerned with a solid material's response to heat. Of special importance is the material's thermal conductivity, which is its ability to transfer heat from a region of high temperature to a region of lower temperature (see Fig. 24.2-7). Because the free electrons in metals transport heat effectively, metals typically exhibit the highest thermal conductivities; ceramics have much

lower conductivities and molecular materials the lowest.

A material's *thermal conductivity* is the number in British thermal units (Btus) (kilojoules [kJ]) conducted in an hour through 1 sq ft (0.093 m²) of 1-in. (25.4-mm)-thick material for each 1°F (−17.2°C) of temperature differential. Conductivity is measured by placing a 1 sq ft (0.093 m²) by 1-in. (25.4-mm)-thick slab of the material to be tested in a laboratory oven with the material's "hot" face 1°F (−17.2°C) hotter than its cold face. The heat energy in Btus (kJ) con-

ducted through the material in 1 hour is the material's thermal conductivity.

### 24.2.2.3 Thermal Expansion

One reason that buildings seem "alive" at times is that temperature changes cause materials to expand and contract. The crackle of ductwork as a furnace starts up is an example of thermal expansion in action. *Thermal expansion* is measured by the coefficient of thermal expansion, which is stated in terms of length change per unit of length per degree of temperature change (in./in./°F [mm/mm/°C]).

Figure 24.2-7 shows that in general, materials with low melting temperatures have high thermal expansion coefficients and vice versa. Ceramics tend to have slightly lower coefficients of thermal expansion than metals, while plastics usually have much higher coefficients than either ceramics or metals.

Structural steel has an expansion coefficient of 0.0000065 in./in./°F, which indicates that each 1 in. length of steel expands 6.5 millionths of an inch for every 1°F temperature rise. A steel bridge 1 mile long expands about 12 in. (305 mm) as its temperature is raised from 70°F (21.1°C) to 100°F (37.8°C).

Thermal expansion is a critical property in applications where several materials are joined. For example, aluminum has twice the thermal expansion of steel. Thus, if aluminum trim is fixed to a steel base, the uneven expansion of the two materials will cause internal stresses and elastic deformation in the assembly. Noises, as well as unattractive buckling effects, may result. The need for properly

**FIGURE 24.2-7** Thermal Properties of Typical Construction Materials

| Category | Material | Melting Temperature (°F) | Thermal Conductivity (Btu/sq ft/°F/in/hr) | Thermal Expansion ($10^{-6}$ in./in./°F) |
|---|---|---|---|---|
| Metals | Steels (carbon or low alloy) | 2700–2800 | 300–500 | 6.5 |
| | Stainless steels | 2700–2800 | 100–150 | 7 |
| | Aluminum alloys | 1100–1200 | 500–1500 | 9 |
| | Copper alloys | 1800–2000 | 500–2500 | 12.5 |
| Ceramics | Brick | 2000–5000 | 3–6 | 5 |
| | Concrete | 3000 | 5–10 | 7 |
| | Plate glass | 1500 | 5–6 | 1 |
| Molecular materials | Wood | 400 (chars) | 1–2 | 10–30 |
| | ABS plastic | 200–300 | 1–2 | 50 |
| | Acrylics (PMMA) | 200–300 | 1 | 30–50 |
| | Vinyls (PV) | 200–300 | 0.5–1 | 30–100 |
| | Polyethylene (PE) | 200 | 2–5 | 60 |
| | Epoxies (EP) | 200–500 | 1–2 | 30 |
| | Silicones (SI) | 400–600 | 1 | 1–100 |

designed expansion joints in such cases is obvious.

### 24.2.3 ELECTRICAL PROPERTIES

The primary electrical property of interest to a building designer is *electrical conductivity*, a characteristic that is closely related to a material's thermal conductivity. The free electrons in metals allow heat and electrical energy to be conducted easily, so conductivity is high. Ceramics have lower electrical conductivities, and molecular materials have the lowest. This property accounts for the extensive use of molecular materials, such as plastics, in electrical insulation products.

Electrical conductivity is usually given in *mho* (ohm spelled backward) per foot of conductor length. It is the reciprocal of electrical resistivity, which is expressed in *ohms per square foot*. For example, metals have *low* electrical *resistivity* and *high* electrical *conductivity*. Figure 24.2-8 summarizes the resistivities of typical construction materials and illustrates the excellent insulating capabilities of ceramics and molecular materials.

### 24.2.4 CHEMICAL PROPERTIES

Air and airborne moisture contain small amounts of active chemical compounds, which, under certain conditions, react with building materials and degrade their properties.

Ceramics resist chemical attack in most normal environments, and plastics resist attack from all but a few organic solvents. The situation is more complicated, however, in the case of metals.

Metals degrade or corrode through the transport of minute amounts of electricity from regions called *anodes* to other regions called *cathodes*. A cathode accepts electrons and remains intact, but an anode is degraded by the chemical reaction.

The essential elements in a corroding system are the anode, the cathode, and the current-carrying medium (*electrolyte*) connecting them. An electrolyte is usually a solution of water and a gas, such as carbon dioxide or sulfur dioxide. Preventing corrosion, therefore, is simply a matter of removing one of the three elements from the system. For example, most paints are molecular materials, which are poor conductors of electricity. Painting metallic surfaces not only shelters them directly from corrosive environments, but also provides a nonconductive barrier to electrical current flow.

Corrosion can occur when an entire material, or a given region in a material, is subjected to a situation conducive to anodic behavior. Such situations develop in the presence of certain impurities in the metal (acting as cathodes) and when metals with different *galvanic potentials* are placed in close proximity. Galvanic tables ranking materials according to their tendency to become anodic can be used to predict potentially corrosive situations (Fig. 24.2-9).

Materials at the top of the list have a strong tendency to become anodic and those toward the bottom cathodic. If a material near the top of the list, such as carbon steel, is placed in contact with a material lower on the list, such as brass, corrosion is likely to result. The steel, acting as an anode, and the brass, acting

**FIGURE 24.2-9** Galvanic Series of Common Metals and Alloys

| Electrolytic Tendency | Metal or Alloy |
|---|---|
| Anodic (read up) | Magnesium alloys |
| | Zinc |
| | Aluminum alloys |
| | Carbon steel |
| | Stainless steel (active) |
| | Lead |
| | Tin |
| | Brass |
| | Copper |
| Cathodic (read down) | Bronze |
| | Stainless steel (passive) |
| | Gold |

as a cathode, will cause an electric current to flow. Corrosion of the steel is inevitable.

Several points regarding galvanic reactions and corrosion problems should be remembered:

1. Intimate contact between materials widely separated in the galvanic table should be avoided.
2. Zinc can be used as a protective coating (galvanizing) on steel. Because it is higher than steel in the galvanic series, zinc serves as a "sacrificial anode" and is intentionally permitted to corrode, thus saving the steel. As long as the anode is present, the cathode (the steel) will not corrode.
3. Tin is cathodic with respect to steel, so, as a coating over steel, it acts as a physical barrier to the corrosive environment. Should the tin coating be scratched or removed in spots, the exposed steel will corrode quickly. Care must be taken, for example, with *terneplate* (a tin and lead coating over steel), as with painted surfaces, to maintain a continuous film of the protective coating.

Surprisingly, aluminum and stainless steel are inherently "actively" anodic and susceptible to corrosion. The corrosion resistance which is normally associated with these metals develops when they are exposed to oxygen. The oxygen forms a thin, impervious, transparent oxide film that protects the underlying metal from further contact with atmospheric oxygen. The anodizing of aluminum alloys is simply an electrochemical process that produces a thicker oxide coating than forms naturally.

**FIGURE 24.2-8** Electrical Resistivity of Typical Construction Materials

| Category | Material | Electrical Resistivity (ohm/ft) |
|---|---|---|
| Metals | Steel (carbon or low alloy) | .000006 |
| | Stainless steel | .00003 |
| | Aluminum alloys | .000001 |
| | Copper alloys | .0000005 |
| Ceramics | Brick | 4,000,000 |
| | Concrete | 4,000,000 |
| | Plate glass | 10,000,000,000,000 |
| Molecular materials | ABS plastic | 1,000,000,000,000,000 |
| | Acrylics (PMMA) | 100,000,000,000,000 |
| | Vinyls (PV) | 100,000,000,000 |
| | Polyethylene (PE) | 1,000,000,000,000 |
| | Epoxies (EP) | 100,000,000,000,000 |
| | Silicones (SI) | 1,000,000,000,000 |

## 24.3 Additional Reading

More information about the subjects discussed in this chapter can be found in the references listed in Section 24.4 and in the following publications.

Kuhn, Karl F. 1996. *Basic Physics: A Self-Teaching Guide*, 2nd ed. Hoboken, NJ: John Wiley & Sons, Inc. (www.wiley.com).

Monteiro, J. M., B. Williamson, and C. Ostertag. "CEE 60, Concrete Materials and Construction,

Atomic Structure and Bonding." Berkeley: University of California (www.ce.berkeley.edu).

Tilley, Richard J. D. 2001. *Colour and Optical Properties of Materials: An Exploration of the Relationship Between Light, the Optical Properties of Materials and Colour*. Hoboken, NJ: John Wiley & Sons, Inc. (www.wiley.com).

White, Mary Anne. 1999. *Properties of Materials*. Cary, NC: Oxford University Press (www.oup.com/us/catalog).

## 24.4 Acknowledgments and References

### 24.4.1 ACKNOWLEDGMENTS

We gratefully acknowledge the assistance of the following organizations and individuals in preparing this chapter. We are also indebted to them for permission to use their illustrations when requested and for the use of their publications as references.

ASTM International (ASTM) (www.astm.org)
Professor N. F. Fiore
Professor Lawrence H. Van Vlack

### REFERENCES

We would also like to thank the authors and publishers of the publications in the following list for their contribution to our research for this chapter.

Monteiro, J. M., B. Williamson, and C. Ostertag. "CEE 60, Concrete Materials and Construction, Atomic Structure and Bonding." Berkeley: University of California (www.ce.berkeley.edu).

Van Vlack, Lawrence H. *Elements of Materials*, 2nd ed. Reading, PA: Addison-Wesley. (This book is out of print. Try libraries and used-book sellers on the Internet.)

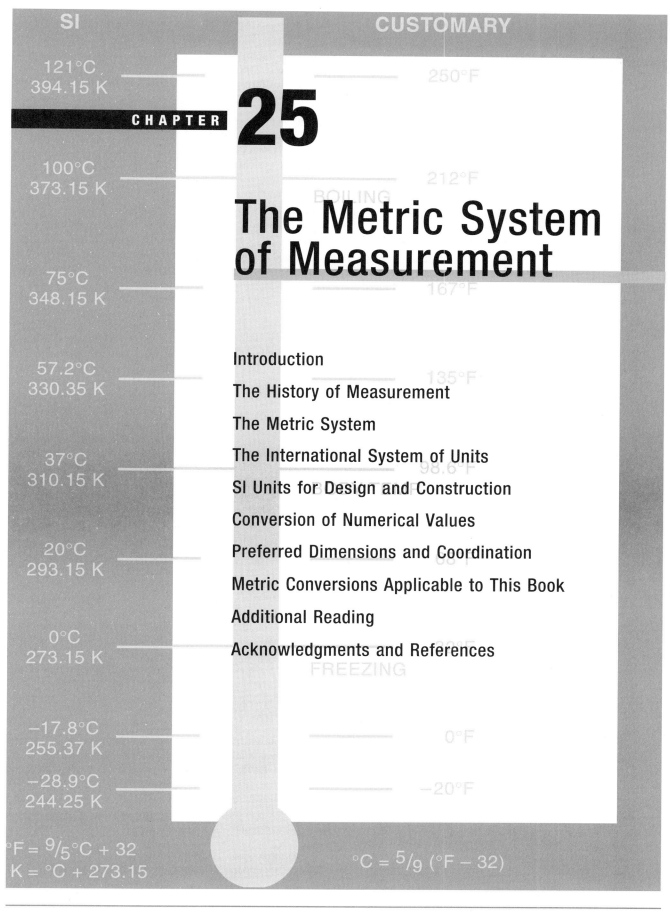

CHAPTER

# 25

# The Metric System of Measurement

121°C
394.15 K

100°C
373.15 K   BOILING   212°F

75°C
348.15 K   167°F

57.2°C
330.35 K   135°F

37°C
310.15 K   98.6°F

20°C
293.15 K   68°F

0°C
273.15 K   FREEZING

−17.8°C
255.37 K   0°F

−28.9°C
244.25 K   −20°F

°F = 9/5°C + 32
K = °C + 273.15

°C = 5/9 (°F − 32)

# Introduction

*Metrication* is a term coined in Britain to describe metric conversion, which is the process of changing from the *customary* to the *metric* system of measurement, including the planning and coordination necessary for the change.

The *customary system* of measurement referred to here is the collection of English and other nonmetric units currently used in the United States. The *metric system* is a system originally developed in France that has been adopted by more than 90% of the nations of the world. It is based on the *metre* and six other standardized *base units*. *Metric units* already in general use include, but are not limited to, the *second, ampere, candela, watt, volt, ohm, farad, coulomb,* and *lumen*.

Much of the material in this section is based on two current metric standards:

1. IEEE/ASTM SI-10, which is a replacement for American Society for Testing and Materials (ASTM) E380-89a, "Standard for Metric Practices."
2. STM E621, originally a supplement to ASTM E380, was based on the now withdrawn National Bureau of Standards (NBS) (now National Institute of Standards and Technology [NIST]) publication 938, "Recommended Practice for the Use of Metric (SI) Units in Building Design and Construction." NBS 938 was authored by the world-renowned metrication expert Hans J. Milton while he was on loan to the NBS by the Australian government.

This section describes the history of measurement and the metric system, the properties of the *International System of Units* (SI) version of the metric system, and its applications and proper usage. It also includes tables that contain:

- Lists of units in the SI system.
- Fraction, decimal, and *metric equivalents* tables.
- Conversion tables for selected customary and SI measurement units. One of these contains conversions for English units that are not converted in the tables in this book.

This section also includes text and tables showing rules and recommendations for:

- Presentation of SI units and symbols
- Presentation of numerical values in conjunction with SI units
- Application for SI quantities and associated names and symbols

If the events in the rest of the world can be used as a guide, conversion to the metric system of measure in this country will probably change the entire measurement base of the construction industry, including guidelines, codes, standards, drawings, specifications, and associated documents. The most extensive use of metric measures in the building industry will be for linear measurement, area, volume, and mass.

Metrication disturbs some people. Manufacturers worry about the cost of retooling and introducing new product sizes. People in general worry about learning many new names and numbers.

As its critics point out, metrication will force a complete revision in the accepted units of measurement and calculation, but at the same time it will offer a logical opportunity to revise standard product sizes, performance criteria, and methods of testing. This, in turn, will reduce inventory, rationalize criteria, and simplify methods of testing and reporting test results. Experience in other countries has shown that metrication takes less time and costs less than estimated. The cost of metrication occurs once; the benefits last forever. Approximately one-third of the contracts awarded to the top U.S. construction firms have been for work outside the United States. American know-how is one of the exports not usually figured into foreign trade discussions. The change to the metric system of measure will help the United States to maintain a role of technical leadership and keep it from lagging behind the metric world.

# 25.1 The History of Measurement

*Measurement* is a comparison between an unknown quantity and a known quantity. It is also a ratio of how many times a known thing can be divided into something unknown. Measurement tells how large, how many, how fast, and how much.

A unified system of measurement is a means of communication. Throughout history, various number systems and representational symbols have been used. Although there are existing remnants of systems based on the numbers 5, 12, and 20, only those systems based on the numbers 2, 10, and 60 are still in wide use. The metric system uses the base-10 (*decimal*) system of numbers.

It took many centuries to develop accurate units of measurement. The first were based on parts of the body. People used the human hand or the human foot; they also used the length of a furrow made by a plow or the distance a man could walk in a day. Because it is harder and takes longer to walk uphill than downhill, the Chinese even made an uphill mile shorter.

All of these forms of measurement were inexact. The size of a human hand or foot varied among individuals. Different people plowed different furrows and walked different distances in a day. There was a need for a unified standard system of weights and measures—one that everybody could agree on.

The Romans established a unified system of weights and measures. They made a bronze rod of specific length and called it a *pes*, Latin for "foot"; they made a bronze weight called a *libra*, Latin for "pound." Both the *pes* and the *libra* were kept under guard in a temple and used as standards of measurement throughout the Roman Empire. The U.S. customary abbreviation for pound, lb, is derived from the Latin *libra*.

After the fall of the Roman Empire, units of measurement once again became localized, and there was confusion when different regions tried to communicate numerically.

## 25.1.1 NOSES, THUMBS, AND BARLEYCORNS

Because local measurements were often based on the king's body measurements, they differed from country to country and depended on the size of the current king.

The U.S. customary system uses the yard as a standard unit of measurement. Legend says that King Henry I of England established by royal decree the length of the yard as 36 in., which was the distance from Henry's nose to his thumb.

Another arbitrary and unrelated unit of the customary system is the system of thirteens, used to measure shoes. It is based on the size of a barleycorn, which is about one-third of an inch. Shoe sizes vary about one-third of an inch from size to size. A size 7 shoe, for example, is 1 in. smaller than a size 10.

### 25.1.2 NUMBER SYSTEMS IN CURRENT USE

There are several numbering systems in current use around the world.

#### 25.1.2.1 Sexagesimal System

The *sexagesimal* system is based on the number 60. First used by the Babylonians, it survives in the subdivision of time and angular measurement in the customary system. For instance, an hour has 60 minutes and a circle has 360 degrees. Both minutes and degrees can be further subdivided into 60 seconds.

#### 25.1.2.2 Binary System

The *binary* system is based only on the numerals 0 and 1; it is used in electronic computers because it can represent yes/no or true/false decision logic.

#### 25.1.2.3 Decimal System

The *decimal* system, which is based on the number 10, is the system on which the metric system is founded. It was probably derived from the early use of human fingers to count. The Greeks and Romans first used a decimally based word structure for numbers, then adopted letter numerals for their representation. Roman numerals, which were used exclusively in Europe until about A.D. 1100, were gradually replaced by Arab numerals of Hindu-Arabic origin, which were much simpler to use.

The *Roman numeral* system consists of *quinary* (base-five) symbols: I equals 1, V equals 5, X equals 10, L equals 50, C equals 100, D equals 500, and M equals 1000. In-between numbers are formed by addition to or subtraction from the seven basic symbols. When a smaller number precedes a larger one, it is subtracted; when the smaller follows the larger number, it is added. Thus, 1978 is written MCMLXXVIII: C before M means 900; all other numbers retain their absolute values and are added.

The *Hindu-Arabic* system we use today expresses all numbers by combinations of 10 symbols, each having an *absolute* value and a *position* value (Fig. 25.1-1).

*Decimal position notation* relates each number's position value as a power of 10. For example, $10^1$ equals 10, $10^2$ equals 100, and $10^3$ equals 1000. Numbers to the right of the decimal point are expressed in negative powers of 10. For example, $10^{-1}$ equals 0.1, $10^{-2}$ equals 0.01, and $10^{-3}$ equals 0.001. Each multidigit number is written as a linear combination of powers of 10. For example:

$$456 = 4 \times 10^2 + 5 \times 10^1 + 6$$
$$37.5 = 3 \times 10^1 + 7 + 5 \ (\times 10^{-1})$$

## 25.2 The Metric System

The scientific evolution of the eighteenth and nineteenth centuries, the Industrial Revolution of the nineteenth century, and the desire for more international trade accelerated the movement toward a universal system of weights and measures. In 1795, the French National Assembly adopted the metric system of measure.

The French metric system was based on a standard unit of length called a *metre* (not meter). The word was derived from the Greek *metron*, meaning "measure." The French metre was expressed in geodetic terms, that is, related to the size and shape of the earth. A metre was the 10-millionth part of an imaginary line running through Paris from the equator to the North Pole. Eventually, that distance was marked on a platinum/iridium bar, which became the international standard (prototype). It was kept at the International Bureau of Weights and Measures in France.

This system, like its modern equivalent, established reference values, called *units*, for each measurable attribute of a physical phenomenon, which is called a *quantity*. A unit, then, was the reference value of a given quantity. For example, the metre was the unit for the given length (quantity) established by the prototype.

Additional units in the system included those for the quantities of *area*, *volume*, *capacity*, and *mass*. These were related decimally to the metre through the properties of water. For example, 1 *litre* represented $1/1000$th of 1 cubic *metre*. When filled with water at its maximum density, a litre had a mass (weight) of 1000 grams or 1 *kilogram*.

The system also contained a set of decimal prefixes that could be attached

FIGURE 25.1-1 Preferred and Nonpreferred Multiples, Submultiples, and Prefix Names for Some Powers of 10

| Multiplication Factor | | | Prefix | | |
| --- | --- | --- | --- | --- | --- |
| | | | Name | Symbol | Pronunciation |
| **Preferred** | | | | | |
| $10^{12}$ | or | 1 000 000 000 000 | tera | T | as in *terrace* |
| $10^9$ | or | 1 000 000 000 | giga | G | *jig'a* |
| $10^6$ | or | 1 000 000 | mega | M | as in *megaphone* |
| $10^3$ | or | 1 000 | kilo | k | as in *kilowatt* |
| $10^{-3}$ | or | 0.001 | milli | m | as in *military* |
| $10^{-6}$ | | 0.000 001 | micro | $\mu$ | as in *microphone* |
| $10^{-9}$ | or | 0.000 000 001 | nano | n | *nan'oh* |
| $10^{-12}$ | or | 0.000 000 000 001 | pico | p | *peek'oh* |
| **Nonpreferred** | | | | | |
| $10^2$ | or | 100 | hecto | h | *heck'toe* |
| $10^1$ | or | 10 | deka | da | *deck'a* |
| $10^{-1}$ | or | 0.1 | deci | d | as in *decimal* |
| $10^{-2}$ | or | 0.01 | centi | c | as in *sentiment* |

(Copyright ASTM; reprinted with permission.)

to reference units to alter their magnitude. For example, *milli* means $^1/_{1000}$ and *kilo* means 1000 times.

In 1881, the *International Electrical Congress* adopted a unit of time to produce the *centimetre-gram-second* (cgs) system. At about 1900, *metric measurements* began to be based on the *metre-kilogram-second* (MKS) system. In 1935, the ampere became the fourth *base unit*, resulting in the metre-kilogram-second-ampere (MKSA) system. The *General Conference on Weights and Measures* (CGPM), an international treaty organization, added the *kelvin* as a unit of temperature and the *candela* as the unit of *luminous intensity*.

All three metric systems were then in use in different countries, making international trade and conversation difficult, even though many nations were using a form of the metric system of measurement.

# 25.3 The International System of Units

In 1960, to help further unify the system of measurement internationally, the CGPM formally introduced a new metric system, called the *International System of Units*, for which the abbreviation is SI in all languages. The abbreviation represents the French *Système International d'Unités*. SI was derived from earlier decimal metric systems but supersedes them. In response to demands by scientists and technologists for a means of precisely defining the metre in a way that could be accurately reproduced without having to refer to the prototype, CGPM introduced in SI a new definition based on the electromagnetic emission of a certain *krypton* atom.

## 25.3.1 ADOPTION OF SI

SI has been adopted by every country that has changed to metric measurement since 1960. Countries that were using the cgs, MKS, or MKSA systems before 1960 are changing those units that have been superseded by the SI system. The decimal metric system is now used in more than 90% of the world's nations. Paradoxically, the United States, the first nation to introduce a system of decimal currency in 1785, and an original party of the 1875

*Treaty and International Metric Convention* that established the CGPM, is one of the last nations to adopt SI. It shares this dubious distinction with only a few small countries.

The metric system was legalized by Congress in 1866, but it was not made mandatory. Renewed interest at about the time SI appeared in the 1960s led to publication of the *ASTM Metric Practice Guide*, which in 1976 became ASTM E380, "Standards for Metric Practice." Meanwhile, the government, responding to Public Law 90-472, signed in 1968, conducted a survey to determine the impact, desirability, feasibility, implications, and difficulties of converting the United States to the metric system of measurement.

At about the same time, ANSI established a *Metric Advisory Committee*, and in 1973, in response to the newfound interest in conversion, ANSI established an independent group called the *American National Metric Council* (ANMC) to provide assistance in the conversion to metric measurement through coordination, planning, and information services to its membership and all segments of society in the United States. Three years later, ANMC separated itself from ANSI, becoming a totally independent, private, and self-supporting national organization.

In 1975, in response to the completed study required by Public Law 90-472, Congress passed Public Law 94-168, the *Metric Conversion Act*. It stated that voluntary metrication was to be national policy and established the *U.S. Metric Board* to plan, publicize, and implement a voluntary change to the metric system based on SI, but the results of this watered-down law were predictably less than were needed to make conversion a reality. A timetable for the changeover to SI units was established by the construction industry. The target date for *metric conversion*, called M Day, was January 1, 1985, which came and went with little accomplished.

Then, with enactment of the *Omnibus Trade and Competitiveness Act* of 1988, metrication was again off and running. This law stated that the metric system is the "preferred system of weights and measures for United States trade and commerce," and required that when it was practicable and economically feasible, federal agencies would use the metric system when making grants and procurements and in other business-related activities by the end of fiscal year 1992. Implementation is the responsibility of the Department of Commerce. Citing the Omnibus Trade and Competitiveness Act

FIGURE 25.3-1 Some Metric Recognition Points

| Quantity | Unit | Description |
|---|---|---|
| Linear measurement | 1 mm | Approximate diameter of a paper clip wire |
| | 25 mm | Vertical dimension of an ordinary U.S. postage stamp; nearest equivalent to 1 in. |
| | 100 mm | International cigarette length; basic metric building module; nearest metric preferred dimension to 4 in. (1.6% less) |
| | 600 mm | Height of three courses of concrete blocks including mortar joints |
| | 2 m | Approximate height of a standard door opening (6 ft. 8 in.) |
| Area measurement | 500 mm$^2$ | Face area of an ordinary U.S. postage stamp [20 mm × 25 mm] |
| | 1 m$^2$ | 1 square metre; approximate area of a shower base |
| Volume measurement | 1 L | 1 litre: 5.7% more than a U.S. quart; new soft drink bottle size |
| Mass [weight] | 1 g | Approximate weight of a paper clip or a dollar bill; artificial sweetener package size |
| | 5 g | mass [weight] of a nickel [5¢ coin] |
| | 1 kg | 1 kilogram: the base unit of mass in SI; mass of water in a cube with 100 mm sides (approximately 2.2 lb) |
| Temperature | 0°C | Freezing point of water (32°F) |
| | 20°C | Comfortable thermostat setting in winter (68°F) |
| | 37°C | Normal body temperature (98.6°F) |
| | 100°C | Boiling point of water (212°F) |

of 1988, the Department of Commerce announced on January 2, 1991, that metrication was no longer voluntary for federal agencies.

Accordingly, by the middle of 1991, the various federal agencies began to make plans to convert. It remains to be seen whether states and local governments will follow, but their response to other federal government initiatives indicates that they will. The private sector will not be far behind.

### 25.3.2 STRUCTURE OF SI

A lifetime of using the customary system gives people a feel for the length of an inch or a foot. New associations or recognition points must be developed in order for them to be comfortable using SI (Fig. 25.3-1). This conversion in thinking happens rather more quickly than most people expected. It can be equated with the relative ease of learning a foreign language when living in the nation of its origin, as opposed to learning the same language in a classroom far removed from its general use. In addition, many SI and related units are already in our vocabulary—for example, *volt*, *ampere*, *watt*, *ohm*, *second*, and *lumen*.

The application of SI units and new *preferred* (see Section 25.6) numeric values will simplify measurements in construction and reduce errors in calculations.

SI symbols have an agreed-upon form and the same meaning worldwide. There is only one recognized SI unit for each physical quantity. All units are derived from seven *base* and two *supplementary* units (Fig. 25.3-2). The entire system consists of *base units*, *supplementary units*, and *derived units*, as summarized in Figure 25.3-3.

SI is a universal, rational, coherent, and preferred measurement system that uses the base-10 (decimal) system of numbers. Relations between the units in the system contain as a numerical factor only the number 1, or "unity," because all derived units have a unity relationship to the constituent base and supplementary units. For example, one litre is equal to $^{1}/_{1000}$th of 1 cubic metre (0.001 m$^3$) and (when filled with water) has a mass of 1 *kilogram*.

### 25.3.2.1 Base Units

SI has fewer units than the customary system. A single unit of length, the *metre* (m), and its decimally related multiples and submultiples, such as *kilometre* (km) and *millimetre* (mm), will replace a variety of customary units, including the *mile*, *furlong*, *chain*, *link*, *rod*, *fathom*, *yard*, *foot*, *hand*, *inch*, and *mil*. In addition, changing to the metre will correct the difference between the *standard foot* and the *survey foot* (which is two parts per million longer) by replacing both. SI uses the French spelling of *metre*.

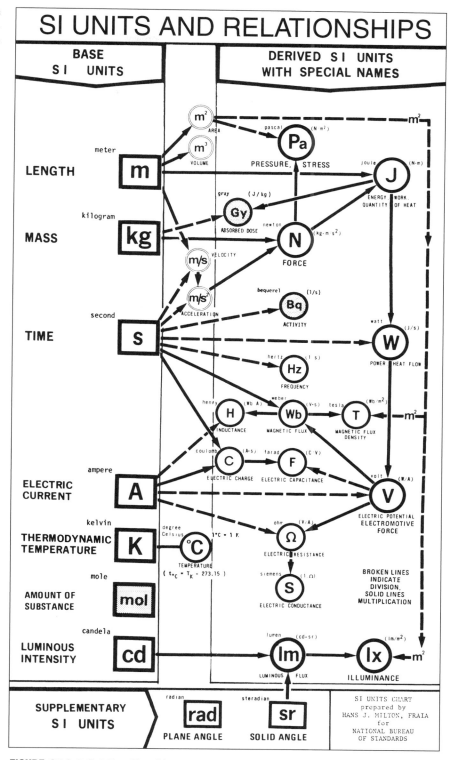

**FIGURE 25.3-2** Relationship of base, supplementary, and derived SI units. (Copyright ASTM; reprinted with permission.)

**FIGURE 25.3-3** Units in the International System (SI)

| Unit Group Quantity | Unit Name | Symbol | Formula | Unit Derivation | Remarks |
|---|---|---|---|---|---|
| **Base Units** | | | | | |
| Length | metre | m | | | An alternative spelling is *meter*. |
| Mass | kilogram | kg | | | |
| Time | second | s | | | Already in common use. |
| Electric current | ampere | A | | | Already in common use. |
| Thermodynamic temperature | kelvin | K | | | The customary unit for temperature is the degree Celsius (°C). |
| Amount of substance | mole | mol | | | The mole has no application in construction. |
| Luminous intensity | candela | cd | | | Already in common use. |
| **Supplementary Units** | | | | | |
| Plane angle | radian | rad | | | Already in common use. |
| Solid angle | steradian | sr | | | Already in common use. |
| **Derived Units with Special Names** | | | | | |
| Frequency (of a periodic phenomenon) | hertz | Hz | 1/s | $s^{-1}$ | The hertz replaces *cycle per second*. |
| Force | newton | N | $kg \cdot m/s^2$ | $m \cdot kg \cdot s^{-2}$ | |
| Pressure, stress, elastic modulus | pascal | Pa | $N/m^2$ | $m^{-1} \cdot kg \cdot s^{-2}$ | |
| Energy, work, quantity of heat | joule | J | $N \cdot m$ | $m^2 \cdot kg \cdot s^{-2}$ | |
| Power, radiant flux | watt | W | J/s | $m^2 \cdot kg \cdot s^{-3}$ | Already in common use. |
| Quantity of electricity, electric charge | coulomb | C | $A \cdot s$ | $s \cdot A$ | Already in common use. |
| Electric potential, potential difference, electromotive force | volt | V | J/C or W/A | $m^2 \cdot kg \cdot s^{-3} \cdot A^{-1}$ | Already in common use. |
| Electric capacitance | farad | F | C/V | $m^{-2} \cdot kg^{-1} \cdot s^4 \cdot A^2$ | Already in common use. |
| Electric resistance | ohm | $\Omega$ | V/A | $m^2 \cdot kg \cdot s^{-3} \cdot A^{-2}$ | Already in common use. |
| Electric conductance | siemens | S | A/V or $1/\Omega$ | $m^{-2} \cdot kg^{-1} \cdot s^3 \cdot A^2$ | The *siemens* was formerly referred to as *mho*. |
| Magnetic flux | weber | Wb | $V \cdot s$ | $m^2 \cdot kg \cdot s^{-2} \cdot A^{-1}$ | Already in common use. |
| Magnetic flux density | tesla | T | $Wb/m^2$ | $kg \cdot s^{-2} \cdot A^{-1}$ | Already in common use. |
| Electric inductance | henry | H | Wb/A | $m^2 \cdot kg \cdot s^{-2} \cdot A^{-2}$ | Already in common use. |
| Luminous flux | lumen | lm | $cd \cdot sr$ | $cd \cdot sr$ | Already in common use. |
| Illuminance | lux | lx | $lm/m^2$ | $m^{-2} \cdot cd \cdot sr$ | |
| Absorbed dose | gray | Gy | J/kg | $m^2 \cdot s^{-2}$ (a) | (a) kg is canceled out. No application in construction. |
| **Derived Units with Generic Names** | | | | | |
| a. Units Expressed in Terms of One Base Unit | | | | | |
| Area | square metre | $m^2$ | | $m^2$ | |
| Volume, capacity | cubic metre | $m^3$ | | $m^3$ | (1 $m^3$ = 1000 L) |
| Section modulus | metre to third power | | $m^3$ | $m^3$ | |
| Second moment of area | metre to fourth power | | $m^4$ | $m^4$ | |
| Curvature | reciprocal (of) metre | | 1/m | $m^{-1}$ | |
| Rotational frequency | reciprocal (of) second | | 1/s | $s^{-1}$ | Revolution per second (r/s) is used in specifications for rotating machinery. |
| Coefficient of linear thermal expansion | reciprocal (of) kelvin | | 1/K | $K^{-1}$ | |
| b. Units Expressed in Terms of Two or More Base Units | | | | | |
| Linear velocity | metre per second | | m/s | $m \cdot s^{-1}$ | |
| Linear acceleration | metre per second squared | | $m/s^2$ | $m \cdot s^{-2}$ | |
| Kinematic viscosity | square metre per second | | $m^2/s$ | $m^2 \cdot s^{-1}$ | |
| Volume rate of flow | cubic metre per second | | $m^3/s$ | $m^3 \cdot s^{-1}$ | |
| Specific volume | cubic metre per kilogram | | $m^3/kg$ | $m^3 \cdot kg^{-1}$ | |
| Mass per unit length | kilogram per metre | | kg/m | $m^{-1} \cdot kg$ | |
| Mass per unit area | kilogram per square metre | | $kg/m^2$ | $m^{-2} \cdot kg$ | |
| Density (mass per unit volume) | kilogram per cubic metre | | $kg/m^3$ | $m^{-3} \cdot kg$ | In this SI form, mass density is conveniently 1000 times specific gravity. |
| Moment of inertia | kilogram metre squared | | $kg \cdot m^2$ | $m^2 \cdot kg$ | |
| Mass flow rate | kilogram per second | | kg/s | $kg \cdot s^{-1}$ | |
| Momentum | kilogram metre per second | | $kg \cdot m/s$ | $m \cdot kg \cdot s^{-1}$ | |

**FIGURE 25.3-3** (Continued)

| Unit Group Quantity | Unit Name | Symbol | Formula | Unit Derivation | Remarks |
|---|---|---|---|---|---|
| **Derived Units with Generic Names (Continued)** | | | | | |
| Angular momentum | kilogram metre squared per second | kg·m²/s | $m^2 \cdot kg \cdot s^{-1}$ | |
| Magnetic field strength | ampere per metre | A/m | $m^{-1} \cdot A$ | |
| Current density | ampere per square metre | A/m² | $m^{-2} \cdot A$ | |
| Luminance | candela per square metre | cd/m² | $m^{-2} \cdot cd$ | |
| **c. Units Expressed in Terms of Base Units and/or Derived Units with Special Names** | | | | | |
| Moment of force, torque | newton metre | N·m | $m^2 \cdot kg \cdot s^{-2}$ | |
| Flexural rigidity | newton square metre | n·m² | $m^3 \cdot kg \cdot s^{-2}$ | |
| Force per unit length, surface tension | newton per metre | N/m | $kg \cdot s^{-2}$ (b) | (b) m is canceled out. |
| Dynamic viscosity | pascal second | Pa·s | $m^{-1} \cdot kg \cdot s^{-1}$ | |
| Impact ductility | joule per square metre | J/m² | $kg \cdot s^{-2}$ (c) | (c) m² is canceled out. |
| Combustion heat (per unit volume) | joule per cubic metre | J/m³ | $m^{-1} \cdot kg \cdot s^{-2}$ | |
| Combustion heat (per unit mass), specific energy, specific latent heat | joule per kilogram | J/kg | $m^2 \cdot s^{-2}$ (d) | (d) kg is canceled out. |
| Heat capacity, entropy | joule per kelvin | J/K | $m^2 \cdot kg \cdot s^{-2} \cdot K^{-1}$ | |
| Specific heat capacity, specific entropy | joule per kilogram kelvin | J/(kg·K) | $m^2 \cdot s^{-2} \cdot K^{-1}$ (e) | (e) kg is canceled out. |
| Heat flux density, irradiance; sound intensity | watt per square metre | W/m² | $kg \cdot s^{-3}$ (f) | (f) m² is canceled out. |
| Thermal conductivity | watt per metre kelvin | W/(m·K) | $m \cdot kg \cdot s^{-3} \cdot K^{-1}$ | |
| Coefficient of heat transfer | watt per square metre kelvin | W/(m²·K) | $kg \cdot s^{-3} \cdot K^{-1}$ (g) | (g) m² is canceled out. |
| Thermal resistance, thermal insulance | square metre kelvin per watt | m²·K/W | $kg^{-1} \cdot s^3 \cdot K$ (h) | (h) m² is canceled out. |
| Electric field strength | volt per metre | V/m | $m \cdot kg \cdot s^{-3} \cdot A^{-1}$ | |
| Electric flux density | coulomb per square metre | C/m² | $m^{-2} \cdot s \cdot A$ | |
| Electric charge density | coulomb per cubic metre | C/m³ | $m^{-3} \cdot s \cdot A$ | |
| Electric permittivity | farad per metre | F/m | $m^{-3} \cdot kg^{-1} \cdot s^4 \cdot A^2$ | |
| Electric permeability | henry per metre | H/m | $m \cdot kg \cdot s^{-2} \cdot A^{-2}$ | |
| Electric resistivity | ohm metre | Ω·m | $m^3 \cdot kg \cdot s^{-3} \cdot A^{-2}$ | |
| Electric conductivity | siemens per metre | S/m | $m^{-3} \cdot kg^{-1} \cdot s^3 \cdot A^2$ | |
| Light exposure | lux second | lx·s | $m^{-2} \cdot s \cdot cd \cdot sr$ | |
| Luminous efficacy | lumen per watt | lm/W | $m^{-2} \cdot kg^{-1} \cdot s^3 \cdot cd \cdot sr$ | |
| **d. Units Expressed in Terms of Supplementary Units and Base and/or Derived Units** | | | | | |
| Angular velocity | radian per second | rad/s | $s^{-1} \cdot rad$ | |
| Angular acceleration | radian per second squared | rad/s² | $m^{-2} \cdot rad$ | |
| Radiant intensity | watt per steradian | W/sr | $m^2 \cdot kg \cdot s^{-3} \cdot sr^{-1}$ | |
| Radiance | watt per square metre steradian | W/m²·sr | $kg \cdot s^{-2} \cdot sr^{-1}$ (i) | (i) m² is canceled out. |

*Note: Metre* is the preferred international spelling applied to this base unit and all its multiples and submultiples.

(Copyright ASTM; reprinted with permission.)

However, the use of the SI spelling is controversial in the United States, and the incorrect spelling *meter* may actually win out over the correct SI spelling and become the U.S. standard.

SI originally addressed the base units for only six physical quantities: *length*, *mass* (which is closely related to weight), *time*, *temperature*, *electrical current*, and *luminous intensity*. In 1971, because of the growth of thermal dynamics, nuclear physics, and electronic science, a seventh quantity, *amount of substance*, was added. Figure 25.3-4 compares the names for the base units for these seven quantities in SI and in the customary system.

The base units shown in Figure 25.3-5 are the building blocks of SI. Except for the *kilogram*, they are reproducible under controlled conditions anywhere (Fig. 25.3-5). The kilogram is a platinum-iridium *prototype* kept by the International Bureau of Weights and Measures in Sevres, France.

### 25.3.2.2 Supplementary Units

In addition to the seven base units, SI has two supplementary units, the *radian* (plain angle) and the *steradian* (solid angle).

### 25.3.2.3 Derived Units

All other quantities are derived from the seven base and two supplementary units. *Derived units* are formed by combining base and supplementary units and other derived units according to algebraic relations linking corresponding quantities (Fig. 25.3-6). For example, the metre is a base unit, the radian is a supplementary unit, and the pascal is a derived unit.

*Standard prefixes* allow a change in magnitude from the subatomic scale $(10^{-18})$ to the astronomic scale $(10^{18})$

**FIGURE 25.3-4** Customary and SI Base Units

| Quantity | Customary | SI |
|---|---|---|
| Length | foot | metre (m) |
| Mass (weight) | pound | kilogram (kg) |
| Time | second | second (s) |
| Temperature | degree Fahrenheit | kelvin (K) |
| Electric current | ampere | ampere (A) |
| Luminous intensity | candela | candela (cd) |
| Amount of a substance | | mole (mol) |

**FIGURE 25.3-5** SI Technical Definitions

**Base Units**

**Metre**
  The length equal to 1 650 763.73 wavelengths in vacuum of the radiation corresponding to the transition between the levels $2p_{10}$ and $5d_5$ of the krypton-86 atom.

**Kilogram**
  The unit of mass, it is equal to the mass of the international prototype of the kilogram. The kilogram is the only base unit whose name for historic reasons contains a prefix.

**Second**
  The measure of time with a duration of 9 192 631 770 periods of the radiation corresponding to the transition between the two hyperfine levels of the ground state of the cesium-133 atom.

**Ampere**
  The measure of current flow equal to that flow that, if maintained in two straight parallel conductors, would produce between those conductors a force equal to $2 \times 10^7$ newtons per metre of length. The conductors are assumed to be of infinite length, of negligible circular cross section, and placed 1 metre apart in vacuum.

**Kelvin**
  The unit of thermodynamic temperature equal to the fraction 1/273.16 of the thermodynamic temperature of the triple point of water.
  The temperature intervals of the degree Celsius and the kelvin are identical and the scales using these two units are related by a difference of exactly 273.15 kelvins.

**Mole**
  The amount of substance of a system that contains as many elementary entities as there are atoms in 0.012 kilogram of carbon-12. When the mole is used, the elementary entities must be specified and may be atoms, molecules, ions, electrons, other particles, or specified groups of such particles.

**Candela**
  The luminous intensity in the perpendicular direction of a surface of 1/600 000 square metre of blackbody at the temperature of freezing platinum under a pressure of 101 325 newtons per square metre.

**Supplementary Units**

**Radian**
  The plane angle between two radii of a circle that cut off on the circumference an arc equal in length to the radius. The radian is equal to $360/2\pi$ and is an angle of approximately 57° 17′ 44.6″.

**Steradian**
  The solid angle that, having its vertex in the center of a sphere, cuts off an area of the surface of the sphere equal to that of a square with sides of length equal to the radius of the sphere. The total solid angle of a sphere is $4\pi$ steradians.

(see Fig. 25.1-1). The use of multiples of SI units that cannot be handled by using the SI prefixes is discouraged.

### 25.3.2.4 Non-SI Units for Use with SI

To preserve the advantages of SI as a *coherent* system, the use of units from other systems should be minimized. There are, however, acceptable but noncoherent traditional units retained for use with SI. Non-SI units accepted for general use with SI include units for volume (*litre*), mass (*metric ton*), time (*minute, hour, day, year*), plane angle (*degree of arc*), and velocity (*kilometre per hour*). Three other non-SI units are accepted for limited use. These include units for area (*hectare*), energy (*kilowatt-hour*), and speed of rotation (*revolutions per minute*). These quantities and their associated units, along with their symbols and their relationship to SI *units*, are given in Figure 25.3-7.

### 25.3.2.5 Abandoned Metric Units

Many metric units in use before SI was adopted have been abandoned. SI supersedes all previous metric systems. Abandoned units include *centimetre-gram-second* (cgs) *units* such as *abampere* and *statvolt*. Decimal multiples of SI units, such as the *angstrom* (0.1 mm), that cannot be handled by using the SI *prefixes* should not be used.

### 25.3.2.6 Special Names

Except for *litre, metric ton*, and *hectare*, special names or multiples and submultiples of SI units should be avoided.

### 25.3.3 RECOMMENDATIONS FOR THE USE OF SI

SI symbols and many rules for their proper use are internationally agreed upon and should be used (Figs. 25.3-8 and 25.3-9).

### 25.3.3.1 Writing Unit Symbols

In general, unit symbols should be:

1. Printed in upright type, regardless of the type style of the surrounding text
2. Unaltered in the plural
3. Not followed by a period except when appearing at the end of a sentence
4. Written in lowercase unless the unit is derived from a proper name
5. Written with a space between the unit symbol and the numerical value

**FIGURE 25.3-6** Derived SI Units with Special Names

| Physical Quantity | Unit | Symbol | Definition |
|---|---|---|---|
| Absorbed dose | gray | Gy | The energy imparted by ionizing radiation to a mass of matter corresponding to 1 joule per kilogram. |
| Activity | becquerel | Bq | The activity of a radionuclide having one spontaneous nuclear transition per second. |
| Electric capacitance | farad | F | The capacitance of a capacitor between the plates of which there appears a difference of potential of 1 when it is charged by a means of electricity equal to one coulomb. |
| Electric conductance | siemens | S | The electric conductance of a conductor in which a current of 1 ampere is produced by an electric potential difference of 1 volt. |
| Electric inductance | henry | H | The inductance of a closed circuit in which an electromotive force of 1 volt is produced when the electric current in the circuit varies uniformly at a rate of 1 ampere per second. |
| Electric potential difference, electromotive force | volt | V | The unit of electric potential difference and electromotive force that is the difference of electric potential between two points of a conductor carrying a constant current of 1 ampere, when the power dissipated between these points is equal to 1 watt. |
| Electric resistance | ohm | Ω | The electric resistance between two points of a conductor when a constant difference of potential of 1 volt, applied between these two points, produces in this conductor a current of 1 ampere, this conductor not being the source of any electromotive force. |
| Energy | joule | J | The work that is done when the point of application of a force of 1 newton is displaced a distance of 1 metre in the direction of the force. |
| Force | newton | N | The force that, when applied to a body having a mass of 1 kilogram, gives it an acceleration of 1 metre per second squared. |
| Frequency | hertz | Hz | The frequency of a periodic phenomenon of which the period is 1 second. |
| Illuminance | lux | lx | The illuminance produced by a luminous flux of 1 lumen uniformly distributed over a surface of 1 square metre. |
| Luminous flux | lumen | lm | The luminous flux emitted in a solid angle of 1 steradian by a point source having a uniform intensity of 1 candela. |
| Magnetic flux | weber | Wb | The magnetic flux that, linking a circuit of one turn, produces in it an electromotive force of 1 volt as it is reduced to zero at a uniform rate in 1 second. |
| Magnetic flux density | tesla | T | The magnetic flux density given by a magnetic flux of 1 weber per square metre. |
| Power | watt | W | The power that gives rise to the production of energy at the rate of 1 joule per second. |
| Pressure or stress | pascal | Pa | The pressure or stress of 1 newton per square metre. |
| Quantity of electricity | coulomb | C | The quantity of electricity transported in 1 second by a current of 1 ampere. |

Because a comma is used in some countries as a decimal marker, it should not be used to separate digits into groups of three. Instead, digits should be divided into groups of three by using a space at least equal to the width of the letter I to separate these groups. A space is not necessary in four-digit numbers. Examples are:

0.133 47
2.141 596
73 722
7372

The word *billion* means a *thousand million* (prefix *giga*) in the United States but a *million million* (prefix *tera*) in many other countries. Therefore, terms such as *billion* and *trillion* should not be used in technical writing.

### 25.3.3.2 Preferred Range of Values

Whole numbers between 1 and 1000 should be used with the proper prefix wherever possible. For example, 725 m is preferred over 0.725 km or 725 000 mm. However, the prefix that adequately covers the whole quantity range should be used. For example, if 725 m is part of a group of numbers shown in kilometres, it should be expressed as 0.725 km.

*Exponential notation*, instead of prefixes, is preferred in calculations. For example, 900 mm² equals $0.9 \times 10^{-3}$ m².

One measurement unit should be used throughout a set of drawings so that numerical values can be represented by numbers only. In a drawing dimensioned in millimetres, five-digit numbers are acceptable—for example, 32 845 mm.

## 25.4 SI Units for Design and Construction

There are two principal factors in any measurement statement: a *number*, which is always written first, and a *measurement unit*. Correct selection of units for use in building construction is essential to minimize errors and optimize coordination. Figure 25.4-1 lists SI units needed for building construction in terms of

**FIGURE 25.3-7** Other Units Whose Use Is Permitted with SI

| Quantity | Unit Name | Symbol | Relationship to SI Unit | Remarks |
|---|---|---|---|---|
| **Units for General Use** | | | | |
| Volume | litre[a] | L[a] | $1 \text{ L} = 0.001 \text{ m}^3 = 10^6 \text{ mm}^3$ | An alternative spelling is *liter*. The litre may only be used with the SI prefix milli-. |
| Mass | metric ton[b] | t[b] | $1 \text{ t} = 1000 \text{ kg} = (1 \text{ Mg})$ | |
| Time | minute | min | $1 \text{ min} = 60 \text{ s}$ | |
| | hour | h | $1 \text{ h} = 3600 \text{ s} = (60 \text{ min})$ | |
| | day (mean solar) | d | $1 \text{ d} = 86\,400 \text{ s} = (24 \text{ h})$ | |
| | year (calendar) | a | $1 \text{ a} = 31\,536\,000 \text{ s} = (365 \text{ d})$ | |
| Temperature interval | degree Celsius | °C | $1°\text{C} = 1 \text{ K}$ | The Celsius temperature 0°C corresponds to 273.15 K exactly. $(t_{°C} = T_K - 273.15)$ |
| Plane angle | degree (of arc) | ° | $1° = 0.017453 \text{ rad} = 17.453 \text{ mrad}$ | $1° = (\pi/180) \text{ rad}$ |
| Velocity | kilometre per hour | km/h | $1 \text{ km/h} = 0.278 \text{ m/s}$ | $1 \text{ m/s} = 3.6 \text{ km/h}$ |
| **Units Accepted for Limited Use** | | | | |
| Area | hectare | ha | $1 \text{ ha} = 10\,000 \text{ m}^2$ | For use in land measurement. |
| Energy | kilowatthour | kWh | $1 \text{ kWh} = 3.6 \text{ MJ}$ | For measurement of electrical energy consumption only. |
| Speed of rotation | revolution per minute | r/min | $1 \text{ r/min} = \frac{1}{60} \text{ r/s} = \frac{2\pi}{60} \text{ rad/s}$ | To measure rotational speed in slow-moving equipment only. |

[a]The international symbol for litre is the lowercase l (ell), which can be easily confused with the numeral 1. Several English-speaking countries have adopted the script l as a symbol for litre to avoid any misinterpretation. The symbol L (capital ell) is recommended for U.S. use to prevent confusion.

[b]The international name for metric ton is tonne. The metric ton is equal to the megogram (Mg).

(Copyright ASTM; reprinted with permission.)

**FIGURE 25.3-8** Rules and Recommendations for the Presentation of SI Units and Symbols

| | Typical Examples | Remarks |
|---|---|---|
| **General** | | |
| 1. All unit names should either be denoted by correct symbols or be written in full. In the interest of simplification and to reduce the amount of writing, use unit symbols rather than fully written forms. | | |
| 2. Do not use mixtures of names and symbols. | Use: J/kg *or* joule per kilogram | Not: joule per kg Not: J/kilogram |
| **Symbols for Unit Quantities and Prefixes** | | |
| 1. SI symbols are internationally agreed upon and there is only *one* symbol for each unit quantity. Multiples and submultiples are formed by using the unit symbol and attaching a prefix symbol in front of it. | m, kg, s, A, cd, K | See also Nos. 5–7, below. |
| 2. All unit symbols are shown in upright letters, and can be produced by a normal typewriter keyboard, with the exceptions of the symbols for the SI unit ohm and the prefix micro-, which are represented by Greek letters $\Omega$ and $\mu$, respectively. | | Exceptions: $\Omega$, $\mu$ |
| 3. Unit symbols are never followed by a period (full stop) except at the end of a sentence. | 60 kg/m | Not: 60 kg./m. |
| 4. Unit symbols are normally written in lowercase, except for unit names derived from a proper name, in which case the initial is capitalized. Some units have symbols consisting of two letters from a proper name, of which *only* the first letter is capitalized. (The symbol for the unit name ohm is the capital Greek letter $\Omega$.) | m, kg, s, mol, cd, etc. A, K, N, J, W, V, etc. Pa, Hz, Wb, etc. | Exception: L |
| 5. Prefixes for magnitudes from $10^6$ to $10^{18}$ have capital upright letter symbols. | M, G, T, etc. | See also cl |
| 6. Prefixes for magnitudes from $10^{-18}$ through $10^3$ have lowercase upright letter symbols. (The symbol for $10^{-6}$ or micro is the lowercase Greek letter $\mu$.) | p, n, $\mu$, m, k, etc. | See also cl |
| 7. Prefix symbols are directly attached to the unit symbol *without a space* between them. | mm, kW, MN, etc. | Not: m m, k W, M N |
| 8. Do not use compound prefixes to form a multiple or submultiple of a unit (e.g., use nanometre, do not use micromillimetre *or* millimicrometre). | nm | Not: $\mu$mm or m$\mu$m |
| 9. In the case of the base unit kilogram, prefixes are attached to the gram (e.g., milligram, not microkilogram). | mg | Not: $\mu$kg |
| 10. Use only one prefix when forming a multiple or a submultiple of a compound unit. Normally, the prefix should be attached to a unit in the numerator. An exception to this rule is made for the base unit kilogram. | km/s, mV/m | Not: mm/$\mu$s, $\mu$V/mm Exception: MJ/kg not: kJ/g |

**FIGURE 25.3-8** *(Continued)*

| | Typical Examples | Remarks |
|---|---|---|
| **Areas of Possible Confusion Requiring Special Care** | | |
| 1. The symbols for SI units and the conventions that govern their use should be strictly followed. A number of prefix and unit symbols use the same letter, but in different form. Exercise care to present the correct symbol for each quantity. | g (gram), G (giga), k (kilo), K (kelvin), m (milli), m (metre), M (mega), n (nano), N (newton) | Others: c (centi), C (coulomb), °C (degree Celsius), s (second), S (siemens), t (metric ton), T (tesla), T (tera) |
| 2. All prefix and unit symbols retain their prescribed form regardless of the surrounding typography. In printouts from limited character sets (telex, computer printers) special considerations apply to symbols for mega, micro, ohm, and siemens. Where confusion is likely to arise, write units in full. | | |
| **Unit Names Written Out in Full** | | |
| 1. Unit names, including prefixes, are treated as common names and are *not capitalized*, except at the beginning of sentences or in titles. (The only exception is Celsius in degree Celsius, where degree is considered as the unit name and is shown in lowercase, whereas Celsius represents an adjective and is capitalized.) | metre, newton, etc. | Not: Metre, Newton Exception: degree Celsius |
| 2. Where a prefix is attached to an SI unit to form a multiple or submultiple, the combination is written as one word. (There are three cases where the final vowel of the prefix is omitted in the combination: megohm, kilohm, and hectare.) | millimetre, kilowatt | Not: milli-metre Not: kilo watt |
| 3. Where a compound unit is formed by multiplication of two units, the use of a space between units is preferred, but a hyphen is acceptable and in some situations more appropriate, to avoid any risk of misinterpretation. | newton metre *or* newton-metre | Not: newtonmetre |
| 4. Where a compound unit is formed by division of the two units, this is expressed by inserting the word *per* between the numerator and the denominator. | metre per second joule per kelvin | Not: metre/second Not: joule/kelvin |
| 5. Where the numerical value of a unit is written in full, the unit should also be written in full. | seven metres | Not: seven m |
| **Plurals** | | |
| 1. Units written in full are subject to the normal rules of grammar. For any unit with a numerical value greater than one (1), an *s* is added to the written unit to denote the plural. | 1.2 metre*s* 2.3 newton*s* 33.2 kilogram*s* | But: 0.8 metre |
| 2. The following units have the same plural as singular when written out in full: hertz, lux, siemens. | 350 kilohertz 12.5 lux | |
| 3. Symbols never change in the plural. | 2.3 N, 33.2 kg | Not: 2.3 N*s*, 33.2 kg*s* |
| **Compound Unit Symbols—Products and Quotients** | | |
| 1. The product of two units is indicated by a dot placed at midheight between the unit symbols. | kN·m, Pa·s | Not: kNm, Pas Not: kN m, Pa s |
| 2. To express a derived unit formed by division, any one of the following methods may be used: | | See also f3 and f5 |
|   a. a solidus (slash, /) | kg/m³, W/(m·K) | See also f3 and f5 |
|   b. a horizontal line between numerator and denominator | $\dfrac{kg}{m^3}, \dfrac{W}{m·K}$ | |
|   c. a negative index (or negative power). | kg·m⁻³, W·m⁻¹·K⁻¹ | |
| 3. *Only one* solidus may be used in any combination. | m/s², m·kg/(s³·A) | Not: m/s/s Not: m·kg/s³/A |
| 4. Do not use the abbreviation *p* for per in the expression of a division. | km/h | Not: kph *or* k.p.h. |
| 5. Where the denominator is a product, this should be shown in parentheses. | W/(m²·K) | |

*Note: Metre* is the preferred international spelling applied to this base unit and all its multiples and submultiples.

(Copyright ASTM; reprinted with permission.)

activity. Figures 25.4-2 to 25.4-7 contain lists of recommended units for use in building design and construction in terms of quantity and application.

## 25.4.1 LINEAR MEASUREMENT

The preferred SI units for linear measurement in building design and construction are the *millimetre* (mm) and the *metre* (m). These have inherently greater precision than units in the customary system, which uses decimalized fractions in sitework but uses feet, inches, and fractions in construction work. Using the millimetre as the principal unit of construction measurement makes fractions unnecessary. One millimetre provides an accuracy of ¹/₂₅ in.; five digits can express any length up to 328 ft. The use of whole numbers greatly speeds calculations and helps to avoid errors; calculations can be done twice as fast with only one-fifth the rate of error (Fig. 25.4-8). In the customary system, fractions have to be reconciled to a common denominator and then added, inches have to be

**FIGURE 25.3-9** Presentation of Numerical Values in Conjunction with SI

| | Typical Examples | Remarks |
|---|---|---|
| **Decimal Marker** | | |
| 1. Whereas most European countries use the comma on the line as the decimal marker and this practice is advocated by ISO, a special exception is made for documents in the English language, which have traditionally used the point (dot) on the line, or period, as the decimal marker. | | See also under g |
| 2. The recommended decimal marker for use in the United States is the point on the line (period), and the comma should not be used.<br>    In handwritten documents the decimal marker may be shown slightly above the line for clear identification. | 9.9; 15.375 | Not: 9,9; 15,375 |
| 3. *Always* show a zero before the decimal point for all numbers smaller than 1.0 (one). | 0.1; 0.725 | Not: .1; .725 |
| **Spacing** | | |
| 1. *Always* leave a gap between the numerical value associated with a symbol and the symbol of at least half a space in width.<br>    In the case of the symbol for degree Celsius this space is optional, but the degree symbol must always be attached to Celsius. | 900 MHz; 200 mg; $10^6$ mm$^2$<br>20°C or 20 °C | Not: 900MHz; 200mg<br>Not: $10^6$mm$^2$<br>Not: 20° C |
| 2. In non-SI expressions of plane angle (°, ′, ″), do not leave a space between the numerical value and the symbol. | 27°30′ (of arc) | Not: 27 ° 30 ′ |
| 3. *Always* leave a space on each side of signs for multiplication, division, addition, and subtraction. | 100 mm × 100 mm;<br>36 MPa + 8 MPa | Not: 100 mm×100 mm<br>Not: 36 MPa+ 8 MPa |
| **Fractions** | | |
| 1. *Avoid* common fractions in connection with SI units. | Write: 0.5 kPa | Not: ½ kPa |
| 2. *Always* use decimal notation to express fractions of any number larger than 1.0 (one). | 1.5; 16.375 | Not: Not: 1½; 16⅜ |
| 3. Although the most common fractions such as half, third, quarter, and fifth will remain in speech, *always* show decimal notation in written, typed, or printed material. | 0.5; 0.33; 0.25; 0.2 | Not: ½; ⅓; ¼; ⅕ |
| **Powers of Units and Exponential Notation** | | |
| 1. When writing unit names with a modifier *squared* or *cubed*, the following rules should be applied: | | |
|    a. In the case of area and volume, the modifier is written before the unit name as *square* and *cubic*. | cubic metre; square millimetre | Not: metre cubed; millimetre squared |
|    b. In all other cases, the modifier is shown after the unit name as *squared*, *cubed*, *to the fourth power*, etc. | metre per second squared | Not: metre per square second (or metre per second per second) |
|    c. The abbreviations *sq.* for square, and *cu.* for cubic should *not* be used. | | Not: sq. millimetre<br>Not: cu. metre |
| 2. For unit symbols with modifiers (such as square, cubic, fourth power, etc.) *always* show the superscript immediately after the symbol. | m$^2$; mm$^3$; s$^4$ | Not: m $^2$; mm $^3$; s $^4$ |
| 3. Show the superscript as a numeral reduced in size raised half a line space. Where a typewriter without superscript numerals is used, the full-sized numeral should be raised half a line space, provided that this does not encroach on the print in the line above. | mm$^3$; m/s$^2$ | Permitted: mm$^3$; m/s$^2$ |
| 4. Where an exponent is attached to a prefixed symbol, it indicates that that multiple (or submultiple) is raised to the power expressed by the exponent. | 1 mm$^3$ = $(10^{-3}$ m$)^3$<br>= $10^{-9}$ m$^3$<br>1 km$^2$ = $(10^3$ m$)^2$<br>= $10^6$ m$^2$ | Not: 1 mm$^3$<br>= $10^{-3}$ m$^3$ |
| **Ratios** | | |
| 1. Do not mix units in expressing a ratio of like unit quantities. | 0.01 m/m<br>0.03 m$^2$/m$^2$ | Not: 10 mm/m<br>Not: 30,000 mm$^2$/m$^2$ |
| 2. Wherever possible, use a nonquantitative expression (ratio or percentage) to indicate measurement of slopes, deflections, etc. | | Preferred:<br>1:100; 0.01; 1%<br>1:33; 0.03; 3% |
| **Range** | | |
| 1. The choice of the appropriate prefix to indicate a multiple or submultiple of an SI unit is governed by convenience to obtain numerical values within a practical range and to eliminate nonsignificant digits. | | |
| 2. In preference, use prefixes representing ternary powers of 10 (10 raised to a power that is a multiple of 3). | milli; kilo; mega | Avoid: centi; deci; deka; hecto |
| 3. Select prefixes so that the numerical value or values occur in a common range between 0.1 and 1000. | 120 kN<br>3.94 mm<br>14.5 MPa | In lieu of: 120 000 N<br>0.003 94 m<br>14 500 kPA |
| 4. Compatibility with the general range must be a consideration; e.g., if all dimensions on a drawing are shown in millimeters (mm), a range from 1 to 99 999 (a maximum of five numerals) would be acceptable to avoid mixing of units. | | Note: Drawings should include the notation<br>*All dimensions in millimetres.* |

**FIGURE 25.3-9** *(Continued)*

| | Typical Examples | Remarks |
|---|---|---|
| **Presentation and Tabulation of Numbers** | | |
| 1. In numbers with many digits it has been common practice in the United States to separate digits into groups of three by means of commas. This practice must *not* be used with SI to avoid confusion. It is recommended international practice to arrange digits in long numbers in groups of three from the decimal marker, with a gap of not less than half a space, and not more than a full space, separating each group. | 54 375.260 55 | Not: 54,375.260,55 |
| 2. For individual numbers with four digits before (or after) the decimal marker, this space is not necessary. | 4500; 0.0355 | |
| 3. In all tabulations of numbers with five or more digits before and/or after the decimal marker, group digits into groups of three: e.g., 12.5255; 5735; 98 300; 0.425 75 | 12.525 5<br>5 735<br>98 300<br>0.425 75<br>———————<br>104 047.951 25 | |
| **Use of Unprefixed Units in Calculations** | | |
| Errors in calculations involving compound units can be minimized if all prefixed units are reverted to coherent base or derived units, with numerical values expressed in powers-of-ten notation. | Preferred: 136 kJ =<br>$136 \times 10^3$ J<br>20 MPa =<br>$20 \times 10^6$ Pa<br>1.5 t (Mg) =<br>$1.5 \times 10^3$ kg | Also acceptable:<br>(or $1.36 \times 10^5$ J)<br><br>(or $2 \times 10^7$ Pa) |

*Note: Metre* is the preferred international spelling applied to this base unit and all its multiples and submultiples.

(Copyright ASTM; reprinted with permission.)

---

added and then converted to feet and inches, and, finally, the feet must be added.

Sizes and dimensions should not be listed in centimetres, because (1) the order of magnitude between millimetres and centimetres is only 10 and using both units causes confusion, and (2) using millimetres permits the use of whole numbers rather than fractions within appropriate tolerances for all building and product dimensions.

Unit symbols may be deleted on a drawing if a note such as "Dimensions are shown in millimetres (or metres), unless otherwise noted" appears on the drawing.

Whole numbers always indicate millimetres. A five-digit number can indicate a length up to 328 ft; 327 ft $10^{11}/_{16}$ in. equals 99 941 mm. A length up to 32 ft 9 in. can be shown by a four-digit number; a length of 3 ft $3^5/_{16}$ in. by a three-digit number; $^1/_{25}$ in. is 1 millimetre.

Numbers carried to three decimal places always indicate metres; for example: 3.600, 0.300, and 0.025.

Using millimetres and metres saves time and space in drawing, typing, and computer applications. It also improves clarity in drawings that have many dimensions.

The drawing scales used by architects and engineers have traditionally differed. Architects have generally used scales such as $^1/_4$ in. equals 1 ft, which resulted in nondecimal scale ratios; even when engineers used seemingly metric scales such as 1 in. equals 100 ft, the ratio was nondecimal (Fig. 25.4-9). Metrication will eliminate these differences.

The consistent use of fewer scales has definite benefits: (1) drawings prepared on transparent media drawn to the same scale may be easily superimposed, which makes drawing faster and easier and reduces the possibility of errors and omissions; (2) drawings lend themselves more to computer-aided reduction and enlargement techniques; and (3) consistently sized drawings are easier to understand and speed construction.

## 25.4.2 AREA

In general, the preferred unit for area is the derived SI unit *square metre* ($m^2$). Large areas can be expressed in square *kilometres* ($km^2$) or *hectares* (ha). *Hectare* is an alternative name for square *hectometre*, which is equal to 10 000 $m^2$. However, the *hectare* (ha)

should be used only for land and water surface area measurements. Small areas should be expressed in square *millimetres* ($mm^2$). The *square centimetre* ($cm^2$) should be avoided to minimize confusion. Sometimes the area of buildings or products is indicated by linear dimensions (e.g., 40 by 90 mm); generally, width is indicated first and height second.

## 25.4.3 VOLUME AND FLUID CAPACITY

The derived SI unit *cubic metre* ($m^3$) is the preferred measurement unit for volume in construction and for large storage tank capacities.

The non-SI units *litre* (L) and *millilitre* (mL) are the preferred measurement units of *fluid capacity* (liquid volume) and gas volume. A *litre* is equal to $^1/_{1000}$th of a cubic metre or 1 cubic *decimetre* ($dm^3$) and can be expressed as 1 L = $10^{-3}$ $m^3$; 1 L = 1 $dm^3$; or 1 $m^3$ = 1000 L.

A cubic metre contains 1 billion ($10^9$) cubic millimetres. Cubic centimetres and decimetres should be converted to the preferred cubic metres or millimetres whenever possible (see Fig. 25.4-2).

**FIGURE 25.4-1** SI Units Used in Building Construction

| Activity | Quantity | Unit | Symbol |
|---|---|---|---|
| Land surveying | Linear measure | kilometre, metre | km, m |
| | Area | square kilometre | km² |
| | | hectare (10 000 m²) | ha |
| | | square metre | m² |
| Excavating | Linear measure | metre, millimetre | m, mm |
| | Volume | cubic metre | m³ |
| Concreting | Linear measure | metre, millimetre | m, mm |
| | Area | square metre | m² |
| | Volume | cubic metre | m³ |
| | Temperature | degree Celsius | °C |
| | Water capacity | litre | L |
| Constituents | Mass (weight) | kilogram, gram | kg, g |
| Reinforcement | Cross section | square millimetre | mm² |
| Trucking | Distance | kilometre | km |
| | Mass (weight) | metric ton (1000 kg) | t |
| Paving and | Linear measure | metre, millimetre | m, mm |
| plastering | Area | square metre | m² |
| Bricklaying | Linear measure | metre, millimetre | m, mm |
| | Area | square metre | m² |
| | Mortar-volume | cubic metre | m³ |
| Carpentry/joinery | Linear measure | metre, millimetre | m, mm |
| Steelworking | Linear measure | metre, millimetre | m, mm |
| | Mass (weight) | metric ton (1000 kg) | t |
| | | kilogram, gram | kg, g |
| Roofing | Linear measure | metre, millimetre | m, mm |
| | Area | square metre | m² |
| | Slope | millimetre/metre | mm/m |
| Painting | Linear measure | metre, millimetre | m, mm |
| | Area | square metre | m² |
| | Capacity | litre, millilitre | L, mL |
| Glazing | Linear measure | metre, millimetre | m, mm |
| | Area | square metre | m² |
| Plumbing | Linear measure | metre, millimetre | m, mm |
| | Mass (weight) | kilogram, gram | kg, g |
| | Capacity | litre | L |
| | Pressure | kilopascal | kPa |
| Drainage | Linear measure | metre, millimetre | m, mm |
| | Area | hectare (10 000 m²) | ha |
| | | square metre | m² |
| | Volume | cubic metre | m³ |
| | Slope | millimetre/metre | mm/m |
| Electrical | Linear measure | metre, millimetre | m, mm |
| services | Frequency | hertz | Hz |
| | Power | watt, kilowatt | W, kW |
| | Energy | megajoule (1 kWh = 3.6 MJ) | MJ |
| | Electric current | ampere | A |
| | Electric potential | volt, kilovolt | V, kV |
| | Resistance | ohm | Ω |
| Mechanical | Linear measure | metre, millimetre | m, mm |
| services | Volume | cubic metre | m³ |
| | Capacity | litre | L |
| | Airflow | metre/second | m/s |
| | Volume flow | cubic metre/second | m³/s |
| | | litre/second | L/s |
| | Temperature | degree Celsius | °C |
| | Force | newton, kilonewton | N, kN |
| | Pressure | kilopascal | kPa |
| | Energy, work | kilojoule, megajoule | kJ, MJ |

## 25.4.4 GEOMETRIC CROSS-SECTIONAL PROPERTIES

The expression of *geometric cross-sectional* properties of structural sections involves raising the unit of length to the third, fourth, or sixth power with exponential notation (see Fig. 25.1-1) as follows:

- *Modulus of section* (S) is written in units of mm³ or m³. For example, 1 mm³ = 0.000 000 001 = $10^{-9}$ m³.
- *Second moment of area* (I) and *torsional constant* (J) are written in units of mm⁴ or m⁴. For example, 1 mm⁴ = $10^{-12}$ m⁴.
- *Warping constant* (C) is written in units of mm⁶ or m⁶. For example, 1 mm⁶ = $10^{-18}$ m⁶).

Figure 25.4-10 shows the *cross-sectional properties* of a wide-flange beam 460 mm deep weighing 82 kg/m per unit of length.

## 25.4.5 PLANE ANGLE

The preferred unit for expressing *plain angles* is the SI supplementary unit *radian* (rad). The radian should be used in engineering calculations when possible, because it is a *coherent unit*. However, use of the customary *degree of arc* and its decimal submultiples (as in 26.250) is permissible when using the radian is not convenient, especially on maps and engineering drawings, and in surveying and construction operations. The use of *minutes* and *seconds* is discouraged, with the recognition that they will probably continue in use for some time.

## 25.4.6 TIME INTERVALS

The *second* (s) is the SI base unit for time measurement. The *day* (d), *hour* (h), and *minute* (min) are non-SI alternatives. The minute should be avoided as much as possible, however, to minimize the variety of units in which time is a dimension.

*Flow rates* should be expressed in *cubic metres per second, litres per second,* or *cubic metres per hour:*

1 m³/s = 1000 L/s (do not use 60 m³/min)

1 L/s = 3.6 m³/h (do not use 60 L/min)

1 m³/h = 1000 L/h (do not use 1667 L/min)

| Quantity and SI Unit Symbol | Preferred Units (Symbols) | Other Acceptable Units | Unit Name | Typical Applications | Remarks |
|---|---|---|---|---|---|
| Length (m) | m | | metre | **Architecture and General Engineering** Levels, overall dimensions, spans, column heights, etc., in engineering computations. **Estimating and Specification** Trenches, curbs, fences, lumber lengths, pipes and conduits; lengths of building materials generally. **Land Surveying** Boundary and cadastral surveys; survey plans; heights, geodetic surveys, contours. **Hydraulic Engineering** Pipe and channel lengths, depth of storage tanks or reservoirs, height of potentiometric head, hydraulic head, piezometric head. | Use metres on all drawings with scale ratios between 1:200 and 1:2000. Where required for purposes of accuracy, show dimensions to three decimal places. |
| | mm | | millimetre | **Architecture and General Engineering** Spans, dimensions in buildings, dimensions of building products; depth and width of sections; displacement, settlement, deflection, elongation; slump of concrete, size of aggregate; radius of gyration, eccentricity; detailed dimensions generally; rainfall. **Estimating and Specification** Lumber cross sections; thicknesses, diameters, sheet metal gauges, fasteners; all other building product dimensions. **Hydraulic Engineering** Pipe diameters; radii of groundwater wells; height of capillary rise; precipitation, evaporation. | Use millimetres on drawings with scale ratios between 1:1 and 1:200. Avoid the use of centimetres (cm). Where cm is shown in documents, such as for snow depth, body dimensions, or carpet sizes, etc. convert to mm or m. |
| | km | | kilometre | Distances for transportation purposes, geographical, or statistical applications in surveying; long pipes and channels. | |
| | $\mu$m | | micrometre | Thickness of coatings (paint, galvanizing, etc.), thin sheet materials, size of find aggregate. | |
| Area (m$^2$) | m$^2$ | | square metre | **General Applications** Small land areas; area of cross section of earthworks, channels, and larger pipes; surface area of tanks and small reservoirs; areas in general. **Estimating and Specification** Site clearing; floor areas; paving, masonry construction, roofing, wall and floor finishes, plastering, paintwork, glass areas, membranes, lining materials, insulation, reinforcing mesh, formwork; areas of all building components. | $(1\ m^2 = 10^6\ mm^2)$  Replaces sq. ft.; sq. yd.; and square. Specify masonry construction by wall area $\times$ wall thickness. |
| | mm$^2$ | | square millimetre | Area of cross section for structural and other sections, bars, pipes, rolled and pressed shapes, etc. | Avoid the use of cm$^2$ (square centimetre) by conversion to mm$^2$ $(1\ cm^2 = 10^2\ mm^2$ $= 100\ mm^2).$ |
| | km$^2$ | | square kilometre | Large catchment areas or land areas. | |
| | | ha | hectare | Land areas; irrigation areas; areas on boundary and other survey plans. | $(1\ ha = (10^2\ m)^2$ $= 10^4\ m^2$ $= 10\ 000\ m^2)$ |

**FIGURE 25.4-2** *(Continued)*

| Quantity and SI Unit Symbol | Preferred Units (Symbols) | Other Acceptable Units | Unit Name | Typical Applications | Remarks |
|---|---|---|---|---|---|
| **Volume, capacity (m³)** | m³ | | cubic metre | **General Applications** Volume, capacity (large quantities); volume of earthworks, excavations, filling, waste removal; concrete, sand, all bulk materials supplied by volume, and large quantities of lumber. **Hydraulic Engineering** Water distribution, irrigation, diversions, sewage, storage capacity, underground basins. | $1 \text{ m}^3 = 1000 \text{ L}$ As far as possible, use the cubic metre as the preferred unit of volume for all engineering purposes. |
| | mm³ | | cubic millimetre | Volume, capacity (small quantities) | |
| | | L | litre | Volume of fluids and containers for fluids; liquid materials, domestic water supply, consumption; volume/capacity of fuel tanks | The litre and its multiples or submultiples may be used for domestic and industrial supplies of liquids $1 \text{ L} = 1 \text{ dm}^3$ $= 1000 \text{ cm}^3$ $1 \text{ mL} = 1 \text{ cm}^3$ |
| | | mL | millilitre | Volume of fluids and containers for fluids (limited application only) | |
| | | cm³ | cubic centimetre | Limited application only (small quantities) | $1 \text{ cm}^3 = 1000 \text{ mm}^3$ $= 10^{-6} \text{ m}^3$ |
| **Modulus of section (m³)** | mm³ | | millimetre to third power | Geometric properties of structural sections, such as plastic section modulus, elastic section modulus, etc. | |
| | m³ | | metre to third power | | |
| **Second moment of area (m⁴)** | mm⁴ | | millimetre to fourth power | Geometric properties of structural sections, such as moment of inertia of a section, torsional constant of a cross section. | |
| | m⁴ | | metre to fourth power | | |
| **Plane angle (rad)** | rad | | radian | Generally used in calculations only to preserve coherence. | Slopes and gradients may be expressed as a ratio or as a percentage: 26.57 = 1:2 = 50% = 0.4637 rad (1 rad = 57.2958°) |
| | mrad | | milliradian | | |
| | | ( _._ °) | degree (of arc) | **General Applications** Angular measurement in construction (generally using decimalized degrees); angle of rotation, torsion, shear resistance, friction, internal friction, etc. **Land Surveying** Bearings shown on boundary and cadastral survey plans; geodetic surveying | |
| **Time, time interval (s)** | s | | second | Time used in methods of test; all calculations involving derived units with a time component, in order to preserve coherence. | Avoid the use of minute (min) as often as possible |
| | | h | hour | Time used in methods of test; all calculations involving labor time, plant hire, maintenance periods, etc. | (1 h = 3600 s 1 d = 86 400 s = 86.4 ks) |
| | | d | day | | |
| | | a | annum (year) | | |
| **Frequency (Hz)** | Hz | | hertz | Frequency of sound, vibration, shock; frequency of electromagnetic waves | $(1 \text{ Hz} = 1/s = s^{-1})$ Replaces cycle(s) per second (c/s or cps) |
| | kHz | | kilohertz | | |
| | MHz | | megahertz | | |

**FIGURE 25.4-2** *(Continued)*

| Quantity and SI Unit Symbol | Preferred Units (Symbols) | Other Acceptable Units | Unit Name | Typical Applications | Remarks |
|---|---|---|---|---|---|
| Rotational frequency, speed of rotation ($s^{-1}$) | | r/s | revolution per second | Widely used in the specification of rotational speed of machinery; use r/min (revolutions per minute) only for slow-moving machinery. | (1 r/s = $2\pi$ rad/s = 60 r/min) |
| Velocity, speed (m/s) | m/s | | metre per second | Calculations involving rectilinear motion, velocity and speed in general; wind velocity; velocity of fluids; pipe flow velocity | (1 m/s = 3.6 km/h) |
| | | km/h | kilometre per hour | Wind speed; speed used in transportation; speed limits | |
| | | mm/h | millimetre per hour | Rainfall intensity | |
| Angular velocity (rad/s) | rad/s | | radian per second | Calculations involving rotational motion | |
| Linear acceleration ($m/s^2$) | $m/s^2$ | | metre per second squared | Kinematics, and calculations of dynamic forces | Recommended value of acceleration of gravity for use in U.S.: $g_{us} = 9.8$ m/s$^2$ |
| Volume rate of flow ($m^3/s$) | $m^3/s$ | | cubic metre per second | Volumetric flow in general; flow in pipes, ducts, channels, rivers; irrigation spray demand | (1 m$^3$/s = 1000 L/s) |
| | | $m^3/h$ | cubic metre per hour | | |
| | | $m^3/d$ | cubic metre per day | | |
| | L/s | | litre per second | Volumetric flow of fluids only | |
| | L/d | | litre per day | | |

*Note: Metre* and *litre* are preferred international spellings applied to these base units and all their multiples and submultiples, such as kilometre and millilitre. (Copyright ASTM; reprinted with permission.)

**FIGURE 25.4-3** Mechanics, Statics, and Dynamics

| Quantity and SI Unit Symbol | Preferred Units (Symbols) | Other Acceptable Units | Unit Name | Typical Applications | Remarks |
|---|---|---|---|---|---|
| Mass (kg) | kg | | kilogram | Mass of materials in general, mass of structural elements and machinery | Use kilograms (kg) in calculations and specifications |
| | g | | gram | Mass of samples of material for testing | Masses greater than 10$^4$ kg (10 000 kg) may be conveniently expressed in metric tons (t): 1 t = 10$^3$ kg = 1 Mg = 1000 kg |
| | | t | metric ton | Mass of large quantities of materials, such as structural steel, reinforcement, aggregates, concrete, etc.; ratings of lifting equipment | |
| Mass per unit length (kg/m) | kg/m | | kilogram per metre | Mass per unit length of sections, bars, and similar items of uniform cross section | Also known as Linear Density |
| | | g/m | gram per metre | Mass per unit length of wire and similar material of uniform cross section | |

**FIGURE 25.4-3** *(Continued)*

| Quantity and SI Unit Symbol | Preferred Units (Symbols) | Other Acceptable Units | Unit Name | Typical Applications | Remarks |
|---|---|---|---|---|---|
| Mass density, concentration ($kg/m^3$) | $kg/m^3$ | | kilogram per cubic metre | Density of materials in general; mass per unit volume of materials in a concrete mix; evaluation of masses of structures and materials | Also known as Mass per Unit Volume ($1\ kg/m^3 = 1\ g/L$) ($1\ g/m^3 = 1\ mg/L$) |
| | | $g/m^3$ | gram per cubic metre | Mass per unit volume (concentration) in pollution control | |
| | | $\mu g/m^3$ | microgram per cubic metre | | |
| Momentum ($kg \cdot m/s$) | $kg \cdot m/s$ | | kilogram metre per second | Used in applied mechanics; evaluation of impact and dynamic forces | |
| Moment of inertia ($kg \cdot m^2$) | $kg \cdot m^2$ | | kilogram square metre | Rotational dynamics; evaluation of the restraining forces required for propellers, windmills, etc. | |
| Mass per unit time ($kg/s$) | $kg/s$ | | kilogram per second | Rate of transport of material on conveyors and other materials handling equipment | $1\ kg/s = 3.6\ t/h$ |
| | | $t/h$ | metric ton per hour | | |
| Force (N) | N | | newton | Unit of force for use in calculations | $1\ N = 1\ kg \cdot m/s^2$ |
| | kN | | kilonewton | Forces in structural elements, such as columns, piles, ties, prestressing tendons, etc.; concentrated forces; axial forces; reactions; shear force; gravitational force | |
| Force per unit length (N/m) | N/m | | newton per metre | Unit for use in calculations | |
| | kN/m | | kilonewton per metre | Transverse force per unit length on a beam, column, etc.; force distribution in a linear direction | |
| Moment of force, torsional, or bending moment ($N \cdot m$) | $N \cdot m$ | | newton metre | Bending moments (in structural sections), torsional moment; overturning moment; tightening tension for high strength bolts; torque in engine drive shafts, axles, etc. | |
| | $kN \cdot m$ | | kilonewton metre | | |
| | $MN \cdot m$ | | meganewton metre | | |
| Pressure, stress, modulus of elasticity (Pa) | Pa | | pascal | Unit for use in calculations; low differential pressure in fluids | ($1\ Pa = 1\ N/m^2$) |
| | kPa | | kilopascal | Uniformly distributed pressure (loads) on floors; soil bearing pressure; wind pressure (loads), snow loads, dead and live loads; pressure in fluids; differential pressure (e.g., in ventilating systems) | Where wind pressure, snow loads, dead and live loads are shown in $kN/m^2$ change units to kPa |
| | MPa | | megapascal | Modulus of elasticity; stress (ultimate, proof, yield, permissible, calculated, etc.) in structural materials; concrete and steel strength grades | $1\ MPa = 1\ MN/m^2$ $= 1\ N/mm^2$ |
| | GPa | | gigapascal | Modulus of elasticity in high strength materials | |
| | $\mu Pa$ | | micropascal | Sound pressure ($20\ \mu Pa$ is the reference quantity for sound pressure level) | |
| Compressibility ($Pa^{-1}$) | 1/Pa | | reciprocal (of) pascal | Settlement analysis (coefficient of compressibility), bulk compressibility | ($1/Pa = 1\ m^2/N$) |
| | 1/kPa | | reciprocal (of) kilopascal | | |

**FIGURE 25.4-3** *(Continued)*

| Quantity and SI Unit Symbol | Preferred Units (Symbols) | Other Acceptable Units | Unit Name | Typical Applications | Remarks |
|---|---|---|---|---|---|
| Dynamic viscosity (Pa·s) | Pa·s | | pascal second | Shear stresses in fluids | $(1\ \text{Pa·s} = 1\ \text{N·s/m}^2)$ The centipoise (cP) = $10^{-3}$ Pa·s will not be used |
| | mPa·s | | millipascal second | | |
| Kinematic viscosity (m²/s) | m²/s | | square metre per second | | The centistokes (cSt) = $10^{-6}$ m²/s will not be used $1\ \text{cSt} = 1\ \text{mm}^2/\text{s}$ |
| | mm²/s | | square millimetre per second | Computation of Reynolds' number, settlement analysis (coefficient of consolidation) | |
| Work energy (J) | J | | joule | Energy absorbed in impact testing of materials; energy in general; calculations involving mechanical and electrical energy | |
| | kJ | | kilojoule | | |
| | MJ | | megajoule | | |
| | | kWh | kilowatthour | Electrical energy applications only | $1\ \text{kWh} = 3.6\ \text{MJ}$ |
| Impact strength (J/m²) | J/m² | | joule per square metre | Impact strength; impact ductility | |
| | kJ/m² | | kilojoule per square metre | | |
| Power (W) | W | | watt | Power in general (mechanical, electrical, thermal); input/output rating, etc. of motors, engines, heating and ventilating plant and other equipment in general | |
| | kW | | kilowatt | | |
| | MW | | megawatt | Power input/output rating, etc. of heavy power plant | |
| | pW | | picowatt | Sound power level (1 pW is the reference quantity for sound power level) | |

*Note: Metre* is the preferred international spelling applied to this base unit and all its multiples and submultiples.
(Copyright ASTM; reprinted with permission.)

**FIGURE 25.4-4** Thermal Effects and Heat Transfer

| Quantity and SI Unit Symbol | Preferred Units (Symbols) | Other Acceptable Units | Unit Name | Typical Applications | Remarks |
|---|---|---|---|---|---|
| Temperature value (K) | K | | kelvin | Expression of thermodynamic temperature; calculations involving units of temperature | $(t_{°\text{C}} = T_\text{K} - 273.15)$ |
| | | °C | degree Celsius | Common temperature scale for use in meteorology and general applications; ambient temperature values | Temperature values will normally be measured in °C (degree Celsius) |
| Temperature interval (K) | K | | kelvin | Heat transfer calculations; temperature intervals in test methods; etc. | $(1\ \text{K} = 1°\text{C})$ The use of K (kelvin) in compound units is recommended |
| | | °C | degree Celsius | | |

**FIGURE 25.4-4** *(Continued)*

| Quantity and SI Unit Symbol | Preferred Units (Symbols) | Other Acceptable Units | Unit Name | Typical Applications | Remarks |
|---|---|---|---|---|---|
| Coefficient of linear thermal expansion (1/K) | 1/K | | reciprocal (of) kelvin | Expansion of materials subject to a change in temperature (generally expressed as a ratio per kelvin or degree Celsius) | |
| | | 1/°C | reciprocal (of) degree Celsius | | |
| Heat, quantity of heat (J) | J | | joule | Thermal energy calculations. Enthalpy, latent heat, sensible heat | |
| | kJ | | kilojoule | | |
| | MJ | | megajoule | | |
| Specific energy, specific latent heat, combustion heat (mass basis) (J/kg) | J/kg | | joule per kilogram | Heat of transition; heat and energy contained in materials; combustion heat per unit mass; calorific value of fuels (mass basis); specific sensible heat; specific latent heat in psychrometric calculations | |
| | kJ/kg | | kilojoule per kilogram | | |
| | MJ/kg | | megajoule per kilogram | | |
| Energy density, combustion heat (volume basis) (J/m³) | J/m³ | | joule per cubic metre | Combustion heat per unit volume | |
| | kJ/m³ | | kilojoule per cubic metre | | $(1 \text{ kJ/m}^3 = 1 \text{ J/L})$ |
| | MJ/m³ | | megajoule per cubic metre | Calorific value of fuels (volume basis) | $(1 \text{ MJ/m}^3 = 1 \text{ kJ/L})$ |
| Heat capacity, entropy (J/K) | J/K | | joule per kelvin | Thermal behavior of materials, heat transmission calculations, entropy | |
| | kJ/K | | kilojoule per kelvin | | |
| Specific heat capacity, specific entropy (J/(kg·K)) | J/(kg·K) | | joule per kilogram kelvin | Thermal behavior of materials, heat transmission calculations | |
| | kJ/(kg·K) | | kilojoule per kilogram kelvin | | |
| Heat flow rate (W) | W | | watt | Heat flow rate through walls, windows, etc.; heat demand | $(1 \text{ W} = 1 \text{ J/s})$ |
| | kW | | kilowatt | | |
| Power density, heat flux density, irradiance (W/m²) | W/m² | | watt per square metre | Density of power or heat flow through building walls and other heat transfer surfaces; heat transmission calculations | |
| | kW/m² | | kilowatt per square metre | | |
| Heat release rate (W/m³) | W/m³ | | watt per cubic metre | Rate of heat release per unit volume over time (for gases and liquids) | $W/m^3 = J/(m^3 \cdot s)$ |
| | kW/m³ | | kilowatt per cubic metre | | |
| Thermal conductivity (W/(m·K)) | W/(m·K) | | watt per metre kelvin | Estimation of thermal behavior of materials and systems; heat transmission calculations. Thermal conductivity of structural and building materials in fire-resistance testing, insulation, etc. | $1 \text{ W/(m·K)} = 1 \text{ W/(m·°C)}$ (k = value) |

**FIGURE 25.4-4** *(Continued)*

| Quantity and SI Unit Symbol | Preferred Units (Symbols) | Other Acceptable Units | Unit Name | Typical Applications | Remarks |
|---|---|---|---|---|---|
| Coefficient of heat transfer (thermal conductance) ($W/(m^2 \cdot K)$) | $W/(m^2 \cdot K)$ | | watt per square metre kelvin | Heat transfer calculations for buildings, building components, and equipment. Transmittance of construction elements | (U = value) |
| | $kW/(m^2 \cdot K)$ | | kilowatt per square metre kelvin | | |
| Thermal resistivity (($m \cdot K)/W$) | $(m \cdot K)/W$ | | metre kelvin per watt | Heat transmission calculations (reciprocal of thermal conductivity) | |
| Thermal insulance (thermal resistance (($m^2 \cdot K)/W$)) | $(m^2 \cdot K)/W$ | | square metre kelvin per watt | Heat transmission calculations (reciprocal of thermal conductance) | (R = value) |

*Note: Metre* and *litre* are the preferred international spellings applied to these units and all their multiples and submultiples.
(Copyright ASTM; reprinted with permission.)

**FIGURE 25.4-5** Electricity and Magnetism

| Quantity and SI Unit Symbol | Preferred Units (Symbols) | Other Acceptable Units | Unit Name | Typical Applications | Remarks |
|---|---|---|---|---|---|
| Electric current (A) | A | | ampere | Maintenance rating of an electrical installation. Leakage current | |
| | kA | | kiloampere | | |
| | mA | | milliampere | | |
| | $\mu A$ | | microampere | | |
| Magnetomotive force, magnetic potential difference (A) | | | | Used in the calculations involved in magnetic circuits | |
| Magnetic field strength, magnetization (A/m) | A/m | | ampere per metre | Magnetic field strength used in calculation of magnetic circuitry such as transformers, magnetic amplifiers, and general cores | (1 kA/m = 1 A/mm) |
| | kA/m | | kiloampere per metre | | |
| Current density ($A/m^2$) | $A/m^2$ | | ampere per square metre | Design of cross-sectional area of electrical conductor | |
| | $kA/m^2$ | | kiloampere per square metre | | |
| | | $A/mm^2$ | ampere per square millimetre | | ($1 A/mm^2 = 1 MA/m^2$) |
| Electric charge, quantity of electricity (C) | C | | coulomb | The voltage on a unit with capacitive characteristics may be related to the amount of charge present (e.g., electrostatic precipitators). Storage battery capacities | $1 C = 1 A \cdot s$ Do not use ampere hour: $1 A \cdot h = 3.6 kC$ |
| | kC | | kilocoulomb | | |
| | $\mu C$ | | microcoulomb | | |
| | nC | | nanocoulomb | | |
| | pC | | picocoulomb | | |
| Electric potential, potential difference, electromotive force (V) | V | | volt | | 1 V = 1 W/A |
| | MV | | megavolt | | |
| | kV | | kilovolt | | |
| | mV | | millivolt | | |
| | $\mu V$ | | microvolt | | |

**FIGURE 25.4-5** *(Continued)*

| Quantity and SI Unit Symbol | Preferred Units (Symbols) | Other Acceptable Units | Unit Name | Typical Applications | Remarks |
|---|---|---|---|---|---|
| **Electric field strength (V/m)** | V/m | | volt per metre | The electric field strength gives the potential gradient at points in space. This may be used to calculate or test electrical parameters such as dielectric strength | |
| | MV/m | | megavolt per metre | | |
| | kV/m | | kilovolt per metre | | |
| | mV/m | | millivolt per metre | | |
| | μV/m | | microvolt per metre | | |
| **Active power (W)** | W | | watt | The useful power of an electrical circuit is expressed in watts (W). (The apparent power in an electrical circuit is expressed in volt-amperes, (V·A).) | 1 W = 1 V·A |
| | GW | | gigawatt | | |
| | MW | | megawatt | | |
| | kW | | kilowatt | | |
| | mW | | milliwatt | | |
| | μW | | microwatt | | |
| **Capacitance (F)** | F | | farad | Electronic components. Electrical design and performance calculators | 1 F = 1 C/V |
| | mF | | millifarad | | |
| | μF | | microfarad | | |
| | nF | | nanofarad | | |
| | pF | | picofarad | | |
| **Resistance (Ω)** | Ω | | ohm | The design of electrical devices with resistance, such as motors, generators, heaters, electrical distribution systems, etc. | 1 Ω = 1 V/A |
| | GΩ | | gigaohm | | |
| | MΩ | | megaohm | | |
| | kΩ | | kiloohm | | |
| | mΩ | | milliohm | | |
| **Conductance, admittance, susceptance (S)** | S | | siemens | | The siemens (S) was formerly known as mho. |
| | MS | | megasiemens | | |
| | kS | | kilosiemens | | |
| | mS | | millisiemens | | |
| | μS | | microsiemens | | |
| **Resistivity (Ω·m)** | Ω·m | | ohm metre | | |
| | GΩ·m | | gigaohm metre | | |
| | MΩ·m | | megaohm metre | | |
| | kΩ·m | | kiloohm metre | | |
| | mΩ·m | | milliohm metre | | |
| | μΩ·m | | microohm metre | | |
| | nΩ·m | | nanoohm metre | | |

**FIGURE 25.4-5** *(Continued)*

| Quantity and SI Unit Symbol | Preferred Units (Symbols) | Other Acceptable Units | Unit Name | Typical Applications | Remarks |
|---|---|---|---|---|---|
| Electrical conductivity (S/m) | S/m | | siemens per metre | A parameter for measuring water quality | |
| | MS/m | | megasiemens per metre | | |
| | kS/m | | kilosiemens per metre | | |
| | $\mu$S/m | | microsiemens per metre | | |
| Magnetic flux, flux of magnetic induction (Wb) | mWb | | milliweber | Used in the calculations involved in magnetic circuits | 1 Wb = 1 V·s |
| Magnetic flux density, magnetic induction (T) | T | | tesla | Used in the calculations involved in magnetic circuits | $1\ T = 1\ Wb/m^2$ |
| | mT | | millitesla | | |
| | $\mu$T | | microtesla | | |
| | nT | | nanotesla | | |
| Magnetic vector potential (Wb/m²) | kWb/m² | | kiloweber per square metre | Used in the calculations involved in magnetic circuits | |
| Self-inductance, mutual inductance, permeance (H) | H | | henry | Used in analysis and calculations involving transformers | 1 H = 1 Wb/A |
| | mH | | millihenry | | |
| | $\mu$H | | microhenry | | |
| | nH | | nanohenry | | |
| | pH | | picohenry | | |
| Reluctance (1/H) | 1/H | | reciprocal of henry | Design of motors and generators | |
| Permeability (H/m) | H/m | | henry per metre | Permeability gives the relationship between the magnetic flux density and the magnetic field strength | |
| | $\mu$H/m | | microhenry per metre | | |
| | nH/m | | nanohenry per metre | | |

*Note: Metre* is the preferred international spelling applied to this base unit and all its multiples and submultiples.
(Copyright ASTM; reprinted with permission.)

**FIGURE 25.4-6** Lighting

| Quantity and SI Unit Symbol | Preferred Units (Symbols) | Other Acceptable Units | Unit Name | Typical Applications | Remarks |
|---|---|---|---|---|---|
| Luminous intensity (cd) | cd | | candela | | |
| Solid angle (sr) | sr | | steradian | | |
| Luminous flux (lm) | lm | | lumen | Luminous flux of light sources, lamps, and lightbulbs | 1 lm = 1 cd·sr |
| | klm | | kilolumen | | Already in general use |
| Quantity of light (lm·s) | lm·s | | lumen second | | 1 lm·h = 3600 lm/s |
| | | lm·h | lumen hour | | |

**FIGURE 25.4-6** *(Continued)*

| Quantity and SI Unit Symbol | Preferred Units (Symbols) | Other Acceptable Units | Unit Name | Typical Applications | Remarks |
|---|---|---|---|---|---|
| Luminance $(cd/m^2)$ | $cd/m^2$ | | candela per square metre | Assessment of surface brightness; luminance of light sources, lamps, and lightbulbs; calculation of glare in lighting layouts | Replaces stilb $(1\ sb = 10^4\ cd/m^2)$ and apostilb $(1\ apostilb = cd/\pi m^2)$ |
| | $kcd/m^2$ | | kilocandela per square metre | | |
| | | $cd/mm^2$ | candela per square millimetre | | |
| Illuminance (lx) | lx | | lux | Luminous flux per unit area used in determination of illumination levels and design and evaluation of interior lighting layouts. (Outdoor daylight illumination on a horizontal plane ranges up to 100 klx.) | (a) Formerly referred to as illumination $1\ lx = 1\ lm/m^2$ (b) Replaces $(1\ ph = 10^4\ lx)$ (c) Luminous exitance is described in $lm/m^2$ |
| | klx | | kilolux | | |
| Light exposure (lx·s) | lx·s | | lux second | | |
| | klx·s | | kilolux second | | |
| Luminous efficacy (lm/W) | lm/W | | lumen per watt | Rating of luminous efficacy of artificial light sources | |

*Note: Metre* is the preferred international spelling applied to this base unit and all its multiples and submultiples.

(Copyright ASTM; reprinted with permission.)

**FIGURE 25.4-7** Acoustics

| Quantity and SI Unit Symbol | Preferred Units (Symbols) | Other Acceptable Units | Unit Name | Typical Applications | Remarks |
|---|---|---|---|---|---|
| Wavelength (m) | m | | metre | Definition of sound wave pitch | |
| | mm | | millimetre | | |
| Area of absorptive surface $(m^2)$ | $m^2$ | | square metre | Calculations of room absorption | |
| Period, periodic time (s) | s | | second | Measurement of time and reverberation time | |
| | ms | | millisecond | | |
| Frequency (Hz) | Hz | | hertz | Frequency ranges in sound absorption calculations and sound pressure measurement | 1 Hz = 1 cycle per second (cps) |
| | kHz | | kilohertz | | |
| Instantaneous sound pressure (Pa) | Pa | | pascal | Measurement of sound pressure; reference level for sound pressure is $20\ \mu Pa$, but sound pressure is shown in decibels (dB) based on a logarithmic scale. Sound pressure level $$L_p = 20 \log_{10} \frac{\text{actual pressure (Pa)}}{20 \times 10^{-6}\ \text{(Pa)}}$$ | Do not use dyne $(1\ dyn = 10\ \mu Pa)$ |
| | mPa | | millipascal | | |
| | $\mu Pa$ | | micropascal | | |
| Sound power, sound energy flux (W) | W | | watt | Measurement of sound power; reference level for sound power is 1 pW. Sound power level $$L = 10 \log_{10} \frac{\text{actual power (W)}}{10^{-12}\ \text{(W)}}$$ | |
| | mW | | milliwatt | | |
| | $\mu W$ | | microwatt | | |
| | pW | | picowatt | | |

**FIGURE 25.4-7** *(Continued)*

| Quantity and SI Unit Symbol | Preferred Units (Symbols) | Other Acceptable Units | Unit Name | Typical Applications | Remarks |
|---|---|---|---|---|---|
| Sound intensity (W/m²) | W/m² | | watt per square metre | Measurement of sound intensity; reference level for sound intensity is 1 pW/m² | |
| | pW/m² | | picowatt per square metre | Sound intensity level $$L_1 = 10 \log_{10} \frac{\text{actual intensity (W/m}^2)}{10^{-12} \text{ (W/m}^2)} \text{ dB}$$ | |
| Specific acoustic impedance ((Pa·s)/m) | (Pa·s)/m | | pascal second per metre | Sound impedance measurement | (1 (Pa·s)/m = 1 (N·s)/m³) |
| Acoustic impedance, resistance ((Pa·s)/m³) | (Pa·s)/m³ | | pascal second per cubic metre | Sound impedance measurement | |

*Note: Metre* is the preferred international spelling applied to this base unit and all its multiples and submultiples.

(Copyright ASTM; reprinted with permission.)

**FIGURE 25.4-8** Comparison of Calculations in Customary Units and SI Units

**Division of a Linear Dimension into Equal Parts**

A detailed drawing shows a height of 8′ 10″ between two floors connected by a stair with 14 risers, but not the height of each riser. What is the target height of risers so that none of them varies by more than ¹⁄₁₆″?

$$8′ \ 10″ = 106″$$
$$106 \div 14 = 7.57″$$
$$7.57″ \cong 7\tfrac{9}{16}″$$

(Check) 7⁹⁄₁₆″ × 14 = 105⅞″ (−⅛″)
Target height of riser: 7⁹⁄₁₆″
with 2 risers at 7⅝″ (+¹⁄₁₆″ each)

A detail drawing shows a height of 2690 mm between two floors connected by a stair with 14 risers, but not the height of each riser. What is the target height of risers so that none of them varies by more than 1 mm?

$$2690 \div 14 = 192.1 \text{ mm}$$
(Check)   192 × 14 = 2688 mm (+2 mm)
Target height of riser: 192 mm with 2 risers at 193 mm each

*Comment: Division of measurement in customary units requires decimalization and change to and from decimal fractions.*

**Area and Volume Calculations to Determine Order Quantity**

A flat concrete slab for an industrial structure has the following dimensions:

Length: 25′0″; width: 26′6″; thickness: 5″

Determine the order quantity of concrete to the nearest half of a cubic yard.

$$\frac{25 \times 26.5 \times 5}{12} = 276 \text{ cu. ft.}$$

276 ÷ 27 = 10.22 cu. yd.
Order: 10½ cu. yd.

A flat concrete slab for an industrial structure has the following dimensions:

Length: 7.5 m; width: 8.4 m; thickness: 125 mm (0.125 m)

Determine the order quantity of concrete to the nearest half of a cubic meter.

7.5 × 8.4 × 0.125 = 7.85 m³
Order: 8 m³

*Comment: Customary units need to be decimalized, and factors are required with inches and conversion of cubic feet to cubic yards. Metric volume is derived directly by multiplication of dimensions.*

The *month*, which varies in length, should not be used, unless a specific calendar month is indicated.

The *calendar year* (a for annum) represents 365 days, equal to 31 536 000 seconds.

### 25.4.7 TEMPERATURE

The *kelvin* (K) is the SI base unit of temperature. The derived SI unit, *degree Celsius* (°C), however, is acceptable and will replace the customary unit, *degree Fahrenheit* (°F), in everyday use (Fig. 25.4-11). The earlier name for Celsius, *centigrade*, is no longer used. The temperature interval of 1 degree Celsius exactly equals 1 kelvin. A temperature expressed in degrees Celsius is equal to the temperature in kelvins less 273.15: K = −273.15°C; 0°C = 273.15 K.

The zero point of the kelvin scale is *absolute zero*, the point at which all molecular movement stops. Therefore, there are no negative (minus) temperatures in the kelvin scale. The lack of negative numbers makes the kelvin scale useful for scientific applications, but the large numeric values make it cumbersome for everyday use. For example, room temperature may be expressed as 68 *degrees Fahrenheit*, 20 *degrees Celsius*, or 293.15 *kelvin*; body temperature as 98.6

**FIGURE 25.4-9** Metric and Nonmetric Drawing Scales

| Traditional Scales (Expressed as Ratios) | | Metric Scales | |
| --- | --- | --- | --- |
| | | Preferred | Other |
| Full size | (1:1) | 1:1 | |
| Half full size | (1:2) | | 1:2 |
| 4″ = 1′0″ | (1:3) | | |
| 3″ = 1′0″ | (1:4) | | |
| | | 1:5 | |
| 2″ = 1′0″ | (1:6) | | |
| 1½″ = 1′0″ | (1:8) | | |
| | | 1:10 | |
| 1″ = 1′0″ | (1:12) | | |
| ¾″ = 1′0″ | (1:16) | | |
| | | 1:20 | |
| ½″ = 1′0″ | (1:24) | | |
| | | | 1:25* |
| ⅜″ = 1′0″ | (1:32) | | |
| ¼″ = 1′0″ | (1:48) | | |
| | | 1:50 | |
| 1″ = 5′0″ | (1:60) | | |
| 3/16″ = 1′0″ | (1:64) | | |
| ⅛″ = 1′0″ | (1:96) | | |
| | | 1:100 | |
| 1″ = 10′0″ | (1:120) | | |
| 3/32″ = 1′0″ | (1:128) | | |
| 1/16″ = 1′0″ | (1:196) | | |
| | | 1:200 | |

*Limited use

degrees Fahrenheit, 37 degrees Celsius, or 310.15 kelvin; boiling water as 212 degrees Fahrenheit, 100 *degrees Celsius*, or 373.15 *Kelvin*.

## 25.4.8  MASS, WEIGHT, AND FORCE

SI differs from both the customary system and other forms of metric measurement in the use of explicit and distinctly separate units for *mass* and *force*. There is considerable confusion in the use of the term *weight* to mean either *force* or *mass*.

### 25.4.8.1  Mass

In everyday use, *weight* generally means *mass*. When one speaks of a person's weight, the quantity referred to is really mass.

*Mass* is the amount of matter in a body. An iron ball and a baseball of exactly the same size take up the same amount of space, but there is more matter (greater density or mass) per unit volume in the iron ball; therefore, the iron ball has more mass. *Density* is the ratio of mass to volume.

The *kilogram* (kg) is the SI base unit for mass, the amount of matter in an object that is constant and independent of gravitational attraction. Multiples of the kilogram are formed by attaching an SI prefix to *gram* (*milligram, megagram*). A *megagram* (Mg), which is equal to 1 000 000 grams, should be used to measure large masses formerly expressed in tons.

The name *ton* has been given to several large mass units that are widely used in commerce and technology: the *long ton* of 2240 lb, the *short ton* of 2000 lb and the *metric ton* of 1000 kg, which is also called a *tonne*. None of these terms is applicable to SI. The term *metric ton* may be used, but it should be restricted to commercial usage and never used with prefixes. Use of the term *tonne* is discouraged.

### 25.4.8.2  Force

In technical papers, *weight* has often been used to mean *force of gravity*, which is a particular force related to *gravitational acceleration*. The force of gravity varies from place to place on the earth and on other planets. A body of the same mass will weigh more or have more force on the earth than on the moon (Fig. 25.4-12).

To alleviate confusion, the term *weight* should be avoided in technical practice. *Force of gravity* should denote *force*. The *newton* (N) is the derived SI unit of force. It equals mass times acceleration: $F = kg \times m/s^2$.

Previous metric systems have used the term *kilogram-force of newton*, often dropping the suffix *force*, which caused confusion. The use of *kilogram-force* should be avoided. The *newton* also supersedes the term *kilopond* (kp). The *newton* can also be used to designate *pressure, stress, energy, work*, and *quantity of heat, power*, and many *electrical units*.

## 25.4.9  PRESSURE, STRESS, AND MODULUS OF ELASTICITY

The derived unit *pascal* (Pa), which is the equivalent of *newtons per square metre*, is the unit for pressure, stress, and elastic modulus.

**FIGURE 25.4-10** Properties of a 460-mm-Deep Wide-Flange Beam Weighing 82 kg/m

| Property | Designation and Units |
| --- | --- |
| Plastic modulus | $Z_x = 1.835 \times 10^6$ mm$^3$ ($1.835 \times 10^{-3}$ m$^3$) |
| Second moment of area | $I_{x-x} = 0.371 \times 10^9$ mm$^4$ ($0.371 \times 10^3$ m$^4$) |
| Torsional constant | $J = 0.691 \times 10^6$ mm$^4$ ($0.691 \times 10^{-3}$ m$^4$) |
| Warping constant | $C_w = 0.924 \times 10^{12}$ mm$^6$ ($0.924 \times 10^{-6}$ m$^6$) |

**FIGURE 24.4-11** Temperature equivalents.

SI | CUSTOMARY

121°C
394.15 K — 250°F

100°C
373.15 K — 212°F
BOILING

75°C
348.15 K — 167°F

57.2°C
330.35 K — 135°F

37°C
310.15 K — 98.6°F
BODY TEMP

20°C
293.15 K — 68°F

0°C
273.15 K — 32°F
FREEZING

−17.8°C
255.37 K — 0°F

−28.9°C
244.25 K — −20°F

$°F = {}^9/_5 °C + 32$
$K = °C + 273.15$

$°C = {}^5/_9 (°F − 32)$

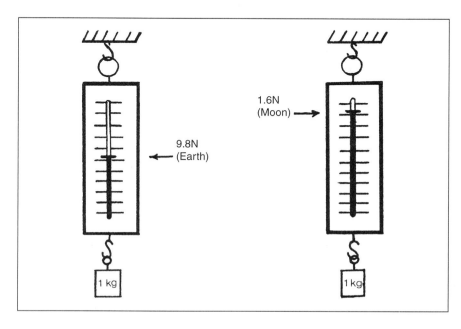

9.8N
(Earth)

1.6N
(Moon)

1 kg

1 kg

*Pressure* is a force applied uniformly over a surface and measured as a force per unit of area.

*Stress* is the effect of a force on a body, such as squeezing (compression), pulling (tension), and bending (compression plus tension). Stress is generally accompanied by *strain*, a physical change in the body's shape or dimension.

The pascal and its prefixed multiples—*megapascal* (MPa), *kilopascal* (kPa), and *gigapascal* (GPa)—replace a large number of customary units such as pounds/square inch (lb/in.² or psi). It also supersedes a few non-SI metric units such as the *bar* (equal to 100 kPa) and the *millibar* (equal to 100 Pa). Old *metric gravitational units* for pressure and stress such as *kilogram-force per square centimetre* (kgf/cm²) should not be used; neither should units such as *bar* and *torr*. Use of the *millibar* will continue in order to permit meteorologists to communicate within their profession, but even they should use the *kilopascal* (kPa) to present data to the public.

Though it may sometimes be useful to report test results in N/mm² (identical to MN/m²), calculations and results should be shown in MPa or kPa.

## 25.4.10 ENERGY, WORK, AND QUANTITY OF HEAT

The *joule* (J) with its multiples is the derived SI unit of *energy*, *work*, and *quantity of heat*. It is equal to a *newton metre* (N-m) and a *watt second* (W-s). The joule supersedes many traditional units, such as *Btu*, *therm*, *calorie*, *kilocalorie*, and *foot pound-force*.

The *kilowatthour* (kWh) has been the customary unit of *energy consumption*; the preferred unit is the *megajoule* (MJ). Since it would be needlessly costly to recalibrate electric meters, the kWh will continue to be used in electrical applications, but it should not be introduced into new areas.

The term *joule* should never be used for *torque*; for this property the designated unit is *newton metre* (N-m).

For dimensional consistency in rotational dynamics, *torque* should be ex-

**FIGURE 24.4-12** Although the mass is the same, 1 kg of matter weighs more than six times as much on the earth as it does on the moon pressed as *newton metre per radian* (N-m/rad) and *moment of inertia* as *kilogram square metre per radian squared* (kg-m$^2$/rad$^2$).

### 25.4.11 POWER AND HEAT FLOW RATE

The *watt* (W), the derived SI unit for power and *heat flow rate*, is already in worldwide use as a unit for *electrical power*. It will replace some traditional units of power and heat flow rate: (1) *horsepower* (electric, boiler), (2) *foot pound-force per hour* (or *minute* or *second*), (3) *Btu per hour*, (4) *calorie* (or *kilocalorie*) *per minute* (*or second*), and (5) *ton of refrigeration*.

### 24.4.12 ELECTRICAL ENGINEERING

The only changes in electrical engineering will be the renaming of the unit of conductance to *siemens* (S) instead of *mho* and the use of *hertz* (Hz) for frequency instead of *cycles per second* (cps).

### 25.4.13 LIGHTING UNITS

The *candela* (cd), the SI base unit for *luminous intensity*, is already in common use and replaces *candle* and *candlepower*.

*Luminance* will be expressed in *candelas per square metre* (cd/m$^2$), which replaces *candelas per square foot*, *foot lambert*, and *lambert*.

The derived SI unit *lumen* (lm) is also currently in common use. *Illuminance* is expressed in lux (lx), which is equal to *lumens per square metre* (lm/m$^2$) and replaces *lumens per square foot* and *footcandle*.

### 25.4.14 VELOCITY

There is no base, supplementary, or derived SI unit expressing *velocity*. The accepted non-SI unit is *kilometre per hour*.

### 25.4.15 SPEED OF ROTATION

There is no SI unit for *speed of rotation*. The accepted non-SI unit is *revolutions per minute*.

### 25.4.16 DIMENSIONLESS QUANTITIES

Ratios such as *relative humidity* (RH), *specific gravity* (SG), and *decibel* (dB)

remain unchanged in the conversion to SI.

# 25.5 Conversion of Numerical Values

*Metrication* is the process of change to a new set of reference units. It involves a change in numerical values by the ratio between customary and SI reference units. This ratio is known as a *conversion factor*.

Types of conversions are (1) *exact*, (2) *soft*, (3) *hard*, and (4) *rationalization*.

### 25.5.1 EXACT CONVERSION

An *exact*, also called *direct*, *conversion* is a change from a customary value to its precise SI equivalent, generally expressed to a number of decimal places. For example, 24 in. is exactly 609.6 mm (Fig. 25.5-1).

*Exact conversion* is often needed in the early transitional period of a change to SI units but should be thought of as only the first step to the other types of conversions. Exact conversions will rarely, if ever, be needed in design and construction, but they are an essential step before rounding or *rationalization* of traditional values can begin. Figure 25.5-2 shows exact conversion of fractions and decimals to three places. Figure 25.5-3 shows conversion tables for the most common SI units used in building design and construction.

Conversion of quantities should be handled carefully so that accuracy is neither sacrificed nor exaggerated. *Conversion factors* and quantities should not be rounded prior to multiplication or division.

Calculated values should be rounded to the minimum number of digits that will maintain the appropriate accuracy. For example, if a metric tape measure shows values to the nearest millimetre, showing lineal values more precisely would be unnecessary if the final dimensions are to be applied using a tape measure.

For calculations using numbers of differing precision, the final answer should be no more precise than the least precise number (Fig. 25.5-4).

### 25.5.2 SOFT CONVERSION

*Soft conversion* is a change in the way a product's size is described without significant change in its actual size. In general, a soft conversion is the rounding of an exact conversion to a more workable numerical value. For example, a piece of 24-in.-wide gypsum lath would become 609.6 mm wide if exact conversion were desired. A soft conversion would simply round off the number to 610 mm. Such a practice would give the appearance of metric conversion but would not result in the most ideal product width, 600 mm, which would be attained through hard conversion (see Fig. 25.5-1).

Since soft conversion is only a metric veneer, a change made only on paper, soft converted products are in production now. Some are already labeled with both customary and metric measures. Although soft conversion may be of some value in specifications, standards, codes, and other technical data, it is inappropriate in product sizing because it perpetuates customary practices and diminishes the advantages of SI.

### 25.5.3 HARD CONVERSION

In a *hard conversion* there is a definite change in physical size or magnitude from customary values to preferred SI values (see Fig. 25.5-1). Such a conver-

**FIGURE 25.5-1** Exact, Soft, and Hard Conversions of Customary Values

| Customary Value | Exact Conversion | Soft Conversion | Hard Conversion | % Change for Hard Conversion |
|---|---|---|---|---|
| 24 in. (2 ft) | 609.6 mm | 610 mm | 600 mm | –1.6 |
| 1 square (100 sq ft) | 9.2903 m$^2$ | 9.3 m$^2$ | 10 m$^2$ | +7.6 |
| 4 cu yd | 3.058 m$^3$ | 3.1 m$^3$ | 3 m$^3$ | –1.9 |
| 1 quart | 0.946 L | 0.95 L | 1 L | +5.7 |

**FIGURE 25.5-2** Fraction, Decimal, and Metric Equivalents

| Fractional Inch | Decimal Equivalent* | Millimetres | Fractional Inch | Decimal Equivalent* | Millimetres |
|---|---|---|---|---|---|
| 1/64 | 0.0156 | 0.397 | 23/64 | 0.3594 | 9.128 |
| 1/32 | 0.0313 | 0.794 | 3/8 | 0.3750 | 9.525 |
| 3/64 | 0.0469 | 1.191 | 25/64 | 0.3906 | 9.922 |
| 1/16 | 0.0625 | 1.588 | 2/5 | 0.4000 | 10.160 |
| 5/64 | 0.0781 | 1.984 | 13/32 | 0.4063 | 10.319 |
| 1/12 | 0.0833 | 2.117 | 5/12 | 0.4167 | 10.583 |
| 3/32 | 0.0938 | 2.381 | 27/64 | 0.4219 | 10.716 |
| 1/10 | 0.1000 | 2.540 | 7/16 | 0.4375 | 11.112 |
| 7/64 | 0.1094 | 2.778 | 29/64 | 0.4531 | 11.509 |
| 1/8 | 0.1250 | 3.175 | 15/32 | 0.4688 | 11.906 |
| 9/64 | 0.1406 | 3.572 | 31/64 | 0.4844 | 12.303 |
| 1/6 | 0.1667 | 4.233 | 1/2 | 0.5000 | 12.700 |
| 11/64 | 0.1719 | 4.366 | 33/64 | 0.5156 | 13.097 |
| 3/16 | 0.1875 | 4.700 | 17/32 | 0.5313 | 13.494 |
| 1/5 | 0.2000 | 5.080 | 35/64 | 0.5469 | 13.891 |
| 13/64 | 0.2031 | 5.159 | 9/16 | 0.5625 | 14.288 |
| 7/32 | 0.2188 | 5.556 | 37/64 | 0.5781 | 14.684 |
| 15/64 | 0.2344 | 5.953 | 7/12 | 0.5833 | 14.817 |
| 1/4 | 0.2500 | 6.350 | 19/32 | 0.5938 | 15.081 |
| 17/64 | 0.2656 | 6.747 | 3/5 | 0.6000 | 15.240 |
| 9/32 | 0.2813 | 7.144 | 39/64 | 0.6094 | 15.478 |
| 19/64 | 0.2969 | 7.541 | 5/8 | 0.6250 | 15.875 |
| 3/10 | 0.3000 | 7.620 | 41/64 | 0.6406 | 16.272 |
| 5/16 | 0.3125 | 7.937 | 21/32 | 0.6563 | 16.669 |
| 1/3 | 0.3333 | 8.467 | 2/3 | 0.6667 | 16.933 |
| 11/32 | 0.3438 | 8.731 | 43/64 | 0.6719 | 17.066 |
| 11/16 | 0.6875 | 17.463 | 27/32 | 0.8438 | 21.431 |
| 7/10 | 0.7000 | 17.780 | 55/64 | 0.8594 | 21.828 |
| 45/64 | 0.7031 | 17.859 | 7/8 | 0.8750 | 22.225 |
| 23/32 | 0.7188 | 18.256 | 57/64 | 0.8906 | 22.622 |
| 47/64 | 0.7344 | 18.653 | 9/10 | 0.9000 | 22.860 |
| 3/4 | 0.7500 | 19.050 | 29/32 | 0.9063 | 23.019 |
| 49/64 | 0.7656 | 19.447 | 11/12 | 0.9167 | 23.283 |
| 25/32 | 0.7813 | 19.844 | 59/64 | 0.9219 | 23.416 |
| 51/64 | 0.7969 | 20.241 | 15/16 | 0.9375 | 23.813 |
| 4/5 | 0.8000 | 20.320 | 61/64 | 0.9531 | 24.209 |
| 13/16 | 0.8125 | 20.638 | 31/32 | 0.9688 | 24.606 |
| 53/64 | 0.8281 | 21.034 | 63/64 | 0.9844 | 25.003 |
| 5/6 | 0.8333 | 21.167 | 1″ | 1.000 | 25.400 |

*To convert a decimal to percentage, carry the decimal point two places to the right. Thus, 63/64, or 0.9844, equals 98.44%.

sion often leads to incompatibility between products made using the SI system and those earlier ones produced in customary measure. For example, in a hard conversion, the 24-in.-wide gypsum lath mentioned in Section 25.5.5.2 might actually be changed to 610 mm wide instead of just being called that. This would be within normal tolerances and would create few interface problems. More likely, however, the width would be changed to an even 600 mm wide, which is 1.6% smaller. Additional examples include the following: a 1-qt container would be replaced by a 1-*litre* container, which is a 5.7% increase in volume; a 12-ft length would become a preferred 3600 mm, which is a 1.6% decrease in length.

Though the changes are more difficult to make, hard conversion is better than the other means of conversion because it creates a simpler SI system. Figure 25.5-1 shows the differences in converting customary values using exact, soft, and hard conversions.

Unfortunately, hard conversion is unrealistic and unacceptable in some cases, such as in changing the standard railroad gauge. The U.S. railroad gauge currently has a nonpreferred dimension of 4 ft 8 1/2 in. The soft conversion is 1435 mm. Since railroad tracks must conform to the large number of existing railroad cars, the soft conversion to 1435 mm is acceptable and would cost little compared with a hard change to 1450 mm or 1500 mm.

In general, hard conversion should be used when possible because it results in preferred values, which are easier to work with. However, the decision to make a hard conversion must always be weighed against the cost of changing the many products or components that are intended to work together.

### 25.5.4 RATIONALIZATION

*Rationalization*, which often occurs during the conversion from customary to SI units of measure, is the development of standards for greater product compatibility and cost effectiveness. It results from selection of the most rational, preferred, and economical alternatives after research into the technical implications and an appraisal of the economic impact. Rationalization usually results in changes in product sizes and often in a reduction in the number of a particular product's available sizes. For instance, during metrication, the number of reinforcing bar

**FIGURE 25.5-3** Conversion Factors for the Most Common Units Used in Building Design and Construction

| Metric to Customary | | | Customary to Metric | | |
|---|---|---|---|---|---|
| **Length** | | | | | |
| 1 km | = 0.621 371 | mile (international) | 1 mile (international) | = *1.609 344* | km |
| | = 49.7096 | chain | 1 chain | = 20.1168 | m |
| 1 m | = 1.093 61 | yd | 1 yd | = *0.9144* | m |
| | = 3.280 84 | ft | 1 ft | = *0.3048* | m |
| 1 mm | = 0.039 370 1 | in | | = *304.8* | mm |
| | | | 1 in | = *25.4* | mm |
| | | | (1 U.S. survey foot | = 0.304 800 6 | m) |
| **Area** | | | | | |
| 1 km$^2$ | = 0.386 101 | mile$^2$ (U.S. survey) | 1 mile$^2$ (U.S. survey) | = 2.590 00 | km$^2$ |
| 1 ha | = 2.471 04 | acre (U.S. survey) | 1 acre (U.S. survey) | = 0.404 687 | ha |
| 1 m$^2$ | = 1.195 99 | yd$^2$ | | = 4046.87 | m$^2$ |
| | = 10.7639 | ft$^2$ | 1 yd$^2$ | = 0.836 127 | m$^2$ |
| 1 mm$^2$ | = 0.001 550 | in$^2$ | 1 ft$^2$ | = 0.092 903 | m$^2$ |
| | | | 1 in$^2$ | = *645.16* | mm$^2$ |
| **Volume, Modulus of Section** | | | | | |
| 1 m$^3$ | = 0.810 709 × 10$^{-3}$ | acre ft | 1 acre ft | = 1233.49 | m$^3$ |
| | = 1.307 95 | yd$^3$ | 1 yd$^3$ | = 0.764 555 | m$^3$ |
| | = 35.3147 | ft$^3$ | 100 board ft | = 0.235 974 | m$^3$ |
| | = 423.776 | board ft | 1 ft$^3$ | = 0.028 316 8 | m$^3$ |
| 1 mm$^3$ | = 61.0237 × 10$^{-6}$ | in$^3$ | | = 28.3168 | L (dm$^3$) |
| | | | 1 in$^3$ | = 16 387.1 | mm$^3$ |
| | | | | = 16.3871 | mL (cm$^3$) |
| **(Fluid) Capacity** | | | | | |
| 1 L | = 0.035 314 7 | ft$^3$ | 1 gal (U.S. liquid)* | = 3.785 41 | L |
| | = 0.264 172 | gal (U.S.) | 1 qt (U.S. liquid) | = 946.353 | mL |
| | = 1.056 69 | qt (U.S.) | 1 pt (U.S. liquid) | = 473.177 | mL |
| 1 mL | = 0.061 023 7 | in$^3$ | 1 fl oz (U.S.) | = 29.5735 | mL |
| | = 0.033 814 | fl oz (U.S.) | | | |
| | | | *1 gal (U.K.) approximately 1.2 gal (U.S.) | | |
| **Second Moment of Area** | | | | | |
| 1 mm$^4$ | = 2.402 51 × 10$^{-6}$ | in$^4$ | 1 in$^4$ | = 416 231 | mm$^4$ |
| | | | | = 0.416 231 × 10$^{-6}$ | m$^4$ |
| **Plane Angle** | | | | | |
| 1 rad | = 57° 17′ 45″ | (degree) | 1° (degree) | = 0.017 453 3 | rad |
| | = 57.2958° | (degree) | | = 17.4533 | mrad |
| | = 3437.75′ | (minute) | 1′ (minute) | = 290.888 | $\mu$rad |
| | = 206 265″ | (second) | 1″ (second) | = 4.848 14 | $\mu$rad |
| **Velocity, Speed** | | | | | |
| 1 m/s | = 3.280 84 | ft/s | 1 ft/s | = *0.3048* | m/s |
| | = 2.236 94 | mile/h | 1 mile/h | = *1.609 344* | km/h |
| 1 km/h | = 0.621 371 | mile/h | | = *0.447 04* | m/s |
| **Acceleration** | | | | | |
| 1 m/s$^2$ | = 3.280 84 | ft/s$^2$ | 1 ft/s$^2$ | = *0.3048* | m/s$^2$ |
| **Volume Rate of Flow** | | | | | |
| 1 m$^3$/s | = 35.3147 | ft$^3$/s | 1 ft$^3$/s | = 0.028 316 8 | m$^3$/s |
| | = 22.8245 | million gal/d | 1 ft$^3$/min | = 0.471 947 | L/s |
| | = 0.810 709 × 10$^{-3}$ | acre ft/s | 1 gal/min | = 0.063 090 2 | L/s |
| 1 L/s | = 2.118 88 | ft$^3$/min | 1 gal/h | = 1.051 50 | mL/s |
| | = 15.850 3 | gal/min | 1 million gal/d | = 43.8126 | L/s |
| | = 951.022 | gal/h | 1 acre ft/s | = 1233.49 | m$^3$/s |
| **Temperature Interval** | | | | | |
| 1°C | = *1* K | = *1.8*°F | 1°F | = 0.555 556 | °C or K |
| | | | | = ⁵⁄₉°C | = ⁵⁄₉ K |
| **Equivalent Temperature Value** ($t_{°C} = T_K - 273.15$) | | | | | |
| $t_{°C}$ | = ⁵⁄₉ ($t_F$ − 32) | | $t_F$ | | = ⁹⁄₅ $t_{°C}$ + 32 |

**FIGURE 25.5-3** *(Continued)*

| Metric to Customary | | | Customary to Metric | | |
|---|---|---|---|---|---|
| **Mass** | | | | | |
| 1 kg | = 2.204 62 | lb (avoirdupois) | 1 ton (short)* | = 0.907 185 | metric ton |
| | = 35.2740 | oz (avoirdupois) | | = 907.185 | kg |
| 1 metric | = 1.102 31 | ton (short, 2000 lb) | 1 lb | = 0.453 592 | kg |
| ton | = 2204.62 | lb | 1 oz | = 28.3495 | g |
| 1 g | = 0.035 274 | oz | 1 pennyweight | = 1.555 17 | g |
| | = 0.643 015 | pennyweight | | | |
| | | | *(1 long ton (2240 lb) = 1016.05 kg) | | |
| **Mass per Unit Length** | | | | | |
| 1 kg/m | = 0.671 969 | lb/ft | 1 lb/ft | = 1.488 16 | kg/m |
| 1 g/m | = 3.547 99 | lb/mile | 1 lb/mile | = 0.281 849 | g/m |
| **Mass per Unit Area** | | | | | |
| 1 kg/m$^2$ | = 0.204 816 | lb/ft$^2$ | 1 lb/ft$^2$ | = 4.882 43 | kg/m$^2$ |
| 1 g/m$^2$ | = 0.029 494 | oz/yd$^2$ | 1 oz/yd$^2$ | = 33.9057 | g/m$^2$ |
| | = 3.277 06 × 10$^{-3}$ | oz/ft$^2$ | 1 oz/ft$^2$ | = 305.152 | g/m$^2$ |
| **Density (Mass per Unit Volume)** | | | | | |
| 1 kg/m$^3$ | = 0.062 428 | lb/ft$^3$ | 1 lb/ft$^3$ | = 16.0185 | kg/m$^3$ |
| | = 1.685 56 | lb/yd$^3$ | 1 lb/yd$^3$ | = 0.593 276 | kg/m$^3$ |
| 1 t/m$^3$ | = 0.842 778 | ton/yd$^3$ | 1 ton/yd$^3$ | = 1.186 55 | t/m$^3$ |
| **Moment of Inertia** | | | | | |
| 1 kg·m$^2$ | = 23.7304 | lb·ft$^2$ | 1 lb·ft$^2$ | = 0.042 140 1 | kg·m$^2$ |
| | = 3417.17 | lb·in$^2$ | 1 lb·in$^2$ | = 292.640 | kg·mm$^2$ |
| **Mass per Unit Time** | | | | | |
| 1 kg/s | = 2.204 62 | lb/s | 1 lb/s | = 0.453 592 | kg/s |
| 1 t/h | = 0.984 207 | ton/h | 1 ton/h | = 1.016 05 | t/h |
| **Force** | | | | | |
| 1 MN | = 112.404 | tonf (ton-force) | 1 tonf (ton-force) | = 8.896 44 | kN |
| 1 kN | = 0.112 404 | tonf | 1 kip (1000 lbf) | = 4.448 22 | kN |
| | = 224.809 | lbf (pound-force) | 1 lbf (pound-force) | = 4.448 22 | N |
| 1 N | = 0.224 809 | | | | |
| **Moment of Force, Torque** | | | | | |
| 1 N·m | = 0.737 562 | lbf·ft | 1 lfb·ft | = 1.355 82 | N·m |
| | = 8.850 75 | lbf·in | 1 lbf·in | = 0.112 985 | N·m |
| 1 kN·m | = 0.368 781 | tonf·ft | 1 tonf·ft | = 2.711 64 | kN·m |
| | = 0.737 562 | kip·ft | 1 kip·ft | = 1.355 82 | kN·m |
| **Force per Unit Length** | | | | | |
| 1 N/m | = 0.068 521 8 | lbf/ft | 1 lbf/ft | = 14.5939 | N/m |
| 1 kN/m | = 0.034 260 9 | tonf/ft | 1 lbf/in | = 175.127 | N/m |
| | | | 1 tonf/ft | = 29.187 8 | kN/m |
| **Pressure, Stress, Modulus of Elasticity (Force per Unit Area) (1 Pa = 1 N/m$^2$)** | | | | | |
| 1 MPa | = 0.072 518 8 | tonf/in$^2$ | 1 tonf/in$^2$ | = 13.7895 | MPa |
| | = 10.4427 | tonf/ft$^2$ | 1 ton/ft$^2$ | = 95.7605 | kPa |
| | = 145.038 | lbf/in$^2$ | 1 kip/in$^2$ | = 6.894 76 | MPa |
| 1 kPa | = 20.8854 | lbf/ft$^2$ | 1 lbf/in$^2$ | = 6.894 76 | kPa |
| | | | 1 lbf/ft$^2$ | = 47.8803 | Pa |
| **Work Energy, Heat (1 J = 1 N·m = 1 W·s)** | | | | | |
| 1 MJ | = 0.277 778 | kWh | 1 kWh (550 ft·lbf/s) | = *3.6* | MJ |
| 1 kJ | = 0.947 817 | Btu | 1 Btu (Int. Table) | = 1.055 06 | kJ |
| 1 J | = 0.737 562 | ft·lbf | | = 1055.06 | J |
| | | | 1 ft·lbf | = 1.355 82 | J |
| **Power, Heat Flow Rate** | | | | | |
| 1 kW | = 1.341 02 | hp (horsepower) | 1 hp | = 0.745 700 | kW |
| 1 W | = 3.412 14 | Btu/h | | = 745.700 | W |
| | = 0.737 562 | ft·lbf/s | 1 Btu/h | = 0.293 071 | W |
| | | | 1 ft·lbf/s | = 1.355 82 | W |

**FIGURE 25.5-3** *(Continued)*

| Metric to Customary | | | Customary to Metric | | |
|---|---|---|---|---|---|
| **Heat Flux Density** | | | | | |
| 1 W/m$^2$ | = 0.316 998 | Btu/(ft$^2$·h) | 1 Btu/(ft$^2$·h) | = 3.154 59 | W/m$^2$ |
| **Coefficient of Heat Transfer** | | | | | |
| 1 W/(m$^2$·K) | = 0.176 100 | Btu/(ft$^2$·h·°F) | 1 Btu/(ft$^2$·h·°F) | = 5.678 26 | W/(m$^2$·K) |
| **Thermal Conductivity** | | | | | |
| 1 W/(m·K) | = 0.577 789 | Btu/(ft·h·°F) | 1 Btu/(ft·h·°F) | = 1.730 73 | W/(m·K) |
| **Calorific Value (Mass and Volume Basis)** | | | | | |
| 1 kJ/kg | = 0.429 923 | Btu/lb | 1 Btu/lb | = *2.326* | kJ/kg |
| (1 J/g) | = 0.429 923 | Btu/lb | 1 Btu/lb | = *2.326* | (J/g) |
| 1 kJ/m$^3$ | = 0.026 839 2 | Btu/ft$^3$ | 1 Btu/ft$^3$ | = 37.2589 | kJ/m$^3$ |
| **Thermal Capacity (Mass and Volume Basis)** | | | | | |
| 1 kJ/(kg·K) | = 0.238 846 | Btu/(lb·°F) | 1 Btu/(lb·°F) | = *4.1868* | kJ/(kg·K) |
| 1 kJ/(m$^3$·K) | = 0.014 910 7 | Btu/(ft$^3$·°F) | 1 Btu/(ft$^3$·°F) | = 67.0661 | kJ/(m$^3$·K) |
| **Illuminance** | | | | | |
| 1 lx (lux) | = 0.092 903 | lm/ft$^2$ (footcandle) | 1 lm/ft$^2$ (footcandle) | = 10.7639 | lx (lux) |
| **Luminance** | | | | | |
| 1 cd/m$^2$ | = 0.092 903 | cd/ft$^2$ | 1 cd/ft$^2$ | = 10.7639 | cd/m$^2$ |
| | = 0.291 864 | footlambert | 1 footlambert | = 3.426 26 | cd/m$^2$ |
| 1 kcd/m$^2$ | = 0.314 159 | lambert | 1 lambert | = 3.183 01 | kcd/m$^2$ |

*Note:* Conversion factors are taken to six significant figures, where appropriate. Values in italics are exact conversions.

sizes has been reduced from 11 to 10 in Australia, from 11 to 9 in Britain, and from 11 to 8 in Canada.

Rationalization can also be applied to product standards, combining them and reducing their number.

The objectives of rationalization are (1) to reduce the variety of products by eliminating unnecessary or uneconomical sizes, (2) to gain optimum steps in a new product range, (3) to harmonize various technical data, standards, and codes by using preferred values, and (4) to simplify practices and procedures through less complicated numerical description.

# 25.6 Preferred Dimensions and Coordination

The internationally recommended basic building module is 100 mm. It is small enough to provide flexibility in design, large enough to promote simplification of the number of component sizes, and acceptable worldwide. It is slightly smaller than the most common module of the customary system, 4 in. For example, the width of a wood stud is nominally 4 in.; the width of a brick and the height of a block are nominally 8 in.; the spacing of studs and the nominal length of a block are 16 in.; the width of gypsum lath and insulating sheathing is 24 in.

Preferred SI sizes are those that are preferred over others for building components and assemblies. These are usually selected multiples of the basic 100 mm building module. During the change to SI, it is important to select preferred dimensions for building products to en-

sure that they will work together when combined into walls, floors, and other building assemblies.

*Dimensional coordination* is the systematic application of preferred and related dimensions in the design of buildings and the manufacture and positioning of building components, assemblies, and elements.

*Modular coordination* is dimensional coordination using multiples of the international 100 mm building module.

## 25.6.1 ADVANTAGES OF PREFERRED DIMENSIONS

Selecting preferred building dimensions, such as floor-to-floor heights, provides an incentive to produce components of a standardized height for use in high-rise buildings, such as precast concrete stairs, ducts, shaft walls, and cladding panels.

Preferred building dimensions will also (1) simplify distributors' inventory by reducing product-size variety, (2) make it easier to combine interrelated components into a building, and (3) reduce the time and waste normally associated with assembling uncoordinated products. In addition to reducing construction cost and time, the application of uniform preferred dimensions will simplify communications and give additional

**FIGURE 25.5-4** Calculating Exact Conversion

**Addition or Subtraction**

| Original numbers: | |
|---|---|
| | 153 787 400 |
| | 647 200 000 |
| | 21 000 000 |
| | 821 987 400 |

Rounded to the least precise and added:

| | |
|---|---|
| | 154 000 000 |
| | 647 000 000 |
| | 21 000 000 |
| | 822 000 000 |

**Multiplication or Division**

175.5 × 3.475 = 609.8625

Round the answer to 609.9 because the least precise number, 175.5, has the decimal carried to only one place.

## Temperature

| °F | °C | °F | °C | °F | °C | °F | °C | °F | °C |
|---|---|---|---|---|---|---|---|---|---|
| −40 | −40 | 18.6 | −7.4 | 39 | 3.9 | 72.2 | 22.3 | 110 | 43.3 |
| −30 | −34.4 | 19 | −7.2 | 40 | 4.4 | 73 | 22.8 | 200 | 93.3 |
| −20 | −28.9 | 20 | −6.7 | 41 | 5 | 73.1 | 22.83 | 204.4 | 95.8 |
| −10 | −23.3 | 20.3 | −6.5 | 43 | 6.1 | 73.5 | 23.1 | 300 | 148.9 |
| 0 | −17.8 | 21 | −6.1 | 44 | 6.7 | 74 | 23.3 | 400 | 204.4 |
| 1 | −17.2 | 21.3 | −5.9 | 44.6 | 7 | 74.3 | 23.5 | 426.7 | 219.3 |
| 5 | −15 | 22 | −5.6 | 45 | 7.2 | 74.8 | 23.8 | 500 | 260 |
| 6 | −14.4 | 22.6 | −5.2 | 46 | 7.8 | 75 | 23.9 | 600 | 315.6 |
| 6.3 | −14.3 | 23 | −5 | 47 | 8.3 | 75.6 | 24.2 | 648.9 | 342.7 |
| 7 | −13.9 | 23.6 | −4.7 | 48 | 8.9 | 76.1 | 24.5 | 800 | 426.7 |
| 8 | −13.3 | 24 | −4.4 | 49 | 9.4 | 76.9 | 24.9 | 871.1 | 466.2 |
| 9 | −12.8 | 25 | −3.9 | 50 | 10 | 77 | 25 | 1000 | 537.8 |
| 10 | −12.2 | 26 | −3.3 | 51 | 10.6 | 77.5 | 25.3 | 1093.3 | 589.6 |
| 10.3 | −12.1 | 26.3 | −3.2 | 55 | 12.8 | 78 | 25.6 | 1100 | 593.3 |
| 11.3 | −11.5 | 27 | −2.8 | 58 | 14.4 | 78.3 | 25.7 | 1200 | 648.9 |
| 12 | −11.1 | 27.6 | −2.4 | 60 | 15.6 | 79.7 | 26.5 | 1315.6 | 713.1 |
| 13 | −10.6 | 28 | −2.2 | 61 | 16.1 | 80 | 26.7 | 1500 | 815.6 |
| 13.6 | −10.2 | 28.6 | −1.9 | 62 | 16.7 | 81.2 | 27.3 | 1600 | 871.1 |
| 14 | −10 | 30 | −1.1 | 65 | 18.3 | 82 | 27.8 | 1800 | 982.2 |
| 15 | −9.4 | 31 | −0.6 | 66 | 18.9 | 83 | 28.3 | 2000 | 1093.3 |
| 15.3 | −9.3 | 32 | 0 | 68 | 20 | 85 | 29.4 | 2400 | 1315.6 |
| 16 | −8.9 | 33.6 | 0.9 | 69 | 20.6 | 90 | 32.2 | 2700 | 1482.2 |
| 16.3 | −8.7 | 34 | 1.1 | 70 | 21.1 | 95 | 35 | 2800 | 1537.8 |
| 17 | −8.3 | 35 | 1.7 | 71 | 21.7 | 100 | 37.8 | 3000 | 1648.9 |
| 17.6 | −8 | 36 | 2.2 | 71.1 | 21.72 | 105 | 40.6 | 5000 | 2760 |
| 18 | −7.8 | 38 | 3.3 | 72 | 22.2 | 108 | 42.2 | | |

## Mass

| grains/lb | kg/kg | lb | kg | lb | kg | lb | kg | lb | kg |
|---|---|---|---|---|---|---|---|---|---|
| 24 | 0.0034 | 5 | 2.27 | 35 | 15.88 | 80 | 36.29 | 144 | 65.32 |
| 36 | 0.0051 | 6 | 2.72 | 40 | 18.14 | 81 | 36.74 | 158 | 71.67 |
| 40 | 0.0057 | 10 | 4.54 | 48 | 21.77 | 82 | 37.19 | 180 | 81.65 |
| 54 | 0.0077 | 15 | 6.80 | 50 | 22.68 | 90 | 40.82 | 192 | 87.09 |
| 77 | 0.0110 | 20 | 9.07 | 55 | 24.95 | 94 | 42.64 | 200 | 90.72 |
| 80 | 0.0114 | 21 | 9.53 | 60 | 27.22 | 100 | 45.36 | 240 | 108.86 |
| 108 | 0.0154 | 22.5 | 10.21 | 64 | 29.03 | 105 | 47.63 | 250 | 113.40 |
| 120 | 0.0171 | 24 | 10.89 | 65 | 29.48 | 110 | 49.90 | 300 | 136.08 |
| 160 | 0.0229 | 25 | 11.34 | 70 | 31.75 | 120 | 54.43 | 360 | 163.29 |
| 200 | 0.0286 | 30 | 13.61 | 72 | 32.66 | 125 | 56.70 | | |
| | | 33 | 14.97 | 75 | 34.02 | 140 | 63.50 | | |

## Mass per Unit Area

| oz/sq ft | g/m² | psi | MPa | psi | MPa | psi | MPa | psi | MPa |
|---|---|---|---|---|---|---|---|---|---|
| 2 | 610 | 360 | 2.48 | 1375 | 9.5 | 2100 | 14.5 | 5000 | 35 |
| 4 | 1220 | 400 | 2.76 | 1400 | 9.6 | 2150 | 14.8 | 6000 | 41 |
| | | 450 | 3.1 | 1500 | 10 | 2200 | 15 | 7000 | 48 |

| psi | MPa | psi | MPa | psi | MPa | psi | MPa | psi | MPa |
|---|---|---|---|---|---|---|---|---|---|
| 55 | 0.38 | 500 | 3.4 | 1575 | 10.9 | 2250 | 15.5 | 8000 | 55 |
| 60 | 0.41 | 600 | 4 | 1600 | 11 | 2300 | 15.8 | 10,000 | 69 |
| 70 | 0.48 | 640 | 4.4 | 1650 | 11.4 | 2325 | 16 | 19,000 | 131 |
| 75 | 0.52 | 700 | 4.8 | 1675 | 11.5 | 2400 | 16.5 | 20,000 | 138 |
| 100 | 0.69 | 720 | 4.96 | 1700 | 11.7 | 2500 | 17 | 30,000 | 207 |
| 115 | 0.79 | 750 | 5.2 | 1750 | 12.1 | 2550 | 17.6 | 40,000 | 276 |
| 120 | 0.83 | 800 | 5.5 | 1800 | 12.4 | 2650 | 18.3 | 50,000 | 345 |
| 140 | 0.96 | 900 | 6.2 | 1825 | 12.6 | 2700 | 18.6 | 60,000 | 413 |
| 160 | 1.1 | 950 | 6.6 | 1875 | 12.9 | 2850 | 19.6 | 70,000 | 482 |
| 200 | 1.4 | 1000 | 6.9 | 1925 | 13.3 | 3000 | 21 | 75,000 | 517 |
| 225 | 1.6 | 1020 | 7 | 1950 | 13.4 | 3150 | 21.7 | 80,000 | 551 |
| 300 | 2 | 1175 | 8.1 | 1975 | 13.6 | 3300 | 22.7 | 100,000 | 689 |
| 350 | 2.4 | 1200 | 8.3 | 2000 | 13.8 | 3500 | 24 | 120,000 | 827 |
| 320 | 2.2 | 1250 | 8.6 | 2035 | 14 | 4000 | 28 | 140,000 | 965 |
| | | 1300 | 8.96 | 2050 | 14.1 | 4500 | 31 | | |
| | | 1350 | 9.4 | | | | | | |

# FIGURE 25.7-1 (Continued)

## Mass per Unit Area (Continued)

| E | MPa | E | MPa | psf | kg/m² | psf | MPa | lb/square | kg/square |
|---|---|---|---|---|---|---|---|---|---|
| 1 | 6890 | 2.3 | 15847 | 40 | 195.3 | 3000 | 143.64 | 350 | 158.75 |
| 1.2 | 8268 | 2.4 | 16536 | 45 | 219.71 | 6000 | 287.28 | 450 | 204.11 |
| 1.4 | 9646 | 2.5 | 17225 | 50 | 244.12 | | | | |

| | | | | | | lb/square | kg/square | tons/sq ft | Mg/m² |
|---|---|---|---|---|---|---|---|---|---|
| 1.5 | 10335 | 2.6 | 17914 | | | 200 | 90.72 | 1 | 9.77 |
| 1.6 | 11024 | | | | | 220 | 99.79 | 2 | 19.53 |
| 1.7 | 11713 | psf | kg/m² | psf | MPa | 250 | 113.39 | 2.5 | 24.61 |
| 1.8 | 12402 | 20 | 97.65 | 100 | 4.79 | 260 | 117.93 | 3 | 29.29 |
| 1.9 | 13091 | 25 | 122.06 | 150 | 7.18 | 280 | 127 | 5 | 48.82 |
| 2 | 13780 | 30 | 146.47 | 200 | 9.58 | | | 10 | 97.65 |
| 2.2 | 15158 | 35 | 170.88 | 400 | 19.15 | | | | |

## Mass per Unit Length

| lb/ft | kg/m | lb/ft | kg/m | lb/ft | kg/m | lb/ft | kg/m | lb/ft | kg/m |
|---|---|---|---|---|---|---|---|---|---|
| 0.376 | 0.5595 | 1.502 | 2.2352 | 2.67 | 3.9734 | 4.303 | 6.4036 | 7.65 | 11.384 |
| 0.668 | 0.9941 | 2.044 | 3.0418 | 3.4 | 5.0597 | 5.313 | 7.9066 | 13.6 | 20.239 |
| 1.043 | 1.5522 | | | | | | | | |

## Mass per Unit Volume

| lb/cu ft | kg/m³ | lb/cu ft | kg/m³ | lb/cu ft | kg/m³ | lb/cu ft | kg/m³ | lb/cu yd | kg/m³ |
|---|---|---|---|---|---|---|---|---|---|
| 10 | 160.19 | 15 | 240.28 | 20 | 320.37 | 105 | 1681.94 | 520 | 308.5 |
| 13 | 208.24 | 18 | 288.33 | 85 | 1361.57 | 125 | 2002.31 | 564 | 334.61 |

## Length

| mil | mm | in. | mm | in. | mm | in. | mm | in. | mm |
|---|---|---|---|---|---|---|---|---|---|
| .35 | 0.0089 | 0.188 (3/16) | 4.76 | 0.7 | 17.78 | 3.2 (3 1/5) | 81.28 | 13.75 (13 1/4) | 336.55 |
| 1 | 0.0254 | 0.191 | 4.83 | 0.719 (23/32) | 18.3 | 3.5 (3 1/2) | 88.90 | 13 | 330.20 |
| 2 | 0.0508 | 0.192 | 4.88 | 0.750 (3/4) | 19.10 | 3.544 | 90.02 | 14 | 355.60 |
| 4 | 0.1016 | 0.203 | 5.16 | 0.79 | 20.07 | 3.625 (3 5/8) | 92.08 | 15 | 381.00 |
| 6 | 0.1524 | 0.2031 | 5.16 | 0.864 | 21.95 | 3.75 (3 3/4) | 95.30 | 15.5 (15 1/2) | 393.70 |
| in. | mm | 0.207 | 5.19 | 0.875 (7/8) | 22.2 | 3.990 | 101.35 | 15.625 (15 5/8) | 396.88 |
| 0.0141 | 0.36 | 0.219 (7/32) | 5.60 | 0.889 | 22.58 | 4 | 101.60 | 16 | 406.40 |
| 0.0142 | 0.36 | 0.225 | 5.72 | 0.987 | 25.07 | 4.25 (4 1/4) | 107.95 | 17 | 431.80 |
| 0.015 | 0.38 | 0.2299 | 5.839 | 1 | 25.40 | 4.43 | 112.52 | 18 | 457.20 |
| 0.020 | 0.51 | 0.2300 | 5.842 | 1.128 | 28.65 | 4.5 (4 1/2) | 114.30 | 19 | 482.60 |
| 0.0255 | 0.65 | 0.239 | 6.07 | 1.178 | 29.92 | 5 | 127.00 | 20 | 508.00 |
| 0.028 | 0.71 | 0.244 | 6.073 | 1.185 | 30.10 | 5.32 | 135.13 | 22 | 558.80 |
| 0.0343 | 0.87 | 0.2499 | 6.35 | 1.25 (1 1/4) | 31.80 | 5.375 (5 3/8) | 136.53 | 22.25 (22 1/4) | 565.15 |
| 0.038 | 0.97 | 0.25 (1/4) | 6.40 | 1.270 | 32.26 | 5.5 (5 1/2) | 139.70 | 22.5 (22 1/2) | 571.50 |
| 0.044 | 1.12 | 0.262 | 6.65 | 1.410 | 35.81 | 5.625 (5 5/8) | 142.88 | 24 | 609.60 |
| 0.0449 | 1.14 | 0.263 | 6.68 | 1.5 (1 1/2) | 38.10 | 5.75 (5 3/4) | 146.10 | 25.9375 (25 15/16) | 658.81 |
| 0.050 | 1.27 | 0.286 | 7.26 | 1.571 | 39.90 | 6 | 152.40 | 30 | 762.00 |
| 0.056 | 1.42 | 0.313 (5/16) | 7.9 | 1.58 | 40.13 | 6.25 (6 1/4) | 171.50 | 32 | 812.80 |
| 0.0625 (1/16) | 1.60 | 0.334 | 8.43 | 1.693 | 43.00 | 7 | 177.80 | 32.75 (32 3/4) | 831.85 |
| 0.064 | 1.63 | 0.344 (11/32) | 8.73 | 1.75 (1 3/4) | 44.45 | 7.09 | 180.09 | 34 | 863.60 |
| 0.071 | 1.80 | 0.350 | 8.89 | 1.963 | 49.86 | 7.5 (7 1/2) | 190.50 | 35 | 889.00 |
| 0.0821 | 2.09 | 0.375 (3/8) | 9.50 | 2 | 50.80 | 7.625 (7 5/8) | 193.68 | 36 | 914.40 |
| 0.085 | 2.16 | 0.383 | 9.73 | 2.2 (2 1/5) | 55.88 | 8 | 203.20 | 38 | 965.20 |
| 0.098 | 2.49 | 0.431 | 10.95 | 2.25 (2 1/4) | 57.20 | 8.5 (8 1/2) | 215.90 | 39.75 (39 3/4) | 1009.65 |
| 0.102 | 2.56 | 0.437 | 11.10 | 2.257 | 57.33 | 8.625 (8 5/8) | 219.08 | 40 | 1016.00 |
| 0.113 | 2.87 | 0.469 (15/32) | 11.91 | 2.356 | 59.84 | 9 | 228.60 | 48 | 1219.20 |
| 0.125 (1/8) | 3.20 | 0.487 | 12.37 | 2.5 (2 1/2) | 63.50 | 9.625 (9 5/8) | 244.48 | 50 | 1270.00 |
| 0.131 | 3.33 | 0.500 (1/2) | 12.70 | 2.625 (2 5/8) | 66.70 | 10 | 254.00 | 53 | 1346.20 |
| 0.143 | 3.63 | 0.525 | 13.34 | 2.666 (2 2/3) | 67.72 | 10.5 (10 1/2) | 266.7 | 54 | 1371.60 |
| 0.148 | 3.76 | 0.540 | 13.72 | 2.749 | 69.82 | 11 | 279.40 | 71 | 1803.40 |
| 0.162 | 4.11 | 0.594 (19/32) | 15.1 | 2.75 (2 3/4) | 69.85 | 11.5 (11 1/2) | 292.10 | 72 | 1828.80 |
| 0.1799 | 4.57 | 0.612 | 15.54 | 2.813 (2 13/16) | 71.44 | 11.625 (11 5/8) | 295.26 | 75 | 1905.00 |
| 0.180 | 4.58 | 0.625 (5/8) | 15.90 | 3 | 76.20 | 12 | 304.80 | 108 | 2743.20 |
| | | 0.648 | 16.46 | 3.142 | 79.81 | 12.25 (12 1/4) | 361.95 | 288 | 7315.20 |

**FIGURE 25.7-1** *(Continued)*

## Length *(Continued)*

| ft & in. | m | ft & in. | m | ft & in. | m | ft & in. | m | ft & in. | m |
|---|---|---|---|---|---|---|---|---|---|
| 1 | 0.305 | 5-5 | 1.651 | 7-2 | 2.184 | 9-10 | 2.997 | 24 | 7.315 |
| 2 | 0.609 | 5-6 | 1.676 | 7-3 | 2.210 | 9-11 | 3.023 | 25 | 7.62 |
| 3 | 0.914 | 5-7 | 1.702 | 7-4 | 2.235 | 9-11¾ | 3.042 | 28 | 8.534 |
| 4 | 1.219 | 5-8 | 1.727 | 7-5 | 2.261 | 10 | 3.048 | 30 | 9.144 |
| 4-1 | 1.245 | 5-9 | 1.753 | 7-6 | 2.286 | 10-2 | 3.099 | 32 | 9.754 |
| 4-2 | 1.270 | 5-10 | 1.778 | 7-7 | 2.311 | 10-3 | 3.124 | 33 | 10.058 |
| 4-3 | 1.295 | 5-11 | 1.803 | 7-8 | 2.337 | 10-9 | 3.277 | 34 | 10.363 |
| 4-4 | 1.321 | 6 | 1.829 | 7-9 | 2.362 | 10-10 | 3.302 | 34.6 | 10.546 |
| 4-5 | 1.346 | 6-1 | 1.854 | 7-10 | 2.388 | 11 | 3.352 | 35 | 10.668 |
| 4-6 | 1.371 | 6-2 | 1.880 | 7-11 | 2.413 | 11-3 | 3.429 | 36 | 10.973 |
| 4-7 | 1.397 | 6-3 | 1.905 | 7-11½ | 2.426 | 11-7 | 3.531 | 50 | 15.240 |
| 4-8 | 1.422 | 6-4 | 1.930 | 8 | 2.438 | 12 | 3.658 | 84 | 25.603 |
| 4-9 | 1.448 | 6-5 | 1.956 | 8-1 | 2.464 | 12-1 | 3.683 | 100 | 30.480 |
| 4-9¼ | 1.454 | 6-6 | 1.981 | 8-2 | 2.489 | 12-7 | 3.835 | 123 | 37.490 |
| 4-10 | 1.473 | 6-7 | 2.007 | 8-4 | 2.540 | 12-11⅝ | 3.953 | 125 | 38.100 |
| 4-11 | 1.499 | 6-8 | 2.032 | 8-6 | 2.591 | 13-2 | 4.013 | 150 | 45.720 |
| 5 | 1.524 | 6-9 | 2.057 | 8-8 | 2.692 | 13-9 | 4.191 | 156 | 47.548 |
| 5-1 | 1.549 | 6-10 | 2.083 | 8-10 | 2.718 | 14-11½ | 4.559 | 200 | 60.96 |
| 5-2 | 1.575 | 6-11 | 2.108 | 9 | 2.743 | 20 | 6.096 | 250 | 76.200 |
| 5-3 | 1.600 | 7 | 2.134 | 9-3 | 2.819 | 22 | 6.706 | 300 | 91.440 |
| 5-4 | 1.626 | 7-1 | 2.159 | 9-6 | 2.896 | | | | |

## Area

| sq in. | mm² | sq ft | m² | sq ft | m² | sq ft | m² | sq ft | m² |
|---|---|---|---|---|---|---|---|---|---|
| 0.11 | 70.97 | 9 | 0.836 | 54 | 5.017 | 109 | 10.126 | 181.5 | 16.862 |
| 0.2 | 129.03 | 9.4 | 0.873 | 55 | 7.110 | 110 | 10.219 | 188 | 17.466 |
| 0.26 | 167.74 | 10 | 0.929 | 57 | 5.295 | 112 | 10.405 | 198 | 18.395 |
| 0.31 | 199.99 | 11 | 1.022 | 59 | 5.481 | 113 | 10.498 | 200 | 18.581 |
| 0.44 | 283.87 | 11.5 | 1.068 | 61 | 5.667 | 115 | 10.684 | 201 | 18.674 |
| 0.6 | 387.10 | 12 | 1.115 | 63 | 5.853 | 120 | 11.148 | 206.5 | 19.184 |
| 0.79 | 509.68 | 12.6 | 1.171 | 64 | 5.946 | 126 | 11.706 | 207 | 19.231 |
| 1 | 645.16 | 13 | 1.208 | 65 | 6.039 | 127 | 11.799 | 209 | 19.417 |
| 1.27 | 819.35 | 14 | 1.301 | 66 | 6.132 | 130 | 12.077 | 213 | 19.788 |
| 1.56 | 1006.45 | 15 | 1.394 | 67 | 6.225 | 132 | 12.263 | 218 | 20.253 |
| 2 | 1290.32 | 15.5 | 1.44 | 69 | 6.410 | 133 | 12.356 | 220 | 20.439 |
| 2.25 | 1451.61 | 15.7 | 1.459 | 70 | 6.503 | 136 | 12.635 | 226 | 20.996 |
| 4 | 2580.64 | 16 | 1.486 | 72 | 6.689 | 138 | 12.821 | 230 | 21.368 |

| sq ft | m² | | | | | | | | |
|---|---|---|---|---|---|---|---|---|---|
| | | 18.8 | 1.747 | 72.5 | 6.735 | 140 | 13.006 | 236 | 21.925 |
| | | 19 | 1.765 | 75 | 6.968 | 141 | 13.099 | 240 | 22.297 |
| 1 | 0.093 | 22 | 2.044 | 79 | 7.339 | 145.5 | 13.517 | 242 | 22.483 |
| 1.25 | 0.126 | 25 | 2.323 | 80 | 7.432 | 146.5 | 13.610 | 251 | 23.319 |
| 2 | 0.186 | 25.1 | 2.332 | 84 | 7.804 | 150 | 13.935 | 254 | 23.597 |
| 2.25 | 0.209 | 28 | 2.601 | 85 | 7.897 | 151 | 14.028 | 254.5 | 23.644 |
| 3 | 0.279 | 28.3 | 2.629 | 86 | 7.990 | 153 | 14.214 | 264 | 24.526 |
| 3.5 | 0.325 | 31 | 2.880 | 88 | 8.175 | 154 | 14.307 | 276 | 25.641 |
| 4 | 0.372 | 31.4 | 2.917 | 90.5 | 8.408 | 154.5 | 14.354 | 283 | 26.292 |
| 5 | 0.465 | 34.6 | 3.214 | 93 | 8.640 | 157 | 14.586 | 302 | 28.057 |
| 5.5 | 0.511 | 37.7 | 3.502 | 94 | 8.733 | 160 | 14.864 | 311 | 28.893 |
| 6 | 0.557 | 38 | 3.530 | 100 | 9.290 | 163.5 | 15.190 | 314 | 29.172 |
| 7 | 0.65 | 43 | 3.995 | 101 | 9.383 | 170 | 15.794 | 339 | 31.494 |
| 7.5 | 0.697 | 44 | 4.088 | 104 | 9.662 | 173 | 16.072 | 346 | 32.144 |
| 8 | 0.743 | 47 | 4.366 | 106 | 9.848 | 176 | 16.351 | 377 | 35.024 |
| 8.5 | 0.79 | 50 | 4.645 | 106.5 | 9.894 | 180 | 16.723 | 500 | 46.45 |

## Volume

| gal | L | gal | L | gal | L | gal | L |
|---|---|---|---|---|---|---|---|
| 0.24 | 0.908 | 10.3 | 38.990 | 14.8 | 56.024 | 16.25 | 61.513 |

**FIGURE 25.7-1** *(Continued)*

### Volume Rate of Flow

| cu ft/hr | m³/hr | cfm/in.² | m³/smm² | cfm/in.² | m³/smm² | cfm/in.² | m³/smm² | cfm/in.² | m³/smm² |
|---|---|---|---|---|---|---|---|---|---|
| 1000 | 28.317 | 1 | 7.314E-07 | 7.3 | 5.339E-06 | 12 | 8.777E-06 | 20 | 1.463E-05 |
| 2000 | 56.634 | 3.5 | 2.560E-06 | 7.8 | 5.705E-06 | 13 | 9.509E-06 | 24 | 1.755E-05 |
| 3000 | 84.950 | 4.4 | 3.218E-06 | 8.4 | 6.144E-06 | 14 | 1.024E-05 | 28 | 2.048E-05 |
| 4000 | 113.267 | 4.7 | 3.438E-06 | 8.5 | 6.217E-06 | 15 | 1.097E-05 | 31 | 2.267E-05 |
| 5000 | 141.584 | 5.8 | 4.242E-06 | 9.4 | 6.876E-06 | 17 | 1.243E-05 | | |
| | | 6.8 | 4.974E-06 | 10 | 7.314E-06 | 19 | 1.390E-05 | | |

### Moisture Penetration

| perms | ng/(Pa·s·m²) | perms | ng/(Pa·s·m²) | perms | ng/(Pa·s·m²) | perms | ng/(Pa·s·m²) | perms | ng/(Pa·s·m²) |
|---|---|---|---|---|---|---|---|---|---|
| 0.002 | 0.11 | 0.2 | 11.49 | 1.6 | 91.92 | 5 | 287.26 | 20 | 1149.05 |
| 0.02 | 1.15 | 0.3 | 17.24 | 1.9 | 109.16 | 5.4 | 310.25 | 30 | 1723.58 |
| 0.05 | 2.87 | 0.4 | 22.98 | 2 | 114.91 | 5.5 | 315.99 | 31 | 1781.03 |
| 0.06 | 3.45 | 0.5 | 28.73 | 2.1 | 120.65 | 5.8 | 333.22 | 50 | 2872.64 |
| 0.08 | 4.60 | 0.8 | 45.96 | 2.4 | 137.89 | 9.5 | 545.80 | 85 | 4883.46 |
| 0.1 | 5.75 | 1 | 57.45 | 3 | 172.36 | 10 | 574.53 | 90 | 5170.73 |
| 0.14 | 8.04 | 1.2 | 68.94 | 3.3 | 189.59 | 11 | 631.98 | 116 | 6664.49 |
| 0.15 | 8.62 | 1.5 | 86.18 | 4 | 229.81 | 15 | 861.79 | 120 | 6894.3 |
| 0.16 | 9.19 | | | | | | | | |

### Lumber

| 1 × 4 | 25 × 100 | 2 × 4 | 50 × 100 | 2 × 8 | 50 × 200 | 2 × 10 | 50 × 250 | 2 × 12 | 50 × 300 |
|---|---|---|---|---|---|---|---|---|---|
| 1 × 6 | 25 × 150 | 2 × 6 | 50 × 150 | | | | | | |

### Metal Gauges

| gauge | mm |
|---|---|
| 22 | .85 |
| 24 | .70 |

### Modulus of Subgrade Reaction

| lb/sq in./in. | kPa/mm |
|---|---|
| 50 | 184.2 |
| 100 | 368.4 |
| 200 | 736.8 |
| 300 | 1105.2 |

### Heat

| Btu | kJ | Btu | kJ | Btu | kJ | Btu | kJ | Btu | kJ |
|---|---|---|---|---|---|---|---|---|---|
| 0.8 | 0.84 | 100 | 105.5 | 240 | 253.2 | 3711 | 3915 | 40,000 | 42202.4 |
| 1 | 1.055 | 140 | 147.7 | 400 | 422 | 5000 | 5275 | 56,000 | 58083 |
| 60 | 63.3 | 160 | 168.8 | 1000 | 1055.1 | | | | |

### Thermal Conductivity (k-value)

| Btu/sq ft/°F/in./hr | W/mK | Btu/sq ft/°F/in./hr | W/mK | Btu/sq ft/°F/in./hr | W/mK |
|---|---|---|---|---|---|
| 0.5 | 0.865 | 5 | 8.654 | 300 | 519.219 |
| 1 | 1.731 | 6 | 10.384 | 500 | 865.365 |
| 2 | 3.461 | 10 | 17.307 | 1500 | 2596.095 |
| 2.5 | 4.327 | 100 | 173.073 | 2500 | 4326.825 |
| 3 | 5.192 | 150 | 259.610 | | |

### Thermal Resistivity (1/k)

| R-value | R-value |
|---|---|
| 11 | 4.68 |
| 19 | 8.09 |

impetus to the use of computers for design and detailing.

### 25.6.2 THE 100 MM MODULE

The 100 mm *building module* has many compelling advantages:

1. It is 1.6% smaller than the common 4 in. module of the customary system, which means that existing equipment can often be used to manufacture new metric products and that new products will often fit into existing openings.
2. The use of the *millimetre* as the working unit in building design and construction makes multimodular dimensions visible immediately. For example, 6400 mm is exactly 64 basic 100 mm modules; it is also 16 modules of 400 mm each, as opposed to sixteen 16 in. stud spaces. In customary units, the use of feet and inches hides the multiplier in modular dimensions. For example, it is not obvious that 21 ft 4 in. is 64 whole 4 in. modules or that 26 ft 8 in. is a whole multiple of 16 in. stud spacings.
3. The modular square, 100 mm × 100 mm, has an area of 0.01 $m^2$. This relationship is useful in cost comparisons because the unit cost per modular square, in cents, has the same number as the unit cost per square metre in dollars. For example, carpet costing 30 cents a modular square would cost $30 a square metre.
4. The modular cube, 100 mm × 100 mm × 100 mm, represents a volume of exactly 1 litre, or 0.001 $m^3$. One litre of water has a specific gravity of 1 and a mass of exactly 1 kg, a relationship that is useful in many calculations. For example, a modular cube of concrete, with a specific gravity of 2.4, has a mass of 2.4 kg. The calculation of the mass of a 100-mm-wide concrete wall 2800 mm high and 7200 mm long is 28 × 72 × 2.4 = 4834.4 kg. The mass per unit length is 28 × 10 (1 metre) × 2.4 = 672 kg/m.
5. One dimension, the *millimetre*, expresses all building dimensions, including tolerances and limitations, which will enable much greater accuracy in construction.

## 25.7 Metric Conversions Applicable to This Book

Early editions of this book did not contain metric measure equivalents for the customary system units shown. Metric units were added for the seventh edition and continued for this eighth edition. However, in order to avoid altering copyrighted documents, metric units have not been added to some figures in the seventh and eighth editions that are copyrighted and reprinted by permission of the copyright holders. Figure 25.7-1 is a list of the customary units and their metric equivalents that are included in the figures for which metric units have not been indicated.

## 25.8 Additional Reading

More information about the subjects discussed in this chapter can be found in the references listed in Section 25.9 and in the following publication.

Ramsey, Charles George, Harold Reeve Sleeper, and John Ray Hoke, Jr. 2000. *Architectural Graphic Standards*, 10th ed. Hoboken, NJ: John Wiley & Sons, Inc. (www.wiley.com).

## 25.9 Acknowledgments and References

### ACKNOWLEDGMENTS

We gratefully acknowledge the assistance of the following organizations and individuals in preparing this chapter. We are also indebted to them for permission to use their illustrations when requested and for the use of their publications as references.

American National Standards Institute (ANSI) (www.ansi.org)
ASTM International (ASTM) (www.astm.org)
National Institute of Building Sciences (www.nibs.org)
National Institute of Standards and Technology (NIST) of the U.S. Department of Commerce (www.nist.gov)

### REFERENCES

We would also like to thank the authors and publishers of the publications in the following list for their contribution to our research for this chapter.

*Architectural Graphic Standards. See* Ramsey et al. (2000).
ASTM International (ASTM). West Conshohocken, PA (www.astm.org). Standards:
  E621, "Standard Practice for the Use of Metric (SI) Units in Building Design and Construction."
  E713, "Standard Guide for Selection of Scales for Metric Building Drawings."
Freund, Eric C. 1990. "U.S. Construction and Practice in Europe." *The Construction Specifier* (September): 68 (www.csinet.org).
Institute of Electrical and Electronics Engineers, Inc. (IEEE) (www.ieee.org). 2002.
IEEE/ASTM SI 10, *American National Standard for Use of the International System of Units (SI): The Modern Metric System.* Piscataway, NJ: IEEE/ASTM.
National Institute of Standards and Technology (NIST). Gaithersburg, MD (www.nist.gov).
  Special Publication 504, *Metric Dimensional Coordination—the Issues and Precedents.*
  Special Publication 530, *Metrication in Building Design.*
  Technical Note 990, *The Selection of Preferred Metric Values for Design and Construction.*
Ramsey, Charles George, Harold Reeve Sleeper, and John Ray Hoke, Jr. 2000. *Architectural Graphic Standards*, 10th ed. Hoboken, NJ: John Wiley & Sons, Inc. (www.wiley.com).

# Glossary

Note: Boldfaced words in the definitions are also glossary entries.

**abrasion**   Wearing away by friction.

**absorption bed**   A wide trench exceeding 36 in. (910 mm) in width containing a minimum of 12 in. (305 mm) of clean, coarse **aggregate** and a system of two or more distribution pipes through which treated sewage may seep into the surrounding soil. Also called *seepage bed*.

**absorption coefficient**   The ratio of the sound-absorbing effectiveness of 1 sq ft (0.09 m$^2$) of a material to 1 sq ft (0.09 m$^2$) of a perfectly absorptive material at a specific frequency; usually expressed as a decimal value (such as .70) or as a percentage.

**absorption field**   An arrangement of absorption trenches through which treated sewage is absorbed into the soil. Also called *disposal field*.

**absorption of sound**   The ability of a material to absorb rather than reflect sound waves striking it by converting sound energy to heat energy within the material.

**absorption of water by clay masonry**   The weight of water a **brick** or other clay **masonry unit** absorbs when immersed in either cold or boiling water for a stated length of time, expressed as a percentage of the weight of the dry unit.

**absorption of water by concrete masonry units (CMU)**   The weight of water a concrete masonry unit absorbs when immersed in water, expressed in pounds of water per cubic foot of concrete.

**absorption rate**   The weight of water absorbed when a clay brick is partially immersed for 1 minute, usually expressed in either grams or ounces per minute. Also called *suction* or *initial rate of absorption*.

**absorption trench**   A trench not more than 36 in. (910 mm) in width, containing a minimum of 12 in. (305 mm) of clean, coarse aggregate and a distribution pipe, through which treated sewage is allowed to seep into the soil.

**absorption unit**   *See* **sabin**.

**accelerator**   An **admixture** used in **concrete** to hasten its set and increase the rate of strength gain (the opposite of a **retarder**).

**accessible**   Describes a site, building, facility, or portion thereof that complies with current standards and can be approached, entered, and used by a **physically handicapped person**.

**accessible route**   A continuous, unobstructed path connecting all **accessible** elements and spaces in a building or facility that can be negotiated by a person with a severe **disability** using a wheelchair and that is also safe for and usable by people with other disabilities.

**acid**   Corrosive chemical substance that attacks many common building materials, decorative finishes, **coatings**, **paints**, and **transparent finishes**.

**acid-resistant brick**   *See under* **brick masonry unit**.

**acoustical correction**   The planning, shaping, and equipping of a space to establish the best possible hearing conditions for faithful reproduction of wanted sound within the space. *See also* **acoustics**.

**acoustics**   The science of sound; the production, transmission, and effect of sound. *See also* **acoustical correction**, **architectural acoustics**, and **room acoustics**.

**acrylic**   (1) In **carpet**, a generic term including acrylic and modified acrylic (modacrylic) fibers. Acrylic is a **polymer** composed of at least 85% by weight of acrylonitrile; modacrylic is a polymer composed of less than 85% but at least 35% by weight of acrylonitrile. (2) In glazing applications, a transparent plastic material.

**actual dimension**  The actual measured dimension of a **masonry unit**, piece of lumber, or other construction material or assembly. *See also* **nominal dimension**.

**actual size**  *See* **size**.

**adaptability**  The capability of certain building spaces and elements, such as kitchen counters, sinks, and grab bars, to be altered or added so as to accommodate the needs of persons with and without disabilities, or to accommodate the needs of persons with different types or degrees of **disability**.

**adhesion**  The property of a **paint** film that enables it to stick to a surface.

**adhesive, drywall**  Adhesives specifically intended for the application of gypsum board. A *contact adhesive* is an adhesive used to bond layers of gypsum board or for bonding gypsum board to metal studs. A *laminating adhesive* is an adhesive used to bond layers of gypsum board. A *stud adhesive* is an adhesive used to attach gypsum board to wood supports.

**adhesive, tile**  Prepared organic material, ready for use with no further addition of liquid or powder, which cures or sets by evaporation; distinguished from mortars by the absence of siliceous fillers (sand) that are included in mortars either at the plant or in the field.

**admixture**  A material (other than **portland cement**, water, or **aggregate**) used in **concrete** to alter its properties (such as **accelerators**, **retarders**, and **air-entraining agents**).

**adobe brick**  *See under* **brick masonry unit**.

**age hardening**  The continuing increase in strength for long periods of time of aluminum **alloys** after **heat treatment**.

**agglomeration**  Process for increasing the particle size of iron ores to make them suitable for ironworking and steelmaking.

**aggregate**  A hard, inert material mixed with **portland cement** and water to form **concrete**. *Fine aggregate* has pieces $1/4$ in. (6.4 mm) in diameter and smaller. *Coarse aggregate* has pieces larger than $1/4$ in. (6.4 mm) in diameter.

**aging**  The period of time in which a **heat-treatable** aluminum **alloy** is al-lowed to remain at room temperature, after heat treatment (heating and quenching) to reach a stable state of increased strength. *See also* **artificial aging**.

**airborne sound transmission**  *See* **sound transmission**.

**air chamber**  A piece of pipe about 10 in. (250 mm) long installed above the hot and cold valves of fixtures such as sinks, lavatories, and clothes washers to cushion the rush of water as the valve is closed and prevent **water hammer**.

**air drying**  (1) In **coatings**, **paints**, and **transparent finishes**, capable of forming a solid film when exposed to air at moderate atmospheric temperatures. (2) Of wood, *see* **seasoning**.

**air-entrained concrete**  Concrete containing minute bubbles of air up to about 7% by volume.

**air-entraining agent**  An **admixture** used to produce **air-entrained concrete**.

**air gap**  The unobstructed vertical distance between the mouth of a water outlet and the flood level rim of the water receptacle. The water outlet may be a faucet, spout, or other outlet; the receptacle may be a **plumbing fixture**, tank, or other receptacle.

**alclad sheet**  A clad product with an aluminum or aluminum **alloy** coating having high resistance to **corrosion**. The coating is anodic to the core alloy it covers, thus protecting it physically and electrolytically against corrosion. *See also* **clad alloy**.

**alkali**  A soluble mineral salt present in some soils. Alkalis are chemical substances characterized by their ability to combine with acids to form neutral salts. They are damaging to many **coatings**, **paints**, and **transparent finishes**.

**allowable unit stress**  *See under* **stress**.

**alloy**  The material formed when two or more metals, or metals and nonmetallic substances, are joined by being dissolved into one another while molten. *See also* **brazing alloy** and **clad alloy**.

**alloy designation**  A numerical system used in designating the various **alloys** of aluminum.

**alloying element**  Element added in steelmaking to achieve desired properties.

**alloy steel**  *See under* **steel**.

**alumina (Al$_2$O$_3$)**  A hydrated form of aluminum oxide found in **bauxite** and in ordinary clays.

**ambient sound**  A continuous background sound that is a composite of individual sounds coming from exterior sources, such as street traffic, and interior sources, such as ventilating equipment and appliances, none of which can be identified individually by a listener. Also called *background noise*.

**ambulatory**  Able to walk without assistance or difficulty.

**amperage**  Electrical rate of flow, measured in amperes (amps) and comparable to gallons per minute (gpm) in a fluid medium. The strength of a current of electricity.

**ampere (amp)**  A unit of electrical current equivalent to that produced by 1 volt applied across a resistance of 1 ohm. One coulomb of electricity in every second.

**anchor**  A piece or assemblage, usually metal, used to attach parts (e.g., plates, joists, trusses, studs, sills, masonry, windows, doors, and other building elements) to wood, concrete, or masonry.

**angular course**  *See under* **course**.

**angular measure**  The deviation between two lines that meet at a point, expressed in **degrees**, **minutes**, and **seconds**.

**annealing**  *See* **heat treatment**.

**annual growth ring**  The growth layer put on by a tree in a single growth year.

**anodic coating**  A surface coating applied to an aluminum **alloy** by **anodizing**.

**anodizing**  Applying an electrolytic oxide coating to an aluminum **alloy** by building up the natural surface film using an electrical current (usually dc) through an oxygen-yielding electrolyte with the alloy serving as the anode.

**antioxidant**  A compound added to other substances to retard oxidation, which deteriorates plastics.

**antique finish**  A finish usually applied to furniture or woodwork to give the appearance of age.

**arch**  A usually curved compressive structural member, spanning openings or recesses; also built flat. A *back arch* is a concealed arch carrying the backing of a wall where the exterior face is carried by a lintel. A *jack arch* is an arch having horizontal or nearly horizontal upper and lower surfaces. Also called a *flat* or *straight arch*. A *major arch* is an arch with a span greater than 6 ft (1800 mm) that carries a load that is equivalent to a uniform load greater than 1000 psf (4882 kg/m²). Typically known as a *Tudor arch, semicircular arch, gothic arch,* or *parabolic arch.* Has a rise-to-span ratio greater than 0.15. A *minor arch* is an arch with a maximum span of 6 ft (1800 mm) carrying a load that does not exceed 1000 psf (4882 kg/m²). Typically known as a *jack arch, segmented arch,* or *multicentered arch.* Has a rise-to-span ratio less than or equal to 0.15. A *relieving arch* is an arch that is built over a lintel, flat arch, or smaller arch to divert loads, thus relieving the lower member from excessive loading. Also known as a *discharging* or *safety arch.* A *trimmer arch* is an arch, usually a low-rise type, of **brick masonry unit** used for supporting a fireplace hearth.

**architectural acoustics** The **acoustics** of buildings and other structures. *See also* **acoustics** and **room acoustics**.

**architectural barrier** A physical condition in a building or facility that creates unsafe or confusing conditions or prevents accessibility and free mobility.

**architectural terra cotta** Custommade, hard-burned, glazed or unglazed clay building units, plain or ornamental, that are machine extruded or hand molded. *See also* **ceramic veneer**.

**area wall** A **retaining wall** around below-grade basement windows.

**aromatic solvents** Group of organic compounds derived from coal or petroleum, such as benzene and toluene.

**artificial aging** The heating of an aluminum **alloy** for a controlled time at an elevated temperature to accelerate and increase its strength gain after heat treatment. *See also* **aging**.

**ashlar masonry** Masonry composed of rectangular units usually larger in size than **brick** with sawed, dressed, or square beds bonded with **mortar**. Ashlar masonry is also described according to its **pattern bond**, which may be coursed, random, or patterned.

**atmospheric pressure steam curing** *See* **curing**.

**atmospheric vacuum breaker** A simple mechanical device consisting essentially of a check valve in a supply line and a valve member (on the discharge side of the check valve) opening to the atmosphere when the pressure in a line drops to atmospheric. Also called a *siphon breaker.*

**attenuation** Reduction of the energy or intensity of sound.

**audible cue** *See* **cue**.

**austenitic steel** *See* **grain structure**.

**autoclave** *See* **curing**.

**average transmission loss (tl)** The numerical average of the **transmission loss** values of an assembly measured at nine frequencies. It is a single-number rating for comparing the **airborne sound transmission** through walls and floors.

**Axminster carpet** *See* **woven carpet**.

**back** The side opposite the face. The poorer side of a **plywood** panel. The surface of **gypsum board** that will be placed toward the supports.

**back arch** *See under* **arch**.

**back blocking** A single-ply **gypsum board** installation procedure for reinforcing butt-end or edge joints to minimize surface imperfections such as cracking and ridging.

**backflow** The unintentional flow of water into the supply pipes of a plumbing system from a nonsupply source.

**background noise** *See* **ambient sound**.

**backing** The carpet foundation of jute, kraftcord, cotton, rayon, or polypropylene yarn that secures the pile yarns and provides stiffness, strength, and dimensional stability.

**backing board** *See under* **gypsum board**.

**backing, tile** A suitable surface for the application of tile, such as a structural subfloor or wall surface.

**backparging** *See* **parging**.

**backplate** *See under* **finished mill products**.

**back-siphonage** A type of **backflow**, usually caused by a temporary occurrence of negative pressure (suction) in pipes.

**backup** The part of a masonry wall behind the exterior facing.

**baking finish** A **coating** that is baked at temperatures above 150°F (65.6°C) to dry and develop desired properties.

**bark** The outer corky layer of a tree, composed of dry, dead tissue.

**bars** *See under* **finished mill products**.

**base coat** The first **plaster** layer in **two-coat work**; or either the **scratch** or **brown coat** in **three-coat work**.

**base course** The lowest course of masonry in a wall or pier.

**base line** A **parallel** of specified **latitude**, used in the **rectangular survey system**, serving as the main east-west reference line, with a **principal meridian** for a particular state or area.

**basic stress** *See* **stress**.

**batter** Recessing or sloping a wall back in successive courses; the opposite of **corbel**.

**bauxite** A raw ore of aluminum consisting of 45 to 60% aluminum oxide, 3 to 25% iron oxide, 2.5 to 18% silicon oxide, 2 to 5% titanium oxide, other impurities, and 12 to 30% water. This ore varies greatly in the proportions of its constituents, color, and consistency.

**Bayer process** The process generally employed to refine **alumina** from **bauxite**.

**beam** A structural member transversely supporting a load.

**bearding** Long-fiber fuzz occurring on some loop pile fabrics, caused by fibers snagging and loosening due to inadequate anchorage.

**bearing wall** A wall that supports a vertical load in addition to its own weight. *See also* **non-load-bearing wall**.

**bed joint** Horizontal layer of mortar in which a masonry unit is laid.

**beneficiation** Concentrating process used to increase the iron content of ores prior to use. *See also* **agglomeration**.

**binder** (1) In paint and coatings, the **vehicle** ingredient with adhesive qualities (linseed oil, resins, etc.) that binds the **pigment** and other ingredients of a **coating, paint,** or **transparent finish** into a cohesive film and facilitates bonding with the underlying surfaces. (2) In **terrazzo**, a cementitious or resinous material that gives the matrix adhesive and other important physical properties.

**bleaching** (1) The process of lightening raw wood. (2) The process of restoring discolored or stained wood to its normal color or making it lighter.

**bleeding** (1) In unit masonry, (a) the loss of water from a **masonry unit** having low **suction** when it comes in contact with mortar or (b) the loss of water from mortar due to low **water retention** when it contacts a masonry unit. Bleeding causes masonry units to float (*see* **floating**). (2) In **concrete**, the appearance of excess water rising to the surface shortly after placing of concrete. (3) In **coatings**, **paints**, and **transparent finishes**, discoloration of a finish coat by coloring matter from the underlying surface or coat of finishing material. (4) In **plastics**, the diffusion of a colorant out of a plastic part into adjacent materials.

**blend to a common level** The meeting of two or more surfaces so that there is no abrupt rise or drop in the surface.

**blistering** The formation of bubbles or pimples on a coated, painted, or transparent finished surface caused by moisture in the underlying material (wood, masonry, concrete, etc.); caused by adding a coat of **coating**, **paint**, or **transparent finish** before the previous coat has dried thoroughly or caused by excessive heat or grease under a coating, paint, or transparent finish.

**blocking** (1) In masonry construction, a method of bonding two adjoining or intersecting walls, not built at the same time, by means of offsets whose vertical dimensions are not less than 8 in. (200 mm). (2) In wood construction, short pieces of wood used as nailers, spacers, or fillers between wood members, between wood members and other construction, or between other materials.

**blow molding** Shaping thermoplastic materials into hollow form by air pressure and heat; usually performed on sheets or tubes.

**blushing** Describes opaque lacquer that loses its gloss and becomes flat or clear lacquer that turns white or milky.

**board** *See under* **lumber**.

**board foot** A measure of lumber. One board foot is the equivalent of a piece of **lumber** whose nominal dimensions are 1 in. (25 mm) thick, 12 in. (305 mm) wide, and 12 in. (305 mm) long.

**body** Used to indicate thickness or thinness of a liquid **coating**, **paint**, or **transparent finish** material.

**bond beam** Course or courses of a masonry wall grouted and usually reinforced in the horizontal direction. Alternatively, may be made of reinforced concrete. Serves as a horizontal tie of wall-bearing courses for structural members or itself as a flexural member.

**bond course** A masonry **course** in which the units overlap more than one **wythe** of masonry.

**bonded rubber cushioning** Rubber or latex cushioning adhered to a carpet at the mill.

**bonder** *See* **header**.

**bonderizing** Process to improve **paint** adhesion on steel by dipping lightly galvanized (see **galvanizing**) objects in a hot phosphate solution to form a surface film of zinc phosphate.

**bond, pattern** *See* **pattern bond**.

**bond, structural** Tying **wythes** of a masonry wall together by lapping units over one another or by connecting them with metal ties.

**bond, tensile** Adhesion between **mortar** and **masonry units** or reinforcement.

**border** The resilient flooring at the perimeter of a room adjacent to the walls, which is installed separately from the **field**.

**bow** *See* **warp**.

**boxing** Mixing a **coating**, **paint**, or **transparent finish** material by pouring it from one container to another several times. Boxing is not recommended for most such materials.

**braille** A special raised-touch alphabet for the blind using a cell of six dots.

**brazing** A welding process in which the filler metal is a nonferrous metal or **alloy** with a melting point higher than 800°F (426.67°C) but lower than that of the metals joined. A *brazing alloy* is an alloy used as filler metal for brazing. In aluminum the brazing alloy is usually in the 4000 Series of alloys. A *brazing sheet* is an unclad or specially clad sheet for brazing purposes, the surface of which has a lower melting point than the core. Brazing sheet of the clad type may be clad on either one or two surfaces.

**breathe** The ability of a **coating**, **paint**, or **transparent finish** film to permit the passage of moisture vapor without causing **blistering**, **cracking**, or **peeling**.

**brick grade** Designation for the durability of a unit, expressed as SW for severe weathering, MW for moderate weathering, and NW for negligible weathering.

**brick masonry unit** A unit that is formed into a rectangular prism while plastic. An *acid-resistant brick* is a clay brick suitable for use in contact with chemicals; designed primarily for use in the chemical industry. Usually used with acid-resistant mortars. An *adobe brick* is a large clay brick of varying size, roughly molded and sun dried. A *building (common) brick* is a clay brick for building purposes, not especially treated for texture or color. Formerly called *common brick*. A *clay brick* is a brick made of clay or shale, formed into a rectangular prism while plastic and burned (fired) in a kiln. A *concrete brick* is a brick made from **portland cement** and an **aggregate** and cured either in normal atmosphere or using either low- or high-pressure steam. It is formed as a rectangular prism, usually not larger than 4 × 4 × 12 in. (100 × 100 × 305 mm). An *economy brick* is a clay brick whose nominal dimensions are 4 × 4 × 8 in. (100 × 100 × 305 mm). An *engineered brick* is a clay brick whose nominal dimensions are 4 × 3.2 × 8 in. (100 × 81.28 × 305 mm). A *facing brick* is a clay brick made especially for facing purposes, often with a finished surface texture. Such units are made of selected clays or treated to produce desired color. A *fire brick* is a brick made of refractory ceramic material that will resist high temperatures. A *floor brick* is a smooth, dense brick that is highly resistant to abrasion; used as a finished floor surface. A *hollow brick* is a clay or shale **masonry unit** whose **net cross-sectional area** in any plane parallel to the bearing surface is not less than 60% of its **gross cross-sectional area** measured in the same plane. *Jumbo brick* is a generic term indicating that a clay brick is larger than the standard. Some producers use this term to describe oversized brick of specific dimensions manufactured by them. A *Norman brick* is a clay brick whose nominal dimensions are 4 × 2²/₃ × 12 in. (100 × 67.7 × 305 mm). A *paving brick* is a clay brick especially suitable for use in pavements where resistance to abrasion is important. A *Roman brick* is a clay brick whose nominal dimensions are 4 × 2 × 12 in. (100 × 50 × 305 mm). An *SCR brick* is a clay brick whose nominal dimensions are 2²/₃ × 6 × 12 in. (67.7 × 150 × 305 mm). It lays up three courses to 8 in. (203.2 mm) and produces a nominal 6-in. (150-mm)-thick wall. Developed by the Structural Clay Products Research Foundation. A *sewer* *brick* is a low-absorption, abrasive-resistant clay brick intended for use in drainage structures.

**brick type** Designation for **facing brick** that controls tolerance, chippage, and distortion. Expressed as FBS, FBX, and FBA for solid brick and HBS, HBX, HBA, and HBB for hollow brick.

**British thermal unit (Btu)** A measure of heat quantity equal to the amount of heat required to raise 1 lb of water 1°F (0.5556°C) at sea level.

**broadloom** Carpet woven on a broad loom in widths of 6 ft (1800 mm) or more.

**brown coat** The second **plaster** layer in **three-coat work**, which provides additional strength and a suitably true and plane surface for the application of the **finish coat**.

**brush** A tool composed of bristles set into a handle; often used to apply **paint** and **transparent finish** materials. Sometimes used to apply **coatings**. Bristles may be synthetic (needed for water-thinned paints) or natural, such as hog hair.

**brushability** Ease with which a paint or **transparent finish** material can be brushed.

**brush marks** Marks of a brush that remain in a dried **coating, paint**, or **transparent finish** film.

**Btu** *See* **British thermal unit**.

**buckling** Wrinkling or ridging of a carpet after installation, caused by insufficient stretching, dimensional instability, or manufacturing defects.

**build** The apparent thickness or depth of a **coating, paint**, or **transparent finish** film after it has dried.

**building brick** *See under* **brick masonry unit**.

**building drain** The lowest part of a building's drainage system. It receives the discharge from soil, waste, and other drainage pipes inside the walls of a building and conveys it to a point 3 ft (900 mm) outside the building walls, where it joins with the **building sewer.**

**building sewer** The horizontal piping of a **drainage system**, which extends from the end of a **building drain** to a public sewer, private sewer, **individual sewage disposal system**, or other point of disposal. A **building sewer** begins at a point 3 ft (900 mm) outside the building wall, where the building drain ends. Also called a **house sewer**.

**building storm drain** A type of **building drain** used for conveying rainwater, surface water, groundwater, subsurface water, and cooling condensate; or a combined **building sewer**, extending to a point not less than 3 ft (900 mm) outside the building wall, where it joins with a storm sewer or combined storm and sewage sewer.

**built environment** The collection of buildings, facilities, transportation systems, and other structures and spaces created for the purpose of providing convenient places for work, play, living, and related human activities; contrasted with the natural environment.

**bulked continuous filament (BFC)** Continuous strands of synthetic fiber made into yarn without spinning; often extruded in modified cross section such as multilocal, mushroom, or bean shape, or textured to increase bulk and covering power.

**burling** Removing surface defects such as knots, loose threads, and high spots to produce acceptable quality after weaving; also, filling in omissions in weaving.

**burnishing** Shiny or lustrous spots on a **coating, paint**, or **transparent finish** surface caused by rubbing.

**butt-end joint** Joint in which mill- or job-cut (exposed core) **gypsum board** ends or edges are butted together.

**buttering** Placing **mortar** on a **masonry unit** with a trowel.

**calcimine** Water-based **paint** generally consisting of animal glue, zinc white, and calcium carbonate or clay; now seldom used.

**calcium chloride** An accelerator added to **concrete** to hasten setting (not to be considered an antifreeze).

**calendering** A process for producing plastic film or sheeting by passing the material between revolving heated rolls.

**call** In surveying, a statement or mention of a **course** or distance.

**cambium** A thin layer of tissue that lies between a tree's bark and its wood. The cambium subdivides to form the new wood and bark cells of each year's growth.

**cant** A log **slabbed** on one or more sides.

**cant strip** A triangular filler between a roof and a **parapet wall**.

**capacity insulation** The ability of masonry to store heat as a result of its mass, density, and specific heat.

**capillaries** (1) Thin-walled tubes or vessels found in wood. (2) In concrete, channels that absorb water and are interrupted by entrained air bubbles.

**carbon steel** *See under* **steel**.

**cardinal points** The four major compass headings of north, east, south, and west.

**carpet** General designation of fabric constructions that serve as soft floor coverings, especially those that cover an entire floor and are fastened to it, as opposed to rugs. *See also* **flocked carpet, fusion-bonded carpet, knitted carpet, needle-punched carpet, tufted carpet,** and **woven carpet**.

**carved carpet** *See sculptured* under **pile**.

**case hardening** Hardening of the outer skin of an iron-based alloy by promoting surface absorption of carbon, nitrogen, or cyanide, generally accomplished by heating the alloy in contact with materials containing these elements and rapid cooling.

**casting** (1) In masonry and **concrete**, pouring a mix into a mold and permitting it to set to form the desired shaped object. (2) In metalwork, pouring molten metal into molds to form desired shapes. (3) In **plastics**, the shaping of plastic objects by pouring the material into molds and allowing it to harden without the use of pressure.

**cast iron** High-carbon iron made by melting **pig iron** with other iron-bearing materials and **casting** in sand or loam molds; characterized by hardness, brittleness, and high compressive and low tensile strengths.

**catalyst** (1) A substance that speeds up or slows down the rate of a chemical reaction without itself undergoing permanent change in composition. (2) A substance used to initiate the polymerization of monomers to form polymers.

**caulking compound** Semidrying or slow-drying plastic material used to seal joints or fill crevices such as those around windows and chimneys. Either elastomeric or acrylic joint sealants are generally used today where caulking compounds are called for. *See also* **sealant, joint**.

**cavity wall**   A wall built of **masonry units** so arranged as to provide a continuous air space within the wall (with or without insulating material) and in which the inner and outer **wythes** of the wall are tied together with metal **wall ties** or continuous metal **joint reinforcement**.

**cavity wall tie**   *See under* **wall tie**.

**C/B ratio**   The ratio of the weight of water absorbed by a clay **masonry unit** during immersion in cold water to the weight absorbed during immersion in boiling water. An identification of the probable resistance of clay **brick** to the action of freezing and thawing.

**cellular plastic**   A plastic whose apparent density is decreased substantially by the presence of numerous cells distributed throughout its mass. *Closed-cell plastic* is cellular plastic in which there is a predominance of nonconnecting cells. *Open-cell plastic* is a cellular plastic in which there is a predominance of interconnecting cells.

**cellulose**   The principal constituent of wood that forms the framework of the wood cells.

**cement**   A binding agent capable of uniting dissimilar materials into a composite whole. *See also* **portland cement**.

**centering**   Temporary formwork for the support of masonry **arches** or **lintels** during construction. Also called *center(s)*.

**ceramic tile**   A thin surfacing unit having either a glazed or unglazed face made from clay or a mixture of clay and other ceramic materials and fired to a temperature sufficiently high to produce specific physical properties and characteristics.

**ceramic tile finish**   All of the elements normally installed by a tile contractor, from the backing to the face of the tile. This may include just a tile-setting product, tile, and **grout** or, in addition, a **mortar bed**, reinforcement, and **cleavage membrane**. Also called *tilework*.

**ceramic veneer**   Architectural terra cotta, characterized by large face dimensions and thick sections. An *adhesion-type ceramic veneer* is a thin section of ceramic veneer held in place without metal **anchors** by adhesion of **mortar** backing. An *anchored-type ceramic ve-*neer is a thicker section of ceramic veneer held in place by **grout** and wire **anchors** connected to backing.

**cesspool**   A covered and lined underground pit used as a holding tank for domestic **sewage** and designed to retain the organic matter and solids, but to permit the liquids to seep through the bottom and sides. Cesspools are almost universally prohibited in this country and are not acceptable as a means of **sewage** disposal.

**chain warp**   *See under* **warp**.

**chalking**   Formation of a loose powder on the surface of a **coating** or **paint** after exposure to the elements.

**chase**   A groove or continuous recess built in a masonry or concrete wall to accommodate pipes, ducts, or conduits.

**check**   A lengthwise separation of wood, the greater part of which occurs across the **annual growth rings**. A check that passes entirely through a piece of wood is called a *split*.

**checking**   Type of **coating**, **paint**, or **transparent finish** failure in which many small cracks appear in the surface.

**circuit vent**   A branch vent that serves two or more traps and extends from in front of the last fixture connection of a horizontal branch to a **vent stack**. *See also* **loop vent**.

**circulation path**   An exterior or interior passageway from one place to another for pedestrians, including, but not limited to, walks, hallways, courtyards, stairways, and stair landings.

**clad alloys**   Alloys having one or both surfaces of a metallurgically bonded coating, the composition of which may or may not be the same as that of the core, and which is applied for such purposes as corrosion protection, surface appearance, or brazing. *See also* **alclad sheet** and **alloy**.

**cladding**   Bonding thin sheets of a coating metal with desirable properties (such as corrosion resistance or chemical inertness) over a less expensive metallic core not possessing these properties. Copper cladding over steel may be applied by hot dipping; stainless steel and aluminum cladding by hot rolling.

**clay brick**   *See under* **brick masonry unit**.

**cleavage (isolation) membrane**   A membrane such as saturated roofing felt, building paper, or 4 mil (0.1 mm) polyethylene film, installed between the **backing** and a **mortar bed** to permit independent movement of a tile finish.

**closer**   The last **masonry unit** laid in a **course**.

**CMU**   *See* **concrete masonry unit**.

**coarse aggregate**   *See* **aggregate**.

**coating**   A mastic or liquid-applied surface finish, regardless of whether a protective film is formed or only decorative treatment results. There are two categories of coatings: coatings and special coatings. Materials called *coatings* in this book are those that are usually applied in the shop or factory on metal, glass, porcelain, wood, and other materials. They include such products as Kynar coatings on metal, anodizing on aluminum, and liquid-applied colored finishes on wood doors and cabinets. Some coatings are virtually identical to **paint**, the major differences being that they are applied in a shop or factory and are usually sprayed on rather than being brushed or rolled on. *Special coatings* are relatively thick, high-performance architectural coatings, such as high-build glaze coatings; fire-retardant coatings; industrial coatings, such as those used in sewage disposal plants; and cementitious coatings. Special coatings are usually applied in the field. These are beyond the scope of this book. *See also* **fire-retardant coating** or **paint**, **paints**, and **transparent finishes**.

**coke**   Processed form of bituminous coal used as a fuel, a reducing agent, and a source of carbon in making pig iron.

**cold flow**   Permanent change in dimension due to stress over time without heat.

**cold forming**   Forming thin sheets and strips to desired shapes at room temperature, generally with little change in the mechanical properties of the metal; includes roll, stretch, shear, and brake forming. *See also* **cold working**.

**cold rolling**   *See under* **cold working**.

**cold working**   (1) In aluminum, forming a metal product at room temperature by means of **rolling**, **drawing**, **forging**, or other mechanical methods of forming or shaping. *Cold rolling* is the forming of sheet metal by rolling at room temperature metal that has been previously hot rolled to a thickness of about 0.125 in. (3.2 mm). (2) In steel, shaping by cold drawing, cold reduction, or cold rolling at room temperature; generally accompanied by an increase in strength and hardness. *Cold drawing* is shaping by pulling through a die to reduce the cross-sectional area and impart the desired shape; generally accompanied by an increase in strength, hardness, closer dimensional tolerances, and a smoother finish. *Cold finishing* is cold working that results in **finished mill products**. *Cold reduction* is cold rolling that drastically reduces sheet and strip thickness with each pass through the rolls; generally accompanied by an increase in hardness, stiffness, and strength and resulting in a smoother finish and improved flatness. *Cold rolling* is a gradual shaping between rolls to reduce the cross-sectional area or impart the desired shape; generally accompanied by an increase in strength and hardness. *See also* **cold forming** and **hot working**.

**collar joint**  Interior longitudinal vertical joint between two **wythes** of masonry.

**colorant**  Concentrated color added to **coatings** and **paints** to make specific colors.

**colorfast**  Fade resistant.

**color uniformity**  Ability of a **coating** or **paint** to maintain a consistent color across its entire surface, particularly during weathering.

**column**  A vertical structural member acting primarily in compression, whose horizontal dimension measured at right angles to its thickness does not exceed three times its thickness.

**combination process**  A process used to retrieve additional **alumina** and soda from the red mud impurities of the **Bayer process**.

**common alloy**  An alloy that does not increase in strength when heat-treated (*non-heat-treatable*). Common alloys may be strengthened by **strain hardening**.

**common brick**  *See* **brick masonry unit**.

**common wall**  A wall that separates adjacent dwelling units within an apartment building or adjacent tenants in townhouses and other buildings; also called a *party wall*.

**composite wall**  A multiple-**wythe** wall in which at least one of the **wythes** is dissimilar to the other wythe or wythes with respect to type or grade of **masonry unit** or **mortar**.

**compounding**  The thorough mixing of a **polymer** or **polymers** with other ingredients such as **fillers**, **plasticizers**, **catalysts**, **pigments**, **dyes**, or curing agents.

**compression molding**  Forming **plastic** in a mold by applying pressure and, usually, heat.

**compressive strength**  A material's ability to resist compressive forces.

**concrete**  A composite material made of **portland cement**, water, and **aggregates**, and sometimes **admixtures**.

**concrete block**  *See* **concrete masonry unit**.

**concrete brick**  *See under* **brick masonry unit**.

**concrete masonry unit (CMU)**  A **masonry unit** having **portland cement** as its primary cementitious material. A *decorative CMU* is one of various available types of concrete masonry units with beveled face shell recesses or other articulation or texture. A *faced CMU* is one that has a special ceramic, glazed, plastic, polished, or ground face. A *slump block* is a CMU produced so that it will slump or sag before it hardens; for use in masonry wall construction. A *split-face block* is a solid or hollow CMU that is machine fractured (split) lengthwise after hardening to produce a rough, varying surface texture.

**condensation**  The change of water from its gaseous form (water vapor) to liquid water; the liquid water so collected.

**condensation polymerization**  A chemical reaction in which the molecules of two substances combine, giving off water or some other simple substance (*see* **polymerization**).

**consistency**  The relative ability of freshly mixed **concrete** to flow, as measured by a **slump test**.

**construction**  The method by which a **carpet** is made (**loom** or machine) and other identifying characteristics, including the number of **pile** rows per inch, **pitch**, **wire** height, number of **shots**, yarn count and plies, total pile yarn weight, and pile **yarn density**.

**construction joint**  A joint placed in **concrete** to permit practical placement of the work in section with predefined boundaries.

**contact adhesive**  *See* **adhesive, drywall**.

**continuous casting**  A process in which molten metal is used directly to produce semifinished products such as slabs or billets, bypassing ingot teaming, stripping, soaking, and **rolling**.

**control joint**  (1) A joint placed in **concrete** to form a plane of weakness to prevent random cracks from forming due to drying shrinkage and temperature changes. (2) A prefabricated metal accessory intended to relieve shrinkage, temperature, or structural stresses in **plaster**, thus minimizing cracking.

**coping**  **Masonry units** forming a finished cap on top of an exposed pier, wall, pilaster, chimney, etc., to protect the masonry below from penetration of water from above.

**copolymer**  A **polymer** formed by the combination of two or more different **monomers**.

**copper staining**  A stain usually caused by the corrosion products of copper screens, gutters, or downspouts washing down on a finished surface. Can be prevented by painting the copper or applying a **transparent finish** to it.

**corbel**  A shelf or ledge formed by projecting successive courses of masonry out from the face of a wall.

**core**  The innermost portion of **plywood**, consisting of either **hardwood** or **softwood** sawed **lumber**, **veneer**, or composition board.

**corner cracking**  Cracks occurring in the apex of inside corners of **gypsum board** surfaces, such as between adjacent walls or at walls and ceilings.

**corner floating**  **Gypsum board** installation procedure that eliminates some mechanical fasteners at interior corners and permits sufficient movement of boards to eliminate **corner cracking**.

**correction lines**  East-west reference lines used in the **rectangular survey system**, located at 24-mile intervals to the north and south of a **base line**.

**corrosion**  Physical deterioration, decomposition, or loss of the cross section of a metal due to **weathering**, galvanic action, or **direct chemical attack**. *Galvanic action* is corrosion produced by electrolytic action between two dissimilar metals in the presence of an **electrolyte**. *Direct chemical attack* is corrosion caused by a chemical dissolving of the metal. *Weathering* is galvanic or chemical corrosion produced by atmospheric conditions.

**coulomb** A unit of electrical charge equal to the number of electrons conducted past a point in 1 second.

**count** A number identifying yarn size or weight per unit of length (or length per unit of weight), depending on the **carpet** spinning system used (such as **denier**, woolen, worsted, cotton, or jute system).

**course** (1) Compass direction from one reference point to the next for each leg of a **metes and bounds** survey. An **angular course** is a compass direction in **degrees**, **minutes**, and **seconds**, stated as a deviation eastward or westward from due north or south; used in metes and bounds descriptions and **surveys**. (2) One of the continuous horizontal layers of **masonry units**, bonded with **mortar**. One course is equal to the thickness of the masonry unit plus the thickness of one **mortar joint**.

**coverage** Area over which a given amount of **coating**, **paint**, or **transparent finish** will spread and hide the previous surface; usually expressed in sq ft/gal (m²/L).

**covering power** *See* **hiding power**.

**cracking** Type of **coating**, **paint**, or **transparent finish** failure characterized by breaks in irregular lines wide enough to expose the underlying surface.

**crazing** (1) Numerous hairline cracks in the surface of newly hardened **concrete**. (2) Similar cracks in a **paint** or **coating**.

**creep** *Same as* **cold flow**.

**crimping** Method of texturing staple and continuous filament yarn to produce irregular alignment of fibers in **carpet** and increase their bulk and covering power; also facilitates interlocking of fibers, which is necessary for spinning staple fibers into yarn.

**crook** *See* **warp**.

**crossband** A layer of **veneer** in a **plywood** panel whose grain direction is at right angles to that of the face plies.

**cross furring** Furring members installed perpendicular to framing members.

**cross grain** *See under* **grain**.

**crosslinking** A chemical reaction in which adjacent **polymer** molecules unite to form a strong three-dimensional network; usually occurs during the curing of **thermoset**ting **plastics**.

**cross slope** The slope of a pedestrian way that is perpendicular to the direction of travel. *See also* **running slope**.

**crown, joint** The maximum height to which joint compound is applied over a **gypsum board** joint.

**cryolite sodium** Aluminum fluoride used with **alumina** in the final electrolytic reduction of aluminum. Found naturally in Greenland; generally produced synthetically from alum, soda, and hydrofluoric acid.

**cubing** The assembling of **concrete masonry units** into cubes after curing for storage and delivery. A cube normally contains six layers of 15 to 18 blocks (8 × 8 × 16 in. [200 × 200 × 400 mm]) or an equivalent volume of other size units.

**cue** A device that alerts a user to an upcoming condition; includes audible, visual, and textural signals. See also **detectable warning**. An *audible cue* is a sound or a verbal alert. A *tactile cue* is one that can be detected by touch. A *visual cue* is one that can be seen.

**cup** *See* **warp**.

**curb ramp** A short ramp cutting through a curb or built into it.

**cure** To change the properties of a polymeric system into a final, more stable condition by the use of heat, radiation, or reaction with chemical additives. Sometimes referred to as *set*.

**curing** (1) The hardening of a **concrete masonry unit (CMU)**. *Atmospheric pressure steam curing* is a method of curing **CMU**s, using steam at atmospheric pressure usually at temperatures of 120°F (48.89°C) to 180°F (82.2°C). Also called *low-pressure steam* curing. *High-pressure steam curing* is a method of curing **CMU**s, using saturated steam (365°F [185°C]) under pressure, usually 125 to 150 psi (87.88 to 105.46 Mg/m²). Also referred to as *autoclave curing*. *Moist curing* is a method of curing **CMU**s using moisture at atmospheric pressure and temperature of approximately 70°F (21.1°C). (2) The process of keeping **concrete** moist for an extended period after placement to ensure proper **hydration** and subsequent strength and quality. (3) Final conversion or drying of a **coating**, **paint**, or **transparent finish** material.

**curtain wall** An exterior non-load-bearing wall. Such walls may be anchored to columns, spandrel beams, structural walls, or floors. *See also* **panel wall**.

**cushioning** Soft, resilient layer provided under **carpet** to increase underfoot comfort, to absorb pile-crushing forces, and to reduce impact sound transmission. Also called *underlay* or *lining*; *see also* **padding**.

**cut and loop** *See* **multilevel loop**.

**cut pile** A **carpet** face construction in which the pile is cut level so that it stands erect in a low, dense, plush, even surface.

**cutting in** Painting of an edge, such as wall color at the ceiling line or at the edge of woodwork.

**cylinder test** A laboratory test for compressive stress of a field sample of **concrete** (6 in. [152.4 mm] in diameter by 12 in. [305 mm] in length).

**daily degree day** The numerical difference between 65°F (18.3°C) and the average of all recorded temperatures on a given day that are lower than 65°F (18.3°C).

**damp course** A course or layer of impervious material that prevents capillary entrance of moisture from the ground or a lower course. Often called *damp check*.

**darby** A tool used to level freshly placed **concrete**.

**darbying** Smoothing the surface of freshly placed **concrete** with a **darby** to level any raised spots and fill depressions.

**dead load** The weight of all permanent and stationary construction or equipment included in a building. *See also* **live load**.

**decay** The decomposition of wood substances by certain fungi.

**decibel** A logarithmic unit expressing the ratio between a **sound** being measured and a reference point.

**decorative CMU** *See under* **concrete masonry unit**.

**degradation** A permanent change in the physical or chemical properties of a **plastic** evidenced by impairment of these properties.

**degree** A unit of **angular measure** equal to the angle contained within two radii of a circle that describe an arc equal to ¹/₃₆₀th of the circumference of the circle; also used to define an arc equal to $1/360$th of the circumference of a circle.

**degree days heating (DDH)** The sum of the **daily degree days** when the temperature dropped below 65°F (18.3°C).

**delustered nylon** Nylon on which the normally high sheen has been reduced by surface treatment.

**denier** System of yarn count used for synthetic **carpet** fibers: number of grams per 9000 meters of yarn length; one denier equals 4,464,528 yards per pound or 279,033 yards per ounce.

**density** *See* **pile yarn density**.

**detectable warning** A standardized surface texture applied to or built into walking surfaces or other elements to warn visually impaired people of hazards in the path of travel. *See also* **cue**.

**deterioration** *See* **degradation**.

**dew point** The temperature above freezing at which air becomes **saturated** and **condensation** occurs. *See also* **frost point**.

**dimensional stability** The ability of a material to retain its dimensions in service.

**dimension lumber** *See under* **lumber**.

**direct chemical attack** Corrosion caused by a chemical dissolving a metal.

**disability** A limitation or loss of use of a physical, mental, or sensory body part or function.

**discontinuous construction** A construction method used to separate a continuous path through which **sound** may be transmitted. Examples include the use of staggered studs, double walls, and the resilient mounting of surfaces.

**dispersion** The distribution of a finely divided solid in a liquid or a solid.

**disposal field** *See* **absorption field**.

**distribution line** Open joint or perforated pipe intended to permit soil absorption of **effluent**.

**divider strips** All-metal or plastic-top metal strips provided in a **terrazzo finish** to control cracking due to drying shrinkage, temperature variations, and minor structural movements; are also used for decorative purposes and convenience in placing a **terrazzo** topping.

**drain** A pipe that carries wastewater or waterborne wastes in a building **drainage system**.

**drainage system** That piping that conveys **sewage**, rainwater, or other liquid wastes up to a point of disposal, such as the mains of a public **sewer** system or a private **septic disposal system**. *See also* **vent system** and **drain, waste, and vent (DWV) system**.

**drain, waste, and vent (DWV) system** The collection of pipes that facilitates the removal of liquid and solid wastes and dissipates **sewer** gases.

**drawing** The process of pulling material through a die to reduce the size, to change the cross section or shape, or to harden the material.

**dressed lumber** **Lumber** that has been surfaced with a planing machine.

**dressed size** *See* **size**.

**drier** An ingredient included to speed the drying of **coatings**, **paints**, and **transparent finishes**.

**drip** A projection shaped to cause water to flow away from a lower surface, thus preventing it from running down the face of the lower surface.

**drying** The various stages of **curing** in a **coating**, **paint**, or **transparent finish** film. *Dust-free* is the stage of drying when particles of dust that settle on the surface do not stick to it. *Tack-free* is the stage of drying when the surface no longer feels sticky when lightly touched. *Dry enough to handle* is the stage of drying when the film has hardened sufficiently so that the object or surface may be used without marring. *Dry enough to recoat* is that stage of drying when the next coat can be applied. *Dry enough to sand* is the stage of drying when the film can be sanded without the sandpaper sticking or clogging.

**drying oil** A **coating**, **paint**, or **transparent finish vehicle** ingredient, such as linseed oil, which, when exposed to the air in a thin layer, oxidizes and hardens to a relatively tough elastic film.

**dry lumber** Under Product Standard PS 20–70, **lumber** with a moisture content of 19% or less.

**dry rot** A term that is loosely applied to many types of decay, which in the advanced state permit wood to be easily crushed to a dry powder. This term is a misnomer since all fungi require moisture.

**drywall nail** *See* **nail, drywall**.

**drywall screw** *See* **screws, drywall**.

**dry well** A covered and lined underground pit, similar to a seepage pit but intended to receive water free of organic matter, such as from roof drains, floor drains, or laundry tubs. Used as an auxiliary to a **septic disposal system** to avoid overloading the **septic tank** absorption system. (Also called *leaching well* or *leaching pit*.)

**ductile** Capable of being drawn out or hammered; able to undergo cold **plastic deformation** without breaking.

**durability** The ability of a **coating**, **paint**, or **transparent finish** to retain its desirable properties for a long time under expected service conditions.

**dusting** The appearance of a powdery material at the surface of a hardened **concrete**.

**dye** or **dyestuff** Colored material used to change the color of a **coating** or **paint** with little or no hiding of the underlying surface.

**dyne** A unit of force, which when acting on a mass of 1 gram accelerates it 1 centimeter per second per second.

**echo** A reflected **sound** loud enough and received late enough to be heard as distinct from the source.

**economy brick** *See under* **brick masonry unit**.

**edge grain** *See under* **grain**.

**edges, gypsum board** *See* **gypsum board**.

**edging** The finishing operation of rounding off the edge of a **concrete** slab to prevent chipping or damage.

**efflorescence** Deposit of soluble salts, usually white in color, appearing on the exposed surface of masonry, **concrete**, or **plaster**.

**effluent** Partially treated liquid sewage flowing from any part of a **septic disposal system**, **septic tank**, or absorption system.

**eggshell finish** Surface sheen midway between **flat** and **semigloss**.

**elastic deformation** Deformation caused by a **stress** small enough that when the stress is removed, the material returns to its original shape. *See also* **plastic deformation**.

**elasticity** The ability to recover the original size and shape after deformation.

**elastic limit** The amount of **stress** that, if exceeded, will cause a given material to deform or set permanently.

**elastomer** (1) A material that at room temperature can be stretched repeatedly to at least twice its original length and that, upon release of the **stress**, will return instantly and with force to its approximate original length. (2) A rubber-like substance.

**electric induction furnace** *See under* **furnace, steel**.

**electrogalvanizing** Electroplating with zinc to provide greater corrosion resistance.

**electrolysis** Also called **galvanic corrosion**. An electrochemical decomposition that results when dissimilar metals are each contacted by the same **electrolyte**, such as water. The process is similar to that which takes place in an automobile battery. One metal acts as a cathode, the other as an anode. When the electrolyte causes an electrical current to flow from one metal to the other, the anodic metal dissolves and hydrogen ions accumulate on the cathodic metal. Electrolysis can also take place in a single metal when one portion of it is cathodic and another portion is anodic if an **electrolyte** makes a bridge between the two portions.

**electrolyte** A nonmetallic substance in which electricity is conducted by the movement of ions.

**electromotive force** Something that moves or tends to move electricity.

**electroplating** A process that employs an electric current to coat a base metal (cathode) with another metal (anode) in an electrolytic solution.

**emulsion** Mixture of liquids (or a liquid and a solid) not soluble in each other, one liquid (or solid) being dispersed as minute particles in the other, base liquid, with the help of an emulsifying agent.

**enamel** A **coating** or **paint** capable of forming a very smooth, hard film, sometimes using varnish as the **vehicle**; may be **flat**, **gloss**, or **semigloss**.

**enamel holdout** Property of producing a tight film that prevents the penetration of subsequent **enamel** coats to underlying surfaces; prevents unequal absorption and uneven gloss.

**ends, gypsum board** *See* **gypsum board**.

**end wall** The wall along the short dimension of a room.

**engineered brick** *See under* **brick masonry unit**.

**engineered brick masonry** Masonry in which the design is based on a rational, accepted structural engineering analysis.

**equilibrium moisture content** The **moisture content** at which wood neither gains nor loses moisture when surrounded by air at a given **relative humidity** and temperature.

**erosion** Wearing away of a **coating**, **paint**, or **transparent finish** film caused by exposure.

**etch** A surface preparation for a coating, paint, or transparent finish by chemical means to improve adhesion.

**evaporation** The change of water from a liquid to a gas.

**expandable plastic** A **plastic** suitable for expansion into cellular form by thermal, chemical, or mechanical means.

**expansion strips** Double **divider strips** in a **terrazzo finish** separated by resilient material and provided generally for the same purpose as **divider strips**, but where a greater degree of structural movement is expected.

**extender** (1) In **plastics**, a low-cost material used to dilute or extend high-cost resins without appreciably lessening the properties of the original resin. (2) In **coatings**, **paints**, and **transparent finishes**, an inexpensive but compatible substance that can be added to a more valuable substance to increase the volume of material without substantially diminishing its desirable properties; in **coatings** and **paints**, extender pigments improve storage and application properties.

**exterior wall** Any outside wall of a building other than a **party wall**.

**external corner** A projecting angle formed by abutting walls or a vertical surface and soffit (not to be confused with *exterior*, meaning exposed to the weather).

**extrude** To form lengths of shaped sections by forcing a **plastic** material through a shaped hole in a die.

**extrusion** (1) In aluminum, a product formed by extruding. (a) An *extrusion billet* is a solid, wrought, semifinished product intended for further extrusion into rods, bars, or shapes. (b) An *extru-sion ingot* is a solid or hollow cylindrical casting used for extrusion into bars, rods, shapes, or tubes. (2) In plastic, forcing plastic material through a shaped orifice to make rod, tubing, or sheeting.

**faced CMU** *See* **concrete masonry unit**.

**faced wall** A wall in which the facing and backing are of different materials and are bonded together to exert common action under load.

**face shell** The side wall of a hollow **masonry unit** or clay tile.

**face size** *See under* **size**.

**facing** A part of a wall that is used as a finished surface.

**facing brick** *See under* **brick masonry unit**.

**factory and shop lumber** *See* **lumber**.

**fading** Loss of color due to exposure to light, heat, or weathering.

**fastener treatment** Method of concealing **gypsum board** fasteners by successive applications of compound until a smooth surface is achieved.

**featheredging (feathering)** Tapering **gypsum board joint compound** to a very thin edge to ensure inconspicuous blending with adjacent gypsum board surfaces.

**feather sanding** Tapering the edge of a dried **coating**, **paint**, or **transparent finish** film with sandpaper.

**ferritic steel** *See* **grain structure**.

**ferroalloys** Iron-based alloys used in steelmaking as a source of desired alloying elements.

**ferrous alloys** Composite metals whose chief ingredient is iron (ferrum), metallurgically combined with one or more alloying elements.

**fiber saturation point** The stage in drying (or wetting) of wood at which the cell walls are saturated with water but the cell cavities are free of water, being approximately 30% **moisture content** in most species.

**fiber, wood** A comparatively long ($1/25$ in. [1.016 mm] or less to $1/3$ in. [8.47 mm]), narrow, tapering unit closed at both ends.

**field** (1) In **ceramic tile**, the general area of the tile excluding trim. (2) In **gypsum board**, the surface of the board exclusive of the perimeter. (3) In masonry walls, the expanse of wall between openings, corners, etc., principally composed of **stretchers**. (4) In resilient flooring, the area of a floor within the **borders**.

**figure** The pattern produced in a wood surface by **annual growth rings**, **rays**, **knots**, and deviations from the regular **grain**.

**fill** The sand, gravel, or compacted earth used to bring a **subgrade** up to a desired level.

**filler** (1) In **plastics**, a relatively inert material added to modify the strength, permanence, or working properties or to lower the cost of a **resin**. (2) In **coatings**, **paints**, and **transparent finishes**, a pigmented composition for filling the pores or irregularities in a surface in preparation for finishing.

**filling** *See* **weft**.

**film** (1) In **plastics**, plastic sheeting having a nominal thickness not greater than 0.010 in. (0.254 mm). (2) In **coatings**, **paints**, and **transparent finishes**, a thin application generally not thicker than 0.010 in. (0.254 mm).

**fine aggregate** *See* **aggregate**.

**fineness modulus** A measure of the average size of an **aggregate**, calculated by passing the **aggregate** through a series of screens of decreasing size.

**finish coat** The final decorative plaster layer in either **two-coat** or **three-coat work**.

**finished mill products** Steel shapes that can be used directly in construction. *Bars* are hot-rolled or cold-drawn round, square, hexagonal, or multifaceted long shapes, generally larger than wire in cross section. Also hot- or cold-rolled rectangular flat shapes (flats) generally narrower than sheets and strip. *Backplate* is a cold-rolled, flat, carbon-steel product that is thinner than sheet and wider than strip, generally coated with zinc, tin, or terne metal. *Foil* is a cold-rolled flat product less than 0.005 in. (0.127 mm) thick. *Plate* is a hot-rolled flat product generally thicker than sheet and wider than strip. *Sheet* is a hot- or cold-rolled flat product generally thinner than plate and wider than strip. *Strip* is a hot- or cold-rolled flat product generally narrower than sheet and thinner than plate. *Structurals* are hot-rolled steel shapes of special design (such as H beams, I beams, channels, angles, and tees) used in construction. *Terneplate* is backplate or sheet

metal that has been coated with terne metal. *Tinplate* is backplate that has been coated with tin. *Tubular products* are hollow products of round, oval, square, rectangular, or multifaceted cross sections. In construction, round products are generally called *pipe*; square or rectangular products with thinner wall sections are called *tube* or *tubing*. *Wire* is a cold-finished product of round, square, or multifaceted cross section, generally smaller than bars; round wire is cold drawn, 0.005 in. (0.127 mm) to less than 1 in. (25 mm) in diameter; flat wire is cold rolled, generally narrower than bar.

**fire brick** *See under* **brick masonry unit**.

**fire division wall** A wall that subdivides a building to resist the spread of fire. It is not necessarily continuous through all stories to and above the roof. *See also* **fire wall**.

**fireproofing** A material or combination of materials built to protect structural members to increase their fire resistance.

**fire-resistive material** *See* **noncombustible material**.

**fire-retardant coating** or **paint** A **coating** or **paint** that will significantly (1) reduce the rate of flame spread, (2) resist ignition at high temperatures, and (3) insulate the underlying material to prolong the time required for the material to reach its ignition, melting, or structural weakening temperature.

**fire wall** A wall that subdivides a building to resist the spread of fire and that extends continuously from the foundation through the roof. *See also* **fire division wall**.

**fixture-unit** A mathematical factor used by engineers to estimate the probable demand on a **drainage** or **water supply system** (volume, duration of flow, and intervals between operations) by various **plumbing fixtures**.

**flaking** Form of **coating**, **paint**, or **transparent finish** failure characterized by the detachment of small pieces of the film from the surface or previous coat; usually preceded by cracking or blistering.

**flanking path** A wall or floor and ceiling assembly that permits **sound** to be transmitted along its surface. Also, an opening that permits the direct transmission of sound through the air.

**flashing** (1) In masonry construction, a thin, relatively impervious sheet material placed in **mortar joints** and across air spaces in masonry walls to collect water that may penetrate the wall and to direct it to the exterior. (2) In masonry manufacture, the step during the burning process of clay **masonry units** that produces varying shades and colors in the units. (3) In resilient flooring, the bending up of resilient sheet material against a wall or a projection, either temporarily for the purpose of fitting or permanently to form a one-piece resilient base.

**flash point** Temperature at which a **coating**, **paint**, **transparent finish**, or **solvent** will ignite; the lower the flash point, the greater the hazard.

**flash set** Undesirable rapid setting of **cement** in **concrete** or **mortar**.

**flat** Dull, nonreflective; opposite of **gloss**.

**flat applicator** Rectangular flat pad with an attached handle that is used to paint shingles, shakes, and other special surfaces.

**flat finish** Finish having no gloss or luster.

**flat grain** *See under* **grain**.

**flats** Term applied to **flat coatings** and **paints**.

**flatting agent** Ingredient added to **coatings** and **paint** to reduce the gloss of the dried film.

**flitch** (1) A portion of a log sawed on two or more sides and intended for manufacture into **lumber** or sliced or sawed **veneer**. (2) A complete bundle of veneers laid together in sequence as they were sliced or sawed.

**floating** (1) A **concrete** slab finishing operation that embeds **aggregate**, removes slight imperfections, humps, and voids to produce a level surface, and consolidates mortar at the surface. (2) A condition in which a layer of water occurs between a mortar bed and a masonry unit, usually due to **bleeding**, causing the unit and the mortar to fail to bond with each other. In this condition, the unit is said to *float*. (3) Separation of **pigment** colors on the surface of applied **paint**.

**flocked carpet** Single-level velvety pile carpet composed of short fibers embedded on an adhesive-coated backing.

**floor brick** *See under* **brick masonry unit**.

**flow** The ability of a **coating**, **paint**, or **transparent finish** to level out and spread into a smooth film; materials that have good flow usually level out uniformly and exhibit few brush or roller marks.

**fluffing** *See* **shedding**.

**flux** A mineral that, due to its affinity to the impurities in iron ores, is used in ironworking and steelmaking to separate impurities in the form of molten slag. *Basic flux* is a mineral, such as limestone or dolomite, used in basic furnaces to make basic (low-phosphorus) steel. *Neutral flux* is a mineral (fluorspar) used to make slag more fluid.

**foamed plastic** *See* **cellular plastic**.

**foil** *See under* **finished mill products**.

**footing** The base of a foundation, column, or wall used to distribute the load over the subgrade.

**forging** The working (shaping) of metal parts by forcing between shaped dies. *Press forging* is shaping by applying pressure in a press. *Hammer forging* is shaping by application of repeated blows, as in a forging hammer. *See also* **hot working**.

**form** A temporary structure erected to contain **concrete** during **placing** and initial hardening.

**foundation wall** A load-bearing wall below the floor nearest to exterior grade serving as a support for a wall, pier, column, floor, or other structural part of a building.

**fractional section** *See under* **section**.

**frames** Racks at the back of a **jacquard loom**, each holding pile yarn of a different color. In **Wilton carpets**, two to six frames may be used. The number is a measure of quality as well as an indication of the number of colors in the pattern, unless some of the yarns are buried in the **backing**.

**freezing cycle day** A day in which the air temperature passes either above or below 32°F (0°C). The average number of freezing cycle days in a year equals the difference between the mean number of days when the minimum temperature was 32°F (0°C) or below and when the maximum temperature was 32°F (0°C) or below.

**frequency** The number of complete cycles of a vibration occurring in each sec-

ond, measured in cycles per second (cps) and expressed in **hertz (Hz)**.

**frieze carpet** *See under* **pile**.

**frost point** That temperature below freezing at which **condensation** occurs. *See also* **dew point**.

**fungicide** Agent that helps prevent mold or mildew growth on **paint**.

**furnace, blast** Tall, cylindrical masonry structure lined with refractory materials, used to smelt iron ores in combination with fluxes, coke, and air into **pig iron**.

**furnace, steel** Masonry or steel structure lined with refractory materials, used to melt **pig iron**, scrap metal, and sometimes agglomerated ores, ferroalloys, and fluxes into steel. A *basic oxygen furnace* is a suspended, tilting vessel that uses high-purity oxygen to oxidize impurities in hot **pig iron** and other iron-bearing materials to produce low-phosphorus (basic) steel. An *electric arc suspended furnace* is a kettle that melts scrap metal, ore, and sometimes ferroalloys with the heat of an electric arc to produce steels of controlled chemical composition. An electric induction furnace is a steel-encased, insulated magnesia pot in which metal, scrap, and ferroalloys are melted with the heat of an electric current induced by windings of electric tubing; used chiefly to produce small quantities of high-grade steels such as alloy, stainless, and heat-resisting steels. An *open hearth furnace* is a masonry structure with a hearth exposed to the sweep of flames in which hot pig iron, scrap metal, and fluxes are melted and oxidized by a mixture of fuel and air to produce basic or acid steel.

**furring** A method of finishing the interior face of a **concrete** or masonry wall to provide space for insulation, to prevent moisture penetration, or to provide a level, plumb, and straight surface for finishing. Furring consists of metal channels or studs or of wood strips or studs.

**fusion-bonded carpet** **Carpet** made by fusing carpet yarn and a backing, then cutting the substrates in two, making two pieces of carpet.

**fuzzing** Temporary condition on new **carpet** consisting of an irregular appearance caused by slack yarn twist, "snagging" of fibers, or breaking of yarn. Can be remedied by spot **shearing**.

**galvanic corrosion** *See* **electrolysis**.

**galvanizing** Zinc coating by electroplating or hot dipping, which produces a characteristic bright, spangled finish and protects the base metal from atmospheric **corrosion**.

**gang grooved** **Plywood** panels produced by passing them under a machine with grooving knives set at certain intervals.

**gauge** The distance between **tufts** across the width of **knitted** and **tufted carpets**, expressed in fractions of an inch.

**glaze** Used to describe several types of finishing materials. *Glazing putty* is a compound of creamy consistency applied to fill surface imperfections. A *glazing stain* is very thin, semitransparent, and usually pigmented with Vandyke brown or burnt sienna, applied over a previously stained, filled, or painted surface to soften or blend the original color without obscuring it. A *glaze coat* is a clear finish applied over previously coated surfaces to create a gloss finish.

**glazing compound** Dough-like material, consisting of **vehicle** and **pigment**, that retains its plasticity over a wide range of temperatures and for an extended period of time. Either elastomeric or acrylic joint sealants are generally used today where glazing compounds are called for. *See also* **sealant, joint**.

**gloss** Shiny, reflective surface quality; term sometimes used broadly to include **coatings**, **paints**, and **transparent finishes** with these surface properties.

**glueline** The line of glue visible on the edge of a **plywood** panel. Also applies to the layer of glue itself.

**grade** The designation of the quality of wood, steel, and other materials and products made from them, such as **plywood**.

**graded aggregate** An **aggregate** containing particles of uniformly graduated size from the finest fine **aggregate** size to the maximum size of coarse aggregate.

**graded sand** A sand containing particles of uniformly graduated size from very fine up to $1/4$ in. (6.4 mm).

**grain** The direction, size, arrangement, appearance, or quality of the **fibers** in wood. **Cross-grained wood** is sawed with the fibers not parallel with the longitudinal axis of the piece. This grain may be diagonal, in a spiral pattern, or a combination of both. *Diagonal-grained wood* is sawn at an angle with the **bark** of the tree such that the **annual growth rings** are at an angle with the axis of the piece. *Edge-grained wood* is sawn parallel with the **pith** of a log and at nearly right angles to the annual growth rings, making an angle of 45 to 90 degrees with the wide surface of the piece. Also called *quarter-sawn wood* and *vertical-grained wood*. *Flat-grained wood* is sawn parallel with the pith of the log and nearly tangent to the annual growth rings, making an angle of less than 45 degrees with the surface of the piece. Also called *plain-sawn wood*. *Open-grained wood* is a common designation for wood with large pores, such as oak, ash, chestnut, and walnut.

**graining**   Simulating the grain of wood by means of specially prepared colors or stains and the use of graining tools or special brushing techniques.

**grain raising**   Swelling and standing up of wood grain caused by absorbed water or solvents.

**grain structure**   The microscopic internal crystalline structure (size and distribution of particles) of a metal that affects its properties, known as *austenitic*, *ferritic*, and *martensitic*. *Austenitic steels* are tough, strong, and nonmagnetic. Austenitic stainless steels have a chromium content of up to 25% and a nickel content of up to 22% and can be hardened by cold working. *Ferritic steels* are soft, ductile, and strongly magnetic. Ferritic stainless steels usually have a chromium content of 12 to 27% and are not hardenable by heat treatment. *Martensitic steels* can be made very hard and tough by heat treatment and rapid cooling. Martensitic stainless steels have a chromium content of 4 to 12%.

**great circle**   A line described on a sphere by a plane bisecting the sphere into equal parts. The **equator** is a great circle, as are pairs of opposing **meridians**.

**green concrete**   Freshly placed **concrete**.

**green lumber**   Under Product Standard PS 20–70, lumber with a moisture content of more than 19%. Unseasoned lumber that has not been exposed to air or kiln drying.

**Greenwich meridian**   *See under* **meridian**.

**grin**   Condition in which the **backing** shows through sparsely spaced **pile tufts**. A **carpet** may be grinned (bent back) deliberately to reveal its **construction**.

**gross cross-sectional area**   In masonry, the total area of a section perpendicular to the direction of the load, including areas within cores, cellular spaces, and other openings in the material.

**ground coat**   Base coat in an antiquing system; applied before graining colors, glazing, or other finish coat.

**grounds**   Nailing strips placed in **concrete** and masonry walls as a means of attaching trim, furring, cabinetry, or equipment.

**grout, masonry**   (1) **Mortar** of a consistency that will flow or pour easily without segregation of the ingredients. (2) A liquid mixture of **cement**, water, and sand of pouring consistency.

**grout, tile**   A formulation used to fill the joints between tiles; may be cementitious, resinous, or a combination of both.

**guide meridian**   *See under* **meridian**.

**gusset plate**   Wood or metal plate used as a means of joining coplanar structural members in trusses. Gusset plates lap the butt joints between members.

**gypsum board**   A panel consisting of a noncombustible core of calcined gypsum, surfaced on both sides with a covering material specifically designed for various uses with respect to performance, application, location, and appearance. *Wallboard* is a class of gypsum board used primarily as an interior finished surface. *Lath* is a class of gypsum board used as a base for gypsum plaster. *Backing boards* are gypsum boards that serve as a base to which gypsum wallboard or tile is applied. *Sheathing* is a class of gypsum board used as a base for exterior finishes. *Edges* are gypsum board extremities that are paperbound and run the long dimension of the board as manufactured. *Ends* are gypsum board extremities that are mill- or job-cut, exposing the gypsum core, and run the short dimension of the board as manufactured.

**hardboard**   A dense panelboard manufactured of wood fibers with the natural lignin in the wood reactivated to serve as a binder for the wood fibers.

**hard-burned**   Clay products that have been fired at high temperatures. They have relatively low absorptions and high compressive strengths.

**hardness**   Cohesion of particles on the surface of a **coating**, **paint**, or **transparent finish** as determined by the ability to resist scratching or indentation.

**hardwood**   The botanical group of trees that are broad-leaved and deciduous. The term does not refer to the actual hardness of the wood.

**header**   A **masonry unit** that overlaps two or more adjacent **wythes** of masonry to provide structural bond. Also called *bonder*.

**header course**   A continuous bonding **course** of header **brick**.

**head joint**   The vertical **mortar joint** between ends of **masonry units**.

**heartwood**   The wood that extends from the **pith** to the **sapwood**, the cells of which no longer participate in the growth process of the tree. Heartwood may be impregnated with gums, resins, and other materials, which usually make it darker and more decay resistant than sapwood.

**heat capacity**   The amount of heat required to raise the temperature of 1 cu ft of a material 1°F (0.5556°C).

**heat-set nylon**   Nylon fiber that has been heat treated to retain a desired shape.

**heat sink**   A material or system that collects and stores heat.

**heat-treatable alloys**   Aluminum alloys capable of gaining strength by being heat treated. The alloying elements show increasingly solid solubility in aluminum with increasing temperature, resulting in pronounced strengthening.

**heat treatment**   (1) Of aluminum (*see* **solution heat treatment**): *Annealing* is a heat treatment process in which an **alloy** is heated to a temperature between 600°F (315.56°C) and 800°F (426.67°C) and then slowly cooled to relieve internal **stresses** and return the **alloy** to its softest and most **ductile** condition. (2) Of steel: Controlled heating and cooling of steels in the solid state for the purpose of obtaining certain desirable mechanical or physical properties. *Annealing* is heating metal to high temperatures (1350°F [732.2°C] to 1600°F [871.1°C] for steel), followed by controlled cooling, to make the metal softer or change its ductility and toughness. *Quenching* is rapid cooling by immersion in oil, water, or another cooling medium to increase hardness. *Tempering* is reheating (in steel to less than 1350°F [732.2°C]) after hardening (as by quenching) and slow cooling to restore ductility.

**hertz (Hz)**   The unit of frequency of a periodic process equal to one cycle per second.

**hiding power**   Ability of a **coating** or **paint** to obscure the surface to which it is applied, generally expressed as the number of square feet that can be covered by 1 gal of material or the number of gallons required to cover 1000 sq ft. Also called *covering power*.

**high-density plywood**   *See under* **plywood**.

**high-early-strength cement**   Cement used to produce a **concrete** that develops strength more rapidly than normal.

**high-low loop**   *See* **multilevel loop**.

**high polymer**   A **polymer** of high molecular weight.

**high-pressure laminate**   A **laminate** molded and cured at pressures not lower than 1000 psi (703.07 mg/m²) and usually between 1200 psi (843.68 mg/m²) and 2000 psi (1406.14 mg/m²).

**high-pressure steam curing**   *See under* **curing**.

**hollow brick**   *See under* **brick masonry unit**.

**hollow masonry unit**   *See* **masonry unit**.

**hollow wall**   A wall built of **masonry units** arranged to provide an air space within the wall between the facing and backing **wythes**.

**horizontal application**   Applying **gypsum board** with the **edges** perpendicular to supporting members such as studs, joists, channels, or furring strips. See also **vertical application**.

**hot-dip process**   Coating a metal with another metal by immersing it in a bath of the molten coating metal. The coating metals most commonly used on steel are zinc, terne metal, and aluminum.

**hot rolling**   *See* **rolling** and **hot working**.

**hot working**   Forming metal at elevated temperatures at which metals can be more easily worked. (1) For aluminum alloys: (a) Hot working temperatures are usually in the 300°F (148.9°C) to 400°F (204.4°C) range. (b) *Hot rolling* is the shaping of aluminum plate by rolling heated slabs of metal. Hot rolling is usually used for work down to about 0.125 in. (3.175 mm) thick. (2) For steel: Shaping hot plastic metal by hot rolling, extruding, or forging. (a) Hot working temperatures are above 1500°F (815.6°C). Hot working is generally accompanied by increases in strength, hardness, and toughness. (b) *Hot forming* is the forming of a hot plastic metal into desired shapes, with little change in the mechanical properties of the metal. (c) *Extruding* is shaping lengths of hot metal by forcing them through a die of the desired profile. (d) *Forging* is shaping hot metal between dies with compression force or impact. (e) *Hot rolling* is gradual shaping by squeezing hot metal between rolls.

**house sewer**   *See* **building sewer**.

**hydration**   The chemical reaction of water and **cement** that produces a hardened **concrete**.

**Hz**   *See* **hertz**.

**impact insulation class (IIC)**   A single-number rating developed by the Federal Housing Administration to estimate the impact **sound isolation** performance of floor and ceiling assemblies.

**impact noise rating (INR)**   A single-number rating used to compare and evaluate the ability of floor and ceiling assemblies to isolate impact sound transmission (*see* **sound isolation**).

**impact sound pressure level (ISPL)**   The **sound** level in **decibels** measured in the receiving room resulting from the transmission of sound produced by a standard tapping machine through an adjacent floor and ceiling assembly.

**impervious soil**   A tight, cohesive soil, such as clay, that does not allow the ready passage of water.

**incident sound**   A **sound** striking a surface, as contrasted with a sound reflected from the surface. The angle of incidence equals the angle of reflection.

**individual sewage disposal system**   A combination of a **sewage** treatment plant (package plant or **septic tank**) and method of **effluent** disposal (soil absorption system or stream discharge) serving a single dwelling.

**inert**   Having inactive chemical properties.

**ingot**   (1) A mass of aluminum cast into a convenient shape for storage or transportation, to be later remelted for casting or finishing by rolling, forging, or another process. (2) A cast **pig iron** or steel shape made by pouring hot metal into a mold.

**inhibitor**   A substance that prevents or retards a chemical reaction; inhibitors are often added to plastic resins to prolong their storage life.

**injection molding**   Forming **plastic** by fusing it in a chamber with heat and pressure and then forcing part of the mass into a cooler chamber, where it solidifies.

**insolation (incident solar radiation)**   The solar radiation that strikes a surface.

**intercoat adhesion**   Adhesion between two coats of **paint**.

**internal corner**   An enclosed angle of less than 180 degrees formed by abutting walls or the juncture of walls and the ceiling (not to be confused with *interior*, meaning protected from the weather).

**isolation**   *See* **sound isolation**.

**isolation joint**   A joint placed to separate a **concrete** slab into individual panels or from adjacent surfaces.

**isolation membrane**   (1) In tile floors, see **cleavage (isolation) membrane**. (2) In a **terrazzo finish**, a membrane such as asphalt-saturated roofing felt, building paper, or polyethylene film installed between the subfloor and the terrazzo underbed to prevent bonding and permit independent movement of each.

**jacquard**   Mechanism for a **Wilton loom** that uses punched cards to produce the desired color design.

**Jaspe carpet**   *See under* **pile**.

**joint**   *See* **mortar joint**.

**joint beading**   *See* **ridging**.

**joint compound**   A material used for finishing joints in **gypsum board**.

**joint crown**   *See* **crown, joint**.

**joint reinforcement**   Steel wire, bar, or fabricated reinforcement that is placed in horizontal **mortar joints**.

**joints**   *See* **control joint**, **construction joint**, and **isolation joint**.

**joint tape**   Paper or paper-faced cotton tape used over joints between **wallboard**

to conceal the joints and provide a smooth surface for painting.

**joint treatment**  Method of reinforcing and concealing **gypsum board** joints with tape and successive layers of **joint compound**.

**joist**  One of a series of parallel beams used to support floor, ceiling, and roof loads. Joists are supported in turn by bigger **beams**, girders, or **bearing walls**.

**joists and planks**  See under **lumber**.

**jumbo brick**  See under **brick masonry unit**.

**jute**  Strong, durable yarn spun from fibers of the jute plant, native to India and the Far East; used in the **backings** of many **carpets**.

**kiln drying**  See under **seasoning**.

**knitted carpet**  **Carpet** made on a knitting machine by looping together backing, stitching, and pile yarns with three sets of needles, as in hand knitting.

**knot**  The portion of a branch or limb that has been surrounded by subsequent growth of wood.

**kraftcord**  Tightly twisted yarn made from wood pulp fiber, used as an alternative for cotton or jute in carpet backing.

**lacquer**  Coating that dries quickly by evaporation of its volatile **solvent** and forms a **film** from its nonvolatile constituent, usually nitrocellulose; may be pigmented or clear.

**laitance**  A soft, weak layer of mortar appearing on a horizontal surface of **concrete** due to **segregation** or **bleeding**.

**laminate**  A product made by bonding together two or more layers of materials.

**laminated wood**  A piece of wood built up of laminations that have been joined with glue, mechanical fastenings, or both.

**laminating adhesive**  See **adhesive, drywall**.

**langley**  The measure of **isolation**; equal to 1 calorie per square centimeter or 3.69 **Btu**s per square foot.

**lap**  To lay or place one coat so that its edge extends over and covers the edge of a previous coat.

**lateral**  A branch of an **absorption field**, consisting of either (1) the length of the distribution line between overflow pipes or (2) the length of the distribution line between the tee or cross fitting and the farthest point in a closed-loop field.

**lateral support of walls**  Means whereby walls are braced either horizontally by **columns**, **pilasters**, or cross walls or vertically by floor and roof construction.

**latex**  A water suspension of fine particles of rubber or rubber-like plastics.

**latex paint**  A paint containing **latex** and thinned with water.

**lath, gypsum**  See **gypsum board**.

**latitude**  The position of a point on the earth's surface north or south of the equator, stated as an angular measure (**degrees**, **minutes**, and **seconds**) of the **meridian** arc contained between that point and the equator.

**leaching pit**  See **dry well**.

**leaching well**  See **dry well**.

**lead**  The section of a masonry wall built up and racked back on successive courses. A line is attached to leads as a guide for building a wall between them.

**legal description**  A written, legally recorded identification of the location and boundaries of a parcel of land. A legal description may be based on a **metes and bounds survey** or the rectangular survey system, or it may make reference to a **recorded plat** or **survey**.

**leveling**  Ability of a **film** to flow out free from ripples, pockmarks, and brush marks after application.

**level loop pile**  A woven or **tufted carpet** style in which the tufts are in loop form and of the same height.

**level tip shear**  Randomly patterned **cut and loop pile carpet**.

**lifting**  Softening and penetration of a previous film by solvents in the **paint** being applied over it, resulting in raising and wrinkling.

**light-fastness**  Ability to resist **fading**.

**light framing lumber**  See **lumber**.

**lignin**  The second most abundant constituent of wood, making up 12 to 28% in most species. It encrusts the cell walls and cements the cells together.

**linear polymer**  A **polymer** that is arranged in a long, continuous chain with a minimum number of side chains or branches.

**lining felt**  A mineral fiber or asphalt felt specifically manufactured for use under resilient flooring.

**lintel**  A structural member used to carry the load over an opening in a wall.

**lip**  When related to barrier-free design, an abrupt vertical change in level.

**live load**  The total of all moving and variable loads that may be placed on or in a building. See also **dead load**.

**longitude**  The position of a point on the earth's surface east or west of the **Greenwich meridian**, stated as an **angular measure (degrees, minutes, and seconds)** of the arc on the equator contained between a meridian passing through that point and the Greenwich meridian.

**long oil**  Varnish or **paint** vehicle with more than 55% of the resin consisting of oil or fatty acid.

**loom**  Machine on which **carpet** is woven, as distinguished from other machines on which **carpets** may be tufted, flocked, or punched.

**loomed carpet**  See **woven carpet**.

**loop pile**  A woven or tufted carpet style in which the pile surface consists of uncut loops; also called **round wire**.

**loop vent**  A branch vent that serves two or more traps and extends from in front of the last fixture connection of a horizontal branch to the **stack vent**. See also **circuit vent**.

**loudness**  A subjective response to **sound**, indicating the magnitude of the hearing sensation, depending on the listener's ear.

**lumber**  The wood product of a saw and planing mill not further manufactured than by sawing, resawing, and passing lengthwise through a standard planing machine, cross-cut to length, and worked. *Beams and stringers* are large pieces of lumber measuring 5 in. (127 mm) or more in thickness and 8 in. (203.2 mm) or more in width. These are graded with respect to their strength in bending when loaded on the narrow face. *Boards* are lumber up to 2 in. (50 mm) thick and 2 in. (50 mm) or more in width. *Dimension lumber* is lumber 2 to 4 in. (50 to 100 mm) thick and 2 in. (50 mm) or more in width. *Factory and shop lumber* is intended to be cut again for use in further manufacture. *Joists and planks* are pieces of lumber 2 to 4 in. (50 to 100 mm) in nominal thickness and 4 in. (100 mm) or more in width. These are graded with respect to strength in bending when loaded either on the narrow face, as a joist, or on the wide face, as a plank. *Light framing* is lumber 2 to 4 in. (50 to 100 mm) thick and 2 to 4 in. (50 to 100 mm) wide. *Matched lumber* is lumber that has been edge dressed and shaped to make a close tongue-and-groove joint at the edges or ends when laid edge to edge or end to end. *Patterned lumber* is lumber that is shaped to a pattern or to a molded form in addition to being dressed, matched, shiplapped, or any combination of these. *Posts and timbers* are lumber that is approximately square, 5 in. (127 mm) and thicker, and having a width not more than 2 in. (50.8 mm) greater than its thickness. *Stress grade lumber* is the same as *structural lumber*. *Structural lumber* is lumber that has been machine rated or visually graded into grades with assigned working stresses. Includes most yard lumber grades except boards. *Yard lumber* is intended for general building purposes; includes boards, dimension lumber, and timbers.

**machinability**  The ability to be milled, sawed, tapped, drilled, and reamed without excessive tool wear, with ease of chip metal removal and surface finishing.

**main**  The principal artery of a **drainage** or **water supply system** to which branches may be connected. It is usually the lowest horizontal piping. In a drainage system the main is the building drain; in a water system it is the distributing main, whether located in the basement or on the top floor.

**mainstreaming**  A process that seeks to integrate **physically handicapped persons** into society by (1) equipping them with the personal devices and adaptive skills needed to function effectively in the built environment and (2) removing physical barriers that prevent them from functioning like able-bodied individuals.

**main vent**  The principal artery of a **vent system** to which vent branches may be connected. It may be either a vertical pipe that collects one or more branch vents or a horizontal pipe to which a number of **vent stacks** are connected before they go through the roof.

**major arch**  *See under* **arch**.

**malleability**  The ability to be shaped without fracture by either hot or cold working.

**marine varnish**  Varnish specially designed for immersion in water and exposure to a marine atmosphere.

**marl**  A soil or rock containing calcium carbonate (limestone).

**martensitic steel**  *See under* **grain structure**.

**masking**  (1) The effect produced by **ambient sound** that seems to diminish the loudness of transmitted noise. (2) Temporary covering of areas not to be painted.

**masking tape**  A strip of paper or cloth tape, easily removable and used temporarily to cover areas that are not to be painted.

**masonry bonded hollow wall**  A **hollow wall** in which the **facing** and **backup** are bonded together with solid **masonry units**.

**masonry cement**  A packaged premixture of **portland cement**, lime, and other ingredients to which sand and water are added in the field to make masonry cement **mortar**.

**masonry unit**  A manufactured building unit of burned clay, shale, concrete, stone, glass, gypsum, or other material. A *hollow masonry unit* is one whose **net cross-sectional area** in any plane parallel to the bearing surface is less than 75% of the **gross cross-sectional area**. A *modular masonry unit* is one whose **nominal dimensions** are based on a 4 in. (100 mm) module. A *solid masonry unit* is one whose net cross-sectional area in every plane parallel to the bearing surface is 75% or more of the gross cross-sectional area.

**mastic**  Heavy-bodied paste-like coating of high build often applied with a trowel.

**matched lumber**  *See under* **lumber**.

**matrix, terrazzo**  Topping mortar consisting of binders, and sometimes **pigments** and inert fillers, that fills the spaces between chips and binds them into a homogeneous mass.

**medium-density plywood**  *See* **plywood**.

**meridian**  Imaginary north-south line on the earth's surface described by a great circle arc from the North Pole to the South Pole. All points on a meridian are of the same longitude. *See also* **great circle**. The **Greenwich meridian**, also called the *prime meridian*, is the meridian passing through the Royal Observatory at Greenwich, England. It is designated as the starting line (zero degrees) for measuring east and west longitude. A *principal meridian* is a meridian of specified longitude, used in the *rectangular survey system*, serving as the main north-south reference line for a particular state or area. *Guide meridians* are north-to-south reference lines located at 24-mile intervals east and west of a *principal meridian*.

**metallics**  Class of **paints** that include metal flakes.

**metes and bounds**  A system of land **survey** and description based on starting from a known reference point and tracing the boundary lines around an area.

**mil**  One one-thousandth of an inch (0.0254 mm).

**mildewcide**  *See* **fungicide**.

**mildew resistance**  Ability of a **coating**, **paint**, or **transparent finish** to resist the growth of molds and mildew; mildew is particularly prevalent in moist, humid, and warm climates.

**mild steel**  *See under* **steel**.

**millwork**  Lumber that is shaped to a pattern or to a molded form in addition to being **dressed**, **matched**, shiplapped, or any combination of these. Millwork includes most finished wood products such as doors, windows, interior trim, and stairways, but not flooring or siding products.

**mineral spirits**  Common **solvent** for **coatings**, **paints**, and **transparent finishes**; derived from the distillation of petroleum.

**minor arch**  *See under* **arch**.

**minute**  A unit of angular measure equal to $1/60$th of a **degree**.

**modacrylic**  *See under* **acrylic**.

**modifier**  An ingredient added to a **plastic** to improve or modify its properties.

**modular masonry unit**  *See* **masonry unit**.

**modulus of elasticity**  The ratio of **stresses** lower than the **elastic limit** to their respective **strains**.

**moist curing**  *See under* **curing**.

**moisture content of a concrete masonry unit**  The amount of water contained in a unit, expressed as a percentage of the total **absorption** (e.g., a concrete masonry unit at 40% moisture content contains 40% of the water it can absorb).

**moisture content of wood** The amount of water contained in wood at the time it is tested, expressed as a percentage of the weight of the wood when **oven-dry**.

**monolithic concrete** Concrete placed in one continuous pour without **construction joints**.

**monomer** A relatively simple compound that can react to form a **polymer**.

**monument** A permanent reference point for land **survey**ing whose location is recorded; either a man-made marker or a natural landmark.

**moresque** Multicolored yarn made by twisting together two or more strands of different shades or colors.

**mortar** Substance used to join **masonry units**, tile, stone, and other materials.

**mortar bed, tile** A $^3/_4$-in. (19.1-mm)- to $1^1/_4$-in. (31.8-mm)-thick bed of **portland cement** and sand mortar onto which a layer of neat **portland cement** (water-cement paste) is applied to receive tile. Also called *thickbed*.

**mortar joint** The joint between masonry units. A *tooled joint* is one that is compressed and shaped with a special concave or V-shaped tool. A *troweled joint* is one finished with a trowel to form a struck joint or a weathered joint. In a *raked joint* the mortar is raked out to a specified depth while the mortar is still green.

**mortar, masonry** A plastic mixture of one or more cementitious materials, sand, and water.

**mortar, tile** A mixture of **portland cement** and other ingredients used to install tile. A *leveling or setting-type mortar* is a mixture of portland cement, sand, and sometimes lime, used as a bed in which tile is installed (setting bed), or as a coat to produce a plumb and level surface (leveling coat) so that subsequent coats can be applied in a uniform thickness. A *bonding-type mortar* is any one of a variety of formulations used to bond tile to a **backing** or **mortar bed**. Formulations may be mainly cementitious, such as commercial cement mortar; resinous, such as epoxy mortar; or a combination of both, as in latex—portland cement mortar. *Also see **adhesive, tile**.

**multilevel loop** A **carpet** style in which the yarns are looped at several levels; also called *high-low loop* and *cut and loop*.

**nail, drywall** A nail suitable for **gypsum board** application. Such nails are typically bright, coated, or chemically treated low-carbon steel nails with flat, thin, slightly filleted and countersunk heads approximately $^1/_4$ in. (6.4 mm) in diameter and medium or long diamond points. Annularly threaded nails (GWB-54) and smooth or deformed shank nails should conform to ASTM C514.

**nail gluing** A method of gluing wood in which the nails hold the wood members until the glue sets.

**nail popping** Surface defect in **gypsum board** resulting in a conspicuous protrusion of **joint compound** directly over a nail head.

**nail spotting** *See* **fastener treatment**.

**nap** Length of fibers on a paint roller cover.

**natural gray yarn** Unbleached and undyed yarn spun from a blend of black, brown, or gray wools.

**natural resin** *See under* **resin**.

**neat cement** A mixture of **cement** and water (no **aggregates**).

**neat portland cement** Unsanded mixture of **portland cement** and water used as a bond coat in conventional thickbed installation of tile.

**needle-punched carpet** Carpet made by punching loose, unspun fibers through a woven sheet, which results in a pileless carpet similar to a heavy felt; usually consists entirely of synthetic fibers.

**net cross-sectional area** The **gross cross-sectional area** of a section minus the area of cores, cellular spaces, and other voids it contains.

**noise** Unwanted **sound**.

**noise reduction** Reducing the level of unwanted **sound** by means of acoustical treatment.

**noise reduction coefficient (NRC)** A single-number index of the noise-reducing efficiency of acoustical materials. Found by averaging a material's sound absorption coefficients at 250, 500, 1000, and 2000 cycles per second.

**nominal dimension** A dimension greater than the **actual dimension** that is used to identify the size of a material. In masonry, the nominal dimension is greater than the actual dimension by the amount of the thickness of a **mortar joint**, but not more than $^1/_2$ in. (12.7 mm).

In wood, the nominal dimension (also called *nominal size*) is greater than the actual dimension by an amount roughly equivalent to the size lost due to wood shrinkage during drying, plus the amount planed off to square and smooth the piece. *See also* **size**.

**nominal size** *See* **nominal dimension**; **size**.

**noncombustible material** A material that will neither ignite nor actively support combustion when exposed to fire in air at a temperature of 1200°F (648.9°C).

**nondrying oil** Oil that does not readily oxidize and harden when exposed to air.

**non-heat-treatable alloy** An **alloy** that is not capable of gaining strength by **heat treatment** and that depends on the initial strength of the **alloy** or **cold working** for additional strength. Also called a *common alloy*.

**non-load-bearing wall** A wall that supports no vertical load other than its own weight. *See also* **bearing wall**.

**nonvolatile vehicle** Liquid portion of **coatings**, **paints**, and **transparent finishes** except for their volatile **thinners** and water.

**normal portland cement** Portland cement, Type I.

**Norman brick** *See under* **brick masonry unit**.

**octave** The interval between two sounds with a frequency ratio of 2:1.

**ohm** A measure of electrical resistance equal to that of a circuit that permits 1 volt to cause 1 amp to flow through it.

**oil color** Single **pigment** dispersion in linseed oil used for tinting **coatings** and **paints**.

**oil paint** *See under* **paint**.

**oil stains** May be penetrating or nonpenetrating. *Penetrating* oil stains contain dyes and resins that penetrate a surface; *nonpenetrating* oil stains contain larger amounts of pigments and are usually opaque or translucent.

**oil varnish** *See under* **transparent finish**.

**olefins** Long-chain synthetic **polymers** composed of at least 85% by weight of ethylene, propylene, or other olefin units.

**one-package (one-part) formulation** Coating or paint formulated to contain all the necessary ingredients in one package and generally not requiring any field additions except **pigment** and **thinner**.

**opacity**  Degree of obstruction to the passage of visible light.

**opaque coating**  A **coating** that hides the previous surface or coating.

**open-grained**  *See under* **grain**.

**open hearth furnace**  *See under* **furnace, steel**.

**oven-dry wood**  Wood that has been dried in an oven to a consistent moisture content and, for all practical purposes, no longer holds any water.

**overlaid plywood**  *See under* **plywood**.

**oxidation**  The chemical combination of a substance with oxygen.

**package dyeing**  Placing spun and wound yarn on large, perforated forms and forcing the dye through the perforations.

**padding**  Cellular rubber, felted animal hair, jute fibers, or plastic foams in sheet form, used as cushioning under **carpet**. *See also* **cushioning**.

**paint**  A liquid, pigmented material applied in the field and producing an opaque film. **Primers** may, however, be applied by hand or machine in a factory or shop, but are part of the **paint system** and are therefore also called paint. In addition, all parts of a paint system are called paint, collectively and individually. Such parts include **primers, emulsions, enamels,** opaque stains, **sealers, fillers,** and other applied materials used as prime, intermediate, or finish coats. *See also* **coating** and **transparent finish**. An *emulsion* is a paint with a vehicle consisting of an oil, oleoresinous varnish, or resin binder dispersed in water (see also **emulsion**). A *latex paint* is a paint with latex resin as the chief binder. An *oil paint* is a paint with drying oil or oil varnish as the basic **vehicle** ingredient. A *paste* is a paint with sufficiently concentrated pigment to permit substantial thinning before use. *Water-based paint* is paint with a vehicle that is a water emulsion or water dispersion, or that has ingredients that react chemically with water.

**paint remover**  Compound that softens old **paint** and **transparent finishes** and permits scraping off the loosened material.

**paint system**  Collectively, the several coats that are necessary to produce a complete paint coating. Materials that are used in the various coats are called **paint**.

**panel wall**  An exterior **non-load-bearing wall** in skeleton frame construction, wholly supported at each story. *See also* **curtain wall**.

**parallel**  An imaginary east-west line on the earth's surface, consisting of a circle on which all points are equidistant from one of the poles. All points on a parallel are at the same **latitude**.

**parapet wall**  That part of a wall entirely above the roof line.

**parging**  (1) The application of mortar to the back of the facing material or the face of the backing material within a masonry wall; the mortar so applied. Also called *back-plastering, backparging,* or *pargeting.* (2) The application of **portland cement** mortar or **plaster** to the exterior face of masonry, often below or near the ground; the material so applied.

**particleboard**  A composition board consisting of distinct particles of wood bonded together with a synthetic resin or other binder.

**partition**  An interior wall, one story or less in height.

**party wall**  A wall used for joint service by adjoining buildings. *See also* **common wall**.

**pattern bond**  Patterns formed by exposed faces of **masonry units** and their joints.

**patterned lumber**  *See under* **lumber**.

**paving brick**  *See under* **brick masonry unit**.

**peeling**  Detachment of a dried **coating, paint,** or **transparent finish** film in relatively large pieces, usually caused by moisture or grease under the finish.

**perceived noise level (PNdB)**  A single-number rating of aircraft **noise** in decibels, used to describe the acceptability or noisiness of aircraft sound. PNdB is calculated from measured interior and exterior noise levels and correlates well with subjective responses to various kinds of aircraft noise.

**perforated wall**  A wall that contains a considerable number of relatively small openings. Often called a *pierced wall* or *screen wall.*

**perimeter**  In **gypsum board** work, the surface (as opposed to the edges) of a **gypsum board** panel near the edges and ends.

**pervious soil**  Usually a granular soil, such as sand or gravel, that allows water to pass readily through it.

**physically handicapped person**  An individual who has a physical impairment, including impaired sensory, manual, or speaking abilities, that results in a functional limitation in gaining access to and using a building or facility.

**pickling**  Removing the oxide scale formed on hot metal as it air cools by dipping it in a solution of sulfuric or hydrochloric acid.

**piece dyeing**  Immersing an entire **carpet** in a dye bath to produce single-color or multicolor pattern effects. *See also* **resist printing**.

**pier**  An isolated column of masonry.

**pig**  A **pig iron** ingot.

**pig iron**  High-carbon crude iron from a blast **furnace** used as the main raw material for ironworking and steelmaking. *Basic pig iron* is a high-phosphorus iron used in basic steelmaking furnaces.

**pigment**  Fine, solid particles suspended in the **vehicle** of a **coating** or **paint** that provide color (**hiding power**) as well as other properties.

**pilaster**  A thickened wall section or column built as an integral part of a wall.

**pile**  The raised yarn tufts of **woven, tufted,** and **knitted carpets** that provide the wearing surface and desired color, design, or texture. In *flocked carpets,* the upstanding, nonwoven fibers. *Cut loop pile* is a pile surface in which tufts have been cut to reveal the fiber ends. A *frieze carpet* is a rough, nubby-textured carpet using tightly twisted yarns; same as twist carpet. *Jaspe* is a carpet surface characterized by irregular stripes produced by varying textures or shades of the same color. *Multilevel carpet* has a texture or design created by different heights of tufts of either cut or uncut loop. *Plush carpet* has a cut pile surface that does not show any yarn texture. *Sculptured (carved) carpet* has surface designs created by combinations of cut and loop pile and variations in pile height. *Shag carpet* has a surface consisting of long, twisted loops. *Stria (striped) carpet* has a striped surface effect obtained by loosely twisting two strands of one shade of yarn with one strand of a lighter or darker shade. *Twist carpet* is the same as *frieze carpet.* *Uncut loop* is pile in which the yarns are continuous from tuft to tuft, forming visible loops.

**pile crushing**  Bending of pile due to foot traffic or the pressure of furniture.

**pile height**  The height of pile measured from the top surface of the backing to the top surface of the pile. Also called *pile wire height*.

**pile setting**  Brushing after shampooing to restore damp pile to its original height.

**pile yarn density**  The weight of pile yarn per unit of volume in **carpet**, usually stated in ounces per cubic yard (grams per cubic meter).

**pilling**  Appearance defect associated with some staple fibers in which balls of tangled fibers are formed on the carpet surface, which are not removed readily by vacuuming or foot traffic; pills can be removed by periodic clipping.

**pipe**  *See under* **finished mill products**.

**pitch**  The number of **tufts** or pile warp yarns in a 27 in. (685.8 mm) width of woven carpet.

**pitch pocket**  An opening extending parallel to the **annual growth rings** of a tree. These contain or have contained either solid or liquid pitch.

**pith**  The small, soft core occurring in the center of the growth rings of a tree, branch, twig, or log.

**placing**  The act of putting **concrete** in position (sometimes incorrectly called *pouring*).

**plain masonry**  Masonry without reinforcement, or reinforced only to resist shrinkage or temperature changes.

**plain-sawn wood**  *See under* **grain**.

**plaster**  All elements of a gypsum or **portland cement** plaster membrane, such as the several coats of plaster, metal reinforcement, accessories, and backing paper, when required.

**plaster mix**  Mortar consisting of properly proportioned quantities of cementitious materials, **aggregate**, water, and sometimes **pigments**, **plasticizers**, and other **admixtures**.

**plastic**  A material that contains as an essential ingredient an organic substance of large molecular weight, or a synthetic or processed substance. A plastic is solid in its finished state but at some stage in its manufacture can be shaped by flow (*see* **polymer**).

**plastic deformation**  Deformation caused by a **stress** exceeding the **elastic limit** of a material so that it does not return to its original shape when the stress is removed. *See also* **elastic deformation**.

**plasticizer**  (1) In **plastics**, a material added to increase the workability and flexibility of a resin. (2) In **coatings**, **paints**, and **transparent finishes**, a substance added in the liquid state to impart flexibility to a hardened film.

**plat**  A map of **surveyed** land showing the location and the boundaries and dimensions of a parcel. A *recorded plat* is a plat that is recorded at an appropriate government office, usually the county recorder's office. In addition to location notes and a boundary line layout, a recorded plat may contain information such as restrictions, easements, approvals by zoning boards and planning commission, and lot and block numbers for a subdivision.

**plate**  *See under* **finished mill products**.

**plumbing fixtures**  Devices or appliances that are supplied with water and receive or discharge liquids or wastes.

**ply**  When preceded by a number, the number of single strands used in a finished yarn.

**plywood**  A crossbanded assembly made of layers of **veneer** or of veneer in combination with a core of **lumber**, **particleboard**, or another composition core, all joined with an adhesive. *Standard plywood* is hardwood plywood produced to meet governing industry standards. Also, unsanded interior softwood plywood for sheathing and flooring. *High-density overlaid plywood* is overlaid plywood in which the overlay sheet contains 40% resin by weight. *Medium-density overlaid plywood* is overlaid plywood in which the overlay sheet contains 20% resin by weight. In *overlaid plywood* the face veneer is bonded on one or both sides with paper, resin-impregnated paper, or metal. *Specialty plywood* is hardwood plywood that is not necessarily manufactured to meet industry standards.

**P.O.B. (place of beginning)**  The starting point of a **metes and bounds** description or **survey**.

**pointing**  Troweling mortar into a joint after a **masonry unit** has been laid.

**polymer**  A chemical compound formed by the reaction of simple molecules to form more complex molecules of higher molecular weight.

**polymerization**  A chemical reaction in which the molecules of a **monomer** are linked together to form large molecules whose molecular weight is a multiple of that of the original substance.

**polypropylene**  *See* **olefins**.

**polyurethane**  *See* **resin**; **transparent finish**.

**polyvinyl acetate**  *See under* **resin**.

**ponding**  A curing method used on flat **concrete** surfaces whereby a small earth dam or other water-retaining material is placed around the perimeter of the surface and the enclosed area is flooded with water.

**portland blast furnace cement**  Cement made by grinding not more than 65% of granulated blast furnace **slag** with at least 35% of **portland cement**.

**portland cement**  A hydraulic cement produced by pulverizing clinker consisting essentially of hydraulic calcium silicates and usually containing one or more of the forms of calcium sulfate as an interground addition. *Types I and IA portland cement* are used in general construction when the special properties of other types are not required. *Types II and IIA portland cement* are used in general construction where moderate heat of hydration is required. *Types III and IIIA portland cement* are used when a high early strength is required. *Type IV portland cement* is used when low heat of hydration is required. *Type V portland cement* is used when high sulfate resistance is required.

**portland-pozzolan cement**  Cement made by blending not more than 50% pozzolan (a material consisting of siliceous or siliceous and aluminous material) with at least 50% of **portland cement**.

**postforming**  The forming of **laminates** that have been cured into simple shapes by heat and pressure.

**posts and timbers**  *See under* **lumber**.

**potable water**  Water of sufficient quality, either through treatment or natural phenomena, to be drinkable.

**potential difference**  The voltage difference between two points.

**pot life** Time period after mixing reactive components, in a two-component **coating**, **paint**, or **transparent finish** system, during which material can be satisfactorily used.

**pots** Carbon-lined vessels used in the reduction processing of aluminum. Molten aluminum is collected and siphoned off from the bottom of pots.

**precast concrete** **Concrete** components that are cast and cured off-site or on-site in a location other than where they will be finally placed.

**prefabricated brick masonry** Masonry panels fabricated other than in their final location in the structure. Also known as *preassembled*, *panelized*, and *sectionalized brick masonry*.

**preframed panels** Panels fabricated with precut **lumber** and **plywood**.

**preservative** A substance that will prevent, for a finite time that varies with quantity, the action of wood-destroying fungi, insects of various kinds, and other destructive life in wood that has been properly impregnated with that substance.

**pressure gluing** A method of gluing that places the wood members under high pressure until the glue sets.

**prestressed concrete** **Concrete** that has been subjected to compressive **stresses** before external loads have been applied by the prestretching (or stressing) of high-strength steel reinforcement within the concrete. This prestressing may be done by *pretensioning* (stretching the steel before the concrete is placed around it) or by *post-tensioning* (stretching the steel after the concrete has been cured).

**prime meridian** *See under* **meridian**.

**primer, paint** First of several coats, intended to prepare a surface for the succeeding coat(s); sometimes a special product, but may be the same as the finish coat.

**primer, resilient flooring** A brushable **solvent-based** asphaltic preparation recommended as a first coat over porous or dusty concrete floors and panel **underlayments**; intended to seal the pores and improve the bond with asphaltic adhesives used for the installation of asphalt and vinyl composition tile.

**primer-sealer** Product formulated to possess properties of both a **primer** and a **sealer**.

**principal meridian** *See under* **meridian**.

**print dyeing** Screen printing a pattern on **carpet** by successive applications of premetalized dyes, which are driven into the pile construction by an electromagnetic charge.

**puddling** The compacting or consolidating of **concrete** with a rod or other tool.

**putty** Dough-like material consisting of **pigment** and **vehicle**; used for sealing glass in sash or frames and for filling imperfections in wood and metal surfaces; does not retain its plasticity for extended periods, as does a **glazing compound** or **joint sealant**. Both elastomeric and acrylic joint sealants are superior to putty and are generally used today where putty is called for. *See also* **sealant, joint**.

**quarter-sawn wood** *See under* **grain**.

**quenching** *See under* **heat treatment**.

**quoin** A right-angled masonry corner that is usually projected when of the same material as the surrounding material, but may be flush with the surrounding masonry when of a different material. A typical example of a quoin consists of squared stones set into a rubble stone or brick wall. Quoins are often larger than the surrounding **masonry units**.

**ramp** When related to accessibility for handicapped persons, a walking surface in an accessible space that has a **running slope** greater than 1:20. The maximum slope is limited by the applicable code or law.

**random shear** A **carpet** style created by shearing only the higher loops of either level loop or multilevel loop carpet.

**range lines** North-south reference lines used in the **rectangular survey system**, located at 6-mile intervals between **guide meridians**.

**rays** Strips of cellulose extending radially within a tree. These store food and transport it horizontally in the tree.

**ready-mixed concrete** **Concrete** mixed at a plant or in trucks en route to the job and delivered ready for **placing**.

**recorded plat** *See under* **plat**.

**rectangular survey system** A land **survey** system based on geographical coordinates of **longitude** and **latitude**; originally established by acts of Congress to survey the lands of public domain and

now used in most states. Also called *Government Survey System*.

**red mud** Solid impurities collected by either filtering or gravity settling during the **Bayer process** of aluminum refining.

**reduction** (1) An electrolytic process used to separate aluminum from aluminum oxide. (2) Separation of iron from its oxide by smelting ores in a blast furnace.

**refining of steel** Melting **pig iron** and other iron-bearing materials in steel furnaces to achieve desired contents of residual and alloying elements.

**reflection** The return from surfaces of **sound** not absorbed on contact with the surfaces or transmitted through the material contacted.

**refractories** When related to steel production, nonmetallic materials with superior heat and impact resistance used for lining furnaces, flues, and vessels for ironworking and steelmaking.

**reinforced masonry** **Masonry** containing embedded steel so that the two materials act together to resist applied forces.

**reinforced plastic** A **plastic** material whose strength and, to a lesser degree, stiffness have been upgraded by the addition of high-strength fillers such as glass fiber.

**reinforcing bars** Steel placed in **concrete** or **masonry** to take tensile, compressive, and shear **stresses**.

**relative humidity** The amount of water vapor in air stated as a percentage. The ratio of the amount of water in air at a given temperature and atmospheric pressure to the amount that same air can contain when it is **saturated**.

**relief vent** A vent whose primary function is to provide circulation of air between **drainage** and **vent systems**.

**relieving arch** *See under* **arch**.

**repeat** The distance along the length of a **carpet** from one point in a figure or pattern to the same point at which it again occurs.

**residual elements** Nonferrous elements (such as carbon, sulfur, phosphorus, manganese, and silicon) that occur naturally in raw materials and are controlled in steelmaking.

**resin** (1) In **plastics**, the essential ingredients of a plastic mix before final processing and fabrication of the plastic object. (2) In **coatings**, **paints**, and **transparent finishes**, a mixture of organic or synthetic compounds with no sharply defined melting point, no tendency to crystallize, soluble in certain organic solvents but not in water. It is the main ingredient of most coatings, paints, and transparent finishes, binds the other ingredients together, and aids adhesion to the surface. Includes acrylic, alkyd, epoxy, nitrocellulose, polyurethane, polyvinyl acetate, silicone, styrene-butadiene, and vinyl. A *natural resin* may be a fossil of ancient origin, such as amber, or an extract of certain pine trees, such as rosin, copal, and damar. A *synthetic resin* is a manmade substance exhibiting properties similar to those of natural resins; typical synthetic resins used in coatings, paints, and transparent finishes include alkyd, acrylic, latex, phenolic, urea, and others.

**resist printing** Placing a dye-resist agent on **carpet** before **piece dyeing** so that the **pile** will absorb color according to a predetermined design.

**resonance** The sympathetic vibration, resounding, or ringing of such things as enclosures, room surfaces, and panels when they are excited at their natural frequencies.

**retaining wall** A wall that is subjected to lateral pressure other than wind pressure, such as a wall built to support a bank of earth.

**retarder** An admixture added to **concrete** to retard its set.

**retempering** Restoring **workability** to **mortar** that has stiffened due to evaporation by adding water and remixing.

**retentivity** *See* **water retention**.

**reverberation** The continuing travel of sound waves between reflective surfaces after the original source has stopped transmitting.

**reverberation time (T)** The time in seconds required for a sound to diminish 60 decibels (dB) after the source has stopped transmitting.

**ridging** A surface defect in a **gypsum board** surface resulting in conspicuous wrinkling of the **joint tape** at treated joints.

**riser** In plumbing systems, a water-supply pipe that extends vertically one full story or more to convey water to branches or fixtures.

**roller** Paint applicator having a revolving cylinder covered with lambswool, fabric, foamed plastic, or other material.

**rolling** Shaping plate and sheet metal, blooms, and billets by passing them through steel rollers. *See also* **cold working** and **hot working**.

**Roman brick** *See under* **brick masonry unit**.

**room acoustics** A branch of **architectural acoustics** dealing with both **acoustical correction** and **noise reduction**. *See also* **acoustics**.

**rosin** Natural **resin** obtained from various pine trees; an ingredient of varnishes and some man-made resins.

**rotary-cut veneer** *See under* **veneer**.

**rough-in** The installation of those parts of a plumbing system that can be completed before the **plumbing fixtures** are installed. This includes drainage, water-supply, and vent piping and the necessary fixture supports. Also, the plumbing system parts so installed.

**round wire** A pile wire that does not cut the pile loop; or **woven carpet** with an **uncut loop pile**.

**rowlock** A brick unit laid on its face edge. Usually laid in a wall with its long dimension perpendicular to the wall face. Also spelled *rolok*.

**running slope** The slope of a pedestrian way that is parallel to the direction of travel. *See also* **cross slope**.

**rust preventive paint or primer** First coat of **paint** applied directly to iron or steel structures to slow down or prevent rusting.

**sabin** The measure of sound absorption of a surface, equivalent to 1 sq ft (0.093 m²) of a perfectly absorptive surface.

**sags** Excessive flow, causing runs or sagging in a **coating**, **paint**, or **transparent finish** film during application; usually caused by applying too heavy a coat or thinning too much.

**salamander** A portable stove used to heat the surrounding air.

**sand finish** *See* **texture**.

**sanding surfacer** Heavily pigmented finishing material used for building a surface to a smooth condition; it is sanded after drying.

**sandwich panel** A composite structural panel made of two thin, strong, hard

**facings** bonded firmly to a core of relatively lightweight, weaker material with insulating properties.

**sapwood** The living wood of pale color near the outside of a tree.

**satin finish** *See* **semigloss**.

**saturated air** Air that contains 100% of the amount of water vapor it can contain. *See also* **relative humidity**.

**saturated molecule** A molecule that will not unite readily with another element or compound.

**saturation coefficient** *See* **C/B ratio**.

**sawed veneer** *See under* **veneer**.

**scaling** The breaking away of a hardened **concrete** surface (to a depth of about $^1/_{16}$ in. [1.6 mm] to $^3/_{16}$ in. [4.76 mm]), usually occurring at an early age of the concrete.

**scarf jointing** A joint in which the ends of **plywood** panels are beveled and glued together.

**scoring** Partial cutting of **concrete** flat work for the control of shrinkage cracking. Also used to denote the roughening of a slab to develop a mechanical bond.

**scrap metal** Source of iron for iron-working and steelmaking, consisting of rolled product croppings, rejects, and obsolete equipment from steel mills and foundries and waste ferrous material from industrial and consumer products.

**scratch coat** The first **plaster** layer in **three-coat work**, which embeds the reinforcement and provides a suitably rigid and roughened (scratched) surface for the following coat.

**SCR brick** *See under* **brick masonry unit**.

**screed** A wood or metal template to which a **concrete** surface is leveled.

**screeding** Striking off excess **concrete** in the finishing operation of concrete slab work.

**screws, drywall** Screws developed for **gypsum board** application, usually with self-tapping, self-threading points, special-contour flat heads, and deep Phillips recesses for use with a power screwdriver. *Type S* drywall screws are used for sheet metal studs and **furring**. *Type W* drywall screws are used for wood framing and furring. *Type G* drywall screws are used for attaching gypsum board to gypsum board.

**scribing** A method of transferring the profile of an obstruction, projection, or material edge to a piece of material, such as resilient flooring or wood trim, so that it can be accurately cut and fitted.

**scrim** Rough, loosely woven fabric often used as a secondary **backing** on **tufted carpets**.

**scrubbability** Ability of a **coating**, **paint**, or **transparent finish** film to withstand scrubbing and cleaning with water, soap, and other household cleaning agents.

**sculptured carpet** *See under* **pile**.

**scum** A mass of sewage matter that floats on the surface of the **sewage** in a **septic tank**.

**scum clear space** In a septic tank, the distance between the bottom of the scum mat and the bottom of the outlet device (tee, ell, or baffle).

**sealant, joint** Any one of a number of plastic materials, including rubber, formulated to fill and seal stationary and moving joints. *See also* **caulking compound**, **glazing compound**, and **putty**, which are types of joint sealants.

**sealer** (1) In **paint** and **coatings**, a formulation intended to prevent excessive absorption of a finish coat of coating, paint, or **transparent finish** into a porous surface or to prevent bleeding through the finish coat. (2) In resilient flooring, a solution of equal parts of wax-free **shellac** and denatured alcohol; recommended as a first coat over existing wood strip floors from which the finish has been removed; intended to seal wood pores, prevent excessive moisture absorption, and provide a dimensionally stable base for direct application of **lining felt** or resilient flooring.

**seasoning** Removal of moisture from green wood. *Air drying* is seasoning by exposure to air, usually in a yard. *Kiln drying* is seasoning in a kiln (oven) under controlled conditions of heat, humidity, and air circulation.

**second** A unit of angular measure equal to $1/60$th of a minute.

**section** An area of land used in the **rectangular survey system**, approximately 1 mile square, bounded by **section lines**. The rectangular survey system provides for the further subdivision of sections into halves, quarters, and quarter-quarters. A **fractional section** is an adjusted

section of land generally containing less (sometimes more) than 1 sq mi. The deficiency (or excess) may be the result of the convergence of **meridians**, the presence of bodies of water, or uncertainties in **surveying**.

**section lines** North-south reference lines used in the **rectangular survey system**, parallel to the nearest **range line** to the east, and east-west lines parallel to the nearest **township line** to the south; these lines divide **townships** into 36 approximately equal squares called **sections**.

**seeds** Small, undesirable particles or granules other than dust found in a **coating**, **paint**, or **transparent finish**.

**seepage bed** *See* **absorption bed**.

**seepage pit** A covered underground pit with a **concrete** or **masonry** lining designed to permit partially treated **sewage** to seep into the surrounding soil. Also called *dry well* in some parts of the country.

**segregation** Separation of the heavier coarse **aggregate** from the **mortar**, or of water from the other ingredients of a **concrete** mix, during handling or **placing**.

**self-cleaning** Controlled chalking of a paint film so that dirt does not adhere to the surface.

**selvage** The finished long edge of **woven carpet** that will not unravel and will not require binding or **serging**.

**semigloss** Degree of surface reflectance midway between **gloss** and **eggshell**; also **coatings** and **paints** displaying these properties.

**semitransparent** Degree of hiding greater than transparent but less than opaque.

**separated construction** *See* **discontinuous construction**.

**septic (sewage) disposal system** A system for the treatment and disposal of **sewage** by means of a **septic tank** and a **soil absorption system**.

**septic tank** A watertight, covered receptacle that receives the discharge of **sewage** from a **building sewer** and is designed and constructed to separate solids from liquids, to digest organic matter during a period of retention, to store digested solids through a period of retention, and to allow the clarified liquids to discharge for final disposal or additional treatment.

**serging** A method of finishing a cut long edge of **carpet** to prevent unraveling; distinguished from finishing a cut end, which may require binding.

**serial distribution** A combination of several **absorption trenches**, **seepage pits**, or **absorption beds** arranged in sequence so that each is forced to utilize the total effective absorption area before liquid flows into the succeeding component.

**service temperature** The maximum temperature at which a material can be continuously employed without noticeable reduction in strength or other properties.

**setting bed** *See* **mortar bed, tile**.

**settling** Coating, paint, or **transparent finish** material separation in which **pigments** and other solids accumulate at the bottom of the container.

**sewage** The liquid and solid waste matter carried off by **sewers**. It contains organic (animal or vegetable) matter in suspension or solution, as well as liquids containing chemicals in solution.

**sewer** An underground conduit for carrying off **sewage** and rainwater.

**sewerage** A system of **sewers**; the removal and disposal of liquid and solid wastes by sewers.

**sewer brick** *See under* **brick masonry unit**.

**shading** Bending or crushing a **pile** surface so that the fibers reflect light unevenly. This is not a defect but rather an inherent characteristic of some pile fabrics.

**shake** (1) A separation along the grain in a tree, the greater part of which occurs between the **annual growth rings**. (2) A wood shingle-like roofing material.

**shake painter** *See* **flat applicator**.

**shale** Laminated clay or silt compressed by earth overburden. Unlike slate, shale splits along its bedding planes.

**shearing** A carpet finishing operation that removes stray fibers and fuzz from **loop pile** and produces a smooth, level surface on **cut loop pile**.

**shear wall** A wall that resists horizontal forces applied in the plane of the wall.

**sheathing, gypsum** *See under* **gypsum board**.

**shedding** A normal temporary condition of dislodged, loose short fibers in new carpet after initial exposure to traffic and sweeping.

**sheen** Degree of luster of a dried **coating**, **paint**, or **transparent finish** film.

**sheen uniformity** Even distribution of luster over the entire surface of an applied finish.

**sheet** *See under* **finished mill products**.

**sheeting** A form of **plastic** in which the thickness is very small in proportion to the length and width; usually refers to a product with a thickness greater than 0.010 in. (0.254 mm) (*see also* **film**).

**sheet lamination** In multi-ply **gypsum board** construction, method of applying adhesive to the entire surface to be bonded. *See also* **strip lamination**.

**shellac** A **transparent finish** material made from **resins** dissolved in alcohol.

**shooting** See **sprouting**.

**short circuit** Related to **sound** control, a bypassing connection or transmission path that tends to nullify or reduce the sound-isolating performance of a construction assembly.

**shot** *See* **weft**.

**shrinkage** Decrease in initial volume due to removal of moisture from fresh **concrete** or **green lumber**. May also refer to a decrease in volume due to subsequent decreases in temperature or moisture content in concrete.

**side wall** The wall along the long dimension of a room.

**signage** Verbal, symbolic, and pictorial information.

**silica** Silicon dioxide ($SiO_2$), occurring as quartz, a major constituent of sand, sandstone, and quartzite.

**silicone** Any of a number of polymeric organic silicon compounds used in **paint**, **sealants**, roof membranes, and water repellents. *See also* **resin**.

**siphon breaker** See **atmospheric vacuum breaker**.

**size** (1) Water-based formulation, with glue or starch binders, intended as a sealer over existing wall **paint** or **plaster**; now seldom used. (2) The size of lumber. The *actual size* of lumber is its measured size after **seasoning** and **dressing**. The *dressed size* of lumber is its size

after surfacing. The *face size* of lumber is the exposed width of a piece of lumber when installed. The *nominal size* of a piece of lumber is its approximate rough-sawn commercial size by which lumber is known and sold.

**skein dyeing** Immersing batches of yarn (skeins) in vats of hot dye.

**skin** Tough covering that forms on liquid **coating**, **paint**, and **transparent finish** materials when left exposed to air.

**slab** (1) In **concrete**, a flat, thin (as compared with its other dimensions) structure. A structural slab is a suspended, self-supporting, reinforced concrete floor or roof slab. A *slab on grade* is a nonsuspended, ground-supported concrete slab, usually but not always having some temperature reinforcement. An *edge-supported slab on grade* rests atop the perimeter foundation wall. A *floating slab on grade* terminates at the inside face of the perimeter foundation wall and is said to "float" independently of the foundation wall. A *monolithic slab on grade* is a combination slab and foundation wall formed into one integral mass of concrete. Also called a *thickened-edge slab*. (2) In wood, the outside piece cut from a log in squaring it.

**slabbed** To have removed an outer slab from a log.

**slag** A glass-like waste product, generally from an ironworking or steelmaking furnace. Molten mass composed of fluxes in combination with unwanted elements, which floats to the surface of the hot metal in the furnace and thus can be removed.

**slag cement** *See* **portland blast furnace cement**.

**sliced veneer** *See under* **veneer**.

**sloshing** Attempting to fill vertical joints after units are laid by throwing mortar into the joint with a trowel from above.

**sludge** The accumulated solids that settle out of **sewage**, forming a semiliquid mass on the bottom of a **septic tank**.

**sludge clear space** In a **septic tank**, the distance between the top of the **sludge** and the bottom of the outlet device.

**slump** A measure of the consistency of a **concrete** mix (in inches).

**slump block** *See under* **concrete masonry units**.

**slump test** A method of measuring **slump** by filling a conical mold, removing it, and measuring the sag or slump of the sample.

**smelting** Melting of iron-bearing materials in a blast furnace to separate iron from impurities with which it is chemically combined or mechanically mixed.

**soap** A brick or tile of normal face dimensions having a nominal 2 in. (50 mm) thickness.

**soft-burned clay** Clay products that have been fired at low temperature ranges. They have relatively high absorptions and low **compressive strengths**.

**softening range** The range of temperature in which a **plastic** changes from a rigid to a soft state. Sometimes referred to as *softening point*.

**softness** Film property displaying low resistance to scratching or indentation; opposite of **hardness**.

**soft temper** The state of maximum **workability** of aluminum obtained by **annealing**.

**softwood** The botanical group of trees that have needle- or scale-like leaves and are evergreen for the most part. Some exceptions are cypress, larch, and tamarack. This term does not relate to the actual hardness of the wood.

**soil absorption field** *See* **absorption field**.

**soil absorption system** Any system that utilizes the soil for subsurface absorption of treated **sewage**, such as an **absorption trench**, **absorption bed**, or **seepage pit**.

**soil pipe** A pipe that conveys the discharge of water closets, urinals, or fixtures having functions similar to those of a **building drain** or **building sewer** (with or without the discharge from other fixtures).

**solar screen** A perforated masonry wall used as a sunshade.

**soldier** Masonry unit set vertically on end with face showing on the masonry surface.

**solid masonry unit** *See under* **masonry unit**.

**solid masonry wall** A wall built of solid or hollow **masonry units** laid continuously, with joints between units completely filled with **mortar**.

**solution dyeing**   Adding dye or colored pigments to synthetic material while in liquid solution before **extrusion** into fiber.

**solution heat treatment**   The first temperature-raising step in the thermal treating of a heat-treatable aluminum alloy. Also called **heat treatment**.

**solvent**   The volatile portion of the **vehicle** of a **coating**, **paint**, or **transparent finish**, such as **turpentine** or **mineral spirits**, which evaporates during the drying process.

**solvent-based**   Adhesives and **primers** consisting of cementitious **binders** and **fillers** dissolved in a volatile hydrocarbon carrier such as alcohol or cutback.

**solvent-thinned**   Formulation of a **coating**, **paint**, or **transparent finish** in which the **binder** is dissolved in the **thinner** (as in **oil paint**), rather than emulsified (as in **latex paint**).

**sound**   A vibration in an elastic medium in a frequency range capable of producing the sensation of hearing.

**sound attenuation**   *See* **attenuation**.

**sound isolation**   Materials or methods of construction designed to resist the transmission of airborne and structure-borne **sound** through walls, floors, and ceilings; the effect of such materials or methods.

**sound pressure**   The instantaneous pressure at a point as a result of the sound vibration minus the static pressure at that point; the change in pressure resulting from sound vibration. It is measured in **dynes** per square centimeter ($cm^2$). A sound pressure of 1 dyne/$cm^2$ is about that of conversational speech at close range and is approximately equal to 1 millionth of atmospheric pressure.

**sound pressure level (SPL)**   **Sound pressure** measured on the decibel scale; the ratio in **decibels** between a measured pressure and a reference pressure.

**sound transmission**   The passage of **sound** through a material or assembly. *Airborne sound transmission* is the transmission of sound through the air as a medium rather than through a solid, such as the structure of a building. Airborne sound is produced when a surface is caused to vibrate, thus producing alternating air pressures adjacent to the surface. The alternating pressures then radiate through the air in waves of higher and

lower pressures. *Structure-borne sound transmission* is the transmission of sound through a solid or assembly of solids, such as a wall or the structure of a building. It occurs when the vibrations from equipment, or the impact of footsteps or dropped objects come in contact with a solid.

**sound transmission class (STC)**   A single-number rating for evaluating the efficiency of assemblies in isolating airborne sound transmission (*see* **sound isolation**). The higher the STC rating, the more efficient the assembly.

**space dyeing**   Alternating bands of color applied to yarn by rollers at predetermined intervals before tufting.

**spackling compound**   Material used as a crack filler for preparing surfaces before painting.

**spall**   A small fragment removed from the face of **concrete** or of a **masonry unit** by a blow or by action of the elements.

**spandrel wall**   That part of a panel **curtain wall** above the head of a window; in a multistory building, includes the panel below the sill of the window in the story above.

**spar varnish**   Very durable varnish designed for service on exterior surfaces.

**spatter**   Small particles or drips of liquid **coating**, **paint**, or **transparent finish** materials thrown or expelled when applying these materials.

**specialty plywood**   *See under* **plywood**.

**split-face block**   *See under* **concrete masonry unit**.

**spot priming**   Method for protecting localized spots. The only areas primed are those that require additional protection due to rusting or **peeling** of the former coat of **paint** or **transparent finish**.

**spreading rate**   Measure of the area that can be covered by a unit volume of **coating**, **paint**, or **transparent finish** material, generally expressed as square feet per gallon ($m^2$/L).

**springwood**   The portion of the **annual growth ring** of a tree that is formed during the early part of the season's growth. This is usually less dense and weaker mechanically than **summerwood**.

**sprouting**   Temporary condition on new **carpets** in which strands of yarn work loose and project above the **pile**.

Can be remedied by careful clipping or spot **shearing**.

**stabilizer**   A material added to prevent or retard degradation of a **plastic** when exposed to sunlight or other environmental conditions.

**stack**   (1) A structure or part thereof that contains a flue or flues for the discharge of gases. (2) The vertical main of a system of soil, waste, or vent piping.

**stack vent**   The extension of a soil or waste **stack** above the highest horizontal drain connected to the **stack**. It is sometimes called a *waste vent* or *soil vent*.

**stain**   (1) A discoloration of wood that may be caused by such diverse agencies as microorganisms, metals, or chemicals. (2) A penetrating formulation intended primarily for wood surfaces. An *opaque stain* is classified as **paint**; a *transparent stain* is a **transparent finish** material. An opaque stain is a stain that hides the **substrate** and previous surface finish materials. A *transparent stain* changes the color of a wood without obscuring the grain; depending on the amount of **pigment**, transparent stains may be more or less transparent and may leave little or no surface film.

**staple fibers**   Relatively short natural (wool) or synthetic fibers ranging from about $1^1/_2$ in. (38 mm) to 7 in. (180 mm) long, which are spun into yarn.

**steam curing**   *See under* **curing**.

**steel**   Iron-based alloy containing manganese, carbon residual, and often other alloying elements, characterized by its strength and toughness; distinguished from gray **cast iron** by its ability to be shaped by **hot working** or **cold working** as initially cast. *Alloy steel* is steel in which residual elements exceed the limits prescribed for carbon steel or to which alloying elements are added within specified ranges. *Carbon steel* is steel in which the residual elements are controlled but to which alloying elements are not usually added. *Heat-resisting steel* is a low-chromium steel with at least 4% chromium which retains its essential mechanical properties at elevated temperatures. *High-strength low-alloy steel* is steel with less than 1% of an alloying element, manufactured to high standards for strength, ductility, and partial chemical specifications. *Mild steel* is carbon steel with a carbon content between 0.15 and 0.25%. *Stainless steel* is steel that contains at least 10% chromium. It has excellent corrosion resistance, strength, and chemical inertness at high and low temperatures.

**stock dyeing**  Dyeing raw fibers before they are carded (combed) or spun.

**storm sewer**  A **sewer** used for conveying rainwater, surface water, condensate, cooling water, or similar liquid wastes, exclusive of **sewage** and industrial waste.

**story**  That portion of a building included between the upper surface of a floor and the upper surface of the floor directly above, except that the topmost story is that portion of a building included between the upper surface of the topmost floor and the ceiling or roof above. Where a finished floor level directly above a basement or cellar is more than 6 ft (1800 mm) above grade, such a basement or cellar is considered a story.

**story pole**  Marked pole for measuring vertical masonry courses during construction.

**strain**  The change in cross-sectional area of a body produced by stress. Measured in inches per inch (millimeter per millimeter) of length.

**strain hardening**  A method of strengthening **non-heat-treatable** aluminum **alloys** by either **cold rolling** or other physical or mechanical working.

**strand caster**  A machine that continuously casts **steel** slabs and billets.

**streaking**  Irregular occurrence of lines or streaks of various lengths and colors in an applied film; usually caused by some form of contamination.

**strength**  A term used to describe all the properties of wood that enable it to resist forces or loads.

**stress**  The intensity of a mechanical force acting on a body; either tensile, compressive, shear, or the combination of compressive and tension forces known as bending, or the twisting force, torsion. Tensile, compressive, and shear forces are measured by dividing the total force by the area over which the force acts (force per unit of area), such as pound per square inch (kilogram per square meter), for example. In wood, the *allowable unit*, or *working, stress* is the stress used in designing wood members. It is appropriate to the species and grade of the wood. Values for each type of stress are obtained

by multiplying the *basic stress* for that species by the strength ratio assigned to each grade. Most codes include the allowable working stress for each grade. In wood, the basic stress is the design stress for a clear wood specimen free from strength-reducing features such as **knots**, **checks**, and **cross grain**. Such a specimen is assumed to have all the factors appropriate to the nature of **structural lumber** and the conditions under which it is used, except those that are accounted for in the strength ratio.

**stretcher**  **Masonry unit** laid with its length horizontal and parallel with the face of the masonry and with its smallest dimension vertical.

**stretch forming**  The stretching of metal for the purpose of flattening it or for forming it to a predetermined shape.

**stria (striped)**  *See under* **pile**.

**stringer**  *See under* **lumber**.

**strip**  *See under* **finished mill products**.

**strip lamination**  In multi-ply **gypsum board** construction, a method of applying **adhesive** in parallel strips spaced 16 in. (406 mm) to 24 in. (610 mm) apart. *See also* **sheet lamination**.

**strip reinforcing**  In single-ply **gypsum board** construction, an installation procedure in which strips of gypsum board are applied to the framing members to reinforce gypsum board joints and provide a base for adhesive application.

**struck joint**  *See* **mortar joint**.

**structural clay tile**  Hollow **masonry units** composed of burned clay, shale, fire clay, or mixtures of these. *End-construction* tile is designed to be laid with the axis of its cells vertical. *Facing tile* is made for exterior and interior use with its face exposed. *Side-construction tile* is intended for placement with the axis of its cells horizontal.

**structural lumber**  See under **lumber**.

**structurals**  *See under* **finished mill products**.

**stucco**  **Portland cement**, water, sand, and possibly a small quantity of lime (**portland cement plaster**), along with, perhaps, other **aggregates**; used on exterior surfaces.

**stucco finish**  A factory-prepared mix of **stucco** for application as finish coats.

**stud adhesive**  *See* **adhesive, drywall**.

**stuffer yarn**  *See under* **warp**.

**styrene-butadiene**  *See under* **resin**.

**subfloor**  The structural material or surface that supports a finish floor and floor loads and serves as a working platform during construction.

**subgrade**  An earth surface on which another material, such as **concrete**, is placed.

**sublimation**  The transition of ice directly into water vapor.

**substrate**  Surface to be covered with a **coating**, **paint**, or **transparent finish**.

**suction**  The initial rate of water **absorption** by a clay **masonry unit**. *See* **absorption rate**.

**summerwood**  The portion of the **annual growth ring** of a tree that is formed after **springwood** formation has ceased. It is usually denser and stronger mechanically than springwood.

**sump**  A tank or pit that receives **sewage** or other liquid waste, located below the normal grade of a building's gravity sewer system, and that must be emptied by mechanical means.

**sump pump**  A mechanical device used for pumping out **sewage**, liquid, or industrial wastes from locations below a building's gravity drain.

**surfacer**  Pigmented formulation for filling minor irregularities before a finish coat of **paint** is applied; usually applied over a **primer** and sanded for smoothness.

**survey**  The measure and marking of land, accompanied by maps and field notes that describe the measures and marks made in the field.

**surveying**  The process of making a **survey**.

**swale**  A low, flat depression in the exterior grade, used to drain away storm water.

**system, paint**  *See* **paint system**.

**tack rag**  Piece of loosely woven cloth that has been dipped into a varnish oil and wrung out. When it becomes tacky or sticky, it is used to wipe a surface to remove small particles of dust.

**tactile**  Describes an object that can be perceived with the sense of touch.

**tactile cue**  *See under* **cue**.

**taping** In **gypsum board** construction, applying **joint tape** over embedding compound in the process of **joint treatment**.

**temper designation** Designation (following an **alloy designation** number) that denotes the temper of an aluminum **alloy**.

**tempering** *See* **heat treatment**.

**terne** A lead-tin alloy usually consisting of 4 parts lead and 1 part tin.

**terneplate** *See under* **finished mill products**.

**terrazzo** A floor topping made of marble or other stone chips set in cement mortar, ground smooth and polished.

**terrazzo finish** All elements installed by a **terrazzo** contractor from subfloor to finished surface, such as a sand bed, **isolation membrane**, **underbed**, and **topping**.

**texture** In wood, a term often used interchangeably with **grain**. It refers to the structure of wood.

**texture paint** A paint that may be manipulated by brush, roller, trowel, or another tool to produce various effects.

**thermal mass** The heat storage capacity of a material.

**thermoplastic** A **plastic** that can be repeatedly softened by heating and hardened by cooling.

**thermoset** A **plastic** that, after curing, forms a permanently hardened product that cannot be softened again by reheating.

**thickbed** *See* **mortar bed, tile**.

**thinners** **Solvents** used to thin **coating**, **paint**, or **transparent finish** materials.

**thin-set** (1) In **terrazzo**, a method of installing relatively thin **toppings** ($^1/_4$ in. [6.4 mm] to $^1/_2$ in. [12.7 mm] thick) directly over a suitable **subfloor**; generally possible with resinous **binders** only. (2) In **tile**, a method of bonding tile with a thin layer ($^1/_{16}$ in. [1.6 mm] to $^1/_4$ in. [6.4 mm] thick) of special mortar or adhesive to a suitable **backing** or to a properly cured **mortar bed**.

**thixotropy** Property of a material that causes it to change from a thick, pasty consistency to a fluid consistency upon agitation, brushing, or rolling.

**three-coat work** Application of plaster in three separate layers (**scratch**, **brown**,

and **finish coats**), totaling at least $^7/_8$ in. (22.2 mm) in thickness.

**tie** *See* **wall tie**.

**tinplate** *See under* **finished mill products**.

**tint base** Basic **paint** in a custom color system to which **colorants** are added to make a wide range of colors.

**tone-on-tone** **Carpet** pattern made by using two or more shades of the same hue.

**tooled joint** *See* **mortar joint**.

**toothing** Projecting **brick** or **concrete masonry unit** in alternate courses to provide for bond with adjoining masonry that will be laid later.

**topping** In **terrazzo finishes**, a decorative wear layer consisting of marble chips embedded in a suitable matrix and requiring grinding, polishing, or washing to form finished **terrazzo**.

**touch-up** (1) Ability of a **coating**, **paint**, or **transparent finish** film to be spot repaired (usually within a few months of the initial application) without showing color or gloss differences. (2) The repair of a coating, paint, or transparent finish film by selectively adding finishing material to damaged or missed areas after the earlier coats have dried.

**toughness** Maximum ability of a material to absorb energy without breaking, as from a sudden shock or impact.

**township** An area of land used in the **rectangular survey system**, approximately 6 miles square, bounded by **range lines** and **township lines**.

**township lines** East-west reference lines used in the **rectangular survey system**, located at 6-mile intervals between **correction lines**.

**transfer molding** Forming **plastic** by fusing it in one chamber with heat and then forcing it into another chamber, where it solidifies; commonly used with **thermoset**ting plastics.

**transmission loss** The decrease in or attenuation of **sound** energy, expressed in **decibels**, of airborne sound as it passes through building construction.

**transparent finish** A system of materials that are applied as liquids or by hand in the field or in a factory or shop to form a finish through which the substrate or previous finish is visible. *Lacquer* is a transparent finish material that is em-

ployed mostly as a shop or factory finish on furniture and casework. It is not usable over existing finish films. In fact, it is an effective paint remover. *Polyurethane* is a hard, highly abrasion- and chemical-resistant transparent finish, often used as a floor finish and as a bar-top finish. Some manufacturers call their polyurethanes varnishes; others call them lacquers. In fact, although they have some characteristics of each, they are neither. Polyurethanes are available in both oil-modified and moist-curing formulations. *See also* **resin**. *Shellac* is a fast-drying, transparent finish material consisting of lac resins, produced by the lac insect, dissolved in alcohol. Shellac has a relatively short shelf life. *Spar varnish* is an exterior, weather-resistant, transparent finish material based generally on long-oil phenolic resin; the term originated from the use of this material on spars of ships. *Varnish* is a finish material that dries to a transparent or translucent film when exposed to air. Varnish is a mixture of resin, **drying oil**, **drier**, and a **solvent**.

**transparent finish system** The several coats and materials necessary to produce a **transparent finish**. The various materials may include **dyes**, bleaches, or **stains** to change the color of substrates such as wood, and **undercoats** and **finish coats** of one of the materials listed under **transparent finish** or another transparent finish material, such as oil.

**trap** A fitting or device designed to provide, when properly vented, a liquid seal that will prevent the back passage of air without materially affecting the flow of **sewage** or wastewater through it.

**trimmer arch** *See under* **arch**.

**troweled joint** *See* **mortar joint**.

**troweling** A **concrete slab** finishing operation that produces a smooth, hard surface.

**tubular steel products** *See under* **finished mill products**.

**tuck pointing** Refilling defective **mortar joints** that have been cut out in existing masonry.

**tufted carpet** **Carpet** made by inserting the pile yarns through a prewoven fabric **backing** on a machine with hundreds of needles (similar to a huge sewing machine).

**tufts** Surface loops of pile fabric.

**turpentine** Colorless liquid used as **thinner** for **oil paints** and **varnishes**, distilled from products of the pine tree.

**twist** *See* **pile, warp**.

**two-coat work** Application of **plaster** in two layers (**base coat** and **finish coat**), totaling at least $^5/_8$ in. (15.9 mm) in thickness.

**two-package (two-part) formulation** **Coating**, **paint**, or **transparent finish** material formulated in two separate packages and requiring that the two ingredients be mixed before characteristic properties can be obtained and the material can be applied.

**underbed** Layer of nonstructural **portland cement mortar** sometimes used over a **subfloor** to provide a suitable base for **portland cement terrazzo** and to minimize cracking.

**undercoat** **Primer** or intermediate coat in a multicoat system.

**underlay** *See* **cushioning**.

**underlayment** A mastic or panelboard material installed over a **subfloor** to provide a suitable base for resilient flooring when the subfloor does not possess the necessary properties for direct application of the flooring.

**uniformity** In **coating**, **paint**, or **transparent finish**: not varying in **gloss**, **sheen**, color, hiding, or other property.

**unsaturated molecule** A molecule that is capable of uniting with certain other elements or compounds without creating any side products.

**vacuum forming** Shaping a heated **plastic** sheet by causing it to flow in the direction of reduced air pressure.

**vehicle** Liquid portion of a **coating**, **paint**, or **transparent finish**, including ingredients dissolved in it.

**velvet carpet** *See* **woven carpet**.

**veneer** (1) A single-facing **withe** of **masonry units** or similar materials attached to a wall for the purpose of providing ornamentation, protection, or insulation, but not bonded or attached to intentionally exert common action under load. (2) A thin sheet of wood. *Rotary-cut veneer* is veneer cut in a continuous strip by rotating a log against the edge of a knife in a lathe. *Sawed veneer* is veneer produced by sawing. *Sliced veneer* is veneer that is sliced by moving a log or **flitch** against a large knife.

**veneered wall** A wall having a face of **masonry units** or other weather-resisting materials attached to the backing, but not so bonded as to intentionally exert common action under load.

**veneer wall tie** *See* **wall tie**.

**vent stack** A vertical vent pipe installed primarily for the purpose of providing circulation of air to and from a part of a **drainage system**.

**vent system** A collection of pipes installed to provide a flow of air to or from a **drainage system**, or to provide circulation of air within such a system to protect trap seals from siphonage and back pressure.

**vertical application** Gypsum boards applied with the **edges** parallel to supporting members and borders. *See also* **horizontal application**.

**vertical grain** *See under* **grain**.

**vibrating** A mechanical method of compacting **concrete**.

**vibrator** A mechanical device that vibrates at a speed of 3000 to 10,000 revolutions per minute and is inserted into wet **concrete** or applied to its **forms** to compact concrete.

**vinyl** A **polymer** derived from ethylene; used in **paint**, **coatings**, and fabric. *See also* **resin**.

**vinyl foam cushioning** Carpet cushioning made from a combination of foamed synthetic materials.

**visual cue** *See under* **cue**.

**voids** Air spaces between pieces of **aggregate** within a **cement** paste.

**volt** The **potential difference** between two points in a wire carrying a current of 1 **ampere** when the power dissipated is 1 **watt**; equivalent to the **potential difference** across 1 **ohm** of resistance when 1 ampere is flowing through it.

**voltage** Electrical pressure, measured in **volts** and comparable to pounds per square inch (psi) in a fluid medium. Reference to 120 and 240 nominal voltages includes typical operating ranges of 115 to 125 and 230 to 250, respectively. Same as **potential difference**.

**wallboard, gypsum** *See under* **gypsum board**.

**wall tie** A **header** (*bonder*) or metal anchor that connects **wythes** of masonry

to each other or to other materials. A *cavity wall tie* is a rigid, corrosion-resistant metal tie that bonds two wythes of a cavity wall. It is usually steel, $^3/_{16}$ in. (4.76 mm) in diameter, and formed in a Z shape or a rectangle. A *veneer wall tie* is a strip or piece of bent metal used to tie a facing veneer to the backing; sometimes in two pieces to permit movement.

**wane** **Bark** or lack of wood, from any cause, on the edge or corner of a piece of **lumber**.

**warp** (1) In wood: variation from a true or plane surface. *Bow* is distortion of a board in which the face is convex or concave longitudinally. *Crook* is distortion of a board in which the edge is convex or concave. *Cup* is distortion of a board in which the face is convex or concave transversely. *Twist* is distortion caused by the turning of the edges of a board so that the four corners of any face are no longer in the same plane. (2) In **carpet**: **backing** yarns running lengthwise in the carpet. In *chain warp*, zigzag warp yarn works over and under the **shot** yarns of the carpet, binding the backing yarns together. *Pile warp* is lengthwise **pile** yarns in Wilton **woven carpets** that form part of the backing. *Stuffer warp* is yarn that runs lengthwise in a **carpet** but does not intertwine with any filling (**weft** shot) yarn; serves to give weight, thickness, and stability to the fabric.

**washability** Ability of a **coating**, **paint**, or **transparent finish** to be easily cleaned without wearing away during cleaning.

**water absorption** *See* **absorption rate**.

**water-cement ratio** The ratio, by weight, of water to **cement**, or the amount of water, in gallons, used per 94 lb sack of cement to make **concrete**. It is an index of strength, durability, watertightness, and consistency.

**water hammer** Pounding noises and vibration that sometimes develop in a piping system when its air chambers become filled with water or have been omitted entirely; the problem may also be due to worn washers in faucets.

**waterproofing** Prevention of moisture flow through **concrete** or masonry due to water pressure; an impervious liquid or sheet material used to waterproof.

**water retention** A property of **mortar** that prevents the loss of water to **ma-**

sonry units having high suction. It also prevents bleeding or water gain when mortar is in contact with units having low suction. Also called *water retentivity*.

water spotting Coating, paint, or transparent finish appearance defect caused by water droplets.

water supply system In a building, consists of the water service pipe, water distributing pipes, branches to plumbing fixtures and appliances, and necessary fittings and control valves.

watt (1) In acoustics, a unit of sound power equal to $1 \times 10^7$ dynes per centimeter per second, which is the basic expression of the flow of sound energy. (2) In electricity, a unit of electrical power equal to the work done by a current of 1 ampere under the pressure of 1 volt.

wattage The amount of electrical power measured in watts; a single unit combining the effect of both voltage (pressure) and rate of flow (amperage) by multiplying these quantities (volts times amperes equals watts).

watt-hour Unit of energy consumed, consisting of watts multiplied by time in hours; the result is often expressed in thousands of watt-hours, called *kilowatt-hours*.

weathering (1) Corrosion (galvanic or chemical) produced by atmospheric conditions. (2) Effect of exposure to weather on a coating, paint, or transparent finish film, raw wood, and other materials.

weathering index The product of the average annual number of freezing cycle days and the average annual winter rainfall.

weaving Process of forming carpet on a loom by interlacing the warp and weft yarns.

weep holes Openings placed in mortar joints of facing materials at the level of a flashing to divert to the exterior any moisture collected on the flashing.

weft Backing yarns that run across the width of a carpet. In woven carpets, the weft shot (filling) yarns and the warp chain (binder) yarns interlock and bind the pile tufts to the backing. In tufted carpets, pile yarns that run across the carpet are also considered weft yarns.

welding Creating a metallurgical bond between metals with heat and sometimes with the use of pressure and filler metal. *Arc welding* employs an electric arc as the source of heat. *Gas welding* employs a fuel gas (acetylene, hydrogen) and oxygen as the sources of heat. *Shielded welding* is a process in which gases or fusible granular materials are used to shield the weld area from the damaging effects of oxygen and nitrogen in the air. *Shielded metal-arc welding* is arc welding in which a flux-coated metal electrode is consumed to form a pool of filler metal and a gas shield around the weld area. Also known as *manual metal-arc welding* and *stick electrode welding*. *Inert-gas-shielded arc welding* is arc welding in which shielding is provided by an inert gas envelope (such as argon, helium, a combination of argon and helium, or carbon dioxide) from an external supply. Filler metal is supplied either by a consumable metal electrode, as in inert-gas-shielded metal arc welding (MIG), or by a separate filler rod used with a nonconsumable tungsten electrode, as in inert-gas-shielded tungsten arc welding (TIG). *Submerged arc welding* is arc welding in which the weld area is shielded by a fusible granular material that melts to protect the weld area. Filler metal is obtained from either a consumable electrode or a separate filler rod.

white lead Oldest white pigment, chemically known as *lead sulfate* or *lead carbonate*.

**Wilton carpet** *See* **woven carpet**.

winter rainfall The sum in inches of the mean monthly corrected precipitation occurring between the first killing frost in the fall and the last killing frost in the spring.

wire, steel *See under* finished mill products.

wires (pile wire, gauge wire, standing wire) Metal strips over which the pile tufts are formed in woven carpets. *See also* round wire.

woolen yarn Soft, bulky yarn spun from both long and short wool fibers that are not combed straight but lie in all directions so that they will interlock to produce a felt-like texture.

workability Relative ease or difficulty with which concrete can be placed and worked into its final position within forms and around reinforcing bars. However, workability is contingent on the absence of segregation of the concrete. If the aggregate segregates, workability is considered to diminish (or terminate), regardless of how easily the concrete flows into place.

working stress *See* stress.

worsted yarn Strong, dense yarn made from long staple fibers that are combed to align the fibers and remove extremely short fibers.

woven carpet Carpet made by simultaneously interweaving backing and pile yarns on one of several types of looms from which the carpets derive their names. *Axminster carpet* is carpet made on an Axminster loom, which is capable of intricate color designs, usually with a level cut-pile surface. *Loomed carpet* is carpet made on a modified upholstery loom with a characteristic dense, low-level loop pile, generally bonded to cellular rubber cushioning. *Velvet carpet* is carpet made on a simple loom, usually of a solid color or moresque, with cut or loop pile of either soft or hard-twisted yarns. *Wilton carpet* is carpet made on a loom employing a jacquard mechanism, which selects two or more colored yarns to create the pile pattern.

wrinkling Development of ridges and furrows in a coating, paint, or transparent finish film when it dries.

wrought aluminum products Products formed by rolling, drawing, extruding, or forging. A *bar* is a solid section that is long in relation to its cross-sectional dimensions and has a completely symmetrical cross section that is square or rectangular (excluding flattened wire) with sharp or rounded corners or edges, or is a regular hexagon or octagon. An aluminum bar has a width or greatest distance between parallel faces of $3/_8$ in. (9.5 mm) or more. An *extruded section* is a rod, bar, tube, or any other shape produced by the extrusion process. *Foil* is a solid sheet section rolled to a thickness of less than 0.006 in. (0.1524 mm). *Forging stock* is a rod, bar, or other section suitable for subsequent change in cross section by forging. A *pipe* is a tube having certain standardized combinations of outside diameters and wall thicknesses. A *plate* is a solid section rolled to a thickness of 0.250 in. (6.4 mm) or more in rectangular form and with either sheared or sawed edges. A *rod* is a solid, round aluminum section $3/_8$ in. (9.5 mm) or greater in diameter, whose length is greater than its diameter. A *sheet* is a solid section rolled to a thickness ranging from 0.006 in. (0.1524 mm) to 0.249 in. (6.3246 mm) inclusive, supplied with sheared, slit, or sawed edges. *Structural shapes* are solid shapes used as load-bearing members. They include angles, channels, W shapes, tees, zees, and others. A *tube* is a hollow product whose cross section is completely symmetrical and is round, square, rectangular, hexagonal, octagonal, or elliptical, with sharp or rounded corners, and whose wall is of uniform thickness except as affected by corner radii. A *wire* is a solid section that is long in relation to its cross-sectional diameter, having a completely symmetrical cross section that is square or rectangular (excluding flattened wire) with sharp or rounded corners or edges, or is round or a rectangular hexagon or octagon, and whose diameter, width, or greatest distance between parallel faces is less than $^3/_8$ in. (9.5 mm).

**wrought iron** Relatively pure iron, mechanically mixed with a small amount of iron-silicate slag; characterized by good corrosion resistance, weldability, toughness, and high ductility.

**wrought steel products** Products formed by rolling, **drawing**, **extruding**, or **forging**. *See also* **finished mill products**.

**wythe** A continuous vertical section of masonry one unit in thickness. Also called *withe* and *tier*.

**yard lumber** *See under* **lumber**.

**yellowing** Development of a yellow color or cast in white, pastel, colored, or clear finishes.

# Index

Nail popping, 694–696
National Association of Home Builders (NAHB), 11
National Building Code, 19, 20
National Conference of States on Building Codes and Standards (NCSBCS), 22
National Electrical Code, 21
National energy codes, 21–22
National Fire Protection Association (NFPA), 11
National Institute of Building Sciences (NIBS), 22
National Standard Plumbing Code, 21
Natural decay resistance, 355
Natural finishes, 300, 301
NCSBCS (National Conference of States on Building Codes and Standards), 22
NC wired systems, 1061
Neat plaster, 656, 660
Neat portland cement, 703
Needlepunched carpet, 790
Negotiation, contract, 43, 44
NIBS (National Institute of Building Sciences), 22
Nickel, 251, 298
Noise, 861, 870–873
Noise barriers, 872, 873
Noise reduction, 862–865
Nondirectional textured mechanical finishes, 302
Nonetched cleaned aluminum, 303
Nonetched cleaned copper and copper alloys, 304
Nonferrous metals:
  adhesive bonding, 285, 286
  mechanical finishes on, 300–302
  welding, 284–285
Non-heat-treatable aluminum alloys, 272
Non-load-bearing walls, 199
Nonmetallic conduit, 1032
Nonmetallic sheathed cable, 1031
Non-pressure-treated wood, 356, 357
Nonproblem soils, 1071–1072
Nonveneer panels, 353
Nonvulcanized elastomers, 529, 531
Nosings, 24, 756, 770, 771
NO wired systems, 1061
Number systems, 1119
Nylons, 421, 422, 792

Oak, 747, 748
Occupant health considerations, 3–4
Octaves, 860
Odor, unpleasant, 939
Ohm's law, 1023
Oil, 1009
Oil and oleoresinous flat paints, 803
Oil-based caulking compound, 551
Oil-base paints, 803–804
Oil-borne wood preservative, 355–356
"Oil canning," 544, 545
Oil hydraulic elevators, 910–911
Oil/oleoresinous flat paints, 803
Oil primers, 801, 802

Oleoresinous paints, 804
One- and Two-Family Dwelling Code, 21
100-mm building module, 1153
One-way slabs, 129–131
Opaque coatings, 305, 805
Open-die forging, 261, 262
Open hearth process, 253–254
Openings:
  floor, 373
  wall, 538
Operative temperature, 981
Organic adhesive, 705
Organic coatings, 304–305
Organic fiber insulation, 447
Organization of spaces, 26, 27, 30
Outdoor design temperatures, 984–985
Outlets, electrical, 1034–1037
Overlaid plywood, 344–345
Overlay drafting, 40
Overload protection devices, 1028, 1030
Overturning forces, 360

Packaged direct refrigeration units, 1011
Padlocks, 612
Paint(s):
  alkali-resistant, 805
  on aluminum, 305
  on ferrous metal, 304–305
  fire-retardant, 805
  as vapor retarders, 437
Painting, 800–818
  application, 811–818
    conditioning/mixing, 811–812
    materials handling, 811
    paint application, 817–818
    surface preparation, 812–816
    surface repair, 816, 817
  materials, 800–806
    exterior primers, 801–802
    exterior topcoat, 803–805
    interior primers, 801
    interior topcoat, 802–803
    opaque coatings, 805
    transparent finishes, 805–806
  surfaces, 806–811
    exterior finishing, 810–811
    exterior priming, 809–810
    interior floors, 808
    interior trim/paneling, 808
    interior walls/ceilings, 806–808
  sustainability considerations for, 643
Paint removers, 815
Panel grades of plywood, 344
Paneling, 345, 413, 414, 808
Panelized assemblies, 384–386
Panel systems, curtain-wall, 587, 589
Parging, 220
Parking, 25
Parquet flooring, 749–750
Partitions:
  in conventional wood framing, 375–376
  plaster, 662, 663
  self-supporting gypsum, 690–692
Passageways, 27, 30, 31

Passenger elevators, 911–912
Passive solar cooling systems, 1008
Passive solar heating systems, 1005–1009
  building envelope, 1006
  collection, 1006–1007
  direct gain systems, 1007
  history, 1005–1006
  indirect gain systems, 1007–1008
  isolated gain systems, –1008
  properties, 1006
  storage, 1007
Pattern bonds, masonry, 190–192
Patterned mechanical finishes, 302
Paver tile, 702, 703
Paving, 734–737
Paving joint sealants, 551
Penetrating sealers, 820–821
Perceptual impairment, 30
Percolation tests, 948, 949
Perforated lath, 652
Performance-rated panels, 344
Perimeter intrusion detection, 1060–1061
Perimeter systems, 996
Perlite, 655
Perlite insulation, 447, 448
Permanent forms, 76, 133
Permanent-mold castings, 277
Permanent wood foundation systems, 357–360
Permeability, vapor, 430, 459
Permeance, vapor, 430–431
Phenolic-based foam, 449
Phenolic insulation, 447
Phenolics, 423
Photoelectric detectors, 1064
Photoelectric/infrared systems, 1062
Pickling, 815
Piers, 126, 127, 200, 201
Pigtailing, 1035
Pilaster concrete blocks, 182
Pilasters, 126, 127, 200, 201
Piles, 117, 118, 1085–1086
Pile fibers, 791–793, 797
Pipes:
  central HVAC, 1011
  for DWV system, 944–945
  roof drainage, 959–960
  sound control for, 967–969
  for water supply, 931–934
Pitch (sound), 860
Pitch of roof, 460, 461
Pitch pockets, 318, 331
Plain lath, 652
Plane angle, 1130
Planing mill, 328, 329
Plank-and-beam systems, 400–405
Plank flooring, 748–749
Plants, 545–546
Plant-cast units, 135
Planting strips, 26
Plaster:
  for fire containment/control, 547
  gypsum, 660–663
  portland cement, 663–670
  sustainability considerations for, 642

Traffic noise, 871–873
Translation forces, 360, 361
Transmission of sound, *see* Sound transmission
Transmittance (heat transfer), 986
Transparent finishes, 805–806, 809
Transverse strength, 196, 197
Traps, DWV system, 940, 942–943
Treatment, water, *see* Water treatment
Trees, 317, 318
    for construction plywood, 339, 340, 343
    hardwoods, 317
    in landscaping, 1090–1093
    lightning protection for, 1044–1045
    softwoods, 317, 318
Trim:
    exterior finishing of, 808
    in finish carpentry, 412–413
    interior, 808
Trim enamels, 804
Trim paints, 804
Triple-track storm windows, 611
Troweled joints, 216
Troweling, 106, 107
Trusses:
    cold-formed metal framing, 295
    steel, 292
    wood, 386–388
Tubular products, 267–268
Tuck pointing, 155–156
Tufted carpet, 789, 790
Tungsten, 251
Two-coat work, 661, 667
Two-pipe systems, 1003
Two-track storm windows, 611
Two-way slabs, 130–132

UL, *see* Underwriters Laboratories
Ultrasonic motion detector, 1062, 1063
Unauthorized entry, 615
Unauthorized entry prevention, 1061, 1062
Unbonded concrete toppings, 141
Uncertainties in legal descriptions of land, 59–60
Uncontrolled natural finishes, 300
Underfloor raceway, 1033
Underlayment:
    and asphalt shingles, 471, 480
    concrete, 143–144
    for metal roofing, 542
    for mineral-fiber-cement shingles, 488
    plywood, 392–394
    for resilient flooring, 756–758
    for roll roofing, 525
    roofing, 462, 463
    for slate roofing, 481–482
    for tile roofing, 485–486
    for wood shingles/shakes, 499
Underpinning, 1073–1075
Underwriters Laboratories (UL), 10–11
Unified Numbering System (UNS), 257
*Uniformat,* 41–42
Uniform Building Code, 20–21

Uniform Plumbing and Uniform Mechanical Codes, 21
Unit-and-mullion systems, 587, 588
Unit curtain-wall systems, 587, 588
Unit design, 183–205
    cavity walls, 187–188
    crack control, 192–195
    fireplaces/chimneys, 201–205
    fire resistance, 201
    lintels/arches, 199–200
    moisture control, 192
    pattern bonds, 190–192
    piers/pilasters, 200, 201
    reinforced walls, 190
    solid walls, 185–187
    veneered walls, 188–190
    wall structure, 195–199
Unit heaters, 1013
Unit kitchens, 841–843
Unit masonry walls, 882–884
Unit reinforcement, 157–160
Unit ventilators, 1013
Unseasoned lumber, 328
Unsheathed construction, 644–645
UNS (Unified Numbering System), 257
Upfeed water supply systems, 930–931
Ureas, 424
Urethane enamels, 803
Urethane foam, 448, 798
Urethane sealants, 550
Urinals, 964
Utilities, 1096–1101
    electric power distribution, 1100–1101
    sanitary sewage utilities, 1097–1098
    subdrainage systems/storm drainage utilities, 1098–1100
    sustainability considerations for, 1096
    water supply sources, 1097

Valley flashing, 463, 480, 500–501
Valves, water-supply, 935
Vanadium, 251
Vapor permeability and condensation, 459
Vapor retarders, 436–442, 1084–1085
    design, 437–441
    foil-backed boards, 436–437
    installation, 441–442
    selection, 436–437
Varnish, 806
Velocity, 1144
Veneer, brick:
    bonding, 211, 212
    thermal insulation of, 456, 457
Veneered walls, 188–190
Veneered wall ties, 159
Veneer grades of plywood, 343–344
Veneer plaster, 657, 662
Veneer plaster bases, 653
Vents, DWV system, 941, 943–944
Ventilating, *see* Heating, ventilating, and air conditioning
Ventilation, interior, 979, 980
Ventilation design, 437–441
    crawl spaces, 438–439

roof and ceiling assemblies, 439–441
    slabs on grade, 441
    walls, 437, 438
Venting, DWV system, 942
Vent stacks, 1009
Vent systems, 940
Vermiculite, 655
Vermiculite insulation, 448
Vertical lift doors, 580, 581
Vertical wheelchair lifts, 915
Vibrating (of concrete), 104–105
Viewing devices, burglary-protection, 615
Vinyls, 422–423
Vinyl-coated fabric wall coverings, 643, 799–800
Vinyl composition tile, 751
Vinyl sheet and tile, 751, 753
Vinyl siding, 515, 517–519
Vinyl tile, 753
Vinyl toluene acrylic resin systems, 820
Vinyl toluene resin coatings, 820
Vinyl wall base, 770
Visual impairment, accessible design for, 29–30
Vitreous finishes on metal, 308
Voltage, 1022
Volume, 1129
Volume change, 151, 179–180
Voluntary Product Standards (Voluntary PS), 11–12
Vulcanized elastomers, 529

Waffle slabs, 132, 133
Walks, 26
Walls:
    cast-in-place structural concrete, 125–126
    concealed condensation in, 433–434
    flashing at base of, 538
    forming, 78, 79
    and HVAC, 988
    masonry, *see* Masonry walls
    painting interior, 806–808
    reinforced, 190, 211
    and sound control, 879–884
    structural precast concrete, 136
    thermal insulation of, 451–454
    and tile backing, 707–708
    vapor retarder installation in, 442
    vapor retarder/ventilation design for, 437, 438
Wall assemblies:
    conventional wood framing, 375–376
    panelized, 385
    and sound control, 890–896, 904
Wall base cement, 761
Wall bases, 755–756, 770
Wallboard, fire-resistant, 672
Wall coverings, vinyl-coated fabric, *see* Vinyl-coated fabric wall coverings
Wall flashing, 537, 538
    at base of walls, 538
    at copings, 538
    at expansion joints, 538
    at floor lines, 538